Flowering Plants

Armen Takhtajan

Flowering Plants

Second Edition

Springer

Armen Takhtajan
Komarov Botanical Institute
St. Petersburg
Russia

ISBN: 978-1-4020-9608-2 e-ISBN: 978-1-4020-9609-9
DOI: 10.1007/978-1-4020-9609-9

Library of Congress Control Number: 2009929400

Cover illustration: Front Cover: Photograph of Painting "Spring Flowers" by Martiros Saryan
Back Cover: Photograph of Armen Takhtajan

Printed on acid-free paper

Springer is part of Springer Science+Business Media (www.springer.com)

Contents

Foreword

Professor Armen Takhtajan, a giant among botanists, has spent a lifetime in the service of his science and of humanity. As a thoroughgoing internationalist, he promoted close relationships between botanists and people of all nations through the most difficult times imaginable, and succeeded with his strong and persistent personal warmth. He also has stood for excellent modern science throughout this life, and taught hundreds of students to appreciate the highest values of civilization whatever their particular pursuits or views, or the problems they encountered.

Takhtajan has made multiple contributions to our understanding of plant evolution, particularly concerning angiosperms and their classification. As early as 1943, in his paper "Correlations of Ontogenesis and Phylogenesis in Higher Plants," he put forward a theory of the macroevolution of many groups of plants through neoteny; he elaborated this theory in later publications. Takhtajan's ideas on macroevolution as a result of changes in developmental timing (heterochrony or heterobatmy) has been viewed favorably by a number of outstanding biologists, including Agnes Arber (in "The Natural Philosophy of Plant Form", 1950) and Stephen Gould (in "Ontogeny and Phylogeny", 1977). His principal ideas were that the origin of herbaceous angiosperms was the result of neoteny and that the origin of some arborescent forms was secondary. He also offered hypotheses about the way in which monocot leaves, with their characteristic parallel venation, and discussed well the patterns involved in the origin of stomata. Takhtajan produced a novel classification for the structural types of gynoecium and of their placentation. He also wrote on the evolution of inflorescences, the evolution of pollen grains, and the evolutionary classification of fruit types. His theory of the evolution of inflorescences, in which he postulated that a leafy cyme was the original type, was accepted by Stebbins ("Flowering Plants, Evolution Above the Species Level," 1974: 263). One of the his most important contributions was the idea that the origin and evolution of male and female gametophytes of the angiosperms came about through evolutionary changes in developmental timing accompanied by drastic modifications of the ontogenetic processes involved.

Takhtajan's most important achievement has been the development of his phylogenetic system of the flowering plants, a system that has greatly influenced all other recent systems of classification; in turn, Takhtajan was inspired by Hans Hallier's earlier theories. He published a preliminary phyletic diagram of the orders of angiosperms as early as in 1942, and this diagram was mentioned by Gundersen in his "Families of Dicotyledons" (1950). Later in his large book, "A System and Phylogeny of the Flowering Plants" (1966) and in his "Systema Magnoliophytorum" (1987), both in Russian, as well as in "Diversity and Classification of Flowering Plants" (1997), in English, Takhtajan provided a detailed exposition of his system as well as the reasons for his delimitation and arrangement of families and orders. One of his main innovations was the subdivision of both the dicots and monocots into subclasses, which was widely accepted as a major advance in angiosperm classification and introduced into some textbooks, including the last edition of Strasburger's "Lehrbuch der Botanik".

Takhtajan's system of classification is a synthetic, integrated one based on all available data, including recent studies in embryology, palynology, comparative anatomy, cytology, phytochemistry, and molecular data, as well as on cladistic analyses of many taxa. This new book, as well as "Diversity and Classification of Flowering Plants", includes also intrafamilial classification (subfamilies and tribes).

Armen Takhtajan has worked for many years at the Komarov Botanical Institute, St. Petersburg, Russia (LE), where he had access to its great herbarium collections and library. He used these rich resources to supplement his field experience in many regions in the world. As a result of his studies and the observations he was able to make during the course of his travels, he prepared a book entitled "Floristic Regions of the World," in which he presented not only floristic divisions for the whole world, but also listed endemic families and genera and provided examples of endemic species for each province.

At the present, the classification of angiosperm families and our ideas of their relationships are moving forward rapidly; current studies have led and are leading to many significant changes in our interpretations, largely following the important clues about relationship that have come from molecular comparisons between taxa. Because of the numerous examples of parallelism and evolutionary convergence among the angiosperms and their individual structures, some of the ideas gained by earlier, often meticulous analyses of morphological, anatomical, and even chemical features. The classification presented in the current book should be understood as a summary of a life's study of plants and the system that his insights support – the work of a very great botanist that takes into account not only his own meticulous studies but as much of the contemporary information as he was able to assimilate and take into account. Although future classifications will clearly go beyond the stage of development represented here, it is important to be able to benefit from Armen Takhtajan's insights into the features of flowering plants and the ways in which the suites of characteristics they present can be viewed in an evolutionary context.

Takhtajan is a botanist of the 20th century, and the views developed from his vast experience – he is nearly 100 years of age – richly deserve publication. Younger research workers and students will appreciate the opportunity to be informed of Armen Takhtajan's ideas, and to be acquainted with the wide ranging data on where they are based. This book naturally draws extensively on the rich Russian literature in the field of plant classification, and many readers will find ideas expressed that are of interest to them. The new insights and ideas in the book likewise will inspire new levels of thinking about the relationships between the families of angiosperms and their evolutionary history, including the convergent and parallel evolution of particular features.

Peter Stevens, one of the reviewers, has pointed out that additional evidence has accumulated regarding the relationships of many angiosperm families, and that comparisons of their DNA have revealed unsuspected similarities. Armen Takhtajan has taken into consideration some, but not all, of this evidence, and future treatments will result in major revisions of some of the concepts presented here. Importantly, he brings to our attention the pertinent Russian botanical literature, which is poorly known in the West. This book presents challenging new ideas and insights clearly, and it is very important to publish for its demonstrated value as the final work of a great scientist, representing the culmination of his experience and study.

It is also important to mention that this book summarizes the ideas and understanding of a lifetime of investigation and thought by one of the most able and influential botanists of our time. Considering his age, it will probably be the last one. – Peter H. Raven, President and Director, Missouri Botanical Garden, St. Louis, Missouri, USA.

Short Biography

Armen Takhtajan was born on June 10, 1910 in Shusha, Nagorny Karabakh. He was graduated from the Institute of Subtropical Cultivation in Tbilisi (1929–1932). He got his Ph.D. (candidate of sciences) in Leningrad, 1938; and his Dr. Sci. (Doctor of Sciences) at the Yerevan State University, Armenia in 1943. He worked as Research fellow, at the Natural History Museum in Yerevan (1932–1937), and as Senior Botanist, at the Biological Institute in Armenia (1935–1943). He was also Lecturer and later Professor of Botany at the Yerevan State University (1936–1948); Director of Botanical Institute, Armenian Academy of Sciences (1943–1948); and Professor of Botany, Leningrad State University (1949–1960). He has been a member of the staff at the Komarov Botanical Institute, Russian Academy of Sciences since 1955, first as a Chief of the Laboratory of Palaeobotany (1955–1987) and also as Director of the Institute (1976–1986). Now he is an Advisor of the Komarov Botanical Institute.

He is a full member (Academician) of the Armenian Academy of Sciences and of the Russian Academy of Sciences, foreign associate of the National Academy of Sciences of the United States of America, foreign member of Finnish Academy of Sciences and Letters, German Academy of Naturalists (Leopoldina), Polish Academy of Sciences, Norwegian Academy of Sciences, foreign member of the Linnaean Society of London. For many years he was President of the Soviet Union Botanical Society.

Armen Takhtajan was awarded the A.L. Komarov Award (1969), Russian State Award (1981), the Allerton Medal (1990) and Henry Shaw Medal (1997) for Botany, and he has 20 books and more than 300 scientific papers to his name. He served as editor for many books and series published in Russia, including the Botanisky Zhurnal. As an editor, he read and corrected the entire text line by line. The recent publication of this kind was the 6-volume series "Plant Life," which serves for many high school and university students as a wonderful textbook.

Armen Takhtajan has been and is an individual of outstanding accomplishment and influence on the biological sciences both in Russia and throughout the world.

Preface

This book is a result of my almost half-century study of the morphology and systematics of flowering plants. It continues my work published in several of my previous books, especially "Systema Magnoliophytorum" (1987), published in Russian, and its continuation and expansion – "Diversity and Classification of Flowering Plants" (1997), published in English. However, when writing this book of mine, I have inevitably analyzed and considered the matter again and in many cases considerably changed the former conclusions. Here I present an essentially new version of my system.

My new revision of the system is based on a great amount of new information published in the last decade as well as on discussions and consultations with many of my colleagues. New taxonomic revisions of large groups, including families, and new comparative-morphological studies of various groups, including an increasing number of micromorphological (ultrastructural) studies, were especially important for phylogenetic inferences.

No less important was a rapidly increasing number of molecular taxonomic studies, provided that they did not contradict the totality of other evidence.

I would like to thank Dr. Peter Stevens and Dr. James Reveal for reading the manuscript. Both of them made valuable suggestions that were very helpful during the preparation of the final version of the book.

My work on this book would be impossible without the great help of Tatiana Wielgorskaya. She has helped me not only in all kinds of computer work but also in the search of literature.

November, 2008 Armen Takhtajan

Introduction

Main Vectors of Evolution in Flowering Plants (The Criteria Used in Evaluating the Relative Degree of Their Advancement)

The vegetative characters there are many easily reversible characters, such as growth habit, arrangement, size and form of leaves, but there are also many trends which either can be reversible with great difficulty or are completely irreversible. In general, vegetative organs are characterized by more reversibility than reproductive organs. However, even the most reversible characters usually reveal more or less definite evolutionary trends.

Growth habit: The most primitive magnoliophytes are woody plants, and the herbaceous growth habit is always secondary (Jeffrey 1899, 1917; Hallier 1905, 1912; Sinnott and Bailey 1914, and many subsequent authors including Eames 1961, and Stebbins 1974). The evolution of flowering plants most probably begins with small, relatively weakly branched woody forms. According to Hallier (1912) the early angiosperms were small trees with a weak crown of relatively few thick branches, like the fossil bennettitaceous genus *Wielandiella* or some living cycads. Stebbins (1974), on the other hand, visualizes the earliest angiosperms as low-growing shrubby plants, having a continuous ring of secondary vascular tissue, and no single well-developed trunk. Amongst the living primitive flowering plants there are both trees (the majority) and shrubs (*Eupomatia laurina*, for example, is a shrubby plant with several trunks). It is difficult to say whether the earliest magnoliophytes were small trees or shrubs. The only thing we can say is that they were small woody plants, which occupied only a modest and insignificant position in the Early Cretaceous vegetation. Big stately trees of tropical rain forest are derived, having originated from primitive, small, woody angiosperms. Trees with numerous slender branches evolved from sparingly branched trees. Deciduous woody plants evolved from evergreen ones.

The evolutionary trend from woody plants to herbs is not irreversible. In some phyletically distant taxa of flowering plants the reverse process of the transformation of herbaceous plants into arborescent plants took place, for example, in Ranunculaceae, Berberidaceae, Papaveraceae, Phytolaccaceae, Nyctaginaceae, Chenopodiaceae, Polygonaceae, Cucurbitaceae, Campanulaceae-Lobelioideae, Asteraceae, and many liliopsids (including Agavaceae, Dracaenaceae, Philesiaceae, Smilacaceae, Poaceae – Bambusoideae, Arecaceae, Pandanaceae). But usually these secondary arborescent plants, especially arborescent liliopsids, strikingly differ from the primary woody plants. As Stebbins (1974: 150) aptly remarks, "Palms and bamboos are as different from primitive preangio- spermous shrubs and trees as whales and seals are from fishes".

Branching: There are two main morphological types of branching in flowering plants – monopodial and sympodial. Both these types are met in many families and even within one and the same genus and change from one to the other with great ease. This makes the determination of the main direction of evolution of the branching in flowering plants somewhat difficult. The study of the most archaic extant magnoliophytes indicates that perhaps the original type has a combination of monopodial and sympodial branching – well expressed, for example, in *Magnolia*. The vegetative branches of *Magnolia* are monopodial, but the short branches carrying the terminal flowers develop in a strictly sympodial manner, and the apparently simple axis of such a branch is in fact a sympode of a certain number of shoots of an ascending series. The sympodial nature of a reproductive branch is determined by the fact that each of the component axes ends in a terminal flower, arresting its subsequent development. So the sympodial nature here is primary and not secondary as in the evolution of the vegetative branches. Monopodial branching is characteristic of many trees of the humid subtropical and particularly the humid tropical forest (Serebryakov 1955: 75). This is explained by the fact that the conditions of humid tropical and subtropical climates help in prolonged preservation of the terminal meristems of the stems so that the growth of the vegetative shoot occurs all the time through a continuously operating apical meristem, which leads to a vigorous development of the main axis and to a greater or lesser suppression of the lateral shoots. But in the extratropical regions as well as in the mountains of tropics and under the conditions of a dry tropical climate, the sympodial branching arises out of monopodial (Takhtajan 1948, 1964; Serebryakov 1955). The growth of the annual shoots ends in the disappearance of their terminal bud, which inevitably leads to the development of a large number of lateral buds and the formation of a larger number of lateral shoots. The main axis ceases to hinder the development of the lateral shoots, the intensity of branching is amplified, and the crown becomes denser. The process of the origin of sympodial branching out of the monopodial type is realized in the most diverse phyletic lines and at various levels of specialization. Sympodial branching is very widespread in the herbaceous angiosperms. It is observed in almost all monocotyledons, where it is a direct result of the reduction of the cambium (Holttum 1955), and quite typical of the herbaceous dicotyledons as well. The biological advantages of sympodial branching is emphasized by Zhukovsky (1964: 125), who thinks that the successive dying off of the terminal buds should be considered as a very useful adaptation. According to Serebryakov, sympodial renewal was in addition a vigorous tool for intensifying vegetative reproduction (1952: 278). Lastly, in his opinion, the dying off of the shoot apex or the terminal buds under sympodial growth provides for an earlier "maturing" of the shoots, their transition to the state of dormancy, and an intensification of the hardiness of the trees and shrubs.

Leaves and leaf arrangement: The leaves of primitive living flowering plants are mostly simple, entire, pinnately nerved, coriaceous and glabrous. This indicates that the simple entire leaf with pinnate venation is primitive (Parkin 1953; Takhtajan 1959, 1964; Eames 1961; Cronquist 1968; Hickey 1971; Stebbins 1974), and it is very likely that the leaves of the earliest angiosperms were more or less similar. But this is not certain – they may have been of a still more primitive type. In Stebbins's (1974: 331) opinion, "The leaves of the original angiosperms are believed to have been elliptical, obovate, or spatulate in outline, and tapered at the base to an indistinct petiole."

Simple, pinnately-nerved leaves are ancestral to pinnately-lobed, pinnatifid, and pinnatisect leaves with pinnate venation. Both pinnatisect and palmatisect leaves gave rise to compound leaves – pinnately compound in one case and palmately compound in the other. These trends in leaf evolution are reversible. Such reversal is well documented in some instances, such as the genera *Berberis* and *Citrus*.

The most primitive type of venation is pinnate venation with brochidodromous secondaries, especially leaves which are characterized by the general irregularity of their venation, expressed in such features as the highly irregular size and shape of areas between secondary veins, the irregularly ramifying courses and poor differentiation of the tertiary and higher vein orders (Hickey 1971; Hickey and Doyle 1972; Doyle and Hickey 1976). Among the living flowering plants this primitive type occurs in some members of Winteraceae, Canellaceae, Magnoliaceae, and Himantandraceae. All other types of pinnate venation are derived.

Palmate (actinodromous) venation evolved from pinnate venation, and in its turn gave rise to various types of campylodromous and acrodromous venation. The most advanced type is parallel (parallelodromous), which is characteristic for the majority of liliopsids and for some magnoliopsids. But parallel venation is not a climax type, and in some taxa of liliopsids, such as the Smilacaceae, Dioscoreaceae and Stemonaceae, it gave rise to reticulate venation with free vein-endings.

Among the various types of leaf vernation (ptyxis) the most primitive is conduplicate vernation with lamina folded once adaxially along midrib (Takhtajan 1948), which is characteristic for some primitive taxa including Magnoliales.

In the evolution of leaf arrangement (phyllotaxy), the most primitive is alternate arrangement. Both the opposite and verticillate types are derived from the alternate arrangement. But as Cronquist (1968) points out, the origin of opposite leaves from alternate leaves is not immutable and is subject to reversal. In his opinion, among the family Asteraceae it is perfectly clear that opposite leaves are primitive and alternate leaves are advanced. As regards verticillate leaves, they are probably less reversible.

Stomatal apparatus: The stomatal apparatus of flowering plants is characterized by diversity of structure. Stomata may be surrounded either by ordinary epidermal cells (the anomocytic type characteristic of Ranunculaceae, Berberidaceae, Liliaceae, and many other families), or by two or more subsidiary cells morphologically distinct from the other epidermal cells (paracytic, tetracytic, anisocytic, diacytic, actinocytic, and other types).

There are two basic types of development of stomata with subsidiary cells – perigenous and mesogenous. There is also an intermediate mesoperigenous (Pant 1965). In the evolution of seed plants the perigenous type preceded the mesogenous type (Florin 1933, 1958), but the flowering plants most probably began with the mesogenous type. This is supported by the occurrence of the mesogenous (and mesoperigenous) type in such archaic families as Degeneriaceae, Himantandraceae, Magnoliaceae, Eupomatiaceae, Annonaceae, Canellaceae, Winteraceae, and Illiciaceae. Moreover, the stomatal apparatus of the mesogenous and mesoperigenous Magnoliidae is of the paracytic type (accompanied on either side by one or more subsidiary cells parallel to the long axis of the pore and guard cells). Mesogeneous paracytic stomata are the most primitive and initial type of the magnoliophyte stomatal apparatus (Takhtajan 1966, 1969; Baranova 1972, 1985, 1987a, b). All other types of stomata, including the anomocytic type which is devoid of subsidiary cells, are derived.

As regards stomatal ontogeny, most of the morphological types of stomatal complexes with subsidiary cells are in fact ontogenetically heterogenous (Baranova 1987a).

Nodal structure: It is generally agreed that in gymnosperms the unilacunar node structure is more primitive, and the multilacunar nodes of cycads and *Gnetum* are derived. But the evolutionary trend in nodal structure of angiosperms is much more debatable. In addition to unilacunar and multilacunar nodal types in flowering plants there is a third type, the trilacunar, unknown in gymnosperms. The presence of three different types of nodal structures complicates the situation and makes more difficult the ascertainment of the evolutionary trends in angiosperms.

At different times and by different authors each of these three types has been accepted as the most primitive and basic nodal structure in angiosperms. The study of all the available data accumulated in literature brings me to the conclusion, that Sinnott's (1914) theory of the primitiveness of the trilacunar type, based on the extensive reconnaissance of 164 families of dicotyledons, is nearest to the truth. It also much better corresponds to the widely accepted theory of the primitiveness of the magnolialian stock. The presence of trilacunar nodes in such an archaic family as the Winteraceae, as well as in Himantandraceae, Annonaceae, Canellaceae, Myristicaceae, Tetracentraceae, Cercidiphyllaceae, and in the orders Ranunculales, Hamamelidales, Caryophyllales, Dilleniales and Violales is very suggestive. But some members of the Magnoliales are penta- or multilacunar. Such an extremely primitive genus as *Degeneria* has pentalacunar nodes (Swamy 1949; Benzing 1967) and in the genus *Eupomatia*, which in its vegetative anatomy is one of the most primitive among the vessel-bearing angiosperms, the nodes are multilacunar (Eames 1961; Benzing 1967). The nodal structure of the Magnoliaceae is usually also multilacunar (6–17 gaps), except in the relatively primitive genus *Michelia,* which is tri-pentalacunar (see Ozenda 1949). This distribution of tri-, penta- and multilacunar types most probably indicates that tri- and pentalacunar nodes are more primitive and multilacunar nodes are derived. But it is much more difficult to decide which of these two types, trilacunar and pentalacunar, is the basic one. In my opinion it is quite possible that the earliest angiosperms were tri-pentalacunar, like the living genus *Michelia*.

The unilacunar nodal structure, which Sinnott (1914) considered as having arisen by reduction from the trilacunar, is according to Marsden and Bailey (1955) the most primitive and basic nodal type in all seed plants, including angiosperms. They considered the primitive node to be the unilacunar type with two discrete leaf traces. This new concept of nodal evolution was based on the fact that the unilacunar node with two distinct traces is characteristic not only for some ferns and gymnosperms (as was well-known earlier), but also occurs in certain dicots (Laurales, certain Verbenaceae, Lamiaceae and Solanaceae). Also it is repeatedly found in the cotyledonary node of various flowering plants. Bailey (1956) concluded that we could no longer think of the unilacunar node of dicotyledons as having arisen by reduction from the trilacunar; in his opinion, "during early stages of the evolution and diversification of the dicotyledons, or of their ancestors, certain of the plants developed trilacunar nodes, whereas others retained the primitive unilacunar structure." Canright (1955), Eames (1961), Fahn (1974) and several other anatomists have even more strongly favored the primitiveness of the unilacunar node with two traces, which they consider the basic type in the evolution of angiosperm nodal structure. But there are also objections. Thus Benzing (1967) has pointed out that the occurrence of plants with two-trace unilacunar nodal structure proposed as primitive by Marsden and Bailey (1955) is limited to a few families characterized by derived decussate phyllotaxy and many specialized floral characters. He also correctly points out that the anatomy of cotyledonary nodes does not necessarily reflect ancestral conditions in the mature stem. "The unique seedling morphology and decussate insertion of the cotyledons make this unlikely," says Benzing. He comes to the conclusion that either the unilacunar node with one trace or the trilacunar node with three traces is more likely to be primitive in the angiosperms than the unilacunar node with two traces. Bierhorst (1971) is also very skeptical about the theory of primitiveness of two-traces unilacunar type and says that "the issue is far from settled".

In my opinion neither of the two types of unilacunar nodes is primitive and basic in flowering plants. The unilacunar nodal structure is characteristic mostly for the advanced taxa. In the Magnolianae the unilacunar node is present only in orders Laurales and Illiciales, which are considerably more advanced than the Magnoliales. The only unilacunar members of the whole subclass Hamamelididae are *Euptelea* and *Casuarina*. On the other hand it is significant that the unilacunar node is characteristic for such advanced orders as Ericales, Ebenales, Primulales, Myrtales, Polygalales, Gentianales, Polemoniales, Scrophulariales, Lamiales and Campanulales. Among the gamopetalous dicotyledons only Plan-taginaceae and Asteraceae are exceptions. In some orders, such as Celastrales and Santalales, it is possible to follow the transition from the trilacunar to the unilacunar type, which occurs along with general specialization of the vegetative organs. It is particularly well shown in the family Icacinaceae (see Bailey and Howard 1941). One may see the same evolutionary trend in the series Dilleniales – Theales. All these facts lead to the conclusion that the unilacunar type of nodal structure is secondary in flowering plants, having originated from the basic tri-pentalacunar type.

Wood anatomy: One of the most reliable and well documented evolutionary trends thus far revealed among the flowering plants is the derivation of vessel members (elements) from tracheids with scalariform bordered pits. And what is more, "this particular phylogenetic sequence clearly is a unidirectional and irreversible one, and cannot be read in reverse" (Bailey 1956: 271). Vessels evolved entirely independently in diverse lines of evolution of angiosperms. They originated independently not only in dicotyledons and monocotyledons, but even independently in some major taxa of these two classes. But in all the cases the evolution of vessels was unidirectional and irreversible from vessel members with scalariform perforations to vessel members with simple perforations. With this main trend in the evolution of vessels are more or less correlated (but not always synchronized) other trends in specialization of vessel members (see any modern textbook of plant anatomy).

As comparative anatomical studies of the phloem from Hemenway (1913) onwards have shown, the sieve elements of primitive angiosperms are long and narrow with very oblique end wall, as, for example, in *Drimys*. This is in agreement with the finding that the sieve elements in ferns and gymnosperms are long and pointed with no pronounced differences between the side and end walls. The absence of companion cells in the phloem of gymnosperms and ferns gives us good reason to suspect that the earliest angiosperms were also devoid of them.

Wood parenchyma (occurring as longitudinal parenchyma strands) in early angiosperms was either very scanty (Hallier 1908, 1912) and apotracheal (independent of the tracheal elements in distribution) or, more probably, was absent. Carlquist (1962) considers absence of parenchyma as primitive. The most primitive type of ray tissue system is a heterogenous ray system which consists of two kinds of rays: one heterocellular-multiseriate composed of elongated or nearly isodiametric cells in the multiseriate part and upright cells in the uniseriate marginal parts which are longer than the multiseriate part; the other homocellular-uniseriate composed entirely of upright (vertically elongated) or of upright and square cells. Such rays are met with in many living angiosperms with relatively primitive wood (Kribs 1935; Metcalf and Chalk 1950; Eames 1961; Esau 1965).

Extensive comparative anatomical studies have revealed trends in evolution of xylem fibers (from tracheids, through fiber-tracheids, to libriform fibers), in radial and axial parenchyma, sieve tubes, plastids in sieve elements, and other structures. All these trends are important as criteria which one can use in evaluating the relative degree of specialization of the conducting system.

Inflorescences: Among living flowering plants solitary flowers, both terminal and axillary, probably represent the surviving members of reduced inflorescences (Eames 1961; Stebbins 1974). In the Winteraceae, for example, the solitary terminal flower of *Zygogynum* represents "the end of a reduction series" (Bailey and Nast 1945).

The various forms of inflorescence are divided into two major categories – cymose, determinate or "closed" and racemose, indeterminate or "open." The boundary between these two basic groups is not sharp and there are many intermediate and combined forms. Nevertheless for phylogenetic purposes this traditional classification is much more suitable than Troll's (1928) typological classification which is based on Aristotelian logic and the tenets of methodological essentialism rooted in Plato's idealistic philosophy.

Of two basic groups of inflorescences, the cymose inflorescence is more primitive and the racemose inflorescence is derived (Parkin 1914). Weberling (1965) also comes to the conclusion tat in general the polytelic type is more highly evolved than and perhaps derived from the monotelic type. The most primitive form of cymose inflorescence is probably a simple, few-flowered terminal leafy cyme (Takhtajan 1948, 1959, 1964; Stebbins 1974). Such a leafy cyme one can see for example in *Paeonia delavayi* or in some primitive ranunculaceous genera. In various evolutionary lines the primitive leafy cyme has given rise to more specialized forms.

By means of repeated branching the simple cyme gives rise to compound cymes – pleiochasium, compound dichasium, and cymose panicle. In some evolutionary lines the compound cymes undergo drastic transformations and give rise to very specialized types such as the capitate inflorescences of some species of *Cornus*, of Dipsacaceae and of certain Valerianaceae and Rubiaceae and especially the inflorescences of Urticaceae, Moraceae, Betulaceae, Fagaceae and Leitneriaceae.

In some genera and even families, for example in Caryophyllaceae, the compound monochasium results by the suppression of one of the two branches of each ramification of the compound cyme.

From the compound cyme evolved the raceme, which is the most primitive form of the racemose inflorescence. The transitions from pleiochasium to raceme may be observed in the genera *Aconitum* and *Thalictrum* or in the Papaveraceae-Fumarioideae and in the Campanulaceae (Parkin 1914; Takhtajan 1948). The simple raceme gives rise to the compound raceme, the spike, and the umbel. The umbel in its turn gives rise to a still more specialized form of racemose inflorescence – the capitulum s.str. or calathidium. It characterizes certain Apiaceae, as *Eryngium* and *Sanicula*. The ancestry of the capitulum in the Calyceraceae and Asteraceae is more debatable, and no opinion is offered here.

The diversity of the types of inflorescences is strengthened by the presence of different and sometimes very complex combinations of their basic types. Examples of such secondary or composite inflorescences (inflorescentiae compositae) are compound umbels of Apiaceae or catkinlike compound inflorescences of *Betula*, *Alnus*, or *Corylus*.

It is most interesting that frequently the ways and trends of evolution of secondary inflorescences repeat those of primary inflorescences. In many cases, the secondary inflorescences imitate the architecture of the primary one. Such are, for example, the catkinlike inflorescences of Betulaceae, which are so similar to aments of *Salix*. Even more remarkable are the secondary capitula of some Asteraceae, for example those of *Echinops*, which are externally almost indistinguishable from the simple (elementary) capitula. It is also interesting that there is a remarkable parallelism in evolution of composite and elementary capitula of Asteraceae.

General floral structure: The most primitive and archaic flowers, like those of *Degeneria* and Winteraceae, are of moderate size with a moderately elongated receptacle. Stebbins (1974) concluded that the original angiosperms had flowers of moderate size, which is in harmony with the hypothesis that they were small woody plants inhabiting pioneer habitats that were exposed to seasonal drought. It is also in harmony with my hypothesis of the neotenous origin of flowering plants, according to which they arose under environmental stress, probably as a result of adaptation to moderate seasonal drought on rocky, mountain slopes in an area with monsoon climate (Takhtajan 1976). Under such conditions flowers of moderate (or even less than moderate) size would be better adapted than the large flowers postulated by Hallier (1912) and Parkin (1914).

Large flowers, like those of some Magnoliaceae and Nymphaeaceae, of Peruvian ranunculaceous *Laccopetalum giganteum*, and especially very large flowers (*Rafflesia arnoldii*) are of secondary origin and evolved in response to selection pressure for different methods of pollination. Small and especially very small flowers are also derived and their origin is usually correlated either with the specialization of inflorescences or with the reduction of the whole plant.

The most primitive flowers have a more or less indefinite and variable number (but not necessarily a large number) of separate parts arranged spirally upon a moderately elongated floral axis. The progressive shortening of the floral axis brings floral parts closer together and gives rise to the gradual transition from spiral to cyclic arrangement and to the fixation of the number of parts. At its earlier evolutionary stages this progressive shortening is reversible, and in some relatively archaic taxa, such as Magnoliaceae (especially *Magnolia pterocarpa*), *Schisandra* or *Myosurus*, the elongated receptacle is of secondary origin. Another result of shortening of the floral axis is a gradual fusion of floral parts – their connation and adnation. Partial or overall reduction of the flower occurs in many evolutionary lines.

Although in the original flowering plants there probably was no corolla yet (Hallier 1912) and the perianth consisted entirely of modified bracts (sepals), in modern angiosperms the presence of petals is a primitive condition and their absence is derived. Petals are a later evolutionary acquisition. It is almost generally agreed that they are of dual origin – in some groups, such as Magnoliales, Illiciales, and Paeoniales, they are of bract origin, whereas in the majority of flowering plants, including Nymphaeales, Ranunculales, Papaverales, Caryophyllales and Alismatales, they are modified stamens. To designate these two types of petals Kozo-Poljanski (1922) aptly coined the terms "bracteopetals" and "andropetals". Bracteopetals occur in more archaic taxa and evidently appeared earlier, they also connected with generally more primitive pollination mechanisms and with less specialized pollinators. Andropetals, on the contrary, are usually connected with more advanced types of pollination.

Among the living angiosperms there are probably no primary apetalous plants. Flowers with vestigial petals, with petals transformed into glands, or devoid of petals are secondary, derived from flowers with normally developed and functioning petals.

Androecium: Comparative studies of the stamens of flowering plants leads to the conclusion that within living angiosperms the most primitive type of stamen is a broad, laminar, three-veined organ not differentiated into filament and connective, and produced beyond the microsporangia; it develops four slender elongated microsporangia embedded in its abaxial or adaxial surface between the lateral veins and the midvein (see especially Bailey and Smith 1942; Ozenda 1949, 1952; Canright 1952; Moseley 1958; Eames 1961; Foster and Gifford 1974). Canright (1952) regards the stamen of *Degeneria,* as "the closest of all *known* types to a primitive angiosperm stamen." It is important to note, however, that in *Degeneria*, *Galbulimima*, *Lactoris*, Annonaceae, *Belliolum* (Winteraceae) and *Liriodendron* the microsporangia occupy the abaxial surface (and therefore the stamens are extrorse), whereas in the Magnoliaceae (except *Liriodendron),* Austrobaileyaceae and Nymphaeaceae they are situated on the adaxial surface (the stamens being introrse). In my opinion both the abaxial and adaxial position have been derived from a common ancestral type, which could only have been the marginal. Thus we must come to the logically inescapable conclusion that in the ancestors of living Magnoliales the microsporangia were marginally situated on the microsporophylls (Takhtajan 1948, 1959, 1964, 1969). Were the original microsporophylls of angiosperms flattened organs, entire or pinnate, or were they branched three-dimensional structures? In my opinion the stamens of the earliest angiosperms or of their immediate ancestor were leaf-like pinnate microsporophylls with marginally situated microsporangia, which in their turn originated from the branched and three-dimensional structures of the more remote ancestors.

Many authors, among them Ozenda (1952), Canright (1952), Moseley (1958), Eames (1961) and Cronquist (1968) consider that the immersion of the microsporangia in the tissue of the stamen is a primitive feature. In *Degeneria* and *Galbulimima* the microsporangia are deeply sunk in the tissue of the stamen, as they are in the Magnoliaceae (except *Liriodendron*) and *Victoria amazonica*. This immersion of the microsporangia is probably a result of the neotenous origins of stamens and the flower as a whole (Takhtajan 1976).

All the accumulated evidence indicates that the stamen is not a surviving solitary branch of the ancestral compound organ, but an individual organ which is homologous to an entire microsporophyll. As regards the stamen fascicles and the branched system like that of *Ricinus,* these are of secondary origin and are not homologous to the ancestral compound microsporangiate organ (see Eames 1961).

During evolution changed not only the number and arrangement of stamens but also the mode of their sequence of ontogenetic development (Payer 1857; Corner 1946). The initial and most widespread type of development is the centripetal (acropetal), when the development of androecium follows the development of the perianth in the normal sequence, spiral or cyclic. The first to develop in this case are the outermost (lowermost) stamens and then, successively, the inner ones. This type is characteristic for all spiral androecia (like those of Magnoliaceae, Annonaceae, Nymphaeaceae, Nelumbonaceae, Ranunculaceae), for cyclic oligomerous androecia, such as those of the Papaveraceae, Rosaceae, Fabaceae – Mimosoideaea, or Myrtaceae. In the centrifugal androecium, there is a break between the order of development of perianth and androecium caused by the intercalation of new stamens. The centrifugal development arose from the centripetal (Corner 1946; Ronse Decraene and Smets 1987). It is characteristic of the Glau-cidiaceae, Paeoniaceae, probably some Phyto-laccaceae with numerous stamens, Aizoaceae, Cactaceae, Dilleniaceae, Actinidiaceae, Theaceae, Clusiaceae, Lecythidaceae, many Violales, some Capparaceae, Bixaceae, Colchlospermaceae, Cistaceae, Tiliaceae, Bombacaceae, Malvaceae, the genus *Lagerstoemia* (Lythraceae), Punicaceae, Loasaceae, Limnocharitaceae, and some other taxa. In some families such as Ochnaceae, Begoniaceae, Lythaceae, and Loasaceae, there are both types of stamen development. Therefore, the distinction between centrifugal and centripetal types of development is by no means clear-cut and there are some transitional forms (Sattler 1972; Philipson 1975; Sattler and Pauzé 1978; Ronse Decraene and Smets 1987). According to Leins (1964, 1975), the difference between centripetal and centrifugal development depends on the shape of the receptacle: a concave receptacle would give rise to a centripetal development, while on a convex receptacle only a centrifugal development would be possible. But this is not a general rule (Hiepko 1964; Mayr 1969; Ronse Decraene and Smets 1987).

Microsporangia, microsporogenesis and pollen grains: Stamens most commonly contain four nicrosporangia arranged in two pairs. Only in some taxa, such as Circaeasteraceae, Epacridaceae, certain Diapensiaceae, Bombacaceae, Malvaceae, Adoxaceae, Philydraceae, Restionaceae, the stamens contain only two microsporangia. Very rarely, as in *Arceuthobium* (Viscaceae) there is only one microsporangium. Multisporangiate stamens of some taxa, e.g., in Rhizophoraceae, result from partition of the sporogeneous tissue by sterile plates.

There are two structural and functional types of tapeta, distinguished on the basis of cell behavior during microsporogenesis: the secretory or glandular tapetum, the cells of which remain intact and persist in situ but, after meiosis at the tetrad stage, or at the beginning of the free microspore stage, and sometimes as late as at the stage of two-celled pollen grains, become disorganized and obliterated, and the plasmodial or amoeboid tapetum, characterized by the breakdown of the cell walls before meiosis and protrusion of the protoplasts into the locule and fusion to form a multinucleate plasmodium. Besides, unusual cyclic-invasive type of tapetum has been found lately (Rowley et al. 1992; Gabarayeva and El-Ghazaly 1997). The overwhelming majority of families of flowering plants, including the majority of the most archaic taxa, is characterized by the secretory tapetum. In additions, some primitive characters are correlated with a secretory tapetum (Sporne 1973; Pacini et al. 1985). On the other hand, the plasmodial type usually occurs in relatively more advanced groups. As Schürhoff (1926) pointed out, the presence of plasmodial tapetum is closely correlated with an advanced character such as tricelled pollen grains.

The ways of dehiscence of the mature anther has also some systematic and evolutionary significance. The commonest and the most primitive dehiscence is the longitudinal dehiscence along the fissure (stomium), situated between a pair of microsporangia. The longitudinal dehiscence is of two types: by one simple longitudinal slit or by two longitudinal valves. The second type is characterized by additional, transverse slits usually at both ends of the longitudinal slit, which results in two windowlike lateral valves (see Endress and Hufford 1989; Hufford and Endress 1989). Whereas the dehiscence by simple longitudinal slit is very common, the second type is characteristic of many Magnoliidae and Hamamelididae with more or less massive anthers and evidently derived from the first type. "Possibly, only the predisposition for easily developing valvate dehiscence was present in the original angiosperm stamen that dehisced via simple longitudinal slits. This predisposition would have been lost in more advanced angiosperms" (Endress and Hufford 1989: 79). More specialized is a valvate dehiscence in Laurales and Berberidaceae, which typically arises by the opening of the thecal wall outward producing apically hinged flaps that lift upward at dehiscence. One of the most advanced types of dehiscence is the poricidal dehiscence, when pollen is released from a small opening situated at one end (distal or proximal). Examples of the latter are: Ochnaceae, Ericaceae, Myrsinaceae, some Fabaceae, the majority of Melastomaceae, Tremandraceae, Solanaceae. There are also other specialized modes of dehiscence including transverse dehiscence (e.g., *Alchemilla*, *Hibiscus*, *Euphorbia*, *Chrysosplenium*).

The microspore tetrads are formed by two patterns determined by the mechanism of cytokenesis in microspore mother cells. In the successive type, the developing cell plate is formed at the end of meiosis I, dividing the microsporocyte into two cells; in each of these two cells, the second meiotic division takes place, followed again by centrifugal formation of cell plates. In the simultaneous type, on the other hand, no wall is formed after meiosis I; division occurs by centripetally advancing constriction furrows, which usually first appear after the second meiotic division, meet in the center, and divide the mother cell into four parts. The constriction furrows originate at the surface of the mother cell and develop inwardly, resulting in the formation of walls that divide the microsporocyte into four microspores.

It is difficult to say which of the two types of microsporogenesis is more primitive. Although some authors (including Schürhoff 1926 and Davis 1966) consider the successive type as the more primitive, there is no definite correlation between this type and archaic Magnoliidae and Ranunculidae. The majority of Magnoliidae and Ranunculidae are characterized by simultaneous microsporogenesis.

The pollen wall, as a rule, consists of two main layers – the inner one, called intine, and the outer one, called exine. The exine typically consists of two layers – the inner layer endexine and the outer layer ectexine. Endexine may be found as a continuous layer (sometimes very thick, as in Lauraceae) or only in apertural regions, in some taxa it is absent.

In an overwhelming majority of flowering plants the ectexine is well developed and stratified. The exine structure and ornamentation (sculpturing) is extremely varied and, at the same time, very constant within the taxonomic groups and has a large systematic and evolutionary significance. The ectexine consist of two basic layers – a roof-like outer layer or tectum and an infratectal layer. The latter is of two main types – granular and columellar. Granular structure is characterized by an infratectal layer consisting of more or less densely aggregated, equidimensional granules of sporopollenin. The tectum, which is not always noticeable, is composed of more densely aggregated granules. Doyle et al. (1975: 436) suspect that at least some of the apparently homogenous "atectate" exine of Walker and Skvarla (Walker and Skvarla 1975), revealed in some of the most archaic Magnoliidae such as *Degeneria* and *Eupomatia*, are extreme members of the granular category, with very closely aggregated granules. The predominant type of infratectal structure is columellar, which characterized by radially directed rods of lineary fused sporopollenin granules, the columellae. Comparative studies of the ectexine ultrastructure suggest an evolutionary trend from granular ectexine to incipient rudimentary columellae and from the incipient columellae to fully developed columellar structure. The great majority of flowering plants have tectate columellate pollen (the heads of the columellae extend laterally over the intercolumellar spaces forming tectum). In the most primitive type of collumellar ectexine the tectum is devoid of any kind of holes or perforations (Walker 1974a). The tectate-imperforate (Walker 1974a) or completely tectate ectexine (Hideux and Ferguson 1976) is found in various groups of flowering plants both archaic and advanced. The next evolutionary stage of the tectum structure is the perforate (Walker 1974a, Hideux and Ferguson 1976). In the perforate tectum, the holes or tectal perforations (lumina) are always small (e.g., in some Annonaceae and Myristicaceae) and the columellae are invisible through them. When perforations enlarge so that their diameter becomes greater than the width of the pollen wall between them (muri), e.g., in Winteraceae, Illiciaceae, and Schisandraceae, the exine becomes semitectate (Walker 1974a). For this partial tectum, the visibility of columellae in oblique view through the lumina is characteristic (Hideux and Ferguson 1976). When the tectum is completely lost, e.g., in some Annonaceae, Myristicaceae, and Salicaceae, and there are only free, exposed columellae or their modified derivatives, we have intectate exine (Walker 1974a). The culmination of an evolutionary trend is the origin of the almost exinless pollen with a much expanded and highly structured intine.

Most pollen grains have specially delimited apertures – generally thin-walled areas or openings in the exine which serve as exits through which the pollen tubes usually emerge. The apertures of flowering plants pollen grains are characterized by a great diversity and are of various types. Various types of apertures correspond to different levels of specialization, and the significance of these types is very important in determining the general level of organization of some taxon or other. The apertural arrangement in the angiosperm pollen grains evolved from distal through zonal to global.

As long ago as 1912 Hallier concluded that the most primitive type of pollen grain is characterized "par une seui pore germinal," by which he apparently meant aperture and not a pore in the strict sense of the word. Later it was shown that the most primitive angiosperm pollen grain is a type with one distal germinal furrow (distal colpus or "sulcus") in the sporoderm (Wodehouse 1936; Bailey and Nast 1943; Takhtajan 1948, 1959, 1964; Eames 1961; Cronquist 1968; Doyle 1969; Muller 1970; Sporne 1972; Stebbins 1974; Walker 1974b, 1976a, b; Walker and Doyle 1975; Straka 1975; Meyer 1977). Such monocolpate ("unisulcate") pollen grains still have a continuous aperture membrane devoid of special openings (ora) in the exine for the emergence of the pollen tube. The distal furrow has given rise to a few other types of distal apertures.

In some taxa, there are two parallel, morphologically distal furrows instead of one (dicolpate or "bisulcate" pollen grains) or even three parallel furrows. In some other taxa, including both dicotyledons and monocotyledons, the distal colpus has been transformed into a peculiar three-armed (very rarely four-armed) distal aperture (trichotomocolpate pollen grains). In some primitive angiosperms, including *Eupomatia* and Nymphaeaceae, the distal aperture has changed its polar position and forms one more or less continuous subequatorial or equatorial ring-like or band-like, encircling aperture, or several apertures parallel to each other (zonacolpate or "zonasulculate" pollen grains). Intermediate stages in the evolution of the zonacolpate type may be observed in the pollen of *Nymphaea* (Walker 1974b). More frequently, as a result of complete reduction of the aperture, monocolpate grains give rise to inaperturate ones. In the inaperturate type the whole exine, which is thin, is a kind of global aperture. But the main trend in distal aperture evolution is the transformation of the distal colpus into a distal pore, which is characteristic for many monocotyledons. In monocotyledons monocolpate pollen grains have also given rise to two-polyporate pollen grains, like those in the Alismatales. In some dicotyledons (Chloranthaceae) monocolpate pollen grains give rise to polycolpate pollen, but the main trend of evolution of sporoderm apertures in dicotyledons is from monocolpate to tricolpate and from tricolpate to tricolporate. According to Straka (1963, 1975) and Wilson (1964) the trichotomocolpate aperture, characteristic of some of the pollen of members of the Winteraceae and Canellaceae, represents an intermediate stage between the monocolpate and tricolpate condition. But nobody has seen any intermediate stage between the trichotomocolpate and tricolpate types, and as Cronquist (1968) has pointed out, several families of monocotyledons including the palms, have trichotomous furrows in the pollen of some species, but here this has not led to the typical tricolpate grains so commonly seen in the dicotyledons.

According to Walker (1974b; Walker and Doyle 1975), the tricolpate aperture, as well as distally dicolpate ("disulculate"), polycolpate and forate apertures are derived de novo from inaperturate pollen grains. I agree that all these apertures types originated de novo, but I can not accept their derivation from the inaperturate type. Typical inaperturate pollen grains have a specialized sporoderm with a more or less reduced, thin exine and a usually thick intine. Functions of the aperture are transferred to the whole of the exine which is transformed into a global aperture. The inaperturate sporoderm is a climax type which hardly can give rise to any type of aperturate pollen grain.

In my opinion the tricolpate condition arose not as a result of the gradual transformation of the monocolpate aperture, but rather as a result of evolutionary deviation of the earlier stages of sporoderm development from their previous course (Takhtajan 1948, 1959, 1964). It originated de novo from monocolpate pollen grains. The sporoderm of monocolpate pollen is less specialized than that of the inaperturate type and therefore is more liable to radical changes in the number and position of apertures. In some cases (in the Canellaceae, for example) polycolpate pollen grains have also evolved the same way.

Tricolpate pollen grains have given rise independently in a number of major taxa of flowering plants to polycolpate pollen, as well as to polyrugate, triporate and polyporate (including pantoporate) types.

The next grade of tricolpate and tricolpate-derived pollen is the origin of composite apertures – tricolporate, polycolporate, tripororate, polypororate (including panpororate). The highest stage of the evolution of the pollen grains in dicotyledons is tri-multiaperturate pollen with composite apertures.

Carpels, gynoecium and placentation: The most primitive carpels are unsealed, conduplicate and more or less stipitate structures (resembling young petiolate leaves lying still in the adaxially folded state inside the bud), containing a relatively large number of ovules (Bailey and Swamy 1951; Eames 1961, and many others). Such primitive conduplicate carpels are especially characteristic of such archaic genera as *Tasmannia* and *Degeneria* (Bailey and Nast 1943; Bailey and Swamy 1951) and to a lesser degree of some other primitive taxa including some primitive monocotyledons.

A very important characteristic of the most primitive carpels is the absence of styles, the stigmas being decurrent along the margins of the carpels (Hallier 1912; Takhtajan 1948; Parkin 1955; Eames 1961). Such stigmatic margins (approximated but not fused at the time of pollination) are the prototypes of the stigma. As Kozo-Poljanski (1922: 121) first pointed out in his commentary on Hallier's codex of characters of the primitive angiosperms, "the stigma developed from the sutures." In the course of evolution the primitive decurrent stigma was transformed into a more localized subapical and then apical stigma. As the stigma is localized in the upper part of the carpel, the latter is usually elongated into a style (stylode), which raises the stigma above the fertile portion of the carpel. During earlier evolutionary stages of the development of the style it is conspicuously conduplicate (Bailey and Swamy 1951).

The most primitive taxa of the flowering plants are characterized by an apocarpous gynoecium. But already in the most primitive families a tendency is observed towards a greater or lesser union of carpels, which leads to the formation of the syncarpous (coenocarpous) gynoecium. As a result, forms with more or less syncarpous gynoecia appear even in such families as Winteraceae, Magnoliaceae, Annonaceae, etc. The overwhelming majority of the magnoliophytes has one or another type of syncarpous gynoecium. I distinguish three main types of syncarpous gynoecium: eusyncarpous, paracarpous, and lysicarpous. An eusyncarpous gynoecium emerged independently in many lines of evolution from an apocarpous gynoecium by lateral concrescence of closely connivent carpels. The eusyncarpous gynoecium usually originates from a more advanced cyclic apocarpous gynoecium. The most primitive forms of eusyncarpous gynoecium still have free upper portions of the fertile regions of the carpels. With specialization of the eusyncarpous gynoecium the concrescence extends also to the individual styles, which finally coalesce completely into one compound style with one apical compound stigma. The union of carpels leads also to anatomical changes: with close fusion of carpel margins, the epidermal layers on the surface of contact are lost and the two ventral bundles form a single bundle (Eames 1931).

The paracarpous gynoecium evolved in many lines of dicotyledons as well as in certain groups of monocotyledons. Usually the paracarpous gynoecium denotes a unilocular gynoecium, consisting of several carpels and having parietal or free-central placentation. But I prefer to limit the concept of paracarpous gynoecium to only the form of unilocular syncarpous gynoecium that has a parietal arrangement of ovules (Takhtajan 1942, 1948, 1959, 1980). A paracarpous gynoecium is characterized by unfolded individual carpels. Their margins are disconnected, while the connection of the borders of the adjoining carpels is maintained.

The paracarpous gynoecium is already found among Magnoliales where it is present in *Takhtajania* (Winteraceae), *Isolona* and *Monodora* (Annonaceae) and the whole family Canellaceae. In these cases, as in many others, including Saururaceae, Cactaceae, Alismatales etc., the paracarpous gynoecium evolved directly from the apocarpous one. The possibility of such an origin of the paracarpous gynoecium is based not only on the existence of apocarpous gynoecia with open conduplicate carpels, but also on the well known fact that the carpels in an apocarpous gynoecium begin development as open structures. If a whorl of such open carpels remained so and became coherent, as is presumed by Parkin (1955: 55), the paracarpous gynoecium originated directly from the apocarpous one (see also Cronquist 1968: 101).

In many other cases, e.g. in the genus *Hypericum* and within the superorder Lilianae, the paracarpous gynoecium arises from the primitive type of eusyncarpous gynoecium in which the margins of individual carpels are not fused yet. As a result of unfolding of these unsealed carpels the eusyncarpous gynoecium gives rise to the paracarpous one.

In many cases the placentae in the paracarpous gynoecium grow thick, expand and intrude inside the ovarian cavity where they meet and often coalesce, forming false septa and pseudoaxile placentation, as for example in the family Campanulaceae. Puri (1952) is quite right in inclining to the conviction, that the multilocular character of this type, i.e. which appeared due to the concrescence of the placentae and not the carpellary margins, is more common than was earlier thought. In many cases, e.g. in the family Campanulaceae, the intruded placentae meet in the center of the ovary and coalesce among themselves; as a result the ovary is subdivided into loculi or rather chambers (pseudoloculi). Thus a typical unilocular paracarpous gynoecium gives rise to the multilocular paracarpous one.

In several lines of evolution of dicotyledons, for example in Primulales, the eusyncarpous gynoecium gave rise to a special type of gynoecium with a unilocular ovary which I named lysicarpous (Takhtajan 1942, 1948, 1959). Like the paracarpous gynoecium, the lysicarpous type is also unilocular but it originates in a completely different manner and is characterized by free-central ("columnar") placentation instead of parietal. The unilocular ovary of the lysicarpous gynoecium is due to the disappearance of the septa of the multilocular ovary, which takes place either during ontogeny, as in Portulacaceae and some Caryophyllaceae, or during evolution, as in Primulaceae. In this context, the carpellary sutures themselves remain entire and the ovules continue to be perched on them as earlier (for literature see Puri 1952). Thus the sutural portion of the carpels together with the placentae is transformed into a column freely rising at the center of the locule and not reaching the top of the ovary.

Specialization of the syncarpous gynoecium as well as that of the apocarpous is usually (but not always) accompanied by greater or lesser reduction in the number of carpels and in most cases also by reduction in the number of ovules. An extreme form of reduction in the number of carpels in the syncarpous gynoecium is the so-called pseudomonomerous gynoecium (Eckardt 1937, 1938), where only one of the carpels is fertile. The sterile carpels (or carpel, if the gynoecium is dimerous) in the pseudomonomerous gynoecium attain often such a degree of reduction that their presence can be detected only through an anatomical study of the vascular system and ontogeny. The pseudomonomerous gynoecium is characteristic for such taxa as Eucommiales, Urticales, Casuarinales, a majority of Thymelaeaceae, Gunneraceae, Garryaceae, Valerianaceae, etc.

The main directions of evolution of the gynoecium determine the main trends of evolution of placentation.

The types of placentation in the flowering plants may be classified as follows (see Takhtajan 1942, 1948, 1959, 1964, 1991):

A. Laminar (superficial) placentation. The ovules occupy the side portions of the inner face of the carpel or are scattered over almost the entire surface, rarely occupy only its back side.
   1. Laminar-lateral placentation. The ovules occupy the side portions of the adaxial surface of the carpel between the median and the lateral veins. Examples: *Tasmannia, Degeneria.*
   2. Laminar-diffuse placentation. The ovules are scattered over almost the entire adaxial surface of the carpel. Examples: *Exospermum,* Nymphaeaceae, Butomaceae, Limnocharitaceae.
   3. Laminar-dorsal placentation. The ovules are attached pseudo-medially, occupying the back of the carpel. Examples: *Nelumbo, Ceratophyllum,* Cabombaceae.

B. Submarginal (sutural) placentation. The ovules occupy morphologically sutural areas of the carpel.
   4. Axile placentation. The ovules are attached along the sutures of the closed carpel i.e. in the corner formed by the ventral area of the carpel in an apocarpous or syncarpous gynoecium. Examples: Ranunculaceae, Dilleniaceae, Rosaceae, Liliaceae.
   5. Parietal placentation. The ovules are situated along the sutures in a paracarpous gynoecium or on the intrusive placentae which in their turn are attached to the sutures. Examples: Violales, Capparales, Juncales.
   6. Free-central or columnar placentation. The ovules are situated along the central column of the lysicarpous gynoecium. Examples: Portulacaceae, Myrsinaceae, Primulaceae. The most primitive type of placentation is laminar-lateral

(Takhtajan 1942, 1948, 1959, 1964; Stebbins 1974). It characterizes such archaic genera as *Degeneria* and *Tasmannia* and certain species of the genus *Zygogynum*. The ovules of these plants are rather far away from carpellary margins and are arranged in the space between the median and lateral veins. Such an arrangement of ovules is most probably an initial one in the evolution of angiosperm placentation. Both the laminar-diffuse and the laminar-dorsal types of placentation are derived from the laminar-lateral (Takhtajan 1942, 1964).

In the course of evolution laminar placentation evolved into submarginal. This is the most widespread type of placentation in flowering plants and it is found already in a majority of taxa with an apocarpous gynoecium, as Magnoliaceae, Annonaceae, Ranunculaceae, etc. But the largest variety of forms of submarginal placentation can be found in syncarpous gynoecia. Two basic types of submarginal placentation are the axile and the parietal types. Their origin and evolution is correlated with the origin and evolution of eusyncarpous and paracarpous gynoecia.

Lastly, free-central or columnar placentation is characteristic for the lysicarpous gynoecium.

Ovules: The ovule is a solitary megasporangium surrounded by a protective cover – the integument. In the most primitive Palaeozoic seeds the integument was segmented (as in *Lagenostoma*), lobed (as in *Archaeosperma*, *Eurystoma*, and *Physostoma*) or even consisted of more or less separate elongated structures (as in *Genomosperma*) (completely separate in *G. kidstonii* and partially fused around the very base of the megasporangium in *G. latens* – see Long 1960). These and other facts suggest that the integument evolved from a distal truss of separate structures (sterilized telomes) which once immediately subtended and surrounded the megasporangium, later became fused together, and eventually more or less fused with the megasporangium, which became almost completely enclosed by the integument (except the terminal micropyle) (see Walton 1953; Kozo-Poljanski 1948; Zimmermann 1959; Andrews 1961, 1963; Camp and Hubbard 1963; Long 1966; Pettit 1970). This telomic theory of the origin of the ovule is a modernized version of Margaret Benson's (1904) "synangial hypothesis."

The morphological interpretation of the integument in the magnoliophytes is complicated by the fact that many dicotyledons and a majority of monocotyledons are bitegmic, that is have two integuments. In all probability the outer integument of the angiosperm ovule emerged from the cupule of the ancient gymnospermous ancestor. The cupule is known to have emerged first in the Lyginopteridaceae, but it is not found in these primitive gymnosperms only. In a modified form it was preserved both in several later gymnosperms and in angiosperms. Already Mary Stopes (1905) considered the outer layer of the seed of Cycadaceae or the sarcotesta as a structure homologous to the "outer integument" (i.e. cupule) of *Lagenostoma*. This homology of the "outer integument" and the cupule is still more clearly visible in the Medullosaceae (Takhtajan 1950; Walton 1953). The cupule gave rise not only to the outer layer of the ovular envelope in a number of gymnosperms but also to the outer integument of the magnoliophytes. Some confirmation of this conjecture mentioned by Stebbins (1974: 232) is the fact that in many families of flowering plants – including the relatively archaic groups – the outer and inner integuments of the ovule differ greatly from each other in their morphology and their histological structure. In these forms, the outer integument is thicker than inner one and has specialized epidermal cells, in some cases including stomata. Moreover, the micropyle may be differently shaped in the two integuments. Stebbins mentions also the lobed distal portion of the outer integument in a few genera. Lobed integuments have been observed in Berberidaceae, Juglandaceae, Rosaceae, and Flacourtiaceae (van Heel 1970, 1976). Distal lobing may involve either the outer or the inner integument, or both. "The lobing suggests that the integuments are compound organs," states Bouman (1984: 144). The cupular origin of the outer integument of the angiosperm ovule was suggested by Gaussen (1946), Takhtajan (1950, 1959, 1964), and Walton (1953).

Unitegmic ovules arose from the bitegmic ones in various lines of flowering plants evolution. As the single integument of the sympetalous magnoliopsids (except for Plumbaginales, Primulales, and Cucurbitales) and some choripetalous ones is usually as massive or even more massive than the double, a suggestion was made (Coulter and Chamberlain 1903), that the single massive envelope has a dual character and resulted from the complete fusion of two integuments at the earliest stages of the differentiation of the integumentary primordia. Presumably in many cases the unitegmic ovule resulted from the congenital fusion of both the envelopes, but in certain taxa it was formed due to the abortion of the inner or the outer integument. Thus, in *Filipendula*, *Rubus*, *Rosa*, *Potentilla*, *Fragaria*, *Alchemilla*, and some other Rosaceae, the envelope resulted from the incomplete development of the inner integument; while in *Peperomia*, Hydnoraceae, *Rafflesia*, *Mitrastemon*, *Cytinus*, and some others, it resulted from the underdevelopment of the outer integument. Another pathway of the origin of unitegmy, integumentary shifting, has been described in Ranunculaceae (Bouman and Calis 1977). According to Bouman (1984: 140), integumentary shifting is a complicated ontogenetic process involving (1) a fusion of primordia, in the sense that the initials of the two integuments give rise to a common structure; (2) a shifting of the inner integument; and (3) an arrested growth of the latter. In some families, like the Piperaceae, Ranunculaceae, Ericaceae, Salicaceae, Rosaceae, Fabaceae and others, even quite close genera are often distinguished by the number of integuments. This shows that the unitegmic condition arose from the bitegmic independently and heterochronously in different evolutionary branches of the flowering plants.

In some taxa, as a result of reduction, the ovular envelope disappears completely, and as the result the megasporangium is naked. This is quite typical in the order Santalales, where in many genera and even entire families ovules are ategmic, the integuments completely lacking. In the Balanophorales this process of reduction goes even further and more or less involves the megasporangium wall.

It is also generally accepted that crassinucellate ovules are more primitive and tenuinucellate ovules evolved from crassinucellate by reduction of the megasporangial wall. Thus the most primitive ovules of the flowering plants are bitegmic and crassinucellate and the most advanced ones are unitegmic and tenuinucellate. But these two types of angiosperm ovules are not always strictly exclusive of each other and there are intermediate types–bitegmic-tenuinucellate (e.g. Theaceae and Primulaceae) and unitegmic-wcrassinucellate (e.g. Cornaceae and Araliaceae) (see Philipson 1974, 1977).

There is also a definite evolutionary trend in the form and orientation of the ovule. Anatropous ovules characterize a majority of flowering plants, including Magnoliales, and are presumably the initial type (Netolizky 1926; Takhtajan 1959; Eames 1961; Cronquist 1968, 1988; Corner 1976). The orthotropous type, as well as campylotropous and amphitropous ones, arose from the anatropous ovule.

Pollination: Long ago the idea was expressed that in angiosperm evolution, entomophily preceded anemophily (Henslow 1888; Bessey 1897; Robertson 1904, and others). The initial agents of cross pollination were undoubtedly animals, insects in the beginning and later small birds, bats and some other animals as well. The original pollinators were most probably beetles (Diels 1916; Faegri and van der Pijl 1979). The original attractant in insect pollination was the pollen (Darwin, 1876; Faegri and van der Pijl 1979). But the necessity for pollen economy leads to a course of evolution in which the flower starts producing a cheaper foodstuff, nectar, as its alternative. For the production of nectar special structures are formed as nectaries. They originated independently in the most diverse lines of angiosperm evolution and on a most widely varying morphological basis. With the emergence of nectaries the plant gets an opportunity for producing pollen in more limited quantities and using it only for transport to other flowers. But the less the pollen production, the more effective should be the utilization of both the pollen and the pollinators. This inevitably leads to the perfection of pollination mechanisms.

In some evolutionary lines of flowering plants a transition takes place from entomophily to anemophily and more rarely to hydrophily. Anemophily arose from entomophily in completely different lines of evolution of both dicotyledons and monocotyledons. As Cronquist (1968: 97) says "Wind-pollination and insect-pollination are not necessarily mutually exclusive. The change from one to the other can take place gradually, without any sudden jumps, especially if the adaptation to insect-pollination is generalized and does not involve complex or unusual structure associated with a particular kind of pollinator".

In all the known cases of anemophily the more primitive entomophilous relatives of anemophilous forms have relatively "generalized" entomophilous flowers. The evolutionary trend from entomophily to anemophily is reversible and in a number of cases there is a return to entomophily in groups earlier adapted to wind pollination (e.g. *Ficus*, *Castanea*). Again the secondary entomophily evolves from the less specialized types of anemophily.

Self-pollination emerged only as a secondary phenomenon, and it is a sort of blind alley for the further evolution "and rarely if ever contributes to major evolutionary trends" (Stebbins 1974). Morphological and phylogenetic analysis indicates that self-pollinating taxa emerged in all cases from cross-pollinating ancestors.

The evolution of pollination was of exceptionally great importance in angiosperm evolution. It had a decisive role in the evolution of flowers and inflorescences and determined many important directions in the evolution of flowering plants.

Gametophytes and fertilization: In the course of evolution both the male and female gametophytes of flowering plants reached a very high degree of simplification and specialization. Gametogenesis occurs in them at such an early stage of an extremely abbreviated ontogeny of the gametophyte that gametangia cannot even be formed, and the gametes are formed without them. Moreover, the development of the gametes themselves is also cut short, and they became extremely simplified. Due to a sharp abbreviation and acceleration of their ontogeny, the gametophytes of angiosperms completely lost their gametangia. As I have suggested in my previous works (beginning with 1948) these drastic changes in the gametophyte structure and development resulted from neoteny and subsequent specialization (see Takhtajan 1976, 1983, 1991).

The entire male gametophyte of the flowering plants consists only of two cells – a small generative cell and a large tube cell ("vegetative" cell). It has neither the prothallial cells, nor the stalk cell ("dislocator" according to terminology of Goebel 1933) and the true spermatogenous cell ("body cell"). The function of the spermatogenous cells has been transferred to the generative cell, which divides to form two nonmotile male gametes, and the function of the stalk cell became unnecessary. Thus the magnoliophytes male gametophyte reached the climax of simplification and miniaturization, which precluded any further major structural changes.

In the majority of flowering plants, including the primitive taxa, the pollen is released from the anther in the two-celled stage in the development of the gametophyte. But in many other flowering plants, including some advanced taxa, the generative cells divide before the pollen grain is shed and the male gametophyte is therefore three-celled. The two-celled condition is primitive and the three-celled type is derived and originated independently in many lines of angiosperm evolution (see Brewbaker 1967; Cronquist 1968; Stebbins 1974).

The female gametophyte of the flowering plants resembles the early stages of the female gametophytes of archegoniate gymnosperms, possessing a peripheral layer of free nuclei arranged around a large central vacuole. It is therefore quite possible that the nonarchegoniate angiosperm gametophyte originated by way of progressive acceleration of gametogenesis and retardation of all other developmental processes (see Coulter 1914; Takhtajan 1976). Even the angiosperm egg is not the former egg of the archegonium but one of the very first cells of the gametophyte which is transformed into a female gamete (see Gerasimova-Navashina 1958).

The female gametophyte is considerably less simplified than the male gametophyte and therefore is more liable to evolutionary modifications. But the evolutionary modifications of the female gametophyte take place within the bounds of some limits. The different types of female gametophytes are distinguished mainly on the basis of the number of megaspores or megaspore nuclei that participate in their formation, on the number of mitotic divisions during gametogenesis, and on the number and arrangement of the cells and free nuclei present in the mature gametophyte (see Johri 1963; Romanov 1971). It is generally agreed that the monosporic eight-nucleate female gametophyte of the *Polygonum*-type, which characterizes the majority of angiosperms, is the basic and the most primitive type. All other types of the development and organization of the female gametophyte are derived. The tetrasporic types of female gametophyte are considered as the most specialized.

In a vast majority of cases the pollen tube penetrates into the female gametophyte through the micropyle (porogamy). Porogamy is the basic and primitive condition. Aporogamy (mesogamy and chalazogamy) is derived.

As it is well known, syngamy in flowering plants is accompanied by triple fusion of one of the two male gametes with the two polar nuclei. Triple fusion, which is one of the most characteristic features of the flowering plants, originated as a result of neotenic simplification of the female gametophyte. It triggers the formation of an entirely new structure, the triploid endosperm, which compensates for the extreme scarcity of nutrient materials in the simplified and miniaturized female gametophyte.

Following triple fusion, both the zygote and the primary nucleus of the endosperm develop further.

The development of the seed begins with the divisions of the primary endosperm nucleus, followed by the emergence of embryo. Three major types of the endosperm development are recognized – cellular, nuclear, and helobial.

In the cellular type, at least the first few divisions of the primary endosperm nucleus are accompanied by wall formation. The cellular pattern of development is found in many families of magnoliopsids (both archaic and advanced) and occurs only in four monocotyledonous families – Hydatellaceae, Araceae, Lemnaceae, Acoraceae.

In the helobial type, which is usually considered as somewhat intermediate between the cellular and nuclear type, the primary endosperm nucleus is always found at the chalazal end of the gametophyte; and, therefore, when it divides, two unequal cells or chambers are produced – a small chalazal cell and a much larger micropylar cell. The nucleus in the chalazal cell either does not divide further (a basic type, according to Swamy and Parameswaran 1963) or undergoes a usually restricted number of free nuclear divisions, whereas the larger micropylar cell undergoes numerous free nuclear divisions. Commonly, the cell-wall formation ultimately takes place in the micropylar chamber. The helobial type is common in liliopsids and is much less frequent in magnoliopsids.

In the nuclear type, the division of the primary endosperm nucleus is followed by a series of free nuclear divisions, resulting in the formation of a large multinucleate cell, which usually becomes cellular in a later phase of development. The nuclear type is widespread in both magnoliopsids and liliopsids.

The helobial type of endosperm development is probably apomorphic, derived either from nuclear or, more likely, from cellular type. But it is much more difficult to decide which of the two types – nuclear or cellular – is the more primitive. The main reason for this difficulty is that the formation of the endosperm is subject to reversal and that there are also many intermediates (Schnarf 1929, 1931; Brink and Cooper 1947). But, in spite of the reversibility of the types of endosperm development, the first flowering plants must have had either cellular or nuclear endosperm.

Seeds: The seeds of primitive flowering plants are of medium-size, 5–10 mm long (Corner 1976). Both small and large seeds are derived. The more primitive seeds are characterized by abundant endosperm and a minute and undifferentiated embryo (Pritzel 1898; Hallier 1912; Martin 1946; Eames 1961, and many others). In advanced seeds, on the contrary, the embryo is large and well differentiated, and the endosperm is more or less reduced or even wanting. Here we observe something analogous to what happens in the animal world where the embryo in the mother's body attains greater development in the higher forms (Nägeli 1884; Hallier 1912).

There are two main types of the specialization of endosperm – rumination and the development of haustoria. The outer surface of the ruminate endosperm tissue is irregularly ridged and furrowed to varying degrees, often very deeply. This furrowing occurs in a number of magnoliopsid families, especially in Magnolianae, and in some liliopsids (some genera of Dioscoreales, Cyclanthaceae, and Arecaceae) (Tamamschian 1951; Periasamy 1962; Corner 1976). Rumination is due to irregular growth activity of the seed coat or endosperm itself, during later stages of seed development (Boesewinkel and Bouman 1984). According to Vijayaraghavan and Prabhaker (1984: 343), ruminate endosperm could be an ancestral character still occurring in present-day seeds, belonging to both primitive and advanced taxa.

Another and more remarkable type of endosperm specialization is the formation of endosperm haustoria. The haustoria may arise at the chalazal or micropylar end, or at both ends of the developing endosperm. Endosperm haustoria are especially characteristic for taxa that develop the cellular type of endosperm. In the most archaic groups of flowering plants, including Magnoliaceae, endosperm haustoria are usually absent. In those rare cases, when haustoria are present in them, as in *Magnolia obovata* and in Saururaceae, they are chalazal.

Endosperm haustoria evolved independently in various lines of angiosperm evolution. The presence of absence of haustoria is a taxonomically useful embryological character, but the evolutionary trends in endosperm haustoria are not yet well known.

The period of dormancy is very weakly expressed or even absent in seeds of some tropical angiosperms. Since there is a long-continuing after-ripening development in some primitive families, absence of dormancy is considered by Eames (1961) as a survival of primitive condition. Dormancy, on the contrary, is considered as an advanced stage in the evolution of the seed.

It is almost universally accepted that the monocotyledonous embryo arose from the dicotyledonous embryo.

The basic primitive type of seed-coat is one with "multiplicative" integuments (Corner 1976), exarillate (Eames 1961), and probably with well developed pinkish or reddish sarcotesta (Zazhurilo 1940; Takhtajan 1948, 1959; van der Pijl 1955, 1969), like those of *Degeneria* and *Magnolia*. The presence of a sarcotesta in some primitive families suggests that endozoochory (possibly at first saurochory and later ornithochory) was probably characteristic of the earliest angiosperms (Zazhurilo 1940; Takhtajan 1948; van der Pijl 1969).

Considering the seed-coat structure of *Degeneria* and *Magnolia* as the initial one for the primitive flowering plants, the derivation of all other types can be easily imagined as a result of reconstructions connected with a transition towards some other non-endozoochorous mode of dissemination. This transition determined the development of the outer layer of sclerenchyma and the reduction of parenchyma, which have become superfluous.

In many lines of angiosperm evolution a gradual simplification of the seed-coat is observed. The maximum simplification of the spermaderm is attained in those cases where the seed adjoins closely or is fused with the pericarp. The role of the protection of the embryo as well as the function of dissemination passes over to the pericarp and so the seed-coat is strongly reduced. In some cases the reduction of the seed-coat goes very far. At times only the outer epidermis is retained in the mature seed.

During the evolution of zoochory, starting from the primitive endosaurochory and ending in the most highly specialized forms of myrmecochory, various types of the succulent nutritive tissue of the outer portion of the seed-coat play a big role. At first presumably the sarcotesta served as the bait for attracting arboreal reptiles and later birds. The aril is a more specialized type of nutritive tissue than sarcotesta. In some cases the aril possibly results from a decrease in the area of sarcotesta and its localization in a definite (usually basal) part of the seed (see van der Pijl 1955, 1969). But in the vast majority of cases arils arise as new structures on very different parts of the ovule and independently in many unrelated taxa. I therefore agree with Eames (1961) that it seems unlikely to consider the arillate seed as a primitive angiosperm character (but see Corner 1976).

Fruits: The most primitive and basic fruit type is a fruit consisting of many-seeded distinct follicles (Hallier 1912; Harvey-Gibson 1909; Bessey 1915; Gobi, 1921 and many others). Such a fruit, developing from a multicarpellate apocarpous gynoecium, was called "multifolliculus" (follicetum) by Gobi. The multifolliculus gave rise to unifolliculus by reduction in the number of carpels (e.g. *Degeneria, Consolida*).

From follicular fruits arose many other types of apocarpous fruits, which in their turn gave rise to numerous syncarpous fruit types (see Takhtajan 1991).

Karyotype: As was proposed as long ago as 1931 by Lewitsky (1931), the morphologically more primitive and basic type of chromosome of the flowering plants was one which had equally developed arms and a median or submedian centromere. Such symmetrical (or, in contemporary terminology, 'metacentric') chromosomes were those from which in course of evolution arose asymmetrical (acrocentric) chromosomes, in which the centromere is situated very close to one end.

The original karyotype of flowering plants was probably characterized by a comparatively small number of medium-sized chromosomes. But it is very difficult to say what in fact was the ancestral basic number of chromosomes in the flowering plants.

In the Magnoliaceae $2n = 38$, 76 and 114; consequently, in this family $n = 19$. In the Eupomatiaceae $n = 10$, i.e. the basic number differs both from that of the Magnoliaceae and from that of the Degeneriaceae and Himantandraceae. In the Annonaceae $n = 7, 8, 9$; in the Canellaceae $n = 11, 13, 14$.

It appears highly probable that the basic number of chromosomes of the early flowering plants was a low one. Darlington (Darlington and Mather 1949: 324) came to the conclusion that the basic number relationships of the chief families of woody flowering plants showed 7 as the common ancestral chromosome number of flowering plants. 'From this origin, 8, 9 and an increasing series have arisen on only a few occasions, whereas 14, with its diminishing series, has arisen very frequently. In this series 12 has often been stabilized, and, from its addition to 7, 19 has appeared several times'. Later, Raven and Kyhos (1965) and Ehrendorfer et al. (1968) reached a similar conclusion. The probability is very high that Darlington's proposal was correct and that the basic number was indeed 7. Yet on the other hand, in such extremely primitive families as Himantandraceae ($n = 6$), in some Winteraceae, Degeneriaceae, Cetarophyllaceae, in some Hernandiaceae n = 12, in the primitive family Eupomatiaceae n = 10. Moreover, in the ancient family Lauraceae is characterized by a basic number of 12, These data occasion some doubt that the basic number of chromosomes in the evolution of the karyotype of the flowering plants was in fact 7; a figure of 5, or better 6 is just as likely. Stebbins (1966) has suggested that the original number was x = 7 or x = 6.

The Significance of Primitive Characters

Evolutionary systematists consider both derived (apomorphic) and primitive (plesiomorphic) characters. For evolutionary taxonomists the retention of a large number of ancestral characters is just as important an indicator of phylogenetic relationships as the joint acquisition of a few apomorphic characters. Plesiomorphic characters are often among the most evident key characters of a taxon and are used for the delimitation and ranking of taxa.

Evidently the information content of primitive and derived characters is different and of different significance. Whereas the ancestral characters cannot be used in cladistic analysis to locate branching points, they have great importance in classification when they represent the dominant characters of a taxon. Often they are even more important in classification than rapidly changing derived characters. The conservative ancestral characters may actually tell us more about the total genotype. Often the discovery of ancestral, plesiomorphic features in some generally advanced groups indicates their phyletic links with less specialized taxa. The presence of distally monocolpate or distally dicolpate pollen grains in such specialized groups as Piperales, Aristolochiales, Hydnorales, and Rafflesiales provides important confirmation of their affinities with the orders of the archaic subclass Magnoliidae. The very definition of the Magnoliidae is based on the fact that they have the greatest number of primitive characters.

For the classification of taxa within individual lineages a systematist uses both the primitive and derived characters. Every new stage of evolution, and consequently every new taxon, differs from the ancestral taxon by an acquisition of some new, derived characters. The ancestral taxon, on the other hand, will differ from its descendants by the absence of these derived characters. For example, prokaryotes differ from the eukaryotes just in their fundamental plesiomorphic character, namely, the absence of the nucleus. The situation is different in the case of cladogenesis. Sister groups differ from each other essentially in derived characters acquired during the process of divergence.

Primitive characters are no less important for the hypothetical reconstruction of the ancestral groups, which was especially emphasized by the paleontologist Tatarinov (1984: 11). Thus knowing the basic evolutionary trends in flowering plants we may by extrapolation extend the transformation series into the past to the lowest level of specialization. In the absence of reliable fossil records this is the only way to reconstruct ancestral types. Out-group comparisons are especially important in these cases.

Convergence and Parallelism

It is generally accepted that a convergence is the origin of apparently similar apomorphies from different plesiomorphic states while parallelism is the independent development of similar apomorphies from the same plesiomorphic state.

The term *convergence* is usually applied to the occasional tendency for distantly related taxa to acquire some similar characters and thus in one or more respects to come to simulate one another and be more nearly alike than were their ancestors. Convergence is an independent analogous adaptation of different organisms to similar environmental conditions in response to similar functional needs. Therefore convergence is usually limited to the organs and tissues that are directly connected with similar environmental factors. However, convergent similarity is never very deep and mostly concerns only a few organs. There is only a very low probability of an independent development of similarity in the total morphological pattern in different lineages. As a rule, the more organisms are differentiated, the less probable is their complete convergence. It is also quite natural that the chance of partial convergence is much higher than that of overall convergence. But even so, any kind of convergent similarity is more or less easily distinguishable by detailed comparative studies. For example, the similarity in general habit between *Cuscuta* and *Cassytha* does not obscure the fact that in the basic characters of their flowers, fruit, pollen morphology, and vegetative anatomy as well as in their chemistry they belong to different subclasses of flowering plants.

The problem of parallelism is much more difficult than that of convergence. In flowering plants parallel evolutionary changes are very common. Similarities due to parallel evolution occur not only in the structure of the vegetative organs but also in flowers, inflorescences, fruits, seeds, pollen grains, and even chemical characters. The phenomenon of parallelism is usually defined as the development of similar characters separately in two or more lineages of common ancestry and on the basis of, or channeled by, characters of that ancestry (see Simpson 1961: 78). In contrast to convergence, parallelism is based on an initial similarity of structure and an initial homology of mutations and common evolutionary trends. It is highly improbable that two unrelated taxa will have a large number of parallel characters. This means that the parallel development of characters implies an evolutionary relationship. Thus the closeness of parallelism in the evolution of given taxa depends on the closeness of their affinity (Takhtajan 1947: 106; Cronquist 1968, 1987, 1988), and therefore we come to the general principle that evolutionary parallelism tends to indicate relationships, and accordingly it should be given due weight, along with other factors, in arriving at conclusions (Cronquist 1988: 32).

Weighting of Taxonomic Characters and Heterobathmy

Every practicing systematist knows the importance of detection, selection, and weighting of characters. He or she knows that some characters have a greater diagnostic value and therefore have greater weight than others. For a systematist different characters hold different information content, and many of them are merely "noise" (Mayr 1969: 208). It is therefore more important to deal with a small number of carefully selected and weighted characters than to deal with a large number of random characters. In addition, the same characters can have a different weight in various related taxa and especially in unrelated ones. It is well known that weighting can be only a posteriori, that is, based on experience. The relative weight of the characters is determined by the trial-and-error method, based on the personal experience of a systematist as well as on the experience of his or her predecessors and colleagues. Weighting is a specific taxonomic problem that can be solved only by the systematist. It is also well known that the higher the rank of a taxon, the more important weighting becomes (Mayr 1969: 211). Unfortunately, the weighting of characters and the selection of characters to use is not completely free of subjective elements. No technique or procedure (including the cladistic one) is immune to the necessity for subjective decisions. I agree with Cronquist (1988: 39) that "complete objectivity in taxonomy or any other complex subject is an unattainable will-o'-the-wisp." There is always room for subjective and even intuitive decisions, but we should always strive for objectivity.

The problem of weighting would not be so difficult if all the characters of an organism evolved harmoniously, at an equal rate and the same level of evolutionary development. But as is well known, the rates of evolution of different organs and parts of an organism are different, often drastically so, and the evolution of one structure may be faster or slower compared to that of other structures in the same taxon. This phenomenon of unequal rates of evolution of different features within one lineage is known under different names, of which the best known is mosaic evolution (De Beer 1954; Simpson 1961; Mayr 1963; Takhtajan 1991). As a result of mosaic evolution an organism may represent a disharmonious combination of characters of quite different evolutionary grade.

These differences in the grades of characters within the same taxon have been designated heterobathmy (from Greek *bathmos* – "step," "grade") (Takhtajan 1959, 1991). Heterobathmy may be expressed not only within the organism as a whole but also within its parts, such as flowers or xylem and their components, such as perianth, stamens, carpels, vessels, rays, and axial parenchyma.

The concept of heterobathmy is of the greatest importance for phylogenetical reconstructions. Thus if two taxa differ from each other by a pair of characters *A* and *B*, in one taxon character A may be primitive (plesiomorphic) while character B may be comparatively advanced (apomorphic), whereas in the other taxon the situation may be the reverse. In such a case of heterobathmy, the phyletic interrelationship between the two taxa can be reconstructed only through a third taxon (real or hypothetical), in which both of these two characters are on the primitive level of evolution. That is why long ago Dollo (1893) emphasized the importance of the *chevauchement des specialisations* ("crossing of specializations") when establishing cladistic relationships of recent organisms.

The more strongly heterobathmy is expressed, the more contradictory is the taxonomic information provided by different sets of characters, and the more difficult it is to pass from the evolutionary series of separate characters to the phyletic sequences of the organisms themselves. Thus in the overwhelming majority of cases, especially in the more archaic groups of the flowering plants, where heterobathmy is most clearly expressed (as in the Magnoliaceae and Winteraceae, and especially in *Amborella*), we cannot establish phyletic relationships and construct phyletic lineages using only floral characters. It is all the more impossible to reconstruct phyletic lineages on the basis of the characters of vegetative organs only, as, for example, on the basis of wood anatomy or leaf architecture. In such cases, instead of phyletic lineages we usually obtain only comparative-morphological series of forms arranged according to evolutionary trends of certain characters. Such series of forms illustrate the gradual evolutionary changes of these or other structures, but they do not express the phyletic interrelationships between organisms. The greater the number of properly chosen high-weight characters used for phylogenetic reconstructions, the closer we shall approach phyletic interrelationships. Studying the evolution of an adequate number of independent noncorrelated characters belonging to a sufficient number of different high-weight character complexes enables us to establish the basic trends of the evolution of a given taxon, to discover those of its members nearest to the phylogenetically initial forms and those more distant, and to deduce the cladistic relationships among orders, families, and subfamilies of the flowering plants. But in doing this we must always reckon with the phenomenon of heterobathmy.

The more heterobathmic the taxon, the more complete and all-around must be its study. Only the application of various methods and techniques can reveal those "critical characters" and "critical tendencies" (Wernham 1912) that are reliable phyletic markers. Correct weighting of the characters and their evolutionary trends takes on special significance in such cases and largely depends on the experience of a systematist and his or her erudition.

In many different lines of the evolution of the flowering plants there occurs a simplification of various structures accompanied by the loss of characters. During this regressive evolution many structures were not only reduced but even completely disappeared, which often represents an irretrievable loss of information. Extreme simplification and loss of information are typical of many aquatic plants and especially of parasites. A considerable loss of information is also characteristic of many specialized anemophilous flowers.

The Linnean Hierarchy

Any evolutionary classification is inevitably a more or less simplified representation of the phylogram in the form of nested hierarchies of increasingly more inclusive monophyletic taxa of different categorical rank. The hierarchical arrangement of taxa is a special information storage system that facilitates information retrieval by permitting an easy survey of taxa. However, from any given phylogenetic tree a number of different classifications can be derived. Nevertheless, there are some requirements essential to any practical and applicable hierarchical classification system. The most important requirement is the manageable number of ranks. A form of hierarchy that has been adopted for usual botanical classifications represents a sequence of basically seven levels (kingdom, phylum or division, class, order, family, genus, and species) with additional levels designated as super- (above the basic levels) and sub- (below them).

While taxa and the branches of phylogenetic trees are realities, taxonomic categories and their hierarchy are based on concepts. The use of any particular number of levels is therefore arbitrary. "It does not correspond with anything in nature but is an artifice imposed by practical necessity in the use of any hierarchy" (Simpson 1961: 17–18). To arrange the system in the form of the Linnean hierarchy is the only way to make it manageable and to achieve economy of memory.

The hierarchical classification is based on the gaps between taxa and the size of the gaps. Evolutionary systematists follow Mayr's recommendation (1969: 92) that the size of the gap be in inverse ratio to the size of the taxon. This is especially important in the recognition of higher taxa, particularly of monotypic higher taxa. According to Ashlock's definition (1979: 446), "A higher taxon is a monophyletic group of species (or a single species) separated from each phylogenetically adjacent taxon of the same rank by a gap greater than any found within these groups".

Bibliography

Andrews HN. 1961. Studies in paleobotany. Wiley, New York/London.

Andrews HN. 1963. Early seed plants. Science 142: 925–931.

Ashlock PD. 1979. An evolutionary systematist's view of classification. Syst. Zool. 28: 441–450.

Avetisyan EM and AM Hayrapetyan. 2002. On the terminology of compound apertures of pollen grains. Flora, vegetation and plant resources of Armenia, 14: 64–67.

Bailey IW. 1956. Nodal anatomy in retrospect. J. Arnold Arbor. 37: 269–287.

Bailey IW and RA Howard. 1941. The comparative morphology of the Icacinaceae. II. Vessels. IV. Rays of the secondary xylem. J. Arnold Arbor. 22: 171–187, 556–568.

Bailey IW and CG Nast. 1943. The comparative morphology of the Winteraceae. I. Pollen and stamens. II. Carpels. J. Arnold Arbor. 24: 340–346, 472–481.

Bailey IW and CG Nast. 1945. The comparative morphology of the Winteraceae. VII. Summary and conclusions. J. Arnold Arbor. 26: 37–47.

Bailey IW and AC Smith. 1942. Degeneriaceae, a new family of flowering plants from Fiji. J. Arnold Arbor. 23: 356–365.

Bailey IW and BGL Swamy. 1951. The conduplicate carpel of dicotyledons and its initial trends of specialization. Am. J. Bot. 38: 373–379.

Baranova MA. 1972. Systematic anatomy of the leaf epidermis in the Magnoliaceae and some related families. Taxon 21: 447–469.

Baranova MA. 1985. Classification of the morphological types of stomata. Bot. Zhurn. (Leningrad) 70 (12): 1585–1594 (in Russian).

Baranova MA. 1987a. On the stephanocytic type of the stomatal apparatus in Angiospermae. Bot. Zhurn. (Leningrad) 72 (1): 59–62 (in Russian).

Baranova MA. 1987b. Historical development of the present classification of morphological types of stomates. Bot. Rev. 53: 53–79.

Benson MJ. 1904. *Telangium scotti*, a new species of *Telangium* (*Calymmatotheca*) showing structure. Ann. Bot. 13: 161–177.

Benzing DH. 1967. Developmental patterns in stem primary xylem of woody Ranales. Am. J. Bot. 54: 805–820.

Bessey CE. 1897. Phylogeny and taxonomy of the angiosperms. Bot. Gaz. 24: 145–178.

Bessey CE. 1915. The phylogenetic taxonomy of flowering plants. Ann. Missouri Bot. Gard. 2: 109–164.

Bierhorst DW. 1971. Morphology of vascular plants. Macmillan, New York.

Boesewinkel FD and FG Bouman. 1984. The seed: structure. In: BM Johri, ed. Embryology of angiosperms, pp. 567–610. Springer, Berlin.

Bouman F. 1984. The ovule. In: BM Johri ed. Embryology of angiosperms, pp. 123–157. Springer, Berlin.

Bouman F and JIM Calis. 1977. Integumentary shifting – A third way to unitegmy. Ber. Deutsch. Bot. Ges. 90: 15–28.

Brewbaker JL. 1967. The distribution and phylogenetic significance of binucleate and trinucleate pollen grains in the angiosperms. Am. J. Bot. 54: 1069–1083.

Brink RA and DC Cooper. 1947. The endosperm and seed development. Bot. Rev. 13(8): 423–477, 13(9): 479–541.

Camp WH and MM Hubbard. 1963. On the origin of the ovule and cupule in Lyginopterid pteridosperms. Am. J. Bot. 50: 235–243.

Canright JE. 1952. The comparative morphology and relationships of the Magnoliaceae. I. Trends of specialization in the stamens. Am. J. Bot. 39: 484–497.

Canright JE. 1955. The comparative morphology and relationships of the Magnoliaceae. IV. Wood and nodal anatomy. J. Arnold Arbor. 36: 119–140.

Carlquist S. 1962. A theory of paedomorphosis in dicotyledonous woods. Phytomorphology 12: 30–45.

Corner EJH. 1946. Centrifugal stamens. J. Arnold Arbor. 27: 423–437.

Corner EJH. 1976. The seeds of dicotyledons, vols. I., II. Cambridge University Press, Cambridge.

Coulter JM. 1914. The evolution of sex in plants. University of Chicago Press, Chicago.

Coulter JM and CJ Chamberlain. 1903. Morphology of the angiosperms. Appleton, New York.

Cronquist A. 1968. The evolution and classification of flowering plants. Nelson, London.

Cronquist A. 1987. A botanical critique of cladism. Bot. Rev. 53: 1–53.

Cronquist A. 1988. The evolution and classification of flowering plants, 2nd ed. Bronx, New York.

Darlington CD and K Mather. 1949. The elements of genetics. Allen & Unwin, London.

Darwin C. 1876. The effect of cross and self fertilization in the vegetable kingdom. John Murray, London.

Davis GL. 1966. Systematic embryology of the angiosperms. Wiley, New York/London/Sydney.

De Beer GR. 1954. Archaeopteryx lithographica: A study based upon the British Museum specimen, British Museum, London.

Diels L. 1916. Käferblümen bei den Ranales and ihre Bedeutung für die Phylogenie der Angiospermen. Ber. Dt. Bot. Ges. 34: 758–774.

Dollo L. 1893. Les lois de l'evolution. Bull. Soc. Belg. Geol. 7: 164–166.

Doyle JA. 1969. Cretaceous angiosperms pollen of the Atlantic Coastal Plain and its evolutionary significance. J. Arnold Arbor. 50: 1–35.

Doyle JA and LJ Hickey. 1976. Pollen and leaves from the Mid-Cretaceous Potomac Group and their bearing on early angiosperm evolution. In: CB Beck, ed. Origin and early evolution of angiosperms, pp. 139–206. Columbia University Press, New York.

Doyle JA, M Van Campo, and B Lugardon. 1975. Observations on exine structure of *Eucommidites* and Lower Cretaceous angiosperm Pollen. Pollen Spores 17: 429–486.

Eames AJ. 1931. The vascular anatomy of the flower with refutation of the theory of carpel polymorphysm. Am. J. Bot. 18: 147–188.

Eames AJ. 1961. Morphology of the angiosperms. McGraw-Hill, New York/Toronto/London.

Eckardt T. 1937. Untersuchungen über Morphologie, Entwicklungsgeschichte und systematische Bedeutung des pseudomonomeren Gynoeceums. Nova Acta Leopold. Carol., N.F. 5: 1–112.

Eckardt T. 1938. Das pseudomonomeren Gynoeceums. Chron. Bot. 4: 206–208.

Ehrendorfer F, F Krendl, E Habeler, and W Sauer. 1968. Chromosome numbers and evolution in primitive angiosperms. Taxon 17: 337–353.

Endress PK and LD Hufford. 1989. The diversity of stamen structure and dehiscence patterns among Magnoliidae. Bot. J. Linn. Soc. 100: 45–85.

Esau K. 1965. Plant anatomy, 2nd ed. Wiley, New York.

Faegri K and L van der Pijl. 1979. The principles of pollination ecology, 3rd revised ed. Pergamon, Oxford.

Fahn A. 1974. Plant anatomy, 2nd ed. Pergamon, Oxford.

Florin R. 1933. Studien über die Cycadales des Mesozoicum Erörterungen über die Spaltöffnungsapparate der Bennititales. K. Svensk. Akad. Handl. 12: 1–134.

Florin R. 1958. On Jurassic taxads and conifers from North-Western Europe and Eastern Greenland. Acta Horti Bergian 17(10): 257–402.

Foster AS and EM Gifford. 1974. Comparative morphology of vascular plants, 2nd ed. Freeman, San Francisco.

Gabarayeva NI and El-Ghazaly G. 1997. Sporoderm development in *Nymphaea mexicana* (Nymphaeaceae). Plant Syst. Evol. 204: 1–19.

Gaussen H. 1946. Les gymnospermes actuelles et fossiles. Fac. Sci., Toulouse.

Gerasimova-Navashina EN. 1958. On the gametophyte and on the salient features of development and functioning of reproducing elements in angiospermous plants. In: Problemy Botaniki 3: 125–167. Leningrad (in Russian).

Gobi C. 1921. Classification génétique des fruits des plantes angiospermes. Ann. Inst. d'essais de semences au Jard. Impér. Princ. bot. Pierre le Grand 4, 4: 5–30 (in Russian with French summary).

Goebel K. 1933. Organographie der Pflanzen. III. Fischer, Jena.

Hallier H. 1905. Provisional sheme of the natural (phylogenetic) system of flowering plants. New Phytol. 4: 151–162.

Hallier H. 1908. Über Juliania, eine Terebinthaceen-Gattung mit Cupula. Beih. bot. Zbl., 23(2): 81–265.

Hallier H. 1912. L'origine et le système phylétique des Angiosperms exposés à l'aide de leur arbre généalogique. Arch. Néerl. Sci. Exact. Nat. Ser. 3,1: 146–234.

Harvey-Gibson RI. 1909. A classification of fruits on a physiological basis. Trans. Liverpool Bot. Soc. 1: 1–15.

Heel WA, van. 1970. Distally lobed integuments in some angiosperm ovules. Blumea 18: 67–70.

Heel WA, van. 1976. Distally-lobed integuments in *Exodhorda*, *Juglans*, *Leontice*, and *Bongardia*. Phytomorphology 26: 1–4.

Hemenway AF. 1913. Studies on the phloem of the dicotyledons. II. The evolution of the Sieve-tube. Bot. Gaz. 55: 236–243.

Hennig W. 1966. Phylogenetic systematics. University of Illinois Press, Urbana/Chicago/London.

Henslow G. 1888. The origin of floral structures through insect and other agencies. Kegan Paul, Trench, Trübner. London.

Hickey LJ. 1971. Evolutionary significance of leaf architectural features in the woody dicots. Am. J. Bot. 58: 469 (Abstract).

Hickey LJ and JA Doyle. 1972. Fossil evidence on evolution of angiosperm leaf venation. Am. J. Bot. 59: 661 (Abstrtact).

Hideux MJ and IK Ferguson. 1976. The stereostructure of the exine and its evolutionary significance in Saxifragaceae sensu lato. In: IK Ferguson and J Muller, eds. The evolutionary significance of the exine, pp. 327–377. Linn. Soc. Symp. ser. 1. London.

Hiepko P. 1964. Das zentrifugale Androecium der Paeoniaceae. Ber. Deutsch. Bot. Ges. 77: 427–435.

Holttum RE. 1955. Growth-habits of monocotyledons – variations on the theme. Phytomorphology 5(4): 399–413.

Hufford LD and PK Endress. 1989. The diversity of anther structure and dehiscence patterns among Hamamelididae. Bot. J. Linn. Soc. 99: 301–346.

Huxley JS. 1942. Evolution: The modern synthesis. Allen & Unwin, London.

Huxley JS. 1958. Evolutionary processes and taxonomy with special reference to grades. Uppsala Univ. Arssks. 1958: 21–39.

Jeffrey EC. 1899. The development, structure and affinities of the genus *Equisetum*. Mem. Boston Soc. Nat. Hist. 5: 157–190.

Jeffrey EC. 1917. The anatomy of plants. University of Chicago Press, Chicago.

Johri BM. 1963. Embryology and taxonomy. In: P Maheshwari, ed. Recent advances in the embryology of angiosperms. pp. 395–444. University of Delhi, Delhi.

Kozo-Poljanski BM. 1922. An introduction to phylogenetic systematics of the higher plants. Voronezh (in Russian).

Kozo-Poljanski BM. 1948. To the modernization of the system of the plant world. Trudy Voronezh State Univ. 15: 76–129 (in Russian).

Kribs DA. 1935. Salient lines of structural specialisation in the wood parenchyma of dicotyledons. Bot. Gaz. 96: 547–557.

Leins P. 1964. Das zentripetale und zentrifugale Androeceum. Ber. Deutsch. Bot. Ges. 77 (71 Gen.-Vers): 22–26.

Leins P. 1975. Die Bezienhungen zwischen multistaminaten und einfuchen Androeceen. Bot. Jahrb. Syst. 96: 231–237.

Lewitsky GA. 1931. Chromosome morphology and the karyotype in systematics. Trudy prikladnoi Bot. Genet. Selek. 27: 187–239 (in Russian and English).

Long AG. 1960. On the structure of *Calymmatotheca kidstonii* Calder (emended) and *Genomosperma latens* gen. et sp. nov. from the calciferous sandstone series of Berwickshire. Trans. Roy. Soc. Edinb. 64: 29–44.

Long AG. 1966. Some lower carboniferous fructifications from Berwickshire, together with a theoretical account of the evolution of ovules, cupules and carpels. Trans. Roy. Soc. Edinb. 66: 345–375.

Marsden MPF and IW Bailey. 1955. A fourth type of nodal anatomy in dicotyledons, illustrated by *Clerodendron trichotomum* Thunb. J. Arnold Arbor. 36: 1–50.

Martin AC. 1946. The comparative internal morphology of seeds. Am. Midl. Nat. 36: 513–660.

Mayr E. 1963. Animal species and evolution. Harvard University Press, Cambridge, MA.

Mayr E. 1969. Principles of systematic zoology. McGraw-Hill, New York.

Mayr E. 1976. Evolution and the diversity of life: Selected essays. Belknap, Cambridge, MA/London.

Mayr E. 1982. The growth of biological thought. Harvard University Press, Cambridge, MA.

Mayr E. 1988. Toward a new philosophy of biology. Harvard University Press, Cambridge, MA/London.

Mayr E and PD Ashlock. 1991. Principles of systematic zoology, 2nd ed. McGraw-Hill, New York.

Metcalf CR and L Chalk. 1950. Anatomy of the dicotyledones. I. Oxford University Press, Oxford.

Meyer NR. 1977. Comparative morphological studies of the development and ultrastructure of the gymnosperms and primitive angiosperms. Thesis of doctoral dissertation, Leningrad, (in Russian).

Moseley MF. 1958. Morphological studies in the Nymphaeaeceae. I. The nature of the stamens. Phytomorphology 8: 1–29.

Muller J. 1970. Palynological evidence on early differentiation of angiosperms. Biol. Rev. 45: 417–450.

Nägeli C von. 1884. Mechanisch-physiologische Theorie der Abstammungslehre. München/Leipzig.

Netolizky F. 1926. Anatomie der Angiospermem-Samen. Boerntraeger. Berlin.

Olmstead RG and JD Palmer. 1994. Chloroplast DNA systematics: A review of methods and data analysis. Am. J. Bot. 81:1205–1224.

Ozenda P. 1949. Recherches sur le Dicotylédones apocarpiques. Contribution à l'étude des Angiospermes dites primitives. Masson, Paris.

Ozenda P. 1952. Remarques sur quelques interprétations de l'étamine. Phytomorphology 2: 225–231.

Pacini E, GG Franchi, and M Hesse. 1985. The tapetum: its form, function, and possible phylogeny in Embryophyta. Plant. Syst. Evol. 149: 155–185.

Pant DD. 1965. On the ontogeny of stomata and other homologous structures. Plant Science Series, Allahabad 1: 1–24.

Parkin J. 1914. The evolution of the inflorescence. J. Linn. Soc., Bot. 42: 511–553.

Parkin J. 1953. The durian theory – a criticism. Phytomorphology 3: 80–88.

Parkin J. 1955. A plea for a simpler gynoecium. Phytomorphology 5: 46–57.

Payer JB. 1857. Traité d'organogénie comparée de la fleur. Masson, Paris.

Periasamy K. 1962. The ruminate endosperm. Development and types of rumination. In: Plant embryology: a symposium, pp. 62–74. CSIR, New Delhi.

Pettit J. 1970. Heterospory and the origin of the seed habit. Biol. Rev. 45: 401–415.

Philipson WR. 1974. Ovular morphology and the major classification of the dicotyledons. Bot. J. Linn. Soc. 68: 89–108.

Philipson WR. 1975. Evolutionary lines within the dicotyledons. New Zealand J. Bot. 13: 73–91.

Philipson WR. 1977. Ovular morphology and the classification of dicotyledons. Plant Syst. Evol., Suppl. 1: 123–140.

Pijl L van der. 1955. Sarcotesta, aril, pulpa and evolution of the angiosperm fruit. I, II. Proc. Ned. Acad. Wet. C 58(2): 154–161: 58(3): 307–312.

Pijl L van der. 1969. Principles of dispersal in higher plants. Springer, New York.

Pritzel E. 1898. Die Systematische Wert der Samenanatomie, insbesondere des Endosperms bei den Parietales. Bot. Jahrb. 24: 345–394.

Puri V. 1952. Placentation in angiosperms. Bot. Rev. 18: 603–651.

Raven PH and DW Kyhos. 1965. New evidence concerning the original basic chromosome number of Angiosperms. Evolution 19: 244–248.

Robertson C. 1904. The structure of the flowers and the mode of pollination of the primitive angiosperms. Bot. Gaz. 37: 294–298.

Romanov ID. 1971. Types of development of embryo sac of the angiosperms. In: VP Zosimovich, ed. Problems in embryology, pp. 72–113. Kiev (in Russian).

Ronse Decraene LP and E. Smets. 1987. The distribution and systematic relevance of the androecial character oligomery and polymery in the Magnoliophytina. Nord. J. Bot. 7: 239–253.

Rowley JR, NI Gabarayeva, and B Walles. 1992. Cyclic invasion of tapetal cells into loculi during microspore development in *Nymphaea colorata* (Nymphaeaceae). Am. J. Bot. 79: 801–808.

Sattler R and P. Sarita 1972. Centrifugal primordia inception in floral development. In: YS Murty et al., eds. Advances in plant morphology, pp. 170–178. Meerut.

Sattler R and F Pauzé. 1978. L'androcée centripète *d'Ochna atropurpurea*. Canadian J. Bot. 56: 2500–2511.

Schürhoff PN. 1926. Die Zytologie der Blütenpflanzen. Stuttgart.

Serebryakov IG. 1952. Morphology of vegetative organs of higher plants. Sovetskaya Nauka, Moscow (in Russian).

Serebryakov IG. 1955. Main directions of the evolution of life forms in angiospermous plants. Bull. Mosc. Obshch. Ispyt. Prirody, Otd. Biol. 60(3): 71–91 (in Russian).

Simpson GG. 1961. Principles of animal taxonomy. Columbia University Press, New York.

Sinnott EW. 1914. Investigations on the phylogeny of the angiosperm. I. The anatomy of the node as an aid in the classification of angiosperms. Am. J. Bot. 1: 303–322.

Sinnott EW and IW Bailey. 1914. Investigations on the phylogeny of the angiosperms: No. 4. The origin and dispersal of herbaceous angiosperms. Ann. Bot. 28: 547–600.

Schnarf K. 1929. Embryologie der Angiospermen. In: K Lindsbauer, ed. Handbuch der Pflancenanatomie, II Abt., 2 Teil. Band 10. Berlin.

Schnarf K. 1931. Vergleichende Embryologie der Angiospermen. Gebrüder Bountraeger, Berlin.

Sporne KR. 1972. Some observations on the evolution of pollen types in dicotyledons. New Phytol. 71: 181–185.

Sporne KR. 1973. A note on the evolutionary status of tapetal types in dicotyledons. New Phytol. 72: 1173–1174.

Stebbins GL. 1966. Chromosomal variation and evolution. Science 152. 1463–1469.

Stebbins GL. 1971. Chromosomal evolution in higher plants. Addison-Wesley, London.

Stebbins GL. 1974. Flowering plants. Evolution above the species level. E. Arnold. London.

Stopes M. 1905. On the double nature of the cycadean integument. Ann. Bot. 19: 561–566.

Straka H. 1975. Pollen- und Sporenkunde. Fischer, Stuttgart.

Swamy BGL. 1949. Further contributions to the morphology of the Degeneriaceae. J. Arnold. Arbor. 30: 10–38.

Swamy BGL and N Parameswaran. 1963. The helobial endosperm. Biol. Rev. 38(1): 1–50.

Takhtajan A. 1942. The structural types of gynoecium and placentation. Bull. Armen. Branch Acad. Sci. USSR, 3–4 (17–18): 91–112 (in Russian with English summary).

Takhtajan A. 1947. On the principles, methods, and symbols of the phylogenetic constructions in botany. Bull. Mosc. Obshch. Ispyt. Prirody, Otd. Biol. 52 (5): 95–120 (in Russian).

Takhtajan A. 1948. Morphological evolution of angiosperms. Nauka, Moscow (in Russian).
Takhtajan A. 1950. Phylogenetic principles of the system of higher plants. Bot. Zhurn. (Leningrad) 35: 113–135 (in Russian, English translation in: Bot. Rev. 1953, 19: 1–45).
Takhtajan A. 1959. Die Evolution der Angiospermen. Fischer, Jena.
Takhtajan A. 1964. Foundations of the evolutionary morphology of angiosperms. Nauka, Moscow/Leningrad (in Russian).
Takhtajan A. 1966. A system and phylogeny of the flowering plants. Moscow/Leningrad (in Russian.)
Takhtajan A. 1969. Flowering plants. Origin and dispersal. Oliver & Boyd, Edinburgh.
Takhtajan A. 1976. Neoteny and the origin of flowering plants. In: CB Beck, ed. Origin and early evolution of angiosperms, pp. 207–219. Columbia University Press, New York.
Takhtajan A. 1980. Outline of the classification of flowering plants (Magnoliophyta). Bot. Rev. 46: 225–359.
Takhtajan A. 1983. Macroevolutionary processes in the history of plant world. Bot. Zhurn. (Leningrad). 68(12): 1593–1603 (in Russian with English summary).
Takhtajan A. 1991. Evolutionary trends in flowering plants. Columbia University Press, New York.
Tamamschian SG. 1951. Rumination of endosperm in angiosperms. Bot. Zhurn. 36(5): 497–514 (in Russian).
Tatarinov LP. 1984. Cladistic analysis and phylogenetics. Palaeontol. Zhurn. 3: 3–16 (in Russian).
Troll W. 1928. Organisation und Gestalt in Bereich der Blute. Berlin.
Vijayaraghavan MR and K Prabhakar. 1984. The endosperm. In: BM Johri, ed. Embryology of angiosperms, pp. 319–376. Springer, Berlin.
Walker JW. 1974a. Evolution of exine structure in the pollen of primitive angiosperms. Am. J. Bot. 61: 891–902.
Walker JW. 1974b. Aperture evolution in the pollen of primitive angiosperms. Am. J. Bot. 61: 1112–1137.
Walker JW. 1976a. Comparative pollen morphology and phylogeny of Ranalean complex. In: CB Beck, ed. Origin and early evolution of angiosperms, pp. 241–299. Columbia University Press, New York.
Walker JW. 1976b. Evolutionary significance of exine in the pollen of primitive angiosperms. In: K Ferguson and J Muller, eds. The evolutionary significance of the exine. Linn. Soc. Symp. Series 1: 251–308.
Walker JW and JA Doyle. 1975. The bases of angiosperm phylogeny: Palynology. Ann. Missouri Bot. Gard. 62: 664–723.
Walker JW and JJ Skvarla. 1975. Primitively columellales pollen: a new concept in the evolutionary morphology of angiosperms. Science 187: 445–447.
Walton J. 1953. The evolution of the ovule in the pteridosperms. Adv. Sci. 10: 223–230 (British Association Adv. Sci., No. 38).
Weberling F. 1965. Typology of inflorescence. J. Linn. Soc. Bot. 59: 215–221.
Wernham HF. 1912. Floral evolution: With particular reference to the sympetalous dicotyledons. New Phytol. 11: 373–397.
Wilson TK. 1964. Comparative morphology of the Canellaceae. III. Pollen. Bot. Gaz. 125: 192–197.
Wodehouse RP. 1936. Evolution of pollen grains. Bot. Rev. 2: 67–89.
Zazhurilo KK. 1940. On the anatomy of the seed coats of Magnoliaceae (*Liriodendron tulipifera* L.). Bull. Soc. Nat. Voronezh Univ. 4, 1: 32–40 (in Russian).
Zimmermann W. 1959. Die Phylogenie der Pflanzen. 2 Aufl. Fischer, Stuttgart.
Zhukovsky PM. 1964. Botany. Nauka, Moscow (in Russian).

SYNOPSIS of families and suprafamilial taxa of the Magnoliophyta

Phylum MAGNOLIOPHYTA (FLOWERING PLANTS)

Class MAGNOLIOPSIDA (DICOTYLEDONS)

Dicotyledonous family of incertae position
1. HAPTANTHACEAE

Subclass I. Magnoliidae

SUPERORDER NYMPHAEANAE

Order 1. Amborellales

Family 1. *Amborellaceae*

Order 2. Nymphaeales

Family 1. *Hydropeltidaceae*

2. Cabombaceaes
3. Nymphaeaceae
4. *Barclayaceae*

Order 3. Austrobaileyales

Family 1. *Austrobaileyaceae*

Order 4. Illiciales

Family 1. *Illiciaceae*

2. *Schisandraceae*

Order 5. Trimeniales

Family 1. *Trimeniaceae*

Order 6. Chloranthales

Family 1. *Chloranthaceae*

Order 7. Ceratophyllales

Family 1. *Ceratophyllaceae*

SUPERORDER MAGNOLIANAE

Order 8. Canellales

Family 1. *Winteraceae*

2. *Canellaceae*

Order 9. Magnoliales

Family 1. *Degeneriaceae*

2. *Magnoliaceae*

Order 10. Himantandrales

Family 1. *Himantandraceae*

Order 11. Annanales

Family 1. *Eupomatiaceae*

2. *Annonaceae*

Order 12. Myristicales

Family 1. *Myristicaceae*

SUPERORDER LAURANAE

Order 13. Laurales

Family 1. *Monimiaceae*

2. *Idiosper*maceae
3. *Calycanthaceae*
4. *Atherospermataceae*
5. *Siparunaceae*
6. *Comortegaceae*
7. *Hernandiaceae*
8. *Lauraceae*

SUPERORDER PIPERANAE

Order 14. Piperales

Family 1. *Lactoridaceae*

2. *Sauruaceae*
3. *Piperacaea*
4. *Peperomiaceae*
5. *Aristolochiaceae*

Order 15. Hydnorales

Family 1. *Hydnoraceae*

SUPERORDER RAFFLESIANAE

Order 16. Mitrostemonales

Family 1. *Mitrastemonaceae*

Order 17. Rafflesiales (Cytinales)

Family 1. *Apodanthaceae*

2. *Rafflesiaceae*
3. *Cytinaceae*

Subclass II. Ranunculidae

SUPERORDER PROTEANAE
Order 18. Platanales
Family 1. *Platanaceae*
Order 19. Proteales
Family 1. *Proteaceae*
Order 20. Nelumbonales
Family 1. *Nelumbonaceae*
SUPERORDER RANUNCULANAE
Order 21. Eupteleales
Family 1. *Eupteleaceae*
Order 22. Lardizabalales
Family 1. *Lardizabalaceae*
Order 23. Menispermales
Family 1. *Menispermaceae*
Order 24. Berberidales
Family 1. *Nandinaceae*
2. *Berberidaceae*
3. *Ranzaniaceae*
4. *Podophyllaceae*
Order 25. Ranunculales
Family 1. *Hydrastidaceae*
2. *Ranunculaceae*
Order 26. Circaeasterales
Family 1. *Kingdoniaceae*
2. *Circaeasteraceae*
Order 27. Papaverales
Family 1. *Papaveruceae*
2. *Pteridophyllaceae*
3. *Hypecoaeceae*
4. *Fumariaceae*
Order 28. Glaucidiales
Family 1. *Glaucidiaceae*
Order 29. Paeoniales
Family 1. *Paeoniaceae*

Subclass III. Hamamelidae

SUPERORDER TROCHODENDRANAE
Order 30. Trochodendrales
Family 1. *Trochodendraceae*
2. *Tetracentraceae*
Order 31. Cercidiphyllales
Family 1. *Cercidiphyllaceae*
SUPERORDER MYROTHAMNANAE
Order 32. Myrothamnales
Family 1. *Myrothamnaceae*
SUPERORDER HAMAMELIDANAE
Order 33. Hamamelidales
Family 1. *Hamamelidaceae*
2. *Altingiaceae*
Order 34. Daphniphyllales
Family 1. *Daphniphyllaceae*
Order 35. Balanopales
Family 1. *Balanopaceae*
Order 36. Buxales
Family 1. *Buxaceae*
2. *Didymelaceae*
Order 37. Fagales
Family 1. *Fagaceae*
2. *Nothofagaceae*
Order 38. Betulales (Corylales)
Family 1. *Betulaceae*
2. *Ticodendraceae*
Order 39. Casuarinales
Family 1. *Casuarinaceae*
SUPERORDER JUGLANDANAE
Order 40. Myricales
Family 1. *Myricaceae*
Order 41. Juglandales
Family 1. *Rhoipteleaceae*
2. *Juglandaceae*

Subclass IV. Caryophyllidae

SUPERORDER CARYOPHYLLANAE
Order 42. Caryophyllales
Family 1. *Phytolaccaceae*
2. *Rhabdodendraceae*
3. *Gisekilaceae*
4. *Sarcobataceae*
5. *Barbeuiaceae*
6. *Achatocarpaceae*
7. *Nyctaginaceae*
8. *Aizoaceae*
9. *Stegnospermaceae*
10. *Portulacaceae*
11. *Hectorellaceae*
12. *Basellaceae*
13. *Halophytaceae*
14. *Cactaceae*
15. *Didiereaceae*
16. *Molluginaceae*
17. *Limeaceae*
18. *Caryophyllaceae*

19. *Simmondsiaceae*
20. *Amaranthaceae*
21. *Chenopodiaceae*

Order 43. Physenales

Family 1. *Asteropeiaceae*
2. *Physenaceae*

SUPERORDER POLYGONANAE

Order 44. Polygonales

Family 1. *Polygonaceae*

Order 45. Plumbaginales

Family 1. *Plumbaginaceae*

Order 46. Tamaricales

Family 1. *Tamaricaceae*
2. *Frankeniaceae*

SUPERORDER NEPENTHANAE

Order 47. Nepenthales

Family 1. *Ancistrocladaceae*
2. *Dioncophyllaceae*
3. *Nepenthaceae*
4. *Drosophyllaceae*
5. *Droseraceae*

Subclass V. Dilleniidae

SUPERORDER DILLENIANAE

Order 48. Dilleniales

Family 1. *Dilleniaceae*

SUPERORDER ERICANAE

Order 49. Paracryphiales

Family 1. *Paracryphiaceae*

Order 50. Theales

Family 1. *Stachyuraceae*
2. *Sladeniaceae*
3. *Pentaphylacaceae*
4. *Ternstroemiaceae*
5. *Theaceae*
6. *Oncothecaceae*
7. *Caryocaraceae*

Order 51. Balsaminales

Family 1. *Balsaminaceae*
2. *Marcgraviaceae*
3. *Tetrameristaceae*
4. *Pellicieraceae*

Order 52. Hypericales

Family 1. *Bonnetiaceae*
2. *Clusiaceae*
3. *Hypericaceae*
4. *Elatinaceae*

Orders 53. Ochnales

Family 1. *Strasburgeriaceae*
2. *Sauvagesiacea*
3. *Lophiraceae*
4. *Ochnaceae*
5. *Medusagynaceae*
6. *Quiinaceae*

Order 54. Ericales

Family 1. *Actinidiaceae*
2. *Clethraceae*
3. *Cyrillaceae*
4. *Ericaceae*

Order 55. Diapensiales

Family 1. *Diapensiaceae*

Order 56. Fouquieriales

Family 1. *Fouquieriaceae*

Order 57. Polemoniales

Family 1. *Polemoniaceae*

Order 58. Lecythidales

Family 1. *Scytopetalaceae*
2. *Lecythidaceae*
3. *Napoleonaeaceae*

Order 59. Sarraceniales

Family 1. *Sarraceniaceae*

Order 60. Roridulales

Family 1. *Roridulaceae*

SUPERORDER PRIMULANAE

Order 61. Styracales (Ebenales)

Family 1. *Styracaceae*
2. *Symplocaceae*
3. *Ebenaceae*

Order 62. Sapotales

Family 1. *Sapotaceae*

Order 63. Primulales

Family 1. *Myrsinaceae*
2. *Maesaceae*
3. *Aegicerataceae*
4. *Theophrastaceae*
5. *Primulaceae*
6. *Coridaceae*

SUPERORDER VIOLANAE

Order 64. Berberidopsidales

Family 1. *Berberidopsidaceae*

Order 65. Aextoxicales

Family 1. *Aextoxicaceae*

Order 66. Violales (Passiflorales)

Family 1. *Flacourtiaceae*
2. *Salicaceae*
3. *Lacistemataceae*

4. *Peridiscaceae*
5. *Violaceae*
6. *Dipentodontaceae*
7. *Scyphostegiaceae*
8. *Passifloraceae*
9. *Turneraceae*
10. *Malesherbiaceae*
11. *Achariaceae*
12. *Aphloiaceae*

Order 67. Elaeocarpales
Family 1. *Elaeocarpaceae*

Order 68. Cucurbitales
Family 1. *Cucurbitaceae*
2. *Datiscaceae*
3. *Tetramelaceae*
4. *Begoniaceae*

SUPERORDER CAPPARANAE

Order 69. Acaniales
Family 1. *Akaniaceae*
2. *Bretschneideraceae*

Order 70. Tropaeolales
Family 1. *Tropaeolaceae*

Order 71. Limnanthales
Family 1. *Limnanthaceae*

Order 72. Caricales
Family 1. *Caricaceae*

Order 73. Moringales
Family 1. *Moringaceae*

Order 74. Capparales (Resedales, Brassicales)
Family 1. *Capparaceae*
2. *Cleomaceae*
3. *Brassiaceae*
4. *Resedaceae*
5. *Gyrostemonaceae*
6. *Pentadiplandraceae*
7. *Koeberliniaceae*
8. *Bataceae*
9. *Salvadoraceae*
10. *Tovariaceae*
11. *Setchellanthaceae*
12. *Emblingiaceae*

SUPERORDER MALVANAE

Order 75. Malvales
Family 1. *Muntingiaceae*
2. *Tiliaceae*
3. *Dipterocarpaceae*
4. *Monotaceae*
5. *Sarcolaenaceae*
6. *Neuradaceae*
7. *Sterculiaceae*
8. *Diegodendraceae*
9. *Sphaerosepalaceae*
10. *Bombacaceae*
11. *Malvaceae*
12. *Bixaceae*
13. *Cochlospermaceae*
14. *Cistaceae*
15. *Tepuianthaceae*
16. *Thymelaeaceae*

Order 76. Urticales
Family 1. *Ulmaceae*
2. *Moraceae*
3. *Cannabaceae*
4. *Cecropiaceae*
5. *Urticaceae*

SUPERORDER EUPHORBIANAE

Order 77. Euphorbiales
Family 1. *Phyllanthaceae*
2. *Putranjivaceae*
3. *Picrodendraceae*
4. *Euphorbiaceae*
5. *Pandaceae*

Subclass VI. Rosidae

SUPERORDER ROSANAE

Order 78. Cunoniales
Family 1. *Cunoniaceae*
2. *Brunelliaceae*

Order 79. Anisophylleales
Family 1. *Anisophylleaceae*

Order 80. Cephalotales
Family 1. *Cephalotaceae*

Order 81. Saxifragales
Family 1. *Tetracarpaeaceae*
2. *Aphanopetalaceae*
3. *Penthoraceae*
4. *Crassulaceae*
5. *Haloragaceae*
6. *Gunneraceae*
7. *Saxifragaceae*
8. *Grossulariaceae*
9. *Pterostemonaceae*
10. *Iteaceae*

Order 82. Podostemales
Family 1. *Podostemaceae*

Order 83. Vitales
Family 1. *Vitaceae*
2. *Leeaceae*
Order 84. Rosales
Family 1. *Rosaceae*
2. *Quillajaceae*
Order 85. Crossosomatales
Family 1. *Crossosomataceae*
Order 86. Chrysobalanales
Family 1. *Chrysobalanaceae*
2. *Dichapetalaceae*
3. *Trigoniaceae*
4. *Euphroniaceae*
SUPERORDER MYRTANAE
Order 87. Myrtales
Family 1. *Alzateaceae*
2. *Rhynchocalycaceae*
3. *Geissolomataceae*
4. *Penaeaceae*
5. *Oliniaceae*
6. *Combretaceae*
7. *Crypteroniaceae*
8. *Memecylaceae*
9. *Melastomataceae*
10. *Lythraceae*
11. *Trapaceae*
12. *Onagracea*
13. *Myrtaceae*
14. *Vochysiaceae*
SUPERORDER FABANAE
Order 88. Fabales
Family 1. *Fabaceae*
Order 89. Polygalales
Family 1. *Polygalaceae*
SUPERORDER RUTANAE
Order 90. Oxalidales (Connarales)
Family 1. *Connaraceae*
2. *Oxalidaceae*
Order 91. Sapindales
Family 1. *Staphyleaceae*
2. *Tapisciaceae*
3. *Sapindaceae*
4. *Hippocastanaceae*
5. *Aceraceae*
Order 92. Sabiales
Famely 1. *Sabiaceae*
Order 93. Biebersteiniales
Family 1. *Biebersteiniaceae*
Order 94. Rutales
Family 1. *Rutaceae*
2. *Cneoraceae*
3. *Simaroubaceae*
4. *Picramniaceae*
5. *Leitneriaceae*
6. *Surianaceae*
7. *Irvingiaceae*
8. *Kirkiaceae*
9. *Pteroxylaceae*
10. *Meliaceae*
11. *Burseraceae*
12. *Anacardiaceae*
13. *Podoaceae*
Order 95. Coriariales
Family 1. *Coriariaceae*
Order 96. Corynocarpales
Family 1. *Corynocarpaceae*
Order 97. Geraniales
Family 1. *Hypseocharitaceae*
2. *Geraniaceae*
3. *Vivianiaceae*
4. *Ledocarpaceae*
5. *Melianthaceae*
6. *Greyiaceae*
7. *Francoaceae*
Order 98. Zygophyllales
Family 1. *Zygophyllaceae*
2. *Balanitaceae*
3. *Peganaceae*
4. *Nitrariaceae*
5. *Tetradiclidaceae*
Order 99. Linales
Family 1. *Hugoniaceae*
2. *Linaceae*
3. *Ctenolophonaceae*
4. *lxonanthaceae*
5. *Humiriaceae*
6. *Erythroxylaceae*
7. *Rhizophoraceae*
Order 100. Malpighiales
Family 1. *Malpighiaceae*
2. *Krameriaceae*
SUPERORDER CELASTRANAE
Order 101. Celastrales
Family 1. *Goupiaceae*
2. *Lepidobotryaceae*
3. *Brexiaceae*
4. *Celastraceae*
5. *Plagiopteraceae*
6. *Lophopyxidaceae*
7. *Stackhousiaceae*

8. *Parnassiaceae*
9. *Huaceae*

SUPERORDER SANTALANAE

Order 102. Santalales

Family 1. *Olacaceae*
2. *Opiliaceae*
3. *Aptandraceae*
4. *Octoknemaceae*
5. *Medusandraceae*
6. *Santalaceae*
7. *Misodendraceae*
8. *Loranthaceae*
9. *Viscaceae*
10. *Eremolepidaceae*

SUPERORDER BALANOPHORANAE

Order 103. Cynomoriales

Family 1. *Cynomoriaceae*

Order 104. Balanophorales

Family 1. *Balanophoraceae*

SUPERORDER RHAMNANAE

Order 105. Rhamnales

Family 1. *Rhamnaceae*
2. *Elaeagnaceae*
3. *Barbeyaceae*
4. *Dirachmaceae*

Subclass VII. Asteridae

SUPERORDER CORNANAE

Order 106. Desfontainiales

Family 1. *Escalloniaceae*
2. *Eremosynaceae*
3. *Abrophyllaceae*
4. *Ixerbaceae*
5. *Tribelaceae*
6. *Kaliphoraceae*
7. *Montiniaceae*
8. *Columelliaceae*
9. *Desfontainiaceae*
10. *Vahliaceae*

Order 107. Bruniales

Family 1. *Bruniaceae*

Order 108. Loasales (Hydrangeales)

Family 1. *Hydrangeaceae*
2. *Loasaceae*

Order 109. Cornales

Family 1. *Davidiaceae*
2. *Nyssaceae*
3. *Mastixiaceae*
4. *Curtisiaceae*
5. *Grubbiaceae*
6. *Cornaceae*
7. *Alangiaceae*

Order 110. Garryales

Family 1. *Garryaceae*
2. *Aucubaceae*

Order 111. Eucommiales

Family 1. *Eucommiaceae*

Order 112. Dipsacales

Family 1. *Viburnaceae*
2. *Sambucaceae*
3. *Adoxaceae*
4. *Caprifoliaceae*
5. *Valerianaceae*
6. *Triplostegiaceae*
7. *Dipsacaceae*
8. *Morinaceae*

Order 113. Aquifoliales

Family 1. *Aquifoliaceae*
2. *Icacinaceae*
3. *Helwingiaceae*
4. *Phyllonomaceae*
5. *Sphenostemonaceae*
6. *Cardiopteridaceae*

Order 114. Apiales

Family 1. *Pennantiaceae*
2. *Griseliniaceae*
3. *Aralidiaceae*
4. *Toricelliaceae*
5. *Melanophyllaceae*
6. *Pittosporaceae*
7. *Araliaceae*
8. *Myodocarpaceae*
9. *Apiaceae*

SUPERORDER ASTERANAE

Order 115. Rousseales

Family 1. *Rousseaceae*
2. *Carpodetaceae*

Order 116. Campanulales

Family 1. *Pentaphragmataceae*
2. *Sphenocleaceae*
3. *Campanulaceae*

Order 117. Stylidiales

Family 1. *Donatiaceae*
2. *Stylidiaceae*

Order 118. Phellinales (Alseuosmiales)

Family 1. *Phellinaceae*

2. *Argophyllaceae*
3. *Corokiaceae*
4. *Alseuosmiaceae*

Order 119. Asterales

Family 1. *Goodeniaceae*
2. *Brunoniaceae*
3. *Menyanthaceae*
4. *Calyceraceae*
5. *Asteraceae*

Subclass VIII. Lamiidae

SUPERORDER LAMIANAE

Order 120. Rubiales (Gentianales)

Family 1. *Gelsemiaceae*
2. *Loganiaceae*
3. *Strychnaceae*
4. *Antoniaceae*
5. *Spigeliaceae*
6. *Dialypetalanthaceae*
7. *Rubiaceae*
8. *Theligonaceae*
9. *Gentianaceae*
10. *Geniostomaceae*
11. *Apocynaceae*

Order 121. Solanales

Family 1. *Solanaceae*
2. *Nolanaceae*
3. *Sclerophylacaceae*
4. *Duckeodendraceae*
5. *Goetzeaceae*
6. *Hydroleaceae*
7. *Convolvulaceae*
8. *Cuscutaceae*
9. *Humbertiaceae*

Order 122. Boraginales

Family 1. *Hydrophyllaceae*
2. *Boraginaceae*
3. *Hoplestigmataceae*
4. *Lennoaceae*

Order 123. Oleales

Family 1. *Oleaceae*

Order 124. Lamiales

Family 1. *Buddlejaceae*
2. *Polypremaceae*
3. *Tetrachondraceae*
4. *Calceolariaceae*
5. *Stilbaceae*
6. *Scrophulariaceae*
7. *Bignoniaceae*
8. *Gesneriaceae*
9. *Plocospermataceae*
10. *Carlemanniaceae*
11. *Globulariaceae*
12. *Plantaginaceae*
13. *Callitrichaceae*
14. *Hippuridaceae*
15. *Pedaliaceae*
16. *Martyniaceae*
17. *Trapellaceae*
18. *Myoporaceae*
19. *Oftiaceae*
20. *Acanthaceae*
21. *Avicenniaceae*
22. *Byblidaceae*
23. *Lentibulariaceae*
24. *Verbenaceae*
25. *Phrymataceae*
26. *Cyclocheilaceae*
27. *Nesogenaceae*
28. *Symphoremataceae*
29. *Lamiaceae*

Order 125. Hydrostachyales

Family 1. *Hydrostachyaceae*

Class LILIOPSIDA
(MONOCOTYLEDONS)

Subclass I. Alismatidae

SUPERORDER PETROSAVIANAE

Order 1. Petrosaviales

Family 1. *Japonoliriaceae*
2. *Petrosaviaceae*
3. *Tofieldiaceae*
4. *Nartheciaceae*

SUPERORDER ALISMATANAE

Order 2. Hydrocharitales

Family 1. *Aponogetonaceae*
2. *Butomaceae*
3. *Najadaceae*
4. *Hydrocharitaceae*

Order 3. Alismatales

Family 1. *Limnocharitaceae*
2. *Alismataceae*

Order 4. Potamogetonales
Family 1. *Scheuchzeriaceae*
2. *Juncaginaceae*
3. *Potamogetonaceae*
4. *Posidoniaceae*
5. *Ruppiaceae*
6. *Zannichelliaceae*
7. *Cymodoceaceae*
8. *Zosteraceae*

SUPERORDER ARANAE
Order 5. Arales
Family 1. *Acoraceae*
2. *Araceae*
3. *Pistaceae*
4. *Lemnaceae*

Subclass II. Liliidae

SUPERORDER LILIANAE
Order 6. Melanthiales
Family 1. *Melanthiaceae*
Order 7. Trilliales
Family 1. *Trilliaceae*
Order 8. Liliales
Family 1. *Campynemataceae*
2. *Colchicaceae*
3. *Tricyrtidaceae*
4. *Scoliopaceae*
5. *Calochortaceae*
6. *Liliaceae*
7. *Medeolaceae*
Order 9. Burmanniales
Family 1. *Burmanniaceae*
2. *Thismiaceae*
3. *Corsiaceae*
Order 10. Alstroemeriales
Family 1. *Luzuriagaceae*
2. *Behniaceae*
3. *Alstroemeriaceae*
4. *Petermanniaceae*
Order 11. Smilacales
Family 1. *Philesiaceae*
2. *Ripogonaceae*
3. *Smilacaceae*
Order 12. Orchidales
Family 1. *Blandfordiaceae*
2. *Asteliaceae*
3. *Lanariaceae*
4. *Hypoxidaceae*
5. *Orchidaceae*
Order 13. Iridales
Family 1. *Ixioliriaceae*
2. *Walleriaceae*
3. *Tecophilaeaceae*
4. *Cyanastraceae*
5. *Doryanthaceae*
6. *Geosiridaceae*
7. *Iridaceae*
Order 14. Amaryllidales
Family 1. *Hemerocallidaceae*
2. *Phormiaceae*
3. *Xeronemataceae*
4. *Asphodelaceae*
5. *Xanthorrhoeaceae*
6. *Anthericaceae*
7. *Anemarrhenaceae*
8. *Hyacinthaceae*
9. *Agavaceae*
10. *Themidaceae*
11. *Agapantaceae*
12. *Alliaceae*
13. *Amaryllidaceae*
14. *Herreriaceae*
15. *Aphyllanthaceae*
Order 15. Asparagales
Family 1. *Convallariaceae*
2. *Dracaenaceae*
3. *Nolinaceae*
4. *Ruscaceae*
5. *Asparagaceae*
6. *Eriospermaceae*

SUPERORDER PANDANANAE
Order 16. Pandanales
Family 1. *Pandanaceae*
Order 17. Cyclanthales
Family 1. *Cyclanthaceae*
Order 18. Triuridales
Family 1. *Triuridaceae*
Order 19. Velloziales
Family 1. *Velloziaceae*
2. *Acanthochlamydaceae*
Order 20. Stemonales
Family 1. *Stemonaceae*
2. *Pentastemonaceae*

SUPERORDER DIOSCOREANAE
Order 21. Dioscoreales

Family 1. *Taccaceae*
2. *Stenomeridaceae*
3. *Dioscoreaceae*
4. *Trichopodaceae*

Subclass III. Arecidae

SUPERORDER ARECANAE
Order 22. Arecales
Family 1. *Arecaceae*

Subclass IV. Commelinidae

SUPERORDER BROMELIANAE
Order 23. Bromeliales
Family 1. *Bromeliaceae*
SUPERORDER ZINGIBERANAE
Order 24. Zingiberales (Cannales)
Family 1. *Strelitziaceae*
2. *Lowiaceae*
3. *Musaceae*
4. *Heliconiaceae*
5. *Zingiberaceae*
6. *Costaceae*
7. *Cannaceae*
8. *Marantaceae*
SUPERORDER COMMELINANAE
Order 25. Commelinales
Family 1. *Hanguanaceae*
2. *Commelinaceae*
3. *Philydraceae*
4. *Pontederiaceae*
5. *Haemodoraceae*
Order 26. Xyridales
Family 1. *Mayacaceae*
2. *Xyridaceae*
3. *Rapateaceae*
4. *Eriocaulaceae*
5. *Hydatellaceae*
Order 27. Dasypogonales
Family 1. *Baxteriaceae*
2. *Lomandraceae*
3. *Dasypogonaceae*
4. *Calectasiaceae*
SUPERORDER JUNCANAE
Order 28. Juncales
Family 1. *Thurniaceae*
2. *Juncaceae*
3. *Cyperaceae*
SUPERORDER POANAE
Order 29. Typhales
Family 1. *Sparganiaceae*
2. *Typhaceae*
Order 30. Restionales
Family 1. *Flagellariaceae*
2. *Joinvilleaceae*
3. *Restionaceae*
4. *Anarthriaceae*
5. *Ecdeiocoleaceae*
6. *Centrolepidaceae*
Order 31. Poales
Family 1. *Poaceae*

Phylum MAGNOLIOPHYTA (FLOWERING PLANTS)

The characteristic features of the phylum (division) of Magnoliophyta are: (1) ovules are enclosed in a more or less completely closed cavity formed by one or by several to many separate or united carpels; (2) pollen grains are deposited not directly on the micropyle of the ovule but on the stigmatic surface, which is capable not only of receiving the pollen but also of assisting in the development of the pollen tube (initially the stigmatic surface stretches along more or less free carpellary margins, but in the majority of flowering plants it is localized and forms a specialized stigma); (3) both male and female gametophytes are devoid of gametangia (antheridia and archegonia) and are extremely simplified and specialized; (4) triple fusion: one of the two male gametes fuses with the egg cell (syngamy), and the other with two free or already fused polar nuclei of the central cell of the female gametophyte; as a result of triple fusion a diploid zygote and a triploid primary nucleus of the endosperm are formed; although in the Podostemaceae and Orchidaceae the endosperm fails to develop or degenerates early in ontogeny, the triple fusion is a characteristic feature of magnoliophytes, sharply distinguishing them from all other plants; (5) sieve elements with companion cells.

The flowering plants, also known as the angiosperms, constitute the largest and most conspicuous group of modern plants. They predominate in number of species and number individuals, and constitute the dominant vegetation of the Earth's surface today. Flowering plants inhabit the most diversified environments and may be found wherever higher plants can survive. They have evolved a remarkable variety of life forms, a much greater variety than is to be found in the other phyla of the plant kingdom. They range in life span from a few weeks to several thousand years.

Bibliography

Agardh CA. 1822. Aphorismi botanici. Lund Unice. Diss. Lund.

Agardh JG. 1858. Theoria systematis plantarum. C.W.K. Gleerup, Lund.

Airy Shaw HK. 1965. Diagnoses of new families, new names, etc., for the seventh edition of Willis's Dictionary. Kew Bull. 18: 249–273.

Ambrose BA, DR Lerner, P *Ciceri, CM Padilla, MF Yanofsky and RJ Schmidt. 2000. Molecular and genetic analyses of the Silky1 gene reveal conservation in floral organ specification between eudicots and monocots. Molecular Cell 5: 569–579.

Antonov A.S. 2006. Plant genotypes systematics. Nauka, Moscow.

APG I (The Angiosperm Phylogeny Group I). 1998. An ordinal classification for the families of flowering plants. Ann. Miss. Bot. Gard. 85: 531–553.

APG II (The Angiosperm Phylogeny Group II). 2003. An update of the Angiosperm Phylogeny Group classification for the order and families of flowering plants. APG II. Bot. J. Linn. Soc. 141: 399–436.

Baillon HE. 1866–1895. Histoire des plantes. 13 vols. Paris.

Bartling FG. 1830. Ordines naturales plantarum. Dieterich, Gottingen.

Barthlott W. 1981. Epidermal and seed surface characters of plants: Systematic applicability and some evolutionary aspects. Nord. J. Bot. 1: 345–355.

Barthlott W, I Theisen, T Borsch, and C Neinhuis. 2003. Epicuticular waxes and vascular plant systematics: Integrating micromorphological and chemical data. In: TF Stuessy, V Mayer, and E Hörandl, eds. Deep morphology: toward a renaissance of morphology in plant systematics, pp. 189–206. Ruggell, Liechtenstein.

Bedell HG and JL Reveal. 1982. Amended outlines and indices for six recently published systems of angiosperm classification. Phytologia 51: 65–156.

Behnke H-D. 1972. Sieve-tube plastids in relation to angiosperm systematics: An attempt towards a classification by ultrastructural analysis. Bot. Rev. 38: 155–197.

Behnke H-D. 1977. Transmission electron microscopy and systematics of flowering plants. Plant Syst. Evol., Suppl. 1: 155–178.

Behnke H-D. 1981. Sieve-element characters. Nord. J. Bot. 1: 381–400.

Bennett MD. 2000. Genomic organization and systematics in the 21st century (Nancy T. Burbidge Memorial Lecture). In: KL Wilson and DA Morrison, eds. Monocots: systematics and Evolution. Collingwood, Australia.

Benson L. 1957. Plant classification. DC Heath, Boston.

Bentham G and D Hooker. 1862–1883. Genera plantarum. 3 vols. Lovell Reeve, London.

Bessey CE. 1893. Evolution and classification. Proc. Am. Assoc. Adv. Sci. August: 237–251.

Bessey CE. 1897. Phylogeny and taxonomy of the angiosperms. Bot. Gaz. 24: 145–178.

Bessy CE. 1907. A synopsis of plant phyla. Univ. Nebr. Stud. 7: 275–373.

Bessey CE. 1915. The phylogenetic taxonomy of flowering plants. Ann. Missouri Bot. Gard. 2: 109–164.

Bews JW. 1927. Studies in the ecological evolution of the angiosperms. New Phytologist reprint 16. London.

Boivin BB. 1956. Les families des Tracheophytes. Bull. Soc. Bot. France 103: 490–505.

Brown WH. 1938. The bearing of nectaries on the phylogeny of flowering plants. Proc. Am. Phil. Soc. 79: 549–595.

Brummitt RK. 1992. Vascular plants families and genera. Royal Botanical Garden, Kew.

Burleigh JG and S Mathews. 2007a. Assessing among-locus variation in the inference of seed plant phylogeny. Int. J. Plant Sci. 168: 111–124.

Burleigh JG and S Mathews. 2007b. Assessing systematic error in the inference of seed plant phylogeny. Int. J. Plant Sci. 168: 125–135.

Candolle AP de. 1813. Théorie élémentaire de la bo-tanique. Deterville, Paris.

Candolle A. P. de. 1824–1873. Prodromus systematis naturalis regni vegetabilis. 17 vols. Paris.

Chase MW. 2005. Relationships between families of flowering plants. In: RJ Henry, ed. Plant diversity and evolution: genotypic and phenotypic variation in higher plants, pp. 7–23. CAB International, Wallingford.

Chase MW et al. 1993. Phylogenetics of seed plants: An analysis of nucleotide sequences from the plastid gene *rbc*L. Ann. Missouri Bot. Gard. 80: 528–580.

Cho Y, JP Mower, Y-L Qiu, and JD Palmer. 2004. Mitochondrial substitution rates are extraordinarily elevated and variable in a genus of flowering plants. Proc. National Acad. Sci. 101: 17741–17746.

Cocucci AE. 2005. Morphogenetic seed types of Spermatophyta. Plant Syst. Evol. 250: 1–6.

Cook CDK. 1990. The aquatic plant book. SPB Academic, The Hague.

Crane PR. 1985. Phylogenetic analysis of seed plants and the origin of angiosperms. Ann. Missouri Bot. Gard. 72: 384–402.

Croizat L. 1960. Principia Botanica. Codicote.

Cronquist A. 1957. Outline of a new system of families and orders of dicotyledons. Bull. Jard. Bot. L'Etat 27: 13–40.

Cronquist A. 1965. The status of the general system of classification of flowering plants. Ann. Missouri Bot. Gard. 52: 281–303.

Cronquist A. 1968. The evolution and classification of flowering plants. London.

Cronquist A. 1969. Broad features of the system of an-giosperms. Taxon 18: 188–193.

Cronquist A. 1981. An integrated system of classification of flowering plants. New York.

Cronquist A. 1983. Some realignments in the dicotyledons. Nord. J. Bot. 3: 75–83.

Cronquist A. 1988. The evolution and classification of flowering plants. 2nd ed. The New York Botanical Garden, New York.

Cronquist A. et al. 1993. Angiosperms. Encyclopaedia Britannica 13: 603–765.

Cronquist A, A Takhtajan, and W Zimmermann. 1966. On the higher taxa of Embryobionta. Taxon 15: 129–134.

Czaja AT. 1978. Structure of starch grains and the classification of vascular plant families. Taxon 27: 463–470.

Dahlgren G. 1989. An updated angiosperm classification. Bot. J. Linn. Soc. 100: 197–203.

Dahlgren RMT. 1975. A system of classification of the angiosperms to be used to demonstrate the distribution of characters. Bot. Not. 128: 119–147.

Dahlgren RMT. 1977a. A commentary on a diagrammatic presentation of the angiosperms in relation to the distribution of character states. Plant Syst. Evol., Suppl. 1: 253–283.

Dahlgren RMT. 1977b. A note on the taxonomy of the "Sympetalae" and related groups. Publ. Cairo Univ. Herb. 7–8: 83–102.

Dahlgren RMT. 1980. A revised system of classification of the angiosperms. Bot. J. Linn. Soc. 80: 91–124.

Dahlgren RMT. 1983a. General aspects of angiosperm evolution and macrosystematics. Nord. J. Bot. 3: 119–149.

Dahlgren RMT. 1983b. The importance of modern serological research for angiosperm classification. In: U Jensen and DE Fairbrothers, eds. Proteins and nucleic acids in plant systematics, pp. 371–374. Springer, New York.

Dahlgren RMT and K Bremen 1985. Major clades of the angiosperms. Cladistics 1: 349–368.

Dahlgren RMT and HT Clifford. 1982. The monocotyledons: A comparative study. Academic Press, London/New York.

Dahlgren RMT, HT Clifford, and PE Yeo. 1985. The families of the monocotyledons. Springer, New York.

Dahlgren RMT, S Rosendal-Jensen, and BJ Nielsen. 1981. A revised classification of the angiosperms with comments on correlation between chemical and other characters. In: DA Joung and DS Seigler, eds. Phytochemistry and angiosperm phytogeny, pp. 149–204. Praeger, New York.

Davis GL. 1966. Systematic embryology of the angiosperms. Wiley, New York.

De Bodt S, S Maere, and Y van de Peer. 2005. Gene duplication and the evolution of angiosperms. Trends Ecol. Evol. 20: 591–597.

Delpino F. 1888–1890, 1893, 1896, 1905. Applicazione di nuovi criterii per la classificazione delle piante. Mem. R. Accad. Sci. Istituto di Bologna. Prima Me-moria. ser. 4a, vol. 9 (1988); Seconda Memoria. Ibid., ser. 4a, vol. 10 (1889); Terza Memoria. Ibid., ser. 4a, vol. 10 (1889); Quarta Memoria. Ibid., ser. 5a, vol. 1 (1890); Quinta Memoria. Ibid., ser. 5a, vol. 3 (1893); Sesta Memoria. Ibid., ser. 5a, vol. 6 (1896); Settima Memoria. Ibid., (1905).

Dickison WC. 1975. The bases of angiosperm phylogeny: Vegetative anatomy. Ann. Missouri Bot. Gard. 62: 590–620.

Dixon RA, D-Y Xie, and SB Sharma. 2005. Proanthocyanidins – a final frontier in flavonoid research. New Phytol. 165: 9–28.

Donoghue MJ and JA Doyle. 1989. Phylogenetic studies of seed plants and angiosperms based on morphological characters.

In: B Fernholm, K Bremer, and H Jornvall, eds. The hierarchy of life: Molecules and morphology in phylogenetic analysis, pp. 181–193. Elsevier Science, Amsterdam.

Doweld A. 2001. Prosyllabus Tracheophytorum. Tentamen systematic plantarum vascularium (Tracheophyta). Geos, Moscow.

Doyle JA and MJ Donogue. 1993. Phylogenesis and angiosperm diversification. Paleobiol. 19: 141–167.

Dumortier B-C. 1829. Analyse des families des plantes. J. Casterman Aîné, Tournay.

Eames AJ. 1961. Morphology of the angiosperms. McGraw-Hill, New York.

Ehrendorfer F. 1977. New ideas about the early differentiation of angiosperms. Plant Syst. Evol., Suppl. 1: 227–234.

Eichler AW. 1875, 1878. Blütendiagramme, 2 vols. Engelmann, Leipzig.

El-Naggar LJ and JL Beal. 1980. Iridoids. A review. J. Natur. Products, 43: 649–707.

Endlicher S. 1836–1840. Genera plantarum secundum ordines naturales disposita. Vindobonae.

Endlicher S. 1841. Enchiridion botanicum. Englemann, Lipsiae/Viennae.

Endress PK. 1994. Diversity and evolutionary biology of tropical flowers. Cambridge University Press, Cambridge.

Endress PK. 1997. Evolutionary biology of flowers in the next century. In: K Iwatsuki and PH Raven, eds. Evolution and diversification of land plants, pp. 99–119. Springer, Tokyo.

Endress PK. 2003a. What should a "complete" morphological phylogenetic analysis entail? In: TF Stuessy, E Hörandl, and V Mayer, eds. Deep morphology: towards a renaissance of morphology in plant systematics, pp. 131–164. Gantner, Ruggell.

Endress PK. 2003b. Morphology and angiosperm systematics in the molecular era. Bot. Rev. 68: 545–570.

Endress PK. 2005. The role of morphology in angiosperm evolutionary studies. Nova Acta Leopold. 92.

Endress PK. 2006. Angiosperm floral evolution: Morphological developmental framework. Adv. Bot. Res. 44: 1–61.

Endress, PK and ML Matthews. 2006. Elaborate petals and staminodes in eudicots: Diversity, function, and evolution. Organisms Divers. Evol. 6: 257–293.

Endress PK and S Stumpf. 1990. Non-tetrasporangiate stamens in the angiosperms: Structure, systematic distribution, and evolutionary aspects. Bot. Jahrb. Syst. 112: 193–240.

Engler A., ed. 1900–1953. Das Pflanzenreich. W. Engelmann/Akademie-Verlag Leipzig/Berlin.

Engler A and L Diels. 1936. Syllabus der Pflanzenfamilien. 2nd ed. Akademie-Verlag Berlin.

Engler A and K Prantl. 1895–1915. Die natürlichen Pflanzenfamilien, 20 vols. W. Engelmann, Leipzig.

Engler A and K Prantl. 1924. Die natürlichen Pflanzenfamilien, 2nd ed. Berlin.

Erbar C. 1991. Sympetaly - a systematic character? Bot. Jahrb. Syst. 112: 417–451.

Erbar C and P Leins. 1994. Flowers in Magnoliidae and the origin of flowers in other subclasses of the angiosperms. I. The relationships between flowers of Magnoliidae and Alismatidae. Plant Syst. Evol., Suppl. 8: 193–208.

Erdtman G. 1952. Pollen morphology and plant taxonomy: An introduction to palynology, vol. 1. Almqvist & Wiksell, Stockholm.

Evert RF. 2006. Esau's plant anatomy. Meristems, cells, and tissues of the plant body: their structure, function, and development, 3rd ed. Wiley, Hoboken, NJ.

Eyde RH. 1971. Evolutionary morphology: Distinguishing ancestral structure from derived structure in flowering plants. Taxon 20: 63–73.

Eyde RH. 1975. The bases of angiosperm phylogeny: Floral anatomy. Ann. Missouri Bot. Gard. 62: 521–537.

Fahey JW, AT Zalcmann, and P Talalay. 2001. The chemical diversity and distribution of glucosinolates and isothiocyanates among plants. Phytochem. 56: 5–51.

Fairbrothers DE. 1977. Perspectives in plant serotaxonomy. Ann. Missouri Bot. Gard. 64: 147–160.

Falser BF. 1975. The bases of the angiosperm phylogeny: Embryology. Ann. Missouri Bot. Gard. 62: 621–646.

Frohlich MW. 2006. Recent developments regarding the evolutionary origin of flowers. Adv. Bot. Res. 44: 63–127.

Frohne D und U Jensen. 1985. Systematik des Pflanze-nreichs. 3rd ed. Fischer, Stuttgart.

Garcia-Cruz J and V Sosa. 2006. Coding quantitative character data for phylogenetic analysis: a comparison of five methods. Syst. Bot. 31: 302–309.

Gaussen H, J-F Leroy, and P Ozenda. 1982. Précis de botanique, vol. 2, Vegetaux superieurs, 2nd ed. Masson, Paris.

Gibbs RD. 1974. Chemotaxonomy of flowering plants, 4 vols. McGiîl Queen's University Press, Montreal/London.

Gobi Chr. 1916. A review of the system of plants. Petrograd (in Russian with French summary.)

Good R. 1956. Features of evolution in flowering plants. Longmans, London.

Grayer RJ, MW Chase, and MSJ Simmonds. 1999. A comparison between the chemical and molecular characters for the determination of phylogenetic relationships among plant families: An appreciation of Hegnauer's "Chemotaxonomie der Pflanzen". Biochem. Syst. Ecol. 27: 369–393.

Gunn CR, JH Wiersema, CA Ritchie, and JH Kirkbride, Jr. 1992. Families and genera of spermatophytes recognized by the Agricultural Research Service. U.S.D.A. Techn. Bull. 1796.

Hallier H. 1901. Über die Verwandtschaftsverhaltnisse der Tubifloren und Ebenalen, den polyphyletischen Ursprung der Sympetalen und Apetalen, und die An-ordnung der Angiospermen über haupt: Vorstudien zum Entwurf eines Stammbaums der Blütenpflanzen. Abh. Naturw. Ver. Hamburg. 16: 1–112.

Hallier H. 1903. Vorläufiger Entwurf des natürlichen (phylogenetischen) Systems der Blütenpflanzen. Bull. Herb. Boiss. 11: 306–317.

Hallier H. 1905a. Ein zweiter Entwurf des natürlichen (phylogenetischen) Systems der Blütenpflanzen. Ber. Deutsch. Bot. Ges. 23: 85–91.

Hallier H. 1905b. Provisional scheme of the natural (phylogenetic) system of flowering plants. New Phytol. 4: 151–162.

Hallier H. 1912. L'origine et le systeme phyletique des Angiospermes exposees a l'aide de leur arbre genealogique. Arch. Neerl., 2nd ser., B (Sci. nat.), 1: 146–234.

Hallier H. 1923. Beitrage zur Kenntnis der Linaceae. Beih. Bot. Centralbl. 39 (2): 1–178.

Hasebe M and JA Banks. 1997. Evolution of MADS gene family of plants. In: K Kwatsuki and PH Raven, eds. Evolution and diversification of land plants, pp. 179–197. Springer, Tokyo.

Hegnauer R. 1962–1991. Chemotaxonomie der Pflanzen, 10 vols. Birkhaeuser, Basel.

Hegnauer R. 1986. Comparative phytochemistry and plant taxonomy. Giorn. Bot. Ital. 120: 15–26.

Hegnauer R. 1988. Biochemistry, distribution and taxonomic relevance of higher plant alkaloids. Phytochemistry 27: 2423–2427.

Heslop-Harrison Y. 1981. Stigma characteristics and angiosperm taxonomy. Nord. J. Bot. 1: 401–420.

Heslop-Harrison Y and KR Shivanna. 1977. The receptive surface of the angiosperm stigma. Ann. Bot. 41: 1233–1258.

Heintze A. 1927. Cormofyternas fylogenis. H. Ohlssons Boktryckeri, Lund.

Heywood VH, ed. 1978. Flowering plants of the world. Oxford University Press, Oxford.

Heywood VH, RK Brummitt, A Culham, and O Seberg. 2007. Flowering plant families of the world. Royal Botanical Gardens, Kew.

Hilu KW, T Borsh, K Müller, DE Soltis, PS Soltis, V Savolainen, MW Chase, MP Powell, LA Alice, R Evans, H Sauquet, C Neinhuis, AB Slotta, JG Rohwer, CS Campbell, and LW Chatrou. 2003. Angiosperm phylogeny based on *mat*K sequence information. Amer. J. Bot. 90: 1758–1776.

Hoogland RD and JL Reveal. 2005. Index nominum familiarum plantarum vascularium. Bot. Rev. 71: 1–291.

Hörandl E. 2006. Paraphyletic versus monophyletic taxa – evolutionary versus cladistic classifications. Taxon 55: 564–570.

Hu Hsien-hsu. 1950. A polyphyletic system of classification ofangiosperms. Sci. Rec. (Peking) 3: 221–230.

Huber H. 1990. Angiospermen. Fischer, Stuttgart.

Hutchinson J. 1924. Contributions towards a phylogenetic classification of flowering plants. IV. Kew Bull. 1924: 114–134.

Hutchinson J. 1926, 1934. The families of flowering plants, 2 vols. Macmillan, London.

Hutchinson J. 1927. Contributions towards a phylogenetic classification of flowering plants. Kew Bull. 6: 100–118.

Hutchinson J. 1959. The families of flowering plants, 3 vols, 2nd ed. Clarendon Press, Oxford.

Hutchinson J. 1969. Evolution and phylogeny of flowering plants. Academic Press, London/New York.

Hutchinson J. 1973. The families of flowering plants arranged according to a new system based on their probable phylogeny. 3rd ed. Clarendon Press, Oxford.

Huysmans S, G El-Ghazaly, and E Smets. 1998. Orbicules in angiosperms: Morphology, function, distribution and relation with tapetum types. Bot. Rev. 64: 240–272.

Jensen SR, BJ Nielsen, and R Dahlgren. 1975. Iridoid compounds, their occurrence, and systematic importance in the angiosperms. Bot. Not. 128: 148–180.

Johri BM, KB Ambegaokar, and PS Srivastava. 1992. Comparative embryology of angiosperms, 2 vols. Springer, Berlin.

Judd WS, RW Sanders, and MJ Donoghue. 1994. Angiosperm family pairs: Preliminary phylogenetic analyses. Harvard Papers Bot. 5: 1–51.

Judd WS, CS Campbell, EA Kellogg, and PF Stevens. 1999. Plant systematics: A phylogenetic approach. Sinauer, Sunderland, MA.

Judd WS, CS Campbell, EA Kellogg, PF Stevens, and MJ Donoghue. 2002. Plant systematics: A phylogenetic approach. 2nd ed. Sinauer, Sunderland, MA.

Jussieu AL de. 1789. Genera plantarum. Herissant, Paris.

Kamelina OP. 2000. Embryological features in phylogenetic systematics of flowering plants. Bot. Zhurn. 85(7): 22–33 (in Russian with English summary).

Kapil RN. 1962. Some recent examples of the value of embryology in relation to taxonomy. Bull. Bot. Surv. India 4: 57–66.

Kapil RN and AK Bhatnagar. 1991. Embryological evidence in angiosperm classification and phylogeny. Bot. Jahrb. Syst. 113: 309–338.

Kimura Y. 1953. The system and phylogenetic tree of plants. J. Jap. Bot. 28: 97–104.

Kozo-Poljanski BM. 1922. An introduction to the phylogenetic systematics of the higher plants. Nature and Culture, Voronezh (in Russian).

Kozo-Poljanski BM. 1928. The ancestors of the angiosperms. Timiryazev Institute Press, Moscow (in Russian).

Kubitzki K. 1972. Probleme der Grossystematik der Blütenpflanzen. Ber. Deutsch. Bot. Ges. 85: 259–277.

Kubitzki K. 1977. Some aspects of the classification and evolution of higher plants. Plant Syst Evol., Suppl. 1: 21–31.

Kubitzki K, ed. 1993, 1998, 2003, 2004, 2007. The families and genera of vascular plants, vols. 2 (1993), 3, 4 (1998), 5 (2003), 6, 7 (2004), 8, 9 (2007). Springer, Berlin/Heidelberg/ New York.

Kubitzki K and OR Gottlieb. 1984a. Micromolecular patterns and the evolution and major classification of angiosperms. Taxon 33: 375–391.

Kubitzki K and OR Gottlieb. 1984b. Phytochemical aspects of angiosperm origin and evolution. Acta Bot. Neerl. 33: 457–468.

Kuijt J. 1969. The biology of parasitic flowering plants. University of California, Berkeley.

Kuznetsov NI. 1936. An introduction to the systematics of flowering plants. Biomedgiz, Leningrad (in Russian).

Kuzoff RK and CS Gasser. 2000. Recent progress in reconstructing angiosperm phylogeny. Trends Plant Sci. 5: 330–336.

Ladd PG. 1994. Pollen presenters in the flowering plants: Form and function. Bot. J. Linn. Soc. 116: 165–195.

Lawrence GHW. 1951. Taxonomy of vascular plants. Macmillan, New York.

Leins P. 2000. Blüte und Frucht. Aspekte der Morphologie, Entwicklungsgeschichte, Phylogenie, Funktion und Ökologie. E. Schwezerbart, Stuttgart.

Leroy J-F. 1993. Origine et évolution des plantes á fleurs. Masson, Paris.

Lersten NR. 2004. Flowering plant embryology. Blackwell, Ames.

Lersten NR, AR Czlapinski, JD Curtis, R Freckmann, and HT Horner. 2006. Oil bodies in leaf mesophyll cells of angiosperms: overview and a selected survey. Am. J. Bot. 93: 1731–1739.

Levin DA. 2006. Flowering phenology in relation to adaptive radiation. Syst. Bot. 31: 239–246.

Lindley J. 1830. An introduction to the natural system of botany. Longman, London.

Lindley J. 1833. Nixus plantarum. Longman, London.

Lindley J. 1836. A natural system of botany. Longman, London.

Lindley J. 1839. An introduction to botany. Longman, London.

Lindley J. 1853. The vegetable kingdom, 3rd ed. Longman, London.

Liscombe DK, BP MacLeod, N Loukanina, OI Nandi, and PJ Facchini. 2005. Evidence for the monophyletic evolution of benzylisoquinoline alkaloid biosynthesis in angiosperms. Phytochemistry 66: 2500–2520.

Lloyd FE. 1942. The carnivorous plants. Chronica Botanica, Waltham, MA.

Loconte H. 1996. Comparison of alternative hypotheses for the origin of the angiosperms. In: DW Taylor, LJ Hickey, eds.

Flowering plant origin, evolution and phylogeny, pp. 267–285. Chapman & Hall, New York.
Lotsy JP. 1911. Vorträge über botanische Stammesg-eschichte. Bd. 3, Heft 1. Cormophyta siphonogamia. Fischer, Jena.
Maas PJM and LYTh Westra. 1993. Neotropical plant families. Koeltz Scientific Books, Germany.
Mabberley DJ. 1997. The plant book: A portable dictionary of the higher plants. Cambridge University Press, Cambridge.
Magallón, SA and MJ Sanderson. 2001. Absolute diversification rates in angiosperm clades. Evolution 55: 1762–1780.
Maheshwari P. 1964. Embryology in relation to taxonomy Vistas Bot. 4: 55–98.
Martin HA. 1946. The comparative internal morphology of seeds. Am. Midi. Nat. 36: 513–660.
Martin PG and JM Dowd. 1991. Studies of angiosperm phylogeny using protein sequences. Ann. Missouri Bot. Gard. 78: 296–337.
Mathews S and MJ Donoghue. 1999. The root of angiosperm phylogeny inferred from duplicate phytochrome genes. Science 286: 947–950.
Mauritzon J. 1939. Die Bedeutung der embryologischen Forschung für das natürliche System der Pflanzen. Lunds Univ. Arsskr. II, 35 (15): 1–70.
Meacham CA. 1994. Phylogenetic relationships at the basal radiation of angiosperms: further study by probability of character compatibility. Syst. Bot. 19: 506–522.
Meeuse ADJ. 1992. Angiosperm evolution: No abominable mystery. Eburon-Delft.
Melchior H, ed. 1964. A. Englers Syllabus der Pflan-zenfamilien. 12th ed, vol. 2, Angiospermen. Borntraeger, Berlin.
Nakai T. 1943. Ordines, familiae, tribi, genera, sections, species, varietates, formae, et combinationes novae a prof. Nakai Takenoschin adhuc ut novis edita. Appendix. Qyestiones acterium naturalium plantarum, etc. Imperial University of Tokyo, Tokyo.
Nakai T. 1952. A synoptical sketch of Korean flora: Or the vascular plants indigenous to Korea arranged in a new natural order. Bull. Nat. Sci. Mus. 31: 1–152.
Nandi O, MW Chase, and PK Endress. 1998. A combined cladistic analysis of angiosperms using *rbc*L and non-molecular data sets. Ann. Missouri Bot. Gard. 85: 137–212.
Nemejc F. 1956. On the problem of the origin and phylogenetic development of the angiosperms. Act Mus. Nat. Prague 12b (23): 59–143.
Netolizky F. 1926. Anatomie der Angiospermen-Samen. Gebrüder Borntraeger, Berlin.
Novak FA. 1954. System Angiosperm. Preslia 26: 337–364.
Novak FA. 1972. Vyssi rostliny, 2 vols. Academia, Praha.
Okada H. 1997. Chromosomal evolution of angiosperms. In: K Iwatsuki and PH Raven, eds. Evolution and diversification of land plants, pp. 209–220. Springer, Tokyo.
Parkin J. 1945. The classification of flowering plants. North West Nat. 23: 18–27.
Parkinson CL, KL Adams, JD Palmer. 1999. Multigene analyses identify the earliest lineages of extant flowering plants. Curr. Biol. 9: 1485–1488.
Poddubnaya-Arnoldi VA. 1982. Characteristic of the families of angiospermous plants according to their cytoembryological features. Springer, Moscow (in Russian).
Pulle AA. 1938, 1952. Compendium van de terminologie, nomenclatuur, en systematiek der zaadplanten. Utrecht.
Qui Y-L, J Lee, F Bernasconi-Quadroni, DE Soltis, PS Soltis, M Zanis, Z Chen, V Savolainen, and MW Chase. 1999. The earliest angiosperms: evidence from mitochondrial, plastid and nuclear genomes. Nature 402: 404–407.
Qui Y-L, O Dombrovska, J Lee, L Li, BA Whitlock, F Bernasconi-Quadroni, JS Rest, CC Davis, T Borsch, KW Hilu, SS Renner, DE Soltis, PS Soltis, MJ Zanis, JJ Cannone, RR Gutell, M Powell, V Savolainen, LW Chatrou, and M Chase. 2005. Phylogenetic analyses of basal angiosperms based on nine plastid, mitochondrial, and nuclear genes. Int. J. Plant Sci. 166: 815–842.
Raven PH. 1975. The bases of angiosperm phylogeny: Cytology. Ann. Missouri Bot. Gard. 62: 724–764.
Rendle AB. 1930, 1938. The classification of flowering plants, 2 vols. Cambridge University Press, Cambridge.
Rohweder O and PK Endress. 1983. Samenpflanzen. Thieme, New York.
Sanderson MJ, JL Thorne, N Wikström, and K Bremer. 2004. Molecular evidence on plant divergence times. Am. J. Bot. 91: 1656–1665.
Savolainen V, MW Chase, SB Hoot, CM Morton, DE Soltis, C Bayer, MF Fay, AY de Bruijn, S Sullivan, and Y-L Qui. 2000. Phylogenetics of flowering plants based upon a combined analysis of plastid *atp*B and *rbc*L gene sequences. Syst. Biol. 49: 306–362.
Schaffner JH. 1929. Principles of plant taxonomy. VII. Ohio J. Sci. 29: 243–252.
Schaffner JH. 1934. Phylogenetic taxonomy of plants. Quart. Rev. Biol. 9: 129–160.
Serna L and C Martin. 2006. Trichomes: Different regulatory networks lead to convergent structures. Trends Plant Sci. 11: 274–280.
Shamrov II. 1998. Ovule classification in flowering plants — new approaches and concepts. Bot. Jahrb. Syst. 120: 377–407.
Shamrov II. 2000. The integument of flowering plants: developmental patterns and evolutionary trends. Acta Biol. Cracoviensis, Ser. Bot. 42: 9–20.
Shipunov AB. 2003. The system of flowering plants: synthesis of classical and molecular approaches. J. Gen. Biol. 64: 499–507 (in Russian, with English summary).
Smith N, SA Mori, A Henderson, DW Stevenson, and SV Heald. eds. 2004. Flowering plants of the Neotropics. Princeton University Press, Princeton, NJ.
Sokoloff, D, PJ Rudall, and M Remizowa. 2006. Flower-like terminal structures in racemose inflorescences: A tool in morphogenetic and evolutionary research. J. Experim. Bot. 57: 3517–3530.
Soltis DE, PA Soltis, MJ Zanis. 2002. Phylogeny of seed plants based on evidence from eight genes. Am. J. Bot. 89: 1670–1681.
Soltis DE, C Hibsch-Jetter, PM Soltis, MW Chase, and J Farris. 1997. Molecular phylogenetic relationships among angiosperms: an overview based on *rbc*L and 18S rDNA sequences. In: K Iwatsuki and PH Raven, eds. Evolution and diversification of land plants, pp. 157–178. Springer, Tokyo.
Soltis DE, PS Soltis, DL Nickrent, LA Johnson, WJ Hahn, SB Hoot, JA Sweere, RK Kuzoff, KA Kron, MW Chase, SM Swenson, EA Zimmer, SM Chaw, LJ Gillespie, WJ Kress, and KJ Systma. 1997. Angiosperm phylogeny inferred from 18S ribosomal DNA sequences. Ann. Missouri Bot. Gard. 84: 1–49.

Soltis DE, PS Soltis, MW Chase, ME Mort, DC Albach, M Zanis, V Savolainen, WH Hahn, SB Hoot, MF Fay, M Axtell, SM Swensen, LM Prince, WJ Kress, KC Nixon, and JS Farris. 2000. Angiosperm phylogeny inferred from a combined data set of 18S rDNA, *rbc*L, and *atp*B sequences. Bot. J. Linn. Soc. 133: 381–461.

Soltis DE, PS Soltis, PK Endress, and MW Chase. 2005. Phylogeny and evolution of angiosperms. Sinauer, Sunderland, MA.

Soltis PS, DE Soltis, MW Chase. 1999. Angiosperm phylogeny inferred from multiple genes as a tool for comparative biology. Nature 402: 402–404.

Soo R. 1967. Die modernen Systeme der Angiospermen. Acta Biol. Acad. Sci. Hungar. 13: 201–233.

Sporne KR. 1956. The phylogenetic classification of the angiosperms. Biol. Rev. Cambridge Philos. Soc. 31: 1–29.

Stebbins GL. 1974. Flowering plants: evolution above the species level. Belknap Press, Cambridge, MA.

Stevens PF. Angiosperm Phylogeny Website [Electronic resource]: Version 7. 2006. Mode of access: http://www.mobot.org/MOBOT/research/APweb/.

Stuessy TF, V Mayer, and E Hörandl, eds. 2003. Deep Morphology: toward a renaissance of morphology in plant systematics. Ruggell, Liechtenstein.

Takhtajan A. 1959. Die Evolution der Angiospermen. Fischer, Jena.

Takhtajan A. 1966 (1967). A system and phylogeny of the flowering plants. Nauka, Moscow/Leningrad (in Russian).

Takhtajan A. 1969. Flowering plants: Origin and dispersal. Oliver & Boyd, Edinburgh.

Takhtajan A. 1980. Outline of the classification of flowering plants (Magnoliophyta). Bot. Rev. 46: 225–359.

Takhtajan A, ed. 1980, 1981, 1982. Life of plants: flowering plants, vols. 5(1), 5(2), 6. Prosveshchenie, Moscow (in Russian).

Takhtajan A, ed. 1985, 1988, 1991, 1992, 1996, 2000. Comparative seed anatomy. Nauka, Vols. 1–6 + (in Russian).

Takhtajan A. 1986. Floristic regions of the world. University of California Press, Berkeley.

Takhtajan A. 1987. Systema Magnoliophytorum. Nauka, Leningrad (in Russian).

Takhtajan A. 1997. Diversity and classification of flowering plants. Columbia University Press, New York.

Thorne RF. 1958. Some guiding principles of angiosperm phylogeny. Brittonia 10: 72–77.

Thorne RF. 1968. Synopsis of a putative phylogenetic classification of the flowering plants. Aliso 6 (4): 57–66.

Thorne RF. 1976. A phylogenetic classification of the Angiospermae. Evol. Biol. 9: 35–106.

Thorne RF. 1977. Some realignments in the Angiospermae. Plant Syst. Evol., Suppl. 1: 299–319.

Thorne RF. 1981. Phytochemistry and angiosperm phylogeny: A summary statement. In: DA Young and DS Seigler, eds. Phytochemistry and angiosperm phylogeny, pp. 233–295. Praeger, New York.

Thorne RF. 1992a. An updated phylogenetic classification of the flowering plants. Aliso 13: 365–389.

Thorne RF. 1992b. Classification and geography of the flowering plants. Bot. Rev. 58: 225–348.

Thorne RE. 2000. The classification and geography of the flowering plants: Dicotyledons of the class Angiospermae. Bot. Rev. 66: 441–647.

Thorne RF. 2006. An updated classification of the class Magnoliopsida ("Angiospermae"). http://rsabg.org/angiosperms.angiosperms.pdf

Tieghem P van. 1891. Traité de botanique. Deuxieme partie, 2nd ed. Libririe F. Savy, Paris.

Tieghem P van. 1906. Elements de botanique. Deuxieme partie, 4th ed. Masson, Paris.

Tucker SC and J Grimes. 1999. The inflorescence: Introduction. Bot. Rev. 65: 303–316.

Turland NJ and FR Barrie. 2001. Family name listings modified in Appendix IIB of the Saint Louis Code. Taxon 50: 897–903.

Tyree MT and MH Zimmermann. 2002. Xylem structure and the ascent of sap, 2nd ed. Springer, New York.

Verdú M. 2006. Tempo, mode and phylogenetic associations of relative embryo size evolution in angiosperms. J. Evol. Biol. 19: 625–634

Wagenitz G. 1975. Blütenreduktion als ein zentrales Problem der Angiospermen-Systematik. Bot. Jahrb. Syst. 96: 448–470.

Wagenitz G. 1997. The impact of molecular methods on the systematics of angiosperms. Bot. Acta 110: 274–281.

Walker JW and JA Doyle. 1975. The bases of angiosperm phylogeny: Palynology. Ann. Missouri Bot. Gard. 62: 664–723.

Warming E. 1912. Froplanterne. Blanco Lunus, Copenhagen.

Warming E. 1913. Observations sur la valeur systématique de l'ovule. Blanco Lunus, Copenhagen.

Watson L and MJ Dallwitz. 2006. The Families of Flowering Plants [Electronic resource]. 2006. http://biodiversity.uno.edu/delta/angio/index

Weberling F. 1989. Morphology of flowers and inflorescences. Cambridge University Press, Cambridge.

Werker E. 1997. Seed anatomy. Gebrüder Borntraeger, Berlin.

Wernham HF. 1912. Floral evolution: With particular reference to the sympetalous dicotyledons. New Phytol. 11: 373–397.

Wettstein R. 1935. Handbuch der systematischen Botanik. Franz Deuticke, Leipzig/Wien.

Wielgorskaya T. 1995. Dictionary of generic names of seed plants. Columbia University Press, New York.

Wikström N, V Savolainen, and MW Chase. 2001. Evolution of the angiosperms: Calibrating the family tree. Proc. Roy. Soc. London B, 268: 2211–2220.

Wikström N, V Savolainen, and MW Chase. 2003. Angiosperm divergence times: Congruence and incongruence between fossils and sequence divergence estimates. In: PCJ Donoghue and MP Smith, eds. Telling the evolutionary time: molecular clocks and the fossil record, pp. 142–165. CRC Press Boca Raton, FL.

Willis JC. 1973. A dictionary of the flowering plants and ferns, 8th ed. revised by HK. Airy Shaw. Cambridge University Press, Cambridge.

Wu ZY, LU Anmin, YC Tang, ZD Chen, and LI Dezhu. 2003. The families and genera of Angiosperms in China. A comprehensive analysis. Science Press, Beijing (in Chinese).

Wu ZY, YC Tang, AM Lu, ZD Chen. 1998. On primary subdivisions of the Magnoliophyta – towards a new scheme for an eight-class system of classification of the angiosperms. Acta Phytotax. Sin. 36(5): 385–402.

Yakovlev MS, ed. 1981–1990. Comparative embryology of flowering plants, 5 vols. Nauka, Leningrad (in Russian).

Yakovlev MS and GY Zhukova. 1980. Chlorophyll in embryos of angiosperm seeds, a review. Bot. Notiser 133: 323–336.

Zimmermann W. 1959. Die Phylogenie der Pflanzen. Fischer, Stuttgart.

Class MAGNOLIOPSIDA (DICOTYLEDONS)

Embryo usually with two cotyledons, sometimes one, rarely with three or four. Cotyledons commonly with three vascular bundles. Leaves mostly petioled. Venation typically reticulate, either pinnate or palmate, and mostly not closed (with free vein endings). Leaf traces 1–3, seldom more. Prophylls and bracteoles are usually paired. Plumule terminal. Vascular bundles usually arranged in a ring, seldom in two or more rings or scattered. The root system mostly that of a tap root from which side branches arise. The root cap and epidermis mostly of a common ontogenetical origin. Plants woody or herbaceous, frequently secondarily arborescent. Epicuticular wax ultrastructure mostly of *Aristolochia*-type, but in Winteraceae and Canellaceae they are of *Berberis*-type. Plastids of the sieve elements of S-type or less often of P-type. Flowers mostly 5- or (less frequently) 4-merous and only in some mainly archaic groups 3-merous. Floral nectaries of various types but never septal or nectaries absent. Pollen grains typically triaperturate or of triap-erturate-derived type, except in some archaic families, where they are distally uniaperturate or rarely biaperturate. The class Magnoliopsida includes 8 subclasses, 125 orders, c. 440 families, almost 10,500 genera, and no less than 195,000 species.

Bibliography

Baas P, E Wheeler, and M Chase. 2000. Dicotyledonous wood anatomy and the APG system of angiosperm classification. Bot. J. Linn. Soc. 134: 3–17.

Baas P, S Jansen, E Smets, and EA Wheeler. 2003. Ecological adaptations and deep phylogenetic splits – evidence and questions from the secondary xylem. In: TF Stuessy, V Mayer, and E Hörandl, eds. Deep morphology: toward a renaissance of morphology in plant systematics, pp. 221–240. A.R.G. Gantner Verlag, Liechtenstein.

Bate-Smith EC. 1962. The phenolic constituents of plants and their taxonomic significance: I. Dicotyledons. Bot. J. Linn. Soc. 58: 95–173.

Behnke H-D. 1991. Distribution and evolution of forms and types of sieve-element plastids in the dicotyledons. Aliso 13: 167–182.

Boros CA and FR Stermitz. 1990. Iridoids. An updated review, I. J. Nat. Prod. 53: 1055–1147.

Boros CA and FR Stermitz. 1991. Iridoids. An updated review, II. J. Nat. Prod. 54: 1172–1246.

Burr B und W Barthlott. 1993. Untersuchungen zur Ultraviolettreflexion von Angiospermenblüten II. Magnoliidae, Ranunculidae, Hamamelididae, Caryophyllidae, Rosidae. Akad. Wiss. Lit. (Münz). Trop. Subtrop. Pflanzenwelt, vol. 87. Stuttgart.

Carlquist S. 1984. Vessel grouping in dicotyledon woods: significance and relationship to imperforate tracheary elements. Aliso 10: 505–525.

Carlquist S. 1992. Wood anatomy of sympetalous dicotyledon families: a summary with comments on systematic relationships and evolution of the woody habit. Ann. Missouri Bot. Gard. 79: 303–332.

Carlquist S. 2001. Comparative wood anatomy. Systematic, ecological, and evolutionary aspects of dicotyledon wood, 2nd ed. Springer-Verlag, Berlin.

Carpenter KJ. 2005. Stomatal architecture and evolution in basal angiosperms. Am. J. Bot. 92: 1595–1615.

Corner EJH. 1976. The seeds of dicotyledons, 2 vols. Cambridge University Press, Cambridge.

Cutler DF and M Gregory. 1998. Anatomy of the dicotyledons, vol. IV. Saxifragales (sensu Armen Takhtajan 1983), 2nd edition. Clarendon, Oxford.

Dahlgren G. 1989. The last Dahlgrenogram: system of classification of the dicotyledons. In: K Tan, RR Mill, and TS Elias, eds. Plant taxonomy, phytogeography, and related subjects, pp. 249–260. University Press Edinburgh.

Dahlgren G. 1991. Steps toward a natural system of the dicotyledons. Aliso 13: 107–165.

Davis CC and KJ Wurdack. 2004. Host-to-parasite gene transfer in flowering plants: phylogenetic evidence from Malpighiales. Science 305: 676–678.

Ditsch F and W Barthlott. 1997. Mikromorphologie der Epicuticularwachse und das System der Dilleniidae und

A. Takhtajan, *Flowering Plants*,

Rosidae. Akad. Wiss. Lit. (Münz). Trop. Subtrop. Pflanzenwelt, vol. 97. Stuttgart.

Doyle JA and PK Endress. 2000. Morphological phylogenetic analyses of basal angiosperms: comparison and combination with molecular data. Int. J. Plant Sci. 161(6, Suppl.): S121–S153.

Drinnan AN, PR Crane, and SB Hoot. 1994. Patterns of floral evolution in the early diversification of non-magnoliid dicotyledons (eudicots). In: PK Endress and EM Friis, eds. Early evolution of flowers, pp. 93–122 Vienna (Plant Syst. Evol. Suppl. 8).

Endress PK. 1994. Floral structure and evolution of primitive angiosperms: recent advance. Plant. Syst. Evol. 192: 79–97.

Endress PK. 2004. Structure and relationships of basal relictual angiosperms. Aust. Syst. Bot. J. 17: 343–366.

Endress PK. 2005. The role of morphology in angiosperm evolutionary studies. Nova Acta Leopold. 92.

Endress PK and A Igersheim. 1999. Gynoecium diversity and systematics of the basal eudicots. Bot. J. Linn. Soc. 130: 305–393.

Friedman WE and JH Williams. 2003. Modularity of the angiosperm female gametophyte and its bearing on the early evolution of endosperm in flowering plants. Evolution 57: 216–230.

Friedman WE and JH Williams. 2004. Developmental evolution of the sexual process in ancient flowering plant lineages. Plant Cell 16: S119–S132.

Friedman WE, RC Moore, and MD Purugganan. 2004. The evolution of plant development. Am. J. Bot. 91: 1726–1741.

Friis EM, PR Crane, and KP Pedersen. 1997. Fossil history of magnoliid angiosperms. In: K. Iwatsuki and PH Raven, eds. Evolution and diversification of land plants, pp. 121–156. Springer, Tokyo.

Furness CA and PJ Rudall. 2004. Pollen aperture evolution – a crucial factor for eudicot success? Trends Plant Sci. 9: 154–158.

Furness CA, PJ Rudall, and FB Sampson. 2002. Evolution of microsporogenesis in angiosperms. Int. J. Plant Sci. 163: 235–260.

Goldberg A. 1986. Classification, evolution, and phylogeny of the families of dicotyledons. Smithsonian Contr. Bot. 58: 1–314.

Gregory M and P Baas. 1989. A survey of mucilage cells in vegetative organs of the dicotyledons. Israel J. Bot. 38: 125–174.

Gundersen A. 1950. Families of Dicotyledons. Chronica Botanica, Waltham, MA.

Hallier H. 1908. Über *Juliania*, eine Terebinthaceen-Gattung mit Cupula und die wahren Stammeltern der Kätzchenblüthler. Neue Beiträge zur Stammesgeschichte nebst einer Übersicht über das natürliche System der Dicotyledonen. Beih. Bot. Centralbl. Bd. 23, Heft 2: 81–265.

Hennig S, W Barthlott, I Meusel, und I Theisen. 1994. Mikromorphologie der Epicuticularwachse und die Systematik der Magnoliidae, Ranunculidae und Hamamelididae. Trop. Subtrop. Pflanzenwelt 90: 1–60.

Hoot SB, S Magallón, and PR Crane. 1999. Phylogeny of basal eudicots based on three molecular data sets: *atp*B, *rbc*L, and 18S nuclear ribosomal DNA sequences. Ann. Missouri Bot. Gard. 86: 1–32.

Hutchinson J. 1964. The genera of flowering plants: Dicotyledons, 2 vols. Clarendon, Oxford.

Igersheim A and PK Endress. 1998. Gynoecium diversity and systematics of the paleoherbs. Bot. J. Linn. Soc. 127: 289–370.

Jensen U. 1991. Stages towards the natural system of the Dicotyledons: serological characters. Aliso 13: 183–190.

Metcaife CR and L Chalk. 1950. Anatomy of the dicotyledons, 2 vols. Clarendon, Oxford.

Metcaife CR and L Chalk. 1979, 1983, 1987. Anatomy of the dicotyledons, 2d ed, 3 vols. (to be continued). Clarendon, Oxford.

Philipson WR. 1961. Relationship and convergence in angiosperms. Phytomorphology 10: 367–376.

Philipson WR. 1974. Ovular morphology and the major classification of the dicotyledons. Bot. J. Linn. Soc. 68: 89–108.

Philipson WR. 1975. Evolutionary lines within the dicotyledons. New Zealand J. Bot. 13: 73–91.

Philipson WR and EE Balfour. 1963. Vascular patterns in dicotyledons. Bot. Rev. 29: 382–404.

Ronse Decraene LP and EF Smets. 1996. The morphological variation and systematic value of stamen pairs in the Magnoliatae. Feddes Repert. 107: 1–17.

Savolainen V, MF Fay, DC Albach, A Backlund, M van der Bank, KM Cameron, SA Johnson, MD Lledó, J-C Pintaud, M Powell, MC Sheahan, DE Soltis, PS Soltis, P Weston, WM Whitten, KJ Wurdack, and MW Chase. 2000. Phylogeny of the eudicots: a nearly complete familial analysis based on *rbc*L gene sequences. Kew Bull. 55: 257–309.

Smets EF. 1986. Localization and systematic importance of the floral nectaries in the Magnoliatae (dicotyledons). Bull. Jard. Bot. Nat. Belg. 56: 51–76.

Soares GLG and Kaplan MAC. 2001. Analysis of flavone-flavonol ratio in Dicotyledoneae. Bot. J. Linn. Soc. 135: 61–66.

Takhtajan A. 1983. The systematic arrangement of dicotyledonous families. In: CR Metcaife and L Chalk, eds. Anatomy of the dicotyledons, vol. 2, 2nd ed, pp. 180–201. Clarendon, Oxford.

Theisen I and W Barthlott. 1994. Mikromorphologie der Epicuticularwachse und die Systematik der Gentianales, Rubiales, Dipsacales und Calycerales. Trop. Subtrop. Pflanzenwelt 89: 1–62.

Thomas V. 1991. Structural, functional and phylogenetic aspects of the colleter. Ann. Bot. 68: 287–305.

Vogel S. 1997. Remarkable nectaries: structure, ecology, organophyletic perspectives. I. Substitutive nectaries. Flora 192: 305–333.

Vogel S. 1998a. Remarkable nectaries: structure, ecology, organophyletic perspectives. II. Nectarioles. Flora 193: 1–29.

Von Teichman I and AE van Wyk. 1991. Trends in the evolution of dicotyledonous seeds based on character associations, with special reference to pachychalazy and recalcitrance. Bot. J. Linn. Soc. 105: 211–237.

Walker JW and JA Doyle. 1975. The bases of angiosperm phylogeny: palynology. Ann. Missouri Bot. Gard. 62: 664–723.

Zahur MS. 1959. Comparative study of secondary phloem of 423 species of woody dicotyledons belonging to 85 families. Mem. Cornell Univ. Agric. Exp. Sta., no. 358.

Zanis MJ, PS Soltis, Y-L Qui, EA Zimmer, and DE Soltis. 2003. Phylogenetic analyses and perianth evolution in basal angiosperms. Ann. Missouri Bot. Gard. 90: 129–150.

Dicotyledonous Family of Incertae Position

1. *HAPTANTHACEAE*

C. Nelson 2002. (Isonym: Haptanthaceae Shipunov in Zhurn. Obshchei Biol. 64: 504. 2003, validated by a diagnosis in Latin). 1/1. Honduras (from 5 km southeast of Mataras, Alantida).

Evergreen glabrous tree. Vessels with scalariform perforations or scalariform and reticulate; scalariform perforations with 30–35 bars. Axial parenchyma apotracheal; wood rays uniseriate and heterocellular. Tracheids with circular pits; fibers storied. Leaves opposite, estipulate. Flowers in axillary cymose, unisexual (plant monoecious), 2-merous. Male flowers: petal absent, stamen inserted at different levels on androphore; each stamen with minute bracteole; filaments flattened; anthers 2-locular, basifixed. Pollen grains small, tricolpate and reticulate. Female flowers: petals ? four, staminodes absent. Gynoecium of three carpels, stipitate; styles three, thick, linear, arched, united at the base; ovary superior, unilacunar, with three parietal placentas, each with 8–15 two-ranked ovules; ovules bitegmic; placentas and ovary wall deeply intruded into the locule for part of the ovary. Fruit unknown, perhaps a berry.
Haptanthus

Bibliography

Doust AN and PF Stevens. 2005. A reinterpretation of the staminate flowers of *Haptanthus*. Syst. Bot. 30: 779–785.
Goldberg A and HA Alden. 2005. Taxonomy of *Haptanthus* Goldberg & C. Nelson. Syst. Bot. 30: 773–778.
Goldberg A and C Nelson. 1989. *Haptanthus*, a new dicotyledonous genus from Honduras. Syst. Bot. 14: 16–19.
Nelson CH. 2002. Haptanthaceae C. Nelson, fam. nov. Ceiba 42: 33.

Subclass I. MAGNOLIIDAE

Woody or herbaceous plants. Secretory cells with oily contents ordinarily present in the parenchymatous tissues. Vessels with scalariform or simple perforations or vessels wanting. Sieve-element plastids usually containing protein crystalloid (Pc-type) or filaments (Pf-type), often also starch, in some families only starch (S-type). Stomata commonly paracytic. Flowers bisexual or less often unisexual, frequently spiral or spirocyclic, actinomorphic. Stamens mostly numerous. Tapetum usually secretory. Microsporogenesis successive or simultaneous. Pollen grains 2-celled or less often 3-celled, 1-colpate, 2-colpate, 3–6-colpate, rugate, porate, or often inaperturate. Gynoecium mostly apocarpous. Ovules bitegmic or much less often unitegmic, usually crassinucellate. Endosperm cellular or nuclear. Seeds mostly with small or minute embryo and copious endosperm, sometimes accompanied or largely replaced by perisperm. Cotyledons typically 2, but occasionally 3 or 4 (Degeneriaceae, Idiospermaceae). Commonly producing neolignans and/or benzyl isoquinoline alkaloids, but without ellagic acid and iridoid compounds.

The subclass Magnoliidae includes a number of relatively very archaic orders and families of flowering plants. All of them are extremely heterobathmic, that is, they have a very disharmonious combination of both primitive and derived characters.

Different families of the magnoliids developed in different directions. Although all of them most probably evolved from a common ancestral stock.

The basal group of flowering plants are superorder Nymphaeanae, which include the most archaic families, beginning with Amborellaceae and ending with Ceratophyllaceae.

Bibliography

Agababian VS. 1973. Pollen grains of primitive angiosperms. University Press Yerevan (in Russian).

Bailey IW and BGL Swamy. 1951. The conduplicate carpel of dicotyledons and its initial trend of specialization. Am. J. Bot. 38: 373–379.

Behnke H-D. 1971. Sieve-tube plastids in Magnoliidae and Ranunculidae in relation to systematics. Taxon 20: 723–730.

Behnke H-D. 1988. Sieve-element plastids, phloem protein, and evolution of flowering plants: III. Magnoliidae. Taxon 37: 699–732.

Carlquist S. 1996. Wood anatomy of primitive angiosperms: new perspectives and syntheses. In: DW Taylor and LJ Hickey, eds. Flowering plant origin, evolution, and phylogeny, pp. 68–90. Chapman & Hall, New York.

Doyle JA and PK Endress. 2000. Morphological phylogenetic analysis of basal angiosperms: comparison and combination with molecular data. Int. J. Plant Sci. 161(Suppl. 6): 121–153.

Ehrendorfer E. 1976. Evolutionary significance of chromosomal differentiation patterns in gymnosperms and primitive angiosperms. In: CB Beck, ed. Origin and early evolution of angiosperms, pp. 220–240. Columbia University Press, New York.

Ehrendorfer F. 1988. Affinities of the African dendroflora: Suggestions from karyo- and chemosystematics. In: P Goldblatt and PP Lowry II., eds. Modern systematic studies in African botany. Monographs Syst. Bot. Missouri Bot. Gard. 25: 105–127.

Ehrendorfer E, K Krendl, E Habeller, and W Sauer. 1968. Chromosome numbers and evolution in primitive angiosperms. Taxon 17: 337–353.

Endress PK. 1973. Arils and aril-like structures in woody Ranales. New. Phytol. 72: 1159–1171.

Endress PK. 1983. Dispersal and distribution in some small archaic relic families (Austrobaileyaceae, Eupomatiaceae, Himantandraceae,Idiospermoideae-Calycanthaceae). Sonderh. Naturwiss. Vereins Hamburg 7: 201–217.

Endress PK. 1986. Reproductive structures and phylogenetic significance of extinct primitive angiosperms. Plant Syst. Evol. 152: 1–18.

Endress PK. 1994. Shapes, sizes, and evolutionary trends in stamens of Magnoliidae. Bot. Jahrb. Syst. 115: 429–460.
Endress PK. 1996. Evolutionary aspects of fruits in basal flowering plants. Det Norske Vidensk. Akademi, I. Mat.-Naturv. Klasse, Achandlinger, n.s. 18: 21–32.
Endress PK. 2001. The flowers in extant basal angiosperms and inferences on ancestral flowers. Int. J. Plant Sci. 162: 1111–1140.
Endress PK and D Hufford. 1989. The diversity of stamen structures and dehiscence patterns among Magnoliidae. Bot. J. Linn. Soc. 100: 45–85.
Endress PK and A Igersheim. 2000. Gynoecium structure and evolution in basal angiosperms. Int. J. Plant Sci. 161(Suppl. 6): 211–223.
Erbar C. 1983. Zum Karpellbau einiger Magnoliiden. Bot. Jahrb. Syst. 104: 3–31.
Erbar C. 1994. Flowers in Magnoliidae and the origin of flowers in other subclasses of the angiosperms. II. The relationships between flowers of Magnoliidae, Dilleniidae, and Caryophyllidae. Plant Syst. Evol., Suppl. 8: 209–218.
Erbar C and P Leins. 1983. Zur Sequenz von Blütenorganen bei einigen Magnoliiden. Bot. Jahrb. Syst. 103: 433–449.
Floyd SF and WE Friedman. 2000. Evolution of endosperm developmental patterns among basal flowering plants. Int. J. Plant Sci. 161: S57–S81.
Floyd SK and WE Friedman. 2001. Developmental evolution of endosperm in basal angiosperms: evidence from *Amborella* (Amborellaceae), *Nuphar* (Nymphaeaceae), and *Illicium* (Illiaceae). Plant Syst. Evol. 228: 153–169.
Furness CA and PJ Rudall. 2001. The tapetum in basal angiosperms: Early diversity. Int. J. Plant Sci. 162: 375–392.
Gabarayeva NI. 1991. Patterns of development in primitive angiosperm pollen. In: S Blackmore and SH Barnes, eds. Pollen and spores: Patterns of diversification, pp. 257–268. Claredon Press, Oxford.
Goldblatt P. 1974. A contribution to the knowledge of cytology in Magnoliidae. J. Arnold Arbor. 55: 453–457.
Gottlieb OR, MAC Kaplan, K Kubitzki, and JR Toledo Barros. 1989. Chemical dichotomies in the Magnolealean complex. Nord. J. Bot. 8: 437–444.
Gottsberger G. 1977. Some aspects of beetle pollination in the evolution of flowering plants. Plant Syst. Evol., Suppl. 1: 211–226.
Gottwald HPJ. 1977. The anatomy of secondary xylem and the classification of ancient dicotyledons. Plant Syst. Evol., Suppl. 1:111–121.
Graham SW and RG Olmstead. 2000. Utility of 17 chloroplast genes for inferring the phylogeny of the basal angiosperms. Am. J. Bot. 87: 1712–1730.
Groot EP, JA Doyle, SA Nichol, and TL Rost. 2004. Phylogenetic distribution and evolution of root apical meristem organization in dicotyledonous angiosperms. Int. J. Plant Sci. 165: 97–105.
Hiepko P. 1965. Vergleichend-morphologische und entwicklungsgeschichdiche Untersuchungen über das Perianth bei den Polycarpicae. Bot. Jahrb. Syst. 84: 359–508.
Hoot SB, S Megallon, and PR Crane. 1999. Phylogeny of basal eudicots based on three molecular datasets: *atp*B, *rbc*L and 18S nuclear ribosomal DNA sequences. Ann. Miss. Bot. Gard. 86: 1–32.
Igersheim A and PK Endress. 1997. Gynoecium diversity and systematics of Magnoliales and winteroids. Bot. J. Linn. Soc. 124: 213–271.
Jensen U and G Greven. 1984. Serological aspects and phylogenetic relationships of the Magnoliidae. Taxon 33: 563–577.
Judd WS and RG Olmstead. 2004. A survey of tricolpate (eudicot) phylogenetic relationships. Am. J. Bot. 91: 1627–1644.
Kramer EM and VF Irish. 2000. Evolution of the petal and stamen developmental programs: evidence from comparative studies of the lower eudicots and basal angiosperms. Int. J. Plant Sci. 161(Suppl.): 29–40.
Kubitzki K and H Reznik. 1966. Flavonoid-Muster der Polycarpicae als systemadsches Merkmal: I. Übersichtuber die Fammen. Beitr. Biol. Pfl. 42: 445–470.
Lemesle R. 1955. Contribution a l'etude de quelques families dedicotyledonesconsiderescommeprimitives. Phytomorphology 5: 11–45.
Le Thomas A. 1988. Les structures reproductives des Magnoliales Africaines et Malgashes: Significations phylogeniques. Monographs Syst. Bot. Missouri Bot. Gard. 25: 161–174.
Loconte H and DW Stevenson. 1991. Cladistics of the Magnoliidae. Cladistics 7: 267–296.
Metcaife CR. 1987. Anatomy of the dicotyledons. Vol. 3, Magnoliales, Illiciales, and Laurales, 2nd ed. Claredon Press, Oxford.
Okada H. 1975. Karyomorphological studies of woody Polycarpicae. J. Sci. Hiroshima Univ., ser. B, Div. 2, Bot. 15: 115–200.
Ozenda P. 1949. Recherches sur les dicotyledones apo-carpiques: Contribudon a l'etude des angiospermes dites primidves. Publ. Lab. de l'Ecole Norm. Sup., Ser. Biol. Fasc. 2. Paris.
Qiu Y-L, MW Chase, DH Les, and CR Parks. 1993. Molecular phylogenedcs of the Magnoliidae: Cladistic analyses of nucleotide sequences of the plastid gene *rbc*L. Ann. Missouri Bot. Gard. 80: 587–606.
Raven PH, DW Kyhos, and MS Cave. 1971. Chromosome numbers and relationships in Annoniflorae. Taxon 20: 479–483.
Ronse Decreane LP and EF Smets. 1992. Complex polyandry in the Magnoliadae: Definition, distribution, and systematic value. Nord. J. Bot. 12: 621–649.
Sampson 2000. Pollen diversity in some modern magnoliids. Int. J. Plant Sci. 161(Suppl.): 193–210.
Sastri RLN. 1969. Comparative morphology and phylogeny of the Ranales. Biol. Rev. Cambridge Philos. Soc. 44: 291–319.
Smith AC. 1971 (1972). An appraisal of the order and families of primitive extant angiosperms. J. Indian Bot. Soc., Golden Jubilee volume, 50A: 215–226.
Sugiyama M. 1979. A comparative study of nodal anatomy in the Magnoliales based on the vascular system in the node-leaf continuum. J. Fac. Sci. Univ. Tokyo, Sect. 3, Bot. 12(5): 199–279.
Takhtajan AL. 1948. Morphological evolution of the angiosperms. Nauka, Moscow (in Russian).
Tucker SC and AW Douglas. 1996. Floral structure, development, and relationships of paleoherbs: *Saruma, Cabomba, Lactoris*, and selected Piperales, pp. 141–175. In: DW Taylor, LJ Hickey, eds. Flowering plant origin, evolution and phylogeny. Chapman & Hall, New York.
Thien LB. 1980. Patterns of pollination in the primitive angiosperms. Biotropica 12: 1–13.

Thorne RB. 1974. A phylogenetic classification of the Annoniflorae. Aliso 8: 147–209.
Walker JW. 1974a. Evolution of exine structure in the pollen of primitive angiosperms. Am. J. Bot. 61: 891–902.
Walker JW. 1974b. Aperture evolution in the pollen of primitive angiosperms. Am. J. Bot. 61: 1112–1137.
Walker JW. 1976a. Comparative pollen morphology and phylogeny of the ranalean complex. In: CB Beck, ed. Origin and early evolution of angiosperms, pp. 241–299. Columbia University Press, New York.
Walker JW. 1976b. Evolutionary significance of the exine in the pollen of primitive angiosperms. In: IK Ferguson and J Muller, eds. The evolutionary significance of the exine, pp. 251–308. Linn. Soc. Symposium, No. 1. Columbia University Press, London/New York.
Wu Z-Y, A-M Lu, and Y-C Tang. 1998. A comprehensive study of "Magnoliidae" sensu lato. – With special consideration on the possibility and the necessity for proposing a new "polyphyletic-polychronic-polytopic" system of angiosperms. In: A-L Zhang and S-G Wu, eds. Floristic characteristics and diversity of East Asian plants, pp. 269–334. China Higher Education Press/Springer, Beijing/Berlin.
Zhu Min and Xiao Pei-gen. 1991. Distribution of benzyl isoquinolines in Magnoliidae and other taxa. Acta Phytotax. Sin. 29: 142–155 (in Chinese with English summary).

Superorder NYMPHAEANAE

Order 1. AMBORELLALES

Evergreen shrubs with several stems from common base up to 8 m; young parts tomentose or glabrous. Nodes unilacunar with a single trace. Xylem vesselless. Axial parenchyma scanty diffuse; pericycle with hippocrepiform sclereids. Rays 1-2-(5)-seriate; ray cells have lignified secondary walls and contains droplets or massive yellowish to brownish deposits. Piths homogeneous, composed of thick-walled, pitted parenchyma, but devoid of sclerenchymatous cells. Sieve-element plastids of Ss-type (Behnke 2002). Oil cells and mucilage cells absent. Leaves alternate, at first spirally arranged, later distichous, simple, entire, or repand and obscurely dentate. Stomata paracytic to anomocytic. Flowers in small, botryoids or poorly ramified panicles (Endress and Igersheim 2000), unisexual, dioecious, with a slightly convex receptacle bearing bracteoles that grade into the larger perianth segments; floral members spirally arranged. Male flowers with 9–11, inconspicuous perianth segments; approximately 5 perianth segments are the largest ones. They are reflexed in the open flowers. Toward the base of the flower these organs are gradually smaller and bractlike, but there is no clear separation into a zone of bracts and a zone of perianth segments (Endress and Igersheim 2000). Stamens 12–21 with short filament, inserted spirally on the inner surface of the receptacle; anthers triangular, tetrasporangiate, two pollen sacs somewhat separated, opening longitudinally; connective slightly prolonged. Pollen grains oblate to spheroidal and monoaperturate (ulcerate) at the distal pole, with a poorly defined aperture margin; the exine consists in the nonapertural region of ektexinous unique tecta elements ("cupules"), a smooth foot layer, and a thin, nonlamellate endexine (Hesse 2001). Microsporogenesis successive. Female flowers with 7–8 perianth segments, 1–2 staminodia (rarely 3 or none); gynoecium of 5, obovoid, shortly stipitate, free carpels in a single cycle at the center of the receptacle; commonly only 3 carpels develop into a fruitlets (Endress and Igersheim 2000), carpels unsealed at the tip, with single ovule attached to the lower ventral wall; stigma large, cup-like, sessile, with 2 conspicuous flanges, shortly decurrent. Ovule pendant, typically orthotropous (the chalaza and micropyle are in a line although the funiculus is curved – Yamada et al. 2001), crassinucellate, bitegmic, outer integument 4–5 cells across, micropyle endostomal. Female gametophyte of *Polygonum*-type. Endosperm formation is if *ab initio* cellular (Tobe et al. 2000). Fruits of oblique, fleshy drupaceous, its wall is formed by both endocarp and mesocarp. Seeds with weakly differentiated membranous testa; seed exsotestal-endotegmic; embryo basal, minute, endosperm copious, fleshy, contains protein. The Amborellaceae are weak accumulators of aluminium (Metcalfe 1987). n = 13.

1. AMBORELLACEAE

Pichon 1948. 1/1. New Caledonia.

Amborella.

The shrubby *Amborella trichopoda* is very heterobathmic and combines both very primitive and a highly specialized features. According to Hesse (2001), "the unique tectal elements (cupules) and the thick homogeneous endexine, which form an odd operculum in the apertural area, are unusual angiosperm pollen characters". According to Carlquist and Schneider (2001) "*Amborella* shares stomatal configurations nodal type (in part), ray types, and porose pit membranes in tracheary elements with Illiciales s.l., but differs from that order in lacking oil cells, vessels and grouped axial parenchyma cells." However it is very isolated genus and one of the most archaic taxon within the Magnoliidae. It differs from all other Magnolliids in its many respects, including structure of unique type of pericarp, and deserve the ordered status (Melikian et al. 1999). *Amborella* is one of the most remarkable living fossils which most probably originated directly from the earlier angiosperm.

Bibliography

Bailey IW. 1957. Additional notes on the vesselless dicotyledon, *Amborella trichopoda* Baill. J. Arnold Arbor. 38: 374–378.

Bailey IW and BGL Swamy. 1948a. *Amborella trichopoda* Baill., a new type of vesselless dicotyledons. J. Arnold Arbor. 29: 215.

Bailey IW and BGL Swamy. 1948b. *Amborella trichopoda* Baill.: A new morphological type of vesselless dicotyledon. J. Arnold Arbor. 29: 245–254.

Bobrov AV, PK Endress, AP Melikian, MS Romanov, AN Sorokin, and AP Bejerano. 2005. Fruit structure of *Amborella trichopoda* (Amborellaceae). Bot. J. Linn. Soc. 148: 265–274.

Buzgo M, PS Soltis, and DE Soltis. 2004. Floral developmental morphology of *Amborella trichopoda* (Amborellaceae). Int. J. Plant Sci. 165: 925–947.

Carlquist S and EL Schneider. 2001. Vegetative anatomy of the New Caledonian endemic *Amborella trichopoda*: relationships with the Illiciales and implications for vessel origin. Pac. Sci. 55: 305–312.

Degtjareva G, TH Samigullin, DD Sokoloff, and CM Valiejo-Roman. 2004. Gene sampling versus taxon sampling: Is *Amborella* (Amborellaceae) a sister group to all other extant angiosperms? Bot. Zhurn. 89: 896–907.

Endress PK and A Igersheim. 2000. The reproductive structures of the basal angiosperm *Amborella trichopoda* (Amborellaceae). Int. J. Plant Sci. 161(Suppl. 6): 237–248.

Field TS, T Brodribb, T Jaffré, and NM Holbrook. 2001. Acclimation of leaf anamoty, photosynthetic light use, and xylem hydraulies to light in *Amborella trichopoda*. Int. J. Plant Sci. 162: 999–1008.

Field TS, MA Zweiniecki, T Brodribb, T Jaffré, MJ Donoghue, and NM Holbrook. 2000. Structure and function of tracheary elements in *Amborella trichopoda*. Int. J. Plant Sci. 161: 705–712.

Floyd SK and WE Friedman. 2001. Developmental evolution of endosperm in basal angiosperms: evidence from *Amborella* (Amborellaceae), *Nuphar* (Nymphaeaceae), and *Illicium* (Illiaceae). Plant Syst. Evol. 228: 153–169.

Goremykin VV, KI Hirsch-Ernst, S Wölfl, and FH Hellwig. 2003. Analysis of the *Amborella trichopoda* chloroplast genome sequence suggests that *Amborella* is not a basal angiosperm. Molec. Biol. Evol. 20: 1499–1505.

Hesse M. 2001. Pollen characters of *Amborella trichopoda* (Amborellaceae): A reinvestigation. Bot. J. Plant Sci. 162: 201–208.

Lockhart PJ and D Penny. 2005. The place of *Amborella* within the radiation of angiosperms. Trends Plant Sci. 10: 201–202.

Metcalfe CR. 1987. Anatomy of the Dicotyledons. Amborellaceae, pp. 97–100. Claredon Press, Oxford.

Melikian AP, AV Bobrov, and ES Zaytzeva. 1999. A new fruit type in *Amborella trichopoda* Baill. (Amborellaceae Pichon). 14 Symp. Biodiversität und Evolutionsbiol. Jena.

Oginuma K, T Jaffré, and H Tobe. 2000. The karyotype analysis of somatic chromosomes in *Amborella trichopoda* (Amborellaceae). J. Plant Res. 113: 281–283.

Philipson WR. 1993. Amborellaceae. In: K Kubitzki, ed. The families and genera of vascular plants, vol. 2, pp. 92–93. Springer, Berlin/Heidelberg/New York.

Plisko MA. 1988. Amborellaceae. In: A Takhtajan, ed. Comparative seed anatomy, vol. 2, pp. 54–55. Nauka, Leningrad (in Russian).

Posluszny U and PB Tomlinson. 2003. Aspects of inflorescence and floral development in the putative basal angiosperm *Amborella trichopoda* (Amborellaceae). Canad. J. Bot. 81: 28–39.

Sampson FB. 1993. Pollen morphology of the Amborellaceae and Hortoniaceae (Hortonioideae: Monimiaceae). Grana 32: 154–162.

Soltis DE and PS Soltis. 2004. *Amborella* not a "basal angiosperm"? Not so fast. Am. J. Bot. 91: 997–1001.

Stefanovic S, DW Rice, and JD Palmer. 2004. Long branch attraction, taxon sampling, and the earliest angiosperms: *Amborella* or monocots. BMC Evol. Biol. (4): 35 (online).

Thien LB, TL Sage, T Jaffré, P Bernhardt, V Pontieri, PH Weston, D Malloch, H Azuma, SW Graham, MA McPherson, HS Rai, RF Sage, and J-L Dupre. 2003. The population structure and floral biology of *Amborella trichopoda* (Amborellaceae). Ann. Missouri Bot. Gard. 90: 466–490.

Tobe H, T Jaffré, and PH Raven. 2000. Embryology of *Amborella* (Amborellaceae): Description and polarity of character states. J. Plant Res. 113: 271–280.

Yamada T, H Tobe, R Imaichi, and M Kato. 2001. Developmental morphology of the ovules of *Amborella trichopoda* (Amborellaceae) and *Chloranthus serratus* (Chloranthaceae). Bot. J. Linn. Soc. 137: 277–290.

Young DA. 1982. Leaf flavonoids of *Amborella tricho-poda*. Biochem. Syst. Ecol. 10: 21–22.

Order 2. NYMPHAEALES

Herbaceous aquatic plants with generally submerged or floating leaves; 4-celled uniseriate secretory trichomes with a large terminal cell. Parenchymatous tissues without ethereal oil cells and with conspicuous air-passages and with laticifers. Vascular bundles closed, scattered, or less commonly in one or more rings. Vessels absent. Sieve-element plastids of S-type. Leaves spiral, involute, peltate (divided), simple, entire or margin toothed, stipulate (the stipules median-axillary), or estipulate. Stomata anomocytic. Flowers solitary, often large, bisexual, actinomorphic, mostly spirocyclic. Perianth basically 3-merous. Stamens usually numerous; filaments free; anthers extrorse or latrorse, tetrasporangiate, opening longitudinally. Tapetum secretory. Microsporogenesis mostly simultaneous. Pollen grains 3-celled or (*Cabomba*) 2-celled, 1-colpate, zonocolpate, or inaperturate, mostly in monads. Gynoecium apocarpous or syncarpous. Carpels 1–50, large to very large. Ovules 3–400, usually anatropous with cup-shaped outer integument, bitegmic, crassinucellate. Endosperm usually cellular. Female gametophyte of *Polygonum*-type or rarely also both *Polygonum*- and *Oenothera*-type (Orban and Bouharmont 1998). Fruits of various types. Seeds operculate, exotesta palisade, mostly with small embryo; endosperm scanty, perisperm present or absent, with or without arillus. Without benzyl isoquinoline alkaloids, but with pseudoalkaloids based on sesquiterpenes and ellagitanninses.

There is every reason for considering the superorder Nymphaeanae as a hydrophilous derivative of some ancient vesselless arborescent or subherbaceous magnoliid, most likely of inhabitants of moist or marshy places, whose vegetative organs became modified in an aquatic environment. According to Arber (1920: 309), the nymphaeids "descended from a stock closely related to that which gave rise to the monocotyledons." The nymphaeids and some liliopsids, particularly the Alismatidae, do indeed have much in common. Besides the apocarpous gynoecium with laminar-diffuse placentation of the ovules, pollen grain morphology, atactostelic vascular cylinder, more or less reduced primary root, and similar arrangement of the prophylls, there are also striking similarities in the "liorhizous" type of root development (because of the presence of secondary dermatogen, their root-cap is sharply cut off from the rhizodermis) (van Tieghem 1886; Voronin 1966; Voronkina 1974), and similar formation of the plumule (Meyer 1960). It is therefore not surprising that some botanists regarded Nymphaeaceae s. l. as monocots (Trecul 1845; Seidel 1869; Schaffner 1904, 1934; Cook 1906; Guttenberg and Müller-Schroeder 1958). But in spite of many similarities with the monocots, nymphaeids are dicotyledonous, and their affinity with the basal Magnoliidaes, especially with Amborellales, is supported also by molecular data (see Qiu et al. 1993, 1999; D. Soltis et al. 2000; Zanis et al. 2002, and others).

Key to Families

1 Plants with leafy stems coated with a mucilaginous secretion. Sclerenchymatous idioblasts lacking. Flowers relatively small. Microsporogenesis simultaneous or successive. Female gametophyte of *Schisandra*-subtype of *Oenothera*-type (at least in Cabombaceae). Endosperm helobial with chalazal haustorium. Fruits coriaceous, follicle like or achenelike. Seeds exarillate.

2 Submerged parts with thick, transparent, gelatinous sheath; stems encased in a thick layer of mucilage; there are paired, glandular patches at the nodes. Leaves monomorphic, all alternate, estipulate, with floating entire and peltate laminas. Flowers without nectariferous spurs, anemophilous. Perianth only slightly differentiated into sepals and petals, dull-purple. Sepals and petals 3 or 4 each. Stamens 18–36. Anthers latrorse. Microsporogenesis simultaneous. Pollen grains 3-celled, the exine scabrate. Carpels 4–22, large; stylodium short, with linear, decurrent, and extremely papillate stigma. Ovules 1 or 2 per carpel, large to very large, dorsal or near dorsal. Antipodals ephemeral. Fruits achenelike, containing 1–2 seeds, n = 40.. . . . 1. HYDROPELTIDACEAE.

2 Plants rhizomatous, without a gelatinous sheath. Leaves dimorphic, submerged opposite or verticillate, palmately dissected into linear segments, estipulate, floating ones few, peltate, usually alternate. Flowers with nectariferous spurs at the adaxial base of petals, myophilous. Perianth clearly differentiated into sepals and petals, white to yellow or purple. Sepals and petals (2)3(-4) each. Stamens 3 or 6. Microsporogenesis successive. Anthers extrorse to latrorse. Pollen grains 2-celled, with striate ornamentation composed of parallel rods. Carpels (1-)2-18, small to medium-sized, whorled, occasionally connate at the base; stylodium long, with a spherical papillate stigma. Ovules (1)2-3 per carpel, the upper ovule ventral, the two lower ones lateral. micropyle endostomal. Antipodals lacking. Fruits follicle like or achenes, dehiscing along the dorsal side, containing 2–4 seeds. Alkaloids absent, n = 40, 48, 52. . . . . 2. CABOMBACEAE.

1 Acaulescent perennial or rarely annual plants with leaves arising directly from the horizontal or vertical rhizomes. Branched sclerenchymatous idioblasts common. Nonbranched articulated laticifers occur throughout the ground tissue and usually associated with the vascular tissue. Flowers large,

generally emergent. Fruits spongy berries. Seeds arillate or exarillate.

3 Petals distinct, very rarely absent. Stamens free. Pollen grains 3-celled, I-colpate or more often zonocolpate. Ovules anatropous. Endosperm cellular. Seeds arillate or exarillate, not spiny. Aquatic, rhizomatous herbs. Leaves medium-sized to large, alternate, simple, petiolate, usually more or less peltate, stipulate or estipulate. Flowers solitary, large. Calyx imbricate; corolla 0–70, imbricate. Stamens 14–200, petaloid, or laminar, of filantherous; anthers adnate, non-versatile, introrse. Gynoecium of 5–35 carpels; ovary 5–35-locular, superior to partly inferior. Ovules 10–100 per locule, orthotropous or anatropous, arillate or nonarillate. Fruit fleshy, dehiscent or indehiscent, or a schizocarp. Seed numerous; embryo green or white, straight; endosperm and perisperm present. Proanthocyanidins (cyanidin or delphinidin) present, flavonols (kaempferol, quercetin, or myricetin) present or (*Nuphar*) absent, n = 10–12, 14–18. . . . . 3. NYMPHAEACEAE.

3 Petals connate into a lobed tube. Stamens 20–30, attached to the corolla tube, free of one another. Pollen grains 2-celled, inaperturate, with a reduced exine. Ovary inferior. Ovules orthotropous. Seeds exarillate, spiny. Stemless, laticiferous herbs. Laticifers in leaves, stems, roots, in flowers, and in the fruits. Leaves floating, alternate, membranous, simple, long petiolate, estipulate. Flowers solitary, long-pedunculate, but often not reaching the surface of the water, at least sometimes cleistogamous. Corolla 12–50 lobes; Staminodes 15–20, external to the fertile stamens. Filaments short; anthers basifixed, non-versatile, 2-locular. Gynoecium of 8–14 carpels; ovary 8–14-locular, ovules 30–50 per locule. Fruit a berry; seeds numerous, densely rather softly spinulose; embryo minute, well differentiated; perisperm and endosperm present, n = 17, 18. 4. BARCLAYACEAE

1. HYDROPELTIDACEAE

Dumortier 1822. 1/1. Eastern Himalayas (Bhutan), Assam, eastern Asia, eastern Australia, tropical Africa, North America (from Alaska to Florida), Central America, West Indies.

Brasenia (Hydropeltis).

2. CABOMBACEAE

Richard ex A. Richard 1822. 1/5. Warm-temperate and tropical regions of America.

Cabomba.

Closely related to the Hydropeltidaceae, which are usually included in the Cabombaceae. However, Moseley et al. (1984), comparing anatomical and morphological features of both vegetative organs and flowers in *Brasenia* and *Cabomba*) came to the conclusion that the two genera are more distant taxonomically than commonly assumed. This view is also shared by Collinson (1980). According to Osborn et al. (1991: 1367), "there is little argument that *Brasenia* and *Cabomba* are closely linked." The differences are so many that I find it necessary to accept Dumortier's family Hydropeltidaceae.

3. NYMPHAEACEAE

R. A. Salisbury 1805 (including Euryalaceae J. Agardh 1858; Nupharaceae Kerner 1891). 5/75. Almost cosmopolitan, in fresh water habitats. *Nuphar* (20) temperate and cold regions of the Northern Hemisphere, *Victoria* (2) in tropical South America, *Euryale* (1) in northern India, China, and Japan, *Ondinea* (1) in northern western Australia (sandstone streams of the northern Kimberley region).

3.1 NUPHAROIDEAE

Gynoecium superior, its distal part forming a flattened to somewhat concave disc with rays of stigmatic tissue. The stigmatic area is displayed on a flat plate. Pollen spiny. Carpels 5-23(-30), glabrous. Sepals 5–17. Petals small, each with an abaxial nectary. Nucellar cap present. Seeds exarillate. Aquatic herbs with stout rhizomes. Leaves large, entire, with basal sinus, submersed, floating or standing erect above water. n = 17. –*Nuphar.*

3.2 NYMPHAEOIDEAE

Gynoecium inferior to semi-inferior, its distal part cup-shaped and covered with stigmatic tissue. Leaf veins, peduncles, sepals, and petals not aculeate. The stigmatic area is displayed on a concave groove. Pollen grains in monads. Carpels 3–47, covered by pluricellular-uniseriate hairs. Sepals 4, rarely 5. Petals (0)4-5 to numerous, not small, without nectaries (?). Seeds arillate. n = 10–12, 14–18. – NYMPHAEEAE: *Nymphaea*; ONDINEAE: *Ondinea.*

3.3 EURYALOIDEAE
Leaf veins, peduncles, sepals, and petals aculeate. Pollen grains in tetrads. – *Euryale, Victoria.*

4. BARCLAYACEAE

H. L. Li 1955. 1/4. Thailand and Burma to East Malesia. *Barclaya (Hydrostemma).*

Close to the Nymphaeaceae, but differ from them in some important features, including congenital carpel fusion encompasses a considerable part of these "stylar processes", and are more advanced.

Bibliography

Arber A. 1920. Water plants. Cambridge University Press, Cambridge.

Batygina TB. 1981. Cabombaceae, Nymphaeaceae. In: MS Yakovlev, ed. Comparative embryology of flowering plants: Winteraceae-Juglandaceae, pp. 101–110. Nauka, Leningrad (in Russian).

Batygina TB and II Shamrov. 1983. Embryology of the Nelumbonaceae and Nymphaeaceae: Pollen grain structure (some features of correlated development of the pollen grain and anther wall). Bot. Zhurn. 68: 1177–1184 (in Russian).

Batygina TB and II Shamrov. 1985. Comparative embryology of the orders Nymphaeales and Nelumbonales and the problems in their systematics and phylogeny. Bot. Zhurn. 70: 368–373 (in Russian).

Batygina TB, TI. Kravtsova, and II Shamrov. 1980. The comparative embryology of some representatives of the order Nymphaeales and Nelumbonales. Bot. Zhurn. 65: 1071–1087 (in Russian with English summary).

Batygina TB, II Shamrov, and GE Kolesova. 1982. Embryology of the Nymphaeales and Nelumbonales: II. The development of the female embryonic structures. Bot. Zhum. 67: 1179–1195 (in Russian with English summary).

Behnke HD. 1996. Endoplasmic reticulum derived decorated tubules in the sieve elements of *Nymphaea*. Protoplasma 193: 213–221.

Bonilla Barbosa J, A Novelo, YH Orozco, J Marquez Guzman. 2000. Comparative seed morphology of Mexican *Nymphaea* species. Aquatic Bot. 68(3): 189–204.

Bukowiecki H, M Furmanova, and H Oledzka. 1972. The numerical taxonomy of Nymphaeaceae: 1. Estimation of taxonomic distance. Acta Pol. Pharm. 29: 319–327.

Carlquist S and EL Schneider. 1996. Vessels in *Brasenia* and *Cabomba* (Cabombaceae). Am. J. Bot. 83(6): 144–145 (Abstract).

Chassat J-R. 1962. Recherches sur la ramification chez les Nymphaeacees. Bull. Soc. Bot. France, Mem., 1962: 72–95.

Chen W-P and S-M Zhang. 1992. Comparative leaf anatomy of Nymphaeaceae (s.l.) Acta Phytotax. Sin. 30: 415–422.

Chen I, SR Manchester, and Z Chen. 2004. Anatomically-preserved seeds of *Nuphar* from the early Eocene of Wutu, Shandong Province, China. Am. J. Bot. 91: 1265–1272.

Chifflot JBJ. 1902. Contribution a Fetude de la classe des Nympheinees. Ann. Univ. Lyon Sei. Med., N. S. 10: 18–38.

Chrysler MA. 1938. The winter buds of *Brasenia*. Bull. Torrey Bot. Club 65: 277–283.

Clarke GCS and MR Jones. 1981. Cabombaceae. Rev. Palaebot. Palynol. 33: 51–55.

Collinson MF. 1980. Recent and tertiary seeds of the Nymphaeaceae sensu lato with a revision of *Brasenia ovula* (Brong.) Reid and Chandler. Ann. Bot. 46: 603–632.

Conard JS. 1905. The waterlilies. A monograph of the genus *Nymphaea*. Washington.

Cook MT. 1902. Development of the embryo sac and embryo of *Castalia odorata* and *Nymphaea advena*. Bull. Torrey Bot. Club 29: 211–220.

Cook MT. 1906. The embryology of some Cuban Nymphaeaceae. Bot. Gaz. 42: 376–396.

Cutter EG. 1957. Studies of morphogenesis in the Nymphaeaceae: 1. Introduction: Some aspects of the morphology of *Nuphar lutea* (L.) Sm. and *Nymphaea alba* L. Phytomorphology 7: 45–46: 57–73.

Cutter EG. 1958. Studies of morphogenesis in the Nymphaeaceae: II. Floral development in *Nuphar* and *Nymphaea*: Bracts and calyx. Phytomorphology 8: 74–95.

Cutter EG. 1961. The inception and distribution of flowers in the Nymphaeaceae. Proc. Linn. Soc. London 172: 93–100.

Dorofeev PI. 1973. Systematics of ancestral forms of *Brasenia*. Palaeontol. J. 7: 219–227.

Dorofeev PI. 1984. The taxonomy and history of the genus *Brasenia* (Cabombaceae). Bot. Zhurn. 69: 137–148 (in Russian).

Earle TT. 1938. Embryology of certain Ranales. Bot. Gaz. 100: 257–275.

El Ghazali GEB and WE Abd Alla. 2001. Pollen morphological study on *Nymphaea lotus* L. (*Nymphaeaceae*) with emphasis on zonisulculate apertures. Jpn. J. Hist. Bot. 9: 79–84.

Endress PK. 2005. Carpels in *Brasenia* (Cabombaceae) are completely ascidiate despite a long stigmatic crest. Ann. Bot. N.S. 96: 209–215.

Fassett NC. 1953. A monograph of *Cabomba*. Castanea 18: 116–128.

Floyd SK and WE Friedman. 2001. Developmental evolution of endosperm in basal angiosperms: evidence from *Amborella* (Amborellaceae), *Nuphar* (Nymphaeaceae), and *Illicium* (Illiaceae). Plant Syst. Evol. 228: 153–169.

Friis EM, KR Pedersen, and PR Crane. 2001. Fossil evidence of water lilies (Nymphaeales) in the Early Cretaceous. Nature 410: 357–360.

Gabarayeva NI and G El-Ghazaly. 1997. Sporoderm development in *Nymphaea mexicana* (Nymphaeaceae). Plant Syst. Evol. 204: 1–19.

Gabarayeva NI and JR Rowley. 1994. Exine development in *Nymphaea colorata* (Nymphaeaceae). Nord. J. Bot. 14: 671–691.

Gabarayeva NI, VV Grigorjeva, and JR Rowley. 2003. Sporoderm ontogeny in *Cabomba aquatica* (Cabombaceae). Rev. Paleobot. Palynol. 127: 147–173.

Gabarayeva N, B Walles, G El Ghazaly, and JR Rowley. 2001. Exine and tapetum development in *Nymphaea capensis*

(Nymphaeaceae): a comparative study. Nord. J. Bot. 21: 529–548.

Galati B. 1981. The ontogeny of hairs and stomata of *Cabomba australis* (Nymphaeaceae). Lilloa 35: 149–158.

Galati BG. 1985. Estudios embriologicos en *Cabomba australis* (Nymphaeaceae). I. La esporagenesis y las grneraciones sexuadas. Bol. Soc. Argentina Bot. 24: 29–47.

Galati BG. 1987. Estudios embriologicos en *Cabomba australis* (Nymphaeaceae). II. Ontogenia de la semilla. Bol. Soc Argentina Bot. 25: 187–196.

Gandolfo MA, KC Nixon, and WL Crepet. 2004. Cretaceous flowers of Nymphaeaceae and implications for complex insect entrapment pollination mechanisms in early angiosperms. Proc. National Acad. Sci. 101: 8056–8060.

Goleniewska-Furmanova M. 1970. Comparative leaf anatomy and alkaloid content in the Nymphaeaceae. Monogr. Bot. 31: 1–55.

Grob V, P Moline, E Pfeifer, AR Novelo, and R Rutsihauser. 2006. Developmental morphology of branching flowers in *Nymphaea prolifera*. J. Plant Res. 119: 561–570.

Gwynne-Vaughan DT. 1897. On some points in the morphology and anatomy of the Nymphaeaceae. Trans. Linn. Soc. London Bot. 5: 287–299.

Haines RW and KA Lye. 1975. Seedlings of Nymphaeaceae. Bot. J. Linn. Soc. 70: 255–265.

Hartog C den. 1970. *Ondinea*: A new genus of Nymphaeaceae. Blumea 18: 413–416.

Heinsbroek PG and WA Van Heel. 1969. Note on the bearing of the pattern of vascular bundles on the morphology of the stamens of *Victoria amazonica* (Poep.) Sowerby. K. Nederl. Akad. Wet. Proc., ser. C, 72: 431–444.

Hiepko P. 1965. Vergleichend-morphologische und entwicklungsgeschichtliche Untersuchungen über das Perianth bei den Polycarpicae. Bot. Jahrb. Syst. 84: 359–508.

Hu G-W, K-M Liu, and L-G Lei. 2003. Comparative study on leaf anatomy of three genera in Nymphaeaceae. Life Sci. Res. 7: 243–248.

Inambar JA and KM Aleykutty. 1979. Studies on *Cabomba aquatica* (Cabombaceae). Plant Syst. Evol. 132: 161–166.

Ishimatsu M, T Tanaka, G Nonaka, I Nishioka, M Nishizawa, and T Yamagishi. 1989. Tannins and related compounds. LXXIX. Isolation and characterisation of novel dimeric and trimeric hydrolyzable tannins, nuphrins C, D, E and F, from *Nuphar japonicum* DC. Chem. Pharmac. Bull. 37: 1735–1743.

Ito M. 1982. On the embryos and the seedlings of the Nymphaeaceae. Acta Phytotax. Geobot. 33: 143–148.

Ito M. 1983. Studies in the floral morphology and anatomy of Nymphaeales: 1. The morphology of vascular bundles in the flower of *Nymphaea tetragona* George. Acta Phytotax. Geobot. 34: 18–26.

Ito M. 1984. Studies in the floral morphology and anatomy of Nymphaeales: II. The floral anatomy of *Nymphaea tetragona* George. Acta Phytotax. Geobot. 35: 94–102.

Ito M. 1986. Studies in the floral morphology and anatomy of Nymphaeales: III. Floral anatomy of *Brasenia schreberi* Gmel. and *Cabomba caroliniana* A. Gray. Bot. Mag. Tokyo 99: 169–184.

Ito M. 1987. Phylogenetic systematics of the Nymphaeales. Bot. Mag. Tokyo 100: 17–35.

Kadono Y and EL Schneider. 1987. The life history of *Euryale ferox* Salisb. in southwestern Japan with special reference to reproductive ecology. Plant Species Biol. 2: 109–115.

Kenneally KF and EL Schneider. 1983. On the genus *Ondinea* (Nymphaeaceae) including a new subspecies from the Kimberley region, Western Australia. Nuytsia 4: 359–365.

Khanna P. 1964. Morphological and embryological studies in Nymphaeaceae: 1. *Euryale ferox*. Proc. Indian Acad. Sci. 59B: 237–243.

Khanna P. 1965. Morphological and embryological studies in Nymphaeaceae: II. *Brasenia schreberi* Gmel. and *Nelumbo nucifera* Gaertn. Austral. J. Bot. 13: 379–387.

Khanna P. 1967. Morphological and embryological studies in Nymphaeaceae: III. *Victoria cruziana* D'Orb. and *Nymphaea stellata* Wilid. Bot. Mag. Tokyo 80: 305–312.

Knoch E. 1899. Untersuchungen über die Blüte von *Victoria regia*. Biol. Bot. 47: 1–60.

Kolesova GE and TB Batygina. 1988. Cabombaceae. In: A Takhtajan, ed. Comparative seed anatomy, 2: 136–140. Nauka, Leningrad (in Russian).

Les DH DK Garvin, and CF Wimpee. 1991. Molecular evolutionary history of ancient aquatic angiosperms. Proc. Nad. Acad. Sci. USA 88: 10119–10123.

Les DH, EL Shneider, and DJ Padgett. 1997. Phylogemy of the Nymphaeaceae: on the verge of a synthesis. Am. J. Bot. 84(Suppl. 6): 219–211.

Les DH, EL Shneider, DJ Padgett, PS Soltis, DE Soltis, and M Zanis. 1999. Phylogeny, classification and floral evolution of water lilies (Nymphaeaceae; Nymphaeales): A synthesis of non-molecular, *rbcL*, *matK*, and 18S rDNA Data. Syst. Bot. 24: 24–46.

Li H-L. 1955. Classification and phylogeny of Nymphaeaceae and allied families. Am. Midi. Naturalist 54: 33–41.

Lippok B and SS Renner. 1997. Polination of *Nuphar* (Nymphaeaceae) in Europe: flies and bees rather than *Donacia* beetles. Plant Syst. Evol. 207: 273–283.

Liu Y-L, L-M Xu, X-M Ni, and J-R Zhao. 2005. Phylogeny of Nymphaeaceae inferred from ITS sequences. Acta Phytotax. Sinica 43: 22–30 (in Chinese with English summary).

Lodkina MM. 1988. Evolutionary relations between mono- and dicotyledons based on embryo and seedling investigation. Bot. Zhurn. 73: 617–630 (in Russian).

Lu P-H and W-P Chen. 1993. Study on the peroxidase isozymes of Nymphaeaceae with analysis of polar ordination. J. Nanjing Normal University (Natural Sci.) 16: 52–56 (in Chinese with English summary).

Lu P-H, W-P Chen, H-X Xu, and M Liu. 1994. Study on the esterase isozymes of Nymphaeaceae with numerical taxonomy. J. Nanjing Normal University (Natural Sci.) 17: 92–97 (in Chinese with English summary).

Meeuse BJD and EL Schneider. 1979/80. *Nymphaea* revisited: A preliminary communication. Israel J. Bot. 28: 65–79.

Melikian AP. 1964. Comparative anatomy of the sporoderm of some representatives of the family Nymphaeaceae. Bot. Zhurn. 49: 432–436 (in Russian).

Melikian AP and EN Nemirovich-Danchenko. 1988. Nymphaeaceae. In: A Takhtajan, ed. Comparative seed anatomy, vol. 2, pp. 140–147. Nauka, Leningrad (in Russian).

Meyer NR. 1964. Palynological studies of the family Nymphaeaceae. Bot. Zhurn. 49: 1421–1429 (in Russian).

Meyer-Melikian NR and N Diamandopulu. 1996. Ultrastructure of pollen grains of the order Nymphaeales. Bot. Zhurn. 81(7): 1–9 (in Russian with English summary).

Moseley MF. 1958. Morphological studies of the Nymphaeaceae: 1. The nature of the stamens. Phytomorphology 8: 1–29.

Moseley MF. 1961. Morphological studies of the Nymphaeaceae: II. The flower of *Nymphaea*, Bot. Gaz. 122: 233–259.
Moseley MF. 1965. Morphological studies of the Nymphaeaceae: III. The floral anatomy of *Nuphar*. Phytomorphology 15: 54–84.
Moseley MF. 1972. Morphological studies of the Nymphaeaceae. IV, Development of flower of *Nuphar*. Phytomorphology 21: 253–283.
Moseley MF, IJ Mehta, PS Williamson, and H Kosakai. 1984. 13th Morphological studies of the Nymphaeaceae (sensu lato): Contributions to the vegetative and floral structure of *Cabomba*. Am. J. Bot. 71: 902–924.
Moseley MF, EL Schneider, and PS Williamson. 1993. Phylogenetic interpretations from selected floral vasculature characters in the Nymphaeaceae sensu lato. Aquatic Bot. 44: 325–342.
Müller J. 1970. Description of pollen grains of *Ondinea purpurea* Den Hartog. Blumea 18: 416–417.
Murthy GVS. 2000. Pollen morphology of Nymphaeaceae (s.l.). Bull. Bot. Surv. India 42: 73–80.
Ni X-M, B Yu, Y-J Zhou, and J-R Zhao. 1994. Studies on the phylogenetic relationships among the Nymphaeaceae. J. Wuhan Bot. Res. 12: 311–320.
Okada H and M Tamura. 1981. Karyomorphological study on the Nymphaeales. J. Jpn. Bot. 56: 367–375.
Orban I and J Bouharmont. 1998. Megagametophyte development of *Nymphaea nouchali* Burm. f. (Nymphaeaceae). Bot. J. Linn. Soc. 126: 339–348.
Orgaard M. 1991. The genus *Cabomba* (Cabombaceae): A taxonomic study. Nord. J. Bot. 11: 179–203.
Orgaard M, HWE Bruggen, van, PJ Vlugt, van der. 1992. Die Familie Cabombaceae (*Cabomba* und *Brasenia*.). Aqua Planta Sonderheft 3.
Osborn JM and EL Schneider. 1988. Morphological studies of the Nymphaeaceae sensu lato: XVI. The floral biology of *Brasenia schreberi*. Ann. Missouri Bot. Card. 75: 778–794.
Osborn JM, TN Taylor, and EL Schneider. 1991. Pollen morphology and ultrastructure of the Cabombaceae: Correlations with pollination biology. Am. J. Bot. 78: 1367–1378.
Padgett DJ, DH Les, and GE Crow. 1999. Phylogenetic relationships in *Nuphar* (Nymphaeaceae): Evidence from morphology, chloroplast DNA, and nuclear ribosomal DNA. Am. J. Bot. 86: 1316–1324.
Padmanabhan D and MY Ramji. 1966. Developmental studies on *Cabomba caroliniana* Gray: II. Floral anatomy and microsporogenesis. Proc. Indian Acad. Sei., Sec. B., 64: 216–223.
Prance GT and AB Anderson. 1976. Studies of the floral biology of Neotropical Nymphaeaceae. 3. Acta Amazonica 6: 163–170.
Prance GT and JR Arias. 1975. A Study of the floral biology of *Victoria amazonica* (Poepp.) Sowerby (Nymphaeaceae). Acta Amazonica 5: 109–139.
Raciborski M. 1894. Die Morphologic der Cabombeen und Nymphaeaceen. Flora 78: 244–299.
Ramji MY and D Padmanabhan. 1965. Developmental studies on *Cabomba caroliniana* Gray: 1. Ovule and carpel. Proc. Indian Acad. Sei., Sec. B, 62: 215–223.
Rao TA and BC Banerjee. 1978. On foliar sclereids in the Nymphaeaceae sensu lato and their use in familial classification. Proc. Indian Acad. Sci. 88B: 413–422.
Richardson FC. 1969. Morphological studies of the Nymphaeaceae: IV. Structure and development of the flower of *Brasenia schreberi* Gmel. Univ. Calif. Publ. Bot. 47: 1–101.
Richardson FC and M Moseley. 1967. The vegetative morphology and nodal structure of *Brasenia schreberi*. Am. J. Bot. 54: 645.
Rowley JR, NI Gabaraeva, and B Walles. 1992. Cyclic invasion of tapetal cells into loculi during microspore development in *Nymphaea colorata* (Nymphaeaceae). Am. J. Bot. 79: 801–808.
Schaffner JH. 1904. Some morphological peculiarities of the Nymphaeaceae and Helobiae. Ohio Nat. 4: 83–92.
Schaffner JH. 1929. Principles of plant taxonomy: VII. Ohio J. Sci. 29: 243–252.
Schaffner JH. 1934. Phylogenetic taxonomy of plants. Quart. Rev. Biol. 9: 129–160.
Schamrov II. 1998. Formation of hypostase, podium and postament in the ovule of *Nuphar lutea* (Nymphaeaceae) and *Ribes aureum* (Grossulariaceae). Bot. Zhurn. 83: 3–14 (in Russian with English summary).
Schmucker Th. 1932. Physiologische und ökologische Untersuchungen an Blüten tropischer *Nymphaea*-Arten. Planta 16: 376–412.
Schneider EL. 1976. Morphological studies of the Nymphaeaceae: VIII. The floral anatomy of *Victoria* Schromb. (Nymphaeaceae). Bot. J. Linn. Soc. 72: 115–148.
Schneider EL. 1978. Morphological studies of the Nymphaeaceae: IX. The seed of *Barclaya longifolia* Wall. Bot. Gaz. 139: 223–230.
Schneider EL. 1979. Pollination biology of the Nymphaeaceae. In: DM Caron, ed., Proc. 4th Internat. Symp. Pollination Md. Agr. Exp. Sta. Spec. Misc. Publ. 1: 419–430.
Schneider EL. 1982. Notes on the floral biology of *Nymphaea elegans* (Nymphaeaceae) in Texas. Aquatic Bot. 12: 197–200.
Schneider EL. 1983. Gross morphology and floral biology of *Ondine apurpurea*. Austral. J. Bot. 31: 371–382.
Schneider EL and S Carlquist. 1995a. Vessels in the roots of *Barclaya rotundifolia* (Nymphaeaceae). Am. J. Bot. 82: 1343–1349.
Schneider EL and S Carlquist. 1995b. Vessels origins in Nymphaeaceae: *Euryale* and *Victoria*. Bot. J. Linn. Soc. 119: 185–193.
Schneider EL and S Carlquist. 1996a. Vessels in *Brasenia* (Cabombaceae): New perspective on vessel origin in primary xylem of angiosperms. Am. J. Bot. 83: 1236–1240.
Schneider EL and S Carlquist. 1996b. Vessel origin in *Cabomba*. Nord. J. Bot. 16: 637–641.
Schneider EL, S Carlquist, K Beamer, and A Kohn. 1995. Vessels in Nymphaeaceae: *Nuphar*, *Nymphaea*, and *Ondinea*. Int. J. Plant Sci. 156: 857–862.
Schneider EL and T Chancy. 1981. The floral biology of *Nymphaea odorata* (Nymphaeaceae). Southw. Naturalist 26: 159–165.
Schneider EL and EG Ford. 1978. Morphological studies of the Nymphaeaceae: X. The seed of *Ondinea purpurea* Den Hartog. Bull. Torrey Bot. Club 105: 192–200.
Schneider EL and JM Jetter. 1982. Morphological studies of the Nymphaeaceae: XII. The floral biology of *Cabomba caroliniana*. Am. J. Bot. 69: 1410–1419.

Schneider EL and LA Moore. 1977. Morphological studies of the Nymphaeaceae: VII. The floral biology of *Nuphar lutea* subsp. *Macrophylla*. Brittonia 29: 88–99.

Schneider EL and PS Williamson. 1993. Nymphaeaceae. In: K Kubitzki, ed., The families and genera of vascular plants, vol. 2, pp. 486–493. Springer, Berlin/Heidelberg/New York.

Schneider EL, SC Tucker, and PS Williamson. 2003. Floral development in the Nymphaeales. Int. J. Plant Sci. 164(Suppl. 5): S279–S292.

Seago JL. 2002. The root cortex of the Nymphaeaceae, Cabombaceae, and Nelumbonaceae. J. Torrey Bot. Soc. 129: 1–9.

Seidel CF. 1869. Zur Entwicklungsgeschichte der *Victoria regia* Lindl. Nov. Act. Acad. Caes.-Leopold-Carol. Nat. Cur. 35(6): 26.

Shamrov II and AN Winter. 1991. The ovule development in representatives of the genera *Nymphaea* and *Victoria* (Nymphaeaceae). Bot. Zhurn. 76: 1072–1083 (in Russian with English summary).

Simon JP. 1971. Comparative serology of the order Nymphaeales: II. Relationships of Nymphaeaceae and Nelumbonaceae. Aliso 7: 325–350.

Snigirevskaya NS. 1955. On the morphology of pollen of Nymphaeales. Bot. Zhurn. 40: 108–115 (in Russian).

Sokolovskaya AP and AP Melikian. 1964. On the karyotype of *Barclaya longifolia* Wall. Bot. Zhurn. 49: 585–586 (in Russian).

Takahashi M. 1992. Development of spinous exine in *Nuphar japonicum* De Candolle (Nymphaeaceae). Rev. Paleobot. Palynol. 75: 317–322.

Tamura M. 1982. Relationship of *Barclaya* and classification of Nymphaeales. Acta Phytotax. Geobot. 33: 336–345.

Taylor ML and JM Osborn. 2006. Pollen ontogeny in *Brasenia* (Cabombaceae, Nympheales). Am. J. Bot. 93: 344–356.

Thorne RF. 1974. A phylogenetic classification of the Annoniflorae. Aliso 8 (2): 147–209.

Tillich H-J. 1990. Die Keimpflanzen der Nymphaeaceae: Monocotyl oder dicotyl. Flora 184: 169–176.

Titova GE. 1990. The development of the female generative structures in *Cabomba caroliniana* A. Gray (Cabombaceae). Abstracts 11th Internat. Symposium Embryology and Seed Reproduction, p. 210. Leningrad.

Titova GE and TB Batygina. 1996. Is the embryo of Nymphaealean plants (Nymphaeales s.l.) a dicotyledonous? Phytomorphology 46 (2): 171–190.

Trecul A. 1845. Recherches sur la structure et le devel-oppement du *Nuphar lutea*. Ann. Sei. Nat. Bot., ser. 3, 4: 286–345.

Troll W. 1933. Beiträge zur Morphologie des Gynaeceums: IV. Über das Gynaeceum der Nymphaeaceen. Planta 21: 447–485.

Ueno J and S Kitaguchi. 1961. On the fine structure of the pollen walls in angiosperms: 1. Nymphaeaceae. J. Biol. Osaka City Univ. 12: 83–89.

Valla JJ and DR Cirino. 1972. Biologia floral del irupe, *Victoria cruziana* Orb. (Nymphaeaceae). Darwiniana 17: 477–500.

Valtseva OV and EI Cavich. 1965. On development of the embryo of *Nymphaea candida* Presl and *N. tetragona* Georgi. Bot. Zhurn. 50: 1323–1326 (in Russian).

Van Heel WA. 1977. The pattern of vascular bundles in the stamens of *Nymphaea lotus* L. and its bearing on the stamen morphology. Blumea 23: 345–348.

Van Meigroet F and M Dujardin. 1992. Cytologie et histology de la reproduction chez *Nymphaea heudelotii*. Canad. J. Bot. 70: 1991–1996.

Van Tieghem P. 1866. Sur la croissance terminale de la racine dans les Nympheacees. Bull. Soc. Bot. France 33: 264–265.

Vinogradov IS. 1967. System of the family Nymphaeaceae on the basis of the analysis of morphological structure. Zapiski Tsentr.-Kavkaz. Otd. Vses. Bot. Obschestva 2: 5–11 (in Russian).

Voronin NS. 1964. Evolution of the primary structure of the roots of plants. Scient. Papers Tsialkovsky Pedagog. Inst. Kaluga 13: 3–179 (in Russian).

Voronkina NV. 1974. The anatomical structure of root apex in Nymphaeales J. Schaffner. Bot. Zhurn. 59: 1417–1424 (in Russian).

Walker JW. 1976a. Comparative pollen morphology and phylogeny of the ranalean complex. In: CB Beck, ed. Origin and early evolution of angiosperms, pp. 241–299. Columbia University Press, New York.

Walker JW. 1976b. Evolutionary significance of the exine in the pollen of primitive angiosperms. In: K Ferguson and J Müller, eds. The evolutionary significance of the exine, pp. 251–308. Linn. Soc. Symposium, No. 1. Columbia University Press, London/New York.

Weberbauer A. 1894. Beiträge zur Samenanatomie der Nymphaeaceen. Engler's Jahrb. 18(3): 213–258.

Wei P-H, W-P Chen, and R-Y Chen. 1993. Chromosome number of the members in the family Nymphaeaceae. J. Nanjing Normal University (Natural Sci.) 16: 52–55 (in Chinese with English summary).

Wei P-H, W-P Chen, and R-Y Chen. 1994. Study on the karyotype analysis of Nymphaeaceae and its taxonomic position. Acta Phytotax. Sin. 32: 293–300.

Weidlich WH. 1976. The organization of the vascular system in the stems of Nymphaeaceae: 1. *Nymphaea* subgenera *Castalia* and *Hydrocallis*; and II. *Nymphaea* subgenera *Anecphya, Lotos*, and *Brachyeeras*. Am. J. Bot. 63: 499–509,1365–1379.

Weidlich WH. 1980. The organization of the vascular system in the stems of Nymphaeaceae: III. *Victoria* and *Euryale*. Am. J. Bot. 67: 790–803.

Williamson PS and MF Moseley. 1989. Morphological studies of the Nymphaeaceae sensu lato: XVII. Floral anatomy of *Ondinea purpurea* subspecies *purpurea* (Nymphaeaceae). Am. J. Bot. 76: 1779–1794.

Williamson PS and EL Schneider. 1993. Cabombaceae. In: K Kubitzki, ed., The families and genera of vascular plants, vol. 2, pp. 157–160. Springer, Berlin/Heidelberg/New York.

Williamson PS and EL Schneider. 1994a. Floral aspects of *Barclaya* (Nymphaeaceae): Pollination, ontogeny, and structure. Plant Syst. Evol., Suppl. 8: 159–173.

Williamson PS and EL Schneider. 1994b. *Ondinea purpurea* den Hartog. Aqua Plants 1: 3–10.

Winter AN. 1990. Peculiarities of the female gametophyte development in the families Nymphaeaceae and Barclayaceae. Abstracts 11th Internal. Symp. Embryology and Seed Reproduction, p. 189. Leningrad.

Winter AN and TB Batygina. 1988. Barclayaceae. In: A Takhtajan, ed. Comparative seed anatomy, vol. 2, pp. 147–150. Nauka, Leningrad (in Russian).

Winter AN and II Shamrov. 1991a. The development of the ovule and embryo sac in *Nuphar lutea* (Nymphaeaceae).

Bot. Zhurn. 76: 378–390 (in Russian with English summary).
Winter AN and II Shamrov. 1991b. Megasporogenesis and embryo sac development in representatives of the genera *Nymphaea* and *Victoria* (Nymphaeaceae). Bot. Zhurn. 76: 1716–1728 (in Russian with English summary).
Wolf M. 1991. Blütenphyllotaxis von Nymphaeaceae: ist das Androecium von *Nymphaea, Nuphar*, etc. spiralig? Symposium Morphologie, Anatomie und Systematik. Göttingen.
Wood CE. 1959. The genera of the Nymphaeaceae and Ceratophyllaceae in the southeastern United States. J. Arnold Arbor. 40: 94–112.
Yamada T, R Imaichi, and M Kato. 2001. Developmental morphology of ovules and seeds of Nympheales. Am. J. Bot. 88: 963–974.

Order 3. AUSTROBAILEYALES

Large scandent shrubs to high climbing lianas with loosely twining main stem and straight, leafy lateral branches, containing more or less spherical oil cells in the leaf mesophyll and the outer cortex and surrounding phloem of young stem and calcium oxalate crystals in the form of crystal sand in the cortex. Nodes unilacunar, with two traces. Vessels with scalariform perforations; lateral pits rare, scalariform to opposite or alternate. Fibers with conspicuous bordered pits on thick-walled tracheids devoid of living contents when mature and vestigial bordered pits on septate fiber-tracheids containing several nuclei. Rays heterogeneous, tall, mixed uniseriate and multiseriate. Axial parenchyma paratracheal. Phloem of primitive type with often extremely inclined compound sieve plates; sieve-element plastids of S-type with about 20 globular starch grains, which are the largest among Magnolianae. Leaves opposite or subopposite, simple, entire, pinnately veined, estipulate. Stomata laterocytic (Baranova 2004). Flowers rather large, ca. 5 cm in diameter, solitary in the axils of foliage leaves or sometimes terminating a longer shoot, bisexual, actinomorphic, spiral, with conical receptacle, more or less pendant when open, entomophilous, emitting a strong smell of decaying fish. Perianth of 19–24 free, imbricate segments in compacted spiral, gradually larger from the outer glossy green sepaloid ones to the inner petaloid ones, which are yellowish-green with red dots. The transition between bracteoles and tepals is gradual as well. Stamens 7–11, spirally arranged, with a complex vascular bundle, with markedly papillose surface and irregular dark purple spots concentrated mainly on the apical and basal parts on both surfaces, the outer 6–11 fertile, the inner 9–16 gradually reduced and sterile; fertile stamens laminar, not differentiated into filament and connective, change from relatively flat to strongly boat-shaped, and the innermost stamens approach in outline the narrow, plicate, and irregularly ridged inner staminodia; anthers of four elongate microsporangia borne in two pairs on the adaxial side, dehiscing by longitudinal slits. Tapetum secretory. Microsporogenesis simultaneous. Pollen grains 2-celled, globose, monocolpate, tectate-columellate, markedly rugulate, with very marked aperture margin and gemmate aperture surface. Gynoecium of (6-)9(-14) large, free and spirally arranged, stipitate, completely sealed and extremely ascidate carpels, each with an excentric, adaxially displaced, elongate, and strongly bilobed stylodium and decurrent, papillate stigma. The carpel wall has cells with oxalate druses and crystals and tanniferous cells. Ovules large, (4)6-8(-10) in each carpel, arranged alternately in two parallel series at the adaxial side, anatropous with hood-shaped outer integument (Yamada et al. 2001), bitegmic, crassinucellate, micropyle endostomal. Female gametophyte of Polygonum-type. Fruitlets up to 8 cm long, ellipsoid-globose, orange, fleshy, berrylike, with a long stalk, with several seeds, resembling certain Annonaceae (e.g., Asinine). Seeds large, lenticular, the inner layer of the outer integument forming a protective inner part with lignified cell wall and a parenchymatous outer part forming a thin mealy-pulpy sarcotesta; embryo very small, straight; endosperm copious, starchy, ruminate. Producing several neolignans and lignans, but lacking alkaloids, n = 22.

The monotypic order Austrobaileyales is very distinct and taxonomically isolated within the Magnoliidae and its taxonomic position is a matter of dispute. According to the Angiosperm Phylogeny Group (2003), Austrobaileyaceae are grouped with the vessel-bearing families Illiciaceae and Schisandraceae into an order that could be called an expanded version of Illiciales. Carlquist (2001) concludes that vegetative anatomy as a whole supports the concept of an expanded Illiciales. The Austrobaileyaceae represent one of the most ancient independent group and are one the most remarkable "living fossils" among the archaic flowering plants.

1. AUSTROBAILEYACEAE

Croizat 1943. 1/1. Northern Queensland.
Austrobaileya.

Bibliography

Bailey IW and BGL Swamy. 1949. The morphology and relationships of *Austrobaileya*. J. Arnold Arbor. 30: 211–220.
Baranova MA. 1992. The epidermal structures and systematic position of the Austrobaileyaceae. Bot. Zhurn. 77: 1–17 (in Russian with English summary).
Baranova MA. 2004. The epidermal structure of *Austrobaileya* (Austrobaileyaceae) – a further comment. Kew Bull. 59: 489–491.
Behnke H-D. 1986. Sieve element characters and the systematic position of *Austrobaileya*, Austrobaileyaceae, with comments on the distinction and definition of sieve cells and sieve tube members. Plant Syst. Evol. 152: 101–121.
Carlquist S. 2001. Observations on the vegetative anatomy of *Austrobaileya*: habital, organographic and phylogenetic conclusions. Bot. J. Linn. Soc. 135: 1–11.
Croizat L. 1940. Notes on the Dilleniaceae and their allies: Austrobaileyeae subfam. nov. J. Arnold Arbor. 21: 397–404.
Croizat L. 1943. New families. Cact. Succ. J. (Los Angeles) 15: 64.
Dickison WC and PK Endress. 1983. Ontogeny of the stem-node-leaf vascular continuum of *Austrobaileya*. Am. J. Bot. 70: 906–911.
Endress PK. 1980. The reproductive structures and systematic position of the Austrobaileyaceae. Bot. Jahrb. Syst. 101: 393–433.
Endress PK. 1983a. The early floral development of *Austrobaileya* Bot. Jahrb. Syst. 103: 481–497.
Endress PK. 1983b. Dispersal and distribution in some small archaic relic angiosperm families (Austrobaileyaceae, Eupomatiaceae, Himantandraceae, Idiospermoideae-Calycanthaceae). Sonderbd. Naturwiss. Verh. Hamburg 7: 201–217.
Endress PK. 1993. Austrobaileyaceae. In: K Kubitzki, ed. The families and genera of vascular plants, vol. 2, pp. 138–140. Springer, Berlin/Heidelberg/New York.
Endress PK and R Honegger. 1980. The pollen of the Austrobaileyaceae and its phylogenetic significance. Grana 19: 177–182.
Morawetz W. 1988. Karyosystematics and evolution of Australia Annonaceae as compared with Eupomatiaceae, Himantandraceae, Austrobaileyaceae. Plant Syst. Evol. 159: 49–79.
Rudenberg L. 1967. The chromosomes of *Austrobaileya*. J. Arnold Arbor. 48: 241–244.
Srivastava LM. 1970. The secondary phloem of *Austrobaileya scandens*. Canad. J. Bot. 48: 341–359.
Zavada MS. 1984. Pollen wall development of *Austrobaileya maculata*. Bot. Gaz. 145: 11–21.

Order 4. ILLICIALES

Small trees, shrubs, or woody lianas. Nodes unilacunar. Vessels with scalariform, simple or mixed perforations; lateral pitting scalariform to opposite. Fibers with conspicuous bordered piths in both radial and tangential walls. Axial parenchyma scanty or very scanty, usually paratracheal, accidentally terminal or diffuse. Rays heterogeneous. Sieve-element plastids of S-type with high starch content. Leaves alternate, simple, entire or toothed, without stipules. Stomata paracytic, mixed paracytic and laterocytic, sometimes laterocytic. Flowers solitary or sometimes two or three (rarely more) together, usually axillary or supra-axillary, bisexual or unisexual, spiral or spirocyclic, actinomorphic. Perianth mostly of numerous (5–33) spirally arranged segments, usually not clearly differentiated into sepals and petals. Stamens mostly numerous (4–50, rarely to 80), more or less spirally arranged; filaments free or more or less connate; anthers basifixed, tetrasporangiate, extrorse to introrse, opening longitudinally. Tapetum secretory. Microsporogenesis simultaneous. Pollen grains 2-celled, globose-oblate, 3-colpate or 6-colpate, or 3-colporate, semitectate, reticulate. Gynoecium of (5-)7-many distinct carpels arranged spirally or in a single cycle; carpels conduplicate, at anthesis not completely sealed (in Illicium only in stylodial portion), with a decurrent stigma. In each carpel 1 or 2-5(-11) ovules; placentation submarginal or subbasal (near the bottom of the ovary). Ovules usually anatropous with cup-shaped outer integument (Yamada et al. 2001), bitegmic, crassinucellate, micropyle exostomal or endostomal. Female gametophyte of Polygonum-type. Endosperm cellular. Fruiting folliculate or baccate. Seeds laterally flat tened; seed coat formed by the outer integument; embryo very small or minute; endosperm copious, oily. Producing proanthocyanins and esters of angelic and tiglid acids, but not ellagic acid; n = 13, 14.

Key to Families

1 Shrubs or small trees, evergreen, glabrous, aromatic with scattered ethereal oil cells. Neolignans and crystalliferous sclereids lacking. Pits vestured. Nodes with a single trace. Vessels narrow, with numerous (up to 150) bars in each perforation plate. Leaves entire, ovate to elliptic, papyraceous or coriaceous, decurrent. Stomata generally paracytic. Flowers mostly axillary, sometimes cauliflorous, bisexual. Perianth segments numerous (7–33), free, spirally arranged. Stamens free, with short and thick filaments. Pollen grains 3-colporate. Carpels (5-)7-15(-21) in a single whorl, each with a single, large near-basal ovule. The carpel wall has cells with oxalate druses and crystals, tanniferous and oil cells. Ovary unilocular, with solitary, near-basal ovule. Fruiting flattened, woody follicles, arranged in a star-shaped fruit and each contains a single seed; seed exotestal, seed coat formed mainly by the testa. Contain glycosides of the flavonolos

kaempferol and quercetin, the toxic dilactonic sesquiterpenes (anisatin, majucin, etc.), the essential oils and shikimic acid. . . . . 1. ILLICIACEAE

1 Clambering or twining woody lianas with toothed or entire leaves. Neolignans and crystalliferous sclereids present. Nodes with three traces. Vessels in late metaxylem and early secondary xylem with scalariform perforations with numerous bars, but in late secondary xylem they are mostly simple, and only occasionally with a few bars (Carlquist 1999); vessels relatively wide with only a few (1–15, rarely to 30) bars in each perforation plate. Flowers unisexual, monoecious or dioecious. Perianth segments free, outermost and innermost sometimes reduced. Stamens partially or wholly united into a fleshy globose mass. Pollen grains 3-colpate or 6-colpate. Carpels 12–120 (*Schisandra*) or 17–300 (*Kadsura*), small, spirally arranged, free, ascidiate, with dorsal bulge, each with 2-5(-11), anatropous to campylotropous, ventrally attached or pendulous ovules. Fruits aggregate of many fruitlets with carnose pericarp, with enlarged subglobose or ellipsoid receptacle in *Kadsura* and elongated slender cylindrical receptacle in *Schisandra* (Saunders 1998, 2000). Seeds exotestal (Johri 1992), the testa with well delimited exotesta, mesotesta and endotesta (Denk and Oh 2006) Producing neolignans and myricetin. . . . . 2. SCHISANDRACEAE

1. ILLICIACEAE

A. Berchtold et J. Presl 1825. 1/42. Bhutan, Assam (Khasi Hills), eastern Asia and Southeast Asia, southeastern North America, eastern Mexico, West Indies.

Illicium.

2. SCHISANDRACEAE

Blume 1830 (including Kadsuraceae Radogizky 1849). 2/c.40. India and Sri Lanka, from Simla to Bhutan, Assam, northern Burma, South and Southeast Asia to western Malesia; Schisandra glabra is endemic to southeastern USA.

Schisandra, Kadsura.

Bibliography

Bailey IW and CG Nast. 1948. Morphology and relationships of *Illicium, Schisandra,* and *Kadsura*: I. Stem and leaf. J. Arnold Arbor. 29: 77–89.

Baranova MA. 1983. On the laterocytic stomatotype in angiosperms. Brittonia 35: 93–102.

Battaglia E. 1986. Embryological questions. 7. Do new types of embryo sac occur in *Schisandra*? Ann. Bot. 44: 69–82.

Behnke H-D. 1988. Sieve-element plastids, phloem protein, and evolution of flowering plants: III. Magnoliidae. Taxon 37: 699–732.

Bhandari NN. 1971. Embryology of the Magnoliales and comments on their relationships. J. Arnold Arbor. 52: 1–39, 285–304.

Carlquist S. 1982. Wood anatomy of *Illicium* (Illiciaceae): Phylogenetic, ecological, and functional interpretations. Am. J. Bot. 69: 1587–1598.

Carlquist S. 1999. Wood and bark anatomy of Schisandraceae: implications for phylogeny, habit, and vessel evolution. Aliso 18: 45–55.

Denk T and I-C Oh. 2006. Phylogeny of Schisandraceae based on morphological data: evidence from modern plants and the fossil record. Plant Syst. Evol. 256: 113–145.

Earle TT. 1941. Embryo and endosperm development in *Illicium floridanum* Ellis. Am. J. Bot. 28: 25 (Abstract).

Floyd SK and WE Friedman. 2001. Developmental evolution of endosperm in basal angiosperms: evidence from *Amborella* (Amborellaceae), *Nuphar* (Nymphaeaceae), and *Illicium* (Illiaceae). Plant Syst. Evol. 228: 153–169.

Friedman WE, WN Gallup, and JH Williams. 2003. Female gametophyte development in *Kadsura*: implications for Schisandraceae, Austrobaileyales, and the early evolution of flowering plants. Int. J. Plant Sci. 164(Suppl. 5): 293–305.

Gabarayeva NI and VV Grigorjeva. 2003. Comparative study of the pollen wall development in *Illicium floridanum* (Illiciaceae) and *Schisandra chinensis* (Schisandraceae). Taiwania 48(3): 147–167.

Hao G, ML Chye and RMK Saunders. 2001. A phylogenetic analysis of Schizandraceae based on morphology and nuclear ribosomal ITS sequences. Bot. J. Linn. Soc. 135: 401–411.

Hao G, RMK Saunders, and M-L Chye. 2000. A phylogenetic analysis of the Illiciaceae based on sequences of internal transcribed spacers (ITM) of nuclear ribosomal DNA. Plant Syst. Evol. 223: 81–90.

Hayashi Y. 1963a. The embryology of the family Magnoliaceae sensu lat. 1. Megasporogenesis, female gametophyte and embryogyne of *Illicium anisatum* L. Sci. Rep. Tohoku Univ. Ser. Biol. 29: 27–33.

Hayashi Y. 1963b. The embryology of the family Magnoliaceae sensu lato: I. Megasporogenesis, female gametophyte, and embryogeny of *Schisandra repanda* Radlkofer and *Kadsura japonica* Dunal. Sci. Rep. Tohoku Univ. Ser. Biol. 29: 403–411.

Hegnauer R. 1997. Phytochemistry and chemotaxonomy of the Illiciaceae. In: C Kalkman et al., eds. Flora Malesiana, ser. 1, 13: 175–177. Leiden.

Huynh K-L. 1976. L'arrangement du pollen du genre *Schisandra* (Schisandraceae) et sa significance phylo-genique chez les angiospermes. Beitr. Biol. Pfl. 52: 227–253.

Jalan S. 1962. The ontogeny of the stomata in *Schisandra grandtfiora* Hook. f. et Thorns. Phytomorphology 12: 239–242.

Jalan S. 1968a. Contribution to the nodal structure of *Schisandra* Michaux. Bot. Jahrb. Syst. 88: 311–316.

Jalan S. 1968b. Observations on the crystalliferous sclereids of some Schisandraceae. Beitr. Biol. Pfl. 44: 277–288.

Jalan S and RN Kapil. 1964. Pollen grains of *Schisandra* Michaux. Grana Palynol. 5: 216–221.
Kapil RN and S Jalan. 1964. *Schisandra* Michaux: Its embryology and systematic position. Bot. Notis. 117: 285–306.
Keng H. 1965. Observations on the flowers of *Illicium*. Bot. Bull. Acad. Sin., 2nd ser., 6: 61–73.
Keng H. 1993a. Illiciaceae. In: K Kubitzki, ed. The families and genera of vascular plants, vol. 2, pp. 344–347. Springer, Berlin/Heidelberg/New York.
Keng H. 1993b. Schisandraceae. In: K Kubitzki, ed. The families and genera of vascular plants, vol. 2, pp. 589–592. Springer, Berlin/Heidelberg/New York.
Kolbasina EI. 1967. Organogenesis of *Schisandra chinensis* (Turcz.) Baill. Bot. Zhurn. 52: 377–378 (in Russian).
Lan S-F. 1984. Pollen morphology of the genus *Kadsura* in China. J. South China Agric. College 5: 83–92 (in Chinese).
Lin Q. 1989. A study of the pollen morphology of genus *Illicium* L. Bull. Bot. Res. 9: 115–124 (in Chinese with English summary).
Lin Q. 1997. Systematics and evolution of the family Illiciaceae. Ph.D. thesis. Forestry College, Guangzhou (in Chinese).
Lin Q. 2000. Taxonomic notes on the genus *Schisandra* Michx. Acta Phytotax. Sinica 38: 532–550.
Liu H and C-S Yang. 1989. Pollen morphology of Illiciaceae and its significance in systematics. China J. Bot. 1: 104–115.
Liu Z, G Hao, Y-B Luo, LB Thien, SW Rosso, A-M Lu, and Z-D Chen. 2006. Phylogeny and androecial evolution in Schisandraceae, inferred from sequences of nuclear ribosomal DNA ITS and chloroplast DNA *trn*L-F regions. Int. J. Plant Sci. 167: 539–550.
Liu Z, X-Q Wang, Z-D Chen, Q Lin, and A-M Lu. 2000. The phylogeny of Schizandraceae inferred from sequence analysis of the nrDNA ITS region. Acta Bot. Sinica 42: 758–761.
Melikian AP. 1988. Illiciales. In: A Takhtajan, ed. Comparative seed anatomy, 2: 48–50. Nauka, Leningrad (in Russian).
Metcalfe CR. 1987. Illiciales. In: CR Metcalfe, ed. Anatomy of dicotyledons, 2nd ed., vol. 3, pp. 73–89. Claredon Press, Oxford.
Oh I-C and T Denk. 2001. Seed and leaf character evolution in the monogeneric basal angiosperm family Illiciaceae. Intern. Symposium Deep Morphology, p. 61. Vienna.
Oh I-C, T Denk, and EM Friis. 2003. Evolution of *Illicium* (Illiciaceae): Mapping morphological characters on the molecular tree. Plant Syst. Evol. 204: 175–209.
Praglowski J. 1976. Schisandraceae Bl. In: Nilsson S, ed. World pollen and spore flora, vol. 5, pp. 1–36. Almqvist & Wiksell, Stockholm.
Roberts ML and RR Haynes. 1983. Ballistic seed dispersal in *Illicium* (Illiciaceae). Plant Syst. Evol. 143: 227–232.
Robertson RE and SC Tucker. 1979. Floral ontogeny of *Illicium fioridanum*, with emphasis on stamen and carpel development. Am. J. Bot. 66: 605–617.
Saunders RMKS. 1995. Systematics of the genus *Illicium* L. (Illiciaceae) in Malesia. Bot. J. Linn. Soc. 117: 333–352.
Saunders RMKS. 1997a. Illiciaceae. In: C Kalkman et al., eds. Flora Malesiana, ser. I, 13: 169–184. Leiden.
Saunders RMKS. 1997b. Schisandraceae. In: C Kalkman et al., eds. Flora Malesiana, ser. I, 13: 185–207. Leiden.
Saunders RMKS. 1998. Monograph of *Kadsura* (Schisandraceae). Syst. Bot. Monogr. 54: 1–106.
Saunders RMKS. 2000. Monograph of *Schisandra* (Schisandraceae). Syst. Bot. Monogr. 58: 1–146.
Smith AC. 1947. The families Illiciaceae and Schisandraceae. Sargentia 7: 1–224.
Stone DE. 1968. Cytological and morphological notes on the southeastern endemic, *Schisandra glabra* (Schisandraceae). J. Elisha Mitchell Sci. Soc. 84: 351–356.
Stone DE and JL Freeman. 1968. Cytotaxonomy of *Illicium fioridanum* and *I. farviflorum* (Illiciaceae). J. Arnold Arbor 49: 41–51.
Sun CR. 2002. Micromorphological features of the seed surface of Schisandraceae and their systematic significance. Acta Phytotax. Sinica 40(2): 97–109.
Swamy BGL. 1965. Macrogametophytic ontogeny in *Schisandra chinensis*. J. Indian Bot. Soc. 43: 391–396.
Sy LK, RMK Saunders, and Brown GD. 1997. Phytochemistry of *Illicium dunnianum* and the systematic position of the Illiciaceae. Phytochemistry. 44: 1099–1108.
Thien LB, DA White, and LA Yatsu. 1983. The reproductive biology of a relic: *Illicium floridanum* Ellis. Am. J. Bot. 70: 719–727.
Van der Ham RWHM. 1997. Pollen morphology of Iilliciaceae. In: C Kalkman et al., eds. Flora Malesiana, ser. I, 13: 173–174. Leiden.
Vijayaraghavan MK and U Dhar. 1975. *Kadsura heteroclicta*: Microsporangium and pollen. J. Arnold Arbor. 56: 176–182.
White DA and LB Thien. 1983. The pollination of *Illicium parviflorum* (Illiciaceae). J. Elisha Mitchell Sci. Soc. 101: 5–18.
Williams JH and WE Friedman. 2004. The four-celled female gametophyte of *Illicium* (Illiciaceae; Austrobaileyales): implications for understanding the origin and early evolution of monocots, eumagnoliids and eudicots. Am. J. Bot 91: 332–351.
Yang Z-R and Qi Lin 2005. Comparative morphology of the leaf epidermis in *Schisandra* (Schisandraceae). Bot. J. Linn. Soc. 148: 39–56.
Yoshida O. 1962. Embryologische Studien uber *Schisandra chinensis* Baillon. J. Coil. Arts and Sci., Chiba Univ., 3: 459–462.

Order 5. TRIMENIALES

Trees, shrubs or woody lianas, young parts tomentose or glabrous. Nodes unilacunar, with two traces. Vessels with scalariform perforations with many bars. Secondary phloem with broad rays. Oil and mucilage cells occur in the mesophyll of the leaf and in the axis. Sieve-element plastids of S-type. Leaves opposite, margins entire or toothed, serrate to entire, often with pellucid glands and widely spreading and quite close secondary veins, estipulate. Stomata paracytic. Flowers small, in cymose inflorescences, bisexual or male (andromonoecious). Receptacle only slightly convex and continuous with the pedicel. Perianth segments, stamens, and carpel(s) are initiated in a spiral phyllotaxis. Perianth segments not differentiated into sepals and petals, caducous before or at anthesis, imbricate, 2–38. Stamens 7-16(-25); connective produced at apex; anthers tetrasporangiate, extrorse or latrorse, opening longitudinally. Pollen grains 2-colpate in monads or tetrads, inaperturate or polyporate, tectate-columellate. Microsporogenesis successive. Carpels solitary or very rarely 2, glabrous or hairy, extremely utriculate, topped by a capitate tufted-papillose stigma. Ovule 1, very large, anatropous with hood-shaped outer integument (Yamada et al. 2001), pendulous, bitegmic, crassinucellate. Fruits are small, spherical one-seeded berries. Seeds hard; seed coat formed mainly by the outer integument and has very stony exotesta consisting of cells with thick and lignified walls; endosperm copious, contains starch and oil; embryo small, straight, with rudimentary cotyledons. Hippocrepiform sclereids absent. Oil cells and mucilage cells present. 5-O-methyl flavonols present. n = 9.

Havesomecommonfeatureswiththe Amborellaceae, including convex floral base, capitate stigma and the gynoecium is strongly ascidiate up to the stigmatic region (Endress et al. 1997). Soltis et al. (2006) are including the Trimeniaceae in the Austrobaileyales sensu lato. According to Hao et al. (2000), *Trimenia* is a sister to well-supported to Schisandraceae. The Trimeniaceae also approaches the Chloranthaceae.

1. TRIMENIACEAE

Gibbs 1917. 1/6–8. Eastern Australia (New South Wales), Central Celebes, Moluccas, New Guinea, New Britain, New Ireland, Bougainville, New Caledonia, Fiji, Samoa, Marquises.

Trimenia (incl. *Piptocalyx*).

Bibliography

Carlquist S. 1984. Wood anatomy of Trimeniaceae. Plant Syst. Evol. 144: 103–118.

Endress PK and EL Sampson. 1983. Floral structure and relationships of the Trimeniaceae (Laurales). J. Arnold Arbor. 64: 447–473.

Goldblatt P. 1979. Chromosome number in two primitive dicots, *Xymalos monospora* (Monimiaceae) and *Piptocalyx moorei* (Trimeniaceae). Ann. Missouri Bot. Gard. 66: 898–899.

Morat P and HS MacKee. 1977. Quelques precisions sur le *Trimenia neocaledonica* Bak. f. et la famille des Trimeniacees en Nouvelle-Caledonie. Adansonia, ser. 2, 17: 205–213.

Petrova LP. 1988. Trimeniaceae. In: A Takhtajan, ed. Comparative seed anatomy, vol. 2, pp. 56. Nauka, Leningrad (in Russian).

Philipson WR. 1986. Trimeniaceae. In: CGGJ van Steenis, ed. Flora Malesiana, ser. 1, 10: 327–333. Nijhoff, Dordrecht.

Philipson WR. 1993. Trimeniaceae. In: K Kubitzki, ed. The families and genera of vascular plants, vol. 2, pp. 596–598. Springer, Berlin/Heidelberg/New York.

Renner SS. 1999. Circumscription and phylogeny of the Laurales: evidence from molecular and morphological data. Am. J. Bot. 86: 1301–1315.

Rodenburg WF. 1971. A revision of the genus *Trimenia* (Trimeniaceae). Blumea 19: 3–15.

Sampson FB. 1987. Short communications: disulculate pollen in the Trimeniaceae (Laurales). Grana 26: 239–241.

Sampson FB and PK Endress. 1984. Pollen morphology in the Trimeniaceae. Grana 23: 129–137.

Wagner WL and DH Lorence. 1999. A revision of *Trimenia* Seem. (Trimeniaceae) in the Marquises Islands with description of a new species, *Trimenia nukuhivensis*. Adansonia, sér. 3, 21: 225–230.

Order 6. CHLORANTHALES

Small trees, shrubs, or suffruticose or perennial, rarely annual herbs. Branches sometimes jointed at the nodes. Ethereal oil cells are generally scattered in the mesophyll and in the cortex and pith of the stem (and also in the testa and tegmen of the seeds). Mucilage ducts present in *Hedyosmum* and in some specimens of *Chloranthus*. Nodes often swollen, unilacunar, trilacunar, or unilacunar with split laterals. Vessels of very primitive type, usually only slightly wider than the tracheids, with very oblique perforation plates that have 50–200 bars; lateral pits mostly rare, ranging from scalariform to opposite; vessels in *Sarcandra* only in roots. Fibers are tracheids and fiber-tracheids. Rays heterogeneous, mixed multiseriate and uniseriate. Axial parenchyma usually scanty, apotracheal or paratracheal (*Hedyosmum*). Sieve-element plastids of S-type and characterized by 10–20 starch grains of different sizes Leaves opposite, decussate, or sometimes subverticillate in whorls of four, simple, serrate, crenate, or dentate, pinnately veined; stipules interpetiolar, minute to fairly conspicuous, threadlike or subulate, occasionally pectinate, borne on the petiole bases or emerging from the margin of the more or less developed vaginate sheath on either side of the petiole (formed by the connate bases of the opposite leaves and supporting the stem mechanically during the period of intercalary growth). Stomata variable-paracytic, laterocytic, encyclocytic, and their variants. Flowers very small and inconspicuous, spicate, paniculate, or capitate axillary or terminal compound or simple inflorescences, zygomorphic, of three distinct types: bisexual and without perianth, with one stamen or a lobed androecium (*Chloranthus, Sarcandra* and rarely in *Ascarina*); or unisexual, with tree short, scalelike organs that are probably tepals (female *Hedyosmum*), and unisexual without perianth (*Ascarina*, male *Hedyosmum*), entomophilous or sometimes anemophilous. Stamens 1–3 (up to 5 in *Ascarina*), not distinctly differentiated into anther and filament, in bisexual flowers adnate to the ovary, usually more or less connate laterally, the lateral ones usually with only disporangiate half-anthers. In the archaic genus *Sarcandra* the solitary stamen are laminar, with separated half-anthers; anthers linear to oblong, tetrasporangiate or bisporangiate, introrse, opening longitudinally by slits or by valves; connective often expanded or extended. Tapetum secretory. Microsporogenesis simultaneous. Pollen grains 2-celled, from globose to oblate, medium-sized to small, monocolpate (*Ascarina*), polyporate *(Sarcandra)* or polycolpate, tectate-perforate to semi-tectate with columellae well developed, irregularly reticulate with the reticulum usually verrucate. Gynoecium of a single, ascidate, barrel-shaped, glabrous carpel with a short stylodium or sessile apical, smooth or unicellular papillate stigma. Ovary superior (*Ascarina*) or inferior; stigma small, hemispherical and smooth (*Sarcandra*) or lobed (in some *Chloranthus* sp.). Ovule solitary, pendulous, orthotropous with cup-shaped outer integument (Yamada et al. 2001), bitegmic, crassinucellate. Female gametophyte of *Polygonum*-type. Endosperm cellular. Fruits in *Sarcandra, Chloranthus*, and *Ascarina* are drupelike berries (not drupes, since the hard protective layer is formed by the integument, not by the ovary wall, Endress 1987). In *Hedyosmum* fruits are drupes with thin exocarp, fleshy mesocarp, and stony endocarp, or (*H. mexicanum*) nuts with soft aril-like outgrowths. Seeds subglobose or ovoid; seed coat formed by both integuments; endotesta palisade, lignified, crystalliferous; embryo minute, weakly differentiated; endosperm copious, oily (in *Sarcandra* also with starch and proteins). Perisperm absent as a storage tissue, but sometimes present as a trace of crushed nucellar tissue next to chalaza and thin hypostase (see Corner 1976; Lodkina 1988). Producing noelignans and sesquiterpenes; n = 8, 13, 14, 15 (Kong 2000).

The order Chloranthales shows some relationships with the Trimeniaceae (Endress 1986, 1987; Todzia 1993). They originated from some very ancient vesselless protrimeniaceous ancestor by reduction and further specialization.

1. CHLORANTHACEAE

R. Brown ex Sims 1820 (including Hedyosmaceae Caruel 1881). 4/75 or more. Madagascar (subgenus *Ascarinopsis* of the genus *Ascarina*), tropical Himalayas, southern and eastern Asia and Southeast Asia, Malesia (Borneo), New Guinea, and Melanesia to the Marquises on the east, New Hebrides, Fiji, New Caledonia, North Island of New Zealand, and tropical America.

Sarcandra, Chloranthus, Hedyosmum, Ascarina.

Bibliography

Armour HM. 1906. On the morphology of *Chloranthus*. New Phytol. 5: 49–55.

Baillon H. 1871. Sur la position des Chloranthacees. Adansonia 10: 138–146.

Baranova M. 1983. On the laterocytic stomatotype in angiosperms. Brittonia 35: 93–102.

Baranova M. 1986. Comparative stomatographic investigations in the family Chloranthaceae. In: Problems of paleobotany, pp. 12–19. Soviet Science Press, Leningrad (in Russian).

Behnke HD. 1988. Sieve-element plastids, phloem protein, and evolution of flowering plants: III. Magnoliidae. Taxon 37: 699–732.

Burger WC. 1977. Flora Costaricensis: Chloranthaceae. Fieldiana Bot. 40: 1–10.

Carlquist S. 1987. Presence of vessels in wood of *Sarcandra* (Chloranthaceae): Comments on vessel origins in angiosperms. Am. J. Bot. 74: 1765–1771.

Carlquist S. 1990. Wood anatomy of *Ascarina* (Chloranthaceae) and the tracheid-vessel element transition. Aliso 12: 667–684.

Carlquist S. 1992a. Wood anatomy of *Hedyosmum* (Chloranthaceae) and the tracheid-vessel element transition. Aliso 13: 447–462.

Carlquist S. 1992b. Wood anatomy and stem of *Chloranthus*: Summary of wood anatomy of Chloranthaceae, with comments on relationships, vessellessness, and the origin of monocotyledons. IAWA Bull. 2, 13: 3–16.

Chan H-S and Y-Q Cheng. 1994. The origin, differentiation and geography of Chloranthaceae. J. Trop. Subtrop. Bot. 2: 31–44 (in Chinese, with English summary).

Cordemoy CJ de. 1863. Monographic du groupe des Chloranthacees. Adansonia 3: 280–310.

Crane PR, EM Friis, and KR Pedersen. 1989. Reproductive structure and function in Cretaceous Chloranthaceae. Plant Syst. Evol. 165: 211–226.

Doyle JA, H Eklund, and PS Herendeen. 2003. Floral evolution in Chloranthaceae: implications of a morphological phylogenetic analysis. Int. J. Plant Sci. 164(Suppl. 5): 365–382.

Edwards JG. 1920. Flower and seed of *Hedyosmum nutans*. Bot. Gaz. 70: 409–424.

Eklund H. 1999. Phylogeny of living and fossil Chloranthaceae. In: H Eklund. Big survivors with small flowers: fossil history and evolution of Laurales and Chloranthaceae. Uppsala University, Uppsala, Sweden.

Eklund H, JA Doyle, and PS Herendeen. 2004. Morphological phylogenetic analysis of living and fossil Chloranthaceae. Int. J. Plant Sci. 165: 107–151.

Endress PK. 1971. Bau der weiblichen Blüten von *Hedyosmum mexicanum* Cordemoy (Chloranthaceae). Bot. Jahrb. Syst. 91: 39–60.

Endress PK. 1986. Reproductive structures and phylo-genetic significance of extant primitive angiosperms. Plant Syst. Evol. 152: 1–28.

Endress PK. 1987. The Chloranthaceae: Reproductive structures and phylogenetic position. Bot. Jahrb. Syst. 109: 153–226.

Endress PK, FLS Igersheim, and A Igersheim. 1997. Gynoecium diversity and systematics of Laurales. Bot. J. Linn. Soc. 125: 93–168.

Jeremie J. 1980. Notes sur le genre *Ascarina* (Chloranthaceae) en Nouvelle-Caledonie et a Madagascar. Adansonia, ser. 2, 20: 273–285.

Kong HZ. 2000a. Karyotypes of *Sarcandra* Gardn. and *Chloranthus* Swartz (Chloranthaceae) from China. Bot. J. Linn. Soc. 133: 327–342.

Kong HZ. 2000b. Taxonomic notes on *Chloranthus henryi* Hemsl. And its allies. Acta Phytotax. Sinica 38: 355–366 (in Chinese with English summary).

Kong HZ. 2001. Comparative morphology of leaf epidermis in the Chloranthaceae. Bot. J. Linn. Soc. 136: 279–294.

Kong HZ and Z-D Chen. 2000. Phylogeny in *Chloranthus* Swartz (Chloranthaceae) inferred from sequence analysis of nrDNA ITS region. Acta Bot. Sinica 42: 762–764.

Kong HZ, Z-D Chen, and AM Lu. 2002. Phylogeny of *Chloranthus* (Chloranthaceae) based on nuclear ribosomal ITS and plastid *trn*L-F sequence data. Am. J. Bot. 89: 940–946.

Kong HZ, AM Lu, and PK Endress. 2002. Floral organogenesis of *Chloranthus sessilifolius*, with special emphasis on the morphological nature of the androecium of *Chloranthus* (Chloranthaceae). Plant Syst. Evol. 232: 181–188.

Kuprianova LA. 1967. Palynological data for the history of the Chloranthaceae. Pollen et Spores 9: 95–100.

Kuprianova LA. 1981. Palynological data on the family Chloranthaceae, its relationships, and the history of distribution. Bot. Zhurn. 66: 3–15 (in Russian).

Leroy JF. 1981. An unrecognized ancestral dicotyledon with a strobiloid flower is living today: *Hedyosmum*. 13th Internal. Bot. Congress, Sydney, Abstr. 136.

Leroy JF. 1983a. Interpretation nouvelle des appareils sexuels chez les Chloranthacees (Chloranthales, Magnoliidees). C. R. Acad. Sci. Paris, ser. 53, 296: 747–752.

Leroy JF. 1983b. The origin of angiosperms: An unrecognized ancestral dicotyledon, *Hedyosmum* (Chloranthales), with a strobiloid flower is living today. Taxon 32: 169–175.

Liu H. 1992. The role of palynology in modern plant taxonomy based on pollen morphology of Chloranthaceae. In: Z-Y Yu, X-Y Li, W-Z Di, eds. Advances in plant taxonomy in northwest China, pp. 69–70. Beijing.

Lodkina MM. 1988. Chloranthaceae. In: A Takhtajan, ed. Comparative anatomy of seeds, pp. 89–92. Nauka, Leningrad (in Russian).

Maekawa F. 1970. Notes on the stamens of *Chloranthus japonicus*. J. Jpn. Bot. 45: 289–294.

Maekawa F. 1971. Further notes on the stamens of *Chloranthus japonicus*. J. Jpn. Bot. 46: 198.

Moore LB. 1977. The flowers of *Ascarina lucida* Hook. f. (Chloranthaceae). New Zealand J. Bot. 15: 491–494.

Nakazawa K. 1956. Vascular course of Piperales: I. Chloranthaceae. Jpn. J. Bot. 15: 199–207.

Occhioni P. 1954. Contribuicao ao estudo da famflia Chloranthaceae corn especial referencia ao genus *Hedyosmum* Sw. Rio de Janeiro: Universidade do Brasil.

Okada H. 1995. Karyological studies of four genera of the Chloranthaceae. Plant Syst. Evol. 195: 177–185.

Patel RN. 1975. Wood anatomy of the dicotyledons indigenous to New Zealand. New Zealand J. Bot. 13: 141–148.

Smith AC. 1976. Studies of Pacific Island plants: XXXIII. The genus *Ascarina* (Chloranthaceae) in the Southern Pacific. J. Arnold Arbor. 57: 405–425.

Stuchlick L. 1984. Morphologia de los granos de polen de las Chloranthaceae y Canellaceae Cubanas. Acta Bot. Hung. 30: 321–328.

Swamy BGL. 1953a. The morphology and relationships of the Chloranthaceae. J. Arnold Arbor. 34: 375–411.

Swamy BGL. 1953b. *Sarcandra irvingbaileyi*: A new species of vesselless dicotyledon from South India. Proc. Natl. Inst. Sci. India 19B: 301–306.

Swamy BGL. 1953c. A taxonomic revision of the genus *Ascarina* Forst. Proc. Nad. Inst. Sci. India 19B: 371–388.

Swamy BGL and IW Bailey. 1950. *Sarcandra*: A vesselless genus of Chloranthaceae. J. Arnold Arbor. 31: 117–129.

Takahashi H. 1988. Morphology and ontogeny of stem xylem elements in *Sarcandra glabra* (Thunb.) Nakai (Chloranthaceae): additional evidence for the occurrence of vessels. Bot. Mag. Tokyo 101: 387–395.

Takahasi H and M Tamura. 1990. Occurrence of vessel elements in the stem of *Sarcandra glabra*. J. Jpn. Bot. 65: 81–86.

Thierry R. 1913. Contribution a l'etude anatomique des Chloranthacees. Thesis, University of Paris. Also: Trav. Lab. Mat. Med., Paris, 9: 1–158.

Todzia CA. 1988. Chloranthaceae: *Hedyosmum*. Flora Neotropica 48: 1–139.

Todzia CA. 1993. Chloranthaceae. In: K Kubitzki, ed., The families and genera of vascular plants, vol. 2, pp. 281–287. Springer, Berlin/Heidelberg/New York.

Todzia CA and RC Keating. 1991. Leaf architecture of the Chloranthaceae. Ann. Missouri Bot. Gard. 78: 476–496.

Verdcourt B. 1985. Notes on Malesian Chloranthaceae. Kew Bull. 40: 213–224.

Verdcourt B. 1986. Chloranthaceae. In: CGGJ van Steenis, ed. Flora Malesiana, ser. 1, 10(2): 123–149. Dodrecht.

Vijayaraghavan MR. 1964. Morphology and embryology of a vesselless dicotyledon: *Sarcandra irvingbaileyi* Swamy and systematic position of the Chloranthaceae. Phytomorphology 14: 429–441.

Von Balthazar M and PK Endress. 1999. Floral bract function, flowering process and breeding systems of *Sarcandra* and

Chloranthus (Chloranthaceae). Plant Syst. Evol. 218: 161–178.

Wang DQ, SH Huang, and ZF Wu. 1984. A preliminary study of the genus *Chloranthus* in Anhui. Bull. Bot. Res. 4: 173–182 (in Chinese with English summary).

Wang YH and K Yang. 2000. The comparation of the heteromorphosic anthers of *Chloranthus henryi* (Chloranthaceae) and its phylogenetic meanings. Bull. Bot. Res. 20: 379–384.

Yamada T, H Tobe, R Imaichi, and M Kato. 2001. Developmental morphology of the ovules of *Amborella trichopoda* (Amborellaceae) and *Chloranthus serratus* (Chloranthaceae). Bot. J. Linn. Soc. 137: 277–290.

Yamazaki T. 1992. Floral morphology of *Hedyosmum orientale* Merr. Et Chun (Chloranthaceae) and phylogenetic significance of its perianth. J. Jpn. Bot. 67: 257–269.

Yamazaki T. 1998. Embryogeny of *Sarcandra glabra* (Thunb.) Nakai (Chloranthaceae) and phylogenetic position of the Chloranthaceae. J. Jpn Bot. 73: 22–25.

Yoshida O. 1957. Embryologische Studien über die Ordnung Piperales: I. Embryologie von *Chloranthus japonicus*. J. Coll. Arts Chiba Univ. 2: 172–178.

Yoshida O. 1959. Embryologische Studien über die Ordnung Piperales: II. Embryologie von *Chloranthus serratus*. J. Coll. Arts Chiba Univ. 2: 295–303.

Yoshida O. 1960. Embryologische Studien über die Ordnung Piperales: III. Embryologie von *Sarcandra glabra*. *J*. Coll. Arts Chiba Univ. 3: 55–60.

Zhang L-B and S Renner. 2003. The deepest splits in Chloranthaceae as resolved by chloroplast sequences. Int. J. Plant Sci. 164(Suppl. 5): 383–392.

Zhang SS and LG Lei. 1991. Characteristics of leaf epidermis of Chloranthaceae from China. Acta Bot. Bor.-Occident. Sinica 11: 17–22 (in Chinese, with English summary).

Zhang SS, LG Lei, HQ Liu, and QY Su. 1990. A preliminary study on tracheary elements in the endemic species *Sarcandra hainanensis* from China – evidence for the occurrence of vessels in *Sarcandra*. Acta Bot. Bor.-Occid. Sinica 10: 95–98.

Zhang SS and Y Wang. 1983. A preliminary study on the anatomy of the Chloranthaceae in China. Acta Bot. Bor-Occid. Sinica 3: 117–121 (in Chinese with English summary).

Zhang S, W Zhang, and Q Su. 1991. Discussion of systematic position of Chloranthaceae by leaf architecture of plants. Acta Bot. Bor.-Occid. Sinica 11: 226–232.

Zhou ZK. 1993. Origin, systematics and distribution of Chloranthaceae. Acta Bot. Yunn. 15: 321–331.

Zhou ZK and HM Li. 1994. Implication of the leaf architecture for systematic studies of Chloranthaceae from China. Chinese J. Bot. 6: 12–18.

Order 7. CERATOPHYLLALES

Submersed herbs with leafy stems, perennating by dormant terminal buds. Stems usually branched, freely suspended or sometimes anchored in bottom sediments by slender rhizoidlike branches. Roots absent. Idioblasts lacking. Xylem very reduced, vesselless, with tracheids modified into unlignified elongate starch-bearing cells. Sieve-element plastids of Ss-type (Behnke 2002). Leaves verticillate, almost sessile, rather rigid, dichotomously dissected into filiform or linear segments that bear two rows of minute denticles and are tipped by a medial multicellular mucilaginous appendage. Stomata absent. Flowers minute (ca. 0.5–1.5 mm), axillary, alternating with leaves, usually solitary (or occasionally in vestigial inflorescences), spirocyclic, monoecious (the males and females commonly on alternate nodes), apetalous, actinomorphic, hydrophilous. Sepals (bracts, according to many authors) 12(8-15) in male flowers and 9–10 in females, bractlike, connate at the base, often dentate or lacerate at the apex. Stamens three to numerous, free, spirally arranged on a convex receptacle and developing centripetally; the innermost stamens are retarded and sterile; filaments very short and broad or almost absent; anthers linear-oblong, tetrasporangiate, extrorse-latrorse, opening longitudinally; connective laminar and thickened, often colored, prolonged apically into short spurs, flanked by two to several small denticles. Tapetum secretory. Microsporogenesis successive (*Ceratophyllum demersum* and *C. submersum*) or simultaneous (*C. pentacanthum*). Pollen grains 2-celled, globose, with very reduced exine and thick intine, indistinctly 1-colpate, in monads, medium sized. Gynoecium of one carpel (in rare cases of two free carpels) tapering into a long slender spinescent or short and awliform stylodium with more or less well-developed decurrent stigmatic groove above the mouth of the stylodial canal. Ovule solitary, dorsally pendulous near the top of the locule, orthotropous, unitegmic (integument reduced, four cells thick at the base, thinning distally to the thickness of a single cell), crassinucellate, with well-developed nucellar cap and tanniniferous hypostase, without funicle. Female gametophyte of *Polygonum*-type. Endosperm cellular, its four large lower cells perform haustorial function. Proembryo without suspensor. Fruit an achene crowned by the persistent stylodium and mostly with basal, basal-lateral, or lateral horns. Seeds minute, elliptical; integument obliterated and very thin, and transparent seed coat formed by the outer epidermis of nucellus; endosperm present as a thin layer only in the chalazal part; perisperm lacking; embryo large, with thick fleshy cotyledons, conspicuous and highly developed greenish plumule consisting of 8–10 whorls of leaves and a few lateral buds and very short and weakly differentiated vestigial radicle, $n = 12$.

In spite of some embryological similarities with the Cabombaceae (Johri et al. 1992), *Ceratophyllum* is not closely related to the Nympheales (Iwamoto, Shimizu and Ohba 2003). *Ceratophyllum* is very isolated and probably derived from some early flowering plants. I therefore tentatively include Ceratophyllales in the superorder Nymphaeanae.

1. CERATOPHYLLACEAE

Gray 1822. 1/11. Worldwide in fresh water, except Arctica.

Ceratophyllum.

Bibliography

Aboy HE. 1936. A study of the anatomy and morphology of *Ceratophyllum demersum*. Thesis, Cornell University.

Batygina TB and II Shamrov. 1981. Ceratophyllaceae. In: MS Yakoviev, ed. Comparative embryology of flowering plants: Winteraceae-Juglandaceae, pp. 110–115. Nauka, Leningrad (in Russian).

Endress PK. 1994. Evolutionary aspects of the floral structure in *Ceratophyllum*. In: PK Endress and EM Friis, eds. Early evolution of flowers, pp. 175–183. Plant Syst. Evol., Suppl.: 8. Wien.

Gray A. 1848. Remarks on the structure and affinities of the order Ceratophyllaceae. Ann. Lyceum Nat. Hist. New York 4: 41–50.

Iwamoto A, A Shimizu, and H Ohba. 2003. Floral development and phyllotaxis variation in *Ceratophyllum demersum* (Ceratophyllaceae). Am. J. Bot. 90: 1124–1130.

Jedrychowska A and A Sroczynska. 1934. On the cytology and embryology of *Ceratophyllum submersum*. Acta Soc. Bot. Polon. 11: 423–441.

Jones EN. 1931. The morphology and biology of *Ceratophyllum demersum*. Stud. Nat. Hist. Iowa Univ. 13: 11–55.

Kaden NN. 1953. Fruits and seeds of the Middle Russian Hornworts. Bull. Moscow Soc. Naturalists, Biol. 38(3): 82–85.

Klercker JEF de. 1885. Sur Fanatomie et le developpe-ment de *Ceratophyllum*. Kgl. Svensk. Vet.-Akad. Handl. 9(10): 1–22.

Les DH. 1985. The taxonomic significance of plumule morphology in *Ceratophyllum* (Ceratophyllaceae). Syst. Bot. 10: 338–346.

Les DH. 1986. Systematic and evolution of *Ceratophyllum* L. (Ceratophyllaceae): a monograph. Ph.D. dissertation, Ohia State University, Ohia.

Les DH. 1988a. The evolution of achene morphology in *Ceratophyllum* (Ceratophyllaceae): II. Fruit variation and systematics of the "spiny-margined" group. Syst. Bot. 13: 73–86.

Les DH. 1988b. The evolution of achene morphology in *Ceratophyllum* (Ceratophyllaceae): III. Relationships of the "facially-spined" group. Syst. Bot. 13: 509–518.

Les DH. 1988c. The origin and affinities of the Ceratophyllaceae. Taxon 37: 326–345.

Les DH. 1989. The evolution of achene morphology in *Ceratophyllum* (Ceratophyllaceae): IV. Summary of proposed relationships and evolutionary trends. Syst. Bot. 14: 254–262.

Les DH. 1993. Ceratophyllaceae. In: K Kubitzki, ed. The families and genera of vascular plants, vol. 2, pp. 246–249. Springer, Berlin/Heidelberg/New York.

Lowden RM. 1978. Studies on the submerged genus *Ceratophyllum* L. in the Neotropics. Aquatic Bot. 4: 127–142.

Mouraviev I. 1945. Recherches sur la microspore du genre *Ceratophyllum*. Bull. Mens. Soc. Linn. Lyon 14(l): 73–82.

Muenscher WC. 1940. Fruits and seedlings of *Ceratophyllum*. Am. J. Bot. 27: 231–233.

Oganezova EP and RM Nalbandyan. 1976. Purification and properties of plastoganin and ferredoxin from *Ceratophyllum demersum* L. Biokhimia (Moscow) 41(5): 794–800 (in Russian).

Sastri RLN. 1955. Embryology of *Ceratophyllum demersum* L. Proc. Indian Sci. Congr. 3: 226.

Schieiden MJ. 1837. Beiträge zur Kenntniss der Cerato-phylleen. Linnaea 11: 513–542.

Schneider EL and S Carlquist. 1996. Conducting tissue in *Ceratophyllum demersum* (Ceratophyllaceae). SIDA 17: 437–443.

Sehgal A and HY Mohan Ram. 1981. Comparative developmental morphology of two populations of *Ceratophyllum* L. (Ceratophyllaceae) and their taxonomy. Bot. J. Linn. Soc. 82: 343–356.

Shamrov II. 1980. Some data on the flower ecology of *Ceratophyllum*. Bot. Zhurn. 65: 703–706 (in Russian).

Shamrov II. 1981. Some peculiar features of the development of the anther in *Ceratophyllum demersum* and *C. pentacanthum* (Ceratophyllaceae). Bot. Zhurn. 66: 1464–1472 (in Russian).

Shamrov II. 1983a. Antecological investigation of three species of the genus *Ceratophyllum* (Ceratophyllaceae). Bot. Zhurn. 68: 1357–1366 (in Russian).

Shamrov II. 1983b. The structure of the anther and some peculiar features of the microsporogenesis and pollen grain development in the representatives of the genus *Ceratophyllum* (Ceratophyllaceae). Bot. Zhurn. 68: 1662–1667 (in Russian).

Shamrov II. 1997. Ovule and seed development in *Ceratophyllum demersum* (Ceratophyllaceae). Bot. Zhurn. 82: 1–13 (in Russian with English summary).

Shamrov II and TB Batygina. 1984. The development of the embryo and endosperm in representatives of the family Ceratophyllaceae. Bot. Zhurn. 69: 1328–1335 (in Russian).

Shamrov II and TB Batygina. 1988. Ceratophyllaceae. In: A Takhtajan, ed. Comparative seed anatomy, vol. 2, pp. 153–156. Nauka, Leningrad (in Russian).

Strasburger E. 1902. Ein Beitrag zur Kenntnis von *Ceratophyllum submersum* und phylogenetische Erörterungen. Jahrb. Wiss. Bot. 34: 477–524.

Sundari KT, M Radhakrishnaiah, and LL Narayana. 1982. Chemotaxonomy of *Ceratophyllum*. Acta Bot. Indica, 10: 304–305.

Wilmot-Dear M. 1985. *Ceratophyllum* revised: A study in fruit and leaf variation. Kew Bull. 40: 243–271.

Wood CE, Jr. 1959. The genera of the Nymphaeaceae and Ceratophyllaceae in the southeastern United States. J. Arnold Arbor. 40: 94–112.

Superorder MAGNOLIANAE

Order 8. CANNELALES

Large to small trees, shrubs or shrublets. Xylem vesselless (Winteraceae) or vessels with scalariform perforations (Cannelaceae). Axial parenchyma scanty, diffuse to tangentially banded (Winteraceae) or apotracheal diffuse to paratracheal. Sieve-element plastids of Ss-, Psc-, and Pfs-types. Nodes trilacunar. Leaves alternate, simple, entire, pinnately veined, gland-dotted, estipulate. Stomata usually paracytic. Flowers terminal or axillary cymose inflorescences or solitary, bisexual or rarely unisexual, actinomorphic. Sepals two, less often three, rarely up to six (Winteraceae) or three. Petals two to many, imbricate, free or rarely united into a tube. Stamens three to numerous (very rarely up to 370); anthers introrse or extrorse, tetrasporangiate, 2-locular, usually opening longitudinally. Tapetum secretory. Microsporogenesis simultaneous. Pollen grains in tetrads or rarely in monads. Gynoecium of several to many (up to 50), rarely two (*Takhtajania*) or even one carpels, more or less 1-seriate, free and remaining so in fruit; ovary superior. Ovules one to many (up to 100), anatropous, descending apotropous, campylotropous, bitegmic, crassinucellate. Female gametophyte of *Polygonum*-type. Endosperm cellular. Fruits follicles, multifollicular, capsular, berries or consisting of free berry-like fruitlets. Seeds with a hard, brittle exotesta formed by the epidermis of the outer integument; embryo minute, usually ovoid and apically slightly bilobed; endosperm copious, oily.

The order Canellales is rather heterobathmic: alongside the extremely primitive xylem and also very primitive carpels and stamens in some genera, its pollen grains and seeds are considerably advanced. Evidently one of the most ancient members of the flowering plants which is supported by its very disjunct geographical distribution. Evidently related to the Magnoliales, especially to Degeneriaceae.

Key to Families

1 Xylem vesselless, with long cambial initials and long slender tracheids. Tracheids usually with 2–3 rows of large bordered pits, but the overlapping radially oriented end walls of *Bubbia* and *Zygogynum* bear scalariform pitting. Axial parenchyma scanty, diffuse to tangentially banded. Rays narrow, heterogeneous with long ends. Sieve-element plastids of Ss- and Pcs-types. Small trees to shrublets. Leaves pinnately veined, gland-dotted. Stomata paracytic, rarely (*Bubbia perrieri*) anomocytic (Baranova 2004). Flowers small to rather large, in terminal or axillary cymose inflorescences or solitary (terminal or axillary), bisexual or (*Tasmannia*) unisexual (dioecious), with short receptacle, spiral or more or less cyclic, variously entomophilous or more or less anemophilous, seldom autogamous. Sepals two, less often three, rarely up to six, connate into calyptra rupturing at very early stages and then persisting or rupturing upon anthesis and then usually dropping. Petals two to many, from small and chaffy to larger and petaloid, all free or the outer ones connate and rupturing upon anthesis, very rarely wanting. According to Vink (1970) a continuation of the spiral of bracteoles through the sepals and the petals has been observed occasionally in *Pseudowintera*. Stamens three to numerous (very rarely up to 370), with more or less irregular spiral arrangement; filaments short, and thick to long-cylindrical, 1-veined; anthers sometimes apically prolonged, usually with submarginal, apical, or subapical bisporangiate half-anthers. Pollen grains in tetrads or rarely (in two species of *Zygogynum*) in monads, monoporate or rarely (*Takhtajania*) ranging from monoporate to trichotomocolpate, semitectate or rarely tectate-perforate, with distinct columellae, reticulate to microreticulate. Carpels several to many (up to 50), rarely two (*Takhtajania*) or even one, more or less 1-seriate, free and remaining so in fruit, partially united and loosely connate in fruit or closely united into a eusyncarpous (*Zygogynum*) or unilocular (*Takhtajania*) gynoecium, conduplicate, stipitate or sessile, varying from unsealed and with a decurrent stigma (as in *Tasmannia* and *Bubbia*) to fully sealed and with a short stylodium and localized stigma; placentation laminar-lateral or submarginal. Ovules one to many (up to 100), with massive parietal tissue. Fruits multifollicular, capsular, berrylike or consisting of free berry-like fruitlets. Seeds with a hard, brittle exotesta formed by the epidermis of the outer integument; embryo minute, usually ovoid and apically slightly bilobed; endosperm copious, oily; seed coat exotestal; cells of exotesta lignified, palisade. Producing proanthocyanins but not alkaloids, plants Aluminium accumulator; n = 13 (*Tasmannia*), 18 (*Takhtajania*), 43 (*Drimys, Pseudowintera, Bubbia, Belliolum, Exospermum, Zygogynum*). . . . . . . 1. WINTERACEAE

1 Vessels with scalariform perforations. Trees or rarely shrubs. Vessel elements usually very long, with oblique ends and typically scalariform perforation plates mostly with numerous fine bars (50–100 in *Cinnamodendron)*; anastomoses between the bars sometimes produce reticulate perforations; lateral pitting rare but, where present, opposite. Fibers mostly long to very long, with evidently bordered pith. Rays narrow, heterogeneous to homogeneous. Axial parenchyma apotracheal diffuse to paratracheal. Sieve elements of Psc- or (*Canella*) of Pfs-type. Leaves leathery, commonly pellucidly dotted. Stomata paracytic or anomocytic. Flowers in terminal or axillary cymes or racemes or solitary in leaf axils, bisexual, basically 3-merous. Sepals three, thick, leathery, basally connate, imbricate. Petals (3-4)5-12, free or (*Cinnamosma*) united high up into a tube with lobes reflexed after anthesis, in 1, 2 or 4 cycles or in an indistinct spiral, often fleshy. Stamens 6–12 (up to 40); anthers adnate to the outer surface of the staminal tube, elongate, opening longitudinally, often apically prolonged. Pollen grains 1-colpate or sometimes trichotomocolpate, tectate, columellate or granular, or (*Cinnamosma*) intectate and reticulate, in *Warburgia* psilate. Gynoecium of 2–6 united carpels; style short, thick, with usually 2–6-lobed stigma; ovary unilocular, with 2–6 parietal placentas and 2-20(-30) campylotropous ovules on each placenta; ovules with massive parietal tissue; micropyle zig-zag. Fruits berries with two or more seeds. Seeds small, black, with shiny crustaceous testa; embryo small to moderately well developed; endosperm copious, oily, ruminate in *Canella* and *Cinnamosma*. Aporphine alkaloids, flavonols and drimane sesquiterpenoids present; n = 11, 13, 14.......................... 2. CANNELACEAE

1. WINTERACEAE

R. Brown ex Lindley 1830 (including Takhtajaniaceae J.-F. Leroy 1980). 8/70–90. Madagascar, Philippines, Malay Archipelago (except Sumatra, Java, and Timor), New Guinea, Moluccas to Solomon Islands, eastern Australia, Tasmania, New Zealand, Lord Howe Island, New Caledonia, America from Mexico to the Straits of Magellan, and Juan Fernandez.

TASMANNEAE: *Tasmannia* (including *Austrodrimys, Pseudodrimys*); TAKHTAJANEAE: *Takhtajania*; WINTEREAE: *Drimys, Pseudowintera, Bubbia, Belliolum, Zygogynum, Exospermum.*

2. CANELLACEAE

C. Martius 1832. 6/16. Madagascar *(Cinnamosma)*, tropical East Africa and northern Transvaal (*Warburgia*), southern end of the Florida Keys and Cape Sable, West Indies, and South America.

Canella, Warburgia, Capsicodendron, Cinnamodendron, Pleodendron, Cinnamosma.

According to Corner (1976), seeds of the Canellaceae appear exotestal and indicate alliance with Winteraceae. Canellaceae and Winteraceae share sesquiterpenoids of the drimane and rearranged drimane type, which occur only in these two families (Gottlieb et al. 1989).

Bibliography

Agababian VS. 1972. Ultrastructure of sporoderm of some primitive Angiospermae. Bot. Zhurn. 57: 955–959 (in Russian).

Bailey IW and CG Nast. 1943a. The comparative morphology of the Winteraceae. I. Pollen and stamens. J. Arnold Arbor. 24: 340–346.

Bailey IW and CG Nast. 1943b. The comparative morphology of the Winteraceae. II. Carpels. J. Arnold Arbor. 24: 472–481.

Bailey IW and CG Nast. 1944a. The comparative morphology of the Winteraceae. III. Wood. J. Arnold Arbor. 25: 97–103.

Bailey IW and CG Nast. 1944b. The comparative morphology of the Winteraceae. IV. Anatomy of the node and vascularization of the leaf. J. Arnold Arbor. 25: 215–221.

Bailey IW and CG Nast. 1944c. The comparative morphology of the Winteraceae. V. Foliar epidermis and sclerenchyma. J. Arnold Arbor. 25: 342–348.

Bailey IW and CG Nast. 1945. The comparative morphology of the Winteraceae. VII. Summary and conclusions. J. Arnold Arbor. 26: 37–47.

Baranova M. 1972. Systematic anatomy of the leaf epidermis in the Magnoliaceae and some related families. Taxon 21: 446–469.

Baranova M. 2004. The stomatal apparatus of *Takhtajania perrieri* (Capuron) M. Baranova et J.-F.Leroy (Winteraceae). Kew Bull. 59: 141–144.

Behnke H-D and U Kiritis. 1983. Ultrastructure and differentiation of sieve elements in primitive angio-sperms: I. Winteraceae. Protoplasma 118: 148–156.

Bhagavathi Kutti Amma PR. 1938. Microsporogenesis in *Drimys*. Presidency College Bot. Mag. Madras, 5: 22–25.

Bhandari NN. 1963. Embryology of *Pseudowintera colorata*, a vesselless dicotyledon. Phytomorphology, 13: 303–316.

Bhandari NN. 1971. Embryology of the Magnoliales and comments on their relationships. J. Arnold Arbor. 52: 1–39, 285–304.

Bhandari NN and R Venkataraman. 1968. Embryology of *Drimys winteri*. J. Arnold Arbor. 49: 509–524.

Birkinshaw C, D Ravelonarivo, R Andriamparany, S Rapanarivo, E Rabakonandriana, GE Schatz, and LB Thien. 1999a.

Risque d'extinction du *Takhtajania perrieri*. Rapport Final I., September, Antananarivo.

Birkinshaw C, D Ravelonarivo, R Andriamparany, S Rapanarivo, E Rabakonandriana, GE Schatz, and LB Thien. 1999b. L'habitat du *Takhtajania perrieri*. Rapport Final II., October, Antananarivo.

Boer R de and F Bouman. 1974. Integumentary studies in the Polycarpicae. III. *Drimys winteri* (Winteraceae). Acta Bot. Neerl. 23: 19–27.

Bongers JM. 1973. Epidermal leaf characters of the Winteraceae. Blumea 21: 381–411.

Burtt BL. 1936. *Bubbia haplopus* B.L. Burtt, Winteraceae. Hooker's Icon. Plant. 34, t.3315, 1–3.

Burtt BL. 1938. The taxonomic position of *Tetrathalamus*. Bull. Misc. Int. Kew 1938: 458–460.

Capuron R. 1963. Contributions á l'étude de la flore de Madagascar. XII. Présence á Madagascar d'un nouveau représentant (*Bubbia perrieri* R. Capuron) de la famille des Wintéracées. Adansonia n.s. 3: 373–378.

Carlquist S. 1981. Wood anatomy of *Zygogynum* (Winteraceae): Field observations. Bull. Mus. Natur. Hist. Nat. Paris, ser. 4, Adansonia 3: 281–292.

Carlquist S. 1982. *Exospermum stipitatum* (Winteraceae): Observations on wood, leaves, flowers, and fruit. Aliso 10: 277–289.

Carlquist S. 1983a. Wood anatomy of *Belliolum* (Winteraceae) and note on flowering. J. Arnold Arbor. 64: 161–169.

Carlquist S. 1983b. Wood anatomy of *Bubbia* (Winteraceae) with comments on origin of vessels in dicotyledons. Am. J. Bot. 70: 578–590.

Carlquist S. 1988. Wood anatomy of *Drimys s.* s. (Winteraceae). Aliso 12: 81–95.

Carlquist S. 1989. Wood anatomy of *Tasmannia*: Summary of wood anatomy of Winteraceae. Aliso 12: 257–275.

Carlquist S. 2000. Wood and bark anatomy of *Takhtajania* (Winteraceae); phylogenetic and ecological implications. Ann. Missouri Bot. Gard. 87: 317–322.

Coetzee JA and J Praglowski. 1988. Winteraceae pollen from the Miocene of the southwestern Cape (South Africa): Relationship to modern taxa and phytogeographical significance. Grana 27: 27–37.

Dandy JE. 1933. The Winteraceae of New Zealand. J. Bot. 71: 119–122.

De Boer R and F Bouman. 1974. Integumentary studies in the Polycarpicae: III. *Drimys winteri* (Winteraceae). Acta Bot. Neerl. 23: 19–27.

Dehay C and A Ghestem 1969. Caractéres de l'appareil libéroligneux foliare chez quelques Wintéracées. Bull. Soc. Bot. France 116: 165–169.

Deroin T. 2000. Notes on the vascular anatomy of the fruit of *Takhtajania* (Winteraceae) and its interpretation. Ann. Missouri Bot. Gard. 87: 398–406.

Deroin T and J-F Leroy. 1993. Sur l'interprétation de la vascularisation ovarienne de *Takhtajania* (Winteracees). C. R. Acad. Sci. Paris 316: 725–729.

Doust AN. 1997. Variability and pattern in the flowers of the Winteraceae (Magnoliidae). Am. J. Bot. 84 (6, Abstract): 40.

Doust AN. 2000. Comparative floral ontogeny in Winteraceae. Ann. Missouri Bot. Gard. 87: 366–379.

Doust AN 2001. The developmental basis of floral variation in *Drimys winteri* (Winteraceae). Int. J. Pland Sci. 162: 697–717.

Doust AN and AN Drinnan. 2004. Floral development and molecular phylogeny support the generic status of *Tasmannia* (Winteraceae). Am. J. Bot. 91: 321–331.

Doweld A. 2000. De genere *Tasmannia* R.Br. ex DC. (Winteraceae). In: TV Egorova ed. Novitates Systematicae Plant. Vascularium, 32: 36–40 (in Russian).

Doyle JA. 2000. Paleobotany, relationships, and geographic history of Winteraceae. Ann. Missouri Bot. Gard. 87: 303–316.

Doyle JA, CL Hotton, and JV Ward. 1990a. Early Cretaceous tetrads, zonasulculate pollen, and Winteraceae. I. Taxonomy, morphology and ultrastructure. Am. J. Bot. 77: 1544–1557.

Doyle JA, CL Hotton, and JV Ward. 1990b. Early Cretaceous tetrads, zonasulculate pollen, and Winteraceae. II. Cladistic analysis and implications. Am. J. Bot. 77: 1558–1568.

Ehrendorfer F and M Lambrou. 2000. Chromosomes of *Takhtajania*, other Winteraceae, and Canellaceae: phylogenetic implications. Ann. Missouri Bot. Gard. 87: 407–413.

Ehrendorfer F, I Silberbauer-Gottsberger, and G Gottsberger. 1979. Variation on the population, racial, and species level in the primitive relic angiosperm genus *Drimys* (Winteraceae) in South America. Plant Syst. Evol. 132: 53–83.

Endress PK, A Igersheim, FB Sampson, and GE Schatz. 2000. Floral structure of *Takhtajania* and its systematic position in Winteraceae. Ann. Missouri Bot. Gard. 87: 347–365.

Erbar C and P Leins. 1983. Zur Sequenzvon Blütenorganen bei einigen Magnoliiden. Bot. Jahrb. Syst. 103: 433–449.

Esau K and VI Cheadle. 1984. Anatomy of the secondary phloem in Winteraceae. JAWA Bull., N.S. 5: 13–43.

Field TS, MA Zwieniecki, and NM Holbrook. 2000. Winteraceae evolution: an ecophysiological perspective. Ann. Missouri Bot. Gard. 87: 323–334.

Fiser J and D Walker 1967. Notes on the pollen morphology of *Drimys* Forst., section *Tasmannia* (R.Br.) F. Muell. Pollen et Spores 9: 229–239.

Frame D. 1996. Carpel development in *Tasmannia insipida* (Winteraceae). Int. J. Plant Sci. 157: 698–702.

Freiberg TE. 1981. Winteraceae. In: MS Yakovlev, ed. Comparative embryology of flowering plants. Winteraceae – Juglandaceae, pp. 26–30. Nauka, Leningrad (in Russian).

Gifford EM. 1951a. Early ontogeny of the foliage leaf in *Drimys winteri* var. *chinensis*. Am. J. Bot. 38: 93–105.

Gifford EM. 1951b. Ontogeny of the vegetative axillary bud in *Drimys winteri* var. *chinensis*. Am. J. Bot. 38: 234–243.

Godley EJ and DH Smith. 1981. Breeding system in New Zealand plants: 5. *Pseudowintera colorata* (Winteraceae). New Zealand J. Bot. 19: 151–156.

Gottsberger G, I Silberbauer-Gottsberger, and F Ehrendorfer. 1980. Reproductive biology in the primitive relic angiosperm *Drimys brasiliensis* (Winteraceae). Plant Syst. Evol. 135: 11–39.

Ham van der R and BJ van Heuven. 2002. Evolutionary trends in Winteraceae pollen. Grana 41: 4–9.

Hiepko P. 1966. Das Blütendiagramm von *Drimys winteri* J. R. et G. Forst. (Winteraceae). Willdenowia 4: 221–226.

Hotchkiss AT. 1955. Chromosome numbers and pollen tetrad size in the Winteraceae. Proc. Linn. Soc. N. S. W. 80: 47–53.

Karol KG, Y Suh, GE Schatz, and EA Zimmer. 2000. Molecular evidence for the phylogenetic position of *Takhtajania* in the Winteraceae: inference from nuclear ribosomal and chloroplast gene spacer sequences. Ann. Missouri Bot. Gard. 87: 414–432.

Keating RC. 2000. Anatomy of the young vegetative shoot of *Takhtajania perrieri* (Winteraceae). Ann. Missouri Bot. Gard. 87: 335–346.

Kubitzki K. 1993. Canellaceae. In: K Kubitzki, ed. The families and genera of vascular plants, vol. 2, pp. 200–203. Springer, Berlin/Heidelberg/New York.

Leinfellner W. 1965, 1966. Wie sind die Winteraceen-Karpelle tatsachlich gebaut?: I. Die Karpelle von *Drimys*, Sektion *Tasmannia*. II. Über das Vorkommen einer ringformigen Plazenta in den Karpellen von *Drimys*, Sektion *Wintera*. III. Die Karpelle von *Bubbia, Belliolum, Pseudowintera, Exospermum*, und *Zygogynum*. Oesterr. Bot. Z. 112: 554–575; 113: 84–95, 245–264.

Leinfellner W. 1967. Über die Karpelle verschiedener Magnoliales. V. *Pleodendron* (Canellaceae). Oesterr. Bot. Z. 114: 502–507.

Lemesle A. 1950. Persistance de caracteres archaïques dubois secondaire chez les Canellacees. C. R. Acad. Sci. Paris 231: 455–456.

Lemesle A. 1951. Nouvelles remarques histologiques et phylogenetiques sur la famille des Canellacees. Rev. Gen. Bot. 58: 193–202.

Leroy J-F. 1977. A compound ovary with open carpels in Winteraceae (Magnoliales): Evolutionary implications. Science 196: 977–978.

Leroy J-F. 1978. Une sous-famille monotypique de Winteraceae endémique á Madagascar: Takhtajanioideae. Adansonia, ser. 2, 17: 385–395.

Leroy J-F. 1980. Nouvelles remarques sur le genre *Takhtajania* (Winteraceae – Takhtajanioideae). Adansonia, n.s. 20: 9–20.

Lloyd DG and MS Wells. 1992. Reproductive biology of a primitive angiosperm, *Pseudowintera colorata* (Winteraceae), and the evolution of pollination systems in the Anthophyta. Plant Syst. Evol. 181: 77–95.

Lobreau-Callen D. 1977. Le pollen de *Bubbia perieri* R. Cap.: Rapports palynologiques avec les autres genres de Winteracees. Adansonia, ser. 2, 16: 445–460.

Melchior H and W Schultze-Motel. 1959. Canellaceae. In: A Engler and K Prantl, eds. Die natürlichen Pflanzenfamilien, 2nd ed., 17a: 221–224. Ducker und Humboldt Berlin.

Meylan BA and BG Butterfield. 1982. Pit membranes structure in the vessel-less woods of *Pseudowintera* Dandy (Winteraceae). IAWA Bull. 3: 167–175.

Miers J. 1858. On the Canellaceae. Ann. Mag. Nat. Hist. 3, 1: 349–353.

Miers J. 1861. On the Winteraceae. Contributions to Botany 1: 123–138. London.

Morawetz W. 1984. How stable are genomes of tropical woody plants? Heterozygosity in C-banded karyotypes of *Porcelia* as compared with *Annona* (Annonaceae) and *Drimys* (Winteraceae). Plant Syst. Evol. 145: 29–39.

Nast CG. 1944. The comparative morphology of the Winteraceae: VI. Vascular anatomy of the flowering shoot. J. Arnold Arbor. 25: 454–466.

Occhioni P. 1948. Contribuicao ao estudo do familia "Canellaceae." Arq. Jard. Bot. Rio de Janeiro 8: 3–165.

Occhioni P. 1949. Contribuicao ao estudo anatomico de *Cinnamodendron sampaioanum* Occh. Arq. Jard. Bot. Rio de Janeiro 9: 101–108.

Parameswaran N. 1961a. Foliar vascularisation and histology in the Canellaceae. Proc. Indian Acad. Sci. 54: 306–317.

Parameswaran N. 1961b. Ruminate endosperm in the Canellaceae. Curr. Sci. 30: 344–345.

Parameswaran N. 1962. Floral morphology and embryology in some taxa of the Canellaceae. Proc. Indian Acad. Sci. 55B: 167–182.

Patel RN. 1974. Wood anatomy of the dicotyledons indigenous to New Zealand: 4. Winteraceae. New Zealand J. Bot. 12: 19–32.

Pellmyr O, LB Thien, G Bergstrom, and I Groth. 1990. Pollination of New Caledonian Winteraceae: Opportunistic shifts or parallel radiation with their pollinators. Plant Syst. Evol. 173: 143–157.

Praglowski J. 1979. Winteraceae Lindl. In: S Nilsson, ed. World pollen and spore flora, vol. 8. Almqvist & Wiksell, Stockholm.

Prakash N, AL Lim, and FB Sampson. 1992. Anther and ovule development in *Tasmannia* (Winteraceae). Austral. J. Bot. 40: 877–885.

Raleigh RE, PY Ladige, TJ Entwisle and AN Drinnan. 1994. Morphometric studies of the genus *Tasmannia* (Winteraceae) in Victoria, Australia. Mulleria 38: 235–256.

Sampson FB. 1963. The floral morphology of *Pseudowintera*, the New Zealand member of the vesselless Winteraceae. Phytomorphology 13: 403–423.

Sampson FB. 1970. Unusual features of cytokinesis in meiosis of pollen mother cells of *Pseudowintera traversii* (Buchan.) Dandy (Winteraceae). Beitr. Biol. Pfl. 47: 71–77.

Sampson FB. 1974. A new pollen type in the Winteraceae. Grana 14: 11–15.

Sampson FB. 1978. Placentation in *Exospermum stipitatum* (Winteraceae). Bot. Gaz. 139: 215–222.

Sampson FB. 1980. Natural hybridism in *Pseudowintera* (Winteraceae). New Zealand J. Bot. 18: 43–51.

Sampson FB. 1981. Synchronous versus asynchronous mitosis within permanent pollen tetrads of the Winteraceae. Grana 20: 19–23.

Sampson FB. 1987. Stamen venation in the Winteraceae. Blumea 32: 79–89.

Sampson FB. 2000. The pollen of *Takhtajania perrieri* (Winteraceae). Ann. Missouri Bot. Gard. 87: 380–388.

Sampson FB and DR Kaplan. 1970. Origin and development of the terminal carpel in *Pseudowintera traversii*. Am. J. Bot. 57: 1185–1196.

Sampson FB and SC Tucker. 1978. Placentation in *Exospermum stipitatum* (Winteraceae). Bot. Gaz. 139: 215–222.

Sampson FB, JB Williams, and PS Woodland. 1988. The morphology and taxonomic position of *Tasmannia glaucifolia* (Winteraceae), a new Australian species. Austral. J. Bot. 36: 395–413.

Schatz GE. 1989. The search for *Takhtajania* (Winteraceae). Bull. Natl. Trop. Bot. Gard. 19(4): 117–118.

Schatz GE. 2000. The rediscovery of a Malagasy endemic: *Takhtajania perrieri* (Winteraceae). Ann. Missouri Bot. Gard. 87: 297–302.

Schatz GE, PP Lowry II, and A Ramisamihantanirina. 1998. *Takhtajania perrieri*: Rediscovered. Nature 391: 133–134.

Smissen R. 1993. Some aspects of the embryology, morphology and anatomy of *Exospermum stipitatum*. B.Sc. (Hons) Project, Victoria University of Wellington, New Zealand.

Smith AC. 1943a. The American species of *Drimys*. J. Arnold Arbor. 24: 1–33.

Smith AC. 1943b. Taxonomic notes on the Old World species of Winteraceae. J. Arnold Arbor. 24: 119–164.

Smith AC. 1945. Geographical distribution of the Winteraceae. J. Arnold Arbor. 26: 48–59.

Smith AC. 1969. A reconsideration of the genus *Tasmannia* (Winteraceae). Taxon 18: 286–290.

Straka H. 1963. Über die mogliche phylogenetische Be-deutung der Pollenmorphologie dcr madagascarischen *Bubbia perieri* R. Cap. (Winteraceae). Grana Palynol. 4: 355–360.

Strasburger E. 1905. Die Samenanlage von *Drimys winteri* und die Endospermbildung bei Angiospermen. Flora 95: 215–231.

Stuchlick L. 1984. Morfologia de los granos de polen de las Chlorantaceae y Canellaceae Cubanas. Acta Bot. Hung. 30: 321–328.

Suh Y, LB Thien, and EA Zimmer. 1992. Nucleotide sequences of the internal transcribed spacers and 5.8S *r*RNA gene in *Canella winterana* (Magnoliales; Canellaceae). Nucl. Acids Res. 20: 6101–6102.

Suh Y, LB Thien, HE Reeve, and EA Zimmer. 1993. Molecular evolution and phylogenetic implications of internal transcribed spacer sequences of ribosomal DNA in Winteraceae. Am. J. Bot. 80: 1042–1055.

Svoma E. 1998. Studies on the embryology and gynoecium structure in *Drimys winteri* (Winteraceae) and some Annonaceae. Plant Syst. Evol. 209: 205–229.

Swamy BGL. 1952. Some aspects of the embryology of *Zygogynum baillonii*. Proc. Natl. Inst. Sci. India 18: 399–406.

Thien LB. 1982. Fly pollination in *Drimys* (Winteraceae), a primitive angiosperm. In: JL Gressitt, ed. Biogeography and ecology of New Guinea, Monographiae Biologicae 42: 529–533. The Hague.

Thien LB, P Bernhardt, GW Gibbs, O Pellmyr, G Bergstrom, I Groth, and G McPherson. 1985. The pollinadon of *Zygogynum* (Winteraceae) by a moth, *Sabatinca* (Micropteridae): An ancient association? Science 227: 540–543.

Thien LB, O Pellmyr, LY Yatsu, G Bergstrom, and G McPherson. 1990. Polysaccharide food-bodies as pollinator rewards in *Exospermum stipitatum* and other Winteraceae. Adansonia, n.s. 12: 191–197.

Thompson WP and IW Bailey. 1916. Are *Tetracentron, Trochodendron*, and *Drimys* specialized or primitive types? Mem. New York Bot. Gard. 6: 27–32.

Tieghem P van. 1900. Sur les dicotyledones du groupe des Homoxylees. J. Bot. (Paris) 14: 259–297, 330–361.

Tobe H and B Sampson. 2000. Embryology of *Takhtajania* (Winteraceae) and a summary statement of embryological features for the family. Ann. Missouri Bot. Gard. 87: 389–397.

Tucker SC. 1959. Ontogeny of the inflorescence and the flower in *Drimys winteri* var. *chinensis*. Univ. Calif. Publ. Bot. 30: 257–336.

Tucker SC. 1975. Carpellary vasculature and the ovular vascular supply in *Drimys*. Am. J. Bot. 62: 191–197.

Tucker SC and EM Gifford. 1964. Carpel vascularization of *Drimys lanceolata*. Phytomorphology 14: 197–203.

Tucker SC and EM Gifford. 1966a. Organogenesis in the carpellate flower of *Drimys lanceolata*. Am. J. Bot. 53: 433–442.

Tucker SC and EM Gifford. 1966b. Carpel development in *Drimys lanceolata*. Am. J. Bot. 53: 671–678.

Tucker SC and FB Sampson. 1979. The gynoecium of winteraceous plants. Science 203: 920–921.

Ueda K. 1977. Floral morphology of *Belliolum pancheri* (Winteraceae). Proc. Jpn Soc. Plant Taxon. 3: 10.

Ueda K. 1978a. Floral morphology of the Winteraceae: 2. *Drimys confertifolia*. Proc. Jpn Soc. Plant Taxon. 4: 11–12.

Ueda K. 1978b. Vasculature in the carpels of *Belliolum pancheri* (Winteraceae). Acta Phytotax. Geobot. 29: 119–125.

Vink W. 1970. The Winteraceae of the Old World: I. *Pseudowintera* and *Drimys*, morphology and taxonomy. Blumea 18: 225–354.

Vink W. 1977. The Winteraceae of the Old World: II. *Zygogynum*, morphology and taxonomy. Blumea 23: 219–250.

Vink W. 1978. The Winteraceae of the Old World: III. Notes on the ovary of *Takhtajania*. Blumea 24: 521–525.

Vink W. 1983. The Winteraceae of the Old World: IV. The Australian species of *Bubbia*. Blumea 28: 311–328.

Vink W. 1985. The Winteraceae of the Old World: V. *Exospermum* links *Bubbia* to *Zygogynum*. Blumea 31: 39–55.

Vink W. 1988. Taxonomy in Winteraceae. Taxon 37: 691–698.

Vink W. 1993. Winteraceae. In: K Kubitzki, ed. The genera and families of vascular plants, vol. 2, pp. 630–638. Springer, Berlin/Heidelberg/New York.

Walker JW, GJ Brenner, and AG Walker. 1983. Winteraceae pollen in the Lower Cretaceous of Israel: Early evidence of a Magnolealean angiosperm family. Science 220: 1273–1275.

Williams CA and WJ Harvey. 1982. Leaf flavonoid patterns in the Winteraceae. Phytochemistry 21: 329–337.

Wilson TK. 1960. The comparative morphology of the Canellaceae: I. Synopsis of genera and wood anatomy. Trop. Woods 112: 1–27.

Wilson TK. 1964. The comparative morphology of the Canellaceae: III. Pollen. Bot. Gaz. 125: 192–197.

Wilson TK. 1965. The comparative morphology of the Canellaceae: II. Anatomy of the young stem and node. Am. J. Bot. 52: 369–378.

Wilson TK. 1966. The comparative morphology of the Canellaceae: IV. Floral morphology and conclusions. Am. J. Bot. 53: 336–343.

Order 9. MAGNOLIALES

Trees or shrubs. Nodes multilacunar. Vessels with scalariform or less frequently simple perforations; lateral pitting scalariform, opposite (*Magnolia fraseri* and *Liriodendron*). Fibers with bordered pits. Rays heterogeneous or rarely homogeneous (in some temperate species of *Magnolia*). Axial parenchyma apotracheal. Sieve-element plastids of S- and Psc-types. Leaves alternate, simple, entire or seldom 2–10-lobed, pinnately veined, with stipules or estipulate. Stomata paracytic or rarely anomocytic. Flowers usually solitary, terminal or axillary, bisexual or very rarely unisexual (*Kmeria*). Perianth well developed, with free and imbricate segments, these variously spiral to cyclic in three or more series, more or less similar and all petaloid or less often clearly differentiated into sepals and petals arranged in whorls of three. Stamens numerous, free, spirally arranged, originating centripetally, more or less ribbon-shaped (laminar), mostly three-veined and typically not clearly differentiated into filament and anther; anthers tetrasporangiate, the microsporangia paired, Tapetum secretory. Microsporogenesis simultaneous. Pollen grains 2-celled, boat-shaped, psilate or more or less rugulose, sulcate, aperturate. Gynoecium sessile or stipitate, apocarpous or, less frequently, more or less syncarpous. Carpels solitary (in *Degeneria* and sometimes in *Michelia montana)* or more frequently several to numerous and spirally arranged (except in *Pachylarnax*), conduplicate, semisealed (*Manglletia* and *Elmerillia*) or more commonly fully sealed. Ovules two to numerous, large to very large, anatropous, bitegmic, crassinucellate, with massive parietal tissue. Female gametophyte of *Polygonum*-type. Endosperm cellular. Fruits carpels, or indehiscent. Seeds large, with sarcotesta; embryo small to minute; endosperm copious and oily, sometimes ruminate. Present sesquiterpene lactones; n = 12, 19.

Key to Families

1 Leaves estipulate. Basic chromosome number n = 12. Large trees, bearing essential oils. Vessels with scalariform perforations. Intravascular pitting scalariform. Rays multiseriate or less often uniseriate. Sieve-element plastids of Psc-type. Nodes 5-lacunar. Leaves entire, non-sheathing, gland-dotted, aromatic. Flowers solitary, pendulous on long peduncles, bisexual. Sepals 3, rarely 4, not calyptrate. Petals 12–18, fleshy, imbricate. Stamens 20–30; anthers adnate, extrorse. Pollen boat-shaped, aperturate, sulcate. Carpel 1(-2), unsealed at anthesis, not differentiated into an ovary and stylodium. The carpel wall has cells with oxalate druses and crystals, sclereids, oil cells. Ovules 22–30, with conspicuous funicular obturator. Fruit a follicle, leathery, with a hard exocarp. Seeds flattened, with orange-red sarcotesta; exotestal cells palisate, thin-walled; endotesta with lignified internal fibrils; embryo very small, but well differentiated, with 3(4) cotyledons; endosperm oily, ruminate.................... 1. DEGENERIACEAE

1 Leaves with large stipules enclosing the young leaves. Basic chromosome number n = 19. Trees or shrubs. Nodes multilacunar. Vessels present, with scalariform or less frequently simple perforations; lateral pitting scalariform, opposite (*Magnolia fraseri* and *Liriodendron*). Fibers with bordered pits. Rays heterogeneous or rarely homogeneous (in some temperate species of *Magnolia*). Sieve-element plastids of S- and Psc-types. Leaves entire or seldom 2–10-lobed, pinnately veined. Flowers often with an elongate receptacle, Perianth with free and imbricate segments, these variously spiral to cyclic in three or more series, more or less similar and all petaloid or less often clearly differentiated into sepals and petals arranged in whorls of three. Stamens numerous, free, anthers often more or less embedded in the adaxial (abaxial in *Liriodendron*) surface, the connective more or less prolonged into a distinct appendage (supracon-nectivum). Pollen grains psilate or more or less rugulose, monocolpate, atectate (primitively columellaless) or with incipient or rarely well-developed columellae. Carpels of several to numerous and spirally arranged (except in *Pachylarnax*), semisealed (*Manglletia* and *Elmerillia*) or more commonly fully sealed. Ovules 2–12(-16). When the fruit is apocarpous, the carpels open primarily dorsally, rarely *(Kmeria)* ventrally and partly dorsally or (*Talauma* and *Tsoongiodendron*) transversely (circumscissile) orindehiscent,samaroidandcaducous(*Liriodendron*), otherwise the carpels are united into a fleshy berry-like syncarp (*Aromadendron, Paramichella* and some species of *Elmerillia*) or into a loculicidal capsule (*Pachylarnax*). Seeds large, with sarcotesta, and usually hanging from the elongated spiral vessels of the funicle in dehiscent carpels, or, when the fruit or individual carpels are indehiscent, then they are without sarcotesta and usually adherent to the endocarp. Embryo minute; endosperm copious and oily, not ruminate..................... 2. MAGNOLIACEAE

2. DEGENERIACEA

I. W. Bailey and A. C. Smith 1942. 1/2. Fiji.
Degeneria.

1. MAGNOLIACEAE

A.L. de Jussieu 1789 (including Liriodendraceae F. A. Barkley 1975). 15/240. Southern India, Sri Lanka, eastern Himalayas, Assam, eastern Asia and Southeast Asia, New Guinea, southeastern North America, Central America, West Indies, Venezuela, Brazil,

Ecuador. Concentrated mainly in East and Southeast Asia, especially in continental China.

1.1 MAGNOLIOIDEAE

Cyanogenic compounds absent. Leaves entire or very rarely 2-lobed. Stipules at first united, surrounding the stem and bud and fused adaxially to part or most of the petiole, then rupturing longitudinally and falling, leaving a circular scar around the twig. Intravascular pitting scalariform. Stomata paracytic. Anthers introrse or latrorse. Fruiting carpels dehiscent or indehiscent, not samaroid. Seeds with colored sarcotesta and scleroendotesta with crystals and lignified fibrils in the cells. – MAGNOLIEAE: *Manglietia, Manglietiastrum, Pachylarnax, Magnolia, Parakmeria, Talauma* (including ? *Dugandiodendron), Aromadendron, Kmeria, Woonyoungia, Alcimandra*; MICHELIEAE: *Elmerrillia, Michelia, Paramichelia, Tsoongiodendron.*

1.2 LIRIODENDROIDEAE

Cyanoid compounds present. Leaves two- or more lobed with apex truncate or widely emarginate. Stipules free, foliaceous, fused laterally to the petiole base. Intervascular pitting opposite. Stomata both paracytic and anomocytic. Anthers extrorse. Seeds without sarcotesta and adherent to endocarp. Fruiting carpels indehiscent, produced at the apex into a long winglike beak, caducous. Seeds without sarcotesta; endosperm slight and not ruminate. – *Liriodendron.*

Related to the Degeneriaceae, but leaves stipulate, and stamens and carpels are generally more specialized.

Bibliography

Agababian VS. 1972. Pollen morphology of the family Magnoliaceae. Grana 12: 166–176.

Azuma H, JG Garcia-Franco, V Rico-Gray, and LB Thien. 2001. Molecular phylogeny of the Magnoliaceae: the biogeography of tropical and temperate disjunctions. Am. J. Bot. 88: 2275–2285.

Azuma H, LB Thien, and S Kawano. 1999a. Floral scents, leaf volatiles and thermogenic flowers in Magnoliaceae. Plant Species Biol. 14: 121–127.

Azuma H, LB Thien, and S Kawano. 1999b. Molecular phylogeny of *Magnolia* (Magnoliaceae) inferred from cdDNA sequences and evolutionary divergence of the floral scents. J. Plant Res. 112: 291–306.

Bailey IW and AC Smith. 1942. Degeneriaceae: A new family of flowering plants from Fiji. J. Arnold Arbor. 23: 356–365.

Baillon H. 1866. Memoire sur la famille de Magnoliacees. Adansonia 7: 1–16.

Baranova M. 1969. A comparative stomatographic investigation of the genus *Manglietia* Bl. Bot. Zhurn. 54: 1952–1964 (in Russian).

Baranova M. 1972. Systematic anatomy of the leaf epidermis in the Magnoliaceae and some related families. Taxon 21: 447–469.

Baranova MA and C Jeffrey. 2000. Stomatographical features in the systematics of the Magnoliaceae. Bot. Zhurn. 85(6): 35–49.

Baranova M, C Jeffrey, QG Wu, and JP Liao. 2000. Leaf epidermis features of Magnoliaceae and their systematic significance, with special reference to the genera *Parakmeria, Manglietiastrum* and *Woonyoungia*. Proc. Internat. Symp. Fam. Magnoliaceae 2000: 143–152. Beijing.

Barkley FA. 1975. Liriodendraceae fam. n. order Mag-noliales. Phytologia 32(4): 304.

Behnke H-D. 1988. Sieve-element plastids, phloem protein, and evolution of flowering plants: III. Magnoliidae. Taxon 37: 699–732.

Bhandari NN. 1971. Embryology of the Magnoliales and comments on their relationships. J. Arnold Arbor. 52: 1–39, 285–304.

Biswas BK and AK Sharma. 1984. Chromosome studies in the family Magnoliaceae. Cytologia 49: 193–200.

Bouman F. 1977. Integumentary studies in the Polycar-picae: IV. *Liriodendron tulipifera* L. Acta Bot. Neerl. 26: 213–223.

Brandza M. 1891. Developpement des teguments de la graine. (Magnoliacees). Rev. Gen. Bot. 3: 124–126.

Cai X and Z-H Hu. 2000a. Studies on the development of oil cells in *Liriodendron chinense*. Acta Bot. Bor.-Occid. Sinica 20: 309–312.

Cai X, ZH Hu. 2000b. Comparative studies on leaf structure and oil cells of the Magnoliaceae in China. Acta Phytotax. Sinica 38(3): 218–230.

Canright JE. 1952. The comparative morphology and relationships of the Magnoliaceae: I. Trend of specialization in the stamens. Am. J. Bot. 39: 484–492.

Canright JE. 1953. The comparative morphology and relationships of the Magnoliaceae: II. Significance of the pollen. Phytomorphology 3: 355–365.

Canright JE. 1955. The comparative morphology and relationships of the Magnoliaceae: IV. Wood and nodal anatomy. J. Arnold Arbor. 36: 119–140.

Canright JE. 1960. The comparative morphology and relationships of the Magnoliaceae: III. Carpels. Am. J. Bot. 47: 145–155.

Carlquist S. 1989 (1990). Wood and bark anatomy of *Degeneria*. Aliso 12: 485–495.

Cheng B-L and HP Nooteboom. 1993. Notes on Magnoliaceae. III. The Magnoliaceae of China. Ann. Missouri Bot. Gard. 80: 999–1104.

Cheng BL, X Huang, R Wang, and S Cheng. 2000. Chromosome data of Magnoliaceae. In: Y Liu et al., eds. Proc. Internat. Symp. Fam. Magnoliaceae 2000, pp. 192–201. Beijing.

Dahl A and JR Rowley. 1965. Pollen of *Degeneria vitiensis*. J. Arnold Arbor. 46: 308–329.

Dandy JE. 1927. The genera of Magnoliaceae. Kew Bull. 1927: 257–265.

Dandy JE. 1971. The classification of the Magnoliaceae. Newslett. Am. Magnolia Soc. 8: 3–6.

Dandy JE. 1974. Magnoliaceae taxonomy. In: J Praglowski, ed. World pollen and spore flora. Magnoliaceae, vol. 3, pp. 1–5. Almqvist & Wiksell, Stockholm.

Dandy JE. 1978. A revised survey of the genus *Magnolia* together with *Manglietia* and *Michelia*. In NG Treseder, ed. Magnolias, pp. 29–37. London.

De Boer R and F Bouman. 1972. Integument studies in the Polycarpicae: II. *Magnolia stellata* and *Magnolia virginiana*. Acta Bot. Neerl. 21: 617–629.

Deroin T. 1991. La vascularisation florale des Magnoliales: Premiere approche experimentale de son role au cours de la pollinisation. C. R. Acad. Sci. Paris, 3rd ser., 312: 355–360.

Doweld AB. 2003. On the structure of *Degeneria* seeds (Degeneriaceae). Bull. Mosc. Ob. Ispyt. Prirody, 108(3): 64–73 (in Russian with English summary).

Earle TT. 1938. Origin of the seed coats in *Magnolia*. Am. J. Bot. 25: 221–223.

Ehrendorfer E, F Krendl, E Habeller, and W Sauer. 1968. Chromosome numbers and evolution in primitive angiosperms. Taxon 17: 337–353.

Endress PK. 1977. Über Blütenbau und Verwandtschaft der Eupomatiaceae und Himantandraceae (Magnoliales). Ber. Deutsch. Bot. Ges. 90: 83–103.

Endress PK. 1984. The role of inner staminodes in the floral display of some relic Magnoliales. Plant Syst. Evol. 146: 269–282.

Endress PK and LD Hufford. 1989. The diversity of stamen structures and dehiscence patterns among Magnoliales. Bot. J. Linn. Soc. 100: 45–85.

Erbar C and P Leins. 1981. Zur Spirale in Magnolien-Bluten. Beitr. Biol. Pfl. 56: 225–241.

Freiberg TE and OP Kamelina. 1981. Magnoliaceae. In: MS Yakovlev, ed. Comparative embryology of flowering plants: Winteraceae-Juglandaceae, pp. 36–41. Nauka, Leningrad (in Russian).

Gabarayeva NI. 1986a. The development of the exine in *Michelia fuscata* (Magnoliaceae) in connection with the change in cytoplasmic organelles of microspores and tapetum. Bot. Zhurn. 71: 311–322 (in Russian with English summary).

Gabarayeva NI. 1986b. Ultrastructure analysis of the intine development of *Michelia fuscata* (Magnoliaceae) in connection with the changes of cytoplasmic organelles of microspores and tapetum. Bot. Zhurn. 71: 416–428 (in Russian with English summary).

Gabarayeva NI. 1987a. Ultrastructure and development of sporoderm in *Manglietia tenuipes* (Magnoliaceae) during the tetrad period: The ptimexine formation in connection with cytoplasmic organelle activity. Bot. Zhurn. 72: 281–290 (in Russian with English summary).

Gabarayeva NI. 1987b. Ultrastructure and development of lamellae of endexine in *Manglietia tenuipes* (Magnoliaceae) in connection with the question of endexine existence in primitive angiosperms. Bot. Zhurn. 72: 1310–1317 (in Russian with English summary).

Gabarayeva NI. 1987c. Ultrastructure and development of pollen grain wall in *Manglietia tenuipes* (Magnoliaceae): The formation of intine in connection with the activity of cytoplasmic organelles. Bot. Zhurn. 72: 1470–1478 (in Russian with English summary).

Gabarayeva NI. 1988. The significance of ontogenetic investigation of the sporoderm for elucidation of the structure and phylogeny of the mature sporoderm of some species of Magnoliaceae and Annonaceae. In: AF Chlonova, ed. Palynology in the USSR, pp. 48–52. Nauka, Novosibirsk (in Russian with English summary).

Gabarayeva NI. 1990. On the site of sporopollenin precursors synthesis in the developing pollen grains in Magnoliaceae. Bot. Zhrun.75: 783–791 (in Russian with English summary).

Gabarayeva NI. 1991a. The ultrastructure and development of exine and orbicules of *Magnolia delavayi* (Magnoliaceae) in the tetrad and the beginning of post-tetrad periods. Bot. Zhurn. 76: 10–19 (in Russian with English summary).

Gabarayeva NI. 1995. Sporoderm development in *Liriodendron chinense* (Magnoliaceae): a probable role of the endoplasmic reticulum. Nord. J. Bot. 16: 307–323.

Goldblatt P. 1974. A contribution to the knowledge of cytology in Magnoliales. J. Arnold Arbor. 55: 453–457.

Guedes M. 1968. Le carpelle du tulipier (*Liriodendron tulipifera*). Oesterr. Bot. Z. 115: 372–378.

Guillaumin A. 1942. Materiaux pour la flore de la Nouvelle Caledonie: 69 Notes sur les Magnoliacees. Bull. Soc. Bot. France 89: 1–3.

Hayashi Y. 1963. The embryology of the Magnoliaceae sensu lato: I. Megasporogenesis, female gametophyte, and embryology. Sci. Rep. Tohoku Imp. Univ., 4th ser. (Biol.), 29: 27–33.

Hayashi Y. 1964. The embryology of the Magnoliaceae sensu lato: III. *Magnolia liltflora* and *Michelia fuscata*. Sci. Rep. Tohoku Imp. Univ., 4th ser. (Biol.), 30: 89–98.

Hayashi Y. 1965. The comparative embryology of the Magnoliaceae s. 1. in relation to the systematic consideration of the family. Sci. Rep. Tohoku Imp. Univ., 4th ser. (Biol.), 31: 29–44.

Hayashi Y. 1966. The embryology of the Magnoliaceae sensu lato: IV. Microsporogenesis and development of the male gametophyte in *Michelia figo* Spreng. Sci. Rep. Tohoku Imp. Univ., 4th ser. (Biol.), 32: 11–118.

Hayashi Y. 1984. Embryology of *Magnolia salicifolia*. J. Jpn. Bot. 59: 289–307.

Heiser Ch B. 1962. Some observations on pollination and compatibility in *Magnolia*. Proc. Indiana Acad. Sci. 72: 259–266.

Hu HH. 1940. A new genus of Magnoliaceae. Sunyatsenia 4: 142–145.

Hu HH and WY Cheng. 1951. *Parakmeria*, a new genus of Magnoliaceae of southwestern China. Acta Phytotax. Sinica 1: 1–2.

Igersheim A, PK Endress, and FLS Endress. 1997. Gynoecium diversity and systematics of the Magnoliales and winteroids. Bot. J. Linn. Soc. 124: 213–271.

Johnson MA and D Fairbrothers 1965. Comparison and interpretation of serological data in the Magnoliaceae. Bot. Gaz. 126: 260–269.

Kaeiser M and SG Boyce. 1962. Embryology of *Liriodendron tulipifera* L. Phytomorphology 12: 103–109.

Kapil RN and NN Bhandari. 1964. Morphology and embryology of *Magnolia*. Proc. Nad. Inst. Sci. India 30B: 245–262.

Keng H. 1978. The delimitation of the genus *Magnolia*. Gard. Bull. Singapore 31: 127–131.

Kim S, C-W Park, Y-D Kim, and Y Suh. 2001. Phylogenetic relationships in family Magnoliaceae inferred from *ndh*F sequences. Am. J. Bot. 88: 717–728.

Korobova SN. 1981. Degeneriaceae. In: MS Yakovlev, ed. Comparative embryology of flowering plants: Winteraceae-Juglandaceae, pp. 30–33. Nauka, Leningrad (in Russian).

Kubitzki K. 1993. Degeneriaceae. In: K Kubitzki, ed., The families and genera of vascular plants, vol. 2, pp. 290–291. Springer, Berlin/Heidelberg/New York.

Law YH. 1984. A preliminary study on the taxonomy of the family Magnoliaceae. Acta Phytotax. Sinica 22: 89–109 (in Chinese with English summary).

Law YH. 2000. Studies on the phylogeny of the Magnoliaceae. In: YH Law, HM Fan, ZY Chen, QG Wu and QW Zeng, eds. Proc. Internat. Symp. Fam. Magnoliaceae 2000, pp. 3–13. Beijing.

Leinfellner W. 1967. Über die Karpelle verschiedener Magnoliales: IV. *Magnolia* und *Michelia* (Magnoliaceae). Oesterr. Bot. Z. 114: 73–84.

Leinfellner W. 1969. Über die Karpelle verschiedener Magnoliales: VIII. Überblick über alle Familien der Ordnung. Oesterr. Bot. Z. 117: 107–127.

Lemesle R. 1953. Les caracteres histologiques du bois secondaire des Magnoliales. Phytomorphology 3: 430–446.

Lemesle R and A Duchaigne. 1955. Contribution a l'etude histologique et phylogenetique du *Degeneria vitiensis* I. W. Bailey et A. C. Smith. Rev. Gen. Bot. E. E. 62: 708–719.

Leppik EE. 1975. Morphogenic stagnation in the evolution of *Magnolia* flowers. Phytomorphology 25: 451–464.

Li J and JG Conran. 2003. Phylogenetic relationships in Magnoliaceae subfam. Magnolioideae: a morphorogical cladistic analysis. Plant Syst. Evol. 242: 33–47.

Li SS, NH Tan, J Zhou, and SX Zhao. 2001. Phytochemical and chemotaxonomic studies on *Liriodendron chinense* and *Paramichelia baillonii* (Magnoliaceae). Acta Bot. Yunnanica. 23: 115–120.

Li Z-L, W-Q Song, Z-P An, and R-Y Chen. 1998. Karyotype comparison between genera in Magnoliaceae. Acta Phytotax. Sinica 36: 232–237 (in Chinese with English summary).

Lozano-Contreras G. 1975. Contribucion a las Magnoliaceae de Colombia. *Dugandiodendron*. Caldesia 11: 27–50.

Maneval WE. 1914. The development of *Magnolia* and *Liriodendron*, including a discussion of the primitive-ness of the Magnoliaceae. Bot. Gaz. 57: 1–31.

Matsui M, R Imaichi, and M Kato. 1993. Ovular development and morphology in some Magnoliaceae species. J Plant Res. 106: 297–304.

McLaughlin RP. 1933. Systematic anatomy of the woods of the Magnoliales. Trop. Woods 34: 3–39.

Melikian AP and MA Plisko. 1988a. Degeneriaceae. In: Takhtajan, ed. Comparative seed anatomy, vol. 2, pp. 8–10. Nauka, Leningrad (in Russian).

Melikian AP and MA Plisko. 1988b. Magnoliaceae. In: A Takhtajan, ed. Comparative seed anatomy, 2: 11–17. Nauka, Leningrad (in Russian).

Melville R. 1969. Studies in floral structure and evolution: I. The Magnoliales. Kew Bull. 23:133–180.

Millington WE and JE Gunkel. 1950. Structure and development of the vegetative shoot tip of *Liriodendron tulipifera* L. Am. J. Bot. 37: 326–335.

Mohana Rao PR. 1975. Seed anatomy in *Michelia champaca* Linn. with comments on the primitiveness of the Magnoliaceous seed. Phytomorphology 25: 81–90.

Morawetz W. 1981. C-banding in *Liriodendron tulipifera* (Magnoliaceae): Some karyological and systematic implications. Plant Syst. Evol. 138: 209–216.

Morawetz W. 1988. Karyosystematics of Australian Annonaceae as compared with Eupomatiaceae, Himantandraceae, and Austrobaileyaceae. Plant Syst. Evol. 159: 49–79.

Nong Van Tiep. 1980. Beitrage zur Sippenstruktur der Gattung *Manglietia*. Feddes Repert. 91: 497–576.

Nooteboom HP. 1985. Notes on Magnoliaceae with a revision of *Pachylarnax* and *Elmerrillia* and the Malesian species of *Manglietia* and *Michelia*. Blumea 31: 65–121.

Nooteboom HP. 1987. Notes on Magnoliaceae: II. Revision of *Magnolia* sections *Maingoia* (Malesian species), *Aromadendron*, and *Blumiana*. Blumea 32: 343–382.

Nooteboom HP. 1993. Magnoliaceae. In: K Kubitzki, ed., The families and genera of vascular plants, vol. 2, pp. 391–401. Springer, Berlin/Heidelberg/New York.

Nooteboom HP. 1998. The tropical Magnoliaceae and their classification. In: D Hung, ed. Magnolias and their allies, pp. 71–80. International Dendrology Society and Mangolia Society, Sherbone.

Nooteboom HP. 2000. Different looks at the classification of the Magnoliaceae. In: YH Law, HM Fan, ZY Chen, QG Wu and QW Zeng, eds. Proc. Internat. Symp. Magnoliaceae 2000, pp. 26–37. Beijing.

Ozenda P. 1947a. Anatomie des genres *Aromadendron, Elmerrillia, Alcimandra, Pachylarnax*, et *Kmeria*. C. R. Acad. Sci. Paris 225: 1360–1362.

Ozenda P. 1947b. Structure du noeud foliare des Magnoliacees et des Anonacees. C. R. Acad. Sci. Paris 224: 1521–1523.

Padmanabhan D. 1960. A contribution to the embryology of *Michelia champaca*. J. Madras Univ. 30B: 155–165.

Pan H, S Shi, H Jin, Y Huang, and H Zhang. 1999. Phylogeny of *Parakmeria* (Magnoliaceae) and its related taxa inferred from the *mat*K gene sequence. Acta Sci. Natur. univ. Sunyatseni 38: 63–67.

Parks CR, NG Miller, JF Wendel, and KM McDougal. 1983. Genetic divergence within the genus *Liriodendron* (Magnoliaceae). Ann. Missouri Bot. Gard. 70: 658–666.

Praglowski J. 1974. Magnoliaceae. World Pollen and Spore Flora, vol. 3. Almqvist & Wiksell, Stockholm.

Qiu Y-L, MW Chase, HL Donald, and RP Clifford. 1993. Molecular phylogenetics of the Magnoliidae: cladistic analysed of nucleotide sequences of the plastid gene *rbc*L. Ann. Missouri Bot. Gard. 80: 587–606.

Raven PH and DW Kyhos. 1965. New evidence concerning the original basic chromosome number of angiosperms. Evolution 19: 244–248.

Sauquet H, JA Doyle, T Scharaschkin, T Borsch, KW Hilu, LW Chatrou, and A Le Thomas. 2003. Phylogenetic analysis of Magnoliales and Myristicaceae based on multiple data sets: implications for character evolution. Bot. J. Linn. Soc. 142: 125–186.

Scvortsova NT. 1953. On the anatomy of the flower of *Magnolia grandtfiora* L. Bot. Zhurn. (Leningrad) 43: 401–408 (in Russian).

Shaparenko KK. 1937. *Liriodendron*: Tulip tree. Trudy Bot. Inst. Akad. Nauk SSSR, 1st ser., 4: 93–170.

Skipworth JP. 1970 (1971). Development of floral vasculature in the Magnoliaceae. Phytomorphology 20: 228–236.

Smith AC. 1949. Additional notes on *Degeneria vitiensis*. J. Arnold Arbor. 30: 1–9.

Sugiyama M. 1979. A comparative study of nodal anatomy in the Magnoliales base on the vascular system in the nodeleaf continuum. J. Fac. Sci. Univ. Tokyo, sec. Botany 12: 199–279.

Swamy BGL. 1949. Further contribution to the morphology of the Degeneriaceae. J. Arnold Arbor. 30: 3–30.

Takhtajan AL. 1980. Degeneriaceae. In: AL Takhtajan, ed. Life of plants, 5(2): 121–125. Moscow (in Russian).

Takhtajan AL and NR Meyer. 1976. Some additional data on the morphology of pollen grains of *Degeneria vitiensis* (Degeneriaceae). Bot. Zhurn. 61: 1531–1535 (in Russian with English summary).

Thien LB. 1974. Floral biology of *Magnolia*. Am. J. Bot. 61: 1037–1045.

Thien LB, WH Heimermann, and RT Holman. 1975. Floral odors and quantitative taxonomy of *Magnolia* and *Liriodendron*. Taxon 24 (5/6): 557–568.

Thorne RE. 1974. A phylogenetic classification of the Annoniflorae. Aliso 8: 147–209.

Tiffney BH. 1977. Fruits and seeds of the Brandon Lignite: Magnoliaceae. Bot. J. Linn. Soc. 75: 299–323.

Tucker SC. 1960. Ontogeny of the floral apex of *Michelia fuscata*. Am. J. Bot. 47: 266–277.

Tucker SC. 1961. Phyllotaxis and vascular organization of the carpels in *Michelia fuscata*. Am. J. Bot. 48: 60–71.

Tucker SC. 1963. Development and phyllotaxis of the vegetative axillary bud of *Michelia fuscata*. Am. J. Bot. 50: 661–668.

Tucker SC. 1977. Foliar sclereids in the Magnoliaceae. Bot. J. Linn. Soc. 75: 325–356.

Ueda Kunihiko. 1984. Vascular systems in Magnoliaceae. Fac. Sci. Kyoto Univ.

Ueda K, J Yamashita, and MN Tamura. 2000. Molecular phylogeny of the Magnoliaceae. In: YH Law, HM Fan, ZY Chen, QG Wu and QW Zeng, eds. Proc. Internat. Symp. Fam. Magnoliaceae, pp. 205–209. Beijing.

Umeda A, R Imaichi, and M Kato. 1994. Ovular development and morphology of the outer integument of *Magnolia grandtflora* (Magnoliaceae). Am. J. Bot 81: 361–367.

Vazquez-Garcia JA. 1994. *Magnolia* (Magnoliaceae) in Mexico and Central America: a synopsis. Brittonia 46: 1–23.

Wei Z-X and Z-Y Wu. 1993. Pollen ultrastructure of *Liriodendron* and its systematic implications. Acta Bot. Yunn. 15 (2): 163–166 (n Chinese with English summary).

Whitaker TW. 1933. Chromosome number and relationships in the Magnoliales. J. Arnold Arbor. 14: 376–385.

Xu F-X. 2000a. Morphology of chalazal region on endotesta in the seeds of Magnoliaceae and those of related families. Subtrop. Plant Res. Commun. 29: 5–10.

Xu F-X. 2000b. A cladistic analysis of Magnoliaceae. J. Trop. and Subtrop. Bot. 8: 207–214.

Xu F-X. 2002. Study on pollen morphology of *Parakmeria lotungensis*). Guihaia. 22: 157–159.

Xu F-X. 2003. Sclerotesta morphology and its systematic implications in magnoliaceous seeds. Bot. J. Linn. Soc. 142: 407–424.

Xu F-X and PJ Rudall. 2006. Comparative floral anatomy and ontogeny in Magnoliaceae. Plant Syst. Evol. 258: 1–15.

Xu F-X and QG Wu. 2000. Morphology of the chalazal region on the endotesta of seeds in genera *Magnolia, Parakmeria* and *Kmeria*. In: YH Law, HM Fan, ZY Chen, QG Wu and QW Zeng, eds. Proc. Internat. Symp. Fam. Magnoliaceae 2000: 129–142.

Xu F-X and Q-G Wu. 2002. Chalazal region morphology on the endotesta of Magnoliaceous seeds and its systematic significance. Acta Phytotax. Sinica 40: 260–270.

Yasui K. 1937. Karyological studies in *Magnolia*, with special reference to the cytokinesis in the pollen mother cell. Bot. Mag. Tokyo 51: 539–564.

Young DA and RW Sterner. 1981. Leaf flavonoids of primitive dicotyledonous angiosperms: *Degeneria vitiensis* and *Idiospermum australense*. Biochem. Syst. Ecol. 9 (2–3): 185–187.

Zazhurilo KK. 1940. On the anatomy of the seed coats of Magnoliaceae *(Liriodendron tulipifera* L.). Bull. Soc. Nat. Voronezh University 4 (I): 32–40 (in Russian).

Zhang B. 2001. Floral analysis of Magnoliaceae. Guihaia. 21(4): 315–320.

Zhang X-J, Y-Q Chen, F-R Chai, and S-M Shu. 2000. Anatomical characteristics and phylogenetic relationships of secondary xylem of Magnoliaceae in China. Bull. Bot. Res. 20: 318–323 (in Chinese with English summary).

Order 10. HIMANTANDRALES

Large trees. Peltate scaly indumentum present. Nodes trilacunar. Vessels mainly with simple perforations, but with some vestigial scalariform plates; intervascular pitting alternate in fully mature wood. Rays markedly heterogenous. Axial parenchyma apotracheal. Sieve-element plastids of S-type. Leaves alternate, distichous, margin entire; blade simple, stipules absent; coppery indumentum on leaf undersurface, young shoots and floral buds consisting of peltate scales. Stomata paracytic. Flowers bisexual, actinomorphic, solitary or paired on short axillary branches, at first enclosed by two calyptriform leathery deciduous sepals (Hutchinson 1964); petals 7–9, spirally arranged, lanceolate, very similar in size and shape to the stamens. Stamens 13–40 (mostly 25–30), not differentiated into filament and anther, one pair of sporangia on each side, each with single longitudinally opening slits. Outer staminodes 3–23, inner staminodes 13–20; inner staminodes and part of the stamens with marginal and laminar glands. Pollen grains subglobose. Microsporogenesis successive. Carpels 7–28, closed, differentiated into an ovary and a short stylodium. The stigmata of all carpels form a common plumose mass. Ovules solitary, rarely two, pendulous, anatropous, apotropous, bitegmic, crassinucellate. Fruits drupes with several flat pyrenes. Seeds flat, without elongated funicle; seed coat mesotestal with testal-tegmic ruminations and unspecialized 2–3-layered tegmen; a single-layered exotesta represented by thin-walled tanniniferous cells; a 2–3-layered mesotesta, composed of thick-walled lignified longitudinal fibers; and an endotesta composed of two or three layers of unspecialized aerenchymatous parenchyma (Doweld and Shevyryova 1998); embryo straight, very small; endosperm oily. Unusually rich in a variety of hallucinogenic pyridine alkaloids of the *Galbulimima* type. n = 6.

Probably related to the Magnoliales, especially to the Degeneriaceae, but differs in so many important features, including 3-locular nodes, calyptriform calyx and especially seed coat structure, that deserves an ordinal rang (Doweld and Shevyryova 1998).

1. HIMANTANDRACEAE

Diels 1917. 1/2. New Guinea, Moluccas, Celebes, New Britain, and northeastern Australia.

Galbulimima (Himantandra).

Bibliography

Bailey IW, CG Nast, and AC Smith. 1943. The family Himantandraceae. J. Arnold Arbor. 24: 190–206.
Buchheim G. 1962. Beobachtungen über den Bau der Frucht der Familie Himantandraceae. Sitzungsber. Ges. Naturf. Freunde Berlin (N. F.) 2: 78–92.
Diels L. 1917. Über die Gattung *Himantandra*: Ihre Verbreitung und ihre systematische Stellung. Bot. Jahrb. Syst. 55: 126–134.
Doweld AB, and NA Shevyryova. 1998. Carpology, seed anatomy and taxonomic relationships of *Calbulimima* (Himantandraceae). Ann. Bot. 81: 337–347.
Endress PK. 1993. Himantandraceae. In: K Kubitzki, ed., The families and genera of vascular plants, vol. 2, pp. 338–341. Springer, Berlin/Heidelberg/New York.
Melikian AP and EF Cherniakovskaya. 1988. Hi-mantandraceae. In: A Takhtajan, ed. Comparative seed anatomy, vol. 2, pp. 10–11. Nauka, Leningrad (in Russian).
Prakash N, DB Foreman, and S Griffith. 1984. Ga-metogenesis in *Galbulimima belgraveana* (Himantan-draceae). Austral. J. Bot. 32: 605–612.
Ritchie E and WC Taylor. 1967. The *Galbulimima* alkaloids. In: RHF Manske, ed. The alkaloids, vol. 9, pp. 529–543. Academic Press, New York.
Royen P van and WA van Heel. 1962. Sertulum Papuanum 6. Himantandraceae. Nova Guinea, Bot., 8–10: 127–135.
Sauer W and F Ehrendorfer. 1970. Chromosomen, Verwandtschaft, und Evolution tropischer Holzpflanzen: II. Himantandraceae. Oesterr. Bot. Z. 118: 38–54.
Smith AC. 1942. A nomenclatural note on the Himantandraceae. J. Arnold Arbor. 23: 366–368.

Order 11. ANNONALES

Evergreen trees, shrubs, or woody lianas. Vessels with scalariform (Eupomatiaceae) or with simple perforations. Fibers with simple or narrowly bordered or rarely distinctly bordered pits. Rays homogeneous or heterogeneous. Axial parenchyma apotracheal or sometimes scanty paratracheal. Sieve-element plastids of Psc- or (*Xylopia*) of Pcsf-types; according to Behnke (1988). Nodes mostly trilacunar. Leaves alternate, distichous or often spirally (Johnson 2003) arranged, simple, entire, pinnately veined, often pellucid-dotted, often with a pulvinus that is swollen and elongate or contorted for climbing, without stipules. Stomata nearly always paracytic. Flowers in various types of basically cymose inflorescences or solitary in the axils, bisexual or rarely unisexual, with 3-merous perianth, or without perianth. Sepals (2)3(4), imbricate or valvate, free or partly connate. Stamens numerous (up to 100) and spirally arranged or rarely 3 or 6 (up to 15) and cyclic. Anthers linear, tetrasporangiate, extrorse or rarely latrorse, very rarely introrse, opening longitudinally; connective prolonged and often expanded above the anther so that the anther appears to be peltate. Tapetum secretory. Microsporogenesis successive or simultaneous. Pollen grains 2-celled, in monads or seldom in tetrads or even polyads. Gynoecium of more or less numerous, several or rarely only one carpel, spiral and apocarpous or, less frequently, cyclic and paracarpous. Carpels 1–400, stylodia free, short, and thick. Ovary superior, 1-locular, or 2–15-locular. Ovules 1–2 – several, ascending, anatropous or apotropous, bitegmic, crassinucellate. Female gametophyte of *Polygonum*-type. Endosperm cellular. Fruits of various types Seeds with usually small embryo and copious, and ruminate, oily endosperm and seldom (some Annonaceae) with perisperm, arillate or not. Frequently producing alkaloids, n = 7, 8, 9, mostly 8, 10.

There are many similarities in the anatomy of the young stem of Annonaceae and Magnoliales: sclerified pith diaphragms, stratification of the secondary phloem into tiers of hard and soft bast, V-shaped phloem rays, cortical scleroids, superficial origin of periderm (see Van der Wyk and Canright 1956). The phylogenetic relationships between these groups have been much discussed in the recent literature. Annonaceae originated from a magnolialean ancestor.

Key to Families

1 Vessels with scalariform perforations with numerous (20–150 or more) bars; vessel elements are very long and long-tapering, with very oblique end walls; lateral pitting scalariform or opposite. Large shrubs to dwarf shrublets with fleshy, starchy tuberous roots. Axial

parenchyma scanty paratracheal or diffuse. Rays heterogeneous with elongate ends. Fibers with simple or slightly bordered pits, sometimes septate. Sieve elements long and slender, with slowly tapering ends; sieve-element plastids of Pcs-type with several protein crystals, one being rod-shaped, which is family specific. Nodes multilacunar. Leaves alternate, distichous (but spirally arranged in young plants), simple, entire, pinnately veined, and without stipules. Prophylls in a median position. Stomata paracytic or occasionally actinocytic. Flowers showy, solitary, bisexual, without perianth, enclosed when young in a deciduous calyptra (a single bract, according to Baillon 1868; Eames 1961; Endress 1977) attached to the rim of an enlarged, somewhat concave, turbinate receptacle and falling off as a conical lid. Stamens 20–100, staminodia numerous, inserted into the receptacle near its rim; the innermost stamens are sterile (staminodia), somewhat fleshy, more or less petaloid, the others fertile, ribbon-shaped, with a short, broad, laminar base. Anthers long, basifixed, tetrasporangiate, latrorse-introrse, opening longitudinally, with prolonged thickened connective. Pollen grains globose-oblate, psilate, with structureless, atectate-amorphous exine, with two parallel, equatorially encircling colpi that are differentiated from the nonapertural areas of the exine by their fossulate-rugulate surface. Gynoecium of numerous (ca. 13–70) small carpels spirally arranged in the hollow receptacle, all appressed and connate, but each has a decurrent stigma with projecting papillae and a slit-like opening. Each carpel with two to several (up to ca. 11) ovules arranged laminar-laterally. Ovules anatropous with hood-shaped outer integument. Endosperm cellular. Fruit an urceolate-turbinate berry with a truncated, flattened, or convex apex. Seeds 1–3 in each carpel, subangular and coarsely pitted; seed coat composed of a single-layered thin-walled exotesta or large tanniniferous cells, but has 3–4 layers of longitudinal thin-walled lignified fibers in the mesotesta and 2 layers of thin-walled parenchymatous endotesta; embryo small, straight; endosperm copious, ruminate, oily. Producing various types of lignans and alkaloids, n = 10. . . . . 1. EUPOMATIACEAE

1 Vessels with simple perforations; lateral pitting alternate or sometimes transitional to opposite. Evergreen trees, shrubs, or woody lianas, bearing essential oils, resinous or not resinous. Pith generally septate. Fibers usually with numerous simple or narrowly bordered or rarely distinctly bordered pits. Rays typically wide and high, homogeneous or weakly heterogeneous, occasionally distinctly heterogeneous. Axial parenchyma apotracheal or sometimes scanty paratracheal. Sieve-element plastids of Psc- or (*Xylopia*) of Pcsf-types. Nodes mostly trilacunar. Leaves entire, pinnately veined, often pellucid-dotted, often with a pulvinus that is swollen and elongate or contorted for climbing, without stipules. Stomata nearly always paracytic. Prophylls in some genera – e.g., in *Annona* and *Asimina* – paired and lateral (as usual in dicotyledons), but usually they are solitary and median, in dorsal position, as in the monocotyledons. Flowers in various types of basically cymose inflorescences or solitary in the axils, mostly entomophilous, bisexual or rarely unisexual, with cyclic and usually 3-merous perianth. Sepals (2)3(4), imbricate or valvate, free or partly connate. Petals 3-6(-12), usually 6 in 2 cycles of 3, free or very rarely basally connate, imbricate or valvate. Stamens numerous (up to 100) and spirally arranged or rarely 3 or 6 (up to 15) and cyclic, generally free, rarely basally connate, filaments short and stout, with single vascular trace. Anthers linear; connective prolonged and often expanded above the anther so that the anther appears to be peltate. Staminodia present only in a few genera. Pollen grains very diverse in size and structure, monocolpate to inaperturate, inconspicuously granular to columellate, atectate, tectate-imperforate, tectate-perforate, semitectate, or intectate. Gynoecium of more or less numerous, several or rarely only one carpel, spiral and apocarpous or, less frequently, cyclic and paracarpous (Monodoroideae). Carpels 1–400 (ca. 400 in *Annona muricata* – Igersheim and Endress 1997), conduplicate, sometimes unsealed, whorled or irregularly arranged, free in most genera, with 1–30 (up to 60 or more in *Monodora myristica*) ovules on submarginal, laminar-lateral, parietal or nearly basal placentas; stylodia free, short, and thick. Ovary superior, 1-locular, or 2–15-locular. Ovules 1–10, ascending, with hood-shaped outer integument. Fruits of various types, but mostly more or less fleshy, distinct, stipitate, indehiscent, or, rarely (*Anaxagorea*), dehiscent fruitlets. Seeds with small embryo and copious, and ruminate, oily endosperm and seldom with perisperm; n = 7, 8, 9, mostly 8. . . . . 2. ANNONACEAE

1. EUPOMATIACEAE

Endlicher 1841. 1/3 Eastern and southeastern Australia (from temperate Victoria to subtropical

New South Wales and tropical Queensland) and New Guinea (from sea level up to 1,300 m altitude).

Eupomatia

The Eupomatiaceae have some features in common with both the Himandandarceae and Annonaceae but differ from them in many respects, including complete absence of perianth, circular aperture of pollen grains, and the basic chromosome number.

2. ANNONACEAE

A.L. de Jussieu 1789 (including Hornschuchiaceae J. Agardh 1858; Monodoraceae J. Agardh 1858). 128/2400. Except for the temperate North American genus *Asimina*, the family has a pantropical distribution, but is concentrated in the Paleotropics. Low and more stable chromosome numbers dominate in the Paleotropics; higher and more diversified ones in the Neotropics (Morawetz 1988).

2.1 ANNONOIDEAE

Carpels spirally arranged (rarely few in I cycle), free from one another from the beginning (rarely only 1), or if at length united, then forming a many-locular syncarp. Stigmas more or less erect, rarely radiating. – UVARIEAE: *Uvaria, Balonga, Tetrapetalum, Ellipeia, Ellipeiopsis, Sapranthus, Stenanona, Afroguatteria, Rauwenhoffia, Sageraea, Dendrokingstonia, Stelechocarpus, Dasoclema, Hexalobus, Cleistopholis, Greenwayodendron, Mkilua, Toussaintia, Enicosanthum, Enicosanthellum*; MILIUSIEAE: *Fenerivia, Heteropetalum, Neostenanthera, Boutiquea, Marsypopetalum, Meiogyne, Phaeanthus, Trivalvaria, Anomianthus, Piptostigma, Cymbopetalum, Miliusa, Mezzettiopsis, Orophea, Phoenicanthus, Alphonsea, Platymitra, Mezzettia*; XYLOPIEAE: *Xylopia, Cardiopetalum, Meiocarpidium, Polyceratocarpus, Dielsiothamnus, Drepananthus, Anaxagorea, Artabotrys, Pseudoartabotrys, Cyathocalyx, Diclinanona*; ANNONEAE: *Desmos, Dasymaschalon, Cyathostemma, Polyalthia, Polyaulax, Oncodostigma, Monocarpia, Exellia, Unonopsis, Uvariodendron, Uvariastrum, Dennettia, Asteranthe, Desmopsis, Guamia, Haplostichanthus, Monocyclanthus, Chieniodendron, Enantia, Woodiellantha, Cleistochlamys, Disepalum, Pseuduvaria, Mitrephora, Goniothalamus, Richella, Schefferomitra, Melodorum, Friesodielsia, Oreomitra, Petalolophus, Popowia, Neo-uvaria, Papualthia, Mitrella, Pyramidanthe, Fissistigma, Cananga, Bocageopsis, Onychopetalum, Monanthotaxis, Mischogyne, Atopostema, Gilbertiella, Uvariopsis, Ambavia, Bocagea, Ophrypetalum, Trigynaea, Porcelia, Hornschuchia, Froesiodendron, Oxandra, Pseudoxandra, Crematosperma, Ephedranthus, Ruizodendrom, Malmea, Guatteria, Guatteriella, Guatteriopsis, Asimina, Deeringothamnus, Fitzalania, Tridimeris, Lettowianthus, Annona, Anonidium, Raimondia, Rollinia, Rolliniopsis, Pachypodanthium, Lestestudoxa, Duckeanthus, Duguetia, Fusaea*; TETRAMERANTHEAE: *Tetrameranthus*.

2.2 MONODOROIDEAE

Carpels cyclically arranged, united from the beginning into I-locular ovary with parietal placentas. Tropical Africa and Madagascar. – *Monodora, Isolona*.

Bibliography

Baillon H. 1868. Recherches organogenique sur les *Eupomatia*. Adansonia 9: 22–28.

Behnke H-D. 1988. Sieve-element plastids, phloem protein, and evolution of flowering plants: III. Magnoliidae. Taxon 37: 699–732.

Bergstrom G, I Groth, O Pellmyr, PK Endress, LB Thien, A Hilbener, and W Francke. 1991. Chemical basis of a highly specific mutualism: Chiral esters attract pollinating beetles in Eupomatiaceae. Phytochemistry 30: 3221–3225.

Bygrave P. 1998. Molecular systematics of the Annonaceae. Annonaceae Newslett. 12: 15–17.

Bhandari NN. 1971. Embryology of the Magnoliales and comments on their relationships. J. Arnold Arbor. 52: 1–39, 285–304.

Carlquist S. 1992. Vegetative anatomy and relationships of Eupomatiaceae. Bull. Torrey Bot. Club 119: 167–180.

Cave A. 1989. Chemical research in Annonaceae. Annonaceae Newslett. 6: 24–36.

Christmann M. 1986. Beiträge zur Histologie der An-nonaceen Samen. Bot. Jahrb. Syst. 106: 379–390.

Christmann M. 1989a. Genera and species of Annonaceae with tritegmic seeds. Annonaceae Newslett. 6: 11–13.

Christmann M. 1989b. Die tritegmischen Annonaceen-Samen. Bot. Jahrb. Syst. 110: 433–439.

Corner EJH. 1949. The annonaceous seed and its integument. New Phytol. 48: 332–364.

Deroin T. 1985. Contribution a la morphologic comparee du gynecee des Annonaceae-Monodoroideae. Bull. Mus. Natn. Hist. Nat. Paris, ser. 4, 7, sect. B, 2: 167–176.

Deroin T. 1987. Anatomie florale de *Meiocarpidium* Engler et Diels (Annonaceae-Unoneae). Bull. Mus. Natn. Hist. Nat., Paris, ser. 4, 9, sect. B, 1: 81–93.

Deroin T. 1989. Definition et signification phylogenique des systemes corticaux floraux: L'exemple des Annonacees. C. R. Acad. Sei. Paris, 3rd ser., 308: 71–75.

Deroin T. 1991. La repartition des modeies de plateaux stigmatiques et l'evolution des Annonacees. C. R. Acad. Sei. Paris, 3rd ser., 312: 561–566.
Deroin T. 1997. Conformation and origin of paracarpy in Annonaceae, with comments on some methodological aspect. Candollea 82: 45–58.
Deroin T. 2000. Floral anatomy of *Toussaintia hallei* Le Thomas, a case of convergence of Annonaceae with Magnoliaceae. In: YH Law, HM Fan, ZY Chen, QG Wu and QW Zeng, eds. Proc. Internat. Symp. Fam. Magnoliaceae 2000: 168–176.
Diels L. 1912. Über primitive Ranales der australischen Flora. Bot. Jahrb. Syst. 48: 7–13.
Doyle JA, P Bygrave, and A Le Thomas. 2000. Implications of molecular data for pollen evolution in Annonaceae. In: MM Harley, CM Morton, and S Blackmore, eds. Pollen and spores: morphology and biology, pp. 259–284. Royal Botanic Gardens, Kew.
Doyle JA and A Le Thomas. 1994. Cladistic analysis and pollen evolution in Annonaceae. Acta Botanica Gallica 141: 149–170.
Doyle JA and A Le Thomas. 1996. Phylogenetic analysis and character evolution in Annonaceae. Bull. Mus. nat. Hist. Nat., Paris, ser.4, 18, sect. B: 279–334.
Doyle JA and A Le Thomas. 1997. Phylogeny and geographic history of Annonaceae. Géogr. Phys. Quatern. 51: 353–361.
Eames A. 1961. Morphology of the angiosperms. McGraw-Hill, New York.
Ehrendorfer F, F Krendl, E Habeier, and W Sauer. 1968. Chromosome numbers and evolution in primitive angiosperms. Taxon 17: 337–353.
Endress PK. 1977. Über Blütenbau und Verwandtschaft der Eupomatiaceae und Himantandraceae (Magnoliales). Ber. Deutsch. Bot. Ges. 90: 83–103.
Endress PK. 1983. Dispersal and distribution in some small archaic relic angiosperm families (Austrobaileyaceae, Eupomatiaceae, Himantandraceae, Idiospermoideae – Calycanthaceae). Sonderbd. Naturwiss. Verh. Hamburg 7: 201–217.
Endress PK. 1984a. The role of inner staminodia in the floral display of some relic Magnoliales. Plant Syst. Evol. 146: 269–282.
Endress PK. 1984b. The flowering process in the Eupomatiaceae (Magnoliales). Bot. Jahrb. Syst. 104: 297–319.
Endress PK. 1993. Eupomatiaceae. In: K Kubitzki ed. The families and genera of vascular plants, vol. 2, pp. 296–298. Springer, Berlin/Heidelberg/New York.
Endress PK. 2003. Early floral development and nature of the calyptra in Eupomatiaceae (Magnoliales). Int. J. Plant Sci. 164: 489–503.
Erbar C. 1996. Early floral developments studies in Annonaceae. In: W Morawetz, H Winkler (eds.). Reproductive morphology in Annonaceae. Biosyst. Ecol. ser. 10: 1–27.
Fournier G, M Leboeuf, and A Cave. 1999. Annonaceae essential oils: a review. J. Essent. Oil Res. 11: 131–142.
Fries RE. 1919. Studien über die Blütenstandsverhältnisse bei der Familie Annonaceae. Acta Horti Berg. 6: 3–48.
Fries RE. 1939. Revision der Arten einiger Annonaceen-Gattungen. Acta Horti Berg. 12: 289–577.
Fries RE. 1959. Annonaceae. In: A. Engler and K. Pranti, eds. Die natürlichen Pflanzenfamilien, 2nd ed., 17a:l–171. Ducker & Humboldt, Berlin.
Gabarayeva NI. 1992. Sporoderm development in *Asimina triloba*: I. The development events before callose dissolution. Grana 31: 213–222.
Gabarayeva NI. 1993. Sporoderm development in *Asimina triloba*: II. The development events after callose dissolution. Grana 32: 210–220.
Gabarayeva NI. 1995. Pollen wall and tapetum development in *Anaxagorea brevipes* (Annonaceae): sporoderm substructure, cytoskeleton, sporopollenin precursor particles, and the endexine problem. Rev. Palaeobot. Palynol. 85: 123–152.
Garwood NC. 1995. Studies in Annonaceae. XX. Morphology and ecology of seedlings, fruits and seeds of selected Panamanian species. Bot. Jahrb. Syst. 117: 1–152.
Gottsberger G. 1970. Beiträge zur Biologic der Annonaceenblüten. Oesterr. Bot. Z. 118: 237–279.
Gottsberger G. 1999. Pollination and evolution in Neotropical Annonaceae. Plant Species Biol. 14: 143–152.
Gottsberger G and I Gottsberger. 1985. Pollen units, pollen shape, and apertural position in the Annonaceae: A reassessment. Beitr. Biol. Pfl. 59: 465–473.
Guedes M and A Le Thomas. 1981. Le gynecee de *Monodora* (Annonacees-Monodoroidees). C. R. Acad. Sci. (Paris) 292 (3): 1025–1028.
Hamilton AG. 1897. On the fertilization of *Eupomatia laurina* R. Br. Proc. Linn. Soc. N. S. W. 22: 48–55.
Heijden E van der, and F Bouman. 1988. Studies in Annonaceae. X. Seed anatomy of the *Annona* group. Bot. Jahrb. Syst. 110: 117–135.
Hesse M, W Morawetz, and F Ehrendorfer. 1985. Pollen ultrastructure and systematic affinities of *Anaxagorea* (Annonaceae). Plant Syst. Evol. 148: 253–285.
Hesse M and M Waha. 1984. Sporoderm characters of *Tetrameranthus duckei* (Annonaceae) and their systematic implications. Plant Syst. Evol. 147: 323–326.
Hooker WJ. 1855. *Eupomatia laurina*. Curtis's Bot. Mag. 81, t. 4848.
Hotchkiss AT. 1955. Geographical distribution of the Eupomatiaceae. J. Arnold Arbor. 36: 385–396.
Hotchkiss AT. 1958. Pollen and pollination in the Eupomatiaceae. Proc. Linn. Soc. N. S. W. 83: 86–91.
Johnson DM. 2003. Phylogenetic significance of spiral and distichous architecture in the Annonaceae. Syst. Bot. 28: 503–511.
Johnson DM and NA Murray. 1995. Synopsis of the tribe Bocageeae (Annonaceae) with revisions of *Cardiopetalum, Froesiodendron, Trigynaea, Boccagea*, and *Hornschuchia*. Brittonia 47: 248–319.
Kamelina OP. 1981. On the embryology of the non-investigated taxa: I. Some data on the embryology of Eupomatiaceae. Bot. Zhurn. 66: 854–859 (in Russian).
Kessler PJA. 1989. Some interesting distribution patterns in Annonaceae. Annonaceae Newslett. 6: 14–23.
Kessler PJA. 1993. Annonaceae. In: K Kubitzki, ed. The families and genera of vascular plants, vol. 2, pp. 93–128. Springer, Berlin/Heidelberg/New York.
Koek-Noorman J. 1989. Multidisciplinary approach to the systematics of Neotropical Annonaceae. Annonaceae Newslett. 6: 3–10.
Koek-Noorman J, LY Th Westra, and PJM Maas. 1990. Studies in Annonaceae: XIII. The role of morphological characters in subsequent classification of Annonaceae: A comparative survey. Taxon 39: 16–32.

Koek-Noorman J, AK van Setten, and CM van Zullen. 1997. Studies in Annonaceae. XXVI: Flower and fruit morphology in Annonaceae: Their contribution to patterns in cluster analysis. Bot. Jahrb. Syst. 119: 213–230.

Klucking EP. 1986. Leaf venation pattern, vol. 1, Annonaceae. Cramer, Berlin/Stuttgart.

Leboeuf M, A Cave, PK Bhaurnik, B Bhaurnik, B Mukherjee, and R Mukherjee. 1982. The phytochem-istry of the Annonaceae. Phytochemistry 21: 2783–2813.

Leins P and C Erbar. 1980. Zur Entwicklung der Blüten von *Monodora crispata* (Annonaceae). Beitr. Biol. Pfl. 55: 11–22.

Leins P and C Erbar. 1982. Das monokarpellate Gynoeceum von *Monodora crispata* (Annonaceae). Beitr. Biol. Pfl. 57: 1–13.

Leins P and C Erbar. 1996. Early floral developmental studies in Annonaceae. Biosyst. Ecol. ser. 10: 1–27.

Lemesle R. 1936. Les vaissaux a perforation scalariformes de *l'Eupomatia* et leur importance dans la phylogenie des Polycarpes. C. R. Acad. Sci. Paris 203: 1538–1540.

Lemesle R. 1938. Contribution a l'etude de *l'Eupomatia* R. Br. Rev. Gen. Bot. 50: 692–712.

Le Thomas A. 1980/1981. Ultrastructural characters of the pollen grains of African Annonaceae and their significance for the phylogeny of primitive Angiosperms. Parts I and 2. Pollen et Spores 22: 267–342; 23: 5–36.

Le Thomas A. 1988. Variation de la region aperturale dans le pollen des Annonacees. Taxon 37: 644–650.

Le Thomas A and B Lugardon. 1975. Ultrastructure d'un pollen original parmi les Annonacees. Bull. Soc. Bot. France 122: 109–111.

Le Thomas A and B Lugardon. 1976. De la structure grenue a la structure columellaire dans le pollen des Annonacees. Adansonia, ser. 2, 15: 543–572.

Le Thomas A, B Lugardon, and JA Doyle. 1994. Pollen ultrastructure and relationships of *Fusaea* (Baillon) Safford and *Duguetia* A.Saint-Hilaire (Annonaceae). Rev. Paleobot. Palynol. 83: 55–64.

Le Thomas A., W. Morawetz, and M. Waha. 1986. Pollen of Palaeo- and Neotropical Annonaceae: Definition of the aperture by morphological and functional characters. In: S Blackmore and IK Ferguson, eds. Pollen and spore: Form and function, pp. 375–388. London.

Melikian AP and MA Plisko. 1988. Eupomatiaceae. In: A. Takhtajan, ed. Comparative seed anatomy, vol. 2, pp. 18–19. Nauka, Leningrad (in Russian).

Mohana Rao PR. 1975. Seed anatomy of *Artabotrys odoratissintus* with discussion on chalaza, integumentary bundles, and ruminate endosperm. Phytomorphology 25: 215–228.

Mohana Rao PR. 1983. Seed and fruit anatomy in *Eupomatia laurina* with a discussion of the affinities of Eupomatiaceae. Flora B 173: 311–319.

Mols JB, B Gravendeel, LW Chatrou, MD Pirie, PC Bygrave, MW Chase, and PJA Keßler. 2004. Identifying clades in Asian Annonaceae: monophyletic genera in the polyphyletic Miliuseae. Am. J. Bot. 91: 590–600.

Morawetz W. 1986. Systematics and karyoevoluton in Magnoliidae: *Tetrameranthus* as compared with other Annonaceae genera of the same chromosome number. Plant Syst. Evol. 154: 145–177.

Morawetz W. 1988. Karyosystematics and evolution of Australian Annonaceae as compared with Eupomatia-ceae, Himantandraceae, and Austrobaileyaceae. Plant Syst. Evol. 159: 49–70.

Okada H and K Ueda. 1984. Cytotaxonomical studies on Asian Annonaceae. Plant Syst. Evol. 144: 165–177.

Periasamy K and BGL Swamy. 1959 (1960). Studies in the Annonaceae: 1. Microsporogenesis in *Cananga odorata* and *Miliusa wightiana*. Phytomorphology 9: 251–263.

Periasamy K and BGL Swamy. 1961. Studies in the Annonaceae: II. The development of ovule and seed in *Cananga odorata* and *Miliusa wightiana*. J. Indian Bot. Soc. 40: 206–216.

Read RW and WC Taylor. 1979. Constituents of *Eupomatia* species: V. The isolation of eupomatenoid-13 (a new neolignan), (±)-trans-Dehydrodiisoeu-genol, and other extractives from the bark of *Eupomatia laurina*. Aust. J. Chem. 32: 2317–2321.

Ronse Decraene LP and E Smets. 1990. The floral development of *Popowia whitei* (Annonaceae). Nord J. Bot. 10: 411–420.

Samuelsson G. 1914. Über die Pollenentwicklung von *Annona* and *Aristolochia* and ihre systematische Bedeutung. Svensk Bot. Tidskr. 8: 181–189.

Sauer W and F Ehrendorfer. 1984. Notes on the karyo-systematics of Annonaceae. Plant Syst. Evol. 146: 47–55.

Setten AK van and J Koek-Noorman. 1992. Fruits and seeds of Annonaceae: Morphology and its significance for classification and identification. Bibl. Bot. 142: 1–101.

Su YCF and RMK Saunders. 2003. Pollen structure, tetrad cohesion and pollen-connecting threads in *Pseuduvaria* (Annonaceae). Bot. J. Linn. Soc. 143: 69–78.

Sugiyama M. 1976. Comparative studies of the vascular system of node-leaf continuum in woody Ranales: II. Node-leaf vascular system of *Eupomatia laurina* R. Br. J. Jpn. Bot. 51: 169–174.

Svoma E. 1993. Contribution to the ontogenetic development and organization of Annonaceae seed. Annonaceae Newslett. 9: 67–70.

Svoma E. 1998. Seed morphology and anatomy in some Annonaceae. Plant Syst. Evol. 209: 177–204.

Taylor WC. 1985. *Eupomatia* alkaloids. Alkaloids 24: 1–23.

Tsou C-H and DM Johnson. 2003. Comparative development of aseptate and septate anthers of Annonaceae. Am. J. Bot. 90: 832–848.

Uphof JCT. 1959. Eupomatiaceae, In: A Engler and K Prantl, eds. Die natürlichen Pflanzenfamilien, 2nd ed., 17a: 173–176. Ducker & Humboldt, Berlin.

Van der Wyk RW and JE Canright. 1956. The anatomy and relationships of the Annonaceae. Trop. Woods 104: 1–24.

Van Heusden ECH. 1992. Flowers of Annonaceae: morphology, classification, and evolution. Blumea, Suppl. 7: 1–128.

Waha M. 1987a. Sporoderm development of pollen tetrads in *Asimina triloba* (Annonaceae). Pollen et Spores 29(1): 31–44.

Waha M. 1987b. Different origins of fragile exines within the Annonaceae. Plant Syst. Evol. 158: 23–27.

Waha M and W Morawetz. 1988. Pollen evolution and systematics in Annonaceae with special reference to the disulcate Australian endemic genera. Plant Syst. Evol. 161: 1–12.

Walker JW. 1971a. Pollen morphology, phytogeography, and phylogeny of the Annonaceae. Contr. Gray Herb. 202: 1–131.

Walker JW. 1971b. Contribution to the pollen morphology and phylogeny of the Annonaceae. I. Grana 11: 45–54.

Walker JW. 1971c. Unique type of angiosperm pollen from the family Annonaceae. Science 172: 565–567.

Walker JW. 1972a. Contributions to the pollen morphology and phylogeny of the Annonaceae, II. Bot. J. Linn. Soc. 65: 173–178.
Walker JW. 1972b. Chromosome numbers, phylogeny, phytogeography of the Annonaceae and their bearing on the (original) basic chromosome number of angiosperms. Taxon 21: 57–65.
Walker JW. 1976. Evolutionary significance of the exine in the pollen of primitive angiosperms. In: K Ferguson and J Müller, eds. The evolutionary significance of the exine, pp. 251–308. Linn. Soc. Symposium, No. 1. Academic Press, London/ New York.
Woodland PS and PR Garlick. 1982. The fine structure of the pollen of Eupomatiaceae. Austral. J. Bot. 30: 297–301.
Woodland PS and N Prakash. 1993. Floral morphology, embryology and relationships of the Eupomatiaceae. Abstr. XV Inter. Bot. Congr., p. 95. Yokohama.
Young DA. 1983. Leaf flavonoids of Eupomatiaceae. Biochem. Syst. Ecol. 11(3): 209–210.

Order 12. MYRISTICALES

Trees or seldom shrubs, usually aromatic and resinous. Trichomes of various kinds, eglandular ones elongate, unicellular and colourless, glandular ones with a multicellular, basally embedded stalk and a multicellular peltate, later balloon-shaped head with golden-yellow content. Nodes trilacunar with three traces. Pith septate. Vessels typically with both simple and scalariform perforations, ranging from mostly simple to all scalariform, with occasional reticulate perforations in most genera; scalariform plates moderately oblique and usually with 1–10 (occasionally up to 20) broad bars; lateral pitting from scalariform to alternate. Fibers with simple or sometimes narrowly bordered pits. Rays uniseriate or biseriate, heterogeneous. Axial parenchyma apotracheal banded and diffuse to paratracheal. In the axial parenchyma and the rays as well as in the phloem there are tanniniferous tubes containing pale or red resin. Sieve-element plastids of S-type and Psc-type, the latter with tiny protein crystals. Leaves alternate, often distichous, sometimes falsely verticillate, simple, entire, pinnately veined, occasionally pellucid-dotted, estipulate. Stomata paracytic. Flowers rather small, in axillary or rarely terminal, cymose or racemose inflorescences, unisexual, dioecious or rarely (*Endocomia*) monoecious, actinomorphic, apetalous, basically 3-merous. Calyx cupulate or campanulate, (2)3(-5)-lobed, the lobes valvate. Male flowers lacking even a vestigial gynoecium; stamens 2–45, with filaments partially or completely fused; anthers free or more often laterally connate, tetrasporangiate, extrorse or seldom latrorse, opening longitudinally, mutinous or shortly apiculate. Tapetum secretory. Microsporogenesis successive. Pollen grains 2-celled, boat-shaped to subglobose, 1-colpate to inaperturate, tectate to intectate, granular to columellate. Female flowers without staminodia; gynoecium of a single conduplicate, unsealed, sometimes short stipitate carpel with solitary, subbasal to basal ovule; stigma sessile or subsessile or rarely (*Brocboneura*) on a long stylodium, small, simple, or more or less 2-lobed. Ovule solitary, anatropous or rarely hemitropous or suborthotropous, bitegmic, crassinucellate. Female gametophyte of *Polygonum*-type. Endosperm nuclear. Fruits fleshy to coriaceous or ligneous capsule, generally dehiscent both ventrally and dorsally. Seeds usually ellipsoid to subglobose, 1–5 cm long, covered mostly with crustaceous to fleshy, laciniate or entire funicular aril; seed coat formed by both integuments, generally multiplicative, with well-developed vascular system and has the most elaborate construction among the flowering plants; inner epidermis of the outer integument strongly developed, transformed into the palisade mechanical layer of the testa. Derivatives of the inner integument transformed into the complex tegmen that is also vascularized and participates in the rumination of the endosperm; chalaza large; endosperm copious, ruminate, oily, and also contains large protein crystals and starch grains; embryo small, straight, nearly basal, differentiated into very short radicle, weakly developed hypocotyl and partly or completely connate cotyledons. Produce essential oils, flavonoids, legnans, polyketides (acetogenins), tryptamine alkaloids. $n = 19, 21, 22, 25, 26$.

The Myristicaceae definitely belong to the Magnolianae, but differ markedly from all of the other families in their very elaborate seed structure and tanniniferous tubes in rays, axial parenchyma, and phloem. The seed structure somewhat resembles the Aristolochiaceae, but the latter has starch-free and not ruminate endosperm and lacks an aril. The Myristicaceae resemble Annonaceae in their ruminate endosperm. According to Hegnauer (2000) Myristicaceae resemble Lauraceae in their seed fats and in their secondary metabolism, in their producing of noelignans and lignans, and of polyketides.

1. MYRISTICACEAE

R. Brown 1810. 21/500. Widely distributed in tropical countries, especially in Asia.

Mauloutchia, Brochoneura, Cephalosphaera, Haematodendron, Scyphocephalium, Pycnanthus, Staudtia, Coelocaryon, Compsoneura, Osteophloeum, Otoba, Virola, Bicuiba, Iryanthera, Knema, Horsfieldia, Endocomia, Gymnacranthera, Myristica, Paramyristica, Doyleanthus.

Bibliography

Armstrong JE and BA Drummond III. 1986. Floral biology of *Myristica fragrans* Houtt.: The nutmeg of commerce. Biotropica 18: 32–38.

Armstrong JE and AK Irvine. 1989. Floral biology of *Myristica insipida* (Myristicaceae): A distinctive beetle pollination syndrome. Am. J. Bot. 78: 86–94.

Armstrong JE and SC Tucker. 1986. Floral development in *Myristica* (Myristicaceae). Am. J. Bot. 73: 1131–1141.

Armstrong JE and TK Wilson. 1978. Floral morphology of *Horsfieldia*. Am. J. Bot. 65: 441–449.

Baas P. 2000. Wood anatomy of Myristicaceae. In: PF Stevens, ed. Flora Malesiana, ser. I, 14: 19. Noordhoff, Leiden.

Behnke D-H. 1991. Sieve-element characters of Myristicaceae: Nuclear crystals, S- and P-type plastids, nacrous walls. Nord. J. Bot. 11: 333–344.

Bhandari NN. 1971. Embryology of the Magnoliales and comments on their relationships. J. Arnold Arbor. 52: 1–39, 285–304.

De Wilde WJJO. 1994a. Taxonomic review of *Myristica* (Myristicaceae) in the Pacific. Blumea 38: 349–406.

De Wilde WJJO. 1994b. *Paramyristica*, a new genus of Myristicaceae. Blumea 39: 341–350.

De Wilde WJJO. 2000. Myristicaceae. In: PF Stevens, ed. Flora Malesiana, ser. I, 14: 1–632. Noordhoff, Leiden.

Garratt GA. 1933a. Systematic anatomy of the woods of the Myristicaceae. Trop. Woods 35: 6–48.

Garratt GA. 1933b. Bearing of wood anatomy on the relationships of the Myristicaceae. Trop. Woods 36: 20–44.

Gottlieb OR. 1979. Chemical studies on medicinal Myristicaceae from Amazonia. J. Ethnopharmacol. 1: 309–323.

Hegnauer R. 2000. Phytocyhemisty and chemotaxonomy of Myristicaceae. In: PF Stevens, ed. Flora Malesiana, ser. I, 14: 21–27. Leiden.

Holmstedt B, JE Lindgren, T Plowman, L River, RE Schultes, and O Tovar. 1980. Indole alkaloids in Amazonian Myristicaceae. Bot. Mus. Leafl. Harv. Univ. 28(3): 215–234.

Joshi AC. 1946. A note on the development of pollen of *Myristica fragrans* Van Houtten and the affinities of the family Myristicaceae. J. Indian Bot. Soc. 25: 139–143.

Koster J and P Baas. 1981. Comparative leaf anatomy of the Asiatic Myristicaceae. Blumea 27: 115–173.

Kühn U and K Kubitzki. 1993. Myristicaceae. In: K Kubitzki, ed. The families and genera of vascular plants, vol. 2, pp. 457–467. Springer, Berlin/Heidelberg/New York.

Nair NC and P. Sarita 1972 (publ. 1975). Floral morphology of *Myristica malabarica* Lamk. with a discussion of certain aspects of the systematics of *Myristica*. In: Murty et al., eds. Advances in plant morphology, pp. 264–277. Meerat

Nair NC and PN Bahl. 1956. Vascular anatomy of the flower of *Myristica malabarica* Lamk. Phytomorphology 6: 127–134.

Periasamy K. 1961. Studies on seeds with ruminate endosperm: 1. Morphology of ruminating tissue in *Myristica fragrans*. J. Madras Univ. 31B: 53–58.

Sauquet H, JA Doyle, T Scharaschkin, T Borsch, KW Hilu, LW Chatrou, and A Le Thomas. 2003. Phylogenetic analysis of Magnoliales and Myristicaceae based on multiple data sets: implications for character evolution. Bot. J. Linn. Soc. 142: 125–186.

Sastri RLN. 1954. On the vascular anatomy of the female flower of *Myristica fragrans*. Proc. Indian Sci. Cong. 3: 172–173 (Abstract).

Sastri RLN. 1955. Structure and development of nutmeg seed. Curr. Sci. 24: 172.

Sastri RLN. 1959. Vascularization of the carpel of *Myristica fragrans*. Bot. Gaz. 121: 92–95.

Sauquet H. 2003. Androecium diversity and evolution in Myristicaceae (Magnoliales), with the description of a new Malagasy genus, *Doyleanthus*, gen. nov. Am. J. Bot. 90: 1293–1305.

Sauquet H and A Le Thomas. 2003. Pollen diversity and evolution in Myristicaceae (Magnoliales). Int. J. Plant Sci. 164: 613–628.

Siddiqi MR and TK Wilson. 1974. Wood anatomy of the genus *Knema* (Myristicaeae). Bull. Torrey Bot. Club 101: 354–362.

Siddiqi MR and TK Wilson. 1975a. Leaf anatomy of the genus *Knema*. Biologia (Pakistan) 21: 167–175.

Siddiqi MR and TK Wilson. 1975b. Pollen of the genus *Knema* (Myristicaceae). Pak. J. Bot. 7: 197–200.

Siddiqi MR and TK Wilson. 1976a. Floral anatomy of the genus *Knema* (Myristicaceae). Biologia (Pakistan) 22: 127–141.

Siddiqi MR and TK Wilson. 1976b. Comparative study of the genus *Knema* (Myristicaceae). Biologia (Pakistan) 22: 305–308.

Van Heel WA. 1982. Note on the structure of developing seeds of *Knema* and *Horsfieldia* (Myristicaceae). Blumea 28: 53–60.

Van der Ham RWJM. 2000. Palynology of Myristicaceae. In: PF Stevens, ed. Flora Malesiana, ser. I, 14: 19–21. Leiden.

Walker JW and AG Walker. 1979. Comparative pollen morphology of the American myristicaceous genera *Compsoneura* and *Virola*. Ann. Missouri Bot. Card. 66: 731–755.

Walker JW and AG Walker. 1980. Comparative pollen morphology of the mainland African genera of Myristicaceae (*Cephalosphaera, Coelocaryon, Pycnanthus*, and *Scyphocephalium*). Am. J. Bot. 67: 603–611.

Walker JW and AG Walker. 1981. Comparative pollen morphology of the Madagascan genera of Myristicaceae (*Mauloutchia, Brochoneura*, and *Haematodendron*). Grana 20: 1–17.

Walker JW and AG Walker. 1983. Comparative pollen morphology of the American myristicaceous genera *Otoba, Iryanthera*, and *Osteophloeum*. Am. J. Bot. 70: 315–326.

Warburg O. 1897. Monographic der Myristicaceae. Nov. Act. Acad. Caes.-Leop.-Carol. Nat. Cur. 68: 1–680.

Wilson TK and LM Maculans. 1967. The morphology of the Myristicaceae: 1. Flowers of *Myristica fragrans* and *M. malabarica*. Am. J. Bot 54: 214–220.

Suborder LAURANAE

Order 13. LAURALES

Trees or shrubs, sometimes woody lianas, rarely twining parasitic herbs (*Cassytha*), usually with oil or mucilage cells in the parenchymatous tissues. Nodes unilacunar, with 1–7 traces. Xylem usually with well-developed vessels. Vessels with scalariform or simple perforations; lateral pitting scalariform to alternate or mostly alternate. Fibers ranging from tracheids through fiber-tracheids to libriform-fibers. Rays mostly heterogeneous. Axial parenchyma apotracheal or paratracheal, rarely absent. Sieve-element plastids of S-, Psc-, or Pest-type. Leaves alternate or opposite, sometimes verticillate, simple, entire or sometimes lobed or dentate, pinnately or rarely palmately compound, without stipules. Stomata paracytic or less often anomocytic. Flowers usually in cymose or racemose inflorescences, rarely solitary, actinomorphic or very rarely oblique, spiral, spirocyclic or more often cyclic, bisexual or unisexual, mostly with more or less developed and usually cup-shaped hypanthium (floral cup) of receptacular origin. Perianth spiral or cyclic (often 3-merous), usually not differentiated into sepals and petals. Androecium of (3-) 5 to many stamens, cyclic or spiral; stamens more or less ribbon-shaped or much more often differentiated into filament and anther; filaments frequently with two lateral glands; anthers tetrasporangiate or disporangiate, opening longitudinally or more often by valves, rarely by pores or very rarely circumscissile (*Hennecartia*). Tapetum secretory or amoeboid. Microsporogenesis successive or simultaneous (Calycanthaceae). Pollen grains 2-celled or rarely 3-celled, mostly inaperturate, frequently 2-colpate, rarely with 2–12 slightly irregular porate apertures, mostly tectate. Gynoecium apocarpous, spiral (Gomortegaceae, Calycanthaceae and some Monimiaceae) or whorled (polymerous or more often monomerous) or less frequently syncarpous; carpels 1–2,000, glabrous or rarely hairy, fully sealed and with a localized stigma on a distinct stylodium or rarely unsealed above and with shortly decurrent stigma. Ovule solitary in each carpel or in each locule, anatropous, rarely orthotropous or hemitropous, bitegmic or rarely unitegmic, crassinucellate. Female gametophyte usually of *Polygonum*-type. Endosperm cellular or nuclear (Lauraceae-Lauroideae). Fruits of various types. Seeds with endotestal seed coat, embryo small to large; endosperm well developed, scanty or absent. Produces proanthocyanins but not ellagic acid, and very often with benzyl isoquinoline or aporphine alkaloids.

Close to the Magnoliales but more advanced. Evidently derived from some vesselless ancestors.

Key to Families

1 Seeds with well-developed endosperm and small or medium sized embryo.

2 Gynoecium apocarpous.

3 Anthers opening by slits. Trees or shrubs, rarely woody lianas. Spherical ethereal cells present; mucilage cells absent. Nodes unilacunar with 1–7 traces. Nodes unilacunar with 1–7 traces. Vessels with simple perforation. Sieve-element plastids of S-, Psc-, or Pcsf-type. Leaves alternate or opposite. Stomata paracytic or anomocytic. Flowers in cymose inflorescences or rarely solitary, unisexual in most genera, bisexual in *Hortonia*, actinomorphic or very rarely oblique. Receptacle transformed into more or less well-developed floral cup, more or less globose or urceolate to widely campanulate. Perianth segments spiral or in cycles, often 4-merous or of decussate pairs, sometimes absent. Stamens usually numerous and dispersed over the inner surface of the receptacle or fewer and sometimes in one or more 4-merous cycles; filaments in *Hortonia, Peumus*, and *Monimia* with two nectariferous glands; anthers tetrasporangiate or disporangiate, opening longitudinally or by valves. Pollen grains inaperturate or dicolpate. Carpels 1–2,000 (1 in *Xymalos*, 1–2 in *Hennecartia*, 1,000 in *Decaryodendron* and up to 2,000 in *Tambourissa ficus*, the largest in Magnoliidae), small, whorled or spiral or irregular, free (in *Tambourissa* congenitally united and congenitally immersed in the floral cup (Endress et al. 1997); the carpel is covered with unicellular, often lignified or stellate hairs, or glabrous; ovary superior (inferior in *Tambourissa* – Monimiaceae); stylodia short or elongate, with terminal stigma. Ovules solitary, anatropous or (*Kibara, Kibaropsis, Xymalos*) hemitropous, erect or pendulous, bitegmic. Fruits drupes. Seeds endotestal, the tegmen eventually crushed; endosperm

copious, oily; embryo very small to moderately large, straight, differentiated, with two or very rarely four (*Kibaropsis*) cotyledons. Hippocrepiform sclereids present. Benzyl isoquinolines present. n = 19–22, 39, 43, 57, etc.. . . . . 1. MONIMIACEAE.

3 Anthers opening by valves.

4 Fruits of long-stylodiated plumose achenes. Trees or shrubs. Nodes with 1 trace. Sieve-tube plastids P-type. Leaves opposite, petiolate, gland-dotted or not, simple, entire or dentate, estipulate. Stomata anomocytic. Flowers solitary or in racemes or cymes, bisexual or monoecious or polygamo-dioecious, actimonorphic or somewhat zygomorphic. Perianth segments with distinct calyx and corolla, or sepaloid, or vestigial, or absent. Corolla when present 7–20, or more. Stamens (4-)6-12(-100). Filaments appendaged, each with a pair of glandular scales at the base. Anthers adnate, extrorse, disporangiate, opening by two valves. Pollen grains reticulate. Gynoecium of 3–100 carpels with a lateral style, or with a gynobasic style. Ovary superior to inferior; ovule 1, anatropous, or rarely orthotropous, micropyle endostomal, directed downward. Fruits plumose, with woody hypanthium. Seeds with small, well differentiated, straight embryo; endosperm oily. Flavonols present; kaempferol and quercetin, n = 22, 57. . . . . 4. ATHEROSPERMATACEAE.

4 Fruitlets drupes. Shrubs or small trees bearing essential oils. Nodes with one trace. Vessels with simple or scalariform and simple perforations. Xylem without fibre tracheids. Axial parenchyma apotracheal. Leaves evergreen, opposite or verticillate, simple, entire, gland-dotted, or not gland-dotted, estipulate. Stomata paracytic. Flowers in cymes, spikes, or in panicles, monoecious or dioecious, actinomorphic to zygomorphic. Free hypanthium present. Perianth with distinct calyx and corolla, perianth segments 4-6(-8), or obscure, calyptrate. Stamens (1)2-100, filaments free, without nectariferous appendages. Anthers often with one flap, introrse, bisporangiate (*Siparuna*), or tetrasporangiate (*Glossocalyx*). Pollen grains inaperturate. Gynoecium of (3-)4-100 carpels; style free or connate at the apex; ovule solitary in each carpel, ascending, anatropous, unitegmic or bitegmic (*Glossocalyx* – Philipson 1993). Fruits druplets, with fleshy appendage, enclosed in the fleshy hypanthium. Seeds erect, endosperm not oily. Plants accumulate Aluminum; n = 22. . . . . 5. SIPARUNACEAE

2 Gynoecium syncarpous. Aromatic trees with secretory cells containing a yellowish, resinous substance. Nodes with two traces. Vessels with scalariform perforations with 9–19 bars. Sieve-element plastids of Pcsf-type. Leaves opposite, decussate, simple, entire, pinnately veined, estipulate. Stomata paracytic. Flowers rather small, in axillary or terminal racemes, subtended by two inconspicuous, caducous bracteoles, bisexual. Perianth segments 7–10, more or less spirally arranged, intergrading with the stamens. Stamens 7–13, the outer 1–3 petaloid, the next 5–10 differentiated into filament and anther, basally provided with two short-stalked nectariferous glands; anthers disporangiate, the outer stamens introrse, the middle ones latrorse, opening from the base upward by two valves. Pollen grains globose, inaperturate, tectate, with a delicate exine and a thicker intine. Gynoecium of (2)3(-5) glabrous carpels; style very short, with 2(3) stigmatic branches; ovary deeply inferior, (2)3(-5)-locular, each locule with a single orthotropous to hemianatropous (Endress et al. 1997) ovule pendulous from the apex. Fruits drupaceous, mostly one-seeded. Seeds thin-walled; endosperm copious, oily; embryo rather large, well differentiated, n = 21. . . . . 6. GOMORTEGACEAE.

1 Seeds without endosperm and with large or medium-sized embryo.

5 Leaves alternate.

6 Ovary superior, sometimes semi-inferior, very rarely inferior (*Hypodaphnis*). Aromatic trees or shrubs or rarely (*Cassytha*) twining parasitic herb. Vessels mostly with simple perforations, but simple and scalariform in some genera and sometimes scalariform (commonly with few bars, rarely up to 25 in some species of *Actinodaphne, Aiouea*, and *Litsea*), lateral pitting alternate. Fibers predominantly libriform. Rays heterogeneous to homogeneous, in *Hypodaphnis* exclusively homogeneous (composed solely of procumbent cells). Axial parenchyma basically paratracheal. Oil cells and

mucilage cells present. Nodes unilacunar, with 1–3 traces. Leaves alternate to opposite or apparently verticillate, simple, entire or sometimes (*Sassafras*) lobed, often coriaceous, gland-dotted, aromatic; venation pinnate or rarely palmate. Stomata paracytic or sometimes anomocytic. Flowers usually in axillary and commonly cymose inflorescences, rarely solitary, bisexual or less often unisexual, generally actinomorphic, mostly 3-merous. Perianth segments commonly six and mostly in two usually equal cycles. Stamens usually in four cycles of three, of which the innermost cycle is sterile or often lacking; sometimes also one or two of the outer three cycles sterile or absent; usually the third cycle with a pair of nectariferous glands at the base; increased number of stamens are found in some genera (up to 32 in *Cinnadenia*); anthers 2- or 4-locular, basifixed, tetrasporangiate or disporangiate, opening by two or four valves, or rarely (*Hexapora*) by pores, mostly from the base upward, in the two outer cycles mostly introrse and in the third cycle introrse to extrorse. Pollen grains 2-celled, more or less spheroidal, inaperturate, with rather thick intine and extremely thin and entirely ectexinous exine, mostly echinate. Gynoecium monocarpelate. The stylodium often shows a deep furrow at the ventral side; stigma papillate, capitate to disciform, lobed, or decurrent. Ovule solitary, large to very large, anatropous, pendulous, apical or subapical, with dorsal raphe; outer integument not contributing to the micropyle; endosperm haustoria present or absent. Fruits baccate (drupes with weakly developed endocarp), 1-seeded; fruit enclosed in the fleshy receptacle, or enclosed in the fleshy hypanthium, or without fleshy investment. Seed with thin testa; embryo well differentiated, straight, with very large, occasionally ruminate cotyledons; endosperm reduced. Plants producing flavones, 5-O-methyl flavonols, polyketides (acetogenins), tryptamine alkaloids, kaempferol and quercetin. n = 12(15)...... 8. LAURACEAE

6 Ovary inferior. Fruits dry and indehiscent. Trees, shrubs, or woody lianas. Vessels with simple perforations; lateral pitting mostly alternate. Fibers with simple or bordered pits. Rays heterogeneous to homogeneous. Axial parenchyma usually para-tracheal. Nodes unilacunar, with several traces. Leaves simple (sometimes 3-lobed) or palmately compound, pinnately, palmately, or pedately veined; glandular hairs in leaf epidermis. Stomata paracytic or anomocytic. Flowers in cymose inflorescences, bisexual or unisexual, more or less actinomorphic. Perianth segments in two cycles of 3-4(-6) or in one cycle of 4–8, imbricate or valvate. Stamens 3-5(-7) in a single cycle; filaments commonly with a pair of dorso-basal or basilateral nectariferous glands; anthers disporangiate, opening by valves. Pollen grains spheroidal, inaperturate, with massive intine composed of unstratified inner and thick, radially channeled outer layer, and very thin exine ornamented by spines and globules. Gynoecium monocarpellate; stylodium with a ventral furrow decurrent from the peltate stigma. Ovule solitary, large to very large, anatropous, pendulous. Seeds ruminate (*Hernandia*), testa vascularised, spongy, tanniniferous; n = 10, 12, 15, 20.................... 7. HERNANDIACEAE.

5 Leaves opposite. Fibers with minute bordered pits. Rays predominantly uniseriate or predominantly multiseriate. Sieve-element plastids P-type. Leaves opposite, simple, entire, pinnately veined, without stipules. Stomata paracytic. Stamens spirally arranged at the rim of the receptacle, usually more or less ribbon-shaped, with short or no filament, with prolonged connective, the inner ones staminodial; anthers extrorse. Pollen grains more or less globose, with two distal colpi, tectate-columellate, psilate. Carpels with usually elongate stylodium; ovules solitary, basal or two superposed ovules, the upper one abortive. Fruits usually of numerous achenes enclosed in the enlarged, fleshy receptacle. Very rich in common mono- and sesquiterpenes. Characterized by the presence of calycanthidine and calycanthine types of alkaloids.

7 Carpels 5–35; cotyledons 2, spirally twisted. Deciduous or sometimes evergreen small trees or shrubs. Nodes with two traces. Vessels with simple perforation. Parenchyma scanty vasicentric with some diffuse. Flowers, bisexual, solitary at the ends of specialized short leafy branches. Perianth sequentially intergrading from sepals to petals; perianth segments 15–30,

free. Stamens 5–30. Anthers adnate, extrorse, tetrasporangiate, appendaged. Pollen grains aperturate; 2(–3) colpate, sulculate, 2-celled. Ovules two, ascending, anatropous or apotropous, crassinucellate or subcrassinucellate. Fruits achenes, enclosed in the fleshy hypanthium; seeds with well differentiated, large embryo; endosperm wanting. Producing tryptamine (calycanthidine) and benzylisoquinoline alkaloids; n = 11. . . . . 3. CALYCANTHACEAE

7 Carpels 1 or 2(3); cotyledons 3 or 4, fleshy, peltate. Evergreen trees. Vessels with scalariform perforation. Axial parenchyma abundant, predominantly apotracheal. Flowers, bisexual and male, solitary (3) on bracteate, axillary peduncles, or in few-flowered terminal inflorescences. Perianth sequentially intergrading from sepals to petals, or petaloid; perianth segments 30–40, free. Stamens 13–15. Anthers adnate, extrorse, appendages. Pollen grains 2 colpate. Gynoecium of 1 or 2(3) carpels. Ovules 1–2, anatropous. Stigma multiseriate-multicellular-papillate, fleshy, subsessile, broad, obliquely terminal. Fruits strictly non-fleshy, but enclosed in the fleshy receptacle. Seeds without endosperm. Embryo with three or more often four massive, fleshy-firm, cotyledons attached in a single whorl. Producing luteolin; n = 11.. . . . . 2. IDIOSPERMACEAE

1. MONIMIACEAE

A. L. de Jussieu 1809 (Hortoniaceae A. C. Smith 1971). 23/200. Tropical and subtropical regions, especially of Southern Hemisphere, where they reach temperate zones; concentrated in Malesia, in the islands of the southwestern Indian Ocean, including Madagascar, and in South America; poorly represented in Africa and absent in India.

1.1 HORTONIOIDEAE

Stomata paracytic. Flowers bisexual. Receptacle concave. Perianth spiral, with no clear distinction between outer sepaloid and inner petaloid members. Stamens with two appendages on the filament. Anthers tetrasporangiate, opening longitudinally. Staminodia present. Pollen grains inaperturate, with semihelical bands. Carpels free. Ovules bitegmic, the micropyle directed upward. Hippocrepiform sclereids present. – *Hortonia* (3, Sri Lanka).

1.2 MOLLINEDIOIDEAE

Stomata paracytic. Flowers unisexual (monoecious or dioecious). Receptacle flat, concave, urceolate or globose, sometimes pollinated at the upper margin of the globose floral cup, at a "hyperstigma" (Endress 1979, 1980b, 1983). Perianth segments sepaloid or absent. Stamens without appendages on the filaments. Anthers tetrasporangiate, opening longitudinally or rarely circumscissile (*Hennecartia*). Staminodia absent in female flowers, but sometimes the inner stamens sterile in male flowers. Pollen grains inaperturate, rarely in tetrads or dyads. Carpels free, or inferior in *Tambourissa*, solitary in *Xymalos*. Stylodia subulate or stigmas sessile. Ovules bitegmic, with the micropyle directed upward (sideways in *Kibaropsis*). – HEDYCARYEAE: *Decarydendron, Ephippiandra, Hedycarya, Kibaropsis, Levieria, Tambourissa, Xymalos*; MOLLINEDIEAE: *Austromatthaea, Faika, Kairoa, Kibara, Macropeplus, Matthaea, Mollinedia, Parakibara, Steganthera, Tetrasynandra, Wilkiea*; HENNECARTIEAE: *Hennecartia*.

1.3 MONIMIOIDEAE

Stomata anomocytic. Flowers unisexual (dioecious). Receptacle a shallow cup or urceolate. Perianth segments sepaloid. Filaments appendaged (except in *Palmeria*). Anthers tetrasporangiate or disporangiate, opening longitudinally. Staminodia absent. Pollen grains inaperturate. Carpels free, stylodia elongate, projecting through the ostiole. Ovules bitegmic, the micropyle directed upward. – PALMERYEAE: *Palmeria*; MONIMIEAE: *Monimia*; PEUMEAE: *Peumus*.

The most archaic member of the family is the genus *Hortonia*. The Monimiaceae are divided into a number of well-defined subfamilies (Schodde 1970; Thorne 1974, 1992, 2000).

2. IDIOSPERMACEAE

S. T. Blake 1972. 1/1. Northeastern Queensland.

Idiospermum.

Very closely related to the Calycanthaceae, but distinctive nature of the flavonoids of *Idiospermum*, together with the morphological and anatomical differences between Calycanthaceae and *Idiospermum* support removing *Idiospermum* from the Calycanthaceae and placing it in a separate family (Sterner and Young 1980).

3. CALYCANTHACEAE

Lindley 1819. 2/7 (including Chimonanthaceae Perleb 1838). Continental China, North America.

Calycanthus (including *Sinocalycanthus), Chimonanthus.*

Related to the Monimiaceae, but markedly differ in the presence of four inverted vascular bundles in the cortex and pericycle of the young stem, and the absence of triple fusion, and autonomous development of the endosperm.

4. ATHEROSPERMATACEAE

R. Brown 1814. 7/16. Temperate to tropical. Australia, New Guinea, New Zealand, New Caledonia, Chile.

ATHEROSPERMEAE: *Atherosperma, Laureliopsis*; LAURELIEAE: *Daphnandra, Doryphora, Dryadodaphne, Laurelia, Nemuaron.*

5. SIPARUNACEAE

Schodde 1970. 2/150. Tropical America, West Indies, West Africa.

Siparuna (including *Bracteanthus), Glossocalyx.*

6. GOMORTEGACEAE

Reiche 1896 1/1. Central Chile.

Gomortega.

Rather closely related to the Monimiaceae and especially to the Atherospermataceae (Goldblatt 1976), differing from them mainly by syncarpous pseudomonomerous gynoecium and inferior ovary. According to Stern (1955), the Gomortegaceae are most likely closely allied to Monimiaceae through a *Hortonia*-like ancestor. The Pcsf-type sieve-element plastids of *Gomortega* are much like those of the Atherospermataceae and the Calycanthaceae but have one large starch grain (Behnke 1988). Gomortegaceae are very heterobathmic (primitive wood anatomy, undifferentiated tepals, and spiral androecium together with syncarpous ovary with single pendulous ovule, valvular dehiscence of the anthers, and large embryo).

7. HERNANDIACEAE

Blume 1826 (including Gyrocarpaceae Dumortier 1829; Illigeraceae Blume 1833). 4/58. Tropics and partly subtropics of both hemispheres, mainly in coastal areas, especially on oceanic islands, reaches north to southern China, Taiwan, Ryukyu Islands, and Mexico.

7.1 HERNANDIOIDEAE

Inflorescences thyrsoid, bracteate, with cincinnate partial inflorescences. Cotyledons more or less crumpled. Cystoliths absent. Evergreen trees with simple leaves (*Hernandia*) or lianas with 3–5-foliolate leaves (*Illigera*). – *Hernandia, Illigera.*

7.2 GYROCARPOIDEAE.

Inflorescences dichasial, ebracteate. Cotyledons foliaceous and folded around the radicle or spirally twisted. Cystoliths present. Deciduous trees with entire or 3–5-lobed leaves (*Gyrocarpus*) or shrubs or lianas with simple, mostly triplin-erved leaves (*Sparattanthelium*). – *Gyrocarpus, Sparattanthelium.*

Closely related to the Lauraceae, but have also some similarities to the Monimiaceae.

8. LAURACEAE

A. L. de Jussieu 1789 (including Cassythaceae Bartling ex J. Lindley 1833; PerseaceaeHoraninow 1834). 54/2500–3500. Tropical and subtropical regions of both hemispheres, centered in Southeast Asia and tropical America. Some genera, including *Laurus, Lindera, Litsea, Persea* and *Sassafras*, reach warm-temperate areas.

8.1 LAUROIDEAE

Trees and shrubs with well-developed green leaves. Tapetum secretory or amoeboid. Endosperm nuclear or rarely (*Umbellularia*) cellular. Micropyle formed by the inner integument alone. – PERSEEA: *Persea, Phoebe, Apollonias, Nothaphoebe, Alseodaphne, Dehaasia, Caryodaphnopsis, Neocinnamomum, Nectandra, Pleurothyrium, Rhodostemonodaphne, Urbanodendron, Dicypellium, Phyllostemonodaphne, Systemonodaphne (Kubitzkia), Paraia, Gamanthera, Povedadaphne, Williamodendron, Mezilaurus* (including *Clinostemon), Anaueria, Beilschmiedia, Brassiodendron, Endiandra, Triadodaphne, Hexapora, Potameia, Syndiclis, Dahlgrenodendron, Aspidostemon, Potoxylon, Cinnadenia, Chlorocardium*; CINNAMOMEAE: *Ocotea,*

Cinnamomum, Actinodaphne, Aiouea, Aniba, Endlicheria, Licaria; LAUREAE: *Umbellularia, Dodecadenia, Litsea, Adenodaphne, Neolitsea, Lindera, Iteadaphne, Laurus, Parasassafras, Sassafras*; CRYPTOCARYEAE: *Cryptocarya, Ravensara, Eusideroxylon*; HYPODAPHNIDEAE: *Hypodaphnis*.

8.2 CASSYTHOIDEAE

Semiparasitic climbing or twining herbs with scalelike leaves attached to host plants by haustoria (modified roots). Tapetum secretory. Endosperm cellular. Micropyle formed by both integuments. Inner integument does not grow beyond nucellus so that overarching funiculus is in close contact with nucellus. – *Cassytha*.

Lauraceae are related to the Monimiaceae and probably derived from *Hortonia*-like ancestors. The taxonomic subdivision of the family as well as its phylogeny have been subjects of controversy and conjecture for a long time and have been discussed by many systematists (Rohwer 1993). There is no satisfactory subdivision of the family, and even its subdivision into two subfamilies, which is apparently well supported by embryological data, creates some doubts. According to Rohwer (1993), there is increasing evidence (from flower and fruit structure, pollen morphology, and alkaloid chemistry) that *Cassytha* is an offshoot of the branch leading to *Cryptocarya*.

Bibliography

Ablett EM, J Playford, and S Mills. 1997. The use of ribisco DNA sequences to examine the systematic position of *Hernandia albiflora* (C.T.White) Kubitzki (Hernandiaceae), and relationships among the Laurales. Austrobaileya 4: 601–607.

Allen CK. 1938. Studies in the Lauraceae: 1. Chinese and Indo-Chinese species of *Litsea, Neolitsea, and Actinodaphne*. Ann. Missouri Bot. Card. 25: 361–434.

Baillon H. 1868. Observations sur les Monimiacees. Adansonia 9: 111–134.

Bello MA, F Gonzalez, G Romero de Perez. 2002. Morfologia del androceo, tapete y ultraestructura del polen de Siparuna aspera (Ruiz et Pavon) A. DC. (Siparunaceae). Rev. Acad. Colomb. Cienc. Exact. Fis. Nat. 26(99): 155–167.

Bhandari NN. 1967. Monimiaceae, Calycanthaceae, Lauraceae. In: BM Johri et al., eds. Seminar on comparative embryology of angiosperms, pp. 19–21. Department of Botany, Univ. Press, Delhi.

Blake ST. 1972. *Idiospermum* (Idiospermaceae): A new genus and family for *Calycanthus australiensis*. Contr. Queensland Herb. 12: 1–37.

Boyle EM. 1980. Vascular anatomy of the flower, seed, and fruit of *Lindera benzoin*. Bull. Torrey Bot. Club 107: 409–417.

Brizicky GK. 1959. Variability in the floral parts of *Gomortega* (Gomortegaceae). Willdenowia 2: 200–207.

Brofferio I. 1930. Osservazioni sullo sviluppo delle Calycanthaceae. Ann. Bot. Roma 18: 387–394.

Carlquist S. 1983. Wood anatomy of Calycanthaceae: Ecological and systematic implications. Aliso 10: 427–441

Chanderbali AS, H van der Werff, and SS Renner. 2001. Phylogeny and historical biogeography of Lauraceae: evidence from the chloroplast and nuclear genomes. Ann. Missouri Bot. Gard. 88: 104–134.

Chang R-H and C-S Ding. 1980. The seedling characters of Chinese Calycanthaceae with a new species of *Chimonanthus* Lindl. Acta Phytotax. Sinica 18: 328–332.

Cheadle CL and K Esau. 1958. Secondary phloem of Calycanthaceae. Univ. Calif. Publ. Bot. 29: 397–510.

Christophel DC, R Kerrigan, and AI Rowett. 1996. The use of cuticular features in the taxonomy of the Lauraceae. Ann. Missouri Bot. Gard. 83: 419–432.

Collins RPN, N Chang, and LE Knaak. 1969. Anthocyanins in *Calycanthus floridus*. Am. Midi. Nat. 82: 633–637.

Coy GV. 1928. Morphology of *Sassafras* in relation to phylogeny of angiosperms. Bot. Gaz. 86: 149–171.

Cummings K and CA Schroeder. 1943. Anatomy of the avocado fruit. Yearb. Calif. Avocado Assoc. 1942: 56–64.

Datta K and S Chanda. 1980. Pollen morphology of a few members of the order Laurales (sensu Takhtajan) with reference to taxonomy and phylogeny. Trans. Böse Res. Inst. Calcutta 43 (3–4): 73–79.

Daumann E. 1930. Blütennektarium von *Magnolia* und die Futterkorper in der Blüte von *Calycanthus*. Planta 11: 108–116.

Dengler NG. 1972. Ontogeny of the vegetative and floral apex of *Calycanthus occidentalis*. Canad. J. Bot. 50: 1349–1356.

Doweld AB. 2001. Carpology and phermatology of *Gomortega* (Gomortegaceae): systematic and evolutionary implications. Acta Bot. Malacitana 26: 19–37.

Endress PK. 1972. Zur vergleichenden Entwicklungsmorphologie, Embryologie, und Systematik bei Laurales. Bot. Jahrb. Syst. 92: 331–428.

Endress PK. 1979. Noncarpellary pollination and "hyperstigma" in an angiosperm *(Tambourissa religiosa*, Monimiaceae). Experientia 35: 45.

Endress PK. 1980a. Floral structure and relationships of *Hortonia* (Monimiaceae). Plant Syst. Evol. 133: 199–221.

Endress PK. 1980b. Ontogeny, function, and evolution of extreme floral construction in Monimiaceae. Plant Syst. Evol. 134: 79–120.

Endress PK. 1983. Dispersal and distribution in some small archaic relic angiosperm families (Austrobaileyaceae, Eupomatiaceae, Himantandraceae, Idiospermaceae-Calycanthaceae). Sonderb. Nat. Wiss. Habmurg 7: 201–217.

Endress PK. 1992. Protogynous flowers in Monimiaceae. Plant Syst. Evol. 181: 227–232.

Endress PK, FLS Igersheim, and A Igersheim. 1997. Gynoecium diversity and systematics of the Laurales. Bot. J. Linn. Soc. 125: 93–168.

Endress PK and DH Lorence. 1983. Diversity and evolutionary trends in the floral structure of *Tambourissa* (Monimiaceae). Plant Syst. Evol. 143: 53–81.

Esenbeck N., von. 1836. Systema Laurinarum. Berlin. Fahn A and IW Bailey. 1957. Nodal anatomy and primary vascular cylinder of Calycanthaceae. J. Arnold Arbor. 38: 107–117.

Foreman DB. 1984. The morphology and phylogeny of the Monimiaceae (sensu lato) in Australia. Ph.D. Thesis, University of New England, Armidale.

Foreman DB. 1987. Notes on the wood anatomy of *Idiospermum australiense* (Idiospermaceae). Muelleria 6: 329–333.

Foreman DB and FB Sampson. 1987. Pollen morphology of *Palmeria scandens* and *Wilkiea huegeliana* (Monimiaceae). Grana 26: 127–133.

Gardner RO. 1974. Trinucleate pollen in *Beilschmiedia* Nees (Lauraceae). New Zealand J. Bot. 12: 243–244.

Garratt GA. 1934. Systematic anatomy of the woods of the Monimiaceae. Trop. Woods 39: 18–44.

Goldblatt P. 1976. Chromosome number in *Gomortega keule*. Ann. Missouri Bot. Gard. 63: 207–208.

Goldblatt P. 1979. Chromosome number in two primitive dicots, *Xymalos monospora* (Monimiaceae) and *Piptocalyx moorei* (Trimeniaceae). Ann. Missouri Bot. Gard. 66: 898–899.

Gottlieb OR. 1972. Chemosystematics of the Lauraceae. Phytochemistry 11: 1537–1570.

Grant V. 1950. The pollination of *Calycanthus occidentalis*. Am. J. Bot. 37: 294–297.

Heilborn O. 1931. Studies on the taxonomy, geographical distribution, and embryology of the genus *Siparuna*. Svensk Bot. Tidskr. 25: 202–228.

Heo K, Y Kinoto, M Riveros, and H Tobe. 2004. Embryology of Gomortegaceae (Laurales): Characteristics and character evolution. J. Plant Res. 117: 221–228.

Heo K and H Tobe. 1995. Embryology and relationships of *Gyrocarpus* and *Hernandia* (Hernandiaceae). J. Plant Res. 108: 327–341.

Heo K, H van der Werff and H Tobe. 1998. Embryology and relationships of Lauraceae. Bot. J. Linn Soc. 126: 295–322.

Hesse M and K Kubitzki. 1983. The sporoderm ultra-structure in *Persea, Nectandra, Hernandia, Gomortega*, and some other Lauralean genera. Plant Syst. Evol. 141: 299–311.

Hiepko P. 1965. Vergleichend-morphologische und en-twicklungsgeschichtliche Untersuchungen über das Perianth bei den Polycarpicae. Bot. Jahrb. Syst. 84: 359–508.

Hyland B. 1989. A revision of Lauraceae in Australia (excluding *Cassytha*). Austral. Syst. Bot. 2: 135–267.

Kamelina OP. 1981a. Monimiaceae. In: MS Yakovlev, ed. Comparative embryology of flowering plants: Winteraceae-Juglandaceae, pp. 65–69. Nauka, Leningrad (in Russian).

Kamelina OP. 1981b. Calycanthaceae. In: MS Yakovlev, ed. Comparative embryology of flowering plants: Winteraceae-Juglandaceae, pp. 69–74. Nauka, Leningrad (in Russian).

Kasapligil B. 1951. Morphological and ontogenetic studies on *Umbellularia californica* Nutt. and *Laurus nobilis* L. Univ. Calif. Publ. Bot. 25: 115–240.

Kimoto Y and H Tobe. 2001. Embryology of Laurales: a review and perspectives. J. Plant Res. 114: 247–267.

Klucking EP. 1987. Leaf venation patterns: Lauraceae. Cramer, Berlin.

Kostermans AJG. 1957. Lauraceae. Reinwardtia 4: 193–256.

Kostermans AJGH. 1988. Materials for a revision of Lauraceae: 5. Reinwardtia 10(5): 439–469.

Kubitzki K. 1969. Monographie der Hernandiaceen. Bot. Jahrb. Syst. 89: 78–209.

Kubitzki K. 1981. The tubular exine of Lauraceae and Hernandiaceae: A novel type of exine structure in seed plants. Plant Syst. Evol. 138: 139–146.

Kubitzki K. 1993a. Calycanthaceae. In: K. Kubitzki, ed. The families and genera of vascular plants, vol. 2, pp. 197–200. Springer, Berlin/Heidelberg/New York.

Kubitzki K. 1993b. Gomortegaceae. In: K Kubitzki, ed. The families and genera of vascular plants, vol. 2, pp. 318–320. Springer, Berlin/Heidelberg/New York.

Kubitzki K. 1993c. Hernandiaceae. In: K Kubitzki, ed. The families and genera of vascular plants, vol. 2, pp. 334–338. Springer, Berlin/Heidelberg/New York.

Kubitzki K and H Reznik. 1966. Flavonoid-Muster der Polycarpicae als systematisches Merkmal: I. Übersicht über die Familien. Beitr. Biol. Pfl. 42: 445–470.

Leinfellner W. 1966. Über die Karpelle verschiedener Magnoliales: II. *Xymalos, Hedicarya*, und *Siparuna* (Monimiaceae). Oesterr. Bot. Z. 113: 448–458.

Leinfellner W. 1968. Über die Karpelle verschiedener Magnoliales: VI. *Gomortega keule* (Gomortegaceae). Oesterr. Bot. Z. 115: 113–119.

Lemesle R and Y Pichard. 1954. Les caracteres histo-logiques du bois des Monimiacees. Rev. Gen. Bot. 61: 69–95.

Li J, J Ledger, T Ward, and P del Tredici. 2004. Phylogenetics of Calycanthaceae based on molecular and morphological data, with a special reference to divergent paralogues of the nrDNA its region. Harvard Papers Bot. 9: 69–82.

Li Y and PT Li. 1999. Epidermal features of the leaves of Calycanthaceae. J. Trop. Subtrop. Bot. 7: 202–206.

Li Y and PT Li. 2000. Cladistic analysis of Calycanthaceae. J. Trop. Subtrop. Bot. 8: 275–281.

Li Y and PT Li. 2000. Origin, evolution and distribution of the Calycanthaceae. Guihaia. 20: 295–300.

Liu L, RH Chang, HE Liu, YQ Zhu, C Zhou, and SF Ye. 1995. Essential oil components in leaves of seven species in Calycanthaceae and their significance for taxonomy. Acta Phytotax. Sinica 33: 171–174.

Longo B. 1899. Osservazioni sulle Calycanthaceae. Ann. R. 1st. Bot. Roma 9 (I): 1–16.

Lorence DH. 1985. A Monograph of the Monimiaceae (Laurales) in the Malagasy Region (SW Indian Ocean). Ann. Missouri Bot. Gard. 72: 142–210.

Lorence DH. 1987. The fruits of *Decarydendron* (Monimiaceae). Ann. Missouri Bot. Gard. 74: 445–446.

Lorence DH, VE Zenger, and P Vinay. 1984. Pollen morphological studies on the Monimiaceae of the Malagasy Region. Grana 23: 11–22.

Ly Thi Ba. 1962. Embryogénie des Calycanthacées. Développment d'embryon chez le *Chimonanthus fragrans* Lidl. Compt. Rend. Hebd. Séances Acad. Sci. 254: 1323–1325.

Martinez-Laborde J. 1988. Some comments on a recent classification of the Monimiaceae. Taxon 37: 834–837.

Mathur SL. 1968. Development of female gametophyte of *Calycanthus fertilis* Walt. Proc. Natl. Inst. Sci. India 34B(6): 323–329.

Mauritzon J. 1935. Zur Embryologie von *Peumus boldus*. Arch. Bot. 11: 317–327.

Meeuse ADJ. 1993. Evolutionary history and classification of the Laurales, especially of the Monimiaceae: deductions based on fossil records and on the Anthocorm theory. Rheedea 3: 35–49.

Mez C. 1888. Morphologische Studien über die Familie der Lauraceen. Verh. Bot. Ver. Prov. Brandenburg 30: 1–31.

Mirande M. 1905. Recherches sur le developpement et l'anatomie des Cassythacees. Ann. Sci. Nat. Bot., ser. 9, 1–2: 181–285.

Mohana Rao PR. 1986. Seed and fruit anatomy in *Gyrocarpus americanus* with a discussion on the affinities of Hernandiaceae. Israel J. Bot. 35: 133–152.

Money LL, IW Bailey, and BGL Swamy 1950. The morphology and relationships of the Monimiaceae. J. Arnold Arbor. 31: 372–404.

Nemirovich-Danchenko EN. 1988. Lauraceae. In: A Takhtajan, ed. Comparative seed anatomy, vol. 2, pp. 75–84. Nauka, Leningrad (in Russian).

Nicely KA. 1965. A monographic study of the Calycanthaceae. Castanea 30: 38–81.

Ning JC. 1993. A palynological study of Calycanthaceae. Cathaya 5: 179–188.

Nozeran R and L Bancilhon. 1960. La structure florale de *Laurus nobilis* L. Naturalia Monspel. Bot. 12: 41–48.

Oginuma K and H Tobe. 2006. Chromosome evolution in the Laurales based on analyses of original and published data. J. Plant Res. 119: 309–320.

Pal S. 1975. Studies in Lauraceae: II. Some aspects of embryology of *Cinnamomum cecidodaphne* Meissn. Geobios 2: 83–84.

Pal S. 1976. Pollen grains of some Lauraceae. J. Palynol. 12: 55–62.

Patel RN. 1973. Wood anatomy of the Dicotyledons indigenous to New Zealand: 3. Monimiaceae and Atherospermataceae. New Zealand J. Bot. 11: 587–598.

Perkins J. 1925. Übersicht über die Gattungen der Monimiaceae. Leipzig.

Peter J. 1920. Zur Entwicklungsgeschichte einiger Calycanthaceen. Beitr. Biol. Pfl. 14: 59–84.

Philipson WR. 1986. Monimiaceae. In: CGGJ van Steenis, ed. Flora Malesiana, ser. 1, 10: 255–326. Noordhoff, Leyden.

Philipson WR. 1987. A classification of the Monimiaceae. Nord. J. Bot. 7: 25–29.

Philipson WR. 1988. A classification of the Monimiaceae: An additional note. Nord. J. Bot. 8: 24.

Philipson WR. 1993. Amborellaceae, Monimiaceae. In: K Kubitzki, ed. The families and genera of vascular plants, vol. 2, pp. 92–93, 426–437. Springer, Berlin/Heidelberg/New York.

Pichon P. 1948. Les Monimiacees: Famille heterogene. Bull. Mus. Hist. Nat. Paris 2(20): 383–384.

Pignal M, B Lugardon, J Jeremie, and A le Thomas. 1999. Morphologie et ultrastructure du pollen des Siparunaceae (Laurales). Grana 38: 210–217.

Poole I and H Gottwald. 2001. Monimiaceae *sensu lato*, an element of gondwanan polar forests: Evidence from late Creaceous-Early Tertiary wood flora of Antarctica. Austral. Syst. Bot. 14: 207–230.

Quinlan CE. 1919. Contributions toward a knowledge of the anatomy of the lower dicotyledons: III. The anatomy of the stem of the Calycanthaceae. Trans. R. Soc. Edinburgh 52: 517–530.

Raj B and H van der Werff. 1988. A contribution to the pollen morphology of Neotropical Lauraceae. Ann. Missouri Bot. Gard. 75: 130–167.

Raven PH, DW Kyhos, and S Marion. 1971. Chromosome number and relationships in Annoniflorae. Taxon 20: 479–483.

Record SJ and RW Hess. 1942. American timbers of the family Lauraceae. Trop. Woods 69: 7–35.

Reece Ph C. 1939. The floral anatomy of the avocado (*Persea americana*). Am. J. Bot. 26: 429–433.

Reiche K. 1896. Zur Kenntniss von *Gomortega nitida* R. et Pav. Ber. Deutsch. Bot. Ges. 14: 225–233.

Renner SS. 1998. Phylogenetic affinities of Monimiaceae based on cpDNA gene and spacer sequences. Perspectives in Plant Ecol., Evol. and Syst. 1: 61–77.

Renner SS. 1999. Circumscription and phylogeney of the Laurales: evidence from molecular and morphological data. Am. J. Bot. 86: 1301–1315.

Renner SS and A Chanderbali. 2000. What is the relationship among Hernandiaceae, Lauraceae, and Monimiaceae, and why is this question so difficult to answer? Int. J. Plant Sci. 161(Suppl.): 109–119.

Renner SS, AE Schwarzbach, and L Lohmann. 1997. Phylogenetic position and floral function of *Siparuna* (Siparunaceae: Laurales). Int. J. Plant Sci. 158(Suppl.): 89–98.

Richter HG. 1980. On the occurrence, morphology, and taxonomic implications of crystalline and siliceous inclusions in the secondary xylem of Lauraceae and related families. Wood Sci. Techn. 14: 35–44.

Richter HG. 1981. Anatomic des sekundaren Xylems und der Rinde der Lauraceae. Sonderbd. Naturwiss. Verh. Hamburg 5: 1–148.

Richter HG. 1985. Wood and bark anatomy of Lauraceae II. *Licaria* Aublet. IAWA Bull. n. s. 6: 187–199.

Rickson FR. 1979. Ultrastructural development of the beetle food tissue of *Calycanthus* flowers. Am. J. Bot. 66: 80–86.

Rohwer JG. 1993. Lauraceae. In: K Kubitzki, ed. The families and genera of vascular plants, vol. 2, pp. 366–391. Springer, Berlin/Heidelberg/New York.

Rohwer JG. 1994. A note on the evolution of stamens in the Laurales, with emphasis on the Lauraceae. Bot. Acta 107: 103–110.

Rohwer JG. 2000. Toward a phylogenetic classification of the Lauraceae: evidence from *mat*K sequences. Syst. Bot. 25: 60–71.

Rohwer JG, HG Richter, and H van der Werff. 1991. Two new genera of neotropical Lauraceae and critical on the generic delimitation. Ann. Miss. Bot. Gard. 78: 388–400.

Romanov MS, PK Endress, AVFCh Bobrov, AP Melikian, and AP Bejerano. 2007. Fruit structure and systematics of Monimiaceae (Laurales). Bot. J. Linn. Soc. 153: 265–285.

Sampson FB. 1969. Studies on the Monimiaceae: I. Floral morphology and gametophyte development of *Hedycarya arborea* J. R. et G. Forst. (subfamily Monimioideae). Austral. J. Bot. 17: 403–424. II. Floral morphology of *Laurelia novae-zelandiae A*. Cunn. (subfamily Atherospermoideae). New Zealand J. Bot. 7: 214–240. III. Gametophyte development of *Laurelia novae-zelandiae* A. Cunn. (subfamily Atherospermoideae). Austral. J. Bot. 17: 425–439.

Sampson FB. 1976. Aperture orientation in *Laurelia* pollen (Atherospermataceae syn. subfamily Atherospermoideae of Monimiaceae). Grana 15: 153–157.

Sampson FB. 1977. Pollen tetrads of *Hedycarya arborea J*. R. et G. Forst. (Monimiaceae). Grana 16: 61–73.

Sampson FB. 1982. Variation in position of the nascent generative cell in pollen of *Hedycarya* (Monimiaceae). Grana 21: 9–14.

Sampson FB. 1996. Pollen morphology and ultrastructure of *Laurelia, Laureliopsis and Dryadodaphne* (Atherospermataceae [Monimiaceae]). Grana 35: 257–265.
Sampson FB. 1997. Pollen morphology and ultrastructure of Australian Monimiaceae – *Austromatthaea, Hedycarya, Kibara, Leviera, Steganthera* and *Tetrasynandra*. Grana 36: 135–145.
Sampson FB and DB Foreman. 1988. Pollen morphology of *Atherosperma, Daphnandra*, and *Doryphora* (Atherospermataceae [Monimiaceae]). Grana 27: 17–25.
Sastri RLN. 1952. Studies in Lauraceae: I. Floral anatomy of *Cinnamomum iners* Reinw. and *Cassytha filiformis* Linn. J. Indian Bot. Soc. 31: 240–246.
Sastri RLN. 1958. Studies in Lauraceae: II. Embryology of *Cinnamomum* and *Litsea*. J. Indian Bot. Soc. 37: 266–278.
Sastri RLN. 1962. Studies in Lauraceae: III. Embryology of *Cassytha*. Bot. Gaz. 123: 197–206.
Sastri RLN. 1963. Studies in Lauraceae: IV. Comparative embryology and phylogeny. Ann. Bot. 27: 425–433.
Sastri RLN. 1965. Studies in Lauraceae: V. Comparative morphology of the flower. Ann. Bot. 29: 39–44.
Schaeppi H. 1953. Morphologische Untersuchungen an den Karpellen der Calycanthaceaen. Phytomorphology 3: 112–118.
Schaeppi H and F Steindl. 1950. Vergleichend-mor-phologische Untersuchungen am Gynoeceum der Ro-soideen. Ber. Schweiz. Bot. Ges. 60: 15–50.
Schaffner JH. 1904. The jacket layer in *Sassafras*. Ohio Naturalist 4: 191–193.
Schodde R. 1970. Two new suprageneric taxa in the Monimiaceae alliance (Laurales). Taxon 19: 324–328.
Schroeder CA. 1940. Floral abnormality in the Avocado. Yearb. Calif. Avocado Assoc. 1940: 36–39.
Schroeder CA. 1952. Floral development, sporogenesis, and embryology in the Avocado, *Persea americana*. Bot. Gaz. 113: 270–278.
Schurhoff PN. 1923. Zur Apogamie von *Calycanthus*. Flora 116: 73–84.
Shutts CF. 1960. Wood anatomy of Hernandiaceae and Gyrocarpaceae. Trop. Woods 113: 85–123.
Smith GH. 1928. Vascular anatomy of Ranakan flowers: II. Menispermaceae, Calycanthaceae, Annonaceae. Bot. Gaz. 85: 152–177.
Sohma K. 1985. Ultrastructure of pollen wall of *Lindera umbellata* Thunb. var. *membranacea* (Maxim.) Momiyama (Lauraceae). Sci. Rep. Tohoku Imp. Univ., 4th ser. (Biol.), 39: 13–19.
Staedler YM, PH Weston, and PK Endress. 2007. Floral phyllotaxis and floral architecture in Calycanthaceae (Laurales). Int. J. Pant Sci. 168: 285–306.
Stern WL. 1954. Comparative anatomy of xylem and phylogeny of Lauraceae. Trop. Woods 100: 1–72.
Stern WL. 1955. Xylem anatomy and relationships of Gomortegaceae. Am. J. Bot. 42: 874–885.
Sterner RW and DA Young. 1980. Flavonoid chemistry and phylogenetic relationships of the Idiospermaceae. Syst. Bot. 5: 432–437.
Tackholm G and E Soderberg. 1917. Über die Pollen-entwicklung bei *Cinnamomum* nebst Erorterungen tiber die phylogenetische Bedeutung des Pollentyps. Arkiv Bot. 15: 1–14.
Thorne RF. 1974. A phylogenetic classification of the Annoniflorae. Aliso 8: 147–209.
Tiagi YD. 1963. Vascular anatomy of the flower of certain species of the Calycanthaceae. Proc. Indian Acad. Sci. 58: 224–234.
Ueda KA, A Nakano, R Rodriguez, C Ramirez, and H Hishida. 1997. Molecular phylogeny of the Gomortegaceae, a Chilean endemic monotypic, and endangered family. Notic. Biol. 5: 124.
Van der Merwe JJM, AE van Wyk, and PDF Kok. 1988. *Dahlgrenodendron*: A remarkable new genus from Natal and Pondoland. S. Afr. J. Bot. 54: 80–88.
Van der Merwe JJM, AE van Wyk, and PDF Kok. 1990. Pollen types in the Lauraceae. Grana 29: 185–196.
Van der Werff H. 1991. A key to the genera of Lauraceae in the New World. Ann. Missouri Bot. Gard. 78: 377–387.
Van der Werff H and PK Endress. 1991. *Gamanthera* (Lauraceae): A new genus from Costa Rica. Ann. Missouri Bot. Gard. 78: 401–408.
Van der Werff H and HG Richter. 1996. Toward an improved classification of Lauraceae. Ann. Missouri Bot. Gard. 83: 409–418.
Van Heel WA. 1971a. The labyrinth seed of *Hernandia feltata* Meissn. In: DC. K. Nederl. Akad. Wet. Proc., ser. C., 74: 46–51.
Van Heel WA. 1971b. The distally lobed inner integument of *Hernandia peltata* Meissn. in DC. (Hernandiaceae). Blumea 19: 147–148.
Vyshenskaya TD. 1988. Monimiaceae, Atherospermataceae, Siparunaceae. In: A Takhtajan, ed. Comparative seed anatomy, vol. 2, pp. 56–68. Nauka, Leningrad (in Russian).
Walker JW. 1976. Evolutionary significance of the exine in the pollen of primitive angiosperms. In: IL Ferguson and J Muller, eds. The evolutionary significance of the exine. Linn. Soc. Symposium ser. 1, pp. 251–308. London.
Weber J. 1981. A taxonomic revision of *Cassytha* (Lauraceae) in Australia. J. Adelaide Bot. Gard. 3: 187–262.
Weberling F. 1985. Zur Infloreszenzmorphologie der Lauraceae. Bot. Jahrb. Syst. 107: 395–414.
Wilson CL. 1976. Floral anatomy of *Idiospermum aus-traliense* (Idiospermaceae). Am. J. Bot. 63: 987–996.
Wilson CL. 1979. *Idiospermum australiense* (Idiospermaceae): Aspects of vegetative anatomy. Am. J. Bot. 66: 280–289.
Young DA and RW Sterner. 1981. Leaf flavonoids of primitive dicotyledonous angiosperms: *Degeneria vitiensis* and *Idiospermum australiense*. Biochem. Syst. Ecol. 9: 185–187.
Zhou S, SS Renner, and J Wen. 2006. Molecular phylogeny and intra- and intercontinental biogeography of Calycanthaceae. Molec. Phylog. Evol. 39: 1–15.

Superorder PIPERANAE

Order 14. PIPERALES

Mostly herbs (rhizomatous or tuberous perennials or delicate annuals), less often shrubs and subshrubs, rarely small trees, sometimes lianas or epiphytes, frequently succulent when herbaceous. Nodes multilacunar or trilacunar, rarely unilacunar, frequently jointed or swollen. Vascular bundles widely spaced, arranged in a single circle (Saururaceae) or in two or more circles with others often scattered inside the rings. Vessels with simple or sometimes scalariform perforations; lateral pitting alternate, opposite, or scalariform. Rays wide and very high, heterogeneous. Axial parenchyma paratracheal. Sieve-element plastids of S-type. Leaves alternate, rarely opposite or verticillate, simple, entire, palmately or pinnately veined, with stipules usually more or less adnate to the petiole or without stipules, often pellucid-dotted. Stomata of various types, mostly encyclocytic, sometimes anomocytic or helicocytic. Prophylls in Saururaceae and many Piperaceae median, in adaxial position. Flowers numerous, small or minute, in terminal or axillary and more or less dense, elongate spikes (generally fleshy in *Peperomia*) or less often racemose, each in the axil of a small and mostly peltate bract but without bracteoles, mostly bisexual, without perianth, zygomorphic, anemophilous or entomophilous. Androecium of six stamens in two cycles in the Saururaceae, or of two stamens in the Peperomiaceae and (1)2-7(-10), often three in the Piperaceae; filaments commonly free; anthers tetrasporangiate or disporangiate (*Peperomia*), basifixed, opening longitudinally. Tapetum secretory. Microsporogenesis usually simultaneous. Pollen grains mostly 2-celled, small to minute, boat-shaped to globose-spherical, monocolpate, trichotomocolpate, inaperturate or almost inaperturate, tectate. Gynoecium of (3)4(5) basally connate, conduplicate carpels (*Saururus*), monocarpellate or, according to Sastrapradja (1968), pseudomonomerous (Peperomiaceae) or syncarpous and unilocular (paracarpous) of (2)3-4(5) carpels (Piperaceae and most of the Saururaceae). Ovary mostly superior, but partly inferior in some species of *Piper* and inferior in *Anemopsis, Gymnotheca* and *Macropiper*. Stigma double-crested and decurrent along the distinct and not completely sealed stylodia (Saururaceae) or sessile or nearly sessile, apical, subapical or lateral. Ovules 1-10(13) to numerous, in laminar-lateral position in *Saururus*, parietal in other genera of the Saururaceae, and solitary and subbasal in the Piperaceae and Peperomiaceae, orthotropous, bitegmic or (Peperomiaceae) unitegmic, crassinucellate or rarely subtenuinucellate (*Houttuynia*), or (*Zippelia* – Lei et al. 2002) tenuinucellate. Female gametophyte monosporic of *Polygonum*-type or tetrasporic. Endosperm cellular or (*Piper*) nuclear. Fruits of somewhat fleshy I-seeded and indehiscent fruitlets in *Saururus*, fleshy capsules with few seeds, or more frequently more or less drupaceous or sometimes berrylike. Seeds small, with small to minute, undifferentiated to weakly differentiated embryo, scanty or copious endosperm, and copious starchy perisperm. Often contain alkaloids and flavonols.

Key to Families

1 Pollen grains 3-celled. Small shrubs with sympodial branching and storied wood. Nodes swollen, unilacunar, with two traces. Vessel elements very short, with simple perforations, but a few plates partly or completely traversed by a single wide bar are located in vessels near the pith (Carlquist 1990). Lateral pitting mostly alternate, rarely scalariform and opposite. Fiber tracheids with vestigial bordered pits. Rays mainly restricted to the nodal regions, wide and tall, cells mostly upright. Axial parenchyma scanty, vasicentric. Fibers, vessels, and axial parenchyma storied. Sieve-element plastids of Ss-type with about ten globular starch grains. Leaves small, alternate, simple, entire, emarginate, with converging main veins, pellucid-punctate, with a conspicuous, membranous, sheathing, ochrealike stipules adnate to the petiole and formed by the fusion of two distinct outgrowth that develop from the flanks of the base of the leaf primordium and partially encircle the shoot apex (Gonzáles and Rudall 2001). Epidermis on the abaxial surface papillate. Stomata anomocytic. Camphor oil and coumaric acid present in leaves. Flowers small, solitary in the axils or more often in 2–4-flowered axillary monochasia, bisexual and unisexual (gynomonoecious), cyclic, 3-merous, apetalous. Sepals three, free, imbricate, membranous. Stamens six in two cycles, very short and wide, narrowly laminar, tetrasporangiate, with abaxial, nearly marginal disporangiate anther halves and shortly prolonged connective; in some flowers inner stamens transformed into staminodia. Anthers basifixed. Pollen grains in permanent tetrahedral tetrads of calymmate type, monocolpate with poorly defined lenticular or ovoid aperture (Carlquist 1964); the tectum has minute perforations and an infratectal region of granules which are sometimes fused to form incipient columellae. The ektexine is continuous across the external walls of neighbouring pollen grains, but only tenuous links occur between the internal walls. There is a poorly developed foot layer, which is absent, or has a granular appearance, in some regions. Endexine is lacking and the intine

has an outer region containing tubular material (Sampson 1995). Gynoecium of three carpels adaxially more or less connate at the base and gradually narrowing above into stylodium with short decurrent stigma consisting of two lines of papillae lining the ventral slit. Ovules (4)5-6(-8) in each carpel, on a long funiculus, arranged alongside the carpel margins, anatropous and with the micropyle directed upward, bitegmic (with both integuments lobed), almost tenuinucellate (but with 2–4-layered nucellar cup), with an endothelium, cup-like hypostase and active integumentary tapetum. Endosperm cellular, with chalazal haustorium. Fruits trifollicles. Seeds small, elongate, with membranous and reticulate testa, very small embryo with two minute cotyledons; endosperm copious, oily with chalazal haustorium. Exhibit flavonoid constituents consisting of 6, 3-0-diglycosides of the flavonols kaempferol and isorhamnetin. n = 20 (probably tetraploid of x = 10, Tobe et al. (1993)... 1. LACTORIDACEAE.

1 Pollen grains 2-celled.

2 Seeds with copious perisperm.

3 Pollen grains monocolpate. Flowers bisexual (in *Houttuynia* bisexual and male). Gynoecium of (3)4(5) carpels, these conduplicate and free above the connate base in *Saururus*, united into a paracarpous gynoecium in the other genera; ovary superior (*Saururus, Houttuynia*) or inferior (*Anemopsis, Gymnotheca*), stylodia free, not wholly closed, with decurrent stigma. Ovules (1)2(3) and laminal-lateral in *Saururus*, 6-10(-13) on each placenta in the other genera; micropyle zig-zag. Endosperm cellular, with chalazal haustorium. Fruits of four basally connate indehiscent fruitlets (*Saururus*) or apically dehiscent capsules. Seed coat formed by both integuments, but the main mechanical layer derived from the inner one; embryo minute, weakly differentiated. Vascular bundles arranged in one or sometimes two circles. Vessels with scalariform perforations and numerous bars, or less often (*Anemopsis*) with simple perforations; lateral pitting opposite or scalariform. Rhizomatous, aromatic or pungent, often stoloniferous herbs with secondary growth. Producing leucanthocyanins, alkaloids lacking, n = 11, 12 (*Saururus and Anemopsis*) and 9 (*Gymnotheca*). . . . . .2. SAURURACEAE

3 Pollen grains monocolpate or inaperturate. Gynoecium of several carpels or monocarpellate. Ovules solitary on each placenta or in each carpel and subbasal. Female gametophyte tetrasporic. Endosperm without chalazal haustorium. Vascular bundles in two or more circles. Lateral pitting of the vessels alternate.

4 Herbs, shrubs, woody lianas, or small trees, sometimes epiphytic, often with adhesive roots when juvenile, often aromatic. The outer vascular bundles united into a cylinder, the inner ones scattered in one or two cycles. Leaves usually alternate, often distichous or spirally arranged, sometimes opposite or verticillate, often pellucid-punctate, pinnately or palmatelyeined; stipules commonly present, free or adnate to the petiole or variously modified, rarely absent; petioles winged, often sheathing the mostly jointed stem. Flowers bisexual or rarely unisexual (*Macropiper*). Subtending bracts peltate, subpeltate, or laterally attached. Stamens 2–6, often 3. Anthers tetrasporangiate, with discrete locules. Pollen grains monocolpate, smooth and slightly rough (*Zippelia*), continuously verrucate (*Piper*). Gynoecium of (2-)3-5(-7) carpels, stigmas essentially apical. Ovules bitegmic. Female gametophyte of *Fritillaria*-type. Endosperm nuclear. Fruits drupaceous or baccate. Seed coat formed by the inner integument; embryo small, globular, undifferentiated or with small primordia of cotyledons, n = 13 (*Micropiper* and *Piper*), 19 (*Zippelia*). . . . . .3. PIPERACEAE.

4 Perennial or annual herbs, often epiphytic, commonly succulent. Vascular bundles separate and wholly scattered or in some species organized into more or less distinct circles. Leaves alternate, opposite, or verticillate, without stipules; petioles not winged, stem not jointed. Flowers bisexual. Subtending bracts usually orbicular. Stamens two. Anthers disporangiate and with eventually confluent locules. Pollen grains inaperturate or almost inaperturate. Gynoecium of 1-3(-4) carpels, stigma simple, apical or subapical, often penicillate. Ovules unitegmic. Female gametophyte of *Peperomia*-type.

Endosperm cellular. Fruits drupaceous. Seed coat formed by single integument; embryo small, in many species globular, undifferentiated, n = 11.......... 4. PEPEROMIACEAE

2 Seeds without perisperm. Rhizomatous herbs, subshrubs, or shrubs, erect, scrambling or scandent, sometimes tall lianas. Branches often slightly swollen and jointed at nodes. Nodes trilacunar. Vessels with simple perforations; lateral pitting alternate. Fibers with bordered pits. Rays in woody lianas limited to only interfascicular and wide and very high, dissecting the stem into discrete bundles; rays in *Thottea* heterogeneous and with short ends. Axial parenchyma usually paratracheal and rather scanty, sometimes apotracheal. Sieve-element plastids of S-type (species of *Thottea*) or more often of several different forms of P1/2c-type. Leaves alternate, distichous, simple, entire, often more or less cordate or reniform, sometimes 3-lobed, 2-lobed, or palmately to pedately lobed, pinnately or more often palmately veined, sometimes pellucid-punctate, estipulate. Stomata anomocytic. Prophylls solitary and median, in adaxial position. Flowers solitary or in rhipidia, terminal or axillary, sometimes on the older wood, bisexual, zygomorphic or seldom actinomorphic, basically 3-merous, entomophilous and often smelling of carrion. Calyx gamosepalous (in *Asarum caulescens* only slightly united at base), actinomorphic, 3-lobed and campanulate, urceolate or cup-shaped or often zygomorphic, tubular or pitcherlike, straight, curved, S-shaped or pipe-shaped, 3 (-6)-lobed or I- lipped and entire, often large and bizarrely colored, fetid. Petals usually absent, very rarely well developed (*Saruma*) or vestigial (three minute teeth alternating with the calyx lobes in some species of *Asarum)*. Stamens (4 or 5)6-12(-36, rarely more than 40), in one or two cycles or very rarely (in some species of *Thottea*) in three or four cycles, often with more or less prolonged connective, free or slightly mutually united at the base, and/or almost completely adnate to the style to form a gynostemium; filaments short, thick; anthers free or dorsally united to the style, tetrasporangiate, extrorse or those of the outer cycle nearly latrorse, opening longitudinally. Microsporogenesis simultaneous or successive. Pollen grains 2-celled, more or less globose, large to medium-sized, monocolpate (*Saruma*) or more commonly inaperturate, sometimes polyporate or, more rarely, polycolpate, semitectate-reticulate (*Saruma*) or tectate. Gynoecium of four (*Thottea*) or six carpels (other genera), commonly syncarpous or rarely (*Saruma*) nearly apocarpous; carpels medium to very large; stylodia free or connate to form a short and stout style with 3–6 lobes; stigmas decurrent in *Saruma* and *Asarum*, but more specialized in the other genera; ovary semiinferior (*Saruma*) or more often inferior, 4–6-locular or sometimes imperfectly locular, with several to usually many horizontal or pendulous ovules arranged submarginally in each locule or on each intruded placenta. Ovules anatropous (very rarely circinotropous), bitegmic (integuments mostly unlobed), crassinucellate, number of ovules 2-50(-150+) in each carpels. Female gametophyte of *Polygonum*-type. Fruits usually capsular, septicidal or rarely (in some species of *Aristolochia*) dehiscing apically toward the base (e.g., *Thottea*) or basally toward the apex (e.g., *most Aristolochia*), sometimes bursting irregularly (some species of *Asarum*). The fruit is rarely a semiapocarpous multifollicle (*Saruma*) or hard and indehiscent or consists of indehiscent I-seeded cocci (*Euglypha*). Seeds usually many in each locule or on each placenta, small to medium-sized, variously shaped, often coated with remains of placental tissue (membranous when dry); seed coat with the mechanical tissue derived from both integuments. The most characteristic layer of the seed coat is the inner epidermis of the outer integument, which consists of cells with often large, solitary or twin crystals of calcium oxalate and proximally thickened cell walls; embryo basal, minute, and sometimes undifferentiated; endosperm copious, fleshy, oily, and in some cases also starchy, sometimes weakly ruminate. Tend to deposit silica and calcium oxalate in their tissues, accumulating essential oils of taxon-specific composition in idioblasts and benzyl isoquinoline alkaloids and their degradation products, but lacking protoanthocyanins and ellagic acid....... 5. ARISTOLOCHIACEAE.

1. LACTORIDACEAE

Engler 1888. I/I. Masatierra of the Juan Fernández Islands.

Lactoris

The only living and highly endangered representative of the order – *Lactoris fernandeziana* – is one of the most remarkable archaic flowering plants. *Lactoris* is characterized by one unique character – saccate pollen grains (Carlquist 1964; Zavada and Taylor 1986), which together with the whole constellation of its traits makes it taxonomically rather isolated. However, some botanists have suggested a definite relationship between the Lactoridaceae and the Piperaceae. According to Weberling (1970), the leaf base and stipules of the Lactoridaceae are similar to those of the Piperaceae-Saururaceae alliance. According to Carlquist (1990), the wood of *Lactoris* is virtually identical with that of Piperaceae. In his opinion evidence available to date supports the placement of Lactoridaceae in Piperales, in which it would be more primitive than Piperaceae and Saururaceae.

2. SAURURACEAE

Martynov 1820. 4/7. From the Himalayas to Japan, Philippines and Indochina, North America and Mexico (*Anemopsis*). The most archaic genus *Saururus* (2) is disjunct between eastern Asia and eastern USA.

Saururus, Anemopsis, Houttuynia, Gymnotheca.

According to Meng et al. (2001), *Anemopsis* departs from the rest of Saururaceae.

3. PIPERACEAE

Giseke 1792. 7/1100. Pantropical.

Zippelia, Macropiper, Piper, Lindeniopiper, Trianaeopiper, Pothomorphe, Sarcorhachis.

Probably related to the Lactoridaceae. Carlquist (1993) listen ten distinctive wood features that unite Piperaceae with Lactoridaceae and Arisotlochiaceae. However, Piperaceae differ from Lactoridaceae markedly in vascular structure of the node and petiole, sieve-element plastids, stomata, absence of perianth, orthotropous and mostly crassinucellate ovules, presence of the copious perisperm, and in the anatomy of the seed coat.

4. PEPEROMIACEAE

A.C. Smith 1981. 4/1000. Pantropical.

Peperomia, Verbuellia, Manekia, Piperanthera.

Usually included in the Piperaceae, but clearly differ in many respects, including estipulate leaves, disporangiate anthers, monocarpellate gynoecium, unitegmic ovules, *Peperomia*-type female gametophyte, cellular endosperm, and basic chromosome number. Probably both Piperaceae and Peperomiaceae had a common origin from a *Saururus*-type ancestor.

5. ARISTOLOCHIACEAE

A.L. de Jussieu 1789 (including Asaraceae Ventenat 1799). 9/c 600. Widely distributed throughout tropical and temperate Eurasia, Africa, America, Australasia.

5.1 ASAROIDEAE

Herbs without twining habit and without uncinate hairs and lacking silicified cells. Flowers not constricted between perianth and ovary. Calyx actinomorphic. Petals present, vestigial, or absent. Stamens in two cycles, free or united with style. Ovary semi-inferior or inferior. Fruits multifollicles (*Saruma*), or capsules (*Asarum*). Elaiosome extending along the raphe. Rich in lignoids. n = 6, 12, 13, 18, 20, 26. – SARUMEAE: *Saruma*, ASAREAE: *Asarum* (including *Hexastylis*).

5.2 ARISTOLOCHIOIDEAE

Frequently twining woody or herbaceous plants, usually with uncinate hairs. Flowers constricted between perianth and ovary. Calyx mostly zygomorphic. Petals absent. Stamens mostly in one cycle, united with style. Ovary inferior. Fruits mostly dehiscent capsules. Rich in alkaloids based on benzyl isoquinoline. n = (4-)6-7(8+). – BRAGANTIEAE: *Asiphonia, Thottea*; ARISTOLOCHIEAE: *Isotrema, Pararistolochia, Euglypha, Holostylis, Aristolochia.*

Closely related to the Annonaceae (Wagner 1907; Wettstein 1924; Lang 1924; Thorne 1968, 1974, 1981, 1992a, b; Dahlgren 1980; Takhtajan 1980, 1987; Rohweder and Endress 1983; Ding Hou 1984; John et al. 1992), which is supported by adaxial prophyll, 3-merous flowers, prolonged connective, embryology (Samuelson 1914; Johri and Bhatnagar 1955; Wyatt 1955), ruminate endosperm, ultrastructure of sieve-element plastids (Behnke 1971, 1988), ethereal oils in the parenchymatous tissues, chemistry (Hegnauer 1960) and also karyology. According to Morawetz (1985), chromosome size and condensing behavior and chromatin and interphase nucleus structure in *Thottea* are very similar to those of Annonaceae. Both the Aristolochiaceae and Annonaceae possess a considerable number of monocotyledonous features and most probably derived from a common ancestor that gave rise to the Liliopsida. However, according to Huber (1990, 1993), they are nearer to the Myristicaceae although the relationship is not a particularly close one.

Bibliography

Baldacci A. 1894. Affinita delle Aristolochiaceae e dei genera aristolochiacei. Bull. Soc. Bot. Ital. 1894: 49–54.

Balfour E. 1957. The development of the vascular systems in *Macropiper excelsum* Forst.: 1. The embryo and seedling. Phytomorphology 7: 354–364.

Balfour E. 1958. The development of the vascular systems in *Macropiper excelsum* Forst.: II. The nature stem. Phytomorphology 8: 224–233.

Behnke H-D. 1971. Zum Feinbau der Siebröhrenplasti-den von *Aristolochia* und *Asarum* (Aristolochiaceae). Planta 97: 62–69.

Behnke H-D. 2002. Sieve-element plastids and evolution of Monocotyledons, with emphasis on Melanthiaceae sensu lato and Aristolochiaceae-Asaroideae, a putative Dicotyledon sister group. Bot. Rev. 68: 524–544.

Bernardello G, GJ Anderson, P Lopez, MA Cleland, TF Stuessy, and DK Crawford. 1999. Reproductive biology of *Lactoris fernandeziana* (Lactoridaceae). Am. J. Bot. 86: 829–840.

Blot J. 1960. Contribution a l'etude cytologique du genre *Peperomia*. Rev. Gen. Bot. 67: 522–535.

Bornstein AJ. 1989. Taxonomic studies in the Piperaceae – 1. The pedicellate Pipers of Mexico and central America (*Piper* subg. *Arctottonia*). J. Arnold Arbor. 70: 1–55.

Bornstein AJ. 1991. The Piperaceae in the southeastern United States. J. Arnold Arbor., Suppl. ser., 1: 349–366.

Bouman F. 1971. Integumentary studies in the Polycarpicae: 1. Lactoridaceae. Acta Bot. Neerl. 20: 565–569

Bowman TC. 1973. Comparative morphological investigations on the Aristolochiaceae. Ph.D. Thesis, Arizona State University.

Brantjes NBM. 1980. Flower morphology of *Aristolochia* species and the consequences of pollination. Acta Bot. Neerl. 29: 212–213.

Brauner S, DJ Crawford, and TF Stuessy. 1992. Ribosomal DNA and RAPD variation in the rare plant family Lactoridaceae. Am. J. Bot. 79: 1436–1439.

Brown WH. 1908. The nature of the embryo sac of *Peperomia*. Bot. Gaz. 46: 445–460.

Burger W. 1977. The Piperales and the monocots. Bot. Rev. 43: 345–393.

Campbell DH. 1901. The embryo sac of *Peperomia*. Ann. Bot. 15: 103–118.

Cariquist S. 1964. Morphology and relationships of Lactoridaceae. Aliso 4: 421–435.

Cariquist S. 1990. Wood anatomy and relationships of Lactoridaceae. Am. J. Bot. 77: 1498–1505.

Carlquist S. 1993. Wood and bark anatomy of Aristolochiaceae – systematic and habital correlations.. IAWA Bull. 14: 341–357.

Carlquist S, K Dauer, and SY Nishimura. 1995. Wood and stem anatomy of Saururaceae with reference to ecology, phylogeny, and origin of the monocotyledons. IAWA Bull. 16: 133–150.

Cammerloher H. 1923. Zur Biologic der Blüte *von. Aristolochia grandiflora* Swartz. Ber. Deutsch. Bot. Ges. 40: 385–393.

Correns C. 1891. Beiträge zur biologischen Anatomie der *Aristolochia*-Blüte. Jahrb. Wiss. Bot. 22: 161–189.

Crawford DJ, TF Stuessy, MB Cosner, D Haines, D Wiens, and P Penalillo. 1994. *Lactoris fernandeziana* (Lactoridaceae) on the Juan Fernandez Islands: allozyme uniformity and field observations. Conserv. Biol. 8: 277–280.

Crawford DJ, TF Stuessy, and MO Silva 1986. Leaf flavonoid chemistry and the relationships of the Lactoridaceae. Plant Syst. Evol. 153: 133–139.

Dasgupta A and PC Datta. 1976. Cytotaxonomy of Piperaceae. Cytologia 41: 697–706.

Datta PC and A Dasgupta. 1977. Comparison of vegetative anatomy of Piperales. Parts I and 2. Acta Biol. Acad. Sci. Hungar. 28: 81–96, 97–110.

Daumann E. 1959. Zur Kentniss der Blütennektarien *von Aristolochia*. Preslia 31:359–372.

De Figuereido RA and M Sazima. 2000. Pollination biology of Piperaceae species in southeastern Brazil. Ann. Bot. 85: 455–460.

Dickison WC. 1992. Morphology and anatomy of the flower and pollen of *Saruma henryi* Oliv.: A phylogenetic relict of the Aristolochiaceae. Bull. Torrey Bot. Club 119: 392–400.

Ding Hou. 1983. Florae Malesianae Precursores LXV: Notes on Aristolochiaceae. Blumea 29: 229–249.

Ding Hou. 1984. Aristolochiaceae. In: CGGJ van Steenis, ed. Flora Malesiana, ser. 1, 10 (1): 53–108. Dordrecht.

Edwards JG. 1920. Flower and seed of *Hediosmum nutans*. Bot. Gaz. 70: 409–424.

Engler A. 1886. Über Familien der Lactoridaceae. Bot. Jahrb. Syst. 8: 53–56.

ErdtmanG. 1964. Ein Beitrag zur Kenntnis den Pollenmorphologie von *Lactoris fernandeziana* und *Drimys winteri*. Grana Palynol. 5: 33–39.

Fagerlind F. 1940. Die Entwicklung des Embryosackes bei *Peperomia pellucida*. Arkiv Bot. 29 (17): 1–15.

Figueiredo RA de and M Sazima. 2000. Pollination biology of Piperaceae species in southeastern Brazil. Ann. Bot. 85: 455–460.

Fisher GC. 1914. Seed development in the genus *Peperomia*. Bull. Torrey Bot. Club 41: 137–156.

González F. 1997. Phylogenetic relationships of the subfamily Aristolochioideae (Aristolochiaceae). Am. J. Bot. 84(6): 198 (Abstract).

González F. 1999a. A phylogenetic analysis of the Aristolochioideae (Aristolochiaceae). Ph.D. thesis. The City University of New York.

González F. 1999b. Inflorescence morphology and the systematics of Aristolochiaceae. Syst. Geogr. Plant 68: 159–172.

Gonzáles F and P Rudall. 2001. The questionable affinities of *Lactoris*: evidence from branching pattern, inflorescence morphology, and stipule development. Am. J. Bot. 88: 214–2150.

González F and DW Stevenson. 2000a. Gynostemium development in *Aristolochia* (Aristolochiaceae). Bot. Jahrb. Syst. 122: 249–291.

González F and DW Stevenson. 2000b. Perianth development and systematics of *Aristolochia*. Flora 195: 370–391.

González FA and Stevenson DW. 2002. A phylogenetic analysis of the subfamily Aristolochioideae (Aristolochiaceae). Rev. Acad. Colomb. Cienc. Exact. Fis. Nat. 26(98): 25–57.

González F, PJ Rudall, and CA Furness. 2001. Microsporogenesis and systematics of Aristolochiaceae. Bot. J. Linn. Soc. 137: 221–242.

Gregory MP. 1956. A phyletic rearrangement in the Aristolochiaceae. Am. J. Bot. 43: 110–122.

Guedes M. 1968. La feuille vegetative et perianthe dequelques *Aristolochia*. Flora B 158: 167–179.

Hagerup O. 1961. The perianthium of *Aristolochia elegans* Mast. Bull. Res. Council Israel IOD: 348–351.

Hegnauer R. 1960. Chemotaxonomische Betrachtungen: II. Phytochemische Hinweise für die Stellung der Aristolochiaceae im System der Dicotyledonen. Die Pharmazie 15: 634–642.

Holm T. 1926. *Saururus cernuas* L.: A morphological study. Am. J. Sei., 5th ser., 12: 162–168.

Huber H. 1985. Samenmerkmale und Gliederung der Aristolochiaceen. Bot. Jahrb. Syst. 107: 277–320.

Huber H. 1993. Aristolochiaceae. In: K Kubitzki, ed., The families and genera of vascular plants, vol. 2, pp. 129–140. Springer, Berlin/Heidelberg/New York.

Jacobsson-Stiasny E. 1918. Zur Embryologie der Aristolochiaceae. Denkschr. Akad. Wiss. Wien, Math.-Naturw. Kl., 95: 1–13.

Jaramillo MA and PS Manos. 2001. Phylogeny and patterns of floral diversity in the genus *Piper* (Piperaceae). Am. J. Bot. 88: 706–716.

Jaramillo MA, PS Manos, and EA Zimmer. 2004. Phylogenetic relationships of the perianthless Piperales: Reconstructing the evolution of floral development. Int. J. Plant Sci. 165: 403–416.

Johnson DS. 1900. On the endosperm and embryo of *Peperomia pellucida*. Bot. Gaz. 30: 1–11.

Johnson DS. 1902. On the development of certain Piperaceae (*Piper adunca, P medium, Heckeria umbellata*). Bot. Gaz. 34: 321–340.

Johnson DS. 1914. Studies on the development of the Piperaceae: II. The structure and seed development of *Peperomia hispidula*. Am. J. Bot. 1: 323–329, 357–397.

Johri BM and SP Bhatnagar. 1955. A contribution to the morphology and life history of *Aristolochia*. Phyto-morphology 5: 123–137.

Kamelina OP. 1997. An addition to the embryology of Lactoridaceae and Fouquieriaceae. Bot. Zhurn. 82: 25–29 (in Russian with English summary).

Kanta K. 1962. Morphology and embryology of *Piper nigrum* L. Phytomorphology 12: 207–221.

Kelly LM. 1997a. Floral morphology, homology, and phylogenetic relationships in Aristolochiaceae. Am. J. Bot. 84(6): 207 (Abstract).

Kelly LM. 1997b. A cladistic analysis of *Asarum* (Aristolochiaceae) and implications for the evolution of herkogamy. Am. J. Bot. 84: 1752–1765.

Kelly LM. 1998. Phylogenetic relationships in *Asarum* (Aristolochiaceae) based on morphology and ITS sequences. Am. J. Bot. 85: 1454–1467.

Kelly LM and F González. 2003. Phylogenetic relationships in Aristolochiaceae. Syst. Bot. 28: 236–249.

Kubitzki K. 1993. Lactoridaceae. In: K Kubitzki, ed. The families and genera of vascular plants, vol. 2, pp. 359–361. Springer, Berlin/Heidelberg/New York.

Lammers TG, TF Stuessy, and MO Silva. 1986. Systematic relationships of the Lactoridaceae: An endemic family of the Juan Fernandez Islands, Chile. Plant Syst. Evol. 152: 3–4.

Lebot V and J Levesque. 1989. The origin and distribution of Kava (*Piper methysticum* Forst.f., Piperaceae): A phytochemical approach. Allertonia 5: 223–281.

Lei L-G and H-X Liang. 1998a. Floral development in dioecious species and trends of floral evolution in *Piper sensu lato*. Bot. J. Linn. Soc. 127: 225–237.

Lei L-G and H-X Liang. 1998b. Pollen morphology and its taxonomic significance of Piperaceae. Acta Bot. Yunn. 20: 427–433.

Lei L-G and H-X Liang. 1999. Variations in floral development in *Peperomia* (Piperaceae) and their taxonomic implications. Bot. J. Linn. Soc. 131: 423–431.

Lei L-G, Z-Y Wu, and H-X Liang. 2002. Embryology of *Zippelia begoniaefolia* (Piperaceae) and it systematic relationships. Bot. J. Linn. Soc. 140: 49–64.

Leinfellner W. 1953. Die hypopeltaten Brakteen von *Peperomia*. Oesterr. Bot. Z. 100: 601–615.

Leins P and C Erbar. 1985. Ein Beitrag zur Blütenent-wicklung der Aristolochiaceen, einer Vermittlergruppe zu den Monokotylen. Bot. Jahrb. Syst. 107: 343–368.

Leins P and C Erbar. 1995. Das frühe Differenzierungsmuster in den Blüten von *Saruma henryi* Oliv. (Aristolochiaceae). Bot. Jahrb. Syst. 117: 365–376.

Leins P, C Erbar, and WA van Heel. 1988. Note on the floral development of *Thottea* (Aristolochiaceae). Blumea 33: 357–370.

Liang H-X. 1991. Karyomorphology of *Gymnotheca* and phylogeny of four genera in Saururaceae. Acta Bot. Yunn. 13: 303–307.

Liang H-X. 1992. Study on the pollen morphology of Saururaceae. Acta Bot. Yunn. 14: 401–404.

Liang H-X. 1994. On the systematic significance of floral organogenesis in Saururaceae. Acta Phytotax. Sin. 32: 425–432.

Liang H-X. 1995. On the evolution and distribution in Saururaceae. Acta Bot. Yunn. 17: 255–267.

Liang H-X and SC Tucker. 1989. Floral development in *Gymnotheca chinensis* (Saururaceae). Am. J. Bot. 76: 806–819.

Liang H-X and SC Tucker. 1990. Comparative studies of the floral vasculature in Saururaceae. Am. J. Bot. 77: 607–623.

Liang H-X and SC Tucker. 1995. Floral ontogeny of *Zippelia begoniaefolia* and its familial affinity: Saururaceae or Piperaceae? Am. J. Bot. 82: 681–687.

Ling HX, KY Pan, and ZD Chen. 1996. Floral organogenesis in *Saururus chinensis* (Saururaceae). Acta Phytotax. Sinica 34: 565–568.

Lorch JW. 1959. The perianth of *Aristolochia*: A new interpretation. Evolution 13: 415–416.

Majumdar GP and P Pal. 1961. Developmental studies: VI. The morphology of the so-called stipule of *Piper*, etc. Proc. Nad. Inst. Sci. India 27: 26–39.

Ma Jin-shuang. 1990. The geographical distribution and the system of Aristolochiaceae. Acta Phytotax. Sinica 28(5): 345–355.

Mathew PJ and Mathew PM. 2001. Pollen morphology of some members of *Piperaceae* and its bearing on the systematics and phylogeny of the family. Rheedea 11(2): 65–78.

Mathew PJ, PM Mathew, and P Pushpangadan. 1999. Cytology and its bearing on the systematics and phylogeny of the Piperaceae. Cytologia (Japan). 64: 301–307.

Meeuse ADJ. 1971. Interpretative gynoecial morphology of the Lactoridaceae and the Winteraceae: a re-assessment. Acta Bot. Neerl. 20: 221–238.

Meeuse ADJ. 1972. Taxonomic affinities between Piperales and Polycarpicae and their implications in interpretative floral morphology. Adv. Plant Morph. 1972: 3–27.

Meng SW, ZD Chen, DZ Li, and HX Liang. 2002. Phylogeny of Saururaceae based on mitochondrial *mat*R gene sequence data. J. Plant Res. 115(1118): 71–76

Meng S-W, D-Z Li, H-Z Liang. 2001. The phylogeny of Saururaceae based on 5.8S rDNA sequences. Acta Bot. Yunn. 23: 309–312.

Meng SW, and H Liang. 1997. Comparative embryology on Saururaceae. Acta Bot. Yunn. 19(1): 67–74.

Metcalfe CR. 1987. Lactoridaceae. In: CR Metcalfe, ed. Anatomy of the dicotyledons, 2nd ed., vol. 3, pp. 147–151. Claredon Press, Oxford.

Mi Qiu-wen and Yang Chun-shu. 1991. Pollen morphology of *Asarum* in China. Acta Phytotax. Sinica 29(2): 164–171.

Miyoshi N and H Kato. 1982. Pollen morphology by means of scanning electron microscope: 5. Angiospermae (Piperales, Podostemonales). Jpn. J. Palynol. 28: 7–11.

Mohana Rao PR. 1989. Seed and fruit anatomy in *Aristolochia* and *Asarum* with a discussion of the affinities of Aristolochiaceae. Swamy Bot. Club 6(3–4): 105–119.

Morawetz W. 1985. Beiträge zur Karyologie und Systematik der Gattung *Thottea* (Aristolochiaceae). Bot. Jahrb. Syst. 107: 1–4.

Murty YS. 1958. Studies in the order Piperales: II. A contribution to the study of vascular anatomy of the flower of *Peperomia*. J. Indian Bot. Soc. 37: 474–491.

Murty YS. 1959a. Studies in the order Piperales: III. A contribution to the study of floral morphology of some species of *Peperomia*. J. Indian Bot. Soc. 38: 120–139.

Murty YS. 1959b. Studies in the order Piperales: V. A contribution to the study of floral morphology of some species of *Piper*. VI. A contribution to the study of floral morphology of *Pothomorphe umbellata* (L.) Miq. Proc. Indian Acad. Sci. 49B: 52–65, 82–85.

Murty YS. 1959c. Studies in the order Piperales: VII. A contribution to the study of morphology of *Saururus cernuus* L. J. Indian Bot. Soc. 38: 195–203.

Murty YS. 1960a. Studies in the order Piperales: 1. A contribution to the study of vegetative anatomy of some species of *Peperomia*. Phytomorphology 10: 50–59.

Murty YS. 1960b. Studies in the order Piperales: VIII. A contribution to the morphology of *Houttuynia cordata* Thunb. Phytomorphology 10: 329–341.

Nair NC and KR Narayanan. 1962. Studies on the Aristolochiaceae: I. Nodal and floral anatomy. Proc. Natl. Inst. Sci. India 28B: 211–227.

Neinhuis C, KW Hilu, and T Borsch. 2000. Systematics of Aristolochiaceae: Molecular evidence. Am. J. Bot. 87(Suppl. 6): 146.

Neinhuis C, S Wanke, KW Hilu, K Müller, and T Borsch. 2005. Phylogeny of Aristolochiaceae based on parsimony, likelihood, and Bayesian analyses of *trn*L-*trn*F sequences. Plant Syst. Evol. 250: 7–26.

Nickrent D, A Blarer, Y-L Qiu, DE Soltis, PS Soltis, and M Zanis. 2002. Molecular data place Hydnoraceae with Aristolochiaceae. Am. J. Bot. 89: 1809–1817.

Nikiticheva ZI. 1981. Embryological features of some Piperales. Acta Soc. Bot. Polon. 50: 329–332.

Nikiticheva ZI. 1988. Piperaceae, Peperomiaceae. In: A Takhtajan, ed. Comparative seed anatomy, vol. 2, pp. 97–104. Nauka, Leningrad (in Russian).

Nikiticheva ZI, MS Yakovlev, and TA Plyushch. 1981. The development of ovule, embryo sac, embryo, and endosperm in some species of the genus *Peperomia* (Piperaceae). Bot. Zhurn. 66: 513–523 (in Russian).

Nishida R, JD Weintraub, P Feeny and H Fukami. 1993. Aristolochic acids from *Thottea* spp. (Aristolochiaceae) and osmeterial secretions of *Thottea*-feeding troidine swallowtail larvae (Papilionidae). J. Chem. Ecol. 19: 1587–1594,

Okada H. 1986. Karyomorphology and relationships in some genera of Saururaceae and Piperaceae. Bot. Mag. Tokyo 99: 289–299.

Pant DD and R Banerji. 1965. Structure and ontogeny of stomata in some Piperaceae. Bot. J. Linn. Soc. 59: 223–228.

Parmer VS, et al. 1998. Polyphenols and alkaloids from *Piper* species. Phytochemistry 49: 1069–1078.

Pfluge GM. 1956. Phyletic rearrangement in the Aristolochiaceae. Am. J. Bot. 43: 110–122.

Plisko MA. 1988. Saururaceae. In: A Takhtajan, ed. Comparative seed anatomy, vol. 2,pp. 93–96. Nauka, Leningrad (in Russian).

Plyushch TA. 1982a. Ultrastructure of *Peperomia blanda* L. (Piperaceae) embryo sac. Ukrain. Bot. Zhur. 39(4): 88–91 (in Russian with English summary).

Plyushch TA. 1982b. Ultrastructure of *Peperomia blanda* (Piperaceae) embryo sac in the process of fertilization. Ukrain. Bot. Zhurn. 39 (6): 30–36 (in Russian with English summary).

Pontieri V and TL Sage. 1997. Characterization of pollen/carpel interaction following self and crosspollination in the paleoherb family: Saururaceae. Am. J. Bot. 84(Suppl. 6): 65.

Prakash N, JF Brown, and Y-H Wang. 1994. An embryological study of kava, *Piper metjysticum*. Austral. J. Bot. 42: 231–237.

Quibell CH. 1941. Floral anatomy and morphology of *Anemopsis californica*. Bot. Gaz. 102: 749–758.

Raju MVS. 1961. Morphology and anatomy of the Saururaceae: 1. Floral anatomy and embryology. Ann. Missouri Bot. Card. 48: 107–124.

Remizowa M, PJ Rudall, and D Sokoloff. 2005. Evolutionary transitions among flowers of perianthless Piperales: inferences from inflorescence and flower development in the anomalous species *Peperomia fraseri* (Piperaceae). Int. J. Plant Sci. 166: 925–943.

Renuka C and K Swarupanandan. 1986. Morphology of the flower in *Thottea siliquosa* and the existence of staminodes in Aristolochiaceae. Blumea 31: 313–318.

Rohweder O and E Treu-Koene. 1971. Bau und morphologische Bedeutung der Infloreszenz von *Houttuynia cordata* Thunb. (Saururaceae). Vierteljahrsschr. Naturf. Ges. Zürich 116: 195–212.

Sampson FB. 1995. Pollen morphology of Lactoridaceae: A re-examination. Grana 34: 100–107.
Samuel R and W Morawetz. 1989. Chromosome evolution within Piperaceae. Plant Syst. Evol. 166(1–2): 105–117.
Samuelson G. 1914. Über die Pollenentwicklung von *Annona* und Aristolochia und ihre systematische Bedeutung. Svensk Bot. Tidskr. 8: 181–189.
Sastrapradja S. 1968. On the morphology of the flower in *Peperomia* (Piperaceae) species. Ann. Bogor. 4: 235–244.
Schmidt OC. 1935. Aristolochiaceae. In: A Engler und K Prantl, eds. Die natürlichen Pflanzenfamilien, 2nd ed., vol. 16b, pp. 204–242. Engelmann. Prantl. Lepipzig..
Schmitz F. 1872. Die Blüten-Entwicklung der Piperaceen. Bot. Abhand. Morphol. Physiol. 2(8): 1–74.
Semple KS. 1974. Pollination in Piperaceae. Ann. Missouri Bot. Gard. 61: 868–871.
Skottsberg C. 1953. The vegetation of the Juan Fernandez Islands. The Nat. Hist. of the Juan Fernandez and Easter Is. 2: 793–960.
Small JK. 1931. The wild pepper plants of continental United States. J. New York Bot. Gard. 32: 210–223.
Solereder H. 1889. Beiträge zur vergleichenden Anatomie der Aristolochiaceen. Engl. Bot. Jahrb. 10: 410–524.
Stuessy TF, DJ Crawford, GJ Anderson, and RJ Jenner. 1998. Systematics, biogeography and conservation of Lactoridaceae. Perspectives in Plant Evol. Syst. 1–2: 267–290.
Sugawara T. 1987. Chromosome number of *Saruma henryi* Oliver (Aristolochiaceae). Bot Mag. Tokyo 100: 99–102.
Täckholm G and E Söderberg. 1918. Neue Beispiele der simultanen und successiven Wandbbildung in den Pollenmutterzellen. Svensk Bot. Tidskr. 12: 189–201.
Tanaka H. 1979. Pollination in *Saururus chinensis* (Lour.) Baill. J. Jpn. Bot. 54: 221–224.
Tebbs MC. 1993. Piperaceae. In: K Kubitzki, ed. The families and genera of vascular plants, vol. 2, pp. 516–520. Springer, Berlin/Heidelberg/New York.
Tobe H, TF Stuessy, PH Raven, and K Oginuma. 1993. Embryology and karyomorphology of Lactoridaceae. Am. J. Bot. 80: 933–946.
Trelease W. 1922. The peltate *Peperomia* of North America. Bot. Gaz. 73: 133–146.
Trelease W and TG Yunker. 1950. The Piperaceae of northern South America. Urbana.
Tucker SC. 1975. Floral development in *Saururus cernuus* (Saururaceae): 1. Floral initiation and stamen development. Am. J. Bot. 62: 993–1007.
Tucker SC. 1976. Floral development in *Saururus cernuus* (Saururaceae): II. Carpel initiation and floral vasculature. Am. J. Bot. 63: 289–301.
Tucker SC. 1979. Ontogeny of the inflorescence of *Saururus cernuus* (Saururaceae). Am. J. Bot. 66: 227–236.
Tucker SC. 1980. Inflorescence and flower development in the Piperaceae: 1. *Peperomia*. Am. J. Bot. 67: 686–702.
Tucker SC. 1981. Inflorescence and development in *Houttuynia cordata* (Saururaceae). Am. J. Bot. 68: 1017–1032.
Tucker SC. 1982a. Inflorescence and flower development in the Piperaceae: II. Floral ontogeny of *Piper*. Am. J. Bot. 69: 1389–1401.
Tucker SC. 1982b. Inflorescence and floral ontogeny in the Piperaceae: III. Floral ontogeny of *Piper*. Am. J. Bot. 69: 743–752.
Tucker SC. 1985. Initiation and development of inflorescence and flower in *Anemopsis californica* (Saururaceae). Am. J. Bot. 72: 20–31.
Tucker SC and AW Douglas. 1996. Floral structure, development, and relationships of paleoherbs: *Saruma, Cabomba, Lactoris*, and selected Piperales, pp. 141–175. In: DW Taylor, LJ Hickey, eds. Flowering plant origin, evolution and phylogeny. Chapman & Hall, New York.
Tucker SC, AW Douglas, and H-X Liang. 1993. Utility of ontogenetic and conventional characters in determining phylogenetic relationships of Saururaceae and Piperaceae (Piperales). Syst. Bot. 18 (4): 614–641.
Vovk AG and GA Komar. 1988. Aristolochiaceae. In: A Takhtajan, ed. Comparative seed anatomy, vol. 2, pp. 105–111. Nauka, Leningrad (in Russian).
Wagner R. 1907. Zur Kentnis des *Saruma henryi* Oliv. Oesterr. Bot. Z. 57: 265–271.
Wanke S and C Neinhuis. 2006. Systematics of pipevines: Combining morphological and fast-evolving molecular characters to investigate the relationships within subfamily Aristolochioideae (Aristolochiaceae). Int. J. Plant Sci. 167: 1215–1227.
Wanke S, M-S Samain, L Vanderschaeve, G Mathieu, P Goetghebeur, and C Neinhuis. 2006. Phylogeny of the genus *Peperomia* (Piperaceae) inferred from the *trn*K/*mat*K region (cpDNA). Plant Biol. 8: 93–102.
Wanke S, MA Jaramillo, T Borsch, M-S Samain, D Quandt, and C Neinhuis. 2007. Evolution of Piperales – *mat*K gene and *trn*K intron sequence data reveal lineage specific resolution contrast. Mol. Phyl. Evol. 42: 477–497.
Weberling F. 1970. Weitere Untersuchungen zur Morphologie des Unterblattes bei den Dikotylen: V. Piperales. Beitr. Biol. Pfl. 46: 403–434.
Weisse A. 1927. Zur Kenntnis von Blattstellung und Blütensstand der Aristolochiaceen. Ber. Deutsch. Bot. Ges. 45: 229–244.
Wood CE. 1971. The Saururaceae in the southeastern United States. J. Arnold Arbor. 52: 479–485.
Wu C-Y and K Kubitzki. 1993. Saururaceae. In: K Kubitzki, ed. The families and genera of vascular plants, vol. 2, pp. 586–588. Springer, Berlin/Heidelberg/New York.
Wyatt RI. 1955. An embryological study of four species of *Asarum*. J. Elisha Mitchell Sei. Soc. 71: 64–82.
Yoshida O. 1957, 1959, 1960, 1961. Embryologische Studien über die Ordnung Piperales. Parts 1–5. J. Coll. Arts Chiba Univ. 2: 172–178, 1957; 295–303, 1959; 3: 56–60, 155–162, 1960; 311–316, 1961.
Yuncker TG. 1958. The Piperaceae: A family profile. Brittonia 10: 1–7.
Zavada MS and JM Benson. 1987. First fossil evidence for the primitive angiosperm family Lactoridaceae. Am. J. Bot. 74: 1590–1594.
Zavada MS and TN Taylor. 1986. Pollen morphology of Lactoridaceae. Plant Syst. Evol. 154: 31–39.
Zhang S-S, S-B HO, and Y Wang. 1984. A study on the anatomy of vegetative organs of the *Gymnotheca* Decne. (Saururaceae) in relation to its systematic position. Acta Phytotax. Sinica 22: 49–52 (in Chinese).
Zhitkov VS. 1977. Forms of phyllotaxis in the genus *Peperomia* Ruiz et Pav. And their morphogenesis. Bull. Mosc. Soc. Nat., Biol. ser. 82: 103–119 (in Russian with English summary).

Order 15. HYDNORALES

Terrestrial parasitic herbs living on the roots of various trees and shrubs. Vegetative body consists of two kinds of roots: the coarse rhizomelike pilot-roots, which traverse the soil and appear to have a root cup, and endogenously emerging from the pilot-root small unbranched, slender, lateral haustorial roots of limited growth, which have a definite root cup. Vessels, when present, with simple perforations. Parenchymatous tissues of the pilot-root with scattered tanniniferous mucilage cells or sometimes lysigenous mucilage canals or cavities containing catechin. Sieve-element plastids of So-type, i.e., devoid of any prominent contents, difficult to recognize, and of different sizes. Leaves absent. Flowers solitary, arising endogenously from the pilot-roots, rather large, usually bisexual (functionally unisexual in *Hydnora esculenta*), actinomorphic, apetalous, malodorous, cantharophilous. Sepals 3-4(5), thick and fleshy, with a coarse, cracked, brown exterior, valvate, connate below. Stamens as many as sepals and opposite to sepals in *Hydnora sessile*; anthers forming a lobed ringlike synandrium on the calyx tube, the lobes opposite the sepals. In *Prosopanche* the very short filaments arising from the calyx tube and the anthers connate to form a dome or cap with a small central opening; anthers elongate, extrorse, opening longitudinally. Tapetum secretory. Microsporogenesis successive. Pollen grains 2-celled, 1-colpate *(Hydnora)* or 2–3-colpate or trichotomocolpate (*Prosopanche*), with homogeneous ectexine, nearly psilate. Small, fleshy staminodia present between androecium and gynoecium and alternate with the stamens in *Prosopanche*. Gynoecium of 3(4-5) carpels; ovary inferior, unilocular, with numerous ovules embedded in much-branched placentas suspended from the top of the ovary (in *Hydnora* individual placentas are deeply intruded; in *Prosopanche* radially arranged placental lamellae fill the cavity of the ovary). Stigma is flat with numerous ridges in parallel series. Ovules numerous (ca. 35,000 in each ovary in *Prosopanche americana*), orthotropous, unitegmic (single massive integument in *Hydnora*, but scarcely differentiated from the placenta and recognizable only in the micropylar area in *Prosopanche*), tenuinucellate. Female gametophyte bisporic, *Allium*-type (*Prosopanche*) or tetrasporic, *Adoxa*-type (*Hydnora*). Endosperm cellular. Fruits rather massive, with thick, leathery pericarp and fleshy interior, edible (the placental tissues contain much starch), in *Prosopanche* bursting in circumscissile fashion when ripe. Seeds very numerous, minute, with extremely hard, exotestal seed coat and minute, undifferentiated embryo surrounded by copious endosperm with polysaccharide food reserves and single layer of perisperm.

Closely related to the Aristolochiaceae, especially to the Asaroideae, and have probably originated directly from their immediate ancestors. The affinity with the Aristolochiaceae has been recently supported by molecular data (Nickrent et al. 2002).

1. HYDNORACEAE

C. Agardh 1821. 2/18. Drier parts of Africa, Reunion, Madagascar, and Saudi Arabia *(Hydnora)* as well as dry regions of Central and South America (*Prosopanche*).

Hydnora, Prosopanche.

Bibliography

Bruch C. 1923. Coleopteros fertilizadores de *Prosopanche burmeisteri* De Bary. Physis 7: 82–88.

Burkart A. 1963. Nota sobre *Prosopanche bonacinae* Speg. (Hydnoraceae) su area y parasitismo sobre algodon. Darwiniana 12: 633–638.

Chodat R. 1915. Les especes du genre *Prosopanche*. Bull. Soc. Bot. Geneve, ser. 2, 7: 65–66.

Chodat R. 1916. Hydnoraceae. Bull. Soc. Bot. Geneve, ser. 2, 8: 186–201.

Cocucci AE. 1965. Estudios en el genero *Prosopanche* (Hydnoraceae): I. Kurtziana 2: 53–74.

Cocucci AE. 1975. Estudios en el genero *Prosopanche* (Hydnoraceae): II. Kurtziana 8: 7–15.

Cocucci AE. 1976. Estudios en el genero *Prosopanche* (Hydnoraceae): III. Kurtziana 9: 19–39.

Cocucci AE. 1983. New evidence from embryology in angiosperm classification. Nord. J. Bot. 3: 67–73.

Cocucci AE. 1996. 1996. *Prosopanche* (Hydnoraceae): somatic and reproductive structures, biology, systematics, phylogeny and potentiality as a parasitic weed. In: MT Moreno, JC Cubero, eds. Advances in parasitic plant research, pp. 178–193. Junta de Andalucia, Sevilla.

Dastur RH. 1922. Notes on the development of the ovule, embryo sac, and embryo of *Hydnora ajricana*.Trans. Roy Soc. South Afr. 10: 27–31.

De Bary A. 1868. *Prosopanche burmeuteri*: Eine neue Hydnoreae aus Südamerika. Abhandl. Naturf. Ges. Halle 10: 243–272.

Harms H. 1935. Hydnoraceae. In: A Engler und K Prantl, eds. Die natürlichen Pflanzenfamilien, vol. 16b, pp. 282–295. Engelmann, Leipzig.

Kuijt J. 1969. The biology of parasitic flowering plants. University of California Press, Berkeley.

Meijer W. 1993. Hydnoraceae. In: K Kubitzki, ed. The families and genera of vascular plants, vol. 2, pp. 341–343. Springer, Berlin/Heidelberg/New York.
Musselman LJ and JH Visser. 1989. Taxonomy and natural history of *Hydnora* (Hydnoraceae). Aliso12(2): 317–326.
Nickrent DL, A Blarer, Y-L Qiu, DE Soltis, PS Soltis, and M Zanis. 2002. Molecular data place Hydnoraceae with Aristolochiaceae. Am. J. Bot. 89: 1809–1817.
Schimper AFW. 1880. Die Vegetationsorgane von *Prosopanche burmeuteri*. Abhandl. Naturf. Ges. Halle 15: 21–47.
Solms-Laubach HG. 1874. Über den Bau des Samens in den Familien der Rafflesiaceae und Hydnoraceae. Bot. Z. 32 (I): 337–342, 353–358, 369–374, 385–389.
Takhtajan AL, NR Meyer, and VN Kosenko. 1979. Morphology of pollen grains of the family Hydnoraceae in relation to its systematic position. Bot. Zhurn. 64: 1774–1777 (in Russian with English summary).
Terekhin ES. 1981. Hydnoraceae. In: MS Yakovlev, ed. Comparative embryology of flowering plants, pp. 95–96. Nauka, Leningrad (in Russian).
Terekhin ES. 1988. Hydnoraceae. In: A Takhtajan, ed. Comparative seed anatomy, vol. 2, pp. 112–113. Nauka, Leningrad (in Russian).
Tieghem P van. 1897. Sur la structure de l'ovule et de la graine chez les Hydnoracees. Jour. de Bot. 11: 233–238.
Visser J. 1981. South African parasitic plants. Juta, Capetown.

Superorder RAFFLESIANAE

Order 16. MITRASTEMONALES

Plants without chlorophyll, endoparasitic in the roots of other plants, often causing broomlike overgrowth of the host-root. Vegetative body much-dissected and largely filamentous, resembling a fungal mycelium, permeating the tissues of the root of the host, but not extending into the apical meristem. Leaves opposite, scale-like but sometimes fleshy and leaf waxes hummocky, the uppermost leaves tending to be somewhat cupped and accumulating nectar. Stomata absent. Flowers of medium size, terminal, solitary, bisexual, protandrous. Nectaries in the axils of the upper bracts. Perianth much reduced, collar-shaped, connate below to form a cup; the rim of the cap 4-lobed. Stamens connate into a tube (androphore); the staminal tube, open at the top by a small hole, circumscissaly separated for the flower as it is pushed up by the glowing gynoecium. The apical portion of the staminal tube is sterile. Anthers numerous, extrorse, sessile in several series of rings shortly below the stamen tube summit, 2-locular, opening longitudinally. Tapetum secretory. Microsporogenesis simultaneous. Pollen grains 2-celled, 3(4)-porate, ectexine reduced to tuberculae. Gynoecium with 9–15 intrusive parietal placentae; style short, stout, the stigma thick, depressed-conical. Ovary superior, 1-locular. Ovules numerous, anatropous, unitegmic (but integument with two cells layers), tenuinucellate, with a funicular obturator. Female gametophyte of *Polygonum*-type. Endosperm cellular. Fruits baccate or capsular and tardily opening by a horizontal slit. Seeds very numerous, minute; exotestal cells with U thickening; embryo undifferentiated, 4-celled, surrounded by endosperm; n = 20.

Usually including in the Rafflesiales, but differ from them in many respects, especially in superior ovary and cellular endosperm.

1. MITRASTEMONACEAE

Makino 1911. 1/2. Eastern Asia and Southeast Asia, Malesia, and America from Mexico to Guatemala and northwestern Colombia.

Mitrastemon.

Bibliography

Jochems SCJ. 1928. Die Verbreitung der Rafflesiaceengattung *Mitrastemon*. Rec. Trav. Bot. Neerl. 25A: 203–207.
Makino T. 1911. *Mitrastemon* – Mitrastemonaceae – Mitrastemonales. Bot. Mag. Tokyo 25: 251–257.
Matuda E. 1947. On the genus *Mitrastemon*. Bull. Torrey Bot. Club 74: 133–141.
Meijer W and JF Veldkamp. 1993. A revision of *Mitrastema* (Rafflesiaceae). Blumea 38: 221–229.
Mitra K, M Mondal. 1982. A note on the pollen morphology of *Mitrastemon yamamotoi* (Makino) Makino (Rafflesiaceae). Bangladesh J. Bot. 11(2): 179–181.
Nickrent DL, A Blarer, Q Yin-Long, R Vidal-Russell, and FE Anderson. 2004. Phylogenetic inference in Rafflesiales: the influence of rate heterogeneity and horizontal gene transfer. BMC Evol. Biol. 4: 40 – http://www.biomedcentral.com/14711-2148/4/40
Watanabe K. 1936. Morphologisch-biologische Studien liber die Gattung *Mitrastemon*: IV. J. Jpn. Bot. 12: 848–858.
Watanabe K. 1937. Morphologisch-biologische Studien über die Gattung *Mitrastemon*: VII. J. Jpn. Bot. 13: 154–162.

Order 17. RAFFLESIALES (CYTINALES)

Parasitic herbs, often fleshy, living on the roots and less often the stems of various trees and shrubs.

External roots and rhizomelike structures absent. The vegetative body is much branched, and the myceliumlike endophytic system is inside the tissues of the host. A very reduced, vestigial vascular system exists only in the fairly massive haustorial system and in the flower-bearing axis. The conducting elements are short, broad, spirally thickened tracheids, but the older elements are mostly devoid of transverse walls. So-plastids difficult to recognize and of varying sizes. Reduced scalelike leaves present on the emergent flowering shoot or around the base of the solitary flower, usually verticillate, sometimes opposite or even alternate. Stomata occur very rarely and are abnormal, mostly absent. Flowers from very small (Apodanthaceae) to very large or even gigantic (Rafflesiaceae), solitary and sessile or less frequently in short racemes or spikes, bisexual or more often unisexual (monoecious or dioecious), apetalous, actinomorphic, often malodorous. Sepals 4–5, sometimes up to 10 or more, in 1 or 2 whorls, free or more often connate below, sometimes petaloid. Stamens five to many (up to 100), connate by their filaments into a tube surrounding the stylar column, or more often adnate to the column, from which the anthers originate in one to several cycles; anthers unilocular or bilocular, tetrasporangiate, opening longitudinally or by transverse slits or by apical pore. Tapetum secretory. Microsporogenesis simultaneous. Pollen grains 2-celled, monocolpate, 3-colpate, or 1-polyporate or inaperturate; ectexine from homogeneous to columellate. Nectary sometimes present, adnate to the base of the style and base of the staminal column. Gynoecium of 4-10(-14) carpels; ovary inferior or semi-inferior, 1-locular with more or less intruded parietal placentas or irregularly multilocular with the ovules covering the surfaces of the placental partitions; stigma on a short style or sessile, distally expanded into an often large disc, capitate or multilobed, more or less papillose. Ovules very numerous, hemianatropous to anatropous or orthotropous (Cytinaceae), mostly bitegmic, usually with reduced short outer integument, tenuinucellate. Female gametophyte of *Polygonum*-type. Endosperm nuclear. Fruits usually more or less berry-like, indehiscent, dehiscent by horizontal slit, or irregularly dehiscent. Seeds very numerous, tiny, with an undifferentiated embryo surrounded by 1(-3)-layered endosperm; seed-coat hard, of a single layer. Perisperm absent.

Related to the Aristolochiaceae. The relationship with the Aristolochiaceae has been accepted by Robert Brown (1821), Solms-Laubach (1901), Hallier (1905b, 1908, 1912), Endress (1988, 1994), Bouman and Meijer (1994), and many others. According to Hallier, the family Rafflesiaceae is closest to the section *Heterotropa* of the genus *Asarum* and originated from the Aristolochiaceae as a result of parasitic reduction.

Key to Families

1 Ovules orthotropous. Flowers of medium size, in clusters or spike-like inflorescences, unisexual (monoecious or dioecious). Stems stout with bract-like leaves. Flowers with bracts and bracteoles, with more or less horseshoe-shaped nectary glands (*Cytinus*), Stamens 8–10, with united filaments, often with appendages above the anthers. Anthers 2-locular. Pollen grains free or in tetrads, 2–4-porate or 1–5-colpate, rugate or polyporate. Nectary disc present (*Cytinus*). Ovary inferior, with 6–8 parietal placentas. Style 1 with punctuate-radiate stigma or stigma disk-like (*Bdallophyton*) Ovules unitegmic, but with reduced outer integument (Tereckin 1988), micropyle endostomal. Contain pelargonidin 3-galactoside and petunidin 3-glucoside (Hegnauer 1997), n = 16. . . . . 3. CYTINACEAE

1 Ovules hemianatropous to anatropous.

2 Pollen grains triporate. Ovules bitegmic, anatropous, covering the whole of inner wall of the ovary or on 4–5 wide placentas. Endoparasitic, achlorophyllous, monoecious or dioecious herbs; their vegetative endophyte, often compared to a fungal mycelium, residents in the host; *Apodanthes* is known to parasitize Flacourtiaceae *Pilostyles* and *Berlinianche* parasitize a wide range of Fabaceae. Flowers small, not more than 4–5 mm in diameter, unisexual. Staminal structures laterally on apex of the central column. Stamens ca. 15, in 2–4 cycles. Anthers 1-locular, extrorse, opening by transverse slits. Pollen grains 3-colpate. Nectary disc present. Female flowers cone-like and broader than the male ones, with annular stigma below the apex of the column. Gynoecium of 4–5 united carpels; carpels opposite inner petals; Stigma hemispherical without separate stigmatic lobes and is almost completely covered with elongate unicellular papillae, each with a rounded apex, which are heavily secretory (*Berlinianche* – Blarer, Nickrent, and Endress

2004). Ovary inferior or semi-inferior, 1-locular, placentation parietal; ovules many, anatropous, micropyle bistomal or none. Fruits baccate, testa thin-walled, exotegmen massively lignified; endosperm present, embryo undifferentiated; n = 16, 30–31 + (*Pilostyles*). 1. APODANTHACEAE

2 Pollen grains unisulcate or uniporate. Flowers very large, solitary or in inflorescences, unisexual or (*Rhizanthes*) bisexual. Staminal structures under the rim of the expanded apex of the central column. Stamens in one cycle. Anthers with two to many locules. Pollen grains 1-porate or with reduced short colpi. *Rafflesia* and *Sapria* have no nectarines, but *Rhizanthes* has a nectary on the distal part of the perianth organs (Bänziger and Hansen 2000). Female flowers with stigma on or under the rim of apical disc. Ovary inferior. Ovules basically anatropous, unitegmic. Seeds with basal chalazal appendage. n = 11, 12. . . . . 2. RAFFLESIACEAE

1. APODANTHACEAE

Tieghem ex Takhtajan 1987. 3/23–30. America from southern California and southeastern USA to the Straits of Magellan, East Africa, western Asia (Asia Minor, Iraq, Iran), southwestern Australia.

Apodanthes, Pilostyles, Berlinianche.

Related to the Rafflesiaceae, but differ in pollen grains and bitegmic ovules.

2. RAFFLESIACEAE

Dumortier 1829. 3/18–20. Southern China, Assam, Bhutan, Thailand, and Indochina (*Sapria himalayana*) and Malesia.

Rafflesia, Sapria, Rhizanthes.

3. CYTINACEAE

A. Richard 1824. 2/8. South Africa, Madagascar, Mediterranean, Asia Minor, and western Caucasus (*Cytinus*), Central America (*Bdallophytori*).

Bdallophyton. Cytinus

Related to the Rafflesiaceae, but differ in orthotropous ovules.

Bibliography

Bänziger H. 1995. Ecological, morphological and taxonomic studies on Thailand fifth species of Rafflesiaceae: *Rhizanthes zippelii* (Blume) Spach. Nat. Hist. Bull. Siam Soc. 43: 337–365.

Bänziger H and B Hansen. 2000. A new taxonomic revision of a deceptive flower, *Rhizanthes* Dumortier (Rafflesiaceae). Nat. Hist. Bull. Siam Soc. 48: 117–143.

Beaman RS, PJ Decker, and JH Beaman. 1988. Pollination in *Rafflesia* (Rafflesiaceae). Am. J. Bot. 75: 1148–1162.

Blarer A, DL Nickrent, and PK Endress. 2004. Comparative floral structure and systematics in Apodanthaceae (Rafflesiales). Plant Syst. Evol. 245: 119–142.

Blarer A., DL Nickrent, H Bänziger, PK Endress, and Y-L Qui. 2000. Phylogenetic relationships among genera of the parasitic family Rafflesiaceae s.l. nased pm mic;ear ITS and SSU rDNA, mitochondrial LSU and SSU rDNA, *atp*1, and *mat*R sequences. Am. J. Bot. 87(Suppl. 6): 503.

Blarer A, DL Nickrent, and PK Endress. 2004. Comparative floral structure and systematics in Apodanthaceae (Rafflesiales). Plant Syst. Evol. 245: 119–142.

Bouman F and W Meijer. 1986. Comparative seed morphology in Rafflesiaceae. Acta Bot. Neerl. 35: 521–521.

Bouman F and W Meijer. 1994. Comparative structure of ovules and seeds in Rafflesiaceae. Plant Syst. Evol. 193: 187–212.

Brown R. 1821. An account of a new genus of plants named *Raffiesia*. Trans. Linn. Soc. London 13: 231–234.

Brown R. 1834. Description of the female flower and fruit of *Rafflesia arnoldi*, with remarks on its affinities and an illustration of the structure of *Hydnora africana*. Trans. Linn. Soc. London 19: 221–247.

Cartellieri E von. 1926. Das Absorptionssystem der Rafflesiaceae *Brugmansia*. Bot. Arch. 14: 284–311.

Chodat R and C Bernard. 1902. Embryologie du *Cytinus hypocistis*. Arch. Sci. Phys. et Nat. Geneve, 13: 1–6.

Davis CC, M Latvis, DL Nickrent, KJ Wurdack, and DA Baum. 2007. Floral gigantism in Rafflesiaceae. Science 315: 1812.

Dell B, J Kuo, and A Brium. 1982. Anatomy of *Pilostyles hamiltonii* C. L. Gardner (Rafflesiaceae) in stems of *Daviesia*. Austral. J. Bot. 30: 1–9.

De Vattimo I. 1955. Notice sur la tribu Apodantheae R.Br. (Rafflesiaceae). Taxon 4: 211–212.

De Vattimo I. 1956. Notes on *Apodanthes caseariae* Poit. and *Pilostyles calliandrae* (Gardn.) R.Br. (Rafflesiaceae – Apodantheae). Not. Syst. 15: 225–229.

De Vattimo I. 1971. Contribução ao conhecimento da tribu Apodantheae R.Br. Parte 1 – Conspecto das especies (Rafflesiaceae). Rodriguesia 26: 37–62.

De Vattimo I. 1978. Uma nova especie de *Apodanthes* Poit. (Rafflesiaceae). Rodriguesia 29: 269–306.

Emmons LH, J Nias, and A Brium. 1991. The fruit and consumers of *Rafflesia keithii* (Rafflesiaceae). Biotropica 23: 197–199.

Ernst A and E Schmid. 1913. Über Blüte und Frucht von *Rafflesia*. Ann. Jard. Bot. Buitenz. 2, 12: 1–58.

Forstmeier L, F Weberling, and HC Weber. 1983. Zum Parasitismus von *Cytinus hypocistis* L. (Rafflesiaceae). Beitr. Biol. Pfl. 58: 299–311.

Garcia Franco JG, V Rico Gray. 1997. Reproductive biology of the holoparasitic endophyte *Bdallophyton bambusarum* (Rafflesiaceae). Bot. J. Linn. Soc. 123(3): 237–247.

Griffith W. 1845. On the root-parasites referred by authors to Rhizantheae: and on various plants related to them. Trans. Linn. Soc. London 19: 303–347.

Gusowska I. 1964. Reinvestigation of embryo sac development, fertilization, and early embryogeny in *Cytinus hypocistis*. Acta Soc. Bot. Polon. 33: 157–166.

Gusowska I. 1966. Microsporogenesis and chromosome number in *Cytinus hypocistis L.* Acta Soc. Bot. Polon. 35: 445–454.

Harms H. 1945. Rafflesiaceae. In: A Engler and K Prantl, eds. Die natürlichen Pflanzenfamilien, S. 243–281. 2 Aufl. Leipzig, Berlin.

Hegnauer R. 1997. Phytochemistry of Rafflesiaceae. In: C Kalkman et al., eds. Flora Malesiana, ser. I, 13: 10, Leiden.

Heinricher E. 1905. Beiträge zur Kenntnis der Rafflesiaceae: I. Denkschr. Akad. Wiss. Wien, Math.-Naturw. Kl., 78: 1–25, 57–81.

Heinricher E. 1917. Zur Kenntnis der Blüte von *Cytinus hypocistis*. Ber. Deutsch. Bot. Ges. 35(3): 513–517.

Heinricher E. 1934. Zur Frage der Artbildung bei *Cytinus hypocistis* nebst anderen Bemerkungen. Ber. Deutsch. Bot. Ges. 52(I): 48–53.

Hosseus CC. 1907. Eine neue Rafflesiaceengattung aus Siam. Bot. Jahrb. Syst. 41:55–61.

Justesen PT. 1922. Morphological and biological notes on *Rafflesia* flowers observed in the Highlands of Mid-Sumatra (Padangsche Bovenlanden). Ann. Jard. Bot. Buitenz. 32: 64–87.

Kuijt J, D Bray, and AR Olson. 1985. Anatomy and ultras tructure of the endophytic system of *Pilostyles thurberi* (Rafflesiaceae). Canad. J. Bot. 63: 1231–1240.

Mat Salleh K, A Latiff. 1995. On the morphology of the female flower of *Rafflesia tengku-adlinii* and notes on the status of *R.borneensis* (Rafflesiaceae). Flora Males. Bull. 11(6): 425–428.

Meijer W. 1958. A contribution to the taxonomy and biology of *Rafflesia arnoldi* in West Sumatra. Ann. Bogor. 3: 33–44.

Meijer W. 1985. Saving the world's largest flower. Natl. Geogr. Mag. 168(I): 136–140.

Meijer W. 1993. Rafflesiaceae. In: K Kubitzki, ed. The families and genera of flowering plants, vol. 2, pp. 557–563. Springer, Berlin/Heidelberg/New York.

Meijer W. 1997. Rafflesiaceae. In: C Kalkman et al., eds. Flora Malesiana, ser. 1, 13: 1–42. Leiden.

Meijer W and JF Veldkamp. 1988. A revision of *Rhizanthes* (Rafflesiaceae). Blumea 31: 329–342.

Nickrent DL, A Blarer, Q Yin-Long, R Vidal-Russell, and FE Anderson. 2004. Phylogenetic inference in Rafflesiales: the influence of rate heterogeneity and horizontal gene transfer. BMC Evol. Biol. 4: 40 – http://www.biomedcentral.com/14711-148/4/40

Olah L. 1960. Cytological and morphological investigations in *Rafflesia arnoldi* R. Br. Bull. Torrey Bot. Club 87: 406–416.

Ponzi R and P Pizzolongo. 1976. *Cytinus hypocistis* L. Embryogenesis: Ultrastructural aspect of megasporo-genesis and megagametogenesis. J. Submicroscop. Cytol. 8: 327–336.

Ponzi R and P Pizzolongo. 1982. *Cytinus hypocistis* em-bryogenesis: Some biological and ultrastructural aspects of fertilization and embryo development. Nuovo Giorn. Bot. Ital. 116: 149–166.

Royen P van. 1963. Sertulum Papuanum: 8. Rafflesiaceae. Nova Guinea Bot. 14: 243–245.

Rutherford RJ. 1970. The anatomy and cytology of *Pilostyles thurberi* Gray (Rafflesiaceae). Aliso 7: 263–288.

Schaar F. 1898. Über den Bau des Thallus von *Rafflesia rochussenii* Teijsm et Binn. Sitzungsber. K. Akad. Wiss. Wien, Math.-Naturw. Kl., 107: 1039–1056.

Solms-Laubach HG. 1874a. Über den Thallus von *Pilostyles haussknechtii*. Bot. Z. 32: 44–59, 65–74.

Solms-Laubach HG. 1874b. Über den Bau des Samens in den Familien der Rafflesiaceae und Hydnoraceae. Bot. Z. 32: 337–342, 353–358, 369–374, 385–389.

Solms-Laubach HG. 1875. Das Haustorium der Loranthaceen und der Thallus der Rafflesiaceen und Balanophoreen. Abhandl. Naturf. Ges. Halle 13(I): 1–40.

Solms-Laubach HG. 1876. Die Entwicklung der Blüten bei *Brugmansia zippelii* und *Aristolochia clematitis*. Bot. Z. 34: 448–461, 464–470, 480–489,479–503.

Solms-Laubach HG. 1898. Die Entwicklung des Ovulum und des Samens bei *Rafflesia* und *Brugmansia*. Ann. Jard. Bot. Buitenz., Suppl. 2: 11–22.

Solms-Laubach HG. 1901. Rafflesiaceae. In: A Engler und K Prantl, eds. Das Pflanzenreich. IV, 75: 1–19. Berlin.

Takhtajan AL, NR Meyer, and VN Kosenko. 1985. Pollen morphology and classification in Rafflesiaceae s. 1. Bot. Zhurn. 70: 153–162 (in Russian).

Terekhin ES. 1988. Rafflesiales. In: A Takhtajan, ed. Comparative seed anatomy, vol. 2, pp. 114–119. Nauka, Leningrad (in Russian).

Terekhin ES and GM Anisimova. 1981. Rafflesiaceae. In: MS Yakovlev, ed. Comparative embryology of flowering plants, pp. 96–100. Nauka, Leningrad (in Russian).

Winlder H. 1927. Über eine *Rafflesia* aus Zentralborneo. Planta 4: 1–79.

Subclass II. RANUNCULIDAE

Mostly herbaceous plants. Parenchymatous tissues usually without ethereal cells. Stomata mostly anomocytic. Vessels usually with simple perforations. Sieve-element plastids of S-type. Flowers actinomorphic or zygomorphic, bisexual or less often unisexual, spirocyclic or more often cyclic. Tapetum secretory, rarely amoeboid (*Kingdonia* and some species of *Mahonia* and *Berberis*). Microsporogenesis simultaneous or rarely successive (some species of *Mahonia*). Pollen grains 2-celled or 3-celled, 3-colpate or of derived types. Gynoecium apocarpous, syncarpous, or paracarpous. Ovules mostly anatropous, bitegmic or sometimes unitegmic, crassinucellate or seldom tenuinucellate. Female gametophyte commonly of *Polygonum*-type. Endosperm cellular or more often nuclear. Fruits of various types. Seeds with small embryo and copious endosperm or less often with large embryo and little or no endosperm. Commonly producing diverse kinds of isoquinoline alkaloids, but without ellagic acid and iridoid compounds.

Most probably derived from some archaic Magnoliidae.

Bibliography

Barthlott W and I Theisen. 1995. Epicuticular wax ultrastructure and classification of Ranunculiflorae. In: U Jensen and JW Kadereit, eds. Systematics and evolution of the Ranunculiflorae. Plant Syst. Evol., Suppl. 9: 39–45.

Behnke H-D 1971. Sieve-tube plastids in Magnoliidae and Ranunculidae in relation to systematics. Taxon 20: 723–730.

Behnke H-D. 1995. Sieve-element plastids, phloem proteins, and the evolution of the Ranunculanae. In: U Jensen and JW Kadereit, eds. Systematics and evolution of the Ranunculiflorae. Plant Syst Evol., Suppl. 9: 25–37.

Blackmore S, P Stafford, and V Persson. 1995. Palynology and systematics of Ranunculiflorae. In: U Jensen and JW Kadereit, eds. Systematics and evolution of the Ranunculiflorae. Plant Syst. Evol., Suppl. 9: 71–82.

Brückner C. 2000. Clarification of the carpel. Number in Papaverales, Capparales, and Berberidaceae. Bot. Rev. 66: 155–309.

Carlquist S. 1995. Wood anatomy of Ranunculiflorae: a summary. In: U Jensen and JW Kadereit, eds. Systematics and evolution of the Ranunculiflorae. Plant Syst. Evol., Suppl. 9: 11–24.

Endress PK. 1995. Floral structure and evolution in Ranunculanae. In: U Jensen and JW Kadereit, eds. Systematics and evolution of the Ranunculiflorae. Plant Syst. Evol., Suppl. 9: 47–61.

Hiepko P. 1965. Vergleichend-morphologische und entwicklungsgeschichtliche Untersuchungen über das Perianth bei den Polycarpicae. Bot. Jahrb. Syst. 84: 359–508.

Hoot SB and PR Crane. 1995. Inter-familial relationships in the Ranunculidae based on molecular systematics. In: U Jensen and JW Kadereit, eds. Systematics and evolution of the Ranunculiflorae. Plant Syst. Evol., Suppl. 9: 119–131.

Jensen U. 1995. Secondary compounds of the Ranunculiflorae. Plant Syst. Evol., Suppl. 9: 85–97.

Jensen U and JW Kadereit, eds. 1995. Systematics and evolution of the Ranunculiflorae. Plant Syst. Evol., Suppl. 9: 133–145.

Kubitzki K. 1995. Ranunculiflorae – delimitation, phylogeny, diversification. In: U Jensen and JW Kadereit, eds. Systematics and evolution of the Ranunculiflorae. Plant Syst. Evol., Suppl. 9: 1–10.

Loconte H, LM Campbell, and SW Stevenson. 1995. Original and familial relationships of the Ranunculiflorae. In: U Jensen and JW Kadereit, eds. Systematics and evolution of the Ranunculiflorae. Plant Syst. Evol., Suppl. 9: 99–118.

Smith AC. 1971 (1972). An appraisal of the orders and families of primitive extant angiosperms. J. Indian Bot. Soc. (Golden Jubilee) 50A: 215–226.

Walker JW. 1976. Comparative pollen morphology and phylogeny of the ranalean complex, pp. 241–299. In: CB Beck, ed. Origin and early evolution of angiosperms. New York.

Superorder PROTEANAE

Order 18. PLATANALES

Large, monoecious, deciduous trees, often with mottled bark falling off in large flakes; hairs candelabriform. Vessels with scalariform and simple perforations; vessels without vestured pits. Axial parenchyma apotracheal, diffuse or in uniseriate bands. Rays homogeneous. Sieve-element plastids of S-type. Nodes multilacunar (7). Leaves alternate, medium-sized to large, palmately veined and lobed or seldom (*Platanus kerrii*) pinnately veined and entire. Leaves without a persistent basal meristem (Watson and Dallwitz 2006). The petiole usually mitriform at the base and enclosing the axillary bud or seldom (*P. kerrii*) buds free, teeth glandular, with a cavity. Candlabriform trichomes with whorls of arms cover the abaxial surface of leaves (Carpenter et al. 2005). Stipules concrescent (around the stem), ochreate, scaly, caducous. Stomata anomocytic, laterocytic or paracytic. Inflorescences monoecious, long pedunculate, pendulous, of 1 or more globular heads of numerous flowers, each head subtended by a circular bract, smaller bracts interspersed among the flowers. Individual flowers small, inconspicuous, basically 4-merous; sepals 3-4(-7) free or basally connate, not vascularized; corolla in male flowers 3-4(-7), tiny or vestigial. Stamens as many as and opposite the sepals; filaments very short. Anthers basifixed or adnate, latrorse, 2-locular, tetrasporangiate, with a connective enlarged at the apex into a peltate appendage. Microsporogenesis simultaneous. Tapetum secretory. Pollen grains 2-celled, 3-colpate or 6-rugate. Male flowers sometimes with vestigial carpels. Female flowers with 3–4 staminodia, or without staminodes. Gynoecium apocarpous, of 5–8 or less often 3–4 or 9 carpels in 2–3 cycles; stylodia long, linear, with decurrent stigma; carpels are not completely sealed distally. Ovules orthotropous or slightly hemitropous, usually solitary, rarely two, pendulous, bitegmic, crassinucellate. Placentation apical to marginal. Endosperm nuclear. Fruits of achenes, rarely of almost follicles, with tuft of basal hairs. Seeds small, with thin testa and tegmen completely disappearing in the ripe seed; embryo straight, rather large; endosperm scanty. Contain cyanogenic (cynogenic constituents tyrosine-derived), proanthocyanidins (cyanidin and delphinidin), flavonols (kaempferol, quercetin, and myricetin), n = 7, 8, 16–21.

1. PLATANACEAE

T.G. Lestiboudois 1826. 1/10. From the Balkan Peninsula to western Himalayas, Indochina, and America from Canada to Mexico.

Platanus.

Bibliography

Baas P. 1969. Comparative anatomy of *Platanus kerrii* Gagenep. Bot. J. Linn. Soc. 62: 413–421.

Boothroyd LE. 1930. The morphology and anatomy of the inflorescence and flower of the Platanaceae. Am. J. Bot. 17: 678–693.

Brett DW. 1979. Ontogeny and classification of the stomatal complex of *Platanus* L. Ann. Bot. 44: 249–251.

Bretzler E. 1938. Bau der Platanenblüte und systematische Stellung der Platanen. Bot. Jahrb. Syst. 62: 305–309.

Carpenter RJ, RS Hill, and GJ Jordan. 2005. Leaf cuticular morphology links Platanaceae and Proteaceae. Int. J. Plant Sci. 166: 843–855.

Ernst WR. 1963. The genera of Hamamelidaceae and Platanaceae in the southeastern United States. J. Arnold Arbor. 44: 193–210.

Feng Y, S-H Oh, and PS Manos. 2005. Phylogeny and historical biogeography of the genus *Platanus* as inferred from nuclear and chloroplast DNA. Syst. Bot. 30: 786–799.

Floyd SK, VT Lerner, and WE Friedman. 1999. A developmental and evolutionary analysis of embryology in *Platanus* (Platanaceae), a basal eudicot. Am. J. Bot. 86: 1523–1537.

Hsiao J-Y. 1973. A numerical taxonomic study of the genus *Platanus* based on morphological and phenolic characters. Am. J. Bot. 60: 678–684.

Kubitzki K. 1993. Platanaceae. In: K Kubitzki, ed. The families and genera of vascular plants, 2: 521–522. Springer, Berlin/Heidelberg/New York.

Leroy J-F. 1982. Origine et evolution du genre *Platanus* (Platanaceae). C. R. Acad. Sci. Paris, ser. 3, 295: 251–254.

Maslova NP. 2003. Extinct and extant Platanaceae and Hamamelidaceae: morphology, systematics, and phylogeny. Paleontol. J. 37: 467–590.

Mindell RA, RA Stockey, and G Beard. 2006. Anatomically preserved staminate inflorescences of *Gynoplatananthus oysterbayensis* gen. et sp. nov. (Platanaceae) and associated pistillate frutifications from the Eocene of Vancouver Island, British Columbia. Int. J. Plant Sci. 167: 591–600.

Oginuma K and H Tobe. 1991. Karyomorphology and evolution in some Hamamelidaceae and Platanaceae (Hamamelididae: Hamamelidales). Bot. Mag. Tokyo 104: 115–135.

Rieseberg LH and DE Soltis. 1987. Flavonoids of fossil Miocene *Platanus* and its extant relatives. Biochem. Syst. Ecol. 15: 109–112.

Schwarzwalder RN. 1986. Systematics and early evolution of the Platanaceae. Ph.D. dissert. Indiana University.

Schwarzwalder RN and DL Dilcher. 1991. Systematic placement of the Platanaceae in the Hamamelididae. Ann. Missouri Bot. Gard. 78: 962–969.

Stevenson DW, and AW Douglas. 1998. Hulles, ocreas and metamers: vegetative architecture of *Platanus* revisited. Am. J. Bot. 85(Suppl.): 21 (Abstract).

Wheeler EA. 1995. Wood of *Platanus kerii*. IAWA Bull. 16: 127–132.

Zemskova EA. 1993. On the karyotype of *Platanus kerrii* (Platanaceae). Bot Zhurn. 78 (9): 15–17 (in Russian).

Order 19. PROTEALES

Trees or more often shrubs, or often lignotuberous subshrubs, sometimes geoflorous, very rarely (in *Stirlingia*) almost herbaceous, glabrous or more often provided with characteristic uniseriate and 3-celled trichomes, mostly more or less xeromorphic. Lateral roots of limited growth, forming clusters, proteoid roots (*Persoonia* lacks proteoid roots), rarely mycorrhizal. Vessels nearly always with simple perforations, but some rare with simple and scalariform plates with few bars sometimes occur, usually with small bordered pits. Rays nearly or quite homogeneous. Axial parenchyma scanty paratracheal, most commonly as narrow bands, occasionally with most of the bands independent of the vessels (apotracheal). Sieve-element plastids commonly of S-type. Nodes mostly trilacunar with three traces, very rarely (*Panopsis*) pentalacunar. Leaves evergreen, small to very large, alternate or sometimes opposite or verticillate, simple and entire to variously divided, less often compound, mostly very coriaceous, leathery or fleshy, or modified into spines; petiolate to sessile, often aromatic, estipulate. Leaves without a persistent basal meristem (Watson and Dallwitz 2000). Stomata paracytic or rarely (in *Bellendena*) laterocytic. Flowers solitary or paired (seldom several) in the axils of bracts (which sometimes absent), arranged in racemes, umbels, or conelike inflorescences, often in involucrate heads, or the primary inflorescences often reduced to a pair of flowers arranged into secondary racemes, bisexual or rarely unisexual (monoecious or more often dioecious), actinomorphic or zygomorphic, mainly 4-merous, protandrous, pollinated by various animals including small marsupials. Perianth with distinct calyx and corolla (there sometimes being 'glands' or 'scales', perhaps representing petals, internal to and alternating with the conspicuous tepals), or sepaline (the conspicuous perianth component seemingly representing the calyx, though 'petaloid'). Sepals 4, valvate, usually petaloid, free or more or less connate below into a calyx tube sometimes cleft on one side, or three sepals connate and one free. Intrastaminal annular or horseshoe-shaped nectary disc represented by (2-)4 free or variously connate nonvasculated alternisepalous scales or glands, sometimes considered vestigial petals (Haber 1960, 1961, 1966) although there are more probably enations (Venkata Rao 1967a). Stamens 4, rarely (*Grevillea, Petrophile*) 3–4, antesepalous, usually with broad filaments adnate to sepals in varying lengths (in *Conospermum, Synaphea*) or rarely (*Bellendena*) free from the calyx; filaments all equal or markedly unequal; anthers tetrasporangiate or rarely (*Conospermum* and *Synaphea*) bisporangiate, erect, basifixed, usually introrse, rarely latrorse, opening longitudinally, appendaged (connective often distally prolonged), or unappendaged. Staminodes when present (in *Petrophile* 1), in the same series as the fertile stamens. Tapeturn secretory. Microsporogenesis simultaneous. Pollen grains 2-celled, (2)3(-8)-porate or sometimes (*Beauprea*) 3-colpoidate, more or less triangular in polar view, pores broadly operculate, arranged according to Garside's Rule – three pores at four points in the tetrad. Gynoecium of 1 conduplicate and often stipitate carpel sealed to a varying degree (in Grevilleoideae mostly open up to the summit of ovary), sometimes carpels 2; the carpel is slightly deflected abaxially, as in *Didymeles*; stylodium elongate, thickened distally, with terminal or lateral dry stigma, often modified into a "pollen-presenter" serving in pollen-presentation. Ovary superior, sessile to stipitate. Ovules one or two or less often several to more or less numerous, from anatropous to othrotropous, but mostly orthotropous, bitegmic, crassinucellate; placentation mostly marginal, or apical. Female gametophyte of *Polygonum*-type. Endosperm nuclear. Endosperm haustoria chalazal. Fruits tardily dehiscent or indehiscena, follicles, achenes, or drupes, often 1-seeded. Seeds endotestal-tegmen, sometimes with specialized exotesta (Vyshenskaya 2000), small to large (*Macadamia*), commonly winged in follicles, with internally fibrillar endotestal cells and fibrous tegmen; exotestal cells thick-walled, crystalliferous, embryo straight, oily, achlorophyllous, with 2 or very rarely (*Persoonia*) 3–9 cotyledons; endosperm wanting (except *Bellendena*). Contain cyanogenic glycosides, proanthocyanidins, (delphinidin or cyanidin and delphinidin), flavonols (kaempferol, quercetin and myricetin), arbutin, sometimes saponins n = 5 (*Bellendena*), 7 (Persoonioideae), 10–14; the usually large (7–17 μm) chromosomes of the Persoonieae are comparable in size with the largest known in plants (Johnson and Briggs 1975).

Proteaceae are related to both Platanaceae and Nelumbonaceae which is well supported by molecular data. According to Soltis et al. (2006), "Although the composition of Proteales has to be one of the major surprises of molecular phylogenetics given the diverse habits and morphologies of its members, analyses of coimbined DNA datasets provide strong support for this clade". They put into one order Proteales these families. But in my opinion, taxonomically they will be more reasonable the consider them as three separate orders of superorder Proteanae.

1. PROTEACEAE

A. L. de Jussieu 1789 (including Lepidocarpaceae Schultz-Schultzenstein 1832). 80/1700. Tropical and South Africa, Madagascar, trop. Asia, Malesia, Australia, Tasmania, Vanuatu, New Zealand, Fiji, Central and South America southward to Chile. The centre of diversity are Australia and southern Africa.

1.1 BELLENDENOIDEAE
Shrubs; cluster roots present. Leaves simple, lobed or entire. Stomata brachyparacytic. Flowers actinomorphic. Pollen grains triporate. Carpel shortly stipitate. Stylodium not a pollen-presenter. Ovules two, orthotropous. Fruits thin, indehiscent, surrounded by a very narrow wing, emdosperm slight, n = 5. – *Bellendena.*

1.2 PERSOONIOIDEAE
Shrubs or small trees. Cluster roots absent. Leaves simple. Flowers solitary in axils, actinomorphic or zygomorphic. Stylodium not a pollen-presenter. Ovules 1–2 or many, orthotropous. Fruits drupes or follicles (*Placospermum*). Seed winged (*Placospermum*) or wingless. Chromosomes very large, n = 7. – PLACOSPERMEAE: *Placospermum*; PERSOONIEAE: *Acidonia, Toronia, Garnieria, Persoonia.*

1.3 SYMPHIONEMATOIDEAE
Plants bisexual. Cluster roots absent. Leaves alternate. Flowers sessile. Carpel sessile to very shortly stipitate. Fruits dry, indehiscent, 1-seeded, n = 10, 13. – *Agastachys, Symphionema.*

1.3 EIDOTHEOIDEAE
Leaves simple. Flowers actinomorphic, andromonoecious, solitary in each bract axil. Stylodium not a pollen-presenter. Ovule 1, orthotropous, pendulous, longitudinally ribbed. Fruits achenes. – *Eidothea.*

1.4 PROTEOIDEAE
Mostly sclerophyllous shrubs. Leaves simple, entire, toothed or variously divided. Flowers solitary in axils, sessile, actinomorphic or zygomorphic. Pollen-presenter present or absent. Ovules orthotropous. Fruits dry, indehiscent, achenes, rarely drupes. Chromosomes small at 4-x and higher levels, n = (10)11-13(14). – CONOSPERMEAE: *Cenarrhenes, Beauprea, Beaupreopsis, Dilobeia, Stirlingia, Petrophile, Isopogon, Conospermum, Synaphea*; FRANKLANDIEAE: *Franklandia, Adenanthos*; PETROPHILEAE: *Petrophile, Aulax*; PROTEEAE: *Faurea, Sorocephalus, Orothamnus, Spatalla, Diastella, Paranomus, Protea, Vexatorella, Leucospermum, Mimetes, Serruria, Leucodendron.*

1.5 SPHALMIOIDEAE
Leaves simple or pinnate on sapling and coppice growth. Flowers solitary in axils of bracts, arranged in racemose inflorescences, andromonoecious, actinomorphic. Stylodium not a pollen-presenter. Ovules two. Fruits follicles. Seeds longitudinal, winged. Chromosomes small; n = 12. – *Sphalmium.*

1.6 CARNARVONIOIDEAE
Leaves partly digitate, partly pinnately divided. Flowers loosely grouped but not paired, bisexual, actinomorphic. Stylodium not a pollen presenter. Ovules two, orthotropous. Fruits follicles. Seeds winged at one end. Chromosomes small; n = 14. – *Carnarvonia.*

1.7 GREVILLEOIDEAE
Flowers paired, actinomorphic or more often zygomorphic. Stylodia are usually pollen presenters. Ovules two to many, anatropous, hemitropous, orthotropous. Fruits usually follicles or less often indehiscent. Seeds winged or wingless, flattened. Chromosomes small to medium-sized, n = (10-) 14(15). – ORITEAE: *Orites, Neorites, Megahertzia*; KNIGHTIEAE: *Darlingia, Eucarpha, Knightia, Cardwellia*; EMBOTHRIEAE: *Stranges, Stenocarpus, Buckinghamia, Opisthiolepis, Lomatia, Alloxylon, Telopea, Embothrium, Oreocallis*; HELICIEAE: *Hollandaea, Helicia, Xylomelum, Triunia*;

MACADAMIEAE: *Euplassa, Sleumerodendron, Turrillia, Gevuina, Malagasia, Heliciopsis, Hicksbeachia, Virotia, Athertonia, Catalepidia, Floydia, Macadamia, Barbejum, Panopsis, Lambertia, Roupala, Kermadecia*; GREVILLEEAE: *Grevillea, Hakea, Finschia*; BANKSIEAE: *Musgravea, Austromuellera, Banksia, Dryandra*.

Bibliography

Barker NP, PH Weston, JP Rourke, and G Reeves. 2002. The relationships of the southern African Proteaceae as elucidated by internal transcribed spacer (ITS) DNA sequence data. Kew Bull. 57: 867–883.

Barraclough TG and G Reeves. 2005. The causes of speciation in plant lineages: Species-level DNA trees in the African genus *Protea*. In: FT Bakker, LW Chatrou, B Gravendeel, and PB Pelser, eds. Plant species-level systematics: new perspectives on pattern and process, pp. 31–46. A. R. G. Gantner, Ruggel, Liechtenstein.

Behnke H-D. 1995. Sieve-element characters of the Proteaceae and Elaeagnaceae: nuclear crystals, phloem proteins and sieve-element plastids. Bot. Acta 108: 514–524.

Bieleski RL, and BG Briggs. 2005. Taxonomic patterns in the distribution of polyols within the Proteaceae. Austral. J. Bot. 53: 205–217.

Briggs BG. 1998. Proteaceae: diverse approaches toward new syntheses. Austral. Syst. Bot. 11: 631–633.

Carolin R. 1961. Pollination of the Proteaceae. Austral. Mus. Mag., 13: 371–374.

Carpenter RJ, RS Hill, and GJ Jordan. 2005. Leaf cuticular morphology links Platanaceae and Proteaceae. Int. J. Plant Sci. 166: 843–855.

Catling DM and PJ Gates. 1998. Nodal and leaf anatomy in *Grevillea* R. Br. (Proteaceae). Bot. Jahrb. Syst. 120: 187–227.

Chattaway MM. 1948. The wood anatomy of the Proteaceae. Austral. J. Sci. Res. B, 1: 279–302.

Cowling RM and BB Lamont. 1998. On the nature of Gondwanan species flocks: diversity of Proteaceae in mediterranenan south-western Australia and South Africa. Austral. J. Bot. 46: 335–355.

Dettmann ME. 1998. Pollen morphology of Eidotheoideae: implications for phylogeny in the Proteaceae. Austral. Syst. Bot. 11: 605–612.

Dettmann ME and DM Jarzen. 1998. The early history of the Proteaceae in Australia: the pollen record. Austral. Syst. Bot. 11: 401–438.

Drinnan AN, PR Crane, and SB Hoot. 1994. Patterns of floral evolution in the early diversification of non-magniliid dicotyledons (eudicots). In: PK Endress and EM Friis, eds. Early evolution of flowers. Plant Syst. Evol., Suppl. 8: 93–122.

Douglas AW and SC Tucker. 1996a. Inflorescence ontogeny and floral organogenesis in Grevilleoideae (Proteaceae), with emphasis on the nature of the flower pairs. Int. J. Plant Sci. 157: 341–372.

Douglas AW and SC Tucker. 1996b. The developmental basis of diverse carpel orientations in Grevilleoideae (Proteaceae). Int. J. Plant Sci. 157: 373–397.

Elsworth JF and KR Martin. 1971. Flavonoids of the Proteaceae: 1. A chemical contribution to studies on the evolutionary relationships in the South African Proteoideae. J. South Afr. Bot. 37: 199–212.

Faegri K. 1965. Reflections on the development of pollination systems in African Proteaceae. J. South Afr. Bot. 31: 133–136.

Feuer S. 1986. Pollen morphology and evolution in the Persoonioideae, Sphalmioideae, and Carnarvonioideae (Proteaceae). Pollen et Spores 28: 123–155.

Feuer S. 1990. Pollen aperture evolution among the subfamilies Persoonioideae, Sphalmioideae and Carnarvonioideae (Proteaceae). Am. J. Bot. 77: 783–794.

Filla F. 1926. Das Perikarp der Proteaceae. Flora 120: 99–142.

George AS. 1998. Proteus in Australia. An overview of the current state of taxonomy of the Australian Proteaceae. Austral. Syst. Bot. 11: 257–266.

Grams AJ. 2002. Proteaceae, 160 p. Eugen Ulmer, Stuttgart.

Haber JM. 1959, 1961, 1966. The comparative anatomy and morphology of the flowers and inflorescences of the Proteaceae: I. Some Australian taxa. II. Some American taxa. III. Some African taxa. Phytomorphology 9: 325–358, 1959 (1960); 11: 1–16, 1961; 16: 490–527, 1966.

Hoot SB and AW Douglas. 1998. Phylogeny of the Proteaceae based on *atp*B and *atp*B-*rbc*L intergenic spacer region dequences. Austral. Syst. Bot. 11: 301–329.

Johnson LAS. 1998. Proteaceae – where are we? Austral. Syst. Bot. 11: 251–255.

Johnson LAS and BG. Briggs. 1963. Evolution in the Proteaceae. Austral. J. Bot. 11: 21–61.

Johnson LAS and BG Briggs. 1975. On the Proteaceae: the evolution and classification of a southern family. Bot. J. Linn. Soc. 70: 83–182.

Johnson LAS and BG Briggs. 1981. Three old southern families: Myrtaceae, Proteaceae, and Restionaceae. In: A Keast, ed. Ecological biogeography of Australia, vol, 3, pp. 429–469. W. Junk, The Hague.

Jordan GJ, RA Dillon, and PH Weston. 2005. Solar radiation as a factor in the evolution of scleromorphic leaf anatomy in Proteaceae. Am. J. Bot. 92: 789–796.

Kausik SB. 1940. Vascular anatomy of the flower of *Macadamia tenuifolia* F. Muell. (Proteaceae). Curr. Sci. 9: 22–25.

Kausik SB. 1941. Studies in the Proteaceae: V. Vascular anatomy of the flower of *Grevillea robusta* Cunn. Proc. Nad. Inst. Sci. India 7: 257–266.

Ladd PG, I Nanni, and GJ Thomson. 1998. Unique stigmatic structure in three genera of Proteaceae. Austral. J. Bot. 46: 479–488.

Lanyon JW. 1979. The wood anatomy of three proteaceous timbers: *Placospermum coriaceum, Dilobeia thouarsii* and *Garnieria spathulaefolia*. IAWA Bull. (2–3): 27–33.

Levyns MR. 1958. The phytogeography of members of Proteaceae in Africa. J. South Afr. Bot. 24: 1–9.

Liu H, G-J Yan, F Shan, and R Sedgley. 2006. Karyotypes in *Leucadendron* (Proteaceae): evidence of the primitiveness of the genus. Bot. J. Linn. Soc. 151: 387–394.

Matthews ML, J Gardner, and M Sedgley. 1999. The proteaceous pistil: morphological and anatomical aspects of the

pollen presenter and style of eight species across five genera. Ann. Bot. 83: 385–399.
Milne LA and ARH Martin. 1998. Conospermeae (Proteaceae) pollen morphology and its phylogenetic implications. Austral. Syst. Bot. 11: 503–552.
Nicolson SW and B-E Van Wyk. 1998. Nectar sugars in Proteaceae: patterns and processes. Austral. J. Bot. 46: 489–504.
Orchard AE, ed. 1995. Flora of Australia. Vol.16. Elaeagnaceae. Proteaceae. 1. Melbourne.
Pole M. 1998. The Proteaceae record in New Zealand. Austral. Syst. Bot. 11: 343–372.
Prance GT and V Plana. 1998. The American Proteaceae. Austral. Syst. Bot. 11: 287–299.
Purnell HM. 1960. Studies of the family Proteaceae, I. Anatomy and morphology of the roots of some Victorian species. Austral. J. Bot. 8: 38–50.
Ramsay HP. 1963. Chromosome numbers in the Proteaceae. Austral. J. Bot. 11: 1–20.
Rourke JP. 1998. A review of the systematics and phylogeny of the African Proteaceae. Austral. Syst. Bot. 11: 267–285.
Rourke J and D Wiens. 1977. Convergent floral evolution in South African and Australian Proteaceae and its possible bearing on pollination by nonflying mammals. Ann. Missouri Bot. Gard. 64: 1–17.
Sauquet H, D Cantrill, PH Weston, N Barker, A Mast, and V Savolainen. 2006. A phylogenetic approach to the evolution of pollen morphology in Proteaceae (Proteales). In Botany 2006 – Looking to the future, conserving the past, pp. 254–255. Abstracts. Chico State University Press, California.
Sleumer H. 1956. Proteaceae. In: CGGJ Steenis van, ed. Flora Malesiana, ser. 1, 5: 147–206. Noordhoff, Groningen.
Stace HM, AW Douglas, and JF Sampson. 1998. Did 'Paleo-polyploidy' really occur in Proteaceae? Austral. Syst. Bot. 11: 613–629.
Staden J van. 1966. A paper chromatographic survey of free amino acids and related compounds in the Proteaceae. J. South Afr. Bot. 32: 77–86.
Swenson WK, JE Dunn, and EE Conn. 1989. Cyanogenesis in the Proteaceae. Phytochemistry 28: 821–823.
Torres F. 1998. Comparative foliar anatomy of South American Proteaceae. Austral. J. Bot. 46: 357–365.
Venkata Rao C. 1957. Cytotaxonomy of the Proteaceae. Proc. Linn. Soc. N. S. W. 82: 257–271.
Venkata Rao C. 1960–1971. Studies in the Proteaceae, parts 1–6, 13. Proc. Nad. Inst. Sci. India 26B: 300–337, 1960; 27B: 126–151, 1961; 29B: 489–510, 1963; 30B: 197–244, 1964; 35B: 471–486, 1969. Parts 5 and 6. J. Indian Bot. Soc. 44: 244–270; 479–494, 1965. Part 14. Proc. Indian Acad. Sci. 36B: 345–363, 1971.
Venkata Rao C. 1963. On the morphology of the calyculus. J. Indian Bot. Soc. 42: 618–628.
Venkata Rao C. 1965. Pollen grains of Proteaceae. J. Palynol. 1: 1–9.
Venkata Rao C. 1967a. Morphology of the nectary in Proteaceae. New Phytol. 66: 99–107.
Venkata Rao C. 1967b. Origin and spread of the Proteaceae. Proc. Indian Acad. Sci. 33B: 219–251.
Venkata Rao C. 1971. Proteaceae. Bot. Monogr. 6. Council of Sci. and Indus. Res. New Delhi.
Vickery JR. 1971. The fatty acid composition of the seed oils of Proteaceae: A chemotaxonomic study. Phytochemistry 10: 123–130.
Vyshenskaya TD. 2000. Proteaceae. In: A Takhtajan, ed. Comparative seed anatomy, vol. 6, pp. 197–217. Nauka, St. Petersburg (in Russian).
Weston PH. 2007. Proteaceae. In: K Kubitzki, ed. The families and genera of vascular plants, vol. IX, pp. 364–404. Springer, Berlin/Heidelberg/New York.
Weston PH and NP Barker. 2006. A new generic classification of the Proteaceae with an annotated checklist of genera. Telopea 11: 314–344.

Order 20. NELUMBONALES

Large acaulescent aquatic perennial herbs developing thick, creeping rhizomes with distinct nodes and fleshy elongate internodes; the internodes near the growing tip form fleshy, banana-like storage tubers. The rhizome bears scaly leaves (cataphylls) and foliage leaves; after each long internode comes a triad of phyllomes – a scale leaf on the lower side of the rhizome and another one on its upper side, immediately followed by a normal foliage leaf; from the axil of the second scale leaf arises the flower, and from that of the foliage leaf develops a new branch. Roots adventitious, produced at nodes. Laticifers in form of thin-walled, much elongated cells occur in ground tissue and, more prominently, in the vascular bundles. Sclerenchymatous sclereids lacking. In the parenchyma of the leaves and the rhizome, especially lining the walls of the aerenchymatic caverns present druses containing calcium oxalate crystals. Vessels occur both in the roots and in rhizomes, very primitive (elongate vessel-elements with very oblique end walls and perforations scalariform with numerous bars). Sieve-elements with transverse end walls with simple sieve plates. Sieve-element plastids of S-type with relatively large starch grains of highly different sizes. Leaves long-petiolate, with 2-keeled petiole and distally free ochreate base; laminas large, orbicular, concave, centrally peltate, some floating, others raised well above the water surface; some submersed, more or less lanceolate, parallel-veined leaves are also present. Stomata anomocytic. Flowers large, solitary, axillary, elevated above the water on long terete peduncles, actinomorphic, spirocyclic, bisexual, entomophilous. Sepals 2–5, often 4 in 2 decussate pairs, greenish, caducous. Petals numerous, usually 20–30, not very clearly differentiated from the sepals, free, spirally arranged. The 5–8 outer petals are transitional and smaller than the more numerous inner true petals; inner petals have a papillose epidermis, but this epidermis is not glandular, and only inner petals are scented (Vogel and Hadacek 2004). Stamens numerous (100–200, sometimes up to 400), spirally arranged, with slender, elongate filaments; anthers elongate, tetrasporangiate, with narrowly laminar connective that is conspicuously prolonged beyond the anther into a slender incurved appendage; the bodies of appendage are covered throughoput with a glandulae epithelium; anthers of the outer stamens extrorse; those of the inner ones latrorse or introrse, opening longitudinally. Tapetum secretory. Microsporogenesis simultaneous. Pollen grains 3-celled, often 3-colpate (sometimes or other times, but never with distal apertura – Kreunen and Osborn 1999; Borsch and Wilde 2000; Pozhedaev 2005, personal communication), and only very rarely with 1 distal aperture. It is most interesting that in some tetrads one can see different types of distal apertures including the circular apertures of Nymphaeaceae-type (Titova and Vasilyeva 2000: Pl.28, Fig. 44). Gynoecium apocarpous, of (2-)8–30(-40) carpels in 2–4 more or less distinct whorls, immersed separately in the broad truncate top of the enlarged turbinate, spongy receptacle; stigma sessile, papillate, circular rim, with a small depression in the center that leads into an ovarian canal occluded with interlocking papillae and serving as passage for the pollen tube. Ovules solitary or rarely two, very large, dorsally pendulous, anatropous, bitegmic (integuments slightly lobed), crassinucellate, strongly vascularized, with an obturator of funicular origin, nucellar cap, and hypostase. Female gametophyte of *Polygonum*-type. Endosperm cellular (Davis 1966; Batygina 1981; nuclear according to Khanna 1965 and Johri et al. 1992; helobial according to Yan 1986), with multinucleate cells in the chalazal part, which probably perform haustorial function. Fruits hard-walled achenes immersed in the dry, accrescent receptacle, each with a small respiratory pore (remnant of the ovarian canal) near the stigma. Seeds large, ovoid, exarillate, with thin seed coat consisting of pressed remnants of both integuments. Very large embryo, inconspicuous vestigial endosperm represented by thin film in the chalazal part and around the tips of cotyledons and plumule, and without perisperm; cotyledons large, fleshy, connate by their margins for most of their length, forming a thick cup-shaped structure surrounding the well-developed green plumule of three or four dark green young leaves; radicle abortive. Producing benzyl isoquinoline and aporphine alkaloids, n = 8.

"Nelumbonaceae are sister to the strongly supported (93%) clade of Platanaceae and Proteaceae" (Soltis et al. 2006), but Burthlott et al. (1996) noted that the cuticule waxes of Platanaceae and Nelumbonaceae were very different.

1. NELUMBONACEAE

Berchtold et J. Presl 1820. 1/2. Eastern Asia, Southeast Asia, and southern Asia, through Malesia to northeastern Australia, and America from eastern North America south through Mexico and the Greater Antilles to Colombia.

Nelumbo.

Bibliography

Baillon H. 1871. Recueil d'observations botaniques.Traite de developpement de la fleur et du fruit: IV. Nelumbees. Adansonia 10: 1–9.

Barthlott W, C Neihhuis, R Jetter, T Bouraul, and M Riederer. 1996. Waterlily, poppy, or sycamore: on the systematic position of *Nelumbo*. Flora 191: 169–174.

Batygina TB. 1981. Nelumbonaceae. In: MS Yakovlev, ed. Comparative embryology of flowering plants, pp. 116–120. Nauka, Leningrad (in Russian).

Batygina TB and II Shamrov. 1985. Comparative embryology of Nymphaeales and Nelumbonales orders and the problems of their taxonomy and phylogeny. Bot. Zhurn. 70: 368–373 (in Russian).

Batygina TB, TI Kravtsova, and II Shamrov. 1980. Comparative embryology of some representatives of the orders Nymphaeales and Nelumbonales. Bot. Zhurn. 65: 1071–1087 (in Russian).

Batygina TB, II Shamrov, and GE Kolesova 1982. Embryology of Nymphaeales and Nelumbonales: 2. Development of female embryonic structures Bot. Zhurn. 67: 1179–1195 (in Russian).

Borsch T and W Barthlott. 1994. Classification and distribution of the genus *Nelumbo* Adans. (Nelumbonaceae). Beitr. Biol. Pfl. 68: 421–450.

Borsch T and V Wilde. 2000. Pollen variability within species, populations, and individuals, with particular reference to *Nelumbo*. In: MM Harley, CM Morton, and S Blackmore, eds. Pollen and spores: morphology and biology, pp. 285–299. Royal Botanic Gardens, Kew.

Borsch T, C Neinhuis, and W Barthlott. 1998. *Nelumbo*: biology and systematics of an exceptional plant. In: AL Zhang, SG Wu, et al., eds. Floristic characteristics and diversity of East Asian plants: proceedings of the first international symposium of floristic characteristics and diversity of East Asian plants, pp. 408–416. China Higher Education Press, Beijing/ Springer, Berlin/Heidelberg/New York/Tokyo.

Comi C. 1939. Ricerche sull'appendice clavata nell'anthera del *Nelumbo nucifera*. Nuovo G. Bot. Ital. 46: 600–510.

Esau K and H Kosakai. 1975a. Laticifers in *Nelumbo nucifera* Gaertn: distribution and structure. Ann Bot. 39: 713–719.

Esau K and H Kosakai. 1975b. The phloem of *Nelumbo nucifera* Gaertn. Ann. Bot. 39: 901–913.

Esau K and H Kosakai. 1975c. Leaf arrangement in *Nelumbo*: a reexamination of a unique phyllotaxy. Phytomorphology 25: 100–112.

Farr CH. 1922. The meiotic cytokinesis in *Nelumbo*. Am. J. Bot. 9: 296–306.

Gupta SC and R Ahluwalia. 1977. The carpel of *Nelumbo nucifera*. Phytomorphology 27: 274–282.

Gupta SC and R Anuja. 1967. Is *Nelumbo* a monocot? Naturwissenschaften 54: 498.

Gupta SC, GS Paliwal, and R Ahuga. 1968. The stomata of *Nelumbo nucifera*: Formation, distribution, and degeneration. Am. J. Bot. 55: 295–301.

Hall TF and WT Penfound. 1944. The biology of the American lotus, *Nelumbo lutea* (Wild.) Pers. Am Midi. Nat. 31: 744–758.

Hayes V, EL Schneider, and S Carlquist. 2000. Floral development of *Nelumbo nucifera* (Nelumbonaceae). Int. J. Plant Sci. 16(6): 183–191.

Ito M. 1986. Studies in the floral morphology and anatomy of Nymphaeales: IV. Floral anatomy of *Nelumbo nicifera*. Acta Phytotax. Geobot. 37: 82–96.

Kak AM and S Durani. 1986. A contribution to the seed anatomy of *Nelumbium nuciferum* Gaertn. J. Plant Anat. Morph. 3: 59–64.

Khanna P. 1965. Morphology and embryological studies in Nymphaeaceae: II. *Brasenia schreberi* Gmel. and *Nelumbo nucifera* Gaertn. Austral. J. Bot. 13: 379–387.

Kolesova GE and TB Batygina. 1988. Nelumbonaceae. In: AL. Takhtajan, ed. Comparative seed anatomy, vol. 2, pp. 157–163. Nauka, Leningrad (in Russian).

Kosakai H, MF Moseley, and VI Cheadle. 1970. Morphological studies of the Nymphaeaceae: V. Does *Nelumbo* have vessels? Am J. Bot. 57: 487–494.

Kreunen SS and JM Osborn. 1999. Pollen and anther development in *Nelumbo* (Nelumbonaceae). Am. J. Bot. 86: 1662–1676.

Kristen U. 1971. Licht- und Elektronenmikroskopische Untersuchungen zur Entwicklung der Hydropoten von *Nelumbo nucifera*. Ber. Deutsch. Bot. Ges. 84: 211–214.

Kuo Huang L-L. 1990. Calcium oxalate crystals in the leaves of *Nelumbo nucifera* and *Nymphaea tetragona*. Taiwania 35: 178–190.

Kuprianova LA. 1979. On the possibility of the development of tricolpate pollen from monosulcate. Grana 18: 1–4.

Kuprianova LA and VF Tarasevich. 1983. Pollen morphology of recent and fossil species of the genus *Nelumbo* (Nelumbonaceae). Bot. Zhurn. 68: 137–146 (in Russian).

Leenwen WAM van. 1963. A study of the structure of the gynoecium of *Nelumbo lutea* (Wilid.) Pers. Acta Bot. Neerl. 12: 84–97.

Les DH, DK Garvin, and CF Wimpee. 1991. Molecular evolutionary history of ancient aquatic angiosperms. Proc. Nad. Acad. Sci. USA 88: 10119–10123.

Moseley MF and NW Uhl. 1985. Morphological studies of the Nymphaeaceae sensu lato: XV. The anatomy of the flower of *Nelumbo*. Bot. Jahrb. Syst. 106: 61–98.

Ni X-M. 1987. Chinese Lotus. Wuhan Research Inst. Bot. Acad. Sinica.

Schneider EL and JD Buchanan. 1980a. Morphological studies of the Nymphaeaceae: XI. The floral biology of *Nelumbo pentapetala*. Am. J. Bot. 67: 182–193.

Schnaider EL and JD Buchanan. 1980b. Morphological studies of *Nelumbo pentapetala* (Walter) Fernald: the American lotus on the Upper Mississippi. Trans. Wisconsin Acad. Sci. 65: 258–273.

Schneider EL and S Carlquist. 1996. Vessels in *Nelumbo* (Nelumbonaceae). Am. J. Bot. 83: 1101–1106.

Seago JL. 2002. The root cortex of the Nymphaeaceae, Cabombaceae, and Nelumbonaceae. J. Torrey Bot. Soc. 129: 1–9.

Seymour RS and AJ Blaylock. 2000. Stigma peroxidase activity in association with thermogenesis in *Nelumbo nucifera*. Aquatic Bot. 67: 155–157.

Seymour R and P Schultze-Motel. 1996. Thermoregulating lotus flowers. Nature 383: 305.

Seymour R and P Schultze-Motel. 1998. Physiological temperature regulation by flowers of the sacred lotus. Philos. Trans. Ser. B, 353: 935–943.

Seymour RS, P Schultze-Motel, and I Lamprecht. 1998. Heat production by sacred lotus depends on ambient temperature, not light cycle. J. Exp. Bot. 49(324): 1213–1217.

Shen-Miller J, JW Schopf, G Harbottle, R Cao, S Ouyang, K-S Shou, JR Southon, and G Liu. 2002. Long-living lotus: germination and soil γ-radiation of centuries old fruits, and cultivation, growth, and phenotypic abnormalities of offspring. Am. J. Bot. 89: 236–247.

Simon J-P. 1970. Comparative serology of the order Nymphaeales: 1. Preliminary survey on the relationships of *Nelumbo*. Aliso 7: 243–261.

Simon J-P. 1971. Comparative serology of the order Nymphaeales: II. Relationships of Nymphaeaceae and Nelumbonaceae. Aliso 7: 325–350.

Snigirevskaya NS. 1964. Contributions to the morphology of the genus *Nelumbo* Adans. Trudy Bot. Inst. Akad. Nauk SSSR, ser. l, 13: 104–172 (in Russian).

Snigirevskaya NS. 1992. The leaf dimorphism and the cotyledon nature in *Nelumbo* (Nelumbonaceae). Bot. Zhurn. 77 (1): 1–6 (in Russian with English summary).

Sohmer SH and DF Sefton. 1978. The reproductive biology of *Nelumbo pentapetala* (Nelumbonaceae) on the Upper Mississippi River: II. The insects associated with the transfer of pollen. Brittonia 30: 355–364.

Tang PH. 1988. Scanning electronmicroscopic observations on development and structure of pericarp of *Nelumbo nucifera* Gaertn. Acta Bot. Sinica 30: 248–253 (in Chinese with English summary).

Tang PH, DI Sun, J Zhao and XZ Su. 1989. Light and electron microscopic observation of the developing pattern of cotyledons of *Nelumbo nucifera*. Acta Bot. Sinica 31: 750–756 (in Chinese with English summary).

Titova GE and VE Vasilyeva. 2000. Reproductive strategy in Nelumbonaceae. In: TB Batygina, ed. Embryology of flowering plants. Terminology and concepts. Reproductive systems, vol. 3, pp. 479–499. Nauka, St. Petersburg.

Trecul A. 1854. Vegetation du Nelumbium codophyllum et disposition anomale de ses feuilles et de ses stipules. Ann. Sci. Nat. Bot. ser. 4, 1: 291–298.

Vasilyeva VE, TB Batygina, and GE Titova. 1987. Morphophysiological correlations in the development of the reproductive structures of *Nelumbo nucifera* Gaertn. Phytomorphology 37: 349–357.

Vogel S. 2004a. Contributions to the functional anatomy and biology of *Nelumbo nucifera* (Nelumbonaceae). I. Pathways of air circulation. Plant Syst. Evol. 249: 9–25.

Vogel S. 2004b. Contributions to the functional anatomy and biology of *Nelumbo nucifera* (Nelumbonaceae). II. Unique emergent druses on the floral receptacle. Plant Syst. Evol. 249: 27–35.

Vogel S and F Hadacek. 2004. Contributions to the functional anatomy and biology of *Nelumbo nucifera* (Nelumbonaceae). III. An ecological reappraisal of floral organs. Plant Syst. Evol. 240: 173–189.

Watling JR, SA Robinson, and RS Seymour. 2006. Contribution of the alternative pathway to respiration during thermogenesis in flowers of the Sacred Lotus. Plant. Physiol. 140: 1367–1373.

Wigand A and E Dennert. 1888. Nelumbium speciosum W.: Eine morphologische Studie. Bibl. Bot. 11: 1–68.

Williamson PS and EL Schneider. 1993. Nelumbonaceae. In: K Kubitzki, ed. The families and genera of vascular plants, vol. 2, pp 470–473. Springer, Berlin/Heidelberg/New York.

Willis JC. 1951. *Nelumbium*. In: A dictionary of the flowering plants and ferns, p. 445. Cambridge University Press, Cambridge.

Wood CE, Jr. 1959. The genera of the Nymphaeaceae and Ceratophyllaceae in the southeastern United States. J. Arnold Arbor. 40: 94–112.

Yan S-Z. 1986. The development of embryo and endosperm of *Nelumbo nucifera*. Acta Bot. Sinica 28: 355–360 (in Chinese with English summary).

Superorder RANUNCULANAE

Order 21. EUPTELEALES

Deciduous trees. Secretory cells scattered in petioles and in pith of the stem. Vessel elements with oblique, scalariform, reticulate, and reticulate-scalariform perforation plates that have 20–90 bars; lateral pitting from opposite to intermediate, rarely alternate or scalariform. Fibers with small, mostly bordered pits. Rays heterogeneous. Axial parenchyma diffuse or in short, tangential aggregates. Multiseriate phloem rays strongly sclerified. Sieve-element plastids of S-type with about ten globular starch grains. Nodes unilacunar, with 5–11 traces. Leaves alternate, subplicate-conduplicate, simple, margins gland-toothed, with pinnate venation, without stipules. Stomata anomocytic. Flowers in axillary racemes appearing before leaves and composed of 6–12 flowers borne in axils of bracts around growing point (the main axis of the raceme transforms into a vegetative shoot after production of fruits), long pedicillate, bisexual or sometimes some of them male, without perianth, disymmetric, curved downward at anthesis, anemophilous or partly entomophilous. Stamens commonly 6–19 in a single cycle on an expanded flattened receptacle; filaments short, filiform or slightly expanded. Anthers elongate, red, tetrasporangiate, basifixed, latrorse, with a prolonged connective, opening by longitudinal slits, and very short horizontal extensions of the dehiscence line, mainly at its lower end, which results in two narrow valves on each theca. Tapetum secretory. Microsporogenesis simultaneous. Pollen grains 2-celled, 3-colpate or 5–7-colpate, tectate-columellate, with minute tectal perforations, finely reticulate and apertural exine is coarsely structured. Gynoecium apocarpous, cyclic, of 6–18 (up to 31) free carpels. Carpels small, long-stipitate, oblique, conduplicate, incompletely sealed, with sessile, decurrent stigma papillate by long unicellular papillae, becoming deformed by asymmetric growth after anthesis, with 1-2(-4) submarginal ovules. Ovules anatropous, apotropous, bitegmic, crassinucellate, outer integument 4–5 cells across; micropyle exo-endotegmic. Female gametophyte of *Polygonum*-type. Endosperm cellular. Fruits of small, long, stipitate, and winged achenes with a papery pericarp. Seeds flat, with large, tanniniferous epidermal cells; seed coat formed mainly by the outer integument, exotestal cells enlarged (sclerotic mesotesta), endotesta lignified subpalisade. Embryo minute, but differentiated; endosperm copious, oily, and proteinaceous. Producing quercetin and kaempferol, but not myricetin. n = 14.

Very archaic order with such pleisiomorphic characters as vessels with scalariform perforation with numerous bars, heterogeneous rays, parenchyma diffuse, in short, tangential aggregates, very primitive, long-stipitate, incompletely seales carpels, sessile, decurrent stigma, minute embryo, and abundent endosperm.

1. EUPTELEACEAE

K. Wilhelm 1910. 1/2. Eastern Himalayas, northeastern India, southwestern and central China, Japan.

Euptelea.

Bibliography

Endress PK. 1969. Gesichtspunkte zur systematischen Stellung der Eupteleaceen (Magnoliales). Bot. Schweiz. Bot. Ges. 79: 229–278.

Endress PK. 1993. Eupteleaceae. In: K Kubitzki, ed. The families and genera of vascular plants, vol. 2, pp. 299–300. Springer, Berlin/Heidelberg/New York.

Leinfellner W. 1969. Über die Karpelle verschiedener Magnoliales: VII. *Euptelea* (Eupteleaceae). Oesterr. Bot. Z. 116: 159–166.

Li H-F and Y Ren. 2005. The variation of perforation plates of vessels in the secondary xylem of *Euptelea pleiosperma* (Eupteleaceae). Acta Phytotax. Sinica 43: 1–11 (in Chinese with English summary).

Nast CG and IW Bailey 1946. Morphology of *Euptelea* and comparison with *Trochodendron*. J. Arnold Arbor. 27: 186–192.

Pan K-Y, A-M Lu, and J Wen. 1991. Chromosome number and development of gametophytes in *Euptelea pleiospermum* (Eupteleaceae). Acta Phytotax. Sinica 29: 439–444.

Smith AC. 1946. A taxonomic review of *Euptelea*. J. Arnold Arbor. 27: 175–185.

Wang FH, NF Chien, and YL Zhang. 1984. A study on the pollen morphology in *Trochodendron, Tetracentron*, and *Euptelea*. Acta Phytotax. Sinica 22: 456–460 (in Chinese).

Order 22. LARDIZABALALES

Twining woody lianas or sometimes (*Decaisnea*) erect arborescent shrubs, glabrous or with uniseriate hairs. Wood with broad primary medullary rays. Vessels usually with simple perforations or seldom (*Decaisnea*) with scalariform perforations with few bars. Fibers with bordered or seldom (*Holboellia*) simple pits. Axial parenchyma sparse or none. Nodes trilacunar. Leaves alternate, palmately or (*Decaisnea*) pinnately compound, or trifoliolate, usually without stipules. Petiolules usually swollen at the base. Stomata anomocytic, or (*Parvatia*) encyclocytic, or (*Sinofranchetia*) actinocytic. Hairs simple. Flowers in racemes arising from perulate buds, unisexual (monoecious or dioecious) or rarely polygamous, actinomorphic, 3-merous, cyclic or spirocyclic. Sepals six to many, rarely (*Akebia*) only three, petaloid. Petals six, in two cycles. Stamens six, opposite the petals, with a broad connective prolonged into a terminal appendage; filaments more or less connate or free; anthers tetrasporangiate, extrorse, opening longitudinally. Pollen grains 2-celled, 3-colpate, with foveolate to reticulate ornamentation. Gynoecium of numerous free carpels arranged spirally or in cycles; carpels with a terminal, often oblique, usually sessile or subsessile, nonpapillate stigma, sometimes (*Akebia*) conduplicate and unsealed, with numerous or seldom few, laminar or laminal-lateral ovules, rarely (*Sargentodoxa*) with solitary, pendulous, subapical ovule. Ovules anatropous to hemitropous, bitegmic, crassinucellate, micropyle endostomal. Female gametophyte of *Polygonum*-type. Endosperm cellular. Fruits fleshy follicles or berries. Seeds exotestal, with small or minute embryo; endosperm copious, starchy or with hemicellulose. Plant Aluminum-accumulators, n = 11, 14–16, ?17, 18.

Lardizabalales are evidently the most archaic member of the ranunculids. In many respects, especially in their cellular endosperm, the Lardizabalaceae differ from the typical Ranunculales and show rather close relationships with some magnoliids.

1. LARDIZABALACEAE

R. Brown 1821 (including Decaisneaceae Loconte 1995, Sargentodoxaceae Stapf 1926, Sinofranchetiaceae Doweld 2000). 10/38. From western Himalayas to Japan, Korean peninsula, continental China, Hainan, Taiwan, and North Vietnam and occurs disjunctively in Chile between the Andes and the Pacific (*Lardizabala, Boquila*).

1.1 DECAISNEOIDEAE

Flowers polygamo-monoecious. Stem erect. Leaves imparipinnate with several pairs of opposite leaflets. Vessels with scalariform perforations. – *Decaisnea.*

1.2 LARDIZABALOIDEAE

Flowers unisexual. Stems climbing. Leaves digitate, pinnately 3-foliate or 2- to 3-ternate. Vessels almost always with simple perforations. – SINOFRANCHETIEAE: *Sinofranchetia*; STAUNTONIEAE: *Akebia, Archakebia, Holboellia, Stauntonia, Parvatia*; LARDIZABALEAE: *Boquila, Lardizabala.*

1.3 SARGENTODOXOIDEAE

1. Flowers spirocyclic, unisexual, dioecious. Woody lianas. Vessels with simple perforation. Leaves trifoliolate. Sepals six, imbricate. Petals very small and scalelike, green, nectariferous. Carpels ca. 100, spirally arranged on an enlarged receptacle; ovules solitary, pendulous, subapical. Fruits of numerous stipitate berries. n = 11. – *Sargentodoxa.*

Bibliography

Bhatnagar SP. 1965. Some observations on the embryology of *Holboellia latifolia* Wall. Curr. Sci. 54: 28–29.

Carlquist S. 1984. Wood and stem anatomy of Lardizabalaceae: with comments on the vining habit, ecology, and systemarics. Bot. J. Linn. Soc. 88: 257–277.

Decaisne J. 1839. Memoire sur la famille des Lardizaba-lacees. Arch. Mus. Hist. Nat. Paris 1: 143–213.

Henderson EM. 1924. The stem structure of *Sargentodoxa cuneata* Rehd. et Wils. Trans. and Proc. Bot. Soc. Edinburgh 29: 57–62.

Hoot SB, A Culham, and PR Crane. 1995. The utility of *atp*B gene sequences in resolving phylogenetic relationships: comparison with *rbc*L and 18S ribosomal DNA sequences in the Lardizabalaceae. Ann. Missouri Bot. Gard. 82: 194–207.

Hoot SB, A Culham, and PR Crane. 1995. Phylogenetic relationships of the Lardizabalaceae and Sargentodoxaceae. Chloroplast and nuclear DNA sequence evidence. Plant Syst. Evol., Suppl. 9: 195–199.

Kofuji R, K Ueda, K Yamaguchi, and T Shimizu. 1994. Molecular phylogeny in the Lardizabalaceae. J. Plant Res. 107: 339–348.

Lemesle R. 1943. Les tracheides a ponctuations areolees de *Sargentodoxa cuneata* Rehd. et Wils. et leur importance dans la phylogenie des Sargentodoxacees. Bull. Soc. Bot. France 90: 104–107.

Melikian AP. 1988. Sargentodoxaceae. In: A Takhtajan, ed. Comparative seed anatomy, vol. 2, pp. 165–167. Nauka, Leningrad (in Russian).

Melikian AP and GA Komar. 1988. Lardizabalaceae. In: A Takhtajan, ed. Comparative seed anatomy, vol. 2, pp. 163–165. Nauka, Leningrad (in Russian).

Nickol MG. 1996. Inflorescence morphology of the Lardizabalaceae. Am. J. Bot. 83(6): 182 (Abstract).

Payne WW and JL Seago. 1968. The open conduplicate carpel of *Akebia quinata* (Berberidales: Lardizabalaceae). Am. J. Bot. 55: 575–581.

Qin H-N. 1989. An investigation on carpels of Lardizabalaceae in relation to taxonomy and phylogeny. Cathaya 1: 61–82.

Qin H-N. 1995. *Archakebia*: a new genus of Lardizabalaceae from China. Acta Phytotax. Sinica 33: 240–243.

Qin H-N. 1997. A taxonomic revision of the Lardizabalaceae. Cathaya 8–9: 1–214.

Stapf O. 1925. *Sargentodoxa cuneata*. Bot. Mag. 151: t. 9111, 9112.

Sugiyama M. 1981. Comparative studies of vascular system of node-leaf continuum in the Ranalian complex: 2. *Sargentodoxa cuneata* Rehd. et Wils. Jpn. J. Bot. 58: 252–258.

Sugiyama M. 1984 (1985). Comparative studies of vascular system in node-leaf continuum in Ranalian complex: 3. Lardizabalaceae. Phytomorphology 34(1–4): 99–109.

Swamy BGL. 1953. Some observations on the embryology of *Decaisnea insignis* Hook. et Thoms. Proc. Nad. Inst. Sci. India 19(2): 307–310.

Taylor BAS. 1967. The comparative morphology and phylogeny of the Lardizabalaceae. Ph.D. thesis, Indiana University, Bloomington.

Wang Y and QY Su. 1993. Researches of the anatomy of vegetative organs in relation to the systematic position of the *Sinofranchetia*. Acta Bot. Bor.-Occident. Sinica 13: 57–59 (n Chinese with English summary).

Wei ZX and DZ Li. 1995. Pollen morphology and wall ultrastructure of Sargentodoxaceae. Acta Bot. Yunn. 17(2): 197–200.

Wu C-Y and K Kubitzki. 1993. Lardizabalaceae. In: K. Kubitzki, ed. The families and genera of vascular plants, vol. 2, pp. 361–365. Springer, Berlin/Heidelberg/New York.

Xia Q and J Kong. 1990. A study on the leaf morphology and anatomy of the Lardizabalaceae, Sargentodoxaceae, and their significance in taxonomy. Bull. Bot. Res. (Harbin) 10: 113–128 (in Chinese with English summary).

Xia Q and J Kong. 1991. A study of the ultrastructure of pollen wall of Lardizabalaceae and Sargentodoxaceae. Bull. Bot. Res. North-East Forest. Univ. 11: 93–98 (in Chinese with English summary).

Xia Q and Z-X Peng. 1989a. A study of the seed of Lardizabalaceae and Sargentodoxaceae: 1. A SEM examination of testa. Acta Phytotax. Sinica 27(4): 273–276 (in Chinese with English summary).

Xia Q and Z-X Peng. 1989b. A study of the pollen morphology of Lardizabalaceae, Sargentodoxaceae, and its significance in taxonomy. Bull. Bot. Res. (Harbin) 9: 99–114.

Yoshida O and A Michikawa. 1973. Embryological studies of genus *Akebia* Decaisne. J. Coil. Arts Chiba Univ. 6B: 25–37.

Yoshida O and A Nakajima. 1978. Embryological study of *Stauntonia hexaphylla* Decne. J. Coil. Arts Chiba Univ. 11B: 45–57 (in Japanese with English summary).

Zhang X-H, Y Ren, X-H Tian, and L-Z Pan. 2005. Anatomical studies on *Sinofranchetia chinensis* (Lardizabalaceae) and their systematic significance. Bot. J. Linn. Soc. 149: 271–281.

Zheng QA and Yang CR. 2001. Chemotaxonomic study on the family of *Lardizabalaceae*. Chinese Bull. Bot. 18: 332–339.

Order 23. MENISPERMALES

Twining or rarely erect shrubs or small trees. Secretory sacs with various sorts of contents often occur in the leaves and stems. Stems mostly with anomalous secondary thickening. Vessels with simple perforations; lateral pitting alternate. Fibers with numerous, distinctly bordered pits. Rays interfascicular only, very wide and usually extremely high. Axial parenchyma apotracheal, diffuse or in short bands. Nodes trilacunar. Leaves alternate, simple or rarely trifoliolate, entire or sometimes deeply 3–5-lobed, often palmati-veined. Stomata of various types. Flowers small, unisexual, always dioecious, actinomorphic or rarely slightly zygomorphic. Sepals 3–12 or more, in cycles of 3, rarely 1, free or slightly connate, imbricate or valvate.

Petals mostly 6 in 2 cycles but sometimes more or fewer or even lacking, free or sometimes more or less connate, usually smaller than the sepals, usually imbricate. Stamens 3, 6, 12 or more (in *Hypserpa* up to 40), rarely 2 or even 1, free or more or less connate; connective rarely prolonged; anthers tetrasporangiate or sometimes disporangiate, introrse or extrorse, opening longitudinally or rarely by transverse slits. Pollen grains 2-celled, mostly 3-colpate or of various derived types. Staminodia present or absent. Gynoecium of (1)3(6-40) free carpels in one or more cycles, often borne on a gynophore; stigma sessile or on a very short stylodium. Ovules in each carpel 2 (1 soon aborting), pendulous, hemianatropous at anthesis, later campylotropous-amphitropous, submarginal, bitegmic or unitegmic, crassinucellate. Female gametophyte of *Poygonum*-type. Endosperm nuclear. Fruits more or less stipitate drupes; exocarp membranous to coriaceous, mesocarp fleshy; endocarp usually bony, often curved and usually variously sculptured. Seeds often horseshoe-shaped or subannular, sometimes straight and more or less broadly ellipsoidal or deeply cup-shaped; embryo small to large, frequently more or less curved; endosperm mostly copious and sometimes ruminate, often scanty or none. Contain poisonous sesquiterpenoids and diterpenoids and usually benzyl isoquinoline alkaloids and/or aporphine alkaloids, n = 11–13, mostly 13.

Related to the Lardizabalales and probably had a common origin.

1. MENISPERMACEAE

A. L. de Jussieu 1789 (including Pseliaceae Rafinesque 1838). 71/450. Almost entirely in tropical and subtropical regions with a few species in temperate regions (*Menispermum* with two disjunct species in North America and northern Asia and a few species of *Cocculus* that extend into North America and temperate Asia).

PACHYGONEAE: *Albertisia, Macrococculus, Eleutharrhena, Penianthus, Sphenocentrum, Synclisia, Ungulipetalum, Syrrheonema, Sciadotenia, Chondrodendron, Anisocycla, Pycnarrhena, Triclisia, Carronia, Pleogyne, Haematocarpus, Curarea, Beirnaertia, Cionomene, Hyperbaena*; ANOMOSPERMEAE: *Tiliacora, Orthomene, Elephantomene, Caryomene, Abuta, Anomospermum, Telitoxicum*; TINOSPOREAE: *Calycocarpum, Arcangelisia, Disciphania, Synsndropus, Leichhardtia, Syntriandrium, Dialytheca, Odontocarya, Burasaia, Aspidocarya, Platytinospora, Rhigiocarya, Jateorhiza, Tinospora, Sarcolophium, Dioscoreophyllum, Chlaenandra, Chasmanthera, Borismene, Kolobopetalum, Leptoterantha, Parabaena*; FIBRAUREAE: *Anamirta, Fibraurea, Coscinium, Tinomisclum*; MENISPERMEAE: *Stephania, Cyclea, Cissampelos, Antizoma, Rhaptonema, Sarcopetalum, Strychnopsis, Legnephora, Diploclisia, Cocculus, Limaciopsis, Pericampylus, Limacia, Hypserpa, Menispermum, Sinomenium.*

Bibliography

Carlquist S. 1996. Wood and stem anatomy of Menispermaceae. Aliso 14: 155–170.

Fedotova TA. 1988. Menispermaceae. In: A Takhtajan, ed. Comparative seed anatomy, 2: 167–176. Nauka, Leningrad (in Russian).

Ferguson IK. 1975. Pollen morphology of the tribe Triclisieae of the Menispermaceae in relation to its taxonomy. Kew Bull. 30: 49–75.

Ferguson IK. 1978. Pollen morphology of the tribe Coscinieae of the Menispermaceae in relation to its taxonomy. Kew Bull. 32: 339–340.

Forman LL. 1986. Menispermaceae. In: CGGJ van Steenis, ed. Flora Malesiana, ser. 1, 10: 157–253. Kluwer, Dordrecht.

Forman LL. 1988. A synopsis of Thai Menispermaceae. Kew Bull. 43: 369–407.

Hong YP, KY Pan, ZD Chen, and AM Lu. 2001. Characters of leaf epidermis and their systematic significance in Menispermaceae. Acta Bot. Sinica 43(6): 615–623.

Joshi AC. 1937. Contribution to the embryology of the Menispermaceae: 1. *Cocculus villosus* DC. Proc. Indian Acad. Sci. 5B: 57–63.

Joshi AC. 1939. Morphology of *Tinospora cordifolia*, with some observations on the origin of the single integument, nature of synergidae, and affinities of the Menispermaceae. Am. J. Bot. 26: 433–439.

Kessler PJA. 1993. Menispermaceae. In: K Kubitzki, ed. The families and genera of vascular plants, vol. 2, pp. 402–418. Springer, Berlin/Heidelberg/New York.

Mennega AMW. 1982. Stem structure of New World Menispermaceae. J. Arnold Arbor. 63: 145–171.

Mohana Rao PR. 1981. Seed and fruit anatomy of *Cocculus hirsutus* (Menispermaceae). Plant Syst. Evol.139: 95–102.

Schaeppi H. 1976. Über die männlichen Blüten einiger Menispermaceen. Beitr. Biol. Pfl. 52: 207–215.

Thanikaimoni G. 1984. Menispermacees: Palynologie et systematique. Inst. Franc. Pondichery, Trav. Sect. Sci.Techn. 18: 1–135.

Thanikaimoni G. 1986. Evolution of Menispermaceae. Canad. J. Bot. 64: 3130–3133.

Thornber CW. 1970. Alkaloids of the Menispermaceae. Phytochemistry 9: 157–187.

Wang H, A Meng, J Li, M Feng, Z Chen, and W Wang. 2006. Floral organogenesis of *Cocculus orbiculatus* and *Stephania*

diesliana (Menispermaceae). Int. J. Plant Sci. 167: 951–960.
Wilkinson HP. 1989. Leaf anatomy of the Menispermaceae, tribe Tiliacoreae Miers. J. Linn. Soc. Bot. 99: 125–174.

Order 24. BERBERIDALES

Perennial herbs, shrubs, or rarely small trees. Vessels with simple perforations or sometimes in *Berberis* also with scalariform perforation plates. Fibers with simple or (*Nandina*) bordered pits. Rays broad, usually homogeneous or nearly so. Sieve-element plastids of S-type. Leaves usually alternate, sometimes all basal, simple to pinnately or ternately compound or decompound (in *Berberis* unifoliolate); stipules vestigial or wanting. Stomata anomocytic. Flowers mostly small, in various kinds of inflorescences or less often solitary, bisexual (in *Achlys* bisexual and male with a rudimentary gynoecium), actinomorphic, (2)3(4)-merous. Perianth (absent in *Achlys*) usually of 6 or 7(-9) series of free members, typically the outer 2 series relatively small and sepaloid, often caducous, the next 2 series larger and more petaloid, and the inner 2 or 3 (often interpreted as staminodia) usually also petaloid and nectariferous (not in *Diphylleia* and *Podophyllum*). Stamens (4-)6 (-18), usually of the same number as nectariferous petals and opposite them (or opposite the nectarless petals when nectariferous petals are wanting), but sometimes twice as many; filaments short; anthers tetrasporangiate, opening by two valves that lift up from the base or less often (*Nandina, Podophyllum, Dysosna* and *Sinopodophyllum*) by longitudinal slits. Pollen grains 2-celled, 3-colpate, 6–12-colpate or spiraperturate. Gynoecium pseudomonomerous, derived from two or three fused carpels; stigma sessile or on a short style, often 3-lobed. Ovules commonly numerous on a thickened marginal placenta or only two or even (*Achlys*) solitary and basal, anatropous or hemitropous, bitegmic, crassinucellate. Female gametophyte of *Polygonum*-type. Endosperm nuclear. Fruits usually a berry, seldom dry and indehiscent or irregularly dehiscent. Seeds often arillate, with small embryo; endosperm copious, with hemicellulose. Produced alkaloids: berberin, a benzyl-isoquinoline, protopine (*Nandina*). n = 6–8, 10.

Related to both the Lardizabalales and Menispermales, but markedly differs in the pseudomonomerous gynoecium.

Key to Families

1 Perianth segments numerous on elongate receptacle, white, arranged in very compressed spiral (nearly in 3-merous cycles) the inner ones gradually larger, becoming petal-like. Nectaries lacking. Stamens six, opposite the inner perianth segments; anthers opening longitudinally. Pollen grains 3-colpate with deeply punctate thick tectum, reduced diminutive columellae and very well-developed massive lamellate endexine. Carpels with 3(-4)-lobed truncate stigma on a short stylodium. Ovules two, rarely three (1 soon aborting), pendulous, submarginally attached one above the other. Fruit generally 1-seeded berries. Seed endotegmic; embryo minute, weakly differentiated; endosperm copious, oily. Lateral pitting of vessels opposite. Erect shrubs with bi- or tripinnate leaves; main joints of pinnae and pinnules bulbously swollen at base. n = 10. . . . . 1. NANDINACEAE.

1 Perianth cyclic, 3-merous or rarely 2-merous, consisting of small sepals and mostly larger petals, rarely perianth absent. Nectaries present. Stamens 4–18, mostly 6; anthers opening by two valves or rarely longitudinally. Pollen grains with thin endexine. Ovules usually numerous or only two, rarely solitary, basal. Fruits of various types. Seed coat exotestal. Nodes trilacunar or multilacunar. Lateral pitting of vessels alternate. Herbs or shrubs with simple or compound leaves.

2 Shrubs or rarely small trees. Leaves imparipinnate to trifoliolate (*Mahonia*) or unifoliolate (*Berberis*), articulate. Nodes trilacunar. Petals (often interpreted as staminodia) with two nectariferous glands on the inner side. Stamens six, opposite the petals; anthers opening by two apical valves. Pollen grains with irregular or spiral aperture and with nearly amorphous ectexine, which shows only an early stage of columellae development. Ovules anatropous, basal. Fruits baccate. n = 7. . . . . 2. BERBERIDACEAE

2 Herbs. Leaves simple or compound, not articulate. Anthers mostly opening by two apical valves or rarely longitudinally. Pollen grains 3-colpate, ectexine amorphous or columellate.

3 Pollen grains 6-pantocolpate, psilate-punctate, with nearly amorphous unstratified ectexine. Rhizomes creeping. Stem simple, with two terminal leaves, long-petiolate, trifoliate; leaflets petiolulate, 3-lobed or 3-incised. Flowers in terminal

fascicle, nodding, with three caducous, sepal-like bracteoles. Sepals six, petaloid, purple, 15 mm long, spreading. Petals six, small, with two nectariferous glands at base. Stamens six; anthers opening by small valves. Carpels with nearly sessile large stigma. Ovules numerous (20–40), lateral, hemitropous, with a long funicle. Fruit a berry. n = 7.. . . . . 3. RANZANIACEAE

3 Pollen grains 3-colpate, columellate. Sepals 5–18, usually petaloid. Petals 4–9, often nectariferous. Stamens 4–18; anthers opening longitudinally (*Podophyllum, Dysosma*) or by two apical valves. Ovules numerous or few, rarely (*Achlys*) solitary, lateral to basal. Fruits berries, capsules, or berry-like. Rhizomes creeping or tuberous, n = 6 or (Leonticoideae) 8.. . 4. PODOPHYLLACEAE

1. NANDINACEAE

Horaninow 1834. 1/1. Continental China and Japan.

Nandina.

The relatively most archaic member of the Berberidales, which is related to the Lardizabalaceae and Menispermaceae, but stands closer to the Berberidaceae.

2. BERBERIDACEAE

A. L. de Jussieu 1789. 2/600 or more. Eurasia, North Africa, mountains of tropical Africa, North, Central, and South America.

Mahonia, Berberis.

Closely related to the Nandinaceae, but differ in many respects, including chromosome number, nonstratified and almost amorphous ectexine, mode of anther dehiscence, nectariferous petals, basal ovules.

3. RANZANIACEAE

Takhtajan 1994. 1/1. Japan (Honshu, Japan Sea side).

Ranzania.

Ranzania, a remarkable and rare Japanese endemic, is included either in the Berberidaceae s. l. or in the Podophyllaceae. Although *Ranzania* resembles the Berberidaceae s. str. in its basic chromosome number, amorphous ectexine, sensitive stamens, nectariferous petals, valvate dehiscence of the anthers, and baccate fruits, it markedly differs in its herbaceous habit, flowers arranged in terminal fascicles subtended by three deciduous bracts, anther valves that roll up on the adaxial side (Kumazawa 1937b), most of the chromosomes with centromeres situated in the median portion (Kuroki 1970), lateral ovules, cortical vascular bundles in the receptacle, recurving ventral and dorsal bundles in the stigmatic region, and the fascicular branching of the ovular traces (Terabayashi 1977) as well as 6(-12) pantocolpate (rugate) apertures (according to Nowicke and Skvaria 1981: 22, "the difference in aperture type may be more significant than the similarity in exine structure"). In any case the pantocolpate aperture is unique in the suborder Berberidineae. I agree with Nowicke and Skvarla (1981: 21) that "the degree to which *Ranzania* is separated from the remaining genera (of the Berberidaceae s. l.) may be exceeded only by *Nandina*." Therefore, they consider *Ranzania* "a taxon incertae." I prefer to consider it a representative of a separate family, the Ranzaniaceae.

4. PODOPHYLLACEAE

A. P. de Candolle 1817 (including Diphylleiaceae Schulz-Schulzenstein 1832, Epimediaceae Menge 1839, Leonticaceae Berchtold et J. Presl 1820). 12/60. Northern temperate regions, especially eastern Asia and northeastern North America.

4.1 LEONTICOIDEAE

Leaves mainly basal, simply or 2–3-pinnate, often fleshy, glaucous. Sepals six in two cycles, petaloid. Petals six in two cycles, clawed, nectariferous. Anthers opening by two valves. Pollen grains reticulate. Carpel with short stylodium and small, plicate-dilated stigma. Ovules 2–4, basal. Fruits with accrescent funicules; pericarp evanescent, exposing the large seeds or an inflated bladder. Seed coat consists of 13–32 layers, weakly differentiated. – *Caulophyllum, Gymnospermium, Leontice.*

4.2 EPIMEDIOIDEAE

Leaves simply pinnate, pinnately 1–3-ternate, more or less bilobed (*Jeffersonia*) or 3-foliate (*Achlys*). Sepals 5–18, absent in *Achlys*. Petals 4, 6, or 8, hooded, spurred, or modified into small glands, with spurs, nectariferous, absent in *Achlys*. Anthers opening by two valves. Pollen grains with striate or striate-reticulate tectum. Ovules many or few (only 1 in *Achlys*), lateral or basal. Fruits capsular (obliquely or transversely dehiscing) or apically dehiscent bladder (*Bongardia*), rarely (*Achlys*) berry-like. Seed coat consists of 6–12 layers of cells, exomesotestal. – EPIMEDIEAE: *Epimedium* (perhaps including *Aceranthus*), *Vancouveria, Jeffersonia, Plagiorhegma*; ACHLYDEAE: *Achlys*; BONGARDIEAE: *Bongardia*.

4.3 PODOPHYLLOIDEAE

Leaves 2(3), simple, peltate, palmately-lobed or 2-lobed. Sepals 6–9. Petals 6–9, without nectaries. Anthers opening by two longitudinal slits. Pollen grains in monads or rarely (*Podophyllum hexandrum*) in tetrads that have a tectum of ripple-surfaced gemmae. Pollen grains of *Diphyllea* have a tectum composed of irregularly placed rods that form the base of stout, blunt spines. Ovules numerous, lateral. Fruits baccate. Seed coat consists of 7–9 layers of cells, exotestal. – *Podophyllum* (perhaps including *Sinopodophyllum*), *Dysosma, Diphylleia*.

Close to Berberidaceae and especially to the Ranzaniaceae. In some respects the most archaic group are Leonticoideae, which have the least specialized seed coat (Takhtajan and Melikian 1972) and the most primitive karyotype (Kosenko 1979) with basic chromosome number n = 8 and long, metacentric, and submetacentric chromosomes. Epimedioideae and Podophylloideae are more advanced.

Bibliography

Adachi J, K Kosuge, T Denda and K Watanabe. 1995. Phylogenetic relationships of the Berberidaceae based on partial sequences of the *gap*A gene. Plant Syst. Evol., Suppl. 9: 351–353.

Ahrendt LWA. 1961. *Berberis* and *Mahonia*: a taxonomic revision. J. Linn. Soc. London Bot. 57: 1–40.

Archangelsky DB and AL Takhtajan. 1972. Morphology of *Leontice* L., *Gymnospermium* Spach, and allied genera of the family Berberidaceae. Bot. Zhurn. 57: 921–926 (in Russian).

Archarova KB and IG Zubkova. 1969. Anatomical structure of the petiole in the family Berberidaceae Juss. Bot. Zhurn. 54: 98–103 (in Russian).

Baillon H. 1862. Remarques sur l'οργανισατιον des Berberidees. Adansonia 2: 268–291.

Barykina RP and NV Chubatova. 1985. Morphology and anatomy of the flower of *Nandina domestica* Thunb. (Berberidaceae). Vestnik Moscow Univ., ser. 16, Biol. 3: 62–68 (in Russian with English summary).

Berg RY. 1972. Dispersal ecology of *Vancouveria* (Berberidaceae). Am. J. Bot. 59: 109–122.

Brett JF and U Posluszny. 1982. Floral development in *Caulophyllum thalictroides* (Berberidaceae). Canad. J. Bot. 60: 2133–2141.

Brückner C. 2000. Clarification of the carpel number in Papaverales, Capparales, and Berberidaceae. Bot. Rev. 66(2): 155–307.

Butters FK. 1909. The seeds and seedling of *Caulophyllum thalictroides*. Minn. Bot. Stud. 4: 11–32.

Carlquist S. 1995. Wood anatomy of Berberidaceae: ecological and phylogenetic consideration. Aliso 14: 85–103.

Chapman M. 1936. Carpel anatomy of the Berberidaceae. Am. J. Bot. 23: 340–348.

De Maggio AE and CL Wilson. 1986. Floral structure and organogenesis in *Podophyllum peltatum* L. (Berberidaceae). Am. J. Bot. 73: 21–32.

Dermen H. 1931. A study of chromosome number in two genera of Berberidaceae: *Mahonia* and *Berberis*. J. Arnold Arbor. 12: 281–287.

Ehdaie M and SD Russell. 1984. Megagametophyte development of *Nandina domestica* and its taxonomic implications. Phytomorphology 34: 221–225.

Endress PK. 1989. Chaotic floral phyllotaxis and reduced perianth in *Achlys* (Berberidaceae). Bot. Acta 102: 159–163.

Ernst WR. 1964. The genera of Berberidaceae, Lardizabalaceae, and Menispermaceae in the southeastern United States. J. Arnold Arbor. 45: 1–35.

Feng M and A-M Lu. 1998. Floral organogenesis and its systematic significance of the genus *Nandina* (Berberidaceae). Acta Bot. Sinica 40: 102–108.

Guedes M. 1977. Le gynecee de *Podophyllum* (Berberidaceae). C. R. Acad. Sci. Paris 285D: 755–758.

Harvey-Gibson RJ and E Horsman. 1919. The anatomy of the stem of the Berberidaceae. Trans. Roy. Soc. Edinb. 52: 501–515.

Himmelbauer W. 1913. Die Berberidaceen und ihre Stellung im System. Denkschr. Akad. Wiss. Wien, Math.-Naturw. KL. 89: 733–796.

Holm Th. 1899. *Podophyllum peltatum*: a morphological study. Bot. Gaz. 27: 419–433.

Ito T. 1888. *Ranzania*. A new genus of Berberidaceae. J. Bot. 26: 302–303.

Kamelina OP and NG Turchina. 1982. A contribution to the embryology of the noninvestigated taxa: II. The development of the anther and pollen grains in *Nandina domestica* (Berberidaceae). Bot. Zhurn. 67: 1459–1468 (in Russian).

Kaute U. 1963. Beitrage zur Morphologie des Gynoeceums der Berberidaceen mit einem Anhang über Rhi-zomknospe von *Plagiorhegma dubium*. Ph.D. dissetration, University of Berlin.

Kawano S and M lhara. 1967. Chromosome morphology of *Caulophyllum robustum* (Podophyllaceae) and its systematic implications. J. Jap. Bot. 42: 129–135.

Kim Y-D and RK Jansen. 1995. Phylogenetic implications of chloroplast DNA variation in the Berberidaceae. Plant Syst. Evol., Suppl. 9: 341–349.

Kim Y-D and RK Jansen. 1996. Phylogenetic implications of *rbc*L and ITS sequence variation in the Berberidaceae. Syst. Botany 21: 381–396.

Kim Y-D and RK Jansen. 1998. Chloroplast DNA restriction site variation and phylogeny of the Berberidaceae. Am. J. Bot. 85: 1766–1778.

Kosenko VN. 1977a. Comparative karyological study of three genera of the family Berberidaceae. Bot. Zhurn. 62: 235–240 (in Russian).
Kosenko VN. 1977b. Comparative karyological study of *Leontice eversmannii* Bunge and *L. incerta* Pall. Bot. Zhurn. 62: 1042–1049 (in Russian).
Kosenko VN. 1978. Comparative karyological study of *Gymnospermium altaicum* (Pall.) Spach and G. *darvasi-cum* (Regel) Takht. (Berberidaceae). Bot. Zhurn. 62: 1206–1212 (in Russian).
Kosenko VN. 1979. Comparative karyological study of representatives of the family Berberidaceae s. 1. Bot. Zhurn. 64: 1539–1552 (in Russian).
Kosenko VN. 1980a. Comparative palynomorphological study of the family Berberidaceae: I. Morphology of pollen grains of the genera *Diphylleia, Podophyllum, Nandina, Berberis, Mahonia, Ranzania*. Bot. Zhurn. 65: 198–205 (in Russian with English summary).
Kosenko VN. 1980b. Comparative palynomorphological study of the family Berberidaceae: II. Morphology of the pollen grains of the genera *Gymnospermium, Leontice, Caulophyllum, Bongardia, Epimedium, Vancouveria, Achlys, Jeffersonia*. Bot. Zhurn. 65: 1412–1421 (in Russian with English summary).
Kumazawa M. 1936. *Podophyllum pleianthum* Hance: A morphological study with supplementary notes on allied plants. Bot. Mag. (Tokyo) 50: 268–276.
Kumazawa M. 1937a. On the morphology and anatomy of *Achlys japonica* Maxim. Bot. Mag. Tokyo 51: 660–668 (in Japanese).
Kumazawa M. 1937b. *Ranzania japonica*: Its morphology, biology, and systematic affinities. Jap. J. Bot. 9: 55–70.
Kumazawa M. 1938. Systematic and phylogenetic consideration of the Ranunculaceae and Berberidaceae. Bot. Mag. (Tokyo) 52: 9–15, 52–53.
Kuroki Y. 1970. Chromosome study in four species of Berberidaceae. Mem. Ehime Univ. Sect. 2, Biol. 6: 215–221.
Kurita M. 1956. Karyotype studies in Berberidaceae: I. Mem. Ehime Univ. Sect. 2. Biol. 2: 247–252.
Langlet O. 1928. Einige Beobachtungen über die Zytologie der Berberidaceen. Svensk Bot. Tidskr. 22: 169–184.
Loconte H. 1993. Berberidaceae. In: K Kubitzki, ed. The families and genera of vascular plants, 2: 147–152. Springer, Berlin/Heidelberg/New York.
Loconte H and JR Estes. 1989a. Genetic relationships within Leonticeae (Berberidaceae). Canad. J. Bot. 67: 2310–2316.
Loconte H and JR Estes. 1989b. Phylogenetic systematics of Berberidaceae and Ranunculales (Magnoliidae). Syst. Bot. 14: 565–579.
Mauritzon J. 1936. Zur Embryologie der Berberidaceen. Acta Horti Goteborg 11: 1–18.
Meacham CA. 1980. Phylogeny of the Berberidaceae with an evaluation of classification. Syst. Bot. 5: 149–172.
Melikyan AP and AA Chernykh. 1998. Comparative carpology of some Epimedioideae (Berberidaceae). Byull. Mosk. Obshch. Ispyt. Prir., Biol. 103(5): 42–48 (in Russian).
Melikian AP and TD Vyshenskaya. 1988. Berberidaceae, Nandinaceae. In: A Takhtajan, ed. Comparative seed anatomy, vol. 2, pp. 185–192. Nauka, Leningrad (in Russian).
Miyaji Y. 1930. Beiträge zur Chromosomenphylogenie der Berberidaceen. Planta 11: 650–659.
Nickol MG. 1995. Phylogeny and inflorescence of Berberidaceae – a morphological survey. Plant Syst. Evol., Suppl. 9: 327–340.
Nowicke JW and JJ Skvarla. 1981. Pollen morphology and phylogenetic relationships of the Berberidaceae. Smithsonian Contr. Bot. 50: 1–83.
Oganezova GG. 1974. Anatomical structure of leaf in Berberidaceae s. 1. in relation to the systematics of the family Bot. Zhurn. 59: 1780–1794 (in Russian).
Oganezova GG. 1975. On the evolution of the life forms in the family Berberidaceae s. 1. Bot. Zhurn. 60: 1665–1675 (in Russian).
Panov PP, MM Mollov, and LN Panova. 1971. Alkaloids from plants of the Berberidaceae family. C. R. Acad. Bulgare Sci. 24: 675–677.
Rao RR, T Hussain, B Datt, and A Garg. 1999. Palynology of *Berberis* (Berberidaceae). Rheedea. 9(2): 115–146.
Sastri RLN. 1969b. Floral morphology, embryology, and relationships of the Berberidaceae. Austral. J. Bot. 17: 60–79.
Schmidt E. 1928. Untersuchungen über Berberidaceen. Beih. Bot. Centralbl., Abt. 2, 45: 329–396.
Shen Y-F. 1954. Phylogeny and wood anatomy of *Nandina*. Taiwania 5: 85–92.
Stearn WT. 2002. The genus *Epimedium* and other herbaceous Berberidaceae (including the genus *Podophyllum*), by Julian M. H. Shaw. Timber Press, Portland, OR.
Takeda H. 1915. On the genus *Achlys*: A morphological study. Bot. Mag. (Tokyo) 29: 169–185.
Takhtajan AL. 1970. On the genus *Gymnospermium* Spach. Bot. Zhurn. 55: 1191–1193 (in Russian).
Takhtajan AL and AP Melikian. 1972. Comparative anatomical study of seed coat anatomy of *Leontice, Gymnospermium, Caulophyllum*, and allied genera in relation to their systematics. Bot. Zhurn. 57: 1271–1278 (in Russian).
Tanaka R and C Takahashi. 1981. Comparative karyotype analysis in *Epimedium* species by C-banding: 1. *E.sempervirens* and *E.perralderianum*. J. Jpn. Bot. 56: 1–24.
Terabayashi S. 1977. Studies in the morphology and systematics of Berberidaceae: I. Floral anatomy of *Ranzania japonica*. Acta Phytotax. Geobot. 28: 45–57.
Terabayashi S. 1978. Studies in the morphology and systematics of Berberidaceae: II. Floral anatomy of *Mahonia japonica* (Thunb.) DC. and *Berberis thunbergii* DC. Acta Phytotax. Geobot. 29: 106–118.
Terabayashi S. 1979. Studies in the morphology and systematics of Berberidaceae: III. Floral anatomy of *Epimedium grandiflorum* Morr. et Decne. ssp. *sempervirens* (Nakai) Kitam. and *Vanconveria hexandra* (Hook.) Morr. et Decne. Acta Phytotax. Geobot. 30: 153–168.
Terabayashi S. 1981. Studies in the morphology and systematics of Berberidaceae: IV. Floral anatomy of *Plagiorhegma dubia* Maxim., *Jeffersonia diphylla* (L.) Pers., and *Achlys triphylla* (Smith) DC. ssp. *japonica* (Maxim.) Kitam. Bot. Mag. (Tokyo) 94: 141–157.
Terabayashi S. 1982. Systematic consideration of the Berberidaceae. Acta Phytotax. Geobot. 33: 355–370 (in Japanese).
Terabayashi S. 1983a. Studies in the morphology and systematics of Berberidaceae: V. Floral anatomy of *Caulophyllum* Michx., *Leontice* L., *Gymnospermium* Spach., and *Bongardia* Mey. Mem. Fac. Sci. Kyoto Univ., Ser. Biol., 8: 197–217.

Terabayashi S. 1983b. Studies on the morphology and systematics of Berberidaceae: VI. Floral anatomy of *Diphylleia* Michx., *Podophyllum* L., and *Dysosma* Woodson. Acta Phytotax. Geobot. 34: 27–47.
Terabayashi S. 1983c. Studies in the morphology and systematics of Berberidaceae: VII. Floral anatomy of *Nandina domestica* Thunb. J. Phytogeogr. Taxon. 31: 16–21.
Terabayashi S. 1985a. The comparative floral anatomy and systematics of the Berberidaceae: I. Morphology. Mem. Fac. Sci. Kyoto Univ., Ser. Biol., 10: 73–90.
Terabayashi S. 1985b. The comparative floral anatomy of the Berberidaceae: II. Systematic considerations. Acta Phytotax. Geobot. 36: 1–13.
Tischler G. 1902. Die Berberidaceen und Podophyllaceen: Versuch einer morphologisch-biologischen Monographic. Bot. Jahrb. Syst. 31: 596–727.
Toren J. 1950. Les caracteres morphologiques, anatom-iques, et cytologiques des *Bongardia chrysogonum* Boiss. Rev. Fac. Sci. Univ. Istanbul, ser. B, 15: 239–263.
Tören J. 1961. Recherches sur les Berberidaceae de la Turquie: I. Morphologic et anatomic du *Leontice leontopetalum* L. Rev. Fac. Sci. Univ. Istanbul, ser. B., 26: 125–162.
Tören J. 1962. Recherches sur les Berberidaceae de la Turquie: II. Caracteres cytologiques du *Leontice leontopetalum* L. Rev. Fac. Sci. Univ. Istanbul, ser. B., 27: 229–250.
Tören J. 1971. Investigation of the Berberidaceae of Turkey: VII. *Bongardia chrysogonum* (L.) Boiss. Rev. Fac. Sci. Univ. Istanbul, ser. B., 36: 81–88.
Toyokuni H and Y Toyokuni. 1964. Ein neuer Anhalt für die Teilung der Podophyllaceae in zwei Unterfamilien. Bot. Mag. (Tokyo) 77: 197–198.
Ying T-S. 1979. On *Dysosma* Woodson and *Sinopodo-phyllum* Ying, gen. nov. of the Berberidaceae. Acta Phytotax. Sinica 17: 15–23 (in Chinese).
Ying T-S, S. Terabayashi, and DE. Boufford. 1984. A monograph of *Diphylleia* (Berberidaceae). J. Arnold Arbor. 65: 57–94.
Zhang K-T and P-L Wang. 1983. Pollen morphology study of Berberidaceae. Acta Phytotax. Sinica 21: 130–140 (in Chinese).
Zhang M-L, CH Uhink, and JW Kadereit. 2007. Phylogeny and biogeography of *Epimedium/Vancouveria* (Berberidaceae): Western North American—East Asian Disjunctions, the origin of European mountain plant taxa, and East Asian species diversity. Syst. Bot. 32: 81–92.
Zhukova GY. 1981. Berberidaceae. In: MS Yakovlev, ed. Comparative embryology of flowering plants: Winteraceae-Juglandaceae, pp. 125–130. Nauka, Leningrad (in Russian).

Order 25. RANUNCULALES

Herbs, sometimes lianas, or erect subshrubs or shrubs. Vascular bundles usually amphivasal and the xylem straight, not V-shaped in cross section. Vessels mostly with simple perforations; lateral pitting mostly alternate. Fibers commonly with simple pits. Woody forms with broad medullary rays. Nodes trilacunar or more often multilacunar. Leaves alternate or rarely opposite, compound or simple, mostly estipulate. Stomata anomocytic or paracytic. Flowers solitary or in various basically cymose inflorescences, bisexual or sometimes unisexual, actinomorphic or zygomorphic. Perianth double or simple, rarely completely wanting. Stamens usually numerous, spirally arranged and centripetal; anthers basifixed, extrorse or introrse, usually tetrasporangiate, opening longitudinally. Tapetum usually secretory. Microsporogenesis simultaneous or (*Thalictrum*) successive. Pollen grains 2-celled or rarely 3-celled, tectate-columellate; tectum spinulose and punctate/perforate or rarely reticulate or striate. Gynoecium mostly apocarpous. Ovary superior. Ovules one to many, mostly anatropous, bitegmic or sometimes unitegmic, crassinucellate or rarely tenuinucellate. Female gametophyte commonly of *Polygonum*-type, but in *Adonis, Trollius acaulis* and *Ceratocephala falcata* of *Allium*-type, in *Callianthemum* of both types and in *Thalictrum angustifolium* of *Pyrethrum*-type. Endosperm nuclear. Fruits of various types. Seeds exotesta, with small to large embryo and with copious or scanty endosperm, sometimes without endosperm. n = 6–9, 7 or 8, or 13 (*Hydrastis*). Chromosomes are of two types – R-type (*Ranunculus*-type) or long chromosomes generally winding or curved several times and with irregularly distributed heterochromatic regions, and T-type (*Thalictrum*-type) or small chromosomes generally simply curved and with well marked heterochromatic chromocentres (Langlet 1932; Gregory 1941; Kurita 1960). According to Tamura (1995), the values of chromosome length of R-type and T-type chromosomes are not clearly distinct. It is thus evident that the distinction between R-type and T-type is not always clear as far as the size of the chromosomes is concerned. Does *Myosurus minimus* has comparatively small metaphase chromosomes, interphase nuclei of the T-type, and prophase chromosomes of the R-type.

Closely related to the Berberidales and probably have a common origin with them.

Key to Families

1 Vascular bundles amphicribral. Vessels in metaxylem with scalariform perforations (the number of bars usually 3–10, rarely up to 18); lateral pitting scalariform. Fibers with bordered pits. Perennial herbs with thick, creeping, knotty, yellow rhizome. Nodes multilacunar. Leaves alternate, palmately

lobed, usually 1 basal and 2 cauline, near the summit of stem. Flowers solitary, terminal, bisexual, actinomorphic, apetalous, without nectaries. Sepals three, petaloid, caducous. Stamens numerous, spirally arranged. Pollen grains 2-celled, 3-colpate, with striate-reticulate tectum of two layers of striae. Gynoecium apocarpous; carpels 8–15, with more or less fleshy wall, short stylodium and flat, 2-lobed stigma with flaring borders. Ovules 2(-4), anatropous, bitegmic, crassinucellate; micropyle zig-zag, formed by both integuments. Fruits of red berrylike follicles. Seeds small, globose, black, with minute, weakly differentiated embryo and copious endosperm. Rhizome contains benzyl isoquinoline alkaloid combination of D-galactose and a ribitol-like substance, n = 13, chromosomes small, of T-type. 1. HYDRASTIDACEAE

1 Vascular bundles usually amphivasal. Vessels in metaxylem usually with simple perforations; lateral pitting scalariform. Fibers with simple pits. Perennial or annual herbs, often rhizomatous or tuberous, herbaceous or woody lianas, or erect subshrubs or small shrubs. Nodes trilacunar or multilacunar, rarely unilacunar. Leaves alternate or rarely opposite, seldom all basal, simple or compound, estipulate or with vestigial stipules. Flowers solitary or more often in various basically cymose inflorescences, bisexual or sometimes unisexual (monoceous or dioecious), actinomorphic to strongly zygomorphic, usually with more or less elongate receptacle and spirally arranged perianth, entomophilous or sometimes ornitophilous, anemophilous, or autogamous. Perianth double or simple, rarely completely wanting. Sepals five or less often fewer or more, imbricate or sometimes valvate, persistent or sometimes caducous. Petals few to numerous, mostly five, apparently modified stamens (andropetals), free, flat, and sessile or more often peltate and usually stipitate, often with basal nectaries. Stamens 5–10, up to 100, spirally arranged and centripetal; anthers adnate, extrorse or latrorse, tetrasporangiate, opening longitudinally. Pollen grains 2-celled or rarely 3-celled, 3-colpate to pantocolpate or pantoporate, very rarely inaperturate, tectate-columellate; tectum spinulose and punctate/perforate or rarely reticulate or striate. Gynoecium apocarpous or rarely more or less syncarpous; carpels single to numerous, up to 10,000 in *Laccopetalum* (largest in Ranunculidae), small to large, multiovulate to uniovulate; stylodia short to long and plumose. Ovary 3–5-locular. Ovules one to many, mostly anatropous, bitegmic or sometimes unitegmic (as for example *Anemone, Clematis, Helleborus*, and *Ranunculus*), crassinucellate or rarely tenuinucellate, with ventral raphe to with dorsal raphe; micropyle nearly always formed by the inner integument alone. Fruits aggregate, sometimes simple, follicular, achenial, rarely baccate, sometimes capsulate. Seeds with minute, small and weakly differentiated (as in Thalictreae and Actaeae) to large embryo and with copious or scanty endosperm, sometimes without endosperm; cotyledons usually two, their petioles usually fused downward to various degree, rarely one. Commonly with various kinds of benzylisoquinoline or/and aporphine alkaloids, n = 6–9, mostly 7 or 8. 2. RANUNCULACEAE

1. HYDRASTIDACEAE

Martynov 1820. 1/1. Central and eastern North America.

Hydrastis.

Hydrastis is usually regarded as a member of the Ranunculaceae or less often as a separate family near to the Ranunculaceae. It differs from the Ranunculaceae in many respects, including the amphicribral vascular bundles, the presence of vessels with scalariform perforations and scalariform lateral pitting (Eames 1961; Tobe and Keating 1985), vasculature and anatomy of the flowers (Tobe and Keating 1985), morphology of pollen grains (Nowicke and Skvarla 1980, 1981, 1982), the micropyle formed by both the outer and inner integuments (Kumazawa 1938; Tobe and Keating 1985), the presence of D-galactose (Iriki and Minamisawa 1983), and the basic number of chromosomes. Hydrastidaceae are rather "a relictual primitive group which very early diverged from a common ancestral stock of the Ranunculaceae, Berberidaceae [s.l.] and probably of the Circaeasteraceae" (Tobe and Keating 1985: 313).

2. RANUNCULACEAE

A.L. de Jussieu 1789 (including Actaeaceae Rafinesque 1828, Anemonaceae Vest 1818, Aquilegiaceae Lilja 1870, Calthaceae Martynov 1820, Cimicifugaceae Bromhead 1840, Clematidaceae Martynov 1820, Delphiniaceae Brenner 1886, Helleboraceae Vest 1818, Nigellaceae J. Agardh 1858, Thalictraceae

Rafinesque 1815). 65–71/2500. Distributed throughout the world but centered in temperate and cold regions of the Northern and Southern Hemispheres.

2.1 COPTIDOIDEAE

Perennial rhizomatous herbs or (*Xanthorhiza*) small shrub. Leaves ternately or pinnately compound, pedate. Flowers terminal to stem, or in a racemiform cyme or cymes, bisexual, sometimes unisexual. Petals stalked, shorter than sepals, blades basically thick. Pollen grains pantocolpate or pantoporate. Carpels 5–15, stalked and sessile, open (*Coptis*) or closed. Ovules several to two in a carpel, bitegmic. Follicles several, free, without, rarely with a longitudinal vein, without distinct transverse veins on lateral faces. Accumulating large amount of benzylisoquinoline alkaloids, the major is berberin; ranunculin absent. According to Yang et al. (1993) the metaphase chromosomes, were smaller than the R-type ones but obviously larger than the T-type ones. In his opinion it seems to be somewhat reasonable that Gregory (1941) classified the chromosomes of *Coptis* as C-type. n = 9 – COPTIDEAE: *Coptis*; XANTHORHIZEAE: *Xanthorhiza*.

2.2 THALICTROIDEAE (ISOPYROIDEAE)

Perennial or annual herbs. Leaves ternately or pinnately compound, pedate. Flowers actinomorphic or more or less zygomorphic, bisexual or rarely polygamous. Petals present or (*Enemion* and *Thalictrum*) absent. Pollen grains 3-colpate to pantocolpate and pantoporate. Carpels sessile, with or without transverse veins, multiovulate to uniovulate. Ovules usually bitegmic. Fruitlets follicles or (*Thalictrum*) achenes. Accumulating large amount of benzylisoquinoline alkaloids like berberine and magnoflorine; cyanogenic compounds and columbinic acid are also common; ranunculin absent. Chromosomes usually of T-type. n = 6 (*Dichocarpum*), 7 (mostly). – ISOPYREAE: *Enemion, Isopyrum, Leptopyrum* (including ? *Neoleptopyrum), Paropyrum, Paraquilegia, Semiaquilegia, Urophysa, Aquilegia*; THALICTREAE: *Thalictrum, Anemonella*; DICHOCARPEAE: *Dichocarpum*.

2.3 CIMICIFUGOIDEAE

Geophilous perennials with sympodial rhizome. Vessels in *Asteropyrum* with scalariform perforation with over 30 bars (Chen and Li 1990). Leaves ternately compound. Flowers actinomorphic or sometimes more or less irregular, bisexual or rarely unisexual. Petals present. Staminodia flat or (*Eranthis*) peltate. Ovules bitegmic. Fruitlets follicles or (*Actaea*) berries. Embryo relatively small. Producing isoprenylated 2-methyl chromones, and in *Eranthis* also small amounts of benzylisoquinoline alkaloids, but not accumulating. Ranunculin absent. Chromosomes of R-type, but in *Actaea* two chromosomes of T-type (Yang 1998), n = 8. – ASTEROPYREAE: *Asteropyrum*; CIMICIFUGEAE (ACTAEEAE): *Beesia, Anemonopsis, Souliea, Cimicifuga, Actaea*; ERANTHIDEAE: *Eranthis, Shibateranthis*.

2.4 TROLLIOIDEAE (ADONIDOIDEAE)

Perenial or seldom annual herbs. Leaves compound or simple. Flowers actinomorphic, bisexual. Petals present or absent. Pollen grains tricolpate. Ovules mostly bitegmic. Fruits of follicles or achenes. Producing, but not accumulating small amount of benzylisoquinoline alkaloids. Ranunculin absent. Chromosomes of R-type, n = 8. – CALTAEAE: *Caltha* (including ? *Thacla*); TROLLIEAE: *Trollius, Megaleranthis, Calathodes*; ADONIDEAE: *Callianthemum, Adonis*.

2.5 ACONITOIDEAE (DELPHINIOIDEAE)

Perennial or annual herbs with simple or compound leaves. Flowers actinomorphic (*Nigelleae*) or more often zygomorphic, bisexual. Petals present, spured or not spured. Pollen grains sometimes striate. Ovules bitegmic. Fruitlets follicles, free or more or less connate, seldom capsulate. Producing benzylisoquinoline alkaloids in small amounts, but not accumulating them, and in *Delphineae* also producing toxic diterpene alkaloids and mannitol. Ranunculin absent. Chromosomes of R-type, n = 8 or less often 6 (*Nigella* and *Garidella*) or 7 (*Komaroffia*). – DELPHINEAE: *Aconitum, Delphinium, Consolida, Aconitella*; NIGELLEAE: *Komaroffia, Nigella, Garidella*.

2.6 HELLEBOROIDEAE

Perennial herbs with pedately, palmately of ternately divided leaves. Sepals persistent. Petals tubular or cup-shaped. Pollen grains 3-colpate and have a finely or coarsely reticulate tectum in which the muri are colliculate, or, more rarely, a complete tectum which is sparsely colliculate and sparsely punctate (Nowicke and Skvarla 1983). Ovules unitegmic. Follicles free or connate, or forming a capsule. Ranunculin-type glucosides generating acrid protoanemonin, bufadienolides, steroidal saponins, phytoecdysones present; benzylisoquinoline alkaloids absent. Unique in contain cardioactive bufadienolides. Chromosomes of R-type, n = 8. – *Helleborus*.

2.7 RANUNCULOIDEAE

Perennial or annual herbs, less often subshrubs or woody lianas. Flowers actinomorphic, mostly bisexual. Petals present or lacking. Pollen grains 3-colpate to pancolpate and pantoporate. Ovules unitegmic. Fruitlets achenes or rarely (*Knowltonia*) 1-seeded berries. Ranunculin-type glucosides generating acrid protoanemonin usually present (Hegnauer 1987); triterpenoid saponins often present; cyanogenesis in some species of *Ranunculus*; benzylisoquinoline alkaloids absent, except the presence of small acount of magnoflorine in *Clematis*. Chromosomes of R-type, n = 8 or rarely (some species of *Anemone* and *Raninculus*) 7, or rarely 9. – ANEMONEAE: *Anemone* (including ? *Anemonastrum, Anemonidium, Anemonoides, Arsenjevia*), *Anemoclema, Pulsatilla, Miyakea, Metanemone, Hepatica, Barneoudia, Oreithales, Knowtonia, Clematis* (including *Archiclematis, Naravelia, Clematopsis*); *Atragene*; RANUNCULEAE: *Myosurus, Arcteranthis, Halerpestes, Cyrtorhyncba, Oxygraphis, Callianthemoides, Paroxygrapis, Hamadryas, Aphanostemma, Ranunculus, Buschia, Ficaria, Batrachium, Ceratocephala, Beckwithia, Krapfia, Laccopetalum*; TRAUTVETTERIEAE: *Trautvetteria, Kumlienia*.

The relatively most primitive subfamily is probably *Coptidoideae*, which includes only three rather isolated genera. The less specialized of them is *Coptis* (16 species in the Himalayas, eastern Asia and Pacific North America). Its carpels stipitate and not completely fused in the flowering stage, their venation without transverse veins and in an early stage they conduplicate. The nodal structure of *Coptis* is also primitive – trilacunar or pentalacunar (Tamura 1981). Relatively primitive are also Cimicifugoideae and Thalictroideae, while Ranunculoideae are the most advanced and diversified.

Bibliography

Aitzetmüller K. 1995. Fatty acid patterns of Ranunculaceae seed oils: phylogenetic relationships. Plant Syst Evol., Suppl. 9: 229–240.

Avita S and JA Inamdar. 1980. Structure and ontogeny of stomata in Ranunculaceae and Paeoniaceae. Flora 171: 354–370.

Avita S and JA Inamdar. 1981. Diversity in the vessel elements of Ranunculaceae-Paeoniaceae complex. Feddes Repert. 92: 397–411.

Bhandari NN. 1966. Studies in the family Ranunculaceae: IX. Embryology of *Adonis*. Phytomorphology 16: 578–587.

Bhandari NN and S Asnani. 1966. Studies in the family Ranunculaceae: XI. Morphology and embryology of *Ceratocephalus falcatus* Per. Beitr. Biol. Pfl. 45: 271–290.

Bhandari NN and MR Vijayaraghavan. 1970. Studies in the family Ranunculaceae: XII. Embryology of *Aquilegia vulgaris*. Beitr. Biol. Pfl. 46: 337–354.

Bowers H. 1891. A contribution to the life-history of *Hydrastis canadensis*. Bot. Gaz. 16: 73–82.

Brouland M. 1935. Recherches sur l'ανατομιε florale des Ranunculacees. Botanist 27: 1–278.

Carlquist 1995. Wood and bark anatomy of Ranunculaceae (including *Hydrastis*) and Glaucidiaceae. Aliso 14: 65–84.

Chen Y-Z and Z-L Li. 1990. Comparative studies of perforation plate structures of vessels in Ranunculaceae. Acta Bot. Sinica 32: 245–251.

Chen Y-Z and Z-L Li. 1993. Comparative anatomical studies on root structure of Ranunculaceae. Acta Bot. Sinica 35: 27–34.

Compton JA and A Culham. 2002. Phylogeny and circumscription of the tribe Actaeeae (Ranunculaceae). Syst. Bot. 27: 502–511.

Compton JA, A Culham, JG Gibbings, and SL Jury. 1998. Phylogeny of *Actaea* including *Cimicifuga* (Ranunculaceae) inferred from *nr*DNA ITS sequence variation. Biochem. Syst. Ecol. 26: 185–197.

Compton JA, A Culham, and SL Jury. 1998. Reclassification of *Actaea* to include *Cimicifuga* and *Souliea* (Ranunculaceae): phylogeny inferred from morphology, *nr*DNA ITS, and *cp*DNA *trn*L-F sequence variation. Taxon. 47: 593–634.

Daumann E. 1969. Blütenmorphologie und Bestäubungsökologie einiger Ranunculaceen (*Cimicifuga* L., *Actaea* L., *Thalictrum* L.). Preslia 41: 213–219.

Dinan L, T Savchenko, and P Whiting. 2002. Chemotaxonomic significance of ecdysteroid agonists and antagonists in the Ranunculaceae: phytoecdysteroids in the genera *Helleborus* and *Hepatica*. Biochem. Syst. Ecol. 30: 171–182.

Doroszewska A. 1974. The genus *Trollius* L.: A taxonomical study. Monogr. Bot. 41: 1–167.

Duncan T and CS Keener. 1991. A classification of the Ranunculaceae with special reference to the Western Hemisphere. Phytologia 70: 24–27.

Ehrendorfer F and R Samuel. 2001. Contributions to a molecular phylogeny and systematics of *Anemone* and related genera (Ranunculaceae-Anemoninae). Acta Phytotax. Sinica 39: 293–307.

Eichler H. 1958. Revision der Ranunculaceen Malesiens. Bibl. Bot. 124: 1–110.

Engell K. 1995. Embryo morphology of the Ranunculaceae. Plant Syst. Evol., Suppl. 9: 207–216.

Erbar C, S Kusma, and P Leins. 1999. Development and interpretation of nectary organs in *Ranunculaceae*. Flora (Germany) 194: 317–332.

Ezelarab GE and KJ Dormer. 1963. The organization of the primary vascular system in Ranunculaceae. Ann. Bot. 27: 23–38.

Forster P. 1997. Die Keimpflanzen der Tribus Ranunculeae DC. und der Tribus Adonideae Kunth (Ranunculaceae). Flora (Germany). 192(2): 133–142.

Fu D-Z. 1990. Phylogenetic consideration of the subfamily Thalictroideae (Ranunculaceae). Cathaya 2: 181–190.

Grey-Wilson C. 2000. *Clematis*, the genus. Timber Press, Portland, OR.

Gregory WC. 1941. Phylogenetic and cytological studies in the Ranunculaceae. Trans. Amer. Philos. Soc., n.s., 31: 443–521.

Hammond HD. 1955. Systematic serological studies in Ranunculaceae. Serol. Mus. Bull. 14: 1–3.

Hegnauer R. 1986. Comparative phytochemistry and plant taxonomy. G. Bot. Ital. 120: 15–26.

Hiepko P, ed. 1995. Ranunculaceae. Die natürlichen Pflanzenfamilien, ed.2, 17a, IV: 555 S. Ducker & Humboldt, Berlin.

Hoot SB. 1991. Phylogeny of the Ranunculaceae based on epidermal microcharacters and macromorphology. Syst. Bot. 16: 741–755.

Hoot SB. 1995. Phylogeny of the Ranunculaceae based on preliminary *atp*B, *rbc*L and 18S nuclear ribosomal DNA sequence data. Plant Syst. Evol., Suppl. 9: 241–251.

Hoot SB, AA Reznicek, and JD Palmer. 1994. Phylogenetic relationships in *Anemone* (Ranunculaceae) based on morphology and chloroplast DNA. Syst. Bot. 19: 169–200.

Hsiao P-K. 1980. A preliminary study of the correlation between phylogeny, chemical constituents and pharmaceutical aspects in the taxa of Chinese Ranunculaceae. Acta Phytotax. Sinica 18: 142–153.

Hutchinson J. 1923. Contributions towards a phylogenetic classification of flowering plants. I. Kew Bull. 1923: 65–89.

Ichinohe Y and M Tamura. 1977. The characteristic components and phylogenetic relationships of genus *Aconitum* and its allies. Bull. Dept. Gen. Educ., Coil. Sci. and Tech., Nihon Univ. 22: 71–81; 23: 27–36.

Iriki Y and H Minamisawa. 1983. D-galactose and a ribitol-like substance in *Hydrastis canadensis* L. Nippon Nogeikagaku Kaishi 57: 319–321.

Ivashina T and S Ootani. 1990. Three flavonol allosides from *Glaucidium palmatum*. Phytochemistry 29: 3639–3641.

Jalan S. 1963. Studies in the family Ranunculaceae – IV. The embryology of *Actaea spicata* Linn. Phytomorphology, 13: 338–347.

Janchen E. 1949. Die systematische Gliederung der Ranunculaceen und Berberidaceen. Oesterr. Akad. Wiss., Math.-Naturwiss. Kl., Denkschr. 108: 1–82.

Jensen U. 1968. Serologische Beiträge zur Systematik der Ranunculaceae. Bot. Jahrb. Syst. 88: 204–268, 269–310.

Jensen U. 1971. Zur systemarischen Stellung der Helle-borinae (Ranunculaceae). Taxon 20: 747–758.

Jensen U. 1973. The interpretation of comparative sero-logical results. In: G Lendz and J Lantesson, eds. Chemistry in botanical classification. Nobel Symposium 25, pp. 17–227. Academic Press, New York/London.

Jensen U. 1995a. Secondary compounds of the Ranunculiflorae. Plant Syst. Evol., Suppl. 9: 85–97.

Jensen U. 1995b. Serological legumin data and the phylogeny of the Ranunculaceae. Plant Syst. Evol., Suppl. 9: 217–227.

Jensen U, SB Hoot, JT Johansson, and K Kosuge. 1995. Systematics and phylogeny of the Ranunculaceae – a revised family concept on the basis of molecular data. Plant Syst. Evol., Suppl. 9: 273–280.

Johansson JT. 1995. A revised chloroplast DNA phylogeny of the Ranunculaceae. Plant Syst. Evol., Suppl. 9: 253–251.

Johansson JT and RK Jansen. 1993. Chloroplast DNA variation and phylogeny of the Ranunculaceae. Plant Syst. Evol. 187: 29–49.

Keener CS. 1993. A review of the classification of the genus *Hydrastis* (Ranunculaceae). Aliso 13: 551–558.

Khan HA. 1991 [1992]. Palynotaxonomy and phylogeny of Ranunculaceae. Geophytology 21: 207–210.

Kordyum EL. 1959. Comparative embryological investigation of the family Ranunculaceae DC. Ukrainsk. Bot. Zhurn. 16: 32–43 (in Ukrainian, with Russian and English summaries).

Kosuge K. 1994. Petal evolution in Ranunculaceae. Plant Syst. Evol., Suppl. 8: 185–191.

Kosuge K, K Sawada, T Denda, J Adacgi, and K Watanabe. 1995. Phylogenetic relationships of some genera in the Ranunculaceae based on alcohol dehydrogenase genes. Plant Syst. Evol., Suppl. 9: 263–271.

Kosuge K and M Tamura. 1989. Ontogenetic studies on petals of the Ranunculaceae. J. Jap. Bot. 64: 65–67.

Kumazawa A. 1930. Morphology and biology of *Glau-cidium palmatum* Sieb. et Zucc. with notes of affinities to the allied genera *Hydrastis, Podophyllum*, and *Diphylleia*. J. Fac. Sci. Univ. Tokyo, Sect. 3, Bot. 2: 345–380.

Kumazawa M. 1936. Pollen grain morphology in Ranunculaceae, Lardizabalaceae, and Berberidaceae. Jpn. J. Bot. 8: 19–46.

Kumazawa M. 1937. Comparative studies on the vernation in the Ranunculaceae and Berberidaceae. Jpn. J. Bot. 13: 573–586, 659–667, 713–726 (in Japanese).

Kumazawa M. 1938a. On the ovular structure in the Ranunculaceae and Berberidaceae. Jpn. J. Bot. 14: 10–25.

Kumazawa M. 1938b. Systematic and phylogenetic consideration of the Ranunculaceae and Berberidaceae. Bot. Mag. (Tokyo) 52: 9–15.

Kurita M. 1957. Chromosome studies in Ranunculaceae: II. Karyotype of subtribe Cimicifuginae. Rep. Biol. Inst. Ehime Univ. 1: 11–17.

Kurita M. 1958a. Chromosome studies in Ranunculaceae: VIII. Karyotype and phylogeny. Rep. Biol. Inst. Ehime Univ. 5: 1–14.

Kurita M. 1958b. Chromosome studies in Ranunculaceae: IX. Comparison of chromosome volume between a 14- and a 16-chromosome species in *Anemone* and in *Ranunculus*. Rep. Biol. Inst. Ehime Univ. 6: 1–7.

Kurita M. 1960. Chromosome studies in Ranunculaceae. 16. Comparison of an aspect of nucleus and chromosome between several genera. Mem. Ehime Univ., sect. 2, ser. B, 4: 53–58.

Kurita M. 1963. Chromosome studies in Ranunculaceae. 21. Karyotype of *Myosurus* and *Adonis*. Mem. Ehime Univ., sect. 2, ser. B, 4: 487–492.

Kurita M. 1965. Chromosome studies in Ranunculaceae. 23. Karyotypes and chromosome numbers of some. Species. Mem. Ehime Univ., sect. 2, ser. B, 5: 89–95.

Langlet O. 1932. Über Chromosomenverhaltnisse und Systematik der Ranunculaceae. Svensk Bot. Tidskr. 26: 381–400.

Leconte H, LM Campbell, and DW Stevenson. 1995. Ordinal and familial relationships of Ranunculid genera. Plant Syst. Evol., Suppl. 9: 99–118.

Lee ST. 1992. Palynological relationships among *Calathodes* and its relative genera. Korean J. Plant. Taxon. 22: 23–31.

Lemesle R. 1948. Position phylogenetique de *l'Hydrastis canadensis* L. et du *Crossosoma californicum* Nutt., d'a-pres les particularites histologiques du xyleme. Compt. Rend. Hebd. Seances Acad. Sci. 227: 221–223.

Lemesle R. 1950. *L'Hydrastis canadensis* L. et ses princi-pales falsifications. Rev. Gen. Bot. 57: 5—23.

Lemesle R. 1955. Contribution a l'ετυδε de quelques families de dicotyledones considerees comme primitives. Phytomorphology 5: 11–45.

Leppik EE. 1964. Floral evolution in the Ranunculaceae. Iowa State Coll. J. Sci. 39: 1–101.

Lewitsky GA. 1931. The karyotype in systematics, on the base of karyology of the subfamily Helleboreae. Trudy Prikl. Bot. 27: 187–240 (in Russian).

Luo Y, F-M Zhang, and Q-E Yang. 2005. Phylogeny of *Aconitum* subgenus *Aconitum* (Ranunculaceae) inferred from ITS sequences. Plant Syst. Evol. 252: 11–25.

Mathew B. 1989. Helleborus. Alpine Garden Society, Woking.

Mikeda O, K Kita, T Handa, and T Yukawa. 2006. Phylogenetic relationships of *Clematis* (Ranunculaceae) based on chloroplast and nuclear DNA sequences. Bot. J. Linn. Soc. 152: 153–168.

Nowicke JW and JJ Skvarla. 1980. Pollen morphology: the potential influence in higher order systematics. Ann. Missouri Bot. Gard. 66: 633–700.

Nowicke JW and JJ Skvarla. 1983. A palynological study of the genus *Helleborus* (Ranunculaceae). Grana 22: 129–140.

Nowicke JW and JJ Skvarla. 1995. Ranunculaceae. In: P Hiepko, ed. Die natürlichen Pflanzenfamilien, ed. 2, 17a: 129–159. Ducker & Humboldt, Berlin.

Okada H and M Tamura. 1979. Karyomorphology and relationship in the Ranunculaceae. J. Jap. Bot. 54: 65–77.

Park CW, and HW Lee. 1995. Trichome morphology of *Cimicifuga* L. (Ranunculaceae) and its taxonomic significance. J. Plant Biol. 38(3): 289–295.

Pohl J. 1894. Botanische Mitteilung über *Hydrastis canadensis*. Bibl. Bot. 29: 1–12.

Prantl K. 1887. Beiträge zur Morphologie und Systematik der Ranunculaceen. Bot. Jahrb. Syst. 9: 225–273.

Qi W-Q, F Yun, X-Z Chenzhu, Z-L Li, and R-L You. 1997. Studies on the reproductive characteristics of *Cimicifuga nanchuanensis* (Ranunculaceae), an endemic endangered species to China. Acta Bot. Sinica 39: 7–10.

Qiong Y and Q-E Yang. 2006. Tribal relationships of *Beesia, Eranthis* and seven other genera of Ranunculaceae: evidence from cytological characters. Bot. J. Linn. Soc. 150: 267–289.

Ro K-E, CS Keener, and BA McPherson. 1997. Molecular phylogenetic study of the Ranunculaceae: utility of the nuclear 26S ribosomal DNA in inferring intrafamilial relationships. Molecular Phylogen. Evol. 8: 117–127.

Ro KE, and BA Mepheron. 1997. Molecular phylogeny of the *Aquilegia* group (Ranunculaceae) based on internal transcribed spacers and 5.8S nuclear ribosomal DNA. Biochem. Syst. Ecol. 25: 445–461.

Rohweder O. 1967. Karpellbau und Synkarpie bei Ranunculaceen. Ber. Schweiz. Bot. Ges. 77: 376–432.

Ruijgrok HWL. 1966. The distribution of ranunculin and cyanogenetic compounds in the Ranunculaceae. In: T Swain, ed. Comparative phytochemistry, pp. 175–186. London.

Santisuk T. 1979. A palynological study of the tribe Ranunculeae. Opera Bot. 48: 1–76.

Sastri RLN. 1969a. Comparative morphology and phylogeny of the Ranales. Biol. Rev. Cambridge Philos. Soc. 44: 291–319.

Savitsky VD. 1982. Morphology, classification, and evolution of pollen in buttercup family. Naukova Dumka Kiev (in Russian).

Schrodinger R. 1909. Der Blütenbau der zygomorphen Ranunculaceen und seine Bedeutung für die Stammesgeschichte der Helleboreen. Abh. K. K. Zool.-Bot. Ges. Wien 4–5: 1–63.

Slavikovaa Z. 1971. Zur Blutenmorphologie von *Adonis vernalis* L. Oesterr. Bot. Z. 119: 447–453.

Smith GH. 1926. Vascular anatomy of *Ranalian* flowers: I. Ranunculaceae. Bot. Gaz. 82: 1–29.

Smith PG. 1973. A revision of *Caltha* (Ranunculaceae). Blumea 21: 119–130.

Sun A-C and F-X Wang. 1983. Contribution to the morphology and embryology of *Asteropyrum peltatum*. Bot. Res. 1: 85–90.

Tamura M. 1962. Petiolar anatomy in the Ranunculaceae. Sci. Rep. Osaka Univ. 11: 19–47.

Tamura M. 1963–1968. Morphology, ecology, and phylogeny of the Ranunculaceae. Parts 1–8. Sci. Rep. Osaka Univ. 11: 115–126; 12: 141–156; 13: 25–35; 14 (I): 53–71; 14 (2): 27–48; 15 (I): 13–35; 16 (2): 21–43; 17 (I): 41–56.

Tamura M. 1981. Morphology of *Coptis japonica* and its meaning in phylogeny. Bot. Mag. (Tokyo) 94: 165–176.

Tamura M. 1984. Phylogenetical consideration of the Ranunculaceae. Korean J. Plant Taxon. 14: 33–42.

Tamura M. 1987. A classification of genus *Clematis*. Acta Phytotax. Geobot. 38: 33–44.

Tamura M. 1990–1992. A new classification of the family Ranunculaceae. Parts 1–2. Acta Phytotax. Geobot. 41: 93–101; 42:177–188; 43: 53–58.

Tamura M. 1993. Ranunculaceae. In: K Kubitzki, ed. The families and genera of vascular plants, vol. 2, pp. 563–583. Springer, Berlin/Heidelberg/New York.

Tamura M. 1995. Phylogeny and classification of the Ranunculaceae. Plant Syst. Evol., Suppl. 9: 201–206.

Tamura M. 1995. Ranunculaceae. In: Hiepko P. ed. Die Natürlichen Pflanzenfamilien, Zwei Aufl., 17a(4). Ducker & Humboldt, Berlin.

Tamura M and K Kosuge. 1989. Classification of the Isopyroideae (Ranunculaceae). Acta Phytotax. Geobot. 40: 31–35.

Tamura M and AL Lauener. 1968. A revision of *Isopyrum, Dichocarpum*, and their allies. Notes Roy. Bot. Gard. Edinb. 37: 431–466.

Tamura M and Y Mizumoto. 1972. Stages of embryo development in ripe seeds or achenes of the Ranunculaceae. J. Jpn. Bot. 47: 225–237.

Tobe H. 1995. Ranunculaceae – embryology. In: P Hiepko, ed. Die natürlichen Pflanzenfamilien. 2 Aufl. Band 17a IV, pp. 106–128. Ducker & Humboldt, Berlin.

Tobe H. 2003. Hydrastidaceae. In: K Kubitzki, ed. The families and genera of vascular plants. Vol. 5: 405–409. Springer, Berlin/Heidelberg/New York.

Tobe H and RC Keating. 1985. The morphology and anatomy of *Hydrastis* (Ranunculaceae): systematic reevaluation of the genus. Bot. Mag. (Tokyo) 98: 291–316.

Trifonova VI. 1988. Ranunculaceae. In: A. Takhtajan, ed. Comparative seed anatomy, vol. 2, pp. 176–181. Nauka, Leningrad (in Russian).

Troll W. 1933. Beiträge zur Morphologic des Gynoe- ceums: III. Über das Gynoeceum von *Nigella* und einiger anderer Helleboreen. Planta 21: 266–291.
Tucker SC. 1966. The gynoecial vascular supply in *Caltha*. Phytomorphology 16: 339–342.
Tucker SC and SA Hodges. 2005. Floral ontogeny of *Aquilegia, Semiaquilegia* and *Enemion* (Ranunculaceae). Int. J. Plant Sci. 166: 557–574.
Vishnu-Mittre and BD Sharma. 1963. Studies of Indian pollen grains 2. Ranunculaceae. Pollen et spores 5: 285–296.
Wang W, R-Q Li, and Z-D Chen. 2005. Systematic position of *Asteropyrum* (Ranunculaceae) inferred from chloroplast and nuclear sequences. Plant Syst. Evol. 255: 41–54.
Wang XQ, DY Hong, and ZY Li. 1993. A study on pollen and seed coat in the tribe Cimicifugeae and some allied genera (Ranunculaceae). Cathaya 5: 131–149.
Wang XQ, ZR Deng, and DY Hong. 1998. The systematic position of *Beesia*: evidence from ITS (*nr*DNA) sequence analysis. Acta Phytotax. Sinica 36: 403–410.
Wang XQ, ZY Li, and DY Hong. 1994. A karyomorphological studie of nine species in four genera of Ranunculaceae. Cathaya 6: 43–56.
Werth E. 1941. Die Blütennektarien der Ranunculaceen und ihre phylogenetische Bedeutung. Ber. Deutsch. Bot. Ges. 59: 246–256.
Xi Y-Z, J-C Ning, and X-P. Fu. 1993. Pollen morphology of the tribe Trollieae and its taxonomic significance. Cathaya 5: 115–130.
Xiao PG. 1980. A preliminary study of correlation between phylogeny, chemical constituents and pharmaceutical aspects in the taxa of Chinese Ranunculaceae. Acta Phytotax. Sinica 18: 142–153.
Xuhan X, and AAM Van Lammeren. 1997. Structural analysis of embryogenesis and endosperm formation in celery-leafed buttercup (*Ranunculus sceleratus* L.). Acta Bot. Neerl. 46: 291–301.
Yang Q-E. 1998. Does *Actaea asiatica* have the most symmetric and primitive karyotype in the Ranunculaceae? Acta. Phytotax. Sinica 36: 490–495.
Yang Q-E. 1999. Karyomorphology of four species in *Cimicifuga* (Ranunculaceae) from China, with some cyto-geographical notes on *C. foetida*. Acta Phytotax. Sinica 37(5): 433–444.
Yang Q-E. 2000. Karyomorphology of the genus *Oxygraphis* Bunge (Ranunculaceae). Acta Phytotax. Sinica 38: 350–354.
Yang Q-E. 2002. Cytology of the tribe Trollieae and of the tribe Cimicifugeae in the Ranunculaceae: a comparative study. Acta Phytotax. Sinica 40: 52–65.
Yang Q-E, X Gong, Z-J Gu, and Q-A Wu. 1993. A karyomorphological study of five species in the Ranunculaceae from Yunnan, with a special consideration on systematic positions of *Asteropyrum* and *Calathodes*. Acta Bot. Yunn. 15: 179–190.
Yang Q-E, ZJ Gu, and H Sun. 1995. The karyotype of *Beesia deltophylla* and its systematic significance. Acta. Phytotax. Sinica 32: 225–229.
Yuan Q and Q-E Yang. 2006a. Tribal relationships of *Beesia, Eranthis* and seven other genera of Ranunculaceae: evidence from cytological characters. Bot. J. Linn. Soc. 150: 267–289.
Yuan Q and Q-E Yang. 2006b. Cytology, palynology and taxonomy of *Asteropyrum* and four other genera of Ranunculaceae. Bot. J. Linn. Soc. 152: 15–26.
Zhang ZY. 1982. Chromosome observation of three Ranunculaceous genera in relation to their systematic position. Acta Phytotax. Sinica 20: 402–409.
Ziman SN and CS Keener. 1989. A geographical analysis of the family Ranunculaceae. Ann. Missouri Bot. Gard. 76: 1012–1049.

Order 26. CIRCAEASTERALES

Small perennial or annual herbs. Vessels usually with simple perforations. Nodes unilacunar. Leaves alternate or opposite, palmate or simple, 2-ranked, petiolate, with an open dichotomous venation, the veins ending in teeth of the lamina, estipulate. Stomata anomocytic. Flowers very small, in terminal cyme (*Circaeaster*) or solitary, devoid of bracts and bracteoles, bisexual, actinomorphic, apetalous. Sepals 2–3 or 5–9, scalelike or petaloid, valvate, persistent. Floral nectaries present (*Kingdonia*) or absent (*Circaeaster*). Stamens 1(2)3 or 14–17; anthers tetrasporangiate or disporangiate, introrse or latrorse, opening longitudinally; connective not prolonged. Tapetum glandular. Pollen grains 2-celled, 3-colpate, with striate tectum. Gynoecium apocarpous, of (1)2-9 conduplicate carpels with 1 or 2 pendulous ovules per carpel. Ovule orthotropous (hemitropous according to Hu et al. 1990), unitegmic, tenuinucellate. Before fertilization integument covers 1/3 part of nucellus, after that, elongates and then disappers (Ren Y. et al. 1998). Female gametophyte tetrasporic, 4- or 8-nucleate. Fertilization porogamous (*Kingdonia*) or mesogamous (*Cicraeaster*) Endosperm helobial or cellular. The first division of primary endosperm nucleus longitudinal. The embryogeny of *Kingdonia* subtype of Caryophyllad-type. Fruits achenes. Seeds small, without testa, seed coat degenerating, embryo relatively large. Endosperm differentiated into two parts – storage micropylar part and absorbing chalazal part, which are separated by an intermediate suberized layer; the cells of endosperm surface with thikend outer. n = 9, 15.

A rather isolated group that shows some similarities with Lardizabalales, Berberidales and Ranunculales. They often included in Ranunculales, but markedly differ from them, testaless seeds and some embryological features such as: Caryophyllad type of embryogeny, the first division of primary endosperm nucleus, differentiating of endosperm on two parts, thickened outer walls of cells of endosperm surface, integument covering 1/3 part of nucellus before fertilization and disappearing after that (Ren Y. et al. 1998). However, according to Nowicke and Skvarla (1982), palynologically *Circaeaster* and *Kingdonia* are related to *Hydrastis* and to the tribe Epimedieae in the Podophyllaceae than to any remaining taxa examined in the Ranunculanae. They base the relationships on the similarity of the tectum, a compound layer of striae. However, there are also many dissimilarities that make Circaeasterales a very distinct group (Takhtajan 1997).

Key to Families

1 Perennial herbs with long, creeping, slender, branching, scaly rhizome with endotrophic mycorrhizae, a solitary leaf and flower arising from each bud. Unilacunar node with four traces in foliage leaves and a double trace in both cataphylls and scale leaves. Leaves usually single, basal, long-petiolate, with membranous, suborbicular lamina, 5–7-palmatisect, segments cuneate, variously lobed and dentate on the upper margin. Flowers solitary, with a long pedicel. Sepals 5–6, rarely 8–9 in two series with 5–6 in the outer and 3 in the inner series, pale yellow-green, ovate, acuminate, with dichotomous venation. Androecium consists of 9–11 outer sterile and 5–6 inner fertile stamens; each sterile stamen differentiates into a stalk and a broadened head with a groove in the centre; fertile stamens differentiate into costate filaments and 2-locular, tetrasporangiate, extrorse anthers (Ren et al. 2005). Pollen grains and the tectum consisting of two layers of striae or mostly striate. Gynoecium of 5–6, rarely 9 carpels, opened in the upper third, with subulate stylodium, which gradually becomes recurved over the dorsal edge of the carpel, and 1 ventral pendulous, hemitropous ovule per carpel. Inner epidermal cells of integument near the micropyle elongated radically to form a palisade-like structure. Endosperm helobial. Achenes 1–5, glabrous, with thick stipe, beaks elongate, strongly bent outward at base. Seeds narrow, ellipsoid; endosperm cell walls thin, n = 9. . . . . 1. KINGDONIACEAE

1 Annual herbs with an elongated stemlike hypocotyl, short stem, and linear persistent cotyledons and leaf-rosette. Unilacunar nodes with a single trace in foliage leaf. Leaves more or less opposite, petiolate, cuneate-spatulate, dentate at apex. Inflorescences are thyrsoid with extremely short main axis and 10-30 (-40) flowers; 2–3 flowers in the center of the inflorescence have longer pedicels and bloom first (Tian et al. 2006). Flowers bisexual, nearly actinomorphic. Sepals 2–3, scalelike, valvate, persistent, without vascular bundles. Stamens (1-)2(-3), alternisepalous; anthers basifixed, disporangiate, latrorse. Pollen grains with the irregularly striate tectum. Gynoecium of 1-2(-4) fusiform, shortly stipitate, closed carpels with very short stylodium and slightly elongate and slightly oblique papillate stigma. Ovule solitary, small to very small, orthotropous, pendulous, submarginal, the uppermost ovule aborting. Integument without palisade-like structure. Endosperm cellular. Fruits long elliptic green achenes covered by uncinate hairs. Seeds small; embryo well differentiated, small, straight, with short cotyledons; endosperm copious; endosperm cell wall thick. n = 15. . . . . 2. CIRCAEASTERACEAE

1. KINGDONIACEAE

A. S. Foster ex Airy Shaw 1964. 1/1. High mountains of southwestern China, primarily in western Sichuan.

Kingdonia.

2. CIRCAEASTERACEAE

Hutchinson 1926. 1/1. Northwestern Himalayas to western China.

Circaeaster.

Despite some common features with the Kingdoniaceae, such as dichotomous leaf venation, pollen morphology, pendulous, unitegmic, tenuinucellate ovules. Circaeasteraceae have many unique characters that make them rather distinct.

Bibliography

Balfour IB and WW Smith. 1914. *Kingdonia uniflora.* In: Diagnoses specierum novarum LI-CII (Species Chinenses). Notes Roy. Bot. Gard. Edinb. 8: 191–192.

Diels L. 1932. *Circaeaster*: Eine hochgradig reduzierte Ranunculaceae. Beih. Bot. Centralbl. 49: 55–60.

Fedotova TA. 1988. Circaeasteraceae. In: A Takhtajan, ed. Comparative seed anatomy, 2: 181–184. Nauka, Leningrad (in Russian).

Foster AS. 1959. The morphological and taxonomic significance of dichotomous venation in *Kingdonia uniflora* Balfour f. et W.W. Smith. Notes Roy. Bot. Gard. Edinb. 23: 1–12.

Foster AS. 1961. The floral morphology and relationships of *Kingdonia uniflora*. J. Arnold Arbor. 41: 397–410.
Foster AS. 1963. The morphology and relationships of Circaeasteraceae. J. Arnold Arbor. 44: 299–321.
Foster AS. 1966. Morphology of anastomoses in the dichotomous venation of *Circaeaster*. Am. J. Bot. 53: 588–599.
Foster AS. 1968. Further morphological studies on anastomoses in the dichotomous venation of *Circaeaster*. J. Arnold Arbor. 49: 52–67.
Foster AS. 1970. Types of blind vein-endings in the dichotomous venation of *Circaeaster*. J. Arnold Arbor. 51: 70–80.
Foster AS. 1971. Additional studies on the morphology of blind vein-endings in the leaf of *Circaeaster agrestis*. Am. J. Bot. 58: 263–272.
Foster AS and HJ Arnott. 1960. Morphology and dichotomous vasculature of the leaf of *Kingdonia uniflora*. Am. J. Bot. 47: 684–698.
Hoot SB, PS Herendeen, and PR Crane. 1995. Phylogenetic relationships and floral morphology in the family Circaeasteraceae (*Circaeaster* and *Kingdonia*). Am. J. Bot. 82(6): 136 (Abstract).
Hu A. 1987. Studies on the morphology of *Kingdonia uniflora* Balf. F. et W.W. Smith I. and *Circaeaster agrestis* Maxim. Intern. Bot. Cong. Abstract 5–162b-3. Berlin.
Hu Z-H and K-M Lee. 1979. Morphological studies of *Kingdonia uniflora* F.Balf. et W.W.Smith. II. The anatomy of rhizome. Acta Phytotax. Sinica 17: 23–29.
Hu Z-H and L-X Tian. 1985. Studies on morphology of *Kingdonia uniflora* F.Balf. et W.W.Smith. III. The morphology and anatomy of flowers, fruits and seeds. Acta Phytotax. Sinica 23: 170–178.
Hu Z-H and J Yang. 1987. Morphological studies of *Circaeaster agrestis* Maxim: I. Process of embryological development. Acta Phytotax. Sinica 25: 350–356 (in Chinese).
Hu Z-H, K-M Li, and X-L Lee. 1964. Distribution and general morphology in *Kingdonia uniflora*. Acta Bot. Sinica 12: 351–358.
Hu ZH, J Yang, RQ Jing, and ZM Dong. 1990. Morphological studies on *Circaeaster agrestis*. II. Morphology and anatomy of flower, fruit and seed. Cathaya 2: 77–88.
Junell S. 1931. Die Entwicklungsgeschichte von *Circaeaster agrestis*. Svensk Bot. Tidskr. 25: 238–270.
Kong H-Z and Q-E Yang. 1997. Karyomorphology and relationships of the genus *Circaeaster* Maxim. Acta Phyt. Sinica 35: 494–499.
Kosuge K, F-D Pu, and M Tamura. 1989. Floral morphology and relationships of *Kingdonia*. Acta Phytotax. Geobot. 40: 61–67.
Li Z-L and Z-M Dong. 1987. Anatomical studies on hypocotyls of *Circaeaster*. Acta Bot. Sinica 29: 132–137.
Mu X-J. 1983. Ovule, female and male gametophyte, and fertilization of *Kingdonia uniflora* Balfour F. et W.W. Smith. Acta Bot. Sinica 25: 297–504 (in Chinese).
Mu X-J. 1984. Early development of the endosperm of *Kingdonia uniflora*. Acta Bot. Sinica 26: 668–671.
Nowicke JW and JJ Skvarla. 1982. Pollen morphology and the relationships of *Circaeaster*, of *Kingdonia*, and of *Sargentodoxa* to the Ranunculales. Am. J. Bot. 69: 990–998.
Oliver D. 1895. *Circaeaster agrestis* Maxim. Hooker's Icones Plantarum 4 (4): pl. 2366.
Oxelman B and M Lidén. 1995. The position of *Circaeaster* – evidence from nuclear rinosomal DNA. Plant Syst. Evol., Suppl. 9: 189–193.
Ren Y and ZH Hu. 1995. The morphology of the vegetative organs of *Circaeaster agrestis* (Ranunculaceae) and its taxonomic significance. Cathaya 7: 177–187.
Ren Y and ZH Hu. 1996. Morphological studies on anastomoses and blind veins in dichotomous venation of the leaf in *Kingdonia uniflora*. Acta Phytotax. Sinica 34: 569–576.
Ren Y and Z-H Hu. 1997. The morphology of the dichotomous leaf venation of *Circaeaster agrestis* and its systematic implication. Acta Phytotax. Sinica 35: 219–224.
Ren Y and Z Hu. 1998. Anatomical studies on root, node and leaf of *Kingdonia uniflora*. Acta Bot. Bor. Occid. Sinica 18: 72–77.
Ren Y, ZH Hu, and ZJ Li. 1997. The morphology of the dichotomous leaf venation of *Cicaeaster agrestis* and its systematic implication. Acta Phytotax. Sinica 35: 219–224.
Ren Y, Z-J Li, H-L Chang, Y-J Lei, and A-M Lu. 2004. Floral development of *Kingdonia* (Ranunculaceae s.l., Ranunculales). Plant Syst. Evol. 247: 145–153.
Ren Y, M-L Wang, and Z-H Hu. 1998. *Kingdonia*, embryology and its systematic significance. Acta Phytotax. Sinica 36: 423–427.
Ren Y, Y-P Xiao, and Z-H Hu. 1998. The morphological nature of the open dichotomous leaf venation of *Kingdonia* and *Circaeaster* and its systematic implication. J. Plant Res. 111: 225–230.
Tamura MN, H Okada, Y Ichinohe, MA Take, FD Pu, and M Tamura. 1995. Relationship of *Kingdonia* based on karyomorphology and alkaloid components. J. Jpn. Bot. 70: 118–121.
Tian X, L Zhang, and Y Ren. 2006. Development of flowers and inflorescences of Circaeaster (Circaeasteraceae, Ranunculales). Plant Syst. Evol. 256: 89–96.
Wu C-Y and K Kubitzki. 1993. Circaeasteraceae. In: K Kubitzki, ed. The families and genera of vascular plants, vol. 2, pp. 288–289. Springer, Berlin/Heidelberg/New York.
Yi R, Y-P Xiao, and Z-H Hu. 1998. The morphological nature of the open dichotomous leaf venation of *Kingdonia* and *Circaeaster* and its systematic implication. J. Plant Res. 111: 225–230.
Zhang Y-L. 1983. Pollen morphology of *Kingdonia un-iflora* and its taxonomic significance. Acta Phytotax. Sinica 21: 441–444 (in Chinese).
Zhang Z-Y. 1982. Observation on the chromosome of *Asteropyrum, Kingdonia*, and *Calathodes* and its systematic significance. Acta Phytotax. Sinica 20: 402–409.

Order 27. PAPAVERALES

Perennial or annual herbs or rarely shrubs (*Dendromecon* and species of *Bocconia*) or small soft-wooded trees up to 6 m tall. Latex present throughout the plant in either laticiferous tubes or laticiferous sacks or plants without latex and laticifers, but with elongate idioblasts containing juice (except

Pteridophyllum, which lacks both idioblasts and laticifers). The transverse section of the stem mostly exhibits a single ring of widely spaced collateral vascular bundles, which are separated by broad multiseriate medullary rays (markedly heterogeneous in *Bocconia*). Vessels with simple perforations. Fibers with numerous, small, simple pits. Axial parenchyma paratracheal. Nodes unilacunar to multilacunar. Leaves alternate (sometimes all basal) or sometimes opposite or almost verticillate, entire or more often variously divided, without stipules. Stomata anomocytic. Flowers solitary or in various types of cymose or racemose inflorescences, bisexual, actinomorphic or zygomorphic, 2-merous or less often 3-merous, usually entomophilous. Sepals 2, sometimes 3 or 4, mostly caducous. Petals usually twice as many as sepals and in 2 whorls, mostly 4, less often 6, sometimes 8–12 (up to 16), free or apically connate or connivent, rarely absent (*Macleaya* and *Bocconia*). Stamens numerous or 6–12, rarely only 4 (*Hypecoum* and *Pteridophyllum*), free or diadelphous (Fumariaceae), when numerous developing in centripetal sequence; anthers opening longitudinally. Tapetum secretory or amoeboid. Microsporogenesis simultaneous. Pollen grains 2-celled, 3-colpate to pantocolpate and pantoporate, seldom 2-colpate, rarely (*Meconopsis, Cathcartia*) inaperturate. Gynoecium paracarpous, of 2 or less often 3–25 carpels; ovary superior, rarely almost semi-inferior, mostly with numerous (up to 180 in *Glaucium*) ovules, rarely two or only one (*Bocconia, Macleaya*) ovule. Ovules anatropous, hemicampylotropous or hemiamphitropous, or campylotropous, bitegmic, crassinucellate, with the micropyle formed by both integuments. Female gametophyte monosporic of *Polygonum*-type or rarely (*Platystemon*) tetrasporic of *Fritillaria*-type. Endosperm nuclear. Fruits typically capsular. Seeds small, arillate (except Hypecoaceae and Pterydophyllaceae); seed coat formed by both integuments; embryo minute and undifferentiated to linear and elongate and more or less differentiated; endosperm copious, oily. Producing various types of isoquinoline alkaloids (including benzyl isoquinoline and aporphine types and most notably protopine). Chromosomes large to very small, n = 5–10, 14, 19, initially probably 6 (Safronova 1988a, b).

Close to Ranunculales and Berberidales (especially to Hydrastidaceae and Podophyllaceae) and Glaucidiales and shares a common origin with them. The Papaverales differ from the other Ranunculanae mainly in their paracarpous gynoecium with parietal placentation.

Key to Families

1 Plants producing milky or colored latex in articulated laticifers or less often in elongate latex cells. Flowers actinomorphic. Sepals 2 or 3, rarely 4, free or sometimes more or less connate, fully enclosing the bud before anthesis. Petals usually in two cycles of two or three, usually caducous, imbricate and often crumpled in bud, rarely lacking or in more than two cycles, not spurred. Stamens mostly numerous, free. Gynoecium of 2–25 carpels. Perennial or annual herbs, rarely evergreen shrubs or small trees. Nodes unilacunar. Leaves alternate, rarely opposite or verticillate, mostly variously divided, n = 5–10, 14 (*Argemone*), 19 (*Romneya*). . . . . . 1. PAPAVERACEAE

1 Plants without laticifers. Secretory idioblasts present or absent. Flowers actinomorphic or zygomorphic. Sepals two, small, not enclosing the developing bud. Petals four, spurred or not. Stamens four or six. Gynoecium of two carpels.

2 Flowers nearly actinomorphic. Outer petals not spurred. Stamens four, free.

3 Stemless perennial herbs with short thick rhizomes. Leaves basal, oblanceolate, gradually narrowed below, pectinate, with 10–20 pairs of dentate segments, fern-like. Inflorescence an open thyrsus with small dentate bracts subtending cymules of 1–4 pedicellate flowers. Individual flowers small, white, 2-merous. Sepals two, small, caducous. Petals four, entire, oblong, inner narrower than outer. Stamens four, alternating the inner petals; filaments short, not winged; anthers linear. Pollen grains (2)3(4)-colpate, with perforate, subechinulate tectum. Gynoecium with a long, slender style and densely papillate bifid stigma. Ovules two or rarely four, anatropous to subcampylotropous. Fruits sili-culiform, 2-seeded capsules dehiscing by two valves. Seeds ellipsoidal, without elaiosomes, with small embryo, n = 9. . . . . 2. PTERIDOPHYLLACEAE

3 Annual herbs with slender stems. Leaves lanceolate, 2–4 pinnatisect, the lobes linear to narrowly obovate. Inflorescence cymose, many-flowered, with foliose bracts. Flowers actinomorphic, 2-merous. Sepals 2, ovate to lanceolate, herbaceous, with membranous, often eroded margin. Petals four in two cycles, small, outer petals entire to 3-lobed, inner petals trifid, with linear lateral lobes and stipitate,

cochleariform, fimbriate to denticulate median lobe. Stamens four; filaments winged, basally connate with the glandular tissue along the margins; anthers oblong. Pollen grains 2-colpate. Gynoecium with two long and narrow stigmatic branches. Ovules several, campylotropous. Fruit a many-seeded lomentum or dehiscing by two valves, n = 7, 8. . . . . .3. HYPECOACEAE

2 Flowers bilaterally symmetrical or more often transversely zygomorphic. One or both of the outer petals usually basally saccate or spurred, and the inner ones narrower and more or less connate or sticky-connivent over the anthers. Stamens six, diadelphous in a pair of lateral phalanges usually with one or two nectary glands at base. Fruits transversely septate elongated or bladdery capsules or seldom 1-seeded nuts. Perennial or annual herbs. Leaves alternate or rarely (some *Corydalis* spp.) opposite, pinnately to ternately divided. In some parts of the plant, particularly below the leaf epidermis, unbranched idioblasts present; their juice contains alkaloids and fatty oils, and may be reddish or yellowish, but never milky, n = 8 or very rarely 7. . . . . 4. FUMARIACEAE

1. PAPAVERACEAE

A. L. de Jussieu 1789 (including Chelidoniaceae Martynov 1820, Eschscholziaceae Seringe 1847, Platystemonaceae Lilja 1870). 26/250. Mostly temperate and subtropical regions of the Northern Hemisphere, especially the Mediterranean, western, central, and eastern Asia and southwestern parts of the USA; extending to northern South America; 1 species of *Papaver* in South Africa and 1 species in Cape Verde Islands.

1.1 CHELIDONIOIDEAE

Rhizomatous perennials or arborescent plants. Latex orange, yellow or red. Hairs multicellular and terminally uniseriate. Perianth 2-merous. Gynoecium 2-carpellate. Fruits opening by valves. Seeds mostly arillate (except *Glaucium* and *Dicranostigma*). n = 5, 6, 7, 9, 10, chromosomes relatively large to very small. – CHELIDONIEAE: *Sanguinaria, Eomecon, Stylophorum, Hylomecon, Coreanomecon, Chelidonium*; GLAUCIEAE: *Glaucium, Dicranostigma*; BOCCONIEAE: *Macleaya, Bocconia*.

1.2 ESCHSCHOLZIOIDEAE

Perennial or annual herbs or arborescent plants. Hairs unicellular. Perianth 2(rarely 3)-merous. Gynoecium 2-carpellate. Fruits with ten conspicuous longitudinal ribs, opening by valves and usually explosively. Seeds nonarillate or (*Dendromecon*) arillate. n = 6, 7, 8, chromosomes small (especially in *Hunnemannia*), metacentric. – *Dendromecon, Hunnemannia, Eschscholzia*.

1.3 PAPAVEROIDEAE

Large suffruticose herbs to tiny annuals. Hairs multicellular and multiseriate. Perianth 2- or 3-merous. Gynoecium 2–24-carpellate. Stigma laterally confluent. Fruits opening by three or usually more valves. Seeds usually nonarillate (except *Arctomecon*). n = 7, less often 6, 8 or (*Argemone*) 14 and (*Romneya*) 19, chromosomes rather large or (*Argemone*) small, generally submetacentric and intercentric. – PAPAVEREAE: *Papaver, Roemeria, Stylomecon, Meconopsis, Cathcartia, Argemone, Canbya*; ROMNNEYEAE: *Romneya, Arctomecon*.

1.4 PLATYSTEMONOIDEAE

Annual herbs with narrow and entire, often opposite or verticillate leaves. Hairs multicellular and multiseriate. Perianth 3-merous. Gynoecium 3-carpellate or (*Platystemon*) 6–25-carpellate. Carpel tips and stigmas free. Fruits opening by basipetal separation of carpels along placentas without formation of valves or (*Platystemon*) breaking transversely into 1-seeded segments. n = 6, 7, 8, chromosomes metacentric, rather large. – *Platystemon, Hesperomecon, Meconella*.

Chelidonioideae are the most archaic subfamily of the Papaveraceae. However, they are rather heterogeneous and the Glaucieae differ in some respects from the rest of the subfamily. According to Kadereit et al. (1994), they differ by the possession of frondose bracts, the absence of a seed appendage, and the absence of chelidonin and stylopin as characteristic alkaloids of the Chelidoniodeae s. str. Perhaps the best solution will be the new subfamily Glaucioideae, as suggested by Stevenson and Loconte (personal communication). As regards Platystemonoideae, they represent simplified rather than primitive forms (Ernst 1962, 1967). They are close to the Papaveroideae, particularly to *Canbya* (Ernst 1962: 1067; Kadereit et al. 1994), but the valvular dehiscence within the carpel wall of the fruit, which at maturity leaves behind the persistent placental bundles and the fused stigmatic mechanism, clearly isolates *Canbya* from the Platystemonoideae (Ernst 1967: 61). I agree with Ernst "that the similarities between *Canbya* and Platystemonoideae are manifestations of highly advanced features in Papaveraceae and that *Canbya* probably can have played no direct role in the evolution of Platystemonoideae."

2. PTERIDOPHYLLACEAE

Nakai ex Reveal and Hoogland 1991. 1/1. Japan (Honshu).

Pteridophyllum.

Pteridophyllum racemosum is a rare plant that grows in coniferous woods of central and northern Honshu. It is usually included in the Papaveraceae s. str., from which it differs markedly by the absence of latex, fernlike pectinately pinnate leaves surrounded at the base by round scales, and constantly 4-merous androecium. Somewhat intermediate between Papaveraceae and Hypecoaceae.

3. HYPECOACEAE

H. M. Wilkomm and J. M. C. Lange 1880. 1/18. From the western Mediterranean to Mongolia and northwestern China.

Hypecoum.

Hypecoum is usually included in the Fumariaceae, from which it differs by the morphology of corolla and androecium as well as by embryological characters (Iljina 1981) and seed coat anatomy (Iljina 1988). In contrast to all other members of the Papaverales, in *Hypecoum* the pollen is deposited not directly onto the stigma but onto the central lobes of the inner petals, which clasp the stamens and the style (Dahl 1989). It differs from the Papaveraceae and Fumariaceae no less than these two families from each other.

4. FUMARIACEAE

Berchtold et J. Presl 1820 (including Corydalaceae Vest 1818). 18/530. Mostly northern temperate regions; only a few are found south of the equator (a few species of *Corydalis* on mountains in East Africa and the small genera *Phacocapnos, Cysticapnos, Trigonocapnos,* and *Discocapnos* in southern Africa).

CORYDALEAE: *Dicentra, Capnoides, Adlumia, Dactylicapnos, Corydalis, Roborowskia*; FUMARIEAE: *Pseudofu-maria, Sarcocapnos, Ceratocapnos, Platycapnos, Fumari-ola, Fumaria, Cryptocapnos, Rupicapnos, Discocapnos, Trigonocapnos, Phacocapnos, Cysticapnos.*

Close to the Papaveraceae, especially to the Chelidonioideae and Eschscholzioideae, with which they share a common origin.

Bibliography

Berg RY. 1966. Seed dispersal of *Dendromecon*: Its ecologic, evolutionary, and taxonomic significance. Am. J. Bot. 53: 61–71.

Berg RY. 1967. Megagametogenesis and seed development in *Dendromecon rigida* (Papaveraceae). Phyto-morphology 17 (1–4): 223–232.

Berg RY. 1969. Adaptation and evolution in *Dicentra* (Fumariaceae) with special reference to seed, fruit, and dispersal mechanism. Nytt. Mag. Bot. 16: 49–75.

Bernath J, ed. 1998. Poppy: the genus *Papaver*. Amsterdam.

Bersillon G. 1955. Recherches sur les Papaveracees: Contribution a l'ετυδε du development des dicotyled-ones herbacees. Ann. Sci. Nat. Bot., ser. 11., 16: 225–447.

Blattner FR and JW Kadereit. 1995. Three intercontinental disjunctions in Papaveraceae subfamily Chelidonioideae: evidence from chloroplast DNA. Plant Syst. Evol., Suppl. 9: 147–157.

Blattner FR and JW Kadereit. 1999. Morphological evolution and ecological diversification of the forest-dwelling poppies (Papaveraceae: Chelidonioideae) as deduced from a molecular phylogeny of the ITS region. Plant Syst. Evol. 219: 181–197.

Brückner C. 1982. Zur Kenntnis der Fruchtmorphologie der Papaveraceae Juss. s. str. und der Hypecoaeae (Prantl et Kündig) Nak. Feddes Repert. 93: 153–212.

Brückner C. 1983. Zur Morphologie der Samenschale in den Papaveraceae Juss. s. str. und Hypecoaceae (Prantl et Kündig) Nak. Feddes Repert. 94: 361–405.

Brückner C. 1984. Zur Narbenform und zur karpelmorphologischen Stellung der Fumariaceae DC. in den Papaverales. Gleditschia 11: 5–16.

Brückner C. 1985. Frucht- und Samenanatomie von *Pteridophyllum racemosum* Sieb. et Zucc. und die Position der monotypischen Gattung in den Papaverales. Feddes Repert. 96(3): 199–213.

Brückner C. 1992. Gynoecium morphology and fruit anatomy in *Pseudofumaria* Medik. (Fumariaceae), with discussion of carpellary composition. Bot. Jahrb. Syst. 114: 251–274.

Brückner C. 2000. Clarification of the carpel number in Papaverales, Capparales, and Berberidaceae. Bot. Rev. 66: 155–307.

Carlquist S and S Zona. 1988. Wood anatomy of Papaveraceae, with comments on vessel restriction patterns. IAWA Bull., n.s., 9 (3): 253–267.
Carolan JC, ILI Hook, MW Chase, JW Kadereit, and TR Hodkinson. 2006. Phylogenetics of *Papaver* and related genera based on DNA sequences from ITS nuclear ribosomal DNA and plastid *trn*L intron and *trn*L-F intergenic spacers. Ann. Bot. 98: 141–155.
Cresson RA and EL Schneider. 1988. Ovule and seed structure in *Argemone aurantiaca* (Papaveraceae). Bull. Torrey Bot. Club 115: 108–112.
Cumbie BG. 1983. Developmental changes in the wood of *Bocconia vulcanica* Donn. Smith. IAWA Bull., n.s., 4: 131–140.
Dahl A. 1989. Taxonomic and morphological studies in *Hypecoum* sect. *Hypecoum* (Papaveraceae). Plant Syst. Evol. 163: 227–280.
Dahl A. 1990. Infrageneric division of *Hypecoum* (Papaveraceae). Nord. J. Bot. 10: 129–140.
Dahl AE. 1990. Biosystematics of *Hypecoum* L. (Papaveraceae). Goteborg.
Dahl A, A-B Wassgren, and G Bergstrom. 1990. Floral scents in *Hypecoum* Sect. *Hypecoum* (Papaveraceae): Chemical composition and relevance to taxonomy and mating system. Biochem. Syst. Ecol. 18: 157–158.
Dickson J. 1935. Studies in floral anatomy: II. The floral anatomy of *Glaucium flavum* with reference to other members of the Papaveraceae. Bot. J. Linn. Soc. 50: 175–224.
Ernst WR. 1962. The genera of Papaveraceae and Fumariaceae in the southeastern United States. J. Arnold Arbor. 43: 315–343.
Ernst WR. 1967. Floral morphology and systematics of *Platystemon* and its allies *Hesperomecon* and *Meconella* (Papaveraceae: Platystemonoideae). Univ. Kansas Sci. Bull. 47: 25–70.
Fairbairn JW and EM Williamson. 1978. Meconic acid as a chemotaxonomic marker in the Papaveraceae. Phytochemistry 17: 2087–2089.
Fedde FF. 1909. Papaveraceae-Hypecoideae et Papaver aceae-Papaveroideae. In: A Engler, ed. Das Pflanzenreich, IV, 104, pp. 1–430. W. Engelmann, Leipzig.
Feng R-Z, W-Y Lian, G-X Fu, and P-G Xiao. 1985. Chemotaxonomy and resource utilization of the tribe Chelidonieae (Papaveraceae). Acta Phytotax. Sinica 23: 36–42.
Friedel J. 1938a. Note sur la structure anatomique du *Pteridophyllum racemosum* Sieb. et Zucc. Bull. Soc. Bot. France 85: 406–408.
Friedel J. 1938b. Anatomic comparee du *Pteridophyllum racemosum* Sieb. et Zucc. et du *Platystemon californicum* Benth. Bull. Soc. Bot. France 85: 482–486.
Fukuhara T. 1992. Seed coat anatomy of Japanese species of *Corydalis* and *Dicentra* (Papaveraceae; Fumarioideae). Bot. Mag. (Tokyo) 105: 303–321.
Fukuhara T. 1999. Seed and funicle morphology of Fumariaceae-Fumarioideae: systematic implications and evolutionary patterns. Int. J. Plant Sci. 160(1): 151–180.
Fukuhara T. 1999. Morphology and phylogeny of Fumariaceae-Fumarioideae. Acta Phytotax. Geobot. 49: 153–170.
Gleissberg S. 1998. Comparative analysis of leaf shape development in Papaveraceae-Chelidonioideae. Flora (Germany) 193(4): 387–409.
Gleissberg S and JW Kadereit. 1999. Evolution of leaf morphogenesis: evidence from developmental and phylogenetic data in Papaveraceae. Int. J. Plant Sci. 160: 787–794.
Gonnermann C. 1980. Beiträge zur Kenntnis der Gynoeceumsstruktur der Papaveraceae Juss. s. str. Feddes Repert. 91: 593–613.
Gonnermann C. 1982. Überblick über die Testa-Morphologie der Papaveraceae Juss. s. str. Gleditschia 9: 17–25.
Gunn CR. 1980. Seeds and fruits of Papaveraceae and Fumariaceae. Seed Sci. Technol. 8: 3–58.
Gunther K-F. 1975a. Beiträge zur Morphologie und Verbreitung der Papaveraceae, 1. Teil: Infloreszenz-morphologie der Papaveraceae, Wuchsformen der Chelidonieae. Flora 164: 185–234.
Gunther K-F. 1975b. Beiträge zur Morphologie und Verbreitung der Papaveraceae, 2. Teil: Die Wuchsformen der Papaveraceae, Eschscholzieae, und Platystemonoideae. Flora 164: 393–436.
Hegnauer R. 1961. Die Gliederung der Rhoeadales sensu Wettstein im Lichte der Inhaltsstoffe. Planta Med. 9: 37–46.
Hoot SB, JW Kadereit, FR Blattner, KB Jork, AE Schwarzbach, and PR Crane. 1997. Data congruence and phylogeny of the Papaveraceae s.l. based on four data sets: *atp*B and *rbc*L sequences, *trn*K restriction sites, and morphological characters. Syst. Bot. 22: 575–590.
Hutchinson J. 1921. The genera of Fumariaceae and their distribution. Kew Bull. Misc. Inform. 3: 97–115.
Iljina GM. 1981. Papaveraceae. In: MS Yakovlev, ed. Comparative embryology of flowering plants: Win-teraceae-Juglandaceae, pp. 142–150 (in Russian).
Iljina GM. 1988. Papaverales. In: AL Takhtajan, ed. Comparative seed anatomy, 2: 208–235. Nauka, Leningrad (in Russian).
Jensen U. 1967 (1968). Serologische Beiträge zur Frage der Verwandtschaft zwischen Ranunculaceen und Papaveraceen. Ber. Deutsch. Bot. Ges. 80: 621–624.
Kadereit JW. 1993. Papaveraceae. In: K Kubitzki, ed. The families and genera of vascular plants, vol. 2, pp. 494–515. Springer, Berlin/Heidelberg/New York.
Kadereit JW and P Leins. 1988. A wind tunnel experi- ment on seed dispersal in *Papaver* L. sect. *Argemonidium* Spach and *Rhoeadium* Spach (Papaveraceae). Flora 181: 189–203.
Kadereit JW and KJ Sytsma. 1992. Disassembling *Papaver*: A restriction site analysis of chloroplast DNA. Nord. J. Bot. 12: 205–217.
Kadereit JW, FR Blattner, KB Jork, and A Schwarzbach. 1994. Phylogenetic analysis of the Papaveraceae s. 1. (incl. Fumariaceae, Hypecoaceae, and *Pteridophyllum*) based on morphological characters. Bot. Jahrb. Syst. 116: 361–390.
Kadereit JW, FR Blattner, KB Jork, and A Schwarzbach. 1995. The phylogeny of the Papaveraceae sensu lato: morphological, geographical and ecological implications. Plant Syst. Evol., Suppl. 9: 133–145.
Kaul MLH. 1972. Studies on *Argemone mexicana* Linn.: VI. Pollen morphology, floral biology, and pollination mechanism. Proc. Indian Acad. Sci. Sect. B. 35: 86–93.
Kölsch A and S Gleissberg. 2006. Diversification of *CYCLOIDEA*-like TCP genes in the basal eudicot families Fumariaceae and Papaveraceae. Plant Biol. 8: 680–687.
Liden M. 1986. Synopsis of Fumarioideae (Papaveraceae) with a monograph of the tribe Fumarieae. Opera Bot. 88: 1–133.

Liden M. 1993a. Fumariaceae. In: K Kubitzki, ed. The families and genera of vascular plants, vol. 2, pp. 310–318. Springer, Berlin/Heidelberg/New York.
Liden M. 1993b. Pteridophyllaceae. In: K Kubitzki, ed. The families and genera of vascular plants, vol. 2, pp. 556–557. Springer, Berlin/Heidelberg/New York.
Liden M, T Fukuhara, J Rylander, and B Oxelman. 1997. Phylogeny and classification of Fumariaceae, with emphasis on *Dicentra* s.l., based on the plastid gene rps16 intron. Plant Syst. Evol. 206: 411–420.
Meyer NR and GM Iljina. 1986. Palynomorphological data on the system of family Papaveraceae. Vestnik Moscow Univ. 16, Biol. 1: 16–21.
Murbeck S. 1912. Untersuchungen über den Blütenbau der Papaveraceen. Kgl. Svensk. Vet.-Akad. Handl. 5: 1–168.
Norris T. 1941. Torus anatomy and nectary characteristics as phylogenetic criteria in the Rhoeadales. Am. J. Bot. 28: 101–113.
Preininger V. 1986. Chemotaxonomy of Papaveraceae and Fumariaceae. In: RHF. Manske, ed. The Alkaloids, 29: 1–98. New York.
Röder L. 1958. Anatomische und fluoreszenzoptische Untersuchungen an Samen von Papaveraceae. Oesterr. Bot. Z. 104: 370–381.
Ronse Decraene LP and EF Smets. 1992. An updated interpretation of the androecium of the Fumariaceae. Canad. J. Bot. 70: 1765–1776.
Ryberg M. 1960. A morphological study of the Fumariaceae and the taxonomic significance of the characters examined. Acta Horti Berg. 19: 122–248.
Sachar RC. 1955. The embryology of *Argemone mexicana*: a reinvestigation. Phytomorphology 5: 200–218.
Sachar RC and HY Mohan Ram. 1958. The embryology of *Eschscholzia californica* Cham. Phytomorphology 8: 114–124.
Safronova IN. 1988a. Chromosome numbers of some representatives of the family Papaveraceae. Bot. Zhurn. 73: 741 (in Russian with English summary).
Safronova IN. 1988b. Karyosystematic study of the tribes Platystemoneae and Eschscholzieae of the family Papaveraceae. Bot. Zhurn. 73: 1126–1130 (in Russian with English summary).
Safronova IN. 1994. Karyotypical analysis of the genera *Dicranostigma, Hylomecon, Macleaya, Sanguinaria, Stylophorum* (Chelidonioideae, Papaveraceae). Bot. Zhurn. 79: 70–76 (in Russian with English summary).
Sagdulaeva AL. 1959. Pollen morphology of the family Papaveraceae. Problemy Botaniki 4: 11–50 (in Russian).
Saksena HB. 1954. Floral morphology and embryology of *Fumaria parvifiora* Lamk. Phytomorphology 4: 409–417.
Sands MJS. 1973. New aspects on the floral vascular anatomy in some members of the Rhoeadales sensu Hutchinson. Kew Bull. 28: 211–256.
Schneider EL and DM Nichols. 1984. Floral biology of *Argemone aurantiaca* (Papaveraceae). Bull. Torrey Bot. Club 111: 1–7.
Sugiura T. 1940. Chromosome studies on Papaveraceae with special reference to the phylogeny. Cytologia 10: 558–576.
Wheeler MJ, VE Franklin-Tong, and FCH Franklin. 2001. The molecular and genetic basis of pollen-pistil interactions. New Phytol. 151: 565–584.

Order 28. GLAUCIDIALES

Perennial herbs with well-developed and creeping rhizome producing a single long-petiolate leaf on its end. Stem solitary or sometimes two together, simple, rather stout, minutely poberulous, eventually glabrescent above. Vascular bundles amphicribral; xylem barely concave at the outer edge, but not V-shaped. Vessels in the rhizome mostly with simple perforations, but sometimes with scalariform or, rarely, reticulate perforations; lateral pitting scalariform. Fibers with bordered pits. Nodes multilacunar. Leaves few, alternate, apparently arranged in 1/2 divergence, simple, palmately lobed, without stipules; basal leaves long-petiolate, cauline leaves usually two, rarely three, borne on the uppermost part of the stem, reniform or cordate-orbicular; vernation conduplicate. Stomata anomocytic. Flowers large, solitary and terminal, bisexual, spirocyclic, actinomorphic, apetalous. Bract foliaceous, sessile, reniform to suborbicular, cordate, sharply incised-dentate. Sepals four, large, decussate, broadly obovate-rhomboid, petaloid, spreading, caducous. Stamens very numerous, spirally arranged, fasciculate, initiated centrifugally; filaments filiform, slightly enlarged and abruptly constricted below the anther; anthers tetrasporangiate, opening longitudinally. Pollen grains 2-celled, 3-colpate, with punctate/spinulose tectum in which spinules are more numerous. Carpels usually two (rarely solitary or 3), shortly connate at the base by the intervention of receptacular tissue, conduplicate, with a bipartite stigma on a very short stylodium and with distinct furrows both along the ventral suture and along the dorsal side, with 15–30 ovules in each carpel. Ovules, anatropous, bitegmic, pseudo-crassinucellate; the outer integument vascularized. Endosperm nuclear. Female gametophyte of *Polygonum*-type. Fruiting carpels laterally compressed, subquadrate, dehiscing both ventricidally and dorsicidally. Seeds pendulous in two rows from upper edge of fruiting carpels, the members of the two rows strictly alternating but forming a single line, compressed, broadly winged, obovate, about 1.5 cm long, brown, exotestal, with minute embryo and copious endosperm. Cotyledons have a tendency to fuse. Producing glaupalol, a substance with a coumarin skeleton. No alkaloids is detected, $n = 10$, chromosomes small and short.

Glaucidiaceae definitely belong to Ranunculidae, but from all of them differ by the assemblage of characters. The nearest are probably the Ranunculaceae or maybe the Hydrastidaceae (see Tobe 2003) but from them they differ in conduplicate leaf vernation, vascular bundles (as in Hydrastidaceae), scalariform lateral pitting (as in Hydrastidaceae), vascular anatomy of the flower, morphology of the gynoecium, micropyle formation, vascularized outer integument, the absence of nectar discs, fasciculate stamens, initiated centrifugally (as in Paeoniaceae), pollen morphology, fruiting carpels dehiscining both ventricidally and dorsicidally, broadly winged seeds, absence of alkaloids and presence of glaupalol, a substance with a coumarin skeleton, the basic number of chromosomes (see Tobe and Keating 1985).

1. GLAUCIDIACEAE

Tamura 1972. 1/1. High mountains of central and northern Japan (Hokkaido and northern and central districts of Honshu).

Glaucidium.

Bibliography

Irie H, S Uyeo, K Yamamoto, and K Kinoshta. 1967. The structure of glaupalol: a novel furanocoumarin from *Glaucidium palmatum* Sieb. et Zucc. Chem. Commun. 1967: 547–548.

Kosuge K, K Doi, and M Tamura. 1994. Immunological investigation of the phylogenetic relationships of *Glaucidium*. Acta Phytotax. Geobot. 45: 139–150.

Kumazawa W. 1930a. Morphology and biology of *Glaucidium palmatum* Sieb. et Zucc. with notes of affinities to the allied genera *Hydrastis, Podophyllum*, and *Diphylleia*. J. Fac. Sci. Univ. Tokyo, Sect. 3, Bot. 2: 345–380.

Kumazawa W. 1930b. Structure and affinities of *Glaucidium* and its allied genera. Bot. Mag. (Tokyo) 44: 479–490.

Melville R. 1983. The affinity of *Paeonia* and a second genus of Paeoniaceae. Kew Bull. 38: 87–105.

Murakami T, Y Mikami, and H Itokawa. 1967. Die Struktur des neu isolierten Glykosids aus den Rhizomen von *Glaucidium palmatum*. Chem. Pharm. Bull. 15: 1817–1818.

Takhtajan AL and VI Trifonova. 1987. Seed anatomy of *Glaucidium palmatum* Sieb. et Zucc. in relation to its taxonomic position. Bot. Zhurn. 72: 191–196 (in Russian with English summary).

Tamura A. 1972. Morphology and phyletic relationship of the Glaucidiaceae. Bot. Mag. (Tokyo) 85: 29–41.

Tobe H. 1981. Embryological studies in *Glaucidium pal-matum* Sieb. et Zucc. with a discussion on the taxonomy of the genus. Bot. Mag. (Tokyo) 94: 207–224.

Tobe H. 2003. Hydrastidaceae. In: K Kubitzki, ed. The families and genera of vascular plants, vol. 5, pp. 405–409. Springer, Berlin/Heidelberg/New York.

Trifonova VI. 1988. Glaucidiaceae. In: A Takhtajan, ed. Comparative seed anatomy, vol. 2, pp. 193–194. Nauka, Leningrad (in Russian).

Order 29. PAEONIALES

Perennial herbs or less often small shrubs or subshrubs with scattered secretory cells in the parenchymatous tissues of the stems. Rhizome and roots thickened, napiform or bulbous. Vascular bundles amphicribral (as in Hydrastidaceae and Glaucidiaceae). Vessels very small, angular, with scalariform or simple perforations that have 1-3(-7) bars (in metaxylem 4–12 bars); lateral pitting alternate. Fibers with bordered pits. Rays heterocellular to homocellular. Wood parenchyma diffuse. Nodes pentalacunar with five traces. Leaves large, alternate, ternately or ternate-pinnately compound or dissected; petioles dilated at base; leaflets entire, 3–5-lobed to – parted, or dissected into linear segments, cuneate to rounded at base; stipules absent. Stomata anomocytic. Flowers large, terminal, solitary or sometimes a few together, spiral or spirocyclic, bisexual, actinomorphic. Sepals (3-)5(-7), persistent. Petals morphologically equivalent to sepals, mostly 5, less often up to 10 (rarely up to 13), large, free. Stamens numerous, with thin filament united into 5 fascicles, the members of which almost always develop in centrifugal sequence (except *Paeonia japonica*, Sawada 1971); anthers tetrasporangiate, extrorse, opening longitudinally. Tapetum secretory. Microsporogenesis simultaneous. Pollen grains 2-celled, 3-colpate/colporoidate/colporate, with finely punctate and rugulose or perforate tectum. Gynoecium apocarpous, of (2-)5, rarely up to 8 and even to 15 carpels; carpels coarse and thick-walled, with almost sessile expanded stigma; gynoecium surrounded by a fleshy, lobed nectary disc that is probably a modified part of the androecium (Eames 1953, 1961). Ovules several or many in each carpel, anatropous, bitegmic (with massive, slightly vascularized outer integument, which is longer than the inner one), crassinucellate, with nucellar cup. Female gametophyte of *Polygonum*-type. Endosperm nuclear. Embryogenesis of unique type: the zygote nucleus undergoes repeated free nuclear divisions to form a coenocyte with the nuclei restricted to a peripheral layer around a large central vacuole; wall formation proceeds centripetally and the proembryo becomes cellular. Fruits of follicles. Seeds medium-sized to large, with more or less well-developed dark red aril; seed coat formed by the outer integument (testal); embryo minute, differentiated; endosperm copious. Producing phenolic substance and glycosides, such as paeonol, paeonoside, paeoniflorin, etc., ethereal oils, flavones, seed polysaccharides (amyloids) and alkaloids nearly lacking; n = 5, somewhat n = 10.

The Paeoniales are closely related to the Glaucidiales (Melville 1982), from which they differ by tricolporate pollen grains, persistent and nondecussate sepals, not sharply limited perianth (the phyllotactic spiral in *Paeonia* is continuous from leaves through bracts, sepals, petals, stamen trunks, and carpels, Eames 1961: 433), somewhat convex receptacle, lobed nectary disc surrounding the gynoecium, the longer outer integument, arillate seeds, presence of ethereal cells in parenchymatous tissues, long chromosomes, absence of glaupalol, and presence of amyloids in seeds. Both orders share a common origin from some ranunculalean ancestor. Affinities with the Dilleniaceae seem less probable.

1. PAEONIACEAE

Rafinesque 1815. 1/35–40. North Africa (north of Sahara), Europe, subtropical, temperate, and partly cold regions of Asia, western regions of North America.

Paeonia.

Bibliography

Barber HN. 1941. Evolution in genus *Paeonia*. Nature 148 (3747): 227–228.

Camp WH and MM Hubbard. 1963. Vascular supply and structure of the ovule and aril in peony and of the aril in nutmeg. Am. J. Bot. 50: 174–178.

Carniel K. 1967. Über die Embryobildung in der Gattang *Paeonia*. Oesterr. Bot. Z. 114: 4–19.

Cave MS, HG Arnott, and SA Cook. 1961. Embryogeny in the California paeonies with reference to their taxonomic position. Am. J. Bot. 48: 397–404.

Davezac Th. 1957. La place systematique du genre *Paeonia* et forme de jeunesse de *P. lusitanica* Mill. Bull. Soc. Hist. Natur. Toulouse 92 (3–4): 197–201.

Fedotova TA. 1988. Paeoniaceae. In: A Takhtajan, ed. Comparative seed anatomy, 2: 195–207. Nauka, Leningrad (in Russian).

Halda JJ. 1998. Notes on the observations upon the structure of the *Paeonia* seeds, fruit and roots. Acta Mus. Richnov. Sect. Nat. 5: 1–11.

Hiepko P. 1965. Das zentrifugale Androecium der Paeoniaceae. Ber. Deutsch. Bot. Ges. 77: 427–435.

Hiepko P. 1966. Zur Morphologie, Anatomie und Funktion des Diskus der Paeoniaceae. Ber. Deutsch. Bot. Gesell. 79: 233–245.

Hong D-Y. 1989. Studies on the genus *Paeonia*. 2. The characters of leaf epidermis and their systematic significance. Chinese J. Bot. 1: 145–154.

Hong D-Y, Z-X Zhang, and X-Y Zhu. 1988. Studies on the genus *Paeonia*. 1. Report of karyotypes of some wild species in China. Acta Phytotax. Sinica 26: 33–43 (in Chinese).

Kartashova NN. 1962. On the nature of the nectaries in flowers of *Paeonia*. Trudy Bot. Inst. Akad. Nauk SSSR, 7th ser., 5: 77–85 (in Russian).

Keefe JM and MF Moseley 1978. Wood anatomy and phylogeny of *Paeonia* section Moutan. J. Arnold Arbor. 59: 274–297.

Kumazawa M. 1935. The structure and affinities of *Paeonia*. Bot. Mag. (Tokyo) 49: 306–315.

Mattiessen A. 1962. A contribution to the embryology of *Paeonia*. Acta Horn Berg. 20: 57–61.

Melville R. 1983. The affinity of *Paeonia* and a second genus of Paeoniaceae. Kew Bull. 38: 87–105.

Murgai P. 1962. Embryology of *Paeonia*, together with discussion of its systematic position. In Plant embryology: a symposium, pp. 215–223. New Delhi: *csir*.

Nowicke JW, JL Bittner, and J Skvarla. 1986. *Paeonia*: Exine substructure and plasma ashing. In: S Blackmore and IK Ferguson, eds. Pollen and spores: form and function, pp. 81–95. Academic Press, London.

Sawada M. 1971. Floral vascularization of *Paeonia japonica* with some consideration of systematic position of Paeoniaceae. Bot. Mag. (Tokyo) 84: 51–60.

Schmitt E. 2000. Etude systematique de genre *Paeonia* L. (suite). Plant Montagne Rocaille 13(193): 5–6.

Serov VP. 1991. The fruit structure of the genus *Clematis* (Ranunculaceae). Bot. Zhurn. 76: 1090–1099 (in Russian with English summary).

Shamrov II. 1997. Ovule and seed development in *Paeonia lactiflora* (Paeoniaceae). Bot. Zhurn. 82: 24–46 (in Russian with English summary).

Stearn WT. 1946. A study of the genus *Paeonia*. Roy Hort. Soc. London.

Stebbins GL. 1938. Cytogenetic studies in *Paeonia: 2* the cytology of the diploid species and hybrids. Genetics 23: 83–110.

Tamura M. 2006. Paeoniaceae. In: K Kubitzki, ed. The families and genera of flowering plants, vol. 9, pp. 265–269. Springer, Berlin/Heidelberg/New York.

Tank DC and T Sang. 2001. Phylogenetic utility of the glycerol-3-phosphate acyltransferase gene: evolution and implications in *Paeonia* (Paeoniaceae). Molec. Phylogenet. Evol.19: 421–429.

Tzanoudakis D. 1983. Karyotypes of four wild *Paeonia* species from Greece. Nord. J. Bot. 3: 307–318.

Ulubelen A. et al. 1968. Phytochemical investigation of *Paeonia decora*. Lloydia 31: 249–251.

Worsdell WC. 1908. The affinities of *Paeonia*. *J*. Bot. 46: 114–116.

Xi Y-Z. 1984. The pollen morphology and exine ultrastructure of *Paeonia* in China. Acta Bot. Sinica 26: 241–246 (in Chinese).

Yakovlev MS. 1983. Paeoniaceae. In: MS Yakovlev, ed. Comparative embryology of flowering plants: Phytolaccaceae-Thymelaeaceae, pp. 70–77. Nauka, Leningrad (in Russian).

Yakovlev MS and MD Yoffe. 1957. On some peculiar features in the embryology of *Paeonia* L. Phytomorphology 7: 74–82.

Yu J and P-G Xiao. 1987. A preliminary study of the chemistry and systematics of Paeoniaceae. Acta Bot. Sinica 25: 172–179.

Subclass III. HAMAMELIDIDAE

Trees or shrubs, very rarely suffruticose herbs (*Pachysandra*). Vessels mostly with scalariform perforations. Sieve-element plastids of S-type or rarely (Buxales) of PCS- or Pc-type. Leaves alternate or sometimes opposite or verticillate, simple or less often pinnately compound. Stomata of various kinds. Flowers mostly small, in various kinds of inflorescences, cyclic, bisexual or unisexual, commonly apetalous and often wholly without perianth, typically anemophilous. Stamens (1)2-several or sometimes numerous, often with a prolonged connective, free, mostly with elongate filaments. Pollen grains mostly 2-celled, 3-colpate to often porate. Gynoecium apocarpous or more often syncarpous, sometimes pseudomonomerous. Ovules anatropous to orthotropous, bitegmic or less often unitegmic, crassinucellate. Endosperm cellular or nuclear. Fruits dry (dehiscent or indehiscent) or drupaceous. Seeds with small (sometimes very small) or more often well-developed embryo; endosperm present or absent. Tannins, proanthocyanidins, ellagic acid, and especially myricetin flavonoids are common; iridoids are very rare and occur only in *Liquidambar, Daphniphyllum*, and *Didymeles* (Giannasi 1986).

The subclass Hamamelididae represents a very ancient line and one of the main branches of the phylogenetic tree of magnoliopsids. The lower Hamamelididae are linked to Magnoliidae via Trochodendrales (Takhtajan 1966, 1987; Endress 1986, 1993). There are also some links between the lower Hamamelididae and the lower Dilleniidae and Rosidae (see Ehrendorfer 1989; Dickison 1989). However, higher members of the subclass are so highly specialized that they show no evident links with either the Dilleniidae or Rosidae.

Bibliography

Abbe EC. 1974. Flowers and inflorescences of the "Amenriferae." Bot. Rev. 40: 159–261.

Barabe D. 1984. Application du cladisme a la systematique des Angiospermes: Cas de Hamamelidales. Candollea 39: 51–70.

Barabe D, Y Bergeron, and GA Vincent. 1982. La position de Daphniphyllaceae, Buxaceae, Simmondsiaceae, et Cecropiaceae dans la sous-classe des Hamamelididae: Etude numerique. C. R. Acad. Sci. Paris, 3rd ser., 294: 891–896.

Barabe D, Y Bergeron, and GA Vincent. 1987. La repartition des caracteres dans la classification des Hamamelididae (Angiospermae). Canad. J. Bot. 65: 1756–1767.

Behnke HD. 1973. Sieve-tube plastids of Hamamelidae. Taxon 22: 205210.

Behnke HD. 1989. Sieve-element plastids, phloem proteins, and the evolution of flowering plants: IV. Hamamelidae. In: PR Crane and S Blackmore, eds. Evolution, systematics, and fossil history of the Hamamelidae, vol. 1, pp. 105–128. Claredon Press, Oxford.

Chupov VS. 1978. Comparative immunoelectrophoretic study of proteins of pollen of some Amentiferae. Bot. Zhurn. 63: 1579–1585 (in Russian).

Clarke B. 1858. On the structure and affinities of Myricaceae, Platanaceae, Altingiaceae, and Chloranthaceae. Ann. Mag. Nat. Hist., 3rd ser., 1: 100–109.

Crane PR and S Blackmore, eds. 1989. Evolution, systematics, and fossil history of the Hamamelidae, 2 vols. Claredon Press, Oxford.

Cronquist A. 1986. Commentary on the status of the Hamamelidae. Ann. Missouri Bot. Card. 73: 227.

Dickison WC. 1989. Comparisons of primitive Rosidae and Hamamelidae. In: PR Crane and S Blackmore, eds. Evolution, systematics, and fossil history of the Hamamelidae, vol. 1, pp. 47–73. Claredon Press, Oxford.

Dilcher DL and S Zavada. 1986. Phytogeny of the Hamamelidae: An introduction. Ann. Missouri Bot. Gard. 73: 225–226.

Donaghue MJ and JA Doyle. 1989. Phylogenetic analysis of angiosperms and the relationships of Hamamelidae. In: PR Crane and S Blackmore, eds. Evolution, systematics, and fossil history of the Hamamelidae, vol. 1, pp. 17–45. Claredon Press, Oxford.

Ehrendorfer F. 1989. The phylogenetic position of the Hamamelidae. In: PR Crane and S Blackmore, eds., Evolution, systematics, and fossil history of the Hamamelidae, vol. 1, pp. 1–7. Claredon Press, Oxford.
Endress PK. 1967. Systematische Studied über die verwandtschaftlichen Beziehungen zwischen den Hamamelidaceen und Betulaceen. Bot. Jahrb. Syst. 87: 431–525.
Endress PK. 1977. Evolutionary trends in the Hamamelidales-Fagales-group. In: K Kubitzki, ed. Flowering plants: Evolution and classification of Higher Categories. Plant Syst. Evol., Suppl. 1: 321–347.
Endress PK and A Igersheim. 1999. Gynoecium diversity and systematics of the basal eudicots. Bot. J. Linn. Soc. 130: 305–393.
Feng Y-X, X-Q Wang, K-Y Pan, D-Y Hong. 1998. A revaluation of the systematic positions of the Cercidiphyllaceae and Daphniphyllaceae based on *rbc*L gene sequence analysis, with reference to the relationship in the "lower" Hamamelidae. Acta Phytotax. Sinica 36: 411–422.
Friis EM and PR Crane. 1989. Reproductive structures of Cretaceous Hamamelidae. In: PR Crane and S Blackmore, eds., Evolution, systematics, and fossil history of the Hamamelidae, vol. 1, pp. 155–174. Claredon Press, Oxford.
Giannasi DE. 1986. Phytochemical aspects of phylogeny in Hamamelidae. Ann. Missouri Bot. Gard. 73: 417–437.
Hallier H. 1904. Über die Gattung *Daphniphyllum*: Ein Übergangsglied von den Magnoliaceen und Hamamelidaceen zu den Kätzchenblütlern. Bot. Mag. (Tokyo) 18: 55–69.
Hjelmqvist H. 1948. Studies on the floral morphology and phylogeny of the Amentiferae. Bot. Not., Suppl. 2(1): 1–171.
Hufford LD and PR Crane. 1989. A preliminary phylogenetic analysis of the "lower" Hamamelidae. In: PR Crane and S Blackmore, eds. Evolution, systematics, and fossil history of the Hamamelidae, vol. 1, pp. 175–192. Claredon Press, Oxford.
Hufford LD and PK Endress. 1989. The diversity of anther structures and dehiscent patterns among Hamamelidae. Bot. J. Linn. Soc. 99: 301–346.
Janchen E. 1950. Die Herkunft der Angiospermen-Blüte und die systematische Stellung der Apetalen. Oesterr. Bot. Z. 97: 129–167.
Jay M. 1968. Distribution des flavonoides chez les Hamamelidacees et families affines. Taxon 17: 136–147.
Kolobkova EV. 1972. Comparative study of albumins and globulins in seeds of Amentiferae. In: EV Kolobkova, ed. Biochemistry and phylogeny, pp. 37–48. Nauka, Moscow (in Russian).
Kuprianova LA. 1965. Palynology of amentifers. The Academy of Sciences of the USSR/Nauka, Moscow/Leningrad (in Russian).
Manos PS and KP Steele. 1997. Phylogenetic analyses of "higher" Hamamelididae based on plastid sequence data. Am. J. Bot. 84: 1407–1419.
Manos PS, KC Nixon, and JJ Doyle. 1993. Cladistic analysis of restriction site variation within the chloroplast DNA inverted repeat region of selected Hamamelididae. Syst. Bot. 18: 551–562.
Mears JA. 1973. Chemical constituents and systematics of Amentiferae. Brittonia 25: 385–394.
Meeuse ADJ. 1975. Floral evolution in the Hamamelidae: III. Hamamelidales and associated groups including Urticales and final conclusions. Acta Bot. Neerl. 24: 181–191.
Melikian AP. 1973. Seed coat of Hamamelidaceae and allied families in relation to their systematics. Bot. Zhurn. 58: 350–359 (in Russian.)
Mohana Rao PR. 1974. Seed anatomy in some Hamamelidaceae and phylogeny. Phytomorphology 24: 113–139.
Moseley MF. 1973 (1974). Vegetative anatomy and morphology of Amentiferae. Brittonia 25: 356–370.
Morawetz W and MRA Samuel. 1989. Karyological patterns in the Hamamelidae. In: PR Crane and S Blackmore, eds. Evolution, systematics, and fossil history of the Hamamelidae, vol. 1, pp. 129–154. Claredon Press, Oxford.
Petersen AE. 1953. A comparison of the secondary xylem elements of certain species of the Amentiferae and Ranales. Bull. Torrey Bot. Club 80: 365–384.
Petersen FP and DE Fairbrothers. 1979. Serological investigations of selected Amentiferous taxa. Syst. Bot. 4: 230–241.
Petersen FP and DE Fairbrothers. 1985. A serotaxonomic appraisal of the "Amentiferae." Bull. Torrey Bot. Club 112: 43–52.
Praglowski J. 1974. Pollen morphology of the Trochodendraceae, Tetracentraceae, Cercidiphyllaceae, and Eupteleaceae with reference to taxonomy. Pollen Spores 16: 449–467.
Qui Y-L, MW Chase, SB Hoot, E Conti, PR Crane, KJ Systma, and CR Parks. 1998. Phylogenetics of the Hamamelidae and their allies: parsimony analyses of nucleotide sequences of the plastid gene *rbc*L. Int. J. Plant Sci. 159: 891–905.
Stern WL. 1973. Development of the Amentiferous concept. Brittonia 25: 316–333.
Stone DE. 1973 (1974). Patterns in the evolution of Amentiferous fruits. Brittonia 25: 371–384.
Thorne RF. 1973 (1974). The "Amentiferae" or Hamamelidae as an artificial group: A summary statement. Brittonia 25: 395–405.
Thorne RF. 1989. "Hamamelididae": A commentary. In: PR Crane and S Blackmore, eds. Evolution, systematics, and fossil history of the Hamamelidae, vol. 1, pp. 9–16. Claredon Press, Oxford.
Tiffney BH. 1986. Fruit and seed dispersal and the evolution of the Hamamelidae. Ann. Missouri Bot. Garden 73: 394–416.
Wolfe JA. 1973 (1974). Fossil forms of Amentiferae. Brittonia 25:334–355.
Wolfe JA. 1989. Leaf architectural analysis of the Hamamelididae. In: PR Crane and S Blackmore, eds. Evolution, systematics, and fossil history of the Hamamelidae, vol. 1, pp. 75–104. Claredon Press, Oxford.
Zavada MS and DL Dilcher. 1986. Comparative pollen morphology and its relationship to phylogeny of pollen in the Hamamelidae. Ann. Missouri Bot. Garden 73: 348–381.
Zhou R-H and Z-H Jiang. 1990. Chemical constituents of the Hamamelidae and their systematic significance. Cathaya 2: 63–76.

Superorder TROCHODENDRANAE

Order 30. TROCHODENDRALES

Evergreen or deciduous trees. Elongate, often branched idioblasts occur in the leaves and in at least some of the parenchymatous tissues of the stem. Secondary xylem mostly of tracheids and only partly of primitive vessels with scalariform perforations with many bars (Snigirevskaya, 2000). Tracheids very long, in early wood with boarded scalariform pits. In late wood, where tracheids are narrower, they have rounded boarded pits. Rays heterogeneous. Axial parenchyma diffuse and in short tangential lines. Nodes multilacunar to unilacunar (*Trochodendron*) or trilacunar (*Tetracentron*). Leaves alternate, simple, margin serrate or serrulate, estipulate (Trochodendraceae) or with stipules adnate to slender petioles (Tetracentraceae). Stomata encyclocytic. Flowers small, in racemelike aggregates or spikes, bisexual or polygamous, actinomorphic or slightly disymmetric, apetalous. Sepals present or absent. Stamens about 40–70 or only 4; anthers basifixed, tetrasporangiate, more or less latrorse, opening by longitudinal slits extending into two transversal lines at both ends, which results in two valves resembling transparent wings (Endress 1986). Tapetum secretory. Microsporogenesis simultaneous. Pollen grains 2-celled, small, spheroidal, 3-colpate, tectate-columellate, rugulate-reticulate between the apertures but striate near the apertures, the apertures coarsely granulate. Gynoecium of 4–17 laterally connate, conduplicate carpels. Stylodia free, with decurrent stigma, which bears unicellular papillae. Ovary semiinferior; dorsal part of ovary horizontally extended at anthesis, with many sunken stomata and nectariferous. Ovules 5–30 per carpel, pendulous, anatropous, apotropous, bitegmic, with long integuments and a chalazal protrusion, crassinucellate; micropyle endostomal. Female gametophyte of *Polygonum*-type. Endosperm cellular or nuclear. Fruits syncarpous multifollicles (follicetum), ventricidal and slightly loculicidal. Seeds small, flat, with chalazal, apical, and lateral winglike extensions. The raphe is in the form of a ridge along the whole length of the seed, its vascular bundle forming a hairpin loop in the chalazal extension and terminating below the hypostase. Seed coat thin, formed by both integuments (endotestal and exotegmic). Embryo straight, small, differentiated into cotyledons, radicle, and plumule; endosperm copious, oily, and proteinaceous. Producing chalkons (Tetracentraceae), proanthocyanidins (Trochodendraceae), flavonols (quercetin and kaempferol); n = 20, 24.

In many respects Trochodendrales occupy, as it were, an intermediate position between Magnoliidae and Hamamelididae, but in the totality of their characters they stand nearer to the latter (Takhtajan 1966, 1969, 1980, 1987; Cronquist 1981, 1988; Endress 1986, 1993). In *Tetracentron* all floral organs contain scattered enlarged cells resembling the "oil" cells that are generally present in many Magnoliidae, but absent in Hamamelididae. They are often more or less spherical as in Magnoliidae; in *Trochodendron* similar cells differentiate into sclerified, branched idioblasts. However, in the gynoecium, at least at anthesis, many of these cells are not sclerified and presumably also correspond to "oil" cells (Endress 1986: 315).

Key to Families

1 Evergreen trees forming long shoots only. Buds large, with many imbricate scales. Leaves closely clustered near the branch tips in pseudowhorls, obovate to broadly lanceolate, crenate serrate, coriaceous, pinnately veined, estipulate; petiole with an arc of three vascular bundles. Idioblasts sclerenchymatous, not secretory. Flowers bisexual or occasionally polygamous, with long pedicels; in short, initially terminal but soon appearing axillary, upright racemelike structure with a terminal flower. Pedicels with bract and several bracteoles, expanded above into subconical torus. Sepals essentially absent (rudimentary tepals are recognizable in early ontogenetic stages of the flower, but they are obliterated at anthesis and never function as protective organs, Endress 1993). Stamens 40–70, spirally arranged; anthers mucronate. Carpels (4-)6-11(17); stylodia conduplicate and deeply canaliculate ventrally, recurved. Ovules 15–30 per carpel. Endosperm cellular. Seeds 15–25 per locule, pendulous, smooth, linear; seed coat endotestal-endotegmic, exotesta consists of one tanniniferous layer. n = 20. . . . . 1. TROCHODENDRACEAE.

1 Deciduous trees with slender branches bearing short shoots. Short shoots alternately arranged, marked with crowded concentric scars of fallen leaves and bud scales and bearing a single subterminal leaf and an inflorescence each year. They form sympodial systems with adaxial prophylls. Buds slender, with two outer sheathing scales. Leaves cordate-ovate, closely crenate, palmately curviveined from base, with stipules adnate at the base of petiole; petiole with a single arcuate or almost cylindrical vascular strand. Idioblasts secretory, not sclerenchymatous. Flowers bisexual, actinomorphic, sessile in the axil of minute bract, in alternate groups of four on

slender, catkinlike spike. Sepals four in two decussate pairs, thin and have only a rudimentary vascular bundle. Stamens four, in two decussate pairs, each opposite a sepal; anthers truncate-rounded at the apex. Carpels four, in a cycle, alternate with stamens; stylodia subulate, at first connivent, later becoming sharply recurved and subbasal by great overgrowth of ventral region of each carpel. Fruits of four laterally coalescent carpels, these loculicidally dehiscent along the entire exposed portion of the ventral face to base of persistent deflexed stylodia. Ovules 5-8(-10) per carpel. Endosperm nuclear. Seeds 5-8(-10) per locule, faceted, linear-oblong; seed coat exotegmic, exotesta consists of one colourless layer. n = 24. . . . 2. TETRACENTRACEAE.

1. TROCHODENDRACEAE

Eichler 1865. 1/1. From Korea to Japan, Ryukyu Islands, and Taiwan.

Trochodendron.

2. TETRACENTRACEAE

A.C. Smith 1945. 1/1. Eastern Himalayas (from eastern Nepal to Arunchal-Pradesh), northern Burma, southwestern and central China.

Tetracentron.

Close related to the Trochodendraceae.

Bibliography

Bailey IW and CG Nast. 1945. Morphology and relationships of *Trochodendron* and *Tetracentron*: I. Stem, root, and leaf. J. Arnold Arbor. 26: 143–153.

Baranova M. 1983. On the laterocytic stomatotype in angiosperms. Brittonia 35: 93–102.

Bonderson W. 1952. Entwicklungsgeschichte und Bau der Spaltöffnungen bei den Gattungen *Trochodendron* Sieb. et Zucc., *Tetracentron* Oliv., und *Drimys* J. R. et G. Forst. Acta Horti Berg. 16: 169–217.

Chaw SM. 1992. Pollination, breeding syndromes, and systematics of *Trochodendron aralioides* Sieb. et Zucc. (Trochodendraceae), a relictual species in eastern Asia. In: Ching-I Peng, ed. Phytogeography and botanical inventory of Taiwan, pp. 63–77, Bot. ser. 12. Taipei.

Chen GZ. 1989. A study on plant Trochodendrales. Acta Sci. Nat. Univ. Sunyat. 28: 73–79.

Crane P, SR Manchester and DL Dilcher. 1991. Reproductive and vegetative structure of *Nordenskioldia* (Trochodendraceae), a vesselless dicotyledon from the early Tertiary of the Northern Hemisphere. Am. J. Bot. 78: 1311–1334.

Croizat L. 1947. *Trochodendron, Tetracentron*, and their meaning in phylogeny. Bull. Torrey Bot. Club 74: 60–76.

Doweld AB. 1998. Carpology, seed anatomy and taxonomic relationships of *Tetracentron* (Tetracentraceae) and *Trochodendron* (Trochodendraceae). Ann. Bot. 82: 412–443.

Endress PK. 1986. Floral structure, systematics, and phylogeny in Trochodendrales. Ann. Missouri Bot. Gard. 73: 297–324.

Endress PK. 1993. Trochodendraceae. In: K Kubitzki, ed., The families and genera of vascular plants, vol. 2, pp. 599–602. Springer, Berlin/Heidelberg/New York.

Foster AS. 1945. The foliar sclereids of *Trochodendron aralioides* Sieb. et Zucc. J. Arnold Arbor. 26: 155–162.

Harms H. 1897. Über die Stellung der Gattung *Tetracentron* und die Familie der Trochodendraceen. Ber. Deutsch. Bot. Ges. 15: 350–360.

Jergensen LB, JD Meller, and P Wagner. 1975. Secondary phloem of *Trochodendron aralioides*. Svensk Bot. Tidskr. 69: 217–238.

Kai-yu P, L Jian-hua, L An-ming, and W Jie. 1993. The embryology of *Tetracentron chinense* Oliver and its systematic significance. Cathaya 5: 49–58.

Keng H. 1959. Androdioecism in the flowers of *Trochodendron aralioides*. J. Arnold Arbor. 40: 158–160.

Leroy J-F. 1988. L'organisation fondamentale d'une Angiosperme archaique: *Tetracentron* (Tetracentracées). Compt. Rend. Acad. Sci. Paris. ser. 3, 306: 229–235.

Mohana Rao PR. 1981. Seed and fruit anatomy of *Trochodendron aralioides*. Phytomorphology 31: 18–23.

Nast CG and IW Bailey 1945. Morphology and relationships of *Trochodendron* and *Tetracentron*: II. Inflorescence, flower, and fruit. J. Arnold Arbor. 26: 267–276.

Oliver D. 1889. *Tetracentron sinense* Oliv. Hooker's Icones Plantarum 19: 1892.

Pan KY et al. 1993. The embryology of *Tetracentron sinense* Oliver and its systematic significance. Cathaya 5: 49–58.

Pervukhina NV. 1962. Interesting peculiarity of the ovary of *Trochodendron aralioides*. Bot. Zhurn. 47: 993–995 (in Russian).

Pervukhina NV. 1963. On the position *of Trochodendron* in the phylogenetic system of angiosperms. Bot. Zhurn. 48: 939–948 (in Russian).

Pervukhina NV and MD Yoffe. 1962. The morphology of *Trochodendron* flower: A contribution to the phylogeny of angiosperms. Bot. Zhurn. 47: 1709–1730 (in Russian).

Praglowski J. 1974. The pollen morphology of the Rochodendraceae, Tetracentraceae, Cercidiphyllaceae, and Eupteleacese with reference to taxonomy. Pollen Spores 16: 449–467.

Rao PRM. 1983. Seed and fruit anatomy of *Trochodendron* aralioides. Phytomorphology. 31(1–2): 18–23.

Smith AC. 1945. A taxonomic review of *Trochodendron* and *Tetracentron*. J. Arnold Arbor. 26: 123–142.

Tsuji SI, M Matsushita. 1991. Notes on pollen morphology of *Trochodendron aralioides*. J. Phytogeogr. Taxon. 39(1): 27–30.

Valen F van. 1978. Contribution to the knowledge of cyanogenesis in angiosperms, 4th Communication: Cyanogenesis in *Trochodendron aralioides* Sieb. and Zucc. Proc. Kon. Nederl. Akad. Wetensch., ser. C., 81: 198–203.

Wagner R. 1903. Beitrage zur Kenntnis der Gattung *Trochodendron* Sieb. et Zucc. Ann. Naturhist. Hofmus. 18: 409–422.

Wang FH, NF Chien, and YL Zhang. 1984. A study on the pollen morphology in *Trochodendron, Tetracentron, Euptelea.* Acta Phytotax. Sinica 22: 456–460 (in Chinese).
Yoffe MD. 1962. On the embryology of *Trochodendron aralioides* Sieb. et Zucc.: Development of pollen and embryo sac. Trudy Bot. Inst. Akad. Nauk SSSR, 7th sen, 5: 250–259 (in Russian).
Yoffe MD. 1965. On the embryology of *Trochodendron aralioides* Sieb. et Zucc.: Embryo and endosperm development. In: MS Yakovlev, ed. Flower morphology and reproductive process of angiosperms, pp. 177–188. The Academy of Sciences of the USSR/Nauka, Moscow/Leningrad (in Russian).

Order 31. CERCIDIPHYLLALES

Deciduous trees with branches differentiated into long and short shoots. Some of the cells of the parenchymatous tissues containing crystals of calcium oxalate. Vessel elements very long, angular, with oblique scalariform perforation plates that have 20–50 bars; lateral pitting scanty, from opposite to scalariform. Fibers with conspicuously bordered pits. Rays heterogeneous. Axial parenchyma scanty, apotracheal. Sieve-element plastids of S-type, large, often surrounded by endoplasmic reticulum membranes. Nodes on long shoots trilacunar, those on short shoots unilacunar. Leaves simple, dimorphic, those on long shoots pinnately veined, elliptic or broadly ovate, entire or finely serrate, opposite, subopposite, or occasionally in irregular whorls of three. However, those on sympodially growing short shoots, which develop from the axillary buds of long shoot leaves and have an adaxial prophyll, are palmately veined, broadly cordate or reniform, crenate, alternate and solitary (each short shoot bears a single leaf in each growing season). In buds leaves are involute and have well-developed marginal glands; stipules present, small, adnate to the petiole and caducous. Stomata anomocytic. Flowers dioecious, without perianth, arranged in small, dense, racemose heads terminal on the short shoot and produced before or with the leaves. Male inflorescences contain 16–35 stamens, but the individual flowers difficult to delimit since perianth and often subtending bracts are absent. Stamens with long, slender, pendulous filaments and long pink apiculate anthers; anthers tetrasporangiate, basifixed, latrorse, opening by simple longitudinal slits. Tapetum secretory. Microsporogenesis simultaneous. Pollen grains 2–2-celled, 3-colpate, tectate-columellate, finely reticulate, with minute tectal perforations; the apertural exine is coarsely structured. Female inflorescences of 2–8 unicarpellate flowers; carpels slightly stipitate, gradually narrowed to a slender stylodium with a long, 2-ridged, decurrent red stigma that bears unicellular papillae. The ventral suture of each carpel is oriented abaxially with respect to the inflorescence axis, which is unusual. According to Solereder (1899) the abaxial orientation of the suture is due to loss of one of a pair of opposite carpels, which is confirmed by paleobotanical data (Crane and Stockey 1986). The ovary contains 15–30 ovules with the laminar lateral placentation. Ovules anatropous, apotropous, bitegmic, crassinucellate. Female gametophyte of *Polygonum*-type Endosperm cellular. Fruits follicles. Seeds small, flattened, with a long chalazal extension forming a wing that contains the ovular vascular bundles forming a hairpin loop (as in Trochodendrales). Seed coat formed mainly by the outer integument; exotestal cells enlarged, slightly thickened; embryo large, well-differentiated into cotyledons and long hypocotyl; endosperm scanty, oily. Tanniniferous and producing both phenolics and flavonols (quercetin and kaempferol). n = 19.

Although Cercidiphyllales are related to the Trochodendrales, particularly to Tetracentraceae, they differ from them markedly in several characters, such as dimorphic leaves, extremely congested inflorescences, much more reduced dioecious and unicarpellate flowers, and the presence of vessels.

1. CERCIDIPHYLLACEAE

Engler 1907. 1/2. Japan and China (Hubei, Sichuan, and Shaanxi).

Cercidiphyllum.

Bibliography

Brown RW. 1939. Fossil leaves, fruits, and seeds of *Cercidiphyllum.* J. Palaeontol. 13: 485–499.
Crane PR. 1984. A reevaluation of *Cercidiphyllum*-like plant fossils from the British early Tertiary. Bot. J. Linn. Soc. 89: 199–230.
Crane PR and RA Stockey. 1985. Growth and reproductive biology of *Joffrea speirsii* gen. et sp. nov., a *Cercidiphyllum*-like plant from the Late Paleocene of Alberta, Canada. Canad. J. Bot. 63: 340–364.
Crane PR and RA Stockey 1986. Morphology and development of pistillate inflorescences in extant and fossil Cercidiphyllaceae. Ann. Missouri Bot. Gard. 73: 382–393.
Endress PK. 1986. Floral structure, systematics, and phylogeny in Trochodendrales. Ann. Missouri Bot. Gard. 73: 297–324.

Endress PK. 1993. Cercidiphyllaceae. In: K Kubitzki, ed., The families and genera of vascular plants, vol. 2, pp. 250–252. Springer, Berlin/Heidelberg/New York.

Feng Y-X, X-Q Wang, K-Y Pan, D-Y Hong. 1998. A reevaluation of the systematic positions of the Cercidiphyllaceae and Daphniphyllaceae based on *rbc*L gene sequence analysis, with reference to the relationship in the 'Lower' Hamamelidae. Acta Phytotax. Sinica 36: 411–422.

Harms H. 1916. Über die Blütenverhältnisse und die systematische Stellung der Gattung *Cercidiphyllum* Sieb. and Zucc. Ber. Deutsch. Bot. Ges. 34: 272–283.

Heel WAV. 1987. Note on the morphology of the male inflorescences in *Cercidiphyllum* (Cercidiphyllaceae). Blumea 32: 303–309.

Jähnichen H, DH Mai, and H Walther. 1980. Blätter und Früchte von *Cercidiphyllum* Siebold et Zuccarini im mitteleuropaischen Tertiär. Schriftenreihe Geol. Wiss. 16: 357–399.

Leroy JE. 1980. Dẻveloppement et organogenese chez le *Cercidiphyllum japonicum*: Un cas semblant unique chez les Angiospermes. C. R. Acad. Sci. Paris., ser. D, 290: 679–682.

Mohana Rao PR. 1986. Seed and fruit anatomy in *Cercidiphyllum japonicum* with a discussion on the affinities of Cercidiphyllaceae. Flora 178: 243–249.

Rowley JR. 1992. Pollen of *Cercidiphyllum* (Cercidiphyllaceae). Bot. Zhurn. 77(11): 13.

Solereder H. 1899. Zur Morphologic und Systematik der Gattung *Cercidiphyllum* Sieb. and Zucc. mit Berücksichtigung der Gattung *Eucommia* Oliv. Ber. Deutsch. Bot. Ges. 17: 387–406.

Spongberg SA. 1979. Cercidiphyllaceae hardy in temperate North America. J. Arnold Arbor. 60: 367–376.

Swamy BGL and IW Bailey. 1949. The morphology and relationships of *Cercidiphyllum*. J. Arnold Arbor. 30: 187–210.

Van Heel WA. 1986. A deviating female flower of *Cercidiphyllum japonicum* (Cercidiphyllaceae). Blumea 31: 273–276.

Van Heel WA. 1987. Note on the morphology of the male inflorescences in *Cercidiphyllum* (Cercidiphyllaceae). Blumea 32: 303–309.

Wang D and SZ Gao. 1990. A study on the relationships of Cercidiphyllaceae. I. Leaf architecture and vasculature changes of petiole. Acta Bot. Bor.-Occid. Sinica 10: 37–41.

Superorder MYROTHAMNANAE

Order 32. MYROTHAMNALES

Small, sympodially branching, ascending, rigid, xerophytic shrubs with an extreme poikilohydrous capacity ("resurrection plants") and characterized by the production of balsam (resin) in large, inflated, thin-walled epidermal cells. Branches clearly differentiated into long and short shoots, and lateral branches largely developed as short shoots. Vessel elements long, narrow, angular, with oblique scalariform or reticulate scalariform perforation plates, with numerous (ca. 45) bars; lateral pitting alternate. Fibers with bordered pits. Rays exclusively uniseriate and composed mostly of upright cells (and a few square cells), without crystals. Axial parenchyma absent. Sieve-element plastids of S-type. Ultrastructurally the chloroplasts possess a unique type of granum (Welburn and Welburn 1976). Nodes trilacunar with three traces. Leaves small, opposite-decussate, with broad sheathing base, cuneate, with distal part rounded and dentate, plicate and folding up like a fan during the dry season, with palmately-flabellate venation. The leaf bases of opposite leaves are fused and form a sheath that remains on the shoot after the leaf has abscised. Stipules intrapetiolar, small, toothlike. Stomata encyclocytic. Inflorescences terminal on short shoots, erect, spikelike, with or without terminal flower, the flowers (or floral triads at the base of the inflorescence) opposite and subtended by foliose (the lowest) or bracteose, cucullate leaves, unisexual, with either male or female flowers (Dahlgren and van Wyk 1988). Flowers sessile, bracteolate or without bracteoles, unisexual, actinomorphic or somewhat zygomorphic, apetalous, with 0–4 (or more) lanceolate scales interpreted as "tepals" (Dahlgren and van Wyk 1988) (vestigial sepals?). Stamens 3–8, when 4 in terminal flowers with 4 "tepals" alternating with them. Filaments cylindrical, distinct or connate at the base. Anthers large, tetrasporangiate, basifixed, with a connective prolonged into a beak, latrorse, opening longitudinally almost slitlike but with a stomial bifurcation at the proximal end (Hufford and Endress 1989). Tapetum secretory. Microsporogenesis simultaneous. Pollen grains 2-celled, in tetrahedral tetrads, each with three ill-defined pores and intectate. The exine sculpturing consists of clavate columellae that are ornamented with minute papillae (Zavada and Dilcher 1986). Gynoecium of 3–4 united carpels with free, recurved, short, broad stylodia and enlarged, winglike, ventrally decurrent, marginally more or less ungulate stigma bearing unicellular papillae. Ovary 3–4 -locular, with 28–32 submarginal ovules in each locule. Ovules anatropous, bitegmic, crassinucellate, micropyle bistomal. Female gametophyte of *Allium*-type. Endosperm nuclear. Fruits small, leathery capsules; the carpels separating above and opening ventrally. Seeds numerous, small, ovoid, with thin, reticulate seed coat formed mainly by the inner integument, exotestal cells with somewhat thickened outer walls. Embryo small, weakly differentiated; endosperm copious, oily. Plants strongly tanniniferous and containing ellagic acid, chalcones, flavonols (quercetin and kaempferol), p-coumaric acid, sinapinic acid, and p-hydrobenzoic acid as well as vanillinic and syringic acids and cyanidin. Crystal druses of calcium oxalate are abundant in the leaves, n = 10.

Although *Myrothamnus* somewhat resembles the genus *Distylium* of the Hamamelidaceae (Endress 1977, 1989), it differs from Hamamelidaceae and related families in many respects, including its vegetative morphology (particularly leaf morphology and anatomy), uniseriate rays, absence of the axial parenchyma, pollen morphology, morphology of the gynoecium, female gametophyte of *Allium*-type, and copious endosperm.

It is therefore more correct to separate the Myrothamnaceae into an order of their own (Nakai 1943; Takhtajan 1987, 1997). Endress (1989: 199) has come to the conclusion that features *Myrothamnus* shares with *Cercidiphyllum* and Trochodendrales s. 1. "strongly point to a closer phylogenetic relationship of the Myrothamnaceae with the Trochodendrales than with the Hamamelidales, and especially with the Cercidiphyllaceae." I agree that Myrothamnaceae are relatively closer to the Trochodendranae (Trochodendrales s.1.) than to the hamamelides. However, taking into consideration so many important differences between Trochodendranae (including Cercidiphyllales) and the Myrothamnaceae, it seems more appropriate to consider Myrothamnales a separate superorder. The Myrothamnales differ from the Trochodendranae by their poikilohydrous capacity, leaf morphology (including very peculiar and unique flabellate venation), uniseriate rays, absence of axial parenchyma, tetradinous and intectate 3-porate pollen grains, *Allium*-type female gametophyte, nuclear endosperm, and septicidal capsules. I consider Myrothamnanae a separate evolutionary lineage and a very advanced side branch of the phylogenetic tree of the Hamamelidae (1997).

1. MYROTHAMNACEAE

Niedenzu 1891. 1/2. Tropical Africa and South Africa and Madagascar.

Myrothamnus.

Bibliography

Carlquist S. 1976. Wood anatomy of *Myrothamnus flabellifolia* (Myrothamnaceae) and the problem of multiperforiate perforation plates. J. Arnold Arbor. 57: 119–126.

Carlquist S. 1990. Leaf anatomy of Geissolomataceae and Myrothamnaceae as a possible indicator of relationship to Bruniaceae. Bull. Torrey Bot. Club 117: 420–428.

Dahlgren R and AE van Wyk. 1988. Structures and relationships of families endemic to or centered in southern Africa. Monogr. Syst. Bot. Missouri Bot. Gard. 25: 1–94.

Endress PK. 1989. The systematic position of the Myrothamnaceae. In: PR Crane and S Blackmore, eds. Evolution, systematics, and fossil history of the Hamamelidaceae, vol. 1, pp. 193–200. Claredon Press, Oxford.

Grundell R. 1933. Zur Anatomic von *Myrothamnus fla-bellifolia* Welw. Symbolae Botanicae Upsaliensis 2: 1–17.

Jäger-Zurn I. 1966. Infloreszenz- und blütenmorpholog-ische, sowie embryologische Untersuchungen an *Myrothamnus* Welw. Beitr. Biol. Pfl. 42: 241–271.

Kubitzki K. 1993. Myrothamnaceae. In: K Kubitzki, ed. The families and genera of vascular plants, vol. 2, pp. 468–469. Springer, Berlin/Heidelberg/New York.

Puff C. 1978a. The nodal anatomy of *Myrothamnus fla-bellifolius* (Myrothamnaceae): Another example of a 'split-lateral' condition. J. Arnold Arbor. 59: 192–196.

Puff C. 1978b. Zur Biologic von *Myrothamnus fiabelli-folius* Welw. (Myrothamnaceae). Dinteria 14: 1–20.

Vieweg GH and H Ziegler. 1969. Zur Physiologic von *Myrothamnus flabellifolia*. Ber. Deutsch. Bot. Ges. 82: 29–36.

Weimarck H. 1936. *Myrothamnus flabellifolius* Welw.: Eine polymorphe Pflanzenart. Bot. Not. 1936: 451–462.

Wellburn FAM. and AR Wellburn. 1976. Novel chloroplasts and unusual cellular ultrastructure in the "resurrection" plant *Myrothamnus flabellifolius* Welw. (Myrothamnaceae). Bot. J. Linn. Soc. 72: 51–54.

Superorder HAMAMELIDANAE

Order 33. HAMAMELIDALES

Evergreen or deciduous trees and shrubs. Crystals of calcium oxalate usually present in some of the cells of the parenchymatous tissues. Vessel elements narrow and sometimes very long, with very oblique end walls, mostly with scalariform perforations that have few to numerous bars (more than 50 in *Corylopsis* and *Exbucklandia* and 50–100, sometimes 130 or even 160 in *Ostrearia* lateral pitting from opposite to scalariform, rarely alternate. Fibers with bordered pits. Rays heterogeneous. Axial parenchyma apotracheal and typically diffuse. Sieve-element plastids of S-type with 5–10 starch grains. Nodes trilacunar to pentalacunar. Leaves alternate, simple, often palmately lobed, pinnately or palmately veined, stipulate. Stomata paracytic. Flowers in inflorescences of various types, of medium size or small, bisexual, polygamous or unisexual, actinomorphic or rarely zygomorphic, with double perianth or seldom apetalous, sometimes without perianth or with perianth extremely reduced. Sepals (3)4-5(-10), small, distinct or more often more or less connate, sometimes very reduced or lacking. Petals usually 4–5, commonly small or narrow, sometimes wanting. Stamens (3)4-5 (up to 10), in one cycle, alternating with petals or staminodia, or sometimes more numerous (up to 32) and initiated in centripetal or centrifugal sequence. Anthers frequently with a shortly prolonged connective, mostly tetrasporangiate, latrorse, opening by two valves like two window shutters (Endress 1986, 1989a, 1993; Hufford and Endress 1989). Only in five genera of tribe Hamamelideae do the thecae open by a single valve that serves both pollen sacs (Endress 1989b). Tapetum secretory. Microsporogenesis simultaneous. Pollen grains 2-celled, from 3-colpate to pantoporate and pantocolpate, tectatecolumellate, reticulate. Gynoecium of 2 or less often 3-8(9) carpels, or more frequently united at the base; stylodia always free, with decurrent stigma. Each locule and each free carpel mostly with one ovule, less often two ovules, and in each locule of syncarpous gynoecia sometimes more. Ovules anatropous, more or less pendulous, bitegmic, crassinucellate. Endosperm cellular (*Parrotiopsis*) or nuclear. Fruits of small achenes, or woody capsules with a leathery exocarp and a bony endocarp. Seed coat formed mainly by the outer integument; embryo long, straight; endosperm copious to scanty, oily, and proteinaceous. More or less tanniniferous, producing phenolics flavonols (quercetin, kaempferol, and frequently also myricetin) and also iridoid compounds.

This order in many ways serves as a connecting link between Trochodendrales on the one hand and the "amentiferous" orders Casuarinales, Fagales, etc., on the other. The order Hamamelidales is evidently derived from an immediate ancestor of Trochodendrales.

Key to Families

1 Pollen grains commonly 3-colpate, seldom pantocolpate (*Matudaea*), 4-colpate to pantoporate (*Distyilum*) or pantoporate (*Sycopsis*). Petioles with a trough-shaped or cylindrical vascular strand. Resin ducts absent (except *Mytilaria*, which has resin canals in the pith of stem). Flowers small, bisexual, in some genera bisexual and male, rarely unisexual (*Fortunearia, Sinowilsonia)*. Sepals usually imbricate. Petals present or absent, when present (2-)4, or 5, gamopetalous, valvate. Stamens 4–32; anthers basifixed, introrse, tetrasporangiate or rarely (*Hamamelis*) disporangiate, opening by pores, or by short or longitudinal slits, of by longitudinal valves, usually appendages. Carpels two or rarely three, small or medium; ovary more or less inferior, ovules mostly solitary, sometimes 6–15; outer integument 6–7 cells across, micropyle zig-zag. Fruits loculicidal and septicidal capsules with woody exocapr and horny endocarp. Seeds winged or wingless, with thick, hard, and conspicuous testa, consisting of fiberlike cells and forming the main protective part of the seed coat; hilum large. The tegmen does not contribute to the construction of the mature seed coat; only the inner epidermis persists; the nucellus persists in ripe seed; endosperm oily; embryo straight. Trees or shrubs, often with stellate indumentum. Vessels with scalariform or scalariform and reticulate perforation. Vestured pits present. Nodes pentalacunar with five traces. Leaves dissected or entire, often palmatifid. Contain flavonols kaempferol, quercetin, and myricetin, producing C-glycosylflavones, sometimes (*Corylopsis*) ellagic acid.. . . . . 1. HAMAMELIDACEAE

1 Pollen grains always pantoporate, the pores circular, often with islands of ectexine on the pore membrane. Petioles with 3–5 vascular bundles. Resin ducts present in roots, stem, leaves, and floral organs. Flowers in terminal racemes (male) or in globular heads. Perianth sepaline or vestigial, absent in male flowers, or 12–50 minute lobes or scales in female flowers. Stamens 12–50, anthers opening by longitudinal slits or valves. Staminodes 3–10. Carpels 2; ovary inferior, 2-locular. Ovules 20–50 per locule, horizontal. Fruit a septicidal capsule. Seeds small, winged; the wing formed by the longitudinal elongation and flattening of the integument around the micropyle; the ripe seed with the 2-celled thick testa, but the protective part of the seed coat contributed by the outer epidermis of the inner integument with lignified and fibrous cells; the nucellus does not persist in ripe seed; endosperm slight, embryo long. Large, monoecious, deciduous trees. Vessels without vestured pits. Leaves ovate, palmatifid or tricuspidate, entire, serrate, or dentate; domatia (*Liquidambar*) presented by pockets, or hair tufts. Iridoids detected (*Liquidambar*), present proanthocyanidins (cyanidin and delphinidin), flavonols (quercetin and myricetin), and ellagic acid, n = 15, 16 . . . . 2. ALTINGIACEAE

1. HAMAMELIDACEAE

R. Brown 1818 (including Disanthaceae Nakai 1943, Exbucklandiaceae Reveal et Doweld 1999, Fothergillaceae Nuttall 1818, Parrotiaceae Horaninow 1834, Rhodoleiaceae Nakai 1943). 28/100-110. Tropical and South Africa, Madagascar, southeastern Transcaucasia, northern Iran, western and eastern Himalayas, Assam, Manipur, eastern Asia and Southeast Asia, New Guinea, northeastern Australia, eastern North America, and Central America.

1.1 DISANTHOIDEAE
Leaves suborbiculate, cordate or truncate at the base, entire, palmately 5-nerved. Stipules scarious, linear, caducous. Flowers two in a head, with short bracts at the base, bisexual, entomophilous. Perianth present, petals five, circinate in bud, flabellate, stellately spreading. Anthers opening by simple longitudinal slits. Each locule with about five ovules; ripe locule with 4–5 seeds. Seeds not ejected out of capsule, n = 8, chromosomes the largest in the family and nuclei similar to the *Tetrameranthus* type (Morawetz and Samuel 1989). – *Disanthus* (1 sp. in eastern China and Japan).

1.2 HAMAMELIDOIDEAE
Leaves mostly ovate, sometimes cordate, pinnately veined, entire or toothed. Stipules mostly small, commonly caducous. Stellate hairs present. Flowers usually in spikes or heads, commonly bisexual entomophilous or anemophilous. Perianth mostly present, but petals often lacking. Each locule mostly with 1 ovule, rarely with one fertile and two sterile; ripe locule with one seed. Fruit with ballistic dispersal. The seeds ejected at maturity, n = 12. – HAMAMELIDEAE: *Hamamelis, Loropetalum, Tetrathyrium, Maingaya, Embolanthera, Dicoryphe, Trichocladus, Ostrearia, Neostrearia, Noahdendron*; CORYLOPSIDEAE: *Corylopsis*; EUSTIGMATEAE: *Eustigma, Fortunearia, Sinowilsonia*; FOTHERGILLEAE: *Molinodendron, Fothergilla, Parrotiopsis, Shaniodendron, Parrotia, Sycopsis, Distyliopsis, Distylium, Matudaea*.

I.3 EXBUCKLANDIOIDEAE
Leaves ovate, often tricuspidate, mostly cordate, palmately veined. Stipules mostly large and covering subsequent bud. Flowers aggregated in dense spikes, bisexual. Perianth present or absent. Each locule with 5–8 ovules. The seed not ejected, n = 8–12. — *Exbucklandia, Mytilaria, Chunia*.

1.4 RHODOLEIOIDEAE
Leaves ovate, rarely cordate, entire, pinnately veined, vestigially stipulate. Inflorescence a pseudanthium of strongly zygomorphic flowers. Flowers bisexual, ornitophilous. Perianth present. Each locule with 1–2 fertile ovules plus 10–20 sterile ones; ripe locule with 2 or more seeds. Seeds not ejected, n = 12. — *Rhodoleia* (1–10 spp. in eastern Asia and Malesia).

Disanthoideae are in general the least specialized subfamily (Takhtajan 1980, 1987, 1997; Cronquist 1981; Wolfe 1989; Pan et al. 1991) while Rhodoleioideae are the most specialized.

2. ALTINGIACEAE

Horaninow 1841. 3/17. Eastern Mediterranean (islands of the Aegean Sea-Rhodes and Cyprus), eastern Himalayas (Bhutan and Assam), continental China, Taiwan, Southeast Asia, eastern and northeastern North America, northeastern Mexico, and Central America to Costa Rica.

Liquidambar, Altingia, Semiliquidambar.

Usually included in the Hamamelidaceae, but the combination of characteristics "sets the subfamily quite apart from the rest of the family" (Bogle 1986: 328). Although specific characteristics, including resin ducts and periporate pollen grains, do occur in certain genera of the other subfamilies (Bogle 1986), in general the Altingiaceae are so distinct that they deserve the status of a separate family. Chemistry also underscores Altingiaceae's distinctive nature (Ferguson 1989).

Bibliography

Bogle AL. 1970. Floral morphology and vascular anatomy of the Hamamelidaceae: The apetalous genera of Hamamelidoideae. J. Arnold Arbor. 51: 310–366.

Bogle AL. 1984. Floral morphology and vascular anatomy of *Mamgaya* Oliv. (Hamamelidaceae, Hamamelidoideae, Hamamelideae). Am. J. Bot. 71: 19.

Bogle AL. 1986. The floral morphology and vascular structure of the Hamamelidaceae: Subfamily Liquidambaroideae. Ann. Missouri Bot. Gard. 73: 325–347.

Bogle AL. 1989. The floral morphology, vascular anatomy, and ontogeny of the Rhodoleioideae (Hamamelidaceae) and their significance in relation to the 'lower' hamamelids. In: PR Crane and S Blackmore, eds. Evolution, systematics, and fossil history of the Hamamelidae, vol. 1, pp. 201–226. Claredon Press, Oxford.

Bogle AL. 1990. Multilacunar nodal anatomy in *Mytilaria* (Hamamelidaceae). J. Arnold Arbor. 71: 111–118.
Bogle AL and CT Philbrick. 1980. A genera atlas of hamamelidaceous pollens. Contr. Gray Herb. 210: 29–103.
Britton EG. 1887. Elongation of the inflorescence of *Liquidambar*. Bull. Torrey Bot. Club 14: 95–96.
Chang CT. 1959. The pollen morphology of *Liquidambar* L. and *Altingia* Nor. Bot. Zhurn. 44: 1375–1380 (in Russian with English summary).
Chang CT. 1964. The pollen morphology of the families Hamamelidaceae and Altingiaceae. Trudy Bot. Inst. Akad. Nauk SSSR, 1st ser., 13: 173–232 (in Russian).
Chang HT. 1973. A revision of the hamamelidaceous flora of China. Bull. Dept. Biol. Sun Yatsen Univ. 1: 54–71 (in Chinese).
Deng MB, HT Wei, XQ Wang. 1992a. *Shaniodendron*, a new genus of Hamamelidaceae from China. Acta Phytotax. Sinica 30: 57–61.
Deng MB, HT Wei, XQ Wang et al. 1992b. On the significance of the discovery of Fothergilleae in China. J. Plant Resour. Environ. 1: 30–35.
Egger K and H Reznik. 1961. Die Flavonolglykoside der Hamamelidaceen. Planta 57: 239–249.
Endress PK. 1967. Systematische Studie über die verwandschaftlichen Beziehungen zwischen den Hamamelidaceen und Betulaceen. Bot. Jahrb. Syst. 87: 431–525.
Endress PK. 1969. Untersuchungen über den phylogenetischen Anschluss der Betulaceen an die Hamamelidaceen. Verh. Schweiz. Naturf. Ges. 1968: 113–114.
Endress PK. 1970. Die Infloreszenzen der apetalen Hamamelidaceen: Ihre grundsatzliche morphologische und systematische Bedeutung. Bot. Jahrb. Syst. 90: 1–54.
Endress PK. 1971. Blütenstände und morphologische Interpretation der Blüten bei apetalen Hamamelidaceen. Ber. Deutsch. Bot. Ges. 84: 183–185.
Endress PK. 1977. Evolutionary trends in the Hamamelidales-Fagales-group. Plant Syst. Evol., Suppl. 1: 321–347.
Endress PK. 1978a. Stipules in *Rhodoleia* (Hamamelidaceae). Plant Syst. Evol. 130: 157–160.
Endress PK. 1978b. Blütenontogenese, Blütenabgrenzung, und systematische Stellung der perianthlosen Hamamelidoideae. Bot. Jahrb. Syst. 100: 249–317.
Endress PK. 1989a. Aspects of evolutionary differentiation of the Hamamelidaceae and the Lower Hamamelididae. Plant Syst. Evol. 162: 193–211.
Endress PK. 1989b. A suprageneric taxonomic classification of the Hamamelidaceae. Taxon 38: 371–376.
Endress PK. 1989c. Phylogenetic relationships in the Hamamelidoideae. In: PR Crane and S Blackmore, eds. Evolution, systematics, and fossil history of the Hamamelidae, vol. 1, pp. 227–240. Claredon Press, Oxford.
Endress PK. 1993. Hamamelidaceae. In: K Kubitzki, ed. The families and genera of vascular plants, vol. 2, pp. 322–331. Springer, Berlin/Heidelberg/New York.
Ernst WR. 1963. The genera of Hamamelidaceae and Platanaceae in the southeastern United States. J. Arnold Arbor. 44: 193–210.
Fang YM, MB Deng. 1996. Wood anatomy of *Shaniodendron subaequale* (Hamamelidaceae) and its systematic implications. J. Plant Resour. Environ. 5(2): 50–54.
Fang YM, RW Fan. 1993. Variation and evolution of leaf trichomes in the Chinese Hamamelidaceae. Acta Phytotax. Sinica 31(2): 147–152.
Ferguson DK. 1989. A survey of the Liquidambaroideae (Hamamelidaceae) with a view to elucidating its fossil record. In: PR Crane and S Blackmore, eds. Evolution, systematics, and fossil history of the Hamamelidaceae, vol. 1, pp. 249–272. Claredon Press, Oxford.
Goldblatt P and PK Endress. 1977. Cytology and evolution in Hamamelidaceae. J. Arnold Arbor. 58: 67–71.
Hao RM, HT Wei, and WG Liu. 1996. Floral morphology of *Shaniodendron* (Hamamelidaceae) and its taxonomic significance. J. Plant Resour. Environ. 5(1): 38–42.
Home AL. 1914. A contribution to the study of the evolution of the flower, with special reference to the Hamamelidaceae, Caprifoliaceae, and Cornaceae. Trans. Linn. Soc. London, Bot., 2nd ser., 8: 239–309.
Huang GL. 1986. Comparative anatomical studies on the woods of Hamamelidaceae in China. Acta Sci. Nat. Univ. Sunjatseni 1: 22–28 (in Chinese with English summary).
Ickert-Bond SM, KB Pigg, and J Wen. 2005. Comparative infructescence morphology in *Liquidambar* (Altingiaceae) and its evolutionary significance. Am. J. Bot. 92: 1234–1255.
Ickert-Bond SM and J Wen. 2006. Phylogeny and biogeography of Altingiaceae: Evidence from combined analysis of five non-coding chloroplast regions. Molec. Phylog. Evol. 39: 512–528.
Jha UN. 1978. Chemotaxonomy of the Hamamelidaceae. J. Indian Bot. Soc. 56: 44–48.
Kapil RN and U Kaul. 1972. Embryologically little known taxon – *Parrotiopsis jacquemontiana*. Phytomorphology 22: 234–245.
Kaul U and RN Kapil. 1974. *Exbucklandia fofulnea*: From flower to fruit. Phytomorphology 24: 217–228.
Li H-M, and LJ Hickey. 1988. Leaf architecture and systematics of the Hamamelidaceae sensu lato. Acta Phytotax. Sinica 26: 96–110.
Li J-H and AL Bogle. 1998. Comparative embryology and floral ontogeny of the *Corylopsis* complex (Hamamelidaceae). Am. J. Bot. 85(Suppl. 6): 173 (Abstract).
Li J, and AL Bogle. 2001. A new suprageneric classification system of Hamamelidoideae based on morphology and sequences of nuclear and chloroplast DNA. Harvard Pap. Bot. 5: 499–515.
Li J-H, AL Bogle, and AS Klein. 1997. A phylogenetic analysis of the Hamamelidaceae using DNA sequences of nuclear ribosomal ITS and chloroplast gene *mat*K. Am. J. Bot. 84(Suppl. 6): 212.
Li J-H, AL Bogle, and AS Klein. 1998. Phylogenetic relationships in the *Corylopsis* complex (Hamamelidaceae) based on sequences of the internal transcribed spacers of nuclear ribosomal DNA and morphology. Rhodora 99: 302–318.
Li J-H, AL Bogle, and AS Klein. 1999a. Phylogenetic relationships of the Hamamelidaceae inferred from sequences of internal transcribed spacers (ITS) of nuclear ribosomal DNA. Am. J. Bot. 86: 1027–1037.
Li J-H, AL Bogle, and AS Klein. 1999b. Phylogenetic relationships in the Hamamelidaceae: evidence from the nucleotide sequences of the plastid gene matK. Plant Syst. Evol. 218(3–4): 205–219.
Li J-H, AL Bogle, AS Klein, and KY Pan. 1997. Close relationship between *Shaniodendron* and *Parrotia* (Hamamelidaceae), evidence from ITS sequences of nuclear ribosomal DNA. Acta Phytotax. Sinica 35: 481–493.

Li J, AL Bogle, AS Klein, and MJ Donoghue. 2000. Phylogeny and biogeography of *Hamamelis* (Hamamelidaceae). Harvard Pap. Bot. 5(1): 171–178.

Melikian AP. 1973. Seed coat types of Hamamelidaceae and allied families in relation to their systematics. Bot. Zhurn. 58: 350–359 (in Russian).

Melikian AP. 1991. Hamamelidales. In: A Takhtajan, ed. Comparative seed anatomy, vol. 3, pp. 105–112. Nauka, Leningrad (in Russian).

Mione T and AL Bogle. 1987. Comparative ontogeny of the flowers *Hamamelis virginiana* and *Loropetalum chinense* (Hamamelidaceae). Am. J. Bot. 74: 620–621.

Mione T and AL Bogle. 1990. Comparative ontogeny of the inflorescence and flower of *Hamamelis virginiana* and *Loropetalum chinense* (Hamamelidaceae). Am. J. Bot. 77: 77–91.

Mizushima M. 1968. On the flower of *Disanthus cercidifolius* Maxim. J. Jpn. Bot. 43: 522–544.

Mohana Rao PR. 1974. Seed anatomy in some Hamamelidaceae and phylogeny. Phytomorphology 24: 113–139.

Oginuma K and H Tobe. 1991. Karyomorphology and evolution in some Hamamelidaceae and Platanaceae (Hamamelididae: Hamamelidales). Bot. Mag. (Tokyo) 104: 115–135.

Pan KY, Lu Anming, and Wen Jie. 1990. Characters of leaf epidermis in Hamamelidaceae (s. 1.). Acta Phytotax. Sinica 28(I): 10–26.

Pan KY, Lu An-ming, and Wen Jie. 1991. A systematic study on the genus *Disanthus* Maxim. (Hamamelidaceae). Cathaya 3: 1–28.

Pan KY and QE Yang. 1994. Karyotypes of Disanthus and Mytilaria (Hamamelidaceae). Acta Phytotax. Sinica 32: 235–239.

Qiu Y-L, MW Chase, SB Hoot, E Conti, PR Crane, KJ Sytsma, and CR Parks. 1998. Phylogenetics of the Hamamelidae and their allies: parsimony analyses of nucleoitide sequences of the plastid gene *rbc*L. Int. J. Plant Sci. 159: 891–905.

Rao TA and OP Bhupal. 1974. Typology of foliar sclereids in various taxa of Hamamelidaceae. Proc. Indian Acad. Sci., Sect. B, 79: 127–138.

Reinsch A. 1889. Über die anatomischen Verhaltnisse der Hamamelidaceae mit Rücksicht auf ihre systematische Grüppierung. Bot. Jahrb. Syst. 2: 347–395.

Shi S-H, H-T Chang, Y-Q Chen, L-H Qu, and J Wen. 1998. Phylogeny of the Hamamelidaceae based on the ITS sequences of nuclear ribosomal DNA. Biochem. Syst. Ecol. 26: 55–69.

Skvortsova NT. 1960a. The structure of epidermis in representatives of the family Hamamelidaceae. Bot. Zhurn. 45: 712–717 (in Russian).

Skvortsova NT. 1960b. The anatomical structure of the conducting system of leaf petioles of representatives of families Hamamelidaceae and Altingiaceae. Doklady Akad. Nauk SSSR 133: 1231–1234 (in Russian).

Skvortsova NT. 1960c. On the types of leaf venation in representatives of the family Hamamelidaceae. Trudy Leningrad Chemico-Pharmaceutical Inst. 12: 75–83 (in Russian).

Skvortsova NT. 1965. On the morphology of the genus *Hamamelis* L. Bot. Zhurn. 50: 1143–1148 (in Russian).

Skvortsova NT. 1975. Comparative morphological studies in representatives of the family Hamamelidaceae and their phylogenetic relationships. In: LY Budantsev, ed. Problems of comparative morphology of the seed plants, pp. 724. Nauka, Leningrad (in Russian).

Tang Y. 1943. Systematic anatomy of the woods of the Hamamelidaceae. Bull. Fan Memorial Inst. Biol., n.s., 1: 8–63.

Tattje DHE, R Bos, and AP Bruins. 1980. Constituents of essential oil from leaves of *Liquidambar styraciflua* L. Planta Medica 38: 79–85.

Tong K. 1930. Studien über die Familie der Hamamelidaceae, mit besonderer Berücksichtigung der Systematik und Entwicklungsgeschichte von *Corylopsis*. Bull. Dept. Biol. Sun Yatsen Univ. 2: 1–72.

Wen J, and S Shi. 1998. A phylogenetic and biogeographic study of *Hamamelis* (Hamamelidaceae), an eastern Asian and eastern North American disjunct genus. Biochem. Syst. Ecol. 27: 55–66.

Wisniewski M and AL Bogle. 1982. The ontogeny of the inflorescence and flower of *Liquidambar styraciflua* L. (Hamamelidaceae). Am. J. Bot. 69: 1612–1624.

Zhang Z-Y. 2001. Pollen morphology and variation of the genus *Fortunearia* (Hamamelidaceae) endemic to China. Israel J. Plant Sci. 49(1): 61–66.

Zhang Z-Y and A-M Lu. 1995. Hamamelidaceae: Geographic distribution, fossil history and origin. Acta Phytotax. Sinica 33: 313–339 (in Chinese with English summary).

Zhang Q, SH Shi, YL Huang, FX Tan, and H Jin. 2000. Phylogeny of the tribe *Hamamelideae (Hamamelidaceae)* based on the ITS sequences. Acta Sci. Nat. Univ. Sunyatseni. 39: 72–76.

Zhang J-T and D-W Zhang. 1991. Studies on pollen morphology of the genus *Corylopsis*. Acta Phytotax. Sinica 29: 347–351.

Order 34. DAPHNIPHYLLALES

Evergreen, dioecious trees and shrubs, all parts glabrous. Vessels with scalariform perforations that have 20–100 or more thin bars; lateral pitting from opposite to scalariform. Fibers with bordered pits. Rays heterogeneous, 1–2 cells wide. Axial parenchyma scanty paratracheal and diffuse. Sieve-element plastids of S-type, containing about five globular starch grains. Leaves alternate, rarely opposite, long-petiolate, sometimes densely crowded at the end of branches and appearing almost verticillate, simple, entire or revolute, usually glaucous and papillate beneath, with pinnate venation, estipulate. Stomata paracytic. Flowers small, in axillary or rarely subterminal racemes, unisexual, actinomorphic, apetalous, subtended by a deciduous bract. Sepals (2)3–6, small, more or less connate, imbricate, caduceus or persistent, sometimes wanting. Stamens (5)6-12(14), free or coherent at apex, subsessile or with long filaments. Anthers large, oblong, tetrasporangiate, basifixed, latrorse, opening longitudinally, commonly with shortly prolonged connective. Tapetum secretory. Microsporogenesis simultaneous. Pollen grains 2-celled, 3–4-colpate, tectate-columellate, psilate to verrucate, microperforate. Female flowers often with staminodia. Gynoecium of two or rarely four (very rarely more) united carpels; stylodia connate only at the base, short, broad, divaricate, recurved or circinate, with papillate, decurrent stigma. Ovary imperfectly 2(-4)-locular, with two more or less apical, pendulous ovules in each locule. Ovules anatropous, epitropous, with ventral raphe, bitegmic, crassinucellate, micropyle zig-zag. Female gametophyte of *Polygonum*-type. Endosperm cellular. Fruits 1(2)-seeded drupes with stony endocarp, often with persistent stylodia and staminodia. Seeds endotegmic (Bhatnagar and Kapil 1982), seed coat persistent but thin-walled and crushed, or endotegmen tanniniferous, walls thickened; embryo minute, straight, apical; endosperm copious, oily, and proteinaceous, and perisperm containing protein crystals. Flavonols are common to *Daphniphyllum* (apigenin and luteolin), as well as the iridoids glucosides – asperuloside and daphniphyloside, and unique diterpene alkaloids (daphniphylline), myricetin, and probably proanthocyanidins (Hegnauer 1997), n = 16.

Recent studies of wood anatomy and phytochemestry indicate that Dapyhniphyllaceae have a close affinity with Hamamelidaceae and Buxaceae.

1. DAPHNIPHYLLACEAE

Müller Aargau 1869. 1/c.30. Southwestern India (western Gats), Sri Lanka, Himalayas, southern Tibet, Assam (Khassia Hills), eastern Asia and Southeast Asia.

Daphniphyllum.

Bibliography

Baas P. 1997. Vegetative anatomy of Daphniphyllaceae. Flora Malesiana. In: C Kalkman et al., eds. Flora Malesiana, ser. 1, 13: 146–147. Leiden.

Bhatnagar AK and M Garg. 1977. Affinities of *Daphniphyllum*: Palynological approach. Phytomorphology 27: 92–97.

Bhatnagar AK and RN Kapil. 1982. Seed development in *Daphniphyllum himalayense* with a discussion on taxonomic position of Daphniphyllaceae. Phytomorphology 32: 66–81.

Carlquist S. 1982. Wood anatomy of Daphniphyllaceae: Ecological and phylogenetic considerations, review of Pittosporalean families. Brittonia 34: 252–266.

Croizat L. 1941. On the systematic position of *Daphniphyllum* and its allies. Lingnan Sci. J. 20: 79–103.

Feng Y-X, X-Q Wang, K-Y Pan, D-Y Hong. 1998. A reevaluation of the systematic positions of the Cercidiphyllaceae and Daphniphyllaceae based on *rbc*L gene sequence analysis, with reference to the relationship in the 'Lower' Hamamelidae. Acta Phytotax. Sinica 36: 411–422.

Hegnauer R. 1997. Phytochemistry and chemotaxonomy of Daphniphyllaceae. In: C Kalkman et al., eds. Flora Malesiana, ser. 1, 13: 148–149. Leiden.

Huang T-C. 1965, 1966. Monograph of *Daphniphyllum*. Part 1–2. Taiwania 11: 57–98; 12: 137–234.

Huang T-C. 1996. Notes on taxonomy and pollen of Malesian *Daphniphyllum* (Daphniphyllaceae). Blumea 41: 231–244.

Huang T-C. 1997. Daphniphyllaceae. In: C Kalkman et al., eds. Flora Malesiana, ser. 1, 13: 145–168. Leiden.

Kamelina OP. 1984. The anther and pollen grain development in *Daphniphyllum macropodum* (Daphniphyllaceae). Bot. Zhurn. 69(3): 376–383 (in Russian with English summary).

Sato Y. 1972. Development of the embryo sac of *Daphniphyllum macropodum* var. *humile* (Maxim.) Rosenth. Sci. Rep. Tohoku Imp. Univ., 4th ser. (Biol.), 36: 129–133.

Sutton DA. 1989. The Daphniphyllaceae: A systematic review. In: PR Crane and S Blackmore, eds. Evolution, systematics, and fossil history of the Hamamelidae, vol. 1, pp. 285–291. Claredon Press, Oxford.

Van der Ham RWJM. 1997. Pollen morphology of Daphniphyllaceae. In: C Kalkman et al., eds. Flora Malesiana I, 13: 147–148.

Zhang ZY and A-M Lu. 1989. On the systematic position of Daphniphyllaceae. Acta Phytotax. Sinica 27: 17–26 (in Chinese, with English summary).

Order 35. BALANOPALES

Evergreen trees or shrubs. Young shoots often with caducous unicellular hairs. Vessel elements elongate, with scalariform reticulate or scalariform perforations that have 10–33 thin bars, or some of them with simple perforations; lateral pitting from alternate to nearly opposite. Fibers thick-walled, pits with small or vestigial borders. Rays heterogeneous. Axial parenchyma usually scanty, diffuse. Sieve-element plastids of S-type with up to ten medium-sized, globular starch grains. Each shoot with minute scale leaves proximally and normal leaves distally. Nodes trilacunar. Leaves alternate, the normal leaves scattered or crowded distally and subverticillate, simple, pinnately veined, the margins recurved and toothed to subentire; stipules vestigial, represented by a pair of minute teeth. Stomata laterocytic or laterocytic and encyclocytic. Flowers small, inconspicuous, dioecious, with vestigial or no perianth, anemophilous. Male flowers in small axillary catkinlike inflorescences, much reduced, but often with pistillodia. Stamens (1-) 5-6(-14); anthers nearly sessile,

2-locular, opening longitudinally by lateral slits, sometimes with shortly protruded connective. Pollen grains 3–5-colpate, tectate-granular to columellate, microperforate, ornamented with small spinules. Female flowers solitary, subsessile or short-pedicellate in the axils of scale leaves, without staminodia. Gynoecium surrounded by numerous, spirally arranged bracts, of two or three united carpels; stylodia connate at the base and deeply bipartite and sometimes forking again. Ovary 2–3-locular, frequently imperfectly (partitions not reaching the top of the ovary), with two nearly basal ovules in each locule or on each placenta. Ovules anatropous and intermediate between apotropous and epitropous, bitegmic, outer integument 5–7 cells across, weakly crssinucellate with a funicle forming an obturatorlike outgrowth toward the micropyle; the two ovules of a carpel are curved away from each other (Sutter and Endress 2003). Fruits drupaceous, with thin, fleshy pericarp, subtended at the base by a persistent involucre of numerous imbricate, concrescent bracts, and containing two or three oneseeded pyrenes. Seeds large, with a fairly large, straight, green embryo surrounded by a very thin layer of endosperm, n = 20(21).

Hallier (1908, 1912) included *Balanops* in his Hamamelidaceae s.l. The Balanopales are much nearer the Hamamelidales and Daphniphyllales and probably have common origin with the Daphniphyllales from a hamamelidalean stock.

1. BALANOPACEAE

Bentham et J. W. Hooker 1880. 1/9. Tropical Australia, New Caledonia, Vanuatu, Fiji.

Balanops.

Bibliography

Carlquist S. 1980. Anatomy and systematics of Balanopaceae. Allertonia 2: 191–246.

Guillaumin A. 1925. Recherches sur l'anatomie et la classification des Balanopsidacées. Rev. Gen. Bot. 37: 433–449.

Litt A and MW Chase. 1999. The systematic position of *Euphronia*, with comments on the position of *Balanops*: an analysis based on *rbc*L sequence data. Syst. Bot. 23: 401–409.

Nemirovich-Danchenko EN. 1991. Balanopaceae. In: A Takhtajan, ed. Comparative seed anatomy, vol. 3, pp. 115–116. Nauka, Leningrad (in Russian).

Sutter DM and PL Endress. 2003. Female flower and cupule structure in Balanopaceae, an enigmatic risid family. Ann. Bot. 92: 459–469.

Order 36. BUXALES

Evergreen, glabrous or pubescent trees, shrubs, subshrubs, or perennial herbs. Vessel with scalariform perforations, with 6–25, rarely 30 bars. Lateral pitting opposite to alternate. Fibers with large or small bordered pits. Rays heterogeneous. Axial parenchyma apotracheal, commonly diffuse, or lacking. Leaves alternate or opposite simple, entire or coarsely toothed, mostly pinnately veined, without stipules. Stomata laterocytic or less often encyclocytic (and also of intermediate type), sometimes anomocytic. Flowers usually in spikes, racemes, panicles, or heads, bracteate, nearly always unisexual, apetalous. Sepals mostly 4 or wanting. Stamens two, or four, seldom six or more; filaments free or basally connate, often broad, long or short. Anthers large, basifixed, tetrasporangiate, sometimes sessile, introrse or extrorse, opening longitudinally. Tapetum secretory. Microsporogenesis simultaneous. Pollen grains 2-celled, reticulate. Vestigial gynoecium in male flowers present (Buxaceae) or absent. Gynoecium of 1 (Didymelaceae), 2 or 2–3 united carpels with free or rarely basally shortly connate stigma, which is decurrent or nearly so, more or less broad, commonly with a median furrow, and papillose. Ovary 1-locular (Didymelaceae), 3- or 2-locular, with one or two pendulous ovules in each locule. Ovules anatropous, hemitropous, or campylotropous, apotropous (with a dorsal raphe), bitegmic, crassinucellate. Female gametophyte of *Polygonum*-type. Endosperm cellular (Naumova 1981). Fruits loculicidal, elastically dehiscent capsules or drupaceous. Seeds carunculate or not, embryo straight or curved, short to long, with thin and flat cotyledons. Endosperm copious. Present steroidal alkaloids and iridoids.

The Buxales in many respects resemble Hamamelidales and related orders.

Key to Families

1 Stamens four, rarely six or more (6–30). Trees, shrubs, or (*Pachysandra*) rhizomatous, procumbent perennial herbs or subshrubs. Hairs simple, usually unicellular; cuticle waxes as curled rodlets or irregular platelets. Secretory cells are scattered in the parenchymatous tissues or arranged in long, branching rows. Vessel elements very short or rarely very long (*Styloceras*), nearly always with scalariform perforations, with not less than 20 bars, but usually

with 30 or more in *Styloceras, Pachysandra*, and *Sarcococca*. Lateral pitting commonly opposite, less often intermediate between opposite and scalariform with a tendency to scalariform. Fibers with large and evidently bordered pits (*Styloceras*) or more often with obscurely bordered pits. Rays with elongate (*Styloceras*) or more often short ends. Axial parenchyma apotracheal, commonly diffuse. Sieve-element plastids of P-type with rather peculiar characteristics (PC or Pcs), which contains a central globular protein crystal (Behnke 1982, 1989). Nodes trilacunar with three traces. Leaves alternate (*Pachysandra, Sarcococca*, and *Styloceras*) or opposite (*Buxus* and *Notobuxus*), simple, entire or coarsely toothed (*Pachysandra*), mostly pinnately veined; leaf mesophyll cells contain oil bodies. Stomata laterocytic or less often encyclocytic (and also of intermediate type), sometimes anomocytic. Flowers in axillary or terminal spikes or dense racemes or heads, bracteate, small, actinomorphic, nearly always uni-sexual, monoecious or seldom dioecious (partly in *Styloceras)* or rarely a few of them bisexual. Sepals mostly four, basally connate, seldom wanting. Stamens typically four and opposite calyx lobes, seldom six (*Notobuxus*) or more (6–30 in *Styloceras*); filaments free, often broad, long or short. Anthers large, basifixed, seldom sessile or nearly sessile (*Notobuxus, Styloceras*), introrse. Pollen grains 2-celled, pantoporate, tripolycolpate (Buxeae), reticulate. Gynoecium of 2 or 2–3 (*Sarcococca* and *Styloceras*) united carpels with free or rarely basally shortly connate stigma, which is decurrent, more or less broad, commonly with a median furrow, and papillose. Ovary 3- or 2-locular, with two ovules in each locule, but only one ovule develops into a seed (von Balthazar and Endress 2002); primary locules divided into uniovulate locelli in *Pachysandra* and *Styloceras*. Ovules anatropous or campylotropous, apotropous (with a dorsal raphe), provided with an obturator at least in *Buxus* and *Sarcococca*. Fruits loculicidal, elastically dehiscent capsules or drupes (*Sarcococca* and *Styloceras*). Seeds black or dark brown, shining, carunculate or not, in *Buxus* and *Notobuxus* with small funicular aril. Seed coat formed by the outer integument (exotestal-mesotestal), exotesta lignified, palisade, hypodermis often lignified; embryo straight or curved, short to long, with thin and flat cotyledons. Endosperm copious, firm, oily; mature seeds of *Sarcococca* also have perisperm. Contain steroidal alkaloids, which are derived from triterpenoids n = 10, 12–14, 27. .... 1. BUXACEAE

1 Stamens two. Dioecious, trees; hairs small, peltate. Calcium oxalate druses present. Vessels with scalariform perforations with 6–25 bars; thin lateral pitting opposite to alternate. Fibers with small bordered pits. Axial parenchyma lacking, replaced by lignified cells. Non-xylem tissues of bark, stem, and leaves with abundant coarse, short fibers. Leaves alternate, leathery, simple, entire, petiolate, pinnately veined, estipulate, but petiole urceolate at the base. Stomata encyclocytic. Flowers in axillary racemes or panicles, small, pedicellate, bracteate, without perianth. Male flowers in open panicles, subtended by 0–2 scales. Stamens with very short filaments, connate; anthers cuneate, 2-locular, extrorse. Pollen grains 3-zonocolporate with 2-orate, operculate colpi, tectatecolumellate, reticulate. Female flowers in racemes, paired, solitary or in threes, apetalous, usually with 1–4 scale-like sepals, interpretable as bracts or sepals. Gynoecium of one carpel, the carpel is slightly deflected abaxially; ovary superior, 1-locular. Stigma sessile or elevated on a very short stylodium, broad, oblique (subdecurrent), with a median furrow, very papillose. Ovule solitary, pendulous from the adaxial wall, hemitropous, epitropous, with the integuments prolonged at the apex into a more or less elongate collar; occasionally an additional, rudimentary ovule is present in lateral position (von Balthazar et al. 2003); a little scale presents between the ovule and the bract. Fruits large, fleshy, globous drupaceous, with a lateral groove and with persistent stigma and style. Seed pendulous, reniform; endosperm copious; embryo large, straight, with thick cotyledons. Produce iridoids and highly distinctive steroidal alkaloids (Sutton 1989)................ 2. DIDYMELACEAE

1. BUXACEAE

Dumortier 1822 (including Pachysandraceae J. Agardh 1858, Stylocerataceae Baillon ex Reveal et Hoogland 1990). 5/80-100. Tropical and southern Africa and Madagascar (*Notobuxus*); Socotra; northeastern Africa; from the western Mediterranean to the Caucasus and northern Iran; Afghanistan; Himalayas; southern, eastern, and Southeast Asia; and North, Central, and South

America. *Sarcococca* (11) in Asia from Afghanistan, south and east through India, Sri Lanka, China, and Thailand to Sumatra and Java. *Pachysandra* has one North American and three eastern Asian species. *Styloceras* (4) occurs in western tropical America.

STYLOCEREAE: *Styloceras*; SARCOCOCCEAE: *Sarcococca*; PACHYSANDREAE: *Pachysandra*; BUXEAE: *Buxus, Notobuxus*.

2. DIDYMELACEAE

Leandri 1937. 1/2. Madagascar.

Didymeles.

Close related to Buxaceae. Presence of steroidal alkaloids in both *Buxus* and *Didymeles* suggests a close affinity between Buxaceae and Didymelaceae. Twelve of these aminopregnane and conane alkaloids have been isolated from the Buxaceae and a similar number from *Didymeles* (Hardman 1987).

Bibliography

Ahond A, MM Debray, F Picot, C Poupat, V Sanchez, and P Potier. 1980. Alkaloids steroidique de *Didymeles* cf. *madagascariensis*. Planta Medica 39: 204.

Baillon H. 1859. Monographic des Buxucées et Stylocerées. Masson, Paris.

Balthazar M von, PK Endress. 2002a. Reproductive structures and systematics of Buxaceae. Bot. J. Linn. Soc. 140: 193–228.

Balthazar M von, PK Endress. 2002b. Development of the inflorescences and flowers in Buxaceae and the problem of perianth interpretation. Int. J. Plant Sci. 163: 847–876.

Balthazar M von, PK Endress, Y-L Qiu. 2000. Phylogenetic relationships in Buxaceae based on nuclear internal transcribed spacers and plastid *ndh*F sequences. Int. J. Plant Sci. 161: 785–792.

Balthazar M von, GE Schatz, and PK Endress. 2003. Female flowers and inflorescences of Didymelaceae. Plant Syst. Evol. 237: 199–208.

Baranova MA. 1980. Comparative stomatographic studies in the families Buxaceae and Simmondsiaceae. In: SG Zhilin, ed. Systematics and evolution of higher plants, pp. 68–75. Nauka, Leningrad (in Russian).

Behnke H-D. 1982. Sieve-element plastids, exine sculpturing, and the systematic affinities of the Buxaceae. Plant Syst. Evol. 139: 257–266.

Brückner P. 1993. Pollen morphology and taxonomy of Eurasiatic species of the genus *Buxus* (Buxaceae). Grana 32: 65–78.

Carlquist S. 1982. Wood anatomy of Buxaceae: Correlations with ecology and phylogeny. Flora 172: 463–491.

Castilho RO, AA da Silva Bulhoes, MAC Kaplan. 1999. Controversy in Buxales systematic positioning. Nord. J. Bot. 19: 541–546.

Cerny V and F Sorm. 1967. Steroid alkaloids: Alkaloids of Apocynaceae and Buxaceae. In: RHF Manske, ed. The Alkaloids, 9: 305–426.

Channell RB and CE Wood. 1987. The Buxaceae in the southeastern United States. J. Arnold Arbor. 68: 241–257.

Dang-van-Liem. 1959. Embryogénie des Buxacées: Développement de l'embryon chez le *Buxus sempervirens* L. C. R. Acad. Sci. Paris 248: 1844–1847.

Friis I. 1988. A synopsis of the Buxaceae in Africa south of the Sahara. Kew Bull. 44: 293–299.

Hardman R. 1987. Recent developments in our knowledge of steroids. Planta Medica 53: 233–238.

Jarvis CE. 1989. A review of the family Buxaceae Dumortier. In: PR Crane and S Blackmore, eds. Evolution, systematics, and history of the Hamamelidae, vol. 1, pp. 273–278. Claredon Press, Oxford.

Köhler E. 1980. Zur Pollenmorphologie und systemadschen Stellung der Didymelaceae Leandri. Feddes Repert. 91: 581–591.

Köhler E. 1981. Pollen morphology of the West Indian Central American species of the genus *Buxus* (Buxaceae) with reference to taxonomy. Pollen Spores 23: 37–91.

Köhler E. 1984. Zur Blattnervatur der neotropischen *Buxus*-Arten und ihre Bedeutung für die Systematic (Buxaceae). Flora 175: 345–374.

Köhler E. 1990. Zur Blattnervatur der afrikanischen *Buxus*- und *Notobuxus*-Arten. Feddes Repert. 101: 243–255.

Köhler E. 1993. Blattnervatur-Muster der Buxaceae Dumortier und Simmondsiaceae van Tieghem. Feddes Repert. 104: 145–167.

Köhler E. 2007a. Buxaceae. In: K Kubitzki, ed. The families and genera of vascular plants, vol. 9, pp. 40–47. Springer, Berlin/Heidelberg/New York.

Köhler E. 2007b. Didymelaceae. In: K Kubitzki, ed. The families and genera of vascular plants, vol. 9, pp. 129–131. Springer, Berlin/Heidelberg/New York.

Köhler E and P Brückner. 1982. Die Pollenmorphologie der afrikanischen *Buxus*-Arten und *Notobuxus*-Arten (Buxaceae): Ihre systematische Bedeutung. Grana 21: 71–82.

Köhler E and P Brückner. 1989. The genus *Buxus* (Buxaceae): Aspects of its differentiation in space and time. Plant Syst. Evol. 162: 267–283.

Köhler E and P Brückner. 1990. Considerations on the evolution and chorogenesis of the genus *Buxus* (Buxaceae). Mem. New York Bot. Gard. 55: 153–168.

Leandri J. 1937. Sur l'aire et la position systematique du genre malgache *Didymeles* Thouars. Ann. Sci. Nat. Bot., ser. 10, 19: 304–318.

Martin-Sans E. 1930. Généralité de la présence d'alkaloides chez les Buxacées. C. R. Acad. Sci. Paris 191: 625–626.

Martin-Sans E and J Ponchet. 1930. Sur l'appareil sécréteur des *Buxus*. Bull. Soc. Hist. Nat. Toulouse 60: 231–232.

Mathou T. 1939. Recherches sur la famille des Buxacées: Étude anatomique, microchimique, et systématique. Ph.D. thesis, University of Toulouse, Paris.

Melikian AP. 1968. On the position of the families Buxaceae and Simmondsiaceae in the system. Bot. Zhurn. 53: 1043–1047 (in Russian with English summary).

Naumova TN. 1980. Nucellar polyembryony in *Sarcococca* (Buxaceae). Bot. Zhurn. 65: 230–240 (in Russian).

Naumova TN. 1981. On the embryology of the representatives of the family Buxaceae. Bot. Zhurn. 66: 1135–1145 iIn Russian with English summary).

Outer RW Den. 1985. Wood anatomy of *Buxus madagascarica* Baill. Acta Bot. Neerl. 34: 111–113.
Qiu M-H and D-Z Li. 2002. Study on chemotaxonomy of Buxaceae. Chinese J. Appl. Environ. Biol. 8(4): 387–391 (in Chinese with English summary).
Radcliffe-Smith A. 1981. A remarkable new species of *Notobuxus* (Buxaceae) from Tanzania. Kew Bull. 36: 39–41.
Robins HC. 1968. The genus *Pachysandra* (Buxaceae). SIDA 3: 211–248.
Sánchez V, A Ahond, M-M Debray, F Picot, and C Poupat. 1984. Alcaloïdes des écorces de tronc de Didymeles cf. madagascariensis (Didymélacées). Bull. Soc. Chim. France 2: 71–76.
Sealy JR. 1986. A revision of the genus *Sarcococca* (Buxaceae). Bot. J. Linn. Soc. 92: 117–159.
Smith DA. 1988. The relationships of the Didymelales. In: PJ Stafford PJ, ed. Evolution, systematics and fossil history of the Hamamelidae. Intern. Symp. Univ. of Reading 22nd–25th March 1988, p. 20..
Straka H. 1966. Palynologia Madagassica et Mascarenica: Didymelaceae. Pollen Spores 8: 242–247.
Sutton DA. 1989. The Didymelales: A systematic review. In: PR Crane and S Blackmore, eds. Evoludon, systemadcs, and fossil history of the Hamamelidae, vol. 1, pp. 279–284. Claredon Press, Oxford.
Takhtajan AL, IA Shilkina, and AA Yatsenko-Khmelevsky. 1986. Wood anatomy of *Didymeles madagascariensis* in connection with the systematic status of die family Didymelaceae. Bot. Zhurn. 71: 1203–1206 (in Russian).
Tieghem P van. 1897. Sur les Buxacees. Ann. Sci. Nat., ser. 10, 5: 289–338.
Webster GL. 1987. The saga of the spurges: A review of classification and relationships in the Euphorbiales. Bot. J. Linn. Soc. 94: 3–46.
Wiger J. 1935. Embryological studies on the families Buxaceae, Meliaceae, Simarubaceae, and Burseraceae. Thesis, University of Lund, Lund.

Order 37. FAGALES

Trees or rarely shrubs, evergreen or deciduous. Buds covered with imbricate scales or very rarely naked. Vessels with scalariform and simple or more often only simple perforations; lateral pitting opposite, intermediate or alternate. Fibers with bordered or simple pits. Rays homogeneous or slightly heterogeneous. Axial parenchyma diffuse or in thin bands. Sieve-element plastids of S-type and contain nondispersive P-protein bodies. Nodes trilacunar. Leaves alternate or very rarely verticillate in whorls of three, simple, pinnately veined, entire to deeply lobed, crenate; stipules present, commonly caducous. Stomata anomocytic and/or encyclocytic. Flowers small, inconspicuous, usually unisexual, monoecious or rarely dioecious, apetalous, anemophilous or secondary entomophilous, in variously modified and frequently very reduced dichasia, which are often arranged in "catkins" with the dichasia sometimes reduced to solitary flowers. Male flowers in more or less reduced dichasia that are arranged into catkins or sometimes small heads (1–3 axillary dichasia in *Nothofagus*). Sepals (4-)6(-9), scalelike, imbricate, more or less connate or calyx very reduced and sometimes completely absent. Stamens (4)6–12(-90); filaments filiform, mostly free; anthers tetrasporangiate, dorsifixed or basifixed, opening longitudinally. Tapetum secretory. Microsporogenesis simultaneous. Pollen grains 2-celled, from 3-colpate to 2–7-porate. Pistilodia sometimes present. Female flowers 1–7-(-15) at base of male inflorescences or in separate axils, individually or collectively subtended, surrounded or completely covered by an involucre that develops into a cupule, an integrated dichasial inflorescence, in which the outer (ultimate order) axes of the dichasium are modified into cupule valves. In a few species of *Nothofagus* from New Guinea the cupule is reduced to two minute loose flaps or may even be entirely absent. Staminodia present or absent; gynoecium of 2(3) or sometimes 6–9 or more (up to 15 in *Lithocarpus*) united carpels, with distinct or nearly distinct stylodia. Ovary inferior, with as many locules (at least in the lower part) as carpels and two pendulous ovules in each locule. Ovules anatropous to hemitropous, bitegmic or seldom unitegmic, crassinucellate. Female gametophyte of *Polygonum*-type. Fertilization chalazogamous. Endosperm nuclear. Fruits 1-seeded nuts subtended by accrescent cupule. Seeds with large embryo and without endosperm. Strongly tanniniferous, with gallic acid and also proanthocyanins and flavonols (quercetin and less frequently kaempferol). Crystals of calcium oxalate commonly present in some of the cells of the parenchymatous tissues, n = 11, 12, 13, 14.

Most probably derived directly from some ancient hamamelidalean stock. From the Hamamelidales and related orders, this order differs sharply in its highly specialized dichasial inflorescences, its very reduced flowers, and in its chalazogamy.

Key to Families

1 Ovules bitegmic. Sclereid nests with rhomboidal crystals in bark. Male flowers borne in aments or small dichasial heads, or the flower clusters distributed along a branching axis. Anthers dorsifixed or subbasifixed, without a well-developed connective extension. Pollen grains 3-colporate, tectate-columellate, striate, scabrate, or rugulate. Gynoecium of (2)3–6(-15) carpels. Fruits nut-like;

endocarp hairy inside. Leaves without large multicellular peltate foliar glands; stipules never peltately attached, insertion not surrounded by sausage-shaped resinous colleters. n = 12, 13 and 11 (*Trigonobalanus*). . . . . 1. FAGACEAE

1 Ovules unitegmic. Male flowers in axillar clusters of (2)3 or solitary in the axils. Anthers basifixed with a well developed connective extension. Pollen grains (3)4-9(10)-colpate, tectate-granular, spinulate. Gynoecium of 2–3 carpels. Leaves with large multicellular glands (missing in two species); stipules mostly peltately attached, very rarely ligulate, insertion surrounded by sausageshaped resinous colleters. n = 13. . . . . 2. NOTHOFAGACEAE

1. FAGACEAE

Dumortier 1829 (including Castaneaceae Adanson 1763, Quercaceae Martynov 1820). 9/c.1000. Temperate to tropical regions of the Northern Hemisphere, in Southeast Asia crossing the equator.

1.1 QUERCOIDEAE
Male inflorescences spikelike, composed of 2- or 3-flowered dichasial clusters or of solitary flowers; female flowers solitary or in few flowered spikes. Stylodia generally flattened, often dilated above, with shortly decurrent or subcapitate stigma. Male flowers without pistillodia. Stamens six; anthers large, more or less basifixed. Fruit rounded in transverse section. Cupule open, containing 1 fruit, not lobed. – *Quercus*.

1.2 TRIGONOBALANOIDEAE
Inflorescence spikelike; male flowers in dichasial clusters of (1)3-7(-12), female flowers in dichasial clusters of (1-)3-7(-15). Stylodia recurved or connate at the base with the subcapitate bilobed stigma. Male flowers with a pistillodium. Stamens six, anthers large, ovoid, cordate at the base, subdorsifixed near the base. Fruit trigonous in transverse section. Cupules open, containing 1–7 fruits, 3–12-lobed. – *Trigonobalanus, Colombobalanus, Formanodendren*.

1.3 CASTANEOIDEAE
Inflorescence spikelike, composed of condensed dichasial clusters or of solitary flowers. Stylodia with minute and punctiform stigma. Male flowers usually with pistillodia. Stamens usually 12; anthers minute, dorsifixed or versatile. Cupule open or enclosing the fruits, containing one or several fruits, lobed or not. – *Lithocarpus, Castanopsis, Chrysolepis, Castanea*.

1.4 FAGOIDEAE
Male inflorescences spikelike, composed of solitary flowers or dichasial clusters of two or three flowers. Male flowers without pistillodia. Female flowers in 2–4-flowered clusters in axils of leaves. Stylodia slender, recurved, stigmatic toward the apex. Stamens 8–16 (up to 40), anthers large, basifixed. Fruits sharply triangular. Cupule open, containing (1-)2-4 fruits, 4-lobed. – *Fagus*.

2. NOTHOFAGACEAE

Kuprianova 1962. 1/37. From southern South America (including Staten Island) beyond 33° S to New Zealand, Tasmania, eastern Australia, New Caledonia, New Guinea, and the adjacent D'Entrecasteaux Islands of Goodenough and Normandy, and New Britain.

Nothofagus.

Share important apomorphic features with both Fagaceae-Fagoideae and Corylaceae. Derived similarities between *Nothofagus* and the Corylaceae include unitegmic ovules, glabrous locule cavity, compound leaf teeth, perorate pollen, and distichous phyllotaxy (Crane 1989). But *Nothofagus* differs from both of them in its unique chromosome number of n = 13, peltate stipules (in most species), stipular colleters, and pollen morphology. According to Jones (1989: 252), the known fossil record suggests that the Nothofagaceae evolved from the primitive fagalean complex in the Southern Hemisphere (probably in southern Australia or Antarctica) while the Fagaceae sensu stricto evolved from representatives of the same complex in the Northern Hemisphere.

Bibliography

Baas P. 1982. Comparative leaf anatomy of *Trigonobalanus* Forman (Fagaceae). Blumea 28: 171–175.

Berridge EM. 1914. The structure of the flower of the Fagaceae and its bearing on the affinities of the group. Ann. Bot. 28: 509–526.

Brett DW. 1964. The inflorescence of *Fagus* and *Castanea* and the evolution of the cupules of the Fagaceae. New Phytol. 63: 96–118.

Cao M and Z-K Zhou. 2002. Pollen morphology and its systematic significance of the *Quercus* from China. Guihaia 22: 14–18.

Codaccioni M. 1962. Recherches morphologiques et ontogenetiques sur quelques cupuliferes. Rev. Cytol. Biol. Veg. 25: 1–208.

Corner EJH. 1990. On *Trigonobalanus* (Fagaceae). Bot. J. Linn. Soc. 102: 219–223.

Crepet WL. 1989. History and implications of the early North American fossil record of Fagaceae. In: PR Crane and S Blackmore, eds. Evolution, systematics, and fossil history of the Hamamelidae, vol. 2, pp. 45–66. Claredon Press, Oxford.

Crepet WL and CP Daghlian. 1980. Castaneoid inflorescences from the Middle Eocene of Tennessee and the diagnostic value of pollen (at the subfamily level) in the Fagaceae. Am. J. Bot. 67: 739–757.

Crepet WL and KC Nixon. 1989. Earlies megafossil evidence of Fagaceae: Phylogenetic and biogeographic implications. Am. J. Bot. 76: 842–855.

Cutler DF. 1964. Anatomy of vegetative organs of *Trigonobalanus* Forman (Fagaceae). Kew Bull. 17: 401–409.

Dadswell HE and HD Ingle. 1954. The wood anatomy of the New Guinea *Nothofagus* Bl. Austral. J. Bot. 70: 639–649.

Denk T, G Grimm, K Stogerer, M Langer, and V Hemleben. 2002. The evolutionary history of *Fagus* in western Eurasia: evidence from genes, morphology and the fossil record. Plant Syst. Evol. 232(3–4): 213–236.

Denk T, and B Meller. 2001. Systematic significance of the cupule/nut complex in living and fossil *Fagus*. Int. J. Plant Sci. 162: 869–897.

Elias TS. 1971. The genera of Fagaceae in the southeastern United States. J. Arnold Arbor. 52: 159–195.

Fey BS. 1981. Untersuchungen über Bau und Ontogenese der Cupula, Infloreszenzen, und Blüten sowie zur Embryologie bei Vertretern der Fagaceae und ihre Bedeutung für die Systematik. Ph.D. dissertration, University of Zurich.

Fey BS and PK Endress. 1983. Development and morphological interpretation of the cupule in Fagaceae. Flora 173: 451–468.

Forman LL. 1964. *Trigonobalanus* and its importance in the taxonomy of the Fagaceae. Proc. Roy. Soc. London, Ser. B, Biol. Sci. 161: 48–49.

Forman LL. 1966a. On the evolution of cupules in the Fagaceae. Kew Bull. 18: 385–419.

Forman LL. 1966b. Generic delimitation in the Castaneoideae. Kew Bull. 18: 421–426.

Gottlieb M. 2001. Zur Phylogenie und Phyto-/Palaophytogeographie der Sudbuchen, Gattung *Nothofagus*. Mitt. Deutsch. Dendrol. Ges. 86: 85–106.

Hanks SL and DE Fairbrothers. 1976. Palynotaxonomic investigation of *Fagus* L. and *Nothofagus* Bl.: Light microscopy, scanning electron microscopy, and computer analysis. Syst. Bot. 1: 1–142.

Hill RS and GJ Jordan. 1993. The evolutionary history of *Nothofagus* (Nothofagaceae). Austral. Syst. Bot. 6: 111–126.

Hill RS and J Read. 1991. A revised infrageneric classification of *Nothofagus* (Fagaceae). Bot. J. Linn. Soc. 105: 37–72.

Hjelmquist H. 1957. Some notes on the endosperm and embryo development in Fagales and related orders. Bot. Not. 110: 173–195.

Hjelmquist H. 1963. Some notes on *Nothofagus* from New Guinea and New Caledonia. Bot. Not. 116: 225–237.

Jenkins R. 1993. The origin of the fagaceous cupule. Bot. Rev. 59: 81–111.

Jones JH. 1986. Evolution of the Fagaceae: The implications of foliar features. Ann. Missouri Bot. Gard. 73: 228–275.

Kaul RB. 1985. Reproductive morphology of *Quercus* (Fagaceae). Am. J. Bot. 72: 1962–1977.

Kaul RB. 1986. Evolution and reproductive biology of inflorescence in *Lithocarpus, Castanopsis, Castanea*, and *Quercus* (Fagaceae). Ann. Missouri Bot. Gard. 73: 284–296.

Kaul RB. 1987. Reproductive structure of *Lithocarpus* sensu lato (Fagaceae): Cymules and fruits. J. Arnold Arbor. 68: 73–104.

Kaul RB. 1988. Cupular structure in paleotropical *Castanopsis* (Fagaceae). Ann. Missouri Bot. Gard. 75: 1480–1498.

Kaul RB. 1989. Fruit structure and ecology in paleotropical *Lithocarpus* (Fagaceae). In: PR Crane and S Backmore, eds. Evolution, systematics, and fossil history of the Hamamelidae, vol. 2, pp. 67–86. Claredon Press, Oxford.

Kaul RB and EC Abbe. 1984. Inflorescence architecture and evolution in the Fagaceae. J. Arnold Arbor. 65: 375–401.

Kubitzki K. 1993. Fagaceae. In: K Kubitzki, ed. The families and genera of vascular plants, vol. 2, pp. 301–309. Springer, Berlin/Heidelberg/New York.

Kuprianova LA. 1962. Palynological data concerning the systematics of the Fagales and Urticales. In Papers of the Soviet palynologists for 1st International Palynological Conference, pp. 17–25 (in Russian).

Kuprianova LA. 1963. On a hitherto undescribed family belonging to the Amentiferae. Taxon 12: 12–13.

Langdon LM. 1939. Ontogenetic and anatomical studies of the flower and fruit of the Fagaceae and Juglandaceae. Bot. Gaz. 101: 301–327.

Langdon LM. 1947. The comparative morphology of the Fagaceae: I. The genus *Nothofagus*. Bot. Gaz. 108: 350–371.

Li J-Q. 1996a. The origin and distribution of the family Fagaceae. Acta Phytotax. Sinica 34: 376–396.

Li J-Q. 1996b. On the phylogeny of the Fagaceae. Acta Phytotax. Sinica 34: 597–609.

Li J-Q, H-C Wang, X-D Li, and X-W Li. 2003. Molecular phylogenetic inference in *Fagus* (Fagaceae) based on ITS (internal transcribed spacer) sequence of nuclear ribosomal DNA. J. Wuhan Bot. Res. 21: 31–36.

Liao JC. 1969. Morphological studies on the flowers and fruits of the genus *Lithocarpus* in Taiwan. National Taiwan Univ. Mem. Agric. 10: 1–113.

Liao H, G Gou, and N Ye. 1998. A study on seedling morphology and anatomy and systematic position of *Trigonobalanus doichangensis* Forman. J. Wuhan Bot. Res. 16: 223–226.

Manos PS. 1997. Systematics of *Nothofagus* (Nothofagaceae) based on rDNA spacer sequences (ITS): Taxonomic congruence with morphology and plastid sequences. Am. J. Bot. 84: 1137–1155.

Manos PS, ZK Zhou, and CH Cannon. 2001. Systematics of Fagaceae: phylogenetic tests of reproductive trait evolution. Int. J. Plant Sci. 162: 1361–1379.

Melville R. 1982. The geography *of Nothofagus* and *Trigonobalanus* and the origin of the Fagaceae. Bot. J. Linn. Soc. 85: 75–88.

Middleton TM. 1988. Intervessel pits in the stem wood of New Zealand *Nothofagus* (Fagaceae). IAWA Bull., n.s., 9(4): 327–331.

Nixon KC. 1989. Origins of Fagaceae. In: PR Crane and S Blackmore, eds. Evolution, systematics, and fossil history of the Hamamelidae, vol. 2, pp. 23–43. Clarendon Press, Oxford.

Nixon KC and WL Crepet. 1989. *Trigonobalanus* (Fagaceae): Taxonomic status and phylogenetic relationships. Am. J. Bot. 76: 826–841.

Palibin JW. 1935. Sur la morphologic florale des Fagacees. Bull. Acad. Sci. URSS. Classe de Sciences Math. et Nat. 1935: 349–381.

Philipson WR and MN Philipson. 1979. Leaf vernation in *Nothofagus*. New Zealand J. Bot. 17: 417–421.

Philipson WR and MN Philipson. 1988. A classification of the genus *Nothofagus* (Fagaceae). J. Linn. Soc. Bot. 98: 27–36.

Puntieri JG, MS Souza, D Barthelemy, C Brion, M Nunez, and C Mazzini. 2000. Preformation, neoformation, and shoot structure in *Nothofagus dombeyi* (Nothofagaceae). Canad. J. Bot. 78: 1044–1054.

Poole AL. 1950. Studies of New Zealand *Nothofagus: 2*. Nut and cupule development. Trans. and Proc. Roy. Soc. New Zealand 78: 502–508.

Poole AL. 1952. The development of *Nothofagus* seed. Trans. and Proc. Roy Soc. New Zealand 80: 207–212.

Praglowski J. 1982. Fagaceae L. (Fagoideae). World Pollen and Spore Flora 11. Almqvist & Wiksell, Stockholm.

Romero EJ. 1986. Fossil evidence regarding the evolution of *Nothofagus* Blume. Ann. Missouri Bot. Gard. 73: 276–283.

Rozefelds AC. 1998. Stamen morphology in *Nothofagus* (Nothofagaceae). Int. J. Plant Sci. 159: 655–667.

Rozefelds AC and AN Drinnan1998. Ontogeny and diversity in staminate flowers of *Nothofagus* (Nothofagaceae). Int. J. Plant Sci. 159: 906–922.

Rozefelds AC and AN Drinnan. 2002. Ontogeny of pistillate flowers and inflorescences in *Nothofagus* subgenus *Lophozonia* (Nothofagaceae). Pland Syst. Evol. 233: 105–126.

Shimaji K. 1962. Anatomical studies on the phylogenetic interrelationships of the genera in the Fagaceae. Bull. Tokyo Univ. Forest 57: 1–64.

Soepagmo E. 1972. Fagaceae. Flora Malesiana 2: 272–294.

Sogo A and H Tobe. 2006. The evolution of fertilization modes independent of the micropyle in Fagales and 'pseudoporogamy'. Plant Syst. Evol. 259: 73–80.

Steenis CGGJ van. 1971. *Nothofagus*: Key genus of plant geography, in time and space, living and fossil, ecology and phylogeny. Blumea 19: 65–98.

Swenson U, RS Hill, and S McLoughlin. 2001. Biography of *Nothofagus* supports the sequence of Gondwana break-up. Taxon. 50: 1025–1041.

Wang P-L and K-T Chang. 1991. The pollen morphology in relation to the taxonomy and phylogeny of Fagaceae. Acta Phytotax. Sinica 29: 60–66.

Wang P-L, F-T Pu, and Z-H Zheng. 1998. Palynological evidence for taxonomy of *Trigonobalanus* (Fagaceae). Acta Phytotax. Sinica 36: 238–241.

Wang PL, FT Pu, ZH Zheng. 2000. Pollen morphology of the genus *Nothofagus* and its taxonomic significance. Acta Phytotax. Sinica 38: 452–461.

Yatsenko-Khmelevsky AA. 1945. Wood structure of some species of the genus *Nothofagus* in relation to its systematic position. Doklady Armenian Acad. Sci 2: 109–113 (in Russian).

Zheng Z-H, P-L Wang, and F-D Pu. 1999. A comparative study on pollen exine ultrastructure of *Nothofagus* and the other genera of Fagaceae. Acta Phytotax. Sinica 37: 253–258.

Order 38. BETULALES (CORYLALES)

Deciduous or (*Ticodendron*) evergreen trees and shrubs with branchlets sometimes differentiated into long and short shoots. Bark close or exfoliating in thin layers, often marked with prominent lenticels. Buds naked or covered with two valvate, stipular scales or with few to many imbricate scales. Vessels with scalariform or simple perforations; lateral pitting scalariform (*Ticodendron*), opposite (*Alnus*) or alternate. Fibers with bordered pits. Rays heterogeneous to homogeneous. Axial parenchyma diffuse or terminal. Sieve-element plastids of S-type. Nodes trilacunar. Leaves alternate, spirally arranged or distichous, simple, petiolate, serrate, abaxially sometimes covered with resinous glands, pinnately veined, with free, deciduous stipules. Stomata anomocytic. Flowers small, inconspicuous, unisexual, monoecious or dioecious, with reduced perianth or naked. Stamens (1)4-10; anthers tetrasporangiate, basifixed (*Ticodendron*) or dorsifixed, opening longitudinally. Tapetum secretory. Microsporogenesis simultaneous. Pollen grains 2-celled, 3(-7)-porate, with microperforate tectum. Gynoecium of 2(3) united carpels with free or nearly free stylodia; ovary inferior, 2(3)-locular, with 1–2 pendulous ovules per locule. Ovules anatropous, hemitropous (*Duschekia* and *Ticodendron*) or campylotropous (*Corylus*), epitropous, unitegmic or (*Carpinus*) bitegmic, crassinucellate. Female gametophyte of (*Polygonum*-type). Fertilization chalazogamous (at least in Betulaceae and Corylaceae). Endosperm nuclear. Fruits nuts or (*Ticodendron*) drupes. Seeds with very thin seed coat, large embryo, and thin layer of endosperm.

Very closely related to the Fagales, especially to the Nothofagaceae and probably had a common origin with them.

Key to Families

1 Deciduous trees and shrubs. Flowers monoecious, in pendulous or erect catkins, consisting of reduced 3-flowered cymules, or reduced to compact clusters of several minute flowers. Male flowers with very small scalelike sepals or without sepals. Anthers dorsifixed, extrorse. Pollen grains 3(-7)-porate, minutely scabrate to slightly rugulate, tectate-granular, although columella-like structures often present in the infratectal layer. Female flowers without sepals with 2–3 bracts. Ovary 2(-3)-locular below, 1-locular

above; stylodia stigmatic above. Ovules 1–2 per locule. Infructescences with woody or scaly bracts or with leafy bracts. Fruits nuts, small, compressed, laterally winged (the wings sometimes reduced to ridges), or not winged, not enclosed in the foliaceous involucre. Hairs peltate glandular. Producing flavones, n = 8, 11, 14. . . . . 1. BETULACEAE

1 Evergreen trees. Hairs T-shaped, unicellular. Stipules encircling the stem. Flowers dioecious or less often polygamodioecious. Male inflorescences long, simple or branched, sometimes crowned by a solitary female flower. Stamens many, arranged in 2–4 verticels surrounded by three deciduous bracts; filaments 2–3 mm long. Anthers oblong, geminate, basifixed, the connective with an apiculate appendage, opening longitudinally. Pollen grains 3-porate with slightly aspidote pores, tectate-columellate with spherical coarse sporadic granules. The tectum traversed by narrow channels, minutely spinulate. Female cymule of solitary flower, surrounded by three early deciduous bracts. Gynoecium of two carpels; ovary included in the calyx tube, 4-locular, with one ovule per locule; ovules hemitropous, unitegmic. Stigmas two, long. Fruits drupaceous. Exotestal cells initially radially elongated, all cells more or less thick-walled and tanniniferous. Vessels with scalariform perforations with mean number per perforation 62; lateral pitting scalariform, less commonly transitional or opposite. Fibers with fully bordered pits. Rays heterogeneous. Axial parenchyma diffuse, n = 13. . . . . . . 2. TICODENDRACEAE

1. BETULACEAE

Gray 1822 (including Carpinaceae Vest 1818, Corylaceae Mirbel 1815). 7/140. Northern temperate regions and Andes of South America, *Carpinus* (35) reaches Central America, *Ostryopsis* (2) is endemic to China.

1.1 BETULOIDEAE

Male flowers with very small scalelike sepals, female flowers without sepals, 2–3 per bract. Inflorescences composed of small, crowded, woody, or coriaceous scales. Infructescences with woody or scaly bracts. Fruits small, compressed, laterally winged (the wings sometimes reduced to ridges), not enclosed in the foliaceous involucre. Hairs peltate glandular. Producing flavones, n = 11, 14. – *Betula, Alnus, Duschekia* (closely related to *Alnus*).

1.2 CARPINOIDEAE

Male flowers without bracteoles. Female flowers spicate. Nuts small subtended by a large 3-lobed or coarsely dentate bract (*Carpinus*) or enclosed in tubular or bladderlike involucres, n = 8. – *Carpinus, Ostrya, Ostryopsis*.

1.3 CORYLOIDEAE

Male flowers with two bracteoles united to bract. Female flowers in pairs in axil of each bract. Nuts large, with lignified mesocarp, surrounded by leaflike involucre (accrescent bract and bracteoles), n = 14. – *Corylus*.

2. TICODENDRACEAE

J. Gomez-Laurito et L. D. Gomez P. 1991. 1/1. Central America (from southern Mexico to central Panama). *Ticodendron*.

Related to the Betulaceae, which is supported also by *rbc*L sequence data (Conti et al. 1994).

Bibliography

Abbe EC. 1935. Studies in the phylogeny of the Betulaceae: I. Floral and inflorescence anatomy and morphology. Bot. Gaz. 97: 1–67.

Abbe EC. 1938. Studies in the phylogeny of the Betulaceae: II. Extremes in the range of variation of floral and inflorescence morphology. Bot. Gaz. 99: 431–469.

Behnke HD. 1991. Sieve-element characters in *Ticodendron*. Ann. Missouri Bot. Gard. 78: 131–134.

Bousquet J, SH Strauss, P Li. 1992. Complete congruence between morphological and *rbc*L based molecular phylogenies in birches and related species (Betulaceae). Mol. Biol. Evol. 9: 1076–1088.

Brunner F and DE Fairbrothers. 1979. Serological investigation of the Corylaceae. Bull. Torrey Bot. Club 106(2): 97–103.

Carlquist S. 1991. Wood and bark anatomy of *Ticodendron*: Comments on relationships. Ann. Missouri Bot. Gard. 78: 96–104.

Chen Z-D. 1991. Pollen morphology of the Betulaceae. Acta Phytotax. Sinica 29: 464–475 (in Chinese with English summary).

Chen Z-D. 1994. Phylogeny and phytogeography of the Betulaceae. Acta Phytotax. Sinica 32: 1–32, 101–153 (in Chinese with English summary).

Chen Z-D, A-M Li, and K-Y Pan. 1990. The embryology of the genus *Ostryopsis* (Betulaceae). Cathaya 2: 53–62.

Chen Z-D, SR Manchester, and H-Y Sun. 1999. Phylogeny and evolution of the Betulaceae as inferred from DNA sequences, morphology, and paleobotany. Am. J. Bot. 86: 1168–1181.

Chen Z-D and Z-Y Zhang. 1991. A study on foliar epidermis in the Betulaceae. Acta Phytotax. Sinica 29: 156–165 (in Chinese with English summary).

Conti E, KJ Sytsma, and WS Alverson. 1994. Neither oak nor alder, but nearly: The relationships of *Ticodendron* based on rbcL data. Am. J. Bot. 81(6): 149 (Abstract).
Erdogan V and SA Mehlenbacher. 2000. Phylogenetic relationships of *Corylus* species (Betulaceae) based on nuclear ribosomal DNA ITS region and chloroplast *mat*K gene sequences. Syst. Bot. 25: 727–737.
Feuer S. 1991. Pollen morphology and the systematic relationships of *Ticodendron incognitum*. Ann. Missouri Bot. Gard. 78: 143–151.
Furlow JJ. 1990. The genera of Betulaceae in the southeastern United States. J. Arnold Arbor. 71: 1–67.
Gomez-Laurito J and PL Gomez. 1989. *Ticodendron*: A new tree from Central America. Ann. Missouri Bot. Gard. 76: 1148–1151.
Gomez-Laurito J and PL Gomez. 1991. Ticodendraceae: A new family of flowering plants. Ann. Missouri Bot. Gard. 78: 87–88.
Hall JW. 1952. The comparative anatomy and phylogeny of the Betulaceae. Bot. Gaz. 113: 235–270.
Hammel B and W Burger. 1991. Neither oak nor alder, but nearly: The history of Ticodendraceae. Ann. Missouri Bot. Gard. 78: 89–95.
Hardin JW and JM Bell. 1986. Atlas of foliar surface features in woody plants: IX. Betulaceae of eastern United States. Brittonia 38: 133–144.
Hickey LJ and DW Taylor. 1991. The leaf architecture of *Ticodendron* and the application of foliar characters in discerning its relationships. Ann. Missouri Bot. Gard. 78: 105–130.
Jager EJ. 1980. Progressionen im Synfloreszenzbau und in der Verbreitung bei den Betulaceae. Flora 170: 91–113.
Kato H, K Oginuma, Z Gu, B Hammel and H Tobe. 1999. Phylogenetic relationships of Betulaceae based on matK sequences with particular reference to the position of *Ostryopsis*. Acta Phytotax. Geobot. 49: 89–97.
Khramova EL. 1996. Palynomorphology of the genus *Alnus* (Betulaceae). Bot. Zhurn. 81(3): 42–53 (in Russian with English summary).
Korchagina IA. 1991. Betulaceae. In: A Takhtajan, ed. Comparative seed anatomy, vol. 3, pp. 134–140. Nauka, Leningrad (in Russian).
Kubitzki K. 1993a. Betulaceae. In: K Kubitzki, ed. The families and genera of vascular plants, vol. 2, pp. 152–157. Springer, Berlin/Heidelberg/New York.
Kubitzki K. 1993b. Ticodendraceae. In: K Kubitzki, ed. The families and genera of vascular plants, vol. 2, pp. 594–596. Springer, Berlin/Heidelberg/New York.
Oginuma K, Kato H, Tobe H. 1999. Chromosome base number of *Corylus* (Betulaceae): correction, and evolution. Acta Phytotax. Geobot. 49(2): 99–104.
Snow N and P Goldblatt. 1992. Chromosome number in *Ticodendron* (Fagales, Ticodendraceae). Ann. Missouri Bot. Gard. 79: 906–907.
Schneider H. 1994. Sprossorganisation und Blattarchitektur bei Betulaceae und extratropischen Fagaceae. Cramer, Berlin.
Tobe H. 1991. Reproductive morphology, anatomy, and relationships of *Ticodendron*. Ann. Missouri Bot. Gard. 78: 135–142.
Whitcher IN, Wen J. 2001. Phylogeny and biogeography of *Corylus* (Betulaceae): inferences from ITS sequences. Syst. Bot. 26(2): 283–298.
Yoo K-O and J Wen. 2002. Phylogeny and biogeography of *Carpinus* and subfamily Coryloideae (Betulaceae). Int. J. Plant Sci. 163: 641–650.
Xing S-P, Z-D Chen, and A-M Lu. 1998. Development of ovule and embryo sac in *Ostrya virginiana* (Betulaceae) and its systematic significance. Acta Phytotax. Sinica 36: 428–435 (in Chinese with English summary).

Order 39. CASUARINALES

Trees and shrubs with slender, green, wiry, articulate branchlets. Roots usually with nodules harboring nitrogen-fixing filamentous soil bacteria of the genus *Frankia*. Solitary or clustered crystals of calcium oxalate usually present in some of the cells of parenchymatous tissues. Vessels with mainly simple or sometimes scalariform perforations with up to 30 bars; lateral pitting opposite or alternate, very rarely intermediate. Fibers with distinctly bordered pits. Rays heterogeneous. Axial parenchyma apotracheal, diffuse and in short bands. Sieve-element plastids of S-type with up to five starch grains; they also incorporate nondispersive P-protein bodies. Nodes unilacunar with one trace. Leaves small, scalelike, verticillate, more or less connate to form a toothed sheath at each node; stipules wanting. Stomata generally tetracytic, sometimes penta- or hexacytic. Inflorescences consist of alternating whorls of toothlike, basally connate bracts; within each bract two lateral, keeled, scalelike bracteoles are found. Male inflorescences from short to elongated catkinlike spikes; female inflorescences from small globular to ovoid heads. Flowers small and inconspicuous, unisexual, monoecious or rarely dioecious, apetalous, anemophilous. Male flowers consist of one or two anteroposterior, hooded, scalelike sepals (deciduous at anthesis) and a single (sometimes more or less bifurcate) stamen. Anthers tetrasporangiate, basifixed, opening longitudinally. Tapetum secretory. Microsporogenesis simultaneous(?). Pollen grains 2-celled, 3-porate, tectate-granular, with thick tectum traversed by minute channels; exine ornamentation consists of small spinules or scabrae and sometimes appears rugulate. Female flowers without perianth; gynoecium pseudomonomerous, consists of two median (anteroposterior) carpels with very long, threadlike, reddish stylodia united at the base. Stigma very long, decurrent; ovary superior, laterally flattened, initially 2-locular, but only the anterior locule is fertile and contains two (rarely 3 or 4) ascending ovules; the posterior one is usually empty or more or less reduced or obsolete. Ovules orthotropous to hemitropous(?), epitropous, bitegmic, crassinucellate, micropyle endostomal. Female gametophytes several to numerous, but only one fully developed and fertile, of *Polygonum*-type. Fertilization chalazogamous. Endosperm nuclear. Female inflorescence develops into a more or less woody infrutescence with the two enlarged bracteoles of each flower forming two tightly appressed lateral valves, which separate when the fruit is mature. Fruits 1-seeded, compressed, terminally winged, samaroid nuts. Seeds small with seed coats adnate to the pericarp, without endosperm, and often with more than one embryo. Embryo large, straight, oily, with two large cotyledons and a superior radicle. Plants tanniniferous, producing proanthocyanins, ellagic acid, and flavonoid monomers consisting of a number of glycosidic variations based on a conservative aglycone component of myricetin, quercetin, and kaempferol. $n = 8$ (*Gymnostoma*) – 14. There is a remarkable differentiation in chromosome size (Morawetz and Samuel 1989).

Exhibits many features in common with Hamamelidales and to a lesser degree also with Fagales and Betulales (especially pollen morphology, which resembles betulaceous pollen, Erdtman 1965; Kuprianova 1965; Zavada and Dilcher 1986). Bessey (1915) derived Casuarinaceae from Hamamelidaceae, which is corroborated by the totality of morphological data (Moseley 1948) including seed anatomy (Melikian 1973). The wood anatomy of the Casuarinaceae is more advanced than that of the Hamamelidaceae, and their flowers originated as a result of reduction and simplification of the flowers of a hamamelidaceous type. The similarities between the flowers of Casuarinaceae and those of Betulaceae (Eames 1961; Flores and Moseley 1982) are a result of the evolutionary parallelism. In both cases the lines of floral reduction seen in the Hamamelidales have been carried to extremes. All of the most important morphological features of the Casuarinaceae are found in one or another representative of Hamamelidales. But at the same time Casuarinaceae diverged so much from the hamamelidalean stock, that "there is a considerable gap between any known living or fossil Casuarinaceae and any other known family" (Johnson and Wilson 1989: 185).

1. CASUARINACEAE

R. Brown 1814. 4/96. Australia, New Caledonia, Fiji, New Guinea, Malesia; concentrated in Australia and to a lesser extent in Malesia (especially the eastern regions) and New Caledonia. One species (*Casuarina equisetifolia*) has wide littoral distribution from the Bay of Bengal to northeastern Australia and Polynesia (but the occurrence of the species in the Mascarenes is regarded as a probable introduction, Johnson and Wilson 1989, 1993).

Gymnostoma, Ceuthostoma, Casuarina, Allocasuarina.

Bibliography

Barlow BA. 1959. Chromosome numbers in the Casuarinaceae. Austral. J. Bot. 7: 230–237.

Boodle LA and WC Worsdell. 1894. On the comparative anatomy of the Casuarinaceae, with special reference to the Gnetaceae and Cupuliferae. Ann. Bot. 8: 231–264.

Dilcher DL, DC Christophel, HO Bhagwandin, Jr., and LJ Scriven. 1990. Evolution of the Casuarinaceae: Morphological comparisons of some extant species. Am. J. Bot. 77: 338–355.

Flores EM. 1980. Shoot vascular system and phyllotaxis of *Casuarina* (Casuarinaceae). Am. J. Bot. 67: 131–140.

Flores EM and MF Moseley. 1982. The anatomy of the pistillate inflorescence and flower of *Casuarina verticillata* Lamarck (Casuarinaceae). Am. J. Bot. 69: 1673–1684.

Flores EM and MF Moseley 1990. Anatomy and aspects of development of the staminate inflorescences and florets of seven species of *Allocasuarina* (Casuarinaceae). Am. J. Bot. 77: 795–808.

Hwang R and JG Conran. 2000. Seedling characteristics in the Casuarinaceae. Telopea 8: 429–439.

Johnson LAS and KL Wilson. 1989. Casuarinaceae: A synopsis. In: PR Crane and S Blackmore, eds. Evolution, systematics, and fossil history of the Hamamelidae, vol. 2, pp. 167–188. Claredon Press, Oxford.

Johnson LAS and KL Wilson. 1993. Casuarinaceae. In: K Kubitzki, ed. The families and genera of vascular plants, vol. 2, pp. 237–242. Springer, Berlin/Heidelberg/New York.

Kershaw AP. 1970. Pollen morphological variation within the Casuarinaceae. Pollen Spores 12: 145–161.

Moseley MF. 1948. Comparative anatomy and phylogeny of the Casuarinaceae. Bot. Gaz. 110: 231–280.

Natarajan S, VVS Murty, and TR Seshadri. 1971. Chemotaxonomical studies of some *Casuarina* species. Phytochemistry 10: 1083–1085.

National Research Council. 1984. Casuarinas: Nitrogenfixing trees for adverse sites. Academic Press, Washington, DC.

Okuda T, T Yoshida, T Hatano, K Yazaki, and M Ashida. 1982. Ellagitannins of the Casuarinaceae, Stachyuraceae, and Myrtaceae. Phytochemistry 21: 2871–2874.

Pant DD, DD Nautiyal, and S Singh. 1974. The cuticle, epidermis, and stomatal ontogeny of *Casuarina equisetifrlia* Forst. Ann. Bot. 39: 1117–1123.

Redell P, GD Bowen, and AD Robson. 1986. Noduladon of Casuarinaceae in relation to host species and soil properties. Austral. J. Bot. 34: 435–444.

Sogo A, H Setoguchi, J Noguchi, T Jaffre, and H Tobe. 2001. Molecular phylogeny of Casuarinaceae based on rbcL and matK gene sequences. J. Plant Res. 114(116): 459–464.

Swamy BGL. 1948. A contribution to the life history of *Casuarina*. Proc. Am. Acad. Arts and Sci. 77 (I): 132.
Torrey JG. 1983. Root development and root nodulation in *Casuarina*. In: SJ Midgley, JW Turnbull, and RD Johnston, eds. *Casuarina* ecology, management, and utilization, pp. 180–192. CSIRO Melbourne.
Torrey JG and RH Berg. 1988. Some morphological features for generic characterization among the Casuarinaceae. Am. J. Bot. 75: 864–874.
Treub M. 1891. Sur les Casuarinees et leur place dans le systeme naturel. Ann. Jard. Bot. Buitenz. 10: 145–231.
Ueno J. 1963. On the fine structure of the pollen walls of Angiospermae: III. *Casuarina*. Grana Palynol. 4: 189–193.

Superorder JUGLANDANAE

Order 40. MYRICALES

Evergreen or deciduous shrubs or small trees. Trichomes when present are distinctive because of their 2–3-celled bases and include regular balloon glands, irregular bowl-shaped glands, and unicellular trichomes. Roots usually bearing nodules with nitrogen fixing bacteria (except perhaps for *Canacomyrica)*. Vessels usually with scalariform perforations with a range of 1–32 bars; *Camptonia* has simple perforation, with a very small proportion of the plates bearing 1–3 bars. Fibers with bordered or both bordered and simple pits. Rays heterogeneous. Axial parenchyma diffuse. Sieve-element plastids of S-type. Nodes trilacunar or unilacunar. Leaves alternate, simple, entire or serrate to irregularly dentate or lobed, rarely (*Comptonia*) deeply pinnatifid, pinnately veined, without stipules or rarely (*Comptonia*) with foliaceous, semicordate stipules, often with aromatic peltate resinous glands. Stomata anomocytic. Flowers small, inconspicuous, in axillary, dense, spikelike racemose inflorescences, unisexual (monoecious or more often dioecious), mostly without perianth, usually subtended by a bract and two transversal bracteoles, anemophilous. Male flowers have 2–8 (very rarely up to 20) stamens, progressively fewer in the more distal flowers. Filaments free or sometimes connate at the base. Anthers ovoid, tetrasporangiate, basifixed, extrorse, opening longitudinally. Pistillodia often present. Tapetum secretory. Microsporogenesis simultaneous. Pollen grains 2-celled, 3-porate, tectate-granular with stout columellae often discernible in the infrastructural layer; the tectum is traversed by minute channels, minutely scabrate. Gynoecium of two united carpels, very rarely of three carpels. Stylodia mostly long and slender, nearly free or shortly connate at the base. Ovary superior *(Comptonia)* or more or less inferior, 1-locular, with one basal erect ovule born in the anterior carpel. Ovules orthotropous, unitegmic (in *Canacomyrica* with elongate, reflexed micropylar tube), crassinucellate. Female gametophyte of *Polygonum*-type. Endosperm nuclear. Fruits drupes with a hard endocarp and wax-secreting exocarp or almost a nut, sometimes enveloped by the persistent, accrescent bracteoles, in *Canacomyrica* completely enclosed in an enlarged, 6-lobed perianth. Seeds with very reduced, exotestal seed coat, straight, large embryo, and nearly or quite without endosperm. Tanniniferous and accumulating triterpenes and sesquiterpenes, also with protoanthocyanins, ellagic acid, and gallic acid. n = 8, 12.

Although Myricales have some features in common with the Betulaceae and Casuarinaceae, they are much nearer to the Juglandales (Chevalier 1901; Kershaw 1909; Stokes 1937; Hjelmquist 1948; Takhtajan 1966, 1987; Dahlgren 1980, 1983; Cronquist 1981; Thorne 1989; Macdonald 1989; Kubitzki 1933), which is supported by the fossil record (Friis 1983). According to Zavada and Dilcher (1986), *Myrica* exhibits particular similarity to *Rhoiptelea* and *Platycarya*, but *Canacomyrica* is probably the closest to the Rhoipteleaceae (see Friis 1983; Macdonald 1989).

1. MYRICACEAE

A. Richard ex Kunth 1817 (including Canacomyricaceae Doweld 2000). 4/57. Widely distributed in America, Africa, and Asia, but absent in Australia and New Zealand. *Comptonia* (1) is endemic to eastern North America, and *Canacomyrica* (1) to New Caledonia.

1.1 CANACOMYRICOIDEAE
Flowers with 6-lobed perianth. Male flowers with six stamens. Female flowers with (5)6 staminodia. Ovules with micropylar tube. – *Canacomyrica.*

1.2 MYRICOIDEAE
Flowers without perianth. Male flowers with 2–8 (very rarely up to 20) stamens. Female flowers without staminodia. Ovules without micropylar tube. – *Myrica* (including *Gale*?), *Morella, Comptonia.*

Bibliography

Abbe EC. 1972. The inflorescence and flower in male *Myrica esculenta* var. *farquhariana*. Bot. Gaz. 133: 206–213.

Abbe LB and EC Abbe. 1971. The vessel member of *Myrica esculenta* Buch. Ham. J. Minnesota Acad. Sci. 37: 72–76.

Carlquist S. 2002. Wood and bark anatomy of Myricaceae: relationships, generic definitions, and ecological interpretations. Aliso 21: 7–29.

Chen Z-D and A-M Lu. 1998. Systematic position of the Rhoipteleaceae: evidence from nucleotide sequences of *rbc*L gene. Acta Phytotax. Sinica 36: 1–7.

Chevalier A. 1901. Monographic des Myricacees: Anatomic et histologie, organographie, classification, et description des especes, distribution geographique. Mem. Soc. Sci. Nat. Cherbourg 32: 85–340.

Collins RP and AF Halim. 1973. Chemotaxonomy of the Myricaceae; II. Essential oil analysis of three Central American species of *Myrica*. Lloydia 36: 320–325.

Davey AJ and CM Gibson 1917. Note on the distribution of sexes in *Myrica gale*. New Phytol. 16: 147–151.

Diboll AG. 1959. Comparative anatomy of staminate catkins of Myricaceae. M.A. thesis, Claremont Grad. School, Claremont.

Elias TS. 1971. The genera of Myricaceae in southeastern United States. J. Arnold Arbor. 52: 305–318.

Ferguson DK. 1998. The contribution of micromorphology to the taxonomy and fossil record of the Myricaceae. Taxon 47: 333–335.

Fletcher WW. 1955. The development and structure of the root-nodules of *Myrica gale* with special reference to the nature of the endophyte. Ann. Bot. 19: 501–573.

Friis EM. 1983. Upper Cretaceous (Senonian) floral structures of juglandalean affinity containing Normapolles pollen. Rev. Palaeobot. Palynol. 39: 161–188.

Gladkova VN. 1962. Fragments of the history of the Myricaceae family. Pollen Spores 4: 345.

Hakansson A and RP Collins. 1955. Endosperm formation in *Myrica gale* L. Bot. Not. 108: 6.

Halim AF and RP Collins. 1973. Essential oil analysis of the Myricaceae of the eastern United States. Phytochemistry 12: 1077–1083.

Kershaw EM. 1909. The structure and development of the ovule of *Myrica gale*. Ann. Bot. 23: 353–362.

Killick DJB. 1969. The South African species of *Myrica*. Bothalia 10: 517.

Kubitzki K. 1993. Myricaceae. In: K Kubitzki, ed. The families and genera of vascular plants, vol. 2, pp. 453–457. Springer, Berlin/Heidelberg/New York.

Leroy JF. 1949. De la morphologic florale et de la classification des Myricaceae. C. R. Acad. Sci. Paris 229: 1162–1163.

Leroy JF. 1957. Sur deux Amentiferes remarquables de la Flore Asiatico Pacifique et Pacifique. Proc. 8th Pacific Science Congress 4: 459–464.

Macdonald AD. 1974. Floral development of *Comptonia peregrina* (Myricaceae). Canad. J. Bot. 52: 2165–2169.

Macdonald AD. 1977. Myricaceae: Floral hypothesis for *Gale* and *Comptonia*. Canad. J. Bot. 55: 2636–2651.

Macdonald AD. 1978. Organogenesis of the male inflorescences and flowers of *Myrica esculenta*. Canad. J. Bot. 56: 2415–2423.

Macdonald AD. 1979. Development of the female flower and gynecandrous partial inflorescence of *Myrica californica*. Canad. J. Bot. 57: 141–151.

Macdonald AD. 1980. Organogenesis of the female reproductive structure of *Myrica pensylvanica*. Canad. J. Bot. 58: 2001–2006.

Macdonald AD. 1989. The morphology and relationships of the Myricaceae. In: PR Crane and S Blackmore, eds., Evolution, systematics, and fossil history of the Hamamelidae, vol. 2, pp. 147–165. Claredon Press, Oxford.

Macdonald AD and R Sattler. 1973. Floral development of *Myrica gale* L. and the controversy over floral concepts. Canad. J. Bot. 51: 1965–1975.

Stokes J. 1937. Cytological studies in the Myricaceae. Bot. Gaz. 99: 387–399.

Sundberg MD. 1985. Pollen of the Myricaceae. Pollen Spores 27: 15–28.

Vikhireva VV. 1957. Anatomical structure and development of the female flower of the common bog myrtle *Myrica gale* L. Trudy Komarov Bot. Inst. Acad. Sci. USSR, ser. 7, Morf. Anat. Rast. 4: 270–287 (in Russian).

Vyshenskaya TD. 1991. Myricaceae. In: A Takhtajan, ed. Comparative seed anatomy, vol. 3, pp. 141–148. Nauka, Leningrad (in Russian).

Yen TK. 1950. Structure and development of the flower and the fruit of *Myrica rubra*. Peking Natural Hist. Bull. 19: 1–20.

Order 41. JUGLANDALES

Deciduous or evergreen, aromatic trees or rarely large shrubs with naked or scaly buds. Vessel elements elongate, with scalariform or more often simple perforations, but in *Alfaroa* and *Engelhardtia* some of them scalariform with a few bars; lateral pitting alternate. Fibers with bordered pits; some gelatinous fibers usually also present. Rays heterogeneous or sometimes (in some Juglandaceae) almost homogeneous. Axial parenchyma apotracheal to paratracheal. Sieve-element plastids of S-type. Nodes trilacunar or sometimes pentalacunar with five traces. Leaves alternate or less often opposite, stipulate or estipulate, pinnately compound (or trifoliolate), with leaflets entire or serrate. Leaves provided with aromatic, resinous, several-celled, peltate, basally sunken glandular scales; hairs mainly nonglandular. Stomata anomocytic. Flowers small, borne in aments, apetalous, unisexual (monoecious or rarely dioecious) or gynomonoecious (the central flower of dichasium in *Rhoiptelea*), borne in elongate, drooping, and rarely erect, catkins or spikelike inflorescences, that are sometimes grouped into panicles. Each flower of *Rhoiptelea* has two lateral bracteoles and four small, scarious, persistent on the fruit sepals in two cycles. In Juglandaceae flowers more or less reduced. The less reduced flowers (e.g., *Juglans)* consist of two lateral bracteoles and four sepals, but mostly there are only three, two, or even one sepals. *Platycarya* lost both the sepals and bracteoles and thus reached the end of the reduction line. The bract and two bracteoles (when present) are fused with the floral receptacle and appear as a part of the calyx, the stamens thus appearing to be on the involucre and calyx in male flowers; modifications in female flowers led to the reduction of the calyx and the adnation of the bract and bracteoles with the entire ovary to form a cupular involucre that ripens with the fruit to form a husk. Stamens 3–105 (mostly 5–40) in Juglandaceae, in *Rhoiptelea* – 6; filaments short, anthers tetrasporangiate, basifixed, opening longitudinally. Tapetum secretory. Microsporogenesis simultaneous, rarely successive. Pollen grains 2-celled or 3-celled, 3-porate or pantoporate, 3-colporate or seldom 4-colporate, tectate-granular. Gynoecium regularly of two united carpels, rarely individual flowers with three or four carpels; stylodia free or united at the base, seldom stigmas sessile. Ovary superior or inferior, incompletely 2-locular (2-locular below and 1-locular above), in Juglandaceae often falsely 4- to 8-locular below because of secondary partitions (sometimes 5-locular at the very apex due to intrusions from the wall). Ovules solitary, attached to the partial partition, campylotropous, or orthotropous, bitegmic or unitegmic, crassinucellate. Female gametophyte of *Polygonum*-type. Fertilization chalazogamous, very rarely porogamous. Endosperm nuclear. Fruits are nuts (often samaroid), or drupe-like but rather dry. Seeds large, obovoid (*Rhoiptelea*) or more often 2- to 4- to 8-lobed. Seed coat reduced to thin exotestal layer (*Rhoiptelea*) or outer epidermis as a layer of rather large, thin-walled cells, highly vascularized. Embryo massive, with 2- (*Rhoiptelea*) or 4-lobed oily cotyledons; endosperm wanting or very scanty.

Has much in common with Myricales (including serolological similarities between Juglandaceae and Myricaceae – Chupov 1978; Petersen and Fairbrothers 1979). Especially many common features with Myricaceae in the structure of flower, pollen grains, ovules, fruit, basic chromosome number and chemistry. Both orders derived from the hamamelidalean type ancestor.

Key to Families

1 Leaves with leafy, thick, papery and asymmetrically caducous stipules. Ovary superior, 2-locular; stigmas two, lamelliform; ovules one per locule, campylotropous, bitegmic, but only one ovule develops. Flowers borne in elongate spikes consisting of 3-flowered dichasial glomerules that are clustered into large, nodding, terminal panicles look like a horse tail (the species is named the "horse-tail tree" – Wu and Kubitzki 1993). Each dichasium is suspended by a large bract. The lateral flowers of each dichasium are female, but more or less reduced and frequently abortive, each flower with two bracteoles. Sepals four, in two cycles, small, scarious, brown, persistent on the fruit. Stamens six. Pollen grains 3-colporate or seldom 4-colporate with very short colpi appearing to approximate the porate condition, scabrate. The female flowers has no staminodes. Fruits small, 2-winged nuts. Vessel elements elongate, with scalariform perforations that have 1–2 (up to 20 bars). Axial parenchyma abundant, mostly vasicentric. Produce protoanthocyanins, ellagic acid, and flavonols (myricetin, quercetin and kaempferol), n = 16. . . . . .1. RHOIPTELEACEAE

1 Leaves estipulate. Ovary inferior; ovules orthotropous, unitegmic. Flowers unisexual, borne in elongate, drooping, and rarely erect, catkins or spikelike inflorescences, that are sometimes grouped into panicles. Male and female inflorescences separate or combined into an androgynous panicle; individual flowers in the axils of entire or 3-lobed bracts. The calyx adnate to the bract or wanting. Stamens 3–105, mostly 5–40. Pollen grains 3-porate or pantoporate, spinulose. Fruits are nuts enclosed in an adherent, thin or thick, fibrous indehiscent or dehiscent husk, or nuts with a thin dry skin, or thin-skinned nutlets with wings, which are formed by the bract or by two bracteoles. The whole fruit is drupelike, but the husk is derived from the involucre and calyx, and the skin from the calyx alone, neither one from the pericarp. Hence the fruit is not a true drupe; it is sometimes called a tryma (Manning 1978); nut loculicidal in germination. Vessel elements elongate, mostly with simple perforations, but in *Alfaroa* and *Engelhardtia* some of them scalariform with a few bars. Axial parenchyma mostly apotracheal, but sometimes intermediate between apotracheal and paratracheal. Mostly with abundant tannin and producing proanthocyanins, ellagic acid, flavonol glycosides, including those of myrcetin, quercetin, kaempferol, and napthoquinones, n = 15, 16, 28. . . . . 2. JUGLANDACEAE

1. RHOIPTELEACEAE

Handel-Mazzetti 1932. 1/1. Southwestern China (Yunnan, Guangxi, Guizhou) and North Vietnam.

Rhoiptelea.

In many respects this unigeneric family approaches closely to the hypothetical intermediate between Hamamelidales and Juglandaceae.

2. JUGLANDACEAE

A.P. de Candolle ex Perleb 1818 (including Engelhardtiaceae Reveal et Doweld 1999, Platycaryaceae Nakai ex Doweld 2000, Pterocaryaceae Nakai 1930). 9/60. Mainly northern temperate and subtropical regions with extensions southward to South and Southeast Asia, New Guinea, West Indies, and Central America and the western mountains of South America as far as northern Argentina.

Juglandaceae are closely related to the Rhoipteleaceae, but are somewhat more advanced. Each flower of the reduced inflorescence of Juglandaceae morphologically corresponds to the central flower of the 3-flowered dichasium of *Rhoiptelea.*

2.1 PLATYCARIOIDEAE
Bracts of female flowers persisting on the axis of inflorescence (not dispersed with the fruit). Pollen grains small, 3-porate, with pseudocolpi. Vessels with simple perforations. – *Platycarya.*

2.2 ENGELHARDTIOIDEAE
Bracts of the female flowers fall together with the fruit, 3-lobed. Pollen grains small, 3- to 4-porate. Vessels with scalariform or simple perforations. Leaf parenchyma with druses. – *Alfaroa, Alfaropsis, Oreomunnea, Engelhardtia.*

2.3 JUGLANDOIDEAE
Bracts of the female flowers fall together with the fruit, entire. Pollen grains large, 3-polyporate. Vessels with simple perforations. – JUGLANDEAE: *Cyclocarya, Pterocarya, Juglans*; HICORIEAE: *Carya.*

Manning (1978) recognized two subfamilies (Platycaryoideae and Juglandoideae), the former with one monotypic east Asian genus (*Platycarya*) and the latter with three tribes (Engelhardieae, Juglandeae, and Hicorieae). Recently Iljinskaya (1990) reviewed previous classifications and came to the conclusion that Manning's tribe Engelhardieae is so distinctive that it deserves the rank of a subfamily.

Bibliography

Berquam DL. 1975. Floral Morphology and Anatomy of Staminate Juglandaceae. Ph.D. Thesis, University of Minnesota, St. Paul.

Bos JAA and W Punt. 1991. Juglandaceae. Rev. Paleobot. Palynol. 69: 79–95.

Chang C-Y. 1981. Morphology of the family Rhoipteleaceae in relation to its systematic position. Acta Phytotax. Sinica 19: 168–178 (in Chinese with English summary).

Chen Z-D, X-Q Wang, H-Y Sun, Y Han, Z-X Zhang, Y-P Zou, and A-M Lu. 1998. Systematic position of the Rhoipteleaceae: Evidence from nucleotide sequences of *rbc*L gene. Acta Phytotax. Sinica 36: 1–7.

Chupov VS. 1978. The comparative immunoelectrophoretic investigations of pollen proteins of some Amentiferous taxa. Bot. Zhurn. 63(11): 1579–1584 Leningrad (in Russian).

Conde LF and DE Stone. 1970. Seedling morphology in the Juglandaceae: The cotyledonary node. J. Arnold Arbor. 51: 463–477.

Elias TS. 1972. The genera of Juglandaceae in the southeastern United States. J. Arnold Arbor. 53: 26–51.

Endress PK. 1986. An entomophily syndrome in Juglandaceae: *Platycarya strobilacea.* Veröffentlichungen des Geobotanischen Institutes. Stiftung Rübel, Zurich 87: 100–111.

Gunter LE, G Kochert, and DE Giannasi. 1994. Phylogenetic relationship of the Juglandaceae. Plant Syst. Evol. 192: 11–29.

Handel-Mazzetti H. 1932. Rhoipteleaceae: Eine neue Familie der Monochlamydeen. Feddes Repert. 30: 75–80.

Heimsch C and RH Wetmore. 1939. The significance of wood anatomy in the taxonomy of the Juglandaceae. Am. J. Bot. 26: 651–660.

Iljinskaya IA. 1953. A monograph of the genus *Pterocarya.* Trudy Bot. Inst. Akad. Nauk SSSR, 1st ser., 10: 7123 (in Russian).

Iljinskaja IA. 1990. On the taxonomy and phylogeny of the family Juglandaceae. Bot. Zhurn. 75: 792–803 (in Russian with English summary).

Iljinskaja IA. 1993. *Alfaropsis*: A new genus of the Juglandaceae. Bot. Zhurn. 78 (10): 79–83 (in Russian with English summary).

Jacobs M. 1960. Juglandaceae. In: CGGJ van Steenis, ed. Flora Malesiana, ser. I, 6: 143–154. Groningen.

Kribs DA. 1927. Comparative anatomy of the woods of the Juglandaceae. Trop. Woods 12: 16–21.

Kuang Ko Zen. 1960. De familis monotypica Rhoipteleaceae. Acta Bot. Sinica 9: 43–47.

Leroy JF. 1951. La théorie généralisee des carpelles sporophylles et la fleur des Juglandales: III. Discussion et conclusions. Compt. Rend. Hebd. Séances Acad. Sci. 233: 1214–1216.

Leroy JF. 1955. Étude sur les Juglandaceae. Mem. Mus. Natl. Hist. Nat, sér. B, Bot. 6: 1–246.

Litvak AI. 1981. Juglandaceae. In: MS Yakovlev, ed. Comparative embryology of flowering plants: Winteraceae Juglandaceae, pp. 216–225. Leningrad (in Russian).

Liu C. 1987. Studies of pollen morphology in the Rhoipteaceae (sic) and the relative families. Acta Bot. Yunn. 9(2): 181–186

Lu AM and Z-Y Zhang. 1990. The differentiation, evolution and systematic relationship of Juglandales. Acta Phytotax. Sinica 28: 96–102 (in Chinese with English summary).

Manchester SR. 1987. The fossil history of the Juglan-dacaea. Monographs Syst. Bot. Missouri Bot. Gard. 21: 1–137.

Manchester SR. 1989. Early history of the Juglandaceae. Plant Syst. Evol. 162: 231–250.

Manning WE. 1938. The morphology of the flowers of the Juglandaceae: I. The inflorescence. Am. J. Bot. 25: 407–419.

Manning WE. 1940. The morphology of the flowers of the Juglandaceae: II. The pistillate flowers and fruits. Am. J. Bot. 27: 830–852.

Manning WE. 1948. The morphology of the flowers of the Juglandaceae: III. The staminate flowers. Am. J. Bot. 35: 606–621.

Manning WE. 1949. The genus *Alfaroa*. Bull. Torrey Bot. Club 76: 196–209.

Manning WE. 1959. *Alfaroa* and *Engelhardtia* in the New World. Bull. Torrey Bot. Club 86: 190–198.

Manning WE. 1974. Juglandales. In: Encyclopaedia Britannica, 15th ed., vol., 10, pp. 329–332. Chicago.

Manning WE. 1975. An analysis of the genus *Cyclocarya* Iljinskaya (Juglandaceae). Bull. Torrey Bot. Club 102: 157–166.

Manning WE. 1978. The classification within the Juglandaceae. Ann. Missouri Bot. Gard. 65: 1058–1087.

Manos PS and ES Donald. 2001. Evolution, phylogeny and systematics of the Juglandaceae. Ann. Missouri Bot. Gard. 88: 231–269.

Nagel K. 1914. Studien über die Familie der Juglandaceen. Bot. Jahrb. Syst. 50: 459–530.

Nast CG. 1935. Morphological development of the fruit of *Juglans regia*. Hilgardia 9: 345–381.

Nast CG. 1941. The embryology and seedling morphology of *Juglans regia* L. Lilloa 6: 163–205.

Oginuma K. 1999. Karyomorphology and evolution in the Juglandales: A Review. Acta Phytotax. Geobot. 50: 229–241 (in Japanese with English summary).

Oginuma K and H Tobe. 1992. Karyomorphology of Juglandaceae. In: R Tanaka, ed. Plant Chromosome Research. Proc. Second Sin-Jap. Symp. Plant Chromosomes, pp. 171–179.

Oginuma K, JZ Gu, ZS Yue. 1996. Karyomorphology of *Rhoiptelea* (Rhoipteleaceae). Acta Phytotax. Geobot. 46: 147–151.

Petersen FP and DE Fairbrothers. 1979. Serological investigation of selected amentiferous taxa. Syst. Bot. 4: 230–241.

Polechko MA and RB Clarkson. 1984. A serological study of the systematics of the Juglandaceae. Biochem. Syst. Ecol. 14: 33–39.

Schaarschmidt H. 1985. Zur Verwandschaft von *Carya* Nutt. und *Pterocarya* Sieb. and Zucc. (Juglandaceae) und zur natürlichen Gliederung der Familie. Feddes Repert. 96: 345–361.

Schaarschmidt H. 1987. Zur Position der Juglandaceae A. Rich. ex Kunth im aktuellen System der Magnoliatae. Folia Geobot. Phytotax. 22: 271–286.

Smith JF and JJ Doyle. 1995. A cladistic analysis of chloroplast DNA restriction site variation and morphology for the genera of the Juglandaceae. Am. J. Bot. 82: 1163–1172.

Stachurska A. 1961. Morphology of pollen grains of the Juglandaceae. Monogr. Bot. 7: 121–143.

Stanford AM, R Harden and CR Parks. 2000. Phylogeny and biogeography of *Juglans* (Juglandaceae) based on *mat*K and ITS sequence data. Am. J. Bot. 87: 872–882.

Stone DE. 1962. Affinities of a Mexican endemic, *Carya palmeri*, with American and Asian hickories. Am. J. Bot. 49: 199–212.

Stone DE. 1970. Evolution of cotyledonary and nodal vasculature in the Juglandaceae. Am. J. Bot. 57: 1219–1225.

Stone DE. 1972. New World Juglandaceae: III. A new perspective of the tropical members with winged fruits. Ann. Missouri Bot. Gard. 59: 297–321.

Stone DE. 1977. Juglandaceae. In: W Burger, ed. Flora Costaricensis. Fieldiana (Botany) 40: 28–53.

Stone DE. 1989. Biology and evolution of temperate and tropical Juglandaceae. In: PR Crane and S Blackmore, eds. Evolution, systematics, and fossil history of the Hamamelidae, vol. 2, pp. 117–145. Claredon Press, Oxford.

Stone DE. 1993. Juglandaceae. In: K Kubitzki, ed. The families and genera of vascular plants, vol. 2, pp. 348–359. Springer, Berlin/Heidelberg/New York.

Stone DE and CR Broome. 1971. Pollen ultrastructure: Evidence for relationship of the Juglandaceae and Rhoipteleaceae. Pollen Spores 13: 5–14.

Stone DE and CR Broome. 1975. Juglandaceae A. Rich. ex Kunth. World Pollen Spore Flora 4: 1–35.

Sun S-G, Y Lu, and S-Q Huang. 2006. Floral phenology and sex expression in functionally monoecious *Rhoiptelea chiliantha* (Rhoipteleaceae). Bot. J. Linn. Soc. 152: 145–151.

Tang Y. 1932. Timber studies of Chinese trees: 1. Timber anatomy of Rhoipteleaceae. Bull. Fan. Inst. Biol. Peking 3: 127–131.

Vyshenskaya TD. 1991. Rhoipteleaceae. In: A Takhtajan, ed. Comparative seed anatomy, vol. 2, pp. 149–151. Nauka, Leningrad (in Russian).

Vyshenskaya TD. 1991. Juglandaceae. In: A Takhtajan, ed. Comparative seed anatomy, vol. 3, pp. 152–162. Nauka, Leningrad (in Russian).

Wing SL and LJ Hickey 1984. The *Platycarya* perplex and the evolution of the Juglandaceae. Am. J. Bot. 71: 388–411.

Whitehead DR. 1965. Pollen morphology in the Juglandaceae: II. Survey of the family. J. Arnold Arbor. 46: 369–410.

Withner CL. 1941. Stem anatomy and phylogeny of the Rhoipteleaceae. Am. J. Bot. 28: 872–878.

Wu CY and K Kubitzki. 1993. Rhoipteleaceae. In: K Kubitzki, ed. The families and genera of vascular plants, vol. 2, pp. 584–585. Springer, Berlin/Heidelberg/ New York.

Zhang ZY, AM Lu, and J Wen. 1994. Embryology of *Rhoiptelea chiliantha* (Rhoipteleaceae) and its systematic relationship. Cathaya 6: 57–66.

Subclass IV. CARYOPHYLLIDAE

Mostly herbaceous plants, undershrubs, or low shrubs, less often trees; succulents and halophytes common, woody species generally with anomalous secondary growth or otherwise anomalous stem structure. Vessels with simple perforations. Sieve-element plastids of P-type (Caryophyllales, Physenales) or of S-type (Polygonales, and Plumbaginales). Nodes unilacunar or less often trilacunar or multilacunar. Leaves simple, mostly entire, sometimes with stipules. Stomata mostly paracytic, anomocytic, or diacytic. Flowers bisexual or unisexual, cyclic. Tapetum secretory. Microsporogenesis simultaneous. Pollen grains 3-celled or rarely 2-celled, 3-colpate or of derived types. Gynoecium apocarpous (certain Phytolaccaceae and *Gisekia*) or more often syncarpous, frequently pseudomonomerous. Ovules mostly bitegmic or rarely (in some Nyctaginaceae) unitegmic, crassinucellate. Endosperm nuclear. Fruits of various types. Seeds with mostly curved embryo and usually without endosperm (except in Polygonaceae and Plumbaginaceae), but mostly with perisperm.

Related to the Ranunculidae, especially to the lower Ranunculales. The Phytolaccaceae in particular are linked with the ranunculids, especially with the Lardizabalaceae and Menispermaceac. In all probability both subclasses share a common origin from some Magnoliidae.

The *rbc*L sequences confirm the monophyly of the Caryophyllidae. According to Olmstead et al. (1992), the Caryophyllidae, including Polygonaceae and Plumbaginaceae, appear as a strongly supported, monophyletic group, whose circumscription in traditional classifications coincides completely with the molecular evidence.

Bibliography

Beck E, H Merxmüller, and H Wagner. 1962. Anthocyane bei Plumbaginaceen, Alsinoideen, und Molluginaceen. Planta 58: 220–224.

Beckstrom-Stemberg SM. 1989. Two-dimensional gelelectrophoresis as a taxonomic tool: Evidence from Centrospermae. Biochem. Syst. Ecol. 17: 573–582.

Behnke H-D. 1976a. Ultrastructure of sieve-element plastids in Caryophyllales (Centrospermae): Evidence for the delimitation and classification of the order. Plant Syst. Evol. 126: 31–54.

Behnke H-D. 1976b. A tabulated survey of some characters of systematic importance in centrospermous families. Plant Syst. Evol. 126: 95–98.

Eckardt T. 1976. Classical morphological features of centrospermous families. Plant Syst. Evol. 126: 5–25.

Ehrendorfer E. 1976. Chromosome numbers and differentiation of centrospermous families. Plant Syst. Evol. 126: 27–30.

Engel T and W Barthlott. 1988. Micromorphology of epicuticular waxes in centrosperms. Plant Syst. Evol. 161: 71–85.

Friedrich HC. 1955. Beitrage zur Kenntnis einiger Familien der Centrospermae. Mitt. Bot. Staatssamml. Munchen 2: 56–66.

Friedrich HC. 1956. Studien über die natürliche Ver-wandtschaft der Plumbaginales und Centrospermae. Phyton 6: 220–263.

Gianasi DE, G Zurawski, G Learn, and MT Clegg. 1992. Evolutionary relationships of the Caryophyllidae based on comparative *rbc*L sequences. Syst. Bot. 17: 1–15.

Hershkovitz MA. 1989. Phylogenetic studies in Centrospermae: A brief appraisal. Taxon 38: 602–610.

Nowicke JW. 1975. Preliminary survey of pollen morphology in the order Centrospermae. Grana 15: 51–77.

Nowicke JW and JJ Skvarla. 1977. Pollen morphology and the relationship of the Plumbaginaceae, Polygonaceae, and Primulaceae to the order Centrospermae. Smithsonian Contr. Bot. 37: 1–64.

Olmstead RG, HJ Michaels, KM Scott, and JD Palmer. 1992. Monophyly of the Asteridae and identification of their major lineages inferred from DNA sequence of *rbc*L. Ann. Missouri Bot. Gard. 79: 249–265.

Rodman JE. 1990. The Centrospermae revisited. Part 1. Taxon 39: 383–393.
Rodman JE, MK Oliver, RR Nakamura, JU McClammer, Jr., and AH Bledsoe. 1984. A taxonomic analysis and revised classification of Centrospermae. Syst. Bot. 9: 297–323.
Skvarla JJ and JW Nowicke. 1976. Ultrastructure of pollen exine in the centrospermous families. Plant Syst. Evol. 126: 55–78.
Veselova TD. 1990. On the systematic position of centrosperms in connection with the embryological data. In: VN Tikhomirov and AN Sladkov, eds. Morphology of centrosperms as a source of evolutionary information, pp. 28–48. Nauka, Moscow (in Russian).
Volgin SA. 1990. Comparative morphology of the flower and phylogeny of centrosperms. In: VN Tikhomirov and AN Sladkov, eds. Morphology of centrosperms as a source of evolutionary information, pp. 11–28. Nauka, Moscow (in Russian).
Wohlpart A and TJ Mabry. 1968. The distribution and phylogenetic significance of the betalains with respect to the Centrospermae. Taxon 17: 148–152.

Superorder CARYOPHYLLANAE

Order 42. CARYOPHYLLALES

Perennial or annual herbs, shrubs or trees, rarely climbing, often more or less succulent. Anomalous secondary thickening of the stems fairly frequent. Vessels with simple perforations, very rarely also reticulate. Fibers with simple or seldom bordered pits. Axial parenchyma diffuse (Stegnospermataceae, Barbeuiaceae) or vasicentric (Phytolaccaceae and Achatocarpaceae). Sieve-element plastids of P-type with a peripheral ring of protein filaments and with or without (Amaranthaceae and Chenopodiaceae) a central protein crystal (Behnke 1994). Nodes unilacunar or rarely trilacunar, very rarely (Rhabdodendraceae) multilacunar. Leaves alternate or opposite, seldom verticillate, simple and mostly entire, sometimes reduced to spines, often showing Kranz anatomy, estipulate or rarely with small stipules. Stomata of various types. Flowers in various types of inflorescences or solitary in the axils, bisexual or less often unisexual, usually actinomorphic, mostly 4–5-merous, entomophilous or less often anemophilous. Sepals (1)2–5(-10), free or more or less connate below, sometimes petaloid. Petals two to many or lacking, free or more or less connate. Stamens one to many, often as many or twice as many as the sepals or petals, in multistaminate flowers often developing centrifugally, especially in Aizoaceae and Cactaceae; anthers usually opening longitudinally. Pollen grains 3-celled or rarely 2-celled, 3-colpate to pantocolpate and pantoporate, tectate-columellate, with spinulose-annular perforate/punctate tectum. Tapetum secretory. Microsporogenesis simultaneous. Gynoecium syncarpous (eusyncarpous, paracarpous, or lysicarpous) or less often more or less apocarpous (but with short, basal, distinctly fused carpellary sections (see Rohweder 1965; Hofmann 1977; Hofmann in Behnke and Mabry 1994), of two to many carpels, with distinct stylodia. Ovary superior, half inferior, or inferior. Nectaries present or absent. Ovules hemitropous to campylotropous, rarely nearly anatropous, bitegmic, crassinucellate. Female gametophyte usually of *Polygonum*-type. Endosperm nuclear. Fruits of various types. Seeds with peripheral and mostly curved or coiled embryo, bordering or encircling the more or less copious perisperm or (Didiereaceae) nearly or quite without perisperm. Perisperm completely absent in *Monococcus* of Phytolaccaceae); endosperm lacking or scanty. Seed-coat exotestal-endotegmic or exotestal. Mostly producing betalains (betacyanins and betaxanthins) but not anthocyanins (only in the Molluginaceae and Caryophyllaceae producing anthocyanins and not betalains). At least some species of *Phytolacca* contain glucosilonates (Daxenbichler et al. 1991).

Key to Families

1 Stylodium gynobasic. Evergreen tall shrubs or small trees. Secretory cavities with resinous contents scattered through parenchymatous tissues. Vessels with simple perforations; lateral pitting alternate. Fibers with many small simple or bordered pits. Rays heterogeneous to sometimes homogeneous. Axial parenchyma paratracheal and very scanty diffuse or largely in apotracheal bands. Sieve-element plastids of Pcs-type. Leaves alternate, simple, entire, leathery, with short-stalked, peltate hair, obscurely pellucid-punctate; stipules minute and caducous or obsolete, or absent. Stomata anomocytic. Flowers in axillary or supra-axillary racemes or racemoid cymes, with thick rachis, usually bisexual, actinomorphic. Calyx very short, turbinate, more or less entire or with five imbricate lobes. Petals five, oblong, sepal-like, pellucid-punctate, slightly imbricate below or more or less valvate at apex, caducous. Stamens 25–50 more or less in two cycles; filaments very short, flattened, persistent and recurved; anthers very long, basifixed, caducous. Pollen grains 3-celled, 3-colpate and rarely 4-colpate (Prance 2003). Nectary

disc wanting. Gynoecium of single carpel with long, thickened gynobasic stylodium with unilateral (dorsal) decurrent stigma. Ovule usually solitary (sometimes another aborted), basal, hemicampylotropous and epitropous, bitegmic (bitegmic condition restricted to the micropylar part). Fruits small, drupaceous, with thin exocarp and slightly woody endocarp, shortly stipitate, within the cupular calyx. Seed with curved, green embryo and large, fleshy cotyledons, small radicle bent inward toward the hilum; endosperm wanting; exotestal cells tangentially elongated. Presence triterpenoids and O-alkylated ellagic acid (Walter-Filho et al. 1985, 1989); alkaloids absent, n = 10. . . . . 2. RHABDODENDRACEAE

1 Stylodium not gynobasic.

2 Sieve-element plastids with a central protein crystal.

3 Producing betalains, very rarely anthocyanins.

4 Carpels free or united. Pollen grains 3-colpate. Sepals five, imbricate. Petals absent or present.

5 Fruits of more or less drupelike fruitlets or syncarpous and baccate, or fruits indehiscent, by abortion 1-seeded, with coriaceous pericarp adherent to the seed (*Agdestis*). Herbs, shrubs, woody lianas, or rarely trees, more or less succulent, or non-succulent, mesophytic or xerophytic. Raphides present. Sieve-element plastids of P3a-type. Nodes unilacunar. Leaves alternate, entire, usually estipulate. Flowers in racemes or spikes, bisexual or rarely dioecious, actinomorphic or more or less zygomorphic, apetalous. Perianth segments 4–5, in one whorl. Stamens (2-)4 to many; filaments filiform, free or basally connate, sometimes inserted on a fleshy disc; anthers dorsifixed, introrse, tetrasporangiate. Pollen grains 3-colpate via pantocolpate to pantoporate, 2- or 3-celled. Carpels 3–16, free or more or less connate; ovary superior, rarely semi-inferior (*Agdestis*), 4–12-locular, ovules solitary in each locule, campylotropous or hemitropous (*Agdestis*). Seeds more or less reniform, perisperm copious or moderate, soft or granular and whitish; embryo well differentiated, curved; cotyledons flat or folded; endosperm absent. Betalains present; saponins, alkaloids and flavonols present or absent; *Gallesia* smells of onions; n = 9, rarely 17. . . . . 1. PHYTOLACCACEAE

5 Fruits achenes with membranous epicarp densely covered by warts. Annual herbs. Leaves opposite or subverticillate. Flowers in dichasia, bisexual or rarely unisexual, more or less actinomorphic. Petales five. Stamens 5 or 10–15, rarely 8, sometimes 20 or more, and then they are paired in groups of three, inserted on a flat disc, alternating with petals; filaments dilated at the base; anthers basifixed. Carpels five or less often three or ten, pseudoapocarpoous, laterally compressed, with short stylodia on their ventral side; ovule solitary in each locule, anatropous. Fruit achenial; seeds smooth, pitted. Present anthocyanins and raphides; n = 9. . . . . 3. GISEKIACEAE

4 Carpels always united. Pollen grains of various types. Petals present or absent.

6 Ovary mostly superior.

7 Fruits winged. Shrubs with thorny branches. Wood rayless. Leaves opposite below, alternate above, deciduous, linear, entire, fleshy, glabrous to hairy. Inflorescence spike, cylindric, densely catkin-like. Flowers unisexual. Male flowers numerous, spirally arranged, with peltate bracts, without perianth. Stamens 2–3 or more, with very short filaments and long anthers; pollen grains pantoporate, distinctive by its raised pore margins produced by elongated columellae (crassimarginate pores – Nowicke 1994). Female flowers 1–3, below male flowers, perianth cup-like, connate with ovary; ovule solitary. Carpels two, stigmas two, horizontally expanded. Seed erect with transparent testa; embryo compressed, green, n = 9. . . . 4. SARCOBATACEAE.

7 Fruits not winged.

8 Petioles articulated at the base. Large, branched woody lianas, blackening on drying. Axial parenchyma diffuse. Leaves alternate, entire. Protein crystal in the sieve-element plastids globular. Flowers in short, rigid, axillary racemes. Pedicels long and slender, thickened below the calyx. Receptacle slightly convex. Sepals five imbricate. Stamens 20–25, in several series, inserted on an annular disc (annulus of receptacle); filaments short, free; anthers linear, 2-locular, introrse, sagittate. Pollen grains 3(4–6)-colporoidate, the exine undulated to verrucate. Gynoecium of two united carpels; ovary 2-locular; stylodia linear-oblong,

somewhat flattened, erect-spreading, very shortly connate at the base, the inner side papillate; ovule solitary. Fruits woody, 2-locular and 1–2-seeded capsules. Seeds partially enclosed by a yellow aril. . . . . 5. BARBEUIACEAE.

8 Petioles not articulated. Plants not blackening on drying.

9 Seeds covered by a large red aril. Small, glabrous trees or shrubs, usually with decumbent or scandent branches. Leaves alternate, entire, fleshy, with translucent margins; diffuse axial parenchyma presents (Carlquist 1999). Crystalloids within the sieve-element plastids polygonal. More or less scandent or sprawling glabrous shrubs. Flowers in terminal racemes. Pedicels bracteate and bracteolate at the base. Sepals five, imbricate, persistent in the fruit. Petals free, imbricate, shorter than the sepals. Stamens ten, in one series; filaments connate at the base; anthers dorsifixed, cordate, introrse. Pollen grains 3-colpate, operculate, with spinulose and punctate rectum. Ovary superior, 1-locular, with a very short style and 3–5 subulate recurved stigmas, papillose inside. Ovules as many as the stigmas, amphitropous, basal around a central column. Fruits coriaceous, 3–5-valved and 1–5-seeded capsule. Seeds almost covered by a large red aril, exotesta more or less palisade, unlignified, endotegmen enlarged, persistent, n = 36 (P. Goldblatt, 2003). . . . . 9. STEGNOSPERMATACEAE.

9 Seeds without aril, or when an aril is present, it does not cover the seed.

10 Fruits 1-seeded berries. Shrubs or small trees with normal secondary growth. Leaves alternate, entire, estipulate. Flowers in axillary racemes or panicles, dioecious. Sepals 4–5. Stamens 10–20; filaments slender, free or basally connate. Pollen grains with mostly 4–6 large pores or more or less irregular apertures, which are often poorly defined. Gynoecium of two carpels with short, basally connate stylodia; ovary superior, 1-locular, with 1 (very rarely 2) campylotropous basal ovule. Seeds black, tuberculate, with very small aril at the hilum. Present C-glycosylflavonoids. . . . 6. ACHATOCARPACEAE.

10 Fruits dry.

11 Fruits dehiscent.

12 Stamens opposite the sepals. Annual herbs to small trees with normal secondary growth. Leaves alternate or sometimes opposite, often with hairs, bristles, or scales in the axils; stipules scarious. Flowers bisexual or rarely unisexual, with two or sometimes five or more *(Lewisia)* imbricate bracteoles and (2)5(-18) free or basally connate, more or less petaloid and often caducous sepals. Stamens commonly as many as and opposite the sepals. Pollen grains ranging from 3-colpate to pantocolpate and pantoporate. Gynoecium 2–8-carpellate, with more or less united stylodia and capitate stigmas. Ovary superior to inferior, with two to many or rarely one ovule; ovules anatropous, acampylotropous, or amphitropous, placentation free-central or basal. Fruits circumscissile or valvate capsules, rarely nutlike. Seeds sometimes arillate; embryo curved to annular, n = 8–12 (-17), mostly 8 or 9, in *Claytonia virginiana* n = 6–95 (Bogle 1969). . . . . . . 10. PORTULACACEAE.

12 Stamens alternate with the petals. Densely caespitose, glabrous subshrubs; cuticle waxes as procumbent platelets. Leaves alternate, imbricated all around the stem, without stipules. Flowers solitary, axillary, with (1)2(3) lateral, apical or basal bracteoles and 4–5(6) petaloid sepals. Stamens 3–5(6), free or adnate to calyx tube when it is present. Pollen grains 3-colpate. Gynoecium 2-carpellate, unilocular; ovary with 4–7 ovules borne near the base of central column. Fruits 1–5-seeded capsules, or indehiscent. Seeds with endosperm and curved embryo, n = 48. . . . . . 11. HECTORELLACEAE.

11 Fruits indehiscent.

13 Sepals united to form a lobed and commonly corolloid tube and sometimes subtended by brightly colored sepaloid bracteoles. Trees, shrubs, herbs, and lianas, sometimes thorny. Nodes swollen. Wood storied, rayless. Vessels with simple perforation, occasionally also reticulate; stem with anomalous secondary growth. Leaves opposite or rarely alternate, toothed. Inflorescences various, cymose, very rarely a raceme; flowers bisexual or rarely dioecious (*Cephalotomandra*). Tubular, petaloid perianth with induplicate-valvate or contorted aestivation. Stamens one to many, free or connate at the base, involute in bud; filaments often of unequal length; anthers dorsifixed near the base. Pollen grains 3(4)-colpate, pericolpate, or periporate. Ovary superior, 1-locular, with slender styles. Ovule solitary, bitegmic, or rarely unitegmic (*Abronia umbellate, Boerhaavia diffusa*). Fruits nutlike, surrounded by accrescent (viscid) basal part of the perianth; endotesta thickened (*Mirabilis*), testa multiplicative; embryo peripheral, arcuate or rarely straight, green; perisperm copious to scanty; n = 8–11(13), rarely 17, mostly 10. . . . .7. NYCTAGINACEAE.

13 Sepals not forming a corolloid calyx tube.

14 Perennial herbs with slender, annual twining or scrambling stems from fleshy-thickened or tuber-bearing rhizomes; vascular bundles separate, bicollateral. Leaves opposite, conduplicate (*Anredera*). Flowers in racemose inflorescences, small, mostly bisexual. Bracteoles two, often united to the base of the calyx. Sepals five, often colored, almost free or connate below, persistent. Stamens five, opposite to the sepals and adnate to the base of sepals or to the calyx tube. Pollen grains cuboid, pantocolpate. Gynoecium of three carpels; ovary unilocular when mature, with three stylodia or a style, often 3-lobed. Ovule solitary, basal, anatropous to campylotropous. Fruits indehiscent, surrounded by persistent, often fleshy calyx or by winglike persistent bracteoles; exotesta and endotesta more or less thickened, embryo green, starch grains clustered; n = 12, 22. . . 12. BASELLACEAE.

14 Not twining or scrambling annual herbs. Flowers unisexual, monoecious. Bracteoles two, very unequal. Sepals in male flowers four, free, membranous, in female flowers lacking. Stamens four, alternating with sepals; filaments slender, long. Pollen grains cuboid, 6-porate. Gynoecium of three carpels; ovary unilocular, with three free stylodia. Ovule solitary, basal, evidently campylotropous. Fruits nutlets, several becoming enclosed by inflorescence axis, n = 12. . . 13. HALOPHYTACEAE.

6 Ovary mostly inferior or semi-inferior.

15 Ovary with a single fertile locule and 2(3) empty locules. Very spiny, cactuslike, xerophytic shrubs and trees with soft wood and rather well-developed pith; cuticular waxes as ribbons or rodlets. Leaves alternate, entire, small, sometimes much reduced. Flowers mostly unisexual and usually dioecious. Sepals four, in two series. Petals absent. Stamens 8(10), in two cycles basally connate and with adaxial nectaries. Pollen grains 3-celled, (4)5–7-zonocolpate. Gynoecium of (2)3(4) carpels with a style and a usually irregularly (2)3(4)-lobed, peltate, fringed stigma; ovary with as many locules as carpels, but usually only one locule is fertile and bears a solitary, basal ovule. Fruits dry, indehiscent, usually 3-angled. Seeds with a

small funicular aril, nearly or quite without both perisperm and endosperm. Present methylated flavonoids; n = 8.......... 15. DIDIEREACEAE.

15 Ovary without empty locules.

16 Plants leaf-succulents or rarely stem-succulents, always unarmed. Leaves alternate or opposite, usually entire, rarely toothed. Stomata anisocytic, anomocytic or paracytic. Nodes 3-locular with three traces. Anomalous secondary growth present. Epicuticular waxes present. Inflorescences axillary or terminal pleiochasia, dichasia or flowers solitary. Flowers cyclic, bisexual or rarely unisexual (*Glischrothamnus*). Calyx lobes 5–8; petals ca. 30–40, or absent. Stamens (1-)4–5, or 8–10, or numerous. Pollen grains 3-colpate or rarely 3-colporate. Ovary superior, half-inferior, or inferior, surrounded by nectary ring. Female gametophyte usually *Polygonum*-type. Ovules campylotropous or anacampylotropous, funicle often long. Fruits usually loculicidalcapsules,rarelybaccate(*Carpobrotus*), sometimes nutlike. Seeds often arillate. n = 8, 9, mostly 9......... 8. AIZOACEAE.

16 Plants stem-succulents or with distinctly spiny stem, cork also in cortex; rays wide and tall; nodes often with two or many traces; mucilage copious; cuticular waxes as ribbons or rodlets. Leaves mostly very reduced. Stomata usually paracytic. Flowers usually solitary, rarely in panicles, spiral or spirocyclic, bisexual, rarely unisexual. Perianth segments spirally arranged, usually not differentiated into sepals and petals, Stamens numerous, centrifugal; anthers small, dorsifixed or basifixed. Pollen grains 3-colpate to 6–15-colpate or -porate. Gynoecium of 3-many carpels; stigma wet, lobed; ovary inferior or rarely (in spp. of *Pereskia*) superior, placentation more or less parietal; ovules numerous, anatropous to campylotropous. Fruits usually indehiscent, commonly fleshy and baccate, rarely dry, funicles fleshy in fruit; endotegmic cell walls thickened or not; n = 11................... 14. CACTACEAE.

3 Producing anthocyanins but not betalains.

17 Leaves opposite to alternate or verticillate, slightly or not succulent; stipules scarious, small or absent.

18 Seeds small, dark brown to shiny black. Herbs, subshrubs, rarely shrubs with anomalous secondary growth. Inflorescences pleiochasail or dichasial cymes, or in glomerules to cincinnii, or flowers solitary. Sepals five, mostly free, rarely shortly connate. Petals small and inconspicuous or more often wanting. Stamens 5–10 to many, filaments filiform, free or shortly connate; anthers dorsifixed; nectary disc absent or annular. Ovary (except in *Adenogramma*) with two to several locules, at least below, but in the upper part of the ovary the partitions sometimes incomplete. Placentation axile (*except Adenogramma*). Fruits usually loculicidal or circumscissible capsules, often surrounded by persistent calyx. Seeds sometimes arillate, embryo curved or around the periphery. Plants containing anthocyanins; n = 8 (*Hypertelis*) or 9. .... 16. MOLLUGINACEAE.

18 Seeds rather large, pale yellow. Herbs or subshrubs, glabrous or with glandular hairs. Sieve-element plastids of PIIIc"f-type. Leaves alternate to subopposite, linear, subulate, lanceolate or orbicular, estipulate. Inflorescence cymose, lax or glomerulate. Flowers small, sepals five, petals five or absent, adnate to the staminal tube. Stamens (5)-7(-10). Pollen grains tricolpate. Ovary pseudomonomerous 2-chambered, with one ovule per chamber. Fruit separating into two woody cocci; rarely broadly winged; seeds with rough surface, strictly reniform, slightly winged, periclinals flat, micropapillate, anticlinals highly raised, laciniate or with longitudinal striations (Hassan et al. 2005), testa with cells in rows along the dorsal junction; n = 9.... 17. LIMEACEAE.

17 Leaves opposite or rarely (as in *Telephium* and *Corrigiola*) alternate; stipules absent or present (Illecebroideae), and then often scarious, white or brownish. Stomata often diacytic or anisocytic. Mostly herbs, perennial or annual, or sometimes subshrubs, very rarely (*Sanctambrosia manicata*) small trees; stems often swollen at the nodes due to an anomalous growth of the concentric rings of xylem and phloem. Inflorescences usually terminal, paniculate, racemiform or

capitate dichasial cymes, or flowers solitary. Flowers actinomorphic, or very rarely (Mediterranean *Drypis spinosa*) slightly zygomorphic, bisexual or rarely the plants dioecious. Sepals (4)5–25, rarely spirally arranged, free or connate. Petals mostly more or less developed, in Illecebroideae often absent. Stamens 10 (3–13), filaments filiform, free, anthers dorsifixed. Gynoecium of 2–5(-10) carpels; ovary superior, unilocular above, but usually more or less distinctly partitioned toward the base, at least when young; ovules mostly numerous, but sometimes few or even solitary. Fruits capsules, dehiscent by apical teeth or by valves, rarely circumscissile, or rarely a scarcely fleshy berry (*Cucubalus*). Seeds not arillate (except seeds with funicular elaiosome in *Moehringia*), embryo peripheral, arcuate to spiral (*Dripis*), or straight (*Dianthus, Stellaria*). Produce anthocyanins (batalains) and anthraquinones, saponins, ferula acid and pinitol, n = 5–15, 17–19, mostly 8. . . . . 18. CARYOPHYLLACEAE.

2 Sieve-element plastids without protein crystal.

19 Stylodia three. Xerophytic, evergreen, freely branched, shortly pubescent shrubs or small trees with anomalous secondary growth of concentric type (concentric rings of vascular bundles); hairs uniseriate. Vessels with simple perforations and with tertiary thickenings; lateral pitting alternate. Fibers with large bordered pits. Axial parenchyma apotracheal, very rare, except for conjunctive tissue associated with the anomalous structure. Sieve-element plastids of Ss-type. Nodes unilacunar. Leaves small, opposite, sessile, simple but jointed at the base, entire, pinnately veined, leathery, estipulate; the mesophyll containing calcium oxalate crystals. Stomata laterocytic or cyclocytic, or anomocytic (and also of intermediate type). Flowers unisexual, dioecious, actinomorphic, apetalous. Male flowers small, in cernuous, capitulate, axillary clusters, the female ones larger, and mostly solitary in the axils (more rarely in pendulous 2–7-flowered racemes). Sepals (4)5(6), much imbricate, fimbriate, accrescent in female flowers. Stamens (8)10(12), free, inserted more or less distantly on flat receptacle; filaments short and stout; anthers elongate, tetrasporangiate, basifixed, 2-locular, extrorse, opening longitudinally. Pollen grains 2-celled, with three large, weakly defined, porelike apertures the membranes of which are bordered with irregularly shaped insulae or granula forming an operculoid structure in the central part and grading to the exine pattern. Pistillodia absent. Gynoecium of three united carpels, with three elongate, reflexed, subulate, papillate, hairy, deciduous stylodia clustered at the top. Ovary superior, 3-locular, with one pendulous, apical-axile ovule per locule. Ovules anatropous, or apotropous, pendulous, bitegmic, crassinucellate (but tenuinucellate, according Köhler 2003). Endosperm nuclear and very reduced. Fruits ovoid loculicidal capsules, with shiny brown coriaceous pericarp and usually one seeded by abortation (two of the locules empty), rarely 2–3-seeded. Seeds large, subtriquatrous, red-brown, glandular, and short-haired. Embryo straight, with thick cotyledons; cotyledons and peripheral layers of the hypocotyl contain a cyanogenic glucoside (simmondsin), mono-ethylenic acids, and a high proportion of a unique liquid wax; endosperm scanty or wanting, n = 13.. . . . . .19. SIMMONDSIACEAE.

19 Stylodium one.

20 Sepals usually dry and scarious or membranous, sometimes absent. Herbs, rarely shrubs or small trees. Vessels with simple perforation. Leaves alternate or opposite, entire, estipulate. Inflorescences terminal and axillary spikes, heads, racemes or cymes. Stamens often connate at the base into a tube. Stylodia united into a style with 2–3-fid or capitate stigma. Epicuticular wax structures lacking, n = 7–11, 13, 15, 17, mostly 8 and 9. . . . . .20. AMARANTHACEAE.

20 Sepals mostly green or greenish and usually more or less herbaceous or somewhat membranous, rarely wanting. Herbs and shrubs, rarely small trees or climbers. Vessels with simple perforation or very rarely with some oblique scalariform perforations (*Axirys*). Leaves mostly alternate, rarely opposite, sometimes fleshy or reduced to scales, entire or toothed, estipulate. Stamens free or sometimes connate at the base. Stylodia distinct or more or less connate into a style. Female gametophyte *Polygonum*- or *Allium*-type. Epicuticular wax structures present (more or less lobed platelets), n = 9, rarely 6. . . . . 21. CHENOPODIACEAE.

1. PHYTOLACCACEAE

R. Brown 1818 (including Agdestidaceae Nakai 1942, Hilleriaceae Nakai 1942, Lophiocarpaceae Doweld et Reveal 2005, Petiveriaceae C. Agardh 1824, Rivinaceae C. Agardh 1824, Sarcocaceae Rafinesque 1837, Seguieriaceae Nakai 1942). 15/50. Subcosmopolitan.

1.1. PHYTOLACCOIDEAE
Fibers vasicentric. Perianth segments usually 5. Carpels 3–17, free or united. Fruit a berry. – *Anisomeria, Ercilla, Phytolacca, Nowickea.*

1.2. RIVINIOIDEAE
Shrubs, lianas or herbs. Perianth segments usually 4. Carpel one. Fruits indehiscent, often samaroid, pericarp mostly adherent to the seed. – RIVINEAE (PETIVERIEAE): *Rivina, Trichostigma, Schindleria, Hilleria, Ledenbergia, Petiveria, Monococcus;* SEGUIERIEAE: *Gallesia, Seguieria*; LOPHIOCARPEAE: *Lophiocarpus.*

1.3. AGDESTIDOIDEAE
Semiwoody liana with massive globose rootstock and anomolous secondary growth. True tracheids present. Sepals four, rarely five (in terminal flowers). Carpels (3)4, united. Ovary semi-inferior. Fruits a 1-seeded achene with sepalline wings. – *Agdestis.*

2. RHABDODENDRACEAE

Prance 1968. 1/4. Tropical South America (the Guianas and Brazil).

Rhabdodendron.

Prance (1968, 1972) closed Rhabdodendraceae with Phytolaccaceae. Gadek et al. (1996) suggested a relationship Rhabdodendraceae with the caryophyllid clade in which Phytolaccaceae are also placed based on molecular evidence. Fay et al. (1997) indicated that *Rhabdodendron* belongs within Caryophyllidae, but without a close relationship to other families. APG II (2003) includes Rhabdodendraceae in to Caryophylales sensu lato. The family is very isolated and represent a separate order and even superorder (Doweld 2002).

3. GISEKIACEAE

Nakai 1942. 1/7. From tropical and South Africa to South Vietnam, but mainly African.

Gisekia.

Beginning with Bentham and Hooker (1867), many authors assigned *Gisekia* to the Aizoaceae. Others, including Cronquist (1981), Thorne (1983, 2005) and Rohwer (1993), transfer it to the Phytolaccaceae s. l. Hutchinson (1926, 1959, 1973) includes *Gisekia* in the Molluginaceae, with some members of which, especially *Limeum*, it shares some morphological features. Embryological data also show a close relationship of *Gisekia* with *Mollugo* (Raghavan and Srinivasan 1940). However, *Gisekia* differs from the Molluginaceae not only in its apocarpous gynoecium and indehiscent fruits but also in the presence of betalains. From the Aizoaceae, with which *Gisekia* shares some chemical characters (Narayana and Narayana 1988; Richardson 1981), it differs in its apocarpous gynoecium as well as its free sepals, apetalous flowers, fruit morphology, etc. From the Phytolaccaceae it differs in cymose inflorescences and dry, nutlike fruitlets as well as in the morphology of its vegetative organs. However, the endexine of *Gisekia* suggests a relationship to the Phytolaccaceae (Nowicke in Behnke and Mabry 1994).

4. SARCOBATACEAE

Behnke 1997. 1/2. North America.

Sarcobatus.

The genus *Sarcobatus* are usually including in the Chenopodiaceae. However, according to Behnke (1997), *Sarcobatus* markedly differs from Chenopodiaceae. The important feature that sets off *Sarcobatus* from the remainder of the Chenopodiaceae are specific sieve-elements plastids. The sieve-element plastids of *Sarcobatus* are of the form P3cf with a globular central and an additional rodlike protein crystal not found elsewhere in the Chenopodiaceae (Behnke 1993). "If I had to decide by the sieve-element plastid data where to place *Sarcobatus*" says Behnke, "I would select the suborder Phytolacineae, where some genera of the Phytolaccaceae and Nyctaginaceae have the same characteristics."

5. BARBEUIACEAE

Nakai 1942. 1/1. Madagascar.

Barbeuia.

This large woody vine may be regarded as a very advanced and specialized derivative of the Phytolaccaceae. Sieve-element plastids are similar to those of the Phytolaccaceae (Behnke 1993).

6. ACHATOCARPACEAE

Heimerl 1934. 2/10. From Texas and northwestern Mexico to Paraguay and Argentina.

Achatocarpus, Phaulothamnus.

Probably derived from the Phytolaccaceae. Among the many differences from the Phytolaccaceae are the sieve-element plastids with polygonal central crystals (Behnke in Behnke and Mabry 1994).

7. NYCTAGINACEAE

A. L. de Jussieu 1789 (including Allioniaceae Horaninow 1834, Bougainvilleaceae J. Agardh 1858, Mirabilidaceae W. R. B. Oliver 1936, Pisoniaceae J. Agardh 1858). 32/400. Tropical and subtropical regions, particularly in America. *Phaeoptilum* is endemic to southwestern Africa.

BOLDOEAE: *Boldoa, Salpianthus, Cryptocarpus;* LEUCASTEREAE: *Andradea, Ramisia, Leucaster, Reichenbachia;* NYCTAGINEAE: *Colignoma, Pisoniella, Boerhavia; Anulocaulis, Cyphomeris, Commicarpus, Caribea, Acleisanthes, Selinocarpus, Okenia, Mirabilis, Cuscatlania, Nyctaginia, Allionia, Phaeoptilum;* ABRONIEAE: *Abronia, Tripterocalyx;* BOUGAINVILLEEAE: *Bougainvillea, Belemia;* PISONIEAE: *Pisonia, Guapira, Neea, Neeopsis, Cephalotomandra, Grajalesia.*

An advanced entomophilous family derived from the Phytolaccaceae stock. Stands close to the Petiveriaceae. According to Baillon (1875), the gynoecium of Nyctaginaceae is constructed like that of the Rivineae.

8. AIZOACEAE

Martynov 1820 (including Galeniaceae Rafinesque 1819, Mesembryaceae Dumortier 1829, Mesembryanthemaceae Burnett 1835, Sesuviaceae Horaninow 1834, Tetragoniaceae Link 1831). Circa 130/2500. Dryer parts of tropical and subtropical regions, but centered mainly in South and southwestern Africa; there is also a small center in western and southern Australia.

8.1 AIZOOIDEAE

Annuals to perennials, with herbaceous or woody branches; leaves slightly succulent, mostly alternate, flat or more rarely cylindrical; inflorescence leafy, perianth internally petaloid, externally sepaloid; flowers perigynous, epigynous, or rarely hypogynous, with a holonectary; capsules loculicidal or septicidal, often hygrochastic or a winged nut or horn-shaped, n = 8. – *Gunniopsis, Aizoanthemum, Aizoon, Plinthus, Galenia, Acrosanthes.*

8.2 TETRAGONIOIDEAE

Shrublets or perennials with alternate, entire leaves. Flowers small, inconspicuous, in andromonoecious inflorescence. Fruits nutlike, usually with acute horns, n = 8. – *Tetragonia, Tribulocarpus.*

8.3. SESUVIOIDEAE

Prostrate to erect perennials or annuals; leaves slightly succulent, petiolate, often stipulate, flat or more rarely cylindrical; inflorescence bracteate; perianth internally petaloid, externally sepaloid; flower perigynous, with a holonectary; fruit a circumscissile capsule or compound and fused to spinous bracts; seed arilate, n = 8. – *Cypselea, Sesuvium, Trianthema, Zaleya.*

8.4 MESEMBRYANTHEMOIDEAE

Annual to short-lived or perennial herbs or shrubs, cortex with vascular bundles; leaves succulent, usually flat to almost cylindrical and channeled, bladder cells often large and conspicuous; flowers with petals of staminodial origin, with a central placenta, ovary half-inferior or inferior, nectaries koilomorphic, i.e., always separated and grooved; fruit a hygrochastic capsule with purely septal expanding keels reaching from the central columella to the tip of the valve or very rarely a woody nut; n = 9. – *Mesembryanthemum, Synaptophyllum, Phyllobolus, Aptenia, Aspazoma, Aridaria, Caulipsolon, Prenia, Sceletium, Psilocaulon, Brownanthus.*

8.5 RUSCHIOIDEAE

Perennial shrubs, rarely annual; leaves succulent, usually cylindrical or trigonous, rarely flat, epidermis rarely with conspicuous bladder cells, usually xeromorphic; flowers with petals of staminodial origin, with a basal or pariental placenta, ovary inferior, nectaries lophomorphic, i. e., crested, either separated or in a ring or rarely glands flattened; fruit a hygrochastic capsule with expanding keels of mainly valvar origin with only small septal portion near the outer rim of the locule, never reaching to the center of the fruit, mostly with covering membranes and additional closing devises; n = 9. – APATESIEAE: *Apatesia, Carpanthea, Hymenogyne, Caryotophora, Conicosia, Suphesia, Skiatophytum;* DOROTHEANTHEAE: *Aethephyllum, Cleretum, Dorotheanthus*; RUSCHIAE: *Bergeranthus, Bijlia, Carruanthus, Cerochlamys, Chasmatophyllum, Corpuscularia, Delosperma, Drosanthemum, Ectotropis, Faucaria, Frithia, Gibbaeum, Hereroa, Machairophyllum, Malephora, Mestoklema, Mossia, Muiria, Neohenricia, Orthopterum, Oscularia, Rabiea, Rhinephyllum, Rhombophyllum, Ruschianthus, Trichodiadema, Stomatium, Deilanthe, Smicrostigma, Oophytum, Vanzijlia, Schlechteranthus, Nananthus, Hammeria, Marlothistella, Zeuktophyllum, Acrodon, Didymaotus, Namibia, Wooleya, Aloinopsis, Dinteranthus, Juttadinteria, Ihlenfeldtia, Cheiridopsis, Diplosoma, Jacobsenia, Sarcozona, Octopoma, Vlokia, Pleiospilos, Titanopsis, Osammophora, Amphibolia, Circandra, Antegibbaeum, Disphyma, Cylindrophyllum, Ottosonderia, Scopelogena, Hartmanthus, Dicrocaulon, Leipoldtia, Arenifera, Meyerophytum, Dracophilus, Feenestraria, Eberlanzia, Glottiphyllum, Jordaaniella, Odontophorus, Ebracteola, Argyroderma, Enarganthe, Hallianthus, Stayneria, Khadia, Lapidaria, Nelia, Lithops, Astridia, Mitrophyllum, Braunsia, Jensenobotrya, Conophytum, Lampranthus, Carpobrotus, Erepsia, Ruschianthemum, Ruschia, Monilaria, Namaquanthus, Antimima, Vanheerdea, Schwantesia, Tanquana, Cephalophyllum, Stroeberia.*

Closely related to the Phytolaccaceae, but more advanced.

9. STEGNOSPERMATACEAE

Nakai 1942. 1/4. From Baja California to Central America and West Indies.

Stegnosperma.

Stegnosperma is usually included in the Phytolaccaceae, from which it differs in its flower morphology (Hofmann 1973), particularly in the presence of petaloid staminodia as well as in sieve-element plastids with polygonal protein crystal as in Achatocarpaceae and Caryophyllaceae (Behnke 1976b, 1982; Behnke in Behnke and Mabry 1994), in seeds almost completely covered by a large fleshy funicular aril, in capsular fruits, and in the presence of quercetin and kaempferol glycosides (Richardson 1981). They differ also in wood anatomy. According to Carlquist (1999), diffuse axial parenchyma is common in *Stegnosperma*, but has not been reported in other Caryophyllales. He also agrees with Rodman (1994), that *Stegnosperma* is basal in Caryophyllales.

10. PORTULACACEAE

A.L. de Jussieu 1789 (including Montiaceae Rafinesque 1820, Talinaceae Doweld 2001). 28/450 or more. Very widely distributed, but mainly in warm and temperate regions, particularly well represented in South Africa, western North America and southern South America.

PORTULACEAE: *Portulaca, Grahamia*; TALINEAE: *Talinopsis, Talinaria, Anacampseros, Lewisia, Calandrinia, Montiopsis, Erocallis, Calandrinia* (including ? *Baitaria*), *Cistanthe, Rumicastrum, Monocosmia, Talinum, Talinella, Amphipetalum, Schreiteria, Lenzia, Calyptridium* (including *Spraguea*), *Claytonia, Limnalsine, Naiocrene, Neopaxia, Mona, Montia, Montiastrum, Maxia, Crunocallis.*

Allied to the Aizoaceae and Cactaceae and especially to the Hectorellaceae and Basellaceae. Probably derived directly from the Phytolaccaceae stock.

11. HECTORELLACEAE

Philipson et Skipworth 1961. 2/2. New Zealand (South Island) and Kerguelen Island.

Hectorella, Lyallia.

Very close to the Portulacaceae, from which they differ in stamens alternating with the sepals and attached to the calyx tube and in the densely imbricate arrangement of coriaceous estipulate leaves. Sieve-element plastids with globular protein crystal resemble those of the Portulacaceae (Behnke 1975, 1976b, 1993).

12. BASELLACEAE

Rafinesque 1837 (including Anrederaceae J. Agardh 1858 and Ullucaceae Nakai 1942). 4/20. Mainly tropical and subtropical regions of America, but also tropical Africa, Madagascar, southern Asia, New Guinea, and some Pacific islands.

BASELLEAE: *Basella, Ullucus, Tournonia;* BOUSSINGAULTIEAE: *Anredera* (including *Boussingaultia*).

Very close to the Portulacaceae, especially to *Montia*, with which they share not only similar features of floral morphology but also the presence of bicollateral bundles resulting from the tardy development of internal phloem, the absence of anomalous secondary growth, the presence of mucilaginous cells as well as sieve-element plastids with central globular protein crystal (Behnke 1976b) and the absence of epicuticular wax projections (Engel and Barthlott 1988; Barthlott in Behnke and Mabry 1994). However, from the Portulacaceae the family Basellaceae differs not only in the twining or scrambling habit but also in its considerably larger protein crystals of the sieve-element plastids (Behnke in Behnke and Mabry 1994), solitary ovules (among the Portulacaceae only *Portulacaria* and *Ceraria* have a solitary ovule), and its indehiscent fruit (indehiscent nutlike fruits are characteristic of the somewhat transitional genus *Portulacaria*). The family is divided into two well-defined groups – Baselleae, with filaments erect and straight in the bud and spirally twisted embryo, and Boussingaultieae, with filaments reflexed outward in the bud and an annular embryo.

13. HALOPHYTACEAE

A. Soriano 1984. 1/1. Arid regions of Argentina (from La-Rioja to Santa-Cruz).

Halophytum.

A very specialized annual succulent herb that has usually been included in the Chenopodiaceae and is kept there even in Cronquist's system (1981). However from the Chenopodiaceae it differs markedly in the ultrastructure of the sieve-element plastids, which possess a central globular crystal (Hunziker et al. 1974; Behnke in Behnke and Mabry 1994). According to Hunziker et al. (2000) Halophytaceae closely related with Basellaceae, and perhaps also with the Hectorellaceae and Portulacaceae, this idea also supported by the P-type of sieve-element plastids, the oral structure and pollen morphology (Nowicke and Skvarla 1979) and ontogeny of stomata (De Fulvio, 1975).

14. CACTACEAE

A.L. de Jussieu 1789 (including Cereaceae De Candolle et Sprengel 1821, Nopaleaceae Schmid et Curtman 1856, Opuntiaceae Desvaux 1817). About 100/1500 or more. Arid and semiarid regions of America from British Columbia and Alberta as far south as Patagonia, concentrated in the dryer regions of Mexico, southwestern United States, and southern South America; the only paleotropical member of the family is *Rhipsalis baccifera* in Africa, Madagascar, Mascarenes, Seychelles, and Sri Lanka.

14.1 PERESKIOIDEAE

Leaves large, supervolute, with lamina. Spines present, but glochids absent. Tree-like, shrubby or scandent plants. Ovary from superior to fully inferior. Funicles fleshy in fruit. Seeds black and without an aril, unspecialized, without complicated seed coat structure; they resemble very much the seed coats of many Portulacaceae (e.g., *Claytonia*) and Didiereaceae, e.g., *Alluaudia*) and support the ancestral position of the family (Behnke and Barthlott 1983). – *Pereskia* (including *Rhodocactus*).

14.2 MAIHUENIOIDEAE

Leaves small, terete, persistent; spines usually three. Caespitose shrubs. Stems succulent, globose or short-cylindric. Stomata parallelocytic. Flowers terminal, solitary. Fruits obovoid or oblong, somewhat fleshy. Seeds almost circular. – *Maihuenia.*

14.3 OPUNTIOIDEAE

Leaves and glochids present. Seeds covered by pale, bony aril or winged, exhibiting particularly specialized and unique type. This verifies the isolated position of the subfamily, also indicated by its pollen morphology (Behnke and Barthlott 1983). – *Pereskiopsis, Quiabentia, Tacinga, Opuntia, Pterocactus.*

14.4 CACTOIDEAE

Leaves absent or very minute, glochids absent. Seeds black or brown, not enveloped by a bony aril. The Cactoideae display a great and complex diversity of seed coat characters that reflects their advanced position within the family. – ECHINOCEREEAE: *Leptocereus, Harrisia, Acanthocereus, Peniocereus, Echinocereus;* HYLOCEREEAE: *Weberocereus, Hylocereus, Selenicereus, Epiphyllum, Disocactus, Pseudorhipsalis*; CEREEAE: *Cereus, Cipocereus, Brasilicereus, Pilosocereus, Stephanocereus, Arrojadoa, Micranthocereus, Coleocephalocereus, Melocactus*; TRICHOCEREEAE: *Cleistocactus, Leocereus, Haageocereus, Brachycereus,*

Espostoa, Espostoopsis, Facheiroa, Samaipaticereus, Cleistocactus, Denmoza, Oreocereus, Matucana, Oroya, Echinopsis, Arthrocereus, Rebutia, Mila, Gymnocalycium, Discocactus; NOTOCACTEAE: *Corryocactus, Austrocactus, Eulychnia, Copiapoa, Neowerdermannia, Eriosyce, Neoporteria, Parodia, Blossfeldia, Frailea, Uebelmannia*; RHIPSALIDEAE: *Lepismium, Rhipsalis, Hatiora, Schlumbergera*; BROWNINGIEAE: *Calymmanthium, Armatocereus, Jasminocereus, Neoraimondia, Browningia, Stetsonia*; PACHYCEREEAE: *Bergerocactus, Pachycereus, Carnegiea, Neobuxbaumia, Cephalocereus, Stenocereus, Rathbunia, Polaskia, Escontria, Myrtillocactus*; CACTEAE: *Echinocactus, Geohintonia, Astrophytum, Sclerocactus, Pediocactus, Thelocactus, Coryphantha, Neolloydia, Mammilloydia, Epithelantha, Ariocarpus, Lophophora, Obergonia, Strombocactus, Aztekium, Pelecyphora, Ferocactus, Leuchtenbergia, Stenocactus, Escobaria, Ortegocactus, Mammillaria*.

Morphological studies over the last few decades (and likewise chemotaxonomic data) have confirmed the close affinity of the Cactaceae with the Phytolaccaceae, Portulacaceae, Didiereaceae, and Aizoaceae. As early as 1862 Bentham and Hooker included Cactaceae together with the Ficoideae (Aizoaceae) in the order Ficoidales and thus correctly showed their affinity. In Hauler's and Wettstein's systems the Cactaceae stand between Aizoaceae and Portulacaceae. Numerous morphological data, accumulated during the last decades, fully confirm the closeness of Cactaceae to the Phytolaccaceae, Aizoaceae, and Portulacaceae (see Chorinsky 1931; Mauritzon 1934; Martin 1946; Buxbaum 1962; Erdtman 1952; Barthlott and Hunt 1993). The relationship of the Cactaceae with the other members of Caryophyllales is supported also by phytochemical data (Reznik 1955; Hegnauer 1964, 1989). The subfamily Pereskioideae is in some respects an intermediate group between the apocarpous ancestors and more advanced subfamilies of the Cactaceae.

15. DIDIEREACEAE

Radlkofer 1896 (including Portulacariaceae Doweld 2001). 7/16. South Africa (*Portulacaria* and *Ceraria*), trop. Africa (*Calyptrotheca*), southern and south western Madagascar.

15.1 DIDIEROIDEAE

Very spiny xerophytic shrubs or trees. Leaves alternate, sometimes much reduced. Ovary (2)3–4-locular, but usually only one locule is fertile. Fruit dry, indehiscent. – *Decaryia, Alluaudiopsis, Alluaudia, Didierea*.

15.2 PORTULACARIOIDEAE

Much-branched glabrous shrubs with waxy bark. Leaves opposite, succulent. Flowers in cymose clusters. Sepals persistent. Ovule solitary. Fruit a thin walled nut with a single seed. – *Portulacaria, Ceraria*.

15.3 CALYPTROTHECOIDEAE

Shrubs or small trees, sometimes climbing. Leaves alternate, succulent with crinkled margins. Flowers in panicles of cymes. Stamens up to 60. Ovules six. Capsule circumscissile at the base, splitting upwards into six valves. Seed solitary, arillate. – *Calyptrotheca*.

Have a common origin with the Cactaceae and Portulacaceae and stand close to them. As early as 1908 and 1912 Hallier included Didiereaceae in the Caryophyllales and put them near the Portulacaceae s. l. as their possible derivatives. According to Erdtman (1948, 1952) pollen grains of the Didiereaceae reveal many common features with the Nyctaginaceae (*Phaeoptilum* and others) and stratification of the exine resembles some pollen types of Cactaceae. Close affinity with the Cactaceae is confirmed by the resemblance in the structure of vegetative organs of Didiereaceae and primitive members of the Cactaceae as well as by the successful grafting of stems of Didiereaceae to *Pereskiopsis* and *Trichocereus* (Rauh and Reznik 1961). At the same time Didiereaceae have much in common in floral morphology with the Portulacaceae (see Scholch 1963) and their epicuticular wax ultrastructure (irregular platelets and curled ribbons) looks much the same as that of Aizoaceae and Cactaceae (Engel and Barthlott 1988; Barthlott in Behnke and Mabry 1994). In addition, embryo development of the Didiereaceae is of the *Chenopodium*-type, confirming their placement in the Caryophyllales (Scholch 1963). The formation of betalains instead of anthocyanins (Reznik 1955, 1975) and sieve-element plastids with globular protein crystal (Behnke 1976b, 1978, 1982) as well as serological studies (Jensen 1965) also place the Didiereaceae in the order Caryophyllales.

16. MOLLUGINACEAE

Bartling 1825 (including Adenogrammaceae Nakai 1942, Glinaceae Dumortier 1829, Pharnaceaceae Martynov 1820, Polpodaceae Nakai 1942, Telephiaceae Martynov 1820). 12/120. Mainly tropical and subtropical regions, especially in Africa.

Corbichonia, Macarthuria, Psammotropha, Adenogramma, Glischrothamnus, Mollugo, Glinus, Hypertelis, Pharnaceum, Suessenguthiella, Coelanthum, Polpoda.

Related to the Phytolaccaceae, Aizoaceae, and especially to the Sesuviaceae, differing mainly in the absence of betalains. Also, they differ from the Aizoaceae and Sesuviaceae in the presence of flavonols and of C-glycosyl-flavonoids (viteksin and isoviteksin) (Richardson 1978, 1981). The plastids of *Glinus* include some starch grains (Behnke 1976b; Behnke et al. 1983a, b), which is rare in the order. In all other Molluginaceae the plastids have globular protein crystals.

According to Endress and Bittrich (1993), the inclusion of the poorly known South African genus *Polpoda* in the Molluginaceae is dubious. The stylodia are connate at the base, a condition not found in the Molluginaceae (Hofmann in Behnke and Mabry 1994).

17. LIMEACEAE

Shipunov ex Reveal 2005. 1/21. Southern Africa, to Ethiopia, S. Asia.

Limeum.

Related to Molluginaceae, but differs in a lacking of pigments and type of sieve-element plastids (Behnke et al. 1983).

18. CARYOPHYLLACEAE

A.L. de Jussieu 1789 (including Alsinaceae Bartling 1825, Cerastiaceae Vest 1818, Corrigiolaceae Dumortier 1829, Dianthaceae Vest 1818, Herniariaceae Martynov 1820, Illecebraceae R. Brown 1810, Lychnidaceae Löll 1843, Ortegaceae Martynov 1820, Paronychiaceae A. L de Jussieu 1815, Polycarpaeaceae Martens 1835, Scleranthaceae J. Presl et C. Presl 1820, Silenaceae Bartling 1825, Spergulaceae Bartling 1825, Stellariaceae Berchtold et J. Presl 1820, Telephiaceae Martynov 1820). 104/2400. Mainly temperate regions of the Northern Hemisphere with some species in the Arctic, in the temperate regions of the Southern Hemisphere, and on mountains in the tropics. The center of distribution is in the Mediterranean region and adjoining parts of Europe, and in western and Central Asia. All the larger genera are found in the Northern Hemisphere.

18.1 ILLECEBRIOIDEAE

Arnot 1832 (PARONYCHIOIDEAE A. Saint-Hilaire 1839). Stipules almost always present, mostly scarious. Leaves alternate. Sepals free or slightly connate. Petals mostly inconspicuous or absent. Pollen grains 3-colpate (rarely 3-colporate in *Polycarpaea* spp.) to pantocolpate and pantoporate. – SPERGULEAE: *Spergula, Spergularia;* TELEPHIEAE: *Telephium;* CORRIGIOLEAE: *Corrigiola;* XEROTIEAE: *Xerotia;* POLYCARPAEAE: *Drymaria, Sanctambrosia, Polycarpon, Polycarpaea, Ortegia, Loeflingia, Haya, Krauseola, Polytepalum, Stipulicida, Cerdia, Microphyes, Pirinia*; PARONYCHIEAE: *Cometes, Dicheranthus, Pteranthus, Sphaerocoma, Sclerocephalus, Lochia, Paronychia, Herniaria, Philippiella, Chaetonychia, Achyronychia, Illecebrum, Cardionema, Scopulophila, Pollichia.*

18.2 MINUARTIOIDEAE (ALSINOIDEAE)

Stipules absent. Leaves opposite. Sepals free or rarely connate at the base. Petals present or rarely absent. Fruits mainly capsules, rarely indehiscent. – ALSINEAE: *Arenaria, Dichodon, Eremogone, Thylacospermum, Moehringia, Brachystemma, Thurya, Bufonia, Lepyrodiclis, Cerastium, Stellaria, Mesostemma, Tytthostemma, Fimbripetallum, Pseudostellaria, Moenchia, Holosteum, Myosoton, Minuartia, Queria, Wilhelmsia, Honkenya, Sagina, Colobanthus, Alsinidendron, Schiedea, Reicheella, Plettkea, Pycnophyllopsis*; PYCNOPHYLLEAE: *Pycnophyllum;* SCLERANTHEAE: *Scleranthus, Pentastemonodiscus*; SGEOCARPEAE: *Geocarpon*; HABROSIEAE: *Habrosia.*

18.3 CARYOPHYLLOIDEAE

Stipules absent. Sepals connate. Petals present, often clawed. – CARYOPHYLLEAE: *Acanthophyllum, Kuhitangia, Allochrusa, Scleranthopsis, Ochotonophila, Diaphanoptera, Gypsophila, Psammophiliella, Dichoglottis, Vaccaria, Pleioneura, Saponaria, Spanizium, Bolbosaponaria, Pseudosaponaria,*

Cyathophylla, Bolanthus, Ankyropetalum, Phrynella, Dianthus, Kohlrauschia, Petrorhagia, Velezia; DRYPIDEAE: *Drypis;* SILENEAE: *Silene, Melandrium, Pleconax, Ixoca, Oberna, Petrocoma, Lychnis, Steris, Gastrolychnis, Petrocoptis, Cucubalus, Agrostemma, Uebelinia.*

The Caryophyllaceae are closely related to the Molluginaceae and probably have a common origin with them. They are probably an early branch from the protophytolaccaceous ancestor, which had no betalain yet. A special variety of P-type sieve-element plastids with polygonal central protein crystal is characteristic of the Caryophyllaceae as well as of *Limeum* and *Macarthuria* in the Molluginaceae (Behnke 1969, 1975b, c, 1976b, c, 1993). Besides, the micromorphology of epicuticular waxes in the Caryophyllaceae is very much similar to that of Molluginaceae by the frequent occurrence of irregular platelets and rodlets (Engel and Barthlott 1988; Barthlott in Behnke and Mabry 1994). According to Eichler (1875) the Caryophyllaceae-type of flower originated from the *Phytolacca*-type as a result of transformation of the outer whorl of stamens into petals.

19. SIMMONDSIACEAE

van Tieghem 1899. 1/1. Southern California, western Arizona, northern Baja California.

Simmondsia.

Simmondsiaceae are similar to Caryophyllales in having normal secondary growth, wood anatomy, pollen morphology, and seeds without endosperm.

20. AMARANTHACEAE

A. L. de Jussieu 1789 (including Achyranthaceae Rafinesque 1837, Celosiaceae Martynov 1820, Deeringiaceae J. Agardh 1858, Gomphrenaceae Rafinesque 1837, Polycnemaceae Menge 1839). 69/1000 or more. Widely distributed in tropical, subtropical and warm-temperate regions. Concentrated in Africa south of the Sahara, southwestern North America, and Central and South America.

20.1 POLYCNEMOIDEAE

Embryo annular. Stem with normal secondary growth. Flowers bisexual, solitary, with bracteoles. — *Polycnemum, Hemichroa, Nitrophila.*

20.2 AMARANTHOIDEAE.

Anthers tetrasporangiate and 2-locular. Ovules several or solitary on a basal or short, free-central placenta. – CELOSIEAE: *Deeringia* (including *Dendroportulaca*), *Pleuropetalum, Henonia, Celosia, Hermbstaedtia;* AMARANTHEAE: *Bosea, Chamissoa, Herbstia, Siamosia, Allmania, Charpentiera, Indobanalia, Lagrezia, Amaranthus, Digera, Neocentema, Pleuropterantha, Saltia, Sericostachys, Sericocomopsis, Sericocoma, Pseudosericocoma, Kyphocarpa, Centemopsis, Nelsia, Sericorema, Centema, Eriostylos, Lopriorea, Rosifax, Leucosphaera, Cyathula, Allmaniopsis, Pupalia, Marcelliopsis, Daysphaera, Volkensinia, Arthraerua, Aerva, Polyrhabda, Trichuriella, Nothosaerva, Nototrichium, Calicorema, Chionothrix, Stilbanthus, Mechowia, Nyssanthes, Ptilotus, Psilotrichum, Psilotrichopsis, Achyranthes, Centrostachys, Achyropsis, Pandiaka.*

20.3 GOMPHRENOIDEAE

Anthers disporangiate and 1-locular. Ovules usually solitary and basal, rarely apical and pendulous. – PSEUDOPLANTAGEAE: *Pseudoplantago*; GOMPHRENEAE: *Guilleminea, Tidestromia, Froelichia, Froelichiella, Pfaffia, Alternanthera, Woehleria, Gomphrena, Pseudogomphrena, Iresine, Orenella, Blutaparon, Lithophila.*

The Amaranthaceae reveal some links with the Petiveriaceae and Achatocarpaceae. In the relatively most archaic members of the family, such as *Deeringia* and *Pleuropetalum*, the gynoecium is even more primitive than in Achatocarpaceae and Petiveriaceae.

21. CHENOPODIACEAE

Ventenat 1799 (including Atriplicaceae A.L. de Jussieu 1789, Betaceae Burnett 1835, Blitaceae Post et Kuntze 1903, Dysphaniaceae Pax 1927, Salicorniaceae Martynov 1820, Salsolaceae Menge 1839, Spinaciaceae Menge 1839). 103/1600. Widely distributed in temperate and subtropical regions, mainly in saline habitats; particularly well represented around the Mediterranean and Red Sea, in deserts and semideserts of Irano-Turanian region, in the edge of the Sahara, and in the Karoo of South Africa, the deserts of Australia, and the alkaline areas of the United States of America and Argentina.

21.1 CHENOPODIOIDEAE

Embryo annular or less often curved. Leaves well developed, mostly broad, sometimes narrow and fleshy or scalelike. Fruits surrounded by the persistent perianth or by bracteoles. Perisperm usually present. – BETEAE: *Acroglochin, Hablitzia, Oreobllton, Aphanisma, Beta;* CHENOPODIEAE: *Rhagodia, Holmbergia, Baolia, Chenopodium, Teloxys, Scleroblitum, Dysphania, Monolepis*; ATRIPLICEAE: *Axyris, Endolepis, Atriplex, Exomis, Archiatriplex, Microgynoecium, Suckleya, Krascheninnikovia, Grayia, Ceratocarpus, Spinacia, Zuckia*; CAMPHOROSMEAE: *Kirilowia, Panderia, Camphorosma, Cycloloma, Chenolea, Bassia, Londesia;* SCLEROLAENEAE: *Maireana, Eremophea, Enchylaena, Didymanthus, Neobassia, Malacocera, Dissocarpus, Babbagia, Threlkeldia, Sclerolaena, Roycea*; CORISPERMEAE: *Corispermum, Agriophyllum, Anthochlamys.*

21.2 MICROTEOIDEAE

Embryo annular. Annual herbs, sometimes suffrutescent at the base. Flowers in spike or spike-like racemes, bisexual, actinomorphic. Anthers globose, pollen grains pantoporate. Style absent or nearly so. Fruit an achene, muricate to spiny. – *Microtea.*

21.3 SALICORNIOIDEAE

Embryo annular or curved, rarely straight. Perisperm present. Stems succulent, articulated, with leaves reduced to tubercles or scales. Flowers usually in groups of three sunken into cavities in the axis of spiciform inflorescences. – HALOPEPLIDEAE: *Halopeplis, Kalidium;* SALICORNIEAE: *Allenrolfea, Halocnemum, Halostachys, Heterostachys, Pachycornia, Sclerostegia, Arthrocnemum, Microcnemum, Halosarcia, Tecticornia, Tegicornia, Sarcocornia, Salicornia.*

21.4 SALSOLOIDEAE

Embryo spirally coiled. Perisperm usually lacking. Leaves mostly linear to filiform, terete, sometimes scalelike. Flowers usually 1–3, in the axils of the bracts, usually bracteolate. – SUAEDEAE: *Suaeda, Alexandra, Bienertia, Borsczowia;* SALSOLEAE: *Haloxylon, Sympegma, Noaea, Salsola, Climacoptera, Halothamnus, Cyathobasis, Anabasis, Halogeton, Micropeplis, Seldlitzia, Arthrophytum, Iljinia, Nanophyton, Girgensohnia, Halocharis, Halanthium, Fadenia, Sevada, Choriptera, Rhaphidophyton, Ofaiston, Cornulaca, Traganum, Traganopsis, Nucularia,Horaninowia, Piptoptera, Gamanthus, Halarchon, Petrosimonia, Halimocnemis, Halotis, Lagenantha.*

The Chenopodiaceae stand very close to the Amaranthaceae with which they have a common origin.

The American genus *Microtea,* which is usually included in the Phytolaccaceae s. 1., has many features in common with the Chenopodiaceae and often has been considered a link between these two families. According to Nowicke (Nowicke in Behnke and Mabry 1994:215), the transfer of *Microtea* to either Amaranthaceae or Chenopodiaceae is supported, for the most part, by pollen data. The grains are pantoporate, have a thin endexine, and a thin foot layer but thick tecta, characters common to the Amaranthaceae and Chenopodiaceae. Moreover, the pollen of *Microtea* lacks the unusual endexine (fibrous and netlike under the aperture) found in many Phytolaccaceae. Recently Behnke (1993) has shown that the sieve-element plastids of *Microtea* are characterized by a very broad peripheral ring of protein filaments and a single tiny starch grain. The average diameter of the sieve-element plastids of *Microtea* also corresponds to that of the Chenopodiaceae and negates closer relationships to the Phytolaccaceae.

Bibliography

Aleshina LA. 1963. Morphology of pollen grains of the genus Claytonia Cronov. and related genera. Bot. Zhurn. 48: 1191–1196 (in Russian).

Al-Turku TA, K Swarupanandan and PG Wilson. 2003. Primary vasculature in Chenopodiaceae: a re-interpretation and implications for systematics and evolution. Bot. J. Linn. Soc. 143: 337–374.

Anderson EF. 2001. The Cactus Family. Timber Press, Portland.

Applequist WL and DB Pratt. 2005. The Malagasy endemic *Dendroportulaca* (Portulacaceae) is referable to *Deeringia* (Amaranthaceae): molecular and morphological evidence. Taxon 54: 681–687.

Applequist WL, WL Wagner, EA Zimmer, and M Nepokroeff. 2006. Molecular evidence resolving the systematic position of *Hectorella* (Portulacaceae). Syst. Bot. 31: 310–319.

Applequist WL and RS Wallace. 2000. Phylogeny of the Madagascarian endemic family Didiereaceae. Plant Syst. Evol. 221: 157–166.

Applequist WL and RS Wallace. 2001. Phylogeny of the Portulacaceous cohort based on *ndhF* sequence data. Syst. Bot. 26: 406–419.

Applequist WL and RS Wallace. 2002. Deletions in the plastid *trn*T-*trn*L intergenic spacer define clades within Cactaceae subfamily Cactoideae. Plant Syst. Evol. 231(1–4): 153–162.

Applequist WL and RS Wallace. 2003. Expanded circumscription of Didiereaceae and its division into three subfamilies. Adansonia, sér. 3, 25: 13–16.

Bailey DC. 1980. Anomalous growth and vegetative anatomy of *Simmondsia chinensis*. Am. J. Bot. 67: 147–161.

Balfour E. 1965. Anomalous secondary thickening in Chenopodiaceae, Nyctaginaceae, and Amaranthaceae. Phytomorphology 15: 111–122.

Baranova M. 1980. Comparative-stomatographical study of the families Buxaceae and Simmondsiaceae. In: SG Zhilin, ed. Systematics and evolution of the higher plants, pp. 68–75. Nauka, Leningrad (in Russian).

Barthlott W. 1988. Über die systematischen Gliederungen der Cactaceae. Beitr. Biol. Pfl. 63: 17–40.

Barthlott W. 1994. Epicuticular wax ultrastructure and systematics. In: H-D Behnke and TJ Mabry, eds. Caryophyllales. Evolution and systematics, pp.75–86. Springer, Berlin

Barthlott W and DR Hunt. 1993. Cactaceae. In: K Kubitzki, ed. The families and genera of vascular plants, vol. 2, pp. 161–197. Springer, Berlin/Heidelberg/New York.

Barthlott W and DR Hunt. 2000. Seed-diversity in the Cactaceae, subfamily Cactoideae. Succ. Plant Res. 5: 1–173

Barthlott W and G Voit. 1979. Mikromorphologie der Samenschalen und Taxonomie der Cactaceae. Plant Syst. Evol. 132:205–229.

Bedell HG. 1980. A taxonomic and morphological reevaluation of Stegnospermataceae (Caryophyllales). Syst. Bot. 5: 419–431.

Behnke H-D. 1969. Über Siebröhren-Plastiden und Plastidenfilamente der Caryophyllales. Untersuchungen zum Feinbau und zur Verbreitun eines weiteren spezifischen Plastidentyps. Planta 89: 275–283.

Behnke H-D. 1974. Elektronenmikroskopische Unter-suchungen an Siebrohren-Plastiden und ihre Aussage über die systematische Stellung von *Lophiocarpus*. Bot. Jahrb. Syst. 94:114–119.

Behnke H-D. 1975. *Hectorella caespitosa:* Ultrastructural evidence against its inclusion into Caryophyllaceae. Plant Syst. Evol. 124: 31–34.

Behnke H-D. 1976a. A tabulated survey of some characters of systematic importance in centrospermous families. Plant Syst. Evol. 125: 95–98.

Behnke H-D. 1976b. Ultrastructure of sieve-element plastids in Caryophyllales (Centrospermae), evidence for the delimitation and classification of the order. Plant Syst. Evol. 126: 31–54.

Behnke H-D. 1976c. Sieve element plastids of *Fouquieria, Frankenia* (Tamaricales), and *Rhabdodendron* (Rutaceae), taxa sometimes allied with Centrospermae (Caryophylales). Taxon 25: 265–268.

Behnke H-D. 1976d. Die Siebelement-Plastiden der Caryophyllaceae: Eine weitere spezifische Form der P-Typ-Plastiden bei Centrospermen. Bot. Jahrb. Syst. 95: 327–333.

Behnke H-D. 1977a. Regular occurring massive deposits of phytoferritin in the phloem of succulent Centrospermae. Z. Pflanzenphysiol. 85: 89–92.

Behnke H-D. 1977b. Zur Skulptur der Pollen-Exine bei drei Centrospermen (*Gisekia, Limeum, Hectorella*) bei Gyrostemonaceen, und Rhabdodendraceen. Plant Syst. Evol. 128: 227–235.

Behnke H-D. 1977c. Transmission electron microscopy and systematics of flowering plants. Plant Syst. Evol., Suppl. 1: 155–178.

Behnke H-D. 1978. Elektronenoptische Untersuchungen am Phloem sukkulenter Centrospermen (incl. Didiereaceen). Bot. Jahrb. Syst. 99: 341–352.

Behnke H-D. 1981. Sieve-element characters. Nord. J. Bot. 1: 381–400.

Behnke H-D. 1982. *Geocarpon minimum:* Sieve-element plastids as additional evidence for its inclusion in the Caryophyllaceae. Taxon 31: 45–47.

Behnke H-D. 1993. Further studies of the sieve-element plastids of the Caryophyllales including *Barbeuia, Corrigiola, Lyallia, Microtea, Sarcobatus,* and *Telephium*. Plant Syst. Evol. 186: 231–243.

Behnke H-D. 1994. Sieve-element plastids: their significance for the evolution and systematics of the order. In: H-D Behnke and TJ Mabry, eds. Evolution and systematics of the Caryophyllales, pp. 87–121. Springer, Berlin/New York.

Behnke H-D. 1997. Sarcobataceae — a new family of Caryophyllales. Taxon 46: 495–507.

Behnke H-D and W Barthlott. 1983. New evidence from ultrastructural and micromorphological fields in angiosperm classification. Nord. J. Bot. 3: 43–66.

Behnke H-D and BL Turner. 1971. On specific sieve-tube plastids in Caryophyllales: Further investigation with special reference to the Bataceae. Taxon 20: 731–737.

Behnke H-D, C Chang, IJ Eifert, and TJ Mabry. 1974. Betalains and P-type sieve-tube plastids in *Petiveria* and *Agdestis* (Phytolaccaceae). Taxon 23: 541–542.

Behnke H-D, TJ Mabry, IJ Eifert, and L Pop. 1975. P-type sieve-element plastids and betalains in Portulacaceae (including *Ceraria, Portulacaria, Talinella*). Canad. J. Bot. 53: 2103–2109.

Behnke H-D, TJ Mabry, P Neuman, and W Barthlott. 1983a. Ultrastructural, micromorphological, and phytochemical evidence for a "central position" of *Macarthuria* (Molluginaceae) within the Caryophyllales. Plant Syst. Evol. 143: 151–161.

Behnke H-D, L Pop, and VV Sivarajan. 1983b. Sieve-element plastids of Caryophyllales: Additional investigations with special reference to the Caryophyl-laceae and Molluginaceae. Plant Syst. Evol. 142: 109–115.

Bisalputra T. 1960. Anatomical and morphological studies in the Chenopodiaceae: I. Inflorescence of *Atriplex* and *Bassia*. Austral. J. Bot. 8: 226–242.

Bisalputra T. 1961. Anatomical and morphological studies in the Chenopodiaceae: II. Vascularization of the seedlings. Austral. J. Bot. 9: 1–19.

Bisalputra T. 1962. Anatomical and morphological studies in the Chenopodiaceae: III. The primary vascular system and nodal anatomy. Austral. J. Bot. 10: 13–24.

Bittrich V. 1990. Systematic studies in Aizoaceae. Mitt. Inst. Allg. Bot. Hamb. 23: 491–507.

Bittrich V. 1993. Achatocarpaceae, Caryophyllaceae, Halophytaceae. In: K Kubitzki, ed. The families and genera of vascular plants, vol. 2, pp. 35–36, 206–236, 320–321. Springer, Berlin/Heidelberg/New York.

Bittrich V and HEK Hartmann. 1988. The Aizoaceae: A new approach. Bot. J. Linn. Soc. 97: 239–254.

Bittrich V and U Kühn. 1993. Nyctaginaceae. In: K Kubitzki, ed. The families and genera of vascular plants, vol. 2, pp. 473–486. Springer, Berlin/Heidelberg/New York.

Bittrich V and M Struck. 1989. What is primitive in Mesembryanthemaceae? An analysis of evolutionary polarity of character states. South Afr. J. Bot. 55: 321–331.

Blackwell WH. 1977. The subfamilies of the Chenopodiaceae. Taxon 26: 395–397.

Blunden G, MH Yang, G Janicsak, I Mathe, and A Carabot-Guervo. 1999. Betaine distribution in the Amaranthaceae. Biochem. Syst. Evol. 27: 87–92.

Bocquet G. 1959 (1960). The structure of the placental column in the genus *Melandrium* (Caryophyllaceae). Phytomorphology 9: 217–221.

Bogle AL. 1969. The genera of Portulacaceae and Ba-sellaceae in the southeastern United States. J. Arnold Arbor. 50: 566–598.

Bogle AL. 1970. The genera of Molluginaceae and Aizoaceae in the southeastern United States. J. Arnold Arbor. 51: 431–462.

Bogle AL. 1974. The genera of Nyctaginaceae in the southeastern United States. J. Arnold Arbor. 55: 1–37.

Bogle AL et al. 1971. *Geocarpon:* Aizoaceae or Caryophyllaceae? Taxon 20: 473—477.

Boke NH. 1963. Anatomy and development of the flower and fruit of *Pereskia pititache*. Am. J. Bot. 50: 843–858.

Boke NH. 1964. The cactus gynoecium: A new interpretation. Am. J. Bot. 51: 598–610.

Boke NH. 1966. Ontogeny and structure of the flower and fruit of *Pereskia aculeata*. Am. J. Bot. 53: 534–542.

Boke NH. 1980. Developmental morphology and anatomy in Cactaceae. Bioscience 30: 605–610.

Borsch T. 1998. Pollen types in the Amaranthaceae: morphology and evolutionary significance. Grana 37(3): 129–142.

Bortenschlager S. 1973. Morphologic pollinique des Phytolaccaceae. Pollen Spores 15: 227–253.

Bortenschlager S, A Auimger, J Blaha, and P Simons-burger. 1972.PollenmorphologyofAchatocarpaceae(Centrospermae). Bot. Naturwiss.-Med. Vereins Innsbruck 59: 7–13.

Brown GK and GS Varadarajan. 1985. Studies in Caryophyllales: I. Reevaluation of classification of Phytolaccaceae s. 1. Syst. Bot. 10: 49–63.

Burret F, Z Rabesa, P Zandonella, and B Viorin. 1981. Contribution biochimique a la systemadque de 1'ordre des Centrospermales. Biochem. Syst. Ecol. 9: 257–262.

Butnik AA. 1991. Chenopodiaceae. In: A Takhtajan, ed. Comparative seed anatomy, vol. 3, pp. 77–82. Nauka, Leningrad (in Russian).

Butnik AA and UN Zhapakova. 1991. Amaranthaceae. In: A Takhtajan, ed. Comparative seed anatomy, vol. 3, pp. 74–77. Nauka, Leningrad (in Russian).

Butterworth CA. 2006. Molecular phylogenetics of Cactaceae Jussieu – a review. In: AK Sharma and A Sharma, eds. Plant Genome Biodiversity and Evolution, vol. 1, pt C: 489–524. Science Publishers, Enfield, NH.

Butterworth CA and RS Wallace. 2005. Molecular phylogenetics of the leafy cactus genus *Pereskia* (Cactaceae). Syst. Bot. 30: 800–808.

Buxbaum F. 1944. Untersuchungen zur Morphologic der Kakteenblüte: I. Das Gynoeceum. Bot. Arch. 45:190–247.

Buxbaum F. 1948. Zur Klärung der phylogenetischen Stellung der Aizoaceae und Cactaceae im Pflanzenreich. Jahrb. Schweiz. Kakt. Ges. 2: 3–16.

Buxbaum F. 1949. Vorläufer des Kakteen-Habitus bei den Phytolaccaceen. Oesterr. Bot. Z. 96: 5–14.

Buxbaum F. 1950, 1954, 1955. Morphology of cacti., vols. 1–3. Abbey Garden Press, Pasadena, CA.

Buxbaum F. 1958. The phylogenetic division of the subfamily Cereoideae, Cactaceae. Madroño 14: 177–206.

Buxbaum F. 1961. Vorläufige Untersuchungen über Um-fang, systematische Stellung, und Gliederung der Caryophyllales (Centrospermae). Beitr. Biol. Pfl. 36: 1–56.

Buxbaum F. 1962. Stellung der Kakteen im Pflanzenreich. Kakt. und Sukk. 13: 194–197.

Buchmann SL. 1987. Floral biology of jojoba (*Simmondsia chinensis*), an anemophilous plant. Desert Plants 8: 111–124.

Carlquist S. 1995. Wood anatomy of Caryophyllaceae: ecological, habitual, systematic, and phylogenetic implications. Aliso 14: 1–17.

Carlquist S. 1998a. Wood anatomy of Portulacaceae and Hectorellaceae: ecological, habitual, and systematic implications. Aliso 16: 137–153.

Carlquist S. 1998b. Wood and stem anatomy of *Petiveria* and *Rivina* (Caryophyllales): systematic implications. IAWA J. 19: 383–391.

Carlquist S. 1999a. Wood, stem, and root anatomy of Basellaceae, with a relation to systematics and cambial variants. Flora 194: 1–12.

Carlquist S. 1999b. Wood anatomy of *Agdestis* (Caryophyllales): systematic position and nature of the successive cambia. Aliso 18: 35–43.

Carlquist S. 1999c. Wood and stem anatomy of *Stegnosperma* (Caryophyllales): phylogenetic relationships; nature of lateral meristems and successive cambial activity. IAWA J. 20: 149–163.

Carlquist S. 1999d. Wood anatomy, stem anatomy, and cambial activity of *Barbeuia* (Caryophyllales). IAWA J. 20: 431–440.

Carlquist S. 2000a. Wood and stem anatomy of *Sarcobatus* (Caryophyllales): Systematic and ecological implications. Taxon 49: 27–34.

Carlquist S. 2000b. Wood and stem anatomy of phytolaccoid and rivinoid Phytolaccaceae (Caryophyllales): ecology, systematics, nature of successive cambia. Aliso 19: 13–29.

Carlquist S. 2000c. Wood and bark anatomy of Achatocarpaceae. SIDA 19: 71–78.

Carlquist S. 2001. Wood and stem anatomy of Rhabdodendraceae is consistent with placement in Caryophyllales sensu lato. IAWA J. 22: 171–18.

Carlquist S. 2002. Wood anatomy and successive cambia in *Simmondsia* (Simmondsiaceae): Evidence for inclusion in Caryophyllales s.l. Madroño 49: 158–164.

Carlquist S. 2003. Wood and stem anatomy of woody Amaranthaceae s.s.: ecology, systematics and the problems of defining rays in dicotyledons. Bot. J. Linn. Soc. 143: 1–19.

Carlquist S. 2005. Lateral meristems, successive cambia and their products: a reinterpretation based on roots and stems of Nyctaginaceae. Bot. J. Linn. Soc. 146: 129–143.

Carlquist S. 2007. Successive cambia in Aizoaceae: products and process. Bot. J. Linn. Soc. 153: 141–155.

Carolin RC. 1983. The trichomes of the Chenopodiaceae and Amaranthaceae. Bot. Jahrb. Syst. 103: 451–466.

Carolin RC. 1987. A review of the family Portulacaceae. Austral. J. Bot. 35: 383–412.

Carolin RC. 1993. Portulacaceae. In: K Kubitzki, ed. The families and genera of vascular plants, vol. 2, pp. 544–555. Springer, Berlin/Heidelberg/New York.

Carolin RC, SWL Jacobs, and M Vesk. 1975. Leaf structure in Chenopodiaceae. Bot. Jahrb. Syst. 95: 226–255.

Chesselet P, Mössmer, and GF Smith. 1995. Research priorities in the succulent plant family Mesembryanthaceae Fenzl. South Afr. J. Sci. 91: 192–209.

Chesselet P, GF Smith, PM Burgoyne, C Klak, SA Hammer, HEK Hartmann, H Kurzweil, EJ van Jaarsveld, BE van Wyk and OA Leistner. 2000. Mesembryanthemaceae. In: Leistner OA eds. Seed plants of South Africa: families and genera. Srelitzia 10: 360–410.

Chesselet P, GF Smith and AE van Wyk. 2002. A new tribal classification of Mesebryanthemaceae: evidence from floral nectaries. Taxon 51: 295–308.

Chesselet P, AE van Wyk, and GF Smith. 2004. Mesembryanthemaceae. A new tribe and adjustments to infrafamilial classification. Bothalia 31: 47–51.

Chorinsky F. 1931. Vergleichende morphologische Untersuchungen der Haargebilde bei Portulacaceae and Cactaceae. Osterr. Bot. Z. 80: 308–327.

Choux P. 1934. Les Didiereacees: Xerophytes de Madagascar. Mem. Acad. Malgache 18: 1–69.

Chuong SDX, VR Franceschi, and GE Edwards. 2006. The cytoskeleton maintains organelle partitioning required for single cell C_4 photosynthesis in Chenopodiaceae species. Plant Cell 18: 2207–2223.

Clement JS and TJ Mabry. 1996. Pigment evolution in the Caryophyllales: a systematic overview. Bot. Acta 109: 360–367.

Costea M and DA DeMason. 2001. Stem morphology and anatomy in *Amaranthus* L. (Amaranthaceae) – taxonomic significance. J. Torr. Bot. Soc. 128: 254–281.

Cranwell LM. 1963. The Hectorellaceae: Pollen type and taxonomic speculation. Grana Palynol. 4: 195–202.

Cronquist A and RE Thorne. 1994. Nomenclatural and taxonomic history. In: H-D Behnke and TJ Mabry, eds. Caryophyllales. Evolution and systematics, pp. 5–25. Springer, Berlin/Heidelberg.

Cuénoud Ph. 2002. Introduction to expanded Caryophyllales. In: K Kubitzki, ed. The Families and genera of vascular plants, vol. 5, pp. 1–4. Springer, Berlin/Heidelberg/New York.

Cuénoud P. 2006. Phylogeny, evolution and diversification of Caryophyllales. In: AK Sharma and A Sharma, eds. Plant genome biodiversity and evolution, vol. 1, pp. 187–218. Science Publishers, Enfield, NH.

Cuénoud Ph, V Savolainen, LW Chatrou, M Powell, RJ Grayer, and M Chase. W. 2002. Molecular phylogenetics of Caryophyllales based on nuclear 18S *r*DNA and plastid *rbc*L, *atp*B, and *mat*K DNA sequences. Am. J. Bot. 89: 132–144.

Daxenbichler M, GF Spencer, DG Carlson, GB Rose, AM Brinker, and RG Powell. 1991. Glucosilonate composition of seeds from 297 species of wild plants. Phytochemistry 30: 2623–2638.

Dickie SL. 1996. Phylogeny and evolution in the subfamily Opuntioideae (Cactaceae): insights from rp116 intron sequence variation. M.S. Thesis. Iowa State University.

Di Fulvio TE. 1975. Estomatogenesis en *Halophytum ameghinoi* (Halophytaceae). Kurtziana 8: 17–29.

Dinan L, P Whiting, and AJ Scott. 1998. Taxonomic distribution of phytoecdysteroids in seed of members of the Chenopodiaceae. Biochem. Syst. Ecol. 26: 553–576.

Douglas NA and PS Manos. 2007. Molecular phylogeny of Nyctaginaceae: taxonomy, biogeography, and characters associated with a radiation of xerophytic genera in North America. Am. J. Bot. 94: 856–872.

Doweld AB. 2001. Conspectus systematicus tribus Cacteae (Cactaceae). Genera et taxa supragenerica, 2. Novosti Sist. Vyssh. Rast. 33: 163–171 (in Russian).

Downie SR and JD Palmer. 1994. A chloroplast DNA phylogeny of the Caryophyllales based on structural and inverted repeat restriction site variation. Syst. Bot. 19: 236–252.

Downie SR, DS Katz-Downie, and KJ Cho. 1997. Relationships in the Caryophyllales as suggested by phylogenetic analyses of partial chloroplast DNA ORF2280 homolog sequences. Am. J. Bot. 84: 253–273.

Eckardt Th. 1954. Morphologische und systematische Auswertung der Placentation von Phytolaccaceen. Ber. Deutsch. Bot. Ges. 67: 113–128.

Eckardt Th. 1967a. Vergleich von *Dysphania* mit *Chenopodium* und mit Illecebraceae. Bauhinia 3: 327–344.

Eckardt Th. 1967b. Blütenbau und Blültenentwicklung von *Dysphania myriocephala* Benth. Bot. Jahrb. Syst. 86: 20–37.

Eckardt Th. 1967c (1968). Blütenmorphologie von *Dysphania planta ginella* F. v. M. Phytomorphology 17: 165–172.

Eckardt Th. 1974. Vom Blütenbau der Centrospermen-Gattung *Lophiocarpus* Turcz. Phyton (Horn) 16: 13–27.

Eckardt Th. 1976. Classical morphological features of centrospermous families. Plant Syst. Evol. 126: 5–25.

Edwards EJ. 2006. Correlated evolution of stem and leaf hydraulic traits in *Pereskia* (Cactaceae). New Phytol. 171.

Edwards EJ and M Diaz. 2006. Ecological physiology of Persekia guamacho, a cactus with leaves. Plant, Cell Environ. 29: 247–256.

Edwards EJ and MJ Donoghue. 2006. *Pereskia* and the origin of the cactus life form. Am. Naturalist 167: 777–793.

Edwards EJ, R Nyffeler, and MJ Donoghue. 2005. Basal cacus phylogeny: implications of *Pereskia* (Cactaceae) paraphyly for the transition to the cactus life form. Am. J. Bot. 92: 1177–1188.

Ehrendorfer F. 1976a. Chromosome numbers and differentiation of centrospermous families. Plant Syst. Evol. 126: 27–30.

Ehrendorfer F. 1976b. Closing remarks: Systematics and evolution of centrospermous families. Plant Syst. Evol. 126: 99–105.

Eliasson UH. 1988. Floral morphology and taxonomic relations among the genera of Amaranthaceae in the New World and the Hawaiian Islands. Bot. J. Linn. Soc. 96: 235–283.

Endress ME and V Bittrich. 1993. Molluginaceae. In: K Kubitzki, ed. The families and genera of vascular plants, vol. 2, pp. 419–426. Springer, Berlin/Heidelberg/New York.

Engel T and W Barthlott. 1988. Micromorphology of epicuticular waxes in Centrosperms. Plant Syst. Evol. 161: 71–85.

Erbar C. 1994. Putative origin and relationships of the order from the viewpoint of developmental flower morphology. In: H-D Behnke, TJ Mabry, eds. Caryophyllales. Evolution and systematics, pp. 3–316. Springer, Berlin/New York.

Erbar C and P Leins. 2006. Floral ontogeny and systematic position of the Didiereaceae. Plant Syst. Evol. 261: 165–185.

Erdtman G. 1948. Pollen morphology and plant taxonomy: VIII. Didiereaceae. Bull. Mus. Hist. Nat., Paris, ser. 2, 20: 387–394.

Esau K and VI Cheadle. 1969. Secondary growth in *Bougainvillea*. Ann. Bot. 33: 807–819.

Fahn A and T Arzee. 1959. Vascularization of articulated Chenopodiaceae and the nature of their fleshy cortex. Am. J. Bot. 46: 330–338.

Fahn A and Y Shchori. 1967 (1968). The organization of the secondary conducting tissues in some species of the Chenopodiaceae. Phytomorphology 17: 147–154.

Fay MF, KM Cameron, GT Prance, MD Lledo, and MW Chase. 1997. Familial relationships of *Rhabdodendron* (Rhabdodendraceae): plastid *rbc*L sequences indicate a caryophyllid placement. Kew Bull. 52: 923–932.

Fedotova TA. 1991. Hectorellaceae, Basellaceae, Halophytaceae. In: A Takhtajan, ed. Comparative seed anatomy, vol. 3, pp. 33–41. Nauka, Leningrad (in Russian).

Fior S, PO Karis, G Casazza, L Minuto, and F Sala. 2006. Molecular phylogeny of the Caryophyllaceae (Caryophyllales) inferred from chloroplast *mat*K and nuclear rDNA ITS sequences. Am. J. Bot. 93: 399–411.

Franz E. 1908. Beiträge zur Kenntnis der Portulacaceen und Basellaceen. Bot. Jahrb. 42, Beibl. 97: 1–46.

Gail PA. 1964. *Simmondsia chinensis* (Link) Schneider: anatomy and morphology of flowers. M.A. Thesis. Clermont.

Garcia M, M Lapp, M Castro, and P Torrecilla. 2000. Anatomia foliar comparada de cuatro especies del genero *Pereskia* (Plum.) Miller (Cactaceae). Ernstia 10: 27–41.

Gentry HS. 1958. The natural history of jojoba (*Simmondsia chinensis*) and its cultural aspects. Econ. Bot. 12: 261–295.

Gibson AC. 1977. Comparative anatomy of secondary xylem in Cactoideae (Cactaceae). Biotropica 5 (I): 29–65.

Gibson AC. 1977b. Vegetative anatomy of *Maihuenia* (Cactaceae) with some theoretical discussions of ontogenetic changes in xylem cell types. Bull. Torrey Bot. Club 104: 35–48.

Gibson AC. 1978. Rayless secondary xylem of *Halophytum*. Bull. Torrey Bot. Club 105: 39–44.

Gibson AC and KE Horak. 1979. Systematic anatomy and phylogeny of Mexican columnar cacti. Ann. Missouri Bot. Gard. 65: 999–1057.

Gibson AC and PS Nobel. 1986. The cactus primer. Harvard University Press, Cambridge, MA.

Gilbert MG. 1987. The taxonomic position of the genera *Telephium* and *Corrigiola*. Taxon 36: 47–49.

Gilbert MG. 1993. A review of *Gisekia* (Gisekiaceae). Kew Bull. 48: 343–356.

Gowik U, S Engelmann, OE Bläsing, AS Raghavendra, and P Westhoff. 2006. Evolution of C_4 phosphoenolpyruvate carboxylase in the genus Alternanthera: Gene families and the enzymatic characteristics of the C_4 isozyme and its orthologues in C_3 and C_3/C_4 alternanthera. Planta 359–368.

Griffith MP. 2002. Phylogenetic relationships in the Opuntioideae (Cactaceae) based on *nr*ITS sequences. In Botany 2002: Botany in the Curriculum, p.126. Madison, WisconSinica (Abstracts).

Guaralnick LJ and MD Jackson. 2001. The occurrence and phylogenetics of Crassulacean Acid Metabolism in the Portulacaceae. Int. J. Plant Sci. 162: 257–262.

Gvinianidze ZI and TA Fedotova. 1991. Caryophyllaceae. In: A Takhtajan, ed. Comparative seed anatomy, vol. 3, pp. 59–74. Nauka, Leningrad (in Russian).

Hartmann HEK. 1993. Aizoaceae. In: K Kubitzki, ed. The families and genera of vascular plants, vol. 2, pp. 37–69. Springer, Berlin/Heidelberg/New York.

Hassan NMS, U Meve, and S Liede-Schumann. 2005. Seed coat morphology of Aizoaceae-Sesuvioideae, Gisekiaceae and Molluginaceae and its systematic significance. Bot. J. Linn. Soc. 148: 189–206.

Hassan NMS, J Thiede, and S Liede-Schumann. 2005. Phylogenetic analysis of Sesuvioideae (Aizoaceae) inferred from nrDNA internal transcribed spacer (ITS) sequences and morphological data. Plant Syst. Evol. 255: 121–143.

Haug I, M Weiß, J Homeier, F Oberwinkler, and I Kottke. 2005. Russulaceae and Thelephoraceae form ectomycorrhizas with members of Nyctaginaceae (Caryophyllales) in the tropical mountain rain forest of southern Ecuador. New Phytol. 165: 923–936.

Hershkovitz MA. 1990. Phylogenetic and morphological studies in Portulacaceae. Ph.D. dissertration, University of California, Davis.

Hershkovitz MA. 1993. Revised circumscriptions and subgeneric taxonomies of *Calandrinia* and *Montiopsis* (Portulacaceae) with notes on phylogeny of the portulacaceous alliance. Ann. Missouri Bot. Gard. 80: 333–365.

Hershkovitz MA. 2006. Ribosomal and chloroplast DNA evidence for diversification of western American Portulacaceae in the Andean region. Gayana Bot. 63: 13–74.

Hershkovitz MA and EA Zimmer. 1997. On the evolutionary origins of the cacti. Taxon 46: 217–232.

Heubl GR, G Bringmann, and H Meimberg. 2006. Molecular phylogeny and character evolution of carnivorous plant families in Caryophyllales – revisited. Plant Biol. 8: 821–830.

Hofmann U. 1973. Centrospermen-Studien: 6. Morphologische Untersuchungen zur Umgrenzung und Gliederung der Aizoaceen. Bot. Jahrb. Syst. 93: 247–324.

Hofmann U. 1977. Centrospermen-Studien: 9. Die Stellung von *Stegnosperma* innerhalb der Centrospermen. Ber. Deutsch. Bot. Ges. 90: 39–52.

Horak KE. 1981. Anomalous secondary thickening in *Stegnosperma* (Phytolaccaceae). Bull. Torrey Bot. Club 108:189–197.

Hu Z-H and P-J Yang. 1944. Comparative anatomy of anomalous structures in the axes of 44 species of Chenopodiaceae. Cathaya 6: 145–162.

Hunt D and N Taylor, eds. 1990. The genera of Cactaceae: Progress towards consensus. Report of the IOS Working Party, 1987–1990. Bradleya 8: 85–107.

Hunziker JH, H-D Behnke, IJ Eifert, and TJ Mabry. 1974. *Halophytum ameghinoi:* A betalain-containing and P-type sieve-tube species. Taxon 23: 537–539.

Hunziker JH, R Pozner, and A Escobar. 2000. Chromosome number in *Halophytum ameghinoi* (Halophytaceae). Plant Syst. Evol. 221: 125–127.

Ihlenfeldt H-D. 1975. Some trends in the evolution of the Mesembryanthemaceae. Boissiera 24: 249–254.

Ihlenfeldt H-D, G Schwantes, and H Straka. 1962. Die hoheren Taxa der Mesembryanthemaceae. Taxon 11: 52–56.

Ihlenfeldt H-D and H Straka. 1961 (1962). Über die systematische Stellung und Gliederung der *Mesembryanthemen*. Ber. Deutsch. Bot. Ges. 74: 485–492.

Inamdar JA. 1968. Epidermal structure and ontogeny of stomata in some Nyctaginaceae. Flora 158: 159–166.

Inamdar JA, M Gagadhara, PG Morge, and RM Patel. 1977. Epidermal structure and ontogeny of stomata in some Centrospermae. Feddes Repert. 88: 465–475.

Jacobs SWL. 2001. Review of leaf anatomy and ultrastructure in the Chenopodiaceae (Caryophyllales). J. Torrey Bot. Soc. 128: 236–253.

Jansen S, LP Ronse Decraene, and E Smets. 2000. On the wood and stem anatomy of *Monococcus echinophorus* (Phytolaccaceae s.l.). Syst. Geogr. Plants 70: 171–179.

Jensen U. 1965. Serologische Untersuchungen zur Frageder systematischen Einordnung der Didiereaceae. Bot. Jahrb. Syst. 84: 233–253.

Jia A-Q, N-H Tan, Y-P Yang, S-G Wu, L-Q Wang, and J Zhou. 2004. Cyclopeptides from three arctic Caryophyllaceae plants, chemotaxonomy and distribution significance of Caryophyllaceae cyclopeptides. Acta Bot. Sinica 46: 625–630.

Joshi AC. 1937. Some salient points in the evolution of the secondary vascular cylinder of Amaranthaceae and Chenopodiaceae. Am. J. Bot. 24: 3–9.

Joshi AC and VSR Rao. 1934. Vascular anatomy of flowers of four Nyctaginaceae. J. Indian Bot. Soc. 13: 169–186.

Joshi AC and VSR Rao. 1936. The embryology of *Gisekia pharnaceoides* L. Proc. Indian Acad. Sci. 3B: 71–92.

Judd WS and IK Ferguson. 1999. The genera of Chenopodiaceae in southeastern United States. Harward Pap. Bot. 4: 365–416.

Kadereit G, T Borsch, K Weising, and H Freitag. 2003. Phylogeny of Amaranthaceae and Chenopodiaceae and the evolution of C_4 photosynthesis. Int. J. Plant Sci. 164: 959–986.

Kadereit G, D Gotzek, S Jacobs, and H Freitag. 2005. Origin and age of Australian Chenopodiaceae. Organisms Divers. Evol. 5: 59–80.

Kadereit G, L Mucina, and H Freitag. 2006. Phylogeny of Salicornioideae (Chenopodiaceae): Diversification, biogeography, and evolutionary trends in floral morphology. Taxon 55: 617–642.

Kajale LB. 1954. Contribution to the embryology of the Phytolaccaceae: II. Fertilization and development of embryo, seed, and fruit in *Rivina humilis* Linn. And *Phytolacca dioica* Linn. J. Indian Bot. Soc. 33: 206–225.

Kakkar L. 1973. Leaf vasculature in *Simmondsia californica*. Indian Sci. Congr. Assoc., Chandigarh. Proc. 60th Sess. Part III: 293–294 (Abstract).

Kamelina OP. 2001. Development of embryonic structures of *Hablitzia* (Chenopodiaceae). Bot. Zhurn. 86(10): 1–9 (in Russian with English summary).

Kamelina OP and OB Proskurina. 1985. On the embryology of *Stegnosperma halimifolium* (Stegnospermataceae). Bot. Zhurn. 70: 721–730 (in Russian with English summary).

Kapralov MV, H Akhani, EV Voznesenskaya, G Edwards, V Franceschi, and EH Ropalson. 2006. Phylogenetic relationships in the Salicornioideae/Suaedoideae/Salsoloideae s.l. (Chenopodiaceae) clade and a clarification of the phylogenetic position of *Bienertia* and *Alexandra* using multiple DNA sequence datasets. Syst. Bot. 31: 572–585.

Kapranova NN. 1991. Aizoaceae. In: A Takhtajan, ed. Comparative seed anatomy, vol. 3, pp. 17–22. Nauka, Leningrad (in Russian).

Klak C, A Khunou, G Reeves and T Hedderson. 2003. A phylogenetic hypothesis for the Aizoaceae (Caryophyllales) based on four plastid DNA regions. Am. J. Bot. 90: 1433–1445.

Köhler E. 1993. Blattnervatur-Muster der Buxaceae Dumortier und Simmondsiaceae van Tieghem. Feddes Repert. 104: 145–167.

Köhler E. 2003. Simmondsiaceae. In: K Kubitzki, ed. The families and genera of vascular plants, vol. 5, pp. 355–358. Springer, Berlin/Heidelberg/New York.

Köhler E and R Brückner. 1983. Zur Pollenmorphologie und systematischen Stellung der Gattung *Simmondsia* Nutt. Wiss. Z. Friedrich-Schiller-Univ. Jena, Math. Naturwiss. Reihe, 32: 945–955.

Komar GL. 1991. Nyctaginaceae. In: A. Takhtajan, ed. Comparative seed anatomy, vol. 3, pp. 12–17. Nauka, Leningrad (in Russian).

Kondorskaya VR. 1983. On the inflorescences of the genus *Chenopodium* L. Bull. Moscow Soc. Nat., Biology, 88: 78–86 (in Russian).

Kondorskaya VR. 1984. Peculiarities in the structure of inflorescences of the tribe Atripliceae C. A. Mey. of the family Chenopodiaceae. Bull. Moscow Soc. Nat., Biology, 89: 104–114 (in Russian).

Kowal T. 1961. Studia nad morfologia i anatomia nasion Portulacaceae Rchb. Monogr. Bot. 12: 1–48.

Kozhanchikov VI. 1967. Morphological characters of the seeds of the family Caryophyllaceae and probable ways of their evolution. Bot. Zhurn. 52: 1277–1280.

Kozhanchikov VI. 1975. Variation of morphological features of seeds of the representatives of the family Caryophyllaceae Juss. In: LY Budantsev, ed. Questions of comparative morphology of seed plants, pp. 108–128. Nauka, Leningrad (in Russian).

Kravtsova TI. 1991. Didiereaceae. In: A Takhtajan, ed. Comparative seed anatomy, vol. 3, pp. 57–58. Nauka, Leningrad (in Russian).

Kubitzki K. 1993. Didiereaceae. In: K Kubitzki, ed. The families and genera of vascular plants, vol. 2, pp. 292–295. Springer, Berlin/Heidelberg/New York.

Kühn U, V Bittrich, R Carolin, H Freitag, IC Hedge, P Uotila, and PG Wilson. 1993. Chenopodiaceae. In: K Kubitzki, ed. The families and genera of vascular plants, vol. 2, pp. 253–281. Springer, Berlin/Heidelberg/New York.

Landrum JV. 2001. Wide-band tracheids in leaves of genera in Aizoaceae: the systematic occurrence of a novel cell type and its implications for the monophyly of the subfamily Ruschioideae. Plant Syst. Evol. 227: 49–61.

Landrum JV and JD Mauseth. 1995. Systematic and evolutionary implications of stem and leaf anatomy in portulacaceous genera of South America, North America, and Africa. Am. J. Bot. 82(6): 176 (Abstract).

Landrum JV and JD Mauseth. 1996. Wide-band tracheids in genera of the Portulacaceae: Systematic and evolutionary implications. Am. J. Bot. 83(6): 170 (Abstract).

Larsen K. 2002. Caryophyllaceae. In: HP Nooteboom, ed. Flora Malesiana, ser.I, 16: 1–51. Leiden.

Lee CW and RA Sherman. 1985. Meiosis in jojoba, *Simmondsia chinensis*. Israel J. Bot. 34: 1–6.

Leins P. 1988. Placentation in Cactaceae. In: P Leins, SC Tucker, and PK Endress, eds. Aspects of Floral Development, pp. 191–204. Cramer, Berlin.

Leins P and C Erbar. 1994. Putative origin and relationships of the order Caryophyllales from the viewpoint of developmental flower morphology. In: H-D Behnke and TJ Mabry, eds. Caryophyllales. Evolution and systematics, pp. 303–316. Springer, Berlin/New York.

Leins P and S Schwitalla. 1986. Studien an Cactaceen-Blüten I. Einige Bemerkungen zur Blütenentwicklung von *Pereskia*. Beitr. Biol. Pflanzen 60: 313–323.

Leins P and S Schwitalla. 1988. Placentation in Cactaceae. In: P Leins, SC Tucker, and PK Endress, eds. Aspects of floral development, pp. 57–68. Cramer, Berlin.

Leins P, A Walter und C Erbar. 2001. Eine morphogenetische Interpretation der Caryophyllaceen-Kronblätter. Bot. Jahrb. Syst. 123(3): 355–367.

Leuenberger BE. 1976. Die Pollenmorphologie der Cactaceae und ihre Bedeutung für die Systematik. Diis. Lot. 31. Vaduz.

Leuenberger BE. 1986. *Pereskia* (Cactaceae). Mem. New York Bot. Gard. 41: 1–141.

Leuenberger BE. 1997. *Maihuenia*: Monograph of a Patagonian genus of Cactaceae. Bot. Jahrb. Syst. 119: 1–92.

Levin RA. 2000. Phylogenetic relationships within Nyctaginaceae tribe Nyctagineae: evidence from nuclear and chloroplast genomes. Syst. Bot. 25: 738–750.

Liden M, Popp M, and B Oxelman. 2001. A revised generic classification of the tribe Sileneae (Caryophyllaceae). Nord. J. Bot. 20: 513–518.

Livingstone DA, M Tomlinson, G Friedman, and R Broome. 1973. Stellate pore ornamentation in pollen grains of the Amaranthaceae. Pollen et Spores 15: 345–351.

Mabry TJ. 1977. The order Centrospermae. Ann. Missouri Bot. Gard. 64: 210–220.

Mabry TJ, H-D Behnke, and IJ Eifert. 1976. Betalains and P-type sieve-element plastids in *Gisekia* L. (Centrospermae). Taxon 25: 112–114.

Mabry TJ, L Kimler, and C Chang. 1972. The betalains: Structure, function, and biogenesis and the plant order Centrospermae. In: VC Runeckles and TC Tso, eds. Recent Adv. Phytochem. 4: 105–134. New York.

Mabry TJ, P Neumann, and WP Philipson. 1978. *Hectorella:* A member of the betalain-suborder Chenopodiineae of the order Centrospermae. Plant Syst. Evol. 130: 163–165.

Mauseth JD. 2004. Wide-band tracheids are present in almost all species of Cactaceae. J. Plant Res. 117: 69–76.

Mauseth JD, Y Uozumi, BJ Plemons, and JK Landrum. 1995. Structural and systematic study of an unusual tracheid type in cacti. J. Plant Res. 108: 517–526.

McNeil J. 1974. Synopsis of a revised classification of the Portulacaceae. Taxon 23: 725–728.

McNeil J. 1975. A generic revision of Portulacaceae tribe Montieae using techniques of numerical taxonomy. Canad. J. Bot. 53: 789–809.

Maheshwari P and RN Chopra. 1955. The structure and development of the ovule and seed of *Opuntia dillenii* Haw. Phytomorphology 5: 112–122.

Mauritzon J. 1934. Ein Beitrag zur Embryologie der Phytolaccaceen und Cactaceen. Bot. Not. 1934: 111–135.

Mauseth JD. 1982–1984. Introduction to cactus anatomy, parts 1–11. Cact. Succ. (U.S.) 54: 263–266; 55: 18–21, 42, 84–89, 113–118, 171–175, 272–276; 56: 33–37,131–135, 181–184, 212–216, 250–255.

Melikian AP. 1968. Position of the families Buxaceae and Simmondsiaceae in the system. Bot. Zhurn. 53: 1043–1047 (in Russian).

Melikian AP. 1991. Phytolaccaceae, Achatocarpaceae, Barbeuiaceae, Stegnospermataceae. In: A Takhtajan, ed. Comparative seed anatomy, vol. 3, pp. 8–11. Nauka, Leningrad (in Russian).

Melikian AP and AG Devyatov. 2000. Exocarp surface structure as a taxonomically significant feature in Caryophyllaceae. Byull. Mosk. Obshch. Ispyt. Prir., Biol. 105(1): 54–58 (in Russian with English summary).

Mikesell JE. 1979. Anomalous secondary thickening in *Phytolacca americana* L. (Phytolaccaceae). Am. J. Bot. 66: 997–1005.

Moeliono BM. 1966. De caulomatische Oorsprong van Zaadknoppen bij Caryophyllaceën en Primulaceën (een histogenetische en anatomische Analyse). Rotoype, Amsterdam.

Müller K and T Borsch. 2005a. Phylogenetics of Amaranthaceae based on *matK/trnK* sequence data – evidence from parsimony, likelihood, and Bayesian analyses. Ann. Missouri Bot. Gard. 92: 66–102.

Müller K and T Borsch. 2005b. Multiple origins of a unique pollen feature: Stellate pore ornamentation in Amaranthaceae. Grana 44: 266–281.

Nageshwar G and M Radhakrishnaiah. 1993. Affinities of *Basella L.* Feddes Repert. 104(3–4): 241–244.

Nair NC and VJ Nair. 1961. Studies on the morphology of some members of the Nyctaginaceae: I. Nodal anatomy of *Boerhavia*. Proc. Indian Acad. Sci. 54B: 281–294.

Narayana PS. 1990. The bearing of amino acids on the status and systematic position of Basellaceae. Indian J. Bot. 13: 129–135.

Narayana PS and LL Narayana. 1986. The embryology of Stegnospermataceae, with a discussion on its status, affinities, and systematic position. Plant Syst. Evol. 154: 137–145.

Narayana PS and LL Narayana. 1988. Systematic position of *Gisekia* L.: A numerical assessment. Feddes Repert. 99 (5–6): 189–193.

Nepokroeff M, WL Wagner, RK Rabeler, EA Zimmer, SG Weller, and AK Sakai. 2002. Relationships within Caryophyllaceae inferred from molecular sequence data. In Botany 2002: Botany in the Curriculum, p.105. Madison, WisconSinica (Abstracts).

Ng Siew Yoong, WR Philipson, and JRL Walker. 1975. Hectorellaceae: A member of the Centrospermae. New Zealand J. Bot. 13: 567–570.

Nowicke JW. 1968. Palynotaxonomic study of the Phytolaccaceae. Ann. Missouri Bot. Gard. 55: 294–364.

Nowicke JW. 1975. Pollen morphology in the order Centrospermae. Grana 15: 51–77.

Nowicke JW. 1994. Pollen morphology and exine ultrastructure. In: H-D Behnke and TJ Mabry, eds. Caryophyllales. Evolution and systematics, pp. 167–221. Springer, Berlin/New York.

Nowicke JW. 1996. Pollen morphology, exine structure and the relationships of Basellaceae and Didiereaceae to Portulacaceae. Syst. Bot. 21: 187–208.

Nowicke JW and JJ Skvarla. 1984. Pollen morphology and relationships of *Simmondsia chinensis* to the order Euphorbiales. Am.J. Bot. 71: 210–215.

Nyananyo BL. 1988. The systematic significance of seed morphology and anatomy in the Portulacaceae (Centrospermae). Folia Geobot. Phytotax. 23 (3): 275–279.

Nyananyo BL. 1990. Tribal and generic relationship in the Portulacaceae (Centrospermae). Feddes Repert. 101 (5–6): 237–241.

Nyffeler R. 2002. Phylogenetic relationships in the cactus family (Cactaceae) based on evidence from *trn*K/*mat*K and *trn*L-*trn*F sequences. Am. J. Bot. 89: 312–326.

O'Quinn R and L Hufford. 2005. Molecular systematics of Montieae (Portulacaceae): Implications for taxonomy, biogeography and ecology. Syst. Bot. 30: 314–331.

Oxelman B, M Lidén, RK Rabeler, and M Popp. 2000. A revised generic classification of the tribe Sileneae (Caryophyllaceae). Nord. J. Bot. 20: 743–748.

Oxelman B, M Liden, K Rabeler, and M Popp. 2001. A revised generic classification of the tribe Sileneae (Caryophyllaceae). Nord. J. Bot. 20: 743–748.

Paliwal GS. 1965. The development of stomata in *Basella rubra* Linn. Phytomorphology 15:50–53.

Pant DD and PF Kidwai. 1968. Structure and ontogeny of stomata in some Caryophyllaceae. Bot. J. Linn. Soc. 60: 309–314.

Patterson GW and S Xu. 1990. Sterol composition in five families of the order Caryophyllales. Phytochemistry 29: 3539–3541.

Pax F. 1927. Zur Phylogenie der Caryophyllaceae. Bot. Jahrb. Syst. 61: 223–241.

Payne MA. 1933. The morphology and anatomy of *Mollugo vertuillata* L. Univ. Kansas Sci. Bull. 21: 399–419.

Philipson WR. 1993. Hectorellaceae. In: K Kubitzki, ed. The families and genera of vascular plants, vol. 2, pp. 331–334. Springer, Berlin/Heidelberg/New York.

Philipson WR and JP Skipworth. 1961. Hectorellaceae: A new family of dicotyledons. Trans. Roy. Soc. New Zealand Bot. 1: 31.

Plisko MA. 1991. Tetragoniaceae, Molluginaceae. In: A Takhtajan, ed. Comparative seed anatomy, 3: 22–27. Nauka, Leningrad (in Russian).

Pozner R. and A Cocucci. 2006. Floral structure, anther development, and pollen dispersal of *Halophytum ameghinoi* (Halophytaceae). Int. J. Plant Sci. 167: 1091–1098.

Prakash N. 1967a. Aizoaceae – a study of its embryology ans systematics. Bot. Notis. 120: 305–323.

Prakash N. 1967b. Life history of *Tetragonia tetragonioides* (Pal.) O. Kuntze. Austral. J. Bot. 15: 413–424.

Prance GT. 1968. The systematic position of *Rhabdodendron* Gilg and Pilg. Bull. Jard. Bot. Nat. Belg. 38: 127–146.

Prance GT. 1972. Monograph of Rhabdodendraceae. Flora Neotropica No.11: 1–22.

Prance GT. 2003. Rhabdodendraceae. In: K Kubitzki, ed. The families and genera of vascular plants, vol. 5, pp. 339–341. Springer, Berlin/Heidelberg/New York.

Pratt DB 2003. Phylogeny and morphological evolution of the Chenopodiaceae-Amaranthaceae alliance. Ph.D. Thesis, Iowa State University. Ames.

Pratt DB, LG Clark, and RS Wallace. 2001. A tale of two families: Phylogeny of the Chenopodiaceae-Amaranthaceae. In Botany 2001: Plants and People, p. 135 (Abstracts).

Puff C and A Weber. 1976. Contributions to the morphology, anatomy, and karyology of *Rhabdodendron* and a reconsideration of the systematic position of the Rhabdodendraceae. Plant Syst. Evol. 125: 195–222.

Rabeler RK and V Bittrich. 1993. Suprageneric nomenclature in the Caryophyllaceae. Taxon 42: 857–863.

Raghavan TS and VK Srinivasan. 1940. Studies in the Indian Aizoaceae. Ann. Bot. 4: 651–661.

Raiput KS. 2002. Stem anatomy of Amaranthaceae: rayless nature of xylem. Flora 197: 224–232.

Rajput KS. 2002. Stem anatomy of Amaranthaceae: Rayless nature of xylem. Flora 197: 224–232.

Raipyt KS and KS Rao. 1999. Structural and developmental studies on cambial variant in *Pupalia lappaeae* (Amaranthaceae). Ann. Bot. Fenn. 36: 137–141.

Rajput KS and KS Rao. 2000. Secondary growth in the stem of some species of *Alternanthera* and *Achyranthes aspera* (Amaranthaceae). IAWA J. 21: 417–424.

Rao BS. 1975. Embryo development in five species of *Mollugo*. Current Sci. 44: 712–713.

Rauh W. 1956. Morphologische, entwicklungsgeschich- tliche, histogenetische, und anatomische Untersu-chungen an Didiereaceen. Sitzungsber. Abh. Akad.Wiss. Lit., Math.-Nat. Kl, 6: 1–104.

Rauh W. 1961. Weitere Untersuchungen an Didiereaceen. Part 1. Sitzungsber. Heidelb. Akad. Wiss., Math.-Nat. Kl, 7: 1–118.

Rauh W. 1965. Weitere Untersuchungen an Didieraceen. II. Teil: Infloreszenz, blütenmorphologische und embryologische Untersuchungen mit Ausblick auf die systematische Stellung der Didieraceen. Sitzungsber. Heidelberger Akad. Wiss. Jahresh. 1965: 221–443.

Rauh W and K Ditmar. 1970. Weitere Untersuchungen an Didiereaceen. Part 3. Sitzungsber. Heidelb. Akad.Wiss., Math.-Nat. Kl, 4: 1–88.

Rauh W and H Reznik. 1961. Zur Frage der systematischen Stellung der Didiereaceen. Bot. Jahrb. Syst. 81: 94–105.

Rauh W and HR Schölch. 1965. Weitere Untersuchungen an Didiereaceen. Part 2. Sitzungsber. Heidelb. Akad. Wiss., Math.-Nat. Kl, 3: 1–218.

Record SJ. 1933. The woods of *Rhabdodendron* and *Duckeodendron*. Trop. Woods 33: 6–10.

Rettig JH, HD Wilson, and JR Manhart. 1992. Phylogeny of the Caryophyllales: Gene sequence data. Taxon 41: 201–223.

Reznik H. 1955. Die Pigmente der Centrospermen alssystematisches Element. Z. Bot. 43: 499–530.

Reznik H. 1975. Betalaine. Ber. Deutsch. Bot. Ges. 88: 179–190.

Richardson M. 1978. Flavonols and C-glycosylflavonoids of the Caryophyllales. Biochem. Syst. Ecol. 6: 283–286.

Richardson M. 1981. Flavonoids of some controversialmembers of the Caryophyllales (Centrospermae). Plant Syst. Evol. 138: 227–233.

Robertson KR. 1981. The genera of the Amaranthaceae in the southeastern United States. J. Arnold Arbor. 62: 267–314.

Rodman JE, MK Oliver, RR Nakamura, JU McClammer, Jr., and AH Bledsoe. 1984. A taxonomic analysis and revised classification of Centrospermae. Syst. Bot. 9: 297–323.

Rogers GK. 1985. The genera of Phytolaccaceae in the southeastern United States. J. Arnold Arbor. 66: 1–37.

Rohweder O. 1965. Centrospermen-Studien: 1. Der Blütenbau bei *Uebelinia kiwuensis* T. C. E. Fries (Caryophyllaceae). 2. Entwicklung und morphologische Deutung der Gynöciums bei *Phytolacca*. Bot. Jahrb. Syst. 83: 406–418; 84: 509–526.

Rohweder O. 1967. Centrospermen-Studien; 3. Blütenentwicklung und Blütenbau bei Silenoideen (Caryophyllaceae). Bot. Jahrb. Syst. 86: 130–185.

Rohweder O. 1970. Centrospermen-Studien: 4. Morphologie und Anatomic der Blüten, Früchte, und Sa-men bei Alsinoideen und Paronychioideen s.lat. (Caryophyllaceae). Bot. Jahrb. Syst. 90: 201–271.

Rohweder O and K Huber. 1974. Centrospermen-Studien: 7. Beobachtungen und Anmerkungen zur Morphologie und Entwicklungsgeschichte einiger Nyctaginaceen. Bot. Jahrb. Syst. 94: 327–359.

Rohweder O and K Konig. 1971. Centrospermen-Studien: 5. Bau der Blüten, Früchte, und Samen von *Pteranthus dichotomus* Forsk. (Caryophyllaceae). Bot. Jahrb. Syst. 90: 447–468.

Rohweder O and K Urmi-König. 1975. Centrospermen-Studien: 8. Beiträge zur Morphologic, Anatomic, und systematischen Stellung von *Gymnocarpos* Forsk. und *Paronychia argentea* Lam. (Caryophyllaceae). Bot. Jahrb. Syst. 96: 375–409.

Rohwer JG. 1993. Phytolaccaceae, Stegnospermataceae. In: K Kubitzki, ed. The families and genera of vascular plants, vol. 2, pp. 506–515, 592–594. Springer, Berlin/Heidelberg/ New York.

Ronse Decraene LP and EF Smets. 1991. The floral ontogeny of some members of the Phytolaccaceae (subfamily Rivinoideae) with a discussion on the evolution of the androecium in the Rivinoideae. Biol. Jahrb. Doidonaea 39: 77–99.

Ronse Decraene LP, FF Smets, and P Vanvinckenroye. 1998. Pseudodiplostemony, and its implications for the evolution of the androecium in the Caryophyllaceae. J. Plant Res. 111: 25–43.

Ronse Decraene LP, P Vanvinckenroye, and EF Smets. 1997. A study of floral morphological diversity in *Phytolacca* (Phytolaccaceae) based on early floral ontogeny. Int. J. Plant Sci. 158: 57–72.

Ronse Decraene LP, SA Volgin, EF Smets. 1999. The flora development of *Pleuropetalum darwinii,* an anomalous member of the *Amaranthaceae.* Flora (Germany) 194(2): 189–199.

Ross R. 1981. Chromosome counts, cytology, and reproduction in the Cactaceae. Am. J. Bot. 68: 463–470.

Rost TL, AD Samper, P Schell, and S Alien. 1977. Anatomy of jojoba *(Simmondsia chinensis)* seed and the utilization of liquid wax during germination. Econ. Bot. 31: 140–147.

Sanderson SC. et al. 1988. Evolutionary loss of flavonoids and other chemical characters in the Chenopodiaceae. Biochem. Syst. Ecol. 16: 143–149.

Satyanarayana P and LL Narayana. 1985. Embryology of Stegnospermataceae. Curr. Sci. 54: 882–884.

Schaeppi H. 1936. Zur Morphologic des Gynoeceums der Phytolaccaceen. Flora 131: 41–59.

Schölch H-F. 1963. Die systematische Stellung der Di-diereaceen im Lichte neuer Untersuchungen über ihren Blütenbereich. Ber. Deutsch. Bot. Ges. 76: 49–55.

Schütze P, H Freitag, and K Weising. 2003. An integrated molecular and morphological study of the subfamily Suaedoideae Ulbr. (Chenopodiaceae). Plant Syst. Evol. 239: 257–286.

Scogin R. 1980. Serotaxonomy of *Simmondsia chinensis* (Simmondsiaceae). Aliso 9: 555–559.

Scogin R and S Brown. 1979. Leaf flavonoids of *Simmondsia chinensis* (Simmondsiaceae). Aliso 9: 475–477.

Scott AJ. 1977. Reinstatement and revision of Salicor-niaceae J. Agardh (Caryophyllales). Bot. J. Linn. Soc. 75: 357–374.

Scott AJ. 1978. A revision of the Camphorosmoideae (Chenopodiaceae). Feddes Repert. 89: 101–119.

Sharma HP. 1961. Contribution to the morphology and anatomy of *Basella rubra* Linn. Bull. Bot. Soc. Bengal. 15: 43–48.

Sharma HP. 1962. Contributions to the morphology of the Nyctaginaceae: I. Anatomy of the node and inflorescence of some species. Proc. Indian Acad. Sci. 56B: 35–50.

Sharma HP. 1963. Contributions to the morphology of the Nyctaginaceae: II. Floral anatomy of some species. Proc. Indian Acad. Sci. 57B: 149–163.

Sharma HP. 1962. Studies in the order Centrospermales: III. Vascular anatomy of the flower of some species of the family Ficoidaceae. Proc. Indian Acad. Sci. 56B: 269–285.

Sharma HP. 1963. Studies in the order Centrospermales: II. Vascular anatomy of the flower of certain species of the Molluginaceae, Nyctaginaceae, and Portulacaceae. J. Indian Bot. Soc. 42: 19–32; 637–645.

Shepherd A, TD Macfarlane and TD Colmer. 2005. Morphology, anatomy and histochemistry of fruits and seeds of the Salicornioideae (Chenopodiaceae). Ann. Bot. 95: 917–933.

Shepherd A, TD Macfarlane, and M Waycott. 2005. Phylogeneic analysis of the Australian Salicornioideae (Chenopodiaceae) based on morphology and nuclear DNA. Austral. Syst. Bot. 18: 89–115.

Shepherd A, M Waycott, and A Calladine. 2004. Radiation of the Australian Salicornioideae (Chenopodiaceae) – based on evidence from nuclear and chloroplast DNA sequences. Am. J. Bot. 91: 1387–1397.

Shilkina IA. 1953. Anatomical peculiarities of the family Chenopodiaceae. Bot. Zhurn. 38: 590–598 (in Russian).

Shipunov AB. 2003. The system of flowering plants from a synthetic point of view. J. Gen. Biol. 64: 501–510 (in Russian).

Singh BP. 1989. The status of Aizoaceae and Molluginaceae in Centrospermae. J. Econ. Taxon. Bot. 13: 311–313.

Singh BP. 1992. Chemotaxonomy of Aizoaceae and Molluginaceae. A study of free amino and organic acids. J. Econ. Taxon. Bot. 16: 103–106.

Sivarajan VV. 1988. A preliminary taxonomic revision of Indian Molluginaceae. J. Taiwan Mus. 41(2): 79–93.

Skipworth JP. 1961. The taxonomic position of *Hectorella caespitosa.* Trans. Roy. Soc. New Zealand Bot. 1: 17–30.

Skvaria JJ and JW Nowicke. 1976. Ultrastructure of pollen exine in centrospermous families. Plant Syst. Evol. 126: 55–78.

Skvaria JJ and JW Nowicke. 1982. Pollen fine structure and relationships of *Achatocarpus triana* and *Phaulothamnus* A. Gray. Taxon 31: 224–249.

Smissen RD, JC Clement, PJ Garnock-Jones, and GK Chambers. 2002. Subfamilial relationships within Caryophyllaceae as inferred from 5'*ndh*F sequences. Am. J. Bot. 89: 1336–1341.

Soriano A. 1946. Halophytaceae. Nueva familia del orden Centrospermae. Notas Prelim. Mus. La Plata, Bot. 11: 161–175.

Sperling CR and V Bittrich. 1993. Basellaceae. In: K Kubitzki, ed. The families and genera of vascular plants, vol. 2, pp. 143–152. Springer, Berlin/Heidelberg/New York.

Straka H. 1965. Die Pollenmorphologie der Didiereaceen. Sitzungsber. Heidelb. Akad. Wiss., Math.-Nat. Kl., 3: 219–227.

Straka H. 1975. Palinologie et différentiation systématique d'une famille endémique de Madagascar: Didiereacees. Boissiera 24: 245–248.

Terrazas T and S Arias. 2002. Comparative stem anatomy in the subfamily Cactoiodeae. Bot. Rev. 68: 444–473.

Thiede J, SA Schmidt, B Rudolph. 2007. Phylogenetic implication of the chloroplast *rpo*C1 intron loss in the Aizoaceae (Caryophyllales). Biochem. Syst. Ecol. 35: 372–380.
Thieret JW. 1966. Seeds of some United States Phytolaccaceae and Aizoaceae. SIDA 2: 352–360.
Thomson BF. 1942. The floral morphology of the Caryophyllaceae. Am. J. Bot. 29: 333–349.
Thorne RE. 1985. Phylogenetic relationships of the monotypic family Simmondsiaceae. Jojoba Happen. 13(2): 8.
Tiagi YD. 1960. Embryological and anatomical studies in the Cactaceae in connection with the position of the family in the system of angiosperms. Vestnic Moscow University 4: 21–30 (in Russian).
Tieghem P van. 1898. Sur le genre *Simmondsia* considere comme type d'une famille distincte, les Simmondsiacees. J. Bot. (Paris) 12: 103–112.
Tobe H and P Raven. 1989. The embryology and systematic position of *Rhabdodendron* (Rhabdodendraceae). In: K Tan, ed. The Davis and Hedge Festschrift: Plant taxonomy, phytogeography, and related subjects, pp. 233–248. Edinburgh University Press, Edinburgh.
Tobe H, S Yasuda, and K Oginuma. 1992. Seed coat anatomy, karyomorphology, and relationships of *Simmondsia* (Simmondsiaceae). Bot. Mag. (Tokyo) 105: 529–538.
Townsend CC. 1993. Amaranthaceae. In: K Kubitzki, ed. The families and genera of vascular plants, vol. 2, pp. 70–91. Springer, Berlin/Heidelberg/New York.
Trckova P, Z Slavikova, and J Chrtek. 2000. Evolution of the gynoecium in the Caryophyllaceae family. Novit. Bot. Univ. Carol. no.13: 47–53.
Tsukada M. 1964. Pollen morphology and identification: II. Cactaceae. Pollen Spores 6: 45–84.
Vanvinckenroye P, E Cresens, LP Ronse Decraene, and E Smets. 1993. A comparative floral developmental study in *Pisonia, Bougainvillea* and *Mirabilis* (Nyctaginaceae) with special emphasis on the gynoecium and floral nectaries. Bull. Jard. Bot. Natl. Belg. 62: 69–96.
Volgin SA. 1987. Origin and trends of specialization of flowers in the Amaranthaceae family. Bot. Zhurn. 72: 1338–1345 (in Russian with English summary).
Volgin SA. 1988. Vergleichende Morphologic und Gefassbundelanatomie der Blute bei den Rivinoideae (Phytolaccaceae). Flora 181 (5–6): 325–337.
Volponi CR. 1999. Stomatogenesis in *Caryophyllaceae*. Phytomorphology. 49(3): 253–259.
Vyshenskaya TD. 1991. Cactaceae. In: A Takhtajan, ed. Comparative seed anatomy, vol. 3, pp. 41–57. Nauka, Leningrad (in Russian).
Wagner WL and EM Harris. 2000. An unique Hawaiian *Schiedea* (Caryophyllaceae: Alsinoideae) with only five fertile stamens. Am. J. Bot. 87: 153–160.
Wallace RS. 1998. Molecular evidence for the elucidation of subfamilial relationships within Aizoaceae s.l. Am. J. Bot. 85(6): 163–164 (Abstract).
Wallace RS, and H Cota. 1996. An intron loss in the chloroplast gene *rpo*C1 supports a monophyletic origin for the subfamily Cactoideae of the Cactaceae. Curr. Genet. 29: 275–281.
Wallace RS and AC Gibson. 2002. Evolution and systematics. In: PS Nobel, ed. Cacti: Biology and Uses, pp. 1–21. University of California Press, Berkeley.
Wheat D. 1977. Successive cambia in the stem of *Phytolacca dioica*. Am. J. Bot. 64: 1209–1217.
Wofford BE. 1981. External seed morphology of *Arenaria* (Caryophyllaceae) of the southeastern United States. Syst. Bot. 6: 126–135.
Wolter-Filho W, AI Da Rocha, M Yoshida, and OR Gottlieb. 1985. Ellagic-acid derivatives from *Rhabdodendron macrophyllum*. Phytochemistry 24: 1991–1994.
Walter-Filho W, AL Da Rocha, M Yoshida, and OR Gottlieb. 1989. Chemosystematics of *Rhabdodendron*. Phytochemistry 28: 2355–2357.
Yamazaki T. 1987. The floral anatomy of the genus *Phytolacca*, with reference to the flower of the Caryophyllaceae. Acta Phytotax. Geobot. 38: 21–32 (in Japanese, with English summary).
Yoong NS, WR Philipson, and JRL Walker. 1975. Hectorellaceae: A member of the Centrospermae. New Zealand J. Bot. 13: 567–570.
Zamyatnin BN. 1951. On the inferior ovary of the axial origin. Bot. Zhurn. 36: 89–92 (in Russian).
Zandonella P. 1967. Les nectaires des Alsinoideae: *Stellaria* et *Cerastium* sensu lato. Compt. Rend. Hebd. Seances Acad. Sci., ser. D, 264: 2466–2469.
Zandonella P. 1977. Apports de l'étude comparée desnectaires floraux á la conception phylogénedque del'ordre des Centrospermales. Ber. Deutsch. Bot. Ges. 90:105–125.
Zandonella P and M Lecocq. 1977. Morphologic pol-linique et mode de pollinisation chez les Amaranthaceae. Pollen Spores 19: 119–141.
Zheng H-C, A-M Lu, and Z-H Hu. 2004. Floral organogenesis in *Phytolacca* (Phytolaccaceae). Acta Phytotax. Sinica 42: 352–364 (in Chinese).
Zibareva L, V Volodin, Z Saatov, T Savchenko, P Whiting, R Lafont, and L Dinan. 2003. Distribution of phytoecdysteroids in the Caryophyllaceae. Phytochemistry 64: 499–517.

Order 43. PHYSENALES

Small trees or shrubs. Vessels with simple perforations with slightly inclined end walls; lateral pitting alternate. Rays homogenous or heterogenous. Fibers with simple to bordered pits. Axial parenchyma aliform to confluent and distinctly unilateral, paratracheal. Sieve-element plastids of P-type. Nodes unilacunar. Leaves alternate, entire, estipulate. Stomata anomocytic. Flowers in racemes or panicles, or solitary, unisexual or bisexual. Sepals 5–9, slightly unequal, imbricate, persistent, each sepal vascularized by a single trace; petals 5 or absent. Stamens 8–25 (usually 10–14), essentially in one cycle; filaments free, short, slightly connate at the base; anthers basifixed or dorsifixed, tetrasporangiate, latrorse, opening longitudinally, with endothecal-like connective hypodermis. Pollen grains in monads, 2–3-colpate or sometimes 6-colpate. Tectum traversed by numerous microchannels, appearing on the surface as microperforations; ornamentation is in the form of small spines or spinules; endexine lamellate in both aperture and nonaperturate regions. Gynoecium of 2–3 united carpels. Ovary superior, mainly 2–3-locular, with 4–5 pendulous, campylotropous, epitropous, bitegmic (with the inner integument longer than the outer), crassinucellate ovules. Fruits dry, indehiscent. Seeds large, with straight or curved embryo, without endosperm. Plants producing betalain.

The sister-group relationship of *Physena* and *Asteropeia* was demonstrated by Morton et al. (1997), and the sister relationship of this clade to the remaining core Caryophyllales has been confirmed in several studies (e.g., Morton et al. 1997; Soltis et al. 2000, 2006; Guénoud et al. 2002).

Key to Families

1 Sepals enlarged and winglike in fruit. Fibers with small bordered pits. Rays homogenous. Small to tall, evergreen trees and scrambling shrubs. Leaves simple, shortly petiolate. Flowers bisexual, in axillary or terminal panicles. Petals five, adnate to the calyx, imbricate. Stamens (9)10–15; filaments dilated and connate at the base; anthers small and versatile, short-dorsifixed. Pollen grains 3-colpate or 6-colpate, rarely 2-colpate, finely spinulate. Gynoecium of 3(2) carpels; style usually with 3 stylar branches or with 3-lobed stigma. Ovary usually incompletely 3-locular or rarely 2-locular, with 2 or less often 4–5 (rarely 6) pendulous, epitropous ovules in each locule. Fruits thick-walled, dry, surrounded by persistent stamens and strongly accrescent and winglike sepals. Seeds reniform to hippocratiform with radicle close to large hilum; embryo curved, sometimes sigmoid; cotyledons large and thin, spirally coiled.................. 1. ASTEROPEIACEAE.

1 Sepals not enlarged and winglike in fruit. Shrub or tree. Fibers with simple or minutely bordered pits. Rays homogeneous typically with square and/or upright cells, or heterogeneous with more than four rows of upright and/or square cells. Leaves coriaceous, simple, with pinnate venation, and arch-shaped petiole bundle. Flowers on long pedicels in 5–10-flowered axillary racemes (*Physena madagascariensis*) or sessile or nearly sessile and solitary, or in 2–3-flowered axillary clusters (*P. sessiliflora*), unisexual (dioecious), actinomorphic, apetalous, anemophilous. Stamens (8)10-14(-25); filaments short, slightly connate at the base; anthers long, linear, basifixed. Pollen grains 3-colpate or sometimes with 4–5 short meriodinal colpi; ornamentation is in the form of small spines or spinules. Male flowers with a pistillodium. Gymoecium of two carpels. Stylodia very long, filiform, arising from the depression at the top of the ovary, connate only at the very base, papillate surface nearly along their whole length. Ovary 2-locular at the base and apex and locular at the midlevel, with four ovules (two ovules per locule) on axile placentas, each with a long funicle directed upward. Ovules campylotropous. Fruits dry, indehiscent, 1-seeded, with coriaceous and somewhat inflated pericarp that is separated from the seeds. Seeds large, with straight embryo consisting of two massive cotyledons of unequal size and minute, punctiform radicle. . . . . 2. PHYSENACEAE

1. ASTEROPEIACEAE

Takhtajan ex Reveal et Hoogland 1990. 1/8. Madagascar.
Asteropeia.

2. PHYSENACEAE

Takhtajan 1985. 1/2. Madagascar.
Physena.

According to Miller and Dickison (1992), the wood of *Physena* so much resembles that of *Asteropeia* that "solely on wood anatomy, these two genera could be placed in the same family." It is all the more important, then, that many years ago Erdtman (1952) noticed similarities between pollen grains of *Physena* and *Asteropeia*. Morton, Karol, and Chase (1997) also mentioned many similarities between *Asteropeia* and *Physena*, and consider them other sister group to the Caryophyllales.

Bibliography

Capuron R. 1968. Sur le genre *Physena* Noronh.ex Thouars. Adansonia, ser. 2, 8: 355–357.

Carlquist S. 2006. *Asteropeia* and *Physena* (Caryophyllales): A case study in comparative wood anatomy. Brittonia 58: 301–313.

Dickison WC and RB Miller. 1993. Morphology and anatomy of the Malagasy genus *Physena* (Physenaceae), with a discussion of the relationships of the genus. Adansonia 1–4: 85–106.

Kubitzki K. 2003a. Asteropeiaceae. In: K Kubitzki, ed. The families and genera of vascular plants, 5: 28–29. Springer, Berlin/Heidelberg/New York.
Kubitzki K. 2003b. Physenaceae. In: K Kubitzki, ed. The families and genera of vascular plants, vol. 5, pp. 332–333. Springer, Berlin/Heidelberg/New York.
Miller RB and WC Dickison. 1992. Wood anatomy of *Asteropeia* (Asteropeiaceae) and *Physena* (Physenaceae): Two endemics from Madagascar. Am. J. Bot. 79(6): 83 (Abstract.)
Morton CM, MW Chase, and KG Karol. 1997. Phylogenetic relationships of two anomalous dicot genera *Physena* and *Asteropeia*: evidence *rbc*L plastid DNA sequences. Bot. Rev. 63: 231–239.
Rao TA, MC Cheluviah, and S Chakraborti. 1984. On foliar sclereids in *Asteropeia* Thou. Curr. Sci. 53: 45–48.
Schatz GE, PP Lowry II, and A-E Wolf. 1999. Endemic families of Madagascar, IV. A synoptic revision of *Asteropeia* (Asteropeiaceae). Adansonia, sér.3, 21: 107–123.

Superorder POLYGONANAE

Order 44. POLYGONALES

Perennial or annual herbs, shrubs, lianas, or trees, often with conspicuous swollen nodes. Woody species often with anomalous secondary growth; silica bodies and calcium oxalate crystals present in wood cells. Vessels with simple perforations; lateral pitting alternate; vestured pits mostly absent (except *Brunnichia, Muehlenbeckia, Triplaris*, and *Coccoloba* – Carlquist 2003). Fibers with simple pits, mostly septate. Axial parenchyma scanty paratracheal or vasicentric. Rays from homogeneous to heterogeneous. Sieve-element plastids of S-type or (*Triplaris, Ruprechtia, Coccoloba*) P*fs*-type (Behnke 1999). Nodes multilacunar or less often trilacunar. Leaves alternate, rarely opposite or verticillate, entire or rarely pinnatifid or palmately cleft, membranous to fleshy or leathery, sometimes articulated at the base; stipules interpetiolar, connate into a usually membranous or scarious sheath (ochrea) around the stem and completely or partly covering the nodes, reduced to a row of hairs or wanting in Eriogonoideae and in a few members of Polygonoideae. Stomata commonly anomocytic, sometimes anisocytic, diacytic, or paracytic. Flowers small, seldom solitary, mostly in various types of inflorescences, bisexual or less often unisexual and dioecious, actinomorphic, 3-merous or 5-merous, rarely 2-merous, apetalous. Sepals (2-)5 or 6, rarely 9 (*Rheum*), imbricate or quincuncial in bud, green and herbaceous to often colored and petaloid, more or less connate at the base or forming a short or long floral tube, persistent and sometimes enlarging in fruit. Stamens (3)5–8(9); filaments filiform, free or basally connate; anthers tetrasporangiate, 2-locular, versatile or basifixed, opening longitudinally. Tapetum secretory. Microsporogenesis simultaneous. Pollen grains 3-celled or rarely 2-celled, mostly 3-colporate, rarely panto-hexocolporate (Hong et al. 2005). A nectary disc present at the base of the ovary, which is annular as in Caryophyllaceae, or nectaries are placed between the bases of the filaments. Gynoecium of three or less often two, rarely four, carpels, with free or more or less connate stylodia; ovary superior, subtended by an annular (often lobed) glandular disc, unilocular, sometimes with vestigial partitions at the base. Ovule solitary, basal on more or less well-expressed stalk representing a reduced free-central placenta, orthotropous or rarely anatropous, bitegmic or sometimes more or less unitegmic through connation (Corner 1976), crassinucellate. Female gametophyte of *Polygonum*-type. Endosperm nuclear. Fruits nutlike, very often trigonous, sometimes closely subtended by the persistent calyx or enclosed in a fleshy floral tube. Seeds of medium size, straight; seed coat formed by the outer integument (testal); embryo straight or more often curved, well-differentiated, surrounded by a copious, starchy, and oily, horny, sometimes ruminate endosperm with solitary starch grains; perisperm essentially lacking. Commonly tanniniferous, producing anthocyanins but not betalains, sulphated flavonols, ellagic acid, n = 7–13.

Polygonales are related to the Caryophyllales and probably derived from some of their archaic members with 3-colporate pollen grains, apetalous 3-merous flowers, and stipulate leaves. Bessey (1915) included Polygonaceae in the Caryophyllales between Chenopodiaceae and Nyctaginaceae, while Hallier (1908, 1912) derived the family from Portulacaceae s. l. (including Basellaceae) and Chenopodiaceae. There are many features in common between the Polygonaceae and Portulacaceae, especially in the morphology of the gynoecium, which in both families usually consists of three carpels and is of lysicarpous type. In the seeds of Caryophyllaceae there are vestiges of nucellar tissue with very small starch grains, which are characteristic of the perisperm of Caryophyllales. The amino acid sequence in members of the Polygonaceae and Caryophyllales also confirms their affinity (Boulter 1973). But at the same time the Polygonales (as well as the next order Plumbaginales) differ from the Caryophyllales in their S-type sieve-element plastids and from most of them also in their lack of betalains.

1. POLYGONACEAE

A. L. de Jussieu 1789 (including Calligonaceae Chalkuziev 1985, Eriogonaceae G. Don 1839, Persicariaceae Martynov 1820, Rumicaceae Martynov 1820). 54/c.1100. Very widely distributed, but chiefly in northern temperate regions; quite a few are tropical, including *Antigonon, Coccoloba,* and *Muehlenbeckla.*

1.1 ERIOGONOIDEAE

Leaves without well-defined stipules (ochrea lacking). Branching often sympodial. Inflorescences cymose. Flowers always 3-merous. Anthraquinones probably lacking. — ERIOGONEAE: *Eriogonum, Oxytheca, Dedeckera, Gilmania, Stenogonum, Goodmania, Nemacaulis, Dodecahema, Aristocapsa, Chorizanthe, Mucronea, Systenotheca, Centrostegia, Hollisteria, Lastarriaea*; PTEROSTEGIEAE: *Pterostegia, Harfordia.*

1.2 POLYGONOIDEAE.

Leaves with stipular sheaths (ochrea more or less developed). Branching monopodial. Inflorescences racemose with cymose partial inflorescences. Flowers 3-merous, 2-merous, or 5-merous. Anthraquinones often present. – RUMICIEAE: *Rheum, Rumex, Oxyria, Emex;* PERSICARIEAE: *Persicaria, Antenoron, Bistorta, Aconogonon, Koenigia, Fagopyrum;* POLYGONEAE: *Polygonum, Cephalolophilon, Truellum, Knorringia, Chylocalyx Polygonella, Reynoutria, Fallopia, Oxygonum;* ATRAPHAXIDEAE: *Atraphaxis, Pteropyrum, Calligonum, Physopyrum, Parapteropyrum*; TRIPLARIDEAE: *Gymnopodium, Leptogonum, Ruprechtia, ? Enneatypus, Triplaris, Symmeria;* COCCOLOBEAE: *Muehlenbeckia, Homalocladium, Coccoloba, Antigonon, Brunnichia, Afrobrunnichia, Podopterus, Neomillspaughia.*

Bibliography

Behnke H-D. 1999. P-type sieve element plastids present in members of the tribes Triplareae and Coccolobeae (Polygonaceae) renew the links between the Polygolales and the Caryophyllales. Plant Syst. Evol. 214: 15–27.

Brandbyge J. 1993. Polygonaceae. In: K Kubitzki, ed. The families and genera of vascular plants, vol. 2, pp. 531–544. Springer, Berlin/Heidelberg/New York.

Buchinger M. 1957. Nota sobre la subdivision de la familia de las Polygonaceas. Bot. Soc. Argent. 7: 42–43.

Carlquist S. 2003. Wood anatomy of Polygonaceae: Analysis of a family with exceptional wood diversity. Bot. J. Linn. Soc. 141: 25–51.

Decraene LR and JR Akeroyd. 1988. Generic limits in *Polygonum* and related genera (Polygonaceae) on the basis of floral characters. Bot. J. Linn. Soc. 98: 321–371.

Decraene LR, SP Hong, and E Smets. 2000. Systematic significance of fruit morphology and anatomy in tribes Persicarieae and Polygoneae (Polygonaceae). Bot. J. Linn. Soc. 134: 301–337.

Edman G. 1929. Zur Entwicldungsgeschichte der Gattung *Oxyria* Hill: Nebst zytologischen, embryologischen, und systematischen Bemerkungen über einige andere Polygonaceen. Acta Horti Berg. 9: 165–291.

Emberger L. 1939. La structure de la fleur des Polygona-cees. C. R. Acad. Sci. Paris 208: 370–372.

Fedotova TA. 1991. Polygonaceae. In: A Takhtajan, ed. Comparative seed anatomy, vol. 3, pp. 83–94. Nakua, Leningrad (in Russian).

Galle P. 1977. Untersuchungen zur Blütenentwicklung der Polygonaceen. Bot. Jahrb. Syst. 98: 449–489.

Geitler L. 1929. Zur Morphologic der Blüten von *Polygonum*. Oesterr. Bot. Z. 78: 229–241.

Graham SA and CE Wood. 1965. The genera of Polygonaceae in the southeastern United States. J. Arnold Arbor. 46: 91–121.

Gross H. 1913. Beiträge zur Kenntnis der Polygonaceen. Bot. Jahrb. Syst. 49: 234–339.

Haraldson K. 1978. Anatomy and taxonomy in Polygonaceae subfam. Polygonoideae Meisn. emend. Jaretzky. Symb. Bot. Uppsal. 22 (2): 1–95.

Hedberg O. 1946. Pollen morphology in the genus *Polygonum* L. s.lato and its taxonomical significance. Svensk Bot. Tidskr. 40: 371–404.

Hong S-P, I-C Oh, and LP Ronse Decraene. 2005. Pollen morphology of the genera *Polygonum* s.str. and *Polygonella* (Polygoneae: Polygonaceae). Plant Syst. Evol. 254: 13–30.

Hong S-P, LP Ronse DeCraene, and E Smets. 1998. Systematic significance of tepal surface morphology in tribes Persicarieae and Polygoneae (Polygonaceae). Bot. J. Linn. Soc. 127: 91–116.

Jaretzky R. 1925. Beiträge zur Systematik der Polygonaceen unter Berucksichtigung des Oxymethyl-anthrachinon-Vorkommens. Feddes Repert. Sp. Nov. 22: 49–83.

Jaretzky R. 1928. Histologische und karyologische Studien an Polygonaceen. Jahrb. Wiss. Bot. 69: 357–490.

Joshi AC. 1938. The nature of the ovular stalk in Polygonaceae and some related families. Ann. Bot. 2: 957–959.

Lamb Frye AS and KA Kron. 2003. *rbc*L phylogeny and character evolution in Polygonaceae. Syst. Bot. 28: 326–332.

Laubengayer RA. 1937. Studies in the anatomy and morphology of the polygonaceous flower. Am. J. Bot. 24: 329–343.

Maekawa F. 1964. On the phylogeny in the Polygonaceae. J. Jpn. Bot. 39: 366–379.

Martin PG and JM Dowd. 1984. The study of plant phylogeny using amino acid sequences of ribulose-1,5-bisphosphate carboxylase. V. Magnoliaceae, Polygonaceae and the concept of primitiveness. Austral. J. Bot. 32: 301–400.

Mondal MS. 1997. Pollen morphology and systematic relationship of the family Polygonaceae. Botanical Survey of India, Calcutta.

Nowicke JW and JJ Skvarla. 1977. Pollen morphology and the relationship of the Plumbaginaceae, Polygonaceae, and Primulaceae to the order Centrospermae. Smithson. Contr. Bot. 37: 1–64.

Perdrigeat CA. 1900. Anatomic comparee des Polygonacees et ses rapports avec la morphologie et la classification. Actes Soc. Linn. Bordeaux 55: 1–91.

Reveal JL. 1978. Distribution and phylogeny of Eriogonoideae (Polygonaceae). Great Basin Nat. Mem. 2: 169–190.

Reveal JL. 1989. A checklist of the Oriogonoideae (Polygonaceae). Phytologia 66: 266–294.

Ronse Decraene LP and J Akeroyd. 1988. Generic limits in *Polygonum* and related genera (Polygonaceae) on the basis of floral characters. Bot. J. Linn. Soc. 98: 321–371.

Ronse Decraene LP, S-P Hong, and E Smets. 2000. Systematic significance of fruit morphology and anatomy in tribes Persicarieae and Polygoneae (Polygonaceae). Bot. J. Linn. Soc. 134: 301–337.

Ronse Decraene LP and E Smets. 1991. The floral nectaries of *Polygonum* s.l. and related genera (Persicarieae and Polygoneae): Position, morphological nature and semophylesis. Flora 185: 165–185.

Sirrine E. 1895. Structure of the seed coats of Polygonaceae. Proc. Iowa Acad. Sci. 2: 128–135.

Vautier S. 1949. La vascularisation florale chez les Polygonacees. Candollea 12: 219–343.

Weberling E. 1970. Weitere Untersuchungen zur Morphologie des Unterblattes bei den Dikotylen: VI. Polygonaceae. Beitr. Biol. Pfl. 47: 127–140.

Wodehouse RP. 1931. Pollen grains in the identification of plants: VI. Polygonaceae. Am. J. Bot. 18: 749–764.

Wolf GR, S Xu, GW Patterson, and TA Salt. 1989. Polygonales and Plumbaginales: Sterol composition in relation to the Caryophyllidae. Phytochemistry 28: 143–145.

Wood CE, Jr. and SA Graham. 1965. The genera of Polygonaceae in the southeastern United States. J. Arnold Arbor. 46: 219–224.

Zhou ZZ, RX Xu, YL Zhuang, and ZQ Lin. 2000. Studies on pollen exine ultrastructure of the *Polygonaceae*. Acta Phytotax. Sinica 38: 446–451.

Order 45. PLUMBAGINALES

Perennial or sometimes annual herbs, subshrubs, or low shrubs, rarely lianas. Secondary growth in some genera (including *Acantholimon*, *Aegialitis*, and *Limoniastrum*) anomalous. Vessels with simple perforations; lateral pitting alternate. Fibers with simple pits. Rays homogeneous. Axial parenchyma scanty paratracheal. Sieve-element plastids of S-type. Nodes trilacunar. Leaves alternate, entire, usually estipulate, only rarely with well-developed stipules, often with scattered chalk-glands exuding water and salts, and often also raised mucilage glands in leaf axils and on the surface. Stomata anomocytic, or anisocytic, or paracytic. Flowers in cymose heads or panicles or in racemes or spikes, bisexual, actinomorphic, 5-merous, with both sepals and petals, bracteate (the bracts sometimes forming on involucre), with 2(1) bracteoles, often heterostylic. Sepals connate to form a plicate, often conspicuously 5-ribbed or 10-ribbed tube, the lobes mostly dry and membranous or scarious, often showy, persistent. Petals connate at the very base or less often into a tube, often persistent; the petals or corolla lobes contorted or imbricate. Stamens five, opposite the petals or corolla lobes, or partly adnate to the tube; anthers dorsifixed or (*Aegialitis*) basifixed, tetrasporangiate, introrse, 2-locular, opening longitudinally. Tapetum secretory. Microsporogenesis simultaneous. Pollen grains 2-celled or more often 3-celled, tectate-columellate, 3-colpate or rarely 4–5-colpate, very rarely pantocolpate, usually dimorphic in the Staticoideae. Gynoecium of five united carpels; stylodia more or less distinct (Staticoideae) or connate into an apically lobed style (Plumbaginoideae); stigmas papillate; ovary superior, unilocular, usually 5-lobed or -ribbed, with solitary basal ovule on elongate, slender funicle that almost encircles it. Ovules pendulous, anatropous, sometimes circinotropous, bitegmic, crassinucellate. Female gametophyte tetrasporic, 4–8-nucleate, of *Fritillaria*-, *Adoxa*-, *Plumbagella*-, *Plumbago*- and *Penaea*-type. Endosperm nuclear. Fruits mostly more or less enclosed within the persistent calyx, dry, indehiscent or less often tardily circumscissally dehiscent or splitting upward into valves. Seeds small, winged; seed coat formed by both integuments and consists of exotesta and endotegmen (Vovk 1991). Embryo rather large, green, straight, well-differentiated; endosperm starchy, more or less copious or scanty, sometimes (as in *Aegialitis* and *Psylliostachys*) absent. Perisperm lacking. Plants often salt tolerant; producing various 0-methylated flavonols and anthocyanidins, glycine betaine, choline-O-sulphate, plumbagin, kaempferol, quercetin, and myricetin, but no betalains, generally tanniniferous. n = 6–9.

Agardh (1858) put Plumbaginaceae close to the Nyctaginaceae and Hallier (1901, 1903, 1908, 1912) derived Plumbaginaceae directly from Portulacaceae s. 1. (including Basellaceae). With both the Portulacaceae and Basellaceae they share the important feature of a lysicarpous gynoecium, which in Plumbaginaceae and Basellaceae is characterized by a reduced free central column. Friedrich (1956) brought together many gross morphological, anatomical, palynological, embryological, and cytological data that confirm the affinity of the Plumbaginaceae with the families of "Centrospermae." He derived Plumbaginaceae directly from the Phytolaccaceae, which seems unlikely.

The Plumbaginaceae differ from the Caryophyllales in so many features that they deserve the status of a separate order. They differ from them by S-type plastids, pollen grain morphology, generally sympetalous corolla, the absence of perisperm and presence of endosperm, and straight embryo. The *rbc*L sequence data indicate that the Polygonales and Plumbaginales should be regarded as sister groups perhaps evolved parallel to, rather than directly related to the Caryophyllales (Giannasi et al. 1992).

1. PLUMBAGINACEAE

A. L. de Jussieu 1789 (including Aegialitidaceae Linczevsky 1968, Armeriaceae Horaninow 1834, Limoniaceae Seringe 1851, Staticaceae Cassel 1817). 29/750–850. Almost cosmopolitan but especially well represented in dry and saline habitats, particularly in the Mediterranean and Irano-Turanian regions.

1.1 PLUMBAGINOIDEAE
Flowers in racemes or spikes. Sepals herbaceous, more or less glandular. Corolla sympetalous. Stamens mostly free or partly connate to the corolla. Pollen grains with well-defined foot-layer, highly irregular columellae, and a complete tectum. Stylodia connate into an apically lobed style. Naphtoquinone plumbagin, glycine betaines, 5-O-methylated flavonols always present, n = 6–7. – *Plumbago, Plumbagella, Dyerophytum, Ceratostigma.*

1.2 STATICOIDEAE
Flowers in panicles or heads with monochasial partial inflorescences or rarely (*Aegialltis*) in leafy, often fork-branched racemes. Sepals scarious. Petals connate only at the very base, rarely (*Psylliostachys*) corolla sympetalous. Filaments basally adnate to the petals. Pollen grains with a foot-layer supporting straight, regular columellae, which are distally fused into an incomplete rectum. Stylodia free or rarely (*Limoniastrum*) connate up to the middle. Present betaalanine betaines, glycine betaines rare, naphtoquinone plumbagin lacking, n = 8, 9. – *Aegialitis, Acantholimon, Neogontscharovia, Gladiolimon, Ghaznianthus, Dictyolimon, Cephalorhizum, Bamiania, Popoviolimon, Chaetolimon, Vassilczenkoa, Limoniopsis, Ikonnikovia, Goniolimon, Bukiniczia, Afrolimon, Bakerolimon, Muellerolimon, Limoniastrum* (including *Bubania*), *Limonium, Myriolimon, Saharanthus, Eremolimon, Armeria, Psylliostachys.*

Bibliography

Baker HG. 1948. Relationships in the Plumbaginaceae. Nature 161: 400.

Baker HG. 1948. Dimorphism and monomorphism in the Plumbaginaceae: I. A survey of the family. Ann. Bot. 12: 207–219.

Baker HG. 1953. Dimorphism and monomorphism in the Plumbaginaceae: II. Pollen and stigmata in the genus *Limonium*; and III. Correlation of geographical distribution patterns with dimorphism and monomorphism in *Limonium*. Ann. Bot. 17: 433–445, 615–627.

Boyes JW and E Bataglia. 1951. Embryo sac development in the Plumbaginaceae. Caryologia 3: 305–310.

Carlquist S and CJ Boggs. 1996. Wood anatomy of Plumbaginaceae. Bull. Torrey Bot. Club 123: 135–147.

Channell RB and CE Wood. 1959. The genera of Plumbaginaceae of the southeastern United States. J. Arnold Arbor. 40: 391–397.

Clinckemaillie D and EF Smets. 1992. Floral similarities between Plumbaginaceae and Primulaceae: Systematic significance. Belg. J. Bot. 125: 151–153.

Dahlgren KVO. 1916. Zytologische und embryologische Studien über die Reihen Primulales und Plumbaginales. Kgl. Svensk. Vet.-Akad. Handl. 56(4): 1–80.

De Laet J, D Clinckemaillie, S Jensen, and E Smets. 1995. Floral ontogeny of Plumbaginaceae. J. Plant Res. 108: 289–304.

Dillon MO, AD Hanson and DA Gage. 1996. Osmoprotective compounds and phylogeny of the Plumbaginaceae. Am. J. Bot. 83(6): 131–132 (Abstract).

Harborne JB. 1967. Comparative biochemistry of the flavonoids: IV. Correlations between chemistry, pollen morphology, and systematics in the family Plumbaginaceae. Phytochemistry 6: 1415–1428.

Korobova SN. 1983. Plumbaginaceae. In: MS Yakovlev, ed. Comparative embryology of flowering plants: Phytolaccaceae-Thymelaeaceae, pp. 59–64. Nauka, Leningrad (in Russian).

Kubitzki K. 1993. Plumbaginaceae. In: K Kubitzki, ed. The families and genera of vascular plants, vol. 2, pp. 523–530. Springer, Berlin/Heidelberg/New York.

Labbe A. 1962. Les Plumbaginacées: Structure, dévelopment, répartition, conséquences en systématique. Thesis, University of Grenoble.

Lledó MD, MB Crespo, KM Cameron, MF Fay, and MW Chase. 1998. Systematics of Plumbaginaceae based upon cladistic analysis of *rbc*L sequence data. Syst. Bot. 23: 21–29.

Lledó MD, MB Crespo, MF Fay, and MW Chase. 2005. Molecular phylogenetics of *Limonium* and related genera (Plumbaginaceae): biogeographical and systematic implications. Am. J. Bot. 92: 1189–1198.

Lledó MD, PO Karis, MB Vrespo, MF Fay, and MW Chase. 2001. Phylogenetic position and taxonomic status of the genus *Aegialitis* and subfamilies Staticoideae and Plumbaginoideae (Plumbaginaceae): evidence from plastid DNA sequences and morphology. Plant Syst. Evol. 229: 107–124.

Maury P. 1886. Études sur l'organisation et la distribution géographique de Plombaginacees. Ann. Sci. Nat. Bot., ser. 7, 4: 1–134.

Nowicke JW and JJ Skvarla. 1977. Pollen morphology and the relationship of the Plumbaginaceae, Polygonaceae, and Primulaceae to the order Centrospermae. Smithson. Contr. Bot. 37: 1–64.

Roth I. 1962a. Histogenese und morphologische Deutung der basilären Plazenta von *Armeria*. Oesterr. Bot. Z. 109: 19–40.

Roth I. 1962b. Histogenese und morphologische Deutung der Kronblatter von *Armeria*. Port. Acta Biol, ser. A., 6: 211–230.

Schoute JC. 1935. Observations on the inflorescence in the family of the Plumbaginaceae. Rec. Trav. Bot. Neerl. 32: 406–424.

Vovk AG. 1991. Plumbaginaceae. In: A Takhtajan, ed. Comparative seed anatomy, vol. 3, pp. 95–97. Nauka, Leningrad (in Russian).

Weberling F. 1956. Weitere Untersuchungen zur Morphologie des Unterblattes bei den Dikotylen: I. Balsaminaceae. II. Plumbaginaceae. Beitr. Biol. Pfl. 33: 17–32.

Wolf GR, S Xu, GW Patterson, and TA Salt. 1989. Polygonales and Plumbaginales: Sterol composition in relation to the Caryophyllidae. Phytochemistry 28: 143–145.

Order 46. TAMARICALES

Trees, shrubs, subshrubs or less often perennial herbs, mostly halophytic or xerophytic with storied wood. Vessels with simple perforations; lateral pitting alternate. Fibers mostly with few, simple pits. Rays in Tamaricaceae very broad and high, heterogeneous, but in Frankeniaceae definite medularry rays are lacking. Axial parenchyma in Tamaricaceae scanty-paratracheal to vasicentric. Sieve-element plastids of S-type. Leaves alternate or opposite simple, entire, mostly small, commonly with embedded, multicellular, salt-excreting glands, estipulate. Stomata usually anomocytic. Flowers mostly small or very small, solitary or more often in various kinds of inflorescences, usually bisexual, actinomorphic. Sepals 4–7, free or less often more or less connate (in Frankeniaceae united into a tube with lobes), imbricate or valvate. Petals as many as the sepals, free, imbricate. Nectary disc present or wanting. Stamens as many as petals, free or more or less connate; anthers tetrasporangiate, opening longitudinally. Tapetum secretory. Microsporogenesis simultaneous. Pollen grains 2-celled or 3-celled, (2)3(4)-colpate, rarely rugate. Gynoecium of 2–5 (mostly 3–4) united carpels with stylodia free or only basally connate. Ovary superior, 1-locular, with (1)2-many ovules on each parietal placenta. Ovules anatropous, bitegmic, more or less tenuinucellate (Tamaricaceae) or thinly crassinucellate (Frankeniaceae – Fedotova 1992). Female gametophyte of *Polygonum*-type or tetrasporic of various types (Kamelina 1983). Endosperm nuclear, or rarely (*Tamarix tetrandra*) cellular, without haustoria or with micropylar haustorium (Frankeniaceae). Fruits loculicidal capsules. Seeds small, covered with long unicellular hairs or with a chalazal hair tuft, or seeds uniformly papillate; seed coat formed by the outer integument (exotestal); embryo large, straight; endosperm copious and starchy, scanty and starchy or lacking. Commonly producing bisulphated flavonols, but not cyanogenic.

According to molecular data Tamaricales are closest to the Plumbaginales, which does not contradict morphological data.

Key to Families

1 Leaves alternate, not ericoid or revolute-margined, often scaly or subulate, commonly with external but embedded, multicellular salt glands, simple, estipulate. Stomata anomocytic or paracytic. Flowers small to medium size, in racemes or in spike, or in panicles, or solitary, bisexual or rarely unisexual (dioecious). Sepals 4–5(6), free or less often basally connate, imbricate. Petals 4–5, alternate with sepals with two scalelike appendages (Reaumurieae) or without appendages, free or slightly connate, imbricate, persistent or caduceus. Stamens 4–6, or 8–12 (often twice as many as the petals), or 15–100 (often more or less numerous), seated on a fleshy nectary disc, free or basally connate; anthers 2-locular, dorsifixed, extrorse, latrorse, or introrse, opening by longitudinal slits. Pollen grains 2- or 3-celled. Gynoecium of 3–4, rarely 2 or 5 united carpels; ovules ascending. Female gametophyte tetrasporic, mostly of *Fritillaria*-type. Seeds covered with long unicellular

hairs; exotestal cells periclinally elongated and thick-walled, endotestal cells thin-walled, crystalliferous; endosperm scanty or lacking. Trees, shrubs or subshrubs. Gypsum crystals present. Contain kaempferol, quercetin, or tamatixin, alkaloids present or absent. n = 11 (Reaumurieae) or 12 (Tamariceae). . . . . 1. TAMARICACEAE

1 Leaves opposite, often ericoid and revolute-margined, commonly punctuate with sunken, salt-excreting glands. Medullary rays wanting. Flowers solitary and sessile in the forks of the branches or forming a dense leafy cyme. Sepals 4–7, connate into a tube with short, induplicate-valvate lobes. Petals 4–7, distinct, imbricate, clawed, usually with a scale-lake appendage inside. Nectary disc wanting. Stamens 4–7, or up to 24, most often 6 in 2 cycles, free or shortly connate at the base with filiform or flattened filaments; anthers versatile, opening length-wise. Pollen grains 3-celled, (2)3(4)-colpate. Gynoecium of (1-)3(-4) carpels united to form a 1-locular ovary; style slender end elongate, usually with distinct stylodia; ovules (1)2–6(-many) in each carpel. Female gametophyte of *Polygonum*-type. Seeds uniformly papillate, exotestal cells large, papilla with terminal nail-like thickenings, endotestal cells thin-walled, endotegmen with thick cuticle, tanniniferous; embryo straight, endosperm copious. Halophytic shrubs, subshrubs or herbs. Contain tannins, proanthocyanidins, ellagic acid, flavonols. n = 10, 15. . . . . . 2. FRANKENIACEAE

1. TAMARICACEAE

Link 1821 (including Reaumuriaceae Ehrenberg ex Lindley 1830). 5/90. Africa and Eurasia, mainly Mediterranean and Irano-Turanian regions.

REAUMURIEAE: *Reaumuria, Hololachna*; TAMARICEAE: *Tamarix, Myricaria, Myrtama.*

2. FRANKENIACEAE

Gray 1821 1/90. Distributed throughout warmer dry regions, especially on maritime shores of America, Africa, Eurasia, and Australia, but best developed in the Mediterranean region and western Asia.

Frankenia (including ? *Hypericopsis, Beatsonia, Anthobryum, Niederleinia*).

Related to the Tamaricaceae and especially to the Reaumurieae, which is confirmed also by a great similarity in the ultrastructure of the salt glands (Campbell and Thomson 1976), as well as secondary chemistry, and often petaline scales, and also the molecular analysis (Gaskin et al. 2004).

Bibliography

Baum BR. 1964. On the vernales-aestivales character in *Tamarix* and its diagnostic value. Israel J. Bot. 13: 30–35.

Baum BR. 1966. Monographic revision of the genus *Tamarix*. Jerusalem.

Baum BR, IJ Bassett, and CW Crompton. 1971. Pollen morphology of *Tamarix* species and its relationship to the taxonomy of the genus. Pollen Spores 13: 495–521.

Behnke H-D. 1976. Sieve-element plastids of *Fouquieria*, *Frankenia* (Tamaricales) and *Rhabdodendron* (Rutaceae), taxa sometimes allied with Centrospermae (Caryophyllales). Taxon 25: 265–268.

Brochmann C, W Lobin, P Sunding, and O Stabbetorp. 1995. Parallel ecoclinal evolution and taxonomy of *Frankenia* (Frankeniaceae) in the Cape Verde Islands, W Africa. Nord. J. Bot. 15: 603–623.

Campbell N and WW Thomson. 1976. The ultra-structure of *Frankenia* salt glands. Ann. Bot. 40: 681–686.

Cheng ZM, HX Pan, and LK Yin. 2000. Study on the phytochemistry taxonomy of *Tamarix* L. and *Myricaria* Desv. Acta Bot. Bor. Occid. Sinica 20: 275–282.

Crins WJ. 1989. The Tamaricaceae in the southeastern United States. J. Arnold Arbor. 70: 403–425.

Cui DF, WB Liao, and B Zhang. 2000. Determination of flavonoid compounds of *Reaumuria* L. (Tamaricaceae) and their taxonomical significance. Acta Bot. Bor. Occid. Sinica 20: 283–287.

Fedotova TA. 1992. Tamaricales. In: A Takhtajan, ed. Comparative seed anatomy, vol. 4, pp. 128–140. Nauka, St. Petersburg (in Russian).

Gaskin JF. 2003. Tamaricaceae. In: K Kubitzki, ed. The families and genera of vascular plants, vol. 5, pp. 363–368. Springer, Berlin/Heidelberg/New York.

Gaskin JF, F Ghahremaninejad, D-Y Zhang, and JP Londo. 2004. A systematic overview of Frankeniaceae and Tamaricaceae from nuclear rDNA and plastid sequence data. Ann. Missouri Bot. Gard. 91: 401–409.

Gundersen A. 1927. The Frankeniaceae as a link in the classification of the dicotyledons. Torreya 27: 65–71.

Gupta AK and YS Murty. 1987. Floral anatomy in Tamaricaceae. J. Indian Bot. Soc. 66: 275–282.

Harborne JB. 1976. Flavonoid bisulphates and their co-occurrences with ellagic acid in the Bixaceae, Frankeniaceae and related families. Phytochemistry 14: 1331–1337.

Horton JS. 1957. Inflorescence development in *Tamarix pentandra* Pallas (Tamaricaceae). Southwest. Nat. 2: 135–139.

Johri BM and D Kak. 1954. The embryology of *Tamarix* L. Phytomorphology 4: 230–247.
Kamelina OP. 1983. Tamaricaceae. In: MS Yakovlev, ed. Comparative embryology of flowering plants: Phytolaccaceae-Thymelaeaceae, pp. 179–185. Nauka, Leningrad.
Kubitzki K. 2003. Frankeniaceae. In: K Kubitzki, ed. The families and genera of vascular plants, vol. 5, pp. 209–212. Springer, Berlin/Heidelberg/New York.
Leinfellner W. 1959. Die falschen Rollblätter der Fran-keniaceen, in Vergleich gesetzt mit jenen der Ericaceae. Oesterr. Bot. Z. 106: 325–351.
Leinfellner W. 1965. Über die Kronblätter der Frankeniaceen. Oesterr. Bot. Z. 112: 44–55.
Mauritzon J. 1933. Über die Embryologie der Turneraceae und Frankeniaceae. Bot. Not. 86: 543–554.
Qaiser M. 1987. Studies in the seed morphology of the family Tamaricaceae from Pakistan. Bot. J. Linn. Soc. 94: 469–484.
Ragonese AM. 1966. Anatomia de las Frankeniaceas argentines. Darwiniana 14: 95–129.
Ronse Decraene LP. 1990. Morphological studies in Tamaricales. I. Floral ontogeny and anatomy of *Reaumuria vermiculata* L. Beitr. Biol. Pflanzen 65: 181–203.
Walia K and RN Kapil. 1965. Embryology of *Frankenia* Linn. with some comments on the systematic position of Frankeniaceae. Bot. Not. 118: 412–429.
Waly NM. 1999. Wood anatomical characters of the Egyptian *Tamarix* L. species and its taxonomic significance. Taeckholmia 19: 115–125.
Wei Y, D-Y Tan, and L Yin. 1999. The discussion on the anatomical structure of leaf and its taxonomical relationship of Tamaricaceae in China. Acta Bot. Borr.-Occident. Sinica 19: 113–118 (in Chinese with English summary).
Whalen MA. 1987. Wood anatomy of the American frankenias (Frankeniaceae): Systematic and evolutionary implications. Am. J. Bot. 74: 1211–1223.
Whalen MA. 1989. Systematics of *Frankenia* (Frankeniaceae) in North and South Am. Syst. Bot. Monogr. 17.
Xi YZ. 1988. Study on pollen morphology of Tamaricaceae from China. Bull. Bot. Res. Harbin 8: 23–42.
Zhang DY, ZD Chen, HY Sun, LK Yin, and BR Pan. 2000. Systematic studies on some questions of Tamaricaceae based on ITS sequence. Acta Bot. Bor. Occid. Sinica 20: 421–431.
Zhang PY and YJ Zhang. 1984. A study on the taxonomy of the genus *Myricaria* Desv. in China. Bull. Bot. Res. North-East For. Inst. 4: 67–80.
Zhang YM. 2001. Pollen morphology of the Tamaricaceae from China and its taxonomic signigicance. Acta Bot. Boreal.-Occid. Sinica 21: 857–864.
Zhang YM, BR Pan, and LK Yin. 1998. Seed morphology of Tamaricaceae in China arid areas and its systematic evolution. J. Plant Res. Environ. 7: 22–27.
Zhang YM, BR Pan, and LK Yin. 2001. Pollen morphology of the Tamaricaceae from China and its taxonomic significance. Acta Bot. Bor. Occid. Sinica 21: 857–864.
Zhang YM, BR Pan, and LK Yin. 2003. The photogeographical studies of *Tamarix* (Tamaricaceae). Acta Bot. Yunnan. 25: 415–427.
Zohari M and B Baum. 1965. On the androecium of *Tamarix* flower and its evolutionary trends. Israel J. Bot. 14: 101–111.

Superorder NEPENTHANAE

Order 47. NEPENTHALES

Herbaceous or sometimes more or less woody plants with normal or anomalous secondary growth, erect, prostrate, or often twining or climbing, aquatic, terrestrial or epiphytic, glabrous or hairy, cornivorous or not. Vessels with simple perforation, true tracheids present. Rays homogeneous or heterogeneous. Axial parenchyma paratracheal (vasicentric) and apotracheal, banded and diffuse. Nodes unilacunar, trilacunar to multilacunar (Nepenthaceae). Leaves alternate, simple, estipulate or with small stipules, Stomata anomocytic, tetracytic, actinocytic or encyclocytic. Flowers usually small, terminal or axillary, in racemes, panicles, spikes, or rarely (*Aldrovanda*) solitary, bisexual or rarely (Nepenthaceae) unisexual, actinomorphic. Sepals usually 4–5 rarely 3, free or less often basally connate, imbricate or rarely valvate, sometimes persistent. Petals 4–5, free or shortly connate, or cohering, imbricate or contorted. Stamens 2–8(-25), rarely 4–6; filaments free or united; anthers 2-locular, basifixed or dorsifixed, extrorse, introrse or latrorse, tetrasporangiate, opening longitudinally. Tapetum secretory. Microsporogenesis simultaneous. Pollen grains 2–3-celled, in tetrads or rarely (*Drosophyllum*) in monads, 3-colpate or 3–4-colporate. Gynoecium of 2 or 5, very rarely 3–4 or 6 united carpels; stylodia more or less free or united; stigma capitate, hippocrepiform or puntofid, or discoid, on a very short style, or sessile; ovary superior to inferior, 1-locular, or rarely 3–4-locular; ovules solitary (*Ancistrocladus*) or numerous, anatropous or hemitropous, bitegmic, mostly crassinucellate. Female gametophyte of *Polygonum*-type. Endosperm nuclear or cellular (*Ancistrocladus*). Fruits loculicidal or valvular capsules, or indehiscent (*Aldrovanda* and *Ancistrocladus*). Seeds small to large, sometimes numerous (100–500), often winged (usually, with a narrow or hair-like wing at either end), or wingless; seed coat formed by both the integuments, but mainly exotestal. Embryo well-differentiated, straight, cylindrical or discoid-obconix, surrounded by starchy, proteinaceous and oily endosperm, sometimes endosperm absent. Produced proanthocyanidins, flavonols, acetogenic napthoquinone plumbagin, and isoquinoline alkaloids.

Probably related to the some families of the Polygonanae. "Possible synapomorphies for the noncore

Caryophyllales are scattered secretory cells containing plumbagin, naphthaquinone (which has been lost in several clades), an indumentum of stalked, gland-headed hairs, basal placentation (with shifts to parietal in some Droseraceae and axile in Nepentaceae), and starchy endosperm" (Judd et al. 2002, Soltis et al. 2006)

Key to Families

1 Ovary mostly superior.

2 Flowers bisexual; petals imbricate or contorted.

3 Plants with normal secondary growth.

4 Perennial or annual carnivorous herbs, commonly growing in bogs and other waterlogged soils containing little or no available nitrogen, rarely (*Aldrovanda*) submerged aquatics, sometimes with corms or root tubers; glandular hairs irritable when present. Nodes unilacunar with 1 trace. Leaves spiral or rarely whorled (*Aldrovanda*), often circinate in bud, usually in basal rosettes, mostly with intrapetiolar stipules; the blade either modified into an active trap that shuts when touched (*Dionaea, Aldrovanda*) or provided with sticky tentacles and sessile glands that entrap and digest insects and other small animals (*Drosera*). Stomata anomocytic, tetracytic or actinocytic. Flowers in cymose inflorescences or (*Aldrovanda*) solitary, bisexual, actinomorphic. Sepals (4)5(-8), more or less connate at the base, imbricate, persistent. Petals as many as the sepals, free, imbricate or contorted. Stamens (4)5(10–20), free or (*Dionaea*) basally connate; anthers 2-locular, dorsifixed or basifixed, extrorse, tetrasporangiate, opening longitudinally; anther wall initially with one middle layer. Pollen grains in tetrads, 3-colpate to more often periporate. Gynoecium of three or five united carpels, with more or less free or (*Dionaea*) united stylodia that have terminal stigma. Ovary superior, or sometimes nearly half-inferior, 1-locular, with three parietal placentas (*Dionaea*) or with essentially basal placentas. Ovules anatropous, bitegmic, crassinucellate or tenuinucellate in some *Drosera* species. Endosperm nuclear. Fruits loculicidal or valvular capsules, or rarely indehiscent (*Aldrovandra*). Seeds usually minute, mostly fusiform; seed coat usually formed by both the integuments; every epidermal layer of the seed coat is potentially specialized, the exotesta as a palisade, the endotesta as thick-walled crystal-cells, and the endotegmen as scleroic cuboid cells; embryo short, straight, embedded in the copious, oily endosperm rich in starch. Produced proanthocyanidins (cyanidin and delphinidin), flavonols (including myricetin, kaempferol and quercetin), acetogenic napthoquinones plumbagin, and ellagic acid (*Drosera*, 2 sp.); n = 5–24. . . . . 5. DROSERACEAE

4 Small insectivorous tap-rooted subshrubs, sometimes branched; axes and leaves covered with rows of stalked mucilage, red tentacles and irregularly distributed sessile digesting glands. Leaves spiral, crowded, narrow-linear, reversely circinate in bud, estipulate. Inflorescence terminal, few-flowered, thyrso-paniculate. Flowers large, short-lived, remain open only a single day, bisexual, actinomorphic, 5-merous. Sepals imbricate, fused at base, glandular. Petals free, contorted, deciduous. Stamens ten, free; anthers extrorse. Pollen grains pantoporate, in monads. Gynoecium of five united carpels; stylodia with capitate stigmas; ovary 1-locular; ovules numerous, basal, anatropous, with long funicles attached to a basal, conical placenta. Fruits loculicidal, coriaceous capsules, opening in upper half. Seeds numerous, seed coat with endotestal crystal layer and sclerotised endotegmen. Embryo small, endosperm copious, fleshy and starchy. Produced proanthocyanins, the naphtoquinone plumbagin, flavone luteolin, and inulin; n = 6. . . . . 4. DROSOPHYLLACEAE

3 Plants with anomalous secondary growth. Insectivorous (*Triphyophyllum*), soft-wooded lianas, climbing by pairs of hooks or tendrils representing the forked tips of the leaf midribs, heterophyllous (*Triphyophyllum*, which in addition to the hook-tipped climbing leaves has 'normal' leaves without hooks, and others which are wholly reduced to the midrib, circinnate and beset with numerous stalked and

sessile glands), or not heterophyllous. Leaves and stems are beset with peltate hairs and with characteristic multicellular, stalked, or sessile glands secreting a sticky mucilage that traps insects. Interxylar (included) phloem present. Vessel elements extremely large (except in *Habropetalum*); lateral pitting alternate. Fibers with large, bordered pits. Axial parenchyma paratracheal (vasicentric) and apotracheal. Vasicentric tracheids present. Leaves entire or crenulate, estipulate, pinnately veined; the midrib usually prolonged and forked into two recurved hooks or pigtailed tendrils. Stomata actinocytic or encyclocytic. Flowers of moderate size, in a lax, axillary or supra-axillary, cymose inflorescences, bracteate. Sepals five, small, free or connate at the base into a short tube, valvate or open in bud, persistent. Petals five, free, contorted, caducous. Stamens 10(-30), free or slightly connate; anthers ovoid or oblong, basifixed, introrse, appendaged, opening longitudinally. Pollen grains 3(-4)-colporate. Gynoecium of two or five united carpels; stylodia either two, free, filiform, with capitate stigmas, or five, free, filiform, with minute stigmas, or five, slightly connate at the base and with plumose stigmas. Ovary superior, 1-locular, with numerous ovules on parietal placentas. Ovules anatropous, bitegmic, crassinucellate (but according to Schmid 1964, ovules in *Triphyophyllum* semitenuinucellate). The outer integument forms a characteristic cover. Endosperm probably nuclear. In connection with the development of the integumentary cover, the female gametophyte (and therefore also the endosperm) elongate perpendicularly. Fruits loculicidal capsules of two or five valves, spreading widely, opening well before maturity, and bearing the few seeds peltately attached on greatly elongate, thickened, rigid funicles. Seeds large, discoid, either thick and surrounded by narrower wing, or thin and surrounded by broad satiny wing; embryo large, discoid-obconic, axile, mostly surrounded by the copious starchy endosperm. Produce naphthoquinones like plumbagin, and naphthylisoquinoline alkaloids, e.g. dioncophylline (Bringmann and Pokorny 1995; Porembski and Barthlott 2002), and ellagic acid, n = 12, 18 (*Triphyophyllum peltatum*).. . . . . 2. DIONCOPHYLLACEAE.

2 Flowers unisexual, apetalous. Leaves transformed into pitchers. Herbaceous or sometimes more or less woody plants with a small amount of secondary growth, erect, prostrate, or often climbing by aid of the leaf tendrils, terrestrial or epiphytic. Various kinds of multicellular glands and peltate hydathodes are widely distributed on the stem and leaves. Schmid-Hollinger (1971) recognises five types of non-glandular trichomes: multicellular simple hairs, hairs with teeth, tufted, rosette and arachnoid hairs. Cortical bundles occur in young stem, and medullary vascular bundles in the stem of certain species. Characteristic spiral cells occur in the parenchymatous tissues of both stem and leaf. Vessels dimorphic, including wide, short elements along with fusiform vessel members, both usually with simple perforation, true tracheids present. Rays mostly biseriate or uniseriate; multiseriate rays more than two cells wide are relatively scarce. Axial parenchyma banded and diffuse. Leaves alternate, simple, estipulate, when fully developed composed of a more or less distinct petiole, a winged or expanded portion (blade) followed by a stout and often coiled tendril (a prolongation of the leaf midrib), usually terminated by a large, pendant, often brightly colored, urceolate or cylindrical pitcher (ascidium) that has a recurved rim and a flattened lid (operculum) projecting over the mouth, but not closing it except in the young state. Below the base of the lid on the outside of the pitcher is a spur (an appendage); at the entrance of the pitcher are numerous nectar-glands, below which the interior is slippery, and for some distance below are digestive glands. The second appendage, if developed, also lies on the midline, but at the apex of the lid; the apical appendage is more filiform, and may not always be glandular (Cheek and Jebb 2001). Stomata anomocytic. Flowers small, in racemes or panicles, dioecious. Sepals usually four, rarely three, imbricate, free or less often basally connate, glandular and nectariferous within. Stamens (2-)8–25, rarely 4–6; filaments united into a column; anthers crowded but distinct, 2-locular, extrorse, tetrasporangiate, opening longitudinally by slits. Pollen grains 3-celled, in tetrads, spinulose, with indistinct apertures. Gynoecium syncarpous, of four or very rarely three or six carpels opposite the sepals; ovary superior, 4- or rarely 3-locular, with numerous ovules multiseriate on

axile placentas; stigma discoid, on a very short style or sessile. Ovules numerous, multiseriate on axile or laminar placentas, ascending, anatropous, bitegmic, crassinucellate. Endosperm nuclear. Fruits elongated, sometimes stipitate, leathery loculicidal capsules. Seeds numerous (100–500), mostly filiform, ascending, imbricate, often winged (usually, with a narrow or hair-like wing at either end), or wingless; seed coat formed by both the integuments, but mainly exotestal; exotesta with much thickened inner walls. Embryo minute but well-differentiated, straight, cylindrical, surrounded by starchy, proteinaceous and oily endosperm, sometimes endosperm absent. Tanniniferous, producing acetogenic napthoquinone plumbagin, cyanidin, and also kaempferol and quercetin, n = 40. . . . . . . 3. NEPENTHACEAE.

1 Ovary inferior or partly inferior. Sympodially branched, scandent shrubs or lianas, with hooked or twining shoot ending of each member of the sympodium, "with juvenile growth often heteromorphic, erect, shrubby, monopodial, unbranched or sparsely branched, and without hooks, with adult growth comprising elongated, usually little-branched stems that bear smaller extra-axillary latral stems" (Taylor et al. 2005). Plant externally glabrous, often with calcium oxalate crystals. Primary cortex contains isolated, thick-walled secretory cells. Vessels with simple perforations in very oblique end walls. Sieve-element plastids Ss-type. Fibers with large, evidently bordered pits. Rays homogeneous. Axial parenchyma commonly in tangential bands. Nodes trilacunar with three traces. Leaves pinnately veined, with very small, mostly caducous stipules, or estipulate, and beset with minute, embedded, peltate, multicellular, waxy glands. Stomata actinocytic (surrounded by more or less distinct subsidiary cells). Flowers small, in panicles, in spikes, and in cymes; the terminal inflorescence unit often apparently cymose (axillary or terminal), with strongly articulated pedicels, 5-merous, rarely (*Ancistrocladus grandiflorus*) 4-merous. Sepals free to connate into a short tube, at length adnate to the base of the ovary; lobes quincuincial, imbricate, becoming unequally enlarged and winglike in fruit. Petals more or less fleshy, shortly connate or cohering, imbricate. Stamens (8-)10 in one cycle or rarely 5 (*A hamatus*) or 15 (*A. robertsoniorum*); filaments short, fleshy, connate below and adnate to the bases of the petals; anthers basifixed, introrse or latrorse, tetrasporangiate, opening longitudinally. Pollen grains 2(4)-colporate. Gynoecium of three united carpels; ovary 1-locular, with one basilateral ovule on a short funiculus; stylodia free or connate, articulated at the base and thickened upward, with more or less hippocrepiform or puntofid stigmas. Ovules ascending, hemitropous, bitegmic, on a short funiculus, crassinucellate. Endosperm cellular. Fruits dry, woody, indehiscent, surrounded by a corky hypanthium and crowned by spreading or erect winglike calyx-lobes, floating in water. Seeds large, with a thin, membranous seed coat, strongly ruminate, hard, starchy, sometimes ruminate endosperm, with testa intruding between folds, and fairly small, straight embryo. Produce acetogenic isoquinoline alkaloids, cyanidin and delphinidin; kaempferol, quercetin, myricetin and ellagic acid. . . . . 1. ANCISTROCLADACEAE

1. ANCISTROCLADACEAE

Planchon ex Walpers 1851. 1/16. Tropical Africa, southern Asia, and from eastern Himalayas to West Malesia.

Ancistrocladus.

Numerous molecular studies (particularly Meimberg et al. 2000), provide evidence for the sister group relationship between Ancistrocladaceae and Dioncophyllaceae and their close relationship with Nepenthaceae, Drosophyllaceae, and Droseraceae, confirming the earlier view of Airy Shaw (1952).

2. DIONCOPHYLLACEAE

Airy Shaw 1952. 3/3. Western tropical Africa. *Dioncophyllum, Triphyophyllum, Habropetalum*.

Closely related to the Ancistrocladaceae and Droseraceae. The traps on young stems and leaves, constituted by multicellular, stalked or sessile secretory glands, recalling those of Ancistrocladaceae and Droseraceae, which secrete sticky, acid mucilage. In their alkaloids Dioncophyllaceae strongly resemble Ancistrocladaceae (Hegnauer 1987; Bringmann et al. 1991).

3. NEPENTHACEAE

Berchtold et J. Presl 1820. 1/87. Madagascar (2), Seychelles (1), Sri Lanka (1), northeastern India

(1, Assam), Southeast Asia (6–9), northern Australia (1), New Caledonia (1), but mostly in Malesia (82).

Nepenthes (including *Anurosperma*).

Related to Ancistrocladaceae and Dioncophyllaceae.

4. DROSOPHYLLACEAE

Chrtek, Slaviková et Studnicka 1989. 1/1. Portugal, southern Spain, Morocco.

Drosophyllum.

For a long time *Drosophyllum* is included in Droseraceae, but recent studies have indicated an isolation of this genus within Droseraceae. Metcalfe (1952) showed that *Drosophyllum* shares the stalked, non-irritable glands, and the whole trapping mechanism, with *Triphyophyllum* of Dioncophyllaceae. Strong evidence for the affinity of *Drosophyllum* with other members of the Nepenthales has been provided by cladistic analysis of *mat*K gene sequences (Meimberg et al. 2000).

5. DROSERACEAE

R. A. Salisbury 1808 (including Aldrovandaceae Nakai 1949, Dionaeaceae Rafinesque 1837). 3/112. *Drosera* (110) subcosmopolitan with concentration in Australia and New Zealand, *Dionaea* (1) in southeastern United States, and *Aldrovanda* (1) throughout much of Eurasia – extends south to northeastern Australia.

Drosera, Dionaea, Aldrovanda.

Probably related to the Nepenthaceae, which is supported also by molecular data (Albert et al. 1992). Meimberg et al. (2000) placed Droseraceae sister to a clade comprising Dioncophyllaceae, Ancistrocladaceae and Nepenthaceae.

Bibliography

Airy Shaw HK. 1952. On the Dioncophyllaceae, a remarkable new family of flowering plants. Kew Bull. 1951: 327–347.

Albert VA, SE Williams, and MW Chase. 1992. Carnivorous plants: Phylogeny and structural evolution. Science 257: 1491–1495.

Arber A. 1941. On the morphology of the pitcher-leaves in *Heliamphora, Sarracenia, Darlingtonia, Cephalotus,* and *Nepenthes.* Ann. Bot. 5: 563–578.

Basak RK and K Subramanyan. 1966. Pollen grains of some species of *Nepenthes.* Phytomorphology 16: 334–338.

Batalin A. 1877. Mechanik der Bewegungen der insek-tenfressenden Pflanzen. Flora 35: 54–58; 60: 33–39, 54–58, 65–73, 105–111, 129–154.

Boesewinkel FD. 1989. Ovule and seed development in Droseraceae. Acta. Bot. Neerl. 38: 295–311.

Bringmann G, D Lisch, H Reuscher, LA Assi, and K Günther. 1991. Atrop-diastereomer separation by racemate resolution techniques: N-methyl-dionco-phylline A and its 7-epimer from *Ancistrocladus* abbreviatus. Phytochemistry 30: 1307–1310.

Bringmann G and F Pokorny. 1995. The naphtylisoquinoline alkaloids. Alkaloids 46: 127–217.

Bringmann G, F Pokorny, H Reucher, D Lisch, and L Ake Assi. 1990. Novel Ancistrocladaceae and Dioncophyllaceae type naphtylisoquinoline alkaloids from *Ancistrocladus abbreviatus*: A phylogenetic link between the two families? Planta Med. 56: 496–497.

Bringmann G, M Rubenacker, P Vogt, H Busse, LA Assi, K Peters, and HG von Shering. 1991. Dioncopeltine A and Dioncolactone A: Alkaloids from *Triphyophyllum peltatum.* Phytochemistry 30: 1691–1696.

Cameron KM, MW Chase, and SM Swensen. 1995. Molecular evidence for the relationships of *Triphyophyllum* (Dioncophyllaceae) and *Ancistrocladus* (Ancistrocladaceae). Am. J. Bot. 82: 117–118 (Abstract).

Carlquist S. 1981. Wood anatomy of Nepenthaceae. Bull. Torrey Bot. Club 108: 324–330.

Carlquist S and EJ Wilson. 1995. Wood anatomy of *Drosophyllum* (Droseraceae); ecological and phylogenetic considerations. Bull. Torrey Bot. Club. 122: 185–189.

Chanda S. 1965. The pollen morphology of Droseraceae with special reference to taxonomy. Pollen Spores 7: 509–528.

Cheek M and M Jebb. 2001. Nepenthaceae. In: HP Nooteboom, ed. Flora Melsiana, ser. 1, 15: 1–157, Leiden.

Chrtek J and Z Slavikova. 2000. Genera and families of the Droserales order. Novit. Bot. Univ. Carol. 13: 39–46.

Chrtek J, Z Slavik6va, and M Studnicka. 1989. Beitrag zur Leitbündelanordnung in den Kronblättern ausgewählter Arten der fleischfressenden Pflanzen. Preslia 61: 107–124.

Conran JG, VG Jaudzems, and ND Hallam. 1997. Droseraceae germination patterns and their taxonomic significance. Bot. J. Linn. Soc. 123: 211–223.

Darwin C. 1875. Insectivorous plants. John Murray, London.

Daumann E. 1930. Das Blütennektarium von *Nepenthes.* Beih. Bot. Centralbl. 47: 1–14.

Diels L. 1906. Droseraceae. Das Pflanzenreich IV, 112: 1–136.

Degreef LD. 1988. Evolution of *Aldrovamda* and *Dionaea* traos. Carnivorous Plant Newslett 17: 119–125.

Erdtman G. 1958. A note on the pollen morphology in the Ancistrocladaceae and Dioncophyllaceae. Veroff. Geobot. Inst. Rübel, Zürich 33: 47–49.

Foucher JP, JL Pousset, A Cave, A Bouquet, and RR Paris. 1971. Chimiotaxinomie des Ancistrocladacees: 1. Sur les alcaloides de l'Ancistrocladus ealaensis J. Leonard. Plant Med. Phytother. 5: 16–24.

Gottwald H and N Parameswaran. 1968. Das sekundäre Xylem und die systematische Stellung der Ancistrocladaceae und Dioncophyllaceae. Bot. Jahrb. Syst. 88: 49–69.

Heubl G and A Wistuba. 1995. A cytological study of the genus *Nepenthes* (Nepenthaceae). Sendtnera 4: 169–174.

Higashi S, A Nagashima, H Ozaki, M Abe and T Uchiumi. 1993. Analysis of feeding mechanisms in a pitcher of *Nepenthes hybrida*. J. Plant Res. 106: 47–54.

Hooker JD. 1859. On the origin and development of the pitcher of *Nepenthes,* with an account of some new Bornean plants of the genus. Trans. Linn. Soc. London 22: 415–424.

Hoshi Y and Kondo K. 1998. A chromosome phylogeny of the Droseraceae by using CMA-DAPI fluorescent banding. Cytologia 63: 329–339.

Iones K. 1964. Pollen structure and development in *Drosera*. J. Bot. Linn. Soc. 59:81–87.

Jebb M and M Cheek. 1997. A skeletal revision of *Nepenthes* (Nepenthaceae). Blumea 42: 1–106.

Kaul RB. 1982. Floral and fruit morphology of *Nepenthes lowii* and *N. villosa,* montane carnivores of Borneo. Am. J. Bot. 69: 793–803.

Kaul RB. 1933. Vergleichende entwicklungsgeschichtliche Untersuchungen an der Insektivore *Nepenthes*. Beih. Bot. Centralbl. 51: 311–334.

Keng H. 1967. Observations on *Ancistrocladus*. Gard. Bull. Singapore 22: 113–121.

Kondo K. 1973. The chromosome numbers of *Striga asiatica* and *Triphyophyllum peltatum*. Fyton 31: 1–2.

Kondo K and PS Lavarack. 1984. A cytotaxonomic study of some Australian species of Drosera. Bot. J. Linn. Soc. 88: 317–333.

Korzshinsky S. 1886. Über die Samen der *Aldrovanda vesiculosa* L. Bot. Centralbl. 27: 302–304, 334–335.

Kovácik J and M Repcák. 2006. Napthoquinones content of some sundews (*Drosera* L.). Carniv. Plant Newslett. 35: 49–51.

Kubitzki K. 2003a. Droseraceae. In: K Kubitzki, ed. The families and genera of vascular plants, vol. 5, pp. 198–202. Springer, Berlin/Heidelberg/New York.

Kubitzki K. 2003b. Drosophyllaceae. In: K Kubitzki, ed. The families and genera of vascular plants, vol. 5, pp. 203–205. Springer, Berlin/Heidelberg/New York.

Kubitzki K. 2003c. Nepenthaceae. In: K Kubitzki, ed. The families and genera of vascular plants, vol. 5, pp. 320–324. Springer, Berlin/Heidelberg/New York.

Kühl R. 1933. Vergleichende entwicklungsgeschichtliche Untersuchunged an der Insektivore *Nepentes*. Beih. Bot. Centralbl. 51, 1: 311–334.

Kuprianova LA. 1973. Pollen morphology within the genus *Drosera*. Grana 13: 103–107.

Lloyd FE. 1942. The carnivorous plants. Chronica Botanica, Waltham, MA.

Marburger JE. 1979. Glandular leaf structure of *Triphyophyllum peltatum* (Dioncophyllaceae): A "fly-paper" insect trapper. Am. J. Bot. 66: 404–411.

Meimberg H and G Heubl. 2006. Introduction of a nuclear marker for phylogenetic analysis of Nepenthaceae. Plant Biol. 8: 831–840.

Meimberg H, A Wistuba, P Dittrich and C Heubl. 2000. Molecular phylogeny of Nepenthaceae based on cladistic analysis of plastid *trn*K intron sequence data. Plant Biol. 3: 164–175.

Metcalfe CR. 1952. The anatomical structure of the Dioncophyllaceae in relation to the taxonomic affinities of the family. Kew Bull. 1951: 351–368.

Moran J, WE Booth, JK Charles. 1999. Aspects of pitcher morphology and spectral characteristics of six Bornean *Nepenthes* pitcher plant species: implications for prey capture. Am. J. Bot. 83: 521–528.

Nemirovich-Danchenko EN. 1988. Nepenthaceae. In: A Takhtajan, ed. Comparative seed anatomy, vol. 2, pp. 133–135. Nauka, Leningrad (in Russian).

Owen TP and KA Lennon. 1999. Structure and development of the pitchers from the carnivorous plant *Nepenthes alata* (Nepenthaceae). Am. J. Bot. 86: 1382–1390.

Pant DD and S Bhatnagar. 1977. Morphological studies in *Nepenthes* (Nepenthaceae). Phytomorphology 27: 13–34.

Patankar TBV. 1956. Further contribution to the embryology of *Drosera burmannii* Vahl. Proc. Indian Acad. Sci. 43B: 161–171.

Porembski S. 2003. Ancistrocladaceae. In: K Kubitzki, ed. The families and genera of vascular plants, vol. 5, pp. 25–27. Springer, Berlin/Heidelberg/New York.

Porembski S and W Barthlott. 2003. Dioncophyllaceae. In: K Kubitzki, ed. The families and genera of vascular plants, vol. 5, pp. 178–181. Springer, Berlin/Heidelberg/ New York.

Ragleti HWJ, M Weintraub, and E Lo. 1972. Characteristics of *Drosera tentacles:* I. Anatomical and cytological detail. Canad. J. Bot. 50: 159–168.

Rao VS. 1969. The floral anatomy of *Ancistrocladus*. Proc. Indian Acad. Sci. 70B: 215–222.

Rivadavia F, K Kondo, M Kato, and M Hasebe. 2003. Phylogeny of the sundews, *Drosera* (Droseraceae), based on chloroplast *rbc*L and nuclear 18S ribsomal DNA sequences. Am. J. Bot. 90: 123–130.

Roth I. 1953. Zur Entwicklungsgeschichte and Histogenese der Schlauchblätter von *Nepenthes*. Planta 42: 177–208.

Roth I. 1954. Entwicklung und histogenetischer Vergleich der Nektar- und Verdauungsdrüsen von *Nepenthes*. Planta 43: 361–378.

Sahashi N and M Ikuse. 1973. Pollen morphology of *Aldrovanda vesiculosa* L. J. Jpn Bot. 48: 374–379.

Schmid R. 1964. Die systematische Stellung der Dioncophyllaceen. Bot. Jahrb. Syst. 83: 1–56.

Schmid-Hollinger R. 1970. *Nepenthes*-Studien: I. Ho-mologien von Deckel (operculum, lid) und Spitzchen (calcar, spur). Bot. Jahrb. Syst. 90: 275–296.

Seine R and W Barthlott. 1993. On the morphology of trichomes and tentacles of Droseraceae Salisb. Beitr. Biol. Pflanzen 67: 345–366.

Shabes LK and AA Morozova. 1996 Droseraceae. In: A Takhtajan, ed. Comparative seed anatomy, vol. 5, pp. 93–98. Nauka, St. Petersburg (in Russian).

Smith CM. 1929. Development of *Dionaea muscipula*: I. Flowers and seed. Bot. Gaz. 87: 507–530.

Takahashi H. 1988. Ontogenetic development of pollen tetrads of *Drosera capensis* L. Bot. Gaz. 149: 275–282.

Takahashi H and K Sohma. 1982. Pollen morphology of the *Drosera* and its related taxa. Sci. Rep. Tohoku Imp. Univ., IV (Biol.), 38: 81–156.

Taliev VI. 1903. On the morphology and genesis of insectivorous plants. Kharkov (in Russian).

Taylor CM, RE Gereau, and GM Walters. 2005. Revision of *Ancistrocladus* Wall. (Ancistrocladaceae). Ann. Missouri Bot. Gard. 92: 360–399.

Terekhin ES. 1985. Droseraceae. In: MS Yakovlev, ed. Comparative embryology of vascular plants: Bru--nelliaceae-Tremandraceae, pp. 48–52. Nauka, Leningrad (in Russian).

Terekhin ES. 1986. Seed development and structure in *Aldrovanda vesiculosa* (Droseraceae). Bot. Zhurn. 71: 527–533 (in Russian).

Tieghem P van. 1903. Sur les Ancistrocladacees. J. Bot. (Paris) 17: 151–168.

Toekes ZA, WC Woon, and SM Chambers. 1974. Digestive enzymes secreted by carnivorous plant *Nepenthes macfarlanei* L. Planta 119: 39–46.

Trankovsky DA. 1938. Spermatogenesis and fertilization in *Drosera*. Bull. Mosc. Soc. Nat. 67(I): 104–111 (in Russian).

Venkatasuban KR. 1950. Studies in the Droseraceae: 2. A contribution to the embryology of three species of *Drosera*. Proc. Indian Acad. Sci. 32B: 23–38.

Vines SH. 1901. The proteolytic enzymes of *Nepenthes* III. Ann. Bot. 15: 563–573.

Williams SE. 1976. Comparative sensory physiology of the Droseraceae: The evolution of a plant sensory system. Proc. Am. Phil. Soc. 120: 187–204.

Williams SE, VA Albert, and MW Chase. 1994. Relationships of Droseraceae: A cladistic analysis of *rbc*L sequence and morphological data. Am. J. Bot. 81: 1027–1037.

Wood CE. 1960. The genera of Sarraceniaceae and Droseraceae in the Southeastern United States. J. Arnold Arbor. 41: 152–163.

Subclass V. DILLENIIDAE

Trees, shrubs, and herbs. Ethereal oil cells wanting. Vessels with scalariform or simple perforations. Sieve-element plastids of S-type or rarely of P-type. Nodes tri-multilacunar or unilacunar. Leaves simple or less often compound, with or without stipules. Stomata of various types, mostly anomocytic. Flowers usually with double perianth, mostly bisexual, spiral, spirocyclic or cyclic. Petals free or less often corolla sympetalous. Pollen grains 2-celled or seldom 3-celled, 3-colpate or of tricolpate-derived type. Gynoecium apocarpous or more often carpels variously united; stylodia distinct or more or less connate; ovary superior or inferior. Ovules mostly bitegmic or less often unitegmic, commonly crassinucellate. Endosperm nuclear or less often cellular. Fruits of various types. Seeds mostly with endosperm.

Dilleniidae are one of the largest subclasses of the magnoliopsids and one of their central groups, archaic members of which have many features in common with the Magnoliidae.

Bibliography

Albert VA, SE Williams, and MW Chase. 1992. Carnivorous plants: phylogeny and structural evolution. Science 257: 1491–1495.

Ditsch F and W Barthlott. 1994. Mikromorphologie der Epicudculargewachse und die Systematik der Dilleniales, Lecytidales, Malvales, und Theales. Trop. Suptrop. Pflanzenwelt 88: 7–72.

Leins P and C Erbar. 1991. Fascicled androecia in Dilleniidae and some remarks on the *Garcinia* androecium. Bot. Acta 104: 257–344.

Superorder DILLENIANAE

Order 48. DILLENIALES

Mostly trees (sometimes buttresses, or tortuous to small rosette trees), and shrubs, rarely woody lianas, subshrubs (*Pachynema*), or semiherbaceous from a woody rhizome (*Acrotrema*); raphides common in wood. Vessel elements angular to circular, from medium to extremely long, usually very long with scalariform perforations that have 5–130 bars (mostly 15–40) or scalariform and simple perforations; lateral pitting from scalariform to opposite. Fibers with distinctly bordered pits. Rays heterogeneous. Axial parenchyma apotracheal diffuse and aggregate diffuse, often paratracheal diffuse to incomplete vasicentric. Sieve-element plastids of S-type. Nodes trilacunar, pentalacunar, multilacunar, or sometimes unilacunar with one trace per gap. Leaves usually alternate, very rarely opposite, entire or serrate to dentate, rarely pinnatifid or trilobed, pinnately veined, often with prominent parallel lateral veins, seldom much reduced and needlelike (species of *Hibbertia*) or scalelike (species of *Pachynema* with stems and branches flattened into phylloclades). Stipules usually wanting, but rarely petioles with stipulelike, often wholly or partly caducous wings. Walls of epidermal cells mostly silicified (as in Magnoliaceae and Calycanthaceae). Stomata of various types, but mostly anomocytic. Flowers small to medium-sized, rarely large, in various cymose or racemose inflorescences or sometimes solitary, bisexual or rarely unisexual (monoecious or dioecious), actinomorphic or sometimes zygomorphic (but only with reference to the androecium), with spirocyclic perianth. Receptacle flat or conical between the carpels (*Dillenia*). Sepals mostly five, rarely more (up to 15) or less (3), strongly imbricate, persistent. Petals five or rarely fewer, imbricate, often crumpled in bud, deciduous. Stamens usually numerous (from 50 to 500 in *Tetracera* and to 900 in *Dillenia ovalifolia*), rarely 10 or fewer. Filaments usually filiform, rarely broad and flattened (*Pachynema*), variously united into fascicles at the base; the members of each fascicle originating in centrifugal sequence, or the whole androecium developing centrifugally. The androecium of *Schumacheria* is unique in its placement at one side of the gynoecium with the large number of stamens united basally by their filaments into an oblique cylindric column. Anthers tetrasporangiate, often more or less embedded in the connective tissue, from short to very elongate, sometimes with more or less prolonged connective, opening by longitudinal slits or apical pores. Tapetum secretory. Microsporogenesis simultaneous. Pollen grains 2-celled, 3-colpate (*Acroterma, Dillenia, Hibbertia, Pachynema*), 4-colpate or 3-colporate, tectate-columellate, with an incomplete tectum, finely lunctate to coarsely reticulate. Gynoecium usually of 1–20 carpels, mostly of several carpels; carpels free, less often partly connate at the base, rarely fused to above the middle (*Curatella*) or even to the apex (*Neowormia*); carpels more or less conduplicate and sometimes not fully sealed. Stylodia usually slender and elongate, with an apical, often capitate, minute stigma. Ovules one to many in each carpel, apotropous, anatropous to amphitropous or campylotropous, bitegmic, crassinucellate; micropyle zigzag or exostomal. Female gametophyte of *Polygonum*-type. Endosperm nuclear. Fruits apocarpous or syncarpous, mostly dry and dehiscent, sometimes indehiscent (*Dillenia*). Seeds mostly with a well-developed crested or laciniate funicular aril (very large in *Dellenia sarmentosa* and *Tetracera* spp.), less often with the aril vestigial or even wanting. Seed coat formed by both integuments and with sclerotic endotesta and tracheidal exotegmen (endotestal-exotegmal); embryo usually minute, straight, more or less differentiated; endosperm copious, oily, and proteinaceous, sometimes truminate. Tanniniferous, containing a great diversity of flavonoids, but mostly without alkaloids and always without benzylisoquinoline alkaloids, n = 4, 5, 8–10, 12, 13.

The Dilleniaceae are a family of considerable morphological and ecological diversity (Hutchinson 1969; Dickison 1970b, 1982; Stebbins 1974; Cronquist 1981). They are also very heterobathmic. Although they retain some primitive features in wood anatomy and flower morphology (apocarpous gynoecium, more or less conduplicate carpels and stamens that sometimes have a prolonged connective), they have also many features of specialization, including specialized androecium and specialized stylodia with apical non-decurrent stigma. Although the Dilleniaceae are the most archaic family in the Dilleniidae, they have already lost direct links with the Magnoliidae. However, they occupy an intermediate position between the Magnolianae and the Ericanae and Violanae.

1. DILLENIACEAE

R.A. Salisbury 1807 (including Hibbertiaceae J. Agardh 1858, Soramiaceae Martynov 1820). 12/400. Almost pantropical, but centered in Asia and Australasia, especially in Australia. Comparatively rare in Africa, where it is represented only by members of the pantropical genus *Tetracera*. *Hibbertia* is native to both Australasia and Madagascar.

1.1 DILLENIOIDEAE
Perforations exclusively scalariform; lateral pitting scalariform-transitional. Anthers linear or oblong, loculi very rarely slightly divergent. – DILLENIEAE: *Dillenia, Neowormia, Schumacheria, Didesmandra, Acrotrema;* HIBBERTIEAE: *Hibbertia, Pachynema.*

1.2 DELIMOIDEAE (TETRACEROIDEAE)
Perforations scalariform and simple; lateral pitting opposite-transitional. Anthers short, loculi divergent, separated at the base by the thickened connective. – *Tetreacera, Curatella, Pinzona, Doliocarpus, Davilla.*

Bibliography

Baretta-Kuipers T. 1972. Some remarks on the wood structure of *Pinzona* and allied genera of the subfamily Tetraceroideae (Dilleniaceae). Acta Bot. Neerl. 21: 573–577.

Dickison WC. 1967a. Comparative morphological studies in Dilleniaceae: I. Wood anatomy; II. The pollen. J. Arnold Arbor. 48: 1–29, 231–240.

Dickison WC. 1968. Comparative morphological studies in Dilleniaceae: III. The carpels. J. Arnold Arbor. 49: 317–329.
Dickison WC. 1969. Comparative morphological studies in Dilleniaceae: IV. Anatomy of the node and vascularization of the leaf. J. Arnold Arbor. 50: 384–400.
Dickison WC. 1970. Comparative morphological studies in Dilleniaceae: V. Leaf anatomy; VI. Stamens and young stem. J. Arnold Arbor. 51: 89–113, 403–418.
Dickison WC. 1971. Comparative morphological studies in Dilleniaceae. VII. Additional notes on *Acrotrema*. J. Arnold Arbor. 52: 319–333.
Dickison WC. 1979. A note on the wood anatomy of *Dillenia* (Dilleniaceae). IAWA Bull. 2/3: 57–60.
Dickison WC, JW Nowicke, and JJ Skwarla. 1982. Pollen morphology of the Dilleniaceae and Actinidiaceae. Am. J. Bot. 69: 1055–1073.
Dulci Elizabeth J and KK Lakshmanan. 1984. Further contributions to the embryology of *Dillenia suffruticosa* (Griff.) Martelli. J. Indian Bot. Soc. 63: 353–359.
Endress PK. 1997. Relationships between floral organization, architecture, and pollination made in *Dillenia* (Dilleniaceae). Plant Syst. Evol. 206: 99–118.
Gurni AA and K Kubitzki. 1981. Flavonoid chemistry and systematics of the Dilleniaceae. Biochem. Syst. Ecol. 9: 109–114.
Horn JW. 2002. Phylogenetics of the Dilleniaceae. In Botany 2002: Botany in the Curriculum, Abstracts: 128, Madison.
Horn JW. 2007. Dilleniaceae. In: K Kubitzki, ed. The families and genera of vascular plants, vol. 9, pp. 132–154. Springer, Berlin/Heidelberg/New York.
Kubitzki K. 1968. Flavonoide und Systematik der Dil-leniaceen. Ber. Deutsch. Bot. Ges. 81: 238–251.
Sastri RLN. 1958. Floral morphology and embryology of some Dilleniaceae. Bot. Not. 111: 495–511.
Swamy BGL and K Periasamy. 1955. Contributions to the embryology of *Acrotrema arnottianum*. Phytomorphology 5: 301–314.
Vyshenskaya TD and GG Oganezova. 1991. Dilleniaceae. In: A Takhtajan, ed. Comparative seed anatomy, vol. 3, pp. 163–171. Nauka, Leningrad (in Russian).
Wilson CL. 1965. The floral anatomy of the Dilleniaceae: I. *Hibbertia* Andr. Phytomorphology 15: 248–274.
Wilson CL. 1973 (1974). The floral anatomy of the Dilleniaceae: II. Genera other than *Hibbertia*. Phytomorphology 23: 35–42.

Superorder ERICANAE

Order 49. PARACRYPHIALES

Small to moderate-sized trees with unicellular, unbranched hairs. Styloid crystals of calcium oxalate present in the bark, pith, and xylem and phloem parenchyma. Vessel elements extremely long, angular to rounded in transverse section; perforations scalariform in almost vertical end walls, with numerous (up to 203) bars; lateral pitting scalariform to transitional. Fibers thick-walled, with numerous bordered pits. Rays heterogeneous, of very primitive type. Axial parenchyma diffuse and diffuse-in-aggregates, apotracheal as well as scanty paratracheal. Phloem of primitive type. Sieve-elements of S-type. Nodes trilacunar. Leaves subverricillate, simple, finely serrate, pinnately veined, teeth vascularized by a single medial vein extending to the tooth apex, leathery, estipulate. Stomata anomocytic. Flowers sessile, in terminal, paniclelike, compound inflorescences consisting of many spikes, bisexual or occasionally exclusively male (without pistillodia) and borne at the lower levels on the inflorescence axis. Perianth consists of four caducous, decussate, concave, free segments (sepals?); aestivation of the perianth parts cochleate, having one large helmet-shaped segment that encloses or covers the others. Stamens usually 8 (–11), in a single cycle, free; filaments of the male flowers conspicuously swollen, filaments of bisexual flowers sometimes accrescent. Anthers tetrasporangiate, basifixed, opening longitudinally. Pollen grains 3-colporate, tectate-columellate, reticulate or occasionally transitional between reticulate and rugulate. Gynoecium of 8–15, laterally connate, conduplicate carpels that are also ventrally adnate to the solid column of central tissue. Stigmas sessile and conduplicately folded; ovary superior, 8–15-locular, with four rather small ovules in each locule borne in a single row on axile placentas. Ovules anatropous, unitegmic, crassinucellate. Endosperm cellular. Fruits capsules with mature carpels separating from the central column (except for a distal connection maintained by two strands apparently representing the lignified ventral carpellary vascular bundles, Dickison and Baas 1977), spreading out from the base and opening ventrally. Seeds small, compressed, winged, exarillate, exotestal, with sinuous anticlinal walls, inner walls lignified; embryo straight, well differentiated; endosperm copious.

As Dickison and Baas (1977) point out, *Paracryphia* is distinguished by an unusual combination of characters. "It is remarkable that most characters from vegetative anatomy are distinctly primitive or at least un-specialized, whereas those of reproductive organs can be regarded as advanced or specialized." *Paracryphia* was originally included into the Eucryphiaceae but was later separated as its own family by Airy Shaw (1965). Dahlgren (1983) placed Paracryphiaceae in his suborder Theineae of the order

Theales, whereas according to Cronquist (1981: 336) *Paracryphia* is reasonably at home in the Theales. He mentions that the persistent central column is a common (though by no means consistent) feature of capsular fruits in the Theales, and the manner of dehiscence in *Paracryphia* is reminiscent of *Medusagyne*. In my opinion *Paracryphiaceae* are distinct enough to be entitled to the status of an order (Takhtajan 1987, 1997).

Paracryphiales have many features in common with the Dilleniales, Theales, and Ericales (see Dickison and Baas 1977; Schmid 1978). This includes sieve-element plastids that are extremely similar in size and starch grain characters to those of Dilleniaceae, Theaceae, and Actinidiaceae, and are almost identical to the plastids of *Hibbertia* in the Dilleniaceae (Behnke 1985). Paracryphiales are possibly an early divergence from an ancient dillenialean stock and one of the evolutionary blind ends. *Paracryphia* is one of the most remarkable ancient relicts.

1. PARACRYPHIACEAE

Airy Shaw 1965. 1/1. New Caledonia.
Paracryphia.

Bibliography

Agababian VS and EL Zavarian. 1971. Palynotaxonomy of the genus *Paracryphia* Bak. f. Biol. Zhurn. Armenii 24 (12): 35–40 (in Russian).

Airy Shaw HK. 1965. Diagnoses of new families, new names, etc., for the seventh edition of Willis's Dictionary. Kew Bull. 18: 249–273.

Dickison WC and P Baas. 1977. The morphology and relationships of *Paracryphia* (Paracryphiaceae). Blumea 23: 417–438.

Melikian AP. 1991. Paracryphiaceae. In: A Takhtajan, ed. Comparative seed anatomy, vol. 3, p. 175. Nauka, Leningrad (in Russian).

Schmid R. 1978. Actinidiaceae, Davidiaceae, and Paracryphiaceae: systematic considerations. Bot. Jahrb. Syst. 100: 196–204.

Steenis CGGJ van. 1950. Note on *Paracryphia* Baker (Eucryphiaceae). Bull. Jard. Bot. Buitenzorg 18: 459–460.

Order 50. THEALES

Trees, shrubs, or sometimes woody lianas or herbs. Often with secretory cells, cavities, or channels. Vessel elements from short to very long, with scalariform or simple perforations; lateral pitting from scalariform to alternate. Fibers with bordered or less often simple pores. Rays mostly heterogeneous. Axial parenchyma mostly apotracheal. Young stems often with stratified phloem. Sieve-element plastids of S-type. Nodes unilacunar, less often pentalacunar (Oncothecaceae), or penta- to multilacunar (Caryocaraceae). Leaves alternate or opposite, simple or seldom compound, mostly entire, estipulate, or very rarely (Caryocaraceae) stipulate. Stomata of various types, mostly paracytic. Flowers of medium size or often large, axillary and solitary or in various kinds of inflorescences, usually bisexual, mostly actinomorphic. Perianth spiral or spirocyclic (many Theaceae) or more often cyclic, mostly 5-merous. Calyx often subtended by 2-several bracteoles, these sometimes gradually passing into sepals. Sepals imbricate, free or often basally connate. Petals imbricate or sometimes contorted, free or rarely connate at the base, rarely (*Anthodiscus* in the Caryocaraceae) connate above to form a calyptra, or corolla sympetalous (Oncothecaceae). Stamens numerous to 5-4, free or connate into bundles, when numerous developing in centrifugal sequence. Anthers tetrasporangiate, basifixed or dorsifixed, less often versatile, opening longitudinally or rarely (spp. of *Eurya* and Pentaphylacaceae) by short, apical, porelike slits; connective sometimes more or less prolonged (very long in *Anneslea frapans* of the Theaceae). Tapeturn secretory. Microsporogenesis simultaneous. Pollen grains 2-celled, usually 3-colporate. Gynoecium of 2-many variously united carpels. Stylodia free or more or less connate; ovary superior, rarely semi-inferior or inferior, with as many locules as carpels or unilocular, usually with many ovules on axile placentas or on intruded parietal placentas. Ovules anatropous or less often hemitropous or more or less campylotropous (Pentaphylacaceae), apotropous, bitegmic or (Oncothecaceae) unitegmic, crassinucellate (Pentaphylacaceae) or more often tenuinucellate. Female gametophyte of *Polygonum*- or *Allium*-types, rarely *Adoxa*-type. Endosperm nuclear. Fruits of various types. Seeds exarillate, seed coat formed mainly or only by the outer integument; typically mesotestal with sclerotic cells; endosperm copious or scanty, often very reduced or absent; embryo small to large, straight or curved.

The order Theales is evidently derived from early dillenialean stock. It differs from the Dilleniales mainly in the syncarpous gynoecium and mesotestal and mostly exarillate seeds.

Key to Families

1 Leaves simple. Nodes unilacunar or seldom trilacunar or pentalacunar.

2 Flowers in axillary, pendulous racemes or spikes, 4-merous throughout, closely subtended by two well-developed bracteoles, bisexual or unisexual (plant polygamo-monoecious). Sepals four, decussate, imbricate; petals four, free, imbricate Stamens eight, free, in two cycles. Anthers deeply sagittate, dorsifixed, introrse, and versatile, with a broad connective and a short protrusion, opening longitudinally. Pollen grains 3-colporate, occasionally 4-ruporate. Nectary at base of gynoecium well developed on sepaline radii. Gynoecium of four carpels, they are congenitally united for almost their entire length, free only at the very apex, with short style, stigma large, capitate, the four free carpel tips are postgenitally connected and form a single stigmatic surface (Matthew and Endress 2005). Ovary 4-locular and with axile placentation in the lower part, but in the middle and upper parts the partitions are incomplete and the ovules borne on deeply intruded placentas. Ovules numerous per locule, anatropous, Outer integument contributing to the micropyle zig-zag. Fruits 4-locular, rather firm and dry berry. Seeds arillate, with small, straight embryo, with oily and proteinaceous endosperm and short, fleshy funicle. Evergreen or deciduous shrubs, small trees, or lianas; non laticiferous and without coloured juice. The branchlets with large pith. Cluster crystals and tanniniferous cells are present in parenchymatous tissue. Nodes trilacunar with three traces. Leaves alternate, involute, simple, entire, glabrous or pubescent, petiole bundle arcuate; stiuples free of one another, caduceus, small, linear-lanceolate. Stomata anomocytic. Ellagic acid, cyanidin, flafonols (kaempferol and quercetin) present; n = 12. . . . . 1. STACHYURACEAE.

2 Inflorescence cymose; sclereids absent; anthers opening by apical pores or slits. Evergreen trees; hairs unicellular. Leaves spiral or distichous, estipulate, leaf margins crenate-serrulate. Stomata anomocytic. Flowers in short dichasial cymes, small, 5(6)-merous, sepals free, imbricate; petals free or adnate to corolla, imbricate. Stamens (8-)10(-15), in one cycle; filaments broad, thick; anthers basifixed, setulose. Carpels three; ovules two or many per locule. Female gametophyte of *Adoxa*-type. Fruits dry, ribbed, with papery exocarp and crustaceous endocarp, eventually splitting septicidally into three cocci. Seeds winged, with transparent seed coat; embryo nearly straight, endosperm copious; n = 24. . . . . 2. SLADENIACEAE.

3 Anthers basifixed; corc usually surficial; pseudopollen absent; Fruits berries or drupes. Evergreen trees or shrubs; hairs unicellular. Leaves distichous or spiral, estipulate, leaf margins entire, rarely serrate. Stomata anomocytic. Flowers generally small, solitary, or in clusters, bisexual or unisexual, 5-merous. Sepals connate or distinct, calyx lobes always persistent; petals distinct or connate basally, thick or membranous. Stamens from 5 or 10 up to ca. 30, usually in one or two cycles; filaments shorter or slightly longer than anthers; anthers introrse, opening by longitudinal slits, rarely by pore-like slits. Exine nearly smooth. Carpels mostly (2)3–5(6); ovary superior to inferior; ovules few or numerous per locule. Female gametophyte of *Polygonum*-type. Fruits berries or pomelike with succulent pericarp. Seeds usually numerous, often small, not winged; seed coat usually cartilaginous, very thin; endosperm present, often copious; embryo slender, curved or hippocrepiform; radicle always longer than the narrow cotyledons, incumbent. Contain proanthocyaninidins, flavonols and ellagic acid, n = 12–13, 15, 18, 20–25. . . . . 4. TERNSTROEMIACEAE.

3 Anthers versatile; cork usually deep-seated; pseudopollen produced from connective; capsule with persistent column. Trees or shrubs. Leaves spiral or distichous, simple, usually coriaceous; leaf margins serrate. Stomata mostly paracytic (except in *Franklinia, Pyrenaria* spp.). Flowers generally large and showy, usually solitary, axillary, bisexual; Sepals five or more, imbricate, calyx lobes persistent or deciduous; petals five, rarely numerous, free or connate basally; stamens generally more than 40, in 2–6 cycles, free, rarely connate; filaments several times as long as anthers. Exine mostly finely granulate or finely reticulate. Carpels mostly (3-)5(-10); ovary superior. Female gametophyte of *Polygonum*-, or rarely *Allium*-type. Fruits dry,

capsular or indehiscent. Pericarp usually woody, rarely coriaceous. Seeds often one or few in each locule, fairly large, often winged, seed coat woody; endosperm present or absent, sometimes copious (*Stewartia*); embryo broad, straight or slightly bent; radicle much shorter than the broad cotyledons, accumbent. Contain flavonols, flavones, proanthocyanins and ellagic acid, n = 15, 18........ 5. THEACEAE.

4 Anthers opening by very small apical pore formed by a small, uplifting valve. Ovules two in each locule, pendulous from the inner angle, campylotropous to apotropous, crassinucellate. Seeds more or less winged, with scanty endosperm and horseshoe-shaped embryo. Fruits rather woody, loculicidal capsules with persistent central column. Flowers in axillary or terminal racemes, 5-merous throughout, with two bracteoles appressed to the calyx. Filaments thickened and expanded toward the middle, loosely connivent toward the base. Evergreen shrubs or small trees with mucilage cells in the cortex. Clustered crystals of calcium oxalate present in parenchymatous tissues, but not in the form of raphides. Vessel elements elongate, with scalariform perforations. Buds perulate. Plants Al-accumulators; .....3. PENTAPHYLACACEAE.

4 Anthers opening by longitudinal slits. Ovules tenuinucellate. Vessels with simple perforations or sometimes some of them with scalariform perforations. Corolla sympetalous, 5-lobed. Sepals five, without glandular pits, persistent in fruit. Stamens five, attached to the tube of the corolla; connectives prolonged and abruptly inflexed, forming a roof over the gynoecium. Ovary 5-locular, 5-grooved, apically shortly 5-lobed, each lobe ventrally stigmatic and imperfectly sealed. Ovules two in each locule, unitegmic, apical-axile, pendulous on a long funicle. Flowers small, in terminal panicles or compound spikes. Fruits oblate-compressed drupes with thin flesh and very thick-walled, 5-locular pyrene. Seeds with very reduced seed coat and copious endosperm surrounding straight, cylindric embryo with a long hypocotyl and two very short cotyledons. Clustered crystals of calcium oxalate present in parenchymatous tissues. Small evergreen, glabrous trees. Vessels with oblique perforation plates that have 10–50 bars. Phloem stratified. Nodes pentalacunar with five traces. Leaves simple, leathery, margin with caduceus glands. Stomata more or less anomocytic, n = 25. 6. ONCOTHECACEAE.

1 Leaves 3–5-foliolate, evergreen, alternate (*Anthodiscus*) or opposite (*Caryocar*), toothed or seldom subentire; simple leaves intermixed with palmate leaves in *Anthodiscus*; leaflet blades usually serrulate or serrate in *Caryocar*, crenulate or crenate in *Antho-discus*. Stipules caducous or none. Stomata mostly anomocytic. Small to very large trees, the boles of some large *Caryocar* have conspicuous running buttresses; *C. brasiliense* shrublike. Nodes penta- or multilacunar. Branched sclerenchymatous idioblasts and solitary or clustered crystals of calcium oxalate present in parenchymatous tissues. Vessels usually with simple perforations. Fibers with simple, small pits, often septate. Rays heterogeneous. Axial parenchyma mostly apotracheal (*Caryocar*) or paratracheal (*Anthodiscus*). Flowers large, in terminal, ebracteate racemes, bisexual, nocturnal (*Caryocar*). Sepals 5–6, imbricate or truncate, cuplike with scarcely deve-loped lobes in *Anthodiscus*. Petals 5–6, imbricate, free (*Caryocar*) or connate above into a calyptra (*Anthodiscus*). Stamens numerous (55–750), in five fascicles alternating with the petals or shortly connate at the base into a ring; the innermost stamens shorter and staminodial. Filaments very long, variously bent in bud, often covered with minute, glandlike tubercules toward apex, sometimes the inner ones without anthers; anthers small, versatile, dorsifixed or basifixed. Gynoecium of 4–20 carpels, with free, very long and filiform (*Caryocar*) or short (*Anthodiscus*) stylodia. Ovary superior, 4–6 (*Caryocar*) or 8–20-locular, with a single ascending, anatropous or orthotropous, bitegmic ovule in each locule; placentation axile. Fruits drupaceous, separating into 1-seeded pyrenes at maturity, the endocarp muricate, tuberculate, or spinulose surfaces, breaking up into 1-seeded parts. Seeds 1–4 in *Caryocar*, 8–20 in *Anthodiscus*, reniform, with scanty or no endosperm; embryo straight (*Caryocar*) with a large, oily, and proteinaceous, or spirally twisted hypocotyl and small hooked inflexed cotyledons; n = 23. . . . . 7. CARYOCARACEAE.

1. STACHYURACEAE

J. Agardh 1858. 1/16. Himalayas, southern Tibet, Assam, northern Burma, continental China, Taiwan, Japan, Bonin Islands, northern Indochina.

Stachyurus

One of the most archaic member of Theales.

2. SLADENIACEAE

Airy Shaw 1965. 2/3. South-East Asia, trop. Africa. *Sladenia* (South-East Asia), *Ficalhoa* (trop. Africa).

Sladeniaceae differ from the Ternstroemiaceae and Theacea in many apomorphic characters. The monotypic Southeast-asiatic genus *Sladenia* differs not only in its dichasial cymes, dilated filaments, and lack of foliar sclereids (Keng 1962) but also in the high degree of vessel grouping (vessels in radial multiples of 2–6) and in the opposite to alternate intervessel pits (Liang and Baas 1990).

3. PENTAPHYLACACEAE

Engler 1897. 1/1–2. Southern China, Vietnam, Malay Peninsula, Sumatra.

Pentaphylax.

Very close to the Ternstroemiaceae, especially to Freziereae, but ovules crassinucellate and embryo horseshoe-shaped.

4. TERNSTROEMIACEAE

Mirbel ex A.P. de Candolle 1816. 11/340. Tropical and subtropical regions, few in Africa.

TERNSTROEMIEAE: *Ternstroemia, Anneslea;* FREZIEREAE: *Adinandra, Cleyera, Archboldiodendron, Eurya, Freziera, Symplococarpon, Euryodendron, Visnea, Balthasaria.*

5. THEACEAE

Mirbel ex Ker Gawler 1816 (including Camelliaceae A.P. de Candolle 1816, Gordoniaceae Sprengel 1826). 9/460. Tropical and subtropical regions, but some species reach temperate parts of eastern Asia and eastern North America.

THEEAE: *Camellia* (including *Thea*), *Pyrenaria* (including *Tutcheria*), *Polyspora, Laplacea, Apterosperma*; GORDONIEAE: *Gordonia, Franklinia, Schima*; STEWARTIEAE: *Stewartia* (including *Hartia*).

Theaceae together with Ternstroemiacea are one of the most archaic members of the order Theales.

6. ONCOTHECACEAE

Kobuski ex Airy Shaw 1965. 1/2. New Caledonia.

Oncotheca.

Near to Ternstroemiaceae (Airy Shaw 1965; Baas 1975; Carpenter and Dickison 1976; Shilkina 1977; Dickison 1982), from which they differ mainly in the sympetalous corolla, pentalacunar node, encyclocytic stomata, prolonged and abruptly inflexed anther connectives forming a roof over the gynoecium, unitegmic ovules, drupelike fruits with 5-locular pyrene and exotestal seeds.

7. CARYOCARACEAE

Voigt 1845. 2/25. Tropical America from Costa Rica to Paraguay, but best developed in the Amazon baSinica

Caryocar, Anthodiscus.

Allied to the Theaceae, but differ by compound leaves, more complex structure of the vascular system of petiole, vessels with simple perforations, size and shape of the hypocotyl and cotyledons, drupaceous fruits. Have some common features with the Tetrameristaceae (subepidermal formation of the cork, presence of idioblasts), but differ in many other respects. A very distinct and isolated family, which according to Corner (1976: 89) is "clearly a relict of an ancestral state far removed from modern Theales."

Bibliography

Airy-Shaw HK. 1936. Notes on the genus *Schima* and on the classification of the Theaceae-Camellioideae. Kew Bull. 1936: 496–499.

Anderberg AA, C-I Peng, I Trift, and M Källersjö. 2001. The *Stimpsonia* problem; evidence from DNA sequences of plastid genes *atp*B, *ndh*F and *rbc*L. Bot. Jahrb. Syst. 123: 369–376.

Ao C-Q, G-X Chen, and H-D Zhang. 2002. Leaf epidermis morphology of *Camellia* and its taxonomic significance. Acta Bot. Yunnan. 24: 68–74.

Baas P. 1975. Vegetative anatomy and the affinities of Aquifoliaceae, *Sphenostemon, Phelline,* and *Oncotheca.* Blumea 22: 311–407.

Baretta-Kuipers T. 1976. Comparative wood anatomy of Bonnetiaceae, Theaceae, and Guttiferae. In: P Baas, AJ Bolton, and DM Catling, eds. Wood structure in biological and technological research. Leiden Bot. Ser. 3: 76–101.

Barker WR. 1980. Taxonomic revisions in Theaceae in Papuasia. I. *Gordonia, Ternstroemia, Adinandra* and *Archboldiodendron.* Brunonia 3: 1–60.

Beauvisage L. 1920. Contribution á l'étude anatomique de la famille des Ternstroemiacées. Tours.

Carlquist S. 1984b. Wood anatomy and relationships of Pentaphylacaceae: significance of vessel features. Phytomorphology 34: 84–90.

Carpenter CS and WC Dickison. 1976. The morphology and relationships of *Oncotheca balansae.* Bot. Gaz. 137: 141–153.

Chang HT. 1981. A taxonomy of the genus *Camellia.* Acta Sci. Nat. Univ. Sunyatseni, Monogr. Ser. 1: 1–180 (in Chinese).

Deng L and P Baas. 1991. The wood anatomy of the Theaceae. IAWA Bull. (New Series) 12: 333–353.

Dickison WC. 1982. Vegetative anatomy of *Oncotheca macrocarpa:* a newly described species of Oncothecaceae. Bull. Mus. Nat. Hist. Nat., Paris, ser. 4, sect. B, Adansonia 3(4): 177–181.

Dickison WC. 1986. Further observations on the floral anatomy and pollen morphology of *Oncotheca* (Oncothecaceae). Brittonia 38: 249–259.

Dickison WC. 1990. A study of floral morphology of the Caryocaraceae. Bull. Torrey Bot. Club 117: 123–137.

Ehrendorfer E, W Morawetz, and J Dawe. 1984. The Neotropical angiosperm families Brunelliaceae and Caryocaraceae: first karyosystematical data and affinities. Plant Syst. Evol. 145: 183–191.

Erbar C. 1986. Untersuchungen zur Entwicklung der spiraligen Blute von *Steuartia fseudocamellia* (Theaceae). Bot. Jahrb. Syst. 106: 391–407.

Guillaumin A. 1938. Observations morphologiques et anatomiques sur le genre *Oncotheca.* Rev. Gen. Bot. 50: 629–635.

Han L, T Hatano, T Okuda, and T Yoshida. 1995. Tannins of *Stachyurus* species. 3. Stachyuranins A, B and C, three new complex tannins from *Stachyurus praecox* leaves. Chem. Pharmacol. Bull. (Tokyo) 43: 2109–2114.

Jin Q-J and Z-X Wei. 2002. Studies on pollen morphology of Stachyuraceae and Staphyleaceae. Acta Bot. Yunnan. 24: 57–63.

John J and K-P Kolbe. 1980. The systematic position of the "Theales" from the viewpoint of serology. Biochem. Syst. Ecol. 8: 241–248.

Keng H. 1962. Comparative morphological studies in Theaceae. Univ. Calif. Publ. Bot. 33: 369–384.

Kimoto Y and T Tokuoka. 1999. Embryology and relationships of *Stachyurus* (Stachyuraceae). Acta Phytotax. Geobot. 50: 187–200.

Kobuski CE. 1942. Studies in the Theaceae. XII. Notes on the South American species of *Ternstroemia.* J. Arnold Arbor. 23: 298–343.

Kobuski CE. 1951. Studies in the Theaceae: XXIII. The genus *Pelliciera.* XXIV. The genus *Sladenia.* J. Arnold Arbor. 32: 256–262; 403–408.

Leins P and C Erbar. 1991. Fascicled androecia in Dilleniidae and some remarks on the *Garcinia* androecium. Bot. Acta 104: 336–344.

Li H. 1992. On the origin of Stachyuraceae. Acta Bot. Yunnan. Suppl., 5: 59–64.

Li J, P del Tredici, S Yang, and MJ Donoghue. 2002. Phylogenetic relationships and biogeography of *Stewartia* (Camellioideae, Theaceae) inferred from nuclear ribosomal DNS ITS sequences. Rhodora 104: 117–133.

Li L. 2001. Chromosome number of *Sladenia celastrifolia.* Acta Bot. Yunnan. 23: 223–224 (in Chinese).

Li L, H-X Liang, and H Peng. 2003. Karyotype of *Sladenia* and its systematic insights. Acta Bot. Yunnan. 25: 321–326 (in Chinese with English summary).

Li L, H-X Liang, H Peng, and L-G Lei. 2003. Sporogenesis and gametogenesis in *Sladenia* and their systematic implication. Bot. J. Linn. Soc. 143: 305–314.

Liang D and P Baas. 1990. Wood anatomy of trees and shrubs from China: II. Theaceae. IAWA Bull. n.s., 11: 337–378.

Liang D and P Baas. 1991. The wood anatomy of the Theaceae. IAWA Bull. 12: 335–353.

Luna Vega MI. 1997. Relaciones filogeneticas de los generos de la familia Theaceae D. Don. Mexico.

Luna I and JL Villasenor. 1996. Generos de Theaceae: aspectos taxonomicos y nomenclaturales. Bol. Soc. Bot. Mex. 59: 81–95.

Mathew CJ. 1978. Development of male and female gametophytes in *Camellia susanqua.* Phytomorphology 28: 262–269.

Mathew CJ and M Chaphekar. 1977. Development of female gametophyte and embryogeny in *Stachyurus chinensis.* Phytomorphology 27: 68–78.

McPherson G, P Morat, and JM Veillon. 1982. Existence d'une deuxieme espece appartenant au genre *Oncotheca* endemique de la Nouvelle-Caledonie et nou-velle donnees concernant les Oncothecacees. Bull. Mus. Nat. Hist. Nat., Paris, ser. 4, sect. B, Adansonia 3: 305–311.

Mori SA. 2004. Caryocaraceae. In: N Smith, SA Mori, A Henderson, DW Stevenson, and SV Heald, eds. Flowering plants of the Neotropica, pp. 87–88. New York Botanical Garden. Princeton, NJ/Oxford.

Prakash N and YY Lau. 1976. Morphology of *Ploiarium alternifolium* and the taxonomic position of *Ploiarium.* Bot. Not. 129(3): 425–435.

Prance GT and MF da Silva. 1973. A monograph of the Caryocaraceae. Flora Neotropica 12: 1–75.

Prince LM. 2002. Circumscription and biogeographic patterns in the eastern North American-East Asian genus *Stewartia* (Theaceae: Stewartieae): insight from chloroplast and nuclear DNA sequence data. Castanea 67: 290–301.

Prince LM and CR Parks. 1997. Evolutionary relationships in the tea subfamily Theoideae based on DNA sequence data. Int. Camellia J. 29: 135–144.

Prince LM and C Parks. 2001. Phylogenetic relationships of Theaceae inferred from chloroplast DNA sequence data. Am. J. Bot. 88: 2309–2320.

Record SJ. 1942. American woods of the family Theaceae. Trop. Woods 70: 23–33.

Schneider JV. 2007. Stachyuraceae. In: K Kubitzki, ed. The families and genera of vascular plants, vol. 9, pp. 436–439. Springer Verlag, Berlin/Heidelberg/New York.

Shilkina IA. 1977. Comparative anatomical characteristics of wood of the genus *Oncotheca* Baill. (Theales). Bot. Zhurn. 62: 1273–1275.

Steenis CGGJ van. 1956. Pentaphylacaceae. In: CGGJ van Steenis, ed. Flora malesiana, ser. 1, 5(2): 121–124. Leiden.

Stevens PF and AL Weitzman. 2004. Sladeniaceae. In: K Kubitzki, ed. The families and genera of vascular plants, vol. 6, pp. 431–433. Springer, Berlin/Heidelberg/ New York.

Stevens PF, S Dressler, and AL Weitzman. 2004. Theaceae. In: K Kubitzki, ed. The families and genera of vascular plants, 6: 463–471. Berlin, Heidelberg, New York.

Tang Y-C, Y-L Cao, Y-Z Xi, and J He. 1983. Systematic studies on Chinese Stachyuraceae: 1. Phytogeographical, cytological, palynological. Acta Phytotax. Sinica 21: 236–253.

Tsou C-H. 1995. Embryology of Theaceae – anther and ovule development of *Adinandra, Cleyera*, and *Eurya*. J. Plant Res. 108: 77–86.

Tsou C-H. 1996. The pseudopollen of the Camellioideae of Theaceae: its development and systematic significance. Am. J. Bot. 83(6): 198–199 (abstract).

Tsou C-H. 1997. Embryology of the Theaceae – anther and ovule development of *Camellia, Franklinia*, and *Schima*. Am. J. Bot. 84: 369–381.

Tsou C-H. 1998. Early floral development of Camellioideae (Theaceae). Am. J. Bot. 85: 1531–1547.

Vyshenskaya TD. 1980. The development of the poly-merous androecium in *Stuartia pseudocamellia*. Bot. Zhurn. 65: 948–957 (in Russian).

Vyshenskaya TD. 1991a. Theaceae. In: A Takhtajan, ed. Comparative seed anatomy, vol. 3, pp. 178–187. Nauka, Leningrad (in Russian).

Vyshenskaya TD. 1991b. Oncothecaceae. In: A Takhtajan, ed. Comparative seed anatomy, vol. 3, pp. 188–101. Nauka, Leningrad (in Russian).

Wei Z-X. 1997. Pollen ultrastructure of Theaceae and its systematic significance. Acta Bot. Yunnan. 19: 143–153.

Wei Z-X and Q-J Jin. 2002. The development of male and female gametophytes of *Stachyurus himalaicus* and its systematics enlightenment. Acta Bot. Yunnan. 24: 733–742.

Wei Z-X, D-Z Li, X-K Fan, and X-L Zhang. 1999. Pollen ultrastructure of Pentaphylacaceae and Sladeniaceae and their relationships to the family Theaceae. Acta Bot. Yunnan. 21: 202–206.

Wei Z-X, Q-J Jin, and H Wang. 2002. Pollen morphology of Stachyuraceae and related taxa. Acta Bot. Yunnan. 24: 483–496.

Wijmstra TA. 1968. The identity of *Psilatricolporites* and *Pelliciera*. Acta Bot. Neerl. 17: 114–116.

Wood CE. 1959. The genera of Theaceae of the southeastern United States. J. Arnold Arbor. 40: 413–419.

Yang J-B, S-X Yang, D-Z Li, L-G Lei, T Ikeda, and H Yoshino. 2006. Phylogenetic relationships of Theaceae inferred from mitochondrial *mat*R gene sequence data. Acta Bot. Yunnan. 29: 29–36 [in Chinese].

Yang S-X and TL Ming. 1955a. Embryological studies on genera *Pyrenaria* and *Tutcheria* of family Theaceae. Acta Bot. Yunnan. 17: 67–71 (in Chinese).

Yang S-X and TL Ming. 1955b. Studies on the systematic position of genera *Pyrenaria, Tutcheria* and *Parapyrenaria* of family Theaceae. Acta Bot. Yunnan. 17: 192–196 (in Chinese).

Ye C-X. 1990a. A discussion on relationship among the genera of Theoideae (Theaceae). Acta Sci. Nat. Univ. Sunyatsenia 29: 74–81 (in Chinese).

Ye C-X. 1990b. The range of Gordonieae (Theaceae) and limitation of genera in the tribe. Guihaia 10: 99–103 (in Chinese).

Zhu Y-P, J Wen, Z-Y Zhang, and Z-D Chen. 2006. Evolutionary relationships and diversification of Stachyuraceae based on sequences of four chloroplast markers and the nuclear ribosomal ITS region. Taxon 55: 931–940.

Order 51. BALSAMINALES

Perennial or annual herbs, subshrubs, shrubs or small trees, sometimes lianas or epiphytic shrubs. Vessels generally with simple perforations, but very few with scalariform perforations; lateral pitting alternate or opposite Fibers with simple or small bordered pits. Rays heterogeneous or homogeneous. Axial parenchyma mostly paratracheal. Sieve-element plastids of S-type. Nodes unilacunar or rarely (Tetrameristaceae). Leaves alternate, opposite or verticillate, simple, entire or toothed, pinnately veined; stipules absent or represented by a pair of small petiolar glands. Stomata anomocytic, with some tendency to anisocytic or staurocytic type. Flowers solitary or in small, sometimes umbelliform cymose inflorescences, bisexual, zygomorphic, 4–5-merous. Sepals free or basally connate, imbricate, somewhat petaloid. Petals free or more or less connate at the base. Nectary disc wanting. Stamens 3–5 to numerous (Marcgraviaceae); filaments broad free or apically or basally connate; anthers small, introrse, basifixed or dorsifixed, tetrasporangiate, opening less often longitudinally. Tapetum secretory. Microsporogenesis simultaneous. Pollen grains 2-celled or very rarely 3-celled, 3-colpate or sometimes 4–5-colpate. Gynoecium of 2–8 variously united carpels; stylodia connate into a very short style or stigma almost sessile; ovary superior, 5(4)-locular, or 1–2-locular, sometimes plurilocular; ovules solitary to numerous in each locule, or 2–3 ovules per locule (*Hydrocera*). Ovules anatropous, apotropous, campylotropous, sometimes pendulous, bitegmic or sometimes unitegmic by fusion or intermediate, tenuinucellate, with a characteristic endothelium. Female gametophyte of *Polygonum-* or *Allium*-type. Endosperm cellular, with chalazal and micropylar haustorium, or nuclear. Fruits indehiscent or elastically dehiscent loculicidal capsules (Balsaminaceae), Seeds exotestal, with large straight, or slightly curved embryo; endosperm copious, scanty or wanting; cotyledons sometimes large.

The families included in this order have many similarities, such as: specialized nectaries or glands on leaves, petioles, sepals or petals, tenuinucellate, bitegmic ovules, endostomal micropyle, short style and little expanded stigma, endosperm with mycropylar haustorium, etc. They closely relationships also strongly (100%) supported by molecular date (Soltis et al. 2000; Geuten et al. 2004; Soltis et al. 2006). Formerly these families have never been included in one order.

Key to Families

1 Endosperm cellular.

2 Perennial (rarely subshrubs) or more often annual herbs, sometimes with tubers or rhizomes and with more or less watery and succulent, stems more or less translucent, often with raphide-sacs. Idioblasts with possible mucilage content in stems and leaves. Vessels generally with simple perforations, but very few with scalariform perforations with 1–3 bars. Nodes unilacunar. Leaves alternate, opposite or verticillate, simple, entire or toothed, pinnately veined; stipules absent or represented by a pair of small petiolar glands. Stomata anomocytic, with some tendency to anisocytic type. Flowers solitary or in small, sometimes umbelliform cymose inflorescences, pedicels twisted, bisexual, zygomorphic, often resupinate, 5-merous. Sepals imbricate, the lowest (posterior) one very large, somewhat petaloid, often saccate and gradually prolonged backward into a tubular nectariferous spur whose cavity is lined by a layer of secretory cells, the two lateral are small and pushed forward to the anterior aspect of the flower, while the two upper (anterior) ones are minute or aborted. The upper petal free, large, external in bud, concave and partly sepaloid, the others free in *Hydrocera,* but in *Impatiens* more or less connate in pairs on each side of the flower, each pair suggesting a single bilobed petal. Stamens five, alternating with the corolla members; filaments broad and short, apically connate; anthers small, introrse, tetrasporangiate, cohering and covering the ovary like a cap; ultimately they rupture at the base and are lifted up by the lengthening gynoecium; microsporangia characteristically divided by trabeculae that separate the sporogeneous tissue into islands. Pollen grains 2-celled or very rarely 3-celled, 3-colpate or sometimes 4–5-colpate. Gynoecium of 5(4) united carpels; stylodia connate into a very short style or stigma almost sessile; stigma solitary or five nonpapillate stigmas; ovary superior, 5(4)-locular, with 5–50 ovules per locule (*Impatiens*) or 2–3 ovules per locule (*Hydrocera*). Ovules pendulous, anatropous, apotropous (the micropyle directed upward and inward), bitegmic or sometimes unitegmic by fusion or intermediate, tenuinucellate, with a characteristic endothelium. Female gametophyte of *Polygonum-* or *Allium*-type. Endosperm cellular, with chalazal and in *Hydrocera* also with micropylar haustorium. Fruits in *Impatiens* elastically dehiscent loculicidal capsules, the valves separating from the placentas and becoming elastically coiled while the seeds are shot out on all sides, in *Hydrocera* pentagonal, berrylike drupes, the stone eventually separating into five 1-seeded pyrenes. Seeds pachychalazal, exotesta only thickened, with large straight embryo; endosperm scanty or (*Hydrocera*) none; cotyledons large; seed coat usually contains crystals. Contain raphides of calcium oxalate and large amounts of leucanthocyanins and napthoquinones, present proanthocyanidins (cyanidin and delphinidin), flavonols (kaempferol and quercetin), n = 8 (*Hydrocera*), in *Impatiens* basic numbers n = 7, 8, 9 and 10 (Fischer 2004). . . . . 1. BALSAMINACEAE.

2 Lianas or epiphytic shrubs with climbing aerial roots, seldom erect shrubs or small trees. Stems often slender, more or less scrambling. Raphides and sclerenchymatous idioblasts present in the parenchymatous tissues. Vessels with scalariform perforations (in *Noranthea* with numerous bars) or more often with simple perforations; lateral pitting alternate or less often opposite. Fibers septate, with small bordered pits. Rays heterogeneous, of primitive type. Axial parenchyma apotracheal or paratracheal. Leaves alternate, simple, sometimes dimorphic, fleshy, entire, with marginal to abaxial cavities, estipulate. Stomata staurocytic. Bracts of the sterile flowers modified into pitcherlike, saccate, spurred or hooded nectariferous structures.

Inflorescences terminal, sometimes cauliflorous on short lateral shoots, racemose, forming pseudospikes or pseudoumbels; extrafloral bracteal nectaries present, often conspicuously colored; bracteoles two, generally sepal-like. Flowers actinomorphic, bisexual. Sepals five or four, free or basally connate, much imbricate, persistent. Petals five or four, free or more or less connate at the base, in *Marcgravia* distally connate into a deciduous calyptra. Stamens three to numerous; filaments free or basally connate, sometimes adnate to the base of the petals; anthers tetrasporangiate, introrse, opening longitudinally. Gynoecium of 2–20 carpels; ovary superior, at first 1-locular, but becoming plurilocular by intrusion and fusion of the parietal placentas; stigmas nearly sessile, radiate, lobed or umbonate. Ovules 10–20 to numerous in several rows on thick placentas, anatropous, with a poorly developed endothelium. Endosperm cellular and at least in *Marcgravia* with a micropylar haustorium. Fruits thick and rather fleshy, indehiscent or slightly dehiscent into the loculi at the base. Seeds numerous, small, with very scanty or no endosperm and straight or slightly curved embryo. Flowers in terminal, often pendulous racemes, spikes, or umbels. Contain terpenes, tannins, saponins, alkaloids, and phenolics, n = 18 (*Marcgravia evenia*) (Dressler 2004). . . 2. MARCGRAVIACEAE.

1 Endosperm nuclear.

3 Ovary 4–5-locular, capped by a terminal style with a punctate or minutely lobed stigma. Ovule solitary in each carpel, anatropous, axile-basal, bitegmic. Fruits dry, more or less coriaceous, 4–5-seeded berry. Seeds relatively large, with copious endosperm surrounding a straight, basal embryo with the hypocotyl much longer than the two cotyledons. Flowers small, in axillary, umbelliform or compactly corymbiform raceme, subtended by two persistent or deciduous bracteoles, 4–5-merous throughout; filaments shortly connate at the base. Sepals free, imbricate, with numerous glandular pits on middle of adaxial surface; petals free, imbricate, greenish. Stamens five, alternate with petals; filaments flattened, basally connate; anthers basifixed, introrse. Leaves not decurrent, alternate, simple, entire, crowded at branch tips, blades coriaceous, glossy, the apex rounded to emarginate, with marginal glands; stipules absent. Stomata mostly anomocytic. Trees to 15 m tall, or shrubs. No sclerenchymatous idioblasts. Vessels with simple perforation (*Tetramerista*), predominantly simple but sometimes scalariform perforations in *Pentamerista* with 2–5 bars. Axial parenchyma mainly diffuse to diffuse-in-aggregates, but also scanty paratracheal (Lens et al. 2005). . . . . 3. TETRAMERISTACEAE.

3 Ovary imperfectly 2-locular, occasionally 1-locular by abortion, superior, conic-cylindric, 10-grooved, with gradually tapering style and punctiform stigma. Ovules solitary in each locule, large, campylotropous, bitegmic, pendulous from the inner angle. Fruits dry and leathery, indehiscent, woody, 10-grooved, long-acuminate, turbinate, reddish brown, covered with resinous pustules, 1-seeded. Seeds large, without endosperm; embryo cordate, with broad, large, fleshy cotyledons, well-developed, pointed radicle projecting into the beak of the fruit, and elongate, hooked, reddish plumule. Flowers large, solitary, axillary, closely subtended and enclosed for a time by two long petaloid foliaceous involute bracteoles surpassing petals, with 5-merous perianth and androecium. Sepals free, petaloid, externally rosy, much shorter than the bracteoles and petals, ovate, adaxially concave, proximal half of adaxial surface covered with glands; petals free, white to pinkish red. Stamens five, free, lying within the alternate groves of the ovary; anthers elongate, basally sagittate, with distinctly prolonged connective. Branched, sclerenchymatous idioblasts present in the cortex and pith. Leaves disposed at the ends of the branches, involute in bud, asymmetric, leathery, shortly decurrent, with almost annular structure of the vascular strand in the petiole; pair of extrafloral nectaries often present near the base of leaf. Mangrove trees with trunks buttressed at base; raphides in idioblasts or the parenchymatous tissue. Vessels with simple perforations. Axial parenchyma apotracheal . . . . .4. PELLICIERACEAE.

1. BALSAMINACEAE

Berchtold and J. Presl 1820 (including Hydroceraceae Blume 1825, Impatientaceae Barnhart 1895). 2/c.1000. Mainly tropical Asia and Africa, a few species in temperate regions of Eurasia, Africa, and North America.

Impatiens (including *Semelocardium, Impatientella*), *Hydrocera*.

Related to the Marcgraviaceae.

2. MARCGRAVIACEAE

A.L. de Jussieu ex A.P. de Candolle 1816 (including Noranteaceae A.P. de Candolle 1935). 7/130. Tropical America.

2.1 NORANTEOIDEAE
Leaves alternate, simple. Flowers 5-merous – *Norantea, Sarcopera, Marcgraviastrum, Ruyschia, Souroubea, Schwartzia*.

2.2 MARCGRAVIOIDEAE
Stems dimorphic (the sterile stems creeping or root-climbing, bearing two rows of small, juvenile, cordate leaves, the fertile stems free and pendulous, usually rootless, bearing adult leaves – Dressler 2004). Leaves distichous, dimorphic. Flowers 4-merous with calyptrate corolla. Nectaries adnate to abortive flowers – *Marcgravia*.

Molecular analyses place the Marcgraviaceae as a sister to the Tetrameristaceae and Pellicieraceae (Soltis et al. 2000, 2006; Anderberg et al. 2002). *Norantea* (about 35 species), characterized by scalariform perforations with numerous bars, is relatively the most archaic genus (Dressler 2004).

3. TETRAMERISTACEAE

Hutchinson 1959. 2/4. Malay Peninsula, Borneo, and Sumatra (*Tetramerista*) and Guyana Highlands in southern Venezuela (*Pentamerista*).

Tetramerista, Pentamerista

Already Hallier (1916) recognized the affinity Tetrameristaceae with the Marcgraviaceae, which is supported recent molecular data (Soltis et al. 2006).

4. PELLICIERACEAE

L. Beauvisage ex Bullock 1959. 1/1. Mangroves along Pacific coasts and islands of Nicaragua, Costa Rica, Panama, Colombia, and Ecuador.

Pelliciera.

Recent molecular studies (Soltis et al. 2000; Anderberg et al. 2002) provide also strong support for a clade consisting of *Pelliciera, Tetramerista*, and Marcgraviaceae.

Bibliography

Akiyama S, M Wakabayashi, and H Ohba. 1992. Chromosome evolution in Himalayan *Impatiens* (Balsaminaceae). Bot. J. Linn. Soc. 109: 247–257.

Antlfinger AE. 1986. Field germination and seedling growth of chasmogamous and cleistogamous progeny of *Impatiens capensis* (Balsaminaceae). Am. J. Bot. 73: 1267–1273.

Arisumi T. 1980. Chromosome numbers and breeding behavior of hybrids among Celebes, Java and New Guinea species of *Impatiens* L. Hort. Sci. 9: 478–479.

Bedell HG. 1985. A generic revision of Marcgraviaceae. I. The *Noranthea* complex. Ph.D. dissertation, University of Maryland, College Park.

Boesewinkel FD and F Bouman. 1991. The development of bi- and unitegmic ovules and seeds in *Impatiens*. Bot Jahrb. Syst. 113: 87–104.

Boesewinkel FD and F Bouman. 2000. Balsaminaceae. In: A Takhtajan, ed. Comparative seed anatomy, 6: 40–42. St. Petersburg (in Russian).

Caris P, KP Geuten, SB Janssens, and EF Smets. 2006. Floral development in three species of *Impatiens* (Balsaminaceae). Am. J. Bot. 93: 1–14.

Dressler S. 2004a. Marcgraviaceae. In: N Smith, SA Mori, A Henderson, DW Stevenson, and SV Heald, eds. Flowering plants of the Neotropics, pp. 236–238. The New York Botanical Garden. Princeton University Press, Princeton.

Dressler S. 2004b. Marcgraviaceae. In: K Kubitzki, ed. The families and genera of vascular plants, 6: 258–265. Springer Berlin, Heidelberg, New York.

Fischer E. 2004. Balsaminaceae. In: K Kubitzki, ed. The families and genera of vascular plants. 6: 20–25. Springer, Berlin/Heidelberg/New York.

Frame D. 2004. Pellicieraceae. In: N Smith, SA Mori, A Henderson, DW Stevenson, and SV Heald, eds. Flowering plants of the Neotropics, pp. 289–290. The New York Botanical Garden. Princeton University Press, Princeton.

Geuten K, E Smets, P Schols, Y-M Yuan, S Janssens, P Küpfer, and N Pyck. 2004. Conflicting phylogenies of balsaminoid families and the polytomy in Ericales: combining data in a Bayesian framework. Molec. Phylog. Evol. 31: 711–729.

Govindarajan T and D Subramanian. 1986. Karyotaxonomy of south Indian Balsaminaceae. Cytologia 51: 107–116.

Grey-Wilson C. 1980. Studies in Balsaminaceae: V. *Hydrocera triflora*, its floral morphology and relationships with *Impatiens*. VI. Some observations on the floral vascular anatomy of *Impatiens*. Kew Bull. 35: 213–219, 221–227.

Hallier H. 1916. Beitrage zur Flora von Borneo, Marcgraviaceae. Bot. Centralbl. Beib. 34: 38.

Huynh K-L. 1968. Morphologie du pollen des Tropaeolacées et des Balsaminacées, parts 1, 2. Grana Palynol. 8: 88–184, 277–516.

Huynh K-L. 1970. Quelques caractéres cytologiques, anatomiques, et embryologiques distinctifs du genre *Tropaeolum* et du

genre *Impatiens,* et position taxonomique de la famille des Balsaminacées. Bull. Soc. Neuchateloise Sci. Nat. 93: 165–177.
Janssens S, F Lens, S Dressler, K Geuten, E Smets, and S Vinckier. 2005. Palynological variation in balsaminoid Ericales. II. Balsaminaceae, Tetrameristaceae, Pellicieraceae and general conclusions. Ann. Bot. 96: 1061–1073.
Janssens S, K Geuten, Y-M Yuan, Y Song, P Küpper, and E Smets. 2006. Phylogenetics of *Impatiens* and *Hydrocera* (Balsaminaceae) using chloroplast *apt*B-*rbc*L spacer sequences. Syst. Bot. 31: 171–180.
Kubitzki K. 2004a. Pellicieraceae. In: K Kubitzki, ed. The families and genera of vascular plants, vol. 6, pp. 297–299. Springer, Berlin/Heidelberg/New York.
Kubitzki K. 2004b. Tetrameristaceae. In: K Kubitzki, ed. The families and genera of vascular plants, vol. 6, pp. 461–462. Springer, Berlin/Heidelberg/New York.
Lens F, S Dressler, S Jansen, L van Evelghem, and E Smets. 2005. Relationships within balsaminoid Ericales: a wood anatomical approach. Am. J. Bot. 92: 941–953.
Lu Y-Q and Y-L Chen. 1991. Pollen morphology of *Impatiens* L. (Balsaminaceae) and its taxonomic implications. Acta Phytotax. Sinica 29: 252–257.
Maguire B, C de Zeeuw, Y-C Huang, and CC Clare Jr. 1972. Botany of Guyana Highland: Part IX. Tetrameristaceae. Mem. New York Bot. Gard. 23: 165–192.
Mauritzon J. 1939. Über die Embryologie von *Marcgravia.* Bot. Not. 1939: 249–255.
Narayana LL. 1963. Contributions to the embryology of Balsaminaceae, part 1. J. Indian Bot. Soc. 42: 102–109.
Narayana LL. 1965. Contributions to the embryology of Balsaminaceae, part 2. J. Jpn. Bot. 40: 104–116.
Narayana LL. 1970. Balsaminaceae. In Symposium on comparative embryology of angiosperms. Bull. Indian Natl. Sci. Acad. 41: 158–162.
Narayana LL. 1974. A contribution to the floral anatomy of Balsaminaceae. J. Jpn. Bot. 49: 315–320.
Plisco MA. 1991. Marcgraviaceae. In: A Takhtajan, ed. Comparative seed anatomy, vol. 3, pp. 191–193. Nauka, Leningrad (in Russian).
Punt W. 1971. Pollen morphology of the genera *Norantea, Souroubea,* and *Marcgravia* (Marcgraviaceae). Pollen et Spores 13: 199–232.
Rama Devi D and LL Narayana. 1989. Floral anatomy of Balsaminaceae. In: M Trivedi, B Gill, and S Saini, eds. Plant Science Research in India, pp. 707–413. Today & Tomorrow's Printers & Publishers, New Delhi, India.
Rao RVS, KR Ayyangar, and R Sampathkumar. 1986. On the karyological characteristics of some members of Balsaminaceae. Cytologia 51: 251–260.
Roon AC de. 1967. Foliar sclereids in the Marcgraviaceae. Acta Bot. Neerl. 15: 585–623.
Roon AC de. 1975. Contributions towards a monograph of the Marcgraviaceae. Ph.D. thesis, University of Utrecht, Utrecht, The Netherlands.
Roth LC and A Grijalva. 1991. New record of the mangrove *Pelliciera rhizophorae* (Theaceae) on the Caribbean coast of Nicaragua. Rhodora 93: 183–186.
Saleh NAM and GHN Towers. 1974. Flavonol glycosides of *Norantea guianensis* flowers. Phytochemistry 13: 2012.
Schemske DW. 1978. Evolution of reproductive characteristics in *Impatiens* (Balsaminaceae): the significance of cleistogamy and chasmogamy. Ecology 59: 596–613.
Schürhoff PN. 1931. Die Haploidgeneration der Balsaminaceae und ihre Verbreitung für die Systematik. Bot. Jahrb. Syst. 64: 324–356.
Simon L. 1975. Le gynécée de *L'Impatiens balsamina:* Étude morphologique, ontogénique, et tératologique. Canad. J. Bot. 53: 2361–2383.
Song Y, Y-M Yuan, and P Küpfer. 2003. Chromosomal evolution in Balsaminaceae with cytological observations on 45 species from Southeast Asia. Caryologia 56: 643–481.
Song Y, Y-M Yuan, and P Küpfer. 2005. Seed coat micromorphology of *Impatiens* (Balsaminaceae) from China. Bot. J. Linn. Soc. 149: 195–208.
Steffen K. 1951. Die Embryoentwicklung von *Impatiens glandulifera* Lindl. Flora 139: 175–244.
Stevenson DW. 2004. Tetrameristaceae. In: N Smith, SA Mori, A Henderson, DW Stevenson, and SV Heald, eds. Flowering plants of the Neotropics, pp. 368–369. The New York Botanical Garden. Princeton University Press, Princeton.
Swamy BGL. 1948. A contribution to the embryology of the Marcgraviaceae. Am. J. Bot. 35: 628–633.
Venkateswarlu J and L Lakshminarayana. 1957. A contribution to the embryology of *Hydrocera triflora* W. et A. Phytomorphology 7: 194–203.
Ward NM and RA Price. 2002. Phylogenetic relationships of Marcgraviaceae: insights from three chloroplast genes. Syst. Bot. 27: 149–160.
Wood CE. 1975. The Balsaminaceae in the southeastern United States. J. Arnold Arbor. 56: 413–426.
Utami N and T Shimizu. 2005. Seed morphology and classification of impatiens (Balsaminaceae). Blumea 50: 447–456.
Yuan Y-M, Y Song, K Geuten, E Rahelivololona, S Wohlhauser, E Fisher, E Smets, and P Küpfer. 2004. Phylogeny and biogeography of Balsaminaceae inferred from ITS sequences. Taxon 53: 391–403.
Zinoveve-Stahevitch AE and WF Grant. 1984. Chromosome numbers in *Impatiens* (Balsaminaceae). Canad. J. Bot. 62: 2630–2635.

Order 52. HYPERICALES

Trees, shrubs, woody lianas, or herbs. Mostly with resinous, often colored juice in schizogenous secretory canals or cavities, and commonly with scattered tanniniferous secretory cells as well. Vessel elements with simple or sometimes mixed simple and scalariform perforations; lateral pitting alternate. Fibers with simple or seldom bordered pits, often septate. Rays heterogeneous to homogeneous. Axial parenchyma apotracheal or paratracheal, sometimes (*Hypericum*) wanting. Sieve-element plastids of S-type. Nodes mostly unilacunar, rarely (Bonnetiaceae) trilacunar. Leaves opposite, verticillate, or sometimes alternate, simple and mostly entire, pinnately veined, usually with many slender lateral veins, often with resin-cavities appearing as pellucid dots, estipulate or very rarely (*Mahurea* of Clusiaceae and Elatinaceae) stipulate. Stomata mostly paracytic. Flowers in terminal, cymose inflorescences or sometimes solitary and terminal, bisexual or unisexual, actinomorphic, often with bracteoles (sometimes passing into sepals). Sepals imbricate. Petals (2)3–6(-14), free or basally connate, imbricate or contorted. Stamens usually numerous, free, and developing in centrifugal sequence or more often grouped into 2–3 centrifugal bundles opposite and often adnate to the petals, rarely only five or even three stamens. Anthers tetrasporangiate or disporangiate, opening longitudinally. Tapetum secretory. Microsporogenesis simultaneous. Pollen grains 2-celled, (2)3(-5)-colporate. Gynoecium of 3-5(-20 or more) united carpels; stylodia free to connate into a style with lobed or peltate stigma. Ovary superior, with as many locules as carpels or less often 1-locular (intruded placenta not meeting in the center), with (1)2 to many ovules per carpel. Ovules anatropous, bitegmic, tenuinucellate. Female gametophyte of *Polygonum*-type. Endosperm nuclear. Fruits capsules, berries, or drupes. Seeds small to large, often with funicular or micropylar aril. Seed coat formed by both integuments and usually characterized by large cells with thick and more or less lignifieed walls in the exotegmen. Endosperm lacking or reduced to 1–2 layers of thin-walled cells; embryo straight or curved. Generally producing proanthocyanins, and often accumulating diverse sorts of xanthones; n = 7–10.

Close related to the Theales.

Key to Families

1 Leaves stipulate; petals imbricate. Perennial or annual herbs of aquatic or moist terrestrial habitats, rarely (*Bergia suffruticosa*) suffrutescent. Stems glabrous or (*Bergia*) glandular pubescent throughout. A resinous secretion, brownish when dry, is widely distributed in the tissues or deposited in granular form on the surface of the stem. Secretory cells with taninlike compounds occur in cortex, pith, and xylem parenchyma of *Bergia*. Vessels with simple perforations; lateral pitting alternate. Fibers with fully bordered pits. Vasicentric tracheids present. Nodes unilacunar. Leaves and decussate or very rarely verticillate (*Elatine alsinastrum*), entire or coarsely dentate, with a pair of scarious, interpetiolar stipules. Stomata anomocytic. Flowers small, solitary, or in dichasia in the axils of the leaves, bisexual, actinomorphic, cyclic, 5-merous in *Bergia* (except for 3-merous *B. trimera*) and 2-, 3-, or 4-merous in *Elatine*. Sepals (2)3–5(6), free or more or less united basally, glandular pubescent in *Bergia*. Petals as many as the sepals, small, membranaceous, free, persistent. Stamens from as many to twice as many as the petals, free, in one or two cycles, the outer cycle alternates with the petals; anthers broadly ovoid, dorsifixed, introrse, 2-locular, opening longitudinally. Pollen 3-colporate. Gynoecium syncarpous, of 2–5 carpels, with more or less free and often short stylodia; stigmas capitate; ovary 2–5-locular (the partitions not reaching the summit in some Asian species of *Bergia*), gradually (*Bergia*) or abruptly (*Elatine*) narrowing into the stylodia, with numerous ovules on axile placentas. Ovules numerous, anatropous, weakly crassinucellate, micropyle zig-zag. Fruits thin-walled septicidal capsules. Seeds small, with reticulate surfaces; exotegmen with anticlinal walls sinuous, low, lignified, embryo straight or less often more or less curved filling nearly the entire seed; endosperm reduced to 1–5 cell layers (*Bergia*) or none (*Elatine*). Several phenolic acids (including ellagic acid) occur in *Bergia* and ellagic acid in *Elatine*. n = 6, 9. . . . . 4. ELATINACEAE.

1 Leaves estipulate (except *Mahurea* in Clusiaceae); petals contorted.

2 Leaves alternate. Nectariferous disc well-developed, alternate with the stamen clusters at the base of the ovary. Trees and shrubs, glabrous, sometimes resinous, mucilage cells common; the heartwood is dark reddish brown and heavy. Vessels with simple perforations. Nodes trilacunar (*Bonnetia*) or unilacunar (*Archytaea, Ploiarium*). Leaves alternate, crowded toward tips of branches, with close, ascending, lateral veins, margins minutely toothed by setae, petiole short. Stomata anomacytic. Flowers sometimes axillary, bisexual, cyclic, perianth members free or only slightly united at the base. Sepals five, unequal, free, quincuncial. Petals five, contorted, very thick, free. Stamens numerous; filaments slender, free, or basally connate, sometimes united into five epipetalous bundles. Anthers short, basifixed, subbasifixed, or versatile and dehiscence is introrse or lateral-introrse; connective sometimes with a gland at the apex. Pollen grains 3-colporate, with

circular endoaperture. Ovary 3-, seldom 4-, or 5-locular with numerous ovules; stylodia free or (*Archytaea*) united into style; stigma punctate, papillate. Ovules anatropous, tenuinucellate, many in each locule; micropyle exostomal. Fruits septicidal capsules with a persistent central column. Seeds quite small, endotestal (*Ploiarium*), with scanty (*Ploiarium*), or no endosperm; embryo straight with short cotyledons and long radicle. Contain xanthones.. . . . . .1. BONNETIACEAE.

2 Leaves opposite or very rarely alternate. Nectary absent.

3 Flowers mostly unisexual, polygamous or dioecious, rarely bisexual, actinomorphic. Inflorescences terminal or axillary, rarely ramiflorous. Sepals 2(3)4 or 5-12(-20), free, occasionally fused. Petals 3–8 free, sometimes absent. Stamens numerous, up to 1,000, rarely few (eight in *Oedematopus*), free or all connate basally. Anthers extrorse to introrse, tetrasporangiate or disporangiate, opening by slits, rarely pores, connective often with glands of various types. Pollen grains 3(2-5)-colporate. Staminodes often present in female flowers, pistillodium sometimes presents in male flowers. Gynoecium of 5-3 (1-15) carpels. Stylodia free to connate into a style; stigmas various, sometimes radiating. Ovary 1-several locules; ovules 1-many per locule; placentation axile or basal, rarely parietal or apical. Fruits often septicidal or (*Mesua*) septifragal capsules, or berries, rarely drupes (*Endodesmia*). Seeds small to large, winged or often arillate; embryo straight, sometimes conferruminate; cotyledons massive to almost absent, endosperm often wanting. Leaves opposite, rarely alternate, not appearing gland-dotted. Indumentum when present not stellate. Evergreen trees, sometimes (some *Calophyllum*) with buttresses, or shrubs, sometimes epiphytic, with schizogenous canals or cavities. Plants glabrous or with unicellular or stellate hairs. Nodes unilacunar with one trace. Vessels with simple or sometimes scalariform perforations. Produce flavones, flavonols, biflavonoids, ellagic acid, and especially xanthones. . . . . .2. CLUSIACEAE.

3 Flowers bisexual, in mostly terminal, cymose-paniculate inflorescences, or flowers solitary. Sepals 4–5, free or basally connate. Petals 4–5, free. Stamens numerous (50–200), rarely 10(5); filaments elongate, free or often connate in 3–5 bundles, or basally monadelphous. Anthers tetrasporangiate, extrorse, opening by slits, connective often with glands. Pollen grains 3-colporate. Gynoecium of 3–5 carpels. Stylodia free or very rarely connate; stigmas apical. Ovary 1- or 3–5-locular; ovules numerous, axile or rarely parietal. Fruits loculicidal or septicidal capsules or berries, rarely drupaceous. Seeds minute, not arillate, embryo green or white, endosperm absent, cotyledons moderate-sized. Leaves opposite or verticillate, very rarely the lower leaves alternate (*Psorospermum*), usually gland-dotted, estipulate. Indumentum when present stellate. Evergreen or sometimes deciduous herbs, shrubs or trees; glands or canals in most parts of the plant. Vessels with simple or sometimes with scalariform perforations. Produce flavones, flavonols, biflavonoids, ellagic acid, and xanthones; n = 7–10, 12, 16–20. . . . . 3. HYPERICACEAE.

1. BONNETIACEAE

L. Beauvisage ex Nakai 1948. 3/35. Southeast Asia, West Malesia, Moluccas, and New Guinea (*Ploiarium*) and West Indies, Guyana, Brazil, and Peru.

Ploiarium, Bonnetia (including *Neblinaria* and *Neogleasoma* and probably also *Acopanea*), *Archytaea.*

Bonnetiaceae occupy a somewhat intermediate position between Theaceae (and closely related families, particularly Pellicieraceae) and Clusiaceae. "All genera together form a kind of gradual transition in the structure of their wood from Theaceae to Guttiferae," states Baretta-Kuipers (1976: 99). The intermediate position of the Bonneriaceae is supported also by their stems and leaves anatomy (Dickison and Weitzman 1996) and floral morphology and anatomy (Dickison and Weitzman 1998). The presence of xanthones in the Bonnetiaceae makes them similar to the Clusiaceae. Also, they have secretory cavities in pith and phloem as do *Kielmeyera* and other Clusiaceae (Baretta-Kuipers 1976).

2. CLUSIACEAE

Lindley 1836 or Guttiferae A.L. de Jussieu 1789 (nom. altern.) (including Calophyllaceae J. Agardh 1858, Cambogiaceae Horaninow 1834, Garciniaceae Bartling

1830). 28/470. Widespread in tropical regions, mainly in rain forests.

2.1 CALOPHYLLOIDEAE (including Kielmeyeroideae)
Leaves alternate, with pellucid dots, rarely canals. Flowers bisexual, unisexual, or polygamous. Stamens free or united at base or apex; anthers sometimes with large, apical glands. Stylodia free or united. Fruits drupes or capsules. Seeds small to large, exarillate. Embryo large, with united large cotyledons. – CALOPHYLLEAE: *Neotatea, Marila, Mahurea, Clusiella, Kielmeyera, Caraipa, Haploclathra, Poeciloneuron, Mesua, Kayea, Mammea, Calophyllum,* ENDODESMIEAE: *Endodesmia, Lebrunia.*

2.2 CLUSIOIDEAE
Trees or shrubs, sometimes epiphytic. Leaves opposite, usually with canals. Flowers unisexual, rarely polygamous. Stamens free or in groups. Stylodia free or united. Fruits capsules or drupes. Seeds arillate or exarillate. Embryo with large hypocotyl and vestigial or no cotyledons. – CLUSIEAE: *Clusia* (including *Quapoya, Havetia, Renggeria, Havetiopsis, Oedematopus, Pilosperma, Decaphalangium*), *Dystovomita, Tovomita* (including *Tovomitidium*), *Tovomitopsis, Chrysochlamys*; GARCINIEAE: *Allanblackia, Garcinia* (including *Ochrocarpus*).

2.3 MORONOBEOIDEAE (CHRYSOPIOIDEAE)
Leaves opposite. Flowers bisexual. Stamens in five groups or united into a tube. Style long, 5-lobed above. Fruit mostly a berry. Seeds exarillate. Embryo undifferentiated. – *Pentadesma, Montrouziera, Moronobea, Platonia, Symphonia, Thysanostemon.*

2.4 LOROSTEMONOIDEAE
Flowers bisexual.– *Lorostemon*

Closely related to the Bonnetiaceae. In many respects, on the basis of pollen morphology, Moronobeoideae differs markedly from the other subfamilies and in Erdtman's (1952) opinion possibly deserves a family rank.

3. HYPERICACEAE

A.L. de Jussieu 1789 (including Ascyraceae Plenck 1796). 9/550. Subcosmopolitan, mainly in temperate regions and mountains of the tropics.

VISMIEAE: *Vismia, Harungana* (including *Psorospermum*); HYPERICEAE: *Hypericum, Lianthus, Triadenum, Thornea, Santomasia;* CRATOXYLEAE: *Cratoxylum, Eliea.*

The Hypericaceae differ from Clusiaceae in their constantly bisexual flowers, and only very rarely do the leaves have the numerous close, parallel nerves or the wormlike secretory cells so characteristic of Clusiaceae. These in Hypericaceae take the form of translucent dots so familiar in the type genus *Hypericum* and in this and other genera's black dots. Besides, it is not less important that in the Hypericaceae the stylodia are more or less free. Both the Hypericaceae and Clusiaceae differ markedly from the Bonnetiaceae in the presence of a latex system.

4. ELATINACEAE

Dumortier 1829 (including Cryptaceae Rafinesque 1820). 2/35. Nearly cosmopolitan. *Bergia* primarily in tropical and subtropical regions and mainly of the Old World, while *Elatine* occurs in all continents except Antarctica.

Bergia, Elatine.

Elatinaceae have many features in common with the Clusiaceae and especially with the Hypericaceae, which also contain some aquatic forms (Cambessedes 1829; Takhtajan 1966; Corner 1976; Cronquist 1981; Melikian 1992). The common features include opposite leaves, syncarpous gynoecium with free stylodia and capitate stigmas, septicidal capsules as well as seed anatomy (Corner 1976; Melikian 1992) and stem anatomy (Carlquist 1984). The reticulate sculpturing of the seed coat in the Elatinaceae is very much like that of some Clusiaceae and Bonnetiaceae (especially the genus *Ploiarium*).

Bibliography

Baretta-Kuipers T. 1976. Comparative wood anatomy of Bonnetiaceae, Theaceae, and Guttiferae. In: P Baas, AJ Bolton, and DM Catling, eds. Wood structure in biological and technological research. Bot. Ser. 3: 76–101. Leiden University Press, Leiden.

Bennett GJ, H-H Lee, and TK Lowrey. 1989. Xanthones from Guttiferae. Phytochemistry 28: 967–998.

Cambessedes J. 1829. Note sur les Elatinees: Nouvelle famille des plantes. Paris.

Carlquist S. 1984a. Wood and stem anatomy of *Bergia suffruticosa:* relationships of Elatinaceae and broader significance of vascular tracheids, vasicentric tracheids, and fibriform vessel elements. Ann. Missouri Bot. Gard. 71: 232–242.

Dathan ASR and D Singh. 1971. Embryology and seed-development in *Bergia* L. J. Indian Bot. Soc. 50: 362–370.

Davis CC and MW Chase. 2004. Elatinaceae are sister to Malpighiaceae; Peridiscaceae belong to Saxifragales. Am. J. Bot. 91: 262–273.

Dickison WC and AL Weitzman. 1996. Comparative anatomy of the young stem, node, and leaf of Bonnetiaceae, including observations on a folior endodermis. Am. J. Bot. 83: 405–418.

Dickison WC and AL Weitzman 1998. Floral morphology and anatomy of Bonnetiaceae. J. Torrey Bot. Soc. 125: 268–286.

Gustafsson MHG, V Bittrich, and PF Stevens. 2002. Phylogeny of Clusiaceae based on *rbc*L sequences. Int. J. Plant Sci. 163: 1045–1054.

Kajale LB. 1939. A contribution to the life history of *Bergia ammanioides*. J. Indian Bot. Soc. 18: 157–167.

Kawano S. 1965. Anatomical studies on the androecia of some members of the Guttiferae-Moronoboideae. Bot. Mag. (Tokyo) 78: 97–108.

Kobuski CE. 1948. Studies in the Theaceae. XVII. A review of the genus *Bonnetia*. J. Arnold Arbor. 29: 393–413.

Kubitzki K. 1978. The botany of Guyana Highland: Part X. *Caraipa* and *Mahurea* (Bonnetiaceae). Mem. New York Bot. Gard. 29: 82–128.

Kubitzki K, AAL Mesquita, and OR Gottlieb. 1978. Chemosystematic implications of xanthones in *Bonnetia* and *Archytaea*. Biochem. Syst. Ecol. 6: 185–187.

Leins P. 1964. Die frühe Blütenentwicklung von *Hypericum hookerianum* Wight et Arn. and *H. aegyptiacum* L. Ber. Deutsch. Bot. Ges. 77: 112–123.

Maguire B. 1972. The botany of Guyana Highland: Part IX. Bonnetiaceae. Mem. New York Bot. Gard. 23: 131–165.

Melikian AP. 1992. Elatinaceae. In: A Takhtajan, ed. Comparative seed anatomy, vol. 4, pp. 11. Nauka, St. Petersburg (in Russian).

Melikian AP and BI Dildarian. 1977. Comparative anatomical and palynological study of representatives of Elatinaceae family. Biol. Zhurn. Armenii 30(11): 44–49 (in Russian with Armenian summary).

Mourao KSM and CM Beltrati. 2000. Morphology and anatomy of developing fruits and seeds of *Mammea americana* L. (Clusiaceae). Brazil. J. Biol. 60: 701–711.

Mourao KSM and CM Beltrati. 2001. Morphology and anatomy of developing fruits and seeds of *Vismia guianensis* (Aubl.) Choisy (Clusiaceae). Brazil J. Biol. 61: 147–158.

Prakash N and YY Lau. 1976. Morphology of *Ploiarium alternifolium* and the taxonomic position of *Ploiarium*. Bot. Not. 129: 279–285.

Raghavan TS and VK Srinivasan. 1940. A contribution to the life history of *Bergia capensis* Linn. J. Indian Bot. Soc. 19: 283–291.

Rao AN. 1957. The embryology of *Hypericum patulum* Thunb. and *H. mysorense* Heyne. Phytomorphology 7: 36–45.

Robson NKB. 1981. Studies in the genus *Hypericum* L. (Guttiferae). 2. Characters of the genus. Bull. Brit. Mus. (Nat. Hist.), Bot. 8(2): 55–226.

Robson NKB and PF Stevens. 1987. Towards a phylogenetic understanding of the Bonnetiaceae-Clusiaceae-Hypericaceae. Abstract XIV Int. Bot. Congr., 5-47b-5. Berlin.

Schofield EK. 1968. Petiole anatomy of the Guttiferae and related families. Mem. New York Bot. Gard. 18: 1–55.

Seetharam YN and JK Maheshwari. 1986. Scanning electron microscopic studies on the pollen of some Clusiaceae. Proc. Indian Acad. Sci. 96: 217–226.

Stevens PF. 2007a. Clusiaceae-Guttiferae. In: K Kubitzki, ed. The families and genera of vascular plants, vol. 9, pp. 48–66. Springer, Berlin/Heidelberg/New York.

Stevens PF. 2007b. Hypericaceae. In: K Kubitzki, ed. The families and genera of vascular plants, vol. 9, 194–201. Springer, Berlin/Heidelberg/New York.

Tucker G. 1986. The genera of Elatinaceae in the southeastern United States. J. Arnold Arbor. 67: 471–483.

Tucker G. 2004. Elatinaceae. In: N Smith, SA Mori, A Henderson, DW Stevenson, and SV Heald, eds. Flowering plants of the Neotropics, pp. 138–139. New York Botanical Garden, Princeton University Press, Princeton.

Vestal PA. 1937. The significance of comparative anatomy in establishing the relationship of the Hypericaceae to the Guttiferae and their allies. Philipp. J. Sci. 64: 413–419.

Weitzman AL and PF Stevens. 1997. Notes on the circumscription of Bonnetiaceae and Clusiaceae, with taxa and new combinations. BioLlania, Edicion Esp. 6: 551–564.

Weitzman AL, K Kubitzki, and PF Stevens. 2007. Bonnetiaceae. In: K Kubitzki, ed. The families and genera of vascular plants, vol. 9, pp. 36–39. Springer, Berlin/Heidelberg/New York.

Order 53. OCHNALES

Trees or shrubs, sometimes lianas (some Quiinaceae), less often subshrubs, very rarely herbs (some species of *Sauvagesia*). Cristarque cells and mucilage cells and canals usually present. Vascular bundles of the petiole mostly cylindrical. Young stems of woody members commonly with cortical vascular bundles and sometimes with medullary bundles as well. Vessels mostly with simple perforations, rarely with both simple and scalariform perforations, but Strasburgeriaceae have exclusively scalariform perforations. Pits vestured. Rays heterogeneous or homogeneous. Axial parenchyma usually apotracheal or metatracheal, often scanty. Nodes mostly trilacunar or multilacunar. Leaves alternate or less often opposite or verticillate (Quiinaceae), simple or very rarely pinnately compound (*Rhytidanthera splendida*), pinnately veined, often with numerous parallel secondary veins, stipulate or rarely (Medusagynaceae) estipulate. Stomata paracytic or less often anomocytic. Flowers of medium size or small, in cymose or racemose inflorescences, bisexual, actinomorphic or rarely with a

tendency to zygomorphy (*Luxemburgia*). Perianth usually cyclic, rarely spirocyclic (*Ochna* and *Ouratea* of the Ochnaceae and Strasburgeriaceae). Sepals (3)5(-12), imbricate, free or more or less united. Petals (3)5(-12), free, contorted or imbricate. Stamens few to many, free, sometimes on elongated androphore; filaments free or connate basally, filiform or broad and thick (Strasburgeriaceae); anthers tetrasporangiate, basifixed or dorsifixed, opening by longitudinal slits or apical pores. Tapetum secretory. Microsporogenesis simultaneous. Pollen grains 2-celled or 3-celled, mostly 3-colporate or 3-porate. Sometimes the innermost stamens sterile and transformed into a tube around the ovary, or they form a lobed intrastaminal disc. Gynoecium paracarpous; stylodia free or more often united into a simple style that is sometimes lobed at the end; stigma capitate; in each locule or on each parietal or axial (Medusagynaceae) placenta 1(-2) to many ovules. Ovules anatropous, campylotropous, or more or less hemitropous, bitegmic or rarely (Lophiraceae) unitegmic, tenuinucellate or rarely (Strasburgeriaceae) crassinucellate. Female gametophyte of *Polygonum*- or *Allium*-type. Endosperm nuclear. Fruits of different types. Seeds with or without endosperm; embryo straight or curved. Often producing proanthocyanins. Solitary or clustered crystals of calcium oxalate frequently present in some of the parenchymatous cells.
Related to the Hypericales.

Key to Families

1 Leaves usually alternate. Stylodia united into a style. Cortical bundles present.

2 Stipules intrapetiolar, basally connate. Ovules crassinucellate. Small to medium-sized trees. Leaves alternate, simple, large, leathery, coriaceous, with very fine, widely spaced teeth along the margins of the distal half of the lamina; petioles with narrow, lateral wings. Stomata anomocytic. Mucilaginous cells containing a frothy, red-staining contents present in both vegetative and reproductive organs. Young stem with cortical vascular bundles. Nodes predominantly trilacunar (nodes from older stems infrequently pentalacunar) with three traces. Vessel elements extremely long with exclusively scalariform perforations in almost vertical end walls and with 18–38 narrow and closely spaced bars; lateral pitting opposite. Fibers (fiber-tracheids) extremely long, with numerous bordered pits. Rays heterogeneous. Axial parenchyma apotracheal (diffuse and scanty). Secondary phloem with primitive sieve-elements and without fibers. Flowers large, solitary, axillary, bisexual, actinomorphic. Sepals 8–10, free, imbricate, spirally arranged, gradually increasing in size from outer to inner, persistent. Petals five, free, imbricate. Stamens 10, free, borne in a single cycle; filaments broad and thick; anthers, dorsifixed, latrorse, opening longitudinally. Pollen grains 3-colporate, distinctly brevicolpate, psilate to psilate granular, tectate. Disc intrastaminal, thick, annular at the base, 10-lobed above, lobes alternating with the stamens. Gynoecium of 4–7 laterally completely concrescent carpels that are also adnate with a central column; style subulate, twisted, elongate, terminates in lobed stigma; ovary angled, 4–7-locular, with one large ovule in each locule. Ovules anatropous, epitropous ventral (ovules pendulous, micropyle above, raphe ventral or toward bundle), bitegmic, crassinucellate, with zigzag micropyle. Endosperm cellular. Fruits large, woody, multiloculate, indehiscent, with persistent sepals and style. Seeds somewhat flattened, exotestal, with a straight embryo surrounded by oily endosperm approximately ten cells in thickness; sometimes with a rudimentary aril on funicle; n = 250. . . . . 1. STRASBURGERIACEAE.

2 Stipules extrapetiolar. Petals contorted or seldon imbricate. Ovules tenuinucellate, bitegmic or rarely (Lophiraceae) unitegmic. Vessels with simple or occasionally with both simple and scalariform perforations. Lateral pitting alternate.

3 Style gynobasic, persistent even beyond ftuitfall. Gynoecium is deeply lobed and the receptacle so enlarged that the carpels appear to be free except for their common style (Ochneae), or receptacle is not distinctly enlarged and the carpels do not appear distinct (Elvasieae). Fruits of separate 1-seeded drupes (Ochneae) or coriaceous and deeply lobed, with one seed in each lobe (locule). Seeds without endosperm, with highly vascularized thin-walled, exotestal seed coat. Flowers bisexual. Sepals five (Ochneae) or 3–6 (Elvasieae), persistent. Petals five (Ochneae) or 3–6 (Elvasieae), caducous. Stamens (8)10 to numerous, free, in one or more cycles; anthers basifixed, more or less latrorse, opening by longitudinal slits or apical pores. Pollen grains 3-colporate. Leaves

distichous, with short but distinct petioles. Stomata paracytic. Vasicentric tracheids absent. Rays heterogeneous. Axial parenchyma apotracheal. Crystarque cells commonly present. Trees or shrubs; n = 12, 14, 24. . . . . 4. OCHNACEAE.

3 Style apical. Carpels 2–5, with two to many ovules per carpel. Ovules bitegmic or unitegmic. Fruits berries or capsules. Receptacle not distinctly enlarged. Weeds with or without endosperm. Flowers bisexual or functionally polygamous. Petals five. Stamens five or numerous, free or connate at the base. Leaves not distichous. Vasicentric tracheids present or absent. Crystarque cells usually absent Rays heterogeneous or homogeneous. Axial parenchyma paratracheal of metatrachel.

4 Ovules bitegmic (outer integument of two layers, inner one of three layers), anatropous. Shrubs, subshrubs or herbs. Rays heterogeneous. Axiall parenchyma paratracheal and often scanty. Mucilage cells present. Vasicentric tracheids absent. Leaves simple, conduplicate-flat, rarely (*Rhytidanthera*) compound; stipules persistent. Flowers bisexual, actinomorphic; sepals free, imbricate, usually persistent; petals contorted; anthers extrorse, basifixed, 2-locular. Stamens five, free or connate at the base, surrounded by numerous staminodia (*Sauvagesia*). Carpels 2, 3, 5, with 1–2 (*Euthemis*) or many ovules per carpel. Fruits berries (*Euthemis*) or capsular. Seeds small, winged or not, with fibrous exotegmen, rarely (*Indovethia*) with long funiculus; exotesta with large cells; endosperm fleshy. n = 18. . . . . .2. SAUVAGESIACEAE.

4 Ovules unitegmic (with two completely fused integuments), numerous (10–20) but only one fertile, anatropous. Carpels two, with 5–10 ovules per carpel. Fruits 1-seeded capsules, with two persistent outer sepals enlarged and winglike. Seeds large, without endosperm, with testa reduced to approximately three layers of unspecialized cells, compressed between pericarp and large embryo. Stamens numerous, in 3–5 cycles. Axial parenchyma metatracheal. Rays homogeneous. Mucilage cells absent. Vasicentric tracheids present. Stomata anomocytic. Tall trees; n = 14. . . . . 3. LOPHIRACEAE.

1 Leaves opposite or verticillate. Cortical bundles absent.

5 Petals contorted. Small, glabrous trees up to 10 m with a rounded crown. Vessels with simple perforations in slightly inclined to nearly vertical end wall; lateral pitting from opposite to predominately alternate. Fibers with small bordered pits. Rays heterogeneous. Axial parenchyma diffuse and diffuse-in-aggregates. Secondary phloem stratified into alternating concentric rings of sieve tissue and sclerenchyma. Nodes pentalacunar with five traces. Leaves opposite, leathery, and shiny, simple, often emarginate, margins crenate with fine, widely spaced teeth, pinnately veined, with a pair of colleters in the leaf margins and in the leaf axils external to the axillary bud, with scattered mucilaginous cells in the mesophyll; stipules wanting, petiole bundles arcuate. Stomata anomocytic. Flowers medium-sized in lax, terminal, cymose inflorescences, andromonoecious (bisexual and exclusively male usually within the same inflorescence), actinomorphic with an elongate receptacle foetid. Sepals five, connate at the base, imbricate, reflexed persistent. Petals five, free, basally thickened, spatulate, contorted (with each petal having one margin exterior and the other margin interior), at anthesis reflexed. Stamens very numerous, free, spirally arranged on an elongated receptacle, and characterized by five major stamen fascicle traces (stamen trunk bundles); filaments very slender, shorter than the petals. Anthers basifixed, usually with a small, apical extension of the connective, 2-locular, the locules often set at different heights, opening longitudinally. Pollen grains 3-porate, tectate-columellate, with short and rather wide columellae and the massive foot layer, forming about one half of total exine thickness; the apertures protrude and are formed by a tapering, conelike extension of the exine; ornamentation consists of very fine striae. Gynoecium of 16–19 (up to 25) carpels fused ventrally to a solid parenchymatous central column (a prolonged receptacle). Below the level of ovular insertion the ventral sutures are open although carpels are always fused to the central column. Stylodia rather short, forming a ring around the shoulders of the ovary persistent; stigmas capitate, disclike, with a papillate upper surface; ovary with two ovules in each locule on axile placentas, upper

one ascending or erect (micropyle below), the lower one descending or hanging (micropyle above). Ovules large, with long and curved funiculi, anatropous, bitegmic, with a straight micropyle, vascularized by a single bundle that extends unbranched from the funiculus to the chalaza, with a very thin nucellus composed of one cell layer at maturity. Fruit a slightly woody, septicidal pendulous capsule; each carpel separates acropetally from the central column, with only a distal connection maintained; the carpels thus diverging from the top of the central column like the ribs of an umbrella; dehiscence of the individual carpels is along the entire ventral suture. Seeds two in each locule, small, compressed, winged all around, exarillate, with straight embryo and narrow layer of endosperm consisting of a cellular zone of very thin-walled cells without endosperm. The mature seed coat thin, consisting of four or five layers of cells and exotegmic with fibers. Tanniniferous. Dark-staining phenolic compounds throughout the plant body. . . . . 5. MEDUSAGYNACEAE.

5 Petals imbricate. Evergreen trees, sometimes to 30 m tall, shrubs, or sometimes woody lianas. With scattered secretory cells in the parenchymatous tissues of the young stem and with lysigenous mucilage-channels, especially within the vascular strand of the midrib and petioles of the leaves. Cortical bundles absent. Nodes trilacunar. Vessels with simple or seldom scalariform perforations. Fibers with evidently bordered pits. Rays heterogeneous. Axial parenchyma apotracheal diffuse or sometimes banded. Leaves simple or (*Touroulla* and *Froesia*) pinnately compound; secondary veins numerous, tertiaries feather-veined; veins, especially the smaller ones, sheathed by very thick-walled fibers; stipules interpetiolar, usually conspicuous, foliaceous or setose, entire or divided into several setose parts, persistent, pubescent. Stomata paracytic. Flowers in axillary or terminal panicles or racemes, bisexual or unisexual (plants androdioecious in *Quiina* and *Touroulia* or dioecious in *Lacunaria*), actinomorphic. Sepals 4-4 small, unequal, imbricate. Petals 4–5(-8), imbricate. Stamens 9 to ca. 100 in *Lacunaria*, *Quiina*, or sometimes over 300 in *Froesia* (Zizka et Schneider 2004); filaments filiform, free or connate at the base, and sometimes adnate to the base of the petals; anthers small, basifixed, 2-locular, opening longitudinally. Pollen grains 3-colporate. Gynoecium of three free carpels (*Froesia*) or more often of 2–14 united carpels with 2-14-locular ovary and linear stylodia; stigmas obliquely peltate. Ovules paired, ascending, anatropous. Fruits berrylike or 1–3 follicles in *Froesia*. Seeds one to more than 20 in berrylike fruits of one per follicle, hairy, or glabrous in *Froesia*, without endosperm and with straight embryo. .6. QUIINACEAE.

1. STRASBURGERIACEAE

Solereder 1908. 1/1. New Caledonia.

Strasburgeria.

Strasburgeria robusta is one of the most remarkable New-Caledonian relicts still preserving primitive characters, such as very long vessel elements with 20–30 bars, spirally arranged sepals, crassinucellate ovules, etc. But like many other phylogenetic relicts, it is very heterobathmic and has a number of specialized characters, especially in pollen, gynoecium, and fruit morphology. It is an early side branch from the ochnalean ancestral stock.

2. SAUVAGESIACEAE

Dumortier 1829 (including Euthemidaceae Solereder 1908, Luxemburgiaceae Solereder 1908). 20/120. Pantropical, but mainly represented in the Neotropics.

2.1 LUXEMBURGIOIDEAE

Tree. Flowers yellow; sepals caducous; androecium obliquely zygomorphic in bud, only adaxial; filaments more or less connate; pollen exine with small perforations. Gynoecium of three carpels. – *Luxemburgia.*

2.2 SAUVAGESIOIDEAE

Trees, shrubs, subshrubs, or herbs. Rays heterogeneous. Axial parenchyma paratracheal and often scanty. Secondary veins archingly joined near the leaf margin. Staminodia 10-numerous in one or more cycles. Stamens five, more or less adnate at the base to a short tube of inner staminodia. Anthers opening by two lateral slits. Gynoecium of 2–3 carpels.

Ovary 1-locular with intruding placentas and numerous ovules. Fruits a coriaceous capsule. Seeds usually numerous, small. – *Godoya, Rhytidanthera, Krukoviella, Cespedesia, Testulea, Fleurydora, Philacra, Blastemanthus, Poecilandra, Wallacea, Schuurmansia, Schuurmansiella, Tyleria* (including *Adenanthe?*), *Adenanthe, Adenarake, Indosinia, Sinia, Sauvagesia* (including *Indovethia?*).

2.3 EUTHEMIDOIDEAE

Shrubs. Numerous parallel secondary veins reaching denticulate leaf margin. Staminodia 0(-5), filamentous. Stamens free. Anthers opening by an apical pore. Gynoecium of five carpels. Ovary 5-locular, with 1–2 ovules per locule. Fruits berries with five, usually 1-seeded pyrenes. Seeds one or two per locule. – *Euthemis.*

Sauvagesiaceae are usually included in the Ochnaceae, which makes the later very heterogeneous. They have no close affinity with the Ochnaceae and differ markedly from them in many respects, including seed coat anatomy (Corner 1976; Takhtajan 1987).

3. LOPHIRACEAE

Loudon 1830. 1/2. Tropical Africa.

Lophira.

A specialized unigeneric family that differs markedly from both the Sauvagesiaceae and Ochnaceae, but it is nearer to the Sauvagesiaceae.

4. OCHNACEAE

A.P. de Candolle 1811 (including Gomphiaceae A.P. de Candolle ex Schnizlein 1843–1870). 6/480. Pantropical with only a few representatives in the subtropics (some species of *Ochna* partly cross the Tropic of Cancer in northern India, to ca. 30° N). The tribe Elvasieae is Neotropical.

OCHNEAE: *Ochna, Brackenridgea, Ouratea* (including *Idertia*), *Gomphia* (including *Rhabdophyllum?, Campylospermum?*), *Perissocarpa;* ELVASIEAE: *Elvasia.*

5. MEDUSAGYNACEAE

Engler et Gilg 1924. 1/1. Seychelles Islands. *Medusagyne.*

Related to the Ochnaceae.

6. QUIINACEAE

Engler 1888. 4/50. Tropical America from Belize south to Bolivia and southern Brazil, but mainly Amazon baSinica.

Froesia, Quiina, Touroulia, Lacunaria.

Evidently close to the Ochnaceae (see Gottwald and Parameswaran 1967).

Bibliography

Amaral MCE. 1991. Phylogenetische Systematik der Ochnaceae. Bot. Jahrb. Syst. 113: 105–196.

Amaral MCE and V Bittrich. 1998. Ontogenia inicial do androceu de especies de Ochnaceae subfam. Sauvagesioideae atraves da analise em microscopia eletronica de varredura. Rev. Brasil. Bot. 21: 269–273.

Cameron KM. 2001. On the phylogenetic position of the New Caledonian endemic families Strasburgeriaceae, Oncothecaceae, and Paracryphiaceae: a comparison of molecules and wood anatomy. In Botany 2001: Plants and People, p. 17. Abstracts. Albuquerque.

Cameron KM. 2003. On the phylogenetic position of the New Caledonian endemic families Paracryphiaceae, Oncothecaceae, and Strasburgeriaceae: a comparison of molecules and morphology. Bot. Rev. 68: 428–443.

Decker JM. 1966. Wood anatomy and phylogeny of Luxemburgieae (Ochnaceae). Phytomorphology 16: 39–55.

Decker JM. 1967. Petiole vascularization of Luxemburgieae (Ochnaceae). Am. J. Bot. 54: 1175–1181.

Den Outer RW. 1977. The secondary phloem of some Ochnaceae and the systematic position of *Lophira lanceolata* Tieghem ex Keay. Blumea 23: 439–447.

Dickison WC. 1981. Contributions to the morphology and anatomy of *Strasburgeria* and discussion on the taxonomic position of the Strasburgeriaceae. Brittonia 33: 564–580.

Dickison WC. 1990a. The morphology and relationships of *Medusagyne* (Medusagynaceae). Plant Syst. Evol. 171: 27–55.

Dickison WC. 1990b. An additional note on the floral morphology and affinities of *Medusagyne oppositifolia* (Medusagynaceae). Brittonia 42: 191–196.

Dickison WC. 2007. Strasburgeriaceae. In: K Kubitzki, ed. The families and genera of vascular plants, vol. 9, pp. 446–448. Springer, Berlin/Heidelberg/New York.

Doweld AB. 1998. On the phylogenetic relationships of *Medusagyne* (Medusagynaceae) as evidenced by the structure of its fruits and seeds. Bot. Zhurn. 83: 54–68 (in Russian with English summary).

Engler A and H Melchior. 1925. Medusagynaceae. In: A Engler and K Prantl, eds. Die natürlichen Pflanzenfamilien, 2nd ed., vol. 21, pp. 50–52. W. Englemann, Leipzig.

Fay MF and MW Chase. 1996. Molecular phylogeny of Ochnaceae and related families. Am. J. Bot. 83(6): 155 (abstract).

Fay MF, SM Swensen, and MW Chase. 1997. Taxonomic affinities of *Medusagyne oppositifolia* (Medusagynaceae). Kew Bull. 52: 111–120.

Fay MF, RS Cowan, G Beltran, and B Allen. 2000. Genetic fingerprinting of two endemics from the Seychelles: *Medusagyne oppositifolia* (Medusagynaceae) and *Rothmannia annae* (Rubiaceae). Phelsuma 8: 11–22.

Foster AS. 1950a. Morphology and venation of the leaf in *Quiina acutangula*. Am. J. Bot. 37: 159–171.

Foster AS. 1950b. Venation and histology of the leaflets in *Touroulia guianensis* Aubl. and *Froesia tricarpa* Pires. Am. J. Bot. 37: 848–862.

Gottwald H and N Parameswaran. 1967. Beiträge zur Anatomic und Systematik der Quiinaceae. Bot. Jahrb. Syst. 87: 361–381.

Guedes M and C Sartre. 1981. Morphology of the gynoecium and systematic position of the Ochnaceae. Bot. J. Linn. Soc. 82: 121–138.

Hemsley WB. 1905. *Medusagyne oppositifolia* J.G. Baker. Hooker's Icones Plantarum, ser. 4, 8: 1–3, pl. 2790.

Kanis A. 1968. A revision of the Ochnaceae of the Indo-Pacific area. Blumea 16: 1–82.

Muller J. 1969. Pollen morphological notes on Ochnaceae. Rev. Palaeobot. Palynol. 9: 149–173.

Narayana LL. 1975. Contribution to the floral anatomy and embryology of Ochnaceae. J. Jpn. Bot. 50: 329–336.

Oginuma K, J Munzinger, and H Tobe. 2005. *Strasburgeria robusta* (Strasburgeriaceae) survives as a high-polyploig species in New Caledonia. In Abstracts of XVII International Botanical Congress, 12–16 July, Vienna. Abstract PO668.

Oginuma K, J Munzinger, and H Tobe. 2006. Exceedingly high chromosome number in Strasburgeriaceae, a monotypic family endemic to New Caledonia. Plant Syst. Evol. 262: 97–101.

Pauze F and R Sattler. 1979. La placentation axillaire chez *Ochna atropurpurea*. Canad. J. Bot. 57: 100–107.

Robertson AR, R Wise and F White. 1989. 138. *Medusagyne oppositifolia:* Medusagynaceae. Kew Mag. 6(4): 166–171.

Schneider JV, U Swenson, and G Zizka. 2002. Phylogenetic reconstruction of the neotropical family Quiinaceae (Malpighiales) based on morphology with remarks on the evolution of an androdioecious sex distribution. Ann. Missouri Bot. Gard. 89: 64–76.

Schneider JV, U Swenson, R Samuel, T Stuessy, and G Zizka. 2006. Phylogenetics of Quiinaceae (Malpighiales): evidence from *trn*L-*trn*F sequence data and morphology. Plant Syst. Evol. 257: 189–203.

Tieghem P van. 1901. Sur le genre *Lophira* considere comme type d'une famille distincte, les Lophiracees. J. Bot. (Paris) 15: 169–194.

Tieghem P van. 1902a. L'embryon des Ochnacees et son emploi dans la definition des genres. Bull. Mus. Hist. Nat., Paris, 8: 208–218.

Tieghem P van. 1902b. Le cristarque dans la tige et la famille des Ochnacees. Bull. Mus. Hist. Nat., Paris, 8: 266–273.

Tieghem P van. 1903. Sur le genre *Strasburgeria* considéré comme type d'une famille nouvelle, les Strasburgériacées. J. Bot. (Morot) 17: 198–204.

Zizka G and JV Schneider. 1999. The genus *Touroulia* Aubl. (Quiinaceae). Willdenowia 29: 1–8.

Zizka G and J Schneider. 2004. Quiinaceae. In: N Smith, S Mori, A Henderson, DW Stevenson, and SV Heald, eds. Flowering plants of the Neotropics, pp. 317–318. The New York Botanical Garden, Princeton University Press, Princeton.

Order 54. ERICALES

Shrubs or less often small trees, perennial herbs or woody lianas. Raphide mostly cells absent. Vessels mostly with scalariform perforations, sometimes with many bars (up to 50 in Clethraceae and up to 65–70 in Cyrillaceae), but often with both scalariform and simple perforations or only with simple perforations; lateral pitting from scalariform to alternate. Fibers with bordered pits. Rays typically heterogeneous. Axial parenchyma diffuse, often scanty or wanting. Sieve-element plastids of S- or (Cyrillaceae) Pcf-type. Nodes unilacunar, trilacunar (some Ericaceae and some Styphelioideae) or multilacunar (some Styphelioideae). Leaves alternate, less often opposite or verticillate, simple, entire, estipulate, sometimes without chlorophyll and reduced to mere scales. Stomata anomocytic, paracytic, tetracytic, or of other types. Flowers in various (mostly racemose) inflorescences or sometimes solitary (axillary or terminal), bisexual or rarely unisexual, usually actinomorphic. Sepals (3-)5(-7), free or basally connate, imbricate or valvate. Petals isomerous with sepals, free or more often connate into sympetalous corolla with usually imbricate or contorted, more rarely valvate lobes; rarely petals wanting. Stamens mostly twice as many as the petals or less often the same number and then alternate, rarely only two (*Ceratiola* in the Empetroideae) or up to 20; filaments free or more or less adnate to the corolla; anthers tetrasporangiate or disporangiate, usually inverted, opening longitudinally or much more often by apical (morphologically basal) pores or slits, often with two or more slender appendages, which are rarely borne on the upper part of the filament instead of on the anther. Tapetum secretory. Microsporogenesis simultaneous. Pollen grains 2-celled or sometimes 3-celled, very often in tetrads, mostly 3-colporate. Nectary disc usually present, intrastaminal. Gynoecium of 2–10 united carpels, with usually simple style; ovary superior or inferior, 1–10-locular, with numerous or rarely several or even one ovule per carpel. Ovules anatropous or hemitropous, unitegmic, tenuinucellate. Female gametophyte usually of *Polygonum*-type. Endothelium present. Endosperm usually cellular (nuclear in *Rhododendron* and *Vaccinium*), with terminal haustoria that are usually micropylar and chalazal. Fruits septicidal or loculicidal capsules, berries, or drupes. Seeds mostly small or minute, exotestal, with copious or scanty endosperm and small or minute embryo. Often producing iridoid compounds, alkaloids present (rarely), or absent.

Key to Families

1 Raphide cells present. Trees, shrubs, or woody lianas; hairs often more or less flattened setose. Pith often septate. Vessel elements with spiral thickening, their ends more or less oblique, with scalariform perforations that have 15 and more bars or with mixed scalariform and simple perforations; lateral pitting opposite or less often scalariform or alternate. Fibers mostly long, with distinctly or dared pits. Nodes unilacunar or seldom trilacunar. Leaves dentate or entire. Stomata anomocytic or sometimes nearly paracytic. Flowers small, in axillary or lateral, basically cymose inflorescences (which sometimes reduced to one or a few flowers), often on old wood, bisexual, polygamous or dioecious, often subtended by small bracteoles. Sepals and petals mostly five each, sometimes four, seldom up to seven, usually quincuncial in bud, free or connate at the base. Stamens branched, or unbranched; maturing centrifugally; free of the perianth; all equal; free of one another, or coherent (then variously united at the base, often in five clusters opposite the petals); numerous to 10, often in five antepetalous fascicles, that may adnate to the base of the petals. Anthers erect or inflexed in bud, versatile, inverted, opening by morphologically basal but seemingly apical short slits or pores. Pollen grains 2-celled, in monads or sometimes (*Saurauia elegans*) in tetrads, with complete tectum and reduced columellae, psilate or granulate. Gynoecium of 3–5 or many (up to 30 or more) united carpels, with free or more or less connate stylodia; ovary superior, completely or incompletely trilacunar to multilacunar, with numerous or (*Clematoclethra*) only 10 ovules in each locule on axile or nearly axile placentas. Ovules anatropous, hypostatic (with the distinctive thick-walled nucellar area in the chalaza), with conspicuous endothelium and a well-developed hypostase. Fruits berries, hairy or glabrous, with massive placentae, or sometimes more or less loculicidal capsules. Seeds small, arillate or not, with copious endosperm and well-developed, straight or slightly curved embryo. Iridoid compounds found in *Actinidia*; myricetin was detected in leaves, actinidin, a proteinase similar to papain was detected in kiwifruits (McDowall 1970; Dressler and Bayer 2004). n = 30 (*Saurauia*, South American species – Soejarto 1969, 1970) and n = 20 (Asian species), 29 (*Actinidia* – Yan et al. 1997). . . . . 1. ACTINIDIACEAE.

1 Raphide cells absent.

2 Petals, when present, free or nearly so. Pollen grains usually in monads. Leaves alternate. Nodes only unilacunar.

3 Anthers deeply inverted, opening by slitlike pores at the apparent apex. Intrastaminal disc wanting, but the basal part of the ovary often nectariferous. Sepals imbricate, persistent. Petals five, free, imbricate. Stamens 10(12), free. Anthers not appendages. Pollen grains single or rarely in tetrads, 3-colporate. Gynoecium of three carpels. Ovary superior, 3-locular, ovules numerous on intrusive placentas, anatropous. Fruits loculicidal capsules. Seeds numerous, often winged. Seed coat present, exotestal, consisting of a single layer of cells; embryo straight, short, cylindrical; endosperm moderate, fleshy Small trees or shrubs; indumentum of simple and fasciculate, often stellate hairs. Vessels with scalariform perforation with 20–50 bars, pits vestured. Alkaloids and iridoids not detected. Kestose and isokestose oligosaccharides present, n = 8. . . . . 2. CLETHRACEAE.

3 Anthers not obviously inverted, opening by longitudinal slits or apical pores. Intrastaminal nectary disc present. Sepals 5(6), free or rarely very shortly connate, imbricate or rarely valvate, persistent. Petals 5(6), free or slightly connected below, imbricate or contorted. Stamens 5 or 10 in two whorls, filaments elongate, sometimes broadened, free or adnate to the corolla at the very base (*Purdiaea*). Pollen grains 3(-4)-colporate. Gynoecium of 2–5 carpels. Ovary 2-4(5)-locular; ovules 1–3 per locule, pendulous, apotropous. Funiculus long. Fruits small, indehiscent, the pericarp thin and dry, sometimes a samaroid nutlet. Seeds usually without seed coat; embryo straight, usually cylindrical, with small, slightly expanded cotyledons; endosperm copious, fleshy-hard. Small glabrous trees or shrubs. Vessels with scalariform perforations with 20-50(-70) bars. Leaves short-petioled or sessile. Present myricetin, ellagic acid and procyanidin, n = 10. . . . . 3. CYRILLACEAE.

2 Petals, when present, mostly united into a sympetalous corolla. Pollen grains mostly in tetrads. Mainly shrubs, less often perennial herbs, small trees, or rarely trailing or scrambling lianas. Indumentum unicellular and multicellular, or none. Vessels mostly with scalariform perforation with many bars. Nodes multilacunar, trilacunar, or unilacunar. Leaves mostly with palmate or subparallel venation, rarely with pinnate venation (*Prionotes* and *Lebetanthus*), alternate or opposite. Stomata mostly anomocytic, but in Vaccinioideae mostly paracytic. Flowers mostly bisexual or rarely (Empetroideae) unisexual. Sepals (3)4–7(8), mostly five, usually more or less connate. Petals the same number as sepals, more or less connate to form a short or conspicuous corolla tube. Stamens mostly in two cycles, twice as many as petals, seldom more numerous (up to 20) or less often the same number as petals. Anthers with exothecium or rarely (*Enkianthus*) with endothecium, tetrasporangiate and 2-locular (rarely 4-locular) or disporangiate and usually 1-locular (Styphelioideae), opening by pores at the apparent apex, or by elongate slits, sometimes by longitudinally, often with appendages. Pollen grains in tetrahedral tetrads or rarely (*Enkianthus*) in monads. Intrastaminal nectary disc present. Gynoecium of 2–10, mostly five or, less often, four carpels. Ovary usually superior or inferior, usually with the same number of locules as carpels and rarely (as in *Scyphogyne* and related genera) unilocular, with a solitary pendulous ovule; ovules 1-several or more often numerous on each placenta, anatropous to subcampylotropous. Fruits septicidal or loculicidal capsules, or berries or drupes, seldom a small nut. Seeds numerous, small to minute, sometimes winged, embryo straight, fusiform, rarely embryo minute, undifferentiated; endosperm fleshy, with haustoria at both ends. Present flavonoids, flavones, simple phenols, anthocyanins, ellagic acid, and methyl salicylate; iridoids detected. . . . . 4. ERICACEAE.

1. ACTINIDIACEAE

Gilg et Werdermann 1925 (including Saurauiaceae Grisebach 1854). 3/365. Tropical and subtropical regions of eastern and Southeast Asia, West Malesia, America, and Australia (only one species of *Saurauia* in northeastern Queensland); *Actinidia* (55) is distributed in the Himalayas and eastern and Southeast Asia, *Clematoclethra* (1) in temperate China.

1.1 SAURAUIOIDEAE

Flowers usually bisexual, rarely functionally dioecious, predominantly 5-merous. Petals usually connate at the base. Stamens 13–50 or more, basally connate, and distinctly epipetalous to only partially so. Anthers opening by rimiform pores. Gynoecium of 3–5 (very rarely seven) carpels. Stylodia emerging from a depression at the apex, free or united for three fourths of their length. Ovules variable in number, commonly numerous in each locule. Fruits usually typical berries with numerous small seeds embedded in a mucilaginous pulp, n = 13. – *Saurauia*.

1.2 ACTINIDIOIDEAE

Flowers dioecious or polygamous, with 5-merous or 4-merous perianth. Petals free to basally connate. Stamens numerous, free. Anthers opening by short subapical longitudinal slits or by rimiform pores. Gynoecium of 3–5 or 15 to slightly more than 30 carpels. Stylodia free. Ovules numerous in each locule. Fruits berries, include about 40 locules and 1,500 seeds, n = 29. – *Actinidia*.

1.3 CLEMATOCLETHROIDEAE

Flowers dioecious or polygamous, basally 5-merous. Stamens 10–30, free. Anthers opening longitudinally. Gynoecium mostly of five or four carpels. Stylodia united into a style arising from a conspicuous depression and terminated by a slightly swollen stigmatic surface. Ovules about 10 per locule. Fruits capsules or drupaceous, n = 12. – *Clematoclethra*.

I agree with Dickison (1972: 53) that "the similarities between the Actinidiaceae and Ericales are impressive and indicate that the close affinities of this family may very well lie within the ericalean complex." Although in many respects, particularly in embryology (Crété 1944), Actinidiaceae are closely related to other families of Ericales (especially Clethraceae), they differ in the presence of the raphides, stamen fascicle traces, hypostatic ovules, the absence of terminal haustoria, and some other less important features.

2. CLETHRACEAE

Klotzsch 1851. 1/85. Eastern and Southeast Asia, Malesia, southeastern United States, central and tropical

South America, with one endemic species on the island of Madeira.

Clethra.

Clethraceae are related to the Actinidiaceae, which is confirmed also by embryological data (Veilllet-Bartoszewska 1960). They reveal also definite closeness to the Theaceae, especially in the anatomy of the seed coat and in flower morphology, including the presence of nectariferous tissues in the basal part of the ovary of some species of *Clethra,* which resemble *Eurya japonica* (Brown 1938). Thus, specialized nectariferous tissues have already developed in *Clethra,* which gave rise to the intrastaminal nectary disc of other Ericales (Takhtajan 1966). Although Clethraceae have many features in common with the Theales, especially with the genus *Eurya* in the Theaceae, they are closer to the Ericales, which is confirmed not only by the flower morphology but also by their unitegmic ovules, presence of endothelium and endospermal haustoria (Poddubnaja-Arnoldi 1982), anatomy of the seed coat (Corner 1976) and wood characters (Giebel and Dickison 1976).

3. CYRILLACEAE

Endlicher 1841. 3/14. Central America, northern parts of South America, West Indies, coastal plains of southeastern United States.

Cyrilla, Cliftonia, Purdiaea.

Jussieu (1789) placed *Cyrilla* (the only known cyrillaceous genus at that time) in his Ericeae, which has been followed by some other botanists including Agardh (1858). According to Hallier (1912), Cyrrillaceae originated from the Ericaceae and are so close to them that they could even be considered a tribe of the Ericaceae. This idea of the closeness of the Cyrillaceae to the Ericaceae has been accepted by an increasing number of botanists (including Wettstein [1935] and Copeland [1953]) and by some later authors. Dahlgren (1975, 1989), Thorne (1976, 1992a, b, 2000, 2006), and Cronquist (1981, 1988) place the Cyrillaceae close to the Clethraceae. These two families have many similarities in gross morphology and in embryology, palynology, and chemistry (including the absence of iridoid compounds). *Cyrilla* and *Clethra* are also very similar in aspect (Cronquist 1981). Anderberg (1993) notes that the genus *Purdiaea* includes one taxon that has formerly been placed in the Clethraceae, thus reflecting the symplesiomorphic similarity of the two families.

4. ERICACEAE

A.L. de Jussieu 1789 (including Andromedaceae A.P. de Candolle ex Schnizlein 1843–1870, Arbutaceae Bromhead 1840, Arctostaphylaceae J.G. Agardh 1858, Azaleaceae Vest 1818, Empetraceae Berchtold et J. Presl 1820, Epacridaceae R. Brown 1810, Hypopityaceae Klotzsch 1851, Ledaceae Gmelin 1803, Menzieziaceae Klotzsch 1851, Monotropaceae Nuttall 1818, Oxycoccaceae A. Kerner 1891, Prionotaceae Hutchinson 1969, Pyrolaceae Lindley 1829, Rhododendraceae A.L. de Jussieu 1789, Rhodoraceae Ventenat 1799, Salaxidaceae J.G. Agardh 1858, Styphe-liaceae Horaninow 1834, Vacciniaceae A.P. de Candolle ex Perleb 1818). 128/c.4000. Widely distributed in almost all parts of the world with large concentrations in the Himalayas and southwestern China, New Guinea and southern Africa; absent from most of Australia.

4.1 ENKIANTHOIDEAE

Usually deciduous or sometimes evergreen shrub. Pith heterogeneous (pith with small, thick-walled and lignified and larger and thin walled cells mixed). Leaves flat and fairly large, serrulate or entire. Flowers in raceme- or umbel-like inflorescences. Corolla urceolate to companulate. Stamens with papillose to hairy filaments, anthers with paired awns (fibrous endothecium), which gradually narrow from the thecae. Pollen grains trinucleate, in monads, without viscin threads. Ovary superior, 5-locular. Fruits loculicidal capsules, opening with five valves. Seeds rather large, often winged on the margins. n = 11. – *Enkianthus.*

4.2 MONOTROPOIDEAE

Strongly mycotrophic and more or less fleshy perennial herbs devoid of chlorophyll. Multicellular hairs present. Leaves reduced to scales. Corolla persistent, rarely absent. Anthers sometimes spurred at the morphological base, without viscin threads, opening by longitudinal slits or sometimes by terminal pores or by slits across the broad locular summits. Pollen grains in monads. Ovary 1–6-locular. Fruits loculicidal or septicidal capsules, or baccate and indehiscent; embryo minute, undifferentiated, n = 8, 13, 19, 23, etc. – PTEROSPOREAE: *Pterospora, Sarcodes, Allotropa;* PLEURICOSPOREAE: *Pleuricospora*; MONOTROPEAE: *Cheilotheca, Monotropsis, Monotropa, Monotropastrum*; HEMITOMEAE: *Hemitomes, Pityopus, Hypopitys.*

4.3 PYROLOIDEAE

Strongly mycotrophic perennial herbs from creeping rhizomes. Multicellular hairs absent. Leaves usually green, serrate. Corolla persistent. Petals free. Stamens without appendages and without viscin threads. Anthers opening by terminal pores. Pollen grains usually in tetrads. Ovary superior, imperfectly 5–4-locular. Fruits loculicidal capsules; embryo small, hardly differentiated. – *Pyrola, Moneses, Chimaphila, Orthilia.*

4.4 ARBUTOIDEAE

Evergreen trees or shrubs. Leaves coriaceous, entire or serrate, scattered or rarely whorled. Flowers 4(5)-merous. Corolla urceolate. Anthers almost smooth with paired granular spurs near the top where the filament and anther join, dehiscence by terminal pores or short slits. Ovary 4–10-locular. Fruit usually baccate or drupaceous; testa cells rather thick-walled. Ellagic acid present, n = 13. – *Arbutus, Comarostaphylos, Ornithostaphylos, Arctostaphylos* (including *Xylococcus*), *Arctous.*

4.5 CASSIOPOIDEAE

Low shrubs, stems lacking pericyclic sheath of fibers; pith with large thin walled cells surrounded by smaller thick-walled and lignified cells (*Caluna*-type). Leaves decussate, ericoid, auriculate, revolute in bud. Indumentum of fasciculate branched and small glandular hairs. Inflorescence axillary. Flowers solitary, with 4–6 basal bracteoles. Corolla sympetalous. Anthers dehiscing by a pair of dorso-apical spurse. Pollen grains in tetrads, without viscin threads. Ovary 4–5-locular, superior. Fruit a loculicidal capsule. – *Cassiope* (including *Harrimanella*).

4.6 EPIGAEOIDEAE

Procumbent shrublets. Leaves rather large, entire. Indumentum of long-stalked glandular and long-celled, or unicellular hairs Winter buds without scales. Inflorescence terminal, shortly racemose or spicate. Calyx lobes large; corolla hypocrateriform to infundibular. Stamens 10, without appendages, having viscin threads among the pollen tetrads, opening longitudinally. Ovary superior, 5-locular, densely pubescent; placentae deeply divided; stigma much expanded, sometimes with radiate arms. Fruits septicidal capsules with fleshy placentae. Seeds ovoid, cells of testa thick-walled. – *Epigaea.*

4.7 ERICOIDEAE

Shrubs. Leaves usually verticillate, rarely decussate. No winter buds or scales. Corolla usually persistent or caducous. Stamens usually with appendages represented by flattened spurs; anthers opening by very large pores or slits; no viscin threads among pollen tetrads. Ovary superior, (1-)2–14-locular. Fruits usually loculicidal capsules, rarely septicidal or indehiscent. Seeds winged or not. – ERICEAE: *Erica* (including *Ericinella, Blaeria*), *Bruckenthalia*; CALLUNEAE: *Calluna.*

4.8 RHODODENDROIDEAE

Small to large evergreen or deciduous shrubs. Leaves convolute or revolute in bud. Corolla gamo- or choripetalous, 4–9-lobed. Anthers dehiscing by means of collapse tissue only, forming pores or short slits; viscin threads usually present. Ovary 4–14-locular. Fruit a many-seeded septicidal capsule. Seeds variably shaped and winged, cells of the testa nearly always much elongated. – BEJARIEAE: *Bejaria, Ledothamnus, Bryanthus;* RHODODENDREAE: *Therorhodion, Rhododendron* (including *Ledum?*), *Tsusiophyllum, Menziesia*; CLADOTHAMNEAE: *Elliottia* (including *Tripetaleia* and *Cladothamnus*); PHYLLODOCEAE: *Kalmia* (including *Leiophyllum, Loiseleuria*), *Phyllodoce, Kalmiopsis, Rhodothamnus;* DABOECIEAE: *Daboecia*; DIPLARCHEAE: *Diplarche* (2, eastern Himalayas).

4.9 EMPETROIDEAE

Flowers typically dioecious, seldom monoecious or polygamous. Sepals 1–3, free. Petals 1–3, free. Stamens two or three, rarely four, free. Anthers opening longitudinally, without appendages. Intrastaminal nectary disc absent. Gynoecium of 2–9 (*Empetrum*) *or* two carpels (*Ceratiola* and *Corema*). Ovary superior, 2–9-locular, with solitary ovule in each locule. Fruits fleshy or dry drupes containing two or more 1-seeded pyrenes. Small shrubs. Nodes unilacunar. Vessels perforation simple, scalariform, or with vestigial bars. Alkaloids and iridoids not detected. n = 13. – *Empetrum, Corema, Ceratiola.*

4.10 STYPHELIOIDEAE (EPACRIDOIDEAE)

Shrubs or small trees. Nodes multilacunar, trilacunar, or unilacunar. Leaves mostly with palmate or subparallel venation, rarely with pinnate venation, alternate. Flowers mostly in racemes, usually bisexual, actinomorphic, with two bracteoles. Sepals 5-4, free, imbricate, persistent. Petals 5-4, connate, attached to the nectary disc; corolla tubular, with imbricate or valvate lobes. Stamens mostly of the same number as corolla-lobes, often attached to the corolla tube; anthers disporangiate and usually 1-locular or seldom (*Prionotes* and *Lebetanthus*) 2-locular and opening longitudinally. Pollen grains in tetrads, but 1–3 grains aborting in many Styphelieae so that the mature pollen sometimes appear solitary. Gynoecium of five or less often four carpels, with simple style and capitate stigma; ovary superior, often surrounded at the base by the nectary disc, 1–10-lacunar, with intrusive placentas, with several or many ovules per carpel. Fruits loculicidal capsules or (Styphelieae) drupes with 1–5 pyrenes. Seeds with thin testa (one layer of cells) and cylindrical embryo surrounded by copious, fleshy endosperm. Alkaloids present (rarely), or absent. Iridoids detected, n = 4–14. – PRIONOTEAE: *Prionotes, Lebetanthus*; ARCHERIEAE: *Archeria*; OLIGARRHENEAE: *Oligarrhena, Needhamiella*; COSMELIEAE: *Cosmelia, Sprengelia, Andersonia*; RICHEEAE: *Richea, Dracophyllum, Sphenotoma*; EPACRUDEAE: *Lysinema, Woollsia, Epacris, Rupicola, Budawangia*; STYPHELIEAE: *Styphelia, Androstoma, Planocarpa, Cyathopsis, Cyathodes, Leptecophylla, Astroloma, Leucopogon, Trochocarpa, Decatoca, Lissanthe, Monotoca, Acrotriche, Brachyloma, Melichrus, Conostephium, Croninia, Coleanthera, Choristemon, Pentachondra.*

4.11 VACCINIOIDEAE

Woody plants of variable habits. Leaves usually more or less broad and flat; stomata often paracytic. Winter buds sometimes with scales. Inflorescences usually axillary. Corolla caducous, usually urceolate or tubular. Anthers with spurs, two or four awns, or appendages absent, often dehiscing via terminal tubules; viscin tetrads absent. Stigma truncate. Ovary superior to inferior, 3–10-locular. Fruits berries, drupes, or loculicidal capsules. Seeds wingless. "Some taxa (Vaccinieae) in both the Old and New Worlds have seeds with a mucilaginous testa and a green embryo; plants with such seeds are generally epiphytic or epilithic" (P.F. Stevens Angiosperm Phylogeny Website 2006). n = 12. – OXYDENDREAE: *Oxydendrum*; LYONIEAE: *Lyonia, Pieris, Craibiodendron, Agarista* (incl. *Agauria*), *Arcterica*; ANDROMEDEAE: *Andromeda, Zenobia*; VACCINIEAE: *Gaylussacia, Vaccinium* (including *Oxycoccus*), *Symphysia, Thibaudia, Satyria, Agapetes* (including *Pentapterigium*), *Didonica, Lateropora, Notopora, Utleya, Gonocalyx, Dimorphanthera, Paphia, Cavendishia, Orthaea, Macleania, Psammisia, Mycerinus, Polyclita, Anthopteropsis, Ceratostema, Semiramisia, Oreanthes, Siphonandra, Pellegrinia, Disterigma, Sphyrospermum, Diogenesia, Rusbya, Themistoclesia, Plutarchia, Demosthenesia, Anthopterus, Costera*; GAULTHERIEAE: *Diplycosia, Chamaedaphne, Leucothoe* (including *Eubotryoides*), *Eubotrys, Tepuia, Gaultheria, Pernettyopsis.*

Ericaceae are closely related to the Clethraceae and differ from them mainly in the presence of two bracteoles, more or less connate petals, the presence of the intrastaminal nectary disc and especially in the presence of iridoid compounds.

I agree with Henderson (1919), Copeland (1941, 1947), Veillet-Bartoszewska (1960), Stevens (1971), Anderberg (1992a, b), Judd and Kron (1993), and Kron and Chase (1993) that Pyroloideae and Monotropoideae are not separable from the rest of the Ericaceae.

Bibliography

Abbott CL. 1936. The phylogeny of the Ericales. Trillia 9: 62–69.

An H-X, D-R Cai, J-R Wang, and N-F Qian. 1983. Investigations on early embryogenesis of *Actinidia chinensis* Planchon var. *chinensis*. Acta Bot. Sinica 25: 99–104.

Anderberg AA. 1992. The circumscription of the Ericales and their cladistic relationships to other families of "higher" dicotyledons. Syst. Bot. 17: 660–675.

Anderberg AA. 1993. Cladistic interrelationships and major classification of the Ericales. Plant Syst. Evol. 184: 207–231.

Anderberg AA. 1994a. Phylogeny of Empetraceae with special emphasis on character evolution in the genus *Empetrum*. Syst. Bot. 19: 35–46.

Anderberg AA. 1994b. Cladistic analysis of *Enkianthus* with notes on the early diversification of the Ericaceae. Nord. J. Bot. 14: 385–401.

Anderberg AA and X Zhang. 2002. Phylogenetic relationships of Cyrillaceae and Clethraceae (Ericales) with special emphasis on the genus *Purdiaea* Planch. Organisms Divers. Evol. 2: 127–137.

Anderberg AA, C Rydin, and M Källersjö. 2002. Phylogenetic relationships in the order Ericales s.l.: analyses of molecular data from five genes from the plastid and mitochondrial genomes. Am. J. Bot. 89: 677–687.

Anisimova GM and II Shamrov. 2000. Ovule and seed development in *Vaccinium myrtillus* L. Acta Biol. Cracoviensia, Ser. Bot. 42: 47–54.

Baas P. 1985. Comparative leaf anatomy of *Pernettya* Gaud. (Ericaceae). Bot. Jahrb. Syst. 105: 481–495.

Behnke H-D. 1982. Sieve element plastids of Cyrillaceae, Erythroxylaceae and Rhizophoraceae: description and significance of subtype PV plastids. Plant Syst. Evol. 141: 31–39.

Bell TL and F Ojeda. 1999. Underground storage in *Erica* species of the Cape floristic region – differences between seeders and resprouters. New Phytol. 144: 143–152.

Bell TL, JS Pate, and KW Dixon. 1996. Relationships between fire response, morphology, root anatomy and starch distribution in south-west Australian Epacridaceae. Ann. Bot. II 77: 357–364.

Bidartondo MI. 2005. The evolutionary ecology of myco-heterotrophy. New Phytol. 167: 335–352.

Bidartondo ML and TD Bruns. 2001. Extreme specificity in epiparasitic Monotropoideae (Ericaceae): widespread phylogenetic and geographic structure. Molec. Ecol. 10: 2285–2295.

Bidartondo ML and TD Bruns. 2002. Fine-level mycorrhizal specificity in the Monotropoideae (Ericaceae): specificity for fungal species groups. Molec. Ecol. 11: 557–569.

Böcher TW. 1981. Evolutionary trends in Ericalean leaf structure. Kongel. Danske Vidensk.-Selsk. Biol. Skr. 23(2): 1–64.

Bohm BA and JE Averett. 1989. Flavonoids in some Monotropoideae. Biochem. Syst. Ecol. 17: 399–401.

Brown EGS. 1935. The floral mechanism of *Saurauia subspinosa* Anth. Trans. Proc. Bot. Soc. Edinb. 31: 485–497.

Brown WH. 1938. The bearing of nectaries on the phylogeny of flowering plants. Proc. Am. Phil. Soc. 79: 549–565.

Brundell DJ. 1975. Flower development of the Chinese gooseberry (*Actinidia chinensis* Planch.): I. Development of the flowering shoot. II. Development of the flower bud. New Zealand J. Bot. 13: 473–483, 485–496.

Cairney JWG and AE Ashford. 2002. Biology of mycorrhizal associations of epacrids (Ericaceae). New Phytol. 154: 305–326.

Carlquist S. 1989. Wood and bark anatomy of Empetraceae; comments on paedomorphosis in woods of certain small shrubs. Aliso 12: 497–515.

Carlquist S and EL Schneider. 2005. Vestigial pit membrane remnants in perforation plates and helical thickening in vessels of Ericaceae. Nord. J. Bot. 23: 353–363.

Chernyakovskaya ER. 1992. Ericales. In: A Takhtajan, ed. Comparative seed anatomy, 4: 24–42. Nauka, St. Petersburg (in Russian).

Cherry W, PA Gadek, EA Brown, MM Heslewood, and CJ Quinn. 2001. *Pentachondra dehiscens* sp. nov. – an abberrant new member of Styphelieae. Aust. Syst. Bot. 14: 513–533.

Copeland HE. 1933. The development of seeds in certain Ericales. Am. J. Bot. 20: 513–517.

Copeland HF. 1939. The structure of monotropsis and the classification of the Monotropoideae. Madroño 5: 105–119.

Copeland HE. 1941. Further studies on Monotropoideae. Madroño 6: 97–119.

Copeland HF. 1943. A study, anatomical and taxonomic, of the genera of Rhododendroideae. Am. Midi. Nat. 30: 533–625.

Copeland HF. 1947. Observations on the structure and classification of the Pyroleae. Madroño 9: 65–102.

Copeland HF. 1953. Observations on the Cyrillaceae, particularly on the reproductive structures of the North American species. Phytomorphology 3: 405–411.

Copeland HF. 1954. Observations on certain Epacridaceae. Am. J. Bot. 41: 215–222.

Cox HT. 1948. Studies in the comparative anatomy of the Ericales: I. Ericaceae subfamily Rhododendroideae. II. Ericaceae subfamily Arbutoideae. Am. Midi. Nat. 39: 220–245; 40: 493–516.

Crayn DM and CJ Quinn. 2000. The evolution of the *atp*B-*rbc*L intergenic spacer in the epacrids (Ericales) and its systematic and evolutionary implications. Molec. Phylogen. Evol. 16: 238–252.

Crayn DM, KA Kron, PA Gadek, and CJ Quinn. 1996. Delimitation of Epacridaceae: preliminary molecular evidence. Ann. Bot. 77: 317–321.

Crayn DM, KA Kron, PA Gadek, and CJ Quinn. 1998. Phylogenetics and evolution of epacrids: a molecular analysis using the plastid gene *rbc*L with a reappraisal of the position of *Lebetanthus*. Aust. J. Bot. 46: 187–200.

Crété P. 1944. Recherches anatomiques sur la semino-genese de *Actinidia chinensis* Planch.: Affinites des Actinidiacees. Bull. Soc. Bot. France 91: 153–160.

Cullings K. 1996. Single phylogenetic origin of ericoid mycorrhizae within the Ericaceae. Canad. J. Bot. 74: 1896–1909.

Cullings K. 2000. Reassessment of phylogenetic relationships of some members of the Monotropoideae based on partial 28S ribosomal RNA gene sequencing. Canad. J. Bot. 78: 1–2.

Cullings KW. 1994. Molecular phylogeny of the Monotropoideae (Ericaceae) with a note on the placement of the Pyroloideae. J. Evol. Biol. 7: 501–516.

Cullings KW and TD Bruns. 1992. Phylogenetic origin of the Monotropoideae inferred from a partial 28S ribosomal RNA sequences. Canad. J. Bot. 70: 1703–1708.

Cullings KW and L Hileman. 1997. The Monotropoideae is [are] a monophyletic sister group to the Arbutoideae (Ericaceae): a molecular test of Copeland's hypothesis. Madroño 44: 297–304.

Dawson MI. 2000. Index of chromosome numbers in the Epacridaceae. Proc. Linn. Soc. N. S. W. 73: 37–56.

Dickison WC. 1972. Observations on the floral morphology of some species of *Saurauia, Actinidia,* and *Clematoclethra*. J. Elisha Mitchell Sci. Soc. 88: 43–54.

Dickison WC, JW Nowicke, and JJ Skwarla. 1982. Pollen morphology of the Dilleniaceae and Actinidiaceae. Am. J. Bot. 69: 1055–1073.

Dormer KJ. 1944. Morphology of the vegetative shoot in Epacridaceae. New Phytol. 44: 149–151.

Dressler S and C Bayer. 2004. Actinidiaceae. In: K Kubitzki, ed. The families and genera of vascular plants, vol. 6, pp. 14–19. Springer, Berlin/Heidelberg/New York.

Foss PJ and GJ Doyle. 1988. A palynological study of the Irish Ericaceae and *Empetrum*. Pollen et Spores 30: 151–178.

Franks JW and L Watson. 1963. The pollen morphology of some critical Ericales. Pollen et Spores 5: 51–68.

Freudenstein JV. 1997. Molecular evidence for relationships and character transformation within Pyroloideae (Ericaceae). Am. J. Bot. 84(6): 195 (abstract).

Freudenstein JV. 1999. Relationships and character transformation in Pyroloideae (Ericaceae) based on ITS sequences, morphology, and development. Syst. Bot. 24: 398–408.

Ganapathy PS and BF Falser. 1964. Studies of floral morphology in the Ericales: VII. Embryology in the Phyllodoceae. Bot. Gaz. 125: 280–297.

Gibbs RD. 1958. Biochemistry as an aid in establishing the relationships of some families of dicotyledons. Proc. Linn. Soc. Lond. 69: 216–230.

Giebel KP and WC Dickison. 1975. Wood anatomy of Clethraceae. J. Elisha Mitchell Sci. Soc. 91: 17–26.

Guedes MS and R Schmid. 1978. The peltate (ascidiate) carpel theory and carpel peltation in *Actinidia chinensis* (Actinidiaceae). Flora 167: 525–543.

Hagerup O. 1946. Studies on the Empetraceae. Biol. Meddel. Kongel. Danske Vidensk. Selsk. 20(5): 1–49.

Hagerup O. 1953. The morphology and systematics of the leaves in Ericales. Phytomorphology 3: 459–464.

Hallett IC and PW Sutherland. 2005. Structure and development of kiwifruit skins. Int. J. Plant Sci. 166: 693–704.

Hara N. 1958. Structure of the vegetative shoot apex and development of the leaf in the Ericaceae and their allies. J. Fac. Sci. Univ. Tokyo, Sect. 3, Bot. 7: 367–450.

Harborne JB and CA Williams. 1973. A chemotaxonomic survey of flavonoids and simple phenols in leaves of the Ericaceae. Bot. J. Linn. Soc. 66: 37–54.

Harvey CF and LG Fraser. 1988. Floral biology of two species of *Actinidia* (Actinidiaceae): 2. Early embryology. Bot. Gaz. 149: 37–44.

He Z-C, Y Zhong, H-T Liu, X-H Tang, L Ye, D-S Huang, and L-M Xu. 2000. Quantitative taxonomic analyses of *Actinidia* (Actinidiaceae) in China based on micromorphological characters. Acta Phytotax. Sinica 38: 121–136.

He Z-C, X-Y Zhang, Y Zhang, and L Ye. 2000. Phylogenetic relationships of *Actinidia* and related genera based on micromorphological characters of foliar trichomes. Genetic Res. Crop Evol. 47: 627–639.

He Z-C, J-Q Li, Cai Q, and Q Wang. 2005. The cytology of *Actinidia, Saurauia* and *Clematoclethra* (Actinidiaceae). Bot. J. Linn. Soc. 147: 369–374.

Henderson MW. 1919. A comparative study of the structure and saprophytism of the Pyrolaceae and Monotropaceae, with reference to their derivation from the Ericaceae. Contr. Bot. Lab. Morris Arbor. Univ. Pennsylvania 5: 42–109.

Hermann PM and BF Palser. 2000. Stamen development in the Ericaceae. I. Anther wall, microsporogenesis, inversion, and appendages. Am. J. Bot. 87: 934–957.

Hileman LC, VT Parker and MC Vasey. 1994. Preliminary generic relationships of the Arbuteae (Ericaceae) based on molecular sequence data. Am. J. Bot. 81(6): 161 (abstract).

Hileman LC, MC Vasey, and VT Parker. 2001. Phylogeny and biogeography of the Arbutoideae (Ericaceae): implications for the Madrean-Tethyan Hypothesis. Syst. Bot. 26: 131–143.

Hopping ME. 1976. Structure and development of fruit and seeds in Chinese gooseberry (*Actinidia chinensis* Planch.). N. Z. J. Bot. 14: 63–68.

Hopping ME. 1990. Floral biology, pollination, and fruit set. In: IJ Warrington and GC Weston, eds. Kiwifruit: science and management, pp. 71–96. Ray Richards, Auckland.

Hopping ME and EM Jerram. 1979. Pollination of kiwifruit (*Actinidia chinensis* Planch.): stigma-style structure and pollen tube growth. N. Z. J. Bot. 17: 233–240.

Howpage D, V Vithanage, and R Spooner-Hart. 1998. Pollen tube distribution in the kiwifruit (*Actinidia deliciosa* A. Chev.), pistil in relation to its reproductive process. Ann. Bot. II, 81: 697–703.

Jackes BR. 1968. Floral anatomy of the genus *Oligarrhena* R. Br. (Epacridaceae). Aust. J. Bot. 33(1): 83–103.

Jarman SJ and RK Crowden. 1974. Anthocyanins in the Epacridaceae. Phytochemistry 13: 743–750.

Jarman SJ and RK Crowden. 1977. The occurrence of flavonol arabinosides in the Epacridaceae. Phytochemistry 16: 929–930.

Judd WS. 1979. Generic relationships in the Andromedeae (Ericaceae). J. Arnold Arbor. 60: 477–303.

Judd WS and KA Kron. 1993. Circumscription of Ericaceae (Ericales) as determined by preliminary cladistic analyses based on morphological, anatomical, and embryological features. Brittonia 45: 99–114.

Judd WS, KA Kron, AA Anderberg, DM Crayn, CJ Quinn, and PS Stevens. 1998. Major clades of Ericaceae, I: morphology. Am. J. Bot. 85(6): 138–139 (abstract).

Kang N, S Wang, R Huang, and X Wu. 1993. Studies on the pollen morphology of nine species of genus *Actinidia*. J Wuhan Bot. Res. 11: 111–116.

Kavaljian LG. 1952. The floral morphology of *Clethra alnifolia* with some notes on *C. acuminata* and *C. arborea*. Bot. Gaz. 113: 392–413.

Kim KH, S Nilsson, and J Praglowski. 1988. A note on the pollen morphology of the Empetraceae. Grana 27: 283–290.

Kolbasina ZI. 1969. Ontogenesis of the inflorescence and flower of *Actinidia kolomicta* Maxim. Bot. Zhurn. 54: 1397–1400 (in Russian).

Kron KA. 1996. Phylogenetic relationships of Empetraceae, Epacridaceae, Ericaceae, Monotropaceae, and Pyrolaceae: evidence from nuclear ribosomal 18S sequence data. Ann. Bot. 77: 293–303.

Kron KA. 1997. Phylogenetic relationships of Rhododendroideae (Ericaceae). Am. J. Bot. 84: 973–980.

Kron KA and M Chase. 1993. Systematics of the Ericaceae, Empetraceae, Epacridaceae, and related taxa based upon *rbc*L sequence data. Ann. Missouri Bot. Gard. 80: 735–741.

Kron KA and SL Johnson. 1997. Phylogenetic analysis of the monotropoids and pyroloids (Ericaceae) using *nrITS* and 18S sequence data. Am. J. Bot. 84: 205–206 (abstract).

Kron KA and WS Judd. 1990. Phylogenetic relationships within the Rhodoreae (Ericaceae) with specific comments on the placement of *Ledum*. Syst. Bot. 15: 57–68.

Kron KA and WS Judd. 1997. Systematics of the *Lyonia* group (Andromedeae, Ericaceae) and the use of species as terminals in higher-level cladistic analyses. Syst. Bot. 22: 479–492.

Kron KA and JM King. 1996. Cladistic relationships of *Kalmia, Leiophyllum*, and *Loiseleuria* (Phyllodoceae, Ericaceae) based or *rbc*L and *nr*ITS data. Syst. Bot. 23: 17–29.

Kron KA and JL Luteyn. 2005. Origin and biogeographic patterns in Ericaceae: new insights from recent phylogenetic analyses. In: I Friis and H Balslev, eds. Proceedings of a Symposium on Plant Diversity and Complexity Patterns – Local, Regional and Global Dimensions. Biol. Skrifte 55: 479–500. Copenhagen.

Kron KA, MW Chase, and HG Hills. 1991. Phylogenetic relationships of Ericaceae, Empetraceae, and Epacridaceae based on rbcL sequence data and their position within the Dilleniidae. Am. J. Bot., Suppl, 78(6): 197–198.

Kron KA, WS Judd, AA Anderberg, DM Crayn, R Fuller, PA Gadek, JL Luteyn, CJ Quinn, and PS Stevens. 1998. Major clades of Ericaceae, II. Molecular evidence. Am. J. Bot. 85: 139–140.

Kron KA, WS Judd, and DM Crayn. 1999. Phylogenetic analyses of Andromedeae (Ericaceae subfam. Vaccinioideae). Am. J. Bot. 86: 1290–1300.

Kron KA, R Fuller, DM Crayn, PA Gadek, and CJ Quinn. 1999. Phylogenetic relationships of epacrids and vaccinioids (Ericaceae s.l.) based on *mat*K sequence data. Plant Syst. Evol. 218: 55–65.

Kron KA, WS Judd, PF Stevens, DM Crayn, AA Anderberg, PA Gadek, CJ Quinn, and JL Luteyn. 2002. Phylogenetic classification of Ericaceae: molecular and morphological evidence. Bot. Rev. 68: 335–423.

Kron KA, EA Powell, and JL Luteyn. 2002. Phylogenetic relationships within the blueberry tribe (Vaccinieae, Ericaceae) based on sequence data from *mat*K and nuclear ribosomal ITS regions, with comments on the placement of *Satyria*. Am. J. Bot. 89: 327–336.

Kubitzki K. 2004. Cyrillaceae. In: K Kubitzki, ed. The families and genera of vascular plants, vol. 6, pp. 114–116. Springer, Berlin/Heidelberg/New York.

Kubota M, NP McGonigle, and M Hyakumachi. 2001. *Clethra barbinervis*, a member of the order Ericales, forms arbuscular mycorrhizae. Canad. J. Bot. 79: 300–306.

Leins P. 1964. Entwicklungsgeschichtliche Studien an Ericales Blüten. Bot. Jahrb. Syst. 83: 57–88.

Lems K. 1964. Evolutionary studies in the Ericaceae, II: leaf anatomy as a phylogenetic index in the Andromedeae. Bot. Gaz. 125: 178–186.

Lens F, P Gasson, E Smets, and S Jansen. 2003. Comparative wood anatomy of epacrids (Styphelioideae, Ericaceae s.l.). Ann. Bot. 91: 835–856.

Lens F, KA Kron, JL Luteyn, E Smets, and S Jansen. 2004. Comparative wood anatomy of the blueberry tribe (Vaccinieae, Ericaceae s.l.). Ann. Missouri Bot. Gard. 91: 566–592.

Lens F, JL Luteyn, E Smets, and S Jansen. 2004. Ecological trends in the wood anatomy of Vaccinioideae (Ericaceae s.l.). Flora 199: 309–319.

Lens F, E Smets, and S Jansen. 2004. Comparative wood anatomy of Andromedeae s.s., Gaultherieae, Lyonieae, and Oxydendreae (Vaccinioideae, Ericaceae s.l.). Bot. J. Linn. Soc. 144: 161–179.

Li H-L. 1952. A taxonomic review of the genus *Actinidia*. J. Arnold Arbor. 23: 1–61.

Li J-Q, Q Cai, and HW Huang. 2000. On the phylogeny of the genus *Actinidia* Lindley. In: H-W Huang, ed. Advance in *Actinidia* research, pp. 80–86. Science Press, Beijing (in Chinese with Enlish abstract).

Li J-W, G Rui, MY Liang, and C Pang. 1989. Studies on the pollen morphology of the *Actinidia*. Guihaia 9: 335–339.

Liang C-F and AR Ferguson. 1986. On the distribution of actinidias. Guihaia 3: 229–248.

Luteyn JL. 1991. Key to the subfamilies and genera of Neotropical Ericaceae. Nord. J. Bot. 11: 623–627.

Luteyn JL. 2002. Diversity, adaptation and endemism in neotropical Ericaceae: biogeographical patterns in the Vaccinieae. Bot. Rev. 68: 55–87.

Matthews JR and EM Knox. 1926. The comparative morphology of the stamen in the Ericaceae. Trans. Proc. Bot. Soc. Edinb. 29: 243–281.

McDowall MA. 1970. Anionic proteinase from *Actinidia* chinensis. Preparation and properties of the crystalline enzyme. Eur. J. Biochem. 14: 214–221.

McLean C. 1995. Mycorrhizae of the Epacridaceae and its use in propagation. Combined Proc. Int. Plant Propagators 45: 108–111.

Middleton DJ. 1991. Taxonomic studies in the *Gaultheria* group of genera of the tribe Andromedeae (Ericaceae). Edinb. J. Bot. 48: 283–306.

Moore DM, JB Harborne, and CA Williams. 1970. Chemotaxonomy, variation, and geographical distribution of the Empetraceae. Bot. J. Linn. Soc. 63: 277–293.

Nishino E. 1983. Corolla tube formation in the Primulaceae and Ericales. Bot. Mag. (Tokyo) 96: 319–342.

Nowicke JW. 1966. Pollen morphology and classification of the Pyrolaceae and Monotropaceae. Ann. Missouri Bot. Gard. 53: 213–219.

Oliver EGH. 1989. The Ericoideae and the southern African heathers. Bot. J. Linn. Soc. 101: 319–327.

Oliver EGH. 2000. Systematics of Ericeae (Ericaceae: Ericoideae) species with indehiscent and partially dehiscent fruits. Contrib. Bolus Herb. 19: 1–483.

Olson AR. 1980. Seed morphology of *Monotropa uniflora* L. (Ericaceae). Am. J. Bot. 67: 986–974.

Palser BF. 1951. Studies of floral morphology in the Ericales: I. Organography and vascular anatomy in the Andromedeae. Bot. Gaz. 112: 447–485.

Palser BF. 1952. Studies of floral morphology in the Ericales: II. Megasporogenesis and megagametophyte development in the Andromedeae. Bot. Gaz. 114: 33–52.

Palser BF. 1954. Studies of floral morphology in the Ericales: III. Organography and vascular anatomy in several species of the Arbuteae. Phytomorphology 4: 335–354.

Palser BF. 1958. Studies of floral morphology in the Ericales: IV. Observations on three members of the Gaultherieae. Trans. Illinois Acad. Sci. 51: 24–34.

Palser BF. 1961. Studies of floral morphology in the Ericales: V. Organography and vascular anatomy in several United States species of the Vacciniaceae. Bot. Gaz. 123: 79–111.

Palser BF and YS Murty. 1967. Studies of floral morphology in the Ericales: VIII. Organography and vascular anatomy in *Erica*. Bull. Torrey Bot. Club 94: 243–320.

Paterson BR. 1961. Studies of floral morphology in the Epacridaceae. Bot. Gaz. 122: 259–279.

Philipson WR and MN Philipson. 1968. Diverse nodal types in *Rhododendron*. J. Arnold Arbor. 49: 193–224.

Powell EA and KA Kron. 2001. An analysis of the phylogenetic relationships in the wintergreen group (*Diplycosia, Gaultheria, Pernettya, Tepuia;* Ericaceae). Syst. Bot. 26: 808–817.

Powell JM, AR Chapman, and ANL Doust. 1987. Classification and generic status in the Epacridaceae: a preliminary analysis. Aust. Syst. Bot. Soc. Newslett. 53: 70–78.

Powell JM, DM Crayn, PA Gadek, CJ Quinn, DA Morrison, and AR Chapman. 1996. A reassessment of relationships within Epacridaceae. Ann. Bot. 77: 305–315.

Powell JM, DA Morrison, PA Gadek, DM Crayn, and CJ Quinn. 1997. Relationships and generic concepts within Styphelieae (Epacridaceae). Aust. Syst. Bot. 10: 15–29.

Pyykko M. 1969. Placentation in the Ericales: I. Pyrolaceae and Monotropaceae. Ann. Bot. Fenn. 6: 255–268.

Quinn CJ, DM Crayn, MM Heslewood, EA Brown, and PA Gadek. 2003. A molecular estimate of the phylogeny of Styphelieae (Ericaceae). Aust. Syst. Bot. 16: 581–594.

Quinn CJ, EA Brown, MM Heslewood, and DM Crayn. 2005. Generic concepts in Styphelieae (Ericaceae): The *Cyathodes* group. Aust. Syst. Bot. 18: 439–454.

Read DJ. 1996. The structure and function of the ericoid mycorrhizal root. Ann. Bot. II, 77: 365–374.

Record SJ. 1932. Woods of the Ericales, with particular reference to *Schizocardia*. Trop. Woods 32: 11–14.

Redgwell RJ. 1983. Composition of *Actinidia* mucilage. Phytochemistry 22: 951–956.

Samuelson G. 1913. Studien über Entwicklungs-geschichte der Blüten einiger Bicornes-Typen: Ein Beitrag zur Kenntnis der systematischen Stellung der Diapensiaceen und Empetraceen. Svensk Bot. Tidskr. 7: 97–188.

Schmid R. 1978a. Reproductive anatomy of *Actinidia chinensis* (Actinidiaceae). Bot. Jahrb. Syst. 100: 149–195.

Schmid R. 1978b. Actinidiaceae, Davidiaceae, and Paracryphiaceae: systematic considerations. Bot. Jahrb. Syst. 100: 196–204.

Schnarf K. 1924. Bemerkungen zur Stellung der Gattung *Saurauia* im System. Sitzungsber. K. Akad. Wiss. Wien., Math.-Naturw. Kl. Abt. 1, 133: 17–28.

Schneider JV and C Bayer. 2004. In: K Kubitzki, ed. The families and genera of vascular plants, vol. 6, pp. 69–73. Springer, Berlin/Heidelberg/New York.

Schneider EL and S Carlquist. 2003. Unusual pit membrane remnants in perforation plates of Cyrillaceae. J. Torrey Bot. Soc. 130: 225–230.

Schönenberger J, AA Anderberg, and KJ Sytsma. 2005. Molecular phylogenetics and patterns of floral evolution in the Ericales. Int. J. Plant Sci. 166: 265–288.

Sleumer H. 1941. Vaccinioideen-Studien. Bot. Jahrb. Syst. 71: 375–510.

Sleumer H. 1966, 1967. Ericaceae. In: CGGJ van Steenis, ed. Flora malesiana, ser. 1, 6: 469–668, 669–914. Noordhoff, Groningen.

Sleumer H. 1967. Monographia Clethracearum. Bot. Jahrb. Syst. 87: 36–175.

Smith-White S. 1955. Chromosome numbers and pollen types in the Epacridaceae. Aust. J. Bot. 3: 48–67.

Smith-White S. 1959. Pollen development patterns in the Epacridaceae: a problem of cytoplasm-nucleus interaction. Proc. Linn. Soc. N. S. W. 84: 8–35.

Soejareto DD. 1980. Revision of South American *Saurauia* (Actinidiaceae). Feldiana, Bot. 2: 1–141.

Stace HM, AR Chapman, KL Lemson, and JM Powell. 1997. Cytoevolution, phylogeny and taxonomy in Epacridaceae. Ann. Bot. (London) 79: 283–290.

Stevens PF. 1971. A classification of the Ericaceae: subfamilies and tribes. Bot. J. Linn. Soc. 64: 1–53.

Stevens PF, J Luteyn (most Vaccinieae), EGH Oliver (*Erica*), TL Bell, EA Brown, RK Crowden, AS George, GJ Jordan, P Ladd, K Lemson, CB McLean, Y Menadue, JS Pate, HM Stace, and CM Weiller (Styphelioideae). 2004. In: K Kubitzki, ed. The families and genera of vascular plants, vol. 6, pp. 145–194. Springer, Berlin/Heidelberg/New York.

Stushnoff C and BF Palser. 1969 (1970). Embryology of five *Vaccinium* taxa including diploid, tetraploid, and hexaploid species or cultivars. Phytomorphology 19: 312–331.

Szkudlarz P. 1999. The morphological and anatomical structure of fleshy fruit in the family Ericaceae. Biol. Bull. Poznan 36: 43–56.

Taaffe G, EA Brown, DM Crayn, PA Gadek, and CJ Quinn. 2001. Generic concepts in Styphelieae: resolving the limits of *Leucopogon*. Aust. J. Bot. 49: 107–120.

Takahashi H. 1987. Pollen morphology and its taxonomic significance in the Monotropoideae (Ericaceae). Bot. Mag. (Tokyo) 100: 385–405.

Takahashi H. 1988. Pollen morphology and systematics in two subfamilies of the Ericaceae: Pyroloideae and Monotropoideae. Korean J. Plant Tax. 18: 9–17.

Takahashi H. 1993. Seed morphology and its systematic implications in Pyroloideae (Ericaceae). Int. J. Plant Sci. 154: 175–186.

Takhtajan AL and AA Yatsenko-Khmelevsky. 1946. The morphological characteristics of the genus *Epigaea*. Doklady Armenian Acad. Sci. 4(2): 45–50 (in Russian).

Tanaka R and K Oginuma. 1980. Karyomorphological studies in *Clethra barbinervis* and two allied species. J. Jpn. Bot. 55: 65–72.

Tang YC and QY Xiang. 1989. A reclassification of the genus *Clematoclethra* (Actinidiaceae) and a further note on the methodology of plant taxonomy. Acta Phytotax. Sinica 27: 81–95.

Thomas JL. 1960. A monographic study of the Cyrillaceae. Contr. Gray Herb. 186: 1–114.

Thomas JL. 1961. The genera of the Cyrillaceae and Clethraceae of the southeastern United States. J. Arnold Arbor. 42: 96–106.

Thomas JL. 1961. *Schizocardia belizensis*: a species of Purdiaea (Cyrillaceae) from Central America. J. Arnold Arbor. 42: 110–111.

Van Heel WA. 1987. Androecium development in *Actinidia chinensis* and *A.melanandra* (Actinidiaceae). Bot. Jahrb. Syst. 109: 17–23.

Veillet-Bartoszewska M. 1960. Embryogenie des Clethracees: Developpement de l'embryon chez le *Clethra alnifolia* L. Compt. Rend. Hebd. Seances Acad. Sci. 251: 2572–2574.

Veillet-Bartoszewska M. 1961. Embryogenie des Epacridacees: Developpement de l'embryon chez le *Dracophyllum secundum* R. Br. Compt. Rend. Hebd. Seances Acad. Sci. 253: 1000–1002.

Veillet-Bartoszewska M. 1963. Recherches embryogeniques sur les Ericales: comparaison avec les Primulales. Rev. Gen. Bot. 70: 141–230.

Venkata Rao C. 1961. Pollen types in the Epacridaceae. J. Indian Bot. Soc. 40: 409–423.

Vijayaraghavan MR. 1960. Studies in the family Cyrillaceae: 1. Development of male and female gametophytes in *Cliftonia monophylla* (Lam.) Britton ex Sarg. Bull. Torrey Bot. Club 96: 484–489.

Vijayaraghavan MR. 1965. Morphology and embryology of *Actinidia polygama* Franch. et Sav. and systematic position of the family Actinidiaceae. Phytomorphology 15: 224–235.

Vijayaraghavan MR. 1970. Actinidiaceae. In: BM Johri (convener), Proceedings of the Symposium on Comparative Embryology of Angiosperms, pp. 69–71. Bulletin of Indian National Science Academy No. 41, New Delhi.

Vijayaraghavan MR and U Dhar. 1978. Embryology of *Cyrilla* and *Cliftonia* (Cyrillaceae). Bot. Not. 131: 127–138.

Villamil P and B Palser. 1980. Studies of floral morphology in the Ericales: 9. Organography, vascular anatomy, and

megagametophyte in three species of Gaultherieae. Phytomorphology 30: 250–265.
Wallace GD. 1975. Studies of the Monotropoideae (Ericaceae): taxonomy and distribution. Wasmann J. Biol. 33: 1–88.
Wallace GD. 1976. Interrelationships of the subfamilies of the Ericaceae and derivation of the Monotropoideae. Bot. Not. 128: 286–298.
Wallace GD. 1977. Studies of the Monotropoideae (Ericaceae). Floral nectaries: anatomy and function in pollination ecology. Am. J. Bot. 64: 199–206.
Walton EF, PJ Fowke, K Weis, and PL McLeay. 1997. Shoot axillary bud morphogenesis in Kiwifrui (*Actinidia deliciosa*). Ann. Bot. (London) 80: 13–21.
Wang Z-Y, KS Gould, and KJ Patterson. 1994. Comparative root anatomy of five *Actinidia* species in relation to root-stock effects on kiwifruit flowering,. Ann. Bot. (London) 73: 403–413.
Watson L. 1962. The taxonomic significance of stomatal distribution and morphology in Epacridaceae. New Phytol. 61: 36–40.
Watson L. 1964. The taxonomic significance of certain anatomical observations on Ericaceae: The Ericoideae, *Calluna*, and *Cassiope*. New Phytol. 63: 274–280.
Watson L. 1965. The taxonomic significance of certain anatomical variations among Ericaceae. Bot. J. Linn. Soc. 59: 111–125.
Watson L. 1967. Taxonomic implications of a comparative anatomical study of Epacridaceae. New Phytol. 66: 495–504.
Watson L. 1976. Ericales revisited. Taxon 35: 269–271.
Watson L, WT Williams, and GN Lance. 1967. A mixed data approach to angiosperm taxonomy: the classification of Ericales. Proc. Linn. Soc. Lond. 178: 25–35.
Webby RF, RD Wilson, and AR Ferguson. 1994. Leaf flavonoids of *Actinidia*. Biochem. Syst. Ecol. 22: 227–286.
Weiller CM, RK Crowden, and JM Powell. 1944. Morphology and taxonomic significance of leaf epicuticular waxes in the Epacridaceae. Aust. Syst. Bot. 7: 125–152.
White J. 1986. Ontogeny and morphology of ovarian and fruit hairs in kiwifruit. N. Z. J. Bot. 24: 403–414.
White J. 1990. Pollen development in *Actinidia deliciosa* var. *deliciosa*: histochemistry of the microspore mother cell walls. Ann. Bot. (London) 65: 231–239.
Wood CE, Jr. 1961. The genera of Ericaceae in the southeastern United States. J. Arnold Arbor. 42: 10–80.
Wood CE, Jr. and RB Channell. 1959. The Empetraceae and Diapensiaceae of southeastern United States. J. Arnold Arbor. 40: 161–171.
Xiong Z-T and R-H Huang. 1988. Chromosome numbers of 10 species and 3 varieties in *Actinidia* Lindl. Acta Phytotax. Sinica 26: 245–247.
Yan G, J Yao, AR Ferguson, MA McNeilage, AG Seal, and BG Murray. 1997. New reports of chromosome numbers in *Actinidia* (Actinidiaceae). N. Z. J. Bot. 35: 181–186.
Yang B-Y. 1952. Pollen grain morphology in the Ericaceae. Quart. J. Taiwan Mus. 5: 1–24.
Yatsenko-Khmelevsky AA. 1946. Wood structure of the Caucasian representatives of the family Ericaceae and its systematic and phylogenetic significance. Izvestia Akademii Nauk Armianskoi SSR 9: 33–58 (in Russian).
Zhang X-P and AA Anderberg. 2002. Pollen morphology in the ericoid clade of the order Ericales, with special emphasis on Cyrillaceae. Grana 41: 201–215.
Zhang Z-Y. 1987. A study on the pollen morphology of Actinidiaceae and its systematic position. Acta Phytotax. Sinica 25: 9–23 (in Chinese with English summary).

Order 55. DIAPENSIALES

Evergreen shrublets or stemless perennial herbs with ecto- and endotrophic mycorrhizae. Glandular hairs absent. Vessels commonly with simple, rarely also with scalariform perforations. Fibers with bordered pits. Nodes trilacunar with three traces, or (*Pyxidanthera*) unilacunar with one trace. Leaves alternate, simple, entire or toothed, often basal, estipulate. Stomata mostly anomocytic, rarely anisocytic. Flowers axillary, solitary or in compact racemes, bisexual, actinomorphic. Sepals five, imbricate, free or connate into a 5-lobed tube, persistent. Petals five, connate into a sympetalous corolla (almost free in *Galax*), imbricate or contorted, caducous. Stamens five, attached to the corolla tube and alternate with its lobes, free or basally connate into a ring, sometimes with as many alternating staminodia (staminodia absent in *Diapensia* and *Pyxidanthera*); anthers tetrasporangiate or (*Galax*) disporangiate, opening longitudinally or (*Pyxidanthera*) by transverse slits. Tapetum secretory. Pollen grains in monads, 3-celled, 3-colporate, tectate-columellate. Nectary disc represented by a weakly developed ring at the base of the ovary or wanting. Gynoecium of three carpels, with a simple style terminated by 3-lobed stigma; ovary superior, 3-locular, with several or numerous ovules on more or less intrusive placentas. Ovules anatropous to hemitropous and campylotropous, unitegmic or also bitegmic (?), tenuinucellate. Female gametophyte of *Polygonum*-type. Endosperm cellular, without haustoria. Fruits loculicidal capsules. Seeds small, with thin, exotestal seed coat, copious, fleshy endosperm, and rather large, straight or slightly curved embryo. Plants Al-accumulators, ellagic acid present. n = 6.

Diapensiales are closely related to the Ericales, differing in the absence of endothelium and endospermal haustoria and the presence of fibrous layer (endothecium) in the anther wall (Samuelson 1913). It seems probable that they evolved from some ericalean stock.

1. DIAPENSIACEAE

Lindley 1836. 5/20 (including Galacaceae D. Don 1828). Temperate and cold regions of the Northern Hemisphere; *Diapensia* (4) is circumpolar, extending through North America, Greenland, and northern Eurasia as far south as the Himalayas; *Berneuxia* (1) occurs in eastern Tibet, southwestern China, and Upper

Burma; *Shortia* (6) occur in southwestern China, Taiwan, Japan, and southeastern United States, and *Pyxidanthera* (1) and *Galax* (1) are endemic to eastern United States.

Diapensia, Pyxidanthera, Berneuxia, Shortia (including *Schizocodon*), *Galax.*

Bibliography

Barwin JT. 1939. Chromosomes of the Diapensiaceae: a cytological approach to a phylogenetic problem. J. Hered. 30: 169–171.

Baldwin JT. 1941. *Galax:* the genus and its chromosomes. J. Hered. 32: 249–254.

Diels L. 1914. Diapensiaceen studien. Bot. Jahrb. Syst. Suppl, 50: 304–330.

Mirphy HT and JW Hardin. 1976. A new and unique venation pattern in the Diapensiaceae. Torreya 103: 177–179.

Palser BF. 1963. Studies of floral morphology in the Ericales: 6. Diapensiaceae. Bot. Gaz. 124: 200–219.

Reynolds JD. 1968. Morphological studies in Diapensiaceae. 1. Chromosome number and microspore development in *Pyxidanthera brevifolia* Wells. Torreya 95: 653–656.

Rönblom K and AA Anderberg. 2002. Phylogeny of Diapensiaceae based on molecular data and morphology. Syst. Bot. 27: 383–395.

Samuelson G. 1913. Studien über Entwicklungs-geschichte der Blüten einiger Bicornestypen. Svensk Bot. Tidskr. 7: 97–188.

Scott PJ. 2004. Diapensiaceae. In: K Kubitzki, ed. The families and genera of vascular plants, vol. 6, pp. 117–121. Springer, Berlin/Heidelberg/New York.

Scott PJ and RT Day. 1983. Diapensiaceae: a review of the taxonomy. Taxon 32: 417–423.

Wood CE and RB Channell. 1959. The Empetraceae and Diapensiaceae of southeastern United States. J. Arnold Arbor. 40: 161–171.

Xi Y-Z and Y-C Tang. 1990. Pollen morphology and phylogenetic relationships in the Diapensiaceae. Cathaya 2: 89–112.

Yamazaki T. 1966. The embryology of *Shortia uniflora* with brief review of the systematic position of the Diapensiaceae. J. Japan Bot. 41: 245–251.

Order 56. FOUQUIERIALES

Xeromorphic, spiny, small trees or shrubs with little branched or unbranched stems and differentiated long and short shoots. A layer of chlorenchymatous tissue occurs in the outermost parenchymatous cortex. Vessels with simple perforations. Fibers with bordered pits. Rays mostly heterogeneous to homogeneous. Axial parenchyma diffuse. Sieve-element plastids of S-type. Nodes unilacunar with one trace. Leaves alternate, small, simple, entire, estipulate, quickly deciduous when the soil dries. Leaves of the long shoots each surmounting a decurrent ridge of the stem and persisting as spines formed usually from the full length of the petiole and may continue below the midrib of the leaf blade; short shoots are axillary to the spines and produce fascicles of leaves that do not form spines. Stomata anomocytic. Flowers in various kinds of inflorescences, very rarely solitary or paired, bisexual, actinomorphic, or slightly zygomorphic, 5-merous. Sepals free, strongly imbricate, unequal, persistent. Petals connate into a tube, lobes imbricate. Stamens in one cycle, free or slightly connate, mostly 10, but in some species very variable in number; filaments often hairy at the base; anthers tetrasporangiate, dorsifixed, apiculate, opening longitudinally. Tapetum secretory. Microsporogenesis simultaneous. Pollen grains 2-celled, 3-colporate, pitted or reticulate, semitectate. Gynoecium of three united carpels, with a style branched near or above the middle; stigma apical; ovary superior, 1-locular, with three parietal septiform placentas that extend in the upper four fifths to the center of the ovary, and in the lower portion the deeply intruded parietal placentas become completely united into a central axile placenta (Henrickson 1972). Ovules (6)14–18(-20), ascending, anatropous, bitegmic, tenuinucellate, with an endothelium and integumentary tapetum; micropyle endostomal. Female gametophyte of *Polygonum*-type. Endosperm cellular, with a chalazal haustorium. Fruits loculicidal capsules containing a large parenchymatous central column (united margins of intruded placentas). Seeds small, flattened, with a conspicuous membranous wings; seed coat becoming crushed, testa and tegmen multiplicative; embryo small, straight and thin layer of oily endosperm, or endosperm wanting. Producing triterpenes, triterpenoid, and steroidal saponins, iridoid compounds, ellagic acid, and sometimes proanthocyanins; flowers sometimes containing anthocyanin 3-galactosides, which are also common in the Ericales. $n = 12$.

Van Tieghem (1899), who studied stem and floral morphology of the Fouquieriaceae, concluded that the affinity of this family is with the Ebenales, and Bessey (1897, 1915) included them in the order Ebenales close to the Styracaceae. However, from the Styracales the family Fouquieriaceae differs in the presence of iridoid compounds, anatomy of the seed coat, and a number of other features. Fouquiericeae are much closer to the Ericales, which stand not very far from the Styracales.

Dumortier (1829) was the first who placed the Fouquieriaceae close to the "Rhodoraceae" and the Ericaceae. With the Ericales the Fouquieriales share such important characters as the presence of iridoid compounds, anthers with the prolonged connective, sympetalous corolla, cellular endosperm, and the presence of the endospermal haustoria. They differ from the Ericales in wood anatomy, bitegmic ovules, and scanty endosperm and therefore deserve the status of an order (Dahlgren 1975, 1983, 1989; Takhtajan 1987, 1997).

1. FOUQUIERIACEAE

A.P. de Candolle 1828. 2/11. Arid parts of Mexico and southwestern United States.

Fouquieria (including *Idria?*).

Bibliography

Bate-Smith EC. 1964. Chemistry and taxonomy of *Fouquieria splendens* Engelm.: a new member of the asperuloside group. Phytochemistry 3: 623–625.

Behnke H-D. 1976. Sieve-element plastids of *Fouquieria, Frankenia* (Tamaricales), and *Rhabdodendron* (Rutaceae), taxa sometimes allied with Centrospermae (Caryophyllales). Taxon 25: 265–268.

Carlquist S. 2001. Wood anatomy of Fouquieriaceae in relation to habit, ecology, and systematics; nature of meristems in wood and bark. Aliso 19: 137–163.

Dahlgren R, SR Jensen, and BJ Nielsen. 1976. Iridoid compounds in Fouquieriaceae and notes on its possible affinities. Bot. Not. 129: 207–212.

Henrickson J. 1967. Pollen morphology of the Fouquieriaceae. Aliso 6: 137–160.

Henrickson J. 1968. Vegetative morphology of the Fouquieriaceae. Thesis. Claremont Graduate School. Claremont.

Henrickson J. 1969. Anatomy of periderm and cortex of Fouquieriaceae. Aliso 7: 97–126.

Henrickson J. 1972. A taxonomic revision of the Fouquieriaceae. Aliso 7: 439–537.

Henrickson J. 1973. Fouquieriaceae DC. World Pollen Spore Flora 1: 1–12.

Humphrey RR. 1931. Thorn formation in *Fouquleria splendens* and *Idria columnaris*. Bull. Torrey Bot. Club 58: 263–264.

Humphrey RR. 1935. A study of *Idria columnaris* and *Fouquieria splendens*. Am. J. Bot. 22: 184–207.

Johansen DA. 1936. Morphology and embryology of *Fouquieria*. Am. J. Bot. 23: 95–99.

Kamelina OP. 1997. An addition to the embryology of Lactoridaceae and Fouquieriaceae. Bot. Zhurn. 82: 25–29 (in Russian with English summary).

Khan R. 1943. The ovule and embryo sac of *Fouquieria*. Proc. Natl. Inst. Sci. India 9: 253–256.

Kubitzki K. 2004. Fouquieriaceae. In: K Kubitzki, ed. The families and genera of vascular plants, vol. 6, pp. 195–198. Springer, Berlin/Heidelberg/New York.

Nash GV. 1903. A revision of the family Fouquieriaceae. Bull. Torrey Bot. Club 30: 449–459.

Nedoff JA, IO Ting, and EM Lord. 1985. Structure and function of the green stem tissue in ocotillo (*Fouquieria splendens*). Am. J. Bot. 72: 143–151.

Schönenberger J and Grenhaged. 2005. Early floral development and androecium organization in Fouquieriaceae (Ericales). Plant Syst. Evol. 254: 233–249.

Schultheis LM and BG Baldwin. 1999. Molecular phylogenetics of Fouquieriaceae: evidence from nuclear rDNA ITS studies. Am. J. Bot. 86: 578–589.

Scogin R. 1977. Anthocyans of the Fouquieriaceae. Biochem. Syst. Ecol. 5: 265–267.

Scogin R. 1978. Leaf phenolics of the Fouquieriaceae. Biochem. Syst. Ecol. 6: 297–298.

Scott FM. 1932. Some features of the anatomy of *Fouquieria splendens*. Am. J. Bot. 19: 673–678.

Order 57. POLEMONIALES

Annual or perennial herbs, sometimes vines or woody shrubs, rarely small trees. Plants variously pubescent to (rarely) nearly glabrous. Sieve-element plastids of S-type. Vessels with simple perforation. Rays heterogeneous to homogeneous. Axial parenchyma apotracheal or less often scanty paratracheal. Nodes unilacunar. Leaves alternate to opposite, entire to deeply divided, sometimes compound, petiolate or sessile, estipulate or rarely (*Cobaea*) stipulate. Stomata anomocytic or less often paracytic. Flowers usually in cymose inflorescences, sometimes condensed in involucrate heads, rarely solitary and axillary, bisexual, actinomorphic to zygomorphic. Sepals (4)5(6), more or less connate (in *Cobaea* only connate at the base), rarely free, lobes imbricate or valvate, persistent. Corolla sympetalous, rotate to salverform, contorted, occasionally somewhat bilabiate. Stamens (3-)5(6), attached to the corolla tube, sometimes at differing levels, alternate with the petals; anthers basifixed to dorsifixed, dithecal, tetrasporangiate, opening longitudinally. Pollen grains 2-celled, 4 to many colporate or porate, often periporate. Annular intrastaminal nectary disk usually present. Gynoecium of three or sometimes two or four united carpels; style filiform, terminal, usually with as many stigma lobes as carpels; ovary superior, (2)3(4)-locular, with 1–many ovules per locule in two rows. Ovules anatropous to hemitropous, tenuinucellate unitegmic, with thick integument. Female gametophyte of *Polygonum*-type. Endosperm nuclear, but the tissue later becomes cellular (Johri et al. 1992). Fruits loculicidal or rarely (*Cobaea*, *Acanthogilia*) septicidal capsules, sometimes explosively dehiscent (*Collomia*, *Phlox*), rarely indehiscent. Seeds small, sometimes winged, exotestal, epidermal cells with spiral thickenings and included mucilage, the cells bursting when wetted and forming a mucilaginous coat; embryo more or less straight, usually spatulate, green or white; endosperm copious and oily, or (in tropical genera) endosperm scanty or wanting. Present flavonoids (patuletin, eupalitin, eupatoletin), C-glycosyflavones, pelargonidin glycosides, and cucurbitacins, kestose and isokestose oligosaccharides, cucurbitacins; n = 6–9, 15, 26.

Close to the Fouquieriales (Nash 1903; Soltis et al. 2006; Thorne 2006).

1. POLEMONIACEAE

L. de Jussieu 1789 (including Cobaeaceae D. Don 1824). 20/320. Mainly North America, but also Mexico, Chile, Peru, Europe, and extratropical Asia.

1.1 ACANTHOGILIOIDEAE
Shrubs. Leaves dimorphic, the primary alternate, spinescent and pinnate, the axillary – simple, in fascicles. Corolla salverform. Pollen grains zonocolporate. Fruits septicidal capsules. Seed flat, winged, n = 9. – *Acanthogilia*.

1.2 COBAEOIDEAE
Climbing shrubs, the tendrils much branched with terminal hooks. Leaves alternate, pinnately compound, lateral leaflets 4–8, terminal leaflets a branched tendril, and the basal pair of leaflets is foliaceous-stipuliform. Flowers axillary, large, solitary, pendulous, actinomorphic. Disc large, fleshy, deeply lobed. Ovules pseudocrassinucellate. Fruits coriaceous capsules, septicidally 3-valved to the base. Seeds compressed, broadly winged, embryo large, white, n = 26 (*Cobea scandens*). – *Cobea*.

1.3 POLEMONIOIDEAE
Annual or perennial herbs, subshrubs, sometimes shrubs. Leaves opposite or alternate, simple to variously divided. Corolla campanulate, rotate to salverform, radially or occasionally bilaterally symmetric. Pollen grains pantoporate. Fruits loculicidal capsules. Seeds usually not winged, rarely winged, n = 7–10, 12, 14, 16–18, 25, 36 (*Gilia*). – CANTUEAE: *Cantua, Huthia;* BONPLANDIEAE: *Bonplandia*; LOESELIEAE: *Loeselia*; POLEMONIEAE: *Polemonium, Allophyllum, Navarretia, Collomia,* PHLOCIDEAE: *Gymnosteris, Phlox, Microsteris*; GILIEAE: *Gilia, Ipomopsis, Eriastrum, Langloisia* (including *Loeseliastrum*), *Giliastrum, Aliciella, Linanthus* (including *Leptodactylon, Linanthastrum*).

Bibliography

Arisawa M, J Pezzuto, A Kinghorn, A Douglas, G Cordell, and N Farnsworth. 1984. Plant anticancer agents. 30. Cucurbitacins from *Ipomopsis aggregata* (Polemoniaceae). J. Pharmaceut. Sci. 73: 411–413.

Carlquist S, VM Eckhart, and DC Michener. 1984. Wood anatomy of Polemoniaceae. Aliso 10: 547–572.

Dawson ML. 1936. The floral morphology of the Polemoniaceae. Am. J. Bot. 23: 501–511.

Day AG and R Moran. 1986. *Acanthogilia,* a new genus of Polemoniaceae from Baja California, Mexico. Proc. California Acad. Sci. 44: 111–126.

Grant V. 1998a. Primary classification and phylogeny of the Polemoniaceae, with comments on molecular cladistics. Am. J. Bot. 85: 741–752.

Grant V. 1998b. Nomenclature of subfamilies and tribes in the Polemoniaceae. Phytologia 83: 385–389 (1997 publ. 1998).

Grant V. 2001. A guide to understanding recent classification of the family Polemoniaceae. Lundellia 4: 12–24.

Grant V and KA Grant. 1965. Flower pollination in the *Phlox* family. Columbia University Press, New York/London.

Johnson LA, DE Soltis, PS Soltis and JM Porter. 1995. Molecular phylogenetic relationships in Polemoniaceae: testing congruence between plastid *mat*K and nuclear rDNA gene trees. Am. J. Bot. Suppl. 82: 138 (abstract).

Johnson LA, JL Schultz, DE Soltis, and PS Soltis. 1996. Monophyly and generic relationships of Polemoniaceae based on *mat*K sequences. Am. J. Bot. 83: 1207–1224.

Johnson LA, DE Soltis and PS Soltis. 1999. Phylogenetic relationships of Polemoniaceae inferred from 18S ribosomal DNA sequences. Plant Syst. Evol. 214: 65–89.

Jurenitsch J, E Haslinger, and W Kubelka. 1979. Structure of sapogenins from *Polemonium reptans*. Pharmazie 34: 445–446.

Kapil RN, PN Rustagi, and R Venkataraman. 1968 (1969). A contribution to the embryology of the Polemoniaceae. Phytomorphology 18: 403–411.

Porter JM. 1997. Phylogeny of Polemoniaceae based on nuclear ribosomal internal transcribed spacer DNA sequences. Aliso 15: 57–77.

Porter JM. 1998. *Aliciella*, a recircumscribed genus of Polemoniaceae. Aliso 17: 23–46.

Porter JM. 1998. Nomenclatural changes in Polemoniaceae. Aliso 17: 83–85.
Porter JM and LA Johnson. 1998. Phylogenetic relationships of Polemoniaceae: inferences from mitochondrial nad1b intron sequences. Aliso 17: 157–188.
Porter JM and LA Johnson. 2000. A phylogenetic classification of Polemoniaceae. Aliso 19: 55–91.
Prather LA and RK Jansen. 1998. Phylogeny of *Cobaea* (Polemoniaceae) based on sequence data from the ITS region of nuclear ribosomal DNA. Syst. Bot. 23: 57–72.
Prather LA, CJ Ferguson, and RK Jansen. 2000. Polemoniaceae phylogeny and classification: implications of sequence data from the chloroplast gene *ndh*F. Am. J. Bot. 87: 1300–1308.
Smith DM, CW Glennie, JB Harborne, and CA Williams. 1977. Flavonoid diversification in the Polemoniaceae. Biochem. Syst. Ecol. 5: 107–115.
Spencer SC and JM Porter. 1998. Evolutionary diversification and adaptation to novel environments in *Navarretia* (Polemoniaceae). Syst. Bot. 22: 649–668.
Steele KP and R Vilgalys. 1994. Phylogenetic analyses of Polemoniaceae using nucleotide sequences of the plastid gene *mat*K. Syst. Bot. 19: 126–142.
Stuchlik L. 1967. Pollen morphology in the Polemoniaceae. Grana Palynol. 7: 146–240.
Taylor TN and DA Levin. 1975 (1976). Pollen morphology of Polemoniaceae in relation to systematics and pollination systems: scanning electron microscopy. Grana 15: 91–112.
Wilken DH. 2004. Polemoniaceae. In: K Kubitzki, ed. The families and genera of vascular plants, vol. 6, pp. 300–312. Springer, Berlin/Heidelberg/New York.
Wilson KA. 1960a. The genera of Hydrophyllaceae and Polemoniaceae in the southeastern United States. J. Arnold Arbor. 41: 197–212.

Order 58. LECYTHIDALES

Small to large trees or shrubs; wood siliceous and/or with SiO_2 grains. Cortical vascular bundles occur in the stem, with internal xylem (normally oriented) or external (inverted). Secondary xylem and primary cortex often with crystalliferous strands. Vessels typically with simple perforations, but scalariform or reticulate perforation plates appear in some genera including *Asteranthos* and *Grias*. Lateral pitting usually alternate, except in *Asteranthos, Grias,* and *Gustavia,* in which it is alternate to opposite. Fibers typically with simple pits, but bordered pits in a few genera. Rays from distinctly heterogeneous to homogeneous (entirely heterogeneous in some genera, including *Asteranthos, Grias, Gustavia, Foetidia, Planchonia,* and others). Axial parenchyma typically abundant and in apotracheal bands with some paratracheal. Secondary phloem generally stratified into fibrous and soft portions. Nodes tri-, penta- or multilacunar. Leaves alternate, simple, entire or dentate, pinnately veined, often crowded at the ends of twigs, estipulate or (Scytopetalaceae and *Gustavia* – Lecythidoideae) with minute, caducous stipules; the structure of petiole is mostly somewhat complex, typically containing numerous vascular bundles, but with reduced number in Napoleonaeaeae (one major bundle and two minor adaxial bundles). Stomata usually anisocytic. Flowers mostly rather large and showy, solitary in the axils of leaves or leaflike bracts, or more often in racemose or paniculate inflorescences, sometimes cauliferous, bisexual, actinomorphic or the corolla and particularly androecium zygomorphic, subtended by a bract and bearing two bracteoles, typically ephemeral, pollinated by insects and bats. Perianth forms a tube more or less adnate to the ovary. Sepals (2)4–6(-12), imbricate or valvate, free or more or less connate, sometimes fully connate and calyptrate. Petals (3)4-6 or more or absent. Stamens commonly numerous (up to about 1,200), in several cycles, developing centripetally or more often centrifugally; filaments mostly connate into a ring that is either actinomorphic or more or less expanded on one side into an incurving flat ligule sometimes curved over the gynoecium as a hood; anthers latrorse, introrse or rarely poricidal, tetrasporangiate or sometimes (as in *Napoleonaea*) disporangiate, opening longitudinally or sometimes (as *Gustavia*) by apical pores. Tapetum secretory. Microsporogenesis simultaneous. Pollen grains 3-celled, 3-colpate or 3-colporate, tectate-columellate, with various types of ornamentation. Intrastaminal disc present or absent. Gynoecium usually of 2–6(-8) united carpels, mostly with a simple style with capitate or lobed stigma; ovary inferior, semi-inferior or (most of Scytopetalaceae) superior, with as many locules as carpels and with 1-many ovules per locule. Ovules anatropous to campylotropous, bitegmic, tenuinucellate, with or without endothelium, often with conspicuous funicle; micropyle endostomal. Female gametophyte of *Polygonum*-type. Endosperm nuclear. Fruits fibrous berries, dry drupes, or capsules, often very large, with one to many seeds. Seeds small to very large, hairy, often with funicular aril; seed coat formed primarily by the outer integument (exotestal-mesotestal); exotestal cells variously thickened, palisade, or low with sinuous anticlinal walls, mesotesta sclerotic or not; endosperm almost or completely absent (well-developed only in Scytopetalaceae); embryo large, straight or curved, oily or proteinaceous. Generally containing riterpenoid saponins and usually also producing proanthocyanins and ellagic acid, n = 11, 13, 16–18, 21.

The Lecythidales have been associated with the Myrtales (Prance and Mori 1979). But many important characters, including the secondary xylem (especially by the absence of vestured pits) and lack of internal phloem, and tenuinucellate (but bitegmic) ovules often with an endothelium, they differ markedly from the Myrtales. Therefore, Cronquist (1957, 1980, 1988) and Dahlgren (1980, 1983) removed the family from its traditional place, and Cronquist placed Lecythidaceae s.l. in a separate order Lecythidales (1957) and suggested placing it somewhere between Malvales and Theales. Thorne (1968) suggested a thealean relationships, and Dahlgren (1980, 1983) included this family in Theales. In my opinion the order Lecythidales are related to the theoid order Ochnales and could have a common origin with it from some remote thealean ancestor.

Key to Families

1 Seeds with endosperm. Evergreen trees or shrubs. Leaves mostly estipulate, papery to coriaceous. Flowers in panicles, thyrsoids, botryoids or racemes on younger branches, often ramiflorous or cauliflorous, rarely reduced to single flowers (*Asteranthos*). Sepals united; calyx copular, thick, leathery, persistent on ripe fruits, occasionally accrescent. Petals lacking. Staminodia 6–16, or 24–28 (*Asteranthos*); apex of staminodia curved inwards or not curved inwards (*Rhaptopetalum, Pierrina, Brazzeia*). Stamens 60–240; filaments adnate to pseudocorolla; anthers basifixed, tetrasporangiate, dehiscing by longitudinal slits or apical and pore-like. Pollen grains 3-colpate or 3-colporoidate (Rhaptopetaloideae). Carpels 3–8. Ovary superior or semi-inferior. Ovules 2–4-many in each locule, pendulous. Fruits monospermous or polyspermous. Seeds glabrous or embedded in a shiny white haircoat, arising from epidermis cells along the raphe (*Rhaptopetalum, Pierrina, Brazzeia*). Endosperm horny, mostly markedly ruminate (Kravtsova 1991); embryo well differentiated; cotyledons large, flat, cordate, rarely reduced (*Asteranthos*). . . . . . . 1. SCYTOPETALACEAE.

1 Seeds without endosperm.

2 Petals free, 3-6(-18); stamens numerous, usually connate at base and often forming a strip-like structure; nectary disc absent; seeds arillate or not; embryos undifferentiated or with fleshy cotyledons. Small to large trees, rarely shrubs. Stipules absent or minute and caduceus. Inflorescences terminal, axillary or cauline, simple racemes, or spicate branches or fascicles. Sepals 2–6 or rarely unlobed. Petals imbricate, free; anthers bilocular, latrorse, introrse or rarely poricidal. Ovary inferior or semi-inferior, 2–4- or 6-locular; ovules 1–115 in each locule. Fruits indehiscent, then dry, fleshy or woody or dehiscent. Seeds exotestal-mesotestal, winged or without wings; endosperm lacking or very scanty. Some species accumulate selenium. . . . . 2. LECYTHIDACEAE.

2 Pseudocorolla of 30–35 fused staminodes. Nectary disc well developed, annular, decagonal, supported by a very complex vascular system. Trees, shrubs or subshrubs. Leaves estipulate. Flowers usually solitary and axillary, occasionally in paniculate, axillary inflorescences, actinomorphic, usually with 4–5 bracts below the flowers. Calyx lobes 3 or 5–6, imbricate or valvate, each with two large glands. Petals absent. Stamens numerous; there are two outer cycles of staminodia, and an inner cycle of 20 fertile stamens, which are aligned into five groups of four stamens in each group, the outer two have very large, disporangiate, extrorse or latrorse anthers; filament ribbonlike, incurved. Pollen grains 3-colporoidate or 3-colporate. Ovary semi-inferior or inferior. Style short, with flat, pentagonal or hexagonal stigma; the entire structure is table-like. Ovules mainly campylotropous. Fruits baccate; seeds exarillate; embryos well-differentiated; cotyledons plano-convex with short radicle and large plumules, n = 16. . . 3 NAPOLEONAEACEAE.

1. SCYTOPETALACEAE

Engler 1897 (including Asteranthaceae Knuth 1939, Rhaptopetalaceae van Thiegem 1908). 6/28. Tropical West Africa and South America.

1.1 ASTERANTHOIDEAE

Flowers solitary in leaf axils. Calyx accrescent. The margin of corolla-like structure is ciliate and curved inward in bud and unfolds like an umbrella at anthesis. Staminodes 24–28. Stamens 200–300. Ovary semi-inferior. Carpels 5–8. Ovules four in each locule. Fruits monospermous. Seeds glabrous. Cotyledons small and strongly reduced. Pollen grains 3-colpate, n = 21. – *Asteranthos*.

1.2 SCYTOPETALOIDEAE
Flowers in few to many-flowered inflorescences in leaf axils of youngest branches. Staminodes 6-12(-16). Stamens 100–240. Carpels 3–5 or (6)7-8. Ovary superior. Ovules 2–4 in each locule. Fruits capsules or drupaceous. – *Scytopetalum, Oubanguias.*

1.3 RHAPTOPETALOIDEAE
Inflorescences ramiflorous or cauliflorous. Staminodes completely united into a thick, leathery pseudocorolla. Stamens 60–100. Anthers dehiscent by short, pore-like longitudinal slits. Ovules many in each locule. Fruits many-seeded. Pollen grains 3-colporoidate. – *Rhaptopetalum, Pierrina, Brazzeia.*

Related to the Lecythidaceae.

2. LECYTHIDACEAE

A. Richard 1825 (including Barringtoniaceae F. Rudolphi 1830, Foetidiaceae Airy Shaw 1965, Gustaviaceae Burnett 1835). 17/210. Tropical, especially America and West Africa.

2.1 PLANCHONIOIDEAE
Flowers in racemose or paniculate inflorescences, actinomorphic. Sepals (3)4(5), free or connate, imbricate (rarely closed). Petals (3)4(5), free, imbricate. Nectary disc usually well developed, annular. Stamens numerous, in several cycles, filaments linear, very long, free or mostly connate at the base, with a basal staminal ring; anthers basifixed, latrorse or less often introrse. Pollen grains syntricolporate, with thick ectexine and the specialized ectexinous ridge along the colpus margin (marginal ridge). Style filiform, often very long, with small stigma; ovary inferior or rarely semi-inferior, 2–4--locular, with two to many anatropous or campylotropous ovules per locule. Fruits one to many-seeded drupes with persistent calyx, sometimes broadly 4-winged. Seeds sometimes very large (up to 6 cm in *Abdulmajidia*), with undifferentiated or differentiated embryo. Cortical bundles in stem reversely oriented. Crystalliferous parenchyma strands absent in secondary xylem. Nodes 3-lacunar with three traces, n = 13. – *Petersianthus, Barringtonia, Chydenanthus, Abdulmajidia, Careya, Planchonia* (*Combretodendron*).

2.2 LECYTHIDOIDEAE
Flowers actinomorphic or zygomorphic. Sepals (4)6, free or more or less connate, valvate or slightly imbricate. Petals 4-6(8), free, imbricate. Stamens numerous, in several cycles, connate at the base into a staminal ring, the ring actinomorphic or more often prolonged on one side and forming a hood over the fertile stamens and gynoecium (the stamens on this sterile); filaments of fertile stamens usually short; anthers basifixed. Pollen grains 3-colporoidate or 3-colporate, smaller, and without marginal ridge. Nectary disc absent. Style usually short, mostly rodlike, with capitate to lobed stigma; ovary inferior or less often semi-inferior, 2–6- or rarely or more-locular. Ovules one to many. Fruits indehiscent (drupaceous or berrylike) or dehiscent with a circumscissile operculum (pyxidia), then often very large and woody. Seeds often arillate; embryo undifferentiated or with fleshy cotyledons. n = 17 (18). – GRIADEAR: *Gustavia, Grias*; COUROUPITEAE: *Allantoma, Cariniana, Couroupita, Corythophora*; LECYTHIDEAE: *Lecythis, Eschweilera*; COURATARIEAE: *Couratari*; BERTHOLETIEAE: *Bertholletia.*

2.3 FOETIDIOIDEAE
Flowers solitary and axillary or in small axillary cymes, actinomorphic, bibracteolate. Calyx of (3)4(5) valvate, thick, woody, deep purple, connate sepals, which enclose the inner floral organs until anthesis. Petals absent. Stamens numerous, free, sometimes in four oppositisepalous groups; filaments filiform, long and upright, or slightly wavy in the bud, spread out in the open flowers; anthers very small, introrse. Intrastaminal disc inconspicuous. Style slender, with usually four short slender divaricate stigmas; ovary inferior, (3)4-locular, with 15–20 ovules per locule; ovules usually campylotropous, arranged in vertical ring around thick peltate placenta. Fruits drupaceous, turbinate, 1- to 4-locular. Seeds with straight or J-shaped embryo. – *Foetidia.*

3. NAPOLEONAEACEAE

A. Richard 1827. 2/10. Tropical West Africa.
Napoleonaea, Crateranthus.
Closely related to Scytopelataceae.

Bibliography

Appel O. 1996. Morphology and systematics of the Scytopetalaceae. Bot. J. Linn. Soc. 121: 207–227.

Appel O. 2004. Scytopetalaceae. In: K Kubitzki, ed. The families and genera of vascular plants, vol. 6, pp. 426–430. Springer, Berlin/Heidelberg/New York.

Breteler FJ. 2002. Scytopetalaceae are stipulate. Kew Bull. 57: 759–761.

Carlquist S. 1988. Wood anatomy of Scytopetalaceae. Aliso 12: 63–76.

Diehl GA. 1935. A study of the Lecythidaceae. Trop. Woods 43: 1–15.

Frame D and S Durou. 2001. Morphology and biology of *Napoleonaea vogelii* (Lecythidaceae) flowers in relation to the natural history of insect visitors. Biotropica 33: 458–471.

Knuth R. 1934. Über die Gattung *Asteranthos*. Notizbl. Bot. Gart. Berlin-Dahlem 11: 1034–1036.

Knuth R. 1939a. Barringtoniaceae. In: A Engler, ed. Pflanzenreich, IV. 219: 1–82. W. Engelmann, Leipzig.

Knuth R. 1939b. Lecythidaceae. In: A Engler, ed. Pflanzenreich, IV. 219a: 1–144. W. Engelmann, Leipzig.

Knuth R. 1939c. Asteranthaceae. In: A Engler, ed. Pflanzenreich, IV. 219b: 1–3. W. Engelmann, Leipzig.

Kowal RR. 1989. Chromosome numbers of *Asteranthos* and the putatively related Lecythidaceae. Brittonia 41: 131–135.

Kowal RR, SA Mori, and JA Kallunki. 1977. Chromosome numbers of Panamanian Lecythidaceae and their use in subfamilial classification. Brittonia 29: 399–410.

Kravtsova TI. 1991. Scytopetalaceae. In: AL Takhtajan, ed. Comparative seed anatomy, 3: 226–228. Nauka, Leningrad (in Russian).

Leins P. 1972. Das zentrifugale Androeceum von *Conroupita guianensis* (Lecythidaceae). Beitr. Biol. Pfl. 48: 313–319.

Lens F, P Baas, S Jansen, and E Smets. 2007. A search for phylogenetically informative wood characters within Lecythidaceae s.l. Am. J. Bot. 94: 483–502.

Letouzey R. 1961. Notes sur les Scytopetalacees (Revision des Scytopetalacees de 1'herbier de Paris). Adansonia, ser. 2, 1: 106–142.

Liben L. 1971. Rĕvision de genre africain *Napoleonaea* P. Beauv. (Lecythidaceae). Bull. Jard. Bot. Natl. Belg. 41: 363–382.

Masters MT. 1869. On the structure of the flower in the genus *Napoleona*. Bot. J. Linn. Soc. 10: 492–504.

Miers J. 1874. On the Lecythidaceae. Trans. Linn. Soc. Lond., Bot. 30: 157–318.

Miers J. 1875a. On *Napoleona, Omphalocarpum,* and *Asteranthos*. Trans. Linn. Soc. Lond., Bot. 2d ser.,1: 17–19.

Miers J. 1875b. On the Barringtoniaceae. Trans. Linn. Soc., 2d ser., 1: 47–118.

Monteiro-Scanavacca WR. 1974. Vascularizay o do gineceu em Lecythidaceae. Bol. Univ. São Paulo, Bot. 2: 53–69.

Monteiro-Scanavacca WR. 1975a. Vascularizac o e natureza de estruturas do androceu em Lecythidaceae. Bol. Bot. Univ. São Paulo 3: 61–74.

Monteiro-Scanavacca WR. 1975b. Estudo da Placentaeo em Lecythidaceae. Bol. Bot. Univ. São Paulo 3: 75–86.

Mori SA, JE Orchard, and GT Prance. 1980. Intrafloral pollen differentiation in the New World Lecythidaceae subfamily Lecythidoideae. Science 209: 400–403.

Mori SA and GT Prance. 1990. Lecythidaceae: Part II. The zygomorphic flowered New World genera (*Couroupita, Corythophora, Bertholletia, Couratari, Eschweilera,* and *Lecythis*). Flora Neotropica 21: 1–376.

Mori SA, GT Prance, and AB Bolten. 1978. Additional notes on the floral biology of Neotropical Lecythidaceae. Brittonia 30: 113–130.

Mori SA, C-H Tsou, C-C Wu, B Crponholm, and AA Anderberg. 2007. Evolution of Lecythidaceae with an emphasis on the circumscription of neotropical genera: information from combined *ndh*F and *trn*L-F sequence data. Am. J. Bot. 94: 289–301.

Morton CM, SA Mori, GT Prance, KG Karol and MW Chase. 1997. Phylogenetic relationships of Lecythidaceae: a cladistic analysis using *rbc*L sequence and morphological data. Am. J. Bot. 84: 530–540.

Morton CM, GT Prance, SA Mori and LG Thorburn. 1998. Recircumscription of the Lecythidaceae. Taxon 47: 817–827.

Muller J. 1972. Pollen morphological evidence for subdivision and affinities of Lecythidaceae. Blumea 20: 350–355.

Muller J. 1973. Pollen morphology of *Barringtonia calyptrocalyx* K. Sch. (Lecythidaceae). Grana 13: 29–44.

Murthy GVS. 1996. Pollen morphology of Barringtoniaceae. J. Palynol. 32: 79–83.

Payens JPDW. 1967. A monograph of the genus *Barringtonia* (Lecythidaceae). Blumea 15: 157–263.

Prance GT. 1976. The pollination and androphore structure of some Amazonian Lecythidaceae. Biotropica 8: 235–241.

Prance GT. 2004. Napoleonaeaceae. In: K Kubitzki, ed. The families and genera of vascular plants, vol. 6, pp. 282–284. Springer, Berlin/Heidelberg/New York.

Prance GT and SA Mori. 1978. Observations of the fruits and seeds of neotropical Lecythidaceae. Brittonia 30: 21–33.

Prance GT and SA Mori. 1979. Lecythidaceae: Part I. The actinomorphic-flowered New World Lecythydaceae (*Asteranthos, Gustavia, Grias, Allantoma,* and *Cariniana*). Flora Neotropica 21: 1–270.

Prance GT and SA Mori. 1983. Dispersal and distribution of Lecythidaceae and Chrysobalanaceae. Sonderb. Naturwiss. Vereins Hamburg 7: 163–186.

Prance GT and SA Mori. 2004. Lecythidaceae. In: K Kubitzki, ed. The families and genera of vascular plants, vol. 6, pp. 221–232. Springer, Berlin/Heidelberg/New York.

Sarkar AK. 1983. Cytology of *Barringtonia* Forst. (Barringtoniaceae) as an aid to their taxonomy. Indian Sci. Congr. Assoc. Proc. 70: 77.

Tsou C-H. 1989. The floral morphology and embryology of *Asteranthos* and its systematic consideration. Am. J. Bot. 76 (Suppl.): 275–276.

Tsou C-H. 1993. Application of comparative embryology to plant taxonomy: taking the Lecythidaceae as an example. Recent Adv. Bot. 13: 293–302.

Tsou C-H. 1994a. The embryology, reproductive morphology, and systematics of Lecythidaceae. Mem. N. Y. Bot. Gard. 71: 1–110.

Tsou C-H. 1994b. The classification and evolution of pollen types of Planchonoideae (Lecythidaceae). Plant Syst. Evol. 189: 15–27.

Tsou C-H and SA Mori. 2007. Floral organogenesis and floral evolution of the Lecythidoideae (Lecythidaceae). Am. J. Bot. 94: 716–736.

Venkateswarlu J. 1952. Embryological studies in Lecythidaceae. Part I. J. Indian Bot. Soc. 31: 103–116.

Vijayaraghavan MR and U Diar. 1976. *Scytopetalum tieghemii:* embryologically unexplored taxon and affinities of the family. Phytomorphology 26: 16–22.

Weberling F. 1958. Über das Vorkommen rudimentarer Stipein bei den Lecythidaceae (s.l.) und Sonneratiaceae. Flora 116: 72–77.
Zeeuw de CH. 1990. Secondary xylem of Neotropical Lecythidaceae. In: I Mori and GT Prance, eds. Lecythidaceae – Part II, chap. II. The zygomorphic-flowered New World genera. Flora Neotropica, Monograph 21(3), pp. 4–59.

Order 59. SARRACENIALES

Perennial, rhizomatous, mostly acaulescent, carnivorous herbs. Rhizomes with vascular bundles arranged in a discontinuous ring interrupted by medullary rays of unequal width. Bundles bounded internally and externally by sclerenchyma. Vessels with oblique, scalariform plates with 3–50 bars; lateral pitting from scalariform to alternate. Fibers with bordered pits. Axial parenchyma diffuse. Leaves alternate, borne in a radical rosette or (species of *Heliamphora*) on an upright stem; all or some of the leaves highly modified, transformed into more or less elongate, often pitcherlike traps partly filled with digestive liquid and with a more or less developed ridge or flat wing on the ventral side and a laminar and relatively small epidermis provided with specialized glands and hairs that serve to entrap insects and other small organisms. Stomata anomocytic. Flowers rather large, solitary on a scape or (*Heliamphora*) in few flowered, sometimes axillary racemes, nodding, bisexual, actinomorphic, subtended by 1–3 bracteoles. Sepals (3)4–5(6), free, imbricate, persistent, often colored and showy. Petals five, free, imbricate, caducous or (*Heliamphora*) wanting. Stamens numerous or 10–15, in *Sarracenia* several, arising from each of a limited number (commonly 10) of primordia; filaments more or less short; anthers basifixed or (*Sarracenia*) versatile, opening longitudinally, introrse. Tapetum secretory. Microsporogenesis simultaneous. Pollen grains 2-celled, (3)4-polycolporate. Gynoecium of five or (*Heliamphora*) three united carpels, with slightly 3-lobed style and more or less truncate at the apex (*Heliamphora*), with five short branches at the apex, each with a terminal stigma (*Darlingtonia*) or apically expanded and peltate, umbrella-shaped, with small stigma under the tip of each of the five lobes (*Sarracenia*); ovary superior, 5-locular or (*Heliamphora*) 3-locular, but the partition of the upper part is often not complete, and the placentation is therefore axile below and parietal above; *Sarracenia* has 10 nectaries on the ovary wall above the stamen fascicles. Ovules numerous, anatropous, unitegmic or (*Darlingtonia*) bitegmic (?), tenuinucellate, with endothelium and a well-developed hypostase. Female gametophyte of *Polygonum*-type. Endosperm cellular. Fruits loculicidal capsules with numerous seeds. Seeds small, often winged, with exotestal seed coat, copious endosperm rich in oil and protein, and small linear embryo. Decarboxylated iridoids and O-methyl flavonols present, n = 13 (*Sarracenia*), 15 (*Darlingtonia*), 21 (*Heliamphora*).

Probably related to the Theales, which is supported by flower morphology, embryology, and wood anatomy (see DeBuhr 1975, 1977).

1. SARRACENIACEAE

Dumortier 1829 (including Heliamphoraceae Chrtek, Slavikova, et Studnicka 1992). 3/15. North America (*Darlingtonia and Sarracenia*) and Guyana Highland.

Heliamphora, Darlingtonia, Sarracenia.

Bibliography

Arber A. 1941. On the morphology of the pitcher-leaves in *Heliamphora, Sarracenia, Darlingtonia, Cephalotus,* and *Nepenthes*. Ann. Bot. II 5: 563–578.
Bayer RJ, L Hufford, and DE Soltis. 1996. Phylogenetic relationships in Sarraceniaceae based on *rbc*L and ITS sequences. Syst. Bot. 21: 121–134.
Bell CR. 1949. A cytotaxonomic study of the Sarraceniaceae of North America. J. Elisha Mitchell Sci. Soc. 65: 137–166.
Chrtek J, Z Slavikova, and M Studnicka. 1992. Beitrag zur Morphologie und Taxonomie der Familie Sarraceniaceae. Preslia 64(1): 1–10.
DeBuhr LE. 1973. Distribution and reproductive biology of *Darlingtonia californica*. M.A. thesis, Claremont Graduate School, Claremont.
DeBuhr LE. 1975. Phylogenetic relationships of the Sarraceniaceae. Taxon 24: 297–306.
DeBuhr LE. 1977. Wood anatomy of the Sarraceniaceae: ecological and evolutionary implications. Plant. Syst. Evol. 128: 159–169.
Juniper BE, RL Robins, and DM Joel. 1989. The carnivorous plants. Academic, London.
Kubitzki K. Sarraceniaceae. 2004. In: K Kubitzki, ed. The families and genera of vascular plants, vol. 6, pp. 422–425. Springer, Berlin/Heidelberg/New York.
Lloyd FE. 1942. The carnivorous plants. Chronica Botanica, Waltham, MA.
Maguire B. 1978. Sarraceniaceae. The botany of Guayana Highland – part X. N. Y. Bot. Gard. 29: 36–62.

Markgraf F. 1955. Über Laubblatt-Homologien und verwandtschaftliche Zusammenhänge bei Sarraceniales. Planta 46: 414–446.
Newman T, S Ibrahim, JW Wheeler, WB McLaughlin, RL Petersen, and RM Duffield. 2000. Identification of sarracenin in four species of *Sarracenia* (Sarraceniaceae). Biochem. Syst. Ecol. 28: 193–195.
Neyland R and M Merchant. 2006. Systematic relationships of Sarrceniaceae inferred from nuclear ribosomal DNA sequences. Madroño 53: 223–232.
Shreve F. 1906. The development and anatomy of *Sarracenia purpurea*. Bot. Gaz. 42: 107–126.
Terekhin ES. 1992. Sarraceniaceae. In: A Takhtajan, ed. Comparative seed anatomy, vol. 4, pp. 21–23. Nauka, St. Petersburg (in Russian).
Thanikaimoni G and G Vasanthy. 1972. Sarraceniaceae: palynology and systematics. Pollen et Spores 14: 143–155.
Wood CR. 1960. The genera of Sarraceniaceae and Droseraceae in the southeastern United States. J. Arnold Arbor. 41: 152–163.

Order 60. RORIDULALES

Sparsely branched shrubs bearing glandular (tentacular) hairs of various lengths and unicellular bristle hairs on young stems, leaves, and calyx; the largest tentacular trichomes are situated on the leaf margins; secretion of tentacular hairs sticky, resinous, without proteolytic enzymes. Vessel elements long and have oblique perforation plates with scalariform perforations that in *Roridula gorgonias* have up to 83 bars per plate; lateral pitting scalariform to alternate. Fibers are tracheids. Rays mostly uniseriate (rarely biseriate) and composed of upright cells. Axial parenchyma sparse, diffuse. Leaves clustered toward the branch ends of adult shrubs, sessile, alternate, linear-lanceolate, entire or with linear lateral lobes, pinnately veined, circinate in vernation, estipulate. Stomata anomocytic. Flowers in few-flowered, axillary panicles terminated by a flower, actinomorphic, 5-merous as to the perianth and androecium, bibracteolate. Sepals free, imbricate, glandular-hairy. Petals free, imbricate, glabrous, purple or crimson to nearly white. Stamens five, alternipetalous; filaments free, slender, thickened at the base; anthers 2-locular, elongate, tetrasporangiate, inverted in bud, with basal thickening containing a nectariferous cavity, irritable (swinging upward and becoming erect and spreading pollen when touched at anthesis), extrorse, basifixed, opening by terminal pores. Tapetum secretory. Microsporogenesis simultaneous. Pollen grains 2-celled, 3-colporate, tectate-perforate and spinuliferous (*R. dentata*) or without distinctly developed apertures and without coherent tectum (*R. gorgonias*). Gynoecium of three carpels; style tapering and with small, capitate stigma (*R. dentata*) or expanding, terminal, apically obconical and with a large stigma (*R. gorgonias*); ovary 3-locular, superior, with solitary (*R. dentata*) or 2–4 (*R. gorgonias*) ovules per lucule. Ovules pendulous, with endothelium. Female gametophyte of *Polygonum*-type, with a conspicuous hypostase at the chalazal end. Endosperm cellular. Fruits 3-valved locultcidal capsules. Seeds exotestal, rather large, reticulate (*R. dentata*) or warty to honeycombed (*R. gorgonias*) with small, straight embryo and copious, fleshy endosperm. Producing iridoid compounds. Tannins present in the seed coat and endosperm cells, n = 6.

1. RORIDULACEAE

Martynov 1820. 1/2. Endemic to the Cape Province.
Roridula.

Hutchinson (1959, 1973) and Cronquist (1981, 1988) place *Roridula* in the Byblidaceae, but as Vani-Hardev (1972) showed there are profound morphological differences between *Roridula* and *Byblis*. Albert et al. (1992) placed the Roridulaceae as sister to the carnivorous pitcher plant family Sarraceniaceae.

Bibliography

Bruce AN. 1907. On the distribution, structure, and function of the tentacles of *Roridula*, Notes Roy. Bot. Gard. Edinb. 17: 85–96.
Carlquist S. 1976. Wood anatomy of Roridulaceae: ecological and phylogenetic implications. Am. J. Bot. 63: 1003–1008.
Conran JG. 2004. Roridulaceae. In: K Kubitzki, ed. The families and genera of vascular plants, vol. 6, pp. 339–342. Springer, Berlin/Heidelberg/New York.
Conran JG and JW Dowd. 1993. The phylogenetic relationships of the *Byblis – Roridula* (Byblidaceae: Roridulaceae) complex inferred from 18S rRNA partial sequences. Plant Syst. Evol. 188: 73–86.
Lobova TA. 2000. Roridulaceae. In: A Takhtajan, ed. Comparative seed anatomy, 6: 289–292. Nauka, St. Petersburg (in Russian).
Vani-Hardev S. 1972. Systematic embryology of *Roridula gorgonias* Planch. Beitr Biol. Pfl. 48: 339–351.
Wilkinson HP. 1998. Roridulaceae. In: DF Cutler and M Gregory, eds. Anatomy of the Dicotyledons, Saxifragales (sensu Armen Takhtajan 1983), vol. 4, pp. 121–130. Clarendon, Oxford.

Superorder PRIMULANAE

Order 61. STYRACALES (EBENALES)

Trees and shrubs. Vessels with scalariform or simple perforations; lateral pitting from scalariform to alternate. Fibers with bordered or simple pits. Rays mostly heterogeneous, mixed uniseriate and pluseriate, or sometimes all uniseriate. Axial parenchyma mostly apotracheal. Nodes unilacunar. Leaves alternate or rarely opposite, simple, entire or dentate to serrate, estipulate. Stomata mostly anomocytic, sometimes paracytic or encyclocytic. Flowers usually in various racemose or less often cymose inflorescences, bisexual or less often unisexual, actinomorphic, with bracteoles or (Styracaceae) without. Calyx synsepalous, usually tubular or copulate, with (2)3-7-9 lobes. Corolla usually sympetalous and 2–7-lobed or very rarely (*Bruinsmia* in Styracaceae) petals free or nearly free. Stamens two times as many as the corolla lobes or less often same number, adnate at different levels upon the petals or corolla tube; filaments free or more or less connate; anthers tetrasporangiate, opening longitudinally or rarely (some species of *Diospyros*) by apical pores. Tapetum secretory. Microsporogenesis simultaneous. Pollen grains 2-celled or less often 3-celled, 3-colporate or 3-porate. Gynoecium of 2–8 united carpels with stylodia more or less completely connate into a style, rarely (some Ebenaceae) nearly free; stigma lobed or capitate; ovary superior to inferior, usually plurilocular, with 1–4 or more ovules per locule; placentation axile (Symplocaceae), axile to basal (Styracaceae), or apical (Ebenaceae). Ovules anatropous to hemitropous, pendulous (Ebenaceae), bitegmic (Ebenaceae) or unitegmic (Symplocaceae and most of the Styracaceae), tenuinucellate, sometimes (Ebenaceae and *Pterostyrax* and *Alniphyllum* in Styracaceae) with an endothelium. Female gametophyte of *Polygonum*-type. Endosperm cellular. Fruits of various types. Seeds large, hard, often shiny; seed coat thin, exotestal; embryo straight or slightly curved; endosperm copious, mostly bony, often ruminate. Without iridoid compounds.

Styracales evidently derived directly from some thealean ancestor. They also resemble the Ericales, but I agree with Cronquist (1981: 493) that "the relationship is collateral rather than ancestral." I also agree that no one family of the Styracales is likely to be directly ancestral to any of the others.

Key to Families

1 Flowers bisexual or seldom polygamo-dioecious. Style with capitate or lobed stigma. Gynoecium of 2–5 carpels. Ovary superior to often inferior. Vessels mostly or always with scalariform perforations; lateral pitting from scalariform to opposite, rarely alternate.

2 Pubescence of stellate or stellate-peltate trichomes. Trees and shrubs, often with resiniferous intercellular canals in the bark and wood but without latex. Leaves alternate, simple, entire to toothed. Stomata anomocytic. Inflorescences axillary or terminal, racemes or panicles, rarely the flowers solitary. Flowers without bracteoles, bisexual or rarely female in gynodioecious; sepals 4–5(-9), connate, the lobes valvate or the calyx open in bud, often truncate; petals 4–5(-9), mostly basally connate. Stamens 4-10(-14), the filaments usually connate and often inserted basally on the corolla, rarely hypogynous. Anthers linear to oblong, basifixed. Pollen grains 3-colporate. Stylodia completely united or (*Huodendron*) united at lower level but free above; ovary superior to inferior, plurilocular below but usually unilocular above; ovules bitegmic (*Styrax*) or unitegmic (*Halesia*); micropyle endostomal. Fruits nut-like, sometimes winged, or a loculicidal capsule, samara, or a drupe, with persistent calyx. Seeds one of few, rarely numerous; embryo large, straight or slightly curved; endosperm copious, oily. Contain quercetin, kaempferol, tannins, a small amount of leucoanthocyanin, saponins, coffeic and ferulic acids, n = 8 (*Styrax*) or 12. . . . . 1. STYRACACEAE.

2 Plants glabrous or with multicellular hairs or stalked glands, but without stellate hairs. Evergreen or rarely deciduous trees or shrubs, commonly accumulating aluminum. Leaves alternate, simple, serrate or entire. Stomata mostly paracytic. Flowers provided with two bracteoles, in racemes, or less often in spikes, panicles, or fascicles, or solitary, bisexual, rarely unisexual. Sepals five, connate, imbricate or valvate, persistent. Petals 5 (3–11), imbricate, connate basally in 1–2 series. Stamens 4–12 to numerous (to 100), the filaments free or connate basally. Anthers short and broad, mostly round ovate. Pollen grains angular, spinuliferous. Ovary wholly or partly inferior, usually 2–5-locular, usually with an

apical 5-glandular, annular, cylindrical, or 5-lobed disc; ovules 2–4 per lacule, pendulous, unitegmic (Corner 1976). Fruits 1-seeded drupes or less often baccate. Seeds with linear, straight or arcuate embryo and copious endosperm; exotestal cells with inner walls thickened or thin. Present gallic and ellagic acids, tannins, leucoanthocyanins, iridoids, n = 11. . . . . 2. SYMPLOCACEAE.

1 Flowers mostly unisexual and commonly dioecious, rarely bisexual. Trees and shrubs, commonly with hard, dark or black wood, without latex. Vessels with simple perforations, or less greauently scalariform; lateral pitting alternate. Nodes unilacunar with three traces. Leaves alternate, rarely opposite or whorled, simple, entire. Stomata usually paracytic. Inflorescences axillary or rarely cauliflorous cymes, or flowers solitary. Sepals 3–5(-8), connate, valvate or imbricate; petals 3–5(-8), connate, usually contorted and imbricate, rarely valvate. Stamens 3-50(-120 ?), usually 2–4 times the number of corolla lobes, rarely equal in number and alternate with them; the filaments free or connate in pairs; anthers elongated, often hairy, adnate or basifixed, dehiscing longitudinally, more rarely by apical pores. Pollen grains 3-colporate or 3-porate. Gynoecium of 2–8 carpels. Ovary superior, 2–16-locunar. Stylodia nearly free or more or less connate below. Ovules 1–2 per locule, bitegmic; micropyle endostomal. Fruits more or less juicy and leathery berries rarely capsular. Seeds pachychalazal, often ruminate, testa multiplicative (not vascularised), exotesta fibriform or mucilaginous, cells cuboid to palisade, endotesta crystalliferous or not, walls thickened or not; embryo straight or slightly curved; endosperm copious, hard. Producing saponins, triterpenes, napthoquinone derivative of seven methyljugone and plumbagin, flavonols, leucodelphinidin, myricetin, n = 15. . . . 3. EBENACEAE.

1. STYRACACEAE

A.P. de Candolle et Sprengel 1821 (including Halesiaceae D. Don 1828). 11/160. Mediterranean, eastern Himalayas (from Nepal to Arunachal-Pradesh), Assam, eastern India, eastern and Southeast Asia to West Malesia, America from United States to Argentina.

Styrax (including *Pamphilia*), *Huodendron*, *Bruinsmia*, *Alniphyllum*, *Changiostyrax*, *Sinojackia*, *Parastyrax*, *Pterostyrax*, *Halesia*, *Rehderodendron*, *Melliodendron*.

Styracaceae (together with the Symplocaceae) are the most primitive members of the order and the nearest to the Theales (see Copeland 1938).

2. SYMPLOCACEAE

Desfontaines 1820. 1/300. Tropical and subtropical regions except Africa, Europe, and western Asia.

Symplocos.

Very close to the Styracaceae and approximately on the same level of specialization.

3. EBENACEAE

Gürke 1891 (including Diospyraceae Durande 1782, Guaiacanaceae Jussieu 1789, Lissocarpaceae Gilg 1924). 4–5/500. The largest genus *Diospyros* (up to 500) is pantropical with the greatest concentration of species in Malesia; only a few species of *Diospyros* occur in western Asia, Japan, and the southeastern United States. *Euclea* (14) occurs in East and South Africa, on the Comoros and Arabian peninsula. *Tetraclis* (3) endemic to Madagascar, *Lissocarpa* (8) tropical South America (from Guyana to Bolivia).

3.1 EBENOIDEAE

Pollen grains 3-colporate. Ovary superior, 2–16-locunar, ovules one or less frequently two; indumentum present; bracteoles alternate. – *Diospyros*, *Royena*, *Tetraclis*, *Euclea*.

3.2 LISSOCARPOIDEAE

Pollen grains 3-porate. Ovary inferior, 4-locular; ovules two; indumentum absent; bracteoles subopposite. – *Lissocarpa*.

Related to the Styracaceae and Symplocaceae, but the affinity is not very close. A very heterobathmic family: ovules bitegmic and stylodia not completely united or sometimes nearly free, but flowers mostly unisexual.

Bibliography

Barth OM. 1982. The sporoderma of Brazilian *Symplocos* pollen types (Symplocaceae). Grana 18: 65–69.

Berry PE. 1999. A synopsis of the family Lissocarpaceae. Brittonia 51: 214–216.

Berry PE, V Savolainen, KJ Sytsma, JC Hall, and MW Chase. 2001. *Lissocarpa* is sister to *Diospyros* (Ebenaceae). Kew Bull. 56: 725–729.

Contreras LS and NR Lersten. 1984. Extrafloral nectaries in Ebenaceae: anatomy, morphology, and distribution. Am. J. Bot. 71: 865–872.

Copeland HF. 1938. The *Styrax* of northern California and the relationships of the Styracaceae. Am. J. Bot. 25: 771–780.

Caris P, LP Ronse Decraene, E Smets, and D Clinckemaillie. 2002. The uncertain systematic position of *Symplocos* (Symplocaceae): evidence from a floral ontogenetic study. Int. J. Plant Sci. 163: 67–74.

Dickison WC. 1993. Floral anatomy of the Styracaceae, including observations on intraovarian trichomes. Bot. J. Linn. Soc. 112: 223–255.

Dickison WC and KD Phend. 1985. Wood anatomy of the Styracaceae: evolutionary and ecological considerations. IAWA Bull. 6: 3–22.

Duangiai S, B Wallnöfer, R Samuel, J Munzinger, and MW Chase. 2006. Generic delimitation and relationships in Ebenaceae sensu lato: evidence from six plastid DNA regions. Am. J. Bot. 93: 1808–1827.

Franceschi, G. de. 1993. Phylogénie des Ebénales: analyse de l'ordre et origin biogéographique des espèces indiennes. Institut Français de Pondichery, Département de Écologie 33: 1–153.

Fritsch P. 1996. Isozyme analysis of intercontinental disjuncts within *Styrax* (Styracaceae): implications for the Madrean-Tethyan hypothesis. Am. J. Bot. 83: 342–355.

Fritsch PW. 1999. Phylogeny of *Styrax* based on morphological characters, with implications for biogeography and infrageneric classification. Syst. Bot. 24: 356–378.

Fritsch PW. 2001. Phylogeny and biogeography of the flowering plant genus *Styrax* (Styracaceae) based on chloroplast DNA restriction sites and DNA sequences of the internal transcribed spacer region. Molec. Phylogen. Evol. 19: 387–408.

Fritsch PW. 2004. Styracaceae. In: K Kubitzki, ed. The families and genera of vascular plants, vol. 6, pp. 434–442. Springer, Berlin/Heidelberg/New York.

Fritsch PW, CM Morton, T Chen, and C Meldrum. 2001. Phylogeny and biogeography of the Styracaceae. Int. J. Plant Sci. 162(6, Suppl.): 95–116.

Fritsch PW, BC Cruz, F Almeda, Y Wang, and S Shi. 2006. Phylogeny of *Symplocos* based on DNA sequences of the chloroplast *trn*C-*trn*D intergenic region. Syst. Bot. 31: 181–192.

Kolbe K-P and J John. 1980. Serology and systematics of the Ebenales and the Theales. Biochem. Syst. Ecol. 8: 249–256.

Komar GA. 1992. Ebenales. In: A Takhtajan, ed. Comparative seed anatomy, 4: 45–48. Nauka, St. Petersburg (in Russian).

Li Y-H and C-H Yu. 1985. Pollen morphology of Styracaceae and its taxonomic significance. Acta Phytotax. Sinica 23: 81–90 (in Chinese with English summary).

Liang YH. 1986. Pollen morphology of the family Symplocaceae from China. Acta Bot. Austro Sinica 2: 111–121.

Machado SR, V Angyalossy-Alfonso, and BL de Morretes. 1997. Comparative woody anatomy of root and stem in *Styrax camporum* (Styracaceae). IAWA J. 18: 13–25.

Mallavadhani UV, AK Panda, and YR Rao. 1998. Pharmacology and chemotaxonomy of *Diospyros*. Phytochemistry 49: 901–951.

Manshard E. 1936. Embryologische Untersuchungen an *Styrax obassia* Sieb. et Zucc. Planta 25: 364–383.

Morawetz W. 1991. The karyology of some Neotropical Styracaceae. Plant Syst. Evol. 177: 111–115.

Morton CM and WC Dickison. 1992. Comparative pollen morphology of the Styracaceae. Grana 31: 1–15.

Morton CM, MW Chase, KA Kron, and SM Swensen. 1996. A molecular evaluation of the monophyly of the order Ebenales based upon *rbc*L sequence data. Syst. Bot. 21: 567–586.

Morton CM, MW Chase, KA Kron, and SA Swensen. 1997. A molecular evaluation of the monophyly of the order Ebenales based upon *rbcL* sequence data. Syst. Bot. 21: 567–586.

Ng ESP. 1971. A taxonomic study of the Ebenaceae with special reference to Malesia. Doctoral thesis, University of Oxford, Oxford.

Nikiticheva ZI. 1983. Ebenales. In: MS Yakovlev, ed. Comparative embryology of flowering plants: Phytolaccaceae-Thymelaeaceae, pp. 221–233. Nauka, Leningrad (in Russian).

Nooteboom HP. 1975. Revision of the Symplocaceae of the Old World (New Caledonia excepted). Leiden University Press, Leiden.

Nooteboom HP. 2004. Symplocaceae. In: K Kubitzki, ed. The families and genera of vascular plants, vol. 6, pp. 443–449. Springer, Berlin/Heidelberg/New York.

Oever L van den, P Baas, and M Zandoe. 1981. Comparative wood anatomy of *Symplocos* and latitude and altitude of provenance. IAWA Bull. 2: 3–24.

Oliver D. 1895. *Lissocarpa benthamii* Gürke. Hooker's Icon. Plant. 25: t.2413.

Periasamy K. 1966. Studies on seeds with ruminate endosperm. VI. Rumination in the Araliaceae, Aristolochiaceae, Caprifoliaceae, and Ebenaceae. Proc. Indian Acad. Sci. 60B: 127–134.

Schadel WE. 1978. Leaf anatomy and venation patterns of the Styracaceae. Thesis, Master of Arts, University of North Carolina at Chapel Hill.

Schadel WE and WC Dickison. 1979. Leaf anatomy and venation patterns of Styracaceae. J. Arnold Arbor. 60: 8–27.

Shyamala R and N Gulati. 1981. Contribution to the embryology of *Symplocos cochinchinensis*. Indian J. Bot. 4: 229–235.

Shyamala R and N Gulati. 1986. Origin and structure of integuments in Ebenales: I. Symplocaceae. Swamy Bot. Club 2: 113–117.

Takasi Y. 1972. Embryological studies in Ebenales, part 4. J. Jpn. Bot. 47: 20–28.

Veillet-Bartoszewska M. 1960. Embryogenie des Styracacees: Developpement de l'embryon chez le *Styrax officinalis* L. Compt. Rend. Hebd. Seances Acad. Sci. 250: 905–907.

Wallnöfer B. 2001. The biology and systematics of Ebenaceae: a review. Ann. Naturhist. Mus. Wien. 103B: 485–512.

Wallnöfer B. 2004a. Ebenaceae. In: K Kubitzki, ed. The families and genera of vascular plants, vol. 6, pp. 125–130. Springer, Berlin/Heidelberg/New York.

Wallnöfer B. 2004b. Lissocarpaceae. In: K Kubitzki, ed. The families and genera of vascular plants, vol. 6, pp. 236–238. Springer, Berlin/Heidelberg/New York.

Wallnöfer B. 2004c. A revision of *Lissocarpa* Benth. (Ebenaceae subfam. Lissocarpoideeae (Gilg in Engler) B. Walln. Ann. Naturhist. Mus. Wien, B, 105: 515–564.

Wang Y, PW Fritsch, S Shi, F Almeda, BC Criz, and LM Kelly. 2004. Phylogeny and infragenetic classification of *Symplocos* (Symplocaceae) inferred from DNA sequence data. Am. J. Bot. 91: 1901–1914.

White EB. 1956–1963. Notes on Ebenaceae: I, II. Bull. Jard. Bot. L'Etat 26: 237–246; 277–307, 1956. III. Bull. Jard. Bot. L'Etat 27: 515–531, 1957. IV. Bol. Soc. Brot. 36: 97–100, 1962. V. Bull. Jard. Bot. L'Etat 33: 345–367, 1963.

Wood CE and RB Channell. 1960. The genera of the Ebenales in the southeastern United States. J. Arnold Arbor. 41: 1–35.

Yamazaki T. 1970, 1972. Embryological studies in Ebenales: 1. Styracaceae. 2. Symplocaceae. 4. Ebenaceae. J. Jpn. Bot. 45: 267–273; 353–358, 1970. 47: 20–28, 1972.

Order 62. SAPOTALES

Trees and shrubs with well-developed, elongate laticiferous sacks in the pith, phloem, cortex, and leaves that contain a refractive white substance resembling resin or guttapercha. Indumentum in all genera, except *Delpydora*, of appressed unicellular hairs with two equal or unequal arms commonly mixed with erect or spreading simple hairs. Vessels with simple perforations; lateral pitting alternate. Fibers typically with simple pits, but pits occasionally bordered. Rays heterogeneous. Axial parenchyma apotracheal. Sieve-element plastids of S-type. Nodes trilacunar or seldom unilacunar. Leaves alternate or seldom opposite, simple, nearly always entire, pinnately veined, with small to large and sometimes persistent stipules (e.g., *Madhuca* spp.) or more often estipulate. Stomata usually anomocytic. Flowers mostly rather small, solitary or more often in small cymes, sometimes (*Sarcosperma*) in large paniculate inflorescences, bisexual, actinomorphic, often nocturnal and bat-pollinated. Sepals (4)5(-12), free or nearly so and imbricate, or sometimes in two cycles of two, three or four. Corolla sympetalous, with 4–8 imbricate lobes, sometimes with paired appendages. Stamens in 1–3 cycles, opposite the corolla lobes, inserted within the corolla tube or occasionally at the base of the corolla lobes, rarely free, often some of them staminodial; anthers tetrasporangiate, opening longitudinally. Tapetum secretory. Microsporogenesis simultaneous. Pollen grains 2-celled, 3–5(6)-colporate, with colpi provided by granular membrane; ectexine nearly always with a thick and usually sparsely punctate or perforate tectum and a very narrow interstitium, either granular or with very reduced columellae. Gynoecium of 2-14(-30) united carpels, rarely (several species of *Pouteria* sect. *Franchetella*) of one carpel, usually hairy; style with capitate or slightly lobed stigma; ovary superior, typically 5-plurilocular (unilocular in *Diploon*), with axile or sometimes basal or axile-basal placentation and one ovule per carpel. Ovules anatropous to hemitropous, apotropous, unitegmic, tenuinucellate, without endothelium. Female gametophyte of *Polygonum*-type. Endosperm nuclear. Fruits typically 1- to many-seeded berries with fleshy, leathery, or rarely woody outer pericarp, rarely (as in *Pradosia*) drupelike or sometimes, as in four genera of Mimusopeae, tardily dehiscing 1-several-seeded loculicidal capsule. Seeds medium-sized to large, ellipsoid, light brown to black, shiny, often with variously expanded hilum (basal, basi-ventral or adaxial); seed coat thick and hard, its outer part (8–25 cells thick or more) often forming a heavily lignified, sclerotic layer of compressed, pitted cells, and its inner part (of similar thickness) thin-walled, aerenchymatous, carrying the vascular bundles, eventually more or less crushed; embryo large, usually with small radicle; endosperm sometimes relatively copious (*Manilkara* and some other genera), but mostly scanty or even wanting. Producing saponins, C-30 oxidised triterpenes, pyrrhlizidine alkaloids, flavonols, leucodelphinidin, myrcetin, but without iridoid compounds, n = (10-)13(-14).

Sapotales are related to the Styracales and share a common origin from the thealean ancestor. However, they differ from the Styracales in many important characters including a well-developed system of latex sacks, nuclear endosperm, absence of the endothelium, and seed coat anatomy.

1. SAPOTACEAE

A.L. de Jussieu 1789 (including Achradaceae Vest 1818, Boerlagellaceae H.J. Lam 1925, Bumeliaceae Barnhardt 1895, Sarcospermataceae H.J. Lam 1925). 53/1200. Pantropical, with a few species extending into temperate regions.

MIMUSOPEAE: *Mimusops, Vitellariopsis, Autranella, Tieghemella, Baillonella, Vitellaria, Manilkara, Labramia, Faucherea, Northia, Labourdonnaisia, Letestua, Inhambanella, Neolemonniera, Lecomtedoxa, Gluema, Eberhardtia;* ISONANDREAE: *Palaquium, Aulandra, Isonandra, Madhuca, Payena, Burckella, Diploknema;* SIDEROXYLEAE: *Sideroxylon, Neohemsleya, Nesoluma, Argania, Sarcosperma, Diploon;* CHRYSOPHYLLEAE: *Pouteria, Aubregrinia, Breviea, Micropholis, Chromolucuma, Chrysophyllum, Ecclinusa, Delpydora, Pichonia, Sarcaulus, Elaeoluma,*

Niemeyera, Pradosia, Leptostylis, Pycnandra, Synsepalum, Englerophytum, Xantolis, Capurodendron; OMPHALOCARPEAE: *Tsebona, Magodendron, Omphalocarpum, Tridesmostemon.*

The delimitation of genera in the Sapotaceae has always been very controversial. Here I follow Pennington's (2004) system of classification.

Bibliography

Anderberg AA and B Ståhl. 1994. Phylogenetic interrelationships in the order Primulales, with special emphasis on the family circumscriptions. Canad. J. Bot. 73: 1699–1730.

Anderberg AA and U Swenson. 2003. Evolutionary lineages in Sapotaceae (Ericales): a cladistic analysis based on *ndh*F sequence data. Int. J. Plant Sci. 164: 763–773.

Aurbeville A. 1964. Système de classification des Sapotacèes. Adansonia, ser. 2, 4: 38–42.

Baehni C. 1938. Memoires sur les Sapotacees: I. Systeme de classification. Candollea 7: 394–508.

Baehni C. 1965. Memoires sur les Sapotacees: III. In-ventaire des genres. Boissiera 11: 1–262.

Bartish IV, U Swenson, J Munzinger, and AA Anderberg. 2005. Phylogenetic relationships among New Caledonian Sapotaceae (Ericales): molecular evidence for generic polyphyly and repeated dispersal. Am. J. Bot. 92: 667–673.

Bokdam J. 1977. Seedling morphology of some African Sapotaceae and its taxonomical significance. Meded. Land. Wagen. 77/20: 1–84.

Cronquist A. 1946a. Studies in the Sapotaceae: II. Survey of the North American genera. Lloydia 9: 241–292.

Cronquist A. 1946b. Studies in the Sapotaceae: VI. Miscellaneous notes. Bull. Torrey Bot. Club 73: 465–471.

Govaerts R, DG Frodin, and TD Pennington. 2001. World checklist and bibliography of Sapotaceae. Royal Botanic Gardens, Kew.

Gunasekera SP, V Kumar, MUS Sultanbawa, and S Balasubramaniam. 1977. Triterpenoids and steroids of some Sapotaceae and their chemotaxonomic significance. Phytochemistry 16: 923–926.

Harley MM. 1986. Distinguishing pollen characters for the Sapotaceae. Canad. J. Bot. 64: 3091–3100.

Harley MM. 1991a. The pollen morphology of the Sapotaceae. Kew Bull. 46: 379–491.

Harley MM. 1991b. Pollen morphology of the Sapotaceae. In: TD Pennington, The genera of Sapotaceae, pp. 23–50. Royal Botanic Gardens, Kew, New York.

Johnson MAT. 1991. Cytology. In: TD Pennington, The genera of Sapotaceae, pp. 15–22. Royal Botanic Gardens, Kew, New York.

Komar GA. 1992. Sapotales. In: A Takhtajan, ed. Comparative seed anatomy, 4: 49–53. Nauka, St. Petersburg (in Russian).

Kukachka BF. 1978, 1979, 1980. Wood anatomy of the Neotropical Sapotaceae: I–XIX. US Forest Service Research Paper. FPL 325–331, 1978; 349–354, 1979; 358–363, 1980.

Lam HJ. 1925. The Sapotaceae, Sarcospermaceae, and Boerlagellaceae of the Dutch East Indies and surrounding countries (Malay Peninsula and Philippine Islands). Bull. Jard. Bot. Buitenzorg 7: 1–289.

Lam HJ. 1939. On the system of the Sapotaceae, with some remarks on taxonomical methods. Rec. Trav. Bot. Neerl. 36: 509–525.

Lam HJ and WW Varossieau. 1938. Revision of the Sarcospermataceae. Blumea 3: 183–200.

Ng FSP. 1991. The relationships of the Sapotaceae within the Ebenales. In TD Pennington. The genera of Sapotaceae, pp. 1–13. Royal Botanic Gardens, Kew.

Pennington TD. 1990. Sapotaceae. Flora Neotrop. Monogr. 52: 1–770.

Pennington TD. 1991. The genera of Sapotaceae. Royal Botanic Gardens, Kew.

Pennington TD. 2004. Sapotaceae. In: K Kubitzki, ed. The families and genera of vascular plants, 6: 390–421. Springer, Berlin/Heidelberg/New York.

Record SJ. 1939. American woods of the family Sapotaceae. Trop. Woods 59: 21–51.

Smedmark JEE, U Swenson, and AA Anderberg. 2006. Accounting for variation in substitution rates through time in Bayesian phylogeny reconstruction of Sapotoideae (Sapotaceae). Molec. Phyl. Evol. 39: 706–721.

Swenson U and AA Anderberg. 2005. Phylogeny, character evolution, and classification of Sapotaceae (Ericales). Cladistics 21: 101–130.

Waterman PG and EN Mahmoud. 1991. Chemical taxonomy of the Sapotaceae: patterns in the distribution of some simple phenolic compounds. In: TD Pennington, ed. The genera of Sapotaceae, pp. 51–74. Royal Botanic Gardens, Kew, New York.

Yamazaki T. 1971. Embryological studies in Ebenales: 1. Sapotaceae. J. Jpn. Bot. 46: 161–165.

Order 63. PRIMULALES

Trees or shrubs, sometimes woody lianas, rarely sub-shrubs, mostly perennial or less often annual herbs, sometimes aquatic (*Hottonia*). In Myrsinaceae, Theophrastaceae and in some genera of Primulaceae, particularly in *Lysimachia,* occur secretory cells or cavities containing red or reddish substance. Vessels with simple perforations or occasionally some of them with scalariform perforations; lateral pitting alternate. Fibers with simple or sometimes more or less indistinctly bordered pits. Rays heterogeneous to homogeneous, typically multiseriate. Axial parenchyma paratracheal and usually scanty, often lacking. Sieve-element plastids of S-type. Nodes unilacunar or sometimes trilacunar. Leaves alternate, opposite, or less often verticillate, often gland-dotted or farinose, simple, entire or variously lobed, rarely (*Hottonia*) pectinately pinnatisect, estipulate. Stomata mostly anomocytic, or anisocytic. Flowers large to small, in various kinds of inflorescences or solitary, bracteate but without bracteoles, bisexual or unisexual, actinomorphic or rarely (*Coris*) somewhat zygomorphic, mostly 5-merous, often heterostylic. Sepals 3–6, mostly five, free or basally connate, imbricate, valvate or contorted, often glandular-punctate, usually persistent. Corolla sympetalous, with imbricate lobes, rarely petals secondarily free (in 3-merous genus *Pelletiera*) or lacking (*Glaux*). Stamens as many as and opposite the corolla lobes, attached to the corolla tube or rarely almost free; sometimes there is also an outer cycle of scalelike staminodia (as in Theophrastaceae and *Samolus* and *Soldanella*) alternating with the functional stamens; anthers tetrasporangiate, sometimes with a prolonged connective, introrse, opening longitudinally, or by apical pores. Tapetum secretory. Microsporogenesis simultaneous. Pollen grains 2-celled, (3-)-5-8-colpate, or 3(-5)-colporate, tectate-columellate, with tectum finely granular to perforate. Gynoecium mostly of five united carpels, with a style terminated by a capitate or punctate, sometimes shallowly lobed stigma; ovary superior, sometimes semi-inferior, unilocular but often with vestigial partitions at the base, mostly with numerous ovules on a free-central placenta. Ovules anatropous or hemitropous, sometimes campylotropous, bitegmic or rarely unitegmic (as in *Aegiceras, Cyclamen* and *Douglassia*), tenuinucellate, with endothelium. Integumentary tapetum present or less often (some Myrsinaceae and Theophrastaceae, *Coris* and *Cyclamen*) absent. Female gametophyte of *Polygonum*-type. Endosperm nuclear. Fruits berries, drupes or capsules. Seeds large to small, with straight embryo and usually copious endosperm; seed coat formed by the outer integument. Commonly containing various saponine compounds, and many genera also produce quinoid compounds, notably primulagenin. n = 5, 8–15, 17, 19, 22.

Primulales have many common features with the Ericales and Sapotales (confirmed by serological data, Frohne and John [1978]), which supports their common ancestry. All this groups originated from the some thealean ancestor. Hallier (1912) derived Primulales from the Ochnaceae. But in Hutchinson's opinion (1959, 1969) the Myrsinaceae (but not the order Primulales) linked with the Sapotaceae.

Key to Families

1 Fruits not viviparous.
  2 Flowers actinomorphic. Calyx not spiny.
    3 Mostly evergreen trees and shrubs, sometimes subshrubs, very rarely subherbaceous.
      4 Plants with secretory cells, cavities, or canals with yellow or reddish-brown contents in the stem and leaves. Staminodia absent.
        5 Ovary superior. Pedicel without bracteoles. Vessels with simple or occasionally scalariform perforations with 2–5 bars. Fibers typically septate. Nodes trilacunar with three traces. Leaves gland-dotted. Sclerenchymatous subepidermal fibers and submarginal strands in leaves wanting. Flowers small, without staminodia. Sepals 5-4(3-6, very rarely 9). Petals 5-4(3-6, very rarely 7). Anthers introrse, sometimes transversely locellate (*Ardisia elliptica*). Gynoecium of 3–5(6) carpels. Fruits usually berries or drupes, rarely capsule, opening with valves. Seeds relatively small to relatively large; seed coat undistinguished, tegmen thickened before becoming crushed, endotegmen crystalliferous; embryo straight or slightly curved; endosperm copious, bony, sometimes ruminate, rarely well preserved only on the funicular and placental side. Mostly evergreen small trees or shrubs, sometimes lianas or epiphytes, rarely subshrubs and very rarely (*Ardisia primulifolia*) subherbaceous. Present benzoquinones, n = 10–13, 15, 17, 23. . . 1. MYRSINACEAE.
        5 Ovary semi-inferior. Pedicel with a single bract and two bracteoles. Evergreen small trees, shrubs or sometimes lianas; secretory canals well developed. Vessels with simple perforation, but sometimes perforation scalariform with 5–7 bars; axial parenchyma scanty paratracheal (Lens et al. 2005). Leaves without immersed glandular hairs, prismatic crystal druses present. Inflorescence terminal or axillary, paniculate or racemose. Flowers small. Corolla campanulate or sometimes urceolate, with a well-demarcated tube. Stamens not exerted, adnate to the corolla at about the middle or higher. Anthers not trans-

versely locellate, dorsifixed. Gynoecium with somatic nectaries, stigma truncate or capitate, entire or 2–5-lobed. Ovules many to numerous, densely set on the placenta and have an inner integument with a tapetum. Fruits many-seeded baccate, formed by the ovary as well as the surrounding tissue, with a somewhat woody endocarp, externally punctate or lineate. Seeds small, testa reticulate, two-layered with rhomboid crystals; embryo with short hypocotyl and short, narrow cotyledons; endosperm copious. n = 10. . . . . 2. MAESACEAE.

4 Plants without secretory system in the stem and leaves. Vessels exclusively with simple perforations; rays broad. Fibers not septate; rays exclusively multiseriate. Nodes unilacunar with one trace (*Jacquinia*) or pentalacunar with five traces (*Clavija*). Leaves commonly crowded at the ends of the branches, with long strands of sclerenchymatous tissue beneath the epidermis of the leaves, estipulate. Flowers rather large, bisexual, or unisexual (dioecious), with 4–5 petaloid or glandular staminodia alternating with the corolla lobes and attached to the corolla tube. Sepals 5(4), free or basally connate (*Clavija*), persistent, imbricate. Petals 5(4), corolla rotate, urceolate or funnel-shaped, somewhat fleshy, gland dotted or streked. Stamens 5(-4), inserted near the base of the corolla tube; anther extrorse or introrse, mostly with a prolonged connective. Staminodia present. Pollen grains 3-colporate or (*Clavija*) 4-colporate. Gynoecium of 5(-4) carpels. Ovary superior or semi-inferior, unilocunar, the cavity filled with mucilage. Ovules numerous, imbedded in mucilage, on central or basal placenta, ascending, anatropous or campylotropous. Fruits berries or rarely drupes, or rarely (*Samolus*) capsules, containing many to few seeds (very rarely only one). Seeds often fairly large; exotestal cells flattened, thick-walled, hypodermal cells often crystalliferous; embryo straight or slightly curved; endosperm copious, bony; testa yellow, orange to red,. n = 9, 12–14, 19. . . . . 4. THEOPHRASTACEAE.

3 Perennial or annual herbs, rarely subshrubs. Nodes unilacunar. Leaves alternate or more often opposite or whorled, often all basal, often glandular-punctate, mostly simple and entire or merely toothed or lobed, in the aquatic genus *Hottonia* pectinately pinnatisect. Stomata anomocytic. Flowers variously in panicles, umbels, racemes, of heads or solitary, individually bracteate but without bracteoles. Sepals 5(4-9), connate, persistent, imbricate, or contorted; petals 5(4-9), (0 in *Glaux*), connate, imbricate or contorted; petals deeply bifid to bilobed, or entire. Stamens 5(4-9), inserted near the base of the corolla tube, or midway down the corolla tube, or in the throat of the corolla tube; filaments long or short, rarely alternating with staminodes (scales); anthers dorsifixed, opening by longitudinal slits or sometimes by terminal pores. Pollen grains 3(2-4)-colporate or colporoidate, or 3–10-colpate. Gynoecium of 5(4-9) carpels. Ovary superior, 1-locular. Ovules numerous, ascending, anatropous to hemianatropous. Fruits capsular, dehiscent by valves or sometimes circumscissile, or rarely indehiscent. Seeds small, usually more or less numerous; seed coat rather thin, formed by both the integuments, endotegmen often crystalliferous; embryo straight, surrounded by usually copious, oily endosperm, rarely (*Soldanella, Hottonia*) endosperm lacking. Cotyledons one (*Cyclamen*) or two, semi-cylindric. Present flavonols (kaempferol, quercetin, myricetin), cucurbitacins, n = 8–12(13). . . . . 5. PRIMULACEAE.

2 Flowers zygomorphic. Calyx spiny. Small, evergreen, thyme-like subshrubs, stem often red-dish-tinged. Leaves alternate, linear, coriaceous, persistent, entire or obscurely dentate, gland-dotted. Flowers in terminal spike-like racemes. Sepals five, membranous, 10-nerved, teeth short, persistent, valvate, each bearing a large black dorsal gland, and an outer ring of 10–15 spreading aculei below the teeth. Corolla campanulate, sub-bilabiate (three upper lobes longer), lobes unequally bifid, bright magenta to rose or white. Stamens five, antepetalous, filaments filiform, exerted, glandular at the base; anthers small, globose, introrse. Pollen grains relatively large and have a unique kind of reticulate pattern in addition to the tectum perforations (Carrion et al. 1993). Gynoecium of five carpels. Ovary superiorn 1-locular, ovules 5–6, hemianatropous. Style filiform. Capsule globose, enclosed by the persistent calyx,

dehiscent by five valves. Seeds 4–6, papillose; embryo well differentiated, straight; endosperm present. Contain C-glycosyl flavonoid in the leaves (Frezet et al. 1975). . . . . . . 6. CORIDACEAE.

1 Fruits viviparous. Mangrove shrubs or small trees. Vessel elements storied (Lens et al. 2005). Leaves alternate, coriaceous, entire, minutely punctuate, estipulate. Flowers in sessile terminal or axillary umbels of very short branches racemes, without bracts and bracteoles, bisexual. Sepals five, strongly and dextrorsely contorted, imbricate, very asymmetric; corolla shortly tubular, lobes five, contorted. Stamens five, opposite the corolla-lobes, inserted in the densely hairy tube of the corolla; filaments shortly connate, villous in the lower part; anthers 2-locular, introrse, opening longitudinally. Ovary superior, 1-locular, fusiform, narrowed into a subulate style, stigma acute. Ovules numerous (30–100), but only one fertile, anatropous, unitegmic; endosperm haustoria present Fruit 1-seeded, capsule, curved and elongate. Seed large, pachychalazal; seed-coat thin; endosperm present (Nemirovich-Danchenko 1992: 71); embryo well differentiated, n = 23. . . . . . . 3. AEGICERATACEAE.

1. MYRSINACEAE

R. Brown 1810 (including Ardisiaceae Jussieu 1810, Embeliaceae J. Agardh 1858). 38/1000. Mainly tropical and subtropical regions, extending to South Africa and New Zealand in the south, and North China, Korea, Japan, and Mexico in the north.

Embelia, Grenacheria, Heberdenia, Pleiomeris, Myrsine, Rapanea, Ardisia, Afrardisia, Synardisia, Hymenandra, Tetrardisia, Solonia, Geissanthus, Emblemantha, Sadiria, Gentlea, Stylogyne, Ctenardisia, Yunckeria, Antistrophe, Parathesis, Amblyanthus, Amblyanthopsis, Elingamita, Wallenia, Loheria, Cybianthus, Grammadenia, Vegaea, Oncostemum, Badula, Tapeinosperma, Discocalyx, Labisia, Systellantha, Monoporus, Fittingia, Conandrium.

2. MAESACEAE

A. Anderberg, B Stähl and M Källersjö 2000. 1/c.150. trop. Africa, Madagascar, India, Sri Lanka, Burma, China, Japan, Indochina, Malesia, Australia, Melanesia eastwards to Fiji, Samoa, and Tonga.

Maesa.

Anderberg and Stahl (1995) came to the definite conclusion that *Maesa* should be raised to the rank of family, which is supported by both morphological and molecular date (see Anderberg et al. 2000).

3. AEGICERATACEAE

Blume 1833. 1/2. Mangroves South and South-East Asia, New Guinea, Bismark Archipelago, Solomon Is., northeastern Australia, Lord Howe I.

Aegiceras.

Aegicerataceae are related to Myrsinaceae but markedly differ from them in the presence of versatile anthers, viviparous fruits with exalbumbinous seeds, and unitegmic ovules (Stähl and Anderberg 2004), and very important are also differences in the wood anatomy, such as "*Aegiceras* can easily be distinguished from other Myrsinaceae by the presence of relatively narrow vessels, a relatively high vessel density, short vessel elements and fibers (non-septate fibers), and multiseriate rays with exclusively procumbent body ray cells and without sheath cells" (Lens et al. 2005).

4. THEOPHRASTACEAE

Link 1829. 7/125. America from Mexico and southern Florida to West Indies and northern Paraguay.

Theophrasta, Neomezia, Deherainia, Votschia, Clavija, Bonellia, Jacquinia.

Related to the Myrsinaceae.

5. PRIMULACEAE

Batsch ex Borkhausen 1797 (including Anagallidaceae Batsch ex Borkhausen 1797, Lysimachiaceae A.L. Jussieu 1789, Samolaceae Rafinesque 1820). 24/1000. Nearly cosmopolitan, but mainly in temperate and cold regions of the Northern Hemisphere.

5.1 PRIMULOIDEAE

Ovary superior, capsule valvular, corolla lobes imbricate orquineuncialinthebud.–PRIMULEAE: *Omphalogramma, Bryocarpum, Primula, Dionysia, Cortusa, Kaufmannia, Hottonia, Sredinskya, Dodecatheon, Soldanella;* ANDROSACEAE: *Androsace* (including *Douglasia*), *Vitaliana, Stimpsonia, Pomatosace;* ARDISIANDREAE: *Ardisiandra*; SAMOLEAE: *Samolus.*

5.2 LYSIMACHIOIDEAE

Ovary superior, capsule valvular or a pyxidium, corolla-lobes contorted in bud. – LYSIMACHIEAE: *Lysimachia* (including *Naumburgia*), *Trientalis, Asterolinon, Pelletiera*; GLAUCEAE: *Glaux;* ANAGALLIDEAE: *Anagallis* (including *Centunculus*).

5.3 CYCLAMINOIDEAE

Ovary superior, capsule valvular. Tuberous plants; flowers with reflexed petal-limbs, twisted in bud. – *Cyclamen.*

Related to the Myrsinaceae, differing mainly in predominantly herbaceous habit, capsular fruits, seed coat formed by both the integuments, and Caryophyllad-type embryogenesis (Onagrad-type in the Myrsinaceae). Although the Primulaceae are in general more advanced than Myrsinaceae and Theophrastaceae, in some respects they are more archaic (Wettstein [1935] considered Myrsinaceae as an advanced family in the Primulales). We may therefore conclude that the Primulaceae are derived not directly from the Myrsinaceae or Theophrastaceae but rather from the common ancestor of these three families (Takhtajan 1959). The genus *Cyclamen* is the most distinctive in the Primulaceae and the systematic position of *Cyclamen* was discussed by Rosvik (1966), who found it isolated in the family. *Cyclamen* differs from all of them by the absence of the integumentary tapetum.

6. CORIDACEAE

J.G. Agardh 1858. 1/2. Mediterranean, Northeastern Africa.

Coris.

Usually included in the Primulaceae, but differs in the zygomorphic flowers, spinose sepals and the majority of them (except for *Cyclamen*) by the absence in the integumentary tapetum.

Bibliography

Airy Shaw HK. 1951 *Coris* (Primulaceae?) in Somaliland. Kew Bull. 1951: 29–31.

Anderberg AA. 1993. Cladistic interrelationships and major clades of the Ericales. Plant Syst. Evol. 184: 207–231.

Anderberg AA. 1994. Phylogeny and character evolution in *Cyclamen* L. (Primulaceae). Kew Bull. 49: 455–467.

Anderberg AA. 2004. Primulaceae. In: K Kubitzki, ed. The families and genera of vascular plants, vol. 6, pp. 313–319. Springer, Berlin/Heidelberg/New York.

Anderberg AA and B Ståhl. 1995. Phylogenetic interrelationships in the order Primulales, with special emphasis on the family circumscriptions. Canad. J. Bot. 73: 1699–1730.

Anderberg AA, B Ståhl, and M Källersjö. 1998. Phylogenetic relationships in the Primulales inferred from *rbc*L sequence data. Plant Syst. Evol. 211: 93–102.

Anderberg AA, I Trift, and M Källersjö. 1998. On the systematic position of the genus *Coris* (Primulaceae). Nord. J. Bot. 18: 203–207.

Anderberg AA, B Ståhl, and M Källersjö. 2000. Maesaceae, a new primuloid family in the order Ericales s.l. Taxon 49: 183–187.

Anderberg AA, I Trift, and M Källersjö. 2000. Phylogeny of *Cyclamen* L. (Primulaceae): evidence from morphology and sequence data from the internal transcribed spacers of nuclear ribosomal DNA. Plant Syst. Evol. 220: 147–160.

Anderberg AA, C-I Peng, I Trift, and M Källersjö. 2001. The *Stimpsonia* problem: evidence from DNA sequences of plastid genes *atp*B, *ndh*F and *rbc*L. Bot. Jahrb. Syst. 123: 369–376.

Carey G and L Fraser. 1932. The embryology and seedling development of *Aegiceras majus* Gaertn. Proc. Linn. Soc. N. S. W. 57: 341–360.

Caris P, LP Ronse Decraene, E Smets, and D Clinckemaillie. 2000 Floral development of three *Maesa* species, with special emphasis on the position of the genus within Primulales. Ann. Bot. 86: 87–97.

Caris PL and EF Smets. 2004. A floral ontogenetic study of the sister group relationship between the genus *Samolus* (Primulaceae) and the Theophrastaceae. Am. J. Bot. 91: 627–643.

Carrasquel N. 1970. Estudios anátomo-morfológicos de las especies del género *Jacquinia* en Venezuela para su interpretación taxonómica. Acta Bot. Venezuel. 4: 303–357.

Carrion JS, MJ Delgado, and M Garcia. 1993. Pollen grain morphology of *Caris* (Primulaceae). Plant Syst. Evol. 184: 89–100.

Channell RB and CE Wood. 1959. The genera of the Primulales of the southeastern United States. J. Arnold Arbor. 40: 268–288.

Clinckemaillie D and EF Smets. 1992. Floral similarities between Plumbaginaceae and Primulaceae: systematic significance. Belg. J. Bot. 125: 151–153.

Conti E, E Suring, D Boyd, J Jorgensen, J Grant, and S Kelso. 2000. Phylogenetic relationships and character evolution in *Primula* L.: the usefulness of ITS sequence data. Plant Biosyst. 134: 385–392.

Dahlgren KVO. 1916. Zytologische und embryologische Studien über die Reihen Primulales und Plumbaginales, Kongl. Svenska Vetenskapsaked. Haandl. 56(4): 1–80.

Decraene Ronse LP, EF Smets, and D Clinckemaillie. 1995. The floral development and floral anatomy of *Coris monspeliensis*. Canad. J. Bot. 73: 1687–1698.

Decrock E. 1901. Anatomie des Primulacées. Ann. Sci. Nat. Bot. Biol. Veg. 13: 1–199.

Dickson J. 1936. Studies in floral anatomy: III. An interpretation of the gynoecium in the Primulaceae. Am. J. Bot. 23: 385–393.

Douglas GE. 1936. Studies in the vascular anatomy of the Primulaceae. Am. J. Bot. 23: 199–212.

Faure P. 1968. Contribution à l'étude caryo-taxonomique des Myrsinacées et des Théophrastacées. Mem. Mus. Hist. Nat. Sér. B Bot. 18: 37–57.

Frezet C, J Raynaud, and M-L Bouillant. 1975. Sur la présence de la di-C-glucosyl-6,8 apigénine chez *Coris monspeliensis*

(Primulaceae). Compte Rendu Acad. Sci. Paris 280: 1079–1081.

Frohne D and J John. 1978. The Primulales: serological contributions to the problem of their systematic position. Biochem. Syst. Ecol. 6: 315–322.

Gardner IC, IM Miller, and A Scott. 1981. The fine structure of the leaf nodules of *Ardisia cripsa* (Thunb.) A. D. C. (Myrsinaceae). Bot. J. Linn. Soc. 83: 93–102.

Geetha K, I Umadevi, and M Daniel. 1993. Primulales: a reassessment of the taxonomy and phylogeny of the group. Feddes Repert. 104: 67–71.

Grey-Wilson C. 1988. The genus *Cyclamen* (Primulaceae). Portland.

Hao F, Y-M Yuan, CM Hu, J-X Ge, and N-X Zhao. 2004. Molecular phylogeny of *Lysimachia* (Myrsinaceae) based on chloroplast *trn*L-F and nuclear ribosomal ITS sequences. Molec. Phylogen. Evol. 31: 323–339.

Källersjö M and B Ståhl. 2003. Phylogeny of Theophrastaceae (Ericales s. lat.). Int. J. Plant Sci. 164: 579–591.

Källersjö M, G Bergqvist, and AA Anderberg. 2000. Generic realignment in primuloid families of the Ericales s.l.: a phylogenetic analysis based on DNA sequences from three chloroplast genes and morphology. Am. J. Bot. 87: 1325–1341.

Knudsen JT and B Stahl. 1994. Floral odours in the Theophrastaceae. Biochem. Ecol. Syst. 22: 259–268.

Lens F, S Jansen, P Caris, L Serlet, and E Smets. 2005. Comparative wood anatomy of the Primuloid clade (Ericales s.l.). Syst. Bot. 30: 163–183.

Lys J. 1955. Sur la nature et 1϶εϖολυτιον biochimique des glucides des Lysimaques. Compt. Rend. Hebd. Seances Acad. Sci. 241: 1842–1844.

Lys J. 1956. Les glucides de quelques Primulacees. Rev. Gen. Bot. 63: 95–100.

Ma OSW and RMK Saunders. 2003. Comparative floral ontogeny of *Maesa* (Maesaceae), *Aegiceras* (Myrsinaceae) and *Embelia* (Myrsinaceae) taxonomic and phylogenetic implications. Plant Syst. Evol. 243: 39–58.

Manns U and AA Anderberg. 2005. Molecular phylogeny of *Anagallis* (Myrsinaceae) based on ITS, *trn*L-F, and *ndh*F sequence data. Int. J. Plant Sci. 166: 1019–1028.

Martins L, C Oberprieler, and FH Hellwig. 2003. A phylogenetic analysis of Primulaceae s.l. based on internal transcribed spacer (ITS) DNA sequence data. Plant Syst. Evol. 237: 75–85.

Mast AR, S Kelso, AJ Richards, DJ Lang, DMS Feller, and E Conti. 2001. Phylogenetic relationships in *Primula* L. and related genera (Primulaceae) based on noncoding chloroplast DNA. Int. J. Plant Sci. 162: 1381–1400.

Mauritzon J. 1936. Embryologische Angaben über Theophrastaceen. Ark. Bot. 28B: 1–4.

Mez C. 1902. Myrsinaceae. In: A. Engler, ed. Das Pflanzenreich, vol. 9 (IV. 236), pp. 1–437. Engelmann, Leipzig.

Moeliono BM. 1966. De caulomatische Oorsprong van Zaadknoppen bij Caryophyllaceën en Primulaceën (een histogenetische en anatomische Analyse). Amsterdam.

Nemirovich-Danchenko EN. 1992. Theophrastaceae, Myrsinaceae, Primulaceae, Aegicerataceae. In: A. Takhtajan, ed. Comparative seed anatomy, vol. 4, pp. 54–77. Nauka, St. Petersburg (in Russian).

Nowicke JW and JJ Skvarla. 1977. Pollen morphology and the relationship of the Plumbaginaceae, Polygonaceae, and Primulaceae to the order Centrospermae. Smithsonian Contr. Bot. 37: 1–64.

Otegui M and A Cocucci. 1999. Flower morphology and biology of *Myrsine laetevirens,* structural and evolutionary implications of anemophily in Myrsinaceae. Nord. J. Bot. 19: 71–85.

Otegui M and S Maldonado. 1998. Embryological features and bacterial transmission to gynoecium and ovule in *Myrsine laetevirens* (Myrsinaceae). Acta Bot. Neerl. 47: 185–194.

Ronse Decraene LP, EF Smets, and D Clinckemaillie. 1995. The floral development and floral anatomy of *Coris monspeliensis*. Canad. J. Bot. 73: 1687–1698.

Røsvik A. 1966. On the taxonomic position of the genera *Dodecatheon* L. and *Cyclamen* L. within Primulaceae. Arb. Univ. Bergen Mat. Nat. Ser. 5: 1–16.

Røsvik A. 1968. Investigations on petal epidermis and its bearings on taxonomy in Primulaceae. Arb. Univ. Bergen Mat. Nat. Ser. 3: 1–32.

Roth I. 1959. Histogenese und morphologische Deutung der Kronblatter von *Primula*. Bot. Jahrb. Syst. 79: 1–16.

Roth I. 1959. Histogenese und morphologische Deutung der Plazenta von *Primula*. Flora 148: 129–152.

Sattler R. 1962. Zur frühen Infloreszenz- und Blütenentwicklung der Primulales sensu lato mit besonderer Berücksichtigung der Stamen-Petalum-Entwicklung. Bot. Jahrb. Syst. 81: 358–396.

Schaeppi H. 1937. Vergleichend-morphologische Untersuchungen am Gynoeceum der Primulaceen. Z. Gesamte Naturwiss. (Brunswick) 3: 239–250.

Schwarz O. 1955. Systematische Monographie der Gattung *Cyclamen* L. 1. Feddes Repert. 58: 234–283.

Schwarz O and L Lepper. 1964. Kritische Revision der Gattung. In: O Schwarz, ed. Systematische Monographie der Gattung *Cyclamen* L. 2. Feddes Repert. 69: 79–92.

Souéges R. 1937. Embryogénie des Primulacées: Developpement de 1϶εμβρψον chez le *Samolus Valerandi* L. Compt. Rend. Hebd. Séances Acad. Sci. 204: 145–147.

Spanowsky W. 1962. Die Bedeutung der Pollenmorpho-logie für die Taxonomie der Primulaceae-Primuloideae. Feddes Repert. Spec. Nov. Regni Veg. 65: 149–214.

Ståhl B. 1987. The genus *Theophrasta* (Theophrastaceae): foliar structures, floral biology, and taxonomy. Nord. J. Bot. 7: 529–538.

Ståhl B. 1989. A synopsis of Central American Theophrastaceae. Nord. J Bot. 9: 15–30.

Ståhl B. 1991. A revision of the *Clavija* (Theophrastaceae). Opera Bot. 107: 1–77.

Ståhl B. 1993. *Votschia*, a new genus of Theophrastaceae from northeastern Panama. Brittonia 45: 204–207.

Ståhl B. 2004a. Samolaceae. In: K Kubitzki, ed. The families and genera of vascular plants, vol. 6, pp. 387–389. Springer, Berlin/Heidelberg/New York.

Ståhl B. 2004b. Theophrastaceae. In: K Kubitzki, ed. The families and genera of vascular plants, vol. 6, pp. 472–478. Springer, Berlin/Heidelberg/New York.

Ståhl B and AA Anderberg. 2004a. Maesaceae. In: K Kubitzki, ed. The families and genera of vascular plants, vol. 6, pp. 255–257. Springer, Berlin/Heidelberg/New York.

Ståhl B and AA Anderberg. 2004b. Myrsinaceae. In: K Kubitzki, ed. The families and genera of vascular plants, 6: 266–281. Springer, Berlin/Heidelberg/New York.

Stone BC. 1992. *Systellantha*: a new genus of Myrsinaceae from Borneo. Malayan Nat. J. 46: 13–24.
Subramanyam K and LL Narayana. 1976. A contribution to the floral anatomy and embryology in certain members of Primulaceae. J. Indian Bot. Soc. 55: 274–282.
Sundberg MD. 1982. Petal-stamen initiation in the genus *Cyclamen* (Primulaceae). Am. J. Bot. 69: 1707–1709.
Thomson HJ. 1953. The biosystematics of *Dodecatheon*. Contr. Dudley Herb. 4: 73–154.
Trift I, M Källersjö, and AA Anderberg. 2002. The monophyly of *Primula* evaluated by analysis of sequences from the chloroplast gene *rbc*L. Syst. Bot. 27: 396–407.
Utteridge TMA and RMK Saunders. 2001. Sexual dimorphism and functional dioecy in *Maesa perlarius* and *M. japonica* (Maesaceae/Myrsinaceae). Biotropica 33: 368–374.
Walker EH. 1940. A revision of the eastern Asiatic Myrsinaceae. Philipp. J. Sci. 73: 1–258.
Wendelbo P. 1961. Studies in Primulaceae. III. On the genera related to *Primula* with special reference to their pollen morphology. Ärbok for Universitetet I Bergen, Matematisk-Naturvitenskapelig Serie 19: 1–31.
Zhang L-B, HP Comes, and JW Kadereit. 2001. Phylogeny and quaternary history of the European montane/alpine endemic *Soldanella* (Primulaceae) based on ITS and AFLP variation. Am. J. Bot. 88: 2331–2345.

Superorder VIOLANAE

Order 64. BERBERIDOPSIDALES

Scandent, evergreen shrubs with sympodial branching and collateral axillary buds. Vessel elements very long to extremely long; perforations exclusively scalariform with 6–15 or more bars (mostly over 15 in *Berberidopsis*), lateral pitting almost absent; vessel-to-ray pitting scalariform, vessel to tracheid pitting composed of axial rows of circular pits. Fibrous elements are nonseptate, with distinctly bordered pits. Rays heterogeneous, over 1 cm in height. Axial parenchyma inconspicuous but present. Sieve-element plastids of S-type. Nodes trilacunar with three traces. Leaves alternate, entire to coarsely dentate, petiolate, coriaceous, estipulate. Stomata cyclocytic and bicyclis. Flowers in terminal racemes (*Berberidopsis*) or axillary, solitary, bisexual, actinomorphic. Sepals and petals (9-)12(-15), and spirally arranged, gradually merging from one to the other (*Berberidopsis*) or the perianth with five petals distinct from the five sepals (*Streptothamnus*). Stamens 6–13 (*Berberidopsis*) or numerous (*Streptothamnus*), free. Filaments very short, connective broad (*Berberidopsis*) or filaments longer than the anthers, filiform, connective inconspicuous (*Streptothamnus*). Anthers basifixed, shortly apiculate, opening longitudinally. Tapetum secretory. Microsporogenesis simultaneous. Pollen grains 3-colpate, tectum imperforate (*Berberidopsis*), or perforate (*Streptothamnus*). Disc dentate (*Berberidopsis*) or absent (*Streptothamnus*). Gynoecium of three, five carpels. Ovary superior, 1-locular, ovules two to many on each of 3–5 parietal placentae. Style stout, stigma punctuate to slightly lobed. Ovules anatropous, bitegmic, crassinucellate. Female gametophyte of *Polygonum*-type. Endosperm nuclear. Fruits berrylike crowned by the persistent style base. Seeds endotestal, with fleshy (*Berberidopsis*) or leathery (*Streptothamnus*) exotesta; endotestal cells crystalliferous, palisade, lignified; embryo minute and endosperm copious. Producing cyclopentenoid cyanogenic glycosides.

Evidently a very ancient family as confirmed by its transpacific disjunction (*Berberidopsis* in Chile and *Streptothamnus* in Australia). Berberidopsidaceae are probably directly related to the Dilleniales and are possible a link between the Dilleniales and Violales.

1. BERBERIDOPSIDACEAE

Takhtajan 1985. 2/3. Chile (from Talca to Osorno) and eastern Australia (Queensland and New South Wales).
Berberidopsis, Streptothamnus.

Bibliography

Baas P. 1984. Vegetative anatomy and taxonomy of *Berberidopsis* and *Streptothamnus* (Flacourtiaceae). Blumea 30: 39–44.
Carlquist S. 2003. Wood anatomy of Aextoxicaceae and Berberidopsidaceae is compatible with their inclusion in Berberidopsidales. Syst. Bot. 28: 317–325.
Heel WA. van. 1984. Flowers and fruits in Flacourtiaceae. V. The seed anatomy and pollen morphology of *Berberidopsis* and *Streptothamnus*. Blumea 30: 31–37.
Jaroszewski JW, PS Jensen, C Cornett, and JR Byberg. 1988. Occurrence of lotaustralin in *Berberidopsis* Beckeri and its relation to the chemical evolution of Flacourtiaceae. Biochem. Syst. Ecol. 16: 23–28.
Kubitzki K. 2007. Berberidopsidaceae. In: K Kubitzki, ed. The families and genera of vascular plants, vol. 9, pp. 33–35. Springer, Berlin/Heidelberg/New York.
Ronse Decraene L-P. 2004. Floral development in *Berberidopsis corallina*: a crucial link in the evolution of flowers in the core eudicots. Ann. Bot. N. S. 94: 741–751.
Trifonova VI. 1992. Berberidopsidaceae. In: AL Takhtajan, ed. Comparative seed anatomy, vol. 4, p. 78. Nauka, St. Petersburg (in Russian).

Van Heel WA. 1979. Flowers and fruits in Flacourtiaceae IV. *Hydnocarpus, Kiggelaria africana* L., *Casearia, Berberidopsis corallina* Hook. f. Blumea 25: 513–529.
Van Heel WA. 1984. Flowers and fruits in Flacourtiaceae. V. The seed anatomy and pollen morphology of *Berberidopsis* and *Streptothamnus*. Blumea 30: 31–37.
Veldkamp JF. 1984. *Berberidopsis* (Flacourtiaceae) in Australia. Blumea 30: 21–29.

Order 65. AEXTOXICALES

Large, evergreen trees. Twigs, leaves (below), and inflorescences lepidote, with peltate scales all over. Vessels with scalariform perforations that have rather numerous bordered bars. Axial parenchyma diffuse, moderately common. Rays heterogeneous, droplets or amorphous deposits of a dark-staining substance common in ray cells (especially procumbent cells). Leaves opposite, simple, conduplicate, entire, without stipules. Stomata encyclocytic. Flowers in axillary, pendulous racemes, dioecious, (4)5(6)-merous, completely enveloped in bud by a calyptrate bracteole. Sepals free, strongly imbricate. Petals broadly clawed, incurved in buds, with a thick midrib, imbricate. Stamens five, antesepalous, alternating with well-developed, fleshy, reniform nectary glands; filaments thick and fleshy; anthers introrse, opening by porelike slits toward the summit. Pollen grains 3-colporate. Tapetum secretory. Microsporogenesis simultaneous. Female flowers with five fleshy staminodia alternating with the nectary glands. Gynoecium of two carpels, with a short style strongly deflexed to one side and appressed to the ovary, shortly bifid; ovary superior, 2-locular, but only one loculus fertile. Ovules two, pendulous, anatropous and apotropous, with a massively beaked nucellus protruding beyond the integuments, bitegmic, crassinucellate. Fruits dry, small drupes with a single 1-seeded pyrene. Seeds with ruminate, copious, fleshy endosperm and a well-developed embryo that has flattened, cordate-orbicular cotyledons. n = 16.

Both molecular data and wood anatomy show the close affinity with the Berberidopsidales. However Aextoxicales markedly differ from the Berberidopsidales in opposite leaves, their dioecious flowers, 3-colporate pollen grains, the 2-carpellate gynoecium, drupaceous fruits, ruminate endosperm, and well-developed embryo.

1. AEXTOXICACEAE

Engler et Gilg 1920. 1/1. Chile.
Aextoxicon.

Bibliography

Carlquist S. 2003. Wood anatomy of Aextoxicaceae and Berberidopsidaceae is compatible with their inclusion in Berberidopsidales. Syst. Bot. 28: 317–325.
Hooker WJ. 1836. *Aextoxicum functatum*. Hooker's Icones Plantarum 1: t. 12.
Kubitzki K. 2006. Aextoxicaceae. In: K Kubitzki, ed. The families and genera of vascular plants, vol. 9, pp. 23–25. Springer, Berlin/Heidelberg/New York.
Komar GA. 1992. Aextroxicaceae. In: AL Takhtajan, ed. Comparative seed anatomy, vol. 4, p. 404. Nauka, St. Petersburg (in Russian).

Order 66. VIOLALES (PASSIFLORALES)

Trees, shrubs, and herbs, often lianas. Vessels mostly with simple perforations, but in some archaic members of the order some or all of them with scalariform perforations, sometimes with many bars; lateral pitting from scalariform to alternate or rarely (*Sayauxia*) scalariform and opposite. Fibers with bordered or simple pits, often septate. Rays from heterogeneous to homogeneous. Axial parenchyma of various types or absent. Sieve-element plastids of S-type. Nodes mostly trilacunar. Leaves alternate or opposite, rarely verticillate, simple, entire to dentate or serrate, sometimes (some Violaceae) lobed or even dissected, commonly pinnately veined; stipules present, often caducous, or absent. Stomata paracytic, anisocytic or anomocytic. Flowers small to large, mostly in various kinds of inflorescences or rarely solitary, bisexual or less often unisexual, actinomorphic or sometimes more or less zygomorphic, cyclic or rarely with a spiral perianth, mostly 5-merous, sometimes apetalous. Sepals 6–3, rarely up to 15, mostly 5, free or more or less connate, usually imbricate. Extrastaminal corona sometimes present. Petals 6–3, rarely up to 15, mostly 5, free or shortly connate at the base (in the Achariaceae connate into a campanulate 5–3-lobed tube), mostly imbricate. Stamens mostly the same number as the number of petals, free or more or less connate; anthers tetrasporangiate, opening longitudinally or rarely by terminal pores (as in *Kiggelaria*). Tapetum secretory. Microsporogenesis simultaneous. Pollen grains 2-celled, 3-colporate to 10–12-colporate, generally reticulate, rarely inaperturate. Gynoecium syncarpous, of 2–12 carpels, rarely of one carpel, with free or more or less united stylodia; ovary superior, rarely inferior (*Bembicia*) with numerous or less often several or only 1–2 ovules mostly on parietal placentas. Ovules mostly anatropous, bitegmic, crassinucellate. Female gametophyte of *Polygonum*-type. Endosperm nuclear. Fruits of various types, mostly capsules or berries. Seed of various sizes and shapes, arilate or not; seed coat formed by both integuments, mostly exotegmic; embryo straight to curved, endosperm copious to scanty. Certain members of the order are cyanogenic (Malesherbiaceae, Passifloraceae, Flacourtiaceae, and Turneraceae) and produce elaborate cyanogens with cyclopentenoid ring structures.

Violates are close to the Theales and share a common origin from a dillenialean ancestor.

Key to Families

1 Gynoecium of one carpel. Evergreen trees and shrubs with pericyclic cork. Vessels usually with exclusively scalariform perforations. Fibers with bordered pits, septate. Axial parenchyma rare to frequent, vasicentric and apotracheal diffuse. Nodes trilacunar with three traces. Leaves alternate, simple, entire, serrate or on the same specimen incised, estipulate. Stomata anisocytic. Flowers axillary, small, sessile or pedicellate, bisexual, actinomorphic. Sepals (4-)5(-6), broadly imbricate. Petals absent. Stamens 65–110, persistent, inserted outside floral cup; filaments are longer than anthers, narrow; anthers very small, basifixed, introrse; the connective broad and without protrusion, opening longitudinally. Nectary is located on the floral cup, between androecium and gynoecium. The carpel is elevated above the floral base on a short stalk; ovary superior, thick-walled, with one short lateral placenta. Stigma large, almost sessile forms two recurved crests and appears capitate-bilobed and somewhat decurrent on the ventral side. Ovules 6–8, biseriate, campylotropous and syntropous; placentation lateral; micropyle formed by the inner integument, which protrudes from the outer integument. Fruits baccate. Seeds incurved, with an hippocrepiform embryo and scanty endosperm. Contain mangiferin (Bate-Smith 1965)…. 12. APHLOIACEAE.

1 Gynoecium mostly of 2–5 carpels.

2 Perianth present.

3 Petals present.

4 Petals more or less free.

5 Flowers with an extrastaminal corona.

6 Seeds exarillate. Xerophitic shrubs up to 3.5 m tall, or small shrublets, perennial or annual herbs with foetid smell, erect or procumbent, bearing branches sparsely to densely pubescent with both simple and glandular hairs. Vessels usually with simple perforations, but sometimes scalariform with a few bars. Fibers rather short, with very small pits. Nodes unilacunar with one trace. Leaves alternate, simple, entire, sessile or petiolated, toothed or sometimes pinnatifid; stipule-like leaves usually present, often 1–5-lobed. Stomata anomocytic. Inflorescences axillary or terminal racemes, panicles or cymes, or flowers solitary. Flowers bisexual, actinomorphic, with a long, slender, straight or curved hypanthium, blue, purple, white, yellow, red to orange. Sepals five, persistent, imbricate or valvate; the margins sometimes lined with glandular hairs. Petals five, valvate or with open aestivation, corolla appendiculate, with a membranous, denticulate corona. Both the androecium and gynoecium borne on a villous, lobed androgynophore. Stamens five, free and arising from the apex of the androgynophore, extending beyond the throat of the floral tube and corona; filaments filiform; anthers 2-locular, dorsifixed. Pollen grains 3-colporate. Gynoecium of 3(-4) carpels; stylodia three, free, filiform, emerging near apex of ovary, generally extending stigmas beyond anthers. Ovary superior, 1-locular, stipitate, with numerous, anatropous ovules on parietal placentas. Fruits loculicidal capsules enclosed by the persistent hypanthium, stipitate. Seeds one to many, crustaceous, pitted seed coat and with straight, medium-sized embryo and

copious, fleshy, dotted endosperm. Producing cyclopentenoid cyanogens; n = 14........ 10. MALESHERBIACEAE.

6 Seeds arillate. Woody or herbaceous lianas, climbing by axillary tendrils that represent modified axes of the inflorescences or part of inflorescences, less often erect shrubs or trees. Woody stems often with anomalous secondary growth. Vessels mostly with simple perforations, but in *Soyauxia* exclusively with scalariform perforations, and in *Androsiphonia* and *Mitostemma glasiovii* with some amount of scalariform plates; lateral pitting usually alternate, but in *Soyauxia* scalariform and opposite. Fibers with bordered pits. Rays of various types. Axial parenchyma usually apotracheal. Nodes trilacunar. Leaves alternate, entire or often palmately lobed, rarely compound, often with nectary glands on the petiole; stipules often small, sometimes foliose, then to 7 cm long, often caducous. Stomata mostly anomocytic. Inflorescences axillary cymes or racemes, or cauliflorous, or the flowers solitary, rarely the middle flower replaced by a hook. Flowers bisexual or rarely unisexual, actinomorphic, mostly with flat or tubular hypanthium, often with elongate androgynophore. Sepals (3-)5(-8), free or sometimes basally connate, imbricate, persistent. Petals as many as sepals, free or basally connate (connate in tube around the gynoecium in *Androsiphonia*), imbricate, rarely wanting or very rudimentary. Extrastaminal corona (poorly developed in *Adenia*) of one or more series of filaments or scales present. Stamens mostly five, less often more (numerous in *Smeathmannia, Soyauxia*, and *Barteria*) or four, usually alternate with the petals; anthers dorsifixed, introrse; anther epidermis persistent. Pollen grains 3–12-colporate. Nectary disc of staminodial origin often present around the ovary. Gynoecium of (2)3(-5) carpels; stylodia mostly free or connate only at the base, each with a capitate to clavate or discoid stigma, rarely (*Barteria* and *Crossostemma*) stylodia united into a simple style; ovary superior, unilocular with more or less numerous, orthotropous to anatropous ovules borne on parietal placentas; micropyle bitegmic or zig-zag. Fruits loculicidal or irregularly decaying capsules or berries, rarely fleshy with irregular, apical dehiscence. Seeds usually much-compressed, with bony, pitted testa and mostly with a fleshy, apical aril; endotesta short-palisade, crystalliferous, lignified or not; embryo large, straight, spatulate, embedded in the copious, oily, rarely ruminate (*Passiflora foetida*) endosperm. Usually producing cyclopentenoid cyanohydrin glycosides dominated by transdioxygenes; alkaloids, ellagic and gallic acids present or absent, n = 6–7, 9–12.... 8. PASSIFLORACEAE.

5 Flowers without extrastaminal corona.

7 Stamens usually 5.

8 Ovules at the top of the columnar. Deciduous shrubs or small trees. Vessel elements narrow and long, with scalariform plates that have numerous (up to 20 or more) bars. Fibers with bordered pits. Lateral pitting scarce. Rays uniseriate, homogeneous. Axial parenchyma absent. Leaves alternate, simple, petiolate, entire, ovate, with small caduceus (lanceolate) stipules. Stomata anomocytic. Flowers in axillary, umbellate, long pedunculate, globose inflorescence, small, actinomorphic. Extrastaminal disc present; 5–7 glands opposite the corolla members. Sepals 5–7, linear, pubescent sepals shortly connate at the base, valvate, persistent. Petals 5–7, very similar to the sepals, valvate, persistent. Stamens as many as and alternate with the sepals, persistent, erect in bud, free;

anthers 2-locular, opening longitudinally. Staminodes 5–7, in the same series as the fertile stamens, or internal to the fertile stamens. Pollen grains 3-colporate. Gynoecium of three carpels with simple style terminated by small capitate stigma. Ovary superior, 1-locular above, perfectly 3-locular at the base, with two, ascending ovules per carpel but only one developing. Fruits tardily dehiscent 1-locular, tomentose, 1-seeded capsules. Seed small, oblong with a persistent style, surrounded by the persistent calyx, corolla and androecium; seed coat with palisade tissue underneath collapsed polygonal cells. . . . . . 6. DIPENTODONTACEAE.

8 Placentation parietal.

9 Petals 5.

10 Anthers basifixed. Shrubs or herbs (mostly perennial), less often small trees or scandent. Vessels with scalariform or (especially in herbs) simple perforations; lateral pitting from opposite to scalariform, sometimes alternate. Fibers with simple or bordered pits, septate. Axial parenchyma absent or very scanty. Leaves alternate or sometimes opposite (some species of *Hybanthus* and *Rinorea*) simple, entire or toothed, rarely lobed, with leafy or small stipules. Stomata paracytic or anisocytic. Inflorescences axillary racemes, spikes, panicles or the flowers solitary. Flowers actinomorphic or more or less zygomorphic, bisexual or rarely polygamous or dioecious. Sepals five, often persistent, imbricate, or open in bud. Petals five, imbricate or contorted, clawed or sessile, in zygomorphic flowers unequal, the lowest one usually prolonged behind into a spur. Stamens five (three in *Leonia triandra*), alternating with the corolla; filaments very short, free or more or less connate; anthers usually more or less connivent in a ring around the ovary, introrse, with a connective often prolonged into a membranous appendage, opening longitudinally. Pollen grains 3(-5)-colporate. Gynoecium of (2)3–5 carpels, with stylodia united into a style; the stigma simple or lobed. Ovary superior, unilocular, with 1–2 or many, anatropous ovules on (2)3–5 parietal placentas. Fruits loculicidal and valvular capsules or sometimes berries, rarely (*Leonia*) nutlike. Seeds winged or wingless, with straight, green (some species of *Viola*) embryo and flat cotyledons surrounded by copious, oily, rarely scanty endosperm, often arillate. Contain flavonols (kaempferol and quercetin), alkaloids present or absent, inulin recorder in *Ionidium*; n = 6–13,17, 21, 23. . . 5. VIOLACEAE.

10 Anthers dorsifixed. Herbs, shrubs, or rarely trees, erect or decumbent, usually with glandular or non-glandular trichomes. Vessels with simple or less often with simple or scalariform perforations; lateral pitting alternate. Rays heterogeneous or homogeneous. Axial parenchyma apotracheal. Leaves alternate, simple, entire or lobed, often with two glands or extrafloral nectaries at the back of the lamina; stipules small (*Erblichia*), or lacking (usually, but often with a pair of glands or extrafloral nectaries at the base of the blade). Stomata of various types. Inflorescences axillary racemes or cymes or flowers solitary. Flowers bisexual, actinomorphic, with a short to more often tubular hypanthium,

often subtended by two bracteoles. Calyx tubular, 5-toothed, teeth imbricate. Petals five, inserted on the calyx tube, free, clawed, contorted in bud, usually ephemeral. Hypanthium sometimes with fringed corona or five glands or lobes between corona and androecium. Stamens five, alternating with the corolla, free or partially attached in throat of hypanthium. Pollen grains in monads, usually 3-colporate. Gynoecium of three carpels; stylodia free, opposite the placentas, sometimes bifid, often with a fringed or brushlike commissural stigma; ovary superior or sometimes semi-inferior, 1-locular; ovules numerous on each parietal placenta, anatropous. Fruits capsules, opening loculicidally into three valves with the placenta in the middle of each. Seeds pitted, arillate, endotestal cells large, exotegmen palisade; embryo large, straight or weakly curved, endosperm copious, oily; cotyledons two, planoconvex. Producing cyclopentenoid cyanohydrins glycosides, linamarin, and sometimes alkaloids, n = 5, 7, 8(10), 13... 9. TURNERACEAE.

9 Petals six. Small dioecious trees with soft wood. Vessels with very oblique end walls but usually with simple perforations. Lateral pitting alternate. Fibers with strongly thickened walls and with large bordered pits and numerous thin cross-partitions. Rays heterogeneous. Axial parenchyma nearly or quite wanting. Nodes trilacunar. Leaves alternate, simple, distichous, serrate or dentate, with very small, caducous stipules. Stomata paracytic. Inflorescences terminal and axillary racemes or panicles. Flowers actinomorphic, 3-merous. Perianth in male flowers tubular, 6-lobed, the lobes in two series, imbricate (the outer series probably represents a calyx, and the inner one a corolla (Swamy 1953). Three large, fleshy nectary glands present between the perianth and the androecium. Stamens united into a column with clavate common prolonged apical connective. Anthers 2-locular, with a prolonged connective, extrorse, opening longitudinally. Pollen grains 3-colpate. Perianth of the female flowers of three free sepals alternating with three free petals. Gynoecium of 8–12 carpels. Stigma sessile, thick, discoid, with as many radiating ridges as carpels and with the aperture in the middle. Ovary superior, 1-locular, more or less septate near the summit but unilocular below, with numerous (30–100), basal, erect ovules on prominent funicles. Fruits fleshy loculicidal capsules, or ultimately dehiscing into 8–12 reflexed segments, enclosing numerous seeds. Seeds arillate, perisperm present in a very thin layer, embryo fairly large, straight; endosperm rather scanty, oily, cotyledons 2, n = 9....... 7. SCYPHOSTEGIACEAE.

7 Stamens usually more than five. Trees and shrubs, rarely with stellate indumentum (*Kiggelaria, Rayania*). Vessels with scalariform or more often simple perforations. Fibers with simple or bordered pits, usually septate. Axial parenchyma absent. Leaves alternate or rarely opposite (*Pseudoscolopia, Abatia, Apherema*), simple, often distichous and coriaceous, serrate or entire, sometimes pellucid-dotted or lined, stipules usually present, often caducous, rarely absent (*Abatia, Bembicia*). Stomata paracytic or anisocytic. Inflorescences axillary, terminal or below the leaves, racemes, cymes, rarely (*Bembicia*) in short axillary conelike heads surrounded by closely

imbricate leathery bracts, the outer ones empty; bracteoles 2-keeled, sometimes adnate to the midrib with the flowers epiphyllous (in *Phyllobotryon, Phylloclinium, Mocquerysia*). Flowers bisexual or unisexual. Perianth parts clearly differentiated into sepals and petals or rarely (Erythrospermeae) poorly differentiated. Sepals imbricate or rarely valvate, persistent. Petals present or absent (Flacourtieae, Casearieae, Bembicieae). Petals with an adaxial scale (in some genera of Erythrospermeae and in Pangieae). Stamens development centrifugally or centripetally, numerous or occasionally equal in number to the petals, the filaments distinct or grouped into fascicles opposite the petals. Anthers linear to about as long as wide, dorsifixed or basifixed, usually latrorse, sometimes the connective produced apically, opening usually longitudinally, or rarely (*Kigellaria*) via pores. Pollen grains 3(-6)-colporate. Prismatic crystals in ray cells usually present (except for *Poliothyrsis*). Hypogenous disc extrastaminal or intrastaminal, or separate members or annular (*Bembicia*). Gynoecium of 2–10 carpels. Ovary superior or rarely (*Bembicia*) inferior, with 2(3)–5(-10) placentas (sometimes appearing almost basal) or seldom placentas axile. Stylodia free or united. Stigma is not peltate, sometimes apical or rarely stigmas numerous (*Dendrostigma, Mayna*). Ovules numerous, pendulous, orthotropous or anatropous, or hemianatropous; outer integument contributing to the micropyle. Fruits various, dry or fleshy, capsule, drupes, or usually berries; in *Flacourtia* the drupes with separable pyrenes. Seeds exotegmic, often with thin or thick sarcotesta (Van Heel 1977), sometimes covered with long woolly hairs (*Calantica*) or with long silky hairs (*Bivinia* and *Gossypiospermum*), winged in *Poliothyrsis, Itoa*, and *Carrierea*; embryo straight, surrounded by oily, copious endosperm. Producing cyanogenic glycosides (Oncobeae, Pangieae, and occasionally Erythrospermeae and several species of *Banara* in Prockieae), cyclopentenyl fatty acids ('chaulmoogra oils') (Oncobeae, Pangieae, *Azara, Dovyalis, Flacouria* in Flacourtieae) (Spencea and Seigler 1985; Lemke 1988), sometimes producing alkaloids, and flavonols, n = 10–12. . . . . 1. FLACOURTIACEAE.

4 Flowers sympetalous, monoecious, solitary or in fascicles or racemes. Acaulescent (*Guthriea*) or climbing (*Ceratosicyos*) glabrous herbs or (*Acharia*) small subherbaceous pubescent shrublets. Vessels mostly with simple perforations. Fibers with simple pits, sometimes septate. Leaves alternate, sometimes radical, palmately lobed, or (*Guthriea*) unlobed, crenate or serrate, without stipules. Inflorescences axillary racemes, fascicles or the flowers solitary. Flowers tardily deciduous to deciduous, small, actinomorphic. Sepals 3–5, free or adnate to the corolla. Petals united into 3–5-lobed tube, valvate. Stamens 3–5, attached to the corolla tube near its base or at the throat; anthers basifixed, with a broadly expanded connective, introrse, opening longitudinally. Staminodes 3–5, short, fleshy, internal to the fertile stamens. Pollen grains 3-colporate. Gynoecium of 3–5 carpels; stylodia united into a more or less deeply lobed style, stigmas 3–5. Ovary superior, 1-locular, sessile, with few to many ovules on parietal placentas. Fruits 3–5-valvate capsules. Seeds with small, straight embryo and copious, oily endosperm; raphe expanded (ridged) in *Acharia* and *Guthriea*. Producing cyclopentenoid cyanogenic glucoside and cyclopentenyl fatty acids, gynocardin, ellagic acid. . . . . 11. ACHARIACEAE.

3 Petals mostly absent.

11 Stamens numerous. Deciduous trees, glabrous or with an indumentum of long simple hairs. Bark fissured, rough, dark brown.

Vessels with exclusively scalariform perforations; lateral pitting opposite. Fibers non-septate. Xylem with fibre tracheids (sometimes septate), without libriform fibres. Axial parenchyma apotracheal. Leaves alternate, large, simple, leathery, distichous, entire, with two large pits at the base of the blade on the dorsal surface; stipules intrapetiolar, free of one another, caducous. Stomata anomocytic. Inflorescences clusters of short racemes (*Peridiscus*), axillary, or at nodes of fallen leaves, covered with short branched hairs. Flowers small, actinomorphic, fragrant, pale yellow, bisexual, with large, persistent bracteoles. Sepals 4–5(6) in *Peridiscus* or seven in *Whittonia*, hairy, imbricate. Petals five (*Soyauxia*), otherwise absent Stamens 30–100, seated on or around the outside of a large, fleshy, multilobate or annular disc; filaments distinct or irregularly connate toward base, flexuous (versus incurved) in *Whittonia*; anthers 1-locular, introrse, opening longitudinally. Pollen grains 3-colporate, reticulate. Gynoecium of 3–4 carpels; styles 3–4, free, short; ovary 1-locular, superior (half immersed in disc in *Peridiscus*), glabrous or lanate (*Whittonia*); ovules 6–8, pendulous from the top of a unilocular ovary. Fruit drupaceous (known only from *Peridiscus*), or (*Soyauxia*) capsular, fleshy, large, 1-seeded. Seeds with a small, straight embryo lying alongside the copious, horny endosperm; the cotyledons two, ovate-lanceolate, membranous. Prismatic crystals absent. . . . . 4. PERIDISCACEAE.

11 Stamen one. Shrubs and small trees. Vessels with scalariform perforation plates that have numerous bars. Fibers with simple or obscurely bordered pits, usually septate. Rays heterogeneous. Axial parenchyma diffuse. Leaves alternate, simple, entire or toothed, distichous, stipules small, caducous. Stomata more or less distinctly anomocytic. Inflorescences often clustered, axillary catkinlike dense spikes or looser racemes. Flowers very small, bisexual or sometimes polygamous, zygomorphic. Sepals absent or (1)2–6, unequal; petals absent. Stamen seated on or within a fleshy disc; anther basifixed, 2-locular, locules well separated on an expanded connective, or sometimes individually stipitate, opening longitudinally. Pollen grains 3-colporate. Gynoecium of 2–3 carpels. Stylodia united into a style with 2–3 distinct stigmas. Ovary 1-locular, superior, with 1–2 ovules on each parietal placenta. Ovules pendulous, anatropous, with rather thick, elongate funicle. Fruits 3-valvate capsules, with three valvular, 1(-3)-seeded. Seeds with brightly colored, oil-rich sarcotesta, straight embryo and rather copious, oily endosperm; cotyledons two, broad and foliar. . . . . 3. LACISTEMATACEAE.

2 Perianth reduced. Trees or shrubs, sometimes shrublets (some arctic and subarctic species of *Salix*). Roots usually forming ectotrophic mycorrhizae. Vessels with simple perforations; lateral pitting alternate. Fibers with simple pits, occasionally pits with vestigial borders. Rays uniseriate-heterogeneous (*Salix* and certain species of the subgenus *Turanga* of *Populus*) or homogeneous (most species of *Populus* and *Chosenia*). Axial parenchyma scanty, terminal, vasicentric, or diffuse. Sieve-element plastids of S-type. Nodes trilacunar. Leaves alternate, simple, entire, dentate, or lobed, with small or sometimes foliaceous stipules. Stomata paracytic. Distal end of the petiole with one or more closed rings of xylem and phloem. Flowers small, axillary to a bract, but without bracteoles, in dense erect (nearly all species of *Salix*) or pendulous (*Populus* and *Chosenia*), unisexual, often precocious catkins (spikes or less often racemes with shortened peduncles), dioecious, anemophilous or (most species of *Salix*) secondarily entomophilous. Calyx vestigial, in *Populus* more or less disclike in male flowers and saucer-shaped or cupulate in female ones, but in *Salix* represented by one or three (rarely 3–5) small nectary glands that may be united into a lobed ring. In *Chosenia*, however, there are usually no glands and female flowers are only sometimes provided by two small lateral glands. Stamens (4)6–40 (rarely up to 60 or even 70) in *Populus* 1–2, rarely three or five (up to 12) in *Salix*, and 3–6 in *Chosenia*; filaments free (*Populus* and some spp. of *Salix*) or basally connate, rarely connate throughout.

Anthers opening longitudinally. Pollen grains 2-celled or sometimes (in some species of *Populus*) 3-celled, usually 3-colporate or (*Populus*) inaperturate; in *Populus* and *Chosenia* pollen dry and dispersed by wind, in *Salix* pollen viscid and dispersed by insects. Gynoecium of two, less often of 3–4 (certain species of *Populus*) united carpels; stigmas sessile or on a short style, often bifid or irregularly lobed, carinal (as in *Populus* and many species of *Salix*) or commissural (as in most species of *Salix*); ovary superior, sessile or on a short or rarely long gynophore, unilocular, with numerous (*Populus*) or 2–10 ovules, in *Salix* rarely only with one ovule. Ovules anatropous, weakly bitegmic (some species of *Populus*) or more often unitegmic, crassinucellate. Fruits 2–4-valvate capsules. Seeds minute, enveloped at the base by a tuft of long hairs (coma) originating from the placenta; seed coat thin, formed by the outer integument; embryo large, straight, with well-developed cotyledons and hypocotyl and very weakly developed radicle; endosperm much reduced or absent. Producing special phenolic heterosides salicin and populin, only seldom cyanogenic. n = 11, 12, 19. . . . . 2 SALICACEAE.

1. FLACOURTIACEAE

Richard ex A.P. de Candolle 1824 (including Bembiciaceae R. Keating et Takhtajan 1996, Erythrospermaceae Doweld 2001, Gerrardinaceae Alford 2006, Homaliaceae R. Brown 1818, Poliothyrsidaceae Doweld 2001, Prockiaceae Bertuch 1801, Samydaceae Ventenat 1808). 82/875. Widespread in the tropics and subtropics, with some species in temperate regions.

ERYTHROSPERMEAE: *Erythrospermum, Camptostylus, Ahernia, Dasylepis, Scottellia, Rawsonia*; ONCOBEAE: *Mayna, Carpotroche, Grandidiera, Xylotheca, Oncoba, Caloncoba, Lindackeria, Poggea, Prockiopsis, Buchnerodendron, Peterodendron*; PANGIEAE: *Chlorocarpa, Hydnocarpus* (*Taraktogenos*), *Eleutherandra, Gynocardia, Pangium, Baileyoxylon, Chiangiodendron, Trichadenia, Scaphocalyx, Ryparosa, Kiggelaria*; HOMALIEAE: *Dissomeria, Calantica, Bartholomaea, Homalium, Byrsanthus, Gerrardina, Neopringlea, Trimeria*; BEMBICIEAE: *Bembicia*; SCOLOPIEAE: *Scolopia, Hemiscolopia, Pseudoscolopia, Phyllobotryon, Phylloclinium, Mocquerysia*; PROCKIEAE: *Banara, Pineda, Prockia, Hasseltia, Hasseltiopsis, Macrohasseltia Neosprucea, Pleuranthodendron*; FLACOURTIEAE: *Bennettiodendron, Flacourtia, Dovyalis, Tisonia, Azara, Ludia, Xylosma, Poliothyrsis, Carrierea, Itoa, Idesia, Olmedlella, Priamosia*; ABATIEAE: *Abatia, Aphaerema*; SAMYDEAE (CASEARIEAE): *Casearia* (including *Synandrina*), *Bivinia, Euceraea, Hecatostemon, Osmelia, Pseudosmelia, Laetia, Neoptychocarpus, Ophiobotrys, Lunania, Ryania, Samyda, Tetrathylacium, Zuelania, Lasiochlamys, Trichostephanus*.

Closely related to the Berberidopsidaceae, especially through the most archaic tribe Erythrospermeae, which is characterized by very primitive wood anatomy as well as by sepals and petals usually spirally arranged and often indistinguishable from one another. There are also similarities with Stachyuraceae (Theales) and Elaeocarpaceae (Elaeocarpales). A rather heterogeneous family whose taxonomic boundaries are not quite clear and have been variously circumscribed. It is especially difficult to draw a clear-cut boundary between the Flacourtiaceae and Passifloraceae (Passiflorales). The tribes Paropsieae and Abatieae occupy a somewhat intermediate position between these two families. They differ from the typical Flacourtiaceae by the presence of the extrastaminal corona (weakly developed in the Abatieae), wood-anatomical (Metcaife and Chalk 1950) and palynological (Keating 1973) features and stand nearer to the Passifloraceae. Lemke (1988) excludes both of these tribes from the Flacourtiaceae. The correct position of the Prockieae is somewhat in doubt. Evidence from xylem anatomy (Miller 1975) and pollen morphology (Keating 1973) suggests that the Prockieae are allied to both the Flacourtiaceae and the Elaeocarpaceae, thus representing a link between these two families (see Lemke 1988). Hutchinson (1967) placed the tribe Prockieae in the Tiliaceae s. 1. (including the Elaeocarpaceae). According to Miller (1975), a comparison of the xylem anatomy of Tiliaceae and Elaeocarpaceae with the secondary xylem of the tribe Prockieae suggests a close alliance between the tribe Prockieae and the Elaeocarpaceae. The family occupies a basal position in the Violales.

2. SALICACEAE

Mirbel 1815. 3/400. Mainly temperate and cold regions of the Northern Hemisphere, but also in southern parts of South America, in South Africa, and in Southeast Asia. Absent in New Guinea and Australia.

Populus, Salix, Chosenia

Very closely related to the Flacourtiaceae, especially to *Idesia* and *Itoa* (Hallier 1908, 1911, 1912; Holden 1912; Gobi 1916; Cronquist 1957, 1981, 1988; Takhtajan 1959, 1966, 1987; Thorne 2001), but differs in unisexual flowers, reduced perianth, inflorescences and exotestal seeds. Affinity with the Flacourtiaceae is confirmed by wood anatomy (Gzyrian 1952, 1955; Miller 1975), pollen morphology (Keating 1973) and uredinological data (Holm 1969) as well as by chemical data (salicin in Salicaceae and *Idesia* and some other flacourtiaceous genera).

3. LACISTEMATACEAE

C. Martius 1826. 2/14. West Indies and tropical America from southern Mexico to southeastern Brazil, Paraguay, Uruguay, and Argentina.

Lacistema, Lozania

Very close to the Flacourtiaceae, especially to the Erythrospermeae. Although wood structure is rather primitive, the floral morphology is highly specialized.

4. PERIDISCACEAE

Kuhlmann 1950. 3/11. Northern Brazil and Venezuela (*Peridiscus*, 1), Guyana (*Whittonia*, 1); *Soyauxia* (9) in West Africa.

Peridiscus, Whittonia, Soyauxia

Related to the Flacourtiaceae, especially to the Erythrospermeae.

5. VIOLACEAE

Batsch 1802 (including Alsodeiaceae J.Agardh 1858, Leoniaceae A.L.P.P. de Candolle 1868). 20/900. Cosmopolitan, mostly in tropical and subtropical regions, but with many representatives of the largest genus *Viola* (500) in temperate regions. The archaic genus *Rinorea* (about 300) is pantropical and especially characteristic of rain forests.

5.1 VIOLIOIDEAE

Flowers actinomorphic to zygomorphic. Petals imbricate. Filaments free or more or less connate. Anthers introrse. Fruits capsules or berries. – RINOREAE: *Rinorea, Allexis, Gloeospermum, Melicytus* (including *Hymenanthera*), *Isodendrion, Rinoreocarpus, Decorsella, Amphirrhox, Hekkingia, Paypayrola*; VIOLEAE: *Hybanthus, Agatea, Anchietea, Corynostylis, Mayanaea* (including *Orthion*), *Schweiggeria, Noisettia, Viola*.

5.2 LEONIOIDEAE

Flowers actinomorphic. Petals quincuncial or imbricate. Filaments united into a tube. Fruits nutlike. – *Leonia*.

5.3 FUSISPERMOIDEAE

Petals convolute. – *Fusispermum*.

6. DIPENTODONTACEAE

Merrill 1941. 1/1. Eastern Himalayas, northeastern India, northern Burma, southeastern Tibet, southwestern China.

Dipentodon.

Despite a free-basal placenta this family is probably related to the Flacourtiaceae as suggested by T.A. Sprague (see Metcaife and Chalk 1950: 126). Pollen grains are flacourtiacean (Lobreau 1969; Lobreau-Callen 1982). The perianth of *Dipentodon* also resembles that of some genera of the tribe Homalieae. Dipentodontaceae are rather isolated within the Violales and Wu et al. (2002) recognized a separate order Dipentodontales in Dilleniidae.

7. SCYPHOSTEGIACEAE

Hutchinson 1926. 1/1. Borneo.

Scyphostegia.

Probably derived from the Flacourtiaceae (Metcalfe 1956; Takhtajan 1966; Van Heel 1967).

8. PASSIFLORACEAE

A.L. de Jussieu ex Roussel 1806 (including Modeccaceae Horaninow 1847, Paropsiaceae Dumortier 1829, Smeathmanniaceae Martius ex Perleb 1838). 17/700–750. Widely distributed in tropical, subtropical, and warm-temperate regions, especially in tropical America and Africa.

PAROPSIEAE: *Smeathmannia, Barteria, Paropsia, Paropsiopsis, Androsiphonia, Viridivia*;. PASSIFLOREAE: *Ancistrothyrsus, Passiflora* (including *Tetrapathaea*), *Hollrungia, Mitostemma, Dilkea, Basananthe* (including *Tryphostemma*), *Schlechterina, Crossostemma, Adenia, Deidamia, Efulensia.*

Derived from the Flacourtiaceae through the archaic tribe Paropsieae and so close to them that it is difficult to draw a clear taxonomic boundary between these two families. The family Passifloraceae or its immediate ancestor was the basal stock from which Turneraceae, Malesherbiaceae, Caricaceae, and Achariaceae as well as the order Cucurbitales arose. In contrast to Flacourtiaceae, which usually produce the dihydroxylated derivative gynocardin, cyclopentenoid cyanohydrin glucosides are typical of Passifloraceae (Jaroszewski and Olafsdottir 1987).

9. TURNERACEAE

Kunth ex A.P. de Candolle 1828 (including Piriquetaceae Martynov 1820). 10/150. Subtropical and tropical parts of America and Africa, Madagascar, and Rodriguez Island (*Mathurina*).

Mathurina, Erblichia, Stapfiella, Loewia, Tricliceras, Hyalocalyx, Streptopetalum, Adenoa, Piriqueta, Turnera.

Very close to the Passifloraceae and Malesherbiaceae and produce structurally related cyanogens. The Turneraceae seem to be similar to the Malesherbiaceae and the primitive members of the Passifloraceae in producing only four basic cyclopentenoid rings (Spencer et al. 1985). But their flowers have no extrastaminal corona (very narrow corona only in *Piriqueta* and *Erblichia*), and their petals are contorted in the bud.

10. MALESHERBIACEAE

D. Don 1827. 1/24. Western South America: Pacific coastal desert and adjacent arid Andes of Chile (including Atacama), Peru, and Argentina.

Malesherbia.

Very close to the Passifloraceae and Turneraceae and have structurally related cyanogens and possess similar hydrolytic enzymes (Spencer and Seigler 1985). Differ mainly in contorted petals. They have both extrastaminal corona and androgynophore, but seeds are without arils.

11. ACHARIACEAE

H. Harms 1897. 3/3. Warm-temperate South Africa (Cape Province, centered in the eastern Cape).

Acharia, Ceratosicyos, Guthriea.

Closely related to the Passifloraceae, which is supported also by chemical data (the presence of the cyclopentenoid cyanogenic glucoside and gynocardin). From the Passifloraceae they differ mainly in sympetalous corolla ("a feature not often associated with a unisexual condition, except in Cucurbitaceae and in a few genera of Euphorbiaceae," according to Hutchinson [1969: 221]).

12. APHLOIACEAE

Takhtajan 1985. 1/1–4–8. Tropical East Africa, Madagascar, Comoros, Seychelles, Mascarene Islands.

Aphloia (*Neumannia*).

Rather isolated family, which probably related to Flacourtiaceae.

Bibliography

Arbo MM. 2007. Turneraceae. In: K Kubitzki, ed. The families and genera of vascular plants, vol. 9, pp. 458–466. Springer, Berlin/Heidelberg/New York.

Argus GW. 1997. Infrageneric classification of *Salix* (Salicaceae) in the New World. Syst. Bot. Monogr. 52.

Ayensu ES and WL Stern. 1964. Systematic anatomy and ontogeny ot the stem in Passifloraceae. Contr. U.S. Natl. Herb. 34: 45–73.

Bate-Smith EC. 1965. Recent progress in the chemical taxonomy of some phenolic constituents of plants. Mem. Soc. Bot. France 1965: 16–28.

Bayer C. 2007. Peridiscaceae. In: K Kubitzki, ed. The families and genera of vascular plants, vol. 9, pp. 297–300. Springer, Berlin/Heidelberg/New York.

Beattie AJ. 1974. Floral evolution in *Viola*. Ann. Missouri Bot. Gard. 61: 781–793.

Belin-Depoux M. 1982. Aspects histologiques des glandes foliaires de l'*Idesia polycarpa* Maxim. (Flacourtiaceae). Rev. Gen. Bot. 89(1055–1057): 111–120.

Bernhard A. 1999a. Floral structure and development of *Ceratiosicyos laevis* (Achariaceae) and its systematic position. Bot. J. Linn. Soc. 131: 103–113.

Bernhard A. 1999b. Flower structure, development, and systematics in Passifloraceae and in *Abatia* (Flacourtiaceae). Int. J. Plant Sci. 160: 135–150.

Bernhard A. 1999c. Paropsieae – the woody Passifloreae: a comparative study of flower morphology and development in

Passifloraceae, In XVI Intern. Bot. Congr. Abstracts, p. 420. St. Louis.

Bernhard A and PK Endress. 1999. Androecial development and systematics in Flacourtiaceae *s.l.* Plant Syst. Evol. 215: 141–155.

Binns WW, G Blumden, and DL Woods. 1968. Distribution of leucoanthocyanidins, phenolic glycosides, and imino-acids in leaves of *Salix* species. Phytochemistry 7: 1577–1581.

Brenan JPM. 1954. *Soyauxia*: a second genus of Medusandraceae. Kew Bull. 8: 507–511.

Brizicky GK. 1961a. The genera of Turneraceae and Passifloraceae in the southeastern United States. J. Arnold Arbor. 42: 204–218.

Brizicky GK. 1961b. The genera of Violaceae in the southeastern United States. J. Arnold Arbor. 42: 321–333.

Carlquist S. 1984. Wood anatomy of Malesherbiaceae. Phytomorphology 34: 180–190.

Chase MW, S Zmarzty, MD Lledo, KJ Wurdack, SM Swensen, and MF Fay. 2002. When in doubt, put it in Flacourtiaceae: a molecular phylogenetic analysis based on plastid *rbc*L DNA sequences. Kew Bull. 57: 141–181.

Chauhan L, RD Raturi, RV Rao, and R Dayal. 1995. Wood anatomy of Indian Flacourtiaceae. Indian Forester 121: 824–837.

Clausen V, K Frydenvang, R Koopmann, LB Jorgensen, DK Abbiw, P Ekpe, and JW Jaroszewski. 2002. Plant analysis by butterflies: occurrence of cyclopentenyl-glycines in Passifloraceae, Flacourtiaceae and Turneraceae and discovery of the novel nonproteinogenic amino acid 2-(3'-cyclopentenyl)glycine in *Rinorea*. J. Nat. Prod. 65: 542–547.

Cusset G. 1968. Les vrilles des Passifloraceae. Bull. Soc. Bot. France 115: 45–61.

Dahlgren R and AE van Wyk. 1988. Structures and relationships of families endemic to or centered in southern Africa. Syst. Bot. Monogr., Missouri Bot. Gard. 25: 1–94.

Daly DC. 2004. Peridiscaceae. In: N Smith, SA Mori, A Henderson, DW Stevenson, and SV Heald, eds. Flowering plants of the Neotropics, pp. 290–291. The New York Botanical Garden, Princeton University Press, Princeton.

Dathan ASR and D Singh. 1973a. Development and structure of seed in *Tacsonia* Juss. and *Passiflora* L. Proc. Indian Acad. Sci. 77B: 5–18.

Dathan ASR and D Singh. 1973b. Structure and development of ovule and seed in *Flacourtia indica* (Burn. f) Menill. Proc. Indian Acad. Sci. 39B: 172–179.

Dathan ASR and D Singh. 1973c. Structure and development of seed coat in *Viola* spp. J. Indian Bot. Soc. 52: 119–126.

Dathan ASR and D Singh. 1979. Structure and development of female gametophyte and seed in *Hydnocarpus laurifolia* (Dennst.) Sleumer. J. Indian Bot. Soc. 59: 256–263.

DeWilde WJJO. 1971. The systematic position of the tribe Paropsieae, in particular the genus *Ancistrothyrsis*, and a key to the genera of Passifloraceae. Blumea 19: 99–104.

DeWilde WJJO. 1974 (1975). The genera of tribe Passifloreae (Passifloraceae) with special reference to flower morphology. Blumea 22: 37–50.

Ding Hou. 1972. Germination, seedling, and chromosome number of *Scyphostegia borneensis* Stapf (Scyphostegiaceae). Blumea 20: 88–92.

Erianson EW and FJ Hermann. 1928. The morphology and cytology of perfect flowers of *Populus tremuloides*. Mich. Acad. Sci. 8: 97–110.

Fedorov Al A. 1949. A case of anomaly in *Salix caprea* L. and its probable interpretation. Bot. Zhurn. 34: 273–279 (in Russian).

Fedotova TA. 1992. Salicales. In: A Takhtajan, ed. Comparative seed anatomy, vol. 4, pp. 146–153. Nauka, St. Petersburg (in Russian).

Feuillet C and JM MacDougal. 2007. Passifloraceae. In: K Kubitzki, ed. The families and genera of vascular plants, vol. 9, pp. 270–281. Springer, Berlin/Heidelberg/New York.

Fisher MJ. 1928. The morphology and anatomy of the flowers of the Salicaceae. Am. J. Bot. 15: 307–326, 372–394.

Gavrilova OA. 1993. Types of pollen structure and their significance for the systematics of the family Flacourtiaceae. Bot. Zhurn. 78(12): 45–52 (in Russian with English summary).

Gavrilova OA. 1997. The significance of palynological characters for systematics and phylogeny in the tribe Flacourtieae (Flacourtiaceae). Bot. Zhurn. 82(2): 74–79 (in Russian with English summary).

Gavrilova OA. 1998. Palynomorphology of the family Kiggelariaceae. Bot. Zhurn. 83: 20–27 (in Russian with English summary).

Gengler KM. 1997. ITS phylogeny and biogeography of Malesherbiaceae, an endemic of the South American Pacific coastal desert. Am. J. Bot. 84(6): 196–197 (abstract).

Gengler-Nowak KM. 2003. Molecular phylogeny and taxonomy of Malesherbiaceae. Syst. Bot. 28: 333–344.

Gengler-Nowak KM. 2004. Malesherbiaceae. In: N Smith, SA Mori, A Henderson, DW Stevenson, and SV Heald, eds. Flowering plants of the Neotropics, pp. 227–228. The New York Botanical Garden, Princeton University Press, Princeton.

Golysheva MD. 1975. Leaf anatomy of *Idesia polycarpa* Maxim, and other Flacourtiaceae in relation to the question of relationships between the families Salicaceae and Flacourtiaceae. Bot. Zhurn. 60: 787–799 (in Russian).

Gonzalez AM. 2001. Nectarios y vascularizacion floral de *Piriqueta* y *Turnera* (Turneraceae). Bol. Soc. Argent. Bot. 36(1–2): 47–68.

Gonzalez AM and MM Arbo. 2004. Trichome complement of *Turnera* and *Piriqueta* (Turneraceae). Bot. J. Linn. Soc. 144: 85–97.

Gzyrian MS. 1952. The family Salicaceae and its place in the system of angiosperms on the basis of the wood anatomy. Candidate's thesis, University of Erevan (in Russian).

Gzyrian MS. 1955. Intrafamilial relationships in the Salicaceae. Doklady USSR Acad. Sci. 105(4): 832–834 (in Russian).

Hagerup O. 1930. Vergleichende morphologische und systematische Studien über die Ranken und andere vegetative Organe der Cucurbitaceen und Passiflora-ceen. Dansk. Bot. Ark. 6: 1–103.

Hansen AK, LE Gilbert, BB Simpson, SR Downie, AC Cervi, and RK Jansen. 2006. Phylogenetic relationships and chromosome number evolution in *Passiflora*. Syst. Bot. 31: 138–150.

Hekking WHA. 1988. Violaceae: Part 1: *Rinorea* and *Rinoreocarpus*. Flora Neotropica 46: 1–208.

Hodges SA, HE Ballard Jr., ML Arnold, and MW Chase. 1995. Generic relationships in the Violaceae: data from morphology,

anatomy, chromosome numbers and *rbc*L sequences. Am. J. Bot. 82(6): 136 (abstract).
Holm L. 1969. An uredinological approach to some problems in angiosperm taxonomy. Nytt. Mag. Bot. 16: 147–150.
Holden R. 1912. (Salicac. Phylog.). Ann. Bot. 20: 161–175.
Hong T, Ma Z-L, and J-S Chen. 1987. Floral morphology of *Populus lasiocarpa* Oliv. and its phylogenetic position in *Populus*. Acta Bot. Sinica 29: 236–241.
Jaroszewski JW and ES Olafsdottir. 1987. Monohy-droxylated cyclopentenoid cyanohydrin glucosides of Flacourtiaceae. Phytochemistry 26: 3348–3349.
Jaroszewski JW et al. 1988. Occurrence of lotaustralin in Berberidopsis beckleri and its relation to chemical evolution of Flacourtiaceae. Biochem Syst. Ecol. 16(1): 23–28.
Judd WS. 1997. The Flacourtiaceae in the Southeastern United States. Harvard Pap. Bot. 1(10): 65–79.
Keating RC. 1973. Pollen morphology and relationships of the Flacourtiaceae. Ann. Missouri Bot. Gard. 60: 273–305.
Keating RC. 1976. Trends of specialization in pollen of Flacourtiaceae with comparative observations of Cochlospermaceae and Bixaceae. Grana 15: 29–49.
Killip EP. 1938. The American species of Passifloraceae. Publ. Field. Mus. Nat. Hist., Bot. Ser., 19: 1–613.
Kimura C. 1963. On the embryo sac in some members of the Salicaceae. Sci. Rep. Tohoku Imp. Univ., 4th ser. (Biol.) 29: 393–398.
Kolbe K-B and J John. 1979. Serologische Untersuchungen zur Systematik der Violales. Bot. Jahrb. Syst. 101: 3–15.
Kubitzki K. 2007. Aphloiaceae. In: K Kubitzki, ed. The families and genera of vascular plants, vol. 9, pp. 31–32. Springer, Berlin/Heidelberg/New York.
Kubitzki K. 2007. Malesherbiaceae. In: K Kubitzki, ed. The families and genera of vascular plants, vol. 9, pp. 247–249. Springer, Berlin/Heidelberg/New York.
Lemke DE. 1987a. Morphology, xylem anatomy, and relationships of *Neopringlea* (Flacourtiaceae). Syst. Bot. 12: 609–616.
Lemke DE. 1987b. Tribal relationships of *Bartholomaea* (Flacourtiaceae). Brittonia 39: 436–439.
Lemke DE. 1988. A synopsis of Flacourtiaceae. Aliso 12: 29–43.
Leskinen E and C Alström-Rapaport. 1999. Molecular phylogeny of Salicaceae and closely related Flacourtiaceae: evidence from 5.8S, ITS 1 and ITS 2 of the rDNA. Plant Syst. Evol. 215: 209–227.
Liu J and C Cheng. 1991. On the systematic position of genus *Dipentodon* Dunn. J. Wuhan Bot. Res. 9: 29–39.
Lobreau-Callen D. 1982. Structure et affinités polliniques des Cardiopterygaceae, Dipentodontaceae, Erythropalaceae et Octoknemataceae. Bot. Jahrb. Syst. 103: 371–412.
Lu L, H Wang, and Z-X Wei. 2005. Eight species of pollen morphology of Clethraceae and Violaceae, with reference to relationships among six families of Dilleniidae and Staphyleaceae. Acta Bot. Yunn. 27: 269–278.
Malyutina ET. 1972. On the morphological nature of the flower parts of certain species of the genus *Salix* L. and possible ways of their evolution. Bot. Zhurn. 57: 524–530.
Mauritzon J. 1933. Über die Embryologie der Turneraceae und Frankeniaceae. Bot. Not. 86: 543–554.
Mauritzon J. 1936. Zur Embryologie einiger Parietales-Familien. Svensk. Bot. Tidskr. 30: 79–113.
Meeuse ADJ. 1975. Taxonomic relationships of Salicaceae and Flacourtiaceae: their bearing on interpretative floral morphology and dilleniid phylogeny. Acta Bot. Neerl. 24: 437–457.
Melchior H. 1925. Die phylogenetische Entwicklung der Violaceen und die natürlichen Verwandtschaftsverhältnisse ihrer Gattungen. Repert. Spec. Nov. Regni. Veg. Beih. 36: 83–125.
Melo NF de, AC Cervi, and M Guerra. 2001. Karyology and cytotaxonomy of the genus *Passiflora* L. (Passifloraceae). Plant Syst. Evol. 226: 69–84.
Metcalfe CR. 1956. *Scyphostegia borneensis* Stapf: anatomy of stem and leaf in relation to its taxonomic position. Reinwardtia 4: 99–104.
Metcaife CR. 1962. Notes on the systematic anatomy of *Whittonia* and *Peridiscus*. Kew Bull. 15: 472–475.
Miller RB. 1975. Systematic anatomy of the xylem and comments on the relationships of the Flacourtiaceae. J. Arnold Arbor. 56: 20–102.
Miyoshi N and R Yamamoto. 1984. Pollen morphology by means of scanning electron microscope: 8. Violales (Angiospermae). Jpn. J. Palynol. 30: 37–42.
Morawetz W. 1981. Zur systematischen Stellung der Gattung *Prockia*: Karyologie und Epidermisskultur im Vergleich zu *Flacourtia* (Flacourtiaceae), *Grewia* (Tiliaceae) and verwandten Gattungen. Plant Syst. Evol. 139: 57–76.
Munzinger J 2000. Systematique des genres *Hybanthus* Jacq. et *Agatea* A. Gray en Nouvelle-Caledonie, implications sur la phylogenie des Violaceae ligneuses tropicales. Museum National d'Histoire Naturelle, Paris.
Munzinger JK and HE Ballard, Jr. 2003. *Hekkingia* (Violaceae), a new arborescent violet genus from French Guiana, with a key to genera in the family. Syst. Bot. 28: 345–351.
Newsholme C. 1992. Willows: the genus *Salix*. Timber Press, Portland, OR.
Olafsdottir ES, JW Jaroszcwski, and MM Arbo. 1990. Cyanohydrin glucosides of Turneraceae. Biochem. Syst. Ecol. 18: 435–438.
Paine AK and AK Sarkar. 1993. Some aspects of anatomy in Flacourtiaceae (tribe Berberidopsideae). J. Plant Anat. Morphol. 6: 34–41.
Paine AK and AK Sarkar. 1997. Some aspects of anatomy in Flacourtiaceae (tribe Scolopieae). J. Plant Anat. Morphol. 7: 1–8.
Peng Y, Z Chen, X Gong, Y Zhong, and S Shi. 2003. Phylogenetic position of *Dipentodon sinicus*: evidence from DNA sequences of chloroplast *rbc*L, nuclear ribosomal 18S, and mitochondria *mat*R genes. Bot Bull. Acad. Sinica 44: 217–222.
Plisko M. Passifloraceae. In: A Takhtajan, ed. Comparative seed anatomy, vol. 4, pp. 112–120. Nauka, Leningrad (in Russian).
Presting D. 1964 (1965). Die Systematik der Passifloraceen aus pollenmorphologischer Sicht. Ber. Deutsch. Bot. Ges. 77: 40–44.
Presting D. 1965. Zur Morphologic der Pollenkörner der Passifloraceen. Pollen et Spores 7: 193–247.
Puri V. 1947. Studies in floral anatomy: IV. Vascular anatomy of the flower of certain species of Passifloraceae. Am. J. Bot. 34: 562–573.
Puri V. 1948. Studies in floral anatomy: V. On the structure and nature of the corona in certain species of Passifloraceae. J. Indian Bot. Soc. 27: 130–149.
Raju MVS. 1954. Pollination mechanism in *Passiflora foetida* Linn. Proc. Nat. Inst. Sci. India 20: 431–436.
Raju MVS. 1956. Embryology of the Passifloraceae: I. Gametogenesis and seed development of *Passiflora calcarata* Mast. J. Indian Bot. Soc. 35: 126–138.

Rao VS. 1949. The morphology of the calyx-tube and the origin of perigyny in Turneraceae. J. Indian Bot. Soc. 28: 198–201.

Record SJ. 1941. American woods of the family Flacourtiaceae. Trop. Woods 68: 40–57.

Rehfeldt AC, E Schulte, and F Spener. 1980. Occurrence and biosynthesis of cyclopentenyl fatty acids in leaves and chloroplasts of Flacourtiaceae. Phytochemistry 19: 1685–1689.

Ricardi SM. 1967. Revisión taxónomica de las Malesherbiaceas. Gayana 16: 1–139.

Risch C. 1960. Die Pollenkorner der Salicaceen. Willdenowia 2: 413–430.

Rowley JR and G Erdtman. 1967. Sporoderm in *Populus* and *Salix*. Grana Palynol. 7: 517–567.

Sandwith NY. 1962. Contribution to the flora of tropical America: LXIX. A new genus of Peridiscaceae. Kew Bull. 16: 467–471.

Shore JS, KL McQueen, and SL Little. 1994. Inheritance of plastid DNA in the *Turnera ulmifolia* complex. Am. J. Bot. 81: 1636–1639.

Shore JS, MM Arbo, and A Fernández. 2006. Breeding system variation, genetics and evolution in the Turneraceae. New Phytol. 171: 539–551.

Schaeffer J. 1972. Pollen morphology of the genus *Hydnocarpus* (Flacourtiaceae) with notes on related genera. Blumea 20: 65–87.

Skvortsov AK and MD Golysheva. 1966. Studies in leaf anatomy of the willows (*Salix*) in connection with systematics of the genus. Acta Biol. Acad. Sci. Hungar. 12: 125–174 (in Russian).

Soltis DE, JW Clayton, CC Davis, MA Gitzendanner, M Cheek, V Savolainen, AM Amorim, and PS Soltis. 2007. Monophyly and relationships of the enigmatic amphitropical family Peridiscaceae. Taxon 56: 65–73.

Sosa V, MW Chase, and C Barcenas. 2003. *Chiangiodendron* (Achariaceae): an example of the Laurasian flora of tropical forests of Central Americas. Taxon 52: 519–524.

Spencer KC. 1987. Cyanogenic glycosides and the systematics of the Passifloraceae. Am. J. Bot. 74: 755–756 (abstract).

Spencer KC and DS Seigler. 1985a. Cyanogenic glycosides of *Malesherbia*. Biochem. Syst. Ecol. 13: 23–24.

Spencer KC and DS Seigler. 1985b. Cyanogenic glycosides and the systematics of the Flacourtiaceae. Biochem. Syst. Ecol. 13: 421–431.

Spencer KC, DS Seigler, and SW Fraly. 1985. Cyanogenic glycosides of the Turneraceae. Biochem. Syst. Ecol. 13: 433–435.

Spirlet ML. 1965. Utilisation taxonomique des graines de pollen de Passifloracees. Pollen et Spores 7: 249–301.

Steyn EMA, AE van Wyk, and GF Smith. 2001. A study of the ovule, embryo sac and young seed of *Guthriea capensis* (Achariaceae). South Afr. J. Bot. 67(2): 206–213.

Steyn EMA, AE van Wyk, and GF Smith. 2002. Ovule, seed and seedling characters in *Acharia* (Achariaceae) with evidence of myrmecochory in the family. South Afr. J. Bot. 68: 143–156.

Suda Y. 1963. The chromosome numbers of salicaceous plants in relation to their taxonomy. Sci. Rep. Tohoku Imp. Univ., 4th ser. (Biol.), 29: 413–430.

Swamy BGL. 1953. On the floral structure of *Scyphostegia*. Proc. Natl. Inst. Sci. India 19: 127–142.

Van Heel WA. 1967. Anatomical and ontogenetic investigations on the morphology of the flowers and the fruit of *Scyphostegia borneensis* Stapf (Scyphostegiaceae). Blumea 15: 107–125.

Van Heel WA. 1973, 1974, 1977, 1979. Flowers and fruits in Flacourtiaceae, parts 1–4. Blumea 21: 259–279; 22: 15–19; 23: 349–369; 25: 513–529.

Vijayaraghaven MR and D Kaur. 1966. Morphology and embryology of *Turnera ulmifolia* L. and affinities of the family Turneraceae. Phytomorphology 19: 539–553.

Wang H. 2004. Pollen morphology of thirteen species of the Flacourtiaceae. Acta Bot. Yunn. 26: 73–82.

Wei M, Liang Y, and Ru T. 1982. Studies on the chemical components of the fruits and seeds of *Idesia polycarpa*. Acta Bot. Bor. Occid. Sinica 2(2): 87–94.

Wendt T. 1988. *Chiangiodendron* (Flacourtiaceae: Pangieae): a new genus from southeastern Mexico representing a new tribe for the New World flora. Syst. Bot. 13: 435–441.

Woodworth RH. 1935. Fibriform vessel members in the Passifloraceae. Trop. Woods 41: 8–16.

Order 67. ELAEOCARPALES

Trees or shrubs, pubescent with unicellular or glandular hairs. Mucilage cavities and canals absent, but mucilage cells are known to occur. Vessels usually with simple and oblique perforation plates, but some vestigial scalariform perforations with a few bars occur in *Aristotelia, Vallea* and *Tremandra*): lateral pitting large, typically opposite, occasionally alternate or transitional to alternate in some genera. Fibers with small bordered pits (the borders sometimes very inconspicuous) or with simple (*Platytheca* and *Tremandra*) pits, mostly septate. Rays markedly heterogeneous, mixed uniseriate and pluriseriate, the latter often with long ends. Axial parenchyma paratracheal or sometimes absent, in Tremandreae diffuse to vasicentric, in narrow sheaths around the vessels, confluent, and diffuse in *Echinocarpus*. Phloem not stratified. Sieve-element plastids of S-type. Nodes trilacunar or unilacunar with one trace. Cells of the epidermis frequently mucilaginous. Leaves alternate or sometimes opposite, simple, entire, crenate, serrate, or denticulate, stipulate or very rarely estipulate; stipules persistent or caducous. Stomata paracytic, encyclocytic, or (Tremandreae) anomocytic. Flowers in racemes, panicles, or in dichasia, rarely solitary in the axils, bisexual or sometimes polygamous, actinomorphic, often apetalous. Sepals 4–5, free or more or less connate, valvate or less often imbricate. Petals 4–5, free or rarely basally connate, rarely united high up into a corolla crenate-dentate at the apex, entire, lobed, dentate, incised, or laciniate, valvate or seldom imbricate, sometimes wanting. Annular extrastaminal nectary disc with lobes opposite petals present in *Tremandra*. Stamens numerous or very 12–8, free, originating centrifugally, often weakly grouped into five antesepalous clusters, inserted on a lobed or annular disc (large and 5-lobed in *Sericolea*) or enlarged receptacle that is sometimes more or less elongated to form an androgynophore; anthers introrse, except *Sloanea*, which has lartorse, extrorse and introrse anthers in the same flower (Matthews and Endress 2002), basifixed, tetrasporangiate, generally relatively large and linear, often with a conspicuously prolonged connective, opening only apically, by two short slits or pores in *Aristotelia, Sloanea* and *Vallea*, or by a single pore in *Elaeocarpus, Crinodendron,* and Tremandreae (Matthews and Endress 2002). Tapetum secretory. Microsporogenesis simultaneous. Pollen grains 2-celled, small, globose, usually 3-colporate. Gynoecium of 2–5 united carpels; stylodia united into a simple or shortly lobed style. Ovary 2–5-loculed (in *Peripentadenia* two of three loculi becoming abortive), with 2-many ovules per locule, rarely (*Platytheca*) with solitary ovule; placentation secondarily axile. Ovules anatropous, bitegmic, crassinucellate, with zig-zag micropyle and endothelium; unicellular hairs on ovules are present in *Elaeocarpus* and *Platytheca*. Endosperm nuclear. Fruits loculicidal, sometimes also septicidal capsules, spiny or nor, or less often drupaceous or baccate. Seeds massive to very small, sometimes (*Peripentadenia, Echinocarpus, Sloanes* spp) with an aril arising from the raphe and chalazal end of the seed or from the chalaza only, the appendage differentiates as an elaiosome in *Tetratheca* and *Tremandra* (Boesewinkel 1999); exotegmen fibrous; embryo straight or strongly curved; endosperm more or less copious, oily or proteinaceous. Often producing pyrrolizidine and tropane alkaloids, ellagic acid, n = 12, 14, 15.

The Elaeocarpaceae are usually included in the Malvales near the Tiliaceae. However, from the Tiliaceae and related families they differ markedly in the absence of stellate and peltate trichomes, the absence of mucilage cavities or canals, in usually septate fibers with bordered pits, usually opposite lateral pitting of the vessels, paratracheal axial parenchyma, unstratified phloem, the absence of nectar hairs at the calyx base, the presence of a nectariferous disc at the base of the stamens and the presence of endothelium, the absence of fatty acids with a cyclopropenyl ring, and important differences in seed anatomy (according to Corner 1976, the fibrous exotegmen denies the immediate affinity with the Tiliaceae, which are characterized by the palisade exotegmen). On the other hand, the Elaeocarpaceae share many features with the Flacourtiaceae, including gross morphology, pollen morphology, and wood anatomy. Fibrous exotegmen is typical both of the Flacourtiaceae and the Elaeocarpaceae (Corner 1976). Both these families have so many external similarities, that "members of the Flacourtiaceae are often confused with *Sloanea*" (Cronquist 1981: 349). Elaeocarpaceae are especially allied to the Prockieae (particularly to *Prockia* and *Hasseltia* according to Schumann [1895]). According to recent molecular and morphological studies Tremandraceae are nested within Elaeocarpaceae (Savolainen et al. 2000; Doweld 2001; APG II 2003). In my opinion, *Platytheca, Tetratheca, Tremandra* constitute a separate tribe in the Elaeocarpaceae.

1. ELAEOCARPACEAE

A.L. de Jussieu 1824 (including Aristoteliaceae Dumortier 1829, Tetrathecaceae R. Brown 1814, Tremandraceae R. Brown ex A.P. de Candolle 1824). 12/625. Madagascar, Mauritius, Socotra, eastern Himalayas, India, eastern Asia and Southeast Asia, Moluccas, New Guinea, New Caledonia, New Hebrides, Fiji, Samoa, Tonga, and some other Pacific Islands, Australia, Tasmania, New Zealand, America from Mexico to Chile, West Indies.

ELAEOCARPEAE: *Peripentadenia, Dubouzetia, Elaeocarpus, Aceratium, Sericolea, Crinodendron;* ARISTOTELIEAE: *Aristotelia;* SLOANEAE: *Vallea, Sloanea*; TREMANDREAE: *Platytheca, Tetratheca, Tremandra.*

Bibliography

Bawa KS and CJ Webb. 1983. Floral variation and sexual differentiation in *Muntingia calabura* (Elaeocarpaceae), a species with hermaphrodite flowers. Evolution 37: 1271–1282.

Biddle JA and DC Christophel. 1978. Intergynoecial development in Tremandraceae. Phytomorphology 28: 411–418.

Boesewinkel FD. 1999. Ovules and seeds of Tremandraceae. Aust. J. Bot. 47: 769–781.

Brizicky GK. 1965. The genera of Tiliaceae and Elaeocarpaceae in the southeastern United States. J. Arnold Arbor. 46: 286–307.

Carlquist 1977. Wood anatomy of the Tremandraceae: phylogenetic and ecological implications. Am. J. Bot. 64: 704–713.
Coode MJE. 1978. A conspectus of Elaeocarpaceae in Papuasia. Brunonia 1: 131–302.
Coode MJE 2004. Elaeocarpaceae. In: K Kubitzki, ed. The families and genera of vascular plants, vol. 6, pp. 135–144. Springer, Berlin/Heidelberg/New York.
Dehay Ch. 1961. Remarques sur 1эανατομιε comparee des Elaeocarpacees. Bull. Soc. Bot. France 14(4): 89–96.
Gasson P. 1996. Wood anatomy of the Elaeocarpaceae. In: LA Donaldson, AP Singh, BG Butterfield, and LJ Whitehous, eds. Resent advances in wood anatomy, pp. 47–71. New Zealand Forest Institute, Rotorua.
Lodkina MM. 1992. Elaeocarpaceae. In: A Takhtajan, ed. Comparative seed anatomy, vol. 4, pp. 249–254. Nauka, St. Petersburg (in Russian).
Matthews ML and PK Endress. 2002. Comparative floral structure and systematics in Oxalidales (Oxalidaceae, Connaraceae, Brunelliaceae, Cephalotaceae, Cunoniaceae, Elaeocarpaceae, Tremandraceae). Bot. J. Linn. Soc. 140: 321–381.
Mauritzon J. 1934. Zur Embryologie der Elaeocarpaceae. Ark. Bot. 26A(10): 1–8.
Plisko MA. 2000. Malpighiaceae, Trigoniaceae, Vochysiaceae, Krameriaceae, Tremandraceae. In: A Takhtajan, ed. Comparative seed anatomy, vol. 6, pp. 49–11. Nauka, St. Petersburg (in Russian).
Sandhya Rani R. 1995. Floral anatomy of Tremandraceae. J. Indian Bot. Soc. 74(1–4): 69–74.
Schumann KM. 1895. Elaeocarpaceae. In: A Engler and K Pranti, eds. Die natürlichen Pflanzenfamilien, 3(6): 1–8.
Smith CE, Jr. 1954. The New World species of *Sloanea* (Elaeocarpaceae). Contr. Gray Herb. 175: 1–114.
Tang Y and Z-Y Wu. 1990. Study on the pollen morphology of Chinese Elaeocarpaceae. Acta Bot. Yunn. 12: 397–403 (in Chinese with English summary).
Thompson J. 1976. A revision of the genus *Tetratheca* (Tremandraceae). Telopea 1: 139–215.
Venkata Rao C. 1953. Floral anatomy and embryology of two species of *Elaeocarpus*. J. Indian Bot. Soc. 32: 21–33.
Weibel R. 1968. Morphologic de 1эεμβρψον et de la graine des *Elaeocarpus*. Candollea 23: 101–108.
West WC, JE Gunkel, and MA Johnson. 1970. Morphology of the shoot apex in Elaeocarpaceae. Phytomorphology 20: 58–67.

Order 68. CUCURBITALES

Perennial or annual herbs, rarely soft-wooded lianas, or less often shrubs and trees. Vessels with scalariform or simple perforations. Rays heterogeneous. Axial parenchyma paratracheal. Sieve-element plastids of S-type. Nodes trilacunar. Leaves alternate, usually palmately lobed, sometimes palmately compound with three or more leaflets, stipulate or less often estipulate. Stomata mostly anomocytic. Flowers in axillary or less often terminal inflorescences, sometimes reduced to a solitary, nearly always unisexual (monoecious or dioecious), rarely bisexual usually actinomorphic. Calyx (3-)5(6–8)-lobed, imbricate or open. Corolla more or less deeply (3-)5(6–8)-lobed, valvate or induplicate-valvate. Stamens three, less often 5(1–5), or numerous, free or sometimes more or less connate; anthers tetrasporangiate or disporangiate, extrorse, opening longitudinally or rarely by pores. Tapetum secretory. Microsporogenesis simultaneous. Pollen grains 2-celled, of various types, the most primitive ones are 3-colpate or 3-colporate. Nectaries in both the male and female flowers commonly present. Gynoecium of (3-)4–5(-8) united carpels or rarely secondary monomerous; stylodia usually more or less united into a style. Ovary inferior or very rarely semi-inferior, basically 1-locular; ovules numerous or rarely (Sicyoeae) solitary, anatropous, bitegmic, crassinucellate. Female gametophyte of *Polygonum*-type or of *Allium*-type. Endosperm nuclear. Fruits berries or pepos, less often dry or fleshy capsules. Seeds usually numerous, rarely solitary, mostly with thick, woody exotesta and papery endotesta, commonly compressed and sometimes even winged; embryo straight, with two large flat cotyledons; endosperm wanting or vestigial. Bitter substances known as cucurbitacins are widespread.

Closely related to the Violales, especially to the Passifloraceae.

Key to Families

1 Fruits usually baccate, rarely (*Zanona*) capsules. Perennial herbs with swollen, tuberous rootstock formed by swelling of hypocotyl, or less often annuals, mostly climbing or prostrate, rarely soft-wooded lianas, very rarely (monotypic Socotran genus *Dendrosicyos*) small trees with soft and juicy stem. Herbaceous species usually with spirally coiled (often branched) tendrils at the side of each leaf base, which represent modified shoots with lower part corresponding to the stem and upper twining portion corresponding to highly modified foliage; the tendrils sometimes modified into spines or wanting. Glandular hairs are very characteristic, but simple unicellular or uniseriate hairs as well as wartlike or spiny trichomes also occur. Calcareous cystoliths and similar bodies occur, usually at the bases of the hairs or in nearby cells. Vascular bundles of the stem are predominantly bicollateral and

frequently arranged in two rings. Anomalous structure is fairly frequent in thick stems and roots. Leaves alternate, usually palmately lobed, sometimes palmately compound with three or more leaflets, estipulate. Petioles usually with a crescent or ring of unequal vascular bundles, the larger ones bicollateral. Flowers in axillary inflorescences that are sometimes reduced to a solitary, nearly always unisexual (monoecious or dioecious), very rarely bisexual (sometimes in the eastern asiatic genera *Actinostemma* and *Schizopepon*), usually actinomorphic, mostly 5-merous. The perianth together with the bases of stamens forms a floral tube (hypanthium) that is shortly or strongly prolonged beyond the ovary. Calyx (3-)5(6)-lobed, imbricate or open. Corolla more or less deeply (3-)5(6)-lobed, valvate or induplicate-valvate. Stamens three or less often 5(1–5), inserted on the lower part of the hypanthium and alternating with the corolla lobes; filaments more or less free, as in *Fevillea,* or more often variously united, as also the anthers; anthers straight or arcuate to triplicate or contorted, when three in number usually two tetrasporangiate and disporangiate, when five in number all disporangiate, extrorse, opening longitudinally. Pollen grains 2-celled. Nectaries in both the male and female flowers commonly present. Gynoecium of 3 (4–5) united carpels or rarely secondary monomerous (*Cyclanthera* and related genera); stylodia usually more or less united into a style; stigmas as many as carpels, usually forked, commissural. Ovary inferior or very rarely semi-inferior, basically 1-locular, mostly almost completely filled with much enlarged placentas, which are sometimes joined in the center making the ovary plurilocular, usually with numerous ovules or rarely (Sicyoeae) ovule solitary. Female gametophyte of *Polygonum*-type. Chalazal end of the female gametophyte usually developing into a tubular aggressive haustorium of various lengths. Fruits berries or pepos (firm-walled berries with hardened pericarp), less often dry or fleshy capsules, or leathery and indehiscent. Seeds usually numerous, rarely solitary, mostly with thick, woody exotesta and papery endotesta, commonly compressed and sometimes even winged; in some genera with a false scarlet aril (placental tissue – Corner 1976); embryo straight, with two large flat cotyledons; endosperm wanting or vestigial, n = 7(8)-14. . . . . . 1. CUCURBITACEAE.

1 Fruits capsules, dehiscing apically, rarely berry-like.

2 Leaves simple, entire, dentate or deeply lobed, or pinnate, often asymmetrical, estipulate. Stomata anomocytic. Vessels with simple perforations; lateral pitting alternate. Fibers with minute simple pits. Rays heterogeneous. Axial parenchyma paratracheal, scanty. Flowers more or less actinomorphic, dioecious or in *Datisca glomerata* functionally androdioecious. Sepals 3–8, mostly unequal. Petals only in male flowers of *Octomeles,* 6–8, small, free, valvate, inserted on the calyx, in the other genera wanting in both the male and female flowers. Staminodia generally wanting, but pistillodia sometimes present. Stamens of the same number as the sepals and opposite them, or more numerous (up to about 25). Anthers 2-locular, opening longitudinally. Gynoecium of 3–8 carpels; ovary inferior, 1-locular. Operculum without collar cells. Fruits capsules. Perennial herbs or very large trees.

3 Perennial herbs engaging in a nitrogen-fixing symbiosis with the actinomycetes *Frankia.* Leaves pinnate or pinnatifid, with long narrow caudate-acuminate pinnae. Inflorescences greatly shortened with flowers appearing axillary. Stamens sessile, indefinite in number, without definite arrangement in regard to the sepals; filaments very short, slender; anthers large, straight, oblong. Styles elongate-subulate, densely papillate from base to apex, with slender bifurcating stigmas. Capsules opening apically between the persistent stylodia. Seeds with lid, exotegmic cells large, cuboid; endosperm slight; Producing unusual B-ring unsubstituted and 2'-hydroxylated flavonols. n = 11. . . . . 1. DATISCACEAE.

3 Large or very large trees, often buttressed, with soft wood; hairs glandular or lepidote. Nodes with two traces from the lateral gaps. Leaves simple, cordate-ovate (sometimes obliquely), entire or dentate, pubescent or lepidote, long-petiolate, palmately 3–5-veined. Flowers in long, pendant axillary spikes or panicles. Stamens isomerous with and opposite to the petals; filaments elongate, somewhat flattened-subulate; anthers relatively short and ovoid, or, if larger and more or less oblong, then conspicuously recurved. Styles shorter or very short, stout, with large, capitate or obliquely oblong-decurrent stigmas. Fruits

septicidal capsules, opening down the sides. Seeds small, numerous, winged (*Tetrameles*). Producing simple flavonols, kaempferol, and quercetin. n = ca. 23. . . .2. TETRAMELACEAE.

2 Leaves simple, often palmately lobed, with mostly large stipules, usually asymmetrical, mostly distichous. Nodes swollen. Stomata surrounded by 3–6 subsidiary cells, these often arranged in two rings. Vessels with simple or (and) scalariform perforations with many bars; lateral pitting scalariform. Flowers often somewhat zygomorphic, monoecious. Sepals in male flowers two, rarely five, free or connate, valvate, in female flowers usually 2–5, rarely 6–8, imbricate; sepals (calyx lobes in female flowers) free, but in *Begoniella* and in female flowers of *Symbegonia* united into a tube. Petals absent or (*Hillebrandia* and *Begoniella*) very small and rudimentary. Stamens four (*Begoniella*) or numerous, in many cycles, free or more or less connate; anthers basifixed, opening by longitudinal slits or rarely by apical pores. Gynoecium of (2)3(6) carpels; stylodia free or sometimes basally connate, usually deeply bifid; stigmas often twisted; ovary inferior or (*Hillebrandia*) semi-inferior, 1-locular (*Hillebrandia* and *Begonia* sect. *Meziera*) or secondary 2–4-locular. Ovules with zig-zag micropyle. Fruits capsules (loculicidal or in *Hillebrandia* opening at apex between the stylodia) or rarely berries. With collar cells arranged in a transverse ring around the operculum. Rather succulent herbs or subshrubs with often jointed stems, usually from thick rhizomes or tubers; present sclereids and uncalcified cystoliths, some climb with the aid of aerial roots. Plants tanniniferous, n = 10–21 +. . . . . 3. BEGONIACEAE.

1. CUCURBITACEAE

A.L. de Jussieu 1789 (including Bryoniaceae G. Meyer 1836, Cyclantheraceae Lilja 1870, Fevilleaceae L. Pfeiffer 1879, Nhandirobaceae Lestibudois 1826, Zanoniaceae Dumortier 1829). 123/800–900. Widely distributed in tropics and subtropics, especially in the rain forests of South America and the drier parts of Africa, with relatively few species in Australasia and the temperate regions.

Classification after C. Jeffrey (2005).

1.1 NHANDIROBOIDEAE

Tendrils two-branched from near the apex, spiraling above and below the point of branching. Perennial, dioecious herbs. Gynoecium with two or three stylodia. Pollen grains small, striate, and uniform. Ovules pendulous. Seeds often winged, n = 8. – ZANONIEAE: *Fevillea, Alsomitra, Bayabusua, Zanonia, Siolmatra, Gerrardanthus, Zygosicyos, Xerosicyos, Neoalsomitra, Hemsleya, Gomphogyne, Gynostemma, Bolbostemma, Actinostemma, Sicydium, Chalema, Pteropepon, Pseudosicydium Cyclantheropsis*.

1.2 CUCURBITOIDEAE

Tendrils unbranched or two-to seven-branched from the lower part, spiraling only above the point of branching. Often annual, monoecious herbs. Stylodia united into a style. Pollen grains colpate and (panto)porate. Seeds unwinged, n = (7)8–14. – JOLIFFIEAE: *Baijiania, Indofevillea, Sinobaijiania, Siraitia, Microlagenaria, Thladiantha, Momordica, Telfairia, Odosicyos;* BRYONIEAE: *Bryonia, Ecballium;* TRICHOSANTHEAE: *Hodgsonia, Ampelosicyos, Tricyclandra, Gymnopetalum, Trichosanthes*; HERPETOSPERMEAE: *Biswarea, Herpetospermum, Edgaria;* SCHIZOPEPONEAE: *Schizopepon;* LUFFEAE: *Luffa*; SICYEAE: *Hanburia, Echinopepon, Apatzingania, Marah, Echinocystis, Vaseyanthus, Brandegea, Elateriopsis, Pseudocyclanthera, Cyclanthera* (including *Cremastopus*), *Rytidostylis, Sicyos, Sicyosperma, Parasicyos, Microsechium, Sechium, Sechiopsis* (including *Pterosicyos*); CONIANDREAE: *Kedrostis, Dendrosicyos, Corallocarpus, Ibervillea, Tumamoca, Halosicyos, Ceratosanthes, Doyerea, Trochomeriopsis, Seyrigia, Dieterlea, Cucurbitella, Apodanthera, Guraniopsis, Melothrianthus, Wilbrandia, Helmontia, Psiguria, Gurania;* BENINCASEAE: *Solena, Borneosicyos, Papuasicyos, Trochomeria, Dactyliandra, Ctenolepis, Peponium, Cephalopentandra, Cogniauxia, Ruthalicia, Lagenaria, Benincasa, Praecitrullus, Citrullus, Acanthosicyos, Eureiandra, Bambekea, Nothoalsomitra, Coccinia, Diplocyclos, Raphidiocystis, Lemurosicyos, Zombitsia, Melancium, Cucumeropsis, Posadaea, Melothria, Muellerargia, Zehneria, Cucumella, Cucumis, Oreosyce, Myrmecosicyos, Mukia, Dicoelospermum*; CUCURBITEAE: *Cucurbita, Sicana, Tecunumania, Calycophysum, Peponopsis, Anacaona, Polyclathra, Schizocarpum, Penelopeia, Cionosicyos, Cayaponis, Selysia, Abobra.*

The Cucurbitaceae are more or less closely related to Datiscaceae and Begoniaceae.

2. DATISCACEAE

Berchtold et J. Presl 1820. 1/2. Crete to western Himalayas and from northern Baja California to northern California.

Datisca.

According to Boeswinkel (1984) Datiscaceae (s. l.) are less specialized than Begoniaceae because they have a larger, less reduced nucellus without strikingly swollen epidermal cells and because there is no special rupture layer aiding the lifting of the operculum. In addition, the vegetative characters of the Datiscaceae are less specialized than those of the Begoniaceae.

3. TETRAMELACEAE

Airy Shaw 1965. 2/2. Malay Archipelago, New Guinea, New Britain, and Solomon Islands (*Octomeles*) and Sri Lanka, Andaman Islands, Indochina, and Malesia (*Tetrameles*).

Tetrameles, Octomeles.

Close to the Datiscaceae, but sufficiently well differentiated in morphological characters (Airy Shaw 1965; Davidson 1973, 1976), flavonoid profile (Bohm 1988) and nucleotide sequence data (Swensen et al. 1994) to deserve familial rank.

4. BEGONIACEAE

Berchtold et J. Presl 1820. 4/950. Widespread in tropics and subtropics, especially in northern parts of South America and in tropical Asia, absent in Australia; *Begonia* (about 900 or more) is widespread; very closely related to *Begonia* is South American *Semibegoniella* with two species in Ecuador; the other two genera are Pacific (the monotypic *Hillebrandia* in Hawaii and *Symbegonia* with 13 species in New Guinea).

Begonia, Semibegoniella, Symbegonia, Hillebrandia.

Begoniaceae are rather closely related to the Datiscaceae but are more advanced.

Bibliography

Aleshina LA. 1971. Palynological data on the systematics and phylogeny of the family Cucurbitaceae Juss. In: LA. Kuprianova and MS. Yakovlev, eds. Morphology of pollen grains of Cucurbitaceae, Thymelaeaceae, Cornaceae, pp. 3–103. Nauka, Leningrad (in Russian).

Barabe D. 1981. Vascularisation de la fleur pistillee de *Begonia handelii*. Canad. J. Bot. 59: 819–825.

Barabe D and L Chretien. 1983. Nouvelles donnees sur la vascularisation de la fleur pistillee de *Begonia* (Begoniaceae). Bull. Soc. Bot. France Lett. Bot. 130: 307–316.

Barabe D, L Brouillet, and C Bertrand. 1985. Les Begonias a placentation parietale: Cas de *Begonia masoniana* Irmscher. Bull. Mus. Nat. Hist. Nat., ser. 4, Sect. B, Adansonia 7: 403–414.

Bates DM, RW Robinson and C Jeffrey, eds. 1990. Biology and utilization of the Cucurbitaceae. Cornell University Press, Ithaca, NY.

Boeswinkel FD. 1984. Ovule and seed structure in Datiscaceae. Acta Bot. Neerl. 33: 419–429.

Boeswinkel FD and A De Lange. 1983. Development of ovule and seed in *Begonia squmulosa* Hook. f. Acta Bot. Neerl. 32: 417–425.

Bohm BA. 1988. Flavonoid systematics of the Datiscaceae. Biochem. Syst. Ecol. 16: 151–155.

Bouman F and A De Lange. 1982. Micromorphology of the seed coats in *Begonia* section *Squamibegonia* Warb. Acta Bot. Neerl. 31: 38–54.

Bouman F and A De Lange. 1983. Structure, micro-morphology of *Begonia* seeds. The Begonian 50(70–78): 91.

Bugnon P. 1926. Valeur morphologique de lɜoϖαιρε infere chez les *Begonia*. Bull. Soc. Linn. Normandie, ser. 7, 9: 7–25.

Bugnon P. 1956. Valeur morphologique du complexe axillaires chez les Cucurbitacees. Ann. Sci. Nat. Bot., ser. 4, 11: 313–323.

Carlquist S. 1985. Wood anatomy of Begoniaceae, with comments on raylessness, paedomorphosis, relationships, vessel diameter, and ecology. Bull. Torrey Bot. Club 112: 59–69.

Carlquist S. 1992. Wood anatomy of selected Cucurbitaceae and its relationship to habit and systematics. Nord. J. Bot. 12: 347–355.

Chakravarty HL. 1958. Morphology of the staminate flowers in the Cucurbitaceae with special reference to the evolution of the stamens. Lloydia 21: 49–87.

Charpentier A, L Broullet, and D Barabe. 1989. Or-ganogenese de la fleur pistillee du *Begonia dregei* et de *Hillebrandia sandwicensis* (Begoniaceae). Canad. J. Bot. 67: 3625–3639.

Chopra RN. 1955. Some observations on endosperm development in the Cucurbitaceae. Phytomorphology 5: 219–230.

Clement WL, LL Forrest, and SM Swensen. 2001. Phylogenetic placement of *Hillebrandia sandwichensis* (Begoniaceae). In Botany 2001: Plants and People, abstracts, p. 156. Albuquerque.

Cuerrier A, L Brouillet, and D Barabé. 1991. Micromorphologie foliaires des Begoniaceae. Bull. Mus. Nat. Hist. Nat., sect. B, 4 ser. 12: 297–335.

Davidson C. 1973. An anatomical and morphological study of Datiscaceae. Aliso 8: 49–110.

Davidson C. 1976. Anatomy of xylem and phloem of the Datiscaceae. Los Angeles County Mus. Contr. Sci. 280: 1–28.

Deshpande PK, SM Bhuskute, and VH Makde. 1986. Microsporogenesis and male gametophyte in some Cucurbitaceae. Phytomorphology 36: 145–150.

Devi HM and KC Naidu. 1979. Embryological studies in the family Begoniaceae. Indian J. Bot. 2: 1–7.

De Wilde WJJO and BEE Duyfjes. 1999. *Bayabusua*, a new genus of Cucurbitaceae. Sandakania 13: 1–16.

Dittmer HJ and ML Roser. 1963. The periderm of certain members of the Cucurbitaceae. Southwest. Nat. 8: 1–9.

Doorenbos J, MSM Sosef, and JJFE de Wilde. 1998. The Sections of *Begonia* Including Descriptions, Keys and Species Lists. Wageningen Agricultural University Papers 98–2.

Dzevaltovsky AK. 1963. Cytoembryological studies of certain representatives of the family Cucurbitaceae. Ukrainian Bot. Zhurn. 20(4): 16–29 (in Ukrainian with Russian and English summaries).

Dzevaltovsky AK. 1983. Cucurbitaceae. In: MS Yakovlev, ed. Comparative embryology of flowering plants: Phytolaccaceae–Thymelaeaceae, pp. 127–138. Nauka, Leningrad (in Russian).

Gauthier R. 1959. L'anatomie vasculaire et lэιντερπρετα-tion de la fleur pistillee de l'*Hillebrandia sandwicensis* Oliv. Phytomorphology 9: 72–87.

Gauthier R and J Arros. 1963. L'anatomie de la fleur staminee de l' *Hillebrandia sandwicensis* Oliver et la vascularisation de lэεταμινε. Phytomorphology 13: 115–127.

Golding J and DC Wasshausen. 2002. Begoniaceae, Edition 2. Part 1: annotated species lists. Part II: illustrated key, abridgement and supplement. Contrib. U.S. National Herb. 43: 1–289.

Goulet I, D Barabé, and L Brouillet. 1994. Analyse structurale et architecture de l'inflorescence des Begoniaceae. Canad. J. Bot. 72: 897–914.

Gulyaiev VA. 1963. Comparative embryology of Cucurbitaceae and its significance for the systematics of the family. Bot. Zhurn. 48: 80–85 (in Russian).

Hagerup O. 1930. Vergleichende morphologische und systematische Studien über die Ranken und andere vegetative Organe der Cucurbitaceen und Passiflora-ceen. Dansk. Bot. Ark. 6(8): 1–104.

Inamdar JA and M Gangadhara. 1976. Structure, ontogeny, and taxonomic significance of stomata in some Cucurbitaceae. Feddes Repert. 87: 293–310.

Irmscher E. 1925. Begoniaceae. In: A Engler and K Prantl, eds. Die natürlichen Pflanzenfamilien, vol. 21, pp. 548–588. Engelmann, Leipzig.

Jeffrey C. 1962. Notes on Cucurbitaceae, including a proposed new classification of the family. Kew Bull. 15: 337–371.

Jeffrey C. 1966. On the classification of the Cucurbitaceae. Kew Bull. 20: 417–426.

Jeffrey C. 1978. Cucurbitaceae. In: VH Heywood, ed. Flowering plants of the world, pp. 115–117. Oxford University Press, Oxford.

Jeffrey C. 1990. An outline classification of the Cucurbitaceae. In: DM Bates, RW Robinson, and C Jeffrey, eds. Biology and utilization of the Cucurbitaceae, pp. 449–463. Cornell University Press, London.

Jeffrey C. 2005. A new system of Cucurbitaceae. Bot. Zhurn. 90(3): 332–335.

Jin XB and FH Wang. 1994. Style and ovary anatomy of Chinese *Begonia* and its taxonomic and evolutionary implications. Cathaya 6: 125–144.

Jobst J, K King, and V Hemleben. 1998. Molecular evolution of the internal transcribed spacers (ITS1 and ITS2) and phylogenetic relationships among species of the family Cucurbitaceae. Molec. Phylogenet. Evol. 9: 204–219.

Kamelina OP. 1983. Datiscaceae. In: MS Yakovlev, ed. Comparative embryology of flowering plants: Phytolaccaceae–Thymelaeaceae, pp. 139–144. Nauka, Leningrad (in Russian).

Kamilova FN and EA Mokeeva. 1961. The nature and structure of tendrils of Cucurbitaceae. In Morphogenesis of plants, vol. 2, pp. 56–59. Nauka, Moscow (in Russian).

Kartashova NN and EN Nemirovich-Danchenko. 1968. On the evolution of nectaries in Cucurbitaceae Juss. Bot. Zhurn. 53: 1219–1225 (in Russian).

Komar GA. 1992. Bignoniales. In: AL Takhtajan, ed. Comparative anatomy of seeds, vol. 4, pp. 178–182. Nauka, St. Petersburg (in Russian).

Kumazawa M. 1964. Morphological interpretation of axillary organs in the Cucurbitaceae. Phytomorphology 14: 287–298.

Lassnig P. 1997. Verzweigungsmuster und Rankenbau der Cucurbitaceae. Trop. Subtrop. Pflanzenwelt 98: 1–146.

Leins P and R Bonnery-Brachtendorf. 1977. Ent-wicklungsgeschichtliche Untersuchungen an Blüten von *Datisca cannabina* (Datiscaceae). Beitr. Biol. Pfl. 53: 143–155.

Liston A, LH Rieseberg, and TS Elias. 1989. Morphological stasis and molecular divergence in the intercontinental disjunct genus *Datisca* (Datiscaceae). Aliso 12: 525–542.

Liston A, LH Rieseberg, and TS Elias. 1990. Functional androdioecy in the flowering plant *Datisca glomerata*. Nature 343: 641–642.

Maheswari Devi H and KC Naidu. 1979. Embryological studies in the family Begoniaceae. Indian J. Bot. 2: 1–7.

Matienko BT. 1957. On the anatomo-morphological nature of the flowers and fruits of Cucurbitaceae. Trudy Bot. Inst. Akad. Nauk SSSR, ser. 7, 4: 288–322.

Matthews ML and PK Endress. 2004. Comparative floral structure and systematics in Cucurbitales (Corynocarpaceae, Coriariaceae, Tetramelaceae, Datiscaceae, Begoniaceae, Cucurbitaceae, Anisophylleaceae). Bot. J. Linn. Soc. 145: 129–185.

Mauritzon J. 1936. Zur Embryologie einiger Parietales-Familien. Svensk Bot. Tidskr. 30(I): 79–113.

Merxmuller H and P Leins. 1971. Zur Entwicklungs-geschichte münnlicher Begonienblüten. Flora 160: 333–339.

Mokeeva EA. 1963. Development and structure of the node of stems in Cucurbitaceae. Bot. Zhurn. 48: 1472–1483 (in Russian).

Muller EGO and F Pax. 1889. Cucurbitaceae. In: A Engler and K Prantl, eds. Die natürlichen Pflanzenfamilien, vol. IV (5), pp. 1–39. Engelmann, Leipzig.

Naudin C. 1855. Observations relatives a la nature des vrilles et a la structure de la fleur chez les Cucurbitacees. Ann. Sci. Nat. Bot. ser. 4, 4: 5–19.

Oginuma K and CI Peng. 2002. Karyomorphology of Taiwanese *Begonia* (Begoniaceae): taxonomic implications. J. Plant Res. 115(1119)no. 3: 225–235.

Puri V. 1954. Studies in floral anatomy: VII. On placentation in the Cucurbitaceae. Phytomorphology 4: 278–299.

Radhakrishnaiah M and A Shanta. 2001. Taxonomic significance of flavonoid data in Cucurbitaceae. Rheedea. 11: 79–86.

Rehm S, PR Enslin, ADJ Meeuse, and JH Wessels. 1957. Bitter principles of the Cucurbitaceae: VII. The distribution of bitter principles in this plant family. J. Sci. Food. Agric. 12: 679–680.

Renner SS, A Weerasooriya, and ME Olson. 2002. Phylogeny of Cucurbitaceae inferred from multiple chloroplast loci. In Botany 2002: Botany in the Curriculum, abstracts, p. 169. Madison, WI.

Rieseberg LH, MA Hanson, and CT Philbrick. 1992. Androdioecy is derived from dioecy in Datiscaceae: evidence from restriction site mapping of PCR-amplified chloroplast DNA fragments. Syst. Bot. 17: 324–336.
Ronse Decraene LP and EF Smets. 1990. The systematic relationship between Begoniaceae and Papaveraceae: a comparative study of their floral development. Bull. Jard. Bot. Nat. Belg. 60: 229–273.
Singh A and ASR Dathan. 2001. Development and structure of seed coat in the Cucurbitaceae and its implications in systematics. In SVS Chauhan and SN Chaturvedi, eds. Botanical essays: tribute to Professor Bahadur Singh, pp. 87–111. Printwell Publishers Distributors, Jaipur.
Singh B. 1953. Studies on the structure and development of seeds of Cucurbitales. Phytomorphology 3: 224–239.
Singh D. 1961a. Studies on endosperm and development of seed in the Cucurbitaceae and some related families. Agra Univ. J. Res. Sci. 10: 117–123.
Singh D. 1961b. Development of embryo in the Cucurbitaceae. J. Indian Bot. Soc. 40: 620–623.
Singh D. 1964. A further contribution to the endosperm of the Cucurbitaceae. Proc. Indian Acad. Sci. 60B: 399–413.
Singh D and ASR Dathan. 1990. Seed coat anatomy of the Cucurbitaceae. In: DM Bates, RW Robinson and C Jeffrey, eds. Biology and utilization of the Cucurbitaceae, pp. 225–238. Cornell University Press, Ithaca, NY.
Singh D and ASR Dathan. 1998. Morphology and embryology. In: NM Nayar and TA More, eds. Cucurbits, pp. 67–84. Science Publishers, Enfield, NH.
Smith LB and BC Schubert. 1963. Begoniaceae Lindl. Field Mass. Nat. Hist. Bot. 13(1): 181–202.
Smith LB, DC Wasshausen, J Golding, and C Karegeannes. 1986. Begoniaceae. Part I: illustrated key, Part II: annotated species list. Smithson. Contrib. Bot. 60: 1–584.
Soyfer VN. 1964. Seed anatomy of the family Cucurbitaceae Juss. as a systematic character. Bull. Moscow Soc. Nat., Sect. Biol., new ser., 69(1): 86–101 (in Russian).
Sridhar and D. Singh. 1986. Development of anther and male gametophyte in Cucurbitaceae. J. Indian Bot. Soc. 65: 487–493.
Stafford PJ and DA Sutton. 1994. Pollen morphology of the Cyclantheinae C. Jeffr. (tribe Sicyeae Schrad., Cucurbitaceae); and its taxonomic significance. Acta Bot. Gallica 141: 171–182.
Steenis CGGJ van. 1953. Datiscaceae. In: CGGJ van Steenis, ed. Flora malesiana, ser. 1, 4: 382–387. Noordhoff, Groningen.
Swensen SM and CM Walsh. 1997. A molecular phylogeny of Cucurbitaceae. Am. J. Bot. 84(6): 236–237 (abstract).
Swensen SM, BC Mullin, and MW Chase. 1994. Phylogenetic affinities of Datiscaceae based on an analysis of nucleotide sequences from the plastid *rbc*L gene. Syst. Bot. 19: 157–168.
Swensen SM, JN Luthi, and LH Rieseberg. 1998. Datiscaceae revised: monophyly and the sequence of breeding system evolution. Syst. Bot. 23: 157–169.
Swensen SM, WL Clement, LL Forrest, and MC Tebbitt. 2001. *Hillebrandia sandwichensis*: evolutionary relationships and biogeography. In Botany 2001: Plants and People, p. 95. Abstracts. Albuquerque.
Tebbitt MC and CM Maciver. 1999. The systematic significance of the endothecium in Begoniaceae. Bot. J. Linn. Soc. 131: 203–221.
Teppner H. 2000. *Cucurbita pepo* (Cucurbitaceae) – history, seed coat type, thin coated seeds and their genetics. Phyton 40: 1–42.
Zhan Z-Y and AM Lu. 1989. Pollen morphology of the subtribe Thiadiantinae (Cucurbitaceae) and its taxonomic significance. Cathaya 1: 23–36.
Zhang L-B and SS Renner. 2003. Phylogeny of Cucurbitales inferred from seven chloroplast and mitochondrial loci. Botany 2003. Abstract.
Zhang L-B, MP Simmons, A Kocyan, and SS Renner. 2006. Phylogeny of the Cucurbitales inferred from plastid, nuclear, and mitochondrial sequences: implications for morphological and sexual system evolution. Molec. Phylog. Evol. 39: 305–322.

Superorder CAPPARANAE

Order 69. ACANIALES

Evergreen or deciduous trees. Idioblastic and stomatal myrosin cells present. Vessels mostly with simple perforations, but in mature wood occasionally with scalariform perforations and with numerous bars, often in aberrant arrangements. Fibers with bordered pits. Rays heterogeneous. Axial parenchyma scanty and vasicentric. Sieve-element plastids of S-type (*Bretschneidera*). Leaves alternate, imparipinnate, leaflets entire or serrate; stipules small or reduced. Inflorescences axillary or terminal, simple racemes of paniculate, bracts minute or lacking. Flowers actinomorphic or somewhat zygomorphic, bisexual. Sepals five, imbricate; petals five, longer than sepals, clawed or slightly constricted at base, contorted or imbricate. Stamens 8–10, free, inserted at the base of receptacular cup; filaments pubescent at base. Anthers dorsifixed or sub-basifixed, versatile, tetrasporangiate, opening longitudinally. Tapetum secretory. Microsporogenesis simultaneous. Pollen grains 2–3-colpate. Gynoecium of 3(5) carpels; style simple, slender. Ovary superior, sessile, surrounded by copular tube, 3-locular, pubescent, placentation axile; style simple, slender; ovules 2–3 in each locule, pendulous, anatropous or campylolotropous bitegmic, crassinucellate. Female gametophyte of *Polygonum*-type (Akaniaceae), or *Allium*-type (Bretschneideraceae). Endosperm nuclear. Fruits loculicidal, leathery, pubescent capsules. Seed broadly ellipsoid; seed-coat exo-mesotestal with long palisade cells; testa vascularised, mesotesta and endoteasta thickened; embryo straight, cotyledons thick; endosperm copious or absent, n = 9.

Morphological date support a relationship Akaniales with Tropaeolaceae (Ronse Decraene and Smets 2001; Ronse Decraene et al. 2002; Ronse Decraene and Haston 2006).

Key to Families

1 Flowers actinomorphic; nectary disc absent. Evergreen or deciduous small trees; the freshly cut wood and bark with a turniplike odour. Rays very wide, multiseriate. Leaves large, imparipinnate, coriaceous; petiolules swollen both proximally and distally, leaflets serrate, stipules small, subulate, or reduced. Inflorescences axillary, paniculate, bracts minute or lacking. Flowers fragrang, white to pinkish. Sepals five, imbricate. Petals five, contorted. Stamens usually eight, rarely 9–10, the five outer ones opposite the sepals; filaments free; anthers dorsi- to sub-basifixed, latrorse. Gynoecium of three united carpels, with simple style and a small 3-lobed stigma; ovary 3-locular, with two axile, superposed, pendulous ovules in each locule. Ovules epitropous, anatropous, or hemitropous. Fruits 3–6-seeded capsules. Seeds yellow, exarillate, with a straight, massive embryo and copious, fleshy endosperm smelling of bitter almonds. Present calcium oxalate druses and solitary crystals, mustard-oils, proanthocyanidins (cyanidin and delphinidin), flavonols (kaempferol and quercetin), producing alkaloids. . . . . 1. AKANIACEAE.

1 Flowers slightly zygomorphic, nectary disc present. Myrosin cells and mustard oils present in the bark and inflorescences. Small deciduous plants. Leaves petiolate, leaflets entire, estipulate. Flowers rather large, pinkish, in erect terminal racemes. Calyx of connate sepals, campanulate, 5-toothed. Petals five, free, unequal, basally clawed, imbricate. Stamens eight, in two cycles, with slender, hairy filaments attached to the rather thin, annular nectary disc; anthers dorsifixed, introrse, connective is slightly prolonged, dehiscence is lateral, but starts at the apex. Pollen grains 3-colpate or 2-colpate. Gynoecium of three (sometimes five) carpels, with elongate, curved style and six-lobed stigma; ovary 3-locular or sometimes 5-locular, with 2–3 campylotropous, apical-axile and pendulous ovules per locule. Fruits thick-walled, 3–5-locular pubescent capsules. Seeds orange to red; embryo large, curved, endosperm lacking. Contain glucosinolates.. . . . . .2. BRETSCHNEIDERACEAE.

1. AKANIACEAE

Stapf 1912. 1/1. Eastern Australia from southern coast of Queensland to northeastern New South Wales.

Akania.

2. BRETSCHNEIDERACEAE

Engler et Gilg 1924. 1/1. Western and southwestern China, northern Vietnam, Thailand, Taiwan.

Bretschneidera.

Closely related to Akaniaceae.

Bibliography

Bayer C and O Appel. 2003. Akaniaceae. In: K Kubitzki, ed. The families and genera of vascular plants, vol. 5, pp. 21–24. Springer, Berlin/Heidelberg/New York.

Boufford DE, A Kjaer, JO Madsen, and T Skrydstrop. 1989. Glucosilonates in Bretschneideraceae. Biochem. Syst. Ecol. 17: 75–379.

Carlquist S. 1996. Wood anatomy of Akaniaceae and Bretschneideraceae: a case of near-identity and its systematic implication. Syst. Bot. 21: 607–616.

Chaw S-M and C-I Peng. 1987. Palynological notes on *Bretschneidera sinensis* Hemsl. Bot. Bull. Acad. Sinica 28: 5–60.

Doweld AB. 1996a. Bretschneideraceae. In: A Takhtajan, ed. Comparative seed anatomy, vol. 5, pp. 346–348. Nauka, St. Petersburg (in Russian).

Doweld AB. 1996b. Akaniaceae. In: A Takhtajan, ed. Comparative seed anatomy, vol. 5, pp. 352–354. Nauka, St. Petersburg (in Russian).

Doweld AB. 1996c. The systematic relevance of fruit and seed anatomy and morphology of *Akania* (Akaniaceae). Bot. J. Linn. Soc. 120: 379–389.

Doweld AB. 1996d. The carpology and taxonomic relationships of *Bretschneidera* (Bretschneideraceae). Acta. Bot. Malacitana 21: 79–90.

Gadek PA, CJ Quinn, JE Rodman, KG Karol, E Conti, RA Price and ES Fernando. 1992. Affinities of the Australian endemic Akaniaceae: new evidence from the *rbc*L sequences. Aust. Syst. Bot. 5: 717–724.

Johnson MAT 1979 Chromosome numbers in *Akania* and *Cephalotus*. Kew Bull. 34: 37–38.

Liu C. 1986. Studies of pollen morphology in the Bretschneideraceae and the relative families. Acta Bot. Yunn. 8: 441–450 (in Chinese with English summary).

Ma ZW and GF He. 1992. Studies on the chemical constituents of *Bretschneidera sinensis* Hemsl. endemic in China. Acta Bot. Sinica 34: 483–484.

Romero EJ and LJ Hickey. 1976. A fossil leaf of Akaniaceae from Paleocene beds in Argentina. Bull. Torrey Bot. Club 103: 126–131.

Ronse Decraene LP and E Haston. 2006. The systematic relationships of glucosinolate-producing plants and related families: a cladistic investigation based on morphological and molecular characters. Bot. J. Linn. Soc. 151: 453–494.
Ronse Decraene LP, TYA Yang, P Schols, and EF Smets. 2002. Floral anatomy and systematics of *Bretschneidera* (Bretschneideraceae). Bot. J. Linn. Soc. 139: 29–45.
Stapf O. 1912. Akaniaceae. A new family of Sapindales. Kew Bull. 1912: 378–380.
Tang Y. 1935. Notes on the systematic position of Bretschneideraceae as shown by its timber anatomy. Bull. Fan Mem. Inst. Biol. 6: 153–157.
Tobe H and C-I Peng. 1990. The embryology and taxonomic relationships of *Bretschneidera* (Bretschneideraceae). Bot. J. Linn. Soc. 103: 139–152.
Tobe H and PH Raven. 1995. Embryology and relationships of *Akania* (Akaniaceae). Bot. J. Linn. Soc. 118: 261–274.
Yang D-Q and C-M Hu. 1985. The chromosomes of *Bretschneidera*. Hems. Notes Roy. Bot. Gard. Edinb. 42: 347–349.

Order 70. TROPAEOLALES

More or less succulent perennial or annual herbs, mostly climbing (petioles twining); perennial species with a rhizome, a tuber or corms. Leaves and stems are often covered by waxes. Stem with a ring of vascular bundles surrounding the large pith and eventually becoming connected through the activity of intravascular cambium. Vessels with simple perforations, but sometimes also with reticulate or intermediate perforations. Sieve-element plasdtids of S-type. Nodes trilacunar. Leaves alternate (or the lower opposite), more or less peltate, entire, more or less lobed to palmately divided, with palmate venation, stipules present or absent, sometimes present only in the seedling stage, stipules mostly small and caducous, but large and leaflike in a few species; petioles usually longer than the lamina, more or less cirrhose in climbing species. Flowers large, solitary, and axillary or rarely in umbellate or umbelliform inflorescences, usually with long and commonly pendant or subpendant, or rarely erect (*Magallana*) pedicels, usually without bracteoles, bisexual, usually strongly zygomorphic, 5-merous. Calyx 5-lobed, imbricate or valvate, the adaxial one or the adaxial and two lateral ones extended into usually long nectariferous spur (much reduced or absent in *Trophaeastrum*). Petals clawed, the three abaxial (lower) ones usually different from the other two, often with a hairy claw, rarely wanting. Stamens eight in a single whorl; filaments free; anthers small, basifixed, introrse, opening longitudinally. Tapetum secretory. Microsporogenesis simultaneous. Pollen grains 3-celled, 3-colporate or less often 2-colporate, with reticulate ornamentation. Gynoecium of three united carpels, with slender, terminal, apically trifid style; ovary superior, 3-locular, with a solitary, apical-axile, pendulous ovule per locule. Ovules anatropous or epitropous, bitegmic, tenuinucellate. Female gametophyte or *Polygonum*-type. Endosperm nuclear. The derivatives of the two uppermost cells of the proembryo divide repeatedly to produce the basal mass of cell that gives rise to very aggressive haustorial system, at least in *Tropaeolum majus*. The peripheral cells of the basal cell mass, close to the suspensor and on the opposite side of the funicle, give rise to a rootlike haustorium that penetrates the integuments below the micropyle and grows around the ovule in the carpel cavity and into the carpel wall; shortly another haustorium arises from the peripheral cells of the basal cell masses, close to the funicle, grows through the integument and funicle, reaches the vascular bundle of the placenta and grows along its curvature (Walker 1947; Bhalla et al. 1979). Fruits separating into 1-seeded fleshy or dry indehiscent mericarps or (*Magellana*) not separating, broadly 3-winged, with only one mericarp maturing; a short carpophore is found in *Magellana* and a rather conspicuous one in *Tropaeolum* sect. *Schizotrophaeum*. Seeds endotestal; embryo large, straight, rich in oil, amyloid and protein, with thick cotyledons; endosperm lacking. Contain myrosin cells and tard oil; present glucosinolates, erucic and eicosenic acids, n = 12–15, 21.

Related to the Capparales.

1. TROPAEOLACEAE

Berchtold et J.Presl 1820. 3/95. Mountains of America from Mexico to cool-temper-Chile (Tierra del Fuego).

Tropaeolum, Tropaeastrum, Magellana.

On the basis of the analysis of the elementary cycles of the flower of *Tropaeolum majus*, Chadefaud (1974) concluded that Tropaeolaceae belong to the Capparales. Dahlgren (1989) placed his Tropaeolales (Tropaeolaceae and Limnanthaceae). Tropaeolaceae shares with Capparales such common characters as the absence of bracteoles, racemose inflorescences, the presence of a gynophore, strongly developed marginal bundles extending into the style, and a few other, more generalized characteristics (Ronse Decraene and Smets 2001). However, strongly differ from them in the axile placentation of the tenuinucellate ovules, fruit morphology and dehiscence, the absence of parietal cells, and very aggressive haustoria.

Bibliography

Andersson L and S Andersson. 2000. A molecular phylogeny of Tropaeolaceae and its systematic implication. Taxon 49: 721–736.

Arnal C and J Loiseau. 1946. L'eperon de la fleur du *Tropaeolum majus*. C. R. Acad. Sci. Paris 223: 361–364.

Batygina TB. 1985. Tropaeolaceae. In: MS Yakovlev, ed. Comparative embryology of flowering plants: Brunelliaceae–Tremandraceae, pp. 215–219. Nauka, Leningrad (in Russian).

Bayer C and O Appel. 2003. Tropaeolaceae. In: K Kubitzki, ed. The families and genera of vascular plants, vol. 5, pp. 400–404. Springer, Berlin/Heidelberg/New York.

Bersier JD. 1960. L'ovule et la placentation dans le genre *Tropaeolum*. Arch. Sci. Geneve. 13: 566–567.

Bhalla PL and CP Malik. 1982. Localization and activity of some glycosidases during early embryogenesis in *Tropaeolum majus* L. J. Indian Bot. Soc. 61: 91–94.

Bhalla PL, MB Singh, and CP Malik. 1979. Physiology and sexual reproduction: VI. Embryogenesis in *Tropaeolum majus* L.: enzyme changes. Acta Bot. Indica 7: 72–86.

Bhalla PL, MB Singh, and CP Malik. 1982. Post-fertilization developmental time table for *Tropaeolum majus*. Acta Bot. Indica 10: 201–205.

Boesewinket FD and F Bouman. 2000. Tropaeolaceae. In: A Takhtajan, ed. Comparative seed anatomy, vol. 6, pp. 43–45. Nauka, St. Petersburg (in Russian).

Bolenbach A. 1928. Microsporogenesis in *Tropaeolum majus* with special reference to the cleavage process in tetrad formation. Bull. Torrey Bot. Club 55: 105–115.

Brunotte C. 1900. Recherches embryogeniques et anatomiques sur quelques especes des genres *Impatiens* (L.) et *Tropaeolum* (L.). Thesis, University of Paris.

Buchenau F. 1878. Bildungsabweichungen der Blüte von *Tropaeolum majus*. Abh. Nat. Ver. Bremen. 5: 599–641.

Buchenau F. 1896. Der Blütenbau von *Tropaeolum*. Abh. Nat. Ver. Bremen. 13: 383–407.

Carlquist S and CJ Donald. 1996. Wood anatomy of Limnanthaceae and Tropaeolaceae in relation to habit and phylogeny. SIDA 17: 333–342.

Chadefaud M. 1974. Sur la formule florale de la Capucine (*Tropaeolum majus* L.). Bull. Bot. France 121: 347–355.

Devi DR and LL Narayana. 1994. Floral anatomy of Tropaeolaceae. Feddes Repert. 105: 437–443.

Dickson A. 1872. On embryogeny of *Tropaeolum peregrinum* and *T. speciosum*. Trans. Roy. Soc. Edinb. 27: 222–235.

Fuchs C. 1975. Ontogenèse foliaire et acquisition de la forme chez le *Tropaeolum peregrinum* L. I. Les premiers stades de l'ontogenèse du lobe median. Ann. Sci. Nat., Bot. XII Biol. Vég. 16: 321–389.

Fuchs C. 1976. Ontogenèse foliaire et acquisition de la forme chez le *Tropaeolum peregrinum* L. II. Le développement du lobe après la formation des lobules. Ann. Sci. Nat., Bot. XII Biol. Vég. 17: 121–158.

Gibbs PE, D Marshall, and D Brunton. 1978. Studies on the cytology of *Oxalis tuberose* and *Tropaeolum tuberosum*. Notes Bot. Gard. Edinb. 37: 215–220.

Hoffman U. 1987. Der Bau des Gynoeceums von *Tropaeolum* und *Pelargonium*. Bot. Jahrb. Syst. 108: 439–448.

Huynh K-L. 1968. Morphologic du pollennes Tropaeolacees et des Balsaminacees, part 1. Grana Palynol. 8: 88–184.

Kjaer A, JO Madsen, and Y Mael. 1978. Seed volatiles within the family Tropaeolaceae. Phytochemistry 17: 1285–1287.

Loiseau JE. 1947. Sur ιεοργανισατιον du gynecee chez les Tropaeolacees. Ann. Fac. Sc. 26: 125–147.

Rachmilevitz T. and A. Fahn. 1975. The floral nectary of *Tropaeolum majus* – the nature of the secretory cells and the cells and the manner of nectar secretion. Ann. Bot. 39: 721–728.

Rama Devi D. 1990. Chemotaxonomy of Tropaeolaceae. Indian J. Bot. 13: 136–141.

Rama Devi D and LL Narayana. 1994. Floral anatomy of Tropaeolaceae. Feddes Repert. 105: 7–8.

Ricardi M, C Marticorena and F Torres. 1957. Nota preliminary de los pólenes de Tropaeolaceae chilenas. Bol. Soc. Biol. Concepción 32: 17–19.

Ronse Decraene LP and EF Smets. 2001. Floral developmental evidence for the systematic relationships of *Tropaeolum* (Tropaeolaceae). Ann. Bot. 88, 5: 879–892.

Schacht H. 1855. Über die Entstehung des Keimes von *Tropaeolum majus*. Bot. Z. 13: 188–208.

Sparre B and L Andersson. 1991. A taxonomic revision of the Tropaeolaceae. Opera Bot. 108: 1–139.

Tiwari SC, F Bouman, and RN Kapil. 1977. Ovule ontogeny in *Tropaeolum majus*. Phytomorphology 27: 350–358.

Walker RI. 1947. Megasporogenesis and embryo development in *Tropaeolum majus* L. Bull. Torrey Bot. Club. 74: 240–249.

Weberling F and L Müller. 1989. Persistierende Blütensporne bei *Tropaeolum*. Flora 169: 295–298.

Woycicki M Z. 1907. Über den Bau des Embryosackes bei *Tropaeolum majus*. Bull. Internat. Acad. Sci. Cracovie 550–557.

Order 71. LIMNANTHALES

Small, delicate annuals of damp places. Vascular system of the stem consists of a ring of 8–10 separate bundles. Vessels with simple perforation. Phloem weakly developed. Sieve-element plastids of S-type. Nodes unilacunar. Leaves alternate, simple or compound (pinnatisect to pinnate), estipulate. Stomata anomocytic. Flowers solitary on long, axillary pedicels, bisexual, actinomorphic, (4)5-merous (*Limnanthes*) or 3-merous (*Floerkea*), Sepals more or less free, valvate (*Limnanthes*), or slightly imbricate (*Floerkea*), persistent, sometimes accrescent. Petals free, contorted, persistent. Stamens 3, or 6, or 8, or 10; twice as many as petals and in two cycles or sometimes in *Floerkea* as many as and alternate with petals; filaments free, opposite sepals with nectary gland outside the base; anthers tetrasporangiate, introrse, dorsifixed, opening longitudinally. Floral nectaries present and adnate to the antesepalous stamens. Tapetum secretory. Microsporogenesis simultaneous. Pollen grains 2-celled, 2–4-colporate, heteropolar (with a smaller distal and a larger proximal area), of very unique zono-colpate type – with two zonocolpi running parallel to the equator. Gynoecium of 5(4) carpels in *Limnanthes* and 2–3 carpels in *Floerkea;* placentation basal-parietal, style gynobasic, more or less deeply cleft or entire and with lobed or capitate stigma; ovary superior to partly inferior, deeply lobed into globular segments each with one basal or ascending ovule. Ovules ascending, anatropous, apotropous, unitegmic, tenuinucellate; the integument is massive and shows a conspicuous vascular supply. Female gametophyte tetrasporic, but pseudo-monosporic (only one of the four megaspore nuclei contributes to the formation of the female gametophyte), a modified and extremely reduced variation of the *Drusa*-type. in which the activity of the chalazal nucleus is very much restricted (Mathur 1956; Maheshwari and John 1956; John 1970); antipodals usually absent in the mature female gametophyte. Endosperm nuclear. Fruits separating into indehiscent, 1-seeded mericarps that remain adherent to the seed coat; the epidermis of the pericarp develops numerous pyramidal thickenings that give the pericarp a specific pattern. Seeds small, with straight green embryo, with large cordate, fleshy cotyledons and small radicle, without endosperm. Seed coat is weakly differentiated testa, and in the later stage its cells are full of starch. Contains glucosinolate (m-methoxybenzyl isothiocyanate), erucic acids (cis-13-docosenoic, cis-5-eicosinoic, cis-5-docosenoic, cis-5, cis-13-docosadienoic and other fatty acids in small amounts), tannins, elagic acid derivatives (myricetins and proanthocyanidins), myrosinase and myrosin cells and mustard oils (Bayer and Appel 2003). n = 5.

Usually placed in the Geraniales, from which the Limnanthaceae differ in their apotropous and tenuinucellate ovules without endothelium, unique types of pollen grains and female gametophyte, usually diplostemonous androecium, valvate sepals, persistent petals, unilacunar nodes, as well as in chemical characters. Many years ago Richard Wettstein in his *Handbuch der systematischen Botanik* (1907, 1935) suggested that Limnanthaceae with their tenuinucellate ovules belong to the Sympetalae rather than to the Choripetalae. On the other hand, in 1991 Rosalie Wunderlich in her posthumously published work on the systematic position of the Limnanthaceae concluded that they are closely related to the Boraginaceae sensu stricto or to the subfamily Boragoideae. According to Wunderlich, the Limnanthaceae show essential embryological characters that are found in all Boraginaceae s. str., such as tenuinucellate ovule with a single multilayered integument, lack of an endothelium, and presence of branched vascular bundles in the integument. The development of the seed coat and pericarp also shows similar features in both families. Wunderlich mentions also some other characters of the Limnanthaceae that are found in almost all Boraginaceae s. str., such as a single gynobasic style, ovary deeply divided into 1-seeded mericarps, a persistent calyx, etc. However, she also mentions differences between the two families, such as pinnatifid to pinnate leaves of the Limnanthaceae, their distinct petals, the diplostemonous androecium, the number of carpels with only one ovule in each carpel. The free petals of the Limnanthaceae, according to Wunderlich, are perhaps a very reduced variant of the corolla tubes of the Boraginaceae s. str., which differ very much in length (from long to quite short). Wunderlich concluded, "There is no doubt that the Limnanthaceae represent a separate family which, however, is closely related to the Boraginaceae s. str. and ought to be placed immediately beside these in the same order." Unfortunately, Wunderlich does not include in her list of differences between the two families such an important character as the presence or absence of myrosin cells and mustard oils.

Dahlgren (1975), Kubitzki (2003), Thorne (2006) include Limnanthaceae in the Capparales (Brassicales). The affinity with the brassicalian families is supported mainly by the occurrence of myrosin cells, glucosinolates and erucic acid. However from the Brassicales they markedly differ in unique type of pollen grains, unitegmic ovules and female gametophyte.

1. LIMNANTHACEAE

R. Brown 1833. 2/11. Temperate North America.
Limnanthes, Floerkea.

Bibliography

Bayer C and O Appel. 2003. Limnanthaceae. In: K Kubitzki, ed. The families and genera of vascular plants, vol. 5, pp. 220–224. Springer, Berlin/Heidelberg/New York.

Buchner R, H Halbritter, G Pfunder, and M Hesse. 1990. Pollen of *Limnanthes douglasii*: a reinvestigation. Grana 29: 207–211.

Carlquist S and CJ Donald. 1996. Wood anatomy of Limnanthaceae and Tropaeolaceae in relation to habit and phylogeny. SIDA 17: 333–342.

Eckert G. 1966. Entwicklungsgeschichtliche blütenanatomische Untersuchungen zum Problem der Obdiplostemonie. Bot. Jahrb. Syst. 85: 523–604.

Ettlinger MG, AJ Lundeen. 1956. The mustard oil of *Limnanthes douglasii* seed, m-methoxybenzyl isothiocyananthe. J. Am. Chem. Soc. 78: 1952–1954.

Fries M. 1936. Über die Chromosomenzahl bei zwei *Limnanthes*-Arten. Svensk Bot. Tidskr. 30: 440–442.

Hauptli H, BD Webster, S Jain. 1978. Variation in nutlet morphology of *Limnanthes*. Am. J. Bot. 65: 615–624.

Hofmann U and J Ludewig. 1985. Morphologische und systematische Stellung von *Limnanthes douglasii* R. Brown, einem reprasentativen Vertreter der Limnanthaceae. Bot. Jahrb. Syst. 105: 401–431.

Huynh K-L. 1971. The morphological development of the pollen of *Limnanthes douglasii* (Limnanthaceae). Grana 11: 58–61.

Huynh K-L. 1972. The original position of the generative nucleus in the pollen tetrads of *Agropyron, Itea, Limnanthes,* and *Onosma*, and its phylogenetic significance in the angiosperms. Grana 12: 105–112.

Huynh K-L. 1982. Le pollen du *Limnanthes douglasii* (Limnanthaceae) en microscopie électronique. Pollen et Spores 24: 211–234.

Johri BM. 1970. Limnanthaceae. Bull. Indian Natl. Sci. Acad. 41: 110–113.

Johri BM and P Maheshwari. 1951. The embryo sac of *Floerkea proserpinacoides* Willd. Curr. Sci. 20: 44–46.

Kesseli RV and SK Jain. 1984. New variation and biosystematic patterns detected by allozyme and morphological comparisons in Limnanthaceae sect. Reflexae (Limnanthaceae). Plant. Syst. Evol. 147: 133–165.

Link DA. 1992. The floral nectaries in the Limnanthaceae. Plant Syst. Evol. 179: 235–243.

Maheshwari P and BM Johri. 1956. The morphology and embryology of *Floerkea proserpinacoides* Willd. with a discussion of the systematic position of the family Limnanthaceae. Bot. Mag. Tokyo 69: 410–423.

Mason Ch T, Jr. 1951. Development of the embryo sac in the genus *Limnanthes*. Am. J. Bot. 38: 17–22.

Mason Ch T, Jr. 1952. A systematic study of the genus *Limnanthes* R. Br. Univ. Calif. Publ. Bot. 25: 455–512.

Mathur N. 1956. The embryology *of Limnanthes*. Phytomorphology 6: 41–51.

Ornduff R. 1971. Systematic studies of Limnanthaceae. Madroño 21: 103–111.

Ornduff R and TJ Crovello. 1968. Numerical taxonomy of Limnanthaceae. Am. J. Bot. 55: 173–182.

Parker WH. 1976. Comparison of numerical taxonomic methods used to estimate flavonoid similarities in the Limnanthaceae. Brittonia 28: 390–399.

Parker WH and BA Bohm. 1979. Flavonoids and taxonomy of the Limnanthaceae. Am. J. Bot. 66: 191–197.

Phillips BE, CR Smith Jr., and WH Tallent. 1971. Glycerides of *Limnanthes douglasii* seed oil. Lipids 6: 93–99.

Propach H. 1934. Cytologische Untersuchungen an *Limnanthes douglasii* R. Br. Zeitschr. Zellforsch. 21: 357–375.

Rama Devi D. 1990. Floral anatomy of Limnanthaceae. J. Indian Bot. Soc. 69: 271–274.

Russell AM. 1919. A comparative study of *Floerkea proserpinacoides* and allies. Contrib. Bot. Lab. Univ. Pennsylvania 4: 401–418.

Stenar H. 1925. Embryologische und zytologische Studien über *Limnanthes douglasii* R. Br. Svensk Bot. Tidskr. 19: 133–152.

Wunderlich R. 1991. Zur Frage nach der systematischen Stellung der Limnanthaceae. Stapfia 25: 1–59.

Zhukova GY. 1985. Limnanthaceae. In: MS Yakovlev, ed. Comparative embryology of flowering plants: III. Brunelliaceae–Tremandraceae, pp. 219–221. Nauka, Leningrad (in Russian).

Order 72. CARICALES

Small, palmlike trees or shrubs with soft wood and usually unbranched, sometimes succulent stems with an apical cluster of leaves, rarely perennial, prostrate (*Jarilla*) herbs, usually prickly. Well-developed anastomosing, articulated laticifers present in the ground tissue of all parts of the plants; they are cellular in origin (Kubitzki 2003). Stomatal myrosin cells present in the epidermis of all green parts (leaves, stems and immature fruits) and also in cotyledons from germinating seeds (Jorgensen 1995). Vessels with simple perforations. Except for the vessels, the xylem consists of unlignified cells; only in a *Calicomorpha* the paratrachelal parenchyma was found to be lignified (Carlquist 1998). Vessels without vestured pits. Nodes trilacunar or multilacunar. Leaves alternate, usually large and long-petiolate, mostly palmately lobed or palmately compound, estipulate or rarely stipules present and spinelike. Stomata anomocytic. Inflorescences axillary, cymose thyrses, 1- to many-flowered. Flowers usually unisexual, mostly dioecious, or more rarely monoecious or polygamous, but sometimes (*Carica*) some flowers bisexual, actinomorphic, 5(4)-merous. Sepals, small, united into a toothed or lobed calyx. Petals contort or valvate, united into an elongate, slender tube in male flowers and short or very short tube in female flowers. Stamens 10, in two cycles, or less often five, inserted in the throat of the corolla tube; filaments free (*Carica*) of one another, or basally connate into a short tube; anthers sometimes with shortly prolonged connective, 2-locular, introrse, dorsifixed to basifixed, tetrasporangiate, opening longitudinally. Tapetum glandular. Microsporogenesis simultaneous. Pollen grains 2-celled, 3-colporate. Gynoecium of 5(4) carpels; stylodia free; ovary superior, 1-locular with deeply intrusive parietal placentas or the placentas meet and join in the center to form a plurilocular ovary. Ovules numerous, with more or less enlarged funicle, anatropous, bitegmic, crassinucellate. Female gametophyte *Polygonum*-type. Endosperm nuclear. Fruits fleshy, large, melonlike berries. Seeds with juicy mucilaginous sarcotesta that surrounds the hard sclerotesta; seed coat formed by both integuments (Komar 1992), mesotesta tanniniferous, with lignified ribs; endotesta crystalliferous, exotegmen fibrous; embryo straight, spatulate, embedded in the fleshy and containing both oil and protein endosperm. Producing myrosinases and glucosilonates, and also contain cyanogenic glycosides (Spencer and Seigler 1984). Often present mustard-oils, sometimes alkaloids carpaine. The latex contains large amounts of the proteolytic enzyme, papain, n = 9.

Related to Moringales, but strongly differ in many characters, including actinomorphic and unisexual flowers, in baccate fruits, presence of endosperm. The close similarity between the two orders in wood anatomy and as well as molecular analyses (e.g., Rodman et al. 1996).

1. CARICACEAE

Dumortier 1829. 5/35. West Indies, tropics and subtropics of America from Mexico to northern Chile and Argentina, and tropical West Africa, centered in South America; *Cylicomorpha* (2) is endemic to tropical West Africa.

Jacaratia, Cylicomorpha, Carica (including and *Horovitzia*), *Vasconcellea, Jarilla.*

Bibliography

Aradhya MK, RM Manshardt, F Zee, and CW Morden. 1999. A phylogenetic analysis of the genus *Carica* L. (Caricaceae) based on fragment length variation in a cpDNA intergenic spacer region. Genet. Resour. Crop Evol. 46: 579–586.

Badillo VM. 1971. Monographia de la familia Caricaceae. Publ. Assoc. Prof., Universidad Central de Venezuela, Maracay.

Badillo VM. 1993. Caricaceae. Segundo esquema. Rev. Fac. Agron. Univ. Centr. Venezuela. Maracay.

Baker HG. 1976. "Mistake" pollination as a reproductive system with special reference to the Caricaceae. In: J Burley and BT Styles, eds. Tropical trees – variation, breeding and conservation. Linn. Soc. Symp. Ser. 2, pp. 161–169. Academic, London.

Bawa KS. 1980. Mimicry of male by female flowers and intrasexual competition for pollinators in *Jacaratia dolichaula* (D. Smith) Woodson (Caricaceae). Evolution 34: 467–474.

Benitez de Rojas CE. 1974. Characteres microscopicos de la epidermis foliar en Caricaceae. Genero *Carica*. Bot. Fac. Agron. (Maracay) 7: 195–274.

Carlquist S. 1998. Wood and bark anatomy of Caricaceae: correlations with systematics and habit. IAWA J. 19: 191–206.

Dathan ASR and D Singh. 1970. Female gametophyte and seed of *Carica candamarensis* Hook. f. Plant (Lucknow, India) 2: 52–60.

Devi S. 1952. Studies in the order Parietales: III. Vascular anatomy of the flower of *Carica papaya* L., with special reference to the structure of the gynoecium. Proc. Indian Acad. Sci. 36B: 59–69.

Fischer JB. 1980. The vegetative and reproductive structure of papaya (*Carica papaya*). Lyonia 1: 191–207.

Gmelin R and A Kjaer. 1970. Glucosilonates in the Caricaceae. Phytochemistry 9: 591–593.

Jorgensen LB. 1981. Guard cell myrosin cells in *Carica papaya*. In: DJ Carr, ed., Abstract XIII Int. Bot. Congr., p. 242. Sydney.

Jorgensen LB. 1995. Stomatal myrosin cells in Caricaceae. Taxonomic implications for a glucosinolate-containing family. Nord. J. Bot. 15: 523–540.

Komar GA. 1992. Caricaceae. In: AL Takhtajan, ed. Comparative seed anatomy, vol. 4, pp. 124–127. Nauka, St. Petersburg (in Russian).

Kubitzki K. 2003. Caricaceae. In: K Kubitzki, ed. The families and genera of vascular plants, vol. 5, pp. 57–61. Springer, Berlin/Heidelberg/New York.

Kyndt T, B van Droogenbroeck, E Romeijn-Peeters, JP Romero-Motochi, X Scheldeman, P Goetghebeur, P van Damme, and G Gheysen. 2005. Molecular phylogeny and evolution of Caricaceae based on rDNA internal transcribed spacers and chloroplast sequence data. Molec. Phylog. Evol. 37: 442–459.

Kyndt T, E Romeijn-Peeters, B van Droogenbroeck, JP Romero-Motochi, G Gheysen, and P Goetghebeur. 2005. Species relationships in the genus *Vasconcellea* (Caricaceae) based on molecular and morphological evidence. Am. J. Bot. 92: 1033–1044.

Miller NG. 1982. The Caricaceae in the southeastern United States. J. Arnold Arbor. 63: 411–427.

Olson ME. 2002. Intergeneric relationships within the Caricaceae-Moringaceae clade (Brassicales) and potential morphological synapomorphies of the clade and its families. Int. J. Plant Sci. 163(1): 51–65.

Ronse Decraene LP and E Smets. 1999. The floral development and anatomy of *Carica papaya* (Caricaceae). Canad. J. Bot. 77: 582–598.

Roth I. 1972. Desarollo y anatomia del fruto y de la semilla de *Carica papaya* L. (Lechosa). Acta Bot. Venez. 7: 1–4.

Spencer KC and DS Seigler. 1984. Cyanogenic glycosides of *Carica papaya* and its phylogenetic position with respect to the Violales and Capparales. Am. J. Bot. 71: 1444–1447.

Sprecher A. 1943. Beitrag zur Morphologie von *Carica papaya* L. Ber. Schweiz. Bot. Ges. 53A: 517–549.

Tang C-S. 1971. Benzyl isothiocyanate of papaya fruit. Phytochemistry 10: 117–131.
Van Droogenbroeck, BT Kyndt, I Maertens, E Romeijn-Peeters, X Scheldeman, JP Romero-Motochi, P van Damme, P Goetghebeur, and G Gyheysen. 2004. Phylogenetic analysis of the highland papayas (*Vasconcellea*) and allied genera (Caricaceae) using PCR-RFLP. Theoret. Appl. Genet. 108: 1473–1486.

Order 73. MORINGALES

Deciduous trees of very different habitus (bottle trees, sarcorhizal trees, slender trees) or tuberous shrubs, exuding a gum resembling tragacanth from bark wounds. Contain myrosin cells and cells with protein-rich, dilated cisternae of the endoplasmic reticulum. Hairs unicellular. Vessels with simple perforations; lateral pitting alternate. Fibers libriform, with large lenticular, simple pits. Rays homogeneous. Axial parenchyma paratracheal of various types. Vessels with vestured pits. Sieve-element plastids of S-type. Leaves alternate, two to three times pinnately compound, with opposite leaflets, estipulate or stipules represented by stipitate glands. Epidermis often mucilaginous. Stomata anomocytic. Flowers in axillary panicles, bisexual, usually more or less zygomorphic, with a saucer-shaped or shortly tubular, often oblique hypanthium, formed partly by an invaginated receptacle. Sepals five, imbricate, equal or unequal, petaloid. Petals five, imbricate, the two posterior ones smaller and reflexed, the two lateral ones ascending, and the anterior one the largest. The hypanthium is lined below with nectary disc. Functional stamens five, inserted on the hypanthium around the margin of the disc, free, declinate, of unequal length, alternating with the outer cycle of 3–5 staminodia; anthers 1-locular, usually disporangiate, seldom tetrasporangiate, dorsifixed, opening longitudinally, forming a head through which the style protrudes at anthesis. Tapetum secretory. Microsporogenesis simultaneous. Pollen grains 2-celled or rarely 3-celled, 3-colporate with costate colpi, the tectum thick and the interstitium granular, the ornamentation psilate, very sparsely and finely perforate with irregularly shaped microslits and puncta, the endexine very thick at the apertures but almost absent from the mesocolpial and polar areas, the foot layer thick (Ferguson 1985). Gynoecium of three united carpels with slender, tubular style terminated by truncate stigma; ovary superior, on a short gynophore, villous, curved, 1-locular, with numerous ovules in two rows on each parietal placenta. Ovules pendulous, anatropous, bitegmic, crassinucellate. Female gametophyte of *Polygonum*-type. Endosperm nuclear. Fruits large, beaked, 3–6-angled, torulose, but without a replum, elongate, explosively dehiscent capsules. Seeds numerous, mostly 3(4)-winged; seed coat formed by the outer integument (mesotestal); testa vascularised, mesotesta thick, outer and inner parts with helical thickenings, tegmen thin (multiplicative); embryo large, green, straight, oily, with well-developed plumule; endosperm scanty or lacking. Producing glucosilonates and leucine; mustard-oils present in myrosin cells; present flavonols (kaempferol and quercetin); n = 11, 14.

Related to the Capparales, with which they have some similarities, including myrosin cells strikingly similar to those in Capparaceae, Brassicaceae, and Tovariaceae (Jørgensen 1981). Moringaceae and Capparaceae also share similarities in carpel vasculature (Ronse Decraene et al. 1998), but differ in pinnately compound leaves, disporangiate anthers, rather distinctive combination (including ornamentation) of pollen characters (Ferguson 1985), and in chemical characters (Dutt et al. 1984).

1. MORINGACEAE

Martynov 1820. 1/14. From northern and northeastern Africa and across Arabian Peninsula and Iran to Pakistan, India and Bangladesh, as well as southern Angola, southwestern Africa, and Madagascar.

Moringa.

Bibliography

Datta RM and IN Mitra. 1947. The systematic position of the family Moringaceae based on a study of *Moringa pterigosperma* Gaerth. (*M. oleifera* Lam.). J. Bot. Nat. Hist. Soc. 47: 355–357.
Dave YS, ND Patel, and SV Desai. 1974. Pericarpal studies in the developing fruit of Horse-dadish (*Moringa oleifera* Lamk.). Flora 163: 398–404.
Dutt BSM. 1978a. Anther in *Moringa concanensis* Nimmo. Curr. Sci. 47: 589.
Dutt BSM. 1978b. Embryo development in *Moringa concanensis* Nimmo. Curr. Sci. 48: 693.

Dutt BSM. 1979. Ovule and seed of *Moringa concanensis* Nimmo. Curr. Sci. 48: 652–654.
Dutt BSM, LL Narayana, and A Parvathi. 1978. Floral anatomy of *Moringa concanensis* Nimmo. Indian J. Bot. 1: 35–39.
Dutt BSM, LL Narayana, M Radhakrishnaiah, and G Nageshwar. 1984. Systematic position of *Moringa*. J. Econ. Taxon. Bot. 5(3): 577–580.
Ernst WR. 1963. The genera of Capparaceae and Mor-ingaceae in the southeastern United States. J. Arnold Arbor. 44: 81–95.
Ferguson IK. 1985. The pollen morphology of Moringaceae. Kew Bull. 40: 25–34.
Fisel KJ. 1989. Untersuchungen zur Morphologie und Ontogenie der Blüten von *Tovaria pendula* Ruiz et Pavon und *Moringa oleifera* Lam. Diploma thesis, Faculty Nat. Sci. Math., University of Ulm, Germany.
Ghasi S, E Nwodobo, and JO Ofili. 2000. Hypocholesterolemic effects of crude extract of leaf of *Moringa oleifera* Lam. in highfat diet fed wistar rats. J. Ethnopharm. 69: 21–25.
Gill LS, YY Karatela, BL Lamina, and SWH Husaini. 1985. Cytology and histomorphology of *Moringa oleifera* Lam. (Moringaceae). Feddes Repert. 96: 299–305.
Komar GA. 1992. Moringaceae. In: AL Takhtajan, ed. Comparative seed anatomy, vol. 4, pp. 209–212. Nauka, St. Petersburg (in Russian).
Kubitzki K. 2003. Moringaceae. In: K Kubitzki, ed. The families and genera or vascular plants, vol. 5, pp. 312–314. Springer, Berlin/Heidelberg/New York.
Narayana HS. 1970. Moringaceae. Bull. Indian Nat. Sci. Acad. 41: 78–83.
Olson ME. 2002a. Intergeneric relationships within the Caricaceae-Moringaceae clade (Brassicales) and potential morphological synapomorphies of the clade and its families. Int. J. Plant Sci. 163: 51–65.
Olson ME. 2002b. Combining data from DNA sequences and morphology for a phylogeny of Moringaceae (Brassicales). Syst. Bot. 27: 55–73.
Olson ME. 2003a. Wood and bark anatomy in *Moringa* (Moringaceae). Haseltonia 8: 85–121.
Olson ME. 2003b. Ontogenetic origins of floral bilateral symmetry in Moringaceae (Brassicales). Am. J. Bot. 90: 49–71.
Olson ME and S Carlquist. 2001. Stem and root anatomical correlations with life form diversity, ecology, and systematics in *Moringa* (Moringaceae). Bot. J. Linn. Soc. 135: 315–348.
Olson ME and JA Rosell. 2006. Using heterochrony to detect modularity in the evolution of stem diversity in the plant family Moringaceae. Evolution 2006: 724–734.
Periasamy K and C Indira. 1986. The carpel of *Moringa*. Ann. Bot. 58: 897–901.
Puri V. 1941. Life-history of *Moringa oleifera* Lam. J. Indian Bot. Soc. 20: 263–284.
Rao NV, S Avita, and JA Inambar. 1983. Studies on the Moringaceae. Feddes Repert. 94: 213–223.
Ronse Decraene LP, J De Laet, and E Smets. 1998. Floral development and anatomy of *Moringa oleifera* (Moringaceae): what is the evidence for a capparalean or sapindalean affinity? Ann. Bot. 82: 273–284.
Verdcourt B. 1985. A synopsis of the Moringaceae. Kew Bull. 48: 1–23.

Order 74. CAPPARALES (RESEDALES, BRASSICALES)

Trees, shrubs, or most often herbs, containing myrosin cells and cells with protein-rich, dilated cisternae of the endoplasmic reticulum. Vessels with simple perforations; lateral pitting alternate. Fibers with simple or bordered pits. Rays heterogeneous or homogeneous. Axial parenchyma apotracheal or paratracheal, rarely diffuse (some Gyrostemonaceae). Sieve-element plastids of S-type or sometimes (some members of Capparaceae and Brassicaceae) of Pfcs-type. Nodes unilacunar or trilacunar. Leaves alternate or rarely opposite, simple or often variously compound, with or without stipules. Stomata anomocytic, anisocytic, or paracytic. Flowers mostly in terminal racemes, bisexual or sometimes unisexual, actinomorphic or zygomorphic, with double perianth or sometimes apetalous, with nectaries of receptacular origin. Perianth 4-merous or less often 5-merous, with free segments. Stamens (2-)4-many, arising from a limited number (most often four) of primordia and centrifugal when in more than one series. In some Capparaceae and in most of Brassicaceae two median primordia divide, and therefore the number of stamens reaches six; in other cases all the four primordia branch once or more and as a result the androecium consists of eight or more stamens. Anthers tetrasporangiate, opening longitudinally. Tapetum secretory. Microsporogenesis simultaneous. Pollen grains 2-celled or 3-celled, mostly 3-colpate or 3-colporate. The extrastaminal nectaries present. Gynoecium paracarpous, of two or less often 3–6 to 12, rarely (Gyrostemonaceae) up to 60 united carpels with stylodia united into a shorter elongated style with a simple or lobed stigma or (Resedaceae) stigmas separate and sessile. Ovary superior, basically 1-locular or partitioned by replum (a thin, unvasculated frame-like partition connecting the two parietal placentas), with usually more or less numerous ovules, sometimes only with one ovule. Ovules anatropous or more often campylotropous, bitegmic, crassinucellate or less often tenuinucellate (part of Brassicaceae and Resedaceae). Female gametophyte usually of *Polygonum*-type. Endosperm nuclear. Fruits of various types. Seeds with large, oily, more or less curved or folded embryo, mostly without endosperm or with scanty endosperm, rarely (Tovariaceae) with well-developed endosperm. Mostly producing glucosilonates (mustard-oil glucosides, thioglucosides), sometimes with alkaloidal amines and sometimes cyanogenic, several genera are accumulators of heavy metals, lacking ellagic acid.

Derived from the archaic representatives of Violales, most likely from a flacourtiaceous stock.

Key to Families

1 Endosperm copious. Petals absent. Trees or shrubs with normal secondary growth, or biennial or annual herbs (*Tersonia*). Axial parenchyma mostly paratracheal, but sometimes also diffuse. Leaves entire, simple, often succulent, sessile or petiolate, stipules very small or absent. Flowers small, actinomorphic or almost so, solitary and axillary or in terminal or axillary racemes or spikes, bracteolate, unisexual and mostly dioecious or monoecious, apetalous. Calyx broadly cupular, more or less lobed or truncate, persistent in fruit, the lobes imbricate in bud. Stamens seven or more (up to 100), in one or several (*Gyrostemon*) concentric series around the edge of the expanded receptacle or (*Gyrostemon*) covering the receptacle; anthers almost sessile, quadrangular. Pollen grains 2-celled, 3-colpate, occasionally 2- or 4-colpate, with thick, granular, undifferentiated ectexine. Gynoecium of 2-many (up to 60) carpels, either connate around a central column or fused, sometimes solitary (*Cypselocarpus*); stigmas decurrent, large and spreading or not (marginal). Ovary with as many locules as carpels (carpel solitary in the genus *Cypselocarpus*). Ovules solitary in each locule, campylotropous, with a thickened funicle. Fruits dry or sometimes succulent, dehiscing septicidally, loculicidally, or both or less often indehiscent, carpels often falling before shedding seeds. Seeds small, hippocrepiform, usually rugose, red-brown, with prominent or small (in *Cypselocarpus* very small), translucent exostomal and funicular aril around the hilum; seed surface with both large and minute ribs; seed coat exotegmic, with cells of exotegmen longitudinally elongate and unevenly radially enlarged, thick-walled, pitted and thus fibrous. Embryo curved around the copious, oily, and fleshy endosperm. Containing glucosilonates (mustard oil glucosides) and lacking both betalains and anthocyanins, n = 14. . . 5. GYROSTEMONACEAE.

1 Endosperm scanty or absent.

2 Ovary usually open at the top from very early stage and bears small, sessile, well-separated stigmas around the rim; it is unilocular, usually with numerous, anatropous to campylotropous ovules, rarely (*Caylusea*) ovules 10–18 on united basal placenta, or (*Sesamoides*) only one or rarely two ovules on laminar placentas of secondarily free carpels. Perennial, biennial or annual herbs, subshrubs, or shrubs (*Ochradenus* and *Randonia*); stems erect, rarely winding. Elongate myrosine cells present in the phloem of the roots, in the stems, in the leaf epidermis and embryo. Nodes unilocular. Fibers with simple pits. Endoplasmic reticulum with ER-dependent vacuoles. Leaves alternate, simple, entire or deeply pinnatifid, with small stipules modified into glands. Stomata anomocytic. Flowers in racemes or spikes, without bracteoles and often without bracts (as in Brassicaceae), bisexual or less often unisexual, more or less strongly zygomorphic, with a short androgynophore or gynophore; androgynophore usually with an exstrastaminal nectary disc often dilated at the adaxial side. Sepals 4–8, valvate or slightly imbricate, inserted below or rarely on the disc. Petals 4–8 or sometimes two, or absent, free, unequal, valvate, posterior ones usually much larger and often incised or laciniate. Stamens (3-)-16–22 or more; anthers introrse, bithecate. Pollen grains 2-celled, 3-colpate or 3-colporate. Gynoecium of (2)3–6(-8) free carpels. Fruits usually an apically open capsule, sometimes baccate (*Ochradenus*), or (*Sesamoides*) of separate spreading carpels. Seeds reniform, often arillate, endotestal cells cuboid, more or less thickened, unlignified, exotegmic cells fibrous, lignified; embryo large, curved, or folded; endosperm little or wanting, n = 6–15. . . . . 4. RESEDACEAE.

2 Ovary closed.

3 Ovary 4–8-locular.

4 Perianth 5-merous. Large glabrous shrubs or lianas; mucilage cells present. Nodes trilacunar with three traces. Leaves alternate, simple, entire; stipules minute. Flowers in short axillary racemes, long-pedicellate, bisexual or unisexual. Sepals five, almost free, valvate, sometimes persistent. Petals five, loosely connivent or coherent at thickened scalelike base or claw, with free, lanceolate, thin limb, imbricate, clawed or sessile. Androgynophore distinct, thick. Stamens 9–13, inserted within a thick, fleshy cupular disc; filaments fused at base,

filiform; anthers small, basifixed, latrorse; connective rounded and developed into a knob at the top. Pistillodium present in the male flowers. Pollen grains 3-colporate. Female flowers with about 10 filiform staminodia within the disc. Gynoecium of 5–3 carpels; style elongate, 5–4-fid at the apex; ovary 3–5-locular, shortly stipitate, with 3–10 axile ovules per locule. Fruits globose, brownish-scaly berry with numerous small, pubescent seeds immersed in pulp; embryo strongly curved, white. Fruits contain sweet-tasting proteins, brazzein and pentadin (van der Wel et al. 1989; Ming and Hellekant 1994; Caldwell et al. 1998)..... 6. PENTADIPLANDRACEAE.

4 Perianth 6–8–9-merous. Coarse herbs or soft shrubs, sometimes somewhat scrambling or tree-like, glabrous or pubescent, with pungent smelling parts. Leaves alternate, trifoliolate, stipules absent. Stomata anomocytic. Flowers in lax terminal racemes, bisexual, actinomorphic. Sepals and petals mostly eight, less often seven or nine; sepals imbricate, caducous; petals shortly clawed, imbricate. Stamens mostly eight, free, borne internally on a lobed nectary disc; filaments long-papillate, dilated near the base; anthers sagittate, basifixed. Pollen grains 2-celled, 2(3)-colporate. Gynoecium of (5)6–8 carpels, with short, beaklike style and 6–8-rayed stigma; ovary (4-)6(-8)-locular, on a very short gynophore, with numerous ovules on deeply intruded parietal placentas that form false partitions. Anthers and stigmatic lobes covered by short unicellular hairs with longitudinal cuticular folds (similarly hairs are also known from *Forchhammeria*–Capparaceae). Ovules anatropous, but become campylotropous by the post-fertilisation development of the exotegmen (Boesewinkel 1990); placentation parietal. Fruits small, globose berries with membranous pericarp. Seeds small, shiny, very numerous, exotestal cells more or less enlarged, walls thickened, endotestal cells small, exotegmic cells fibrous; embryo curved, endosperm thin, n = 14............... 10. TOVARIACEAE.

3 Ovary 1–2-, rarely 3-locular.

5 Ovary 1–2-locular.

6 Leaves well-developed.

7 Stomata paracytic.

8 Small maritime shrubs. Branches often (3)4(15)-angled. Contain myrosin cells and protein-rich, dilated cisternae of the endoplasmic reticulum. Vessels with simple perforations, perforation plate borders vestigial; pits vestured. Fibers small, thick-walled, with minute bordered pits. Rays somewhat heterogeneous. Axial parenchyma mainly paratracheal, but also apotracheal in bands. Nodes unilocular. Leaves opposite, sessile, entire, slender, succulent, with a saccate colorless base. "The stipules arise after the initiation of the leaf or bract and remain always smaller than the leaves. They are reminiscent of colleters and have an obvious function in keeping the buds sufficiently moist during early development" (Ronse Decraene 2005). Flowers small and very reduced, unisexual, monoecious or dioecious in axillary conelike spikes. In monoecious *Batis argillicola* male and female flowers are intermingled in the same spike; in dioecious *B. maritima* flowers are congested in short, dense spikes. Male flowers subtended by bracts and enclosed in a membranous, saccate, gamosepalous calyx, which eventually splits near the top into two or four lobes (*B.maritima*) or along one side only (*B. argillicola*). Petals (or staminodia?) four, free, spatulate. Stamens four, alternating with four whitish spathulate appendages (Ronse Decraene 2005); filaments free; anthers basifixed, latrorse. Pollen grains 2-celled, 3(4)-colporate, with solid undifferentiated exine. Sometimes a vestigial gynoecium present in the male flowers. Female flowers without perianth. Gynoecium of two united carpels, paracarpous, with two sessile styles with a mass of

papillate trichomes; ovary divided by false partitions into four loculcelles with one parietal-basal ovule in each locelle. Ovules anatropous, epitropous, crassinucellate. Female gametophyte of a unique bisporic 4-nucleate type, in which only the upper cell of the dyad is functional (the female gametophyte never develops into an 8-nucleate stage) (Tobe and Raven 1992). Fruits drupaceous with four woody pyrenes. Seeds straight and cylindrical, though more or less bilaterally flattened; seed surface scalariform with fine markings; the mature seed coat extremely thin and composed only of the thin, persistent outer cell walls of the exotesta and the thick cuticle of the nucellar epidermal cells, with straight or slightly curved embryo, lacking both endosperm and perisperm. Myrosinase and benzylglucosinolate is present (Schraudolf et al. 1971; Mabry 1976) betacyanins and betaxanthins are absent (Mabry and Turner 1964). n = 11. . . 8. BATACEAE.

8 Shrubs (sometimes scandent) or rather small trees of dry hot regions, some with odour of mustard oils, sometimes with axillary spines (*Azima*) and storied wood. Interxylary phloem present in *Dobera* and *Salvadora*. Vessels with simple perforations; lateral pitting alternate and very small. Fibers with simple pits, very short. Rays homogeneous or slightly heterogeneous. Axial parenchyma scanty paratracheal or vasicentric. Leaves opposite, simple, entire, with very small stipules; leaf ptyxis flat-curved (*Salvadora*). Flowers in terminal or axillary spikes, racemes, panicles, or sometimes dense fascicles, small, bisexual or variously polygamous, actinomorphic. Sepals 2–4(5), connate into a lobed tube, lobes imbricate or more or less valvate. Petals 4(5), free or (*Salvadora*) basally shortly connate, imbricate, usually with teeth or glands on inner side. Stamens four or rarely five, alternipetalous, filaments free (*Azima*), connate below into a tube (*Dobera*), or adnate at the base of corolla tube (*Salvadora*); anthers dorsifixed. Pollen grains 2-celled, usually 3-colporoidate. Nectary disc absent or represented by glands alternating with stamens. Gynoecium of two united carpels; style short or very short, with entire or 2-lobed stigma; ovary 2-locular in *Azima*, 1-locular in *Dobera* and *Salvadora*, with one or two basal or axile-basal, erect ovules in each locule; ovules apotropous, bitegmic, crassinucellate. Fruits mostly 1-seeded berries or drupes. Seeds exotestal (Melikian and Savinov 2000); exotestal cells palisade, slightly thickened, inner walls mucilaginous, crystalliferous, tegmen becoming crushed, exotegmic cells fibrous, not lignified (Corner 1976), endosperm absent; embryo straight, with thick, oily, cordate cotyledons. Plants producing mustard oils and piperidine alkaloids, leucoanthocyanins and tannins are lacking (Hegnauer 1973; 1990). n = 12 (*Salvadora*), n = 11 (*Azima*).. . . . . 9. SALVADORACEAE.

7 Stomata anomocytic or anisocytic.

9 Fruits mostly baccate or capsules, rarely schizocarps, nuts, or samara, no replum. Perianth mostly 4-merous. Flowers mostly bisexual. Shrubs, herbs, or trees, rarely lianas. Vessels with vestured pits. Leaves alternate or rarely opposite, spiral or rarely distichous, simple or trifoliolate or palmately compound, petiolate, nonsheathing; often with small stipules, sometimes modified into glands or spinules. Stomata anomocytic. Flowers usually in racemes, rarely solitary and axillary, mostly bisexual, actinomorphic or more or less zygomorphic, generally with an evident gynophore or androgynophore. Sepals (2-)4(-6), often decussate, sometimes basally connate, imbricate or valvate. Petals (2-)4(-6), rarely lacking, often clawed. Stamens basically four (derived from four primordia), but often two or all four of staminal primordia developing to produce six to many (up to 250 in *Capparis*) stamens,

some staminodial. Pollen grains 2-celled or seldom 3-celled, (2)4(4)-colporate. Nectary an extrastaminal ring or merely receptacular protrusion. Gynoecium of 2(-12) carpels; style with bilobed or capitate stigma; ovary usually 1-locular, or 2-locular (by false septa), with parietal placentas, sometimes more or less intruded and rarely meeting in the center of the ovary. Ovules many to few on each placenta, generally campylotropous, crassinucellate. Seeds often reniform; seed coat fibrous-exotegmic embryo linear, arcuate; endosperm scanty. Present methyl glucosinolates, flavonols, proanthocyanidins, pyrrolidine alkaloids, sinapic and ferulic acids (Hegnauer 1964; Kers 2003), n = 7–10(-17), mostly 10. . . . . 1. CAPPARACEAE.

9 Fruits usually siliques or silicles divided into two chambers by the replum, dehiscent or indehiscent, rarely transversely jointed, sometimes one to few-seeded nuts.

10 Ovary divided lengthwise by a membranous and spurious septum, or sometimes transversely several- or many-locular. Stamens usually six, four longer and two shorter filaments. Herbs or rarely subshrubs or shrubs. Fibers with small bordered pits. Leaves alternate or rarely subopposite, simple, often more or less pinnately dissected, only rarely with distinct, articulated leaflets, without stipules. Stomata anisocytic. Inflorescences usually a raceme or corymb. Flowers only seldom with a gynophore and never with an androgynophore. Stamens six and usually the two outer ones shorter than the four inner ones (the inner stamens derived from only two primordia and sometimes connate below in pairs), rarely the stamens only 2–4 or up to 16; anthers dorsifixed. At the base of the stamens are nectar-secreting outgrowths of the floral receptacle which vary in form and number in different genera. Pollen grains 3-celled or seldom 2-celled, (2-)3-colpate, 4-rupate, 6-rugate or nonaperturate. Ovary and fruit nearly always partitioned by a septum; ovary 2-locular; ovules 1–300, usually pendulous. Seed folded, but no invagination of the coat, testa often mucilaginous, exotestal cells reticulately thickened on radial walls, endotesta palisade, with U-shaped thickenings or not, tegmen multiplicative; embryo folded or coiled; the cotyledons 2–10 times as wide as the radicle, thick or thin; endosperm absent or rarely very little present. Containing glucosinolates (mustard oil glucosides, thioglucosides), and often cyanogenic; very rarely alkaloids; n = 4(-8(-13). 3. BRASSICACEAE.

10 Ovary not divided by a spurious septum, shortly to long-stipitate. Stamens more or less equal. Perennial or annual herbs, glabrous, pilose or glandular. Leaves simple or more often 3–7-foliolate, alternate, estipulate or with minute stipules (*Oxystylis*). Flowers bisexual or rarely monoecious, often very showy, mostly in racemes or corymbs, or solitary and axillary, often with foliaceous bracts, rarely without bracts. Sepals mostly four, free, or calyx 4-lobed or partite. Petals usually four, rarely reduced to two, free, often clawed, imbricate. Stamens mostly six, rarely four, five, or seven, very rarely reduced to one fertile; filaments free or partly adnate to the gynophore; anthers 2-locular, sometimes some aborted, rarely some stamens transformed into petaloid staminodes. Torus sometimes produced at the back into a tube or gland. Ovary stipitate or rarely sessile, usually 1-locular. Style usually filiform with pointlike or capitate stigma; ovules two (*Oxystylis*) or few to numerous. Fruits 2–3-valved capsule, the valves falling away and leaving a persistent replum, rarely inflated and bladder-like, very rarely fruits inde-

hiscent, rarely compressed contrary to the replus. Seeds often reniform, smooth, rugose or hairy; exotegmen cells radially enlarged, sclerified, endotegmen cells with lignified bands on periclinal walls; embryo conforming to the shape of the seed, with two cotyledons; endosperm wanting; n = 9. . . . . 2. CLEOMACEAE.

6 Leaves scalelike. Shrubs or small trees with minute, estipulate, early deciduous leaves and rigid, interlocking, thorny twigs. Fibers with distinctly bordered pits, sometimes with spiral thickening, very short. Vessels very to extremely small, with spiral thickening in the late wood. Rays nearly homogeneous, with intercellular canals. Axial parenchyma apotracheal. Flowers in very short axillary racemes, bisexual, with minute bracts. Sepals 4(5), small, imbricate, free. Petals 4(5), imbricate, free, slightly clawed. Stamens eight; filaments flat, with a superficial appendage at the base of each; anthers deeply bilobed, slightly apiculate, dorsifixed, introrse or latrorse. Pollen grains 2-celled, 3-colporate. Nectary disc absent. Gynoecium of two or rarely three carpels; style subulate, with minute, slightly bilobed stigma, persistent; ovary on a short gynophore, 2-locular, with several to many axile ovules per locule; ovules anatropous, crassinucellate, with zig-zag micropyles formed by both integuments. Fruits few-seeded, subglobose berries. Seeds black, hard, and circinate; exotesta with massive cuticle, then tanniniferous cells, exotegmen walls very thick, lignified; embryo much curved, endosperm very scanty. Mustard oils present or absent, n = 11. 7. KOEBERLINIACEAE.

5 Ovary 2–3-locular.

11 Leaves alternate. Unarmed, pungent-smelling shrub 30–120 cm tall with rigid stems. Whole plant densely pubescent; the trichomes strigose, appressed, T-shaped. Wood with uniseriate rays of upright cells; vestured pits and myrosin cells lacking (Carlquist et Miller 1999). Leaves entire, simple, ovate to broadly oblong-lanceolate, the apex broadly acute to obtuse or rounded, minutely mucronate-apiculate, estipulate. Stomata anomocytic. Flowers solitary in long-shoot leaf axils, with 1–5 spaced out near the end of the branchlets, bisexual, 5-, 6-, or 7-merous. Calyx in bud spheroid to ovoid and abruptly pointed; sepals 5–7, greatly thickened at the base and permanently fused their whole length into a seamless cap. Petals 5–7, blue, imbricate, obovate-spathulate. Stamens (40-)60–76, with centrifugal maturation, united in (5-)6(-7) fascicles that originate presumably by dedoublement form (5-)6(-7) initials, and axile placentation; anthers basifixed, round, with two introrse slits. Pollen grains 3-colpate, semitectate with a complex strio-regulate exine; the colpi are covered with a colpus membrane that has shallow, irregular verrucate projections (Tomb 1999). Ovary of three carpels, 3-locular with axile placentation but "proto-parietal" vasculature. Styles glabrous, with three short, filiform, stigmatiferous, subcapitate apical branches arching outwards and with down-turned margins (Iltes 1999). Ovules 10–14 per locule, anatropous. Fruit a linear downward-pointing capsule with wulging locules. Seeds 3–10 per carpel (6–30 per fruit), in a narrow double row, each borne on a short, curved, exarillate funicle, covered with a thin, soft and pith-like, exotegmic seed coat; embryo large, flat, straight, spathulate; endosperm of only one cell layer (Tobe et al. 1999). The mustard oils seem to accumulate unequally in different tissues. . . . . . 11. SETCHELLANTHACEAE.

11 Leaves opposite or subopposite. Prostrate herbaceous subshrubs with elongated, hispid branches and short axillary branchlets. Vessels with simple perforations; lateral pitting mainly alternate but with a tendency to be opposite. Fibers with bordered pits. Leaves entire, with catrilagenous margines, scabrous, petiolate, minutely stipulate. Stomata anomocytic. Flowers solitary, axillary, bisexual, strongly zygomorphic. Sepals five, unequal but not bilabiate, connate for about half of their length, except the two abaxial ones, which are separate from each other right down to the base. Corolla is slipper-like, consists of two pet-

als alternating with three adaxial sepals and interconnected with each other by their cuticules; the upper and lowest parts are free from each other; ventrally the lower part of each petal is provided with two narrow, ribbonlike appendages. Receptacle produced into a linear, flattened, incurved androgynophore ascending in the slit of the calyx, bearing a triangular nectary disc at the base between the two petals. Stamens 8–9, interconnected at the base and forming a spreading disc-shape ring at the top of the androgynophore; four stamens on the adaxial part of the flower are fertile, four or five stamens on the abaxial (outer) part are staminodial, connate in their lower part, pubescent; anthers ovoid. Pollen grains 3(4)-colporate, with very short colpi. Gynoecium of three carpels, with a small, sessile stigma; placentation axial, ovary 2–3-locular, with one basal ovule in each locule. Fruits 1-seeded, dry, indehiscent, with thin pericarp adherent to the seed, pendant within the calyx from the apex of the androgynophore. Seeds reniform, arilate, more or less compressed, with rugose hard testa and laciniate funicle; embryo linear, conduplicate; endosperm scanty......12. EMBLINGIACEAE.

1. CAPPARACEAE

Jussieu 1789. 20/480–500. Widely distributed in tropics and subtropics and in warm-temperate regions.

CAPPAREAE: *Capparis, Tirania, Crateva, Eudenia Cladostemon, Dhofaria, Ritchiea, Belencita, Steriphoma, Morisonia*; MAERUEAE: *Maerua*; CADABEAE: *Cadaba, Buchholzia, Atamisquea, Thilachlum, Boscia, Bachmannia*; STIXEAE: *Stixis*; APOPHYLLEAE: *Apophyllum, Forchhammeria.*

2. CLEOMACEAE

Horaninow 1834 (including Oxystylidaceae Hutchinson 1969). 13/300. Tropical and warm temperate regions, esp. America.

2.1 DIPTERYGIOIDEAE

Annual or perennial herb, or subshrub. Leaves simple, stipules very insignificant, setaceous, entire or apically 3-fid, tardily deciduous. Inflorescences terminal racemes. Flowers minute, almost actinomorphic. Sepals four, valvate, free or connate. Petals four. Stamens six. Fruit a samara, 1-seeded, laterally compressed, laterally winged. – *Dipterygium.*

2.2 CLEOMOIDEAE

Perennial or annual herbs, rarely somewhat shrubby, very rarely arborescent. Hairs present, or absent, commonly glandular. Flowers bisexual or rarely monoecious, zygomorphic. Sepals mostly four, free or calyx 4-lobed or partite. Petals usually four, rarely reduced to two, free, often clawed, imbricate. Stamens mostly six, rarely 4, 5, 6, or 7, rarely reduced to one fertile, free or partly adnate to the gynophore. Fruits 2- or 3-valved capsules with the valves falling away and leaving a persistent replum, rarely inflated and bladderlike, very rarely indehiscent, rarely compressed contrary to the replum, sometimes (*Wislizenia* and *Oxystylis*) with a reduced replum and didymous, each lobe falling away from the thickened corky axis and indehiscent and enclosing one or rarely two seeds. – CLEOMEAE: *Cleome, Puccionia, Polanlsia, Cleomella, Isomeris, Buhsia, Haptocarpum, Cristatella, Dactylaena*; PODANDROGYNEAE: *Podandrogyne*; OXYSTYLIDEAE: *Oxystylis, Wislizenia*

3. BRASSICACEAE

Burnett 1835 or Cruciferae A.L. de Jussieu 1789 (nomen altern.) (including Drabaceae Martynov 1820, Erysimaceae Martynov 1820, Raphanaceae Horaninow 1847, Schizopetalaceae A.L. de Jusseau 1848, Sisymbriaceae Martynov 1820, Stanleyaceae Nuttall 1834, Thlaspiaceae Martynov 1820). 337–368/3200. Subcosmopolitan, but mainly in temperate and cold areas, and more especially in the Mediterranean and Irano-Turanian regions; only sparingly represented in the Southern Hemisphere and with very few species in tropical regions.

MACROPODIEAE: *Macropodium*; THELYPODIEAE: *Stanleya, Rollinsia, Warea, Thelypodium, Irenepharsus, Chlorocrambe, Romanschulzia, Thelypodiopsis, Streptanthus* (including *Euklisia, Microsemia, Icianthus*), *Caulanthus* (including *Stanfordia*), *Microsisymbrium, Raphanorhyncha, Streptanthella*; PRINGLEEAE: *Pringlea*; CREMOLOBEAE: *Cremolobus* (including *Urbanodoxa*), *Menonvillea*; MEGADENIEAE:

Megadenia; MEGACARPAEAE: *Megacarpaea, Dimorphocarpa, Dithyrea, Biscutella*; BRASSICEAE: *Conringia, Morisia, Succowia, Quizeliantha, Psychine, Schouwia, Fortuynia, Pseudofortuynia, Calcanthus, Physorhynchus, Diplotaxis, Carrichtera, Vella* (including *Euzomodendron*), *Zilla, Henophyton, Hemicrambe, Foleyola, Moricandia, Oudneya, Douepea* (including *Dolichorhynchus*), *Sinapis, Eruca, Erucastrum, Hirschfeldia, Erucaria* (including *Reboudia*), *Brassica, Sinapidendron, Cordylocarpus, Fezia, Rytidocarpus, Pseuderucaria, Orychophragmus, Raphanus, Enarthrocarpus, Cakile, Muricaria, Didesmus, Ceratocnemum, Crambella, Kremeriella, Guiraoa, Otocarpus, Coincya* (including *Hutera, Rhynchosinapis*), *Rapistrum, Trachystoma, Crambe, Calepina, Schimpera, Eremophyton*; CHAMIREAE: *Chamira*; ISATIDEAE: *Didymophysa, Brachycarpaea, Moriera, Pugionium, Graellsia, Coluteocarpus, Myagrum, Horwoodia, Chartoloma, Tauscheria, Sameraria, Pachypterygium, Isatis*; SISYMBRIEAE: *Streptantella, Caulostramina, Glaucocarpum, Mostacillastrum* (including *Phlebiophragmus*), *Sisymbrium* (including *Schoenocrambe*), *Pachycladon, Cymatocarpus, Lycocarpus, Ammosperma, Trichotolinum, Chilocardamum, Polypsecadium, Werdermannia, Chaunanthus, Neuontobotrys, Descurainia* (including *Robeschia*). *Hugueninia, Sophiopsis, Sinosophiopsis, Redowskia, Yinshania, Ischnocarpus, Alliaria, Eutrema* (including *Wasabia*), *Parlatoria, Sobolewskia, Scambopus, Neotorularia, Dimorphostemon, Sisymbriopsis, Octoceras, Dichasianthus, Atelanthera, Thellungiella, Catenulina, Streptoloma, Berteroella, Smelowskia, Hedinia, Christolea, Lignariella, Taphrospermum* (including *Glaribraya*), *Platycraspedum, Sibaropsis, Aphragmus* (including *Staintoniella*), *Braya, Weberbauera* (including *Alpaminia, Pelagatia, Stenodraba*), *Eremodraba, Catadysia, Lithodraba, Xerodraba, Brayopsis, Aschersoniodoxa, Hollermayera, Englerocharis, Delpinophytum, Eudema, Onuris, Ivania*; HESPERIDEAE: *Erysimum* (including *Syrenia, Syrenopsis*), *Beringia, Eigia, Cheiranthus, Hesperis, Chalcanthus, Litwinowia, Ochtodium, Bunias, Clausia, Pseudoclausia, Malcolmia* (including *Strigosella*), *Paraclausia, Sprygina, Cryptospora, Trichochiton, Gynophorea, Diptychocarpus, Maresia, Zuvanda, Eremobium, Nasturtiopsis, Leptaleum, Leiospora, Geococcus, Parrya* (including *Neuroloma*), *Ballantinia, Solms-Laubachia, Pycnoplinthopsis, Baimashania, Iskandera, Desideria* (including *Ermaniopsis*), *Pycnoplinthus, Cheesemania, Lepidostemon, Oreoblastus, Dielsiocharis, Dontostemon, Blennodia, Oreoloma, Citharelоma, Lachnoloma, Anchonlum, Zerdana, Eremoblastus, Euclidium, Lobularia, Notoceras, Sterigmostemum* (including *Petiniotia*), *Raffenaldia, Chorispora, Spirorhynchus, Matthiola* (including *Lonchophora*), *Parolinia, Diceratella, Microstigma, Veselskya* (including *Pyramidium*), *Anastatica, Morettia, Tetracme* (including *Tetracmidion*), *Synstemon, Alyssum* (including *Meniocus, Gamosepalum*), *Lesquerella, Physocardamum, Physaria, Hormathophylla, Aurinia, Straussiella, Degenia, Takhtajaniella, Physoptychis, Alyssoides, Clastopus, Fibigia, Berteroa, Galitzkya, Bornmuellera, Clypeola* (including *Pseudoanastatica*), *Asperuginoides, Athysanus, Thysanocarpus, Heterodraba*; ARABIDEAE: *Cardamine, Arabidella, Gorodkovia, Loxostemon, Dentaria, Pegaephyton, Neomartinella, Iodanthus, Planodes, Idahoa, Leavenworthia, Sibara, Barbarea, Nasturtium, Rorippa* (including *Rorippiella, Kardanoglyphos, Sisymbrella, Trochiscus*), *Caulostramina, Sphaerocardamum, Pennellia, Phlebolobium, Armoracia, Dictyophragmus, Arabis* (including *Parryodes*), *Turritis, Sandbergia, Crucihimalaya, Olimarabidopsis, Halimolobos, Pachymitus, Harmsiodoxa, Drabopsis, Alyssopsis, Farsetia, Murbeckiella, Prionotrichon, Rhammatophyllum, Phaeonychium, Vvedenskyella, Parryopsis, Eurycarpus, Oreophyton, Aplanodes, Phoenicaulis, Anelsonia, Aubrieta, Botschantzevia, Arcyosperma, Cardaminopsis, Arabidopsis, Pseudoarabidopsis, Pachyneurum, Stevenia, Petroravenis, Borodinia, Draba* (including *Thylacodraba, Coelonema*), *Drabastrum, Ermania, Sarcodraba* (including *Ateixa*), *Menkea, Schivereckia, Erophila, Petrocallis, Pseudovesicaria, Elburzia, Cusickiella*; LUNARIEAE: *Lunaria, Pterygiosperma, Ricotia, Neotchihatchewia, Peltaria, Peltariopsis, Selenia, Schlechteria, Thlaspeocarpa* (including *Palmstruckia*), *Savignya*; LEPIDIEAE: *Lepidium* (including *Stroganowia, Papuzilla*), *Stubendorffia, Cardaria, Iberis, Heldreichia, Winklera, Cyphocardamum, Acanthocardamum, Coronopus, Dilophia, Parodiodoxa, Dipoma, Tropidocarpum* (including *Agallis*), *Dactylocardamum, Grammosperma, Cuphonotus, Cochlearia, Microlepidium, ? Nerisyrenia, Synthlipsis, Kernera, Rhizobotrya, Pseudocamelina, Pseudosempervivum, Chrysochamela, Camelinopsis, Hymenolobus, Polyctenium, Hornungia* (including *Hutchinsiella, Microcardamum*), *Calymmatium, Hemilophia, Asta, Scoliaxon, Mancoa, Phlegmatospermum,*

Subularia, Camelina, Capsella, Notothlaspi, Pachyphragma, Thlaspi (including *Carpoceras, Raparia*), *Carinavalva, Noccaea, Teesdalia* (including *Teesdaliopsis*), *Eunomia, Synthlipsis, Aethionema, Lachnocapsa, Campyloptera, Iberidella, Andrzeiowskya, Ionopsidium* (including *Pastorea, Minaea*), *Iti, Bivonaea, Brossardia, Boreava, Neslia, Glastaria, Goldbachia*; SCHIZOPETALEAE: *Schizopetalon, Dryope-talon, Ornithocarpa, Mathewsia*; STENOPETALEAE: *Stenopetalum*; HELIOPHILEAE: *Heliophila* (including *Carponema*), *Cycloptychis, Silicularia*; LYROCARPEAE: *Lyrocarpa*.

The Brassicaceae are very closely related to Capparaceae and Cleomaceae and linked to them through the tribe Thelypodieae.

4. RESEDACEAE

Berchtold and J. Presl 1820 (including Astrocarpaceae A. Kerner 1891). 6/75–80. From Macaronesia across northern Africa and the Mediterranean and western Asia to northwestern India and central Asia, in Europe and western Siberia, South Africa, Socotra, and in southwestern United States and Mexico.

RESEDEAE: *Oligomeris, Ochradenus, Reseda, Randonia*; CAYLUSEEAE: *Caylusea*; ASTROCARPEAE: *Astrocarpus* (*Sesamoides*).

Nearer to the Cleomaceae, which is confirmed by the presence of the androgynophore, morphology of the nectary disc, and seed anatomy (Orr 1921).

5. GYROSTEMONACEAE

Endlicher 1841. 5/18. Australia (mostly in drier regions); *Gyrostemon thesioldes* reaches northern Tasmania.

Codonocarpus, Gyrostemon, Tersonia, Walteranthus, Cypselocarpus.

Systematic position is not very certain. Many nineteenth-century botanists, including Baillon and Bentham and Hooker, placed *Gyrostemon* and related genera in Phytolaccaceae, but Endlicher (1841) recognized the Gyrostemonaceae as a separate family, which is usually placed in the order Caryophyllales. However, from most of the Caryophyllales the family Gyrostemonaceae differs chemically through the presence of mustard oil, through S-type sieve-element plastids and undifferentiated, solid ectexine (Mabry 1976; Behnke and Turner 1971; Behnke 1976, 1977; Prijanto 1970; Nowicke 1976; Goldblatt et al. 1976). Chromosome number of n = 14 is also very uncommon throughout the order Caryophyllales and occurs only in occasional polyploids. However, this number is fundamental in Gyrostemonaceae (Goldblatt et al. 1976: 204). All these differences are fully sufficient for excluding Gyrostemonaceae from the order Caryophyllales. According to Dahlgren (1975, 1983), Goldblatt et al. (1976), Behnke (1977), Tobe and Raven (1991), and some others, the Gyrostemonaceae belong in the expanded order Brassicales or are related to the Brassicales (Cronquist 1981, 1988). This is supported by the nucleotide sequences of *rbc*L (Rodman et al. 1994)

6. PENTADIPLANDRACEAE

Hutchinson and Dalziel 1928. 1/2. Tropical West Africa.

Pentadiplandra.

Related to the Capparaceae.

7. KOEBERLINIACEAE

Engler 1895. 1/1. Southwestern United States, Mexico, tropical Bolivia.

Koeberlinia.

Related to the Capparaceae (Wettstein 1935; Takhtajan 1959, 1987; Cronquist 1968, 1981, 1988; Gibson 1979; Mehta and Moseley 1984), but also they have similarities with *Batis* and Salvadoraceae (Ronse Decreane and Haston 2006).

8. BATACEAE

Perleb 1838. 1/2. Coasts of tropical and subtropical America and Galapagos Islands (*B. maritima*) and along the coasts of New Guinea and northern Australia.

Batis.

It is now widely accepted that Bataceae (together with the Gyrostemonaceae) are close to the Capparaceae and related families (Dahlgren 1975a, 1977; Cronquist 1981, 1988; Rodman 1991; Rodman et al. 1993; Thorne 2006) with which the Bataceae have many similarities including the guard-cell myrosin cells strikingly similar to those in Resedaceae and Tovariaceae and representing a specific type of myrosin cells (Jargensen 1981). However the Bataceae differ from the other Capparales in so many characters, including distinctive "*Batis*-type" female gametophyte (Tobe and Raven 1992). Although Bataceae reveal some similarities with the Gyrostemonaceae (Kuprianova 1965; Takhtajan 1966; Thorne 1977), they differ markedly in many respects, including the habit, wood anatomy, inflorescence and flower morphology, compound apertures without operculum, anatropous ovules, presence of nucellar cap and absence of hypostase, a unique type of bisporic female gametophyte, absence of endosperm in mature seeds, straight seeds, absence of aril, and extremely thin exotestal seed coat (see especially Tobe and Raven 1992).

9. SALVADORACEAE

Lindley 1836 (including Azimaceae Wight and Gardner 1845). 3/11. From Africa and Madagascar to Pakistan, India, Sri Lanka, and Southeast Asia.

Azima, Dobera, Salvadora.

The family has often been referred to a celastraceous affinity, but Corner (1976) pointed out that this is in conflict with its lack of a fibrous tegmen. Dahlgren (1975) suggested Salvadoraceae to form part of an expanded concept of Capparales. In phyllotaxis and floral merosity, Salvadoraceae agree remarkably with Bataceae. Cladistic studies and various molecular analyses employing plastid and/or nuclear genes or combined morphological/molecular data sets always place Salvadoraceae close to *Batis* (Rodman et al. 1996, 1998).

10. TOVARIACEAE

Pax 1891. 1/2. Tropical America: *Tovaria pendula* – Andes from Mexico to Peru and Bolivia; *T. diffusa* – Jamaica.

Tovaria.

Closely related to the Cleomaceae.

11. SETCHELLANTHACEAE

Iltis 1999. 1/1. Mexico (Chihuahua Desert).

Setchellanthus.

This very specialized family probably belongs to the Capparales.

12. EMBLINGIACEAE

Airy Shaw 1965. 1/1. Western and southwestern Australia.

Emblingia.

Emblingia has many unique morphological features, there are some features in common with members of Capparales, including androgynophore, a conduplicate, curved embryo, scanty endosperm, that "agreeing with Capparidaceae, but differing widely in almost every other character" (Airy Show 1973), but no studies have been conducted to examine the presence of mustard oils (glucosinolates) in *Emblingia.* Including Emblingiaceae in the Capparales is supported by molecular studies (Chandler and Bayer 2000; Soltis et al. 2006; Ronse Decraene et Haston 2006).

Bibliography

Andallah MS. 1967. The Resedaceae. A taxonomic revision of the family. Meded. Land. Wagen. 67: 1–98.

Abdallah MS and HCD de Wit. 1978. The Resedaceae: a taxonomic revision of the family (final installment) Meded Landbouwhogeschool Wageningen 78–14: 99–416.

Abdel Khalik K, RG van den Berg, J van der Maesen, and MN El Hadidi. 2002. Pollen morphology of some tribes of Brassicaceae from Egypt and its systematic implications. Feddes Repert. 113(3–4): 211–223.

Alexander I. 1952. Entwicklungsstudien an Blüten von Cruciferen und Papaveraceen. Planta 41: 125–144.

Aleykutty KM and JA Inamdar. 1978a. Studies in the vessels of some Capparaceae. Flora 167: 103–109.

Aleykutty KM and JA Inamdar. 1978b. Structure, ontogeny, and taxonomic significance of trichomes and stomata in some Capparidaceae. Feddes Repert. 89: 19–30.

Al-Shehbaz IA. 1984. The tribes of Cruciferae (Brassicaceae) in the southeastern United States. J. Arnold Arbor. 65: 343–373.

Al-Shehbaz IA. 1985. The genera of Brassicaceae (Cruciferae) in the southeastern United States. J. Arnold Arbor. 66: 279–351.

Al-Shehbaz IA. 1988a. The genera of Arabideae (Cruciferae: Brassicaceae) in the southeastern United States. J. Arnold Arbor. 69: 85–166.

Al-Shehbaz IA. 1988b. The genera of Anchoniea (Hesperideae) (Cruciferae: Brassicaceae) in the southeastern United States. J. Arnold Arbor. 69: 193–212.

Al-Shehbaz IA. 1988c. The genera of Sisymbrieae in the southeastern United States. J. Arnold Arbor. 69: 213–237.

Al-Shehbaz IA. 1989. *Dactylocardamum* (Brassicaceae): a remarkable new genus from Peru. J. Arnold Arbor. 70: 515–521.

Al-Shehbaz IA and KI Al-Shammary. 1987. Distribution and chemotaxonomic significance of glucosinolates in certain

Middle-Eastern Cruciferae. Biochem. Syst. Ecol. 15: 599–569.

Al-Shehbaz IA, MA Beilstein, and EA Kellogg. 2006. Systematics and phylogeny of the Brassicaceae (Cruciferae): an overview. Plant Syst. Evol. 259: 89–120.

Anuradha SMJ, M Radhakrisnaiah, and LL Narayana. 1988. Chemosystematics of Capparaceae. Feddes Repert. 99: 391–394.

Appel O. 1999. The so-called 'beak', a character in the systematics of Brassicaceae? Bot. Jahrb. Syst. 121: 85–98.

Appel O and C Bayer. 2003. Tovariaceae. In: K Kubitzki, ed. The families and genera of vascular plants, vol. 5, pp. 397–399. Springer, Berlin/Heidelberg/New York.

Appel O and IA Al-Shehbaz. 1997. Re-evaluation of the tribe Heliophileae (Brassicaceae). Mitt. Inst. Allg. Bot. Hamburg 27: 85–92.

Appel O and IA Al-Shehbaz. 2003. Cruciferae. In: K Kubitzki, ed. The families and genera of vascular plants, vol. 5, pp. 75–174. Springer, Berlin/Heidelberg/New York.

Appelqvist LA. 1976. Lipids in the Cruciferae. In: JG Vaughan, AJ Macleod, and BMG Jones, eds. The biology and chemistry of the Cruciferae, pp. 221–277. Academic, London.

Arber A. 1931. Studies in floral morphology: I. On some structural features of the Cruciferous flower. New Phytol. 30: 11–41.

Arber A. 1938. Studies in flower structure: VII. On the gynoecium of *Reseda* with a consideration of paracarpy. Ann. Bot. 6: 43–48.

Arunalakshmi V. 1985. Embryological studies in Capparidaceae. J. Indian Bot. Soc. 64: 17–24.

Arunalakshmi V. 1989. Structure and development of seed coat in *Cleome*. J. Indian Bot. Soc. 68: 116–121.

Avetisian VE. 1976. Some modifications of the system of the family Brassicaceae. Bot. Zhurn. 61: 1198–1203 (in Russian).

Avetisian VE. 1987. Some peculiarities in carpologic evolution of Brassicaceae. Sbornik Nauch. Trudov Arm. Otd. Vsesoyusn. Bot. Ob. 10: 67–69 (in Russian).

Avetisian VE. 1990. A review of the system of Brassicaceae flora of the Caucasus. Bot. Zhurn. 75: 1029–1032 (in Russian).

Balwant Singh. 1944. A contribution to the anatomy of *Salvadora persica* L. with special reference to the origin of the included phloem. J. Indian Bot. Soc. 23: 71–78.

Bayer A. 1905. Beiträge zur systematischen Gliederung der Cruciferen. Bot. Centralbl. 18: 119–180.

Bayer C and O Appel. 2003. Bataceae. In: K Kubitzki, ed. The families and genera of vascular plants, vol. 5, pp. 30–32. Springer, Berlin/Heidelberg/New York.

Bayer C and O Appel. 2003. Pentadiplandraceae. In: K Kubitzki, ed. The families and genera of vascular plants, vol. 5, 329–331. Springer, Berlin/Heidelberg/New York.

Behnke H-D. 1977a. Phloem ultrastructure and systematic position of Gyrostemonaceae. Bot. Not. 130: 255–260.

Behnke H-D. 1977b. Zur Skulptur der Pollen-Exine bei drei Centrospermen (*Gisekia*, *Limeum*, *Hectorella*), bei Gyrostemonaceen und Rhabdodendraceen. Plant Syst. Evol. 128: 227–235.

Behnke H-D. 1977c. Dilatierte ER-Zisternen: Ein mik-romorphologisches Merkmal der Capparales? Ber. Deutsch. Bot. Ges. 90: 241–251.

Behnke HD and G Eschleck. 1978. Dilated cisternae in Capparales – an attempt towards the characterization of a specific endoplasmatic reticulum. Protoplasma 97: 351–363.

Beilstein MA, IA Al-Shehbaz, EA Kellogg. 2006. Brassicaceae phylogeny and trichome evolution. Am. J. Bot. 93: 607–619.

Belyaeva LE and GB Rodionova. 1983. In: MS Yakovlev, ed. Comparative embryology of flowering plants, vol. 2, pp. 154–164. Nauka, Leningrad (in Russian).

Boesewinkel FD. 1990. Ovule and seed development of *Tovaria pendula* Ruiz et Pavon. Bot. Jahrb. Syst. 111: 389–401.

Bouman F. 1975. Integument initiation and testa development in some Cruciferae. Bot. J. Linn. Soc. 70: 213–229.

Bowman JL. 2006. Molecules and morphology: comparative developmental genetics of the Brassicaceae. Plant Syst. Evol. 259: 199–215.

Brock A, T Herzfeld, R Paschke, M Koch, and B Dräger. 2006. Brassicaceae contain nortropane alkaloids. Phytochemistry 67: 2050–2057.

Brown RC, BE Lemmon, and H Nguyen. 2004. Comparative anatomy of the chalazal endosperm cyst in seed of the Brassicaceae. Bot. J. Linn. Soc. 144: 375–394.

Brückner C. 1996. Carpelloid stamens in Papaveraceae Juss. and Brassdicaceae Burnett (Cruciferae Juss.) and their bearing on theories of gynoecium organization. Feddes Repert. 107: 321–337.

Brückner C. 2000. Clarification of the carpel number in Papaverales, Capparales, and Berberidaceae. Bot. Rev. 66: 155–307.

Carlquist S. 1971. Wood anatomy of Macaronesian and other Brassicaceae. Aliso 7: 365–384.

Carlquist S. 1978. Wood anatomy and relationships of Bataceae, Gyrostemonaceae, and Stylobasiaceae. Allertonia 1: 297–330.

Carlquist S. 1985. Wood anatomy and familial placement of *Tovaria*. Aliso 11: 69–76.

Carlquist S. 1998. Wood anatomy of Resedaceae. Aliso 16: 127–135.

Carlquist S. 2002. Wood and bark anatomy of Salvadoraceae: ecology, relationships, histology of interxylary phloem. J. Torrey Bot. Soc. 129: 10–20.

Carlquist S and RB Miller 1999. Vegetative anatomy and relationships of *Setchellanthus caeruleus* (Setchellanthaceae). Taxon 48: 289–302.

Chaban IA. 1983. Resedaceae. In: MS Yakovlev, ed. Comparative embryology of flowering plants, vol. 2, pp. 164–172. Nauka, Leningrad (in Russian).

Chaban IA and MS Yakovlev. 1974. Embryology of *Reseda lutea* L.: I. Megasporogenesis and development of the embryo sac. Bot. Zhurn. 59: 24–37 (in Russian).

Chandler GT and RJ Bayer. 2000. Phylogenetic placement of the enigmatic Western Australian genus *Emblingia* based on *rbc*L sequences. Plant Sp. Biol. 15: 67–72.

Chernyakovskaya EF. 1992. Resedaceae. In: AL Takhtajan, ed. Comparative seed anatomy, vol. 4, pp. 202–209. Nauka, St. Petersburg (in Russian).

Crete R. 1951. Embryogenie des Capparidacees: Developpement de lɜεμβρψον chez le *Cleome graveolens* Raffin. Compt. Rend. Hebd. Seances Acad. Sci. 233: 562–564.

D'Arcy WG. 1979. Family 73A. Capparaceae–Tovarioideae. In Flora of Panama. Ann. Missouri Bot. Gard. 66: 117–121.

David E. 1938. Embryologische Untersuchungen an Myoporaceen, Salvadoraceen, and Hippocrateaceen. Planta 28: 680–703.

Delaveau P, B Koudogbo, and J-L Pousset. 1973. Al-kaloides chez les Capparidaceae. Phytochemistry 12: 2983–2985.

De Leonardis W, G Fichera, N Longhitano, A Zizza. 1997. Pollen and seed morphology of three species of *Reseda* L. in Sicily and relationship with their systematic position. Plant Biosystems 131(1): 43–49.

Deng Y and Z Hu. 1995. The comparative morphology of the floral nectarines of Cruciferae. Acta Phytotax. Sinica 33: 209–220.

Den Outer RW and WLH van Veeneudaal. 1981. Wood and bark anatomy of *Azima tetracantha* Lam. (Salvadoraceae). Acta Bot. Neerl. 30: 199–207.

Dvorak F. 1968. A contribution to the study of the variability of the nectaries. Preslia 40: 13–17.

Dvorak F. 1971. On the evolutionary relationship in the family Brassicaceae. Feddes Repert. 82: 357–372.

Dvorak F. 1973. The importance of the indumentum for the investigation of evolutional relationship in the family Brassicaceae. Oesterr. Bot. Z. 121: 155–164.

Eames AJ. 1930. Crucifer carpels. Am. J. Bot. 17: 638–656.

Eames AJ and CL Wilson. 1928. Carpel morphology in the Cruciferae. Am. J. Bot. 15: 251–270.

Eckardt Th. 1959 (1960). Das Blütendiagramm von *Batis* P. Br. Ber. Deutsch. Bot. Ges. 72: 411–418.

Eckard Th. 1971. Anlegung und Entwicklung der Blüten von *Gyrostemon ramulosus* Desf. Bot. Jahrb. Syst. 90: 434–446.

El Migirab S, Y Berger, and J Jadot. 1977. Isothiocyanates, thiourees et thiocarbamates isolés de *Pentadiplandra brazzeana*. Phytochemistry 16: 1917–1921.

Erbar C and P Leins. 1997a. Different patterns of floral development in whorled flowers, exemplified by Apiaceae and Brassicaceae. Int. J. Plant Sci. 158: 49–64.

Erbar C and P Leins. 1997b. Studies on the early floral development in Cleomoideae (Capparaceae) with emphasis on the androecial development. Plant Syst. Evol. 206: 119–132.

Erdtman G, P Leins, R Melville, and CR Metcalfe. 1969. On the relationships of *Emblingia*. Bot. J. Linn. Soc. 62: 169–186.

Ernst WR. 1963. The genera of Capparaceae and Mor-ingaceae in the southeastern United States. J. Arnold Arbor. 44: 81–95.

Ezelarab GE and KJ Dormer. 1966. The organization of the primary vascular system in the Rhoeadales. Ann. Bot. 30: 123–132.

Fedotova TA. 1992. Gyrostemonaceae, Bataceae. In: A Takhtajan, ed. Comparative seed anatomy, vol. 4, pp. 213–228. Nauka, St. Petersburg (in Russian).

Fisel KJ and F Weberling. 1990. Untersuchungen zur Morphologic und Ontogenie der Blüten von *Tovaria pendula* Ruiz und Pavon und *Tovaria diffusa* (Macfad.) Fawcett and Rendle (Tovariaceae). Bot. Jahrb. Syst. 111: 365–387.

Frohne D. 1962. Das Verhaltnis von vergleichender Ser-obotanik zu vergleichender Phytochemie, dargestellt an serologischen Untersuchungen im Bereich der "Rhoeadales." Planta Medica 10: 283–297.

Fulcher WE. 1972. An anatomical and morphological study of *Batis maritime* L. with systematic implications. Diss. Abstr. Int. B 32: 6965.

Fursa NS, LE Belyaeva, and VE Avetisian. 1986. Natural compounds of the family Brassicaceae as possible chemotaxonomic characters. Report 4. Glucosinolates, alkaloids, sinapine. Rastitel'n. Resursy 22: 449–474 (in Russian).

Galloway GL, RL Malmberg, and RA Price. 1998. Phylogenetic utility of the nuclear gene arginine decarboxylase: an example from Brassicaceae. Molec. Biol. Evol. 15: 1312–1320.

Gazet du Chatelier G. 1946. Le diagramme de la fleur des Cruciferes. Rec. Trav. Inst. Bot. 2: 5–9.

George AS. 1982. Gyrostemonaceae. Flora Australia 8: 362–379. Canberra.

George AS. 2003. Gyrostemonaceae. In: Kubitzki K, ed. The families and genera of vascular plants, vol. 5, pp. 213–217. Springer, Berlin/Heidelberg/New York.

Gibson AC. 1979. Anatomy of *Koeberlinia* and *Canotia* revisited. Madroño 26: 1–12.

Goldblatt P. 1976. Chromosome number and its significance in *Batis maritima* (Bataceae). J. Arnold Arbor. 57: 526–530.

Goldblatt P. 1979. Chromosome number in two cytologically unknown New World families, Tovariaceae and Viviniaceae. Ann. Missouri Bot. Gard. 65: 776–777.

Goldblatt P, JW Nowicke, TJ Mabry, and H-D. Behnke. 1976. Gyrostemonaceae: status and affinity. Bot. Not. 129: 201–206.

Gomez-Campo C. 1980. Morphology and morpho-taxonomy of the tribe Brassicaceae. In: S Tsunoda, K Hinata, and C Gomez-Campo, eds. *Brassica* crops and wild allies, pp. 3–31. Japan Scientific Societies Press, Tokyo.

González-Aguilera JJ and AM Fernández-Peralta. 1984. Phylogenetic relationships in the family Resedaceae L. Genetica 64: 185–197.

Grubb CD and S Abel. 2006. Glucosinolate metabolism and its control. Trends Plant Sci. 11: 89–100.

Guerra M dos Santos. 1989. The chromosome number of *Azima tetracantha* (Salvadoraceae). Plant Syst. Evol. 168: 83–86.

Hall JC and KJ Sytsma. 2000. Solving the riddle of Californian cuisine: phylogenetic relationships of capers and mustards. Am. J. Bot. 87(6, Suppl.): 132.

Hall JC, KJ Sytsma, and HH Iltis. 2002. Phylogeny of Capparaceae and Brassicaceae base on chloroplast sequence data. Am. J. Bot. 89: 1826–1842.

Hall JC, HH Iltis, and KJ Sytsma. 2004. Molecular phylogenetics of core Brassicales, placement of orphan genera *Emblingia, Forchhammeria, Tirania,* and character evolution. Syst. Bot. 29: 654–669.

Hedge IC, A Kjaer, and O Malver. 1980. *Dipterygium* – Cruciferae or Capparaceae? Notes Roy. Bot. Gard. Edinb. 38: 247–250.

Hennig L. 1929. Beiträge zur Kentnis der Resedaceen-Blüte und Frucht. Planta 9: 507–563.

Holl JC, HH Iltis, and KJ Sytsma. 2004. Molecular phylogenetics of core Brassicales, placement of orphan genera *Emblingia, Forchhammeria, Tirania,* and character evolution. Syst. Bot. 29: 654–669.

Hufford L. 1996. Developmental morphology of female flowers of *Gyrostemon* and *Tersonia* and floral evolution among Gyrostemonaceae. Am. J. Bot. 83: 1471–1487.

Hurka H, M Paetsch, W Bleeker, and B Neuffer. 2005. Evolution within the Brassicaceae. Nova Acta Leopoldina NF 92: 113–127.

Iltis HH. 1957. Studies in the Capparidaceae: III. Evolution and phytogeny of the western North American Cleomoideae. Ann. Missouri Bot. Gard. 44: 77–119.

Iltis HH. 1999. Setchellanthaceae (Capparales), a new family for a relictual, glucosinolate-producing endemic of the Mexican deserts. Taxon 48: 257–275.

Inamdar JA. 1969. The stomatal structure and ontogeny in *Azima* and *Salvadora*. Flora 158: 519–525.

Iversen T-H. 1970. The morphology, occurrence, and distribution of dilated cisternae of the endoplasmic reticulum in tissues and plants of the Cruciferae. Protoplasma 71: 467–477.

Janchen E. 1942. Das System der Cruciferen. Oesterr. Bot. Z. 91: 1–28.

Johnson DS. 1935. The development of the shoot, male flower, and seedling of *Batis maritima* L. Bull. Torrey Bot. Club 62: 19–31.

Jørgensen LB. 1981. Myrosin cells and dilated cisternae of the endoplasmic reticulum in the order Capparales. Nord. J. Bot. 1: 341–460.

Jørgensen LB. 1999. Absence of myrosin cells in *Koeberlinia*. In Linnean Society Symposium under the microscope: plant anatomy and systematics. Abstracts: 32.

Jørgensen LB, H-D Behnke, and TJ Mabry. 1977. Protein-accumulating cells and dilated cisternae of the endoplasmic reticulum in three glucosinolate containing genera: *Armoracia, Capparis, Drypetes*. Planta 137: 215–224.

Kamelin RV. 2002. Cruciferae: concise review of the system. Barnaul University, Barnaul. (In Russian).

Karol KG, JE Rodman, E Conti, and KJ Sytsma. 1999. Nucleotide dequence of *rbc*L and phylogenetic relationships of *Setchellanthus caeruleus* (Setchellanthaceae). Taxon 48: 303–315.

Keighery GJ. 1975. Chromosome numbers in the Gyrostemonaceae Endl. and the Phytolaccaceae Lindl.: a comparison. Aust. J. Bot. 23: 335–338.

Keighery GJ. 1981. The breeding system of *Emblingia* (Emblingiaceae). Plant Syst. Evol. 137(1–2): 63–65.

Kers LE. 2003. Capparaceae. In: K Kubitzki, ed. The families and genera of vascular plants, vol. 5, pp. 36–56. Springer, Berlin/Heidelberg/New York.

Khalik KA and LJG van der Maesen. 2002. Seed morphology of some tribes of Brassicaceae (implications for taxonomy and species identification for the flora of Egypt). Blumea 47: 363–383.

Kjaer A. 1968. Glucosilonates in Tovariaceae. Phytochemistry 7: 131–133.

Kjaer A. 1976. Glucosinolates in the Cruciferae. In: JG Vaughan, AJ Macleod, and BMG Jones, eds. The biology and chemistry of the Cruciferae, pp. 207–219. Academic, London.

Kjaer A and O Malver. 1979. Glucosinulates in *Tersonia brevipes* (Gyrostemonaceae). Phytochemistry 18: 1565.

Koch M. 2003. Molecular phylogenetics, evolution and population biology in Brassicaceae. In: AK Sharma and A Sharma, eds. Plant genome: biodiversity and evolution, vol. 1, part A. Phanerogams, pp. 1–35. Science Publishers, Enfield, NH.

Koch M, B Haubold, and T Mitchell-Olds. 2001. Molecular systematics of the Brassicaceae: evidence from coding plastidic *mat*K and nuclear Chs sequences. Am. J. Bot. 88: 534–544.

Koch, M, IA Al-Shehbaz, and K Mummenhoff. 2003. Molecular systematics, evolution, and population biology in the mustard family (Brassicaceae). Ann. Missouri Bot. Gard. 90: 151–171.

Koch M, C Dobes, M Matschinger, W Bleeker, J Vogel, M Kiefer, and T Mitchell-Olds. 2005. Evolution of the *trn*F(GAA) gene in *Arabidopsis* relatives and the Brassicaceae family: monophyletic origin and subsequent diversification of a plastidic pseudogene. Molec. Biol. Evol. 22: 1032–1043.

Kolbe KP. 1978. Serologische Beiträge zur Systematik der Capparales. Bot. Jahrb. Syst. 99: 468–489.

Koudogbo B and P Delaveau. 1974. Chimotaxonomie des Capparidaceae. Plant Med. Phytotherapie 8: 96–103.

Koul KK, N Ranjna, and SN Raina. 2000. Seed coat microsculpturing in *Brassica* and allied genera (subtribes Brassicinae, Raphaninae, Moricandiinae). Ann. Bot. 86: 385–397.

Kozo-Poljanski BM. 1945. The origin of the flowers of Cruciferae in the light of teratology: teratology of the flower and new problems of its theory. Bot. Zhurn. 30: 14–30 (in Russian).

Kshetrapal S. 1970. A contribution to the vascular anatomy of the flower of certain species of the Salvadoraceae. J. Indian Bot. Soc. 49: 92–99.

Kubitzki K. 2003a. Emblingiaceae. In: K Kubitzki, ed. The families and genera of vascular plants, vol. 5, pp. 206–208. Springer, Berlin/Heidelberg/New York.

Kubitzki K. 2003b. Koeberliniaceae. In: K Kubitzki, ed. The families and genera of vascular plants, vol. 5, pp. 218–219. Springer, Berlin/Heidelberg/New York.

Kubitzki K. 2003c. Resedaceae. In: K Kubitzki, ed. The families and genera of vascular plants, vol. 5, pp. 334–338. Springer, Berlin/Heidelberg/New York.

Kubitzki K. 2003d. Salvadoraceae. In: K Kubitzki, ed. The families and genera of vascular plants, vol. 5, pp. 342–344. Springer, Berlin/Heidelberg/New York.

Kubitzki K. 2003e. Setchellanthaceae. In: K Kubitzki, ed. The families and genera of vascular plants, vol. 5, pp. 353–354. Springer, Berlin/Heidelberg/New York.

Kumar A, SR Rao, and TS Rathore. 2002. Cytological investigations in some important tree species of Rajasthan: 4. Male meiosis studies in the genus *Salvadora* L. Cytologia (Japan) 67(2): 105–115.

Kuprianova LA. 1965. Palynology of Amentiferae. Nauka, Leningrad (in Russian).

La-Serna Ramos I. 1996. Pollen characters of Canary Resedaceae with special reference to endemic taxa. Grana 35: 16–23.

Leins P and G Metzenauer. 1979. Entwicklungs-geschichtliche Untersuchungen an Capparis-Bluten. Bot. Jahrb. Syst. 100: 542–554.

Leins P and U Sobick. 1977. Die Blütenentwicklung von *Reseda lutea*. Bot. Jahrb. Syst. 98: 133–149.

Lysack MA, MA Koch, A Pecinka, and I Schubert. 2005. Chromosome triplication across the tribe Brassiceae. Genome Res. 15: 516–525.

Mabry TJ. 1976. Pigment dichotomy and DNA-RNA hybridization data for centrospermous families. Plant Syst. Evol. 126: 79–94.

Mabry TJ and BL Turner. 1964. Chemical investigations of the Batidaceae. Betaxanthins and their systematic implications. Taxon 13: 197–200.

Maheshwari Devi H. 1972. Salvadoraceae: a study of its embryology and systematics. J. Indian Bot. Soc. 51: 56–62.

Mauritzon J. 1934. Die Embryologie einiger Capparidaceen sowie von *Tovaria pendula*. Arkiv Bot. 26A(15): 1–14.

McLaughlin L. 1959. The woods and flora of the Florida Keys: wood anatomy and phylogeny of Batidaceae. Trop. Woods 110: 1–15.

McLean WFH, G Blunden, and K Jewers. 1996. Quaternary ammonium compounds in the Capparaceae. Biochem. Syst. Ecol. 24(5): 427–434.

Mehta IJ and MF Moseley, Jr. 1981. The floral anatomy of *Koeberlinia* Zucc.: systematic implications. Am. J. Bot. 68: 482–497.

Melikian AP. 1990. Comparative studies of fruits and seeds of the representatives of the family Phytolaccaceae R. Br. In: VN Tikhomirov and AN Sladkov, eds. Morphology of the centrosperms as a source of evolutionary information, pp. 151–180. Nauka, Moscow (in Russian).

Melikian AP and IA Savinov. 2000. Salvadoraceae. In: AL Takhtajan, ed. Comparative seed anatomy, vol. 6, pp. 139–140. Nauka, St. Petersburg (in Russian).

Meyer-Melikian NR. 1990. Palynomorphology of the family Phytolaccaceae R. Br. In: VN Tikhomirov and AN Sladkov, eds. Morphology of the centrosperms as a source of evolutionary information, pp. 104–151. Nauka, Moscow (in Russian).

Ming D and G Hellekant. 1994. Brazzein, a new high-potency sweet protein from *Pentadiplandra brazzeana*. B. FEBS Lett. 335: 106–108.

Mitchell-Olds T, IA Al-Shehbaz, MA Koch, and TF Sharbel. 2005. Crucifer evolution in the post-genomic era. In: RJ Henry, ed. Plant diversity and evolution: genotypic and phenotypic variation in higher plants, pp. 119–137. CABI International, Wallingford.

Murty YS. 1953. A contribution to the anatomy and morphology of normal and some abnormal flowers of *Gynandropsis gynandra* (L.) Briq. J. Indian Bot. Soc. 32: 108–122.

Narayana HS. 1962, 1965. Studies in the Capparidaceae: I. The embryology of *Capparis decidua* (Forsk.) Pax. II. Floral morphology and embryology of *Cadaba indica* Lamk. and *Grata evanurvala* Buch.-Ham. Phytomorphology 12: 167–177, 1962; 15: 158–175, 1965.

Narayana LL and A Parvathi. 1980. Chemotaxonomy of Capparidaceae. Phyta 2–3: 87–91.

Norris T. 1941. Torus anatomy and nectary characteristics as phylogenetic criteria in the Rhoeadeles. Am. J. Bot. 28: 101–113.

Oksiyuk PF. 1937. Comparative cyto-embryological study in the Resedaceae. I. *Reseda* and *Astrocarpus*. Zhurn. Inst. Bot. (Kiev) 12: 2–46 (in Russian).

Orr MY. 1919. The occurrence of tracheal tissue enveloping the embryo in certain Capparidaceae. Notes Roy. Bot. Gard. Edinb. 11: 249–257.

Orr MY. 1921. Observations on the structure of the seed in the Capparidaceae and Resedaceae. Notes Roy. Bot. Gard. Edinb. 12: 259–260.

Pant DD and PF Kidwai. 1967. Development of stomata in some Cruciferae. Ann. Bot. (London) 31: 513–521.

Pillai A and SK Pillai. 1977. Some aspects of the anatomy of *Salvadora oleoides* Decne. Flora 166: 211–218.

Polowick PL and VK Sawhney. 1986. A scanning electron microscopic study on the initiation and development of floral organs *Brassica napus* (cv. Westar). Am. J. Bot. 73: 254–263.

Prasad K. 1975. Development and organization of gametophytes in certain species of Cruciferae. Acta Bot. Indica 3: 147–154.

Prasad K. 1979. Morphology and histochemistry of the nucellus and endosperm of Cruciferae. Bot. Jahrb. Syst. 100: 536–541.

Price RA, IA Al-Shehbaz, and JD Palmer. 1994. Systematic relationships of *Arabidopsis:* a molecular and morphological perspective. In: E Meyerowitz and C Somerville, eds. *Arabidopsis*, pp. 7–19. Cold Spring Harbor Laboratory Press, New York.

Prijanto B. 1970. Batidaceae, Gyrostemonaceae. In: G Erdtman, ed. World pollen flora, vol. 3, pp. 1–11. Plenum, New York.

Puri V. 1941. Studies in floral anatomy: I. Gynoecium constitution in the Cruciferae. Proc. Indian Acad. Sci. 14B: 166–187.

Puri V. 1942. Studies in floral anatomy: II. Floral anatomy of the Moringaceae with special reference to gynoedum constitution. Proc. Natl. Inst. Sci. India 8: 71–88.

Puri V. 1950. Studies in floral anatomy: VI. Vascular anatomy of the flower of *Crataeva religiosa* Forst., with special reference to the nature of the carpels in the Capparidaceae. Am. J. Bot. 37: 363–370.

Raghavan TS. 1937, 1939. Studies on the Capparidaceae: I. The life-history of *Cleome cholidonii* Linn. fil. II. Floral anatomy and some structural features of the capparidaceous flower. Bot. J. Linn. Soc. 51: 43–72, 1937; 52: 239–257.

Raghavan TS and KR Venkatasubban. 1941. Studies in the Capparidaceae: V. The floral morphology of *Crataeva religiosa* Forst. VII. Floral ontogeny and anatomy of *Crataeva religiosa* with special reference to the morphology of the carpel. Beih. Bot. Centralbl. 60: 388–396; 397–416.

Record S. 1926. The wood of *Koeberlinia spinosa* Zuccarini. Trop. Woods 8: 15–17.

Rodionova GB. 1982. Tovariaceae and Capparaceae. In: MS Yakovlev, ed. Comparative embryology of flowering plants, vol. 2, pp. 149–154. Nauka, Leningrad (in Russian).

Rodionova GB. 1992. Capparaceae. Brassicaceae. Tovariaceae. In: AL Takhtajan, ed. Comparative seed anatomy, vol. 4, pp. 183–202. Nauka, St. Petersburg (in Russian).

Rodionova GB. 1997. The main directions of evolution of families in the order Capparales on the basis of embryological characteristics. Byull. Glavn. Bot. Sada (Moscow) 175: 142–146 (in Russian with English summary).

Rodman JE, KG Karol, RA Price, E Conti, and KJ Sytsma. 1994. Nucleotide sequences of *rbc*L confirm the capparalean affinity of the Australian endemic Gyrostemonaceae. Aust. Syst. Bot. 7: 57–69.

Rodman JE, KG Karol, RA Price, and KJ Sytsma. 1996. Molecules, morphology, and Dahlgren's expanded order Capparales. Syst. Bot. 21: 289–307.

Rodman JE, PS Soltis, DE Soltis, KJ Sytsma, KG Karol. 1998. Parallel evolution of glucosinolate biosynthesis inferred from congruent nuclear and plastid gene phylogenies. Am. J. Bot. 85: 997–1006.

Rogers GK. 1982. The Bataceae in the southeastern United States. J. Arnold Arbor. 63: 375–386.

Rollins RC. 1993. The Cruciferae of continental North America: systematics of the mustard family from the Arctic to Panama. Standford University Press, Stanford.

Rollins RC and UC Banerjee. 1976. Trichomes in studies of the Cruciferae. In: JG Vaughan, AJ Macleod, and BMG Jones, eds. The biology and chemistry of the Cruciferae, pp. 145–166. Academic, London.

Rollins RC and UC Banerjee. 1979. Pollens of the Cruciferae. The Bussey Institution of Harvard University, Cambridge, MA.

Ronse Decraene LP. 2002. Floral development and anatomy of *Pentadiplandra* (Pentadiplandraceae): a key genus in the identification of floral morphological trends in the core Brassicales. Canad. J. Bot. 80: 443–459.

Ronse Decraene LP. 2005. Floral developmental evidence for the systematic position of *Batis* (Bataceae). Am. J. Bot. 92: 752–760.
Ronse Decraene LP and E Haston. 2006. The systematic relationships of glucosinolate-producing plants and related families: a cladistic investigation based on morphological and molecular characters. Bot. J. Linn. Soc. 151: 453–494.
Ronse Decraene LP and EF Smets. 1997a. A floral ontogenetic study of some species of *Capparis* and *Boscia*, with special emphasis on the androecium. Bot. Jahrb. Syst. 119: 231–255.
Ronse Decraene LP and EF Smets. 1997b. Evidence for carpel multiplications in the Capparaceae. Belg. J. Bot. 130: 59–67.
Rössler W. 1974. Myrosinzellen bei *Tovaria*. Phyton (Horn) 16: 231–522.
Royen P van. 1956. A new Batidaceae, *Batis argillicola*. Nova Guinea 7: 187–195.
Schranz ME and T Mitchell-Olds. 2006. Independent ancient polyploidy events in the sister families Brassicaceae and Cleomaceae. Plant Cell 18: 1152–1165.
Schraudolf H. 1969. Serotonin und Indoglucosinolate in *Tovaria pendula*. Naturwissenschaften 56: 462–463.
Schraudolf H., Schmidt B., Weberling F. 1971. Das Vorkommen von "Myrosinase" als Hinweis auf die systematische Stellung der Batidaceae. Experientia 27: 1090–1091.
Schweingruber FH. 2006. Anatomical characteristics and ecological trends in the xylem and phloem of Brassicaceae and Resedaceae. IAWA J. 27: 419–442.
Singh D and S Gupta. 1968. The seeds of the Violaceae and Resedaceae: a comparison. J. Indian Bot. Soc. 46: 248–256.
Sobick U. 1983. Blütenentwicklungsgeschichtliche Untersuchungen an Resedaceen unter besonderer Berücksichrigung von Androeceum und Gynoeceum. Bot. Jahrb. Syst. 104: 203–248.
Spratt ER. 1932. The gynoecium of the family Cruciferae. J. Bot. 70: 308–314.
Stern WL, GK Brizicky, and FN Tamolang. 1963. The woods and flora of the Florida keys: Capparaceae. Contr. U.S. Natl. Herb., vol. 34, part 2. Washington, DC.
Stoudt HN. 1941. The floral morphology of some of the Capparidaceae. Am. J. Bot. 28: 664–675.
Subramanian D and G Susheela. 1988. Cytotaxonomical studies of South Indian Capparidaceae. Cytologia 53: 679–684.
Tieghem P van. 1903. Sur les Batidacees. J. Bot. (Morot) 17: 363–376.
Tobe H, S Carlquist, HH Iltis. 1999. Reproductive anatomy and relationships of *Setchellanthus caeruleus* (Setchellanthaceae). Taxon 48: 287–283.
Tobe H and PH Raven. 1991. The embryology and relationships of Gyrostemonaceae. Aust. Syst. Bot. 4: 407–420.
Tobe H and PH Raven. 1992. The embryology and relationships of Bataceae. Syst. Bot. 19: 485–496.
Tobe H and M Takahashi. 1995. Pollen morphology of Gyrostemonaceae, Bataceae, and *Koeberlinia*. J. Plant Res. 108(1091): 283–288.
Tomb AS. 1999. Pollen morphology and relationships of *Setchellanthus caeruleus* (Setchellanthaceae). Taxon 48: 285–288.
Uphof JC. 1930. Biologische Beobachtungen an *Batis maritima* L. Oesterr. Bot. Z. 79: 355–367.
Vaughan JG and JM Whitehouse. 1971. Seed structure and the taxonomy of the Cruciferae. Bot. J. Linn. Soc. 64: 383–409.
Vaughan JG, AJ McLeod, and BMG Jones, eds. 1976a. The biology and chemistry of the Cruciferae. Academic, London/New York/San Francisco.
Vaughan JG, JR Phelan, and KE Denford. 1976b. Seeds studies in the Cruciferae. In: JG Vaughan JG, AJ McLeod, and BMG Jones, eds. The biology and chemistry of the Cruciferae, pp. 119–144. Academic, London.
Verdcourt B. 1968. Salvadoraceae. In: E Milne-Redhead and RM Polhill, eds. Flora of tropical East Africa. Government Printer, London.
Warwick SI and IA Al-Shehbaz. 2006. Brassicaceae: Chromosome number index and database on CD-Rom. Plant Syst. Evol. 259: 237–248.
Warwick SI and LD Black. 1997. Phylogenetic implications of chloroplast DNA restriction site variation in subtribes Raphaninae and Cakilinae (Brassicaceae, tribe Brassiceae). Canad. J. Bot. 72: 1692–1701.
Warwick SI and CA Saunder. 2005. Phylogeny of the tribe Brassiceae (Brassicaceae) based on chloroplast restriction site polymorphisms and nuclear ribosomal internal transcribed spacer and chloroplast *trn*L intron sequences. Canad. J. Bot. 83: 467–483.
Warwick SI, A Francis, and IA Al-Shehbaz. 2006. Brassicaceae: Species checklist and database on CD-Rom. Plant Syst. Evol. 259: 249–258.
Wel H van der, G Larson, A Hladik, G Hellekant, and D Glaser. 1989. Isolation and characterization of pentadin, the sweet principle of *Pentadiplandra brazzeana* Baillon. Chem. Senses 14: 75–79.
Wittstock U and BA Halkier. 2002. Glucosinolate research in the *Arabidopsis* era. Trends in Plant Science 7: 263–270.
Zohary M. 1948. Carpological studies in Cruciferae. Palestine J. Bot. Jerusalem 4: 158–165.
Zunk K, K Mummenhoff, and H Hurka. 1999. Phylogenetic relationships in tribe Lepidieae (Brassicaceae) based on chloroplast DNA restriction site variation. Canad. J. Bot. 77: 1504–1512.

Superorder MALVANAE

Order 75. MALVALES

Trees, shrubs, woody lianas, or herbs. Mostly with tufted or stellate hairs or peltate scales, or less often with unicellular hairs, sometimes also with glands, more rarely lepidote, exceptionally with prickles or spines. Mucilage cells, cavities, or canals usually present in the parenchymatous tissues and sometimes in the wood and roots. Vessels with simple perforations lateral pitting usually alternate. Fibers with simple pits. Rays heterogeneous to homogeneous. Axial parenchyma apotracheal or paratracheal. Some cells of the rays and the axial parenchyma usually contain prismatic, mostly solitary crystals. Phloem in young stem stratified tangentially into hard and soft layers and generally with wedge-shaped rays Sieve-element plastids of S-type. Nodes trilacunar, pentalacunar, multilacunar, or rarely unilacunar. Leaves alternate or opposite, or rarely verticillate, often palmately veined, simple to palmately compound, usually stipulate, petiolar vascular system usually complex. Stomata anomocytic or paracytic, sometimes anisocytic or encyclocytic. Flowers in various types of basically cymose inflorescences, or less often solitary, mostly bisexual, usually actinomorphic, mostly 5-merous, commonly with a double perianth, often provided with an epicalyx. Nectaries usually consisting of tufts of glandular hairs and mostly borne at the base of sepals. Sepals free or basally connate, valvate or rarely (Monotaceae except *Pseudomonotes*) imbricate. Petals free or sometimes basally adnate to the filament tube, contorted, imbricate or valvate, sometimes reduced or wanting. Nectary disc, if present, of glandular trichomes on ventral face of sepals, more rarely on petals or androgynophore. Stamens basically in two cycles, mostly five to very numerous (usually more or less numerous), rarely (*Pimelea*) 1–2, developing in centrifugal sequence when numerous, free or often the filaments connate into a tube or 5–10(15) fascicles; members of the outer cycle often wanting or transformed into staminodia. Anthers basifixed or dorsifixed, tetrasporangiate and 2-locular or disporangiate and 1-locular, opening longitudinally or less often by apical pores. Tapetum secretory or amoeboid. Microsporogenesis simultaneous. Pollen grains 2-celled or 3-celled, from 3-colpate or 3-colporate to pantoporate. Gynoecium of two to many (often five) united carpels (secondarily apocarpous or partly apocarpous in some Sterculiaceae) with free or more or less united stylodia; ovary superior or rarely inferior (*Neotessmannia*), with as many locules as the carpels or seldom unilocular by failure of the intruded parietal placentas to meet in the center. Ovules 1-many in each locule or on each placenta, ascending or pendulous, anatropous to campylotropous and hemitropous, rarely orthotropous, bitegmic, crassinucellate. Female gametophyte of *Polygonum*-type. Endosperm mostly nuclear. Fruits usually dry, dehiscent or indehiscent, less often fleshy, rarely winged. Seed coat generally formed by both the integuments, and usually exotegmic; embryo straight or less often curved; endosperm present or wanting. Often produce fatty acids containing a cyclopropenyl ring.

Malvales have many features in common with the Violales, particularly with Flacourtiaceae (especially with the Prockieae). Both the anatomy of vegetative organs and the palynological data confirm the origin of Malvales from Flacourtiaceae through the intermediate group Scolopieae s.l. (including Prockieae and Banareae) (Keating 1973; Miller 1975). It is also confirmed by chemotaxonomical data (Alston and Turner 1963).

Key to Families

1 Anthers tetrasporangiate and 2-locular.
  2 Nodes 3-lacunar.
    3 Anthers opening longitudinally.
      4 Style not gynobasic.
        5 Stamens free or shortly connate at the base.
          6 Mucilage cells or cavities present.
            7 Pollen grains usually in monads.
              8 Seeds with copious or scanty, oily endosperm, often with fatty acids that contain cyclopropene. Trees, shrubs, or sometimes herbs, provided with simple or more often stellate hairs or peltate scales. Nodes trilacunar. Leaves alternate or rarely opposite, entire or often dentate or lobed, mostly palmately or pinnipalmately veined, often asymmetric. Stomata usually anomocytic. Flowers in various types of cymose inflorescences, paired or solitary, mostly bisexual, sometimes with basally connate. Petals as an epicalyx. Sepals (3-)5, valvate, free or sometimes many as sepals, free, contorted, imbricate or valvate, sometimes wanting. Nectaries usually consist of variously located tufts of glandular trichomes. Stamens mostly numerous, free or shortly basally connate or in 5–10 fascicles; anthers opening longitudinally or sometimes by apical pores. Pollen grains in monads, 3-colporate to 3-porate. Androgynophore present or absent. Gynoecium of two

to many carpels; style with a lobed or capitate stigma; ovary superior, plurilocular or (*Goethalsia, Mollia*) unilocular, with (1)2-several ovules on axile or intruded parietal placentas in each locule. Fruits dry or fleshy, dehiscent or not. Embryo mostly straight, n = 7–41. 2. TILIACEAE.

8 Endosperm scanty or none.

9 Stem with characteristic vertical branching system of intercellular resin canals in the pith, wood,andbark.Ectomycorrhizal trees. Secretory cavities in pith absent. Rays heterogeneous to homogeneous, larger rays multiseriate. Flowers without androgynophore. Sepals valvate, often two or more of them conspicuously accrescent and wing-like in fruit. Anthers basifixed. Pollen grains 3-colpate, the exine convoluted, without clear columellae. Ovary (2)3-locular, each locule with two ovules; micropyle endostomal. Produced terpenoids, sesquiterpene oleoresins and dipterocarpol. n = 6, 7, 10, 11(-13). . . . 3. DIPTEROCARPACEAE.

9 Stem without resin canals. Secretory cavities in pith present. Rays heterogeneous, uniseriate or sometimes biseriate. Flowers with or without androgynophore. Sepals imbricate or less often valvate, equally accrescent, sometimes winglike. Anthers basiversatile. Pollen grains 3-colporate, the exine tectate-columellate. Ovary 2–5-locular, each locule with 1–2 or 4 ovules. Fruits 1-seeded, nutlike, or 3–5 locular capsules. Produced sesqui- and triterpenoids. . . . . .4. MONOTACEAE.

7 Pollen grains in tetrads, 3–6-colpate. Stamens usually very numerous, rarely (spp. of *Leptolaena*) 5–10, usually persistent, generally inserted inside an entire or toothed, more or less cupular nectary disc of staminodial origin; filaments free or weakly connate at the base into 5–10 fascicles; anthers basi- or dorsifixed, opening longitudinally. Gynoecium of 3–4(5) carpels with a terminal style and expanded, usually lobed stigma; ovary with as many locules as carpels, with (1)2-many ovules per carpel; ovules apical-descendent, basal-ascendant, or axile in the middle of the ovary, anatropous. Petals contorted, 5(-6), large, free or slightly united at the base. Sepals mostly 3(4–5), imbricate-contorted, when more than three, the outer ones smaller than the inner three. Flowers in umbel-like or more or less paniculiform inflorescence or rarely solitary, bisexual, usually subtended by an involucel (epicalyx) of bracteoles more or less connate to form a cup; in *Eremolaena* epicalyx lacking. Fruits loculicidal capsules or indehiscent and with only one or a few seeds. Seeds hairy or not, with straight embryo; seed coat exotegmic; endosperm usually copious and starchy, sometimes ruminate or reduced; cotyledons cordate. Leaves simple, entire, usually persistent, with usually caducous stipules. Nodes trilacunar. Trees or shrubs, often with stellate hairs. Presence of sapinins, glucosides of myrcetin and other flavonols, phenolic acids of the cinnamic and the benzoic series, n = 11. 5. SARCOLAENACEAE.

6 Mucilage cells and cavities absent. Phloem not stratified.

10 Trichomes glandular with globular heads. Small to medium sized trees or shrubs; pubescent of stellate, long unbranched, and glandular hairs, especially on young growth. Perforation plates non-bordered or

minimally bordered. Axial parenchyma diffuse, scanty vasicentric, diffuse-in-aggregated. Leaves alternate, distichous, petiolate, blades asymmetrically cordate at base, palminerved, their margin serrate; stipule-like appendages filiform or foliaceous and peltate. Flowers in supra-axillary position, solitary of in few-flowered clusters, pedicellate, actinomorphic, usually bisexual. Sepals (4-)5(-7), valvate in bud, with more or less spreading tips, basally fused, forming a saucer- to cup-like tube. Petals (4)5(-7), more or less crumpled in bud, imbricate, free, longer than the calyx, thin, caducous, the outer margin irregular. Stamens numerous; filaments filiform, free or almost so; anthers dithecal, sometimes versatile, fixed at or near the base of the connective, dehiscing by longitudinal slits, sometimes restricted to their apical portion. Pollen grains 3-colporate. Ovary superior to inferior, syncarpous, 5-many locular, sometimes 1-locular in the upper part, lobed, pendulous; style thick, sometimes almost lacking; stigma thick, lobed-sulcate to decurrent. Ovules numerous. Fruits indehiscent, baccate. Seeds numerous, small; embryo short, straight. Endosperm composed of large, thin-walled cells, oily. Contain gallic and ellagic acids, n = 14, 15. . . . 1. MUNTINGIACEAE.

10 Trichomes stellate or simple, not glandular. Prostrate, tomentose, perennial and annual herbs with stellate trichomes; stems sympodial, more or less woody below. Vessels with simple perforations; lateral pitting opposite, sometimes nearly alternate. Have lysigenous mucilage ducts in the pith. Sieve-element plastids of Pcs-type. Leaves alternate, simple, toothed, lobed to pinnatifid; estipulate. Stomata anomocytic. Flowers solitary on axillary peduncles, bisexual, actinomorphic, sometimes with an epicalyx of five bracteoles, 5-merous. Sepals and petals united into a floral tube that is dry at maturity. Calyx lobes more or less valvate, corolla lobes imbricate or contorted. Stamens 10, borne on the floral tube; filaments elongate, subulate, glabrous or rarely pubescent, more or less persistent; anthers small, introrse, dorsifixed, opening longitudinally. Pollen grains 3-celled, 3-colporate, of a unique bipolar type, with 3(4) pores at each end. Gynoecium of 10 or (*Grielum*) 5–10 united carpels, with free subulate or filiform stylodia that eventually become indurated; stigmas small, capitate; ovary more or less inferior, multilocular, but 2–4 of the locules on the side toward the peduncle more or less reduced, or their ovules not maturing, with one ovule in each of the fully developed locules. Ovules pendulous, anatropous, apotropous. Fruits dry, laterally membranous-winged or spinose-muricate, carpels dehiscing ventrally, stylodia sometimes spinescent. Seeds six to several; exotegmic cells tangentially elongated, crystalliferous, other tegmic cells persistent; embryo contains protein and oil, but lacks starch; endosperm lacking. Cyclopropenoid fatty acids present; n = 7. . .6. NEURADACEAE.

5 Stamens generally connate into a tube around the ovary, often seated on an androgynophore, basically in two cycles, the outer usually staminodial or sometimes suppressed, the inner often branched. Anthers opening by longitudinal slits or rarely by apical pores. Pollen grains from 3-colporate and smooth or reticulate to periporate and spinulose.

Gynoecium usually of five carpels, with stylodia free or more or less connate into a style. Ovary with as many locules as carpels or rarely unilocular through the failure of the parietal placentas to meet in the center; in the tribe Sterculieae the carpels united only by their stylodia and becoming free at maturity, or wholly distinct from the beginning (as in *Cola*), but this kind of apocarpy is most probably of secondary origin (Takhtajan 1966). Ovules (1)2 to many per locule, free carpel or free placenta. Fruits of various types. Seeds sometimes arillate, with straight or curved embryo and usually copious endosperm or rarely (as in *Cola*) without endosperm. Flowers bisexual or less often unisexual, usually actinomorphic. Sepals valvate. Petals often small and reduced, contorted, sometimes absent, as in Sterculieae. Leaves alternate, from simple and entire to palmately lobed and palmately compound, stipulate. Stomata mostly anomocytic. Trees, shrubs (sometimes lianoid), or rarely herbs. Vessels with simple perforation of rarely (*Lasiopetalum*) with rudimentarily scalariform, n = 15, 16, 18, 20, 21 +. . . . . .7. STERCULIACEAE.

4 Style more or less gynobasic.

11 Sepals and petals 5(6). Stipules broad and coiled around the terminal bud, caducous, and leaving a horizontal scare. Leaves simple, entire, pinnately veined, densely minutely pellucid-punctate, smelling of camphor when crushed. Stomata cyclocytic. Flowers in terminal, few-flowered, paniculiform inflorescences, large, bisexual, fragrant. Sepals free, widely imbricate, unequal (the outer smaller), persistent until the beginning of fruit development. Petals free, large, slightly unequal, widely imbricate, caducous. Stamens very numerous (up to 435 – Capuron 1963), free; filaments filiform; anthers basifixed, opening laterally by longitudinal slits. Pollen grains small, spheroidal, 3-colporate. Gynoecium of 2–4 free carpels with common central elongate gynobasic style with punctiform stigma; carpels ovoid, very verrucose and with numerous small, peltate glands. Ovules two in each carpel, basal, ascending, with micropyle facing outward. Fruits of 1–3 free 1-seeded mericarps with thin, coriaceous pericarps and with many small glands; seed coat thin, no palisade layer, inwardly-curving exotegmen or hypostase plug; embryo with thick cotyledons; endosperm absent. Shrubs or small trees. . . . . . 8. DIEGODENDRACEAE.

11 Sepals and petals four. Stipules large, caducous. Leaves simple, entire, pubescent or glabrous, pinnately veined or 3-veined, not pellucid-punctate and not smelling of camphor when crushed. Stomata anomocytic or anisocytic. Flowers in axillary or terminal cymose inflorescences, bisexual. Sepals, unequal, much imbricate. Petals unequal, imbricate, densely streaked with short, resinous lines, caducous. Stamens numerous (more than 100), in 2–4 series, free or inserted below a large, thick, gynophore-like, cupular, wrinkled, denticulate nectary disc; filaments slender, elongate, more or less connate at the base, resinducted; anthers small, introrse, loculi short and widely separated by a broad, glandular connective, opening longitudinally. Pollen grains more or less spinose, 3–5-, rarely 6-colpate. Gynoecium of two or four, or rarely three or five carpels; ovary superior, but partly immersed in the nectary disc, usually 2-locular and vertically more or less deeply lobed (in *Dialyceras* resulting separation of carpels) with simple style inserted between the lobes; stigma more or less capitate. Ovules usually three in each locule, subbasal, anatropous, epitropous and ascending. Fruits more or less deeply lobed or globose, densely muricate, in *Dialyceras* divided into mericarps. Seeds large, with straight embryo; endosperm ruminate or not, copious, oily, contains elliptic starch grains (Bayer 2003);

cotyledons cordate and bilobed apically. Trees or shrubs. Leaves and seeds of *Rhopalocarpus*) contain saponins, n = 19 (*Rhopalocarpus lucidus* – Sarkar et al. 1982). . . 9. SPHAEROSEPALACEAE.

3 Anthers opening by pores or pore-like slits.

12 Leaves simple and entire, long-petiolate, stipulate. Flowers in terminal thyrsoids, large, actinomorphic. Sepals five, free, imbricate, caduceus. Petals five, free, imbricate, alternating with sepals. Stamens numerous, free; anthers dithecal, horseshoe-shaped opening with short apical slits. Gynoecium of two, rarely three carpels; ovary superior; ovules many per carpel, anatropous, pseudo-crassinucellate; micropyle zig-zag. Fruits 2–4-valved setose capsules. Seeds more or less straight, with a vestigial funicular aril; endosperm starchy; testa pulpy, eventually drying up and minutely papillate from the bixin cells. Small trees or shrubs, The leaves contains several sesquiterpenes, elagic acid and flavonoids, n = 7. . . . .12. BIXACEAE.

12 Leaves palmatisect, palmately lobed or subentire-reniform, glabrous or pubescent. Inflorescence terminal or very rarely axillary, thyrsoid. Anthers with one or two apical and sometimes also with two small basal pores. Gynoecium of 3–5 carpels; ovules many in each carpel; micropyle zig-zag. Fruits 3–5-valved capsules, the inner and outer layers separating and forming alternative valves. Seeds large, usually cochleate-reniform, without an aril; endosperm oily; testa densely pilose with white hairs along the curved dorsal side of the seed. Seeds of *Cochlospermum* possess cotton-like whitish or reddish hairs, whereas those of *Amoreuxia* glabrous, pilose or echinate. Trees, shrubs, or occasionally rhizomatous perennial herbs with woody subterranean trunks. Present ellagic acid, n = 6 (*Cochlospermum*) . . . . 13. COCHLOSPERMACEAE.

2 Nodes unilacunar.

13 Flowers with extrastaminal disc. Trees or large shrubs (*Tepuianthus auyantepuiensis*) with resinous cells, and usually densely pubescent with simple thrichomes. Vessels with simple perforations. Fibers with distinct bordered pits. Rays homogeneous, mostly uniseriate. Axial parenchyma scanty, vasicentric. Leaves alternate or opposite, small, simple, entire, pinnately veined, estipulate. Flowers small, in cymose inflorescences (terminal or in the uppermost axils), androdioecious or rarely bisexual, actinomorphic, receptacle thickened. Sepals five, free, imbricate, densely sericeous abaxially, glabrous adaxially. Petals five, free, imbricate, tend to be clawed, glabrous, yellowish. Nectary disc consisting of 5–10 contiguous, suborbicular, fleshy glands. Stamens 12–16 or 5, in 1–3 cycles, when unicyclic alternipetalous; anthers 2-locular, sagittate, versatile, introrse, dorsifixed, with more or less appendaged connective. Pollen grains 3–6-colporate, with prominent ornamentation. Gynoecium of three united carpels with free, forked stylodia; ovary superior, 3-locular, with one pendulous ovule per locule; placentation apical. Ovules anatropous with ventral raphe, the micropyle directed upward and outward. Fruits densely sericeous, 3-locular, bony, loculicidal capsules. Seeds with conspicuously ridged raphe, small embryo and copious endosperm; the cotyledons poorly differentiated (Boom and Stevenson 2004). . . . . 15. TEPUIANTHACEAE.

13 Flowers without extrastaminal disc.

14 Pollen grains 3-colporate. Plants with glandular, or simple, clustered, or stellate hairs and colorless juice. Nodes with one trace. Leaves opposite or less often alternate or verticillate, often more or less reduced, simple, entire, petiolate to sessile, stipulate or estipulate, often possess cystoliths. Phloem not stratified. Axial parenchyma absent or very scanty. Flowers solitary or in cymose inflorescences; sepals 3 or (4)5, when five the inner contorted, often persistent. Anthers basifixed. Gynoecium of 5 (6–10)

carpels in *Cistus,* in all other genera of three carpels. Ovules mostly orthotropous, rarely hemitropous or anatropous. Seeds very small; seed coat of two integuments, the outer thin, the inner very hard; embryo usually curved or bent into a hook or ring, or folded or circinately more or less coiled, rarely almost straight; endosperm starchy. Shrubs, subshrubs, or herbs. Contain glycosides of common flavonoids, including myricetin and other flavonols, and condensed and hydrolysable tannins, n = 5, 7, 9–12, 16, 18, 20, 24................. 14. CISTACEAE.

14 Pollen grains mostly 8–60-porate. Trees, shrubs, lianas, or rarely herbs, evergreen or deciduous, usually poisonous; wood often fluoresces. Stems usually with intraxylary phloem (except in the some ericoid or mosslike species of *Drapetes,* and the anomalous *Lethedon* and *Solmsia*) and in nine genera (including *Aquilaria*) with interxylary (included) phloem of "foraminate" type. Leaves alternate or opposite, not glandular-punctate, petiolate or sessile, simple, entire, estipulate. Stomata cyclocytic, rarely (*Octolepis*) anomocytic. Flowers bisexual or polygamous or dioecious, actinomorphic, but only some species of *Lachnaea* zygomorphic. Calyx 4–5-lobed, tube campanulate or cylindric; lobes imbricate to valvate. Corolla represented by free or united petaloid appendages, isomerous and alternating with the calyx lobes, less often double or treble in number, entire or lobed, rarely united into a ring, inserted at the throat of calyx tube or at the middle or near the base, or corolla absent. Stamens equal or twice as many as the calyx lobes or very rarely reduced to two or even one, inserted below the petals; in two or in one cycle, if in two cycles then at two different levels (the upper cycle antesepalous); filaments filiform or slightly flattened, entirely or partly adnate to the calyx tube; anthers basifixed or rarely dorsifixed, introrse or rarely extrorse. Pollen grains with more or less clearly expressed "crotonoid" ornamentation or rarely (*Octolepis*) 3–6-porate with reticulate ornamentation, somewhat resembling those of some Tiliaceae (Erdtman 1952; Archangelsky 1966, 1971). Nectary disc mostly present, annular, cupular, membranous, or of 4–5 scales. Gynoecium of (2)3–5, rarely 6–8, or 10–12 carpels (*Lethedon*), 3–5 carpels (*Solmsia, Deltaria,* and *Octolepis*), or more often of two carpels; style elongate or short, sometimes very short, not contorted, terminal or sublateral. Stigma usually capitate, sometimes papillose. Ovary sessile or shortly stipitate, 1–2-locular (rarely 3–12-locular), with one pendulous ovule per locule. Fruits loculicidal capsules or more often indehiscent, nut-like, baccate, or drupaceous, usually enclosed in the base of the persistent calyx tube. Seeds usually with a caruncle-like or tail-like appendage, without an aril; exotegmen and endotegmen with brown contents, reticulately thickened and lignified; embryo straight, white; endosperm scanty or absent, rarely copious; cotyledons large. Thymelaeoidea contain highly toxic diterpene esters, which often occur in the whole plants, n = 7(-9)-10 (Thymelaeoideae). .... 16. THYMELAEACEAE.

1 Anthers disporangiate and 1-locular.

15 Pollen grains generally smooth or merely rugose, only rarely minutely spinose. Fruits loculicidal capsules, rarely fleshy or indehiscent. Trees, sometimes spiny, buttressed or with swollen trunk, often with verticillate branches, or rarely shrubs or epiphytic stranglers. Indumentum of stellate, exceptionally lepidote trichomes, or absent. Leaves simple, rarely lobed, or digitate, stipulate. Flowers actinomorphic or rarely more or less zygomorphic, solitary of in axillary clusters, rarely in cymes or paniculate,

or terminal or leaf-opposed. Epicalyx of 3(2) distinct bracts usually present; sepals commonly five, usually fused to more than half their length, forming copular to tubular calyx, sometimes persistent or even accrescent, exceptionally imbricate in bud; petals usually contorted, generally fused at base with staminal column, often many times longer than sepals and exceeding calyx before anthesis; stamens usually numerous (up to more than 1,000) or sometimes only five, generally forming a staminal tube with antepetalous (or more or less laterally shifted) lobes or phalanges and/or distally free filament portions; anthers usually monothecal, sometimes septate, straight or reniform or spirally contorted; staminodes absent or staminal tube exceptionally with sterile appendages or collar-like thickening; ovary (8-)5–2-locular, superior to rarely semiinferior; ovules numerous to two per locule; style usually simple or with short stylar branches, stigma sometimes thickened, capitate and/or lobed. Seeds usually glabrous, rarely winged or arillate; endosperm reduced to absent; cotyledons folded. n = 36–46.. . . . . 10. BOMBACACEAE.

15 Pollen grains generally spiny, only rarely smooth. Fruits schizocarpic or loculicidal or septicidal capsules, rarely baccate or samaroid. Herbs or shrubs, only rarely trees; vestitute principally of stellate, lepidote, simple, and unbranched glandular hairs, sometimes spiny. Leaves simple or lobed, exceptionally digitate, mostly stipulate. Flowers solitary or in cymes, collectively forming spicate to paniculate or even head-like inflorescences, actinomorphic or more or less zygomorphic, bisexual or sometimes unisexual; epicalyx present or absent, of (2)3 or more bracts, occasionally persistent; calyx lobes valvate in bud, usually nectariferous within at base; corolla tubular to rotate or reflexed; petals five, convolute in bud; stamens monadelphous; filaments few to many, terminating the staminal column; anthers reniform, 1-celled, dorsifixed. Pollen grains often spiny. Carpels (1-)3 to many, ovary superior; style exceeding the staminal column, single with lobed, decurrent or divergent stigmas as many as the carpels; ovules 1–9 or many in each carpels, ascending to pendulous. Seeds reniform, trigonous or turbinate, rarely arillate; embryo curved, endosperm present or absent, cotyledons folded, n = 5–13, 15–17, 19–23, 29, 33, 36–39. . . . . 11. MALVACEAE.

1. MUNTINGIACEAE

Bayer, M.W. Chase and F. Fay 1998. 3/3. Central and tropical South America, West Indies.

1.1 MUNTINGIOIDEAE
Leaves serrate. Ovary superior, surrounded by a disc-like elevation of the receptacle, 7–5-locular. – *Muntingia.*

1.2 NEOTESSMANNIOIDEAE
Leaves cordate, stipulate. Calyx deeply lobed. Anthers loculi not confluent at the apex, opening by longitudinal slits or apical pores. Pollen grains in tetrads. Ovary inferior, 5-locular (*Dicraspidia*) or multilocular in the lower part and 1-locular in the upper part (*Neotessmannia*). – *Neotessmannia, Dicraspidia.*

The basal family in the Malvales.

2. TILIACEAE

A.L. de Jussieu 1789 (including Berryaceae Doweld 2001, Grewiaceae Doweld et Reveal 2005, Sparmanniaceae J.G. Agardh 1858). 39/450+. Widely distributed in tropical regions, especially in tropical America, Africa, and Southeast Asia. The genus *Tilia* (80) extends into the temperate regions of the Northern Hemisphere. The largest genus *Grewia* (150) occurs in tropical Africa, Asia, and Australasia. *Corchorus* (about 40 species in Africa and Asia) is one of the few herbaceous genera in the family.

2.1 TILIOIDEAE
Sepals free. Anthers with the separate locules. Ovary superior, with 2-many anatropous ovules in each locule. – APEIBEAE: *Glyphaea, Apeiba, Ancistrocarpus;* ENTELEAE: *Entelea, Mortoniodendron, Burretiodendron, Sicrea;* CORCHOREAE: *Corchorus* (? including *Oceanopapaver* – Tirel et al. 1996; Whitlock et al.

2000), *Pseudocorchorus;* SPARMANNIEAE: *Sparmannia, Clappertonia;* LUEHEEAE: *Luehea, Trichospermum, Mollia;* TILIEAE: *Tilia, Schoutenia;* DUBOSCIEAE: *Duboscia;* DESPLATSIEAE: *Desplatsia, Vasivaea, Hydrogaster;* GREWIEAE: *Grewia, Eleutherostylis, Tetralix, Lueheopsis;* COLONEAE: *Colona, Goethalsia;* TRIUMFETTEAE: *Erinocarpus, Triumfetta, Heliocarpus;* CRAIGIEAE: *Craigia.*

2.2 BROWNLOWIOIDEAE
Sepals united into a 3–5th or lobed campanulate calyx. Anthers with the loculi often confluent at the apex. Ovary superior, 1–5-locular; in Brownlowieae carpels soon free, indumentum often lepidote, n = 10. – BROWNLOWIEAE: *Christiana* (including *Asterophorum* and *Tahitia*), *Brownlowia, Pentaplaris;* DIPLODISCEAE: *Diplodiscus, Pityranthe* (including *Hainania*), *Jarandersonia;* BERRYEAE: *Berrya, Carpodiptera, Pentace.*

Tiliaceae are related to the Elaeocarpaceae but differ markedly in the character of their indumentum, in the presence of usually well-developed mucilage cavities, mucilage cells, or canals, fibers with simple pits, alternate lateral pitting of the vessels, mostly apotracheal axial parenchyma, usually somewhat more specialized rays, stratified phloem in young stems, and, as Corner (1976) notes, also in the anatomy of the seed coat.

3. DIPTEROCARPACEAE

Blume 1825. 13/500. Tropical regions of the Old World, especially the rain forests of Malesia.
SHOREAE: *Hopea, Shorea, Neobalanocarpus, Upuna;* PARASHOREAE: *Parashorea*; DRYOBALANEAE: *Dryobalanops;* DIPTEROCARPEAE: *Cotylelobium, Vateria, Vateriopsis, Vatica, Stemonoporus, Dipterocarpus, Anisoptera.*

Dipterocarpaceae are close to the Tiliaceae and especially to the Monotaceae. According to Hallier (1901, 1912), Dipterocarpaceae (but without *Monotes*) belong to the Malvales and are related to Elaeocarpaceae and Tiliaceae.

4. MONOTACEAE

Maury ex Takhtajan 1987. 4/40. Tropical Africa (*Monotes* and *Marquesia*) and tropical South America (*Pakaraimaea* and *Pseudomonotes*).

4.1 MONOTOIDEAE
Flowers with an androgynophore. Sepals imbricate. Petals longer than the sepals, variously pubescent. Anthers little or deeply basiversatile. Ovary (3)4-locular, each locule with two ovules. Trichomes fasciculate, often glandular. Wood rays uniseriate. – *Monotes, Marquesia.*

4.2 PAKARAIMAEOIDEAE
Flowers without androgynophore. Sepals imbricate. Petals shorter than sepals, glabrous. Anthers deeply basiversatile. Ovary 4(5)-locular, each locule with four ovules. Trichomes fasciculate. Wood rays biseriate. – *Pakaraimaea.*

4.3 PSEUDOMONOTOIDEAE
Flowers without androgynophore. Sepals valvate. Petals longer than sepals, glabrous. Anthers imbedded in thickened connective, with prominent appendage. Ovary 3-locular, each locule with one subbasal ovule. Trichomes glandular. – *Pseudomonotes.*

Monotaceae are closely related to the Tiliaceae from which they differ mainly in imbricate sepals and vestured intervessel pores (in Tiliaceae known only for *Schoutenia*). Heim (1892) transferred the genus *Monotes* to the Tiliaceae, and according to Fries (1914), *Marquesia macroura* is related to *Schoutenia* in Tiliaceae. According to Bancroft (1933, 1935) the subfamily Monotoideae serves as a connecting link with tropical Asiatic members of the Tiliaceae.

Finally, Kostermans (1978) argued for the close relationships between the genus *Pakaraimaea* and *Monotes* and *Marquesia* and concluded that they should be transferred to the Tiliaceae. They are undoubtedly closer to the Tiliaceae than to the Dipterocarpaceae. Based on the wood anatomical data, Gottwald and Parameswaran (1966) suggested raising the subfamily Monotoideae into a separate family Monotaceae, and later this was confirmed by palynological studies (Maury 1981).

5. SARCOLAENACEAE

Caruel 1881 (including Rhodolaenaceae Bullock 1958, Schizolaenaceae Barnhart 1895). 10/35. Madagascar.

Sarcolaena, Leptolaena, Xyloolaena, Perrierodendron, Eremolaena, Schizolaena, Rhodolaena, Pentachlaena, Mediusella, Xerochlamys.

Sarcolaenaceae are related to the Tiliaceae and especially to Dipterocarpaceae and have many common features with them in the structure of stem, petiole, and pollen grains (Metcalfe and Chalk 1950; Dehay 1957; Carlquist 1964; Takhtajan 1997). Hutchinson (1969) draws attention to a remarkable feature of most of the genera – the metamorphosis of the bracteoles, which are united to form an involucre or epicalyx usually enclosing the fruit – as well as to another characteristic of the family, the pollinia, consisting of 4 or 16 pollen grains.

6. NEURADACEAE

Link 1831 (including Grielaceae Martynov 1820). 3/10. Arid regions of North Africa, Arabian Peninsula, Syria, Iraq, northern Iran, Afghanistan, and from Pakistan to India (*Neurada*), southern and southeastern Africa (*Grielum* and *Neuradopsis*).

Neurada, Grielum, Neuradopsis

Neuradaceae related to the Malvaceae (Hallier 1907, 1912; Airy Shaw 1973; Melikian and Bondar 1996; Bayer 2003).

7. STERCULIACEAE

E.P. Ventenat ex Salisbury 1807 (including Byttneriaceae R. Brown 1814, Chiranthodendraceae A. Gray 1887, Dombeyaceae Kunth 1829, Helicteraceae J.G. Agardh 1858, Hermanniaceae Marquis 1820, Lasiopetalaceae Reichenbach 1823, Melochiaceae J.G. Agardh 1858, Pentapetaceae Brechtold et J. Presl 1820, Theobromataceae J.G. Agardh 1858, Triplobaceae Rafinesque 1838, Triplochitonaceae K. Schumann 1900). 62/1500. Pantropical, extending into subtropical and warm-temperate regions.

7.1 BYTTNERIOIDEAE
Flowers bisexual or less often polygamous. Petals mostly present, sometimes very small, scalelike or even absent (in some Lasiopetaleae and in Fremontodendreae). Carpels mostly united until dehiscence, rarely free at maturity. – LASIOPETALEAE: *Hannafordia, Leptonychiopsis, Thomasia, Seringia, Keraudrenia, Maxwellia, Lysiosepalum, Lasiopetalum, Guichenotia;* HERMANNIEAE: *Hermannia, Melochia, Dicarpidium, Waltheria;* HELMIOPSIDEAE: *Helmiopsiella, Helmiopsis, Nesogordonia;* BYTTNERIEAE: *Rulingia, Commersonia, Byttneria, Ayenia;* THEOBROMEAE: *Glossostemon, Scaphopetalum, Leptonychia, Abroma, Theobroma, Herrania, Guazuma;* HELICTERETEAE: *Pterospermum, Helicteres, Neoregnellia, Kleinhovia, Reevesia, Ungeria;* FREMONTODENDREAE: *Fremontodendron, Chiranthodendron;* ERIOLAENEAE: *Eriolaena;* TRIPLOCHITONEAE: *Triplochiton.*

7.2 DOMBEYOIDEAE
Flowers usually in axillary cymes or solitary; petals often persistent, flat, with apical appendage; pollen usually sphaeroidal and spinose. – *Ruizia, Astiria, Cheirolaena, Trochetia, Helmiopsiella, Dombeya, Paradombeya Harmsia, Melhania, Paramelhania, Pentapetes.*

7.3 STERCULIOIDEAE
Flowers mostly unisexual or polygamous, rarely (in some Tarrietieae) bisexual. Petals absent (present in *Mansonia*). Androgynophore present. Carpels free or at least free at maturity, mostly dehiscent and not winged (Sterculieae) or indehiscent and obliquely winged or keeled. – STERCULIEAE: *Sterculia, Brachychiton, Firmiana, Scaphium, Pterocymbium, Pterygota, Acropogon, Octolobus, Cola;* TARRIETIEAE: *Heritiera, Argyrodendron, Tarrietia;* MANSONIEAE: *Mansonia, Hildegardia*

Sterculiaceae stand close to both the Tiliaceae and the Bombacaceae and Malvaceae. The boundaries between these families are however not very clear, and the taxonomic position of some genera, such as the paleotropical genus *Leptonychia* ("The strangely arillate seed has a fibrous exotegmen which is neither Sterculiaceae nor Tiliaceous," states Corner 1976: 261) and of the very isolated New Caledonian genus *Maxwellia* is still not certain.

Sterculiaceae are very near to the Tiliaceae, which is confirmed by the anatomy of vascular system of the petiole (Dehay 1941, 1942). But the family Sterculiaceae is very diverse and is characterized by a great diversity of morphological, anatomical, and palynological features. In particular, there are so many differences between the Byttnerioideae and Sterculioideae that Edlin (1935) proposed to accept them as two separate families "on account of the structure of their wood, vegetative organs, flowers, and fruit" (p. 6). However, the boundary between these two groups is not so very clear-cut. For example, on the basis of the vascular system of the petiole, the tribe Byttnerieae is closer to the Sterculieae than to the other tribes of the Byttnerioideae (Gazet du Chatelier 1940a, b). Besides, as Rao (1952) has shown, the structure of the androecium is similar for the whole family, and therefore there is no reason to separate the Sterculioideae in a family of their own.

8. DIEGODENDRACEAE

Capuron 1964. 1/1. Madagascar.

Diegodendron.

According to Capuron (1962, 1970), Hutchinson (1969, 1973), Straka and Albers (1978), and Cronquist (1980, 1988), the nearest ally of Diegodendraceae is the family Ochnaceae. This conclusion is based mainly on the gynobasic style, which is evidently the result of parallel evolution. The specialized wood anatomy of *Diegodendron* supports a close affinity with the malvalean families, particularly with the Sphaerosepalaceae (Dickison 1988). However, according to Nandi (1998), Diegodendraceae are closely related to the Bixaceae, and Thorne (2006) places them between the Bixaceae and Cochlospermaceae.

9. SPHAEROSEPALACEAE

van Tieghem ex Bullock 1959 (Rhopalocarpaceae Hemsley 1903). 2/17. Madagascar.

Dialyceras, Rhopalocarpus.

Very closely related to the Diegodendraceae. Both of these families exhibit definite affinities to the Sarcolaenaceae.

10. BOMBACACEAE

Kunth 1822. 29/c.300. Pantropical, especially in rain forests of South America, above all in Brazil.

DURIONEAE: *Neesia, Durio, Camptostemon, Coelostegia, Boschia, Kostermansia, Cullenia, Scleronema, Cavanillesia;* MATISIEAE: *Quararibea, Ochroma, Matisia, Patinoa, Bernoullia, Septotheca, Phragmotheca, Huberodendron;* CATOSTEMATEAE: *Aguiaria, Catostemma;* BOMBACEAE: *Bombax, Eriotheca, Adansonia, Pseudobombax, Pachira* (including *Rhodognaphalon, Rhodognaphalopsis,* and *Bombacopsis*) *Gyranthera;* CEIBEAE: *Ceiba, Spirotheca, Chorisia, Neobuchia.*

Very close to the Malvaceae.

11. MALVACEAE

A.L. de Jussieu 1789 (including Hibiscaceae J.G. Agardh 1858, Philippodendraceae Endlicher 1841, Plagianthaceae J.G. Agardh 1858). 111/1500–1600. Essentially cosmopolitan, but best developed in the tropics.

KYDIEAE: *Julostylis, Dicellostyles, Nayariophyton, Kydia;* HIBISCEAE: *Decaschistia, Radyera, Hibiscus, Papuodendron, Talipariti, Kosteletzkya, Fioria, Wercklea, Abelmoschus, Hibiscadelphus, Senra, Symphyochlamys, Megistostegium, Perrierophytum, Humbertiella, Macrostelia, Helicteropsis, Humbertianthus, Cenocentrum, Lagunaria, Urena, Malachra, Peltaea, Phragmocarpidium, Rojasimalva* (including *Lopimia*), *Pavonia, Malvaviscus, Anotea*; GOSSYPIEAE: *Cienfuegosia, Cephalohibiscus, Lebronnecia, Hampea, Thespesia, Gossypioides, Kokia, Gossypium, Alyogyne*; MALVEAE: *Anoda, Periptera, Horsfordia, Bakeridesia, Bastardiastrum, Wissadula, Tetrasida, Pseudabutilon, Bastardia, Bastardiopsis, Abutilon, Herissantia, Neobaclea, Meximalva, Corynabutilon, Briquetia, Hochreutinera, Dirhamphis, Gaya, Billieturnera, Sidastrum, Sida, Krapovickasia, Rhynchosida, Robinsonella, Dendrosida, Fryxellia, Allowissadula, Allosidastrum, Neobrittonia, Batesimalva, Malvella, Lecanophora, Cristaria, Asterotrichion, Plagianthus, Gynatrix, Hoheria, Lawrencia, Sidasodes, Sidalcea, Eremalche, Iliamna, Malvastrum, Modiolastrum, Modiola, Calyculogygas, Callirhoe, Calyptraemalva, Napaea, Monteiroa, Lavatera, Malva, Althaea, Navaea, Malope, Malacothamnus, Phymosia, Alcea, Kitaibelia, Anisodontea, Kearnemalvastrum, Acaulimalva, Urocarpidium, Fuertesimalva, Palaua, Nototriche, Tarasa, Sphaeralcea, Andeimalva.*

Malvaceae are so closely related to the Bombacaceae, that from time to time the proposal arises to merge these two families. Edlin (1935) transferred the tribe Hibisceae together with the genus *Kydia* to the Bombacaceae, but as Hutchinson (1967: 538) noted, "Malvaceae without the great genus *Hibiscus* would be like a horse without a tail." The most archaic tribe is Malopeae, in which the carpels are in two or more superposed and spirally arranged cycles. All other tribes have carpels in a single cycle. The tribe Ureneae, probably the most advanced, has twice as many style branches as carpels, which may be due to a splitting of the styles, just as the anthers are split and unilocular in the whole family (Hutchinson 1967: 538).

12. BIXACEAE

Kunth 1822. 1/5. Tropical America and West Indies.
Bixa.

13. COCHLOSPERMACEAE

Planchon 1847. 2/15–20. Tropical and subtropical regions of America, western and central tropical Africa, southern and southeastern Asia and northern Australia.
Cochlospermum, Amoreuxia.
Very close to the Bixaceae.

14. CISTACEAE

A.L. de Jussieu 1789 (including Helianthemaceae G. Meyer 1836). 8/180–200. Mostly temperate and subtropical regions of the Northern Hemisphere, especially the Mediterranean and eastern North Africa, with some in West Indies and South America.

CISTEAE: *Cistus, Halimium, Crocanthemum, Tuberaria, Helianthemum* (including *Atlanthemum?*), *Fumana;* HUDSONIEAE: *Hudsonia;* LECHIDIEAE: *Lechea.*

15. TEPUIANTHACEAE

Maguire et Steyermark 1981. 1/7. Venezuela, Brazil, and Colombia; lowland savannas and Roraima sandstone.
Tepuianthus.
Very close to the Thymelaeaceae.

16. THYMELAEACEAE

A.L. de Jussieu 1789 (including Aquilariaceae R. Brown ex A.P. de Candolle 1825, Daphnaceae Ventenat 1799, Gonystylaceae van Tieghem 1896, Phaleriaceae Meissner 1841). 48/800. Subcosmopolitan but concentrated mainly in tropical Africa and Australia, with local centers in the Mediterranean region and western, eastern, and Southeast Asia.

Classification after Domke (1934), revised by B.E. Herber (2003).

16.1 OCTOLEPIDOIDEAE (GONYSTYLOIDEAE)

Trees, rarely shrubs. Inflorescences usually determinate; bracts minute or absent. Floral tube not articulated, persistent; sepals (3)4–5(6), valvate or imbricate; petals 4–40, free or sometimes slightly fused at base, rarely forming a fleshy annulus, inserted at receptacle, or absent. Stamens inserted at receptacle; filaments slender, sometimes broader at base; anthers straight, reflexed, peltate or horseshoe-shaped. Pollen grains 8–60-porate, or rarely (*Octolepis*) 3–6-porate, and reticulate. tectate-columellate. Nectary disc wanting. Gynoecium of (2)3–5, rarely 6–8 carpels, with elongate, filiform, wiry, contorted style, occasionally accompanied by a few small processes (parastyles) around the base; stigma small, capitate to punctiform. Fruits 1–9-seeded, usually loculicidally dehiscent or (*Aetoxylon*) 1-seeded, indehiscent. Seeds large, without chalazal fold, usually with thin dorsal aril arising from the fleshy funicle or (*Amyxa* and *Aetoxylon*) not arillate; endosperm wanting. – OCTOLEPIDEAE: *Octolepis*; MICROSEMMATEAE: *Lethedon* (*Microsemma*); SOLMSIEAE: *Deltaria, Solmsia, Arnhemia*; GONYSTYLIDEAE: *Gonystylus, Amyxa, Aetoxylon.*

16.2 THYMELAEOIDEAE

Trees or shrubs (sometimes lianas), rarely perennial herbs, very rarely annuals; interxylar phloem present in some genera. Inflorescences usually indeterminate, occasionally with involuctum. Sepals 4–5(6), connate into a campanulate or cylindric tube, often broader around the ovary, imbricate (valvate in *Lagetta*); petals well-developed, entire or divided, ringlike, scalelike, reduced to small, fleshy glands, obscure or lacking. Stamens equal or double the number of calyx lobes (at most 10) or very rarely (*Pimelea*) reduced to two or even one. Pollen grains 8–60-porate and with more or less clearly expressed crotonoid ornamentation. Nectary disc annular, cupular, composed of scales, or lacking. Gynoecium of two carpels. Ovary 2-locular (*Peddiea, Synandrodaphne, Oreodendron,* and *Phaleria*) or more often 1-locular. Fruits indehiscent, nutlike, baccate, or drupaceous, or more rarely capsule. Seeds sometimes appendiculate, with or without endosperm. – SYNANDRODAPHNEAE: *Synandrodaphne* (*Gilgiodaphne*); AQUILARIEAE: *Aquilaria, Gyrinops*; PHALERIEAE: *Peddiea, Phaleria*; DICRANOLEPIDEAE: *Lachnaea, Lagetta, Goodallia, Funifera, Lophostoma, Gnidia, Enkleia, Linostoma, Stephanodaphne*; DAPHNEAE: *Jedda, Dicranolepis, Synaptolepis, Craterosiphon, Linodendron, Daphnopsis, Schoenobiblus, Ovidia, Dirca, Daphne, Wikstroemia, Rhamnoneuron, Edgeworthia, Thymelaea, Diarthron,*

Dendristellera, Restella, Stelleropsis, Stellera, Dais, Struthiola, Passerina, Drapetes, Kelleria, Pimelea, Lasiadenia.

Evidently related to the Malvalean families (Hallier 1922; Domke 1934). They have many common features with the Tiliaceae. According to Metcalfe and Chalk (1950: 1181), general anatomical evidence "appears to indicate that the Gonystylaceae, Thymelaeaceae, and Tiliaceae have definite affinities with each other. The occurrence of secretory elements containing mucilage is common to *Gonystylus*, the Tiliaceae, and Thymelaeaceae". The occurrence of cyclopropenyl fatty acids (Vickery 1980), and exotegmic seeds (Corner 1976) also support the affinities.

Bibliography

Alexandrov VG and AV Dobrotvorskaya. 1957. On the morphological nature of the stamens, petals, and so-called stamen tube in the flower of Malvaceae. Trudy Bot. Inst. Akad. Nauk SSSR, ser. 7, 4: 83–137 (in Russian).

Alverson WS, KG Karol, DA Baum, MW Chase, SM Swensen, R McCourt, and KJ Sytsma. 1998. Circumscription of the Malvales and relationships to other Rosidae: evidence from *rbc*L sequence data. Am. J. Bot. 85: 876–887.

Alverson WS, BA Whitlock, R Nyffeler, C Bayer, and DA Baum. 1999. Phylogeny of the core Malvales. Evidence from *ndhF* sequence data. Am. J. Bot. 86: 1474–1486.

Anderson GJ. 1976. The pollination biology of *Tilia*. Am. J. Bot. 63: 1203–1212.

Arbo MM. 1972. Estructura y ontogenia de los nectarios foliares del género *Byttneria* (Sterculiaceae). Darwiniana 17: 104–158.

Archangelsky DB. 1966. Pollen grains of the families Thymelaeaceae and Gonystylaceae. Bot. Zhurn. 51: 484–494 (in Russian).

Archangelsky DB. 1971. Palynotaxonomy of Thymelaeaceae s.l. In: LA Kuprianova and MS Yakovlev, eds. Morphology of pollen grains of Cucurbitaceae, Thymelaeaceae, Cornaceae, pp. 104–234. Nauka, Leningrad (in Russian).

Arrington JM and K Kubitzki. 2003. Cistaceae. In: K Kubitzki, ed. The families and genera of vascular plants, vol. 5, pp. 62–70. Springer, Berlin/Heidelberg/New York.

Ashton PS. 1977. Phylogenetic speculations on Dipterocarpaceae. Mém. Mus. Natl. Hist. Nat. B., Bot. 26: 145–149.

Ashton PS. 1979. Phylogenetic speculations on Dipterocarpaceae. Mem. Mus. Natl. Hist. Nat., Ser. B., Bot. 26: 145–149.

Ashton PS. 1982. Dipterocarpaceae. In: CGGJ van Steenis, ed., Flora malesiana, vol. 9, pp. 237–552. Martinus Nijhoff, The Hague.

Ashton PS. 2003. Dipterocarpaceae. In: K Kubitzki, ed. The families and genera of vascular plants, vol. 5, pp. 182–197. Springer, Berlin/Heidelberg/New York.

Baas P and E Werker. 1981. A new record of vestured pits in Cistaceae. IAWA Bull. N.s., 2: 41–42.

Baer DF 1977. Systematics of the genus *Bixa* and geography of the cultivated annatto tree. Diss. Abstr. Int., B 37(10): 4845.

Baker HG and I Baker. 1968. Chromosome numbers in the Bombacaceae, Bot. Gaz. 129: 294–296.

Bakhuizen van der Brink RC. 1924. Revisio Bombacearum. Bull. Jard. Bot. Buitenzorg. III 6: 161–240.

Balasubramanian A, T Sekar, and C Devadoss. 1995. The vascular system of stem-node-leaf in *Bixa orellana* L. Phytomorphology 45: 219–227.

Bancroft H. 1935. The wood anatomy of representative members of the Monotoideae. Am. J. Bot. 22: 717–739.

Bank van der M, MF Fay, and MW Chase. 2002. Molecular phylogenetics of Thymelaeaceae with particular reference to African and Australian genera. Taxon. 51: 329–339.

Bate-Smith EC and TC Whitmore. 1959. Chemistry and taxonomy of the Dipterocarpaceae. Nature 184: 795–796.

Bates DM. 1968. Generic relationships in the Malvaceae tribe Malveae. Gentes Herb. 10: 117–135.

Bates DM. 1976. Chromosome numbers in the Malvales: III. Miscellaneous counts from the Byttneriaceae and Malvaceae. Gentes Herb. 11: 143–150.

Bates DM and OJ Blanchard, Jr. 1970. Chromosome numbers in Malvales. II. New or otherwise noteworthy counts relevant to classification in the Malvaceae, tribe Malveae. Am. J. Bot. 57: 927–934.

Baum DA and K Oginuma. 1994. A review of chromosome numbers in Bombacaceae with new counts for *Adansonia*. Taxon 43: 11–20.

Baum DA, WS Alverson, and R Nyffeler. 1998. A durian by any other name: taxonomy and nomenclature of the core Malvales. Harvard Pap. Bot. 3: 315–330.

Baum DA, S de Witt Smith, A Yen, WS Alverson, R Nyffeler, BA Whitlock, and RL Oldham. 2004. Phylogenetic relationships of *Malvatheca* (Bombacoideae and Malvoideae; Malvaceae sensu lato) as inferred from plastid DNA sequences. Am. J. Bot. 91: 1863–1871.

Bawa KS and CJ Webb. 1983. Floral variation and sexual differentiation in *Muntingia calabura* (Elaeocarpaceae), a species with hermaphrodite flowers. Evolution 37: 1271–1282.

Bayer C. 1994. Zur Infloreszenzmorphologie der Malvales. Diss. Bot. 212: 1–280.

Bayer C. 1995. Zur Verzweigung der vegetativen und blühenden Achsen einiger Bombacaceen. Feddes Repert. 106: 407–413.

Bayer C. 1998. Synflorescences of Malvaceae. Nord. J. Bot. 18: 335–338.

Bayer C. 1999. The bicolor unit-homology and transformation of an inflorescence structure unique to core Malvales. Plant Syst. Evol. 124: 187–198.

Bayer C. 2003a. Diegodendraceae. In: K Kubitzki, ed. The families and genera of vascular plants, vol. 5, pp. 175–177. Springer, Berlin/Heidelberg/New York.

Bayer C. 2003b. Muntingiaceae. In: K Kubitzki, ed. The families and genera of vascular plants, vol. 5, pp. 315–319. Springer, Berlin/Heidelberg/New York.

Bayer C. 2003c. Neuradaceae. In: K Kubitzki, ed. The families and genera of vascular plants, vol. 5, pp. 325–328. Springer, Berlin/Heidelberg/New York.

Bayer C. 2003d. Sarcolaenaceae. In: K Kubitzki, ed. The families and genera of vascular plants, vol. 5, pp. 345–352. Springer, Berlin/Heidelberg/New York.

Bayer C. 2003e. Sphaerosepalaceae. In: K Kubitzki, ed. The families and genera of vascular plants, vol. 5, pp. 359–362. Springer, Berlin/Heidelberg/New York.

Bayer C and LJ Dorr. 1999. A synopsis of the neotropical genus *Pentaplaris*, with remarks on its systematic position within core Malvales. Brittonia 51: 134–148.

Bayer C and K Kubitzki. 1996. Inflorescence morphology of some Australian Lasiopetaleae (Sterculiaceae). Telopea 6: 721–728.

Bayer C and K Kubitzki. 2003. Malvaceae. In: K Kubitzki, ed. The families and genera of vascular plants, vol. 5, pp. 225–311. Springer, Berlin/Heidelberg/New York.

Bayer C, MW Chase, and MF Fay. 1998. Muntingiaceae, a new family of dicotyledons with malvalean affinities. Taxon 47: 37–42.

Bayer C, MF Fay, AY De Bruijn, V Savolainen, CM Morton, K Kubitzki, WS Alverson, and MW Chase. 1999. Support for an expanded family concept of Malvaceae within a recircumscribed order Malvales: a combined analysis of plastid *atp*B and *rbc*L DNA sequences. Bot. J. Linn. Soc. 129: 267–303.

Benn SJ and DE Lembke. 1991. Taxonomy of Neotessmannieae (Tiliaceae). Am. J. Bot. 78: 166–167 (Suppl.).

Beyers JBP and EM Marais. 1998. Palynological studies of the Thymelaeaceae of the Cape flora. Grana 37: 193–202.

Blunden G, AV Patel, NJ Armstrong, and J Gorham. 2001. Betaine distribution in the Malvaceae. Phytochemistry 58: 451–454.

Boom B and DW Stevenson. 2004. Tepuianthaceae. In: N Smith, SA Mori, A Henderson, DW Stevenson, and SV Heald, eds. Flowering plants of the Neotropics, pp. 366–367. The New York Botanical Garden. Princeton University Press, Princeton.

Boureau E. 1958. Contribution a l϶ετυδε anatomiquedes especes actuelles de Rhopalocarpaceae. Bull. Mus. Hist. Nat., Paris, 30: 213–221.

Bredenkamp CL and JBP Beyers. 2000. Thymelaeaceae. In: OA Leistner, ed. Seed plants of southern Africa: families and genera. Strelitzia 10. Pretoria.

Brizicky GK. 1965. The genera of Tiliaceae and Elaeocarpaceae in the southeastern United States. J. Arnold Arbor. 46: 286–307.

Brizicky GK. 1966. The genera of Sterculiaceae in the southeastern United States. J. Arnold Arbor. 47: 60–71.

Bunniger L. 1972. Untersuchungen über die morpholog-ische Natur des Hypanthiums bei Myrtales-und Thymelaeales-Familien: II. Myrtaceae. III. Vergleich mit den Thymelaeaceae. Beitr. Biol. Pfl. 48: 79–156.

Burret M. 1926. Beiträge zur Kenntnis der Tiliaceen. Notizbl. Bot. Gart. Berlin-Dahlem 9: 592–880.

Capuron R. 1962. Revision des Rhopalocarpacees. Adansonia, ser. 2, 2: 228–267.

Capuron R. 1963. Contributions l'étude de la flora de Madagascar. XV. *Diegodendron* Capuron gen. nov. type de la nouvelle famille des Diegodendraceae (Ochnales sensu Hutchinson). Adansonia, ser. 2, 3: 385–392.

Capuron R. 1965. Déscription des fruits du *Diegodendron humberti* R. Capuron (Diegodendracées). Adansonia, ser. 2, 5: 503–505.

Capuron R. 1970. Observations sur les Sarcolaenacees. Adansonia, ser. 2, 10: 247–265.

Carlquist S. 1964. Pollen morphology and evolution of Sarcolaenaceae (Chlaenaceae). Brittonia 16: 231–254.

Carlquist S. 2005. Wood and bark anatomy of Muntingiaceae: a phylogenetic comparison within Malvales s.l. Brittonia 57: 59–67.

Chattaway M. 1932. The wood of Sterculiaceae: I. Specialization of the vertical wood parenchyma within the subfamily Sterculieae. New Phytol. 31: 119–132.

Chattaway MM. 1933a. Ray development in the Sterculiaceae. Forestry 7: 93–108.

Chattaway MM. 1933b. Tile-cells in the rays of the Malvales. New Phytol. 32: 261–273.

Chattaway MM. 1937. The wood anatomy of the Sterculiaceae. Phil. Trans. Roy. Soc. 228: 313–365.

Chattaway MM. 1956. Crystals in woody tissues, part II. Trop. Woods 104: 100–124.

Chiarugi A. 1925. Embriologia delle Cistaceae. Nuovo. G. Bot. Ital. 32: 223–314.

Chopra RN and H Kaur. 1965. Embryology of *Bixa orellana* Linn. Phytomorphology 15: 211–214.

Christensen PB. 1986. Pollen morphological studies in the Malvaceae. Grana 25: 95–117.

Cristóbal CL. 1967. Cromosomas de Malvales. Kurtziana 4: 139–142.

Cristóbal CL. 1968. Estudio morfológico de los granos de pollen de *Byttneria* (Sterculiaceae). Pollen et Spores 10: 57–72.

Cristóbal CL. 1976. Estudio taxonómico del género *Byttneria* Loefling (Sterculiaceae). Bonplandia 4: 1–428.

Cuatrecasas J. 1964. Cacao and its allies: a taxonomic revision of the genus *Theobroma*. Contr. U.S. Natl. Herb. 35: 379–614.

Curtis WF. 1976. Chromosome counts in *Grielum* and *Cercis*. Ann. Missouri Bot. Gard. 63: 379–380.

Dali TM. 1998. Structure and development of fruits of Tiliaceae. Vallabh Vidyanagar, Sardar Patel University, India.

Dathan ASR and D Singh. 1971. Development of embryo sac and seed of *Bixa* L. and *Cochlospermum* Kunth. J. Indian Bot. Soc. 61: 254–266.

Dayanandan S, PS Ashton, SM Williams, and RB Primack. 1999. Phylogeny of the tropical tree family Dipterocarpaceae based on nucleotide sequences of the chloroplast *rbc*L gene. Am. J. Bot. 86: 1182–1190.

Dehay C. 1938. Les affinités entre les Euphorbiales, les Morales et les Malvales, d'aprés l'appareil libéro-ligneux foliare. Bull. Soc. Bot. France 85: 23–31.

Dehay C. 1941. L'appareil libero-ligneux foliaire des Sterculiacees. Ann. Sci. Nat. Bot., ser. 2, 45: 127–128.

Dehay C. 1942. Remarques sur l϶αππαρειλ libero-ligneux foliaire des Sterculiacees. Bull. Soc. Bot. France 89: 76–78.

Dehay C. 1944. L'appareil libero-ligneux foliaire des Tiliacees. Bull. Soc. Bot. France 91: 27–29.

Dehay C. 1957. Anatomic comparée de la feuille des Chlénacées. Mém. Inst. Sci. Madagascar, Sér. B, Biol. Vég. 8: 145–203.

Delpierre X. 1970. Trivalved capsules in *Bixa* L. Taxon 19: 304.

Dempsey RE and NC Garwood. 1994. A study of *Bixa* (Bixaceae), with particular reference to the leaf undersurface indumentum as a diagnostic character. Bull. Nat. Hist. Mus. Lond., Bot. 24: 173–179.

Den Outer RW and PR Schütz. 1981. Wood anatomy of some Sarcolaenaceae and Rhopalocarpaceae and their systematic position. Meded. Land. Wagen. 81(8): 1–25.

Den Outer RW and AP Vooren. 1980. Bark anatomy of some Sarcolaenaceae and Rhopalocarpaceae and their systematic position. Meded. Land. Wagen. 80(6): 1–15.

Dickison WC. 1988. Xylem anatomy of *Diegodendron humbertii*. IAWA Bull., n.s., 9(4): 332–336.

Ding Hou. 1960. Thymelaeaceae. In: CGGJ van Steenis, ed. Flora Malesiana, ser.1, 6: 1–59. Noordhoff-Kolff N.V., Djakarta.

Domke W. 1934. Untersuchungen über die systematische und geographische Gliederung der Thymelaeaceen nebst einer Neubeschreibung ihrer Gattungen. Bibl. Bot. 111: 1–151.

Ducousso M, G Béna, C Bourgeois, B Buyck, G Eyssartier, M Vincelette, R Rabevohitra, L Randrihasipara, B Dreyfus, and T Prin. 2004. The last common ancestor of Sarcolaenaceae and Asian dipterocarp trees was ectomycorrhizal before the India-Madagascar separation, about 88 million years ago. Molec. Ecol. 13: 231–236.

Edlin HL. 1935. A critical revision of certain taxonomic groups of the Malvales. New Phytol. 34: 1–20, 122–143.

Endress PK, M Jenny, and M Fallen. 1983. Convergent elaboration of apocarpous gynoecia in higher advanced dicotyledons (Sapindales, Malvales, Genrianales). Nord. J. Bot. 3: 293–300.

Fagerlind F. 1940. Zytologie und Gametophytenbildung in der Gattung *Wikstroemia*. Hereditas 26: 23–50.

Fay MF, C Bayer, WS Alverson, AY de Bruijn, and MW Chase. 1998. Plastid *rbc*L sequence data indicate a close affinity between *Diegodendron* and *Bixa*. Taxon 47: 43–50.

Fedalto LC. 1983. Estudo anatomico do lenho de *Bixa arborea* Huber. Acta Amazon. 12: 389–399.

Fleming TH, CF Williams, FJ Bonaccorso, and LH Herbst. 1985. Phenology, seed dispersal and colonization in *Muntingia calabura*, a neotropical pioneer tree. Am. J. Bot. 72: 383–391.

Fryxell PA. 1988. Malvaceae of Mexico. Syst. Bot. Monogr. 25: 1–522.

Fryxell PA. 1997. The American genera of Malvaceae. II. Brittonia 49: 204–269.

Fuchs A. 1938. Beiträge zur Embryologie der Thymelaeaceae. Oesterr. Bot. Z. 87: 1–41.

Fuchs HP. 1967. Pollen morphology of the family Bombacaceae. Rev. Palaeobot. Palynol. 3: 119–132.

Gallego MJ and A Aparicio. 1993. Karyological studies in the genus *Tuberaria* Sect. *Scorpioides* (Cistaceae): taxonomic and evolutionary inferences. Plant Syst. Evol. 184: 11–25.

Garwood NC. 1994. Morphology and ecology of seedlings, fruits and seeds of Panama: Bixaceae and Cochlospermaceae. Bull. Nat. Hist. Mus. (London), Bot. 24: 161–171.

Gasson P. 1996. Wood anatomy of the Elaeocarpaceae. In: L Donaldson, AP Singh, BG Butterfield, and LJ Whitehous, eds. Recent advances in wood anatomy, pp. 47–71. New Zealand Forest Institute, Rotorua.

Gaydou EM and ARP Ramanoelina. 1983. A survey of the Sarcolaenaceae for cyclopropene fatty acids. Phytochemistry 22: 1725–1728.

Gazet du Chatelier G. 1940a. Recherches sur les Sterculiacees. Rev. Gen. Bot. 52: 174–191, 211–233, 257–284, 305–332.

Gazet du Chatelier G. 1940b. La structure florale des Sterculiacees. C. R. Acad. Sci. Paris 210: 57–59.

Giannasi DE and KJ Niklas. 1977. Pakaraimoideae, Dipterocarpaceae of the Western Hemisphere, IV. Phytochemistry. Taxon 26: 380–385.

Goldblatt P. 1986. Chromosome number in Sarcolaenaceae Ann. Missouri Bot. Gard. 73: 828–829.

Gottwald H and N Parameswaran. 1966. Das sekundäre Xylem der Familie Dipterocarpaceae, anatomische Untersuchungen zur Taxonomie und Phylogenie. Bot. Jahrb. Syst. 85: 410–508.

Gottsberger G. 1972. Blütenmorphologische Beobachtungen an brasilianischen Malvaceen. II. Oesterr. Bot. Z. 120: 439–509.

Hallier H. 1922. Beiträge zur Kenntnis der Thymelaeaceen und ihrer natürlichen Umgrenzung. Mededebingen van's Rijks Herbarium 44: 1–31.

Harborne JB. 1975. Flavonoid bisulphates and their co- occurrences with ellagic acid in the Bixaceae, Frankeniaceae, and related families. Phytochemistry 14: 1331–1337.

Heel WA van. 1966. Morphology of the androecium in Malvales. Blumea 13: 177–394.

Heinig KH. 1951. Studies in the floral morphology of the Thymelaeaceae. Am. J. Bot. 38: 113–132.

Herber BE. 2002. Pollen morphology of the Thymelaeaceae in relation to its taxonomy. Plant Syst. Evol. 232: 107–121.

Herber BE. 2003. Thymelaeaceae. In: K Kubitzki, ed. The families and genera of vascular plants, vol. 5, pp. 373–396. Springer, Berlin/Heidelberg/New York.

Hochreutiner BPG. 1937. La valeur relative de groupes systematique. Boissiera 2: 1–7.

Horn JW. 2004. The morphology and relationships of Sphaerosepalaceae (Malvales). Bot. J. Linn. Soc. 144: 1–40.

Horn JW and WC Dickison. 1997. Structural biology and phylogenetics of the Sphaerosepalaceae and Diegodendraceae. Am. J. Bot. 84(6): 44.

Huard J. 1965a. Anatomie des Rhopalocarpacees. Adansonia, ser. 2, 5: 103–123.

Huard J. 1965b. Palynologia Madagassica et Mascarenica. 127. Rhopalocarpaceae. Pollen et Spores 7: 303–312.

Huard J. 1965c. Remarques sur la position systématique des Ropalocarpacées d'aprés leur anatomie et leur Morphologie pollinique. Bull. Soc. Bot. France 112: 252–254.

Huber H. 1993. *Neurada* – eine Gattung der Malvales. Sendtnera 1: 7–10.

Inamdar JA and AJ Chohan. 1969. Epidermal structure and stomatal development in some Malvaceae and Bombacaceae. Ann. Bot., ser. 2, 33: 865–878.

Inamdar JA, Bhat R Balakrishna, and Rao TV Ramana. 1983. Structure, ontogeny, classification, and taxonomic significance of trichomes in Malvales. Korean J. Bot. 26: 151–160.

Jabeen F, M Prabhakar, and P Leelavathi. 1994. Structure and distribution of mucilage cells in leaf epidermis of Malvales. Acta Bot. Hung. 38(1–4): 345–352.

Jabeen F, M Prabhakar, and P Leelavathi. 1995. Cyrstalliferous cells in leaf epidermis of Malvales in relation to taxonomy. Geophytology 24: 213–217.

Jansen S, P Baas, and E Smets. 2000. Vestured pits in the Malvales s.l.: a character with taxonomic significance hidden in the secondary xylem. Taxon 48: 169–182.

Jenny M. 1985. Struktur, Funktion und systematische Bedeutung des Gynoeciums bei Sterculiaceen. Ph.D. thesis, University of Zürich, Zürich.

Jenny M. 1988. Different gynoecium types in Sterculiaceae: ontogeny and functional aspects. In: P Leins, SC Tucker, and

PK Endress, eds. Aspects of floral development, pp. 225–236. J. Cramer, Berlin.

Jong K and A Lethbridge. 1967. Cytological studies in the Dipterocarpaceae. 1. Chromosome numbers of certain Malaysian genera. Notes Roy. Bot. Gard., Edinb. 27: 175–184.

Joshi AC. 1936. Anatomy of the flowers of *Stellera chamaejasme*. J. Indian Bot. Soc. 15: 77–85.

Judd WS and SR Manchester. 1997. Circumscription of Malvaceae (Malvales) as determined by a preliminary cladistic analysis of morphological, anatomical, palynological, and chemical characters. Brittonia 49: 384–405.

Kajita T, K Kamiya, K Nakamura, T Tachida, R Wickneswari, Y Tsumura, H Yoshimaru, and T Yamazaki. 1998. Molecular phylogeny of Dipterocarpaceae in Southeast Asia based on nucleotide sequences of *mat*K, *trn*L intron, and *trn*L-*trn*F intergenic spacer region in chloroplast DNA. Molec. Phylogen. Evol. 10: 202–209.

Kamiya K, K Harada, K Ogino, T Kayita, T Yamazaki, HS Lee, and PS Ashton. 1998. Molecular phylogeny of Dipterocarp species using nucleotide sequences of two non-coding regions in chloroplast DNA. Tropics 7: 195–207.

Kamiya K, K Harada, H Tachida, and PS Ashton. 2005. Phylogeny of *Pgi*C gene in *Shorea* and its closely related genera (Dipterocarpaceae), the dominant trees in Southeast Asian tropical rain forests. Am. J. Bot. 92: 775–788.

Kapil RN and R Maheshwari. 1964. Embryology of *Helianthemum vulgare* Gaertn. Phytomorphology 14: 547–557.

Kaur H. 1969. Embryological investigations on *Bixa oreliana* Linn. Proc. Natl. Inst. Sci. India 35: 487–506.

Kausik SB. 1940. Structure and development of the ovule and embryo sac of *Lasiosiphon eriocephalus* Decne. Proc. Natl. Inst. Sci. India 6: 117–132.

Keating RC. 1968. Comparative morphology of Cochlospermaceae: I. Synopsis of the family and wood anatomy. Phytomorphology 18: 379–392.

Keating RC. 1970. Comparative morphology of Cochlospermaceae: II. Anatomy of the young vegetative shoot. Am. J. Bot. 57: 889–898.

Keating RC. 1972. Comparative morphology of Cochlospermaceae: III. The flower and pollen. Ann. Missouri Bot. Gard. 59: 282–296.

Keating RC. 1976. Trends of specialization in pollen of Flacourtiaceae with comparative observations of Cochlospermaceae and Bixaceae. Grana 15: 29–49.

Koechlin J. 1972. L'appareil floral des Sarcolaenacées et la notion de l'angiocarpie. Candollea 27: 171–179.

Kostermans AJGH. 1978. *Pakaraimaea dipterocarpacea* belongs to Tiliaceae. Taxon 27: 357–359.

Kostermans AJGH. 1985. Family status for the Monotoideae Gilg and the Pakaraimoideae Ashton, Maguire, and Zeeuw (Dipterocarpaceae). Taxon 34: 426–435.

Kostermans AJGH. 1992. A handbook of the Dipterocarpaceae of Sri Lanka. PT Gramedia, Jakarta.

Kubitzki K. 2003. Tepuianthaceae. In: K Kubitzki, ed. The families and genera of vascular plants, vol. 5, pp. 371–372. Springer, Berlin/Heidelberg/New York.

Kukachka BF and LW Rees. 1943. Systematic anatomy of the woods of the Tiliaceae. Tech. Bull. Minn. Agric. Exp. Sta. 158: 1–70.

Kumar P and D Singh. 1990. Development and structure of seed coat in *Lavetera* (sic) L. and *Anoda* Cav. J. Indian Bot. Soc. 69(1–2): 121–124.

La Duke JC and J Doebley. 1995. A chloroplast DNA based phylogeny of the Malvaceae. Syst. Bot. 20: 259–271.

Leandri J. 1930. Recherches anatomiques surles Thymé-léacées. Ann. Sci. Nat. Bot., ser. 10, 12: 125–237.

Leinfellner W. 1960. Zur Entwicklungsgeschichte der Kronblätter der Sterculiaceae-Byttnerieae. Oesterr. Bot. Z. 107: 153–176.

Londono C, E Alvarez, E Forero, and C Morton. 1995. A new genus and species of Dipterocarpaceae from the Neotropics: I. Introduction, taxonomy, ecology, and distribution. Brittonia 47: 225–236.

Lowry PP, II, GE Schatz, J-F Leroy, and A-E Wolf. 1999. Endemic families of Madagascar, III. A synoptic revision of *Schizolaena* (Sarcolaenaceae). Adansonia, sér. 3, 21: 183–212.

Lowry PP, II, T Haevermans, J-N Labat, GE Schatz, J-F Leroy, and A-E Wolf. 2000. Endemic families of Madagascar. V. A synoptic revision of *Eremolaena, Pentachlaena,* and *Perrierodendron* (Sarcolaenaceae). Adansonia, sér. 3, 22: 11–31.

Maguire B and PS Ashton. 1980. *Pakaraimaea dipterocarpacea:* II. Taxon 29: 225–231.

Maguire B and JA Steyermark. 1981. Tepuianthaceae, Sapindales. In: B Maguire et al. The botany of the Guayana Highland. Part XI. Mem. N. Y. Bot. Gard. 32: 4–6, 18–21.

Maguire B, PS Ashton, DE Giannasi, KJ Niklas, and C de Zeeuw. 1977. Pakaraimoideae: Dipterocarpaceae of the Western Hemisphere. Taxon 26: 341–385.

Martinez-Hernandez E, P Fernandez, and S Lozano. 1978. Pollen of tropical trees: I. Tiliaceae. J. Arnold Arbor 59: 299–309.

Maury G. 1981. Dipterocarpacees du fruit a la plan tule. Thesis, Montpellier.

Maury G, J Muller, and B Lugardon. 1975. Notes on the morphology and fine structure of the exine of some pollen types in Dipterocarpaceae. Rev. Palaeobot. Palynol. 19: 241–289.

Melikian AP and NA Bondar. 1996. Rosaceae. Neuradaceae. In: A Takhtajan, ed. Comparative seed anatomy, vol. 5, pp. 103–127. Nauka, St. Petersburg (in Russian).

Mohana Rao PR. 1976. Seed and fruit anatomy of *Pterospermumacerifolium*(Sterculiaceae).Phytomorphology 26: 363–369.

Morawetz W. 1986. Remarks on karyological differentiation patterns in tropical woody plants. Plant Syst. Evol. 152: 49–100.

Morton C. 1995. A new genus and species of Dipterocarpaceae from the Neotropics: II. Stem anatomy. Brittonia 47: 237–247.

Murbeck S. 1916. Über die Organisation, Biologic, und verwandtschaftlichen Beziehungen der Neuradoideen. Acta Univ. Lund., II 12: 1–28.

Nandi OI. 1998a. Ovule and seed anatomy of Cistaceae and related Malvanae. Plant Syst. Evol. 209: 239–264.

Nandi OI. 1998b. Floral development and systematics of Cistaceae. Plant Syst. Evol. 212: 107–134.

Nemirovich-Danchenko EN. 1992. Thymelaeales. In: A Takhtajan, ed. Comparative seed anatomy, vol. 4, pp. 405–418. Nauka, St. Petersburg (in Russian).

Nevling LI. 1962. The Thymelaeaceae in the southeastern United States. J. Arnold Arbor. 43: 428–434.

Nilsson S and A Randrianasolo. 1999. Morphology and functional aspects of pollen in the Sarcolaenaceae. Palaeocol. Africa 26: 191–200.

Nilsson S and A Robins. 1986. Bombacaceae. World Pollen Spore Flora 14: 1–59.

Nilsson S, J Coetzee, and E Grafström. 1996. On the origin of the Sarcolaenaceae with reference to pollen morphological evidence. Grana 35: 321–334.

Nowicke JW, V Patel, and J Skvarla. 1985. Pollen morphology and relationships of *Aetoxylon, Amyxa,* and *Gonostylus* to the Thymelaeaceae. Am. J. Bot. 72: 1106–1113.

Nyffeler R and DA Baum. 2000. Phylogenetic relationships of the durians (Bombacaceae-Durioneae or /Malvaceae/ Helicteroideae/Durioneae) based on chloroplast and nuclear ribosomal DNA sequences. Plant Syst. Evol. 224: 55–82.

Nyffeler R and DA Baum 2001. Systematics and character evolution in *Durio* s.l. (/Malvaceae?Helicteroideae/Durioneae or Bimbacaceae-Durioneae). Org. Divers. Evol. 1: 165–178.

Oginuma K and K Fujita. 1997. Karyomorphology of *Neurada procumbens* L. (Neuradaceae). Acta Phytotax. Geobot. 48: 69–71.

Oginuma K, WS Alverson, and DA Baum. 1999. A cytological study of three genera of neotropical Bombacaceae (clades Bombacoideae and Malvoideae). Acta Phytotax. Geobot. 50: 173–178.

Ourisson G. 1979. Chimietaxonomie des Dipterocarpacees. In: G Maury-Lechon, ed. Dipterocarpacees: Taxonomie-Phylogenie-Ecologie. Mem. Mus. Nat. Hist. Nat., ser. B. Bot. 26: 57–67.

Outer RW den and PR Schütz. 1981. Wood anatomy of some Sarcolaenaceae and Rhopalocarpaceae and their systematic position. Meded. Land. Wagen. 81: 1–25.

Outer RE den and AP Vooren. 1980. Bark anatomy of some Sarcolaenaceae and Rhopalocarpaceae and their systematic position. Meded. Land. Wagen. 80(6): 1–15.

Parameswaran N. 1979. Relationships of the family Dipterocarpaceae. Mem. Mus. Natl. Hist. Nat., ser. B, Bot. 26: 155.

Parameswaran N and H Gotwald. 1979. Problematic taxa in the Dipterocarpaceae. Their anatomy and taxonomy. In: G Maury-Lechon, ed. Dipterocarpacees: Taxonomie-Phylogenie-Ecologie. Mem. Mus. Nat. Hist. Nat., ser. B. Bot. 26: 69–75.

Pfeil BE, CL Brubaker, LA Craven, and MD Crisp. 2002. Phylogeny of *Hibiscus* and the tribe Hibisceae (Malvaceae) using chloroplast DNA sequences of *ndh*F and the rpl16 intron. Syst. Bot. 27: 333–350.

Pire SM and CL Cristobal. 2001. El polen de *Helicteres* (Sterculiaceae) y su comparacion con generos vecinos. Bonplandia. 11: 207–230.

Plisko MA. 1996. Tepuianthaceae. In: A Takhtajan, ed. Comparative seed anatomy, vol. 5, pp. 429–430. Nauka, St. Petersburg (in Russian).

Poppendieck H-H. 1980. A monograph of the Cochlospermaceae. Bot. Jahrb. Syst. 101: 191–265.

Poppendieck H-H. 2003. Bixaceae. In: K Kubitzki, ed. The families and genera of vascular plants, vol. 5, pp. 33–35. Springer, Berlin/Heidelberg/New York.

Qi Y-D and Y-Z Wang. 2002. Floral morphogenesis of *Wikstroemia delavayi* (Thymelaeaceae) and its phylogenetic implication. Acta Phytotax. Sinica 40: 271–282.

Ramayya N and SR Shanmukha Rao. 1976. Morphology, phylesis and biology of the peltate scale stellate and tufted hair in some Malvaceae. J. Indian Bot. 55: 75–79.

Randrianasolo A and JS Miller. 1999. Taxonomic revision of the genus *Sarcolaena* (Sarcolaenaceae). Ann. Missouri Bot. Gard. 86: 702–722.

Rao AN. 1953. Embryology of *Shorea talura*. Phytomorphology 3: 476–484.

Rao AN. 1955. A contribution to the embryology of *Valeria indica*. Proc. Natl. Inst. Sci. India 21: 247–255.

Rao AN. 1956. Life history of *Shorea robusta*. Curr. Sci. 25: 128–129.

Record SJ. 1939. American woods of the family Bombacaceae. Trop. Woods 59: 1–20.

Robyns A. 1963. Essai de monographie du genre *Bombax* s. l. (Bombacaceae). Bull. Jard. Bot. L'Etat 33: 1–311.

Rohweder O. 1972. Das Androcium der Malvales und der "Konservatismus" des Leitgewebes. Bot. Jahrb. Syst. 92: 155–167.

Ronse Decraene LP. 1989. Floral development of *Cochlospermum tinctorium* and *Bixa orellana* with special emphasis on the androecium. Am. J. Bot. 76: 1344–1359.

Ronse Decraene LP and EF Smets. 1995. The floral development of *Neurada procumbens* L. (Neuradaceae). Acta Bot. Neerl. 44(4): 439–451.

Ronse Decraene LP and EF Smets. 1996. The floral development of *Neurada procumbens* L. (Neuradaceae). Acta Bot. Neerl. 45: 229–241.

Roth I and H Lindore. 1990. Blatt- und Rindenstruktur von *Tepuianthus auyantepuiensis*, einer neueren Familie aus Venezuela. Bot. Jahrb. Syst. 111: 403–421.

Rye BL. 1988. A revision of western Australian Thymelaeaceae. Nuytsia 6: 129–278.

Saad SI. 1960. The sporoderm stratification in the Malvaceae. Pollen et Spores 2: 13–41.

Schatz GE, P Lowry II, and A-E Wolf. 1999. Endemic families of Madagascar. II. A synoptic revision of Sphaerosepalaceae. Adansonia, sér. 3, 21: 107–123.

Schmid R, S Carlquist, LD Hufford, and GL Webster. 1984. Systematic anatomy of *Oceanopapaver*: a monotypic genus of the Capparaceae from New Caledonia. Bot. J. Linn. Soc. 89: 119–152.

Sensarma P. 1957. On the vascularization of the leaf and its associated structures in *Muntingia calabura*. Bot. Gaz. 119: 116–119.

Shanmukha Rao SR. 1987. Structure, distribution and classification of plant trichomes in relation to taxonomy: Sterculiaceae. Feddes Repert. 98: 127–135.

Shanmukha Rao SR. 1990. Trichome ontogenesis in some Tiliaceae. Beirt. Biol. Pflanz. 65: 363–375.

Shanmukha Rao SR and N Ramayya. 1984. Structure and taxonomic distribution of the epidermal idioblasts in the Malvales. Indian J. Bot. 7: 117–123.

Shanmukha Rao SR and N Ramayya. 1987. Trichome types and their taxonomic importance in the Tiliaceae. Indian J. Bot. 10: 65–73.

Sharma BD. 1969a. Studies of Indian pollen grains in relation to plant taxonomy – Sterculiaceae. Proc. Natl. Inst. Sci. India, part B, Biol. Sci. 35B: 320–359.

Sharma BD. 1969b. Pollen morphology of Tiliaceae in relation to plant taxonomy. J. Palyn. (Lucknow) 5: 7–29.

Sharma BD. 1970. Contribution to the pollen morphology and plant taxonomy of the family Bombacaceae. Proc. Indian Acad. Sci. 36B: 175–191.

Sharma BD. 1990. Trichomes in some Tiliaceae. J. Indian Bot. Soc. 69: 11–14.

Schatz GE, PP Lowry II, and A-E Wolf. 1999. Endemic families of Madagascar, II. A synoptic revision of Sphaerosepalaceae. Adansonia, sér. 3, 21(1): 107–123.

Schnarf K. 1931. Ein Beitrag zur Kenntnis der Samenentwicklung der Gattung *Cochlospermum*. Oesterr. Bot. Z. 80: 45–50.

Shenstone FS and JR Vickery. 1961. Occurrence of cyclo-propene acids in some plants of the order Malvales. Nature 190: 168–169.

Singh HB and VP Dube. 1993. Taxonomic significance of foliar epidermal features of *Muntingia* Linn. (Tilliaceae). J. Plant Anat. Morphol. 6: 123–128.

Somego M. 1978. Cytogenetical study of Dipterocarpaceae. Malaysian For. 41: 358–366.

Straka H. 1963 (1964). Betrachtungen zur Phylogenie der Sarcolaenaceae (Chlaenaceae). Ber. Deutsch. Bot. Ges. 76: 55–62.

Straka H. 1964. Palynologia Madagassica et Mascarenica. Fam. 126. Sarcolaenaceae (Chlaenaceae). Pollen et Spores 6: 289–301.

Straka H. 1965. Über die Pollenmorphologie der Gattung *Eremolaena* (Sarcolaenaceae). Bietr. Biol. Pflanzen 41: 65–68.

Straka H. 1971. Über das System der madagassischen Sarcolaenaceae. Ber. Deutsch. Bot. Ges. 84: 731–735.

Straka H and F Albers. 1978. Die Pollenmorphologie von *Diegodendrom humbertii* R. Capuron (Diegodendraceae, Ochnales bzw. Theales). Bot. Jahrb. Syst. 99: 363–369.

Suzuki E and PS Ashton. 1996. Sepal and nut size ration of fruits of Asian Dipterocarpaceae and its implications for dispersal. J. Trop. Ecol. 12: 853–870.

Swarupanandan K. 1986. Late embryogenesis and morphology of mature embryos in three species of Dipterocarpaceae. Canad. J. Bot. 64: 2582–2587.

Tan K. 1980. Studies in the Thymelaeaceae: 1. Germination, seedlings, fruits, and seeds. Notes Roy. Bot. Gard. Edinb. 38: 149–164.

Tang Y. 1998. Floral morphology and embryo sac development in *Burretiodendron kydiifolium* Y.C. Hsu et R. Zhuge (Tiliaceae) and their systematic significance. Bot. J. Linn. Soc. 128: 149–158.

Tang Y and X-F Gao. 1993. Pollen morphology of *Burretiodendron* sensu lato (Tiliaceae) and its systematic significance. Cathaya 5: 81–88.

Tang Y and K-Y Pan. 1994. Gametophytic development of *Melhania hamiltoniana* Wall. (Sterculiaceae) and its systematic implications. Cathaya 6: 67–74.

Tate JA, JF Aguilar, SJ Wagstaff, JC La Duke, TA Bodo Slotta, and BB Simpson. 2005. Phylogenetic relationships within the tribe Malveae (Malvaceae, subfamily Malvoideae) as inferred from ITS sequence data. Am. J. Bot. 92: 584–602.

Thanos CA, K Georghiou, C Kadis, and C Pantazi. 1993. Cistaceae: a plant family with hard seeds. Israel J. Bot. 41: 251–263.

Thirumalachar MJ and BA Razi. 1941. Megasporogenesis and endosperm formation in *Eriodendron anfractuosum* DC. Proc. Indian Acad. Sci. B14: 461–465.

Tieghem P van. 1900. Sur les Bixacées, les Cochlospermacées et les Sphérosépalacées. J. Bot. (Morot) 14: 32–54.

Tirel C, J Jérémie, and D Lobreau-Callen. 1996. *Corchorus neocaledonicus* (Tiliaceae), veritable identité de l'enigmatique *Oceanopapaver*. Bull. Mus. Hist. Nat. (Paris), 4th ser. 18: 35–43.

Ukraintseva VV. 1993. Pollen morphology of the family Cistaceae in relation to its taxonomy. Grana. (Suppl. 2): 33–36.

Van Heel WA. 1966. Morphology of the androecium in Malvales. Blumea 13: 177–394.

Van Heel WA. 1978. Morphology of the pistil in Malvaceae-Ureneae. Blumea 24: 123–127.

Van Heel WA. 1995. Morphology of the gynoecium of *Kitaibelia vitifolia* Willd. and *Malope trifida* L. (Malvaceae-Malopeae). Bot. Jahrb. Syst. 117: 485–493.

Venkata Rao C. 1949. Floral anatomy of some Sterculiaceae with special reference to the position of stamens. J. Indian Bot. Soc. 28: 237–245.

Venkata Rao C. 1949–1954. Contributions to the embryology of Sterculiaceae, parts 1–5. J. Indian Bot. Soc. 28: 180–197, 1949; 29: 163–176, 1950; 30: 122–131, 1951; 31: 251–260, 1953; 32: 208–238, 1953 (1954).

Venkata Rao C. 1950. Pollen grains of Sterculiaceae. J. Indian Bot. Soc. 29: 130–137.

Venkata Rao C. 1951. Life history of *Muntingia calabura* L. Curr. Sci. 20: 47–48.

Venkata Rao C. 1952a. The embryology of *Muntingia calabura* L. J. Indian Bot. Soc. 31: 87–101.

Venkata Rao C. 1952b. Floral anatomy of some Malvales and its bearing on the affinities of families included in the order. J. Indian Bot. Soc. 31: 171–203.

Venkata Rao C. 1954, 1955. Embryological studies in Malvaceae: I. Development of gametophytes. II. Fertilization and seed development. Proc. Natl. Inst. Sci. India 20: 127–150, 1954; 21B: 53–67, 1955.

Venkata Rao C and KV Sambasiva Rao. 1952. A contribution to the embryology of *Triumfetta rhamboidea* Jacq. and *Corchorus acutangulus* L. J. Indian Bot. Soc. 31: 56–68.

Venkatesh CS. 1956. The curious anther of *Bixa*: its structure and dehiscence. Am. Midi. Nat. 55: 473–476.

Venkatesvarlu J. 1945. Embryological studies in the Thymelaeaceae: I. *Thymelaea arvensis*. J. Indian Bot. Soc. 24: 45–66.

Venkatesvarlu J. 1947. Embryological studies in the Thymelaeaceae: II. *Daphne cannabina* Wall. and *Wikstroemia canescens* Meisn. J. Indian Bot. Soc. 26: 13–39.

Vickery JR. 1980. The fatty acid composition of seed oils from ten plant families with particular reference to cyclopropene and dihydrosterculic acids. J. Am. Oil. Chem. Soc. 57: 87–91.

Vogel S. 2000. The floral nectaries of Malvaceae sensu lato: a conspectus. Kurtziana. 28: 155–171.

Von Balthazar M, WS Alverson, J Schönenberger, and DA Baum. 2004. Comparative floral development and androecium structure in Malvoideae (Malvaceae s.l.). Int. J. Plant Sci. 165: 445–473.

Von Balthazar M, J Schönenberger, WS Alverson, H Janka, C Bayer, and DA Baum. 2006. Structure and evolution of the androecium in the Malvatheca clade (Malvaceae s.l.) and implications for Malvaceae and Malvales. Plant Syst. Evol. 260: 171–197.

Vyshenskaya TD. 1992. Bixales (Cistales). In: A Takhtajan, ed. Comparative seed anatomy, vol. 4, pp. 229–248. Nauka, St. Petersburg (in Russian).
Webber IE. 1934. Systematic anatomy of the woods of the Malvaceae. Trop. Woods 38: 15–36.
Weberling F and U Herkommer. 1989. Untersuchungen zur Infloreszenzmorphologie der Thymelaeaceae. Akad. Wiss. Abh. Math-Naturwiss. Kl. 68: 1–124.
Weibel R. 1945. La placentation chez les Tiliacees. Candollea 10: 155–177.
Whitlock BA, KG Karol, and WS Alverson. 2000. Chloroplast DNA sequences confirm the placement of *Oceanopapaver* within the Grewioideae (Tiliaceae or Malvaceae s.l.). Am. J. Bot. 87: 184 (abstract).
Whithlock BA, C Bayer, and DA Baum. 2001. Phylogenetic relationships and floral evolution of the Byttherioideae ("Sterculiaceae" or Malvaceae s.l.) based on sequences of the chloroplast gene, *ndh*F. Syst. Bot. 26: 420–437.
Whitmore TC. 1962. Studies in systematic bark morphology. III. Bark taxonomy in Dipterocarpaceae. Gard. Bull. Singapore 19: 321–371.
Wilkie P, A Clark, RT Pennington, M Cheek, C Bayer, and CC Wilcock. 2006. Phylogenetic relationships with the subfamily Sterculioideae (Malvaceae/Sterculiaceae-Sterculieae) using the chloroplast gene *ndh*F. Syst. Bot. 31: 160–170.
Wilkins CF and JA Chappill. 2002a. New chromosome numbers for Lasiopetaleae: Malvaceae s.l. (or Sterculiaceae). Aust. Syst. Bot. 15: 1–8.
Wilkins CF and JA Chappill. 2002b. Seed and seedling morphology and seed anatomy of Lasiopetaleae (Malvaceae s.l. or Sterculiaceae). Aust. Syst. Bot. 15: 545–563.
Woon C and H Keng. 1979. Observations on stamens of the Dipterocarpaceae. Gard. Bull. Singapore 32: 1–55.
Wurdack KJ and JW Horn. 2001. A re-evaluation of the affinities of the Tepuianthaceae: molecular and morphological evidence for placement in the Malvales. In Botany 2001: Plants and People, Abstracts, p. 151. Albuquerque.
Xu SJ and HH Hsue. 2000. Comments on the taxonomic position of some genera in Sterculiaceae. J. Trop. Subtrop. Bot. 8(1): 11–16.
Ya T and K-Y Pan. 1994. Gametophytic development of *Melhania hamiltoniana* Wall. (Sterculiaceae) and its systematic implications. Cathaya 6: 67–74.
Young AM, M Schaller, and M Strand. 1984. Floral nectarines and trichomes in relation to pollination in some species of *Theobroma* and *Herrania*. Am. J. Bot. 71: 466–480.
Zabeen F, M Prabhakar, and P Leelavathi. 1994. Structure, distribution and taxonomic significance of crystalliferous cells in Malvales. J. Indian Bot. Soc. 73(1–2): 89–94.
Zeeuw C de. 1977. Pakaraimoideae, Dipterocarpaceae of the Western Hemisphere: III. Stem anatomy. Taxon 26: 368–380.

Order 76. URTICALES

Trees, shrubs, or herbs, often with cystoliths or laticifers or both. Vessels with simple perforations; lateral pitting alternate. Fibers mostly with simple pits. Rays heterogeneous or sometimes homogeneous. Axial parenchyma usually paratracheal. Sieve-element plastids of S-type or (some Ulmaceae) of Pc-type. Nodes trilacunar or (*Ficus*) pentalacunar. Leaves alternate or less often opposite, almost always stipulate (absent in some Parietarieae), mostly simple, very often oblique at the base. Stomata anomocytic or sometimes of other types. Flowers small and inconspicuous, in cymose or racemose inflorescences of sometimes very complex structure, rarely solitary, unisexual or less often bisexual, cyclic, apetalous, anemophilous or secondarily entomophilous. Sepals 2–8 (mostly 4–5), in one or sometimes two cycles, usually valvate, sometimes more or less connate, sometimes wanting. Stamens mostly as many as sepals and opposite them, rarely fewer or more numerous; filaments free or sometimes connate; anthers usually 2-locular (1-locular in *Ficus* subsect. *Malvanthera*). Tapetum secretory. Microsporogenesis simultaneous. Pollen grains 2-celled or less often 3-celled, 2-pantoporate. Gynoecium usually pseudomonomerous, of two carpels (rarely of three carpels in some Moraceae), with one or two separate stylodia, stigmas de-current. Ovary superior or inferior, usually 1-locular, but occasionally 2-locular (e.g., in *Ulmus*). Ovules solitary (or solitary in each of two locules), anatropous, hemitropous, or (Cecropiaceae and Urticaceae) orthotropous, bitegmic, crassinucellate. Female gametophyte of *Polygonum*- or less often *Allium*-, *Adoxa*-, or *Drusa*-types. Endosperm nuclear. Fruits of various types. Seeds with straight or curved embryo, with or without endosperm.

Weddell (1856) was the first who discussed several characters indicating affinity of the Urticaceae with the Malvales, especially with the Tiliaceae. Hallier (1903, 1905) included Urticaceae s.1. in the Malvales but later (Hallier 1908, 1912) changed his opinion and moved them in his "Terebinthines." Bessey (1915) also included Ulmaceae, Moraceae, and Urticaceae in his Malvales s.l. More recently Thorne (1968, 1974, 1976, 1983, 1992a, b, 2000), Stebbins (1974), Dahlgren (1975, 1980, 1983), and Friis (1993: 617) came to the conclusion that the Urticales are closely related to the Malvales but are distinct enough to merit their own order. The affinity between them is confirmed by anatomical data, including wood anatomy and the occurrence of mucilage cells and canals (Guerin 1923; Holm 1927; Metcalfe and Chalk 1950; Hegnauer 1973), as well as by serological studies (Petersen and Fairbrothers 1985) and phytochemical data (Hegnauer

1964, 1969, 1973; Soepadmo 1977), as well seed anatomy (Kravtsova 2003). According to Berg (1977b:371), "The affinities of the Urticales with the Malvales are closest between the Ulmaceae and Tiliaceae, the least derived families of the orders."

The Urticales have most probably originated from an ancient tiliaceous ancestor.

Key to Families

1 Ovules subapical, pendulous, anatropous to campylotropous. Ovary with two stylodia, unilocular or bilocular, but one of them sometimes more or less reduced. Milky latex present or absent. Fruits dry or drupaceous.

2 Plants without milky latex. Stamens erect in bud. Flowers unisexual, polygamous, or rarely bisexual.

3 Trees and shrubs, usually with scattered mucilage cells or even mucilage canals, and with strong tendency toward mineralization of the cell walls with calcium carbonate or silica. Leaves simple, often oblique at the base, usually distichous, rarely opposite, with typically caducous stipules. Flowers in cymose or fasciculate inflorescences, monoecious or less often bisexual. Pollen grains 4–7-porate, exine rugulose. Ovary unilocular or rarely bilocular. Fruits dry or drupaceous. Seed flattened, coat undistinguished, exotestal cells elongated, unthickened; embryo straight or curved (*Zelkova*); endosperm scanty or absent. Contain sesquiterpene lactones, n = 10, 14. . . . . 1. ULMACEAE.

3 Perennial or annual herbs with secretory canal in the phloem, but without mucilage cells. Stems erect or twining, either scabrous or armed with rigid climbing hairs and often with glandular trichomes. Leaves palmately lobed (*Humulus*) or palmately compound (*Cannabis*), sometimes simple, petiolate, serrate, generally decussate, near the stem apex often alternate, opposite at least below, with triangular, persistent stipules. Flowers in complex, basically cymose inflorescences, unisexual (monoecious or more often dioecious). Ovary unilocular with solitary anatropous ovule, pendent from near the apex of the locule. Fruits nutlike or achenes covered by the perianth. Seeds with curved (*Cannabis*) or coiled (*Humulus*) embryo and a small amount of fleshy, oily endosperm. Accumulating quebrachitol and producing pyridine alkaloids, and sometimes tanniferous, with proanthocyanins, n = 7, 9, 10. . . . . .3. CANNABACEAE.

2 Plants nearly always with milky latex (wanting in *Fatoua*) borne in laticifers widely distributed in the parenchymatous parts of the stem and often in the leaves. Trees, shrubs, woody lianas (including stranglers), subshrubs, succulent or tuberous geophytes, and rarely annuals (*Fatoua*). Leaves simple or rarely compound, alternate or opposite, often with cell walls more or less mineralized with calcium carbonate or silica; stipules present, sometimes much reduced. Stomata anisocytic or cyclocytic. Flowers always unisexual, in compact inflorescences with the axis often thickening to form a head or invaginated common receptacle. Stamens erect or inflexed in bud. Ovary unilocular or less often bilocular. Fruits drupaceous, sometimes with dehiscent exocarp. Seed coat undistinguished; endosperm absent or present; embryo strait or curved, n = 7, 12–14. . . . . . . 2. MORACEAE.

1 Ovules subbasal or basal, erect, more or less orthotropous. Ovary always unilocular, with a single stylodium. Plants usually without milky latex. Fruits dry, nutlike.

4 Trees, shrubs, or woody lianas (sometimes epiphytic), often with stilt roots. Laticifers more or less reduced or absent, but mucilage cells and sacs often present. Leaves alternate, simple and entire to often so deeply palmately or radiately incised as to appear compound, with fused and often fully amplexicaul stipules. Cystoliths wanting. Flowers dioecious. Stamens erect in bud (except for *Poikilospermum* subgenus *Poikilospermum*), not abruptly bending in dehiscence. Fruits nutlike. Seeds large and without endosperm or small and with endosperm, x = 7.. . . . . . . 4. CECROPIACEAE.

4 Herbs or occasionally subshrubs or rarely small, soft-wooded trees, very rarely lianas, in Urticeae provided with stinging hairs. Frequently with mucilage cells, and sometimes also with nonarticulated latex canals, but usually without milky juice. Leaves alternate or opposite, simple, entire or 3–5(-7)-lobed, usually with three subequal veins from the base, with epidermal cells tending to be mineralized with silica or calcium carbonate; stipules usually present (absent in some Parietarieae), lateral or often intrapetiolar, often

fused. Stomata anisocytic or paracytic. Cystoliths usually present in both the stems and leaves, linear to punctiform. Flowers unisexual or rarely bisexual. Stamens inflexed in bud, abruptly bending in dehiscence. Fruits nutlike or seldom drupaceous. Seeds with thin oily or starchy endosperm, but sometimes without endosperm, n = 6–14. . . . . 5. URTICACEAE.

1. ULMACEAE

Mirbel 1815 (including Celtidaceae Link 1831). 15/200. Widely distributed in tropical, subtropical, and temperate regions; Ulmoideae distributed mainly in temperate regions.

1.1 CELTIDOIDEAE

Leaves with three main veins from the base and pairs of secondary veins usually less than six, alternate or (*Lozanella*) opposite, brochidodromous. Rays essentially heterogeneous (in *Gironniera* and *Trema* markedly heterogeneous). Flowers unisexual or polygamous. Pollen grains usually 2–5 porate (pantoporate in *Ampelocera*), ectexine palisade (*Ampelocera* and *Chaetachme*) or granular in the middle part (not recognizable in *Lozanella, Parasponia,* and *Trema*); ornamentation finely scabrate or consisting of spinules or of densely spaced warts and microechinules. Fruits more or less globose, drupaceous, and with thick-walled and well-lignified endocarp, or (*Pteroceltis*) small, nutlike, with spherical endocarps whose reticulate surface sculpture is almost identical to that of *Celtis,* and with an exocarp developed into a pair of wings differing from those of Ulmoideae in lacking venation. Seeds with curved embryo and scanty or no endosperm (endosperm fleshy in *Parasponia* and *Lozanella*); cotyledons mostly variously folded or involute. Produce quebrachitol and glycoflavones (which are absent in *Gironniera* subgenus *Galumpita*) but not lignans, sesquiterpens, and flavonols (the latter present in *Gironniera* subgenus *Galumpita*). x = 10, 14. – *Ampelocera, Chaetachme, Gironniera, Aphananthe, Celtis, Pteroceltis, Trema, Parasponia, Lozanella.*

1.2 ULMOIDEAE

Leaves pinnately veined, with seven or more pairs of secondary veins, the lowest pairs not prominent, craspedodromous, alternate. Rays homogeneous or weakly heterogeneous. Flowers mostly bisexual or polygamous. Pollen grains 4–5-porate, with the exine distinctly and finely granular throughout; ornamentation rugulate, with small spinules. Fruits dry, usually more or less compressed (samaroid) or less often subdrupaceous (*Zelkova* and *Planera*) or small and nutlike with a rigid endocarp similar to that of *Zelkova* and a crestlike wing (*Hemiptelea*). Seeds with straight embryo and usually without endosperm; cotyledons broad, flat or longitudinally folded. Produce lignans, sesquiterpenes, and flavonols, but not quebrachitol and glycoflavons. x = 14. – *Ulmus, Holoptelea, Hemiptelea, Zelkova, Planera, Phyllostylon.*

Celtidoideae were traditionally included in the Ulmaceae and only recently have some authors supported their distinctiveness as a separate family (Grudzinskaya 1967; Chernik 1975, 1980, 1981, 1982; Zavada 1983; Takahashi 1989). However, some genera, including *Ampelocera* and *Aphananthe,* possess character combinations (including floral anatomy) that seem to bridge the gap between typical celtidoids and ulmoids (see Chernik 1975; Gianasi 1978; Manchester 1989; Terabayashi 1991; Zhong et al. 1992). *Ampelocera* and *Aphananthe* are celtidoid in most nonchemical characters (except for special leaf venation of *Ampelocera*) although they contain flavonols (Gianasi 1978). *Zelkova* and *Hemiptelea,* long considered celtidoid, are in fact flavonol producers (Gianasi 1978) and included in the Ulmaceae sensu stricto by Grudzinskaya (1967), which is confirmed also by wood anatomy (Zhong et al. 1992). Besides, *Gironniera* consists of two distinct entities, one of which (subgenus *Gironniera*) is really celtidoid and the other (subgenus *Galumpita*) is in many respects rather different (chemically ulmoid) and was even separated as a genus *Galumpita* Blume. Finally, tropical American *Celtis monoica* (a monotypic genus *Mirandaceltis,* according to Sharp [1958]) is in fact a flavonol producer (Gianasi 1978) and as Sharp (1958) indicates, its leaf venation and drupe size are somewhat intermediate between *Celtis* and *Zelkova*. "The flavonoid data at this level do not suggest any particular Ulmoid relative for *Mirandaceltis monoica,* but do indeed support Sharp's contention that the taxon should be removed from *Celtis* sensu stricto," states Gianasi (1978: 342). All these data show that the differences between celtidoid and ulmoid genera are not clear-cut, and therefore I am reluctant to accept a separate family rank for celtidoids. At best they are two subfamilies.

2. MORACEAE

Link 1831 (including Artocarpaceae Dumortier 1829, Dorsteniaceae Chevalier 1827, Ficaceae Berchtold et J. Pres 1,96720). 37/1200+. Widely distributed in tropics and subtropics with a few species in temperate regions of both hemispheres.

MOREAE: *Morus, Broussonetia, Milicia, Maclura, Trophis* (including *Olmedia*), *Streblus, Bleekrodea, Fatoua;* ARTOCARPEAE: *Artocarpus, Parartocarpus, Treculia, Prainea, Hullettia, Antiaropsis, Sparattosyce, Batocarpus, Bagassa, Sorocea, Clarisia, Poulsenia;* CASTILLEAE: *Perebea, Maquira, Castilla, Helicostylis, Pseudolmedia, Naucleopsis, Antiaris, Mesogyne;* DORSTENIEAE: *Utsetela, Bosqueiopsis, Helianthostylis, Trymatococcus, Brosimum, Trilepislum, Scyphosyce, Dorstenia;* FICEAE: *Ficus.*

The subdivision of the Moraceae follows Rohwer (1993). But as Rohwer notes, the family forms such a closely knit complex that tribes and genera are difficult to delimit. According to Rohwer, the best defined tribes are Castilleae and Ficeae.

Related to the Ulmaceae, especially to the Celtidoideae.

3. CANNABACEAE

Martynov 1820 (including Lupulaceae Link 1831). 2–3/4. Temperate regions of the Northern Hemisphere.

Humulus (including *Humulopsis* ?), *Cannabis*

Cannabaceae are related to Urticaceae and Moraceae, but differ in the absence of laticifers and the erect position of stamens in bud (Kubitzki 1993). 2n = 18 + XX or XY, in *Humulus lupuls* 2n = 20 + XX or XY, while in *Humulus* (*Humulopsis*) *scandens* 2n = 14 + XX (female) or YXY (male).

4. CECROPIACEAE

C.C. Berg 1978. 6/200. Mainly tropical America, but also tropical Africa (*Myrianthus* and *Musanga*) and Asia from eastern Himalayas to Malesia (*Poikilospermum*).

Myrianthus, Pourouma, Cecropia, Musanga, Coussapoa, Poikilospermum.

Cecropiaceae are intermediate between the Moraceae, with which they share possession of laticifers, and the Urticaceae, with which they share orthotropous subbasal or basal ovules (Berg 1978). According to Berg (1989: 215), the delimitation of the family is doubtful only with respect to the genus *Poikilospermum,* which deviates in several features (e.g., the elongate "urticaceous" cystoliths) from the group of African and American genera. He states that anatomical data (Bonsen and ter Welle 1983) also suggest that the position of *Poikilospermum* is separate from the other five genera and closer to the Urticaceae. According to Friis (1993) the position of *Poikilospermum* is still uncertain.

5. URTICACEAE

A.L. de Jussieu 1879 (including Parietariaceae Berchtold et J. Presl 1820). 49/1000. Widely distributed in tropical and subtropical regions, but some species occur in temperate and cold regions. The largest concentration of genera and species is in tropical Asia.

URTICEAE: *Urtica, Hesperocnide, Nanocnide, Obetia, Laportea, Discocnide, Girardinia, Dendrocnide, Urera, Gyrotaenia:* LECANTHEAE: *Elatostema, Meniscogyne, Procris, Pilea, Achudemia, Sarcopilea, Lecanthus, Petelotiella*; BOEHMERIEAE: *Boehmeria, Archiboemeria, Chamabainia, Pouzolzia, Hyrtanandra, Gonostegia, Neodistemon, Cypholophus, Sarcochlamys, Touchardia, Neraudia, Pipturus, Nothocnide, Oreocnide, Debregeasia, Astrothalamus, Leucosyke, Gibbsia, Phenax, Maoutia, Myriocarpa;* PARIETARIEAE: *Gesnouinia, Hemistylus, Parietaria, Soleirolla, Rousselia;* FORSSKAOLEAE: *Forsskaolea, Droguetia, Didymodoxa, Australina.*

Bibliography

Anderson LC. 1974. A study of systematic wood anatomy in *Cannabis*. Bot. Mus. Leafl. Harv. Univ. 24: 29–36.

Banerji I. 1953. A contribution to the life history of *Artocarpus lakooche* Roxb. Proc. Indian Acad. Sci. 39B: 128–132.

Bechtel AR. 1921. The floral anatomy of the Urticales. Am. J. Bot. 8: 386–410.

Behnke H-D. 1973. Sieve-tube plastids of Hamamelidae: electron microscopic investigations with special reference to Urticales. Taxon 22: 205–210.

Berg CC. 1972. Olmedieae and Brosimeae (Moraceae). Flora Neotropica 7: 1–228.

Berg CC. 1973. Some remarks on the classification and differentiation of Moraceae. Meded. Bot. Mus. Herb. Rijks Univ. Utrecht 386: 1–10.

Berg CC. 1977a. The Castilleae, a tribe of the Moraceae, renamed and redefined due to the exclusion of the type genus *Olmedia* from the "Olmedieae." Acta Bot. Neerl. 26: 73–82.

Berg CC. 1977b. Urticales, their differentiation and systematic position. Plant Syst. Evol, Suppl. 1: 349–374.

Berg CC. 1977c. Revisions of African Moraceae (excluding *Dorstenia, Ficus, Musanga,* and *Myrianthus*). Bull. Jard. Bot. Natl. Belgium 47: 267–407.

Berg CC. 1978. Cecropiaceae: a new family of the Urticales. Taxon 27: 39–44.

Berg CC. 1983. Dispersal and distribution in the Urticales. In: K Kubitzki, ed. Dispersal and distribution: an international symposium, pp. 219–229. Hamburg, Berlin.

Berg CC. 1989. Systematics and phylogeny of the Urticales. In: PR Crane and S Blackmore, eds. Evolution, systematics, and fossil history of the Hamamelidae, pp. 193–220. Clarendon, Oxford.

Berg CC. 1990. Differentiation of flowers and inflorescences of Urticales in relation to their protection against breeding insects and to pollination. Sommerfeltia 11: 13–34.

Berg CC. 1998. Phytogeography, systematics and diversification of African Moraceae compared with those of other tropical areas. In: CR Huxley, JM Lock, and DF Cutler, eds. Chorology, taxonomy and ecology of the floras of Africa and Madagascar. Royal Botanic Gardens, Kew.

Berg CC. 2005. Moraceae diversity in a global perspective. In: I Friis and H Balslev, eds. Proceedings of a Symposium on Plant Diversity and Complexity Patterns – Local, Regional and Global Dimensions. Biol. Skrifte 55: 423–440. Copenhagen.

Bernbeck F. 1932. Vergleichende Morphologic der Urticaceen- und Moraceen-Infloreszenzen. Bot. Abb. 19: 1–100.

Bonsen KJ and BJH ter Welle. 1983. Comparative wood and leaf anatomy of the Cecropiaceae (Urticales). Bull. Mus. Natl. Hist. Nat. Paris, 4th ser, 5: 151–177.

Bonsen KJ and BJH ter Welle. 1984. Systematic wood anatomy and affinities of the Urticaceae. Bot. Jahrb. Syst. 105: 49–71.

Chernik VV. 1975. Arrangement and reduction of perianth and androecium parts in representatives of the Ulmaceae Mirbel and Celtidaceae Link. Bot. Zhurn. 66: 1561–1573 (in Russian).

Chernik VV. 1980. Peculiarities of structure and development of the pericarp of the representatives of the families Ulmaceae and Celtidaceae. Bot. Zhurn. 65: 521–531 (in Russian).

Chernik VV. 1981. Pseudomonomerous gynoecium of the representatives of Ulmaceae and Celtidaceae. Bot. Zhurn. 66: 958–962 (in Russian).

Chernik VV. 1982. Characteristics of the structural development of spermoderm in some representatives of Ulmaceae and Celtidaceae. Bot. Zhurn. 67: 1216–1220 (in Russian).

Chew Wee-Lek. 1963. A revision of the genus *Poikilospermum* (Urticaceae). Gardens Bull. Singapore 21: 1–186.

Corner EJH. 1962. The classification of Moraceae. Gard. Bull. Singapore 19: 187–252.

Datwyler SL and G Weiblen. 2004. On the origin of the fig: phylogenetic relationships of Moraceae from *ndh*F sequences. Am. J. Bot. 91: 767–777.

Datwyler SL, W Clement, S Swenson, and G Weiblen. 2003. Where did the figs come from? Phylogenetic analysis of Moraceae based on *ndh*F and 26S sequences. In Botany 2003: Aquatic and Wetland Plants: Wet and Wild. Abstracts, p. 77. Mobile, Alabama.

Fukuoka N. 1982. On pseudomonomerous pistil of the Ulmaceae. Acta Phytotax. Geobot. 32: 84–91.

Elias TS. 1970. The genera of Ulmaceae in the southeastern United States. J. Arnold Arbor. 51: 18–40.

Freisleben R. 1933. Untersuchungen über Bildung und Auflösung von Cystolithen bei den Urticales. Flora 127: 1–45.

Friis I. 1989. The Urticaceae: a systematic review. In: PR Crane and S Blackmore, eds. Evolution, systematics, and fossil history of the Hamamelidae, vol. 2, pp. 285–308. Clarendon, Oxford.

Friis I. 1993. Urticaceae. In: K Kubitzki, ed. The families and genera of vascular plants, vol. 2, pp. 612–630. Springer, Berlin/Heidelberg/New York.

Gangadhara MK and JA Inamdar. 1977. Trichomes and stomata and their taxonomic significance in the Urticales. Plant Syst. Evol. 127: 121–137.

Gentry AH. 1983. *Plagioceltis* (Ulmaceae): a superfluous genus. Taxon 32: 460–461.

Gianasi DE. 1978. Generic relationships in the Ulmaceae based on flavonoid chemistry. Taxon 27: 331–344.

Gianasi DE. 1986. Phytochemical aspects of phylogeny in Hamamelidae. Ann. Missouri Bot. Gard. 73: 417–437.

Grudzinskaya IA. 1967. The Ulmaceae and reasons for distinguishing Celtidoideae as a separate family Celtidaceae Link. Bot. Zhurn. 52: 1723–1748 (in Russian with English summary).

Grudzinskaya IA. 1988. On the taxonomy of Cannabaceae. Bot. Zhurn. 73: 589–596 (in Russian with English summary).

Grudzinskaya IA. 1990. On the evolution of flowers and inflorescences in the Urticaceae family. Bot. Zhurn. 75: 774–782. (in Russian with English summary).

Guerin P. 1923. Les Urticees: cellules a mucilage laticiferes et canaux secreteurs. Bull. Bot. Soc. Fr. 70: 125–136, 207–215, 255–263.

Holm T. 1927. *Boehmeria cylindrica* (L.) Sw.: a morphological study. Am. J. Sci., 5th ser., 13: 115–122.

Humphries J and S Blackmore. 1989. A review of the classification of the Moraceae. In: PR Crane and S Blackmore, eds. Evolution, systematics, and fossil history of the Hamamelidae, vol. 2, pp. 267–277. Clarendon, Oxford.

Jackson AP. 2004. Cophylogeny of the *Ficus microcosm*. Biol. Rev. 79: 751–768.

Johri BM and RN Konar. 1956. The floral morphology and embryology of *Ficus religiosa* Linn. Phytomorphology 6: 97–111.

Kravtsova TI. 1995. Pericarp and seed coat structure in the Cecropiaceae (Urticales). Bot. Jahrb. Syst. 80: 1–14.

Kravtsova TI. 2001a. Pericarp structure in the representatives of the tribe Boehmerieae (Urticaceae). Bot. Zhurn. 86(6): 18–39 (in Russian with English summary).

Kravtsova TI. 2001b. Pericarp structure in the tribe Urticeae (Urticaceae) in relation to its systematics. Bot. Zhurn. 86(11): 49–72 (in Russian with English summary).

Kravtsova TI. 2003. Seed coat structure in the Urticaceae and relations of the Urticales. Bot. Zhurn. 88(11): 11–41 (in Russian with English summary).

Kravtsova TI. 2006. Cells with wall ingrowths in the pericarp and seed coat of the representatives of the Urticaceae. Bot. Zhurn. 91(9): 1369–1378 (in Russian with English summary).

Kravtsova TI. 2007. A system of the family Urticaceae. Bot. Zhurn. 92(1): 3–28 (in Russian with English summary).

Kravtsova TI, I Friis, and CM Wilmot-Dear. 2000. Morphology and anatomy of fruits in New World *Boehmeria* in relation to taxonomy. Kew Bull. 55: 43–62.

Fravtsova TI, I Friis, and CM Wilmot-Dear. 2003. Morphology and anatomy of fruits in *Pouzolzia* (Urticaceae) in relation to taxonomy. Kew Bull. 58: 297–327.

Kubitzki K. 1993a. Cannabaceae. In: K Kubitzki, ed. The families and genera of vascular plants, vol. 2, pp. 204–206. Springer, Berlin/Heidelberg/New York.

Kubitzki K. 1993b. Cecropiaceae. In: K Kubitzki, ed. The families and genera of vascular plants, vol. 2, pp. 243–246. Springer, Berlin/Heidelberg/New York.

Leberton P. 1965. Elements de chimiotaxonomie botanique: 2. Cas de flavonoides chez les Urticales; conclusions generales. Bull. Soc. Bot. France 111: 80–93.

Le Coq C. 1963. Contribution a l϶ετυδε cytotaxonomique de Moracees et des Urticacees. Rev. Gen. Bot. 70: 385–426.

Leins P and C Orth. 1979. Zur Entwicklungsgeschichte männlicher Blüten von *Humulus lupulus* (Cannabaceae). Bot. Jahrb. Syst. 100: 372–378.

Macdonald AD. 1974. Theoretical problems of interpreting floral organogenesis of *Laportea canadensis*. Canad. J. Bot. 52: 639–644.

Manchester SR. 1989. Systematics and fossil history of the Ulmaceae. In: PR Crane and S Blackmore, eds. Evolution, systematics, and fossil history of the Hamamelidae, vol. 2, pp. 221–251. Clarendon, Oxford.

Mehra PN and RS Gill. 1974. Cytological studies in Ulmaceae, Moraceae, and Urticaceae. J. Arnold Arbor. 55: 663–677.

Miller NG. 1970. The genera of the Cannabaceae in the southeastern United States. J. Arnold Arbor. 51: 185–203.

Miller NG. 1971. The genera of the Urticaceae in the southeastern United States. J. Arnold Arbor. 52: 40–68.

Mohan Ram HY and R Nath. 1964. The morphology and embryology of *Cannabis sativa* Linn. Phytomorphology 14: 414–429.

Oginuma K and H Tobe. 1995. Karyomorphology of some Moraceae and Cecropiaceae (Urticales). J. Plant Res. 108(1091): 313–326.

Oginuma K, PH Raven, and H Tobe. 1990. Karyomorphology and relationships of Celtidaceae and Ulmaceae (Urticales). Bot. Mag. (Tokyo) 103: 113–131.

Omori Y and S Tarabayashi. 1993. Gynoecial vascular anatomy and its systematic implications in Celtidaceae and Ulmaceae (Urticales). J. Plant Res. Jpn. 106: 249–258.

Pillay M and ST Kenny. 2006. Structural organization of the nuclear ribosomal RNA genes in *Cannabis* and *Humulus* (Cannabaceae). Plant Syst. Evol. 258: 97–105.

Punt W and E Eetnerink. 1982. On the pollen morphology of some genera of the tribe Moreae (Moraceae). Grana 21: 15–19.

Record SJ and RW Hess. 1940. American woods of the family Moraceae. Trop. Woods 61: 11–54.

Renner O. 1906. Beiträge zur Anatomic und Systematik der Artocarpeen und Conocephaleen insbesondere der Gattung *Ficus*. Bot. Jahrb. Syst. 39: 319–448.

Rohwer JG. 1993. Moraceae. In: K Kubitzki, ed. The families and genera of vascular plants, vol. 2, pp. 438–453. Springer, Berlin/Heidelberg/New York.

Shah AM and P Kachroo. 1975. Comparative anatomy in Urticales: I. The trichomes in Moraceae. J. Indian Bot. Soc. 54: 138 153.

Shanahan M, So S, Compton SG, and R Corlett. 2001. Fig-eating by vertebrate frugivores: a global review. Biol. Rev. 76: 529–572.

Sharp AJ. 1958. *Mirandaceltis*: a new genus from Mexico. Bol. Soc. Bot. Mexico 23: 38–42.

Singh SP. 1956. Floral anatomy of *Cannabis sativa* L. Agra Univ. J. Res. Sci. 5: 155–162.

Soepadmo E. 1977. Ulmaceae. In: CGGJ van Steenis, ed. Flora malesiana, ser. 1, 8: 31–76. Sijthoff & Noordhoff, Alphen aan den Rijn.

Sorsa P and P Huttunen. 1975. On the pollen morphology of the Urticaceae. Ann. Bot. Fenn. 12: 165–182.

Swarupanandan K. 1986. Late embryogenesis and morphology of mature embryos in three species of Dipterocarpaceae. Canad. J. Bot. 64: 2582–2587.

Sweitzer EM. 1971. Comparative anatomy of Ulmaceae. J. Arnold Arbor. 52: 523–585.

Sytsma KJ, E Conti, M Nepokroeff, JC Pires, Y-L Qiu, and MW Chase. 1996. Urticales: *rbc*L sequences clarify placement in Rosidae, composition, and familial relationships. Am. J. Bot. 83(Suppl. 6): 197.

Sytsma KJ, J Morawetz, JC Pires, M Nepokroeff, E Conti, M Zihra, JC Hall, and MW Chase. 2002. Utricalean rosids: circumscription, rosid ancestry, and phylogenetics based on *rbc*L, *trn*L-F, and *ndh*F sequences. Am. J. Bot. 89: 1531–1546.

Takahashi M. 1989. Pollen morphology of Celtidaceae and Ulmaceae: a reinvestigation. In: PR Crane and S Blackmore, eds. Evolution, systematics, and fossil history of the Hamamelidae, 2: 253–265. Clarendon, Oxford.

Takaso T and H Tobe. 1990. Seed coat morphology and evolution in Celtidaceae and Ulmaceae (Urticales). Bot. Mag. (Tokyo) 103: 25–41.

Terabayashi S. 1991. Vernation patterns in Celtidaceae and Ulmaceae (Urticales) and their evolutionary and systematic implications. Bot. Mag. (Tokyo) 104: 1–13.

Tippo O. 1938. Comparative anatomy of the Moraceae and their presumed allies. Bot. Gaz. 100: 1–99.

Tobe H and T Takaso. 1997. Trichome micromorphology in Celtidaceae and Ulmaceae (Urticales). Acta Phytotax. Geobot. 47(2): 153–168.

Todzia CA. 1993. Ulmaceae. In: K Kubitzki, ed. The families and genera of vascular plants, vol. 2, pp. 603–611. Springer, Berlin/Heidelberg/New York.

Todzia CA. 1989. A revision of Ampelocera. Ann. Missouri Bot. Gard. 76: 1087–1102.

Trecul A. 1847. Sur la famille des Artocarpees. Ann. Sci. Nat. Bot., ser. 3, 8: 38–157.

Ueda K, K Kosuge, and H. Tobe. 1997. A molecular phylogeny of Celtidaceae and Ulmaceae (Urticales) based on rbcL nucleotide sequences. J. Plant Res. 110(1098): 171–178.

Venkataraman K. 1972. Wood phenolics in the chemo-taxonomy of the Moraceae. Phytochemistry 11: 1571–1586.

Weddel HA. 1856. Monographic des Urticacees. Arch. Mus. Hist. Nat., Paris 9: 1–592.

Wiegrefe SJ, KJ Sytsma, and RP Guries. 1998. The Ulmaceae, one family or two? Evidence from chloroplast DNA restriction site mapping. Plant Syst. Evol. 210: 249–270.

Yamada H and O Yoshida. 1979. Embryological study of the family Ulmaceae: I. Embryology of *Zelkova serrata* Makino. J. Coil. Arts Chiba Univ. 12B: 27–43.
Zavada MS. 1983. Pollen morphology of Ulmaceae. Grana 22: 23–30.
Zavada MS and M Kim. 1996. Phylogenetic analysis of Ulmaceae. Plant Syst. Evol. 2000: 13–20.
Zhong Y, P Baas, and EA Wheeler. 1992. Wood anatomy of trees and shrubs from China: IV. Ulmaceae. IAWA Bull., n. s., 13: 419–453.

Superorder EUPHORBIANAE

Order 77. EUPHORBIALES

Trees and shrubs, woody lianas, or perennial and annual herbs. Plants often with milky or colored juice. Vessels with scalariform or more often simple perforations; lateral pitting alternate or rarely opposite, very rarely scalariform. Fibers mostly with simple pits. Rays heterogeneous of various types. Axial parenchyma mostly abundant and apotracheal or less often paratracheal, but sometimes scanty or even wanting. Sieve-element plastids of S-type. Nodes uni-, penta-, or multilacunar. Leaves mostly alternate, simple or less often compound, with pinnate or palmate venation, mostly stipulate. Stomata mostly paracytic. Flowers in very divers types of inflorescences, unisexual (monoecious or dioecious) or rarely bisexual, actinomorphic or rarely slightly zygomorphic, often apetalous, sometimes perianthless. Perianth 5-merous, rarely 4-or 3-merous. Sepals and petals free or less often connate at the base. Stamens 2-many, or sometimes fewer or even solitary, free or variously connate; anthers usually 2-locular, opening by longitudinal slits or sometimes transversely, rarely by apical pores. Tapetum secretory. Microsporogenesis simultaneous. Pollen grains 2-celled or 3-celled, 3-colpate, 3–4-colporate, pantoporate or sometimes inaperturate. Nectary disc often present. Gynoecium usually of three or less often two or four and sometimes many united carpels, rarely pseudomonomerous; ovary usually superior, 3-locular, rarely 2- or 4-locular or plurilocular, with free or more or less connate stylodia. Ovules two or one per locule, pendulous, anatropous, epitropous (the raphe ventral) or rarely (*Panda*) orthotropous, mostly bitegmic, crassinucellate or (Pandaceae) tenuinucellate. Female gametophyte monosporic, bisporic, or tetrasporic. Endosperm nuclear. Fruits of diverse types, but mostly capsules. Seed coat usually formed by both integuments; embryo straight or rarely curved; endosperm usually copious, rarely lacking or represented by one layer of cells.

The archaic members of Euphorbiales show obvious links with the Elaeocarpales and Malvales. Evidently the gynoecium was originally typically paracarpous (as in Elaeocarpales and Malvales) and became multilocular as a result of fusion of intrusive parietal placentas in the center of the ovary. On the other hand, the Euphorbiales have much in common with the archaic members of the Violales, particularly with the Flacourtiaceae. Hallier (1912: 217) included the Euphorbiaceae in his "Passionales" close to the Flacourtiaceae and noted that they originated from the latter *"pres de Pangiees et Idesiees."* Earlier (Hallier 1911: 5) he had emphasized the existence of numerous parallel forms between these two families, such as *Pangium* and *Aleurites*, *Hydnocarpus* and *Cyclostemon*, or *Ryparosa* and *Baccaurea*. There are many common features in wood anatomy, flower morphology, and palynology between Flacourtiaceae and Euphorbiaceae. According to Metcaife and Chalk (1950: 1230), "*Antidesma, Bischoffia*, and *Phyllanthus* are very similar to certain genera of the Flacourtiaceae, e.g., *Calancoba, Erythrospermum*, and *Kiggelaria; Acalypha, Aporosella, Glochidion*, and *Hymenocardia*, with small intervascular and vessel ray pitting, suggest the group of genera in the Flacourtiaceae that were formerly included in the Samydaceae. *Bridelia* and *Cleistanthus* closely resemble some genera of the Samydaceae, e.g., *Homalium* (see also Miller 1975). Keating (1973) has observed palynological similarities between the Flacourtiaceae and the Phyllanthaceae. There are also similarities in the leaf morphology, flower morphology, and seed anatomy (fibrous exotegmen of some Flacourtiaceae and some Euphorbiaceae, according to Corner [1976]). In his *Principia botanica* (1960: 436, Fig. 50), Croizat regarded links both with the Flacourtiaceae and the Sterculiaceae. There are indeed some similarities between the Euphorbiaceae and the Sterculiaceae, including petiole anatomy (Dehay 1935), the presence of branched multicellular hairs, the structure of the gynoecium, a uniformity in the exotegmic palisade of Euphorbiaceae-Crotonoideae, that, according to Corner (1976: 130), link them with Sterculiaceae and related families.

One may presume that the Euphorbiales have arisen either from some ancient group intermediate between the Flacourtiaceae and the Elaeocarpales and Malvales or directly from some Flacourtiaceae-like lower Violales. In both cases the Euphorbiales have only collateral affinities with the Elaeocarpales and Malvales.

Key to Families

1 Ovules with obturator.

2 Ovules two per locule.

3 Vessel elements with scalariform perforation, very rarely (*Glochidion*) with simple perforation.

4 Flowers in axillary or terminal spicate, racemose or paniculate inflorescences, or flowers solitary. Monoecious or rarely (*Antidesma*) dioecious herbs to trees. Laticifers absent. Nodes unilacunar with one trace. Leaves usually alternate (very rarely opposite), simple and unlobed, rarely (*Bischofia*) trifoliolate, entire, stipulate. Stomata paracytic or anisocytic. Flowers small. Calyx often basally conntate; petals five or absent, nectary disc present or absent. Stamens 2–35, often more or less connate; anthers extrorse. Pollen grains 2-celled, 3–4-colporate (rarely porate, pantoporate in *Phyllanthus*), semitectate, rarely echinate. Gynoecium of 1(2–5) to 15 carpels; style single or branched, stigmas usually bifid, fringed, wet; ovules hemitropous, bitegmic, micropyle bistomal, outer integument 2-many and inner integument 2–3(-5) cells across, nucellus 10 or more cells across, protruding. Fruits septicidal capsule, berries or drupes. Seeds large, without caruncle or sometimes with rudimentary caruncle; endosperm copious or absent. Present cyanogenesis via the tyrosine pathway, tropane and pyrrolizidine alkaloids, cucurbitacins [triterpenes], nonhydrolysable tannins [geraniin], in *Aleurites* detected inulin, n = 13, 14. . . 1. PHYLLANTHACEAE.

4 Flowers in axillary clusters, or cauliflorous. Evergreen trees or shrubs; hairs unicellular; Leaves alternate, entire or serrulate, 2-ranked, stipulate Flowers axillary, pedicelled, apetalous; male flowers: sepals 3–6-lobed, imbricate; stamens (2-)3–20 (-many); filaments free or subconnate; anthers extrorse or introrse; pollen grains 3-colporate; disc present or absent; female flowers: sepals 3–6-lobed, imbricate, gynoecium of 1[-4(-9)] carpels; ovary usually 2–3-locular; style branches short or absent, stigmas flap-like; ovules two in each locule, bitegmic, rarely (*Drypetes macrostigma*) unitegmic (Corner 1976); fruit drupaceous or dehiscing septicidally (*Lingelsheimia*) into two 3-valved cocci or six valves; seeds with sclereidal exomesotesta, exotegmen cells cuboidal; embryo straight, endosperm copious, cotyledons broad, flat and somewhat longer than the radicle. Contain cucurbitacins glucosinolates, mustard oils present in *Drypetes*, n = (19) 20 (21).. . . . . PUTRANJIVACEAE.

3 Vessels with simple perforations. Monoecious or dioecious trees. Mucilage cells present in epidermis. Axial parenchyma apotracheal and paratracheal; prismatic crystals in wood parenchyma present. Leaves alternate, long petiolate, trifoliolate, leaflets oblong-elliptic, entire; stipules very minute, with colleters, caduceus or persistent. Stomata paracytic. Flowers small, axillary, solitary (the female flowers, each on an apically expanded, long peduncles), or aggregated in axillary panicles or in catkins; perianth in male flowers absent, stamens 3–54, loosely clustered and forming a globose head; filaments free, short; anthers minutely pubescent, 2-locular, slightly extrorse, opening longitudinally. Pollen grains spinulose, 6-colporate. Female flowers on slender pedicels; sepals 4–5, narrowly lanceolate, unequal, valvate or imbricate; nectariferous weak disk around the base of the ovary. Gynoecium of two carpels; ovary superior, 2-locular; stigmas stout, 2-lobed; ovules two per locule, pendulous, anatropous, with obturators. Walls of the ovary, the style, the perianth and pedicel have large cavities with a viscous substance. Fruits globous drupes, with a thin, fleshy, orange pericarp containing numerous vesicles of bitter juice; seeds two or one by abortion. Seeds massive, embryo bent, with rather large and leaf-like cotyledons without a plumule, much corrugated, with very scanty endosperm. Present picrotoxanes. . . . . 3. PICRODENDRACEAE.

2 Ovules one. Trees, shrubs, subshrubs, or herbs (some climbing or twining), very rarely floating aquatics. Indumentum simple, sometimes urticating, amalpighiaceous, dendritic, stellate, lepidote, or absent. Stems sometimes succulent and/ or with latex. Vessels usually with simple perforations, but sometimes with scalariform ones. Fibers with bordered or simple pits. Rays uniseriate and multiseriate. Axial parenchyma apotracheal, sometimes wanting. Nodes mostly pentalacunar with five traces. Leaves usually alternate, sometimes opposite, rarely verticillate, entire to dentate or palmately lobed or compound, with pinnate or palmate venation; stipules usually present, sometimes transformed into glands or spines, or absent. Stomata of various types, but mostly paracytic. Flowers in terminal or axillary, basically cymose inflorescences, always unisexual (monoecious or dioecious), sometimes individually very much reduced and grouped into pseudanthia, often apetalous, actinomorphic (but pseudanthia often zygomorphic). Perianth segments free or connate, valvate, imbricate, or open, sometimes reduced or absent; sepals and petals (1-)3–6(-8), sometimes distinctly colored. Nectary disc mostly present, intrastaminal or extrastaminal, annular, cupular or of separate glands. Stamens one (Euphorbieae), 3–100 (up to 1,000 in *Ricinus*); filaments free or connate; anthers introrse, latrorse or extrorse. Pollen grains 2- or 3-celled, tectate or semitectate, often reticulate, mostly 3-colporate, very diverse in exine ornamentation. Gynoecium of (2)3(4-many) carpels, rarely pseudomonomerous; stylodia free or more or less united, entire, bifid, multifid or laciniate; ovary superior, sessile or rarely stipitate, (1-)2–5(-20)-locular, most commonly 3-locular, with ovule per locule; styles (1)3–4(20), free or connate, erect or spreading, entire, 2-lobed, -fid or –partite to multifid or laciniate; the adaxial surface usually stigmatic throughout, smooth, granulate, papillose, plumose or fimbriate. Ovules anatropous or less often hemitropous or amphitropous, epitropous, bitegmic; the nucellus usually forms a beak that extends beyond the inner integument and comes in contact with the placental obturator, which forms a roof over the exo(bi)stomal, micropyle. Female gametophyte of diverse types, but generally of *Polygonum*-type. Fruits typically capsular schizocarps, primarily septicidal, with carpels (cocci) elastically dehiscent from a persistent columella, but sometimes drupaceous or baccate, very rarely samaroid. Seeds one or ttwo in each locule, exotegmic, mostly with a micropylar caruncle; embryo straight to curved or folded, usually with broad and flat cotyledons; endosperm mostly well developed, less often wanting, Plants Al-accumulators, producing cucurbitacins (triterpenes), ellagitannins (geraniin and mallotussic acid), lectins (hemagglutinins), cocarcinogens (phorbol ester diterpenes), antidesmone and related compounds; n = (5)6–9(-11), 12–14, 19. . .4. EUPHORBIACEAE.

1 Ovules without obturator. Seeds not carunculate. Dioecious trees or shrubs, indumentum simple. Vessels with scalariform or both scalariform and simple perforations. Leaves alternate, distichous, simple, often serrate, entire or toothed; stipules small, generally persistent. Inflorescences terminal or cauliflorous and thyrsiform, or axillary and fasciculate; bracts minute. Sepals free or connate, imbricate or open. Petals slightly imbricate to valvate, often hooded. Stamens 5–15, free or nearly so; anthers usually introrse. Nectary disc very small or absent, rarely developed (*Centroplacus*). Gynoecium of 2–5 carpels with 2–5(10) lobed style. Ovules pendulous, anatropous and epitropous (the raphe ventral) or (*Panda*) orthotropous, two (*Centroplacus*) or one in each locule. Fruits capsular (*Centroplacus*) or drupaceous; exocarp fleshy, tartareous or woody; seeds ecarunculate; exotesta and endotegmen tanniniferous, endosperm copious; cotyledons thin and flat, oily. n = 15. . . . . 5. PANDACEAE.

1. PHYLLANTHACEAE

Martynov 1820 (including Antidesmataceae Loudon 1830, Aporosaceae Lindley ex Planchon 1854, Bischofiaceae Airy Shaw 1965, Hymenocardiaceae Airy Shaw 1965, Porantheraceae Hurusawa 1954, Scepaceae Lindley 1836, Stilaginaceae C.A. Agardh 1824, Uapacaceae Airy Shaw 1965). 59/1745. Pantropical, but mainly in Malesia.

WIELANDIEAE: *Heywoodia, Savia, Gonatogyne, Petalodiscus, Blotia, Actephila, Discocarpus, Lachnostylis, Chonocentrum, Wielandia*; AMANOEAE: *Pentabrachion, Amanoa;* BRIDELIEAE: *Cleistanthus, Bridelia;* PHYLLANTHEAE: *Astrocasia, Leptopus, Chascotheca, Zimmermannia,*

Zimmermanniopsis, Meineckia, Pseudolachnostylis, Keayodendron, Securinega, Andrachne, Flueggea, Richeriella, Margaritaria, Phyllanthus, Reverchonia, Sauropus, Breynia, Glochidion; ANTIDESMEAE: *Spondianthus, Uapaca, Protomegabaria, Maesobotrya, Richeria, Jablonskia, Baccaurea, Ashtonia, Aporusa, Thecacoris, Phyllanoa, Celianella, Leptonema, Antidesma, Aerisilvaea, Meborea, Hyeronima, Poranthera, Oreoporanthera;* TACARCUNEAE: *Tacarcuna*; HYMENOCARDIEAE: *Didymocistus, Hymenocardia;* BISCHOFIEAE: *Bischofia*; MARTRETIEAE: *Martretia*; DICOELIEAE: *Dicoelia*; CROIZATIEAE: *Croizatia.*

The family Phyllanthaceae is the most archaic member of the order.

2. PUTRANJIVACEAE

Endlicher 1841. 4/215. Tropical America, tropical and South Africa, South, Southeast and East Asia, Malesia, Australia.

Lingelsheimia, Drypetes, Sibangea, Putranjiva.

Related to the Phyllanthaceae. Radcliffe-Smith (2001) included all four genera in tribe Drypeteae (Phyllanthoideae).

3. PICRODENDRACEAE

J.K. Small 1917. 1/3. West Indies (Greater Antilles, Bahamas, Cayman and Swan Islands).

Picrodendron.

"The anomalous structure of the fruit and seed, as well as the male inflorescence, seems to indicate that, despite pollen and other similarities, *Picrodendron* does not really sit comfortably in the Oldfieldioideae, nor indeed in the Euphorbiaceae." (Radcliffe-Smith 2001).

4. EUPHORBIACEAE

A.L. de Jussieu 1789 (including Acalyphaceae A.L. de Jussieu ex Menge 1839, Androstachyaceae Airy Shaw 1965, Bertyaceae J.G. Agardh 1858, Cheilosaceae Doweld 2001, Crotonaceae J.G. Agardh 1858, Hippomanaceae J.G. Agardh 1858, Mercurialaceae Martynov 1820, Micrantheaceae J.G. Agardh 1858, Paivaeusaceae A. Meeuse 1990, Peraceae Klotzsch 1859, Pseudanthaceae Endlicher ex Pfeiffer 1873, Ricinaceae Martynov 1820, Ricinocarpaceae Hurusawa 1954, Tithymalaceae Ventenat 1799, Tragiaceae Rafinesque 1838, Trewiaceae Lindley 1836). 276/c.6000. Subcosmopolitan, although predominantly tropical; there are strong local concentrations (particularly of the genus *Euphorbia*) in South Africa, the Mediterranean and Irano-Turanian regions, and southern North America.

4.1 STACHYSTEMONOIDEAE (OLDFIELDIOIDEAE)

Laticifers absent. Leaves stipulate or estipulate, alternate or more often opposite or verticillate, simple or palmately compound, entire or dentate. Petals absent (except in *Croizatia*). Nectary disc present or absent. Pollen grains 2-celled, tectate, brevicolporate or porate, usually distinctly spinulose. Ovules two per locule (except in *Scagea*). Seeds often carunculate. Endosperm usually copious (except in *Hyaenanche*). – PODOCALYCEAE: *Podocalyx, Tetracoccus, Paradrypetes;* OLDFIELDEAE: *Piranhea, Parodiodendron, Oldfieldia, Aristogeitonia, Mischodon, Voatamalo, Androstachys, Stachyandra*; CALETIEAE: *Hyaenanche, Austrobuxus, Myladenia, Canaca, Dissiliaria, Sankowskia, Whyanbeelia, Choriceras, Longetia, Petalostigma, Kairothamnus, Celaenodendron, Scagea, Neoroepera, Micrantheum, Pseudanthus, Stachystemon.*

4.2 PEROIDEAE

Laticifers absent. Trees, shrubs or herbs, with stellate or lepidote or occasionally simple indumentum, more or less glabrous. Leaves alternate or very rately opposite (in *Pera*), sometimes pellucid-punctate, stipulate or estipulate. Inflorescences axillary, without or rarely with elongated axed, surrounded by involucral bracts in *Pera* only. Petals and nectary disc absent or present. Stamens 2–20, free to connate. Pollen grains 3–4-colporate, mostly tectate-perforate, sometimes almost tectate or finely reticulate. Ovary 3(-4)-locular, styles bifid to bipartite. Ovule one per locule. Fruits dehiscent, septa membranous, fragile, valves often remaining attached to the base of the columella after dehiscence. Seeds carunculate or arillate; endosperm copious, rarely (*Trigonopleura*) scanty; cotyledons longer and wider than radicle. – CLUTIEAE: *Clutia;* POGONOPHOREAE: *Pogonophora;* CHAETOCARPEAE: *Trigonopleura, Chaetocarpus;* PEREAE: *Pera.*

4.3 CHEILOSOIDEAE

Laticifers absent. Indumentum simple and/or stellate. Leaves alternate, simple, unlobed, usually glandular at the base of the lamina, stipulate. Inflorescences pseudoterminal or axillary. Petals absent; nectary disc present in male flowers, present or absent in female flowers. Stamens 5–12, free; anthers introrse. Pollen grains 3-colporate, echinate. Ovary 2–4-locular, styles bifid. Ovule one per locule. Fruits tardily loculicidally dehiscent. Seeds not carunculate but with sarcotesta, endosperm present. – *Cheilosa, Neoscortechinia.*

4.4 ACALYPHOIDEAE

Laticifers generally absent (if present they are inarticulate). Leaves usually stipulate, simple and entire, dentate or palmately lobed, rarely compound, alternate or rarely opposite, often with foliar glands. Petals and nectary disc present or absent. Pollen grains 2-celled, mostly 3–4-colporate, semitectate, very rarely echinate. Ovules one per locule. Seeds carunculate or not. Endosperm usually copious. – ERISMANTHEAE: *Erismanthus, Moultonianthus, Syndyophyllum;* AMPEREAE: *Monotaxis, Amperea;* AGROSTISTACHYDEAE: *Agrostistachys, Pseudagrostistachys, Cyttaranthus, Chondrostylis*; SPHYRANTHEREAE: *Sphyranthera*; CHROZOPHOREAE: *Speranskia, Caperonia, Philyra, Ditaxis, Argythamnia, Chiropetalum, Doryxylon, Sumbaviopsis, Thyrsanthera, Melanolepis, Chrozophora*; CARYODENDREAE: *Caryodendron, Discoglypremna, Alchorneopsis*; BERNARDIEAE: *Bernardia, Necepsia, Amyrea, Paranecepsia, Afrotrewia, Discocleidion, Adenophaedra*; PYCNOCOMEAE: *Pycnocoma, Droceloncia, Argomuellera, Blumeodendron, Podadenia, Ptychopyxis, Botryophora*; EPIPRINEAE: *Epiprinus, Symphyllia, Cleidiocarpon, Koilodepas, Cladogynos, Cephalocrotonopsis, Cephalocroton, Adenochlaena, Cephalomappa*; ADELIEAE: *Adelia, Crotonogynopsis, Enriquebeltrania, Lasiocroton, Leucocroton*; ALCHORNEEAE: *Orfilea, Bossera, Alchornea, Caelebogyne, Aparisthmium, Bocquillonia, Conceveiba, Gavarretia, Polyandra*; ACALYPHEAE: *Ricinus, Adriana, Mercurialis, Seidelia, Leidesia, Dysopsis, Wetria, Cleidion, Sampantaea, Macaranga, Erythrococca, Claoxylon, Claoxylopsis, Discoclaoxylon, Micrococca, Lobanilia, Mareya, Mareyopsis, Mallotus, Deuteromallotus, Cordemoya, Coccoceras, Avellanita, Trewia, Neotrewia, Rockinghamia, Octospermum, Acalypha, Lasiococca, Spathiostemon, Homonoia*; PLUKENETIEAE: *Haematostemon, Astrococcus, Angostylis, Romanoa, Eleutherostigma, Plukenetia, Vigia, Cnesmone, Megistostigma, Sphaerostylis, Tragia, Tragiella, Platygyne, Acidoton, Gitara, Pachystylidium, Dalechampia*; OMPHALEAE: *Omphalea.*

4.5 CROTONOIDEAE

Articulated and/or inarticulated laticifers usually present. Leaves stipulate or estipulate, simple and entire, dentate or palmately lobed or compound, alternate, opposite, or rarely verticillate. Petals and nectary disc present or absent. Pollen grains 2-celled or 3-celled, 3-colporate to more commonly porate or inaperturate, mostly with ornamentation of knobs in hexagonal pattern. Ovule one per locule. Seeds often carunculate. Endosperm usually copious, often oily. – MICRANDREAE: *Micrandra, Micrandropsis, Cunuria, Hevea;* MANIHOTEAE: *Manihot, Cnidoscolus;* ADENOCLINEAE: *Glycydendron, Klaineanthus, Tetrorchidium, Adenocline, Ditta, Endospermum*; GELONIEAE: *Suregada, Cladogelonium;* ELATERIOSPERMEAE: *Elateriospermum;* JATROPHEAE: *Jatropha, Vaupesia, Oligoceras, Deutzianthus, Loerzingia, Joannesia, Leeuwenbergia, Annesijoa*; CODIAEAE: *Baloghia, Hylandia, Ostodes, Pausandra, Dodecastigma, Pantadenia, Parapantadenia, Dimorphocalyx, Fontainea, Codiaeum, Acidocroton, Ophellantha, Blachia, Strophioblachia, Sagotia, Baliospermum*; TRIGONOSTEMONEAE: *Trigonostemon;* RICINOCARPEAE: *Ricinocarpos, Alphandia, Beyeria, Bertya, Myricanthe, Cocconerion, Borneodendron*; CROTONEAE: *Mildbraedia, Chylamydojatropha, Paracroton, Moacroton, Apodiscus, Croton, Crotonopsis, Eremocarpus, Julocroton*; RICINODENDREAE: *Givotia, Ricinodendron, Schinziophyton;* ALEURITIDEAE: *Aleurites, Reutealis, Vernicia, Garcia, Cavacoa, Grossera, Tapoides, Anomalocalyx, Sandwithia, Tannodia, Neoholstia, Domohinea, Cyrtogonone, Crotonogyne, Manniophyton, Neoboutonia, Benoistia.*

4.6 EUPHORBIOIDEAE

Laticifers inarticulate. Leaves mostly stipulate (except in *Euphorbia*), simple entire, dentate or lobed, alternate (rarely opposite). Sepals mostly valvate (rarely imbricate). Petals and usually nectary disc absent. Pollen grains 2-celled or 3-celled, 3-colporate. Ovules one per locule. Seeds carunculate or not. Endosperm copious. – STOMATOCALYCEAE: *Plagiostyles, Pimelodendron, Hamilcoa, Nealchornea*; HIPPOMANEAE: *Homalanthus, Dendrocousinsia, Excoecaria, Spirostachys, Sebastiania, Colliguaja, Microstachys, Conosapium, Grimmeodendron, Bonania, Adenopeltis, Stillingia, Spegazziniophytum, Falconeria, Sapium,*

Hippomane, Dendrothrix, Senefelderopsis, Pleradenophora, Sclerocroton, Balakata, Triadica, Shirakiopsis, Mabea, Gymnanthes, Ditrysinia, Anomostachys, Neoshirakia, Pseudosenefeldera, Actinostemon, Senefeldera, Rhodothyrsus, Dalembertia, Maprounea; PACHYSTROMATEAE: *Pachystroma;* HUREAE: *Algernonia, Tetraplandra, Ophthalmoblapton, Hura*; EUPHORBIEAE: *Anthostema, Dichostemma, Neoguillauminia, Calycopeplus, Euphorbia, Elaeophorbia, Chamaesyce, Cubanthus, Monadenium, Synadenium, Endadenium, Pedilanthus.*

The Euphorbiaceae are one of the most diversified angiosperm families.

5. PANDACEAE

Engler et Gilg 1912–1913. 4/18. Tropical Africa, Southeast Asia to Solomon Islands, New Guinea.

CENRTOPLACEAE: *Centroplacus*; GALEARIEAE: *Galearia, Microdesmis, Panda*

It is widely accepted that the Pandaceae belong to the order Euphorbiales and Webster (1994), Thorne (2006) and Radcliffe-Smith (2001) even include them in the Euphorbiaceae in the subfamily Acalyphoideae. However Stuppy (1996) considers that from the ovule- and seed-morphology, the Pandaceae should be re-instated.

Bibliography

Airy Shaw HK. 1965. Diagnoses of new families, new names, etc. for the seventh edition of Willis's "Dictionary." Kew Bull. 18: 249–273.

Baillon H. 1874. Euphorbiaceae: histoire des plantes. 5: 105–256. Paris.

Bentham G. 1878. Notes on Euphorbiaceae. J. Linn. Soc. Bot. (London) 17: 185–267.

Bentham G. 1880. Euphorbiaceae. In: G Bentham and JD Hooker, eds. Genera plantarum, 3: 239–340. L. Reeve, London.

Berry PE, AL Hipp, KJ Wurdack, B van Ee, and R Riina. 2005. Molecular phylogenetics of the giant genus *Croton* and tribe Crotoneae (Eurphorbiaceae sensu stricto) using ITS and *trn*L-*trn*F DNA sequence data. Am. J. Bot. 92: 1520–1534.

Beutler JA, AB Alvarado-Lindner, TG McCloud, and GM Cragg. 1989. Distribution of phorbol ester bioactivity in the Euphorbiaceae. Phytotherapy Res. 3: 188–192.

Beutler JA, AB Alvarado-Lindner, TG McCloud, and GM Cragg. 1996. Further studies on phorbol ester bioactivity in the Euphorbiaceae. Ann. Missouri Bot. Gard. 83: 53–533.

Bhathnagar AK and RN Kapil. 1974. *Bischofia javanica:* its relationship with Euphorbiaceae. Phytomorphology 23: 264–267.

Bhathnagar AK and RN Kapil. 1979. Ontogeny and taxonomic significance of anther in *Bischofia javanica*. Phytomorphology 29: 298–306.

Buske A, J Schmidt, and P Hoffmann. 2002. Chemotaxonomy of the tribe Antidesmeae (Euphorbiaceae): antidesmone and related compounds. Phytochemistry 60: 489–496.

Croizat L. 1940. On the phylogeny of the Euphorbiaceae and some of their presumed allies. Revista Univ. (Universidad Catolica de Chile, Santiago) 25: 205–220.

Croizat L. 1960. Principia botanica. 2 vols. Caracas.

Dang-Van-Liem. 1962. Recherches sur lɘεμβρψoγεvιε des Tricoques. Thesis, University of Paris.

Dehay C. 1935. L'Appareil libero-ligneux foliaire des Euphorbiacees. Ann. Sci. Nat. Bot., 10th ser., 17: 147–295.

Dehay C. 1938. Les affinités entre les Euphorbiales, les Morales et les Malvales, d'aprés l'appareil libéro-ligneux foliare. Bull. Soc. Bot. France 85: 23–31.

Esser H-J. 1999. A partial revision of the Hippomaneae (Euphorbiaceae) in Malesia. Blumea 44: 149–215.

Evans FJ and SE Taylor. 1983. Pro-inflammatory, tumour-promoting and anti-tumour diterpenes of the families Euphorbiaceae and Thymelaeaceae. Prog. Chem. Org. Natural Prod. 44: 1–99.

Fernández-Gonzáles D and D Lobreau-Callen. 1996. Le pollen de la tribu des Acalyphea (Acalyphoideae, Euphorbiaceae). Grana 35: 266–284.

Forman LL. 1966. The reinstatement of *Galearia* Zoll. et Mor. and *Microdesmis* Hook. f. in the Pandaceae. With appendices by CR Metcalfe and N Parameswaran. Kew Bull. 20: 309–321.

Forman LL. 1968. The systematic position of *Panda* Pierre. Proc. Linn. Soc. London 179: 269–270.

Gilbert MC. 1994. The relationships of the Euphorbieae (Euphorbiaceae). Ann. Missouri Bot. Gard. 81: 283–288.

Gill BS, SS Bir, and YS Bedi. 1981. Cytological studies on woody Euphorbiaceae from north and central India. New Botanist 8: 35–44.

Gillespie LJ. 1994. Pollen morphology and phylogeny of the tribe Plukenetieae (Euphorbiaceae). Ann. Missouri Bot. Gard. 81: 317–348.

Govaerts R, DG Frodin, and A Radcliffe-Smith. 2000. World Checklist and Bibliography of Euphorbiaceae. 4 vols. Royal Botanic Gardens, Kew.

Hakki MI. 1985. Studies on West Indian plants: 3. On floral morphology, anatomy, and relationship of *Picrodendron* (L.) Krug and Urban (Euphorbiaceae). Bot. Jahrb. Syst. 107: 379–394.

Hans AS. 1973. Chromosomal conspectus of the Euphorbiaceae. Taxon 22: 591–636.

Hayden WJ. 1977. Comparative anatomy and systematics of *Picrodendron,* genus incertae sedis. J. Arnold Arbor. 58: 257–279.

Hayden WJ. 1987. The identity of the genus *Neowawraea* (Euphorbiaceae). Brittonia 39: 268–277.

Hayden WJ. 1994. Systematic anatomy of Euphorbiaceae subfamily Oldfieldioideae: I. Overview. Ann. Missouri Bot. Gard. 81: 180–202.

Hayden WJ and DS Brandt. 1984. Wood anatomy and relationships of *Neowawraea* (Euphorbiaceae). Syst. Bot. 9: 458–466.

Hayden WJ and SM Hayden. 2000. Wood anatomy of Acalyphoideae (Euphorbiaceae). IAWA J. 21: 213–235.

Hayden WJ, WT Gillis, DE Stone, CR Broome, and GL Webster. 1984. Systematics and phylogeny of *Picrodendron*. Further evidence for relationship with the Oldfieldioideae (Euphorbiaceae). J. Arnold Arbor. 65: 105–127.

Hoffmann P, H Kathriarachchi, and KJ Wurdack. 2006. A phylogenetic classification of Phyllanthaceae (Malpighiales; Euphorbiaceae sensu lato). Kew Bull. 61: 37–53.

Hurusawa I. 1954. Eine nochmalige Durchsicht des her-kommlichen Systems der Euphorbiaceen im weiteren Sinne. J. Fac. Sci. Univ. Tokyo, Sect. 3, Bot. 6: 209–342.

Hutchinson J. 1969. Tribulism in the family Euphorbiaceae. Am. J. Bot. 56: 738–758.

Janssonius HH. 1929. A contribution to the natural classification of the Euphorbiaceae. Trop. Woods 19: 8–11.

Jensen U, I Vogel-Bauer, and M Nitsche. 1994. Leguminlike proteins and the systematics of the Euphorbiaceae. Ann. Missouri Bot. Gard. 81: 160–170.

Jury SL, T Reynolds, DF Cutler, and FJ Evans, eds. 1987. The Eurphorbiales: chemistry, taxonomy, and economic botany. Academic Press, London/Toronto.

Kapil RN and AK Bhatnagar. 1994. The contribution of embryology to the systematics of the Euphorbiaceae. Ann. Missouri Bot. Gard. 81: 145–159.

Kathriarachchi H, P Hoffmann, R Samuel, KJ Wurdack, and MW Chase. 2005. Molecular phylogenetics of Phyllanthaceae inferred from five genes (plastid *atp*B, *mat*K, 3' *ndh*F, *rbc*L, and nuclear *PHYC*). Molec. Phylog. Evol. 36: 112–134.

Kato M, A Takimura, and A Kawakita. 2003. An obligate pollination mutualism and reciprocal diversification in the tree genus *Glochidion* (Euphorbiaceae). Proc. Nation. Acad. Sci. USA. 100: 5264–5267.

Klucking EP. 1997. Leaf venation patterns. Vol. 8. Euphorbiaceae. Part I. Phyllanthoideae and Oldfieldioideae. J. Cramer, Berlin.

Köhler E. 1965. Die Pollenmorphologie der biovulaten Euphorbiaceae und ihre Bedeutung für die Taxonomie. Grana Palynol. 6: 26–120.

Komar GA. 1992. Euphorbiales. In: A Takhtajan, ed. Comparative seed anatomy, vol. 4, pp. 382–404. Nauka, St. Petersburg (in Russian).

Levin GA. 1986a. Systematic foliar morphology of Phyllanthoideae (Euphorbiaceae): I. Conspectus. II. Phenetic analysis. Ann. Missouri Bot. Gard. 73: 29–85, 86–98.

Levin GA. 1986b. Systematic foliar morphology of Phyllanthoideae (Euphorbiaceae). III. Cladistic analysis. Syst. Bot. 11: 515–530.

Levin GA and MG Simpson. 1994a. Phylogenetic implications of pollen ultrastructure in the Oldfieldioideae (Euphorbiaceae). Ann. Missouri Bot. Gard. 81: 203–238.

Levin GA and MG Simpson. 1994b. Phylogenetic relationships of *Didymocistus* and *Hymenocardia* (Euphorbiaceae). Ann. Missouri Bot. Gard. 81: 239–244.

Lobreau-Callen D and M Suarez-Cervera. 1994. Pollen ultrastructure of *Hymenocardia* Wallich ex Lindley and comparison with other Euphorbiaceae. Rev. Paleobot. Palynol. 81: 257–258.

Lobreau-Callen D and M Suarez-Cervera. 1997. Le pollen des Crotonoideae apétales (Euphorbiaceae): ultrastructure de l'exine. (Pollen exine ultrastructure of the apetalous Crotonoideae). Rev. Paleobot. Palynol. 98: 257–291.

Lobreau-Callen D, V Malecot, and M Suarez Cervera. 2000. Comparative study of pollen from apetalous Crotonoideae and some other uniovulate Euphorbiaceae: exine ulstrastructure at the aperture. In: MM Harley, CM Morton, and S Blackmore, eds. Pollen and spores: morphology and biology, pp. 301–324. Royal Botanic Gardens, Kew.

Mahlberg PG, DG Davis, DS Galitz, and GD Manners. 1987. Laticifers and the classification of *Euphorbia:* the chemotaxonomy of *Euphorbia esula* L. Bot. J. Linn. Soc. 94: 165–180.

Meeuse ADJ. 1990. The Euphorbiaceae auct. plur.: an unnatural taxon. Eburon, Delft.

Mennega AMW. 1987. Wood anatomy of the Eurph-orbiaceae, in particular of the subfamily Phyllan-thoideae. Bot. J. Linn. Soc. 94: 111–126.

Mennega AMW. 2005. A comparison with subfamilies Crotonoideae and Acalyphoideae and the implications for the circumscription of the Euphorbiaceae. IAWA J. 26: 1–68.

Michaelis P. 1924. Blütenmorphologische Untersuchungen an den Euphorbiaceen. Bot. Abhandlungen (Jena) 3: 1–150.

Nair N and V Abraham. 1962. Floral morphology of a few species of Euphorbiaceae. Proc. Indian Acad. Sci. 56B: 1–12.

Nowicke JW. 1984. A palynological study of the Pandaceae. Pollen et Spores 26: 31–42.

Nowicke JW. 1994. A palynological study of Crotonoideae (Euphorbiaceae). Ann. Missouri Bot. Gard. 81: 245–269.

Nowicke JW and M Takahashi. 2002. Pollen morphology, exine structure and systematics of Acalyphoideae (Euphorbiaceae), part 4. Tribes Acalypheae *pro parte* (*Erthyrococca, Claoxylon, Claoxylopsis, Mareya, Mareyopsis, Discoclaoxylon, Micrococca, Amyrea, Lobanilia, Mallotus, Deuteromallotus, Cordemoya, Coccoceras, Trewia, Neotrewia, Rockinghamia, Octospermum, Acalypha, Lasiococca, Spathiostemon, Homonoia*), Plukenetieae (*Haematostemon, Astrococcus, Angostyles, Romanoa, Eleutherostigma, Plukenetia, Vigia, Cnesmone, Megistostigma, Sphaerostylis, Tragiella, Platygyna, Tragia, Acidoton, Pachystlidium, Dalechampia*), Omphaleae (*Omphalea*), and discussion and summary of the complete subfamily. Rev. Palaeaobot. Palynol. 121: 231–336.

Nowicke JW, M Takahashi, and GL Webster. 1998, 1999. Pollen morphology, exine structure and systematics of Acalyphoideae (Euphorbiaceae). Part 1. Tribes Clutieae (*Clutia*), Pogonophoreae (*Pogonophora*), Chaetocarpeae (*Chaetocarpus, Trigonopleura*), Pereae (*Pera*), Cheiloseae (*Cheilosa, Neoscortechinia*), Dicoelieae (*Dicoelia*), Galearieae (*Galearia, Microdesmis, Panda*) and Amperea (*Amperea, Monotaxis*). Part II. Tribes Agrostistachydeae, Chrozophoreae, Caryodendreae, Bernardieae, and Pycnocomeae. Rev. Palaeobot. Palynol. 102: 115–152; 105: 1–62.

Park KR and A Backlund. 2002. Origin of the cyathium-bearing Euphorbieae (Euphorbiaceae): phylogenetic study based on morphological characters. Bot. Bull. Acad. Sinica 43: 57–62.

Park KR and WJ Elisens. 2000. A phylogenetic study of tribe Euphorbieae (Euphorbiaceae). Int. J. Plant Sci. 161: 425–434.

Pax E 1924. Die Phylogenie der Euphorbiaceae. Bot. Jahrb. Syst. 59: 129–182.

Punt W. 1962. Pollen morphology of the Euphorbiaceae with special reference to taxonomy. Wentia 7: 1–116.

Punt W. 1987. A survey of pollen morphology in Euphorbiaceae with special reference to *Phyllanthus*. Bot. J. Linn. Soc. 94: 127–142.

Radcliffe-Smith A. 1987. Segregate families from the Euphorbiaceae. Bot. J. Linn. Soc. 94: 47–66.

Radcliffe-Smith A. 2001. Genera Euphorbiacearum. Royal Botanical Gardens, Kew.
Rajn VS and PN Rao. 1977. Variation in the structure and development of foliar stomata in the Euphorbiaceae. Bot. J. Linn. Soc. 75: 69–97.
Rao PN. 1970. Euphorbiaceae. In: BM Johri, ed. Symposium on comparative embryology of angiosperms. Bull. Indian Nad. Acad. 41: 136–141.
Record SJ. 1938. The American woods of the family Euphorbiaceae. Trop. Woods 54: 7–40.
Rizk A-FM. 1987. The chemical constituents and economic plants of the Euphorbiaceae. Bot. J. Linn. Soc. 94: 293–326.
Rudall PJ. 1987. Laticifers in Euphorbiaceae: a conspectus. Bot. J. Linn. Soc. 94: 143–163.
Rudal PJ. 1994. Laticifers in Crotonoideae (Euphorbiaceae): homology and evolution. Ann. Missouri Bot. Gard. 81: 270–282.
Sagun VG, and RWJM van der Ham. 2003. Pollen morphology of the Flueggeinae (Euphorbiaceae, Phyllanthoideae). Grana 42: 193–219.
Samuel R, H Kathriarachchi, P Hoffmann, MHJ Barfuss, KJ Wurdack, CC Davis, and MW Chase. 2005. Molecular phylogenetics of Phyllanthaceae: evidence from plastid *mat*K and nuclear *PHYC* sequences. Am. J. Bot. 92: 132–141.
Seigler DS. 1994. Phytochemistry and systematics of the Euphorbiaceae. Ann Missouri Bot. Gard. 81: 380–401.
Simpson MG and GA Levin. 1994. Pollen ultrastructure of the biovulate Euphorbiaceae. Int. J. Plant Sci. 155: 313–341.
Singh RP. 1954. Structure and development of seeds in Euphorbiaceae: *Ricinus communis* L. Phytomorphology 4: 118–123.
Steinmann V and JM Porter. 2002. Phylogenetic relationships in Euphorbioideae (Euphorbiaceae) based on ITS and *ndh*F sequence data. Ann. Missouri Bot. Gard. 89: 453–490.
Stuppy W. 1996. Systematische Anatomie und Morphologie der Samen der Biovulaten Euphorbiaceen. Dissertation. University of Kaiserslautern, Germany.
Suarez Cervera M, L Gillespie, E Arcalis, A Le Thomas, D Lobreau Callen, and JA Seoane Camba. 2001. Taxonomic significance of sporoderm structure in pollen of Euphorbiaceae: tribes Plukenetieae and Euphorbieae. Grana 40: 78–104.
Sutter D and PE Endress. 1995. Aspects of gynoecium structure and macrosystematics in Euphorbiaceae. Bot. Jahrb. Syst. 116: 517–536.
Sutter DM, PI Forster, and PK Endress. 2006. Female flowers and systematic position of Picrodendraceae (Euphorbiaceae s.l., Malpighiales). Plant Syst. Evol. 261: 187–215.
Takahashi M, JW Nowicke, and GL Webster. 1995. A note on remarkable exines in Acalyphoideae (Euphorbiaceae). Grana 34: 282–290.
Takahashi M, JW Nowicke, GL Webster, SS Orli, and S Yankowski. 2000. Pollen morphology, exine structure, and systematics of Acalyphoideae (Euphorbiaceae), part 3: tribes Epiprineae, Alchorneae, Acalypheae pro parte. Rev. Palaeobot. Palynol. 110: 1–66.
Thakur HA and DA Patil. 2002. Nodal organization in some Euphorbiaceae. J. Swamy Bot. Club 19: 59–62.
Tokuoka T and H Tobe. 1993. Embryology and systematics of Euphorbiaceae sensu lato. A review and perspective. J. Plant Res. 108: 97–106.
Tokuoka T and H Tobe. 1998. Ovules and seeds in Crotonoideae (Euphorbiaceae): structure and systematic implications. Bot. Jahrb. Syst. 120: 165–186.
Tokuoka T and H Tobe. 1999. Embryology of tribe Drypeteae, an enigmatic taxon of Euphorbiaceae. Plant Syst. Evol. 215–189–208.
Tokuoka T and H Tobe. 2001. Ovules and seeds in subfamily Phyllanthoideae (Euphorbiaceae): structure and systematic implications. J. Plant Res. 114: 75–92.
Tokuoka T and H Tobe 2002. Ovules and seeds in Euphorbioideae (Euphorbiaceae): structure and systematic implications. J. Plant. Res. 115: 361–374.
Tokuoka T and H Tobe. 2003. Ovules and seeds in Acalyphoideae (Euphorbiaceae): structure and systematic implications. J. Plant Res. 116: 355–380.
Vaughan JG and JA Rest. 1969. Note on the testa structure of *Panda* Pierre, *Galearia* Zoll. et Mor. and *Microdesmis* Hook. f. (Pandaceae). Kew Bull. 23: 215–218.
Vogel C. 1986. Phytoserologische Untersuchungen zur Systematik der Euphorbiaceae. Diss. Bot. 98: 1–124.
Webster GL. 1967. The genera of Euphorbiaceae in the southeastern United States. J. Arnold Arbor. 48: 303–430.
Webster GL. 1975. Conspectus of a new classification of the Euphorbiaceae. Taxon 34: 593–601.
Webster GL. 1987a. The saga of the spurges: a review of classification and relationships in the Euphorbiaceae. Bot. J. Linn. Soc. 94: 3–46.
Webster GL. 1994a. Classification of the Euphorbiaceae. Ann. Missouri Bot. Gard. 81: 3–32.
Webster GL. 1994b. Synopsis of the genera and suprageneric taxa of Euphorbiaceae. Ann. Missouri Bot. Gard. 81: 33–144.
Webster GL and KJ Carpenter. 2002. Pollen morphology and phylogenetic relationships in neotropical *Phyllanthus* (Euphorbiaceae). Bot. J. Linn. Soc. 138: 325–338.
Webster GL and EA Rupert. 1973. Phylogenetic significance of pollen nuclear number in the Euphorbiaceae. Evolution 27: 524–531.
Welzen PC van and W Stuppy. 1999. Phylogenetic considerations of Euphorbiaceae tribe Aleuritideae. Ann. Missouri Bot. Gard. 86: 894–903.
Westra LYTh and J Koek-Noorman. 2004. Wood altas of the Euphorbiaceae s.l. IAWA J. Suppl. 4.
Wurdack KJ. 2002. Molecular systematics and evolution of Euphorbiaceae sensu lato. Ph.D. dissertation. University of North Carolina, Chapel Hill, NC.
Wurdack KJ and MW Chase. 1999. Spurges split. Molecular systematics and changing concepts of Euphorbiaceae s.l. XVI Intern. Bot. Congr. Abstract No 3334. St. Louis.
Wurdack KJ, P Hoffmann, R Samuel, A de Bruijn, M van der Bank, and MW Chase. 2004. Molecular phylogenetic analysis of Phyllanthaceae (Phyllanthoideae pro parte, Euphorbiaceae sensu lato) using plastic *rbc*L DNA sequences. Am. J. Bot. 91: 1882–1900.
Wurdack KJ, P Hoffmann, and MW Chase. 2005. Molecular phylogenetic analysis of uniovulate Euphorbiaceae (Euphorbiaceae sensu stricto) using plastid *rbc*L and *trn*L-F DNA sequences. Am. J. Bot. 92: 1397–1420.

Subclass VI. ROSIDAE

Trees, shrubs, subshrubs, or perennial or annual herbs. Vessels with simple or less often scalariform perforations (sometimes with many bars). Sieve-element plastids mostly of S-type. Leaves alternate, opposite or verticillate, simple or often compound, stipulate or estipulate. Stomata of various types. Flowers in various types of inflorescences or solitary, bisexual or less often unisexual, actinomorphic or zygomorphic, cyclic, usually with double perianth or sometimes the petals much reduced or wanting. Sepals and petals free or more or less connate. Stamens from numerous to few, rarely solitary. Tapetum secretory, rarely amoeboid. Microsporogenesis simultaneous or very rarely successive. Pollen grains 2-celled or less often 3-celled, mostly 3-colporate. Nectaries of various type, often of staminodial origin, often forming an intrastaminal or extrastaminal disc. Gynoecium apocarpous or more often of united carpels; stylodia free or connate into a style; ovary superior, or sometimes semi-inferior, often with only one or two ovules per carpel or per locule, but sometimes with several or many. Ovules usually anatropous, apotropous or epitropous, bitegmic or sometimes unitegmic or ategmic, crassinucellate or sometimes tenuinucellate. Endosperm nuclear or cellular (not produced in Podostemales). Fruits of various types. Seeds with small to large, straight, or curved embryo, with copious or scanty endosperm or endosperm wanting. Frequently with alkaloids.

The subclass Rosidae demonstrates ambivalent affinities to both the Hamamelididae and Dilleniidae.

Bibliography

Dickison WC. 1989. Comparisons of primitive Rosidae and Hamamelidae. In: P Crane and S Blackmore, eds. Evolution, systematics, and fossil history of Hamamelidae, vol. 1, pp. 47–73. Clarendon Press, Oxford.

Endress PK and ML Matthews. 2006. First steps towards a floral characterization of the major rosid subclades. Plant Syst. Evol. 260: 223–251.

Endress PE and S Stumpf. 1991. The diversity of stamen structures in "Lower" Rosidae (Rosales, Fabales, Proteales, Sapindales). Bot. J. Linn. Soc. 107: 217–293.

Fehrenbach S and W Barthlott. 1988. Mikromorphologie der Epicuticular-Wachse der Rosales s.l. und deren systematische Gliederung. Bot. Jahrb. Syst. 109: 407–428.

Hallier H. 1903. Über Verwandtschaftsverhältnisse bei Engler's Rosalen, Parietalen, Myrtifloren, und in an-deren Ordnungen der Dikotylen. Abh. Naturw. Ver. Hamburg 18: 1–98.

Hallier H. 1908. Über *Juliania,* eine Terebinthaceen-Gattung mit *Cupula,* und die wahren Stammeltern der Kätzchenblütler: Neue Beiträge zur Stammesgeschichte nebst einer Übersicht über das natürliche System der Dicotyledonen. Dresden.

Heimsch CH. 1942. Comparative anatomy of the secondary xylem in the "Gruinales" and "Terebinthales" of Wettstein with reference to taxonomic grouping. Lilloa 8: 83–198.

Hijwegen T. 1981. Fungi as plant taxonomists: II. Affinities of the Rosiflorae. Acta Bot. Neerl. 30: 479–491.

Huber H. 1963. Die Verwandtschaftsverhältnisse der Rosifloren. Mitt. Bot. Staatssamml. München 5: 1–48.

Hufford L. 1992. Rosidae and their relationships to other nonmarginal dicotyledons: A phylogenetic analysis using morphological and chemical data. Ann. Missouri Bot. Gard. 79: 218–248.

Mauritzon J. 1939. Contributions to the embryology of the orders Rosales and Myrtales. Acta Univ. Lund. 2 (35): 1–121.

Superorder ROSANAE

Order 78. CUNONIALES

Trees and shrubs. Vessels with scalariform perforation (sometimes with many bars), mixed scalariform and simple perforations, or rarely only with simple perforations; lateral pitting from scalariform to alternate.

Fibers with bordered or simple pits. Rays heterogeneous. Axial parenchyma apotracheal or (Brunelliaceae) wanting. Sieve-element plastids of S- or Pc-type. Nodes trilacunar or multilacunar, rarely (*Bauera*) unilacunar. Leaves alternate, opposite or verticillate, simple or compound, stipulate. Stomata paracytic, anomocytic, anisocytic or encyclocytic. Flowers small or rarely large, in axillary or terminal inflorescences, rarely solitary in the axils, mostly bisexual, actinomorphic, with double perianth or less often apetalous (some Cunoniaceae and Brunelliaceae). Stamens numerous to few, free; filaments slender; anthers tetrasporangiate, opening longitudinally or (*Davidsonia*) by apical pores that elongate into longitudinal slits. Tapetum secretory. Microsporogenesis simultaneous. Pollen grains 2-celled, small, mostly 3-colporate, 2-colporate, or 2-colpate. Nectary disk mostly present. Gynoecium apocarpous or mostly of more or less united carpels; stylodia free, with capitate or (Brunelliaceae) decurrent stigma. Ovary superior, semi-inferior or inferior, with two to numerous ovules per carpel or per locule. Ovules anatropous or less often campylotropous, apotropous, or epitropous, bitegmic, crassinucellate. Female gametophyte of *Polygonum*-type. Endosperm nuclear. Fruits mostly capsular. Seeds exotegmic, with small or large, straight embryo; endosperm more or less copious or (*Davidsonia*) wanting.

In many respects Cunoniales are relatively the most archaic member of the subclass Rosidae, and they are probably the nearest to the hamamelids. The similarities between the cunonialean families, such as Cunoniaceae, and the archaic members of the Hamamelididae are discussed in detail by Dickison (1989).

Key to Families

1 Axial parenchyma present. Trees and erect shrubs. Indumentum generally of simple hairs, or occasionally stellate and peltate trichomes, or rarely stinging hairs (*Davidsonia*). Vessels with scalariform (*Acsmithia*, *Gillbeea*, *Pullea*, *Spiraeanthemum*, and *Weinmannia* p.p.) and mixed simple and scalariform perforations. Vessel-ray pits coarse and scalariform to opposite. Axial parenchyma mainly diffuse, diffuse-in-aggregates, and scanty paratracheal. Nodes mostly trilacunar with three traces, sometimes (*Codia* and *Pulled*) multilacunar or (*Bauera*) unilacunar. Sieve-element plastids of S-type, or of Pc-type with single polygonal protein crystalloid (*Eucryphia*). Leaves opposite or sometimes verticillate, rarely alternate (*Davidsonia*), imparipinnately compound or trifoliolate, seldom unifoliolate; the leaflets vary from entire or serrate to dentate. Stipules sometimes large and conspicuous, usually united in pairs to form the characteristic interpetiolar stipular condition, commonly with small colleters (multicellular, glandular secretory structures). Flowers small, bisexual or less often unisexual (plants dioecious or polygamous), actinomorphic, mostly both sepals and petals. Sepals (3)-6(-10), valvate or imbricate, in some genera enlarged in fruit, in others persistent or caduceus. Petals as many as sepals and alternate with them, or more numerous than the sepals (*Bauera*). Stamens 8–10 or less often 4–5, sometimes numerous; filaments usually long and thin; anthers introrse, more or less dorsifixed, versatile (except *Bauera*), opening longitudinally. Pollen grains mostly 3-colporate, rarely (*Bauera*) syncolpate, smooth or nearly so. A saucershaped, annular nectary disc usually developed around the gynoecium. Gynoecium of two or less often 3–5 carpels, syncarpous or sometimes (*Acsmithia, Acrophyllum, Pancheria,* and *Spiraeanthemum*) apocarpous; carpellary margins mostly inrolled; stylodia free or more or less connate, usually with stigmas ranging from relatively undifferentiated to capitate or subcapitate, but in some species of *Cunonia* extending downward and in *Vesselowskya* decurrent on the ventral surface of the conspicuously elongate stylodium. Ovary superior or sometimes semi-inferior or inferior. Ovules (1)2-many per locule or per carpel, apotropous or sometimes (*Acsmithia* and *Spiraeanthemum*) epitropous. Fruits dehiscent or indehiscent, mostly capsular, sometimes follicular (*Acsmithia* and *Spiraeanthemum*), drupaceous (*Aistopetalum* and *Schizomeria*), or samara (*Gillbeea*), or pseudosamara (*Ceratopetalum*). Seeds small to relatively large, in dehiscent fruits usually with winges at one or both ends, or hairy without wings; embryo embedded in abundant starchy endosperm (except *Davidsonia*). Contain much tannin, glycosides of kaempferol and quercetin, flavonoids prodelphinidins and myricetin; *Bauera rubioides* contain proanthocyanidins n = 16, 15 (*Weinmannia, Euchriphia*), 12 (*Pancheria*)..... 1. CUNONIACEAE.

1 Axial parenchyma absent. Evergreen, tall trees up to 40 m tall, usually tomentose throughout; hairs unicellular. Twigs angular, with rather large pits.

Vessel elements elongate, thin-walled, some of them with scalariform perforations that have up to 35 bars, others with simple and scalariform perforations. Fibers with small, simple, or obscurely bordered pits, often septate. Nodes trilacunar or pentalacunar. Leaves opposite or ternate, pinnately compound, trifoliolate to unifoliolate or simple, the leaflets opposite, pinnately veined, entire to double-dentate; stipules small, caducous, often more than two. Stomata anomocytic. Flowers small, in axillary or terminal cymes, bisexual to more or less unisexual (commonly dioecious or gynodioecious), apetalous; male flowers with a vestigial gynoecium, and female flowers with a vestigial androecium. Sepals (4)5–6(-8), shortly connate below, valvate, persistent in the fruits. Nectary disc intrastaminal, adnate to the calyx, cupular, 8–10-lobed. Stamens 8-10(-14), in two cycles; filaments slender, hairy, inserted in the notches of the nectary disc; anthers dorsifixed, versatile, introrse, opening longitudinally. Pollen grains 3-colporate. Gynoecium apocarpous, of 2–3 carpels or more often up to same number as sepals, more or less adnate to the disc; each carpel gradually attenuates into long, slender, elongate, curved, or hooked stylodium with a linear, decurrent, sutural stigma that extends along its entire length and is composed of bands of papillae. Ovules two per carpel, collateral, pendulous, epitropous. Fruits of 1–2-seeded follicles, usually densely short reddish to yellowish tomentose and with long, pointed trichomes; endocarp more or less lignified, separating from the exocarp at maturity. Seeds with thick, hard, shiny testa and corky, subarillate raphe, attached by a funicle in dehisced follicle; embryo large, straight, surrounded by copious, carnose, mealy, white endosperm, n = 14. . . . . 2. BRUNELLIACEAE.

1. CUNONIACEAE

R. Brown 1814 (including Baueraceae Lindley 1830, Belangeraceae J. Agardh 1858, Callicomaceae J. Agardh 1858, Davidsoniaceae Bange 1952, Eucryphiaceae Endlicher 1841, Spiraeanthemaceae Doweld 2001). 27/300. Almost exclusively confined to the Southern Hemisphere between 13° and 35° S, mainly Australia, New Caledonia, and New Guinea; a few (*Weinmannia* spp.) north to the Philippines and southern Mexico; there are a few genera in tropical America and South Africa. The largest genus *Weinmannia* (190) distributed through Madagascar, Mascarenes, Malesia, the Pacific, New Zealand, Chile, Mexico, and the West Indies. *Cunonia* has a discontinuous distribution is South Africa (1) and New Caledonia (16). *Bauera* (3) and *Davidsonia* are endemic to Australia, *Eucryphia* (5) Eastern Australia, Tasmania, Chile, Argentina.

1.1 CUNONIOIDEAE

Erect trees and shrubs. Nodes trilacunar or sometimes (*Codia* and *Pullea*) multilacunar. Leaves with interpetiolar stipules. Petals mostly present. Anthers dorsifixed, versatile. Carpels mostly 2, rarely 5–3. – *Spraeanthemum, Acsmithia, Aistopetalum, Hooglandia, Schizomeria, Ceratopetalum, Anodopetalum, Platylophus, Gillbeea, Acrophyllum, Lamanonia, Pseudoweinmannia, Geissois, Caldcluvia, Opocunonia, Ackama, Spiraeopsis, Pullea, Codia, Callicoma, Vesselowskya, Pancheria, Cunonia, Weinmania.*

1.2 BAUEROIDEAE

Erect shrubs. Nodes unilacunar with three traces. Leaves trifoliate and estipulate, or simple with well-developed leaflike stipules. Petals present. Anthers not versatile. Carpels 2. – *Bauera.*

1.3 EUCRYPHIOIDEAE

Trees and shrubs with simple, unicellular hairs, producing gum or mucilage and sticky terminal buds. Sieve-element plastids with protein crystals. Nodes trilacunar. Leaves simple or pinnately compound; stipules interpetiolar, small, caducous, with large colleters. Stomata paracytic. Petals 4(5), imbricate. Athers versatile. Carpels 4–14(-18). Fruits leathery or woody septicidal capsules. n = 15, 16. – *Eucryphia.*

1.4 DAVIDSONIOIDEAE

Small trees; hairs urticating. Leaves alternate, to 1 m long. Stomata paracytic. Flowers in large, axillary or supra-axillary panicles with spikelike branches, bisexual, apetalous. Anthers dorsifixed, introrse. Carpels 2, rarely 3; ovules 5–7, pendulous. Fruits large 2-pyrened drupes, red-velvety when young, pruinose and glaucous at maturity, with fleshy mesocarp and flattened, fimbriate-laciniate pyrenes. Seeds strongly compressed, glabrous, broad-ovate to circular, without endosperm. – *Davidsonia.*

Dickison (1975a, 1980b, 1984) has shown, the genera *Ascmithia* (17, Moluccas to western Pacific) and *Spiraeanthemum* (6, western Pacific) are evidently the most archaic members of the family (primitive wood anatomy, paracytic stomata, and apocarpous gynoecium.). They differ from the other members also in the carpels, which are vascularized by five or four major vascular bundles, and the carpel margins, which are not inrolled (Dickison 1975b). These differences show their rather isolated position.

2. BRUNELLIACEAE

Engler 1897. 1/c.60. America from Mexico, Costa Rica, and West Indies to Venezuela and the Andes of Colombia, Bolivia, and Peru.

Brunellia.

Close to the Cunoniaceae (Hallier 1903a, 1912; Engler 1930; Cuatrecasas 1970; Dickison 1989; Thorne 2000, 2006). According to Dickison (1989: 50), strong similarities exist between the two families in vegetative morphology, similar floral diagram, ovular type (particularly *Acsmithia* and *Spiraeanthemum*), and fruit anatomy. As Cuatrecasas (1970) and Dickison (1989) point out, *Brunellia* is uniquely characterized by a linear, decurrent, sutural stigma and a highly specialized carpel maturation and mature fruit. The Brunelliaceae are a highly heterobathmic family with a mixture of both plesiomorphic and apomorphic characters.

Bibliography

Agababian VS. 1964. Evolution of pollen in the orders Cunoniales and Saxifragales in relation to some question of their systematics and phylogeny. Izvestia of Acad. Sci. of the Armenian SSR. Biol. Sciences 17: 59–72 (in Russian).

Bange GGJ. 1952. A new family of dicotyledons: Davidsoniaceae. Blumea 7: 293–296.

Barnes RW and AC Rozefelds. 2000. Comparative morphology of *Anodopetalum* (Cunoniaceae). Austral. Syst. Bot. 13: 267–282.

Bate-Smith EC. 1977. Chemistry and taxonomy of the Cunoniaceae. Biochem. Syst. Ecol. 5: 95–105.

Bate-Smith EC, SM Davenport, and JB Harborne. 1967. Comparative biochemistry of flavonoids: A correlation between chemistry and plant geography in the genus *Eucryphia*. Phytochemistry 6: 1407–1413.

Bausch J. 1938. A revision of the Eucryphiaceae. Kew Bull. Misc. Inform. 1938: 317–349.

Behnke H-D. 1985. Contributions to the knowledge of P-type sieve-element plastids in dicotyledons: II. Eucryphiaceae. Taxon 34: 607–610.

Bradford JC. 2002. Molecular phylogenetics and morphological evolution in Cunonieae (Cunoniaceae). Ann. Missouri Bot. Gard. 89: 491–503.

Bradford JC and RW Barnes. 2001. Phylogenetics and classification of Cunoniaceae (Oxalidales) using chloroplast DNA sequences and morphology. Syst. Bot. 26: 354–385.

Bradford JC, HC Fortune Hopkins, and RW Barnes. 2004. Cunnoniaceae. In: K Kubitzki, ed. The families and genera of vascular plants, vol. 6, pp. 91–111. Springer, Berlin/Heidelberg/New York.

Cherniakovskaya EF. 1996a. Cunoniaceae, Baueraceae. In: A Takhtajan, ed. Comparative seed anatomy, vol. 5, pp. 7–14. Nauka, St. Petersburg (in Russian).

Cherniakovskaya EF. 1996b. Eucryphiaceae, Brunelliaceae. In: A Takhtajan, ed. Comparative seed anatomy, vol. 5, pp. 16–19. Nauka, St. Petersburg (in Russian).

Cherniakovskaya EF and AB Doweld. 1996. Davidsoniaceae. In: A. Takhtajan, ed. Comparative seed anatomy, vol. 5, pp. 14–16. Nauka, St. Petersburg (in Russian).

Cuatrecasas J. 1970. Brunelliaceae. Flora Neotropica 2: 1–189.

Cuatrecasas J. 1985. Brunelliaceae. Flora Neotropica, 2(Suppl.): 29–103.

Dickison WC. 1975a. Floral morphology and anatomy of *Bauera*. Phytomorphology 25: 69–76.

Dickison WC. 1975b. Studies of the floral anatomy of the Cunoniaceae. Am. J. Bot. 62: 433–447.

Dickison WC. 1975c. Leaf anatomy of Cunoniaceae. Bot. J. Linn. Soc. 71: 275–294.

Dickison WC. 1978. Comparative anatomy of Eucryphiaceae. Am. J. Bot. 65: 722–735.

Dickison WC. 1980a. Diverse nodal anatomy of the Cunoniaceae. Am. J. Bot. 67: 975–981.

Dickison WC. 1980b. Comparative wood anatomy and evolution of the Cunoniaceae. Allertonia 2: 281–321.

Dickison WC. 1984. Fruits and seeds of the Cunoniaceae. J. Arnold Arbor. 65: 149–190.

Dickison WC. 1989. Comparisons of primitive Rosidae and Hamamelidae. In: PR Crane and S Blackmore, eds. Evolution, systematics, and fossil history of the Hamamelidae, vol. 1, pp. 47–73. Clarendon Press, Oxford.

Dickison WC and R Rutishauser. 1990. Developmental morphology of stipules and systematics of the Cunoniceae and presumed al lies. II. Taxa without interpetiolar stipules and conclusions. Bot. Helvetica 100: 75–95.

Doweld AB. 1998. The carpology and taxonomic relationships of *Davidsonia* (Davidsoniaceae). Edinb. J. Bot. 55: 13–25.

Ehrendorfer F, W Morawetz, and J Dawe. 1984. The Neotropical angiosperm families Brunelliaceae and Caryocaraceae: First karyosystematical data and affinities. Plant Syst. Evol. 145: 183–191.

Engler A. 1928. Cunoniaceae. In: A. Engler and K. Prantl, eds. Die natürlichen Pflanzenfamilien, 18a: 229–262. Leipzig.

Fortune Hopkins HCF and RD Hoogland. 2002. Cunoniaceae. In: HP Nooteboom, ed. Flora Malesiana, ser. I, 16: 53–165. Leiden.

Fryns-Claessens E and W Van Cotthem. 1966. L'appareil stomatique des Pandacees et Davidsoniacees. Rev. Gen. Bot. 63: 783–789.

Gregory M. 1998. Cunoniaceae. In: DF Cutler, M Gregory, eds. Anatomy of the dicotyledons. Saxifragales, vol. 4, pp. 10–27. Clarendon Press, Oxford.

Harden GJ and JB Williams. 2000. A revision of *Davidsonia* (Cunoniaceae). Telopea 8: 413–428.
Hoogland RD. 1960. Studies in the Cunoniaceae: I. The genera *Ceratopetalum, Gilleea, Aistopetalum,* and *Calycomis.* Austral. J. Bot. 8: 318–341.
Hoogland RD. 1979. Studies in the Cuoniaceae: II. The genera *Caldcluvia, Pillea, Acsmithia,* and *Spi-raeanthemum.* Blumea 25: 481–505.
Hoogland RD. 1981. Studies in the Cunoniaceae: III. Additional notes on *Ceratopetalum* and *Arophyllum.* Brunonia 4: 213–216.
Hufford L and WC Dickison. 1992. A phylogenetic analysis of Cunoniaceae. Syst. Bot. 17: 181–200.
Ingle HD and HE Dadswell. 1956. The anatomy of the timbers of the Southwest Pacific area: IV. Cunoniaceae, Davidsoniaceae, and Eucryphiaceae. Austral. J. Bot. 4: 125–151.
Jay M. 1968. Distribution des flavonoides chez les Cu-noniacees. Taxon 17: 489–495.
Kubitzki K. 2004. Brunelliaceae. In: K Kubitzki, ed. The families and genera of vascular plants, vol. 6, pp. 26–28. Springer, Berlin/Heidelberg/New York.
Lopez Naranjo H and H Huber. 1971. Anatomia com-parativa de las semillas de *Brunellia y Weinmannia* con respeto a su posicion sistematica. Pittieria 3: 19–28.
Matthews ML and PK Endress. 2002. Comparative floral structure and systematics in Oxalidales (Oxalidaceae, Connaraceae, Brunelliaceae, Cephalotaceae, Cunnoniaceae, Elaeocarpaceae, Tremandraceae). Bot. J. Linn. Soc. 140: 321–381.
Matthews ML, PK Endress, J Schönenberger, and EM Friis. 2001. A comparison of floral structures of Anisophylleaceae and Cunnoniaceae and the problem of their systematic position. Ann. Bot. 88: 439–455.
Miranda-Esquivel DR. 2001. Sobre la posici6n sisteḿatica de *Brunellia* Ruiz & Pavon: Un reanálisis de Orozco (1997). Caldasia 22: 337–340.
Moody M and L Hufford. 2000. Floral development and structure of *Davidsonia* (Cunoniaceae). Canad. J. Bot. 78: 1034–1043.
Orozco CI. 1997. Sobre la posicion sistematica de *Brunellia* Ruiz et Pavon. Caldasia 19: 145–164.
Orozco CI. 2000. Indumento del enves foliar en *Brunellia* Ruiz et Pavon y su relativa importancia en la filogenia. Rev. Acad. Colomb. Cienc. Exact. Fis. Nat. 23(Suppl. Esp.): 97–102.
Orozco CI. 2001b. Pollen morphology of *Brunellia* (Brunelliaceae) and related taxa in the Cunoniaceae. Grana. 40: 245–255.
Orozco CI. 2001a. Reanálisis de *Brunellia* Ruiz & Pavon: Una repuesta a Miranda. Caldasia 22: 341–346.
Orozco CI and F Weberling. 1999. A comparative study of inflorescences in *Brunellia* Ruiz and Pavon (Brunelliaceae) and related taxa. Beitr. Biol. Pfl. 71: 261–279.
Orozco-Pardo CI. 2002. Evolutionary biology of *Brunellia* Ruiz & Pav6n (Brunelliaceae, Oxalidales). Bogotá, Colombia.
Orozco-Prado CI and B Coba. 2003. Leaf anatomy in *Brunellia* Ruiz & Pav6n. In: CI Orozco-Prado, eds. Evolutionary biology of *Brunellia* Ruiz & Pav6n (Brunelliaceae, Oxalidales), pp. 59–79. Bogotá, Colombia.
Patel RN. 1990. Wood anatomy of the dicotyledons indigenous to New Zealand: 20. Cunoniaceae. New Zealand J. Bot. 28: 347–355.
Prakash N and EJ McAlister. 1977. An embryological study of *Bauera capitata* with comments on the systematic position of Bauera. Austral. J. Bot. 25: 615–622.
Rao TA and WC Dickison. 1985a. The veinsheath syndrome in Cunoniaceae: I. *Pancheria* Brong. et Gris. Proc. Indian Acad. Sci. 95: 87–94.
Rao TA and WC Dickison. 1985b. The veinsheath syndrome in Cunoniaceae: II. The genera *Acsmithia, Codia, Cunonia, Geissois,* and *Weinmannia.* Proc. Indian Acad. Sci. 95: 247–261.
Rozefelds AC, RW Barnes, and B Pellow. 2001. A new species and comparative morphology of *Vesselowskya* (Cunoniaceae). Austral. Syst. Bot. 14: 175–192.
Rutishauser R and WC Dickison. 1990. Developmental morphology of stipules and systematics of the Cunoniaceae and presumed allies: I. Taxa with inter-petiolar stipules. Bot. Helvetica 99: 147–169.
Schönenberger J, EM Friis, ML Matthews, and PK Endress. 2001. Cunoniaceae in the Cretaceous of Europe: evidence from fossil flowers. Ann. Bot. 88: 423–437.
Schrödinger R. 1927. Die Stipein der Cunoniaceen. Verh. Zool. Bot. Ges. Wien 77: 5–38.
Sweeney P and JC Bradford. 2004. The phylogenetic position and comparative morphology of *Hooglandia* within Cunoniaceae. Ann. Missouri Bot. Gard. 91: 266–274.
Taylor F and RS Hill. 1996. A phylogenetic analysis of the Eucryphiaceae. Austral. Syst. Bot. 9: 735–748.
Webb CJ and MJA Simpson. 1991. Seed morphology in relation to taxonomy in New Zealand species of *Weinmannia, Ackama,* and the related South American *Caldcluvia paniculata* (Cunoniaceae). New Zealand J. Bot. 29: 451–453.
Weberling F. 1976. Weitere Untersuchungen zur Morphologie des Unterblattes bei den Dikotylen: IX. Saxifragaceae s. 1., Brunelliaceae, and Bruniaceae. Beitr. Biol. Pfl. 52: 163–181.
Wollenweber E, M Dörr, AC Rozefelds, P Minchin, and PI Forster. 2000. Variation in flavonoid exudates in *Eucryphia* species from Australia and South America. Biochem. Syst. Ecol. 28: 111–118.

Order 79. ANISOPHYLLEALES

Large trees or shrubs of wet primary forests. Vessels with simple perforations; lateral pitting alternate, with coalescent apertures. Fibers with distinctly bordered pits. Rays heterogeneous, very broad, multiseriate. Axial parenchyma apotracheal and often banded, varying to paratracheal and irregular. Sieve-element plastids of S-type. Lysigenous secretory cavities are present in the parenchymatous tissues of *Poga.* Nodes unilacunar with one trace. Leaves alternate, simple, entire, often 2-ranked, base asymmetrical, sometimes pellucid-punctate, estipulate or (*Polygonanthus*) with 2–4 minute stipules, very base of the petiole. Stomata mostly paracytic. Flowers small (except for the female flowers of *Polygonanthus*), in axillary spikes, racemes, or panicles of catkin-like spikes, rarely solitary, bisexual (*Combretocarpus* and a few species of *Anisophyllea*) or more often unisexual and monoecious or dioecious, sometimes polygamous, actinomorphic, mostly 4-merous, but in *Combretocarpus* usually 3-merous and in other genera sometimes 5-merous. Calyx and petals valvate. Petals commonly deeply incised, generally with three, five, or seven lobes, which in taxa of *Anisophyllea* and *Poga* may have an enlarged, glandular tip; in *Poga* and *Polygonanthus* petals entire, but sometimes finely fimbriate, in *Combretocarpus* they are lacking or reduced. Stamens (6-)8 (-10), in 2 cycles, incurved in bud, free, with narrow filaments; anthers tetrasporangiate or (*Polygonanthus*) occasionally the pollen sacs divided by tapetal septa, dorsifixed, ovoid, introrse, opening longitudinally. Tapetum secretory. Microsporogenesis simultaneous. Pollen grains 2-celled, 3-colporate, have a rather thin tectum with reticulate-punctate or rarely (*Anisophyllea disticha*) striate surface, endoapertures, when present, are circular and poorly defined. A lobed intra- and interstaminal, discontinuous nectary disc surrounds the top of the ovary. Gynoecium of (3)4(5) united carpels with free, subulate stylodia. Ovary inferior, 3–4-locular, with two (*Combretocarpus*) or one ovule per locule. Ovules pendulous from the axile placenta, anatropous, bitegmic or (*Anisophyllea* and *Combretocarpus*) unitegmic, clearly crassinucellate; no endothelium is formed. Female gametophyte of *Polygonum*-type or (*Combretocarpus*) of *Allium*-type. Endosperm nuclear. Fruits drupaceous (*Anisophyllea*), or dry, woody, strongly winged, crowned by much enlarged and persistent calyx lobes (*Polygonanthus*), samara (*Combretocarpus*), or 3–4-seeded capsules (*Poga*). Seeds winged or wingless; coat testal (consists of the testal epidermis only, as in *Poga*, of multilayered testa, as in *Polygonanthus*, or of multilayered or ultimately thin single integument, as in *Anisophyllea* and *Combretocarpus* respectively); embryo has a long hypocotyl and small to rudimentary cotyledons or no cotyledons at all; no endosperm. Aluminum accumulation is highly characteristic. Clearly tanniniferous plants. Alkaloids are unknown, n = 7, 8.

According to Dahlgren (1988: 1274), in gross morphology Anisophylleaceae are "largely Saxifragalean-Rosalean in nature," and he agrees with Cronquist (1981), who places Anisophylleaceae in Rosales sensu latissimo. According to Dahlgren the floral morphological data as well as embryological conditions in Anisophylleaceae agree completely with those common in Rosaceae sensu stricto. "It is probable that Anisophylleaceae comprise a rather isolated family evolved from ancestors shared between those in Rosales, Cunoniales, and Saxifragales," concludes Dahlgren (1988: 1275). On the other hand, according to Tobe and Raven (1988: 1427), "embryological features suggest strongly that Anisophylleaceae, even though there are some points of similarity to Rosales sensu stricto, do not belong in that order." In their opinion it is more appropriate to regard Anisophylleaceae as constituting a distinct order. Matthews et al. (2001), Schönenberger et al. (2001), Matthews and Endress (2004) showed striking similarities in floral structure of Anisophylleaceae with the Cunoniaceae, which is supported by greater similarity between in flowers of *Anisophyllea* (Anisophylleaceae) and the genus *Ceratopetalum* (Cunoniaceae) (Matthews et al. 2001). According to molecular date the Anisophylleaceae belong in the Cucurbitales, which is not acceptable.

1. ANISOPHYLLEACEAE

Ridley 1922. 4/34 (including Polygonanthaceae Croizat 1943). Tropical Africa, tropical Asia (India to Malesia), South America; *Combretocarpus* (1) is endemic to Borneo, *Polygonanthus* (2) to Brazil (Amazonia).

Anisophyllea, Poga, Polygonanthus, Combretocarpus.

Bibliography

Baehni C and P Dansereau. 1939. La position systématique du genre *Polygonanthus*. Bull. Soc. Bot. Suisse 49: 415–416.

Behnke H-D. 1988. Sieve-element plastids and systematic relationships of Rhizophoraceae, Anisophylleaceae, and allied groups. Ann. Missouri Bot. Gard. 75: 1387–1409.

Dahlgren RMT. 1988. Rhizophoraceae and Anisophylleaceae: summary statement, relationships. Ann. Missouri Bot. Gard. 75: 1259–1277.

Juncosa AM and RB Tomlinson. 1988a. A historical and taxonomic synopsis of Rhizophoraceae and Anisophylleaceae. Ann. Missouri Bot. Gard. 75: 1278–1295.

Juncosa AM and RB Tomlinson. 1988b. Systematic comparison and some biological characteristics of Rhizophoraceae and Anisophylleaceae. Ann. Missouri Bot. Gard. 75: 1296–1318.

Keating RC and V Randrianosola. 1988. The contribution of leaf architecture and wood anatomy to classification of the Rhizophoraceae and Anisophylleaceae. Ann. Missouri Bot. Gard. 75: 1343–1368.

Matthews ML and PK Endress. 2004. Comparative floral structure and systematics in Cucurbitales (Corynocarpaceae, Coriariaceae, Tetramelaceae, Datiscaceae, Begoniaceae,

Cucurbitaceae, Anisophylleaceae). Bot. J. Linn. Soc. 145: 129–185.
Matthews ML, PK Endress, J Schönenberger, and EM Friis. 2001. A comparison of floral structures of Anisophylleaceae and Cunnoniaceae and the problem of their systematic position. Ann. Bot. 88: 439–455.
Plisko MA. 1996. Anisophylleaceae. In: A Takhtajan, ed. Comparative seed anatomy, vol. 5, pp. 151–152. Nauka, St. Petersburg (in Russian).
Schwarzbach AE and RE Ricklefs. 2000. Systematic affinities of Rhizophoraceae and Anisophylleaceae, and intergeneric relationships within Rhizophoraceae, based on chloroplast DNA, nuclear ribosomal DNA, and morphology. Am. J. Bot. 87: 547–564.
Tobe H. and R H. Raven. 1987. Systematic embryology of the Anisophylleaceae. Ann. Missouri Bot. Gard. 74: 1–26.
Tobe H and RH Raven. 1988a. Floral morphology and evolution in Anisophylleaceae. Bot. J. Linn. Soc. 98: 1–25.
Tobe H and RH Raven. 1988b. Additional notes on the embryology of *Polygonanthus* (Anisophylleaceae) and relationships of the family. Ann. Missouri Bot. Gard. 75: 1425–1428.

Order 80. CEPHALOTALES

Small, insectivorous, perennial bog-herbs with a short underground rhizome. Vascular system composed of a ring of xylem and phloem interrupted by broad, foliar rays. Vessels with scalariform perforations. Sieve-element plastids of Ss-type. Leaves alternate, all basal, of two kinds, the lower (outer) ones of the rosette modified into ground level pitchers, the upper ones flat, entire, estipulate; the pitchers probably represent an ascidiate petiole dilated at the top into a lid (the lid is an outgrowth from the upper surface of the petiole below the pitcher proper while the pitcher has been produced by a ventral-dorsal invagination of the upper, more distal region according to Lloyd 1942); the lower surface of the lid and the distal inner surface of the pitcher slippery, coated with overlapping, downwardly directed projections from the epidermal cells; multicellular, embedded glands present on the pitcher surfaces as well as on the petiole and lower surface of the vegetative leaves; flask-shaped embedded glands also present in the interior of the pitcher and especially large on the brightly colored, cushionlike projections from the surface. The mouth of the pitcher is surrounded by a corrugated rim, each corrugation forming a clawlike tooth extending inward and downward. Stomata paracytic. Flowers small, borne on a leafless scape arising from the center of the rosette and bearing racemously arranged dichasia, bracteate, bisexual, actinomorphic, 6-merous, apetalous. Calyx colored, synsepalous, 6-lobed; lobes valvate, hooded. Stamens 12, in two alternating cycles, inserted at the top of the calyx tube on the inner margin of a broad, thick, green, papillate, glandular nectary disc; anthers small, strongly introrse and cruciform, dorsifixed, versatile, tetrasporangiate, opening longitudinally; connective large, globose, swollen at the top and glandular. Pollen grains 2-celled, 3-colpate, smooth. Tapetum secretory. Gynoecium of 6 free carpels each with more or less circinately recurved, subulate stylodium and simple stigma, with one (rarely 2) ovule per carpel; ovary superior, 1-locular. Ovules basal, ascending, anatropous, with dorsal raphe, bitegmic, crassinucellate, with an endothelium. Fruits multifollicles, densely hairy. Seeds with small, straight embryo surrounded by copious, fleshy endosperm; testa and tegmen very thin. Present myricetin and quercetin, ellagic and gallic acids, and tannin cells; $n = 10$.

Evidently related to the Cunoniales, which is supported by the 18S rDNA and *rbc*L sequences data (Sotlis and Soltis 1997; Soltis et al. 2006) and even more to Brunelliaceae, they have an identical floral diagram (Conran 2004).

1. CEPHALOTACEAE

Dumortier 1829. 1/1. Southwestern Australia.
Cephalotus.

Bibliography

Arber A. 1941. On the morphology of the pitcher-leaves in *Heliamphora, Sarracenia, Cephalotus,* and *Nepenthes*. Ann. Bot. 5: 563–578.
Conran JG. 2004. Cephalotaceae. In: K Kubitzki, ed. The families and genera of vascular plants, vol. 6, pp. 65–68. Springer, Berlin/Heidelberg/New York.
Conran JG and MD Denton. 1996. Germination in the Western Australian Pitcher Plant *Cephalotus follicularis* and its unusual early seedling development. W.A. Nat. 21: 37–42.
Dickson A. 1878. The structure of the pitcher of *Cepha-lotus follicularis*. J. Bot. (London) 16: 1–5.
Dickson A. 1882. On the morphology of the pitcher of *Cephalotus follicularis*. Trans. Proc. Bot. Soc. Edinb. 14: 172–181.
Hamelton AC. 1904. Notes on the West Australian pitcher plants (*Cephalotus follicularis* La Bill.). Proc. Linn. Soc. N.S.W. 29: 36–53.

Nicholls KW, BA Bohm, and R Ornduff. 1985. Flavonoids and affinities of Cephalotaceae. Biochem. Syst. Ecol. 13: 261–263.
Peng C-I and P Goldblatt. 1983. Confirmation of the chromosome number in Cephalotaceae and Roridulaceae. Ann. Missouri Bot. Gard. 70: 197–198.

Order 81. SAXIFRAGALES

Trees and shrubs, or perennial and annual herbs. Vessels mostly with scalariform or more often simple perforations. Sieve element plastids of S- or P-types. Nodes trilacunar, multilacunar, or unilacunar. Leaves alternate, opposite or verticillate, simple or compound, stipulate or estipulate. Stomata of various types. Flowers in various kinds of inflorescences or solitary, mostly bisexual, actinomorphic, actinomorphic or less often zygomorphic, usually with double perianth, sometimes apetalous. Perianth forming a more or less well-developed floral tube, free or more or less adnate to the base of the ovary. Sepals mostly five, less often less (3, or 4) or more (up to 10), free or more or less connate, imbricate or valvate. Stamens few to numerous, mostly as many or twice as many as the sepals; anthers basifixed or less often slightly dorsifixed, introrse or latrorse, tetrasporangiate, opening longitudinally. Tapetum secretory. Microsporogenesis simultaneous. Pollen grains 2-celled or 3-celled, mostly 3-colporate to pantoporate. Nectary disc mostly present. Gynoecium of free or more or less united carpels, stylodia free, stigma decurrent to capitate; ovary superior, semi-inferior or inferior, with 1–2 to numerous ovules per carpel or per locule. Ovules usually anatropous, bitegmic or rarely unitegmic (Tetracarpaeaceae, *Darmera,* section *Micranthes* of *Saxifraga, Choristylis*), crassinucellate. Female gametophyte of *Polygonum-* or *Allium*-type, or rarely (Gunneraceae) *Peperomia*-type. Endosperm cellular or less often helobial or nuclear. Fruits multifollicles (Penthoraceae, majority of Crassulaceae), capsular, baccate (Grossulariaceae), drupaceous(Gunneraceae),ornutlike(Aphanopetalaceae). Seeds small or minute, exotestal or rarely (*Astilbe* and *Rodgersia* in Saxifragaceae) exotestal-endotegmic; embryo small or large, mostly straight; endosperm usually copious. Present flavones, flavonoids, proanthocyanidins, alkaloids.

The basal member of the Rosidae.

Key to Families

1 Flowers usually 4-merous.

2 Leaves mostly alternate.

3 Ovules bitegmic.

4 Embryo straight, cylindrical. Perennial or annual aquatic or terrestrial herbs, subshrubs or shrublets, or seldom (*Haloragodendron*) shrubs or small trees up to 3 m tall, glabrous or scabrous with simple uniseriate hairs; often calcium oxalate crystals in hair-like cortical cells. Vessels with simple perforation. Rays heterogeneous to homogeneous. Leaves alternate, opposite or verticillate, simple, entire to more or less deeply dissected, pinnatifid or multifid, with minute vestigial stipules. Stomata usually anomocytic. Flowers in dichasial, often compounded inflorescences, sometimes reduced to solitary axillary flowers, usually small, bisexual or unisexual (monoecious), actinomorphic, basically 4-merous, but often reduced to 3-merous or sometimes 2-merous condition, 2-bracteolate. Sepals valvate, persistent, absent in the female flowers of *Myriophyllum.* Petals imbricate, keeled, hooded, or navicular in the apex and more or less unguiculate in the base, deciduous with stamens, absent in *Proserpinaca* and female flowers of *Myriophyllum* and *Laurembergia.* Stamens typically eight in two cycles, but sometimes the antesepalous or antepetalous cycle is absent or rudimentary or (antesepalous stamens in some species *Gonocarpus*) transformed into staminodia; filaments usually very short, slender; anthers large, oblong (reniform in *Proserpinaca*), basifixed, with shortly apiculate connective, introrse, the antesepalous anthers sometimes more or less longer than antepetalous ones. Pollen grains 3-celled, pertectate, 4–5(6)-colpate or 4–5-porate. Gynoecium usually of four united carpels, but reduction to three or two occurs in some genera and species; stylodia free, clavate, more or less bulbously based, convergent at tips until anthers shed, with more or less capitate, fimbriate stigmas; ovary inferior, (1–3)4-locular, but sometimes septa are weakly developed or completely

lacking (*Laurembergia* and most species of *Glischrocaryon*), with 1–2 pendulous ovules per locule (if two ovules, then one aborts at an early stage). Ovules anatropous or hemitropous, with hypostase and weakly developed funicular obturator. Haustorial suspensor present. Endosperm cellular (*Haloragis* and some species of *Myriophyllum*) or nuclear (*Laurembergia* and some species of *Myriophyllum*). Fruits small, nutlike or drupelike, in *Myriophyllum* splitting septicidally into (2–3)4 mericarps, but in all other genera indehiscent, 1–4 seeded, variously ornamented with wings, ribs and tubercles. Seeds with straight, cylindrical embryo and usually with more or less copious, starchy endosperm; exotesta (and hypodermal payer) persistent, thin-walled. Present flavones, tannins, myriophyllin and ellagic acid; n = 6 (*Gonocarpus*), 7, 9, 21, 29, mostly 7. . . . . 5. HALORAGACEAE.

4 Embryo curved to bent (obcordate). Perennial herbs with very short upright stems approaching acaulescence, and short or prostrate and creeping, usually more or less fleshy rhizomes, and adventitious roots, harboring symbiotic *Nostoc* and *Chlorococcus* colonies in superficial tissues of stem and petiole entered through hydathodes or mucilage glands. The stem has an endodermis. Hairs unicellular. Vascular system of the stem and rhizome consists of anastomosing bundles, each with its own endodermis. Scalariform perforation plates occur in most vessel elements of stolons and roots; vessels with simple perforations are more frequent in the stems of large leafed-species (Wilkinson 2000). Fibers thick-walled, with simple pits. Sieve-element plastids of Pcs-type. Nodes multilacunar. Leaves alternate, all radical, extremely large (sometimes more than 3 m across) to small, long-petiolate, orbicular, reniform or ovate, sometimes peltate, entire, toothed, or variously lobed; the margins of larger-leafed species are crenate-denate with crenations and lobes having protruding hydathodes, up to 8 mm; pinnately or subpalmately veined, with large median axillary scales (squamulae intravaginales), which sometimes have been interpreted as stipular. Stomata anomocytic. Flowers small, in mostly very large, terminal or upper axillary, often much branched, racemose panicles, usually not bracteolate, the basal ones often female, upper ones male and middle ones bisexual, or seldom all the flowers bisexual or the flowers all unisexual and dioecious, monoecious or gynomonoecious. Sepals two, rarely three, valvate, very small, often almost absent. Petals two, miter-shaped, or more often absent. Stamens one or two, rarely three, with short filaments; anthers 2-locular, opening longitudinally. Pollen grains 2-celled, with 3(-5) long, deep colpi (with bulging mesocolpia). Gynoecium of two united carpels, with free subulate, long papillate stylodia; ovary inferior, 1-locular, rarely 2-locular (*Gunnera chilensis*), with 1(2) pendulous ovules. Ovules large, hemianatropous, with broad nucellus; micropyle endostomal, formed by inner integument. Female gametophyte of *Peperomia*-type (tetracyclic, 16-nucleate). Fruits one-seeded drupes. Seeds with a very small, cordate embryo and copious, oily endosperm. Producing caffeic and ellagic acids, pelargonidin, two flavonoids: quercetin 3–0-galactoside, and quercetin 3–0-glucoside and high concentration of an unidentified ellagitannin, n = 12, 17, 27, more often 17. . . . . .6. GUNNERACEAE.

3 Ovules unitegmic. Low, evergreen, glabrous shrubs. Imperforate tracheary elements are exclusively very short tracheids. Vessels scalariform, with 5–18 bars per plate; lateral pitting scalariform to opposite. Rays heterogeneous. Axial parenchyma apotracheal, sparse and either diffuse or diffuse-in-aggregates. Nodes unilacunar and one-trace. Leaves small, simple, serrate, with rounded to acute teeth, pinnately veined with conspicuous secondary veins terminating near the leaf margin. Stomata anomocytic. Flowers in erect bracteate racemes, small, bisexual, actinomorphic. Sepals four, small, essentially free, imbricate, persistent. Petals four, free, spreading, spatulate, slightly

imbricate, caducous. Stamens four or eight, borne in a single cycle and positioned in both sepal- and petal-planes although one or more members of the whorl may be missing; filaments free, filiform; anthers elliptic-oblong, basifixed, latrorse. Pollen grains very small, 3-colporate, rugulate. Carpels 4(5), fusiform, supplied by three veins that all reach the stigma, prominently stipitate, erect, free for most of their length, but sometimes adjacent ones are united to the middle of their ventral surfaces; usually one or more carpels remain unfused; stigma subsessile, small, lobed. Ovules numerous, borne on branched submarginal placentas. Fruits multifollicles; follicles erect, stipitate, coriaceous, many-seeded. Seeds very small, obovoid-subulate, testa membranous, slightly prolonged at each end, with narrow wings extended along their entire length and parallel ridges on the surface; embryo minute, at the base of fleshy copious endosperm. .... 1. TETRACARPAEACEAE.

2 Leaves opposite, simple, ovate-lanceolate or elliptical, serrate to mostly entire; stipules wanting but with minute colleters at each side of the nodes. Stomata anomocytic. Scrambling or viny, thin-stemmed, glabrous shrubs, with endodermis in young stems. Vessels with scalariform perforations; rays homocellular, uniseriate and heterocellular, multiseriate. Axial parenchyma scarce, being apotracheal diffuse with occasional paratracheal scanty; cells of both ray parenchyma and axial parenchyma contain abundant starch grains (Dickison et al. 1994). Nodes unilacunar with one trace. Flowers solitary or in axillary panicles, bisexual, 4-merous. Sepals free, imbricate, greatly enlarged after flowering and persistent. Petals very small or absent. Stamens eight, perigynous; anthers long, almost basifixed, not versatile, with connective protrusion, opening longitudinally. Pollen grains 3-colporate with regulate-stellate sculpture. Gynoecium of four carpels; style 4-lobed apically, with four canals, stigmas terminal, highly papillate; ovary semi-inferior, 4-locular; ovules one per locule, suspended on axile placenta, with long, thick funiculus. Fruits nut-like, involucrate by the enlarged calyx-lobes. Seed rugose; embryo curved, endosperm fleshy. .... 2. APHANOPETALACEAE.

1 Flowers mostly 5-merous.

5 Gynoecium apocarpous or less often semiapocarpous (Penthoraceae and some Crassulaceae).

6 Vessels with scalariform perforations. Vessels have many to numerous bars (9–57, usually 18–34 per plate); lateral pits bordered. Imperforate tracheary elements fiber-tracheids and vascular tracheids. Axial parenchyma absent. Nodes unilacunar. Perennial rhizomatous, rather fleshy herbs; young stem with pseudosiphonostele. Leaves alternate, simple, serrate, pinnately veined, nonsucculent. Stomata anomocytic. Flowers in terminal cymose inflorescences, bisexual, 6–7(8)-merous. Sepals basally connate, valvate, persistent. Petals inconspicuous and devoid of venation or more often wanting. Stamens twice as many as sepals, in two cycles; filaments filiform; anthers basifixed, latrorse. Pollen grains 3-colporate. Gynoecium of 5-6(7-8) carpels united halfway, recurved, each with short stylodium and small capitate stigma; ovules numerous in each carpel, on a thick, axile placenta. Fruits syncarpous multifollicles; follicles circumscissile above their union. Seeds numerous, scobiform, papillate, with an operculum formed by an inner integument in the micropylar part; seed coat consists of one layer of exotesta. Present flavonoids and diglucosides, n = 8, 9. .... 3. PENTHORACEAE.

6 Vessels with simple perforations. Lateral pits simple. Imperforate tracheary elements absent, or libriform fibers. Nodes unilacunar, trilacunar or multilacunar. Mostly perennial herbs with fleshy leaves and stem, sometimes succulent subshrubs and small shrubs, rarely arborescent. Leaves alternate, opposite, or sometimes verticillate, simple and usually entire, often with hydathodes. Stomata anisocytic or diacytic. Flowers in various kinds of cymose inflorescences or sometimes solitary, bisexual or rarely (*Rhodiola* p.p.) unisexual, usually actinomorphic, mainly 5-merous or less, free or nearly so, less often more or less connate, persistent. Sepals usually green, free or more or less connate, much shorter than the corolla. Petals usually free or connate only at the base, rarely corolla sympetalous. Stamens mostly twice as many as petals, in two

obdiplostemonous cycles, less often isomerous and alternate with the petals (Crassuloideae); filaments free or very rarely connate below, in sympetalous flowers borne on the corolla tube (the antepetalous ones a little higher than the antesepalous ones); anthers latrorse or more or less introrse, more or less basifixed. Pollen grains 2-celled, usually 3-colporate. Carpels as many as the sepals or petals, free or connate only at the base, rarely (*Diamorpha, Pagella*) connate nearly to the middle, each tapering into a short or elongate stylodium; near the base of each carpel a small nectariferous scale borne externally, which is larger and petaloid in *Monanthes* and some species of *Sedum*. Ovules usually numerous in each carpel, rarely few or sometimes only one, on submarginal placentas, crassinucellate. Endosperm cellular, sometimes with chalazal haustoria. Fruits mostly many-seeded multifollicles which dehisce xerochastically, rarely (*Diamorpha*) syncarpous multifollicle with the carpels opening dorsally. Seeds very small; "seed coat is 4-layered: the exotestal cells have thickened outer wall, the inner exotegmic cell layer is pigmented, and the two middle layers are completely crushed" (Thiede and Eggle 2007); embryo long, straight; endosperm copious to scanty. Plants often with pyridine alkaloids, flavones, acylated flavonol glycosides; red pigment common, even in roots; n = 4–8 with a reduction to 7 in *Crassula* and n = 6, 7 in the *Leucosedum* clade (Mort et al. 2001). . . . 4. CRASSULACEAE.

5 Gynoecium usually syncarpous, rarely apocarpous or semiapocarpous.

7 Fruits juicy berries crowned by persistent calyx. Seeds usually numerous, rather small, with a funicular aril investing the seed or reduced to crenulate fleshy placental ridge; exotestal cells palisade, mucilaginous, endotestal cells crystalliferous, radial and inner walls lignified; embryo small, short; endosperm oily, with slightly thickened walls. Flowers small, in racemes or solitary, bisexual or unisexual, actinomorphic. Sepals (4)5, basally connate, imbricate or subvalvate, sometimes petaloid, persistent; calyx tube adnate to the ovary. Petals (or staminodia) five, small, squamiform, obovate or subulate, adnate to the calyx tube. Stamens five, alternating with the petals. Pollen grains periporate. Gynoecium of two carpels with free or more or less connate stylodia and simple stigma. Ovary inferior, 1-locular, with few to numerous ovules on lateral placentas. Endosperm cellular, helobial, or nuclear. Leaves alternate, simple, variously lobed, with palmate venation, estipulate but often dilated at base; hydathodes associated with the apiculate marginal teeth. Stomata anamocytic. Nodes trilacunar. Vessels with scalariform perforations or very rarely also with a few simple perforations; lateral pitting mostly alternate. Fibers often with bordered pits. Rays heterogeneous. Axial parenchyma absent or in very rare bands. Erect, arching, trailing, or prostrate shrubs, often with conspicuous 3-forked or simple nodal spines, and smaller, internodal bristles. Present flavonoids and tannins, proanthocyanidins, myricetin glycosides and some acids; n = 8, rarely 16. . . . . 8. GROSSULARIACEAE.

7 Fruits mostly capsular, sometimes multifollicles or syncarpous multifollicles.

8 Vessels usually with simple perforations. Fibers, when present, small, with bordered pits. Nodes usually trilacunar or less often multilacunar (*Astilboides, Astilbe, Bergenia, Mukdenia, Rodgersia, Darmera*). Perennial or rarely annual or biennial, often rhizomatous herbs; hairs uni- to multiseriate with multicellular glandular head; young stem with separate bundles (pseudosiphonostele). Leaves generally basal, usually alternate, rarely opposite, simple, entire or lobed and pinnately or palmately veined, or pinnately or palmately compound or decompound, estipulate. Stomata usually anomocytic. Flowers in various kinds of cymose or racemose inflorescences or rarely solitary, bisexual or rarely unisexual (as in *Tanakaea*), actinomorphic or rarely zygomorphic. Perianth usually 5-merous, less often 3-merous, forming a more or less developed floral tube, which is free or variously adnate to the base of the ovary. Sepals imbricate or valvate. Petals 5(-10), imbricate or valvate, rarely wanting (as in *Chrysosplenium* and *Bensoniella oregona*). Stamens as many as petals and alternate with them or twice that number, in which case there is a stamen opposite each perianth; sometimes eight, four or even (*Tolmiea*) three

stamens; anthers slightly introrse or latrorse, mostly basifixed, sometimes slightly dorsifixed, versatile. Pollen grains 2-celled, 3-colpate, 3-colporate or 6–9-porate. Gynoecium mostly syncarpous, rarely more or less apocarpous (*Astilbe* spp. and *Darmera*), of 2–5 (mostly only 2) carpels, usually with more or less free, apical stylodia; stigmas dorsal, papillate; the carpels sometimes more or less open ventrally in the free distal regions (as in *Huchera, Tiarella, Tolmiea, Mitella, Bensoniella,* and *Tellima*); ovary superior to inferior; ovules 9–30 in several rows, pendulous or ascending, anatropous, bitegmic or rarely (*Saxifraga* section *Micranthes, Darmera*) unitegmic; micropyle zig-zag or not zig-zag. Endosperm cellular, helobial, or nuclear and in some members develops various kinds of haustoria. Fruits mostly septicidal capsules, sometimes multifollicles or syncarpous multifollicles. Seeds small, smooth, exotestal cells with outer wall more or less thickened; embryo straight, minute to medium sized, surrounded by copious, oily endosperm. Present flavonols (quercetin, kaempferol, myricetin), sometimes proanthocyanidins (cyanidin and delphinidin), n = 7–17, mostly 7....... 7. SAXIFRAGACEAE.

8 Vessels usually with scalariform perforations.

9 Ovary 5-locular. Much branched shrubs. Conical to peltate glandular hairs present. Vessels with simple to scalariform perforations. Nodes trilacunar. Leaves alternate, dentate, shining, glandular, and glutinous-resinous above, softly pubescent below; stipules minute, subulate, deciduous. Stomata anomocytic. Flowers in few-flowered, corymbose cymes, bisexual, actinomorphic, 5-merous. Calyx tube turbinate, adnate to ovary, lobes triangular-subulate, erect, valvate. Petals imbricate, conspicuous, at length reflexed, pubescent, white. Stamens ten in two cycles, inserted with the petals, erect, the five outer fertile, with broad filaments denticulate at apex and ovoid, dorsifixed, mucronate, introrse anthers. Pollen grains 3-colporate, with complete tectum and complex endoaperture. Gynoecium of five carpels; style shortly 5-lobed, with short and radiate stigmas, separating more with age; ovary inferior, with 4–6 ascending axile ovules per locule. Fruits few-seeded, septicidal, woody capsules crowned by the erect sepals and reflexed petals. Seeds with cartilaginous testa, attenuated at either end; embryo elongate, surrounded by copious, fleshy endosperm. Plants contain C-glycosyl flavones and 3-O-glycosides of quercetin........ 9. PTEROSTEMONACEAE.

9 Ovary 2-locular. Trees and shrubs with simple, unicellular and glandular hairs; young stem with separate bundles. Branches with lamillate pith. Vessels with scalariform perforations that have numerous slender bars; lateral pitting scalariform. Rays uniseriate, heterocellular. Nodes trilacunar with three traces. Leaves alternate, dentate (spiny-dentate in *Itea ilicifolia*), pinnately veined, estipulate or with minute, linear stipules. Stomata paracytic. Flowers small, in terminal or axillary racemes or panicles, bisexual or polygamous, actinomorphic. Sepals five, basally connate into a short turbinate or obconic tube, lobes valvate or open, persistent. Petals five, valvate, persistent. Stamens five, alternating with the petals, inserted under the edge of the more or less developed annular nectary disc; filaments subulate; anthers small, oblong to ovoid, dorsifixed, introrse. Pollen grains 2-porate, bilateral, ektexine homogeneous. Gynoecium of two united carpels; stylodia connate into a 2-grooved style but become free at maturity although often still united by capitate stigmas; ovary nearly superior to more than three-quarters inferior, longitudinally 2-grooved, with usually numerous ovules on axile placentas. Fruits septicidally 2-valved capsules; fruit valves often attached by the stigma. Seeds few to many; exotestal cells with outer walls thickened; embryo large, curved, surrounded by sparse fleshy endosperm. Present C-gycosyl glavones, an unusually in Saxifragales (Bohm et al. 1988); n = 11. .10. ITEACEAE.

1. TETRACARPAEACEAE

Nakai 1943. 1/1. Mountains of Tasmania.
Tetracarpaea.

Most closely allied with the Penthoraceae, Crassulaceae, and Saxifragaceae, which is supported by comparative morphological studies (Hils et al. 1988) and also ***rbc*L** sequence data (Morgan and Soltis 1993).

2. APHANOPETALACEAE

Doweld 2001. 1/2. Southeastern Australia.

Aphanopetalum.

Probably related to Haloragaceae (see Fishbein et al. 2001).

3. PENTHORACEAE

Rydberg ex Britton 1901. 1/2. Far East of Russia, Japan, Korea, China, northern Vietnam, and eastern parts of the United States from Maine to Ontario and Minnesota and south to Florida and Texas.

Penthorum.

Related to both the Crassulaceae and Saxifragaceae, differing from them in having vessels with scalariform perforations, fibers with bordered pits, in their chemical makeup (Jay 1970; Soltis and Bohm 1982), and especially in the presence of an operculum (Nemirovich-Danchenko 1994a). Grund and Jensen (1981) declined to place *Penthorum* in either Crassulaceae or Saxifragaceae, and according to Haskins and Hayden (1987) *Penthorum* is probably best classified in the monogeneric Penthoraceae. A separate family Penthoraceae has been accepted by van Tieghem (1898), Rydberg (1905), Novak (1961, 1972), Airy Shaw (1973), Stern (1974) and Thorne (2006). The *rbc*L sequence data also suggest that *Penthorum* should not be accommodated in either Saxifragaceae or Crassulaceae although the genus is more closely related to Crassulaceae than to Saxifragaceae (Morgan and Soltis 1993).

4. CRASSULACEAE

J. Saint-Hilaire 1805 (including Cotyledonaceae Martynov 1820, Rhodiolaceae Martynov 1820, Sedaceae Martynov 1820, Sempervivacеae A.L. de Jussieu 1782, Tillaeaceae Martynov 1820). About 41/1550–1600. Widely distributed, mainly in dry and warm regions, but centered in tropical and especially South Africa.

Classification after J.Thiede and U.Eggli (2007).

4.1 SEMPERVIVOIDEAE (including Sedoideae and Echeverioideae)
Androecium usually obdiplostemonous. Flowers 3-12(-32)-merous, mostly 5-merous. Petals free or basally connate, rarely corolla sympetalous. – THELEPHIEAE: *Sinocrassula, Kungia, Meterostachys, Orostachys, Hylotelephium Hylotelephium*; UMBILICEAE: *Umbilicus* (including *Chiasto-phyllum*), *Pseudosedum, Rhodiola, Phedimus*; SEMPERVIVEAE: *Sempervivum, Petrosedum*; AEONIEAE: *Aichryson, Monanthes, Perrierosedum, Aeonium* (including *Greenovia*), SEDEAE: *Pistorinia, Rosularia, Prometheum, Afrovivella, Sedella* (including *Parvisedum*), *Dudleya, Sedum* (including *Diamorpha, Cremnophyla, Mucizonia*), *Villadia, Lenophyllum, Graptopetalum* (including, *Tacitus*), *Thompsonella;* ECHEVERIEAE: *Echeveria, Pachyphytum.*

4.2 KALANCHOIDEAE (including Cotyledonoideae)
Androecium obdiplostemonous. Flowers 5-merous or 4-merous. Petals more or less forming tubes or campanulate corolla. – *Adromischus, Kalanchoe* (including *Bryophyllum*), *Cotyledon, Tylecodon.*

4.3 CRASSULOIDEAE
Androecium haplostemonous. Petals free or basally connate, rarely forming tubular or campanulate corolla. – *Crassula* (including *Tillaea, Rochea, Dinacria, Pagella*), *Hypagophytum.*

Close to the Penthoraceae, differing mainly in succulent leaves, anisocytic stomata, vessels with simple perforations, and the presence of small nectariferous appendages near the base of each carpel.

5. HALORAGACEAE

R. Brown 1814 (including Cercodiaceae A.-L. de Jussieu 1817, Myriophyllaceae Schultz-Schultzenstein 1832). 8/145. Mainly in the Southern Hemisphere, particularly in Australia; *Myriophyllum* is cosmopolitan and *Proserpinaca* is confined to the Northern Hemisphere; *Laurembergia* is endemic to South Africa.

Haloragis, Haloragodendron, Glischrocaryon, Meziella, Gonocarpus, Laurembergia, Proserpinaca, Myriophyllum.

The Haloragaceae have often been considered to be allied to the Myrtales, particularly to the Onagraceae. However, they differ from the Myrtales in lacking internal phloem and vestured pits, in having free stylodia covered by large papillae and cellular endosperm in some of their members, and seeds with well-developed endosperm. They have more features in common with the Saxifragales, particularly free stylodia, cellular endosperm of some members (always nuclear in Myrtales), and seeds with copious endosperm (absent or very scanty in Myrtales). A rather close relationship between Haloragaceae and Saxifragales, particularly between *Myriophyllum* and *Penthorum*. In Thorn's system (2006) the Haloragaceae are including in the Saxifragales.

6. GUNNERACEAE

Meisner 1842. 1/40–60. Southeastern tropical and South Africa, Madagascar, Malesia, New Guinea, Tasmania, New Zealand, Hawaii, America from Mexico to Tierra del Fuego.

Gunnera.

Gunnera is often included in the Haloragaceae (Bentham and Hooker 1865; de Candolle 1868; Engler et Prantl 1893; Schindler 1905; Hutchinson 1973; Cronquist 1981; Haywood 1993), but from that family it differs in sieve-element plastids, inflorescence, pollen morphology, in some embryological features, including *Peperomia*-type female gametophyte, and especially the absence of suspensor hausrotium. They also related to the Saxifragaceae (Huber 1963; Hegnauer 1969; Dahlgren 1975, 1980, 1983; Takhtajan 1980, 1987, 1997; Behnke 1986; Doyle and Scogin 1988). According to Fuller and Hickey (2005) Gunneraceae are closely affinity to the Saxifragaceae, especially to the genus *Chrysosplenium*. However, they differ from the Saxifragaceae in unicellular hairs, 2-merous flowers, pollen morphology, female gametophyte, drupaceous fruits, and cordate embryo.

Crane and Hoot (1994) and also D. Soltis et al. (1999, 2003) placed this family as the sister group to Myrothamnaceae. But this conclusion strongly contradict morphological differences between this families.

7. SAXIFRAGACEAE

A.L. de Jussieu 1789 (including Chrysospleniaceae Berchtold et J. Presl 1820, Pectiantiaceae Rafinesque 1837) 33/525–550. Mainly in northern temperate regions with a few species in southern temperate regions and in tropical mountains.

Cascadia, Saxifragodes, Micranthes, Peltoboykinia, Chrysosplenium, Darmera, Astilboides, Rodgersia, Mukdenia, Oresitrophe, Bergenia, Tellima, Lithophragma, Bensoniella, Heuchera, Conimitella, Elmera, Tolmiea, Mitella, Tiarella, Telesonix, Jepsonia, Bolandra, Sullivantia, Suksdorfia, Hieronymusia, Boykinia, Astilbe, Saxifragopsis, Leptarrhena, Tanakaea, Saxifraga (including, *Saxifragella*), *Saniculiphyllum.*

Close to the Crassulaceae, but somewhat more advanced. The genus *Chrysosplenium* (5 species) is rather isolated within the family by its peculiar inflorescence, 4-merous, apetalous flowers (Bensel and Palser 1975), micromorphology of seeds (Kaplan and Strohschneider 1984), and also by its chemical makeup (Bohm et al. 1977; Bohm and Collins 1979).

8. GROSSULARIACEAE

A. P. de Candolle 1805 (including Ribesiaceae Marquis 1820). 1/150. Temperate regions of Eurasia, northeastern Africa, North and Central America, Pacific South America south to Tierra del Fuego.

Ribes (including *Grossularia*).

Flower morphology of *Ribes* is very close to that of Saxifragaceae (Gelius 1967), and in floral structure are strikingly similar to *Tiarella* (Bensel and Palser 1975). Xylem anatomy and embryological characters (Huber 1963) and seed structure (Corner 1976) of *Ribes* is also similar to that of Saxifragaceae as well as to that of Crassulaceae. The affinity of the Grossulariaceae to the Saxifragaceae is also supported by chemical (Jay 1970), serological (Grund and Jensen 1981), and *rbc*L sequence data (Morgan and Soltis 1993).

9. PTEROSTEMONACEAE

Small 1905. 1/3. High mountains of Mexico.

Pterostemon.

Pterostemonaceae are usually associated with Escalloniaceae, Hydrangeaceae, and related families, but ovules in *Pterostemon* are bitegmic and crassinucellate. Evidence from *rbc*L sequence suggests that there is close alliance between *Pterostemon* and *Itea*. However, *Pterostemon* differs markedly from *Itea* in imbricate petals, ten stamens, 3-colporate pollen grains,

5-locular inferior ovary and a number of other features. Most probably the Pterostemonaceae belong to the Saxifragales but occupy a rather isolated position.

10. ITEACEAE

J. Agardh 1858. 1/27. East, Central, and South Africa, Himalayas, Assam, eastern Asia and Southeast Asia, West Malesia, and one species in Atlantic North America from New Jersey to Florida.

Itea (including *Choristylis*).

According to Bensel and Palser (1975: 675), the flowers of *Itea* appear very "saxifragaceous" in external morphology, and their floral vascular patterns are very similar to those typical of the Saxifragoideae. The *rbc*L sequence data also suggest a close relationship between *Itea* and Saxifragaceae s. str. and especially between *Itea* and *Pterostemon* (Morgan 1993). On the other hand, chemical data suggest that *Itea* differs from Saxifragaceae s. str. and more closely resembles other taxa, particularly *Brexia*. Plouvier (1965) reported the presence of allitol in *Itea* and suggested that this may indicate a relationship with *Brexia*, which accumulates dulcitol, an isomer of allitol. Both allitol and dulcitol are not reported in Saxifragaceae and related families. Also, *Itea* accumulates only flavones but no flavonols that are characteristic of the Saxifragales (Bohm et al. 1988). Thus, the Iteaceae is a rather distinct family that needs further study.

Bibliography

Abe K. 1982. Embryological studies in the family Saxifragaceae (s.l.): I. Development of the ovule and embryo sac in *Saxifraga fortunei* var. *partita* (Maldno) Nakai. Am. J. Bot. 69: 416–420.

Agababian VS. 1960. On the palynosystematics of the family Iteaceae. Bull. Armenian Acad. Sci., Biol., 13: 99–102 (in Russian.)

Agababian VS. 1961. Materials toward the palynological study of the family Saxifragaceae s.l. Izvestia Armenian Acad. Sci., Biol., 14(2): 45–61 (in Russian).

Agababian VS. 1963. On palynomorphology of the genus *Ribes*. Bull. Armenian Acad. Sci. 16 (4): 93–98.

Agababian VS. 1964. Evolution of pollen in the orders Cunoniales and Saxifragales in relation to some question of their systematics and phylogeny. Izvestia of Acad. Sci. of the Armenian SSR. Biol. Sciences 17: 59–72 (in Russian).

Al-Shammary KI 1991. Systematic studies of the Saxifragaceae, chiefly from the Southern Hemisphere. Ph.D. dissertation, University of Leicester.

Al-Shammary KI and RJ Gornall. 1994. Trichome anatomy of the Saxifragaceae s. 1. from the Southern Hemisphere. Bot. J. Linn. Soc. 114: 99–131.

Bader F and J Walter. 1961. Das Areal der Gattung *Gunnera* L. Bot. Jahrb. Syst. 80: 281–293.

Bates JC. 1933. Comparative anatomical research within the genus *Ribes*. Univ. Kansas Sci. Bull. 21: 369–398.

Batham E. 1943. Vascular anatomy of New Zealand species of *Gunnera*. Trans. Roy. Soc. N. Zealand 73: 209–216.

Bawa SB. 1969a. Embryological studies on the Halora-gidaceae: II. *Laurembergia brevipes* Schindl. and a discussion of systematic considerations. Proc. Natl. Inst. Sci. India 353: 273–290.

Bawa SB. 1969b. Embryological studies on the Halora-gidaceae: III. *Myriophyllum intermedium* DC. Beitr. Biol. Pfl. 45: 447–464.

Bawa SB. 1970. Haloragaceae. Bull. Indian Natl. Acad. 41: 226–229.

Behnke H-D. 1986. Contributions to the knowledge of sieve-element plastids in Gunneraceae and allied families. Plant Syst. Evol. 151: 215–222.

Bensel CR and BF Palser. 1975a. Floral anatomy in the Saxifragaceae sensu lato: II. Saxifragoideae and Iteoideae. Am. J. Bot. 62: 661–675.

Bensel CR and BF Palser. 1975b. Floral anatomy in the Saxifragaceae sensu lato: IV. Baueroideae and conclusions. Am. J. Bot. 62: 688–694.

Bergman B, C Johansson, and E Söderbäck. 1992. The *Nostoc–Gunnera* symbiosis. New Phytologist 122: 379–400.

Bohm BA and UG Bhat. 1985. Flavonoids of *Astilbe* and *Rodgersia*. Biochem. Syst. Ecol. 13: 437–440.

Bohm BA and FW Collins. 1979. Flavonoids of some species of *Chrysosplenium*. Biochem. Syst. Ecol. 7: 195–201.

Bohm BA and R Ornduff. 1978. Chemotaxonomic studies in the Saxifragaceae s.1.: 9. Flavonoids of *Jepsonia*. Madroño 25: 39–43.

Bohm BA and CK Wilkins. 1978. The flavonoids of *Heuchera cylindrica*. Canad. J. Bot. 56: 1174–1176.

Bohm BA, FW Collins, and R Bose. 1977. Flavonoids of *Bergenia, Francoa,* and *Parnassia*. Biochem. Syst. Ecol. 13: 221–233.

Bohm BA, LS Donevan, and UC Bhat. 1986. Flavonoids of some species of *Bergenia, Francoa, Parnassia,* and *Lepuropetalon*. Biochem. Syst. Ecol. 14: 75–77.

Bohm BA, G Chalmers, and UG Bhat. 1988. Flavonoids and the relationship of *Itea* to the Saxifragaceae. Phytochemistry 27: 2651–2653.Chapman M. 1933. The ovule and embryo sac of *Saxifraga virginiensis*. Am. J. Bot. 20: 151–158.

Collins FW, BA Bohm, and CK Wilkins. 1975. Flavonol glycoside gallates from *Tellima grandiflora*. Phytochemistry 14: 1099–1102.

Cutler DF and M Gregory. 1998. Anatomy of the Dicotyledons, Saxifragales (sensu Armen Takhtajan 1983), 2nd ed, vol. 4. Clarendon Press, Oxford.

Dahlgren KVO. 1930. Zur Embryologie der Saxifra-goideen. Svensk. Bot. Tidskr. 24: 429–448.

Dandy JE. 1927. The genera of Saxifragaceae. In: J Hutchinson, Contributions towards a phylogenetic classification of flowering plants. Kew Bull. Misc. Inform. 6: 100–118.

Danilova MF. 1996. Crassulaceae. In: A Takhtajan, ed. Comparative seed anatomy, vol. 5, pp. 25–32. Nauka, St. Petersburg (in Russian).

Dawson MI. 1983. Chromosome numbers of three South American species of *Gunnera* (Gunneraceae). New Zealand J. Bot. 21: 457–459.
Dickison WC, MH Hils, TW Lugansky, and WL Stern. 1994. Comparative anatomy and systematics of woody Saxifragaceae *Aphanopetalum* Endl. Bot. J. Linn. Soc. 114: 167–182.
Doyle MF. 1990. The biosystematics of *Gunnera* (Gunneraceae) in the Hawaiian Islands. Ph.D. dissertation, Claremont Graduate School.
Doyle MF and R Scogin. 1988a. A comparative phyto-chemical profile of the Gunneraceae. New Zealand J. Bot. 26: 493–496.
Doyle MF and R Scogin. 1988b. Leaf phenolics of *Gunnera mamcata* (Gunneraceae). Aliso 12: 77–80.
Eggli U, ed. 2003. Illustrated handbook of succulent plants. Crassulaceae. Springer, Berlin.
Elst P van der. 1909. Beiträge zur Kenntnis der Samen-auflage der Saxifragaceae. Phyl. Dissertation, University of Utrecht.
Endress PK and A Igersheim. 1999. Gynoecium diversity and systematics of the basal eudicots. Bot. J. Linn. Soc. 130: 305–393.
Favarger C. 1957. Sur deux criteres nouveaux utilisables dans la taxinomie des Saxifragacées. Rev. Cytol. Biol. Veg. 18: 125–137.
Ferguson IK and DA Webb. 1970. Pollen morphology in the genus *Saxifraga* and its taxonomic significance. Bot. J. Linn. Soc. 63: 295–311.
Fernandes Herrera D. 1984. Estudio anatomico y morfologico de *Gunnera insignis* (Oerst.) A.DC. Revista Biol. Trop. 32: 197–202 (with English summary).
Fishbein M and DE Soltis. 2004. Further resolution of the rapid radiation of Saxifragales (Angiosperms, Eudicots) supported by mixed-model Bayesian analysis. Syst. Bot. 29: 883–891.
Fishbein M, C Hibsch-Jetter, C Soltis, DE Soltis, and L Hufford. 2001. Phylogeny of Saxifragales (Angiosperms, Eudicots): analysis of a rapid ancient radiation. Syst. Biol. 50: 817–847.
Fuller DQ. 1995. Systematics and leaf architecture of the Gunneraceae. Am. J. Bot. 82(6): 130–131 (Abstract).
Fuller DQ and LJ Hickey. 2005. Systematics and leaf architecture of the Gunneraceae. Bot. Rev. 71: 295–353.
Gäumann E. 1919. Studien über die Entwicklungsg-eschichte einiger Saxifragales. Rec. Trav. Bot. Neerl. 16: 285–322.
Ge L-P, A-M Lu, and K-Y Pan. 2002. Floral ontogeny in *Itea yunnanensis* (Iteaceae). Act. Bot. Sinica 44: 1261–1267.
Gelius L. 1967. Studien zur Entwicklungsgeschichte an Blüten der Saxifragales sensu lato mit besonderer Be-rücksichtigung des Androeceums. Bot. Jahrb. Syst. 87: 253–303.
Gontcharova SB. 1999. Ornamentation of the testa of some eastern Asian Sedoideae (Crassulaceae). Bull. Nation. Sci. Mus., B (Tokyo) 25(4): 131–141.
Gornall RJ 1986. Trichome anatomy and the taxonomy of *Saxifraga* (Saxifragaceae). Nord. J. Bot. 6: 257–275.
Gornall RJ. 1987. Foliar crystals in *Saxifraga* and segregate genera (Saxifragaceae). Nord. J. Bot. 7: 233–238.
Gornall RJ. 1989. Anatomical evidence and the taxonomic position of *Darmera* (Saxifragaceae). Bot. J. Linn. Soc. 100: 173–182.
Gornall RJ and BA Bohm. 1980. The use of flavonoids in the taxonomy of *Boykinia* and allies (Saxifragaceae). Canad. J. Bot. 58: 1768–1779.
Gornall RJ and BA Bohm. 1985. A monograph of *Peltoboykinia, Boykinia, Bolandra,* and *Sarsdorfia* (Saxifragaceae). Bot. J. Linn. Soc. 90: 1–71.
Grund C and U Jensen. 1981. Systematic relationships of the Saxifragales revealed by serological characteristics of seed proteins. Plant Syst. Evol. 137: 1–22.
Ham RCHJ van and H t'Hart. 1998. Phylogenetic relationships in the Crassulaceae inferred from chloroplast DNA restriction-site variation. Am. J. Bot. 85: 123–134.
Hamel J. 1953. Contribution á l'étude cytotaxinomique des Saxifragacées. Rev. Cytol. Biol. Vég. 14: 113–311.
Harmsen L. 1939. Studies in the embryology and cytology of *Saxifraga*. Medd. Gronl. 125: 1–15.
Hart H't, RDHJ van Ham, JF Stevens, ET Elema, H van der Klis, and TWJ Gadella. 1999. Biosystematic, molecular and phytochemical evidence for the multiple origin of sympetaly in Eurasian Sedoideae (Crassulaceae). Biochem. Syst. Ecol. 27: 407–426.
Haskins ML and WJ Hayden. 1987. Anatomy and affinities of *Penthorum*. Am. J. Bot. 74: 164–177.
Hayden WJ and JD Lewandowski. 1997. Gynoecium structure in *Penthorum*. Am. J. Bot. 84(6): 201 (Abstract).
Hegnauer R. 1969. Chemical evidence for classification of some plant taxa. In: JB Harborne and T. Swain, eds. Perspectives in phytochemistry, pp. 128–138. Academic Press, London/New York.
Herr JM. 1954. The development of the ovule and female gametophyte in *Tiarella cordifolia*. Am. J. Bot. 41: 333–338.
Hideux MJ. 1979. Le pollen donnees nouvelles de la microscopie electronique et de l'informatique: Structure du sporoderme des rosidae-saxifragales, etude comparative et dynamique. Paris.
Hideux MJ and IK Ferguson. 1976. The stereostructure of the exine and its evolutionary significance in Saxifragaceae sensu lato. In: IK Ferguson and J Muller, eds. The evolutionary significance of the exine, pp. 327–377. Linn. Soc. Symposium, No. 1. London/New York.
Hils MH. 1985. Comparative anatomy and systematics of twelve woody Australasian genera of the Saxifragaceae. Matthew Hils, Florida.
Hils MH, WC Dickison, TW Lucansky, and WL Stern. 1988. Comparative anatomy and systematics of woody Saxifragaceae; *Tetracarpaea*. Am. J. Bot. 75: 1687–1700.
Jay M. 1970 (1971). Quelques problmes taxinomiques et phylogénétique des Saxifragacées vus á la lumiére de la biochimie flavonique. Bull. Mus. Hist. Nat., Paris, sér. 2, 42: 754–775.
Jay M and B Voirin. 1976. Les flavonoides de deux espéces du genre *Chrysosplenium*. Phytochemistry 15: 517–519.
Jensen LCW. 1968. Primary stem vascular patterns in three subfamilies of the Crassulaceae. Am. J. Bot. 55: 553–563.
Johansson C and B Bergman. 1992. Early events during the establishment of the *Gunnera/Nostoc* symbiosis. Planta 188: 403–413.
Johnson LA and DE Soltis. 1994. *Mat*K DNA sequences and phylogenetic reconstruction in Saxifragaceae sensu stricto. Syst. Bot. 19: 143–156.
Johnson LA and DE Soltis. 1995. Phylogenetic inference in Saxifragaceae *sensu stricto* and *Gilia* (Polemoniaceae) using *mat*K sequences. Ann. Missouri Bot. Gard. 82: 149–175.
Juel HO. 1907. Studien über die Entwicklungsgeschichte von *Saxifraga granulata*. Nova Acta Reg. Soc. Sci. Uppsal. 4(1): 1–41.
Kapil RN and SB Bawa. 1968. Embryological studies on the Haloragidaceae: I. *Haloragis colensoi* Skottsb. Bot. Not. 121: 11–28.

Kaplan K. 1981. Embryologische pollen- und samen-morphologische Untersuchungen zur Systematik von *Saxifraga* (Saxifragaceae). Bibl. Bot. 134: 1–54.

Kaplan K and M Strohschneider. 1984. Mikromorpho- logische Untersuchungen an Samenoberflächen in der Gattung *Chrysosplenium* (Saxifragaceae). Bot. Jahrb. Syst. 104: 469–482.

Kellermann WA. 1881. Die Entwicklungsgeschichte der Blüten von *Gunnera chilensis*. Zürich.

Klopfer K. 1968, 1970. Beiträge zur floralen Morpho-genese und Histogenese der Saxifragaceae: 1. Die In-floreszenz-Entwicklung von *Tellima grandiflora*. 2. Die Blütenentwicklung einiger *Saxifragen*-Arten. Flora 157B: 461–476, 1968; 158: 1–21, 1968; 159: 347–365, 1970.

Klopfer K. 1973. Florale Morphogenese und Taxonomie der Saxifragaceae sensu lato. Feddes Repert. 84: 475–516.

Knapp U. 1994. Skulptur der Samenschale und Gliederung der Crassulaceae. Bot. Jahrb. Syst. 116: 157–187.

Komar GA. 1967. On the nature of inferior ovary of the Grossulariaceae. Bot. Zhurn. 52: 1611–1629 (in Russian).

Komar GA. 1970. Development of the flower and inflorescence in some representatives of the family Grossulariaceae. Bot. Zhurn. 55: 954–971 (in Russian).

Komar GA. 1996. Grossulariaceae. In: A Takhtajan, ed. Comparative seed anatomy, vol. 5, pp. 60–67. Nauka, St. Petersburg (in Russian).

Krach JE. 1976. Samenanatomie der Rosifloren: I. Die Samen der Saxifragaceae. Bot. Jahrb. Syst. 97: 1–60.

Krach JE. 1977. Seed characters in and affinities among the Saxifragineae. Plant Syst. Evol., Suppl., 1: 141–153.

Kubitzki K. 2007a. Aphanopetalaceae. In: K Kubitzki, ed. The families and genera of vascular plants, vol. 9, pp. 29–30. Sprinegr. Berlin/Heidelberg/New York.

Kubitzki K. 2007b. Haloragaceae. In: K Kubitzki, ed. The families and genera of vascular plants, vol. 9, pp. 184–190. Springer, Berlin/Heidelberg/New York.

Kubitzki K. 2007c. Iteaceae. In: K Kubitzki, ed. The families and genera of vascular plants, vol. 9, pp. 202–204. Springer, Berlin/Heidelberg/New York.

Kubitzki K. 2007d. Pterostemonaceae. In: K Kubitzki, ed. The families and genera of vascular plants, vol. 9, pp. 405–406. Springer, Berlin/Heidelberg/New York.

Kubitzki K. 2007e. Tetracarpaceae. In: K Kubitzki, ed. The families and genera of vascular plants, vol. 9, pp. 456–457. Springer, Berlin/Heidelberg/New York.

Kulbaeva B Zh. 1992a. Seed anatomy in representatives of Saxifragaceae family. Bot. Zhurn. 77(3): 36–49 (in Russian with English summary).

Kulbaeva B Zh. 1992b. The seed surface structure in representatives of the family Saxifragaceae. Bot. Zhurn. 77(4): 61–68 (in Russian with English summary).

Kulbaeva B Zh. 1992c. Seed surface in representatives of the family Saxifragaceae: Typology and significance for systematics and phylogeny. Bot. Zhurn. 77(8): 98–105 (in Russian with English summary).

Kulbaeva B Zh. 1996. Saxifragaceae. In: A. Takhtajan, ed. Comparative seed anatomy, vol. 5, pp. 34–60. Nauka, St. Petersburg (in Russian).

Lobova TA. 2000. Pterostemonaceae. In: A Takhtajan, ed. Comparative seed anatomy, vol. 6, pp. 296. Nauka, St. Petersburg (in Russian).

Lowrey TK and ER Robinson. 1988. The interaction of gynomonoecy, dichogamy, and wind-pollination in *Gunnera perpensa* L. (Gunneraceae) in South Africa. Monograph Syst. Bot. 25: 237–246.

MacGaughey V. 1917. *Gunnera petaloidea* Gaud., a remarkable plant of the Hawaiian Islands. Am. J. Bot. 4: 33–39.

Mandrik VY and LV Golyshkin. 1973. Embryological studies of some species of the family Crassulaceae. Bot. Zhurn. 58: 263–272 (in Russian).

Mauritzon J. 1933. Studien über die Embryologie der Familien Crassulaceae und Saxifragaceae. Ph.D. dissertation, University of Lund.

Mayuzumi S and H Ohba. 2004. The phylogenetic position of Eastern Asian Sedoideae (Crassulaceae) inferred from chloroplast and nuclear DNA sequences. Syst. Bot. 29: 587–598.

Messinger W, K Hummer and A Liston. 1999. *Ribes* (Grossulariaceae) phylogeny as indicated by restriction site polymorphisms of PCR-amplified chloroplast DNA. Plant Syst. Evol. 217: 185–195.

Miehe H. 1924. Entwicklungsgeschichtliche Untersu-chungen der Algensymbiose bei *Gunnera macrophylla* Bl. Flora 17: 1–15.

Modilewski J. 1908. Zur Embryobildung von *Gunnera chilensis*. Ber. Deutsch. Bot. Ges. 26a: 550–556.

Moody ML and DH Les. 2001. Phylogenetic relationships in Haloragaceae emphasizing the aquatic genus *Myriophyllum*. In Botany 2001: Plants and People, p. 128. Albuquerque (Abstracts).

Moreau F. 1971. Apport des caractéres stomatiques á la taxinomie et á la phylogénie des Saxifragées. Bull. Soc. Bot. France 118: 381–427.

Morf E. 1950. Vergleichend-morphologische Untersu-chungen am Gynoeceum der Saxifragaceen. Ber. Schweiz. Bot. Ges. 60: 516–590.

Morgan DR and DE Soltis. 1993. Phylogenetic relationships among members of Saxifragaceae sensulato based on *rbc*L sequence data. Ann. Missouri Bot. Gard. 80: 631–660.

Mort ME, DE Soltis, PS Soltis, J Franciso-Ortega, and A Santos-Guerra. 2001. Phylogenetic relationships and evolution of Crassulaceae inferred from *mat*K sequence data. Am. J. Bot. 88: 76–91.

Mort ME, DE Soltis, PS Soltis, J Francisco Ortega, and A Santos Guerra. 2002. Phylogenetics and evolution of the Macaronesian clade of Crassulaceae inferred from nuclear and chloroplast sequence data. Syst. Bot. 27: 271–288.

Nagaraj M and BHM Nijalingappa. 1967. Embryological studies in *Myriophyllum intermedium* DC. Proc. Indian Acad. Sci. 65B: 210–220.

Nagaraj M and BHM Nijalingappa. 1974. Embryological studies in *Laurembergia hirsuta*. Bot. Gaz. 135: 19–28.

Nemirovich-Danchenko EN. 1994a. Seed structure in *Penthorum sedoides* and *P. chinense* (Penthoraceae). Bot. Zhurn. 79 (6): 56–64 (in Russian with English summary).

Nemirovich-Danchenko EN. 1994b. Morphology and anatomy of the seeds of Iteaceae. Bot. Zhurn. 79(9): 83–87 (in Russian with English summary).

Nemirovich-Danchenko EN. 1996a. Penthoraceae, Crassulaceae. In: A Takhtajan, ed. Comparative seed anatomy, vol. 5, pp. 22–32. Nauka, St. Petersburg (in Russian).

Nemirovich-Danchenko EN. 1996b. Iteaceae. In: A Takhtajan, ed. Comparative seed anatomy, vol. 5, pp. 67–72. Nauka, St. Petersburg (in Russian).

Nemirovich-Danchenko EN. 2000. Tetracarpaceae. In: A Takhtajan, ed. Comparative seed anatomy, vol. 6, pp. 232–233. Nauka, St. Petersburg (in Russian).

Nicholls KW, BA Bohm, and EF Wells. 1986. The flavonoids of *Mitella, Bensoniella,* and *Conimitella* (Saxifragaceae). Canad. J. Bot. 64: 525–530.

Nikiticheva ZI. 1985. Haloragaceae. In: MS Yakovlev, ed. Comparative embryology of flowering plants: Brunelliaceae-Tremandraceae, pp. 116–120. Nauka, Leningrad (in Russian).

Orchard AE. 1975. Taxonomic revisions in the family Haloragaceae: I. The genera *Haloragis, Haloragodendron, Glischrocaryon, Meziella,* and *Gonocarpus*. Bull. Auckland Inst. and Mus. 10: 1–299.

Orchard AE. 1990. Gunneraceae. Flora of Australia 18: 85–87. Canberra.

Pacheco P, DJ Crawford, TF Stuessy, and MO Silva. 1993. Flavonoid chemistry and evolution of *Gunnera* (Gunneraceae) of the Juan Fernandez Islands, Chile. Pacific Sci. 45: 389–399.

Pastre A and A Pons. 1973. Quelques aspects de la systematique des Saxifragacees a la lumiere des donnees de la palynologie. Pollen Spores 15: 117–133.

Planchon JE. 1854. Affinités et synonymic de quelques genres nouveaux. Ann. Sci. Nat. Bot. ser. 4, 2: 256–266.

Plouvier V. 1965. Etudes chimiotaxinomiques sur les Saxifragacees. Bull. Soc. Bot. France, Mem., 1965: 150–161.

Praglowski J. 1970. The pollen morphology of the Haloragaceae with reference to taxonomy. Grana 10: 159–239.

Ramamonjiarisoa BA. 1980. Comparative anatomy and systematics of African and Malagasy wood Saxifragaceae s.l. Ph.D. dissertation. University of Massachusetts. Amherst, MA.

Reinke J. 1873. Untersuchungen über die Morphologie der Vegetationsorgane von *Gunnera*. In: Morphologische Abhandlungen, S. 47–122. Leipzig.

Rocen T. 1928. Beiträge zur Embryologie der Crassulaceen. Svensk Bot. Tidskr. 22: 368–376.

Roeland CHJ Van Ham and H 't Hart. 1998. Phylogenetic relationships in the Crassulaceae inferred from chloroplast DNA restriction-site variation. Am. J. Bot. 85: 123–134.

Rylova TB. 1989. Morphological features of pollen in some fossil and extant species of *Itea* (Iteaceae). Bot. Zhurn. 74: 694–699 (in Russian with English summary).

Savile DBO. 1953. Splash-cup dispersal mechanism in *Chrysosplenium* and *Mitella*. Science 117: 250–251.

Savile DBO. 1975. Evolution and biogeography of Saxifragaceae with guidance from their rust parasites. Ann. Missouri Bot. Gard. 62: 354–361.

Saxena NP. 1964. Studies in the family Saxifragaceae: I. A contribution to the morphology and embryology of *Saxifraga diversifolia* Wall. Proc. Indian Acad. Sci. B 60: 38–51.

Scherwin PA and RL Wilbur. 1971. The contributions of floral anatomy to the generic placement of *Diamorpha smallii* and *Sedum pusillum*. J. Elisha Mitchell Sci. Soc. 87: 103–114.

Schnegg H. 1902. Beiträge zur Kenntnis der Gattung *Gunnera*. Flora 90: 161–208.

Senters AE and DE Soltis. 2003. Phylogenetic relationships in *Ribes* (Grossulariaceae) inferred from ITS sequence data. Taxon 52: 51–66.

Silvester WB and DR Smith. 1969. Nitrogen fixation by *Gunnera-Nostoc* symbiosis. Nature 224: 1231.

Skottsberg C. 1929. Bemerkungen über die Morphologic von *Gunnera macrophylla* Bl. Meddelanden fran Göteborgs bot. Träd. V: 115–126.

Soltis DE. 1980. Karyotypic relationships among species of *Boykinia, Heuchera, Mitella, Sullivantia, Tiarella,* and *Tolmiea* (Saxifragaceae). Syst. Bot. 5: 17–29.

Soltis DE. 1984a. Karyotypic relationships among *Elmera, Heuchera,* and *Tellima* (Saxifragaceae). Syst. Bot. 9: 6–11.

Soltis DE. 1984b. Karyotypes and relationships of species of *Jepsonia* (Saxifragaceae). Syst. Bot. 9: 137–141.

Soltis DE. 1984c. Karyotypes of *Leptarrhena* and *Tana-kaea* (Saxifragaceae). Canad. J. Bot. 62: 671–673.

Soltis DE. 1986. Karyotypic relationships among *Astil-boides, Bergenia, Darmera,* and *Mukdenia* and their implications for subtribal boundaries in Saxifrageae (Saxifragaceae). Canad. J. Bot. 64: 586–588.

Soltis DE. 1988. Karyotypes of *Bensoniella, Conimitella, Lithophragma,* and *Mitella,* and relationships in Saxifragaceae. Syst. Bot. 13: 64–72.

Soltis DE. 2007. Saxifragaceae. In: K Kubitzki, ed. The families and genera of vascular plants, vol. 9, pp. 418–435. Springer, Berlin/Heidelberg/New York.

Soltis DE and BA Bohm. 1982. Flavonoids of *Penthorum sedoides*. Biochem. Syst. Ecol. 10: 221–224.

Soltis DE and L Hufford. 2002. Ovary position diversity in Saxifragaceae: clarifying the homology of epigny. Int. J. Plant Sci. 163(2): 277–293.

Soltis DE and PS Soltis. 1997. Phylogenetic relationships in Saxifragaceae sensu lato: a comparison of topologies based on 18S rDNA and *rbc*L sequences. Am. J. Bot. 84: 504–522.

Soltis DE, PS Soltis, and KD Bothel. 1990. Chloroplast DNA evidence for the origins of the monotypic *Bensoniella* and *Conimitella* (Saxifragaceae). Syst. Bot. 15: 349–362.

Soltis DE, PS Soltis, MT Clegg, and M Durbin. 1990. *rbc*L sequence divergence and phylogenetic relationships in Saxifragaceae sensu lato. Proc. Nad. Acad. Sci. USA 87: 4640–4644.

Soltis DE, A Crable, D Morgan, PS Soltis, and R Kuzoff. 1993. Molecular systematics of Saxifragaceae sensu stricto. Am. J. Bot. 80: 1056–1081.

Soltis DE, RK Kuzoff, E Conti, R Gornall, and K Ferguson. 1996. *mat*k and *rbc*L gene sequence data indicate that *Saxifraga* (Saxifragaceae) is polyphyletic. Am. J. Bot. 83: 371–382.

Soltis DE, RK Kuzoff, ME Mort, M Zanis, M Fishbein, L Hufforl, J Koontz, and MK Arroyo. 2001a. Elucidating deep-level phylogenetic relationships in Saxifragaceae using sequences for six chloroplastic and nuclear DNA regions. Ann. Missouri Bot. Gard. 88: 669–693.

Soltis DE, M Tago-Nakazawa, Q-Y Xiang, S Kawano, J Murata, M Wakabayashi, and C Hibsch-Jetter. 2001b. Phylogenetic relationships and evolution in *Chrysosplenium* (Saxifragaceae) based on *mat*K sequence data. Am. J. Bot. 88: 883–893.

Soltis DE, AE Senters, MJ Zanis, S Kim, JD Thompson, PS Soltis, LP Ronse Decraene, PK Endress, and JS Farris.

2003. Gunnerales are sister to other core eudicots: implications for the evolution of pentamery. Am. J. Bot. 90: 461–470.

Spongberg SA. 1978. The genera of Crassulaceae in the southeastern United States. J. Arnold Arbor. 59: 197–248.

Stern WL. 1974. Saxifragales. In: Encyclopaedia Britannica, 15th ed., vol. 16, pp. 291–302. Chicago.

Stern WL, EM Sweitzer, and RE Phipps. 1970. Comparative anatomy and systematics of woody Saxifragaceae: *Ribes*. In: NKB Robson, DF Cutler, and M Gregory, eds. New research in plant anatomy, pp. 215–237. Bot. J. Linn. Soc. 63 (Suppl.). London/New York.

Steyn EMA, PJ Robbertse, and AE van Wyk. 1987. Floral development in *Greyia flanaganii* with notes on inflorescence initiation and sympodial branching. S. Afr. Tydskr. Plantk. 53: 194–201.

Stolt KAH. 1928. Die Embryologie von *Myriophyllum alterniflorum* DC. Svensk. Bot. Tidskr. 22: 305–319.

Subramanyam K. 1962. Embryology in relation to systematic botany with particular reference to the Crassulaceae. In: Plant embryology: A symposium, pp. 94–112. New Delhi.

Taneyama M and S Yoshida. 1978. Studies on C-glycosides in higher plants: I. Occurrence of bergenin in Saxifragaceae. Bot. Mag. (Tokyo) 91: 109–112.

Thiede J. 2007. Penthoraceae. In: K Kubitzki, ed. The families and genera of vascular plants, vol. 9, pp. 292–296. Springer, Berlin/Heidelberg/New York.

Thiede J and U Eggli. 2007. Crassulaceae. In: K Kubitzki, ed. The families and genera of vascular plants, vol. 9, pp. 83–118. Springer, Berlin/Heidelberg/New York.

Tillson AH. 1940. The floral anatomy of the Kalan-choideae. Am. J. Bot. 27: 595–600.

Towata EM. 1985a. Mucilage glands and cyanobacterial colonization in *Gunnera kaalensis* (Haloragaceae). Bot. Gaz. 146: 56–62.

Towata EM. 1985b. Morphometric and cytochemical ultrastructural analyses of the *Gunnera kaalensis/Nostoc* symbiosis. Bot. Gaz. 146: 293–301.

Uhl CH. 1963. Chromosomes and phylogeny of the Crassulaceae. Cact. Succ. J. (Los Angeles) 35: 80–84.

Van der Meijden R, and N Caspers. 1971. Haloragaceae. In: CGGJ van Steenis, ed. Flora Malesiana, ser.1, 7: 239–263. Noordhoff, Groningen.

Wakabayashi M. 1970. On the affinity in Saxifragaceae s. lato with special reference to the pollen morphology. Acta Phytotax. Geobot. 24: 128–145 (in Japanese, with English summary).

Wanntorp L, J Praglowski, and E Grafström. 2004. New insights into the pollen morphology of the genus *Gunnera* (Gunneraceae). Grana 43: 15–21.

Wanntorp L and LP Ronse Decraene. 2005. The *Gunnera* flower: Key to eudicot diversification or response to pollination mode? Int. J. Plant Sci. 166: 945–953.

Wanntorp L and H-E Wanntorp. 2003. The biogeography of *Gunnera* L.: Vicariance and dispersal. J. Biogeogr. 30: 979–987.

Wanntorp L, H-E Wanntorp, B Oxelman, and M Källersjö. 2001. Phylogeny of *Gunnera*. Plant Syst. Evol. 226: 85–107.

Wanntorp L, H-E Wanntorp, and M Källersjä. 2002. Phylogenetic relationships of *Gunnera* based on nuclear ribosomal DNA ITS region, *rbc*L and *rps*16 intron sequences. Syst. Bot. 27: 512–521.

Wanntorp L, H-E Wanntorp, and R Rutishauser. 2003. On the homology of the scales in *Gunnera* (Gunneraceae). Bot. J. Linn. Soc. 142: 301–308.

Warming E. 1909. Saxifragaceae: 1. Morphology and biology. Medd. Gronl. 36: 169–236.

Watari S. 1930. Anatomical studies on the leaves of some saxifragaceous plants, with special reference to the vascular system. J. Fac. Sci. Univ. Tokyo, Sect. 3, Bot. 5: 195–316.

Webb DA and RJ Gornall. 1989. Saxifrages of Europe. Christopher Helm, London.

Webb CJ, WR Sykes, and PJ Garnock-Jones. 1988. Gunneraceae. Flora of New Zealand IV: 748–750.

Weigend M. 2007. Grossulariaceae. In: K Kubitzki, ed. The families and genera of vascular plants, vol. 9, pp. 168–176. Springer, Berlin/Heidelberg/ New York.

Weigend M, T Motley, and O Mohr. 2002. Phylogeny and classification in the genus *Ribes* (Grossulariaceae) based on 5S-NTS sequences and morphological and anatomical data. Bot. Jahrb. Syst. 124: 163–182.

Wiggins IL. 1959. Development of the ovule and megagametophyte in *Saxifraga hiericifolia*. Am. J. Bot. 46: 692–697.

Wilkinson HP. 1994. Leaf and twig anatomy of the Pterostemonaceae (Engl.) Small: Ecological and systematic features. Bot. J. Linn. Soc. 115: 115–131.

Wilkinson HP. 1998. Gunneraceae. In: DF Cutler, M Gregory, eds. Anatomy of the Dicotyledons, 2nd ed. Saxifragales, 4: 260–272. Oxford.

Wilkinson HP. 2000. A revision of the anatomy of Gunneraceae. Bot. J. Linn. Soc. 134: 233–266.

Wilkinson HP and L Wanntorp. 2007. Gunneraceae. In: K Kubitzki, ed. The families and genera of vascular plants, vol. 9, pp. 177–183. Springer, Berlin/Heidelberg/New York.

Wu C-Y and T-C Ku. 1992. A new tribe with a new monotypic genus of Saxifragaceae (s.l.) from China. Acta Phytotax. Sinica 30: 193–196.

Yen TK. 1936. Floral development and vascular anatomy of the fruit of *Ribes aureum*. Bot. Gaz. 98: 105–120.

Order 82. PODOSTEMALES

Aquatic, annual or perennial herbs living in rushing water and growing only on rocks and stones in rivers or cataracts, mostly submerged or with some of the parts floating. Plants with more or less confluent stems and leaves, usually more or less thalloid and often resembling bryophytes, lichens, algae, or unlike any other plants in appearance. Primary root absent. The thalloid body usually attached to the substrate by numerous hairs or by specialized rootlike branches (haptera). Shoots nearly always arising as endogenous buds from roots; stems reduced or elongate, simple or branched, sometimes dimorphic, occasionally only present when flowering. Xylem very much reduced, usually represented only by a few tracheids with spiral or annular thickenings or completely wanting. Sieve plates recorded only in a few genera. Sieve-element plastids of S-type and large. Leaves, when discernible, alternate, entire or more or less dissected, without axillary buds; stipules present or absent. Flowers very small, solitary, or in cymose inflorescences, bisexual, actinomorphic or zygomorphic, apetalous, anemophilous, entomophilous, or cleistogamous, subtended or enclosed by two bracteoles (spathe-like and enclosing up to 20 flowers in Podostomoideae). Perianth of 2 or 3(-5) free or more or less connate petaloid sepals or small, annular scale or completely wanting; tepals (in Podostemoideae) 2–20, small, or large and 4–6, imbricate. Stamens 1–40 in one to many cycles or in more reduced flowers there is only a solitary stamen; filaments free or more often basally connate; anthers 2-locular, extrorse, opening longitudinally. Tapetum secretory. Microsporogenesis simultaneous or (*Podostemum*) successive. Pollen grains 2-celled, in monads or (Podostemoideae) in dyads or rarely in tetrads, from 3-colpate and 3-colporate to pantoporate, sometimes inaperturate. Gynoecium of (1)2(3) united carpels with as many locules and free or less often more or less basally connate stylodia. Ovary superior, with numerous or rarely (*Farmeria*) 2–4 ovules on thickened axile placentas. Ovules anatropous, bitegmic, tenuinucellate. As a result of the disintegration of the cells of the nucellus lying immediately below the female gametophyte, the so-called nucellar plasmodium or "pseudoembryo" develops, which serves as nutrition for the developing embryo. Female gametophyte monosporitc or bisporic, 4-celled or sometimes 5- or 6-celled; it contains an egg, one or two synergids, and one or two chalazal cells; polar nuclei absent, and therefore there is no triple fusion. Fruits septicidal capsules (except *Farmeria metzgerioides*, it has indehiscent fruit), with usually numerous seeds. Seeds minute, exotesta thick-walled, usually mucilaginous, endotegmen lignified; endosperm wanting; embryo straight, large, with two massive cotyledons and large suspensor; the large basal cell of the suspensor gives rise to a prominent and usually branched haustorium, which draws nourishment from the external tissues. Present xanthones, mangostin (in *Podostemum ceratophyllum*); n = 8, 10, 12, 14, 15, 17, 20.

In spite of its high specialization and very far-reaching reduction processes, the order Podostemales reveals clearly expressed links with the Saxifragales (Warming 1891 and many others), particularly with the Crassulaceae (Magnus 1913; Mauritzon 1933–1939; Maheshwari 1945; van Royen 1951; Subramanyam 1962; Kapil 1970, Les et al. 1997, Ueda et al. 1997). The presence in both of these families of such a unique structure as well-developed suspensor-haustorium is one of many common features in favor of their affinity.

1. PODOSTEMACEAE

Kunth 1816 (including Marathraceae Dumortier 1829, Philocrenaceae Bongard 1834, Tristichaceae J. C. Willis 1915). 49/280–300. Pantropical, especially in Asia and America, extending into temperate eastern North America and temperate East Asia.

1.1 WEDDELLINOIDEAE

Young flowers not enclosed in spathella or cupule. Stomata anomocytic. Sepals (4)5, imbricate; anthers 5–25; pollen grains in monads, 3-colporate. Capsula not ribbed, opening by two equal valves. Tegmen thick walled. – *Weddellina*.

1.2 TRISTICHOIDEAE

Young flowers subtended or enclosed by two bracteoles, which are not spathelike. Inflorescences terminal. Sepals three, imbricate; anthers (1-)3; pollen grains in monads, pantoporate. Gynoecium of three carpals; stigmas three. Capsule strongly ribbed, opening by three equal valves. – *Tristicha* (including *Malaccotristicha*), *Indotristicha, Dalzellia.*

1.3 PODOSTEMOIDEAE

Young flowers enclosed in spathelike bracteoles (spatella). Inflorescences axillary to leaves or bracts. Sepals 2–20, reduced to small scales or wanting. Anthers 2–40, often free. Pollen grains in monads or dyads, rarely (*Diamantina*) in tetrads. Gynoecium of 3–7 carpals. Capsule ribbed, opening by two equal or unequal valves, or rarely indehiscent. – PODOSTEMEAE: *Apinagia, Castelnavia, Ceratolacis, Cipoia, Crenias, Devillea, Marathrum, Rhyncholacis, Jenmaniella, Lophogyne, Hydrobryum, Leiothylax, Diamantina, Cladopus, Zeylanidium* (including *Hydrobryopsis*), *Podostemum, Farmeria, Macarenia, Oserya, Vanroyenella, Wettsteiniola, Angolaea, Dicraeanthus, Endocaulos, Butumia, Djinga, Ledermanniella, Letestuella, Macropodiella,*

Paleodicraeia, Sphaerothylax, Saxicolla, Saxicolella, Zehnderia, Stonesia, Winklerella, Thelethylax, Diplobryum, Griffithella, Hanseniella, Polypleurum, Thawatchaia, Willisia; mourereae: *Mourera, Tulasneantha, Lonchostephus.*

Bibliography

Ameka KG, E Pfeifer, and R Rutishauser. 2002. Developmental morphology of *Saxicolella amicorum* and *S. submersa* (Podostemaceae: Podostemoideae) from Ghana. Bot. J. Linn. Soc. 139: 255–273.

Arber A. 1920. Water plants. Cambridge University Press, Cambridge.

Arelcal GD and CR Nagendran. 1975. *Hydrobryopsis sessilis* (Podostemaceae): Origin, organization, and significance. Bot. Not. 128: 332–338.

Arekal GD and CR Nagendran. 1977. Female gametophyte of *Zeylanidium* (Podostemaceae): A clarification. Phytomorphology 27: 123–129.

Battaglia E. 1971. The embryo sac of Podostemaceae: An interpretation. Caryologia 24: 403–420.

Battaglia E. 1987. Embryological questions. 11. Has the debated case of Podostemaceae been resolved? Annali Bot. (Roma) 45: 37–64.

Bezuidenhout A. 1964. Pollen of the African Podostemaceae. Pollen Spores 6: 463–478.

Burkhardt G, H Becker, M Grubert, J Thomas, and T Eicher. 1992. Biphenyls and xanthones from the Podostemaceae. Phytochemistry 31: 1593–1597.

Carano E. 1914. Embriologia delle Podostemaceae. Ann. Bot. (Roma) 12: 163–164.

Chiarugi A. 1933. Lo sviluppo del gametofito femineo della *Weddellina squamulosa*. C. R. Acad. Lincei, Roma, 17: 1095–1100.

Chopra RN and AJ Mukkada. 1966. Gametogenesis and pseudoembryosac in *Indotristicha ramosissima* (Wight) van Royen. Phytomorphology 16: 182–188.

Contreras VR, R Scogin, C Philbrick and A Novelo R. 1993. A phytochemical study of selected Podostemaceae: Systematic implications. Aliso 13: 513–520.

Cook CDK and R Rutishauser. 2001. Name changes in the Podostemaceae. Taxon 50: 1163–1167.

Cook CDK and R Rutishauser. 2007. Podostemaceae. In: K Kubitzki, ed. The families and genera of vascular plants, vol. 9, pp. 304–344. Springer, Berlin/Heidelberg/New York.

Cusset G and C Cusset. 1988a. Étude sur les Podostemales. 9. Délimitations taxinomiques dans les Tristichaceae. Bull. Mus. Natl. Hist. Nat. Sect.B Adansonia, 10: 149–177.

Cusset G and C Cusset. 1988b. Étude sur les Podostemales. 10. Structures florales et végétatives des Tristichaceae. Bull. Mus. Natl. Hist. Nat. Adansonia 10: 179–218.

Cusset G and C Cusset. 1988c. Études sur les Podostemales. 11. Répartition et évolution des Tristichaceae. Bull. Mus. Hist. Nat. (Paris), ser. 4, 10, B(3): 223–262.

Graham SA and CE Wood. 1975. The Podostemaceae in the southeastern United States. J. Arnold Arbor. 56: 456–465.

Grubert M. 1976. Podostemaceen-Studien. Teil II. Untersuchungen über die Keimung. Bot. Jahrb. Syst. 95: 455–477.

Hammond BL. 1937. Development of *Podostemon ceratofhyllum*. Bull. Torrey Bot. Club 64: 17–36.

Imaichi R, T Ichiba, and M Kato. 1999. Developmental morphology and anatomy of the vegetative organs in *Malaccotristicha malayana* (Podostemaceae). Int. J. Plant Sci. 160: 253–259.

Jäger-Zürn I. 1967. Embryologische Untersuchungen an vier Podostemaceen. Oesterr. Bot. Z. 114: 20–45.

Jäger-Zürn I. 1970. Morphologie der Podostemaceae: I. *Tristicha trifaria* (Bory ex Willd.) Spreng. Beitr. Biol. Pfl. 47: 11–52.

Jäger-Zürn I. 1997a. Comparative morphology of the vegetative structures of *Tristicha trifaria, Indotristicha ramosissima* and *Dalzellia ceylanica* (Podostemaceae, Tristichoideae): a review. Aquat Bot. 57: 71–96.

Jäger-Zürn I. 1997b. Embryological and floral studies in *Weddellina squamulosa* Tul. (Podostemaceae, Tristichoideae). Aquatic Bot. 57: 151–182.

Jäger-Zürn I. 2003a. The occurrence of apical septum in the ovary of *Rhyncholacis, Apinagia, Marathrum,* and *Mourera* (Podostemoideae – Podostemaceae): Taxonomic implications. Bot. Jahrb. Syst. 124: 303–324.

Jäger-Zürn I. 2003b. Comparative morphology as an approach to reveal the intricate structures of the aquatic flowering plant family Podostemaceae. Recent Res. Dev. Plant Sci. 1: 147–172.

Jäger-Zürn I. 2005a. Morphology and morphogenesis of ensiform leaves, syndesmy of shoots and an understanding of the thalloid plant body in species of *Apinagia, Mourera,* and *Marathrum* (Podostemaceae). Bot. J. Linn. Soc. 147: 47–71.

Jäger-Zürn I. 2005b. Shoot apex and spathella: two problematical structures of Podostemaceae-Podostemoideae. Plant Syst. Evol. 253: 209–218.

Jäger-Zürn I and CJ Mathew CJ. 2002. Cupule structure of *Dalzellia ceylanica* and *Indotristicha ramosissima* (Podostemaceae). Aquatic Bot. 72: 79–91.

Kapil RN. 1970. Podostemaceae. Bull. Indian Sci. Acad. 41: 104–109.

Kato M. 2006. Taxonomic studies of Podostemaceae of Thailand. 2. Subfamily Tristichoideae and subfamily Podostemoideae with ribbon-like roots. Acta Phytotax. Geobot. 57: 1–54.

Kato M, CMS de Oliveira, V Bittrich V, and M do Amaral. 2005. Xanthones from *Weddelina squamulosa* Tul. (Podostemaceae). Biochem. Syst. Ecol. 55: 331–334.

Kita Y. 2002. Molecular phylogeny, morphological evolution, and biogeography of Podostemaceae. Bunrui 2: 19–26 (in Japanese).

Kita Y and M Kato. 2001. Intrafamilial phylogeny of the aquatic angiosperm Podostemaceae inferred from the nucleotide sequences of the *mat*K gene. Plant Biol. 3: 156–163.

Kita Y and M Kato. 2005. Seedling developmental anatomy of an undescribed *Malaccotristicha* species (Podostemaceae, subfamily Tristichoideae) with implications for body plan evolution. Plant Syst. Evol. 254: 221–232.

Koi S and M Kato. 2003. Comparative developmental anatomy of the root in three species of *Cladopus* (Podostemaceae). Ann. Bot. N.S. 91: 927–937.

Koi S, R Imaichi, and M Kato. 2005. Endogeneous leaf initiation in the apical-meristemless shoot of *Cladopus quennslandicus* (Podostemaceae) and implications for evolution of shoot morphology. Int. J. Plant Sci. 166: 199–206.

Koi S, R Fujinami, N Kubo, I Tsukamoto, R Inagawa, R Imaichi, and M Kato. 2006. Comparative anatomy of root meristem and root cap in some species of Podostemaceae and the evolution of root dorsiventrality. Am. J. Bot. 93: 682–692.

Les DH, CT Philbrick, and RA Novelo. 1997. The phylogenetic position of riverweeds (Podostemaceae) insights from *rbc*L sequence data. Aquatic Bot. 57: 5–27.

Lobreau-Callen D, A Le Thomas, and M Suarez-Cuervera. 1998. Caractéres ultrastructuraux du pollen de quelques Podostémales. Affinités avec les Rosidae évoluées. Comtes Rend. Acad. Sci. Paris Sci. Viv. 321: 335–345.

Magnus W. 1913. Die atypische Embryosackent-wicklung der Podostemaceen. Flora 105: 275–336.

Maheshwari P. 1945. The place of angiosperm embryology in research and teaching. J. Indian Bot. Soc. 24: 25–41.

Mauritzon J. 1933. Über die systematische Stellung der Familien Hydrostachyaceae and Podostemonaceae. Bot. Not. 1933: 172–180.

Mohan Ram NY and A Sehgal. 1992. Podostemaceae — the strange family of aquatic angiosperms. Paleobotanist 41: 192–197.

Mohan Ram NY and A Sehgal. 1997. In vitro studies on developmental morphology of Indian podostemaceae. Aquat. Bot. 57: 97–132.

Moline P, M Thiv, GK Ameka, J-P Chogue, E Pfeifer, and R Rutishauser. 2007. Comparative morphology and molecular systematics of African Podostemaceae-Podostemoideae, with emphasis on *Dicraeanthus* and *Ledermanniella* from Cameroon. Int. J. Plant Sci. 168: 159–180.

Mukkada AJ. 1962. Some observations on the embryology of *Dicraea stylosa*. In Plant embryology: A symposium, pp. 139–145. CSIR: New Delhi.

Mukkada AJ. 1964. An addition to the bisporic embryo sacs: The *Dicraea* type. New Phytol. 63: 289–292.

Mukkada AJ and RN Chopra. 1973. Post-fertilization development in *Indotristicha ramosissima* (Wight) van Royen. New Phytol. 72: 639–646.

Murguía-Sánchez G, RA Novelo, CT Philbrick, and GJ Márquez-Guzmán. 2002. Embryo sac development in *Vanroyenella plumosa*, Podostemaceae. Aquat. Bot. 73: 201–210.

Nagendran CR and GD Arekal. 1976. Embryo sac of *Griffithella hookeriana: A* reinvestigation. Phytomorphology 26: 359–363.

Nagendran CR, GD Arekal, and K Subramanyam. 1977. Embryo sac studies in three Indian species of *Polypleurum* (Podostemaceae). Plant Syst. Evol. 128: 215–226.

Nagendran CR, K Subramanyam, and GD Arekal. 1976. Development of the female gametophyte in *Hydrobryum griffithii* (Podostemaceae). Ann. Bot. 40: 511–513.

Nemirovich-Danchenko EN. 1996. Podostemaceae. In: A. Takhtajan, ed. Comparative seed anatomy, vol. 5, pp. 128–133. Nauka, St. Petersburg (in Russian).

Novelo RA and CT Philbrick. 1997. Taxonomy of Mexican Podostemaceae. Aquatic Bot. 57: 275–303.

O'Neill SP, JM Osborn, CT Philbrick, and A Novelo R. 1997. Comparative pollen morphology of five New World genera of Podostemaceae. Aquatic Bot. 57: 133–150.

Passarelli LM, SB Girarde, and NM Tur. 2002. Palynology of South American Podostemaceae 1. *Alpinagia* Tul. Grana 41: 10–15.

Philbrick CT. 2004. Podostemaceae. In: N Smith, SA Mori, A Henderso, DW Stevenson, and SV Heald, eds. Flowering plants of the Neotropics, pp. 302–304. The New York Botanical Garden, Princeton University Press, Princeton.

Philbrick CT and RA Novelo. 1995. New World Podostemaceae: ecological and evolutionary enigmas. Brittonia 47: 210–222.

Philbrick CT, A Novelo R, and BE Irgang. 2004. Two new genera of Podostemaceae from the State of Minas Gerais, Brazil. Syst. Bot. 29: 109–117.

Razi BA. 1949. Embryological studies of the members of Podostemonaceae. Bot. Gaz. 111: 211–218.

Razi BA. 1955. Some aspects of the embryology of *Zeylanidium olivaceum* and *Lawia zeylanica*. Bull. Bot. Soc. Bengal. 9: 36–41.

Romo Comtreras V, R Scogin, CT Philbrick, and RA Novelo. 1993. A phytochemical study of selected Podostemaceae: systematic implications. Aliso 13: 513–520.

Royen P van. 1951–1955. The Podostemaceae of the New World. Med ed. Bot. Mus. Herb. Rijks Univ. Utrecht 107: 1–151, 1951; 115: 2–21, 1953; 119: 215–263. 1955.

Rutishauser R. 1995. Developmental patterns of leaves in Podostemaceae compared with more typical flowering plants: saltational evolution and fuzzy morphology. Canad. J. Bot. 73: 1305–1317.

Rutishauser R. 1997. Structural and developmental diversity in Podostemaceae (river-weeds). Aquatic Bot. 57: 29–70.

Rutishauser R and E Pfeifer. 2002. Comparative morphology of *Cladopus* (including *Torrenticola*, Podostemaceae) from east Asia to northeastern Australia. Austral. J. Bot. 6: 725–739.

Rutishauser R, A Novelo, and CT Philbrick. 1999. Developmental morphology of New World Podostemaceae: *Marathrum* and *Vanroyenella*. Int. J. Plant Sci. 160: 29–45.

Schell R. 1967. Etudes sur l'anatomie et la morphologic des Podostemacees. Candollea 22: 157–225.

Schnell R. 1969. Contribution a l'etude des Podostemacees de Guyane. Adansonia, ser. 2, 9: 249–271.

Schnell RAA. 1998. III. Anatomie des Podostémacées. In: S Carlquist, DF Cutler, S Fink, P Ozenda, I Roth, and H Ziegler, eds. Encyclopedia of plant anatomy: extreme adaptations in angiospermous hydrophytes, pp. 197–290. Gebrüder Borntraeger, Berlin.

Schnell R and G Cusset. 1963. Remarques sur la structure des plantules des Podostemonacees. Adansonia, ser. 2, 3: 358–369.

Sehgal A, M Sethi, and HY Mohan Ram. 2002. Origin, structure and interpretation of thallus in *Hydrobryopsis sessilis* (Podostemaceae). Int. J. Plant Sci. 163: 891–905.

Soltis DE, ME Mort, PS Soltis, C Hibsch-Jetter, EA Zimmer, and D Morgan. 1999. Phylogenetic relationships of the enigmatic angiosperm family Podostemaceae inferred from 18S rDNA and *rbc*L sequence data. Molec. Phylog. Evol. 11: 261–272.

Steude H. 1935. Beiträge zur Morphologic und Anatomie von *Mourera aspera*. Beih. Bot. Centralbl. 53A: 627–650.

Subramanyam K. 1962. Embryology in relation to systematic botany, with particular reference to the Crassulaceae. In: Plant embryology: A symposium, pp. 94–112. CSIR, New Delhi.

Subramanyam K and CP Sreemadhavan. 1969. A conspectus of the families Podostemaceae and Tristichaceae. Bull. Bot. Surv. India 11: 161–168.

Suzuki K, Y Kita, and M Kato. 2002. Comparative developmental anatomy of seedlings in nine species of Podotemaceae (subfamily Podostemoideae). Ann. Bot. N.S. 89: 755–765.

Ueda K, T Hanyunda, A Nakano, T Shiuchi, A Seo, H Okubo, and M Hotta. 1997. Molecular phylogenetic position of Podostemaceae, a marvelous aquatic flowering plant family. J. Plant Res. 110: 87–92.

Warming E. 1891. Familien Podostemaceae. Kongel. Dansk. Viten skab. Selsk. Skrifter, Sjette Raekke Afd. VII, 4: 135–179.

Went FAFC. 1909. The development of the ovule, embryo sac, and egg of Podostemaceae. Rec. Bot. Neerl. 5: 1–16.

Went FAFC. 1910, 1912, 1926. Untersuchungen über Podostemaceen. Verh. Kon. Akad. Wetensch. 2 (16): 1–88, 1910; 17: 5–18, 1912; 25: 3–58, 1926.

Went FAFC. 1929. Morphological and histological peculiarities of the Podostemaceae. Proc. Int. Congr. Plant Sci. Ithaca 1: 351–358.

Willis IC. 1902. Studies in the morphology and ecology of the Podostemaceae of Ceylon and India. Ann. Roy. Bot. Gard. 1: 267–465.

Order 83. VITALES

Small trees, erect shrubs, and herbs or more often woody lianas. Roots sometimes adventitious. Cambium storied. Vessel elements usually rather large, with simple perforations; lateral pitting scalariform or sometimes opposite to alternate. Fibers with simple or sometimes bordered pits, septate. Rays heterogeneous. Axial parenchyma mainly paratracheal. Sieve-element plastids of Pcs-type that contain several polygonal protein crystalloids and similar to the plastids of *Gunnera* (Behnke 1981). Nodes 3–7-lacunar. Leaves alternate or very rarely opposite, simple or compound, of diverse shapes and structures, stipulate or estipulate, often bearing specialized, multicellular, stalked, deciduous "pearl glands" and sometimes also peltate scales. Stomata of diverse kinds. Inflorescences leaf-opposed or terminal, rarely axillary, cymose or racemose, sometimes with 1–2 tendril-like branches; bract and bracteoles small, sometimes nectar secreting, the bract subtending each branch. Flowers rather small, bisexual or (some members of haplostemonous Vitaceae) polygamo-monoecious or dioecious, actinomorphic, (3)4–5(6–7)-merous. Calyx mostly weakly developed, lobed, dentate or truncate, often much reduced. Petals 4–5 (rarely 3 or 6–7), valvate, free or basally connate or sometimes coherent distally and calyptra-like (*Vitis*), rarely persistent. Stamens 5–4 (3–7), opposite the petals; filaments free, or connate into a tube. Anthers dorsifixed, tetrasporangiate or rarely disporangiate, introrse, opening longitudinally. Tapetum secretory. Microsporogenesis simultaneous. Pollen grains 2-celled or 3-celled, 3-colporate, with reticulate ornamentation. Nectary disc usually well developed, annular or cupular or sometimes of five free glands, usually more or less adnate to the ovary. Gynoecium of 2-6(-8) united carpels; stylodia connate into a style with capitate, seldom (*Tetrastigma*) quadrifid, rarely sessile stigma; ovary superior, but sometimes partly sunken in the nectary disc, completely or incompletely divided into locules, with one or two ovules per locule. Ovules erect, axile or basal, anatropous, apotropous, with ventral raphe, bitegmic, crassinucellate. Female gametophyte of *Polygonum*-type. Endosperm nuclear. Haustoria of the female gametophyte present or absent. Fruits berries (sometimes rather dry). Seeds endotestal (with lignified, several-layered endotesta and persistent thin tracheidal tegmen); embryo small to minute, straight, surrounded by copious, horny, generally ruminate, oily endosperm with two main folds, one from each side of the raphe. Alkaloids present (rarely), or absent, ellagic acid present (*Vitis vinifera*), or absent (*Parthenocissus*, *Rhoicissus*), present myricetin.

Two rather closely related families Vitaceae and Leeaceae are usually included in the order Rhamnales. However, from the Rhamnales they differ in the presence of raphide sacs in the parenchymatous tissues, berrylike fruits, seed anatomy (Corner 1976) and P-type sieve-element plastids (Behnke 1974). According to Corner (1976: 36), seed structure of Vitaceae "is scarcely improved on that of *Magnolia* and may, even, be more primitive. Certainly there is no affinity in this regard with Rhamnales." However, any direct links between Magnoliales or Magnoliidae in general and Vitales is very doubtful. Endlicher (1841) placed his Ampelideae between Araliaceae and Corneae. Lindley (1853: 439) placed Vitaceae s.l. between Berberidaceae and Pittosporaceae, but indicated the connection, although no direct, with the Araliaceae. He stated: "If the Vine is compared with *Aralia racemosa*, the relationship of the present Order to it will be too obvious to be mistaken. Suppose that *Aralia racemosa* had an adherent calyx, erect ovules, with stamens opposite the petals, and it would be a *Vitis*".

Order Vitales probably derived from some Rosanae ancestor.

Key to Families

1 Woody lianas usually with leaf-opposite tendrils, rarely succulent small trees or erect herbs, branches often swollen or articulated at nodes. Tendrils

simple, bifid, or 2–3-, or 4–12-branched, usually leaf-opposite. Leaves evergreen, or deciduous, gland-dotted, simple, lobed or unlobed, or compound; when compound ternate, or pinnate, or palmate. Stipules if present petiolar and usually caduceus, rarely absent. Inflorescences leaf-opposed or terminal, infrequently axillary, panicles, corymbs, or rarely spikes. Flowers small, pedicellate, with prophyls, actinomorphic, 4–5(-7)-merous. Calyx small, often indistinctly lobed or dentate or much reduced and truncate. Petals free or basally connate, valvate, sometimes coherent distally and calyptrate. Stamens free, equal in number to and opposite the petals; anthers tetrasporaginate or disporaginate. Nectary disc intrastaminal, ring-shaped, copular, or gland-shaped. Ovary 2(-6)-locular, the ovules two per locule, anatropous or apotropous. Style long or short, stigma inconspicuous or capitate or 4-lobed. Female gametophyte with a haustorium. Endostome forming the micropyle. Fruits 1–4-seeded berries. Seeds endotestal with abaxial chalazal knot and an adaxial raphe; embryo small, straight; endosperm oily and proteinaceous, copious, ruminate. Present proanthocyanins, and catechin, prodelphinisin, gallic, ellagic, coffeic, chlorogenic and tartaric acids; n = 11–16, 19–20. . . . . 1. VITACEAE.

1 Small trees or erect, creeping or scrambling shrubs without tendrils, or herbaceous plants with a woody base; stema unarmed or with rows of prickles. Leaves distichous, uni-trifoliolate, or 1– 4-pinnate. Stipules attached to base of petiole and sheathing, often early caducous. Stomata cyclocytic, actinocytic, or rarely anomocytic. The mesophyll contains mucilage cells sometimes with raphides and usually with calcium oxalate crystals. Inflorescences leaf-opposed cyme, lax or condensed by reduction of branches or peduncle. Calyx campanulate or cupuliform. Petals distally cohering in bud by ventrally apical keels, reflexed at anthesis, the basal portions connate to one another and adnate to androecium forming a shortly tubular structure free from calyx. Stamens connate distally (above the common stricture with the corolla) for some distance into a tube that may bear lobes alternate with the anthers. Nectary disc absent. Pollen grains 3-celled or rarely 2-celled, 3-colporate. Ovary 4-6(-10)-locular (sometimes incompletely), with one ovule per locule. Style short, entire, the stigma slightly thickened. Female gametophyte without a haustorium. Exostome forming the micropyle. Large numbers of raphids are commonly in the tissues of corolla, anthers, staminodial tube. Fruits 4–6-seeded berries. Seeds with ruminate endosperm; seed coat has a thick lignified endotesta and inconspicuous tegmen; raphides in seed coat absent; n = 10–12. . . . . 2. LEEACEAE.

1. VITACEAE

A. L. de Jussieu 1789 (including Ampelopsidaceae Kosteletzky 1835, Cissaceae Horaninow 1847, Pterisanthaceae J. G. Agardh 1858). 15/750. Tropical and subtropical regions, with relatively few species in warm-temperate regions (mostly species of *Vitis, Ampelopsis,* and *Parthenocissus*).

Cissus, Nothocissus, Cyphostemma, Acareosperma, Cayratia, Vitis, Ampeloclssus, Parthenocissus, Yua, Ampelopsis, Pterisanthes, Clematicissus, Tetrastigma, Rhoicissus, Pterocissus.

2. LEEACEAE

Dumortier 1829. 1/34. Paleotropics, mainly southern and southeastern Asia.

Leea.

Very close to Vitaceae and sometimes (e.g., Thorne 1992, 2000, 2006) are considered a subfamily Leeoideae. From Vitaceae they differ in the erect habit, absence of tendrils, absence of nectary disc, the presence of secondary septa in the ovary and absence of raphides in seed coat.

Bibliography

Adkinson J. 1913. Some features of the anatomy of the Vitaceae. Ann. Bot. 27: 133–139.

Angeles G and León-Gómez C. 1997. Bark anatomy of four tropical Vitaceae from Veracruz, Mexico. IAWA J. 18: 215–228.

Arnott HJ and MA Webb. 2000. Twined raphides of calcium oxalate in grape (*Vitis*): implications for crystal stability and function. Int. J. Plant Sci. 161: 133–142.

Behnke H-D. 1974. P- und S-Typ Siebelement-Plasti-den bei Rhamnales. Beitr. Biol. Pfl. 50: 457–464.

Brizicky GK. 1965. The genera of Vitaceae in the southeastern United States. J. Arnold Arbor. 46: 48–67.

Descoings B. 1997. Note sur deux *Cyphostemma* (Vitacees) crassulescentes de Madagascar. J. Bot. Soc. Bot. 1: 69–73. France.

Faure O. 1990. Embryons somatiques de *Vitis rupestis* et embryons zygotiques de *Vitis* sp.: morphologie, histology, histochemie et development. Canad. J. Bot. 2305–2315.

Gerrath JM, CR Lacroix, and U Posluszny 1990. The developmental morphology of *Leea guineensis:* II. Floral development. Bot. Gaz. 151: 210–220.

Gerrath JM and U Posluszny 1988, 1989. Morphological and anatomical development in the Vitaceae: I. Vegetative development in *Vitis riparia*. Canad. J. Bot. 66: 209–224, 1988. II. Floral development in *Vitis riparia*. Canad. J. Bot. 66: 1334–1352, 1988. III. Vegetative development in *Parthenocissus inserta*. Canad. J. Bot. 67: 803–816, 1989. IV. Floral development in *Parthenocissus inserta*. Canad. J. Bot. 67: 1356–1365, 1989. V. Vegetative and floral development in *Ampelocissus brevipedunculata*. Canad. J. Bot. 67: 2371–2386, 1989. VI. Floral development in *Cissus antarctica*. Canad. J. Bot. 72: 635–643, 1993.

Gerrath JM and U Posluszny. 1988. Comparative floral development in some members of the Vitaceae. In: P Leins, SC Tucker, and PK Endress, eds. Aspects of floral development, pp. 121–131. J. Cramer, Berlin.

Gerrath JM, CR Lacroix, and U Posluszny. 1998. Phyllotaxis in the Vitaceae. In: RV Jean and D Barabé, eds. Symmetry in Plants, pp. 89–107. Singapore.

Gerrath JM, U Posluszny, and NG Dengler. 2001. Primary vascular patterns in the Vitaceae. Intern. J. Plant Sci. 162: 729–745.

Gerrath JM, T Wilson, and U Posluszny. 2004. Morphological and anatomical development in the Vitaceae. VII. Floral development in *Rhoicissus digitata* with respect to other genera in the family. Canad. J. Bot. 82: 198–206.

Ingrouille MJ, MW Chase, MF Fay, D. Bowman, M. van der Bank, and A. de Bruijin. 2002. Systematics of Vitaceae from the viewpoint of plastid *rbc*L DNA sequence data. Bot. J. Linn. Soc. 138: 421–432.

Inceoglu O, NM Pinar, and E Oybak Donmez. 2000. Pollen morphology of wild *Vitis sylvestris* Gmelin (Vitaceae). Turk. J. Bot. 24: 147–150.

Jayasankar S, BR Bondada, Z Li, and DJ Gray. 2003. Comparative anatomy and morphology of *Vitis vinifera* (Vitaceae) somatic embryos from solid- and liquid-culture-derived proembryogenic masses. Am. J. Bot. 90: 973–979.

Kashyap G. 1955, 1957, 1958. Studies in the family Vitaceae: I. Floral morphology of *Vitis trifolia* L. Agra Univ. J. Res. Sci., (Suppl. 4): 777–783, 1955. II. Floral anatomy of *Vitis trifolia* Linn., *Vitis latifolia* Roxb., and *Vitis himalayana* Brandis. III. Floral morphology of *Vitis latifolia* Roxb., *Vitis himalayana* Brandis, and *Vitis trifolia* Linn. J. Indian Bot. Soc. 36: 317–323, 1957; 37: 240–248, 1958.

Lacroix CR, JM Gerrath, and U Posluszny 1990. The developmental morphology of *Leea guineensis:* I. Vegetative development. Bot. Gaz. 151: 204–209.

Latiff A. 1984. The morphology and systematics of Vitaceae in the Malay peninsula. In: S. Sahid et al., eds. A collection of working papers, 4: 33–41. Univ. Kebangsaan Malaysia.

Latiff A. 2001a. Studies in Malesian Vitaceae: 12. Taxonomic notes on *Cissus, Ampelocissus, Nothocissus* and *Tetrastigma* and other genera. Folia Malaysiana 2: 179–189.

Latiff A. 2001b. Diversity of the Vitaceae in the Malay Archipelago. Malayan Nat. J. 55: 29–42.

Latiff A and K Mat Salleh. 1999. Studies in Malesian Vitaceae: 11. *Pterisanthes* in Borneo, with additional notes on *P. grandis* and *P. hirtiflora*. Sandakania 13: 67–78.

Lombardi JA. 2000. Flora Neotropica Monogr. 80: 1–250. Vitaceae – Gêneros *Ampelocissus, Ampelopsis* e *Cissus*. New York.

Nair NC. 1968. Contribution to the floral morphology and embryology of two species of *Leea* with a discussion on the taxonomic position of the genus. J. Indian Bot. Soc. 47: 193–205.

Nair NC. 1970. Meliaceae, Rhamnaceae, Vitaceae, Leeaceae. Indian National Sci. Acad. Bull. [Symposium: Comparative Embryology of Angiosperms] 41: 151–155, 168–173, 174–179, 180–184.

Nair NC and KV Muni. 1960. Organography and floral anatomy of some species of Vitaceae. Phytomorphology 10: 138–144.

Nair NC and PNN Nambisan. 1957. Contribution to the floral morphology and embryology of *Leea sambucina* Willd. Bot. Not. 110: 160–172.

Patil SG. 2001. Pollen grains studies in *Leea* L. (fam. Leeaceae) and their taxonomic significance. J. Ecol. Taxon. Bot. 25: 75–80.

Patil SG and BK Honrao. 1998. Chemotaxonomical studies in the genus *Tetrastigma* Planch. (Vitaceae). Acta Bot. Indica 26: 137–142.

Petria E. 1973. Macrosporogenesis, development of female gametophyte and beginning embryo differentiation in some members of the families Vitaceae and Leeaceae, followed by some phylogenetic remarks. Lucr. Grad. Bot. Bucuresti 1972–1973: 285–352.

Reille M. 1967. Contribution a 1∞ετυδε palynologique de la famille des Vitacees. Pollen Spores 9: 279–303.

Ren N, K-Y Pan, Z-D Chen, and R-Q Wang. 2003. Structural characters of leaf epidermis and their systematic significance in Vitaceae. Acta Phytotax. Sinica: 41: 531–544.

Risdale CE. 1974. A revision of the family Leeaceae. Blumea 22: 57–100.

Risdale CE. 1976. Leeaceae. In: GGGJ van Steenis, ed. Flora Malesiana ser. I, 7: 759–782. Leiden.

Rossetto M, BR Jackes, KD Scott, and RJ Henry. 2002. Is the genus *Cissus* (Vitaceae) monophyletic? Evidence from plastid and nuclear ribosomal DNA. Syst. Bot. 27: 522–533.

Shah JJ. 1959. Studies on the stipules of six species of Vitaceae. J. Arnold Arbor. 40: 398–412.

Shah JJ and YS Dave. 1970. Morpho-histogenetic studies on the tendrils of Vitaceae. Am. J. Bot. 57: 363–370.

Soejima A and J Wen. 2006. Phylogenetic analysis of the grape family (Vitaceae) based on three chloroplast markers. Am. J. Bot. 93: 278–287.

Thorne ET, BM Young, CM Young, JF Stevenson, JM Labavitch, MAm Matthews, and TL Rost. 2006. The structure of xylem vessels in grapevine (Vitaceae) and a possible passive mechanism for the systemic spread of bacterial disease. Am. J. Bot. 93: 497–504.

Umadevi I and M Daniel. 1991. Taxonomy of the Vitaceae: a chemical approach. Acta Bot. Indica 19: 168–170.

Vyshenskaya TD. 2000. Vitaceae, Leeaceae. In: A Takhtajan, ed. Comparative seed anatomy, vol. 6, pp. 218–231. Nauka, St. Petersburg (in Russian).

Wang X-D, S-C Chen, and C-L Li. 2000. A study on pollen morphology of the genus *Vitis* L. Acta Phytotax. Sinica 38: 43–52.
Wen J. 2007a. Leeaceae. In: K Kubitzki, ed. The families and genera of vascular plants, vol. 9, pp. 221–225. Springer, Berlin/Heidelberg/New York.
Wen J. 2007b. Vitaceae. In: K Kubitzki, ed. The families and genera of vascular plants, vol. 9, pp. 467–479. Springer, Berlin/Heidelberg/New York.
Wheeler EA and CA LaPasha. 1994. Woods of the Vitaceae – fossil and modern. Rev. Paleobot. Palynol. 80: 175–207.
Zubkova IG. 1963. Leaf epidermis of Vitaceae and its systematic significance. Bot. Zhurn. 51: 278–283 (in Russian).
Zubkova IG. 1965. Anatomical study of the petiole of the family Vitaceae Juss., its taxonomic and evolutionary significance. Bot. Zhurn. 50: 1556–1567 (in Russian).

Order 84. ROSALES

Evergreen or deciduous trees and shrubs or often herbs (usually perennial). Vessels mostly with simple perforations. Axial parenchyma scanty to abundant, differently patterned but mainly diffuse and scanty paratracheal. Sieve-element plastids of S-type. Nodes usually trilacunar, or unilacunar, or pentalacunar to multilacunar. Leaves alternate or rarely opposite, mostly serrate or less often entire, simple or variously compound, persistent or caducous, mostly with stipules that are sometimes adnate to the petiole; leaf mesophyll cells contain oil bodies. Stomata anomocytic. Flowers in various kinds of inflorescences (mostly cymose), less often solitary, bisexual or less often unisexual, actinomorphic or more or less zygomorphic, 5-merous, cyclic, mostly with double perianth. Hypanthium usually well-developed (not evident in some male flowers). Sepals, petals and stamens inserted on its rim; disk sometimes distinct, intrastaminal. Sepals and petals free, usually imbricate. Stamens few to numerous; filaments free; anthers 2-locular, tetrasporangiate, usually introrse, dorsifixed, versatile, opening longitudinally or rarely by apical pores. Pollen grains 2-celled, 3-colporate or rarely 3-porate, pantoporate, or 3-zonocolporate. Tapetum secretory. Microsporogenesis simultaneous. Gynoecium of 1 to many (up to 50), free or more or less united carpels, with free stylodia and usually small capitate stigma; ovary superior to inferior, with one to many, but mostly 1–2 ovules per carpel or per locule. Ovules pendulous to subbasal, anatropous or less often hemitropous or campylotropous, bitegmic or less often unitegmic, crassinucellate. Female gametophyte of *Polygonum*-type. Endosperm nuclear, rarely helobial (*Lyonothamnus*). Fruits various, fleshy or dry, dehiscent or not. Seeds small to fairly large, wingless or rarely winged; embryo small to large, straight or bent, linear or folded, sometimes the radicle invested; cotyledons moderately to markedly thick, rarely convolute; endosperm mostly present, varying between 1–20 cells thick in different genera, sometimes abundant (as in *Physocarpus* and *Rhodotypos*); perisperm often present. Present triterpenes, ellagic acid, cyanogenic glycosides, flavones.

Linked to the Saxifragales through the most archaic members of Rosaceae-Spiraeoideae.

Key to Families

1 Sepals imbricate. Evergreen or deciduous trees and shrubs or often herbs (usually perennial). Lateral pitting alternate or less often opposite. Parenchyma scanty to abundant, differently patterned but mainly diffuse and scanty paratracheal. Lysigenous mucilage ducts in the pith wanting. Nodes usually trilacunar, or unilacunar, or pentalacunar to multilacunar. Leaves alternate or rarely opposite, mostly serrate or less often entire, simple or variously compound, persistent or caducous, mostly with stipules that are sometimes adnate to the petiole. Flowers in various kinds of inflorescences (mostly cymose), less often solitary, bisexual or less often unisexual, actinomorphic or more or less zygomorphic. Sepals (3-)5(-10), free or adnate to the ovary, imbricate, sometimes with an epicalyx of bracteoles. Petals inserted below the margin of the disk, as many as the calyx-lobes, free, equal or rarely unequal, imbricate, deciduous, rarely absent. Stamens two, three, or four times as many as petals, sometimes more numerous (up to 400) or rarely only five or even (*Aphanes*) only one, bent inward in bud; filaments free or more or less connate with the floral tube or between themselves; anthers usually introrse or rarely (*Potentilla*) latrorse, dorsifixed, versatile, opening longitudinally or rarely by apical pores. Pollen grains 2-celled, 3-colporate or rarely 3-porate or (*Sanguisorba*) pantoporate. Gynoecium of one to many (up to 50), free or more or less united carpels, with free stylodia and usually small capitate stigma; ovary with one to many, but mostly 1–2 ovules per locule. Ovules pendulous to subbasal, anatropous or less often hemitropous or campylotropous, bitegmic or less often unitegmic, crassinucellate. Fruits multifollicles, achenes, or pomes, or drupes, rarely capsules (Lindleyieae).

Seeds small to fairly large, wingless or rarely (e.g. *Exochorda*) winged; exotestal cells periclinally elongated, radial walls thickened, or palisade or tubular; outer wall often becoming mucilaginous, endotegmic cells slightly thickened, or seed coat undistinguished; embryo small to large, straight or bent, linear or folded, sometimes the radicle invested; cotyledons moderately to markedly thick, rarely convolute; endosperm mostly present, varying between 1–20 cells thick in different genera, sometimes abundant (as in *Physocarpus* and *Rhodotypos*); perisperm often present. Present triterpenes, ellagic acid, cyanogenic glycosides, flavones, n = 7–10, 15–17, 27. . . . . 1. ROSACEAE.

1 Sepals valvate. Evergreen, small glabrous trees (up to 10 m tall and 1 m in diameter) or shrubs, sometimes with saponaceous bark. Vessels with simple perforations. Axial parenchyma abundant, apotracheally diffuse, diffuse-in-aggregates and scanty paratracheal Rays heterogenous (Zhang 1992). Extremely numerous styloids present in the phloem of stems, petioles and in the leaf lamina (Lersten and Horner 2005). Nodes unilacunar with 3 traces. Leaves alternate, simple, leathery, almost entire or serrate, shortly petioled, margins toothed; stipules small, caduceus, petiolar. Inflorescences axillary or terminal, 3–5-flowered, pedicels 2-bracteolate. Flowers polygamo-dioecious (terminal flowers bisexual, lateral ones staminate), 5-merous, rather large, tomentose. Calyx-tube persistent, leathery, small, subcupular, lobes broadly ovate. Petals white, sessile, spathulate. Disk thick, fleshy, lining the calyx-tube, lobes 5, thick, depressed, emarginated, adnate to the calyx-lobes. Stamens ten, five opposite the petals and inserted at the base of the disk, five alternate and inserted at apex of the disk-lobes; filaments subulate, free; anthers 2-locular, introrse. Pollen grains tri-zonocolporate (Claxton et al. 2005). Gynoecium of five carpels, tomentose, partly connate; styles terminal, free, stigmatic zone elongated along styles; ovules numerous, pleurotropous, flattened, subhorizontal, densely 2-seriate. Fruits multifollicles, strongly asymmetrically lobed, spreading stellately, opening down both surfaces of the lobes. Seeds exotestal, long- and broadly winged in the upper part, outer testa layers thickened; endosperm thin; cotyledons convolute. Present flavone C-glycosides; n = 14. . . . . 2. QUILLAJACEAE.

1. ROSACEAE

A. L. de Jussieu 1789 (including Agrimoniaceae Gray 1822, Alchemillaceae Martynov 1820, Amygdalaceae Marquis 1820, Cercocarpaceae J. G. Agardh 1858, Chamaemoraceae Lilja 1870, Coleogynaceae J. Agardh 1858, Cydoniaceae Schnizlein 1856, Dryadaceae Gray 1822, Fragariaceae Richard ex Nestler 1816, Guamatelaceae S. Oh et D. Potter 2006, Lindleyaceae J. Agardh 1858, Malaceae J. K. Small ex Britton 1903, Mespilaceae Schultz-Schultzenstein 1832, Neilliaceae Miquel 1855, Potentillaceae Berchtold et J. Presl 1820, Poteriaceae Rafinesque 1815, Prunaceae Martynov 1820, Pyraceae Vest 1818, Rhodotypaceae J.Agardh 1858, Sanguis orbaceae Berchtold et J. Presl 1820, Sorbaceae Brenner 1886, Spiraeaceae Bertuch 1801, Tormentillaceae Martynov 1820, Ulmariaceae Gray 1822). 111/3000–3500. Subcosmopolitan, but especially in subtropical and temperate regions of the Northern Hemisphere.

1.1 SPIRAEOIDEAE

Carpels 2–5 (up to 12), rarely (*Stephanandra, Adenostoma,* and spp. of *Neillia*) only one, free or nearly so, less often more or less connate. Ovules 1–2 to several (about 8 in *Chamaebatiaria* and 11 in *Neillia*) per carpel, pendulous or less often ascending, anatropous, hemitropous, or campylotropous, mostly epitropous, bitegmic or unitegmic. Fruits multifollicles or (*Holodiscus* and *Adenostoma*) achenes. Seeds mostly with scanty endosperm, rarely (*Physocarpus*) endosperm copious. Leaves alternate, mostly simple, stipulate or estipulate. Shrubs, subshrubs or rarely herbs, n = 9. – NEILLIEAE: *Physocarpus, Neillia, Stephanandra;* SPIRAEEAE: *Spiraea, Pentactina, Sibiraea, Petrophytum, Kelseya, Luetkea, Aruncus, Pleiosepalum;* GILLENIEAE: *Gillenia, Spiraeanthus;* HOLODISCEAE: *Holodiscus, Xerospiraea;* SORBARIEAE: *Sorbaria, Chamaebatiaria;* ADENOSTOMATEAE: *Adenostoma.*

1.2 LYONOTHAMNOIDEAE

Carpels two, free, glandular, with four ovules in each carpel. Leaves opposite, entire or deeply divided, pinnate and fernlike, with small, caducous stipules. Fruits woody. Seeds apiculate. Trees or shrubs with deciduous bark. n = 27. – *Lyonothamnus.*

1.3 FILIPENDULOIDEAE

Carpels 5–15, free, sessile, or stipitate. Ovules two per carpel, pendulous, campylotropous, epitropous, unitegmic. Fruitlets 1-seeded achenes. Seeds with very scanty endosperm. Leaves alternate, the basal ones imparipinnately compound, cauline leaves similar, but becoming smaller and less compound toward the stem apices; stipules of the basal leaves adnate to the petioles, those of the cauline leaves free, large, asymmetrically reniform, clasping. Tall, rhizomatous perennial herbs, n = 7, 8. – *Filipendula*.

1.4 KERRIOIDEAE

Carpels 2–8, mostly 4–5. Ovules two (Rhodotypeae) or one (Kerrieae), pendulous or (*Kerria*) laterally attached to the middle of the suture, anatropous, epitropous, unitegmic. Fruits achenes (in *Neviusia* achenes are drupaceous). Seeds with copious (*Rhodotypos*) or scanty endosperm. Leaves alternate or (*Rhodotypos*) opposite, simple, stipulate. Deciduous shrubs, n = 9. – RHODOTYPEAE: *Rhodotypos;* KERRIEAE: *Kerria, Neviusia*.

1.5 COLEOGYNOIDEAE

Carpel solitary, sessile in the base of a leathery floral tube; style lateral, longhaired especially below, twisted, persistent. Ovule one. Flowers apetalous. Fruits coriaceous, glabrous, more or less crescent-shaped achenes. Seeds with large embryo and rather scanty endosperm. Leaves opposite, small, entire, with small persistent stipules; hairs medifixed. Much branched spinescent shrubs. – *Coleogyne*.

1.6 RUBOIDEAE

Carpels mostly numerous, free, inserted on a nearly flat to cylindrical receptacle. Ovules two per carpel, collateral or superposed, pendulous, anatropous, epitropous, unitegmic. Fruitlets 1-seeded drupelets crowned on the dry or spongy, often elongated receptacle. Seeds with small embryo and rather abundant endosperm. Leaves alternate, simple, lobed, 3–5-foliolate or imparipinnate, with stipules adnate to the petiole. Mostly erect to trailing shrubs or scramblers, rarely herbaceous, n = 7. – *Rubus* (including *Dalibarda* ?).

1.7 ROSOIDEAE

Carpels numerous, free, borne on the inner surface of an invaginated receptacle near the bottom of the hollow center. Ovules one per carpel, pendulous, anatropous, unitegmic. Fruits numerous achenes enclosed by the enlarged, fleshy receptacle. Seeds with large embryo and scanty endosperm. Leaves alternate, imparipinnate or trifoliolate, rarely 1-foliolate or reduced to connate, leaflike stipules; stipules adnate to the petioles and forming wings. Erect, trailing, or scandent shrubs, mostly armed with prickles, n = 7. – *Rosa, Hulthemia*.

1.8 POTENTILLOIDEAE

Carpels numerous, free, borne on convex, conical, or ellipsoid receptacle. Ovule one per carpel, anatropous, bitegmic (*Dryas* and possibly some others) or more often unitegmic. Fruits achenes. Seeds with endosperm 7–8 cells thick (*Dryas*) or more often endosperm 1-cell thick. Leaves alternate or (*Potaninia*) opposite, simple or compound, stipulate. Shrubs or more often perennial herbs, rarely small trees, n = 7, 8, 9. – DRYADEAE: *Dryas, Fallugia;* PURSHIEAE: *Chamaebatia, Cowania, Purshia;* GEEAE: *Geum* (including *Acomastylis, Oreogeum, Parageum*), *Oreogeum, Orthurus, Taihangia, Novosieversia, Sieversia, Woronowia, Waldsteinia, Coluria;* CERCOCARPEAE: *Cercocarpus;* POTENTILLEAE: *Potentilla* (including *Pentaphylloides*), *Tylosperma, Comarum, Chamaerhodos, Duchesnea, Fragaria, Horkelia, Horkelliella, Ivesia, Sibbaldia* (including *Sibbaldianthe*); ALCHEMILLEAE: *Alchemilla, Aphanes;* SANGUISORBEAE: *Agrimonia, Aremonia, Spenceria, Hagenia, Leucosiaea, Sanguisorba, Poterium, Sarcopoterium, Margyricarpus, Tetraglochin, Acaena, Polylepis, Cliffortia, Bencomia;* POTANINIEAE: *Potaninia*.

1.9 DICHOTOMANTHOIDEAE

Carpel solitary at the base of the calyx tube, with tomentose lateral stylodium. Ovary superior. Ovules two, erect, subbasal, anatropous, bitegmic. Floral tube free, fleshy, and enlarged in fruit. Fruits achenes, often 1-seeded, with coriaceous pericarp surrounded by thick fleshy floral tube. Seeds with endosperm 1 cell thick. Leaves alternate, simple, entire. Trees, n = 17. – *Dichotomanthes*.

1.10 PYROIDEAE (Maloideae)

Carpels 2–5, rarely one, more or less adnate to the floral tube and partly to the invaginated receptacle. Ovules (1)2-many per carpel, anatropous, bitegmic or rarely (*Raphiolepis, Chamaemeles*) unitegmic. Fruits usually consist of more or less well-developed and usually fleshy parenchymatous pericarp (with groups of brachysclereids in *Cydonia* and *Pyrus*) and carpellar endocarps consisting of elongated scleroids. In Maleae the inner walls of the carpels become chartaceous or cartilaginous, rarely bony, and the carpels laterally connate; in Crategeae the inner walls of the carpels become hard and each carpel develops into a separate pyrene. Seeds with endosperm 2–15 cells thick (up to 15 cells in *Crataegus*). Leaves alternate, mostly simple, stipulate. Shrubs or trees, n = 17 or (*Vauquelinia*) 15. – KAGENECKIEAE: *Kageneckia;* LINDLEYIEAE: *Vauquelinia, Lindleya;* MALEAE: *Photinia* (including *Stranvaesia*), *Heteromeles, Eriobotrya, Rhaphiolepis, Sorbus, Chamaemespilus, Aronia, Amelanchier, Pyrus, Malus, Eriolobus, Peraphyllum, Docynia, Cydonia, Pseudocydonia, Chaenomeles;* CRATAEGEAE: *Cotoneaster, Malacomeles, Chamaemeles, Pyracantha, Crataegus, Mespilus, Hesperomeles, Osteomeles.*

1.11 AMYGDALOIDEAE

Carpel solitary or sometimes two (*Maddenia*) or two (*Oemleria* and *Exochorda*) carpels, free from the floral tube; stylodium terminal or (Prinsepieae) lateral or subbasal. Ovules two per carpel, pendulous, anatropous, more or less hemitropous, or (Prinsepieae) campylotropous, bitegmic. Fruits mostly 1-seeded drupes, the mesocarp fleshy and indehiscent to dryish and dehiscent, the endocarp hard or (Prinsepieae) coriaceous, indehiscent. Seeds with endosperm 1 to 15 cells thick. Leaves alternate, simple, stipulate. Trees or shrubs, n = 8. – OSMARONIEAE: *Oemleria;* EXOCHORDEAE: *Exochorda;* AMYGDALEAE: *Maddenia, Pygeum, Laurocerasus, Padus, Cerasus, Padellus, Prunus, Armeniaca, Persica, Amygdalus;* PRINSEPIEAE: *Prinsepia.*

Doubtfully Rosaceous genus: *Guamatela.*

The Spiraeoideae represent the least advanced subfamily and are nearest to the ancestral stock.

It is widely accepted that all other subfamilies are offshoots from a basically spiraeoid-like stock. The nearest to the Spiraeoideae are Lyonothamnoideae and Quillajoideae. Both of them are taxonomically isolated, especially the latter. Based on their study of the *rbc*L sequence, Morgan et al. (1994) even conclude that *Quillaja* is not a member of the family Rosaceae, which has been accepted by some recent authors, including Kalkman (2003). As regards *Lyonothamnus,* I agree with Morgan et al. (1994:899) that "it may be described most accurately as an isolated descendant of the ancestral spiraeoid complex."

Filipenduloideae occupy a somewhat intermediate position between them and Rosoideae although, according to Hegnauer (1973), because of the presence of ellagic acid and the absence of cyanogenetic substances, *Filipendula* is definitely rosoid, not spiraeoid. Nakai (1943), who established the subfamily Filipenduloideae, placed it after his Potentilloideae, Ruboideae, Sanguisorboideae, Cercocarpoideae, and Kerrioideae.

Kerrioideae and the closely related Coleogynoideae (Golubkova 1991) with their achenous fruits and basic chromosome number x = 9 are probably also derived from the spiraeoid complex.

The traditional subfamily Rosoideae has been divided by Nakai (1943) into a number of segregate subfamilies, some of which are fairly distinct. His Filipenduloideae, Kerrioideae, Ruboideae, Rosoideae s.str., and Potentilloideae I find acceptable. However, I accept the Potentilloideae in a somewhat broadened sense and include in them – albeit with some hesitation – also his Sanguisorboideae and Cercocarpoideae while excluding *Coleogyne.* The Potentilloideae are the most heterogeneous subfamily (even sensu Nakai), and there are some isolated genera, such as *Cercocarpus, Purshia,* and other genera with basic chromosome number x = 9.

Dichotomanthoideae are close to the Pyroideae but differ in flower morphology, in which, according to Rohrer et al. (1991), they resemble more the Amygdaloideae.

The Pyroideae have definite affinities with the Spiraeoideae, especially with the Sorbarieae, and are closely linked to them via *Vauquelinia, Kageneckia,* and *Lindleya.* They have been derived from spiraeoid ancestors by autopolyploidy or allopolyploidy (Gladkova 1972; Cronquist 1981; Takhtajan 1987, 1997; Morgan et al. 1994).

The Amygdaloideae have probably also originated directly from the spiraeoidlike ancestor. *Exochorda* belongs to the Amygdaloideae (Juel 1918; Goldblatt 1976; Takhtajan 1987, 1997; Thorne 1992, 2000; Morgan et al. 1994) although it differs in having follicular fruits, winged seeds, and in its chemical makeup. However, according to Challice (1981), the presence of arbutin in *Exochorda* indicates that this genus should perhaps remain in the Spiraeoideae and not be transferred to the Amygdaloideae (where arbutin is absent).

2. QUILLAJACEAE

D. Don 1831. 1/2–3. Peru, Chile, southern Brazil.

Quillaja.

Quillaja had typically been placed in Rosaceae, sometimes as a separate subfamily (Takhtajan 1987, 1997), because of several shared features, including follicular fruits and winged seeds. However Quillajaceae differ from Rosaceae in valvate sepals, pollen grains (Claxton et al. 2005), the base chromosome number (n = 14), which is not found elsewhere in Rosaceae, and in the presence of saponins and trihydroxy-substituted flavonoids. In analyses based on *rbc*L sequence data, *Quillaja* appeared as sister to Polygalaceae, but these relationship was not strongly supported (Savolainen et al. 2000).

Bibliography

Aldasoro JJ, C Aedo, and C Navarro. 2005. Phylogenetic and phytogeographical relationships in Maloideae (Rosaceae) based on morphological and anatomical characters. Blumea 50: 3–32.

Bate-Smith EC. 1961. Chromatography and taxonomy in the Rosaceae, with special reference to *Potentilla* and *Prunus*. Bot. J. Linn. Soc. 58: 39–54.

Bate-Smith EC. 1965. Investigation of the chemistry and taxonomy of subtribe Quillageae of the Rosaceae using comparisons of fresh and herbarium material. Phytochemistry 4: 535–539.

Bate-Smith EC and LL Creasy. 1969. Luteoforol in strawberry leaves. Phytochemistry 8: 1811–1813.

Berggren ST, S Lee, and J Wen. 2000. Phylogenetic studies in the Amygdaloideae (Rosaceae) using ITS and *ndh*F sequences. Am. J. Bot. 87(6): 114 (Abstract).

Bondar N. 1997. Comparative study of seed anatomy in Rosaceae—Spiraeoideae. Scripta Bot. Belgica 15: 31.

Bonne G. 1925. Sur les faiscaux de rebroussement dans la coupe florale de certains Rosacees. C. R. Acad. Sci. Paris 219: 181–191.

Bonne G. 1928. Recherches sur le pedicelle et la fleur des Rosacees. Paris.

Campbell CS, MJ Donoghue, BG Baldwin, and ME Wojciechowski. 1995. Phylogenetic relationships in Maloideae (Rosaceae): evidence from sequences of the internal transcribed spacers of nuclear ribosomal DNA and its congruence with morphology. Am. J. Bot. 82: 903–910.

Challice JS. 1973. Phenolic compounds of the subfamily Pomoideae: A chemotaxonomic survey. Phytochemistry 12: 1095–1101.

Challice JS. 1975. Rosaceae: Chemotaxonomy and the origins of the Pomoideae. Bot. J. Linn. Soc. 69: 239–259.

Challice JS. 1981. Chemotaxonomic studies in the family Rosaceae and the evolutionary origins of the subfamily Maloideae. Preslia 53: 289–304.

Claxton F, H Banks, BB Klitgaard, and PR Crane. 2005. Pollen morphology of families Quillajaceae and Surianaceae (Fabales). Rev. Palaeobot. Palynol. 133: 221–233.

Dickson EE, K Arumuganathan, S Kresovich, and JJ Doyle. 1992. Nuclear DNA content variation within the Rosaceae. Am. J. Bot. 79: 1081–1086.

El-Gazzar A. 1981. Chromosome numbers and rust susceptibility as taxonomic criteria in Rosaceae. Plant Syst. Evol. 137: 23–38.

Ericksson LT, MJ Donoghue, and MS Hibbs. 1998. Phylogenetic analysis of *Potentilla* using DNA sequences of nuclear ribosomal internal transcribed spacers (ITS), and implications for the classification of Rosoideae (Rosaceae). Plant Syst. Evol. 211: 155–179.

Eriksson T, MS Hibbs, AD Yoder, CF Delwiche, and MJ Donoghue. 2003. The phylogeny of Rosoideae (Rosaceae) based on sequences of the internal transcribed spacers (ITS) of nuclear ribosomal DNA and the *trn*L/F region of chloroplast DNA. Int. J. Plant Sci. 164: 197–211.

Evans RC and CS Campbell. 2000a. A multigene tree of the Rosales. Am. J. Bot. 87(6): 125 (Abstract).

Evans RC and CS Campbell. 2000b. The polyploid origin of a large clade: Nuclear Granule-Bound Starch Synthase (GBSSI or waxy) gene sequences support a spiraeoid ancestry of the Maloideae (Rosaceae). Am. J. Bot. 87(6): 125–126 (Abstract).

Evans RC and CS Campbell. 2002. The origin of the apple subfamily (Maloideae; Rosaceae) is clarified by DNA sequence data from duplicated GBSSI genes. Am. J. Bot. 89: 1478–1484.

Evans RC and TA Dickinson. 1999a. Floral ontogeny and morphology in subfamily Amygdaloideae T. & G. (Rosaceae). Int. J. Plant Sci. 160: 955–979.

Evans RC and TA Dickinson. 1999b. Floral ontogeny and morphology in subfamily Spiraeoideae Endl. (Rosaceae). Int. J. Plant Sci. 160: 981–1012.

Evans RC and TA Dickinson. 2002. How do studies of comparative ontogeny and morphology aid in elucidation of relationships within the Rosaceae? In Botany 2002: Botany in the Curriculum, p. 108. Madison, Wisconsin (Abstracts).

Fehrenbach S and W Barthlott. 1988. Mikromorphologie der Epicuticular-Wachse der Rosales s.l. und deren systematische Gliederung. Bot. Jahrb. Syst. 109: 407–428.

Gajewsld W. 1957. A cytological study of the genus *Geum*. Monogr. Bot. Warszawa 4: 1–416.

Gajewsld W. 1959. Evolution in the genus *Geum*. Evolution 13: 378–388.

Gao FY. 1998. *Exochorda:* five species or one? A biosystematic study of the Rosaceous genus *Exochorda*. Wageningen Agricultural University, Wageningen, Netherlands.

Gerstner E, V Matzke, and E Pfeil. 1968. Zur chemischen und biologischen Systematik der Rosaceen: Unter suchungen des Flavynsystems D-Oxynitrilase. Naturwiss enschaften 55: 561–563.

Gladkova VN. 1969. On the systematic position of the genus *Dichotomanthes*. Bot. Zhurn. 54: 431–436 (in Russian).

Gladkova VN. 1972. On the origin of the subfamily Maloideae. Bot. Zhurn. 57: 42–49 (in Russian).

Goldblatt P. 1976. Cytotaxonomic studies in the tribe Quillajeae (Rosaceae). Ann. Missouri Bot. Gard. 63: 200–206.

Golubkova EI. 1991. New subfamily Coleogynoideae (Rosaceae). Sci. Reports Higher School, Biol. Sci., 3 (327): 102–109.

Hess WJ and J Henrickson. 1987. A taxonomic revision of *Vauquelinia* (Rosaceae). SIDA 12: 101–163.

Huber H. 1963. Die Verwandtschaftsverhaltnisse der Rosifloren. Mitt. Bot. Staatssamml. Munchen 5: 1–48.

Iketani H and H Ohashi. 1991. Anatomical structure of fruits and evolution of the tribe Sorbeae in the subfamily Maloideae (Rosaceae). J. Jpn. Bot. 66: 319–351.

Iwatsubo Y and N Naruhashi. 1998. Chromosome study of *Exochorda racemosa* (Rosaceae). J. Phytogeogr. Taxon. 46: 201–203.

Jackson G. 1934. The morphology of the flowers of *Rosa* and certain closely related genera. Am. J. Bot. 21: 453–466.

Jansen S, Y Sano, B Choat, D Rabaey, F Lens, and RR Dute. 2007. Pit membranes om tracheary elements of Rosaceae and related families: new records of tori and pseudotori. Am. J. Bot. 94: 503–514.

Juel HO. 1918. Beiträge zur Blütenanatomie und zur Systematik der Rosaceen. Kongl. Vetensk. Akad. Handl. 58(5): 1–32. Stockholm.

Juel HO. 1927. Über die Blütenanatomie einiger Rosaceen. Nova Acta Regiae Soc. Sci. Upsaliensis, vol. extr. Uppsala.

Kalkman C. 1973. The Malesian species of subfamily Maloideae (Rosaceae). Blumea 21: 413–442.

Kalkman C. 1988. The phylogeny of the Rosaceae. Bot. J. Linn. Soc. 98: 37–59.

Kalkman C. 2004. Rosaceae. In: K Kubitzki, ed. The families and genera of vascular plants, vol. 6, pp. 343–386. Springer, Berlin/Heidelberg/New York.

Kamelin RV. 2006. Rosaceae. Barnaul University, Barnaul.

Kania W. 1973. Entwicklungsgeschichtliche Untersu-chungen an Rosaceenblüten. Bot. Jahrb. Syst. 93: 175–246.

Kubitzki K. Quillajaceae. In: K Kubitzki, ed. The families and genera of vascular plants, vol. 9, pp. 407–408. Springer, Berlin/Heidelberg/New York.

Lersten NR and HT Horner. 2005. Macropattern of styloid and druse crystals in *Quillaja* (Quillajaceae) bark and leaves. Int. J. Plant Sci. 166: 705–711.

Lindenhofer A and A Weber. 1999a. Polyandry in Rosaceae: evidence for a spiral origin of the androecium in Spiraeoideae. Bot. Jahrb. Syst. 121: 553–582.

Lindenhofer A and A Weber. 1999b. The spiraeoid androecium of Pyroideae and Amygdaloideae (Rosaceae). Bot. Jahrb. Syst. 121: 583–605.

Lindenhofer A and A Weber. 2000. Structural and developmental diversity in the androecium of Rosoideae (Rosaceae). Bot. Jahrb. Syst. 122: 63–91.

Lotova LI and AC Timonin. 1998a. Anatomy of cortex and secondary phloem of Rosaceae. 1. Spiraeoideae—Spireae; 2. Spiraeoideae except Spiraeae and Lymno thamneae. Bot. Zhurn. 83: 16–27 (in Russian with English summary).

Lotova LI and AC Timonin. 1999. Anatomy of cortex and secondary phloem in Rosaceae: 3. Quillajoideae. Bot. Zhurn. 84(2): 34–41 (in English with Russian summary).

Lotova LI and AC Timonin. 2002. Anatomy of cortex and secondary phloem of Rosaceae. 12. Prunoideae. Bot. Zhurn. 87(5): 68–82 (in Russian with English summary).

MacDaniels LH. 1940. The morphology of the apple and other pome fruits. Cornell Univ. Agr. Exp. Sta., Mem. 230: 1–32.

Mai DH. 1984. Karpologischc Untersuchungen der Steinkerne fossiler und rezenter Amygdalaceae (Rosales). Feddes Repert. 95: 299–329.

Mauritzon J. 1939. Contributions to the embryology of the orders Rosales and Myrtales. Lunds Universitets Arsskrift, 35(2). Gleerup, Lund, Sweden.

Melikian AP and NA Bondar. 1996a. Rosaceae. Neuradaceae. In: A Takhtajan, ed. Comparative seed anatomy, vol. 5, pp. 102–127. Nauka, St. Petersburg (in Russian).

Moffett AA. 1931. The chromosome constitution of the Pomoideae. Proc. Roy Soc. London, Ser. B, Biol. Sci. 108: 423–446.

Morgan DR, DE Soltis, and KR Robertson. 1994. Systematic and evolutionary implications of *rbc*L sequence variation in Rosaceae. Am. J. Bot. 81: 890–903.

Naruhashi N and Y Toyoshima. 1979. Pollen morphology of Japanese Rosaceae. J. Phytogeogr. Taxon. 27: 46–50.

Oh S-H and D Potter. 2002. Where does *Guamatela* belong? Botany 2002 BSA/ASPT meeting, Madison (Abstract).

Oh S-H and D Potter. 2005. Molecular phylogenetic systematics and biogeography of tribe Neillieae (Rosaceae) using DNA sequences of cpDNA, rDNA, and LEAFY. Am. J. Bot. 92: 179–192.

Oh S-H and D Potter. 2006. Description and phylogenetic position of a new Angiosperm family, Guamatelaceae, inferred from chloroplast *rbc*L, *atp*B, and *Mat*K sequences. Syst. Bot. 31: 730–738.

Phipps JB, KR Robertson, PG Smith, and JR Rohrer. 1990. A checklist of the subfamily Maloideae (Rosaceae). Canad. J. Bot. 68: 2209–2269.

Phipps JB, KR Robertson, JR Rohrer, and PG Smith. 1991. Origin and evolution of subfam. Maloideae (Rosaceae). Syst. Bot. 16: 303–332.

Potter D, F Gao, PE Bortiri, S-H Oh, and S Baggett. 2002. Phylogenetic relationships in Rosaceae inferred from chloroplast *mat*K and *trn*L-*trn*F nucleotide sequence data. Plant Syst. Evol. 231: 77–89.

Robertson KR. 1974. The genera of Rosaceae in the southeastern United States. J. Arnold Arbor. 55: 303–332, 344–401, 611, 662.

Robertson KR, JB Phillis, JR Rohrer, and PG Smith. 1991. A synopsis of genera in Maloideae. Syst. Bot. 16: 376–394.

Rohrer JR, KR Robertson, and JB Phipps. 1991. Variation in structure among fruits of Maloideae (Rosaceae). Am. J. Bot. 78: 1617–1635.

Rohrer JR, KR Robertson, and JB Phipps. 1994. Floral morphology of Maloideae (Rosaceae) and its systematic relevance. Am. J. Bot. 81: 574–581.

Sax K. 1932. The origin and relationships of the Pomoideae. J. Arnold Arbor. 12: 3–22.

Schaeppi H and F Steindl. 1950. Vergleichend-mor-phologische Untersuchungen am Gynoeceum der Rosoideen. Ber. Schweiz. Bot. Ges. 60: 15–50.

Sterling C. 1963. The affinities of *Prinsepia* (Rosaceae). Am. J. Bot. 50: 693–699.

Sterling C. 1964, 1965, 1966a, b, c. Comparative morphology of the carpel in the Rosaceae: I, II. Prunoideae. Oesterr. Bot. Z. 51: 36–44, 354–360, 1964; III, IV, V, VI, VII. Pomoideae. Oesterr. Bot. Z. 52: 705–712, 52: 47–54, 418–426, 938–946, 1965; 53: 225–231, 1966; VIII. Spiraeoideae. Oesterr. Bot. Z. 53: 521–530, 951–960,1966.

Sterling C. 1969. Comparative morphology of the carpel in the Rosaceae: X. Evolution and summary. Oesterr. Bot. Z. 116: 46–54.

Tumanian SA. 1950. Anatomical structure of the wood of the Caucasian representatives of the subfamily Pomoideae of the fam. Rosaceae. Trudy Bot. Inst. Armenian Acad. Sci. 7: 69–103 (in Russian).
Vamosi JC and TA Dickinson. 2006. Polyploidy and diversification: a phylogenetic investigation in Rosaceae. Int. J. Plant Sci. 167: 349–358.
Wallaart RAM. 1980. Distribution of sorbitol in Rosaceae. Phytochemistry 19: 2603–2610.
Webb JE. 1902. A morphological study of flower and embryo of *Spiraea*. Bot. Gaz. 33: 451–460.
Wilkes S and H Glasl. 2001. Isolation, characterization, and systematic significance of 2-pyrone-4, 6-dicarboxylic acid in Rosaceae. Phytochemistry 58: 441–449.
Yu T-T. 1984. Origin and evolution of Rosaceae. Acta Phytotax. Sin. 22: 431–444 (in Chinese with English summary).
Zhang S-Y. 1992. Wood anatomy of the Rosaceae. Blumea 37: 81–158.
Zhang S-Y and P Baas. 1992. Wood anatomy of trees and shrubs from China III. Rosaceae. IAWA Bull., n.s. 13: 21–91.
Zhou LH, ZX Wei, and ZY Wu. 1999a. Pollen morphology of Prunoideae of China (Rosaceae). Acta Bot. Yunn. 21: 207–211.
Zhou LH, ZX Wei, and ZY Wu. 1999b. Pollen morphology of Spiraeoideae in China (Rosaceae). Acta Bot. Yunn. 21: 303–308.
Zhou LH, ZX Wei, and ZY Wu. 1999c. Pollen morphology of Rosoideae (Rosaceae) of China. Acta Bot. Yunn. 21: 455–460.
Zhou LH, ZX Wei, and ZY Wu. 2000a. Pollen morphology and systematic position of the Chinese endemic genus *Dichotomanthes* (Rosaceae). Acta Bot. Yunn. 22: 143–147.
Zhou LH, X Gong, and ZY Wu. 2000b. The karyomorphology and systematic position of the Chinese endemic genus *Dichotomanthes*. Acta Bot. Yunn. 22: 282–285.

Order 85. CROSSOSOMATALES

Xerophytic, glabrous, sometimes spinescent shrubs or small trees, sometimes with hyaline to black trichomes. Vessel elements small and short, with simple perforations; lateral pitting alternate. Fibers short, with bordered pits. Rays heterogeneous, mixed uniseriate and pluriseriate. Axial parenchyma scanty, mostly apotracheal and diffuse. Sieve-element plastids of S-type. Nodes unilacunar with one trace, or trilacunar with three traces. Yellow acicular crystals common in all organs. Leaves deciduous, alternate or (*Apacheria*) opposite, small, simple, entire or some of them apically tridentate, with minute, subulate, linear stipules or estipulate. Stomata anomocytic. Flowers solitary, axillary or terminal, bisexual or some of them unisexual (plant polygamo-monoecious), actinomorphic, with double perianth. Sepals (3-)4-5(-6), imbricate, free, persistent. Petals (3-)4-5(-6), free, imbricate, deciduous or persistent, usually longer than sepals, often short-clawed. Stamens attached to a thickened nectary disc or seated around an annular nectary disc at its base, numerous (up to 50–100) in three or four cycles associated with about ten trunk bundles (arising centrifugally or centripetally), but in two cycles in *Apacheria* and some spp. of *Glossopetalon*, and more or less reduced to one cycle (antepetalous cycle lost) in rest of *Glossopetalon*. Anthers small, tetrasporangiate, slightly extrorse, tetrasporangiate, basifixed to slightly ventrifixed, versatile, opening longitudinally. Tapetum secretory. Pollen grains 2-celled, (2)3-colporate, reticulate. Nectary disc adnate to the floral tube. Gynoecium of 1-5(-9) free carpels, each with a short, stout stylodium and terminal, expanded stigma, or (*Apocheria*) the stigma decurrent on the stylodium; ovary superior; ovules (1)2 to many per carpel, campylotropous, bitegmic, crassinucellate; micropyle zig-zag. Female gametophyte of *Polygonum*-type. Endosperm nuclear. Fruits asymmetric follicles, ventrally dehiscent, surface smooth or rugose. Seeds testal, provided with well-developed, large, whitish or yellowish, and fimbriate or small and entire aril arising around the funicle; embryo slender, slightly curved; endosperm scanty or copious, oily. Containing inulin, ellagic and gallic acids, and cyanidin-3-glucoside; *Glossopetalon* leaves contain acacetin and syringin; n = 6 (*Crossosoma*).

The affinities of the Crossosomataceae are not very clear. They are most probably related to the archaic members of the Rosales and especially to the Saxifragales although in the presence of the aril they also resemble the Dilleniaceae.

1. CROSSOSOMATACEAE

Engler 1897. 4/8 or 9. Arid parts of western United States and adjacent Mexico.

Crossosoma, Glossopetalon (*Forsellesia*), *Apacheria, Velascoa.*

Bibliography

Calderon de Rzedowski G and J Rzedowski. 1997. *Velascoa* (Crossosomataceae), un género nuevo de la Sierra Madre Oriental de México. Acta Bot. Mex. 39: 53–59.

DeBuhr LE. 1978. Wood anatomy of *Forsellesia* (*Glossopetalon*) and *Crossosoma* (Crossosomataceae, Rosales). Aliso 9: 179–184.
Ensign M. 1942. A revision of the celastraceous genus *Forsellesia (Glossopetalon)*. Am. Midl. Nat. 27: 501–511.
Holmgren NH. 1988. *Glossopetalon* (Crossosomataceae) a new variety of *G. spinescens* from the Great Basin, U.S.A. Brittonia 40: 269–274.
Kapil RN and RS Vani. 1963. Embryology and systematic position of *Crossosoma californicum*. Curr. Sci. 32: 493–495.
Lemésle R. 1948. Position phylogénétique de l'*Hydrastis canadensis* L. et du *Crossosoma californicum* Nutt., d'a-pres particularités histologiques du xylème. C. R. Acad. Sci. Paris 227: 221–223.
Mason CT. 1975. *Apacheria chiracahuensis:* A new genus and species from Arizona. Madroño 23: 105–108.
Masson CT. 1992. Crossosomataceae: *Crossosoma* family. J. Arizona Nevada Acad. Sci. 26: 7–9.
Melikian AP and NA Bondar. 1996. Crossosomataceae. In: A Takhtajan, ed. Comparative seed anatomy, vol. 5, pp. 126–127. Nauka, St. Petersburg (in Russian).
Richardson PE. 1970. The morphology of the Crossosomataceae: I. Leaf, stem, and node. Bull. Torrey Bot. Club 97: 34–39.
Sosa V and MW Chase. 2003. Phylogenetics of Crossosomataceae based on *rbc*L sequence data. Syst. Bot. 28: 96–105.
Tatsuno A and R Scogin. 1978. Biochemical profile of Crossosomataceae. Aliso 9: 185–188.
Thorne RF and R Scogin. 1978. *Forsellesia* Greene (*Glossopetalum* Gray): A third genus in the Crossosomataceae, Rosineae, Rosales. Aliso 9: 171–178

Order 86. CHRYSOBALANALES

Trees or shrubs, rarely subshrubs, or woody lianas. Vessels with simple perforations, rarely (in some species of *Trigonia*) with scalariform perforation, or mixed simple and scalariform perforations; lateral pitting usually alternate; vestured pits present. Fibers with bordered pits. Rays mainly heterogeneous or rarely almost homogeneous. Axial parenchyma apotracheal or paratrachel. Sieve-element plastids of S-, or Ss-type. Nodes uni-, or pentalacunar. Leaves alternate or opposite, simple, entire, pinnately veined, sometimes conduplicate, stipulate. Stomata paracytic. Flowers mostly rather small, in various kinds of cymose racemose or racemose, terminal or axillary inflorescence or rarely solitary, bisexual or rarely polygamous or unisexual, actinomorphic or more or less zygomorphic. Sepals five, mostly imbricate. Petals five, free, usually imbricate, rarely absent. Stamens (2-)8-20 or numerous (up to 300), all fertile or some of them staminodial; filaments free or basally connate; anthers small or thickened, tetrasporangiate, introrse, basifixed or dorsifixed, opening longitudinally. Nectaries present or rarely wanting. Tapetum secretory. Microsporogenesis simultaneous. Pollen grains 2-celled, 3(4)-colpate or -colporate. Gynoecium of 2–3 carpels; ovary superior to quite inferior, 2–3-locular; ovules usually 1–2, rarely numerous in each fertile locule, anatropous, bitegmic, mostly tenuinucellate, sometimes with endothelium and micropyle zig-zag. Female gametophyte monosporic, bisporic, or tetrasporic. Endosperm nuclear. Fruits drupes or septicidal capsules. Seeds exotestal, testal-tegmic, or endotegmic; embryo straight, with or without endosperm.

Related to the Rosaceae and were by many authors included in the Rosales, but they differ from them markedly in tenuinucellate ovules, morphology of the gynoecium, 4-celled female gametophyte.

Key to Families

1 Fruits septicidal capsules.
   2 Fertile stamens 4–8. Evergreen trees, shrubs, or woody lianas with T-shaped hairs. Intraxylary phloem absent. Vessels with simple perforations, but in some species of *Trigonia* occur vestigial scalariform perforations; lateral pitting extremely rare. Fibers with distinctly bordered pits, in some species septate. Rays heterogeneous. Axial parenchyma apotracheal or paratracheal. Leaves usually opposite or rarely alternate (*Trigoniastrum*), simple, entire, densely whitish hairy below; stipules interpetiolar, often connate, deciduous. Stomata paracytic. Flowers in terminal or axillary cymes, racemes, or panicles, or thyrses, strongly zygomorphic. Sepals five, unequal, imbricate, united into a cup below. Petals three, or five, imbricate or contorted, or valvate, the two anterior (lower) ones forming a keel, the posterior (upper) ones forming a basally saccate or spurred standard, the laterals spatulate. Androecium unilateral on anterior (lower) side of flower, of 4–13 stamens, 4–8 (often 6) of them fertile (up to 4 staminodial); filaments more or less united. Pollen grains 3–5-porate. Nectary glands sometimes lobed or cleft, borne in front of standard or wanting (*Isidodendron*). Gynoecium of 3(4) carpels; style terminal, with a capitate or slightly trilobed stigma; ovules 1–2 to rather numerous in each locule, pendulous, anatropous, crassinucellate (Mauritzon 1936; Corner 1976) or tenuinucellate

(Boesewinkel 1987), with endothelium; placentation parietal; micropyle zig-zag. Fruits septicidal capsules (*Trigonia*) or more often three winged samaras. Seeds winged, hairy (usually long-pilose), seed coat testal-tegmic; exotesta with thickened outer walls, endotegmic cells tanniniferous; embryo straight, green (*Trigonia*); cotyledons large, flat with very short radicle; endosperm copious and flesh (*Trigonia*) or scarce (*Trigoniastrum*) (Plisko 2000); n = ca. 10.. . . . . 3. TRIGONIACEAE

2 Fertile stamens three. Shrubs or subshrubs (*Euphronia acuminatissima*) or trees; young growth densely covered with white or gray pubescent, the branchlets, petioles and blades may become glabrous with age; mucilaginous hypodermis lacking. Parenchyma more or less aliform-confluent. Leaves alternate, simple, entire, white tomentose below; stipules small. Flowers in terminal or axillary, pubescent racemes, bracteate, bisexual, zygomorphic. Sepals three (5 – according A.Litt 2004), unequal, pubescent abaxially, connate et the base. Petals three, spatulate, clawed, purplish, generally silky pubescent. Stamens four and generally one staminodial; stamens and staminode united at the base into tube; the split opposite staminode; filaments basally connate; anthers reddish brown, basifixed, introrse, opening by a single central slit. Gynoecium of three carpels; the style geniculate near apex, pilose, persistent in fruit, the stigma trilobed; placentation axile. Ovary superior, 3-locular; ovules two in each locule, anatropous. Fruits cylindrical septicidal, trivalvate capsule, exocarp thin, fleshy, pubescent. Seeds one per locule, slightly winged, glabrous, reddish brown. . . . . 4. EUPHRONIACEAE.

1 Fruits drupes.

3 Stylodia free or united into simple styles.

1 Ovules tenuinucellate. Small trees, shrubs, or woody lianas; bark of lianas often smooth, pale, with scattered, dark, lenticellate pustules. Mucilage cells sometimes present in the parenchymatous tissues and in the leaf epidermis and hypodermis. Vessels with simple or mixed simple and scalariform perforations; lateral pitting alternate, rarely (*Tapura*) scalariform or intermediate. Fibers with small bordered pits. Axial parenchyma mostly paratracheal. Indumentum consists of unicellular hairs with warty papillae. Leaves alternate, simple, entire, pinnately veined, stipulate; stipules usually caducous. Flowers small, in axillary to petiolar or epiphyllous cymes, bisexual or rarely unisexual (the plants monoecious), actinomorphic or (*Tapura*) more or less zygomorphic, often with articulated pedicels. Sepals five, imbricate, free or rarely basally connate. Petals five, usually 2-lobed or bifid, imbricate, free or basally connate into a tube. Stamens three or five, free or adnate to the corolla tube, in *Tapura* only three fertile and two staminodia. Anthers thickened, opening by longitudinal slits; the connective often dorsally thickened. A basal nectary gland borne opposite each petal (confluent to form a ring when the corolla is sympetalous). Gynoecium of 2–3 carpels; stylodia free or united into simple or lobed styles; stigma capitate or simple; ovary superior to quite inferior; placentation apical. Ovules two in each locule, pendulous at apex of each locule, anatropous; micropyle endostomal. Fruits drupes, dry or rarely fleshy, with a unilocular or sometimes 2–3-locular pyrene that generally has only one seed per locule; exocarp sometimes splitting. Seeds with large, straight, oily, green or orange embryo and without endosperm, often with caruncle. Containing fluoracetic acid and pyridine alkaloids, n = 10, 12. . . . 2. DICHAPETALACEAE.

3 Style more or less gynobasic, with simple or 3-lobed stigmas. Trees or shrubs, rarely subshrubs; woods siliceous and with SiO_2 grains. Vessels with simple perforations; lateral pitting alternate. Fibers with numerous bordered pits. Rays mainly or wholly uniseriate heterogeneous or rarely almost homogeneous. Axial parenchyma apotracheal, in numerous narrow bands. Sieve-element plastids of Ss-type. parenchyma in apotracheal bands; Nodes pentalacunar with five traces. Leaves alternate, simple, entire, pinnately veined, conduplicate, often with abaxial glands, stipulate. Flowers mostly rather small, in various kinds of cymose racemose or racemose, terminal or axillary inflorescence or rarely solitary, bisexual or rarely polygamous actinomorphic or more or less zygomorphic, with well-developed floral tube that is lined with an annular nectary disc. Sepals five, imbricate. Petals five, free, imbricate, sometimes absent. Stamens (2-)8–20 or numerous (up to 300), all fertile or some of them staminodial; filaments

elongate and curved toward the floral center in bud, free or all connate or connate into groups; in the more zygomorphic flowers all on one side of the floral tube; anthers small, tetrasporangiate, introrse, dorsifixed, opening longitudinally. Tapetum secretory. Microsporogenesis simultaneous. Pollen grains 2-celled, 3(4)-colpate or -colporate. Gynoecium of three carpels united only by the common gynobasic style, but usually two of them more or less reduced, so that the gynoecium may appear to be monomerous with a basal stylodium; two ovules in each fertile locule, or sometimes the single fertile locule divided by a median partition to appear as two uniovulate locules; style more or less gynobasic, with a simple or 3-lobed stigma; ovary superior. Ovules ascending from the base, anatropous, bitegmic, tenuinucellate, with a small nucellus with the tissue soon disintegrating, with endothelium; micropyle zig-zag. Female gametophyte consists of four cells: the egg, two synergids, and the central cell (Tobe and Raven 1984). Endosperm nuclear. Fruits 1-seeded drupes, often hairy inside, medium-sized to large. Seeds exotestal, ruminate, with large embryo and without endosperm. Plants contain trihydroxyflavonoids, n = 10, 11............ 1 CHRYSOBALANACEAE.

1. CHRYSOBALANACEAE

R. Brown 1818 (including Hirtellaceae Horaninow 1847, Licaniaceae Martynov 1820). 17/495. Lowlands of tropical and partly subtropical regions of both hemispheres, but mainly Central and South America and West Indies; reaches subtropics of North America, South Africa, and eastern Asia.

CHRYSOBALANEAE: *Chrysobalanus, Grangeria, Licania, Parastemon;* PARINARIEAE: *Bafodeya, Exellodendron, Hunga, Neocarya, Parinari;* GOUEPIEAE: *Acioa, Couepia, Maranthes;* HIRTELLEAE: *Atuna, Dactyladenia, Hirtella, Kostermanthus, Magnistipula.*

2. DICHAPETALACEAE

Baillon 1886 (including Chailletiaceae R. Brown 1818). 3/180. Pantropical, but *Dichapetalum* reaches southeastern Africa.

Dichapetalum, Stephanopodium, Tapura.

Traditionally are included in the Euphorbiales. Recent molecular data indicates a relationship to the Chrysobalanaceae and Trigoniaceae, a placement suggested as early as 1921 by H.Hallier (Prance 2004).

3. TRIGONIACEAE

Endlicher 1841. 5/35. Madagascar *(Humbertiodendron,* 1), western Malesia *(Trigoniastrum,* 1) and central and tropical South America: *Trigonia* (24) from Mexico to Paraguay, *Trigoniodendron* – southeastern Brazil.

Trigoniastrum, Trigonia, Trigoniodendron, Humbertiodendron, Isidodendron.

Trigoniaceae traditionally placed in the Polygalales, and related to the Vochisiaceae. Molecular data suggested affinities with Chrysobalanaceae and Dichapetalaceae (Soltis et al. 2006).

4. EUPHRONIACEAE

Marcano-Berti 1989. 1/3. Colombia, Venezuela, Guayana, and northwestern Brazil.

Euphronia.

Closely related to the Trigoniaceae.

Bibliography

Boesewinkel FD. 1987. Ovules and seeds of Trigoniaceae. Acta Bot. Neerl. 36: 81–91.

Boeswinkel ED and E Bouman. 1980. Development of ovule and seed coat of *Dichapetalum mombuttense* Engl. with notes on other species. Acta Bot. Neerl. 29: 103–115.

Bonne G. 1926. Sur la constitution du gynécée chez les Chrysobalanées. C. R. Acad. Sci. Paris 182: 1404–1406.

Demchenko NI. 1973. On the morphology of pollen of family Chrysobalanaceae. In: Trudy 3rd Intern. Palynol. Conference, USSR, Novosibirsk, 1971, pp. 69–73. Leningrad (in Russian).

Fritsch C. 1888. Über die Gattung der Chrysobalanaceen. Verh. Zool.-Bot. Ges. Wien 38: 93–95.

Hauman L. 1951. Contribution a l''étude des Chrysobalanoides africaines. Bull. Jard. Bot. L'État 21: 167–198.

Juel HO. 1915. Über den Bau des Gynoeceums bei *Parinari.* Ark. Bot. 14: 1–12.

Litt A. 2004. Euphroniaceae. In: N Smith, SA Mori, A Henderson, DW Stevenson, and SH Heald, eds. Flowering plants of the Neotropics, pp.150–151. Princeton University Press, Princeton.

Litt A and MW Chase. 1999. The systematic position of *Euphronia*, with comments on the position of *Balanops*:

an analysis based on *rbc*L sequence data. Syst. Bot. 23: 401–409.
Lleras E. 1976. Revision and taxonomic position of the genus *Euphronia* Martius ex Martius et Zuccarini (Vochysiaceae). Acta Amazon. 6: 43–47.
Lleras E. 1978. Trigoniaceae. Flora Neotropica Monograph 19: 1–73.
Marcano-Berti L. 1989. Euphroniaceae: una nueva familia. Pittieria 18: 15–19.
Melikian AP and NA Bondar. 1996. Chrysobalanaceae. In: A Takhtajan, ed. Comparative seed anatomy, vol. 5, pp. 123–125. Nauka, St. Petersburg (in Russian).
Nilova MV. 2000. Bark and wood anatomy of stems in two species of Chrysobalanaceae R. Br. Byull. Mosk. Obshch. Ispyt. Prir., Biol. 105(3): 62–65 (in Russian with English summary).
Prance GT. 1970a. The genera of Chrysobalanaceae in the southeastern United States. J. Arnold Arbor. 51: 521–528.
Prance GT. 1970b. Chrysobalanaceae. Flora Neotropica No. 9. New York.
Prance GT. 1972. A monograph of the Neotropical Dichapetalaceae. Flora Neotropica 10: 1–84. New York.
Prance GT. 1989. Chrysobalanaceae. In: CGGJ van Steenis, ed. Flora Malesiana, ser. I, 10: 635–678. Noordhoff, Groningen.
Prance GT. 2004. Dichapetalaceae. In: N Smith, SA Mori, A Henderson, DW Stevenson, and SH Heald, eds. Flowering plants of the Neotropics, pp. 127–128. The New York Botanical Garden, Princeton University Press, Princeton.
Prance GT and SA Mori. 1983. Dispersal and distribution of Lecythidaceae and Chrysobalanaceae. Sonderb. Naturwiss. Vereins Hamburg 7: 163–186.
Prance GT and CA Sothers. 2003a. Species plantarum. Flora of the World. Part 9. Chrysobalanaceae 1: Chrysobalanus to *Parinari*. Austral. Biol. Resources. Canberra.
Prance GT and CA Sothers. 2003b. Species plantarum. Flora of the World. Part 10. Chrysobalanaceae 1: *Acioa* to *Magnistipula*. Austral. Biol. Resources. Canberra.
Prance GT and F White. 1988. The genera of Chrysobalanaceae: A study in practical and theoretical taxonomy and its relevance to evolutionary biology. Philos. Trans. 320B: 1–184.
Prance GT, DJ Rogers, and F White. 1969. A taxonomic study of an angiosperm family: Generic delimitation in the Chrysobalanaceae. New Phytol. 68: 1203–1234.
Punt W. 1975. Pollen morphology of the Dichapetalaceae with special reference to evolutionary trends and mutual relationships of pollen types. Rev. Palaeobot. Palynol. 19: 1–97.
Punt W. 1976. Evolutionary trends in the pollen grains of Dichapetalaceae. In: IK Ferguson and J Muller, eds. The evolutionary significance of the exine, pp. 139–146. Linn. Soc. Symposium, No. 1. London/New York.
Tieghem P. van. 1903. Structure de l'ovule Dichopetalacées et place de cette famille dans la classification. J. Bot. (Paris) 17: 229–233.
Tobe H and PH Raven. 1984. An embryological contribution to systematics of the Chrysobalanaceae: I. Tribe Chrysobalaneae. Bot. Mag. (Tokyo) 97: 397–411.
White F. 1976. The taxonomy, ecology and chorology of African Chrysobalanaceae (excluding *Acioa*). Bull. Jard. Bot. État 46: 265–350.

Superorder MYRTANAE

Order 87. MYRTALES

Trees, shrubs, or herbs. Intraxylary (internal) phloem in woody stems always present. Vessels mostly with simple or rarely (Geissolomataceae and some Myrtaceae) with scalariform perforations and with vestured, bordered pits; lateral pitting almost always alternate. Fibers with simple or bordered pits, septate or nonseptate. Rays heterogeneous to homogeneous. Axial parenchyma most typically scanty paratracheal, but in addition it may be apotracheally diffuse. Sieve-element plastids of Ss-type. Nodes unilacunar or sometimes (Alzateaceae) trilacunar. Leaves opposite or less often alternate or verticillate, simple, mostly entire, usually with vestigial stipules, rarely estipulate. Stomata of various types, mostly anomocytic. Flowers in various cymose or racemose inflorescences or sometimes solitary in the axils, bisexual or rarely unisexual, actinomorphic or sometimes more or less zygomorphic (usually slightly zygomorphic), mostly 4- or 5-merous, sometimes apetalous (some Lythraceae and all Alzateaceae and Penaeaceae). Sepals connate, sometimes much reduced or obsolete. Petals alternisepalous. Stamens numerous, rarely reduced to one; filaments bent inward in the bud or rarely erect (as in *Psiloxylon*); anthers 2-locular or (Heteropyxidoideae in Myrtaceae, Rhynchocalycaceae) 4-locular at anthesis, opening longitudinally or (some Myrtaceae and the majority of Melastomataceae) by apical pores. Tapetum secretory. Microsporogenesis simultaneous. Pollen grains 2-celled or less often 3-celled, in monads or sometimes (as in Onagraceae) in tetrads, basically 3-colporate, but sometimes colpate or porate, basically tectate and frequently characterized by pseudocolpi. Intrastaminal annular nectary disc often present. Gynoecium of 2-many (often 4 or 5) united carpels with stylodia connate into a style (very short in *Psiloxylon*); stigma simple or lobed. Ovary superior to inferior, plurilocular to 1-locular, usually with (1)2 to many ovules per carpel. Ovules anatropous or in various taxa hemitropous, amphitropous or campylotropous, usually bitegmic or very rarely (in some Myrtaceae) unitegmic, crassinucellate. Endothelium absent or rarely (*Axinandra* in the Crypteroniaceae) present. Female gametophyte of various types. Endosperm nuclear. Fruits of various types.

Seeds mostly with very little or no endosperm. Contain flavonols, myricetin, methyulated ellagic acid, n = 5–14, 20, 24.

Evidently derived from the lower Rosidae, most probably from the Saxifragales.

Key to Families

1 Nodes trilacunar with three traces. Rays heterogeneous with long ends. Vessel-ray pits large and simple, reticulate to scalariform. Evergreen small trees or shrubs, sometimes hemiepiphytic; branches opposite or verticillate; the young branches scarcely 4-angled, the stems of the inflorescence distinctly 4-angled to narrowly winged; nodes enlarged, especially knobby when branches are verticillate. Leaves mostly clustered at the ends of the branches, opposite or less often verticillate, simple, entire, thick, leathery, with axillary stipules. Stomata anomocytic to nearly encyclocytic. Flowers in axillary, basically paniculate inflorescences, lacking bracteoles, articulated at the base, bisexual, actinomorphic, apetalous or possibly petals rudimentary, with open campanulate, leathery, floral tube, 5(6)-merous. Calyx lobes fleshy, valvate, persistent. Nectary disc well developed, wide, lobed. Stamens five, situated in the floral tube at the sinuses of the lobes; filaments green, stout, with large cordate connective, as long as the anthers; anthers dorsifixed, introrse, heart-shaped, pinkish white, fleshy, at the anthesis held at right angles to the filament with the apiculate apex extending between the lobes, resembling petals, opening longitudinally. Pollen grains 3-colporate, psilate, except for the slightly depressed middle of the mesocolpia, which is psilate-punctate. Gynoecium of two carpels, slightly bilaterally flattened, with a short style and capitate stigma; ovary superior, 2-locular, with septum divided into two half-septa and with 40–60 anatropous ovules horizontally imbricate in vertical rows, micropyle endostomal, placentation parietal. Female gametophyte of *Allium*-type. Fruits bilaterally compressed dry, loculicidal capsules. Seeds flattened, oblong to lunate, with fragile membranous wing; exotestal cells low; embryo straight; endosperm lacking. Ellagic acid and flavonoid (quercetin); n = 14. . . . . 1. ALZATEACEAE.

1 Nodes unilacunar with one trace.

2 Plants woody or herbaceous, terrestrial or less often aquatic.

3 Secretory cavities lined with epithelium lacking. Calyx and corolla usually valvate.

4 Micropyle is formed only by one integument.

5 Micropyle is formed only by inner integument Evergreen tree up to 12 m tall. Leaves opposite, decussate, simple, entire; stipules rudimentary. Stomata anomocytic and cyclocytic. Flowers in multifloral panicle, small, bisexual, actinomorphic, 6-merous, with a short, saucer-shaped floral tube. Sepals six, white and petal-like, recurved, and distinctly beaked in bud, persistent. Petals six, inserted on the rim of the floral tube, conspicuously and narrowly clawed, recurved, caducous, margin lacerate, hoodlike, and covering the anthers in bud only. Stamens six, oppositipetalous, inserted near the inner rim of the floral tube, incurved in bud; filaments much longer than the anthers; anthers subbasifixed, versatile, introrse, opening longitudinally. Nectary disc wanting. Pollen grains 3-colporate, heterocolpate, apertures alternating with three pseudocolpi. Gynoecium of 2(3) carpels, with simple style and capitate stigma covered with minute papillae; ovary superior, bilaterally compressed, in young flowers 1-locular, but subsequently rendered 2-locular (rarely 3-locular) by an extension of placentas from opposite sides of the ovary wall; ovules ca. 15–20 per locule, anatropous, pleurotropous, superposed in a single vertical series. Female gametophyte of *Polygonum*-type. Fruits dorsiventrally compressed (rarely trigonous) capsule, loculicidal at the apex. Seeds narrowly winged (expanded raphe), the wings is formed by divisions and elongation of cells of the funiculus; seed coat papery, rather smooth; embryo more or less flattened, cotyledons folded inside; endosperm lacking. Contain flavonoids and accumulated Aluminium; n = 10. . . . . . 2. RHYNCHOCALYCACEAE.

5 The micropyle is formed by the outer integument. Evergreen, densely leafy

ascending shrubs. Young branches more or less densely covered with long unicellular two-armed (Malpighian) hairs that have one long and one quite short arm. Branches with four longitudinal ridges and with relatively inconspicuous transverse leaf scars. Young stems somewhat tetragonal. Vessels with scalariform perforations that have rather numerous (12–23) bars; lateral pitting opposite or scalariform. Fibers are tracheids. Rays heterogeneous, mixed uniseriate and pluriseriate. Axial parenchyma scanty and diffuse and occurs in strands of 4–5 very long cells. The stem cortex contains scattered stone cells. Small calcium oxalate crystals present in ray cells. Leaves opposite, subsessile, leathery, simple, entire, with the thickened, slightly recurved margins pinnately veined; epidermis containing some mucilaginous cells; stipules small, subulate, situated on the sides of the short, petiolelike leaf base. Stomata anomocytic. Flowers solitary, terminal (on short axillary branches), or axillary, subtended by three pairs of decussate bracts, these increasingly large and petaloid (pink) above, often also with vestigial floral buds in axils of uppermost bracts, on short, sharply 4-angled pedicel, bisexual actinomorphic, apetalous, 4-merous throughout. Sepals large, imbricate, pointed, rose-colored, persistent. Nectary disc wanting. Stamens in two cycles of four each, the outer cycle alternating with the sepals; filaments slender, long erect, free or inserted at the base of the calyx; anthers dorsifixed, versatile, tetrasporangiate, with short, apical connective, opening longitudinally. Tapetum secretory. Pollen grains usually 3-colporate, but some 6-pantoporate grains also occur; the exine tectate-columellate, perforate. Gynoecium of four united carpels; ovary superior, 4-locular, 4-ridged, and 4-grooved with essentially free stylodia; stylodia basally and centrally separate, but their upper slender parts connivent and twisted, separate apically and covered with stigmatic papillae adaxially; each locule with two pendulous ovules attached centrally in the upper middle part of the ovary. Female gametophyte of *Polygonum*-type. Fruits hard, 4-lobed, loculicidal capsules enclosed in the persistent calyx. Seeds reniform, smooth, with a short, whitish funicle, which is swollen and forms a collarlike caruncle; embryo straight, linear with fleshy, linear cotyledons; endosperm present. Plants Al-accumulators. 3. GEISSOLOMATACEAE.

4 Micropyle is formed by both the integuments.

6 Female gametophyte of *Penaea*-type (tetrasporic, 16-nuclear). Glabrous small, evergreen, ericoid shrubs or shrublets, procumbent to ascending or erect. Leaves opposite, decussate, entire, acuminate to retuse, often coriaceous; stipules wanting or small, each divided into a small row of fingerlike or rarely hairlike, secretory structures in the leaf axil. Stomata anomocytic. Flowers solitary in the upper leaf-axils, bisexual, actinomorphic, apetalous, 4-merous, with a variably long floral tube. Calyx lobes valvate or reduplicate-valvate, persistent, calyx often coloured like a corolla. Stamens four, inserted in throat of calyx and alternating with calyx lobes; filaments very short, narrow, erect; anthers 4-locular, introrse, basifixed, opening longitudinally; connective fleshy, much expanded, often much longer than the frequently well separated locules. Pollen grains 3–6-colporate, always provided with pseudocolpi isomerous with the apertures. Nectary disc lacking. Gynoecium of four carpels; style with capitate or 4-lobed stigma; ovary superior, 4-locular, often 4-lobed, with 2–4 anatropous, pendulous or ascending ovules per locule. Placentation basal to axile, or axile, or axile to apical. Fruits firm, often hard loculicidal capsules, included in the persistent floral tube. Seeds elongate,

arilate, endotestal cells much elongated, endotegmen fibrous, nearly or quite without endosperm; embryo with a large hypocotyl and small cotyledons. Contain non-hydrolysable tannins, n = 10–12. . . . . 4. PENAEACEAE.

6 Female gametophyte of *Polygonum*- or *Oenothera*-type.

7 Female gametophyte of *Polygonum*-type. Oxalate raphids wanting.

8 Each flower is basally subtended by a short internode ending with some blunt teeth, a "calyculus" (which is a stem structure). Flowers in paniculate inflorescences with the branches ending in 3-flowered cymules, small, bisexual, actinomorphic, 4(5)-merous, with floral tube prolonged over the ovary, truncate or obscurely and minutely 5-toothed, ultimately deciduous. Calyx represented by a narrow rim at the throat of the floral tube, usually with (4)5 more or less linear-spatulate or obovate spatulate, white or pinkish, valvate lobes, not enclosing the petals in bud. Petals 4(5), inserted at the throat of the floral tube, scalelike, alternating with end much shorter than the calyx lobes, incurved and closing the tube in bud, spreading and reflexed at anthesis, valvate, usually pink to crimson, pubescent. Stamens (4)5, inserted on the inner rim of the floral tube below the petals, incurved in bud; very short, not exserted and with the anthers almost filling the mouth of the floral tube; anthers dorsifixed, not versatile, introrse, with thickened connective, opening longitudinally. Pollen grains heteropolar with one pole colpate and the other heterocolpate; pseudocolpi restricted to one polar face only. Nectary disc lacking. Gynoecium of (2)4–5 carpels; style with a capitate stigma; ovary inferior, with as many locules as carpels, with (2)3 superposed ovules per carpel. Ovules pendulous, campylotropous, hypotropous. Fruits drupaceous, marked by a circular scar left by the deciduous floral tube; endocarp woody, longitudinally ribbed. Seeds usually one per locule and nearly always only one per fruit, without endosperm; embryo oily, with a short radicle and unfolded, spirally twisted or convolute cotyledons. Evergreen shrubs or trees up to 25 m tall; stem often fluted and buttressed; young leafy branchlets quadrangular. The inner bark and leaves (when crushed) usually with a strong almond scent. Leaves opposite or rarely ternate, entire, coriaceous, with rudimentary stipules. Stomata paracytic to anomocytic, verging on encyclocytic. Fibers with simple pits, septate. Contain non-hydrolysable tannins, cyanogenic glycoside prunasin; n = 15, 20. . . . . 5. OLINIACEAE.

8 Flowers not subtended by "calyculus."

9 Placentation apical. Evergreen or deciduous trees up to 50 m tall, shrubs, subshrubs (in savannas, often do not exceed 20 cm), or often lianous (in forests, often climb above 30 m high), sometimes mangroves, rarely spiny; the bark is often flaking. Roots rarely with pneumatophores (*Laguncularia racemosa*); stems rarely with spines. Fibers with simple pits, frequently septate, or (*Strephonema*) with distinctly bordered pits. Intraxylary (internal) phloem commonly but not always present, interxylary (included) phloem often present as well (restricted to some genera of the tribe Com-breteae); mucilage ducts present in *Terminalia*. Leaves alternate, opposite or less often verticillate, entire, simple, variously pubescent, sometimes glandular (Combretinae), conduplicate or supervolute (revolute in

Laguncularia), often with two gland-containing flask-shaped cavities at base; stipules lacking or vestigial. Stomata anomocytic or sometimes encyclocytic, rarely (*Strephonema*) paracytic. Flowers in racemes, spikes, or heads, bisexual or sometimes unisexual, actinomorphic or seldom slightly zygomorphic, with a usually fairly short floral tube (sometimes absent). Sepals 4–5(-8), often appearing as lobes of floral tube, valvate or imbricate or sometimes very small. Petals 4–5(-8), mostly rather small, imbricate or valvate, often lacking. Stamens usually twice as many as the sepals and mostly in two cycles; filaments inflexed in bud; anthers usually versatile, opening longitudinally. Pollen grains 3(4)-colporate, mostly heterocolpate, with pseudocolpi (absent in *Strephonema*). Well-developed intrastaminal disc is very often present. Gynoecium of 2–5(-8) carpels with elongate style ending in a punctate or seldom capitate stigma; ovary inferior, or rarely (*Strephonema*) semi-inferior, 1-locular, with 2(-6), sometimes up to 7(20), only one developing, anatropous ovules pendulous on an elongate, slender funicle from the top of locule; placentation apical. "Fruits 1-seeded pseudocarps, formed from the ovary and surrounding the lower hypanthium; the fruit wall (pericarp plus hypanthium) may be thin and hard, or become differentiated as spongy or succulent tissue, or develop 2–5 wings" (Stace 2007). Seeds without endosperm; embryo green, oily, with folded or spirally twisted cotyledons; rarely (*Strephonema*) cotyledons massive and hemispheric or (some African spp. of *Combretum*) united, some species of *Terminalia* have three or more cotyledons. Contain tannins, 5-desoxyflavonoids, flavonoid sulphates, myricetin, only few alkaloids and triterpene saponins; n = 7, 11–13. 6. COMBRETACEAE.

9 Placentation not apical, only rarely (Trapaceae) apical-axile.

10 Flowers polygamo-dioecious or bisexual, in axillary racemes or panicles, very small, actinomorphic, 4–5-merous. Sepals 4–5, appearing as valvate lobes on the cupulate or shortly tubular floral tube. Petals absent (*Crypteronia*) or small and connate apically and falling off as a cup when the flower opens (*Axinandra*). Stamens as many as and alternate with the sepals, inserted on the margin of the floral tube, inflexed in bud, and with a wide connective, which is conduplicate in *Axinandra*. "The anther endothecium degenerates early, and the anthers dehisce not via differential shrinking of endothecium cells, but via rupture of walls along their thinnest sections, caused by the shrinking of connective cells" (Renner 2007). Pollen grains 2- or 3-colporate, with apertures alternating with pseudocolpi. Gynoecium of 2–6 carpels; style with capitate stigma; ovary superior to semi-inferior, with numerous, anatropous ovules on axile placentas. Endothelium present at least in *Axinandra*. Fruits chartaceous or woody, loculicidal capsules, the valves often held together apically by persistent style. Seeds usually small, flat,

with membranous, minutely wing, endotesta crystalliferous, endotegmen tanniniferous; endosperm lacking. Fibers with distinctly bordered pits. Internal phloem characteristically present. Nodes trilacunar with three traces. Leaves opposite, glabrous, entire, with a marked midrib and continuous marginal vein; petioles short; small stipules sometimes present. Stomata paracytic or (*Dactylocladus*) anomocytic. Evergreen trees up to 50m tall, often with quadrangular twigs. Plants accumulate aluminium. ....
7. CRYPTERONIACEAE.

10 Flowers not polygamo-dioecious. Pollen grains never 2-colporate. Endothelium always absent.

11 Anthers mostly opening by single apical pore or seldom by two pores or by longitudinal slits; connective usually provided by conspicuous appendage.

12 Interxylary (included) phloem of the foraminate type present. Fibers with distinctly bordered pits. Nodes unilacunar with three traces. Large shrubs or huge trees (up to 35m tall); indumentum absent or consists of uniseriate hairs. Leaves opposite, petiolate, simple, entire; venation pinnate with indistinct lateral and intramarginal veins; stipules absent. Stomata paracytic. Inflorescences at nodes of young wood, sometimes cauliflorous, cymose, umbelloid, fasciculate, or sometimes solitary at nodes of older wood. Flowers generally small, actinomorphic, bisexual or rarely (*Lijndenia*) androdioecious, 4–5-merous. Sepals 4–5, minute, imbricate in bud. Petals imbricate in bud, yellow, pink, or bluish. Stamens 4–5 or 8–10, with a carnose connective generally provided with a terpenoid-producing gland; anthers 2-locular, versatile, with fibrous endothecium, opening by longitudinal slits, sometimes short and functioning as pores. Pollen grains heterocolpate. Gynoecium of 2–5 carpels; style elongate, stigma punctate; placentation basal or axile; ovary inferior, 1–5-locular, with two to numerous, ascending, anatropous, or campylotropous ovules per locule. Fruits fleshy, 1–5-seeded berries. Seeds usually large, without endosperm; seed coat with fibrous exotegmen; embryo green, with thick or convolute cotyledons. n = 7.
8. MEMECYLACEAE.

12 Interxylary phloem lacking. Herbs, shrubs, treelets, or trees up to 45m tall, less frequently scandent shrubs or lianas, self supporting, or epiphytic, or climbing. Young stems often tetrago-

nal. Indumentum very differentiated, the trichomes being generally large, multicellular, and scalelike. Vessels with simple perforation and with vestured pits. Fibers with simple pits, often septate. Wood parenchyma paratracheal, or apotracheal and paratracheal. Nodes unilacunar with three traces. Leaves nearly always opposite, simple, entire to dentate, conduplicate or supervolute, with 3-9(-19) main veins separate from the base of the lamina; stipules lacking. Stomata anomocytic, polycytic, or encyclocytic. Flowers in terminal panicles or cymes, rarely solitary, bisexual, or rarely androdioecious, small and nearly actinomorphic to large and distinctly zygomorphic, 3-5(-7)-merous. Sepals 4, or 5(-7), imbricate, or valvate. Petals imbricate in bud. Stamens 4–5, or 8, or 10(-96), with connectives often prolonged or provided with appendages, but without glands; filaments often twisted, bringing all the anthers to one side of the flowers, free of one another; anthers basifixed, tetrasporangiate, without fibrous endothecium, usually opening poricidal (apically, with one, two, or rarely four pores per anther), or by short slits, or rarely opening longitudinally. Nectaries present or absent. Pollen grains heterocolpate, supplied with pseudocolpi or with intercolpate depressions isomerous with the usually three apertures. Gynoecium generally of 3-5 (-14) carpels; ovary superior to inferior, mostly 3–5-locular, rarely 1-locular when the partitions are dissolved, with axile or rarely basal placentation. Styles one, apical, elongate; stigma punctate to capitate. Ovules (2-)6–50 per locule, usually anatropous, rarely (*Rhexia*) orthotropous, bitegmic, outer integument contributing to the micropyle (in addition to the inner one); mizropyle zigzag. Fruits berries or loculicidal capsules. Seeds numerous, small; seed coat without fibrous exotegmen; embryo minute, with two small unequal cotyledons. Contain acylated anthocyanins (cyanidin and delphinidin), flavonols (kaempferol, quercetin, and myricetin), rarely ellagic acid n = 5(-7)-8–18.
9. MELASTOMATACEAE.

11 Anthers always opening by longitudinal slits. Mostly herbs, less often subshrubs, shrubs, or sometimes trees, rarely very large and tall; bark flaky, mucilage cells common; stems often 4-angled or narrowly 4-winged when young. Hairs uni- or bi-, sometimes multicellular, or (*Cuphea*) with cystolitic, malpighiaceous, or glandular hairs. Branched foliar sclereids mostly absent. Internal phloem usually present. Vestured pits present in the

secondary xylem and internal phloem. Axial parenchyma scanty paratrachel. Leaves usually opposite, less often verticillate, rarely subalternate, generally decussate, simple, commonly entire, flat to conduplicate, stipules vestigial or none. Stomata mostly anomocytic. Inflorescences axillary or terminal, thyrses, cymes, umbelliform clusters, of flowers solitary; epicalyx of small teeth often present, the teeth alternating with calyx lobes. Flowers bisexual, actinomorphic or sometimes more or less zygomorphic, with conspicuous, sometimes spurred floral tube, usually (3)4-6(-8-16)-merous. Sepals rather thick, valvate, appearing as valvate lobes of the floral tube, often alternating with external appendages at the sinuses. Petals free, attached at the summit or within the floral tube, pinnately veined, crumpled in bud, sometimes wanting. Stamens usually twice as many as the sepals or petals, in two cycles, rarely (*Rotala*) solitary or numerous and centrifugal as in *Lagerstroemia*; filaments more or less elongate, straight or incurved (the long filaments of *Lafoensia* inrolled) in bud, usually inserted on the inside of the floral tube. Anthers versatile, introrse, dorsifixed or seldom basifixed. Pollen grains 3-colporate (some heterocolpate with either three or six subsidiary colpi). Nectary often present, annular or unilateral (*Cuphea*). Gynoecium of 2–4(-8), rarely (Punicoideae) 7–9(-15) carpels; style filiform, commonly elongate, sometimes bent in bud, mostly with capitate stigma; ovary superior to inferior, often surrounded at the base by an annular nectary disc (or the nectary unilateral), multilocular with as many locules as carpels, but sometimes the septa not reaching the summit; rarely the ovary pseudomonomerous; placentation axile or rarely free-central. Ovules usually more or less numerous in each locule, anatropous or very rarely amphitropous. Fruits usually capsular, dehiscing variously or seldom indehiscent, rarely berrylike. Seeds generally more or less numerous, flattened, sometimes winged, nearly or quite without endosperm, with straight oily embryo; cotyledons folded. Often producing quinolizidine alkaloids. n = 5, 7, 8, 15, 24, 28. 10. LYTHRACEAE.

7 Female gametophyte of *Oenothera*-type. Annual or perennial herbs or sometimes shrubs, lianas (*Fuchsia*), rarely trees to 30 m tall (*Hauya*, some *Fuchsia*), often with epidermal oil cells. Included phloem of foraminate type often present. Fibers with simple pits, often septate. The vegetative parts are rich in oxalate raphides. Leaves alternate, opposite, or more rarely verticillate, usually entire or dentate, but sometimes lobed or lyrate-pinnatifid; small stipules sometimes present, but in most genera they are vestigial or wanting. Stomata with three or more subsidiary cells, often anisocytic. Flowers solitary in the axils, or in spikes, or panicles, bisexual or seldom (spp. of *Fuchsia*) unisexual, mostly with long

floral tube, which is nectariferous within, actinomorphic or less often more or less zygomorphic (strong zygomorphy in *Lopezia*), mostly 4-merous or sometimes 2-merous (*Circaea*), 5-merous, or 7-merous (*Ludwigia*). Sepals 2–7, valvate, commonly appearing as lobes on the floral tube. Petals 2–7, imbricate, valvate, or contorted, often clawed or stipitate, rarely wanting. Stamens attached within the floral tube, or surrounding the epigynous nectary disc, generally 2-cyclic or 1-cyclic, or rarely reduced to 1–2 (and one of these sterile in *Lopezia*). Staminodes sometimes present in *Lopezia* and *Fuchsia*. Pollen grains borne in monads or sometimes in tetrads, or polyads (*Ludwigia*), (2)3(-6)-aperturate (colpate, colporate, or often porate), with viscin threads on the proximal surface in all but one species (*Circaea alpina*), with dominant circular or triangular central body and usually markedly to slightly protruding apertures; pseudocolpi lacking. Gynoecium generally of as many carpels as the sepals or petals; style with a capitate, cylindrical, clavate, or 4-lobed stigma; ovary inferior or rarely semi-inferior, with as many locules as carpels or the partitions sometimes imperfect, so placentation accordingly axile or parietal. Ovules numerous or several (seldom 1 or 2) in each locule or on each placenta, anatropous. Fruits loculicidal (or both loculicidal and septicidal) capsules or sometimes berries (*Fuchsia*) or small nuts (*Gaura* and *Circaea*). Seeds numerous, several, or only one; exotesta often hairy or papillate, inner walls thickened and lignified; embryo oily, straight or nearly so; endosperm absent. Contain flavonoid sulphates, n = 7, 9, 11. . . . . . 12. ONAGRACEAE.

3 Characteristic, schizogenous secretory cavities lined with epithelium when young and filled with essential oils present in leaves and other aerial organs. Calyx and corolla imbricate.

13 Stamens usually numerous. Trees (often rather large), shrubs or subshrubs, with unicellular or sometimes bicellular or multicellular hairs, usually with ectotrophic mycorrhizae. Bark fairly smooth or rough and fissured, sometimes peeling in large plates (e.g., *Psidium guajava*). Internal phloem characteristically present, next to the pith. Secondary phloem of young stems usually stratified. Vessels usually with simple perforations, rarely (*Neomyrtus, Myrceugenia, Luma, Ugni, Myrteola*, and several other Central and South American genera) with scalariform perforations. Fibers mostly with distinctly bordered pits (fiber-tracheids), very rarely septate. Axial parenchyma typically apotracheally diffuse or diffuse-in-aggregates. Leaves opposite or less often alternate, rarely verticillate, simple, entire, often coriaceous, glandular-punctate, often with a continuous intramarginal vein; stipules vestigial or wanting. Stomata anomocytic or seldom paracytic. Flowers in various types of cymose or racemose, often complex inflorescences, rarely solitary and axillary, bisexual, nearly always actinomorphic, mostly 4- or 5-merous, with a floral tube of variable length often prolonged above the ovary, on the margin of which the sepals, petals, and normally numerous stamens are all inserted. Sepals (3)4–5(-8), usually imbricate, or the calyx sometimes very reduced or splitting at anthesis or forming a deciduous calyptra. Petals similar, sometimes (as in spp. of *Eucalyptus*) connivent to form a calyptra, or sometimes wanting. Stamens usually numerous, developing centripetally, free or basally united into four or five fascicles (nearly always antepetalous) each supplied by

one vascular trunk bundle, usually inflexed in bud; anthers small, versatile, opening by longitudinal slits or (as in certain sections of *Eucalyptus* and in *Chamelaucium* and related genera) by pores, the connectives usually have one or more apical secretory cavities (glands). Pollen grains 2-celled, generally triangular, usually syncolporate, and lack pseudocolpi. Staminodia are rarely present. Nectary disc borne on the summit of the ovary or lining the prolonged floral tube. Gynoecium of 2–5(18) carpels; style generally elongate, with capitate or punctiform stigma; ovary with as many locules as carpels or seldom (some Leptospermoideae) the ovary pseudomonomerous, with a single locule, semi-inferior or inferior, or rarely superior; ovules usually numerous, occasionally few, very rarely solitary, anatropous or campylotropous, bitegmic or (some ssp. of *Eugenia* and *Syzygium*) unitegmic. Female gametophyte of *Polygonum*-type or rarely (Heteropyxideae) *Allium*-type. Fruits baccate or loculicidal (very rarely circumscissile) capsules, sometimes drupes, schizocarps, or nutlike, crowned by persistent sepal-lobes or scars. Seeds one to few, very occasionally many; seed coat usually membranous or chartaceous, sometimes hard and bony; embryo of diverse shape and size, rarely (*Eugenia*) undifferentiated, green or white; endosperm usually absent, occasionally in a scanty amount; cotyledons often connate, intricately folded. Contain ethereal oils (usually terpenes) and polyhydroxyalkaloids (Porter et al. 2000), n = 5–11(12), less often 6–9. . . . . 13. MYRTACEAE.

13 Stamens 1–5(-7), usually only one fertile. Trees, sometimes very large, often with well-developed buttresses, shrubs, or somewhat woody vines, rarely herbs; young parts often densely pubescent with simple or stellate (*Erisma*) hairs. Intraxylary phloem usually present in the premedullary region. Indumentum often brown, hairs T-shaped, unicellular, or stellate. Vessels exclusively with simple perforations; lateral pitting alternate and vestured. Fibers with few, small, simple or sometimes with very small borders in *Vochysia*, a few cells sometimes septate in *Erisma* and *Vochysia*. Rays heterogeneous to homogeneous. Axial parenchyma predominantly paratracheal. Nodes trilacunar or unilacunar. Leaves evergreen, opposite, or sometimes verticillate, leathery; stipules small, sometimes represented by large glands, or colleter-like, or wanting. Stomata paracytic or anomocytic. Inflorescence terminal, axillary racemes (panicles) or cincinni. Flowers obliquely zygomorphic. Sepals five, basally connate, persistent, imbricate, one of them usually the largest and with gibbous or spurred base, rarely (*Korupodendron*) calyx petaloid. Petals 1–3, very rarely 5, or sometimes wanting, more or less unequal, contorted, or imbricate. Stamens 1-5(-7), usually only one fertile; filaments tree or rarely connate in two groups; anthers introrse, sometimes with an expanded connective, opening longitudinally. Pollen grains 3-colporate, with reticulate or striate exine. Nectaries wanting. Gynoecium usually of three carpels, sometimes pseudomonomerous; style with a capitate or lateral stigma, sometimes very short; ovary superior and 3-locular or inferior and 1-locular, ovules (1)2-many in each locule, pendulous, epitropous or anatropous, bitegmic, crassinucellate. Fruits loculicidal capsules or sometimes samaras winged with accrescent calyx. Seeds often winged or hairy, testa thin, endotestal cells

more or less thickened, pectic, mesotegmic cells fibrous, thick-walled or nor, or testa multiplicative, exotesta with thickened hairs; embryo straight (longitudinal to the seed), endosperm wanting or rarely developed; cotyledons folded. Plants 5-deoxyflavonoids, usually accumulate aluminum, n = 11 or 12. . . . 14. VOCHYSIACEAE.

2 Plants annual aquatics, sometimes free-floating. Raphides wanting. Submerged part of the stem with elongate internodes and more or less opposite or ternate, filiform-dissected, leaflike green organs (transformed stipules according to Vasiliev [1960]) and also with filiform adventitious roots from some of the nodes; aerial part of the stem short, densely crowded leaves. Leaves with more or less rhombic laminas, the elongate petiole usually with an inflated, aerenchymatous float near midlength and small, cleft, deciduous stipules; the floating leaves have marginal teeth with a unique double apex. Stomata anomocytic. Flowers solitary on short pubescent pedicels, in the axils, slightly elevated above the water, bisexual, actinomorphic, 4-merous. Sepals basally connate into a short hypanthial tube, lobes valvate, two or four of them persistent and accrescent in fruit as hornlike or spinelike projections. Petals alternate with the sepals, imbricate. Stamens alternate with the petals, with short filaments; anthers dorsifixed, opening longitudinally. Pollen grains 2-celled, triangular, 3-colporate, with three prominent meridional ridges formed by the uplifting and folding of the ectexine and enclosing a cavity; the ridges pass over the colpi as a wide verrucate-granular band; the apertures are prodding and swollen as elongated domes. Gynoecium of two carpels, with subulate style and capitate stigma; ovary partly inferior, surrounded at the base by 8-lobed cupular disc, 2-locular, with usually one, pendulous, apical-axile, anatropous ovule per locule, but one locule and its ovule aborting after anthesis. Endosperm formation hardly takes place at all. Fruits mostly 1-seeded, horned, drupelike, with thin, soon disappearing, fleshy exocarps and persistent, stony endocarp. Seed large, without endosperm; embryo straight, the cotyledons very unequal in size, shape, and behavior, one large, starchy, kept within the fruit, the other small, scalelike, growing out with plumule through the pore left by dehiscence of the style, n = 20, 24. . . . . . . 11. TRAPACEAE.

1. ALZATEACEAE

S. Graham 1985. 1/1. Distributed along the lower slopes of the Andean Cordillera in Colombia, Ecuador, Peru and Bolivia, in humid forests of the upper Amazonian basin, and disjunctly in the cloud forests of Costa Rica and Panama.

Alzatea.

"Comparison of specialized morphological attributes and sequence data from the *rbc*L gene indicate that *Alzatea* belongs near the small myrtalean families Rhynchocalycaceae, Oliniaceae, and Penaeaceae" (Graham 2004). From all other myrtalean families the Alzateaceae differ in their trilacunar node.

2. RHYNCHOCALYCACEAE

L. Johnson et B. Briggs 1985. 1/1. Eastern parts of South Africa (endemic to the sandstone region of southern Natal and Pondoland).

Rhynchocalyx.

In the cladistic analysis of Johnson and Briggs (1984:732), *Rhynchocalyx* comes out on a common stem with *Alzatea*. According to Dahlgren and van Wyk (1988: 47), *Rhynchocalyx* is clearly related to a group of families consisting of Crypteroniaceae, Penaeaceae, Oliniaceae, Alzateaceae, and Lythraceae without, however, particularly close living relatives. They conclude that Oliniaceae, Rhynchocalycaceae, and Penaeaceae seem to have evolved from the same ancestral stock, which is to some extent also reflected in their present geographical distribution.

3. GEISSOLOMATACEAE

Endlicher 1841. 1/1. Restricted to the Cape Province, occurring in the Langeberg Mountains of the Swellendam and Riverside districts, where they grow in fynbos.

Geissoloma.

Related to the Pennaeaceae. Differing mainly in female gametophyte. Lindley (1853) and Gundersen (1950) even included *Geissoloma* in Penaeaceae.

4. PENAEACEAE

Sweet ex Guillemin 1828. 7/25. Confined to the Cape Province and concentrated in the southwest. *Penaea* has the widest distribution toward the east and extends to Port Elizabeth.

Endonema, Glischrocolla, Sonderothamnus, Saltera, Brachysiphon, Stylapterus, Penaea.

Penaeaceae are closely related to the Memecylaceae, Cryptoreniaceae, and Alzateaceae, and especially to Oliniaceae and Rhynchocalycaceae (Dahlgren and van Wyk 1988: 45). According to Dahlgren and Thorne (1984: 670), the ancestors of Penaeaceae could have had a common origin with Rhynchocalycaceae.

5. OLINIACEAE

Arnott 1839. 1/8. Southern and eastern Africa (*Olinia ventosa* on St. Helena is most probably an introduction from South Africa – Dahlgren and van Wyk [1988]).

Olinia.

Probably related to the Penaeaceae (Johnson and Briggs 1984). "Penaeaceae, despite their rather different type of embryo sac, appear to be especially close to Oliniaceae," state Dahlgren and van Wyk (1988: 43). Molecular studies clearly support a placement Oliniaceae in the order Myrtales (Conti et al. 1996).

6. COMBRETACEAE

R. Brown 1810 (including Myrobalanaceae Martynov 1820, Terminaliaceae Jaume Saint-Hilaire 1805). 14/500. Mainly tropical regions, with a few species extending into subtropical areas; particularly common in Africa. *Strephonema* (3) is endemic to tropical West Africa.

6.1 STREPHONEMATOIDEAE
Ovary semi-inferior. Pollen grains without pseudocolpi. Cotyledons massive, hemispheric. Stomata paracytic. Fibers with clearly bordered pits. – *Strephonema.*

6.2 COMBRETOIDEAE
Ovary inferior. Pollen grains with pseudocolpi. Cotyledons folded, spirally twisted. Stomata anomocytic or encyclocytic. Fibers with simple pits. – LAGUNCULARIEAE: *Laguncularia, Lumnitzera, Macropteranthes, Dansiea*; COMBRETEAE: *Combretum* (including *Quisqualis, Calopyxis, Thiloa, Melostemon*), *Guiera, Calycopteris, Pteleopsis, Terminalia* (including *Terminaliopsis, Bucida*), *Buchenavia, Anogeissus, Finetia, Conocarpus.*

Combretaceae occupy a somewhat isolated position within the Myrtales. Johnson and Briggs (1984) place them between Penaeaceae-Oliniaceae and Cryptoreniaceae-Melastomataceae.

7. CRYPTERONIACEAE

A. P. de Candolle 1868 (including Henslowiaceae Lindley 1835). 3/9–10. Sri Lanka, India (Assam and Bengal), southwestern China, Andaman Islands, Malay Peninsula, Philippine Islands, Malay Archipelago.

Crypteronia, Dactylocladus, Axinandra.

Crypteroniaceae exhibit relationships to Alzateaceae-Rhynchocalycaceae and especially to Memecylaceae. Some authors, including van Vliet, Koek-Noorman, and ter Welle (1981) and van Vliet and Baas (1984) even include Crypteronioideae and Memecyloideae in the Melastomataceae sensu lato. But from all other Myrtales the Crypteroniaceae (at least the genus *Axinandra*) differ in the presence of endothelium (Tobe and Raven 1983b).

8. MEMECYLACEAE

A. P. de Candolle 1828 (including Mouririaceae Gardner 1840). 6/450. Pantropical.

Memecylon, Warneckea, Lijndenia, Spathandra, Mouriri, Votomita.

Related to Melastomataceae, but differ in the leaf venation, the number and size of the seeds and the presence of an elliptic terpenoid-producing gland on the staminal connective (Renner 2004).

9. MELASTOMATACEAE

A. L. de Jussieu 1789 (including Blakeaceae Barnhart 1895, Miconiaceae C. Martius 1835, Rhexiaceae

Dumortier 1822). 163/4500–4570. Mainly in tropical and subtropical regions, especially in South America.

KIBESSIEAE: *Pternandra* (including *Kibessia*); ASTRONIEAE: *Astrocalyx, Astronia, Astronidium, Becca-rianthus;* SONERILEAE: *Anerincleistus, Barthea, Blastus, Bredia, Oxyspora, Poikilogyne, Driessenia, Neodriessenia, Ochthocharis, Poilannommia, Sporoxeia, Stussenia, Amphiblemma, Aschistanthera, Bertolonia, Boyania, Brittenia, Calvoa, Centradenia, Cincinobotrys, Cyanandrium, Dicellandra, Diplarpea, Enaulophyton, Fordiophyton, Gravesia, Kerriothyrsus, Macrocentrum, Monolea, Opisthocentra, Phyllagathis, Preussiella, Salpinga, Sarcopyramis, Scorpiothyrsus, Sonerila, Tateanthus, Triolena, Tryssophyton, Maguireanthus*; MERIANIEAE: *Adelobotrys, Axinaea, Behuria, Benevidesia, Bisglaziovia, Centronia, Dolichoura, Graffenrieda, Huberia, Meriania, Merianthera, Neblinanthera, Ochthephilus, Pachyloma, Phainantha, Tessmannianthus*; RHEXIEAE: *Rhexia*; MELASTOMEAE: *Dissotis, Acanthella, Acotis, Acisanthera, Amphorocalyx, Antherotoma, Appendicularia, Brachyotum, Cailliella, Chaetolepis, Comolia, Comoliopsis, Desmoscelis, Dichaetanthera, Dinophora, Dionycha, Ernestia, Fritzschia, Guyonia, Heterocentron, Heterotis, Loricalepis, Macairea, Mallophyton, Marcetia, Melastoma, Melastomastrum, Microlepis, Monochaetum, Nepsera, Nerophila, Otanthera, Osbeckia, Pilocosta, Poteranthera, Pseudoosbeckia, Pterogastra, Pterolepis, Sandemania, Schwackaea, Siphanthera, Svitramia, Tibouchina, Tibouchinopsis, Tristemma*; MICROLICIEAE: *Lavoisiera, Bucquetia, Cambessedesia, Castratella, Chaetostoma, Eriocnema, Lavoisiera, Lithobium, Microlicia, Rhynchanthera, Stenodon, Trembleya*; MICONIEAE: *Medinella, Alloneuron, Allomaieta, Anaectocalyx, Bellucia, Boerlagea, Calycogonium, Catanthera, Catocoryne, Chalybea, Charianthus, Clidemia, Conostegia, Creochiton, Cyphostyla, Diplectria, Dissochaeta, Henriettea, Henriettella, Huilaea, Killipia, Kendrickia, Kirkbridea, Leandra, Llewelynia, Loreya, Macrolenes, Maieta, Mecranium, Medinilla, Miconia, Myriaspora, Myrmidone, Necramium, Ossaea, Pachyanthus, Pachycentria, Pleiochiton, Pleithiandra, Pogonanthera, Tetrazygia, Tococa*; BLAKEEAE: *Blakea, Topobea.*

Closely related to the Memecylaceae.

10. LYTHRACEAE

Jaume Saint-Hilaire 1805 (including Ammaniaceae Horaninow 1834, Blattiaceae Engler 1892, Duabangaceae Takhtajan 1986, Lagerstroemiaceae J. Agardh 1858, Lawsoniaceae J. Agardh 1858, Punicaceae Horaninow 1834, Sonneratiaceae Engler 1897). 31/600. Subcosmopolitan, but mainly tropical and subtropical regions, especially in America; relatively few species (mostly annual or perennial herbs) in temperate and cold areas.

10.1 LYTHROIDEAE

Fruits usually capsular, dehiscing variously or seldom indehiscent. Seeds generally more or less numerous, flattened, sometimes winged, nearly or quite without endosperm, with straight oily embryo; cotyledons folded. Mostly herbs, less often subshrubs, shrubs, or sometimes trees; bark flaky, mucilage cells common. Hairs uni- or bi-, sometimes multicellular. Branched foliar sclereids mostly absent. Internal phloem usually present. Leaves opposite, less often verticillate, rarely alternate, commonly entire, flat to conduplicate, stipules vestigial or none. Stomata mostly anomocytic. Flowers bisexual, actinomorphic or sometimes more or less zygomorphic, with conspicuous, sometimes spurred floral tube, usually (3)4-6(-8-16)-merous. Sepals rather thick, valvate, appearing as valvate lobes of the floral tube, often alternating with external appendages at the sinuses. Petals free, attached at the summit or within the floral tube, pinnately veined, crumpled in bud, sometimes wanting. Stamens usually twice as many as the sepals or petals, in two cycles, rarely (*Rotala*) solitary or numerous and centrifugal as in *Lagerstroemia*; filaments more or less elongate, straight or incurved (the long filaments of *Lafoensia* inrolled) in bud, usually inserted on the inside of the floral tube. Anthers versatile or seldom basifixed. Pollen grains 3-colporate (some heterocolpate with either three or six subsidiary colpi). Gynoecium of 2-4(-6) carpels; style filiform, commonly elongate, sometimes bent in bud, mostly with capitate stigma; ovary superior to inferior, often surrounded at the base by an annular nectary disc (or the nectary unilateral), multilocular with as many locules as carpels, but sometimes the septa not reaching the summit; rarely the ovary pseudomonomerous; placentation axile or rarely free-central. Ovules usually more or less numerous in each locule, anatropous or very rarely amphitropous.

Often producing quinolizidine alkaloids. n = 5, 7, 8, 15, 24, 28. – LYTHREAE: *Rotala, Ammannia, Hionanthera, Peplis, Didiplis, Lythrum, Woodfordia, Cuphea, Pleurophora, Galpinia, Pemphis, Capuronia, Diplusodon, Physocalymma, Lafoensia*; NESAEEAE: *Crenea, Nesaea, Heimia, Decodon, Pehria, Adenaria, Koehneria, Lourtella, Tetrataxis, Ginoria* (including, *Haitia*); LAGERSTROEMIEAE: *Lagerstroemia* (including *Orias*), *Lawsonia*.

10.2 PUNICOIDEAE

Fruits with 2–3 layers of locules and a leathery pericarp and persistent calyx. Seeds embedded in a pulpy mass; endosperm wanting, embryo with spirally rolled, large cotyledons. Small trees or shrubs with quadrangular twigs, often thorny; scattered secretory cells in the cortex and pith. Vessel ray and vessel-parenchyma pits half-bordered. Fibers septate. Axial parenchyma scanty paratracheal to virtually absent. Nodes unilacunar. Leaves alternate to opposite, sometimes crowded at the tips of the twigs, estipulate. Flowers solitary or in fascicles terminating axillary branches, bisexual, actinomorphic, with colored floral tube prolonged well beyond the ovary. Sepals 5–8, appearing as valvate lobes on the floral tube, persistent on the leathery hypanthium. Petals imbricate and crumpled in bud. Stamens numerous at the rim of the floral tube, developing centrifugally; filaments slender; anthers dorsiftxed, versatile. Pollen grains 3-colporate, with slightly angular mesocolpia. Gynoecium of 7-9(-15) carpels. Ovary inferior. Ovules more or less numerous on each placenta. Placentation axile (*Punica protopunica*) or the axile placentation at the base of the ovary differentiates into two or three superposed upper layers that appear to have parietal placentation as a result of asymmetric growth. Ovules 20–50 per locule, anatropous. Produing pyridine alkaloids an ellagic acid, n = 7 in *P. protopunica* and 8 in *P. granatum*. 1/2. From the Balkan Peninsula to the western Himalayas (*Punica granatum*) and on Socotra (*P. protopunica*). – *Punica* (including *Socotria* ?).

10.3 DUABANGOIDEAE

Fruits loculicidal capsules, coriaceous, more or less perfectly 4–8-locular, seated on the spreading leathery calyx. Seeds numerous, ellipsoid, testa prolonged at both ends into longish tails; endosperm lacking. Flowers in terminal umbelliform cymes, rather large, 4-merous (*Duabanga moluccana*) or 5–8-merous (*D. grandiflora*). Petals shortly clawed, broadly obovate, crisped, and undulate. Stamens 12 (*D. moluccana*) or more than 50, biseriate (*D. grandiflora*), inserted on the floral tube in one or two cycles, inflexed in bud; anthers recurved or replicate over one end of the connective. Pollen grains 3-porate. Gynoecium of 4–8 carpels; style curved, with 4–8-lobed stigma; ovary 4–8-locular. Ovules numerous, on diffuse placentas, covering nearly the whole inner walls of locules. Tall, buttressed trees with drooping, 4-angled ultimate branches. Leaves opposite, glaucous beneath, estipulate. Vessel ray and vessel-parenchyma pits large and simple. Fibers not septate. Axial parenchyma paratracheal. n = 12. 1/2. Southeast Asia and New Guinea. – *Duabanga*.

10.4 SONNERATIOIDEAE

Fruits indehiscent berries crowned by the style base. Seeds embedded in evil-smelling pulp, not tailed at the ends; embryo green, endosperm lacking. Flowers 1–3 together at the tips of the ultimate, mostly pendulous branchlets. Petals linear or linear-lanceolate or quite absent. Stamens numerous, inserted on the floral tube in several cycles; anthers reniform. Disc saucer-shaped. Pollen grains 3-colporate. Gynoecium of 10–20 carpels; style with capitate stigma; ovary 10–20-locular. Ovules many, ascending, on axile placentas. Trees with trunk surrounded by aerial roots arising vertically as lateral, negatively geotropic branches upon often very long horizontal roots buried in the substratum. Vessel ray and vessel-parenchyma pits mostly half-bordered. Fibers septate. Axial parenchyma absent, n = 12. 1/5. From coasts of East Africa and Madagascar to Micronesia, New Hebrides, and the Solomons, northern Australia and New Caledonia. – *Sonneratia*.

Lythraceae are a rather heterobathmic family (primitive characters, such as superior ovary and monosporic female gametophyte of *Polygonum*-type, are combined with advanced characters, such as pinnate venation of the petals and the dominance of herbs). Probably are nearest to the Crypteroniaceae and Memecylaceae-Melastomataceae.

11. TRAPACEAE

Dumortier 1829. 1/3 or about 15. Tropical, subtropical, and temperate regions of the Old World, except Australia.

Trapa.

Related to the Onagraceae and especially to the Lythraceae. From the Onagraceae they differ markedly in embryological characters and the absence of the viscin threads on the pollen grains. From the Lyrthraceae they differ in semi-inferior ovary, almost 4-merous flowers, and the endosperm-nucleus degenerating soon after its formation. According to Miki (1959), *Trapa* evolved from *Lythrum* through the sintermediate fossil genus *Hemitrapa*.

12. ONAGRACEAE

A. L. de Jussieu 1789 (including Circaeaceae Ruthe 1827, Epilobiaceae Ventenat 1799, Fuchsiaceae Lilja 1870, Isnardiaceae Martynov 1820, Jussiaeaceae Martynov 1820, Lopeziaceae Lilja 1870, Oenotheraceae C. C. Robin 1807). 17/680. Widespread in tropical, subtropical, and temperate regions, especially in the New World, where they are centered in southwestern North America.

12.1 JUSSIAEAOIDEAE

Flowers 4–5-merous, hypanthium absent. Nectary on gynoecium; style short, stigma capitate. – JUSSIAEEAE: *Ludwigia* (including *Jussiaea*).

12.2 ONAGROIDEAE

Flowers 4-merous (2-merous in *Circaea*), hypanthium long. Style short, stigma divided, dry, n = 9–11. – HAUYEAE: *Hauya*; FUCHSIEAE: *Fuchsia*; CIRCAEEAE: *Circaea*; LOPEZIEAE: *Lopezia*; GONGYLOCARPEAE: *Gongylocarpus*; EPILOBIEAE: *Chamerion, Epilobium* (including *Boisduvalia, Chamaenerion* and *Zauschneria*); ONAGREAE: *Xylonagra, Clarkia, Gayophytum, Camissonia, Oenothera, Calylophus, Stenosiphon, Gaura, Megacorax*.

A very distinctive family (Dahlgren and Thorne 1984) that reveals some similarities with the Lythraceae s. str.

13. MYRTACEAE

A. L. de Jussieu 1789 (including Chamelauciaceae F. Rudolphi 1830; Heteropyxidaceae Engler et Gilg 1920; Kaniaceae Nakai 1943; Leptospermaceae F. Rudolphi 1830; Melaleucaceae Vest 1818; Myrrhiniaceae Arnott 1839; Psiloxylaceae Croizat 1961). 131/4000–4600. Tropical and partly subtropical and warm-temperate regions, mainly in Australia, but also in South America.

Classification after P. G. Wilson et al. (2005).

13.1 PSILOXYLOIDEAE

Plants dioecious. Leaves alternate or pseudo-alternate (disjunct-opposite), secretory cavities present (but not containing essential oils in *Psiloxylon*). Stamens not inflexed in bud, anthers 4-locular. Stigma in female flowers (but not in pistillodia) with fleshy, flattened lobes or capitate. Female gametophyte of *Allium*-type. Fruits punctate, many-seeded berries, seed unwinged (Psiloxylea) or 2(3)-valved loculicidal capsules, seed winged (Heteropyxideae), n = 12. – PSILOXYLEA: *Psiloxylon*; HETEROPYXIDEAE: *Heteropyxis*.

13.2 MYRTOIDEAE

Plants with bisexual flowers, rarely andromonoecious. Leaves opposite or spirally arranged; secretory cavities present, containing essential oils. Stamens inflexed in bud, anthers 2-locular, terpenoid-containing glands in the apex of the connective, stigma dry. Female gametophyte of *Polygonum*-type. Fruits indehiscent, pericarp usually fleshy; seeds usually with sclerotic testa. Polyhydroxyalkaloids present, n = 11. –XANTHOSTEMONEAE: *Xanthostemon, Pleurocalyptus, Purpureostemon*; LOPHOSTEMONEAE: *Lophostemon, Welchiodendron, Kjellbergiodendron, Whiteodendron*; OSBORNIEAE: *Osbornia*; MELALEUCEAE: *Beaufortia, Callistemon, Calothamnus, Eremaea, Lamarchea, Melaleuca, Petraeomyrtus, Phymatocarpus, Regelia*; KANIEAE: *Barongia, Basisperma, Kania, Lysicarpus, Mitrantia, Ristantia, Sphaerantia, Tristaniopsis*; BACKHOUSIEAE: *Backhousia, Choricarpia*; METROSIDEREAE: *Carpolepis, Mearnsia, Metrosideros, Tepualia*; TRISTANIEAE: *Tristania, Thaleropia, Xanthomyrtus*; SYZYGIEAE: *Acmena, Acmenosperma, Anetholea, Syzygium* (including *Aphanomyrtus*), *Waterhousea*; MYRTEAE: *Acca, Feijoa, Accara, Amomyrtella, Amomyrtus, Archirhodomyrtus, Austromyrtus, Blepharoxalyx, Calycolpus, Calycorectes, Calyptranthes, Campomanesia, Chamguava, Decaspermum, Eugenia* (including *Monimiastrum*), *Gomidesia, Gossia, Hexachlamys, Legrandia, Lithomyrtus, Lophomyrtus, Luma, Marlierea, Meteoromyrtus, Mosiera, Myrceugenia, Myrcia, Myrcianthes, Myrciaria,*

Myrrhinium, Myrtastrum, Myrtella, Myrteola, Myrtus, Neomitranthes, Neomyrtus, Octamyrtus, Pilidiostigma, Pimenta, Plinia, Psidium, Rhodamnia, Rhodomyrtus, Siphoneugenia, Stereocaryum, Ugni, Uromyrtus; EUCALYPTEAE: *Allosyncarpia, Angophora, Arillastrum, Corymbia, Eucalyptopsis, Eucalyptus, Stockwellia*; SYNCARPIEAE: *Syncarpia*; LINDSAYOMYRTEAE: *Lindsayomyrtus*; LEPTOSPERMEAE: *Agonis, Angasomyrtus, Asteromyrtus, Homalospermum, Kunzea, Leptospermum, Neofabricia, Pericalymma*; CHAMELAUCIEAE: *Actinodium, Aluta, Astartea, Babingtonia, Baeckea, Balaustion, Calytrix, Chamelaucium, Coryanthera, Darwinia, Euryomyrtus, Homalocalyx, Homoranthus, Rylstonea, Hypocalymma, Malleostemon, Micromytrus, Ochrosperma, Pileanthus, Rinzia, Scholtzia, Thryptomene, Triplarina, Verticordia*.

According to Johnson and Briggs (1984: 752), intra-familial phylogeny in the Myrtaceae "is decidedly inconsistent with recognition of the two traditional subfamilies Myrtoideae and Leptospermoideae, not to mention Chamelaucioideae."

14. VOCHYSIACEAE

A. Saint-Hilaire 1820 7/210. Mostly tropical America and the West Indies, but *Erismadelphus* (3) and *Korupodendron* in tropical West Africa.

14.1 VOCHYSIOIDEAE (*subfam. nov.*: Tribe *Vochysieae* Dumortier, Anal. Fam. Pl.: 41. 1829)

Ovary superior, 3-locular. Ovules two to many. Fruit loculicidal capsule. Seeds winged; embryo with two convolute cotyledons. – *Vochysia, Salvertia, Qualea* (including *Ruizterania*), *Callisthene*.

14.2 ERISMOIDEAE (*subfam. nov.*: Tribe *Erismeae* Dumortier, Anal. Fam. Pl.: 41. 1829)

Ovary inferior, 1-locular. Ovules 1–2, lateral to apical. Fruit winged, samaroid, with persistent enlarged calyx. Seeds not winged; testa undifferentiated, with vascular bundles; embryo with two planoconvex cotyledons. – *Erisma, Erismadelphus, Korupodendron*.

Have many similarities with the Myrtaceae both in wood anatomy and in floral morphology (see Dahlgren and Thorne 1984). The affinity with Myrtaceae is well supported by molecular data (Soltis 2006).

Bibliography

Almeda F. 1997a. Chromosome numbers and their evolutionary significance in some neotropical and paleotropical Melastomataceae. BioLlania 6: 167–190.

Almeda F. 1997b. Chromosomal observations on the Alzateaceae (Myrtales). Ann. Missouri Bot. Gard. 84: 305–308.

Averett JE and SA Graham. 1984. Flavonoids of Rhynchocalycaceae (Myrtales). Ann. Missouri Bot. Gard. 71: 853–854.

Averett JE and PH Raven. 1984. Flavonoids of Onagraceae. Ann. Missouri Bot. Gard. 70: 30–34.

Baas P. 1979. The anatomy of *Alzatea* Ruiz et Pav. (Myrtales). Acta Bot. Neerl. 28: 156–158.

Baas P. 1981. A note on stomatal types and crystals in the leaves of Melastomataceae. Blumea 27: 475–479.

Baas P. 1986. Wood anatomy of Lythraceae: Additional genera (*Capuronia, Haitia, Orias,* and *Pleurophora*). Ann. Missouri Bot. Gard. 73: 810–819.

Baas P and RCVJ Zweypfenning. 1979. Wood anatomy of the Lythraceae. Acta Bot. Neerl. 28: 117–155.

Balthazar M von and J Schönenberger. 2007. Oliniaceae. In: K Kubitzki, ed. The families and genera of vascular plants, vol. 9, pp. 260–264. Springer, Berlin/Heidelberg/New York.

Batygina TB. 1985. Onagraceae. In: MS Yakovlev, ed. Comparative embryology of flowering plants: Brunel-liaceae-Tremandraceae, pp. 104–110. Nauka, Leningrad (in Russian).

Batygina TB and GE Kolesova. 1985. Lythraceae, Trapaceae. In: MS Yakovlev, ed. Comparative embryology of flowering plants: Brunelliaceae-Tremandraceae, pp. 85–88, 110–116. Nauka, Leningrad (in Russian).

Baum DA, KJ Sytsma, and PC Hoch. 1994. A phylogenetic analysis of *Epilobium* (Onagraceae) based on nuclear ribosomal DNA sequences. Syst. Bot. 19: 363–388.

Behnke H-D. 1984 [1985]. Ultrastructure of sieve-element plastids of Myrtales and allied group. Ann. Missouri Bot. Gard. 71: 824–831.

Berry PE and PC Hoch. 2004. Onagraceae. In: N Smith, SA Mori, A Henderson, DW Stevenson, and SV Heald, eds. Flowering plants of the Neotropics, pp. 279–280. The New York Botanical Garden, Princeton University Press, Princeton.

Beusekom-Osinga RJ van. 1977. Crypteroniaceae. In: CGGJ van Steenis, ed. Flora Malesiana ser. I, 8: 187–204. Noordhoff, Groningen.

Beusekom-Osinga RJ van and CF van Beusekom. 1975. Delimitation and subdivision of the Crypteroniaceae (Myrtales). Blumea 22: 255–266.

Biffin E, LA Craven, MD Crisp, and PA Gadek. 2006. Molecular systematics of *Syzygium* and allied genera (Myrtaceae): Evidence from the chloroplast genome. Taxon 555: 79–94.

Boesewinkel FD and M Venturelli. 1987. Ovule and seed structure in Vochysiaceae. Bot. Jahrb. Syst. 108: 547–566.

Bohte A and A Drinnan. 2005a. Floral development and systematic position of *Arillastrum, Allosyncarpia, Stockwellia* and *Eucalyptopsis* (Myrtaceae). Plant Syst. Evol. 251: 53–70.

Bohte A and A Drinnan. 2005b. Ontogeny, anatomy and systematic significance of ovular structures in the 'eucalypt group' (Eucalypteae, Myrtaceae). Plant Syst. Evol. 255: 17–39.

Bremer K. 1981. Seeds and embryos in Sri Lanka (Ceylonese) species of *Memecylon*, with notes on *Spathandra* (Melastomataceae). Nordic J. Bot. 1: 62–65.

Bridgwater SD and P Baas. 1978. Wood anatomy of the Punicaceae. IAWA Bull. 1: 3–6.

Briggs BG and LAS Johnson. 1979. Evolution in the Myrtaceae: Evidence from inflorescence structure. Proc. Linn. Soc. N. S. W. 102: 157–256.

Brown CA. 1967. Pollen morphology of the Onagraceae. Rev. Palaeobot. Palynol. 3: 163–180.

Bult CJ and EA Zimmer. 1993. Nuclear ribosomal RNA sequences for inferring tribal relationships within Onagraceae. Syst. Bot. 18: 48–63.

Bunniger L. 1972. Untersuchungen über die morpholog-ische Natur des Hypanthiums bei Myrtales- und Thymelaeales-Familien: II. Myrtaceae. III. Vergleich mit den Thymelaeaceae. Beitr. Biol. Pfl. 48: 79–156.

Bunniger L and F Weberling. 1968. Untersuchungen über die morphologische Natur des Hypanthiums bei Myrtales-Familien: I. Onagraceae. Beitr. Biol. Pfl. 421: 447–477.

Carlquist S. 1975a. Wood anatomy of Onagraceae, with notes on alternative modes of photosynthate movement in dicotyledon woods. Ann. Missouri Bot. Gard. 62: 386–424.

Carlquist S. 1975b. Wood anatomy and relationships of the Geissolomataceae. Bull. Torrey Bot. Club 102: 128–134.

Carlquist S. 1990. Leaf anatomy of Geissolomataceae and Myrothamnaceae as a possible indicator of relationship to Bruniaceae. Bull. Torrey Bot. Club 117: 420–428.

Carlquist S and L De Buhr. 1977. Wood anatomy of Penaeaceae (Myrtales): Comparative, phylogenetic, and ecological implications. Bot. J. Linn. Soc. 75: 211–227.

Carr SGM and DJ Carr. 1966. Cotyledonary stipules in the Myrtaceae. Nature 210: 185–186.

Carr SGM and DJ Carr. 1969, 1970a. Oil glands and dots in *Eucalyptus* L'Herit: I. The phloem and the pith. II. Development and structure of oil glands in the embryo. Austral. J. Bot. 17: 471–513, 1969. 18: 191–212, 1970.

Carr SGM, DJ Carr, and L Milkovits. 1970b. Oil glands and ducts in *Eucalyptus* L'Herit: III. The flowers of series *Corymbosae* (Benth.) Miden. Austral. J. Bot. 18: 313–333.

Carrugan AE and AN Drinnan. 2000. The ontogenetic basis for floral diversity in the *Baeckea* sub-group (Myrtaceae). Kew Bull. 55: 593–613.

Cellinese N. 1997. Notes on the systematics and biogeography of the *Sonerila* generic alliance (Melastomataceae) with special focus on fruit characters. Trop. Biodiv. 4: 83–93.

Cheung M and R Sattler. 1967. Early floral development of *Lythrum salicaria*. Canad. J. Bot. 45: 1609–1618.

Chen Z. 1996. The morphology and anatomy of Sonneratia Linn. f. in China. J. Trop. Subtrop. Bot. 4(2): 18–24.

Chen ZY, XX Huang and WC Ko. 1994. Chromosome counts in genus *Sonneratia* (Sonneratiaceae). Acta Bot. Sinica 9: 60–63 (in Chinese).

Chernyakovskaya EF. 1996. Geissolomataceae. In: AL Takhtajan, ed. Comparative seed anatomy, vol. 5, pp. 21. Nauka, St. Petersburg (in Russian).

Chrtek J. 1969. Die Kronblattnervatur in der Familie Lythraceae. Preslia 55: 323–326.

Clausing G. 1999. Die Systematik der Siddochaeteae und ihre Stellung innerhalb der Melastomataceae. Ph. D. dissertation. University of Mainz.

Clausing G and SS Renner. 2001a. Molecular phylogenetics of Melastomataceae and Memecylaceae: implications for character evolution. Am. J. Bot. 88: 486–498.

Clausing G and SS Renner. 2001b. Evolution of growth form in epiphytic Dissochaeteae (Melastomataceae). Organisms, Diversity, and Evolution 1: 45–60.

Clausing G, K Meyer, and SS Renner. 2000. Correlations among fruit traits and evolution of different fruits within Melastomataceae. Bot. J. Linn. Soc. 133: 303–326.

Conti E, D Baum, and K Sytsma, 1999. Phylogeny of Crypteroniaceae and related families implications for morphology and biogeography. XVI Intern. Bot. Congress, 250. St. Louis (Abstract).

Conti E, A Fishbach, and K Sytsma. 1993. Tribal relationships in Onagraceae: Implications from *rbc*L sequence data. Ann. Missouri Bot. Gard. 80: 672–685.

Conti E, A Litt and KJ Sytsma. 1996. Circumscription of Myrtales and their relationships to other rosids: evidence from *rbc*L sequence data. Am. J. Bot. 83: 221–233.

Conti E, A Litt, PG Wilson, SA Graham, BG Briggs, LAS Johnson, and KJ Sytsma. 1997. Interfamilial relationships in Myrtales: molecular phylogeny and patterns of morphological evolution. Syst. Bot. 22: 629–647.

Conti E, T Eriksson, J Schönenberger, KJ Sytsma, and DA Baum. 2002. Early Tertiary out-of-India dispersal of Crypteroniaceae: evidence from phylogeny and molecular dating. Evolution 56: 1931–1942.

Cronquist A. 1984 [1985]. A commentary on the definition of the order Myrtales. Ann. Missouri Bot. Gard. 71: 780–782.

Dahlgren R and VS Rao. 1969. A study of the Geissolomataceae. Bot. Not. 122: 207–227.

Dahlgren R and RF Thorne. 1984. The order Myrtales: Circumscription, variation, and relationships. Ann. Missouri Bot. Gard. 71: 633–699.

Dahlgren R and AE van Wyk. 1988. Structures and relationships of families endemic to or centered in southern Africa. Monographs Syst. Bot. Missouri Bot. Gard. 25: 1–94.

Davis GL. 1966. Floral morphology and the development of gametophytes in *Eucalyptus melliodora* A. Cunn. Austral. J. Bot. 16: 19–35.

Davis GL. 1969. Floral morphology and the development of gametophytes in *Eucalyptus stellulata* Sieb. Austral. J. Bot. 17: 177–190.

Dickie JB and PE Gasson. 1999. Comparative leaf anatomy of the Penaeaceae and its ecological implications. Bot. J. Linn. Soc. 131: 327–351.

Drinnan AN and A Carrucan. 2005. The ontogenetic basis for floral diversity in *Agonia*, *Leptospermum* and *Kunzea* (Myrtaceae). Plant Syst. Evol. 251: 71–88.

Duke NC and BR Jackes. 1987. A systematic revision of the mangrove genus *Sonneratia* (Sonneratiaceae) in Australasia. Blumea 32: 277–382.

El Ghazali GEB, S Tsuji, GA El Ghazaly, and S Nilsson. 1998. Combretaceae R. Brown. World Pollen Spore Flora 21: 1–40.

Exell A. W. 1930. The genera of Combretaceae. J. Bot. 69: 113–128.

Exell AW and CA Stace. 1966. Revision of the Combretaceae. Bol. Soc. Brot., ser. 2, 40: 5–25.

Eyde RH. 1977, 1978, 1981. Reproductive structures and evolution in *Ludwigia* (Onagraceae): I. Androecium, placentation, merism. II. Fruit and seed. III. Vasculature, nectaries, conclusions. Ann. Missouri Bot. Gard. 64: 644–655, 1977; 65: 656–675, 1978; 68: 470–503,1981.

Eyde RH and JT Morgan. 1973. Floral structure and evolution in Lopezieae (Onagraceae). Am. J. Bot. 60: 771–787.

Eyde RH and JA Teeri. 1967. Floral anatomy of *Khexia virginica* (Melastomataceae). Rhodora 69: 163–178.

Fagerlind F. 1941. Der Bau der Samenanlage und der Markogametophyten bei *Quisqualis indica*. Bot. Not. 1941: 217–222.

Forest F. 2007. Geissolomataceae. In: K Kubitzki, ed. The families and genera of vascular plants, vol. 9, pp. 155–156. Springer, Berlin/Heidelberg/New York.

Gadek PA, PG Wilson, and CJ Quinn. 1996. Phylogenetic reconstruction in Myrtaceae using *mat*K, with particular reference to the position of *Psiloxylon* and *Heteropyxis*. Austral. Syst. Bot. 9: 283–290.

Graham SA. 1964a. The genera of Lythraceae in the southeastern United States. J. Arnold Arbor. 45: 235–250.

Graham SA. 1964b. The genera of Rhizophoraceae and Combretaceae in the southeastern United States. J. Arnold Arbor. 45: 285–301.

Graham SA. 1984 [1985]. Alzateaceae: A new family of Myrtales in the American tropics. Ann. Missouri Bot. Gard. 71: 759–779.

Graham SA. 1995. Innovative seed morphology in Lythraceae. Am. J. Bot. 82(Suppl.): 132.

Graham SA. 2002. Phylogenetic relationships and biogeography of the endemic Caribbean genera *Crenea*, *Ginoria*, and *Haitia* (Lythraceae). Carib J. Sci. 38: 195–204.

Graham SA. 2004a. Alzateaceae. In: N Smith, SA Mori, A Henderson, DW Stevenson, and SV Heald, eds. Flowering plants of the Neotropics, pp. 11–13. The New York Botanical Garden, Princeton University Press, Princeton.

Graham SA. 2004b. Lythraceae. In: N Smith, SA Mori, A Henderson, DW Stevenson, and SV Heald, eds. Flowering plants of the Neotropics, pp. 223–225. The New York Botanical Garden, Princeton University Press, Princeton.

Graham SA. 2007a. Alzateaceae. In: K Kubitzki, ed. The families and genera of vascular plants, vol. 9, pp. 26–28. Springer, Berlin/Heidelberg/New York.

Graham SA. 2007b. Lythraceae. In: K Kubitzki, ed. The families and genera of vascular plants, vol. 9, pp. 226–246. Springer, Berlin/Heidelberg/New York.

Graham SA and JE Averett. 1984. Flavonoids of Alzateaceae (Myrtales). Ann. Missouri Bot. Gard. 71: 855–957.

Graham SA and TB Cavalcanti. 2001. New chromosome counts in the Lythraceae and a review of chromosome numbers in the family. Syst. Bot. 26: 445–458.

Graham SA and R Kleiman. 1987. Seed lipids of the Lythraceae. Biochem. Syst. Ecol. 15: 433–439.

Graham A, SA Graham, JW Nowicke, V Patel, and S Lee. 1990. Palynology and systematics of the Lythraceae: III. Genera *Physocalymma* through *Woodfordia,* addenda, and conclusions. Am. J. Bot. 77: 159–177.

Graham SA, JV Crisci, and PC Hoch. 1993a. Cladistic analysis of the Lythraceae sensu lato based on morphological characters. Bot. J. Linn. Soc. 113: 1–33.

Graham SA, K Oginuma, PH Raven, and H Tobe. 1993b. Chromosome numbers in *Sonneratia* and *Duabanga* (Lythraceae s. 1.) and their systematic significance. Taxon 42: 35–41.

Graham SA, RF Thorne, and JL Reveal. 1998. Validation of subfamily names in Lythraceae. Taxon 47: 435–436.

Graham SA, J Hall, K Sytsma, and S-H Shi. 2005. Phylogenetic analysis of the Lythraceae based on four gene regions and morphology. Int. J. Plant Sci. 166: 995–1017.

Heslop-Harrison Y. 1990. Stigma form and surface in relation to self-incompatibitily in the Onagraceae. Nord. J. Bot. 10: 1–19.

Hoch PC, JV Crisci, and H Tobe. 1993. A cladistic analysis of the plant family Onagraceae. Syst. Bot. 18: 31–47.

Huang YL and SH Shi. 2002. Phylogenetics of Lythraceae sensu lato: a preliminary analysis based on chloroplast *rbc*L gene, *psa*A-*ycf*3 spacer, and nuclear RDNA internal transcribed spacer (ITS) sequences. Int. J. Plant Sci. 163: 215–225.

Jansen S, T Watanabe, and E Smets. 2002. Aluminium accumulation in leaves of 127 species in Melastomataceae, with comments on the order Myrtales. Ann. Bot. 90: 53–64.

Jayaweera NMA. 1967. The genus *Duabanga*. J. Arnold Arbor. 48: 89–100.

Johansen DA. 1931. Studies on the morphology of the Onagraceae: 4. *Stenosiphon linifolium*. Bull. Torr. Bot. Club 57: 315–326.

Johansen DA. 1934. Studies on the morphology of the Onagraceae: 8. *Circaea pacifica*. Am. J. Bot. 21: 508–510.

Johnson LAS and BG Briggs. 1984 [1985]. Myrtales and Myrtaceae: A phylogenetic analysis. Ann. Missouri Bot. Gard. 71: 700–756.

Joshi AC and J Venkateswarlu. 1935, 1936. Embryo-logical studies in the Lythraceae: I. *Lawsonia inermis* Linn. II. *Lagerstroemia* Linn. III. Proc. Indian Acad. Sci. B 2: 481–493, 523–534, 1935; 3: 377–400, 1936.

Katinas L, J Crisci, WL Wagner, and PC Hoch. 2004. Geographical diversification of tribes Epilobieae, Gongylocarpeae, and Onagreae (Onagraceae) in North America, based on parsimony analysis of endemicity and track compatibility analysis. Ann. Missouri Bot. Gard. 91: 159–185.

Kausel E. 1955. Beitrag zur Systematik der Myrtaceen, parts 1 and 2. Ark. Bot. 3: 491–516, 607–611.

Kawasaki ML. 1998. Systematics of *Erisma* (Vochysiaceae). Mem. New York Bot. Gard. 81: 1–40.

Kawasaki ML. 2007. Vochysiaceae. In: K Kubitzki, ed. The families and genera of vascular plants, vol. 9, pp. 480–487. Springer, Berlin/Heidelberg/New York.

Kawasaki ML and BK Holst. 2004. Myrtaceae. In: N Smith, SA Mori, A Henderson, DW Stevenson, and SV Heald, eds. Flowering plants of the Neotropics, pp. 264–266. The New York Botanical Garden, Princeton University Press, Princeton.

Keating RC. 1982 [1983]. The evolution and systematics of Onagraceae: Leaf anatomy. Ann. Missouri Bot. Gard. 69: 770–803.

Keating RC. 1984 [1985]. Leaf histology and its contribution to relationships in the Myrtales. Ann. Missouri Bot. Gard. 71: 801–823.

Koehne E. 1881. Lythraceae monographice describuntur. Bot. Jahrb. Syst. 1: 142–157.

Koehne E. 1902. Lythraceae. In: A Engler, ed. Das Pflanzenreich, Heft 17 (IV.216), pp. 1–326. Weinheim.

Kolesova GE. 1996. Trapaceae. In: A Takhtajan, ed. Comparative seed anatomy, vol. 5, pp. 242–259. Nauka, St. Petersburg (in Russian).

Kosenko VN. 1985. Palinomorfologiya predstavitelei semeistva Punicaceae. Bot. Zhurn. 70(1): 39–41 (in Russian with English summary).

Kurabayashi M, H Lewis, and PH Raven. 1962. A comparative study of mitosis in the Onagraceae. Am. J. Bot. 49: 1003–1026.

Leins P. 1988. Das zentripetale Androecium von *Punica*. Bot. Jahrb. Syst. 109: 555–561.

Levin GM. 1980. Contributions to the study of the family Punicaceae. Bot. Zhurn. 65: 427–430 (in Russian).

Levin RA, WL Wagner, PC Hoch, M Nepokroeff, JC Pires, EA Zimmer, and KJ Sytsma. 2003. Family-level, relationships of Onagraceae based on chloroplast *rbc*L and *ndh*F data. Am. J. Bot. 90: 107–115.

Levin RA, WL Wagner, PC Hoch, WJ Hahn, A Rodriquez, DA Baum, L Katinas, EA Zimmer, and KJ Sytsma. 2004. Paraphyly in tribe Onagreae: insights into phylogenetic relationships of Onagraceae based on nuclear and chloroplast sequence data. Syst. Bot. 29: 147–164.

Litt A. 1999. Floral morphology and phylogeny of Vochysiaceae. Ph.D. dissertation. New York.

Litt A and DW Stevenson. 2003a. Floral development and morphology of Vochysiaceae. I. The structure of the gynoecium. Am. J. Bot. 90: 1533–1547.

Litt A and DW Stevenson. 2003b. Floral development and morphology of Vochysiaceae. II. The position of the single fertile stamen. Am. J. Bot. 90: 1548–1559.

Lowry JB. 1976. Anthocyanins of the Melastomataceae, Myrtaceae, and some allied families. Phytochemistry 15: 513–516.

Lucas EJ, SR Belsham, EM Nic Lughadha, DA Orlovich, CM Sakuragui, MW Chase, and PG Wilson, 2005. Phylogenetic patterns in fleshy-fruited Myrtaceae–preliminary molecular evidence. Plant Syst. Evol. 251: 33–51.

Martin HA. 2003. The history of the family Onagraceae in Australia and its relevance to biogeography. Austral. J. Bot. 51: 585–598.

Martin PG and JM Dowd. 1986. Phylogenetic studies using protein sequences within the order Myrtales. Ann. Missouri Bot. Gard. 73: 442–448.

Mauritzon J. 1934. Zur Embryologie einiger Lythraceae. Acta Horti Göteborg 9: 1–21.

Mauritzon J. 1939. Contributions to the embryology of the orders Rosales and Myrtales. Acta Univ. Lund. 35: 1–121.

Mayr B. 1969. Ontogenetische Studien an Myrtales Blüten. Bot. Jahrb. Syst. 89: 210–271.

McDonald DJ. 1998. The enigma of the Geissolomataceae. Veld and Flora (Kirstenbosch). 84: 122–123.

Meijer W. 1972. The genus *Axinandra*-Melastomataceae: A missing link in Myrtales? Ceylon J. Sci. Biol. Sci. 10: 72–74.

Mentink H and P Baas. 1992. Leaf anatomy of the Melastomataceae, Menecylaceae, and Crypteroniaceae. Blumea 37: 189–225.

Michelangeli F-A, DS Penneys, J Giza, D Soltis, MH Hils, and JD Skean, Jr. 2004. A preliminary phylogeny of the tribe Miconieae (Melastomataceae) based on nrITS sequence data and its implications on inflorescence position. Taxon 53: 279–290.

Miki S. 1959. Evolution of *Trapa* from ancestral *Lythrum* through *Hemitrapa*. Proc. Imp. Acad. Jpn. 35(6): 289–294.

Morley RJ, CW Dick. 2003. Missing fossils, molecular clocks, and the origin of the Melastomataceae. Am. J. Bot. 90: 1638–1645.

Morley T. 1976. Memecyleae (Melastomataceae). Flora Neotropica Monograph 15: 1–295.

Muller J. 1969. A palynological study of the genus *Sonneratia* (Sonneratiaceae). Pollen Spores 11: 223–298.

Muller J. 1975. Note on the pollen morphology of Crypteroniaceae s. 1. Blumea 22: 275–294.

Muller J. 1978. New observations on pollen morphology and fossil contribution of the genus *Sonneratia* (Sonneratiaceae). Rev. Palaeobot. Palynol. 26: 277–300.

Muller J. 1981. Exine architecture and function in some Lythraceae and Sonneratiaceae. Rev. Paleobot. Palynol. 35: 93–123.

Oliveira PE. 1998. Reproductive biology, evolution and taxonomy of Vochysiaceae in central Brazil. In: SJ Owens and P Rudall, eds. Reproductive biology. pp. 381–393. Royal Botanic Gardens, Kew.

Orlovich DA, AN Drinnan, and PT Ladiges. 1996. Floral development in the *Metrosideros* group with special emphasis on the androecium. Telopea 6: 689–719.

Orlovich DA, AN Drinnan, and PT Lagides. 1999. Floral development in *Melaleuca* and *Callistemon* (Myrtaceae). Austral. Syst. Bot. 11: 689–710.

Outer RW den and JM Fundter. 1976. The secondary phloem of some Combretaceae and the systematic position of *Strephonema pseudocola* A. Chev. Acta Bot. Neerl. 25: 481–493.

Patel VC, JJ Skvarla, and PH Raven. 1983. Half pseudocolpi, a unique feature of *Olinia* (Oliniaceae) pollen. Am. J. Bot. 70: 469–473.

Patel VC, JJ Skvarla, and PH Raven. 1984 [1985]. Pollen characters in relation to the delimitation of Myrtales. Ann. Missouri Bot. Card. 71: 858–969.

Peng C-I, CL Schmidt, PC Hock, and PH Raven. 2005. Systematics and evolution of *Ludwigia* section *Dantia* (Onagraceae). Ann. Missouri Bot. Gard. 92: 307–359.

Porter EA, E Nic Lughadha, and MSJ Simmonds. 2000. Taxonomic significance of polyhydroxyalkaloids in the Myrtaceae. Kew Bull. 55: 615–632.

Quirk JT. 1980. Wood anatomy of the Vochysiaceae. IAWA Bull., n.s., 1: 172–179.

Ram M. 1956. Floral morphology and embryology of *Trapa bispinosa* Roxb. with discussion on the systematic position of the genus. Phytomorphology 6: 312–323.

Rama Devi D, DVL Satyavathi, and LL Narayana. 1991. Floral anatomy of some Melastomataceae. Feddes Repert. 102: 595–599.

Rao VS and R Dahlgren. 1969. The floral anatomy and relationships of Oliniaceae. Bot. Not. 122: 160–171.

Rao RV, B Sharma, L Chauhan, and R Dayal. 1987. Reinvestigation of the wood anatomy of *Duabanga* and *Sonnerata* with particular reference to their systematic position, IAWA Bull. 8: 337–345.

Raven PH. 1964. The generic subdivision of Onagraceae, tribe Onagreae. Brittonia 16: 276–288.

Raven PH. 1976. Generic and sectional delimitation in Onagraceae, tribe Epilobieae. Ann. Missouri Bot. Gard. 63: 326–340.

Raven PH. 1979. A survey of reproductive biology in Onagraceae. New Zealand J. Bot. 17: 575–593.

Renner SS. 1989. Systematic studies in the Melastomataceae: *Bellucia, Loreya,* and *Macairea*. Memor. New York Bot. Gard. 50: 1–112.

Renner SS. 1993. Phylogeny and classification of the Melastomataceae and Memecylaceae. Nord. J. Bot. 13: 519–540.

Renner SS. 2004a. Multiple Miocene Melastomataceae dispersal between Madagascar, Africa and India. Phil. Trans. Roy. Soc. London B 359: 1485–1494.

Renner SS. 2004b. Memecylaceae. In: N Smith, SA Mori, A Henderson, DW Stevenson, and SV Heald, eds. Flowering plants of the Neotropics, pp. 246–247. The New York Botanical Garden, Princeton University Press, Princeton.

Renner SS. 2007. Crypteroniaceae. In: K Kubitzki, ed. The families and genera of vascular plants, vol. 9, pp. 123–126. Springer, Berlin/Heidelberg/New York.

Renner SS, G Clausing, and K Meyer. 2001. Historical biogeography of Melastomataceae: the roles of Tertiary migration and long-distance dispersal. Am. J. Bot. 88: 1290–1300.

Rutschmann F, T. Eriksson, J Schönenberger, and E Conti. 2004. Did Crypteroniaceae really disperse out of India? Molecular dating evidence from *rbc*L, *ndh*F, and *rpl*16 intron sequences. Int. J. Plant Sci. 165(Suppl. 4): 69–83.

Rye BL. 1979. Chromosome number variation in the Myrtaceae and its taxonomic implications. Austral. J. Bot. 27: 547–573.

Sajo MG and P Rudall. 2002. Leaf and stem anatomy of Vochysiaceae in relation to subfamilial and suprafamilial systematics. Bot. J. Linn. Soc. 138: 339–364.

Salywon A, N Snow, and LR Landrum. 2002. Phylogenetic relationships in the berry-fruited Myrtaceae as inferred from ITS sequences. In Botany 2002: Botany in the Curriculum. p. 149. Madison, Wisconsin (Abstract).

Schmid R 1972a. Floral anatomy of Myrtaceae: I. *Sy-zygium* s. 1. Bot. Jahrb. Syst. 92: 433–489.

Schmid R. 1972b. Floral anatomy of Myrtaceae: II. Eu-genia. J. Arnold Arbor. 52: 336–363.

Schmid R. 1980. Comparative anatomy and morphology of *Psiloxylon* and *Heteropyxis,* and the subfamilial and tribal classification of Myrtaceae. Taxon 29: 559–595.

Schmid R. 1984 [1985]. Reproductive anatomy and morphology of Myrtales in relation to systematics. Ann. Missouri Bot. Gard. 71: 832–835.

Schönenberger J. 2007. Rhynchocalycaceae. In: K Kubitzki, ed. The families and genera of vascular plants, vol. 9, pp. 409–412. Springer, Berlin/Heidelberg/New York.

Schönenberger J and E Conti. 2003. Molecular phylogeny and floral evolution of Pennaeaceae, Oliniaceae, Rhynchocalycaceae, and Alzateaceae (Myrtales). Am. J. Bot. 90: 293–309.

Schönenberger J and E Conti, and F Rutschmann. 2007. Penaeaceae. In: K Kubitzki, ed. The families and genera of vascular plants, vol. 9, pp. 282–291. Springer, Berlin/Heidelberg/New York.

Schulman L and J Hyvönen. 2003. A cladistic analysis of *Adelobotrys* (Melastomataceae) basen on morphology with notes on generic limits within the tribe Merianieae. Syst. Bot. 28: 738–756.

Seavey SR, RE Magill, and PH Raven. 1977. Evolution of seed size, shape, and surface architecture in the tribe Epilobieae (Onagraceae). Ann. Missouri Bot. Gard. 64: 18–47.

Shabes LK and AA Morozova. 1996. Onagraceae. In: A Takhtajan, ed. Comparative seed anatomy, vol. 5, pp. 236–242. Nauka, St. Petersburg (in Russian).

Shi S, Y Huang, F-X Tan, X-J He, and DE Boufford. 2000. Phylogenetic analysis of the Sonneratiaceae and its relationship to Lythraceae based on ITS sequences of nrDNA. J. Plant Res. 113: 253–258.

Shilkina IA. 1973. On the wood anatomy of *Punica* L. Bot. Zhurn. 58: 1628–1630 (in Russian).

Silverstone-Sopkin PA and SA Graham. 1986. Alzateaceae, a plant family new to Colombia. Brittonia 38: 340–343.

Sinha SC and BC Joshi. 1959. Vascular anatomy of the flower of *Punica granatum* L. J. Indian Bot. Soc. 38: 35–45.

Skvarla JJ, PH Raven, and J Praglowski. 1975. The evolution of pollen tetrads in Onagraceae. Am. J. Bot. 62: 6–35.

Skvarla JJ, PH Raven, and J Praglowski. 1976. Ultra-structural survey of Onagraceae pollen. In: IK Ferguson and J. Muller, eds. The evolutionary significance of the exine, pp. 447–479. Linn. Soc. Symposium, ser. 1. London.

Skvarla JJ, PH Raven, WF Chissoe, and M Sharp. 1978. An ultrastructural survey of viscin threads in Onagraceae pollen. Pollen Spores 20: 5–143.

Smith BB and JM Herr. 1971. Ovule development, megagametogenesis, and early embryogeny in *Ammonia coecinea* Rothb. J. Elisha Mitchell Sci. Soc. 87: 192–199.

Snow N. 2000. Systematic conspectus of Australasian Myrtinae (Myrtaceae). Kew Bull. 55: 647–654.

Solt ML and JJ Wurdack. 1980. Chromosome numbers in the Melastomataceae. Phytology 47: 199–220.

Sprague TA and CR Metcalfe. 1947. The taxonomic position of *Rhynchocalyx*. Kew Bull. 2: 392–394.

Stace CA. 1965. The significance of leaf epidermis in the taxonomy of the Combretaceae: I. A general review of tribal, generic, and specific characters. Bot. J. Linn. Soc. 59: 229–252.

Stace CA. 1980. The significance of the leaf epidermis in the taxonomy of the Combretaceae: I. A general review of tribal, generic, and specific characters. Bot. J. Linn. Soc. 81: 327–339.

Stace CA. 2004. Combretaceae. In: N Smith, SA Mori, A Henderson, DW Stevenson, and SV Heald, eds. Flowering plants of the Neotropics, pp. 110–111. The New York Botanical Garden, Princeton University Press, Princeton.

Stace CA. 2007. Combretaceae. In: K Kubitzki, ed. The families and genera of vascular plants, vol. 9, pp. 67–82. Springer, Berlin/Heidelberg/New York.

Stafleu FA. 1948–1954. A monograph of the Vochysiaceae, parts 1–4. Rec. Trav. Bot. Neerl. 41: 397–540, 1948; Acta Bot. Neerl. 1: 222–242, 1952; 2: 144–217, 1953; 3: 459–480, 1954.

Stephens EL. 1909. The embryo-sac and embryo of certain Penaeaceae. Ann. Bot. (London) 23: 363–378.

Stern WL and K Brizicky. 1958. Comparative anatomy and taxonomy of *Heteropyxis*. Bull. Torrey Bot. Club 85: 111–123.

Stone RD. 2004. Phylogenetic systematics of Melastomataceae, subfamily Olisbeoideae, a species-rich and geographically widespread group of tropical forest understory trees. Ph.D. dissertation, University of California, Berkeley.

Stone RD. 2006. Phylogeny of major lineages in Melastomataceae, subfamily Olisbeoideae: utility of nuclear glyceraldehydes 3-phsphate dehydrogenase (*Gap*C) gene sequences. Syst. Bot. 31: 107–121.

Strey RJ and OA Leistner. 1968. The rediscovery of *Rhynchocalyx lawsonioides* Oliv. J. South Afr. Bot. 34: 9–13.

Subramanyam K. 1942. Gametogenesis and embryogeny in a few members of the Melastomataceae. J. Indian Bot. Soc. 21: 69–85.

Subramanyam K and LL Narayana. 1969. A contribution to the floral anatomy of some members of Melastomataceae. J. Jpn. Bot. 44: 6–16.

Sytsma KJ, A Litt, ML Zjhra, C Pires, M Nepokroeff, E Conti, J Walker, and PG Wilson. 2004. Clades, clocks, and continents: Historical and biogeographical analysis of Myrtaceae, Vochysiaceae, and relatives in the southern hemisphere. Int. J. Plant Sci. 165(Suppl. 4): S85–S105.

Sytsma KJ, M Nepokroeff, and JC Pires. 1996. The utility of *ndhF* sequence analysis in Myrtales, with emphasis on the relationships within the Myrtaceae and Melastomataceae clade. Am. J. Bot. 83(6): 197 (Abstract).

Sytsma KJ and JF Smith. 1992. Molecular systematics of Onagraceae: Examples from *Clarkia* and *Fuchsia*. In: PM Soltis et al., eds. Molecular systematics of plants, pp. 295–323. Chafman & Hall, New York/London.

Sytsma KJ, ML Zjhra, M Nepokroeff, CJ Quinn, and PG Wilson. 1998. Phylogenetic relationships, morphological evolution, and biogeography in Myrtaceae based on *ndh*F sequence analysis. Am. J. Bot. 85(Suppl. 6): 161.

Tan F, S Shi, Y Zhong, X Gong, and Y Wang. 2002. Phylogenetic relationships of Combretoideae (Combretaceae) inferred from plastid, nuclear gene and spacer sequences. J. Plant Res. 115: 475–481.

Thanikaimoni G and DMA Jayaweera. 1966. Pollen morphology of Sonneratiaceae. Trav. Sect. Sci. Techn. Inst. Franc. Pondichery 5: 1–12.

Tiagi YD. 1969. Vascular anatomy of the flower of certain species of Combretaceae. Bot. Gaz. 130: 150–157.

Tilney PM. 2002. A contribution to the leaf and young stem anatomy of the Combretaceae. Bot. J. Linn. Soc.138: 163–196.

Tischler G. 1917. Über die Entwicklung und phylogene-tische Bedeutung des Embryosackes von *Lythrum salicaria*. Ber. Deutsch. Bot. Ges. 35: 233–246.

Titiva GE, AA Zakharova and II Shamrov. 1997. Ovule and seed development in *Trapa natans* L. (Trapaceae) in connection with the specific of embryo sac structure, absence of endosperm and pseudomonocotyledony. Bull. Pol. Acad. Sci. Biol. Sci. 45: 81–92.

Tobe H and PH Raven. 1983a [1984]. An embryological analysis of Myrtales: Its definition and characteristics. Ann. Missouri Bot. Gard. 70: 71–94.

Tobe H and PH Raven. 1983b. The embryology of *Axinandra zeylanica* (Myrtales) and the relationships of the genus. Bot. Gaz. 144: 426–432.

Tobe H and PH Raven 1984a. The embryology and relationships of Oliniaceae (Myrtales). Plant Syst. Evol. 146: 105–116.

Tobe H and PH Raven. 1984b. The embryology and relationships of Penaeaceae (Myrtales). Plant Syst. Evol. 146: 181–195.

Tobe H and PH Raven. 1984c. The number of cells in the pollen of Melastomataceae. Bot. Mag. (Tokyo) 97: 131–136.

Tobe H and PH Raven. 1984d [1985]. The embryology and relationships of *Rhynchocalyx* Oliv. (Myrtales). Ann. Missouri Bot. Gard. 71: 836–843.

Tobe H and PH Raven. 1984e [1985]. The embryology and relationships of *Alzatea* Ruiz et Pav. (Myrtales). Ann. Missouri Bot. Gard. 71: 844–852.

Tobe H and PH Raven. 1984f. The embryology and relationships of Oliniaceae. Plant Syst. Evol. 146: 105–116.

Tobe H and PH Raven. 1987a. Embryology and systematic position of *Heteropyxis* (Myrtales). Am. J. Bot. 74: 197–208.

Tobe H and PH Raven. 1987b. The embryology and relationships of *Crypteronia* (Crypteroniaceae). Bot. Gaz. 148: 96–102.

Tobe H and PH Raven. 1987c. The embryology and relationships of *Dactylocladus* (Crypteroniaceae) and a discussion of the family. Bot. Gaz. 148: 103–111.

Tobe H and PH Raven. 1990. Embryology and systematic position of *Psiloxylon* (Myrtales). Bot. Bull. Acad. Sinica 31: 119–127.

Tobe H and PH Raven. 1996. Embryology of Onagraceae (Myrtales): Characteristics, variation and relationships. Telopea 6: 667–688.

Tobe H, DS Graham, and PH Raven. 1998. Floral morphology and evolution in Lythraceae *sensu lato*. In: SJ Owens and PJ Rudall, eds. Reproductive biology, pp. 392–344. Royal Botanical Gardens, Kew.

Trela-Sawicka Z. 1978. Embryological studies in *Trapa natans* L. Acta Biol. Cracov., Ser. Bot. 22: 101–108.

Turner GW and NR Lersten. 1983. Apical foliar nectary of pomegranate: Punicaceae. Am. J. Bot. 70: 475–480.

Udovicic F and PY Ladiges. 2000. Informativeness of nuclear and chloroplast DNA regions and the phylogeny of the eucalypts and related genera (Myrtaceae). Kew Bull. 55: 633–645.

Vasiliev VN. 1960. Water chestnut and the prospects of its cultivation in the USSR. Moscow/Leningrad (in Russian).

Venkatesh CS. 1955. The structure and dehiscence of the anther in *Memecylm* and *Mouriria*. Phytomorphology 5: 435–440.

Venkateswarlu FNI and PSP Rao. 1970. The floral anatomy of Combretaceae. Proc. Indian Acad. Sci. B 36: 1–20.

Venkateswarlu JA. 1937. A contribution to the embryology of Sonneratiaceae. Proc. Indian Acad. Sci. B 5: 206–223.

Venkateswarlu JA and PSP Rao. 1971. Wood anatomy and systematic position of *Strephonema*. New Phytol.70: 767–771.

Venkateswarlu JA and PSP Rao. 1972. Embryological studies in some Combretaceae. Bot. Not. 125: 161–179.

Venkateswarlu JA and PS Venkata Rao. 1964. The wood anatomy and the taxonomic position of Sonneratiaceae. Curr. Sci. 33: 6–9.

Verdcourt R. 1994. Lythraceae. In: RM Polhill, ed. Flora of tropical East Africa, pp. 1–63. AA Balkema, Rotterdam.

Vliet GJCM van. 1974. Wood anatomy of the Crypteroniaceae *sensu lato*. Blumea 22: 175–195.

Vliet GJCM van. 1978. Vestured pits of Combretaceae and allied families. Acta Bot. Neerl. 27: 273–285.

Vliet GJCM van. 1979. Wood anatomy of the Combretaceae. Blumea 25: 141–223.

Vliet GJCM van. 1981. Wood anatomy of paleotropical Melastomataceae. Blumea 27: 395–462.

Vliet GJCM van and P Baas. 1975. Comparative anatomy of the Crypteroniaceae *sensu lato*. Blumea 22: 173–195.

Vliet GJCM van and P Baas. 1984 [1985]. Wood anatomy and classification of the Myrtales. Ann. Missouri Bot. Gard. 71: 783–800.

Vliet GJCM van, J Koek-Noorman, and BJH ter Welle. 1981. Wood anatomy, classification, and phylogeny of the Melastomataceae. Blumea 27: 463–473.
Vyshenskaya TD. 1996. Psiloxylaceae, Heteropyxidaceae, Myrtaceae, Alzateaceae, Rhynchocalycaceae, Penaeaceae, Oliniaceae, Combretaceae, Crypteroniaceae, Melastomataceae, Lythraceae, Punicaceae, Duabangaceae, Sonneratiaceae. In: A Takhtajan, ed. Comparative seed anatomy, vol. 5, pp. 153–236. Nauka, St. Petersburg (in Russian).
Weberling F. 1960. Weitere Untersuchungen über das Vorkommen rudimentärer Stipein bei den Myrtales (Combretaceae, Melastomataceae). Flora 149: 189–205.
Weberling F. 1963. Ein Beitrag zur systematischen Stellung der Geissolomataceae, Penaeaceae, und Oliniaceae sowie der Gattung *Heteropyxis* (Myrtaceae). Bot. Jahrb. Syst. 82: 119–128.
Weberling F. 1966. Additional notes on the Myrtalean affinity of *Kania eugenioides* Schltr. Kew Bull. 20: 517–520.
Whiffin T and AS Tomb. 1972. The systematic significance of seed morphology in the neotropical capsular-fruited Melastomataceae. Am. J. Bot. 59: 411–422.
Wilson CL. 1950. Vascularization of the stamens in the Melastomataceae with some phyletic implications. Am. J. Bot. 37: 431–444.
Wilson KA. 1960. The genera of Myrtaceae in the southeastern United States. J. Arnold Arbor. 41: 270–278.
Wilson PG, MM O'Brien, PA Gadek, and CJ Quinn. 2001. Myrtaceae revisited: a reassessment of infrafamilial groups. Am. J. Bot. 88: 2013–2025.
Wilson PG, MM O'Brien, MM Heslewood, and CJ Quinn. 2005. Relationships within Myrtaceae sensu lato based on a *matK* phylogeny. Plant Syst. Evol. 251: 3–19.
Ziegler A. 1925. Beiträge zur Kenntnis des Androeciums und der Samenentwicklung einiger Melastomataceen. Bot. Arch. 9: 398–467.

Superorder FABANAE

Order 88. FABALES

Trees, shrubs, subshrubs, and perennial and annual herbs, with a great variety of habit, including lianas, aquatics, and xerophytes, mostly bearing root nodules containing nitrogen-fixing bacteria; secretory cells common. Vessels with simple perforations; lateral pitting alternate; vessels with vestured pits, or without vestured pits. Fibers mostly or all with small, simple pits, sometimes septate. Rays homogeneous or heterogeneous. Axial parenchyma commonly abundant, mostly paratracheal. Sieve-element plastids mostly of P-type, less often and only in some Faboideae (such as *Phaseolus, Vigna,* and a few other genera) of S-type. Nodes trilacunar or less often pentalacunar. Leaves evergreen or deciduous, minute to very large, alternate or rarely opposite, pinnately or less often palmately compound, seldom trifoliolate or unifoliolate, petiolate to sessile, usually with stipules that are sometimes modified into prickles or spines; leaf mesophyll cells contain oil bodies. Stomata anomocytic, paracytic, anisocytic, tetracytic, or cyclocytic. Flowers in racemes, corymbs, spikes, or heads, bisexual or rarely unisexual, actinomorphic (Mimosoideae) or more often zygomorphic, mostly 5-merous. Sepals (3-)5(6), free or more often more or less connate, valvate (Mimosoideae) or imbricate, seldom very reduced. Petals typically five, rarely reduced to one or even absent, free or two anterior ones basally connate (Mimosoideae), valvate (Mimosoideae) or imbricate. Stamens mostly ten (sometimes numerous in Mimosoideae), less often nine, sometimes fewer (50–100); when stamens 50–100 they free or fused into tube, or when stamens ten, free or fused, or nine fused with darsal stamen distinct (Seigler 2004); anthers tetrasporangiate, introrse or latrorse, basifixed or dorsifixed, appendaged or unappendaged, opening longitudinally or infrequently by apical pores, sometimes with an apical deciduous gland. Nectary often surrounding ovary, ring-shaped. Tapetum secretory. Microsporogenesis simultaneous. Pollen grains in monads, tetrads, or polyads, 2-celled or less often 3-celled, mostly 3-colporate. Gynoecium nearly always of one carpel, rarely of two or more (up to 16) free carpels, with a terminal stylodium that has a capitate stigma and with two to many ovules per carpel. Ovary superior, sessile to stipitate. Ovules mostly anatropous or campylotropous, less often hemitropous or amphitropous, pendulous to ascending, bitegmic (the micropyle mostly formed by both integuments), or very rarely (some species of *Luplnus*) unitegmic, crassinucellate or rarely (e.g., in some species of *Lathyrus, Lupinus,* and *Robinia*) tenuinucellate; funicle often rather long or stout; micropyle zig-zag. Endothelium often present in some genera of the Faboideae. Female gametophyte mostly of *Polygonum*-type, but in some Faboideae it is of *Allium*-type. Endosperm nuclear. Fruits typically legumes, but sometimes of various other types. Seeds small to very large, arillate or not, exotestal (exotesta palisade), sometimes winged; mesotesta of stellate cells; seed coat undistinguished; embryo large, straight or more or less curved, often green; endosperm mostly lacking or varies from a trace adnate to the inner seed coat and adjacent to the radicle to seldom copious one; cotyledons two, usually flat. Generally producing proanthocyanins and sometimes cyanogenic, often producing alkaloids, especially of the pyridine, quinolizidine, and indole groups; flavonols mostly present, n = 7, with n = 14 established early in their evolution with subsequent aneuploid reduction leading to low diploid numbers in specialized tribes (Goldblatt 1981).

Many authors have placed the Fabales (Leguminosae) near Rosales sensu lato or even include them in that group. But Corner (1976:36), rejected the affinity between them on the basis of Comparative seed anatomy. "I see no reason to subordinate Leguminous ancestry to Rosalean," states Corner. According to Dickison (1981: 49), "The legumes differ from Rosalean taxa by possessing vestured pits in the secondary xylem, a character which tends to isolate the family from other rosalean taxa." In his opinion the presence of relictual apocarpy in the Leguminosae "links the family nicely with the Connaraceae." I agree that Fabales and Oxalidales are somewhat related. Both orders had a common origin from the archaic members of the Rosanae but represent two different clades.

1. FABACEAE

Lindley 1836 or Leguminosae A.L. de Jussieu 1789, nom. altern. (including Aspalathaceae Martynov 1820, Astragalaceae Martynov 1820, Bauhiniaceae Martynov 1820, Caesalpiniaceae R. Brown 1814, Cassiaceae Vest 1818, Ciceraceae Steele 1847, Coronilloaceae Martynov 1820, Detariaceae J. Hess 1832, Galedupaceae Martynov 1820, Hedysaraceae Oken 1826, Inocarpaceae Zollinger 1854–1855, Lathyraceae Burnett 1835, Lotaceae Oken 1826, Mimosaceae R. Brown 1814, Papilionaceae Giseke 1792, Phaseolaceae C. Martius 1835, Robiniaceae Vest 1818, Sophoraceae J. Weinmann 1824, Swartziaceae Bartling 1830, Tamarindaceae Berchtold and J. Presl 1820, Viciaceae Berchtold and J. Presl 1820). 760–786/17.500–18.000. Subcosmopolitan; Faboideae is the only one of the 4 subfamilies that is widely distributed in temperate and cold areas.

1.1. CAESALPINIOIDEAE

Mainly in tropical and subtropical regions, with several genera represented in temperate areas. Trees and shrubs, sometimes scandent, or lianas, rarely subshrubs and herbs, N-fixing, often with ectotrophic mycorrhizae. Leaves bipinnate or pinnate, rarely unifoliolate or simple. Stomata mostly paracytic or anomocytic. Flowers usually more or less zygomorphic. Sepals mostly five, generally free (except in Cercideae), imbricate or rarely valvate. Petals four or less, sometimes wanting, imbricate-ascending: the adaxial (uppermost) petal overlapped by the pair of adjacent lateral petals when these are present. Stamens generally ten, rarely more (up to 100 in *Maniltoa*), often less (to solitary); filaments free or less often variously connate, but not usually forming a definite sheath around the gynoecium. Pollen grains usually in monads or very rarely in tetrads. Gynoecium always monomerous, mostly on at least short gynophore, which in some taxa becomes much elongated. Ovules anatropous, or sometimes campylotropous, with a funiculus often much thickened and arilloid. Seeds usually without a hilar groove, and usually with a straight radicle. Endosperm present or wanting. Plants are often rich in tannins, and contain resins produced by secretory cells; n = 7 with tetraploid n = 7, 8, 12, 14. – CERCIDEAE: *Cercis, Adenolobus, Griffonia, Brenierea, Bauhlnia, Gigasiphon, Tylosema, Barklya, Lysiphyllum, Phanera, Lasiobema, Piliostigma*; DETARIEAE: *Neoapaloxylon, Schotia, Barnebydendron (Phyllocarpus), Goniorrhachis, Brandezeia, Oxystigma, Kingiodendron, Gossweilerodendron, Priori, Colophospermum, Hardwickia, Daniellia, Eurypetalum, Eperua, Augouardia, Stemonocoleus, Peltogyne, Hymenaea, Guibourtia, Hylodendron, Gilletiodendron, Baikiaea, Tessmannia, Sindora, Sindoropsis, Copaifera, Pseudosindora, Detarium, Endertia, Lysidice, Saraca, Leucostegane, Talbotiella, Scorodophloeus, Crudia, Lebruniodendron, Plagiosiphon, Micklethwaitia, Maniltoa, Cynometra, Tamarindus, Intsia, Afzelia, Brodriguesia, Loesenera, Neochevalierodendron, Normandiodendron, Zenkerella, Humboldtia, Hymenostegia, Leonardoxa, Amherstia, Ecuadendron, Paloue, Paloveopsis, Brachycylix, Heterostemon, Elizabetha, Brownea, Browniopsis*; MACROLOBIEAE: *Macrolobium, Paramacrolobium, Cryptosepalum, Dicymbe, Polystemonanthus, Pseudomacrolobium, Englerodendron, Anthonotha, Berlinia, Westia, Librevillea, Didelotia, Bathieae, Pellegriniodendron, Gilbertlodendron, Isoberlinia, Oddoniodendron, Microberlinia, Julbernardia* (including *Paraberlinia, Pseudoberlinia*), *Thylacanthus, Brachystegia, Tetraberlinia, Bikinia, Icuria, Aphanocalyx, Michelsonia*; CASSIEAE: *Duparquetia, Poeppigia,*

Baudouinia, Eligmocarpus, Mendoravia, Distemonanthus, Apuleia, Storckiella, Labichea, Petalostylis, Koompassia, Martiodendron, Androcalymma, Kalappia, Zenia, Uittienia, Dialium, Dicorynia, Chamaecrista, Senna, Cassia; CAESALPINIEAE: *Gymnocladus, Gledltsia, Umtiza, Tetrapterocarpon, Arcoa, Acrocarpus, Ceratonia, Pterogyne, Haematoxylum, Cordeauxia, Stuhlmannia, Mezoneuron, Pterolobium, Tara, Coulteria, Caesalpinia, Pomaria, Erythrostemon, Poincianella, Monopetalanthus, Cenostigma, Guilandina, Libidibia, Stahlia, Hoffmannseggia, Stenodrepanum, Zuccagnia, Lophocarpinia, Balsamocarpon, Moullava, Batesia, Recordoxylon, Melanoxylon, Moldenhawera, Tachigali, Sclerolobium, Aparatiella, Jacqueshuberia, Schizolobium, Bussea, Peltophorum, Parkinsonia, Conzattia, Delonix, Colvillea, Lemuropsium, Pachyelasma, Erythrophleum, Dimorphandra, Mora, Burkea, Stachyothyrsus, Sympetalandra, Campsiandra, Chidlowia, Diptychandra, Orphanodendron, Vouacapoua.*

1.2 MIMOSOIDEAE

Mainly in tropical and subtropical regions with several genera penetrating into temperate areas. Trees and shrubs (rarely lianous), shrubs, or rarely herbs, sometimes aquatic (*Neptunia*). Leaves mostly bipinnate, less often pinnate, sometimes transformed into phyllodia. Flowers usually actinomorphic, often small, at least as to the corolla. Sepals imbricate (Parkieae and Mimozygantheae) or more often valvate, free or more often connate at the base. Petals usually five, often connate below to form a tube, valvate or rarely (*Dinizia*) imbricate. Stamens as many or twice as many as the petals or more often numerous, free or more or less connate, often colored and long-exerted. Extrafloral nectarines present. Pollen grains in diads, tetrads or polyads. Gynoecium monomerous or rarely (*Affonsea* and *Archidendron*) of several free carpels. Ovules mostly anatropous, often with long and slender funicle, in some cases with funicular aril. Endosperm with thickened walls or absent. Embryo straight; radicle short, thick; cotyledons usually without starch grains. Often with uncommon amino acids in seeds (but not canavanine). – MIMOSEAE: *Dinizia, Pentaclethra, Aubrevillea, Adenanthera, Tetrapleura, Amblygonocarpus, Pseudoprosopis, Calpocalyx, Xylia, Piptadeniastrum, Entada, Elephantorrhiza, Plathymenia, Indopiptadenia, Lemurodendron, Newtonia, Fillaeopsis, Cylicodiscus, Prosopis, Lagonychium, Xeroclada, Prosopidastrum, Piptadeniopsis, Neptunia, Leucaena, Schleinitzia, Desmanthus, Kanaloa, Calliandropsis, Affonsea, Gagnebina, Dichrostachys, Alantsilodendron, Anadenanthera, Pseudopiptadenia, Piptadenia, Parapiptadenia, Microlobus, Stryphnodendron, Adenopodia, Mimosa*; PARKIEAE: *Parkia;* MIMOZYGANTHEAE: *Mimozyganthus;* ACACIEAE: *Acacia;* INGEAE: *Faidherbia, Zapoteca, Guinetia, Calliandra, Viguieranthus, Macrosamanea, Cojoba, Obolinga, Inga, Cedrelinga, Zygia, Marmaroxylon, Archidendron, Falcataria, Serianthes, Paraserianthes, Goldmania, Schrankia, Archidendropsis, Wallaceodendron, Pararchidendron, Hydrochorea, Abarema, Punjuba, Blanchetiodendron, Leucochloron, Chloroleucon, Cathormion, Thailentadopsis, Sphinga, Havardia, Ebinopsis, Painteria, Pithecelloblum, Hesperalbizia, Pseudosamanea, Samanea, Albizia, Enterolobium, Lysiloma.*

1.3 FABOIDEAE (Papilionoideae)

Widely distributed in tropical, subtropical, temperate, and cold regions. Trees, shrubs, subshrubs, and herbs of widely varying habit. Root nodules are regularly formed as in most Mimosoideae and unlike most Caesalpinioideae. Leaves pinnately or less often palmately once compound or trifoliolate, seldom unifoliolate or simple. Flowers strongly zygomorphic. Sepals mostly connate below, forming a tube, imbricate. Petals five, free, imbricate-ascending: the adaxial petal (vexillum) embrace in bud (except Swartzieae and Sophoreae p. p.), two mostly free lateral petals (alae) in their turn embrace two lower (innermost) petals connated distally to form a keel (carina) enfolding the androecium and gynoecium. Stamens mostly 10, rarely 9–5; filaments usually connate to form an open or closed sheath around the gynoecium, the uppermost one often more or less separate from the others so that the androecium is diadelphous, or the uppermost stamen sometimes obsolete, or sometimes the filaments all free. Pollen grains in monads. Gynoecium monomerous. Ovules mostly more or less campylotropous, with a short funicle. Seeds bean-shaped with round or usually more or less elongate hilum with a median groove, typically with rim aril; embryo mostly curved with long, curved radicle (except Sophoreae), with well-developed suspension, bent; endosperm usually absent, when present thick-walled. Able to produce quinolizidine alkaloids and isoflavones as well as nonprotein amino acids, such as canavanine, not found elsewhere. n = 7–16, mostly 14. – SWARTZIEAE: *Bobgunnia, Bocoa, Swartzia, Candolleodendron, Trischidium, Cyathostegia, Ateleia,*

Amburana, Mildbraediodendron, Cordyla, Aldina, Zollernia, Holocalyx, Lecointea, Harleyodendron, Exostyles, Baphiopsis; SOPHOREAE: *Alexa, Castanospermum, Angylocalyx, Xanthocercis, Dussia, Myrocarpus, Myroxylon, Myrospermum, Monopteryx, Cladrastis, Styphnolobium, Calia, Uribea, Sweetia, Leutzelburgia, Ormosia, Haplormosia, Pericopsis, Acosmium, Bowdichia, Diplotropis, Clathrotropis, Petaladenium, Sakoanala, Neoharmsia, Bolusanthus, Platycelyphium, Dicraeopetalum, Cadia, Ammodendron, Ammothamnus, Maackia, Sophora, Pseudosophora, Goebelia, Keyserlingia, Salweenia, Camoensia, Dalhousiea, Airyantha, Leucomphalos, Bowringia, Baphia, Baphiastrum, Amphimas, Panurea, Spirotropis, Uleanthus*; DIPTERYXEAE: *Dipteryx, Taralea, Pterodon;* BRONGNIARTIEAE: *Cyclolobium, Poecilanthe, Harpalyce, Brongniartia, Plagiocarpus, Templetonia, Hovea, Cristonia, Thinicola, Lamprolobium*; EUCHRESTEAE: *Euchresta;* THERMOPSIDEAE: *Pickeringia, Ammopiptanthus, Anagyris, Piptanthus, Thermopsis, Baptisia;* PODALYIEAE: *Cyclopia, Xiphotheca, Amphithalea, Coelidium, Stirtonanthus, Podalyria, Liparia, Priestleya, Virgilia, Calpurnia;* CROTALARIEAE: *Spartidium, Lebeckia, Wiborgia, Rafnia, Aspalathus, Lotononis* (including *Buchenroedera*), *Bolusia, Crotalaria, Pearsonia, Rothia, Robinsiophyton*; GENISTEAE: *Melolobium, Dichilus, Polhillia, Argyrolobium, Calispepla, Lupinus, Anarthrophyllum, Sellocharis, Adenocarpus, Cytisophyllum, Argyrocytisus, Petteria, Laburnum, Podocytisus, Hesperolaburnum, Cytisus, Chamaecytisus, Corothamnus, Sarothamnus, Lembotropis, Calicotome, Echinospartum, Erinacea, Retama, Gonocytisus, Genista, Genistella, Teline, Spartium, Stauracanthus, Ulex*; AMORPHEAE: *Apoplanesia, Parryella, Amorpha, Errazuriza, Eysenhardtia, Psorothamnus, Marina, Dalea*; DALBERGIEAE: *Vatairea, Vataireopsis, Hymenolobium, Andira, Amicia, Zornia, Poiretia, Nissolia, Chaetocalyx, Riedeliella, Discolobium, Cranocarpus, Brya, Platymiscium, Platypodium, Inocarpus, Maraniona, Tipuana, Ramorinoa, Centrolobium, Paramachaerium, Etaballia, Pterocarpus, Cascaronia, Geoffroea, Fissicalyx, Fiebrigiella, Chapmannia, Arthocarpum, Pachecoa, Stylosanthes, Arachis, Grazielodendron, Dalbergia* (including *Coroya*), *Machaerium*; ADESMIEAE: *Adesmia;* AESCHYNOMENEAE: *Aeschynomene, Cyclocarpa, Soemmeringia, Smithia, Kotschya, Humularia, Bryaspis, Geissaspis, Pictetia, Belairia, Diphysa, Zygocarpum, Ormocarpum, Ormocarpopsis, Peltiera, Weberbauerella;* HYPOCALYPTEAE: *Hypocalyptus*; MIRBELIEAE: *Gompholobium, Burtonia, Sphaerolobium, Daviesia, Erichsenia, Viminaria, Isotropis, Jacksonia, Leptosema, Latrobea, Euchilopsis, Phyllota, Otion, Aotus, Urodon, Stonesiella, Almaleea, Eutaxia, Dillwynia, Pultenaea, Mirbelia, Chorizema, Oxylobium, Podolobium, Callistachys, Gastrolobium, Brachysema, Burgesia, Nemicia, Jansonia, Cupulanthus*; BOSSIAEEAE:, *Goodia, Bossiaea, Platylobium, Muelleranthus, Ptychosema, Aenictophyton*; INDIGOFEREAE: *Phylloxylon, Cyamopsis, Indigastrum, Microcharis, Rhynchotropis, Vaughania, Indigofera;* MILLETTIEAE: *Callerya, Padbruggea, Whitfordiodendron, Antheroporum, Endosamara, Sarcodum, Afgekia, Wisteria, Austrosteenisia, Leptoderris, Dalbergiella, Aganope, Ostryocarpus, Xeroderris, Fordia, Dewevrea, Platysepalum, Sylvichadsia, Schefflerodendron, Craibia, Disynstemon, Platycyamus, Kunstleria, Burkilliodendron, Craspedolobium, Philenoptera, Hesperothamnus, Piscidia, Dahlstedtia, Deguelia, Lonchocarpus, Willardia, Behaimia, Bergeronia, Margaritolobium, Muellera, Derris, Paraderris, Millettia, Pongamia, Neodunnia, Pongamiopsis, Pyranthus, Chadsia, Mundulea, Tephrosia* (including *Cracca*), *Apurimacia, Paratephrosia, Requienia, Ptycholobium*; ABREAE: *Abrus;* PHASEOLEAE: *Dioclea, Luzonia, Macropsychanthus, Canavalia, Cymbosema, Cleobulia, Camptostema, Cratylia, Galactia, Collaea, Lackeya, Rhodopis, Neorudolphia, Cruddasia, Ophrestia, Pseudoeriosema, Clitoria, Barbieria, Centrosema, Periandra, Clitoriopsis, Apios, Cochlianthus, Shuteria, Mastersia, Diphyllarium, Mucuna* (including *Stizolobium*), *Kennedia, Hardenbergia, Vandasina, Spatholobus, Butea, Meizotropis, Adenodolichos, Paracalyx, Bolusafra, Carrissoa, Chrysoscias, Rhynchosia, Baukea, Eriosema, Dunbaria, Cajanus* (including *Atylosia, Endomallus*), *Flemingia, Erythrina, Psophocarpus, Dysolobium, Otoptera, Decorsea, Strongylodon, Colopogonium, Cologania, Pachyrhizus, Herpyza, Neurautanenia, Neonotonia, Teyleria, Dumasia, Pueraria, Nogra, Eminia, Sinodolichos, Pseudeminia, Pseudovigna, Amphicarpaea, Teramnus, Glycine, Phylacium, Neocollettia, Wajira, Sphenostylis, Nesphostylis, Alistilus, Austrodolichos, Dolichos, Macrotyloma, Dipogon, Lablab, Spathionema, Vatovaea, Physostigma, Vigna,*

Oxyrhynchus, Phaseolus, Ramirezella, Strophostyles, Dolichopsis, Macroptilium, Mysanthus, Oryxis; DESMODIEAE: *Campylotropis, Kummerowia, Lespedeza, Dendrolobium, Phyllodium, Ougeinia, Aphyllodium, Dicerma, Ohwia, Hanslia, Arthroclianthus, Nephrodesmus, Tadehagi, Akschindlium, Droogmansia, Monarthrocarpus, Trifidacanthus, Desmodium, Podocarpium, Codariocalyx, Hylodesmum, Hegnera, Pseudarthria, Pycnospora, Mecopus, Uraria, Urariopsis, Christia, Alysicarpus, Desmodiastrum, Melliniella, Leptodesmia, Eleiotis*; PSORALEEAE: *Otholobium, Psoralea, Hallia, Orbexilum, Hoita, Rupertia, Psoralidium, Pediomelum, Bituminaria, Cullen*, SESBANIEAE: *Sesbania, Glottidium*; LOTEAE: *Hippocrepis, Scorpiurus, Securigera, Coronilla, Podolotus, Anthyllis, Hymenocarpos, Pseudolotus, Antopetitia, Hosackia, Ornithopus, Dorycnopsis, Kebirita, Ottleya, Acmispon, Syrmatium, Lotus, Dorycnium, Tetragonolobus, Tripodion, Hammatolobium, Cytisopsis;* ROBINIEAE: *Hebsetigma, Lennea, Gliricidia* (including *Yucaratonia*), *Hybosema, Poitea* (including *Sauvallella*), *Sabinea, Notodon, Corynella, Bembicidium, Olneya, Robinia, Poissonia* (including *Noecracca*), *Coursetia, Peteria, Genistidium, Sphinctospermum*; GALEGEAE: *Glycyrrhiza, Meristotropis, Chesneya, Spongiocarpella, Gueldenstaedtia, Tibetia, Erophaca, Oxytropis, Biserrula, Astragalus* (including *Neodielsia*), *Astracantha, Ophiocarpus, Barnebyella, Colutea, Oreophysa, Smirnowia, Eremosparton, Sphaerophysa, Lessertia, Sutherlandia, Swainsona, Montigena, Clianthus, Streblorrhiza, Galega*; CARMICHAELIEAE: *Carmichaelia, Notospartium, Chordospartium, Corallospartium;* HEDYSAREAE: *Calophaca, Caragana, Halimodendron, Alhagi, Eversmannia, Hedysarum, Corethrodendron, Sulla, Taverniera, Onobrychis, Stracheya, Sartoria, Ebenus;* CICEREAE: *Cicer;* TRIFOLIEAE: *Parochetus, Trifolium, Lupinaster, Amoria, Chrysaspis, Ononis, Melilotus, Trigonella, Factorovskya, Medicago, Radiata*; VICIEAE (FABEAE): *Vicia, Lathyrus, Lens, Pisum, Vavilovia;*

The most archaic genera of the family seem to be diverse extratropical woody Caesalpinioideae – *Gleditsia, Gymnocladus, Ceratonia – Zenia*, and *Cercis* (Polhillet al. 1981: 2; Tucker 2002). It is therefore preferable to begin the system of classification of Fabaceae with the subfamily Caesalpinioideae. The most archaic tribes are Caesalpinieae, Cassieae, and Cercideae. Partial helical order of floral initiation is common among Caesalpinioideae, and *Gleditsia* and *Ceratonia* are known to have helical order throughout organ initiation (Tucker 1991). According to Goldblatt (1981: 455), "x = 7 would seem basic for Caesalpinioideae, with the retention of ancestral diploidy only in *Cercis*. Polyploidy apparently became established early in the evolution of the subfamily with n = 14 basic in one line and n = 12 in a second more specialized and derived line."

Mimosoideae are closely linked to the Caesalpinioideae. According to Elias (1981: 143), the apparent link is between the caesalpinioid genus *Dimorphandra* and the mimosoid *Pentaclethra*. Both of them have similar bipinnate leaves, elongated spikes or often paniculate inflorescences, small, actinomorphic, bisexual flowers with imbricate sepals, five fertile, alternisepalous stamens, five or more staminodia, and pollen grains in monads. According to Goldblatt (1981: 455), Mimosoideae are evidently of tetraploid origin, with x = 14 and 13 frequent. In his opinion the closest ally of Mimosoideae is probably the *Erythrophleum* group of Caesalpinioideae. On the other hand, as Polhill (1981: 191) noticed, there are genera in Swartzieae and Sophoreae with open radial flowers, no apparent hilar groove, and a general similarity to some genera in the Caesalpinieae. The less advanced position in the subfamily is occupied by Swartzieae and Sophoreae.

Bibliography

Ainouche A-K and RJ Bayer. 1999. Phylogenetic relationships in *Lupinus* (Fabaceae: Papilionoideae) based on internal transcribed spacer sequences (ITS) of nuclear ribosomal DNA. Am. J. Bot. 86: 590–607.

Allan GJ. 1999. Molecular systematic and biogeographic studies of the temperate herbaceous papilionoid tribes Loteae and Coronilleae (Fabaceae). Ann Arbor, MI.

Banks H and BB Klitgaard. 2000. Palynological systematics of detarioid legumes (Leguminosae: Caesalpinioideae). In: PS Herendeen and A Bruneau, eds. Advances in legume systematics, vol. 9, pp. 79–106. Royal Botanic Gardens, Kew.

Banks H and BB Klitgaard, GP Lewis, PR Crane, and A Bruneau. 2003. Pollen and the systematics of the tribes Caesalpinieae and Cassieae. In: BB Klitgaard and A Bruneau, eds. Advances in legume systematics, vol. 10, pp. 95–122. Royal Botanic Gardens, Kew.

Baretta-Kuipers T. 1981. Wood anatomy of Leguminosae: its relevance to taxonomy. In: RM Polhill and PH Raven, eds. Advances in legume systematics, vol. 2, pp. 677–705. Royal Botanic Gardens, Kew.

Behnke H-D and L Pop. 1981. Sieve-element plastids and crystallineP(hloem)-proteininLeguminosae:Micromorphological characters as an aid to the circumscription of the family and subfamilies. In: RM Polhill and PH Raven, eds. Advances in legume systematics, vol. 2, pp. 707–715. Royal Botanic Gardens, Kew.

Bell EA, JA Lackey, and RM Polhill. 1978. Systematic significance of canavanine in the Papilionoideae (Fabaceae). Biochem. Syst. Ecol. 6: 201–212.

Breteler FJ. 1995. The boundary between Amherstieae and Detariaea (Caesalpinioideae). In: M Crisp and JJ Doyle, eds. Advances in legume systematics: Phylogeny, vol. 7, pp. 53–61. Royal Botanic Gardens, Kew.

Bruneau A, F Forest, PS Herendeen, BB Klitgaard, and GP Lewis. 2001. Phylogenetic relationships in the Caesalpinioideae (Leguminosae) as inferred from chloroplast *trn*L intron sequences. Syst. Bot. 26: 487–514.

Buss PA and NR Lersten. 1975. Survey of tapetal nuclear number as a taxonomic character in Leguminosae. Bot. Gaz. 136: 388–395.

Chapill JA. 1995. Cladistic analysis of the Leguminosae: The development of an explicit phylogenetic hypothesis, pp. 1–9. In: MD Crisp and JJ Doyle, eds. Advances in legume systematics: Phylogeny, vol. 7, pp. 1–9. Royal Botanic Gardens, Kew.

Chubirko MM and LN Kostrikova. 1985. Fabaceae. In: MS Yakovlev, ed. Comparative embryology of flowering plants: Brunelliaceae-Trimandraceae, pp. 67–77. Nauka, Leningrad (in Russian).

Crepet WL and DW Taylor. 1985. The diversification of the Leguminosae: First fossil evidence of the Mimosoideae and Papilionoideae. Science 228: 1087–1089.

Crepet WL and DW Taylor. 1986. Primitive mimosoid flowers from the Paleocene-Eocene and their systematic and evolutionary implications. Am. J. Bot. 73: 548–563.

Crisp MD and LG Cook. 2003a. Molecular evidence for the definition of genera in the *Oxylobium* group (Fabaceae: Mirbelieae). Syst. Bot. 28: 705–713.

Crisp MD and LG Cook. 2003b. Phylogeny and embryo sac evolution in the endemic Australasian papilionoid tribes Mirbelieae and Bossiaeeae. In: BB Klitgaard and A Bruneau, eds. Advances in legume systematics: Higher level systematics, vol. 10, pp. 253–268. Royal Botanic Gardens, Kew.

Crisp MD and JJ Doyle, eds. 1995. Advances in legume systematics: Phylogeny. Part 7. Royal Botanic Gardens, Kew.

Crisp MD and PH Weston. 1995. Mirbelieae. In: M Crisp and JJ Doyle, eds. Advances in legume systematics: Phylogeny, vol. 7, pp. 245–282. Royal Botanic Gardens, Kew.

De Faria SM and JI Sprent. 1994. Legume nodule development: An evolutionary hypothesis. In: JI Sprent and D McKey, eds. Advances in legume systematics: The nitrogen factor, vol. 5, pp. 33–39. Royal Botanic Gardens, Kew.

Dickison WC. 1981. Evolutionary relationships of the Leguminosae. In: RM Polhill and PH Raven, eds. Advances in legume systematics, vol. 1, pp. 35–54. Royal Botanic Gardens, Kew.

Ditsch F, H Patha, and W Barthlott. 1995. Micromorphology of epicuticular waxes in Fabales *s.l.* and its systematic significance. Beitr. Biol. Pflanzen 68: 297–310.

Dormer KJ. 1946. Vegetative morphology as a guide to the classification of the Papilionatae. New Phytol. 45: 145–161.

Doyle JJ. 1995. DNA data and legume phylogeny: A progress report.. In: MD Crisp and JJ Doyle, eds. Advances in legume systematics: Phylogeny, vol. 7, pp. 11–30. Royal Botanic Gardens, Kew.

Doyle JJ, JL Doyle, JA Ballenger, and JD Palmer. 1996. The distribution and phylogenetic significance of a 5-kb chloroplast DNA inversion in the flowering plant family Leguminosae. Molec. Phylogen. Evol. 5: 429–438.

Doyle JJ, JL Doyle, JA Ballenger, EE Dickson, T Kajita, and H Ohashi. 1997. A phylogeny of the chloroplast gene *rbc*L in the Leguminosae: taxonomic correlations and insights into the evolution of nodulation. Am. J. Bot. 84: 541–554.

El-Gazzar A and MA El-Fiki. 1977. The main subdivisions of Leguminosae. Bot. Not. 129: 371–375.

Elias TS. 1974. The genera of Mimosoideae (Leguminosae) in the southeastern United States. J. Arnold Arbor. 55: 67–118.

Elias TS. 1981. Mimosoideae. In: RM Polhill and PH Raven, eds. Advances in legume systematics, vol.1, pp. 143–151. Royal Botanic Gardens, Kew.

Endo Y and H Ohashi. 1998. The features of cotyledon areoles in Leguminosae and their systematic utility. Am. J. Bot. 85: 753–759.

Evans JA, PE Gasson, and GP Lewis. 2006. Wood Anatomy of the Mimosoideae (Leguminosae). IAWA, Nationaal Herbarium Nederland, Leiden.

Ferguson IK and JJ Skvarla. 1988. Pollen morphology of the tribe Swartzieae (subfamily Papilionoideae: Leguminosae) 1. Introduction and all genera excluding *Aldina* and *Swartzia*. Am. J. Bot. 75: 1884–1897.

Ferguson IK, BD Schrire, and R Shepperson. 1994. Pollen morphology of the tribe Sophoreae and relationships between subfamilies Caesalpinioideae and Papilionoideae. In: IK Ferguson and SC Tucker, eds. Advances in legume systematics: Structural botany, vol. 6, pp. 53–96. Royal Botanic Gardens, Kew.

Forest F, A Bruneau, J Hawkins, T Kajita, JJ Doyle, and PR Crane. 2002. The sister of the Leguminosae revealed: Phylogenetic relationships in the Fabales determined using *trn*L and *rbc*L sequences. In: Botany 2002: Botany in the Curriculum, p. 124. Madison, WI (Abstracts).

Goldblatt P. 1981. Cytology and the phylogeny of Leguminosae. In: RM Polhill and PH Raven, eds. Advances in legume systematics, vol. 2, pp. 427–463. Royal Botanic Gardens, Kew.

Gottlieb OR, AMMS Dan, DHT Zocher, and MRMB Borin. 1994. Micromolecular clues for evolution of the Leguminosae.. In: JI Sprent and D McKey, eds. Advances in legume systematics: The nitrogen factor, vol. 5, pp. 107–128. Royal Botanic Gardens, Kew.

Grimes J. 1995. Generic relationships of Mimosoideae tribe Ingeae, with emphasis on the new world *Pithecellobium*-complex. In: M Crisp and JJ Doyle, eds. Advances in legume systematics: Phylogeny, vol. 7, pp. 101–121. Royal Botanic Gardens, Kew.

Guinet P and IK Ferguson. 1989. Structure, evolution and biology of pollen in Leguminosae. In: CH Stirton and JL Zarucchi, eds. Advances in legume biology. Monogr. Syst. Bot. 29: 77–103.

Gunn CR. 1991. Fruits and seeds of genera in the subfamily Caesalpinioideae (Fabaceae). U.S.A. Department of Agriculture, Tech. Bull.: 1755.

Harborne JBD, D Boulter, and BL Turner, eds. 1971. Chemotaxonomy of the Leguminosae. Academic Press, London/ New York.

Heel WA van. 1993. Floral ontogeny of *Archidendron lucyi* (Mimosaceae), with remarks on *Amherstia nobilis* (Caesalpiniaceae). Bot. Jahrb. Syst. 114: 551–560.

Hegnauer R. 1956. Chemotaxonomische Betrachtung der Leguminosae. Die Pharmazie 11: 2–16.

Herendeen PS. 1995. Phylogenetic relationships of the tribe Swartzieae, pp.123–132. In: MD Crisp et JJ Doyle, eds. Advances in legume systematics: Phylogeny, vol. 7, pp. 123–132. Royal Botanic Gardens, Kew.

Herendeen PS. 2000. Structural evolution in the Caesalpinioideae (Leguminosae). In: PS Herendeen and A Bruneau, eds. Advances in legume systematics, vol. 9, pp. 45–64. Royal Botanic Gardens, Kew

Herendeen PS and A Bruneau, eds. 2000. Advances in legume systematics. Part 9. Royal Botanic Gardens, Kew.

Herendeen PS, WL Crepet, and DL Dilcher. 1992. The fossil history of the Leguminosae: Phylogenetic and biogeographic implications. In: Herendeen PS and DL Dilcher, eds. Advances in legume systematics: The fossil record, vol. 4, pp. 303–316. Royal Botanic Gardens, Kew.

Herendeen PS, GP Lewis, and A Bruneau. 2003a. Floral morphology in Caesalpinioid legumes: Testing the monophyly of the "*Umtiza* clade". Int. J. Plant Sci. 164(Suppl. 5): 393–407.

Herendeen PS, A Bruneau, and GP Lewis. 2003b. Phylogenetic relationships in Caesalpiniod legumes: A preliminary analysis based on morphological and molecular data. In: BB Kligaard and A Bruneau, eds. Advances in legume systematics, vol. 10, pp. 37–62. Royal Botanic Gardens, Kew.

Hou D, K Larsen, and SS Larsen. 1996. Caesalpiniaceae. In: CGGJ van Steenis, ed. Flora Malesiana ser. 1, 12: 409–730. Noordhoff, Groningen.

Hu J-M, M Lavin, MF Wojciechowski, and MJ Sanderson. 2000. Phylogenetic systematics of the tribe Millettieae (Leguminosae) based on chloroplast *trn*K/*mat*K sequences and its implications for evolutionary patterns in Papilionoideae. Am. J. Bot. 87: 418–430.

Hu J-M, M Lavin, MF Wojciechowski, and MJ Sanderson. 2002. Phylogenetic analysis of nuclear ribosomal ITS/5.8S sequences in the tribe Millettieae (Fabaceae): *Poecilanthe-Cyclolobium*, the core Millettieae, and the *Callerya* group. Syst. Bot. 27: 722–733.

Ireland H, RT Pennington, and J Preston. 2000. Molecular systematics of the Swartzieae. In: PS Herendeen and A Bruneau, eds. Advances in legume systematics, vol. 9, pp. 217–232. Royal Botanic Gardens, Kew.

Isely D. 1973, 1975, 1981. Leguminosae of the United States. Respective parts published in Mem. New York Bot. Gard.: I. Subfamily Mimosoideae, 25(1): 1–125; II. Subfamily Caesalpinioideae, 25(2): 1–228; III. Subfamily Papilionoideae: Tribes Sophoreae, Podalyrieae, Loteae, 25(3): 1–264.

Kajita T, H Ohashi, Y Tateishi, CD Bailey and JJ Doyle. 2001. *Rbc*L and legume phylogeny, with particular reference to Phaseoleae, Millettieae, and allies. Syst. Bot. 26: 515–536.

Kass E, and M Wink. 1996. Molecular evolution of the Leguminosae: phylogeny of the three subfamilies based on *rbc*L sequences. Biochem. Syst. Ecol. 24: 365–378.

Kirkbride JH Jr., CR Gunn, and AL Weitzman. 2003. Fruits and seeds of genera in the subfamily Faboideae (Fabaceae). Tech. Bull. USA Department of Agriculture: 1980.

Kopooshian H and D Isely 1966. Seed character relationships in the Leguminosae. Proc. Iowa Acad. Sci. 73: 59–67.

Lackey JA. 1977. A revised classification of the tribe Phaseoleae (Leguminosae: Papilionoideae), and its relation to canavanine distribution. Bot. J. Linn. Soc. 74: 163–178.

Lavin M. 1995. Tribe Robinieae and Allies; model groups for assessing early tertiary northern latitude diversification of tropical legumes. In: M Crisp and JJ Doyle, eds. Advances in legume systematics: Phylogeny, vol. 7, pp. 141–160. Royal Botanic Gardens, Kew.

Lee J and T Hymowitz. 2001. A molecular phylogenetic study of the subtribe Glycininae (Leguminosae) derived from the chloroplast DNA *rps*16 intron sequences. Am. J. Bot. 88: 2064–2073.

Leinfellner W. 1970. Zur Kenntnis der Karpelle der Leguminosen: 2. Caesalpiniaceae and Mimosaceae. Oesterr. Bot. Z. 118: 108–120.

Lewis G, B Schrire, B Mackinder, M Lock, eds. 2005. Legumes of the world. Royal Botanic Gardens, Kew.

Luckow M, JT Miller, DJ Murphy, and T Livshultz. 2003. Phylogenetic analysis of the Mimosoideae (Leguminosae) based on chloroplast DNA sequence data. In: BB Klitgaard and A Bruneau, eds. Advances in legume systematics: Higher level systematics, vol. 10, pp. 197–220. Royal Botanic Gardens, Kew.

Manning JC and CH Stirton. 1994. Endothecal thickenings and phylogeny of the Leguminosae, pp. 141–163. In: IK Ferguson and SC Tucker, eds. Advances in legume systematics: Structural botany, vol. 6, pp. 141–163. Royal Botanic Gardens, Kew.

Martin PG and JM Dowd. 1990. A protein sequence study of the dicotyledons and it relevance to the evolution of the legumes and nitrogen fixation. Austral. Syst. Bot. 3: 91–100.

Miller JT, JW Grimes, DJ Murphy, RJ Bayer, and P.Y Ladiges. 2003. A phylogenetic analysis of the Acacieae and Ingeae (Mimosoideae: Fabaceae) based on *trn*K, *mat*K, *psb*A-*trn*H, and *trn*L/*trn*F sequence data. Syst. Bot. 28: 558–566.

Pennington RT, BB Klitgaard, H Ireland, and M Lavin. 2000. New insight into floral evolution of basal Papilionoideae from molecular phylogenies. In: PS Herendeen and A Bruneau, eds. Advances in legume systematics, vol. 9, pp. 233–248. Royal Botanic Gardens, Kew.

Pennington RT, M Lavin, H Ireland, B Klitgaard, J Preston, and J-M Hu. 2001. Phylogenetic relationships of basal Papilionoid legumes based upon sequences of the chloroplast *trn*L intron. Syst. Bot. 26: 537–556.

Pettigrew CJ and L Watson. 1977. On the classification of Caesalpinioideae. Taxon 26: 57–64.

Pfeil BE, JA Schlueter, RC Shoemaker, and JJ Doyle. 2005. Placing paleopolyploidy in relation to taxon divergence: A phylogenetic analysis in legumes using 39 gene families. Syst. Biol. 54: 441–454.

Polhill RM. 1981. Papilionoideae. In: RM Polhill and PH Raven, eds. Advances in legume systematics, vol. 1, pp. 191–208. Royal Botanic Gardens, Kew.

Polhill RM and PH Raven, eds. 1981. Advances in legume systematics, 2 vols. Royal Botanic Gardens, Kew.

Polhill RM, PH Raven, and CH Stirton. 1981. Evolution and systematics of the Leguminosae, pp. 1–26. In: RM Polhill and PH Raven, eds. Advances in legume systematics, vol. 1, pp. 1–26. Royal Botanic Gardens, Kew.

Ponomarenko SF. 1996. Fabaceae. In: A Takhtajan, ed. Comparative seed anatomy, vol. 5, pp. 264–298. Nauka, St. Petersburg (in Russian).

Ramirez-Domenech J and SC Tucker. 1988. Patterns of organ development in mimosoid legume flowers. In: P Leins, SC Tucker, and PK Endress, eds. Aspects of floral development, pp. 171–180. J. Cramer, Berlin/Stuttgart.

Rao VS, K Sirdeshmukh, and MG Sardar. 1958. The floral anatomy of the Leguminosae. J. Univ. Bombay 26, n.s., Pt 5B: 65–138.

Rau MA. 1953. Some observations on the endosperm in Papilionaceae. Phytomorphology 3: 209–222.

Redden KM and PS Herendeen. 2006. Morphology and phylogenetic analysis of *Paloue* and related genera in the *Brownea* clade (Detarieae, Caesalpiniodeae). Int. J. Plant Sci. 167: 1229–1246.

Robertson KR and YT Lee. 1976. The genera of Caeselpinioideae (Leguminosae) in the southeastern United States. J. Arnold Arbor. 51: 1–53.

Schrire BD. 1995. Evolution of the tribe Indigofereae (Leguminosae-Papilionoideae). In: M Crips and JJ Doyle, eds. Advances in legume systematics: Phylogeny, vol. 7, pp. 161–244. Royal Botanic Gardens, Kew.

Seigler DS. 2004. Fabaceae. In: N Smith, SA Mori, A Henderson, DW Stevenson, SV Heald, eds. Flowering plants of the Neotropics, pp. 151–156. The New York Botanical Garden, Princeton University Press, Princeton.

Thompson IR, PY Ladiges, and JH Ross. 2001. Phylogenetic studies of the tribe Brongniartieae (Fabaceae) using nuclear DNA (ITS-1) and morphological data. Syst. Bot. 26: 557–570.

Tucker SC. 1991. Helical floral organogenesis in *Gledit-sia:* A primitive caesalpinioid legume. Am. J. Bot. 78: 1130–1149.

Tucker SC. 2002. Floral ontogeny of *Cercis* (Leguminosae: Caesalpinoideae: Cercideae): Does it show convergence with Papilionoids? Int. J. Plant Sci. 163: 75–87.

Tucker SC and KE Kantz. 2001. Open carpels with ovules in Fabaceae. Int. J. Plant Sci. 162: 1065–1073.

Tucker SC, BB Klitgaard, M Fougère-Danezan, and A Bruneau. 2002. Phylogenetic analysis of combined floral ontogenetic and molecular characters in the Detarieae s.l. (Caesalpinioideae: Leguminosae). In Botany 2002: Botany in the Curriculum, p. 155. Madison, Wisconsin (Abstracts).

Wojciechowski MF. 2003. Reconstructing the phylogeny of legumes (Leguminosae): An early 21st century perspective. In: BB Klitgaard and A Bruneau, eds. Advances in legume systematics: Higher level systematics, vol. 10, pp. 5–35. Royal Botanic Gardens, Kew.

Wojciechowski MF, M Lavin, and MJ Sanderson. 2003. A phylogeny of legumes based on sequences of the plastid *mat*K gene. In: Botany 2003: Aquatic and wetland plants: wet and wild, p. 99. Mobile, Alabama (Abstracts).

Wojciechowski MF, M Lavin, and MJ Sanderson. 2004. A phylogeny of legumes (Leguminosae) based on analysis of the plastid *mat*K gene resolves many well-supported subclades within the family. Am. J. Bot. 91: 1846–1862.

Wunderlin R, K Larsen, and SS Larsen. 1987. Reorganization of the Cercideae (Fabaceae: Caesalpinoideae). Roy. Danish Acad. Sci. Let. Biol. Skr. 28: 1–40.

Wyk van B-E. 1995. Phylogenetic relationships in the tribes Podalyrieae, Liparieae and Crotalarieae. In: M Crisp and JJ Doyle, eds. Advances in legume systematics: Phylogeny, vol. 7, pp. 283–308. Royal Botanic Gardens, Kew.

Yakovlev GP. 1972. Supplements to the system of the order Fabales Nakai (Leguminosales Jones). Bot. Zhurn. 57: 585–595.

Order 89. POLYGALALES

Trees, sometimes (*Xanthophyllum*) up to 50 tall, or more often shrubs, woody lianas (*Balgoya* up to 30 m length), and perennial or annual herbs, very rarely (*Epirixanthes*) achlorophyllous parasites; plants unarmed, glabrous or with simple unicellular or sometimes uniseriate hairs; bearing essential oils or without essential oils. Xylem with tracheids (e.g. *Diclidanthera*), or without tracheids. Vessels with simple perforations; lateral pitting usually alternate. Fibers with bordered pits. Rays heterogeneous. Axial parenchyma mainly paratracheal, rarely (*Xanthophyllum* and *Moutabea*) apotracheal. Sieve-element plastids of S-type. Nodes unilacunar with one trace. Leaves usually alternate, rarely (some *Polygala*) opposite or verticillate, leathery or membranous, sometimes reduced to scales (*Salomonia* and some spp. of *Polygala*), simple, sometimes gland-dotted, stipulate or estipulate, stipules intrapetiolar, free of one another, often scaly, or spiny, or represented by glands. Stomata anomocytic or seldom paracytic, or anisocytic. Flowers in terminal or axillary racemes, spikes or panicles, or rarely solitary in the axil, bracteate and almost always bibracteolate, bisexual, or (*Balgoya*) functionally unisexual, usually strongly zygomorphic. Sepals five, imbricate free or rarely basally connate (sometimes only lower or upper ones connate); the two lateral (inner) sepals much larger and petaloid. Petals free or more often adnate to stamens to form a common tube, sometimes five, but more often the two lateral ones much reduced or completely aborted, so that there are only three evident petals – two upper ones and one lower median one, the latter (carina) large, boat-shaped, and often apically fringed. Stamens (2-) 3–10, basically ten in two cycles, but the median member of each cycle usually suppressed and the number of stamens is commonly 8, or as a result of further reduction sometimes 7–4 or even only 3; filaments generally more or less connate into a cleft tube, that is more or less adnate to petals; anthers 1–2-locular, basifixed or (*Xanthophyllum*) more or less dorsifixed, tetrasporangiate or disporangiate, opening by one or two apical or subapical pores or very short slits, or less often by long longitudinal slits (as in *Xanthophyllum*), or by longitudinal valves. Tapetum secretory. Microsporogenesis simultaneous. Pollen grains 2–3-celled, 3-colporate, polycolporate, often synorate. Intrastaminal nectaries present or (Moutabeoideae) absent. Gynoecium of 2–5(-8) united carpels; stylodia connate into a usually curved and commonly apically 2-lobed style, one lobe stigmatic, and the other with a tuft of hairs; ovary superior, 2-multilocular with parietal placentas and stipitate (*Xanthophyllum*), sometimes pseudomonomerous (as in *Securidaca*) and 1-locular, each locule with (1)2–5(-8) or (*Xanthophyllum*) 4–40 ovules. Ovules anatropous or hemitropous, epitropous, pendulous, bitegmic, crassinucellate; micropyle zig-zag (exostoma often long). Female gametophyte of *Polygonum*-type. Endosperm nuclear. Fruits mostly loculicidal capsules, less often samara, nuts, drupes, or berries. Seeds often hairy, often with a caruncle, arillate or exarillate; testa multiplicative, exotesta subsclerotic, endotestal cells more or less palisade or not, U-thickened (Plisco 2000); embryo straight, sometimes (*Polygala* sp.) green; cotyledons two, planoconves; endosperm copious, oily, scanty or wanting. Contain commonly alkaloids, flavonols (kaempferol and quercetin), often saponins, commonly accumulating aluminum, n = 5–14(-17).

Evidently related to the Fabales (see Soltis et al. 2006).

1. POLYGALACEAE

Hoffmannsegg et Link 1809 (including Diclidantheraceae J. Agardh 1858, Moutabeaceae Endlicher 1841, Xanthophyllaceae Gagnepain ex Reveal et Hoogland 1990). 21/950–1000. Nearly cosmopolitan, except for the Arctic, Polynesia, and New Zealand.

1.1 MOUTABEOIDEAE

Ovary 2–8 locular, with 1 ovule in each locule. Calyx united with the petals into a tube. Abaxial petals not keeled. Filaments connate into a singular tube (*Diclidanthera, Eriandra*) or either in two bundles (*Moutabea*) or inserted on the petals, not united into a tube (*Barnhartia*). Intrastaminal nectaries absent. Axial parenchyma apotracheal or paratracheal. – XANTHOPHYLLEAE: *Xanthophyllum*; MOUTABEAE: *Balgoya, Eriandra, Barnhartia, Diclindanthera, Moutabea.*

2.2 POLYGALOIDEAE

Ovary 2–3-locular with one ovule on each axile placenta. Calyx not united with the petals into a tube. Abaxial petals keeled. Filaments united into a tube open dorsally. Intrastaminal nectaries present. Axial parenchyma predominantly paratracheal. – CARPOLOBIEAE: *Atroxima, Carpolobia*; POLYGALEAE: *Bredemeyera, Acanthocladus, Badiera, Comesperma, Epirixanthes, Muraltia* (including *Nylandtia*), *Salomonia, Monnina, Ancylotropis, Pseudomonnina, Securidaca, Hualania, Pteromonnina.*

Bibliography

Arreguín-Sánchez ML, R Palacios-Chávez, DL Quiroz-García, and D Ramos-Zamora. 1988. Morfoligia de los granos de pollen de la familia Polygalaceae del Valle de Mexico. Acta Bot. Mex. 4: 21–27.

Bernardi I. 2000. Consideraciones taxonómicas y fitogeográficas acerca de 101 Polygalae americanas. Cavanillesia Altera 1: 1–456.

Bridgwater S and P Baas. 1982. Wood anatomy of *Xanthophyllum* Roxb. IAWA Bull. 3: 115–125.

Detienne P. 1991. Anatomic de bois de *Balgoya pacifica* (Polygalaceae) de Nouvelle Caledonie. Adansonia, ser.4, 13: 17–20.

Dickison WC. 1973. Nodal and leaf anatomy of *Xantophyllum* (Polygalaceae). Bot. J. Linn. Soc. 67: 103–115.

Dube VP. 1962. Morphological and anatomical studies in Polygalaceae and its allied families. Agra Univ. J. Res. Sci. 11: 109–112.

Dube VP and DK Awasthi. 1985. Morphological and anatomical studies of floral nectaries in Polygalacaea. J. Indian Bot. Soc. 64 (2 and 3): 231–235.

Erdtman G. 1944. The systematic position of the genus *Diclidanthera* Mart. Bot. Not. 1944: 80–84.

Eriksen B. 1993a. Floral anatomy and morphology in the Polygalaceae. Plant Syst. Evol. 186: 17–32.

Eriksen B. 1993b. Phylogeny of the Polygalaceae and its taxonomic implications. Plant Syst. Evol. 186: 33–35.

Eriksen B. 1993c. Taxonomical studies in the Polygalaceae and Valerianaceae. Thesis, Fac. Nat. Sc., Univ,. Goteborg.

Eriksen B and C Persson. 2007. Polygalaceae. In: K Kubitzki, ed. The families and genera of vascular plants, vol. 9, pp. 345–363. Springer, Berlin/Heidelberg/New York.

Leinfellner W. 1972. Zur Morphologic des Gynozeums der Polygalaceen. Oesterr. Bot. Z. 120: 51–76.
Meijden R van der. 1982. Systematics and evolution of *Xanthophyllum* (Polygalaceae). Leiden Bot. ser. 7: 1–159.
Meijden R van der. 1986. Polygalaceae. In: CGGJ Steenis van, ed. Flora Malesiana, ser. 1, 10: 455–539. Noordhoff, Groningen.
Miller NG. 1971. The Polygalaceae in the southeastern United States. J. Arnold Arbor. 52: 267–284.
Persson C. 2001. Phylogenetic relationships in Polygalaceae based on plastid DNA sequences from the *trn*-F region. Taxon 50: 763–779.
Plisko MA. 2000. Order Polygalales. In: A Takhtajan, ed. Comparative seed anatomy, vol. 6, pp. 49–111. Nauka, St. Petersburg (in Russian).
Prenner G. 2004. Floral development in *Polygala myrtifolia* (Polygalaceae) and its similarities with Leguminosae. Plant Syst. Evol. 249: 67–76.
Rao AN. 1964. An embryological study of *Salomonia contoniensis* Lour. New Phytol. 63: 281–288.
Styer CH. 1937. Comparative anatomy and systematic of Moutabeae (Polygalaceae). J. Arnold Arbor. 58: 109–145.
Verkerke W. 1984. Ovule and seed of *Xanthophyllum* (Polygalaceae). Blumea 29: 409–421.
Verkerke W. 1985. Ovules and seeds of the Polygalaceae. J. Arnold Arbor. 66: 353–394.
Verkerke W. 1991. Fruits and seeds of *Balgoya pacifica* (Polygalaceae) from New Caledonia. Adansonia, ser. 4, 14: 9–12.
Westerkamp C and A Weber. 1999. Keel flowers of the Polygalaceae and Fabaceae: a functional comparison. Bot. J. Linn. Soc. 129: 207–221.

Superorder RUTANAE

Order 90. OXALIDALES (CONNARALES)

Small trees, shrubs, woody lianas, or, less often, herbs. Commonly with mucilage canals and/or tanniniferous secretory cavities. Indumentum of multicellular or unicellular glandular hairs. Vessels with simple perforations; lateral pitting alternate. Fibers with simple pits, septate. Rays homogeneous or heterogeneous, all uniseriate or occasionally biseriate for part of their length or some of them up to five cells wide. Axial parenchyma typically absent or scanty-paratracheal. Sieve-element plastids of PSC- or Pcfs-type. Nodes trilacunar (Oxalidaceae) or pentalacunar (Connaraceae). Leaves alternate, imparipinnate with two or more pairs of leaflets or less often trifoliolate or unifoliolate, with coriaceous, entire or rarely lobed, very rarely peltate leaflets; commonly estipulate. Stomata paracytic, anisocytic and anomo-cyclocytic. Flowers small, in terminal or axillary panicles or racemes, bisexual or very rarely unisexual (plants dioecious), actinomorphic or rarely slightly zygomorphic, 5-merous or seldom 4-merous. Sepals free or rarely basally connate, imbricate or valvate, usually persistent and often embracing the base of the fruit. Petals free or rarely slightly connate, imbricate or valvate, sometimes loriform and circinate. Stamens ten or eight, in two cycles, the inner sometimes staminodial; filaments slender, free, shortly connate at the base; anthers short, dorsifixed, tetrasporangiate, introrse or extrorse, opening longitudinally. Tapetum secretory. Microsporogenesis simultaneous. Pollen grains 2-celled, mostly 3-colporate, sometimes 3-colpate, rarely (*Jollydora*) 4-colpate. Nectary disc poorly developed or wanting. Gynoecium of 5 or 1 carpels, rarely of 3, 7–8 carpels, free or only basally connate; ovary superior; ovules anatropous, hemitropous, or more often orthotropous, bitegmic, crassinucellate or tenuinucellate, sometimes (*Connarus, Jollydora,* and *Averrhoa*) with an endothelium. Female gametophyte of *Polygonum*-type, or very rarely of *Allium*-type. Endosperm nuclear. Fruits capsules or indehiscent. Seeds mostly arillate, without or with fibrous exotegmen; embryo straight; endosperm copious and oily to scanty or none; n = 5(-7)-14, 16.

Oxalidales have many similarities both with the Rosanae (especially with Cunoniales) and with Rutanae (especially with Sapindales).

Key to Families

1 Shrubs, woody lianas, or, less often, small trees. Commonly with mucilage canals and/or tanniniferous secretory cavities. Indumentum of multicellular or unicellular glandular hairs. Nodes pentalacunar with five traces or septalacunar with 7seven traces. Leaves imparipinnate with two or more pairs of leaflets or less often trifoliolate or unifoliolate, with coriaceous, entire or rarely lobed, very rarely peltate leaflets; stipules wanting. Stomata paracytic, anisocytic and anomo-cyclocytic. Flowers small, in terminal or axillary panicles or racemes, bisexual or very rarely unisexual (plants dioecious), actinomorphic or rarely slightly zygomorphic, 5-merous or seldom 4-merous. Sepals free or rarely basally connate, imbricate or valvate, usually persistent and often embracing the base of the fruit. Petals free or rarely slightly connate, imbricate or valvate, sometimes

loriform and circinate. Stamens ten or eight, in two cycles, the inner sometimes staminodial; filaments slender, free, shortly connate at the base; anthers short, dorsifixed, tetrasporangiate, introrse. Pollen grains 2-celled, mostly 3-colporate, sometimes 3-colpate, rarely (*Jollydora*) 4-colpate. Nectary disc poorly developed and usually extrastaminal, or wanting, but the receptacle sometimes nectariferous at the center. Gynoecium of five or one carpels, rarely of three, seven, or eight carpels (often some of them abortive), free or only basally connate, often not fully sealed, each with terminal stylodium and capitate stigma and containing two collateral ovules ascending from near the base, one of which may abort; ovary superior. Ovules anatropous, hemitropous, or more often orthotropous, crassinucellate, sometimes (*Connarus, Jollydora*) with an endothelium. Fruiting carpels sessile or stipitate, dry, 1-seeded or sometimes (*Jollydora*) 2-seeded, opening along the ventral suture (follicles) or seldom along both sides (legumes), or rarely (*Jollydora* and *Hemandradenia*) indehiscent and nutlike. Seeds mostly arillate, exotestal with more or less developed fibrous exotegmen; embryo straight, with short hypocotyl and minute radicle; endosperm copious and oily to scanty or none; funicle very short or none; according to Corner (1976: 105), aril present to some extent in all cases, primarily as an outgrowth from the short raphe and chalaza, yellow to red, pulpy, typically (*Connarus*) with a thick, often crenulate, free border or short limb and a sarcotestal area along the raphe and chalaza; with extensive free limb (*Rourea* spp.); more or less wholly developed as a sarcotesta in genera with baccate seeds. Bark, fruits, and seeds often highly poisonous, n = 14, 16 . . . . 1. CONNARACEAE.

1 Herbs, sometimes succulent, often with fleshy rhizomes or bulb-like tubers, sometimes subshrubs or arborescent. Leaves alternate, palmately or pinnately compound, often trifoliolate, rarely unifoliolate or phyllodic, stipules small or absent, colleters (*Oxalis*) present; leaflets usually folded back in bud and at night. Stomata paracytic. Flowers in axillary cymes or solitary, bisexual, usually tristylous, sometimes cleistogamous and apetalous. Sepals five, basally connate. Petals five, free or united at the base, contorted or rarely imbricate. Stamens 10 or 15, sometimes staminodal. Filaments united at the base. Anthers extrorse. Pollen grains mostly 3-colporate, often contain starch. Nectaries often glands opposite the petals. Gynoecium of (3-)5 carpels, united to form a 5-locular ovary; ovary superior, with (1)2 to many more or less pendulous, anatropous ovules per locule; ovules mostly tenuinucellate, rarely crassinucellate and with endothelium (*Averrhoa*). Fruits loculicidal, ribbed or angled capsules (*Oxalis, Biophytum*) or fleshy capsules (*Dapania*), or (*Averrhoa, Sarcotheca*) baccate. Seeds with fibrous 2-layered exotegmen, endotesta walls thickened, not palisade; testa and tegmen less differentiated when fruit a berry; embryo large, green (*Oxalis*); ednosperm fleshy, starchy, subruminate. Present tannins, juice acrid, with oxalated, calcium oxalatae as crystals, n = (5-)7–12, mostly 7. . . . . . 2. OXALIDACEAE.

1. CONNARACEAE

R. Brown 1818 (including Cnestiaceae Rafinesque 1830). 13/380. Pantropical, with the greatest diversification in Africa, Southeast Asia, and tropical America.

Cnestis, Rourea (including *Bernardinia, Byrsocarpus, Eichleria, Jaundea, Paxia, Roureopsis, Spiropetalum*), *Jollydora, Connarus, Agelaea, Pseudoconnarus, Burttia, Schellenbergia, Vismianthus, Hemandradenia, Cnestidium, Manotes, Ellipanthus* (including *Pseudoellipanthus*).

The family Connaraceae shares with the Oxalidaceae. According to some botanists, including Hallier (1912), Gundersen (1950), and Airy Shaw (1973), Connaraceae are allied to the Oxalidaceae (according to Hallier related to both the Leguminosae and Oxalidaceae, according to Schellenberg (1938) "perhaps near to Oxalidaceae," and according to Airy Shaw "somewhat intermediate between Leguminosae and Averrhoaceae"). Some perhaps rather remote affinity between Connaraceae and Oxalidaceae is supported by the same Pies-type of sieve-element plastids (Behnke 1983).

2. OXALIDACEAE

R. Brown 1818 (including Averrhoaceae Hutchinson 1959). 5/700. Tropical and subtropical regions with a few species widespread in temperate regions.

Oxalis, Biophytum, Averrhoa, Sarcotheca, Dapania.

Bibliography

Behnke H-D. 1982. Sieve-element plastids of Connaraceae and Oxalidaceae: A contribution to the knowledge of P-type plastids in dicotyledons and their significance. Bot. Jahrb. Syst. 103: 1–8.

Boesewinkel FD. 1985. Development of ovule and seed coat in *Averrhoa* (Oxalidaceae) with notes on some related genera. Acta Bot. Neerl. 34: 413–424.

Boesewinkel FD and F Bouman. 2000. Oxalidaceae. In: A Takhtajan, ed. Comparative seed anatomy, vol. 6, pp. 21–23. Nauka, St. Petersburg (in Russian).

Breteler FJ, ed. 1989. The Connaraceae. A taxonomic study with emphasis on Africa. Agric. Univ. Wageningen Pap. 89–6: 1–403.

Chung RCK, and AL Lim. 1998. The embryology of *Averrhoa* (Oxalidaceae). Sandakania 12: 37–55.

Cocucci AA. 2004. Oxalidaceae. In: K Kubitzki, ed. The Families and genera of vascular plants, vol. 6, pp. 285–290. *Springer,* Berlin/Heidelberg/New York.

Devi DR and LL Narayana 1990. Systematic position of *Averrhoa* (Oxalidaceae). Feddes Repert. 101: 165–170.

Dickison WC. 1971–1974. Anatomical studies in the Connaraceae: I. Carpels. II. Wood anatomy. III. Leaf anatomy. IV. The bark and young stem. J. Elisha Mitchell Sci. Soc. 87: 77–86, 1971; 88: 120–136; 89: 121–138, 1973; 166–171, 1973 (1974).

Dickison WC. 1979. A survey of pollen morphology of the Connaraceae. Pollen Spores 21: 31–79.

Forero E. 1983. Connaraceae. Flora Neotrop. Monograph 36. New York.

Heimsch C. 1942. Comparative anatomy of the secondary xylem in the "Gruinales" and "Terebinthales" of Wettstein with reference to taxonomic grouping. Lilloa 8: 83–198.

Huynh K-L. 1969. Étude du pollen des Oxalidaceae Morphologic générale — Palynotaxonomie des *Oxalis americains*. Bot. Jahrb. Syst. 89: 272–303.

Jongkind CCH, and RHJ Lemmens. 1989. The Connaraceae: A taxonomic study with special emphasis on Africa. Landbouwuniversiteit te Wageningen.

Knuth R. 1930. Oxalidaceae. In: A Engler, ed. Pflanzenreich, 95 (IV. 130): 1–481. Berlin.

Leenhouts PW. 1958. Connaraceae. In: CGGJ van Steenis ed. Flora Malesiana ser. 1, 1 (5): 495–541. Noordhoff, Groningen.

Leinfellner W. 1970. Über die Karpelle der Connaraceen. Oesterr. Bot. Z. 118: 542–559.

Lemmens RHMJ, EJ Breteler, and CCH Jongkind. 2004. Connaraceae. In: K Kubitzki, ed. The families and genera of vascular plants, vol. 6, pp. 74–81. Springer, Berlin/Heidelberg/New York.

Matthews ML and PK Endress. 2002. Comparative floral morphology and systematics in Oxalidales (Oxalidaceae, Connaraceae, Brunelliaceae, Cephalotaceae, Cunoniaceae, Elaeocarpaceae, Tremandraceae). Bot. J. Linn. Soc. 140: 321–381.

Mosebach G. 1934. Die Fruchtstielschwellung der Oxalidaceen und Geraniaceen. Jahrb. Wiss. Bot. 79: 353–384.

Mondal MS. 1990. Pollen morphology and systematic relationships of families Sabiaceae (s.l.) and Connaraceae. Botanical Survey of India, New Delhi.

Narayana LL. 1966. A contribution to the floral anatomy of Oxalidaceae. J. Jpn. Bot. 41: 321–328.

Narayana LL. 1970. Oxalidaceae, Geraniaceae. In Symposium on comparative embryology of angiosperms. Bull. Indian Natl. Sci. Acad. 41: 114–120.

Oltmann O. 1971. Pollenmorphologiscyh-systematische Untersuchungen innerhalb der Geraniales. Diss. bot. 11, 163 + XI pp. Cramer, Lehre.

Radlkofer L. 1886. Über die dursichtigen Punkte und andere anatomische Charakter der Connaraceae. Sitzungsber. Bayer. Akad. Wiss. (M.-Ph. Kl.) 16: 345–378.

Rama Devi D. 1991. Floral anatomy of *Hypseocharis* (Oxalidaceae) with a discussion on its systematic position. Plant Syst. Evol. 177: 161–164.

Reddy B Bal and LL Narayana. 1982. Systematic position of Averrhoaceae. J. Econ. Taxon. Bot. 3: 343–348.

Robertson KR. 1975. The Oxalidaceae in the southeastern United States. J. Arnold Arb. 56: 223–239.

Schellenberg G. 1925. Die phylogenetische Entwiclung und die Wanderungen der Connaraceen. Bot. Jahrb. Syst. 60: 207–251.

Yakovleva OV. 1996. Connaraceae. In: A Takhtajan, ed. Comparative seed anatomy, vol. 5, pp. 299–311. Nauka, St. Petersburg (in Russian).

Order 91. SAPINDALES

Trees, shrubs, woody lianas, or sometimes subshrubs or herbs. Vessels with simple perforations or less often perforations scalariform; lateral pitting alternate or sometimes opposite. Fibers with bordered or simple pits, often septate. Rays heterogeneous or homogeneous. Axial parenchyma mostly paratracheal, sometimes scanty or even wanting. Sieve-element plastids of Ss-type. Nodes usually trilacunar. Leaves opposite or less often alternate, simple or more often compound, estipulate or sometimes with small, caducous stipules. Stomata of diverse types. Flowers in various types of inflorescences, rarely solitary, bisexual or more often unisexual, actinomorphic or less often more or less zygomorphic (strongly zygomorphic in Hippocastanaceae), usually 5–4-merous. Sepals free or more or less connate, imbricate or rarely (some Sapindaceae) valvate. Petals free, equal or unequal, imbricate or rarely valvate (some Sapindaceae) or seldom wanting. Nectary disc of receptacular origin commonly present, extrastaminal or less often (Aceraceae) intrastaminal. Stamens (4)5-10, rarely more; filaments free; anthers tetrasporangiate, introrse, basifixed or more often dorsifixed, opening longitudinally. Tapetum secretory. Microsporogenesis simultaneous. Pollen grains 2-celled or 3-celled mostly 3-colporate. Gynoecium of 3-2(4-5) united carpels; stylodia free or more or less connate; ovary superior, with (1)2(-4-10) ovules per locule. Ovules epitropous or apotropous, anatropous or less often hemitropous, campylotropous or orthotropous, bitegmic, crassinucellate. Female gametophyte of *Polygonum*-type. Endosperm nuclear. Fruits of various types. Seeds exotestal, mesotestal, or rarely (*Alectryon* in Sapindaceae) with fibrous exotegmen; embryo mostly curved; endosperm copious, or very scanty or lacking.

The order Sapindales most probably originated from some cunonialean ancestor. The Staphyleaceae, which are relatively the least advanced member of the order (Takhtajan 1959, 1966, 1987; Cronquist 1981, 1988), have much in common with the Cunoniaceae (and related families) that Hallier (1908, 1912) placed these two families side by side in his order Rosales sensu lato, in which he was later followed by van Steenis (1959). The Staphyleaceae are undoubtedly so closely related to Cunoniaceae and related families (especially to the Brunelliaceae) that there is not much objection to their inclusion in the order Cunoniales. However, it seems to me that the similarities between Staphyleaceae and the sapindalean families are greater than those between Staphyleaceae and Cunoniales. "It is interesting and perhaps significant that the Staphyleaceae have the same basic chromosome number (13) as the Aceraceae," says Cronquist (1981:791). "The characteristic inflated capsules of *Staphylea* are much like those of *Koelreuteria* in the Sapindaceae." In many respects, Staphyleaceae represent a link between the Sapindales and Cunoniales. However, as Doweld (1996) states, exotegmic seeds of Cunoniaceae are more specialized, than exo-mesotestal seeds of Staphyleaceae. It is therefore more probable, that the Staphyleaceae derived not directly from the Cunoniaceae, but rather from their less specialized ancestor.

Key to Families

1 Vessels with scalariform perforation.

2 Intrastaminal nectary disc well developed. Leaves stipulate. Seeds with copious endosperm. Trees and shrubs, evergreen or deciduous, non-laticiferous and without coloured juice; wood often fluorescing; fibers with bordered pits; vessels with scalariform perforation. Druses and frothy mucilage cells widespread in *Dalrympelea*. Axial parenchyma scanty paratracheal. Nodes pentalacunar with five traces. Leaves opposite, with well-developed caducous stipules, imparipinnate or often trifoliolate, occasionally (some species of *Staphylea*) unifoliolate, serrate or dentate; stipules usually present, but sometimes reduced to glands or absent. Stomata anisocytic. Flowers in racemes or in panicles, bisexual or unisexual (plant monoecious, dioecious, or polygamo-monoecious), small, actinomorphic, 5-merous. Sepals more or less free, imbricate; petals imbricate. Stamens five, isomerous with the perianth; filaments glabrous to pubescent; anthers dorsifixed, introrse, tetrasporangiate. Pollen grains 3-colporate or rugate. Carpels 2-3(-4), almost free or partly connate. Ovary superior to semi-inferior, laterally and sometimes also apically lobed, with (1-)6-12 apotropous ovules in two rows in each locule. Fruits multifollicles, inflated capsules opening at the tip, or berries (*Dalrympelea*). Seeds arillate or exarillate, exo-mesotestal; chalaza nonbullate; radicle short; embryo straight, green; endosperm scanty (*Staphylea japonica*) or fleshy; cotyledons large. Present cyanidin, kaempferol and quercetin, n = 11, 13, 14. . . . . 1. STAPHYLEACEAE.

2 Intrastaminal nectary disc poorly developed or wanting. Evergreen trees with reddish wood (*Huertea*) or (*Tapiscia*) deciduous trees; the wood without growth rings. The imperforathe tracheary elements septate fibre-tracheids with reduced borders on the pits, or septate libriform fibers with simple pits. Nodes trilacunar. Leaves 2–10-jugate; leaflets sometimes with pocket-shaped domatia or with hairy domation in axils of and along lateral nerves; bases of petioles, petioluli and blades of the leaflets bearing several small glands. Stipules lacking (*Huertea*) or reduced to glands (*Tapiscia*). Stomata anisocytic. Flowers 5(6)-merous; sepals fused only in low third, copular. Petals covered with multicellular trichomes. Anthers dorsifixed, with (*Tapiscia*) or without a fibrous endothecium. Pollen grains rugate or polyrugate. Nectary disc poorly developed, 5-lobed, or lacking (*Tapiscia*). Ovary 1-locular (*Tapiscia*), style elongate, unbranched, stigma bilobed, or (*Huertea*) basally 2-locular, with two spreading stylar branches; ovules 1–2 in each locule, sessile or attached to an axile placenta. Fruit 1-seeded, without sclerotis tissue, apparently indehiscent (*Tapiscia*) or baccate (*Huertea*), with thin, fleshy or crustaceous pericarp and without sclerotic tissue. Seeds

of *Huertea* subglobose, brownish, smooth, with sclerotic cells in the mesotesta; seed-coat rather thin; vascular bundle spreading into fine rows of tracheids in the bullate chalaza, separated from the endosperm; raphe with hard sclerotic tissue of lignified pitted cells; endosperm thick-walled, oily; embryo with rather long radicle; cotyledons flat; seeds of *Tapiscia* plump, micropype acute at the base of the loculus, the adjacent hilum small, chalazal area slightly depressed; endosperm thin-walled, horny, oily; embryo small with relatively long radicle and small thin cotyledons. . . . . .2. TAPISCIACEAE.

1 Vessels with simple perforation.

3 Leaves alternate or very rarely opposite, often gland-dotted, often pulvinate; usually ternate, or pinnate, or bipinnate, or multiply compound (sometimes biternate), seldom simple; stipules wanting in trees, frequently present in lianas, sometimes conspicuously large. Trees and shrubs, sometimes woody or rarely herbaceous (*Cardiospermum*) lianas, these often with tendrils, which are transformed inflorescence axes, and often their stems with unusual, secondary anomalous vascular structure. Vegetative organs usually contain resinous or latex-like secretions in special cells. Fibers with simple pits, commonly septate. Rays heterogeneous. Axial parenchyma paratracheal, or apotracheal and paratracheal. Nodes trilacunar. Stomata anomocytic or sometimes paracytic. Flowers small, in terminal or axillary cymose inflorescences, rarely solitary and axillary, sometimes cauliflorous in lianas, usually unisexual (plants monoecious or polygamomonoecious), actinomorphic or more often obliquely zygomorphic. Sepals 5(4), free or sometimes basally connate, usually imbricate, sometimes valvate. Petals mostly 5(4), imbricate, sometimes more than 5 or only 3 or wanting, often with basal scalelike appendage concealing a nectary. Nectary disc usually well marked, annular or often unilateral, extrastaminal or rarely (*Dodonaea*) intrastaminal and minute. Stamens 4-5(-6), or 8, or 10, or 11–100; filaments free, often hairy. Anthers dorsifixed, or basifixed, to slightly ventrifixed, more or less versatile, usually apically appendaged, opening longitudinally. Pollen grains 2-celled, usually 3-colporate, sometimes 2-colporate or syncolporate, or rarely 3–4-porate. Gynoecium of (2-)3(-8) united carpels; stylodia more or less free or united into a terminal, often lobed or cleft style; stigma simple or lobed; placentation usually axile, sometimes parietal; ovary usually 3-locular, superior, locules without 'false septa', usually with one ovule, less often (Dodonaeoideae) with two or several ovules per locule. Ovules pendulous or horizontal, or ascending, anatropous to hemitropous or campylotropous, often without a denned funicle, but broadly attached to placental protuberance. Fruits capsules, drupes, schizocarps, or berries, often red, sometimes winged (*Serjania, Thouinia, Paullinia*, and *Diatenopteryx*); capsules septifragal or loculicidal (*Cupania*). Seeds often arillate or with sarcotesta (in dehiscent fruits), usually with curved oily and starchy, often green embryo, often with plicate or twisted cotyledons; endosperm lacking. Present proanthocyanidins, flavonols, toxic saponins, cyclopropane amino acids, sometimes alkaloids, n = 10–16.. .3. SAPINDACEAE.

3 Leaves opposite.

4 Flowers actinomorphic, small, some or all functionally or truly unisexual (andromonoecious, androdioecious, or polygamodioecious), in terminal or sometimes axillary racemes sometimes contracted to corymbs or umbels, rarely in large panicles. Sepals five or rarely four, free or rarely basally connate, imbricate. Petals five, rarely four or wanting, free, often sepal-like, shortly clawed. Nectary disc usually present, extrastaminal, usually flat but sometimes lobed or divided or reduced to teeth, rarely absent. Stamens 4-10(12), mostly 8; anthers slightly dorsifixed or basifixed, versatile. Pollen grains mostly 3-colpate or 3-colporate. Vestigial gynoecium often present in male flowers. Gynoecium of two carpels, stylodia divergent and free or basally connate, each with a terminal stigma; ovary 2-locular, usually compressed at right angels to the septum, with two axile ovules in each locule. Ovules pendulous, anatropous to almost orthotropous. Fruits compressed, splitting when ripe into two usually 1-seeded, nutlike, winged mericarps, sep-

arating from the persistent carpo-phore. Seeds becoming campylotropous; embryo curved, oily or starchy, with elongate radicle and two flat or plicate, green cotyledons; endosperm wanting. Deciduous or evergreen trees or less often shrubs; plants with or without laticifers. Sieve elements plastids with only few, moderately large, globular starch grains. Nodes trilacunar. Leaves simple and entire or more often palmately or pinnately lobed or pinnate (*Dipteronia*, *Acer* sect. *Negundo*). Proanthocyanidins present (cyanidina and delphinidin), flavonols (kaempferol, quercetin, and myricetin), alkaloids and ellagic acid rarely present, n = 13. . . . . 5. ACERACEAE.

4 Flowers zygomorphic, rather large, bisexual (but some apical ones functionally male), in terminal large thyrses or racemes; inflorescences glabrous, or golden tomentose and with bracts (*Billia hippocastanum*). Sepals five, almost free (*Billia*) or more or less connate (*Aesculus*), campanulate or tubular, imbricate. Petals four or five, free, imbricate, clawed. Nectary disc small, extrastaminal and often one-sided. Stamens (5)6-8, free; the filaments red or white; anthers introrse, dorsifixed (near the base), versatile. Pollen grains 3-colporate. Gynoecium of (2)3(4) united carpels; style terminal, often pubescent, with simple or obscurely lobed, papillate stigma; ovary usually 3-locular, superior, sessile, with two superposed, axile ovules per locule. Ovules pendulous to ascending (sometimes the lower ascending, the upper pendulous), anatropous to orthotropous. Fruits usually large, leathery, 1-locular and 1-seeded (often, by abortion), or 2–5-seeded loculicidal capsules. Seeds large, with a large hilum reflecting incorporation of the funicle in the placenta, and adnation of the placental obturator; embryo large, curved, green, often starchy; cotyledons two, thick and fleshy; endosperm wanting. Vessels usually with simple perforations, but occasional scalariform perforations also recorded. Leaves palmately 3–11-foliolate. Nodes trilacunar or sometimes pentalacunar. Evergreen (*Billia*) or deciduous (*Aesculus*) trees or less often shrubs; branches often with yellowish lenticels. Leaves large to medium-sized, compound, palmate (3–11 foliolate), crenate or serrate, estipulate. Proanthocyanidins (cyanidin), flavonols (kaempferon and quercetin) present, n = 20. . . . 4. HIPPOCASTANACEAE.

1. STAPHYLEACEAE

Martynov 1820 (including Ochranthaceae Endlicher 1841). 2/c.40. Northern temperate regions, tropical Asia, West Indies, Central and South America. *Euscaphis* (1) is endemic to Eastasiatic floristic region, *Staphylea* (11) distributed in northern temperate regions, and *Turpinia* (10) from Sri Lanka, peninsular India, and the Himalayas to Japan and New Guinea and in Central and South America.

Dalrympelea (including *Ochranthe*), *Staphylea* (including, *Turpinia, Euscaphis*).

2. TAPISCIACEAE

Takhtajan 1987 (including Huerteaceae Doweld 2001). 2/7. Central and southwestern China (*Tapiscia*), Greater Antilles, trop. America along the Andes from Honduras to Colombia, Ecuador, and Peru (*Huertea*).

Tapiscia, Huertea.

Tapisciaceae are close related to the Staphyleaceae and are included in that family. However, they differ from the Staphyleaceae in so many important characters that their separation into a family of their own is justified. One of the most important differences between them are well developed intrastaminal disc and the seed structure.

3. SAPINDACEAE

A. L. de Jussieu 1789 (including Allophylaceae Martynov 1820, Dodonaeaceae Link 1831, Koelreuteriaceae J. Agardh 1858, Ornithrophaceae Martynov 1820). 130/1500. Widely distributed in tropical and subtropical regions; relatively few species in warm-temperate zones.

3.1 DODONAEOIDEAE

Ovules usually two or several per locule, in the first case erect or pendulous, in the second case horizontal, rarely one pendulous with micropyle up. – Dodonaeeae: *Loxodiscus, Diplopeltis, Dodonaea, Distichostemon;*

Doratoxyleae: *Hypelate, Exothea, Zanha, Doratoxylon, Ganophyllum*, etc.; Harpullieae: *Harpullia, Delavaya, Ungnadia, Eurycorymbus*, etc.; Cossignieae: *Cossinia, Llagwioa*; Koelreuterieae: *Boniodendron, Koelreuteria, Stocksia, Erythrophysa.*

3.2 XANTHOCEROIDEAE
Ovules 7–8 in each carpel; disc with golden, horn-like glands; leaves deciduous. – *Xanthoceras.*

3.3 SAPINDOIDEAE
Ovules solitary in each locule, erect or ascending, micropyle down. – MELICOCCEAE: *Melicoccus, Talma*, etc.; LEPISANTHEAE: *Zollingeria, Lepisanthes, Otophora, Radlkofera, Pancovia, Placodiscus*, etc.; SAPINDEAE: *Atalaya, Thouinidium, Toulicia, Sapindus, Delnbollla, Homea*, etc.; APHANIEAE: *Aphania, Erioglossum*; CUPANIEAE: *Cupania, Matayba, Tina, Mollnaea, Blighia, Guioa, Cupaniopsis, Rhysotoechia, Elattostachys, Arytera, Mischocarpus, Paviaesia, Amesiodendron, Lepidopetalum, Paranephelium*, etc.: SCHLEICHEREAE: *Schleichera, Haplocoelum, Macphersonia*, etc.; NEPHELIEAE: *Litchi, Dimocarpus, Pometia, Xerospermum, Nephelium, Alectryon, Pappea*, etc.; THOUINIEAE: *Thouinia, Allophylus*, etc.; PAULLINIEAE: *Serjania, Paullinia, Urvillea, Cardiospermum, Houssayanthus, Thinouia.*

Together with morphological characters, chemical data show that the Dodonaeoideae are in general less advanced and nearer to the ancestral stock than the Sapindoideae. The genus *Dodonaea* is unique in not containing elaborate flavones 0- and C-glycosides (Umadevi and Daniel 1991).

The Sapindaceae are considerably more advanced than the Staphyleaceae

4. HIPPOCASTANACEAE

A. Richard 1823 (including Aesculaceae Berchtold et J. Presl 1820, Paviaceae Horaninow 1834). 2/15. Balkan Peninsula, from western Himalayas to Japan and northern Indochina, temperate North America (*Aesculus*), and from southern Mexico to northern South America (*Billia*).

Aesculus, Billia.

So closely related to Sapindaceae, especially to the tribe Harpullieae (see Radlkofer 1890, 1931–1944; Muller and Leenhouts 1976; Doweld 1996), that perhaps even do not deserve the family rank. According to Muller and Leenhouts, Hippocastanaceae may be connected with the Harpullieae via *Handellodendron* and *Delavaya*, and possibly could be included in the Harpullieae.

5. ACERACEAE

A. L. de Jussieu 1789. 2/120–150 or more. Northern temperate regions and Southeast Asia. *Dipteronia* (2) is endemic to central and southern China.

Acer, Dipteronia.

Aceraceae are closely related to the Sapindaceae, especially to the tribe Harpullieae (see Radlkofer 1890, 1931–1934; Muller and Leenhouts 1976). Beginning with Bentham and Hooker (1862) and some authors included the Aceraceae in the Sapindaceae.

Bibliography

Acevedo-Rodriquez P. 2003. Meliococceae (Sapindaceae) – *Melicoccus* and *Talisia*. Flora Neotropica Monogr. 87: 1–179.

Adema F, PW Leenhouts, and PC van Welzen. 1994. Sapindaceae. In: CGGJ van Steenis, ed. Flora Malesiana ser. 1, 11: 419–768. Noordhoff, Groningen.

Agarwal M, S Gupta, and V Painuly. 2005. Xyltomic study of the family Sapindaceae: Microstructure, systematics and ecological trends. Indian Forester 131: 1024–1040.

Beck HT. 2004. Sapindaceae. In: N Smith, SA Mori, A Henderson, DW Stevenson, and SV Heald, eds. Flowering plants of the Neotropics, pp. 339–341. The New York Botanical Garden, Princeton University Press, Princeton.

Brizicky GK. 1963. The genera of Sapindales in the southeastern United States. J. Arnold Arbor. 44: 62–501.

Carlquist S and DA Hoekman 1985. Wood anatomy of Staphyleaceae: Ecology, statistical correlations, and systematics. Flora 177: 95–216.

Celakovsky LJ. 1899. Über achtzühlige Cyclen pentamer veranlagter Blüten. Prings. Jahrb. 33: 368–416.

Dahlgren R and AE van Wyk. 1988. Structures and relationships of families endemic to or centered in Southern Africa. Monogr. Syst. Bot. Missouri Bot. Gard. 25: 1–94.

Dickison WC. 1986. Floral morphology and anatomy of Staphyleaceae. Bot. Gaz. 147: 312–326.

Dickison WC. 1987a. Leaf and nodal anatomy and systematics of Staphyleaceae. Bot. Gaz. 148: 475–489.

Dickison WC. 1987b. A palynological study of the Staphyleaceae. Grana 26: 11–24.

Doweld AB. 1996a. Staphyleaceae. In: A Takhtajan, ed. Comparative seed anatomy, vol. 5, pp. 312–315. Nauka, St. Petersburg (in Russian).

Doweld AB. 1996b. Tapisciaceae. In: A Takhtajan, ed. Comparative seed anatomy, vol. 5, pp. 315–317. Nauka, St. Petersburg (in Russian).

Doweld AB. 1996c. Sapindaceae. In: A Takhtajan, ed. Comparative seed anatomy, vol. 5, pp. 317–339. Nauka, St. Petersburg (in Russian).

Doweld AB. 1996d. Hippocastanaceae. In: A Takhtajan, ed. Comparative seed anatomy, vol. 5, pp. 343–346. Nauka, St. Petersburg (in Russian).

Ferrucci MS and LM Anzótegui 1993. Et pollen de Paullinieae tribe (Sapindaceae). Bonplandia 6: 211–243.

Gadek PA, ES Fernando, CJ Cuinn, SB Hoot, T Terrazas, MC Sheahan and MW Case. 1996. Sapindales, molecular delimitation and infraordinal groups. Am. J. Bot. 83: 802–811.

Gelderen DM van, PC de Jong, and HJ Oterdoom. 1994. Maples of the world. Timber Press, Portland, OR.

Gornall RJ, BA Bohm, and R Dahlgren. 1979. The distribution of flavonoids in the angiosperms. Bot. Not. 132: 1–30.

Gulati N and S Mather. 1977. Embryology and taxonomy of *Filicium decipiens* THW. Phytomorphology 27: 261–266.

Hall BA. 1951. The floral anatomy of the genus *Acer*. Am. J. Bot. 38: 793–799.

Hall BA. 1961. The floral anatomy of *Dipteronia*. Am. J. Bot. 48: 918–924.

Ham RWJM van der. 1990. Nephelieae pollen (Sapindaceae): Form, function, and evolution. Leiden Bot. Ser. 13: 1–255.

Ham RWJM van der and A Tomlik. 1994. *Serjania* pollen and the origin of the tribe Paullinieae (Sapindaceae). Rev. Palaeobot. Palynol. 83: 43–53.

Hardin JW. 1957. A revision of the American Hippocastanaceae. Brittonia 9: 145–195.

Hardin JW. 1960. Studies in the Hippocastanaceae. V. Species of the Old World. Brittonia 12: 26–38.

Heimsch C. 1942. Comparative anatomy of the secondary xylem in the "Gruinales" and "Terebinthales" of Wettstein with reference to taxonomic grouping. Lilloa 8: 83–198.

Hideux MJ and IK Ferguson. 1976. The stereostructure of the exine and its evolutionary significance in Saxifragaceae sensu lato. In: IK Ferguson and J Muller, eds. The evolutionary significance of the exine, pp. 327–337. Linn. Soc. Symposium, No. 1. London/New York.

Klaassen R. 1999. Wood anatomy of the Sapindaceae. IAWA J. Suppl. 2: 1–214.

Kubitzki K. 2003. Tapisciaceae. In: K Kubitzki, ed. The families and genera of vascular plants, vol. 5, pp. 369–370. Springer, Berlin/Heidelberg/New York.

Linden BL van der. 1960. Staphyleaceae. In: CGGJ van Steenis, ed. Flora Malesiana ser. 1, 6: 49–59. Noordhoff, Groningen.

Lombello RA and ER Forni Martins. 1998. Chromosomal studies and evolution in *Sapindaceae*. Caryologia. 51): 81–93.

Mauritzon J. 1936. Zur Embryologie und systematischen Abgrenzung der Reihen Terbinthales and Celastrales. Bot. Not. 1936: 161–212.

Muller J and PW Leenhouts. 1976. A general survey of pollen types in Sapindaceae in relation to taxonomy. In: IK Ferguson and J Muller, eds. The evolutionary significance of the exine, pp. 407–445. Linn. Soc. Symposium, No 1. London/ New York.

Narayana LL. 1960. Embryology of Staphyleaceae. Curr. Sci. 10: 403–404.

Pozhidaev AE. 1993. Polymorphism of pollen in the genus *Acer* (Aceraceae). Isomorphism of deviate pollen forms of angiosperm plant. Grana 32: 79–85.

Pozhidaev AE. 1995. Pollen morphology of the genus *Aesculus* (Hippocastanaceae). Grana 34: 10–20.

Radlkofer L. 1890. Über die Gliederung der Familie der Sapindaceen. Sitzungsber. Bayer. Akad. Wiss. (M.-Ph. Kl.) 20: 105–379.

Radlkofer L. 1931–1934. Sapindaceae. In: A Engler ed. Das Pflanzenreich, I, v: 165. Leipzig.

Ramp E. 1987. Funktionelle Anatomic des Gynoeceums bei *Staphylea*. Bot. Helvet. 97: 89–98.

Ronse Decraene LP, E Smets, and D Clinckemaillie. 2000. Floral ontogeny and anatomy in *Koelreuteria* with special emphasis on monosymmetry and septal cavities. Plant Syst. Evol. 223: 91–107.

Ronse Decraene LP, HP Linder, T Dlamini, and E Smets. 2001. Evolution and development of floral diversity of Melianthaceae, an enigmatic Southern African family. Int. J. Plant Sci. 162: 69–82.

Shabes LK and AA Morozova. 1996. Aceraceae. In: A Takhtajan, ed. Comparative seed anatomy, vol. 5, pp. 339–343. Nauka, St. Petersburg (in Russian).

Simmons S.L. 2006. Staphyleaceae. In: K Kubitzki, ed. The families and genera of vascular plants, vol. 9, pp. 440–445. Springer, Berlin/Heidelberg/New York.

Simmons SL and JL Panero. 2000. Phylogeny and biogeography of Staphyleaceae (DC.) Lindl. Am. J. Bot. 87(Suppl. 6): 157.

Solereder H. 1892. Über die Staphyleaceengattung *Tupi-scia*. Ber. Deutsch. Bot. Ges. 10: 545–551.

Spongber S. 1971. The Staphyleaceae in the southeastern United States. J. Arnold Arbor. 52: 196–203.

Tanai T. 1978. Taxonomical investigation of the living species of the genus *Acer* L., based on vein architecture of leaves. J. Fac. Sci., Hokkaido Imp. Univ., Ser. 5, Bot. 18: 243–282.

Tian X, QJ Jin, DZ Li, ZX Wei, and TZ Xu. 2001. Pollen morphology of Aceraceae and its systematic implication. Acta Bot. Yunn. 23: 457–465.

Umadevi I and M Daniel. 1991. Chemosystematics of the Sapindaceae. Feddes Repert. 102: 607–612.

Umadevi I, M Daniel, and SD Sabnis. 1986. Interrelationships among the families Aceraceae, Hippocastanaceae, Melianthaceae, Staphyleaceae. J. Plant Anat. Morph. 3: 169–172.

Vuillemin P. 1915. Differences essentielles entre la Capucine et les Geraniacees. C. R. Acad. Sci. Paris 161: 297–301.

Weberling F. 1976. Die Pseudostipein der Sapindaceae. Akad. Wiss. Abh. Math.-Naturwiss. Kl. 2 1976: 1–27.

Weberling F and PW Leenhouts. 1965. Systematisch-morphologische Studien an Terebinthales-Familien (Burseraceae, Simaroubaceae, Meliaceae, Anacardia-ceae, Sapindaceae). Akad. Wiss. Abh. Math.- Natur-wiss. Kl. 10: 495–584.

Weckerle CS and R Rutishauser. 2005. Gynoecium, fruit and seed structure of Paullinieae (Sapindaceae). Bot. J. Linn. Soc. 147: 159–189.

Wolfe JA and T Tanai. 1987. Systematics, phylogeny, and distribution of *Acer* (maples) in Cenozoic of western North America. J. Fac. Sci., Hokkaido Imp. Univ. 22(1): 1–246.

Order 92. SABIALES

Evergreen or rarely deciduous trees, scandent shrubs or woody climbers, glabrous or pubescent, very rarely armed with short spines (*Sabia japonica*). Vessels with simple, simple and scalariform, or occasionally reticulate perforations; lateral pitting alternate. Bars few to 30; wood with broad rays Fibers with simple or very small bordered pits, sometimes septate. Rays heterogeneous or nearly homogeneous. Aixal parenchyma paratracheal, scanty or rarely wanting. Sieve element plastids of S-type, or rarely (*Meliosma glabrata*) Pcs-type. Nodes unilacunar (*Meliosma*). Leaves alternate, simple (*Sabia*) or imparipinnate, conduplicate (*Meliosma*), sometimes heteromorphic, entire or dentate, pinnately veined, estipulate; petiole bases often subwoody, petiolules often with pulvini. Stomata anomocytic or paracytic. Flowers small, in terminal or axillary panicles or cymes, rarely solitary, bisexual or polygamo-dioecious, actinomorphic or zygomorphic. Sepals (4)5, sometimes basally connate, unequal, imbricate. Petals (4)5; the three outer petals slightly imbricate and completely enclose the stamens and gynoecium; the two inner petals strongly reduced and the only two functional stamens are adnate to them. Stamens (4)5, opposite the petals, all fertile or the three outer staminodial, swollen or collarlike extension formed by connective or filaments present; anthers tetrasporangiate, opening longitudinally. Tapetum secretory. Microsporogenesis simultaneous. Pollen grains 2-celled, 3-colporate. Nectary disc small, annular or cupular, sometimes dentate or lobed, surrounding the base of the ovary. Gynoecium of two or rarely three united carpels; stylodia more or less connate into a subulate style, with minute, rounded or capitate stigma; ovary superior, 2- or rarely 3-locular, with one or two pendulous or horizontal ovules in each locule. Ovules apotropous, unitegmic, crassinucellate. As the integument fails to grow over the dome-shaped nucellus, no micropyle is formed. Female gametophyte of *Polygonum*-type. Endosperm helobial. Fruits drupes or schizocarps breaking transversely into two drupaceous or dry 1-seeded mericarps, sometimes one aborted. Seed with condyle (placental intrusion); embryo large, oily, with a curved radicle and folded or coiled (*Ophiocaryon*) cotyledons; endosperm scanty or more often wanting. Present pentacyclic triterpenoids, tanniniferous alkaloids, n = 12, 16.

Mostly included in the Sapindales, from which Sabiales differ in unitegmic ovules, helobial endosperm, and morphology of the androecium and seeds. Radlkofer (1890), following earlier authors, suggested the affinity with the Menispermaceae rather than the Sapindaceae. Erdtman (1952) pointed out the similarity of pollen grains of Sabiaceae to those of Menispermaceae. Cronquist (1957, 1981, 1988) tentatively included Sabiaceae in the Ranunculales. Recent molecular studies placed Sabiaceae among the "Early-Diverging Eudicots" (Soltis et al. 2006) between the Proteaceae and Buxaceae, but these families, however, are considerably different morphologically.

Taxonomic position of Sabiales is still uncertain.

1. SABIACEAE

Blume 1851 (including Meliosmaceae Endlicher 1841, Wellingtoniaceae Meisner 1840). 3/50. Himalayas, southern Asia, and Southeast Asia, Malesia eastwards to Solomon Islands – *Sabia* (c.20); *Meliosma* (15 sp. in Asia, 10 – in the Western Hemisphere); *Ophiocaryon* (7) is endemic to tropical America (north to Mexico).

1.1 MELIOSMOIDEAE
Evergreen trees and shrubs; winterbuds naked. Flowers zygomorphic. Sepals subequal. Stamens very unequal, free or adnate to the base of petals, two inner and larger fertile and their anthers mostly surrounded at base by the cup-shaped apex of the short filaments, the outer ones sterile and reduced to cup-shaped staminodia; Fruits obliquely subglobose, 1-seeded drupe with bony endocarp (rarely 2-locular), n = 8. – *Meliosma, Ophiocaryon.*

1.2 SABIOIDEAE
Deciduous or evergreen climbing shrubs, more rarely erect; winterbuds small, with pointed persistent scales. Flowers actinomorphic. Stamens (4)5(6), all fertile, attached to the base of petals; filaments subulate. Fruits of two flattened, dorsally gibbous, drupaceous carpels, stylodia becoming adaxially subbasal and persistent; exocarp fleshy, endocarp crustaceous and conspicuously sculptured or pitted, n = 13. – *Sabia.*

Bibliography

Beusekom CF van and TPM van de Water. 1989. Sabiaceae. In: CGGJ van Steenis, ed. Flora Malesiana ser. 1, 10: 679–715. Noordhoff, Groningen.
Carlquist S, PL Morrell, and SR Manchester. 1993. Wood anatomy of Sabiaceae (s.l.): ecological and systematic implications. Aliso 13: 521–549.
Kubitzki K. 2004. Sabiaceae. In: N Smith, SA Mori, A Henderson, DW Stevenson, and SV Heald, eds. Flowering plants of the Neotropics, pp. 335–336. The New York Botanical Garden, Princeton University Press, Princeton.
Kubitzki K. 2007. Sabiaceae. In: K Kubitzki, ed. The families and genera of vascular plants, vol. 9, pp. 413–417. Springer, Berlin/Heidelberg/New York.
Le Renard A. 1908. Recherches anatomiques sur la tige et la famille des Sabiacees. J. Bot. 21: 290–332.
Mauritzon J. 1936. Zur Embryologie und systematischen Abgrenzung der Reihen Terebinthales un Celastrales. Bot. Not. 1936: 161–212.
Mondal MS. 1990. Pollen morphology and systematic relationships of families Sabiaceae (s.l.) and Connaraceae. Botanical Survery of India, New Delhi.
Radlkofer L. 1890. Über die Gliederung der Familie der Sapindaceen. Sitzungsber. Bayer. Akad. Wiss. (M.-Ph. Kl.) 20: 105–379.
Raju MVS. 1952. Embryology of Sabiaceae. Curr. Sci. 21 (4): 107–108.
Sharanina EA. 1996. Sabiaceae. In: A Takhtajan, ed. Comparative seed anatomy, vol. 5, pp. 356–359. Nauka, St. Petersburg (in Russian).
Thanikaimoni G. 1984. Menispermacees: Palynologie et systematique. Inst. Fr. Pondichtery, Trav. Sect. Sci. Tech. 13: 1–135.
Water Th PM van de. 1980. A taxonomic revision of the genus *Sabia* (Sabiaceae). Blumea 26: 1–64.

Order 93. BIEBERSTEINIALES

Perennial herbs, occasionally nearly stemless, with more or less tuberous, underground rhizome. Trichomes of glandular shaggy type with long, multiseriate stalks and multicellular knob-shaped heads. Vessels with simple perforation; Lateral piting alternate, at the end of vessels both mixed alternate or opposite. Rays gomogenous. Axial parenchyma apotracheal. Leaves alternate, pinnate or pinnatipartite, with the stipules adnate to the petiole. Stomata anomocytic. Flowers in pedunculate racemose inflorescences, or in panicles, bisexual, actinomorphic, 2-bracteolate, 5-merous. Sepals free, imbricate, persistent. Petals free, imbricate or sometimes contorted, often denticulate at the apex, sometimes unguiculate, alternate with five fleshy extrastaminal glands. Stamens ten; filaments shortly connate into a ring at the base. Anthers dorsifixed, versatile, introrse, tetrasporangiate, opening longitudinally. Staminodes five, external to the fertile stamens, represented by a whorl of glands, alternating with the petals. Tapetum secretory. Microsporogenesis simultaneous. Pollen grains 3-colporate, tectate-columellate, with striate ornamentation. Gynoecium of five carpels; ovary superior, 5-locular, on a short gynophore, deeply lobed; stylodia arising from the base of the ovary lobes, free, filiform, afterward cohering among themselves in a slender column with capitate stigma. Ovule solitary in each locule, apical, pendulous, anacampylotropous, bitegmic, crassinucellate. Female gametophyte tetrasporic, 16-nucleate, 13-celled, of *Penaea*-type. Endosperm nuclear. Fruits separating into five indehiscent 1-seeded, dry, rugose, crustaceous mericarps. Seeds large, with rounded raphe, rugulose; testa thin-walled, more or less collapsed, tegmen composed of thick-walled, strongly lignified, scarcely pitted cells with outward, radially directed arms; embryo slightly curved, surrounded by scanty endosperm; cotyledons foliaceous, thick (Boesewinkel and Bouman 2000). Present flavones ad methyl ethers, n = 5.

Very few botanists accept a separate family Biebersteiniaceae and the genus *Biebersteinia* is usually included in the Geraniaceae. However, *Biebersteinia* differs from all the geranialean families in its tetrasporic female gametophyte, the absence of suspensor (Kamelina and Konnova 1990), low number of relatively large chromosomes (Aryavand 1975), seed coat anatomy, morphology of pollen grains (Bortenschlager 1967), and gynobasic stylodia, and as well as in the flavonoid patterns (Greenham et al. 2001). The structure of tegmen differs from the Geraniales, and resembles that of Linales (Boesewinkel and Bouman 2000). However, molecular studies reveal close affinity with the Sapindales (Bakker et al. 1998). In my opinion the Biebersteiniaceae deserve the status of a separate order Biebersteiniales that occupies a rather isolated position (Takhtajan 1997).

1. BIEBERSTEINIACEAE

Endlicher 1841. 1/4–5. From Greece to western Siberia and western Tibet.

Biebersteinia.

Bibliography

Aryavand A. 1975. Contribution a l´ετυδε cytotaxonomique de *Biebersteinia multifida* DC. (Geraniacees). Compt. Rend. Sci. Paris. 280: 1551–1554.

Bakker ET, DD Vassillades, C Morton, and V Savolainen. 1998. Phylogenetic relationships of *Biebersteinia* Stephan (Geraniaceae) inferred from *rbc*L and *atp*B sequence comparisons. Bot. J. Linn. Soc. 127: 149–159.

Boesewinkel FD and F Bouman. 2000. Biebersteiniaceae. In: A Takhtajan, ed. Comparative seed anatomy, vol. 6, pp. 28–29. Nauka, St. Petersburg (in Russian).

Bortenschlager S. 1967. Vorlaufige Mitteilungen zur Pollenmorphologie in der Familie der Geraniaceen und ihre systematische Bedeutung. Grana Palynol. 7: 400–468.

Greenham J, DD Vassiliades, JB Harborne, CA Williams, J Eagles, RJ Grayer, and NC Veitch. 2001. A distinctive flavonoid chemistry for the anomalous genus Biebersteinia. Phytochemistry 56: 87–91.

Kamelina OP and VA. Konnova. 1990. Embryological characters of the genus *Biebersteinia* Steph. in relation to its systematic position. Doklady Acad. Sci. Tajik SSR. 33 (3): 193–195.

Liu JQ, TN Ho, SL Chen, and AM Lu. 2001. Karyomorphology of *Biebersteinia* Stephan (Geraniaceae) and its systematic and taxonomic significance. Bot. Bull. Acad. Sin. (Taipei). 42: 61–66.

Shen S and R Huang. 1997. Cytological and morpho-anatomical studies of *Biebersteinia heterostemon* Maxim. Acta Biol. Plat. Sinica. 13: 5–8.

Tzakou O, A Yannitsaros, and DD Vassiliades. 2001. Investigation of the C16:3/C18:3 fatty acid balance in leaf tissues of *Biebersteinia orphanidis* Boiss. (Biebersteiniaceae). Biochem. Syst. Ecol. 29: 765–767.

Tutel B. 1984. Comparison on the taxonomy and leaf anatomy of the genus *Biebersteinia* with the other genera of Geraniaceae in Turkey. Istanbul Univ. Fen Fak. Mecm., B, 47–48: 51–87.

Order 94. RUTALES

Trees, shrubs, woody lianas, sometimes subshrubs, less often herbs. Resinous secretory cells or/and cavities containing ethereal oils or vertical resin canals often present in vegetative parts. Vessels with simple or less often scalariform perforations. Fibers with simple or indistinctly bordered pits, often septate. Rays heterogeneous to homogeneous. Axial parenchyma mostly paratracheal. Sieve-element plastids of Ss- or (in some Rutaceae) So-, or (***Stylobasium***) Pfs-type. Nodes trilacunar or less often unilacunar, rarely multilacunar. Leaves alternate or less often opposite, compound or less often simple, mostly estipulate. Stomata of diverse types, mostly anomocytic. Flowers mostly bisexual, commonly actinomorphic, generally 5- or 4-merous, with double perianth and bicyclic androecium. Anthers tetrasporangiate, introrse or sometimes extrorse, mostly dorsifixed, opening longitudinally. Tapetum secretory. Microsporogenesis simultaneous. Pollen grains commonly 2-celled, usually 3-colporate. Intrastaminal or extrastaminal, annular or sometimes cupular, nectary disc typically present. Gynoecium of free or more often united carpels, mostly with more or less connate stylodia; ovary superior or inferior, with two or less often one or several ovules per carpel or per locule. Ovules mostly anatropous, epitropous, or, less often, apotropous (Anacardiaceae and some Simaroubaceae), bitegmic or sometimes unitegmic, crassinucellate. Female gametophyte of ***Polygonum***-type. Endosperm nuclear. Fruits of diverse types. Seeds testal with fibrous exotegmen (Meliaceae and some representatives of Rutaceae) or predominantly testal; embryo straight or more or less curved; endosperm scanty or none.

Closely related to the Sapindales and share a common origin with them.

Key to Families

1 Plants resinous or not, but without resin-ducts in the bark, wood-rays and veins of the leaves; ovules with or more often without an obturator; plants very often producing triterpenoid substances and/or alkaloids, but only seldom saponiferous.

2 Leaves pellucid-punctate. Plants with secretory cavities scattered through the parenchymatous tissues. Aromatic trees or shrubs, sometimes lianous, rarely subshrubs or perennial herbs, sometimes thorny (*Zanthoxylum*). Secondary growth of stem normal. Secretory cavities are nearly always scattered through the parenchymatous tissue of both leaf and axis and usually also in the pericarp (in the leaf they appear to the naked eye as translucent pellucid dots). The secretory cavities are sometimes replaced or accompanied by secretory cells. Resin cells are common in the pith, primary cortex, and rays of the young stem. Vessels with simple perforations except for rare multiperforate plates in a few species; lateral pitting alternate. Fibers with simple or slightly bordered pits, with occasional septa in a few species. Rays homogeneous or more or less heterogeneous. Axial parenchyma terminal and paratracheal or less often diffuse, rarely

wanting. Nodes trilacunar or sometimes unilacunar. Leaves well developed or sometimes much reduced, usually pellucid-punctate (except for *Leptothyrsa* and *Phellodendron*), alternate or less often opposite, rarely verticillate, usually pinnately compound (sometimes bipinnate) or trifoliolate or unifoliolate, rarely simple or pinnatisect; estipules or rarely stipulate. Stomata of various types. Flowers in cymose inflorescences or less often in racemes, sometimes as in *Diplolaena,* in a dense head with a 3–4-seriate involucre of bracts, the inner of which are petaloid, sometimes solitary, bisexual or rarely unisexual, actinomorphic or sometimes more or less zygomorphic. Sepals (2-)5, free or sometimes more or less connate (connate into a cupular calyx in *Correa* and *Empleurum*), usually imbricate, rarely valvate (e.g. *Boronia, Correa*) or obsolete (*Asterolasia*). Petals as many as and alternate with the sepals, mostly free, sometimes basally connate, rarely (as in *Correa*) united into a sympetalous corolla, imbricate or sometimes valvate, rarely (*Empleurum*) wanting. Androecium mostly obdiplostemonous, rarely diplostemonous or stamens in one cycle and antesepalous; often antepetalous stamens transformed into staminodia; stamens sometimes 3–4 as many as the petals, or even up to 60, rarely only two or three stamens fertile and the others staminodial; filaments free or more or less connate, sometimes adnate to petals, or (*Zieria*) inserted above the disc; anthers 2-locular, basifixed or dorsifixed, introrse, or rarely (*Dictamnus*) latrorse, the connective often with a glandular apex. Pollen grains (2)3–6(-8)-colporate. Nectary disc intrastaminal, annular to cuplike, sometimes unilateral, sometimes elongated into a more or less developed gynophore, rarely obsolete. Gynoecium of (2)4-5(-many) carpels, mostly only 7, slightly united at the base or united only by the stylodia or stigmas, less often completely united and with a simple terminal or sometimes subgynobasic or gynobasic style, very rarely gynoecium reduced to a single carpel; ovary superior, 2-multilocular, rarely (as in *Feronia*) unilocular, with typically parietal placentas, with 2-many or only 1 ovule in each locule. Ovules more or less distinctive epitropous with ventral raphe and upwardly and outwardly directed micropyle, anatropous, hemitropous, or campylotropous, pendulous or ascending, bitegmic or rarely (*Glycosmis arborea*) unitegmic; micropyle zig-zag. Fruits berries (often in the form of a hesperidium), or drupaceous, or samaras or follicles. Seeds with large, straight, or curved, sometimes green embryo, more or less well-developed endosperm or without endosperm; exotesta often mucilaginous, or lignified and fibrous. Usually producing triterpenoid bitter substances and usually with one or more alkaloids of diverse types, proanthocyanidins (cyanidin and delphinidin) sometimes flavonols (kaempferol, quercetin, and myricetin), pyranochromones, furanocoumarins and polyacetate acids, n = 7–11, 18 +. . . . . 1. RUTACEAE.

2 Leaves not pellucid-punctate. Plant without secretory cavities in the parenchymatous tissues.

3 Stamens free.

4 Seeds with copious endosperm. Small, evergreen shrubs containing scattered secretory cells with oily or resinous contents in the parenchymatous tissues in the mesophyll and primary cortex. Stems glabrous or with medifixed hairs. Vessels with simple perforations. Fibers very short, with numerous simple pits. Rays heterogeneous to homogeneous. Axial parenchyma predominantly paratracheal. Leaves small, alternate, simple, entire, gray-green, not punctuate, coriaceous, estipulate. Stomata anomocytic. Flowers solitary and axillary or in small, few-flowered, axillary cymes (the peduncle sometimes adnate to the petiole), actinomorphic, bisexual or unisexual (the plant andromonoecious), yellow. Sepals three or occasionally four (Traveset 1995), small, free or sometimes basally connate, persistent. Petals three or four, elongate, free, imbricate, yellow. Nectary disc modified into a columnar nectariferous androgynophore. Stamens as many as and alternate with the petals, seated in the pits in the androgynophore; filaments free; anthers 2-locular, opening longitudinally. Pollen grains 3-celled, 3- or 4–6-colporate, with a striate-reticulate ornamentation. Gynoecium of three or four united carpels with a terminal style and lobed stigmas; ovary superior,

3–4-locular, with (1)2 pendulous collateral ovules per carpel, when two ovules, they are more or less separated by partition from the carpellary midrib. Ovules epitropous, with ventral raphe, amphitropous, bitegmic. Fruits red schizocarpous, each mericarp drupaceous, with one or two seeds. Seeds with strongly curved embryo and copious endosperm. Plants producing coumarins and triterpenoid bitter substances, n = 9. . . . . 2. CNEORACEAE.

4 Seeds without endosperm or with scanty endosperm.

5 Stylodia connate into terminal style. Flowers with a nectary disc.

6 Stamens 4–5.

7 Stigma lobed. Evergreen or deciduous shrubs or trees. Secretory cells occur in the leaves. Vessels with simple perforations; lateral pitting alternate. Fibers with simple pits, septate. Leaves alternate and imparipinnate (*Cedrelopsis*) or opposite and paripinnate (*Ptaeroxylon*), estipulate, leaflets often with the base asymmetric, entire. Stomata anomocytic or cyclocytic. Flowers in axillary multiflorous inflorescences, polygamous (*Cedrelopsis*) or dioecious (*Ptaeroxylon*), actinomorphic. Sepals four or five, basally slightly connate, imbricate or open. Petals four or five, free, recurved, much longer than the sepals, imbricate or valvate. Stamens four or five, free, alternipetalous, reduced to staminodia in female flowers; anthers dorsifixed, versatile, introrse. Nectary disc well developed, intrastaminal, fused to the base of the ovary to form a thick gynophore in female flowers. Gynoecium of 3–5 (*Cedrelopsis*) or 2 (*Ptaeroxylon*) united carpels, with shortly connate stylodia; stigma 2–5-lobed; ovary 2–5-locular, with one or two ovules per locule. Ovules hypotropous, campylotropous, or subcampylotropous, bitegmic. Fruits loculicidal capsules dehiscing from the central column. Seeds flattened with a long terminal wing; testa papery; embryo bent, with fleshy, oily cotyledons; endosperm wanting. Producing remarkable range of chromones and some unusual coumarins, but not simarouboides, limonoids, or alkaloids. . . . . 9. PTAEROXYLACEAE.

7 Stigma punctiform. Monoecious trees or shrubs. Branchlets marked clearly with scars of old leaves, often viscid around tips. Vessels with simple perforations; lateral pitting alternate. Fibers with simple pores, septate. Rays heterogeneous. Axial parenchyma vasicentric. The wood of *Pleiokirkia* is reported to smell like honey (Schatz 2001). Leaves alternate or more or less opposite, mostly crowded at ends of shoots, imparipinnate, leaflets mostly opposite, toothed, estipulate; Stomata paracytic (?). Flowers in axillar dichasia, polygamous or unisexual, actinomorphic, 4-merous. Sepals 4, very shortly united at base, deltoid. Petals four, valvate or imbricate, much exceeding calyx. Stamens four, alternipetalous, inserted outside and beneath disc, in female flowers much reduced; filaments slender; anthers dorsifixed introrse. Pollen grains 3-colporate, syncolpate. Nectary disc intrastaminal, annular, usually fleshy, quadrangular. Gynoecium of 4–8 united carpels; stylodia slender, coherent; stigmas coherent, capitate. Ovary partly immersed in disc, small, ampuliform, lobed, 4–8-locular, with 1–2 small, pendulous ovule in each locule. Fruits woody, prismatic, angled, splitting longitudinally into linear-oblong, dorsally compressed, indehiscent, 1-seeded mericarps with leathery endocarp, suspended from the top of a central carpophore, each mericarp with the remaining base of one of the previously coherent stylodia reflexed back over its apex. Seeds

with slender, slightly curved embryo, triangular in cross section, the thin testa without endosperm. Producing ellagic acid, quassinoids and limonoids lacking.. . 8. KIRKIACEAE.

3 Stamens mostly connate, less often (Cedreleae) free. Evergreen trees, rarely shrubs or suckering shrublets, sometimes almost herbaceous. Bark often bitter. Secretory cells occur in the leaves, cortex, and pith of the axis, while secretory cavities present in a few genera. Indumentum of uni- or multicellular hairs, less often stellate, malpighiaceous, or with lepidote scales. Nodes mostly pentalacunar with five traces. Vessels with simple perforations; lateral pitting alternate. Fibers with simple or narrowly bordered pits, mostly septate. Rays heterogeneous or less often homogeneous. Axial parenchyma paratracheal. Leaves alternate or rarely opposite, pinnate to bipinnate, unifoliolate or simple, with usually entire leaflets, sometimes spines, estipulate. Stomata anomocytic. Flowers mostly small, in various types of mostly axillary to supra-axillary inflorescences, often cauliflorous or even epiphyllous, bisexual or less often polygamous, sometimes monoecious or dioecious (when unisexual often with the rudiments of opposite sex), actinomorphic. Sepals (2)3-5(-7), sometimes transitional to bracteoles, usually more or less connate, imbricate or rarely open or valvate, sometimes more or less entire, or closed and basally circumscissile. Petals much longer than the sepals, 3-7(-14) in one or rarely two cycles, free or sometimes basally (rarely in half-length) connate, imbricate or contorted or adnate to the filament tube and valvate. Androecium diplostemonous (of 8–10 stamens) or less often haplostemonous (of 3–6 stamens), rarely the stamens numerous (up to 30); filaments free (Cedreleae) or much more often united into a staminal tube with anthers in one or two whorls. Pollen grains 2–3-celled, 2–5-colporate. Nectary disc intrastaminal, annular, sometimes adnate to the ovary, sometimes developed into an androgynophore. Gynoecium of (1)2-6(-20) united carpels, with a terminal style; the stigma often capitate or discoid, lobed or not; ovary multilocular or rarely unilocular with parietal placentas, occasionally more or less sunken in the nectary disc, usually with two ovules, less often one, several or many in each locule (or on each placenta). Ovules anatropous, campylotropous, or orthotropous, generally pendulous and epitropous, with ventral raphe, often with a placental obturator; micropyle (bi) endostomal. Fruits septicidal, loculicidal or septifragal capsules, less often baccate or drupes, very rarely nuts. Seeds winged and then attached to woody columella, or with corky outer layers, or with fleshy sarcotesta or aril or a combination of both, or none of these, with spatulate, white or rarely (*Trichilia*) green embryo; endosperm wanting or rarely well developed. Plants usually producing triterpenoid bitter substances and sometimes coumarins. n = 10–14 + ... 10. MELIACEAE.

6 Stamens usually twice as many as the petals.

8 Leaves usually estipulate.

9 Flowers in axillary or terminal racemes, cymes, panicles, or thyrses.

10 Ovule one. Trees and shrubs, rarely herbaceous with woody rootstocks (sp. of *Simaba*); bark very bitter by simaroubilide quassinoids. Secretory cells containing oil, resin, or mucilage, often occur in the parenchymatous tissues; secretory canals occur in certain genera, especially in the peripheral region of the pith; vertical intercellular canals, probably traumatic, also often present. Vessels with simple perforations, rarely some of them reticulate; lateral pitting alternate. Fibers with simple or bordered pits, sometimes septate. Rays mostly homogeneous. Axial parenchyma of very diverse types. Nodes tri- or multilacunar. Leaves alternate or rarely opposite, pinnate to unifoliolate or rarely simple (*Castela*), estipulate or rarely (*Picrasma*) stipulate. Stomata anomocytic or less often paracytic. Flowers bisexual or more often unisexual (dioecious in *Simarouba*

and *Castela*), and then with the rudiments of the opposite sex, actinomorphic, (3-)4-6(-8)-merous. Sepals mostly five, basally connate, imbricate or valvate. Petals mostly five, free, imbricate, contorted, or valvate, rarely wanting. Stamens usually twice as many as the petals, less often as many and alternate with the petals, rarely more than twice as many as the petals, free; filaments often with ventral appendages near the base; anthers basifixed (*Soulamea*, and more or less ventrifixed in *Ailanthus*) or dorsifixed, introrse (usually), or extrorse to latrorse (*Ailanthus*, *Soulamea*). Pollen grains 2-celled, 3-colporate or colporoidate. Nectary disc usually well developed, intrastaminal, sometimes modified into a short gynophore or androgynophore. Gynoecium of one (*Amaroria*), or 2-5(-8) mostly more or less united carpels, sometimes connate by their stylodia, rarely (*Picrolemma*, species of *Ailanthus*) free; ovary superior, 2-5(-8)-locular; ovule solitary in each carpel or each locule, hemianatropous to anatropous, pendulous; micropyle zig-zag, endostomal in *Brucea*. Fruits of various types. Seeds with straight or curved, green oily embryo; cotyledons large, expanded; endosperm thin or none. Bark, wood and seeds are often very bitter by simaroubilide quassinoids and limonoids, alkaloids and flavonols present or absent, ellagic acid, kaempferol, and quercetin present; n = 8, 13 +. . . . . 3. SIMAROUBACEAE.

10 Ovules two. Dioecious, evergreen small trees or shrubs, rarely subshrubs, non-laticiferous and without coloured juice; bark often very bitter. The indumentum consists of simple hairs. Vessels with simple perforation. Xylem with fibre tracheids Nodes trilacunar with three traces. Leaves alternate, petiolate, conduplicate, imparipinnate, shiny and with wavy margins (many *Picramnia*), estipulate. Stomata anomocytic. Flowers in terminal or axillary racemes or panicles, or cauliflorous, small or minute, actinomorphic, 3-5(-6)-merous. Sepals imbricate or valvate, persistent. Petals 3-5(-6), imbricate, sometimes absent from male flowers or reduced in female flowers. Stamens 3-5(-6), isomerous with the perianth, opposite the petals, sometimes borne on a column; anthers versatile, introrse. Pollen grains 2-locular, 3-colporate. Gynoecium of 2–3 carpels; ovary superior, 1–3-locular; styles shorter than the ovary, stigmas 2(3), divergent, persistent. Placentation axile to apical, or basal. Ovules pendulous (when apical), or ascending (when basal), apotropos or epitropous. Fruits berry or capsular indehiscent and a samara (*Alvaradoa*). Seeds plano-convex to narrowly ellipsoid, with straight or curved embryo, cotyledons large, endosperm lacking (*Alvaradoa*). Arthroquinones detected, polyacetate derived. . .4. PICRAMNIACEAE.

9. Flowers in erect axillary catkins of reduced dichasia, dioecious, rarely with bisexual flowers among the males ones, small, inconspicuous, precocious. Male inflorescences lax, of 40–50 small 3-flowered cymules, the flowers without bracteoles and a perianth. Stamens 1–4(5); filaments short, free; anthers tetrasporangiate, basifixed, opening longitudinally. Pollen grains 3–6-colporate, minutely verrucate, with a microperforate tectum. Female inflorescences stiffly spikelike, with sessile solitary flowers inserted in the axils of the bracts, each flower with two bractlets and a perianth of (3)4(-

8). Gynoecium typically of one carpel, but occasionally gynoecia of two united carpels and with 2-locular ovary occurring; stylodium linear, with long, decurrent ventral stigma; each carpel with one ovule. Ovules anatropous to hemitropous, bitegmic. Fruits dry drupes. Seeds testal, with large, straight, linear embryo, perisperm of a few cell-layers, and thin starchy endosperm. Small deciduous trees and shrubs with very soft, slightly porous, and extremely light wood; bark bitter. Secretory canals with yellow resinous contents present at the margins of the pith and extending into the minor veins of the leaf. Vessels with simple perforations; lateral pitting scalariform to opposite. Fibers with extremely small simple pits. Rays exclusively uniseriate, composed of rather large, oblong to square cells. Axial parenchyma paratracheal, scanty to vasicentric, and in fine terminal bands. Secondary phloem including concentric, tangentially compressed layers of thin-walled fibers. Sieve-element plastids of S-type. Nodes trilacunar. Leaves alternate, simple, entire, pinnately veined, glandular, estipulate. Stomata anomocytic. minute segments. Tanniferous. $n = 16$.. . . . . 5. LEITNERIACEAE.

8 Leaves with well-developed stipules. Trees. Vessels with simple perforations; lateral pitting alternate. Rays homogeneous. Leaves alternate, simple, entire, pinnately veined, coriaceous, with large or very long intrapetiolar stipules folded around the terminal bud, early caducous and leaving a very distinct scar. Stomata paracytic. Flowers small, in axillary or terminal panicles, bisexual, actinomorphic. Sepals five, small, imbricate. Petals five, free, imbricate. Stamens 10(9), free, inserted below the large intrastaminal nectary disc; anthers sub-basifixed, short. Pollen grains 3-colporate. Gynoecium of two (*Desbordesia*) or 5–4 united carpels with terminal short simple style; ovary 2- or 5–4-locular, with one ovule per locule, pendulous from the tip of the central axis. Fruits drupes or (*Desbordesia*) broadly winged samara. Testa thick, much sclerotised; endosperm slight to copious; cotyledons large, cordate. . . . . 7. IRVINGIACEAE.

5 Stylodia free, ventral or ventral-basal. Shrubs or small trees; trichomes simple or glandular or plants glabrous. Vessels with simple perforations. Rays heterogeneous or less often homogeneous. Nodes unilacunar with one traces. Leaves small, alternate, usually simple and entire, petiolate or almost sessile, stipulate or estipulate. Stomata anomocytic or anisocytic. Flowers in large mixed panicles, leafy racemes, or solitary in the axils, actinomorphic, bisexual, sometimes unisexual or polygamous. Sepals 5(-7), free or connate at the base, imbricate, persistent. Petals five, shortly clawed, imbricate, or less often (*Stylobasium*) wanting. Stamens ten, in two cycles, in *Suriana* the inner (antepetalous) shorter or sterile or abortive; filaments subulate or filiform, glabrous or basally pilose, caduceus or persistent; anthers basifixed (*Stylobasium*) or dorsifixed, extrorse or introrse. Pollen grains 3-zonocolporate (Claxton et al. 2005). Gynoecium of one carpel (*Stylobasium*), one to seldom two carpels (*Guilfoylia* and *Recchia*) or five free carpels (*Suriana* and *Cadellia*); each carpel with a ventral-basal filiform stylodium (more or less sygmoid in *Stylobasium*); ovary superior; stigma scarcely capitate or (*Stylobasium*) peltate and glandular. Ovules two in each carpel, anatropous, bitegmic, crassinucellate; outer integument not contributing to the micropyle. Fruits 1-seeded, drupaceous, baccate, or bony nuts. Seeds small, endocarp with outer layer of palisade sclereids, exotestal cells enlarged, cuboidal, tanniniferous, rest crushed; embryo curved or folded; cotyledons usually thickened, endosperm very little (*Stylobasium*) or lacking. Proanthocyanidins present or absent. . . . . .6. SURIANACEAE.

1 Plants usually strongly resinous, with vertical intercellular resin-canals in the bark and often also horizontal ones in the wood-rays, usually also with similar resin-ducts in the phloem of the larger veins of the leaves; plants seldom producing alkaloids and only seldom with triterpenoid compounds.

11 Ovary with two or rarely one epitropous ovules in each locule. Trees or less often shrubs, rarely epiphytes; brown bark often flaky, resin present, often aromatic and smelling like turpentine. Parenchymatous tissues often with scattered mucilage cells. Vessels with simple perforations (or in *Beiselia* predominantly with simple but occasionally with scalariform perforations). Fibers usually septate. Rays heterogeneous or more often homogeneous. Nodes mostly pentalacunar. Leaves alternate, rarely opposite, pinnately compound or decompound, rarely simple (1-foliolate), toothed, usually estipulate. Inflorescences axillary to subterminal, indeterminate panicles of racemes, laxly branched or variously reduced, sometimes short and fasciculate. Flowers small, bisexual or more often unisexual (usually dioecious), actinomorphic. Sepals (3)4-5, mostly basally connate, valvate. Petals (3)4-5, mostly free, induplicate-valvate, usually greenish-yellow, sometimes wine red (*Dacryodes* and *Trattinnickia*), rarely wanting. Some genera (e.g. *Garuga*) have a well-developed hypanthium. Stamens 6–10, in one or more often two cycles, obdiplostemonous, the antesepalous cycle often more or less reduced; filaments free or rarely connate; anthers slightly dorsifixed to basifixed, slightly versatile, introrse. Staminodia often present in the female flowers. Nectary disc intrastaminal or rarely extrastaminal, annular or cupular, usually orange or red. Gynoecium of (2)3-5 united carpels, or rarely (*Beiselia*) of 9–12 carpels; style with 2–5-lobed or capitate stigma; ovary (2)3–5-locular, with 2 or rarely 1 pendulous ovule per locule. Ovules with ventral raphe and upwardly and outwardly directed micropyle, hemitropous to campylotropous, rarely orthotropous, bitegmic or rarely unitegmic. Fruits more or less drupaceous with 1–5 one-seeded pyrene or with one plurilocular pyrene, rarely septicidal capsules. Seeds with straight or curved minute embryo and almost without endosperm; cotyledons folded, usually palmately lobed, n = 11, 13, 23. 11. BURSERACEAE.

11 Ovary with one apotropous or epitropous ovule in each locule.

12 Pedicels of the female flowers not adnate to the bract. Trees, shrubs, less frequently wines, scandent shrubs or woody lianas, rarely subshrubs (some species of *Rhus*). Resin canals present, clear or viscous to milky sap present, often poisonous. Vessels with simple perforations, but sometimes some of the perforations scalariform or reticulate (in *Heeria* with many bars). Nodes mostly trilacunar with 3 traces. Leaves alternate or rarely opposite, imparipinnate or trifoliolate, less often simple, without stipules or very rarely with inconspicuous and vestigial stipules. Flowers in terminal or axillary and usually paniculate inflorescences, mostly unisexual (often with rudiments of other sex), seldom apetalous or even without perianth (e.g., male flowers of *Pistacia*). Sepals and petals each (3)5(-7), valvate or imbricate. Sepals usually connate. Petals free or less often basally connate. Stamens 5–10, rarely more numerous or only 1 fertile stamen; filaments free or rarely basally connate, borne outside or rarely within the usually annular, sometimes 5-lobed nectary disc, or the disc modified into a short gynophore, as in *Anacardium*; anthers dorsifixed, rarely basifixed. Staminodia present in female flowers. Gynoecium sometimes of 5(4–6) or occasionally up to 12 (as in *Pleiogynium*) free carpels and then usually only one fertile (as in *Buchanania* and *Androtium*) or monomerous (as in *Mangifera, Anacardium* and related genera), but more often it consists of 3, less often 4–5 or 2 united carpels; stylodia free or united into a style; ovary superior or rarely (*Drimycarpus, Holigarna*) inferior, in plurilocular ovaries often with only one locule fully developed, with one apical and pendulous to basal and erect ovule in each carpel or each fertile locule. Ovules anatropous, apotropous, unitegmic or bitegmic, often with a placental obturator at the base of the long funiculus; micropyle zig-zag. Fruits usually drupaceous, with more or less resinous mesocarp, occasionally samaroid, rarely subtended by a large fleshy hypocarp (*Anacardium*). Seeds often more or less pachychalazal, exotestal cells and endotegmen thickened, lignified; embryo usually more or less curved, oily; endosperm scanty or none; cotyledons fleshy. Present biflavonyls, 5-deoxyflavonoids, and phenols with unsaturated side chains (allergenic), n = 7–12, 14–16, 21. . . . . 12. ANACARDIACEAE.

12 Pedicels of the female flowers adnate to the large bract. Shrubs or perennial, rhizomatous herbs. Vessels with simple perforations. Leaves alternate or opposite, long petiolate, simple, entire and pinnately veined (*Dobinea*) or 3-lobed and palmately veined (*Campylopetalum*), serrate, estipulate. Flowers in terminal thyrses, unisexual and dioecious; bracts linear or (*Campylopetalum*) some in male inflorescences large, long-stalked and membranous-reticulate. Male flowers: calyx cupular, 4–5-dentate, open in bud; petals 4–5, free, clawed, valvate, sometimes elongate-filiform; stamens 8–10; filaments free or adnate to the pistillodium if one is present; anthers short, extrorse or introrse. Pollen grains 3(4)-colporate, reticulate. Pistillodia sometimes present. Female flowers without perianth, with more or less developed, annular nectary disc. Gynoecium of 1 carpel; ovary superior, 1-locular; style very short or elongate-filiform with lateral stigma; ovule solitary, pendulous from the long basal funicle, anatropous. Fruit a lenticular compressed achene, attached to the midrib of accrescent bract. Seeds without endosperm; cotyledons 2, flat, oval, n = 7 (*Campylopetalum*)......... 13. PODOACEAE.

1. RUTACEAE

A. L. de Jussieu 1789 (including Amyridaceae Kunth 1824, Aurantiaceae A.L. de Jussieu 1789, Boroniaceae J.G. Agardh 1858, Citraceae Roussel 1806, Dictamnaceae Vest 1818, Diosmaceae R. Brown 1814, Diplolaenaceae J.G. Agardh 1858, Flindersiaceae C. T. White ex Airy Shaw 1965, Fraxinellaceae Nees von Esenbeck et Martius 1823, Jamboliferaceae Martynov 1820, Pilocarpaceae J.G. Agardh 1858, Pteleaceae Kunth 1824, Spatheliaceae J. Agardh 1858, Zanthoxylaceae Berchtold et J. Presl 1820). 160/1650–1800. Widely distributed in tropical, subtropical, and warm-temperate regions, especially in South Africa and Australia.

1.1 RUTOIDEAE

Gynoecium usually of 4–5 carpels, rarely of 3–1 or more than 5 separate carpels, often united only by stylodia and stigmas and more or less divided when ripe. Ovary with two or five deep lobes. Fruits with loculicid dehiscence, rarely fleshy drupes. – ZANTHOXYLEAE: *Bosistoa, Bouchardatia, Zanthoxylum* (including *Fagaria*), *Tetradium, Geijera, Comptonella, Euodia, Evodiella, Tetractomia, Boninia, Orixa, Melicope, Pentaceras, Plethadenia, Decazyx, Lunasia, Decatropis, Polyaster, Megastigma, Medicosma, Peltostigma, Brombya, Choisya, Platydesma, Dutaillyea, Pitavia*; RUTEAE: *Boenninghausenia, Psilopeganum, Ruta, Haplophyllum, Thamnosma, Cneoridium, Dictamnus*; BORONIEAE: *Boronia* (including *Cyanothamnus*), *Boronella, Acradenia, Myrtopsis, Zieria, Eriostemon, Crowea, Phebalium, Rhadinothmnus, Nematolepis, Chorilaena, Muiriantha, Asterolasia, Philotheca, Urocarpus, Druimmondita, Microcybe, Geleznowia, Correa, Diplolaena*; DIOSMEAE: *Calodendrum, Agathosma* (including, *Barosma*), *Adenandra, Coleonema, Acmadenia, Diosma, Phyllosma, Macrostylis, Euchaetis, Empleurum*; CUSPARIEAE: *Pilocarpus, Esenbeckia, Metrodorea, Spiranthera, Almeidea, Euxylophora, Adiscanthus, Leptothyrsa, Ticorea, Lubaria, Rauia, Galipea, Raputia, Decagonocarpus, Angostura* (including *Cusparia*), *Erythrochiton, Naudinia, Ravenia, Monnieria*.

1.2 SPATHELIOIDEAE

Gynoecium of (2)3 united carpels, with two pendulous ovules in each locule. Fruits winged drupes. Secretory cells and lysigenous oil glands present at the margins of leaves. – *Spathelia, Harrisonia*.

1.3 DICTYOLOMATOIDEAE

Gynoecium of five carpels united only at base; each carpel with several ovules. Fruits of separating 3–4-seeded carpels. Androecium haplostemonous. – *Dictyoloma*.

1.4 TODDALIOIDEAE

Gynoecium of 5–2 united carpels or of only 1 carpel, each with 2–1 ovules. Fruits drupes or dry and winged. Endosperm present or wanting. Leaves and bark with lysigenous oil glands. – TODDALIEAE: *Phellodendron, Fagaropsis, Helietta, Balfourodendron, Ptelea, Oricia, Diphasia, Araliopsis, Casimiroa, Vepris, Toddalia, Toddaliopsis, Oriciopsis, Acronychia, Halfordia, Hortia, Skimmia, Amyris, Maclurodendron, Sarcomelicope, Teclea, Stauranthus*.

1.5 FLINDERSIOIDEAE

Gynoecium of 5 or 3 united carpels with simple style and 5-lobed peltate stigma and 2–8 biseriate ovules in each locule. Fruits woody capsules, septicidal or loculicidal, with 2–8 seeds per locule. Seeds compressed, winged, without endosperm. – *Flindersia, Chloroxylon*.

1.6 AURANTIOIDEAE (CITROIDEAE)
Fruits hesperidia (berrylike fruits often with periderm and with pulp derived from multicellular juice sacs that fill the locules when the fruit ripens). Seeds without endosperm, often with two or more embryos. Lysigenous oil glands present. – AURANTIEAE: *Poncirus, Citrus, Burkillanthus, Clymenia, Oxanthera, Microcitrus, Fortunella, Citropsis, Eremocitrus, Limonia* (including *Feronia*), *Triphasia* (including *Echinocitrus*), *Monanthocitrus, Wenzelia, Naringi, Balsamocitrus*; CLAUSENEAE: *Murraya, Luvunga, Paramignya, Micromelum, Glycosmis, Merrillia, Pleiospermium, Clausena, Atalantia, Severinia, Pamburus, Merope, Swinglea, Balsamocitrus, Aegle, Afraegle, Aeglopsis, Feroniella*.

Rutaceae are a very diversified and extremely heterobathmic family.

2. CNEORACEAE

Vest 1818. 1/3. Canary Islands (*C. pulverulentum*), Mediterranean (*C. tricoccon*, coasts of Spine, France, Italy, the Balearic Is., Sardinia), Cuba (*C. trimerum*).

Cneorum (including *Neochamaelea* = *C. pulverulentum*).

Close to the Rutaceae (see Heimsch 1942; Straka et al. 1976; Lobreau-Callen et al. 1978; Boesewinkel 1984; Plisko 1996), especially to Ruteae.

3. SIMAROUBACEAE

A. P. de Candolle 1811 (including Ailanthaceae J.G. Agardh 1858, Castelaceae J.G. Agardh 1858, Holacanthaceae Engler 1896, Quassiaceae Bertolini 1827, Simabaceae Horaninow 1847, Soulameaceae Endlicher ex Bullock 1958). 21/150. Tropical and subtropical regions, with some species of *Picrasma* and *Ailanthus* in temperate eastern Asia.

Eurycoma, Quassia (including *Samadera, Odyendea*), *Pierreodendron, Iridosma, Simarouba, Simaba, Hannoa, Castela, Brucea, Laumoniera, Picrasma, Perriera, Picrolemma, Ailanthus, Amaroria, Soulamea, Gymnostemon, Gumillea, Nothospondias*.

According to Simao et al. (1991), the presence of canthinones and beta-carbolines suggests the derivation of the Simaroubaceae from protorutaceous stock.

4. PICRAMNIACEAE

Fernando et Quinn 1995. 2/50. Southeastern U.S.A., Central and South America, West Indies.

4.1 PICRAMNIOIDEAE
Placentation axille to apical, ovules pendulous, epitropous. Fruit berry; seed coat vascularised, unlignified, inner layers crushed, embryo minute. – *Picramnia*.

4.2 ALVARADOIDEAE
Placentation basal to axile; ovules ascending, apotropous. Fruit a samaroid capsule; exotesta resinous. – *Alvaradoa*.

Related to the Simaroubaceae.

5. LEITNERIACEAE

Bentham and J. D. Hooker 1880. 1/1. Coastal plains of the southeastern United States from southern Missouri to eastern Texas and northern Florida.

Leitneria.

Serotaxonomical studies suggest relationships with the Rutales, especially with the Simaroubaceae (Petersen and Fairbrothers 1983, 1985). Thorne (2000, 2006) included *Leitneria* in the Simaroubaceae. However, from Simaroubaceae the Leitneriaceae differ in many important characters and I prefer to consider then other separate family closely related to Simaroubaceae.

6. SURIANACEAE

Arnott 1834 (including Stylobasiaceae J. Agardh 1858). 5/8. Tropical littorals (*Suriana*), subtropical Australia (*Cadellia*), northeastern and eastern central Australia (*Guilfoylia*), northwestern Australia (*Stylobasium*), and Mexico (*Recchia*).

STILOBASIEAE: *Stylobasium*; SURIANEAE: *Suriana, Cadellia, Guilfoylia*; RIGIOSTACHYEAE: *Recchia* (*Rigiostachys*).

Closely related to the Simaroubaceae, and Engler in his Sylabus (1964) and Hutchinson (1969) include they in that family. However from the Simaroubaceae they differs in usually (except two of three species of *Recchia*) simple, entire leaves, the absence of terpenoid lactones or simaroubolides (quassinoids), and especially in the presence of distinctive endocarp with its outer palisade layer of sclereids, characteristic of all five genera including *Stylobasium* and *Recchia* (Fernando and Quinn 1992).

According to Fernando and Quinn (1992: 285), the absence of a sclerified layer in the mature pericarp of *Guilfoylia* "is atypical of the Surianoideae and indeed of all the other subfamilies, and also reinforces the distinction between this genus and *Cadellia*..., with which it has sometimes been combined.... This condition might be interpreted, then, as a specialization of the *Suriana*-type that was probably associated with a change in dispersal mechanism, although the evidence for the inclusion of *Guilfoylia* within the Surianaceae obviously needs to be evaluated in the light of this marked difference in pericarp anatomy."

7. IRVINGIACEAE

Exell et Mendonça 1951. 3/8 Tropical Africa, Madagascar, Southeast Asia, and Malesia. *Irvingia* (4) occurs in tropical Africa, Indochina, Malay Peninsula, and Borneo. *Klainedoxa* (3) and *Desbordesia* (1) are endemic to tropical Africa.

Klainedoxa, Desbordesia, Irvingia.

Closely related to the Simaroubaceae.

8. KIRKIACEAE

Takhtajan 1967. 2/6. Tropical and South Africa (*Kirkia,* 5) and Madagascar (*Pleiokirkia,* 1).

Kirkia, Pleiokirkia.

Closely related to the Simaroubaceae but differing from them in wood anatomy (Webber 1936; Heimsch 1942; Metcaife and Chalk 1950), aestivation of petals ("induplicate-imbricate" – Airy Shaw in Willis [1973:616]), pollen morphology (Erdtman 1952), the broad, strongly lignified endocarp with prolific layers of crystalliferous cells (Fernando and Quinn 1992), mode of dehiscence of fruits, and seed anatomy.

9. PTAEROXYLACEAE

J. Leroy 1960. 2/8. Near the Atlantic coast of tropical West Africa and eastern and southeastern Africa (*Ptaeroxylon,* 1) and Madagascar (*Cedrelopsis*, 7).

Ptaeroxylon, Cedrelopsis.

These two genera have traditionally been included either in Meliaceae or Simaroubaceae. From both of them they differ in many respects, and Leroy (1960), Sender (1960), Airy Shaw (in Willis 1973), Pennington and Styles (1975), Takhtajan (1980, 1987), Dahlgren (1983, 1990), Waterman and Grunden (1983), and Dahlgren and van Wyk (1988) consider them a separate family. However, the Ptaeroxylaceae are closely related to the Simaroubaceae and the Kirkiaceae.

10. MELIACEAE

A. L. de Jussieu 1789 (including Cedrelaceae R. Browne 1814, Swieteniaceae Berchtold et Presi 1820). 52/600–650. Tropical and subtropical regions with a few species, including *Melia azederach* and *Cedrela sinensis,* in warm-temperate areas.

10.1 MELIOIDEAE

Buds naked. Stigma capitate. Fruits fleshy or leathery loculicidal capsules, berries, drupes or nut, usually not winged; seeds usually with arillode or sarcotesta. – TURRAEEAE: *Munronia, Naregamia, Turraea, Humbertioturraea, Calodecaryia, Nymania* (*Aitonia*); MELIEAE: *Melia, Azadirachta, Trichilla, Pseudobrsama, Pterorhachis, Walsura, Lepidotrichilia, Malleastrum, Ekebergia, Astrotrichilia, Owenia, Cipadessa*; VAVAEEAE: *Vavaea;* AGLAIEAE: *Aglaia, Lansium, Aphanamixis, Reinwardtiodendron, Sphaerosacme*; GUAREEAE: *Heckeldora, Ruagea, Turraeanthus, Guarea, Chisocheton, Megaphyllaea, Synoum, Anthocarapa, Pseudocarapa, Dysoxylum*; SANDORICEAE: *Sandoricum.*

10.2 QUIVISIANTHOIDEAE

Loculicidal capsule. Seeds dry, winged. – *Quivisianthe.*

10.3 CAPURONIANTHOIDEAE

Buds naked. Leaves opposite, decussate. Fruits dry. Seeds with corny sarcotesta. n = 29. – *Capuronianthus.*

10.4 SWIETENIOIDEAE

Buds usually with scales. Fruits woody, septifragal capsules with central columella and winged seeds or columella rudimentary and seeds with woody or corky sarcotesta. – CEDRELEAE: *Cedrela, Toona;* SWIETENIEAE: *Khaya, Neobeguea, Soymida, Entandrophragma, Chukrasia, Pseudocedrela, Schmardaea, Swietenia, Lovoa*; XYLOCARPEAE: *Xylocarpus*.

11. BURSERACEAE

Kunth 1824 (including Balsameaceae Dumortier 1829). 17/550. Pantropical, especially America, northeastern Africa, and Malesia.

BURSEREAE: *Aucoumea, Triomma, Beiselia, Boswellia, Bursera, Commiphora;* PROTIEAE: *Protium, Crepidospermum, Tetragastris, Garuga, Tapirocarpus*; CANARIEAE: *Canarium, Dacryodes, Santiria, Haplolobus, Scutinanthe, Trattinnickia*.

12. ANACARDIACEAE

R. Brown 1818 (including Blepharocaryaceae Airy Shaw 1965, Cassuviaceae R Brown 1818, Comocladiaceae Martynov 1820, Julianiaceae Hemsley 1906, Lentiscaceae Horaninow 1847, Pistaciaceae Martius ex Perleb 1838, Rhoaceae Sprengel ex Sadler 1826, Schinaceae Rafinesque 1837, Spondiaceae Martynov 1820, Terebinthaceae A. L. de Jussieu 1789, Vernicaceae Link 1831). 75–78/600–900. Mainly tropical and subtropical regions, but with some species in warm-temperate zones.

12.1 ANACARDIOIDEAE
Gynoecium of five free carpels or of one carpel. Leaves simple, entire. – *Buchanania, Androtium, Mangifera, Fegimanra, Anacardium, Gluta, Swintonia, Bouea*.

12.2 SPONDIOIDEAE
Gynoecium of 3–5 or more united carpels, rarely only 1 fertile. Leaves usually pinnate, rarely simple. – SPONDIEAE: *Spondias, Dracontomelon, Sclerocarya, Choerospondias, Poupartia, Pegia, Lannea, Tapirira*, etc.; RHOEAE (RHOIDEAE): *Sorindela, Trichoscypha, Parishia, Campnosperma, Euroschinus, Schinus, Schinopsis, Blepharocarya, Cotinus, Heeria, Comocladia, Metopium, Rhus, Poupartiopsis, Woodfordia, Quebrachia, Toxicodendron, Astronium*, etc.; SEMECARPEAE: *Nothopegia, Melanochyla, Semecarpus Drimycarpus, Holigarna*.

12.3 JULIANIOIDEAE
Flowers much reduced, the female ones without perianth. Nectary disc and gynophore wanting. Ovary unilocular, with a single unitegmic ovule. – *Orthopterygium, Amphipterygium* (*Juliania*).

12.4 PISTACIOIDEAE
Flowers much reduced, apetalous, dioecious, chalazogamous. Gynoecium of three united carpels, but with two of the carpels represented only by their stylodia; ovary 1-locular, with a single ovule. – *Pistacia*.

13. PODOACEAE

Baillon ex Franchet 1889. 2/3. Eastern Himalayas (Nepal, Sikkim), southern China (Yunnan) (*Dobinea*), and Thailand (*Campylopetalum*).

Dobinea (Podoon), Campylopetalum.

Close to the Anacardiaceae but differing from them in the gynoecium morphology, cytologically (Carlin in Forman 1954), and palynologically (Erdtman 1952; Forman 1954; Kuprianova 1965). Hutchinson (1959, 1973) accepts the separate family Podoaceae but places it near the Sapindaceae. However, both the anatomy of vegetative organs (Heimsch 1942; Metcaife in Forman 1954) and gross morphology (Forman 1954) suggest affinity with the Anacardiaceae.

Bibliography

Abbe EC. 1974. Flowers and inflorescences of the "Amentiferae." Bot. Rev. 40: 159–261.

Abbe EC and TT Earle. 1940. Inflorescence, floral anatomy, and morphology of *Leitneria floridana*. Bull.Torrey Bot. Club 67: 173–193.

Aguilar Sierra CI and TSA Melhem. 1998. Morfologia polinica da tribo Bursereae (Burseraceae) na America do Sul. Rev. Brasil. Bot. 21: 17–63.

Aguilar-Ortigoza CJ and V Sosa. 2004. The evolution of toxic phenolic compounds in a group of Anacardiaceae genera. Taxon 53: 357–364.

Araújo EF de, LP de Queiroz, and MA Machado. 2003. What is *Citrus*? Taxonomic implications from a study of cp-DNA evolution in tribe Citreae (Rutaceae subfamily Aurantioideae). Organism. Divers. Evol. 3: 55–62.

Armstrong J. 1991. Studies on pollination and systematics in the Australian Rutaceae. Ph.D. Thesis. The University of New South Wales, Sydney.

Barkley FA. 1957. Generic key to the Sumac family (Anacardiaceae). Lloydia 20: 255–265.

Behnke H-D, U Kiritsis, SJ Patrick, and KF Kenneally. 1996. Form *Pfs* plastids, stem anatomy and systematic affinities of *Stylobasium* Desf. (Stylobasiaceae). A contribution to the knowledge of sieve-element plastids in the Rutales and Sapindales. Bot. Acta 109: 346–359.

Boesewinkel FD. 1977. Development of ovule and testain Rutaceae: I. *Ruta, Zanthoxylum*, and *Skimmia*. Acta Bot. Neerl. 26: 193–211.

Boesewinkel FD. 1978. Development of ovule and testa in Rutaceae: III. Some representatives of the Aurantioideae. Acta Bot. Neerl. 27: 341–354.

Boesewinkel FD. 1984. Development of ovule and seed coat in *Cneorum tricoccum* L. (Cneoraceae). Acta Bot. Neerl. 33: 61–70.

Boesewinkel FD and F Bouman. 1978. Development of ovule and testa in Rutaceae: II. The unitegmic and pachychalazal seed of *Glycosmis* cf. *arborea* (Roxb.) DC. Acta Bot. Neerl. 27: 69–78.

Brizicky GK. 1962a. The genera of Rutaceae in the southeastern United States. J. Arnold Arbor. 43: 1–22.

Brizicky GK. 1962b. The genera of Simaroubaceae and Burseraceae in the southeastern United States. J. Arnold Arbor. 43: 173–186.

Brizicky GK. 1962c. The genera of Anacardiaceae in the southeastern United States. J. Arnold Arbor. 43: 359–375.

Caris P, E Smets, K De Coster, and LP Ronse Decraene. 2006. Floral ontogeny of *Cneorum tricoccon* L. (Rutaceae). Plant Syst. Evol. 257: 223–232.

Carlquist S. 1988. Wood anatomy of Cneoraceae: Ecology, relationships, and generic definition. Aliso 12: 7–16.

Carmello Guerreiro SM and AAS Paoli. 2000. Estrutura do pericarpo e da semente de *Astronium graveolens* Jacq. *(Anacardiaceae)* com notas taxonomicas. Rev. Brasil. Bot. 23: 87–96.

Channell RB and CE Wood Jr. 1962. The Leitneriaceae in the southeastern United States. J. Arnold Arbor. 43: 435–439.

Chase MW, CM Morton, and JA Kallunki. 1999. Phylogenetic relationships of Rutaceae: a cladistic analysis of the subfamilies using evidence from *rbc*L and *atp*B sequence variation. Am. J. Bot. 86: 1191–1199.

Cheek MR. 1989. The systematic seed anatomy of the Meliaceae. Ph.D. dissertation. Oxford University, Oxford.

Cheek MR. 1990. Systematic seed anatomy of the Turraeeae (Meliaceae); taxonomic and ecological aspects. Mitteil. Inst. Allgem. Bot. Hamburg 23b: 683–706.

Cheek M and A Rakotozafy. 1991. The identity of Leroy's fifth subfamily of the Meliaceae, and a new combination in *Commiphora* (Burseraceae). Taxon 40: 231–237.

Clarkson JJ, MW Chase, and MM Harley. 2002. Phylogenetic relationships in Burseraceae based on plastid *rps*16 intron sequences. Kew Bull. 57: 183–193.

Claxton F, H Banks, BB Klitgaard, and PR Crane. 2005. Pollen morphology of families Quillajaceae and Surianaceae (Fabales). Rev. Paleobot. Palynol. 133: 221–233.

Copeland HF. 1955. The reproductive structures of *Pistacia chinensis* (Anacardiaceae). Phytomorphology 5: 440–449.

Copeland HF. 1959. The reproductive structures of *Schinus molle* (Anacardiaceae). Madroño 15: 14–25.

Copeland HF. 1961. Observations on the reproductive structures of *Anacardium occidentale*. Phytomorphology 11: 315–325.

Crayn DM, ES Fernando, PA Gadek, and CJ Quinn. 1995. A reassessment of the familial affinity of the Mexican genus *Recchia* Moçiño et Sessé ex DC. Brittonia. 47: 397–402.

Cronquist A. 1944. Studies in the Simaroubaceae. IV. Resumé of the American genera. Brittonia 5: 128–147.

Daly DC. 2004. Burseraceae. In: N Smith, SA Mori, A Henderson, DW Stevenson, and SV Heald, eds. Flowering plants of the Neotropics, pp. 67–70. The New York Botanical Garden, Princeton University Press, Princeton.

Da Silva GF, GF Das, OR Gotlieb, and DL Dreyer. 1984. Evolution of limonoids in the Meliaceae. Biochem. Syst. Ecol. 12: 299–310.

Da Silva GF and OR Gotlieb. 1987. Evolution of quassinos and liminoids in the Rutales. Biochem. Syst. Ecol. 15: 85–103.

Da Silva GF, OR Gottlieb, and F Ehrendorfer. 1988. Chemosystematics of the Rutaceae: suggestions for a more natural taxonomy and evolutionary interpretation of the family. Plant Syst. Evol. 161: 97–134.

Das MF GF Da Silva, 0R Gottlieb, and DL Dreyer. 1984. Evolution of liminoids in the Meliaceae. Biochem. Syst. Ecol. 12: 299–310.

Desai S. 1962. Cytology and embryology of the Rutaceae. Phytomorphology 12: 178–184.

Ding Hou. 1978. Anacardiaceae. In: CGGJ van Steenis, ed. Flora malesiana, ser 1, 8(3): 395–548. Noordhoff, Groningen.

Dreyer DL. 1966. Citrus bitter principles, V. Botanical distribution and chemotaxonomy in the Rutaceae. Phytochemistry 5: 367–378.

Fernando ES, PA Gadek and CJ Quinn. 1995. Simaroubaceae, an artificial construct: evidence from *rbc*L sequence variation. Am. J. Bot. 82: 92–103.

Fernando ES and CJ Quinn 1992. Pericarp anatomy and systematics of the Simaroubaceae sensu lato. Austral. J. Bot. 40: 263–289.

Fernando ES and CJ Quinn. 1995. Picramniaceae, a new family, and a recircumscription of Simaroubaceae. Taxon 44: 177–181.

Fish F and PG Waterman. 1973. Chemosystematics in the Rutaceae: II. The chemosystematics of the *Zanthoxylum/ Fagara* complex. Taxon 22: 177–203.

Forman L. 1954. A new genus from Thailand. Kew Bull. 4: 555–564.

Forman LL, PE Brandham, MM Harley, and TJ Lawrence. 1989. *Beiselia mexicana* (Burseraceae) and its affinities. Kew Bull. 44: 1–31.

Fritsch FE. 1908. The anatomy of the Julianiaceae considered from systematic point of view. Trans. Linn. Soc. London, 2nd ser., Bot. 7: 129–152.

Fukuda T, J Yokoyama, and H Tsukaya. H. 2003. Phylogenetic relationships among species in the genera *Chisocheton* and *Guarea* that have unique indeterminate leaves as inferred from sequences of chloroplast data. Int. J. Plant Sci. 164: 13–24.

Godfrey RK and AF Clewell. 1965. Polygamodioecious *Leitneria floridana* (Leitneriaceae). SIDA 2: 172–173.

Grant M, S Blackmore, and C Morton. 2000. Pollen morphology of the subfamilyAuranthioideae (Rutaceae). Grana. 39: 8–20.

Gray AI and PG Waterman. 1978. Coumarins in the Rutaceae. Phytochemistry 17: 845–864.

Gregor HJ. 1989. Aspects of the fossil record and phylogeny of the family Rutaceae (Zanthoxyleae, Toddalioideae). Plant Syst. Evol. 162: 251–265.

Grundwag M. 1976. Embryology and fruit development in four species of *Pistacia* L. (Anacardiaceae). Bot. J. Linn. Soc. 73: 355–370.

Gut BJ. 1966. Beiträge zur Morphologic der Gynoeceums und der Blütenachse einiger Rutaceen. Bot. Jahrb. Syst. 85: 151–247.

Gutzwiller MA. 1961. Die phylogenetische Stellung von *Suriana maritima* L. Bot. Jahrb. Syst. 81: 1–49.

Harley MM and DC Daly. 1995. Burseraceae Kunth, Protieae March. em. Engl. World Pollen Spore Flora 20: 1–44.

Harris BJ. 1996. A revision of the Irvingiaceae of Africa. Bull. Jard. Bot. Nat. Belg. 65: 143–196.

Harris DJ. 1999. Species Plantarum: Flora of the World. Part 1, Irvingiaceae. Australian Biological Resoures Study, Canberra.

Harti D. 1957. Struktur und Herkunft des Endokarps der Rutaceen. Beitr. Biol. Pfl. 34: 35–49.

Hegnauer R. 1983. Chemical characters and the classification of the Rutales. In: PG Waterman and MF Grunden, eds. Chemistry and chemical taxonomy in Rutales, pp. 401–440. Academic Press, London.

Heimsch C. 1940. Wood anatomy and pollen morphology of *Rhus* and allied genera. J. Arnold Arbor. 21: 279–291.

Heimsch C. 1942. Comparative anatomy of the secondary xylem in the "Gruinales" and "Terebinthales" of Wettstein with reference to taxonomic grouping. Lilloa 8: 83–198.

Henderson A. 2004. Cneoraceae. In: N Smith, SA Mori, A Henderson, DW Stevenson, and SV Heald, eds. Flowering plants of the Neotropics, pp. 108–109. The New York Botanical Garden, Princeton University Press, Princeton.

Heo K and H Tobe. 1994. Embryology and relationships of *Suriana maritima* L. (Surianaceae). J. Plant Res. 107 (1085): 29–37.

Jacobs H. 2003. Comparative phytochemistry of *Picramnia* and *Alvaradoa*, genera of the newly established family Picramniaceae. Biochem. Syst. Ecol. 31: 773–783.

Jadin F. 1901. Contribution a l'ετυδε des Simaroubacees. Ann. Sci. Nat. Bot., 13th ser., 8: 201–304.

Jarvis CE. 1989. A review of the order Leitneriales. In: Crane PR and S Blackmore, eds. Evoludon, systematics, and fossil history of Hamamelidae, vol. 2, pp. 189–192. Oxford University Press, Oxford.

Jimenez Reyes N and XM Cuevas Figueroa. 2001. Morfologia del polen de *Amphipterygium* Schiede ex Standley (Julianiaceae). Bol. Inst. Bot. (Guadalajara). 8: 65–73.

Johri BM and MR Ahuja. 1957. A contribution to the floral morphology and embryology of *Aegle marmelos* Correa. Phytomorphology 7: 10–24.

Khalid SA. 1983. Chemistry of the Burseraceae. In: PG Waterman and MF Grundon, eds. Chemistry and chemical taxonomy of the Rutales, pp. 281–299. Academic Press, London.

Khosla PK and BT Styles. 1975. Karyological studies and chromosome evolution in Meliaceae. Silvae Genetica 24: 73–83.

Kribs DA. 1930. Comparative anatomy of the woods of the Meliaceae. Am. J. Bot. 17: 724–738.

Kubitzki K. 2007. Picramniaceae. In: K Kubitzki, ed. The families and genera of vascular plants, vol. 9, pp. 301–304. Springer, Berlin/Heidelberg/New York.

Kuprianova LA. 1965. Palynological data on the sys-tematics of the genus *Pistacia* L. Bot. Zhurn. 46: 803–814 (in Russian).

Lam HJ. 1931. Beiträge zur Morphologie der dreizähligen Burseraceae-Canarieae. Ann. Bot. Jard. Buitenzorg 42: 23–56.

Lam HJ. 1932. Beiträge zur Morphologic der Burseraceae insbesondere der Canarieae. Ann. Jard. Bot. Buitenz. 42: 97–226.

Lam HJ. 1938. Studies in phylogeny: II. On the phylogeny of the Malaysian Burseraceae, Canarieae in general, and of *Haplolobus* in particular. Blumea 3: 126–158.

Leenhouts PW. 1956. Burseraceae. In: CGGJ van Steenis, ed. Flora Malesiana, ser. 1, vol. 5: 209–296. Noordhoff, Groningen.

Leenhouts PW. 1978. The pollen morphology of Burseraceae: A taxonomic comment. Grana 17: 175–177.

Leroy J-F. 1959. Sur une petite famille de Sapindales propre a l'Αφριθυε australe et a Madagascar: Les Ptaeroxylaceae. C. R. Acad. Sci. Paris 248: 1001–1003.

Li H. 1994. Phylogeny, origin and species replacement of Podoaceae. Acta Bot. Yunn. Suppl. 6: 121–126.

Link DA. 1992. The floral nectaries in the Irvingiaceae. Plant Syst. Evol. 180: 235–242.

Liu W-Z and Z-H Hu. 1999. Studies on the secretory cavity of stems in Rutaceae. Acta Bot. Borr.-Occident. Sinica 19: 456–460 (in Chinese with English summary).

Lobreau-Called D, J Jérémie. 1986. L'Espece *Cneorum tricoccon* (Cneoraceae, Rutales) representée a Cuba. Grana 25: 155–158.

Lobreau-Callen D, S Nilsson, F Albers, and H Straka. 1978. Les Cneoraceae (Rutales): Etude taxonomique, palynologique, et systematique. Grana 17: 125–139.

Mabberley DJ. 1998. Australian Citreae with notes on other Aurantioideae (Rutaceae). Telopea. 7: 333–344.

Mabberley DJ, CM Pannell, and AM Sing. 1995. Meliaceae. In: CGGJ van Steenis, ed. Flora Malesiana, ser. 1, 12: 1–407. Noordhoff, Groningen.

Mauritzon J. 1935. Über die Embryologie der Familie Rutaceae. Svensk. Bot. Tidskr. 29: 319–347.

Mauritzon J. 1936. Zur Embryologie und systematischen Abgrenzung der Reihen Terebinthales und Celastrales. Bot. Not. 1936: 161–212.

Mauritzon J. 1939. Contribution to the embryology of the order Rosales and Myrtales. Acta Univ. Lund 35: 1–121.

Milanez FR. 1943. Anatomia das pricipais madeiras brasilieras das Rutaceae. Rodriguesia 7: 5–22.

Mitchell RE and TA, Geissman. 1971. Constituents of *Suriana maritima*: a triterpene diol of novel structure and a new flavonol glycoside. Phytochemistry 10: 1559–1567 (1971)

Mole BJ, F Udovicic, PY Ladiges, and MF Duretto. 2004. Molecular phylogeny of *Phebalium* (Rutaceae: Boronieae) and related genera based on the *nr*DNA regions ITS 1 + 2. Plant Syst. Evol. 249: 197–212.

Moore J. 1936. Floral anatomy and phylogeny in the Rutaceae. New Phytol. 35: 318–322.

Morton CM, MW Chase, and J Kallunki. 1996. Evaluation of the six subfamilies of Rutaceae using evidence from *rbcL* sequence vatiation. Am. J. Bot. 83(6): 180–181 (Abstract).

Morton CM, M Grant, and S Blackmore. 2003. Phylogenetic relationships of the Aurantioideae inferred from chloroplast DNA sequence data. Am. J. Bot. 90: 1463–1469.

Moskaleva GI. 1996a. Rutaceae. In: A Takhtajan, ed. Comparative seed anatomy, vol. 5, pp. 361–376. Nauka, St. Petersburg (in Russian).

Moskaleva GI. 1996b. Simaroubaceae, Irvingiaceae, Surianaceae. In: A Takhtajan, ed. Comparative seed anatomy, vol. 5, pp. 384–392. Nauka, St. Petersburg (in Russian).

Muellner AN, R Samuel, SA Johnson, M Cheek, TD Pennington, and MW Chase. 2003. Molecular phylogenetics of Meliaceae (Sapindales) base on nuclear and plastid DNA sequences. Am. J. Bot. 90: 471–480.

Nair MNB. 1991. Wood anatomy of some members of the Meliaceae. Phytomorphology 41: 63–73.

Nair NC. 1959, 1962, 1963. Studies in Meliaceae: I. Floral morphology and embryology of *Naregamia alata* W. et A. II. Floral morphology and embryology of *Melia azederach* Linn.: A reinvestigation. V. Morphology and anatomy of the flower of the tribes Melieae, Trichileae, and Sietenieae. VI. Morphology and anatomy of the flower of the tribe Cedrelieae and discussion of the floral anatomy of the family. J. Indian Bot. Soc. 38: 353–366, 366–378, 1959; 41: 226–242, 1962; 42: 177–189,1963.

Nair NC and RK Joshi. 1958. Floral morphology of some members of the Simaroubaceae. Bot. Gaz. 120: 88–99.

Narayana LL. 1958, 1959. Floral anatomy of Meliaceae, parts 1 and 2. J. Indian Bot. Soc. 37: 365–374, 1958; 38: 288–295,1959.

Narayana LL. 1960. Studies in Burseraceae, parts 1 and 2. J. Indian Bot. Soc. 39: 204–209, 402–409.

Narayana LL. 1963. A note on the embryology of a few Rutaceae. Curr. Sci. 32: 516–517.

Narayana LL and M Sayeeduddin. 1958. Floral anatomy of Simaroubaceae, part 1. J. Indian Bot. Soc. 37: 517–522.

Nene PM and VD Tilak. 1977. Placentation in the Rutaceae. Proc. Indian Acad. Sci. 85B: 378–383.

Nooteboom HP. 1962a. The generic delimitation in Simaroubaceae tribus Simaroubeae and conspectus of the genus *Quassia* L. Blumea 11: 509–528.

Nooteboom HP. 1962b. Simaroubaceae. In: CGGJ van Steenis, ed. Flora Malesiana, ser. 1, 6: 193–226. Noordhoff, Groningen.

Nooteboom HP. 1966. Flavonols, leuco-anthocyanins, cinnamic adds, and alkaloids in dried leaves of some Asiatic and Malesian Simaroubaceae. Blumea 14: 309–313.

Nooteboom HP. 1967. The taxonomic position of Irvingioideae, *Allantospermum* Forman, and *Cyrillopsis* Kuhlm. Adansonia, ser. 2, 7: 161–168.

Palacios Chavez R, M de la L Arreguin Sanchez, and DL Quiroz Garcia. 1997. Estudio palinologico de las Burseraceae del Estado de Queretaro, Mexico. Phytologia. 83: 58–66.

Panshin AJ. 1933. Comparative anatomy of the woods of the Meliaceae, subfamily Swietenioideae. Am. J. Bot. 20: 639–668.

Pell SK, and L Urbatsch. 2000. Evaluation of evolutionary relationships in Anacardiaceae using *mat*K sequence data. Am. J. Bot. 87(Suppl. 6): 149.

Pell SK and L Urbatsch. 2001. Tribal relationships and character evolution in the cashew family (Anacardiaceae): Inferences from three regions of the chloroplast genome. In: Botany 2001: Plants and people, p. 132. Abstracts [Albuquerque.]

Pennington TD. 2004. Meliaceae. In: N Smith, SA Mori, A Henderson, DW Stevenson, and SV Heald, eds. Flowering plants of the Neotropics, pp. 243–246. The New York Botanical Garden, Princeton University Press, Princeton.

Pennington TD and BT Styles. 1975. A generic monograph of the Meliaceae. Blumea 22: 419–540.

Pernet R. 1972. Phytochemie der Burseraceae. Lloydia 35: 280–287.

Petersen FP and DE Fairbrothers. 1983. A serotaxonomic appraisal of *Amphipterygium* and *Leitneria:* Two amentiferous taxa of Rutiflorae (Rosidae). Syst. Bot. 8: 134–148.

Petersen FP and DE Fairbrothers. 1985. A serotaxonomic appraisal of the "Amentiferae." Bull. Torrey Bot. Club 112: 43–52.

Pfeiffer WM. 1912. The morphology of *Leitneria floridana*. Bot. Gaz. 53: 189–203.

Plisko MA. 1996a. Cneoraceae. In: A Takhtajan, ed. Comparative seed anatomy, vol. 5, pp. 381–384. Nauka, St. Petersburg (in Russian).

Plisko MA. 1996b. Meliaceae, Kirkiaceae, Ptaeroxylaceae. In: A Takhtajan, ed. Comparative seed anatomy, vol. 5, pp. 405–430. Nauka, St. Petersburg (in Russian).

Plisko MA. 1996c. Burseraceae, Anacardiaceae. In: A Takhtajan, ed. Comparative seed anatomy, vol. 5, pp. 431–469 (in Russian).

Prance GT. 1965. The systematic position of *Stylobasium* Desf. Bull. Jard. Bot. Etat, Bruxelles 35: 435–448.

Price JR. 1963. The distribution of alkaloids in the Rutaceae. In: T Swain, ed. Chemical plant taxonomy, pp. 429–452. Academic Press, London/New York.

Ramp E. 1988. Struktur, Funktion, und systematische Bedeutung des Gynoeciums bei den Rutaceae und Simmaroubaceae. Ph.D. dissertration, University of Zurich.

Record SJ. 1939. American woods of the family Anacardiaceae. Trop. Woods 60: 11–45.

Record SJ. 1941. American timbers of the Mahogany family. Trop. Woods 66: 7–33.

Record SJ and RW Hess. 1940. American woods of the family Rutaceae. Trop. Woods 64: 1–28.

Samuel R, F Ehrendorfer, MW Chase, and H Greger. 2001. Phylogenetic analysis of Aurantioideae (Rutaceae) based on non-coding plastid DNA sequences and phytochemical features. Plant Biol. 3: 77–87.

Schatz GE. 2001. Generic flora of the trees of Madagascar. Royal Botanic GardensMissouri Botanical Garden, Kew/ St. Louis.

Schneider JV. 2007. Surianaceae. In: K Kubitzki, ed. The families and genera of vascular plants, vol. 9, pp. 449–455. Springer, Berlin/Heidelberg/New York.

Scott KD, CL McIntyre, and J Playford. 2000. Molecular analyses suggest a need for a significant rearrangement of Rutaceae subfamilies and a minor reassessment of species relationships within *Flindersia*. Plant Syst. Evol. 223: 14–27.

Sharma MR. 1954. Studies in the family Anacardiaceae: I. Vascular anatomy of flowers of *Mangifera indica*. Phytomorphology 4: 201–208.

Silva MF das GF da, OR Gottlieb, and F Ehrendorfer. 1988. Chemosystematics of the Rutaceae: Suggestions for a more natural taxonomy and evolutionary interpretation of the family. Plant Syst. Evol. 161: 97–134.

Simao SM, EL Barreiros, M Fatima das GF da Silva, and OR Gottlieb. 1991. Chemogeographical evolution of quassinoids in Simaroubaceae. Phytochemistry 30: 853–865.

Smith-White S. 1954. Chromosome numbers in the Boronieae (Rutaceae) and their bearing on the evolutionary development of the tribe in the Australian flora. Austral. J. Bot. 2: 287–303.

Stace HM, JA Armstrong, and SH James. 1993. Cytoevolutionary patterns in Rutaceae. Plant Syst. Evol. 187: 1–28.

Stannard B. 1981. A revision of *Kirkia* (Simaroubaceae). Kew. Bull. 35: 829–839.

Stern WL. 1952. The comparative anatomy of the xylem and the phylogeny of the Julianiaceae. Am. J. Bot. 39: 220–229.

Straka H, F Albers, and A Mondon. 1976. Die Stellung und Gliederung der Familie Cneoraceae (Rutales). Beitr. Biol. Pfl. 52: 267–310.

Tieghem P van and H Lecomte. 1886. Structure et affinite du *Leitneria*. Bull. Soc. Bot. France 33: 181–184.

Tilson AH and R Barnford. 1938. The floral anatomy of the Aurantioideae. Am. J. Bot. 25: 780–793.

Traveset A. 1995. Reproductive ecology of *Cneorum tricoccon* L. (Cneoraceae) in the Balearic Islands. Bot. J. Linn. Soc. 117: 221–232.

Trelease W. 1895. *Leitneria floridana*. Missouri Bot. Gard. Sixth Report, pp. 65–90, plates 30–44.

Tschunko AN and NH Nickerson. 1976. The androcium of *Suriana maritima*. Rhodora 78: 162–164.

Van der Ham RWJM, P Baas, ME Bakker, FD Boesewinkel, F Bouman, BJ van Heuven, and RKWM Klaasen. 1995. *Bottegoa* Chiov. transferred to Ptaeroxylaceae. Kew Bull. 50: 243–265.

Venning FD. 1948. The ontogeny of the laticiferous canals in the Anacardiaceae. Am. J. Bot. 35: 637–644.

Victor JE and AE Van Wyk. 1998. Palynology of *Acmadenia* (Diosminae: Rutaceae) and its taxonomic implications. Grana. 37: 143–154.

Victor JE and AE van Wyk. 1999. Pollen morphology of *Adenandra* (Rutaceae: Diosminae) and its taxonomic implications. Grana. 38: 1–11.

Victor JE and AE van Wyk. 1999. Pollen morphology of *Diosma* and *Coleonema* (Rutaceae: Diosminae) and its taxonomic implications. Grana. 38: 12–19.

Vieira PC, AR Lazaro, JB Fernandes, and MF das GF da Silva. 1988. The chemosystematics of *Dictyoloma*. Biochem. Syst. Ecol. 16: 541–544.

Wannan, B. S. 2006. Analysis of generic relationships in Anacardiaceae. Blumea 51: 165–195.

Wannan BS and CJ Quinn. 1988. Biflavonoids in the Julianiaceae. Phytochemistry 27: 3161–3162.

Wannan BS and CJ Quinn. 1990. Pericarp structure and generic affinities in the Anacardiaceae. Bot. J. Linn. Soc. 102: 225–252.

Wannan BS and CJ Quinn. 1991. Floral structure and evolution in the Anacardiaceae. Bot. J. Linn. Soc. 107: 349–385.

Wannan BS and CJ Quinn. 1992. Inflorescence structure and affinities of *Laurophyllus* (Anacardiaceae). J. Linn. Soc., Bot. 109: 235–245.

Waterman PG. 1975. Alkaloids of the Rutaceae: Their distribution and systematic significance. Biochem. Syst. Ecol. 3: 149–180.

Waterman PG. 1983. Phylogenetic implications of the distribution of secondary metabolites within the Rutales. In: PG Waterman and MF Grundon, eds. Chemistry and chemical taxonomy of the Rutales, pp. 377–400. Academic Press, London.

Waterman PG. 1990. Chemosystematics of the Rutaceae: Comments on the interpretation of da Silva et al. Plant Syst. Evol. 173: 39–48.

Waterman PG and MF Grunden, eds. 1983. Chemistry and chemical taxonomy of the Rutales. Academic Press, London.

Webber IE. 1936. Systematic anatomy of the woods of the Simaroubaceae. Am. J. Bot. 23: 577–587.

Webber IE. 1941. Systematic anatomy of the woods of the "Burseraceae." Lilloa 6: 441–465.

Weberling F and PW Leenhouts. 1965. Systematisch-morphologische Studien an Terebinthales-Familien (Burseraceae, Simaroubaceae, Meliaceae, Anacardia-ceae, Sapindaceae). Akad. Wiss. Abh. Math.-Naturwiss. Kl. 10: 495–584.

Webster GL and KI Miller. 1963. The chromosomes and relationships of *Leitneria*. Am. J. Bot. 50: 638. (Abstract).

Weeks A, DC Daly, and BB Simpson. 2005. The phylogenetic history and biogeography of the frankincense and myrrh family (Burseraceae) based on nuclear and chloroplast sequence data. Molec. Phylog. Evol. 35: 85–101.

Wendt T and EJ Lott. 1985. A new simple-leaves species of *Rechia* (Simaroubaceae) from southeastern Mexico. Brittonia 37: 219–225.

White F. 1986. The taxonomy, chorology, and reproductive biology of southern African Meliaceae and Ptaeroxylaceae. Bothalia 16: 143–168.

White F. 1990. *Ptaeroxylon obliquum* (Ptaeroxylaceae), some other disjuncts, and the Quaternary history of African vegetation. Bull. Mus. Natur. Hist. Nat. Paris, ser. 4, 12: 139–185.

White F and BT Styles. 1966. Ptaeroxylaceae. In: AW Exell, A Fernandes, and H Wild, eds. Flora zambesiaca, 2(2): 547–550. Crown Agents, London.

Young DA. 1976. Flavonoid chemistry and the phylo-genetic relationships of the Julianiaceae. Syst. Bot. 1: 149–162.

Order 95. CORIARIALES

Erect or scandent shrubs or less often subshrubs or perennial herbs, usually with scaly buds. Twigs angular, with numerous corky lenticels. Usually with nitrogen-fixing bacteria in nodules in the roots. Young stems with large pith. Vessels with simple perforations; lateral pitting alternate. Fibers with simple pits. Rays broad (up to 15 cells wide), homogeneous. Axial parenchyma paratracheal, scanty. Sieve-element plastids of S-type. Nodes unilacunar. Leaves opposite or less often verticillate, simple, entire, palmately veined, with minute, often caducous stipules. Flowers small, numerous, in terminal or axillary racemes, greenish, bisexual or sometimes polygamous, actinomorphic or nearly so. Plants entomophilous. Sepals five, free, imbricate, persistent. Petals five, free, purplish, valvate, usually shorter than sepals, mostly keeled inside, persistent, becoming more or less fleshy in fruits. Stamens ten, in two cycles; filaments filiform, free or those opposite the petals, adnate to petal keel; anthers large, oblongish, tetrasporangiate, basifixed or slightly dorsifixed, opening longitudinally, sometimes apiculate. Tapetum secretory. Microsporogenesis simultaneous. Pollen grains 3-celled, with (2)3(4)-colporate or 3(4)-pororate. Gynoecium of five or seldom ten carpels; carpels free or sometimes basally united, adnate to the shortly columnar receptacle, each with a long slender ventral stylodium that is papillate, stigmatic all around for most of its length, usually divergent; each carpel with one pendulous ovule. Ovules anatropous, with dorsal raphe, bitegmic, crassinucellate, micropyle endostomal. Female gametophyte or *Polygonum*-type. Endosperm nuclear. Fruits of separate hard-walled 5–10 achene-like mericarps, more or less enclosed by the much accrescent, fleshy petals. Seeds small, rather compressed, red-brown; exotesta of cuboid, thick-walled; embryo straight, oily; cotyledons two, large, plano-convex; endosperm scanty or none. Contain bitter sesquiterpenoid substances and producing kaempferol and quercetin, ellagic and often gallic acid, strongly tanniferous. n = 10, 15.

A rather isolated taxon that is one of the most controversial with respect to its affinities. Shows some similarities with the Rutales, especially with the Simaroubaceae, which is supported also by wood anatomical studies (Carlquist 1985), embryology (Kamelina 1991), and chemical data (Bohm and Ornduff 1981).

1. CORIARIACEAE

A. P. de Candolle 1824. 1/15–20. Western Mediterranean, temperate and subtropical Himalayas to Japan and New Guinea, New Zealand, South Pacific islands (including Solomons, Vanuatu, Fiji, Samoa, Society Islands), America from Mexico to Chile.

Coriaria.

Bibliography

Bohm BA and R Ornduff. 1981. Leaf flavonoids and ordinal affinities of Coriariaceae. Syst. Bot. 6: 15–26.

Carlquist S. 1985. Wood anatomy of Coriariaceae: Phylogenetic and ecological implications. Syst. Bot. 10: 174–183.

Duyjes BEE. 1993. Coriariaceae. In: CGGJ van Steenis, ed. Flora Malesiana ser. 1, 11: 385–391. Noordhoff, Groningen.

Garg M. 1981. Pollen morphology and systematic position of *Coriaria*. Phytomorphology 30: 5–10.

Good RDO. 1930. The geography of the genus *Coriaria*. New Phytol. 29: 170–198.

Kamelina OP. 1991. Comparative-embryological analysis as a method of phylogenetic systematics of flowering plants. Ph.D.. Thesis. Tashkent.

Kravtsova TI. 1996. Coriariaceae. In: A Takhtajan, ed. Comparative seed anatomy, vol. 5, pp. 472–473. Nauka, St. Petersburg (in Russian).

Lobreau D. 1969. Les limites de 1′ "ordre" des Celastrales d'apres le pollen. Pollen Spores 11: 499–555.

Matthews ML and PK Endress. 2004. Comparative floral structure and systematics in Cucurbitales (Corynocarpaceae, Coriariaceae, Tetramelaceae, Datiscaceae, Begoniaceae, Cucurbitaceae, Anisophylleaceae). Bot. J. Linn. Soc. 145: 129–185.

Petseon B. 1970. Coriariaceae: Taxonomy and relationships. World Pollen Flora 1: 23–24.

Praglowski J. 1970. Coriariaceae. World Pollen Flora 1: 17–31.

Sharma VK. 1968. Floral morphology, anatomy, and embryology of *Coriaria nepalensis* Wall. with a discussion of the interrelationships of the family Coriariaceae. Phytomorphology 18: 143–153.

Skog LE. 1972. The genus *Coriaria* in the Western Hemisphere. Rhodora 74: 242–253.

Tobe H and N Suzuki. 1993. Occurrence of uni- and bitegmic ovules in *Coriaria* (Coriariaceae), and evolutionary relationships in the genus. Abstracts XV Intern. Bot. Congr., p. 95. Yokohama.

Tobe H, M Suzuki, and T Fukuhara. 1992. Pericarp anatomy and evolution in *Coriaria* (Coriariaceae). Bot. Mag. (Tokyo) 195: 289–302.

Yoda K and M Suzuki. 1992. Comparative wood anatomy of *Coriaria*. Bot. Mag. (Tokyo) 105, 1078: 235–245.

Yokoyama J and M Hasebe. 2001. Phylogeny and biogeography of *Coriaria* based on molecular data. Did *Coriaria* originate in the southern hemisphere? Proc. Jpn. Soc. Plant Taxon. 16: 43–48.

Order 96. CORYNOCARPALES

Evergreen trees or shrubs; young stems with separate bundles. Vessels with simple perforations, lateral pitting alternate. Fibers with simple or faintly bordered pits. Rays very broad (up to 16 cells wide). Axial parenchyma vasicentric and in broad paratracheal strips. Cambium storied. Sieve-element plastids of S-type. Nodes trilacunar. Leaves alternate, conduplicate, coriaceous, more or less crowded towards the ends of the shoots, shining; stipules intrapetiolar, subtending the axillary bud, deciduous. Stomata paracytic. Flowers in terminal panicles and in racemes, bisexual, actinomorphic, 5-merous. Plants anemophilous. Sepals free, strongly imbricate, fleshy. Petals free, imbricate, inserted at the margin of the disc-like receptacle. Stamens five, incurved in bud, opposite and basally slightly adnate to the petals, alternate with five petaloid staminodes attached to the outer surface of the nectaries; filaments flattened or terete; anthers dorsifixed, tetrasporangiate, introrse, opening longitudinally. Tapetum secretory. Microsporogenesis simultaneous. Pollen grains nearly smooth, 2-colporate, with a short colpus at each end. Large cushionlike five nectaries borne opposite and internal to petaloid scales (staminodes). Gynoecium pseudomonomerous, reduced to one fertile carpel with more or less well-expressed trace of a second carpel and very rarely the rudiment of the second stylodium; styles short, conduplicate, stigma capitate; ovary superior, 1–2-locular, solitary ovule pendulous, anatropous, bitegmic, crassinucellate, with micropyle directed upwards. Female gametophyte of *Polygonum*-type. Endosperm nuclear. Fruits fleshy drupes, with fibrous endocarp. Seeds with large, straight, oily, and starchy embryo, minute radicle and plumule; endosperm starchy; cotyledons two, very large. Plants containing very toxic bitter glycosides, at least in the bark and seed; present kaempferol and ellagic acid. n = 22, 23.

Both structural similarities with Coriariaceae (Matthews and Endress 2004) and molecular studies (Nandi et al. 1998; Savolainen et al. 2000; Soltis et al. 2000; Zhang and Renner 2003) support the affinity with Coriariaceae

1. CORYNOCARPACEAE

Engler 1897. 1/5–7. Aru Islands, New Guinea, Bismarck Archipelago, New Britain, New Ireland, Solomon Islands, eastern Australia, New Caledonia, Vanuatu, New Zealand.

Corynocarpus.

Bibliography

Carlquist S and RB Miller. 2001. Wood anatomy of Corynocarpaceae is consistent with Cucurbitalean placement. Syst. Bot. 26: 54–65.

Dawson MI. 1997. Chromosome numbers in *Corynocarpus* (Corynocarpaceae). New Zealand J. Bot. 35(2): 255–258.

Hemsley WB. 1903. On the genus *Corynocarpus* Forst., with descriptions of two species. Ann. Bot. 17: 179–180.

Ingle HD. 1956. A note on the wood anatomy of the genus *Corynocarpus*. Trop. Woods 105: 8–12.

Matthews ML and PK Endress. 2004. Comparative floral structure and systematics in Cucurbitales (Corynocarpaceae, Coriariaceae, Tetramelaceae, Datiscaceae, Begoniaceae, Cucurbitaceae, Anisophylleaceae). Bot. J. Linn. Soc. 145: 129–185.

Melikian AP and IA Savinov. 2000. Corynocarpaceae. In: A Takhtajan, ed. Comparative seed anatomy, vol. 6, pp. 140–142. Nauka, St. Petersburg (in Russian).

Nandi O, MW Chase and PK Endress. 1998. A combined cladistic analysis of angiosperms using *rbs*L and non-molecular data sets. Ann. Missouri Bot. Gard. 85: 137–212.

Narayana LL, PS Narayana, and SR Vatsavaya. 1986. A contribution to the numerical chemotaxonomy of Corynocarpaceae. J. Econ. Taxon. Bot. 8: 249–254.

Nowicke JW and JJ Skvarla. 1983. Pollen morphology and the relationships of the Corynocarpaceae. Taxon 32: 176–183.

Patel RN. 1975. Wood anatomy of the dicotyledons indigenous to New Zealand: 9. *Corynocarpus*. New Zealand J. Bot. 13: 19–29.

Philipson WR. 1987. *Corynocarpus* J. R. and G. Forst.: An isolated genus. Bot. J. Linn. Soc. 95: 9–18.

Soltis DE, PS Soltis, DR Morgan, SM Swensen, BC Mullin, JM Dowd, and PG Martin. 1995. Chloroplast gene sequence data suggest a single origin of the predisposition for symbiotic nitrogen fixation in angiosperms. Proc. Nat. Acad. Sci. USA 92: 2647–2651.

Steenis van CGGJ. 1952. Corynocarpaceae. In: CGGJ van Steenis, ed. Flora Malesiana, ser.1, vol.4, pt.3: 262–264. Noordhoff, Groningen.

Swensen SM. 1996. The evolution of actinorhizal symbioses: evidence for multiple origins of the symbiotic association. Am. J. Bot. 83: 1503–1512.

Wagstaff SJ and MI Dawson. 2000. Classification, origin, and patterns of diversification of *Corynocarpus* (Corynocarpaceae) inferred from DNA sequences. Syst. Bot. 25: 134–149.

Order 97. GERANIALES

Mostly herbs, less often deciduous and evergreen trees, shrubs, subshrubs. Vessels typically with simple perforations; lateral pitting alternate, scalariform or transitional. Fibers with simple pores, often septate. Rays heterogeneous or seldom nearly homogeneous, often wanting. Axial parenchyma paratracheal and also with a few diffusely scattered cells. Sieve-element plastids of S-type or Plc-type. Nodes trilacunar or rarely unilacunar, pentalacunar or multilacunar (Melianthaceae). Leaves alternate or less often opposite, usually compound or dissected or more or less lobed, rarely simple and entire, stipulate or estipulate. Stomata paracytic or more often anomocytic, or diacytic (*Francoa*). Flowers usually in cymose inflorescences or in terminal racemes or panicles, less often solitary and axillary, bisexual or rarely unisexual, actinomorphic or rarely zygomorphic, 5-merous or seldom 4-merous, rarely 3-merous. Sepals free or basally connate, imbricate, valvate, or contorted, sometimes persistent. Petals free or basally connate, imbricate or contorted, rarely wanting. Nectary disc represented by 5 extrastaminal glands or wanting; in Francoaceae nectary disc interstaminal. Stamens 1–2(3) times as many as the sepals or petals, free or more often with basally connate; filaments sometimes subulate; anthers slightly introrse or latrorse, basifixed or dorsifixed, tetrasporangiate, opening longitudinally. Tapetum secretory or plasmodial. Microsporogenesis simultaneous. Pollen grains 2-celled or 3-celled, 3-colpate or 3-colporate, sometimes (Vivianiaceae and Rhynchothecoideae) pantoporate or (Ledocarpoideae) inaperturate. Gynoecium of (2-)5(6) carpels; stylodia free or more or less connate; stigma small, capitate, shortly 5-lobed, or commisural (Francoaceae). Ovary superior, or (Greyiaceae and Francoaceae), with (1)2-several to numerous, more or less pendulous ovules in each locule. Ovules anatropous to campylotropous, sometimes hemitropous, epitropous, with the micropyle directed upward, bitegmic, crassinucellate. Female gametophyte of *Polygonum*-type or very rarely of *Allium*-type. Endosperm nuclear. Fruits of various types. Seeds usually exotegmic, rarely endotegmic (Ledocarpaceae) or seed coat derived from both integuments and not specialized (Rhynchothecoideae and Vivianiaceae, except seeds of *Caesarea* which are exotegmic – Boesewinkel, Bouman 2000); embryo straight or curved; endosperm copious or scanty, sometimes wanting.

Have a common origin from a rutalean ancestor.

Key to Families

1 Nodes trilacunar or rately unilacunar.

2 Extrastaminal nectary glands usually present.

3 Fruits loculicidal capsules. Leaves estipulate.

4 Androecium of 15 stamens in two cycles (10 short antepetalous stamens and 5 long antesepalous ones); stamens with persistent subulate filaments and dorsifixed, introrse anthers. Pollen grains 3-colporate. Acaulescent perennial herbs from thick taproot. Leaves rosulate, alternate, simple, pinnatifid. Sieve-element plastids of S-type. Inflorescences radical, 1–9-flowered cymes. Flowers bisexual, actinomorphic, 5-merous. Calyx strongly imbricate, persistent. Corolla contorted. Pollen grains tricolpate. Extrastaminal lobed nectary disk well developed. Gynoecium of five carpels attached to central column; stylodia united into a simple filiform style with capitate stigma; ovary 5-locular and 5-lobed, with about 12–50 axile, biseriate ovules in each locule; ovules anatropous to campylotropous (becoming campylotropous after fertilization, via a bulge from the inner integument on the raphal side); micropyle zig-zag. Fruits tardily and irregularly loculicidal capsules or rarely (*H. tridentata*) schizocarp but the mericarps are not beaked. Seeds with cochlear embryo with spirally folded cotyledons and scanty endosperm; seed coat with an exotegmen consisting of stellately lobed cells and a crystalliferous endotesta. . . . . . . 1. HYPSEOCHARITACEAE.

4 Androecium of 10 or 8, sometimes 5 stamens, equal or unequal. Pollen grains periporate. Perennial herbs, sometimes with extensive secondary growth, very rarely annuals. Hairs glandular. Nodes unilacunar with one trace. Leaves opposite, entire, crenate to coarsely dentate, often more or less white-tomentose below. Flowers in terminal or apparently lateral, one to many-flowered cymes, bisexual, actinomorphic, mostly 5-merous. Sepals prominently 3-veined, more or less connate, lobes valvate. Petals free, imbricate. Nectary glands as many as and alternate with the petals, entire or 2-lobed. Gynoecium of 3(2) carpels. Stylodia almost free or connate, the stigmatic surface papillose, often extended to cover all or part of the adaxial surface; ovary 3(2)-lobed, with two superposed, anatropous, axile ovules (1 ascending, 1 pendulous) in each carpel. Fruits 3(2)-lobed loculicidal capsules. Seeds with much curved or circinate green embryo and abundant fleshy endosperm, seed coat derived from both integuments and not specialized (except seeds of *Caesarea* which are exotegmic), n = 7. . . . . 3. VIVIANIACEAE.

3 Fruits schizocarps. Stipules present. Herbs or less often subshrubs or shrublets, sometimes (*Sarcocaulon*) with fleshy stems and bearing essential oils. Hairs glandular. Sieve-element plastids of S-type. Nodes trilacunar with three traces. Leaves alternate (the upper, often) or opposite (the lower, usually), mostly lobed, dissected or compound, petiolate, often gland-dotted and aromatic; stipules well developed, interpetiolar or intrapetiolar, usually twinned at the base of the petiole. Stomata anomacytic. Inflorescences terminal or axillary, or leaf-opposite, often pedunculate, usually consisting of paired flowers or umbels, commonly with involucral bracts. Flowers bisexual, actinomorphic or slightly zygomorphic (*Pelargonium*), 5-merous. Sepals free or connate to the middle, imbricate with valvate tips, persistent. Petals 2–5, free, imbricate or rarely contorted, caduceus, rarely absent. Stamens 5, or 10, or 15, sometimes a few sterile, mostly more or less connate at the base, obdiplostemonous when more than one cycle. Anthers dorsifixed, usually versatile, introrse. Pollen grains 3-colpate (*Sarcocaulon* and *Monsoa*) or more often 3-colporate, tectate-columellate, reticulate to striate. Gynoecium of 5 carpels; style with 5–3 stigmatic branches; ovary 5-locular, 5–3-lobed, with 1–2 pendulous or ascending, anatropous to often campylotropous (becoming campylotropous after fertilization) ovules in each locule; micropyle zig-zag. Fruits of five 1-seeded mericarps that separate elastically from a central beak (the persistent style), the mericarps often opening to discharge the seed. Seeds with usually or more or less curved embryo with green cotyledons; endosperm much reduced or absent; seed coat with exotegmen consisting of thick-walled cells. Containing hydrolysable tannins, flavonols (kaempferol, quercetin, and myricetin), commonly alkaloids, and sometimes ellagic acid, n = mostly 10 (*Erodium*), mostly 11 (*Monsonia, Sarcocaulon, Pelargonium*), or mostly 14 (*Geranium*). . . . . 2. GERANIACEAE.

2 Extrastaminal nectary glands absent. Small shrubs or shrublets. Leaves usually opposite, simple or compound, entire or deeply lobed or dissected, estipulate. Flowers solitary or in few-flowered corymbs or terminal racemes, bisexual, actinomorphic, 5-merous. Sepals free, imbricate. Calyx with an epicalyx of a whorl of bracteoles (Ledocarpoideae), or epicalyx wanting (*Rhynchotheca*). Petals imbricate (*Wendtia*) or contorted (*Balbisia*), or absent (*Rhynchotheca*). Stamens ten, free, alternisepalous; anthers extrorse or introrse. Pollen grains inaperturate, periporate (*Rhynchotheca*). Gynoecium of five (*Balbisia* and *Rhynchotheca*) or three (*Wendtia*) carpels. Stylodia free, short (*Wendtia*), or stigmas sessile (*Balbisia*); ovary 3- or 5-locular, ovules 1–2 (*Wendtia, Rhynchotheca*), or 5–50 (*Balbisia*) per locule, pendulous, anatropous to campylotropous, with ventral raphe. Fruits capsules, valvular ('septifragal') or loculicidal. Seeds endotegmic, embryo straight, with thin endosperm, n = 9. . . . . 4. LEDOCARPACEAE.

1 Nodes pentalacunar or multilacunar.

5 Stipules intrapetiolar, relatively small, papery and caduceus (*Bersama*) or often large, showy and persistent. Capsules loculicidal. Flowers more or less zygomorphic, bisexual or polygamous, in axillary or terminal racemes, often large, resupinate by twisting of the pedicel. Sepals 5 or 4, unequal, free or basally connate, imbricate. Petals 4–5, free, clawed, five and unequal or one abortive. Nectary disc extrastaminal, unilateral, well developed. Stamens 4–5, sometimes 8, often declinate, free or sometimes basally connate; anthers dorsifixed, slightly versatile, introrse. Pollen grains 3-celled, 3-colporate. Gynoecium of 4(5) carpels; style with four to five-lobed stigma; ovary superior, slightly sunken, 4–5-locular, each locule with 1 basal (*Bersama*) or 2–5 axile (*Melianthus*), erect to pendulous ovules; ovule of *Bersama* may smell of mustard, and ovule of *Melianthus* – smells nasty. Fruits capsules, opening loculicidally or only at the apex, in *Melianthus* papery or coriaceous, inflated and longitudinally deeply sulcate and sometimes apically lobed, in *Bersama* tough-walled, often woody, dehiscing longitudinally by 4–5 valves. Seeds exotestal, the tegmen is crushed and unlignified; embryo small or large, straight, endosperm copious, starchy or oily, exarillate in *Melianthus* and subtended by a large yellowish aril in *Bersama*. Characteristic anatomical feature is the presence of styloids in

vegetative parts and sometimes also in the seed coat. Nodes penta- or multilacunar. Evergreen subshrubs (*Melianthus*), or gard-wooded shrubs, or small trees. Leaves alternate, imparipinnate, or occasionally trifoliate, conduplicate, strong odour. Stomata anomocytic. Toxic principles, known as bufadienolides, present in both genera. Producing flavonols (quercetin), ellagic acid; present traces of inulin and elongated prismatic crystals. n = 18, 19. . . . . 5. MELIANTHACEAE.

5 Stipules absent, capsules septicidal.

6 Leaves simple, slightly peltate, crenately or dentately lobed. Soft-wooded shrubs or small trees with leafless older branchlets covered with a thin layer of periderm. Internodes green and photosynthetic, glabrous to densely pubescent, often with minute glandular hairs. Vessel elements very short; lateral pitting scalariform or transitional. Rays composed entirely of square and upright cells containing raphides. Leaves crowded apically on the branchlets, alternate, petiolate, coriaceous, palmately or subpalmately veined, either essentially glabrous or with a dense indumentum of long, multicellular, uniseriately branched hairs when young, often becoming glabrous with age, minute glandular hairs present, producing a shiny layer of glutinous exudate; petiole distinctly expanded basally into a decurrent sheath fused with the preceding internode, forming a pseudocortex; stipulelike outgrowths may occur on the sheath lobes; at the end of the growing season the sheath, petiole, and lamina are detached together, which is initiated by the formation of a periderm at the junction of the pseudocortex and the real cortex. Stomata anomocytic. Flowers conspicuous, in terminal racemes eventually displaced laterally by new leafy shoots emerging below the peduncle, bracteolate, bisexual, actinomorphic to slightly zygomorphic, 5-merous, or occasionally 4-merous, pollinated by sunbirds. The bracts, sepals and petals densely covered with short trichomes with globose glandular head cells. Sepals five, free or shortly connate, imbricate, persistent. Petals five, free, imbricate, much longer than the sepals, minutely ciliate, partly reflexed or (*Greyia flanaganii*) loosely coherent apically, brilliant red or crimson, rarely white, caducous. Extrastaminal nectary disc well developed, connected by five interstaminal ridges with the base of the ovary, fleshy, cupular, with the margin crowned by ten small vascularized staminodia the distal portions of which are subpeltate with a glandular area at the apex or irregularly laceratedentate at the apex, producing copious nectar. Stamens ten, free, seemingly placed in a single series but most probably obdiplostemonous (Schonland 1914; Dahlgren and van Wyk 1988), episepalous stamens maturing first; filaments subulate; anthers fairly large, with very narrow connective protrusion, slightly introrse. Gynoecium of (4)5(6) united carpels with stylodia connate into a slender, subulate style; stigma small, shortly 5-lobed, pappilae on stigmatic cleft. Ovary semi-inferior "because of growth of the peripheral tissue: (Ronse Decraene and Smets 1992: 389), deeply (4)5(6)-lobed, with slightly bifid parietal placentas developing from the intrusive margins of the carpels and eventually meeting and forming the placental column around a central hollow; the borderline between the individual carpel margins remains clearly distinguishable (Steyn 1977), and therefore the ovary is essentially unilocular. Ovules numerous, in two rows on each placenta. Fruits capsules, mericarps five, comprising follicles, the mature carpels separating from the central column or hollow, and opening ventrally. Seeds numerous per locule, minute, elongate; seed coat membranous, formed by both integuments, the exotegmen and next layer is fibrous, but although the mesotegmic cells elongate; embryo small, straight; endosperm copious, fleshy. Producing ellagic acid and quercetin, but not myricetin; present durses and raphids; n = 16 or 17 . . . . .6. GREYIACEAE

6 Leaves lyrate-pinnatifid to pinnate or rounded-cordate. Perennial rhizomatous, scapose herbs, glandular-pilose or tomentose. Nodes trilacunar with five traces. Leaves crowded or subradical, conduplicate-plicate, pinnately veined (*Francoa*) or palmately veined (*Tetilla*), estipulate. Stomata (in *Francoa*) diacytic. Flowers in lax, elongated racemes, bracteate, bisexual, actinomorphic, sometimes zygomorphic, usually

4-merous, but occasionally 5-merous. Sepals (4)5, persistent, valvate, equal (*Francoa*) or (*Tetilla*) unequal with two much smaller or deficient; petals clawed or sessile, inserted at the base of the calyx, imbricate or contorted. Stamens 8 (*Tetilla*) or 4–5 (*Francoa*), alternating with as many staminodia; filaments free; anthers oblong or subglobose, latrorse. Nectary disc interstaminal, covered with nectarostomata. Pollen grains 3-colporate. Gynoecium of four (rarely two) carpels; stigmas four, rarely two, sessile, commissural (stand over the "commissures," lines of fusion of the adjacent carpels). Ovary semi-inferior, tetragonous, 4-lobed at the apex, 4-locular or very rarely 2-locular; stigma papillae on stigmatic cleft; ovules numerous, biseriate. Fruits oblong, erect, 4-locular, septicidal (*Tetilla*) or loculicidal (*Francoa*) capsules with numerous seeds. Seeds very small, striolate; exotestal cells elongated, thickened, endotesta elongated, with thickened anticlinal walls, tegmen of pigmented cells; embryo, minute, straight, in the copious, fleshy endosperm. Present ellagic acid (strongly – *Francoa*) and anthocyanin in roots, and traces of inulin; n = 13. . . . . 7. FRANCOACEAE.

1. HYPSEOCHARITACEAE

Weddell 1861. 1/6. Andes of South America from Peru to Chile and northern Argentina.

Hypseocharis.

Hypseocharitaceae close related to the Geraniaceae and their seeds have many common features (Boesewinkel and Bauman 2000) In habit *Hypseocharis* somewhat resembles *Erodium*, and its staminal arrangement compares well with that of *Monsonia* and *Sarcocaulon* (Rama Devi 1991), but it differs from the Geraniaceae in its capsular fruits and estipulate leaves, as well as in its cochlear embryo with spirally folded cotyledons (Boesewinkel 1988).

2. GERANIACEAE

A. L. de Jussieu 1789 (including Erodiaceae Horaninow 1847). 4/c.830. Subcosmopolitan, but mainly in temperate and subtropical regions.

Geranium, Erodium (including *California*), *Pelargonium, Monsonia* (including *Sarcocaulon*).

Close to both the Hypseocharitaceae and the Vivianiaceae but their fruits are schizocarps.

3. VIVIANIACEAE

Klotzsch 1836. 4/6. Southern Brazil, Uruguay, Argentina, and Chile.

Viviania, Araeoandra, Caesarea, Cissarobryon.

Usually included in the Geraniaceae from which they differ in capsular fruits, estipulate leaves, and abundant fleshy endosperm.

4. LEDOCARPACEAE

Meyen 1834 (including Rhynchothecaceae Endlicher 1841). 3/12. Andian South America.

4.1 LEDOCARPOIDEAE.
Calyx with an epicalyx of a whorl of bracteoles. Small shrubs or shrublets. Leaves usually opposite, simple or compound, entire or deeply lobed or dissected, estipulate. Flowers solitary or in few-flowered corymbs or terminal racemes, bisexual, actinomorphic, 5-merous. Sepals free, imbricate. Petals imbricate (*Wendtia*) or contorted (*Balbisia*). Stamens ten, free, alternisepalous; anthers extrorse or introrse. Pollen grains inaperturate. Gynoecium of five (*Balbisia*) or three (*Wendtia*) carpels. Stylodia free, short (*Wendtia*), or stigmas sessile (*Balbisia*); ovary 3- or 5-locular, ovules 1–2 (*Wendtia*), or 5–50 (*Balbisia*) per locule, pendulous, anatropous to campylotropous, with ventral raphe. Fruits capsules, valvular ('septifragal') or loculicidal. Seeds endotegmic, with thin endosperm, n = 9. – *Balbisia* (*Ledocarpon*), *Wendtia*.

4.2 RHYNCHOTHECOIDEAE
Calyx without an epicalyx. Many branched spiny shrublets with 4-angled branches. Leaves small, entire, 3-fid or 3-partite. Flowers terminal, sometimes fasciculate, apetalous. Sepals acuminate. Anthers large, oblong. Pollen grains periporate. Gynoecium of five carpels; style short, thick, with five elongate stigmas; ovules two per locule, pendulous from the middle of the loculi. Fruits septifragally dehiscent

capsules, beaked (beaks not revolute). Seeds with straight embryo and thin endosperm, seed coat derived from both integuments and not specialized. – *Rhynchotheca*.

Related to the Geraniaceae, however, they clearly differ from the Geraniaceae in absence of nectary glands, free stamens, inaperturate pollen grains, free stylodia, capsular fruits, and also endotegmic seeds.

5. MELIANTHACEAE

Berchtold et J. Presl 1820. 2/16 (including Bersamaceae Doweld 2001). Tropical and South Africa; *Melianthus* (8) is endemic to South Africa, often occurring in arid areas; *Bersama* (8) is usually confined to forest regions in South and tropical Africa.

Melianthus, Bersama.

Related to the Geraniaceae.

6. GREYIACEAE

Hutchinson 1926. 1/3. South Africa, mainly the eastern Transvaal escarpment and the Drakenberg Mountains.

Greyia.

Greyiaceae related to the Melianthaceae.

7. FRANCOACEAE

Adr. Jussieu 1832. 2/2. Mountains of middle Chile.

Francoa, Tetilla.

Very closely related to the Greyiaceae.

Bibliography

Albers F. 1990. Comparative karyological studies in Geraniaceae on family, genus, and section level. In: P Vorster, ed. Proc. Intern. Geraniaceae Symp. Republic of South Africa, pp. 117–122. Stellenbosch.

Albers F. 1996. The taxonomic status of Sarcocaulon (Geraniaceae). South Afr. J. Bot. 62: 345–347.

Albers F and JJA Van der Walt. 2007. Geraniaceae. In: K Kubitzki, ed. The families and genera of vascular plants, vol. 9, pp. 157–167. Springer, Berlin/Heidelberg/New York.

Baker 1907. A revision of *Bersama*. J. Bot. 45: 12–21.

Bate-Smith EC. 1973. Chemotaxonomy of *Geranium*. Bot. J. Linn. Soc. 67: 347–359.

Behnke H-D and TJ Mabry. 1977. S-Type sieve-element plastids and anthocyanins in Vivianiaceae: Evidence against its inclusion into the Centrospermae. Plant Syst. Evol. 126: 371–375.

Boesewinkel FD. 1988. The seed structure and taxonomic relationships of *Hypseocharis* Remy. Acta Bot. Neerl. 37: 111–120.

Boesewinkel FD. 1997. Seed structure and phylogenetic relationships of the Geraniales. Bot. Jahrb. Syst. 119: 277–391.

Boesewinkel FD and W Been. 1979. Development of ovule and testa of *Geranium pratense* L. and some other representatives of the Geraniaceae. Acta. Bot. Neerl. 28: 335–348.

Boesewinkel FD and F Bouman. 2000. Oxalidaceae, Lepidobotryaceae, Hypseocharitaceae, Geraniaceae, Ledocarpaceae, Rhynchothecaceae, Vivianiaceae. In: A Takhtajan, ed. Comparative seed anatomy, vol. 6, pp. 21–39. Nauka, St. Petersburg (in Russian).

Bohm BA and J Chan. 1992. Flavonoids and affinities of Greyaceae with a discussion of the occurrence of B-ring deoxyflavonoids in dicotyledonous families. Syst. Bot. 17: 272–281.

Bohm BA, LS Donevan, and UG Bhat. 1986. Flavonoids of some species of *Bergenia, Francoa, Parnassia,* and *Lepuropetalon*. Biochem. Syst. Ecol. 14: 75–77.

Bortenschlager S. 1967. Vorläufige Mitteilungen zur Pollenmorphologie in der Familie der Geraniaceen und ihre systematische Bedeutung. Grana Palynol. 7: 400–468.

Carlquist S. 1985. Wood anatomy and familial status of *Viviania*. Aliso 11: 159–165.

Dahlgren KVO. 1930. Zur Embryologie der Saxifragaceen. Svensk. Bot. Tidskr. 24: 429–448.

Dahlgren R and AE van Wyk. 1988. Structures and relationships of families endemic to or centered in Southern Africa. Monogr. Syst. Bot. Missouri Bot. Gard. 25: 1–94.

Doweld AB. 1996. Melanthiaceae. In: A Takhtajan, ed. Comparative seed anatomy, vol. 5, pp. 348–351. Nauka, St. Petersburg (in Russian).

Doweld AB. 2001. The systematic relevance of fruit and seed structure in *Bersama* and *Melianthus* (Melianthaceae). Plant Syst. Evol. 227: 75–103.

Gäumann E. 1919. Studien über die Entwicklungsgeschichte einiger Saxifragales. Rec. Trav. Bot. Neerl. 16: 85–322.

Goldblatt P. 1979. Chromosome number in two cytologically unknown New World families, Tovariaceae and Vivianiaceae. Ann. Missouri Bot. Gard. 65(2): 776 – 777.

Gornall RJ and KIA Al-Shammary. 1998. Francoaceae. In: DF Cutler and M Gregory, eds. Anatomy of the Dicotyledons. Saxifragales (sensu Armen Takhtajan 1983), 4: 243–245. Oxford.

Gornall RJ, BA Bohm, and R Dahlgren. 1979. The distribution of flavonoids in the angiosperms. Bot. Not. 132: 1–30.

Gregory M. 1998. Greyiaceae. In: DF Cutler and M Gregory, eds. Anatomy of the Dicotyledons. Saxifragales (sensu Armen Takhtajan 1983), vol. 4, pp. 238–243. Academic Press, Oxford.

Hideux M and IK Ferguson. 1976. The stereostructure of the exine and its evolutionary significance in Saxifragaceae sensu lato. In: IK Ferguson and J Muller, eds. The evolutionary significance of the exine, pp. 327–377. Linn. Soc. Symposium, No 1. London/New York.

Hilger HH. 1978. Der multilacunare Knoten einiger *Melianthus* und *Greyia*-Arten im Vergleich mit anderen Knotentypen. Flora 167: 165–176.

Hooker JD. 1873. On *Melianthus trimenianus* H.F., and the affinities of *Greyia sutherlandi*. J. Bot. 11: 353–358.

Hunziker AT and LA Espinar. 1973. Aporte a la rehabilitación de Ledocarpaceae, familia monotipica. Kurtziana 7: 232–240.

Jackson BP and KR Jethwa. 1973. Morphology and anatomy of the leaves of *Bersama abyssinica* Fresen. from Kenya and Uganda. Bot. J. Linn. Soc. 66: 245–257.

Jay MM. 1967. Recherches chimiotaxinomiques sur les plantes vasculaires: Les flavonoides de *Greyia sutherlandii* Harv. C. R. Acad. Sci. Paris 265: 1086–1088.

Jay MM. 1970 (1971). Quelques problémes taxinomiques et phylogénétiques des Saxifragacées vus á la lumiére de la biochimie flavonique. Bull. Mus. Hist. Nat. (Paris), sér. 2, 42: 754–775.

Kaden NN. 1964. Morphology of fruits of the Geraniaceae. Nauchn. Dokl. Vyssh. Shkoly, Biol. Nauki 2: 97–102 (in Russian).

Khushalani I. 1963. Floral morphology and embryology of *Melianthus major* Linnaeus. Phyton (Buenos Aires) 10: 145–156.

Killick DJB. 1976. *Greyia sutherlandii*. In: DJB Killick, ed. Flowering plants of Africa. Vol. 44. Botanical Research Institute, Pretoria.

Klopfer K. 1972. Beiträge zur floralen Morphogenese und Histogenese der Saxifragaceae: 7. *Parnassia palustris* und *Francoa sonchifolia*. Flora 161B: 320–332.

Kumar A. 1976. Studies in Geraniales: II. The floral anatomy. J. Indian Bot. Soc. 55: 233–253.

Lefor MW. 1975. A taxonomic revision of the Vivianiaceae. Univ. Connecticut Occas. Papers 2: 225–255.

Leifertova I, H Buckova, and L Natherova. 1965. K chemotaxonomii znaku trislovin u rodu *Geranium*. Preslia 37: 413–418.

Linder HP. 2007. Melianthaceae. In: K Kubitzki, ed. The families and genera of vascular plants, vol. 9, pp. 250–259. Springer, Berlin/Heidelberg/New York.

Linder HP, T Dlamini, J Henning, and GA Verboom. 2006. The evolutionary history of *Melianthus* (Melianthaceae). Am. J. Bot. 93: 1052–1064.

Link DA. 1994. The nectaries of Geraniaceae. In: P Vorster, ed. Proc. Intern. Geraniaceae Symp., Republic of South Africa, pp. 217–225. Stellenbosch.

Lis-Balchin M. ed. 2002. *Geranium* and *Pelargonium*: The Genera *Geranium* and *Pelargonium*. Talyor & Francis, London.

Lu ZH, WZ Cao, and YL Zhang. 1995. Seed morphology of Geraniaceae in northeastern China and its taxonomic significance. Bull. Bot. Res. 15: 206–212.

Lu ZH, LP Yang, and WZ Cao. 1996. Pollen morphology and systematics of Geraniaceae in northeastern China. Bull. Bot. Res. 16: 315–321.

Meisert A, D Schulz, and H Lehmann. 2001. The ultrastructure and development of the light line in the Geraniaceae seed coat. Plant Biol. 3: 351–356.

Narayana LL. 1970. Oxalidaceae, Geraniaceae. In Symposium on comparative embryology of angiosperms. Bull. Indian Natl. Sci. Acad. 41: 114–120.

Narayana LL and D Rama Devi. 1995. Floral anatomy and systematic position of Vivianiaceae. Plant Syst. Evol. 196: 123–129.

Nemirovich-Danchenko EN. 1994. Seed structure in species of the genera *Francoa* and *Tetilla* (Francoaceae). Bot. Zhurn. 79: 109–114 (in Russian with English summary).

Nemirovich-Danchenko EN. 1995. Structure of seeds of the family Greyiaceae. Bot. Zhurn. 80(1): 99–104 (in Russian with English summary).

Nemirovich-Danchenko EN. 1996. Greyiaceae, Francoaceae. In: A Takhtajan, ed. Comparative seed anatomy, vol. 5, pp. 79–88. Nauka, St. Petersburg (in Russian).

Oltmann O. 1971. Pollenmorphologisch-systematische Untersuchungen innerhalb der Geraniales. Dissertaci-ones Botanicae, vol. 11.

Phillips EP. 1922. The genus *Bersama*. Bothalia 1: 33–39.

Price RA and JD Palmer. 1993. Phylogenetic relationships of the Geraniaceae and Geraniales from *rbc*L sequence comparisons. Ann. Missouri Bot. Gard. 80: 661–671.

Rama Devi D. 1991. Floral anatomy of *Hypseocharis* (Oxalidaceae) with a discussion of its systematic position. Plant Syst. Evol. 177: 161–164.

Ramamonjiarisoa BA. 1980. Comparative anatomy and systematics of African and Malagasy woody Saxifragaceae sensu lato. Ph.D. dissertration, University of Massachusetts, Amherst.

Robertson KR. 1972. The genera of Geraniaceae in the southeastern United States. J. Arnold Arbor. 53: 182–201.

Ronse Decraene LP and E Smets. 1999. Similarities in floral ontogeny and anatomy between the genera *Francoa* (Francoaceae) and *Greyia* (Greyiaceae). Int. J. Plant Sci. 160(2): 377–393.

Ronse Decraene LP, HP Linder, T Dlamini, and EF Smets. 2001. Evolution and development of floral diversity of Melianthaceae, an enigmatic Southern African family. Int. J. Plant Sci. 162: 59–82.

Sant Prasad Reddy T and LL Narayana. 1986. Chemotaxonomy of Geraniaceae. J. Indian Bot. Soc. 65: 428–235.

Schönland S. 1914. Notes on the genus *Greyia* Hook. and Harv. Rec. Albany Mus. 3: 40–51.

Soltis DE and PS Soltis. 1997. Phylogenetic relationships in Saxifragaceae sensu lato: a comparison of topologies based on 18S rDNA and *rbc*L sequences. Am. J. Bot. 84: 504–522.

Soltis DE, PS Soltis, MT Clegg, and M Durbin. 1990. *rbc*L sequence divergence and phylogenetic relationships in Saxifragaceae sensu lato. Proc. Nac. Acad. Sci. USA 87: 4640–4644.

Steyn E. 1974a. Leaf anatomy of *Greyia* (Greyiaceae). Bot. J. Linn. Soc. 69: 45–51.

Steyn E. 1974b. Abscission of leaves in *Greyia* Hook. and Harv. South Afr. J. Bot. 40: 193–200.

Steyn EMA. 1975. Embriogenie van *Melianthus major* L. South Afr. J. Bot. 41: 199–205.

Steyn E, PJ Robbertse, and HP van der Schijff. 1986. An embryogenetic study of *Bersama transvaalensis* and *Greyia sutherlandii*. South Afr. J. Bot. 52: 25–29.

Steyn E, PJ Robbertse, and AE van Wyk. 1987. Floral development in *Greyia flanaganii* with notes on inflorescence initiation and sympodial branching. South Afr. J. Bot. 53: 194–201.

Steyn EMA and AE van Wyk. 1987. Floral development of *Greyia flanaganii* with notes on inflorescence initiation and sympodial development. South Afr. J. Bot. 53: 194–201.

Takahashi M and S Kawano. 1989. Pollen morphology of the Melanthiaceae and its systematic implications. Ann. Missouri Bot. Gard. 76: 863–876.
Umadevi I, M Daniel, and SD Sabnis. 1986. Inter-relationships among the families Aceraceae, Hippocastanaceae, Melianthaceae and Staphyleaceae. J. Plant Anat. Morphol. 3: 169–172.
Verdcourt B. 1956. Melianthaceae. Further noted on the genus *Bersama* Fres. Kew Bull. 1955: 600–601.
Verhoeven RL and EM Marais. 1994. Pollen morphology of the Geraniaceae. In: P Vorster, ed. Proc. Intern. Geraniaceae Symp., Republic of South Africa, pp. 137–173. Stellenbosch.
Vorster P, ed. 1990. Proceedings of the International Geraniaceae Symposium held at the University of Stellenbosch, Republic of South Africa. Stellenbosch.
Warburg EF. 1938. Taxonomy and relationship in the Geraniales in the light of their cytology. New Phytol. 37: 130–159, 189–210.
Weigend M. 2005. Notes on the floral morphology in Vivianaceae (Geraniales). Plant Syst. Evol. 253: 125–131.
Weigend M. 2007. Ledocarpaceae. In: K Kubitzki, ed. The families and genera of vascular plants, vol. 9, pp. 213–220. Springer, Berlin/Heidelberg/New York.
Yeo PF. 1984. Fruit-discharge-type in *Geranium* (Geraniaceae): Its use in classification and its evolutionary implications. Bot. J. Linn. Soc. 89: 1–36.

Order 98. ZYGOPHYLLALES

Small trees, shrubs, or subshrubs, less often perennial or annual herbs. Stems often swollen or jointed at the nodes. Trichomes usually unicellular and simple, more rarely unicellular and 2-armed, sometimes glandular. Usually without scattered mucilage cells (except for spp. of *Nitraria*) and without secretory cells. All elements of wood are usually storied, except in *Balanites*. Vessel elements very small to moderately large, very to extremely short, exclusively with simple perforations and with vestured pits (Carlquist 2005); lateral pitting alternate and very small and numerous or (Nitrariaceae) intermediate between opposite or alternate; rarely (Balanitaceae) intervascular pits vestured. Fibers with bordered or (Tetradiclidaceae) simple pits. Rays usually narrow and short, homogeneous or rarely slightly heterogeneous. Axial parenchyma predominantly apotracheal, almost entirely paratracheal in *Bulnesia* and *Nitraria*, scanty metatracheal and vasicentric in *Tetraena*. Sieve-element plastids of S- or less often (some Zygophyllaceae) of Pcs-type. Nodes trilacunar. Leaves opposite or less often alternate, mostly paripinnate, often 2-foliolate, less often simple and entire, very rarely (*Pintoa*) gland-dotted, usually with well-developed and commonly slender stipules (sometimes modified into spines). Stomata anomocytic or (Balanitaceae) encyclocytic. Flowers in inflorescences (cymose or seldom racemose) or solitary, usually bisexual, commonly actinomorphic, (4)5(6)-merous. Sepals free or sometimes basally connate, imbricate or valvate. Petals usually free, imbricate or contorted, rarely valvate, sometimes wanting. Stamens in (1)2(3) cycles; filaments often with basal glands; anthers small or (*Peganum*) large, tetrasporangiate, dorsifixed, introrse, opening longitudinally. Tapetum secretory (becoming the false periplasmodium in Tetradiclidaceae). Microsporogenesis simultaneous. Pollen grains 2-celled or rarely 3-celled, of various types, but mostly 3-colporate. Nectary disc usually well developed, more or less distinctly intrastaminal, sometimes modified into a gynophore. Gynoecium of five or less often four, seldom six or only two united carpels, with a slender (sometimes very short) simple style; stigma capitate or sometimes lobed or cleft; ovary superior, multilocular, with one to several (rather numerous in *Peganum*) pendulous or seldom ascending ovules per locule. Ovules epitropous or rarely (*Nitraria*) apotropous, anatropous to sometimes hemitropous, campylotropous, or orthotropous, bitegmic, crassinucellate, mostly with an endothelium (except in *Nitraria, Peganum, Malacocarpus,* and *Seetzenia*), sometimes (*Malacocarpus* and *Seetzenia*) with a funicular-placental obturator. Female gametophyte of *Polygonum*-type. Endosperm nuclear. Fruits often capsules or schizocarps, rarely drupes or berries. Seeds typically exotestal-endotestal, with endotegmic cells elongate tangentially; embryo straight or slightly curved, in hard oily endosperm, which is sometimes absent.

Closely related to the Linales and had a common origin from a rutalean stock.

Key to Families

1 Leaves usually stipulate.
  2 Leaves alternate.
    3 Glabrous or hairy perennial herbs (*Peganum*) or shrubs (*Malacocarpus*) with raphids. Leaves entire or irregularly dissected, leathery or fleshy, sessile, simple, with minute setaceous stipules on some leaves (usually absent in *Malacocarpus*); glands with a clavate or spherical head are characteristic of *Peganum*; hypoderm present beneath the upper epidermis

in *Peganum*. Flowers solitary and axillary, bisexual, actinomorphic, 4–5-merous. Sepals leaflike, entire or pinnately cleft, valvate. Petals free, contorted and imbricate. Stamens 12, or 15, in three cycles, attached to the nectary disc; filament dilated at the base. Pollen grains 3-colporate, with reticulate ornamentation. Gynoecium of 3–2 carpels, with simple cylindrical style continued into long 3–2-keeled stigma papillate along the keels; ovary superior, 3–2-lobed and 3–2-locular, with numerous pendulous, anatropous ovules per locule. Fruits loculicidal capsule (*Peganum*) or 2-locular berry (*Malacocarpus*). Seeds angular, with spongy, mucilage testa bearing small, funicular outgrowth; embryo green, slightly curved, in fleshy endosperm. Present alkaloids, n = 12.. . . . . 3. PEGANACEAE.

3 Succulent shrubs, sometimes spiny, secretary cavities sometimes present, with mucilage cells. Leaves entire or sometimes 2–3 toothed above, petiolate, simple stipules minute, intrapetiolar, free of one another, often caduceus or persistent; very large clusters of crystals occur in the leaves. Stomata anomicytic. Flowers in the axils of small caducous bracts, in scorpioid cymose inflorescences, actinomorphic, 5-merous. Calyx synsepalous, small, fleshy, toothed, imbricate, persistent. Petals free, concave, conduplicate-valvate. Stamens 15–10, five of them antesepalous, the other ones in pairs or solitary antepetalous; filaments subulate, without basal appendages; anthers versatile. Pollen grains 3(4)-colporate, with thick ectexine. Gynoecium of three or rarely of six carpels, with a continuous fleshy style; stigma of three (rarely 6) papillate decurrent ridges; ovary 3(6)-locular, not lobed, with one pendulous, apotropous ovule per carpel. Fruits drupaceous, 1-seeded, derived from single fertile locule, with fleshy exocarp with lignified inner part (named scleromesocarp) and stony thin fovealate endocarp (in *N. sphaerocarpa* the drupe is almost dry). Seeds with a straight, green embryo, with the radicle next to hilum; endosperm wanting, n = 12, 30. . . . . 4. NITRARIACEAE.

2 Leaves opposite or sometimes alternate, usually pinnate, rarely 3-foliolate or 2-foliolate or simple and entire, leathery, fleshy, or modified into spines, usually with stipules (sometimes spiny, or scaly, or leafy, persistent). Stomata mostly anomocytic. Small trees or shrubs, rarely perennial or annual herbs, resinous or not resinous, more or less succulent. Wood often fluorescing; vessels with simple perforation and usually with vestured pits. Axial parenchyma diffuse or diffuse-in-aggregates; rays heterogenerous or homogeneous; crystals one per cell or septate portion of cell in wood or secondary phloem (Carlquist 2005). Nodes often swollen or jointed, trilacunar with three traces. Flowers in cymose inflorescences or solitary, terminal or axillary, or leaf-opposed, bisexual or very rarely unisexual and dioecious (*Neoluederitzia*), actinomorphic or rarely zygomorphic, 5-merous or sometimes 4-merous. Sepals free or less often basally connate, imbricate or valvate. Petals usually free, imbricate or contorted, rarely valvate, sometimes wanting (*Seetzenia, Miltianthus*). Nectary disc usually present, extrastaminal or intrastaminal. Stamens in (1)2(3) cycles, alternisepalous, or oppositisepalous, when the outer whorld is staminodal; filaments often with basal ligular appendages; anthers versatile, introrse, or latrorse. Pollen grains mostly 3-colporate, less often 3-colpate (*Seetzenia, Sisyndite, Viscainoa*) or pantoporate (Tribuloideae). Gynoecium of five or less often four carpels, gradually attenuated into a simple style with lobed or capitate stigma; ovary often angled or winged, 5–6-locular (in *Tribulus* locules secondarily divided by 'false septa'), with one to many axile, pendulous, apotropous to orthotropous or campylotropous ovules per locule; micropyle not zig-zag. Endothelium present (absent in *Seetzenia*). Fruits usually loculicidal and/or septicidal capsules or schizocarps, rarely (*Sericodes*) succulent and berrylike. Fruit elastically dehiscent (when of cocci), or passively dehiscent. Seeds with straight to weakly curved, green embryo, with oily or without endosperm; seed-coat with endotestal crystal layer and a lignified endotemgen. Present anthroquinones, flavonols (kaempferol and quercetin), some species produce steroid and triterpenoid saponins, alkaloids (harman, harmin, and harmol); n = 6, 8–13 or (*Tetraena*) 14. . . . . 1. ZYGOPHYLLACEAE.

1 Leaves estipulate (Balanitaceae) or stipules minute and only visible in the youngest leaves.

4 Small trees and shrubs with very long axillary spines and bitter bark. Leaves alternate, leathery, petiolate, 2-foliolate, leaflets coriaceous, entire; stipules absent. Stomata anomocytic. Flowers in axillary 3–9-flowered racemose inflorescences, small, greenish-yellow, bisexual, actinomorphic, usually 5-merous. Sepals free, imbricate, caducous. Petals free, narrow oblong, spreading, imbricate. Stamens usually ten, inserted in the grooves below the outside of the nectary disc; filaments free, filiform; anthers dorsifixed, introrse. Pollen grains usually 2-cellular, 3-colporate or 6-rugorate. Nectary disc intrastaminal, thick, shortly cupular or cushion-shaped, 10-grooved. Gynoecium of five carpels, with very short, subulate style; stigma minute; ovary usually partly immersed in the disc, 5-locular, with one anatropous ovule per locule, pendulous from the axis below the apex of the loculi; micropyle zig-zag. Endothelium present. Fruits 1-seeded drupes with very thick, bony, 5-angled, 1locular endocarp surrounded by oily mesocarp. Seeds with subfibrous testa, seed-coat consists of isodiametrical parenchyma cells with scattered lignified sclerenchyma elements, green embryo; endosperm wanting, n = 9.. . . . . 2. BALANITACEAE.

4 Small succulent annual herbs. Fibers with rare simple pits. Leaves with two minute, lateral, colorless outgrowths at the base; two lowest pairs of leaves of the basal rosette opposite, decussate, entire, spatulate, cauline ones alternate, usually slightly dissected, often with two basal, earlike lobes, the uppermost ones entire; stipules minute. Flowers very small, subsessile in the axils of the basal leaflike bracts, forming a scorpioid terminal inflorescence, bisexual, actinomorphic, 4-merous or rarely 3-merous. Sepals basally connate, imbricate, persistent. Petals free, very shortly clawed, imbricate, persistent. Stamens attached to the intrastaminal, annular, slightly lobed nectary disc; filaments free, long, slender; anthers subglobose. Microsporogenesis simultaneous. Pollen grains 3-colporate, with striate ornamentation. Gynoecium of four carpels alternating with the sepals; style simple, arising from the apical depression of the ovary; stigma clavate, 4-sulcate, with four short, decurrent, double series of papillae; ovary superior, on very short obconic gynophore, slightly 4-lobed, 4-locular; each locule contains 6 ovules hanging on the 6-branched free placentas attached to the center and directed toward the dorsal sides of the ovary; 4 branches of each placenta are situated along the median line of the locule and bear four ovules, each of the two other branches is directed to the lateral locelli and bears one ovule close to the ovary wall. During the fruit development two lateral ovules of each placenta become included in the tissues of the ovary wall that completely frame the developing seed; as a result, the developing seed turns out to be completely within an endocarplike chamber connected with the middle chamber by a squeezed orifice; thus four ovules of each locule are in the middle chamber and two other ovules are in the lateral one. Ovules hemitropous, bitegmic, crassinucellate; the outer integument contains specific cells that accumulate mucilage. Endothelium absent. Female gametophyte of *Polygonum*-type. Endosperm nuclear. The fruit is at first fleshy and red but after maturing becomes dry, capsulelike; it is of very unique structure, dehiscence, and seed dispersal. At first it is opening loculicidally and discharges four seeds of the middle chamber while the lateral seeds are still in their chambers; the discharge of the lateral seeds of each locule takes place at the later stage of the ripening of the fruit; dehiscence of the lateral chambers occurs as a result of a mechanical rupture of the epicarp tissues around the seed and the seed discharges together with a part of the surrounding tissues including the epidermis of the wall of the locule; part of the epidermis becomes a membranous edging around the seed and serves as an anemochorous adaptation. Thus dispersing seeds are of two kinds: with or without membranous wings. The anemochorous seeds are adapted for more distant dispersal (Bunge 1840; Yanishevsky 1940). Seeds small, oblong, dimorphic; those from the middle chambers of the fruit with tuberculate and mucous surface, whereas those from the lateral chambers are almost glabrous and not mucous; testa thin; embryo straight, in fleshy endosperm, n = 7.. . . . . .5. TETRADICLIDACEAE.

1. ZYGOPHYLLACEAE

R. Brown 1814 (including Tribulaceae Trautvetter 1853). 24/240. Mostly tropical and subtropical regions, but also in warm-temperate regions, mainly in arid and subarid areas.

1.1 ZYGOPHYLLOIDEAE
Shrubs, herbs, or rarely trees. Leaves alternate (*Sericodes* and *Plectrocarpa*) or more often opposite, paripinnate or imparipinnate, often 2-foliolate, sometimes unifoliolate or simple, stipulate. Nectary disc usually present, often small. Pollen grains 3-colporate or rarely (*Seetzenia*) 3-colpate. Gynoecium of 5–2 carpels. Fruits capsular, dehiscing septicidally or breaking into cocci. Seeds usually with endosperm. – SEETZENIEAE: *Seetzenia;* SISYNDITEAE: *Sisyndite;* SERICODEAE: *Sericodes;* ZYGOPHYLLEAE: *Zygophyllum* (including *Halimiphyllum*), *Miltianthus, Pintoa, Roepera, Sarcozygium, Plectrocarpa, Metharme, Fagonia;* LARREAE: *Guaiacum, Bulnesia, Porlieria, Larrea*; MORKILLIEAE: *Morkillia, Viscalnoa.* Widely distributed.

1.2 TETRAENOIDEAE
Shrubs with very tortuous branches and short, lateral branchlets. Young shoots and leaves covered by 2-armed trichomes. Axial parenchyma scanty metatracheal and vasicentric. Leaves alternate, 2-foliolate. Flowers solitary in the leaf axils, 4-merous with a very short androgynophore. Sepals imbricate. Petals valvate. Stipules membranous, lacerate. Stamens eight in two cycles; filaments with membranous scaly appendage at the base. Pollen grains 3-colporate. Nectary disc annular. Gynoecium of 4-carpels with long, simple, persistent style and capitate stigma; ovary deeply lobed, with 3–5 ovules per locule; ovules orthotropous, with long, curved funicles. Fruits deeply 4-lobed, breaking into four one-seeded indehiscent cocci, with the style in the middle, indehiscent; exocarp spongy, densely villose. Seeds one in each locule, oblong, with thick testa; embryo oblong-ovoid, green, with fleshy cotyledons; endosperm wanting, n = 14. – *Tetraena.* Arid regions of China (Inner Mongolia).

1.3 TRIBULOIDEAE
Shrubs (*Kelleronia*) or biennial prostrate herbs. Leaves alternate (*Kelleronia*) or opposite, pinnate, stipulate. Trichomes of two markedly different sizes: short and pointed, or much longer and wider and growing from a basal "plinth" of epidermal cells. Pollen grains pantocolpate. Nectary disc present, lobed. Fruits breaking up into five cocci (*Kelleronia*), dividing by oblique transverse septa into 5–3 one-seeded compartments (*Tribulus*) or separating into 12–5 indehiscent 1- or rarely 2-seeded cocci. Seeds without endosperm. – *Kelleronia, Kallstroemia, Tribulopis, Tribulus.* Widely distributed.

1.4 NEOLUEDERITZIOIDEAE
Shrubs with axillary thorns. Leaves alternate, imparipinnate. Flowers dioecious. Filaments with a deeply 2-lobed hairy scale at the base. Pollen grains 3-colporate. Disc in the female flowers hyaline, embracing the young fruits. Gynoecium of five carpels. Fruits densely silky-villous, breaking up into five cocci. Seeds without endosperm. – *Neoluederitzia.* Southwestern Africa.

1.5 AUGEOIDEAE
Very succulent annual herbs with jointed branches. Leaves opposite, clavate, connate at the base, simple, semiterete, estipulate. Petals 3-lobed. Nectary disc urceolate, stipitate, membranous, with ten subulate teeth. Stamens with 3-fid filaments, the lateral lobes longer than the oblong anthers. Pollen grains 3-colporate. Gynoecium of 10 carpels. Ovary 10-locular. Fruits 10-ribbed capsules. Seeds without endosperm. – *Augea.* Dry regions of South Africa.

The family is very diversified and specialized in various directions. Perhaps the Morkillioideae are relatively the least advanced. Seeds of both *Viscainoa* and *Morkillia* have endosperm and their fruits are septicidal capsules whereas *Sericodes* is devoid of endosperm and its villous fruits are breaking into indehiscent cocci. The Zygophylloideae are also less specialized than the three next subfamilies, especially *Seetzenia, Zygophyllum,* and *Guaiacum.* Palynologically *Seetzenia* and *Sysindite* are the most primitive (3-colpate pollen grains), but *Seetzenia* is an apetalous prostrate herb with brittle branches, whereas *Sysindite* is a *Spartium*-like shrub and its seeds are without endosperm. The genus is rather isolated within the Zygophylloideae. The subfamily Tetraenoideae is close to the Zygophylloideae but very specialized.

Tribuloideae, although differing in many respects from the rest of the family, do not deserve the status of a separate family.

The monotypic African subfamily Neoluederitzioideae is connected with the Tribuloideae, and El Hadidi

(1977) includes both of them in the family Tribulaceae. Besides, the subfamily Neoluederitzioidieae comprises both Neoluedentzioideae and *Sisyndite,* which are not closely related.

Augeoideae are the most specialized subfamily of the Zygophyllaceae. *Augea capensis* is one of the most remarkable African succulent plants.

2. BALANITACEAE

Endlicher 1841. 1/25. Tropical and North Africa (from Mauritania to Egypt) and from western Asia through India to Burma.

Balanites.

Sheahan and Chase (1996, 2000) include *Balanites* in the Zhygophyllaceae – Tribuloideae, but from the Zygophyllaceae it markedly differs in so many respects (including estipulate leaves, wood anatomy, pollen morphology, structure of drupaceous fruits, and seed-coat anatomy) that they definitely constitute a separate family. According to Boesewinkel (1994), in the structure of ovule and seed anatomy there are many similarities with the members of Linales.

3. PEGANACEAE

van Tieghem ex Takhtajan 1987. 2/6. Southern Europe, North Africa, arid and semi-arid regions of Asia and North America (northwestern Texas and northern Mexico); monotypic genus *Malacocarpus* is endemic to northern Iran and central Asia.

Peganum, Malacocarpus.

Both genera are usually included in the Zygophyllaceae. However, they differ from the Zygophyllaceae in many characters, including multicellular, capitate trichomes, elongated leaf epidermal cells alternating with shorter cells, the presence of a hypodermis in the stem, angular, thin-walled pericyclic fibers that sometimes form a complete cylinder, and the absence of bracysclereids, raphid sacs in the leaf tissue (see Sheahan and Cutler 1993), the absence of ligular appendages at the base of stamens, the absence of the endothelium (and other embryological differences – see Kapil and Ahluwalia 1963; Kamelina 1959, 1991), pollen morphology (Erdtman 1959; Agababian 1965), the structure of stigma, and spongy seed coat.

4. NITRARIACEAE

Berthold et J. Presl 1820. 1/10. From North Africa and southeastern Europe to Siberia, Afghanistan, central Asia, and 1 sp. in arid and coastal regions of southern Australia.

Nitraria.

Probably related to the Zygophyllaceae, but differ in the absence of endothelium (Kamelina 1991), leaf morphology, pollen morphology (Agababian 1965), structure of the drupe and seed. The family is relatively isolated within the order.

5. TETRADICLIDACEAE

Takhtajan 1986. 1/1. North Africa and Irano-Turanian region.

Tetradiclis.

Tetradiclis has historically been included in such different families as Crassulaceae, Elatinaceae, Zygophyllaceae, and Rutaceae. Endlicher (1841), Lindley (1853), and Le Maout and Decaisne (1868) placed *Tetradiclis* in the Elatinaceae whereas in Engler's (1896) system it is included in Zygophyllaceae as a monotypic subfamily Tetradiclidoideae. Fenzl (1841), Hallier (1908, 1912) and Takhtajan (1966) spoke in favor of Rutaceae, but the majority of modern systematists follow Engler. In 1987 I still accepted the rutalean affinity of *Tetradiclis* but separated it into a family of its own. Later I came to the conclusion that Tetradiclidaceae differ from the Rutales in so many respects, that it probably must be raised to the ordinal level. The structure of the gynoecium and fruits, a very unique type of placentation, and no less the unique type of fruit dehiscence confirm a very isolated position of the family. No less unique are some embryological characters, including the very specific type of tapetum (Kamelina 1991, 1994). Kamelina also has come to the conclusion that *Tetradiclis* deserves the rank of an order that would be placed between Rutales and Linales.

Bibliography

Agababian VS. 1965. Pollen structure and systematics in the family Zygophyllaceae. Trudy Bot. Inst. Armenian Acad. Sci. 15: 66–90 (in Russian).

Barker RM. 1998. Notes on the genus *Tribulopsis* (Zygophyllaceae) in Australia including the descriptions of five new species and one new subspecies, revised keys and typifications. J. Adelaide Bot. Gard. 18: 43–74

Beier B-A, MW Chase, and M Thulin. 2003. Phylogenetic relationships and taxonomy of subfamily Zygophylloideae (Zygophyllaceae) based on molecular and morphological data. Plant Syst. Evol. 240: 11–39.

Bobrov EG. 1944. On the Asiatic species of the genus *Nitraria* L. Sov. Bot. 14: 19–30 (in Russian).

Bobrov EG. 1965. On the origin of the flora of deserts of the Old World in relation to the revision of the genus *Nitraria* L. Bot. Zhurn. 50: 1053–1067 (in Russian).

Boesewinkel FD. 1994. Ovule and seed characters of *Balanites aegyptiaca* and the classification of the Linales – Geraniales – Polygalales assembly. Acta Bot. Neerl. 43: 15–25.

Bunge A. 1840. Über die Gattung *Tetradiclis* Stev. Linnaea 14: 161–178, Taf. I.

Carlquist S. 2005. Wood anatomy of Krameriaceae with comparisons with Zygophyllaceae: phylesis, ecology and systematics. Bot. J. Linn. Soc. 149: 257–270.

El Hadidi MN. 1977. Tribulaceae as a distinct family. Publ. Cairo Univ. Herb. (7 and 8): 103–108.

Fenzl E. 1841. Die Gattung *Tetradiclis* Stev. und ihre Stellung im naturlichen Systeme. Linnaea 15: 289–299.

Iljin MM. 1951. On the systematic position of the genus *Tetraena* Maxim. Trudy Tomsk State Univ. 116: 143–144 (in Russian).

Inamdar JA. 1969. Epidermal structure, stomatal ontogeny, and relationship of some Zygophyllaceae and Simaroubaceae. Flora 158B: 360–368.

Inamdar JA and RC Patel. 1970. Epidermal structure and development of stomata in vegetative and floral organs of *Fagonia oretica* Linn. Flora 159B: 63–70.

Kamelina OP. 1959. Zygophyllaceae, Nitrariaceae, Ba-lanitaceae. In: MS Yakovlev, ed. Comparative embryology of flowering plants: Brunelliaceae-Tremandraceae, pp. 145–157. Nauka, Leningrad (in Russian).

Kamelina OP. 1991. Comparative morphological analysis as a method of phylogenetic systematics of flowering plants. Ph.D. dissertration, University of Tashkent (in Russian).

Kamelina OP. 1994. Embryology and systematic position of *Tetradiclis* (Tetradiclidaceae). Bot Zhurn. 79(5): 11–27 (in Russian with English summary).

Kamelina OP. 1996. Tetradiclidaceae. In: A. Takhtajan, ed. Comparative seed anatomy, vol. 5, pp. 377–381. St. Petersburg (in Russian).

Kapil RN and K Ahluwalia. 1963. Embryology of *Peganum harmala* Linn. Phytomorphology 13: 127–140.

Lia VV, VA Confalonieri, CI Comas, and JH Hunziker. 2001. Molecular phylogeny of *Larrea* and its allies (Zygophyllaceae): Reticulate evolution and the probable time of the creosote bush arrival to North America. Molecular Phylog. Evol. 21: 309–320.

Ma Yu-chuan and S Zhang. 1990. Study on the systematic position of *Tetraena* Maxim. Acta Phytotax. Sinica 28(2): 89–95.

Maksoud SA and MN El Hadidi. 1988. The flavonoids of *Balanites aegyptiaca* (Balanitaceae) from Egypt. Plant Syst. Evol. 160: 153–158.

Mauritzon J. 1934. Etwas über die Embryologie der Zygophyllaceen sowie einige Fragmente über die der Humiriaceen. Bot. Not. 87: 407–422.

Nag TN et al. 1986. Free endogenous ascorbic acid from Zygophyllaceous plants growing in the arid zone of Rajasthan. J. Indian Bot. Soc. 9: 112–113.

Nair NC and RK Jain. 1956. Floral morphology and embryology of *Balanites roxburghii* Planch. Lloydia 19: 269–279.

Nair NC and KS Nathawat. 1958. Vascular anatomy of the flower or some species of Zygophyllaceae, part l. J. Indian Bot. Soc. 37: 172–180.

Narayana HS and CG Prakasa Rao. 1963. Floral morphology and embryology of *Seetzenia orientalis* Decne. Phytomorphology 13: 197–205.

Narayana L, P Satyauarayana, and H Radhakrishnaiah. 1990. Systematic position of Balanitaceae. In: K Bilgrami and J Dogra, eds. Phytochemistry and plant taxonomy, pp. 157–164. CBS, Delhi.

Nemirovich-Danchenko EN. 1996. Nitrariaceae, Balanitaceae, Peganaceae. In: A Takhtajan, ed. Comparative seed anatomy, vol. 5, pp. 399–405. Nauka, St. Petersburg (in Russian).

Parameswaran N and H Conrad. 1992. Wood and bark anatomy of *Balanites aegyptiaca* in relation to ecology and taxonomy. IAWA Bull. 3: 75–88.

Phatak VG. 1971. Embryology of *Zygophyllum coccineum* L. and *Z. fabago* L. Proc. Kon. Nederl. Akad. Wetensch. 74C: 379–397.

Poggio L. 1978. Estudios cromosómicos en *Bulnesia, Pintoa, Porlieria, Plectrocarpa* y *Sericodes* (Zygophyllaceae). Darwiniana 21: 139–151.

Porter DM. 1972. The genera of Zygophyllaceae in the southeastern United States. J. Arnold Arbor. 53: 531–552.

Porter DM. 1974. Disjunct distributions in the New World Zygophyllaceae. Taxon 23: 339–346.

Praglowski J. 1987. Pollen morphology of Tribulaceae. Grana 26: 193–211.

Ronse Decraene LP and EF Smets. 1991. Morphological studies in Zygophyllaceae: I. The floral development and vascular anatomy of *Nitraria retusa*. Am. J. Bot. 78: 1438–1448.

Ronse Decraene LP, J De Laet, and EF Smets. 1996. Morphological studies in Zygophyllaceae. II. The floral development and vascular anatomy of *Peganum harmala*. Am. J. Bot. 83: 201–215.

Safina LI and EN Nemirovich-Danchenko. 1996. Zygophyllaceae. In: A Takhtajan, ed. Comparative seed anatomy, vol. 5, pp. 392–399. Nauka, St. Petersburg (in Russian).

Saleh NAM and MN El Hadidi. 1977. An approach to the chemosystematics of the Zygophyllaceae. Biochem. Syst. Ecol. 5: 121–128.

Saleh N, MN El Hadidi, and A Ahmed. 1982. The chemosystematics of Tribulaceae. Biochem. Syst. Ecol. 10: 313–317.

Sarma V and RS Rao. 1991. Taxonomic importance of epidermis in Simaroubaceae-Zygophyllaceae with special reference to position of *Balanites*. Feddes Repert. 102: 579–585.

Sheahan MC. 2007. Zygophyllaceae. In: K Kubitzki, ed. The families and genera of vascular plants, vol. 9, pp. 488–500. Springer, Berlin/Heidelberg/New York.

Sheahan MC and MW Chase. 1996. A phylogenetic analysis of Zygophyllaceae R.Br. based on morphological, anatomical and *rbc*L DNA sequence data. Bot. J. Linn. Soc. 122: 279–300.

Sheahan MC and MW Chase. 2000. Phylogenetic relationships within Zygophyllaceae based on DNA sequences of three plastid regions, with special emphasis on Zygophylloideae. Syst. Bot. 25: 371–384.

Sheahan MC and DF Cutler. 1993. Contribution of vegetative anatomy to the systematics of the Zygophyllaceae R. Br. Bot. J. Linn. Soc. 113: 227–262.

Shulda RD. 1955. On the morphology of the two abnormal gynoecia of *Peganum harmala*. J. Indian Bot. Soc. 34 (4): 382–387.

Singh BP and I Kaur. 1999. Systematic position of the genus *Peganum*. J. Econ. Taxon. Bot. 22(3): 705–708.

Singh BP, I Kaur, and DP Gauchan. 2002. Floral anatomy and systematic position of the genus *Balanites*. Acta Bot. Hung. 44(1–2): 137–143.

Thulin M. 1993. Zygophyllaceae (including Tribulaceae). In: M Thulin, ed. Flora of Somalia, vol. 1. Royal Botanic Gardens, Kew.

Vasilevskaya VK and MP Petrov. 1963. Centralasiatic endemic *Tetraena mongolica* Maxim. Bot. Zhurn. 49: 1506–1513 (in Russian).

Vijayalakshmi S and S Raja Shanmukha Rao. 1991. Taxonomic importance of epidermis in Simaroubaceae – Zygophyllaceae with special reference to position of *Balanites*. Feddes Repert. 102: 579–585.

Wei N. 1991. A comparative anatomy on the vegetative organs of *Tetraena mongolica* Maxim. and *Zygophyllum xanthoxylum* (Bunge) Maxim. Acta Sci. Natur. Univ. Intramongolicae 22: 528–533.

Wu S and L Tu. 1990. The embryology of *Tetraena mongolica* Maxim. Acta Sci. Natur. Univ. Intramongolicae 21: 177–183.

Xi Y and S Zhou. 1989. Pollen morphology and its exine ultrastructure of the Zygophyllaceae in China. Bot. Research (China) 4: 75–86.

Xi Y and S Zhou. 1992. A contribution to the pollen morphology of *Tetraena* and Malpighiaceae, with discussion of the affinity and taxonomic position of *Tetraena*. Chinese J. Bot. 4: 6–12.

Yanishevski DE. 1940. *Tetradiclis tenella* (Ehrenb.) Litv. as the example of an ephemeral on the solonchaks of the Mediterranean desert regions. Trudy Bot. Inst. Akad. Nauk SSSR, 4th ser., 4: 236–248 (in Russian).

Order 99. LINALES

Trees, shrubs, woody lianas, or herbs. Vessels with scalariform or simple perforations; lateral pitting usually alternate. Fibers with distinctly bordered pits, rarely with simple pits. Rays heterogeneous or homogeneous. Axial parenchyma of diverse types. Sieve-element plastids of Ps- (Erythroxylaceae and Rhizophoraceae), Pcs- (Humiriaceae), or more often of S-type. Nodes trilacunar or unilacunar, or rarely (Rhizophoraceae) multilacunar. Leaves alternate or less often opposite, simple and often entire, with stipules or estipulate; stipules bearing colleters on the adaxiale face (Erythroxylaceae and some Rhizophoraceae – see Thiebaut and Hoffmann 2005). Stomata mostly paracytic. Flowers in terminal or axillary inflorescences (mostly more or less cymose), bisexual or sometimes unisexual, actinomorphic or with dissimilar sepals, 5-merous or sometimes 4-merous. Sepals free or more or less connate, mostly imbricate. Petals free, often clawed, imbricate or contorted, sometimes appendaged. Nectary disc mostly present, of staminal origin, intra- or extrastaminal. Stamens in one or two cycles or less often in three or four cycles or even numerous; filaments usually connate at the base or into a short tube (except in the Ixonanthaceae); anthers tetrasporangiate, introrse, basifixed or dorsifixed, opening longitudinally. Tapetum secretory. Microsporogenesis simultaneous or rarely (Gynotrocheae – Rhizophoraceae) successive. Pollen grains 3-celled or sometimes 2-celled, 3(-4)-colporate or less often, pantocolpate, pantocolporate or pantoporate. Gynoecium of 2-5(-20) united carpels, with free or more or less connate stylodia; ovary superior, rarely semi-inferior to inferior, mostly 2–5-locular, with 1–2 or more pendulous ovules per locule. Ovules anatropous to hemitropous, rarely campylotropous or epitropous, bitegmic, crassinucellate or sometimes tenuinucellate, often with endothelium (lacking in Humiriaceae and Rhizophorea). Female gametophyte of *Polygonum*-type. Endosperm nuclear. Fruits usually drupaceous or capsular, sometimes schizocarps, rarely baccate. Seeds typically with fibrous exotegmen, sometimes arillate or winged, with straight or slightly curved embryo; endosperm copious, fleshy and oily to scanty or wanting. Plants with ellagic acid and proanthocyanins, sometimes producing alkaloids.

Most probably derived from some relatively primitive members of the Rutales with scalariform perforation of the vessels and free stylodia. The Hugoniaceae are the most archaic member of the order.

Key to Families

1 Petals not appendaged. Fibers always with bordered pits. Ovary locules all fertile. Plants not producing alkaloids.

2 Stylodia more or less free or when they are united into a simple style, it is evidently cleft. Petals contorted, caducous. Nectary disc extrastaminal or wanting.

3 Trees, shrubs, or woody lianas. Fruits drupaceous or capsular. Anthers dorsoversatile. Ovules two per locule.

4 Fruits drupaceous. Trees or more often woody lianas (the lower inflorescence

branches modified into hooks), rarely shrubs. Hairs, if present, simple, uniseriate, or multiseriate with multicellular, glandular heads; tufted hairs absent. Vessels in *Indorouchera* with scalariform perforations that have 15–25 bars, but in all other genera perforations predominantly or (*Hugonia*) exclusively simple. Nodes trilacunar. Leaves alternate, entire, with lateral, caducous stipules. Stomata paracytic, subsidiary cells are lobed underneath the guard cells. Flowers in terminal or axillary racemes, spikes or panicles, small to large, 5-merous. Sepals often more or less unequal, imbricate. Petals free, rarely basally clasping, contorted. Extrastaminal nectary disc absent or some-rimes represented by glands, adnate to staminal tube. Stamens 10 or 15, alternately shorter or longer; filaments basally connate into a tube. Pollen grains 3-colporate. Gynoecium of 3-5(-8) carpels; stylodia free or basally connate, with capitate stigma; ovary 2–5-locular, locules without 'false septa'; ovules two per locule, pendulous, anatropous, with ventral raphe. Fruits fleshy, indehiscent or rarely splitting finally into indehiscent mericarps. Seeds sometimes with an arillode; mesotesta with sclerotic cells, endotesta lignified or tegmen obliterated; embryo straight or slightly curved; cotyledons large; endosperm copious or scanty, n = 6, 12, 13. . . . . 1. HUGONIACEAE.

4 Fruits 1-locular capsules with woody pericarp finally lengthwise splitting into two valves. Trees with stellately tufted and simple hairs on the young shoots, stipules, and the outsides of the sepals and petals. Vessels with scalariform perforations. Leaves opposite, leathery, entire, coriaceous, with interpetiolar, large, caducous stipules. Stomata anomocytic to anisocytic. Flowers in terminal or axillary cymes or racemes, 5-merous. Sepals shortly connate, persistent. Petals free, thick, imbricate, spoon-shaped at the base. Extrastaminal nectary disc well developed, cuplike. Stamens ten, free, adnate at base to inner side of the disc, alternately longer and shorter; anthers with broad connective; anthers introrse. Pollen grains 3–9-colporate. Gynoecium of two carpels; stylodia more or less free or connate, with capitate stigma; ovary 2-locular, ovules two in each locule, pendulous. Fruits 1-seeded, with seed persisting after falling of pericarp and pendulous from the top of a long filiform funiculus. Seeds with papillose arilloid surrounding their lower half; exotestal cells palisade; embryo straight; endosperm copious; cotyledons very large, folded. . . . . 3. CTENOLOPHONACEAE.

3 Mostly herbs, seldom subshrubs, shrublets, or (*Tirpitzia*) small shrubs. Vessels with simple perforations. Nodes unilacunar with one trace (*Linum*), or trilacunar. Leaves alternate or opposite, entire; stipules lateral, small and inconspicuous, sometimes (*Linum* and *Hesperolinon*) modified into glands, or wanting; epidermis often somewhat mucilaginous. Stomata commonly paracytic. Flowers in cymose inflorescences or rarely (*Anisadenia*) racemes, bisexual, actinomorphic, 5-merous or rarely (*Radiola*) 4-merous. Sepals free or basally connate, persistent, imbricate (quincuncial). Petals free, usually clawed, imbricate or contorted, usually caducous. Small nectary glands usually present external to the stamens or at the inner base of the petals. Stamens 5, 10, or 15, or rarely (*Radiola*) 4, usually alternate the petals or rarely (*Anisadenia*) opposite the petals, sometimes alternating with filiform or toothed staminodia; filaments expanded below and more or less connate into a tube; anthers introrse. Pollen grains 3-colpate, pantocolpate or pantoporate. Gynoecium of 3–5 carpels, very rarely (*Linum digynum*) of 2 carpels; stylodia free or shortly connate below, with terminal stigma; ovary (2)3–5-locular, sometimes unilocular at the very top, ovules 2 in each locule, usually (but not in *Anisadenia*) separated by an incomplete secondary septum extending inward from the ovary wall; ovules crassinucellate or tenuinucellate, bitegmic, outer integument not contributing to the micropyle. Fruits septicidal capsules (Lineae) or of two indehiscent, 1-seeded mericarps (*Anisadenia*). Seeds often mucilaginous, exotesta with outer walls massively thickened; embryo a straight, spatulate, oily, green; endosperm scanty, n = 6, 8–12. . . 2. LINACEAE.

2 Stylodia more or less completely united into a style. Petals imbricate or contorted. Nectary disc intrastaminal or rarely (*Ochthocosmus* in the Ixonanthaceae) nectary is represented by interstaminal glands.

5 Fruits capsular. Trees (sometimes to 30 m tall) or shrubs, sometimes with slender stems, wiry, wandlike, with sparse leaves; mucilage cells present. Vessels with simple perforations. Leaves alternate, petiolate, simple, entire or toothed, with small lateral stipules, or stipules wanting. Stomata paracytic. Flowers small, un axillary inflorescences, mostly bisexual, actinomorphic or nearly so, commonly 5-merous. Sepals free or connate only at the base, imbricate or contorted. Petals free, imbricate or contorted, persistent. Stamens 5, or 10, or 20; filaments widened at the base, free or basally adnate to the conspicuous annular or cupular nectary disc. Pollen grains 2-locular and 3-colporate. Gynoecium of (4-)5 or rarely 2 carpels; style with a more or less capitate stigma; ovary superior or semiinferior, 2- or (4-)-5 locular, sometimes apically unilocular and sometimes the locules are divided into locelli by incomplete secondary septa as in the Linaceae; style and filaments folded in the bud. Ovules one (*Allantospermum*) or two per locule, pendulous, anatropous, with ventral raphe, the micropyle directed upward and outward. Fruits septicidal capsules, sometimes also loculicidal by secondary septa, with or without a persistent central column. Seeds with an obvious basal wing (*Ixonanthes* and *Ochthocosmus*) or aril arising between the hilum and micropyle (laciniate and covers the seed in *Phyllocosmus* and 2-lobed and membranous in *Cyrillopsis*); endotegmen with sinuous anticlinal walls. Embryo straight, endosperm scanty or absent; cotyledons large. Proanthocyanidins and ellagic acid present. . . . .4. IXONANTHACEAE.

5 Fruits drupaceous. Evergreen trees and shrubs, often with an aromatic juice, but without secretory structures (except often in the fruits). Vessels usually with scalariform perforations that have mostly 15–25 bars. Leaves alternate, coriaceous, entire or toothed, with tiny and caducous stipules or estipulate; leaves often aromatic, with balsamic juice. Stomata anomacytic or paracytic. Flowers bisexual, actinomorphic or nearly so, 5-merous. Sepals persistent, more or less connate, imbricate, two outer ones often smaller than others, rarely the lobes suppressed. Petals free, thick, usually 3–5 veined, imbricate or contorted. Stamens numerous in fascicles (*Vantanea*) or more often 10–30, sometimes in 5 antesepalous groups of 3, plus 5 antepetalous singles; filaments connate into a tube; some of the stamens sometimes staminodial; anthers basifixed or dorsifixed or attached near the base, with expanded, prolonged connective, versatile. Pollen grains 3(4)-colporate or seldom 3(4)-porate. The intrastaminal nectary disc free or adnate to the base of the ovary or to the base of the filaments, usually cupulate to tubular or toothed or lobed, or sometimes of 10–20 free scales. Gynoecium of (4)5(-7) carpels; style with a lobed or cleft stigma; ovary (4-)5(-7)-locular, sometimes apically 1-locular, with 1(-2) pendulous ovule per locule. Fruits with more or less fleshy, thin exocarp and hard, woody, multilocular, and usually 1–2-seeded stone (pyrene) sometimes containing numerous resinous secretory cavities and adapted to the water dispersal; the stone with as many longitudinal valves as carpels, 1 or more being pushed off at germination. Seeds with nucellar remnants; exotestal cells thick-walled, lignified, tegmen multiplicative, cross layer beneath exotegmen; embryo straight or slightly curved; endosperm copious oily. Ellagic acid present, n = 12. 5. HUMIRIACEAE.

1 Petals appendages. Fibers with simple or bordered pits. Plants contain alkaloids.

6 Nectary disc absent. Evergreen or deciduous, glabrous, small trees and shrubs; wood hard, commonly with silicium grains. Vessels with simple perforations. Fibers with distinctly bordered to sometimes simple pits. Nodes unilacunal. Leaves alternate or rarely (*Aneulophus*) opposite, entire, simple, sometimes accompanied by scale leaves, with intrapetiolar, abaxially bicosytate, often caducous stipules. Stomata paracytic. Flowers small, in axillary fascicles or solitary in the axils, bisexual or seldom unisexual (plant dioecious), actinomorphic, 5-merous, often heterostylous; bracteoles small, scarious. Sepals

persistent, connate below into a tube with imbricate or valvate lobes. Petals free, imbricate or contorted, caducous, mostly with an adaxial more or less basal ligular appendage. Stamens ten, in two whorls of five, the outer whorl alternate with petals; filaments united at least at the base, usually forming a short tube; anthers dorsifixed, versatile, introrse. Pollen grains 3-colporate. Gynoecium of three or less often two carpels; stylodia free or more or less connate into a style, stigmas capitellate, rarely sessile; ovary superior, (2-)-3-locular, usually only with one fertile locule, the others empty. Ovules solitary or less often two in the fertile locule, pendulous, anatropous to hemitropous, with an endothelium. Fruits 1-, or sometimes 2-seeded drupes. Seeds with more or less thickened exotestal cells; embryo straight, green; cotyledons two, flat; endosperm copious, starchy, or rarely lacking. Present alkaloids, proanthocyanidins (cyanidin), flavonols (kaempferol and quercetin) n = 12.. . . . . 6. ERYTHROXYLACEAE.

6 Intrastaminal nectary disc present. Shrubs or trees (up to 50 m tall) of dry to wet forests or mangrove swamps, mostly with aerial roots and unicellular hairs. Vessels with scalariform or less often simple perforations; lateral pitting scalariform to opposite and alternate. Fibers with simple or bordered pits. Rays nearly always heterogeneous, usually mixed uniseriate and pluriseriate with short ends. Axial parenchyma typically scanty paratracheal. Sieve-element plastids of Pc-type with about 20 or more square or polygonal protein bodies. Nodes trilacunar or multilacunar with split-lateral traces. Leaves opposite or sometimes verticillate, simple, entire, crenate or dentate, with well-developed interpetiolar stipules, imbricate (Rhizophoreae), or valvate (Macarisieae), sheathing the terminal bud; stipules bear colleters on the inner surface at the base that secrete gummy substances onto the buds. Stomata paracytic or (Rhizophoreae) encyclocytic. Flowers in axillary and fundamentally cymose few-flowered inflorescence or (in large-flowered species of *Bruguiera*) solitary, clearly articulated at juncture with pedicel, bisexual or rarely unisexual (plants monoecious), actinomorphic, most often 4–5-merous, with well-developed floral tube that is sometimes prolonged beyond the ovary; both the calyx and the ovary have a layer of hypodermal laticiferous cells, which in *Gynotroches* and *Pellacalyx* are represented by many idioblastic secretory cells. Sepals (3)4-5(-16), fused at base, valvate, thick, usually fleshy or leathery. Petals as many as and alternate with the sepals, fused at base, commonly fleshy and often shorter than the sepals, contorted or infolded in bud, generally conduplicate, mostly distally divided or appendaged, each petal individually enclosing 1–5 stamens. Androecium diplostemonous to polyandrous; filaments free or sometimes basally connate, borne around the base of a fleshy, often conspicuously lobed, intrastaminal nectary disc; anthers tetrasporangiate, 2-locular or (*Rhizophora*) multilocellate, cross-partitioned and dehiscent by a separating longitudinal valve. Microsporogenesis simultaneous or also successive in most Gynotrocheae. Pollen grains 2-celled, 3(4)-colporate, with relatively thick tectum and rugulate-punctate to psilate surface; endoapertures with some degree of fusion. Gynoecium of (2) 3-5(-20) united carpels, with a terminal simple style; stigma capitate or with pronounced lobes (in *Gynotroches* the stigmatic lobes may be rather long), generally papillate; ovary superior (Macarisieae) or semi-inferior to inferior, with as many locules as carpels, or the locules often incompletely or not at all separated by septa at anthesis. Ovules 2–6 or more per carpel, usually apical-axial, pendulous, anatropous to hemitropous or rarely campylotropous, bitegmic, crassinucellate or (*Gynotroches* and *Pellacalyx*) tenuinucellate; micropyle zig-zag or endostomal. Endothelium present (except for Rhizophoreae). Fruits capsular (Macarisieae and *Crossostylis*), baccate (Gynotrocheae), or hard-walled and indehiscent (Rhizophoreae). Seeds arillate or winged (in capsular fruits) or nonappendaged (in baccate or indehiscent hard-walled fruits); seed coat in Macarisieae and Gynotrocheae with well-developed exotesta and (fibrous) exotegmen and is either exotestal or exotestal-exotegmic; seed coat in Rhizophoreae is not histologically differentiated and entirely lacking a tegmen; embryo straight, in the mangrove genera viviparous and with an enlarged hypocotyl; endosperm well developed, fleshy, oily. Plants contain pyrrolizidine alkaloids and ellagic acid, n = (13), 14, 16, 18, 21. . . . . 7. RHIZOPHORACEAE.

1. HUGONIACEAE

Arnott 1834. 5/55–60. From tropical Africa and Madagascar to New Caledonia, Fiji, and tropical South America.

Indorouchera, Roucheria, Philbornea, Hebepetalum, Hugonia (including *Durandea).*

Probably the most archaic member of the order.

2. LINACEAE

A. P. de Candolle ex Perleb 1818. 8/250. Widely distributed, especially in temperate and subtropical regions.

LINEAE: *Tirpitzia, Reinwardtia, Linum, Cliococca, Sclerolinon, Hesperolinon, Radiola;* ANISADENIEAE: *Anisadenia.*

Linaceae has probably derived from the nearest ancestor of the Hugoniaceae. The most primitive members of the Linaceae are *Tirpitzia* (southwestern China, northern Vietnam) and *Reinwardtia* (northern India and China).

3. CTENOLOPHONACEAE

Exell et Mendonça 1951. 1/4. Tropical West Africa and Malesia.

Ctenolophon.

Related to the Linaceae and also like Humiriaceae (Link 1992a).

4. IXONANTHACEAE

Planchon ex Miquel 1858. 5/35. Tropical Africa (*Phyllocosmus*), Himalayas, northeastern India, southern China, Southeast Asia, New Guinea (*Ixonanthes*), and tropical America (*Cyrillopsis* and *Ochtocosmus*). *Allantospermum* (2) occurs in Madagascar (1) and Borneo (1).

Cyrillopsis, Ixonanthes, Ochthocosmus, Phyllocosmus, Allantospermum.

Related to the Linaceae and especially to the Humiriaceae. According to Link (1992b), the Ixonanthaceae have more or less prominent discs in common only with the Humiriaceae whereas the nectaries of the Linaceae sensu stricto (and the Erythroxylaceae) are clearly parts of the staminal tubes.

5. HUMIRIACEAE

A. H. L. de Jussieu 1829. 8/65. Tropical South America, north to Costa Rica, but mostly in the Amazon basin and with 1 species (*Sacoglottis gabonensis*) in tropical West Africa.

VANTANEEAE: *Vantanea* (20); HUMIRIEAE: *Duckesia, Endopleura, Hylocarpa, Humiria, Sacoglottis* (11), *Schistostemon, Humiriastrum* (17).

Close to the Linaceae and have also much in common with Ixonanthaceae. According to Boesewinkel (1985), the Humiriaceae seem to be more primitive than the Erythroxylaceae and the Linaceae because of the large persistent nucellus, the lack of an endothelium, the absence of repeated divisions in the original middle layer of the inner integument, and perhaps also because of the strongly lignified exotesta.

6. ERYTHROXYLACEAE

Kunth 1822 (including Nectaropetalaceae Exell et Mendonça 1951). 4/250–260. Pantropical, but most abundant in the Andes, in the Amazon Basin of South America, and in Madagascar.

Erythroxylum, Aneulophus (2), *Nectaropetalum* (8), *Pinacopodium* (2).

Related to the Linaceae, Ixonanthaceae, and Humiriaceae.

7. RHIZOPHORACEAE

C. H. Persoon 1806 (including Cassipoureaceae J. Agardh 1858, Macarisiaceae J. Agardh 1858, Mangiaceae Rafinesque 1837). 15/140. Pantropical, but extend into the subtropics of North America and Asia; the greatest diversity in the paleotropics.

MACARISIEAE: *Anopyxis, Macarisia Blepharostemma, Comiphyton, Dactylopetalum, Cassipourea, Sterigmapetalum;* GYNOTROCHEAE: *Carallia, Gynotroches, Pellacalyx;* CROSSOSTYLIDEAE: *Crossostylis;* RHIZOPHOREAE: *Bruguiera, Ceriops, Kandelia, Rhizophora.*

According to Tobe and Raven (1988), the genera of Macarisieae retain many plesiomorphic features, including a superior ovary. The most specialized are Rhizophoreae: endothelium not formed, outer integument vascularized, endosperm overflowing, viviparous seedlings, testa not differentiated histologically, and tegmen lacking. The cladistic analysis indicates that Rhizophoreae have direct relationships with Gynotrocheae (not including *Crossostylis*) rather than with Macarisieae (Tobe and Raven 1988).

Setoguch et al. (1999) and Schwarzbach and Ricklefs (2000) identify the Rhizophoraceae and Erythroxylaceae as sister group within the Malpighiales sensu the APG II.

Bibliography

Behnke H-D. 1982. Sieve-element plastids of Cyrillaceae, Erythroxylaceae, and Rhizophoraceae: Description and significance of subtype PV plastids. Plant Syst. Evol. 141: 31-39.

Behnke H-D. 1988. Sieve-element plastids and systematic relationships of Rhizophoraceae, Anisophylleaceae and allied groups. Ann. Missouri Bot. Gard. 75: 1387–1409.

Boesewinkel FD. 1980a. Development of ovule and testa of *Linum usitatassimum* L. Acta Bot. Neerl. 29: 17–32.

Boesewinkel FD. 1980b. Development of ovule and seed coat of *Erythroxylum coca* Lamk. Acta Bot. Neerl. 29: 231–241.

Boesewinkel FD. 1985. The ovule and seed of *Humiria balsamifera* (Aubl.) St.Hil. Acta Bot. Neerl. 34: 183–191.

Boesewinkel FD. 1994. Ovules and seed characters of *Balanites aegyptiaca* and the classification of the Linales-Geraniales-Polygalales assembly. Acta Bot. Neerl. 43: 15–25.

Boesewinkel FD and F Bauman. 2000. Hugoniaceae, Linaceae, Ctenolophonaceae, Ixonanthaceae, Humariaceae, Erythroxylaceae. In: A Takhtajan, ed. Comparative seed anatomy, vol. 6, pp. 9–20. Nauka, St. Petersburg (in Russian).

Boesewinkel FD and J Geenen. 1980. Development of ovule and seed coat of *Erythroxylum coca* Lamk. Acta Bot. Neerl. 29: 231–241.

Bove CP. 1997. Phylogenetic analysis of Humiriaceae with notes on the monophyly of Ixonanthaceae. J. Comp. Biol. 2: 19–24.

Bove CP and TS Melhem. 2000. Humiriaceae Juss. World Pollen Spore Flora 22: 1–35.

Chase MW, S Zmarzty, MD Lledó, KJ Wurdack, SM Swensen, and MF Fay. 2002. When in doubt, put it in Flacourtiaceae: A molecular phylogenetic analysis based on plastid *rbc*L DNA sequences. Kew Bull. 57: 141–181.

Cuatrecasas J. 1961. A taxonomic revision of the Humiriaceae. Contr. U.S. Natl. Herb. 35: 25–214.

Dahlgren RMT. 1988. Rhizophoraceae and Anisophylleaceae: summary statement, relationships. Ann. Missouri Bot. Gard. 75: 1259–1277.

Daly DC. 2004. Erythroxylaceae. In: N Smith, SA Mori, A Henderson, DW Stevenson, and SV Heald, eds. Flowering plants of the Neotropics, pp. 143–145. The New York Botanical Garden, Princeton University Press, Princeton.

Ding Hou D. 1958. Rhizophoraceae. In: CGGJ van Steenis, ed. Flora Malesiana, ser. 1, 5: 429–493. Noordhoff, Djakarta.

Dorr LJ. 1994. The identity and neotypification of *Endosteira* Turcz. (Rhizophoraceae). Taxon 43: 639–640.

Floret J-J. 1989. *Cassipourea* Aublet (Rhizophoraceae-Macarisieae): Organisation florale et divisions subgeneriques. Adansonia, ser. 2, 11: 1–83.

Forman LL. 1965. A new genus of Ixonanthaceae with notes on the family. Kew Bull. 19: 517–526.

Graham SA. 1964. The genera of Rhizophoraceae and Combretaceae in the southeastern United States. J. Arnold Arbor. 45: 285–301.

Grigorieva VV. 1990. The pollen grain morphology in members of the Linaceae family. Bot. Zhurn. 75: 1345–1352 (in Russian with English summary).

Gustafsson MHG. 2004. Rhizophoraceae. In: N Smith, SA Mori, A Henderson, DW Stevenson, SV Heald, eds. Flowering plants of the Neotropics, pp. 324–326. The New York Botanical Garden, Princeton University Press, Princeton.

Hallier H. 1923. Beiträge zur Kenntnis der Linaceae (DC. 1819) Dumort. Beih. Bot. Centralbl. 39: 1–178.

Ham RWJM van der. 1989. New observations on the pollen of *Ctenolophon* Oliver (Ctenolophonaceae), with remarks on the evolutionary history of the genus. Rev. Palaeobot. Palynol. 59: 153–160.

Heimsch C and EE Tschabold. 1972. Xylem studies in the Linaceae. Bot. Gaz. 133: 242–253.

Hooren AMN van and HP Nooteboom. 1988. Linaceae, Ctenolophonaceae. In: CGGJ van Steenis, ed. Flora Malesiana, ser. 1, 10: 607–619, 629–634. Noordhoff, Groningen.

Jardim AG. 1999. A revision of *Roucheria* Planch. and *Hebepetalum* Benth. (Hugoniaceae). University of Michigan, Ann Arbor, MI.

Jardim A. 1999. A revision of *Roucheria* Planch. and *Hebepetalum* Benth. (Hugoniaceae). M.Sc. Thesis, University of Missouri, St. Louis.

Jayeola AA, JR Thorpe, TA Adenegan. 2001. Macromorphological and micromorphological studies of the West African *Rhizophora* L. Feddes Repert. 112(5–6): 349–356.

Juncosa AM. 1988. Floral development and character evolution in Rhizophoraceae. In: R Leins, SC Tucker, and RK Endress, eds. Aspects of floral development, pp. 83–101. Springer, Berlin.

Juncosa AM and H Tobe. 1988. Embryology of tribe Gynotrocheae (Rhizophoraceae) and its developmental and systematic implications. Ann. Missouri Bot. Gard. 75: 1410–1424.

Juncosa AM and PB Tomlinson. 1987. Floral development in mangrove Rhizophoraceae. Am. J. Bot. 74: 1263–1279.

Juncosa AM and PB Tomlinson. 1988a. Systematic comparison and some biological characteristics of Rhizophoraceae and Anisophylleaceae. Ann. Missouri Bot. Gard. 75: 1296–1318.

Juncosa AM and PB. Tomlinson. 1988b. A historical and taxonomic synopsis of Rhizophoraceae and Anizophylleaceae. Ann. Missouri Bot. Gard. 75: 1278–1295.

Keating RC and V Randrianosola. 1988. The contribution of leaf architecture and wood anatomy to classification of the Rhizophoraceae and Anisophylleaceae. Ann. Missouri Bot. Gard. 75: 1343–1368.

Kool R. 1980. A taxonomic revision of the genus *Ixonanthes* (Linaceae). Blumea 26: 191–204.

Kool R. 1986. Ixonanthaceae. In: CGGJ van Steenis, ed. Flora Malesiana, ser. 1, 10: 621–627. Kluwer, Dodrecht.

Lersten NR and JD Curtis. 1974. Colleter anatomy in red mangrove, *Rhizophora mangle* (Rhizophoraceae). Canad. J. Bot. 52: 2277–2278.

Link DA. 1992a. The floral nectaries of the Geraniales and their systematic implications: IV. *Ctenolophon* Badre. Flora 187: 103–107.

Link DA. 1992b. The floral nectaries of the Geraniales and their systematic implications: VI. Ixonanthaceae Exell and Mendonca. Bot. Jahrb. Syst. 114(I): 81–90.

Link DA. 1992c. The nectaries of the Geraniales and their systematic implications. VII. Humiriaceae Cuatr. Bot. Jahrb. Syst. 114(2): 211–241.

Link DA. 1993. The floral nectaries of the Geraniales and their systematic implications, VI. Ixonanthaceae Exell et Mendonca. Bot. Jahrb. Syst. 114: 81–90.

Narayana LL. 1964. A contribution to the floral anatomy and embryology of Linaceae. J. Indian Bot. Soc. 43: 343–357.

Narayana LL. 1970. Linaceae, Erythroxylaceae, Indian National Sci. Acad. Bull. [Symposium: Comparative Embryology of Angiosperms] 41: 127–132, 133–135.

Narayana LL and D Rao. 1966. Floral morphology of Linaceae. J. Jpn. Bot. 41: 1–10.

Narayana LL and D Rao. 1969–1977. Contributions to the floral anatomy of Humiriaceae, parts 1–6. J. Jpn. Bot. 44: 328–335, 1969; 48: 143–146, 242–276, 1973; 51: 12–15, 42–44, 1976; 52: 145–153, 1977.

Narayana LL and D Rao. 1969–1978. Contributions to the floral anatomy of the Linaceae. [13 parts] J. Jap. Bot. 44: 289–294, 1969; 48: 205–208, 1973; 51: 92–96, 349–352, 1976; 52: 56–59, 231–234, 315–317, 1977; 53: 12–14, 161–163, 213–218, 1978. Phytomorphology 21: 64–67, 1971 (1972). Curr. Sci. 43: 226–227, 391–393, 1974.

Narayana LL and D Rao. 1978. Systematic position of Humiriaceae, Linaceae, and Erythroxylaceae in the light of their comparative floral morphology and embryology: A discussion. J. Indian Bot. Soc. 57: 258–266.

Nikiticheva ZI and MS Yakovlev. 1985. Rhizophoraceae. In: MS Yakovlev, ed. Comparative embryology of flowering plants: Brunelliaceae-Tremandraceae, pp. 120–125. Nauka, Leningrad (in Russian).

Nooteboom HR 1967. The taxonomic position of Ir-vingioideae, *Allantospermum* Forman, and *Cyrillopsis* Kuhlm. Adansonia, ser. 2, 7: 161–168.

Oltmann O. 1968 (1969). Die Pollenmorphologie der Erythroxylaceae und ihre systematische Bedeutung. Ber. Deutsch. Bot. Ges. 81: 505–511.

Oltmann O. 1971. Pollenmorphologisch-systematische Untersuchungen innerhalb der Geraniales. Dissertationes Botanicae, vol. 11.

Plisko MA. 1996. Rhizophoraceae. In: A Takhtajan, ed. Comparative seed anatomy, vol. 5, pp. 134–150. Nauka, St. Petersburg (in Russian).

Price RA and JD Palmer. 1993. Phylogenetic relationships of the Gcraniaceae and Geraniales from *rbc*L sequence comparisons. Ann. Missouri Bot. Gard. 80: 661–671.

Ramassamy V and B Kannabiran. 1996. Leaf epidermis and taxonomy in Rhizophoraceae. Indian Forester. 122(11): 1049–1061.

Ramírez N. 2004. Ixonanthaceae. In: N Smith, SA Mori, A Henderson, DW Stevenson, and SV Heald, eds. Flowering plants of the Neotropics, pp. 195–196. The New York Botanical Garden, Princeton University Press, Princeton.

Robertson KR. 1971. The Linaceae in the southeastern United States. J. Arnold Arbor. 52: 649–665.

Robson NKB and HK Airy 1962. A note on the taxonomic position of the genus *Cyrillopsis* Kuhlmann. Kew Bull. 15: 387–388.

Rojo JP. 1968. The wood anatomy of *Allantospermum borneense* Forman and *Allantospermum multicaule* (Capuron) Nooteboom. Adansonia, ser. 2, 8: 73–83.

Rury PM. 1985. Systematic and ecological wood anatomy of the Erythroxylaceae. IAWA Bull., n.s., 6: 365–397.

Saad SI. 1961. Pollen morphology and sporoderm stratification in *Linum*. Grana Palynol. 3(I): 109–129.

Saad SI. 1962a. Pollen morphology of *Ctenolophon*. Bot. Not. 115: 49–57.

Saad SI. 1962b. Palynological studies in the Linaceae. Pollen Spores 4: 65–82.

Schwarzbach AE and RE Ricklefs. 2000. Systematic affinities of Rhizophoraceae and Anisophylleaceae, and intergeneric relationships within Rhizophoraceae, based on chloroplast DNA, nuclear ribosomal DNA, and morphology. Am. J. Bot. 87: 547–564.

Setoguchi H, K Kosuge, H Tobe. 1999. Molecular phylogeny of Rhizophoraceae based on *rbc*L gene sequences. J. Plant Res. (Tokyo) 112: 443–455.

Shi S, Y Zhong, Y Huang, YQ Du, XZ Qiu, and HT Chang. 2002. Phylogenetic relationships of the Rhizophoraceae in China based on sequences of the chloroplast gene matK and the internal transcribed spacer regions of nuclear ribosomal DNA and combined data set. Biochem. Syst. Ecol. 30: 309–319.

Steyermark JA and JL Luteyn. 1980. Revision of the genus *Ochthocosmus* (Linaceae). Brittonia 32: 128–143.

Thiebaut LF and P Hoffmann. 2005. Occurrence of colleters in Erythroxylaceae. Kew Bull. 60: 455–459.

Tobe H and RH Raven. 1988. Seed morphology and anatomy of Rhizophoraceae: Inter- and intrafamilial relationships. Ann. Missouri Bot. Gard. 75: 1319–1342.

Tomlinson PB. 1986. The botany of mangrove. Cambridge University Press, Cambridge.

Tomlinson PB and PA Cox. 2000. Systematic and functional anatomy of seedlings in mangrove Rhizophoraceae: vivipary explained? Bot. J. Linn. Soc. 134: 215–231.

Tomlinson PB, RB Primack, and JS Bunt. 1979. Preliminary observations on floral biology in mangrove Rhizophoraceae. Biotropica 11: 256–277.

Van Hooren AMN and HP Nooteboom. 1984. A taxonomic revision of the Malesian Linaceae and Ctenolophonaceae, especially of Malesia, with notes on their demarcation and the relationships with Ixonanthaceae. Blumea 29: 547–563.

Van Hooren AMN and HP Nooteboom. 1988a. Linaceae. In: WJJO de Wilde, ed. Flora Malesiana, ser. 1, vol. 10, pp. 607–619. Kluwer, Dordrecht.

Van Hooren AMN and HP Nooteboom. 1988b. Ctenolophonaceae. In: WJJO de Wilde, ed. Flora Malesiana, ser. 1, vol. 10, pp. 629–634. Kluwer, Dordrecht.

Van Welzen PC and P Baas. 1984. A leaf anatomical contribution to the classification of the Linaceae complex. Blumea 29: 453–479.

Vezey EL, VP Shah, JJ Skvarla, and PH Raven. 1988. Morphology and phenetics of Rhizophoraceae pollen. Ann. Missouri Bot. Gard. 75: 1369–1386

Vliet GJCM van. 1976. Wood anatomy of the Rhizophoraceae. Leiden Bot. Series 3: 20–75.

Wheat DW. 1981. Sylleptic branching in the Rhizophoreae (Rhizophoraceae). Bot. Gaz. 142: 115–123.
Yokhioka H, K Kondo, M Legrand, K Nehira, and S Maxeda. 1984. Karyomorphological studies in five species of mangrove genera in Rhizophoraceae. La Kromosomo, ser. 2, 35–36: 1111–1116.

Order 100. MALPIGHIALES

Trees, shrubs, subshrubs or herbs, sometimes woody lianas. Vessels with simple perforations, lateral pitting alternate. Fibers with simple or bordered pits. Rays narrow, heterogeneous or homogeneous. Axial parenchyma apotracheal or paratracheal, sometimes wanting. Sieve-element plastids of So-types. Nodes unilacunar or sometimes trilacunar. Leaves alternate, opposite or verticillate, simple and usually entire; stipules present or absent, sometimes modified into spines or glands; leaf mesophyll cells contain oil bodies. Stomata mostly paracytic. Flowers in various types of inflorescences, bisexual, actinomorphic or more or less strongly zygomorphic, basically 5-merous or less often 4-merous. Sepals free or nearly free, imbricate or valvate. Petals free and often clawed or basally connate or only some of them connate. Nectary disc present or absent. Stamens 1-10(-15), often some of them staminodial; filaments free or often variously connate; anthers tetrasporangiate or rarely disporangiate, mostly basifixed, introrse, opening in various ways. Tapetum secretory. Microsporogenesis simultaneous. Pollen grains 2-celled or 3-celled, very diverse, but mostly 3-colporate or pantoporate. Gynoecium of 3–5, less often two or up to eight united carpels; stylodia free or only basally connate or more often connate into a terminal style; ovary usually superior, commonly plurilocular with axile placentas, seldom unilocular and with two parietal placentas or with 1 or 2, rarely 2–16 pendulous subapical ovules, sometimes pseudo-monomerous. Ovules epitropous, anatropous to hemitropous, bitegmic, crassinucellate. Female gametophyte of *Polygonum*-type or) of *Penaea*-type, less often of *Allium*-type. Endosperm nuclear. Fruits of diverse types. Seeds testaltegmic, exotegmic, or endotestal, often with fibers; embryo straight or more or less curved; endosperm present (sometimes copious) or lacking.

Malpighiales are closely related to the Linales, especially through the Malpighiaceae, which have much in common with the Erythroxylaceae. The family Malpighiaceae occupies a somewhat intermediate position between the Linales and Malpighiales and could with almost equal reason be included in either of them.

Key to Families

1 Stylodia usually free, only rarely (as in spp. of *Bunchosia*) fully connate. Small trees, shrubs, or more often woody lianas or often stem twiners (thryallis twining anticlockwise), often with anomalous secondary growth. Plants generally with 2-armed (Malpighian) hairs, rarely with basifixed or stellate hairs. Latex rarely present (*Galphimia*, *Lophanthera*, *Spachea*, and *Verrucularia*). Vessels with simple perforations and verstured pits. Fibers with simple pits, often septate. Rays heterogeneous. Axial parenchyma mostly scanty and paratracheal. Nodes trilacunar or sometimes unilacunar. Leaves mostly opposite, rarely subopposite, or verticillate, simple, usually entire (lobed in *Stigmaphyllon*), mostly pinnately veined, very often with two large, fleshy glands on petiole or abaxial surface, stipulate (stipules interpetiolar or intrapetiolar, free of one another or concrescent, often rudimentary, sometimes large, more then 14 cm long – Anderson 2004), or estipulate. Stomata mostly paracytic. Inflorescences terminal or axillary, very diverse, most often racemose or paniculate, often flowers in umbels or corymbs. Flowers mostly large, bisexual or rarely unisexual (dioecious), or polygamous, sometimes dimorphic, usually somewhat zygomorphic, 5-merous, receptacle more or less convex. Sepals free or slightly basally connate, imbricate, often with two large glands at the bases of the outside members. Petals free, imbricate, usually clawed, with ciliate, toothed, or fringed margins. Stamens mostly 10 in 2 obdiplostemonous cycles, or 5, up to 15 in *Lasiocarpus*, often some of them without anthers or with abortive anthers, seldom in 1 or 3 cycles; filaments usually more or less connate into a tube; anthers 4-locular, opening longitudinally along inner edge of each locule, or rarely by apical pores or very short slits, sometimes with an enlarged connective. Pollen grains 2-celled, 3–5-colporate or 4- to polyporate. Gynoecium of (2-)3(-5) carpels, mostly all fertile, or with one or more caples aborting, more or less united or rarely (as in *Coleostachys, Acmanthera,* and *Pterandra*) almost free; stylodia usually free or only basally connate, terminal to ventral; stigmas acute, capitate,

or more or less truncate; ovary inferior, entire or lobed, multilocular with axile placentas and solitary, anatropous to nearly orthotropous, pendulous ovule in each locule. Fruiting carpels 1-seeded, either free and indehiscent or more often united into a fleshy, drupaceous or woody fruit that usually separates into nutlike, samaroid, or variously winged indehiscent or rarely 2-valved mericarps, sometimes nuts or drupes. Seeds with large, oily, straight or sometimes curved or even circinate embryo; nearly or quite without endosperm. Seed coat with exotesta and with sclerofied endotegmen, rarely with fibrous endotegmen. Plants producing proanthocyanins, alkaloids of the indole group and sometimes storing carbohydrates as insulin, flavonols (kaempferol and quercetin), detected iridoids (only from *Stigmaphyllon*), n = 6, 9–12 +. . . . . 1. MALPIGHIACEAE.

1 Stylodia united into a style. Plants seldom with 2-armed hairs. Hemiparasitic shrubs (sometimes arborescent) or perennial rhizomatous herbs with a woody rootstock, mostly pubescent or silky all over, with scatterous, tanniniferous secretory cells in the parenchymatous tissues; cuticle waxed more or less ribbon-like platelets. Vessels with simple perforations, lateral pitting alternate. Fibers with bordered pits. Rays uniseriate, homogeneous. Axial parenchyma diffuse-in-aggregates (Carlquist 2005). Crystals many per cell, of varied sized, rare in wood, but common in axial parenchyma of secondary phloem. Nodes unilacunar with one trace. Leaves alternate, simple entire or rarely 3-foliolate, petiole bundle, deeply arcuate; stipules wanting. Flowers solitary and axillary, or in terminal racemes, showy, pink, purple, yellow, or rose, bibracteolate, bisexual, strongly zygomorphic. Sepals five or sometimes four, unequal, free, imbricate, three outer often larger than two inner and often more or less enclosing flower; the sepals have three principal veins, like the leaves. Petals five or sometimes four, the three adaxial (upper) ones long-clawed, free or connate by their claws, the two abaxial (lower) ones smaller, usually broad thick, sessile, often modified into lipid-secreting glands. Nectary disc wanting. Stamens (3)4, alternating with the upper petals, a fifth sterile stamen rarely found below; filaments thick, free or sometimes basally connate or adnate below to the claws of the upper petals; anthers 4-locular, opening by one or two apical pore or short slit. Pollen grains 3–4-colporate to 3-porate, with striate ornamentation. Gynoecium of two carpels, one of them reduced and empty; style terminal, curved, slightly longer than stamens, protruding from flower, with a discoid or punctiform stigma; ovules two in a single fertile locule, collateral and pendulous, anatropous. Fruits dry, indehiscent, 1-seeded; usually armed with barbed bristles or with spines. Seeds long, with straight embryo and without endosperm; cotyledons large, cordate; seed coat very thin, membranous, developed from the outer integument; the inner integument is represented by one layer of endotegmen (Plisko 2000). Contain tannins, principally those of the catechin type, and fatty acids; n = 6. . . . . .2. KRAMERIACEAE

1. MALPIGHIACEAE

A. L. de Jussieu 1789. 66/1250. Tropical and subtropical regions, especially in South America.

1.1 MALPIGHIOIDEAE

Plants mostly erect, rarely climbing. Receptacle flat or slightly concave. Stamens mostly all fertile. Carpels united in the ovary stage, drupaceous in fruit, neither winged nor covered with long bristles. – *Thryallis, Galphimia, Blepharandra Lophanthera, Byrsonima, Dicella, Spachea, Burdachia, Clonodia, Bunchosia, Heladena, Verrucularia, Glandonia, Mcvaughia, Malpighia, Diacidia, Coleostachys, Acmanthera, Pterandra.*

1.2 HIRAEOIDEAE

Plants mostly climbing. Receptacle pyramidal, usually 3-sided. Stamens all fertile or some stamens sterile. Carpels united in the ovary stage, separating when mature, usually winged or covered with long bristles. – TRICOMARIEAE: *Tricomaria, Ptilochaeta,* etc.; HIRAEEAE: *Aspidopterys, Microsteira, Mascagnia, Hiraea, Tetrapteris, Hiptage,* etc.; RHYNCHOPHOREAE: *Rhynchophora;* BANISTERIEAE: *Barnebya, Acridocarpus, Heteropteris, Sphedamnocarpus, Rhyssopteris, Banisteriopsis, Peixotoa, Stigmaphyllon,* etc.

1.3 GAUDICHAUDIOIDEAE

Erect or climbing shrubs. Flowers often dimorphic. Sepals always glandular. Some anthers sterile. Carpels free in the ovary stage, winged or wingless, often at length hanging by a raphelike thread. – *Gaudichaudia, Janusia, Aspicarpa, Camarea.*

2. KRAMERIACEAE

Dumortier 1829. 1/18. Southwestern United States (3 from Kansas to Arizona and Florida), Mexico (11), Central and South America southwards to Brazil (5), Argentina and Chile; 1 sp. in West Indies.

Krameria.

A rather isolated family that, according to Verkerke (1986), is likely to be related to the Malpighiaceae, which is supported also by wood anatomical data (Heimsch 1942). It is interesting "and perhaps significant" (Cronquist 1981: 780) that *Krameria* is pollinated primarily by bees of the genus *Centris,* which also collect oil from species of Malpighiaceae (Simpson and Neff 1978).

Bibliography

Anderson WR. 1977 (1978). Byrsonimoideae, a new subfamily of the Malpighiaceae. Leandra 7: 5–18.

Anderson WR. 1979. Floral conservatism in Neotropical Malpighiaceae. Biotropica 11: 219–223.

Anderson WR. 1990. The origin of the Mapighiaceae – the evidence from morphology. Mem. New York Bot. Gard 64: 210–224.

Anderson WR. 1993. Chromosome numbers of neotropical Malpighiaceae. Contr. Univ. Michigan Herb. 19: 341–354.

Anderson WR. 2004. Malpighiaceae. In: N Smith, SA Mori, A Henderson, SW Stevenson, and SV Heald, eds. Flowering plants of the Neotropics, pp. 229–232. The New York Botanical Garden, Princeton University Press, Princeton.

Anderson C, WR Anderson, and CC Davies. 2006 onwards. Malpighiaceae (herbarium.lsa.umich.edu/malpigh).

Busse-Jung F. 1979. Phytoserologische Untersuchungen zur Frage der systematischen Stellung von *Krameria triandra* Ruiz et Pav. Ph.D. dissertration, Christian-Albrechts-Universitat, Kiel.

Cameron KM, MW Chase, WR Anderson, and JG Hills. 2001. Molecular systematics of Malpighiaceae: evidence from plastid *rbc*L and *mat*K sequences. Am. J. Bot. 88: 1847–1862.

Cannon WA. 1910. The root habits and parasitism of *Krameria canescens* Gray. In: DT MacDougall and WA Cannon, eds. The conditions of parasitism in plants. Publ. Carnegie Inst. Washington, 129: 5–24.

Carlquist S. 2005. Wood anatomy of Krameriaceae with comparisons with Zygophyllaceae: phylesis, ecology and systematics. Bot. J. Linn. Soc. 149: 257–270.

Davis CC. 2002. *Madagasikaria* (Malpighiaceae): a new genus from Madagascar with implications for floral evolution in Malpighiaceae. Am. J. Bot. 89(4): 699–706.

Davis CC, WR Anderson, MJ Donoghue. 2001. Phylogeny of Malpighiaceae: evidence from chloroplast *ndh*F and *trnl*-F nucleotide sequences. Am. J. Bot. 88: 1830–1846.

Heimsch C. 1942. Comparative anatomy of the secondary xylem in the "Gruinales" and "Terebinthales" of Wettstein with reference to taxonomic grouping. Lilloa 8: 83–198.

Leinfellner W. 1971. Das Gynozeum von *Krameria* und sein Vergleich mit jenem der Leguminosae und der Polygalaceae. Oesterr. Bot. Z. 119: 102–117.

Lobreau-Callen D. 1983. Analyse de la répartition géographique des Malpighiaceae d'après les caractères du pollen et de la pollinisation. Bathalia 14: 871–881.

Lobreau-Callen D. 1984. Pollen et paleobotanique des Malpighiaceae. Rev. Paleobiol., special vol., pp. 131–138.

Lombello RA and ER Forni Martins. 2002. Cytogenetics and evolutionary analysis of *Lophanthera,* and Amazonian arboreal Malpighiaceae. Cytologia (Japan). 67: 41–45.

Lowrie SR. 1982. The palynology of the Malpighiaceae and its contribution to family systematics. Ph.D. dissertation University of Michigan.

Milby TH. 1971. Floral anatomy of *Krameria lanceolata.* Am. J. Bot. 58: 569–576.

Morton CV. 1968. A typification of some subfamily, sectional, and subsectional names in the family Malpighiaceae. Taxon 17: 314–324.

Musselman LJ. 1975. Parasitism and haustorial structure in *Krameria lanceolata* (Krameriaceae). A preliminary study. Phytomorphology 25: 416–422.

Robertson KR. 1972. The Malpighiaceae in the southeastern United States. J. Arnold Arbor. 53: 101–112.

Robertson KR. 1973. The Krameriaceae in the southeastern United States. J. Arnold Arbor. 54: 322–327.

Simpson BB. 1982. *Krameria* (Krameriaceae) flowers: orientation and elaiophore morphology. Taxon 31: 517–528.

Simpson BB. 1989. Krameriaceae. Flora Neotropica Monograph 49: 1–109.

Simpson BB. 2007. Krameriaceae. In: K Kubitzki, ed. The families and genera of vascular plants, vol. 9, pp. 208–212. Springer, Berlin/Heidelberg/New York.

Simpson BB and JL Neff. 1978. Dynamics and derivation of the pollination syndrome of *Krameria* (Krameriaceae). Bot. Soc. Am. Misc. Publ. 156: 14.

Simpson BB and JJ Skvarla. 1981. Pollen morphology and ultrastructure of *Krameria* (Krameriaceae): Utility in questions of intrafamilial and interfamilial classification. Am. J. Bot. 68: 277–294.

Simpson BB, DS Seigler, and JL Neff. 1978. Lepids from the floral glands of *Krameria.* Biochem. Syst. Ecol. 7: 193–194.

Simpson BB, A Weeks, DM Helfgott, and LL Larkin. 2004. Species relationships in *Krameria* (Krameriaceae) based on its sequences and morphology: Implications for character utility and biogeography. Syst. Bot. 29: 97–108.

Singh B. 1959, 1961. Studies in the family Malpighiaceae: I. Morphology of *Thryallis glauca* Kuntze. II. Morphology of *Malpighia glabra* Linn. III. Development and structure of seed and fruit of *Malpighia glabra* Linn. Hort. Advance 3: 1–19, 1959; 5: 83–96, 145–155. 1961.

Stenar H. 1937. Zur Embryosackentwicklung einiger Malpighiaceen. Bot. Not. 1937: 110–118.

Subra Rao AM. 1940. Studies in the Malpighiaceae: I. Embryo sac development and embryogeny in the genera *Hiptage,*

Banistera, and *Stigmatophyllum.* J. Indian Bot. Soc. 18: 145–156.
Subra Rao AM. 1941. Studies in the Malpighiaceae: 2. Structure and development of the ovules and embryo sacs of *Malpighia coccifera* Linn. and *Tristellatera australis* Linn. Proc. Nad. Inst. Sci. India, Pt. B, Biol. Sci. 7: 393–404.
Taylor DW and WL Crepet. 1990. Fossil floral evidence of Malpighiaceae and an early plant-pollinator relationship. Am. J. Bot. 74: 274–286.
Tokuoka T and H Tobe. 2006. Phylogenetic analyses of Malpighiales using plastid and nuclear DNA sequences, with particular reference to the embryology of Euphorbiaceae s. str. J. Plant Res. 119: 599–616.
Turner BL. 1958. Chromosome numbers in genus *Krameria*; Evidence for familial status. Rhodora 60: 101–106.
Vega AS, MA Castro, and WR Anderson. 2002. Occurrence and phylogenetic significance of latex in the Malpighiaceae. Am. J. Bot. 89: 1725–1729.
Verkerke W. 1986. Ovule ontogeny and seed coat development in *Krameria* Loefling (Krameriaceae). Beitr. Biol. Pfl. 60: 341–351.
Vogel S. 1987. History of the Malpighiaceae in the light of pollination ecology. 14th Intern. Bot. Congr. Abstracts 5–47a-4. Berlin.
Yunus D. 1990. Studies in the pollen morphology of Malpighiaceae. Phytomorphology 40: 21–25.

Superorder CELASTRANAE

Order 101. CELASTRALES

Trees, shrubs, or woody lianas, sometimes herbs. Often with laticifers in the stem and leaves. Vessels mostly with simple perforations, less often with scalariform perforations. Fibers nearly always with bordered pits. Rays heterogeneous or homogeneous. Axial parenchyma apotracheal to paratracheal. Sieve-element plastids of S-type. Nodes trilacunar (with tree traces – Brexiaceae), or bilacunar (with two traces – Lepidobortyaceae), or unilacunar. Leaves alternate or opposite, simple, usually entire, with small stipules or estipulate; leaf mesophyll cells contain oil bodies. Stomata of diverse types. Flowers in terminal or axillary cymose or less often racemose inflorescences, rarely solitary in the axils, small to rather large, bisexual or unisexual, actinomorphic or rarely weakly zygomorphic, usually 4-merous or 5-merous. Sepals free or basally connate, usually imbricate. Petals free or more or less connate, imbricate or less often contorted or valvate, rarely wanting. Androecium haplostemonous with antesepalous stamens or (in a few Celastraceae) diplostemonous. Stamens 5 (2–20), free. Anthers dorsifixed or basifixed, commonly introrse, mostly tetrasporangiate, opening longitudinally. Tapetum secretory. Microsporogenesis simultaneous. Pollen grains 2-celled or 3-celled, usually 3-colporate. Gynoecium of (2-)3–5(-7) united carpels, with terminal and usually short style or (Goupiaceae) free stylodia; ovary superior or rarely semi-inferior, (1)2–5-locular, with (1)2 or sometimes up to 6 (numerous in Goupiaceae and Brexiaceae), or numerous ovules in each locule. Ovules anatropous, apotropous, or rarely epitropous, bitegmic or rarely (Lepuropetalaceae) unitegmic, crassinucellate to tenuinucellate, often with endothelium. Female gametophyte of *Polygonum*-type. Endosperm nuclear. Fruits loculicidal, rarely septicidal capsules, samaras, baccate, drupaceous, or indehiscent capsules or dry and separate into indehiscent mericarps. Seeds exotestal, endotegmic, with exotegmic fibers; embryo large to very small, straight; endosperm usually copious and more or less oily or less often scanty or wanting.

The most primitive family of the order is Brexiaceae which probably derived from some woody Rosanae.

Key to Families

1 Nodes trilacunar or bilacunar.
  2 Nodes trilacunar.
    3 Stipules minute, deciduous. Trees or shrubs, glabrous. Vessels almost exclusively with simple perforations; lateral pitting opposite, but occasionally with some tendency to alternate. Rays essentially uniseriate (mostly homogeneous) and, also biseriate (always heterogeneous and consisting mainly of upright and square cells). Fibers usually with bordered pits. Axial parenchyma abundant, several layered (1–6), apotracheal banded. Nodes 3-locunar with three traces. Leaves with minute deciduous stipules, alternate, conduplicate, entire to spinulose-dentate, without marginal glands; marginal venation is characteristically fimbriate. Thin-walled, unicellular hairs are present sporadically. Flowers large, bisexual, 5-merous, in axillary umbel-like cymes or they may occur on the stem. Sepals connate into a short tube. Petals not clawed. Anthers almost basifixed, with a basal pseudopit, sagittate, introrse and versatile; connectives protrude in long acute appendages. Pollen grains 3-colporate. There are 4–5 filiform staminodia with fused bases attached to the thick,

5-lobed nectary disc opposite to each petal. Ovary superior, 5–7-locular, stigma capitate, the short stigmatic lobes are commissural (Matthews and Endress 2005). Ovules numerous (up to 40), weakly crassinucellate, without nucellar cap, funicular obturator, and arillode, but with endothelium. Fruits baccate, large, woody, 5-cornered, marked with numerous small, scarcely elevated papillae like the surface of an orange many-seeded. Seeds smooth and shining, slightly angular, seed coat exotestal-exotegmal; exotestal cells tanniniferous, outer wall thickened, with transverse thickenings in surface view; embryo small, straight, endosperm copious, albuminous; n = 30, 32. . . . . 3. BREXIACEAE

3 Stipules intrapetiolar, caduceus. Trees, sometimes tall, or shrubs, sometimes lianoid, with simple or stellate hairs or peltate scales and with strong odour of garlic. Nodes trilacunar. Vessels mostly with oblique, simple perforations, but some of the first-formed ones of the secondary xylem with scalariform perforations, without vestured pits. Fibers with bordered or simple pits. Phloem with wedge-shaped rays, and phloem in young stem not stratified. Leaves simple, alternate, entire, pinnately veined, stipules free of one another, small in *Hua*, large in *Afrostyrax*. Stomata paracytic. Flowers small, axillary and solitary or in small clusters, bisexual. Sepals 5, free, valvate (*Hua*), or the calyx closed in bud and opening by 3–5 irregular lobes (*Afrostyrax*). Petals (4)5, free, induplicate-valvate, long clawed (*Hua*) or with a very short, broad base beneath the obovate blade (*Afrostyrax*). Stamens (8)10, free, in 1 cycle; filaments flattened; anthers basifixed, 2-locular (*Hua*) or 4-locular, the lower 2 loculi smaller than the upper, appendaged (*Afrostyrax*). Pollen grains 3-porate. Gynoecium of five carpels with a terminal style and small, punctate stigma; ovary 1-locular; ovules solitary *(Hua)* or ovules 4–6 (mostly abortive), anatropous, erect on a basal placenta. Fruits dry, 1(2)-seeded, dehiscent from the apex into five valves *(Hua)* or drupes *(Afrostyrax)*. Seeds large, with a basal hilum, hairy (*Hua*); exotegmen of lignified palisade cells; embryo straight, surrounded by the copious endosperm; cotyledons flattened. Plants with a strong odor of garlic and without cyclopropenoid fatty acids. . . . . . 9. HUACEAE.

2 Nodes bilacunar with two traces. Evergreen trees up to 40 m tall, with very sparse indumentum of unicellular hairs; wood fluoresces. Vessels with simple perforations. Vestured pits present (*Ruptiliocarpon*). Leaves alternate, unifoliolate, petiolate and petiolule pulvinate, not gland-dotted, with caducous stipules and stipelli, stipules adnate to petiole. Stomata paracytic. Flowers in axillary, rarely terminal panicles or spikes, strobiliform when young, bracteolate, 5-merous, unisexual, dioecious, fertile flowers functionally male, or functionally female, female flowers with staminodes. Sepals shortly connate below, persistent, imbricate; petals imbricate. Staminodes and pistillodes present. Stamens ten, in two whorls, on fleshy disc, shortly connate below, the antipetalous anthers sessile, the antisepalous filaments distinct beyond tube; anthers basifixed or dorsifixed, versatile. Pollen grains 3-colporoidate. Nectary disc present, receptacular, protuberant between the filament bases. Gynoecium of 2–3 carpels; stylodia partially joined, attenuate from the ovary, apical, stigma 2-lobed; ovary superior, 2–3-locular, with two collateral ovules per locule; ovules pendulous, anatropous, with an obturator and with ventral raphe, carunculate, the caruncle eventually orange-red. Fruits large, leathery, septicidal and valvular capsules. Seeds covered with a fleshy red-orange aril, integuments multiplicative, with fibrous, strongly thickened exotegmen; embryo straight or oblique; endosperm nearly absent. . 2. LEPIDOBOTRYACEAE.

1 Nodes unilacunar.

4 Perforations always scalariform, with 3–10 thick bars. Evergreen trees or shrubs, cork superficial. Fibers with distinctly bordered pits. Rays heterogeneous, with elongate ends. Axial parenchyma apotracheal and paratracheal. Leaves alternate, shining, coriaceous, entire (but seedlings with dentate leaves), subtriplinerved and transversely venose, petiolate (the petiole with complex and unusual vascularization), with very narrow and rather long, caducous stipules. Stomata anomocytic. Inflorescences axillary, pedunculate, umbel-like clusters of very short, bracteate racemes. Flowers small, bisexual, actinomorphic, 5-merous. Pedicels slender. Bracts short and triangular, hairy. Sepals connate, lobes imbricate. Petals very long and subulate, concave, induplicate-valvate, the upper third sharply inflexed in bud and sometimes geniculate or

sigmoid at anthesis. Stamens five, inserted on the edge of the nectary disc; filaments extremely short; anthers small, ovoid, with a thickened and apically pilose connective and ellipsoidal, separate, introrse loculi. Pollen grains 3-colporate. Nectary disc thin and cupular, sinuate on the margin. Gynoecium of five carpels; ovary superior, partly enclosed by the disc, but free, depressed-globose, 5-locular; stylodia free, short, subulate, very divergent, shorter than the ovary; ovules several to numerous in each locule, basal axile, ascending. Fruits small, hard, globose, 2–3-locular, indehiscent, many seeded. Seeds erect, obovoid, exotestal, with axile straight embryo in copious, fleshy endosperm; testa reticulate, pitted inside. . . . . 1. GOUPIACEAE

4 Perforations simple or (some Celastraceae) partly scalariform. Leaves not transversely-venose. Petals not very long and sharply inflexed in bud.

5 Trees, shrubs, or woody lianas.

6 Petals usually imbricate. Trees, shrubs, or woody lianas, laticiferous, or non-laticiferous. Commonly present yellow triterpene derivatives in the bark. Vessels with simple or rarely (*Bhesa, Elaeodendron, Perrottetia*) scalariform perforation, without vestured pits. Fibers with bordered or simple pits, often septate. Leaves alternate or opposite, rarely much reduced (*Psammomoya* and *Canotia* leaves represented by cataphylls), sometimes leathery or membranous, margins entire, spiny; stipules small, caducous or absent. Flowers in terminal, axillary, or rarely (*Polycardia*) epiphyllous cymose or rarely racemose inflorescences, rarely flowers solitary, usually greenish or white, bisexual or less often unisexual (plants monoecious or dioecious), usually 4–5-merous. Sepals (2-)4-5, free or basally connate, mostly imbricate. Petals usually free, imbricate or rarely contorted or valvate, rarely wanting. Stamens (2)3-5(8-10), alternipetalous; anthers 2-locular or rarely (*Euonymus* spp) 1-locunar, extrorse or introrse, disporangiate or tetrasporangiate. Nectary disc well developed, intrastaminal (most Celastroideae) or extrastaminal, often cupular, often adnate to the ovary, sometimes forming a short androgynophore, very rarely wanting (as in *Campylostemon*). Pollen grains 2- or 3-celled, 3-colporate or sometimes (as in *Siphonodon*) 3-porate, mostly reticulate. Gynoecium of 2–5 carpels (usually with all but one abortive), with commonly short style; stigma capitate or sometimes 2–5-lobed; ovary superior or very rarely semi-inferior, 2–5-locular, rarely only 1 locule developed, with (1)2–10(-15) ovules per locule. Ovules anatropous, apotropous, or very rarely (*Tripterygium*) epitropous, pendulous orascending,crassinucellatetotenuinucellate, with differentiated endothelium. Fruits mostly loculicidal, very rarely (*Canotia*) loculicidal and septicidal, dehiscent or indehiscent capsules, samaras, berries, or drupes. Seeds with a fleshy or submembranous brightly colored aril, or winged with the funicle free from the wing, originating from the integument near the funicle, or wingless, exotestal-exotegmic, with copious and more or less oily endosperm or sometimes without endosperm; embryo sometimes rather large, green, straight, with short or minute radicle; cotyledons very large, connate. Present gutta-percha, pyrrolizidine and sesquiterpene alkaloids, triterpenoids, proanthocyanidins (cyaniding and delphinidin), flavonols (kaempferol, quercetin, or myricetin) present or rarely (*Salacia*) absent, n = 8–10, 12, 14–17, 23, 28. . . . . 4. CELASTRACEAE.

6 Petals noncontiguous. Glabrous lianas with leaf tendrils. Cork superficial, vessels with simple perforation, phloem stratified. Leaves alternate, simple, entire, serrulate or crenulate, stipulate, with watch-spring tendrils (modified inflorescences). Stomata paracytic. Flowers in glomerules on the branches of loose axillary panicles, unisexual, monoecious, actinomorphic, 5-merous. Sepals free or shortly connate, valvate, persistent. Petals free, much smaller. Stamens five, oppositisepalous, alternate with five spreading, opposite, petalous staminodia or glands that in the male flowers are more or less adnate to the subtending p etals and in the female flowers are more or less concrescent into a 5-lobed nectary disc; filaments filiform; anthers introrse. Pollen grains 3-colporate. Gynoecium of 5–4 carpels, ovary superior, 4–5-locular, ovoid-oblong, shallowly ribbed;

pubescent, with 5–4 sessile, subulate stigmas and 2 pendulous, apical-axile, anatropous ovules per locule, each surrounded by an obturator-like appendage. Fruits indehiscent, fusiform, 1-locular, 1-seeded, with five broad, stramineous wings. Seeds oblong, with endosperm; embryo straight with long cotyledons.........6. LOPHOPYXIDACEAE.

5 Perennial herbs with a woody, branched rhizome to nearly subshrubs with moderate wood accumulation or (*Macgregoria*) annuals. Vessels mostly with simple perforations but a few double perforation plates are also found; lateral pitting alternate. Fibers with bordered pits. Rays both multiseriate and uniseriate, cells predominantly erect. Axial parenchyma essentially absent. Leaves alternate, entire, sessile, fleshy or leathery, sometimes much reduced; stipules interpetiolar, scaly, deciduous or persistent. Stomata anomocytic. Flowers rather small, in racemose or cymose (rarely umbellate) inflorescences, bisexual, actinomorphic (except for the usually unequal stamens), 5-merous. Sepals connate into a tube, lobes imbricate. Petals linear or spatulate, unequal but not bilabiate, clawed, claws free, but the blades mostly connate. Stamens five, alternipetalous, usually three long and two short; anthers almost basifixed, with a basal pseudopit, sagittate, introrse, apically appendaged. Nectary disc thin, extends between androecium and gynoecium. Pollen grains 3-celled, 3-colporate. Gynoecium of 2–5 carpels; stylodia more or less connate, mostly sunken between the ovary lobes; ovary laterally and usually also apically lobed, with one erect, axile-basal ovule in each locule; ovules apotropous to sometimes epitropous, anatropous, bitegmic, tenuinucellate. Fruits dry, of 2–5 indehiscent cocci. Seeds small, with straight embryo and rather thick cotyledons, endosperm fleshy, contains starch and oils; testa composed of tangentially elongate tannincells; meso-endotesta thin-walled; exotesta with tanniniferous cells, tegmen absorbed, vascular bundle extending in the antiraphe to the micropyle. n = 9, 10, 15........... 7. STACKHOUSIACEAE.

7 Leaves alternate, all basal or both basal and cauline. Perennial herbs from rhizomes or caudices, or diminutive winter annuals, glabrous or young petiole bases puberulent. Secretory cells with tanniniferous contents present in the unlignified tissues of leaves and stem. Vessels with simple perforation. Leaves alternate or subopposite, flat, long petiolate, the blades palmately veined, ovate, reniform, or orbicular, entire, with cuneate, truncate, or cordate bases; leaf venation campylodromous. Stomata anomocytic, on the abaxial leaf surface. Flowers protandrous, solitary, bisexual, weakly zygomorphic, 5-merous, on a long peduncle, terminal on scapiform stems with one or (in section *Fimbripetalum*) many sessile and reduced leaves, often with a clasping base. Sepals basally connate, forming a more or less denned cup, lobes spreading, imbricate, persistent. Petals showy, larger than sepals, sessile, or clawed, imbricate, persistent. Stamens five, dehiscing over the gynoecium in sequence before moving outwards; anthers ventrifixed, versatile, extrorse, the connective is broad with a short protrusion, tetrasporangiate, opening longitudinally. Staminodia five, candelabra-like, consist of three or many filiform divisions with glandular-appearing apices. Nectar secretion from the bases of the staminodes. Pollen grains 2-celled, 3-colporate, 4-colporate and syncolpate, with thin exine and fine columellae, reticulate. Gynoecium of 3–4(5) carpels; stigmas sessile or subsessile; ovary superior or partially inferior. Ovules bitegmic (Parnassioideae) or unitegmic (Lepuropetaloideae), many to numerous, Capsules loculicidal, many-seeded. Seeds minute (in *Parnassia palustris* about 100–2,000 in capsules – Hultgärd 1987), oblong, thinly, with a winglike testa; endotegmic cells with U-shaped thickening; endosperm scanty or absent, embryo straight, cylindrical, n = 7, 8, 9, or 23 (Lepuropetaloideae)..... 8. PARNASSIACEAE.

7 Leaves opposite, simple, entire, pinnately veined, estipulate, with very short stellate hairs. Veins with adaxial caps of thin-walled fibers containing elastic material. Stomata anomocytic and encyclocytic. Climbing shrubs. Young twigs rectangular in transverse section. Branches and leaves with latex canals. Vascular system sheathed by thin-walled fibers containing elastic material.

Vessels with simple perforations and with vestured pits. Fibers with bordered pits, septate and nonseptate. Flowers small, in axillary, paniculate inflorescences, very fragrant, bisexual, with an epicalyx. Sepals (3)4(5), conspicuous in bud, valvate, covered by the same kind of hairs as the vegetative parts, revolute. Corolla (3)4(5), valvate, externally pubescent, revolute. Stamens numerous (30–100) on small disc, shortly connate at the base; filaments slender, free; anthers small, 4-locular, transversely dehiscent by an apical horizontal slit. Pollen grains 3-celled, 3-colporate, tectate-columellate. Gynoecium of 3 carpels, with elongate, subulate, shortly 3-lobed style; stigma punctate, ovary 3-locular, with two basal, ascending, apotropous, weakly crassinucellate ovules per locule; micropyle zig-zag. Endothelium differentiated. Fruits turbinate, finely septicidal capsules expanded at the apex into two or three divergent, stiff wings. 5. PLAGIOPTERACEAE.

1. GOUPIACEAE

Miers 1862. 1/2–3. Tropical South America (Guyana, Surinam, northern Brazil).

Goupia.

Goupiaceae differ from Celastraceae in many respects, including the peculiar structure of the petiole (Metcaife and Chalk 1950), leaf venation, long stipules, morphology of anthers, free stylodia. In some respects *Goupia* is more archaic than Celastraceae. Goupiaceae are characterized by primitive wood anatomy and have vessels with scalariform perforations with numerous bars. The APG (2003) included this family in Malpighiales. However, the including Goupiaceae in the Celastrales is debatable.

2. LEPIDOBOTRYACEAE

J. Léonard 1950. 2/2–3. Tropical Africa (*Lepidobotrys*), trop. America from Costa Rica to Peru and Suriname (*Ruptiliocarpon*).

Lepidobotrys, Ruptiliocarpon.

The Lepidobotryaceae are related to the Celastraceae (Savolainen et al. 2000). According to Matthews and Endress (2005), some "features are shared by Celastraceae and Lepidobortyaceae", e.g. "stamen bases are united and form a collar around the gynoecium, associated with a pronounced nectariferous disc, and seeds have a conspicuous red or orange aril". In their opinion, "Lepidobotriaceae are more distant and share with Malpighiales several features that deviate from other Celastrales".

3. BREXIACEAE

Loudon 1830. 1/12. Mainly Madagascar (10), with one species extending to East Africa (Tanzania, Mozambique, Zanzibar) and Comoro Islands, and one species restricted to the Seychelles.

Brexia.

Bensel and Palser (1975) noted that both Brixiaceae and Celastraceae have dulcitol, and Ramamonjiarisoa (1980) has found the anatomical similarities between *Brexia* and Celastraceae. According to Simmons et al. (2001), *Brexia* is closely related to *Elaeodendron* and *Pleurostylia*.

4. CELASTRACEAE

R. Brown 1814 (including Canotiaceae Airy Shaw 1965, Chingithamnaceae Handel-Mazzetti 1932, Euonymaceae Berchtold and J. Presl 1820, Hippocrateaceae A. L. de Jussieu 1811, Pottingeriaceae Takhtajan 1987, Salaciaceae Rafinesque 1838, Siphonodontaceae Gagnepain et Tardieu ex Tardieu 1951). 85–90/860. Widely distributed, but chiefly in tropical and subtropical regions.

4.1 POTTINGERIOIDEAE
Fruits septicidal capsules. Seeds mesotestal. Nectary disc large, annular. – *Pottingeria*.

4.2 CELASTROIDEAE
Fruits mostly loculicidal capsules, or drupes, berries. Seeds with endosperm or without (Lophopetaleae), with or without an aril. Nectary disc usually intrastaminal. – EUONYMEAE: *Mommopetalum, Euonymus, Microtropis, Glyptopetalum, Empleuridium*, etc.; CELASTREAE: *Celastrus, Psammomoya, Maytenus, Catha, Polycardia, Bhesa, Xylonymus, Paxistima*, etc.; LOPHOPETALEAE: *Lophopetalum, Kokoona, Peripterygia*. CASSINEAE: *Cassine, Forsellesia, Elaeodendron, Pleurostylia, Gyminda, Crossopetalum, Myginda*,

Acanthothamnus, Canotia, Mortonia, Schaefferia, etc.; PERROTTETIEAE: *Perrottetia*.

4.3 TRIPTERYGOIDEAE
Fruits samaras or nut. Seeds without an aril. – *Ptelidium, Zinowiewia, Plenckia, Tripterygium, Wimmeria, Platypterocarpus*.

4.4 HIPPOCRATEOIDEAE
Fruits drupes, berries, or often strongly 3-lobed capsules. Seeds without endosperm and usually without true aril. – SALACIEAE: *Cheiloclinium, Peritassa, Salacia, Salacighia, Thyrsosalacia, Tontelea;* CAMPYLOSTEMONEAE: *Bequaertia, Campylostemon, Tristemona nthus*; HELICTONEMEAE: *Helictonema*; HIPPOCRATEAE: *Hippocratea, Anthodon, Apodostigma, Arnicratea, Cuervea, Elachyptera, Hylenaea, Pristimera, Prionostemma, Simirestis, Loeseneriella, Reissantia, Semialarium, Simicratea*.

4.5 SARAWAKODENDROIDEAE
Fruits 3-lobed capsules. Seeds with filamentous arils, endosperm thin-walled, oily. – *Sarawakodendron*.

4.6 SIPHONODONTOIDEAE
Fruits drupaceous. Nectary disc very large, almost completely enclosing gynoecium. Seeds with endosperm. Aril wanting. – *Siphonodon*.

Genus *Bhesa* (2 sp. in India, Sri Lanka, Burma, Thailand, Indochina, Malesia) is traditionally including into the Celastraceae (see Kubitzki 2004). However, molecular analysis does not support the including *Bhesa* in Celastraceae and show affinity with families of Malpighiales sensu APG II (2003). The systematic position of *Bhesa* is debatable.

All six subfamilies are closely related. The Hippocrateoideae are connected with the Celastroideae through the tropical African genus *Campylostemon*. Seemingly more isolated is an Indomalesian-Australian genus *Siphonodon,* but as Airy Shaw (in Willis 1973) points out, it is possibly only an extreme modification of Celastraceae. *Perrottetia,* according to Corner (1976), is entirely anomalous in Celastraceae.

5. PLAGIOPTERACEAE

Airy Shaw 1965. 1/2. Southwestern China, Lower Burma, Thailand.

Plagiopteron.

Related to the Celastraceae, especially Hippocrateoideae (see M.P.Simmons 2004).

6. LOPHOPYXIDACEAE

H. Pfeiffer 1951. 1/2. Malay Peninsula, Borneo, East Malesia, Palau, and Solomon Islands.

Lophopyxis.

Closely related to the Celastraceae.

7. STACKHOUSIACEAE

R. Brown 1814. 3/28. Malesia, Micronesia, New Guinea, Australia, Tasmania, New Zealand.

7.1 MACGREGORIOIDEAE
Flowers ebracteate. Petals free. Stamens equal; filaments very short; anthers with apical appendages. Pollen grains smooth. Carpels five; style with discoid collar beneath stigma. – *Macgregoria*.

7.2 STACKHOUSIOIDEAE
Flowers with three transverse bracteoles. Claws of petals free below, connate into a tube above. Stamens unequal, three long, three short; filaments elongate; anthers obtuse or shortly mucronate at apex. Pollen grains lamellate-areolate. Carpels three, rarely two or five, style without collar. – *Stackhousia, Tripterococcus*.

Close to the Celastraceae but much more specialized, especially in seed coat structure.

8. PARNASSIACEAE

Martynov 1820 (including Lepuropetalaceae Nakai 1943). 2/c.50–70. Cold and temperate regions of the Northern Hemisphere, especially in the Himalayas, eastern Asia, and northwestern parts of North America (*Parnassia*, c.70) and Southwestern United States, Mexico, Ecuador, Brazil, Argentina, central Chile, and Uruguay (1, *Lepuropetalon*).

8.1 PARNASSIOIDEAE
Ovules bitegmic. Perennial herbs from rhizomes or caudices, glabrous or young petiole bases puberulent.

Secretory cells with tanniniferous contents present in the unlignified tissues of leaves and stem. Leaves all basal, flat, long petiolate, the blades palmately veined, ovate, reniform, or orbicular, entire, with cuneate, truncate, or cordate bases; leaf venation campylodromous. Stomata on the abaxial leaf surface. Flowers protandrous, solitary, on a long peduncle, terminal on scapiform stems with one or (in section *Fimbripetalum*) many sessile and reduced leaves, often with a clasping base. Sepals persistent. Petals showy, larger than sepals, sessile, or clawed, imbricate, persistent. Stamens dehiscing over the gynoecium in sequence before moving outwards; anthers ventrifixed, versatile, extrorse, the connective is broad with a short protrusion. Staminodia five, candelabra-like, consist of three or many filiform divisions with glandular-appearing apices. Nectar secretion from the bases of the staminodes. Pollen grains 2-celled, 3-colporate, 4-colporate and syncolpate, with thin exine and fine columellae, reticulate. Seeds minute (in *Parnassia palustris* about 100–2,000 in capsules – Hultgärd 1987), oblong, thinly, with a winglike testa; endotegmic cells with U-shaped thickening; endosperm scanty or absent, embryo straight, cylindrical, n = 7, 8, 9. – *Parnassia*.

8.2 LEPUROPETALOIDEAE
Ovules unitegmic. Diminutive annual, more or less succulent herbs, usually forming small hemispherical tufts. Leaves both basal and cauline, alternate or subopposite, spatulate, entire, the blade constricted into a broad petiole; leaves brown-streaked from tannins; leaf venation acrodromous. Stomata on both leaf surface. Flowers small and inconspicuous, either appearing sessile in the basal rosette or terminal at the ends of short, leafy, and angled stems. Petals white, smaller than sepals, sometimes absent or of (4)5 linear, bractlike, submicroscopic structures inserted on the rim of the floral cup. Stamens very short; anthers bisporangiate at anthesis (Simmons 2004), Staminodia external to the stamens, dilated distally. Seeds cylindrical, blackish, with a short beak on one end, the surface reticulate; endotestal cell walls much thickened, endosperm scanty, n = 23. – *Lepuropetalon*.

9. HUACEAE

A. Chevalier 1947. 2/3. Tropical Africa. *Hua, Afrostyrax*.

The affinities of Huaceae have long been obscure (Soltis et al. 2005). According to Zhang and Simmons (2006), "Huaceae are best supported as the sister group of Oxalidales, suggesting that it should be placed within that order.". Here I follow to recent version of Thorne system (2006) in inclusion of Huaceae in the Celastrales.

Bibliography

Adatia RD and SG Gavde. 1962. Embryology of the Celastraceae. In: Plant embryology: A symposium. New Delhi, pp. 1–11.

Airy Shaw HK, DF Cutler, and S Nilsson. 1973. *Pottingeria*, its taxonomic position, anatomy and palynology. Kew Bull. 28: 97–104.

Arber A. 1913. On the structure of the androecium in *Parnassia* and its bearing on the affinities of the genus. Ann. Bot. 27: 491–510.

Arber A. 1915. The anatomy of the stamens in certain Indian species of *Parnassia*. Ann. Bot. 29: 159–160.

Archer RH and AE van Wyk. 1992. Palynology and intergeneric relationships in some southern African species of subfamily Cassinoideae (Celastraceae). Grana 31: 241–252.

Archer RH and AE van Wyk. 1993a. Bark structure and intergeneric relationships of some southern African Cassinoideae (Celastraceae). IAWA J. 14: 35–53.

Archer RH and AE van Wyk. 1993b. Wood structure and generic status of some southern African Cassinoideae (Celastraceae). IAWA J. 14: 373–389.

Baas P. 1972. Anatomical contributions to plant taxonomy: II. The affinities of *Hua* Pierre and *Afrostyrax* Perkins and Gilg. Blumea 20: 161–192.

Baas P, R Geesink, WA Van Heel, and J Muller. 1979. The affinities of *Plagiopteron suaveolens* Griff. (Plagiopteraceae). Grana 18: 69–89.

Baker WR. 1977. Taxonomic studies in *Stackhousia* Sm. (Stackhousiaceae) in South Australia. J. Adelaide Bot. Gard. 1: 69–82.

Barker WR. 1984. Stackhousiaceae. Flora of Australia 22: 186–199. Canberra.

Bayer C. 2007. Huaceae. In: K Kubitzki, ed. The families and genera of vascular plants, vol. 9, pp. 191–193. Springer, Berlin/Heidelberg/New York.

Beijersbergen A. 1972. Notes on the chemotaxonomy of Huaceae. Blumea 20: 160.

Bennett AW. 1871. Note on the structure and affinities of *Parnassia palustris* L. Bot. J. Linn. Soc. 11: 24–31.

Bensel CR and BF Palser. 1975a. Floral anatomy in the Saxifragaceae sensu lato: I. Introduction, Parnasioideae, and Brexioideae. Am. J. Bot. 62: 176–185.

Bensel CR and BF Palser. 1975b. Floral anatomy in the Saxifragaceae sensu lato: II. Saxifragoideae and Iteoideae. Am. J. Bot. 62: 661–675.

Berkeley F. 1953. Morphological studies in the Celastraceae. J. Elisha Mitchell Sci. Soc. 69: 185–206.

Boesewinkel FD and F Bouman. 2000. Lepidobotryaceae. In: A Takhtajan, ed. Comparative seed anatomy, vol. 6, pp. 23–24. Nauka, St. Petersburg (in Russian).

Bohm BA, LS Donevan, and UG Bhat. 1986. Flavonoids of some species of *Bergenia, Francoa, Parnassia,* and *Lepuropetalon*. Biochem. Syst. Ecol. 14: 75–77.

Brizicky GK. 1964. The genera of Celastrales in the southeastern United States. J. Arnold Arbor. 45: 206–234.

Brüning R and H Wagner. 1978. Übersicht über die Celastraceen-Inhaltsstoffe: Chemie, Chemotaxonomie, Biosynthese, Pharmakologie. Phytochemistry 17: 1821–1858.

Carlquist S. 1987. Wood anatomy and relationships of Stackhousiaceae. Bot. Jahrb. Syst. 108: 473–480.

Chevalier A. 1947. La famille des Huacaceae et sesaffinités. Rev. Int. Bot. Appl. Agric. Trop. 27(291–292): 26–29.

Copeland HF. 1966. Morphology and embryology of *Euonymus japonica*. Phytomorphology 16: 326–334.

Croizat L. 1947. A study in the Celastraceae: Siphonodonoideae subf. Nov. Lilloa 13: 31–43.

Daumann E. 1960. Über die Bestäubungsökologie der *Parnassia*-Blüte: Ein weiterer Beitrag zur experimen-tellen Blütenökologie. Biol. Plant 2: 113–125.

David E. 1938. Embryologische Untersuchungen an Myoporaceen, Salvadoraceen, und Hippocrateaceen. Planta 28: 680–703.

Davison JD. 1927. Celastraceae R.Br. Bothalia 2: 289–346.

Dehay C. 1951. Caractéres anatomique des Huacacées. Bull. Soc. Bot. N. France. 4: 14–18.

Ding Hou. 1962, 1964. Celastraceae. In: CGGJ van Steenis, ed. Flora Malesiana, ser. 1, 6(2): 227–291, 6(3): 389–421. Noordhoff, Groningen.

Ding Hou. 1967. *Sarawakodendron*, a new genus of Celastraceae. Blumea 15: 139–143.

Ding Hou. 1969. Pollen of *Sarawakodendron* (Celastraceae) and some related genera, with notes on techniques. Blumea 17: 97–120.

Doweld AB. 2000. Brexiaceae. In: A Takhtajan, ed. Comparative seed anatomy, vol. 6, pp. 121–123. Nauka, St. Petersburg (in Russian).

Drude O. 1875. Über die Blütengestaltung und die Verwandtschaftsverhältnisse des Genus *Parnassia,* nebst einer systematischen Revision seiner Arten. Linnaea 39: 239–324.

Eichinger A. 1908. Beiträge zur Kenntnis und systematischen Stellung der Gattung *Parnassia.* Beih. Bot. Centralbl. 23: 298–317.

Gastony GJ and DE Soltis. 1977. Chromosome studies of *Parnassia* and *Lepurapetalon* (Saxifragaceae) from the eastern United States: A new base number for *Parnassia.* Rhodora 79: 573–578.

Gibson AC. 1979. Anatomy of *Koeberlinia* and *Canotia* revisited. Madroño 26: 1–12.

Goldblatt PH, H Tobe, S Carlquist, and VC Patel. 1985. Familial position of the Cape genus *Empleuridium.* Ann. Missouri Bot. Gard. 72: 167–183.

Gornall RJ and KLA Al-Shammary. 1998. Parnassiaceae. In: DF Cutler and M Gregory, eds. Anatomy of the Dicotyledons: Saxifragales, vol. 4, pp. 245–247. Clarendon Press, Oxford.

Hallé N. 1962. Monographie des Hippocratéacées d'Afrique occidentale. Mém. Inst. Franc. Afrique Noire 64: 1–245.

Hallé N. 1983. Révision des Hippocrateae (Celastreae): 3. Fruits, graines et structures placentaires. Bull. Mus. Nat. Hist. Natur. 4 sér., sect. B, Adansonia 5: 11–26.

Hallier H. 1923. *Lepidobotrys* Engl.: Die Oxalidaceen und die Geraniaceen. Beih. Bot. Centralbl. 39 (2): 163.

Hammel BE and NA Zamora. 1993. *Ruptiliocarpon* (Lepidobotryaceae): A new arborescent genus and tropical American link to Africa, with reconsideration of the family. Novon 3: 408–417.

Hammel BE and N Smith. 2004. Lepidobotryaceae. In: N Smith, SA Mori, A Henderson, DW Stevenson, and SV Heals, eds. Flowering plants of the Neotropics, pp. 213–214. Princeton University Press, Princeton.

Hartog RM née, Van Ter Tholen den, and P Baas. 1978. Epidermal characters of the Celastraceae sensu lato. Acta Bot. Neerl. 27: 355–388.

Hideux MJ and IK Ferguson. 1976. The stereostructure of the exine and its evolutionary significance in Saxifragaceae sensu lato. In: IK Ferguson and J Muller, eds. The evolutionary significance of the exine, pp. 327–377. Linn. Soc. Symposium, No. 1. London/New York.

Hultgärd U-M. 1987. *Parnassia palustris* L. in Scandinavia. Acta Univ. Upsal. Symb. Bot. Upsal. 28: 1–128.

Jordaan M and AE van Wyk. 1999. Systematic studies in subfamily Celastroideae (Celastraceae) in southern Africa: reinstatement of the genus *Gymnosporia.* South Afr. J. Bot. 65: 177–181.

Kamelina OP. 1988a. Embryology of the genus *Brexia* in connections with its systematic position. Bot. Zhurn. 73: 355–366. (In Russian with English summary.)

Kamelina OP. 1988b. Sporo-, gametogenesis, and fertilization of *Escallonia* and *Brexia* with comments on their taxonomy. In Sexual Reprod. Higher Plants, pp. 431–435. Siena.

Klopfer K. 1972. Beiträge zur floralen Morphogenese und Histogenese der Saxifgagaceae: 7. *Parnassia palustris* und *Francoa sonchifolia.* Flora 161B: 320–332.

Koontz JA and DE Soltis. 1999. DNA sequence data reveal polyphyly of Brexioideae (Brexiaceae: Saxifragaceae sensu lato). Plant Syst. Evol. 219: 199–208.

Kozo-Poljanski BM. 1927. On the autecology of *Parnassia palustris* L. Bot. Zhurn. 32: 15–22 (in Russian).

Krach JE. 1976. Die Samen der Saxifragaceae. Bot. Jahrb. Syst. 97: 1–60.

Ku T. 1987. A revision of the genus *Parnassia* (Saxifragaceae) in China. Bull. Bot. Res., Herbin 7: 1–59.

Kubitzki K. 2004. Lepidobotryaceae. In: K Kubitzki, ed. The families and genera of vascular plants, vol. 6, pp. 233–235. Springer, Berlin/Heidelberg/New York.

Kullenberg B. 1953. Some observations on insects visiting and pollinating the flowers of *Parnassia palustris* L. Svensk Bot. Tidskr. 47: 439–448 (in Swedish with English summary.)

Lebgue A. 1953. Embryogénie des Parnassiacees: De-veloppement de l'εμβρψον chez le *Parnassia palustris* L. C. R. Acad. Sci. Paris 236: 1693–1695.

Léonard J. 1950. *Lepidobotrys* Engl., type d'une famille nouvelle de Spermatophytes: Les Lepidobotryaceae Bull. Jard. Bot. Nat. Belg. 20: 31–40.

Link DA. 1991. The floral nectaries of Geraniales: III. Lepidobotryaceae J. Leonard. Bull. Jard. Bot. Nat. Belg. 61: 347–354.

Lobova TA. 2000b. Pottingeriaceae. In: A Takhtajan, ed. Comparative seed anatomy, vol. 6, pp. 292–296. Nauka, St. Petersburg (in Russian).

Lobreau D. 1969. Les limites de l'' "ordre" des Celastrales d'apres le pollen. Pollen et Spores 11: 499–555.

Lobreau-Callen D. 1975a. Les pollens colpes dans les Celastrales: Interpretation nouvelle de l'απερτυρε simple. Compt. Rend. Hebd. Seances Acad. Sci. 280: 2547–2550.

Lobreau-Callen D. 1975b. Les pollens des Celastrales et groupes parentes. Thesis, University of Montpellier, C. N. R. S.: no A.08071.

Lobreau-Callen D. 1977. Les pollens des Celastrales (illustrations, commentaries). Memoires et Travaux de L'lnstitut de Montpellier. N. 3. Montpellier.

Martens P. 1936. Pollination et biologic florale chez *Parnassia palustris*. Bull. Soc. Bot. Belg. 68: 183–221.

Matthew ML and Endress PK. 2005. Comparative floral structure and systematics in Celastrales (Celastraceae, Parnassiaceae, Lepidobotryaceae). Bot. J. Linn. Soc. 149: 129–194.

Mauritzon J. 1936a. Zur Embryologie und systematischen Abgrenzung der Reihen Terebinthales und Celastrales. Bot. Not. 1936: 161–212.

Mauritzon J. 1936b. Embryologische Angaben über Stakhousiaceae, Hippocrateaceae, und Icacinaceae. Svensk. Bot. Tidskr. 30: 541–550.

Melikian AP and IA Savinov. 2000. Celastraceae, Goupiaceae, Lophopyxidaceae, Stackhousiaceae. In: AL Takhtajan, ed. Comparative seed anatomy, vol. 6, pp. 121–139. Nauka, St. Petersburg (in Russian).

Mennega AMW. 1993. Comparative wood anatomy of *Ruptiliocarpon caracolito* (Lepidobotryaceae). Novon 3: 418–422.

Mennega AMW. 1997. Wood anatomy of the Hippocrateoideae (Celastraceae). IAWA J. 18: 331–368.

Metcalfe CR. 1956. The taxonomic affinities of *Sphenostemon* in the light of the anatomy of its stem and leaf. Kew Bull. 1956: 249–253.

Murbeck S. 1918. Über die Organisation und verwandtschaftlichen Beziehungen der Gattung *Lepuropetalon*. Arkiv Bot. 15: 1–12.

Narang N. 1953. The life-history of *Stackhousia linariaefolia* A. Cunn. with a discussion of its systematic position. Phytomorphology 3: 485–493.

Nemirovich-Danchenko EN. 1996. Lepuropetalaceae. In: A Takhtajan, ed. Comparative seed anatomy, vol. 5, pp. 90–92. Nauka, St. Petersburg (in Russian).

Pace L. 1912. *Parnassia* and some allied genera. Bot. Gaz. 54: 306–329.

Pant DD and PF Kidwai. 1966. Epidermal structure and stomatal ontogeny in some Celastraceae. New Phytol. 65: 288–295.

Perrier de la Bathie H. 1942. Au sujet des affinités des *Brexia,* des Celastracées, et de deux *Brexia* nouveaux de Madagascar. Bull. Soc. Bot. France 89: 219–221.

Pfeiffer H. 1951. *Lophopyxis* als Typus einer eigenen Familie. Revista Sudamer. Bot. 10: 3–6.

Plouvier V. 1956. Sur la présence d'aspéruloside chez les *Escallonia* et de dulcitol chez de le *Brexia madagascariensis* Thou. (Saxifragaceae). Comptes Rendus Hebdomad. Seanc. Acad. Sci. 242: 1643–1645.

Ramamonjiarisoa BA. 1980. Comparative anatomy and systematics of African and Malagasy woody Saxifragaceae sensu lato. Ph.D. Thesis, University of Massachusetts.

Record SJ. 1938. The American woods of the orders Celastrales, Olacales, and Santalales. Trop. Woods 53: 11–38.

Robson NKB, N Hallé, B Mathew, and R Blakelock. 1994. Celastraceae. In: RM Polhill, ed. Flora of tropical East Africa, vol. 108, pp. 1–78. AA Balkema, Rotterdam.

Savinov IA. 2002. Sarawakodendroideae – a new subfamily from Celastraceae family. Bot. Zhurn. 87(7): 108–109 (in Russian with English summary).

Savolainen V, J-F Manen, E Douzery, and R Spichiger. 1994. Molecular phylogeny of families related to Celastrales based on *rbc*L 5′ flanking sequences. Molec. Phylogenet. Evol. 3: 27–37.

Savolainen V, R Spichiger, and J-M Manen. 1997. Polyphyletism of Celastrales deduced from a chloroplast noncoding DNA region. Molec. Phylog. Evol. 7: 145–157.

Saxena NP. 1964. Studies in the family Saxifragaceae: II. Development of ovule and megagametophyte in *Parnassia nubicola* Wall. Proc. Indian Acad. Sci, 60B: 196–202.

Schatz GE and PP Lowry II. 2004. A synoptic revision of *Brexia* (Celastraceae) in Madagascar. Adansonia 26: 67–81.

Shabes LK. 1996. Parnassiaceae. In: A Takhtajan, ed. Comparative seed anatomy, vol. 5, pp. 88–90. Nauka, St. Petersburg (in Russian).

Sharma VK. 1968. Morphology, floral anatomy, and embryology of *Parnassia nubicola* Wall. Phytomorphology 18: 193–204.

Simmons MP. 2004. Celastraceae. In: K Kubitzki, ed. The families and genera of vascular plants, vol. 6, pp. 29–64. Springer, Berlin/Heidelberg/New York.

Simmons MP. 2004. Parnassiaceae. In: K Kubitzki, ed. The families and genera of vascular plants, vol .6, pp. 291–296. Springer, Berlin/Heidelberg/New York.

Simmons MP, CC Clevinger, V Savolainen, RH Archer, S Mathews, and JJ Doyle. 2001a. Phylogeny of the Celastraceae inferred from phytochrome B gene sequence and morphology. Am. J. Bot. 88: 313–325.

Simmons MP and JP Hedin. 1999. Relationships and morphological character change among genera of Celastraceae sensu lato (including Hippocrateaceae). Ann. Missouri Bot. Gard. 86: 723–757.

Simmons MP, V Savolainen, CC Clevinger, RH Archer, and JI Davis. 2001b. Phylogeny of the Celastraceae inferred from 26S nuclear ribosomal DNA, phytochrome B, *rbc*L, *atp*B, and morphology. Molec. Phylogenet. Evol. 19: 353–366.

Sleumer H. 1968. The genus *Lophopyxis* Hook.f. (Lophopyxidaceae). Blumea 16: 321–323.

Smith AC. 1940. The American species of Hippocrateaceae. Brittonia 3: 341–555.

Smith AC. 1941. Notes on Old World Hippocrateaceae. Am. J. Bot. 28: 438–443.

Smith AC. 1945. Notes on Hippocrateaceae in Southeastern Asia. J. Arnold Arbor. 26: 169–180.

Stant MY. 1952. Notes on the systematic anatomy of *Stackhousia*. Kew Bull. 1951: 309–318.

Tang Y. 1994. Embryology of *Plagiopteron suaveolens* Griffith (Plagiopteraceae) and its systematic implications. Bot. J. Linn. Soc. 116: 145–157.

Tang Y. 1995. Leaf and stem anatomy of *Plagiopteron* suaveolens and its systematic significance. Acta Bot. Yunn. 17: 439–444.
Tieghem P van. 1899. Sur les Parnassiacees. J. Bot. (Morot) 13: 326–332.
Tobe H and B Hammel. 1993. Floral morphology, embryology, and seed anatomy of *Ruptiliocarpon caracolito* (Lepidobotryaceae). Novon 3: 423–428.
Tobe H and PH Raven. 1993. Embryology of *Acanthothamnus, Brexia,* and *Canotia* (Celastrales): A comparison. Bot. J. Linn. Soc. 112: 17–32.
Verdcourt B. 1968. Brexiaceae. Flora of tropical East Africa. 49: 1–3. London.
Wanders GL, JJ Skvarla, and CC Pyle. 1968. Fine structure of Hippocrateaceae pollen walls. Pollen Spores 10: 189–196.
Wu D, H Wang, J-M Lu, and D-Z Li. 2005. Comparative morphology of leaf epidermis in Parnassia (Parnassiaceae) from China. Acta Phytotax. Sinica 43: 210–224.
Zhang L-B and MP Simmons. 2006. Phylogeny and delimitation of the Celastrales inferred from nuclear and plastid genes. Syst. Bot. 31: 122–137.

Superorder SANTALANAE

Order 102. SANTALALES

Trees, shrubs, woody lianas, or perennial herbs, mostly hemiparasitic and attaching to the roots or stems of other plants. Vessels with simple or sometimes scalariform perforations; lateral pitting scalariform to alternate. Fibers with bordered or simple pits. Rays heterogeneous or homogeneous. Axial parenchyma mostly apotracheal. Sieve-element plastids of Ss- or (Misodendraceae) So-type. Nodes trilacunar or unilacunar, rarely pentalacunar. Leaves alternate or opposite, simple, entire, mostly estipulate. Stomata of various types. Inflorescences axillary or terminal. Flowers mostly rather small, bisexual or unisexual, usually actinomorphic, often apetalous or in the most reduced groups even naked. Calyx often very reduced or absent. Petals free or basally connate, mostly valvate, seldom absent. Stamens usually isomerous and opposite the petals; filaments free or sometimes adnate to the petals (very rarely to the sepals) or connate into a column; anthers basifixed, tetrasporangiate, disporangiate, or monosporangiate, opening longitudinally or by apical pore or transverse slit. Tapetum secretory. Microsporogenesis simultaneous. Pollen grains 2-celled or sometimes 3-celled, with various types of apertures (from 3-colpate to pantoporate). Gynoecium of 2–3(-4) united carpels, with short or rarely long style; ovary superior to inferior, mostly 1-locular or partitioned only at the base, with the ovules pendulous from a columnar placenta that extends upward beyond the basal partitions. Ovules anatropous, epitropous, hemitropous, or orthotropous, bitegmic to unitegmic and usually tenuinucellate, but more often ategmic, sometimes not differentiated from the placenta and mostly characterized by clearly expressed reduction of nucellus. Female gametophyte of *Polygonum*- or sometimes of *Allium*-type, rarely *(Thesium)* of *Adoxa*-type. Endosperm cellular or less often helobial. Fruits of various types. Seeds mostly without seed coat, usually with well-developed endosperm; embryo terete and lug-like. Contain triglycerides with C18 acteylenic acids, triterpenic sapogenins.

Probably related to Celastrales (Hutchinson 1973; Dahlgren 1983).

Key to Families

1 Ovules well differentiated from placenta (except in *Exocarpos* of Santalaceae). Ovary superior or inferior. Endosperm cellular or helobial.
  2 Plants usually terrestrial. Calyx present, at least in female flowers, but usually greatly reduced in *Schoepfia,* and minute or almost absent in Opiliaceae.
    3 Leaves alternate. Ovary superior or inferior. Stamens opposite the petals.
      4 Nectary disc present. Petals usually present.
        5 Stamens not united into a column. Gynoecium of 2–5 carpels.
          6 Nectary disc annular, more or less adherent around the base of ovary. Trees, woody lianas, and shrubs, rarely subshrubs, sometimes root parasites (*Schoepfia* and *Ximenia*), sometimes with schizogenous resin glands and/or branched laticifers. Nodes mostly trilacunar with three traces. Vessels usually with simple perforations, but in some genera perforations scalariform with 6–10 bars (up to 20 or more in *Heisteria*); lateral pitting scalariform to alternate. Fibers with bordered or more often simple pores. Rays heterogeneous with long or short ends. Axial paren-

chyma apotracheal to paratracheal. Flowers usually small, 3–6-merous, in axillary panicles, racemes, heads, or (*Erythropalum*) very slender compound dichasia, sometimes solitary in the axils, usually bisexual or rarely unisexual (dioecious). Calyx small, more or less cupular, mostly 3–6-toothed, often accrescent in fruit. Petals 3–6, usually valvate or rarely imbricate, free or basally connate, seldom forming a long tube. Stamens as many as and opposite the petals or more often 2–3 times as many, but apparently always in one cycle; filaments free or sometimes adnate to the corolla, very rarely adnate to calyx; anthers 4-locular, opening longitudinally or sometimes by terminal valves. Pollen grains 3-colpate, 3-colporate or 3–8-porate. Gynoecium of (2)3(-5) carpels. Stigma 2–5-lobed; ovary free or more or less embedded in nectary disc, 2–5-locular at the base, but mostly 1-locular above, with 1 ovule in each locule or semilocule, pendulous from the top of the free central placenta; several genera, e.g., *Anacolosa*, have a complete placental column attached to the top of ovary; ovules mostly anatropous, bitegmic, unitegmic, or ategmic, tenuinucellate or very rarely (*Ximenia americana*) crassinucellate. Endosperm cellular (as in *Strombosia*) or helobial (as in *Olax*). Fruits usually 1-seeded drupes or nuts, often subtended or enclosed by accrescent and colorful calyx, the pericarp fleshy, thin, the endocarp woody. Seeds with green small or minute embryo near tip of copious, oily (and sometimes also starchy) endosperm; cotyledons 2–6; seed coat thin, crushed. n = 12, 13, 16, 19, 20. . . . . .1. OLACACEAE.

6 Nectary disc of separate glands or parts only united at the base but not adherent to ovary, alternating with stamens. Trees and shrubs, sometimes lianoid, usually or always hemiparasitic. Vessels with simple perforations; lateral pitting alternate. Fibers with bordered or rarely simple pits. Rays homogeneous. Axial parenchyma apotracheal, usually diffuse. Cystoliths present. Flowers small, in axillary or cauliflorous inflorescences (spikes, racemes, umbels, or panicles), bisexual or (*Gjellerupia* and *Agonandra*) dioecious, 4–5-merous. Calyx small and inconspicuous, cupular or with 4–5 small lobes or teeth. Petals free or connate below or less often for more than half of their length, valvate, in female flowers usually wanting. Stamens as many as and opposite petals; filaments free or borne on petals or on corolla tube; anthers 4-locular, opening longitudinally; rudimentary gynoecium present. Pollen grains 3-colporate, microechinate. Stigma small or capitately expanded or stigma sessile; ovary free or half-sunken in nectary disc, superior, 1-locular, with a single (very rarely 2) ovule pendulous from the summit of the columnar free-central placenta or (Agonandreae) the ovule basal and erect. Ovules pendulous, anatropous (or orthotropous ? – Corner 1976), unitegmic and tenuinucellate or not divided into nucellus and integument (*Agonandra*). Endosperm cellular. Fruits drupaceous; the pericarp thin, usually yellow to orange, the mesocarp fleshy, the endocarp woody or crustaceous. Seeds large, with rather small embryo and copious, oily and starchy endosperm; seed coat thin-walled, crushed; cotyledons (2) 3–4, n = 10. . . . . 2. OPILIACEAE.

5 Stamens united into a column around the style with as many sessile anthers in a whorl at the top. Trees. Vessels with simple perforations; lateral pitting alternate. Fibers with simple or (*Ongokea*) distinctly bordered pits. Rays heterogeneous. Axial parenchyma typically apotracheal. Flowers in short, axillary inflorescences racemes (*Harmandia*) or panicles, usually bisexual or dioecious. Calyx 4–5-dentate or -lobed, small at anthesis but very much enlarged and

persistent in fruit. Petals 4–5 in bisexual flowers, or 6–8 in female ones, free, valvate. Nectary disc extrastaminal, annular (*Harmandia*) or lobed between petals. Stamens 4 or (*Ongokea*) 5; anthers 2-locular, opening from apex to base by a membranous finally reflexed valve. Pollen grains (3)4(5)-porate. Gynoecium of 3–2 carpels, with elongated style between the staminal column or (*Harmandia*) stigmas sessile; ovary 2- or 1-locular, with 1–3 ovules pendulous from the top of the free-central placenta. Ovules not divided into nucellus and integument. Fruits drupes, more or less enclosed by the enlarged calyx; pericarp coriaceous, endocarp woody. Seeds with oily endosperm and very small embryo near its tip. *Ongokea* has polyacetylenes. . .3. APTANDRACEAE.

4 Nectary disc absent or present in female flowers.

7 Ovary superior. Trees with scattered secretory canals containing a yellow liquid. Vessel elements very long with oblique, scalariform perforation plates that have 11–43 (mostly 19–24) bars; lateral pitting scalariform. Rays high, uniseriate-heterogeneous. Axial parenchyma scanty paratracheal and diffuse. Leaves alternate, apparently simple, remotely and minutely crenulate, pinnately veined, transverse veins numerous and parallel; the lower surface provided with hairs of a unique, thick-based type; petiole with a pulvinus at the top, suggesting that the leaves may be unifoliolate rather than simple; stipules small, deciduous. Stomata anomocytic. Flowers small, in axillary, pendulous catkinlike racemes, bisexual, actinomorphic. Sepals five, free or nearly so, open in bud, persistent. Petals five, free, small, imbricate. Fertile stamens five, oppositipetalous, short, erect in bud, free from each other but filaments slightly adnate to the base of petals; anthers tetrasporangiate, the outer (abaxial) loculi opening by an outwardly recurved valve, the inner (adaxial) ones opening by an inwardly recurved valve; staminodia five, oppositisepalous, elongate, long-exerted, linear, very densely and shortly pubescent above the glabrous base, in-flexed at the apex and with an abortive anther at the tip. Pollen grains 3-colporate. Nectary disc wanting. Gynoecium of 3 (-4) united carpels; stylodia free, widely separate, short, with a very small, terminal stigma; ovary 1-locular, with a slender free-central placental column and 6(-8) anatropous, epitropous ovules pendulous from the top of the column. Fruits 3(4)-valved coriaceous 1-seeded capsule, silky-fibrous within, subtended by persistent reflexed sepals. Seeds large, pulviniform, often with six radiating ribs above, with small, straight embryo near the edge of the copious, slightly ruminate endosperm. . . . . . . 5. MEDUSANDRACEAE.

7 Ovary inferior. Trees or shrubs with stellate hairs (as in Couleae, Olacaceae). Vessels with scalariform (*Octoknema*) or simple perforations (*Okoubaka*); lateral pitting opposite to alternate. Fibers with simple pores, rarely septate. Rays heterogeneous with long ends (*Octoknema*) or homogeneous (*Okoubaka*). Axial parenchyma apotracheal, diffuse or (*Octoknema*) absent. Stomata anomocytic. Flowers usually in axillary racemes, spines, or panicles, unisexual, 5-merous. Sepals valvate, mostly very reduced, in male flowers obsolete. Petals more or less connate, valvate, persistent, the outer surface of the petals densely covered by hairs. Stamens free, oppositipetalous; filaments short and broad; anthers short, introrse. Pollen grains 3-colporate. Rudimentary gynoecium in male flowers with 5-lobed disc at the base. Staminodia in female flowers without anthers or (*Okoubaka*) with sterile anthers. Nectary disc in female flowers lobed. Gynoecium of three carpels; style very short and thick, with three broad, reflexed, irregularly lobulate lobes; ovary inferior, in lower part 3-locular, more or less 1-locular above with three ovules at the apex of a threadlike, free-central placenta that reaches and adnate to the top of the ovary. Ovules anatropous, unitegmic. Fruits drupaceous, 1-seeded.

Seeds with small embryo and copious, slightly ruminate endosperm; radicle much longer than cotyledons; cotyledons up to 6. n = 36. . . . . 4. OCTOKNEMACEAE.

3 Leaves opposite or less commonly alternate, sometimes reduced to scales, rarely (*Ionidium*) margins with spines. Stomata mostly paracytic. Hemiparasitic terrestrial small trees, shrubs, and perennial herbs usually on roots of other plants or rarely (as *Dendrotrophe*) on branches, sometimes thorny or xeromorphic. Cuticular epithelium common (cuticle waxes as rodlets); epidermal cells sclerified, with druses. Vessels with simple perforations; lateral pitting usually alternate. Fibers with bordered or simple pores. Rays heterogeneous or homogeneous. Axial parenchyma usually rather scanty, apotrachel, or paratracheal. Nodes unilacunar. Flowers small, often greenish, in various kinds of inflorescences (often small dichasium in axil of each bract), rarely solitary in leaf axil (as in *Thesium*), bisexual or unisexual (monoecious or dioecious). Calyx extremely reduced, inconspicuous. Petals (3)4–5(-8), free or forming a valvately lobed, often fleshy cup or tube. Stamens as many as and opposite the petals; filaments often adnate to base of petals; anthers 4-locular or rarely (*Exocarpos*) 2-locular, opening longitudinally of by single apical pore. Pollen grains 3-aperturate, with apertures of various types. Lobed nectary disc mostly present (absent in Thesieae), surrounding or seated on the ovary or lining the lower part of floral tube. Gynoecium of (2)3(-5) carpels, with terminal style and lobed or capitate stigma; ovary superior to inferior, completely 1-locular or with basal partitions, with straight or spirally coiled free-central placenta bearing 1–4, less often 4–5 pendulous ovules. Ovules anatropous or less often hemitropous, or orthotropous, unitegmic and tenuinucellate, ategmic (*Mida* and *Santalum*) or (*Exocarpos*) more or less undifferentiated into nucellus and integument. Endosperm cellular or (*Mida* and *Santalum*) helobial. Fruits drupaceous, nuts, or baccate. Seeds without a differentiated testa, with straight embryo surrounded by copious, fleshy, oily, or (*Thesium*) starchy endosperm; seed coat absent. n = 5–7. . . . . .6. SANTALACEAE.

2 Hemiparasitic shrublets almost always on *Nothofagus* spp., with radially expanded haustorial region and stout twigs; shoot apex aborts annually and one or more lateral branches continue vegetative growth in a sympodial fashion in the next growing season. Vessel elements very short, with simple perforations; lateral pitting alternate to scalariform. Fibers very short and broad, with bordered pits. Rays uniseriate, homogeneous. Axial parenchyma vasicentric. Leaves alternate, small, sometimes scalelike. Stomata mostly paracytic. Flowers very small, in axillary catkinlike compound racemes or spikes, subtended by a bract, dioecious, but very occasionally bisexual (Skottsberg 1913); male flowers without perianth of two or three stamens around small, lobed nectary disc; anthers at least initially 2-locular, monothecous, opening by apical tangential slit. Pollen grains 4–12-colpate. Female flowers apetalous, but with three sepals basally adnate to the ovary and free at the top, they fused with the ovary along their midregion and extended sideways in a winglike fashion, having three deep grooves that rub the length of the ovary; from each groove protrudes a short bristlelike staminodium that is lengthening into long plumose bristle during maturation of the fruit. Gynoecium of three carpels; style very short, stout, with 3-lobed stigma; ovary essentially superior (Kuijt 1969), 1-locular, with three ovules pendulous from the top of short, free, central placental column, each hanging down into a basal pocket of ovarian cavity; ovules anatropous, reduced, not differentiated into nucellus and integument. Female gametophyte *Polygonum*-type. Endosperm cellular, with elongate chalazal haustorium. Fruits small, 1-seeded nuts crowned the strongly accrescent staminodia covered by numerous long, unicellular trichomes and serving as an adaptation for wind dispersal and anchoring while the seeds germinate. Seed without testa or testa rudimentary (*M. punctulatum*), with straight embryo surrounded by oily green endosperm; seed coat absent. . . . . .7. MISODENDRACEAE.

1 Ovules embedded in placenta, placentalike body, or in the base of the ovary and scarcely differentiated from them. Ovary inferior, usually 1-locular. Endosperm cellular. Seeds without testa.

8 Flowers usually bisexual, only rarely unisexual. Hemiparasitic brittle shrublets on tree branches (or on epiphytes), less often terrestrial shrubs, lianas, or even (*Nuytsia*) trees on host roots, with a single or several haustoria at ends of epicortical roots, occasionally almost *Cuscuta-like* in habit. Stem often dichasially branched, but without nodal construction. Vessels with simple perforations; lateral pitting alternate. Fibers short to very short, with simple or clearly bordered pits. Rays slightly to strongly heterogeneous. Axial parenchyma apotracheal. Nodes unilacunar. Leaves mostly opposite, less frequently alternate or verticillate, always simple, sometimes reduced to scales. Stomata mostly paracytic. Flowers rather large and brightly colored to small, in various types of inflorescences the basic unit of which is a triad or simple dichasium, rarely (*Ixocactus*) solitary, bisexual or rarely unisexual (plant dioecious), actinomorphic or (in African members) with a strong tendency to zygomorphic, frequently red or yellow, entomophilous or ornithophilous. Calyx represented by toothed or lobed rim around the summit of ovary. Petals (3-)5–6(-9), valvate, free or often connate below into a tube, sometimes nectariferous basally within. Stamens as many as and opposite to petals; filaments often adnate to petals; anthers 4-locular or sometimes 2-locular or 1-locular opening longitudinally. Pollen grains 3- or seldom 4-(5)-aperturate, rarely (*Atkinsonia*) inaperturate, trilobate or triangular. Nectary disc present or wanting. Gynoecium of 3–4-carpels, with very short to much elongated style and small stigma; ovary 1-locular, with commonly 4–12 ovules embedded in the erect, free-central placenta or the basal tissue of the ovary, rarely (*Lysiana*) ovary 4-locular with an axile placentas; each ovule consists essentially of monosporic, 8-nucleate female gametophyte of *Polygonum*-type. Suspensor highly elongate. Fruits usually laticiferous berries or drupes, mostly 1-seeded, less often 2–3-seeded; rarely fruits dry and indehiscent. Seeds more or less covered with viscous material, embryo green, rather large, at least sometimes without obvious radicle; endosperm copious, starchy, rarely wanting at maturity. Contain triterpenes (loranthin, betulin, etc.), rather large amounts of tannins, alkaloids, saponins, and fatty acids, n = 8–12 (Wilson et Calvin 2006).. . . . . .8. LORANTHACEAE.

8 Flowers unisexual, generally small and inconspicuous. Calyx obsolete. Ovules two. Female gametophyte bisporic.

9 Leaves opposite. Hemiparasitic, brittle shrublets on trees without epicortical roots; haustoria (sometimes almost like mycelium) penetrating and ramifying in host tissues; aerial branding often pseudodichotomous and often with nodal constructions. Vessels with simple perforations; lateral pitting alternate. Fibers very short, with simple or evidently bordered pits. Rays more or less heterogeneous. Axial parenchyma apotracheal. Leaves opposite, often reduced to scales. Stomata mostly paracytic. Flowers very small, not brightly colored (usually yellow or green), more or less sessile, mostly in spikelike often branching inflorescences with (1)3-flowered dichasium in axil of each bract, unisexual (monoecious or dioecious), anemophilous or entomophilous. Petals small, valvate, usually four in male flowers and three in female flowers, often reduced to merely teeth or bumps on ovary rim. Stamens opposite and often adnate to petals; tetrasporangiate and opening longitudinally or reduced to one or two microsporangia and opening by apical pores or slits or transversely partitioned and opening by a number of transverse slits, or the anthers confluent into synandrium (*Viscum album, Korthalsella opuntia*). Pollen grains usually spheroidal, triaperturate or in some African *Viscum* species (3-)4–6-aperturate, in one African *Viscum* species (5-)7–10-rugorate. Gynoecium of 3–4 carpels; style very short with a small stigma; ovary inferior, usually 1-locular, with massive placenta (mamelon) bearing two ategmic ovules usually completely reduced to only a female gametophyte. Female gametophyte bisporic, of *Allium*-type. Suspensor very short or absent. Cleavage of zygote transverse. Fruits 1(2)-seeded berries, sometimes explosive. Seeds with viscid tissue at one end; embryo rather large, green, embedded in starchy green endosperm. Chemistry is similarity of that of Loranthaceae, except for toxic polypeptides (visco-toxins and lectins), which present only in Viscaceae, n = 10–15(-17). . . 9. VISCACEAE.

9 Leaves alternate or sometimes opposite (*Lepidoceras* and juvenile plants of *Eubrachion*). Small hemiparasitic shrubs on woody plants, sometimes (spp. of *Antidaphne*) with epicortical roots. Vessels with simple perforations. Leaves foliaceous or (in mature plants of *Eubrachion*) peltate and scalelike. Flowers small, mostly sessile, ebracteolate, in catkinlike spikes or racemes, often with deciduous scale leaves; sometimes female flowers solitary in leaf axils, unisexual (dioecious or monoecious), 2–4-merous, without calyx. Aborted organs of opposite sex wanting. Petals small, two or four (or rarely none) in male flowers, two or three in female flowers. Stamens free, opposite petals; filaments often very short; anthers tetrasporangiate, 2-locular, opening longitudinally. Pollen grains 3-colporate, echinate or rarely (*Eubrachion*) shallowly reticulate and lacking any spines. Gynoecium of three or five carpels with a short, stout style surrounded by a capitate or crestlike stigma, sometimes shallowly lobed; ovary inferior or semi-inferior, 1-locular and contains a small, central ovarian papilla (mamelon) bearing two extremely reduced ovules without clearly defined nucellus and integument and almost completely reduced to only a female gametophyte. Female gametophyte bisporic of *Allium*-type. Fruits 1-seeded berries. Seeds extremely slippery on account of large amount of associated viscin concentrated at its radicular pole; embryo rather large, with two cotyledons or (*Lepidoceras peruvianum*) undifferentiated and without cotyledons; endosperm green or white (absent in *Lepidoceras*), n = 10, 13. ....10. EREMOLEPIDACEAE.

1. OLACACEAE

A. L. de Jussieu ex R. Brown 1818 (including Cathedraceae van Tieghem 1900, Coulaceae van Tieghem 1900, Erythropalaceae Planchon ex Miquel 1856, Harmandiaceae van Tieghem 1900, Heisteriaceae van Tieghem 1900, Schoepfiaceae van Tieghem 1900, Strombosiaceae van Tieghem 1900, Tetrastylidiaceae van Tieghem 1900, Ximeniaceae Horaninow 1834). 25/200–250. Widespread in tropical and subtropical regions.

1.1 ANACOLOSOIDEAE
Ovules with one or two integuments, rarely (*Ximenia americana*) without integuments, anatropous, with dorsal raphe; micropyle facing upward. Calyx often enlarged with ripening of fruit. Ovary superior. – COULEAE: *Coula, Maburea, Minquartia, Ochano stachys*; HEISTERIEAE: *Heisteria, Chaunochiton;* XIMENIEAE: *Ximenia;* ANACOLOSEAE: *Scorodocarpus, Engomegoma, Phanerodiscus, Brachynema, Diogoa, Tetrastylidium, Cathedra, Anacolosa, Strombosia, Strombosiopsis.*

1.2 ERYTHROPALOIDEAE
Ovules with one integument, anatropous, with dorsal raphe. Calyx not enlarged. Ovary embedded in nectary disc. – *Erythropalum.*

1.3 OLACOIDEAE
Ovules without integument, anatropous, with dorsal raphe; micropyle facing upwards. Calyx more or less enlarged. – *Curupira, Ptychopetalum, Douradoa, Olax, Dulacia, Malania.*

1.4 SCHOEPFIOIDEAE
Ovules with one integument, with dorsal raphe. Calyx inconspicuous, not enlarged. Ovary embedded in nectary disc. Corolla sympetalous. – *Schoepfia.*

2. OPILIACEAE

Valeton 1886 (including Cansjeraceae J.G.Agardh 1858). 10/36–72. Widespread in tropical and subtropical regions, especially in Asia; *Agonandra* (10) the only American genus.

OPILIEAE: *Lepionurus, Cansjera, Melientha, Champereia, Opilia, Urobotrya, Rhopalopilia, Pentarhopalopilia, Gjellerupia;* AGONANDREAE: *Agonandra.*

Very close to the Olacaceae. According to Baas et al. (1982), Opiliaceae might be considered a direct derivative from Olacaceae through *Ximenia.*

3. APTANDRACEAE

Miers 1853. 3/6. Southeast Asia (*Harmandia*), tropical Africa (*Ongokea* and *Aptandra zenkeri*), and tropical South America (three other species of *Aptandra*).
Aptandra, Ongokea, Harmandia.

Very close to Olacaceae.

4. OCTOKNEMACEAE

Solereder 1908. 2/6. Tropical Africa.

Octoknema, Okoubaka.

Very close to Olacaceae.

5. MEDUSANDRACEAE

Brenan 1952. 1/2. Equatorial West Africa.

Medusandra.

Close to Olacaceae and especially to Octoknemaceae.

6. SANTALACEAE

R. Brown 1810 (including Anthobolaceae Dumortier 1829, Arjonaceae van Tieghem 1898, Canopodaceae K. Presl 1851, Osyridaceae Rafinesque 1820, Thesiaceae Vest 1818). 37/450. Widely distributed, but mostly tropical and subtropical regions; best developed in relatively dry areas.

SANTALEAE: *Choretrum, Spirogardnera, Leptomeria, Phacellaria, Dufrenoya, Scleropyrum, Jodina, Cervantesia, Buckleya, Omphacomeria, Pyrularia, Osyris, Geocaulon, Comandra, Cladomyza, Nanodea, Myoschilos, Acanthosyris, Santalum, Rhoiacarpos, Amphorogyne, Daenikera, Colpoon, Dendromyza, Elaphanthera, Mida, Dendrotrophe, Kunkeliella, Nestronia*; THESIEAE: *Osyridocarpos, Thesidium, Thesium, Austroamericium, Arjona, Quinchamalium;* ANTHOBOLEAE: *Anthobolus, Exocarpos.*

Close to the Olacaceae and originated from them (Hallier 1912; Fagerlind 1948; Agarwal 1963; Kuijt 1968, 1969).

7. MISODENDRACEAE

J. Agardh 1858. 1/11. Cool-temperate South America from the Strait of Magellan north to about 33° southern latitude.

Misodendrum.

Close to the Santalaceae, especially to South American genera *Arjona* and *Quinchomalium* (Skottsberg 1935), but differ in the absence of the perianth of the male flowers, and the absence of petals in female flowers, and male flowers with monothecous anthers and a rudimentary testa in the seed, and probably originated from them (Hallier 1912). Kuijt (1968, 1969) derives Misodendraceae from Olacaceae. According to Carlquist (1985), wood anatomy of Misodendraceae is similar to that of *Psittacanthus* and other Loranthaceae, disregarding the unique features found in *Misodendrum.*

8. LORANTHACEAE

A. L. de Jussieu 1808 (including Dendrophthoaceae van Tieghem 1898, Elytranthaceae van Tieghem 1898, Gaiadendraceae van Tieghem 1898, Nuytsiaceae van Tieghem 1898, Psittacanthaceae Nakai 1952). 75/c.1000. Mostly in tropical and subtropical regions, especially in Southern Hemisphere.

NUYTSIEAE: *Nuytsia*; LORANTHEAE: *Gaiadendron, Panamanthus, Atkinsonia, Desmaria, Amylotheca, Decaisnina, Lampas, Lepeostegeres, Thaumasianthes, Cyne, Loxanthera, Peraxilla, Alepis, Lysiana, Trilepidea, Macrosolen, Elytranthe, Lepidaria, Loranthus, Helicanthera, Moquiniella, Tristerix, Dactyliophora, Amyema, Psittacanthus, Amyema, Ileostylus, Dendropthoe, Baratranthus, Scurulla, Emelianthe, Pedistylis, Plicosepalus, Oncella, Kingella, Trithecanthera, Trithecanthera, Taxillus, Vanwykia, Pedistylis, Oliverella, Oncocalyx, Phragmanthera, Septulina, Tapinanthus, Agelanthus, Globimetula, Struthanthus, Benthamina, Erianthemum, Tolypanthus, Papuanthes, Diplatia, Distrianthes, Tetradyas, Bakerella, Helicanthes, Sogerianthe, Englerina, Tupeia, Notanthera, Tripodanthus, Muellerina, Psittacanthus, Oryctanthus, Cladocolea, Oryctina, Phthirusa, Dendropemon, Aetanthus, Ligaria, Actinanthella, Berhautia, Cecarria, Ixocactus, Socratina, Spragueanella.*

Probably derived directly from the Olacaceae (Hallier 1912), possibly from *Chaunochiton*-like ancestor (Kuijt 1968, 1969). According to Kuijt, Loranthaceae, contrary to tradition, cannot have evolved from Santalaceae ("The flower of Loranthaceae is dichlamydous and can scarcely have a monochlamydous ancestor in Santalaceae." [1969: 80]).

9. VISCACEAE

Batsch 1802 (including Arceuthobiaceae van Tieghem 1897, Bifariaceae Nakai 1952, Ginalloaceae van

Tieghem 1899, Phoradendraceae H. Karsten 1860). 7/450. Very widely distributed, but mostly in tropical and subtropical regions; all genera except *Arceuthobium* occur either in the Old or the New World; *Arceuthobium* is restricted to the Northern Hemisphere.

PHORADENDREAE: *Phoradendron, Dendrophthora, Korthalsella, Ginalloa;* ARCEUTHOBIEAE: *Arceuthobium;* VISCEAE: *Notothixos, Viscum.*

Closely allied to the Loranthaceae. According to Kuijt (1968, 1969), Viscaceae may be traceable to Santalaceae plants similar to present-day epiphytic parasites such as *Phacellaria.*

10. EREMOLEPIDACEAE

Nakai 1952 (including Lepidocerataceae Nakai 1952). 3/12. From Isia Chiloe (Chile) north along the Andes and the lower, eastern areas of South America to Venezuela and Colombia; only *Antidaphne viscoldea* reaches through Central America as far as Chiapas (Mexico); *Eubrachion* and *Antidaphne wrightii* are represented on both Hispaniola and Puerto Rico; the former also occurs on Jamaica, the latter in eastern Cuba.

Antidaphne (including *Eremolepis*), *Eubrachion, Lepidoceras.*

Kuijt (1988) recognizes the distinct possibility of an early separation of Eremolepidaceae from a preloranthaceous stock. The affinities to Loranthaceae have also been stated by various authors, including Bhandari and Vohra (1983), who mention the striking similarities between the epicortical roots and secondary haustoria of the more primitive eremolepidaceous genus *Antidaphne* and the loranthacous genus *Notanthera.*

Bibliography

Agarwal S. 1963. Morphological and embryological studies in the family Olacaceae: I. *Olax* L. II. *Strombosia* Blume. Phytomorphology 13: 185–196, 348–356.

Baas P, E van Oosterhoud, and JL Scholtes. 1982. Leaf anatomy and classification of the Olacaceae, *Oc-toknema,* and *Erythropalum.* Allertonia 3: 155–210.

Barlow BA. 1964. Classification of the Loranthaceae and Viscaceae. Proc. Linn. Soc. N. S. W., 2nd ser., 89: 268–272.

Barlow BA. 1983. Biogeography of Loranthaceae and Viscaceae. In: M Calder and P Bernhardt, eds. The biology of mistletoes, pp. 19–46. Academic Press, Sydney.

Barlow BA. 1997. Loranthaceae. In: C Kalkman et al., eds. Flora Malesiana ser. 1, 13: 209–401. Leiden.

Barlow BA and D Wiens. 1971. The cytogeography of the loranthaceous mistletoes. Taxon 20: 291–312.

Bhandari NN and K Indira. 1969. Studies in the Viscaceae: IV. Embryology of *Eubrachion* (Hook. et Am.) Engl. Bot. Not. 122: 183–203.

Bhandari NN and SCA Vohra. 1983. Embryology and affinities of Viscaceae. In: M Calder and P Bernhardt, eds. The biology of mistletoes, pp. 69–86. Academic Press, New York.

Bhatnagar SP. 1960. Morphological and embryological studies in the family Santalaceae: IV. *Mida salicifolia* A. Cunn. Phytomorphology 10: 198–207.

Bhatnagar SP and S Agarwal. 1961. Morphological and embryological studies in the family Santalaceae: VI. Thesium. Phytomorphology 11: 273–282.

Bhatnagar SP and PC Joshi. 1965. Morphological and embryological studies in the family Santalaceae: VII. *Exocarpus bidwellii* Hook. f. Proc. Natl. Inst. Sci. India 31B: 34–44.

Bhandari NN and SCA Vohra. 1983. Embryology and affinities of Viscaceae. In: M Calder and P Bernhardt, eds. The biology of mistletoes, pp. 69–86. Academic Press, Sydney.

Brenan JPM. 1952. Plants of the Cambridge Expedition, 1947–1948: II. A new order of flowering plants from the British Cameroons. Kew Bull. 1952: 227–236.

Calvin CL and CA Wilson. 1998. Comparative morphology of haustoria within African Loranthaceae. In: R Polhill and D Wiens, eds. Mistletoes of Africa, pp. 17–36. Royal Botanic Gardens, Kew.

Calvin CL and CA Wilson. 2006. Comparative morphology of epicortical roots in Old and New World Loranthaceae with reference to root types, origin, growth, patterns of longitudinal extension and potential for clonal growth. Flora 201: 51–64.

Carlquist S. 1985. Wood and stem anatomy of Misodendraceae: Systematic and ecological conclusions. Brittonia 37: 58–75.

Cocucci AE. 1983. New evidence from embryology in angiosperm classification. Nord. J. Bot. 3: 67–73.

Cohen LI. 1968. Development of the staminate flower in the dwarf mistletoe, *Arceuthobium.* Am. J. Bot. 55: 187–193.

Cohen LI. 1970. The development of the pistillate flower in the dwarf mistletoe, *Arceuthobium.* Am. J. Bot. 57: 477–485.

Davis CC, WR Anderson, and KJ Wurdack. 2005. Gene transfer from a parasitic flowering plant to a fern. Proc. Roy. Soc. London B, 272: 2237–2242.

Der J and D Nickrent. 2005. Molecular systematics of Santalaceae: Phylogeny and classification of a paraphyletic family of hemiparasitic plants. In: Botany 2005. Learning from plants, pp. 110. Abstracts, pp. 110: Botanical Society of America.

Dixit SN. 1958, 1961. Morphological and embryological studies of the family Loranthaceae: IV. *Amyema* Van Tiegh. V. *Lepeostegeres gemmiflorus* (Bl.) Bl. VIII. *Tolypantus* Bl. Phytomorphology 8: 346–364, 365–376, 1958; 11: 335–345, 1961.

Fagerlind F. 1947. Gynoceummorphologische und embryologische Studien in der Familie Olacaceae. Bot. Not. 1947: 207–230.

Fagerlind F. 1948. Beiträge zur Kenntnis der Gynoceummorphologie und Phylogenie der Santalales-Familien. Svensk Bot. Tidskr. 42: 195–229.

Fagerlind F. 1959. Development and structure of the flower and gametophytes in the genus *Exocarpus.* Svensk. Bot. Tidskr. 53: 257–282.

Feuer S. 1981. Pollen morphology and relationships of Misodendraceae (Santalales). Nord. J. Bot. 1: 731–734.

Feuer S and J Kuijt. 1978. Fine structure of mistletoe pollen: I. Eremolepidaceae, *Lepidoceras*, and *Tufieia*. Canad. J. Bot. 56: 2853–2864.

Fineran B.A. 1991. Root hemi-parasitism in the Santalales. Bot. Jahrb. Syst. 113: 277–308.

Haron NW and ST Ping. 1997. Distribution and taxonomic significance of flavonoids in the Olacaceae and Icacinaceae. Biochem. Syst. Ecol. 25: 265–263.

Hiepko P. 1979. A revision of Opiliaceae: I. Genera of the eastern Old World, excluding *Opilia*. Willdenowia 9: 13–56.

Hiepko P. 1982. A revision of Opiliaceae: II. *Opilia* Roxb. Willdenowia 12: 161–182.

Hiepko P. 1984. Opiliaceae. In: CCGJ van Steenis, ed. Flora Malesiana, ser.1, 10: 31–52. Nijhoff, The Hague.

Hiepko P. 1985. A revision of Opiliaceae: III. *Urobotrya* Stapf. Bot. Jahrb. Syst. 107: 137–152.

Hiepko P. 1987. A revision of Opiliaceae: IV. *Rhopalopilia* Pierre and *Pentarhopalopilia* (Engler) Hiepko gen. nov. Bot. Jahrb. Syst. 108: 271–291.

Hiepko P. 2000. Opiliaceae. Flora Neotropica Monograph 82. New York.

Johri BM, JS Agrawal, and S Garg. 1957. Morphological and embryological studies in the family Loranthaceae: I. *Helicanthes elastica* (Desr.) Dans. Phytomorphology 7: 336–354.

Johri BM and SP Bhatnagar. 1960. Embryology and taxonomy of the Santalales. Proc. Natl. Inst. Sci. India, Suppl., 26B: 199–220.

Johri BM and B Raj. 1969. Morphological and embryological studies in the family Loranthaceae: XII. *Moquiniella rubra* (Spreng. f.) Balle. Oesterr. Bot. Z. 16: 475–485.

Johri BM and S Agarwal. 1965. Morphological and embryological studies in the family Santalaceae: VIII. *Quinchamalium chilense* Lam. Phytomorphology 15: 360–372.

Johri BM and SP Bhatnagar. 1960. Embryology and taxonomy of the Santalales. Proc. Nad. Inst. Sci. India, Suppl., 26B: 199–220.

Jones BL and CC Gordon. 1965. Embryology and development of the endosperm haustorium of *Arceuthobium douglasii*. Am. J. Bot. 52: 127–132.

Joshi PC. 1960. Morphological and embryological studies in the family Santalaceae: V. *Osyris wightiana* Wall. Phytomorphology 10: 239–248.

Koek-Noorman J and P van Rijckevorsel. 1983. Wood and leaf anatomy of Opiliaceae. Wildenowia 13: 147–174.

Kuijt J. 1968. Mutual affinities of Santalean families. Brittonia 20: 136–147.

Kuijt J. 1969. The biology of parasitic flowering plants. University of California Press, Berkeley.

Kuijt J. 1977. Haustoria of phanerogamic parasites. Ann. Rev. Phytopathol. 17: 91–118.

Kuijt J. 1979. Host selection by parasitic angiosperm. Symb. Bot. Upsali 22: 194–199.

Kuijt J. 1981. Inflorescence morphology of Loranthaceae: An evolutionary synthesis. Blumea 27: 1–73.

Kuijt J. 1982a. The Viscaceae of the southeastern United States. J. Arnold Arbor. 63: 401–410.

Kuijt J. 1982b. Seedling morphology and its systematics significance in Loranthaceae of the World with supplementary comments on Eremolepidaceae. Bot. Jahrb. Syst. 103: 305–342.

Kuijt J. 1985. Morphology, biology, and systematic relationships of *Deimaria* (Loranthaceae). Plant Syst. Evol. 151: 121–130.

Kuijt J. 1988. Monograph of the Eremolepidaceae. Syst. Bot. Monograph 18: 1–60.

Kuijt J. 1991. *Panamanthus:* A new monotypic genus of Neotropical Loranthaceae. Ann. Missouri Bot. Gard. 78: 172–176.

Kuijt J and D Lye. 2005. A preliminary survey of foliar sclerenchyma in neotropical Loranthaceae. Blumea 50: 323–355.

Ling Y-R. 1982. A revision and addition of Olacaceae from China and a primitive discussion for the taxonomy and floristics of the family. Bull. Bot. Res. 2: 7–36.

Lobreau-Callen D. 1980. Caractéres comparés du pollen des Icacinaceae et des Olacaceae. Adansonia 20: 29–89.

Lobreau-Callen D. 1982. Structures et affinites polliniques des Cardiopterydaceae, Dipentodontaceae, Erythropalaceae, et Octoknemataceae. Bot. Jahrb. Syst. 103: 371–412.

Maheshwari P, BM Johri, and SN Dixit. 1957. The floral morphology and embryology of the Loranthoideae (Loranthaceae). J. Madras Univ. 27B: 121–136.

Malécot V. 2002. Histoire, classification et phylogénie des Olacaceae Brown (Santalales). Thèse de Doctorat de l'Université Paris 6.

Malécot V, DL Nickrent, P Baas, L van den Oever, and D Lobreau-Callen. 2004. A morphological cladistic analysis of Olacaceae. Syst. Bot. 29: 569–586.

Metcalfe CR. 1952. *Medusandra richardsiana* Brenan; Anatomy of the leaf, stem, and wood. Kew Bull. 1952: 237–244.

Narayana R. 1958. Morphological and embryological studies in the family Loranthaceae: II. *Lysiana exocarpi* (Behr.) van Tieghem. III. *Nuytsia fioribunda* (Labill.) R. Br. Phytomorphology 8: 146–169, 306–323.

Nickrent DL. 1998-onwards. The parasitic plant connection. http://www.science.siu.edu/parasitic-plants/

Nickrent DL and RJ Duff. 1996. Molecular studies of parasitic plants using ribosomal RNA. In: MT Moreno, JI Cubero, D Berner, D Joel, LJ Musselman., and C Parker, eds. Adv. Parasitic Plant Res., pp. 28–52. Cordoba, Spain.

Nickrent DL and CR Franchina. 1990. Phylogenetic relationships of the Santalales and relatives. J. Molec. Evol. 31: 294–301.

Nickrent DL and V Malécot. 2001. A molecular phylogeny of the Santalales. In: A Fer, P Thalouarn, DM Joel, LJ Musselman, C Parker and IAC Verkleij, eds. 7th Intern. parasitic weed symp., pp. 69–74. Nantes.

Norverto CA. 1993. Wood anatomy and relationships of Santalaceae. I. *Acanthosyris, Jodina,* and *Myoschilos*. Aliso 13: 499–511.

Nyananyo BL. 1987. Systematic survey of the leaf epidermis in the Medusandraceae (Rosidae). Feddes Repert. 98(11–12): 595–598.

Paliwal RL. 1956. Morphological and embryological studies in some Santalaceae. Agra Univ. J. Res. Sci. 5: 193–284.

Piehl MA. 1965. The natural history and taxonomy of *Comandra* (Santalaceae). Mem. Torrey Bot. Club 22: 1–97.

Ping ST. 1997. Root hemi-parasitism in Malayan Olacaceae. Garde. Bull. Singapore 49: 7–13.

Prakash S. 1960, 1963. Morphological and embryological studies in the family Loranthaceae: VI. *Peraxilla tetrapetala*

(Linn. f.) Van Tiegh. X. *Barathranthus axanthus* (Korth.) Miq. Phytomorphology 10: 224–234, 1960; 13: 97–103, 1963.
Ram M. 1957, 1959. Morphological and embryological studies in the family Santalaceae: I. *Comandra umbellata* (L.) Nutt. II. *Exocarpus* with a discussion on its systematic position. III. *Leptomeria* R. Br. Phytomorphology 7: 24–35, 1957; 9: 4–19, 20–33, 1959.
Rao LN. 1942. Studies in the Santalaceae. Ann. Bot. 6: 151–175.
Record SJ. 1938. The American woods of the orders Celastrales, Olacales, and Santalales. Trop. Woods 53: 11–38.
Reed C. 1955. The comparative morphology of the Ola-caceae, Opiliaceae, and Octoknemaceae. Mem. Soc. Brot. 10: 29–79.
Ross CM and MJ Sumner. 2004. Development of the unfertilized embryo sac and pollen tubes in the dwarf mistletoe *Arceuthobium americanum* (Viscaceae). Canad. J. Bot. 82: 1566–1575.
Ross CM and MJ Sumner. 2005. Early endosperm and embryo development of the dwarf mistletoe *Arceuthobium americanum* (Viscaceae). Int. J. Plant Sci. 166: 901–907.
Schaeppi H and F Steindl. 1942. Blütenmorphologische und embryologische Untersuchungen an Loranthoideen. Vierteljahrsschr. Naturf. Ges. Zurich 87: 301–372.
Schaeppi H and F Steindl. 1945. Blütenmorphologische und embryologische Untersuchungen an einigen Viscoideen. Vierteljahrsschr. Naturf. Ges. Zurich 90 (I): 1–46.
Shamrov II, GM Anisimova, TB Batygina, G Lakshmi Sita. 2001. The types and morphological evolution of the ovule in the order Santalales. Bot. Zhurn. 85 (7): 1–14 (in Russian with English summary).
Singh V and G Ratnakar. 1974. Contribution to the floral anatomy of the Loranthaceae: I. Subfamily Loranthoideae. J. Indian Bot. Soc. 53: 162–169.
Skottsberg C. 1913. Morphologische und embryologische Studien über die Myzodendraceae. Kgl. Svensk Vetes.-kapsakad. Handl. Bd. 51: 34–46.
Sleumer HO. 1984. Olacaceae. In: CGGJ van Steenis, ed. Flora Malesiana ser. 1, 10(1): 1–29. Noordhoff, Groningen.
Smith FH and EC Smith. 1943. Floral anatomy of the Santalaceae and some related forms. Oregon State Monogr. Stud. Bot. 5.
Stauffer HU. 1959. Santalales-Studien: IV. Revisio *Anthobolearum,* morphologische Studie mit Einschluss der Geographic, Phylogenie, und Taxonomie. Mitt.Bot. Mus. Univ. Zurich 213: 1–260.
Stauffer HU. 1961. Santalales-Studien: VIII. Zur Morphologic und Taxonomie des Olacaceae, Tribus Coulea. Vierteljahrsschr. Naturf. Ges. Zurich 106: 412–418.
Swamy BGL. 1949. The comparative morphology of the Santalaceae: Node, secondary xylem, and pollen. Am. J. Bot. 36: 661–673.
Tamamshian SG. 1958. Santalales in the system of Spermatophyta. Problems Bot. (Moscow-Leningrad) 3: 67–97 (in Russian).
Van den Oever L. 1984. Comparative wood anatomy of the Olacaceae. In: S Sudo, ed. Proceed. Pacific Regional Wood Anatomy Conference, pp. 177–178.
Van den Oever L. 1990. Phylogenetic wood anatomy of the Olacaceae and related families. IAWA Bull., N.S. 11: 133.
Venkata Rao C. 1963. On the morphology of the calycu-lus. J. Indian Bot. Soc. 42: 618–628.
Vidal-Russell R and D Nickrent. 2005. A molecular phylogeny of the mistletoe family Loranthaceae. In Annual Meeting of the Botanical Society of America 2005, p. 101. Austin, Texas.
Wiens D and BA Barlow. 1971. The cytogeography and relationships of the viscaceous and eremolepidaceous mistletoes. Taxon 20: 313–332.
Wilson CA and CL Calvin. 2003. Development, taxonomic significance and ecological role of the cuticular epithelium in the Santalales. IAWA J. 24: 129–138.
Wilson CA and CL Calvin. 2006a. Character divergences and convergences in canopy-dwelling Loranthaceae. Bot. J. Linn. Soc. 150: 101–113.
Wilson CA and CL Calvin. 2006b. An origin of aerial branch parasitism in the mistletoe family, Loranthaceae. Am. J. Bot. 93: 787–796.
Wulff AF. 1991. Los cromosomas de *Misodendron punctulatum* (Misodendraceae). Darwiniana. 30(1–4): 297–298.
Yakovleva OV. 2000. Olacaceae, Opiliaceae, Octoknemaceae, Medusandraceae, Santalaceae, Misodendraceae, Loranthaceae, Viscaceae, Eremolepidaceae. In: A Takhtajan, ed. Comparative seed anatomy, vol. 6, pp. 143–182. Nauka, St. Petersburg (in Russian).
Yoshida O and H Kawaguchi. 1971. Embryology of *Korthalsella japonica* (Thunb.) Engler. J. Coil. Arts Chiba Univ. 4B: 37–47.
Zheng XF and YL Ding. 2001. Morphological features of *Phacellaria rigidula* Benth. J. Nanjing Forest Univ. 25(5): 33–36.

Superorder BALANOPHORANAE

Order 103. CYNOMORIALES

Terrestrial reddish-brown parasitic herbs with branched, thick, fleshy, underground rhizome. Stem simple, fleshy, bearing numerous, alternate, deltoid, scalelike leaves. Haustorial contacts are made by means of numerous small lateral roots on the rhizome and even with the base of the flowering stem, which fuses with the host root. Food reserve is starch. Vascular system reduced. Stomata vestigial, anomocytic. Flowers minute, numerous, on short peduncles, interspersed with the scalelike leaves in terminal clavate or cylindrical inflorescence with fleshy axis, polygamous, apetalous, entomophilous. Calyx of (1)4–5 (rarely 6–8) strap-shaped, linear-spatulate sepals, free or connate at the base. Stamen single, at the base adnate to the perianth; anthers dorsifixed, versatile, tetrasporangiate, introrse, opening longitudinally. Tapetum secretory. Microsporogenesis simultaneous. Pollen grains 2-celled, 3-colporate, reticulate. Male flowers with

1–2 bright yellow nectaries-pistillodia. Gynoecium monocarpellate (Terekhin and Nikiticheva 1976), with a thick and long stylodium, which forms a long canal lined with conducting tissue and has a conduplicate upper part; ovary inferior, unilocular. with 1 pendulous, submarginal ovule. Ovule pendulous, submarginal, orthohemitropous, unitegmic with thick integument, almost crassinucellate. Female gametophyte of ***Polygonum***-type. Endosperm cellular. Fruits 1-seeded berries (Doweld 2000). Seeds with copious endosperm and minute, lentiform, undifferentiated embryo; seed coat of four or five layered; cells of exotesta covered with thick layer of cuticule (Doweld 2000), n = 12, size strongly bimodal.

Usually included in the Balanophorales, but differ from them markedly in monocarpellate gynoecium with conduplicate stylodium and defined ovarian cavity, not reduced ovule, presence of stomata, pollen grains with reticulate ornamentation, and presence of bisexual flowers.

1. CYNOMORIACEAE

Endlicher ex Lindley 1833. 1/2. From the Canary Islands, the Mediterranean coastal regions, from Morocco to Egypt and western Asia to Mongolia.

Cynomorium.

Bibliography

Doweld AB. 2000. Cynomoriaceae. In: AL Takhtajan, ed. Comparative seed anatomy, vol. 6, pp. 414–416. Nauka, St. Petersburg (in Russian).

Juel O. 1902. Zur Entwicklungsgeschichte des Samens von *Cynomorium*. Beih. Bot. Centralbl. 13: 194–202.

Juel O. 1910. *Cynomorium* und *Hippuris*. Svensk Bot. Tidskr. 4: 151–159.

Lye KA. 1991. Strange flowering plants from the Mediterranean area: *Cynomorium coccineum*. Svensk Bot. Tidskr. 85: 1–6 (in Swedish with English summary).

Terekhin ES. 1987. Cynomoriaceae In: MS Yakovlev, ed. Comparative embryology of flowering plants: Davidiaceae-Asteraceae, pp. 87–89. Nauka, Leningrad (in Russian).

Terekhin ES, ZI Nikiticheva, and MS Yakovlev. 1975a. Development of seed, embryo, and endosperm in *Cynomorium songaricum* Rupr. (Cynomoriaceae). Bot. Zhurn. 60: 1603–1613 (in Russian with English summary).

Terekhin ES, MS Yakovlev, and ZI Nikiticheva. 1975b. Development of microsporangia, pollen grains, ovule, and embryo sac of *Cynomorium songaricum* Rupr. (Cynomoriaceae). Bot. Zhurn. 60: 153—162 (in Russian).

Weddell HA. 1858–1861. Memoire sur le *Cynomorium coccineum:* Parasite de 1'ordre Balanophorees. Arch. Mus. Hist. Nat. 10: 269–308.

Order 104. BALANOPHORALES

Terrestrial herbs with yellowish-white to yellow, orange to red, or brown colors parasiting roots of trees, shrubs, or rarely herbs. Attaching to the roots of the host by an often irregularly lobed or branched, sometimes more or less horizontal and elongated tuberous organ that ranges from several centimeters to the size of a human head. Vascular system very reduced, but sometimes contains vessels with simple or (?) scalariform perforations. Sieve-element plastids of S-type. Stems with or without leaves, unbranched. Leaves much reduced (but subterranean only), or absent. Leaves when present, membranous, scaly, usually spiral, rarely opposite, decussate or distichous, very rarely verticillate (*Balanophora involullata*), without stomata. Inflorescences terminal, racemose, spicate or spadix-like; branches subtended by scaly caducous bracts, sometimes with sterile apical part peltately widened. Flowers usually very small (those of some genera among the smallest in flowering plants), numerous or very numerous, unisexual (monoecious or dioecious), apetalous and frequently without calyx, entomophilous. Male flowers often with 2(3) (4–8) free or basally connate, valvate sepals. Stamens usually 3–4 or more opposite to the sepals or less often only 1–2 stamens (1 stamen in *Dactylanthus* and 2 stamens in *Mystropetalon* and Lophophytoideae), free and with tetrasporangiate anthers or more or less united into a synandrium; synandria opening irregularly; anthers extrorse, usually connate, opening longitudinally or (*Balanophora* subgenus *Balania*) transversally. Tapetum secretory. Microsporogenesis simultaneous (successive – *Corynaea*?). Pollen grains 2-celledor3-celled(Mistropetaloideae,Helosidoideae), 3-colpate, rugate, 3–6-colporate, 3–5-porate or 3(7–8)-pororate or rarely inaperturate (*Balanophora* subgenus *Balania*), smooth or variously sculptured. Female flowers commonly with the calyx tube completely fused with the ovary wall, or the calyx apparently absent, but in *Mystropetalon* there is still 3-lobed upper free part of the calyx, which in some other genera is 2-lobed or irregularly lobed or reduced to an inconspicuous, lobed rim or represented by a few cells at the top of the ovary. Gynoecium of three or two completely fused carpels with more or less distinct stylodia (Helosidoideae and Lophophytoideae), stylodia united into a 1–3-lobed style (Mystropetaloideae, Dactylanthoideae, and Balanophoroideae) or stigma sessile (Sarcophytoideae). Ovary superior to inferior, 1–3-locular (*Mystropetalon*) or more often unilocular, with one ovule per locule, which is sometimes erect on the floor of the cavity or more commonly suspended from the roof by a stalklike cell or (*Langsdorffia* and *Thonningia*) completely united with the wall of the ovary. Ovules ategmic, with reduced nucellus, usually little more than a female gametophyte. In *Balanophora* even the placenta is completely reduced. Female gametophyte bisporic of *Allium*-type or monosporic of *Polygonum*-type. Endosperm cellular, with large chalazal haustorium. Fruits 1-seeded, indehiscent, often very tiny, nutlets or drupelets, in *Mystropetalon* surrounded by the swollen calyx tube and with disclike, whitish elaiosome (modified pedicel) at the base. Seeds with a minute, undifferentiated embryo surrounded by copious endosperm; cotyledons lacking; testa probably always absent. Contain starch or a waxy resinous substance, balanophorin (*Langsdorffia* and *Balanophora*). n = 14, 16, 18.

Both Cynomoreales and Balanophorales are probably related to the Santalales.

1. BALANOPHORACEAE

A. Richard 1822 (including Dactylanthaceae Takhtajan, 1987, Hachetteaceae Doweld 2001, Helosidaceae Bromhead 1840, Langsdorffiaceae van Tieghem ex Pilger et K. Krause 1914, Lophophytaceae Bromhead 1840, Mystropetalaceae J. D. Hooker 1853, Sarcophytaceae A. Kerner 1891, Scybaliaceae A. Kerner 1891). 16/50. Tropical and South (*Mystropetalon*) Africa, Madagascar, Comoros, Tropical Asia, Malesia, Pacific Islands, tropical Australia, New Caledonia (*Hachettea*), New Zealand (*Dactylanthus*), Tropical America.

1.1 MYSTROPETALOIDEAE

Cells with abundant deposits of reddish-brown substance (mystrin) present in leaves, superficial layers of the cortex of axis, interspersed among the vascular tissue and present also in bracts, flowers, and fruits. Calyx of the male flowers 3-merous and zygomorphic, gamosepalous. Stamens two, filaments adnate to lower, narrow part of the adaxial sepals. Calyx of the female flowers extremely small, cuplike, slightly zygomorphic. Ovary inferior, completely adnate to the calyx tube, 3-merous, basally encircled by an irregularly crenate disc; from the base of the calyx cup arises the relatively long filiform style, which ends in a papillate, more or less clearly 3-lobed capitate stigma. Ovules three, pendulous. – *Mystropetalon*.

1.2 DACTYLANTHOIDEAE

Male flowers with or without petals. Stamens 1–2, free or united. Female flowers with a small tubular, 3-lobed calyx or with 2 filiform sepals. Stylodia united into a style. Pollen grains pantoporate. Seed coat lacking. – *Hachettea, Dactylanthus*.

1.3 SARCOPHYTOIDEAE

Inflorescence conspicuously branched, paniculate, with fleshy axis, bracteate (*Sarcophyte*) or (*Chlamydophytum*) ebracteate or minutely bracteate. Stamens 3–4 or 7–10, free. Pollen grains 3-porate. Stigma sessile. Ovary with 3–1 ovules. – *Sarcophyte, Chlamydophytum*.

1.4 SCYBALIOIDEAE

Stem with triangular, scaly leaves. Inflorescence when young covered by triangular, imbricate bracts. Pollen grains mostly 2-celled, pantoporate. – *Scybalium*.

1.5 HELOSIDOIDEAE

Stem without scaly leaves. Inflorescence when young covered by hexagonal, marginally coherent peltate bracts. Pollen grains 3-celled, 3-colpate. – *Helosis* (including *Latraeophila* and *Exorhopala*), *Corynaea, Ditepalanthus, Rhopalocnemis*.

1.6 LOPHOPHYTOIDEAE

Both female and male flowers mostly naked.

Stylodia distinct, 2. Pollen grains 3-colporate. Flowers in clavate panicles, not immersed in a layer of chaffy hairs. – *Lophophytum, Lathrophytum, Ombrophytum* (including *Juelia*).

1.7 LANGSDORFFOIDEAE

Female flowers with tubular irregularly lobed or dentate calyx. The ovule is completely united with the wall of the ovary. The stigmatoid tissue of the style is conspicuously enveloped. – *Langsdorffia, Thonningia.*

1.8 BALANOPHOROIDEAE

Female flowers without calyx. The ovule is pendulous in the cavity of the ovary. The stigmatic tissue is not morphologically distinguished from the rest of the style. – *Balanophora* (including *Balania*?).

Bibliography

Andersen H. 1976. *Ombrophytum peruvianum* (Balano-phoraceae) found in the Galapagos Islands. Bot. Not. 129: 113–117.

Arekal GD and GR Shivamurthy. 1976. "Seed" germination in *Balanophora abbreviata*. Phytomorphology 26: 135–138.

Asplund E. 1928. Eine neue Balanophoraceen-Gattung aus Bolivien. Svensk Bot. Tidskr. 22: 261–277.

Beccari O. 1869. Illustrazione di nuove specie di piante Bornensi: *Balanophora reflexa* e *Brugmansia* Lowi. Nuovo Giorn. Bot. Ital. 1: 65–91.

Chodat R and C Bernard. 1900. Sur le sac embryonnaire l'*Helosis guayanensis*. Jour. de Bot. 14: 72–79.

Eichler AW. 1867. Sur la structure de la fleur de quelques Balanophorees. Act. Congr. Int. Bot. Paris 1867: 137–155.

Eichler AW. 1868. *Lathrophytum:* Ein neues Balanophoraceengeschlecht aus Brasilien. Bot. Z. 26: 513–520, 529–537, 545–551.

Eichler AW. 1869. Balanophoraceae. In: CFP Martius, Flora Brasil, 4(2): 1–74.

Ekambaran T and RR Panje. 1935. Contributions to our knowledge of *Balanophora: 2.* Life history of *B.dioica*. Proc. Indian Acad. Sci. 1B: 522–543.

Engell K. 1979. Morphology and embryology of Scybalioideae (Balanophoraceae): I. *Corynaea crassa* Hook. f. var. *spruces* (Eichl.) B. Hansen. Svensk. Bot. Tidskr. 73 (3–4): 155–166.

Ernst A. 1913. Embryobildung bei *Balanophora*. Flora 106: 129–159.

Fagerlind F. 1938a. *Ditepalanthus:* Eine neue Balanophoraceen-Gattung aus Madagascar. Arkiv f. Bot. 29A: 1–15.

Fagerlind F. 1938b. Bau und Entwicklung der floralen Organe von *Helosis cayannensis*. Svensk Bot. Tidskr. 32: 139–159.

Fagerlind F. 1945a. Blüte und Blütenstand der Gattung *Balanophora*. Bot. Not. 1945: 330–350.

Fagerlind F. 1945b. Bildung und Entwicklung des Emryosacks bei sexuellen und agamospermischen *Balanophora*-Arten. Svensk Bot. Tidskr. 39: 65–82.

Fagerlind F. 1945c. Bau der floralen Organe der Gattung *Langsdorffia*. Svensk Bot. Tidskr. 39: 197–210.

Fagerlind F. 1948. Bau und Entwicklung der vegetativen Organe von *Balanophora*. Kgl. Svensk. Vet.-Akad. Handl. 25 (3): 1–72.

Govindappa DA and GR Shivamurthy. 1975. The pollination mechanism in *Balanophora abbreviata* Blume. Ann. Bot. 39: 977–978.

Goeppert HR. 1847. Zur Kenntnis der Balanophoreninsbesondere der Gattung *Rhopalocnemis* Jungh. Nov. Act. Acad. Caes.-Leopol.-Carol. Nat. Cur. 22: 117–158.

Halle N. 1978. Illustration de deux rares Balanophoracees d'Afrique equatorial appartenant aux genres *Chlamydophytum* Mildbr. et *Balanophora* Forst. Adansonia, ser. 2,17: 249–261.

Hansen B. 1972. The genus *Balanophora* J. R. &. C. Forster: A taxonomic monograph. Dank. Bot. Ark. 28(1): 1–188.

Hansen B. 1976a. Balanophoraceae. In: CGGJ van Steenis, ed. Flora Malesiana, ser. 1, 7: 783–805. Noordhoff, Groningen.

Hansen B. 1976b. Pollen and stigma conditions in the Balanophoraceae s. lat. Bot. Not. 129: 341–345.

Hansen B. 1980. Balanophoraceae. Flora Neotropica 23: 1–80.

Hansen B. 1982. The Balanophoraceae of the Pacific. Acta Phytotax. Geobot. 33: 92–102.

Hansen B. 1984. Balanophoraceae. Flore de Madagascar et de Comores. Famille 61: 1–10. Paris.

Hansen B. 1986. The Balanophoraceae of Continental Africa. Bot. Jahrb. Syst. 106: 359–377.

Hansen B and K Engell. 1978. Inflorescences in Balanophoroideae, Lophophytoideae, and Scybalioideae (Balanophoraceae). Svensk Bot. Tidskr. 72: 177–187.

Harms H. 1935. Balanophoraceae. In: A Engler and K Prantl, eds. Die natürlichen Pflanzenfamilien, 16B:296–339. 2nd ed. Leipzig.

Harvey-Gibson RJ. 1913. Observations on the morphology and anatomy of the genus *Mystropetalon*. Trans. Linn. Soc. London 8: 143–154.

Heinricher E. 1907. Beitrage zur Kenntnis der Gattung *Balanophora*. Sitzungsber. K. Akad. Wiss. Wien, Math.-Naturw. Kl., 116: 439–465.

Heinricher E. 1908. Van Tieghem's Anschauungen über den Bau der *Balanophora*-Knolle. Sitzungsber. K. Akad. Wiss. Wien., Math.-Naturw. Kl., 117: 337–346.

Hill H. 1926. *Dactylanthus taylori:* Order Balanophoreae, tribe Cynomorieae. Trans. New Zealand Inst. 56: 87–90.

Holzapfel S. 2001. Studies of the New Zealand root-parasite *Dactylanthus taylorii* (Balanophoraceae). Englera 22: 1–176.

Hooker JD. 1856. On the structure and affinities of Balanophoreae. Trans. Linn. Soc. London 22: 1–68

Hooker JD. 1859. On a new genus of Balanophoreae from New Zealand, and two new species of *Balanophora*. Trans. Linn. Soc. London 22: 425–427.

Hooker WJ. 1840. *Langsdorffia indica*. Icones plantarum 3: 205–206.

Kuijt J. 1969. The biology of flowering plants. University of California Press, Berkeley.

Kuijt J and Wei-Xiang Dong. 1990. Surface features of the leaves of Balanophoraceae: A family without stomata? Plant Syst. Evol. 170: 29–35.

Kuwada Y. 1928. An occurrence of restitution-nuclei in the formation of the embryo sacs in *Balanophora japonica* Mak. Bot. Mag. (Tokyo) 42: 117–129.

Lotsy JP. 1899. *Balanophora globosa* Jungh: Eine wenigstens örtlich verwitwete Pflanze. Ann. Jard. Bot. Buitenz., ser. 2,1: 174–184.

Lotsy JP. 1901. *Rhopalocnemis phalloides* Jungh.: A morphological-systematical study. Ann. Jard. Bot. Bui-tenz., ser. 2, 2: 73–101.

Mangenot G. 1947. Recherches sur l'organisation d'une Balanophoracee: *Thonningia coccinea*. Rev. Gen. Bot. 54: 201–244, 271–294.

Marloth R. 1913. The flora of South Africa. Vol. 1. Cape Town.
Moore L. B. 1940. The structure and life-history of the root parasite *Dactylanthus taylori* Hook. f. New Zealand Jour. Sci. and Technol. 21: 206B–224B.
Steenis CGGJ van. 1932. Some remarks on the genus *Rhopalocnemis* Junghuhn. Handel, 6th Nederl.-Ind. Natuurwet. Congr. (Bandoeng), 1931: 464–475.
Strigl M. 1907. Der anatomische Bau der Knollenrinde von *Balanophora* und seine mutmassliche funktionelle Bedeutung. Sitzungsber. K. Akad. Wiss. Wien, Math.-Naturw. Kl., 116(I): 1041–1060.
Terekhin ES. 1987. Balanophoraceae. In: MS Yakovlev, ed. Comparative morphology of flowering plants: Davidiaceae-Asteraceae, pp. 89–93. Nauka, Leningrad (in Russian).
Terekhin ES. 1988. Balanophorales. In: A Takhtajan, ed. Comparative seed anatomy, vol. 2, pp. 120–132. Nauka, Leningrad (in Russian).
Terekhin ES and MS Yakovlev. 1967. Embryology of the Balanophoraceae. Bot. Zhurn. 52: 745–758 (in Russian).
Tieghem P van. 1896. Sur l'organisation florale des Balanophoracees et sur la place de cette famille. Bull. Soc. Bot. France 43: 295–310.
Treub M. 1898. L'organe femelle et l'apogamie du *Balanophora elongata* Bl. Ann. Jard. Bot. Buitenz. 15: 1–25.
Ultée AJ. 1926. Über das sogenannte Balanophorin. Bull. Jard. Bot. Buitenz., ser. 3, 8: 32–34.
Umiker O. 1920. Entwicklungsgeschichtlich-cytologische Untersuchungen an *Helosis guyanensis* Rich. Arb. Inst. Allgem. Bot. Pfl. Physiol. Univ. Zurich, no. 23.
Visser J. 1981. South African parasitic flowering plants. Cape Town.
Zweifel R. 1939. Cytologisch-morphologische Untersuchungen an *Balanophora abbreviata* und *B. indica*. Vierteljahrsschr. Naturf. Ges. Zurich 84: 245–306. Ph.D. dissertration, University of Zurich.

Superorder RHAMNANAE

Order 105. RHAMNALES

Trees and shrubs, rarely woody climbers, or subshrubs, very seldom herbs. Branches, leaves and flowers often covered with simple (mostly), stellate or peltate trichomes. Vessels mostly with simple perforations, rarely scalariform to reticulate; lateral pitting alternate. Fibers with simple or bordered pits. Rays from markedly heterogeneous to homogeneous. Axial parenchyma predominantly paratracheal in most species, predominantly apotracheal, sometimes very scanty. Sieve-element plastids of S-type. Nodes unilacunar or trilacunar. Leaves alternate or opposite, simple, pinnately veined or with several main veins from the base, with small stipules or estipulate. Stomata mostly anomocytic, less often paracytic or anisocytic. Flowers in various kinds of terminal or axillary inflorescences, sometimes solitary, bisexual or rarely unisexual (plants dioecious), actinomorphic, 5–6-merous or more rarely 3–4-merous, sometimes apetalous. Sepals 5–6, or 3–4, usually basally connate, valvate, deciduous or seldom persistent. Petals 3–6, rarely 8, often inserted on the calyx tube, sometimes wanting (Elaeagnaceae). Stamens 4–9(-12), free, alternisepalous; filaments thin or very short (Barbeyaceae and Elaeagnaceae), free or adnate to petals; anthers generally 2-locular, basifixed or dorsifixed, introrse, latrorse or extrorse, tetrasporangiate, opening longitudinally. Tapetum secretory. Microsporogenesis simultaneous. Pollen grains 2-celled or 3-celled, usually 3-colporate, often more or less triangular. Nectary disc usually well developed. Gynoecium of (1-)2–3(-5) united carpels, rarely (Dirachmaceae) carpels 8; style long with cylindrical, linear or capitate stigmas; ovary superior to inferior, 1-locular or rarely 5–6(-8)-locular. Ovules solitary, rarely 2 (in *Karwinskia*), anatropous, with inwardly directed micropyle, bitegmic, crassinucellate. Female gametophyte of *Polygonum*- or *Allium*-type. Endosperm nuclear. Fruits drupaceous, dry, and indehiscent or eventually dehiscent, or separating into dehiscent or indehiscent mericarps, baccate, rarely samaroid, or capsular, or achenes. Seeds exotestal, arillate or more often exarillate; mesotesta with a few sclerotic cells, embryo usually straight, endosperm starchy, usually scanty, sometimes wanting. Usually producing glycosides, often with various kinds of alkaloids, flavonols, n = (6–8)10–14(23).

Probably related to the Celastrales, from which they are chiefly distinguished by the antepetalous stamens. Both Rhamnales and Celastrales evidently had a common origin from a diplostemonous rosanean stock, the antepetalous or antesepalous cycles having aborted.

Key to Families

1 Flowers with developed nectaries. Alkaloids present.
   2 Gynoecium of 2–3(-5) united carpels. Trees and shrubs, often lianoid (by aid of hooks in *Ventilago*, tendrils in *Gouania*, twining stems in *Berchemia*), or xerophytic (sometimes with flattened, spiny, assimilatory stems and reduced leaves, such as *Colletia armata*), rarely (*Crumenaria*) subshrubs or even (*Crumenaria decumbens*) herbs. Tendrils climbering or scrambling. Hairs mostly simple,

stellate in Pomaderreae. Some members of the family, such as ssp. of *Ceanothus* and *Colletia*, contain nitrogen-fixing actynomycets in the roots. Mucilage cells are common in the leaf and in the primary cortex of the axis, and mucilage cavities are sometimes present as well in the latter. Vessels mostly with simple perforations, rarely scalariform to reticulate; lateral pitting alternate. Fibers with simple pits. Rays from markedly heterogeneous to homogeneous. Axial parenchyma predominantly paratracheal in most species, predominantly apotracheal in a few species. Nodes trilacunar. Leaves alternate or opposite, simple, pinnately veined or with several main veins from the base, sometimes much reduced, with very small stipules (sometimes modified into spines), wanting in most species of *Phylica*. Stomata mostly anomocytic, less often paracytic or anisocytic. Flowers mostly rather small, greenish, in various kinds of terminal or axillary inflorescences (sometimes reduced to a solitary flower), bisexual or rarely unisexual (plants dioecious), actinomorphic, hypanthium present, 5–6-merous or more rarely 4-merous, with more or less cuplike floral tube that is often circumscissile and deciduous above the middle. Calyx 5–4-lobed, lobes triangular, valvate and are individually deciduous or seldom persistent. Petals usually smaller than sepals and more or less concave or hooded, and holding anthers frequently clawed at the base, rarely wanting. Stamens 5–4, free, alternisepalous; filaments thin, adnate to base of petals; anthers minute, dorsifixed, introrse, generally 2-locular. Pollen grains 2-celled usually 3-colporate, often more or less triangular. Nectary disc usually well developed, intrastaminal, usually adnate to the floral tube and sometimes to the ovary. Gynoecium of 2–3(-5) carpels, with a terminal lobed or often deeply cleft style; in 3-carpellate ovaries the sequence is usually fertile-sterile-fertile: one septum bears 2 ovules, the second – a single ovule, and the third is sterile (Medan and Schirarend 2004); ovary superior to inferior, or fully or imperfectly multilocular (rarely pseudomonomerous), with 1 (2 in *Karwinskia*) basal and erect ovule in each locule. Ovules with inwardly directed micropyle. Fruits drupaceous, dry, and indehiscent or eventually dehiscent, or separating into dehiscent or indehiscent mericarps, baccate, rarely samaroid or capsular. Seeds exotestal, sometimes with dorsal groove, arillate, sometimes with a large aril (*Alphitonia*) or more often exarillate; mesotesta with a few sclerotic cells, endotegmen of cuboid cells, with scalariform thickenings; embryo large, green, oily, usually straight, rarely (*Reynosia*) ruminate; endosperm starchy, usually rather scanty (ruminate in *Reynosia*), sometimes wanting. Usually producing anthraquinone glycosides, often with various kinds of alkaloids (sometimes benzylisoquinoline alkaloids), flavonols, n = (6–8)10–13(23)..................... 1. RHAMNACEAE.

2 Gynoecium of 1 carpel. Shrubs or small trees, rarely woody climbers; shoots often reduced to spines; cambium storied. Leaves, young branches, and calyx tube are covered with peltate and stellate trichomes; mycorrhiza present in the cortex of the rootlets of *Hippophae* and *Shepherdia*, and nodules on the roots of all three genera contain nitrogen-fixing bacteria. Vessels with simple perforations; lateral pitting alternate, rather small. Fibers with conspicuously bordered pits. Rays weakly heterogeneous to homogeneous in some species of *Elaeagnus*. Axial parenchyma diffuse, sometimes very scanty. Phloem usually stratified tangentially into hard and soft layers. Nodes unilacunar. Leaves alternate or (*Shepherdia*) opposite, simple, entire, pinnately veined, estipulate. Stomata anomocytic. Flowers in racemose inflorescences or sometimes solitary in the axils of leaves, bisexual (spp. of *Elaeagnus*) or more rarely polygamous (spp. of *Elaeagnus*) or dioecious (*Shepherdia* and *Hippophae*), actinomorphic, mostly 4-merous, apetalous. Calyx often somewhat corolloid, tubular to saucer-shaped, usually 4-lobed or very rarely 6-lobed, in *Hippophae* with only 2 lobes, valvate. Stamens borne in the throat of the calyx tube, equal in number and alternating with the sepals (*Elaeagnus*), or twice as many and both alternate with and opposite to them (4 in *Hippophae* and 8 in *Shepherdia*), filaments very short; anthers tetrasporangiate, dorsifixed or basifixed. Pollen grains 2-celled or 3-celled, (2)3(4)-colporate. More or less well-developed, often lobed nectary disc usually present on the inner side of the calyx tube. Gynoecium of 1 carpel (probably as a result

of complete abortion of the other carpels), with an elongate, slender stylodium ending in a linear or capitate stigma; ovary superior, 1-locular. Ovule solitary, basal, with a short and broad funicle, with a funicular obturator. Female gametophyte of *Polygonum-* or (*Shepherdia*) *Allium-type*. Endospermal chalazal haustorium develops in *Elaeagnus*. Fruits achenes with thin, membranous pericarp enveloped by the persistent base of the calyx tube that becomes mealy or fleshy, often with a long, inner layer. Seeds have a hard testa and contain a straight embryo with large, thick, plano-convex cotyledons and a short radicle pointing downward; endosperm scanty around the embryo (*Shepherdia*), only near the radicle (*Hippophae*), or wanting (*Elaeagnus*). Strongly tanniniferous, usually producing quebrachitol and sometimes also indole alkaloids, sinapinic and ellagic acids, flavonols, 0-methyl flavonoids, triterpenes, n = 6, 10, 11, 13, 14. 2. ELAEAGNACEAE.

1 Flowers without nectaries. Alkaloids absent.

3 Leaves alternate. Deciduous small trees or shrubs. Young annual shoots and leaves with unicellular hairs. Leaves clustered on the short shoots, rapidly covered with long unicellular trichomes on the abaxial side, simple, dentate-serrate, with persistent subulate stipules. Stomata anomocytic or cyclocytic. Flowers solitary and axillary, pedicellate, bisexual, actinomorphic, with an epicalyx of 4–8 bracteoles. Hypanthium present between the petals and the calyx lobes. Sepals 5–6 or 8, basally connate, externally densely pubescent and in *Dirachma somalensis* also with glandular peltate hairs (as well as on pedicels), valvate. Petals 5–6 or 8, free, inserted on the calyx tube, imbricate, with fleshy and densely haired appendages near base that are covering a separate cavity in which nectar is exuded by a flat epithelial gland or nectar is secreted by the basal part of the appendages. Stamens 5–6 or 8, opposite the petals, free, inserted on calyx tube above the petals; filaments subulate, in *D. somalensis* with two diverging short spurs at the lower ventral part, propping the petal appendages; anthers large, 2-locular, oblong-ellipsoidal, basifixed, latrorse or extrorse. Pollen grains 3-colporate, reticulate. Gynoecium of 8 carpels; style long, with cylindrical or linear stigmas; ovary superior, tomentose, 5–6- or 8-locular; ovule solitary in each locule, ascending from the inner angle, anatropous, hypotropous, bitegmic; micropyle zig-zag. Fruits septicidal and septifragal, 8-locular (*Dirachma socotrana*) or 5-locular (*D. somalensis*), woolly inside with very long, unicellular trichomes and separating from base to apex into ventrally dehiscent segments. Seeds ellipsoidal, with straight embryo and scanty endosperm; seed coat with exotesta and endotegmic pigment layer. Contain flavonoids. . . . . 4. DIRACHMACEAE.

3 Leaves opposite. Small trees; the wood without growth rings and very hard. Vessel elements very short, with oblique end walls provided by simple perforations; lateral pitting alternate. Fibers with simple or bordered pits. Rays somewhat heterogeneous, mixed uniseriate and pluriseriate. Axial parenchyma very scanty, diffuse. Sieve-elements plastids of S-type, with very oblique, compound sieve plate. Nodes unilacunar, with a single trace. Leaves petiolate, decussate, simple, entire, pinnately veined, densely white-tomentose below with curled, unicellular hairs, estipulate. Stomata paracytic. Flowers small, in short 3-flowered dichasia, dioecious, actinomorphic, apetalous, without bracts and bracteoles. Sepals 3 or 4, slightly connate at the base, valvate; those of the female flowers pinnately veined and slightly imbricate, accrescent in fruit. Stamens 6–9 (up to 12), with very short, free filaments; anthers elongate, apiculate, latrorse. Pollen grains 3-colporate, tectate-columellate, with granular layer beneath the thick, perforated tecturn. Exine ornamentation rugulate with spinules, colpi with thickened margins, and pores covered with exinous islets. Gynoecium of 1 (mostly) to 3 free or basally more or less connate carpels, each with linear stylodium; stigma linear, papillous all round; ovary superior, sessile to shortly stipitate, 1-locular; ovules solitary, pendulous, subapical ovule on a short funiculus, anatropous, bitegmic (Bouman and Boesewinkel 1997), crassinucellate. Endosperm development nuclear. Fruits dry, indehiscent, nut-like, with thin pericarp and accrescent, submembranous, prominently veined sepals. Seeds

unspecialized, remain parenchymatic and locally compressed, exotesta perforated, endotegmen tanniniferous; embryo straight, endosperm scanty; cotyledons flat, fleshy. Embryo and endosperm contain aleuron grains. Accumulating ellagic acid and flavonols quercetin and kaempferol. . 3. BARBEYACEAE.

1. RHAMNACEAE

A.L. de Jussieu 1789 (including Frangulaceae A.P. de Candolle 1805, Gouaniaceae Rafinesque 1837, Phylicaceae J.G.Agardh 1858, Ziziphaceae Adanson ex Post et Kuntze 1903). 52/900. Nearly cosmopolitan, but mainly in tropical and subtropical regions.

Classification after D. Medan and C. Schirarend (2004).

PALIUREAE: *Paliurus, Ziziphus* (incl. *Sarcomphalus*), *Hovenia;* COLLETIEAE: *Trevoa* (incl. *Talguenea*), *Retanilla, Adolphia, Kentrothamnus, Colletia, Discaria*; PHYLICEAE: *Phylica, Trichocephalus, Nesiota, Noltea*; GOUANIEAE: *Gouania, Helinus, Reissekia, Alvimiantha, Crumenaria, Pleuranthodes*; POMADERREAE: *Pomaderris, Siegfriedia, Trymalium, Spyridium, Stenanthemum, Cryptandra, Blackallia;* RHAMNEAE: *Rhamnus* (including *Frangula*), *Scutia, Sageretia, Berchemiella, Rhamnella* (including *Chaydaia*), *Dallachya, Berchemia, Rhamnidium, Karwinskia, Condalia, Auerodendron, Reynosia, Krugiodendron;* MAESOPSIDEAE: *Maesopsis;* VENTILAGINEAE: *Smythea, Ventilago*; AMPELOZIZIPHEAE: *Ampeloziziphus*; DOERPFELDIEAE: *Doerpfeldia*; BATHIORHAMNEAE: *Bathiorhamnus*.

Incertae sedis: *Alphitonia, Ceanothus, Colubrina, Emmenosperma, Granitites, Lasiodiscus, Schistocarpaea.*

2. ELAEAGNACEAE

A. L. de Jussieu 1789 (including Hippophaeaceae G. Meyer 1836). 3/50. Subtropical and temperate regions of the Northern Hemisphere and in tropical Asia, with 1 species of *Elaeagnus* in Queensland; *Shepherdia* (3) is endemic to North America.

Elaeagnus, Hippophae, Shepherdia.

Share many features with the Rhamnaceae, including seed anatomy ("The seed-structure is typically Rhamnaceous," states Corner [1976:124]). The affinity with the Rhamnaceae is also confirmed by uredinological data (see Holm 1979 and Savile 1979). A common origin is very probable.

3. BARBEYACEAE

Rendle 1916. 1/1. Northeastern Africa and adjacent parts of the Arabian Peninsula.

Barbeya.

Have many similarities with Rhamnaceae, including wood anatomy and flower morphology.

4. DIRACHMACEAE

Hutchinson 1959. 1/2. Socotra and central Somalia.

Dirachma.

The genus *Dirachma* is usually included in the Ger-aniales with some members of which it has certain similarities, such as beaked fruits dehiscing from below to apex. However, Hutchinson (1959, 1967, 1968, 1973) included his family into Tiliales. Link (1991, 1993), Ronse Decraene and Smets (1995) have found *Dirachma* more closely related to the Malvales than to any other group, including Geraniales. They especially emphasize the location and anatomy of epithelial nectaries, which in some important details resemble trichomatic nectaries of Grewieae, Coloneae, Duboscieae, Lueheeae, Trichospermeae, and Triumpheteae of the Tiliaceae as well as the development of long, unicellular trichomes on the fruit valves' ventral sides known in some Bombacaceae, such as *Ceiba* and *Ochroma*. Also, *Dirachma* has mucilage cells (Yakovleva 1994), stellate hairs, hypocalyx, haplostemonous androecium and hypotropous ovules (epitropous in Geraniaceae). However, Thulin et al. (1998) conclude that Dirachmaceae are closely related to the Barbeyaceae. But *Dirachma* differs from *Barbeya* in many importance characters, including alternate and stipulate leaves, bisexual flowers, 8–9 sepals not accrescent in fruit, pollen morphology, the number of carpels, bitegmic ovules, septicidal 5–8-locular capsules, seed anatomy and also in wood structure such as the thicker vessel walls, smaller vessel pits, scanty or entirely lacking axial parenchyma, absence of vasicentric or vascular tracheids, and lack of prismatic crystals in the wood. Boesewinkel and Bouman (1997) proposed relationships Dirachmaceae to the

Rhamnaceae, and Thorne (2000, 2006) includes Dirachmaceae in the Rhamnales. According to Baas, Jensen and Smets (2001), *Dirachma* has many common anatomical features with Rhamnaceae. However Dirachmaceae markedly differ from the Rhamnaceae and related families in the presence of epicalyx, the lack of a nectary disc, and structure of gynoecium and fruit. The caducous glands of Dirachmaceae are not homologous with the persistent receptacular disc-nectaries of Rhamnaceae and Elaeagnaceae (Smets 1986; Smets and Cresens 1988).

Bibliography

Aagesen L. 1999. Phylogeny of the tribe Colletieae, Rhamnaceae. Bot. J. Linn. Soc. 131: 1–43.

Baas P, S Jensen, and E Smets. 2001. Vegetative anatomy and affinities of *Dirachma socotrana* (Dirachmaceae). Syst. Bot. 26: 231–241.

Bartish IV and U Swenson. 2004. Elaeagnaceae. In: K Kubitzki, ed. The families and genera of vascular plants, vol. 6, pp. 131–134. Springer, Berlin/Heidelberg/New York.

Bayer C. 2004. Dirachmaceae. In: K Kubitzki, ed. The families and genera of vascular plants, vol. 6, pp. 122–124. Springer, Berlin/Heidelberg/New York.

Bazar'a M, L Guarino, A Miller, and N Obadi. 1991. *Dirachma socotrana* – back from the brink? Oryx 25: 229–232.

Behnke H-D. 1974. P- und S-Typ Siebelement-Plastiden bei Rhamnales. Beitr. Biol. Pfl. 50: 457–464.

Behnke H-D. 1995. Sieve-element characters of the Proteaceae and Elaeagnaceae: nuclear crystals, phloem proteins and sieve-element plastids. Bot. Acta 108: 514–524.

Bennek C. 1958. Die morphologische Beurteilung der Staub- und Blümenblatter der Rhamnaceae. Bot. Jahrb. Syst. 77: 423–457.

Boesewinkel FD and F Bouman. 1997. Ovules and seeds of *Dirachma socotrana* (Dirachmaceae). Plant Syst. Evol. 205: 195–204.

Boesewinkel FD and F Bouman. 2000. Dirachmaceae. In: A Takhtajan, ed. Comparative seed anatomy, vol. 6, pp. 33–34. Nauka, St. Petersburg (in Russian).

Bond G, JT MacConnel, and AH McCallum. 1956. The nitrogen nutrition of *Hippophae rhamnoides* L. Ann. Bot. 20: 501–512.

Bouman F and D Boesewinkel. 1997. Ovules and seeds of *Barbeya* with additional arguments for an urticalean affinity of the Barbeyaceae. Acta Bot. Neerl. 46: 255–261.

Brizicky GK. 1964. The genera of Rhamnaceae in the southeastern United States. J. Arnold Arbor. 45: 439–463.

Cooper DC. 1932. The development of the peltate hairs of *Shepherdia canadensis*. Am. J. Bot. 19: 423–428.

Davtian AG. 1950. Comparative-anatomical study of the wood of the Caucasian species of the genus *Elaeagnus*. Trudy Bot. Inst. Armenian Acad. Sci. 7: 133–144 (in Russian).

Dickison WS and EM Sweitzer. 1970. The morphology and relationships *of Barbeya oleoides*. Am.J. Bot. 57: 468–476.

Friis I. 1993. Barbeyaceae. In: K Kubitzki, ed. The families and genera of vascular plants, vol. 2, pp. 141–143. Springer, Berlin/Heidelberg/New York.

Gardner IC. 1958. Nitrogen fixation in *Eiaeagnus* root nodules. Nature 181: 717–718.

Gardner IC and G Bond. 1957. Observations on the root nodules of *Shepherdia*. Canad. J. Bot. 35: 305–314.

Graham SA. 1964. The Elaeagnaceae in the southeastern United States. J. Arnold Arbor. 45: 274–278.

Harrison JE and T Beveridge. 2002. Fruit structure of *Hippophae rhamnoides* cv. Indian Summer (sea buckthorn). Canad. J. Bot. 80: 399–409

Holm L. 1979. Some problems in angiosperm taxonomy in light of the rust data. In: I Hedberg, ed. Parasites as plant taxonomists, pp. 177–181. Upsala.

Jensen S, F Piesschaert, and E Smets. 2000. Wood anatomy of Elaeagnaceae, with comments on vestured pits, helical thickenings, and systematic relationships. Am. J. Bot. 87: 20–28.

Kellermann J, F Udovicic, and PY Ladiges. 2005. Phylogenetic analysis and generic limits of the tribe Pomaderreae (Rhamnaceae) using internal transcribed spacer DNA sequences. Taxon 54: 619–631.

Leins P. 1967. Morphologische Untersuchungen an Elaeagnaceen Pollenkornern. Grana Palynol. 7: 390–399.

Link DA. 1991. *Dirachma somalensis* D. A. Link sp. nov.: A new species of a remarkable and highly endangered monogeneric family. Bull. Jard. Bot. Belg. 61: 3–13.

Link DA. 1993. Dirachmaceae. In: M Thulin, ed. Flora of Somalia, pp. 191–192. Royal Botanic Gardens, Kew.

Massagetov PS. 1946. Alkaloids in plants of the family Elaeagnaceae. Zhurn. General Chem. 16: 139–140 (in Russian).

Medan D. 1985. Fruit morphogenesis and seed dispersal in the Colletieae (Rhamnaceae). I. The genus *Discaria*. Bot. Jahrb. Syst. 105: 205–262.

Medan D. 1988. Gynoecium ontogenesis in the Rhamnaceae. In: P Leins, SC Tucker, and PK Endress, eds. Aspects of floral development, pp. 133–141. J. Cramer, Berlin.

Medan D and L Aagesen. 1995. Comparative flower and fruit structure in the Colletieae (Rhamnaceae). Bot. Jahrb. 117: 531–564.

Medan D and C Schirarend. 2004. Rhamnaceae. In: K Kubitzki, ed. The families and genera of vascular plants, vol. 6, pp. 320–338. Springer, Berlin/Heidelberg/New York.

Nemirovich-Danchenko EN. 1992. Berbeyaceae. In: A Takhtajan, ed. Comparative seed anatomy, vol. 4, pp. 380–381. Nauka, St. Petersburg (in Russian).

Rao VS. 1974. The nature of the perianth in *Eiaeagnus* on the basis of floral anatomy, with comments on the systematic position of Elaeagnaceae. Part 1. J. Indian Bot. Soc. 53: 156–161.

Rendle AB. 1916. Barbeyaceae: Prain. Flora Trop. Africa 6(2): 14–15.

Richardson JE, MF Fay, QCB Cronk, D Bowman, and MW Chase. 2000. A phylogenetic analysis of Rhamnaceae using *rbc*L and *trn*L-F plastid DNA sequences. Am. J. Bot. 87: 1309–1324.

Richardson JE, MF Fay, QCB Cronk, and MW Chase. 2000. A revision of the tribal classification of Rhamnaceae. Kew Bull. 55: 311–340.

Ronse Decraene LP and AG Miller. 2004. Floral development and anatomy of *Dirachma socotrana* (Dirachmaceae): a controversial member of the Rosales. Plant Syst. Evol. 249: 111–127.

Saville PB. 1979. Fungi as aids in higher plant classification. Bot. Rev. 45: 377–503.

Schirarend C and E Kohler. 1993. Ramnaceae Juss. World Pollen Spore Flora, 17–18: 1–53.

Schweinfurth C. 1891. *Barbeya* Schf. gen. nov. Urticacearum. Malpighia 5: 332–340.

Servettaz C. 1909. Monographic des Eleagnacees. Beih. Bot. Centralbl. 25(2): 1–420.

Shabes LK and AA Morozova. 2000. Elaeagnaceae. In: A Takhtajan, ed. Comparative seed anatomy, vol. 6, pp. 192–196. Nauka, St. Petersburg (in Russian).

Sorsa P. 1971. Pollen morphological study of the genus *Hippophae* L., including the new taxa recongized by A.Rousi. Ann. Bot. Fenn. 8: 228–236.

Thulin M, B Bremer, J Richardson, J Niklasson, MF Fay, and MW Chase. 1998. Family relationships of the enigmatic rosid genera *Barbeya* and *Dirachma* from the Horn of Africa region. Plant Syst. Evol. 213: 103–119.

Tobe H and M Takahashi. 1990. Trichome and pollen morphology of *Barbeya* (Barbeyaceae) and its relationships. Taxon 39: 561–567.

Tortosa RD, L Aagesen, and GM Tourn. 1996. Morphological studies in the tribe Colletieae (Rhamnaceae): analyses of architecture and inflorescences. Bot. J. Linn. Soc. 122: 353–367.

Tortosa RD and D Medan. 1992. Rhamnaceae with multiple lateral buds: an architectural analysis. Bot. J. Linn. Soc. 108: 275–286.

Veldkamp JF. 1986. Elaeagnaceae. In: CGGJ van Steenis, ed. Flora Malesiana, ser. 1, 10: 151–156. Noordhoff, Groningen.

Vyshenskaya TD. 2000. Rhamnaceae. In: A Takhtajan, ed. Comparative seed anatomy, vol. 6, pp. 183–191. Nauka, St. Petersburg (in Russian).

Yakovleva OV. 1994. The ultrastructure of mucilage cells in the leaf epidermis of *Dirachma socotrana* (Dirachmaceae). Bot. Zhurn. 79: 52–58 (in Russian with English summary).

Yamazaki T. 1975. Embryology of *Elaeagnus umbellata* Thunb. J. Jpn. Bot. 50: 281–284.

Subclass VII. ASTERIDAE

Trees, shrubs, subshrubs, perennial or annual herbs. Vessels with scalariform or simple perforations. Sieve-element plastids usually of S-type. Nodes trilacunar, multilacunar or unilacunar. Leaves simple or compound, alternate, opposite or verticillate. Stomata of various types. Flowers in various kinds of inflorescences or sometimes solitary, mostly bisexual or less often unisexual, actinomorphic or more or less zygomorphic, often characterized by the secondary pollen presentation. Corolla often sympetalous. Stamens usually five, often attached to the corolla tube. Anthers tetrasporangiate, opening longitudinally. Pollen grains 2-celled or 3-celled, colpate, colporate or porate. Tapetum secretory or (some Dipsacales) amoeboid. Microsporogenesis simultaneous. Pollen grains 2- or 3-celled, 3-colporate or of derived types. Nectary disc mostly present. Gynoecium of 2–5 (up to 12) united carpels; stylodia free or more or less connate; ovary mostly inferior. Ovules anatropous, unitegmic, mostly tenuinucellate. Endosperm cellular or more often nuclear. Fruits of various types. Seeds with small or large embryo; endosperm copious to scanty, rarely absent.

Probably originated from a Cunoniales-like saxifraganaean ancestor.

Bibliography

Albach DC, PS Soltis, and DE Soltis. 2001a. Patterns of embryological and biochemical evolution in the Asterids. Syst. Bot. 26: 242–262.

Albach DC, PS Soltis, DE Soltis, and RG Olmstead. 2001b. Phylogenetic analysis of the Asteridae s.l. using sequences of four genes. Ann. Missouri Bot. Gard. 88: 163–212.

Boros CA and FR Stermitz. 1990. Iridoids. An updated review. I. J. Nat. Prod. 53: 1055–1147.

Boros CA and FR Stermitz. 1990. Iridoids. An updated review. II. J. Nat. Prod. 54: 1172–1246.

Bremer K, EM Friis, and B Bremer. 2004. Molecular phylogenetic dating of asterid flowering plants shows Early Cretaceous diversification. Syst. Biol. 53: 496–505.

Bucklund A and B Bremer. 1997. Phylogeny of the Asteridae s.str. based on *rbc*L sequences, with particular reference to the Dipsacales. Plant Syst. Evol. 207: 225–254.

Carlquist S. 1992. Wood anatomy of sympetalous dicotyledons families: a summary, with comments on systematic relationships and evolution of the woody habit. Ann. Missouri Bot. Gard. 79: 303–332.

Dahlgren R. 1977. A note on the taxonomy of the «Sym-petalae» and related groups. Cairo Univ. Herb. 7 and 8: 83–102.

Downie SR and JD Palmer. 1992. Restriction site mapping of the chloroplast DNA inverted repeat: a molecular phylogeny of the Asteridae. Ann. Missouri Bot. Gard. 79: 266–283.

Erbar C. 1988. Sympetaly: a systematic character? Bot. Jahrb. Syst. 112: 417–451.

Erbar C and P Leins. 1995. Portioned pollen release and the syndromes of secondary pollen presentation in the Campanulales-Asterales-complex. Flora 190: 323–338.

Eyde RH. 1988. Comprehending *Cornus*: puzzles and progress in the systematics of the dogwoods. Bot. Rev. 54: 233–351.

Frohne D and U Jensen. 1985. Systematik des Pflanzenreichs, 3rd ed. Gustav Fischer Verlag, Stuttgart.

Gustafsson MHG. 1995. Petal venation in the Asterales and related orders. Bot. J. Linn. Soc. 118: 1–18.

Gustafsson MHG. 1996. Phylogenetic studies in the Asterales sensu lato. Uppsala.

Gustafsson MHG and K Bremer. 1995. Morphology and phylogenetic interrelationships of the Asteraceae, Calyceraceae, Campanulaceae, Goodeniaceae, and related families (Asterales). Am. J. Bot. 82: 250–265.

Gustafsson MHG, A Backlund, and B Bremer. 1996. Phylogeny of the Asterales sensu lato based on *rbc*L sequences with particular reference to the Goodeniaceae. Plant Syst. Evol. 199: 217–242.

Hegnauer R. 1969. Chemical evidence for the classification of some plant taxa. In: JB Harborne and T Swain, eds. Perspectives in phytochemistry, pp. 121–138. Academic, London.

Jensen SR. 1992. Systematic implications of the distribution of iridoids and other chemical compounds in the Loganiaceae and other families of the Asteridae. Ann. Missouri Bot. Gard. 79: 284–302.

Kårehed J. 2002. Introduction. In: Evolutionary studies in Asterids emphasising Euasterids II, pp. 5–50. Acta Universitatis Upsaliensis, Uppsala.

Leins P and C Erbar. 1990. On the mechanisms of secondary pollen presentation in the Campanulales-Asterales complex. Bot. Acta 103: 87–92.
Lundberg J and K Bremer. 2003. A phylogenetic study of the order Asterales using one morphological and three molecular data sets. Int. J. Plant Sci. 164: 553–578.
Michaels HJ, KM Scott, RG Olmstead, T Szaro, R Jansen, and JD Palmer. 1993. Interfamilial relationships of the Asteraceae: Insights from *rbc*L sequence variation. Ann. Missouri Bot. Gard. 80: 742–751.
Noshiro S and P Baas. 1998. Systematic wood anatomy of Cornaceae and allies. IAWA J. 19: 43–97.
Olmstead RG, HJ Michaels, KM Scott, and JD Palmer. 1992. Monophyly of the Asteridae and identification of their major lineages inferred from DNA sequences of *rbc*L. Ann. Missouri Bot. Gard. 79: 249–265.
Olmstead RG, B Bremer, KM Scott, and JD Palmer. 1993. A parsimony analysis of the Asteridae sensu lato based on *rbc*L sequences. Ann. Missouri Bot. Gard. 80: 700–722.
Robyns W. 1972. Outline of a new system of orders and families of Sympetalae. Bull. Jard. Bot. Nat. Belg. 42: 363–372.
Wagenitz G. 1977. New aspects of the systematics of Asteridae. Plant Syst. Evol. Suppl 1: 375–385.
Wagenitz G. 1992. The Asteridae: Evolution of a concept and its present status. Ann. Missouri Bot. Gard 79: 209–217.
Yamazaki T. 1974. A system of Gamopetalae based on the embryology. J. Fac. Sci. Univ. Tokyo, Sect. 3, Bot. 11: 263–281.

Superoder CORNANAE

Order 106. DESFONTAINIALES

Trees or shrubs, less often subshrubs or herbs. Vessels mostly with scalariform perforations (sometimes with numerous bars), lateral pitting scalariform to alternate. Fibers mostly with bordered pits, often septate. Rays heterogeneous. Axial parenchyma apotracheal to scanty paratracheal, often wanting. Sieve-element plastids of S-type. Nodes trilacunar or less often unilacunar or multilacunar. Leaves alternate, opposite or less often verticillate, simple, entire or less often lobed or dentate, stipulate or more often estipulate. Stomata paracytic or more often anomocytic. Flowers in inflorescences of various types or rarely solitary, bisexual or sometimes unisexual, actinomorphic or sometimes more or less zygomorphic. Sepals (3)4–5(-12), valvate or imbricate, free or more often more or less connate. Petals (3)4–5(-12), valvate, imbricate or contorted, free or connate into a short tube. Stamens (3)4–5 or 8–10 or more, sometimes numerous (even up to 200 in *Carpenteria*); filaments free, basally connate or adnate to the corolla tube; anthers opening longitudinally. Tapetum secretory. Microsporogenesis simultaneous. Pollen grains 2-celled, mostly 3-colporate, sometimes 2–3-colpate. Intrastaminal nectary disc present or less often wanting. Gynoecium of (2)3–5 or less often up to 12 united carpels; stylodia free or more often connate into a style with the stigmatic branches or lobed or capitate stigma; ovary superior or more often semi-inferior or inferior, with 1-several or many ovules per locule. Ovules mostly anatropous, usually unitegmic, tenuinucellate, or rarely crassinucellate, usually with endothelium. Female gametophyte of *Polygonum-* or *Allium*-type. Endosperm cellular, with terminal haustoria. Fruits septicidal or loculicidal capsules or less often baccate or drupaceous. Seeds exotestal, rarely (Vahliaceae) endotestal, with straight embryo surrounded by fleshy endosperm or less often without endosperm.

The Escalloniaceae, and related families are usually included in Rosales sensu amplissimo or in Cunoniales or even in the family Saxifragaceae s. 1. However, this group of families differs from Saxifragaceae, Cunoniaceae, and related families and is much closer to the Cornales (see Huber 1963; Dahlgren 1975, 1980, 1983, 1989 [who even included them in the Cornales]; Takhtajan 1987; Soltis et al. 1990; Thorne 1992a, b, 2006; Xiang et al. 1993). From the saxifraganaean and rosanaean families are differ not only morphologically (unitegmic and tenuinucellate ovules) but also in the presence of iridoid compounds (Jensen et al. 1975; El-Naggar and Beal 1980). Its distinctness is also supported by serological studies (Grund and Jensen 1981).

Key to Families

1 Ovules bitegmic.
  2 Vessel elements long (sometimes very long), with exclusively scalariform perforations that have 16–71 (occasionally up to 111) thin bars; lateral pitting extremely rare and confined to overlapping end walls (when present typically occur in uniseriate files). Fibers with bordered pits. Rays of homocellular, uniseriate rays of upright cells and heterocellular, bi- and multiseriate rays. Axial parenchyma sparse and predominantly apotracheal diffuse and paratracheal scanty. Evergreen tree up to 15 m tall. Nodes trilacunar with three traces. Leaves alternate, opposite or verticillate, coarsely and bluntly glandular-serrate, estipulate. Stomata anomocytic. Marginal veins are not fimbriate. Unicellular T-shaped trichomes occur on both the pedicels

and flowers. Flowers in terminal, subumbellate panicles. Sepals spirally arranged, and the three lower ones, which are larger than the other two, completely enclose the other organs in the bud. Calyx-tube adnate to base of ovary, lobes five, imbricate, deciduous. Petals five, inserted below 5-lobed disc, clawed, imbricate. Connectives are not prolonged. Stamens five, alternate with disc lobes, filaments slender. Pollen grains 2-celled, (3)4–5-colporate. Ovary superior to semi-inferior, conical, 5-lobed, 5-locular, with two ovules per locule; style spirally twisted, 5-grooved. Ovules crassinucellate, with nucellar cap, funicular obturator and ruminate arillode and without endothelium. Fruits broadly ovoid, coriaceous, 5-locular, loculicidal capsules. Seeds large, glossy, with a broad "aril", hilar scar elongated, exotestal; cotyledons large. Contains proanthocyanidins (from cortex), urolic acid; n = 25. . . . . . 4. IXERBACEAE.

2 Vessels with simple perforations. Leaves opposite, simple, entire, ovate to linear, estipulate. Stomata anomocytic. Annual herbs rather densely pubescent by multicellular glandulose hairs. Nodes unilacunar. Flowers small, paired, in axillary cymose inflorescences, bisexual, actinomorphic. Sepals five, connate into hemispherical tube adnate to the ovary; lobes valvate. Petals five, free, shorter than the calyx lobes, obovate-spatulate, imbricate. Stamens five, alternipetalous, inserted on the margin of the epigynous nectary disc; filaments free, subulate; anthers dorsifixed, introrse. Pollen grains 2-celled, 3-colporate. Gynoecium of two or three united carpels, with two or three thick, divergent stylodia terminated by capitellate stigma; ovary inferior, 1-locular, with numerous, anatropous, bitegmic, tenuinucellate ovules attached to the large placentas hanging from the apex of the locule. Fruits more or less globose, loculicidal capsules, 2–3-valved at the top, with persistent calyx. Seeds numerous, minute, oblong, appendaged; exotestal cells more or less elongated; endotesta a layer of cells with well-developed U-thickening; embryo large, straight; endosperm scarce, of two cell layers. Contain iridoids n = 6, 9. . . . . 10. VAHLIACEAE.

1 Ovules unitegmic.

3 Trees, shrubs, subshrubs or perennial herbs.

4 Corolla not sympetalous. Stem and petiole without endodermis.

5 Flowers usually bisexual, rarely polygamo-dioecious.

6 Nectary disc usually present.

7 Pericycle without a ring of lactiferous sacs and without a composite and continuous ring of sclerenchyma. Trees and shrubs. Leaves alternate or rarely (*Polyosma*) opposite or subopposite, estipulate. Hairs unicellular and thick-walled, glandular-shaggy with uniseriate stalks of variable length and aspheroid or (*Quintinia*) peltate glands. Vessels with scalariform perforations that have many bars (up to 125 in some species of *Polyosma*); lateral pitting transitional, opposite or alternate. Fibers with bordered pits. Rays heterogeneous. Axial parenchyma apotracheal, diffuse and/or diffuse-in-aggregates. Nodes (at least in *Escallonia*) unilacunar. Leaves entire, serrate or dentate, often glandular-serrate. Flowers in terminal or axillary cymes, small to rather large (*Anopterus*), actinomorphic. Sepals 4–6, free or connate into a short tube. Petals 4–6, free, imbricate or valvate. Stamens mostly five, free, alternate with the lobes of nectary disc. Pollen grains 3-colporate or (*Quintinia*) 5-colporate. Gynoecium of 2–5 carpels; stylodia connate into a short or elongate style with capitate, 2–5-lobed stigma; ovary superior to inferior, usually with numerous ovules on axile or less often (*Polyosma, Anopterus*) parietal placentas. Fruits septicidal or loculicidal capsules or (*Polyosma, Abrophyllum*) berries. Seeds small or minute; embryo small, surrounded by copious fleshy endosperm. Contain decarboxylated iridoids, flavonols, n = 12 (*Escallonia*). . . . . 1. ESCALLONIACEAE.

7 Pericycle with a ring of laticiferous sacs consisting of vertically elongate cells arranged in longitudinal rows and filled with white, friable

contents and with a composite and continuous ring of sclerenchyma. Trichomes glandular with short sunken stalks and unicellular heads. Vessels with scalariform perforations. Leaves alternate, large, long-acuminate, subserrate. Stomata with a pair of small cells nearly circular in outline. Flowers in terminal or axillary cymes, yellowish. Calyx tube very short, lobes 6–5, small, decieduous. Petals 6–5, valvate, spreading, deciduous. Stamens 6–5, inserted on the margin of the inconspicuous nectary disk; filaments very short. Gynoecium of five carpels; stigma sessile, 5-lobed; ovary superior, 5-locular, with numerous axile ovules. Fruits a small, black, many-seeded berry, corsned by the stigma. Seeds small, with deeply latticed testa; embryo very small. . . . . .3. ABROPHYLLACEAE.

6 Nectary disc absent. Prostrate, glabrous shrublets with robust, flexuous stems emitting short ascending, densely leafy branches. Vessels with scalariform perforations that have 12–14 bars. Leaves alternate, sessile, semiamplexicaul, simple, entire but apically minutely tridentate, more or less glaucous above, obscurely veined, estipulate. Flowers small, solitary at apex of branchlets; subsessile, ebracteate, actinomorphic, 5-merous as to the perianth and androecium. Sepals small, shortly connate bellow, imbricate, persistent. Petals relatively large, slightly clawed, contorted, fleshy or not. Stamens alternipetalous; filaments subulate; anthers small, ovoid, subdidymous, extrorse. Pollen grains 3-colporate, rugulate-reticulate, with fractioned, partial tectum with interrupted muri on the intercolpium. Gynoecium of three carpels; style short, with 3-lobed, capitate stigma; ovary superior, 3-locular, with many (30–50) anatropous ovules per locule. Fruits small, many-seeded, loculicidal and valvular capsules, borne on a short, erect pedicel; the valves finally separating from the axis, to which the seeds long remain attached after the capsule has dehisced. Seeds black, shining, with fleshy and oily endosperm. . . . . .5. TRIBELACEAE.

5 Flowers unisexual, dioecious. Vessels usually with simple perforations, even in the primary xylem; lateral pitting alternate, pits more or less vestured. Fibers with simple pits. Rays mostly heterogeneous. Nodes unilacunar.

8 Petals valvate. Small trees or shrubs. Vessels with simple perforation or rarely with simple and scalariform perforations, with 1–2 bars. Fibers not septate. Rays essentially heterogeneous. Axial parenchyma absent. Leaves alternate, entire, pinnately veined, coriaceous, estipulate. Stomata anomocytic. Flowers small, in small axillary panicles. Male flowers: calyx 4-angled, 4-lobed. Petals four, oblong-ovate, free; stamens four; filaments very short, anthers basifixed, oblong-linear, acuminate. Pollen grains 3-colporate, reticulate. Nectary disc pulvinate, obtusely 4-angled; rudimentary gynoecium subulate. Female flowers: calyx tube obtusely 4-angled, teeth four, remote; petals four, staminodia absent; nectary disc hemispherical. Gynoecium of two carpels; stylodia minute, recurved, with decurrent stigmas; ovary semi-inferior, 2-locular, with one ascending, campylotropous ovule in each locule. Fruits small, compressed, didymous drupes with two compressed, 1-seeded, crustaceous, thin-walled pyrenes. Seeds linear-oblong, with membranous testa and fleshy, orange or brownish endosperm; embryo large, with inferior radicle, n = 16. . . . . .6. KALIPHORACEAE.

8 Petals imbricate. Small trees or shrubs, sometimes (*Grevea bosseri*) lianas. Vessels with simple perforation. Fibers

with bordered pits, septate (*Grevea*). Rays heterogeneous (*Montinia*) or homogeneous (*Grevea*). Axial parenchyma scanty paratracheal. Leaves alternate, subopposite or opposite, entire, coriaceous (*Montinia*) or membranous (*Grevea*), estipulate. Stomata anomocytic (*Montinia*) or anisocytic (*Grevea*). Flowers in terminal or axillary cymes or panicles, or solitary and terminal (female), small, actinomorphic, unisexual, 4–5-merous. Male flowers: sepals 3–5, connate into cupular or flattened, entire or shortly 3–5-lobed tube; petals 3–5, alternipetalous, thick; stamens 3–5, alternipetalous; filaments short, thick; anthers rather large, ellipsoid, dorsifixed, extrorse. Pollen grains 3-colporate reticulate; rudimentary gynoecium absent or minute; nectary disc flat, discoid. Female flowers: sepals 4–5, adnate to ovary, limb shortly tubular, entire or minutely 4–5-toothed; petals 4–5, fleshy, deciduous; staminodia 4–5; nectary disc epigynous, fleshy, 4–5-angled. Gynoecium of two carpels; style short, thick, persistent, stigmas 2-fid or shortly 2-lobed; ovary inferior to semi-inferior, 2-locular, with numerous (*Montinia*) or 4–6 ovules per locule; ovules pendulous, anatropous. Fruits 2-locular and loculicidal, dehiscent (*Montinia*) or indehiscent (*Grevea*) capsules with few seeds and tipped by persistent style. Seeds compressed and winged all around and auriculate at the point of attachment (*Montinia*) or subglobose, and not winged; exotesta lignified, periclinal walls thickened; endosperm more or less developed in *Grevea* (Krach 1976), or copious in *Montinia* (Nemirovich-Danchenko 2000). n = 34 (*Montinia*). 7. MONTINIACEAE.

4 Corolla sympetalous.

9 Ovules tenuinucellate. Evergreen shrubs or small trees; plants usually bitter tasting. Vessels with scalariform perforations; lateral pitting is generally absent owing to the solitary nature of vessels and tends toward alternate when present. Fibers are tracheids. Rays entirely uniseriate and comprise axially elongated or upright cells only. Axial parenchyma vasicentric, scanty with some diffuse. Leaves opposite, rather small, simple, asymmetrical, entire or dentate, with glands on lower surface or stipules present. Stomata anomocytic. Flowers in few-flowered terminal cymes or solitary and terminal, two bracteolate, bisexual, slightly zygomorphic, (4)5(-8)-merous as to the perianth. Sepals connate, lobes valvate or weakly imbricate, persistent. Petals yellow, connate into a very short tube, lobes spreading, imbricate. Stamens two, inserted near the base of corolla, alternate with the adaxial and lateral lobes; filaments short, stout; anthers large, extrorse, with broad connective and two undulately plicate, latrorse, twisted loculi that open longitudinally. Pollen grains 3-colporate. Gynoecium of two carpels; style short, thick, stigma cup-like, with two broad lobes; ovary almost quite inferior, imperfectly 2-locular, with two intruded parietal placentas that are subcontiguous in the center; ovules numerous, ascending, anatropous, with endothelium. Fruits septicidal and valvular capsules, with persistent calyx. Seeds numerous, minute, laterally compressed, smooth, with minute straight embryo and copious, fleshy endosperm. . . . . .8. COLUMELLIACEAE.

9 Ovules crassinucellate. Erect or sprawling evergreen shrubs or rarely small trees with somewhat angular, opposite branches, glabrous or with a few simple hairs. Pith parenchymatous. Vessels with scalariform perforations; perforation plates very oblique, the bars fine and numerous from 15 to 65; lateral pitting partly scalariform, partly small, round, bordered. Fibers with bordered pits. Rays uniseriate, composed of upright cells. Axial parenchyma mainly diffuse, but also some isolated strands associated

with the vessels (Mennega 1980). Sieve-element plastids of S-type. Leaves opposite, simple, petiolate, spinulose-dentate like those of *Ilex aquifolium,* pinnately veined, coriaceous, connected at the base by stipular line. Stomata anomocytic. Flowers solitary in the axils of the leaves near the apices of the branchlets or in few-flowered cymes, large, bisexual, actinomorphic or nearly actinomorphic, 5-merous, with leafy or sepal-like bracts. Sepals basally connate, strongly imbricate, equal or subequal, persistent. Corolla much longer than the calyx, nearly cylindrical, fleshy, lobes imbricate or contorted, scarcely spreading. Stamens five, inserted just below the corolla throat, equal; filaments much shorter than the anthers; anthers basifixed, narrowly triangular to nearly oblong, deeply cordate at the base, introrse. Pollen grains 3-colporate, with a thick aperture membrane, rugulate tectum surface with small perforations, and elongate pores (Hoc and Bravo 1984). Gynoecium of five united carpels; style long, slender, with capitate, scarcely dilated, obscurely 5-lobed stigma, only slightly expanded, persistent; ovary superior, cylindrical, 5-locular at the base and 1-locular at the apex, with five parietal placentas with many ovules; placentas mushroom-shaped on section, united inside in the basal half of the ovary. Ovules anatropous, with endothelium. Fruits berries with a persistent calyx. Seeds many, obliquely ellipsoid or ovoid, slightly angular, outer walls of testa pectic, other walls lignified; embryo straight, minute; endosperm copious, starchy. Producing iridoid glycosides loganin and loganic acid (Houghton and Ming 1985). n = 7. . . . . 9. DESFONTAINIACEAE.

3 Annual herbs. Leaves alternate, simple, somewhat fleshy, the radical leaves spatulate-oblanceolate, entire to dissected, when dissected, basically palmatifid (the upper members stagshornlike), estipulate. Small, densely glandular-hairy annual herbs, much branched from the base, thinly pilose all over. Vessels with simple perforations. Flowers very small, in terminal, cymose, paniculate inflorescences, bisexual, actinomorphic, 5-merous. Sepals narrow, deeply lobed, basally connate into a very short tube, adnate to the lowest third of the ovary; lobes valvate. Petals free, narrowly elliptic, spreading. Stamens five, alternate with the petals, free; filaments subulate; anthers small, rounded-ellipsoidal, dorsifixed, versatile, introrse to latrorse. Pollen grains 3-colporate, reticulate. Epigynous disc absent. Gynoecium of two united carpels, with free slender divergent stylodia terminated by capitate stigma; ovary 2-locular, three-quarters superior, shortly adnate to the calyx tube at the base, with one subbasal axile ascending, anatropous ovule per locule. Fruits semi-inferior, subdidymous, loculicidal capsules. Seeds small, erect, with minute, straight embryo and copious endosperm; seed coat 3–4 layered. . . . . .2. EREMOSYNACEAE.

1. ESCALLONIACEAE

R. Brown ex Dumortier 1829 (including Anopteraceae Doweld 2001; Polyosmaceae Blume 1851). 6/150. The largest genus *Polyosma* (60) is distributed from eastern Himalayas through Southeast Asia to New Guinea, tropical Australia, and New Caledonia, *Escallonia* (about 40) in South America (especially Andes), *Quintinia* (25) from Malesia to Australia, New Zealand, and New Caledonia, *Cuttsia* (1), and *Anopterus* (2) are endemic to Australia, *Forgesia* (1) to Reunion and *Valdivia* (1) to Chile.

ANOPTEREAE: *Anopterus;* FORGESIEAE: *Forgesia, Quintinia;* ESCALLONIEAE: *Escallonia, Valdivia;* POLYOSMEAE: *Polyosma.*

The family, even in its narrow sense, as defined, is rather heterogeneous.

2. EREMOSYNACEAE

Dandy 1959. 1/1. Southwestern Australia.

Eremosyne

Furthermore, Al-Shammary and Gornall (1994) suggested a close relationship between *Eremosyne* and

Escallonia based on trichome anatomy. According to Soltis and Soltis (1997), Eremosynaceae are the sister to the Escalloniaceae.

3. ABROPHYLLACEAE

Nakai 1943. 2/2. Eastern Australia.

Cuttsia, Abrophyllum

Related to the Escalloniaceae, but differing in the structure of pericycle and *Abrophyllum* and *Cuttsia* both have clusters of small, unlignified cells in the mesophyll that look like little white raphide bundles (Hils 1985).

4. IXERBACEAE

Grisebach 1854. 1/1. Northern New Zealand (lowland and lower montane forests from lat. 35°30′ to a little south of 3 8°).

Ixerba

Vegetative anatomy of *Ixerba* is very primitive.

5. TRIBELACEAE

Airy Shaw 1965. 1/1. Temperate South America (from southern Chile to Tierra del Fuego).

Tribeles

Related to the Escalloniaceae, differing from them in the absence of nectary disc, in contorted petals, extrorse anthers, persistent seed-bearing column of the fruits, and the obscurely veined, glaucescent leaves ("very distinctive in appearance, and unlike any others known to the writer," states Airy Shaw [1965: 269]), as well as in seed coat anatomy (Nemirovich-Danchenko and Lobova 1998).

6. KALIPHORACEAE

Takhtajan 1996. 1/1. Madagascar.

Kaliphora

Capuron was the first who noticed the uncornaceous features of *Kaliphora,* which were later confirmed by Eyde (1988). *Kaliphora* is related to both the Montiniaceae. There are many palynological similarities between *Montinia, Grevea,* and *Kaliphora* (Hideux and Ferguson 1976; Ferguson 1977; Ferguson and Hideux 1978). The wood anatomy of *Kaliphora* resembles that of *Montinia* and *Grevea* (Bakolimalala-Ramamonjiarisoa 1989). However, *Kaliphora* differ from the Montiniaceae s. str. in many respects, including free stylodia, solitary pendulous ovule, and drupaceous fruits. According to Eyde (1988: 310), *Kaliphora* «has commissural stigmas: receptive tissue extends downward in the septal radii not in the radii of the dorsal carpel bundles as is the case in the comparable Cornaceae (that is, in the Nyssoideae: only they have well-defined style branches». Also, as Eyde points out «Insertion of the ovule is lower – about midlevel in the cavity – in *Kaliphora* than in true cornads and the cushionlike epigynous nectary of true cornads is absent. Furthermore, the seeds develop with their micropyle turned down».

7. MONTINIACEAE

Nakai 1943. 2/4 Southern Africa (*Montinia*) and tropical West and East Africa and Madagascar (*Grevea*).

Montinia, Grevea

Related to the Escalloniaceae, and in Hutchinson's system (1967, 1969, 1973) they are included in them. However, as early as 1891 Kerner von Marilaun and later Nakai (1943) and independently Takhtajan (1954) and Milne-Redhead (1955) considered both genera a separate family, which was later accepted by Capuron (1969), Airy Shaw (in Willis 1973), Krach (1976), Dahlgren (1980, 1983, 1989), Carlquist (1989) and Thorne (1992a, b). Capuron (1969) also proposed to include in this family *Kaliphora,* which was accepted by Dahlgren (1980, 1983) and later by Carlquist (1989) and Thorne (1992), who also added *Melanophylla.* Palynologically *Montinia, Grevea,* and *Kaliphora* are very similar (Hideux and Ferguson 1976; Ferguson and Hideux 1978). Montiniaceae are considerably more advanced than Escalloniaceae. According to Carlquist (1989), wood of *Montinia* resembles that of *Kaliphora* and *Grevea,* but *Grevea* is more distant from *Montinia* than is *Kaliphora.*

8. COLUMELLIACEAE

D. Don 1828. 1/4. Andes of South America from Colombia to Bolivia.

Columellia

Probably related to the Escalloniaceae and differing from them mainly in absence of the nectary disc, weakly sympetalous corolla, slightly zygomorphic flowers, androecium of two stamens with broad connective and contorted anthers, and thick integument. A great variety of families have been proposed as relatives of *Columellia* (Columelliaceae). They include Oleaceae, Gesneriaceae, Ericales, Ebenales, Gentianales, Scrophulariales, Myrtales, Saxifragales, and even Cucurbitaceae. The most recent treatments favor inclusion in Cornales (Dahlgren 1975, 1983, 1989), Resales s. 1. (Cronquist 1981, 1988; Thorne 1983), Saxifragales s. 1. (Hallier 1908, 1911, 1912; Takhtajan 1980), or Hydrangeales (Takhtajan 1987; Thorne 1992a, b). Hallier (1908, 1911, 1912) included *Columellia* in Saxifragaceae-Philadelpheae. According to Stern et al. (1969: 70), it is clearly impossible to assemble an array of data that would affirm unequivocally the relationships of Columelliaceae with any one of the several families to which it has been allied. In their opinion «Perhaps its nearest relatives are in the Escalloniaceae» (p. 70).

9. DESFONTAINIACEAE

Endlicher 1841. 1/5. The Andes from Costa Rica to Cape Horn.

Desfontainia

Closely related to the Columelliaceae (Hallier 1908, 1911, 1912). Thorne (1992a, b, 2000, 2006) placed the Desfontainiaceae in his Cornales close to the Columelliaceae. A relationship between *Desfontainia* and *Columellia* is also supported by the wood anatomy (Mennega 1980). Backlund and Donoghue (1996) even included *Desfontainia* in Columelliaceae. However, there are also many differences in floral morphology.

10. VAHLIACEAE

Dandy 1959. 1/8. Tropical and South Africa, Madagascar, and from upper Egypt and Iran to northwestern India.

Vahlia (Bistella)

Usually included in the Saxifragales, but Vahliaceae differ from them in their placentas are pendulous, ovules are tenuinucellate, and micropyle is formed by the inner integument. Besides, mature seeds do not show any traces of the raphe (Krach 1976, 1977). And what is even more important, *Vahlia* contains iridoids (Al-Shammary 1991). According to *rbs*L and 18S rDNA sequence data, *Vahlia* is well separated from Saxifragales and included in the same clade which contains *Montinia* (Morgan and Soltis 1993; Soltis and Soltis 1997). According to Al-Shammary and Gornall (1994) *Vahlia* resembles *Montinia* also in thin-walled, eglandular, uniseriate trichomes with a smooth cuticle, opposite, estipulate leaves. Probably belong to the Hydrangeales. However, *Vahlia* differs from the typical members of Desfontainiales in bitegmic ovules. Systematic position of *Vahlia* is still uncertain.

Bibliography

Al-Shammary KIA and RJ Gornall. 1994. Trichome anatomy of the Saxifragaceae s.l. from the southern hemisphere. Bot. J. Linn. Soc. 114: 99–131.

Bakolimalala-Ramamonjiarisoa R. 1989. Wood anatomy of Malagasy and African Montiniaceae. Serv. Bot. ser. B: 906. Antananarivo.

Bremer B, RG Olmstead, L Struwe, and JA Sweere. 1994. *rbc*L sequences support exclusion of *Retzia*, *Desfontainia*, and *Nicodemia* from the Gentianales. Plant Syst. Evol. 190: 213–230.

Brizicky GK. 1961. A synopsis of the genus *Columellia* (Columelliaceae). J. Arnold Arbor. 42: 363–372.

Bridsen DM. 1975. A revision of the family Vahliaceae. Kew Bull. 30: 163–182.

Carlquist S. 1989. Wood anatomy and relationships of *Montinia*. Aliso 12: 369–378.

Dahlgren R and AE van Wyk. 1988. Structures and relationships of families endemic to or centered in southern Africa. Monogr. Syst. Bot. Missouri Bot. Gard. 25: 1–94.

Dahlgren R, SR Jensen, and BJ Nielsen. 1977. Seedling morphology and iridoid occurrence in *Montinia caryophyllaceae* (Montiniaceae). Bot. Not. 130: 329–332.

Fairbrothers DE. 1977. Perspectives in plant serotaxonomy. Ann. Missouri Bot. Gard. 64: 147–160.

Hibsch-Jetter C and TD Macfarlane. 1997. Phylogenetic analysis of *Eremosyne pectinata* (Saxifragaceae s.l.) based on *rbc*L sequence data. Plant Syst. Evol. 204: 225–232.

Hideux MJ and IK Ferguson. 1976. The stereostructure of the exine and its evolutionary significance in Saxifragaceae sensu lato. In: IK Ferguson and J Muller, eds. The evolutionary significance of the exine, pp. 327–377. Linn. Soc. Symposium, No. 1. London/New York.

Hils MH. 1985. Comparative anatomy and systematics of twelve woody Australian genera of the Saxifragaceae. Ph.D. dissertation, University of Florida, Gainesville.

Hoc R and L Bravo. 1984. Estudio palinologico sobre las especies presentes en Argentina de *Spigelia*, *Strychnos*, y *Desfontainia* (Loganiaceae). Kurtziana 17: 71–89.

Houghton PJ and LL Ming. 1985. Iridoids from *Desfontainia spinosa*. Phytochemistry 24: 1841–1842.

Huber H. 1963. Die Verwandtschaftsverhaltnisse der Rosifloren. Mitt. Bot. Staatssamml. Munchen 5: 48.

Kamelina OP. 1984. On the embryology of the genus *Escallonia*. Bot. Zhurn. 69: 1304–1316 (in Russian).
Kamelina OP. 1992. On the embryology of the genus *Ixerba* in relation to its systematic position. Bot. Zhurn. 77(12): 112–117 (in Russian with English summary).
Krach JE. 1976. Die Samen der Saxifragaceae. Bot. Jahrb. Syst. 97: 1–60.
Krach JE. 1977. Seed characters in and affinities among the Saxifragaceae. Plant Syst. Evol Suppl. 1: 141–153.
Leenhouts PW. 1980. Taxonomy. In: AJM Leeuwenberg, ed. Die natürlichen Pflanzenfamilien: Fam. Loganiaceae, 28b (I): 8–96. Berlin.
Leeuwenberg AJM. 1969. Notes on American Loganiaceae: IV. Revision of *Desfontainia* Ruiz et Pav. Acta Bot. Neerl. 18: 669–679.
Lundberg J. 2001a. Phylogenetic studies in the Euasterids II with particular reference to Asterales and Escalloniaceae. Acta Universitatis Upsaliensis, Uppsala.
Lundberg J. 2001b. Polyosmaceae. Chapter III. In: J Lundberg, ed. Phylogenetic studies in the Euasterids II with particular reference to Asterales and Escalloniaceae. Acta Universitatis Upsaliensis, Uppsala.
Lundberg J. 2001c. Escalloniaceae. Chapter IV. In: J Lundberg, ed. Phylogenetic studies in the Euasterids II with particular reference to Asterales and Escalloniaceae. Acta Universitatis Upsaliensis, Uppsala.
Lundberg J. 2001d. A well resolved and supported phylogeny of Euasterids II based on a Bayesian inference, with special emphasis on Escalloniaceae and other incertae sedis. Chapter V. In: J Lundberg, ed. Phylogenetic studies in the Euasterids II with particular reference to Asterales and Escalloniaceae. Acta Universitatis Upsaliensis, Uppsala.
Maldonado de Magnano S. 1986. Estudos embriologicos en *Desfontainia spinosa* (Desfontainiaceae). Darwiniana 27: 207–224.
Mennega AMW. 1980. Anatomy of the secondary xylem. In: AJM Leeuwenberg, ed. Die natürlichen Pflanzenfamilien: Fam. Loganiaceae, 28b (I): 112–161.
Milne-Redhead E. 1955. Montiniaceae. Hooker's Icones Plantarum 36: Tab. 3541–3544.
Nemirovich-Danchenko EN. 1996. Vahliaceae, Eremosynaceae. In: A Takhtajan, ed. Comparative seed anatomy, vol. 5, pp. 72–79. Nauka, St. Petersburg (in Russian).
Nemirovich-Danchenko EN. 2000b. Montiniaceae. In: A Takhtajan, ed. Comparative seed anatomy, vol. 6, pp. 257–262. Nauka, St. Petersburg (in Russian).
Nemirovich-Danchenko EN. 2000c. Columelliaceae. In: A Takhtajan, ed. Comparative seed anatomy, vol. 6, p. 267. Nauka, St. Petersburg (in Russian).
Nemirovich-Danchenko EN and TA Lobova. 2000a. Dulongiaceae, Tribelaceae. In: A Takhtajan, ed. Comparative seed anatomy, vol. 6, pp. 267–276. Nauka, St. Petersburg (in Russian).
Patel RN. 1973. Wood anatomy of the dicotyledons indigenous to New Zealand, part 2: Escalloniaceae. N. Z. J. Bot. 11: 421–434.
Punt W. 1980. Pollen morphology. In: AJM Leeuwenberg, ed. Die natürlichen Pflanzenfamilien: Fam. Loganiaceae, 28b (I): 162–191. Ducker & Humboldt, Berlin.
Raghavan TS and VK Srinivasan. 1942. A contribution to the life history of *Vahlia viscosa* Roxb. and *V. oldenlandioides* Rosb. Proc. Indian Acad. Sci. B 15: 83–105.
Ramamonjiarisoa BA. 1980. Comparative anatomy and systematics of African and Malagasy woody Saxifragaceae sensu lato. Ph.D. dissertation, University of Massachusetts.
Ronse Decraene LP, HP Linder, and EF Smets. 2000. The questionable relationship of *Montinia* (Montiniaceae): evidence from a floral ontogenetic and anatomical study. Am. J. Bot. 87: 1408–1424.
Schneider EL and S Carlquist 2004. Pit membrane remnants in perforation plates and other vessel details of Cornales. Brittonia 56: 275–283.
Schneider JV. Ixerbaceae. In: K Kubitzki, ed. The families and genera of vascular plants, vol. 9, pp. 205–207. Springer, Berlin/Heidelberg/New York.
Schultes RE. 1989. De speciebus varietatibusque *Desfontainia* Columbianae notae. Revista Acad. Colomb. Ci. Exact. 17: 313–319.
Sleumer H. 1968. Die Gattung *Escallonia* (Saxifragaceae). Verh. Kon. Ned. Akad. Wetensch. Afd. Natuurk., Tweede Sect. 58(2): 1–146.
Soltis DE and PS Soltis. 1997. Phylogenetic relationships in Saxifragaceae sensu lato: a comparison of topologies base on 18S rDNA and *rbc*L sequences. Am. J. Bot. 84: 504–522.
Soltis DE, PS Soltis, MT Glegg, and M Durbin. 1990. *rbc*L sequence divergence and phylogenetic relationships in Saxifragaceae sensu lato. Proc. Natl. Acad. Sci. USA 87: 4640–4644.
Stern WL. 1974. Comparative anatomy and systematics of woody Saxifragaceae: *Escallonia*. Bot. J. Linn. Soc. 68: 1–20.
Stern WL, GK Brizicky, and RH Eyde. 1969. Comparative anatomy and relationships of Columelliaceae. J. Arnold Arbor. 50: 36–75.
Swamy BGL. 1954. Morpho-taxonomical notes on the Escallonioideae: part A. Nodal and petiolar vasculature. J. Madras Univ. 24B: 299–306.
Takhtajan AL and VI Trifonova. 1999. Fruit and seed anatomy of the genus *Kaliphora* (Kaliphoraceae) in relation to its taxonomical position. Bot. Zhurn. 84(4): 1–7 (in Russian with English summary).
Weigend M. 2001 *Desfontainia* Ruiz et Pav. (Desfontainiaceae) revisited a first step back towards alpha-diversity. Bot. Jahrb. Syst. 123: 281–301.

Order 107. BRUNIALES

Shrubs or subshrubs, or rarely small trees, mostly with ericoid habit, with subepidermal core. The hairs are mostly long, slender and unicellular, with thick smooth walls and narrow lumina. Resinous deposits observed in cork, phloem, medullary rays and pith of *Audoinia* and *Lonchostoma*. Vessels with scalariform perforations that usually have numerous bars. Fibers with bordered pits. Axial parenchyma scanty paratracheal. Rays heterogeneous, mixed uniseriate and pluriseriate. Nodes unilacunar with one trace. Leaves evergreen, alternate, simple, densely set, mostly subtrigonous and acerose, often imbricate, entire, estipulate or with small, rudimentary stipules, water-storage tissue reported to occur at the apex of the leaf near the terminations of the veins. Stomata anomocytic. Flowers in dense spicate or capitate, terminal or axillary inflorescences, occasionally solitary, sometimes subtended by several imbricate bracts, mostly small, bisexual, actinomorphic. Sepals (4)5, free almost to the base, imbricate, persistent. Petals (4)5, sessile or unguiculate, free or connate into a variably long tube, imbricate, often persistent. Stamens as many as petals and alternating with them, often persistent; filaments free or sometimes adnate to the corolla tube, in *Lonchostoma* anthers subsessile and borne on the corolla tube, incurved in bud; anthers small or very small, tetrasporangiate, dorsifixed, often versatile, introrse, sometimes with a prolonged connective, opening longitudinally. Intrastaminal nectary disc sometimes present. Pollen 3-aperturate, or 6–11 aperturate, 3–10-colporate, tectate-columellate, almost psilate to rugulose and murate-reticulate. Gynoecium of two carpels, in *Mniothamnea* and *Berzelia* seemingly unicarpellate or at least unilocular, in *Audouinia* with three carpels, stylodia almost free to connate nearly to the top into a style with stylar branches standing close together; each stylodium or each stylar branch with capitate stigma; ovary usually semi-inferior or sometimes inferior, rarely almost superior, with (1)2–4(-12) ovules per carpel. Dependent on the superior or almost inferior position of the ovary, nectarostomata are scattered on the ovary flanks, or accumulated on the small hypgeneous part of the ovary (Quint and Classen-Bockhoff 2006b). Ovules pendulous from the summit of the septum or near the top of the central axis of the ovary. Ovules anatropous, epitropous, unitegmic, crassinucellate, with ventral raphe. Female gametophyte of *Polygonum*-type. Fruit dry, commonly crowned by the persistent calyx, often indehiscent and achenelike or nutlike, 1-seeded, or the 1–2-seeded carpels separating and opening along the ventral suture. Seeds with thin, exotestal seed coat, sometimes arillate, with minute, straight well differentiated embryo at the base of copious, fleshy endosperm. Present proanthocyanidins and myricetin. n = 10–11, 16 (*Staavia*), (21, 23).

Evidently belong to the Asteridae, probably related to the Desfontainiales. According to Albach et al. (2001) and Bremer et al. (2001, 2002), Bruniaceae are sister to the Neotropical families Collumelliaceae and Desfontainiaceae.

1. BRUNIACEAE

Berchtold et J. Presl 1820 (including Berzeliaceae Nakai 1943). 12/75. Southern coastal regions of South Africa, almost completely restricted to the Cape Province.

Classification after M. Quint and R. Classen-Bockhoff (2006a).

LINCONIEAE: *Linconia*; AUDOUINIEAE: *Audouinia, Thamnea, Tittmannia*; BRUNIEAE: *Raspalia, Nebelia, Staavia, Pseudobaeckea, Mniothamnea, Brunia, Lonchostoma, Berzelia*.

Bibliography

Carlquist S. 1978. Wood anatomy of Bruniaceae: correlations with ecology, phylogeny, and organography. Aliso 9: 323–364.

Carlquist S. 1990. Leaf anatomy of Geissolomataceae and Myrothamnaceae as a possible indicator of relationship to Bruniaceae. Bull. Torrey Bot. Club 117: 420–428.

Carlquist S. 1991. Leaf anatomy of Bruniaceae: ecological, systematic, and phylogenetic aspects. Bot. J. Linn. Soc. 107: 1–34.

Classen Bockhoff R. 2000. Inflorescences in Bruniaceae: with general comments on inflorescences in woody plants. Opera Bot. Belg. 12: 5–310.

Dahlgren R and AE van Wyk. 1988. Structures and relationships of families endemic to or centered in southern Africa. Monographs Syst. Bot. Missouri Bot. Card. 25: 1–94.

Goldblatt P. 1981. Chromosome cytology of Bruniaceae. Ann. Missouri Bot. Gard. 68: 546–550.

Hall AV. 1987. Evidence of a Cretaceous alliance for the Bruniaceae. South Afr. J. Sci. 83: 58–59.

Hall AV. 1988. Systematic palynology of the Bruniaceae. Bot. J. Linn. Soc. 96: 285–296.

Jay MM. 1968. Distribution des flavonoides chez les Bruniacees. Taxon 17: 484–488.

Leinfellner W. 1964a. Über die falsche Sympetalie bei *Lonchostoma* and anderen Gattungen der Bruniaceae. Oesterr. Bot. Z. 111: 345–353.

Leinfellner W. 1964b. Sind die Krnoblätter der Bruniaceae peltat gebaut? Oesterr. Bot. Z. 111: 500–526.

Pillans NS. 1947. A revision of Bruniaceae. South Afr. J. Bot. 8: 121–206.

Quint M. 2004. Evolution of Bruniaceae: evidence from molecular and morphological studies. Ph.D. dissertation, University of Mainz, Göttingen.

Quint M and R Classen-Bockhoff. 2006a. Phylogeny of Bruniaceae based on *mat*K and its sequence data. Int. J. Plant Sci. 167: 135–146.

Quint M and R Classen-Bockhoff. 2006b. Floral ontogeny, petal diversity and nectary uniformity in Bruniaceae. Bot. J. Linn. Soc. 150: 459–477.

Saxton WT. 1910. The ovule of the Bruniaceae. Trans. Roy. Soc. South Afr. 2: 27–31.
Scott G. 1999. A chemosystematic and cladistic study of the Southern African endemic family Bruniaceae. Ph.D. dissertation, University of Cape Town.
Weberling F. 1976. Weitere Untersuchungen zur Morphologic des Unterblattes bei den Dikotylen: IX. Saxifragaceae s. 1., Brunelliaceae, und Bruniaceae. Beitr. Biol. Pfl. 52: 163–181.

Order 108. LOASALES (HYDRANGEALES)

Trees, shrubs, sometimes lianas, subshrubs, or perennial or annual herbs. Indumentum consists of a wide variety of characteristic trichomes. Woods range from relatively soft to quite succulent. Vessels with scalariform, scalariform and simple, or (Loasaceae) basically with simple perforations, but altered versions of scalariform plates or plates with other aberrant configurations, lateral pitting various, sometimes laterally elongate. Fibers with large or small bordered pits. Rays usually heterogeneous to homogeneous, rarely almost wanting. Axial parenchyma basically diffuse or vasicentric, rarely absent. Sieve-element plastids are of S-type. Nodes mostly trilacunar. Leaves alternate or opposite, simple, entire or variously divided, estipulate or pseudostipulate. Stomata anomocytic or paracytic. Flowers in cymose or rarely racemose inflorescences, spicate or capitate, less often solitary, usually bisexual, actinomorphic. Sepals 4–5 (rarely 6–12), mostly connate into a tube, imbricate or valvate, persistent. Petals 4–5 (rarely 6–12), free or sometimes more or less connate, imbricate, contorted or valvate, Stamens (1)2 to several, or numerous (up to 300), free or basally connate into a short tube or fascicled or adnate to the corolla tube, anthers basifixed or dorsifixed, versatile, introrse or latrorse, tetrasporangiate, opening longitudinally. Tapetum secretory. Microsporogenesis simultaneous. Pollen grains 2-celled or 3-celled, 3-colpate or 3-colporate. Gynoecium of (2)3–5(-12) usually united carpels; stylodia free or shortly connate below; ovary mostly inferior or nearly inferior, plulocular to 1-locular, with one to several, or numerous, anatropous to hemitropous, pendulous or ascending ovules in each locule, or in each intruded parietal placenta. Ovules unitegmic, tenuinucellate, rarely (Loasaceae – Petalonychoideae and Gronovioideae) crassinucellate, with endothelium. Female gametophyte of *Polygonum*-type. Endosperm cellular, with micropylar and chalazal haustoria. Fruits septicidal or loculicidal capsules, often ribbed, or sometimes indehiscent. Seeds often minute, winged or not, exotestal, with straight or curved embryo; endosperm copious, fleshy and oily, or (Gronovioideae) endosperm scanty or wanting. Producing various iridoid compounds (Weigend et al. 2000; Rodriguez et al. 2002), proanthocyanidins, flavonols, and coffeic acids; n = 6–18 to 37.

The Loasales probably originated from the cunonialean stock. However, they are only distantly related to the Cunoniales (see Morgan and Soltis 1993).

Key to Families

1 Tanniniferous. Evergreen or deciduous, small trees or shrubs, sometimes lianas, subshrubs, or rhizomatous herbs. Indumentum of stellate hairs, long, unicellular trichomes or (*Broussaisia*) tufted trichomes. Often with raphid sacs in both leaf and axis. Vessels typically with scalariform perforations that have 20–90 thin bars, but sometimes with both scalariform and simple plates or rarely even all simple; lateral pitting scalariform to opposite or rarely alternate. Fibers typically with large or small bordered pits, but sometimes with simple pits, sometimes septate. Rays heterogeneous or less often homogeneous. Axial parenchyma absent, or more often diffuse and scanty vasicentric. Nodes trilacunar or sometimes multilacunar. Leaves generally opposite, entire or dentate, or sometimes lobed, glandular leaf teeth occur in *Decumaria, Deutzia,* and *Philadelphus,* estipulate. Stomata anomocytic or paracytic. Inflorescences cymose or corymbose, sometimes heads, or racemes by abortation, rarely flowers solitary. Flowers usually conspicuous, all equal or marginal ones sterile and with enlarged sepals, bisexual or rarely (*Broussaisia*) polygamodioecious. Sepals 4–5(-12), mostly connate into a tube, lobes valvate or imbricate. Petals 4–5(-12), free, valvate, imbricate or contorted. Stamens (1)2-several times as many as petals, sometimes as many as 50 or even (*Carpenteria*) up to 200; filaments free or slightly connate at base; anthers basifixed or dorsifixed, versatile, introrse or latrorse. Pollen grains 3-colpate (some species of *Hydrangea*) or 3-colporate. Gynoecium of (2)3–5(-12) carpels; stylodia free or shortly connate below, the branches often stigmatic for much of their length or seldom the style short, with distinct stigmas or a capitate, nearly lobed stigma; ovary inferior or less often

superior or semi-inferior, plurilocular to unilocular, with one to several or numerous, anatropous, pendulous or ascending ovules in each locule or on each intruded parietal placenta. Epigynous disk usually present. Fruits septicidal or more often loculicidal capsules or sometimes (*Dichroa, Broussaisia*) berries. Seeds winged or not; embryo large, straight, surrounded by fleshy endosperm, sometimes with aril (*Whipplea, Fendlerella* – Hufford 1995; Lobova 2000); exotestal cell signified. Present rare iridoid compounds (deutziosideas), seco-iridoids and loganin, proanthocyanidins (cyanidin or delphinidin), flavonols (kaempferol, myricetin, flavonols), n = 13–18.. . . . . 1. HYDRANGEACEAE.

1 Non-tanniniferous perennial or annual herbs, coarse, erect or twining, less often shrubs or very rarely (*Mentzelia arborescens*) small and succulent trees or (*Fuertesia*) woody lianas. Indumentum consists of a wide variety of characteristic trichomes, some of which are elongated, silicified, bristly stinging hairs filled with a yellowish irritant substance in the hairs themselves or the surrounding basal cells; the irritant substance is readily liberated owing to the brittleness of the hairs; glandular hairs with uniseriate stalks and uniseriate heads frequently contain cystoliths. Woods range from relatively soft to quite succulent. Vessels basically with simple perforations, but sometimes (in *Loasa picta* and *Mentzelia humilis*) altered versions of scalariform plates or plates with other aberrant configurations were observed by Carlquist (1984); lateral pitting various, sometimes laterally elongate or even scalariform-like. Fibers with bordered pits, but sometimes with vestigial borders on pits. Rays tall and wide, predominantly multiseriate, heterogeneous to homogeneous, rarely almost wanting. Axial parenchyma basically diffuse, rarely absent (*Presliophytum*), or vasicentric (*Plakothira frutescens*). Leaves alternate or opposite, simple, entire or variously divided, estipulate or pseudostipulate. Stomata anomocytic. Flowers in cymose or rarely racemose inflorescences, spicate or capitate, less often solitary, bisexual, actinomorphic. Sepals 4–5 (rarely 6–7), connate; calyx tube adnate to the ovary, often ribbed, the *ribs* sometimes spirally twisted; lobes imbricate or valvate, persistent. Petals 4–5 (rarely 6–7), inserted on the calyx, sessile or clawed, imbricate, contorted or valvate, free or sometimes more or less connate. Stamens commonly numerous (up to 300), less often 10 in two cycles or only five or only two fertile, free or basally connate into a short tube or fascicled or adnate to the corolla tube, sometimes with nearly sessile anthers on the corolla tube; some of the stamens often modified into petaloid, scalelike, or nectariferous staminodia; anthers basifixed. Pollen grains 3-colporate, tectate-columellate, with striate-reticulate, striate, or, more rarely, shallowly reticulate tectum. Gynoecium of 5–3, rarely of 6–7 united carpels, with simple style; ovary inferior or nearly inferior, 1-locular often with more or less deeply intruded parietal placentas, rarely fully partitioned, with 1-many ovules on each placenta; in Gronovioideae pseudomonomerous, with a solitary pendulous apical ovule. Ovules anatropous to hemitropous, unitegmic with a fairly long micropyle, tenuinucellate, with endothelium; in Gronovioideae and Petalonychoideae ovules crassinucellate, in many genera ovules with cup-shaped funicula-projections. Female gametophyte of *Polygonum*-type. Endosperm cellular, with aggressive micropylar and chalazal haustoria. Fruits septicidal or loculicidal capsules, often ribbed, or sometimes indehiscent. Seeds often minute, exotestal, with straight or curved embryo surrounded by copious, oily, thin-walled endosperm or (Gronovioideae) endosperm scanty or wanting. Producing quercetin, coffeic acid, and verious iridoid compounds (Weigend et al. 2000; Rodriguez et al. 2002), n = 6–15 to 37. . . . . . . 2. LOASACEAE.

1. HYDRANGEACEAE

Dumortier 1829 (including Hortensiaceae Berchtold et J. Presl 1820; Kirengeshomaceae Nakai 1943; Philadelphaceae Martynov 1820). 17/250. Subtropical and warm-temperate regions of the Northern Hemisphere, especially eastern Asia (mainly China) and North America; a few species in Southeast Asia and in the Andes, from Mexico to southern Chile.

1.1 PHILADELPHOIDEAE

Indumentum mostly of stellate hairs. Flowers in terminal cymes or racemes, rarely solitary. Filaments often with elongate lateral teeth at apex. – PHILADELPHEAE: *Carpenteria, Fendlerella, Fendlera, Jamesia, Philadelphus, Whipplea;* DEUTZIEAE: *Deutzia.*

1.2 HYDRANGEOIDEAE
Indumentum of simple hairs or absent. Flowers in cymose inflorescences, sometimes corymbose or capitate, or pseudoracemose by reduction. Filaments not toothed. – HYDRANGEAE: *Hydrangea, Platycrater, Decumaria, Pileostegia, Schizophragma;* DICHROEAE: *Dichroa, Broussaisia*; CARDIANDREAE: *Cardiandra, Deinanthe*.

1.3 KIRENGESHOMOIDEAE
Indumentum of 2-armed hairs. Leaves palmatifid. Flowers in tornately branched terminal and axillary paniculate cymes. Filaments not toothed. – *Kirengeshoma*.

The monotypic genus *Kirengeshoma* (southern Japan, Korea, temperate China) occupies a rather isolated position. Typically, it strongly differs from other members of the family in that its indumentum consists of 2-armed hairs; it has 15 stamens in three cycles, its three antipodal cells become binucleate and function as haustoria, and its seeds are winged and anatomically differ from those of the rest of Hydrangeaceae (Krach 1976). Besides, its pollen grain has a unique complete tectum associated with partial tectum (Hideux and Ferguson 1976). However, according to Hideux and Ferguson, a transitional state between a complete tectum and perforate tectum (a sparsely, perforated, unsculptured tectum) occurs in *Anopterus macleyanus* and *Polyosma*. Also, *Kirengeshoma* has many features in common with the tribe Cardiandreae, and serologically it is very close to *Hydrangea* (Grund and Jensen 1981).

2. LOASACEAE

A.L. de Jussieu 1804 (including Cevalliaceae Grisebach 1854; Gronoviaceae Endlicher 1841). 20/300. Temperate, subtropical, and tropical regions of America from southwestern Canada to Argentina and Chile, and West Indies, Galapagos Islands, with only two genera in the Old World – *Plakothira* (Marquesas) and *Fissenia* (*F. capensis* in southwestern Africa and *F. arabica* in Somalia, northeastern Ethiopia, and southern and southwestern parts of the Arabian peninsula).

Classification after M. Weigend (2004).

2.1 LOASOIDEAE
Leaves opposite below and alternate above or opposite throughout; stinging hairs sometimes present. Fertile stamens numerous, arising with centrifugal maturation, rarely fewer and opposite the petals. Staminodia absent. Ovules numerous to three, on parietal (sometimes subapical) placentas, n = 6. – LOASEAE: *Chichicaste, Huidobria, Presliophytum, Aosa, Blumenbachia, Loasa, Caiophora, Scyphanthus, Nasa*; KISSENIEAE: *Fissenia* (*Kissenia*); KLAPROTHIEAE: *Klaprothia, Plakothira, Xylopodia*.

2.2 MENTZELIOIDEAE
Leaves opposite below and alternate above, with scabrid-glochidiate trichomes, sometimes with stinging hairs. Stamens numerous, arising with centripetal maturation. Staminodia, if present, petaloid and free to base. Ovules three to numerous, n = 7. – *Mentzelia, Eucnide, Schismocarpus*.

2.3 GRONOVIOIDEAE
Leaves always alternate, always with glochidiate trichomes, sometimes with stinging hairs. Inflorescences terminal thyrsoids. Fertile stamens 4–5, alternating with the petals. Staminodia absent. Ovules solitary, pendulous from an apical placenta. – *Cevallia, Gronovia, Fuertesia*.

2.4 PETALONYCHOIDEAE
Leaves alternate with glochidiate hairs with short, spreading branches, without stinging hairs. Inflorescences terminal racemes. Fertile stamens reduced to two. – *Petalonyx*.

Related to the Hydrangeaceae, see Fan and Xiang (2003) and Weigend (2004), Soltis et al. (2006).

Bibliography

Agababian VS. 1961. On palynomorphology of the family Hydrangeaceae Dum. Izvestia Armenian Acad. Sci., Biol. Sci. 14 (II): 17–26 (in Russian).

Ackermann M and M Weigend. 2006. Nectar, floral morphology and pollination syndrome in Loasaceae subfamily Loasoideae (Cornales). Ann. Bot. 98: 503–514.

Avetisian EM. 1975. Palynomorphology of the family Loasaceae. In: Palynology, pp. 5–18. Armenian Academy of Science, Yerevan (in Russian).

Bensel CR and BF Palser. 1975. Floral anatomy in the Saxifragaceae sensu lato: III. Kirengeshomoideae, Hydrangeoideae, and Escallonioideae. Am. J. Bot. 62: 661–675.

Bohm BA, KW Nicholls, and UG Bhat. 1985 Flavonoids of the Hydrangeaceae Dumortier. Biochem. Syst. Ecol. 13: 441–445.

Brown DK. 1971. A study of floral morphology in the Loasaceae with emphasis on relationships among the subfamilies. Diss. Abstr. Int. B 32(6): 3208.

Brown DK and RB Kaul. 1981. Floral structure and mechanism in Loasaceae. Am. J. Bot. 68: 361–372.

Carlquist S. 1984. Wood anatomy of Loasaceae with relation to systematics, habit, and ecology. Aliso 10: 583–602.

Carlquist S. 1987. Wood anatomy of *Plakothira* (Loasaceae). Aliso 11: 563–569.

Carlquist S and EL Schneider. 2004. Pit membrane remnants in perforation plates of Hydrangeales with comments on pit membranes remnant occurrence, physiological significance and phylogenetic distribution in Dicotyledons. Bot. J. Linn. Soc. 146: 41–51.

Christy CM. 1998. Loasaceae stickleaf or blazing-star family. J. Arizona-Nevada Acad. Sci. 30(2): 96–111.

Darlington J. 1934. A monograph of *Mentzelia*. Ann. Missouri Bot. Gard. 21: 103–226.

Davis WS and HJ Thompson. 1967. A revision of *Petalonyx* (Loasaceae) with a consideration of affinities of subfamily Gronovioideae. Madroño 19: 1–18.

Ernst WR and HJ Thompson. 1963. The Loasaceae in the southeastern United States. J. Arnold Arbor. 44: 138–142.

Fan C and Q-Y Xiang. 2003. Phylogenetic analyses of Cornales based on 26S rDNA and combined 26S rDNA-*matK-rbcL* sequence data. Am. J. Bot. 90: 1263–1270.

Florence J. 1985. Sertum polynesicum. I. *Plakothira* Florence (Loasaceae), genre nouveau des Îles Marquises. Adansonia 3: 239–245.

Garcia V. 1962a. Embryological studies of the Loasaceae with special reference to the endosperm haustoria. In: Plant embryology: a symposium, pp. 157–161. New Delhi.

Garcia V. 1962b. Embryological studies in the Loasaceae: development of endosperm in *Blumenbachia hieronymi* Urb. Phytomorphol. 12: 307–312.

Grau J. 1997. *Huidobria,* ein isolierte Gattung der Loasaceae aus Chile. Sendtnera 4: 77–93.

Gregory M. 1998. Hydrangeaceae. In: DF Cutler and M Gregory, eds. Anatomy of dicotyledons, 2nd ed. Saxifragales, vol. 4, pp. 87–108. Clarendon, Oxford.

Hao G and C Hu. 1996a. A study of leaf venation of Hydrangeoideae (Hydrangeaceae). Guihaia 16: 155–160.

Hao G and C Hu. 1996b. A study of pollen morphology of Hydrangeoideae (Hydrangeaceae). J. Trop. Subtrop. Bot. 4: 26–31.

Hempel AL, PA Reeves, RG Olmstead, and RK Jansen. 1995. Implications of *rbc*L sequence data for higher order relationships of the Loasaceae and the anomalous aquatic plant *Hydrostachys* (Hydrostachyaceae). Plant Syst. Evol. 194: 25–37.

Hill RJ. 1976. Taxonomic and phylogenetic significance of seed coat microsculpturing in *Mentzelia* (Loasaceae) in Wyoming and adjacent western States. Brittonia 28: 86–112.

Hufford LD. 1988a. Systematics of Loasaceae: a consideration of familial circumscription and a cladistic analysis of generic relationships. Am. J. Bot. 75: 181.

Hufford LD. 1988b. Seed morphology of *Eucnide* and other Loasaceae. Syst. Bot. 13: 154–167.

Hufford LD. 1989. The structure and potential loasaceous affinities of *Schismocarpus*. Nord. J. Bot. 9: 217–227.

Hufford L. 1990. Androecial development and the problem of monophyly of Loasaceae. Canad. J. Bot. 68: 402–419.

Hufford LD. 1992. Rosidae and their relationships to other nonmagnoliid dicotyledons: a phylogenetic analysis using morphological and chemical data. Ann. Missouri Bot. Gard. 79: 218–248.

Hufford LD. 1995. Seed morphology of Hydrangeaceae and its phylogenetic implication. Int. J. Plant Sci. 156: 555–580.

Hufford LD. 1997. A phylogenetic analysis of Hydrangeaceae based on morphological data. Int. J. Plant Sci. 158: 652–672.

Hufford LD. 1998. Early development of androecia in polystemonous Hydrangeaceae. Am. J. Bot. 85: 1057–1067.

Hufford LD. 2001. Ontogeny and morphology of the fertile flowers of *Hydrangea* and allied genera of tribe Hydrangeeae (Hydrangeaceae). Bot. J. Linn. Soc. 137: 139–187.

Hufford L. 2003. Homology and developmental transformation: models for the origins of the staminodes of Loasaceae subfamily Loasoideae. Int. J. Plant Sci. 164(5 Suppl.): 409–439.

Hufford LD. 2004. Hydrangeaceae. In: K Kubitzki, ed. The families and genera of vascular plants, vol. 6, pp. 202–215. Springer, Berlin/Heidelberg/New York.

Hufford L, ML Moody, and DE Soltis. 2001. A phylogenetic analysis of Hydrangeaceae based on sequences of the plastid gene *mat*K and their combination with *rbc*L and morphological data. Int. J. Plant Sci. 162: 835–846.

Hufford L, MM McMahon, AM Sherwood, G Reeves, and MW Chase. 2003. The major clades of Loasaceae: phylogenetic analysis using the plastid *mat*K and *trn*L –*trn*F regions. Am. J. Bot. 90: 1215–1228.

Hutchinson J. 1927. Contributions towards a phylogenetic classification of flowering plants. VI. A. The genera of Hydrangeaceae. Kew Bull. 1927: 100–107.

Hwang S-M. 1987. A preliminary study on *Dichroa* Lour. Acta Phytotax. Sinica 25: 384–389.

Klopfer K. 1971. Beiträge zur floralen Morphogenese und Histogenese der Saxifragaceae 6. Die Hydrangeoideen. Wissenschaft. Zeitschr. Pädagog. Hochschule Potsdam 15: 77–95.

Kooimann P. 1974. Iridoid glycosides in the Loasaceae and the taxonomic position of the family. Acta Bot. Neerl. 23: 677–679.

Leins P and W Winhard. 1973. Entwicklungsgeschichtliche Studien an Loasaceen-Blüten. Oesterr. Bot. Z. 121/122: 145–165.

McClintock E. 1957. A monograph of the genus *Hydrangea*. Proc. California Acad. Sci. 29: 147–256.

Moody ML and L Hufford. 2000. Floral ontogeny and morphology in *Cevallia, Fuertesia*, and *Gronovia* (Loasaceae subfamily Gronovioideae). Int. J. Plant Sci. 161: 869–883.

Moody ML, L Hufford, DE Soltis, and PS Soltis. 2001. Phylogenetic relationships of Loasaceae subfamily Gronovioideae inferred from *mat*K and ITS sequence data. Am. J. Bot. 88: 326–336.

Müller AA, JK Kufer, KG Dietl, and M Weigend. 1998. A dimeric iridoid from *Loasa acerifolia*. Phytochemistry 49: 1705–1707.

Müller AA, JK Kufer, KG Dietl, and M Weigend. 1999. Iridoid glucosides – chemotaxonomic markers in Loasoideae. Phytochemistry 52: 67–78.

Nemirovich-Danchenko EN, and TA Lobova. 1998. The seed coat structure in some representatives of the order Hydrangeales. Bot. Zhurn. 83: 1–9 (in Russian with English summary).

Poston ME and JW Nowicke. 1990. Pollen ultrastructure of Loasoideae (Loasaceae). Am. J. Bot. 77: 151 (abstract).

Poston ME and JW Nowicke 1993. Pollen morphology, trichome types, and relationships of the Gronovioideae. Am. J. Bot. 80: 689–704.
Poston ME and HJ Thompson. 1977. Cytotaxonomic observations in Loasaceae subfamily Loasoideae. Syst. Bot. 2: 28–42.
Quibell CF. 1972. Comparative and systematic anatomy of Carpenterieae (Philadelphiaceae). Ph.D. dissertation. University of California, Berkeley.
Rodriguez V, J Schripsema, and SR Jensen. 2002. An iridoid glucoside from *Gronovia scandens* (Loasaceae). Biochem. Syst. Ecol. 30: 243–247.
Roels P, LP Ronse Decraene, and E Smets. 1997. A floral ontogenetic investigation of the Hydrangeaceae. Nord. J. Bot. 17: 235–254.
Soltis DE, Q-Y Xiang, and L Hufford. 1995. Relationships and evolution of Hydrangeaceae based on *rbc*L sequence. Am. J. Bot. 82: 504–514.
Stern WL. 1978. Comparative anatomy and systematics of woody Saxifragaceae: *Hydrangea*. Bot. J. Linn. Soc. 76: 83–113.
Styer CH and WL Stern. 1979. Comparative anatomy and systematics of woody Saxifragaceae: *Philadelphus, Deutzia*. Bot. J. Linn. Soc. 79: 267–289, 291–319.
Thompson HJ and WR Ernst. 1967. Floral biology and systematics of *Eucnide* (Loasaceae). J. Arnold Arbor. 48: 56–88.
Weigend M. 1996. A revision of the Loasaceae of Ecuador. Bot. Jahrb. Syst. 118: 229–294.
Weigend M. 1997. Some aspects of the biogeography, morphology and systematics of Loasoideae in northern South America. In: R Valencia and H Balslev, eds. Estudios Sobre Diversidad y Ecologia de Plants, pp. 37–50. PUCE, Quito.
Weigend M. 2004. Loasaceae. In: K Kubitzki, ed. The families and genera of vascular plants, vol. 6, pp. 239–254. Springer, Berlin/Heidelberg/New York.
Weigend M, J Kufer, and AA Müller. 2000. Phytochemistry and the systematics and ecology of Loasaceae and Gronoviaceae (Loasales). Am. J. Bot. 87: 1202–1210.

Order 109. CORNALES

Trees or more often shrubs, sometimes ericoid shrubs (Grubbiaceae), rarely rhizomatous herbs. Hairs simple, unicellular. Vessels mostly with scalariform perforations usually with many bars; lateral pitting scalariform or opposite. Fibers typically with large, often conspicuous bordered pits. Rays markedly heterogeneous. Axial parenchyma typically apotracheal. Sieve-element plastids of Ss-, or less often So-, and Pcs-types. Nodes trilacunar. Leaves alternate or more often opposite, simple, mostly entire, estipulate. Stomata usually anomocytic. Flowers small, in various types of cymose inflorescences, usually bisexual, actinomorphic or very rarely slightly zygomorphic, 4-merous or less often 5-merous. Calyx adnate to ovary, 4–10 lobes or teeth or obsolete. Petals free or rarely basally connate, imbricate or valvate, reduced or wanting in female flowers. Stamens as many as and alternate with petals or less often twice or even three or four times as many as the petals, mostly attached to or around the edge of epigynous nectary disc; filaments free; anthers tetrasporangiate, opening longitudinally. Tapetum secretory. Microsporogenesis simultaneous. Pollen grains 2-celled, 3-colporate or of 3-colporate-derived type. Nectary disc usually present. Gynoecium of 2–9 (mostly two) united carpels; stylodia more or less free or united into a style with a lobed or capitate stigma; ovary inferior, with one pendulous ovule per locule. Ovules anatropous or rarely (some species of *Cornus*) hemitropous, apotropous, or epitropous, unitegmic, crassinucellate or (Nyssaceae and some species of *Cornus*) tenuinucellate, often with endothelium. Female gametophyte usually of *Polygonum*-type, rarely (some species of *Cornus*) tetrasporic and then usually of *Fritillaria*-type. Endosperm cellular (Davidiaceae, *Nyssa aquatica, Alangium chinense,* Cornaceae, Grubbiaceae) or nuclear (Nyssaceae, *Alangium lamarckii*). Fruits drupes. Seeds small to medium-sized, exotestal, with small to rather large embryo; endosperm usually copious. Usually contain iridoid compounds, mostly cornin, and gallic and ellagic acids (and corresponding tannins).

Cornales are so close to the Loasales, that the boundary between them is almost conventional. The most archaic cornalean family is Davidiaceae.

Key to Families

1 Gynoecium of (6)7(-10) carpels. Nectary disc absent. Deciduous trees. Vessel elements very long, with very slanting end walls and scalariform perforations with 60–100 thin bars; lateral pitting opposite or less often transitional. Fibers with distinct bordered pits in both radial and tangential walls. Axial parenchyma diffuse(-in-aggregates). Rays heterogeneous with long ends. Crystals absent. Leaves alternate, cordate at base, acuminate, serrate-dentate, pinnately veined with 3–7 subbasal veins, usually silky-pubescent below when young. Flowers andromonoecious, apetalous, in terminal globose heads subtended by two (rarely three) large, creamy-white, subopposite spreading bracts; male flowers numerous, scarcely distinct in dense head, bisexual flower solitary in the head, laterally placed

obliquely above the middle of the axis. Male flowers completely devoid of perianth and consist of (1-)5–6(-12) stamens with slender, elongate filaments and small introrse anthers. Pollen grains 3-colporate. Bisexual flowers: calyx adnate to the ovary and with much reduced limb; stamens 15–26, mostly 21, arranged around the base of style; style short, thick, conical, with 6–10 spreading stigmas; stigmas sulcate and papillose on their adaxial side; ovary 6–10-locular, with one ovule per carpel. Endothelium present. Fruits drupaceous, with granular mesocarp and bony longitudinally sulcate (1–2) 3–5-seeded endocarp. Seeds with large embryo subequal to fleshy, oily endosperm. Contain gallic and ellagic acids (and corresponding tannins), n = 20, 21. . . . . 1. DAVIDIACEAE.

1 Gynoecium of two or less often (Curtisiaceae) four carpels. Nectary disc present. Petals mostly present.

2 Petals (when present) imbricate, small. Trees and shrubs. Hairs simple, unicellular, sometimes with granular surface, together with variously shaped unicellular glands on leaves. Vessels with scalariform or scalariform-reticulate perforations that have numerous (17–50) thin bars; lateral pitting opposite or scalariform. Fibers with distinctly bordered pits in both radial and tangential walls, occasionally (in some species of *Nyssa*) septate, or with dark purple deposits. Rays heterogeneous, with short ends. Axial parenchyma diffuse, but in *Nyssa* occasionally scanty paratracheal. Prismatic crystals present (Noshiro et Baas 1998). Leaves alternate, entire or denticulate, pinnately veined. Flowers small, greenish-white, polygamous, polygamo-dioecious, or dioecious, in terminal panicles of globose heads (*Camptotheca*) or (*Nyssa*) male flowers crowded in axillary, short racemes, umbels, or heads, bisexual and female flowers in few-flowered heads or solitary by reduction. Calyx represented by a minute, 5-dentate or irregularly toothed rim, or virtually obsolete. Petals five or rarely more, small, imbricate, in female flowers sometimes wanting. Stamens in male flowers (5-) 10(16), often 2-seriate; filaments elongate, subulate; anthers small, introrse or latrorse. Pollen grains 3-colporate. Nectary disc in male flowers fleshy, pulvinate. Rudimentary gynoecium wanting, or less often there is a very small subulate rudiment in the middle of the disc. Stamens in bisexual flowers as many and alternate with the petals or wanting; anthers fertile or sterile. Gynoecium usually pseudomonomerous; 1-locular, with one pendulous ovule, but in *Nyssa* there sometimes occurs a gynoecium with a 2-locular ovary and two free stylodia; ovary is crowned at the top with broad pulvinate or conical nectary disc in the middle of which stands a recurved stylodium of the fertile carpel with decurrent stigma on its inner side (*Nyssa*) or a short, 2-fid style (*Camptotheca*). Endothelium absent. Fruits drupaceous (*Nyssa*) or shortly subsamaroid (*Camptotheca*). Seeds with short, straight embryo and copious endosperm; radicle short. Contain gallic and ellagic acids, alkaloid camptothecin (*Camptotheca*), and corresponding tannins, n = 21 (*Camptotheca*), 22 (*Nyssa*). . . 2. NYSSACEAE.

2 Petals valvate.

3 Leaves strongly serrate, more or less flat. Evergreen trees. Branchlets, petioles, and inflorescences with stellate hairs. Resin ducts absent. Wood diffuse-porous. Vessels with scalariform perforations that have 26–39 bars; lateral pitting opposite, without vestures. Fibers with distinctly bordered pits. Rays heterogeneous. Axial parenchyma diffuse to diffuse-in-aggregates. Prismatic crystals common in ray cells (Noshiro et Baas 1998). Leaves opposite, coarsely dentate, coriaceous, shiny above, more or less ferrugineo-tomentose below. Flowers minute, in many-flowered, terminal trichotomously branched tomentose thyrses, bisexual or rarely (*Afrocrania*) dioecious, bracteolate. Calyx tube turbinate, 4-lobed, lobes open. Petals four, triangular-ovate. Stamens four, alternipetalous; filaments subulate; anthers cordate at the base, introrse. Pollen grains small, 3-colporate, tectum complete with pseudoperforations, or H-shaped. Intrastaminal disc broad, 4-angled, densely barbate. Ovary inferior, 4-locular, ovule one per locule. Style short, subconical, glabrous, with 4-lobed stigma; Fruits small, subglobose, 4-locular, 4-seeded drupes arcolate at the apex, crowned by the calyx. Seeds elongate-oblong, subcylindric, with elongate embryo in copious endosperm. Contain tannins, proanthocyanins, ellagitannins and iridoids, n = 13. . . . . .4. CURTISIACEAE.

3 Leaves entire.

4 Inflorescences capitate or cone-like, compound of dichasia. Evergreen ericoid shrubs; hairs

unicellular. Leaves opposite-decussate, estipulate. Flowers in small axillary 3(2)-flowered dichasia or in many-flowered conelike compound dichasia, the ovaries coherent or connate. Flowers subtended by bracts and provided with two prophylls. Sepals four, free, valvate, hairy on outer surface. Petals absent. Stamens eight, half of them alternate with the sepals, and the other longer ones are opposite and slightly adnate to their bases; anthers medifixed or basifixed, inverted, opening longitudinally along the lateral C-like line; peripheral microsporangia nonfunctional, vestigial. Pollen grains 3-colporate, psilate. Disc epigynous, papillate to shortly hairy. Gynoecium of two carpels; style unbranched, with two short apical lobes; ovary inferior, at first 2-locular, but later becoming incompletely 1-locular, with two ovules pendulous from the summit of the septum. At the cotyledonary stage the massive cellular endosperm shows chalazal and micropylar haustoria (Fagerlind 1947). Fruits 1-seeded drupes that connate forming a dense frutescence. Seeds with thin seed coat and large, straight, cylindrical embryo surrounded by the copious, oily endosperm. Contain tannins but there are negative reports of flavonoids and iridoids (Kubitzki 2004). . . . . . 5. GRUBBIACEAE.

4 Inflorescences thyrso-paniculate.

5 Ovules epitropous (with a ventral raphe and micropyle directed upwards). Evergreen trees. Resin ducts with secretory epithelium present in vegetative organs and in fruits. Hairs unicellular, two-armed. Wood diffuse-porous. Vessels with scalariform perforations that have numerous bars (34–66); lateral pitting opposite-scalariform, without vestures. Fibers with distinctly bordered pits in both radial and tangential walls. Rays heterogeneous with long ends. Axial parenchyma apotracheal in narrow bands. Crystals absent (*Diplopanax*) or (*Mastixia*) prismatic crystals occasionally present in body ray cells (Noshiro et Baas 1998). Leaves alternate or opposite, entire, coriaceous, pinnately veined. Flowers small, usually in terminal many-flowered compound dichasia, bisexual; pedicels 2-bracteolate, articulated. Calyx campanulate, 4–5-dentate, open. Petals 4–5, coriaceous, inflexed at the apex and fimbriate or 2-dentate. Stamens 4–5, alternipetalous, inserted at the base of fleshy, intrastaminal tetragonous or pentagonous nectary disc; filaments short; anthers introrse. Pollen grains 3-colporate, sometimes 4-colporate, with complete or partially perforate tectum. Gynoecium pseudomonomerous; style short, thick, with punctiform or rarely 2-lobed stigma; ovary 1-locular. Fruit drupes with grooved endocarp. Seeds with small embryo in copious endosperm. Contain gallic and ellagic acids (and corresponding tannins), unidentified iridoids reported from leaf tissue (Bate-Smith et al. 1975), n = 11, 13. . . . . 3. MASTIXIACEAE.

5 Ovules apotropous (with a dorsal raphe and micropyle directed downward).

6 Leaves opposite or rarely (in *Swida*) alternate. Evergreen or deciduous trees and shrubs, rarely rhizomatous herbs with annual stems, provided with characteristic two-armed hairs. Plants without laticifers. Vessels with scalariform perforations that usually have 14–71 bars. Fibers with distinctly bordered pits in both radial and tangential walls, without vestures. Rays heterogeneous. Axial parenchyma usually diffuse or diffuse-in-aggregates. Crystals rare: prismatic crystals in *Cornus peruviana*, together with druses in some species (Noshiro et Baas 1998). Leaves entire. Flowers small, in terminal, cymose inflorescences, paniculate, corymbose umbellate or capitate (the latter sometimes subtended by a whorl of large petaloid bracts), bisexual or rarely (*Afrocrania*) dioecious. Calyx with four small teeth around the summit of the ovary or entire. Petals four. Stamens four, alternipetalous, around the edge of pulvinate, intrastaminal nectary disc; anthers oblong or ellipsoid, dorsifixed. Pollen grains 3-colporate, with H-shaped columellae or tectum complete with short supratectal spines, of which at least some are prolongations of the columellae, sometimes perforate. Style filiform to columnar, with capitate or truncate stigma; ovary 2-locular, with one ovule per locule. Fruits drupes, areolate at apex, with bony or crustaceous, 2-locular endocarp. Seeds oblong, compressed, with membranous testa, elongate embryo and fleshy endosperm; seed-coat thin-walled and crushed. With iridoid compounds but without isoquinoline alkaloids and petroselinic acid. Contain gallic and ellagic acids (and

corresponding tannins), n = 9–11, or (*C. canadensis*) 22. . . . . .6. CORNACEAE.

6 Leaves alternate. Small trees or less often shrubs or woody lianas, sometimes thorny. Hairs somewhat diverse, sometimes stellate and glandular; the Madagascan species *Alangium grisolleoides* has 2-armed hairs. Articulate laticifers present in various tissues of the shoot. Vessels mostly with simple perforations, but in section *Constigma* they are scalariform with 7–16 bars; lateral pitting alternate and dense. Fibers mostly with minutely bordered pits, or with simple pits and slit-like abertures in section *Constigma*. Rays heterogeneous with long ends. Axial parenchyma diffuse or diffuse-in-aggregates. Prismatic crystals usually conspicuous in both ray cells and axial parenchyma (Noshiro et Baas 1998). Leaves entire or lobed, pinnately or palmately veined, often some of the cells of the mesophyll containing cluster crystals and appearing as translucent dots. Flowers in axillary cymes, bisexual or rarely unisexual (*A.grisolleoides* dioecious). Calyx with 4–10 lobes or teeth or truncate. Petals 4–10, linear or strap-shaped, valvate or rarely contorted, hairy within, sometimes connate at base, finally recurved or revolute. Stamens 4–40, arranged in a single cycle around a conspicious, pulvinate, or sometimes almost hemispherical nectary disc; filaments free or slightly connate at base, sometimes adnate to the petals at base, more or less barbate or villose on inner side; anthers more or less linear, introrse, basifixed or more rarely dorsifixed and versatile and widely sagittate. Pollen grains 3–4(-8)-colporate or porate. Style elongate, filiform with clavate, 2–4-lobed or more or less capitate stigma or rarely style short and with two elongate, ventrally stigmatic branches; ovary 2-locular or more often 1-locular. Endothelium present. Fruits drupes crowned by calyx and disc and with crustaceous or woody endocarp, 1-locular or less often 2-locular, each locule with one seed or one locule empty. Seeds with large, straight embryo, foliaceous cotyledons, and elongate hypocotyl; endosperm copious, oily, and also with reserves of hemicellulose; seed-coat thin-walled. Without iridoid compounds but commonly producing isoquinoline (emetine) alkaloids (alangiside and tubulosin), n = 8, 9, 11. . . . .7. ALANGIACEAE.

1. DAVIDIACEAE

H.L. Li 1955. 1/1. Southwestern and central China.

Davidia

In spite of very specialized inflorescences and reduced perianth, *Davidia* is characterized by very primitive wood structure and multicarpellate gynoecium. We may therefore agree with Eyde (1967: 167) that «with respect to these characters *Davidia* is the nearest living thing to the common ancestor of *Cornus* and *Nyssa*». Usually *Davidia* is included in Nyssaceae, but from *Nyssa* and *Camptotheca* it differs not only in morphology of inflorescences and flowers (including the absence of the nectary disc), crassinucellate ovules with endothelium and some other embryological characters (Kamelina and Shevchenko 1988) but also palynologically (Eramian 1971), chemically (Bate-Smith et al. 1975), and serologically (Fairbrothers 1977).

2. NYSSACEAE

A.L. de Jussieu ex Dumortier 1829. 2/9. Eastern Himalayas, Assam (Khasi Hills), continental China, Southeast Asia, eastern North America. *Camptotheca* (1) is endemic to eastern Tibet and southwestern and central China.

Nyssa, Camptotheca

Related to the Davidiaceae but in some respects more advanced.

3. MASTIXIACEAE

Calestani 1905. 2/c.20. Sri Lanka, southwestern and northeastern India, eastern Himalayas, northern Burma, southwestern and southern China, Hainan Indochina, Andaman Islands, Malesia, New Guinea New Britain, Solomon Islands. *Diplopanax* (2) is distributed in China (Guangdong, Guangxi, Guizhou and Yunnan) and northern Vietnam.

Diplopanax, Mastixia

Differ from the Nyssaceae and Cornaceae and related families in chemical features (Bate-Smith et al.

1975), presence of secretory canals, presence of fibers in the secondary phloem, complete absence of collenchyma in primary cortex, presence of cortical bundles, and pollen grain morphology. Mastixiaceae, especially the genus *Diplopanax,* have some features in common with the Araliaceae, including epitropous ovules and secretory canals. According to Eyde (1988), the *Nyssa*-like ribbing of fossil *Mastixiopsis* makes it a likely common ancestor of nyssoids and mastixioids.

4. CURTISIACEAE

Takhtajan 1987. 1/1–2. From the Cape peninsula eastward to Natal and from there to the north through the Transvaal into the mountains of eastern Zimbabwe.

Curtisia

Curtisia is usually included in Cornaceae, but it markedly differs from them in pollen morphology (Eramian 1971; Ferguson 1977; Ferguson and Hideux 1978), epitropous ovules, 4-locular ovary and fruit, and pyrenes that have vascular bundles running up the middle («no such bundles occur among the living members of the true Cornaceae» states Eude [1988: 309]). On the basis of chromosome number n = 13, Goldblatt (1978) suggests a common ancestry for *Curtisia* and *Mastixia.*

5. GRUBBIACEAE

Endlicher 1839. 1/3 (including Ophiraceae Arnott 1841). Cape Province and ranges from the Cape Peninsula to the Uniondale District in the east and the Clanwilliam District in the northwest.

Grubbia

Rather closely related to Curtisiaceae (Xiang 1999; Xiang et al. 2002; Kubitzki 2004).

6. CORNACEAE

Dumortier 1829. 4/55–60. Northern temperate regions and the Arctic, southern China, Central America, Peru, Bolivia, and tropical East Africa.

Swida (including *Bothrocaryum*) *Cornus* (including *Chamaepericlymenum*), *Afrocrania, Cynoxylon* (including *Dendrobenthamia*).

The family Cornaceae as here defined is a very natural unit. Morphologically, cytologically, and serologically Cornaceae are nearest to Nyssaceae and Mastixiaceae. According to Eyde (1988: 317), "*Cornus, Mastixia, Nyssa, Camptotheca,* and *Davidia* are tied together by 11 basic chromosomes, by pollen similarities, by shared features of the flower – epigynous disc, transseptal stands – and by fruit traits that square with origins inferred from fossils."

7. ALANGIACEAE

A.P. de Candolle 1828. 1/20. Tropical Africa, Madagascar, eastern and tropical Asia, New Guinea and adjacent islands, eastern Australia, New Caledonia, Fiji.

Alangium

According to Eyde (1988), *Alangium* has flowers, fruits, and ovules much like those of *Cornus.* It also has the same basic chromosome complement. It resembles Cornaceae also in placentation of apotropous ovules. However, there are also important differences, including articulated laticifers, habit, leaf architecture, long, linear petals, pollen morphology, absence of iridoid substances, and presence of emetine alkaloids.

Bibliography

Adams JE. 1949. Studies in the comparative anatomy of the Cornaceae. J. Elisha Mitchell Sci. Soc. 65: 218–244.

Averyanov LV and TH Nguyen. 2002. *Diplopanax vietnamensis*, a new species of Nyssaceae from Vietnam – one more living representative of the Tertiary flora of Eurasia. Novon 12: 433–436.

Bate-Smith EC, IK Ferguson, K Hutson, SR Jensen, BJ Nielsen, and T Swain. 1975. Phytochemical interrelationships in the Cornaceae. Biochem. Syst. Ecol. 3: 79–89.

Bloembergen S. 1939. A revision of the genus *Alangium*. Bull. Jard. Bot. Buitenzorg III. 6: 139–235.

Brunner F and DE Fairbrothers. 1978. A comparative serological investigation within the Cornales. Serol. Mus. Bull. 53: 2–5.

Burckhalter RE. 1992. The genus *Nyssa* (Cornaceae) in North America – a revision. SIDA 15: 323–342.

Carlquist S. 1977a. A revision of Grubbiaceae. South Afr. J. Bot. 43: 115–128.

Carlquist S. 1977b. Wood anatomy of Grubbiaceae. South Afr. J. Bot. 43: 129–144.

Carlquist S. 1978. Vegetative anatomy and systematics of Grubbiaceae. Bot. Not. 131: 117–126.

Chao C-Y. 1954. Comparative pollen morphology of the Cornaceae and allies. Taiwania 5: 93–101.

Chopra RN and H Kaur. 1965. Some aspects of the embryology of *Cornus*. Phytomorphology 15: 353–359.

Eramian EN. 1971. Palynological data on the systematics and phylogeny of Cornaceae Dumort. and related families. In: LA Kuprianova and MS Yakovlev, eds. Pollen morphology of Cucurbitaceae, Thymelaeaceae, Cornaceae, pp. 235–273. Nauka, Leningrad (in Russian).

Eyde RH. 1966. The Nyssaceae in the southeastern United States. J. Arnold Arbor. 47: 117–125.

Eyde RH. 1967 (1968). The peculiar gynoecial vasculature of Cornaceae and its systematic significance. Phytomorphology 17: 172–182.

Eyde RH. 1968. Flowers, fruits, and phylogeny of Alangiaceae. J. Arnold Arbor. 49: 167–192.

Eyde RH. 1972. Pollen of *Alangium:* toward a more satisfactory synthesis. Taxon 21: 471–477.

Eyde RH. 1987. The case for keeping *Cornus* in the broad Linnean sense. Syst. Bot. 12: 505–518.

Eyde RH. 1988. Comprehending *Cornus:* puzzles and progress in the systematics of dogwoods. Bot. Rev. 54: 233–351.

Eyde RH and ES Barghoorn. 1963. Morphological and paleobotanical studies of the Nyssaceae: II. The fossil record. J. Arnold Arbor. 44: 328–376.

Eyde RH, A Bartlett, and ES Barghoorn. 1969. Fossil record of *Alangium*. Bull. Torrey Bot. Club 96: 288–314.

Eyde RH and Xiang Qiuyun. 1990. Fossil mastixioid (Cornaceae) alive in eastern Asia. Am. J. Bot. 77: 689–692.

Fagerlind F. 1947. Die systematische Stellung der Familie Grubbiaceae. Svensk Bot. Tidskr. 41: 315–320.

Fairbrothers DE. 1977. Perspectives in plant serotaxonomy. Ann. Missouri Bot. Gard. 64: 147–160.

Fan C and Q-Y Xiang. 2001a. Phylogenetic relationships within *Cornus* (Cornaceae) based on 26S rDNA sequences. Am. J. Bot. 88: 1131–1138.

Fan C and Q-Y Xiang. 2001b. Systematic affinities of two enigmatic African families, Grubbiaceae and Hydrostachyaceae – evidence from nuclear 26S rDNA sequence data. In: Botany 2001: Plants and People. Abstracts, p. 112. Albuquerque.

Fan C and Q-Y Xiang. 2003. Phylogenetic analyses of Cornales based on 26S rRNA and combined 26S rDNA-*mat*K-*rbc*L sequence data. Am. J. Bot. 90: 1357–1372.

Ferguson IK. 1966. The Cornaceae in the southeastern United States. J. Arnold Arbor. 47: 106–116.

Ferguson IK. 1977. Cornaceae Dumort. World Pollen and Spore Flora 6: 1–34. Stockholm.

Ferguson IK and MJ Hideux. 1978 [1979]. Some aspects of the pollen morphology and its taxonomic significance in Cornaceae sens, lat. 4th Intern. Palynol. Conf., Lucknow, 1976–77, 1: 240–249.

Goldblatt P. 1978. A contribution to cytology in Cornales. Ann. Missouri Bot. Gard. 65: 650–655.

Gopinath DM. 1945. A contribution to the embryology of *Alangium lamarckii* Thw. with a discussion of the systematic position of the family Alangiaceae. Proc. Indian Acad. Sci. 22B: 225–231.

Govindarajalu E. 1961, 1962, 1979. The comparative morphology of the Alangiaceae: I. The anatomy of the node and internode. II. Foliar histology and vascularization. III. Pubescence. IV. Crystals. VI. On the foliar anatomy of two new species of *Alangium*. Proc. Natl. Inst. Sci. 27B: 375–388, 1961; 28B: 100–114, 507–531, 1962; 88B: 283–291, 1979.

Govindarajalu E and BGL Swamy. 1956. Petiolar anatomy and subgeneric classification of the genus *Alangium*. J. Madras Univ. 26B: 583–588.

He Z-C, J-Q Li, and HC Wang. 2004. Karyomorphology of *Davidia involucrata* and *Camptotheca acuminata*, with special reference to their systematic position. Bot. J. Linn. Soc. 144: 193–198.

Hegnauer R. 1969. Chemical evidence for the classification of some plant taxa. In: JB Harborne and T Swain, eds. Perspectives in phytochemistry, pp. 121–138. Academic, London/New York.

Hohn ME and WG Meinshein. 1976. Seed oil fatty acids: Evolutionary significance in the Nyssaceae and Cornaceae. Biochem. Syst. Ecol. 4: 193–199.

Horne AS. 1914. A contribution to the study of the evolution of the flower with special reference to the Hamamelidaceae, Caprifoliaceae, and Cornaceae. Trans. Linn. Soc. London II, 8: 239–309, pl. 28–30.

Jensen SR, A Kjaer, and BJ Nielsen. 1975. The genus *Cornus*: Non-flavonoid glucosides as taxonomic markers. Biochem. Syst. Ecol. 3: 75–78.

Kamelina OP and SV Shevchenko. 1988. On the embryology of *Davidia involucrata* (Davidiaceae). Bot. Zhurn. 73: 203–212 (in Russian with English summary).

Kubitzki K. 1963. Zur Kenntnis des unilokularen Cornaceen-Gynozeums (Cornaceae – Studien I). Ber. Deutsch. Bot. Gas. 76: 33–39.

Kubitzki K. 2004a. Cornaceae. In: K Kubitzki, ed. The families and genera of vascular plants, vol. 6, pp. 82–90. Springer, Berlin/Heidelberg/New York.

Kubitzki K. 2004b. Curtisiaceae. In: K Kubitzki, ed. The families and genera of vascular plants, vol. 6, pp. 112–113. Springer, Berlin/Heidelberg/New York.

Kubitzki K. 2004c. Grubbiaceae. In: K Kubitzki, ed. The families and genera of vascular plants, vol. 6, pp. 199–201. Springer, Berlin/Heidelberg/New York.

Li H-L. 1954 (1955). *Davidia* as the type of a new family Davidiaceae. Lloydia 17: 329–331.

Li H-L and C-Y Chao. 1954. Comparative anatomy of the woods of Cornaceae and allies. Quart. J. Taiwan Mus. 7: 119–136.

Li Y-L, H Zhu, and J-B Yang. 2002. Systematic position of the genus *Mastixia:* evidence from *rbc*L gene sequences. Acta Bot. Yunn. 24: 352–358.

Markgraf F. 1963. Die phylogenetische Stellung der Gat-tung *Davidia*. Ber. Deutsch. Bot. Ges. 76: 63–69.

Matthew KM. 1976. A revision of the genus *Mastixia* (Cornaceae). Blumea 23: 51–93.

Mohana Rao PR. 1972 [1973]. Embryology of *Nyssa sylvatica* and systematic consideration of the family Nyssaceae. Phytomorphology 22: 8–21.

Murrell ZE. 1993. Phylogenetic relationships in *Cornus* (Cornaceae). Syst. Bot. 18: 469–495.

Neubauer HF. 1978. Über Knotenbau und Vaskularisation von Blattgrund und Blattsteile bei einigen Cornaceae und einigen ihnen nahestanden Arten, sowie über Knotenbau in allgemeinen. Bot. Jahrb. Syst. 99: 410–424.

Noshiro S and P Baas. 1998. Systematic wood anatomy of Cornaceae and allies. IAWA J. 19: 43–97.

Purkayastha SK and KN Bahadur. 1997. A note on the taxonomy and wood anatomy of the Indian Cornaceae with special reference to the genus *Cornus*. Indian For. 103: 249–332.

Reitsma T. 1970. Pollen morphology of the Alangiaceae. Rev. Palaeobot. Palynol. 10: 249–332.
Sohma K. 1963, 1967. Pollen morphology of the Nyssaceae: I. *Nyssa* and *Camptotheca*. II. *Nyssa* and *Davidia*. Sci. Rep. Tohoku Imp. Univ., 4th ser., 29: 389–392, 1963; 33: 527–532, 1967.
Tandon SR and JM Herr. 1971. Embryological features of taxonomic significance in the genus *Nyssa*. Canad. J. Bot. 49: 505–514.
Tieghem P van. 1897. Sur les caractéres et les affinités des Grubbiacées. J. Bot. (Morot) 11: 127–138.
Titman PW. 1949. Studies in the woody anatomy of the family Nyssaceae. J. Elisha Mitchell Sci. Soc. 65: 245–261.
Trifonova VI. 2000. Davidiaceae, Curtisiaceae. In: A Takhtajan, ed. Comparative seed anatomy, vol. 6, pp. 297–298, 306. Nauka, St. Petersburg (in Russian).
Trifonova VI and EY Yembaturova. 2000. Nyssaceae, Cornaceae, Mastixiaceae, Alangiaceae. In: A Takhtajan, ed. Comparative seed anatomy, vol. 6, pp. 298–306, 312–314. Nauka, St. Petersburg (in Russian).
Trifonova VI and IG Zubkova. 1990. The comparative-anatomical study of petiole of representatives of the genus *Cornus* s. l. (Cornaceae) in relation to its taxonomy. Bot. Zhurn. 75: 820–830.
Wen J and TF Stuessy. 1993. The phylogeny and biogeography of *Nyssa* (Cornaceae). Syst. Bot. 18: 68–79.
Wilkinson AM. 1944. Floral anatomy of some species of *Cornus*. Bull. Torrey Bot. Club 71: 276–301.
Xiang G-Q and Lu F-S. 1989. Study on chemical components of *Davidia involucrata* Baill. native to China. Acta Bot. Sinica 31: 540–543.
Xiang Q-Y. 1999. Systematic affinities of Grubbiaceae and Hydrostachyaceae within Cornales: insights from *rbc*L sequences. Harvard Pap. Bot. 4: 527–541.
Xiang Q-Y and DE Soltis. 1998. *Rbc*L sequence data define a cornaceous clade and clarify relationship of Cornaceae *sensu lato*. In: DE Bouffard and H Ohba, eds. Sino-Japanise Flora: its characteristics and diversification, Bull. No. 37, pp. 123–137. University of Tokyo, Tokyo.
Xiang Q-Y, DE Soltis, DR Morgan, and PS Soltis. 1993. Phylogenetic relationships of *Cornus* L. sensu lato and putative relatives inferred from *rbc*L sequence data. Annals Missouri Bot. Gard. 80: 723–734.
Xiang Q-Y, Brunsfeld SJ, DE Soltis, and PS Soltis. 1997. Phylogenetic relationships in *Cornus* based on chloroplast DNA restriction sites: implications for biogeography and character evolution. Syst. Bot. 21: 515–534.
Xiang Q-Y, DE Soltis, and PS Soltis. 1998. Phylogenetic relationships of Cornaceae and close relatives inferred from *mat*K and *rbc*L sequences. Am. J. Bot. 85: 285–297.
Xiang Q-Y, ML Moody, DE Soltis, C Fan, and PS Soltis. 2002. Relationships within Cornales and circumscription of Cornaceae: *mat*K and *rbc*L sequence data and effects of outgroups and long branches. Mol. Phylogenet. Evol. 24: 35–57.
Zeng Cangjiang [C-J Tzeng]. 1983. The systematic position of *Diplopanax* Hand.-Mazz. Acta Phytotax. Sinica 21: 151–152.
Zhu W-H and Q-B Xiang 1999. Morphological characters of the genus *Diplopanax* Hand.-Mazz. and its systematic implication. Bull. Bot. Res. 19: 286–291.

Order 110. GARRYALES

Evergreen shrubs or sometimes small trees, usually provided with various unicellular trichomes or rarely glabrous. Nodes trilacunar. Wood diffuse-porous. Vessels with scalariform perforations; lateral pitting alternate-opposite. Fibers with bordered pits. Rays heterogeneous. Axial parenchyma usually diffuse or diffuse-in-aggregates and scanty paratracheal. Crystals absent. Sieve-element plastids of S-type. Leaves opposite, simple, entire or subentire, coriaceous, leathery, estipulate. Stomata laterocytic or anomocytic. Inflorescences terminal, sometimes axillary. Flowers small, unisexual (dioecious). Male flowers: calyx almost obsolete appearing at the early stages of flower development as vestigial structures; petals four, valvate, sometimes apically connate; stamens four, alternipetalous, free; anthers basifixed or dorsifixed, tetrasporangiate, introrse or latrorse, opening longitudinally. Nectary disc development or vestigial. Tapetum secretory (amoeboid – *Garrya*). Pollen grains 2-celled, usually 3-colporate. Female flowers: calyx tube ovoid, petals four, valvate. Gynoecium pseudomonomerous or monomerous, of one (*Aucuba*), two, rarely three carpels; ovary inferior, 1-locular. Ovules 1–2, pendulous, anatropous, unitegmic, crassinucellate. Female gametophyte of *Polygonum*-type. Endosperm nuclear or cellular (*Aucuba*). Fruits baccate. Seeds rather small, with minute linear embryo and copious endosperm; Commonly producing iridoid compounds (notably aucubin) and highly toxic diterpenoid alkaloids. Lack both gallic and ellagic acid, but contain caffeic acid (chlorogenic acid), $n = 8$ (*Aucuba*), 11 (*Garrya*).

Markedly differ from the Cornales s.str. in the inflorescence, anemophilous flowers, the absence of endothelium, suspensor polyembryony, absence of gallic and ellagic acids, and presence of caffeic acid and aucubin. Its minute embryo and the presence in its seeds of petroselinic acid seems, according to Eyde (1988), to confirm the long-postulated kinship of *Garrya* to *Aucuba*, which has been supported by *rbc*L sequence data (Xiang et al. 1993).

Key to Families

1 Flowers in terminal, sometimes axillary, pendulous catlinlike or racemose inflorescences. Evergreen shrubs or sometimes small trees, usually provided with various unicellular trichomes or rarely glabrous (*Carrya glaberrima*). Young stems tetragonal. Leaves flat, petiolate (with the petioles united at the base), entire or dissected, sometimes pinnatifid. Stomata laterocytic. Flowers small, pendulous, silky-hairy, long stalked in the males and sessile in the females. Male flowers: calyx almost obsolete appearing at the early stages of flower development as vestigial structures (a vestigial nectary disc); petals four, valvate, sometimes apically connate and hairy at the apex; stamens four, free; anthers elongate, basifixed, introrse or latrorse. Pollen grains 2-celled, usually 3-colporate. Female flowers: calyx tube completely adnate to the ovary, lobes four or two, very small or obsolete; petals seems wanting, but as the study of vascular system shows (Hallock 1930; Moseley and Becks 1955; Eyde 1964), the reduced corolla together with the calyx tube completely adnate to the ovary; no staminodia. Gynoecium pseudomonomerous, of two (rarely three) carpels, with two free subulate, spreading stylodia; ovules two, unitegmic (integument very massive), with an obturator and rather long funicle. Endosperm nuclear. Fruits 1–2- (rarely 3-)-seeded berries, crowned by persistent stylodia, becoming dry and thin-walled at maturity, but indehiscent or only tardily dehiscent. Seeds rather small, sarcotestal, oblong, compressed, with minute linear embryo near the apex of copious, starchy and horny endosperm; endosperm with high percentage of petroselinic acid and with reserves of hemicellulose. Commonly producing iridoid compounds (notably aucubin) and highly toxic diterpenoid alkaloids, contain caffeic acid (chlorogenic acid), n = 11.. . . . . 1. GARRYACEAE.

1 Flowers in silky terminal or axillary compound dichasia. Small evergreen shrubs or herbs, glabrous or sparsely pilose. Branchlets rather thick, scarred with the fallen petioles. Vessels with 27–70 bars; lateral pitting rare, opposite, without vestures. Leaves petiolate, entire or dentate, shining. Stomata anomocytic. Male flowers: calyx small, 4-dentate. Petals four, ovate to lanceolate, valvate, often with slender inflexed tip; stamens four, filaments very short; anthers oblong, dorsifixed, introrse. Nectary disc 4-angular, fleshy. Pollen grains 3-colporate, intectate, psilate. Female flowers: pedicel articulated and 2-bracteolate; calyx tube ovoid or subcylindrical, 4-dentate; petals as in the male. Gynoecium monomerous (ostensibly), or syncarpous (i.e. supposedly 'pseudomonomeric'), or eusyncarpous; style short, thick, with conspicuous, oblique, capitate stigma; ovule solitary, pendulous, anatropous, with dorsal raphe. Endothelium differentiated. Endosperm cellular. Fruits 1-seeded, ovoid, scarlet baccate, crowned by the calyx teeth. Seeds with minute embryo at apex of copious endosperm; cotyledons short; radicle cylindrical; seed coat coloured, thin, undifferentiated, of 3–4 or six layers. Seeds contain petroselinic and coffeic (chlorogenic) acids; leaves of *Aucuba japonica* contain flavonols and flavone glycosides, producing aucubin and eucommioside, n = 8. . . . . 2. AUCUBACEAE.

1. GARRYACEAE

Lindley 1834. 1/14. Western North America from Washington to Panama, with one species in the Greater Antilles.

Garrya

2. AUCUBACEAE

J. Agardh 1858. 1/5. Eastern Himalayas, northern Burma, continental China, Taiwan, Ryukyu Islands, southern Korea, Japan.

Aucuba

H. Baillon (1879–1895) the first proposed the close relationship between *Garrya* and *Aucuba*, which is strong supported by recent molecular studies (see Soltis et al. 2000, 2006; Savolainen et al. 2000, and others).

Bibliography

Coulter JM and WH Evans. 1890. *Garrya*. Bot. Gaz. 15: 93–97.

Dahling GV. 1978. Systematics and evolution of *Garrya*. Contr. Gray Herb. 209: 1–104.

Eyde RH. 1964. Inferior ovary and generic affinities of *Garrya*. Am. J. Bot. 51: 1083–1092.

Hallock FA. 1930. The relationship of *Garrya*. Ann. Bot. 44: 771–812.

Iwashina T, K Kamenosono, and H Hatta. 1997. Flavonoid glycosides from leaves of *Aucuba japonica* and *Helwingia*

japonica (Cornaceae): phytochemical relationship with the genus *Cornus*. J. Jpn. Bot. 72: 337–346.
Kapil RN and PR Mohana Rao. 1966. Studies of the Garryaceae: II. Embryology and systematic position of *Garrya* Douglas ex Lindley. Phytomorphology 16: 564–578.
Kurosawa S. 1971. Cytotaxonomical studies on the genus *Aucuba*. J. Jpn. Bot. 46: 231–238.
Liston A. 2003a. A new interpretation of floral morphology in *Garrya* (Garryaceae). Taxon 52: 271–276.
Maekawa F. 1965. *Aucuba* and its allies: The phylogenetic consideration on the Cornaceae. J. Jpn. Bot. 40: 41–47.
Moseley MF and RM Beeks. 1955. Studies of the Garryaceae: I. The comparative morphology and phylogeny. Phytomorphology 5: 314–346.
Mulligan BO. 1980. A revision of the genus *Garrya*. Arbor. Bull. Arbor. Found. Univ. Wash., 43(2): 39.
Noshiro S and P Baas. 1998. Systematic wood anatomy of Cornaceae and allies. IAWA J. 19: 43–97.
Trifonova VI. 2000. Garryaceae. In: A Takhtajan, ed. Comparative seed anatomy, vol. 6, pp. 310–312. Nauka, St. Petersburg (in Russian).
Trifonova VI and EY Yembaturova. 2000. Aucubaceae. In: A Takhtajan, ed. Comparative seed anatomy, vol. 6, pp. 308–309. Nauka, St. Petersburg (in Russian).

Order 111. EUCOMMIALES

Small, deciduous trees with sympodial branches; hairs micropapillate. The phloem and cortex of vegetative organs contain articulated (unicellular) laticifers that are particularly numerous in the leaves and produce gutta-percha. Nodes unilacunar. Pith lamellate. Vessels with simple perforation, but in vessels of the earliest secondary xylem the perforation is scalariform; lateral pitting alternate or less often intermediate between alternate and opposite; usually spirally thickenings present. Fibers with bordered pits. Rays homogeneous or slightly heterogeneous. Axial parenchyma diffuse, with some tendency to terminal. Sieve-element plastids of S-type, with about ten starch grains. Buds usually with six scales. Leaves alternate, simple, serrate, pinnately veined (strikingly resembling in basic pattern of secondary venation and extensive looping *Stachyurus* – Wolfe [1989]), estipulate. Stomata anomocytic. Flowers in racemose inflorescences arranged on the proximal, bracteate part of a distally leafy shoot (and thus resembling intercalary inflorescence of *Euptelea* – Eckardt [1963]), short pedicillate in the axils of the bracts, but without bracteoles, dioecious, actinomorphic, without perianth. Male flowers (4)5–10 (up to 12 or sometimes more) stamens; filaments very short, anthers linear, with apically prolonged connective, linear, basifixed, tetrasporangiate, opening longitudinally. Tapetum secretory. Microsporogenesis simultaneous. Pollen grains 2-celled, 3-colpate, with one of them narrowing in the equatorial region, and the other straight; exine atectate (with homogeneous ectexine), rather thick endexine and minutely spinulose to verrucate exine ornamentation (Zavada and Dilcher 1986). Gynoecium pseudomonomerous, flattened, bifid at apex, consisting of two carpels, one of which aborts; style one, very short (shorter than the ovary), recurved, with two unequal, ventrally stigmas. Ovary superior, 1-locular. Ovules two (but only one fertile), collateral, pendulous, anatropous, apotropous, with dorsal raphe, unitegmic, weakly crassinucellate. Female gametophyte of *Polygonum*-type. Endosperm cellular. Fruits samara, flattened and winged, with thin epicarp, 1-seeded. Seeds with membranous testa, copious endosperm and large, straight, central embryo equal in length to the endosperm. Producing flavonols (quercetin and kaempferol), various iridoid compounds (including aucubin, eucommioside), proanthocyanidins (cyanidin), saponins, and storing carbohydrate as inulin. Bark with condensed tannins, $n = 17$.

Close related to the Garryales (see Boros and Stermitz 1990, 1991; Soltis et al. 2000, 2006; Albach et al. 2001; Bremer et al. 2002)

1. EUCOMMIACEAE

Engler 1909. 1/1. Continental China.
Eucommia

Bibliography

Eckardt Th. 1956. Zur systematischen Stellung von *Eucommia ulmoides*. Ber. Deutsch. Bot. Ges. 69: 487–498.
Eckardt Th. 1963. Some observations on the morphology and embryology of *Eucommia ulmoides*. J. Indian Bot. Soc. 42A: 27–34.
Rowley JR, JJ Skvarla, and JM Pettitt. 1992. Pollen wall development in *Eucommia ulmoides* (Eucommiaceae). Rev. Paleobot. Palynol. 70: 297–323.
Sogo A and H Tobe. 2006. Mode of pollen tube growth in pistils of *Eucommia ulmoides* (Eucommiaceae, Garryales). Int. J. Plant Sci. 167: 933–941.
Tang SH. 1962. Sporogenesis and gametophyte development in *Eucommia ulmoides*. Acta Bot. Sinica 10: 29–34.
Tippo O. 1940. The comparative anatomy of the secondary xylem and the phylogeny of the Eucommiaceae. Am. J. Bot. 27: 832–838.

Varossieau WW. 1942. On taxonomic position of *Eucommia ulmoides* Oliv. (Eucommiaceae). Blumea 5: 81–92.
Wolfe J. 1973 (1974). Fossil forms of Amentiferae. Brittonia 25: 334–466.
Zhang Y-L, F-S Wang, and N-F Chien. 1988. A study on pollen morphology of *Eucommia ulmoides*. Acta Phytotax. Sinica 26: 367–370.
Zhang Z-Y, A-M Lu, and K-Y Pan. 1990. The anatomy, embryology, and systematic relationships of Eucommiaceae. Acta Phytotax. Sinica 28: 430–441.

Order 112. DIPSACALES

Small trees, shrubs, more or less woody lianas, or herbs. Vessels with simple or less often scalariform perforations; lateral pitting commonly alternate. Fibers with bordered or simple pits. Sieve-element plastids of S-type. Nodes trilacunar or sometimes unilacunar, pentalacunar, or multilacunar. Leaves opposite or sometimes verticillate, simple or less often compound, estipulate or stipulate, but sometimes with stipulelike appendages; leaf mesophyll cells contain oil bodies. Stomata anomocytic, rarely paracytic or anisocytic. Flowers in various kinds of mostly cymose or mixed inflorescences, usually bracteolate, mostly bisexual, actinomorphic or zygomorphic, 4–5-merous. Calyx mostly small, often variously reduced or modified. Corolla sympetalous, usually 5-lobed or 4-lobed by fusion of two, the lobes imbricate or sometimes valvate; corolla tube usually with a nectary consisting of one-celled epidermal hairs and an underlying nectary tissue. Stamens attached to the corolla tube or at the throat, as many as and alternate with the corolla lobes or fewer than the lobes. Anthers mostly tetrasporangiate, opening longitudinally. Tapetum amoeboid or secretory. Microsporogenesis simultaneous. Pollen grains 3-celled, mostly 3-colporate. Gynoecium of 2–5 united carpels, with more or less connate stylodia; stigma capitate or lobed; ovary inferior or semi-inferior, sometimes more or less pseudomonomerous, with one to many ovules per locule. Ovules anatropous, unitegmic, mostly tenuinucellate, with well developed endothelium. Endosperm cellular, usually without haustoria, rarely (*Centranthus*) with chalazal haustorium. Female gametophyte of *Polygonum*-type, rarely *Allium*- or *Adoxa*-type. Fruits of various types. Seeds with small to large, straight embryo and with copious endosperm or rarely endosperm reduced to single layer of cells (most Valerianaceae). Producing iridoid compounds and sometimes various kinds of alkaloids, sometimes somewhat tanniniferous.

Dipsacales are closely related to the Desfontainiales, especially to the Columelliaceae and Desfontainiaceae, which in some respects occupy in intermediate position between these two orders.

Key to Families

1 Vessels mostly or always with scalariform perforations.

2 Ovary nearly semi-inferior; style short, 3-lobed. Deciduous or evergreen shrubs or small trees; hairs stellate and variants. Winterbuds naked or scaly. Fibers with bordered pits. Rays heterogeneous. Axial parenchyma apotracheal. Leaves opposite or rarely verticillate of three, entire, serrate-dentate or lobed, pinnately or palmately veined, sometimes with stellate hairs, estipulate or sometimes with small stipules adnate to the petiole. Stomata paracytic or anomocytic. Flowers small, usually with small caducous bracts and bracteoles, in terminal cymose inflorescences, bisexual or in some species of *Viburnum* the marginal flowers are neutral, with greatly enlarged, somewhat zygomorphic corolla. Calyx 5-lobed to minutely 5-toothed, persistent. Corolla sympetalous, rotate to campanulate or tubular, 5-lobed, lobes imbricate. Stamens five, inserted on the corolla tube; anthers exerted. Pollen grains 3-celled, 3-colporate, semitectate to intectate, reticulate, with smooth muri. Gynoecium of three united carpels; stylodia connate into a short, 3-lobed style; ovary 3-locular with one fertile locule and two abortive locules, or pseudomonomerous, with one pendulous ovule per locule. Nectary on the ovary roof. Ovules tenuinucellate to nearly crassinucellate (Suneson 1933). Female gametophyte of *Polygonum*- or *Allium*-type. Fruits drupaceous, crowned by persistent calyx and style, endocarp horny or stony, in cross section often undulate or with inflexed edges. Seeds large, oblong, usually compressed, with minute embryo; endosperm copious, subruminate from folded endocarp; seed coat 1-layered of large radially elongated cells, with thin or slightly thickened walls. Producing iridoid compounds, n = 8 or 9, mostly 9. . . . . 1. VIBURNACEAE.

2 Ovary inferior; style elongate, mostly with a capitate stigma. Plants mostly woody, seldom herbaceous. Trees or shrubs, often woody lianas (ssp. of *Lonicera*), sometimes creeping shrublets (*Linnaea*) or rhizomatous herbs (*Triosteum*). Vessels mostly with scalariform perforations that sometimes have numerous bars; lateral pitting alternate to scalariform. Fibers with bordered pits. Rays heterogeneous. Axial parenchyma apotracheal or paratracheal, diffuse. Nodes mostly trilacunar, rarely unilacunar. Leaves opposite or whorled, petiolate, simple, entire, serrate or dentate, stipulate (sometimes large – *Leycesteria*) or estipulate. Stomata usually anomocytic. Flowers in various kinds of inflorescences, rarely (*Triostem*) solitary and axillary, bisexual, nearly actinomorphic (*Weigela*) or more or less zygomorphic. Sepals 4–5, rarely (*Weigela*) polysepalous, imbricate or open in bud. Corolla 4–5-lobed, lobes imbricate, tube often gibbous at the base. Stamens (2-)4–5, attached to the corolla tube or in the throat of the corolla tube; filaments glabrous. Pollen grains (2)3(4)-colporate to perorate, tectate, psilate or spinulose. Gynoecium of 2–5(-8) carpels; ovary with one to many pendulous ovules per locule. Fruit capsules, drupes, or berries. Seed with rather small embryo and copious, fleshy endosperm; seed coat of 1–5 layers, formed by differentiated thick-walled outer epidermis and subepidermal parenchyma (Plisko 2000). Alkaloids present or absent, iridoids detected, mostly present kaempferol and quercetin, n = 8–12, 16, 18, mostly 9. . . . . 4. CAPRIFOLIACEAE.

1 Vessels with simple or sometimes with scalariform perforations.

3 Female gametophyte of *Polygonum*-type.

4 Flowers without epicalyx or involucel. Perennial or annual herbs, sometimes woody at base, with characteristic fetid odor caused by the monoterpenoid and sesquiterpenoid ethereal oil cells, which are especially abundant in the outer part of the cortex and cork of the rhizomes and roots. Hairs include simple and glandular types. Vessel elements extremely small, with simple perforations; lateral pitting alternate. Fibers with simple or narrowly bordered pits. Rays are composed almost entirely of square or upright cells. Axial parenchyma absent. Nodes trilacunar. Leaves opposite, often in basal rosettes, entire to pinnatifid or four pinnately divided, the base often sheathing. Stomata anomocytic, rarely anisocytic. Flowers usually small, in compound dichasial cyme or monochasium with many flowers, sometimes congested into heads, bracteate and usually bracteolate, bisexual or sometimes polygamous or unisexual and then usually dioecious, more or less asymmetrical or less often (*Nardostachys* and *Patrinia*) nearly actinomorphic, usually 5-merous. Calyx in *Nardostachys* relatively well developed and 5-lobed, but in other genera the limb is variously reduced, sometimes represented by inconspicuous teeth or even epigynous rim, often considerably modified in fruit (as in *Valeriana* and *Centranthus*) into a plumose pappus or (as in some species of *Valerianella*) becoming coroniform or inflated. Corolla from nearly actinomorphic to more or less strongly zygomorphic, usually tubular, often with a long tube (particularly strongly developed in *Centranthus*), (3-)5-lobed, imbricate, often basally gibbous or more or less spurred and nectariferous, sometimes (*Centranthus*) bilabiate. Stamens four (*Nardostachys* and most species of *Patrinia*) or more often three, in *Fedia* two of three stamens connate or only two stamens, while *Centranthus* has only one stamen attached to the corolla tube; anthers introrse. Pollen grains 3(4)-colporate, tectate. Gynoecium of three carpels, with slender style and simple or 2–3-lobed stigma; ovary inferior, basically 3-locular but two locules more or less strongly reduced and empty or even obsolete (their presence evident in *Valeriana* and *Fedia*), with one anatropous ovule in fertile locule. Fruits dry cypselalike, often crowned by an accrescent winged, awned, plumose or inflated calyx. Seeds small, with large, straight, oily embryo, differentiated into two round cotyledons and hypocotyle; endosperm reduced to a single layer of cells, or rarely (*Patrinia*) well developed (Plisko 2000); seed coat unspecialized, membranous, crushed. Usually producing iridoid compounds and sometimes monoterpene alkaloids, triterpenoid saponins, and valpotriates, n = 7–12, mostly 9. . . . . 5. VALERIANACEAE.

4 Flowers nearly always with an epicalyx.
5 Flowers in compact heads.
6 Flowers with a double epicalyx below the calyx. Perennial rhizomatous herbs smelling of Valerianaceae. Rhizomes sometimes with fusiform tubers. Vessels with simple perforations. Leaves opposite, dentate to pinnatifid, estipulate. Flowers small, in a more or less few-flowered terminal cymose inflorescences with glandular hairs on branches and epicalyces, bisexual, almost actinomorphic, each subtended by two opposite narrow bracts, an outer epicalyx of four further conspicuous capitate-glandular persistent bracts, connate at base and uncinate cuspidate at apex, and an 8-ribbed, urceolate, persistent inner epicalyx. Calyx minute, 5-dentate, persistent. Corolla funnel-shaped, caducous, lobes imbricate, subequal. Stamens four, epipetalous, anthers dorsifixed, introrse; anther epidermis degenerating. Gynoecium of one or three carpels (only one carpel fertile), with slender style and small capitate stigma; ovary inferior, one locule fertile, with solitary ovule, the other two locules abortive and evanescent. Fruit dry, cypselalike, surrounded by persistent indurated epicalyces. Seeds small, with small embryo and copious endosperm; seed coat of 3–4 layered; outer epidermis of large, vacuolarized cells, cells of the inner layer small and flat (Plisko 2000). Iridoids detected. . . . . 6. TRIPLOSTEGIACEAE.
6 Flowers with a single epicalyx below the calyx. Perennial or annual herbs, sometimes woody at base, rarely shrubby. Hairs unicellular, usually lignified, or often multicellular, glandular or (*Knautia*) capitate-glandular or (*Cephalaria*) tufted. Vessels with simple perforations or sometimes some of them scalariform. Nodes trilacunar. Leaves opposite or sometimes verticillate, entire or dentate to deeply pinnatifid or pinnately dissected. Stomata mostly anomocytic, less often anisocytic. Flowers in dense, essentially cymose, involucrate heads with a conspicuously bracteate to hairy or naked common receptacle, bisexual, mostly zygomorphic, 5- or 4-merous. Each flower enveloped by a more or less cupulate epicalyx formed by connection of two subtending bracteoles, wanting in a few species of *Cephalaria* and *Succisa*. Calyx mostly small and variable, cuplike or the lobes divided into 5–10 pappuslike segments. Corolla 5-lobed or 4-lobed by union of two members, imbricate, often more or less two-lipped or the lobes are unequal. Stamens four (the fifth posterior stamen is always absent and the two anterior ones are often smaller), adnate to the summit of corolla tube; anthers introrse. Pollen grains mostly 3-colporate. Gynoecium of two carpels, but one carpel aborts; style slender and with entire or 2-lobed stigma; ovary 1-locular, with one or very rarely (*Knautia arvensis*) 2–3 ovules, pendulous from the top of locule. Fruits cypselalike, usually enclosed in thickened and leathery epicalyx; pericarp thin and easily ruptured at the apex. Seeds small, with thin membranous testa, large, straight embryo and copious or rarely rather scanty (*Scabiosa*), fleshy, oily endosperm; seed coat membranous; outer thin-walled epidermis covered by cuticule and hypodermal layer (Kamelina and Plisko 2000). Usually producing iridoid compounds and sometimes also alkaloids. n = 5, 7–9(-10). . . . . 7. DIPSACACEAE.
5 Flowers in axillary verticillasters. Perennial robust herbs with simple, hollow stems. Nodes pentalacunar with five traces. Leaves opposite or verticillate, pinnatifid or spinose-dentate, rarely entire, estipulate, with secretory ethereal cells in the epidermis. Each flower enclosed in a tubular-campanulate, spinose-margined epicalyx, bisexual, zygomorphic. Calyx bilabiate, lips entire or 2-lobed, persistent. Corolla bilabiate, imbricate, tube usually slender and exerted, sometimes very long. Stamens two, inserted in the throat of the corolla tube and alternating with the corolla members; staminodes two; filaments long and thick; anthers basifixed (?). Tapetum amoeboid. Pollen grains 3–4-colporate or 3-pororate. Gynoecium of two carpels, but one carpel aborts; style slender, with capitate stigma; ovary inferior, 1-locular, with one pendulous ovule. Fruits of cypsela-type, with thickened

and more or less rugose pericarp. Seeds of middle-sized, with thin testa, large, straight embryo and copious, ruminate endosperm; seed coat thin, membranous, formed by the epidermis of the outer integument (Kamelina and Plisko 2000). Present ethereal oils, n = 17. . . . . 8. MORINACEAE.

3 Female gametophyte *Adoxa*-type.

7 Deciduous or semi-evergreen small to stately trees, shrubs (sometimes rhizomatous) or perennial sub-shrubby rhizomatous herbs. Woody stems lenticellate, with conspicuous pith; cambium storied. Crystal sand present in the cortex and phloem. Leaves imparipinnate or sometimes incompletely bipinnate, rarely laciniate, stipules foliaceous or forming small, linear, sometimes glandular appendages or absent. Flowers numerous in terminal cymose corymbs or panicles, articulated with the pedicel, mostly bractless, bisexual or rarely unisexual, 5-merous or less often 3–4-merous. Calyx small, (3–4)5-lobed or dentate. Corolla rotate or rotate-campanulate, deeply (3–4)-lobed, lobes imbricate or valvate. Trichomatic nectaries at the base of corolla lobes absent. Stamens five, inserted at the base of the corolla; filaments short; anthers oblong, extrorse or latrorse. Style very short, cushionlike, more or less deeply 3–5-lobed, with papillate stigmas; ovary 3–5-locular. Fruits berrylike drupes with 5–3 one-seeded, bony pyrenes. Seeds oblong, with the linear embryo and copious endosperm; seed coat membranous of several rows of flat brown cells. x = 9 (n = 18, 19, 20, 36, mostly 18). . . . . 2. SAMBUCACEAE.

7 Small geophytic perennial herbs with a scaly creeping or erect monopodial rhizome, one to several alternate basal leaves and two opposite cauline leaves. Rhizome with two flattened vascular bundles separated by a pith; stem with two vascular bundles, each serving to supply a leaf, and two smaller ones leading to the flowers. Leaves imparipinnate, ternate or 3-lobed, bearing multicellular, tanniniferous secretory cells when young. Flowers greenish, in terminal or sometimes also axillary compound and spikelike (*Sinadoxa*) or simple and racemelike (*Tetradoxa*) or head-like (*Adoxa*) cymose inflorescences, bisexual, without bracts and bracteoles. Calyx 2–4-lobed, persistent. Corolla more or less rotate, 3–5-lobed, the lobes imbricate; nectar is secreted by multicellular hairs situated in groups at the base of each corolla lobe (Erbar 1994). Stamens alternipetalous, inserted into the top of corolla tube, more or less divided into two disporangiate halves. Stylodia 3–5, free, short, with small capitate, papillate stigma, but in *Sinadoxa* 1-locular ovary attenuates to the top, with inconspicuous apical stigma; ovary 1–5-locular. Fruits small, dry drupes with several pyrenes. Seeds with a very small embryo and copious endosperm; seed coat thin, crushed. Present 0-methylated flavonols, n = 9. . . . . 3. ADOXACEAE.

1. VIBURNACEAE

Rafinesque 1820 (including Tinaceae Martynov 1820). 1/175. Mostly northern temperate regions, also in tropical mountains; best developed in the Himalayas and China; crossing the equator in Malesia.

Viburnum

Probably the most archaic member of the Dipsacales.

2. SAMBUCACEAE

Batsch ex Borckhausen 1797. 1/27. Temperate subtropical regions and partly tropical mountains of Eurasia and America, especially eastern Asia and North America, but also North and tropical East Africa and eastern Australia and Tasmania.

Sambucus

Have some similarities with *Viburnum*, including pollen morphology (Bohnke-Gutlein and Weberling 1981) and chromosomes (Sax and Kribs 1930; Benko-Iseppon and Morawetz 1993), but differ in tetrasporic female gametophyte, dissected leaves, thick and soft pith, the presence of narrow, thick-walled pericycle fibers in the stem, simple perforations of the vessels, paratracheal axial parenchyma, fibers with small simple pits, semitectate pollen grains (intectate in *Viburnum* – see Bassett and Compton 1970), 5–3-pyrenous drupes, and longer embryo. Additionally, *Sambucus* is serologically distinct from *Viburnum* (Hillebrand and Fairbrothers 1970).

3. ADOXACEAE

E. Meyer 1839. 3/5. Temperate and cold regions of the Northern Hemisphere; *Sinadoxa* (1) and *Tetradoxa* (1) are endemic to China.

Adoxa, Tetradoxa, Sinadoxa

Very close to the Sambucaceae, especially to their more archaic herbaceous representatives. Both families have many similarities in floral morphology (Eichler 1875; Hallier 1908, 1912; Erbar 1994), fairly similar karyomorphological patterns (Greilhuber 1979; Benko-Iseppon and Morawetz 1993), and they share the same type of female gametophyte (Moissle 1941), pollen grains (Reitsma and Reuvers 1976), and mode of leaf dissection. Thorne (1983, 1992a, b, 2001) included *Adoxa* and *Sambucus* in the family Adoxaceae. However, in *Adoxa* nectariferous multicellular hairs are situated in groups at the base of each corolla lobe (Erbar 1994), whereas trichomatic nectaries are absent in *Sambucus,* stamens in *Adoxa* are more or less divided, stylodia are free, and druses in the tissues of ovary wanting (Fukuoka 1974).

4. CAPRIFOLIACEAE

A.L. de Jussieu 1789 (including Diervillaceae Pyck 1998; Linnaeaceae Backlund 1998; Loniceraceae Vest 1818). 12/275–300. Mostly northern temperate regions, with a few species in tropical mountains; best developed in the Himalayas and China; crossing the equator only in Malesia.

DIERVILLEAE: *Weigela, Diervilla*; LONICEREAE *Leycesteria, Lonicera, Heptacodium, Symphoricarpos;* TRIOSTOMEAE: *Triosteum;* LINNAEEAE: *Linnaea, Dipelta, Kolkwitzia, Abelia, Zabelia.*

The Eastasian genus *Weigela* seems to be the most archaic member of the family (capsular fruits together with very primitive vessel elements with scalariform perforation that have 10–120 bars, and heterogeneous rays with long ends).

5. VALERIANACEAE

Batsch 1802. 16/400. Nearly cosmopolitan but best developed in the northern temperate regions (especially in the Mediterranean region and in western Asia), and in the Andes of South America, being absent from Australia, New Zealand, and much of Africa.

PATRINIEAE: *Nardostachys, Patrinia;* VALERIANEAE: *Phyllactis, Valerianella, Pseudobetckea, Fedia, Siphonella, Stangea, Valeriana, Belonanthus, Aretiastrum, Phuodendron, Astrephia, Plectritis, Plectritis, Centranthus.*

Very near to the Caprifoliaceae, especially to the Linnaeeae (Wilkinson 1949), to which a relatively archaic Sino-Himalayan genus *Nardostachys* is especially closely related and of which vascular anatomy of the flower «is fundamentally the same as that in *Abelia*" (Wilkinson 1949: 488). According to Wilkinson, «The similarities between the Valerianaceae and the Linnaeeae are too significant to be considered as due simply to parallel development. The two groups are closely related.» (Wilkinson 1949: 488). And a tendency toward the oligomerization of the gynoecium and androecium, clearly expressed within the Caprifoliaceae, becomes a fixed character in the Valerianaceae. Besides, the zygomorphy of the corolla «foreshadowed in Caprifoliaceae is carried through every degree of transition in Valerianaceae» (Wernham 1912: 95).

6. TRIPLOSTEGIACEAE

Bobrov ex Airy Shaw 1965. 1/2. Temperate and alpine Himalayas (Garhwal to Bhutan), Assam, southeastern Tibet, Burma, continental China, Taiwan, Sulawesi, and New Guinea.

Triplostegia

The genus *Triplostegia* occupies a somewhat intermediate position between Valerianaceae and Dipsacaceae. Therefore some authors, such as Lindley (1853), Wagenitz in Englers Syllabus (1964), Cronquist (1981, 1982), include it in the Valerianaceae while some others (Hutchinson 1959, 1973; Burtt 1977; Zhang et al. 2001) include it in the Dipsacaceae. Only a few authors, including Airy Shaw (1965 and in Willis 1973), Dahlgren (1975, 1983), Takhtajan (1987, 1997), Brummitt (1992), and Backlund and Bremer (1998) accept Triplostegiaceae as a distinct family. According to Backlund and Nilsson (1997), *Triplostegia* is best included in the family Valerianaceae, as the sole member of a subfamily Triplostegioideae. The close affinity with the Valerianaceae is supported by many features, ranging from wood anatomy to inflorescence morphology (Hoffmann and Göttman 1990), embryology (Kamelina 1980, 1983). However *Triplostegia* differs from all members of the Valerianaceae in its peculiar double epicalyx (see Airy

Shaw 1965) and as well as in not branched aperture operculum elements (Backlund and Nilsson 1997). I therefore to accept the Triplostegiaceae as a separate family very closely related to the Valerianaceae.

7. DIPSACACEAE

A.L. de Jussieu 1789 (including Scabiosaceae Martynov 1820). 14/300. Temperate Eurasia (especially in Mediterranean region and western Asia) and tropical Africa and South Africa.

Cephalaria, Dipsacus, Succisa, Succisella, Knautia, Pterocephalus, Pterocephalidium, Scabiosa, Lomelosia, Pseudoscabiosa, Scabiosiopsis, Sixalix, Tremastelma, Pycnocomon.

Probably originated from some *Triplostegia*-like ancestor whose open inflorescence condensed into a ca-pitulum (see Burtt 1977: 106).

8. MORINACEAE

Rafinesque 1820. 3/13. From the Balkans (Greece, Bulgaria, Romania) and Lebanon and Israel to continental China.

Morina, Acanthocalyx, Cryptothladia

The distinctness of the family Morinaceae is confirmed by palynological and embryological data (Vinokurova 1959; Vijayaraghavan and Sarveshwari 1968; Verlaque 1977; Kamelina 1980; Blackmore and Cannon 1983) as well as by morphology of the inflorescence and basic chromosome numbers. Morinaceae pollen is morphologically distinct from that of the Dipsacaceae in the lack of columellae and operculae and in the presence of zonorate endoapertures in some of the pollen types (Blackmore and Cannon 1983).

Bibliography

Airy Shaw HK. 1995. Diagnoses of new families, new names, etc., for the seventh edition of Willis's 'Dictionary'. Kew Bull. 18: 249–273.

Artyushenko ZT. 1951. Development of the flowers and fruits in Caprifoliaceae. Trudy Bot. Inst. Akad. Nauk SSSR, 7th ser., 2: 131–169 (in Russian).

Backlund A. 1996. Phylogeny of the Dipsacales. Acta Universitatis Upsaliensis 243: 1–34.

Backlund A and B Bremer. 1997. Phylogeny of the Asteridae s.s. based on *rbc*L sequences, with particular reference to the Dipsacales. Plant Syst. Evol. 207: 225–255.

Backlung A and B Bremer. 1998. To be or not to be – principles of classification and monotypic plant families. Taxon 47: 391–400.

Backlund A and MJ Donoghue. 1996. Morphology and phylogeny of the order Dipsacales. In: A Backlund, ed. Phylogeny of the Dipsacales, pp. 1–27. Acta Universitatis Upsaliensis, Uppsala.

Backlund A and T Moritz. 1998. Phylogenetic implications of an expanded valepotriate distribution in the Valerianaceae. Biochem. Syst. Evol. 26: 309–335.

Backlund A and S Nilsson. 1997. Pollen morphology and the systematic position of *Triplostegia* (Dipsacales). Taxon 46: 21–31.

Backlund A and N Pyck. 1998. Diervillaceae and Linnaeaceae, two new families of caprifolioids. Taxon 47: 657–661.

Backlund A and M Thomas. 1998. Phylogenetic implications of an expanded valepotriate distribution in the Valerianaceae. Biochem. Syst. Ecol. 26: 309–335.

Bassett II and CW Crompton. 1970. Pollen morphology of the family Caprifoliaceae in Canada. Pollen et Spores 12: 365–380.

Bell CD. 2004. Preliminary phylogeny of Valerianaceae (Dipsacales) inferred from nuclear and chloroplast DNA sequence data. Mol. Phylogenet. Evol. 31: 340–350.

Bell CD and MJ Donoghue. 2000. Dipsacales phylogeny based on chloroplast DNA sequences. Am. J. Bot. 87(6, Suppl.): 171.

Bell CD and MJ Donoghue. 2003. Phylogeny and biogeography of Morinaceae (Dipsacales) based on nuclear and chloroplast DNA sequences. Organ. Divers. Evol. 3: 227–237.

Bell CD and MJ Donoghue. 2005a. Dating the diversification of Dipsacales: comparing models, genes, and evolutionary implications. Am. J. Bot. 92: 284–314.

Bell CD and MJ Donoghue. 2005b. Phylogeny and biogeography of Valerianaceae (Dipsacales) with special reference to the South American valerians. Organ. Divers. Evol. 5: 147–159.

Bell CD, EJ Edwards, ST Kim, and MJ Donoghue. 2001. Dipsacales phylogeny based on chloroplast DNA sequences. Harvard Pap. Bot. 6: 481–499.

Benko-Iseppon AM and W Morawetz. 1993. Cold-induced chromosome regions and karyosystematics in *Sambucus* and *Viburnum*. Bot. Acta 106: 183–191.

Benko-Iseppon AM and W Morawetz. 2000a. Viburnales: cytological features and a new circumscription. Taxon 49: 5–16.

Benko-Iseppon AM and W Morawetz. 2000b. Cytological comparison and Calyceraceae and Dipsacaceae with special reference to their taxonomic relationships. Cytologia 65: 123–128.

Blackmore S and MJ Cannon. 1983. Palynology and systematics of Morinaceae. Rev. Palaeobot. Palynol. 40: 207–226.

Bohm BA and CW Glennie. 1971. A chemosystematic study of the Caprifoliaceae. Canad. J. Bot. 49: 1799–1807.

Böhnke-Gütlein E und F Weberling. 1981. Palynologische Untersuchungen an Caprifoliaceae: I. Sambuceae, Viburneae, und Diervilleae. Akad. Wiss. Lit. (Mainz), Trop. Subtrop. Pflanzenwelt 34: 131–189.

Bolli R. 1994. Revision of the genus *Sambucus*. Dissert. Bot. 223: 1–227.

Cannon MJ and JFM Cannon. 1984. A revision of the family Morinaceae (Magnoliophyta/Dipsacales). Bull. Br. Mus. (Nat. Hist), Bot. Ser. 12: 1–35.

Caputo P and S Cozzolino. 1994. A cladistic analysis of Dipsacaceae (Dipsacales). Plant Syst. Evol. 189: 41–61.

Caputo P, S Cozzolino, and A Moretti. 2004. Molecular phylogenetics of Dipsacales reveals parallel trends in seed dispersal syndromes. Plant Syst. Evol. 246: 163–175.

Chang CS. 1997. Flavonoid chemistry of *Weigela* (Caprifoliaceae) in Korea. J. Plant Res. 110(1098): 275–281.

Clarke G. 1978. Pollen morphology and generic relationships in the Valerianaceae. Grana 17: 61–75.

Clarke G. 1981. The northwest European pollen flora 21. Dipsacaceae. Rev. Paleobot. Palynol. 33: 1–26.

Clarke G and MR Jones. 1977. The northwest European pollen flora 16. Valerianaceae. Rev. Palaeobot. Palynol. 24: 155–179.

Donoghue MJ. 1982. Systematic studies in the genus *Viburnum*. Ph.D. dissertation. Harvard University, Cambridge, MA.

Donoghue MJ. 1983a. A preliminary analysis of phylogenetic relationships in *Viburnum* (Caprifoliaceae s.l.). Syst. Bot. 8: 45–58.

Donoghue MJ. 1983b. The phylogenetic relationships of *Viburnum*. Advances in Cladistics 2: 143–166.

Donoghue MJ. 1985. Pollen diversity in *Viburnum* and the Caprifoliaceae *sensu lato*. J. Arnold Arbor. 66: 421–459.

Donoghue MJ and W Friedman. 1988. Development of the ovary in *Viburnum*. Am. J. Bot. 75(6, Suppl.): 27.

Donoghue MJ and G Levin. 1986. Leaf morphology and evolution in the genus *Viburnum*. Am. J. Bot. 73(6, Suppl.): 760.

Donoghue MJ, RG Olmstead, J Smith, and JD Palmer. 1992. Phylogenetic relationships of Dipsacales based on *rbc*L sequences. Ann. Missouri Bot. Gard. 79: 333–345.

Donoghue MJ, T Eriksson, PA Reeves, and RG Olmstead. 2001. Phylogeny and phylogenetic taxonomy of Dipsacales, with special reference to *Sinadoxa* and *Tetradoxa* (Adoxaceae). Harvard Pap. Bot. 6: 459–479.

Donoghue MJ, CD Bell, and RC Winkworth. 2003. The evolution of reproductive characters in Dipsacales. Int. J. Plant Sci. 164(5 Suppl.): S453–S464.

Donoghue MJ, BG Baldwin, J Li, and RC Winkworth. 2004. *Viburnum* phylogeny based on chloroplast *trn*K intron and nuclear ribosomal ITS DNA sequences. Syst. Bot. 29: 188–198.

Ehrendorfer F. 1964a. Über stammesgeschichtliche Differenzierungsmuster bei den Dipsacaceen. Ber. Deutsch. Bot. Ges. 77: 83–94.

Ehrendorfer F. 1964b. Evolution and karyotype differentiation in a family of flowering plants: Dipsacaceae. In: Genetics Today: Proc. llth Int. Cong. of Genetics, The Hague, 1963, pp. 399–407.

Erbar C. 1994. Contributions to the affinities of *Adoxa* from viewpoint of floral development. Bot. Jahrb. Syst. 116: 259–282.

Eriksen B. 1993. Taxonomical studies in the Polygalaceae and Valerianaceae. Thesis, Fac. Nat. Sc., Univ. Goteborg.

Eriksson LT and MJ Donoghue. 1997. Phylogenetic relationships of *Sambucus* and *Adoxa* (Adoxaceae) based on nuclear ribosomal ITS sequence and preliminary morphological data. Syst. Bot. 22: 555–573.

Ernet D. 1977. Blütenbau und Fortpflanzungsbiologie von *Valerianella* und *Fedia* (Valerianaceae). Plant Syst. Evol. 128: 1–22.

Ernet D. 1978. Fruchtbau und Verbreitungsbiologie von *Valerianella* und *Fedia* (Valerianaceae). Plant Syst. Evol. 130: 85–126.

Farmer JB. 1888–1889. On the development of the endocarp in *Sambucus nigra*. Ann. Bot. 2: 389–392.

Ferguson IK. 1965. The genera of Valerianaceae and Dipsacaceae in the southeastern United State. J. Arnold Arbor. 46: 218–231.

Ferguson IK. 1966. The genera of Caprifoliaceae in the southeastern United States. J. Arnold Arbor. 47: 33–59.

Fukuoka N. 1968. Phytogeny of the tribe Linnaeeae. Acta Phytotax. Geobot. 23: 82–94 (in Japanese with English summary).

Fukuoka N. 1972. Taxonomic study of the Caprifoliaceae. Mem. Coil. Sci. Kyoto Univ., Ser, Biol. 6: 15–58.

Fukuoka N. 1974. Floral morphology of *Adoxa moschatellina*. Acta Phytotax. Geobot. 26: 65–76.

Fukuoka N. 1975. Studies in the systematics of Caprifoliaceae, part 2. Acta Phytotax. Geobot. 26: 133–139.

Gregor J and D Ernet. 1971. Flavonoide und Systematik der Valerianaceae. Naturwissenschaften 58: 416–417.

Greilhuber J. 1979. C-band distribution, DNA-content, and base composition in *Adoxa moschatellina* (Adoxaceae): a plant with cold-sensitive chromosome segments. Plant Syst. Evol. 131: 243–259.

Gtitlein R and F Weberling 1982. Fruchtanatomische Untersuchungen an Valerianaceae. Ber. Deutsch. Bot. Ges. 95: 35–43.

Hara H. 1981. A new species of the genus *Adoxa* from Mt. Omei of China. J. Jpn. Bot. 56: 271–274.

Hara H. 1983. A revision of Caprifoliaceae of Japan with reference to allied plants in other districts and to Adoxaceae. Ginkgoana (Tokyo) 5: 1–336.

Hidalgo O, T Garnatje, A Susanna, and J Mathez. 2004. Phylogeny of Valerianaceae based on *mat*K and ITS markers, with reference to *mat*K individual polymorphism. Ann. Bot. N.S. 93: 283–294.

Hillebrand GR. 1969. A serological investigation of intrageneric relationships of *Viburnum* (Caprifoliaceae). Bull. Torrey Bot. Club 96: 556–567.

Hillebrand GR and DE Fairbrothers. 1970a. Serological investigation of the systematic position of the Caprifoliaceae: I. Correspondence with selected Rubiaceae and Cornaceae. Am. J. Bot. 57: 810–815.

Hillebrand GR and DE Fairbrothers. 1970b. Phytoserological systematic survey of the Caprifoliaceae. Brittonia 22: 125–133.

Hock F. 1892. Zur systematischen Stellung von *Sambucus*. Beih. Bot. Centralbl. 51: 233–234.

Hofmann U. 1989. Warum sind *Sambucus* und *Viburnum* keine Dipsacales? In: A Weber, E Vitek and M Kiehn, eds. Symposium Morphologie, Anatomie und Systematik: 24. Wien.

Hofmann U and JG Göttmann. 1990. *Morina* L. und *Triplostegia* Wall. ex DC. im Vergleich mit Valerianaceae und Dipsacaceae. Bot. Jahrb. Syst.111: 499–553.

Hounsel RW. 1968. Cytological studies in *Sambucus*. Can. J. Genet. Cytol. 10: 237–247.
Hsu P-Sh. 1983. A preliminary numerical taxonomy of the family Caprifoliaceae. Acta Phytotax. Sinica 21: 26–33.
Jordheim M, NH Giske, and ØM Andersen. 2006. Anthocyanins in Caprifoliaceae. Biochem. Syst. Ecol. 35: 153–159.
Kamelina OP. 1980. Comparative embryology of the families Dipsacaceae and Morinaceae. Nauka, Leningrad (in Russian).
Kamelina OP. 1983. Basic results of the comparative embryological investigation of Dipsacaceae and Morinaceae. In: Fertilization and embryogenesis in ovulated plants, pp. 343–346. VEDA, Bratislava.
Kamelina OP. 1987. Adoxaceae, Morinaceae, Dipsacaceae. In: MS Yakovlev, ed. Comparative embryology of flowering plants: Davidiaceae-Asteraceae, pp. 168–171, 177–192. Nauka, Leningrad (in Russian).
Kamelina OP. 1991. Comparative-embryological analysis as a method of phylogenetic systematics of flowering plants. Masters thesis, University of Tashkent (in Russian).
Kamelina OP and MA Plisko. 2000. Dipsacaceae, Morinaceae. In: A Takhtajan, ed. Comparative seed anatomy, vol. 6, pp. 400–413. Nauka, St. Petersburg (in Russian).
Kamelina OP and MS Yakovlev. 1974. Development of the embryo sac in the Genus *Morina*. Bot. Zhurn. 59: 1609–1617 (in Russian).
Kamelina OP and MS Yakovlev. 1976. Development of the anther and microgametogenesis in the representatives of Dipsacaceae and Morinaceae. Bot. Zhurn. 61: 932–945 (in Russian).
Kamelina OP and GY Zhukova. 1987. Valerianaceae. In: MS Yakovlev, ed. Comparative embryology of flowering plants: Davidiaceae-Asteraceae, pp. 171–177. Nauka, Leningrad (in Russian).
Kamelina OP, NG Tuchina, and MA Vishnyakova. 1987. Caprifoliaceae. In: MS Yakovlev, ed. Comparative embryology of flowering plants: Davidiaceae-Asteraceae, pp. 158–168. Nauka, Leningrad (in Russian).
Katina ZF. 1953. Anatomical data on the localization of essential oil in some species of *Valeriana*. Bot. Zhurn. Kiev 10: 81–86 (in Ukrainian).
Kern JH and CGGJ van Steenis. 1951. Caprifoliaceae. In: CGGJ van Steenis, ed. Flora malesiana ser. 1, vol. 4, pp. 175–194. Noordhoff-Kolff N.V., Djakarta.
Lagerberg T. 1909. Studien über die Entwicklungsg-eschichte und systematische Stellung von *Adoxa mos-chatellina* L. Kongl. Svenska Vetenskapsakad. Handl., 2nd ser. 44: 1–86.
Liang H-X. 1993. The chromosome numbers of Adoxaceae and their systematic significance. Acta Bot. Yunn. 15: 260–262.
Lobstein A, G Haan Archipoff, J Englert, JG Kuhry, and R Anton. 1999. Chemotaxonomical investigation in the genus *Viburnum*. Phytochemistry 50: 1175–1180.
Maheshwari P. 1946. The *Adoxa* type embryo sac: a critical review. Lloydia 9: 73–113.
Moissle HE. 1941. Vergleichende embryologische Studien über die Familie der Caprifoliaceae. Oesterr. Bot. Z. 90: 153–212.
Mukherjee A, S Gupta, and S Gantait. 1997. A census of Dipsacaceae Juss. in the Eastern Himalaya. J. Nation. Bot. Soc. (India) 51(1–2): 87–91.
Nepomnyashchaya OA. 1984. Structure of the flower and evolutionary trends in species of the genus *Adoxa* (Adoxaceae). Bot. Zhurn. 69: 1030–1039 (in Russian).
Neubauer HF. 1977. Morphologische Beobachtungen an Semlingen von *Sambucus nigra*. Phyton 18: 57–69.
Nilova MV. 2001. Comparative bark anatomy of representatives of the family Caprifoliaceae s.l. Bot. Zhurn. 86(11): 37–48 (in Russian with English summary).
Ogata K. 1988. Wood anatomy of the Caprifoliaceae of Japan. IAWA Bull. n.s. 9: 299–316.
Patel VC and JJ Skvarla. 1979. Valerianaceae pollen morphology. Pollen et Spores 21: 81–104.
Peng C-L, H Tobe, and M Takahashi. 1995. Reproductive morphology and relationships of *Triplostegia* (Dipsacales). Bot. Jahrb. Syst. 116: 505–516.
Perdetzoglou DK, C Kofinas, I Chinou, A Loukis, and C Harvala. 2000. A comparative chemotaxonomic study of eight taxa of the *Dipsacaceae* family. Plant Biosyst. 134: 213–218.
Persidsky D. 1939. Gynoecium evolution in the family Caprifoliaceae. J. Inst. Bot. Acad. Sci. Ukraine 21–22(20–30): 45–75 (in Ukrainian with Russian and English summaries).
Plisko MA. 2000a. Viburnaceae. In: A Takhtajan, ed. Comparative seed anatomy, vol. 6, pp. 378–383. Nauka, St. Petersburg (in Russian).
Plisko MA. 2000b. Adoxaceae. In: A Takhtajan, ed. Comparative seed anatomy, vol. 6, pp. 389–392. Nauka, St. Petersburg (in Russian).
Plisko MA. 2000c. Caprifoliaceae, Valerianaceae, Triplostegiaceae. In: A Takhtajan, ed. Comparative seed anatomy, vol. 6, pp. 367–378, 392–400. Nauka, St. Petersburg (in Russian).
Plisko MA and NG Tuchina. 2000. Sambucaceae. In: A Takhtajan, ed. Comparative seed anatomy, vol. 6, pp. 383–389. Nauka, St. Petersburg (in Russian).
Pyck N and E Smets. 2000. A search for the phylogenetic position of the sevenson flower (*Heptacodium*, Dipsacales): combining molecular and morphological evidence. Plant Syst. Evol. 225: 185–199.
Pyck N and E Smets. 2001. Dipsacales phylogeny: combining chloroplast sequences with morphological evidence. In Botany 2001: Plants and People. Abstracts, p. 162. Albuquerque.
Pyck N and E Smets. 2004. On the systematic position of *Triplostegia* (Dipsacales): a combined molecular and morphological approach. Belg. J. Bot. 137: 125–139.
Pyck N, P Roels, and E Smets. 1999. Tribal relationships in Caprifoliaceae: evidence from a cladistic analysis using *ndh*F sequences. Syst. Geogr. Plant 69: 145–159.
Pyck N, A van Lysebetten, J Stessens, and E Smets. 2002. The phylogeny of Patrinieae sensu Graebner (Valerianaceae) revisited: additional evidence from *ndh*F sequence data. Plant Syst. Evol. 233: 29–46.
Reese Krug H, E Meyer, M Hildenbrand, and F Weberling. 2001. Palynological investigations in Valerianaceae: some elementary aspects and problems. Wulfenia 8: 61–80.
Reidt G and P Leins. 1994. Das Initialstadium der sympetalen Krone bei *Sambucus racemosa* L. und *Viburnum farreri* Steam. Bot. Jahrb. Syst. 116: 1–9.
Reitsma T and AAML Reuvers. 1976. Adoxaceae. In: W. Punt, ed. The northwest European pollen flora, vol. 1, pp. 71–73. Elsevier, Amsterdam.

Roels P and E Smets. 1994. A comparative floral ontogenetical study between *Adoxa moschatellina* and *Sambucus edulus*. Belg. J. Bot. 127: 157–170.
Samutina ML. 1986. Comparative-morphological analysis of the pollen of the genus *Sambucus* (Caprifoliaceae). Bot. Zhurn. 71: 168–174 (in Russian).
Sax K and DA Kribs. 1930. Chromosomes and phylogeny in Caprifoliaceae. J. Arnold Arbor. 11: 147–153.
Sprague TA. 1927. The morphology and taxonomy of the Adoxaceae. Bot. J. Linn. Soc. 47: 471–487.
Stabbetorp OE. 1989. Gynoecial anatomy in *Sambucus callicarpa* (Caprifoliaceae) with emphasis on meiotic divisions in a special tissue. Nord. J. Bot. 9: 73–79.
Suneson S. 1933. Zur Embryologie der Gattung *Viburnum*. Bot. Not. 1933: 181–194.
Toko E. 1980. Embryology of *Sambucus racemosa* L. Acta Biol. Crakov. (Bot.). 2(2): 173–188.
Troll W and F Weberling. 1966. Die Infloreszenzen der Caprifoliaceae und ihre systematische Bedeutung. Akad. Wiss. Abh. Math.-Naturwiss. Kl. 1966(4): 459–605.
Verlaque R. 1977. Rapports entre les Valerianaceae, les Morinaceae, et les Dipsacaceae. Bull. Soc. Bot. France 124: 475–482.
Verlaque R. 1983. Contribution a l'étude du genre *Morina* L. Pollen and Spores 25: 143–162.
Vijayaraghavan MR and GS Sarveshwari. 1968. Embryology and systematic position of *Morina longifolia* Wall. Bot. Not. 121: 383–402.
Vinokurova LV. 1959. Palynological data on the sys-tematics of the families Dipsacaceae and Morinaceae. Problems Bot. 4: 51–67 (in Russian).
Wagenitz G and B Laing. 1984. Die Nektarien der Dipsacales und ihre systematlsche Bedeutung. Bot. Jahrb. Syst. 104: 91–113.
Weberling F. 1957. Morphologische Untersuchungen zur Systematik der Caprifoliaceen. Abh. Akad. Wiss. Lit. Mainz (Math. Nat.) 1957: 1–50.
Weberling F. 1961. Die Infloreszenzen der Valerianaceen und ihre systematische Bedeutung. Akad. Wiss. Abh. Math.-Natunviss. Kl. 5: 151–281.
Weberling F. 1977. Vergleichende und entwicklungsgeschichtliche Untersuchungen über die Haarformen der Dipsacales. Beitr. Biol. Pfl. 53: 61–89.
Weberling F. 1978. Monographic der Gattung *Nardosta- chys* (Valerianaceae). Bot. Jahrb. Syst. 99: 188–221.
Weberling F and M Hildenbrand. 1982. Tapetument-wicklung bei *Triosteum* L., *Leycesteria* Wall., und *Kolkwitzia* Graebn. (Caprifoliaceae). Beitr. Biol. Pfl. 57: 481–486.
Weberling F and M Hildenbrand. 1986. Weitere Untersuchungen der Tapetumentwicklung der Caprifoliaceae. Beitr. Biol. Pfl. 61: 3–20.
Wilkinson AM. 1948a. Floral anatomy and morphology of some species of the tribe Lonicereae of the Caprifoliaceae. Am. J. Bot. 35: 261–271.
Wilkinson AM. 1948b. Floral anatomy and morphology of some species of the tribes Linnaeeae and Sambuceae. Am. J. Bot. 35: 365–371.
Wilkinson AM. 1948c. Floral anatomy and morphology of the genus *Viburnum* of the Caprifoliaceae. Am. J. Bot. 35: 455–465.
Wilkinson AM. 1949. Floral anatomy and morphology of *Triosteum* and of the Caprifoliaceae in general. Am. J. Bot. 36: 481–489.
Winkworth RC and MJ Donoghue. 2004. *Viburnum* phylogeny: evidence from the duplicated nuclear gene GBSSI. Mol. Phylogenet. Evol. 33: 109–126.
Winkworth RC and MJ Donoghue. 2005. *Viburnum* phylogeny based on combined molecular data: implication for taxonomy and biogeography. Am. J. Bot. 92: 653–666.
Wu C-Y. 1981. Another new genus of Adoxaceae, with special reference on the intrafamiliar evolution and the systematic position of the family. Acta Bot. Yunn. 3: 383–388 (in Chinese with English summary).
Wu C-Y, ZL Wu, and RF Huang. 1985. *Sinadoxa* C.-Y. Wu, Z. L. Wu, et R. F. Huang, genus novum familiae Adoxacearum. Acta Phytotax. Sinica 19: 203–210.
Zhang W-H, Z-D Chen, H-B Chen, and Y-C Tang. 2001. Phylogenetic relationships of the disputed genus *Triplostegia* based on *trn*L-F sequences. Acta Phytotax. Sinica 39: 337–344 (in Chinese with English summary).
Zhang W-H, Z-D Chen, J-H Li, H-B Chen, and Y-C Tang 2003. Phylogeny of the Dipsacales s.l. based on chloroplast *trn*L-F and *ndh*F sequences. Mol. Phylogenet. Evol. 26: 176–189.

Order 113. AQUIFOLIALES

Trees and shrubs, sometimes woody lianas. Vessels with scalariform or (Icacinaceae) scalariform and simple perforations, or simple perforation; lateral pitting scalariform to alternate. Fibers usually with distinctly bordered pits. Rays heterogeneous, with elongate or short ends. Axial parenchyma apotracheal to paratracheal. Sieve-element plastids of S-type. Nodes trilacunar or less often unilacunar. Leaves alternate or rarely opposite, simple, estipulate or (most of Aquifoliaceae and Helwingiaceae) with very small and caduceus stipules. Stomata of diverse types. Flowers small, greenish or whitish, in cymose or racemose inflorescences or rarely solitary, bisexual or unisexual, actinomorphic, usually 4-merous or 5-merous. Sepals free or more or less connate. Petals free or less often connate, imbricate or valvate, rarely wanting. Androecium haplostemonous with antesepalous stamens; anthers tetrasporangiate, usually opening longitudinally. Tapetum secretory. Microsporogenesis simultaneous. Pollen grains 2-celled or 3-celled, usually 3-colporate. Gynoecium mostly of 2–5 united carpels, with short style or sessile stigma. Ovary superior or inferior, with 1–2 pendulous ovules in each locule. Ovules commonly anatropous, unitegmic or (Sphenostemonaceae and Cardiopteridaceae) bitegmic, crassinucellate to tenuinucellate, sometimes (Aquifoliaceae) with endothelium. Female gametophyte of *Polygonum*-type. Endosperm cellular (Aquifoliaceae and *Nothapodytes* in Icacinaceae) or nuclear (*Gonocaryum* in Icacinaceae) (the other families not studied for nuclear development). Fruits drupaceous, baccate, or samara. Seeds exotestal, with very small to large embryo; endosperm present or absent. Icacinaceae contain iridoid compounds.

According to molecular date (Olmstead et al. 2000; Soltis et al. 2000, 2006; Bremer et al. 2002) Aquifoliales related to Dipsacales and Apiales.

Key to Families

1 Gynoecium of 3–5 carpels.

2 Ovary superior.

3 All (2)4–6 locules of the ovary fertile. Evergreen or deciduous trees (usually small) or shrubs, sometimes climbers and epiphytic; resiniferous and laticiferous idioblasts present. Vessels with exclusively scalariform perforation, and without vestured pits, usually with numerous bars; lateral pitting usually opposite, rarely scalariform or alternate. Fibers often septate. Rays heterogeneous with elongate ends. Axial parenchyma diffuse. Nodes unilacunar, or trilacunar. Leaves alternate or very rarely opposite, or whorled, entire, leathery, toothed or spinose, often with resiniferous or laticiferous cells in mesophyll; stipules very small and often caducous or wanting. Stomata mostly anomocytic. Flowers very small, in axillary or supraaxillary fascicles or cymes, dioecious or occasionally polygamous or (*Nemopanthus*) polygamo-dioecious, 4-merous or less often 5-merous, rarely 6–9-merous. Sepals small, more or less connate at the base, imbricate, in *Nemopanthus* minute and deciduous in male flowers and wanting from female ones. Petals free (*Nemopanthus*) or shortly connate at base (*Ilex*), imbricate. Stamens usually as many as and alternate with the petals, mostly adnate to the petals at base, sometimes more numerous than the petals in polycarpous species of *Ilex*. Staminodia and pistillodia present. Pollen grains 3-colporate. Gynoecium usually of 4–5 carpels, rarely less (2–3) or more (up to 24); style short or nearly wanting, with lobed or capitate stigma; ovary with solitary or very rarely two ovules in each locule; ovules anatropous, unitegmic, crassinucellate to tenuinucellate (two species of *Ilex*), with endothelium. Fruits drupes with as many pyrenes as carpels. Seeds with cuboid exotestal cells, tangentially elongated, inner walls lignified, rest crushed, endotesta tanniniferous; embryo near micropyle, very small; endosperm copious, oily and proteinaceous, without starch. Plants sometimes producing alkaloids and flavonols (kaempferol and quercetin), n = 9, 10. . . . . 1. AQUIFOLIACEAE.

3 Fertile locule solitary, rarely all three locules fertile. Trees and shrubs, sometimes lianas with non-axillary branch tendrils, laticiferous, or non-laticiferous and without coloured juice. Trichomes unicellular. Stem often with anomalous structure, sometimes with interxylary phloem. Vessels scalariform, scalariform and simple, or (in stems with unilacunar nodes) only simple perforations; vessels without vestured pits; lateral pitting usually intermediate between scalariform and opposite or (in vessels with simple perforations) alternate. Fibers usually with bordered pits. Rays heterogeneous with elongate or short ends. Axial parenchyma apotracheal or less often paratracheal. Nodes trilacunar or less often unilacunar. Leaves alternate or less often opposite (*Iodes*), leathery, petiolate, entire to toothed, estipulate. Stomata anomocytic, anisocytic, or paracytic. Inflorescences axillary, sometimes terminal, or opposite leaves (*Citronella*), racemes, thyrsoids, cymes, pseudospikes (pistilate plants of *Calatola*). Flowers small, bisexual or less often polygamous to dioecious, (3)4–5(6)-merous. Sepals more or less connate, imbricate or rarely valvate, rarely wanting. Petals free or rarely basally connate, usually valvate, rarely wanting. Stamens as many as and alternate petals; filaments free or borne on corolla tube alternate with its lobes; anthers usually introrse, dorsifixed, opening longitudinally or less often by apical pores. Pollen grains 2-celled, aperturate, or nonaperturate (*Stachyanthus*), colpate, or porate, or colporate, or foraminate. Nectary disc mostly absent, but sometimes present or represented by distinct glands. Gynoecium of (2)3(-5) carpels with short style; stigma punctiform, subcapitate, capitate, or lobed; ovary usually 1-locular by abortion of the remaining locules; rarely (*Emmotum*) 3-locular; ovules

two, rarely one per locule, pendulous, anatropous, unitegmic but with free integuments at the micropyle in *Phytocrene,* crassinucellate to tenuinucellate; funicle usually thickened near the micropyle. Fruits usually drupaceous, often flattened and or ribbed, 1-seeded, rarely dry and winged. Seeds with straight or curved, green embryo; endosperm mostly well developed, oily, sometimes wanting. Plants sometimes with iridoid compounds, flavonols (quercetin); alkaloids and proanthocyanidins present or absent; verbascosides present (*Cassinopsis*), n = 10, 12. . . . 2. ICACINACEAE.

2 Ovary inferior. Glabrous dioecious shrubs. Vessels with scalariform perforations that have 17–41 widely spaced bars; lateral pitting scalariform. Fibers with minutely bordered pits in radial walls, all septate. Axial parenchyma scanty paratracheal. Rays heterogeneous. Silica grains present in ray cells, and axial parenchyma, and also in septate fibres of *Helwingia himalaica*. Nodes unilacunar. Leaves alternate to subopposite, spiral, petiolate, simple, remotely serrate or crenate-serrate, ovate to lanceolate, with deciduous stipules often divided into filiform lobes. Stomata anomocytic. Flowers unisexual, in small umbels mostly borne on the leaves through the axillary peduncle being adnate to the upper side of midrib, the males several in an umbel, the female 1–3; bracts absent, male flowers sometimes with more or less long pedicel, female flowers with very short pedicels. Calyx limb obsolete; petals 3–5, valvate, continuous with the calyx tube. Stamens 3–5, alternipetalous, inserted outside flat angled nectary disc; anthers short. Pollen grains 3-colporate, with perforate tectum and diffuse endoaperture. Gynoecium of 3–4 carpels; flat nectary disc forming a cone passing into short style with 3–4 recurved stigmas; ovary inferior, 3–4-locular, with one pendulous, anatropous, with dorsal raphe, unitegmic, tenuinucellate ovule per locule. Fruits ovoid or globose drupes, angular when dry, with 3–4 separate cartilaginous or crustaceous pyrenes. Seeds with minute and nearly globose embryo at the apex of copious endosperm; seed-coat thin, consists of 2–3 layers compressed, thin-walled cells. Contain caffeic acid (chlorogeneic acid), n = 19.. . . . . 3. HELWINGIACEAE.

1 Gynoecium of two carpels.

4 Stipules present. Trees or shrubs, glabrous, terrestrial or epiphytic; young stem with separate bundles. Vessels with scalariform perforations; lateral pitting opposite. Fibers with bordered pits. Rays heterogeneous. Axial parenchyma apotracheal. Leaves alternate, entire or serrate, acuminate; stipules small, fimbriate, caduceus, with glandular hairs on adaxial surface. Inflorescences epiphyllous, flowers borne on the leaf toward the tip of the midrib. Flowers very small, actinomorphic, bisexual, green or white, fasciculate (cymose false umbels), 5-merous. Sepals connate into obconic tube adnate to ovary, open, toothed, persistent. Petals inserted below the margins of the epigynous flat nectary disc, spreading, ovate-triangular, acute, valvate. Stamens five, with short, subulate filaments and small, bilobed anthers. Pollen grains 3-colporate, with complete tectum and granules. Gynoecium of two carpels and two small, recurved stigmas on very short stylodia; ovary inferior, incompletely 2-locular, with usually numerous 2-seriate ovules on two parietal placentas. Fruits small, incompletely 2-locular, 3–6-seeded berries. Seeds oblong, slightly curved, rugose; testa multilayered, exotestal cells large, thick-walled, mucilaginous; embryo very minute at the base of fleshy endosperm. . . . . . . 4. PHYLLONOMACEAE.

4 Leaves estipulate.

5 Ovary 2-locular. Evergreen trees and shrubs. Vessels with scalariform perforations that have 32–185 bars; lateral pitting scalariform; phloem stratified. Leaves alternate or subopposite or pseudo-verticillate, simple, leathery, glandular-dentate to entire. Flowers in subterminal, racemose inflorescences, bracteate, bisexual, actinomorphic, 4-merous. Sepals free, imbricate, caducous. Petals free, fleshy, caducous, or wanting. Stamens 4–6, or 8–12, filaments short and thick or obsolete; anthers with thick connective, elongate and latrorse or short and introrse, petaloid or filantherous. Stamens laminal with embedded microsporangia. Pollen grains 3-perorate, reticulate to scabrate. Gynoecium of two carpels with large sessile capitate stigma. Ovary superior, 2-locular, with 1(2), pendulous ovules in each carpel; ovules anatropous, bitegmic, crassinucellate. Fruits drupelike (berry, according to Bailey 1956,

pseudo-drupe – van Steenis 1986), with separable pyrenes or with one stone (with 1–2 bony pyrenes), usually more or less asymmetrical, ovoid, crowned by the stigma (Watson and Dallwits 2006). Pericarp fleshy, finally black. Seeds more or less ruminate, exotestal or exo-endotestal, endotestal cells with dark contents; embryo small, surrounded by a thick, chalazal envelope, of which the outer layer is hard and bony and whether or not ruminate functioning as and superficially resembling the structure of a pyrene (van Steenis 1986), endosperm copious, oily. . . . . 5. SPHENOSTEMONACEAE.

5 Ovary 1-locular. Climbing herbs with articulated laticifers bearing abundant milky juice in the leaves and stems. Vessel elements short and very broad, with simple perforations; lateral pitting alternate. Fibers with small bordered pits. Nodes trilacunar with three traces. Leaves alternate, thinly membranous, simple or lobed to varying degrees, long-petioled, cordate, 3–7-veined from the base, estipulate. Stomata anomocytic and paracytic. Flowers very small, in axillary branched, more or less scorpioid cymes, bisexual or polygamous, actinomorphic, 4–5-merous. Calyx (4-)5-lobed, lobes imbricate, persistent. Corolla shortly sympetalous, (4-)5-lobed, lobes spreading, thin, imbricate, caduceus. Stamens (4-)5, inserted midway down the corolla tube, or in the throat of the corolla tube and alternate with the sepals; filaments very short, glabrous; anthers introrse, dorsifixed, tetrasporangiate, opening longitudinally. Pollen grains 3-colporate, with striate ornamentation. Nectary disc wanting. Gynoecium of two united carpels; stylodia two, dissimilar, one longer, thicker, cylindrical or subclavate, persistent on fruit, the other very short, thinner, with capitate stigma; ovary superior, 1-locular, with two ovules pendulous from the summit of the locule, one of which often abortive. Ovules anatropous, bitegmic, crassinucellete, with dorsal raphe. Fruits obovate-oblong, dry, indehiscent, flat, 1-seeded, broadly 2-winged samara, more or less stipitate; wings transversely striolate. Seeds linear, sulcate, exotestal, embryo minute, well differentiated, at the apex of very densely granular; endosperm fleshy. . . . . 6. CARDIOPTERIDACEAE.

1. AQUIFOLIACEAE

A. Richard 1828 (including Ilicaceae Berchtold et J. Presl 1820). 2/400–450. *Ilex* is very widely distributed; *Nemopanthus* (2) is found only in eastern North America.

Ilex (including *Byronia*), *Nemopanthus*

One is the most archaic families which is characterized by primitive wood structure, the absence of nectarines, and very small embryo.

2. ICACINACEAE

Mires 1851. 53/400 (including Emmotaceae Van Tieghem 1899; Iodaceae Van Tieghem 1899; Leptaulaceae Van Tieghem 1899; Metteniusaceae H. Karsten ex Schnizlein 1843–1870; Pennantiaceae J.G. Agardh 1858; Phytocrenaceae Arnott ex R. Browne 1852; Pleurisanthaceae Van Tieghem 1899; Sarcostigmataceae Van Tieghem ec Bullock 1958; Stemonuraceae Kårehed 2001). Pantropical, with relatively few species in temperate regions.

ICACINEAE: *Citronella* (including *Villaresia*), *Emmotum, Poraqueiba, Oecopetalum, Pittosporopsis, Pseudobotrys, Platea, Gonocaryum, Medusanthera, Ottoschulazia, Gomphandra, Rhyticaryum, Hartleya, Codiocarpus, Medusanthera, Apodytes, Casimirella, Leretia, Pleurisanthes, Merrilliodendron, Icacina, Mappia, Lasianthera, Cantleya, Nothapodytes, Stemonurus, Whitmorea,* etc.; DISCOPHOREAE: *Discophora*; IODEAE: *Mappianthus, Polyporandra, Iodes, Hosiea, Natsiatum, Natsiatopsis;* PHYTOCRENEAE: *Pyrenacantha* (including *Trematosperma*), *Miquelia, Phytocrene,* etc.; SARCOSTIGMATEAE: *Sarcostigma*; *METTENIUSEAE*: *metteniusa.*

Insertae sedis: *Alsodeiopsis, Chlamydocarya, Desmostachys, Lavigeria, Polycephalium, Stachyanthus, Cassinopsis, Calatola, Raphiostylis, Dendrobangia, Leptaulus, Gastrolepis, Grisollea, Irvingbaileya.*

Share many common features with the Aquifoliaceae.

3. HELWINGIACEAE

Decaisne 1836. 1/5. Eastern Himalayas (from central Nepal to Bhutan), northeastern India (Assam and Mantpur), northern Burma, continental China, northern Vietnam, Taiwan, Ryukyu Islands, Japan.

Helwingia

Helwingiaceae related to Aquifoliaceae and Phyllonomaceae (Soltis et al. 2006).

4. PHYLLONOMACEAE

Small 1905 (Dulongiaceae J. Agardh 1858, nom. illeg). 1/4. Mexico to northwestern Bolivia.

Phyllonoma (*Dulongia*)

According to Soltis and Soltis (1997), "Analyses or *rbcL* sequences similarly revealed that *Phyllonoma* is allied with traditional asterids, with a particularly close relationship between *Phyllonoma* and *Helwingia*".

5. SPHENOSTEMONACEAE

P. Royen and Airy Shaw 1972. 1/7–10. Central Malesia to New Guinea, tropical Australia, and New Caledonia.

Sphenostemon

Close to the Aquifoliaceae and especially to Icacinaceae (Baas 1975).

6. CARDIOPTERIDACEAE

Blume 1847. 1/3 (including Peripterygiaceae G. King 1895). Northeastern India (Assam and Bengal), Burma, Southeast Asia, Malesia from Malay Peninsula to New Guinea, and Queensland.

Cardiopteris (*Peripterygium*)

Related to the Icacinaceae, but differ in bitegmic ovules and fruit type.

Bibliography

Baas P. 1973. The wood anatomical range in *Ilex* (Aquifoliaceae) and its ecological and phylogenetic significance. Blumea 21: 193–258.

Baas P. 1974. Stomatal types in Icacinaceae: additional observations of genera outside Malesia. Acta Bot. Neerl. 23: 193–200.

Baas P. 1975. Vegetative anatomy and the affinities of Aquifoliaceae, *Sphenostemon, Phelline,* and *Oncotheca.* Blumea 22: 311–407.

Baas P. 1984. Vegetative anatomy and the taxonomic status of *Ilex collina* and *Nemopanthus* (Aquifoliaceae). J. Arnold Arbor. 65: 243–250.

Bailey IW. 1956. The relationship between *Sphenostemon* of New Caledonia and *Nouhuysia* of New Guinea. J. Arnold Arbot. 37: 360–365.

Bailey IW and RA Howard. 1941a. The comparative morphology of the Icacinaceae. I. Anatomy of the node and internode. J. Arnold Arbor. 22: 125–132.

Bailey IW and RA Howard. 1941b. The comparative morphology of the Icacinaceae. II. Vessels. J. Arnold Arbor. 22: 171–187.

Bailey IW and RA Howard. 1941c. The comparative morphology of the Icacinaceae. III. Imperforate tracheary elements and xylem parenchyma. J. Arnold Arbor. 22: 432–442.

Bailey IW and RA Howard. 1941d. The comparative morphology of the Icacinaceae. IV. Rays and the secondary xylem. J. Arnold Arbor. 22: 555–568.

Bernardi L. 1964. La position systematique du genre *Sphenostemon* Baillon sensu van Steenis. Candollea 19: 199–205.

Copeland HF. 1963. Structural notes on lollies (*Ilex aquifolium* and *I. cornuta,* family Aquifoliaceae). Phytomorphology 13: 455–464.

Cuénoud P, MA Del Pero-Martinez, P-A Loizeau, R Spichiger, S Andrews, and J-F Manen. 2000. Molecular phylogeny and biogeography of the genus *Ilex* (Aquifoliaceae). Ann. Bot. 85: 111–122.

Dahl O. 1952. The comparative morphology of the Icacinaceae: VI. The pollen. J. Arnold Arbor. 33: 252–295.

Dahl O. 1955. The pollen morphology of several genera excluded from the family Icacinaceae. J. Arnold Arbor. 36: 159–163.

Dickinson TA and R Sattler. 1974. Development of the epiphyllous inflorescence of *Phyllonoma integerrima* (Turcz.) Loes.: implications for comparative morphology. Bot. J. Linn. Soc. 69: 1–13.

Dickinson TA and R Sattler. 1975. Development of the epiphyllous inflorescence of *Helwingia japonica* (Helwingiaceae). Am. J. Bot. 62: 962–973.

Doweld AB. 2000a. Aquifoliaceae. In: A Takhtajan, ed. Comparative seed anatomy, vol. 6, pp. 112–114. Nauka, St. Petersburg (in Russian).

Doweld AB. 2000b Icacinaceae. In: A Takhtajan, ed. Comparative seed anatomy, vol. 6, pp. 114–117. Nauka, St. Petersburg (in Russian).

Doweld AB. 2000b. Sphenostemonaceae. In: A Takhtajan, ed. Comparative seed anatomy, vol. 6, pp. 118–119. Nauka, St. Petersburg (in Russian).

Doweld AB. 2000c. Cardiopteridaceae. In: A Takhtajan, ed. Comparative seed anatomy, vol. 6, pp. 120–121. Nauka, St. Petersburg (in Russian).

Fagerlind F. 1945. Bau des Gynoceums, der Samenanlage, und des Embryosackes bei einigen Reprasentanten der Familie Icacinaceae. Svensk. Bot. Tidskr. 39: 346–364.

Fang WP. 1951. A study on *Helwingia* Willd. Syml. Sinica 7(3): 685–688.

Gottlieb AM, GC Giberti, and L Poggio. 2005. Molecular analyses of the genus *Ilex* (Aquifoliaceae) in southern South America, evidence from AFLP and ITS sequence date. Am. J. Bot. 92: 352–369.

Hara H and S Kurosawa. 1975. A revision of the genus *Helwingia.* Univ. Mus., Univ. Tokyo, Bull. 8: 393–413.

Haron NW and ST Ping. 1997. Distribution and taxonomic significance of flavonoids in the Olacaceae and Icacinaceae. Biochem. Syst. Ecol. 25: 265–263.

Heintzelmann CE and RA Howard. 1948. The comparative morphology of the Icacinaceae: V. The pubescence and the crystals. Am. J. Bot. 35: 42–52.

Herr JM. 1959. The development of the ovule and megagametophyte in the genus *Ilex* L. J. Elisha Mitchell Sci. Soc. 74: 107–128.

Herr JM. 1969. Endosperm development and associated ovule modifications in the genus *Ilex*. J. Elisha Mitchell Sci. Soc. 77: 26–32.

Howard RA. 1940. Studies of the Icacinaceae. I. Preliminary taxonomic notes. J. Arnold Arbor. 21: 461–489.

Howard RA. 1942a. Studies of the Icacinaceae. II. *Humirianthera, Leretia, Mappia* and *Nothapodytes*, valid genera of the Icacineae. J. Arnold Arbor. 23: 55–78.

Howard RA. 1942b. Studies of the Icacinaceae. III. A revision of *Emmotum*. J. Arnold Arbor. 23: 479–494.

Howard RA. 1942c. Studies of the Icacinaceae. IV. Considerations of the New World genera. Contr. Gray Herb., Harvard Univ. 142: 3–60.

Howard RA. 1942d. Studies of the Icacinaceae. V. A revision of the genus *Citronella* D. Don. Contr. Gray Herb., Harvard Univ. 142: 60–89.

Howard RA. 1943a. Studies of the Icacinaceae. VI. *Irvingbaileya* and *Codiocarpus*, two new genera of the Icacineae. Brittonia 5: 47–57.

Howard RA. 1943b. Studies of the Icacinaceae. VII. A revision of the genus *Medusanthera* Seeman. Lloydia 6: 133–143.

Howard RA. 1943c. Studies of the Icacinaceae. VIII. Brief notes of some Old World genera. Lloydia 6: 144–154.

Jérémie J. 1996. Deux nouvelles especes de *Sphenostemon* (Sphenostemonaceae) de Nouvelle-Calédonie. Adansonia 13(3–4): 219–224.

Jérémie J. 1997. Sphenostemonaceae. In: Ph Morat, ed. Flore de la Nouvelle Calédonie, pp. 3–21. Muséum National d'Histoire Naturelle, Paris.

Kaplan MAC, J Ribeiro, and OR Gottlieb. 1991. Chemogeographical evolution of terpenoids in Icacinaceae. Phytochemistry 30: 2671–2676.

Kårehed J. 2001. Multiple origin of the tropical forest tree family Icacinaceae. Am. J. Bot. 88: 2259–2274.

Kårehed J. 2002. Not just hollies – the expansion of Aquifoliales. In Evolutionary Studies in Asterids Emphasising Euasterids II, pp. 1–14. Acta Universitatis Upsaliensis, Uppsala.

Karsten H. 1859. *Metteniusa* Karst. Flora of Colombia 1: 79–80, t. 39.

Kong DR, H Peng, and HX Liang. 2002. A new type of embryo sac in *Cardiopteris* and its systematic implication. Acta Bot. Sinica 44: 496–498.

Lobreau-Callen D. 1972, 1973. Pollen des Icacinaceae: I. Atlas (1). II. Observations en microscopic electronique, correlations, conclusions (1). Pollen et Spores 14: 345–388, 1972; 15: 47–89, 1973.

Lobreau-Callen D. 1980. Caractéres comparés du pollen des Icacinaceae et des Olacaceae. Adansonia 20: 29–89.

Lobreau-Callen D. 1982. Structures et affinites polliniques des Cardiopterydaceae, Dipentodontaceae, Erythropalaceae, et Octocnemataceae. Bot. Jahrb. Syst. 17: 127–139.

Lozano-Contreras G and Nubia B de Lozano. 1988. Metteniusaceae. In: P Pinto and G Lozano, eds. Flora de Colombia, Monografia No. 11. Bogota.

Mauritzon J. 1936. Embryologische Angaben über Stakhousiaceae, Hippocrateaceae, und Icacinaceae. Svensk. Bot. Tidskr. 30: 541–550.

Metcalfe CR. 1956. The taxonomic affinities of *Sphenostemon* in the light of the anatomy of its stem and leaf. Kew Bull. 1956: 249–253.

Mori SA and JA Kallunki. 1977. A revision of the genus *Phyllonoma* (Grossulariaceae). Brittonia 29: 69–84.

Padmanabhan D. 1961. A contribution to the embryology of *Gomphandra polymorpha*. Proc. Natl. Inst. Sci. India 27B: 389–398.

Patel RN and A Bowles. 1978. Wood anatomy of the dicotyledons indigenous to New Zealand. 12. Icacinaceae. N. Z. J. Bot. 16: 7–12.

Pittier H. 1925. Arboles y arbustos nuevos de Venezuela. Bol. Sci. Techn. Mus. Corn. Venez. 1: 45–47.

Potgieter MJ and AE van Wyk. 1994. Fruit structure of the southern African species of *Apodytes* E Meyer ex Arn. (Icacinaceae). Bot. J. Linn. Soc. 115: 221–233.

Potgieter MJ and AE van Wyk. 1999. Leaf anatomy of the southern African Icacinaceae and its taxonomic significance. South Afr. J. Bot. 65: 153–162.

Powell M, V Savolainen V, P Cuénoud, J-F Manen, and S Andrews. 2000. The mountain holly (*Nemopanthus mucronatus:* Aquifoliaceae) revisited with molecular data. Kew Bull. 55: 341–347.

Savinov IA. 2003. Comparative carpology of the genus *Sphenostemon* (Sphenostemonaceae) in the context of its taxonomy and phylogeny. Bot. Zhurn 88(2): 5–16 (in Russian with English summary).

Simmons MP, V Savolainen, CC Clevinger, RH Archer, and JI Davis. 2001b. Phylogeny of the Celastraceae inferred from 26S nuclear ribosomal DNA, phytochrome B, *rbc*L, *atp*B, and morphology. Mol. Phylogenet. Evol. 19: 353–366.

Sleumer H. 1936. Über die Gattung *Metteniusa* Karsten (= *Aveledoa* Pittier). Notizbl. Bot. Gart. Berlin-Dahlem 13(118): 359–361.

Sleumer H. 1969. Materials towards the knowledge of the Icacinaceae of Asia, Malesia, and adjacent areas. Blumea 17: 181–264.

Sleumer H. 1970. The identity of *Plectomirtha* Oliv. with *Pennantia* J.R. & G. Forster (Icacinaceae). Blumea 18: 217–218.

Sleumer H. 1971. Icacinaceae. In: CGGJ van Steenis, ed. Flora malesiana, ser. 1, vol. 7, pp. 1–87. Noordhoff, Groningen.

Sleumer H. 1971. Cardiopteridaceae. In: CGGJ van Steenis, ed. Flora malesiana, ser. 1, vol. 7, pp. 93–96. Noordhoff, Groningen.

Spichiger R, V Savolainen, and J-F Manen. 1993. Systematic affinities of Aquifoliaceae and Icacinaceae from molecular data analysis. Candollea 48: 459–464.

Steenis CGGJ van. 1955. Some notes on the flora of New Caledonia and reduction of *Nouhuysia* to *Sphenostemon*. Svensk Bot. Tidskr. 49: 19–23.

Steenis CGGJ van. 1986. Sphenostemonaceae. In: CGGJ van Steenis, ed. Flora malesiana, ser. 1, vol. 10, pp. 145–149. Kluwert, Leiden.

Van Staveren MGC and P Baas. 1973. Epidermal characters of the malesian Icacinaceae. Acta Bot. Neerl. 22: 329–359.

Order 114. APIALES

Trees, shrubs, woody lianas, or herbs, glabrous or with various types of trichomes. Mostly with schizogenous secretory canals or cavities containing ethereal oils, resins, and gums usually well developed in various parts, especially in the parenchymatous tissues. Pith bundles and cortical bundles often present. Vessels with simple or sometimes with scalariform, or simple and scalariform perforations; lateral pitting scalariform to alternate. Fibers with simple or (Toricelliaceae) bordered pits, often septate. Rays heterogeneous to seldom homogeneous. Axial parenchyma apotracheal (Toricelliaceae, *Myodocarpus, Delarbrea, Porospermum, Pseudosciadium*) or more often paratracheal. Sieve-element plastids of S-type. Nodes multilacunar, sometimes pentalacunar, or rarely trilacunar. Leaves alternate or rarely opposite or verticillate, simple to compound; petioles usually broad and sheathing, sometimes stipulate, commonly with a ring or arc of vascular bundles, and sometimes with medullary bundles as well; leaf mesophyll cells contain oil bodies. Stomata paracytic, anomocytic, or of other types. Flowers usually small, in terminal or axillary umbels, that are very often arranged into secondary umbels or sometimes into other types of compound inflorescences, rarely in racemes or spikes, bisexual or seldom unisexual, actinomorphic or the outermost ones sometimes more or less zygomorphic, mostly 5-merous. Calyx usually represented by small teeth around the top of the ovary, which often much reduced or even wanting. Petals nearly always free, valvate or sometimes (some Araliaceae) more or less imbricate, rarely wanting. Stamens as many as and alternate with the petals, sometimes twice as many as petals or even (some Araliaceae) numerous; anthers mostly tetrasporangiate, opening longitudinally. Tapetum secretory. Microsporogenesis simultaneous. Pollen grains 3-celled or rarely 2-celled, mostly 3-colporate or (Toricelliaceae) 4-colporate. Gynoecium of 2–5(-many) united carpels; stylodia free or more or less connate into a style that is often enlarged at the base into a stylopodium confluent with the nectary disc; ovary inferior, or semi-inferior, rarely superior, crowned by the glandular nectary disc, with one pendulous ovule per carpel (but an abortive ovule sometimes present). Ovules anatropous, pendulous, epitropous, with ventral or dorsal raphe and with micropyle directed upward and outward, unitegmic, crassinucellate to tenuinucellate, usually with endothelium. Female gametophyte of *Polygonum*-type or sometimes of *Oenothera-, Allium-, Penaea-*, or *Drusa*-types. Endosperm nuclear. Fruits drupaceous or baccate or more often schizocarps consisting of two indehiscent, 1-seeded mericarps usually separating in the plane of union of two carpels, before falling the two mericarps usually hang for a time on a commonly forked stalk, the carpophore, which develops from the carpellary tissues. Seeds exotestal, with small (sometimes minute), straight embryo and copious endosperm. Usually containing sesquiterpenes, triterpenoid compounds, caffeic acid derivatives, and polyacetolenes, iridoids and ellagic acid present or lacking. Seeds containing petroselinic acid.

Related to both the Hydrangeales and Cornales, especially to the former. Probably had a common origin.

Key to Families

1 Ovary superior.
  2 Secretory canals absent.
    3 Petiole without sheathing base. Shrubs or trees. Trichomes unicellular, recurved or multicellular, uniseriate. Vessesl with scalariform perforation and with bordered pits. Rays heterogenous. Axial parenchyma apotracheal or scanty paratracheal. Fibers nonseptate. Nodes trilacunar with three traces. Leaves alternate, simple, serrate or entire, petiolate, glabrous, estipulate. Stomata paracytic. Flowers in terminal panicles, with articulated pedicels, unisexual (plants dioecious or polygamous), 5-merous. Calyx minute, imbricate, caducous. Petals glabrous, free, valvate. Stamens five, free; filaments filiform, connate; anthers 2-locular, introrse, dorsifixed. Pollen grains 3-colporate. Nectary disc absent. Gynoecium of two carpels. Ovary superior, 1-locular, without fleshy appendage; stigma 3-lobed, on a very short style, or discoid and sessile. Ovule solitary and even though a second abortive may be present, pendulous. Fruit small drupe, stone obtusely trigonous; seed coat thin; embryo short, minute, endosperm present. Leuco-anthocyanins and caffeic acid present; n = 25. . . . . 1. PENNANTIACEAE.
    3 Petioles with sheathing base. Woody plants. Vessels with simple or almost exclusively scalariform perforations, without helical

thickenings. Axial parenchyma apotracheal. Rays homogeneous, with simple or bordered pits. Leaves alternate, pinnately compound or simple, entire, estipulate. Inflorescences panicles of umbellules. Flowers with articulated pedicels, unisexual or bisexual, 5-merous. Sepals large, imbricate. Petals free, imbricate, caducous. Nectary disc present. Gynoecium of two carpels. Ovary inferior, 2-locular. Fruits fleshy or spongy drupes (*Delarbrea*), or dry schizocarps (*Myodocarpus*); seeds laterally flattened; woody endocarp hard, with large oil ducts; endosperm smooth and has druses, in *Myodocarpus* up to 18 druses in cell (Vyshenskaya 2000).; n = 12. 8. MYODOCARPACEAE.

2 Secretory canals present. Small trees, shrubs, often woody lianas, sometimes subshrubs, glabrous or with simple, uniseriate trichomes with two or three basal cells and a long terminal cell, two-armed hairs with a short stalk, or club-shaped glandular hairs. A constant and characteristic feature is the presence of well-developed, schizogenous secretory canals in the pericycle of the stem, roots, and leaves, and also in the secondary cortex and secondary phloem of older stems. Vessels with simple perforations or very rarely scalariform perforations (in primary xylem always with scalariform perforations); lateral pitting alternate, small to very small. Fibers septate, with small, simple, or very narrowly bordered pits. Rays nearly homogeneous. Axial parenchyma scanty paratracheal. Sieve-element plastids of Ss-type. Nodes trilacunar, rarely unilacunar. Leaves alternate, sometimes very closely crowded at the branch tips, simple, entire, or rarely serrate, sinuate or occasionally more or less lobed (but seedling leaves and juvenile leaves may be pinnatifid), sometimes with wavy margins, coriaceous, estipulate. Stomata paracytic. Flowers in corymbs or cymose panicles or solitary, bisexual or seldom polygamous (in several New Zealand spp. of *Pittosporum*), or unisexual, actinomorphic or rarely (*Cheiranthera*) slightly zygomorphic, 5-merous (except for gynoecium), bibracteolate. Sepals imbricate, free or some times basally connate. Petals usually basally connate forming more or less distinct tube, with imbricate lobes. Stamens alternate with petals, free or weakly connate basally; anthers tetrasporangiate, opening longitudinally or rarely (*Cheiranthera*) by apical pores. Nectary disc absent. Microsporogenesis simultaneous. Pollen grains 2-celled or 3-celled, 3-colporate. Gynoecium of two or rarely (spp. of *Pittosporum*) 3–5 carpels, with simple style and capitate or slightly lobed stigma; ovary superior, 1-locular, with parietal placentas or as a result of protrusion and fusion the placentas become 2- or rarely 3-locular, with several to usually more or less numerous ovules per placenta or per locule. Ovules anatropous or nearly campylotropous, tenuinucellate. Female gametophyte of *Polygonum*-type. Fruits loculicidal (sometimes also septicidal) capsules (Pittosporeae) or berries (Billardiereae). Seeds often immersed in a viscid pulp, with a minute embryo placed near the hilum; endosperm copious, oily, and proteinaceous; seed coat 6–12 cells thick, unspecialized. Lack iridoid compounds and ellagic and gallic acid but contain ethereal oils, resins and mucilages, leucoanthocyanins, caffeic, quinic, and sinapic acids, flavonols kaempferol and quercetin, polyacetylenes, and often triterpenoid saponins, n = 12. . . . . 6. PITTOSPORACEAE.

1 Ovary usually inferior.

4 Fruits nearly always carpels, drupaceous, or baccate. Trees, shrubs, woody lianas, or woody epiphytes, rarely perennial herbs; plants bearing essential oils, or without essential oils. Vessels solitary (*Pentapanax, Arthrophyllum, Myodocarpus, Delarbrea, Boerlagiodendron, Oreopanax, Plerandra, Dendropanax,* some species of *Schefflera* and *Polyscias* – Oskolski 1996), or in short radial multiples, mostly with simple perforations, in a few genera with scalariform perforations (*Myodocarpus* exclusively scalariform) that have few to numerous bars (up to 56 in *Boerlagiodendron*); lateral pitting scalariform to alternate. Fibers with small, simple pits, often septate, but in *Boerlagiodendron* fiber tracheids. Rays mostly heterogeneous with short ends or rarely homogeneous. Axial parenchyma usually scanty paratracheal, rarely (*Myodocarpus, Delarbrea*) diffuse-in-aggregates or (*Reynoldia*) metatracheal. Nodes multilacunar, pentalacunar, or seldom trilacunar. Leaves alternate or very rarely (*Cheirodendron*) opposite or verticillate,

simple, dissected, pinnately or palmately compound, sometimes to second or third degree, aromatic or not, gland-dotted or not; petiole usually broad and sheathing, stipulate or estipulate; stipules when present, intrapetiolar, often adnate to and scarcely distinguishable from the base of the petiole. Stomata mostly paracytic or anisocytic, rarely anomocytic. Flowers usually small, mostly in simple umbels, bisexual or rarely polygamous or dioecious, rarely calyptrate, usually more or less 5-merous. Calyx teeth small, often much reduced or even obsolete. Petals 5–10(-12), rarely three, free or rarely connate at the base or forming a calyptra (as in *Tupidanthus*) or in a tube (as in *Boerlagiodendron*), deciduous, valvate or sometimes slightly imbricate. Stamens usually as many as and alternate with petals or rarely twice as many or numerous (up to 120 in *Tupidanthus*), anthers dorsifixed, tetrasporangiate or sometimes of as many as eight microsporangia. Pollen grains 3-celled, usually 3-colporate, sometimes 2- or 4-colporate, commonly reticulate. Gynoecium of 2–5 to many (up to 200 in *Tupidanthus*), mostly five carpels; stylodia free or sometimes more or less connate into a style, generally swollen at the base to form a more or less definite stylopodium that is confluent with the epigynous nectary disc; ovary inferior or semi-inferior, very rarely superior, 1-locular (*Seemannaralia*) or pseudomonomerous, with one (sometimes the second abortive one) pendulous ovule per locule; ovules crassinucellate to tenuinucellate. Female gametophyte usually monosporic, of *Polygonum*-type, Fruits drupes with as many pyrenes as carpels, or berries, rarely a drupelike schizocarp, which in *Stilbocarpa* and *Astrotricha*, has carpophores, *Porospermum* have oily glands in endocarp, *Mackinlaya* has glands in pericarp. Seeds with small embryo and copious, ruminate or not, oily endosperm containing large amount of petroselinic acid. Endosperm of *Tetraplasandra*, *Arthrophyllum*, and *Aralia* has druses (Vyshenskaya 2000). Alkaloids and flavonols (quercetin and kaempferol) present or absent; n = 11, 12 + (usually) or more. . . . . .7. ARALIACEAE.

4 Fruits usually schizocarps with persistent carpophore. Mostly perennial or less often annual herbs, sometimes subshrubs, rarely shrubs or soft-wooded small trees. Plants with essential oils, or without, resinous, or not. Stems usually with well-developed collenchyma and large pith that shrinks or dries at maturity. Internodes commonly becoming hollow. Secretory cavities with resin, oil, or with mucilage. Vessels almost always with simple perforations, or rarely with scalariform, very rarely with vestigial bars. Nodes multilacunar. Leaves small to large, usually alternate, or alternate and opposite (the upper sometimes more or less opposite), usually much dissected (ternate or variously pinnate) infrequently entire, and sometimes more or less parallelveined, gland-dotted or not, aromatic or foetid, or rarely without marked odour, estipulate, but sometimes with stipular flanges; petioles usually with broad sheathing base. Stomata anomocytic or paracytic. Flowers small, mostly compound umbels or less often in heads or simple umbels, rarely (as in *Petagnia*) in dichasial inflorescence; at the top of the stalk of each individual umbel (umbellule) an involucre of bracts is often found (the bracts of outer flowers), and a similar large involucre often occurs at the top of the main stalk bearing the compound umbel. Flowers mostly bisexual or seldom unisexual and monoecious (as *Echinophora*) or dioecious (as *Acronema* and *Arctopus*) actinomorphic or often peripheral flowers more or less zygomorphic, usually 5-merous except for 2-merous gynoecium. Calyx teeth usually more or less reduced and frequently inconspicuous, but in Saniculeae they are well developed. Petals free, typically inflexed at the tip, valvate, rarely wanting. Stamens five, alternate with the petals, borne on the epigynous nectary disc; anthers dorsifixed or basifixed, introrse, tetrasporangiate; filaments long; anthers introrse. Pollen grains 3-celled, 3-colporate, very rarely 2-colporate, with more or less smooth surface. Gynoecium of two rather reduced carpels with free, short stylodia usually set on variously and characteristically developed swollen base (stylopodia); ovary inferior, 2-locular or rarely 1-locular and pseudomonomerous, with one ovule per carpel (sometimes with an additional abortive ovule); ovules anatropous, tenuinucellate or pseudocrassinucellate. Female gametophyte mostly of *Polygonum*-type. Fruits non-fleshy schizocarps; mericarps two, dry, united facially, 1-seeded, the integument sometimes united with

the pericarp, always separating at maturity. Seeds adherent to the pericarp, with usually small, well differentiated, straight embryo and copious, oily endosperm containing large amount of petroselinic acid. Alkaloids present (and usually toxic via falcorinone) or absent; flavonols (kaempferol and quercetin) present or absent; n = 4–12, most commonly 8 or 11. . . . . 9. APIACEAE.

5 Stomata anomocytic. Small trees or shrubs, glabrous. Secretory cavities absent. Hairs simple, glandular. Vessels mostly with simple perforations with distinct rims or occasionally scalariform, with 5–24 bars. Fibers with minutely bordered pits, some fibers septate. Rays heterogeneous. Axial parenchyma scanty apotracheal. Crystals absent (Noshiro et Baas 1998). Leaves alternate, entire or dentate, long-petiolate, petioles more or less sheathing at base, stipulate or have shortly stalked glandular hairs with spherical heads, blackening on drying. Stomata amonocytic. Flowers in axillary or terminal racemes or panicles, bisexual, 5-merous; pedicels bracteate at the base and mostly 2-bracteolate at the apex. Sepals connate into a minutely 5-dentate tube, imbricate. Petals free, imbricate, reflexed. Stamens five; filaments short, slender; anthers oblongate, basifixed or dorsifixed. Pollen grains 3-colporate. Nectary disc obscure or obsolete. Gynoecium of 2–3-carpels; stylodia 2–3, free, subulate, erect or recurved, with linear or punctiform stigma; ovary inferior, 2–3-locular (only one of which is fertile), with one pendulous, anatropous ovule per locule. Fruits ovoid or oblong drupaceous, with one fertile dorsal and two sterile ventral loculi, the septa marked by external grooves. Seeds light brown, with very thin testa; embryo minute, straight, differentiated; endosperm fleshy, copious. . . . . 5. MELANOPHYLLACEAE.

5 Stomata encyclocytic. Trees or shrubs, often epiphytic or climbing. Secretory canals absent. Hairs all unicellular, provided with granular, nodular, or verrucose prominences. Wood diffuse-porous. Vessels with scalariform perforations of 10–17 (up to 28) bars. Fibers with distinctly bordered pits on both radial and tangential walls. Rays heterogeneous. Axial parenchyma diffuse-in-aggregates. Nodes penta- or trilacunar. Prismatic crystals usually in body ray cells. Leaves alternate, often asymmetrical, coriaceous, simple, entire or angled or spinose-dentate, estipulate; epidermal cells rhomboidal, small; palisade cells entire, cylindric; hypodermis present; petioles dilated and slightly sheathing at base, with arc of bundles. Flowers small, in terminal or axillary panicles or racemes, dioecious, pedicels ebracteolate, articulated. Male flowers: calyx minute, 5-dentate; petals five, imbricate; stamens five, alternipetalous; anthers broadly oblong, tetrasporangiate, dorsifixed, introrse, opening longitudinally. Pollen grains 3-colporate. Nectary disc fleshy, 5-sided. Pistillodia lacking. Female flowers: calyx tube ovoid or turbinate, 5-dentate; petals imbricate or absent; gynoecium of three united carpels; stylodia three, very short, free, and/or shortly connate, stigmatose at the apex or inside; ovary inferior, 1–2-locular (only one of which is fertile), with one pendulous ovule in each locule. Staminodia lacking. Ovules anatropous, unitegmic, crassinucellate, with endothelium. Female gametophyte of *Polygonum*-type. Fruits ovoid, usually 1-locular and 1-seeded berries. Seeds oblong, with small, elongate, straight embryo in copious endosperm. Contain iridoid glucoside griselinoside and caffeic acid (chlorogenic acid), but lack gallic and ellagic acids. Seeds contain petroselinic acid, n = 18. . . . . 2. GRISELIANACEAE.

6 Funiculus thickened to form an obturator. Small trees with thick branches and broad pith. Secretory canals absent. Hairs both simple unicellular (in floral parts sometimes with transverse wall) and characteristic multicellular glandular hairs consisting of three or four cells. Primary cork with typical collenchyma. Vessel elements short, with simple perforations; lateral pitting opposite to alternate, without vestures pits. Fibers with minutely bordered pits, septate. Rays heterogeneous with long ends. Axial parenchyma scanty paratracheal. Nodes multilacunar. Crystals absent. Leaves alternate, simple, more or less palmatilobed, entire or more often coarsely toothed, palmately veined, long-petiolate, broadly

sheathed at base, estipulate. Stomata anomocytic. Flowers small, in lax, multiflowered, axillary panicles (thyrses), dioecious, actinomorphic, bracteolate. Male flowers: pedicel articulate; calyx 5-lobed, with short tube, lobes open, more or less unequal; petals five, induplicate-valvate, concave, apex elongate, inflexed; stamens five, alternipetalous; filaments short; anthers tetrasporangiate, basifixed, opening longitudinally, latero-introrse; pollen grains 4-colporate, with short colpi, simple endoaperture and either granulose or superficially rugulate tectum with processes or projections that are «short capillilike» and in section are denticulate; nectary disc more or less flat; pistillodia of 1–3 subulate processes in the middle of disc. Female flowers: pedicel not articulate; calyx minutely and irregularly 3–5-toothed; petals wanting; staminodia absent; nectary disc inconspicuous or absent; gynoecium of 3–4 carpels, with three short, thick, subulate, erect, or divaricate, papillose free stylodia; ovary inferior, 3–4-locular with one apical, pendulous ovule per locule (often partly sterile); ovules anatropous, apotropous, with micropyle directed inward. Fruits obliquely ovoid 3–4 locular, usually 1-seeded, rarely 2-seeded drupe crowned by persistent stylodia and with more or less circular opening into each empty chamber. Seeds linear with small, curved embryo at the apex of copious, fleshy endosperm. Contains iridoid compound griselinoside; n = 12. . .4. TORICELLIACEAE.

6 Funiculus not thickened. Small trees. Hairs both unicellular and strongly asymmetrical, glandular trichomes. Secretory canals absent. Wood diffuse-porous. Pith with groups of sclereids. Vessels with scalariform perforations with up to 34 bars; lateral pitting opposite or sometimes scalariform. Fibers with minutely bordered pits mostly in radial walls, all septate; walls thin to thick. Rays multiseriate (up to 12 cells wide), heterogeneous. Axial parenchyma mostly paratracheal. Nodes multilacunar. Crystals absent. Leaves large, alternate, entire or more often deeply and irregularly pinnately lobed or partite with an end lobe, estipulate; epidermal cells large, sinuate, palisade cells lobed, tubular; petioles dilated and sheathing at base, with scattered bundles. Stomata anisocytic. Flowers small (buds c 2.5 mm long), creamy or red-tinged, in large terminal much branched panicles or cymules, dioecious, actinomorphic. Pedicels articulated. Calyx with five very small teeth. Petals five, imbricate. Stamens five, alternipetalous; filaments flattened, anthers round, tetrasporangiate, dorsifixed, opening longitudinally. Pollen grains 3-colporate, tectate, smooth, minutely perforate. Stamens in female flowers sterile (staminodia). Nectary disc intrastaminal, represented in male flowers by uniform cushion of tissue and in female flowers by three (rarely four) gibbous stylopodia. Gynoecium of three (rarely four) carpels; stylodia 3(-4), free, short, diverging, subulate from a broad base (stylopodium), with terminal stigma; ovary inferior, with three (rarely four) loculi, of which all but one remain small with no ovular rudiments; the fertile locule with one pendulous ovule. Ovule anatropous, with dorsal raphe, unitegmic (with very massive integument), tenuinucellate. Fruits drupaceous, large, ellipsoid, exocarp with network surface and inconspicuous grooves near the apex, which represent incompletely lignified vestiges of empty locules; mesocarp fleshy; endocarp hard, ridged. Seeds large, with 4–5 grooves; endosperm deeply ruminate, enclosing small embryo at the micropylar end. Plants contain the iridoid glucosides griselinoside and aralioside, and caffeic acid. n = 20. . . . .3. ARALIDIACEAE.

1. PENNANTIACEAE

J.G. Agardh 1858. 1/4. Eastern Australia (1, *P. cunninghamii*), New Zealand (3).

Pennantia

The genus *Pennantia* has been assigned by different authors to Rhamnaceae, Euphorbiaceae, Terebinthaceae, Anacardiaceae, Icacinaceae. Considering these uncertainties, Agardh (1858) placed this genus in a family of its own. According to recent studies Pennantiaceae belong to the Apiales (Kårehed 2001, 2003).

2. GRISELINIACEAE

J.R. Forster et G. Forster ex A. Cunningham 1839. 1/7. New Zealand, Chile, Argentina, Paraguay, southeastern Brazil.

Griselinia

Usually *Griselinia* is included in the Cornaceae (in the tribe Griselinieae Wangerin, which contains two genera – *Griselinia* and *Melanophylla*). However, as Philipson (1967) and others have demonstrated, *Griselinia* differs from Cornaceae in a number of important characters, including alternate leaves, nodal anatomy, the presence of a hypodermis below the upper epidermis of leaves, encyclocytic stomata, the presence of spiral thickenings in both vessels and fibers, imbricate petals, free or only shortly connate stylodia, usually 1-locular ovary, and baccate fruits. Both in wood anatomy (Metcaife and Chalk 1951) and floral vasculature (Philipson 1967) the Griseliniaceae have some similarities with Araliaceae. Molecular data also supported transfer *Griselinia* into Apiales.

3. ARALIDIACEAE

Philipson et B.C. Stone 1980. 1/1. Southern Thailand, peninsular Malaysia, Sumatra, Singapore, the Anambas Islands, and Borneo.

Aralidium

The genus *Aralidium* has been assigned by different authors to both the Araliaceae and the Cornaceae. From the Cornaceae it differs mainly in multilacunar nodes, pinnatifid leaves, leaf sheaths, endocarp morphology, pollen morphology, and ruminate endosperm. From the Araliaceae it differs in the absence of secretory canals, peripheral origin of the ventral bundles of the ovary (they rise through the ovary wall and not the axis, and the central axis of the gynoecium has no vascular bundles), the dorsal position of the raphe, the absence of the rudiment of second ovule, the presence of glandular hairs, and anisocytic stomata. In spite of some features it has in common with the Araliaceae, including sheathing base of the petiole, and some features of wood anatomy, the genus *Aralidium* is not closely related with the Araliaceae and related families. With the latter *Aralidium* shares characters such as dioecious flowers, multilacunar nodes, paratracheal axial parenchyma, functionally unilocular ovary, three free stylodia, and the presence of griselinoside (Philipson et al. 1980).

4. TORICELLIACEAE

H. Hu 1934. 1/3. Eastern Himalayas (from western Nepal to Bhutan), northern Burma, continental western China.

Toricellia

The genus *Toricellia* is usually included in the Cornaceae. However, it differs from them in many characters, including multicellular glandular hairs, vessels with simple perforations, the presence of typical collenchyma, free stylodia, the presence of obturator, and especially in its very distinctive pollen type (Ferguson and Hideux 1978). Most probably related to the Aralidiaceae (Eyde 1988: 313).

5. MELANOPHYLLACEAE

Takhtajan ex Airy Shaw 1972. 1/6–8. Madagascar.

Melanophylla

Diverse opinions have been expressed as to the affinities of this remarkable Madagascan genus. Generally, it is included in Cornaceae. However, the genus *Melanophylla* differs markedly from Cornaceae in alternate leaves with sheathing petioles, glandular hairs with spherical heads, racemose inflorescences, imbricate petals, and free stylodia. Plankett (2001) and Lundberg both suggest that the Melanophyllaceae and Griseliniaceae are successive lineages near the base of Apiales. Chandler and Plunkett (2004) include *Melanophylla* in their "Torricellia group".

6. PITTOSPORACEAE

R. Brown 1814. 12/c.240. Tropical and subtropical regions of the Old World, mostly Australia (nine endemic genera); the largest and the only widely distributed genus *Pittosporum* (ca. 200) occurs from southern Africa to New Zealand and the Pacific Islands with one outlier, *P. coriaceum*, on the island of Madeira and in a single locality in Tenerife.

PITTOSPOREAE: *Pittosporum, Auranticarpa, Rhytidosporum, Bursaria, Hymenosporum*; BILLARDIEREAE: *Citriobatus, Pronaya, Billardiera, Marianthus, Bentleya, Sollya, Cheiranthera*.

Pittosporaceae have many features in common with the Araliaceae and most probably originated from the same protoaralealean stock. In some respects Pittosporaceae are even less specialized than Araliaceae (usually well-developed calyx and superior ovary), but in some other respects, including wood anatomy, they are more advanced. As early as 1884, van Tieghem suggested that Pittosporaceae are probably rather closely related to the Araliaceae and Apiaceae. He especially emphasized the possession of schizogenous secretory canals and the similarity in the origin of lateral roots in both families (secretory canals situated externally to the xylem inhibit the development of lateral roots on the outer side of the xylem groups, and consequently they arise between the secretory passage opposite a xylem group and the neighboring secretory passage opposite a phloem group; see Metcaife and Chalk [1950: 129]). Later Solereder (1899) and Guenot (1906) have suggested on anatomical grounds that Pittosporaceae are related to the Araliaceae and Apiaceae. Schurhoff (1929) studied the embryology of *Pittisporum* and came to the same conclusion. Huber (1963) arrived at a similar conclusion. Finally, Hegnauer (1969) concluded that on phytochemical grounds the family is unlike members of the saxifragaceous stock (accepted by many authors) and that the inclusion of the family in the Araliales «appears much more likely to reflect its natural incorporation in the Resales.» According to Jay (1969), the Pittosporaceae should be placed near the Apiaceae rather than near the Saxifragaceae.

7. ARALIACEAE

A.L. de Jussieu 1789 (including Botryodendraceae J.G. Agardh 1858; Hederaceae Giseke 1792; Hydrocotylaceae N. Hylander 1945). 47–52/1325–1350. Tropical and subtropical regions, with relatively few genera in temperate regions; concentrated mostly in southern and southeastern Asia and Pacific islands (especially New Caledonia); one of the most archaic genus *Myodocarpus* is endemic to New Caledonia.

7.1 HYDROCOTYLOIDEAE

Mostly perennial herbs. Nodes trilacunar with three traces. Leaves rounded-peltate, crenate; stipules cauline. Stomata anisocytic. Gynoecium of two carpels. Carpophore undivided. – *Hydrocotyle, Trachymene, Neosciadium*.

7.2 ARALIOIDEAE

Trees, shrubs, woody lianad, rarely herbaceous plants. Nodes usually multilacunar or pentalacunar. Leaves pinnately or palmately compound, stipulate or estipulate, stipules when present intrapetiolar. Stomata mostly paracytic or anisocytic, rarely anomocytic. Gynoecium mostly of five carpels. Carpophore usually absent. – SCHEFFLEREAE: *Schefflera* (including *Plerandra*), *Tupidanthus, Scheffleropsis, Didymopanax, Eleutherococcus* (*Acanthopanax*), *Kalopanax, Opiopanax, Tetrapanax, Merilliopanax, Dendropanax, Boerlagiodendron, Osmoxylon, Trevesia, Fatsia, Gastonia, Cussonia, Reynoldsia, Tetraplasandra, Munroidendron*; MERYTEAE: *Meryta*; HEDEREAE: *Hedera, Gamblea, Oreopanax, Astrotricha, Trevesia, Kissodendron, Brassaiopsis, Macropanax, Heteropanax, Arthrophyllum, Wardenia*; ARALIEAE: *Polyscias, Pseudopanax, Raukaua, Sciadodendron, Pentapanax, Motherwellia, Sciadodendron, Aralia, Cephalaralia*, etc; PANACEAE: *Harmsiopanax, Panax*. ?? *Anakasia, Cheirodendrom*.

8. MYODOCARPACEAE

Doweld 2001. 2/17. Eastern Malesia, New Caledonia, Vanuatu, Australia (Queensland).

Delarbrea (including *Porospermum, Pseudosciadium*), *Myodocarpus*.

Related to the Pittosporaceae and Araliaceae, but closely related to the Araliaceae.

9. APIACEAE

Lindley 1836 or Umbelliferae A.L. de Jussieu 1789 (nom. altern.) (including Actinotaceae Konstantinova et Melikian 2006; Ammiaceae Barnhard 1895; Angelicaceae Martynov 1820; Bupleuraceae Martynov 1820; Coriandraceae Burnett 1835; Daucaceae Martynov 1820; Eryngiaceae Rafinesque 1838; Ferulaceae Saccardo 1872; Imperatoriaceae Martynov 1820;

Mackinlayaceae Doweld 2001; Saniculaceae A. Löve et Löve 1974; Smyrniaceae Burnett 1835). 435/3100–3200. Cosmopolitan, but mainly temperate regions of the Northern Hemisphere (particularly Mediterranean and Irano-Turanian regions).

Closely related to the Araliaceae.

9.1 APIOIDEAE

Fruits with a soft parenchymatous endocarp, sometimes hardened by woody subepidermal layers. Stylodia on apex of disc. – ECHINOPHOREAE: *Echinophora, Anisosciadium, Dicyclophora, Pycnocycia, Thecocarpus, Ergocarpon*; SCANDICEAE: *Scandix, Physocaulis, Chaerophyllum, Grammosciadium, Anthriscus, Myrrhis, Geocaryum, Rhabdosciadium, Myrrhoides, Osmorhiza, Sphallerocarpus, Kozlovia, Krasnovia, Balansaea, Molopospermum, Tinguarra, Noeconopodium, Athamanta, Conopodium, Oreomyrrhis, Ptychotis, Hladnikia;* HETEROMORPHEAE: *Heteromorpha, Anginon, Dracosciadium, Glia, Polemannia*; CAUCALIDEAE: *Astrodaucus, Caucalis, Daucus, Exoacantha, Lisaea, Orlaya, Torilis, Turgenia,* etc.; CORIANDREAE: *Schrenkia, Coriandrum, Bifora, Kosopoljanskia, Fuernrohria,* etc.; SMYRNIEAE: *Scaligeria, Smyrnium, Oreomyrrhis, Smyrniopsis, Conium, Tauschia, Arracacia, Lecokia, Trachydium, Cachrys, Prangos;* PLEUROSPERMEAE: *Pleurospermum, Eleutherospermum, Aulacospermum, Physospermum*; ERIGENIEAE: *Erigenia*; HOHENACKERIEAE: *Hohenackeria;* PYRAMIDOPTEREAE: *Pyramidoptera;* APIEAE: *Bupleurum, Trinia, Szovitsia, Aphanopleura, Froriepia, Apium, Petroselinum, Cicufa, Trachyspermum, Cryptotaenia, Ammi, Falcaria, Carum, Bunium, Chamaesciadium, Conopodium, Pimpinella, Aegopodium. Slum, Crithmum, Seseli, Oenanthe, Aethusa, Athamanta, Foeniculum, Ligusticum,* etc.; ANGELICEAE: *Angelica, Agasyllis, Chymsydia,* etc.; PEUCEDANEAE: *Myrrhidendron, Diplotaenia, Johrenia, Cymbocarpum, Ferula, Ferulago, Dorema, Opopanax, Peucedanum, Anethum,* etc.; TORDYLIEAE: *Pastinaca, Heracleum, Tordylium, Malabaila, Stenotaenia, Zosima,* etc.: LASERPITIEAE: *Laserpitium, Polylophium, Thapsia, Elaeoselinum,* etc.

9.2 SANICULOIDEAE

Fruits with a soft parenchymatous endocarp, but containing scattered druses. Base of the stylodia surrounded by a ring-like disc. – SANICULEAE: *Sanicula, Polemanniopsis, Steganotaenia, Hacquetia, Astrantia, Actinolema, Alepidea, Eryngium;* LAGOECIEAE: *Lagoecia, Petagnaea.*

9.3 AZORELLOIDEAE

Leaves compound, stipulate. Fruits dorsally flattened, with woody endocarp; n = 8. – *Azorella, Bowlesia, Diplaspis, Oschatzia, Homalocarpus, Drusa, Huanaca, Spananthe, Diposis, Mulinum, Schizeilema, Laretia, Dichosciadium, Dickinsia, Hermas, Choritaenia, Bolax, Pozoa, Asteriscium, Gymnophyton, Eremocharis, Domeykos, Stilbocarpa*; *Homalosciadium, Trachymene, Uldinia, Klotzschia, Naufraga, Schoenolaena, Pentapeltis, Chlaenosciadium, Brachyscias*.

9.4 MACKINLAYOIDEAE

Woody or herbaceous plants. Axial parenchyma apotraceal or paratracheal. Leaves palmately compound (*Mackinlaya*). Inflorescences panicles or racemes, often of umbels; carpophore present. Fruits laterally flattened. – *Apiopetalum, Mackinlaya, Centella, Micropleura, Actinotus, Platysace, Xanthosia.*

Bibliography

Airy Shaw HK. 1972. A new species of *Melanophylla* Baker (Melanophyllaceae). Kew Bull. 26: 491–493.

Baumann MG. 1946. *Myodocarpus* und die Phylogenie der Umbelliferen-Frucht. Ber. Schweiz. Bot. Ges. 56: 13–112.

Baumann-Bodenheim MG. 1955. Ableitung und Bau bicarpellatmonospermer und pseudomonocarpellater Araliaceen- und Umbelliferen-Früchte. Ber. Schweiz. Bot. Ges. 65: 481–510.

Baylis GTS. 1997. *Pennantia baylisiana* New Zealand's rarest tree – its discovery and propagations. N. Z. Gard. J. 2: 12–13.

Bell CR and L Constance. 1957, 1960, 1966. Chromosome numbers in the Umbelliferae. Am. J. Bot. Pt. 1, 44: 565–572, 1957; Pt. 2, 47: 24–32, 1960; Pt. 3, 53: 512–520, 1966.

Berenbaum MR. 2001. Chemical mediation of coevolution: phylogenetic evidence for *Apiaceae* and associates. Ann. Missouri Bot. Gard. 88(1): 45–59.

Bremer G. 1915. Reliquiae Treubianae: 2. The development of the ovule and embryo of *Pittosporum ramiflorum* and *Pittosporum timorense*. Ann. Jard. Bot. Buitenz. 14: 161–164.

Burtt BL and WC Dickison. 1975. The morphology and relationships of *Seemannaralia* (Araliaceae). Notes Roy. Bot. Gard. Edinb. 33: 449–464.

Calviño CI, PM Tilney, B-E van Wyk, and SR Downie. 2006. A molecular phylogenetic study of southern African Apiaceae. Am. J. Bot. 93: 1828–1847.

Carlquist S. 1981. Wood anatomy of the Pittosporaceae. Allertonia 2: 355–392.

Cayzer LW. 1998. Revision of the family Pittosporaceae in Australia. Ph.D. dissertation, Australian National University, Canberra.

Cayzer LW, MD Crips, and Ian RH Telford. 2000. Revision of *Pittosporum* (Pittosporaceae) in Australia. Aust. Syst. Bot. 13: 845–902.

Cerceau-Larrival MT. 1962. Plantules et pollens d'Om-belliferes: Leur interet systematique et phylogenique. Mem. Mus. Natl. Hist. Nat., Ser. B, Bot. 14: 1–166.

Cerceau-Larrival MT. 1965. Involucre et involucelle chezles Ombelliferes. Bull. Soc. Bot. France 112: 252–267.

Chandler GT and GM Plunkett. 2004. Evolution in Apiales: nuclear and chloroplast markers together in (almost) perfect harmony. Bot. J. Linn. Soc. 144: 123–147.

Constance L, T-I Chuang, and CR Bell. 1976. Chromosome numbers in Umbelliferae, part 5. Am. J. Bot. 63: 608–625.

Costello A. 2002. Molecular and morphological systematics of the *Tetraplasandra* group (Araliaceae) and the development of the superior ovary in *Tetraplasandra*. Ph.D. dissertation, New York University, New York.

Costello A and TJ Motley. 2000. Molecular systematics of *Tetraplasandra*, *Munroidendron*, and *Reynoldsia sandwicensis* (Araliaceae) and the evolution of superior ovaries in *Tetraplasandra*. Am. J. Bot. 87(6, Suppl.): 120.

Costello A and TJ Motley. 2001. Molecular systematics of *Tetraplasandra*, *Munroidendron* and *Reynoldsia sandwicensis* (Araliaceae) and the evolution of superior ovaries in *Tetraplasandra*. Edinb. J. Bot. 58: 229–242.

Crowden RK, JB Harborne, and VH Heywood. 1969. Chemosystematics of the Umbelliferae: a general survey. Phytochemistry 8: 1963–1984.

Dillon MO and M Munoz-Schick. 1993. A revision of the dioecious genus *Griselinia* (Griseliniaceae), including a new species from the coastal Atacama Desert of northern Chile. Brittonia 45: 261–274.

Doweld AB. 2000. Aralidiaceae. In: A Takhtajan, ed. Comparative seed anatomy, vol. 6, pp. 315–316. Nauka, St. Petersburg (in Russian).

Doweld AB. Pittosporaceae. In: A Takhtajan, ed. Comparative seed anatomy, vol. 6, pp. 361–363. Nauka, St. Petersburg (in Russian).

Downie SR and DS Katz-Downie. 1996. A molecular phylogeny of Apiaceae subfamily Apioideae: evidence from nuclear ribosomal DNA internal transcribed spacer sequences. Am. J. Bot. 83: 234–251.

Downie SR and DS Katz-Downie. 1999. Phylogenetic analysis of chloroplast *rps*16 intron sequences reveals relationships within the woody southern African Apiaceae subfamily Apioideae. Canad. J. Bot. 77: 1120–1135.

Downie SR, S Ramanath, DS Katz-Downie, and E Llanas. 1998. Molecular systematics of Apiaceae subfamily Apioideae: phylogenetic analyses of nuclear ribosomal DNA internal transcribed spacer and plastid *rpo*C1 intron sequences. Am. J. Bot. 85: 563–591.

Downie SR, MF Watson, K Spalik, and DS Katz-Downie. 2000a. Molecular systematics of Old World Apioideae (Apiaceae): relationships among some members of tribe Peucedaneae sensu lato, the placement of several island-endemic species, and resolution within the apioid superclade. Canad. J. Bot. 78: 506–528.

Downie SR, DS Katz-Downie, and K Spalik. 2000b. A phylogeny of Apiaceae tribe Scandiceae: evidence from ribosomal DNA internal transcribed spacer sequences. Am. J. Bot. 87: 76–95.

Downie SR, DS Katz-Downie, and MF Watson. 2000c. A phylogeny of the flowering plant family Apiaceae based on chloroplast DNA *rpl*16 and *rpo*C1 intron sequences: towards a suprageneric classification of subfamily Apioideae. Am. J. Bot. 87: 273–292.

Downie SR, GM Plunkett, MF Watson, K Spalik, DS Katz-Downie, CM Valiejo-Roman, EI Terentieva, AV Troitsky, B-Y Lee, J Lahham, and A El-Oqlah. 2001. Tribes and clades within Apiaceae subfamily Apioideae: the contribution of molecular data. Edinb. J. Bot. 58: 301–330.

Ducamp L. 1902. Recherches sur l'embryogénie des Araliacées. Annales des Sciences Naturelles, Bot. 15: 311–402.

Erbar C and P Leins. 1985. Studien zur Organsequenz in Apiaceen Blüten. Bot. Jahrb. Syst. 105: 379–400.

Erbar C and P Leins. 1988. Blütenentwicklungsg-eschichtliche Studien an *Aralia* und *Hedera* (Araliaceae). Flora 180: 391–406.

Erber C and P Leins. 1996. An analysis of the early floral development of *Pittosporum tobira* (Thunb.) Aiton and some remarks on the systematic position of the family Pittosporaceae. Feddes Repert. 106: 463–473.

Erber C and P Leins. 2004. Sympetaly in Apiales (Apiaceae, Araliaceae, Pittosporaceae). South Afr. J. Bot. 70: 458–467.

Eyde RH and CC Tseng. 1969. Flower of *Tetraplasandra gymnocarpa:* hypogyny with epigynous ancestry. Science 166: 506–508.

Eyde RH and CC Tseng. 1971. What is the primitive floral structure of Araliaceae? J. Arnold Arbor. 52: 205–239.

Friis I. 1987. A reconsideration of *Pittosporum* in Africa and Arabia. Kew Bull. 42: 319–335.

Frodin DG. 1982. Systematics of Araliaceae and inflorescence morphology. Aust. Syst. Bot. Soc. Newslett. 30: 43–55.

Froebe HA. 1979. Die Infloreszenzen der Hydrocotyloideen (Apiaceae). Trop. Suptrop. Pflanzenwelt 29: 501–679.

Froebe HA. 1980. Randmusterbildung und Synorganisation bei strahlenden Apiaceendolden. Plant Syst. Evol. 133: 223–237.

Gardner RO and PJ de Lange. 2002. Revision of *Pennantia* (Icacinaceae), a small isolated genus of southern hemisphere trees. J. Roy. Soc. New Zealand 32: 669–695.

Gilmartin AJ. 1980. Variations within populations and classification, II. Patterns of variation within Asclepiadaceae and Umbelliferae. Taxon 29: 199–212.

Gopinath DM. 1944. Gametogenesis and embryogenus in a few members of the Araliaceae. Proc. Indian Acad. Sci., ser. B 20(5): 239–309.

Graham SA. 1966. The genera of Araliaceae in the southeastern United States. J. Arnold Arbor. 47: 126–136.

Grushvitsky IV, VN Tikhomirov, ES Axenov, and GV Shibakina. 1969. Fleshy fruit with a carpophore in species of the genus *Stilbocarpa* Decne. et Planch. (Araliaceae). Bull. Moscow Soc. Naturalists, Biol. 74(2): 64–76 (in Russian).

Guenot JF. 1906. Contributions a l'etude anatomique des Pittosporacees. Master's thesis, University of Paris.

Guha S. 1971. Cytotaxonomy of Araliaceae. Proc. Indian Sci. Congr. Assoc., Sect. 6, 58: 471.

Guyot M. 1966. Les types stomatiques des Ombelliferes. Bull. Soc. Bot. France 113: 244–273.

Hansen L and PM Boll. 1986. Polyacetylenes in Araliaceae: their chemistry, biosynthesis and biological significance. Phytochemistry 25: 285–293.

Henwood MJ and JM Hart. 2001. Towards an understanding of the phylogenetic relationships of Australian Hydrocotyloideae (Apiaceae). Edinb. J. Bot. 58: 269–289.

Heywood VH, ed. 1971. The biology and chemistry of the Umbelliferae. Bot. J. Linn. Soc. 64(Suppl. 1). London.

Hoar CS. 1915. A comparison of the stem anatomy of the cohort Umbelliflorae. Ann. Bot. 29: 55–63.

Holub M, J Toman, and V Herout. 1987. The phylogenetic relationships of the Asteraceae and Apiaceae based on phytochemical characters. Biochem. Syst. Ecol. 15: 321–326.

Hoo C. The systematics, relationship and distribution of the Araliaceae of China. Bull. Amoi Univ. (Nat. Sci.) 8: 1–11.

Hoo C and CJ Tseng. 1965. Contributions to the Araliaceae of China. Acta Phytotax. Sinica, add. 1: 129–176.

Hu HH. 1934. Notulae systematicae ad florem sinesium V. (Toricelliaceae). Bull. Fan Mem. Inst. Biol. 5: 305–318.

Hu SY. 1980. *Eleutherococcus* vs. *Acanthopanax*. J. Arnold Arbor. 61: 107–111.

Huber H. 1963. Die Verwandtschaftsverhaltnisse der Rosifloren. Mitt. Bot. Staatssamml. Munchen 5: 1–48.

Jackson G. 1933. A study of the carpophore of the Umbelliferae. Am. J. Bot. 20: 121–144.

Jahnke P. 1986. Der Infloreszenzbau der Cornaceen sensu lato und seine systematischen Konsequenzen. Trop. Subtrop. Pflanzenwelt, vol. 57. Stuttgart.

Jay M. 1969. Chemotaxonomic researches in vascular plants: XIX. Flavonoid distribution in the Pittosporaceae. Bot. J. Linn. Soc. 62: 423–429.

Jensen SR and BJ Nielsen. 1980a. Iridoid glucosides in *Griselinia, Aralidium* and *Toricellia*. Phytochemistry 19: 2685–2688.

Jensen SR and BJ Nielsen. 1980b. The systematic position of *Aralidium* Miq. A multidisciplinary study. 4. Iridoid glucosides. Taxon 29: 409–411.

Judd WS. 1996. The Pittosporaceae in the southeastern United States. Harvard Pap. Bot. 8: 15–26.

Kårehed J. 2002. Introduction. In: Evolutionary Studies in Asterids Emphasising Euasterids II, pp. 5–50. Acta Universitatis Upsaliensis, Uppsala.

Kårehed J. 2003. The family Pennantiaceae and its relationships to Apiales. Bot. J. Linn. Soc. 141: 1–24.

Katz-Downie DS, CM Valiejo-Romano, EI Terentieva, AV Troitsky, MG Pimenov, B Lee, and SR Downie. 1999. Towards a molecular phylogeny of Apiaceae subfamily Apioideae: additional information from nuclear ribosomal DNA ITS sequences. Plant Syst. Evol. 216: 167–195.

Kim CH. 1997. Systematics of *Eleutherococcus* and related genera (Araliaceae). Ph.D. dissertation. Chonbuk National University, Korea.

Kordyum EL. 1967. Cytoembryology of the family Umbelliferae. Naukova Dumka, Kiev (in Russian).

Kozo-Poljanski BM. 1915. Scaidophytorum systematis lineamenta. Bull. Soc. Imp. Natur. Moscow, n.s., 29: 93–221.

Kozo-Poljanski BM. 1924. Hydrocotyloidearum revisio, part 1. Not. Syst. Hort. Petropolitani 5(2): 17–24.

Kozo-Poljanski BM. 1926. On the morphology of the flower of Umbelliferae. Bull. Voronesh Soc. Naturalists 1: 153–168 (in Russian).

Kozo-Poljansky BM. 1938. Carpology of *Exoacantha* Labill. and its relation to the diagnostics and taxonomy of Umbelliferae in general. Bull. Mosc. Soc. Naturalists, ser. biol. 47, 1: 39–55 (in Russian).

Kozo-Poljanski BM. 1943. Morphological significance of *Deutes calycini* in Apioideae: a contribution to the morphology of the flower of the umbellifers. Sov. Bot. 4: 36–41 (in Russian).

Lee B-Y and SR Downie. 1999. A molecular phylogeny of Apiaceae tribe Caucalideae and related taxa: inferences based on ITS sequence data. Syst. Bot. 24: 461–479.

Lee B-Y and SR Downie. 2000. Phylogenetic analysis of cpDNA restriction sites and *rps*16 intron sequences reveals relationships among Apiaceae tribes Caucalideae, Scandiceae and related taxa. Plant Syst. Evol. 221: 35–60.

Leins P and C Erbar. 1985. Zur frühen Entwicklungs-geschichte des Apiaceen-Gynoceums. Bot. Jahrb. Syst. 106: 53–60.

Leins P and C Erbar. 2004. Floral organ sequences in Apiales (Apiaceae, Araliaceae, Pittosporaceae). South Afr. J. Bot. 70: 468–474.

Li HL. 1942. The Araliaceae in China. Sargentia 2: 1–134.

Liu M. 2004. A taxonomic evaluation of fruits structure in the family Apiaceae. Ph.D. Rand Afrikaans University, Johannesburg.

Liu M, B-E Van Wyk, and PM Tilney. 2003a. The taxonomic value of fruit structure in the subfamily Saniculoideae and related African genera (Apiaceae). Taxon 52: 261–270.

Liu M, B-E Van Wyk, and PM Tilney. 2003b. Ontogeny of the fruits of two anomalous African woody genera *Polemanniopsis* and *Steganotaenia* (Apiaceae) and their phylogenetic relationship. Edinb. J. Bot. 60: 249–257.

Liu Mei (Rebecca), GM Plunkett, PP Lowry II, B-E Van Wyk, and PM Tilney. 2006. The taxonomic value of fruit wing types in the order Apiales. Am. J. Bot. 93: 1357–1368.

Lowry PP II. 1986. A systematic study of *Delarbrea* Vieill. (Araliaceae). Allertonia 4: 169–201.

Lowry PP II. 1989. A revision of Araliaceae from Vanuatu. Bull. Mus. Nat. Hist. Natu. Paris, Adansonia 11: 117–155.

Lowry PP II, GM Plunkett, and AA Oskolski. 2001. Early lineages in Apiales: insights from morphology, wood anatomy and molecular data. Edinb. J. Bot. 58: 207–220.

Lowry PP II, GM Plankett, V Raquet, TS Sprenkle, and J Jérémie. 2004a. Inclusion of the endemic New Caledonian genus *Pseudosciadium* Baill. in *Delarbrea* Vieill. (Apiales: Myodocarpaceae). Adansonia, sér. 3, 26: 251–256.

Lowry PP II, GM Plankett, and J Wen. 2004b. Genetic relationships in Araliaceae: looking into the crystal ball. South Afr. J. Bot. 70: 382–392.

Magin N. 1977. Das Gynoecium der Apiaceae: Modell und Ontogenie. Ber. Deutsch. Bot. Ges. 90: 53–66.

Meylan BA and BG Butterfield. 1978. The structure of New Zealand woods. New Zealand Department of Scientific and Industrial Research, Wellington.

Mitchell AD, CD Meurk, and SJ Wagstaff SJ. 1999. Evolution of *Stilbocarpa*, a megaherb from New Zealand's subantarctic islands. N. Z. J. Bot. 37: 205–211.

Mittal SP. 1961. Studies in the Umbellales: II. The vegetative anatomy. J. Indian Bot. Soc. 40: 424–443.

Mohana Rao PR. 1972 (1973). Morphology and embryology of *Tiegemopanax sambucifolius* with comments on the affinities of the family Araliaceae. Phytomorphology 22: 75–87.

Narayana LL and M Radhakrishnaiah. 1976. Floral anatomy of the Pittosporaceae, part 1. J. Jpn. Bot. 51: 278–282.

Narayana LL and M Radhakrishnaiah. 1978. Floral anatomy of the Pittosporaceae, part 2. Acta Bot. Indica 6: 104–107.

Narayana LL and M Radhakrishnaiah. 1979. Floral anatomy of the Pittosporaceae, part 3. J. Jpn. Bot. 54: 324–349.

Narayana LL and M Radhakrishnaiah. 1980. Floral anatomy of the Pittosporaceae, part 4. Indian J. Bot. 3: 6–12.

Narayana LL and M Radhakrishnaiah. 1981. Floral anatomy of the Pittosporaceae, part 5. J. Jpn. Bot. 56: 137–141.

Narayana LL and M Radhakrishnaiah. 1982. Floral anatomy of the Pittosporaceae: VI. Five of *Pittosporum*. Canad. J. Bot. 60: 1859–1867.

Narayana LL and M Radhakrishnaiah. 1984. Floral anatomy of the Pittosporaceae: a discussion. J. Plant Anat. Morph. 1: 45–51, 1984. J. Indian Bot. 63: 53–56.

Narayana LL and KT Sundari. 1977. Embryology of Pittosporaceae, part I. J. Jpn. Bot. 52: 204–209.

Ng N and JRL Walker. 1975. A chemotaxonomic study of the New Zealand Araliaceae: phenolic compounds. Mauri Ora 3: 3–10.

Noshiro S and P Baas. 1998. Systematic wood anatomy of Cornaceae and allies. IAWA J. 19: 43–97.

Oliver D. 1889. *Toricellia angulata* Oliv. Hooker's Icon. Plantarum 19: t.1893.

Oskolski AA. 1994. Wood anatomy of Araliaceae. Komarov Botanical Institute. St. Petersburg (in Russian).

Oskolski AA. 1995. Wood anatomy of *Schefflera* and related taxa (Araliaceae). IAWA J. 16: 191–215.

Oskolski AA. 1996. A survey of the wood anatomy of the Araliaceae. In: LA Donaldson, AP Singh, BG Butterfield, and LJ Whitehouse, eds. Resent advances in wood anatomy. New Zealand Forest Research Institute, Rotorua, pp. 99–119.

Oskolski AA. 2001. Phylogenetic relationships within Apiales: evidence from wood anatomy. Edinb. J. Bot. 58: 201–206.

Oskolski AA and PP Lowry II. 2000a. Systematic wood anatomy of *Mackinlaya* and *Apiopetalum* (Araliaceae). Ann. Missouri Bot. Gard. 87: 171–182.

Oskolski AA and PP Lowry II. 2000b. Systematic wood anatomy of *Myodocarpus, Delabrea,* and *Apiopetalum*. Ann. Missouri Bot. Garden. 80: 700–722.

Oskolski AA and PP Lowry II. 2001. Wood anatomy of *Schefflera* and related taxa (Araliaceae). II. Systematic wood anatomy of New Caledonian *Schefflera*. IAWA J. 22: 301–330.

Oskolski AA, PP Lowry II, and HG Richter. 1997. Systematic wood anatomy of *Myodocarpus, Delarbrea,* and *Pseudosciadium* (Araliaceae). Adansonia 19: 61–75.

Oskolski AA, EL Kotina, IV Fomichev, F Tronchet et PP Lowry II. 2007. Systematic implications of wood and bark anatomy in the Pacific Island genus *Meryta* (Araliaceae). Bot. J. Linn. Soc. 153: 363–379.

Park WC and S Lee. 1989. A palynotaxonomic study of the Korean Araliaceae. Korean J. Plant Taxon. 19: 103–121.

Philipson WR. 1965. The New Zealand genera of the Araliaceae. N. Z. J. Bot. 3: 333–341.

Philipson WR. 1967. *Griselinia* Forst. fil.: anomaly or link. N. Z. J. Bot. 5: 134–165.

Philipson WR. 1970. Constant and variable features of the Araliaceae. J. Linn. Soc. Bot. 63: 87–100.

Philipson WR. 1978. Araliaceae: growth forms and shoot morphology. In: Tropical trees as living systems, pp. 269–284. Cambridge University Press, Cambridge.

Philipson WR. 1979. Araliaceae. In: CGGJ van Steenis, ed. Flora malesiana, ser. 1, 9(1): 1–105.

Philipson WR and BG Butterfield. 1980a. The systematic position of *Aralidium* Miq.: a multidisciplinary study. Wood anatomy. Taxon 29: 404–406.

Philipson WR and BC Stone. 1980b. The systematic position of *Aralidium* Miq.: a multidisciplinary study. Introduction and floral and general anatomy. Taxon 29: 391–403.

Pickering JL and DE Fairbrothers. 1970. A serological comparison of Umbelliferae subfamilies. Am. J. Bot. 57: 988–992.

Pimenov MG and MV Leonov. 1993. The genera of the Umbelliferae: a nomenclator. Royal Botanic Gardens, Kew/ Moscow.

Pimenov MG and LI Sdobnina. 1984. Nodal anatomy as a taxonomic character in the family Umbelliferae. Bot. Zhurn. 69: 283–294 (in Russian with English summary).

Plunkett GM. 1998. Endemic araliad genera from New Caledonia and their implications for phylogeny and taxonomy in Araliales (Apiaceae and Araliaceae). Am. J. Bot. 85: 151 (abstract).

Plunkett GM. 2001. Relationships of the order Apiales to subclass Asteridae: a re-evolution of morphological characters based on insights from molecular data. Edinb. J. Bot. 58: 183–200.

Plunkett GM and SR Downie. 1999. Major lineages within Apiaceae subfamily Apioideae: a comparison of chloroplast restriction site and DNA sequence data. Am. J. Bot. 86: 1014–1026.

Plunkett GM and SR Downie. 2000. Expansion and contraction of the chloroplast inverted repeat in Apiaceae subfamily Apioideae. Syst. Bot. 25: 648–667.

Plunkett GM and PP Lowry II. 2001. Relationships among 'ancient araliads' and their significance of the systematics of Apiales. Mol. Phylogenet. Evol. 19: 259–276.

Plunkett GM, DE Soltis, and PS Soltis. 1996a. Higher level relationships of Apiales (Apiaceae and Araliaceae) based on *rbc*L sequences. Am. J. Bot. 83: 499–515.

Plunkett GM, DE Soltis, and PS Soltis. 1996b. Evolutionary patterns in Apiaceae: inferences base on *mat*K sequence data. Syst. Bot. 21: 477–495.

Plunkett GM, DE Soltis, and PS Soltis. 1997. Clarification of the relationship between Apiacea and Araliaceae based on *mat*K and *rbc*L sequence data. Am. J. Bot. 84: 365–380.

Plankett GM, PP Lowry II, and MK Burke. 2001. The phylogenetic status of *Polyscias* (Araliaceae) based on nuclear ITS sequence data. Ann. Missouri Bot. Gard. 88: 213–230.

Plankett GM, GT Chandler, PP Lowry II, SM Pinney, and TS Sprenkle. 2004a. Recent advances in understanding Apiales and a revised classification. South Afr. J. Bot. 70: 371–381.

Plankett GM, PP Lowry II, and N-V Vu. 2004b. Phylogenetic relationships among *Polyscias* (Araliaceae) and close relatives from the Western Indian Ocean Basin. Int. J. Plant Sci. 165: 861–873.

Plankett GM, J Wen, and PP Lowry II. 2004c. Infrafamilial classification and characters in Araliaceae: insights from the phylogenetic analysis of nuclear (ITS) and plastid (*trn*L-*trn*F) sequence data. Plant Syst. Evol. 245: 1–39.

Plunkett GM, GT Chandler, PP Lowry II, SM Pinney, and TS Sprenkle. 2004d. Recent advances in understanding Apiales with a revised classification. South Afr. J. Bot. 70: 371–381.

Plankett GM, PP Lowry II, DG Frodin, and J Wen. 2005. Phylogeny and geography of *Schefflera*: pervasive polyphyly in the largest genus of Araliaceae. Ann. Missouri Bot. Gard. 92: 202–224.

Rodriguez RL. 1957. Systematic anatomical studies on *Myrrhidendron* and other woody Umbellales. Univ. Calif. Publ. Bot. 29: 145–318.

Rodriguez RL. 1971. The relationships of the Umbellales. In: VH Heywood, ed. The biology and chemistry of the Umbelliferae. Bot. J. Linn. Soc. Suppl. 1: 63–92.

Roland-Heydacker F and M-T Cerceau-Larrival. 1978. Ultrastructure du rectum de pollen d'Ombelliferes. Grana 17: 81–89.

Schaeppi H. 1971. Zur Gestaltung des Gynoeceums von *Pittosporum tobira*. Ber. Schweiz. Bot. Ges. 81: 40–51.

Schilrhoff NN. 1929. Über die systematische Stellung der Pittosporaceae. Beitr. Biol. Pfl. 17: 72–86.

Schlessman MA, DG Lloyd, and PP Lowry II. 1990. Evolution of sexual systems in New Caledonian Araliaceae. Mem. N. Y. Bot. Gard. 55: 105–117.

Schlessman MA, GM Plunkett, PP Lowry II, and DG Lloyd. 2001. Sexual systems in New Caledonian Araliaceae: a preliminary phylogenetic reappraisal. Edinb. J. Bot. 58: 221–228.

Schneyer VC, GP Borshchenko, MG Pimenov, and MV Leonov. 1991. The serological investigation of intergeneric relationships in the subfamily Apioideae (Apiaceae). Bot. Zhurn. 76: 245–257 (in Russian with English summary).

Schodde R. 1972. A review of the family Pittosporaceae in Papuasia. Aust. J. Bot. Suppl. Ser. 3: 1–600.

Schürhoff PN. 1926. Über die systematische Stellung der Pittosporaceae. Beitr. Biol. Pflanzen Berg. V. Cohn 17: 72–86.

Seemann B. 1868. Revision of the natural order Hederaceae. J. Bot. 6: 52–57, 163–165.

Shang C-B. 1983. Révision du genre *Macropanax* Miq. (Araliaceae). Bull. Mus. nat. Hist. Nat. Paris, Adansonia 1: 33–52.

Shang C-B and D Callen. 1988. Pollen morphology of the family Araliaceae in China. Bull. Bot. Res. 8: 13–35 (in Chinese with English summary).

Shneyer VS, GP Borschtschenko, MG Pimenov, and MV Leonov. 1992. The tribe Smyrnieae (Umbelliferae) in the light of serotaxonomical analysis. Plant Syst. Evol. 182: 135–148.

Shoup JR and CC Tseng. 1977. Pollen of *Klotzschia* (Umbelliferae): a possible link to Araliaceae. Am. J. Bot. 64: 461–463.

Singh D. 1954. Floral morphology and embryology of *Hedera nepalensis* K. Koch. Agra Univ. J. Res. Sci. 3: 289–299.

Spalik K, A Wojewodzka, and SR Downie. 2001. The evolution of fruit in Scandiceae subtribe Scandicinae (Apiaceae). Canad. J. Bot. 79: 1358–1374.

Stuhlfauth T, H Fock, H Huber, and K King. 1985. The distribution of fatty acids including petroselinic and tartaric acids in the fruit and seed oils of the Pittosporaceae, Araliaceae, Umbelliferae, Simaroubaceae and Rutaceae. Biochem. Syst. Ecol. 13: 447–453.

Sun BY, CH Kim, and WY Soh. 1988. Chromosome numbers of Araliaceae in Korea. Korean J. Plant Taxon. 18: 291–296.

Sun F-J and SR Downie. 2004. A molecular systematic investigation of *Cymopterus* and its allies (Apiaceae) based on phylogenetic analyses of nuclear (ITS) and plastid (*rps*16 intron) DNA sequences. South Afr. J. Bot. 70: 407–416.

Sun F-J, SR Downie, and RL Hartman. 2004. An ITS-based phylogenetic analysis of the perennial, endemic Apiaceae subfamily Apioideae of western North America. Syst. Bot. 29: 419–431.

Tamamshian SG. 1948. Secondary hypogyny of the flower of the umbellifers and principles of the change of function in plants. Doklady USSR Acad. Sci. 61: 537–540 (in Russian).

Thorne RF. 1973. Inclusion of the Apiaceae (Umbelliferae) in the Araliaceae. Notes Roy. Bot. Gard. Edinb. 32: 161–165.

Tieghem P van. 1884. Sur la structure et les affinites des Pittosporees. Bull. Soc. Bot. France 31: 383–385.

Tikhomirov VN. 1961. On the systematic position of the genus *Hydrocotyle* L. and *Centella* L. emend. Urban. Bot. Zhurn. 46: 584–586 (in Russian).

Tikhomirov VN and AL Konstantinova. 1995. On phylogenetic value of some characters of fruit structure in Umbelliferae-Hydrocotyloideae. Byull. Mosk. Obshch. Ispyt. Prir., Biol. 100(6): 61–73 (in Russian).

Tikhomirov VN and AI Konstantinova. 2000. Apiaceae (Umbelliferae). In: A Takhtajan, ed. Comparative seed anatomy, vol. 6, pp. 342–360. Nauka, St. Petersburg (in Russian).

Ting WS, CC Tseng, and ME Mathias. 1964. A survey of pollen morphology of Hydrocotyloideae (Umbelliferae). Pollen et Spores 6: 479–514.

Trifonova VI. 1998. Fruit and seed anatomy of the genus *Melanophylla* (Melanophyllaceae) in relation to its taxonomic position. Bot. Zhurn. 83: 97–103 (in Russian with English summary).

Trifonova VI. 2000a. Melanophyllaceae. In: A Takhtajan, ed. Comparative seed anatomy, vol. 6, pp. 262–266. Nauka, St. Petersburg (in Russian).

Trifonova VI. 2000b. Toricelliaceae. In: A Takhtajan, ed. Comparative seed anatomy, vol. 6, pp. 317–318. Nauka, St. Petersburg (in Russian).

Trifonova VI and EY Yembaturova. 2000a. Griseliniaceae. In: A Takhtajan, ed. Comparative seed anatomy, vol. 6, pp. 251–253. Nauka, St. Petersburg (in Russian).

Trifonova VI and EY Yembaturova. 2000b. Helwingiaceae. In: A Takhtajan, ed. Comparative seed anatomy, vol. 6, pp. 319–320. Nauka, St. Petersburg (in Russian).

Tsaturian TG. 1948. Palynologica caucasica: II. Pollen of the Caucasian representatives of the family Umbelliferae. Trudy Bot. Inst. Armenian Acad. Sci. 5: 157–198 (in Russian).

Tseng CC. 1967. Anatomical studies of flower and fruit in the Hydrocotyloideae (Umbelliferae). Univ. Calif. Publ. Bot. 42: 1–79.

Tseng CC. 1971. Light and scanning electron microscopic studies on pollen of *Tetraplasandra* (Araliaceae) and relatives. Am. J. Bot. 58: 505–516.

Tseng CC. 1980. The systematic position of *Aralidium* Miq.: a multidisciplinary study. 3. Pollen morphology. Taxon 29: 407–409.

Tseng CC and G Hoo. 1982. A new classification for the family Araliaceae. Acta Phytotax. Sinica 20: 125–129.

Tseng CC and JR Shoup. 1978. Pollen morphology of *Schefflera* (Araliaceae). Am. J. Bot. 65: 384–394.

Valiejo-Roman KM, MG Pimenov, EI Terentieva, SR Downie, DS Katz-Downie, and AV Troitsky. 1998. Molecular systematics of the Umbelliferae: using nuclear ribosomal DNA[1]

internal transcribed space sequences to resolve issues of evolutionary relationships. Bot. Zhurn. (St. Petersburg) 83: 1–22.
Valiejo-Roman KM, EI Terentieva, TH Samigulin, and MG Pimenov. 2002. Relationships among genera in Saniculoideae and selected Apioideae (Umbelliferae) inferred from nrITS sequences. Taxon 51: 91–101.
Van Wyk B-E. 2000. Apiaceae. In: OA Leistner, ed. Seed plants of southern Africa: families and genera. Strelitzia 10: 62–71.
Van Wyk B-E. 2001. A preliminary analysis of evolution of African and Madagascan Apiaceae. Edinb. J. Bot. 58: 291–299.
Viguier R. 1906. Recherche anatomique sur la classification des Araliacees. Ann. Sci. Nat. 4: 1–210.
Vyshenskaya TD. 2000. Araliaceae. In: A Takhtajan, ed. Comparative seed anatomy, vol. 6, pp. 321–342. Nauka, St. Petersburg (in Russian).
Wen J. 2001. Evolution of the *Aralia–Panax* complex (Araliaceae) as inferred from nuclear ribosomal its sequences. Edinb. J. Bot. 58: 243–257.
Wen J and EA Zimmer. 1996. Phylogeny and biogeography of *Panax* L. (the ginseng genus, Araliaceae): inferences from ITS sequences of nuclear ribosomal DNA. Mol. Phylogenet. Evol. 6: 167–177.
Wen J, GM Plunkett, AD Mitchell, and SJ Wagstaff. 2001. The evolution of Araliaceae: a phylogenetic analysis based on ITS sequences of nuclear ribosomal DNA. Syst. Bot. 26: 144–167.
Wilkinson HP. 1992. Leaf anatomy of the Pittosporaceae. Bot. J. Linn. Soc. 110: 1–59.
Winter PJD, B-E Van Wyk, and PM Tilney. 1993. The morphology and development of the fruit of Heteromorpha (Apiaceae). South Afr. J. Bot. 59: 336–341.

Superorder ASTERANAE

Order 115. ROUSSEALES

Small trees or shrubs, or (Rousseaceae) climbing shrubs. Vessels with scalariform perforation; lateral pitting scalariform to alternate, or occasionally transitional. Fibers with bordered pits. Rays mostly heterogeneous. Axial parenchyma scanty paratracheal, or diffuse-in-aggregates. Sieve-elements plastids of S-type. Nodes trilacunar. Leaves alternate, opposite or verticillate, estipulate. Stomata anomocytic. Flowers small, in terminal and axillary panicles, or flowers solitary, actinomorphic, mostly bisexual, rarely unisexual, polygamo-dioecious. Sepals 4–5, or 5–6, valvate, persistent or deciduous. Petals 4–6, valvate, persistent or deciduous. Stamens free or connate; anthers tetrasporangiate, opening longitudinally. Tapetum secretory. Microsporogenesis simultaneous. Pollen grains 3-celled, mostly 3-colporate, rarely (*Roussea*) pantoporate. Nectary disc present. Gynoecium of 5–7 or 3–5 (Carpodetaceae) united carpels; style thick or slender with capitate or lobed stigma. Ovary superior or semi-inferior to inferior, (4-)5–7-locular. Ovules numerous, anatropous, unitegmic, tenuinucellate. Female gametophyte of *Polygonum*-type. Endosperm cellular. Fruits baccate. Seeds numerous, small; embryo small; endosperm copious.

Rousseales belong to Asteranae, and occupied the basal position there (see Soltis et al. 1997a, b; Savolainen et al. 2000; Bremer et al. 2002; Soltis et al. 2006).

Key to Families

1 Climbing shrubs, sometimes strangler, up to 4 m in height. Trichomes glandular-peltate with multicellular heads, or unicellular, borne either solitary or in clusters; resin canals present. Vessels with very oblique end walls and exclusively with scalariform perforations that have an average of 20 (up to 64) bars; lateral pitting scalariform or occasionally transitional. Fibers short, with bordered pits. Rays tri- to multiseriate, heterogeneous and comprise procumbent, upright, and square cells. Axial parenchyma scanty paratracheal. Leaves opposite and verticillate, simple, petiolate, serrate-glandular, with unicellular conic hairs and wartlike cuticle protrusions on both surfaces and peltate, multicellular, and stalked glands on the lower surface and on the petiole, estipulate; petiole with radially elongated, schizogenous resin canals. Flowers solitary or few in the leaf axils, arising from bracteate buds, nodding; pedicels thick, arcuate. Copious amounts of nectaries present in flowers (Koontz et al. 2007). Calyx 4–5, united at the base, calyx lobes rather large, thick, valvate, persistent. Petals thick, hairy, connate into a 4–5-lobed tube, revolute at the apex, villous outside, persistent, valvate in bud. Stamens alternating with petals, inserted within the lobes of the nectary disc; anthers large, oblong-sagittate, basifixed, extrorse. Pollen grains in monads, 5–6- or 8-porate, with smooth and complete tectum. Style thick, persistent, expanding apically; stigma capitate, 4–5-lobed, stigmatic lobes narrower, erect, with revolute margins and like an indusium. Ovary superior, piramidal, (4-)5–7-locular, with numerous ovules many-seriate on thick, axile placentas.

Micropylar haustoria and a filamentous suspensor are developed. Fruits large, pale green, fleshy berries, acutely 5–7-angled, angles dilated at the base, (4-)5–7 locular. Seeds numerous, flattened, ovoid, exotestal. Testa with elongated epidermis cells with a strongly undulating outline and thickened and lignified cells walls (Lundberg 2001). Embryo minute, straight; endosperm copious.. . . . 1. ROUSSEACEAE.

1 Evergreen trees to 20 m tall or shrubs with unicellular haris. Young branchlets, petioles, peduncles, and pedicels pubescent. Vessels with long scalariform perforation plates; lateral pitting scalariform to alternate. Fibers with bordered pits. Rays broad, heterogeneous, with prismatic crystals in cells. Axial parenchyma sparse and predominantly diffuse or diffuse-in-aggregates. Leaves alternate, sharply glandular-serrate, with domatia on the abaxial surface in the axils of the secondary veins. Flowers small, white, in terminal and axillary panicles. Calyx tube turbinate, adnate to ovary, lobes 5–6, short, deciduous. Petals 5–6, inserted under margin of epigynous nectary disc, deciduous, valvate. Stamens (4-)5(-7), alternipetalous; filaments free, short, sometimes hairy; anthers oblong, introrse. Pollen grains in tetrahedral tetrads, 3-colporate. Gynoecium of 3–5 carpels; style slender, with capitate stigma; ovary semi-inferior or inferior; ovules numerous, many-seriate on placentas projecting into the loculi. Fruits 3–5-locular, berry-like, with more or less leathery pericarp, surrounded in the middle by the scar of calyx. Seeds numerous, small, pendulous, ovoid, with elongated funiculus; testa coriaceous, foveolate; embryo small, terete, at the base of copious, fleshy endosperm. Contain leucodelphinidin, quercetin and kaempferol; n = 14, 15. . . 2. CARPODETACEAE.

1. ROUSSEACEAE

A.P. de Candolle 1839. 1/1. Mauritius.

Roussea

Systematic position of *Roussea* is debatable. Hutchinson (1967) put it in the Escalloniaceae. According to Lundberg (2001), *Roussea* is close related to *Carpodetus* and they even combine their Carpodetaceae sensu lato (which includes also *Abrophyllum* and *Cuttsia*) with Rousseaceae. However, I prefer to keep Rousseaceae as a separate family.

2. CARPODETACEAE

Fenzl 1841. 1/10. New Guinea, Solomon Islands, New Zealand, and Stewart Island.

Carpodetus

Bibliography

Brook JP. 1951. Vegetative anatomy of *Carpodetus serratus* Forst. Trans. and Proc. Roy. Soc. N. Z. 79: 276–285.

Gustafsson MHG. 2007. Carpodetaceae. In: K Kubitzki, ed. The families and genera of vascular plants, vol. 8, pp. 57–60. Springer, Berlin/Heidelberg/New York.

Gustafsson MHG and B Bremer. 1997. The circumscription and systematic position of Carpodetaceae. Aus. Syst. Bot. 10: 855–862.

Koontz JA, J Lundberg, and DE Soltis. 2007. Rousseaceae. In: K Kubitzki, ed. The families and genera of vascular plants, vol. 8, pp. 611–613. Springer, Berlin/Heidelberg/New York.

Lundberg J. 2001a. The asteralean affinity of the Mauritian *Roussea* (Rousseaceae). Bot. J. Linn. Soc. 137: 267–276.

Nemirovich-Danchenko EN. 2000a. Carpodetaceae. In: A Takhtajan, ed. Comparative seed anatomy, vol. 6, pp. 253–255. Nauka, St. Petersburg (in Russian).

Praglowski J and E Grafstrom. 1985. The genus *Carpodetus* (Escalloniaceae). Grana 24: 11–21.

Order 116. CAMPANULALES

Mostly herbs (perennial or rarely annual), but sometimes secondarily woody (shrubs or less often small trees). Hairs when present mostly unicellular. In leaves and stems (Campanulaceae) articulated, anastomosing laticifers forming a well-developed system in the phloem and commonly extending into the surrounding tissues. Vessels mostly with simple or less often scalariform perforations, or scalariform and simple; lateral pitting scalariform to alternate. Wood parenchyma absent or indistinct. Fibers with simple or indistinctly bordered pits. Rays commonly 4–8 cells wide. Axial parenchyma scanty paratracheal or often wanting. Nodes usually unilacunar. Sieve-element plastids of S-type. Leaves alternate or sometimes opposite, rarely (*Ostrowskia*) verticillate, simple, entire, estipulate. Stomata commonly anomocytic. Flowers in various cymose or more often racemose inflorescences, sometimes solitary, usually bracteate and often bracteolate, bisexual or rarely unisexual, actinomorphic to zygomorphic, mostly with 5-merous perianth and androecium. Calyx (2-)5(-10)-lobed, lobes imbricate or valvate. Corolla usually 5-lobed, lobes valvate (sometimes induplicate) or less often imbricate (Sphenocleaceae). Stamens as many as petals and alternate with them, or 2–3, seemingly free from the corolla and attached to the annular, epigynous nectary disc or variously attached to the corolla tube; filaments free, or variously coherent or connate, the filament bases often expanded, forming a dome-shaped chamber over the nectary disc (the latter enlarged and glandlike in *Adenophora*), anthers basifixed, introrse or extrorse, tetrasporangiate, opening longitudinally. Tapetum secretory. Microsporogenesis simultaneous. Pollen grains 2-celled or less often 3-celled, colpate, colporate, or porate. Gynoecium of 2–10 carpels; style simple or lobed; ovary inferior or seldom semi-inferior, rarely superior, 1–3-locular with two parietal placentas (as in some Lobelioideae), or often with as many locules as carpels, commonly crowned by nectary disc. Ovules usually numerous, anatropous, unitegmic, tenuinucellate, with well-developed endothelium. Female gametophyte of *Polygonum*-type. Endosperm cellular, with typical terminal haustoria (but only in Pentaphragmataceae with micropylar haustorium). Fruits usually capsules, loculicidal, septicidal, or less often poricidal, in Sphenocleaceae and *Parischella* circumscissile (pyxidia), but sometimes indehiscent. Seeds with small, winged, or wingless, usually straight embryo and usually copious, oily or rarely starchy endosperm. Iridoids wanting, alkaloids usually present, cyanogenic; storing carbohydrate as inulin, n = 5–20, rarely up to 56.

Via the Pentaphragmataceae and archaic genus *Cyananthus* the order Campanulales is connected with the Desfontainiales, especially with Escalloniaceae (Takhtajan 1987).

Key to Families

1 Plants without laticifers. Flowers lacking any specialized mechanism for secondary pollen presentation. Inulin and alkaloids absent.

2 Vessels with scalariform perforations with numerous bars. Fiber-tracheids with prominently bordered pits. Perennial coarse herbs often with fleshy or succulent stems. Leaves alternate, often relatively large, petiolate, simple, entire or dentate, usually strongly asymmetric at base, often with branching hairs. Stomata anomocytic, surrounded by three or four cells somewhat differentiated from the ordinary epidermal cells. Flowers in dense, axillary or extra-axillary, sympodial helicoid cymes with usually conspicuous membranous bracts, bisexual or rarely unisexual (plant dioecious). Sepals five, unequal, imbricate, persistent. Corolla usually fleshy or cartilaginous, rarely delicate, persistent, seemingly choripetalous in some species, with five, rarely (*Pentaphragma tetrapetalum*) four, valvate lobes or petals. Stamens 5(4), alternating with petals, inserted in the throat of the corolla tube, shortly below sinuses or on the margins of the top of ovary in choripetalous species; filaments free, flattened, or anthers sessile (rarely filaments almost wanting); anthers 2-thecal. Pollen grains 2-celled, 3-colpate or 3-colporate, trilobate, smooth. Floral nectaries present and nectar secretion from between the hypanthium and the gynoecium, creating 4–5 nectariferous channels. Gynoecium of 2(3) carpels; style short and thick, with massive glabrous stigma; ovary 2(3)-locular, inferior, adnate to the floral tube only by means of five narrow longitudinal septa formed by continuation of filaments, which leave intervening lacunae or pits in which nectar is produced. Ovules numerous on bifid axile placenta, pendulous. Endosperm only with micropylar haustorium. Fruits berries with the perianth persisting at the tip. Seeds minute, exotestal cells cuboid, inner walls lignified; endosperm copious, starchy; n = 54–56. . . . . 1. PENTAPHRAGMATACEAE.

2 Vessels with simple perforations. Annual herbs. Stem somewhat succulent and with large, vertical, cortical air canals, and pericyclic sclerenchyma. Leaves alternate, entire, papillose; the mesophyll containing calcium oxalate crystals. Stomata tetracytic. Flowers in dense terminal spikes, in the axils of small bracts and bibracteolate, bisexual, actinomorphic, 5-merous. Sepals five, with imbricate lobes, persistent. Corolla small, caducous, urceolate-campanulate, with five imbricate lobes. Stamens as many as and alternate with petals, attached to the base of corolla tube; filaments short, free; anthers dorsifixed, introrse. Pollen grains 3-celled, 3-colporate to colporoidate, reticulate. Nectary disc wanting. Gynoecium of two carpels; stigma sessile or on very short style, capitate; ovary 2-locular, inferior or semi-inferior. Ovules numerous on the large, spongy, pendulous placenta. Endosperm with ter-

minal haustoria. Embryogeny of Onagrad-type. Fruits membranous pyxidia. Seeds small, with straight embryo and very scanty or no endosperm; exotestal cells with inner walls thickened. Producing fructose with isokestose linkages, n = 12, 16, 20, + ...2. SPHENOCLEACEAE.

1 Plants with network of articulated laticifers. Flowers usually with specialized mechanisms for pollen presentation. Anthers free or imperfectly connate. Inulin present. Alkaloids present or absent. Perennial or less often biennial or annual herbs, or less often shrubs, woody lianas (sometimes twining), treelets or trees to 15 m tall, typically terrestrial, rarely aquatic or epiphytic; trichomes typically unicellular or sometimes uniseriate. Vessels commonly with simple perforations, but some genera (*Cyananthus, Platycodon, Canarina, Musschia*) have scalariform perforations with 1–8 bars; or scalariform and simple, lateral pitting scalariform to alternate. Rays wide, consisting mostly of upright cells. Axial parenchyma wanting or indistinct. Nodes unilacunar with one trace. Leaves alternate, sometimes opposite, entire to pinnately divided, petiolate or subsessile; leaves of hydrophytes submerged and emergent. Stomata anomocytic. Flowers mostly in cymose or racemose inflorescences, solitary and axillary, usually bisexual, actinomorphic, or more or less zygomorphic; bracts foliose or reduced, rarely absent; pedicels often bracteolate. Calyx (3-)5(-10)-lobed, the lobes imbricate or valvate; sometimes basally appendaged, the odd lobe is dorsal (posterior). Corolla campanulate to tubular, the lobes valvate or rarely induplicate. Stamens as many as the corolla lobes and alternate, inserted near the base of corolla or on the nectary disc or rarely (*Siphocodon* and *Rhigiophyllum*) partly epipetalous; filaments usually free; anthers basifixed, rarely (*Berenice*) dorsifixed, introrse, usually connivent, but commonly separating after anthesis. Pollen grains in tetrahedral or isobilateral tetrads, 2- or 3-celled, 5–10-colpate (*Cyananthus, Codonopsis, Leptocodon*), 6-colpate (*Parishella*), (2)3(4)-colporate, 5–6-colporate, or more often 3–8-zonoporate, rarely pantoporate, echinate. Gynoecium mostly of 2–10, mostly three carpels; style terminates in a more or less lobed stigma and is provided with collecting hairs below the stigma; the stylar hairs eventually invaginate, dislodging pollen grains. Ovary inferior or rarely semi-inferior or even superior, often crowned by nectary disc, with as many locules as carpels or rarely 1-locular. Ovules mostly numerous, horizontal on axile placentas, rarely placentation apical *Siphocodon*) or (*Merciera*) basal. Fruits septicidal or loculicidal, or valvular capsules, which open by valves, slits or pores, located either at the apex of the ovary (i.e. above the calyx lobes) or on the lateral walls of the hypanthium (i.e. below the calyx lobes), or splitting irregularly, often crowned by the persistent calyx lobes, rarely baccate (Canarinoideae and *Clermontia* in Lobelioideae) or very rarely (*Theodoravia*) schizocarp. Seeds numerous, small, winged or wingless; exotesta of lignified cuboid or fibriform cells, endotestal cells thickened; embryo straight; endosperm fleshy, oily, or rarely starchy endosperm. Alkaloids and flavonols present or absent, contain tyrosine-derived (with triglochinin); producing 14-carbon polyacetelenes, ursolic and caffeic acid; very commonly inulin; iridoids and tannins wanting, n = 6–21, 23–30, 32, 34–36, 40, 45, 48, 51, 52 (the commonest number is 17). . . . . 3. CAMPANULACEAE.

1. PENTAPHRAGMATACEAE

J. Agardh 1858. 1/c.30. Lower Burma, southern China, Indochina, Malesia to New Guinea.

Pentaphragma

It is a relatively less specialized member of the order. However, inflorescences, pollen grains, fruits, and seed coat are characterized by a number of derived characters. Pentaphragmataceae are probably nearest to the ancestral stock of the Campanulales. Rather isolated within the order.

2. SPHENOCLEACEAE

Baskerville 1839. 1/2. One species pantropical, another one in West Africa.

Sphenoclea

An isolated family with a number of unique features, but it definitely belongs to the Campanulales.

3. CAMPANULACEAE

A.L. de Jussieu 1789 (including Cyananthaceae J. Agardh 1858; Cyphiaceae A.P. de Candolle 1839; Cyphocarpaceae Reveal et Hoogland 1996; Dortmannaceae Ruprecht 1856; Jasionaceae Dumortier 1829; Lobeliaceae A.L. de Jussieu 1813; Nemacladaceae Nuttall 1842). 92/2000–2400. Very widely distributed, especially in the temperate regions of the Northern Hemisphere but relatively sparsely developed in the tropics. There are only a few endemic genera in South Africa and adjacent islands of Indian Ocean (including monotypic *Berenice* on Reunion). In Australia, Tasmania, and New Zealand there are only several species of *Wahlenbergia,* a large complex subcosmopolitan widespread in the Southern Hemisphere.

3.1 CYANANTHOIDEAE (subfam. nov. basionym: Cyanantheae Meissner 1840)
Pollen grains 4–10-colpate or 5–6-colporate. Vessels with both simple and scalariform perforations. Ovary superior to semi-inferior. Calyx lobes without appendages between them. Fruits capsular, dehiscing by apical valves.–CYANANTHEAE: *Cyananthus;* CODONOPSIDEAE: *Codonopsis, Numaeacampa, Leptocodon*; PLATYCODONEAE: *Platycodon.*

3.2 CANARINOIDEAE
Pollen grains 3(4)-colporate to 3-porate. Vessels with both simple and scalariform perforations. Ovary inferior, 4–7-locular. Fruits baccate. – *Canarina, Campanumoea.*

3.3 OSTROWSKIOIDEAE
Pollen grains compressed-spheroidal, with strongly perforate tectum and massive endexine, but without footlayer, 5–7-aperturate, apertures poroid, elongate-oval; ornamentation gemmate, with two different types of gemmae – numerous small and sparsely dispersed large ones. Vessels only annular and helical. Vascular bundles widely separated. Pith and primary cortex well developed. Calyx lobes without appendages. Corolla very large (up to 10 cm long). Anthers with a placentoid. Gynoecium of (5-)7(-9) carpels; ovules with massive integument. Fruits large capsules with oblong, longitudinal pores of dehiscence, twice as many as sepals; each pore is covered by translucent tissue that later ruptures by vertical fissure in the middle. Seeds winged. High geophytes with thick stem and tuberous root. – *Ostrowskia,* Central Asia.

3.4 CAMPANULOIDEAE
Pollen grains zonoporate or sometimes pantoporate. Roots often thick. Vessels usually with simple perforations. Ovary inferior or semi-inferior, rarely nearly superior. Calyx with or without appendages between the lobes. Fruits capsules, dehiscing in various ways in different genera. – WAHLENBERGIEAE: *Wahlenbergia* (including *Cephalostigma*), *Berenice, Theilera, Gunillaea, Nesocodon, Heterochaenia, Microcodon;* AZORINEAE: *Azorina;* MUSSCHIEAE: *Musschia;* ECHINOCODONEAE: *Echhinocodon*; CAMPANULEAE: *Campanula* (including *Hyssaria, Annaea, Fedorovia, Mzymtella, Megalocalyx, Sicyocodon*), *Gadellia, Symphyandra, Zeugandra, Brachycodonia, Hanabusaya, Adenophora, Popoviocodonia, Astrocodon, Trachelium, Feeria;* PERACARPEAE: *Peracarpa, Homocodon*; MICHAUXIEAE: *Michauxia;* PHYTEUMATEAE: *Phyteuma, Physoplexis, Asyneuma, Sergia, Petromarula, Cryptocodon, Cylindrocarpa, Legouzia, Triodanis, Heterocodon, Githopsis;* EDRAIANTHEAE: *Edraianthus, Muehlbergella;* JASIONEAE: *Jasione;* PRISMATOCARPEAE: *Prismatocarpus, Namacodon, Roella, Craterocapsa, Treichelia;* SIPHOCODONEAE: *Siphocodon, Rhigiophyllum*; MERCIEREAE: *Merciera.*

3.5 NEMACLADOIDEAE
Pollen grains 6-colpate (*Parishella*) or 3-colporate, with spinules. Stamens inserted a top the ovary. *Nemacladus* has groups of remarkable reflexed pseudonectaries at the bases of two filaments. Small, annual or (*Pseudonemacladus*) perennial herbs with erect or prostrate stems and without tubers. Leaves small, basal. Flowers small, in racemes or capitate clusters. Calyx 5-lobed. Corolla bilabiate, with two ventral and three dorsal lobes and lack any sort of palate. Filaments connate in their upper or middle parts; anthers distinct. Stigma 2-lobed. Ovary inferior, 2-locular, with numerous ovules. Fruits capsules, 2- or 4-valved from the tip or (*Parishella*) its low-conic apex circumscissile just above the calyx lobes. Seeds minute, n = 9. – *Pseudonemacladus, Nemacladus, Parishella.*

3.6 CYPHOCARPOIDEAE
Pollen grains 3-colporate, with characteristic reticulate ornamentation. Stamens epipetalous. Annual herbs. Leaves spinulose-sinuato-dentate, rigid. Flowers sessile, solitary in the axils of the upper leaves, often in leafy spikes, with two leaflike bracteoles. Calyx 5-lobed. Corolla bilabiate, with one-lobed upper (dorsal) lip and four-lobed ventral lip with a gibbous palate; corolla tube curved, long. Stamens attached to the corolla tube under the limb; filaments short, distinct; anthers basifixed, introrse. Style filiform, glabrous,

apically curved, surrounded at the base by fleshy annular nectary disc; stigma divided into two broad lobes with a small, sticky gland between the lobes; lobes of the closed stigma are externally covered with setulose hairs; lobes of the open stigma are glabrous (the externally hairy stigma functions as a collector of pollen); ovary 2-locular, with numerous ovules on very thin septa (fused intrusive placentas), which in very early stage begins to separate from the ovary wall and transforms into free-central placenta. Fruits membranous capsules crowned by the calyx lobes. Seeds numerous, cylindrical, longitudinally costate. – *Cyphocarpus*.

3.7 CYPHIOIDEAE

Pollen grains 3-colporate, nanogranulate, almost smooth. Style in the upper part with a fluid-filled stigmatic cavity communicating with the air through a small lateral aperture. Perennial herbs usually with a subterranean root tuber containing both inulin and starch. Stems erect or twining. Leaves alternate, radical or cauline, entire, dentate or divided. Flowers zygomorphic, in racemes, bracteate and with two bracteoles. Calyx 5-lobed; the odd (unpaired) lobe originates in a ventral (anterior) position. Corolla either split to the base into two lips with the upper lip 3-lobed and the lower lip consisting of two free petals, or all petals united into a tube with a 5-lobed limb. Stamens five, free from petals; filaments free or more or less connate at the base; anthers free or apically loosely united, glabrous or more or less pubescent on the back of the connective. Gynoecium of two carpels; style shorter than the stamens, obliquely clavate, crowned by hairs; ovary semi-inferior to semi-superior, 2-locular, with usually many ovules per locule. Fruits capsules dehiscing loculicidally by two apical, usually bifid valves. Seeds minute, numerous, with reticulate testa, n = 9. – *Cyphia*.

3.8 LOBELIOIDEAE

Flowers resupinate at anthesis (except *Downingia laeta*) and with specialized mechanism for protandrous pollen presentation. Herbs or less often more or less woody, sometimes (*Brighamia*) with succulent stem. Vessels with simple perforations. Fibers with simple pits. Rays heterogeneous. Axial parenchyma scanty paratracheal. Leaves alternate, entire to pinnately divided. Flowers usually in racemose inflorescences, bisexual or rarely unisexual, more or less zygomorphic. Calyx 5-lobed. Corolla bilabiate or unilabiate, the odd lobe originates in a dorsal position. Stamens five, inserted at the base of the corolla tube or at the annular nectary disc; filaments connate for at least part of their length; anthers connate into a tube around the style; the two ventral anthers are shorter than the other three. Pollen grains usually 3-colporate or sometimes 3-colpate, usually reticulate-striate, never echinate. Gynoecium of two carpels; style provided with a brush of collecting hairs below the initially appressed stigmas; ovary inferior or rarely semi-inferior (*Diastatea*), usually 2-locular, with numerous ovules. Fruits capsular and variously dehiscent or rarely baccate (*Clermontia*). Seeds numerous, small, with small, straight embryo and copious endosperm. Producing pyridine alkaloids (which commonly accumulate in the latex), chelidonic acid and 14-carbon polyacetylenes, but no caffeic acid. n = 6–7, commonly 14. – LOBELIEAE: *Siphocampylus, Diastatea, Laurentia, Palmerella, Solenopsis, Hippobroma, Porterella, Isotoma, Lobelia, Trimeris, Wimmerella, Trematolobelia, Grammatotheca, Monopsis, Heterotoma, Dialypetalum, Ruthiella, Legenere, Unigenes, Downingia, Howellia, Brighamia, Sclerotheca, Apetahia, Lysipomia;* DELISSEEAE: *Pratia, Hypsela, Burmeistera, Centropogon, Delissea, Cyanea, Clermontia, Rollandia, Dielsantha.*

Bibliography

Adamson RS. 1951. A revision of the genera *Prismato-carpus* and *Roella*. South Afr. J. Bot. 17: 93–166.

Adamson RS. 1954. The genus *Merciera* A. DC. South Afr. J. Bot. 20: 157–163.

Airy Shaw HK. 1954. Pentaphragmataceae. In: CGGJ van Steenis, ed. Flora malesiana ser. 1, 4(1): 517–528.

Avetisian EM. 1948. Pollen of the Caucasian representatives of the fam. Campanulaceae. Trudy Bot. Inst. Armenian Acad. Sci. 1948: 198–206 (in Russian).

Avetisian EM. 1967. Pollen morphology of the family Campanulaceae and related families (Sphenocleaceae, Lobeliaceae, Cyphiaceae) in relation to their systematics and phylogeny. Trudy Bot. Inst. Armenian Acad. Sci. 16: 5–41 (in Russian).

Avetisian EM. 1973. Palynology of the order Campan-ulales s. 1. In: Spores and pollen morphology of recent plants, pp. 90–93. Proc. 3rd Int. Palynol. Conf., Acad. Sci. USSR. Leningrad (in Russian).

Avetisian EM. 1986. Palynomorphology of the families Campanulaceae, Sphenocleaceae, and Pentaphragmataceae. Bot. Zhurn. 71: 1003–1010 (in Russian with English summary).

Avetisian EM. 1988. Palynology of the superorder Campanulanae. Ph.D. dissertation, University of Erevan (in Russian).

Ayers TJ. 1990. Systematics of *Heterotoma* (Campanulaceae) and the evolution of nectar spurs in the New World Lobelioideae. Syst. Bot. 15: 296–327.

Ayers T and R Haberle. 1999. Systematics of *Cyphocarpus* (Campanulaceae): placement of an evolutionary enigma. In: XVI Int. Bot. Congr. Abstracts, p. 196. Missouri Botanical Garden, St. Louis, MO.

Badre E, Th Cadet, and M Malplanche. 1972. Etude systematique et palynologique du genre *Heterochaenia* (Campanulaceae) endemique des Mascareignes. Adansonia, ser. 1, 12: 267–278.

Batterman MRW and TG Lammers. 2004. Branched foliar trichomes of Lobelioideae (Campanulaceae) and the infrageneric classification of *Centropogon*. Syst. Bot. 29: 448–458.

Belyaev AA. 1984a. Seed anatomy of some representatives of the family Campanulaceae. Bot. Zhurn. 69: 585–594 (in Russian with English summary).

Belyaev AA. 1984b. Ultrastructure of the surface and some morphological characteristics of seeds of some representatives of the family Campanulaceae. Bot. Zhurn. 69: 890–898 (in Russian with English summary).

Belyaev AA. 1985. New data on the anatomical structure of seed coat and ultrastructure of seed-surface of two representatives of the genus *Pentaphragma* (Campanulaceae). Bot. Zhurn. 70: 955–957 (in Russian with English summary).

Belyaev AA. 1986. Features of the anatomy and ultrastructure of the surface of the seed coat in some species of critical genera of the family Campanulaceae. Bot. Zhurn. 71: 1371–1375 (in Russian with English summary).

Bigazzi M. 1986. Ultrastructural and cytochemical observations on fibrillar intranuclear inclusions in the family Campanulaceae. Caryologia 39: 199–210.

Buss CC, TG Lammers, and RR Wise. 2001. Seed coat morphology and its systematic implications in *Cyanea* and other genera of Lobelioideae (Campanulaceae). Am. J. Bot. 88: 1301–1308.

Candolle AP de. 1830. Monographic des Campanulees. Paris.

Carlquist S. 1969. Wood anatomy of Lobelioideae (Campanulaceae). Biotropica 1: 47–72.

Carlquist S. 1997. *Pentaphragma*: a unique wood and its significance. IAWA J. 18: 3–12.

Carolin RC. 1960. The structures involved in the presentation of pollen to visiting insects in the order Campanulales. Proc. Linn. Soc. N. S. W. 85: 197–207.

Carolin RC. 1967. The concept of the inflorescence in the order Campanulales. Proc. Linn. Soc. N. S. W. 92: 7–26.

Chapman JL. 1966. Comparative palynology in Campanulaceae. Trans. Kansas Acad. Sci. 69: 197–204.

Cosner ME, RK Jansen, and TG Lammers. 1994. Phylogenetic relationships in the Campanulales based on *rbc*L sequences. Plant Syst. Evol. 190: 79–95.

Cosner ME, LA Raubeson, and RK Jansen. 2004. Chloroplast DNA rearrangements in Campanulaceae: phylogenetic utility of highly rearranged genomes. Mol. Phylogenet. Evol. 4: 27, see http://www.biomedcentral.com/1471–2148/4/27

Crete P. 1951. Repartition et interet phylogenetique des albumens a formations haustoriales chez les Angi-ospermes et plus particulierement chez Gamopetales. Ann. Sci. Nat. 2nd ser. 12: 131–191.

Dunbar A. 1975a. On pollen of Campanulaceae and related families with special reference to the surface ultrastructure: I. Campanulaceae subfam. Campanuloideae. Bot. Not. 128: 73–101.

Dunbar A. 1975b. On pollen of Campanulaceae and related families with special reference to the surface ultrastructure: II. Campanulaceae subfam. Cyphioideae and subfam. Lobelioideae; Goodeniaceae; Sphenocleaceae. Bot. Not. 128: 102–118.

Dunbar A. 1978. Pollen morphology and taxonomic position of the genus *Pentaphragma* Wall. (Pentaphragmataceae): the use of compound fixatives. Grana 17: 141–147.

Dunbar A. 1981. The preservation of soluble material on the surface and in cavities of the pollen wall of Campanulaceae and Pentaphragmataceae. Micron 12: 47–64.

Dunbar A. 1984. Pollen morphology in Campanulaceae, part 4. Nord. J. Bot. 4: 1–19.

Dunbar A and H-G Wallentinus. 1976. On pollen of Campanulaceae: III. A numerical taxonomic investigation. Bot. Not. 129: 69–72.

Eddie WM, RC Haberle, and RK Jansen. 2002. The phylogeny of the Campanulaceae inferred from DNA sequences of the chloroplast gene *mat*K. In Botany 2002: Botany in the Curriculum. Abstracts, p. 122. Madison, WI.

Eddie WMM, T Shulkina, J Gaskin, RC Haberle, and RK Jansen. 2003. Phylogeny of Campanulaceae s.str. inferred from its sequences of nuclear ribosomal DNA. Ann. Missouri Bot. Gard. 90: 554–575.

Erbar C. 1995. On the floral development of *Sphenoclea zeylanica* (Sphenocleaceae, Campanulales): SEM-investigations on herbarium material. Bot. Jahrb. Syst. 117: 469–483.

Erbar C and P Leins. 1989. On the early floral development and the mechanism of secondary pollen presentation in *Campanula, Jasione,* and *Lobelia*. Bot. Jahrb. Syst. 111: 29–55.

Erdtman G and CR Metcalfe. 1963. Affinities of certain genera *incertae sedis* suggested by pollen morphology and vegetative anatomy: III. The Campanulaceous affinity of *Berenice argenta* Tulasne. Kew Bull. 17: 253–256.

Erickson R. 1958. Triggerplants. Patterson Brokensha, Perth.

Fedorov AA. 1957. Campanulaceae. Flora URSS 24: 126–450 (in Russian).

Gadella TWJ. 1966. Some notes on the delimitation of genera of Campanulaceae, part 1. Proc. Kon. Nederl. Akad. Wetensch., ser. C, 69: 502–521.

Givnish TJ, E Knox, JF Smith, and WJ Hahn. 1994. Thorn-like prickles and heterophylly in *Cyanea*: adaptations to extinct avian browsers on Hawaii? Proc. Nat. Acad. Sci. USA 91: 2810–2814.

Givnish TJ, E Knox, JF Smith, and WJ Hahn. 1995. Molecular evolution, adaptive radiation, and geographic speciation in *Cyanea* (Campanulaceae: Lobelioideae). In: WL Wagner and VA Funk, eds. Hawaiian biogeography: evolution on a hot spot archipelago, pp. 288–337. Smithsonian Institution Press, Washington.

Gudkova IY and GP Borshchenko. 1991. The serological study of the Campanulaceae: The phylogenetic relations in the tribe Phyteumateae. Bot. Zhurn. 76: 809–817 (in Russian with English summary).

Gupta DP. 1959. Vascular anatomy of the flower of *Sphenoclea zeylanica* Gaertn. and some other related species. Proc. Natl. Inst. Sci. India 25B: 55–64.

Gustafsson MHG. 1996. Phylogenetic studies in the Asterales *sensu lato*. Acta Universitatus Upsaliensis, pp. 1–36.

Haberle RC. 1998. Phylogenetic systematics of *Pseudonemacladus* and the North American Cyphioids (Campanulaceae, sensu lato). M.S. thesis, Northern Arizona University.

Haberle RC and TJ Ayers. 1997. Systematics of *Pseudonemacladum* (Nemacladaceae). Am. J. Bot. 84(6): 200 (abstract).

Haridasan VK and PK Mukherjee. 1988. Seed surface features of some members of the Indian Campanulaceae. Phytomorphology 37: 277–285.

Hedberg O. 1961. Monograph of the genus *Canarina* L. (Campanulaceae). Svensk. Bot. Tidskr. 55: 16–62.

Heidenhain B. 1952. Über die Blütenstande der Cam-panulaceen. Akad. Wiss. Abh. Math.-Naturwiss. Kl. 9: 621–650.

Hils MH. 1985. Comparative anatomy and systematics of twelve woody Australasian genera of the Saxifragaceae. Ph.D. dissertation, University of Florida.

Hong D-Y. 1995. The geography of the Campanulaceae: on the distribution centres. Acta Phytotax. Sinica 33: 521–536.

Hong D-Y and L-M Ma. 1991. Systematics of the genus *Cyananthus*. Acta Phytotax. Sinica 29: 25–49.

Hong DY and KY Pan. 1998. The restoration of the genus *Cyclocodon* (Campanulaceae) and its evidence from pollen and seed-coat. Acta Phytotax. Sinica 36: 106–110.

Kamelina OP and NA Zhinkina. 1989. On the embryology of *Ostrowskia magnifica* (Campanulaceae). Development of the male embryological structures. Bot. Zhurn. 74: 1293–1300 (in Russian with English summary).

Kamelina OP and NA Zhinkina. 1998. On the embryology of *Ostrowskia magnifica* (Campanulaceae). The ovule and seed. Bot. Zhurn. 83: 9–20 (in Russian with English summary).

Kapil RN and MR Vijayaraghavan. 1962. Embryology and systematic position of *Pentaphragma horsfieldii* (Miq.) Airy Shaw. Curr. Sci. 31: 270–272.

Kapil RN and MR Vijayaraghavan. 1965. Embryology of *Pentaphragma horsfieldii* (Miq.) Airy Shaw with a discussion of the systematic position. Phytomorphology 15: 93–102.

Kaplan DR. 1967. Floral morphology, organogenesis, and interpretation of the inferior ovary in *Downingia bacigalupii*. Am. J. Bot. 54: 1274–1290.

Kausik SB and K Subramanyam. 1946. A contribution to the life history of *Sphenoclea zeylanica* Gaertn. Proc. Indian Acad. Sci. 23B: 274–280.

Knox EB and AM Muasya. 2001. The phylogeny and biogeography of the Lobeliaceae based on the chloroplast genes *atp*B and *rbc*L and their intergenic spacer sequence. In: Botany 2001: Plants and People. Abstracts, pp. 121–122. Albuquerque.

Knox EB and JD Palmer. 1998. Chloroplast DNA evidence of the origin and radiation of the giant lobelias in eastern Africa. Syst. Bot. 23: 109–149.

Knox EB and JD Palmer. 1999. The chloroplast genome of *Lobelia thuliana* (Lobeliaceae): expansion of the inverted repeat in an ancestor of the Campanulales. Plant Syst. Evol. 214: 49–64.

Knox EB, SR Downie, and JD Palmer. 1993. Chloroplast genome rearrangements and the evolution of giant lobelias from herbaceous ancestors. Mol. Biol. Evol. 10: 414–430.

Kolakovsky AA. 1986. Carpology of the Campanulaceae and problems of taxonomy. Bot. Zhurn. 71: 1155–1168 (in Russian with English summary).

Kolakovsky AA. 1987. System of the Campanulaceae of the Old World. Bot. Zhurn. 72: 1572–1579 (in Russian with English summary).

Kolakovsky AA. 1990. New data on the morphology of the flower and fruit in the family Campanulaceae. Soobschch. Akad. Nauk Gruz. SSR. 139(2): 381–384 (in Russian).

Kolakovsky AA. 1994. The conspectus of the system of the Old World Campanulaceaae. Bot. Zhurn. 79: 109–124 (in Russian with English summary).

Lammers TG. 1988. Chromosome numbers and their systematic implications in the Hawaiian Lobelioideae (Campanulaceae). Am. J. Bot. 75: 1130–1134.

Lammers TG. 1992. Circumscription and phylogeny of the Campanulales. Ann. Missouri Bot. Gard. 79: 388–413.

Lammers TG. 1993. Chromosome numbers of Campanulaceae: III. Review and integration of data for subfamily Lobelioideae. Am. J. Bot. 80: 660–675.

Lammers TG. 1998a. Review of the neotropical endemics *Burmeistera, Centropogon,* and *Siphocampylus* (Campanulaceae: Lobelioideae), with description of 18 new species and a new section. Brittonia 50: 233–262.

Lammers TG. 1998b. Nemacladoideae, a new subfamily of Campanulaceae. Novon 8: 36–37.

Lammers TG. 2007a. Campanulaceae. In: K Kubitzki, ed. The families and genera of vascular plants, vol. 8, pp. 26–56. Springer, Berlin/Heidelberg/New York.

Lammers TG. 2007b. Pentaphragmataceae. In: K Kubitzki, ed. The families and genera of vascular plants, vol. 8, pp. 605–607. Springer, Berlin/Heidelberg/New York.

Lammers TG, TJ Givnish, and KJ Sytsma. 1993. Merger of the endemic Hawaiian genera *Cyanea* and *Rollandia* (Campanulaceae: Lobelioideae). Novon 3: 437–441.

Leins P and C Erbar. 1990. On the mechanisms of secondary pollen presentation in the Campanulales-Asterales complex. Bot. Acta 103: 87–92.

Leins P and C Erbar. 2003. The pollen box in Cyphiaceae (Campanulales). Int. J. Plant Sci. 164(5 Suppl.): 321–328.

Markgraf F. 1950. Die Campanulaceen von Südwestafrika. Bot. Jahrb. Syst. 75: 206–220.

Marloth R. 1932. The flora of South Africa, vol. 3, part 2. Wheldon & Wesley, London.

McVaugh R. 1948. Generic status of *Triodanis* and *Specularia*. Rhodora 50: 38–49.

Miers J. 1848. Contribution to the botany of South America. J. Bot. (London) 7: 57–64.

Monod T. 1980. A propos du *Sphenoclea zeylanica* (Sphenocleaceae). Adansonia, ser., 2, 20: 147–164.

Morin N. 1980. Systematics of the annual California Campanulas (Campanulaceae). Madroño 4: 149–163.

Morin N. 1983. Systematics of *Githopsis* (Campanulaceae). Syst. Bot. 8: 436–468.

Morin N. 1987. Pollen surface morphology in North American *Campanula*, *Triodanis*, and *Heterocodon* (Campanulaceae). Am. J. Bot. 74: 746–747.

Moris KE and TG Lammers. 1997. Circumscription of *Codonopsis* and the allied genera *Campanumoea* and *Leptocodon* (Campanulaceae: Campanuloideae). I. Palynoloical data. Bot. Bull. Acad. Sinica (Taipei) 38: 277–284.

Munz PA. 1924. A revision of the genus *Nemacladus* (Campanulaceae). Am. J. Bot. 11: 233–248 and plates 9 and 10.

Murata J. 1992. Systematic implications of seed coat morphology in *Lobelia* (Campanulaceae – Lobelioideae) J. Fac. Sci. Univ. Tokyo 15: 155–172.

Murthy CVS. 1982. Pollen morphology of Indian *Campanumoea* Bl. J. Palynol. 18: 55–59.
Nowicke JW, SG Shetler, and N Morin. 1992. Exine structure of pantoporate *Campanula* (Campanulaceae) species. Ann. Missouri Bot. Gard. 79: 65–80.
Oganesian ME. 1955. Synopsis of Caucasian Campanulaceae. Candollea 50: 275–308.
Oganesian ME. 1985. Structural features of seeds in *Campanula* and *Symphyandra* (Campanulaceae). Bot. Zhurn. 70: 947–955 (in Russian with English Summary).
Rosatti TJ. 1986. The genera of Sphenocleaceae and Campanulaceae in the southeastern United States. J. Arnold Arbor. 67: 1–64.
Rosen W. 1932. Zur Embryologie der Campanulaceen und Lobeliaceen. Acta Hort. Goteb. 7: 31–42.
Rosen W. 1949. Endosperm development in Campanulaceae and closely related families. Bot. Not. 1949: 137–147.
Schonland S. 1889. Campanulaceae. In: A. Engler, ed. Die natürlichen Pflanzenfamilien, Bd. 4, Heft 5, S. 40–70. Leipzig.
Scottsberg C. 1915. Notes on the relations between the floras of Subantarctic America and New Zealand. Plant World 18: 129–142.
Shamrov II and NA Zhinkina. 1994. Development of the ovule of *Azorina vidalii* (Campanulaceae). Bot. Zhurn. 79: 19–34 (in Russian with English summary).
Shetler SG. 1979. Pollen-collecting hairs of *Campanula* (Campanulaceae). Taxon 28: 205–215.
Shetler SG and NR Morin. 1986. Seed morphology in North American Campanulaceae. Ann. Missouri Bot. Gard. 73: 653–688.
Shresta KK. 1992. Taxonomic revision of the genus *Cyananthus* Wall. ex Benth. (Campanulaceae). Ph.D. dissertation, University of St. Petersburgh.
Shresta KK and TI Kravtsova. 1992. Seed coat anatomy and ultrastructure in the genus *Cyananthus* in relation to its systematics. Bot. Zhurn. 77(6): 18–29 (in Russian with English summary).
Shresta KK and VF Tarasevich. 1992. Comparative pollen morphology of the genus *Cyananthus* in relation to its systematics and its position within the family Campanulaceae. Bot. Zhurn. 77(10): 1–13 (in Russian with English summary).
Shulkina TV. 1978. Life-forms in the family Campanulaceae Juss., their geographical distribution in connection with taxonomy. Bot. Zhurn. 63: 153–169 (in Russian with English summary).
Shulkina TV and SE Zykov. 1980. The anatomical structure of the stem in the family Campanulaceae s. str. in relation to the evolution of life forms. Bot. Zhurn. 65: 627–638 (in Russian with English summary).
Shulkina TV, JF Gaskin, and WMM Eddie. 2003. Morphological studies toward in improved classification of Campanulaceae s.str. Ann. Missouri Bot. Gard. 90: 576–591.
Subramanyam K. 1950a. A contribution to our knowledge of the systematic position of the Sphenocleaceae. Proc. Indian Acad. Sci. 31B: 60–65.
Subramanyam K. 1950b. An embryological study of *Levenhookia dubia*. Proc. Natl. Inst. Sci. India 16B: 245–253.
Subramanyam K. 1951. Interrelationships of Campanulacae. J. Mysore Univ. B. 12: 331–339.
Subramanyam K. 1953. The nutritional mechanism of embryo sac and embryo in the families Campanulaceae, Lobeliaceae, and Stylidiaceae. J. Mysore Univ. B. 13: 1–4.
Tarasevich VF and KK Shresta. 1992. Palynological data on the position of the genus *Ostrowskia* within the family Campanulaceae. Bot. Zhurn. 77(9): 27–30 (in Russian with English summary).
Thulin M. 1974. *Gunillaea* and *Namacodon:* two new genera of Campanulaceae in Africa. Bot. Not. 127: 165–182.
Thulin M. 1975. The genus *Wahlenbergia* s. lat. (Campanulaceae) in tropical Africa and Madagascar. Symb. Bot. Upsal. 21: 1–223.
Thulin M. 1978. *Cyphia* (Lobeliaceae) in tropical Africa. Bot. Not. 131: 455–471.
Thulin M. 1983. Some tropical African Lobeliaceae: Chromosome numbers, new taxa, and comments on taxonomy and nomenclature. Nord. J. Bot. 3: 371–382.
Tjon Sie Fat L. 1978. Contribution to the knowledge of cyanogenesis in angiosperms, 2. Communication: cyanogenesis in Campanulaceae. Proc. Kon. Nederl. Akad. Wetensch., ser. C., 81: 126–131.
Tobe H and NR Morin. 1996. Embryology and circumscription of Campanulaceae and Campanulales: a review of the literature. Int. J. Plant Res. 109: 425–435.
Vasilevskaya VK and TV Schulkina. 1976. Morphological and anatomical structure of the arborescent plant *Azorina vidalii.* Trudy Moskovsk. Ob. Isp. Prirody 42: 131–140 (in Russian).
Weio Z-X. 2001. Pollen morphology of some *Codonopsis* and related species of Campanulaceae. Acta Bot. Yunn. 2001: 335–338.
Wimmer FE. 1943. Campanulaceae-Lobelioideae, part 1. In: H Stubbe and K Noack, eds. Das Pflanzenreich, vol. 4, part 276b, pp. i–viii and 1–260. W. Engelmann, Leipzig.
Wimmer FE 1953. Campanulaceae-Lobelioideae, part 2. In: H Stubbe and K Noack, eds. Das Pflanzenreich, vol. 4, part 276b., pp. i–viii and 261–813. Academie Verlag, Berlin.
Wimmer FE. 1968. Campanulaceae-Lobelioideae: Supplementum et Campanulaceae-Cyphioideae. In: H Stubbe and K Noack, eds. Das Pflanzenreich, vol. 4, part 276c, pp. i–x and 815–1024. Academie Verlag, Berlin.
Yeo PF. 1993. Platycodoneae: a new tribe in Campanulaceae. Taxon 42: 109.
Zhinkina NA and II Shamrov. 1997. Embryology of *Azorina vidalii* and *Gadellia lactiflora* (Campanulaceae). Bull. Pol. Acad. Sci., Biol. Sci. 45: 120–133.

Order 117. STYLIDIALES

Small herbs, seldom subshrubs or shrubs. Secondary growth anomalous (Stylidiaceae). Vessels with simple or (*Donatia*) with scalariform perforations. Sieve-elements plastids of S-type. Leaves alternate, densely crowded, simple, entire, linear, coriaceous, estipulate. Stomata anomocytic or less often paracytic. Flowers in racemes or cymose inflorescences, of less often solitary, bisexual or sometimes unisexual, usually zygomorphic. Calyx (2-)5(-7)-lobed. Corolla 5–10-partite

(Donatiaceae) or 5-lobed, lobes mostly unequal. Stamens two or three (Donatiaceae) or only two (Stylidiaceae); filaments free (*Donatia*) or completely connate with the style into a column; anthers basifixed, extrorse, opening longitudinally. Tapetum secretory. Microsporogenesis simultaneous. Pollen grains 2-celled or 3-celled, 3–8-colpate (Stylidiaceae) or 3(4)-colporate (Donatiaceae). Nectary disc or a pair of nectary glands often present. Gynoecium of 2–3 (Donatiaceae) or two (Stylidiaceae) united carpels; stylodia free (Donatiaceae) or connate between themselves and with the filaments (Stylidiaceae); ovary inferior, 2- or 3-locular, but in Stylidiaceae the posterior carpel sometimes reduced or obsolete and the gynoecium pseudomonomerous with 1-locular ovary. Ovules many in each locule, anatropous, unitegmic, tenuinucellate, with endothelium. Tapetal cells binucleate. Female gametophyte of *Polygonum*-type. Endosperm cellular, with terminal haustoria. Fruits septicidal capsules or sometimes indehiscent. Seeds with minute embryo and copious fleshy endosperm. Storing carbohydrate as inulin, producing carbocyclic iridoids and caffeic acid. Alkaloids absent. Iridoids (Stylidiaceae) present; n = 5–16, 18, 24, 26, 28, and 30.

Related to the Asterales.

Key to Families

1 Stamens and stylodia free. Very densely branched perennial herbs with erect crowded stems forming compact, often broad, cushions. Roots fleshy, clitlike. Vessels with oblique scalariform perforation plates. Secondary thickening absent or developing from a conventional cambial ring. Leaves alternate, imbricate, non-sheating, simple, entire, more or less hairy at base, without a persistent basal meristem; mucilage cells abundant in the mesophyll. Stomata paracytic (*Donatia magellanica*) or anomocytic (D. *novae-zelandiae*), Flowers solitary, sessile, terminal, bisexual, essentially actinomorphic. Calyx 5–7-lobed, lobes equal or unequal, open, with obconical tube adnate to the ovary, imbricate. Petals 5–10, free, oblong-linear or ovate, imbricate, more or less fleshy. Stamens two or three, inserted on top of the ovary within the nectary disc; filaments free, short or filiform; anthers broadly ellipsoidal, extrorse. Pollen grains 3(4)-colporate. Epigynous disc present. Gynoecium of two or three carpels; stylodia subulate, recurved, with capitate stigma; ovary inferior, 2–3-locular, with numerous (15–100) ovules on axile placentas near to top of the locule. Fruits turbinate, indehiscent, 2–3-locular capsules. Seeds few, with minute embryo and copious, oily endosperm. Present inulin, iridoids not detected, n = 24. . . . . 1. DONATIACEAE.

1 Stamens and stylodia not free. Small, rhizomatous or tuberous, mostly erect or rarely climbing herbs or rarely subshrubs, more or less xerophytic, occasionally rush-like, often bearing stalked glands with two to multicellular head; some of glandular cells secrete mucilage. Vessels with simple perforations. Secondary thickening anomalous. Leaves alternate or whorled (*Stylidium scandens*), imbricate, simple, entire, linear, often in basal rosettes with flowers on a scape (successive rosettes may be separated by a slightly leafy piece of stem but sometimes crowded on creeping or shortly erect stems); rosettes sometimes almost bulbous, with aerial roots. Stomata anomocytic. Flowers in terminal racemes or cymes or solitary in the upper axils, usually zygomorphic, bisexual or sometimes unisexual (monoecious, polygamomonoecious, or dioecious ?). Calyx (2-)5(-7)-lobed, often covered with stalked viscid glands, imbricate, persistent. Corolla 5-lobed, lobes imbricate, mostly unequal, the odd lobe originates in a ventral position and in the zygomorphic genera takes the form of small or more or less reduced (*Stylidium*) or large, hooded (*Levenhookia*) labellum, but the flower commonly semiresupinate, so that the labellum is lateral; the remaining four lobes are more or less similar in size and shape or heteromorphic; the corolla throat is often marked by a series of various glandular appendages. Stamens 2(-3); filaments free from the corolla but united with the style to form a column, often irritable; anthers extrorse. Pollen grains 2-celled or 3-celled, 3–8-colpate. Epigynous nectary disc or a pair of nectary glands often present. Gynoecium of two carpels; stigmas papillate, diverging above the anthers; ovary inferior, more or less completely 2-locular or almost 1-locular, with axile to free central placentation, or the posterior locule reduced or suppressed, so that the ovary is pseudomonomerous; ovules numerous in each locule, pendulous or horizontal, or ascending. Fruits fleshy to non-fleshy capsules opening from the summit in two valves, or rarely indehiscent. Seeds mostly numerous, minute; seed coat thin, 2-layered, exo-endotestal (Plisko 2005) with small, dicotyledonous or sometimes monocotyledonous, straight

embryo and copious, fleshy oily endosperm. Iridoids detected; storing inulin, proanthocyanidins, flavonols (kaempferol and quercetin) present or absent, n = 5–16, 18. . . . . 2. STYLIDIACEAE.

1. DONATIACEAE

Chandler 1911. 1/2, New Zealand, Tasmania, subantarctic South America.

Donatia

In spite of some important morphological differences, including free stamens and 3(4)-colporate pollen grains, which are slightly similar to those of *Cyphia* (Erdtman 1952; Avetisian 1973), Donatiaceae are rather closely related to the Stylidiaceae, an affinity that is confirmed by many embryological similarities. According to Philipson and Philipson (1973: 458), the embryological details of *Donatia* are so similar to those of *Forstera* and the other members of the Stylidiaceae described in the literature that the position of this genus close to the Stylidiaceae is confirmed.

2. STYLIDIACEAE

R. Brown 1810. 5/170. Mainly extratropical regions of Australia, also Tasmania, New Zealand, southern and especially Southeast Asia, and the southernmost part of South America.

PHYLLACHNEAE: *Phyllachne, Forstera;* STYLIDIEAE: *Oreostylidium, Levenhookia, Stylidium.*

Related to Campanulaceae, but differs in the presence of carbocyclic iridoids, glandular hairs with two- to multicellular stalks (Stylidiaceae), anomalous secondary growth, reduced number of stamens that in Donatiaceae are free from the style, and extrorse anthers. However, there are many similarities between the Stylidiaceae and the Campanulaceae, including pollen morphology (Erdtman 1952; Bronckers and Stainer 1972; Avetisian 1973), floral development (Erbar 1992), the presence of inulin, and some embryological features. According to Crete (1951), the haustoria in the Stylidiaceae arise from endospermal cells by the same sequence of divisions as in *Codonopsis* in Campanulaceae. The embryological similarities between Stylidiaceae and some members of Campanulales (especially the Lobelioideae) have been further substantiated by Rosen (1935, 1949), Subramanyam (1950a, b, 1951a, b, 1953, 1970), and Philipson and Philipson (1973).

Bibliography

Bronckers F and F Stainer. 1972. A l'étude morphologique du pollen de la famille des Stylidiaceae. Grana 12: 1–22.

Burns GP. 1900. Beiträge zur Kenntnis der Stylidiaceae. Flora 87: 313–354.

Carlquist S. 1969. Studies in Stylidiaceae: new taxa, field observations, evolutionary tendencies. Aliso 7: 13–64.

Carlquist S. 1981a. Studies in Stylidiaceae: monocotyly in the family; nomenclatural change. Aliso 10: 35–38.

Carlquist S. 1981b. Types of cambial activity and wood anatomy of *Stylidium* (Stylidiaceae). Am. J. Bot. 68: 778–785.

Carolin RC. 1960a. Floral structure and anatomy in the family Stylidiaceae Swartz. Proc. Linn. Soc. N. S. W. 85: 189–196.

Carolin RC. 1960b. The structures involved in the presentation of pollen to visiting insects in the order Campanales. Proc. Linn. Soc. N. S. W. 85: 197–207.

Carolin RC. 1967. The concept of the inflorescence in the order Campanulales. Proc. Linn. Soc. N. S. W. 92: 7–26.

Carolin RC. 2007. Stylidiaceae. In: K Kubitzki, ed. The families and genera of vascular plants, vol. 8, 614–619. Springer, Berlin/Heidelberg/New York.

Chandler B. 1911. Note on *Donatia novae-zelandiae* Hook. f. Notes Roy. Bot. Gard. Edinb. 6: 43–48.

Darnowski DW, DM Carroll, B Plachno, E Kabanoff, and E Cinnamon. 2006. Evidence of protocarnivory in triggerplants (*Stylidium* spp.; Stylidiaceae). Plant Biol. 8: 805–812.

Dunbar A. 1975b. On pollen of Campanulaceae and related families with special reference to the surface ultrastructure: II. Campanulaceae subfam. Cyphioideae and subfam. Lobelioideae; Goodeniaceae; Sphenocleaceae. Bot. Not. 128: 102–118.

Erbar C. 1992. Floral development of two species of *Stylidium* (Stylidiaceae) and some remarks on the systematic position of the family Stylidiaceae. Canad. J. Bot. 70: 258–271.

Findley N and GP Findley 1989. The structure of the column in *Stylidium*. Aust. J. Bot. 37: 81–101.

Gustafsson MHG. 1996. Phylogenetic studies in the Asterales *sensu lato*. Acta Universitatus Upsaliensis, pp. 1–36. Uppsala.

James SH. 1979. Chromosome numbers and genetic systems in the triggerplants of western Australia (*Stylidium;* Stylidiaceae). Aust. J. Bot. 27: 17–25.

Laurent NB, B Bremer, and K Bremer. 1999. Phylogeny and generic interrelationships of the Stylidiaceae (Asterales), with a possible extreme case of floral paedomorphosis. Syst. Bot. 23: 289–304.

Leins P and C Erbar. 1990. On the mechanisms of secondary pollen presentation in the Campanulales-Asterales complex. Bot. Acta 103: 87–92.

Philipson WR and MN Philipson. 1973. A comparison of the embryology of *Forstera* L. and *Donatia* J. R. et G. Forst. N. Z. J. Bot. 11: 449–460.

Plisko MA. 2005. Fruit and seed structure of *Levenhookia stipitata* and *Stylidium* species (Stylidiaceae). Bot. Zhurn. 90(4): 575–582 (in Russian with English summary).

Rapson LJ. 1953. Vegetative anatomy of *Donatia, Phyllachne, Forstera,* and *Oreostylidium* and its taxonomic significance. Trans. Proc. Roy. Soc. N. Z. 80: 399–402.

Rauling EJ and PY Ladiges. 2001. Morphological variation and speciation in *Stylidium graminifolium* (Stylidiaceae): description of *S. montanum* and reinstatement of *S. armeria*. Aust. Syst. Bot. 14: 901–935.
Rosen W. 1935. Beiträge zur Embryologie der Stylidiaceen. Bot. Not. 1935: 273–278.
Subramanyam K. 1950. Development of embryo sac and endosperm in *Stylidium lenellum*. Curr. Sci. 19: 294.
Subramanyam K. 1951. A morphological study of *Stylidium graminifolium* Swartz. Lloydia 14: 65–81.
Subramanyam K. 1953. The nutritional mechanism of embryo sac and embryo in the families Campanulaceae, Lobeliaceae, and Stylidiaceae. J. Mysore Univ. B. 13: 1–4.
Subramanyam K. 1970. Stylidiaceae. Bull. Indian Natl. Sci. Acad. 41: 317–320.
Wagstaff SJ and J Wege. 2002. Patterns of diversification in New Zealand Stylidiaceae. Am. J. Bot. 89: 865–874.
Wege JA. 2001a. Corolla venation in Stylidiaceae. J. Roy. Soc. West. Aust. 84: 97–101.
Wege JA. 2001b. Scape anatomy in *Stylidium* (Stylidiaceae). Kew Bull. 56: 955–963.

Order 118. PHELLINALES (ALSEUOSMIALES)

Small trees or shrubs. Vessels with scalariform or scalariform and simple perforation, with numerous bars; lateral pitting opposite to alternate, sometimes scalariform. Fibers with bordered or simple pits, often septate. Rays heterogeneous and homogeneous, rarely (Alseuosmiaceae) absent. Axial parenchyma scanty paratracheal, or even absent. Sieve-element plastids of S-type. Nodes mostly trilacunar with three traces, rarely unilacunar or pentalacunar. Leaves alternate to subopposte, entire or dentate; lamina serrate, gland-toothed; estipulate. Stomata anomocytic. Flowers in axillary cymes, paniculate or corymbs, or solitary, bisexual or rarely unisexual, usually actinomorphic, 4–6-merous. Sepals free or basally connate, valvate. Petals free or basally connate, valvate. Stamens as many and alternate with petals; filaments free or basally connate or adnate to the corolla tube; anthers opening longitudinally. Tapetum secretory. Microsporogenesis simultaneous. Pollen grains 2-celled or 3-celled, mostly 3-colporate. Gynoecium of 2–5 united carpels; stylodia free or connate into a style; style slender or stout, with capitate or clavate, 2–5-lobed stigma, or stigma sessile. Ovary semi-inferior or inferior, rarely (Phellinaceae) superior, 2–6-locular; ovules one to many per locule, mostly anatropous, unitegmic, tenuinucellate or crassinucellate (Phellinaceae). Female gametophyte of *Polygonum*-type. Endosperm cellular. Fruits loculicidal capsules, drupes or berries. Seeds with small straight or slightly curved embryo and with copious endosperm. Contain gallic and ellagic acids, ellagitanins, and alkaloids (Phellinaceae).

The monophyly of the order Phellinales has very high support values in both the molecular and combined analyses (see also Soltis et al. 2006).

Key to Families

1 Ovary superior. Evergreen dioecious small trees or shrubs with uniseriate hairs; cuticle waxes as platelets and rodlets. Vessels with scalariform perforations and with numerous bars; lateral pitting scalariform or opposite. Rays heterogeneous with short ends. Axial parenchyma scanty paratracheal. Leaves alternate or subopposite, simple, usually papyraceous, more or less crowded towards tips of branched, estipulate. Stomata anomocytic. Flowers in axillary spicate or paniculate inflorescences, unisexual, 4–6-merous. Sepals small, free, more or less persistent. Petals free, fleshy, valvate with a small inflexed apiculus. Stamens as many and alternate with petals, free; anthers onlong and introrse. Pollen grains 3-colporate. Nectary disc absent. Staminodia present in female flowers. Gynoecium of 2–5 carpels, with sessile lobed stigma. Ovary with solitary (rarely two – Doweld 2000), pendulous, hemitropous or slightly campylotropous, ovule in each locule. Pistilodia present in female flowers. Fruits drupes with 2–5 pyrenes. Seeds with small embryo and copious endosperm. Contain benzylisoquinoline alkaloids, irdoid ?; n = 17. . . . 1. PELLINACEAE

1 Ovary inferior or semi-inferior.

2 Small trees or shrubs with characteristic two-armed trichomes consisting of a very elongate terminal «T-cell» that tapers to a fine point at each end and is attached at its middle to multicellular, uniseriate stalk. Vessels with scalariform perforations; lateral pitting commonly alternate. Fibers with bordered and bordered and simple pits, septate. Rays heterogeneous and homogeneous. Axial parenchyma absent or very scarce. Nodes trilacunar, unilacunar (*Corokia virgata*) or pentalacunar (*Argophyllum laxum*). Leaves alternate; petioles subvaginate. Stomata anomocytic. Flowers small. Petals mostly 4–6, basally connate, valvate, silky outside. Petals usually (except *Corokia macrocarpa*) with ligulate, nonvascularized

appendages on their inner side near the base of their free parts just above the tube. Pollen grains 3-colporate.

3 Fruits loculicidal capsules. Shrubs, often silky pubescent. Leaves often snow-white below, entire or coarsely dentate, ovate. Flowers 4–5(-8)-merous, whitish or greenish, in terminal and axillary panicles or corymbs. Corolla basally connate. Anther disporangiate. Style short, with capitate, 5–2-lobed stigma. Ovary semisuperior, 6–2-locular; ovules many, many-seriate on tumid placentas in the axils of the locules. Seeds very small, with large, elongate, nearly straight embryo in the middle of the copious, fleshy endosperm; exotestal cells with inner walls massively thickened and lignified. Ellagic acid present. . .2. ARGOPHYLLACEAE.

3 Fruits drupes crowned by the persistent calyx teeth and style and with 2–1-locular and 2–1-seeded bony endocarp with characteristic germination plug. Small trees or shrubs with branchlets, petioles, leaves below, peduncles and pedicels densely white, tomentose. Wood diffuse-porous. Lateral pitting opposite-alternate. Perforations scalariform with 16–23 bars. Rays heterogeneous. Vascular tracheids present, except in *C. collenettei* (Noshiro et Baas 1998). Crystals absent. Leaves entire. Flowers bright yellow, in axillary or terminal panicles, racemes, fascicles, or solitary. Anthers tetrasporangiate. Nectary disc bright orange, epigynous, pulvinate. Style short or subelongate, with 2-lobed stigma. Ovary inferior, 1(-2)-locular; ovules solitary in the locule; micropylar and chalazal haustoria present. Seeds linear-oblong, with large, elongate, slightly curved embryo in fleshy endosperm. Contain gallic and ellagic acids, n = 9. 3. COROKIACEAE.

2 Shrubs (sometimes up to 6 m tall – *Periomphale*), or creeping or epiphytic subshrubs, with unicellular, bicellular, or multicellular, unbranched trichomes. Stems and usually also petioles have a morphologically differentiated uniseriate endodermis of cells with casparian strips. Secondary thickening developing from a conventional cambial ring. Vessels with scalariform, or scalariform and simple perforations in almost vertical wall and composed of 20–45 bars; lateral pitting opposite to alternate, sometimes scalariform. Fibers with indistinctly bordered or simple pits, often septate, mostly living and with stored starch at maturity. Rays absent. Axial parenchyma scanty or even absent. Nodes trilacunar with three traces. Leaves alternate or subopposite, entire or dentate, simple, pinnately veined, with multicellular and uniseriate hairs in the axils containing dark, reddish pigment, estipulate. Stomata anomocytic. Flowers in axillary cymes or in terminal umbel-like inflorescences, or flowers solitary in the axils or on old wood, bisexual, often including cleistogamous ones, heavily scented, actinomorphic, perianth and androecium (4)5(-7)-merous. Sepals free, more or less triangular, valvate, persistent or caduceus (*Alseuosmia*). Petals connate, with valvate lobes; corolla margins with wings. Stamens alternipetalous, attached to the throat of the corolla tube or free; anthers dorsifixed, introrse; in *Platyspermation* the anthers sessile and extrorse, brown hairy below, and with a large flat connective appendage at the apex (van Steenis 1982). Pollen grains 3-colporate. Gynoecium of two carpels; style slender or stout, with more or less 2-lobed, clavate or capitate stigma; ovary inferior or sometimes semi-inferior, 2-, rarely (*Wittsteinia vacciniaceae*) 3-locular, crowned by a more or less developed nectary disc, with 1-many ovules in each locule. Fruits 2-locular berries, 2–3-locular (*Wittsteinia*), or 2-locular (*Platyspermation*) capsules, with persistent calyx and one to many seeds. Seeds with small, straight embryo and copious endosperm; exotesta little thickened, lignified, mesotesta persistent. Contain quercetin and kaempferol, caffeic and p-coumaric, stearic acids, ellagitanins and triterpenoid saponins; n = 9 (*Alseuosmia*). . . . . .4. ALSEUOSMIACEAE.

1. PHELLINACEAE

Takhtajan 1967. 1/12. New Caledonia.

Phelline

Wood structure of *Phelline* is very primitive and as Baas (1975: 339) points out, "Amongst the dicotyledons with vessels, the wood anatomy of *Phelline,* belongs to the most primitive."

2. ARGOPHYLLACEAE

Takhtajan 1987. 1/11. Tropical Australia and New Caledonia.

Argophyllum

3. COROKIACEAE

Kapil ex Takhtajan 1997. 1/6. Australia (1, NSW), New Zealand (2), Lord Howe Island (1), Chatham Is. (1), and Rapa I. (1)

Corokia

Corokia is usually placed in the Cornaceae s. l. However Hallier (1908, 1923) came to the conclusion that *Corokia* is closely related to *Argophyllum* ("*Corokia* ist durch Reduktion aus *Argophyllum* enstanden und bildet mit ihm und *Cuttsia* die Sippe der Argophylleen» (1908: 258). Engler (1930) included *Corokia* in his Escallionioideae-Argophylleae. The affinity of *Argophyllum* and *Corokia* was convincingly shown by Eyde (1966). Both Hallier and Eyde emphasized the remarkable similarity between the peculiar multicellular T-shaped trichomes of these genera, which are known only in few groups of flowering plants. They also emphasized the similarity of corolline appendages in the two genera. Moreover, *Corokia* and *Argophyllum* are similar in gross vegetative characters, wood anatomy (including septate tracheids, which are absent in Escalloniaceae [see Patel 1973; Hils 1985]) and pollen grain structure. Also, both *Corokia* and *Argophyllum* have a lamellate endexine in the region of the endoaperture, which resembles Cornaceae (Ferguson 1977; Ferguson and Hideux 1978). Thus, palynologically *Corokia* and *Argophyllum* occupy a somewhat intermediate position between Escalloniaceae and Cornaceae. There are also important differences between *Argophyllum* and *Corokia,* including placentation, the number of ovules, and the structure of fruits and seeds. It is appropriate to consider both these related genera members of two separate families. According to Kapil (1992) embryologically *Corokia* is distinctive enough to deserve family status. However, in his opinion the Corokiaceae resemble the Stylidiaceae more rather than the Cornaceae and the Escalloniceae. According to Gustafsson and Bremer (1995), *Corokia* is the most likely sister group of the Asterales s. l.

4. ALSEUOSMIACEAE

Airy Shaw 1965 (including Platyspermatiaceae Doweld 2001). 5/10. Southwestern Pacific. *Wittsteinia* (2), New Guinae and southeastern Australia; *Periomphale* (1) New Caledonia; *Platyspermation* (1) New Caledonia; *Alseuosmia* (5) is endemic to New Zealand, and *Crispiloba* (1) is endemic to Queensland.

Wittsteinia, Periomphale, Alseuosmia, Crispiloba, Platyspermation

Bibliography

Baas P. 1975. Vegetative anatomy and the affinities of Aquifoliaceae, *Sphenostemon, Phelline,* and *Oncotheca.* Blumea 22: 311–407.

Barriera G, V Savolainen, and R Spighiger. 2007. Phellinaceae. In: K Kubitzki, ed. The families and genera of vascular plants, vol. 8, pp. 608–610. Springer, Berlin/Heidelberg/New York.

Bhatnagar AK. 1973. Morphological and embryological studies in *Corokia.* Botanica (Delhi) 23(4): 149.

Dickison WC. 1986. Wood anatomy and affinities of the Alseuosmiaceae. Bot. J. Linn. Soc. 11: 214–221.

Dickison WC. 1989. Stem and leaf anatomy of the Alseuosmiaceae. Aliso 12: 567–578.

Doweld AB. 2000. Phellinaceae. In: A Takhtajan, ed. Comparative seed anatomy, vol. 6, pp. 114–115. Nauka, St. Petersburg (in Russian).

Eyde RH. 1966. Systematic anatomy of the flower and fruit of *Corokia.* Am. J. Bot. 53: 833–847.

Gardner RO. 1978. Systematic notes on the Alseuosmiaceae. Blumea 24: 138–142.

Kapil RN. 1992. Embryology and systematic position of *Corokia.* In: TB Batygina, ed. XI International Symposium on embryology and seed reproduction, pp. 246–247. St. Petersburg.

Kapil RN and AK Bhatnagar. 1974. Stomata on leaves and floral parts of *Corokia.* Bot. Jahrb. Syst. 94: 257–266.

Kårehed J. 2007a. Alseuosmiaceae. In: K Kubitzki, ed. The families and genera of vascular plants, vol. 8, pp. 7–12. Springer, Berlin/Heidelberg/New York.

Kårehed J. 2007b. Argophyllaceae. In: K Kubitzki, ed. The families and genera of vascular plants, vol. 8, pp. 13–18. Springer, Berlin/Heidelberg/New York.

Kårehed J, J Lundberg, B Bremer, and K Bremer. 1999. Evolution of the Australasian families Alseuosmiaceae, Argophyllaceae, and Phellinaceae. Syst. Bot. 24: 660–682.

Lobova TA. 1997. Seed morphology and anatomy in the genera *Argophyllum* and *Corokia* (Argophyllaceae). Bot. Zhurn. 82: 68–77 (in Russian with English summary).

Lobova TA. 2000. Escalloniaceae. Argophyllaceae. In: A Takhtajan, ed. Comparative seed anatomy, vol. 6, pp. 233–251. Nauka, St. Petersburg (in Russian).

Nemirovich-Danchenko EN and TA Lobova. 2000a. Alseuosmiaceae. In: A Takhtajan, ed. Comparative seed anatomy, vol. 6, pp. 267–269. Nauka, St. Petersburg (in Russian).

Paliwal GS and LM Srivastava. 1969 (1970). The cambium of *Alseuosmia*. Phytomorphology 19: 5–8.
Tirel C. 1996. Rétablissement de *Periomphale* Baill. (Alseuosmiaceae) genre endémique de Nouvelle Caledonie. Bull. Mus. Natl. Hist. Nat. Paris 4e sér. sect. B Adansonia 18: 155–160.
Zeman M. 1907. Studien zu einer Monographie der Gattung *Argophyllum* Forst. Ann. K. K. Naturalist. Hofmuns. 22: 270–292.

Order 119. ASTERALES

Perennial or annual herbs and subshrubs, less often shrubs, lianas or small or medium-sized trees, usually with various types of glandular or nonglandular trichomes. Lactifers mostly present. Vessels with simple or sometimes scalariform perforations. Fibers with simple pits, or less often (Goodeniaceae) with bordered pits. Rays mostly heterogeneous. Axial parenchyma scanty paratracheal. Sieve-element plastids of S-type. Nodes trilacunar to multilacunar, rarely (some Goodeniaceae) unilacunar. Leaves alternate or less often opposite, rarely verticillate, simple to variously divided, gland-dotted or not gland-dotted, estipulate or rarely with stipules; leaf mesophyll cells contain oil bodies. Stomata usually anomocytic. Inflorescences of various types, or less often flowers solitary. Flowers bisexual or unisexual, actinomorphic to zygomorphic, mostly 5-merous. Sepals usually connate at the base, sometimes adnate to the ovary. Petals connate, valvate. Nectary disc usually present. Stamens usually five, alternate with the corolla lobes; filaments free or sometimes connate; anthers basifixed, introrse, tetrasporangiate, or occasionally disporangiate, opening longitudinally. Tapetum secretory or amoeboidal. Microsporogenesis simultaneous. Pollen grains mostly 3-celled, 3-colporate. Gynoecium of two united carpels; stylodia united into terminal style; ovary mostly inferior, 1-locular; ovule mostly solitary, usually anatropous, unitegmic, tenuinucellate, with endothelium. Female gametophyte of various types, but mostly of *Polygonum*-type. Endosperm cellular or less often nuclear, without haustoria. Fruits of various types, but mostly cypselas. Seeds with straight embryo; endosperm usually copious, rarely wanting. Iridoids present or absent.

Very close to the Campanulales.

Key to Families

1 Iridoids present.

2 Fruits cylindrical or prismatic achenes. Perennial, biennial or annual herbs, rarely suffrutescens (typically forming only 1–2 mm thick of secondary xylem at stem base and adjacent root area), rarely pubescent. Laticifers and secretory cavities wanting, but in *Boopis gracilis* present droplets of a resin-like substance (Hellwig 2007). Vessels with simple perforations. Fibers with thick walls and simple pits. Leaves alternate, often rosulate, simple, entire or pinnately lobed, estipulate. Stomata anomocytic. Inflorescences terminal on scapes or aillary, capitulate, these solitary or in cymose panicles, surrounded by 1–2 series of bracts. Flowers numerous, opening centripetally (*Acicarpha*) or in a more complex manner, on short peduncles, or peduncles absent (*Acicarpha* and some species of *Boopis*), bisexual or seldom the central flowers of a head functionally unisexual, actinomorphic or zygomorphic. Calyx with 4–6 small lobes or teeth, aerenchymatous or spine-like. Corolla tubular, with 4–6 valvate lobes, the outer layer of the tubular corolla is photosynthetic. Stamens as many as and alternate with the corolla lobes, attached near the summit of the tube; filaments often basally or wholly connate; gland-like nectaries generally situated basally in the filament tube extending to the top of the stamen-corolla tube; anthers free or basally coherent to connate; secondarily presenting pollen by depositing it on extending style (Heald 2004). Pollen grains 2-celled, 3-colporate, spinulate to nearly smooth. Gynoecium of two carpels; style slender, glabrous, growing up and pushing out the pollen; stigma capitate, indistinctly papillate; ovary inferior, essentially 1-locular (incompletely 5-locular due to reduced septa), with a solitary ovule pendulous from the top of the locule. Ovules anatropous, with (?) endothelium. Fruits with persistent, spine-like calyx (this lignescent and spiny in *Acicarpha* and *Calycera*), ribbed and receptacle small in *Boopis*. Seeds with straight embryo and oily, copious, or scanty endosperm. Storing carbohydrate as inulin. Producing iridoid compounds (secologanin); n = 7–9, 13, 17, 21, 22. . . . . . .4. CALYCERACEAE

2 Fruits capsules, baccate, drupaceous or nutlike.

3 Terrestrial perennial herbs, subshrubs or shrubs, rarely arborescent with trichomes of both glandular and nonglandular types.

Laticifers wanting. Sclerenchymatous idioblasts, which are sometimes branched, are common in the mesophyll of the leaf and sometimes occur in the cortex of the stem as well. Vessels with simple or sometimes with scalariform perforations. Fibers with bordered pits. Rays heterogeneous to homogeneous. Axial parenchyma vasicentric or diffuse or both or wanting. Secondary thickening absent or developing from a conventional cambial ring. Nodes unilacunar or sometimes trilacunar to pentalacunar. Leaves well developed or much reduced, alternate or rarely opposite or verticillate, nearly always spiral, petiolate to sessile, simple, entire, serrate or dentate, estipulate. Stomata anomocytic. Flowers zygomorphic, variously in cymes, racemes, or heads, or solitary in the axils. Calyx tube adnate to the ovary or rarely free. Corolla bilabiate (with two dorsal and three ventral lobes) or sometimes unilabiate (with five ventral lobes), the lobes valvate, often induplicate, spurred or not (*Selliera*). Stamens five, attached to the base of corolla tube or free; anthers separate from one another (*Velleia*, *Scaevola*), or connivent around the style, or (*Dampiera*, *Leschenaultia*) connate into a tube. Pollen grains 2-celled, 3-colporate or (*Lechenaultia*) porate. Gynoecium of two carpels; style one, with a 'pollen cup' close beneath to stigma, bearing an 'indusium' beneath the stigma. Ovary inferior or semi-inferior or rarely (*Velleia*) superior, 2-locular or seldom 1-locular, or (some other spp. of *Scaevola*) with two vestigial lateral locules in addition to the fertile median ones, rarely (*Scaevola porocarya*) 4-locular, with one to many ovules per locule mostly erect or ascending on axile or basal-axile placentas (in *Verreauxia* the locule is solitary and basal). Fruits mostly capsular, less often drupaceous, or nutlike. Seeds usually flat, winged or wingless, exotestal cells usually palisade, all walls thickened; embryo straight, endosperm copious, oily. Storing carbohydrate as inulin and producing 13-carbon polyacetylenes, caffeic acid, sometimes iridoids (secologanin), O-metyl flavonols, and alkaloids, n = 7–9. . . . . . .1. GOODENIACEAE.

3 Aquatic or wetland herbs, mostly perennial and with tufted rootstocks or horizontal creeping rhizomes, but some species of *Nymphoides* and *Villarsia* are annual. Vascular system is composed of isolated or scattered bundles. A well-developed system of intercellular canals and spaces is also characteristic. Branched, sclerenchymatous idioblasts, similar to those of the Nympheaceae, often present in the parenchyma of the stem. Vessels usually with simple perforations, but scalariform and simple plates may occur in *Menyanthes*. Nodes trilacunar or pentalacunar. Leaves alternate, simple or rarely (*Menyanthes*) 3-foliolate with sheating petioles, linear to cordate or reniform; stipules wanting or represented by the expanded wing margins of the petiole. Stomata anomocytic. Flowers solitary or in simple or branched cymes or racemes or in dense heads or clusters, bisexual, sometimes functionally unisexual, actinomorphic, usually 5-merous (except gynoecium), often heterostylous. Sepals more or less connate at base, sometimes adnate to the ovary below, persistent. Corolla sympetalous, appendiculate (with intrastaminal scales), or not appendiculate, valvate or induplicate-valvate; the margins or the inner surface of the lobes often fimbriate or crested. Stamens attached to the base of the corolla tube alternate with the lobes, or midway down the corolla tube, or in the throat of the corolla tube; anthers 2-locular, dorsifixed, mostly sagittate, versatile. Sometimes fringed scales (staminodia?) alternate with stamens. Pollen grains 3-celled, 3-colpate or 3-colporate, or rugate (exceptionally 6-rugate). Nectary disc usually present around the base of the ovary. Gynoecium of two united carpels; stylodia connate into a terminal, 2-lobed style, with papillate stigma; ovary superior to semiinferior, 1-locular, with numerous ovules on two parietal and often intruded placentas. Ovules horizontal, with short funicle and with endothelium. Fruits septicidal, loculicidal, or valvular capsules, or rarely (*Liparophyllum*) fleshy berries. Seeds 4–100, glabrous or with trichomes of different shape, winged or wingless, in some species surrounded by a caruncle which is rich in lipids (Kadereit 2007);

exotestal cells with outer walls thickened, often with a variety of projections; embryo straight, linear; endosperm copious, oily. Producing iridoid compounds (loganin, secologanin, sweroside), flavonols, caffeic acid (except *Menyanthes*), and usually storing inulin. n = 9, 17. . . . . . . 3. MENYANTHACEAE.

1 Iridoids absent.

4 Stomata paracytic. Perennial scapose herbs with characteristic hairs of short, thin-walled, erect basal cell and two to several-celled arm attached at right angles to the basal cell and appressed to the surface. Secondary thickening developing from a conventional cambial ring. Leaves all basal or alternate towards the base of the stems, more or less petiolate, simple, estipulate. Flowers actinomorphic or very nearly so, bisexual, borne in involucrate, cymose heads or spikes, each flower with several hyaline bracts. Calyx tube free from the ovary and calyx lobes markedly longer than the tube. Corolla with five spreading the lobes valvate, blue, markedly longer than the tube. Stamens five, attached to the base of the corolla tube; anthers connate into a tube around the style. Gynoecium of two carpels; style one, bearing an 'indusium' beneath the stigma; stigma surrounded by a collar or indusium; ovary 1-locular, superior, with septum only at base and solitary, ascending ovule inserted on one side of the septum. Fruits dry and indehiscent, enclosed by persistent calyx tube. Seeds ellipsoid, with straight embryo and without endosperm. Iridoids wanting, n = 9. . . . . 2. BRUNONIACEAE.

4 Stomata anomocytic or occasionally anisocytic. Perennial or annual herbs and subshrubs, less often shrubs, lianas or small or medium-sized trees (up to 30 m tall or more), sometimes nearly succulents, halophytes, marsh plants, lianas, epiphytes and aquatics, usually with various types of glandular or nonglandular trichomes. Often with a system of articulated laticifers (mostly Chicorioideae) in the phloem, containing latex rich in triterpene, or with a more or less well-developed system of schizogenous resin canals that are often lined with an epithelium (the two secretory systems largely alternative rather than coexistive); scattered latex cells occasionally present in addition to resin canals. Secondary thickening well developed even in many herbaceous species, or absent, or anomalous. Vessels commonly with simple perforations, but sometimes with scalariform and simple, or simple and reticulate perforations; lateral pitting alternate. Fibers typically with small simple pits, or the pits with very small borders. Rays mostly heterogeneous, mostly 4–10(-18) cells wide. Axial parenchyma scanty paratracheal. Nodes trilacunar to multilacunar. Leaves alternate or less often opposite, rarely verticillate, simple or variously divided, gland-dotted or not gland-dotted, aromatic or foetid, or without marked odour; leaves estipulate or rarely with stipules. Flowers in dense racemose heads (capitula) with one to many sessile flowers on common receptacle, the head nearly always subtended by an involucre of one to several series of bracts; capitula occasionally aggregated into cymose secondary heads sometimes with a secondary involucre; common receptacle of the head flat to conical or cylindrical, sometimes with a bract subtending each flower (mainly in Heliantheae) or densely bristly (mainly in Cynareae); heads homogamous (all flowers alike) or heterogamous; in homogamous heads all flowers are usually bisexual; in heterogamous heads the peripheral (radiate or filiform) flowers female or neutral and the central (disc) flowers bisexual or unisexual (monoecious or dioecious, or androdioecious (rarely), or gynodioecious (rarely), or polygamomonoecious. Perianth and androecium 5-merous or rarely 4–3-merous. Calyx tube together with the base of corolla tube completely fused with the ovary and its free lobes usually greatly modified and transformed into a pappus, which is often strongly reduced or obsolete. Corolla of three basic types: (1) tubular or discoid (actinomorphic, 5-lobed or 5-dentate, with a conspicuous tube and usually a short limb), ligulate (stap-shaped with five apical teeth) or bilabiate in bisexual (or functionally male) flowers; (2) radiate (strap-shaped with 0–4, usually three, apical teeth) in peripheral female or sterile flowers; (3) filiform, narrow, apically truncate, minutely lobed or minutely radiate, in peripheral female flowers; rarely corolla absent (in peripheral female flowers). Stamens usually five, epipetalous, alternate with the corolla lobes; filaments usually free, but sometimes connate into tube (e.g., in *Silybum, Galactites,* and

Dipterocome); anthers usually tetrasporangiate, or occasionally disporangiate, often with short apical appendage and often basal tails. Tapetum amoeboidal or less often secretory. Microsporogenesis simultaneous. Pollen grains 3-celled, mostly 3-colporate, psilate, spinulose or lophate, often caveate. Gynoecium of two carpels; style terminal, usually papillate, bifid (branches commonly separating after passage through the anther tube), with annular or short tubular nectary disc at the base; ovary inferior, 1-locular, with one basal, erect ovule. Female gametophyte of various types, but mostly of *Polygonum*-type. Endosperm cellular or less often nuclear, without haustoria. Fruits cypselas (achenes), usually with persistent pappus, very rarely drupaceous with fleshy pericarp (*Chrysanthemoides*, tribe Calenduleae, and *Wulffia*, tribe Heliantheae). Seeds with straight, oily embryo; endosperm wanting or sometimes present as a very thin vestigial layer; exotestal cells thickened, palisade or flattened, or undistinguished. Usually storing carbohydrate as inulin, and producing polyacetylenes, terpenoid essential oils, alkaloids and flavonols present or absent; producing diverse array of polyacetylenes, bitter sesquiterpene lactones, caffeic and ursolic acids, and others; n = 2–19 + ... 5. ASTERACEAE.

1. GOODENIACEAE

R. Brown 1810 (including Scaevolaceae Lindley 1830). 12/420. Widely dispersed on tropical shores of the Atlantic and Indian oceans while the remainder extend north and east in the Pacific as far as China, Japan, the Philippines, the Hawaiian Islands, New Zealand, and Chile. Over 95% of the species are endemic to Australia (especially western Australia) and Tasmania.

Velleia, Goodenia (including *Calogyne*), *Selliera, Scaevola, Leschenaultia, Anthotium, Pentaptilon, Diaspasis, Verreauxia, Dampiera, Coopernookia.*

Morphological and molecular data support the relationships the Goodeniaceae, Calyceraceae, and Asteraceae.

2. BRUNONIACEAE

Dumortier 1829. 1/1. Australia and Tasmania.

Brunonia

A rather heterobathmic family that differs from the Goodeniaccae mainly in actinomorphic or nearly actinomorphic flowers borne in involucrate heads and the endospermless seeds.

3. MENYANTHACEAE

Dumortier 1829. 5/60. Subcosmopolitan.

Menyanthes, Villarsia, Nymphoides, Liparophyllum, Nephrophyllidium

Belong to the Asterales, but occupies somewhat isolated position. The affinity with the Asterales is confirmed by molecular data (Michaels et al. 1993; Gustafsson and Bremer 1995).

4. CALYCERACEAE

Richard 1820. 5/60 (including Boopidaceae Cassini 1816). South America, mainly in the southern Andes and Patagonia.

Boopis, Calycera, Gamocarpha, Moschopsis, Acicarpha (including *Nastanthus*).

Calyceraceae have similarities with Goodeniaceae, but differ from them mainly in the absence of pollen-collecting cups, epipetalous stamens, in multinucleate tapetal cells, and fruit-type. Calyceraceae also related to Asteraceae: they have similar second pollen presentation, and also Calyceraceae and Barnadesioideae have a similar simple flavonoid profile. However they markedly differ from Asteraceae in having en entire stigma, glabrous terminal style, and in lacking a pappus, as well as the presence of iridoids.

5. ASTERACEAE

Martynov 1820 or Compositae Giseke 1792 (nom. altern.) (including Ambrosiaceae Marynov 1820; Anthemidaceae Martynov 1820; Aposeridaceae Rafinesque 1838; Arctotidaceae Bessey 1914; Artemisiaceae Martynov 1820; Athanasiaceae Martynov 1820; Calendulaceae Link 1829; Carduaceae Dumortier 1822; Cassiniaceae C.H. Schultz 1852; Centauraceae Martynov 1820; Cichoriaceae A.L. de Jussieu 1789; Cnicaceae Vest 1818; Cynaraceae Durande 1782; Echinopaceae Dimortier 1822; Eupatoriaceae

Martynov 1820; Gnaphaliaceae Link ex F. Rudolphi 1830; Heleniaceae Rafinesque 1824; Helianthaceae Dumortier 1822; Inulaceae Bessey 1914; Lactucaceae Drude 1979; Matricariaceae J.Voigt 1845; Mutisiaceae Burnett 1835; Nassauviaceae Burmeister 1836; Picridaceae Martynov 1820; Santolinaceae Martynov 1820; Senecionaceae Berchtold et J. Presl 1820; Serratulaceae Martynov 1820; Syngenetiaceae Horaninow 1847; Tanacetaceae Vest 1818; Vernoniaceae Burmeister 1837; Xanthiaceae Vest 1818). C.1600/ 23000 or more. Cosmopolitan, but mostly in temperate and subtropical regions.

Delpino (1871), Small (1919), Rosen (1946, 1949), and many others derive the Asteraceae from the Lobeliaceae or Lobeliaceae-like ancestor. In the Englerian system, they occupy a position near and after the Calyceraceae at the top of Campanulales sensu lato. Recent morphological and molecular data suggest that the Asteraceae form a sister group to the clade Calyceraceae and Goodeniaceae (Jansen et al. 1991b; Harris 1995). The Asteraceae differ from them mainly in producing various types of sesquiterpene lactones, in the absence of iridoid compounds, in calyx transformed into pappus, and in mostly amoeboid tapetum. The similarities between Asteraceae and Calyceraceae include involucrate capitula, specialized pollen presentation mechanism, epipetalous stamens, uniovulate ovary, and the presence of unique intercolpar concavities. Both families evidently share a common origin from some Goodeniaceae-like ancestor.

Classification after C. Jeffrey 2007.

5.1 BARNADESIOIDEAE

Shrubs, trees, or perennial or annual herbs, usually with fascicled nodal spines. Capitula homogamous or heterogamous, discoid or pseudoradiate or ligulate, sessile or pedunculate. Floral parts having 3-cellular hairs with swollen basal cell (barnadesioid hairs), often with axillary spines; corolla tubular, often villous; filaments free or rarely fused, inserted at different levels; style shortly bilobed or bifid, glabrous or papillose below bifurcation; cypselas densely villous, with straight simple hairs, rarely glabrous. Pappus uniseriate or rarely absent. Poor in flavonoids, flavonols absent; n = 8, 24, 25, 27. – *Schlechtendalia, Doniophyton, Duseniella, Fulcaldea, Barnadesia, Huarpea, Dasyphyllum, Arnaldoa, Chuquiraga.*

5.2 MUTISIOIDEAE

Shrubs, trees or perennial or rarely annual herbs, rarely scandent. Leaves usually alternate, denticulate or lobulate. Capitula homogamous, bilabiate or discoid, rarely ligulate. Florets 1-many, 5-merous. Corolla lobes long, with an apical tuft of minute hairs; style with or without hairs, without thickening on shaft below branches; stigma lobes short, pollen spinulose; cypselas usually with twin hairs. Pappus usually present, commonly of bristles, often uniseriate; n = (6-)9. – STIFFTIEAE: *Stifftia*, etc.; MUTISIEAE: *Mutisia, Acourtia, Chaptalia, Anislaea, Trixis*, etc.

5.3 CARDUOIDEAE

Perennial, biennial or less often annual herbs, shrubs or rarely trees, rarely scandent. Leaves alternate, denticulate or lobulate, especially in herbaceous members often spiny. Capitula homogamous or heterogamous, discoid or discoid with marginal florets sterile and radiant, rarely bilabiate-radiate or ligulate. Florets one to many, 5-merous; corolla lobes long; anthers calcarate and caudate; pollen spinulose or spiny; plant often spiny; style usually with a papillose or hairy thickening on the shaft below the style branches; cypselas with twin hairs, simple hairs or glabrous. Pappus usually present, of bristles or scales; n = 12. – GOCHNATIEAE: *Ainsliaea, Gochnatia, Brachylaena*; HECASTOCLEIDEAE: *Hecastocleis*; TARCHONANTHEAE: *Tarchonanthus*; DICOMEAE: *Dicoma*; CYNAREAE-CARLININAE: *Atractylodes, Thevenotia, Carlina, Atractylis, etc.;* PERTYEAE: *Ainsliaea*, CYNAREAE-ECHINOPSINAE: *Echinops, Acantholepis, Amphoricarpos, Xeranthemum, Chardinia*, etc.; CYNAREAE-CARDUINAE: *Arctium, Cousinia, Onopordum, Saussurea, Jurinea, Carduus, Cirsium, Silybum, Cynara, Ptilostemon, Lamyropsis, Alfredia, Olgaea, Galactites, Picnomon*, etc.; CYNAREAE-CENTAUREINAE: *Serratula, Stemmacantha, Leuzea, Tricholepis, Acroptilon, Callicephalus, Centaurea, Centaurodendron, Chartolepis, Stizolophus, Zoegea, Cnicus, Carthamus, Carduncellus, Amberboa, Oligochaeta, Volutaria, Crupina*, etc.

5.4 CICHORIOIDEAE (Lactucoideae)

Perennial, biennial or annual herbs, shrubs or trees, rarely scandent, very rarely aquatic. Leaves alternate or opposite, entire to deeply lobed, sometimes spiny. Capitula homogamous, ligulate, radiate or discoid, less often heterogamous, radiate or discoid. Disc florets usually with long, narrow lobes. Anthers dorsifixed, mostly calcarate and caudate. Pollen grains globose, mostly spiny, usually excavate, sometimes honeycombed. Style arms usually long, acute, with single stigmatic area on inner surface; style hairs usually evently distributed along the style-branches and on the shaft below. Cypselas with twin hairs. Pappus usually present, sometimes heteromorphic. Plants mostly unarmed. Laticifers often present; n = (7-)9–10(-13). – GYMNARRHENEAE: *Gymnarrhena*; MOQUINIEAE: *Moquinia, Pseudostifftia;* VERNONIEAE: *Vernonia, Piptocarpha, Stokesia, Lychnophora, Elephantopus,* etc.; LIABEAE: *Munnozia, Chrysactinium, Liabum, Oligactis,* etc.; CICHORIEAE (LACTUCEAE): *Scolymus, Cichorium, Tolpis, Amoseris, Andryala, Hieracium, Catananche, Krigia, Malacothrix, Stephanomeria, Hyoseris, Hypochaeris, Leontodon, Picris, Urospermum, Hedypnois, Rhagadiolus, Scorzonera, Epilasia, Tragopogon, Koelpinia, Dubyaea, Soroseris, Prenanthes, Lactuca, Steptorhamphus, Scariola, Cephalorrhynchus, Mycelis, Cicerbita, Lapsana, Crepis, Ixeris, Youngia, Taraxacum, Chondrilla, Heteracia, Launaea, Reichardia, Sonchus,* etc.; GUNDELIEAE: *Gundelia;* ARCTOTIDEAE-ARCTOTIDINAE: *Arctotis, Berkheya,* etc.; ARCTOTIDEAE-GORTERIINAE: *Gorteria, Gazania, Cullumia, Hirpicium,* etc.; ARCTOTIDEAE-EREMOTHAMNINAE: *Eremothamnus, Hoplophyllum.*

5.5 ASTEROIDEAE

Perennial to annual herbs or shrubs, less often trees, sometimes scandent, epiphytic or aquatic, sometimes succulent. Leaves alternate or opposite, not spiny. Florets one to many, (3–4-)5(-6)-merous; marginal flowers often differentiated, radiate and 3-dentate or outer flowers female, with filiform corolla, central flowers actinomorphic and usually shallowly lobed; anthers basifixed, mostly ecalcarate and often ecaudate, usually slender. Pollen grains mostly spiny and caveate, never honeycombed. Styles generally pilose mainly at the tips of the branches, mostly with stigmatic areas in two separate lines. Cypselas with twin hairs. Pappus present or absent, usually of bristles or scales or coroniform, sometimes auriculiform or of awns, sometimes heteromorphic. Laticifers usually absent; n = (4-)9–10(-19). – CORYMBIEAE: *Corymbium*; SENECIONEAE: *Ligularia, Farfugium, Doronicum, Tussilago, Petasites, Adenostyles, Dendrocacalia, Paragynoxys, Tetradymia, Robinsonia, Brachyglottis, Werneria, Gynoxys, Cineraria, Gynura, Crassocephalum, Senecio, Emilia, Othonna, Lopholaena, Kleinia, Euryops,* etc.; CALENDULEAE: *Dimorphotheca, Osteospermum, Oligocarpus, Tripteris, Calendula, Gibbaria, Chrysanthemoides, Garuleum,* etc.; GNAPHALIEAE: *Phagnalon, Athrixia, Filago, Micropus, Evax, Bombycilaena, Cymbolaena, Lucilia, Metalasia, Relhania, Loricaria, Cassinia, Helichrysum, Ifloga, Lasiopogon, Gnaphalium, Antennaria, Leontopodium, Angianthus,* etc.; ASTEREAE: *Grangea, Chrysopsis, Engleria, Grindelia, Gutierrezia, Solidago, Haplopappus, Chrysothamnus, Pteronia, Dichrocephala, Bellis, Myriactis, Calotis, Callistephus, Aster, Erigeron, Lachnophyllum, Psychrogeton, Melanodendron, Commidendrum, Diplostephium, Olearia, Felicia, Microglossa, Conyza, Nolletia, Chrysocoma, Baccharis,* etc.; ANTHEMIDEAE: *Santolina, Anthemis, Achillea, Chamaemelum, Matricaria, Chamomilla, Cladanthus, Anacyclus, Chrysanthemum, Tanacetum, Microcephala, Tridactylina, Cancrinia, Lepidolopha, Leucanthemum, Cotula, Soliva, Artemisia, Lasiospermum, Eriocephalus, Ursinia* etc.; INULEAE: *Inula, Pulicaria, Blumea, Jasonia, Pegolettia, Nauplius, Telekia, Buphthalmum, Carpesium, Amblyocarpum, Pluchea, Laggera, Nicolasia, Pterocaulon, Epaltes, Blumeopsis, Adelostigma, Karelinia, Streptoglossa, Stenachaenium, Cylindrocline,* etc.; HELIANTHEAE: *Blepharispermum, Arnica, Bahia, Flaveria, Hymenopapus, Gaillardia, Helenium, Hymenoxys, Dyssodia, Tagetes, Pectis, Porophyllum, Bidens, Chrysanthellum, Coreopsis, Cosmos, Dahlia, Fitchia, Echinacea, Rudbeckia, Zinnia, Verbesina, Wedelia, Helianthus, Simsia, Galinsoga, Tridax, Melampodium, Sigesbeckia, Silphium, Ambrosia, Iva, Parthenium, Xanthium, Adenostemma, Eupatorium, Trichogonia, Campuloclinium, Ayapana, Critonia, Koanophyllon, Chromolaena, Bartlettina, Neomirandea, Symphyopappus, Ageratum, Stevia, Brickellia, Helogyne, Liatris, Fleischmannia, Micania, Ageratina,* etc.

Bibliography

Alvarenga SAV, MJP Ferreira, GV Rodrigues, and VP Emerenciano. 2005. A general survey and some taxonomic implications of diterpenes in the Asteraceae. Bot. J. Linn. Soc. 147: 291–308.

Anderberg A. 1989. Phylogeny and reclassification of the tribe Inuleae (Asteraceae). Canad. J. Bot. 67: 2277–2296.

Anderberg A. 1991a. Taxonomy and phylogeny of the tribe Gnaphalieae. Opera Bot. 104: 1–195.

Anderberg A. 1991b. Taxonomy and phylogeny of the tribe Inuleae (Asteraceae). Plant Syst. Evol. 176: 75–123.

Anderberg A. 1991c. Taxonomy and phylogeny of the tribe Plucheeae (Benth.) A. Anderb. (Asteraceae). Plant Syst. Evol. 176: 145–177.

Anderberg AA, P Eldenäs, RJ Bayer, and M Englund. 2005. Evolutionary relationships in the Asteraceae tribe Inuleae (incl. Plucheae) evidenced by DNA sequences of *ndh*F; with notes on the systematic positions of some aberrant genera. Organ. Divers. Evol. 5: 135–146.

Anderberg AA, BG Baldwin, RG Bayer, J Breitwieser, C Jeffrey, MO Dillon, P Eldenäs, V Funk, N Garcia-Jacas, DJN Hind, PO Karis, HW Lack, G Nesom, B Nordenstam, Ch Oberprieler, JL Panero, C Puttock, H Robinson, TF Stuessy, A Susanna, E Urturey, R Vogt, J Ward, and LE Watson. 2007. Compositae. In: K Kubitzki, ed. The families and genera of vascular plants, vol. 8, pp. 61–575. Springer, Berlin/Heidelberg/New York.

Augier J and M-L Rubat du Merac. 1951. La phylogenie des Composees. Rev. Sci. 89: 167–182.

Avetisian EM. 1980. Palynomorphology of the family Calyceraceae. In: S Zhilin, ed. Systematics and evolution of higher plants, pp. 57–64. Nauka, Leningrad (in Russian).

Baagoe J. 1977a. Microcharacters in the ligules of the Compositae. In: VH Heywood, JB Harborne, and BL Turner, eds. The biology and chemistry of the Compositae, pp. 119–139. Academic, London.

Baagoe J. 1977b. Taxonomical application of the ligule microcharacters in Compositae: I. Anthemideae, Heliantheae, and Tageteae. Bot. Tidskr. 71: 193–224.

Baagoe J. 1978. Taxonomical application of the ligule microcharacters in Compositae: II. Arctoideae, Astereae, Calenduleae, Eremothamneae, Inuleae, Liabeae, Mutisieae, and Senecioneae. Bot. Tidskr. 72: 125–147.

Baldwin BG, BL Wessa, and JL Panero. 2002. Nuclear rDNA evidence for major lineages of Helenioid Heliantheae (Compositae). Syst. Bot. 27: 161–198.

Bayer RJ and JR Starr. 1998. Tribal phylogeny of the Asteraceae based on two non-coding chloroplast sequences, the *trn*L intron and *trn*L-*trn*F intergenene spacer. Ann. Missouri Bot. Gard. 85: 242–256.

Bayer RJ, CF Puttock, and SA Kelchner. 2000. Phylogeny of South African Gnaphalieae (Asteraceae) based on two noncoding chloroplast sequences. Am. J. Bot. 87: 259–272.

Bayer RJ, DG Greber, and NH Bagnall. 2002. Phylogeny of Australian Gnaphalieae (Asteraceae) based on chloroplast and nuclear sequences, the *trn*L intron, *trn*L/*trn*F intergenic spacer, *mat*K and ETS. Syst. Bot. 27: 801–814.

Bentham G. 1873. Notes on the classification, history, and geographical distribution of the Compositae. Bot. J. Linn. Soc. 13: 335–577.

Bergquist G, B Bremer, and K Bremer. 1992. Chloroplast DNA restriction site variation and phylogenetic interrelationships of some genera of the Heliantheae *sensu lato* (Asteraceae). Nord. J. Bot. 12: 149–154.

Binns SE, BR Baum, and JT Arnason. 2002. A taxonomic revision of Echinaceae (Asteraceae: Heliantheae). Syst. Bot. 27: 610–632.

Blackmore S. 1986. The identification and taxonomic significance of lophate pollen in the Compositae. J. Canad. Bot. 64: 3101–3112.

Bohm BA and TF Stuessy. 1995. Flavonoid chemistry of Barnadesioideae (Asteraceae). Syst. Bot. 20: 22–27.

Bohm BA and TF Stuessy. 2001. Flavonoids of the sunflower family (Asteraceae). Springer, Berlin/Heidelberg/New York.

Bohm BA, KW Nicholls, and R Ornduff. 1986. Flavonoids of the Menyanthaceae: intra- and interfamilial relationships. Am. J. Bot. 73: 204–213.

Bohm BA, A Reid, ML DeVore, and TF Stuessy. 1995. Flavonoid chemistry of Calyceraceae. Canad. J. Bot. 73(12): 1962–1965.

Bolick MR. 1978. Taxonomic, evolutionary, and functional considerations of Compositae pollen ultrastructure and sculpture. Plant Syst. Evol. 130: 209–218.

Boulter D, JT Gleaves, BG Haslett, D Peacock, and U Jensen. 1978. The relationships of 8 tribes of the Compositae as suggested by plastocyanin ammo acid sequence data. Phytochemistry 17: 1585–1589.

Brant AJC. 2003. Flavonóides, cumarinas e benzofuranos como marcadores quimiotaxonômicos em Asteraceae. MSc dissertation, Chemistry Institute, University of São Paulo.

Breitwieser I and JM Ward. 2003. Phylogenetic relationships and character evolution in New Zealand and selected Australian Gnaphalieae (Compositae) inferred from morphological and anatomical data. Bot. J. Linn. Soc. 141: 183–203.

Bremer B. 1987. Tribal interrelationships of the Asteraceae. Cladistics 3: 210–253.

Bremer K. 1994. Asteraceae: cladistics and classification. Timber Press, Portland, OR.

Bremer K. 1996. Major clades and grades of the Asteraceae. In: JN Hind and HJ Beentie, eds. Proceedings of the International Compositae Conference, Kew, 1994, vol. 1. Compositae: Systematics, pp.1–7. Kew.

Bremer K and MHG Gustafsson. 1997. East-Gondwana ancestry of the sunflower alliance of families. Proc. Natl. Acad. Sci. USA 94: 9188–9190.

Bremer K and CJ Humphries. 1993. Generic monograph of the Asteraceae-Anthemideae. Bull. Brit. Mus. (Nat. Hist.) Lond., Bot. 23: 71–177.

Bremer K and RK Jansen. 1992. A new subfamily of the Asteraceae. Ann. Missouri Bot. Gard. 79: 414–415.

Bremer K, RK Jansen, PO Karis, M Kallersjo, SC Keeley, K-J Kirn, HJ Michaels, JD Palmer, and RS Wallace. 1992. A review of the phylogeny and classification of the Asteraceae. Nord. J. Bot. 12: 141–148.

Cabrera AL. 1977. Mutiseae: Systematic review. In: VH Heywood, JB Harborne, and BL Turner, eds. The biology and chemistry of the Compositae, pp. 1039–1066. Academic, London.

Carlquist S. 1959. Studies on Madinae: anatomy, cytology, and evolutionary relationships. Aliso 4: 171–236.

Carlquist S. 1966. Wood anatomy of Compositae: a summary, with comments on factors controlling wood evolution. Aliso 6: 25–44.

Carlquist S. 1969. Wood anatomy of Goodeniaceae and the problem of insular woodiness. Ann. Missouri Bot. Gard. 56: 358–390.

Carlquist S. 1976. Tribal interrelationships and phylogeny of the Asteraceae. Aliso 8: 465–492.

Carlquist S and ML DeVore. 1998. Wood anatomy of Calyceraceae. Ecology, habit, and systematic relationships. Am. J. Bot. 85(6): 118 (abstract).

Carlquist S, BG Baldwin, and GD Carr, eds. 2004. Tarweeds and Silverswords: evolution of the Madiinae (Asteraceae). Missouri Botanical Garden, St. Louis (see also Madiinae Showcase: http://www.botany.hawaii.edu/faculty/carr/tarweeds.htm)

Carolin RC. 1959. Floral structure and anatomy in the family Goodeniaccae Dumort. Proc. Linn. Soc. N. S. W. 84: 242–255.

Carolin RC. 1960. The structures involved in the presentation of pollen to visiting insects in the order Campanulales. Proc. Linn. Soc. N. S. W. 85: 197–207.

Carolin RC. 1966. Seeds and fruit of the Goodeniaceae. Proc. Linn. Soc. N. S. W. 91: 9–18.

Carolin RC. 1967. The concept of the inflorescence in the order Campanulales. Proc. Linn. Soc. N. S. W. 92: 7–26.

Carolin RC. 1971. The trichomes of the Goodeniaceae. Proc. Linn. Soc. N. S. W. 96: 8–22.

Carolin RC. 1978. The systematic relationships of *Brunonia*. Brunonia 1: 9–29.

Carolin RC. 1980. Pattern of the seed surface of *Goodenia* and related genera. Aust. J. Bot. 28: 123–137.

Carolin RC. 2007. Goodeniaceae. In: K Kubitzki, ed. The families and genera of vascular plants, vol. 8, pp. 589–598. Springer, Berlin/Heidelberg/New York.

Carr GD, RM King, AM Powell, and H Robinson. 1999. Chromosome numbers in Compositae. XVIII. Am. J. Bot. 86: 1003–1013.

Christensen LP. 1992. Acetylenes and related compounds in Anthemideae. Phytochemistry 31: 7–49.

Christensen LP and J Lam. 1991. Acetylenes and related compounds in Astereae. Phytochemistry 30: 2453–2476.

Chuang TI and R Ornduff. 1992. Seed morphology and systematics of Menyanthaceae. Am. J. Bot. 79: 1396–1406.

Col A. 1899–1901. Quelques recherches sur 1эαππαρειλ secreteur des Composees. J. Bot. (Morot) 13: 234–252; 15: 166–168.

Cronquist A. 1955. Phylogeny and taxonomy of the Compositae. Am. Midl. Nat. 53: 478–511.

Cronquist A. 1977. The Compositae revisited. Brittonia 29: 137–153.

Dahlgren KVO. 1915. Über die Embryologie von *Acicarpha tribuloides* Juss. Svensk Bot. Tidskr. 9: 184–191.

DeVore ML. 1994. Systematic studies of Calyceraceae. Ph.D. thesis, Ohio State University.

DeVore ML and TF Stuessy. 1995. The place and time of origin of the Asteraceae, with additional comments on Calyceraceae and Goodeniaceae. In: DJN Hind, C Jeffrey, and GV Pope, eds. Advances in compositae systematics, pp. 23–40. Royal Botanic Gardens, Kew.

DeVore ML, Z Zhao, J Skvarla, and R Jansen. 1997. Pollen morphology und ultrastructure of Calyceraceae. Am. J. Bot. 84(6): 185–186 (abstract).

Dittrich M. 1977. Cynareae: Systematic review. In: VH Heywood, JB Harborne, and BL Turner, eds. The biology and chemistry of the Compositae, pp. 999–1015. Academic, London.

Drury DG and L Watson. 1966. Taxonomic implications of a comparative anatomical study of Inuloideae-Compositae. Am. J. Bot. 53: 828–833.

Duigan SL. 1961. Studies of the pollen grains of plants native to Victoria, Australia. I. Goodeniaceae (including *Brunonia*). Proc. Roy. Soc. Victoria 74: 87–109.

Dunbar A. 1975. On pollen of Campanulaceae and related families with special reference to the surface ultrastructure: II. Campanulaceae subfam. Cyphioideae and subfam. Lobelioideae; Goodeniaceae; Sphenocleaceae. Bot. Not. 128: 102–118.

Eldenäs P, AA Anderberg, and M Källersjö. 1998. Molecular phylogenetics of the tribe Inuleae *s.str.* (Asteraceae), based on ITS sequences of nuclear ribosomal DNA. Plant Syst. Evol. 210: 159–173.

Eldenäs P, M Källersjö, and AA Anderberg. 1999. Phylogenetic placement and circumscription of tribes Inuleae s. str. and Plucheeae (Asteraceae): evidence from sequences of chloroplast gene *ndhF*. Mol. Phylogenet. Evol. 13: 50–58.

Emerenciano VP, JSLT Militao, CC Campos, P Romoff, MAC Kaplan, M Zambon, and AJC Brant. 2001. Flavonoids as chemotaxonomic markers for Asteraceae. Biochem. Syst. Ecol. 29: 947–957.

Erbar C. 1993. Studies on the floral development and pollen presentation in *Acicarpha tribuloides* with a discussion of the systematic position of the family Calyceraceae. Bot. Jahrb. Syst. 115: 325–350.

Erbar C. 1997. Zur Blütenentwicklung und systematischen Stellung der Menyanthaceae. Bot. Jahrb. Syst. 119: 115–135.

Erbar C and P Leins. 1988. Studien zur Blütenentwicklung und Pollenpräsentation bei *Brunonia australis* Smith (Brunoniaceae). Bot. Jahrb. Syst. 110: 263–282.

Erbar C and P Leins. 2000. Some interesting features in the capitulum and flower of *Arnoldoa macbrideana* Ferreyra (Asteraceae, Barnadesioideae). Bot. Jahrb. Syst. 122: 517–537.

Eriksson T. 1991. The systematic position of the *Blepharispermum* group (Asteraceae, Heliantheae). Taxon 40: 33–39.

Ezcurra C. 2002. Phylogeny, morphology, and biogeography of *Chuquiraga*, an Andean-Patagonian genus of Asteraceae-Barnadesioideae. Bot. Rev. 68: 153–170.

Fischer H and U Jensen. 1990. Phytoserological investigation of the tribe Cardueae s.1. (Compositae). Plant Syst. Evol. Suppl. 4: 99–111.

Fischer H and U Jensen. 1992. Utilization of proteins to estimate relationships in plants: serology; a discussion based on the Asteraceae-Cichorioideae. Belg. J. Bot. 125: 243–255.

Funk VA, RJ Bayer, S Keeley, R Chan, L Watson, B Gemeinholzer, E Schilling, JL Panero, BG Baldwin, NT Garcia-Jacas, A Susanna, RK Jansen. 2005. Everywhere but Antarctica: using a supertree to understand the diversity and distribution of the Compositae. In: I Friis and H Balslev, eds. Proc. Symp. Plant Diversity and Complexity Patterns – Local, Regional and Global Dimensions, pp. 343–374. Copenhagen.

Garcia-Jacas N, A Susanna, A Mozaffarian, and R Harslan. 2000. The natural delimitation of *Centaurea*: *ITS* analysis of the *Centaurea jacea* group. Plant Syst. Evol. 223, 3–4: 185–199.

Garcia-Jacas N, T Garnatje, A Susanna, and R Vilatersana. 2002. Tribal and subtribal delimitation and phylogeny of the

Cardueae (Asteraceae): a combined nuclear and chloroplast DNA analysis. Mol. Phylogenet. Evol. 22: 51–64.

Gattuso M and S Gattuso. 1989. Exomorfologia y anatomia de *Nymphoides indica* (L.) O. Kuntze (Menyanthaceae). Parodiana 5(2): 249–259.

Goertzen LR, J Francisco-Ortega, A Santos-Guerra, JP Mower, C Randal Linder, and RK Jansen. 2002. Molecular systematics of *Asteriscus* alliance (Asteraceae: Inuleae) II: combined nuclear and chloroplast data. Syst. Bot. 27: 815–823.

Grau J. 1980. Die Testa der Mutisieae und ihre systematische Bedeutung. Mitt. Bot. Staatssamml. München, 16: 269–332.

Grau J and H Hopf. 1985. Das Endosperm der Compositae. Bot. Jahrb. Syst. 107: 251–268.

Gustafsson MHG. 1995. Petal venation in the Asterales and related orders. Bot. J. Linn. Soc. 118: 1–18.

Gustafsson MHG. 1996a. Phylogeny of Asterales. In MHG Gustafsson, Phylogenetic studies in the Asterales sensu lato, pp. 1–36. Acta Universitatis Upsaliensis, Uppsala.

Gustafsson MHG. 1996b. Inter- and intrafamilial relationships of the Goodeniaceae as evidenced by morphology and *rbc*L sequences. In: Phylogenetic Studies in the Asterales *sensu lato*. Uppsala.

Gustafsson MHG. 1996c. Phylogenetic hypotheses for Asteraceae relationships. In: DJN Hind, H Beentje, and GV Pope, eds. Compositae: Systematics. Proc. Intern. Compositae Conference, Kew, 1994, vol. 1, pp. 9–19. Kew.

Gustafsson MHG and K Bremer. 1995. Morphology and phylogenetic relationships of the Asteraceae, Calyceraceae, Campanulaceae, Goodeniaceae and related families (Asterales). Am. J. Bot. 82: 250–265.

Gustafsson MHG, A Bachlund, and B Bremer. 1996a. Phylogeny of the Asterales *sensu lato* based on *rbc*L sequences with particular reference to the Goodeniaceae. Plant Syst. Evol. 199: 217–242.

Gustafsson MHG, S Nilsson, and E Grafström. 1996b. Pollen morphology of the Goodeniaceae and comparisons with related families. In: MHG Gustafsson, ed. Phylogenetic studies in the Asterales *sensu lato*, part V, pp. 1–30. Uppsala.

Gustafsson MHG, E Grafström, and S Nilsson. 1997. Pollen morphology of the Goodeniaceae and comparisons with related families. Grana 36: 185–207.

Gustafsson MHG, ASR Pepper, VA Albert, and M Kallersjo. 2001. Molecular phylogeny of the Barnadesioideae (Asteraceae). Nord. J. Bot. 21: 149–160.

Häffner E and FH Hellwig. 1999. Phylogeny of the tribe Cardueae (Compositae) with emphasis on the subtribe Carduinae: an analysis based on ITS sequence data. Willdenowia 29: 27–39.

Hansen HV. 1991. Phylogenetic studies in Compositae tribe Mutisieae. Opera Bot. 109: 1–50.

Hansen HV. 1992. Studies in the Calyceraceae with a discussion of its relationship to Compositae. Nord. J. Bot. 12: 63–75.

Hansen HV. 1997. Studies in the Goodeniaceae and the Brunoniaceae with a discussion of their relationship to Asteraceae and Calyceraceae. Nord. J. Bot. 17: 495–510.

Haque MZ and MB Godward. 1984. New records of the carpopodium in Compositae and its taxonomic use. Bot. J. Linn. Soc. 89: 321–340.

Harling G. 1950, 1951. Embryological studies in the Compositae: I. Anthemideae-Anthemidinae. II. An-themideae-Chrysantheminae. III. Astereae. Acta Horti. Berg. 15: 135–168, 1950; 16: 1–56, 73–120, 1951.

Harris EM. 1995. Inflorescence and floral ontogeny in Asteraceae: a synthesis of historical and current concepts. Bot. Rev. 61: 93–278.

Harris EM. 1999. Capitula in the Asteridae: a widespread and varied phenomenon. Bot. Rev. 65: 348–369.

Heald SV. 2004. Calyceraceae. In: N Amith, SA Mori, A Henderson, SW Stevenson, and SV Heald, eds. Flowering plants of the Neotropics, pp. 77–78. Princeton University Press, Princeton.

Hegnauer R. 1977. The chemistry of the Compositae. In: VH Heywood, JB Harborne, and BL Turner, eds. The biology and chemistry of the Compositae, pp. 283–335. Academic, London.

Hellwig FH. 2007. Calyceraceae. In: K Kubitzki, ed. The families and genera of vascular plants, vol. 8, pp. 19–25. Springer, Berlin/Heidelberg/New York.

Heywood VH, JB Harborne, and BL Turner, eds. 1977. The biology and chemistry of the Compositae. Academic, London.

Hind DJN, C Jeffrey, and GV Pope, eds. 1995. Advances in Compositae systematics. Royal Botanic Gardens, Kew.

Holub M, J Toman, and V Herout. 1987. The phylogenetic relationships of the Asteraceae and Apiaceae based on phytochemical characters. Biochem. Syst. Ecol. 15: 321–326.

Inoue N and H Tobe. 1999. Integumentary studies in Menyanthaceae (Campanulales sensu lato). Acta Phytotax. Geobot. 50: 75–79.

Jansen RK and JD Palmer. 1987. A chloroplast DNA inversion marks an ancient evolutionary split in the sunflower family (Asteraceae). Proc. Natl. Acad. Sci. USA 84: 5818–5822.

Jansen RK and JD Palmer. 1988. Phylogenetic implications of chloroplast DNA restriction site variation in the Mutisieae (Asteraceae). Am. J. Bot. 75: 753–766.

Jansen RK, KE Holsinger, HJ Michaels, and JD Palmer. 1990. Phylogenetic analysis of chloroplast DNA restriction site data at higher taxonomic levels: an example from the Asteraceae. Evolution 44: 2089–2105.

Jansen RK, HJ Michaels, and JD Palmer. 1991a. Phylogeny and character evolution in the Asteraceae based on chloroplast DNA restriction site mapping. Syst. Bot. 16: 98–115.

Jansen RK, HJ Michaels, RS Wallace, K-J Kirn, SC Keeley, LE Watson, and JD Palmer. 1991b. Chloroplast DNA variation in the Asteraceae: phylogenetic and evolutionary implications. In: D Soltis, P Soltis, and J Doyle, eds. Molecular systematics of plants, pp. 252–279. Chapman & Hall, New York.

Jeffrey C. 1977. Corolla forms in the Compositae: Some evolutionary and taxonomic speculations. In: VH Heywood, JB Harborne, and BL Turner, eds. The biology and chemistry of the Compositae, pp. 111–118. Academic, London.

Jeffrey C. 1978. Compositae. In: VH Heywood, ed., Flowering plants of the world, pp. 263–268. Mayflower Books, New York.

Jeffrey C. 1995. Compositae systematics 1975–1993 developments and desiderata. In: DJN Hind, C Jeffrey, and GV Pope, eds. Advances in compositae systematics, pp. 3–21. Royal Botanic Gardens, Kew.

Jeffrey C. 2002. Systematics of Compositae at the beginning of the 21st century. Bot. Zhurn. 87(11) 1–15 (in Russian with English summary).

Jeffrey C. 2004. Systema Compositarum (Asteracearum) nova. Bot. Zhurn. 89(12): 1817–1822.

Jeffrey C and YL Chen. 1984. Taxonomic studies on the tribe Senecioneae (Compositae) of eastern Asia. Kew Bull. 39: 205–446.

Jones SB. 1977. Vernonieae: systematic review. In: VH Heywood, JB Harborne, and BL Turner, eds. The biology and chemistry of the Compositae, pp. 503–521. Academic, London.

Kadereit G. 2007. Menyanthaceae. In: K Kubitzki, ed. The family and genera of vascular plants, vol. 8, pp. 599–604. Springer, Berlin/Heidelberg/New York.

Kadereit JW. 1989. Chloroplast DNA, cladistics and the phylogeny of the Asteraceae. Bot. Acta 102: 7–10.

Kamelin RV. 2000. Compositae (short review of the system). Barnaul (in Russian).

Karis PO. 1992. *Haplophyllum* DC, the sister group to *Eremothamnus* O. Hoffm. (Asteraceae)? Taxon 41: 193–198.

Karis PO. 1993a. Morphological phylogenetics of the Asteraceae-Asteroideae, with notes on character evolution. Plant. Syst. Evol. 186: 69–93.

Karis PO. 1993b. Heliantheae *sensu lato* (Asteraceae), clades and classification. Plant Syst. Evol. 188: 139–195.

Karis PO. 1996. Phylogeny of the Asteraceae-Asteroideae revised. In: DJN Hind and HJ Beentje, eds. Proceedings of the International Compositae Conference, Kew, 1994, vol. 1. Compositae: Systematics, pp. 41–47. Kew.

Karis PO, M Källersjö, and K Bremer. 1992. Phylogenetic analysis of the Cichorioideae (Asteraceae), with emphasis on the Mutisieae. Ann. Missouri Bot. Gard. 79: 416–427.

Keeley SC and RK Jansen. 1991. Evidence from chloroplast DNA for the recognition of a new tribe, the Tarchonantheae, and the tribal placement of *Pluchea* (Asteraceae). Syst. Bot. 16: 173–181.

Keywood VH, JB Harborne, and BL Turner. 1977. The biology and chemistry of the Compositae, vol. 1. Academic, London.

Khanjian NS. 1991. On the position of the genus *Ursinia* in the system of the family Asteraceae. Bot. Zhurn. 76: 1728–1733 (in Russian with English summary).

Kim HG, DJ Loockerman, and RK Jansen. 2002. Systematic implications of ndhF sequence variation in the Mutisieae (Asteraceae). Syst. Bot. 27: 598–609.

Kim K-J, BL Turner, and RK Jansen. 1990. Chloroplast DNA evidence for the phylogenetic relationships among Coreopsideae, Eupatorieae, Gaillardieae, Heliantheae, and Tagateae. Am. J. Bot. Suppl. 77: 140–141 (abstract).

Kim K-J, RK Jansen, RS Wallace, HJ Michaels, and JD Palmer. 1992. Phylogenetic implications of *rbc*L sequence variation in the Asteraceae. Ann. Missouri Bot. Gard. 79: 428–445.

Kim Y-D and RK Jansen. 1995. *Ndh*F sequence evolution and the major clades in the sunflower family. Proc. Natl. Acad. Sci. USA 92: 10379–10383.

King RM and H Robinson. 1987. The genera of the Eupatorieae (Asteraceae). Allen Press, Lawrence, KS.

Knox EB and HD Palmer. 1995. The origin of *Dendrosenecio* within the Senecioneae (Asteraceae) based on chloroplast DNA evidence. Am. J. Bot. 82: 1567–1573.

Kozo-Poljanski BM. 1923. On the systematic position of the family Compositae. Zhurn. Russian Bot. Soc. 8: 167–191 (in Russian).

Lammers TG. 1992. Circumscription and phytogeny of the Campanulales. Ann. Missouri Bot. Gard. 79: 388–413.

Lavialle P. 1912. Recherches sur le développement de 1϶οϖαιρε en fruit chez les Composées. Ann. Sci. Nat. Bot., 9th ser. 15: 39–152.

Leins P. 1971. Pollensystematische Studien an Inuleen: I. Tarchonanthinae, Plucheinae, Buphtalminae. Engl. Bot. Jahrb. 91: 91–146.

Leins P and C Erbar. 1989. Zur Blütenentwicklung und sekundaren Pollenprasentation bei *Selliera radicans* Cav. (Goodeniaceae). Flora 182: 43–56.

Leins P and C Erbar. 1990. On the mechanisms of secondary pollen presentation in the Campanulales-Asterales-complex. Bot. Acta 103: 87–92.

Leins P and C Erbar. 2000. Die frühesten Entwicklungsstadien der Blüten bei den Asteraceae. Bot. Jahrb. Syst. 122: 503–515.

Lindsey AA. 1938. Anatomical evidence for the Menyanthaceae. Am. J. Bot. 25: 480–485.

Loockerman DJ, BL Turner, and RK Jansen. 2003. Phylogenetic relationships within the Tageteae (Asteraceae) based on nuclear ribosomal ITS and chloroplast *ndh*F gene sequences. Syst. Bot. 28: 191–207.

Lundberg J and K Bremer. 2003. A phylogenetic study of the order Asterales using one morphological and three molecular data sets. Int. J. Plant Sci. 164: 553–578.

Mani MS and JM Saravanan. 1999. Pollination ecology and evolution in Compositae (Asteraceae). Science Publishers, Enfield, NH.

Manilal KS. 1971. Vascularization of corolla of the Compositae. J. Indian Bot. Soc. 50: 189–196.

McKenzie RJ, NP Barker, J Samuel, EM Muller, and AK Skinner. 2005. Morphology of cypselae in subtribe Arctotidinae (Compositae-Arctotideae) and its taxonomic implications. Ann. Missouri Bot. Gard. 92: 569–594.

Miao B, BL Turner, and TJ Mabry. 1955. Systematic implications of chloroplast DNA variation in the subtribe Ambrosiinae (Asteraceae: Heliantheae). Am. J. Bot. 82: 924–932.

Michaels HJ, KM Scott, RG Olmstead, T Szaro, RK Jansen, and JD Palmer. 1993. Interfamilial relationships of the Asteraceae: insights from *rbcL* sequence variation. Ann. Missouri Bot. Gard. 80: 742–751.

Mishler BD, VA Albert, MW Chase, PO Karis, and K Bremer. 1996. Character-state weighting for DNA restriction site data: asymmetry, ancestors and the Asteraceae. Cladistics 12: 11–19.

Morales E. 2001. The comparative method and trait evolution in the Asteraceae. Compositae Newslett. 36: 14–25.

Mukherjee SK and AK Sarkar. 2001. Comparative morpho-anatomical study of cypselas in some species of the tribe Heliantheae (Asteraceae). Bull. Bot. Surv. India 40: 34–46 (1998 publ. 2001).

Nic Lughadha EM and JAN Parnell. 1989. Heterostyly and geneflow in *Menyanthes trifoliata* L. (Menyanthaceae). Bot. J. Linn. Soc. 100: 337–354.

Nillson S and R Ornduff. 1973. Menyanthaceae. In: S Nilsson, ed. World pollen and spore flora, vol. 2, pp. 1–19. Almquist & Wiksell, Stockholm.

Nishino E. 1983. Corolla tube formation in the Tubiflorae and Gentianales. Bot. Mag. (Tokyo) 96: 223–243.

Nordenstam B. 1977. Senecioneae and Liabeae – systematic review. In: VH Heywood, JB Harborne, and BL Turner, eds.

The biology and chemistry of the Compositae, pp. 799–830. Academic, London.
Nordenstam B. 1978. Taxonomic studies in the tribe Senecioneae (Compositae). Opera Bot. 44: 1–84.
Olmstead RG, HJ Michaels, KM Scott, and JD Palmer. 1992. Monophyly of the Asteridae and identification of its major lineages inferred from *rbcL* sequences. Ann. Missouri Bot. Gard. 79: 249–265.
Ornduff R. 1988. Distyly and monomorphism in *Villarsia* (Menyanthaceae): some evolutionary considerations. Ann. Missouri Bot. Gard. 75: 761–767.
Panero JL and VA Funk. 2002. Toward a phylogenetic classification for the Compositae (Asteraceae). Proc. Biol. Soc. Washington 115: 909–922.
Patel RC, JA Inamdar, and NV Rao. 1981. Structure and ontogeny of stomata in some Gentianaceae and Menianthaceae complex. Feddes Repert. 92: 535–550.
Peacock WJ. 1963. Chromosome numbers and cytoevolution in the Goodeniaceae. Proc. Linn. Soc. N. S. W. 88: 8–27.
Pesacreta TC and TF Stuessy. 1996. Autofluorescent walls of connective bases in anthers of Barnadesioideae (Asteraceae), and systematic implication. Taxon 45: 473–485.
Petit DP. 1997. Generic interrelationships of the Cardueae (Compositae): a cladistic analysis of morphological data. Plant Syst. Evol. 207: 173–203.
Philipson WR. 1953. The relationships of the Compositae, particularly as illustrated by the morphology of the inflorescence in the Rubiales and the Campanulatae. Phytomorphology 3: 391–404.
Poljakov PP. 1967. Systematics and origin of the Compositae. Nauka, Alma-Ata (in Russian).
Pontiroli A. 1963. Flora Argentina. Calyceraceae. Rev. Mus. La Plata 9: 175–241.
Proksch P and E Rodriquez. 1983. Chromenes and benzofurans of the Asteraceae, their chemistry and biological significance. Phytochemistry 22: 2335–2348.
Reiche K. 1900. Beiträge zur Systematik der Calyceraceen. Bot. Jahrb. Syst. 29: 107–119.
Robinson H. 1977. An analysis of the characters and relationships of the tribes Eupatorieae and Vernonieae (Asteraceae). Syst. Bot. 2: 199–208.
Robinson H. 1981. A revision of the tribal and subtribal limits of the Heliantheae (Asteraceae). Smithsonian Contr. Bot. 51: 1–102.
Robinson H. 1983. A generic review of the tribe Liabeae (Asteraceae). Smithsonian Contr. Bot. 54: 1–69.
Robinson H. 1992. Observations on the unique form of sweeping hairs on the styles of the Eremothamnaea (Asteraceae). Taxon 41: 199–208.
Robinson H. 1994. Notes on the tribes Eremothamneae, Gundelieae, and Moquinieae, with comparisons of their pollen. Taxon 43: 33–44.
Robinson H and RD Brettell. 1973a. Tribal revisions in the Asteraceae: III. A new tribe, Liabeae. Phytologia 25: 104–107.
Robinson H and RD Brettell. 1973b. Tribal revisions in the Asteraceae: VIII. A new tribe, Ursinieae. Phytologia 26: 76–86.
Robinson H and L Brouillet. 1994. Notes on the tribes Eremothamneae, Gundelieae, and Moquinieae, with comparisons of their pollen. Taxon 43: 33–44.
Rosen W. 1938. Beiträge zur Kenntnis der Embryologie der Goodeniaceen. Acta Horti. Goteborg. 12: 1–10.
Rosen W. 1946. Further notes on the embryology of the Goodeniaceae. Acta Horti. Goteborg. 16: 235–249.
Ryding O and K Bremer. 1992. Phylogeny, distribution, and classification of the Coreopsideae (Asteraceae). Syst. Bot. 17: 649–659.
Sancho G and L Katinas. 2002. Are the trichomes in corollas of Mutisieae (Asteraceae) really twin hairs? Bot. J. Linn. Soc. 140: 427–433.
Schilling EE and JL Panero. 2002. A revised classification of subtribe Helianthinae (Asteraceae; Heliantheae). 1. Basal lineages. Bot. J. Linn. Soc. 140: 65–76.
Seaman F, F Bohlmann, C Zdero, and TJ Mabry. 1990. Diterpenes of flowering plants – Compositae (Asteraceae). Springer, New York.
Senianinova-Korchagina MV. 1952. On the inferior ovary in Compositae. Bull. Moscow Soc. Nat., Biol. Sect., n.s. 57(4): 63–75 (in Russian).
Singh BP. 2000. Morphological nature and trends of evolution in the pappus of the Asteraceae. Acta Bot. Hung. 42(1–4): 285–293 (1999–2000).
Singh BP. 2002. Diversity of vascular system in the wall of inferior ovary in the family Asteraceae. J. Jpn. Bot. 77(3): 171–176.
Singh BP and I Kaur. 1998. Structure, vascular anatomy and evolution of naked pistillate flower in the Asteraceae. Acta Bot. Hung. 41(1–4): 287–292.
Skvarla JJ and BL Turner. 1966. Systematic implications from electron microscopic studies of Compositae pollen: a review. Ann. Missouri Bot. Gard. 53: 220–256.
Skvarla JJ, BL Turner, VC Patel, and AS Tomb. 1977. Pollen morphology in the Compositae and morphologically related families. In: VH Heywood, JB Harborne, and BL Turner, eds. The biology and chemistry of the Compositae, pp. 141–248. Academic, London.
Small J. 1919. The origin and development of the Compositae. New Phytologist Reprint No. 11.
Soltis P, DE Soltis, and JJ Doyle, eds. 1992. Molecular systematics of plants. Chapman & Hall, New York.
Stebbins GL, Jr. 1953. A new classification of the tribe Cichorieae, family Compositae. Madroño 12: 33–64.
Stebbins GL, Jr. 1977. Developmental and comparative anatomy of the Compositae. In: VH Heywood, JB Harborne, and BL Turner, eds. The biology and chemistry of the Compositae, pp. 92–109. Academic, London.
Stix H. 1960. Pollenmorphologische Untersuchungen an Compositen. Grana Palynol. 2(2): 41–114.
Strother JL 1977. Tageteae: systematic review. In: VH Heywood, JB Harborne, and BL Turner, eds. The biology and chemistry of the Compositae, pp. 769–783. Academic, London.
Stuessy TF. 1977. Heliantheae: systematic review. In: VH Heywood, JB Harborne, and BL Turner, eds. The biology and chemistry of the Compositae, pp. 621–671. Academic, London.
Stuessy TF and DM Spooner. 1988. The adaptive and phylogenetic significance of receptacular bracts in the Compositae. Taxon 37: 114–126.
Susanna A, NG Jacas, DE Soltis, and PS Soltis. 1995. Phylogenetic relationships in tribe Cardueae (Asteraceae) based on ITS sequences. Am. J. Bot. 82: 1056–1068.
Susanna A, T Garnatje, N Garcia-Jacas, and R Vilatersana. 2002. On the correct subtribal placement of the genera *Syreitschikovia*

and *Nikitinia* (Asteraceae, Cardueae): Carduinae or Centaureinae? Bot. J. Linn. Soc. 140: 313–319.

Tamamshian SG. 1956. On the origin of the pappus in the family Asteraceae (Compositae). Bot. Zhurn. 41: 634–651 (in Russian).

Tamamshian SG. 1965. «Superevolutionary» form of the calyx and its significance for phylogenetic problems in Asteraceae Link. In Problems in plant phylogeny. Trudy Moscow Soc. Nat. 13: 161–174 (in Russian).

Thiele E-M. 1988. Bau und Funktion des Antheren-Griffel-Komplexes der Compositen. Diss. Bot. 117: 1–169.

Turner BL. 1977a. Fossil history and geography. In: VH Heywood, JB Harborne, and BL Turner, eds. The biology and chemistry of the Compositae, pp. 21–39. Academic, London.

Turner BL. 1977b. Summary of the biology and chemistry of the Compositae. In: VH Heywood, JB Harborne, and BL Turner, eds. The biology and chemistry of the Compositae, pp. 1105–1118. Academic, London.

Turner BL and AM Powell. 1977. Helenieae: systematic review. In: VH Heywood, JB Harborne, and BL Turner, eds. The biology and chemistry of the Compositae, pp. 699–737. Academic, London.

Urbatsch LE, BG Baldwin, and MJ Donoghue. 2000. Phylogeny of the coneflowers and relatives (Heliantheae: Asteraceae) based on nuclear rDNA internal transcribed spacer (ITS) sequences and chloroplast DNA restriction site data. Syst. Bot. 25: 539–565.

Urtubey E and TF Stuessy. 2001. New hypotheses of phylogenetic relationships in Barnadesioideae (Asteraceae) based on morphology. Taxon 50: 1043–1066.

Urtubey E and MC Tellaria. 1998. Pollen morphology of the subfamily Barnadesioideae (Asteraceae) and its phylogenetic and taxonomic significance. Rev. Palaeobot. Palynol. 104: 19–37.

Vezey EL, LE Watson, JJ Skvarla, and JR Estes. 1994. Plesiomorphic and apomorphic pollen structure characteristics of Anthemideae (Asteroideae: Asteraceae). Am. J. Bot. 81: 648–657.

Vijayaraghavan MR and U Malik. 1972. Morphology and embryology of *Scaevola frutescens* K. and affinities of the family Goodeniaceae. Bot. Not. 125: 241–254.

Wagenitz G. 1976. Systematics and phytogeny of the Compositae (Asteraceae). Plant Syst. Evol. 125: 29–46.

Watanabe K, RM King, T Yahara, M Ito, J Yokoyama, T Suzuki, and DJ Crawford. 1995. Chromosomal cytology and evolution in Eupatorieae (Asteraceae). Ann. Missouri Bot. Gard. 82: 581–592.

Watson LE, TM Evans, and T Boluarte. 2000. Molecular phylogeny and biogeography of the tribe Anthemideae (Asteraceae) based on chloroplast gene *ndh*F. Mol. Phylogenet. Evol. 15: 59–69.

Whitton J, RS Wallace, and RK Jansen. 1995. Phylogenetic relationships and patterns of character change in the tribe Lactuceae (Asteraceae) based on chloroplast DNA restriction site variation. Canad. J. Bot. 73: 1058–1073.

Wood CE, Jr. 1983. The genera of Menyanthaceae in the southeastern United States. J. Arnold Arbor. 64: 431–445.

Yamazaki T. 1974. A system of Gamopetalae based on the embryology. J. Fac. Sci. Univ. Tokyo, Sect. 3, Bot. 8–9: 263–281.

Zdero C and F Bohlmann. 1990. Systematics and evolution within the Compositae, seen with the eyes of a chemist. Plant Syst. Evol. 171: 1–14.

Zhang X-P and K Bremer. 1993. A cladistic analysis of the tribe Astereae (Asteraceae) with notes on their evolution and subtribal classification. Plant Syst. Evol. 184: 259–283.

Zhao Z, JJ Skvarla, RK Jansen, and M DeVore. 2000. Phylogenetic significance of pollen morphology and ultrastructure in the Barnadesioideae (Asteraceae). Lundellia, 3: 26–40.

Zohary M. 1950. Evolutionary trends in the fruiting head of Compositae. Evolution 4: 103–109.

Subclass VIII. LAMIIDAE

Trees, shrubs, subshrubs, and herbs. Vessels mostly with simple perforations. Sieve-element plastids of S-type. Nodes unilacunar, trilacunar, or multilacunar. Leaves alternate or more often opposite, sometimes verticillate, estipulate or with stipules. Flowers nearly always sympetalous. Tapetum usually secretory. Microsporogenesis simultaneous. Gynoecium mostly of two united carpels; ovary superior to inferior. Ovules unitegmic and usually tenuinucellate. Endosperm nuclear or cellular, often with haustoria. Fruits of various types. Seeds with or without endosperm.

The subclass Lamiidae is most probably derived from the most primitive Cornanae – Desfontainiales. According to Olmstead et al. (1992: 259; 1993), this lineages is one of the most strongly supported clades in the higher dicots.

Bibliography

Soltis DE, PS Soltis, PK Endress, and MW Chase. 2006. Phylogeny and evolution of Angiosperms. Sinauer, Sanderland, MA.

Superorder LAMIANAE

Order 120. RUBIALES (GENTIANALES)

Trees, shrubs, subshrubs, and perennial or annual herbs. Intraxylary (internal) phloem present in all investigated members; interxylary (included) phloem of the foraminate type present in Antoniaceae, *Strychnos* and some species of *Logania*. Vessels with simple perforations or rarely scalariform perforations; lateral pitting mostly alternate. Fibers with bordered or more often simple pits, sometimes septate. Vestured pits present or rarely (*Gelsemium* and *Logania*) absent. Rays heterogeneous to homogeneous or wanting. Axial parenchyma of diverse types or absent. Sieve-element plastids of S-type or Ss-type. Nodes trilacunar, unilacunar, or multilacunar. Leaves opposite or sometimes verticillate, rarely alternate, simple and usually entire, estipulate or less often with interpetiolar stipules; leaf mesophyll cells sometimes contain oil bodies. Colleters often present. Stomata mostly paracytic or anomocytic. Flowers in cymose or less often racemose inflorescences, sometimes solitary, bisexual or rarely unisexual, mostly actinomorphic or nearly actinomorphic, with double perianth. Corolla sympetalous, lobes usually contorted, less often imbricate or valvate. Stamens usually as many as and alternate with the corolla lobes, in a single whorl, or very rarely (*Dialypetalanthus*) stamens 16–26, in two whorls, usually attached to the corolla tube; anthers tetrasporangiate, opening longitudinally or very rarely (as in *Exacum*) by apical pores. Tapetum secretory. Microsporogenesis simultaneous or (some Apocynaceae) successive. Pollen grains 2-celled or 3-celled, mostly 3-colporate. Intrastaminal nectary disc present or absent. Gynoecium of two or less often 3–5 (up to eight) united carpels; stylodia more or less united into a terminal style; ovary superior to inferior, mostly 1-locular, with (1-) many ovules in each locule or on each placenta. Ovules anatropous to hemitropous, unitegmic, tenuinucellate, sometimes with endothelium. Female gametophyte usually of *Polygonum*-type. Endosperm nuclear or rarely (*Mitrasacme*, *Voyriella*) cellular (Johri et al. 1992), very rarely (in some Apocynaceae) with chalazal haustorium. Fruits capsules or of diverse other types. Seeds exotestal, with more or less straight embryo and copious endosperm, rarely endosperm nearly wanting. Usually producing iridoid

A. Takhtajan, *Flowering Plants*,

compounds and often also alkaloids, but without cardiotonic glycosides.

The basal group within the Lamiidae, probably directly connected with the Hydrangeales and have a common origin from the saxifragalean stock.

Key to Families

1 Plant without a latex-system, and without cardiotonic glycosides.

2 Intraxylary (internal) phloem present.

3 Stigma twice dichotomously branched. Shrubs, subshrubs, or lianas with simple hairs or almost glabrous. Vessels mostly with simple perforations, except for a few scalariform perforations, plates with few bars close to the pith in *Mostuea*; intervascular pits small, vestured (*Mostuea*) or not vestured (*Gelsemium*). Fibers with small bordered pits, nonseptate (*Gelsemium*) or some septate ones present (*Mostuea*). Rays heterogeneous (*Mostuea*) or homogeneous (*Gelsemium*). Axial parenchyma scanty vasicentric (*Gelsemium*) or absent (*Mostuea*). Secondary thickening anomalous from a single cambial ring. Leaves opposite or whorled, entire or obscurely sinuate-dentate; stipulate (stipeles interpetiolar to intrapetiolar, sometimes reduced to a stipular sheath joining the opposite leaf bases) or estipulate. Colleters present in the axils of the leaves, bracts, and sepals, or only in the axils of the leaves. Flowers in axillary or terminal cymes or sometimes solitary, bracteate, nearly always heterostylous, actinomorphic or nearly so, bisexual, with variously shaped bracts. Perianth and androecium 5-merous, occasionally (*Mostuea*) 4-merous. Sepals free or connate up to the middle, equal or unequal, imbricate, persistent. Corolla funnel-shaped, tube much longer than the lobes, lobes imbricate, equal or subequal. Stamens four, rarely five, inserted near the base of the corolla tube, or midway down the corolla tube, equal or unequal; filament very short; anthers dorsifixed, extrorse (*Gelsemium*) or latrorse. Pollen grains 3-colporate, with long, unbranched columellae and granulate colpus membrane (*Gelsemium*) or finely suprastriate, with long and branched columellae and more or less smooth colpus membrane. Gynoecium of two carpels; styles 1–2, free to partially joined, stigmas two, or four (each bifurcated). Ovary superior, 2-locular, glabrous or hairy, with two (*Mostuea*) or 2–8 (*Gelsemium*) anatropous ovules in each locule. Fruits septicidal (*Gelsemium*) or loculicidal (*Mostuea*) capsules. Seeds rather large, flattened, winged (*Gelsemium*) or wingless, with rather thin testa, small, straight or curved embryo and copious, starchy (*Gelsemium*), or bony (*Mostuea*) endosperm. Producing seco-iridoids, C-17 indole alkaloids, quercetin, kaempferol; n = 4, 8 (*Gelsemium*) or 10 (*Mostuea*).............. 1. GELSEMIACEAE.

3 Stigma not branched.

4 Mostly trees and shrubs, sometimes woody lianas or herbs.

5 Fruits capsules.

6 Seeds free (not embedded in the pulp).

7 Plants producing iridoids and small amount of alkaloids. Shrubs, subshrubs, or herbs, *Logania tortuosa* is switch-plant, glabrous or with simple hairs. Vessels usually with simple, or rarely scalariform perforations, without vestured pits. Fibers with simple pits. Nodes unilacunar (with one to several traces), trilacunar or multilacunar. Secondary thickening developing from a conventional cambial ring, or anomalous. Leaves opposite, entire, simple, sometimes reduced to scales, connected at the base by more or less distinct stipular line or a short ochrea. Flowers in terminal or axillary cymose inflorescences or solitary, bisexual or unisexual (dioecious), actinomorphic. Perianth and androecium 5-merous, rarely (only *Logania micrntha*) 4-merous. Sepals subequal, basally connate, imbricate, with colleters inside or not. Corolla rotated or campanulate, lobed, lobes imbricate. Stamens attached to the corolla tube or corolla throat, in female flowers smaller or absent; anthers dorsifixed, introrse. Pollen grains 3-colporate. Gynoecium of two carpels; style thick, with large, subglobose, or

oblong stigma; ovary superior, 2-locular, with several (3–50) ovules per locule; ovules anatropous or hemianatropous. Fruits septicidal or loculicidal (dehiscing only in the upper half) capsules, or schizocarp (Watson and Dallwitz 1992). Seeds more or less ellipsoid, wingless, with a honeycomblike reticulation; embryo straight, cylindrical, endosperm starchy, n = 16. . . . 2. LOGANIACEAE.

7 Plants not producing either iridoids or alkaloids. Erect or climbing shrubs or small to large trees, glabrous or with simple hairs. Wood diffuse porous, not storied, secondary thickening anomalous present. Vessels with simple perforations, with vestured pits; xylem with tracheids (abundant in *Usteria*) or without tracheids. Fibers with small bordered pits. Leaves opposite, simple, entire or dentate, connected at the base by an interpetiolar line, short sheath, or ciliolate rim; colleters often present in the axils of leaves and bracts and at the base of sepals. Stomata anomocytic, paracytic, or encyclocytic. Flowers in usually terminal cymes or in panicles, actinomorphic or (*Usteria*) zygomorphic, 5-merous or (*Usteria*) 4-merous. Calyx sometimes (*Antonia*) surrounded by numerous imbricate, scalelike bracts, 5–4-lobed or partite, lobes or segments equal or (*Usteria*) outer lobe enlarged and petaloid. Corolla tubular, tube often slender; lobes five or four, valvate. Hypogynous disc small or absent. Stamens five, four or (*Usteria*) one, inserted in the throat of the corolla tube;. anthers dorsifixed, latrorse or introrse. Pollen grains 3-colporate. Gynoecium of two carpels; style deciduous or persistent, with capitate or shortly 2-lobed stigma; ovary superior or rarely inferior, 2-locular, with numerous (15–50), anatropous or amphitropous ovules. Fruits septicidal, 2-valved capsules. Seeds winged, the wing diaphanous, reticulate-veined; embryo relatively large, straight, or incurved in fleshy or starchy endosperm. Iridoids not detected, n = 11 (*Usteria*). . 4. ANTONIACEAE.

6 Seeds embedded in pulp. Small, glabrous shrubs to small trees, rarely scrambling to semiclimbing. Branches terete, quadrangular, or 4-winged. Wood diffuse porous; not storied. Vessels mostly with simple perforations, rarely (a few spp. *Labordia*), with scalariform perforations, with vestured pits. Leaves opposite, petiolate to sessile, simple, entire, stipulate (stipules interpetiolar or intrapetiolar) or estipulate. Colleters present or absent. Flowers in terminal (*Labordia*) or axillary, ramiflorous or cauliflorous (*Geniostoma*) inflorescences or occasionally solitary, unisexual or gynodioecious, actinomorphic, 5-merous. Sepals nearly free or connate, imbricate, inside mostly with colleters at the base, margin mostly fimbriate. Corolla campanulate to rotate, lobed, lobes imbricate or contorted. Stamens five, inserted at the corolla mouth; filaments mostly short, not appendiculate; anthers dorsifixed, introrse; often with a distinct apical appendage. Pollen grains 3-pororate or with more than three pores in zonoporate or pantoporate position, psilate or coarsely reticulate. Gynoecium of two or three (*Labordia*) carpels. Ovary superior, 2- or rarely 3-locular, with many anatropous or amphitropous ovules in each locule; style mostly short; stigma about as large as the ovary, clavate, ellipsoid or subglobose, sometimes more or less oblong. Capsules with thick and woody exocarp, septicidal to subseptifragal, with two (or three) valves breaking away from and exposing the persistent and pulpy placentas. Seeds wingless, ellipsoid, ovoid, or subglobose, intruded

on the hilar side, mostly minutely papillate to areolate, with small straight embryo and fleshy endosperm. Alkaloids present or absent, iridoids detected, n = 10. . . . . 10. GENIOSTOMACEAE.

5 Fruits drupes or berries. Trees, shrubs, or woody lianas, often with tendrils and/or axillary spines or spreading prickles, glabrous or with simple hairs. Vessels with simple perforations and with vestured pits. Fibers with bordered pits. Leaves opposite or on the main axis sometimes ternate, entire, pinnately veined or 3–5-veined from or above the base, estipulate or with persistent or caducous stipules. Flowers in terminal or/and axillary cymose inflorescences, rarely solitary, bisexual, actinomorphic or nearly so, with scalelike to leafy bracts, perianth and androecium 4- or 5-merous. Sepals free or up to the middle connate, imbricate, persistent. Corolla rotated to salver-shaped, more or less fleshy, lobes valvate. Stamens 4–5, inserted on the corolla tube or in the throat of the corolla tube; filaments shorter or longer than the anthers; anthers introrse. Pollen grains 3-colporate or rarely 4- or 2-colporate, or syncolpate, usually psilate with widely distributed perforations. Ovary superior, 2-locular, with an axile placenta with few to many ovules or, if 1-locular, with a basal placenta; ovules anatropous; style usually short, with capitate to 2-lobed stigma. Fruits berries or (*Neuburgia*) drupes. Seeds variously shaped, with thin or thick testa; embryo minute to very large; endosperm thick, horny, with a large central cavity. Colleters in the axils of the leaves and bracts and at the base of the sepals or not. Producing loganin and related iridoid glycosides and indole alkaloids n = 11. . . . . 3. STRYCHNACEAE.

4 Mostly herbs, rarely shrublets, shrubs, or small trees.

8 Corolla lobes valvate. Herbs, rarely subshrubs, sometimes swith-plants. Vessels with simple perforations, but in *Mitreola* sometimes occur scalariform perforations with a few bars. Wood diffuse porous, not storied. Axial parenchyma absent. Secondary thickening anomalous from a single cambial ring. Leaves well developed or much reduced, opposite, or opposite and verticillate, simple, entire, petiolate to subsessile, with stipular lines or small, interpetiolar stipules. Colleters in the axils of leaves and bracts and at the base of inner side of sepals sometimes present. Flowers in terminal or axillary cymes or more or less unilateral spikes, rarely solitary or fasciculate or subumbellate, 5-merous. Calyx persistent, sometimes glandular within the base. Corolla valvate (the lobes pointed, erect), funnel-shaped or tubular. Stamens five, attached to the corolla tube or in the throat of the corolla tube; anthers dorsifixed, introrse. Pollen grains 2(3)4-colpate (*Spigelia*) or 3-colporate. Gynoecium of two carpels; style bifid or deeply 2-lobed or 2-partite, with truncate, clavate, or capitate stigmas; ovary superior, sessile, 2-locular, with numerous ovules densely arranged on peltate placentas; ovules anatropous, or campylotropous, or hemianatropous. Fruits capsules, septicidal or (*Spigelia*) circumscissile at the base. Seeds wingless and not particularly compressed, sometimes polyhedral; embryo straight; endosperm fleshy, ruminate. Producing iridoid compounds asperuloside, geniposide, and geniposidic acid. n = 8 (*Spigelia*), 10 (*Mitreola*). . . . . 5. SPIGELIACEAE.

8 Corolla lobes contorted or imbricate. Perennial or annual herbs, often mycorrhizal and sometimes strongly mycotrophic and without chlorophyll, rarely shrubs or trees. Glabrous or with various types of trichomes, sometimes glandular. Vessels with simple or sometimes (*Saccifolium*) scalariform perforations. Rays when present predominantly uniseriate. Axial parenchyma scanty vasicentric or (*Saccifolium*) wanting. Leaves opposite or sometimes verticillate, very rarely alternate, usually entire, simple, saccate (*Saccifolium*), in

mycotrophic genera reduced to scales. Stomata anomocytic or sometimes anisocytic. Flowers in cymose or rarely racemose inflorescences, sometimes solitary, bisexual or rarely (*Veratrilla*) unisexual, actinomorphic or rarely slightly zygomorphic, 5–4(rarely 6–12)-merous. Calyx mostly synsepalous, usually with a short to well-developed tube and imbricate or sometimes valvate or open lobes, sometimes the lobes much reduced or suppressed, rarely sepals nearly free. Corolla often with scales or nectary pits within the tube. Stamens attached to the corolla tube or throat, free or rarely connate by anthers; anthers basifixed, introrse or rarely (*Saccifolium*), latrorse, opening longitudinally or rarely (as in *Exacum* and *Cotylanthera*) by apical pores that extend halfway to the base. Pollen grains 2-celled or 3-celled, (2)3(4)-colporate or 2–3-porate, sometimes monoporate (as in *Voyriella* and spp. of *Leiphamos*). Intrastaminal nectary disc or nectary glands usually present. Gynoecium of two carpels; style with an entire or 2-lobed, papillate stigma; ovary 1-locular, with two parietal placentas that usually project more or less deeply into the cavity and sometimes connate in its center, forming a false partition as in Exacinae; in *Cotylanthera* as a result of suppression of the false partition the parietal placentation transformed into free central. Ovules more or less numerous, anatropous or sometimes hemitropous or orthotropous, rarely (in some parasitic species) ategmic. Fruits septicidal capsules or rarely baccate. Seeds numerous, small, winged or not, with small embryo and copious endosperm, very simplified in mycotrophic genera. Usually producing bitter substances consisting of iridoid compounds, especially gentiopicroside and related substances, n = 5–13. . . . . 9. GENTIANACEAE.

2 Intraxylary (internal) phloem absent.

9 Flowers bisexual or rarely unisexual.

10 Stamens (8)16–18(-25) in two cycles, free. Trees up to 30 m tall, with dense white-tomentose young branches. The wood is light but extremely hard. Vessels with simple perforations; lateral pitting alternate. Fibers septate. Rays heterogeneous. Axial parenchyma scanty paratracheal strands. Leaves opposite, entire, simple, with large, intrapetiolar and interpetiolar stipules that are laterally connate basally in pairs. Stomata paracytic. Flowers in terminal thyrses or cymose panicles, bisexual, actinomorphic, subtended by three bracteoles. Calyx of four free sepals decussate in two cycles of two, imbricate, persistent. Petals four, fleshy, white, free, in two cycles, imbricate. Stamens 16–25 (usually 16–17), united at their base into a short androecial ring atop the ovary, free from the corolla; filaments short, basally connate into a tube; anthers elongate, 2-locular (the locules possess two horn-like appendages at apex), basifixed, introrse, opening by apical pores. Pollen grains 3-colporate, reticulate. Nectary disc forming a fimbriate ring at apex of ovary. Gynoecium of two fused carpels, they are flattened, grooved and densely covered with uniseriate, white hairs; style elongate, with very shortly 2-lobed stigma; ovary inferior, 2-locular, with numerous, ascending, anatropous (Piesschaert et al. 1997) ovules, on axile placentas. Fruits septifragal capsules crowned by persistent calyx and sparsely covered with hairs, after maturation fruits glabrous. Seeds exotestal, numerous, elongate, flat, winged, with straight, large embryo and scanty, oily endosperm. Iridoids not detected. . . 6. DIALYPETALANTHACEAE.

10 Stamens as many as corolla lobes, epipetalous. Trees and shrubs, often lianous or epiphytic, or subshrubs and perennial or annual herbs, mostly terrestrial, rarely epiphytic or aquatic (*Limnosipanea* and species of *Spermacoce*). Secretory cells or cavities of various types often occur in parenchymatous tissues. Calcium oxalate crystals are found as raphides, styloids, druses, and crystal sand, and can be stores in various plant parts (Delprete 2004). Vessels almost always with simple perforations; lateral pitting alternate, pits

vestured. Rays heterogeneous with long ends to homogeneous. Axial parenchyma apotracheal, paratracheal, or wanting. Nodes unilacunar or less often trilacunar, or more. Leaves opposite or rarely verticillate (*Henriquezia*), entire or serrate, very rarely alternate by suppression of one of a pair at each node, or dentate, very rarely sinuate-lobate or pinnatifid, petiolate to sessile, gland-dotted, or not, sometimes containing small galls with symbiotic bacterial colonies; stipules present, entire, divided, or fimbriate, connate, or sheathing at base, often caduceus and only represented by scars, usually interpetiolar or less often intrapetiolar, rarely with four stipules in *Condaminea*, or leaflike in *Galium*; often bearing colleters inside that produce mucilage and protect the meristematic tissues of the growing buds from herbivory (Delprete 2004). The mesophyll containing calcium oxalate crystals. Stomata usually paracytic. Inflorescences terminal or axillary, cymes, panicles, umbels, racemes, spikes, heads inflorescences, or rarely uniflorous. Flowers bisexual or rarely unisexual (then plants usually dioecious), mostly actinomorphic, often heterostylous, in many groups (e.g., the Gardenieae and related tribes) protandrous with specialized stylar pollen presentation mechanism similar to one in Campanulaceae, Goodeniaceae, and Asteraceae, mostly with bracts and bracteoles. Calyx 5–4-lobed, lobes open and often small or sometimes obsolete (as in *Neogaillonia*) or sometimes one or more of them enlarged and brightly colored (e.g., in *Mussaenda*), frequently with colleters inside the base of the calyx. Corolla sympetalous (very rarely with petals free to base, as in *Mastixiodendron* and *Hedstromia*), (3)4–5(8–10)-lobed, rarely 8–11-, or 12–15-lobed, with valvate, imbricate or contorted lobes, actinomorphic to bilabiate. Stamens 4–5, attached to corolla tube or at its throat, sometimes at its base; anthers sometimes slightly appendaged, basifixed or dorsifixed, introrse or extrorse, opening by longitudinal slits or by two apical pores (*Rustia*), or by one common lateral pore (in *Tresanthera*). Pollen grains 2-celled or 3-celled, of various types, mostly 3-colporate or 3(-4)-zonocolporate, in monads or in tetrads (Dessein et al. 2005). Nectary disc often present at the top of ovary. Gynoecium of two or seldom 3–5 or even more carpels; style slender, with lobed or capitate stigma, rarely stylodia free to base (as in *Galium, Hedstromia,* and *Coprosma*); ovary inferior or seldom semi-inferior (as in *Mastixiodendron*) rarely almost completely superior (*Gaertnera* and *Pagamea*), with as many locules as carpels, rarely 1-locular with parietal placentas (as in *Gardenia,* etc.), each locule with one to many ovules (commonly one ovule in Rubioideae). Ovules pendulous, horizontal or ascending, anatropous to hemitropous, very rarely circinotropous or hypertropous, often with funicular obturator, nearly always without endothelium. Endosperm nuclear or very rarely cellular. Fruits of diverse types, but mostly capsules (septicidal or loculicidal), fleshy or leathery berries, or drupes. Seeds wingless or rarely winged, exotesta alone persisting, papillate hairy or not, mesotestal cells thickened; embryo usually straight to curved, embedded in usually copious, oily, ruminate endosperm, or sometimes endosperm scanty or wanting, in certain groups without seed coat. Producing iridoids, various types of alkaloids, flavonols, kaempferol and quercetin, ursolic acid, inulin (*Cinchona*), n = 6–17 (mostly 11, less often 9). . . . . 7. RUBIACEAE.

9 Flowers unisexual. Perennial or annual herbs, sometimes succulent with raphide sacs. Xylem forming a continuous cylinder. Vessels very numerous, small, with simple perforations. Mechanical tissues lacking. Leaves opposite below, alternate above by suppression of one of each pair, simple, entire, fleshy, with large, club-shaped glands at the apex; stipules interpetiolar, united, membranous, bearing colleters near the tip. Stomata paracytic. The mesophyll containing calcium oxalate crystals (raphides). Flowers monoecious, anemophil-

ous, and evidently apetalous. Male flowers solitary or paired opposite the leaves of upper nodes, female flowers mostly in simple axillary dichasia at lower nodes. Calyx of male flowers inconspicuous, closed in bud but soon valvately splitting to the base into 2–5 broad, recurved lobes. Calyx of female flowers membranous, very oblique, tubular, shortly 2–4-dentate at the apex. Corolla of female flowers 2–4-toothed, valvate, unequal, but not bilabiate. Stamens (2)7–12(-30), sometimes basally united into groups of two, four, or six; filaments slender; anthers erect in bud, but pendulous later, versatile, 2-locular. Pollen grains 3-celled, (3)4–6(7)-zonoporate. Gynoecium pseudomonomerous; style gynobasic, slender, exerted from mouth of calyx; ovary superior, 1-locular, with one basal and more or less campylotropous or amphitropous, unitegmic (the integument massive) ovule. Fruits subglobose, nutlike drupes, pericarp with elaiosome. Seeds hippocrepiform, with strongly curved, large embryo in copious, oily endosperm. Producing iridoid compounds, n = 11. . . . . 8. THELIGONACEAE.

1 Plant with a well-developed latex-system (except in *Nerium*), and commonly producing cardiotonic glycosides. steroidal alkaloids, cardenolides, group II decarboxylated iridoids present. Trees, shrubs, herbs, very often lianas, sometimes succulents, glabrous or with various kinds of trichomes.). Stem often of anomalous structure. Vessels with scalariform or simple perforations; Leaves opposite or sometimes verticillate, rarely alternate, simple, entire or rarely lobed or dentate; stipules wanting or rarely small and interpetiolar or vestigial. Flowers in cymose or racemose inflorescences, or solitary, bisexual or very rarely functionally unisexual, actinomorphic or nearly so, commonly 5-merous. Sepals (4)5, more or less connate, imbricate or valvate. Corolla sympetalous, lobes contorted or rarely imbricate or valvate. Stamens as many as and alternate with the corolla lobes, inserted on the corolla tube, free or more often connate; filaments short; anthers free or more or less closely coherent or connate around the style head, basifixed or dorsifixed, tetrasporangiate, bisporangiate or disporangiate, introrse, opening longitudinally or apically. Pollen grains 3-celled, in monads, tetrads or pollinia, 3-colporate or porate. Nectar secreted in alternistaminal troughs on staminal tube or staminal feet or from disclike nectary around base of ovary, more rarely from sides of ovary or absent. Gynoecium of 2(-8) carpels, these connate in varying degrees; ovary superior to semiinferior, stigma mostly on underside of style head, often restricted to five chambers behind guide rails. Ovules two to many, commonly pendulous, anatropous, amphitropous, or hemitropous, very rarely orthotropous, with or without endothelium. Fruits of diverse types. Seeds small, exotestal, often with a terminal coma of long hairs, naked, winged or arillate, with straight embryo and oily, copious to scanty endosperm. n = 8–12. . . . . 11. APOCYNACEAE.

1. GELSEMIACEAE

L. Struwe et V.A. Albert 1995. 2/11. Tropical Africa, Madagascar, Southeast Asia, southern North America, and Central and northern South America.

Gelsemium, Mostuea

Probably the most archaic member of the order.

2. LOGANIACEAE

R. Brown ex C. Martius 1827. 1/15. New Caledonia (1), New Zealand (1), and Australia (13).

Logania

3. STRYCHNACEAE

Perleb 1818 (including Gardneriaceae Perleb 1838). 4/250. Eastern and Southeast Asia, East Malesia, Melanesia, Fiji.

Strychnos, Scyphostrychnos, Neuburgia, Gardneria

4. ANTONIACEAE

Hutchinson 1959. 4/8. West Malesia (*Norrisia*), tropical South America, and West and tropical Africa (*Usteria*).

Norrisia, Bonyunia, Antonia, Usteria

Connected with the Loganiaceae and Gelsemiaceae, but lacking both iridoids and alkaloids (Bisset 1975; Jensen 1992).

5. SPIGELIACEAE

C. Martius 1827. 3/96. Madagascar, southern, eastern, and Southeast Asia, New Guinea, Australia, Tasmania, New Zealand, New Caledonia, Caroline Islands, North and South America.

Spigelia, Mitreola, Mitrasacme

Spigeliaceae are a rather heterogeneous and perhaps not a natural family. Pollen grains of the Spigeliaceae are of different types (Punt 1980), and this together with diverse types of fruits and different basic numbers of chromosomes does not indicate any great coherence within the group (Bisset et al. 1980). There are affinities with Loganiaceae but even more with Rubiaceae, especially with the Hedyotideae in habit, stipules, flowers, and wood anatomy (Bisset et al. 1980). According to Mennega (1980: 119), "striking is the similarity in wood structure between Spigelieae and *Hedyotis*" Here I am accepting the family Spigeliaceae only tentatively.

6. DIALYPETALANTHACEAE

Rizzini et Occhioni 1948. 1/1. Amazonian Brazil and Peru.

Dialypetalanthus

Very closed to the Rubiaceae (Dahlgren and Thorne 1984 ["*Dialypetalanthus* is probably an aberrant early off-shoot of the Rubiaceae or a relict family closely related to the Rubiaceae", p. 690]). It is important to mention that *Dialypetalanthus* has the same type of bifid stipules and many-seeded capsules as in the Rondeletieae complex (Delprete 2004), but differs in dimerous origin of calyx and corolla, free corolla lobes, and numerous stamens.

7. RUBIACEAE

A.L. de Jussieu 1789 (including Aparinaceae Hoffmannsegg et Link 1813–1829, Asperulaceae Chamisso ex Spenner 1835, Catesbaeaceae Martynov 1820, Cephalanthaceae Rafinesque 1820, Cinchonaceae Batsch 1802, Coffeaceae Batsch 1802, Coutareaceae Martynov 1820, Cynocrambaceae Endlicher 1841, Galiaceae Lindley 1836, Gardeniaceae Dumortier 1829, Guettardaceae Batsch 1802, Hedyotidaceae Dumortier 1829, Henriqueziaceae Bremekamp 1957, Houstoniaceae Rafinesque 1840, Hydrophylacaceae Martynov 1820, Lippayaceae Meisner 1836–1843, Lygodysodeaceae Bartling 1830, Naucleaceae Wernham 1912, Nonateliaceae Martynov 1820, Operculariaceae A.L. de Jussieu ex Perleb 1818, Pagamaeaceae Martynov 1820, Psychotriaceae F. Rudolphi 1830, Randicaceae Martynov 1820, Sabiceaceae Martynov 1820). 650/13000. Cosmopolitan, but mainly in tropical and subtropical regions with some genera in temperate and cold regions (even in the Arctic and Antarctic).

7.1 RUBIOIDEAE

Small trees, shrubs, lianas, epiphytes, or more often herbs; hairs articulated. Raphides generally present. Stipules entire, frequently bifid or fimbriate (Spermacoceea). Corolla mostly valvate, rarely imbricate or contorted. Stamens inserted at middle or near corolla mouth. Heterostyly very common. Placentas pluriovulate or with a single (rarely two) erect ovules. Fruits dry (capsules, dehiscent into mericarps, opening with operculum, or indehiscent) or fleshy. Exotestal cells often parenchymalike. – OPHIORRHIZEAE: *Colletoecema, Ophiorrhiza, Neurocalyx, Xanthophytum, Lerchea, Coptophyllum, Spiradiclis*; PAURIDIANTHEAE: *Pauridiantha, Poecilocalyx, Rhipidantha, Stelechantha*; UROPHYLLEAE: *Urophyllum, Amphidasya, Commitheca, Maschalocorymus, Pravinaria, Praravinia*; LASIANTHEAE: *Lasianthus, Saldinia, Perama, Trichostachys, Metabolus*; CRUCKSHANKSIEAE: *Cruckshanksia*; COUSSAREEAE: *Coussarea, Faramea, Oreopolus, Heterophyllaea*; COCCOCYPSELEA: *Coccocypselum, Declieuxia, Hindsia*; CRATERISPERMEAE: *Craterispermum*; PSYCHOTRIEAE: *Psychotria, Amaracarpus, Cephaelis, Chasallia, Chazaliella, Geophila, Hydnophytum, Hymenocoleus, Myrmecodia, Palicourea, Rudgea, Uragoga, Anthorrhiza, Myrmephytum, Squamellaria, Readea, Streblosa, Margaretopsis*; MORINDEAE: *Morinda, Appunia, Coelospermum, Damnacanthus, Gentingia, Gynochthodes, Pogonolobus, Prismatometis, Renellia, Motleyia, Mitchella,*; SCHRADEREAE: *Schradera, Lecananthus, Leucocodon*; GAERTNEREAE: *Gaertnera, Pagamea*; DANAIDEAE: *Danais, Schismatoclada, Payera*; HEDYOTIDEAE: *Pentas, Batopedina, Carphalea, Otiophora, Chaemepentas, Parapentas, Otomeria, Knoxia, Calanda, Pentanisia, Neopentanisia, Chlorochorion, Pseudohedyotis, Triainolepis, Agathisanthemum, Amphiasma, Arcytophyllum,*

Conostomium, Pentodon, Dentella, Scleromitrion, Phylohydrax, Dibrachionostylus, Exallage, Gomphocalyx, Gouldia, Hedyotis, Hedythyrsus, Houstonia, Kohautia, Lelya, Manettia, Manostachya, Mitrasacmopsis, Neohymenopogon, Neanotis, Oldenlandia; SPERMACOCEAE: *Spermacoce, Bouvardia, Diodella, Diodia, Emmeorhiza, Ernodea, Scandentia, Galianthe, Hemidiodia, Mitracarpus, Paratriaina, Pentanopsis, Placopoda, Psyllocarpus, Tobagoa, Richardia, Crusea, Dolichometra, Hydrophylax, Lucya, Nodocapaea, Schwendera, Synaptantha, Staelia, Thecorchis, Thyridocalyx*; ANTHOSPERMEAE: *Anthospermum, Coprosma, Carpacoce, Opercularia, Galopina, Nenax, Nertera* (including *Peratanthe*), *Phyllis, Durringtonia, Eleutheranthus, Leptostigma, Normandia, Pomax*; ARGOSTEMMATEAE: *Argostemma, Cyaneuron, Mouretia, Mycetia*; PAEDERIEAE: *Paederia, Gaillonia, Saprosma, Neogaillonia, Plocama, Aitchisonia, Putoria, Serissa, Spermadictyon, Choulettia, Jaubertia, Kelloggia, Leptodermis, Pseudogaillonia, Pseudopyxis, Pterogaillonia*; RUBIEAE: *Rubia, Asperula, Crucianella, Cruciata, Galium, Phuopsis, Sherardia, Valantia, Callipeltis, Relbunium, Didymaea, Mericarpaea, Microphysa, Warburgia.*

7.2 IXOROIDEAE
Shrubs or less often trees or woody lianas. Raphides nearly always absent. Stipules mostly entire, rarely bifid. Corolla contorted. Stylar pollen presentation almost general. Stamens inserted at middle of tube or near corolla mouth. Placentas mostly pluriovulate. Fruits fleshy. Exotestal cells rarely parenchymalike, generally with thickenings along the walls. – CONDAMINEAE: *Alseis, Bathysa, Calycophyllum, Capirona, Chimarrhis, Condaminea, Dolichodelphys, Elaeagia, Emmenopterys, Ferdinandusa, Hippotis, Macrocnemum, Mastixiodendron, Parachimarrhis, Pentagonia, Picardaea, Pinckneya, Pogonopus, Rustia, Semaphyllanthe, Simira, Sommera, Tammsia, Tresanthera, Wittmackanthus*; MUSSAENDEAE: *Mussaenda, Neomussaenda, Pseudomussaenda, Schizomussaenda, Aphaenandra, Heinsia, Landiopsis*; SABICEEAE: *Sabicea, Ecpoma, Pseudosabicea, Stipularia*; VIRECTARIEAE: *Virectaria, Hekistocarpa*; SIPANEEAE: *Sipanea, Sipaneopsis, Limnosipanea*; HENRIQUEZIEAE: *Gleasonia, Henriquezia, Platycarpum, ?Monopanthera, Posoqueria*; RETINIPHYLLEAE: *Retiniphyllum*; IXOREAE: *Ixora, Captaincookia, Doricera, Myonima, Versteegia, ? Scyphiphora*; VANGUERIEAE: *Canthium, Cuviera, Fadogia, Fadogiella, Hutchinsonia, Keetia, Meyna, Multidentia, Peponidium, Perakanthus, Psydrax, Pygmaeothamnus, Pyrostria, Rytigynia, Scyphochlamya, Vangueria, Vangueriella, Vangueriopsis*; ALBERTEAE: *Alberta, Nematostylis*; COFFEEAE: *Coffea, Psilanthus* (including *Paracoffea*), *Diplospora, Discospermum, Tricalysia, Bertiera, Sericanthe*; OCTOTROPIDEAE: *Jovetia, Lamprothamnus, Octotropis, Paragenipa, Polysphaeria, Ramosmania*; CREMASPOREAE: *Cremaspora*; PAVETTEAE: *Pavetta, Rutidea, Tarrena, Robbrechtia, Coptosperma, Leptactina, Paracephaelis, Schizenterospermum, Dictyandra, Cladoceras, Nichallea, Pachystylus, Tennantia, Homollea, Homolliella, Coleactina.* GARDENIEAE: *Aidia, Aidiopsis, Alibertia, Alleizettella, Amaioua, Atractocarpus, Atractogyne, Aulacocalyx, Benkara, Borojoa, Brachytome, Brenania, Burchellia, Byrsophyllum, Calochone, Casasia, Catunaregam, Ceriscoides, Coddia, Deccania, Didymosalpinx, Dioecrescis, Duperrea, Duroia, Euclinia, Fagerlindia, Ganguelia, Gardenia, Genipa, Adenorandia, Glossostipula, Heinsenia, Himalrandia, Hyperacanthus, Kailarsenia, Kochummenia, Kutchubaea, Larsenaikia, Macrosphyra, Mantalania, Massularia, Melanopsidium, Mitriostigma, Monosalpinx, Morelia, Oligocodon, Oxyanthus, Oxyceros, Pelagodendron, Phellocalyx, Pleiocoryne, Porterandia, Preussiodora, Pseudomantalania, Randia, Riodocea, Rosenbergiodendron, Rothmannia, Rubovietnamia, Schumanniophyton, Sherbournia, Sphinctanthus, Stachyarrhena, Sukunia, Sulitia, Tamilnadia, Tarennoidea, Tocoyena, Trukia, Vidalasia, ? Argocoffeopsis.*

7.3 CINCHONOIDEAE
Trees, shrubs, woody lianas, or sometimes epiphytic shrubs, very rarely herbaceous. Raphides mostly absent (present in *Hillia* and *Hamelia*). Stipules mostly entire, rarely bifid. Corolla imbricate or valvate, rarely contorted. Stamens inserted at base of tube, or near corolla mouth. Heterostyly frequent. Stylar pollen presentation rare. Placentas with numerous ovules. Fruits mostly capsular, sometimes fleshy, or drupaceous. Seeds numerous, often winged. Exotestal cells always with thickenings along tangential walls. – CINCHONEAE: *Cinchona, Dolicholobium, Joosia, Ladenbergia, Remijia, Stilpnophyllum*; ISERTIEAE: *Isertia, Kerianthera, Raritebe, Schizostigma, Temnopteryx;* CATESBAEAE: *Asemnantha, Badusa,*

Bikkia, Catesbaea, Ceratopyxis, Chiococca, Coutaportla, Coutarea, Erithalis, Exostema, Hintonia, Isidorea, Morierina, Nernstia, Osa, Portlandia, Salzmannia, Schmidtottia, Scolosanthus, Syringantha; HYMENODICTYEAE: *Hymenodictyon, Paracorynanthe*; NAUCLEEAE: *Adina, Adinauclea, Breonadia, Breonia, Burttdavya, Cephalanthus, Corynanthe, Cubanola, Gyrostipula, Haldina, Hallea, Janotia, Ludekia, Metadina, Mitragyna, Myrmeconauclea, Nauclea, Neolamarckia, Neonauclea, Ochreinauclea, Pausinystalia, Pertusadina, Sarcocephalus, Sinoadina, Uncaria*; HAMELIEAE: *Hillia, Hamelia, Cosmibuena, Hoffmannia,* ? *Balmea, Chione*; RONDELETIEAE: *Rondeletia, Blepharidium, Roigella, Suberanthus,* ? *Acunaeanthus, Glionnetia, Habroneuron, Rogiera* (pro parte), *Stevensia*; GUETTARDEAE: *Arachnothryx, Chomelia, Gonzalagunia, Guettarda, Javorkaea, Timonius,* ? *Dichilanthe.*

Uncertain Position: *Acranthera, Airosperma, Aleisanthia, Aleisanthiopsis, Aoranthe, Aphanocarpus, Ariadne, Augusta, Calycosiphonia, Coryphothamnus, Crossopteryx, Dunnia, Greeniopsis, Jackiopsis, Kajewskiella, Lathraeocarpa, Leucolophus, Lindenia, Mitchella, Mussaendopsis, Myrioneuron, Pagameopsis, Petitiocoden, Placocarpa, Sarcosperma, Stephanococcus, Trailliaedoxa, Wendlandia.*

Very closely related to the Loganiaceae and even more to the Spigeliaceae. It is also interesting that *Gaertnera* and *Pagamea,* which are characterized by superior ovary, are sometimes transferred to the Loganiaceae. All the available evidence leads to the conclusion that the Rubiaceae share a common origin with the lower Rubiales from the hydrangealean ancestor.

8. THELIGONACEAE

Dumortier 1829. 1/3. Canary Islands, Mediterranean, Crimea, western Asia, southwestern China, Japan.

Theligonum

Very closely related to the Rubiaceae, especially to the Rubioideae (see Wunderlich 1971; Kooiman 1971; Cronquist 1981). Robbrecht (1993) and Thorne (1992b, 2006) even include *Theligonum* in Rubiaceae-Rubioideae. However, *Theligonum* is remarkable in having a large number of stamens arranged in bundles and pseudomonomerous gynoecium and 1-locular ovary without a trace of second locule (a unique feature of *Theligonum* according to Robbrecht [1993: 90]) as well as highly specialized and reduced anemophilous flowers, the absence of mechanical tissues and campylotropous ovules).

9. GENTIANACEAE

A.L. de Jussieu 1789 (including Chironiaceae Horaninow 1847, Coutoubeaceae Martynov 1820, Obolariaceae Martynov 1829, Potaliaceae C. Martius 1827, Saccifoliaceae Maguire and Pires 1978). 88/1600–1690. Cosmopolitan, but mainly in temperate and subtropical regions and also in tropical mountains.

SACCIFOLIEAE: *Curtia, Hockinia, Saccifolium, Tapeinostemon, Voyriella*; EXACEAE: *Cotylanthera, Exacum, Gentianothamnus, Ornichia, Sebaea, Tachiadenus*; CHIRONIEAE: *Bisgoeppertia, Blackstonia, Centaurium, Chironia, Cicendia, Eustoma, Exaculum, Geniostemon, Ixanthus, Orphium, Sabatia, Zygostigma*; *Canscora, Cracosna, Hoppea, Microphium, Phyllocyclus, Schinziella*; *Coutoubea, Deianira, Schultesia, Symphyllophyton, Xestaea*; HELIEAE: *Adenolisianthus, Aripuana, Calolisianthus, Celiantha, Chelonanthus, Chorisepalum, Helia, Irlbachia, Lagenanthus, Lehmanniella, Macrocarpaea, Neblinantha, Prepusa, Purdieanthus, Rogersonanthus, Senaea, Sipapoantha, Symbolanthus, Tachia, Tetrapollinia, Wurdackanthus, Zonanthus*; POTALIEAE: *Congolanthus, Djaloniella, Enicostema, Faroa, Karina, Neurotheca, Oreonesion, Pycnosphaera, Urogentias, Lisianthus, Anthocleista, Fagraea, Potalia,* GENTIANEAE: *Crawfurdia, Gentiana, Tripterospermum, Anagallidium, Bartonia, Comastoma, Frasera, Gentianella, Gentianopsis, Halenia, Jaeschkea, Lotouchea, Lomatogonium, Megacodon, Obolaria, Pterygocalyx, Swertia, Veratrilla*; VOYRIEAE: *Voyria.*

Gentianaceae are closely connected with the Gelsemiaceae and evidently derived from the gelsemiaceous stock.

10. GENIOSTOMACEAE

L. Struwe et V.A. Albert 1995. 2/75. Mascarene Islands, Malesia (except for the Malay Peninsula), Kyushu, Bonin Islands, Micronesia (Mariana Islands and Caroline Islands), New Guinea, Solomon Islands,

Australia (eastern Queensland), New Caledonia, New Hebrides, Fiji, Lord Howe Island, northern New Zealand, Polynesia (including Hawaii).

Geniostoma, Labordia

A rather isolated family that is probably related to the Gentianaceae.

11. APOCYNACEAE

A.L. de Jussieu 1789 (including Asclepiadaceae Borkhausen 1797, Cerberaceae Martynov 1820, Cryptostegiaceae Schlechter 1905, Cynanchaceae G. Meyer 1836, Ophioxylaceae Perleb 1838, Pacouriaceae Martynov 1820, Periplocaceae Schlechter 1905, Plumeriaceae Horaninow 1834, Stapeliaceae Horaninow 1834, Vincaceae Vest 1818, Willughbeiaceae J.G. Agardh 1858). 432/5100. Tropical and subtropical regions with relatively few members in temperate areas.

Classification after M.E. Endress and P.V. Bruyns (2000) and M.E. Endress and W.D. Stevens (2001).

11.1 RAUVOLFIOIDEAE

Anthers not adherent to style head by a reticulum of viscid exudates, full of pollen, rarely with spines. Fruits dehiscent or indehiscent, berries, drupes, follicles, or capsules. Seeds naked, with wings or arils, but almost never with coma at one end; holmindole alkaloids present or absent, n = 9 (Alyxieae), 10, 11, (23). – ALSTONIEAE: *Alstonia, Aspidosperma, Geissospermum, Haplophyton, Laxoplumeria, Microplumeria, Strempeliopsis, Tonduzia, Vallesia;* VINCEAE: *Amsonia, Catharanthus, Kopsia, Neisosperma, Ochrosia, Petchia, Rauvolfia, Vinca;* WULLUGHBEIEAE: *Ancylobotrys, Bousigonia, Chamaeclitandra, Clitandra, Couma, Cyclocotyla, Cylindropsis, Dictyophleba, Hancornia, Lacmellea, Landolphia, Willughbeia, Leuconotis, Orthopichonia, Paucouria, Parahancornia, Saba, Vahadenia;* TABERNAEMONTANEAE: *Ambelania, Bonafousia, Callichilia, Calocrater, Carvalhoa, Crioceras, Macoubea, Mucoa, Molongum, Neocouma, Rhigospira, Schizozygia, Spongiosperma, Stenosolen, Tabernaemontana, Stemmadenia, Tabernanthe, Voacanga, Woytkowskia;* MELODINEAE: *Craspidospermum, Diplorhynchus, Dyera, Gonioma, Kamettia, Melodinus, Pycnobotrya, Stephanostegia;* HUNTERIEAE: *Hunteria, Picralima, Pleiocarpa;* PLUMERIEAE: *Allamanda, Anechites, Cerbera, Cerberiopsis, Cameraria, Thevetia, Himatanthus, Mortoniella, Plumeria, Skytanthus;* CARISSEAE: *Carissa, Acokanthera;* ALYXIEAE: *Chilocarpus, Alyxia, Pteralyxia, Condylocarpon, Lepinia, Lepiniopsis, Plectaneia.*

11.2 APOCYNOIDEAE

Corolla left-contorted, valvate. Anthers tetrasporangiate, adherent to style head by a retinaculum of viscoid exudates, empty at base and with spines. Nectaries, if present, in ring around base of ovary. Seeds ecomose or comose, n = (6-) 10, 11 (12). – WRIGHTIEAE: *Nerium, Isonema, Pleioceras, Stephanostema, Strophanthus, Wrightia, Adenium;* MALOUETIEAE: *Mascarenhasia, Funtumia Kibatalia, Holarrhena, Malouetia, Malouetiella, Alafia, Allowoodsonia, Carruthersia, Farquharia, Pachypodium, Spirolobium;* APOCYNEAE: *Beaumontia, Aganonerion, Anodendron, Urceola, Ecdysanthera, Odontadenia, Baissea, Apocynum, Trachomitum, Aganosma, Baharuia, Chonemorpha, Epigynum, Forsteronia, Ichnocarpus, Vallaris, Vallariopsis, Trachelospermum, Cleghornia, Dewevrella, Elytropus, Epigynum, Eucorymbia, Ixodonerium, Montandra, Oncinotis, Papuechites, Parameria, Parepigynum, Sindechites, Baissea;* MESECHITEAE: *Allomarkgrafia, Galactophora, Mandevilla, Secondatia, Macrosiphonia, Mesechites, Tintinnabularia, Quiotania, Telosiphonia;* ECHITEAE: *Prestonia, Echites, Laubertia, Cycladenia, Parsonsia, Pottsia, Pentalinon, Thenardia, Angadenia, Temnadenia, Rhabdadenia, Stipecoma, Amalocalyx, Neobracea, Macropharynx, Salpinctes, Asketanthera, Fernaldia, Peltastes, Artia, Ecua, Hylaea.*

11.3 PERIPLOCOIDEAE

Corolla valvate, tube formation intermediate. Filaments free from each other. Anthers tetrasporangiate. Two pollinia (consisting of loosely coherent tetrads) in each locule, transferred by spoon-shaped translator ending in a sticky disc, n = 11 (mostly). – *Cryptostegia, Telectadium, Pentopetia, Stomatostemma, Ectadium, Maclaudia, Omphalogonus, Cryptolepis* (including *Mangenotia*), *Schlechterella, Gymnanthera, Streptocaulon, Hemidesmus, Decalepis* (including *Utleria*), *Epistemma, Finlaysonia* (including *Atherolepis, Stelmacrypton, Gongylosperma, Hanghomia, Meladerma*), *Sarcorhiza, Phyllanthera* (including *Streptomanes, Pentanura*), *Baroniella, Sacleuxia,*

Batesanthus, Baseonema, Rhaphionacme, Camptocarpus, Mondia, Myriopteron, Periploca, Tacazze (including *Zacateza*), *Petopentia, Atherandra, Ischnolepis* (*Peptopentia*), *Buckollia, Zygostelma*.

11.4 SECAMONOIDEAE
Corolla left-contorted. Filaments connate. Anthers tetrasporangiate. Two very small pollinia in apical part of each locule, transferred by translator with arm that has two clefts. – *Secamone, Secamonopsis, Toxocarpus, Goniostemma, Pervillea, Trichosandra, Rhynchostigma, Calyptranthera, Genianthus*.

11.5 ASCLEPIADOIDEAE
Filaments connate. Anthers bisporangiate, with one solid pollinium in each locule (pollen in tetrads – Fockeae), transferred by translator, orbicules present or (*Riocreuxia*) absent, granular layer of exine thin; n = 9–14. – FOCKEEAE: *Fockea*: MARSDENIEAE: *Gymnema, Dischidia, Hoya, Marsdenia, Sarcolobus, Cosmostigma, Asterostemma, Rhyssolobium, Gongronema, Cathetostemma, Pseusmagennetus, Pycnorhachis, Lygisma, Treutlera, Absolmsia, Anisopus, Telosma, Anatropanthus, Micholitzia, Campestigma, Spirella, Stigmatorhynchus, Clemensiella, Oreosparte, Heynella, Cibirhiza, Gunnessia, Madangia*; CEROPEGIEAE: *Leptadenia, Brachystelma, Ceropegia, Macropetalum, Hoodia, Piaranthus, Caralluma, Stapella, Huernia, Duvalia, Orbea, Anisotoma, Conomitra, Orthanthera, Pectinaria, Pentasachme, Riocreuxia, Sisyranthus, Heterostemma, Desmidorchis, Frerea, Tridentea, Tromotriche, Tenaris, Tavaresia, Quaqua, Echidnopsis, Edithcolea, Emplectanthus, Dittoceras, Neoschumannia, Stapeliopsis, Stapelianthus, Whitesloanea, Rhytidocaulon, Pseudolithos, Duvaliandra, Lavrania, Ophionella, Notechidnopsis*; ASCLEPIADEAE: *Tylophora, Adelostemma, Matelea, Oxystelma, Sarcostemma, Cynanchum, Vincetoxicum, Vincetoxicopsis, Calotropis, Lachnostoma, Macroscepis, Metastelma, Microloma, Gonolobus, Ditassa, Macroditassa, Pergularia, Schubertia, Funastrum, Asclepias, Gomphocarpus, Kanahia, Morrenia, Pentastelma, Merrillanthus, Biondia, Pentatropis, Seutera, Ampelamus, Calostigma, Cyathella, Cordylogyne, Glossonema, Pachyglossum, Schizostemma, Tweedia, Urostelma, Odontanthera, Pentarrhinum, Pachycarpus, Metaplexis, Pherotrichis, Solenostemma, Tassadia, Turrigera, Nematostemma, Raphistemma, Philibertia, Schizoglossum, Schistogyne, Trichosacme, Aspidoglossum, Holostemma, Graphistemma, Xysmalobium, Oxypetalum, Diplolepis, Oncinema, Stenomeria, Eustegia, Astephanus, Acrocoryne, Blyttia, Dictyanthus, Diploglossum, Fischeria, Araujia, Parapodium, Nautonia, Pycnostelma, Rhyssostelma, Orthosia, Mitostigma, Peplonia, Melinia, Hemipogon, Ibatia, Barjonia, Blepharodon, Polystemma, Nematuris, Rhyncharrhena, Nephradenia, Amblistigma, Sphaerocodon, Fanninia, Lugonia, Margaretta, Metalepis, Calathostelma, Lagoa, Glossostelma, Husnotia, Telminostelma, Cyathostelma, Lorostelma, Stelmation, Sattadia, Bustelma, Kerbera, Lagenia, Anomotassa, Dactylostelma, Hypolobus, Jobinia, Odontostelma, Petalostelma, Diplostigma, Grisebachiella, Pentabothra, Podandra, Pleurostelma, Stelmagonum, Platykeleba, Scyphostelma, Steleostemma, Stenostelma, Prosthecidiscus, Stathmostelma, Woodia, Vailia, Widgrenia, Tetraphysa, Trachycalymma, Meresaldia, Karimbolea, Schistonema, Stelmatocodon, Labidostelma, Mahawoa, Rajasia, Stuckertia, Rojasia, Miraglossum, Seshagiria, Emicocarpus, Folotsia, Gonioanthela, Amblyopetalum, Dicarpophora, Corollonema, Hickenia, Goydera, Aidomene, Aspidonepsis, Stephanotis*.

Here I follow Hallier (1905, 1912), Stebbins (1974), Thorne (1983, 1992a, b, 2000, 2006), Endress and Bruyns (2000), and Endress and Stevins (2001) in combining Apocynaceae and Asclepiadaceae into one family. They are connected by a series of transitional forms, and as Cronquist (1981: 861) states, "There is a fairly straight-line evolutionary series in floral morphology within the Apocynaceae and Asclepiadaceae collectively, from the Plumerioideae to the Apocynoideae to the Periplocoideae to the Secamoneae and thence to the other tribes of the Asclepiadoideae."

Closely related to the Strychnaceae, Gentianaceae and Geniostomaceae clade (see Struwe et al. 1995). Some genera, such as *Tabernaemontana*, are quite similar in their wood anatomy to the genus *Neuburgia* (Mennega 1980). It is interesting that the fruit morphology of *Neuburgia* also resembles that of some apocynaceous genera, such as *Cerbera, Ochrosia*, and *Rauvolfia* (Bisset et al. 1980). Formerly *Neuburgia* was even included in Apocynaceae.

Bibliography

Achille F, TJ Motley, PP Lowry II, and J Jérémie. 2006. Polyphyly in *Guettarda* L. (Rubiaceae, Guettardeae) based on *nr*DNA ITS sequence data. Ann. Missouri Bot. Gard. 93: 103–121.

Agababian VS and KT Tumanian. 1977. Contributions to the palynomorphological study of the family Gentianaceae, part 4. Biol. Zhurn. Armenia 30(8): 43–53 (in Russian).

Albers F and U Meve. 2001. A karyological survey of Asclepiadoideae, Periplocoideae, and Secamonoideae, and evolutionary considerations within Apocynaceae s.l. Ann. Missouri Bot. Gard. 88: 624–656.

Albert VA and L Struwe. 1996. Morphological cladistics of Gentianaceae. III. Phylogeny and biogeography of the saprophytic tribe Voyrieae. Am. J. Bot. 83(6): 135–136 (abstract).

Albert VA and L Struwe. 2002. Gentianaceae in context. In: L Struwe and VA Albert, eds. Gentianaceae: systematics and natural history, pp. 1–20. Cambridge University Press, Cambridge.

Anderson CL, JHE Rova, and L Andersson. 2001. Molecular phylogeny of the tribe Anthospermeae (Rubiaceae): systematic and biogeographic implications. Aust. Syst. Bot. 14: 231–244.

Andersson L. 1995. Tribes and genera of the Cinchoneae comples (Rubiaceae). Ann. Missouri Bot. Gard. 82: 409–427.

Andersson L. 1996. Circumscription of the tribe Isertieae (Rubiaceae). Opera Bot. Belg. 7: 139–164.

Andersson L and A Antonelli. 2005. Phylogeny of the tribe Cinchoneae (Rubiaceae), its position in Cinchonoideae, and description of a new genus, *Ciliosema*. Taxon 54: 17–28.

Andersson L and C Persson. 1991. Circumscription of the tribe Cinchoneae (Rubiaceae) – a cladistic approach. Plant Syst. Evol. 178: 65–94.

Andersson L and JHE Rova. 1999. The *rps*16 intron and the phylogeny of the Rubioideae. Plant Syst. Evol. 214: 161–186.

Andreasen K and B Bremer. 1996. Phylogeny of the subfamily Ixoroideae (Rubiaceae). Opera Bot. Belgica 7: 119–138.

Andreasen K and B Bremer. 1999. Phylogenetic utility of the nuclear rDNA ITS region in the subfamily Ixoroideae (Rubiaceae): comparisons with cpDNA *rbc*L sequence data. Plant Syst. Evol. 217: 119–135.

Andreasen K and B Bremer. 2000. Combined phylogenetic analysis in the Rubiaceae-Ixoroideae: morphology, nuclear and chloroplast DNA data. Am. J. Bot. 87: 1731–1748.

Andronova NN. 1977. Structure of the ovule in Rubiaceae Juss. Bot. Zhurn. 62: 1461–1469 (in Russian).

Andronova NN. 1988. Comparative embryology of Loganiaceae and Rubiaceae. Bot. Zhurn. 73: 937–951 (in Russian with English summary).

Backlund M, B Oxelman, and B Bremer. 2000. Phylogenetic relationships within the Gentianales based on *ndh*F and *rbc*L sequences, with particular reference to the Loganiaceae. Am. J. Bot. 87: 1029–1043.

Behnke H-D. 1975. Elektronenmikroskopische Untersuchungen zur Frage der verwandtschaftlichen Beziehungen zwischen *Theligonum* und Rubiaceae: Feinbau der Siebelement-Plastiden und Anmerkungenzur Strucktur der Pollenexine. Plant Syst. Evol. 123: 317–326.

Bendre AM. 1973. Studies in the family Loganiaceae. I. Trichomes. J. Indian Bot. Soc. 52: 225–234.

Bendre AM. 1975. Studies in the family Loganiaceae: II. Embryology of *Buddleia* and *Strychnos*. J. Indian Bot. Soc. 54: 272–279.

Bisset NG. 1958, 1961. The occurrence of alkaloids in the Apocynaceae. Ann. Bogor. 3: 105–236, 1958; 4: 65–144, 1961.

Bisset NG. 1975. Chemical structures and biosynthesis of Loganiaceae alkaloids. Pharm. Weekbl. 110: 425–441.

Bisset NG. 1980a. Alkaloids of the Loganiaceae. In: JD Phillipson and MH Zenk, eds. Indole and biogenetically related alkaloids, pp. 27–61. Academic, London.

Bisset NG. 1980b. Phytochemistry of the Loganiaceae. In: A Engler and K Prantl, eds. Die natürlichen Pflanzenfamilien, 28b: 211–237.

Bisset NG and AK Choudhury. 1974. Alkaloids and iridoids from *Strychnos nux-vomica* fruits. Phytochemistry 13: 265–269.

Bisset NG, Th WJ Gadella, AJM Leeuwenberg, AMW Mennega, and W Punt. 1980. General discussion of relationships between taxa inside and with taxa outside the family Loganiaceae. In: A. Engler and K. Prantl, eds. Die natürlichen Pflanzenfamilien, 28b: 3–7.

Boppré M. 2005. Pyrrolizidine alkaloids in Apocynaceae. In: XVII International Botanical Congress, Vienna, Abstracts, p. 398.

Bouman F and S Schier. 1979. Ovule ontogeny and seed coat development in *Gentiana,* with a discussion on the evolutionary origin of the single integument. Acta Bot. Neerl. 28: 467–478.

Bouman F, L Cobb, N Devente, V Goethals, PJM Maas, and E Smets. 2002. The seeds of Gentianaceae. In: L Struwe and VA Albert, eds. Gentianaceae: systematics and natural history, pp. 498–572. Cambridge University Press, Cambridge.

Bremer B. 1992. Phylogeny of the Rubiaceae (Chiococceae) based on molecular and morphological data: useful approaches for classification and comparative ecology. Ann. Missouri Bot. Gard. 79: 380–387.

Bremer B. 1996a. Combined and separate analyses of morphological and molecular data in the plant family Rubiaceae. Cladistics 12: 21–40.

Bremer B. 1996b. Phylogenetic studies within Rubiaceae and relationships to other families based on molecular data. In: E Robbrecht, C Puff and E Smets, eds. Second International Rubiaceae Conference Proceedings, pp. 33–50. Opera Bot. Belgica 7.

Bremer B. 1997. Combined and separate analyses of morphological and molecular data in the plant family Rubiaceae. Cladistics 12: 21–40.

Bremer B and O Eriksson. 1992. Evolution of fruit characters and dispersal modes in the tropical family Rubiaceae. Biol. J. Linn. Soc. 47: 79–95.

Bremer B and RK Jansen. 1991. Comparative restriction site mapping of chloroplast DNA implies new phylogenetic relationships within Rubiaceae. Am. J. Bot. 78: 198–213.

Bremer B and J-F Manen. 2000. Phylogeny and classification of the subfamily Rubioideae (Rubiaceae). Plant Syst. Evol. 225: 43–72.

Bremer B and L Struwe. 1992. Phylogeny of the Rubiaceae and the Loganiaceae: congruence or conflict between morphological and molecular data? Am. J. Bot. 79: 1171–1184.

Bremer B and M Thulin. 1998. Collapse of Isertieae, reestablishment of Mussaendeae, and a new genus of Sabiceeae (Rubiaceae): phylogenetic relationships based on *rbc*L data. Plant Syst. Evol. 211: 71–92.

Bremer B, RG Olmstead, L Struwe, and JA Sweere. 1994. *rbcL* sequences support exclusion of *Retzia*, *Desfontainia*, and *Nicodemia* from the Gentianales. Plant Syst. Evol. 190: 213–230.

Bremer B, K Andreasen, and D Olsson. 1995. Subfamilial and tribal relationships in the Rubiaceae based on *rbc*L sequence data. Ann. Missouri Bot. Gard. 82: 383–397.

Bremer B, RK Jansen, B Oxelman, M Backlund, H Lantz, and K-J Kim. 1999. More characters or more taxa for a robust phylogeny – a case study from the coffee family (Rubiaceae). Syst. Biol. 48: 413–435.

Bruyns PV. 2000. Phylogeny and biogeography of the Stapeliads. Plant Syst. Evol. 221: 199–226.

Carlquist S. 1984. Wood anatomy of some Gentianaceae: systematic and ecological conclusions. Aliso 10: 573–582.

Carlquist S and JR Grant. 2005. Wood anatomy of Gentianaceae, tribe Helieae, in relation to ecology, habit, systematics, and sample diameter. Brittonia 57: 276–291.

Civeyrel L and N Rowe. 2001. Phylogenetic relationships of Secamonoideae based on the plastid gene *mat*K morphology, and biomechanics. Ann. Missouri Bot. Gard. 88: 583–602.

Civeyrel L, A Le Thomas, K Ferguson, and MW Chase. 1998. Critical reexamination of palynological characters used to delimit Asclepiadaceae in comparison to the molecular phylogeny obtained from plastid *matK* sequences. Molec. Phylogenet. Evol. 9: 517–527.

Classen Bockhoff R. 1996. A survey of flower-like inflorescences in the Rubiaceae. Opera Bot. Belgica 7: 329–367.

Cockrell RA. 1941. A comparative study of the wood structure of several South American species of *Strychnos*. Am. J. Bot. 28: 32–41.

Conn BJ. 1995. Description of inflorescence axes in the genus *Logania* R. Br. (Loganiaceae). Kew Bull. 50: 777–783.

Coulaud J. 1989. Comparaison de quelques Loganiacées á k'aide des caractéres quantitatifs et qualitatifs de l'anatomie du bois. Canad. J. Bot. 67: 872–878.

Daniels M and SD Sabris. 1990. The chemical phylogeny of the order Gentianales. In: KS Bilgrami and VV Dogra, eds. Phytochemistry and plant taxonomy, pp. 151–156. CBS Publishers and Distributors, Delhi.

Darwin SR 1976. The subfamilial, tribal, and subtribal nomenclature of the Rubiaceae. Taxon 27: 159–185.

De Block P. 1997. Biosystematic studies in the tribe Pavetteae (Rubiaceae – Ixoroideae). Universitaire Instelling Antwerpen.

De Block P and E Robbrecht. 1998. Pollen morphology of the Pavetteae (Rubiaceae, Ixoroideae) and its taxonomic significance. Grana 37: 260–275.

De Laet J and E Smets. 1996. A commentary on the circumscription and evolution of the order Gentianales, with special emphasis on the position of the Rubiaceae. Opera Bot. Belg. 7: 11–18.

Delprete PG. 1996. Evaluation of the tribes Chiococceae, Condamineeae and Catesbaeeae (Rubiaceae) based on morphological characters. Opera Bot. Belg. 7: 165–192.

Delprete PG. 2004. Rubiaceae. In: N Smith, SA Mori, A Henderson, DW Stevenson, and SV Heald, eds. Flowering plants of the neotropics, pp. 328–333. Princeton University Press, Princeton.

Demarco D, LS Kinoshita, and M de Casrom. 2006. Laticíferos articulados anastomosados – novos registros para Apocynaceae. Rev. Brasil. Bot. 29: 133–144.

Dessein S, H Ochoterena, P de Block, F Lens, E Robbrecht, P Schols, E Smets, S Vinckier, and S Huysmans. 2005. Palynological characters and their phylogenetic signal in Rubiaceae. Bot. Rev. 71: 354–414.

Devi HM. 1962. Embryological studies in the Gentianaceae (Gentianoideae and Menyanthoideae). Proc. Indian Acad. Sci. 56B: 195–216.

Devi HM. 1964. Embryological studies in Asclepiadaceae. Proc. Indian Acad. Sci. 60B: 54–65.

Devi HM. 1971. Embryology of Apocynaceae: I. Plumerieae. J. Indian Bot. Soc. 50: 74–85.

Devi HM. 1974. Embryology of Apocynaceae: II. Arduineae (*Carissa spinarum* Linn.). Plant Sci. 6: 24–29.

Devi HM and K Lakshminarayana. 1977. Embryological studies in Gentianaceae. J. Indian Bot. Soc. 56: 182–188.

Devi HM and K Lakshminarayana. 1980. Embryological studies in Loganiaceae. J. Indian Bot. Soc. 59: 370–375.

D'Hondt C, P Schols, S Huysmans, and E Smets. 2004. Systematic relevance of pollen and orbicule characters in the tribe Hillieae (Rubiaceae). Bot. J. Linn. Soc. 146: 303–321.

El-Gazzar A and MK Hamza. 1980. The subdivision of Asclepiadaceae. Phytologia 45: 1–16.

Endress ME. 1997. Key characters at the interface of the apocynacs and the asclepiads. Am. J. Bot. 84(6, Abstracts): 189.

Endress ME. 2004. Apocynaceae: brown and now. Telopea 10: 525–541.

Endress ME and VA Albert. 1995. A morphological cladistic study of Apocynaceae: trends in character evolution within a broadened familial circumscription. Am. J. Bot. 82(6, Abstracts): 127.

Endress ME and PV Bruyns. 2000. A revised classification of the Apocynaceae s.l. Bot. Rev. 66: 1–56.

Endress ME and WD Stevens. 2001. The renaissance of the Apocynaceae s.l.: recent advances in systematics, phylogeny, and evolution: introduction. Ann. Missouri Bot. Gard. 88: 517–522.

Endress ME, M Hesse, S Nilsson, A Guggisberg, and J-P Zhu. 1990. The systematic position of the Holarrheninae (Apocynaceae). Plant Syst. Evol. 171: 157–185.

Endress ME, B Sennblad, S Nilsson, L Civeyrel, MW Chase, S Huysmans, E Grafström, and B Bremer. 1996. A phylogenetic analysis of Apocynaceae s.str. and some related taxa in Gentianales. A multidisciplinary approach. Opera Bot. Belg. 7: 59–102.

Erbar C and P Leins. 1996. The formation of corolla tubes in Rubiaceae and presumably related families. Opera Bot. Belg. 7: 103–112.

Erbar C and P Leins. 1999. Secondary pollen presentation and a curious rupture of the style in *Spigelia* (Spigeliaceae, Gentianales). Plant Biol. 1: 389–402.

Fallen ME. 1986. Floral structure in the Apocynaceae: Morphological, functional, and evolutionary aspects. Bot. Jahrb. Syst. 106: 245–286.

Fay MF, B Bremer, GT Prance, M van der Bank, D Bridson, and MW Chase. 2000. Plastid *rbc*L sequence data show *Dialypetalanthus* to be a member of Rubiaceae. Kew Bull. 55: 853–864.

Fedotova TA. 1999. Structural, organization of seed cover in the Rubiaceae s. l.: diversity, general patterns, taxonomical implication, specialization. Bot. Zhurn. 84(11): 17–35 (in Russian with English summary).

Fishbein M. 2001. Evolutionary innovation and diversification in the flowers of Asclepiadaceae. Ann. Missouri Bot. Gard. 88: 603–623.

Fosberg FR and M-H Sachet. 1974. A new variety of *Fagraea berteriana* (Gentianaceae). Phytologia 28: 470–472.

Fosberg FR and M-H Sachet. 1980. Systematic studies in Micronesian plants. Smithsonian Contr. Bot. 45: 1–40.

Ge XJ. 1996. Seed coat micromorphology of Gentianaceae in Xinjiang. Bull. Bot. Res. 16: 455–458.

Gilmartin AJ. 1980. Variations within populations and classification, II. Patterns of variation within Asclepiadaceae and Umbelliferae. Taxon 29: 199–212.

Gopal Krishna G and V Puri. 1962. Morphology of the flower of some Gentianaceae with special reference to placentation. Bot. Gaz. 124: 42–57.

Hagen KB von and JW Kadereit. 2002. Phylogeny and flower evolution of the Swertiinae (Gentianaceae-Gentianeae): homoplasy and the principle of variable proportions. Syst. Bot. 27: 548–572.

Hakki MI. 1998. On the floral morphology and embryology of *Usteria guineensis* Willd. (Loganiaceae). Bot. Jahrb. Syst. 120: 275–293.

Ham van der R, YM Zimmermann, S Nilsson, and A Igersheim. 2001. Pollen morphology and phylogeny of the Alyxieae (Apocynaceae). Grana 40: 169–191.

Hayden MV and JD Dwyer. 1969. Seed morphology in the tribe Morindeae (Rubiaceae). Bull. Torrey Bot. Club 96: 704–710.

Henrickson J. 1996. Notes on *Spigelia* (Loganiaceae). SIDA 17: 89–103.

Hoc PS and LD Bravo. 1984. Estudio palinologico sobre las especies presentes en Argentina de *Spigelia, Strychnos,* y *Desfontainia* (Loganiaceae). Kurtziana 17: 71–89.

Houghton PJ and LL Ming. 1985. Iridoids from *Desfontainia spinosa*. Phytochemistry 24: 1841–1842.

Huysmans S, G El-Ghazaly, S Nilsson, and E Smets. 1997. Systematic value of tapetal orbicules: a preliminary survey of the Cinchonioideae (Rubiaceae). Canad. J. Bot. 75: 815–826.

Igersheim A and E Robbrecht. 1993. The character status and relationships of the Prismatomerideae (Rubiaceae-Rubioideae). Opera Bot. Belg. 6: 61–80.

Igersheim A and U Rohrhofer. 1993. The tribal position of *Otiophora* (Rubiaceae): new evidence from gynoecium structure and development. South Afr. J. Bot. 59: 431–441.

Igersheim A, C Puff, P Leint, and C Erbar. 1994. Gynoecial development of *Gaertnera* Lam. and of presumably allied taxa of the Psychotrieae (Rubiaceae): secondarily "superior" vs. inferior ovaries. Bot. Jahrb. Syst. 116: 401–414.

Inouye H, Y Takeda, H Nishimura, A Kanomi, T Okuda, and C Puff. 1988. Chemotaxonomic studies of rubiaceous plants containing iridoid glycosides. Phytochemistry 27: 2591–2598.

Jansen S and E Smets. 1998. Vestured pits in some woody Gentianaceae. IAWA J. 19: 35–42.

Jansen S and E Smets. 2000. Morphology, distribution, and systematics of vestures in the Gentianales. In: E Nordenstam, G El-Ghazaly and M Kassas, eds. Plant systematics for the 21st century, pp. 227–296. Portland Press, London.

Jansen S, E Robbrecht, H Beeckman, and E Smets. 1996. *Gaertnera* and *Pagamea*: genera within the Psychotrieae or constituting the tribe Gaertnereae? A wood anatomical and palynological approach. Bot. Acta 109: 466–476.

Jansen S, P De Block, and E Smets. 1997. Wood anatomy of the Pavetteae (Rubiaceae-Ixoroideae). Scripta Bot. Belg. 15: 86.

Jansen S, P De Block, H Beeckman, E Smets. 1999. Systematic wood anatomy of the Pavetteae (Rubiaceae, Ixoroideae). Syst. Geogr. Plant 68(1–2): 113–133.

Jansen S, S Dessein, F Piesschaert, E Robbrecht, and E Smets. 2000. Aluminium accumulation in leaves of Rubiaceae: systematic and phylogenetic implications. Ann. Bot. 85: 91–101.

Jansen S, F Lens, S Ntore, F Piesschaert, E Robbrecht, and E Smets. 2001. Contribution to the wood anatomy of the Rubioideae (Rubiaceae). J. Plant Res. 114: 269–289.

Jansen S, E Robbrecht, H Beeckman, and E Smets. 2002a. Aluminium accumulation in Rubiaceae: an additional character for the delimitation of subfamily Rubioideae? IAWA J. 21: 197–212.

Jansen S, E Robbrecht, H Beeckman, E Smets. 2002b. A survey of the systematic wood anatomy of the Rubiaceae. IAWA J. 23: 1–67.

Jansen S, T Watanabe, S Dessein, E Smets, and E Robbrecht. 2003. A comparative study of metal levels in some Al-accumulating Rubiaceae. Ann. Bot. N. S. 91: 657–663.

Jansen SR. 1992. Systematic implications of the distribution of iridoids and other compounds in the Loganiaceae and other families of the Asteridae. Ann. Missouri Bot. Gard. 79: 284–302.

Jansen SR and J Schripsema. 2002. Chemotaxonomy and pharmacology of Gentianaceae. In: L Struwe and VA Albert, eds. Gentianaceae: systematics and natural history, pp. 573–631. Cambridge University Press, Cambridge.

Kern JH and CGGJ van Steenis. 1951. Caprifoliaceae. In: CGGJ van Steenis, ed. Flora malesiana, 1st ser., 4(3): 176. Leiden.

Kiehn M. 1986. Karyosystematic studies on Rubiaceae: Chromosome counts from Sri Lanka. Plant Syst. Evol. 154: 213–223.

Kiehn M. 1995. Chromosome survey of the Rubiaceae. Ann. Missouri Bot. Gard. 82: 398–408.

Kisakurek MV and M Hesse. 1980. Chemotaxonomic studies of the Apocynaceae, Loganiaceae, and Rubiaceae with reference to indole alkaloids. In: JD Phillipson and MH Zenk, eds. Indole and biogenetically related alkaloids, pp. 11–26. Academic, London.

Koek-Noorman J. 1969a. A contribution to the wood anatomy of South American (chiefly Suriname) Rubiaceae, part 1. Acta Bot. Neerl. 18: 108–123.

Koek-Noorman J. 1969b. A contribution to the wood anatomy of South American (chiefly Suriname) Rubiaceae, part 2. Acta Bot. Neerl. 18: 377–395.

Koek-Noorman J. 1970. A contribution to the wood anatomy of the Cinchoneae, Coptosapelteae, and Naucleeae (Rubiaceae). Acta Bot. Neerl. 19: 154–164.

Koek-Noorman J. 1972. The wood anatomy of Gardenieae, Ixoreae, and Mussaendeae (Rubiaceae). Acta Bot. Neerl. 21: 301–320.

Koek-Noorman J. 1977. Systematische Holzanatomie einiger Rubiaceae. Ber. Deutsch. Bot. Ges. 90: 183–190.

Koek-Noorman J and P Hogeweg. 1974. The wood anatomy of Vanguerieae, Cinchoneae, Condamineae, and Rondeletieae (Rubiaceae). Acta Bot. Neerl. 23: 627–653.

Koek-Noorman J and C Puff. 1983. The wood anatomy of Rubiaceae tribes Anthospermeae and Paederieae. Plant Syst. Evol. 147: 17–45.

Kooiman P. 1971. Ein phytochemischer Beitrag zur Lösung des Verwndtschaftsproblems der Theligonaceae. Oesterr. Bot. Z. 119: 395–398.

Kunze H. 1990. Morphology and evolution of the corona in Asclepiadaceae and related families. Trop. Subtrop. Pflanzenwelt 76: 1–51.

Kunze H. 1991. Structure and function in asclepiad pollination. Plant Syst. Evol. 176: 227–253.

Kunze H. 1993. Evolution of the translator in Periplocaceae and Asclepiadaceae. Plant Syst. Evol. 185: 99–122.

Kunze H. 1995. Floral morphology of some Gonolobeae (Asclepiadaceae). Bot. Jahrb. Syst. 117: 211–238.

Kunze H. 1996. Morphology of the stamen in the Asclepiadaceae and its systematic relevance. Bot. Jahrb. Syst. 118: 547–579.

Kunze H. 2006. Morphology and evolution of the corolla and corona in the Apocynaceae s.l. Bot. Jahrb. Syst. 126: 347–393.

Lakshminarayana K and HM Devi. 1985. Embryological studies in Gentianaceae. Proc. Indian Acad. Sci. 95B (2): 213–219.

Lantz H and B Bremer. 2004. Phylogeny inferred from morphology and DNA data: characterizing well-supported groups in Vanguerieae (Rubiaceae). Bot. J. Linn. Soc. 146: 257–283.

Lantz H and B Bremer. 2005. Phylogeny of the complex Vanguerieae (Rubiaceae) genera *Fadogia, Rytigynia*, and *Vangueria* with close relatives and a new circumscription of *Vangueria*. Plant Syst. Evol. 253: 159–183.

Lantz H, K Andreasen, and B Bremer. 2002. Nuclear rDNA ITS sequence data used to construct the first phylogeny of Vanguerieae (Rubiaceae). Plant Syst. Evol. 230: 173–187.

Lee YS and DE Fairbrothers. 1978. Serological approach to the systematics of the Rubiaceae and related families. Taxon 27: 159–185.

Leenhouts PW. 1962. Loganiaceae. In: CGGJ van Steenis, ed. Flora malesiana, ser. 1, 6: 293–387. Groningen.

Leeuwenberg AJM. 1983. Some remarks on the taxonomy of the Plumerioideae (Apocynaceae). Bothalia 14: 799–801.

Leeuwenberg AJM and PW Leenhouts. 1980. Taxonomy of the Loganiaceae. In: A Engler and K Prantl, eds. Die natürlichen Pflanzenfamilien, 28b: 8–92. Berlin.

Lens F, S Jansen, S Huysmans, E Robbrecht, and E Smets. 2000. Pollen morphological variation in Vanguerieae (Ixoroideae: Rubiaceae). Grana 39: 1–13.

Lersten NR. 1975. Colleter types in Rubiaceae, especially in relation to the bacterial leaf nodule symbiosis. Bot. J. Linn. Soc. 71: 311–319.

Liede S. 1997. Subtribes and genera of the tribe Asclepiadeae (Apocynaceae – Asclepioideae) – a synopsis. Taxon 46: 233–247.

Liede S and F Albers. 1994. Tribal disposition of genera in the Asclepiadaceae. Taxon 43: 201–231.

Liede S and H Kunze. 2002. *Cynanchum* and Cynanchinae (Apocynaceae – Aclepiadoideae): a molecular, anatomical and latex triterpenoid study. Org. Divers. Evol. 2: 239–269.

Liede S, and A Täuber. 2000. *Sarcostemma* R. Br. (Apocynaceae – Asclepiadoideae) – a controversial generic circumscription reconsidered: Evidence from *trn*L-F spacers. Plant Syst. Evol. 225: 133–140.

Liede S, A Täuber, and J Schneidt. 2002. Molecular considerations in the Tylophorinae K. Schum. (Apocynaceae – Aclepiadoideae). Edinb. J. Bot. 59: 377–403.

Liede-Schumann S, A Rapini, DJ Goyder, and MW Chase. 2005. Phylogenetics of the New World subtribes of Asclepiadeae (Apocynaceae – Asclepiadoideae): Metastelmatinae, Oxypetalinae, and Gonolobinae. Syst. Bot. 30: 184–195.

Lindsey AA. 1940. Floral anatomy in the Gentianaceae. Am. J. Bot. 27: 640–652.

Liu JQ, ZD Chen, AM Lu. 2001. A preliminary analysis of the phylogeny of the Swertiinae (Gentianaceae) based on ITS data. Israel J. Plant Sci. 49: 301–308.

Lorence DH. 1990. A phylogenetic list of the genera of Rubiaceae in Mexico. Acta Bot. Mex. 12: 1–7.

Lorence DH. 1999. A nomenclator of Mexican and Central American Rubiaceae. Missouri Botanical Garden Press, St. Louis, MO.

Maguire B and JM Pires. 1978. Saccifoliaceae – a new monotypic family of the Gentianales. Mem. New York Bot. Gard. 29: 230–245.

Majumdar S and P Pal. 1958. The stipules of the Rubiaceae: a review. Trans. Bose Res. Inst. Calcutta 22: 57–68.

Maldonado de Magnano S. 1989. Embryological studies in *Spigelia humboldtiana* (Loganiaceae). Phytomorphology 39: 299–309.

Manen J-F, A Natali, and F Ehrendorfer. 1994. Phylogeny of Rubiaceae-Rubieae inferred from the sequence of a cpDNA intergene regions. Plant Syst. Evol. 190: 195–211.

Mathew PM and Philip Omana. 1986. The distribution and systematic significance of pollen nuclear number in the Rubiaceae. Cytologia 51: 117–124.

Mennega AMW. 1980. Anatomy of the secondary xylem of the Loganiaceae. In: A Engler and K Prantl, eds. Die natürlichen Pflanzenfamilien, 28b: 112–161. Berlin.

Mészáros S, J De Laet, and E Smets. 1996. Phylogeny of temperate Gentianaceae: a morphological approach. Syst. Bot. 21: 153–168.

Mészéras S, J De Laet, V Goethals, E Smets, and S Nilsson. 2002. Cladistics of Gentianaceae: a morphological approach. In: L Struwe and VA Albert, eds. Gentianaceae: systematics and natural history, pp. 310–376. Cambridge University Press, Cambridge.

Meve U and S Liede. 2002. A molecular phylogeny and generic rearrangement of the stapelioid Ceropegieae (Apocynaceae – Asclepidoideae). Plant Syst. Evol. 234: 171–209.

Mohrbutter C. 1936. Embryologische Studien an Loganiaceen. Planta 26: 64–80.

Moore RJ. 1947. Cytotaxonomic studies in the Loganiaceae: I. Chromosome numbers and phylogeny in the Loganiaceae. Am. J. Bot. 34: 527–538.

Morales F, MG Hahn, MR Braga. 2001. Comparative analysis of leaf cell-wall polysaccharides of *Dialypetalanthus fuscescens* and *Bathysa meridionalis:* evidence of biochemical similarities between Dialypetalanthaceae and Rubiaceae-Cinchonoideae. Rev. Brasil. Bot. 24: 289–294.

Motley TJ, KJ Wurdack, and PG Delprete. 2005. Molecular systematics of the Catesbaeeae-Chiococceae complex (Rubiaceae): flower and fruit evolution and biogeographic implications. Am. J. Bot. 92: 316–329.

Mulay BN, BD Deshpande, and U Tolani. 1965. Studies in Asclepiadaceae: II. Floral morphology and gametogenesis in certain members of the Asclepiadaceae. J. Indian Bot. Soc. 44: 95–104.

Natali A, J-F Manen, and F Ehrendorfer. 1995. Phylogeny of the Rubiaceae-Rubioideae, in particular the tribe Rubieae: evidence from a non-coding chloroplast DNA sequence. Ann. Missouri Bot. Gard. 82: 428–439.

Natali A, JF Manen, M Kiehn, and F Ehrendorfer. 1996. Tribal, generic and specific relationships in the Rubioideae-Rubieae (Rubiaceae) based on sequence data of a cpDNA intergene region. Opera Bot. Belg. 7: 193–203.

Nicholas A and H Baijnath. 1994. A consensus classification for the order Gentianales and additional details on the suborder Apocynineae. Bot. Rev. 60: 440–482.

Nilsson S. 2002. Gentianaceae: a review of palynology. In: L Struwe and VA Albert, eds. Gentianaceae: systematics and natural history, pp. 377–497. Cambridge University Press, Cambridge.

Nilsson S and JJ Skvarla. 1969. Pollen morphology of saprophytic taxa in the Gentianaceae. Ann. Missouri Bot. Gard. 56: 420–438.

Nilsson S, ME Endress, and E Grafström. 1993. On the relationship of the Apocynaceae and Periplocaceae. Grana. Suppl. 2: 3–20.

Nilsson S, M Hellbom, and W Smolenski. 2002. A reappraisal of the significance of pollen in classifications of the Gentianaceae. Grana 41: 90–106.

Pares Y and L Ruat. 1953. Observations sur le trichome des Rubiacees et des Loganiacees. Rec. Trav. Laborat. Bot. Geol. Zool. Montpellier, ser. Bot., 6: 127–133.

Patel RC, JA Inamdar, and NV Rao. 1981. Structure and ontogeny of stomata in some Gentianaceae and Menyanthaceae complex. Feddes Repert. 92: 535–550.

Persson C. 1996. Phylogeny of the Gardenieae (Rubiaceae). Bot. J. Linn. Soc. 121: 91–109.

Pichon M. 1948, 1950. Classification des Apocynacees: I. Carissees et Ambelaniees. IX. Rauvolfieae, Alstoniees, Allamandees, et Tabernemontanodees. XXV. Echitoidees. Mem. Mus. Natl. Hist. Nat., n.s., 24: 111–181, 1948; 27: 153–251, 1948; ser. B, 1: 1–174, 1950.

Piesschaert F, E Robbrecht, and E Smets. 1997. *Dialypetalanthus fuscescens* Kuhlm. (Dialypetalanthaceae): the problematic taxonomic position of an Amazonian endemic. Ann. Missouri Bot. Gard. 84: 201–223.

Piesschaert F, E Robbrecht, and E Smets. 2001. Patterns in pyrenes: the systematic significance of pyrene morphology in *Chassalia* (Rubiaceae-Psychotrieae) and related genera. Flora (Germany) 196(2): 121–131.

Post DM. 1958. Studies in Gentianaceae: I. Nodal anatomy of *Frasera* and *Swertia perennis*. Bot. Gaz. 120: 1–14.

Potgieter K and VA Albert. 2001. Phylogenetic relationships within Apocynaceae s.l. based on *trn*L intron and *trn*L-F spacer sequences and propagule characters. Ann. Missouri Bot. Gard. 88: 523–549.

Puff C. 1982. The delimitation of the tribe Anthospermeae and its affinities to the Paederieae (Rubiaceae). Biol. J. Linn. Soc. 84: 355–377.

Puff C. 1986. A biosystematic study of the African and Madagascan Rubiaceae-Anthospermeae. Plant Syst. Evol. Suppl. 3: 1–535.

Puff C. 2001. The significance of gynoecium and fruit and seed characters for the classification of the Rubiaceae. Malayan Nat. J. 55(1–2): 133–146.

Puff C and E Robbrecht. 1989. A survey of the Knoxieae (Rubiaceae-Antirheoideae). Bot. Jahrb. Syst. 110: 511–558.

Puff C, L Andersson, U Rohrhofer, and A Igersheim. 1993. The tribe Schradereae (Rubiaceae) reexamined. Bot. Jahrb. Syst. 114: 449–479.

Puff C, E Robbrecht, R Buchner, and P De Block. 1996. A survey of secondary pollen presentation in the Rubiaceae. Opera Bot. Belg. 7: 369–402.

Punt W. 1978. Evolutionary trends in the Potalieae (Loganiaceae). Rev. Palaeobot. Palynol. 26: 313–335.

Punt W. 1980. Pollen morphology of the Loganiaceae. In: A Engler and K Prantl, eds. Die natürlichen Pflanzenfamilien, 28b: 162–191. Berlin.

Punt W and PW Leenhouts. 1967. Pollen morphology and taxonomy in the Loganiaceae. Grana Palynol. 7: 469–516.

Rao KS and CC Chinnappa. 1983. Studies in Gentianaceae: microsporangium and Pollen. Canad. J. Bot. 61: 324–336.

Rao KS and M Nagaraj. 1982. Studies in Gentianaceae: embryology of *Swertia minor*. Canad. J. Bot. 60: 141–151.

Rao VS and A Ganguli. 1963. The floral anatomy of some Asclepiadaceae. Proc. Indian Acad. Sci. 57: 15–44.

Rapini A, MW Chase, DJ Goyder, and J Griffiths. 2003. Asclepiadeae classification: evaluating the phylogenetic relationships of New World Asclepiadoideae (Apocynaceae). Taxon 52: 33–50.

Razafimandimbison SG and B Bremer. 2001. Tribal delimitation of Naucleeae (Rubiaceae): inference from molecular and morphological data. Syst. Geogr. Plant 71: 515–538.

Razafimandimbison SG and B Bremer. 2002. Phylogeny and classification of Naucleeae (Rubiaceae) inferred from molecular (ITS, *rbc*L, and *trn*T-F) and morphological data. Ams. J. Bot. 89: 1027–1041.

Razafimandimbison SG and B Bremer. 2006. Taxonomic revision of the tribe Hymenodictyeae (Rubiaceae, Cinchonoideae). Bot. J. Linn. Soc. 152: 331–386.

Robbrecht E. 1982. Pollen morphology of the tribes Anthospermeae and Paederieae (Rubiaceae) in relation to taxonomy. Bull. Jard. Bot. Nat. Belg. 52: 349–366.

Robbrecht E. 1988. Tropical woody Rubiaceae. Opera Bot. Belg. 1: 1–271.

Robbrecht E. 1993a. Seventy years of systematics of tropical Rubiaceae at the National Botanic Garden. Bull. Jard. Bot. Nat. Belg. 62: 7–34.

Robbrecht E. 1993b. Advances Rubiaceae macrosystematics. Introduction. Opera Bot. Belg. 6: 7–18.

Robbrecht E. 1993c. Advances in Rubiaceae macrosystematics. In the delimitation of the Rubiaceae. A review. Opera Bot. Belg. 6: 19–30.

Robbrecht E and JF Manen. 2006. The major evolutionary lineages of the coffee family (Rubiaceae, angiosperms). Combined analysis (*n*DNA and *cp*DNA) to infer the position of *Coptosapelta* and *Luculia*, and supertree construction based on *rbc*L, *rps*16, *trn*L-*trn*F and *atp*B-*rbc*L data. A new classification in two subfamilies, Cinchonioideae and Rubioideae. Syst. Geogr. Plant 76: 85–146.

Robbrecht E. C Puff, and E Smets, eds. 1996. Second International Rubiaceae Conference Proceedings. Opera Bot. Belg. 7: 1–432.

Rodriguez P. 1976. Study of fruits and seeds of 11 tribes of Rubiaceae. Acta Bot. Venez. 11: 283–384.

Rogers GK. 1986. The genera of the Loganiaceae in the southeastern United States. J. Arnold Arbor. 67: 143–185.

Rogers GK. 2005. The genera of Rubiaceae in the southeastern United States, part II. Subfamily Rubioideae, and subfamily Cinchonoideae revisited (*Chiococca*, *Erithalis*, and *Guettarda*). Harvard Pap. Bot. 10: 1–45.

Rosatti TJ. 1989. The genera of suborder Apocynineae (Apocynaceae and Asclepiadaceae) in the southeastern United States. J. Arnold Arbor. 70: 307–401, 443–514.

Roth I and H Lindorf. 1974. The morphological interpretation of the seed of the Rubiaceae and especially that of *Coffea*. Acta Bot. Venez. 9: 141–147.

Rova JHE and L Andersson. 1995. A reevaluation of the tribes Hippotideae and Tammsieae (Rubiaceae). Nord. J. Bot. 15: 269–284.

Rova JHE, PG Delprete, L Andersson, and VA Albert. 2002. A *trn*L-F cpDNA sequence study of the Condamineeae-Rondeletieae-Sipaneeae complex with implications on the phylogeny of the Rubiaceae. Am. J. Bot. 89: 145–159.

Rutishauser R. 1984. Blattquirele, Stipeln und Kolleteren bei der Rubieae (Rubiaceae) im Vergleich mit anderen Angiospermen. Beitr. Biol. Pfl. 59: 375–424.

Rutishauser R, LP Ronse Decraene, E Smets, and I Mendoza-Heuer. 1998. *Theligonum cynocrambe*: developmental morphology of a peculiar rubiaceous herb. Plant Syst. Evol. 210: 1–24.

Safwat FM. 1962. The floral morphology of *Secamone* and the evolution of the pollinating apparatus in Asclepiadaceae. Ann. Missouri Bot. Gard. 49: 95–129.

Schick B. 1980. Untersuchungen über die Biotechnik der Apocynaceenblute: I. Morphologic und Funktion der Narbenkopfes. Flora 170: 394–432.

Schick B. 1982. Zur Morphologic, Entwicklung, Feins-truktur, und Funktion des Translators von *Periploca* L. (Asclepiadaceae). Trop. Subtrop. Pflanzenwelt 40: 513–553.

Sennblad B. 1997. Phylogeny of the Apocynaceae s.l. Acta Universitatis Upsaliensis, p. 295. Uppsala.

Sennblad B and B Bremer. 1996. The familial and subfamilial relationships of Apocynaceae and Asclepiadaceae evaluated with *rbc*L data. Plant Syst. Evol. 202: 153–175.

Sennblad B and B Bremer. 2002. Classification of Apocynaceae s.l. according to a new approach combining Linnaean and phylogenetic taxonomy. Syst. Biol. 51: 389–409.

Shamrov II. 1988. Ovule development and structural characteristics of embryo sac in some members of the Gentianaceae family. Bot. Zhurn. 73: 213–222 (in Russian).

Shamrov II. 1996. Ovule development and significance of its features for Gentianaceae systematics. Opera Bot. Belgica 7: 113–118.

Simões AO, ME Endress, T van der Niet, LS Konoshita, and E Conti. 2004. Tribal and intergenetic relationships of Mesechiteae (Apocynoideae, Apocynaceae): evidence from three noncoding plastid DNA regions and morphology. Am. J. Bot. 91: 1409–1418.

Simões AO, MM Castro de, and LS Konoshita. 2006. Calycine colleters of seven species of Apocynaceae (Apocynoideae) from Brazil. Bot. J. Linn. Soc. 152: 387–398.

Solereder H. 1893. Ein Beitrag zur anatomischen Charakteristik und zur Systematik der Rubiaceen. Bull. Herb. Boiss. 1: 167–183.

Struwe L and VA Albert 1997. Floristics, cladistics, and classification: three case studies in Gentianales. In: J Dransfield, MJE Coode, and DA Simpson, eds. Plant diversity in Malesia, vol. 3, pp. 321–352. Royal Botanic Gardens, Kew.

Struwe L and VA Albert. 2002. Gentianaceae. Systematics and natural history. Cambridge University Press, Cambridge.

Struwe L, VA Albert, and B Bremen 1995. Cladistics and family level classification of the Gentianales. Cladistics 10: 175–206.

Struwe L, JW Kadereit, J Klackenberg, S Nilsson, M Thiv, KB von Hagen, and VA Albert. 2002. Systematics, character evolution and biogeography of Gentianaceae, including a new tribal and subtribal classification. In: L Struwe and VA Albert, eds. Gentianaceae: systematics and natural history, pp. 21–309. Cambridge University Press, Cambridge.

Sundell E. 1980. The subfamilial, tribal, and subtribal nomenclature of the Asclepiadaceae. Taxon 29: 257–265.

Swarupanandan K, JK Mangaly, TK Sonny, K Kishorekumar, and S Chand Basha. 1996. The subfamilial and tribal classification of the family Asclepiadaceae. Bot. J. Linn. Soc. 120: 327–369.

Taylor CM. 1996. Overview of the Psychotrieae (Rubiaceae) in the Neotropics. Opera Bot. Belg. 7: 261–270.

Thiv M and JW Kadereit. 2002. A morphological cladistic analysis of Gentianaceae-Canscorinae and the evolution of anisomorphic androecia in the subtribe. Syst. Bot. 27: 780–788.

Thiv M, L Struwe, VA Albert, and JW Kadereit. 1999. The phylogenetic relationships of *Saccifolium bandeirae* (Gentianaceae) reconsidered. Harvard Pap. Bot. 4: 519–526.

Tiagi YC and S Kshetrapal. 1974. Studies on the floral anatomy, evolution of the gynoecium, and relationships of the family Loganiaceae. Adv. Plant Morph. 1972: 408–416.

Venter HJT and RL Verhoeven. 1997. A tribal classification of the Periplocoideae (Apocynaceae). Taxon 46: 705–720.

Venter HJT and RL Verhoeven. 2001. Diversity and relationships within the Periplocoideae (Apocynaceae). Ann. Missouri Bot. Gard. 88: 550–568.

Verdcourt B. 1958. Remarks on the classification of the Rubiaceae. Bull. Rijksplantentuin, Bruss. 28: 209–281.

Verellen J, E Smets, and S Huysmans. 2004. The remarkable genus *Coptosapelta* (Rubiaceae): pollen and orbicule morphology and systematic implications. J. Plant Res. 117: 57–68.

Verellen J, S Dessein, SG Razafimandimbison, E Smets, and S Huysmans. 2007. Pollen morphology of the tribes Naucleeae and Hymenodictyeae (Rubiaceae – Cinchonoideae) and its phylogenetic significance. Bot. J. Linn. Soc. 153: 329–341.

Verhoeven RL and HJT Venter. 1998. Pollinium structure in Periplocoideae (Apocynaceae). Grana 37: 1–14.

Verhoeven RL and HJT Venter. 2001. Pollen morphology of the Periplocoideae, Secamonoideae, and Asclepiadoideae (Apocynaceae). Ann. Missouri Bot. Gard. 88: 569–582.

Verhoeven RL, S Liede, and M Endress. 2003. The tribal position of *Fockea* and *Cibirhiza* (Apocynaceae – Asclepioideae): evidence from pollinium structure and cpDNA sequence data. Grana 42: 70–81.

Vijayaraghavan MR and U Padmanabhan. 1969. Morphology and embryology of *Centaurium ramosissimum* Druce and affinities of the family Gentianaceae. Beitr. Biol. Pfl. 46: 15–37.

Vinckier S and E Smets. 2002a. Morphology, ultrastructure and typology of orbicules in Loganiaceae s.l. and related genera, in relation to systematics. Rev. Palaeobot. Palynol. 119: 161–189.

Vinckier S and E Smets. 2002b. Morphological and ultrastructural diversity of orbicules in relation to evolutionary tendencies in Apocynaceae s.l. Ann. Bot. N. S. 90: 647–662.

Vinckier S and E Smets. 2002c. Systematic importance of orbicule diversity in Gentianales. Grana 41: 158–182.
Vinckier S, S Huysmans, and E Smets. 2000. Morphology and ultrastrucutre or orbicules in the subfamily Ixoroideae (Rubiaceae). Rev. Palaeobot. Palynol. 108: 151–174.
Wagenitz G. 1959. Die systematische Stellung der Rubiaceae: Ein Beitrag zum System der Sympetalen. Bot. Jahrb. Syst. 79: 17–35.
Weberling F. 1977. Beiträge zur Morphologic der Rubiaceen-Infloreszenzen. Ber. Deutsch. Bot. Ges. 90: 191–209.
Woodson RE. 1930. Studies in the Apocynaceae: I. A critical study of the Apocynoideae (with special reference to the genus *Apocynum*). Ann. Missouri Bot. Gard. 17: 1–212.
Woodson RE and JA Moore. 1938. The vascular anatomy and comparative morphology of apocynaceous flowers. Bull. Torrey Bot. Club 65: 135–166.
Wunderlich R. 1971. Die systematische Stellung von *Theligonum*. Oesterr. Bot. Z. 119: 491–518.
Xue C-Y and D-Z Li. 2005. Embryology of *Megacodon stylophorus* and *Veratrilla baillonii* (Gentianaceae): descriptions and systematic implications. Bot. J. Linn. Soc. 147: 317–331.
Xue C-Y, T-N He, and D-Z Li. 2002. Floral nectaries in *Swertia:* anatomy and morphology. Acta Bot. Yunn. 24: 359–369.
Young MCM, MR Braga, SMC Dietrich, VS Bolzani, LMV Trevisan, and OR Gottlieb. 1996. Chemosystematic markers in Rubiaceae. Opera Bot. Belgica 7: 205–212.
Yuan Y-M. 1993. Seed coat micromorphology and its systematic implications for Gentianaceae of western China. Bot. Helv. 103: 73–62.
Yuan Y-M, S Wohlhauser, M Möller, P Chassot, G Mansion, J Grant, P Kupfer, and J Klackenberg. 2003. Monophyly and relationships of the tribe Exaceae (Gentianaceae) inferred drom nuclear ribosomal and chloroplast DNA sequences. Mol. Phylog. Evol. 28: 500–517.

Order 121. SOLANALES

Herbs (erect or climbing), shrubs, or trees, sometimes parasitic. Indumentum of diverse types of trichomes, often stellate or otherwise branched. Intraxylar phloem usually present in stem and petioles. Vessels with simple perforations. Fibers mostly with simple pores. Rays heterogeneous to homogeneous. Axial parenchyma apotracheal or less often scanty paratracheal. Sieve-element plastids of S-type. Nodes unilacunar. Leaves alternate, or rarely opposite, simple or sometimes compound, sometimes succulent, rarely scale-like, estipulate. Stomata of diverse types. Flowers in various types of inflorescences (cymose or derived from the cymose type) or solitary and axillary, usually bisexual, commonly 5-merous, actinomorphic to more or less zygomorphic. Calyx synsepalous, usually 5-lobed, persistent. Corolla sympetalous, usually 5-lobed, lobes mostly plicate in bud, contorted, imbricate, or valvate. Stamens usually five, attached to the corolla tube and alternate with its lobes, but in Solanaceae-Salpiglossideae one or even three of them transformed into staminodia; anthers usually introrse, tetrasporangiate, opening longitudinally or by apical pores. Tapetum secretory. Microsporogenesis simultaneous. Pollen grains 2-celled or sometimes 3-celled, 3-colporate, 6-polycolpate or pantoporate. Well-developed intrastaminal nectary disc usually present. Gynoecium usually of two, sometimes five, rarely three united carpels; style usually terminal or sometimes gynobasic, with usually 2-lobed or sometimes peltate stigma; ovary superior, (1)2(3–5)-locular, with one or more often several or numerous ovules in each locule or on each placenta. Ovules anatropous, hemitropous, campylotropous, or amphitropous, unitegmic, commonly tenuinucellate, usually with endothelium. Female gametophyte of *Polygonum*- or *Allium*-types. Endosperm cellular, sometimes nuclear, or helobial. Fruits of diverse types, mostly baccate or septicidal capsules. Seeds exotestal, with straight or curved or sometimes spiral embryo and usually copious endosperm. Commonly producing various kinds of alkaloids, flavonol and flavone glycosides, coumarins, acylated anthocyanins, and caffeic acid; iridoid compound absent, n = 7–15.

Derived from the Rubiales, probably from the lower Rubiales. The origin from the Rubiales is confirmed by both the flower and seed structure as well as by the presence of intraxylary phloem.

Key to Families

1 Tall tree.

2 Secretory cavities present. Fibers with numerous small bordered pits. Rays closely spaced, heterogeneous, mostly uniseriate. Axial parenchyma occasionally diffuse, but most commonly in apotracheal sheaths 1–3 cells wide or in vasicentric sheaths 1–3 cells wide. Internal phloem present. Leaves alternate, simple, entire. Flowers in few-flowered terminal or subterminal cymes, bisexual, actinomorphic, 5-merous except for the gynoecium. Calyx imbricate, persistent, tubular, with terminal teeth. Corolla funnelform, with long tube and short, very imbricate lobes, greenish-white. Stamens five, alternating with the corolla members; filaments attached to lower part of corolla tube; anthers exserted, oblong,

introrse, basally deeply sagittate, with a thick connective, opening longitudinally. Pollen grains with striate exine (Hunziker 2001). Well-developed intrastaminal nectary disc present around the base of ovary. Gynoecium of two carpels, more or less immersed in disc; style terminal, elongate, with dilated, shortly 2-lobed stigma; ovary 2-locular, but only one locule fertile; ovule solitary, anatropous. Fruits large shining bright orange drupes; endocarp surrounded by a though persistent fibrous mesocarp; 2-locular, with more or less straight sterile locule and U-shaped fertile; seed U-shaped, with curved (U-shaped) embryo, cotyledons small; endosperm scanty, oily. . . . . 4. DUCKEODENDRACEAE.

2 Secretory cavities absent. Stems without internal phloem. Leaves alternate, leathery, simple, entire. Stomata paracytic. Flowers solitary, axillary, bibracteolate, 5-merous. Floral receptacle developing a short and thick gynophore. Hypogynous disc absent. Flowers solitary, axillary, zygomorphic. Sepals persistent, much imbricate, petals contorted and unilaterally induplicate, campanulate, with short lobes. unequal but not bilabiate, densely adpressed-ferrugineous-pubescent externally. Stamens five, inserted near the base of the corolla tube; anthers long exserted, basifixed. Gynoecium of two carpels; style one, deflexed in bud, apical; stigmas small, truncate; ovary 2-locular, stipitate. Ovules 30–50 per locule, anatropous. Placentation axile. Fruit baccate, crustaceous, 1–4-seeded. Endosperm copious. . . . . 9. HUMBERTIACEAE.

1 Herbs, shrubs, vines, rarely small trees.

3 Stems with internal phloem.

4 Articulated latex canals or latex cells absent.

5 Seeds with endosperm.

6 Leaves alternate or alternate to opposite.

7 Fruits mostly berries or capsules. Trees, shrubs, perennial to annual herbs, and lianas, prickly or not.

8 One sepal lies in the upper median plane of the flowers. Small trees, shrubs, perennial to annual herbs, and lianas, often prickly, resinous, or not resinous. Trichomes simple, glandular, stellate, echinoid or peltate. Cork superficial or deep-seated, or pericycle. Plants with or without crystal sand. Nodes with 2–3 traces. Leaves alternate or alternate to opposite (usually alternate below, but often becoming opposite towards the inflorescence), sometimes leathery or modified into spines, entire, simple to pinnately compound or 3-foliolate, rarely gland-dotted (e.g. *Anthocercis*). Stomata anomocytic or anisocytic. Flowers in terminal or axillary inflorescences, or solitary, occasionally leaf-opposite, small, to medium sized, usually actinomorphic, mostly bisexual, rarely unisexual (monoecious or rarely dioecious, e.g. sometimes in *Solanum* and *Symonanthus*). Calyx 5-lobed or -parted, sometimes 4- or 6–7-lobed, often enlarging in fruit. Corolla rotated to tubular, with (4-)5(-7) plicate and sometimes also contorted or rarely merely contorted, imbricate or valvate lobes, actinomorphic to (some *Salpiglossis*) distinctly zygomorphic. Androecium exclusively of fertile stamens or rarely including staminodia (Salpiglossideae). Stamens usually five, rarely four or even (*Schizanthus*) only two fertile; inserted near the base of the corolla tube, or midway down the corolla tube, or in the throat of the corolla tube; anthers dorsifixed or basifixed, extrorse (e.g. *Anthocercis*) or introrse, opening longitudinally or by apical pores or short slits. Pollen grains 2-celled, (2)3–5(6)-colpate or colporate, or colporoidate, or rugate, sometimes inaperturate, rarely (species of *Salpiglossis, Bouchetia, Nierembergia, Reyesia*) in tetrads. Gynoecium mostly of two carpels generally oriented obliquely to the median plane of the flower or rarely three carpels; style terminal, with mostly 2-lobed stigma; ovary 2-locular [but sometimes complicated by secondary divisions; locules secondarily divided by 'false septa' (Datureae and Nicandreae), or without 'false

septa'], or 1-locular apically (as in *Capsicum*). Ovules (1-) more or less numerous on often thickened placentas, anatropous or hemianatropous. Endosperm haustoria usually present. Fruits capsules (commonly septicidal, or loculicidal, or valvular, or circumscissile – *Hyoscyamus*), of berries, or drupes. Seeds with straight or curved embryo, two semi-cylindric cotyledons, and mostly oily or proteinaceous, rarely starchy endosperm, is some *Datura* species seeds with elaiosomes. Containing alkaloids, arthroquinones, ursolic acid; flavonols (kaempferol and quercetin) and saponins present or absent, n = 7–12 +.. . . . . 1. SOLANACEAE.

8 The upper median plane passes more or less between two adaxial sepals. Annual to perennial herbs, or shrubs to 2 m tall (*Hydrolea spinosa*). Leaves alternate, entire. Flowers in short terminal raceme of cymes, 5-merous. Calyx basally connate. Anthers versatile, basifixed. Nectary disc annular, five-humped. Gynoecium of 2–4 carpels; styles 3–5, glandular-pubescent; stigma slightly tunneliform or capitate; ovary 2(-3)-locular, ovules numerous (200–300), on thickened, axile placentae. Endosperm cellular, with branching micropylar and chalazal haustoria. Fruits globose or ovoid, septicidal or irregularly 4-valved. Seeds longitudinally ridged and ruminate, exotestal cells thin-walled, endotestal cells tanniniferous, with a cuticle; n = (9)10 (12). . . . . .6. HYDROLEACEAE.

7 Fruits schizocarps. Herbs or shrublets, sometimes diffuse or prostrate, commonly more or less succulent. Plants with "crystal sand". Leaves alternate to opposite (alternate below, those towards the inflorescence sometimes in alternate pairs of unequal members towards the same side of the stem), sessile or petiolate, simple, entire, flat or terrete, leathery or fleshy, sometimes small and ericoid. Stomata anisocytic. Flowers axillary, subsessile or pedicellate, solitary, actinomorphic or somewhat zygomorphic, 5-merous. Calyx tubular-campanulate, lobes imbricate, rarely valvate, persistent. Corolla actinomorphic to obscurely bilabiate, plicate (between the lobes). Stamens inserted low in the corolla tube, unequal, three longer than the other two; anthers ovoid or oblong, introrse. Pollen grains 2-celled. Gynoecium of (3-)5 carpels; the carpels of *Alona* are united to form a conventional 5-locular ovary and style terminal; in *Nolana* style gynobasic, with peltate or capitate stigma; ovary superior, divided into mericarps or 5-locular. Locules consist of 3–30 locelli, each locules divided horizontally into one-ovulate locelli, or not horizontally divided (Watson and Dallwitz 1992), with 1–7 ascending or appendiculate, hemianatropous ovules in each, uniseriate or in superimposed series. Fruits separating into small 1–7-seeded 5–25 mericarps provided with germination plugs formed by the funicule, mericarps five, or 10–30 (in 1–3 series). Seeds with curved or spiral embryo, two cotyledons, and copious, oily endosperm. Contain flavonols (kaempferol and quercetin), n = 12. . . . . 2. NOLANACEAE.

6 Leaves opposite, more or less entire, thick and fleshy, usually asymmetrical. Stomata diacytic. Plants with "crystal sand". Flowers solitary, sessile, in the axils. Calyx asymmetrical, spinescent, unequally 5-lobed, with short tube, persistent. Corolla slightly zygomorphic, conduplicate-contorted. Stamens five, attached at different levels to the corolla tube; anthers opening longitudinally. Pollen grains 3-colporate, reticulate, colpi with granulate membrane, tectum thick and perforate, columellae well defined and some of them bifurcate. Gynoecium of two carpels; style terminal, with simple

papillate stigma; ovary 2-locular, inferior, with the septum oblique in relation to the median plane of the flower. Ovules 2–3, apotropous, pendulous from the upper part of the locules. Fruits dry, indehiscent, 2-locular, usually 2-seeded, with thin membranous pericarp, usually concrescent with the stem. Seeds with endosperm and straight or more or less curved embryo, n = 12. 3. SCLEROPHYLACACEAE.

5 Seeds without endosperm. Shrubs or small trees with many branches. Young parts and flowers often densely ferrugineo-velutinous. Fibers with bordered pits. Rays 1–3 cells wide, composed primarily of procumbent cells with upright cells only at the tips of multiseriate rays and in uniseriate rays. Axial parenchyma diffuse plus diffuse-in-aggregate plus narrow bands. Leaves alternate, leathery, simple, entire, coriaceous, closely parallel-ascending, striate-veined, midrib sometimes fusiform-thickened. Stomata anomocytic. Flowers solitary and extra-axillary or in few-flowered, axillary racemes or fascicles, bisexual, actinomorphic or zygomorphic, 4–6-merous. Sepals connate or rarely almost free, valvate, persistent. Corolla infundibulate or campanulate, 4–6-lobed, lobes mostly short, equal or unequal, valvate. Stamens 4–6, epipetalous; filaments long-exserted, equal or unequal; anthers exserted, versatile, sagittate, dorsifixed near base. Pollen grains 3-colporate, with thick perforate tectum and prominent columellae, echinate. Nectary disc usually large, fleshy, lobed. Gynoecium of 1–2 carpels; style terminal, long, with 2-lobed or capitate stigma; ovary 1–2-locular, with two ovules per locule. Fruits baccate (drupoid?), fleshy, or chartaceous, 1–2-seeded. Seeds with plicate testa; embryo straight, well differentiated and large, cotyledons two or sometimes four. The berry of *Espadaea* edible, with the flavour of apricot. . . . . 5. GOETZEACEAE.

4 Articulated latex canals or latex cells usually present, either scattered or in vertical rows, are fairly widespread in both leaf and stem. Autotrophic plants: herbs, shrubs, or lianas, rarely small trees, usually with twining stems, occasionally prostrate or creeping or erect, sometimes succulent; some thorny xerophytes, some with tuberous roots or stems, others rhizomatous. *Convolvulus, Ipomoea, Rivea* twining anticlockwise. Hairs both glandular and eglandular. Stem with intraxylary phloem. Leaves alternate, usually petiolate, simple, cordate, hastate, or sagittate, entire to variously divided, sometimes with extrafloral nectaries. Stomata usually paracytic. Flowers in terminale cymes or solitary, bracteate, often bracteolate, bisexual (except *Hildebrandtia* – plants dioecious), actinomorphic, 5-merous. Hypogynous disc present. Sepals imbricate, persistent; Corolla valvate and plicate, or contorted and plicate, tubular or campanulate, or urceolate, without scale staminodia. Stamens five, inserted near the base of the corolla tube, or midway down the corolla tube. Pollen grains 2-, or 3-celled, 3-colpate to pantoporate. Intrastaminal nectary disc usually present, annular or copular. Gynoecium of 2–5 carpels. Sty les one or two, without an indusium, often filiform, simple or forked, rarely very short or absent, stigma entire or 2-lobed, rarely 3-lobed or stigmas 2–4, of various shapes, globular or ellipsoid to filiform, sometimes aplanate, rarely peltate, reniform, conical, or funnel-shaped. Ovules anatropous, with the micropyle directed downward and outward, unitegmic, with a massive integument, tenuninucellate or sometimes crassinucellate. Placentation basal. Fruits usually loculicidal capsules, sometimes circumscissile or irregularly dehiscing, rarely baccate or nutlike. Seeds conspicuously hairy, or not; embryo large, green, straight or curved, with folded emarginate or bifid cotyledons, embedded in a hard, often cartilaginous endosperm. Often producing alkaloids of the indole and other groups, sometimes flavonols (kaempferol and quercetin), n = 7, 14, 15. 7. CONVOLVULACEAE.

3 Stems without internal phloem. Herbaceous rootless (the normal root system is short-lived) parasites with slender, often filiform, twining, yellow, or reddish stem, attaching on aerial parts of the host by haustoria. Hairs when present mostly unicellular or bicellular, not glandular. Intraxylary phloem absent. Leaves alternate, reduced to

minute scales. Stem without intraxylary phloem. Flowers small, 3–5-merous. Calyx imbricate. Corolla valvate, with crenulate or fimbriate scale-like five staminodia inside. Stamens five, inserted in the throat of the corolla tube. Pollen grains 3-celled, 3–6-aperturate, colpate. Nectary disc commonly present around the base of the ovary. Gynoecium of two carpels. Stylodia two, free or united into a simple style, stigmas globose, subglobose, or elongated. Ovary 2-locular; ovules two per locule, ascending, tenuinucellate, unitegmic, with a massive integument. Fruits capsules, opening circumscissilely, irregularly, or indehiscent and somewhat fleshy. Embryo slender, filiform, green, nearly or quite acotyledonous, but sometimes with an enlargement at one end, peripheral and strongly curved or spirally wound around the starchy, oily endosperm. Present proanthocyanidins, flavonols (kaempferol and quercetin), n mostly = 7, sometimes 15. . . . . .8. CUSCUTACEAE.

1. SOLANACEAE

A.L. de Jussieu 1789 (including Atropaceae Martynov 1820, Cestraceae Schlechtendal 1833, Daturaceae Rafinesque 1828, Hyoscyamaceae Vest 1818, Lyciaceae Rafinesque 1840, Nicotianaceae Martynov 1820, Salpiglossidaceae Hutchinson 1969). 92/2300. Subcosmopolitan, but especially diversified in South America.

Classification based on Hunziker (2001).

1.1 BROWALLIOIDEAE (CESTROIDEAE)
Cork superficial or deep-seated; bordered pits present; pericyclic fibers present. Corolla often zygomorphic. Stamens inserted high in the corolla tube. Fruits usually capsular. Seeds prismatic to subglobose, with straight or slightly curved embryo. Alkaloids mostly pyridinic and tropane types, also steroidal, quinolinic and phenilethylaminic types (Hunziker 2001), n = 7–14, rarely 12. – CESTREAE: *Cestrum, Vestia, Sessea* (including *Sesseopsis*); METTERNICHIEAE: *Metternichia*; LATUEAE: *Latua*; NICOTIANEAE: *Nicotiana, Petunia* (including *Calibrachoa*), *Fabiana, Nierembergia, Bouchetia, Leptoglossis, Hunzikeria, Plowmania*; BENTHAMIELLEAE: *Benthamiella, Pantacantha, Combera*; FRANCISCEAE: *Brunfelsia*; BROWALLIEAE: *Browallia, Streptosolen*; SCHWENCKIEAE: *Schwenckia, Melananthus, Protoschwenckia, Heteranthia.*

1.2 JUANULLOIDEAE
Corolla actinomorphic or zygomorphic. Stigma usually saddle-shaped. Fruits generally baccate, multiseeded. Seeds reniform, boomerang-shaped, endosperm scanty, n = 12. – *Juanulloa, Dyssochroma, Ectozoma, Hawkesiophyton, Markea, Merinthopodium, Rahowardiana, Schultesianthus, Trianaea.*

1.3 SOLANOIDEAE
Corolla actinomorphic, aestivation varied: imbricate, vavate, contorted. Stamens five, inserted low in the corolla tube. Stigma small, unmodified. Fruits usually baccate. Seeds flattened, discoid, with coiled embryo, endosperm usually abundant, n = 10, 12, 14, 17. – NICANDREAE: *Nicandra;* MANDRAGOREAE: *Mandragora*; DATUREAE: *Datura, Brugmansia;* LYCIEAE: *Lyceum, Phrodus, Grabowskia*; SOLANEAE: *Witheringia, Brachistus, Cuatresia, Deprea, Discopodium, Exodeconus, Jaltomata, Nothocestrum, Physalis* (including *Margaranthus*), *Tzeltalia, Quincula, Leucophisalis* (*Physaliastrum*), *Chamaesaracha, Oryctes, Tubocapsicum, Capsicum, Aureliana, Athenaea, Darcyanthus, Larnax, Withania* (including *Mellisia, Archiphysalis*), *Solanum, Cyphomandra, Lycopersicon, Lycianthes, Triguera, Normania*; IOCHROMINAE: *Iochroma, Saracha, Acnistus, Eriolarynx, Vassobia, Dunalia*; ATROPEAE: *Atropa* (including *Pauia*); JABOROSEAE: *Jaborosa, Salpichroa, Nectouxia*; SOLANDREAE: *Solandra*; HYOSCYAMEAE: *Hyoscyamus, Anisodus, Atropanthe, Physochlaina, Przewalskia, Scopolia.*

1.4 SALPIGLOSSOIDEAE
Corolla actinomorphic or zygomorphic, imbricate. Pollen grains in monads or in tetrads. Capsules septicide-loculicidal, with two or four valves, many seeded (up to 300 in *Salpiglossis*, or 2–25 in *Reyesia*). Endosperm cellular. Present pyridinic alkaloids, n = 11. – *Salpiglossis, Reyesia.*

1.5 SCHIZANTHOIDEAE
Cork pericycle; pericycle fibers absent. Flowers zygomorphic, stalk curved and the two really upper petals form the lower lip which is 3–4-lobed, and the lowest petal forms the simple or slightly 2-lobed upper lip. Stamens four: two fertile and two staminodial;

endosperm nuclear; embryo curved. Present unique tropane alkaloids with senecioic, mesaconic and itaconic acids, n = 10. – *Schizanthus*.

1.6 ANTHOCERCIDOIDEAE
Woody shrubs or trees. Flowers with non-accrescent calyx, inflexo-valvate aestivation of the corolla lobes, a short relatively broad actinomorphic corolla tube, extrorsely dehiscing stamens inserted low in the corolla tube, and an oblong to ellipsoid, slightly curved seed with reticulate testa, embryo slightly curved. Fruits capsules or rarely (*Duboisia*) berries. Endosperm copious, sometimes including a small oily sector (Hunziker 2001) Produce nicotinic and tropane alkaloids, n = 9, 10. – *Anthocercis, Anthotroche, Cyphanthera, Crenidium, Duboisia, Gramnosolen, Symonanthus*.

2. NOLANACEAE

Dumortier 1829. 1–2/60. According to Mesa (1981) *Nolana* comprises 18 species of which 15 are in tropical America, other modern authors have suggested that there may be as many as 83 American species: Southern Peru, northern Chile, Galspogos Is.

Nolana (? including *Alona*)

3. SCLEROPHYLACACEAE

Miers 1848. 1/15. Argentina, Paraguay, Uruguay.

Sclerophylax

Close to Solanaceae.

4. DUCKEODENDRACEAE

Kuhlmann 1950. 1/1. Amazon Basin in Brazil.

Duckeodendron

Related to the Solanaceae (see Kuhlmann 1950; Carlquist 1988).

5. GOETZEACEAE

Miers ex Airy Shaw 1965. 4/5. Greater Antilles (Cuba, Hispaniola, Puerto Rico).

Coeloneurum, Goetzea, Henoonia, Espadaea

Close to the Solanaceae (Radlkofer 1888; Hunziker 1979; Carlquist 1988; Zona 1989), but differ from them markedly in many respects, including pollen morphology (see Gentry 1986).

6. HYDROLEACEAE

Berchtold et J. Presl 1820. 1/12. Tropical.

Hydrolea

Related to the Solanaceae (APG II 2003; Erbar et al. 2005).

7. CONVOLVULACEAE

A.L. de Jussieu 1789 (including Cressaceae Rafinesque 1821, Dichondraceae Dumortier 1829, Erycibaceae Endlicher ex Meisner 1840, Poranaceae J.G. Agardh 1858). 56/1700. Cosmopolitan, but mainly in subtropical regions of Asia and America.

IPOMOEEAE: *Argyreia, Astripomoea, Blinkworthia, Ipomoea* (including *Pentacrostigma*), *Lepistemon, Lepistemonopsis, Paralepistemon, Stictocardia, Rivea, Turbina,* MERREMIEAE: *Merremia, Hewittia, Hyalocystis, Decalobanthus, Xenostegia, Remirema, Operculina*; CONVOLVULEAE: *Convolvulus, Calystegia, Polymeria;* ANISEIEAE: *Aniseia, Hewettia, Iseia, Odonellia, Tetralocularia*; JACQUEMONTIEAE: *Jacquemontia*; CRESSEAE: *Cressa, Bonamia, Cladostigma, Evolvulus, Seddera, Hildebrandtia, Sabaudiella, Stylisma, Itzaea, Neuropeltis, Neuropeltopsis*; WILSONIEAE: *Wilsonia*; MARIPEAE: *Dicranostyles, Maripa, Lysiostyles,* DICHONDREAE: *Dichondra, Falkia, Nephrophyllum, Petrogenia, Porana, Metaporana, Calycobolus, Dipteropeltis, Rapona*; ERYCIBEAE: *Erycibe*; CARDIOCHAMYEAE: *Cordisepalum, Poranopsis, Cardiochlamys, Tridynamia, Dinetus*.

8. CUSCUTACEAE

Berchtold and J. Presl. 1820. 1/170. Nearly cosmopolitan, but best represented in America, especially in warmer regions.

Cuscuta

Very close to the Convolvulaceae, but differing from them markedly in parasitic habit, corolla morphology, seeds with more copious endosperm, the absence of intraxylary phloem, and embryological features (Tiagi 1951; Johri and Tiagi 1952).

9. HUMBERTIACEAE

Pichon 1947. 1/1. Madagascar
Humbertia
Closely related to the Convolvulaceae.

Bibliography

Acosta MC, A del V Ordóñez, AA Cocucci, and EA Moscone. 2006. Chromosome reports in South American Nicotaneae (Solanaceae), with particular reference to *Nierembergia*. Ann. Missouri Bot. Gard. 93: 634–646.

Alfaro ME and A Mesa. 1979. El origen morfologico del floema in traxylar en Nolanaceae y la posicion systermática de esta familia. Bol. Soc. Argent. Bot. 18(3–4): 123–126.

Al Nowaihi AES and MM Mourad. 1999. Morphological and anatomical characters of the spermoderm of certain taxa of the tribe Solaneae (Solanaceae). Taeckholmia 19: 157–181.

Amporpan L and JE Armstrong. 2002. Floral ontogeny of *Salpiglossis* (Solanaceae) and the oblique gynoecium. J. Torrey Bot. Soc. 129: 85–95.

Armstrong JE. 1986. Comparative floral anatomy of Solanaceae: a preliminary survey. In: WG D'Arcy, ed. Solanaceae: Biology and systematics, pp. 101–113. Columbia University Press, New York.

Austin DF. 1973. The American Ericibeae (Convolvulaceae): *Maripa, Dicranostyles,* and *Lysiostyles:* II. Palynology. Pollen et Spores 15: 203–226.

Austin DF. 1998. Parallel and convergent evolution in the Convolvulaceae. In: P Mathew and M Sivadasan, eds. Diversity and taxonomy of tropical flowering plants, pp. 201–234. Mentor Books, Calicut, India.

Axelius B. 1996. The phylogenetic relationships of the physaloid genera (Solanaceae) based on morphological data. Am. J. Bot. 83: 118–124.

Baehni C. 1946. L'ouverture du bouton chez les fleurs des Solanees. Candollea 10: 399–492.

Badr A, SF Khalifa, Aboel-Atta, and Abou-El-Enain. 1997. Chromosome criteria and taxonomic relationships in the Solanaceae. Cytologia 62: 103–113.

Basak RK. 1967. The pollen grains of Solanaceae. Bull. Bot. Soc. Bengal 21: 49–58.

Bell AD and TD Dines. 1995. Branching patterns in the Solanaceae. In: PC Hoch and AG Stephenson, eds. Experimental and molecular approaches to plant biosystematics, pp. 157–172. Missouri Botanical Garden, St. Louis, MO.

Berg van den RG, GWM Barendse, GM Weerden van der, and C Marinni, eds. 2001. Solanaceae V: advances in taxonomy and utilization. Nijmegen, The Netherlands.

Bernardello L and F Chiang-Cabrera. 1998. A cladistic study on the American species of *Lycium* (Solanaceae) based on morphological variation. Monogr. Syst. Bot. 68: 33–46.

Bernardello L and MC Lujan. 1997. Pollen morphology of tribe Lycieae: *Grabowskia, Lycium, Phrodus* (Solanaceae). Rev. Palaeobot. Palynol. 96: 305–315.

Bohs L and RG Olmstead. 2001. A reassessment of *Normania* and *Triguera* (Solanaceae). Plant Syst. Evol. 228: 33–48.

Bonderson WE. 1986. Gynoecial morphology and funicular plugs in the Nolanaceae. Nord. J. Bot. 6: 183–198.

Carlquist S. 1987. Wood anatomy of Nolanaceae. Aliso 11: 473–483.

Carlquist S. 1988. Wood anatomy and relationships of Duckeodendraceae and Goetzeaceae. IAWA Bull., n.s., 9: 3–12.

Carlquist S and MA Hanson. 1991. Wood and stem anatomy of Convolvulaceae. A survey. Aliso 13: 51–94.

Carrizo García C. 2002. An approach to the diversity of endothecial thickenings in Solanaceae. Flora 197: 214–223.

Carrizo García C. 2003. Combination of sequences of cell divisions in the anther wall formation in Solanaceae species. Flora 198: 243–246.

Cocucci AA. 1989. El mecanismo floral de *Schizanthus* (Solanaceae). Kurtziana 20: 113–132.

Cocucci A. 1995. Floral mechanisms in the tribe Salpiglossidae (Solanaceae). Plant Syst. Evol. 194: 207–230.

Cocucci A. 1999. Evolutionary radiation in Neotropical Solanaceae. In: M Nee, DE Symon, RN Lester, and JP Jessop, eds. Solanaceae IV: advances in biology and utilization, pp. 9–22. Royal Botanic Gardens, Kew.

D'Arcy WG. 1978. A preliminary synopsis of *Salpiglossis* and other Cestreae (Solanaceae). Ann. Missouri Bot. Gard. 65: 698–724.

D'Arcy WG. 1979. The classification of the Solanaceae. In: JG Hawkes, RN Lester, and AD Skelding, eds. The biology and taxonomy of the Solanaceae, pp. 3–47. Finn. Soc. Symposium, No. 7. Academic, London.

D'Arcy WG. 1991. The Solanaceae since 1976, with a review of its biogeography. In: JG Hawkes, RN Lester, and AD Skelding, eds. Biology and taxonomy of the Solanaceae. Royal Botanic Gardens, Kew.

D'Arcy WG and ZY Zhang. 1992. Notes on the Solanaceae of China and neighboring areas. Novon 2: 124–28.

D'Arcy WG, RC Keating, ZY Zhang, and CI Peng. 2001. The genus *Tubocapsicum* (Solanaceae). Bot. Bull. Acad. Sinica (Taipei) 42(1): 67–84.

Davenport LJ. 1988. A monograph of *Hydrolea* (Hydrophyllaceae). Rhodora 90: 169–204.

Demissew S and DF Austin. 1996. Generic delimitation and relationships in the tribe Hildebrandtieae (Convolvulaceae). In: LJC van der Maesen, XM van der Burgt, and JM Medenbach de Rooy, eds. The biodiversity of African plants, pp. 409–420. Kluwer, Dordrecht.

Deroin T. 1992 (publ. 1993). Anatomie florale de *Humbertia madagascariensis* Lam. Contribution á la morphologie comparée de la fleur et du fruit des Convolvulaceae. Bull. Mus. natl. Hist. Nat., Paris, 4 sér., 14: 235–255.

Deroin T. 1999. Ontogeny and phylogeny and Convolvulaceae-Ipomoeae: preliminary comparative remarks on ovary morphology. Syst. Geogr. Plant 68: 225–232.

DeWitt Smith S and DA Baum 2006. Phylogenetics of the florally diverse Andean clade Iochrominae (Solanaceae). Am. J. Bot. 93: 1140–1153.

Diez MJ and IK Ferguson. 1984. Pollen morphology of *Mandragora autumnalis* Bertol. (Solanaceae). Pollen et Spores 26: 151–160.

Di Fulvio TE. 1961. El genero *Sclerophylax* (Solanaceae). Estudio anatomico, embriologico, y caryologico con especial referenda a la taxonomica. Kurtziana 1: 9–103.

Di Fulvio TE. 1969. Embriologi'a de *Nolana paradoxa* (Nolanaceae). Kurtziana 5: 39–54.

Di Fulvio TE. 1971. Morfologi'a de *Nolana paradoxa* (Nolanaceae), con especial referencia a la organizacion del gineceo. Kurtziana 6: 41–51.

Di Fulvio TE. 1989. Observaciones embriológicas en especies Argentinas de *Hydrolea* (Hydrophyllaceae) con especial referencia a la endospermogénesis. Kurtziana 20: 33–64.

Estrada E and M Martinez. 1998. *Physalis* (Solanaceae) and allied genera: *Tzeltalia,* a new genus from the highlands of southern Mexico and northwestern Guatemala. Brittonia 50: 285–295.

Evans WC. 1979. Tropane alkaloids of the Solanaceae. In: JG Hawkes, RN Lester and AD Skelding, eds. The biology and taxonomy of the Solanaceae, ser. 7, pp. 241–254. Academic, London.

Fay MF, RG Olmstead, JE Richardson, E Santiago, GT Prance, and MW Chase. 1998. Molecular data support the inclusion of *Duckeodendron cestroides* in Solanaceae. Kew Bull. 53: 203–212.

Finn VV. 1937. Vergleichende Embryologie und Karyologie einiger *Cuscuta*-Arten. Zhurn. Inst. Bot. Vseukrainsk. Akad. Nauk 12(20): 83–99 (in Ukrainian with German and Russian summary).

Fuentes V and N Rodriguez. 1984. Estudios en el genero *Henoonia* Griseb: I. Morfologia y biometrica de las hojas. Rev. Jard. Bot. Nac. 5(3): 29–40.

Fukuda T, J Yokoyama, and H Ohashi. 2001. Phylogeny and biogeography of the genus *Lycium* (Solanaceae): inferences from chloroplast DNA sequences. Mol. Phylogenet. Evol. 19: 246–258.

Garcia VF and RG Olmstead. 2003. Phylogenetics of tribe Anthocercideae (Solanaceae) based on *ndh*F and *trn*L/F sequence data. Syst. Bot. 28: 609–615.

Gemeinholzer B and M Wink. 2001. Solanaceae: occurrence of secondary compounds versus molecular phylogeny. In: RG van den Berg, GWM Barendse, GM van der Weerden, C Marinni, eds. Solanaceae V: advances in taxonomy and utilization, pp. 165–178. Nijmegen, The Netherlands.

Gentry JL. 1979. Pollen morphology of the Salpiglossideae (Solanaceae). In: JG Hawkes, JG Lester, and AD Skelding, eds. The biology and taxonomy of the Solanaceae, pp. 241–254. Academic, London.

Gentry JL. 1986. Pollen studies in the Cestreae (Solanaceae). In: WD'Arcy, ed. Solanaceae: biology and systematics, pp. 138–158. Columbia University Press, New York.

Govil CM. 1971. Morphological studies in the family Convolvulaceae: I. Development and structure of the seed coat. J. Indian Bot. Soc. 50: 32–38.

Govil CM. 1972. Morphological studies in the family Convolvulaceae: IV. Vascular anatomy of the flower. Proc. Indian Acad. Sci. 758: 271–282.

Grant V. 1959. Natural history of the *Phlox* family: I. Systematic Botany. M. Nijhoff, The Hague.

Haegi L. 1986. The affinities of *Anthocercis* (Solanaceae) and related genera. In: W D'Arcy, ed. Solanaceae: biology and systematics, pp. 27–40. Columbia University Press, New York.

Haegi L. 1994. Inflorescence structure in tribe Anthocercideae. Solanaceae Newslett. 4(1): 19.

Haider Ali SN, MS Ramanna, E Jacobsen, and R Visser. 2001. Alien chromosome additions indicate taxonomic distance in Solanaceae. In: RG van den Berg, GWM Barendse, GM van der Weerden, C Marinni, eds. Solanaceae V: advances in taxonomy and utilization, pp. 209–215. Nijmegen, The Netherlands.

Hawkes JG and WG Tucker. 1968. Serological assessment of relationships in a flowering plant family (Solanaceae). In: JG Hawkes, ed. Chemotaxonomy and serotaxonomy. Syst. All. Special 2: 77–88.

Hawkes JG, RN Lester, and AD Skelding. 1979. The biology and taxonomy of the Solanaceae. Linn. Soc. Symposium, No. 7. London.

Hawkes JG, M Nee, KN Lester, N Estrada, WG D'Arcy, D Symon, and S Dickerson, eds. 1988. Solanaceae. III. Taxonomy, chemistry, evolution. Royal Botanic Gardens, Kew.

Hayrapetian AM. 1992. Palynomorphology of the family Solanaceae Juss. Cand. Sc. Thesis, University of Yerevan (in Russian).

Hayrapetian AM. 1995. The aperture types of pollen and possible ways of their evolution in the family Solanaceae. Bot. Zhurn. 80(8): 1–10 (in Russian with English summary).

Hayrapetyan AM. 2002. The palynomorphological data of the subfamily Solanoideae (Solanaceae Juss.): tribe Solaneae. Flora, vegetation and plant resources of Armenia, 14: 118–130.

Hayrapetian AM. 2004. Pollen morphology of the family Goetzeaceae Miers ex Airy Shaw. Flora, vegetation and plant resources of Armenia, 15: 66–69.

Hoare AL and S Knapp. 1997. A phylogenetic conspectus of the tribe Hyoscyameae (Solanaceae). Bull. Nat. Hist. Mus. (Bot. Ser.) 27: 11–29.

Huber KA. 1980. Morphologische und entwicklungsgeschichtliche Untersuchungen an Blüten und Blütenstanden von Solanaceen und von *Nolana paradoxa* Lindl. Nolanaceae. Dissertationes Botanicae 55. Fl. Vaduz.

Hunziker AT. 1979. South American Solanaceae: a synoptic survey. In: JG Hawkes, RN Lester, and AD Skelding, eds. The biology and taxonomy of the Solanaceae, pp. 49–86. Linn. Soc. Symposium, No. 7. London.

Hunziker AT. 2000a. The tribe Solaneae (Solanaceae): key for its genera and description of *Darcya* gen. nov. Bol. Soc. Argent. Bot. 35(1–2): 163–169.

Hunziker AT. 2000b. *Darcyanthus* nom. nov. substitutes *Darcya* (Solanaceae). Bol. Soc. Argent. Bot. 35(3–4): 345.

Hunziker AT. 2001. Genera Solanacearum: the genera of Solanaceae illustrated, arranged according to a new system. Ruggell, Liechtenstein.

Inamdar JA and SR Murthy. 1977. Vessels in some Solanaceae. Flora 166B: 441–447.

Inamdar JA and RC Patel. 1969. Development of stomata in some Solanaceae. Flora 158B: 462–472.

Jazewitch W, von. 1959. Contribution á l'étude de *Humbertia madagascariensis* Lamk. 1. Anatomie de l'écorce. J. Agric. Trop. 6: 609–615.

Johnston IM. 1936. A study of the Nolanaceae. Proc. Am. Acad. Arts 71: 1–87.

Johri BM. 1934. The development of the male and female gametophytes in *Cuscuta reflexa* Roxb. Proc. Indian Acad. Sci. 1B: 283–289.

Johri BM and B Tiagi. 1952. Floral morphology and seed formation in *Cuscuta reflexa* Roxb. Phytomorphology 2: 162–180.

Keeler KH. 1977. The extrafloral nectaries of *Ipomoea cornea* (Convolvulaceae). Am. J. Bot. 64: 1182–1188.

Kennedy PB and AS Crafts. 1931. The anatomy of *Convolvulus arvensis,* wild morning-glory or field bindweed. Hilgardia 4: 591–622.

Knapp S. 2002a. Floral diversity and evolution in the Solanaceae. In: QC Cronk, RM Bateman, and JA Hawkins eds. Developmental Genetics and Plant Evolution, pp. 267–297. Taylor & Francis, London.

Knapp S. 2002b. Tobacco to tomatoes: a phylogenetic perspective on fruit diversity in the Solanaceae. J. Exp. Bot. 53: 2001–2022.

Knapp S, V Persson, and S Blackmore. 1997. A phylogenetic conspectus of the tribe Juanulloeae (Solanaceae). Ann. Missouri Bot. Gard. 84: 67–89.

Knapp S, P Stafford, and V Persson. 2000. Pollen morphology in the Anthocercideae (Solanaceae). Kurtziana 28: 7–18.

Knapp S, L Bohs, M Nee, and DM Spooner. 2004. Solanaceae – a model for linking genomics with biodiversity. Comp. Funct. Genom. 5: 285–291.

Kothari S, S Gaur, and S Sharma. 1997. Pollen morphology of Convolvulaceae. Acta Bot. Indica 25: 125–127.

Kuhlmann JG. 1950. Duckeodendraceae Kuhlmann (Nova familia). Arquiv. Serv. Florest. 3: 7–8.

Lee KB. 2007. Structure and development of the upper haustorium in the parasitic flowering plant *Cuscuta japonica* (Convolvulaceae). Am. J. Bot. 94: 737–745.

Lesyer RN and PA Roberts. 1986. Serotaxonomy of *Solanum, Capsicum, Dunalia* and other selected Solanaceae. Ann. Missouri Bot. Gard. 73: 128–133.

Levin RA and JS Miller. 2005. Relationships within tribe Lycieae (Solanaceae): paraphyly of *Lycium* and multiple origins of gender dimorphism. Am. J. Bot. 92: 2044–2053.

Luna-Cavazos M and E García-Moya. 2002. Morphological and pollen differentiation in *Solanum cardiophyllum* ssp. *cardiophyllum* and *S. cardiophyllum* ssp. *ehrenbergii*. Bot. J. Linn. Soc. 140: 415–426.

Maestri DM and CA Guzman. 1995. A comparative study of seed lipid components of Nicotianeae (Solanaceae). Biochem. Syst. Ecol. 23(2): 201–207.

Mann P, B Tofern, M Kaloga, and E Eich. 1999. Flavonoid sulfates from the Convolvulaceae. Phytochemistry 50: 267–271.

Mariaux A. 1959. Contribution á l'étude de *Humbertia madagascariensis* Lamk., 2. Note sur le bois. J. Agric. Trop. 6: 616–619.

Martins TR and TJ Barkman. 2005. Reconstruction of Solanaceae phylogeny using the nuclear gene SAMT. Syst. Bot. 30: 435–447.

Mesa A. 1981. Nolanaceae. Flora Neotrop. Monogr. 16: 1–197.

Mesa A. 1986. The classification of the Nolanaceae. In: WG D'Arcy, ed. Solanaceae: Biology and systematics, pp. 86–90. Columbia University Press, New York.

Momin AR. 1975a. Bearing of embryological data on taxonomy of Convolvulaceae. J. Univ. Bombay 44: 50–65.

Momin AR. 1975b. On embryological formulae of some Convolvulaceae members. Sci. Cult. 41: 318.

Moncada M and V Fuentes. 1991. Palinologia de Goetzeaceae. (Palynology of Goetzeaceae.). Rev. Jard. Bot. Nacion. Univ. Habana 12: 75–79.

Miers J. 1869. On the genera *Goetzia* and *Espadaea*. Trans. Linn. Soc. London 27: 187–195.

Miller JS. 2002. Phylogenetic relationships and the evolution of gender dimorphism in *Lycium* (Solanaceae). Syst. Bot. 27: 416–428.

Murray MA. 1945. Capellary and placental structure in the Solanaceae. Bot. Gaz. 107: 243–260.

Nee M. 1986. Placentation patterns in the Solanaceae. In: W D'Arcy, ed. Solanaceae: Biology and systematics, pp. 169–175. Columbia University Press, New York.

Nee M. 2001a. Solanaceae systematics for the 21st century. In: RG van den Berg, GWM Barendse, GM van der Weerden, and C Marinni, eds. Solanaceae V: advances in taxonomy and utilization, pp. 3–22. Nijmegen, The Netherlands.

Nee M. 2001b. An overview of Cestrum. In: RG van den Berg, GWM Barendse GM van der Weerden, and C Marinni, eds. Solanaceae V: advances in taxonomy and utilization, pp. 109–136. Nijmegen, The Netherlands.

Neyland R. 2001. A phylogeny inferred from large ribosomal subunit (26S) rDNA sequences suggests that *Cuscuta* is a derived member of Convolvulaceae. Brittonia 53: 108–115.

Olmstead RG. 1994. Phylogeny of the Solanaceae based on chloroplast DNA. Solanaceae Newslett. 4: 33–34.

Olmstead RG and JD Palmer. 1991. Chloroplast DNA and systematics of the Solanaceae. In: JG Hawkes, RN Lester, M Nee, and N Estrada, eds. Solanaceae III: taxonomy, chemistry, evolution, pp. 161–168. Royal Botanic Gardens, Kew.

Olmstead RG and JD Palmer. 1992. A chloroplast DNA phylogeny of the Solanaceae: subfamilial relationships and character evolution. Ann. Missouri Bot. Gard. 79: 346–360.

Olmstead RG and JA Sweere. 1994. Combining data in phylogenetic systematics: an empirical approach using three molecular data sets in the Solanaceae. Syst. Biol. 43: 467–481.

Olmstead RG, JA Sweere, RE Spangler, L Bohs, and JD Palmer. 2000. Phylogeny and provisional classification of the Solanaceae based on chloroplast DNA. In: M Nee, DE Symon, RN Lester, and JP Jessop, eds. Solanaceae IV: advances in biology and utilization, pp. 111–137. Royal Botanic Gardens, Kew.

Pena RC and O Munoz. 2002. Cladistic relationships in the genus *Schizanthus* (Solanaceae). Biochem. Syst. Ecol. 30: 45–53.

Persson V, S Knapp, and S Blackmore. 1994. Pollen morphology and systematics of tribe Juanulloeae A.T. Hunziker (Solanaceae). Rev. Palaeobot. Palynol. 83(1–3): 1–30.

Persson V, S Knapp, and S Blackmore. 2000. Pollen morphology and the phylogenetic analysis of *Datura* and *Brugmansia*. In: M Nee, DE Symon, RN Lester, and JP Jessop, eds. Solanaceae IV: advances in biology and utilization, pp. 171–187. Royal Botanic Gardens, Kew.

Philomina K. 1980. Cytotaxonomic notes on the Solanaceae. J. Indian Bot. Soc. 59: 173–176.

Prashar G and V Singh. 1990. Development and evolution of inflorescence in the Solanaceae. Indian J. Bot. 13: 29–35.

Prenner G, G Deutsch, and P Harvey. 2002. Floral development and morphology in *Cuscuta reflexa* Roxb. (Convolvulaceae). Stapfia 80: 311–322.

Radlkofer L. 1888. Über die Versetzung der Gattung *Henoonia* von den Sapotaceen zu den Solanaceen. Sitzungsber. Math.-Phys. Cl. Konigl. Bayer. Akad. Wiss. München 18: 405–421.

Roberty G. 1964. Les genres de Convolvulacees (esquisse). Boissiera 10: 129–156.

Robyns W. 1930. L'organisation florale des Solanacees zygomorphes. Mem. Acad. Roy. Belgique, Cl. Set. 11(8): 1–82.

Santiago-Valentin E and RG Olmstead. 2001. Biogeography of the Goetzeoideae (Solanaceae). In Botany 2001: Plants and People. Abstracts, p. 93. Albuquerque.

Santiago-Valentin E and RG Olmstead. 2003. Phylogenetics of Antillean Goetzeoideae (Solanaceae) and their relationships with the Solanaceae based on chloroplast and ITS sequence data. Syst. Bot. 28: 452–460.

Sandina IB. 1980. A critical analysis of the genus *Scopolia* (Solanaceae). Bot. Zhurn. 65: 485–496 (in Russian with English summary).

Sandina IB and VF Tarasevich. 1982. Some palynological data on the study of the genera *Whitleya, Atropanthe* and *Scopolia* s.st. (Solanaceae). Bot. Zhurn. 67: 146–154 (in Russian with English summary).

Sayeedud Din M. 1953. Observations on the anatomy of some of the Convolvulaceae. Proc. Indian Acad. Sci. 37B: 106–109.

Schimming T, B Tofern, P Mann, A Richter, K Jenett Siems, B Drager, N Asano, MP Gupta, MD Correa, and E Eich. 1998. Distribution and taxonomic significance of calystegines in the Convolvulaceae. Phytochemistry 49: 1989–1995.

Sengupta S. 1972. On the pollen morphology of Convolvulaceae, with special reference to taxonomy. Rev. Palaeobot. Palynol. 13: 157–212.

Sripleng A and FH Smith. 1960. Anatomy of the seed of *Convolvulus arvensis*. Am. J. Bot. 47: 386–392.

Stafford P and S Knapp. 2006. Pollen morphology and systematics of the zygomorphic-flowered nightshades (Solanaceae; Salpiglossideae sensu D'Arcy, 1978 and Cestroideae sensu D'Arcy, 1991, pro parte): a review. Syst. Biodiversity 4: 173–201.

Stefanovic S and RC Olmstead. 2000. Molecular systematics of Convolvulaceae inferred from *cp*DNA sequences. Am. J. Bot. 67(6, Suppl.): 160.

Stefanovic S and RC Olmstead. 2001. Molecular systematics of Convolvulaceae inferred from multiple chloroplast loci. In Botany 2001: Plants and People, p. 143, Abstracts. Albuquerque.

Stefanovic S and RC Olmstead. 2004. Testing the phylogenetic position of a parasitic plant (*Cuscuta*, Convolvulaceae, asteridae): Bayesian inference and the parametric bootstrap on data drawn from three genomes. Syst. Biol. 53: 384–399.

Stefanovic S, L Krueger, and RG Olmstead. 2002. Monophyly of the Convolvulaceae and circumscription of their major lineages based on DNA sequences of multiple chloroplast loci. Am. J. Bot. 89: 1510–1522.

Stefanovic S, DF Austin, and RG Olmstead. 2003. Classification of Convolvulaceae: a phylogenetic approach. Syst. Bot. 28: 791–806.

Terekhin ES and VA Kotov. 1988. Embryology of *Cuscuta japonica* (Cuscutaceae). Bot. Zhurn. 73: 222–230 (in Russian with English summary).

Tétényi P. 1987. A chemotaxonomic classification of the Solanaceae. Ann. Missouri Bot. Gard. 74: 600–608.

Tiagi B. 1951. A contribution to the morphology of *Cuscuta hyalina* Roth and C. *planiflora* Tenore. Phytomorphology 1: 9–21.

Tiagi B. 1966. Floral morphology of *Cuscuta reflexa* Roxb. and *C. lupuliformis* Krocker with a brief review of the literature on the genus *Cuscuta*. Bot. Mag. Tokyo 79: 89–97.

Tu T-Y, H Sun, Z-J Gu, and J-P Yue. 2005. Cytological studies on the Sino-Himalayan endemic *Anisodus* and four related genera from the tribe Hyoscyameae (Solanaceae) and their systematic and evolutionary implications. Bot. J. Linn. Soc. 147: 457–468.

Tucker WG. 1969. Serotaxonomy of the Solanaceae: a preliminary survey. Ann. Bot. 33: 1–23.

Ungricht S, S Knapp, and JR Press. 1998. A revision of the genus *Mandragora* (Solanaceae). Bull. Nat. Hist. Mus. (London) Bot. 28: 17–40.

Vales MA and V Fuentes. 1991. Caracteristicas de la epidermis foliar de Goetzeaceae. Acta Bot. Hung. 36(1–4): 255–265 (1990–1991 publ. 1991).

Walsh BM and SB Hoot. 2001. Phylogenetic relationships of *Capsicum* (Solanaceae) using DNA sequences from two noncoding regions: the chloroplast *atp*B-*rbc*L spacer region and nuclear waxy introns. Int. J. Plant Sci. 162: 1409–1418.

Weberling F. 1956. Weitere Untersuchungen zur Morphologic des Unterblattes bei den Dikotylen: III. Convolvulaceae. IV. Zygophyllaceae. Beitr. Biol. Pfl. 33: 149–161.

Whitson M and PS Manos. 2005. Untangling *Physalis* (Solanaceae) from the Physaloids: a two-gene phylogeny of the Physalinae. Syst. Bot. 30: 216–230.

Wilkin P. 2000. A morphological cladistic analysis of the Ipomoeeae (Convolvulaceae). Kew Bull. 54: 853–876.

Wilson KA. 1960. The genera of Convolvulaceae in the southeastern United States. J. Arnold Arbor. 41: 298–317.

Wojciechowska B. 1972. Systematic studies on the seeds of the Solanaceae family. Monogr. Bot. 36: 117–179.

Wollenweber E and M Dorr. 1995. Exudate flavonoids in some Solanaceae. Biochem. Syst. Ecol. 23(4): 457–458.

Yang D-Z. 2002. Tribe Hyoscyameae of the Solanaceae: structure, differentiation and phylogenetic relationship. Ph.D. thesis, Institute of Botany, The Chinese Academy of Sciences.

Yang D-Z, ZY Zhang, and J Wen. 2000. Structural characters of leaf epidermis in Hyoscyameae (Solanaceae) and their systematic significance. Acta Bot. Sinica 42: 133–142.

Zhang Z-Y and A-M Lu. 1984. Pollen morphology of the subtribe Hyoscyaminae (Solanaceae). Acta Phytotax. Sinica 22: 175–180.

Zhang Z-Y and A-M Lu. 2000. A comparative study of *Physalis, Capsicum* and *Tubocapsicum;* three genera of Solanaceae. In: M Nee, DE Symon, RN Lester, and JP Jessop, eds. Solanaceae IV: advances in biology and utilization, pp. 81–96. Royal Botanic Gardens, Kew.

Zona S. 1989. Leaf anatomy of the Goetzeaceae. Aliso 12: 303–312.

Order 122. BORAGINALES

Herbs, shrubs, or less often trees, sometimes lianas, rarely herbaceous parasites. Stems and leaves of the herbaceous forms and less frequent in the woody genera provided with characteristic harsh unicellular hairs that usually have a basal cystolith and often calcified or silicified walls; less often also with gland-tipped hairs or (Lennoaceae) provided only with stalked, glandular hairs. Vessels with simple or very rarely reticulate perforations; lateral pitting alternate. Fibers with bordered or more often simple pits, often septate. Rays heterogeneous to homogeneous. Axial parenchyma of various types, but mostly apotracheal. Vestured pits present or absent. Sieve-element plastids of S-type. Nodes unilacunar, very rarely (*Cordia*) trilacunar with three traces. Leaves alternate or sometimes partly or wholly opposite, simple, usually entire, sometimes lobed, pinnatisect or pinnate, rarely palmate, estipulate. Stomata usually anomocytic. Flowers in scorpioid cymose inflorescences (Buys and Hilger 2003), rarely solitary, mostly bisexual, actinomorphic or slightly zygomorphic, perianth and androecium usually 5-merous. Sepals more or less connate, imbricate or rarely valvate. Corolla sympetalous, imbricate or contorted, rarely valvate. Stamens as many as and alternate with the corolla lobes, attached to the tube; anthers tetrasporangiate, introrse, or rarely (*Hoplestigma*) latrorse, opening longitudinally. Pollen grains (2)3-celled, 3-colpate, 3-colporate, 6-colpate, pericolpate, pericolporate or 3-porate. Nectary disc present around the base of the ovary or wanting. Gynoecium of two median carpels, rarely of 4–5 (in *Zoelleria,* Boraginaceae), 6–14 (Lennoaceae) carpels; stylodia usually more or less united into terminal or gynobasic style; ovary superior or rarely semi-inferior (as in *Nama,* Hydrophyllaceae), with two to many ovules on each placenta or in each locule. Ovules anatropous, rarely hemitropous (some Boraginaceae) or campylotropous (Lennoaceae), unitegmic, tenuinucellate or rarely (Ehretioideae) pseudocrassinucellate (crassinucellate according to Gottschling 2004), sometimes with endothelium (Hydrophyllaceae, *Heliotropium* in Boraginaceae). Female gametophyte usually of *Polygonum*-type, rarely (some Boraginaceae) of *Allium*-type. Endosperm cellular or less often (species of *Phacelia* in Hydrophyllaceae and some Boraginoideae) nuclear, rarely (species of *Echium* and *Pulmonaria*) helobial. Often micropylar and/or chalazal haustoria present. Fruits of various types. Seeds exotestal, with straight or nearly straight embryo and copious (many Hydrophyllaceae) or scanty endosperm or often (most of Boraginaceae) without endosperm; in seed coat, funicule and placenta of most member or the order (except Boraginoideae) present specialized transfer cells (Diane, Hilger and Gottschling 2002).

Boraginales are connected with the Solanales and most probably share a common origin.

Key to Families

1 Autotrophic plants.

2 Petals usually five.

3 Ovary basically 1-locular, with two, often more or less intruded, parietal placentas, which are sometimes meeting and joined, dividing the ovary into two secondary locules (locelli). Ovules many to two on each placenta. Annual, biennial, and perennial herbs, subshrubs, shrubs, sometimes spiny, rarely small trees (to 5 m in *Wigandia*). Stems, leaves and flowers often variously pubescent, sometimes with stringing hairs (as in *Wigandia*). Leaves alternate or sometimes partly or wholly opposite, basal leaves sometimes in rosettes, petiolate, simple to pinnately or rarely palmately divided, entire or lobed. Stomata anomocytic. Flowers in terminal or axillary scorpioid cymes, rarely solitary or paired in axils, bisexual, 5-merous or rarely (South African genus *Codon*) 10–12-merous. Calyx usually divided sometimes to the base or nearly so, lobes imbricate, sometimes with appendages between the calyx lobes. Corolla sympetalous, actinomorphic or nearly so, with imbricate or less often contorted lobes, often with scales inside the tube, alternating with the stamens. Stamens (4-)5(-10), inserted near the base of the corolla tube; filaments variously, basally appendiculate; anthers dorsifixed, usually with small scales at each side of the base, introrse. Pollen grains 3-colpate, 3-colporate, or 5–6-colporate, with tectate-perforate to semitectate exine structure, basically reticulate. Nectaries present at base of ovary (Hofmann 2004). Gynoecium of two carpels; stylodia more or less connate, sometimes free essentially to the base, mostly with capitate stigmas; ovary usually superior, rarely

semi-inferior, 1–2-locular. Ovules two to many in each locule, pendulous, anatropous, usually tenuinucellate, crassinucellate (*Nama* – see di Fulvio 1989a), with an endothelium. Endosperm cellular or less often nuclear. Fruits usually loculicidal (sometimes also septicidal) capsules or sometimes irregularly dehiscent or indehiscent. Seeds small, ruminate by inpushings of the exotestal cells or not, sometimes carunculate; exotestal cells thickened on inner and radial walls, endotestal cells persistent; embryo small to large, green or white, straight, surrounded by copious or scanty, oily endosperm. Flavonols present (kaempferol and quercetin), n = 5–9, 11–13, 19, 38 (*Wigandia kunthii*).. . . . . 1. HYDROPHYLLACEAE.

3 Ovary at maturity with twice as many secondary locules (locelli) as carpels, each with one ovule, rarely one carpel suppressed in subfamily Cordioideae and the ovary with only two locelli. Herbs, subshrubs, less often shrubs or trees, seldom lianas (*Cordia, Tournefortia* spp.); taproots or creeping rhizomes sometimes present, rarely present tubers (in *Lithospermum* and *Heliotropium*); plant often roughly hairy, hairs usually strigose to hirsute, with a basal cystolith or cystolith-like body. Leaves alternate, very rarely opposite (in *Tournefortia*), or sometimes the lowest of them or all opposite, mostly entire, in a very few species serrate. Inflorescences of one or more terminal or lateral helicoids or scorpioid cymes, or flowers contracted into globose or club-shaped heads, or flowers solitary; bracts sometimes present. Flowers mostly bisexual, usually 5-merous, actinomorphic or rarely zygomorphic. Sepals free or basally connate, imbricate or rarely valvate. Corolla imbricate, rarely valvate, or contorted (in *Myosotis*), with faucal, hairy or scale-shaped appendages. Stamens of the same number as corolla lobes, alternating with them; anthers 2-locular, sessile or on distinct filaments, sometimes with a sterile pit of connective tissue or produced to an awn-like structure, awns of one flower twisted of not, basifixed or dorsifixed. Pollen grains from 3-colporate or 3-porate to pericolpate or pericolporate, sometimes with six colpi, the alternate ones with and without a pore. Annular nectary disc present around the base of the ovary, often covered by basal appendages formed by corolla-tube, or absent. Gynoecium of two or rarely (*Zoelleria*) 4–5 carpels; stylodia connate into terminal or gynobasic style; ovary usually 2-locular, undivided, entire or with four deep furrows, breaking into two biovulate halves or into four uniovulate mericarps, sometimes part of the locules aborted, rarely a greater number of carpels developed (e.g. in *Trigonotis procumbens*), placentation axillary (Riedl 1997). Ovules anatropous to hemitropous, erect or nearly horizontal, rarely pendulous, sometimes (*Heliotropium*) with an endothelium. Fruits drupes with 1–4 pyrenes (Cordiodeae, Ehretroideae, and part of Heliotropioideae), separated into four or rarely two (*Cerinthe*) 1-seeded, nutlike mericarps (Boraginoideae), or capsular (Wellstedioideae). Seeds with straight or sometimes more or less curved embryo; endosperm of several cell layers (Heliotropoideae) or reduced to a single cell layer. Producing alkaloids of pyrolizidine group and flavonols (kaempferol and quercetin), flavonol glycosides astragalin and nicotiflorin, isokestose and higher inulin oligosaccharids as storage, n = 4–14, 16, 21.. . 2. BORAGINACEAE.

2 Petals 11–14. Deciduous trees; cork superficial, secondary thickening developing from a conventional cambial ring. Leaves alternate, very large, petiolate, chartaceous, simple, entire. Flowers large, in terminal bractless, brown-hirsute subscropioid cymes, bisexual, actinomorphic. Calyx globose, persistent, irregularly splitting into 2–4 lobes. Corolla with short tube, lobes imbricate. Stamens about 20–35, in about three irregular series attached to the base of the corolla tube; filaments filiform; anthers dorsifixed, latrorse. Pollen grains 3-colporate with meridional ridges alternating with intercolpar concavities. Gynoecium of two carpels; ovary superior, 1-locular with two intruded, forked, parietal placentas; each placenta with two pendulous, anatropous ovules; stylodia basally connate, elongate, slender, curved, or bent in the middle and each with a disclike horseshoe-shaped stigma. Fruits drupaceous, 4-seeded, surrounded at the base by the calyx, with leathery pericarp and hard endocarp. Seeds with rather large, nearly straight embryo with elongate

hypocotyl and expanded cotyledons and scanty endosperm. . . . . 3. HOPLESTIGMATACEAE.

1 Fleshy perennial (*Pholisma*) or annual (*Lennoa*) parasitic herbs. Stems subterranean or nearly so. Roots spreading horizontally and initiating new haustoria by short, lateral branches. Leaves alternate, simple, entire, reduced to mere scales, linear to broadly deltoid. Flowers in cymose-thyroid, cymose-paniculate, spicate, or capitulate inflorescences, bisexual, actinomorphic or nearly so, (4)5–9(10)-merous as to the perianth and androecium. Calyx persistent, tubular or divided nearly to base into narrow lobes, puberulent to tomentose with stalked glandular trichomes. Petals united nearly to apex, limb variously lobed, induplicate valvate or imbricate, persistent. Stamens 5–10, inserted in the throat of the corolla tube, 1–2-seriate; filaments short; anthers introrse. Pollen grains 2-celled, 3–4(5)-colporate or colpate and colporate, often with alternating aperturate and inaperturate colpi. Nectary disc wanting. Gynoecium of 5–16 carpels; style terminal, with capitate or slightly 5–9-lobed stigma; ovary superior, entire or shallowly lobed; each primary locule secondarily divided by false septum into two locelli; ovules two per locule (solitary in each locellus), horizontal, anatropous, epitropous. Endosperm cellular. Fruits slightly depressed capsules, somewhat fleshy, but drying at maturity; dehiscence circumscissile by an irregular ring below the middle of capsule. Seeds small, ovate to slightly reniform, laterally flattened, with reticulate surface; embryo undifferentiated, globose, embedded in the copious, starchy endosperm, n = 9. . . . . .4. LENNOACEAE.

1. HYDROPHYLLACEAE

R. Brown 1817 (including Ellisiaceae Berchtold et J. Presl 1820, Eutocaceae Horaninow 1847, Sagoneaceae Martynov 1820). 16/300. Mainly America, especially western parts of the U.S.A., but also Hawaiian Islands (*Nama sandwicensis*), and South Africa (*Codon*).

1.1 HYDROPHYLLOIDEAE

Flowers (4-)5-merous; stigma capitate. – HYDROPHYLLEAE: *Hydrophyllum, Pholistoma, Ellisia, Nemophila;* PHACELIEAE: *Eucrypta, Draperia, Phacelia, Emmenanthe, Eriodictyon, Turricula, Tricardia, Hesperochiron;* NAMEAE: *Nama, Wigandia, Romanzoffia.*

1.2 CODONOIDEAE

Flowers 10–12-merous. Calyx deeply linear-lobed. Corolla campanulate, 10–12-lobed. Style terminal, stigma dentate. – *Codon.*

Related to the Boraginaceae.

2. BORAGINACEAE

A.L. de Jussieu 1789 (including Anchusaceae Vest 1818, Buglossaceae Hoffmannsegg et Link 1809, Cerinthaceae Martynov 1820, Cordiaceae R. Brown ex Dumortier 1829, Ehretiaceae C. Martius 1827, Heliotropiaceae Schrader 1820, Onosmataceae Martynov 1820, Wellstediaceae Novak 1943). Circa 130/2500. Widely distributed in tropical and especially subtropical and temperate regions; the largest subfamily, the Boraginoideae, is centered in the Mediterranean and Irano-Turanian regions.

2.1 EHRETIOIDEAE

Trees or shrubs, seldom herbs (*Coldenia*). Flowers (4-)5-merous. Corolla imbricate or inwards-folded. Pollen 3-porate. Style terminal, simple or more or less cleft, or two terminal nearly free stylodia (*Rochefortia, Coldenia, Pteleocarpa*). Ovary entire or 4-lobed. Endosperm cellular. Fruits more or less drupaceous or dry, not breaking at maturity (as in *Carmona*) or breaking into four one-seeded or into two two-seeded mericarps. Seeds with copious to scanty endosperm or endosperm absent. Cotyledons flat. Tropical and subtropical regions, n = 5, 7–11, 13, 16 21. – *Coldenia, Tiquilia, Pteleocarpa, Lepidocordia* (including *Antrophora*), *Rochefortia, Rotula, Ehretia, Carmona, Pteleocarpa, Bourreria, Cortesia, Halgania.*

2.2 CORDIOIDEAE

Odoriferous trees, lianes, or shrubs. Corolla contorted (imbricate). Style terminal, once or twice forked. Ovary undivided. Endosperm cellular. Fruits variable, dry with a fibrous or bony pericarp, or thin-fleshed drupes with a thin exocarp, juicy to mucilaginous mesocarp, and bony endocarp, usually 1-locular and 1-seeded. Seeds without endosperm. Cotyledons plicate, toothed. Present terpenoid-based quinones. Tropical and subtropical, n = 9, 14–16, 19. – *Cordia* (including *Varronia, Gerascanthus,* ? *Saccellium, Patagonula, Auxemma*).

2.3 HELIOTROPIOIDEAE
Odoriferous trees, shrubs, lianes, or herbs. Corolla imbricate or with involute margins. Stigma receptive only basi-laterally, discoid, then conical and more or less 2-lobed at sterile apex, or hemispherical, with a ring of hairs, wet. Ovary entire or 4-lobed. Endosperm cellular. Fruits fleshy or dry, lobed or unlobed, at maturity breaking into two or four bony, 1- or 2-seeded mericarps. Seeds exotestal, without endosperm. Cotyledons not plicate. Contain pyrrolizidine alkaloids and salicylic acid (*Tournefortia*). Tropical, subtropical, and warm-temperate regions. n = 5, 7–9, 11–14. – *Tournefortia, Argusia, Heliotropium* (including *Parabouchetia*), *Valentiniella, Mallotonia, Beruniella, Ixorhea, Nogalia.*

2.4 BORAGINOIDEAE
Mainly herbs. Corolla often rotate. Pollen grains 4–6 colporate. Style gynobasic, entire or lobed, arising from the base of usually deeply 4-lobed ovary. Endosperm cellular or nuclear, sometimes helobial. Fruits of four one-seeded mericarps (8–10 in *Zoelleria*). Endosperm scanty or absent. Mostly subtropical and temperate regions. Containing pyrrolizidine alkaloids, lakannin, n = 4–13. – TRIGONOTIDEAE: *Trigonotis* (including *Pedinogyne, Zoelleria*), *Omphalotrigonotis, Sinojohnstonia, Brachybotrys, Bothriospermum, Mertensia, Pseudomertensia,* (including *Scapicephalus*), etc; ERITRICHEAE: *Plagiobothrys, Allocarya, Amblynotus, Antiphytum, Echiochilon, Ogastemma, Sericostoma, Chamissoniophila, Amsinckia, Microula* (including *Schistocaryum*), *Anoplocaryum, Trigonocaryum, Microcaryum, Chionocharis, Actinocarya, Hackelia, Asperugo,, Eritrichium, Metaeritrichium, Tianschaniella, Lepechiniella, Heterocaryum, Craniospermum, Lappula, Lasiocaryum, Stephanocaryum, Cryptantha;* ROCHELIEAE: *Rochelia*; CYNOGLOSEAE: *Bothriospermum, Antiotrema, Cynoglossum, Pardoglossum, Ivanjohnstonia, Austrocynoglossum, Paracynoglossum, Harpagonella, Lindelofia, Adelocaryum, Rindera, Trichodesma, Lacaitaea, Omphalodes, Solenanthus, Kuschakewiczia, Trachelanthus, Mattiastrum, Paracaryum, Thyrocarpus, Suchtelena, Caccinia, Heliocarya,* etc. LITHOSPERMEAE: *Arnebia, Huynhia, Macrotomia, Lithospermum* (including *Aegonychon*), *Ulugbekia, Neatostema, Buglossoides, Lithodora, Onosma, Maharanga, Cerinthe, Moltkia, Moltkiopsis, Stenosolenium, Alkanna, Halacsya, Echium, Lobostemon, Echiostachys, Ancistrocarya,* etc.; BORAGINEAE: *Pulmonaria, Mertensia, Nonea, Nephrocarya, Paraskevia, Elizaldia, Symphytum, Brunnera, Anchusa, Lycopsis, Anchusella, Phyllocara, Cynoglottis, Gastrocotyle, Hormuzakia, Pentaglottis, Pectocarya, Borago, Trachystemon,* etc.; MYOSOTIDEAE: *Myosotis.*

2.5 WELLSTEDIOIDEAE
Low woody herbs or shrublets. Leaves grey-strigose. Flowers bisexual, solitary, 4-merous. Calyx imbricate or open in bud; corolla imbricate. Pollen grains 3-colporate. Nectary disc absent. Style terminal, shortly bind; stigmas two; ovary 2-locular, with solitary, pendulous ovule per locule. Fruits compressed, broadly obcordate, 1–2-seeded loculicidal capsules. Seeds comose, without endosperm, embryo curved, cotyledons accumbent. Southwestern Africa, Somalia, Socotra. – *Wellstedia.*

Ehretioideae are the most archaic group within the family (free or nearly free stylodia and copious endosperm in some genera). Cordioideae are more advanced (plicate cotyledons and usually more specialized pollen grains). Heliotropioideae are closely connected with the Ehretioideae and probably derived from them. Boraginoideae, which are by far the largest and the most diverse subfamily of the Boraginaceae, have probably a common origin with the Heliotropioideae. Wellstedioideae with their 4-numerous flowers and capsular fruits are the most isolated group within the family.

3. HOPLESTIGMATACEAE

Engler et Gilg 1924. 1/2. Tropical West Africa from Cameroon to Gabon.

Hoplestigma

Closely related to the Boraginaceae (Hallier [1911, 1912] even included *Hoplestigma* in Boraginaceae), especially to the Ehretioideae (Takhtajan 1987), which is supported by palynological data (Erdtman 1952; Nowicke and Miller 1989).

4. LENNOACEAE

Solms-Laubach 1870. 3/6–7. California, Arizona, Mexico, Central America, northernmost Colombia, and Venezuela.

Lennoa, Pholisma, Ammobroma

Related to the Hydrophyllaceae and Boraginaceae (Hallier 1912, 1923; Suessenguth 1927; Copeland 1935; Wettstein 1935; Avetisian 1956; Drugg 1962; Takhtajan 1966, 1987; Yatskievych and Mason 1986; Thorne 1992a, b, 2000), in particular to the Ehretioideae. However, the Lennoaceae are very advanced and rather isolated within the order.

Bibliography

Al Nowaihi, SF Khalifa, and K Hamed. 1987. A contribution to the taxonomy of Boraginaceae. Phytologia 62: 107–125.

Al Shehbaz IA. 1991. The genera of Boraginaceae in the southeastern United States. J. Arnold Arbor. Suppl. 1: 1–169.

Avetisian EM. 1956. Morphology of pollen of Boraginaceae. Trudy Bot. Inst. Armenian Acad. Sci. 13(1): 99–102 (in Russian).

Baas P. 1997. Vegetative anatomy of Boraginaceae. Flora malesiana, ser. 1, 13: 46–48.

Berg RY. 1985, Gynoecium and development of embryo sac, endosperm, and seed in *Pholistoma* (Hydrophyllaceae) relative to taxonomy. Am. J. Bot. 72: 1775–1787.

Bigazzi M and F Selvi. 1998. Pollen morphology in the Boragineae (Boraginaceae) in relation to the taxonomy of the tribe. Plant Syst. Evol. 213: 121–151.

Bigazzi M and F Selvi. 2000. Stigma form and surface in the tribe Boragineae (Boraginaceae): micromorphological diversity, relationships with pollen, and systematic relevance. Canad. J. Bot. 78: 388–408.

Bigazzi M and F Selvi. 2001. Karyotype morphology and cytogeography in *Brunnera* and *Cynoglottis* (Boraginaceae). Bot. J. Linn. Soc. 136: 365–378.

Bigazzi M, F Selvi, and G Fiorini. 1999. A reappraisal of the generic status of *Gastrocotyle, Hormuzakia* and *Phyllocara* (Boraginaceae) in the light of micromorphological and karyological evidence. Edinb. J. Bot. 56: 229–251.

Buys MH and HH Hilger. 2003. Boraginaceae cymes are exclusively scorpioid and not helicoid. Taxon 52: 719–724.

Carlquist S and VM Eckhard. 1984. Wood anatomy of Hydrophyllaceae. II. Genera other than *Eriodictyon*, with comments on parenchyma bands containing vessels with large pits. Aliso 10: 27–46.

Chuang TI and L Constance. 1992. Seeds and systematics in Hydrophyllaceae: tribe Hydrophylleae. Am. J. Bot. 79: 257–264.

Constance L. 1963. Chromosome number and classification in Hydrophyllaceae. Brittonia 15: 273–285.

Constance L and TI Chuang. 1982. SEM survey of pollen morphology and classification in Hydrophyllaceae (waterleaf family). Am. J. Bot. 69: 40–53.

Copeland HF. 1935. The structure of the flower of *Pholisma arenarium*. Am. J. Bot. 22: 366–383.

Craven LA. 2005. Malesian and Australian *Tournefortia* transferred to *Heliotropium* and notes on delimitation of Boraginaceae. Blumea 50: 375–381.

Diane N, H Forther, and HH Hilger. 2002a. A systematic analysis of *Heliotropium, Tournefortia,* and allied taxa of the Heliotropiaceae (Boraginales) based on ITS1 sequences and morphological data. Am. J. Bot. 89: 287–295.

Diane N, HH Hilger, and M Gottschling. 2002b. Transfer cells in the seeds of Boraginales. Bot. J. Linn. Soc. 140: 155–164.

Diane N, C Jacob, and HH Hilger. 2003. Leaf anatomy and foliar trichomes in Heliotropiaceae and their systematic relevance. Flora 198: 468–485.

Di Fulvio TE. 1978. Sobre la vasculatura floral, embriología y cromosomas de *Ixorhea tschudiana* (Heliotropiaceae). Kurtziana 11: 75–105.

Di Fulvio TE. 1979. El endosperma y el embrion en el sistema de Tubiflorae, con especial referenda a Boraginaceae e Hydrophyllaceae. Kurtziana 12–13: 101–112.

Di Fulvio TE. 1987. La endospermogenesis en Hydrophylleae (Hydrophyllaceae) con la relacion a la taxonomia. Kurtziana 19: 13–34.

Di Fulvio TE. 1989. Embriolgía de *Nama jamaicense* (Phacelieae, Hydrophyllaceae). Kurziana 20: 9–31.

Di Fulvio TE. 1990. Endospermogenesis y taxonomia de la familia Hydrophyllaceae y su relacion con las demas Gamopetales. Monografia 5: 73–82. Buenos Aires.

Drugg WS. 1962. Pollen morphology of the Lennoaceae. Am. J. Bot. 49: 1027–1032.

Ferguson DM. 1999. Phylogenetic analysis and relationships in Hydrophyllaceae base on *ndhF* sequence data. Syst. Bot. 23: 253–268.

Förther H. 1998. Die infragenerische Gliederung der Gattung *Heliotropium* L. und ihre Stellung innerhalb der subfam. Heliotropioideae (Schrad.) Arn. (Boraginaceae). Sendtnera 5: 35–241.

Gottschling M. 2004. Floral ontogeny in *Bourreria* (Ehretiaceae, Boraginales). Flora 199: 409–423.

Gottschling M and HH Hilger. 2001. Phylogenetic analysis and character evolution of *Ehretia* and *Bourreria* (Ehretiaceae, Boraginales) and their allies based on ITS1 sequences. Bot. Jahrb. Syst. 123: 249–268.

Gottschling M and JS Miller. 2006. Clarification of the taxonomic position of Auxemma, Patagonula, and Saccellium (Cordiaceae, Boraginales). Syst. Bot. 31: 361–367.

Gottschling M, HH Hilger, M Wolf, and N Diane. 2001. Secondary structure of the ITS1 transcript and its application in a reconstruction of the phylogeny of Boraginales. Plant Biol 3: 629–636.

Gottschling M, M Weigend, JS Miller, and HH Hilger. 2003. Phylogeny of Cordiaceae (Boraginales) inferred from ITS1 sequence data. In Botany 2003: Aquatic and Wetland Plants: Wet and Wild. Abstracts, p. 113. Mobile, Alabama.

Gottschling M, N Diane, HH Hilger, and M Weigend, M. 2004. Testing hypotheses on disjunctions present in the primarily woody Boraginales: Ehretiaceae, Cordiaceae, and Heliotropaceae, inferred from ITS1 sequence data. Int. J. Plant Sci. 165(4 Suppl.): S123–S135.

Gottschling M, JS Miller, M Weigend, and HH Hilger. 2005. Congruence of a phylogeny of Cordiaceae (Boraginales) inferred from ITS1 sequence data with morphology, ecology, and biogeography. Ann. Missouri Bot. Gard. 92: 425–437.

Gottwald H. 1982. First description of the wood anatomy of *Antrophora*, *Lepidocordia* and *Pteleocarpa* (Boraginaceae). IAWA. Bull. n.s. 3: 161–165.

Gottwald H. 1984. Wood anatomical studies of Boraginaceae (s. 1.): I. Cordioideae. IAWA Bull., n.s., 4(2–3): 161–178.

Hallier H. 1923. Über die Lennoeen: Eine zu Linne's Bicornes verirrte Sippe der Boraginaceen. Beih. Bot. Centralbl. 40(2): 1–19.

Hegnauer R. 1997. Phytochemistry and chemotaxonomy of Boraginaceae. In: C Kalkman et al., eds, Flora malesiana ser. 1, 13: 52–58. Leiden.

Heubl GR, JC Gaviria, and G Wanner. 1990. A contribution to the taxonomy and evolution of *Cordia* (Boraginaceae) and allied genera: chromosome numbers, pollen morphology, and crystal pattern in wood. Bot. Jahrb. Syst. 112: 129–165.

Hilger HH. 1985. Ontogenie, Morphologie und systematische Bedeutung geflügelter un glochidientragender Cynoglosseae- und Eritrichieae-Fruchte (Boraginaceae). Bot. Jahrb. Syst. 105: 323–378.

Hilger HH. 1987. Fruchtbiologische Untersuchungen an Heliotropioideae (Boraginaceae). I. Die Ontogenie der monospermen Fruchte von *Heliotropium supinum* L. Flora 179: 291–303.

Hilger HH and N Diane. 2003. A systematic analysis of Heliotropiaceae (Boraginales) based on *trn*L and ITS1 sequence data. Bot. Jahrb. Syst. 125: 19–51.

Hilger HH and H Förther. 2004. Boraginaceae. In: N Smith, SA Mori, A Henderson, DW Stevenson, and SV Heald, eds. Flowering plants of the neotropics, pp. 59–62. Princeton University Press, Princeton.

Hilger HH, JR Hoppe, and M Hofmann. 1993. Energiedispersive Rontgenmikroanalyse (EDX) von Boraginaceae subfam. Boraginoideae – Klausenoberflachen (Sind Silicium- und Calcium – Einlagerungen in die Fruchtwand systematische verwertbare Merkmale?) Flora 188: 387–398.

Hofmann M. 1994. Untersuchungen zur Gynoecealentwicklung in der Tribus Eritrichieae (Boraginaceae – Boraginoideae). Inaugural dissertation. Freie Universitat. Berlin.

Hofmann M. 1999. Flower and fruit development in the genus *Phacelia* (Phacelieae, Hydrophyllaceae): characters of systematic value. Syst. Geogr. Plant 68: 203–212.

Hofmann M. 2004. Hydrophyllaceae. In: N Smith, SA Mori, A Henderson, DW Stevenson, and SV Heald, eds. Flowering plants of the neotropics, pp. 190–191. Princeton University Press, Princeton.

John BM and IK Vasil. 1956. The embryology of *Ehretia laevis* Roxb. Phytomorphology 6: 134–143.

Khaleel TE 1974. Embryology of *Cynoglossum denticulatum* DC. Bot. Not. 127: 193–210.

Khaleel TE 1975. Embryology of *Cordia*. Bot. Gaz. 136: 380–387.

Khaleel TE 1977a. Embryology of *Trichodesma* (Boraginaceae). Bot. Not. 130: 441–452.

Khaleel TE. 1977b. Embryology of *Ehretia acuminata* R. Br. Proc. Montana Acad. Sci. 37: 35–53.

Khaleel TE. 1985. A review of endosperm and the taxonomic position of Boraginaceae. J. Plant Sci. Res. 1: 117–133.

Khaleel TE. 1978. Embryology of *Heliotropium scabrum* and *H. strigosum* (Boraginaceae). Plant Syst. Evol. 129: 45–62.

Khaleel TE. 1982. Embryology of *Cordia sebestena* (Boraginaceae). Plant Syst. Evol. 139: 303–311.

Khanna P. 1964. Embryology of *Mertensia*. J. Indian Bot. Soc. 43: 192–202.

Khatamsaz M. 2001. Pollen morphology of Iranian Boraginaceae family and its taxonomic significance. Iranian J. Bot. 9: 27–40.

Kuijt J. 1969. The biology of parasitic flowering plants. University of California Press, Berkeley, CA.

Långström E. 2002. Systematics of *Echiochilon* and *Ogastemma* (Boraginaceae), and the phylogeny of Boraginoideae. Ph.D. thesis, Uppsala University, Uppsala.

Långström E and MW Chase. 2002. Tribes of Boraginoideae (Boraginaceae) and placement of *Antiphytum, Echiochilon, Ogastemma*, and *Sericostoma*: a phylogenetic analysis based on *atp*B plastid DNA sequence data. Plant Syst. Evol. 234: 137–153.

Lawrence R. 1937. A correlation of the taxonomy and the floral anatomy of certain of the Boraginaceae. Am. J. Bot. 24: 333–444.

Liu JX, YZ Xi, JC Ning, JM Zhang, YX Li, YY Zhao, and XH Sun. 2001a. Pollen morphology and exine ultrastructure of genus *Cordia* in Boraginaceae in China. Acta Bot. Sinica 43: 893–898.

Liu JX, YL Zhang, JC Ning, YY Zhao, YX Li, JM Zhang, and XH Sun. 2001b. Pollen morphology of the tribe Cynoglosseae of Boraginoideae (Boraginaceae) in China. Acta Phytotax. Sinica 39: 515–522.

Liu J-X, Y-Z Xi, J-C Ning, X-H Sun, J-M Zhang and Q Li. 2003. Pollen morphology of the subfamily Ehretioideae in the Boraginaceae from China and its systematic significance. Act. Phytotax. Sinica 41: 209–219 (in Chinese).

Miller RB. 1977. Vestured pits in Boraginaceae. IAWA Bull. 3: 43–48.

Miller JS. 1988. A revision of the New World species of *Ehretia* (Boraginaceae). Ann. Missouri Bot. Gard. 76: 1050–1076.

Miller JS. 2003. Classification of Boraginaceae subfam. Ehretioideae: resurrection of the genus *Hilsenbergia* Tausch ex Meisn. Adansonia 25: 151–189.

Miller JS and M Gottschling. 2007. Generic classification in the Cordiaceae (Boraginales): resurrection of the genus *Varronia* P. Br.. Taxon 56: 163–169.

Moore MJ, and RK Jansen. 2006. Molecular evidence for the age, origin, and evolutionary history of the American desert plant *Tiquilia* (Boraginaceae). Mol. Phylogenet. Evol. 39: 668–687.

Nagaraj M. and E Tasneem. 1967. Embryological studies in *Rotula aquatica* Lour. Proc. Indian Acad Sci. 66B: 106–116.

Nagaraj M and E Tasneem. 1971. Studies on the structure and development of pericarp and seed coat of *Rotula aquatica* Lour. Proc. Indian Acad. Sci. 74B: 314–318.

Nikiforova OD. 2001. System of the genus *Myosotis* (Boraginaceae). Bot. Zhurn. 86(12): 77–86 (in Russian with English summary).

Nowicke JW and JS Miller. 1987. Pollen morphology of the Cordioideae (Boraginaceae): *Auxemma, Cordia,* and *Patagonula*. In 14th Int. Congr. Abstracts, 5–36a-3. Berlin.

Nowicke JW and JS Miller. 1989. Pollen morphology and the relationships of Hoplestigmataceae. Taxon 38: 12–16.

Nowike JW and JE Ridgway 1973. Pollen morphology in the genus *Cordia*. Am. J. Bot. 60: 584–591.

Olmstead RG and D Ferguson. 2001. A molecular phylogeny of the Boraginaceae/Hydrophyllaceae. In Botany 2001: Plants and People. Abstracts, p. 131. Albuquerque.

Ovchinnikova SV. 2001. Palynomorphology of the genus *Craniospermum* (Boraginaceae). Bot. Zhurn. 86(12): 44–50 (in Russian with English summary).

Pal PK. 1963. Comparative studies in four species of *Heliotropium*. Proc. Natl. Inst. Sci. India 29: 1–40.
Paria N, P Kamilya, and B Bhattacharyya. 1995. Seedling morphology in taxonomic study of some members of the Boraginaceae. J. Nation. Bot. Soc. (India) 52(1–2): 75–81.
Records SJ and RW Hess. 1941. American woods of the family Boraginaceae. Trop. Woods 67: 19–33.
Retief E, and Van Wyk AE. 1997. Palynology of southern African Boraginaceae; the genera *Lobostemon, Echiostachys* and *Echium*. Grana 36: 271–278.
Retief E and AE van Wyk. 2005. Boraginaceae. Codonoideae, a new subfamily based on *Codon*. Bothalia 35: 78–80.
Richardson AT. 1977. Monograph of the genus *Tiquilia* (*Coldenia*, sensu lato), Boraginaceae: Ehretioideae. Rhodora 79: 467–572.
Riedl H. 1968. Der neue Tribus Trigonotideae und das System der Boraginoideae. Oesterr. Bot. Z. 115: 291–321.
Riedl H. 1997. Boraginaceae. In: C Kalkman et al., eds, Flora malesiana, ser. 1, 13: 43–144. Leiden.
Sahay SK. 1978. On the pollen morphology of Ehretiaceae with reference to taxonomy. In Fourth International Palynological Conference. Proceedings: volume 1. Birbal Sahni Institute of Palaeobotany, 1976–77, pp. 471–479. Lucknow.
Seibert J. 1978. Fruchtanatomische Untersuchungen an Lithospermeae (Boraginaceae). Dissertationes Botanicae 44: 1–207.
Selvi F and M Bigazzi. 2001. Leaf surface and anatomy in Boraginaceae tribe Boragineae with respect to ecology and taxonomy. Flora 196: 269–285.
Smith RA and CW DePamphilis. 1998. Phylogenetic placement of the holoparasistic family Lennoaceae: preliminary molecular evidence. Am. J. Bot. 65: 157.
Smith RA, DM Ferguson, TJ Barkman, and CW DePamphilis. 2000. Molecular phylogenetic evidence for the origin of Lennoaceae: A case of adelphoparasitism in the angiosperms? Am. J. Bot. 87(Suppl.): 158.
Steyermark JA. 1968. El genero *Lennoa*. Acta Bot. Venez. 3: 213–237.
Suessenguth K. 1927. Über die Gattung *Lennoa*. Flora 122: 264–301.
Svenson HG. 1925. Zur Embryologie der Hydrophyllaceen, Boraginaceen, und Heliotropiaceen. Uppsala Univ. Arsskr., 2d ser.
Thulin M and ANB Johansson. 1996. Taxonomy and biogeography of the anomalous genus *Wellstedia*. In: LJG van der Maesen, XM van der Burgt, JM van Medenbach de Rooy, eds. The biodiversity of African plants, pp. 73–86. Kluwer, Dordrecht.
Trivedi ML, V Khanna, and J Shailja. 1976. Nodal anatomy of certain members of Boraginaceae. Proc. Indian Acad. Sci. 84B: 31–36.
Van der Ham RWJM. 1997. Palynology of Boraginaceae. In: C Kalkman et al., eds. Flora malesiana ser. 1, 13: 48–51. Leiden.
Venkateswarlu J and B Atchutaramamurtu. 1955. Embryological studies in Boraginaceae: I. *Coldenia procumbens* L. J. Indian Bot. Soc. 34: 235–247.
Wilson KA. 1960. The genera of Hydrophyllaceae and Polemoniaceae in the southeastern United States. J. Arnold Arbor. 41: 197–212.
Yatskievych G. 1985. Notes on the biology of the Lennoaceae. Cact. Succ. J. (Los Angeles) 57: 73–79.
Yatskievych G and C Mason, Jr. 1986. A revision of the Lennoaceae. Syst. Bot. 11: 531–548.
Yatskievych G and M Zavada. 1984. Pollen morphology of Lennoaceae. Pollen et Spores 26: 19–30.

Order 123. OLEALES

Trees or shrubs, sometimes woody climbers or scramblers; *Jasminum* twining anticlockwise. The indumentum characteristically consists of scale-like, peltate trichomes, sometimes glandular or appearing as transparent or sunken dots in the leaf; groups of secretory type of trichomes frequently form extrafloral nectaries that are common in some of the genera. Small acicular or prismatic crystals of calcium oxalate occur in both stem and leaf. Vessels with scalariform or scalariform and simple perforations (in *Forsythia* infrequently occur scalariform perforations, and vestigial scalariform plates present in *Syringa* and *Fontanesia*); pits vestured, lateral pitting alternate, often more or less distinctly vestured (Baas et al. 1988). Fibers with bordered or simple pits, sometimes septate. Rays heterogeneous to homogeneous. Axial parenchyma apotracheally diffuse or more often paratracheal. Sieve-element plastids of S-type. Nodes unilacunar. Leaves often deciduous, or evergreen, opposite or rarely (some species of *Jasminum*) alternate, petiolate, simple and often entire or lobed to pinnately compound or trifoliolate or unifoliolate, conduplicate (*Chionanthus*), estipulate. Stomata usually anomocytic. Domatia occurring in six genera and numerous species and manifested as pits, or pockets, or hair tufts. The nuclei of mesophyll parenchyma cells contain characteristic crystalline inclusions. Flowers mostly rather small, basically in dichasial cymes but usually modified so as to give the appearance of racemes, panicles or fascicles, sometimes solitary, bisexual or rarely polygamous or dioecious, actinomorphic, usually 4-merous. Calyx mostly small, 4(-15)-lobed, valvate, rarely absent (some spp. of *Fraxinus*). Corolla sympetalous, 4(-15)-lobed, but sometimes, as in spp. of *Fraxinus* very deeply lobed or divided and seemingly choripetalous, lobes imbricate, valvate (or induplicate-valvate), or contorted, sometimes (as in *Forestiera* and some spp. of *Fraxinus*) obsolete, rarely corolla absent. Stamens usually two, rarely four (*Hesperelaea, Tessarandra*), very rarely three or five, attached to the corolla tube (when the tube is expressed); filantherous, or with sessile anthers;

anthers dorsifixed or basifixed, introrse, tetrasporangiate, with locules usually back to back, often apiculate, opening longitudinally; endothecium developing fibrous thickenings. Tapetum secretory. Microsporogenesis simultaneous. Pollen grains 2-celled or very rarely 3-celled, tricolpate, or tricolporate, occasionally rupate, with reticulate ornamentation. A nectary intrastaminal disc sometimes present around the base of the ovary. Gynoecium of two carpels, mostly median; style terminal, with entire, 2-lobed or bifid stigma or sometimes stigma sessile; ovary superior, 2-locular, with two, one or four to many ovules per locule, attached to the apex, side, or base of the partition. Ovules pendulous or ascending, with dorsal raphe, usually anatropous, unitegmic, with usually massive integument, tenuinucellate, with persistent endothelium around the lower part of female gametophyte. Female gametophyte monosporic, usually of *Polygonum*-type, rarely (*Olea*) bisporic of *Allium*-type. Endosperm cellular or sometimes (*Jasminum*) nuclear. Fruits of various types: usually 1-seeded berries, drupes, sometimes samaras (*Fraxinus*), woody loculicidal or circumscissile (*Menodora*) capsule. Seeds small, exotestal (exotesta palisade or not, endotesta fibrous), embryo straight, cotyledons flat, endosperm copious, oily, thick-walled, or cotyledons storage and endosperm almost absent. Usually producing mannitol, flavones and flavonols, carboxycyclic iridoid compounds, myricetin, and the phenolic glycosides orobanchin and syringin as well as verbascosides (Jensen et al. 2002), ursolic acid. Triterpenes and saponins are also present. n = 11, 13–14, 18, 20, 22–24.

Some authors, including Dahlgren, consider Oleaceae a separate order Oleales related to the Rubiales (Gentianales), especially to the Loganiaceae, and others (Cronquist 1981, 1988) placed them in the Scrophulariales, and Thorne (1992a, b, 2000, 2006) – in the Lamiales. Wood anatomy supports the treatment of Oleaceae in or near the Rubiales, closest to the Loganiaceae s. 1. (Baas et al. 1988: 168). According to Hegnauer (1969), chemical evidence also supports affinities of the Oleaceae with Rubiales. However, recent molecular data (*rbc*L sequences, Olmstead et al. [1992]) and restriction site mapping of the chloroplast DNA (Downie and Palmer 1992) seem to agree with the position of the Oleaceae near the Scrophulariaceae (Jensen 1992). "It may have been the presence of seco-iridoids, which never occur in Lamianae, that was decisive for Dahlgren's removal of Oleaceae from the superorder. However, the type of seco-iridoids in Oleaceae is unique when compared to those found in other seco-iridoid-containing taxa.... Thus, the presence of seco-iridoids alone may not be justification for keeping Oleaceae in Gentianales" (Jensen 1992: 293). Serological studies support affinities with both Rubiales and Scrophulariales (Piechura and Fairbrothers 1979). On the other hand, the chloroplast DNA data suggest that the Oleaceae are more closely related to the Scrophulariales than to the Rubiales (Downie and Palmer 1992). In some respects the Oleales occupy an intermediate position between Gentianales and Lamiales assemblage.

1. OLEACEAE

Hoffmannsegg et Link 1809 (including Bolivariaceae Grisebach 1838, Forestieraceae Endlicher 1841, Fraxinaceae Vest 1818, Jasminaceae A.L. de Jussieu 1789, Ligustraceae G. Meyer 1836, Nyctanthaceae J. Agardh 1858, Schreberaceae Schnitzlein 1857–1870, Syringaceae Horaninow 1847). 25/615–900. Nearly cosmopolitan, but best developed in Southeast Asia and Australia; the largest genus *Jasminum* (450) is confined to the Old World, mainly to tropical and subtropical regions.

JASMINEAE: *Menodora, Jasminum;* FORSYTHIEAE: *Abeliophyllum, Forsythia;* FONTANESIEAE: *Fontanesia;* MYXOPYREAE: *Myxopyrum, Nyctanthes, Dimetra*; FRAXINEAE: *Fraxinus*; SCHREBEREAE: *Comoranthus, Schrebera;* OLEEAE: *Syringa* (including *Ligustrina*), *Ligustrum, Olea, Chionanthus, Forestiera, Haenianthus, Tessarandra, Noronhia, Notelaea, Nestegis, Osmanthus, Phyllyrea, Picconia, Hesperelaea.*

Bibliography

Airy Shaw HK. 1952. Notes on the taxonomic position of *Nyctanthes* and *Dimetra*. Kew Bull. 1952: 271–272.

Altamura Betti MM, G Pasqua, and G Mazzolani. 1982. Embryogenesis in *Olea europaea* L. Ann. Bot. (Rome) 40: 141–152.

Baas P, PM Esser, MET van der Westen, and M Zandee. 1988. Wood anatomy of the Oleaceae. IAWA Bull., n.s., 9: 103–182.

Bedi YS and SS Bir. 1986. Cytological observations on certain woody members of family Oleaceae. New Bot. 12(2–4): 135–141.

Bigazzi M. 1989. Ultrastructure of nuclear inclusions and the separation of Verbenaceae and Oleaceae (incl. *Nyctanthes*). Plant Syst. Evol. 163: 1–12.

Devi HM. 1975. Embryology of *Jasminum* and its bearing on the position of Oleaceae. Acta Bot. Indica 3: 52–61.
George K, VA Albert, and CA Ninan. 1989. Chromosome evolution in Oleaceae. J. Cytol. Genet. 24: 71–77.
Green PS. 2004. Oleaceae. In: K Kubitzki, ed. The families and genera of vascular plants, vol. 7, pp. 296–307. Springer, Berlin/Heidelberg/New York.
Harborne JB and PS Green. 1980. A chemotaxonomic survey of flavonoids in leaves of the Oleaceae. Bot. J. Linn. Soc. 81: 155–167.
Inamdar JA, RC Patel, and JSS Mohan. 1986. Structure and ontogeny of stomata in some Oleaceae. Feddes Repert. 97: 147–160.
Jensen SR, H Franzyk, and E Wallander. 2002. Chemotaxonomy of the *Oleaceae:* iridoids as taxonomic markers. Phytochemistry 60: 213–231.
Johnson LAS. 1957. Review of the family Oleaceae. Contr. N. S. W. Natl. Herb. 2: 395–418.
Kapil RN and RS Vani. 1966. *Nyctanthes arbortristis* Linn.: embryology and relationships. Phytomorphology 16: 553–563.
Kiew R and P Baas. 1984. *Nyctanthes* is a member of the Oleaceae. Proc. Indian Acad. Sci. (Plant Sciences) 93: 349–358.
Kim K-J and RK Jansen. 1998. Paraphyly of Jasminoideae and monophyly of Oleoideae in Oleaceae. Am. J. Bot. 85(6): 139 (abstract).
Kshetrapal S and YG Tiagi. 1970. Structure, vascular anatomy, and evolution of the gynoecium in family Oleaceae and their bearing on the systematic position of the genus *Nyctanthes* L. Acta Biol. Acad. Sci. Hungar. 16: 143–151.
Kuriachen PM and YS Dave. 1989. Structural studies in the fruits of Oleaceae with discussion on the systematic position of *Nyctanthes* L. Phytomorphology 39: 51–60.
Lakshminarayana K and HM Devi. 1985. Embryology of *Linociera intermedia* (Oleaceae). J. Jpn. Bot. 95: 213–219.
Li J, JH Alexander, and D Zhang. 2002. Paraphyletic *Syringa* (Oleaceae): evidence from sequences of nuclear ribosomal DNA ITS and ETS region. Syst. Bot. 27: 592–597.
Litvinenko NM. 1987. Oleaceae. In: MS Yakovlev, ed. Comparative embryology of flowering plants: Davidiaceae–Asteraceae, pp. 154–158. Nauka, Leningrad (in Russian).
Mohan JSS and JA Inamdar. 1983. Studies of the leaf architecture of the Oleaceae with a note on the systematic position of the genus *Nyctanthes*. Feddes Repert. 94: 201–211.
Murthy GSR, KN Aleykutty, VS Rao, and JA Inamdar. 1978. Vessels of Oleaceae and Verbenaceae. Feddes Repert. 89: 359–368.
Nilsson S. 2000. Fragrance glands (osmophores) in the family Oleaceae. In: G Nordenstam, El-Ghazaly, M Kasas, eds. Plant systematics for the 21st century, pp. 305–320. Portland Press, London.
Piechura JE and DE Fairbrothers. 1979. Serological investigation of the Oleaceae and putative relatives. Bot. Soc. Am. Misc. Ser. Publ. 157: 65.
Piechura JE and DE Fairbrothers. 1983. The use of protein-serological characters in the systematics of the family Oleaceae. Am. J. Bot. 70: 780–789.
Qin X-K. 1996. The use of peroxidases in the systematics of Oleaceae. Acta Bot. Yunn. 18: 159–166.
Rohwer JG. 1993. A preliminary survey of the fruits and seeds of the Oleaceae. Bot. Jahrb. Syst. 115: 271–291.
Rohwer JG. 1994. Fruits and seeds of *Nyctanthes arbor-tristis* L. (Oleaceae): a comparison with some Verbenaceae. Bot. Jahrb. Syst. 115: 461–473.
Rohwer JG. 1996. Die Frücht- und Samenstrukturen der Oleaceae. Bibliogr. Bot. 148: 1–177.
Taylor H. 1945. Cytotaxonomy and phylogeny of the Oleaceae. Brittonia 5: 337–367.
Vetlugina TG. 2000. Bark anatomy and taxonomic relationships between *Ligustrina* Rupr. and *Syringa* L. (Oleaceae). Byull. Mosk. Obshch. Ispyt. Prir., Biol. 105(2): 35–39 (in Russian).
Wallander E and VA Albert. 2000. Phylogeny and classification of Oleaceae based on *rps*16 and *trn*L-F sequence data. Am. J. Bot. 87: 1827–1841.
Weng R-F and M-Z Zhang. 1992. Chromosome numbers in Chinese Oleaceae I. Investigation et Studium Naturae 12: 66–77.
Wilson KA and CE Wood. 1959. The genera of Oleaceae in the southeastern United States. J. Arnold Arbor. 40: 369–384.

Order 124. LAMIALES

Herbs and subshrubs, less often shrubs and trees, sometimes lianous. Young twigs often quadrangular. Stems sometimes with anomalous structure. Vessels with simple or rarely scalariform perforations; lateral pitting alternate. Fibers usually with simple pits, often septate. Rays mostly heterogeneous of various types to homogeneous. Axial parenchyma paratracheal, very scanty or absent. Intraxylary phloem rare. Sieve-element plastids usually of S-type, but in Acanthaceae also of Pcs- and Pc-types. Nodes unilacunar or rarely trilacunar or multilacunar. Leaves alternate or opposite, sometimes verticillate, simple or sometimes compound, almost always estipulate; leaf mesophyll cells sometimes contain oil bodies. Stomata of various types. Flowers in cymose or racemose inflorescences or solitary, bisexual or rarely unisexual, usually zygomorphic. Perianth 5-merous or less often 4-merous. Sepals more or less connate, lobed or toothed, sometimes bilabiate, sometimes persistent. Corolla sympetalous, often bilabiate, sometimes more or less actinomorphic, lobes imbricate. Stamens attached to the corolla tube and alternate with the corolla lobes, in flowers with 4-merous perianth as many as the lobes or only two, in flowers with 5-merous perianth usually only four fertile stamens and the upper (adaxial) sterile or wanting or staminodial, or only two fertile stamens and the others are wanting or staminodial. Anthers tetrasporangiate or less often disporangiate, opening longitudinally. Tapetum secretory or rarely (Bignoniaceae) amoeboid. Microsporogenesis simultaneous. Pollen grains 2-celled or 3-celled, 3-aperturate or of various derived types. Intrastaminal nectary disc usually present. Gynoecium mostly of two united median carpels, sometimes pseudomonomerous, rarely of three or four carpels; style terminal or gynobasic, with 2-lobed stigma (style sometimes very short and the stigma is essentially sessile); ovary superior or less often semi-inferior or inferior, with (1)2 or mostly many ovules per locule. Ovules anatropous, hemitropous, amphitropous, or less often orthotropous, unitegmic, tenuinucellate, usually with endothelium. Female gametophyte usually of *Polygonum*-type. Endosperm cellular. Endosperm haustoria developed. Fruits capsules or capsulelike, or sometimes indehiscent and then drupaceous or baccate, sometimes separating into drupelike mericarps. Seeds exotestal, with copious endosperm or endosperm scanty or wanting. Usually producing various iridoid compounds (but not secoiridoids) and the phenolic glycoside orobanchin, flavones, but only rarely alkaloids.

Lamiales are related to the Oleales. In some respects the Buddlejaceae are a connecting link between these orders.

Key to Families

1 Flowers essentially actinomorphic.
  2 Perianth usually 4-merous.
    3 Trees or shrubs, rarely lianas or suffrutescent herbs, often with lepidote or stellate or branching trichomes and/or glandular hairs. Vessels with simple perforation, without vestured pits. Fibers with simple or obscurely bordered pits, septate (in some *Buddleja, Peltanthera, Nuxia*). Rays heterogeneous. Axial parenchyma sparse, paratracheal, or absent. Leaves opposite or less often verticillate, rarely (a few spp. of *Buddleja*) alternate, entire or more often dentate or lobed, with interpetiolar stipules usually represented by a line, rarely well developed. Stomata paracytic, anomocytic, or anisocytic, and these types often mixed on one plant (Oxelman, Kornhall and Norman 2004). Flowers in various types of inflorescences, bisexual or often functionally dioecious, bracteate and bracteolate, 4-merous, rarely (*Peltanthera*) 5-merous. Calyx lobed or very deeply cleft, persistent, imbricate. Corolla with imbricate or valvate lobes, tubular, funnelform. Stamens four, inserted midway down the corolla tube, or in the throat of the corolla tube; anthers bilocular, or 4-locular (*Androya*), dorsifixed or peltate, introrse. Pollen grains 2-celled, 3(4)-colporate. Nectary disc absent, except for *Gomphostigma*. Gynoecium of two carpels, or of four carpels (in some *Buddleja* species); style with a capitate or lobed stigma; ovary 2-, or 4-locular, superior, locules secondarily divided by 'false septa', or without 'false septa'; ovules numerous, hemitropous, pendulous or ascending, on thickened, axile placentas. Fruits septicidal capsules, rarely barriers (*Nicodemia*). Seeds often winged, sometimes with pockets between the exotesta and the endothelium (Norman 2000); embryo straight, small to large, endosperm oily or starchy, copious to scanty. Produce iridoids aucubin and catalpol, and verbascosides (Jensen 2000). n = 7, 14, 15, 19. . . . . 1. BUDDLEJACEAE.
    3 Small herbs, terrestrial, or subaquatic to aquatic.
      4 Fruits capsular, with numerous seeds. Small perennial, rarely annual herbs arising from a central taproot with numerous sympodial branches; branches procumbent to ascending, fluted, often with four sharp flanges and 2- or 4-rounded ridges. Leaves opposite, sessile, connate, simple, entire, linear, slightly scabrid along the margins and stem ridges, stipulate or estipulate, stipules, if present, interpetiolar. Young stems tetragonal. Flowers axillary, solitary, bracteolate (the bracteoles sepal-like), small, sessile. Calyx divided almost to the base, persistent. Corolla campanulate, imbricate, with a ring of hairs in the throat. Stamens usually four, included, filaments about as long as the anthers, alternating with the corolla members; anthers 2-locular, introrse. Pollen grains 3-colporate. Gynoecium of two carpels; style short, persistent with a capitate stigma; ovary 2-locular, partly inferior. Ovules 20–50 per locule, anatropous, on peltate placentae that are attached to the base of the septum. Endosperm cellular. Endosperm haustoria chalazal or micropylar. Fruits loculicidal and valvular capsules, laterally compressed, obovoid to ovoid. Seeds small, wingless, cuboidal, with a smooth surface, translucent;

embryo straight; endosperm copious. Detected verbascosides and cornoside, n = 11. . . . . 2. POLYPREMACEAE.

4 Fruits separate into four basally attached, 1-seeded, setulose, dry mericarps. Mostly subaquatic to aquatic, succulent, small creeping herbs with prostrate rooting stems emitting short, leafy branches, bearing essential oils. Leaves opposite, leathery, with flat connate petioles, ciliate toward base, minutely denticulate distally, more or less fleshy, entire, simple, obscurely gland-dotted, aromatic. Flowers solitary, terminal or axillary, minute. Sepals connate, imbricate, persistent. Petals connate, imbricate, corolla subrotate, without a ring of hairs in the throat. Stamens four, inserted to the corolla tube, alternating with the corolla members; anthers 2-locular, dorsifixed, introrse. Pollen grains 3-colporate, rhomboidal, exine regularly dimorphic, the polar regions and borders of colpi psilate and tectate-perforate, the mesocolpal regions intectate and baculate. Gynoecium of two carpels, with long or short, slender gynobasic style and inconspicuous stigma; ovary superior, 4-locular (ostensibly), or 2-locular with four locelae via false septa, with one erect anatropous ovule per locule; placentation basal. Mericarps comprising nutlets. Seeds with copious endosperm. Contain verbascoside. . . . 3. TETRACHONDRACEAE.

2 Flowers apetalous.

5 Gynoecium of two united carpels; style gynobasic, deeply cleft or often divided to the base. Slender perennial or annual herbs; some of them entirely submerged aquatics with underwater pollination, others are amphibious, and few are terrestrial in wet habitats with aerial pollination. Glandular, fan-like scale hairs in the leaf axils, in terrestrial and amphibious species additionally peltate, disc-like hairs. Vascular system of the stem very reduced, xylem with or without vessels. Leaves small, opposite, often in rosettes at branch tips, usually linear and bifid when submerged, linear to spatulate when floating and aerial, epulvinate. Stomata present, sporadically on submerged leaves, mainly confined to one surface (adaxial, when floating), or on both surfaces. Flowers without perianth, minute, unisexual, monoecious or dioecious, usually solitary in axils or rarely one flower of each sex in the same axil, usually with two, minute, hornlike bracteoles. Male flower consists of one stamen, rarely of 2–3 stamens; anthers reniform. Pollen grains small, spheroidal, intectate, with three, not well-defined apertures or inaperturate, 3-coplate. Ovary superior, somewhat compressed, laterally 4-lobed, apically somewhat indented; each carpel longitudinally divided by an intrusive partition from the carpellary midrib into two locelli each with one pendulous, anatropous ovule with ventral raphe. Endosperm with terminal, chalazal and micropylar (the latter the more aggressive) haustoria. Fruits dry, 4-lobed, with each lobe winged or keeled, splitting into four one-seeded mericarps. Seeds minute, with membranous testa, straight or slightly curved embryo and fleshy, oily endosperm; cotyledons short. Contains carbocyclic iridoid aucubin (Wieffering 1966), verbascosides, flavones and caffeic acid, n = 3–6, rarely 10, 14, 19–20. . . . . 13. CALLITRICHACEAE

5 Gynoecium monocarpellate (derived from a pseudomonomerous one), with long subulate stylodium stigmatic throughout. Perennial water herbs with creeping, sympodial rhizome and erect stems, whose upper parts usually project above the water. Vascular system of the axial organs very reduced. Sieve-element plastids of S-type. Leaves verticillate, entire, linear, estipulate, provided with minute, deciduous, peltate, glandular trichomes; submersed leaves long, pale green and flaccid, the emergent leaves obovate to linear, dark green and rigid. Flowers small and inconspicuous, solitary and sessile in the axils of upper leaves, bisexual or sometimes some or all of them unisexual, actinomorphic, apetalous, anemophilous. Calyx reduced to 2–4-lobed or subentire rim around the top of the ovary. Stamens solitary, abaxial, with slender filament and large, red, tetrasporangiate, introrse anther opening longitudinally. Tapetum secretory. Microsporogenesis simultaneous. Pollen grains 3-celled, 4–6-colporate. Ovary inferior,

1-locular, with one apical and pendulous, anatropous, unitegmic to ategmic, tenuinucellate ovule with an endothelium. Endosperm cellular, with inconspicuous terminal haustoria. Female gametophyte of *Polygonum*-type. Fruits drupaceous, with very thin, fleshy exocarp and hard endocarp. Seeds with large, elongate, straight embryo and very thin, proteinaceous endosperm. Producing the carbocyclic iridoid aucubin and catalpol, n = 8, 15, 16. . . . . 14. HIPPURIDACEAE.

1 Flowers mostly more or less zygomorphic. Perianth mostly 5-lobed.

6 Plants with ericoid habit and densely crowded, verticillate leaves and glandular trichomes. Trees, small shrubs or shrublets; shrubs single-stemmed, or multistemmed, young branches often angular and puberulous. Vessels simple, rarely scalariform. Axial parenchyma vasicentric, scarce or lacking. Leaves opposite or rarely alternate, entire, narrowly lanceolate to linear, with recurved margins, usually estipulate. Inflorescences terminal or rarely on axillary branches (*Retzia*). Flowers bisexual, each flower subtended by leafy bract and two bracteoles. Calyx of 4–5 sepals, usually tubular basally, with 4–5 free, imbricate to valvate lobes. Corolla actinomorphic or zygomorphic, with narrowly funnel-shaped tube and 5- or less often 4-lobed, imbricate. Stamens usually four or five, equal; filaments free, short to long, slender or stout, inserted in the throat of the corolla tube, or between the lobes; anthers 2-locular, dorsifixed. Pollen grains 3-colporate, with punctate tectum and granular membrane of colpi. Small, inconspicuous nectary disc usually present. Gynoecium of two carpels; ovary superior, 2-locular, often becoming 1-locular by abortation or loss of septum. Ovules 1–4(-6) in each locule; ovules anatropous, ascending. Style terminal, filiform, erect, stigma capitate or minutely bilobed. Fruits loculicidal and valvular capsules. Seeds reticulate, with straight embryo and oily or starchy endosperm. Present C-8–C-11 iridioid glucosides (Frederiksen et al. 1999), and verbascosides; n = 12 (*Retzia*), 19 (*Nuxia*). . . . . . 5. STILBACEAE.

6 Plants not ericoid. Vessels with simple perforations (except for a few members of Bignoniaceae).

7 Plants not insectivorous. Placentation not free-central.

8 Seeds mostly with copious endosperm (except some Gesneriaceae).

9 Corolla not scarious. Flowers mostly entomophilous or ornithophilous.

10 Leaves mostly alternate.

11 Ovules (2-) more or less numerous on each placenta. Herbs, subshrubs, or shrubs, sometimes climbers, rarely trees (*Paulownia, Wightia*); plants sometimes succulent, rarely hydrophytic (*Limnophila*), sometimes totally parasitic, or partially parasitic, or autotrophic. Often provided with hairs that commonly have a basal cystolith, often also with other types of trichomes including glandular hairs. Vessels with simple perforation, without vestured pits; wood partially storied or not; axial parenchyma apotracheal or paratracheal, usually very sparse, or absent. Nodes unilacunar. Leaves alternate or less often opposite, rarely verticillate, entire or sometimes dissected, often reduced to scales, or absent (e.g. *Lathraea*). Stomata usually anomocytic or anysocytic. Flowers in various types of cymose or racemose inflorescences or sometimes solitary, bisexual, usually zygomorphic. Calyx deeply (1–2)4–5-lobed, persistent, lobes imbricate or valvate, rarely open. Corolla (4)5(-8)-lobed, imbricate or valvate, often bilabiate, sometimes spurred, persistent or deciduous, rarely (spp. of *Besseya*) wanting. Stamens usually four (with adaxial one staminodial or staminodium absent) or sometimes five (as in *Verbascum*) or only 2(3) (with the abaxial pair reduced or wanting); filaments sometimes appendiculate; anthers bilocular, or rarely (Selagineae) unilocular, tetrasporangiate or rarely disporangiate, opening by longitudinally. Pollen grains

2-celled, (2)3(-6)-colporate or panto-colpate, rarely inaperturate, mostly more or less reticulate. A unilateral or annular intrastaminal nectary disc usually present. Gynoecium of two or rarely three carpels; style with 2(-4)-lobed to capitate or crateriform stigma; ovary 2(3)-locular, rarely (Orobanchoideae) 1-locular, with more or less numerous (rarely solitary), pendulous to ascending, anatropous or hemitropous, seldom amphitropous or campylotropous ovules in each locule. Fruits usually capsules, septicidal or less often loculicidal, or occasionally poricidal, or circumscissile, or splitting irregularly, rarely baccate, very rarely schizocarps. Seeds small to minute, exotestal, mostly angular, winged or wingless, without amyloid; embryo straight or slightly curved, in parasitic taxa usually undifferentiated; endosperm oily, very rarely (*Monttea*) lacking. Contain iridoid glycosides (especially aucubin and catapol), pyrrolizidine alkaloids, flavonoids and flavones, saponins. In *Digitalis* containing glycosides (digitalin and lanoxin) and lacking iridoids (Fischer 2004), n = 6–9, 11, 13, 15–17, 20, 21, 23–28. . . . . 6. SCROPHULARIACEAE.

11 Ovules usually solitary in locule. Small shrubs, subshrubs, or rarely perennial herbs. Imperforate tracheary elements are all tracheids. Axial parenchyma scanty, vasicentric, less commonly diffuse. Leaves alternate, simple, entire, coriaceous, estipulate. Flowers in heads (*Globularia*) or spikes (*Poskea*) with or without involucre of bracts, bisexual, zygomorphic. Calyx persistent, tubular, 5-lobed, actinomorphic or rarely slightly bilabiate. Corolla 4–5-lobed, bilabiate (the upper lip of two petals more or less shorter than the 3-petaled lower one), lobes imbricate. Stamens four, inserted at the top of the corolla tube; anthers reniform, at first 2-locular, at length the loculi confluent and opening by a single slit. Pollen grains 3(4)-colporate, tectate-perforate (*Globularia*) or semitectate (*Poskea*), with granular to spinulose colpus membranes. Intrastaminal nectary disc annular, semicircular, or glandlike or absent. Gynoecium pseudomonomerous, with a slender style and capitate or shortly 2-lobed stigma; ovary superior, 1-locular, with anatropous, pendulous ovule. Fruits nutlike, enveloped by the persistent calyx. Seeds with straight long embryo surrounded by scanty endosperm. Present iridoids: globularin, catalpol, globularicosin, globularidin, asperulin, and lytanthosalin (Hegnauer 1989b), n = 8, 16. . . . . 11. GLOBULARIACEAE.

10 Leaves mostly opposite.

12 Ovary bilocular.

13 Ovary superior to semi-inferior. Annual or perennial herbs or shrubs (*Stemotria*), glandular pubescent to densely villose, rarely glabrous. Stems erect, prostrate or ascending. Leaves cauline or in a basal rosette, sessile to petiolate, linear lanceolate to ovoid-orbicular, acuminate, entire or pinnatifid to pinnatisect. Inflorescence thyrsic, with accessory flowers, paired, each pair composed of a terminal flower and the flower from its reduced prophyll. Flowers distinctly zygomorphic, 4-merous, bright yellow; sepals valvate, free to the base, persistent. Corolla with whorl tube, usually with shoe-like

invagination of lower lit, with trichome nectaries. Stamens two or three (*Stemotria triandra*), filaments short, inserted in a distinct, dark purple ring at the base of the corolla tube; anthers 2-locular. Gynoecium of two carpels; style terminal, with inconspicuous or slightly capitate stigma. Capsule septicidal and loculicidal. Seeds numerous, testa with anticlinal walls sinuous (straight); endosperm present, n = (8)9. . . . . 4. CALCEOLARIACEAE.

13 Ovary inferior. Perennial herbs or subshrubs. Leaves opposite, sometimes oblique, membranous, serrate or irregularly dentate, estipulate, petioles connected by a raised line. Stomata anomocytic. Flowers in dense terminal or axillary cymes, bisexual, slightly zygomorphic or nearly actinomorphic. Calyx adnate to the ovary, with four or five more or less unequal lobes, open, persistent. Corolla campanulate or funnel-shaped, lobes imbricate or induplicate-valvate. Stamens two, epipetalous; filaments short; anthers linear-oblong, sometimes latrorse, cohering round style. Pollen grains 3-colpate (*Silvianthus*) or 5–6-colpate (*Carlemannia*) intectate, broadly reticulate. Nectary disc conspicuous, shortly cylindrical or conical. Gynoecium of two carpels; style elongate, with clavate or fusiform, bifid stigma; ovary 2-locular, with many (30–100) ovules on axile or subbasal placenta. Fruits loculicidal capsules (*Carlemannia*) or fruits open by five irregular, longitudinal valves (*Silvianthus*), calyx persistent. Seeds many, exotesta cells narrow, radial walls thickened, endothelium persistent (Silvianthus) or polygonal, all walls thickened (*Carlemannia*); endosperm ruminate (*Silvianthus*), more or less fleshy, copious; embryo small. 10. CARLEMANNIACEAE.

12 Ovary 1-locular, or with two parietal placentas. Herbs or subshrubs, rarely shrubs or small trees, sometimes lianas or epiphytes. Usually provided with uniseriate hairs or stalked glands or both. Secretory cavities present or absent, when present – with oil, or with resin. Fibers with simple pits and septate. Axial parenchyma paratracheal, or very sparse, or absent. Nodes unilacunar, trilacunar, or pentalacunar. Leaves mostly opposite, rarely verticillate or alternate, entire or dentate, rarely pinnatifid, commonly estipulate. Stomata usually anisocytic (often very large). Flowers in various types of cymose or seldom racemose inflorescences or solitary in the leaf axils, sometimes epiphyllous, bisexual or rarely monoecious, mostly zygomorphic. Sepals five, free or usually connate into a lobed tube, valvate or rarely imbricate. Corolla 5-lobed, usually bilabiate and often spurred, sometimes nearly or quite actinomorphic, lobes imbricate, adaxial (posterior) ones usually internal. Stamens commonly four, or two, with the anthers connivent in pairs or all together (posterior reduced or absent), less often a single pair of stamens developed, rarely (*Ramonda,* some *Sinningia* spp.) all five stamens developed; often 1–3 staminodia present in place of missing stamens, anthers

basifixed or rarely dorsifixed, opening by longitudinal slits or by apical (rarely basal) poles. Pollen grains (2)3(4)-colporate. Intrastaminal nectary disc usually present, usually annular or cupular, sometimes unilateral or of discrete glands. Gynoecium of two carpels, with slender style and usually 2-lobed stigma; ovary superior or more or less inferior, 1-locular, or with two parietal placentas that are usually more or less intruded and bifurcate, sometimes more or less joined and dividing the ovary into two or even four chambers. Ovules numerous, anatropous or very rarely orthotropous. Fruits usually loculicidal or rarely septicidal, or rarely valvular capsules, rarely berries. Seeds small, with straight embryo and oily endosperm (Gesnerideae) or no endosperm (Cyrtandroideae). Iridoids absent. Contain flavonoids, anthocyanins (pelargonidin, malvidin, cyanidin, gesnerin, delphinidin, and etc.), cornosids and some genera also verbascosides and orobanchine, n = (4)8–18(64); n = 30 + in Coronanthereae (Smith 1996)......... 8. GESNERIACEAE.

9 Corolla scarious. Perennial or annual herbs, rarely small shrubs or subshrubs, terrestrial to aquatic, sometimes succulent. Trichomes non-glandular or glandular with a head of one or a few cells. Fibers very short, with simple or bordered pits. Nodes unilacunar, trilacunar, or multilacunar. Leaves usually all basal and alternate, rarely cauline and alternate or opposite, mostly sheathing at the base, with more or less curvi-palmate or nearly parallel venation, or leaves much reduced. Stomata mostly diacytic, sometimes anomocytic. Flowers small, mostly in bracteate spikes or heads, or solitary, without bracteoles, usually bisexual, but monoecious in *Littorella* and gynomonoecious in *Bougueria,* actinomorphic, anemophilous or partly entomophilous, sometimes cleistogamous, 4-merous or rarely 3-mcrous. Calyx membranous, lobed or cleft, sometimes abaxial segments more or less connate, imbricate. Corolla 4-lobed or 4-dentate, rarely 3-lobed (*Bougueria*), lobes imbricate. Stamens usually four or rarely 1–2 (*Bougueria*); filaments usually very long; anthers long-exerted, dorsifixed, versatile. Pollen grains 2–3-celled, 4–14-porate. Gynoecium of two carpels, with slender style and usually 2-lobed stigma; ovary superior, 2-locular, with 5–50, usually anatropous to hemitropous ovules in each locule, or (*Bougueria* and *Littorella*) 1-locular with usually one basal ovule. Fruits membranous circumscissile capsule (*Plantago*) or nutlike in persistent calyx. Seeds with straight or (*Bougueria*) curved embryo in copious translucent endosperm. Present aucubin and related iridoid glucosides, phenylethanoid glycosides (Ronsted et al. 2003), alkaloids, terpenes and saponins, n = 4–6, 12. .... 12. PLANTAGINACEAE.

8 Seeds mostly with scanty endosperm (except some Acanthaceae-Nelsonioideae and Nesogenaceae) or endospermless.

14 Seeds flat, often alate with membranous or corky wing. Trees, shrubs, or often woody lianas, rarely herbaceous vines (*Tourretia*) or erect perennial herbs (*Argylia, Incarvillea, Niedzwedzkia*); the twiners twining anticlockwise (*Tecoma*). Vessels usually with simple perforations, but in some genera some of them with reticulate or scalariform perforation plates that have numerous bars. Fibers with simple or bordered pits. Rays homogeneous to sometimes heterogeneous. Axial parenchyma mostly paratracheal. Nodes unilacunar, with three to several traces. Leaves

opposite or sometimes verticillate, rarely alternate, simple or more often pinnately compound or trifoliolate or palmately compound. Stomata anomocytic, paracytic, or diacytic (*Kigelia*). Inflorescences terminal to axillary, ramiflorous to cauliflorous cymes, or reduced to a single flower. Flowers mostly large, bracteate and bracteolate, 5-merous. Calyx campanulate, lobed or toothed, sometimes bilabiate (spathelike in *Spathodea*) or truncate, rarely calyptrate (*Lundia* spp.). Corolla campanulate or funnelform, usually more or less strongly zygomorphic, sometimes bilabiate, lobes imbricate or rarely valvate. Stamens inserted near the base of the corolla tube, or midway down the corolla tube, typically four in two pairs, the fifth (adaxial) staminodial or absent, rarely all five fertile (*Oroxylum*) or two fertile and three staminodial (*Catalpa*); anthers tetrasporangiate or disporangiate. Tapetum amoeboid. Pollen grains 2-celled, in monads, tetrads, or rarely polyads, 3(-4)-colpate, 3(-4)-colporate, 3-porate, zonocolpate, pericolpate and syncolpate. Annular or sometimes cupular intrastaminal nectary disc usually present. Gynoecium of two carpels; style with 2-lobed stigma; ovary superior, typically 2-locular with two placentas per locule or sometimes 1-locular with two or four more or less intruded parietal placentas or rarely (*Tourrettia*) 4-locular with numerous and usually anatropous ovules uniseriate in each locule. Ovules numerous, ascending, anatropous, hemitropous, or orthotropous. Fruits septicidal or loculicidal, bivalved capsules, very often with a replum, or berries (in Crescentieae and Coleae). Seeds in capsular fruits usually flat, winged; wings can be formed by fused capillary hairs or usually by a thin membrance (Hesse and Morawetz 1980); seed coat exotestal (Corner 1976); embryo straight, with usually enlarged and more or less foliaceous cotyledons; endosperm wanting. Present iridoid glycosides, iridoid alkaloids, flavonols, and flavones, and also ursolic and shikimic acids, n = 11, 14, 15, 18, 20. .... 7. BIGNONIACEAE.

14 Seeds not flat and not winged. Herbs, subshrubs, or shrubs, rarely trees.

15 Style terminal or nearly so.

16 Calyx well developed.

17 Seeds with endosperm. Perennial or annual herbs, rarely shrubs or subshrubs, with both simple and multicellular and usually also glandular hairs. Leaves decussate, simple, entire to dentate or rarely deeply pinnatifid, estipulate. Inflorescence axillary of one or more superposed cymes, the latter 1-several-flowered or flowers collected in spicate bracteate inflorescences; calyx (4-)5-lobed, accrescent; corolla sympetalous, slightly zygomorphic, unequ-ally 5-lobed, the lobes imbricate; stamens didynamous, inserted low down in the tube, included; anthers with small apical connective, opening longitudinally. Pollen grains 3-colpate. Ovary superior, 2-locular, ovule one per locule, anatropous, axile at the base of the ovary. Fruit indehiscent, oblong-ovoid, compressed, shallowly grooved between the two halves, pubescent around the thickened persistent base of the style; testa of the seeds foveolate, the foveoles arranged in longitudinal lines...... 27. NESOGENACEAE.

17 Seeds without endosperm. Large scandent shrubs, sometimes stellate-tomentose. Young twigs quadrangular or terete. Leaves opposite, simple, entire or serrate, or dentate, estipulate. Flowers sessile, in 3–7(-9)-flowered capitate cymes, often aggregate into large terminal or axillary panicles, each cyme with an involucre of more or less membranous, colored, persistent bracts and bracteoles, bisexual, zygomorphic. Calyx 4–5–8-toothed, more or less accrescent or inflated in fruit, teeth open or scarcely valvate. Corolla 6–16(-18)-lobed, lobes subequal or bilabiate, imbricate. Stamens 4–16(-18), alternating with the corolla members. Pollen grains 2-celled, 3-colpate, tectate-perforate or rugulose, rarely operculicolpate. Disc absent. Gynoecium of two carpels, with filiform style and bifid stigma; ovary superior, 2-locular (morphologically), or 1-locular (above); locules partially secondarily divided by 'false septa', with two ovules pendulous from the free-central placenta. Fruits small, dry, indehiscent, 1–4-seeded drupe, more or less included in the persistent calyx; embryo straight. n = 12, 14, 17, 18. . . . 8. SYMPHOREMATACEAE.

18 Plants usually woody.

19 Stamens usually four (the fifth one staminodial or absent), inserted on the corolla, or reduced to two. Trees or shrubs, herbs, sometimes lianas, glabrous or pubescent, nonglandular trichomes usually unicellular. Vessels with simple, or occasionally scalariform perforation. Nodes unilacunar, with 1-several traces. Leaves opposite or rarely verticillate or alternate, entire, simple or pinnately or palmately compound, lobed to dissected, sometimes much reduced, scale-like, or spinose, estipulate. Stomata anomocytic or diacytic, rarely paracytic. Flowers in various types of racemose inflorescences, mainly cymose in the Viticoideae, often with involucre of colored bracts, mostly bracteolate, bisexual or rarely some of them unisexual (dioecious). Calyx 5(4)-lobed or toothed, sometimes zygomorphic. Corolla mostly 5-lobed, with imbricate lobes, mostly more or less zygomorphic, sometimes bilabiate, often with narrow tube, rarely campanulate. Anthers 2-locular, basifixed or dorsifixed. Pollen grains with the thickening of the exine immediately adjacent to the colpi or pores, 3-colpate or more often 3-colporate, sometimes 4-colporate, rarely 3-pororate or 3-porate, tectate-perforate and without supratectal processes, psilate or irregularly rough. Gynoecium of two or rarely (*Duranta*) four carpels; style with usually lobed stigma and with prominent stigmatoid tissue; ovary usually initially 2-locular but later commonly subdivided by intrusion of partitions from the carpellary midribs into four uniovulate locelli; rarely the anterior or superior carpel suppressed. Ovules 1–2 per locule, erect or rarely pendulous, mostly anatropous. Fruits drupaceous with two or four pyrenes or often of 1-seeded, separating, nutlike mericarps. Seeds with straight, oily embryo; endosperm mostly wanting; seed coat consists of several layers of thin-walled cells. Usually

producing iridoid compounds and often the phenolic glycoside orobanchin, less often alkaloids, contain ethereal oils, n = 5, 7, 10–12, 15–20, 22–26, 30. . . . . 24. VERBENACEAE.

19 Stamens 5–6, attached to the corolla tube and alternating with corolla lobes, unequal. Evergreen small trees or shrubs with two types of hairs: simple, unicellular trichomes filled with calcium carbonate and accompanied by cystoliths in the adjacent epidermal cells at the base, and bicellular, club-shaped glandular hairs. Vessels with simple perforations. Leaves opposite or subverticillate, small, leathery, simple entire, and sometimes emarginate, estipulate. Stomata anomocytic. Flowers axillary, in 1–7-flowered congested racemes or dichasia, often reduced to only 1–2 flowers, bisexual, but functionally unisexual, dioecious, slightly zygomorphic. Calyx small, 5–6-partite, imbricate or open, persistent. Corolla 5–6-lobed, tube broadly funnel-shaped, lobes imbricate. Anthers basifixed, introrse to extrorse, 2- or 4-celled. Pollen grains 3-colporate, reticulate, with long columellae and smooth colpus membrane. Nectariferous disc presents at the base of ovary in female flowers. Gynoecium of two united carpels; ovary superior, 1-locular, shortly stipitate, with two parietal placentas, the placentas bearing either one basal erect ovule each, two basal ovules each, or two basal ovules on one and two subapical, ascending or pendulous ovules on the other. Fruits elongated, many-ribbed, and sulcate, 1–4-seeded capsule, 2-valved from the apex. Seeds long-linear, with a dense tuft of hairs at the apex; embryo straight, linear, almost filling the seed; endosperm very thin, fleshy. Colleters absent. Producing quinol glycoside cornoside and two compounds (lugrandoside and echinacoside) belonging to verbascoside group, but lack iridoids. 9. PLOCOSPERMATACEAE.

18 Perennial, erect herbs from short, vertical rhizome, non-aromatic, with simple hairs. Young stems tetragonal. Nodes unilacunar, with three traces. Leaves opposite, entire, coarsely dentate. Stomata anomocytic, sometimes anisocytic. Flowers in elongate, narrow terminal and axillary slender spikes, bracteate, small, zygomorphic, bisexual, bracteate, 5-merous. Calyx tubular, 5-toothed, the dorsal teeth becoming spinose and hooked in fruit, two ventral teeth unchanged. Corolla tubular, bilabiate, purple or blue. Stamens four, inserted midway down the corolla tube to in the throat of the corolla tude, didynamous; filaments glabrous; anthers subreniform. Pollen grains 3-celled, tricolpate. Gynoecium of two carpels, pseudo-monomerous (the abaxial carpel is suppressed); style with 2-lobed stigma, without persistent stigmatoid tissue; ovary oblique, 1-locular, with subbasal, erect, orthotropous ovule; endosperm haustoria chalazal. Fruits 1-seeded achenes enclosed in the persistent accrescent calyx, the testa adnate to the membranous pericarp; seed coat vestigial, embryo oblong, endosperm scanty, consists of only two layers of cells, cotyledons convo-

lute. Iridoids detected, n = 7, 14. ....25. PHRYMATACEAE.

16 Calyx absent or reduced to a minute disc-like rim (in *Cyclocheilon physocalyx*), its function taken over by two bracteoles. Shrubs or subshrubs with red roots (?) and simple and glandular hairs. Leaves opposite or subopposite, simple, entire, estipulate. Flowers solitary on bracteate axillary pedicels, bisexual; bracteoles lateral, membranous, more or less free or joined, flat or convex and winged, their margins closely adpressed and enclosing the bud, accrescent. Corolla obliquely funnel-shaped, 5-lobed, lobes broad, imbricate. Stamens four, didynamous; filaments long-pilose, inserted on corolla tube; anthers with a short connective. Pollen grains 3-, rarely 4-colpate, tectum very finely perforate-rugulate. Gynoecium of two carpels; style curved, with lingulate stigma; ovary compressed, completely or incompletely 2-locular or 1-locular, with 2–10 ovules on long funicles; placentation axile or parietal. Fruits very strongly compressed discoid 1-locular capsules with loculicidal dehiscence (*Cyclocheilon*) or breaking up into two 1-seeded mericarps (*Asepalum*). Seeds without endosperm......... 26. CYCLOCHEILACEAE.

15 Style mostly gynobasic. Perennial or annual herbs, subshrubs, shrubs (sometimes ericoid), or less often trees, rarely woody climbers, mostly aromatic. Stems erect to prostrate, sometimes forming stolons or large or slender rhizomes; young stems and twigs usually quadrangular. Indumentum of glandular and non-glandular trichomes, rarely scale-like, dendroid or stellate. Leaves opposite or sometimes verticillate, very rarely alternate, dissected or entire, simple or occasionally pinnately compound, stipulate or estipulate. Stomata mostly diacytic. Inflorescences of various types, mostly composed of small, compact, axillary cymes that form a verticillaster at each node, or sometimes the axillary cymes reduced to single flowers; flowers mostly bracteolate, bisexual or sometimes unisexual, more or less strongly zygomorphic. Calyx 2- (e.g. *Prostanthera*), or 3- (in *Melittis*), or 4- (e.g. *Preslia*) lobed, but basically 5-lobed or dentate, tubular or more or less bilabiate, seldom more or less actinomorphic with four lobes, as in *Mentha* (one lobe representing two connate lobes), imbricate, persistent. Corolla limb 5- or rarely 4-lobed, mostly bilabiate, imbricate. Stamens two or four, usually didynamous, sometimes the upper (posterior) pair staminodial, rarely the lower pair sterile (*Mosla*); filaments inserted near the base of the corolla tube, or midway down the corolla tube, or in the throat of the corolla tube, and sometimes the pairs attached at different levels; anthers 1–2-locular, basifixed or dorsifixed, introrse, opening longitudinally or rarely by apical pores; the connective often much developed. Pollen grains 2- or 3-celled, 3(4)-colpate or 6(-12)-colpate,

mostly suprareticulate, colpal membrane mostly without ectexine elements. Intrastaminal nectary disc usually prominent, often developed on anterior side only. Gynoecium of two carpels, sometimes on gynophore; style with 2-lobed stigma; ovary 4-lobed, fundamentally 2-locular, but early in development a constriction appears in the ovary in the antero-posterior plane, longitudinally dividing each carpel in half, so that the ovary becomes divided into four more or less distinct segments united by gynobasic style. Ovules two per locule, or solitary in each of four ovary lobes, basal-axile, usually anatropous to hemianatropous, erect, often with funicular obturator and usually with endothelium; endosperm haustoria usually chalazal and micropylar, the latter aggressive. Fruits of (1-)4 one-seeded, nutlike mericarps with a hard pericarp or rarely mericarps with a fleshy pericarp, rarely druplets (*Stenogyne, Gomphostemma, Prasium*). Seeds with straight or rarely bent (*Catoferia, Scutellaria*), or slightly curved embryo; the radicle directed downward; endosperm wanting or scant, fleshy endosperm often absorbed by the developing embryo. Characterized by very broad spectrum of isoprenoid compounds, caffeic acids derivatives and apigenin- and luteolin-derived flavonoids and by the replacement of starch accumulation in perennial parts by oligo-saccharidcs; store mainly oils rich in linolic or linoleic acids in their starch-free seeds; true tannins and alkaloids are lacking; iridoids present or absent, n = 5–11, 13, 15, 17, 19, rarely up to 120 (*Salvia ombrophila* – Harley et al. 2004). . . 29. LAMIACEAE.

20 Plants characterized by mucilage trichomes with very shortly stalked heads composed of four or more cells, the latter being convexly arched outward from the apex of the stalks, and the lamina almost completely filled with mucilage. Fruits very often with hooks or horns or prickles, or sometimes winged.

21 Plants terrestrial. Ovary superior.

22 Flowers mostly with characteristic glands (extrafloral nectaries) representing abortive flowers axillary to as many bracts at the base of the pedicel. Perennial or annular herbs, rarely shrubs or small trees, covered by mucilage-producing glandular hairs. Leaves opposite, the upper ones alternate, simple, entire, dentate or lobed, estipulate. Stomata anomocytic, anisocytic, or dicytic and paracytic. Flowers solitary or in terminal cymes (usually 3-flowered), bisexual, zygomorphic, 5-merous. Sepals more or less united. Corolla often with oblique limb, lobes imbricate. Stamens four, inserted near the base of the corolla tube, and generally paired, with small, subulate staminodium in place of the fifth (posterior) one; anthers often connivent or contiguous in pairs. Pollen grains 2-celled, 5–13 zonocolpate. Intrastaminal, fleshy nectary

disc present. Gynoecium of two carpels, with long style and 2-lobed stigma; ovary 2–4-locular or apparently 1-locular, often with false septa; ovules one (*Josephinia*), or two to many in each locule, pendulous, or horizontal, or ascending, anatropous. Fruits loculicidal carpels or nuts, often with horns or hooks. Seeds winged or not, testa multiplicative, exotestal cells palisade or otherwise thickened, mesotesta with crystals; embryo straight, endosperm scanty or absent. Contain iridoids, flavone glycoside, verbascoisede-type compounds, n = 8, 13, 16. .15. PEDALIACEAE.

22 Flowers without extrafloral nectaries. Perennial or annual herbs, erect to prostrate, often with tuberous roots. Arial parts covered with glandular-viscid trichomes. Leaves opposite or alternate, petiolate, simple, sticky, entire to sinuate or almost lobed. Stomata anomocytic or anisocytic, paracytic and diacytic. Flowers in terminal racemes, bisexual, zygomorphic. Sepals five, nearly free or partly united, sometimes spathaceous. Corolla 5-lobed, lobes imbricate. Fertile stamens four in two pairs (the fifth reduced to a staminodium) or less often (*Martynia*) two (anterior ones); anthers connivent, tetrasporangiate, opening longitudinally. Pollen grains not colpate, reticulate (*Martynia*), areolate (*Proboscidea* and *Ibicella*), and "ringlike" (*Craniolaria*); exine dissected into 20–40 platelets. Intrastaminal nectary disc glandular. Gynoecium of two carpels, with long style and 2-lobed stigma; ovary 1-locular with two massive parietal T-shaped placentas divided into two divaricate lobes ovules few to numerous, anatropous. Fruits usually hooked at the end or horned, loculicidal after the separation of a soft, deciduous exocarp, mesocarp more or less fleshy, falling off, endocarp wood. Seeds with straight embryo and scanty endosperm. Contain iridoids, flavone glycosides and verbascoside-type compounds, n = 15, 16. . . . . 16. MARTYNIACEAE.

21 Plants aquatic. Ovary inferior. Perennial herbs with creeping rhizomes and more or less floating, elongate slender stems. Leaves opposite, dimorphic, the submerged ones oblong, remotely serrate, the upper (floating) deltoid-rotundate or reniform-cordate, crenate. Both side of the leaves covered by glandular hairs. Stomata anomocytic, exclusively occur on the upper face of the floating leaves. Flowers solitary, axillary, on slender pedicels, bisexual, zygomorphic, chasmogamous from floating leaves, cleistogamous from submerged leaves. Calyx 5-lobed, imbricate, persistent, with five hornlike appendages near the apex, these elongate after anthesis. Corolla tubular, 5-lobed, slightly bilabiate, imbricate. Two posterior stamens fertile, the other two staminodial; filaments filiform; anthers on large peltate connective, included. Pollen grains 3-zonocolpate. Gynoecium of two carpels, with slender style and dilated, shortly 2-lobed

stigma; ovary 2-locular, posterior locule fertile, with two pendulous, anatropous ovules, the anterior one abortive and empty. Fruits narrow-elongate, 1-seeded, indehiscent, crowned with five spreading, rigid appendages below the calyx, subulate and spinose. Seeds linear, 4-ridged; seed coat exotestal (Singh 1960); embryo straight, endosperm scanty. n = ca. 25. . . . 17. TRAPELLACEAE.

20 Plants without mucilage trichomes.

23 Plants characterized by scattered secretory cavities lined with epithelium, which occur in the leaves and in the primary cortex and/ or pith of the axis. Shrubs or sometimes small trees, often with lepidote or plumose indumentum or with gland-tipped hairs. Nodes unilacunar, internal phloem absent; primary vascular tissue centrifugal. Leaves alternate or seldom opposite, entire or dentate, often more or less leathery, petiolate to sessile, often (when secretory cavities large) with pellucid or raised dots. Flowers in small axillary cymes or solitary in the axils, bisexual. Sepals five, connate, lobes imbricate or open. Corolla from essentially actinomorphic (as in *Myoporum*) to strongly zygomorphic (as in *Bontia*), often bilabiate, lobes imbricate. Stamens mostly four (the fifth posterior represented by a staminodium or absent) or rarely five. Anthers connivent or separate, introrse, confluent at apex, opening with longitudinal slits or transversely. Pollen grains 2–4-colporate, often with two pores in each colpus. Gynoecium of two carpels, with terminal, sometimes impressed style and simple stigma; ovary superior, 2-locular, rarely 4-locular, with (1)2 anatropous ovules pendulous near the summit of the partition in each locule, or 4–8 ovules per locule, superposed in pairs, or the locules subdivided into 4–10 uniovulate compartments. Fruits indehiscent drupaceous or separating into 1-seeded, drupe-like segments, the exocarp dry and papery and separable from the thin cartilaginous or thickened endocarp by mesocarp (Henrickson and Flyr 1985). Seeds unwinged or rarely (*Androya*) winged, with more or less straight embryo; endosperm scanty or lacking. Contain iridoid and verbascosides, n = 17, 18. . . . . .18. MYOPORACEAE.

23 Plants without scattered secretory cells. Lamina without secretory cavities. Internal phloem present, or absent.

24 Flowers ebracteate and mostly without bracteoles. Perennial suffrutescent herbs or shrubs, decumbent to ascending, or erect, glabrous or more often leaves and branchlets provided with short, glandular, and long, multicellular and usually nonglandular hairs. Internal phloem present; primary vascular tissue bicollateral. Leaves opposite, sometimes alternate above, simple, lanceolate to ovoid, acuminate, dentate to serrate, not gland-dotted. Stomata anomocytic. Flowers in the axil of the upper leaves, subsessile or with short pedicels, solitary, bisexual, zygomorphic. Calyx

deeply and almost regularly 5-lobed, persistent, lobes are covered with glandular hairs. Corolla with rather narrow cylindrical tube and five lobes that overlap in bud; corolla tube glabrous or often shortly glandular and puberulous on the outer side, and closely covered with hairs pointing forward on the upper half of the inner side. Stamens four inserted on the corolla tube; filaments short and slender; anthers comparatively large. Pollen grains 4-colporate, foveolate, with small perforations. Gynoecium of two carpels; style short, with the stigma directed upward; ovary 2-locular, with 4(-6) ovules per locule, pendulous from the upper part of the septum. Fruits 2-locular (or by abortion 1-locular) globose drupes with one pyrene and extremely hard, thick inner walls. Seeds with stout, clavate funicle and hard, black testa; endosperm wanting, n = 19. . . .19. OFTIACEAE.

24 Flowers bracteate and bracteolate.

25 Ovules anatropous to campylotropous, each with a more or less strongly modified funiculus. Mostly perennial herbs, subshrubs, and shrubs, rarely trees, very diverse in habit, including lianas, xerophytes, and aquatics (even mangrove *Acanthus ilicifolius*). Provided with various types of trichomes. Internal phloem present, or absent. Leaves usually well developed or much reduced, opposite or (Nelsonioideae) alternate, entire or pinnately lobed, gland-dotted or not gland-dotted, simple. Cystoliths very commonly present, showing as streaks in the lamina. Stomata always almost diacytic. Flowers in various kinds of cymose or less often racemose inflorescences or sometimes solitary, the bracts and bracteoles often showy, bisexual. Calyx more or less deeply (4)5–16-lobed, or lobes sometimes much reduced (as in *Thunbergia*), lobes imbricate or valvate, or contorted, or open in bud. Corolla more or less zygomorphic, usually typically 5-lobed and usually bilabiate (upper lip sometimes not developed, as in *Acanthus*), lobes imbricate or contorted. Stamens usually four or two and paired (upper stamens represented by staminodia), rarely all five stamens fertile (*Pentstemona-canthus*). Anthers separate from one another, or connivent, dorsifixed, often with one lobe reduced or abortive, unilocular to bilocular. Pollen monosiphonous, shed as single grains. Pollen grains 3-colpate, or porate, or colporate, or foraminate; 2-celled, or 3-celled. Annular or glandular intrastaminal disc usually present. Gynoecium of two carpels, with slender style and 2-lobed and variously shaped (often funnelform) stigma; ovary superior, usually 2-locular, with two or more (in Nelsonioideae more or less numerous) ovules per locule on each intruded parietal placenta. Fruits usually loculicidal, explosive capsules or (Mendoncieae) drupes. Seeds with large, straight, or more or less curved embryo and usually without endosperm, but in Nelsonioideae endosperm more or less well developed, oily, and

ruminate; in many genera seeds with often hook-like funicular projection (retinaculum or jaculator). Iridoids detected, alkaloids present or absent; flavonols, when present, kaempferol and quercetin; n = 7–21. . . . . 20. ACANTHACEAE.

25 Ovules orthotropous to hemianatropous. Trees or shrubs of tropical coasts with pneumatophores. Branchlets and twigs with jointed appearance from swollen nodes. Secondary thickening anomalous. Intraxylary phloem present. Axial parenchyma apotracheal. Nodes trilacunar with three traces. Leaves opposite (decussate), entire, coriaceous, leathery; simple, estipulate, with salt glands on both sides. Stomata diacytic. Flowers sessile, in terminal or axillary pubescent spicate or subcapitate inflorescences, subtended by a single bracts and a pair of bracteoles, small, bisexual, actinomorphic. Calyx cup-shaped, deeply 5-lobed, lobes imbricate. Corolla actinomorphic or variously zygomorphic, campanulate-rotate, 4(5–6)-lobed, rotate or campanulate, yellow, lobes imbricate. Nectary disc inconspicuous. Stamens mostly four, equal or subdidynamous, inserted in the throat of the corolla tube; anthers sessile. Pollen grains 3-colporate, tectate-perforate to microreticulate, psilate, colpal membrane granular. Gynoecium of two carpels; style with bifid stigma; ovary superior, 2-locular; locules partially secondarily divided by false septa, with a free-central more or less winged placenta; ovules two per locule (one per locellus), hanging from tip of central placenta, orthotropous to hemianatropous, unitegmic, but with incomplete growth of the integument, which does not form a micropyle; endosperm haustoria chalazal and micropylar. Endothelium wanting. Fruits compressed capsule dehiscent by two thick, leathery valves, by abortion regularly 1-seeded, bracteoles and sepals persistent. Seeds imperfectly covered by its testa, with viviparous, green embryo; cotyledons folded around the plumular axis, radicle fully or partly hairy, endosperm lacking. Contain iridoid glycosides, n = 18, 32. . . .21. AVICENNIACEAE.

7 Plants insectivorous.

26 Flowers actinomorphic, nectary spur absent. Evergreen shrubs, subshrubs or perennial insectivorous herbs with woody rhizome bearing in any one season the dying parts of the previous and the growing ones of the present season. Trapping mechanism passive. Indumentum consists of numerous glands of two types – sessile and stalked; the stalked glands, which are found even on the ovary wall, have a mucilaginous head of usually 32 cells radiating from the center and standing out like an umbrella top; the apex of the stalked head traps small insects, while the sessile glands digest them. The glands closely resemble those of *Pinguicula* in the Lentibulariaceae (Conran and Carolin 2004). Vessels mostly with simple perforations, but scalariform perforations with 1–3 bars are also found; *Byblis gigantean* has unusual,

narrow, tracheid-like vessels; lateral pitting alternate. Fibers consisting of true tracheids. Rays mostly biseriate or triseriate, composed of erect cells and very few are square. Axial parenchyma sparse, diffuse. Sieve-element plastids of Ss-type. Nodes unilacunar with one or three traces. Leaves alternate, elongate linear, terete or filiform, crowded, circinate in vernation, estipulate. Stomata paracytic. Flowers solitary in the axils of leaves, bisexual, ebracteolate, 5-merous (except gynoecium). Sepals imbricate, basally connate, persistent. Petals imbricate or contorted, basally shortly connate, broad-cuneate, apically fimbriate, or entire (*B. gigantea*). Stamens five, alternipetalous, inserted at the base of petals or free, often declinate; filaments short, subulate; anthers basifixed, tetrasporangiate, introrse, bilocular, opening by apical pore or very short, porelike slit, with a glandular binucleate tapetum. Pollen grains 2-celled, 3–4-colpate, 3-colporate, or 4-, or 6-rugate. Gynoecium of two united carpels, with elongate, filiform style and capitate or oblong stigma; ovary superior and as a result of the fusion of two parietal placentas 2-locular, with 10–50 ovules per locule. Ovules pendulous, anatropous, unitegmic, tenuinucellate, with endothelium. Endosperm cellular, with both chalazal and micropylar haustoria. Fruits 2-locular, 2–4-valved, loculicidal, smooth capsules with many seeds. Seeds small, with coarsely verrucose testa, straight, elongate, and thin, linear embryo with small but differentiated cotyledons, surrounded by copious, starchy endosperm, occasionally with deposits of calcium oxalate; seed coat consists of 2–3 layers of cells of pachytesta. Iridoids present, but iridoid type is unknown, n = 7, 8, ... 9. 22. BYBLIDACEAE.

26 Flowers zygomorphic, nectary spur present. Insectivorous herbs found in water or other wet habitats, occasionally epiphytic, rooted in the substrate or rootless and free-floating with the photosynthetic organs submerged. Trapping mechanism active or passive. Provided with stalked and/or sessile glands. Stem in *Utricularia* bearing bladders with a trap mechanism capturing small animals. Nodes unilacunar. Leaves alternate, linear to broadly ovate, entire, estipulate, in basal rosettes in *Pinguicula* and *Genlisea,* which has also tubular trap-leaves arising separately from the rhizome. Stomata diacytic or rarely anomocytic or anisocytic, sometimes absent. Flowers in a bracteate raceme (rarely 1-flowered) or solitary on a bractless scape (*Pinguicula*), bisexual. Calyx equally 4–5-lobed or more or less 2-cleft, persistent, imbricate or open in bud. Corolla bilabiate or more or less 5-lobed, lobes imbricate, the lower lip basally more or less spurred or saccate. Stamens two (anterior pair), inserted near the base of the corolla tube; anthers connivent, dorsifixed, unilocular or 2-locular. Pollen grains 3–4-colporate or multicolporate, in tetrads or in spiroapertutate monads. Nectary disc absent. Gynoecium of two carpels with

more or less sessile, unequally 2-lobed stigma; ovary 1-locular, superior, with free-central placenta and usually anatropous, usually numerous (two in *Utricularia* subgenus *Biovularia*) ovules. Fruits usually capsules opening by 2–4 valves or irregularly or circumscissile, in subgenus *Biovularia* of the genus *Utricularia* indehiscent and 1-seeded. Seeds small, exotestal cells variously thickened; embryo green, scarcely differentiated, endosperm absent. Contain iridoid glycosids (globularin, globulariscin, scutellariosid, aucubin, etc.), flavones, p-cumarid and rarely caffeic acid, n = 7–16, 21–24, 32 (*Pinguicula*)........................22. LENTIBULARIACEAE.

1. BUDDLEJACEAE

K. Wilhelm 1910. 5/c.100. Tropical, subtropical, and warm-temperate regions of Africa, America, and Asia. *Emorya* (1) from Texas to northern Mexico; *Nicodemia* – Madagascar.

Gomphostigma, Buddleja, Nicodemia, Emorya, Chilianthus

Show many similarities with the Loganiaceae, but differ from them in the absence of true stipules (only "auriculae") and of intraxylary phloem, in glandular, stellate, or lepidite indumentum, and chemically (*Buddleja* has been reported to contain an iridoid group typical of Lamiales, but not of Rubiales, see Jensen [2000]). In general the Buddlejaceae are nearer to the Scrophulariaceae than to the Rubiales (Wagenitz 1959; Takhtajan 1966, 1980, 1987, 1997; Punt and Leenhouts 1967; Cronquist 1981, 1988; Dahlgren 1980, 1983, 1989). This is supported by serological studies (Piechura *fide* Cronquist 1981: 947), embryology, anatomy, and molecular date.

2. POLYPREMACEAE

L. Watson ex Doweld et Reveal 2007. 1/1. Southern USA, Mexico, Central and South America, West Indies.

Polypremum

Related to the Buddlejaceae, but differ in regard to the absence of endosperm haustoria, early development of the proembryo, and a chromosome number (Wagstaff 2004).

3. TETRACHONDRACEAE

Wettstein 1924. 1/2. New Zealand and Patagonia and Tierra del Fuego.

Tetrachondra

Tetrachondra is a reduced and rather isolated plant with uncertain affinity.

4. CALCEOLARIACEAE

Olmstead 2001. 3/307. Tropical and Western temperate South America, Brazil, also New Zealand (some *Jovellana*).

Calceolaria, Jovellana, Stemotria (including *Porodittia*)

Related to the Gesneriaceae (see Weber 1972, 1982, Andersson and Molau 1980).

5. STILBACEAE

Kunth 1831 (including Retziaceae Bartling 1830). 7/28. Cape Province of South Africa; *Nuxia* (15): southern Arabia to tropical Africa, Mascarenes, and South Africa.

5.1 NUXIOIDEAE

Stamens four. Shrubs or trees. Leaves opposite, rarely alternate, with stipular ring. Flowers 4-merous, in thyrsoid inflorescence, either solitary on terminal branches or aggregated into compact heads. Ovary 2-locular; ovules numerous in each locule, n = 19. – *Nuxia.*

5.2 STILBOIDEAE

Stamens four. Shrubs or shrublets. Flowers 5-merous, entomophilous, in dense terminal spikes. Ovary 2-locular, but one locule always empty, septum between locules sometimes incomplete; fertile locule with two basal-axile, erect ovules. – *Campylostachys, Euthystachys, Kogelbergia, Tesmophora, Stilbe* (including *Xeroplana* and *Eurylobium*).

5.3 RETZIOIDEAE

Stamens five, rarely six or seven. Monopodially branched shrubs. Flowers ornithophilous, on very short axillary shoots. Ovary basally 2-locular but apically 1-locular, the septum not reaching beyond the placenta, this situated centrally a little more than halfway up the locules. Ovules 2–3, each locule with one ascending and 1–2 pendulous ovules, n = 12. – *Retzia*.

Have similarities with both Scrophulariaceae – Manuleae and also with Buddlejaceae (*Nuxia* is sometimes included in Buddlejaceae).

6. SCROPHULARIACEAE

A.L. de Jussieu 1789 (including Aeginetiaceae Livera 1927, Antirrhinaceae Persoon 1807, Aragoaceae D. Don 1835, Buchneraceae Lilja 1870, Calceolariaceae Raf. ex Olmstead 2001, Caprariaceae Martynov 1820, Chelonaceae Martynov 1820, Digitalidaceae Martynov 1820, Ellisiophyllaceae Honda 1930, Erinaceae Duvau ex Pfeiffer 1874, Euphrasiaceae Martynov 1820, Gratiolaceae Martynov 1820, Hebenstreitiaceae Horaninow 1834, Hemimeridaceae Doweld 2001, Limosellaceae J.G. Agardh 1858, Linariaceae Martynov 1820, Lindenbergiaceae Doweld 2001, Linderniaceae Borsch, K Müller, et Eb. Fischer 2004, Melampyraceae Hooker and Lindley 1821, Orobanchaceae Ventenat 1799, Oxycladaceae Schnizlein 1857–1870, Paulowniaceae Nakai 1949, Pediculariaceae A.L. de Jussieu 1789, Phelypaeaceae Horaninow 1834, Rhinanthaceae Ventenat 1799, Schlegeliaceae Reveal 1996, Selaginaceae Choisy 1823, Sibthorpiaceae D. Don 1835, Verbascaceae Rafinesque 1821, Veronicaceae Cassel 1817). c.300–310/5000. Cosmopolitan, but more abundant in temperate regions (especially in the Northern Hemisphere) and in tropical mountains.

6.1 SCROPHULARIOIDEAE

The two posterior corolla lobes covering the lateral lobes in bud. Placentation basically axile. Mostly autotrophic plants. – VERBASCEAE: *Verbascum* (including *Celsia, Staurophragma*), *Rhabdotosperma;* SCROPHULARIEAE: *Nathalliella, Oreosolen, Scrophularia* (including *Tuerckheimocharis*), *Antherothamnus;* ALONSOEAE: *Alonsoa, Charadrophila;* HEMIMERIDEAE: *Basistemon, Diascia, Hemimeris, Diclis, Nemesia, Colpias*; RUSSELIEAE: *Russelia, Ameroglossum, Dermatobotrys*; TEEDIEAE: *Ranopisoa, Teedia*; FREYLINIEAE: *Phygelius, Freylinia*; LEUCOPHYLLEAE: *Eremogeton, Leucophyllum* (including *Faxonanthus*); APTOSIMEAE: *Anticharis, Aptosimum, Peliostomum*; MANULEEAE: *Barthlottia, Manuleopsis, Jamesbrittenia, Lyperia, Sutera* (including *Chaenostoma, Sphenandra*), *Manulea, Melanospermum, Polycarena, Glekia, Trieenea, Phyllopodium, Zaluzianskya, Reyemia, Glumicalyx, Strobilopsis, Tetraselago, Globulariopsis, Microdon* (including *Agathelpis*), *Gosela, Cromidon, Selago* (including *Walafrida*), *Pseudoselago, Chenopodiopsis, Dischisma, Hebenstretia, ? Camptoloma*; SCHLEGELIEAE: *Gibsoniothamnus, Schlegelia, Synapsis, Exarata*; PAULOWNIEAE: *Paulownia* (including *Shiuyinghua*); WIGHTIEAE: *Brandisia, Wightia*; CHELONEAE: *Chelone, Nothochelone, Chionophila* (including *Penstemonopsis*), *Keckiella, Pennellianthus, Penstemon, Tetranema, Uroskinnera*; COLLINSIEAE: *Collinsia, Tonella;* ANTIRRHINEAE: *Anarrhinum, Kickxia* (including *Elatinoides*), *Linaria, Cymbalaria, Asarina, Maurandella, Rhodochiton, Lophospermum, Pseudorontium, Schweinfurthia, Antirrhinum, Pseudomisopates, Misopates, Acanthorrhinum, Howelliella, Neogarrhinum, Sairocarpus, Mohavea, Galvezia, Chaenorrhinum, Albraunia, Holzneria, Nanorrhinum* (including *Pogonorrhinum*), *Nuttallanthus, Maurandya, Epixiphium, Mabrya, Holmgrenanthe, Gambelia*; ANGELONIEAE: *Angelonia, Monopera*; GRATIOLEAE: *Amphianthus, Bacopa* (including *Moniera, Herpestis, Hydranthelium, Ildefonsia, Geochorda, Monocardia, Ancistrostylis, Sinobacopa*), *Maeviella, Boelckea, Benjaminia, Braunblanquetia, Sophronanthe, Gratiola* (including *Fonkia, Tragiola*), *Mecardonia, Scoparia, Capraria, Deinostema, Dopatrium, Hydrotriche, Limnophila, Philcoxia*; LIMOSELLEAE: *Limosella*; STEMODIEAE: *Lindenbergia* (including *Omania*), *Adenosma, Otacanthus, Achetaria, Tetraulacium, Dizygostemon, Stemodia* (including *Poarium*), *Cheilophyllum, Morgania, Schistophragma, Conobea, Leucospora, Schizosepala, Stemodiopsis*; LINDERNIEAE: *Artanema, Picria, Pierranthus, Schizotorenia, Legazpia, Torenia, Scolophyllum, Hemiarrhena, Hamaegigas, Lindernia* (including *Vandellia, Ilysanthes, Trichotaenia*), *Craterostigma, Crepidorhopalon, Hartliella, Bampsia*; MELOSPERMEAE: *Monttea, Melosperma*. MICROCARPEAE: *Micranthemum* (including *Amphyolanthus, Hemisiphonia*), *Hemianthus, Microcarpaea, Elacholoma, Peplidium, Encopella, Psammetes, Dintera, Bryodes, Bythophyton, Glossostigma*; LEUCOCARPEAE:

Hemichaena (including *Berendtia, Berendtiella*), *Leucocarpus*; MIMULEAE: *Dodartia, Lancea, Mazus, Mimetanthe, Mimulus, Mimulicalyx.*

6.2 RHINANTHOIDEAE

The posterior corolla lobes covered in the bud by one or both of the lateral lobes. Placentation basically axile. Autotrophic, hemiparasitic, or seldom wholly parasitic plants (*Lathraea*). – DIGITALEAE: *Digitalis, Isoplexis*; VERONICEAE: *Erinus, Lafuentea, Wulfenia, Wulfeniopsis, Kashmiria, Picrorhiza, Neopicrorhiza, Veronicastrum, Scrofella, Synthyris, Besseya, Paederota, Paederotella, Pseudolysimachion, Veronica* (including *Cochlidiosperma, Odicardis*), *Hebe, Heliohebe, Parahebe, Chionohebe, Derwentia, Detzneria, Lagotis,* ? *Campylanthus, Ourisia*; SIBTHORPIEAE: *Sibthorpia, Ellisiophyllum*; HEMIPHRAGMEAE: *Hemiphragma;* ARAGOEAE: *Aragoa*; GERARDIEAE: *Agalinis* (*Gerardia*), *Anisantherina, Brachystigma, Aureolaria, Tomanthera, Macranthera, Seymeria, Seymeriopsis, Dasistoma, Silviella, Esterhazya, Lamourouxia*; ESCOBEDIEAE: *Alectra* (including *Pseudorobanche*), *Pseudomelasma, Melasma, Magdalenaea, Velloziella, Nothochilus, Physocalyx, Escobedia, Leptorhabdos, Gerardiina, Micrargeria, Micrargeriella*; BUCHNEREAE: *Sopubia, Parasopubia, Baumia, Petitmenginia, Sieversandreas, Pseudosopubia, Hiernia, Graderia, Ghikaea, Centranthera, Buchnera* (including *Benthamistella, Stellularia*), *Tetraspidium, Striga, Pseudostriga, Parastriga, Cycniopsis, Cycnium, Rhamphicarpa, Rhaphispermum, Radamaea, Leucosalpa, Buttonia, Thunbergianthus, Xylocalyx*; CYMBARIEAE: *Schwalbea, Cymbaria* (including *Cymbochasma*), *Siphonostegia, Lesquereuxia, Bungea, Monochasma, Castilleja, Orthocarpus, Triphysaria, Clevelandia, Gentrya, Cordylanthus, Ophiocephalus*; RHINANTHEAE: *Hedbergia, Bartsia* (including *Bellardia*), *Bornmuellerantha, Macrosyringion, Odontitella, Bartsiella, Odontites* (including *Orthanthella*), *Nothobartsia, Euphrasia, Omphalothrix, Pterygiella, Xizangia, Phtheirospermum, Parentucellia, Pseudobartsia, Pedicularis, Melampyrum, Rhinanthus, Rhynchocorys, Tozzia, Lathraea*; REHMANNIAE: *Rehmannia, Titanotrichum, Triaenophora* (including *Spirostegia*); BOWKERIEAE: *Anastrabe, Bowkeria, Brookea, Halleria, Ixianthes.*

6.3 OROBANCHOIDEAE

The two posterior corolla lobes covered in bud by lateral lobes. Placentation parietal. Parasitic herbs. – *Harveya, Paraharveya, Hyobanche, Aeginetia, Boschniakia* (including *Kopsiopsis*), *Conopholis, Christisonia* (including *Campbellia*), *Cistanche, Phelypaea* (including *Diphelypaea*), *Gleadovia, Mannagettaea, Orobanche* (*Aphyllon*), *Phacellanthus* (including *Tienmuia*), *Platypholis, Xylanche, Epifagus, Phelipanche, Necranthus.*

The Scrophulariaceae stand close to the Buddlejaceae, an affinity that is supported both by morphology (including embryology) and chemistry.

The subfamily Orobanchoideae is very closely related to the Rhinantoideae. It is an advanced group that "represents the final stage of the parasitic tendency exhibited in Rhinanthoideae" (Wernham 1912) and can be regarded as a direct continuation of the line of parasitism developed in the Scrophulariaceae (Boeshore 1920). It probably derived from the Rhinanthoideae through forms like *Striga orobanchoides* (Tiagi 1956, 1963, 1970). There are important embryological similarities between these two subfamilies (Tiagi 1950; Terekhin and Nikiticheva 1981).

7. BIGNONIACEAE

A.L. de Jussieu 1789 (including Crescentiaceae Dumortier 1829). 105/800. Mainly tropical regions, especially in South America, with a few species in subtropical and warm-temperate Asia.

Classification based on E. Fischer, I.Theisen and L.G. Lohmann (2004).

TECOMEAE: *Incarvillea, Niedzwedzkia, Perianthomega, Perichlaena, Pandorea, Tecomanthe, Podranea, Dinklageodoxa, Campsidium, Campsis, Markhamia, Spathodea, Newbouldia, Dolichandrone, Digomphia, Rhigozum, Heterophragma, Santisukia, Pajanelia, Tecoma, Lamiodendron, Fernandoa, Radermachera* (including *Mayodendron*), *Stereospermum, Pauldopia, Catophractes, Deplanchea, Tecomella, Astianthus, Chilopsis, Catalpa, Jacaranda, Spirotecoma, Romeroa, Tabebuia, Delostoma, Argylia, Neosepicea, Cybistax, Zeyhria, Paratecoma, Godmania, Sparattosperma, Ekmanianthe;* OROXYLEAE: *Oroxylum, Millingtonia, Nyctocalos, Hieris;* BIGNONIEAE: *Spingiphila, Pseudocatalpa, Macfadyena, Melloa, Dolichandra, Parabignonia, Phryganocydia, Callichlamys, Manaosella, Gardnerodoxa, Neojobertia, Pleonotoma, Memora, Pyrostegia, Bignonia, Macranthisiphon, Piriadacus, Glaziovia, Amphilophium, Haplolophium, Stizophyllum, Paragonia,*

Roentgenia, Clytostoma, Mansoa, Cydista, Fridericia, Martinella, Cuspidaria, Distictis, Saritaea, Potamoganos, Arrabidaea, Xylophragma, Mussatia, Adenocalymma, Anemopaegma, Tanaecium, Leucocalantha, Lundia, Tynanthus, Pithecoctenium, Spathicalyx, Distictella, Ceratophytum, Periarrabidaea; ECCREMOCARPEAE: *Eccremocarpus;* TOURRETTIEAE: *Tourrettia;* COLEEAE: *Ophiocolea, Colea, Kigelia, Rhodocolea, Phylloctenium, Phyllarthron;* CRESCENTIEAE: *Parmentiera, Amphitecna, Crescentia.*

Very near to the Scrophulariaceae, especially to Scrophularieae (Cheloneae), and probably had a common origin with them. They differ from Scrophulariaceae mainly in endospermless seeds and also in vascular anatomy of the flowers (see Armstrong 1985).

8. GESNERIACEAE

Richard et A.L. de Jussieu ex A.P. de Candolle 1816 (including Belloniaceae Martynov 1820, Besleriaceae Rafinesque 1838, Cyrtandraceae Jack 1823, Didymocarpaceae D. Don 1822, Ramondaceae Godron 1853). c. 145–150/2500–3500. Pantropical, with only a few species in the Pyrenees and the Balkan Peninsula.

Classification after A. Weber (2004).

8.1 CORONATHEROIDEAE

Trees, shrubs or subshrubs. Stomata anomocytic. Nectary adnate to the ovary. Ovary superior. Fruits capsules or berries, n = 37–45. – *Coronanthera, Depanthus, Fieldia, Lenbrassia, Negria, Rhabdothamnus, Mitraria, Sarmienta, Asteranthera.*

8.2 GESNERIOIDEAE

Perennial herbs, subshrubs, shrubs, small trees, vines, climbers or epiphytes. Stomata anisocytic, scattered or in groups. Cotyledons remaining equal after germination. Ovary superior or more or less inferior. Inflorescences often racemose. Nectary disc annular or more often represented by 1–5 separate glands, sometimes wanting, n = 6–16. – BESLERIEAE: *Besleria, Gasteranthus, Cremosperma, Anetanthus, Resia, Reldia, Tylopsacas*; NEPEANTHEAE: *Napeanthus;* GLOXINIEAE: *Gloxinia, Anodiscus, Koellikeria, Monopyle, Diastema, Kohleria, Pearcea, Capanea, Achimenes, Eucodonia, Smithiantha, Niphaea, Moussonia, Solenophora, Phinaea, Heppiella, Goyazia*; GESNERIEAE: *Pheidonocarpa, Gesneria, Rytidophyllum*; SINNINGIEAE: *Sinningia, Paliavana, Vanhouttea*; EPISCIEAE: *Nautilocalyx, Chrysothemis, Paradrymonia, Alloplectus, Drymonia, Columnea, Dalbergaria, Pentadenia, Trichantha, Bucinellina, Corytoplectus, Neomortonia, Episcia, Alsobia, Rufodorsia, Oerstedina, Cobananthus, Nematanthus, Codonanthe, Codonanthopsis;* BELLONIEAE: *Bellonia, Lembocarpus, Rhoogeton, Lampadaria, Cremersia, Cremospermopsis.*

8.3 EPITHEMATOIDEAE

Plants usually fleshy-succulent, sometimes with strongly asymmetrical leaves. Ovary short, abruptly narrowed into the style. Fruit more or less globose. Endosperm absent, n = (8-)10(-12). – *Rhynchoglossum, Gyrogyne, Stauranthera, Loxonia, Epithema, Whytockia, Monophyllaea* (including *Moultonia*).

8.4 DIDYMOCARPOIDEAE (CYTRTANDROIDEAE)

Perennial or rarely annual herbs. Cotyledons becoming unequal after germination. Ovary superior. Inflorescences various, usually cymose with two flowers at each dichotomy, never a simple raceme. Nectary disc annular, rarely divided or split on one side, n = (4) 8–13, 14–17, etc. – *Corallodiscus, Leptoboea, Boeica, Rhynchotechum, Tetraphyllum, Platystemma, Championia, Haberlea, Ramonda, Jankaea, Streptocarpus, Saintpaulia, Linnaeopsis, Acanthonema, Trachystigma, Nodonema, Schizoboea, Hovanella, Colpogyne, Oreocharis* (including *Dasydesmus* and *Perantha*), *Tremacron, Isometrum, Paraisometrum, Ancylostemon, Petrocosmea, Metapetrocosmea, Deinocheilos, Lagarosolen, Petrocodon, Dayaoshania, Dolicholoma, Calcareoboea, Allocheilos, Opithandra, Thamnocharis, Bournea, Tengia, Conandron, Briggsia, Briggsiopsis, Raphiocarpus, Loxostigma, Anna, Lysionotus, Cathayanthe, Beccarinda, Chirita, Chiritopsis, Primulina, Metabriggsia, Didymocarpus, Gyrocheilos, Didymostigma, Deinostigma, Pseudochirita, Allostigma, Phylloboeae, Hemiboeopsis, Hemiboea, Didissandra, Ridleyandra, Hexatheca, Aeschynanthus* (including *Euthamnus, Oxychlamys*), *Micraeschynanthus, Agalmyla* (including *Dichrotrichum*), *Orchadocarpa, Henckelia* (including *Loxocarpus, Codonoboea, Platyadenia*), *Boea, Paraboea* (including *Chlamydoboea*), *Trisepalum* (including *Dichiloboea*), *Rhabdothamnopsis, Ornithoboea, Kaisupeea, Senyumia, Spelaeanthus, Emarhendia, Cyrtandra* (including *Protocyrtandra*), *Sepikaea.*

Genera of uncertain familial affiliation: *Sanango, Cubitanthus, Jerdonia.*

Very near to the Scrophulariaceae, especially to the tribe Scrophularieae. According to Burtt (1977), "a highly advanced, and presumably recent family."

9. PLOCOSPERMATACEAE

Hutchinson 1973. 1/1. Southern Mexico and Guatemala.

Plocosperma (including *Lithophytum*)

Have similarities, including two parietal placentas, with the Gesneriaceae.

10. CARLEMANNIACEAE

Airy Shaw 1965. 1/6. Himalayas (from eastern Nepal to Bhutan), northeastern India, northern Burma, southwestern China, Laos, northern Vietnam, Sumatra.

Carlemannia, Silvianthus

Hooker (in Bentham et Hooker 1973, Genera Plantarum 2: 63–64) included *Carlemannia* and *Silvianthus* in Rubiaceae-Hedyotideae. However, later Solireder (1893) proposed to transfer them into the Caprifoliaceae. Hallier (1910a, b), however, suggested that Carlemanniaceae are closely related to Gesneriaceae. Savolainen et al. (2000) and Bremer et al. (2001) indicated an inclusion of Carlemanniaceae into Lamiales.

11. GLOBULARIACEAE

A.P. de Candolle 1805. 2/25. Macaronesia, Mediterranean, Europe, western Asia, and also Somalia and Socotra (*Poskea*, 2).

Globularia, Poskea

Globulariaceae are closely related to the Scrophulariaceae-Selagineae, differing from them especially in their inflorescence as well as in pseudomonomerous gynoecium and fruit morphology. Probably derived from the Scrophulariaceae.

12. PLANTAGINACEAE

A.L. de Jussieu 1789. 3/270. Temperate regions and in tropical mountains; *Bougueria* (1): Andes.

Plantago, Littorella, Bougueria

Near to and probably derived from the Scrophulariaceae. Hallier (1903, 1912) included them in Scrophulariaceae s. l. and placed them near Selagineae and Manuleae.

13. CALLITRICHACEAE

Berchtold et J. Presl 1820. 1/45–50. Nearly cosmopolitan, but mainly in temperate regions.

Callitriche

According to Olmstead et al. (2001), the clade Callitrichaceae, Globulariaceae, Plantaginaceae, and Hippuridaceae is support very strong.

14. HIPPURIDACEAE

Vest 1818. 1/1. Temperate and cold regions of the Northern Hemisphere, Antarctic America and southern Australia.

Hippuris

Hippuridaceae are the closest relative of the Callitrichaceae (Leins and Erbar 2004).

15. PEDALIACEAE

R. Brown 1810 (including Sesamaceae Berchtold et J. Presl 1820). 13/75. From tropical and southern Africa and Madagascar to Malesia, New Guinea, and northern Australia and adjacent islands.

PEDALIEAE: *Pedalium, Pterodiscus, Pedaliodiscus, Harpagophytum, Holubia, Uncarina;* SESAMEAE: *Sesamothamnus, Rogeria, Sesamum, Ceratotheca;* PRETREEAE: *Dicerocaryum* (*Pretrea*), *Linariopsis, Josephinia.*

Close to the Bignoniaceae and Scrophulariaceae.

16. MARTYNIACEAE

Stapf 1895. 4/17. Subtropical and especially tropical regions of America.

Martynia, Holoregmia, Craniolaria, Proboscidea (including *Ibicella*).

Have many similarities with Pedaliaceae, including wood anatomy, but comparison of pollen suggested that Martyniaceae and Pedaliaceae should be maintained as separate families (Stapf 1895; Bretting and Nilsson 1988).

17. TRAPELLACEAE

Honda et Sakisake 1930. 1/2. Eastern Asia.

Trapella

Very near to the Pedaliaceae. Previously only a few botanists, including Airy Shaw (in Willis 1973) and Dahlgren (1980, 1983, 1989), accepted Trapellaceae as a distinct family. However, the Trapellaceae differ not only in pollen, ovule, ovary, and fruit structure but also in such embryological features as persistence of antipodals, simultaneous differentiation of micropylar and chalazal haustorium, and a four-celled micropylar haustorium and chalazal haustorium of two elongated cells (John et al. 1992: 774).

18. MYOPORACEAE

R. Brown 1810 (including Bontiaceae Horaninow 1834). 4/210. Mainly Australia, but *Myoporum* (32) also extends to New Guinea, eastern Asia, New Zealand, Pacific islands, and Mascarene Islands; *Bontia* (1) occurs in the West Indies and northern South America, and *Androya* (1) is endemic to Madagascar.

Eremophila (*Pholidia, Stenochilus*), *Myoporum, Bontia, Androya*

Closely related to the Scrophulariaceae – Scrophularioideae. According to palynological data (Niezgoda and Tomb 1975) Myoporaceae related to the scrophularien tribe Leucophylleae. In addition Karrfalt and Tomb (1983) found that air spaces of *Leucophyllum* are homologous with the secretory cavities of *Bontia* (Myoporaceae).

19. OFTIACEAE

Takhtajan et Reveal 1993 (Spielmanniaceae J. Agardh 1858, nom.illeg.). 1/3. South Africa

Oftia

Oftia (*Spielmannia*) has been placed in various families, such as Verbenaceae (Endlicher 1841; Lindley 1853; and some others), Scrophulariaceae (Baillon 1888; Thorne 1992, 2000), and Myoporaceae (Bentham and Hooker 1876; Takhtajan 1966; Cronquist 1981; and many others), or separated into a family of its own (Takhtajan 1987, 1997). From the Myoporaceae the genus *Oftia* differs in the presence of internal phloem and in primary vascular tissue. Besides, it differs from the Myoporaceae in the absence of the secretory cavities and not gland-dotted leaves, and from the Scrophulariaceae in absent endosperm. However, the Oftiaceae have some links to both the Scrophulariaceae and Myoporaceae.

20. ACANTHACEAE

A.L. de Jussieu 1789 (including Justiciaceae Rafinesque 1838, Mendonciaceae Bremekamp 1954, Nelsoniaceae Sreemadhavan 1977, Thomandersiaceae Sreemadhavan 1977, Thunbergiaceae Lilja 1870). 220–240/3500–4000. Mainly tropical regions, especially southern and Southeast Asia, Africa, Brazil, and Central America; a few genera extend into warm-temperate regions; some species of *Acanthus* occur in the Mediterranean region and in western Asia.

20.1 NELSONIOIDEAE

Ovary 2-locular with numerous ovules or (*Elytraria*) 1-locular with 6–10 ovules on parietal placentas; ovules with endothelium. Jaculators usually wanting. Fruits capsular. Seeds with more or less developed oily and ruminate endosperm. Leaves alternate. Cystoliths wanting; descending cochlear aestivation. – *Staurogyne, Nelsonia, Elytraria, Anisosepalum, Gynocraterium, Ophiorrhiziphyllon, Saintpauliopsis*.

20.2 THUNBERGIOIDEAE. (including Mendoncioideae)

Ovary 2-locular with two ovules per locule or less often one of the locules aborted. Jaculators wanting. Fruits capsules or drupes. Seeds without endosperm. Leaves usually opposite. Cystoliths present. – THUNBERGIEAE: *Thunbergia, Meyenia, Pseudocalyx;* MENDONCIEAE: *Mendoncia, Anomacanthus*

20.3 ACANTHOIDEAE

Ovary 2-locular with two to many ovules per locule. Jaculators usually present. Seeds without endosperm. Leaves opposite. Cystoliths present or absent (Acantheae). – ACANTHEAE: *Acanthus, Acanthopsis, Achyrocalyx, Blepharis, Crossandra, Crossndrella,*

Gynarospermum, Cyphacanthus, Encephalosphaera, Geissomeria, Holographis, Neriacanthus, Orophochilus, Rhombochlamys, Salpixantha, Sclerochiton, Streptosiphon, Strobilacanthus, Xantheranthemum; APHELANDREAE: *Stenandrium, Aphelandra;* RUELLIEAE: *Blechum, Acanthopale, Apassalus, Benoicanthus, Bravaisia, Brunoniella, Calacanthus, Clarkeasia, Dischistocalyx, Duosperma, Sanchezia, Ruellia, Dyschoriste, Echinacanthus, Epiclastopelma, Eranthemum, Eremomastax, Eusiphon, Hemigraphis, Heteradelphia, Hygrophila, Ionacanthus, Kosmosiphon, Leptosiphonium, Louteridium, Lychniothyrsus, Mellera. Mimulopsis, Pararuellia, Petalidium, Phaulopsis, Physacanthus, Polylychnis, Pseudoruellia, Ruelliospis, Satanocrater, Sautiera, Spirostigma, Stenosiphonium, Stenothyrsus, Strobilanthes, Strobilanthopsis, Suessenguthia, Trichanthera, Trichosanchezia, Zygoruellia;* LEPIDAGATHIDAE: *Lepidagathis;* ANDROGRAPHINAE: *Andrographis, Cystacanthus, Diotacanthus, Graphandra, Gymnostachyum, Haplanthodes, Indonesiella, Phlogacanthus; Afrofittonia, Ambongia, Ancistranthus, Angkalanthus, Anisacanthus, Aphanosperma, Asotheca, Asystasia, Ballochia, Brachystephanus, Calycacanthus, Carlowrightia, Celerina, Centrilla, Cephalocanthus, Chalarothyrsus, Chamaeranthemum, Chileranthemum, Chlamydocardia, Chlamydostachya, Chorisochora, Clinacanthus, Clistax, Codonacanthus, Conocalyx, Cosmianthemum, Cyclacanthus, Cylindrosolenium, Danguya, Dasytropis, Dichazothece, Dicladanthera, Dicliptera, Ecbolium, Filetia, Fittonia, Forcipella, Glossochilus, Graptophyllum, Gypsacanthus, Harpochilus, Henrya, Herpetacanthus, Hoverdenia, Hypoestes, Ichtyostoma, Isoglossa, Isotheca, Jadunia, Juruasia, Justicia, Kalbreyeriella, Linariantha, Mackaya, Marcania, Megalochlamys, Megalostoma, Megaskepasma, Mellitacanthus, Metarungia, Mexacanthus, Mirandea, Monechma, Monothecium, Odontonema, Oplonia, Oreacanthus, Pachystachys, Pelecostemon, Peristrophe, Phialacanthus, Podorungia, Poikilacanthus, Polulina, Pranceacanthus, Pseuderanthemum, Pseudodicliptera, Psilanthele, Ptyssiglottis, Pulchranthus, Razisea, Rhinacanthus, Ritonia, Rungia, Ruspolia, Ruttya, Samuelssonia, Sapphoa, Schaueria, Sebastiano-Schaueria, Spathacanthus, Sphinctacanthus, Stenostephanus, Streblacanthus, Tessmanniacanthus, Tetramerium, Thysanostigma, Trichaulax, Trichocalyx, Xerothamnella, Yeatesia; Barleria, Barleriola, Borneacanthus, Boutonia, Chroesthes, Crabbea, Hulemacanthus, Lepidagathis, Lophostachys.* WHITFIELDIEAE: *Whitfieldia, Chlamydacanthus, Lankesteria.*

INSERTAE SEDIS: *Acanthostelma, Acanthura, Aphelandrella, Camarotea, Dolichostachys, Golaea, Idiacanthus, Kudoacanthus, Lasiocladus, Leandriella, Morsacanthus, Neuracanthus, Perenideboles, Pericalypta, Sphacanthus, Vavara, Vindasia.*

Near to the Scrophulariaceae. Nelsonioideae are in some respects a connecting link between the two families.

21. AVICENNIACEAE

Endlicher 1841. 1/8. Mostly tropical coasts mainly within the mangrove zone of tropical sheltered coastlines, but also in subtropical and even warm-temperate tidal habitats where *Avicennia* occurs as the exclusive tree or shrub form, making it the most widely distributed of any mangrove. Three species of *Avicennia* occur in the Atlantic, Caribbean, and eastern Pacific, while five other species occur in the Old World (Tom-Unson 1986; Duke 1991).

Avicennia

According to Schwarzbach and McDade (2002), "sequence data from these two chloroplast and one nuclear regions consistently place *Avicenia* with Acanthaceae s.l. They analyzed that resolve relationships further place *Avicennia* as sister to Thunbergioideae". However, from the Acanthaceae they differ in the secondary anomalous growth, in leaves with salt glands on both sides, orthotropous ovules and fruit morphology.

22. BYBLIDACEAE

Domin 1922. 1/6. The southernmost part of New Guinea, northern and southwestern Australia.

Byblis

According to recent studies, both molecular and morphological, the Byblidaceae are related to the Lentibulariaceae (see especially Albert et al. 1992; Conran and Carolin 2004).

23. LENTIBULARIACEAE

Richard 1808 (including Pinguiculaceae Dumortier 1829, Utriculariaceae Hoffmannsegg and Link 1809). 3/320. Cosmopolitan.

Pinguicula, Genlisea, Ultricularia (including *Polypompholyx* and *Biovularia*)

Near to and derived from the Scrophulariceae. Hallier (1903a, b, 1905, 1908, 1912) included Lentibulariaceae in Scrophulariaceae s.l., deriving them from Gratioleae (see also Casper 1963). Embryological data suggest the derivation from the Scrophulariaceae (Khan 1970; Johri et al. 1992). Floral morphology and phytochemistry (iridoid glycosids) indicate a close relationship to Scrophulariaceae.

24. VERBENACEAE

J. Saint-Hilaire 1805 (including Durantaceae J.G. Agardh 1858, Lantanaceae Martynov 1820, Petreaceae J.G. Agardh 1858, Viticaceae A.L. de Jussieu 1789). 50/1800–2500. Tropical and subtropical regions with a few species in warm-temperate countries.

24.1 VITICOIDEAE
Leaves opposite or rarely verticillate, simple or sometimes 3–7-foliolate. Inflorescence cymose, corymbose, umbellate, or occasionally flowers solitary. – TEJSMANNIODENDREAE: *Tejsmanniodendron, Garrettia, Hymenopyramis;* CALLICARPEAE: *Callicarpa* (including *Geunsia*); TECTONEAE: *Petitia, Tectona;* VITICEAE: *Pseudocarpidium, Cornutia, Adelosa, Viticipremna, Premna* (including *Pygmaeopremna*), *Tsoongia, Paravitex, Petraeovitex, Vitex* (including *Neorapinia*), *Gmelina.*

24.2 VERBENOIDEAE
Leaves opposite or rarely alternate, entire, lobed to dissected, sometimes much reduced, scale-like, or spinose. Flowers in various types of racemose inflorescences, often with involucre of colored bracts. – VERBENEAE: *Verbena* (including *Stylodon*), *Glandularia, Junellia, Urbania, Hierobotana*; LANTANEAE: *Lantana, Neosparton, Xeroaloysia, Nashia, Lippia* (including *Burroughsia*), *Phyla, Aloysia, Acantholippia, Stachytarpheta* (including *Ubochea*), *Bouchea, Chascanum* (including *Svensonia*), *Diostea, Lampaya*; PRIVEAE: *Priva, Pitraea* (including *Castelia*), *Dipyrena*; PETREAE: *Petrea, Xolocotzia*; CASSELIEAE: *Casselia, Parodianthus, Tamonea*; DURANTEAE (CITHAREXYLEAE): *Citharexylum, Rehdera, Verbenoxylum, Rhaphithamnus, Baillonia, Recordia, Duranta.*

Insertae sedis: *Coelocarpum.*

25. PHRYMATACEAE

Schauer 1847. 1/1. Himalayas, eastern Asia, southeastern North America.

Phryma

Closely related to the Verbenaceae-Verbeneae, especially to *Verbeneae* (see Whipple 1972; Chadwell et al. 1992; Wagstaff 1992; Cantino 1992a, b; Ramana et al. 2000). In overall morphology closely resembles *Stachytarpheta,* but differs in having trichomes and in pollen morphology.

26. CYCLOCHEILACEAE

Marais 1981. 2/4. Eastern and northeastern tropical Africa and Yemen.

Cyclocheilon, Asepalum

A distinct family probably derived from the verbenaceous stock.

27. NESOGENACEAE

Marais 1981. 1/9. Tanzania, Madagascar, Seychelles and Mascarene Is., Malesia, Oceania (Tuamotu Arch.).

Nesogenes

Close related to the Cyclocheilaceae.

28. SYMPHOREMATACEAE

Wight 1849. 3/35. Southern, Southeast, and eastern Asia, Malesia.

Symphorema, Sphenodesme, Congea

Related to the Verbenaceae.

29. LAMIACEAE

Martynov 1820 or Labiatae A.L. de Jussieu 1789, nom. altern. (including Aegiphilaceae Rafinesque 1838, Chloanthaceae Hutchinson 1959, Dicrastylidiaceae Drummond ex Harvey 1855, Glechomaceae Martynov 1820, Melittdaaceae Martynov 1820, Menthaceae Burnett 1835, Nepetaceae Horaninow 1834, Salazariaceae F.A. Barkley 1975, Salviaceae Rafinesque 1837, Scutellariaceae Caruel 1894, Siphonanthaceae Rafinesque 1838). Over

230–234/c.7000. Cosmopolitan, but especially diversified in the Mediterranean, Irano-Turanian, and Eastern Asiatic Regions.

29.1 AJUGOIDEAE
Shrubs, subshrubs, or herbs, sometimes aromatic or ill-scented. Leaves opposite. Flowers more or less zygomorphic. Calyx 5-lobed or dentate. Corolla often with deeply cleft tube (*Teucrium, Teucridium, Monochilus,* and some species of *Spartothamnella*), the corolla limb therefore seemingly unilabiate. Pollen grains 3-colpate or rarely 3-porate, often operculate (*Teucrium* and related genera), with granular exine structure and often more or less branched columella (profusely branched in *Teucrium*), suprareticulate, tectate-perforate to microreticulate, with supratectal spinules or spines or verrucate. Style terminal to gynobasic; ovary mostly shallowly lobed, completely or incompletely 4-locular, with one ovule per locule (locellus). Nutlets with lateral-ventral attachment, the usually large surface of contact often more than half the height of the ovary. Seeds with straight embryo and straight and short radicle; endosperm scanty or none. Terpenoids and rosmarinic acid absent, iridoid glycosides and acteosides present, n = 7, 10, 13, 14, 16 +. – CLERODENDREAE: *Clerodendrum* (including *Kalaharia, Tetraclea*), *Rotheca, Faradaya, Teucridium, Oncinocalyx*; AJUGEAE: *Ajuga*; MONOCHILEAE: *Aegiphila, Amasonia, Monochilus, Amethystea;* TEUCRIEAE: *Teucrium, Spartothamnella*; CARYOPTERIDEAE: *Caryopteris, Glossocarya, Hosea, Huxleya, Karomia, Oxera, Rubiteucris, Discretitheca, Pseudocaryopteris, Tripora, Schnabelia, Trichostema.*

29.2 PROSTANTHEROIDEAE
Shrubs, subshrubs, or less often trees, non-aromatic or occasionally with distal branches and leaves aromatic, usually provided with various types of trichomes, often tomentose. Leaves opposite or verticillate, rarely alternate, entire or dentate, often densely clothed with stellate hairs. Flowers in terminal or axillary inflorescences, often spicate, capitate or corymbose-paniculate. Calyx lobed or dentate. Corolla actinomorphic or zygomorphic, lobed to truncate. Stamens 2–8, mostly four. Pollen grains 3-colpate or sometimes 6(-8)-colpate, tectate-perforate to microreticulate, with or without spinulose ornamentation, often with ectexine elements on the colpi membrane. Nectary disc wanting. Style terminal or gynobasic. Ovary 2-locular, often shallowly 4-lobed, with usually two ovules per carpel. Fruits indehiscent, often 1-seeded or separated into four mericarps. Seeds with straight embryo and usually with copious or scanty endosperm. – CHLOANTHAE: *Chloanthes, Hemiphora, Cyanostegia, Dicrastylis, Mallophora, Physopsis, Newcastelia, Lachnostachys, Pityrodia, Brachysola;* WESTRINGIEAE (PROSTANTHEREAE): *Hemiandra, Hemigenia, Microcorys, Prostanthera* (including *Eichlerago*), *Wrixonia, Westringia.*

29.3 WENCHENGIOIDEAE
Subshrubs. Leaves alternate, entire. Flowers zygomorphic. Calyx 2-lipped, 5-dentate. Corolla 5-lobed, three lower lobes much larger. Stamens four, subequal. Pollen grains 3-colpate, suprareticulate, with simple columellae. Nectary disc absent. Style subterminal, shortly 2-lobed. Ovary shallowly lobed, 4-locular, with one ovule per locule. Fruit of nutlike mericarps attached to the receptacle by means of slender stalks. Seeds with spatulate embryo, endosperm present. – *Wenchengia.*

29.4 SCUTELLARIOIDEAE
Herbs, subshrubs, or rarely shrubs. Leaves opposite, entire or dentate. Calyx 2-labiate with entire, rounded lips that become tightly appressed after anthests, enclosing the developing fruit inside. Corolla with 4-lobed upper lip and 1-lobed lower lip. Nectary disc tubular, elongate. Anterior stamens dimidiate. Pollen grains mostly inoperculate, suprareticulate to psilate, tectate-perforate to microreticulate, with simple columellae. Style mostly attached to the ovary lobes above the apex of the disc extension, rarely (some species of *Scutellaria*) truly gynobasic. Nutlets with basal attachment, the surface of contact small. Seeds more or less transverse. Embryo with a bent radicle on one cotyledon. – *Scutellaria* (including *Perilomia, Salazaria, Cruzia, Harlanlewisia*), *Renschia, Tinnea, Holmskioldia.*

29.5 LAMIOIDEAE
Herbs, subshrubs, or rarely shrubs, mostly non-aromatic or slightly aromatic. Leaves opposite, entire, dentate or lobed. Inflorescence thyrsoid or racemelike, rarely flowers solitary. Calyx of various shape. Corolla bilabiate to nearly actinomorphic. Stamens four, ascending or spreading and projecting straight

forward, often long exserted from the corolla; anthers opening longitudinally, rarely by valves (*Galeopsis*). Pollen grains mostly 3-colpate, but with four colpi arranged in two pairs in *Haplostachys*, tetrazonocolpate in one species of *Marrubium* and many species of *Sideritis* and hexapantocolpate in *Sideritis* sect. *Empedoclia* (Abu-Asab and Cantino 1992), usually tectate-perforate to microreticulate, mostly suprareticulate, usually with simple columellae. Disc lobes, when present, alternate with ovary lobes. Style gynobasic. Ovary 4-locular, deeply 4-lobed; ovule solitary in each locule. Nutlets with small basal attachment, sometimes (Prasieae) with fleshy exocarp. Seeds with spatulate embryo with short, straight superior radicle; endosperm several-layered. Often present iridoid glycosides and laballenic fatty acid, n = 6. – POGOSTEMEAE: *Pogostemon* (including *Eusteralis, Anuragia*), *Dysophylla, Colebrookea, Eurysolen, Leucosceptrum, Comanthosphace, Rostrinucula, Anisomeles, Suzukia;* PRASIEAE: *Gomphostemma, Bostrychanthera, Stenogyne, Phyllostegia, Haplostachys, Prasium*; MARRUBIEAE: *Acrotome, Paralamium, Thuspeinantha, Hypogomphia, Marrubium, Lagopsis, Sideritis*; LAMIEAE: *Physostegia, Chelonopsis, Synandra, Macbridea, Brazoria, Warnockia, Melittis, Alajja, Eriophyton, Loxocalyx, Ajugoides, Matsumurella, Lamium* (including *Galeobdolon*), *Orvala, Wiedemannia, Galeopsis, Paraphlomis, Lamiophlomis, Phlomis, Phlomoides, Eremostachys, Paraeremostachys, Pseuderemostachys, Pseudomarrubium, Stachyopsis, Leonurus, Chaiturus, Panzerina* (*Panzeria*), *Lagochilus, Moluccella, Otostegia, Isoleucas, Ballota, Sulaimania, Roylea, Metastachydium* (*Metastachys*), *Stachys, Epimeredi, Colquhounia, Craniotome, Achyrospermum, Microtoena, Phlomidoschema, Chamaesphacos, Notochaete, Leonotis, Leucas* (*Physoleucas*).

29.6 NEPETOIDEAE

Shrubs and herbs, rarely trees, commonly aromatic. Leaves simple, entire or lobed, sometimes compound. Corolla zygomorphic, sometimes weakly so, often strongly 2-lipped. Stamens four or two, ascending or declinate. Pollen grains usually 3-celled and typically 6(8–12)-colpare, predominantly suprareticulate. Disc usually well developed, often 4-lobed. Style gynobasic. Seeds without endosperm (except *Bystropogon*). Pericarp usually dry, rarely fleshy; exocarp with mucilaginous cells producing hygroscopic spiral fibrils. Embryo erect, investing (except *Catoferia* which has bent embryo). Rich in volatile terpenoids and rosmarinic acid; nepetoidin A and B (caffeic acid esters) present. Iridoid glycosides and acteosides usually absent. Seed oils highly unsaturated, n = 6 +. – ELSHOLTZIEAE: *Elsholtzia, Perilla, Perillula, Collinsonia* (including *Micheliella, Keiskea*), *Mosla;* SATUREIEAE: *Mentha* (including *Preslia*), *Lycopus, Hyssopus, Pycnanthemum, Origanum* (including *Majorana*), *Amaracus, Bystropogon, Minthostachys, Thymus, Zataria, Monardella, Cyclotrichium, Obtegomeria, Thymbra* (including *Coridothymus*), *Cunila, Piloblephis, Ziziphora, Melissa, Heterolamium, Conradina, Micromeria, Gontscharovia, Cuminia, Satureja* (including *Euhesperida*), *Ceratominthe, Saccocalyx, Kurzamra, Dicerandra, Clinopodium, Calamintha, Acinos, Antonina, Hedeoma, Poliomintha, Hesperozygis, Rhododon, Stachydeoma, Pogogyne, Eriothymus, Pentapleura*; PRUNELLEAE: *Prunella, Cleonia*; MERIANDREAE: *Zhumeria, Meriandra, Dorystoechas, Perovskia;* LEPECHINIEAE: *Sphacele, Lepechinia*; HORMINEAE: *Horminum*; NEPETEAE: *Dracocephalum, Lophanthus, Hymenocrater, Agastache, Brittonastrum, Meehania, Glechoma, Schizonepeta, Cedronella, Nepeta* (including *Pitardia*), *Kudrjaschevia, Drepanocaryum, Lallemantia, Monarda, Marmoritis* (including *Phyllophyton*), *Blephilia, Chaunostoma, Neoeplingia, Rhabdocaulon, Hoehnea;* GLECHONEAE: *Glechon, Acanthomintha*; SALVIEAE: *Salvia* (including *Ramona, Schraderia, Arischrada*); ROSMARINEAE: *Rosmarinus*; CATOFERIEAE: *Catoferia*; OCIMEAE: *Eriope* (including *Eriopidion*), *Hypenia, Hyptis, Hyptidendron, Marsypianthes, Peltodon, Rhaphiodon, Asterohyptis, Aeollanthus* (including *Icomum, Bovonia*), *Alvesia* (including. *Plectranthastrum*), *Pycnostachys, Anisochilus* (including *Stiptanthus*), *Leocus, Plectranthus* (including *Rabdosia, Rabdosiella, Neohyptis, Solenostemon, Coleus, Englerastrum, Ascocaridion, Neomuellera, Symphostemon, Perrierastrum, Holostylon, Capitanya,*), *Thorncroftia, Tetradenia, Isodon* (including *Amethystanthus, Skapanthus*),*Siphocranion, Hanceola, Isodictyophorus, Hoslundia, Syncolostemon, Platostoma* (including *Mesona, Ceratanthus, Octomeron*), *Basilicum, Benguellia, Endostemon* (including *Puntia*), *Hemizygia, Ocimum* (including *Becium, Erythrochlamis, Nautochilus*), *Orthosiphon, Fuerstia, Geniosporum, Haumaniastrum* (including

Acrocephalus), *Dauphinea, Capitanopsis, Madlabium,* LAVANDULEAE: *Lavandula.*

Insertae sedis: *Acrymia, Holocheila, Peronema, Cymaria*

Very near to Verbenaceae. The taxonomic boundary between these two families is not clear-cut. Until recently it was supposed that Verbenaceae and Lamiaceae are sister clades and that Lamiaceae are derived from Verbenaceae.

Bibliography

Aagaard JE, RG Olmstead, JH Willis, and PC Phillips. 2005. Duplication of floral regulatory genes in the Lamiales. Am. J. Bot. 92: 1284–1293.

Abid MA. 1967. A revision of *Symphorema* (Verbenaceae). Gard. Bull. Singapore 22: 153–171.

Abu-Asab MS. 1990. Phylogenetic implications of pollen morphology in subfamily Lamioideae (Labiatae) and related taxa. Ph.D. dissertation, Ohio University, Athens.

Abu-Asab MS and PD Cantino. 1992. Pollen morphology in subfamily Lamioideae (Labiatae) and its phylogenetic implications. In: RM Hartley and T Reynolds, eds. Advances in Labiate Science, pp. 97–112. Royal Botanic Gardens, Kew.

Abu-Asab MS and PD Cantino. 1993. Phylogenetic implications of pollen morphology in tribe Ajugeae (Labiatae). Syst. Bot. 18: 100–122.

Abu-Asab MS and PD Cantino. 1994. Systematic implications of pollen morphology in subfamilies Lamioideae and Pogostemonoideae (Labiatae). Ann. Missouri Bot. Gard. 81: 653–686.

Abu-Asab MS, PD Cantino, JW Nowicke, and Tao Sang. 1993. Systematic implications of pollen morphology in *Cartopteris* (Labiatae). Syst. Bot. 18: 502–515.

Abu Sbaih HA, DM Keith-Lucas, and SL Jury. 1994. Pollen morphology of the genus *Orobanche* L. (Orobanchaceae). Bot. J. Linn. Soc. 116: 305–313.

Ahmad KJ. 1974a. Cuticular studies in some Nelsonioideae (Acanthaceae). Bot. J. Linn. Soc. 68: 73–80.

Ahmad KJ. 1974b. Cuticular studies in some species of *Mendoncia* and *Thunbergia* (Acanthaceae). Bot. J. Linn. Soc. 69: 53–63.

Ahmad KJ. 1978. Epidermal hairs of Acanthaceae. Blumea 24: 101–117.

Airy Shaw HK. 1965. On a new species of the *Silvianthus* Hook. f. and on the family Carlemanniaceae. Kew Bull. 19: 507–512.

Albach DC 2002. Biosystematics of *Veronica*. Ph.D.thesis, University of Vienna, Austria.

Albach DC and MW Chase. 2001. Paraphyly of *Veronica* (Veroniceae, Scrophulariaceae): evidence from internal transcribed spacer (ITS) sequences of nuclear ribosomal DNA. J. Plant Res. 114: 9–18.

Albach DC, MM Martinez-Ortega, MA Fischer, and MW Chase 2004a. Evolution of Veroniceae: a phylogenetic perspective. Ann. Missouri Bot. Gard. 91: 275–302.

Albach DC, MM Martinez-Ortega, MA Fischer, and MW Chase. 2004b. A new classification of the tribe Veroniceae – problems and a possible solution. Taxon 53: 429–452.

Albach DC, HM Meudt, and B Oxelman. 2005. Pieding together the "new" Plantaginaceae. Am. J. Bot. 92: 297–315.

Alimova GK. 1987. Gesneriaceae. In: MS Yakovlev, ed. Comparative embryology of flowering plants: Davidiaceae–Asteraceae, pp. 273–278. Nauka, Leningrad (in Russian).

Anderson F. 1922. The development of the flowers and embryogeny of *Martynia louisiana*. Bull. Torrey Bot. Club 49: 141–157.

Andersson S. 2006. On the phylogeny of the genus *Calceolaria* (Calceolariaceae) as inferred from ITS and plastid *mat*K sequences. Taxon 55: 125–137.

Andrzejewska Golec E. 1997. Taxonomic aspects of the iridoid glucosides occurring in the genus *Plantago* L. Acta Soc. Bot. Pol. 66(2): 201–205.

Arekal GD. 1963. Contribution to the embryology of *Chelone glabra* L. Phytomorphology 13: 376–388.

Argue C. 1993. Pollen morphology in the Selagineae, Manuleae (Scrophulariaceae), and selected Globulariaceae, and its taxonomic significance. Am. J. Bot. 80: 723–733.

Argue CL. 2000. Pollen morphology in *Lyperia* (Scrophulariaceae) and related taxa with a multivariate analysis of the major patterns of pollen character variation in tribe Manuleae. Canad. J. Bot. 78: 208–220.

Armbruster WS, CPH Mulder, BG Baldwin, S Kalisz, B Wessa, and H Nute. 2002. Comparative analysis of late floral development and mating-system evolution in tribe Collinsieae (Scrophulariaceae s.l.). Am. J. Bot. 89: 37–49.

Armstrong JE. 1985. The delimitation of Bignoniaceae and Scrophulariaceae based on floral anatomy, and placement of problem genera. Am. J. Bot. 72: 755–766.

Armstrong JE and AW Douglas. 1989. The ontogenetic basis for corolla aestivation in Scrophulariaceae. Bull. Torrey Bot. Club 116: 378–389.

Atkins S. 2004. Verbenaceae. In: K Kubitzki, ed. The families and genera of vascular plants, vol. 7, pp. 449–468. Springer, Berlin/Heidelberg/New York.

Attawi F. 1977. Morphologisch-anatomische Untersu-chungen an den Haustorien einiger *Orobanche*-Arten. Ber. Deutsch. Bot. Ges. 90: 173–182.

Baikova EV. 2001. Structure of trichomes in the genus *Salvia* (Lamiaceae). Byull. Mosk. Obshch. Ispyt. Prir., Biol. 106(4): 58–70 (in Russian).

Bakker FT, F Breman, and V Merckx. 2006. DNA sequence evolution in fast-evolving mitochondrial DNA *nad*1 exons in Geraniaceae and Plantaginaceae. Taxon 55: 887–896.

Balkwill K and F Getliefe Norris. 1988. Classification of the Acanthaceae: a southern African perspective. Missouri Bot. Gard. Monograph Syst. Bot. 25: 503–516.

Barber JC, J Francisco Ortega, A Santos Guerra, KG Turner, and RK Jansen. 2002. Origin of Macaronesian *Sideritis* L. (Lamioideae: Lamiaceae) inferred from nuclear and chloroplast sequence datasets. Mol. Phylogenet. Evol. 23: 293–306.

Barringer K. 1984. Seed morphology and the classification of the Scrophulariaceae. (Abstr.). Am. J. Bot. 71(5, Part 2): 156.

Barringer K. 1993. Five new tribes in the Scrophulariaceae. Novon 3: 15–17.

Barthlott W. 1980. Morphogenese und Mikromorphologie komplexer Cuticularfaltungsmuster an Blütentrichomen von *Antirrhinum* L. (Scrophulariaceae. Ber. Deutsch. Bot. Ges. 93: 379–390.

Beardsley PM and WR Barker. 2005. Patterns of evolution in Australian *Mimulus* and related genera (Phrymaceae ~ Scrophulariaceae): a molecular phylogeny using chloroplast and nuclear sequence data. Aust. Syst. Bot. 18: 61–73.

Beardsley PM and RG Olmstead. 2002. Redefining Phrymaceae: the placement of *Mimulus*, tribe Mimuleae, and *Phryma*. Am. J. Bot. 89: 1093–1102.

Beardsley PM, SE Schoenig, JB Whittall, and RG Olmstead. 2004. Patterns of evolution in western North American *Mimulus* (Phrymaceae). Am. J. Bot. 91: 474–489.

Beaufort-Murphy HT. 1983. The seed surface morphology of the Gesneriaceae utilizing the scanning electron microscope and a new system for diagnosing seed morphology. Selbyana 6: 220–422.

Behnke H-D. 1986. Contributions to the knowledge of P-type sieve-element plastids in dicotyledons: IV. Acanthaceae. Bot. Jahrb. Syst. 106: 499–510.

Belleni R. 1907. Criteri per una nuova classificazione della Personatae Scrophulariaceae et Rhinanthaceae). Ann. Bot. 6: 131–145.

Bello MA, MW Chase, R Olmstead, N Rønsted, and D Albach. 2002. The páramo endemic *Aragoa* is the sister genus of *Plantago* (Plantaginaceae): Evidence from plastid *rbc*L and nuclear ribosomal ITS sequence data. Kew Bull. 57: 585–597.

Bello MA, PJ Rudall, F González, and JJL Fernández-Alonso. 2004. Floral morphology and development in *Aragoa* (Plantaginaceae) and related members of the order Lamiales. Int. J. Plant Sci. 165: 723–738.

Bendre AM. 1975. Studies in the family Loganiaceae: II. Embryology of *Buddleia* and *Strychnos*. J. Indian Bot. Soc. 54: 272–279.

Bennett JR and S Mathews. 2006. Phylogeny of the parasitic plant family Orobanchaceae inferred from phytochorme A. Am. J. Bot. 93: 1039–1051.

Bentham G. 1832–1836. Labiatarum genera et species. London.

Bigazzi M. 1989. Ultrastructure of nuclear inclusions and the separation of Verbenaceae and Oleaceae (incl. *Nyctanthes*). Plant Syst. Evol. 163: 1–12.

Bigazzi M. 1993. A survey of the intranuclear inclusions in the Scrophulariaceae and their systematic significance. Nord. J. Bot. 13: 19–31.

Bigazzi M. 1995. Investigation on occurrence and ultrastructure of the proteinaceous nuclear inclusions (PNIs) in the Bignoniaceae, with special reference to geographic distribution patterns. Caryologia 48: 211–223.

Bigazzi M and M Tardelli. 1990. Pollen morphology and ultrastructure of the Old World Antirrhineae (Scrophulariaceae). Grana 29: 257–275.

Bilimovich OE 1935. Significance of the anatomy of pericarp of Labiatae for their systematics. Trudy Voronezh State Univ. 7: 21–84 (in Russian).

Blatt CTT, MD dos Santos, and A Salatino. 1998. Flavonoids of Bignoniaceae from the "cerrado" and their possible taxonomic significance. Plant Syst. Evol. 210: 289–292.

Blunden G, M-H Yang, Z-X Yuan, BE Smith, A Patel, JA Cegarra, I Máthé Jr., and G Janisak. 1996. Betaine distribution in the Labiatae. Biochem. Syst. Ecol. 24: 71–81.

Bocquilon H-T. 1861. Observations sur le genre *Oftia* Adans. Adansonia 2: 5–12.

Boeshore I. 1920. The morphological continuity of Scrophulariaceae and Orobanchaceae. Contr. Bot. Lab. Morris Arbor. Univ. Pennsylvania 5: 139–177.

Borzova IA. 1960. The question of the origin of six-grooved type of pollen of the mints. Dokl. USSR Acad. Sci. 133: 1456–1467. For English translation see Trans. Doklady (Bot. Sci. Sect.) 133: 132–133.

Bousquet Melou A and MT Fauvel. 1998. Inter-specific variation in the concentration of two iridoid glucosides in *Avicennia* L. (Avicenniaceae Endl.). Biochem. Syst. Ecol. 26: 935–940.

Bremekamp CEB. 1939. On the position of the genera *Carlemannia* Benth. and *Silvianthus* Hook.f. Recueil Trav. Bot. Neerl. 36: 372.

Bremekamp CEB. 1953. The delimitation of Acanthaceae. Kl. Nederl. Akad. Wet. Proc., ser. C., 56: 533–546.

Bremekamp CEB. 1965. Delimitation and subdivision of the Acanthaceae. Bull. Bot. Survey India 7: 21–30.

Bremer B, RG Olmstead, L Struwe, and JA Sweere. 1994. *rbc*L sequences support exclusion of *Retzia, Desfontainia,* and *Nicodemia* from the Gentianales. Plant Syst. Evol. 190: 213–230.

Bretting PK and S Nilsson. 1988. Pollen morphology of the Martiniaceae and its systematic implications. Syst. Bot. 13: 51–59.

Briquet J. 1895–1897. Labiatae. In: A Engler and K Prantl, eds. Die natürlichen Pflanzenfamilien, 4, 3a: 183–375. W. Engelmann, Leipzig.

Brown JL. 2004. Orobanchaceae. In: N Smith, SA Mori, A Henderson, SW Stevenson, and SV Heald, eds. Flowering plants of the neotropics, p. 282. Princeton University Press, Princeton.

Bruce AN. 1905, 1907. On the activity of the glands of *Byblis gigantea*. Notes Roy. Bot. Gard. Edinb. 16: 9–14, 1905; 17: 83, 1907.

Brugger J and R Rutishauser. 1989. Bau und Entiwcklung landbewohnender *Utricularia*-Arten. Bot. Helvetica 99: 91–146.

Brummitt RK. 1989. Against separating Mendonciaceae from Acanthaceae. Acanthus 5: 1–3.

Budantsev AL and TA Lobova. 1997. Fruit morphology, anatomy and taxonomy of tribe Nepeteae (Labiatae). Edinb. J. Bot. 54: 183–216.

Burtt BL. 1962. Studies in the Gesneriaceae of the old world XXIV: tentative keys to the tribes and genera. Not. Roy. Bot. Gard. Edinb. 24: 205–220.

Burtt BL. 1965. The transfer of *Cyrtandromoea* from Gesneriaceae to Scrophulariaceae, with notes on the classification of that family. Bull. Bot. Surv. India 7: 73–88.

Burtt BL. 1970. Studies on the Gesneriaceae of the Old World: XXXI. Some aspects of functional evolution. Notes Roy Bot. Gard. Edinb. 30: 1–9.

Burtt BL. 1977. Classification above the genus, as exemplified by Gesneriaceae, with parallels from other groups. Plant Syst. Evol. Suppl. 1: 97–109.

Burtt BL and H Wiehler. 1995. Classification of the family Gesneriaceae. Gesneriana 1: 1–4.

Buurman J. 1978. Contribution to the pollen morphology of the Bignoniaceae with special reference to the tricolpate type. Pollen et Spores 19: 447–519.
Campbell DH. 1930. The relationships of *Paulownia*. Bull. Torrey Bot. Club 57: 47–50.
Cantino P. 1982. Affinities of the Lamiales: a cladistic analysis. Syst. Bot. 7: 237–248.
Cantino P. 1990. The phylogenetic significance of stomata and trichomes in the Labiatae and Verbenaceae. J. Arnold Arbor. 71: 323–370.
Cantino P. 1992a. Toward a phylogenetic classification of the Labiatae. In: RM Hartley and T Reynolds, eds. Advances in Labiate Science, pp. 27–37. Royal Botanic Gardens, Kew.
Cantino P. 1992b. Evidence for polyphyletic origin of the Labiatae. Ann. Missouri Bot. Gard. 79: 361–379.
Cantino PD. 2004. Phrymaceae. In: K Kubitzki, ed. The families and genera of vascular plants, vol. 7, pp. 323–326. Springer, Berlin/Heidelberg/New York.
Cantino PD and MS Abu-Asab. 1993. A new look at the enigmatic genus *Wenchengia* (Labiatae). Taxon 42: 339–344.
Cantino PD and RW Sanders. 1986. Subfamilial classification of Labiatae. Syst. Bot. 11: 163–183.
Cantino PD, RM Harley, and SJ Wagstaff. 1992. Genera of Labiatae: status and classification. In: RM Harley and T Reynolds, eds. Advances in Labiate Science, pp. 511–522. Royal Botanic Gardens, Kew.
Cantino PD, SJ Wagstaff, and RG Olmstead. 1999. *Cryopteris* (Lamiaceae) and the conflict between phylogenetic and pragmatic considerations in botanical nomenclature. Syst. Bot. 23: 369–386.
Cao Y-F, Q-G Chu, H Xin, SR Tao, and RY Lu. 2000. Developmental and anatomical studies on the floral nectary in *Vitex chinensii*. J. Wuhan Bot. Res. 18: 363–367.
Capuron R. 1972. Myoporacées, famille nouvelle pour Madagascar. Adansonia, ser. 2, 12: 39–43.
Carine MA and RW Scotland. 2002. Classification of Strobilanthinae (Acanthaceae): trying to classify the unclassifiable? Taxon 51: 259–279.
Carlquist S. 1970. Wood anatomy of insular species of *Plantago* and the problem of insular woodiness. Bull. Torrey Bot. Club 97: 353–361.
Carlquist S. 1976. Wood anatomy of Byblidaceae. Bot. Gaz. 137: 35–38.
Carlquist S. 1981. Wood anatomy of Chloanthaceae (Di-crastylidaceae). Aliso 10: 19–34.
Carlquist S. 1986. Wood anatomy of Stilbaceae and Retziaceae: ecological and systematic implications. Aliso 11: 299–316.
Carlquist S. 1987. Wood anatomy of Martyniaceae and Pedaliaceae. Aliso 11: 473–483.
Carlquist S. 1992a. Wood anatomy of sympetalous dicotyledon families: a summary, with comments on systematic relationships and evolution of the wood habit. Ann. Missouri Bot. Gard. 79: 303–332.
Carlquist S. 1992b. Wood anatomy of Lamiaceae. A survey, with comments on vascular and vasicentric tracheids. Aliso 13: 309–338.
Carlquist S. 1997. Wood anatomy of Buddlejaceae. Aliso 15: 41–56.
Carlquist S and DA Hoekman. 1986a. Wood anatomy of Gesneriaceae. Aliso 11: 279–297.
Carlquist S and DA Hoekman. 1986b. Wood anatomy of Myoporaceae: ecological and systematic considerations. Aliso 11: 317–334.
Carlquist S and S Zona. 1988. Wood anatomy of Acanthaceae: a survey. Aliso 12: 201–227.
Casper SJ. 1963. "Systematisch massgebende" Merkmale für die Einordnung der Lentibulariaceen in das System. Oesterr. Bot. Z. 110: 108–131.
Chadwell TB, SJ Wagstaff, and PO Cantino. 1992. Pollen morphology of *Phryma* and some putative relatives. Syst. Bot. 17: 210–219.
Chen S-T, K Guan, Z Zhou, R Olmstead, and Q Cronk. 2005. Molecular phylogeny of *Incarvillea* (Bignoniaceae) based on ITS and *trn*L-F sequences. Am. J. Bot. 92: 625–633.
Chen Z. 1983. Preliminary study on the pollen morphology of *Paulownia*. Wuhan Bot. Res. 1: 143–146.
Chiang F and D Frame. 1987. The identity of *Lithophytum* (Loganiaceae–Plocospermeae). Brittonia 39: 260–262.
Chuang TI and LR Heckard. 1991. Generic realignment and synopsis of subtribe Castillejinae (Scrophulariaceae, tribe Pediculareae). Syst. Bot. 16: 644–666.
Clark JL, PS Herendeen, LE Skog, and EA Zimmer. 2006. Phylogenetic relationships and generic boundaries in the Episceae (Gesneriaceae) inferred from nuclear, chloroplast, and morphological data. Taxon 55: 313–336.
Cole MD. 1992. The significance of the terpenoids in the Labiatae. In: RM Harley and T Reynolds, eds. Advances in Labiate science, pp. 315–324. Royal Botanic Gardens, Kew.
Conn BJ. 1992. Relationships within the tribe Prostanthereae (Labiatae). In: RM Harley and T Reynolds, eds. Advances in Labiate Science, pp. 55–64. Royal Botanic Gardens, Kew.
Conran JG. 1996. The embryology and relationships of the Byblidaceae. Aust. Syst. Bot. 9: 243–254.
Conran JG and R Carolin. 2004. Byblicaceae. In: K Kubitzki, ed. The families and genera of vascular plants, vol. 7, pp. 45–49. Springer, Berlin/Heidelberg/New York.
Conran JG and JM Dowd. 1993. The phylogenetic relationships of *Byblis* and *Roridula* (Byblidaceae: Roridulaceae) inferred from partial 18S ribosomal RNA sequences. Plant Syst. Evol. 188: 73–86.
Conran JG, A Houben, and A Lowrie. 2002a. Chromosome numbers in Byblidaceae. Aust. J. Bot. 50: 583–586.
Conran JG, A Lowrie, and J Moyle-Croft. 2002b. A revision of *Byblis* (Byblicaceae) in south-western Australia. Nuytsia 15: 11–20.
Cooper RL, JM Osborn, and CT Philbrick. 2000. Comparative pollen morphology and ultrastructure of the Callitrichaceae. Am. J. Bot. 87: 161–175.
D'Arcy WG and RC Keating. 1973. The affinities of *Lithophytum*: a transfer from Solanaceae to Verbenaceae. Brittonia 25: 213–225.
Dahlgren R and VS Rao. 1971. The genus *Oftia* Adans. and its systematic position. Bot. Not. 124: 451–472.
Dahlgren R and AE van Wyk. 1988. Structure and relationships of families endemic to or centered in Southern Africa. In: P Goldblatt and PP Lowry, eds. Modern systematic studies in African botany, pp. 1–94. Monographs Syst. Bot. Missouri Bot. Gard. 25.
Dahlgren R, BJ Nielsen, P Goldblatt, and JP Rourke. 1979. Further notes on Retziaceae: Its chemical contents and affinities. Ann. Missouri Bot. Gard. 66: 545–556.

Damtoft S, H Franzyk, SR Jensen, and BJ Nielsen. 1993. Iridoids and verbascosides in *Retzia*. Phytochemistry 34: 239–243.

Daniel TF. 1993. Chromosome numbers of New World Acanthaceae. Syst. Bot. 25: 15–25.

Daniel TF. 1998. Pollen morphology of Mexican Acanthaceae: diversity and systematic significance. Proc. Calif. Acad. Sci. 50: 217–256.

Daniel TF. 1999. Intraspecific seed heteromorphism in Acanthaceae. Acanthus 7: 8–9.

Daniel TF, K Balkwill, and MJ Balkwill. 2000. Chromosome numbers of South African Acanthaceae. Proc. Calif. Acad. Sci. 52: 143–158.

Das S and M Ghose. 2003. Seed structure and germination pattern of some Indian mangroves with taxonomic relevance. Taiwania 48: 287–298.

De A. 1966, 1967, 1968. Cytological, anatomical, and palynological studies as an aid in tracing affinity and phytogeny in the family Acanthaceae: I. Cytological studies. II. Floral anatomy. III. General anatomy. IV. Palynology and final conclusion. Trans. Bose Res. Inst. Calcutta 29: 139–175, 1966; 30: 27–43; 51–65, 1967; 31: 17–29, 1968.

Degtjareva G, J Casper, F Hellwig, and D Sokoloff. 2004. Seed morphology in the genus *Pinguicula* (Lentibulariaceae) and its relation to taxonomy and phylogeny. Bot. Jahrb. Syst. 125: 431–452.

Demissew S. 2004. Cyclocheilaceae. In: K Kubitzki, ed. The families and genera of vascular plants, vol. 7, pp. 60–62. Springer, Berlin/Heidelberg/New York.

Demissew S and RM Harley 1992. Trichome, seed-surface, and pollen characters in *Stachys* (Lamioideae: Labiatae) in tropical Africa. In: RM Harley and T Reynolds, eds. Advances in Labiate Science, pp. 149–166. Royal Botanic Gardens, Kew.

Depamphilis CW, TN Atkinson, and WJ Elisens. 1994. Tribal relationships in the Scrophulariaceae subfamily Antirrhinoideae: insights from sequence variation of the plastid-encoded gene *rps*2. Am. J. Bot. 81(6): 152 (abstract).

De Vos MP. 1947. Die ontwikkeling van die saadknop en saad by die Myoporaceae en die systematiese posisie van *Oftia* Adans. South Afr. J. Sci. 43: 171–187.

Dickison WC. 1994. A re-examination of *Sanango racemosum*. 2. Vegetative and floral anatomy. Taxon 43: 601–618.

Diels L. 1930. Byblidaceae. In: A. Engler and K. Prantl, eds. Die natürlichen Pflanzenfamilien, 2nd ed., pp. 286–288. Engelmann, Leipzig.

Dietrich H. 1969. Über die Aussagekraft pollenmorphologischer Fakten bei stenopalynen Verwandtschaftsbereichen am Beispiel der Plantaginaceae. Feddes Repert. 79: 347–353.

Domin K. 1922. Byblidaceae: a new archychlamydous family. Acta Bot. Bohem. 1: 3–4.

Dong ZM, ZH Hu, and J Yang. 1992. Anatomic studies of the nectaries of *Paulownia tomentosa* (Thunb.) Steud. Chinese J. Bot. 4: 13–19.

Doweld AB. 2000. Byblidaceae. In: A Takhtajan, ed. Comparative seed anatomy, vol. 6, pp. 364–366. Nauka, St. Petersburg (in Russian).

Drewes SI and S Martínez. 1999. Morfologia de la inflorescencias en Verbenaceae-Verbenoideae: 2. Tribu Petreeae. Darwiniana 37: 209–218.

Duke NC. 1991. A systematic revision of the mangrove genus *Avicennia* (Avicenniaceae) in Australasia. Aust. Syst. Bot. 4: 299–324.

Duletiae Lauseviae S and PD Marin. 1999. Pericarp structure and myxocarpy in selected genera of Nepetoideae (Lamiaceae). Nord. J. Bot. 19: 435–446.

El-Gazzar A. 1974. Numerical taxonomy of the Verbenaceae: a reassessment. Egypt. J. Bot. 17: 69–83.

El-Gazzar A and L Watson. 1970. A taxonomic study of Labiatae and related genera. New Phytol. 69: 451–486.

Elisens WJ. 1985. Monograph of the Maurandyinae (Scrophulariaceae-Antirrhineae). Monogr. Syst. Bot., 5.

Elisens WJ. 1986. Pollen morphology and systematic relationships among New World species of the tribe Antirrhineae (Scrophulariaceae). Am. J. Bot. 73: 1298–1311.

Elisens WJ and AS Tomb. 1983. Seed morphology in New World Antirrhineae (Scrophulariaceae). Systematic and phylogenetic implications. Plant Syst. Evol. 142: 47.

Endress PK. 1998. *Antirrhinum* and Asteridae – evolutionary changes of floral symmetry. Soc. Exp. Biol. Symp. Ser. 51: 133–140.

Engell K. 1987. Embryology and taxonomical position of *Retzia capensis* (Retziaceae). Nord. J. Bot. 7: 117–124.

Erbar C and P Leins. 2004. Callitrichaceae. In: K Kubitzki, ed. The families and genera of vascular plants, vol. 7, pp. 50–56. Springer, Berlin/Heidelberg/New York.

Erdtman G. 1945. Pollen morphology and plant taxonomy: IV. Labiatae, Verbenaceae, and Avicenniaceae. Svensk Bot. Tidskr. 39: 279–285.

Fahn A and C Shimony. 1977. Development of the glandular and non-glandular leaf hairs of *Avicennia marina* (Forsskal) Vierh. Bot. J. Linn. Soc. 74: 37–46.

Fedotova TA. 1996. Morphology of the fruit and seed of *Carlemannia* species (Carlemanniaceae). Bot. Zhurn. 81(5): 24–34 (in Russian with English summary).

Ferguson IK and T Santisuk. 1973. Notes on the pollen morphology of some Asiatic Bignoniaceae. Kew Bull. 28: 187–194.

Finn V. 1930. On the history of development of male gametophyte in Labiatae. Zhurn. Inst. Bot. Acad. Sci. Ukraine 20: 77–96 (in Russian).

Fischer E. 1989. Contributions for the flora of Central Africa II – *Crepidorhopalon*, a new genus within the relationship of *Craterostigma*, *Torenia* and *Lindernia* (Scrophulariaceae) with two new or noteworthy species from Central and South Central Africa (Zaire, Zambia). Feddes Repert. 100: 439–450.

Fischer E. 1992. Systematik der afrikanischen Lindernieae (Scrophulariaceae). Trop. Subtrop. Pflanzenwelt 81: 1–365.

Fischer E. 2004. Scrophulariaceae. In: K Kubitzki, ed. The families and genera of vascular plants, vol. 7, pp. 333–432. Springer, Berlin/Heidelberg/New York.

Fischer E, W Barthlott, R Seine, and I Theisen. 2004a. Lentibulariaceae. In: K Kubitzki, ed. The families and genera of vascular plants, vol. 7, pp. 276–282. Springer, Berlin/Heidelberg/New York.

Fischer E, I Theisen, and LG Lohmann. 2004b. Bignoniaceae. In: K Kubitzki, ed. The families and genera of vascular plants, vol. 7, pp. 9–38. Springer, Berlin/Heidelberg/New York.

Flyr LD. 1970. A systematic study of the tribe Leucophylleae (Scrophulariaceae). Doctoral dissertation, University of Texas, Austin.

Fosberg FR and D Herbst. 1983 (1984). A *Nesogenes* (Chloanthaceae) from Micronesia. Micronesica 19(1–2): 11–15.

Frederiksen LB, S Damtoft, and SR Jensen. 1999. Biosynthesis of iridoid lacking C-10 and the chemotaxonomic implications of their distribution. Phytochemistry 52: 1409–1420.
Freeman CE and R Scogin. 1999. Potential utility of chloroplast *trn*L (UAA) gene intron sequences for inferring phylogeny in Scrophulariaceae. Asilo 18: 141–159.
Galati BG and LI Strittmatter. 1999. Microsporogenesis and microgametogenesis in *Jacaranda mimosifolia* (Bignoniaceae). Phytomorphology 49: 147–155.
Galetto L. 1995. Nectary structure and nectar characteristics in some Bignoniaceae. Plant Syst. Evol. 196: 99–121.
Gasson P and DR Dobbins. 1991. Wood anatomy of the Bignoniaceae, with a comparison of trees and lianas. IAWA Bull. 12: 389–417.
Gentry A. 1974. Coevolutionary patterns in Central American Bignoniaceae. Ann. Missouri Bot. Gard. 61: 728–759.
Gentry A. 1976. Relationships of the Madagascar Bignoniaceae: a striking case of convergent evolution. Plant Syst. Evol. 126: 255–266.
Gentry AH. 1980. Bignoniaceae: Part 1. (Crescentieae and Tourrettieae). Flora Neotropica Monograph 25(1): 1–150.
Gentry AN. 1990. Evolutionary patterns in neotropical Bignoniaceae. Mem. New York Bot. Gard. 55: 118–129.
Gentry AN and AS Tomb. 1979. Taxonomic implications of Bignoniaceae palynology. Ann. Missouri Bot. Gard. 66: 756–777.
Ghebrehiwet M. 2000. Taxonomy, phylogeny and biogeography of *Kickxia* and *Nanorrhinum* (Scrophulariaceae). Nord. J. Bot. 20: 655–689.
Ghebrehiwet M, B Bremer, and M Thulin. 2000. Phylogeny of the tribe Antirrhineae (Scrophulariaceae) based on morphological and *ndh*F sequence data. Plant Syst. Evol. 220: 223–239.
Glück H. 1940. Die Gattung *Trapella*. Bot. Jahrb. Syst. 71(1941): 267–336.
Goldblatt P and AH Gentry. 1979. Cytology of Bignoniaceae. Bot. Not. 132: 475–482.
Goldblatt P and RC Keating. 1976. Chromosome cytology, pollen structure, and relationships of *Retzia capensis*. Ann. Missouri Bot. Gard. 63: 321–325.
Govindu HC. 1950. Studiea in the embryology of some members of the Bignoniaceae Proc. Indian Acad. Sci. B. 32(4): 164–178.
Grant WF. 1955. A cytogenetic study in the Acanthaceae. Brittonia 8: 121–149.
Grayer RJ and RPJ de Kok. 1998. Flavonoids and verbascoside as chemotaxonomic characters in the genera *Oxera* and *Faradaya* (Labiatae). Biochem. Syst. Ecol. 26: 729–741.
Grayer RJ, NC Veitch, GC Kite, AJ Paton, and PJ Garnock Jones. 2002. Scutellarein 4 -methyl ether glycosides as taxonomic markers in *Teucridium* and *Tripora* (Lamiaceae, Ajugoideae). Phytochemistry 60: 727–731.
Guedes M. 1974. Le gynecee de *Paulownia* et *Schlegelia* et le probleme de la delimitation des Scrofulariacees et Bignoniacees. Compt. R. Hebd. Seances Acad. Sci., D (Paris) 278, 21: 2629–2632.
Hallier H. 1903. Über die Abgrenzung und Verwandtschaft der einzelnen Sippen bei den Scrophularineen. Bull. Herb. Boiss. 3: 181–207.
Hamilton AG. 1903. Notes on *Byblis gigantea*. Proc. Linn. Soc. N. S. W. 28: 680–684.
Harborne JB. 1967. Comparative biochemistry of the flavonoids. VI. Flavonoid patterns in the Bignoniaceae and the Gesneriaceae. Phytochemistry 6: 1643–1651.
Harley RM. 1992. The potential value of pollen morphology as an additional taxonomic character in subtribe Ociminae (Ocimeae: Nepetoideae: Labiatae). In: RM Harley and T Reynolds, eds. Advances in Labiate Science, pp. 125–138. Royal Botanic Gardens, Kew.
Harley RM. 2004. Nesogenaceae. In: K Kubitzki, ed. The families and genera of vascular plants, vol. 7, pp. 293–295. Springer, Berlin/Heidelberg/New York.
Harley RM and CA Heywood. 1992. Chromosome numbers in tropical American Labiatae. In: RM Harley and T Reynolds, eds. Advances in Labiate Science, pp. 211–246. Royal Botanic Gardens, Kew.
Harley RM and T Reynolds, eds. 1992. Advances in Labiate Science. Royal Botanic Gardens, Kew.
Harley RM, AJ Paton, and O Ryding. 2003. New synonyme and taxonomic changes in the Labiatae. Kew Bull. 58: 485–489.
Harley RM, AM Giulietti, and FR Dos Santos. 2003. *Holoregmia* Nees, a recently rediscovered genus of Martyniaceae from Bahia, Brazil. Kew Bull. 58: 205–212.
Harley RM, S Atkins, AL Budantsev, PD Cantino, BJ Conn, R Grayer, MM Harley, R de Kok, T Krestovskaja, R Morales, AJ Paton, O Ryding, and T Upson. 2004. Labiatae. In: K Kubitzki, ed. The families and genera of vascular plants, vol. 7, pp. 167–275. Springer, Berlin/Heidelberg/New York.
Harrison CJ, M Möller, and QCB Cronk. 1999. Evolution and development of floral diversity in *Streptocarpus*. Ann. Bot. 84: 49–60.
Hartl D. 1957. Die stellung von *Lindenbergia* Lehmann im system der Schrophulariaceen. Beit. Biol. Pflanzen 33: 265–277.
Hartl D. 1959. Das alveolierte Endosperm bei Scrophulariaceen, seine Entstehung, Anatomie und taxonomische Bedeutung. Beitr. Biol. Pflanzen 35: 95–110.
Hartley IH and K Balkwill. 1990. A taxonomic account of *Agathelpis, Globulariopsis,* and *Gosela* (Scrophulariaceae). South Afr. J. Bot. 56: 471–481.
Hartmeyer S. 1998. Carnivory in *Byblis* revisited II: the phenomenon of symbiosis on insect trapping plants. Carniv. Plant Newslett. 27: 110–113.
Hauk WD. 2002. A review of the genera *Roentgenia* and *Potamoganos* (Bignoniaceae). Ann. Missouri Bot. Gard. 89: 77–87.
Hedrén M, MW Chase, and RG Olmstead. 1995. Relationships in the Acanthaceae and related families as suggested by cladistic analysis or *rbc*L nucleotide sequences. Plant Syst. Evol. 194: 93–109.
Hedge IC. 1992. A global survey of the biogeography of the Labiatae. In: RM Harley and T Reynolds, eds. Advances in Labiate Science, pp. 7–18. Royal Botanic Gardens, Kew.
Hegnauer R. 1969. Chemical evidence for the classification of some plant taxa. In: JB Harborne and T Swain, eds. Perspectives in phytochemistry, pp. 121–138. Academic, London/New York.
Hegnauer R. 1989a. Bignoniaceae. Chemotaxonomie der Pflanzen 8: 128–138.

Hegnauer R. 1989b. Globulariaceae. Chemotaxonomie der Pflanzen 8: 520–522.
Hegnauer R. 1989c. Myoporaceae. Chemotaxonomie der Pflanzen 8: 90–96.
Hegnauer R. 1990. Scrophulariaceae. Chemotaxonomie der Pflanzen 9: 527–556.
Henrickson J and LD Flyr. 1985. Systematics of *Leucophyllum* and *Eremogeton* (Scrophulariaceae). SIDA 11: 107–172.
Hesse M und W Morawetz. 1980. Skulptur und systematischer Wert der Samenoberfläche bei *Jacaranda* und anderen Bignoniaceae. Plant Syst. Evol. 135: 1–10.
Hillard OM. 1990. A brief survey of Scrophulariaceae-Selagineae. Edinb. J. Bot. 47: 315–343.
Hilliard OM. 1999. The tribe Selagineae (Scrophulariaceae). Royal Botanic Gardens, Kew.
Hillson CJ. 1959. Comparative studies of floral morphology of the Labiatae. Am. J. Bot. 46: 451–459.
Hjertson M. 1997. Systematics of *Lindenbergia* and *Campylanthus* (Scrophulariaceae). Uppsala.
Hoggard RK, PJ Kores, M Molvray, GD Hoggard, and DA Broughton. 2003. Molecular systematics and biogeography of the amphibious genus *Littorella* (Plantaginaceae). Am. J. Bot. 90: 429–435.
Holm T. 1913. *Phryma leptostachya* L., a morphological study. Bot. Gaz. 56: 306–318.
Holmgren N. 2004. Scrophulariaceae. In: N Smith, SA Mori, A Henderson, SW Stevenson, and SV Heald, eds. Flowering plants of the neotropics, pp. 348–350. Princeton University Press, Princeton.
Hong DY. 1984. Taxonomy and evolution of the Veroniceae (Scrophulariaceae) with special reference to palynology. Opera Bot. 75: 1–60.
Hu TW and TP Lin. 1975. The trichomes of *Paulownia* of Taiwan. Quart. J. Chinese Forest. 8(2): 1–9.
Hu C-C, H-P Tsue, and Y-L Zhang. 2005a. Pollen morphology of the tribe Ruellieae (Acanthaceae) from China. Acta Phytotax. Sinica 43: 123–150 (in Chinese with English summary).
Hu C-C, H-P Tsui, Y-Z Zhen, and Y-L Zhang. 2005b. Pollen morphology of one genus in Lepidagathideae, two in Andrographideae and eight in Justicieae (Acanthaceae) from China. Acta Phytotax. Sinica 43: 151–162 (in Chinese with English summary).
Hu S-Y. 1959. A monograph of the genus *Paulownia*. Quart. J. Taiwan Mus. 12: 1–54.
Ihlenfeldt H-D. 2004. Martyniaceae. In: K Kubitzki, ed. The families and genera of vascular plants, vol. 7, pp. 283–288. Springer, Berlin/Heidelberg/New York.
Ihlenfeldt H-D. 1967. Über die Abgrenzung und die natürliche Gliederung der Pedaliaceae R.Br. Mitt. Staatsinst. Allg. Bot. Hamburg 12: 43–128.
Ihlenfeldt H-D. 2004a. Pedaliaceae. In: K Kubitzki, ed. The families and genera of vascular plants, vol. 7, pp. 307–322. Springer, Berlin/Heidelberg/New York.
Ihlenfeldt H-D. 2004b. Trapellaceae. In: K Kubitzki, ed. The families and genera of vascular plants, vol. 7, pp. 445–448. Springer, Berlin/Heidelberg/New York.
Inamdar JA. 1969a. Epidermal structure and ontogeny of stomata in some Verbenaceae. Ann. Bot. 33: 55–66.
Inamdar JA. 1969b. Epidermal structure and stomatal ontogeny in vegetative and floral organs of *Martynia annua* L., *Pedalium murex* L. and *Sesamum indicum* L. Flora Abt. B 158(1968/1969): 526–537.
Inamdar JA and DC Bhat. 1972. Structure and development of stomata in some Labiatae. Ann. Bot. 36: 335–344.
Ivanina LI. 1965. Application of the carpological method to the taxonomy of Gesneriaceae. Notes Roy. Bot. Gard. Edinb. 26: 383–402.
Ivanina LI. 1967. The family Gesneriaceae (The carpological review). Nauka, Leningrad (in Russian).
Jamzad Z. 2001. A phylogenetic study of *Nepeta* L. (Lamiaceae). University of London, London.
Jensen HFW, SR Jensen, and BJ Nielsen. 1988. Chemotaxonomy of the Acanthaceae: iridoids and quaternary amines. Phytochemistry 27: 2581–2589.
Jensen SR. 1994. A re-examination of *Sanango racemosum*. 3. Chemotaxonomy. Taxon 43: 619–623.
Jensen SR. 1999. Chemical relationships of *Polypremum procumbens, Tetrachondra hamiltonii* and *Peltanthera floribunda*. Biochem. Syst. Ecol. 28: 45–51.
Jensen SR. 2000. Chemistry of the Buddlejaceae. In: E Norman, ed. Buddlejaceae. Flora Neotrop. Monogr. 81: 1–225.
Jensen SR, BJ Nielsen, and R Dahlgren. 1975. Iridoid compounds, their occurrence and systematic importance in the angiosperms. Bot. Not. 128: 148–173.
Jobson RW, J Playford, KM Cameron, and VA Albert. 2003. Molecular phylogenetics of Lentibulariaceae inferred from plastid *rps*16 intron and *trn*L-F DNA sequences: implications for character evolution and biogeography. Syst. Bot. 28: 157–171.
Jobson RW, R Nielsen, L Laakkonen, M Wikström, and VA Albert. 2004. Adaptive evolution of cytochrome *c* oxidase: Infrastructure for a carnivorous plant radiation. Proc. Natl. Acad. Sci. 101: 18064–18068.
Jørgensen CA. 1923. Studies on Callitrichaceae. Svensk Bot. Tidskr. 38: 81–122.
Jørgensen CA. 1925. Frage der systematischen Stellung der Callitrichaceae. Jahrb. Wiss. Bot. 64: 440–442.
Junell S. 1934. Zur Gynäceummorphologie und Systematik der Verbenaceen und Labiaten, nebst Bemerkungen über ihre Samenentwicklung. Symb. Bot. Upsal. 1(4): 1–219.
Junell S. 1937. Die Samenentwicklung bei einigen Labiaten. Svensk Bot. Tidskr. 31: 67–110.
Junell S. 1961. Ovarian morphology and taxonomic position of Selagineae. Svensk Bot. Tidskr. 55: 168–192.
Kampany CM. 1995. Pollination and flower diversity in Scrophulariaceae. Bot. Rev. 61: 350–366.
Kampany CM and JM Canne-Hilliker. 1988. Aspects of floral development in Scrophulariaceae: striking early differences in three tribes. In: P Leins, SC Tucker, and PK Endress, eds. Aspects of floral development, pp. 147–157. J. Cramer, Berlin.
Kampany CM and NG Dengler. 1997. Evolution of flower shape in Veroniceae (Scrophulariaceae). Plant Syst. Evol. 205: 1–25.
Karatela YY and LS Gill. 1984. Phytodermology and ontogeny of stomata in some Pedaliaceae. J. Econ. Tax. Bot. 5: 237–240.
Karrfalt EE and AS Tomb. 1983. Air spaces, secretory cavities, and the relationship between Leucophylleae (Scrophulariaceae) and Myoporaceae. Syst. Bot. 8: 29–32.

Keenan J. 1969. Notes on *Buddleja*. II. Pollen. Notes Roy. Bot. Gard. Edinb. 29: 199–203.

Kok de RPJ. 1997. The biology and systematics of *Oxera, Faradaya* and *Hosea* (Labiatae). Oxford.

Khan R. 1970 Lentibulariaceae. In Symp. Comparative embryology of angiosperms. Bull. Indian Natl. Sci. Acad. 41: 290–297.

Kiehn M and A Weber. 1998. Chromosome numbers of Malayan and other palaeotropical Gesneriaceae. II. Tribes Trichosporeae, Cyrtandreae and Epithemateae. Beitr. Biol. Pflanzen 70: 445–470.

Kiehn M, E Hellmayer, and A Weber. 1998. Chromosome numbers of Malayan and other palaeotropical Gesneriaceae. I. Tribe Didymocarpeae. Beitr. Biol. Pflanzen 70: 407–444.

Kiel CA, LA McDade, TF Daniel, and D Champluvier. 2006. Phylogenetic delimitation of Isoglossinae (Acanthaceae: Justicieae) and relationships among constituent genera. Taxon 55: 683–694.

Kondo K, M Segawa, and K Nehira. 1978. Anatomical studies on seeds and seedlings of some *Utricularia* (Lentibulariaceae). Brittonia 30: 89–95.

Kooiman P. 1970. The occurrence of iridoid glycosides in the Scrophulariaceae. Acta Bot. Neerl. 19: 329–340.

Kooiman PT. 1972. The occurrence of iridoid glycosides in the Labiatae. Acta Bot. Neerl. 21: 417–427.

Kornhall P. 2004. Phylogenetic studies in the Lamiales with special focus on Scrophulariaceae and Stilbaceae. Acta Universitatis Upsaliensis, Upsala.

Kornhall P and B Bremer. 2004. New circumscription of the tribe Limoselleae (Scrophulariaceae) that includes taxa of the tribe Manuleeae. Bot. J. Linn. Soc. 146: 453–467, see also ibid, 2005, Bot. J. Linn. Soc. 147: 385–386. Corrigenda.

Kornhall P, N Heidari, and B Bremer. 2001. Selagineae and Manuleeae, two tribes or one? Phylogenetic studies in the Scrophulariaceae. Plant Syst. Evol. 228: 199–218.

Krishnamurthy KV and K Sigamani. 1987. Wood anatomy of two South Indian species of *Avicennia*. Feddes Repert. 98: 537–542.

Kumar Ravi KV and BH Rao. 1988. Wood anatomy of three species of *Avicennia* L. Swamy Bot. Cl. 5: 83–88.

Kundu BC and S Bose. 1971. Taxonomic position of the genus *Phryma*. Proc. Indian Sci. Congr. 58: 437.

Kvist LP and JA Pedersen. 1986. Distribution and taxonomic implications of some phenolics in the family Gesneriaceae determined by EPR spectrosopy. Biochem. Syst. Ecol. 14: 385–405.

Lang FX. 1901. Untersuchungen über Morphologic, Anatomic, und Samenentwicklung von Polypompholyx und *Byblis gigantea*. Flora 88: 3–60.

Large MF and DJ Mabberley. 1995. An assessment of pollen morphology in the genus *Vitex* L. (Labiatae). Grana 34: 291–299.

Lawrence RM. 1992. Chemical components of Labiatae oils and their exploitation. In: RM Harley and T Reynolds, eds. Advances in Labiate science, pp. 399–436. Royal Botanic Gardens, Kew.

Leeuwenberg AJM. 1967. Notes on American Loganiaceae I. Revision of *Plocosperma* Benth. Acta Bot. Neerl. 16: 56–61.

Legendre L. 2000. The genus *Pinguicula* L. (Lentibulariaceae): an overview. Acta Bot. Gallica 147: 77–95.

Leins P and C Erbar. 1988. Einige Bemerkungen zur Blütenentwicklung und systematischen Stellung der Wasserpflanzen *Callitriche, Hippuris,* und *Hydrostachys*. Beitr. Biol. Pfl. 63: 157–178.

Leins P and C Erbar. 2004. Hippuridaceae. In: K Kubitzki, ed. The families and genera of vascular plants, vol. 7, pp. 163–166. Springer, Berlin/Heidelberg/New York.

Leitner J. 1942. Ein Beitrag zur Kenntnis der Pollenkorner der Labiatae. Oesterr. Bot. Z. 91: 29–40.

Leroy J-F. 1977. Les Myoporaceae á Madagascar: niveau d'endémisme. Adansonia, ser. 2, 17: 113–118.

Lersten NR and JM Beaman. 1998. First report of oil cavities in Scrophulariaceae and renvestigation of air spaces in leaves of *Leucophyllum frutescens*. Am. J. Bot. 85: 1646–1649.

Lersten NR and JD Curtis. 1997. Anatomy and distribution of foliar idioblasts in *Scrophularia* and *Verbascum* (Scrophulariaceae). Am. J. Bot. 84: 1638–1645.

Lersten NR and JD Curtis. 2001. Idioblasts and other unusual internal foliar secretory structures in Scrophulariaceae. Plant Syst. Evol. 227: 63–73.

Lersten NR, L Krueger, and JD Curtis. 2002. Tracheoid vatiation among Bignoniaceae seed wings, with emphasis on *Campsis radicans*. Int. J. Plant Sci. 163: 369–378.

Lewis DQ. 2000. A revision of the New World species of *Lindernia* (Scrophulariaceae). Castanea 65: 93–122.

Li H-L. 1954. Trapellaceae: a familial segregate from the Asiatic flora. J. Washington Acad. Sci. 44: 11–13.

Li ZY. 2000. Taxonomic notes on the genus *Phryma* L. from Asia. Acta Phytotax. Sinica 38: 386–391.

Linder HP. 2004. Stilbaceae. In: K Kubitzki, ed. The families and genera of vascular plants, vol. 7, pp. 433–440. Springer, Berlin/Heidelberg/New York.

Litvinenko VI, TP Popova, AV Simonian, IG Zoz, and VS Sokolov. 1975. "Gerbstoffe" und Oxyzimtsäureabkommlinge in Labiaten. Plant Med. 27: 372–380.

Lloyd FW. 1942. The carnivorous plants. Ronald Press, New York.

Lobreau-Callen D. 1978. Les caracteres polliniques du genre *Ranopisoa* Leroy (Myoporaceae). Adansonia, ser. 2, 18: 237–241.

Lobreau-Callen D, J Jérémie and M Suarez-Cervera. 1999. Morphologie et ultrastructure du pollen dans le genre *Utricularia* L. (Lentibulariaceae). Canad. J. Bot. 77: 744–767.

Lohmann LG. 2006. Untangling the phylogeny of neotropical lianas (Bignonieae, Bignoniaceae). Am. J. Bot. 93: 304–318.

Long RW. 1970. The genera of Acanthaceae in the southeastern United States. J. Arnold Arbor. 51: 257–309.

Lopes AV, S Vogel, and IC Machado. 2002. Secretory trichomes, a substitutive floral nectar source in *Lundia* A. DC. (*Bignoniaceae*), a genus lacking a functional disc. Ann. Bot. 90: 169–174.

Lowrie A and JG Conran. 1998. A taxonomic revision of the genus *Byblis* (Byblidaceae) in northern Australia. Nuytsia 12: 59–74.

Lu A. 1990. A preliminary cladistic study of the families of the superorder Lamiiflorae. Bot. J. Linn. Soc. 103: 39–57.

Luegmayr E. 1993a. Pollen characters of Old World Gesneriaceae (Cyrtandroideae) with special reference to SE Asian taxa. Grana 32: 221–232.

Luegmayr E. 1993b. Pollen of Hawaiian *Cyrtandra* (Gesneriaceae), including notes on Southeast Asian taxa. Blumea 38: 25–38.

Machado dos A and G dos Santos. 1990. Systematic wood anatomy of Tecomeae (Bignoniaceae). St. Louis, MO.

Maffei M. 1994. Discriminant analysis of leaf wax alkanes in the Lamiaceae and four other plant families. Biochem. Syst. Ecol. 22: 711–728.

Maheshwari JK. 1961. The genus *Wightia* Wall. in India with a discussion on its systematic position. Bull. Bot. Surv. India 3: 31–35.

Majumdar S and S Chanda. 1978. Pollen morphology and taxonomy of *Carlemannia* and *Silvianthus* of the family Rubiaceae sensu lato. Trans. Bose Res. Inst. Calcutta 41(3–4): 99–105.

Maldonado de Magnano S. 1986. Estudios embriologicos en *Buddleja* (Buddlejaceae): I. Endosperma y episperma. Darwiniana 27: 225–236.

Maldonado de Magnano S. 1987. Estudios embriologicos en *Buddleja* (Buddlejaceae): II. Embriogenesis. Darwiniana 28: 391–395.

Manktelow M. 2000. The filament curtain – a structure important to systematics and pollination biology in the Acanthaceae. Bot. J. Linn. Soc. 133: 129–160.

Manning SD. 1991. The genera of Pedaliaceae in the southeastern United States. J. Arnold Arbor. Suppl. Ser. 1: 313–347.

Manning SD. 2000. The genera of Bignoniaceae in the southeastern United States. Harvard Pap. Bot. 5: 1–77.

Marais W. 1981. Two new gamopetalous families, Cyclocheilaceae and Nesogenaceae, for extra Australian 'Dicrastylidaceae'. Kew Bull. 35: 797–812.

Marais W. 1984. Cyclocheilaceae, Nesogenaceae. In: RM Polhill, ed. Flora tropical East Aftica. A.A. Balkema, Rotterdam.

Marin PD. 1994. Nutlet sculpturing of selected species from Ajugoideae, Scutellarioideae and Stachyoideae (Lamiaceae). Bull. Inst. Jard. Bot. Univ. Beograd 26–27: 21–29 (1992–1993 publ. 1994).

Marin PD, V Sajdl, S Kapor, B Tatic, and B Petkovic. 1991. Fatty acids of the Saturejoideae, Ajugoideae and Scutellarioideae (Lamiaceae) Phytochemistry 30: 2979–2982.

Martinez S, S Botta, and ME Mulgura. 1996. Morfologia de las inflorescencias en Verbenaceae-Verbenoideae: 1. Tribu Verbeneae. Darwiniana 34: 1–17.

Martinsson K. 1993. The pollen of Swedish *Callitriche* (Callitrichaceae) – trends toward submergence. Grana 32: 198–209.

Mayer V, M Möller, M Perret, and A Weber. 2003. Phylogenetic position and generic differentiation of Epithemateae (Gesneriaceae) inferred from plastid DNA sequence data. Am. J. Bot. 90: 321–329.

Mayr EM and A Weber. 2006. Calceolariaceae: floral development and systematic implications. Am. J. Bot. 93: 327–343.

McDade LA and ML Moody. 1999. Phylogenetic relationships among Acanthaceae: evidence from noncoding *trn*L–*trn*F chloroplast DNA sequences. Am. J. Bot. 86: 70–80.

McDade LA, TF Daniel, SE Masta, and KM Riley. 2000. Phylogenetic relationships within the tribe Justicieae (Acanthaceae): evidence from molecular sequences, morphology, and cytology. Ann. Missouri Bot. Gard. 87: 435–458.

McDade LA, Masta SE, ML Moody, and E Waters. 2000. Phylogenetic relationships among Acanthaceae: evidence from two genomes. Syst. Bot. 25: 106–121.

Mehra KR and AR Kulkarni. 1985. Embryological studies in Bignoniaceae. phytomorphology 35(3, 4): 235–251.

Melhelm T and Mauro C. 1973. Pollen morphological studies in Gesneriaceae. Hoehnea 3: 13–27.

Millsaps V. 1936. The structure and development of the seed of *Paulownia tomentosa* Steud. J. Elisha Mitchell Soc. 52: 56–75.

Minkin JP and WH Eshbaugh. 1989. Pollen morphology of the Orobanchaceae and rhinanthoid Scrophulariaceae. Grana 28: 1–18.

Mitra K. 1968. Pollen morphology in Bignoniaceae in relation to taxonomy. Bull. Bot. Surv. India, 10: 319–326.

Moar NT. 1960. Studies in pollen morphology. 2. The pollen of New Zealand and two Australian species of *Callitriche* L. N. Z. J. Sci. 3: 415–421.

Mohan Ram HY and M Wadhi. 1964. Endosperm in Acanthaceae. Phytomorphology 14: 388–413.

Mohan Ram HY and M Wadhi. 1965. Embryology and delimitation of the Acanthaceae. Phytomorphology 15: 201–205.

Molau U. 1988. Scrophulariaceae – Part I. Calceolarieae. Flora Neotrop. Monogr. 47: 1–325.

Moldenke HN. 1980. Notes on the genus *Symphorema*. Phytologia 45: 396–414.

Moldenke HN. 1983. A sixth summary of the Verbenaceae, Avicenniaceae, Stilbaceae, Chloanthaceae, Symphoremaceae, Nyctanthaceae, and Eriocaulaceae of the world as to valid taxa, geographic distribution and synonymy. Supplement 3. Phytologia 54: 228–245.

Moller M and QCB Cronk. 2001. Evolution of morphological novelty: a phylogenetic analysis of growth pattern in *Streptocarpus* (Gesneriaceae). Evolution 55: 918–929.

Moller M and M Kiehn. 2004. A synopsis of cytological studies in Gesneriaceae. Edinb. J. Bot. 60: 425–457.

Moore RJ. 1948. Cytotaxonomic studies in the Loganiaceae. II. Embryology of *Polypremum procumbens* L. Am. J. Bot. 35: 404–410.

Mukherjee J. 1974. Role of palynology in the taxonomy and phylogeny of Myoporaceae. Sci. Cult. 40: 331–332.

Mukherjee J. 1976. The use of pollen morphology in the taxonomy of the Chloantoideae Briq. (Verbenaceae). Trans. Bose Res. Inst. Calcutta 39: 37–46.

Mukherjee J and S Chanda. 1974. Biosynthesis of *Avicennia* L. in relation to taxonomy. Geophytology 3: 85–88 (1973 publ. 1974).

Müller K and T Borsch. 2005. Phylogenetics of *Utricularia* (Lentibulariaceae) and molecular evolution of the *trn*K intron in a lineage with high substitutional rates. Plant Syst. Evol. 250: 39–67.

Müller K, T Borsch, L Legendre, S Porembski, and W Barthlott. 2000. A phylogeny of Lentibulariaceae based on sequences of *mat*K and adjacent noncoding regions. Am. J. Bot. Suppl. 87(6): 133–134.

Mueller K, T Borsch, L Legendre, E Fischer, S Porembski, and W Barthlott. 2001. The evolution of carnivory in the Lamiales: Evidence from *mat*K and adjacent noncoding

regions. In Botany 2001: Plants and People. Abstracts, p. 129. Albuquerque.
Müller K, T Borsch, L Legrende, S Porembski, I Theisen, and W Barthlott. 2004. Evolution of carnivory in Lentibulariaceae and the Lamiales. Plant Biol. 6: 477–490.
Munir AA. 1978. Taxonomic revision of Chloanthaceae trib. Physopsideae. Brunonia 1: 407–692.
Musselman LJ and WC Dickison. 1975. The structure and development of the haustorium in parasitic Scrophulariaceae. Bot. J. Linn. Soc. 70: 183–212.
Musselman LJ and WF Mann. 1976. A survey of surface characteristics of seeds of Scrophulariaceae and Orobanchaceae using scanning electron microscopy. Phytomorphology 26: 370–378.
Nabli MA. 1976. Etude ultrastructurale comparee de lɜεξιvε chez quelques genres de Labiatae. In: IK Ferguson and I Muller, eds. The evolutionary significance of the exine, pp. 499–525. Academic, London.
Narayana LL and DVL Satyavathi. 1987. Floral anatomy of two Lentibulariaceae. Indian Sci. Congr. Assoc. Proc. 3: 138.
Neizgoda GJ and AS Tomb. 1976. Systematic palynology of tribe Leucophylleae (Scrophulariaceae) and selected Myoporaceae. Pollen Spores 17: 495–516 (1975 publ. 1976).
Nickrent DL. 2006. The parasitic plant connection: parasitic plant genera. Department of Plant Biology. Website http://www.parasiticplants.siu.edu/ListParasites.html
Nicoletti M, M Serafina, JA Garbarino and V Gambaro. 1988. A chemosystematic study of Scrophulariaceae: Iridoid glycosides. Giorn. Bot. Ital. 122: 13–24.
Nie Z-L, H Sun, PM Beardsley, RG Olmstead, and J Wen. 2006. Evolution of biogeographic disjunction between eastern Asia and eastern North America in *Phryma* (Phrymaceae). Am. J. Bot. 93: 1343–1356.
Niezgoda CJ and AS Tomb. 1975. Systematic palynology of tribe Leucophylleae (Scrophulariaceae) and selected Myoporaceae. Pollen et Spores 17: 495–516.
Nikiticheva ZI. 1987. Scrophulariaceae and Lentibula- riaceae. In: MS Yakovlev, ed. Comparative embryology of flowering plants: Davidiaceae-Asteraceae, pp. 251–259, 283–288. Nauka, Leningrad (in Russian).
Norman EM. 1994. A re-examination of *Sanango racemosum*. 1. Morphology and distribution. Taxon 43: 591–600.
Norman E. 2000. Buddlejaceae. Flora Neotrop. Monogr. 81: 1–225.
Novitskaya GV and VI Krishtopa. 1971. Composition of the fatty acids of Labiatae oils in connection with their taxonomic position. J. Rast. Resour. 7: 32–40 (in Russian).
Oxelman B, P Kornhall, RG Olmstead, and B Bremer. 2005. Further disintegration of Scrophulariaceae. Taxon 54: 411–425.
Olmstead RG and PA Reeves. 1995. Evidence for the polyphyly of the Scrophulariaceae based on chloroplast *rbcL* and *ndhF* sequences. Ann. Missouri Bot. Gard. 82: 176–193.
Olmstead RG, KM Scott, and JD Palmer. 1992. A chloroplast DNA phylogeny for the Asteridae: Implications for the Lamiales. In: RM Harley and T Reynolds, eds. Advances in Labiate Science, pp. 19–26. Royal Botanic Gardens, Kew.
Olmstead RG, PA Reeves, and BJ Lepschiz. 1998. Confirmation of a monophyletic Chloanthoideae (Lamiaceae) comprising tribes Chloantheae and Prostanthereae. Lamiales Newslett. 6: 7–10.
Olmsted RG et al. 1999. A Synoptical Classification of the Lamiales Bromhead. www.inform.umd.edu
Olmstead RG, CW de Pamphilis, AD Wolfe, ND Young, WJ Elisons, and PA Reeves. 2001. Disintegration of the Scrophulariaceae. Am. J. Bot. 88: 348–361.
Osborn JM, and CT Philbrick. 1994. Comparative pollen structure and pollination biology in the Callitrichaceae. Acta Bot. Gallica 141: 257–266.
Oxelman B, M Backlund, and B Bremer. 1999. Relationships of the Buddlejaceae s.l. investigated using parsimony jackknife and branch support analysis of chloroplast *ndhF* and *rbcL* sequence data. Syst. Bot. 24: 164–182.
Oxelman B, P Kornhall, and EM Norman. 2004. Buddlejaceae. In: K Kubitzki, ed. The families and genera of vascular plants, vol. 7, pp. 39–44. Berlin/Heidelberg/New York.
Oxelman B, P Kornhall, RG Olmstead, and B Bremer. 2005. Further disintegration of Scrophulariaceae. Taxon 54: 411–425.
Padhye PM, M Daniel, and SD Sabnis. 1983. Chemistry and taxonomy of genus *Paulownia*. Indian Bot. Rep. 2: 137–138.
Pal N. 1951. Studies in the embryology of some Verbenaceae. J. Indian. Bot. Soc. 30: 59–74.
Parvati A and LL Narayana. 1978. Chemotaxonomy of a few taxa of Pedaliaceae. Curr. Sci. 47: 282–283.
Paton A. 1992. The adaptive significance of calyx and nutlet morphology in *Scutellaria*. In: RM Harley and T Reynolds, eds. Advances in Labiate Science, pp. 203–210. Royal Botanic Gardens, Kew.
Pazy B. 1998. Diploidization failure and apomixes in Orobanchaceae. Bot. J. Linn. Soc. 128: 99–103.
Pedersen JA. 2000. Distribution and taxonomic implications of some phenolics in the family Lamiaceae determined by ESR spectroscopy. Biochem. Syst. Ecol. 28: 229–253.
Pennell FW. 1935. The Scrophulariaceae of eastern temperate North America. Acad. Nat. Sci. Philadelphia Monogr. 1: 1–650.
Perret M, A Chautems, R Spichiger, G Kite, and V Savolainen. 2003. Systematics and evolution of tribe Sinningieae (Gesneriaceae): evidence from phylogenetic analyses of six plastid DNA regions and nuclear *ncp*GS. Am. J. Bot. 90: 445–460.
Philbrick CT and GJ Anderson. 1992. Pollination biology in the Callitrichaceae. Syst. Bot. 17: 282–292.
Philbrick CT and RK Jansen. 1991. Phylogenetic studies of North American *Callitriche* (Callitrichaceae) using chloroplast DNA restriction fragment analysis. Syst. Bot. 16: 476–491.
Philbrick CT and DH Les. 1995. Systematics of North American and European *Callitriche* (Callitrichaceae). Am. J. Bot. 82(6): 156 (abstract).
Philbrick CT and DH Les. 2000. Phylogenetic studies in *Callitriche* implications for interpretation of ecological, karyological and pollination system evolution. Aquat. Bot. 68: 123–141.
Piazzano M. 1998. Numerous cromosomicos en Bignoniaceae de Argentina. Kurtziana 26: 179–189.
Plisko MA. 2002. Seed-coat structure of alate seeds of some Bignoniaceae members. Bot. Zhurn. 87(5): 82–90 (in Russian with English summary).

Plisko MA. 2003. Fruit and seed structure of some Verbenaceae members. Bot. Zhurn. 88(5): 57–63 (in Russian with English summary).

Poser von GL, ME Toffoli, M Sobral, and AT Henriques. 1997. Iridoid glucosides substitution patterns in Verbenaceae and their taxonomic implication. Plant Syst. Evol. 205: 265–287.

Poser von GL, J Schripsema, AT Henriques, and SR Jensen. 2000. The distribution of iridoids in Bignoniaceae. Biochem. Syst. Ecol. 28: 351–366.

Pozhidaev AE. 1989. Exine structure in pollen grains of the Lamiaceae family. Bot. Zhurn. 74: 1410–1422 (in Russian with English summary).

Pozhidaev AE. 1990. On the problem of the origin of three- and six-colpate pollen grains in the family Lamiaceae. Bot. Zhurn. 75: 832–835 (in Russian with English summary).

Pozhidaev AE. 1992. The origin of three- and six-colpate pollen grains in the Lamiaceae, Grana 31: 49–52.

Praglowski J and K Gyllander. 1973. Globulariaceae. World Pollen Flora 4: 3–21.

Prather LA, AK Monfils, AL Posto, and RA Williams. 2002. Monophyly and phylogeny of *Monarda* (Lamiaceae) evicence from the internal transcribed spacer (ITS) region of nuclear ribosomal DNA. Syst. Bot. 27: 127–137.

Press JR. 1982. Taxonomic studies in the Labiatae tribe Pogostemoneae. Bull. Brit. Mus. (Nat. Hist.), Bot. 10(1): 1–89.

Punt W. 1978. Evolutionary trends in the Buddlieae (Loganiaceae). 4th Int. Palynol. Conf. (1976–1977), 1: 285–290. Lucknow.

Rahmanzadeh R, K Müller, E Fischer, D Bartels, and T Borsch. 2004. The Linderniaceae and Gratiolaceae are further lineages distinct from the Scrophulariaceae (Lamiales). Plant Biol. 7: 67–78.

Rahn K. 1992. Trichomes within Plantaginaceae. Nord. J. Bot. 12: 3–12.

Rahn K. 1996. A phylogenetic study of the Plantaginaceae. Bot. J. Linn. Soc. 120: 145–198.

Raj B. 1961. Pollen morphological studies in the Acanthaceae. Grana Palynol. 3: 3–108.

Raj B. 1983a. A contribution to the pollen morphology of Stilbaceae Kunth. Pollen et Spores 25: 395–408.

Raj B. 1983b. A contribution to the pollen morphology of Verbenaceae. Rev. Palaeobot. Palynol. 39: 343–422.

Raj B. 1984. A contribution to the pollen morphology of Symphoremaceae. Pollen et Spores 26: 363–384.

Raj B. 1985. A contribution to the pollen morphology of Nesogenaceae and Cyclocheilaceae. Pollen and Spores 27: 295–306.

Raj B. 1987. Pollen morphology of three monotypic genera of Verbenaceae. Pollen et Spores 29: 295–306.

Raj B and G El-Ghazaly. 1987. Morphology and taxonomic application of orbicules (Ubisch bodies) in Chloanthaceae. Pollen et Spores 29: 151–166.

Raj B and E Graftson. 1984. A contribution to the pollen morphology of Chloanthaceae. Grana 23: 139–156.

Ramana RV, PSP Rao, and LL Narayana. 1983. A contribution to the floral anatomy of *Phryma leptostachya* L. Curr. Sci. 52: 922–924.

Ramana RV, PSP Rao, BSM Dutt, and LL Narayana. 2000. Embryology of *Phryma leptostchya* L. (Verbenaceae) with consideration of its systematic status and affinities. Feddes Repert. 111: 231–248.

Raman S. 1987. A code proposed for the classification of trichomes as applied to the Scrophulariaceae. Beitr. Biol. Pflanzen 62: 349–367.

Raman S. 1989–1990. The trichomes on the corolla of the Scrophulariaceae. I. Tribes Aposimeae and Scrophularieae. II. Tribes Hemimerideae and Callceolarieae. III. Tribes Digitalieae. V. Tribe Antirrhineae Chavannes. VI. Tribe Pedicularieae D. Don 1825 (Rhinantheae Benth. 1835). VII. Tribe Cheloneae. Beitr. Biol. Pflanzen 64: 127–140; 64: 141–155; 64: 199–212; 64: 357–375; 64: 377–390; 65: 223–234.

Rao VS. 1952. The floral anatomy of some Verbenaceae with special reference to the gynoecium. J. Indian Bot. Soc. 31: 297–315.

Rao VS. 1953, 1954. The floral anatomy of some Bicar-pellatae: I. Acanthaceae. II. Bignoniaceae. J. Univ. Bombay, n.s., 21(5B): 1–34, 1953; 22(5B): 55–70, 1954.

Ratter JA. 1975. A survey of chromosome numbers in the Gesneriaceae of the Old World. Notes Roy. Bot. Gard. Edinb. 33: 527–543.

Ravi Kumar KV and B Hanumantha Rao. 1988. Wood anatomy of three species of *Avicennia* L. Swamy Bot. Club 5(2): 83–88.

Record SJ and RW Hess. 1941. American woods of the family Verbenaceae. Trop. Woods 67: 19–33.

Reddy MS, C Vitaya Kumari, and M Radhakrishnaiah. 1993. Systematic position of Avicenniaceae. Feddes Repert. 104: 237–239.

Ree RH. 2005. Phylogeny and the evolution of floral diversity in *Pedicularis* (Orobanchaceae). Int. J. Plant Sci. 166: 595–613.

Reifenrath K, I Theisen, J Schnitzler, S Porembski, and W Barthlott. 2006. Trap architecture in carnivorous *Utricularia* (Lentibulariaceae). Flora 201: 597–605.

Richardson PM. 1992. The chemistry of the Labiatae: an introduction and overview. In: RM Harley and T Reynolds, eds. Advances in Labiate Science, pp. 291–297. Royal Botanic Gardens, Kew.

Rimpler H, C Winterhalter, and U Falk. 1992. Cladistic analysis of the subfamily Caryopteridoideae Briq. and related taxa of Verbenaceae and Lamiaceae using morphological and chemical characters. In: RM Harley and T Reynolds, eds. Advances in Labiate Science, pp. 39–54. Royal Botanic Gardens, Kew.

Risch C. 1956. Die Pollenkorner der Labiaten. Willdenowia 1: 617–641.

Rivera G. 1996. Nectarios y tricomas florales en cuatro especies de Tecomeae (Bignoniaceae). Darwiniana 34: 19–26.

Rivera GL. 2000. Nuptial nectary structure of Bignoniaceae from Argentina. Darwiniana 38: 227–239.

Roalson EH, AE Senters, LE Skog, and EA Zimmer. 2002. A morphological cladistic analysis of the neotropical flowering plant genus *Gasteranthus* (Gesneriaceae). Syst. Bot. 27: 573–591.

Roalson EH, JK Boggan, LE Skog, and EA Zimmer. 2005. Untangling Gloxinieae (Gesneriaceae). I. Phylogenetic patterns and generic boundaries inferred from nuclear, chloroplast, and morphological cladistic datasets. Taxon 54: 389–410.

Rønsted N, E Gobel, H Franzyk, SR Jensen, and CE Olsen. 2000. Chemotaxonomy of *Plantago:* iridoid glucosides and caffeoyl phenylethanoid glycosides. Phytochemistry 55: 337–348.

Rønsted N, H Franzyk, P Molgaard, JW Jaroszewski, and SR Jensen. 2003. Chemotaxonomy and evolution of *Plantago* L. Plant Syst. Evol. 242: 63–82.

Rosatti TJ. 1984. The Plantaginaceae in the southeastern United States. J. Arnold Arbor. 65: 533–562.

Rosen W. 1940. Notes on the embryology of *Globularia vulgaris* L. Bot. Not. 1940: 253–261.

Ross H. 1902. *Byblis gigantea*. Gartenflora 51: 337–339 (pi. 1500).

Rourke JP. 1993. *Thesmophora:* a new genus of Stilbaceae from South Africa. Edinb. J. Bot. 50: 89–95.

Rourke JP. 2000. A review of generic concepts in the Stilbaceae. Bothalia 30: 9–15.

Rudall PJ. 1980a. Leaf anatomy of the subtribe Hyptidinae (Labiatae). Bot. J. Linn. Soc. 80: 319–340.

Rudall PJ. 1980b. Pollen morphology in the subtribe Hyptidinae (Labiatae). Kew Bull. 35: 452–458.

Rudall PJ. 1981. Wood anatomy in the Hyptidinae (Labiatae). Kew Bull. 35: 735–741.

Rudall PJ and RM Bateman. 2003. Evolutionary changes in flowers: evidence from naturally-occurring angiosperm tetra. Trends Plant Sci. 8: 76–82.

Rudall PJ and L Clark. 1992. The megagametophyte in Labiatae. In: RM Harley and T Reynolds, eds. Advances in Labiate Science, pp. 65–84. Royal Botanic Gardens, Kew.

Rutishauser R and B Isler. 2001. Developmental genetics and morphological evolution of flowering plants, especially bladderworts (*Utricularia*): fuzzy arberian morphology complements classical morphology. Ann. Bot. 88: 1173–1202.

Ryding O. 1992a. Pericarp structure and phylogeny within Lamiaceae subfamily Nepetoideae tribe Ocimeae. Nord. J. Bot. 12: 273–298.

Ryding O. 1992b. The distribution and evolution of myxocarpy in Lamiaceae. In: RM Harley and T Reynolds, eds. Advances in Labiate Science, pp. 85–96. Royal Botanic Gardens, Kew.

Ryding O. 1993. Pericarp structure of *Leucas* and related genera (Lamiaceae subfam. Lamioideae). Nord. J. Bot. 13: 637–646.

Ryding O. 1994a. Pericarp structure and phylogeny of Lamiaceae subfamily Pogostemoideae. Nord. J. Bot. 14: 59–63.

Ryding O. 1994b. Pericarp structure in the subtribe Melittidinae (Lamiaceae: Lamioideae) and its systematic implications. Bot. Jahrb. 115: 547–555.

Ryding O. 1994c. Pericarp structure in the tribe Prasieae (Lamiaceae: Lamioideae) and its systematic implications. Bot. Jahrb. 116: 391–399.

Ryding O. 1995. Pericarp structure and phylogeny of the Lamiaceae-Verbenaceae complex. Plant Syst. Evol. 198: 101–141.

Ryding O. 1996. Pericarp structure and phylogenetic position of the genus *Wenchengia* (Lamiaceae). Bot. Jahrb. Syst. 118: 153–158.

Ryding O. 1998. Phylogeny of the *Leucas* Group (Lamiaceae). Syst. Bot. 23: 235–247.

Ryding O. 2000. Notes on the Lamiaceae-Verbenaceae boundary. In: JR Timberlake and S Kativu, eds. African plants: biodiversity, taxonomy and uses: proceedings of the 1997 AETFAT congress, Harare, Zimbabwe, pp. 329–337. Kew.

Ryding O. 2002. Myxocarpy in the Nepetoideae (Lamiaceae) with notes on myxodiaspory in general. Syst. Geogr. Plants 71: 503–514.

Sahay SK. 1973. Trend of apertural evolution in pollen of some members of Myoporaceae. Proc. Indian Sci. Congr. 60(3): 369.

Sahay SK. 1980. Palynotaxonomy of Boraginaceae and some other families of Tubiflorae. Biol. Mem. 4: 117–205.

Sanders RW. 1997. The Avicenniaceae in the Southeastern United States. Harvard Pap. Bot. 5: 81–92.

Sanders RW. 2001. The genera of Verbenaceae in the southeastern United States. Harvard Pap. Bot. 5: 303–358.

Sanders RW and PD Cantino. 1984. Nomenclature of the subdivisions of the Lamiaceae. Taxon 33: 64–72.

Sandhya Rani R. 1994. Floral anatomy and the affinities of Byblidaceae. Rheedea 4: 144–150.

Santos GM and A Dos. 1995. Wood anatomy, chloroplast DNA and flavonoids of the tribe Bignonieae (Bignoniaceae). Reading.

Saueregger J and A Weber. 2004. Factors controlling initiation and orientation of the macrocotyl in anisocotylous Gesneriaceae. Edinb. J. Bot. 60: 467–482.

Saxena MR. 1981a. Palynotaxonomy of Symphoremaceae Moldenke. Indian J. Bot. 4: 163–167.

Saxena MR. 1981b. Contribution to the palynotaxonomy of Avicenniaceae Endl. J. Indian Bot. Soc. 60: 28–32.

Schmid E. 1906. Beiträge zur Entwicklungsgeschichte der Scrophulariaceae. Beih. Bot. Centralb. 20: 175–299.

Schneeweiss GM, A Colwell, J-M Park, C-J Jang, and TF Stuessy. 2004a. Phylogeny of holoparasitic *Orobanche* (Orobanchaceae) inferred from nuclear ITS sequences. Mol. Phylogenet. Evol. 30: 465–478.

Schneeweiss GM, T Palomeque, AE Colwell, and H Weiss-Schneeweiss. 2004b. Chromosome number and karyotype evolution in holoparasitic *Orobanche* (Orobanchaceae) and related genera. Am. J. Bot. 91: 439–448.

Schönenberger J. 1999. Floral structure, development and diversity in *Thunbergia* (Acanthaceae). Bot. J. Linn. Soc. 130: 1–36.

Schönenberger J and PK Endress. 1998. Structure and development of the flowers in *Mendoncia*, *Pseudocalyx* and *Thunbergia* (Acanthaceae). Int. J. Plant Sci. 159: 446–465.

Schwarzbach AE. 2004. Plantaginaceae. In: K Kubitzki, ed. The families and genera of vascular plants, vol. 7, pp. 327–329. Springer, Berlin/Heidelberg/New York.

Schwarzbach AE and LA McDade. 2002. Phylogenetic relationships of the mangrove family Avicenniaceae based on chloroplast and nuclear ribosomal DNA sequences. Syst. Bot. 27: 84–98.

Scotland RW. 1991. A systematic analysis of pollen morphology of Acanthaceae genera with contorted corollas. In: S Blackmore and SH Barnes, eds. Pollen and spores: patterns of diversification, pp. 269–289. Oxford University Press, Oxford.

Scotland RW. 1992a. Pollen morphology of Andrographideae (Acanthaceae). Rev. Palaeobot. Palynol. 72: 229–243.

Scotland RW. 1992b. Systematics, similarity and Acanthaceae pollen morphology. Bot. J. Linn. Soc. 109: 529–541.

Scotland RW. 1993. Pollen morphology of Contortae (Acanthaceae). Bot. J. Linn. Soc. 111: 471–504.

Scotland RW and K Vollesen. 2000. Classification of Acanthaceae. Kew Bull. 55: 513–589.

Scotland RW, PK Endress, and TJ Lawrence. 1994. Corolla ontogeny and aestivation in the Acanthaceae. Bot. J. Linn. Soc. 114: 49–65.

Scotland RW, JS Sweere, PA Reeves, and RG Olmstead. 1995. Higher level systematics of Acanthaceae determined by chloroplast DNA sequences. Am. J. Bot. 82: 266–275.

Shivaramiah G. 1998. Endosperm development in Bignoniaceae. Phytomorphology 48: 45–50.

Singh J and SVS Chauhan. 1999. Presence of glandular and non-glandular trichomes on anthers of *Tecoma stans* L. Phytomorphology 49: 469–472.

Singh SP. 1960. Morphological studies in some members of the family Pedaliaceae. Agra Univ. J. Res., Sci. 9: 218–220.

Singh TP. 1995. Alterations in the basic chromosome number as a means of speciation in Labiatae. Feddes Repert. 106: 39–47.

Singh V and DK Jain. 1975. Trichomes in Acanthaceae: I. General structure. J. Indian Bot. Soc. 54: 116–127.

Skog LE. 1976. A study of the tribe Gesnerieae, with a revision of *Gesneria* (Gesneriaceae: Gesnerioideae). Smithsonian Contr. Bot. 29: 1–182.

Skog LE. 1984. A review of the chromosome numbers in the Gesneriaceae. Selbyana 7: 252–273.

Skog LE and JK Boggan. 2005. The World Checklist of Gesneriaceae. Department of Botany, Smithsonian Institution, Washington, DC. http://persoon.si.edu/gesneriaceae/checklist

Skottsberg C. 1912. *Tetrachondra patagonica* n.sp und die systematische Stellung der Gattung. Bot. Jahrb. Syst. Beiblatt 107: 17–26.

Smith JF. 1996. Tribal relationships within the Gesneriaceae: a cladistic analysis of morphological data. Syst. Bot. 21: 497–513.

Smith JF. 2000a. A phylogenetic analysis of tribes Beslerieae and Napeantheae (Gesneriaceae) and evolution of fruit types: parsimony and maximum likelihood analyses of *ndh*F sequences. Syst. Bot. 25: 72–81.

Smith JF. 2000b. Phylogenetic resolution within the tribe Episcieae (Gesneriaceae): congruence of ITS and *ndh*F sequences from parsimony and maximum-likelihood analyses. Am. J. Bot. 87: 883–897.

Smith JF. 2001. The phylogenetic relationships of *Lembocarpus* and *Goyazia* (Gesneriaceae) based on *ndh*F sequences. Ann. Missouri Bot. Gard. 88: 135–143.

Smith JF and S Atkinson. 1998. Phylogenetic analysis of the tribes Gloxinieae and Gesnerieae (Gesneriaceae): Data from *mdh*F sequences. Selbyana 19: 122–131.

Smith JF and CL Carroll. 1997. Phylogenetic relationships of the Episcieae (Gesneriaceae) based on *ndh*F sequences. Syst. Bot. 22: 713–724.

Smith JF, KD Brown, CL Carroll, and DS Denton. 1997a. Familial placement of *Cyrtandromoea*, *Titanotrichum*, and *Sanango*, three problematic genera of the Lamiales. Taxon 46: 65–74.

Smith JF, JC Wolfram, KD Brown, CL Carroll, and DS Denton. 1997b. Tribal relationships in the Gesneriaceae: evidence from DNA sequences of the chloroplast gene *nfhF*. Ann. Miss. Bot. Gard. 84: 50–66.

Smith JF, SB Draper, LC Hileman, and DA Baum. 2004b. A phylogenetic analysis within tribes Gloxinieae and Gesnerieae (Gesnerioideae: Gesneriaceae). Syst. Bot. 29: 947–958.

Smith JF, LC Hileman, MP Powell, and DA Baum. 2004c. Evolution of *GCYC*, a Gesneriaceae homolog of *CYCLOIDEA*, within Gesnerioideae (Gesneriaceae). Mol. Phylogenet. Evol. 31: 765–779.

Spangler RE and RG Olmstead. 1999. Phylogenetic analysis of Bignoniaceae based on the *cpDA* gene sequences *rbcL* and *ndhF*. Ann. Missouri Bot. Gard. 86: 33–46.

Steane DA, RW Scotland, and DJ Mabberley. 1997. Phylogenetic relationships of *Clerodendrum* s.l. (Lamiaceae) inferred from chloroplast DNA. Syst. Bot. 22: 229–243.

Steane DA, RW Scotland, DJ Mabberley, and RG Olmstead. 1999. Molecular systematics of *Clerodendrum* (Lamiaceae): ITS sequences and total evidence. Am. J. Bot. 86: 98–107.

Steenis CGGJ van. 1949. Notes on the genus *Wightia* (Scrophulariaceae). Bull. Jard. Bot. Buitenzorg, III. 18: 213–227.

Steenis CGGJ van. 1971. Byblidaceae. In: CGGJ van Steenis, ed. Flora malesiana, ser 1, 7: 135–137. Groningen.

Steiner K. 1996. Chromosome numbers and relationships in tribe Hemimerideae (Scrophulariaceae). Syst. Bot. 21: 63–76.

Straka H and H-D Ihlenfeldt. 1965. Pollenmorphologie und Systematik der Pedaliaceae R. Br. Beitr. Biol. Pfl. 41: 175–207.

Straw RM. 1966. A redefinition of *Pentastemon* (Scrophulariaceae). Brittonia 18: 80–95.

Struwe L and SR Jensen. 2004. Plocospermataceae. In: K Kubitzki, ed. The families and genera of vascular plants, vol. 7, pp. 330–331. Springer, Berlin/Heidelberg/New York.

Sutton DA. 1988. A revision of the tribe Antirrhineae. Oxford University Press, London.

Tange C. 1998. *Silvianthus* (Carlemanniaceae) a genus and family new to Thailand. Thai For. Bull. Bot. 26: 59–65.

Taniguchi E and Y Oshima. 1972. Phrymarolin-I, a novel lignan from *Phryma leptostachya* L. Agric. Biol. Chem. 36: 1013–1025.

Tank DC, PM Beardsley, SA Kelchner, and RG Olmstead. 2006. Review of the systematics of Scrophulariaceae s.l. and their current disposition. Aust. Syst. Bot. 19: 289–307.

Taskova R, L Evstatieva, N Handjieva, and N Popov. 2002. Iridoid patterns of genus *Plantago* L. and their systematic significance. Z. Naturforsch. 57: 42–50.

Taskova R, D Peev, and N Handjieva. 2002. Iridoid glucosides of the genus *Veronica* s.l. and their systematic significance. Plant Syst. Evol. 231: 1–17.

Taskova RM, CH Gotfredsen, and Sr Jensen. 2006. Chemotaxonomy of Veroniceae and its allies in the Plantaginaceae. Phytochem. 67: 286–301.

Taylor P. 1989. The genus *Utricularia*: a taxonomic monograph. Kew Bull. Add. Ser. 14. London.

Terekhin ES and ZI Nikiticheva. 1981. The family Orobanchaceae: ontogeny and phylogeny. Nauka, Leningrad (in Russian).

Thanikaimoni G. 1966. Pollen morphology of the genus *Utricularia*. Pollen et Spores 8: 265–284.

Theisen I and E Fischer. 2004. Myoporaceae. In: K Kubitzki, ed. The families and genera of vascular plants, vol. 7, pp. 289–292. Springer, Berlin/Heidelberg/New York.

Thieret JW. 1967. Supraspecific classification in the Scrophulariaceae: a review. SIDA 3: 87–106.

Thieret JW. 1971. The genera of Orobanchaceae in the southeastern United States. J. Arnold Arbor. 52: 404–434.

Thieret JW. 1972. The Phrymaceae in the southeastern United States. J. Arnold Arbor. 53: 226–233.

Thieret JW. 1977. The Martyniaceae in the southeastern United States. J. Arnold Arbor. 58: 25–39.

Thiv M. 2004. Carlemanniaceae. In: K Kubitzki, ed. The families and genera of vascular plants, vol. 7, pp. 57–59. Springer, Berlin/Heidelberg/New York.

Tiagi B. 1963. Studies in the family Orobanchaceae: IV. Embryology of *Boschniakia himalaica* Hook. and *B. tuberosa* (Hook.) Jepson, with remarks on the evolution of the family. Bot. Not. 116: 81–93.

Tomás-Barberán FA and MI Gil. 1992. Chemistry and natural distribution of flavonoids in the Labiatae. In: RM Harley and T Reynolds, eds. Advances in Labiate science, pp. 299–305. Royal Botanics Gardens, Kew.

Tomlinson PB. 1986. Avicenniaceae. In The botany of mangroves, pp. 186–207. Cambridge University Press, Cambridge.

Trudel MC and JK Morton. 1992. Pollen morphology and taxonomy in North American Labiatae. Canad. J. Bot. 70: 975–995.

Upson TM, RJ Grayer, JR Greenham, CA Williams, F Al-Ghamdi, and F-H Chen. 2000. Leaf flavonoids as systematic characters in the genera *Lavandula* and *Sabaudia*. Biochem. Syst. Ecol. 28: 991–1007.

Vargas P, JA Rosselló, R Oyama, and J Güemes. 2004. Molecular evidence for naturalness of genera in the tribe Antirrhineae (Scrophulariaceae) and three independent evolutionary lineages from the New World and the Old. Plant Syst. Evol. 249: 151–172.

Vasilchenko IT. 1947. Morphology of germination in the mint family (fam. Lamiaceae) in relation to its systematics. Flora and systematics of higher plants, 1st ser., 6: 72–104 (in Russian).

Venkata Ramana R, PS Prakasa Rao, BSM Dutt, and LL Narayana. 2000. Embryology of *Phryma leptostachya* L. (Verbenaceae) with considerations of its systematic status and affinities. Feddes Repert. 111: 231–248.

Verdcourt B. 1992. Verbenaceae. In: RM Polhill, ed. Flora of tropical East Africa, pp. 1–155. Balkema, Rotterdam.

Vestri Alvarenga SA, JP Gastmans, G do V Rodrigues, PRH Moreno, and V de Emerenciano. 2001. A computer-assisted approach for chemataxonomic studies – diterpenes in Lamiaceae. Phytochemistry 56: 583–595.

Vij DP and SK Kashyap. 1975. Pollen grain studies in some Labiatae. J. Palynol. 11: 29–42.

Vintejoux C and A Shoar-Ghafar. 1997. Secretion of mucilages by an aquatic carnivorous plant: *Utricularia neglecta* L. (Lentibulariaceae). Acta Bot. Gallica 144: 347–351.

Vujicic R, D Grubisic, and R Konjevic. 1993. Scanningelectron microscopy of the seed coat in the genus *Paulownia*. Bot. J. Linn. Soc. 111: 505–511.

Wagenitz G. 1977. New aspects of the systematics of Asteridae. Plant Syst. Evol. 1: 375–395.

Wagenitz G. 2004. Globulariaceae. In: K Kubitzki, ed. The families and genera of vascular plants, vol. 7, pp. 159–162. Springer, Berlin/Heidelberg/New York.

Wagstaff SJ. 1992. A phylogenetic interpretation of pollen morphology in tribe Mentheae (Labiatae). In: RM Harley and T Reynolds, eds., Advances in Labiate Science, pp. 113–124. Royal Botanic Gardens, Kew.

Wagstaff SJ. 2004. Tetrachondraceae. In: K Kubitzki, ed. The families and genera of vascular plants, vol. 7, pp. 441–444. Springer, Berlin/Heidelberg/New York.

Wagstaff SJ and RG Olmstead. 1997. Phylogeny of Labiatae and Verbenaceae inferred from *rbc*L sequences. Syst. Bot. 22: 165–179.

Wagstaff SJ, RG Olmstead, PD Cantino. 1995. A phylogenetic analysis of restriction site variation in subfamily Nepetoideae (Lapiatae). Am. J. Bot. 82: 886–892.

Wagstaff SJ, L Hicherson, RE Spangler, PA Reeves, and RG Olmstead. 1998. Phylogeny of Labiatae s.l. inferred from cpDNA sequences. Plant Syst. Evol. 209: 265–274.

Wagstaff SJ, K Martinsson, U Swenson. 2000. Divergence estimates of *Tetrachondra hamiltonii* and *T. patagonica* (Tetrachondraceae) and their implications for austral biogeography. N. Z. J. Bot. 38: 595–606.

Wallick KP, M Molvray, PJ Kores, and WJ Elisens. 2000. A phylogenetic analysis of Buddlejaceae s.s. inferred from the *trn*L region of cpDNA. Am. J. Bot. Suppl. 87(6): 183 (abstracts).

Wang C-N, M Möller, and QCB Cronk. 2004. Phylogenetic position of *Titanotrichum oldhamii* (Gesneriaceae) inferred from four different gene regions. Syst. Bot. 29: 407–418.

Wang L and Y-Z Wang. 2005. Floral development of *Triaenophora* (Veronicaceae) and phylogenetic implications. Plant Syst. Evol. 250: 69–79.

Weber HC. 1980. Zur Evolution des Parasitismus bei den Scrophulariaceae und Orobanchaceae. Plant Syst. Evol. 136: 217–232.

Weber HC. 2004a. Gesneriaceae. In: K Kubitzki, ed. The families and genera of vascular plants, vol. 7, pp. 276–282. Springer, Berlin/Heidelberg/New York.

Weber HC. 2004b. Gesneriaceae and Scrophulariaceae: Robert Brown and now. Telopea 10: 543–571.

Wei Z-X, H Wang, and Q-J Jin. 2001. Pollen morphology of Bignoniaceae in China. Acta Bot. Yunn. 23: 225–230.

Whipple HL. 1972. Structure and systematics of *Phryma leptostachys* L. J. Elisha Mitchell Sci. Soc. 88: 1–17.

Wiehler H. 1983. A synopsis of the neotropical Gesneriaceae. Selbyana 6: 1–219.

Wiehler H. 1994. A re-examination of *Sanango racemosum*. 4. Its new systematic position in Gesneriaceae. Taxon 43: 625–632.

Williams NH. 1978. Pollen structure and the systematics of the neotropical Gesneriaceae. Selbyana 2: 310–322.

Wilson CL. 1974. Floral anatomy in Gesneriaceae: I. Cyrtandroideae. II. Gesnerioideae. Bot. Gaz. 135: 247–256, 256–268.

Wolfe AD and CW dePamphilis. 1997. Alternate paths of evolution for the photosynthetic gene *rbc*L in four nonphotosynthetic species of *Orobanche*. Plant Mol. Biol. 33: 965–977.

Wolfe AD, WJ Elisens, LE Watson, and CW dePamphilis. 1997. Using restriction-site variation of PCR-amplified cpDNA genes from phylogenetic analysis of tribe Cheloneae (Scrophulariaceae). Am. J. Bot. 84: 555–564.

Wolfe AD, SL Datwyler, and CP Randle. 2002. A phylogenetic and biogeographic analysis of the Cheloneae (Scrophylariaceae) based on ITS and *mat*k sequence data. Syst. Bot. 27: 138–148.

Wolfe AD, CP Randle, L Liu, and KE Steiner. 2005. Phylogeny and biogeography of Orobanchaceae. Folia Geobot. 40: 115–134.

Wood J. 1999. Flowering patterns in Acanthaceae. Acanthus 7: 5.

Wu CY and H Li. 1982. On the evolution and distribution in Labiatae. Acta Bot. Yunn. 4: 97–117.

Wunderlich R. 1967. Ein Vorschlag zu einer natürlichen Gliederung der Labiaten auf Grund der Pollenkörner, Samenentwicklung, und des reifen Samens. Oesterr. Bot. Z. 114: 383–483.

Xiong JQ and ZY Chen. 1992. A study on numerical taxonomy of the genus *Paulownia*. Bull. Bot. Res. North-East. Forest. Univ. 12: 185–188.

Yashodhara K and SR Shanmukha Rao. 1993. Epidermis in relation to the taxonomy of *Phryma* L. Indian Bot. Rep. 11(1–2): 23–26 (1992 publ. 1993).

Yashodhara K and SR Shanmukha Rao. 1995. Epidermis in relation to taxonomy: Symphoremataceae Van Tiegh. Beitr. Biol. Pfl. 68: 157–167 (1993–1994 publ. 1995).

Yashodhara K, SR Shanmukha Rao, and JV Subba Rao. 2002. Structure, distribution and taxonomic importance of trichomes in the tribe Verbeneae (Verbenaceae). Beitr. Biol. Pfl. 72: 265–285 (2000–2001 publ. 2002).

Young ND, KE Steiner, and CW DePamphilis. 1999. The evolution of parasitism in Scrophulariaceae/Orobanchaceae: plastid gene sequences refute an evolutionary transition series. Ann. Missouri Bot. Gard. 86: 876–893.

Yu Z, S Li, C Hsu, and Y Hsu. 1987. A preliminary study on the number and morphology of chromosomes in *Paulownia*. Acta Bot. Bor. Occid. Sinica 7: 127–132.

Zanoni TA. 2004. Martyniaceae. In: N Smith, SA Mori, A Henderson, DW Stevenson, and SV Heald, eds. Flowering plants of the Neotropics, pp. 239–240. Princeton University Press, Princeton.

Zijra ML, KJ Sytsma, and RG Olmstead. 2004. Delimitation of Malagasy tribe Coleeae and implications for fruit evolution in Bignoniaceae inferred from a chloroplast DNA phylogeny. Plant Syst. Evol. 245: 55–67.

Zimmer EA, EH Roalson, LE Skog, JK Boggan, and A Idnurm. 2002. Phylogenetic relationships in the Gesnerioideae (Gesneriaceae) based on nrDNA ITS and cpDNA *trn*L-F and *trn*E-T spacer region sequences. Am. J. Bot. 89: 296–311.

Zona S. 1998. The Myoporaceae in the southeastern United States. Harvard Pap. Bot. 3: 171–179.

Zoz IG and VI Litvinenko. 1979. On the division of the family Lamiaceae Juss. into natural groups. Bot. Zhurn. 64: 989–997 (in Russian with English summary).

Order 125. HYDROSTACHYALES

Submerged fresh-water perennial or annual herbs with short, thick, tuberlike stems and fibrous roots; stems, leaves, and adventitious roots provided with numerous small, scale-like or fringed emergences. Vascular system much reduced, in the disc-like stem procambial tissue with annular- and spiral vessels is present (Erbar and Leins 2004). Leaves alternate, basal, elongate, entire to two or three times pinnatisect, dilated and ligulate at the base, with inconspicuous, membranous, intrapetiolar stipule. Stomata wanting. Flowers small and inconspicuous, sessile in the axils of the bracts of the dense spike, unisexual, dioecious or seldom monoecious, without a perianth. The male flower usually is interpreted as consisting of a single stamen, which is deeply divided into two theca (Rauh and Jager-Zurn 1966; Hutchinson 1973; Cronquist 1981; Takhtajan 1987). However, according to Leins and Erbar (1990), it consists of two independent disporangiate stamens that can be derived from originally tetrasporangiate stamens in which the sporangia of the neighboring thecae have fused apically (as in some Scrophulariaceae, e.g., *Verbascum*) and shifted into a lateral position (as in some Scrophulariaceae, e.g., in *Hebenstretia* and *Zaluzianskya*) microsporangia opening longitudinally. Tapetum secretory. Microsporogenesis simultaneous. Pollen grains in permanent tetrahedral or hexahedral to rhomboidal tetrads (Erbar and Leins 2004), 2-celled, inaperturate. Gynoecium of two united carpels arranged in transversal position and with two elongate, persistent, subulate stylodia that are sometimes connate at the base; ovary superior, zygomorphic, 1-locular, with numerous ovules on two parietal placentas. Ovules anatropous, unitegmic, tenuinucellate, with a narrow, elongated nucellus with a tiny gametophyte located at its apex, with endothelium. Female gametophyte of *Polygonum*-type, Endosperm cellular, with micropylar haustorium. Fruits septicidal capsules with numerous tiny, exotestal seeds, outer cell walls much thickened, mucilaginous (contain pectin); endosperm scanty or lacking. Contain kaempferol, mono- and di-glucosides (Scogin 1992), and caffeic acid (Rønsted et al. 2002), n = 10–12 (*Hydrostachys imbricata* – Palm 1915).

Probably derived from Scrophulariaceae (see Jäger-Zürn 1965; Rauh and Jäger-Zürn 1966; Takhtajan 1966; Leins and Erbar 1990) and probably have a common origin with Plantaginaceae.

Taxonomic position is unknown.

1. HYDROSTACHYACEAE

Engler 1894. 1/22. Madagascar and tropical and South Africa.

Hydrostachys

Bibliography

Albach DC, DE Soltis, MW Chase, and PS Soltis. 2001. Phylogenetic placement of the enigmatic angiosperm *Hydrostachys*. Taxon 50: 781–805.

Cusset C. 1973. Revision des Hydrostachyaceae. Adan-sonia, ser. 2, 13: 75–119.

Erbar C and P Leins. 2004. Hydrostachyaceae. In: K Kubitzki, ed. The families and genera of vascular plants, vol. 6, pp. 216–220. Springer, Berlin/Heidelberg/New York.

Fan C and Q-Y Xiang. 2001. Systematic affinities of two enigmatic African families, Grubbiaceae and Hydrostachyaceae – evidence from nuclear 26S rDNA sequence data. In Botany 2001: Plants and People. Abstracts, p. 112. Albuquerque.

Hempel AL, PA Reeves, RG Olmstead, and RK Jansen. 1995. Implications of *rbc*L sequence data for higher order relationships of the Loasaceae and the anomalous aquatic plant *Hydrostachys* (Hydrostachyaceae). Plant Syst. Evol. 194: 25–37.

Jäger-Zürn I. 1965. Zur Fragen der systematischen Stellung der Hydrostachyaceae auf Grund ihrer Embryologie, Blüten, und Infloreszenzmorphologie. Oesterr. Bot. Z. 112: 621–639.

Jäger-Zürn I. 1998. Anatomy of the Hydrostachyaceae. In: E Landolt, I Jäger-Zürn, and RAA Schnell, eds. Handbuch der Pflanzenanatomie. Extreme adaptations in angiospermous hydrophytes, vol. 13, pp. 129–196. Borntraeger, Berlin.

Leins P and C Erbar. 1988. Einige Benerkungen zur Blütenentwicklung und systematischen Stellung der Wasserpflanzen *Callitriche*, *Hippuris* und *Hydrostachys*. Beitr. Biol. Pflanzen 63: 157–178.

Leins P and C Erbar. 1990. The possible relationship of Hydrostachyaceae based on comparative ontogenetical flower studies. Mitt. Inst. Allg. Bot. Hamburg 23b: 723–729.

Palm B. 1915. Studien über Konstruktionstypen und Entwicklungswege des Embryosacks der Angiospermen. Akad. Afhandl. Stockholm 1915: 1–259.

Rauh W and I Jäger-Zürn. 1966. Zur Kenntnis der Hydrostachyaceae: I. Blütenmorphologische und embryologische Untersuchungen an Hydrostachyaceen unter besonderer Berucksichtigung ihrer systematischen Stellung. Sitzungsber. Heidelb. Akad. Wiss., Math.-Nat. KL. 1966: 1–177.

Rauh W and I Jäger-Zürn. 1967. Le probleme de la position systematique des Hydrostachyacees. Adansonia 6: 515–523.

Rønsted N, H Strandgaard, SR Jensen, and P Mølgaard. 2002. Chlorogenic acid from three species of *Hydrostachys*. Biochem. Syst. Ecol. 30: 1105–1108.

Scogin R. 1992. Phytochemical profile of *Hydrostachys insignis* (Hydrostachyaceae). Aliso 123: 471–474.

Straka H. 1988. Die Pollenmorphologie von *Hydrostachys*, Hydrostachyaceae, und ihre Bedeutung für die systematische Einreihung der Familie. Beitr. Biol. Pflanzen 63: 413–419.

Xiang Q-Y. 1999. Systematic affinities of Grubbiaceae and Hydrostachyaceae within Cornales: insights from *rbcL* sequences. Harvard Pap. Bot. 4: 527–542.

Class LILIOPSIDA (MONOCOTYLEDONS)

Embryo, when differentiated, always with one cotyledon. The cotyledon usually with two main vascular bundles. Leaf venation striate or of derived types, mostly arcuate-striate or longitudinally striate (parallel), less often palmate-striate or pinnate-striate, almost always more or less closed at the apex (the veins emerging from the leaf base usually run together again at their apices). Leaves usually not clearly divided into petiole and lamina, less often more or less differentiated, but in these cases the "petiole" and the "lamina" are not homologous to those of magnoliopsids (are of secondary origin), often with sheathing base. Leaf traces usually numerous. Prophylls (including bracteoles) usually solitary and nearly always adaxial. Vascular bundles usually without cambium or rarely with vestigial cambium only. Vascular system of the stem usually consists of many separate scattered bundles or sometimes of two or more rings of vascular bundles, and the axis mostly attains its full diameter early, after which no increase in thickness takes place; only in some groups does thickening of the axis occur by means of division and enlargement of ground parenchyma cells (so-called diffuse secondary growth), as in palms, or by means of special kind of cambium that arises in the parenchyma outside the primary vascular system, as in some herbaceous and woody Lilianae. Sieve-element plastids of P-type with several to numerous cuneate (triangular) crystalloid bodies (lacking in all magnoliopsids studied except *Saruma* and *Asarum* in Aristolochiaceae). Phloem without parenchyma. Usually without clearly differentiated bark and pith. The primary root is usually ephemeral, dries out early in the growth of the plant, and is replaced by an adventitious root system that develops from the stem or (as in grasses) directly from the hypocotyl. Ontogenetically root cap and root epidermis are of different origin. Usually herbs, but often secondarily arborescent plants (primary woody plants are absent among the monocots). Flowers usually 3-merous, sometimes 4- or 2-merous, very rarely 5-merous. Nectaries predominantly septal. Pollen grains mostly 1-colpate (sulcate) or of derived types, often 1-porate.

The Liliopsida most probably originated from some very ancient vesselless herbaceous member of Magnoliopsida that had atactostelic vascular system, P-type sieve-element plastids, 3-merous flowers, apocarpous gynoecium with laminar-diffuse (scattered) placentation, bitegmic and crassinucellate ovules (with parietal tissue between the female gametophyte) and the nucellar epidermis, and primitive 2-celled and 1-colpate pollen grains.

Unfortunately there is no convincing dicotylidonous sister group to the monocotyledons. According to some authors the nearest group are nymphaeids. Some of the relatively most archaic monocots have some similarities with the nymphaeids (Hallier 1905; Schaffner 1929, 1934; Eber 1934; Takhtajan 1954, 1959, 1969, 1987; Kaul 1967; Cronquist 1968, 1981, 1988). As long ago as 1905, Hallier suggested that the Nymphaeaceae (s.l.) were the "ancestors of the whole division of monocotyledons" though later (1912) he changed his opinion. According to Arber (1920: 309), the Nymphaeaceae "descended from a stock closely related to that which gave rise to the monocotyledons." Similar ideas have also been expressed by some other botanists.

The nymphaeoids and some archaic monocots do indeed have some important characters in common. In the morphology of their gynoecia the families Butomaceae and Limnocharitaceae resemble the Cabombaceae, and in their laminar-diffuse placentation they recall the Nymphaeaceae. There are also some other important similarities, including atactostelic vascular cylinder and especially root ontogeny (see Voronin 1964) as well as the development of female gametophytes, stomatal patterns, seed anatomy, and the arrangement of the first leaves (prophylls) on lateral axes. However, the sieve-element plastids of the nymphaeids are of S-type, and they are too specialized to be considered the ancestors of monocots. It is much more probable that they evolved from some remote common ancestor that was already more or less adapted to a relatively wet (but not yet aquatic) habitat.

Henslow (1911) considered the distinctive features of monocots the result of the primary adaptation to an aquatic habitat while Jeffrey (1917: 415) in his classical "*Anatomy of Woody Plants*" suggested an aquatic or amphibious way of life might have led to the loss of cambial activity. Henslow's hypothesis has been criticized by Sargant (1903, 1904), who concluded that many of the characteristic features of the monocots may be easier explained as having arisen as a result of adaptation to a geophilous habit. But apparently nearer to the truth was Parkin (1923: 59), who suggested "the golden mean" between the two hypotheses. He writes: "Respecting the relative merits of an aquatic or geophilous ancestry of monocotyledons, the two views may be somewhat reconciled by regarding the earliest ones as neither markedly aquatic nor extremely geophilous—in fact, marsh plants with stout rhizomes. Some of their descendants have become completely hydrophytic, others sharply geophytic, while others have returned to the arborescent habit by fresh means." Apparently the ancient common ancestor of both the Nymphaeidae and monocots was a hygrophilous or perhaps even amphibious geophyte in which geophyly arose under wet terrestrial conditions – most probably under the forest canopy or in the forest margin. But as is well known, underground storage is usually a response to a resting season, and geophytes are abundant and highly diversified in areas with a pronounced resting season (see Bews 1927). Therefore, they could originate in a climate having a marked dry season (Sargant 1903; Stebbins 1974).

The class Liliopsida includes 4 subclasses, 31 orders, 120 families, more than 3,000 genera, and about 65,000 species.

Bibliography

Arber A. 1920. Water plants: a study of aquatic angiosperms. Cambridge University Press, Cambridge.

Arber A. 1925. Monocotyledons: a morphological study. Cambridge University Press, Cambridge.

Bancroft N. 1914. A review of the literature concerning the evolution of Monocotyledons. New Phytol. 13: 285–303.

Batygina TB and MS Yakovlev, eds. 1990. Comparative embryology of flowering plants: monocotyledons. Nauka, Leningrad (in Russian).

Behnke H-D. 1981. Siebelement-plastiden, phloem-protein, und evolution der blütenpflanzen: II. Monocotyledonen. Ber. Deutsch. Bot. Ges. 94: 647–662.

Behnke H-D. 1998. Forms and sizes of sieve-element plastids and evolution of the monocotyledons. In Monocots II, pp. 8–9 (abstract). Sydney.

Behnke H-D. 2000. Forms and sized of sieve-element plastids and evolution of the monocotyledons. In: KL Wilson and DA Morrison, eds. Monocots: systematics and evolution, pp.163–188. CSIRO, Collingwood.

Behnke H-D. 2002(2003). Sieve-element plastids and evolution of Monocotyledons, with emphasis on Melanthiaceae sensu lato and Aristolochiaceae-Asaroideae, a putative Dicotyledon sister group. Bot. Rev. 68: 524–544.

Bennett MD and IJ Leitch. 2000. Variation in nuclear DNA amount (C-value) in monocots and its significance. In: KL Wilson and DA Morrison, eds. Monocots: systematics and evolution, pp.137–146. CSIRO, Collingwood.

Bews JW. 1927. Studies in the ecological evolution of the angiosperms. New Phytologist reprint 16. Wheldon & Wesley, London.

Bharathan G, G Lambert, and DW Galbraith. 1944. Nuclear DNA content of monocotyledons and related taxa. Am. J. Bot. 81: 381–386.

Boyd L. 1932. Monocotyledons seedlings: morphological studies in the post-seminal development of the embryo. Trans. Proc. Bot. Soc. Edinb. 31: 5–224.

Brickell CD et al., eds. 1980. Petaloid monocotyledons: horticultural and botanical research. Linn. Soc. Symposium, No. 8. London/New York.

Burger WC. 1977. The Piperales and the monocots: alternate hypotheses for the origin of monocotyledonous flowers. Bot. Rev. 43: 345–393.

Burger WC. 1981. Heresy revised: the monocot theory of angiosperm origin. Evol. Theory (Chicago) 3: 189–225.

Charlton WA. 1999. Morphological traffic between the inflorescence and the vegetative shoot in Helobial Monocotyledons. Bot. Rev. 65: 370–384.

Chase MW. 2004. Monocot relationships: an overview. Am. J. Bot. 91: 1645–1655.

Chase MW, DE Soltis, PS Soltis, PJ Rudall, MF Fay, WH Hahn, S Sullivan, J Joseph, M Molvray, PJ Kores, TJ Givnish, KJ Sytsma, and JC Pires. 2000. Higher-level systematics of the monocotyledons: an assessment of current knowledge and a new classification. In: KL Wilson and DA Morrison, eds. Monocots: systematics and evolution, pp. 3–16. CSIRO, Sydney.

Cheadle VI. 1942. The occurrence and types of vessels in the various organs of the plant in the Monocotyledoneae. Am. J. Bot. 29: 441–450.

Cheadle VI. 1943a. The origin and certain trends of specialization of the vessels in the Monocotyledoneae. Am. J. Bot. 30: 11–17.

Cheadle VI. 1943b. Vessel specialization in the late metaxylem of the various organs in the Monocotyledoneae. Am. J. Bot. 30: 484–490.

Cheadle VI. 1944. Specialization of vessels within the xylem of each organ in the Monocotyledoneae. Am. J. Bot. 31: 81–92.
Cheadle VI and JM Tucker. 1961. Vessels and phylogeny of Monocotyledoneae. In Recent advances in botany, pp. 161–165. University of Toronto Press, Toronto.
Clifford HT. 1977. Quantitative studies of interrelationships amongst the Liliatae. Plant Syst. Evol. Suppl. 1: 77–95.
Clifford HT and WT Williams. 1980. Interrelationships amongst the Liliatae: a graph theory approach. Aust. J. Bot. 28: 261–268.
Conran JG. 2000. Biogeographic studies in the monocotyledons: an overview of methods and literature. In: KL Wilson and DA Morrison, eds. Monocots: systematics and evolution, pp. 35–43. CSIRO, Collingwood.
Daghlian CP. 1981. A review of the fossil record of monocotyledons. Bot. Rev. 47: 517–555.
Dahlgren RMT and FN Rasmussen. 1983. Monocotyledon evolution: characters and phylogenetic estimation. In: MK Hecht, B Wallace, and GT Prance, eds. Evolutionary biology, vol. 16, pp. 255–395. Plenum, New York.
Dahlgren RMT, HT Clifford, and PF Yeo. 1985. The families of the monocotyledons: structure, evolution, and taxonomy. Springer, Berlin.
Danilova MF, EN Nemirovich-Danchenko, GA Komar, and MM Lodkina. 1990. Some trends of structural evolution of seeds in monocotyledons. Bot. Zhurn. 75: 755–773 (in Russian with English summary).
Danilova MF, EN Nemirovich-Danchekno, GA Komar, and MM. Lodkina. 1995. The seed structure of monocotyledons. In: P Rudal, PJ Cribb, DF Cutler, and CJ Humphries, eds. Monocotyledons: systematics and evolution, pp. 461–472. Royal Botanic Gardens. Kew.
Davis JI, DW Stevenson, G Petesen, O Seberg, LM Campbell, JV Freudenstein, DH Goldman, CR Hardy, FA Michelangeli, MP Simmons, CD Specht, F Vergara-Silva, and M Gandolfo. 2004. A phylogeny of the monocots, as inferred from *rbc*L and *atp*A sequence variation, and a comparison of methods for calculating jackknife and bootstrap values. Syst. Bot. 29: 467–510.
Delpino F. 1903. Aggiunte alla teoria della classificazione della Monocotyledoni. Mem. Acad. Bologna, ser. V, 10: 569–584.
Daumann E. 1970. Das Blütennektarium der Monocotyledonen unter besonderer Berucksichtigung seiner systematischen und phylogenetischen Bedeutung. Feddes Repert. 80: 463–590.
Davis JI, DW Stevenson, CD Specht, JV Freudenstein, and R DeSalle. 1998. A phylogenetic analysis of the monocotyledons: based on morphological and molecular character sets. In Monocots II, p.18 (abstract). Sydney.
Deyl M. 1955. The evolution of the plants and the taxonomy of the monocotyledons. Acta Mus. Natl. Prag., ser. 11B, 3(6): 1–143.
Doyle JA. 1973. The monocotyledons: their evolution and comparative biology. Quart. Rev. Biol. 48: 399–413.
Duval MR et al. 1993. Phylogenetic hypotheses for the monocotyledons constructed from *rbc*L sequence data. Ann. Missouri Bot. Gard. 80: 607–619.
Duvall MR. 2000. Seeking the dicot sister group of the monocots. In: KL Wilson and DA Morrison, eds. Monocots: systematics and evolution, pp. 25–32. CSIRO, Collingwood.
Eber E. 1934. Karpellbau und Pflanzenverhaltnisse in dem Reiche der Helobiae. Flora 127: 273–330.
El-Gazzar A and MK Hamza. 1975. On the monocots-dicots distinction. Publ. Cairo Univ. Herb. 6: 15–28.
Endress PK 1995. Major evolutionary traits of monocot flowers. In: PJ Rudall, PJ Cribb, DF Cutler, and CJ Humphries, eds. Monocotyledons: systematics and evolution, pp. 43–79. Royal Botanic Gardens, Kew.
Erbar C and P Leins. 1994. Flowers in Magnoliidae and the origin of flowers in other subclasses of the angiosperms. I. The relationships between flowers of Magnoliidae and Alismatidae. Plant Syst. Evol. Suppl. 8: 193–208.
Fisher JB and JC French. 1978. Internodal meristems of monocotyledons: further studies and general taxonomic summary. Ann. Bot. 42: 41–50.
French JC, MC Chung, and YK Hur. 1995. Chloroplast DNA phylogeny of the Ariflorae. In: PJ Rudall, PJ Cribb, DF Cutler, and CJ Humphries, eds. Monocotyledons: systematics and evolution, pp. 255–275. Royal Botanic Gardens, Kew.
Frolich D and W Barthlott. 1988. Micromorphologie der epicuticularen Wachse und das System der Monocotylen. Trop. Subtrop. Pflanzenwelt, vol. 63. Stuttgart.
Furness CA and PJ Rudall. 1998a. The tapetum and systematics in monocotiledons. Bot. Rev. 64: 201–239.
Furness CA and PJ Rudall. 1998b. Microsporogenesis in monocotyledons. In Monocots II, p. 21 (abstract). Sydney.
Furness CA and PJ Rudall. 1999. Inaperturate pollen in Monocotyledons. Int. J. Plant. Sci. 160 (2): 195–414.
Furness CA and PJ Rudall. 2000a. The systematic significance of simultaneous cytokinesis during microsporogenesis in monocotyledons. In: KL Wilson and DA Morrison, eds. Monocots: systematics and evolution, pp. 189–193. CSIRO, Collingwood.
Furness CA and PJ Rudall. 2000b. Aperture absence in pollen of monocotyledons. In: MM Harley, CM Morton, and S Blackmore, eds. Pollen and spores: morphology and biology, pp. 249–257. Royal Botanic Gardens, Kew.
Furness CA and PJ Rudall. 2001. Pollen and anther characters in monocot systematics. Grana 40: 17–25.
Furness CA and PJ Rudall. 2003. Apertures with lids: distribution and significance of operculate pollen in monocotyledons. Int. J. Plant Sci. 164: 835–854.
Furness CA and PJ Rudall. 2006. The operculum in pollen of monocotyledons. In: JT Columbus, EA Friar, JM Porter, LM Prince, and MG Simpson, eds. Monocots: comparative biology and evolution (excluding Poales), pp. 191–196, Rancho Santa Ana Botanical Garden, Claremont (Aliso 22: 191–196).
Gandolfo MA, KC Nixon, and WL Crepet. 1998. Monocotyledons and their fossil record: a review. In Monocots II, p. 21 (abstract). Sydney.
Gandolfo MA, KC Nixon, and WL Crepet. 2000. Monocotyledons: a review of their Early Cretaceous record. In: KL Wilson and DA Morrison, eds. Monocots: systematics and evolution, pp. 44–51. CSIRO, Collingwood.
Gaut B, S Muse, WD Clark, and M Clegg. 1992. Relative rates of nucleotide substitution at the *rbc*L locus of monocotyledonous plants. J. Molec. Evol. 35: 292–303.
Givnish TJ, JC Pires JC, SW Graham, MA McPherson, LM Prince, TB Patterson, HS Rai, EH Roalson, TM Evans, HJ Hahn, KC Millam, AW Meerow, M Molvray, PJ Kores, HE O'Brien, JC Hall, WJ Kress, and KJ Sytsma. 2006. Phylogeny of the monocots based on *ndh*F: evidence for

widespread concerted convergence. In: JT Columbus, EA Friar, CW Hamilton, JM Porter, LM Prince, and MG Simpson, eds. Monocots: comparative biology and evolution (excluding Poales), vol. 2, pp. 28–51. Rancho Santa Ana Botanical Garden, Claremont.

Gluck H. 1901. Die Stipulargebilde der Monocotyledonen. Verhandl. Naturhist. Med. Vereins zu Heidelberg, N. F, 7: 1–96.

Goldberg A. 1989. Classification, evolution, and phylogeny of the families of monocotyledons. Smithsonian Contr. Bot. 71.

Graham SW, JM Zgurski, MA McPherson, DM Cherniawsky, JM Saarela, ESC Horne, SY Smith, WA Wong, HE O'Brien, VL Biron, JC Pires, RG Olmstead, MW Chase, and HS Rai. 2006. Robust inference of monocot deep phylogeny using an expanded multigene plastid data set. In: JT Columbus, EA Friar, CW Hamilton, JM Porter, LM Prince, and MG Simpson, eds. Monocots: comparative biology and evolution, vol. 2, pp. 3–21. Rancho Santa Ana Botanical Garden, Claremont.

Greilhuber J. 1995. Chromosomes of the monocotyledons (general aspects). In: PJ Rudall, PJ Cribb, DF Cutler, and CJ Humphries, eds. Monocotyledons: systematics and evolution, vol. 2, pp. 379–414. Royal Botanic Gardens, Kew.

Guerra M. 2000. Chromosome number variation and evolution in monocots. In: KL Wilson and DA Morrison, eds. Monocots: systematics and evolution, pp. 127–136. CSIRO, Collingwood.

Guignard J-L. 1975. Du cotyledon des monocotyledones. Phytomorphology 25: 193–200.

Gunawardena A and NG Dengler. 2006. Alternative modes of leaf dissection in monocotyledons. Bot. J. Linn. Soc. 150: 25–44.

Haines RW and KA Lye. 1979. Monocotylar seedlings: a review of evidence supporting origin by fusion. Bot. J. Linn. Soc. 78: 123–140.

Halbitter H and M Hesse. 1993. Sulcus morphology in some monocot families. Grana 32: 87–99.

Harley MM and MS Zavada. 2000. Pollen of the monocotyledons: selecting characters for cladistic analysis. In: KL Wilson and DA Morrison, eds. Monocots: systematics and evolution, pp. 194–213. CSIRO, Collingwood.

Harris PJ. Compositions of monocotyledon cell walls: implications for biosystematics. In: KL Wilson and DA Morrison, eds. Monocots: systematics and evolution, pp. 114–126. CSIRO, Collingwood.

Harris PJ and RD Hartley. 1980. Phenolic constituents of the cell walls of monocotyledons. Biochem. Syst. Ecol. 8: 153–160.

Heel van WA. 1988. On the development of some gynoecia with septal nectaries. Blumea 33: 477–504.

Hegnauer R. 1963. Chemotaxonomie der Pflanzen: 2. Monocotyledoneae. Basel.

Henslow G. 1893. A theoretical origin of endogens from exogens through self-adaptation to an aquatic habit. Bot. J. Linn. Soc. 29: 485–528.

Henslow G. 1911. The origin of monocotyledons from dicotyledons through self-adaptation to a moist or aquatic habit. Ann. Bot. 26: 717–744.

Hofmeister W. 1861. Neue Beiträge zur Erkenntnis der Embryobildung der Phanerogamen: 2. Monokotyledonen. Abh. Kongl. Sachs. Ges. Wiss. 5: 629–760.

Holttum RE. 1955. Growth-habits of Monocotyledons: Variation on a theme. Phytomorphology 5: 399–413.

Huber H. 1969. Die Samenmerkmale und Verwandtschaftsverhaltnisse der Liliifloren. Mitt. Bot. Staatssamml. München 8: 219–538.

Huber H. 1977. The treatment of the monocotyledons in an evolutionary system of classification. Plant Syst. Evol., Suppl., 1: 285–298.

Igersheim A, M Buzgo and PK Endress. 2001. Gynoecium diversity and systematics in basal monocots. Bot. J. Linn. Soc. 136: 1–65.

Janssen T and K Bremer. 2004. The age of major monocot groups inferred from 800 + *rbc*L sequences. Bot. J. Linn. Soc. 146: 385–398.

Johnson KA. 2000. Development of non-zygotic embryos from callus in three Australian monocots. In: KL Wilson and DA Morrison, eds. Monocots: systematics and evolution, pp. 218–220. CSIRO, Collingwood.

Kaplan DR. 1973. The monocotyledons: their evolution and comparative biology: VII. The problem of leaf morphology and evolution in the monocotyledons. Quart. Rev. Biol. 48: 437–457.

Kaplan DR. 1975. Comparative developmental evaluation of the morphology of unifacial leaves in the monocotyledons. Bot. Jahrb. Syst. 95: 1–105.

Kellog EA. 2000. A model of inflorescence development. In: KL Wilson and DA Morrison, eds. Monocots: systematics and evolution, pp. 84–88. CSIRO, Collingwood.

Khokhrjakov AP. 1975. Somatic evolution of the monocotyledons. Nauka, Moscow (in Russian).

Kimura Y. 1956. Systeme et phylogenie des monocotyledones. Notul. Syst. (Paris) 15: 137–159.

Kite GC, RJ Grayer, PJ Rudall, and MSJ Simmonds. 2000. The potential for chemical characters in monocotyledon systematics. In: KL Wilson and DA Morrison, eds. Monocots: systematics and evolution, pp. 101–113. CSIRO, Collingwood.

Kubitzki K, ed. 1998. The families and genera of vascular plants, vols. 3 and 4. Monocotyledons. Springer, Berlin/Heidelberg/New York.

Les DH and EL Schneider. 1995. The Nymphaeales, Alismatiodae, and the theory of an aquatic monocotyledon origin. In: PJ Rudall, PJ Cribb, DF Cutler, and CJ Humphries, eds. Monocotyledons: systematics and evolution, vol. 1, pp. 23–42. Royal Botanic Gardens, Kew.

Li X-X and Z-K Zhou. 2006. A cladistic analysis of monocotyledons at the family level based on morphological data. Acta Bot. Yunn. 28: 241–249 (in Chinese).

Li X-X and Z-K Zhou. 2007. The higher-level phylogeny of monocots based on matK, rbcL and 18S rDNA sequences. Acta Phytotax. Sinica 45: 113–133 (in Chinese).

Lodkina MM. 1988. Evolutionary relations between mono- and dicotyledons based on embryo and seedling investigation. Bot. Zhurn. 73: 617–630 (in Russian with English summary).

Metcaife CR. 1961. The anatomical approach to systematics: general introduction with special reference to recent work on monocotyledons. In recent advances in botany, pp. 146–150. University of Toronto Press, Toronto.

Metcalfe CR, ed. 1960–1982. Anatomy of the monocotyledons, 7 vols. Clarendon, Oxford.

Meusel I, E Leistner, and W Barthlott. 1994. Chemistry and micromorphology of compound epicuticular wax crystalloids (*Strelitzia* type). Plant Syst. Evol. 193: 115–123.

Nadot S, G Bittar, L Carter, R Lacroix, and B Lejune. 1995. A phylogenetic analysis of monocotyledons based on the chloroplast gene rps4, using parsimony and a new numerical phenetics method. Mol. Phylogenet. Evol. 4: 257–282.

Paliwal GS. 1969. Stomatal ontogeny and phylogeny: 1. Monocotyledons. Acta Bot. Neerl. 18: 654–668.

Parkin J. 1923. The strobilus theory of angiospermous descent. Proc. Linn. Soc. Lond. 153: 51–64.

Prychid CJ and PJ Rudall. 1999. Calcium oxalate crystals in monocotyledons: structure and systematics. Ann. Bot. 84: 725–739.

Prychid CJ and PJ Rudall. 2000. Distribution of calcium oxalate crystals in monocotyledons. In: KL Wilson and DA Morrison, eds. Monocots: systematics and evolution, pp. 159–162. CSIRO, Collingwood.

Puri V. 1989. Monocotyledons: some comments on their morphology and evolution. Professor Panchanan Maheshwari Memorial Lecture. New Delhi.

Ronse Decraene LP and EF Smets. 1995. The androecium of monocotyledons. In: P Rudall, PJ Cribb, DF Cutler, and CJ Humphries, eds. Monocotyledons: systematic and evolution, pp. 243–254. Royal Botanic Gardens. Kew.

Rudall P. 1991. Lateral meristems and stem thickening growth in monocotyledons. Bot. Rev. 57: 150–163.

Rudall P. 1997. The nucellus and chalaza in Monocotyledons: structure and systematics. Bot. Rev. 63: 140–181.

Rudall PJ. 2000. 'Cryptic' characters in monocotyledons: homology and coding. In: Scotland R and RT Pennington, eds. Homology and systematics: coding characters for phylogenetic analysis, pp. 114–123. Taylor & Francis, London.

Rudall PJ. 2002a. Homologies of inferior ovaries and septal nectaries in monocotyledons. Int. J. Plant Sci. 163: 261–276.

Rudall PJ. 2002b. Unique floral structures and iterative evolutionary themes in Asparagales: insights from a morphological cladistic analysis. Bot. Rev. 68: 488–509.

Rudall PJ. and R Caddick. 1994. Investigation on the presence of phenolic compounds in monocotyledonous cell walls, using UV fluorescence microscopy. Ann. Bot. 74: 483–491.

Rudall P, PJ Cribb, DF Cutler, and CJ Humphries, eds. 1995. Monocotyledons: systematics and evolution, 2 vols. Royal Botanic Gardens. Kew.

Rudall PJ, CJ Prychid, and CJ Jones. 1998. Intra-ovarian trichomes in monocotyledons. In: SJ Owens and PJ Rudall, eds. Reproductive biology, pp. 219–230. Royal Botanic Gardens, Kew.

Sargant E. 1903. A theory of the origin of monocotyledons, founded on the structure of their seedlings. Ann. Bot. 17: 1–92.

Sargant E. 1904. The evolution of Monocotyledons. Bot. Gaz. 37: 325–345.

Scotland R and RT Pennington, eds. 2000. Homology and systematics: coding characters for phylogenetic analysis. Taylor and Francis, London.

Sharma AK 1969. Evolution and taxonomy of monocotyledons. In: CD Darlington and KR Lewis, eds. Chromosomes today, vol. 2, pp. 241–249. Plenum, New York.

Smets EF, L-P Ronse Decraene, P Caris, and PJ Rudall. 2000. Floral nectaries in monocotyledons: distribution and evolution. In: KL Wilson and DA Morrison, eds. Monocots: systematics and evolution, pp. 221–229. CSIRO, Collingwood.

Stebbins GL and GS Khush. 1961. Variation in the organization of the stomatal complex in the leaf epidermis of monocotyledons and its bearing on their phylogeny. Am. J. Bot. 48: 51–59.

Stevenson DW, JI Davis, JV Freudenstein, CR Hardy, MP Simmons, and CD Specht. 2000. A phylogenetic analysis of the monocotyledons based on morphological and molecular character sets, with comments on the placement of *Acorus* and Hydatellaceae. In: KL Wilson and DA Morrison, eds. Monocots: systematics and evolution, pp. 17–24. CSIRO, Collingwood.

Stevenson DW and H Loconte. 1995. Cladistic analysis of monocotyledons. In: PJ Rudall, PJ Cribb, DF Cutler, and CJ Humphries, eds. Monocotyledons: systematics and evolution, vol. 2, pp. 543–576. Royal Botanic Gardens, Kew.

Takhtajan A, ed. 1982. Plant life: 6. Liliopsids or monocotyledons. Nauka, Moscow (in Russian).

Takhtajan A, ed. 1985. Comparative seed anatomy: 1. Monocotyledons. Nauka, Leningrad (in Russian).

Tamura MN, J Yamashita, S Fuse, and M Haraguchi. 2004. Molecular phylogeny of monocotyledons inferred from combined analysis of plastid *mat*K and *rbc*L gene sequences. J. Plant Res. 117: 109–120.

Thorne RF. 2000. The classification and geography of the monocotyledon subclasses Alismatidae, Liliidae and Commelinidae. In: Nordenstam et al., eds. Plant systematics for the 21st century, pp. 75–122. Portland, London.

Tieghem R van and H Duliot. 1888. Recherches comparatives sur l'origine des membres endogenes dans les plantes vasculaires. Ann. Sci. Nat., ser. 7, 8: 1–666.

Tillich H-J. 1992. Bauprinzipien und Evolutionslinien bei monocotylen Keimpflanzen. Bot. Jahrb. Syst. 114: 91–132.

Tillich H-J. 1995. Seedlings and systematics in monocotyledons. In: PJ Rudall, PJ Cribb, DF Cutler, and CJ Humphries, eds. Monocotyledons: systematics and evolution, 1: 303–352. Royal Botanic Gardens, Kew.

Tillich H-J. 1998. Plesiomorphies and apomorphies in seedlings of monocotyledons. In Monocots II, p. 53 (abstract). Sydney.

Tillich H-J. 2000. Ancestral and derived character states in seedlings of monocotyledons. In: KL Wilson and DA Morrison, eds. Monocots: systematics and evolution, pp. 221–229. CSIRO, Collingwood.

Tomlinson RB. 1970. Monocotyledons: towards an understanding of their morphology and anatomy. In: RD Preston, ed. Advances in botanical research, pp. 207–292. Academic, New York.

Tomlinson PB 1974. Development of the stomatal complex as a taxonomic character in the monocotyledons. Taxon 23: 109–128.

Tomlinson PB. 1995. Non-homology of vascular organization in monocotyledons and dicotyledons. In: PJ Rudall, PJ Cribb, DF Cutler, CJ Humphries, eds. Monocotyledons: systematics and evolution, pp. 489–622. Royal Botanic Gardens. Kew.

Tomlinson PB and JB Fisher. 2000. Stem vasculature in climbing monocotyledons: a comparative approach. In: KL Wilson and DA Morrison, eds. Monocots: systematics and evolution, pp. 89–100. CSIRO, Collingwood.

Van Heel WA. 1988. On the development of some gynoecia with septal nectaries. Blumea 33: 477–504.

Von Staudermann W. 1924. Die Haare der Monocotyledonen. Bot. Arch. 8: 105–184.

Wagner A. 1977. Vessel types of the monocotyledons: a survey Bot. Not. 130: 383–402.

Weber A. 1980. Die Homologie des Perigons der Zingiberaceen: Ein Beitrag zur Morphologic und Phylogenie des Monocotylen-Perigons. Plant Syst. Evol. 133: 149–179.
Williams CA, JB Harborne, and B Mathew. 1988. A chemical appraisal via leaf flavonoids of Dahlgren's Liliiflorae. Phytochemistry 27: 2609–2629.
Wilson KL and DA Morison, eds. 2000. Monocots: systematics and evolution. CSIRO, Collingwood.
Yeo PE. 1989. What is happening to the monocotyledons? Plant Syst. Evol. 167: 75–86.
Zavada M. 1983. Comparative morphology of monocot pollen and evolutionary trends of apertures and wall structures. Bot. Rev. 49: 331–379.
Zimmerman MH and PB Tomlinson. 1972. The vascular system of Monocotyledonous stems. Bot. Gaz. 133: 141–155.

Subclass I. ALISMATIDAE

Aquatic or marsh herbs. Leaves alternate or less often opposite or verticillate, with parallel or arcuate venation, usually with vaginate base. Stems with axillary, multiseriate hairs (intravaginal squamules) at the nodes (replaced by uniseriate, threadlike hairs in *Scheuchzeria*). Vessels absent or only in roots. Flowers in various types of inflorescences or solitary, bisexual or unisexual, actinomorphic or zygomorphic. Tapetum always plasmodial. Pollen grains almost always 3-celled, 1-colpate or inaperturate, less often pantoporate. Gynoecium apocarpous or syncarpous (mostly paracarpous), sometimes pseudomonomerous. Ovules bitegmic, crassinucellate, or sometimes almost tenuinucellate. Endosperm helobial or less often nuclear. Fruits of various types, in most archaic members multifollicles. Seeds usually without endosperm or sometimes only with vestigial endosperm.

Alismatidae share a common origin with Liliidae and some genera (e.g. *Tofieldia*) occupied the somewhat intermediate position.

Alismatidae are one of the most archaic groups of monocotyledons. However, they are one of the blind branches of the archaic liliopsids rather than a basal, ancestral group (Les and Haynes 1995; Takhtajan 1997). They are extremely heterobathmic and in general very specialized. Cladistically the Alismatidae are very diversified and consist of fairly isolated groups, which I prefer to consider as separate orders (Takhtajan 1987).

Derived from some archaic herbaceous members of early liliopsids.

Bibliography

Arber A. 1923. On the "squamulae intravaginales" of the Helobiae. Ann. Bot. 37: 31–41.

Buchenau F. 1882. Beiträge zur Kenntnis der Butomaceen, Alismaceen, und Juncaginaceen. Engler's Bot. Jahrb. 2: 465–510.

Chen J-M, D Chen, GW Robert, Q-F Wang, and Y-H Guo. 2004a. Evolution of apocarpy in Alismatidae using phylogenetic evidence from chloroplast *rbc*L sequence data. Bot. Bull. Acad. Sinica 45: 33–40.

Chen J-M, GW Robert, Q-F Wang. 2004b. Evolution of aquatic life forms in Alismatidae: phylogenetic estimation from chloroplast *rbc*L sequence data. Israel J. Plant Sci. 52: 323–329.

Eber E. 1934. Karpellbau und Plazentationsverhaltnisse in der Reihe der Helobiae. Flora 127: 273–330.

Gibson RJH. 1905. The axillary scales of aquatic monocotyledons. Bot. J. Linn. Soc. 37: 228–236.

Harada I. 1956. Cytological studies in Helobiae: I. Chromosome idiograms and a list of chromosome numbers in seven families. Cytologia 21: 306–328.

Haynes RR and LB Holm-Nielsen. 1985 (1987). A generic treatment of Alismatidae in the Neotropics with special reference to Brazil. Acta Amazonica Suppl. 15: 153–193.

Haynes RR and LB Holm-Nielsen. 1989. Speciation of Alismatidae in the Neotropics. In: LB Holm-Nielsen, IC Nielsen, and H Balslev, eds. Tropical forests: botanical dynamics, speciation, and diversity, pp. 211–219. Academic, London.

Lakshmanan KK. 1970. Hydrocharitaceae, Juncaginaceae, Scheuchzeriaceae, Potamogetonaceae, Zannichelliaceae, Najadaceae. Bull. Ind. Natl. Sci. Acad. 41: 336–357.

Les DH and MA Cleland. 1997. Phylogenetic studies in Alismatidae, II. Evolution of marine angiosperms (sea grasses) and hydrophily. Syst. Bot. 22: 443–463.

Les DH and RR Haynes. 1995. Systematics of subclass Alismatidae: a synthesis of approaches. In: PJ Rudall,

A. Takhtajan, *Flowering Plants*,

PJ Cribb, DF Cutler, and CJ Humphries, eds. Monocotyledons: systematics and evolution 2: 353–377. Royal Botanic Gardens, Kew.
Les DH, DK Garvin, and CF Wimpee. 1993. Phylogenetic studies in the monocot subclass Alismatidae, evidence for a reappraisal of the aquatic order Najadales. Mol. Phylogeb. Evol. 2: 304–314.
Markgraf F. 1936. Blütenbau und Verwandtschaft bei den einfachsten Helobiae. Ber. Deutsch. Bot. Ges. 54: 191–229.
Mavrodiev EV and DD Sokolov. 1998. On the morphology of European species of Zannichelliaceae, Ruppiaceae, Potamogetonaceae and Zosteraceae. Byull. Mosk. Obshch. Ispyt. Prir., Biol. 103(5): 49–60 (in Russian).
Pettitt JM and AC Jenny 1975. Pollen in hydrophilous angiosperms. Micron 5: 377–405.
Posluszny U and WA Charlton. 1993. Evolution of the helobial flower. Aquatic Bot. 44: 303–324.
Salisbury EJ 1926. Floral construction in Helobiales. Ann. Bot. 40: 419–455.
Sharma AK and T Chaterjee. 1967. Cytotaxonomy of Helobiae with special reference to the mode of evolution. Cytologia 32: 286–307.
Singh V. 1966. Morphological and anatomical studies of the flower of Helobiae. Agra Univ. J. Res. (Sci.) 15: 147–150.
Tomlinson PB. 1982. Helobiae (Alismatidae), including the sea grasses. In: CR Metcalfe, ed. Anatomy of monocotyledons, vol. 3. Clarendon, Oxford.
Wilder GJ. 1974. Symmetry and development of *Butomus umbellatus* (Butomaceae) and *Limnocharis flava* (Limnocharitaceae). Am. J. Bot. 61: 379–394.
Wilder GJ. 1975. Phylogenetic trends in the Alismatidae (Monocotyledoneae). Bot. Gaz. 136: 159–170.

Superorder PETROSAVIANAE

Order 1. PETROSAVIALES

Small, chlorophyllous or achlorophyllous, rhizomatous herbs with slender, usually simple, erect to creeping stems; stems with a ring of bundles. Calcium oxalate crystals present (Tofieldiaceae) or absent. Raphides absent or present. Vascular system weakly developed, with vascular bundles in singular ring. Some vessels only in roots, with scalariform perforations. Sieve-element plastids of P1/2c-type (Behnke 2002). Leaves alternate, spirally arranged, basal or distichous, linear or reduced to scales, unifacial of bifacial (Japonoliriaceae). Stomata anomocytic. Flowers small, in terminal or axillary (Japonoliriaceae) corymbs or racemes, rarely solitary, bracteate, with or without bracteoles, 3-merous, bisexual, actinomorphic. Perianth segments of six, in two cycles, persistent, free or basally connate. Stamens six, in two cycles, free or adnate to the base of perianth segments; filaments subulate; anthers basifixed or dorsifixed, introrse or latrorse, sometimes appendaged, opening longitudinally. Tapetum secretory. Microsporogenesis simultaneous. Pollen grains 1-colpate, gemmate or reticulate. Septal nectaries present. Gynoecium of three, rarely 4–6, carpels; stylodia free or rarely connate into a style, short, stigma capitate, subcapitate or styles with decurrent stigma. Ovary superior or semi-inferior, with numerous or only 4–5 (Japonoliriaceae) ovules in each carpel. Ovules anatropous to campylotropous, bitegmic or very rarely (*Tofieldia iridaceae*, some sp. *Narthecium*) unitegmic, crassinucellate to semicrassinucellate. Female gametophyte of *Polygonum*-type or sometimes of *Allium*-type. Endosperm helobial or nuclear (*Petrosavia*). Fruits follicular or capsular, septicidal or ventrally dehiscent in upper part. Seeds small, more or less winged, often with appendages, testal; embryo small to minute and undifferentiated; endosperm copious, containing oil, proteins, or starch (*Japonolirion*). Steroidal saponins and chelidonic acid present.

Key to Families

1 Plants mycotrophic. Flowers in terminal corymbs or racemes; anthers dorsifixed. Small herbs with slender, usually simple, erect stem and thin rhizome covered with sheathing scale leaves. Roots with a cortex consisting of 4–6 layers of parenchyma cells containing mycorrhizal hyphae. Raphide idioblasta present. Vascular system weakly developed, with vascular bundles in singular ring. Some vessels in roots only, with scalariform perforations. Leaves reduced to alternate scales, the leaf primordial formed in distichous succession (Stant 1970). Flowers small, in terminal corymbs or racemes, each subtended by a well-developed linear bract and bears a small bracteole at the base of the pedicel, bisexual, actinomorphic. Perianth of six imbricate, basally connate, persistent, colorless segments in two cycles, the outer narrow, the inner broad-ovate, with a small gland at base. Stamens six in two cycles; filaments subulate, adnate to the base of the perianth segments; anthers ovate, dorsifixed, introrse, opening longitudinally; connective very shortly prolonged. Pollen grains gemmate. Gynoecium of three very shortly basally connate carpels with sutures closed at anthesis; stylodia short, with slightly bifid, subcapitate stigma; ovary semi-inferior, with numerous ovules on basal

submarginal placentas. Septal nectary present as septal glands, which extend from just below the lowermost ovular insertion up to the level of carpellary separation, where they open on the lateral surfaces of the carpels and produce nectar drops (Sterling 1978). Ovules anatropous, bitegmic, semicrassinucellate, with nucellar cap. Endosperm nuclear. Fruits recurved-spreading, basally united follicles. Seeds numerous, minute, brown-ribbed, winged all around, with minute, undifferentiated embryo consisting of embryo proper and a suspensor; endosperm copious, containing oil and protein; seed coat formed by the outer layer of inner integument. Raphide idioblasts present; sieve tube plastids also with polygonal protein crystalloids. n = 15, 30.... 2. PETROSAVIACEAE.

1 Plants not mycotrophic.

2 Capsules septicidal.

3 Calcium oxalate crystals lacking. Flowers in axillary racemes; anthers basifixed. Inflorescences axillary. Pollen grains gemmate or reticulate with roundish brochi. Herbs with slender, short to creeping, scaly rhizome. Leaves spiral, basal, tufted, bifacial, linear, with few parallel veins, scabrous on margin, sometimes longer than scape. Flowers small, in simple, long axillary racemes, bracteate, without bracteoles, with short pedicels, 3-merous, bisexual, actinomorphic, without nectaries. Perianth segments six in two cycles, free, membranous, persistent, outer oblong, inner obovate to spatulate. Stamens six in two cycles; filaments subulate; anthers ovate, introrse. Gynoecium of three carpels, 3-lobed, essentially apocarpous – carpels loosely connected only with the interdigitating papillae along the septal faces; stylodia short, recurved, with decurrent and papillate stigma; ovary usually with 4–5 ovules per locule; ovules anatropous, attached to the ovary wall by their micropyles and completely fill the locule (Remizowa et al. 2006a). The septal nectaries are located in the lower part of ovary. Fruits separate septicidally through the weakly connate zone between the follicules (Utech 1984). Seeds small, broadly elliptical, without appendages; embryo small, straight; endosperm copious, starchy, weakly ruminate (Plisko 2004, personal communication); seed coat 1-layered, n = 12 (three chromosomes long and with subterminal or submedian constrictions and nine chromosomes short and with median or submedian constrictions – Satô 1942)....... 1. JAPONOLIRIACEAE.

3 Calcium oxalate crystals (druses, cuboidal crystals) present (unique crystalline inclusions, druses, in parenchymatous tissues, and prismatic crystals in the bundle sheaths (Ambrose 1980). Perennial herbs with creeping rhizome and simple or branched sympodial stem. Raphides lacking. Roots fibrous. Vessels only in roots, with scalariform perforations. Sieve-element plastids with polygonal crystals. Leaves spirally arranged, mostly basal, reduced upwards to bracts, laterally flattened (unifacial), sheathing at base, linear; venation parallel, sometimes with midrib. Flowers small, in terminal, sometimes glandular, racemose inflorescences, rarely (*Harperocallis*) solitary, bisexual, actinomorphic, pedicillate or sessile, typically subtended by a bract and a calyx-like involucre (clyculus) of three (occasionally 2, 4, or 0) distinct or connate bracteoles. Perianth segments six in two cycles, free, petaloid, persistent, with outer cycle sometimes slightly wider or longer than inner. Stamens usually six in two cycles or (*Pleea*) nine or rarely 10 or more; filaments free, more or less broadened at base; anthers basifixed or dorsifixed, introrse or latrorse, occasionally with apical appendage, dehiscing longitudinally. Pollen grains 2-celled, tectate or semitectate, with two distal colpi, reticulate. Gynoecium of three (rarely 4–6) carpels, nearly apocarpous or syncarpous below and apocarpous above or (*Isidrogalvia*) totally syncarpous, often (*Tofieldia, Pleea, Harperocallis*) stipitate (probably a plesiomorphic condition); stylodia free or (*Isidrogalvia*) connate into a style with capitate stigma. Septal nectaries (basal intercarpellary glands – a unique nectary type, which may be nonhomologous with the septal nectaries of other monocots – Utech 1978; Zomlefer 1997) present. Ovules few to numerous in each carpel or in each locule, anatropous to campylotropous, bitegmic or more rarely unitegmic (*Tofildia iridaceae*), crassinucellate to tenuinucellate (?). The archesporial cell cuts off a primary parietal cell. Female gametophyte of *Polygonum*-type or sometimes

of *Allium*-type. Endosperm helobial. Fruits follicular or septicidal capsules. Seeds elliptical to fusiform, often with terminal appendages on one or both ends (micropylar and chalazal); seed coat testal, phytomelan lacking, but in *Tofieldia* contains phlobaphene deposits; embryo small; endosperm copious. Steroidal saponins, tannins, chelidonic acid, flavonoids (kaempferol, quercetin), and sometimes, as in *Tofieldia,* also glycosides and alkaloid absent (L.S. Teslov 2000, personal communication); n = 15, rarely 14, 16. . . . . . 3. TOFIELDIACEAE.

2 Capsules loculicidal. Perennial herbs with short or creeping rhizome. Raphides present. Scattered fibres present in the phloem. Root cortex aerenchyma present. Vessels only in roots, with scalariform perforations. Leaves mostly basal, spiral, bifacial or unifacial, linear to ovate, sessile, narrowed to sheathing base, parallel veined. Flowers small, in terminal spikes or simple or compound racemes, bracteate and usually bracteolate, 3-merous, bisexual, actinomorphic. Perianth segments six in two cycles, petaloid, free or more or less connate into a tube, adnate to the lower part of ovary, persistent. Stamens six in two cycles; filaments filiform, inserted at base of perianth segments, slender, glabrous, or wooly pubescent; anthers basifixed to dorsifixed, linear to cordate-orbicular, tetrasporangiate, introrse. Pollen grains ellipsoid, tectate or semitectate, reticulate, foveolate, or gemmate. Gynoecium of three united carpels; style with 3-lobed to punctiform stigma; ovary superior to semi-inferior, 3-locular or 3-locular below and 1-locular above, sometimes with septal nectaries. Ovules numerous, anatropous, according to Remizowa et al. (2006), all species examined ovules bitegmic, (unitegmic according to Sterling 1979 at least in *Narthecium*), crassinucellate, with parietal cell. Female gametophyte of *Polygonum*-type. Endosperm helobial. Fruits loculicidal capsules subtended by persistent perianth Seeds ellipsoid to fusiform, appendaged at both ends (*Narthecium*) or without appendages (*Aletris*). Accumulate steroidal saponins and sometimes glycosides and chelidonic acid, no flavonols. n = 12, 13 (common), 21, 22, 23, × more probably 7 (Zomlefer 1997). . . . . 4. NARTHECIACEAE.

1. JAPONOLIRIACEAE

Takhtajan 1996. 1/1. Serpentine swamps of northern and central Japan (Hokkaido, Honshu).

Japonolirion

Most probably the basal family in the monocotyledons.

2. PETROSAVIACEAE

Hutchinson 1934. 1/2. Eastern Asia, West Malesia (Malaya, Borneo).

Petrosavia (including *Miyoshia* and *Protolirion*).

Related to the Japonoliriaceae, but differ in mycotrophic habitus, in having a semi-inferior ovary, in the structure of the septal nectaries, stigma, number of ovules, nuclear endosperm, fruits and oily and protein endosperm.

3. TOFIELDIACEAE

Takhtajan 1995. 4/25. Temperate Eurasia (especially eastern Asia), North America, South America (Venezuela, Guyana, Colombia, Ecuador, Peru). *Harperocallis* (1) is endemic to Florida, *Isidrogalvia* (5) to northern South America.

Harperocallis, Tofieldia, Pleea (sometimes included in *Tofieldia*), *Isidrogalvia.*

The Tofieldiaceae is one of the archaic family, which has been considered by many authors as the basal in monocotyledons (Lotsy 1911; Takhtajan 1959, 1966, 1980, 1987, 1997; Eames 1961; Radulescu 1973; Goldblatt 1995; Tamura 1998).

4. NARTHECIACEAE

E.M. Fries ex J. Bjurzon 1846 (including Lophiolaceae Nakai 1943). 4/40. Temperate Eurasia to Malesia, North America, and (monotypic genus *Nietneria*) Venezuela and Guyana.

Narthecium, Nietneria, Aletris (including *Metanarthecium*), *Lophiola.*

Related to the Tofieldiaceae, but stylodia connate into a style, anthers mostly extrorse, raphides are present, and capsule loculicidal. No alkaloids are known.

Bibliography

Ambrose JD. 1985. *Lophiola*, familial affinity with the Liliaceae. Taxon 34: 149–150.

Beccari O. 1871. *Petrosavia*: Nuovo genere di piante parasite della famiglia delle Melanthiaceae. Nuovo Giorn. Bot. Ital. 3: 7–11.

Browne ET, Jr. 1961. Morphological studies in *Aletris*. I. Development of the ovule, megaspores and megagametophyte of *A. aurea* and their connection with the systematics of the genus. Am. J. Bot. 48: 143–147.

Cameron KM. 1998. Systematics of heteromycotrophic Petrosaviaceae. In Monocots II, p. 64 (abstract). Sydney.

Cameron KM, MW Chase, and PJ Rudall. 2003. Recircumscription of the monocotyledonous family Petrosaviaceae to include *Japonolirion*. Brittonia 55: 214–225.

Eie S. 1972. Floral anatomy in *Tofieldia fusilla* (Michx.) Pers. with special reference to the gynoecium. Norweg. J. Bot. 19: 31–36.

Groom P. 1892. On the embryo of *Petrosavia* Beccari. Ann. Bot. 6: 380–382.

Groom P. 1895. On a new saprophytic monocotyledon. Ann. Bot. 9: 45–58.

Hara H. 1967. The status of the genus *Metanarthecium* Maxim. Jpn. J. Bot. 42: 312–316.

Kosenko VN. 1987. Pollen morphology of Tofieldieae, Narthecieae, Xerophylleae, Melanthieae (Melanthiaceae). Bot. Zhurn. 72: 1318–1330 (in Russian with English summary).

Lersten NR and JD Curtis. 1977. Anatomy and distribution of secretory glands and other emergences in *Tofieldia* (Liliaceae). Ann. Bot. (UK) 41(174): 879–882.

McDaniel S. 1968. Harperocallis. A new genus of the Liliaceae from Florida. J. Arnold Arbor. 49: 35–40.

Ohba H 1984. A review of *Petrosavia* (Liliaceae), with special reference to the floral features. J. Jpn. Bot. 59: 106–109.

Remizova M and D Sokoloff. 2003. Inflorescence and floral morphology in *Tofieldia* (Tofieldiaceae) compared with Araceae, Acoraceae and Alismatales s. str. Bot. Jahrb. Syst. 124: 255–271.

Remizowa M, D Sokoloff, and PJ Rudall. 2006a. Evolution of the monocot gynoecium: evidence from comparative morphology and development in *Tofieldia*, *Japonolirion*, *Petrosavia* and *Narthecium*. Plant Syst. Evol. 258: 183–209.

Remizowa M, D Sokoloff, and PJ Rudall. 2006b. Comparative patterns of floral orientation, bracts and bracteoles in *Tofieldia*, *Japonolirion*, and *Narthecium*. Aliso 24: 157–169.

Sokolowska-Kulczycka A. 1980. Embryological studies of *Tofieldia calyculata* (l.) Whlb. Acta Biol. Cracov. Ser. Bot. 22: 113–128.

Stant MY. 1970. Anatomy of *Petrosavia stellaris* Becc., a saprophytic monocotyledon. Bot. J. Linn Soc. 63(Suppl. 1): 147–161.

Sterling C. 1978. Comparative morphology of the carpel of the Liliaceae: Hewardieae, Petrosavieae, and Tricyrteae. Bot. J. Linn. Soc. 77: 95–106.

Sterling C. 1979. Comparative morphology of the carpel in the Liliaceae: Tofieldieae. Bot. J. Linn. Soc. 79: 321–332.

Takahashi HR, E Nishio, and H Hayashi. 1993. Pollination biology of the saprophytic species *Petrosavia sakuraii* (Makino) J. J. Smith ex van Steenis in central Japan. J. Plant Res. 106: 213–217.

Takhtajan AL. 1994. Six new families of flowering plants. Bot. Zhurn. 79(1): 96–97 (in Russian).

Takhtajan AL. 1994 (1995). New families of the monocotyledons. Bot. Zhurn. 79(12): 65–66 (in Russian).

Takhtajan AL. 1996. Validization of some formerly established families of flowering plants. Bot. Zhurn. 81(2): 85–86.

Tamura MN. 1998b. Nartheciaceae. In: K Kubitzki, ed. The families and genera of vascular plants, vol. 3, pp. 381–392. Springer, Berlin/Heidelberg/New York.

Tamura MN and H Takahashi. 1998. Karyotype analysis of the saprophyte *Petrosavia sakuraii* (Makino) J. J. Smith ex van Steenis. And its systematic implications. Acta Phytotax. Geobot. 49.

Tanaka R and N Tanaka. 1976. Karyomorphological studies in *Tofieldia* of Japan. Chromosome Inform. Serv., no. 19: 9–11.

Tomimatsu H, A Hoya, H Takahashi, and M Ohara. 2004. Genetic diversity and multilocus genetic structure in the relictual endemic herb *Japonolirion osense* (Petrosaviaceae). J. Plant Res. 117: 13–18.

Utech FH. 1978a. Floral vascular anatomy of monotypic Japanese *Metanarthecium luteoviride* Maxim. (Liliaceae-Melanthioideae). Ann. Carnegie Mus. 47: 455–477.

Utech FH. 1978b. Floral vascular anatomy of *Pleea tenuifolia* Michx. (Liliaceae-Tofieldieae) and its reassignment to *Tofieldia*. Ann. Carnegie Mus. 47: 423–454.

Utech FH. 1979. Karyotype analysis, palynology, and external seed morphology of *Tofieldia tenuifolia* Michx. Utech (Liliaceae-Tofieldieae). Ann. Carnegie Mus. 48: 161–174.

Utech FH. 1984. Floral vascular anatomy of *Japonolirion osense* Nakai (Liliaceae) and its tribal relationship. Ann. Carnegie Mus. 53: 447–461.

Zomlefer WB. 1997a. The genera of Tofieldiaceae in the southeastern United States. Harvard Pap. Bot. 2: 179–194.

Zomlefer WB. 1997b. The genera of Nartheciaceae in southeastern United States. Harvard Pap. Bot. 2: 195–211.

Superorder ALISMATANAE

Order 2. HYDROCHARITALES

Perennial or rarely annual, aquatic herbs often with an elongate rhizome or stolon, frequently partly or wholly submerged and rarely free-floating. Vessels absent or only in roots, with scalariform (Hydrocharitaceae) or with simple (Butomaceae) perforations. Leaves radical or cauline, alternate, sometimes distichous, or opposite or verticillate, often more or less sheathing or rarely stipular-expanded at the base, sometimes differentiated into petiole and lanceolate, elliptic or cordate-ovate lamina. Stomata usually paracytic or more often lacking. Flowers from rather large to very small and inconspicuous, solitary, paired, or in few-flowered cymose inflorescences subtended by two distinct or more or less connate bracts forming a sessile or long-pedunculate (sometimes spirally twisted) spathe, bisexual or more often unisexual and dioecious, actinomorphic or sometimes slightly zygomorphic, usually 3-merous. Perianth segments free, mostly six in two cycles or three in one cycle, rarely perianth absent. Stamens 2–3 to numerous; anthers basifixed or dorsifixed, tetrasporangiate or seldom disporangiate, opening longitudinally, introrse to generally extrorse. Tapetum plasmodial. Microsporogenesis successive or rarely (Aponogetonaceae) simultaneous. Pollen grains usually 3-celled. Nectaries present or absent. Gynoecium of (2)3–6 (up to 15–20) united or nearly free carpels, rarely (Najadaceae) of one carpel; stylodia often bilobed or bifid, sometimes shortly connate at the base; stigmas papillate, sometimes (Butomaceae, Aponogetonaceae) decurrent; ovary superior or inferior, unilocular, often with more or less deeply intruded, free carpellary margins. Ovules numerous to few, or solitary (Najadaceae), anatropous or orthotropous, bitegmic, or rarely (*Aponogeton distachyos*) unitegmic, crassinucellate. Female gametophyte of *Polygonum*-type. Endosperm helobial or nuclear (Najadaceae). Fruits submerged, multifollicles, achene-like, or fleshy, berrylike, or less often dehiscent. Seeds usually few to numerous, ellipsoidal to cylindrical or fusiform, smooth or with highly complex testa; endosperm absent or rarely scanty; embryo usually more or less straight, relatively large, with a massive hypocotyls, rich of starch. Present proanthocyanidins (cyanidin) n = 6, 7, 8–15.

The Hydrocharitales are the basal group of the Alismatidae.

Key to Families

1 Vessels absent. Annual or rarely perennial submerged herbs of fresh or brackish water Stems slender, more or less branched, with a much reduced conducting system. Roots fibrous. Leaves spiral, sub-opposite (essentially spirodistichous) or pseudoverticillate, with sheathing base, linear to linear-lanceolate, 1-veined, the margins serrulate to denticulate, the apex acute or acuminate, with 1–3 teeth on each side; sheaths open, and commonly enclosing two tiny, intravaginal squamules. Stomata lacking. Flowers very small and inconspicuous, axillary, solitary, or in small clusters, unisexual (monoecious or rarely dioecious), hydrophilous (pollination takes place under water). Male flowers are subsessile and consist of a single stamen and mostly subtended by one or two very thin hyaline envelopes of unknown origin; the anther subsessile or on a very short filament, tetrasporangiate or rarely with one microsporangium, opening by an apical slit. Microsporogenesis successive. Pollen grains 3-celled, globose to ellipsoidal, inaperturate or rarely slightly 1-colpate (Bolkhovskikh 1983), with thick intine and very thin exine. Female flowers are sessile; they are naked and composed of one ovule surrounded by a very thin (two cell layers thick) transparent carpellary wall. The gynoecium is either monocarpellary or pseudomonomerous; the stylodium divided some distance from the base into two stigmatic branches and sometimes into some shorter, nonstigmatic lobes as well; ovary superior, with one basal ovule. Ovules anatropous, crassinucellate. A parietal cell is cut off from the primary archesporial cell. Female gametophyte usually of *Polygonum*-type, rarely bisporic. Endosperm nuclear. Fruits indehiscent, with the endocarp tightly enclosing the seed and dehiscing by decay. Seeds fusiform to elliptic, more or less areolate, without endosperm, mesotestal; exotesta more or less deeply pitted with areolae which are arranged in longitudinal rows embryo cylindrical, straight or rarely slightly curved; the mesotesta consists of stone cells; the lower periclinal wall of endotegmen tuberculate. n = 6, 7, 8, 17 +. . . . . 3. NAJADACEAE.

1 Vessels present only in roots.

2 Vessels with scalariform perforations.

3 Flowers bracteate. Perennial or rarely annual, fresh-water or salt-water herbs often with an elongate rhizome or stolon, frequently partly or wholly submerged and rarely free-floating. Vessels absent or only in roots, with scalariform perforations. Leaves radical or cauline, alternate, sometimes distichous, or opposite or verticillate, often more or less sheathing or rarely (as in *Hydrocharis*) stipular-expanded at the base, sometimes differentiated into petiole and lanceolate, elliptic or cordate-ovate lamina, with various kinds of venation; intravaginal scales present, usually 2–10 or more, or rarely only one in each leaf axil. Stomata usually paracytic or more often lacking. Flowers from rather large to very small and

inconspicuous, solitary, paired, or in few-flowered cymose inflorescences subtended by two distinct or more or less connate bracts forming a sessile or long-pedunculate (sometimes spirally twisted) spathe, bisexual or more often unisexual and dioecious, actinomorphic or sometimes slightly zygomorphic, usually 3-merous. Perianth segments free to the base, mostly six in two cycles or three in one cycle, rarely only two; when six the outer may be sepaloid, green and valvate, and the inner petaloid, white or variously colored and imbricate, sometimes with a basal gland, rarely perianth absent. Stamens numerous to 2–3 (in some plants even to one), when numerous in up to six cycles of three; anthers basifixed or dorsifixed, tetrasporangiate or seldom disporangiate, introrse to extrorse. Microsporogenesis successive. Pollen grains usually 3-celled, inaperturate; the exine is very thin or reduced; pollen transferred directly from the male flower to the female, through the air, or (in *Thalassia* and *Halophila*) through the water. Gynoecium of (2)3–6 (up to 15–20) united carpels; stylodia often bilobed or bifid, sometimes shortly connate at the base; stigmas papillate; ovary inferior, unilocular, often with more or less deeply intruded, free carpellary margins; placentation laminar-diffuse (ovules scattered over the surfaces of the partial partitions) or parietal when the partitions are not intruded. Nectaries, when present, usually three, rarely more, borne on the bases of the styles (probably androecial in origin – Cook 1998). Ovules numerous to few, anatropous or orthotropous, bitegmic, crassinucellate, with the parietal cell usually cut off from the primary archesporial cell. Endosperm helobial. Fruits submerged, fleshy, berrylike, but generally dehiscent, splitting up irregularly or sometimes (as in some species of *Ottelia*) regularly. Seeds few to numerous, ellipsoidal to cylindrical or fusiform, smooth or with highly complex testa; seed coat formed by the outer integument; endosperm absent or rarely (*Ottelia*) scanty; embryo usually more or less straight, relatively large, with a massive hypocotyl (often with unicellular hairs) and well developed radicle and in some genera has several leaves already developed before germination the lower periclinal wall of the inner or less often upper layer of the tegmen more or less tuberculate. n = ? 6, 7–15. . . . . 4. HYDROCHARITACEAE.

3 Flowers without bracts. Perennial freshwater herbs with starch-rich, sympodial, tuberlike rhizomes or corms, rooted in the substrate. Articulated laticifers consisting of long cells containing tannins (proan-thocyanins) or oil are present in leaves and the inflorescence axis. Vessels wanting or only in roots, with scalariform perforations. Leaves all basal, usually distinctly petiolate, loosely sheathing at the base, with intravaginal squamules in the leaf axils; lamina linear to oblong-elliptic, sometimes fenestrate by loss of the mesophyll between the veins, submerged or floating, usually with a distinct midrib and one or more pairs of apically converging parallel or arcuate primary veins connected by numerous transverse secondary veinlets. Stomata only on the floating laminas, paracytic. Calcium oxalate crystals are abundant in the mesophyll of floating leaves. Inflorescence is a simple, bifurcate or sometimes up to ten times forked spike exerted from the water on a leafless scape and in bud enclosed by a caducous or sometimes persistent spathelike bract; rarely (*Aponogeton ranunculiflorus*) the inflorescence is condensed into a few-flowered pseudanthium. Flowers usually small, bisexual or sometimes unisexual and dioecious, actinomorphic or sometimes zygomorphic, 3-merous, without bracts. Perianth segments (1-)2(-6), absent in female flowers of dioecious species, free, usually petaloid and mostly persistent. Stamens free, mostly six in two cycles, but in *A. distachyos* 8–16 in several cycles; staminodia present; filaments elongate, anthers small, tetrasporangiate, extrorse. Microsporogenesis successive or simultaneous. Pollen grains 3-celled, more or less ellipsoid, 1-colpate, with long colpi; the tectum perreticulate with fine supratectal spines; the aperture membrane may bear small spinules or granules. Gynoecium of (2)3(-9) carpels; carpels nearly free or more or less basally and adaxially

connate, but separating at maturity, with septal nectaries; stylodia short, with a ventral stigmatic groove (stigmas decurrent); in each carpel 2–12 basal-submarginal ovules. Ovules anatropous, bitegmic or rarely (*A.distachyos*) unitegmic, crassinucellate; a parietal cell is cut off from the primary archesporial cell. Endosperm helobial. Fruits of nearly free follicles. Seeds smooth, ridged or winged, testal, with straight embryo and without endosperm. Testa protective or not, endotegmen tanniniferous, or undifferentiated and translucent. Leucoanthocyanins and flavonols present, bur C-glycosylflavons absent. n = 8, 12 +. . . . . 1. APONOGETONACEAE.

2 Vessels with simple perforations. Emergent aquatic perennial herbs with stout, creeping, monopodial, dorsiventral, aerenchymatous rhizome, rich in starch and provided with scattered tanniniferous cells containing proanthocyanins. Vessels only in the roots, with simple perforations. Leaves all radical and arise in two rows along the rhizome, nearly distichous or nearly distichous, linear, erect, and more or less triangular in transverse section, not differentiated into petiole and lamina, parallel-veined, sheathing at the base, and provided with numerous intravaginal squamules. Stomata mostly paracytic. Flowers in scapose, axillary, umbel-like complex of cymes subtended by (2)3(4) bracts, bisexual, actinomorphic, 3-merous. Perianth segments six in two cycles, free, the outer three a little smaller, greenish and sepal-like, the inner pink and petaloid. Stamens nine in two cycles, the outer consisting of three pairs of obliquely antesepalous members, the inner of three separate antepetalous members; filaments long, erect, flattened; anthers basifixed, tetra-sporangiate, latrorse. Microsporogenesis successive. Pollen grains 3-celled, 1-colpate, boat-shaped, tectate-columellate, reticulate. Gynoecium of 6(-9) nearly free conduplicate carpels with short stylodia ending in a shortly bilobed and shortly decurrent stigma; the margins of each carpel are never fused but are held together in the distal portion by interlocking hairs; the basal lateral sides of the carpels nectariferous (septal nectaries); each carpel with numerous ovules scattered over their inner surface, except on the midrib and edges. Ovules anatropous, bitegmic, nearly crassinucellate, with parietal cell cut off from the archesporial cell. Female gametophyte of *Polygonum*-type. Endosperm helobial. Fruits multifollicles. Seeds numerous, small, cylindrical, longitudinally ribbed; seed coat formed by both integuments, but mainly by testa; embryo straight, rich of starch; endosperm vestigial, of thin layer of thick-walled obliterated cells, n = 8, 10–13. . . . . 2. BUTOMACEAE.

1. APONOGETONACEAE

J. Agardh 1858. 1/50. Tropical and subtropical regions of the Old World, mostly in Africa and Madagascar.

Aponogeton.

In spite of their highly specialized habit, the Aponogetonaceae retain such plesiomorphic characters, as apocarpous gynoecium composed of primitive carpels with decurrent stigma and 1-colpate pollen grains.

2. BUTOMACEAE

Mirbel 1804. 1/1. Temperate regions of Eurasia and North Africa.

Butomus.

Butomus umbellatus, the only member of this monotypic order, is extremely heterobathmic. In regard to its apocarpous gynoecium, decurrent stigma, laminar-diffuse placentation, and 1-colpate pollen grains it is a rather archaic monocot. However, a rather specialized inflorescence, simple perforations of the vessels, and very reduced, vestigial endosperm in the mature seeds makes it a considerably advanced member of the monocots.

3. NAJADACEAE

A.L. de Jussieu 1789. 1/40–50. Subcosmopolitan.

Najas.

Closely related and could have originated from the submerged Hydrocharitaceae (Shaffer-Fehre 1991b; Haynes et al. 1998), especially to Hydrocharitoideae, but differ markedly in superior ovary and nuclear endosperm.

4. HYDROCHARITACEAE

A.L. de Jussieu 1789 (including Blyxaceae Nakai 1949; Elodeaceae Dumortier 1829; Enhalaceae Nakai 1943; Halophilaceae J. Agardh 1858; Hydrillaceae Prantl 1879; Stratiotaceae Link 1829; Thalassiaceae Nakai 1943; Vallisneriaceae Link 1829). 17/c.75. Widely distributed, but mainly in tropics and subtropics.

4.1 HYDROCHARITOIDEAE
Freshwater or rarely marine plants, pollinated at or above the surface of water. Pollen grains in monads. Perianth mostly double. – *Ottelia, Stratiotes, Hydrocharis, Limnobium, Blyxa, Apalanthe, Egeria, Elodea, Hydrilla, Appertiella, Lagarosiphon, Nechamandra, Maidenia, Vallisneria, Enhalus.*

4.2 THALASSIOIDEAE
Marine plants, pollinated beneath the surface of the water. Pollen grains cohere in moniliform chains; they form pollen tubes before reaching the stigma. Leaf-bearing branches arising from the rhizome at distances of several internodes. Leaves alternate, distichous, with longitudinal venation. Spathial leaves partly connate. Inflorescences pedunculate, uniflorous. Flowers dioecious, with three uncolored perianth segments. Stamens 3–13, all fertile; anthers latrorse. Gynoecium of 6–8 carpels; each stylodium divided into two filiform stigmas that are 2–6 times as long as the stylodium; ovary 1-locular or imperfectly 2–3-locular. Fruits with fleshy pericarp bursting into a number of irregular, stellately spreading valves. Seeds conical with a thickened basal portion. Embryo straight. Cotyledon not coiled. Tannin cells present. – *Thalassia.*

4.3 HALOPHILOIDEAE
Marine plants pollinated beneath the water or at the water surface. Pollen grains cohere in moniliform chains; they form pollen tube before reaching the stigma. Leaf-bearing branches arising from the rhizome at each internode. Leaves opposite, in pseudowhorls or distichous, sessile or petiolate, linear to ovate, pinnately veined. Spathial leaves free. Inflorescences sessile, uniflorous or rarely having one male flower and one female flower on one spathe. Flowers monoecious or dioecious. Male flowers have three tepals and three stamens; anthers extrorse. Female flowers with a vestigial perianth. Gynoecium of 3–5 carpels; stylodia linear, 3–5, undivided; ovary 1-locular. Fruits ovoid or globose, with membranous pericarp, dehiscent by decay of the pericarp. Embryo somewhat curved, cotyledon spirally coiled upon itself. Tannin cells absent. – *Halophila.*

A considerably advanced taxon, which is clearly closely related to the Butomaceae with which shares a common origin.

Bibliography

Ancibor E. 1979. Systematic anatomy of vegetative organs of the Hydrocharitaceae. Bot. J. Linn. Soc. 78: 237–266.

Argue CL. 1971. Pollen of the Butomaceae and Alismataceae: I. Development of the pollen wall in *Butomus umbellatus* L. Grana 11: 131–144.

Balfour IB. 1870. On the genus *Halophila*. Trans. Proc. Bot. Soc. Edinb. 13: 290–343.

Baude E. 1956. Die Embryoentwicklung von *Stratiotes aloides* L. Planta 46: 649–671.

Bercu R and M Fagaras. 2002. Anatomical features of the root, stem and leaf blade of *Potamogeton pectinatus* L. and *Vallisneria spiralis* L. Contrib. Bot. Univ. Babes Rolyal Gard. Bot. (Cluj Napoca) 37: 41–47.

Bolkhovskikh ZV. 1983. On the morphology of pollen grains of *Najas major* (Najadaceae). Bot. Zhurn. 68: 448–452 (in Russian with English summary).

Bouman F. 1985. Embryology. In: HWE van Bruggen, ed. Monograph of the genus *Aponogeton* (Aponogetonaceae). Bibl. Bot. 137: 4–9.

Bruggen HWE von, ed. 1985. Monograph of the genus *Aponogeton* (Aponogetonaceae). Bibl. Bot. 137: 1–76.

Bruggen HWE von. 1998. *Aponogetonaceae*. In: K Kubitzki, ed. The families and genera of vascular plants vol. 4, pp. 21–25. Springer, Berlin/Heidelberg/New York.

Brunaud A. 1976, 1977. Ramification chez les Hydrocharitaceae: I. Ontogenie du systeme des pousses. II. Organisation des rameaux lateraux. Rev. Gen. Bot. 83: 397–413, 1976; 84: 137–157, 1977.

Campbell DH. 1897. A morphological study of *Najas* and *Zannichellia*. Proc. Calif. Acad. Sci., 3d ser., 1: 1–61.

Caspary R. 1858. Die Hydrilleen. Jahrb. Wiss. Bot. 1: 377–513.

Chanda S, S Nusson, and S Blackmore. 1988. Phylogenetic trends in the Alismatales with reference to pollen grains. Grana 27: 257–272.

Charlton WA and A Ahmed. 1973. Studies in the Alismataceae: IV. Developmental morphology of *Ra-nalisma humile* and comparisons with two members of the Butomaceae, *Hydrocleis nymphoides* and *Butomus umbellatus*. Canad. J. Bot. 51: 899–910.

Cook CDK. 1982. Pollination mechanisms in the Hydrocharitaceae. In: JJ Symoens, SS Hooper, and F Compere, eds. Studies on aquatic vascular plants, pp. 1–15. Royal Botanical Society of Belgium, Brussels.

Cook CDK. 1998a. Butomaceae. In: K Kubitzki, ed. The families and genera of vascular plants, vol 4, pp. 100–102. Springer, Berlin/Heidelberg/New York.

Cook CDK. 1998b. Hydrocharitaceae. In: K Kubitzki, ed. The families and genera of vascular plants vol 4, pp. 234–248. Springer, Berlin/Heidelberg/New York.

Govindappa DA and TRB Najdu. 1956. The embryo sac and endosperm *Blyxa oryzetorum* Hook. f. J. Indian Bot. Soc. 35: 417–422.

Guo YH and SQ Huang. 1999. Evolution of pollination system and characters of stigmas in Najadales. Acta Phytotax Sinica 37(2): 131–136.

Haynes RR. 1977. The Najadaceae in the Southeastern United states. J. Arnold Arbor. 58: 161–170.

Haynes RR. 1979. Revision of north and central American *Najas* (Najadaceae). SIDA 8: 34–56.

Haynes RR and LB Holm-Nielsen. 2001. The genera of Hydrocharitaceae in the southeastern United States. Harvard Pap. Bot. 5: 201–275.

Haynes RR, LB Holm-Nielsen, and DH Les. 1998. Najadaceae. In: K Kubitzki, ed. The families and genera of vascular plants, vol 4, pp. 301–306. Springer, Berlin/Heidelberg/New York.

Islam AS. 1950. A contribution to the life history of *Ottelia alismoides* Pers. J. Indian Bot. Soc. 29: 79–91.

Kaul RB. 1965. Development and vasculature of the androecium in the Butomaceae. Am. J. Bot. 52: 624 (Abstract).

Kaul RB. 1968. Floral morphology and phylogeny in the Hydrocharitaceae. Phytomorphology 18: 13–35.

Kaul RB. 1969. Morphology and development of the flowers of *Boottia cordata, Ottelia alismoides,* and their synthetic hybrid (Hydrocharitaceae). Am. J. Bot. 58: 951–959.

Kaul RB. 1970. Evolution and adaptation of inflorescences in the Hydrocharitaceae. Am. J. Bot. 57: 708–715.

Kaul RB. 1976. Conduplicate and specialized carpels in the Alismatales. Am. J. Bot. 63: 175–182.

Kun S, QF Wang, JK Chen. 1997. Micromorphological characters of seed coats of Chinese Najadaceae and their systematic significance. Acta Phytotax. Sinica 35: 521–526.

Lakshmanan KK. 1961. Embryological studies in the Hydrocharitaceae: I. *Blyxa octandra* Planch. J. Madras Univ. 31B: 133–142.

Lakshmanan KK. 1963. Embryological studies in the Hydrocharitaceae: II. *Halophila ovata* Gaudich. J. Indian Bot. Soc. 42: 15–18.

Lakshmanan KK. 1965. Embryological studies in the Hydrocharitaceae: III. *Nechamandra alternifolia.* Phyton (Buenos Aires) 20: 49–58. IV. Post-fertilization development in the *Hydrilla verticillata* Royle. Phyton (Buenos Aires) 22: 13–14.

Les DH, DK Garvin, and CF Wimpee. 1993. Phylogenetic studies in the monocot subclass Alismatidae: evidence for a reappraisal of the aquatic order Najadales. Mol. Phylogenet. Evol. 2: 304–314.

Les DH, ML Moody, and SWL Jacobs. 2005. Phylogeny and systematics of *Aponogeton* (Aponogetonaceae): the Australian species. Syst. Bot. 30: 503–519.

Les DH, ML Moodly, and CL Soros. 2006. A reappraisal of phylogenetic relationships in the monocotyledon family Hydrocharitaceae (Alismatidae). In: JT Columbus, EA Friar, JM Porter, LM Prince, and MG Simpson, eds. Monocots: comparative biology and evolution (excluding Poales), pp. 211–230. Rancho Santa Ana Botanical Garden, Claremont (Aliso 22: 211–230).

Lowden RM. 1986. Taxonomy of the genus *Najas* L. (Najadaceae) in the neotropics. Aquat. Bot. 24: 147–187.

Magnus P. 1894. Uber die Gattung *Najas*. Ber. Deutsch. Bot. Ges. 12: 214–224.

Miki S. 1937. The origin of *Najas* and *Potamogeton*. Bot. Mag. (Tokyo) 51: 472–480.

Naidoo Y, JR Lawton, AD Barnabas, and J Goetzee. 1990. Ultrastructure and cytochemistry of squamulae intravaginales of the marine angiosperm, *Halophila ovalis*. South Afr. Tydskr. Plantk. 56: 546–553.

Pettitt JM. 1980. Reproduction in sea grasses: nature of the pollen and receptive surface of the stigma in the Hydrocharitaceae. Ann. Bot. 45: 257–271.

Pettitt JM. 1981. Reproduction in sea grasses: pollen development in *Thalassia hemprichii, Halophila stipulacea,* and *Thalassodendron ciliatum*. Ann. Bot. 48: 609–622.

Posluszny U and R Sattler. 1976. Floral development of *Najas flexilis*. Canad. J. Bot. 54: 1140–1151.

Rangasamy K. 1941. A morphological study of the flower of *Blyxa echinosperma* Hook. f. J. Indian Bot. Soc. 20: 123–133.

Rao YS. 1953. Karyosystematic studies of Helobiales: I. Butomaceae. Proc. Natl. Inst. Sci. India 19: 563–581.

Rendle AE. 1899. A systematic revision of the genus *Najas*. Trans. Linn. Soc., Bot., ser. 2, 5: 379–444.

Roper RB. 1952. The embryo sac of *Butomus umbellatus* L. Phytomorphology 2: 61–74.

Sane YK. 1939. A contribution to the embryology of the *Aponogetonaceae*. J. Indian Bot. Soc. 18: 79–91.

Sattler R and V Singh. 1978. Floral development of *Hydrocleis nymphoides*. Canad. J. Bot. 51: 2455–2458.

Scribailo RW and U Posluszny. 1985. Floral development of *Hydrocharis morsus-ranae* (Hydrocharitaceae). Am. J. Bot. 72: 1678–1589.

Shaffer-Fehre M. 1991a. The endotegmen tuberculae: An account of little-known structures from the seed coat of the Hydrocharitoideae (Hydrocharitaceae) and *Najas* (Najadaceae). Bot. J. Linn. Soc. 107: 169–188.

Shaffer-Fehre M. 1991b. The position of *Najas* within the subclass Alismatidae (Monocotyledones) in the light of new evidence from seed coat structures in the Hydrocharitoideae (Hydrocharitales). Bot. J. Linn. Soc. 107: 189–209.

Singh V. 1965. Morphological and anatomical studies in Helobiae: III. Vascular anatomy of the node and flower of Najadaceae. Proc. Indian Acad. Sci. 61B: 98–108.

Singh V. 1966. Morphological and anatomical studies in Helobiae: VII. Vascular anatomy of the flower of *Butomus umbellatus* Linn. Proc. Indian Acad. Sci. 63B: 313–320.

Singh V and R Sattler. 1974. Floral development of *Butomus umbellatus*. Canad. J. Bot. 52: 223–230.

Singh V and R Sattler. 1977. Floral development of *Aponogeton natans* and *A. undulatus*. Canad. J. Bot. 55: 1106–1120.

Soros CL and DH Les. 2002. Phylogenetic relationships in the Alismataceae. In Botany 2002: Botany in the Curriculum. Abstracts, p. 152. Madison, WI.

Stant MY. 1967. Anatomy of the Butomaceae. Bot. J. Linn. Soc. 60: 31–60.

Sun K, Q-F Wang, and J-K Chen. 1997. Micromorphological characters of seed coats of Chinese Najadaceae and their systematic significance. Acta Phytotax. Sinica 35: 521–526 (in Chinese with English summary).

Sun K, J-K Chen, and Z-Y Zhang. 2001. Pollen morphology of Najadaceae and Zannichelliaceae. Acta Phytotax. Sinica 39: 31–37 (in Chinese with English summary).

Sun K, J-K Chen, and Z-Y Zhang. 2002. Studies on pollen morphology of Aponogetonaceae. Bull. Bot. Res. (China) 22: 33–36.

Swamy BGL and KK Lakshmanan. 1962. Contributions to the embryology of the Najadaceae. J. Indian Bot. Soc. 41: 247–267.

Tanaka N, H Setoguchi, and J Murata. 1997. Phylogeny of the family Hydrocharitaceae inferred from *rbc*L and *mat*K gene sequence data. J. Plant Res. 110: 329–337.

Tanaka N, K Uehara, and J Murata. 2004. Correlation between pollen morphology and pollination mechanisms in the Hydrocharitaceae. J. Plant Res. 117: 265–276.

Terekhin EC. 1985. Hydrocharitaceae. In: A Takhtajan, ed. Comparative seed anatomy vol 1, pp. 38–43. Nauka, Leningrad (in Russian).

Thanikaimoni G. 1985. Palynology and phylogeny. In: HWE van Bruggen, ed. Monograph of the genus *Aponogeton* (Aponogetonaceae). Bibl. Bot. 137: 11–14.

Tomlinson PB. 1969, 1972. On the morphology and anatomy of turtle grass, *Thalassia testudinum* (Hydrocharitaceae): II. Anatomy and development of the root in relation to function. IV. Leaf anatomy and development. Bull. Marine Sci. 19 (I): 57–71, 1969; 22(1): 75–93, 1972.

Troll W. 1931. Beitrage zur Morphologic des Gynaeceums: I. Über das Gynaeceum der Hydrocharitaceen. Planta 14: 1–18.

Vijayaraghavan MR and T Kapoor. 1985. Embryogenesis in *Najas marina* L.: Structural and histochemical approach. Aquatic Bot. 22: 45–60.

Order 3. ALISMATALES

Perennial or rarely annual emergent aquatic or marsh herbs with strongly aerenchymatous and commonly creeping rhizomes. Roots fibrous. Plants usually with schizogenous secretory canals that are lined with an epithelium. Vessels only in roots, with scalariform, or simple, or scalariform and simple perforations. Sieve-element plastids of P2c-type (Behnke 2002). Leaves alternate, more or less crowded near the tip of the rhizome or stolen, usually somewhat expanded and sheathing at the base, sometimes differentiated into lamina and petiole. Venation arcuate or parallel. Stomata paracytic or (some Alismataceae) tetracytic. Flowers in various kinds of inflorescences or solitary, bisexual or unisexual, usually actinomorphic, spirocyclic or cyclic, 3-merous, with double or rarely simple perianth, sometimes naked. Stamens 6–9 to many, rarely three; filaments free; anthers basifixed or versatile, extrorse or latrorse, tetrasporangiate, opening longitudinally. Tapetum plasmodial. Microsporogenesis successive. Pollen grains tectate-columellate, 3-celled, 2-pantoporate (up to 29 pores) or sometimes inaperturate, spinose. Gynoecium of three, six, nine, or more (up to 15–20) free or basally connate carpels; stylodia terminal or (as in many Alismataceae) gynobasic, with slightly decurrent or apical stigma; each carpel with more or less numerous ovules with laminar-diffuse placentation or in each carpel one or several basal or subbasal ovules. Ovules anatropous or amphitropous, bitegmic, weakly crassinucellate, sometimes with weakly developed endothelium, without a parietal cell cut off from the primary archesporial cell. Female gametophyte of *Allium*-type. Endosperm helobial or (in some Alismataceae) nuclear. Fruits multi-follicles or achenes. Seeds horshoe-shaped, testal, without endosperm or only with vestigial layer (Alismataceae).

Close to the Hydrocharitales. In Engler's system the Limnocharitoideae are even included in the Butomaceae s.l. However, they differ from the Butomaceae s. str. in anatomy (secretory canals), pollen morphology, embryology, and seed shape, "whereas in all these features the Limnocharitaceae approach very closely the Alismataceae" (Dahlgren et al. 1985: 301).

Key to Families

1 Each carpel with numerous ovules scattered over its inner surface. Perennial aquatic herbs rooted in the substrate or free-floating; rhizome with endodermis. Vessels with more or less oblique, scalariform and simple perforations. Leaves spiral to distichous (spirodistichous), petiolate, the laminas orbicular to lanceolate, the base cordate to attenuate; axillary scales present. Stomata paracytic. Flowers terminal, solitary or aggregated in pseudoumbels, bracteate, actinomorphic, bisexual. Periant segments six, free, in two cycles. Stamens 3, or 6, or 7–100; anthers extrorse or latrorse. Carpels 3, or 5–9, or 12–20, conduplicate and distally unsealed, each with a short, terminal stylodium ending in a shortly decurrent stigma or the stigma sessile. Ovary superior, ovules 12–100 per carpel, anatropous to campylotropous, pseudocrassinucellate. Lateral surface of the carpels nectariferous toward the base. Fruits follicular, aggregated in a head. Seeds with curved, horseshoe shaped or bent; without endosperm. Flavone and phenolic sulphates and tannins present. n = 7, 8, 10, chromosomes large. .... 1. LIMNOCHARITACEAE.

1 Each carpel with solitary or rarely several ventral-basal ovules. Perennial or rarely annual, rhizomatous herbs, generally rooted in the substrate. Vessels only in roots, mostly with simple perforations. Leaves submerged and emergent, alternate, distichous, spirodistichuous or spirally arranged, sessile or petiolate; the laminas linear, lanceolate, ovate to rhomboid, the base attenuate, truncate, cordate, sagittate, or hastate; axillary scales present. Stomata paracytic or rarely tetracytic. Flowers bisexual or unisexual, in terminal inflorescences, bracteate. Perianth segments six, free, in two cycles, white, red or pink. Stamens 3–6 in one cycle, or 18–100 (to many) in many cycles; anthers extrorse. Gynoecium of three carpels or of 6–100 (or more); carpels closed, with a terminal, lateral, or gynobasic stylodium and often with slightly decurrent stigma. Ovules 1–2, rarely more per carpel, ascending, anatropous or amphitropous, weakly crassinucellate. Nectaries generally borne at the basal margins of the carpels. Fruits achenes or rarely, as in *Damasonium,* basally dehiscent follicles. Seeds with starch; embryo curved (horsesho-shaped), achlorophyllous, endosperm lacking; testa without phytomela, usually membranous, sometimes thinly leather, then brown. Flavone and phenolic sulphates and tannins present; Alkaloids and proanthocyanidins (cyanidin), flavonols (kaempferol and quercetin) present or absent; n = (5-)7–11(-13). .... 2. ALISMATACEAE.

1. LIMNOCHARITACEAE

Takhtajan ex Cronquist 1981. 3/12. Tropical and subtropical regions of the Old and New World.

Limnocharis, Hydrocleys, Butomopsis.

The most archaic member of the family is *Limnocharis* (Takhtajan 1966; Haynes and Holm-Nielsen 1992), which is also the least specialized genus in the order Alismatales.

2. ALISMATACEAE

Ventenat 1799 (including Damasoniaceae Nakai 1943). 12/80–90. Subcosmopolitans.

Damasonium (including *Machaerocarpus*), *Baldellia, Alisma, Luronium, Ranalisma, Echinodorus* (? including *Helianthium*), *Caldesia, Limnophyton, Astonia, Sagittaria, Wiesneria, Burnatia.*

Very closely related to the Limnocharitaceae. The genus *Ranalisma* forms a connecting link between the Alismataceae and Limnocharitaceae (Posluszny and Charlton 1993).

Bibliography

Argue CL. 1973. The pollen of *Limnocharis flava* Buch., *Hydrocleis nymphoides* (Willd.) Buch., and *Tenogacharis latifolia* (Don) Buch. (Limnocharitaceae). Grana 13: 108–113.

Argue CL. 1976. Pollen studies in the Alismataceae with special reference to taxonomy. Pollen et Spores 18: 161–201.

Chanda S, S Nilsson, and S Blackmore. 1988. Phylogenetic trends in the Alismatales with reference to pollen grains. Grana 27: 257–272.

Charlton WA. 1968, 1973, 1991. Studies in the Alismataceae: I. Developmental morphology of *Echinodorus tenellus.* II. Inflorescences of Alismataceae. IX. Development of the flower of *Ranalisma humile.* Canad. J. Bot. 46: 1345–1360, 1968; 51: 775–789, 1973; 69: 2790–2796, 1991.

Charlton WA. 2004. Studies in the Alismataceae. XII. Floral organogenesis in *Damasonium alisma* and *Baldellia ranunculoides,* and comparisons with *Butomus umbellatus.* Canad. J. Bot. 82: 528–539.

Charlton WA and A Ahmed. 1973. Studies in the Alismataceae: III. Floral anatomy of *Ranalisma humile.* IV. Developmental morphology of *Ranalisma humile* and comparison with two members of the Butomaceae, *Hydrocleis nymphoides* and *Butomus umbellatus.* Canad. J. Bot. 51: 891–897, 899–910.

Chen J-M, D Chen, GW Robert, Q-F Wang, and Y-H Guo. 2004a. Evolution of apocarpy in Alismatidae using phylogenetic evidence from chloroplast *rbc*L sequence data. Bot. Bull. Acad. Sinica 45: 33–40.

Chen J-M, GW Robert, and Q-F Wang. 2004b. Evolution of aquatic life forms in Alismatidae: phylogenetic estimation from chloroplast *rbc*L sequence data. Israel J. Plant Sci. 52: 323–329.

Daumann E. 1964. Zur Morphologic der Blüte von *Alisma plantago-aquatica* L. Preslia 36: 226–239.

Forni Martins ER and KP Calligaris. 2002. Chromosomal studies on neotropical Limnocharitaceae (*Alismatales*). Aquatic Bot. 74(1): 33–41.

Harley MM. 1982. Palynological evidence of a close association between *Butomopsis* Kunth and *Hydrocleys* L. C. Rich. (Limnocharitaceae). In: JJ Symoens, SS Hooper, and P Compere, eds. Studies on aquatic vascular plants, pp. 61–65. Botanical Society of Belgium, Brussels.

Haynes RR and LB Holm-Nielsen. 1992. Limnocharitaceae. Flora Neotropica 56: 1–34.

Haynes RR and LB Holm-Nielsen. 1994. Alismataceae. Flora Neotropica 64: 1–112.

Haynes RR, DH Les, and LB Holm-Nielsen. 1998a. Alismataceae. In: K Kubitzki, ed. The families and genera of vascular plants, vol 4, pp 11–18. Springer, Berlin/Heidelberg/New York.

Haynes RR, DH Les, and LB Holm-Nielsen. 1998b. Limnocharitaceae. In: K. Kubitzki, ed. The families and genera of vascular plants vol 4, pp 271–275. Springer, Berlin/Heidelberg/New York.

Hooper SS and JJ Symoens. 1982. Observations on the family Limnocharitaceae Takhtajan ex Hooper and Symoens. In: JJ Symoens, SS Hooper, and P Compere, eds. Studies on aquatic vascular plants, pp. 56–60. Royal Botanical Society of Belgium, Brussels.

Johri BM. 1936. The life-history of *Butomopsis lanceolata* Kunth. Proc. Indian Acad. Sci. 4B: 139–162.

Johri BM. 1938a. The embryo sac of *Hydrocleis nymphoides* Buchen. Beih. Bot. Centralbl. 58A: 165–172.

Johri BM. 1938b. The embryo sac of *Limnocharis emarginata* L. New Phytol. 37: 279–285.

Kak AM and S Durani. 1989. Seed morphology of the family Alismataceae. J. Econ. Taxon, Bot. 13: 501–509.

Kaul RB. 1967a. Development and vasculature of the flowers of *Lophotocarpus calycinus* and *Sagittaria latifolia* (Alismataceae). Am. J. Bot. 54: 914–920.

Kaul RB. 1967b. Ontogeny and anatomy of the flower of *Limmnocharis flava* (Butomaceae). Am. J. Bot. 54: 1223–1230.

Kaul RB. 1968. Floral development and vasculature in *Hydrocleis nymphoides* (Butomaceae). Am. J. Bot. 55: 236–242.

Kaul RB. 1976. Conduplicate and specialized carpels in the Alismatales. Am. J. Bot. 63: 175–182.

Kudryashov LV and EI Savich. 1963. Some data on the embryology of *Alisma plantago-aquatica* L. Bull. Moscow Soc. Naturalists Div. Biol. 68(4): 50–63 (in Russian).

Leins P and P Stadler. 1973. Entwicklungsgeschichtliche Untersuchungen am Androeceum der Alismatales. Oesterr. Bot. Z. 121: 51–63.

Liu K-M, L-G Lei, and G-W Hu. 2002. Developmental study on the inflorescence and flower of *Caldesia grandis* Samuel (Alismataceae). Bot. J. Linn. Soc. 140: 39–47.

Markgraf F. 1936. Blütenbau und Verwandtschaft bei den einfachsten Helobiae. Ber. Deutsch. Bot. Ges. 54: 191–229.

Mayr F. 1943. Beiträge zur Anatomic der Alismataceen: Die Blattanatomie von *Caldesia parnassifolia* (Bassi) Parl. Beih. Bot. Centralbl. 62: 61–77.

Meyer FJ 1932. Beiträge zur Anatomic der Alismataceen. Beih. Bot. Centralbl. 49(I): 54–63, 272–91, 309–68; 50 (I): 54–63; 52B: 96–111; 54A: 156–69.

Pichon M. 1946. Sur les Alismatacees et les Butomacees. Notul. Syst. (Paris) 12: 170–183.

Rogers GK. 1983. The genera of Alismataceae in the southeastern United States. J. Arnold Arbor. 64: 383–420.

Sattler R and V Singh. 1973. Floral development of *Hydrocleis nymphoides*. Canad. J. Bot. 51: 2455–2458.

Sattler R and V Singh. 1977. Floral organogenesis of *Limnocharis flava*. Canad. J. Bot. 55: 1076–1086.

Sattler R and V Singh. 1978. Floral organogenesis of *Echinodorus amazonicus* Rataj and floral construction of the Alismatales. Bot. J. Linn. Soc. 77: 141–156.

Singh V. 1966. Morphological and anatomical studies in Helobiae: VI. Vascular anatomy of the flower of Alismaceae. Proc. Natl. Acad. Sci. India B 36: 329–344.

Singh V and R Sattler. 1972. Floral development of *Alisma triviak*. Canad. J. Bot. 50: 619–627.

Singh V and R Sattler, 1973. Nonspiral androecium and gynoecium of *Sagittaria latifolia*. Canad. J. Bot. 51: 1093–1095.

Singh V and R Sattler. 1977. Development of the inflorescence and flower of *Sagittaria cuneata*. Canad. J. Bot. 55: 1087–1105.

Stant MY. 1964. Anatomy of the Alismataceae. Bot. J. Linn. Soc. 59: 1–42.

Troll W. 1932. Beiträge zur Morphologic des Gynaeceums: II. Über das Gynaeceum von *Limnocharis* Humb. and Bonpl. Planta 17: 453–460.

Wilder GJ. 1974. Symmetry and development of *Butomus umbellatus* (Butomaceae) and *Limnocharis flava* (Limnocharitaceae). Am. J. Bot. 61: 379–394.

Wodehouse RP. 1936. Pollen grains in the identification and classification of plants: VIII. The Alismataceae. Am. J. Bot. 23: 535–539.

Order 4. POTAMOGETONALES

Perennial or rarely annual plants growing in marshes or other very wet places, or aquatic or sub-aquatic and mostly freshwater herbs with rhizome rooted in substrate. Roots fibrous. Vessels only in roots and rarely (Scheuchzeriaceae) in rhizomes, with scalariform perforations, or less often vessels absent. Sieve-element plastids of P2c-type (Potamogetonaceae – Behnke 2002). Leaves basal and cauline, alternate to opposite or rarely verticillate, mostly linear or capillary, with well-developed, open sheath, mostly ligulate, with intravaginal squamules, submerged or the upper ones floating and with expanded, lanceolate to elliptic, ovate or orbicular lamina and a slender petiole; veins when more than one parallel or arcuate, often connected by cross-veins. Stomata present or absent. Tannin cells present or absent. Flowers small to minute, in axillary or terminal raceme-like inflorescences, bisexual or unisexual, actinomorphic or less often slightly zygomorphic, mostly 6–4-merous. Perianth segments in one or two cycles or wanting, when free, or adnate to the filaments. Stamens 6, 4 or rarely 2–3, or 1. Anthers nearly sessile, basifixed, mostly tetrasporangiate, extrorse, opening longitudinally. Tapetum plasmodial. Microsporogenesis successive. Pollen grains 2-celled or more often 3-celled, tectate-columellate, inaperturate, usually reticulate. Gynoecium of (1–3)4–6(-16) free or only partly united carpels; stigmas sessile or on short stylodia, sometimes slightly decurrent ventrally. Ovary superior, ovules 1–2, pendulous from the ventral margin of the carpel, anatropous or orthotropous, or ortho-campylotropous, but after fertilization becoming amphitropous, bitegmic, crassinucellate, with a parietal cell cut off from the archesporial cell. Female gametophyte of *Polygonum*-type, or rarely (Zannichelliaceae) of *Allium*-type. Endosperm helobial or nuclear. Fruits indehiscent or rarely dehiscent, follicles, achenes, drupe-like or achene-like, nutlets. Seeds without endosperm or endosperm scanty; embryo straight or slightly curved, sometimes with large hypocotyl. Seeds exotestal, or coat crushed. Flavone aglycones, flavone glycosides, and C-glycosylated flavones are common; n = 6, 8–11.

Related to the Alismatales and have a common origin from the alismatalean ancestor.

Key to Families

1 Vessels with scalariform perforations.

2 Stamens mostly six in two cycles.

3 Fruits follicles. Perennial marsh herbs with sympodial rhizome covered with persistent fibrous remains of older leaves and leafy stem. Adventitious roots inserted below the nodes. Vessels both in roots and in rhizomes. Leaves basal and cauline, alternate, distichous, linear, compressed-terete, parallel-veined, with open sheathing base, the apex obtuse with a conspicuous apical pore; sheath marginally hyaline, with two delicate auricles, at the base with a dense axillary row of numerous long hairs that replace intravaginal squamules, with well-developed air lacunae in the mesophyll. Stomata tetracytic. Tannin cells present, and some calcium oxalate crystals occur in the mesophyll. Flowers in terminal raceme-like inflorescence in the axils of more or less bract-like leaves, bisexual, actinomorphic. Perianth segments six in two cycles, all similar, relatively small, yellow-green. Stamens free; the filaments short, anthers linear, basifixed, with a short connective tip. Pollen grains 3-celled, in dyads, globose, inaperturate, reticulate, the reticulum is continuous between the two grains. Gynoecium of three or rarely six carpels shortly connate at the base; carpel margins not completely fused ("The carpel walls retain an open ventral suture throughout the early and mid stages of development and even in late developmental stages the suture does not fully fuse" – Posluszny 1983); stigmas sessile, decurrent on the upper part papillate. Each carpel with two (rarely one) basal-axile, anatropous ovules. Endosperm helobial. Fruits of divaricate follicles. Seeds oblong, testal, with large straight embryo and scanty vestigial endosperm consisting of several layers of strongly compressed cells. Testa smooth, thick, derived mainly from the outer integument, the outer epidermis of the outer integument transforms into exotesta. The plants are rich in the cyanogenic glucoside triglochinin. n = 11. . . . . 1. SCHEUCHZERIACEAE.

3 Fruits achenes. Perennial or rarely annual plants growing in marshes or other very wet, often saline, places, sometimes aquatic or subaquatic. Perennial species have short or elongated rhizome or stolone. Roots fibrous or sometimes distally tuberous. Vessels only in roots. Leaves mostly or wholly basal, alternate, linear, with open sheathing base; sheath adnate to the lamina, with two auricles; intravaginal squamules membranous. Stomata mostly paracytic. Crystals of calcium oxalate few and tannin cells lacking. Flowers small and inconspicuous, in terminal, mostly bractless spikes or spikelike racemes, bisexual or unisexual (in *Tetroncium* dioecious), actinomorphic or less often slightly zygomorphic, 3-merous, 2-merous, or 1-merous, generally ebracteate, anemophilous. Perianth mostly of six free and more or less similar segments in two cycles (*Triglochin*) *or* of four, three, two, or even one segments in one or two cycles, rarely wanting (unisexual flowers of *Lilaea*). Stamens mostly six in two cycles, rarely three, four or eight, or (in *Lilaea*) even one; filaments very short. Pollen grains 2-celled or 3-celled, in monads, globose, inaperturate, tectate-columellate, reticulate. Gynoecium of 3, 4–10, or commonly six carpels, nearly apocarpous (*Cyanogeton*) to weakly connate (*Tetroncium, Maundia*), but becoming separate in the fruit stage; in (*Lilaea*) the gynoecium is either of only of one carpel (Cronquist 1981) or pseudomonomerous, initially consisted of three carpels (Singh 1965), which is probably supported by its three dorsal vascular bundles; stigma mostly sessile or on short or rarely very long stylodium; in each carpel or in each ovary locule one or two, basal or less often (*Maundia*) subapically pendulous, ovules. Ovules anatro-

pous or less often (*Maundia*) orthotropous. with a parietal cell cut off from the archesporial cell. Endosperm nuclear. Fruits of achenes, in *Lilaea* of a single achene. Seeds with straight or nearly straight embryo and vestigial endosperm consisting of thin, inconspicuous membrane. Cyanogenic, containing the cyanogenic glucoside triglochinin. n = 6, 8, 9, 11. . . . . 2. JUNCAGINACEAE.

2 Stamens four. Perennial or rarely annual aquatic and mostly freshwater herbs with creeping rhizome rooted in substrate. Roots fibrous. Vessels only in roots. Leaves alternate to opposite or rarely verticillate, mostly linear or capillary, with well-developed, open sheath, mostly ligulate, with intravaginal squamules, submerged or (in some species of *Potamogeton*) the upper ones floating and with expanded, lanceolate to elliptic, ovate or orbicular lamina and a slender petiole; veins when more than one parallel or arcuate, often connected by cross-veins; leaves stipulate (commonly, the stipular appendages adhering to the sheath or not), or estipulate. Axillary scales present. The floating and sometimes also the upper submerged leaves with paracytic stomata. Silica bodies and calcium oxalate crystals are lacking. Flowers small to minute, in axillary or terminal, bractless spikes that are more or less elevated above the water, bisexual, actinomorphic, mostly 4-merous. Perianth segments four, in one cycle, free, bractlike, short-clawed, adnate to the filaments or connectives of the opposite stamens. Stamens four or rarely 2–3; anthers nearly sessile. Pollen grains 3-celled, ellipsoid to spheroid, semitectate, nearly inaperturate, with an inconspicuous vestigial colpus under which the intine is somewhat thickened. Gynoecium of (1–3)4(5–16) free or only partly united carpels; carpels sometimes shortly stipitate; stigmas sessile or on short stylodia, slightly decurrent ventrally. Ovary superior, ovule solitary, pendulous from the ventral margin of the carpel, orthotropous or ortho-campylotropous, but after fertilization becoming amphitropous, bitegmic, crassinucellate, with a parietal cell cut off from the archesporial cell. Endosperm helobial. Fruits drupe-like or rarely (*Groenlandia*) achene-like. Seeds without endosperm; embryo slightly curved, with large hypocotyl. Seeds exotestal, or coat crushed. Present flavonoids (apigenin, luteolin), flavone glycosides, and C-glycosylated. The basic chromosome number (x) is 7 (Haynes et al. 1998), n ranging from 7 to 52. . 3. POTAMOGETONACEAE.

1 Plants without vessels.

4 Flowers bisexual or rarely male by abortion of gynoecium. Stamens three. Submerged marine perennial herbs with thick, laterally flattened, creeping, monopodially branching rhizome covered with fibers of old leaf sheaths and erect lateral stems. All vegetative parts with scattered tanniniferous cells containing proanthocyanins. Xylem poorly developed and without vessels. Leaves alternate, distichous on the rhizome, linear, parallel-veined or with only a mid-vein, basally with open, biauriculate, ligulate sheath that is persistent after the lamina has been shed. Intravaginal squamules present at the nodes. Stomata absent. Flowers in long-pedunculate modified cymose inflorescences subtended by 2–4 reduced leaves with broadened sheaths; according to Hartog (1970: 123), the inflorescence "has to be regarded as an anthela of rhipidia, of which the rhipidia distally become reduced to a 'spike' of flowers. Such a 'spike' has lost its terminal flowers as well as its bracts; its cymose characters is still apparent from the fact that the axis is not straight but proceeds more or less zigzag." Individual flowers small, sessile, bractless, bisexual or rarely male by abortion of gynoecium, hydrophilous. Perianth absent; anthers large, sessile, with a much expanded, shield-like connective produced beyond the loculi, the latter widely separated. Pollen grains 3-celled, spheroidal when released but soon become filamentous (pollen tubes), inaperturate, without exine. Gynoecium of one carpel with sessile, obliquely placed, irregularly lobed, discoid stigma, and one orthotropous ovule pendulous from the carpellary margin. Fruits slightly fleshy, with spongy pericarp, freely floating when shed. Seeds with very thin, membranous seed coat and without endosperm embryo straight, large consisting for the greater part of a large, fleshy hypocotyl; hypocotyl with numerous tannin cells. n = 10, dimorphic. 4. POSIDONIACEAE.

4 Flowers unisexual.

5 Stamens two.

6 Fruits long-stipitate or sessile drupes. Annual or rarely perennial submersed

glabrous herbs growing in fresh or brackish waters. Roots fibrous, few, non-septate, from lower nodes of stems. Stems slender, branched or unbranched, often dimorphic, the lower stems rhizomatous, the upper erect, leafy, the tips not modified into turions. Leaves alternate to subopposite, linear, divided into blade and stipular sheath encircling the stem; infravaginal scales present. Flowers in 1-few-flowered capitate axillary or terminal spike, the first enclosed by two supopposite foliage leaves, bisexual, actinomorphic, without perianth segments. Stamens two, sessile with an expanded connective; anthers with two bisporangiate thecae dehiscing by longitudinal slits, connective broad. Pollen grains elongate-arcuate, inaperturate, intectate at both ends, with distinctly separated columellae. Carpels (2)4(-16), free; stigma sessile, peltate, lobed; ovary 1-locular, ovule solitary, more or less pendulous. Endosperm helobial. Seed coat consists of a thin, 2-layered testa and a tegmen of which only the outer epidermis persists; endosperm wanting, embryo with enlarged hypocotyls. Present leucoanthocyanins, flavonols, and G-glycosyl flavones, n = 8, 10–12, 20, dimorphic.. . . . . .5. RUPPIACEAE.

6 Fruits nutlets with stony pericarp or viviparous. Marine perennial herbs with creeping monopodially or sympodially branched rhizomes. Roots often branched, with few to many root hairs. Tannin cells present. Leaves alternate, distichously arranged, sheathing base and a ligule between sheath and blade, linear, flat or terete, with three to several veins. Flowers usually solitary and terminal on a short branch or (*Syringodium*) in small cymes. No vestige of perianth except for a transient ridge that forms below the stamens in the male flower of *Syringodium filiforme* (Tomlinson and Posluszny 1978). Male flowers consist of two united stamens; anthers on a common filament or virtually sessile, fused back to back, tetrasporangiate; connective with an apical prolongation (except in *Syringodium*). Pollen grains without exine, become filamentous (up to about 1 mm long). Gynoecium of two free carpels, each with long, slender, and simple (*Halodule*) or short and bibrachiate or tribrachiate stylodium. Endosperm nuclear. Fruits indehiscent, either with stony pericarp (*Halodule, Cymodocea, Syringodium*) or viviparous (*Amphibolis*). Testa absent. Embryo consists either for the larger part of the plumule with a lateral radicle and a cylindrical hypocotyl or of a long hypocotyl and a short plumule without radicle. Sulphated phenolic compounds are present in all genera; n = 7, 8, 10, 14, 15. . . . . 7. CYMODOCEACEAE.

5 Stamen solitary.

7 Ovule anatropous. Plants of fresh or alkaline or brackish water. Not saponiferous. Leaves 1-veined or incompletely 3-veined. Flowers in small, axillary, usually complex, cymose inflorescences. *Althenia* and *Lepilaena* have 3-membered perianth in flowers of both sexes, while *Zannichellia* have tiny, spathelike structure that may represent a perianth only in the female flower. Male flowers with a single filament with 4, 8, or 12 microsporangia, and probably in the last two cases represent two or three fully connate stamens (Cronquist 1981; Tomlinson 1982; Dahlgren et al. 1985). Pollen grains inaperturate, subspheroidal or subprolate, minutely spinose, without exine or sometimes with thin exine and 1-colpate. Female flowers with small, cupule-like or 3-lobed perianth. Gynoecium of (1-)3–4(-9) free carpels, each with short or long stylodium and symmetrically or asymmetrically funnel-form stigma, which is fimbriate in *Lepilaena*. Ovule solitary, ventral-pendulous. Endosperm helobial. Fruits drupaceous, with membranous exocarp, fleshy mesocarp and stony endocarp, but in *Althenia* they are dehiscent. Seeds with spirally coiled cotyledon, swollen hypocotyl, and very small radicle. Flavone sulphates present. n = 6–8. . . . . 6. ZANNICHELLIACEAE.

7 Ovule orthotropous. Submerged perennial, monoecious or dioecious, marine herbs with creeping or sometimes tuberous-thickened,

monopodial, or sympodial rhizomes and slender, often flattened stems. Tannin cells absent. Vessels and stomata wanting. Leaves alternate, usually decussate, linear or filiform, parallel-veined, sheathing at the base, sheaths open or closed, commonly with stipulelike margins, ligulate. Inflorescence terminal or lateral, sympodial, erect, consisting of a panicle of rhipidia, but often reduced to a single rhipidium; each rhipidium consisting of 2–5 spathes, but sometimes reduced to single one; peduncle of each spathe partially coalescent with the axis from which it springs or completely free; spathe consisting of a sheath and a leaf blade; spathial sheath ligulate, enclosing a sessile or stalked spadix on the dorsal side of which the flowers are arranged (Hartog 1970). Individual flowers small, sessile, bractless, unisexual (monoecious or dioecious), without perianth. Stamen solitary; anther sessile, extrorse, tetrasporangiate (but an alternative explanation is that a flower with two widely separated unithecate (bisporangiate) anthers – Tomlinson and Posluszny 2001), opening longitudinally. Pollen grains 2-celled or 3-celled, inaperturate, filamentous, up to about 2 mm long, with very thin simple exine. Gynoecium of one carpel with a short, bibrachiate stylodium and one apical, pendulous ovule. Ovule orthotropous; a parietal cell is not cut off, but there are periclinal divisions in the nucellar epidermis. Endosperm nuclear. Fruits small, with scarious, dehiscent pericarp or less often (*Phyllospadix*) with soft exocarp and a hard fibrous endocarp. Testa derived from the outer integument. Tegmen degenerates. Seeds without endosperm, with slightly curved embryo consisting for the most part of the hypocotyl and lacking a radicle; n = 6, 9, 10.. . . . . 8. ZOSTERACEAE.

1. SCHEUCHZERIACEAE

F. Rudolphi 1830. 1/1. In bogs of cold and temperate regions of the Northern Hemisphere.

Scheuchzeria.

2. JUNCAGINACEAE

Richard 1808 (including Heterostylaceae Hutchinson 1934; Lilaeaceae Dumortier 1829; Maundiaceae Nakai 1943) Triglochinaceae Chevalier 1827). 5/17. Widely distributed in cold and temperate regions of Northern and Southern Hemispheres; monotypic genus *Cycnogeton,* very closely related to *Triglochin,* is endemic to Australia and Tasmania, and another monotypic genus *Tetroncium* is restricted to the Holantarctic South America (Strait of Magellan); *Lilaea* in mountains of Pacific America from British Columbia to Chile and Argentina, and *Maundia* – Northeastern Australia, southern Queensland, northern New South Wales.

Triglochin, Cycnogeton, Tetroncium, Lilaea, Maundia.

Related to the Scheuchzeriaceae. With the Scheuchzeriaceae they share such a unique character as the presence of triglochinin, but markedly differ in many other respects, including nuclear endosperm and pollen grains.

3. POTAMOGETONACEAE

Reichenbach 1828 (including Hydrogetonaceae Link 1829). 2/80–100. Subcosmopolitan.

Potamogeton (including *Coleogeton*), *Groenlandia.*

Related to the Juncaginaceae.

4. POSIDONIACEAE

Hutchinson 1934. 1/9. Along the coasts of the Mediterranean, locally on the Atlantic coast of southwestern Europe, and along the coasts of southern, southwestern and southwestern Australia and Tasmania.

Posidonia.

Molecular data indicate a close related to Ruppiaceae and Cymodoceaceae.

5. RUPPIACEAE

Horaninow 1834. 1/7–8. Subcosmopolitan.

Ruppia.

Closely related to the Cymodoceaceae.

6. ZANNICHELLIACEAE

Chevalier 1827. 4/12. Subcosmopolitan.

Zannichellia, (including *Pseudalthenia*) *Althenia, Lepilaena, Vieisia.*

Recent molecular studies indicate that Zannichelliaceae are most closely related to Potamogetonaceae.

7. CYMODOCEACEAE

N. Taylor 1909. 5/16. Mainly tropical and subtropical seas with a few species in warm-temperate seas.

Cymodocea, Halodule, Syringodium, Amphibolis, Thalassodendron.

Related to the Zannichelliaceae, but much more specialized.

8. ZOSTERACEAE

Dumortier 1829. 3/14–20. Temperate seas of both hemispheres, with a few species extending into tropical seas.

Zostera, Heterozostera, Phyllospadix.

A very specialized group of marine alismatids that probably derived from the juncaginalean ancestor.

Bibliography

Aalto M. 1970. Potamogetonaceae fruits. I. Recent and subfossil endocarps of the Fennoscandian species. Acta Bot. Fenn. 88: 1–85.

Agrawal JS 1952. The embryology of *Lilaea subulata* H. B. K. with a discussion on its systematic position. Phytomorphology 2: 15–29.

Albergoni FG, B Basso, and G Tedesco. 1978. Considerations sur l'anatomie de *Posidonia oceanica* (Zosteraceae). Plant Syst. Evol. 130: 191–210.

Arber A. 1940. Studies in flower structure: VI. On the residual vascular tissue in the apices of reproductive shoots, with special reference to *Lilaea* and *Amherstia*. Ann. Bot. 2(4): 617–627.

Barnabas AD. 1982. Fine structure of the leaf epidermis of *Thalassodendron ciliatum* (Forsk.) den Hartog. Aquatic Bot. 12: 41–55.

Barnabas AD. 1983. Composition and fine structural features of longitudinal veins in leaves of *Thalassodendron ciliatum*. South Afr. J. Bot. 2: 317–325.

Barnabas AD. 1994. Anatomical, histochemical and ultrastructural features of the seagrass *Phyllospadix scouleri* Hook. Aquatic Bot. 49: 167–182.

Barnabas AD and HJ Arnott. 1987. *Zostera capensis* Setchell: root structure in relation to function. Aquatic Bot. 27: 309–322.

Barnabas AD and S Kasavan. 1983. Structural features of the leaf epidermis of *Halodule uninervis*. South Afr. J. Bot. 2: 311–316.

Black JM. 1913. The flowering and fruiting of *Pectinella antarctica* (*Cymodocea antarctica*). Trans. Proc. Roy. Soc. South Australia 37: 1–5.

Bowes G, SK Rao, GM Estavillo, and JB Reiskind. 2002. C_4 mechanisms in aquatic angiosperms: Comparisons with terrestrial C_4 systems. Funct. Plant Biol. 29: 379–392.

Buzgo M and PK Endress. 1999. The gynoecium of *Gymnostachys* (Araceae) and Potamogeton (Potomagetonaceae). In XVI Int. Bot. Congr. Abstracts, p. 240. Missouri Botanical Garden, St. Louis, MO.

Buzgo M, DE Soltis, PS Soltis, S Kim, H Ma, BA Hauser, J Leebens-Mackl, and B Johansen. 2006. Perianth development in the basal monocot *Triglochin maritime* (Juncginaceae). In: JT Columbus, EA Friar, JM Prince, MG Simpson, eds. Monocots: comparative biology and evolution (excluding Poales), pp. 107–125. Clermont (Aliso 22: 107–125).

Cambridge ML and J Kuo. 1982. Morphology, anatomy and histochemistry of the Australian seagrasses genus *Posidonia sinuosa* Cambridge & Kuo. Aquatic Bot. 14: 1–14.

Campbell DH. 1897. A morphological study of *Nolas* and *Zannichellia*. Proc. Calif. Acad. Set., ser. 3, 1: 1–71.

Campbell DH. 1898. Development of the flower and embryo of *Lilaea subulata* H. B. K. Ann. Bot. 12: 1–12.

Campbell GKG. 1936. The anatomy of *Potamogeton pectinatus*. Trans. Proc. Bot. Soc. Edinb. 32: 179–186.

Charlton WA. 1981. Features of the inflorescence of *Triglochin maritimum*. Canad. J. Bot. 59: 2108–2115.

Chrysler MA. 1907. The structure and relationships in Potamogetonaceae and allied families. Bot. Gaz. 44: 161–188.

Cook MT. 1908. The development of the embryo sac and embryo of *Potamogeton lucens*. Bull. Torrey Bot. Club 35: 209–218.

Cox PA and CJ Humphries. 1993. Hydrophilous pollination and breeding system evolution in sea grasses: a phylogenetic approach to the evolutionary ecology of the Cymodoceaceae. Bot. J. Linn. Soc. 113: 217–226.

Cox PA, PB Tomlinson, and K Nieznanski. 1992. Hydrophilous pollination and reproductive morphology in the seagrass *Phyllospadix scouleri* (Zosteraceae). Plant Syst. Evol. 180: 65–75.

Dahlgren KVO. 1939. Endosperm- und Embryobildung bei *Zostera marina*. Bot. Not. 1939: 607–615.

De Cock AWAM. 1978. Germination of the thread like pollen grains of the seagrass *Zostera marina* L. Bull. Soc. Bot. France Act. Bot. 1–2: 145–148.

De Cock AWAM. 1980. Flowering pollination and fruiting in *Zostera marina* L. Aquatic Bot. 9: 201–220.

Ducker SC, NJ Foord, and RB Knox. 1977. Biology of Australian sea grasses: the genus *Amphibolis* C. Agardh (Cymodoeaceae). Aust. J. Bot. 25: 67–95.

Ducker SC, JM Pettitt, and RB Knox. 1978. Biology of Australian sea grasses: Pollen development and submarine pollination in *Amphibolis antarctica* and *Thalassodendron ciliatum* (Cymodoceaceae). Aust. J. Bot. 26: 265–285.

Gardner RO. 1976. Binucleate pollen in *Triglochin* L. N. Z. J. Bot. 14: 115–116.

Graves AH. 1908. The morphology of *Ruppia maritima*. Conn. Acad. Arts. Sci. 14: 59–170.

Grönland J. 1851. Beitrag zur Kenntnis der *Zostera marina*. Bot. Z. 9: 185–192.

Guo Y-H and CDK Cook. 1990. The floral biology of *Groenlandia densa* (L.) Fourreau (Potamogetonaceae). Aquatic Bot. 38: 283–288.

Gupta BL. 1934. Contribution to the life history of *Potamogeton crispus*. J. Indian Bot. Soc. 13: 51–65.

Hagstrom ML. 1916. Critical researches on the *Potamogeton*. Kongl. Svenska Vetenskapsakad. Handl. 55: 1–281.

Hartog C den. 1970. Sea grasses of the World. Verh. Kon. Ned. Akad. Wetensch. Afd. Natuurk. Tweede Sect. 59(I): 1–275.

Haynes RR. 1978. The Potamogetonaceae in the southeastern United States. J. Arnold Arbor. 59: 170–191.

Haynes RR and LB Holm-Nielsen. 1987. The Zannichelliaceae in the southeastern United States. J. Arnold Arbor. 68: 259–268.

Haynes RR, DH Les, and LB Holm-Nielsen. 1998a. Juncaginaceae. In: K Kubitzki, ed. The families and genera of vascular plants, vol 4, 260–263. Springer, Berlin/Heidelberg/New York.

Haynes RR, DH Les, and LB Holm-Nielsen. 1998b. Potamogetonaceae. In: K Kubitzki, ed. The families and genera of vascular plants, vol 4, pp. 408–415. Springer, Berlin/Heidelberg/New York.

Haynes RR, LB Holm-Nielsen, and DH Les. 1998c. Ruppiaceae. In: K Kubitzki, ed. The families and genera of vascular plants, vol 4, pp. 445–448. Springer, Berlin/Heidelberg/New York.

Haynes RR, DH Les, and LB Holm-Nielsen. 1998d. Scheuchzeriaceae. In: K Kubitzki ed., The families and genera of vascular plants, vol 4, pp. 449–451. Springer, Berlin/Heidelberg/New York.

Haynes RR, DH Les, and LB Holm-Nielsen. 1998e. Zannichelliaceae. In: K Kubitzki, ed. The families and genera of vascular plants, vol 4, pp. 470–474. Springer, Berlin/Heidelberg/New York.

Hegelmaier F. 1870. Über die Entwicklung der Blütenteile von *Potamogeton*. Bot. Z. 18: 283–320.

Hofmeister W. 1852. Zur Entwicklungsgeschichte der *Zostera*. Bot. Z. 10: 121–131, 137–149, 157–158.

Holferty GM. 1901. Ovule and embryo of *Potamogeton natans*. Bot. Gaz. 31: 339–346.

Isaac FM. 1969 (1970). Floral structure and germination in *Cymodocea ciliata*. Phytomorphology 19: 44–51.

Jacobs SWL and MA Brock. 1982. A revision of the genus *Ruppia* (Potamogetonaceae) in Australia. Aquatic Bot. 14: 325–337.

Kamelina OP. 1990. Potamogetonaceae. In: TB Batygina and MS Yakovlev, eds. Comparative embryology of flowering plants. Monocotyledons. Butomaceae – Lemnaceae, vol 1, pp 34–39. Nauka, Leningrad (in Russian).

Kamelina OP and ES Terekhin. 1990. Ruppiaceae. In: TB Batygina and MS Yakovlev, eds. Comparative embryology of flowering plants. Monocotyledons. Butomaceae – Lemnaceae, vol 1, pp. 39–44. Nauka, Leningrad (in Russian).

Kato Y, K Aioi, Y Omori, N Takahata, and Y Satta. 2003. Phylogenetic analyses of *Zostera* species based on *rbc*L and *mat*K sequences: implications for the origin and diversification of seagrasses in Japanese waters. Genes Genet. Syst. 78: 329–342.

Kay QON. 1971. Floral structure in the marine angiosperms *Cymodocea serrulata* and *Thalassodendron ciliatum* (*Cymodocea ciliata*). Bot. J. Linn. Soc. 64: 423–429.

Keighery GJ and DJ Coates. 1981. Chromosome counts in *Posidonia* (Posidoniaceae). Plant Syst. Evol. 137: 221–222.

Kirkman H. 1975. Male floral structure in the marine angiosperm *Cymodocea serrulata* (R. Br.) Ascherson and Magnus (Zannichelliaceae). Bot. J. Linn. Soc. 79: 267–268.

Kuo J. 1978. Morphology, anatomy and histochemistry of the Australian sea grasses genus *Posidonia* Konig (Posidoniaceae). I. Leaf blade and leaf sheath of *Posidonia australis* Hook.f. Aquatic Bot. 5: 171–190.

Kuo J. 1983. The nacreous walls of sieve elements in sea grasses. Am. J. Bot. 70: 159–164.

Kuo J. 1993a. Functional leaf anatomy and ultrastructure in a marine angiosperm, *Syringodium isoetifolium* (Aschers.) Dandy (Cymodoceaceae). Aust. J. Mar. Freshwater Res. 44: 59–73.

Kuo J. 1993b. Root anatomy and rhizosphere ultrastructure in tropical sea grasses. Aust. J. Mar. Freshwater Res. 44: 75–84.

Kuo J. 2001. Chromosome numbers of the Australian *Zosteraceae*. Plant Syst. Ecol. 226(3–4): 155–163.

Kuo J and ML Cambridge. 1978a. Morphology, anatomy, and histochemistry of the Australian sea grasses of the genus *Posidonia* Konig (Posidoniaceae): I. Leaf blade and leaf sheath of *Posidonia australis* Hook. f. Aquatic Bot. 5: 163–170.

Kuo J and ML Cambridge. 1978b. Morphology, anatomy, and histochemistry of the Australian species of the genus *Posidonia* Konig (Posidoniaceae): rhizome and root of *Posidonia australis* Hook. f. Aquatic Bot. 5: 191–206.

Kuo J and H Kirkman. 1987. Floral and seeding morphology and anatomy of *Thalassodendron pachyrhizum* den Hartog (Cymodoceaceae). Aquatic Bot. 29: 1–17.

Kuo J and H Kirkman. 1990. Anatomy of vipiparous sea grasses of *Amphibolis* and *Thalassodendron* and their nutrient supply. Bot. Mar. 33: 117–126.

Kuo J and AJ McComb. 1989. Sea grass taxonomy, structure and development. In: AWD Larkum, AJ McComb, SA Shephard, eds. Biology of seagrasses. A treatise on the biology of seagrasses with special reference to the Australian region, pp. 6–73. Elsevier Science, Amsterdam.

Kuo J and AJ McComb. 1998a. Cymodoceaceae. In: K Kubitzki, ed. The families and genera of vascular plants, vol 4, pp 133–140. Springer, Berlin/Heidelberg/New York.

Kuo J and AJ McComb. 1998b. Posidoniaceae. In: K Kubitzki, ed., The families and genera of vascular plants, vol 4, pp 404–408. Springer, Berlin/Heidelberg/New York.

Kuo J and AJ McComb. 1998c. Zosteraceae. In: K Kubitzki, ed. The families and genera of vascular plants, vol 4, pp. 496–502. Springer, Berlin/Heidelberg/New York.

Kuo J, K Aioi, and H Iizumi. 1988. Comparative leaf structure and its functional significance in *Phyllospadix iwatensis* Makino and *Phyllospadix japonicus* Makino (Zosteraceae). Aquatic Bot. 30: 169–187.

Kuo J, K Seto, T.Nasu, H Iizumi, and K Aioi. 1989. Notes on *Archaeozostera* in relation to the Zosteraceae. Aquatic Bot. 34: 317–328.

Kuo J, RW Ridge, and S Lewis. 1990a. Leaf internal morphology and ultrastructure of *Zostera moelleri* Irmisch ex Aschers.: a comparative study of intertidal and subtidal forms. Aquatic Bot. 36: 217–236.

Kuo J, K Aioi, and H Iizumi. 1990b. Chromosome numbers and their systematic implications in Australian marine angiosperms: The Posidoniaceae. Plant Syst. Evol. 171: 199–204.

Lakshmanan KK. 1965. Note on the endosperm formation in *Zannichellia palustris* L. Phyton 22: 13–14.

Larkum AWD, AJ McComb, and SA Shepherd, eds. 1989. Biology of sea grasses: a treatise on the biology of sea grasses with special reference to the Australian region. Elsevier, Amsterdam.

Larsen K. 1966. Cytotaxonomical note on *Lilaea*. Bot. Not. 119: 496–497.

Les DH and RR Haynes. 1996. *Coleogeton* (Potamogetonaceae), a new genus of pondweeds. Novon 6: 389–391.

Les DH and DJ Sheridan. 1990a. Hagstrom's concept of phylogenetic relationships in *Potamogeton* L. (Potamogetonaceae). Taxon 39: 41–58.

Les DH and DJ Sheridan. 1990b. Biochemical heterophylly and flavonoid evolution in North American *Potamogeton* (Potamogetonaceae). Am. J. Bot. 77: 453–465.

Les DH, ML Moody, SWL Jacobs, and RJ Bayer. 2002. Systematics of Seagrasses (Zosteraceae) in Australia and New Zealand. Syst. Bot. 27: 468–484.

Lieu SM. 1979. Organogenesis in *Triglochin striata*. Canad. J. Bot. 57: 1418–1438.

Lindqvist C, J de Laet, RR Haynes, L Aagesen, BR Keener, and VA Albert. 2006. Molecular phylogenetics of an aquatic plant lineage, Potamogetonaceae. Cladistics 22: 568–588.

Lupnitz D. 1969. Histogenese and Anatomie von Primarwurzeln und sprossburtigen Wurzeln einer Potamogetonaceae. Beitr. Biol. Pflanz. 46 : 247–313.

McConchie CA, SC Ducker, and RB Knox. 1982a Biology of Australian seagrasses: floral development and morphology in *Amphibolis* (Cymodoceaceae). Aust. J. Bot. 30: 251–264.

McConchie CA, RB Knox, and SC Ducker. 1982b. Pollen wall structure and cytochemistry in the seagrass *Amphibolis griffithii* (Cymodoceaceae). Ann Bot. 50: 792–732.

McMillan C. 1983. Seed germination in *Halodule wrightii* and *Syringodium filiforme* from Texas and the US Virgin Islands, Aquatic Bot. 15: 217–220.

McMillan C and LH Bragg. 1987. Comparison of fruits of *Syringodium* (Cymodoceaceae) from Texas, the US Virgin Islands and the Philippines. Aquatic Bot. 28: 97–100.

Muenscher WC. 1936. The germination of seeds of *Potamogeton* (pondweeds). Mich. Bot. 23: 35–38.

Murbeck S. 1902. Über die Embryologie von *Ruppia rostellata* Koch. Koninkl. Sven. Vetensk. Ak. Handl. 36(5): 1–21.

Nikiticheva ZI and OB Proskurina. 1992. Embryology of *Scheuchzeria palustris* (Scheuchzeriaceae). Bot. Zhurn. 77: 3–18 (in Russian with English summary).

Pettitt JM and AC Jenny. 1975. Pollen in hydrophilous angiosperms. Micron 5: 377–405.

Plisco MA. 1985. Scheuchzeriaceae. In: A Takhtajan, ed. Comparative seed anatomy, vol 1, pp. 47–49. Nauka, Leningrad (in Russian).

Posluszny U. 1981. Unicarpellate floral development in *Potamogeton zosteriformis*. Canad. J. Bot. 59: 495–504.

Posluszny U. 1983. Re-evaluation of certain key relationships in the Alismatidae: floral organogenesis of *Scheuchzeria palustris*. Am. J. Bot. 70: 925–933.

Posluszny U and R Sattler. 1973. Floral development of *Potamogeton densus*. Canad. J. Bot. 51: 647–656.

Posluszny U and R Sattler. 1974a. Floral development of *Potamogeton richardsonii*. Am. J. Bot. 61: 209–216.

Posluszny U and R Sattler. 1974b. Floral development of *Ruppia maritima* var. *maritima*. Canad. J. Bot. 52: 1607–1612.

Posluszny U and R Sattler. 1976. Floral development of *Zannichellia palustris*. Canad. J. Bot. 54: 651–662.

Posluszny U and PB Tomlinson. 1977. Morphology and development of floral shoots and organs in certain Zannichelliaceae. Bot. J. Linn. Soc. 75: 21–46.

Posluszny U, WA Charlton, and DK Jain. 1986. Morphology and development of the reproductive shoots of *Lilaea scilloides* (Poir.) Hauman (Alismatidae). Bot. J. Linn. Soc. 92: 323–342.

Reinecke P. 1964. A contribution to the morphology of *Zannichellia aschersoniana* Graebn. South Afr. J. Bot. 30: 93–101.

Roth I. 1961. Histogenese der Laubblatter von *Zostera nana*. Bot. Jahrb. Syst. 80: 500–507.

Sattler R. 1965. Perianth development of *Potamogeton richardsonii*. Am. J. Bot. 52: 35–41.

Schneider EL and S Carlquist. 1997. Origins and nature of vessels in monocotyledons. 2. Juncaginaceae and Scheuchzeriaceae. Nord. J. Bot. 17: 397–401.

Schwantz G. 1967. Untersuchungen zur postmeiotischen Mikrosporogenese. I. Morphogenese des *Ruppia*-Pollens. Pollen et Spores 9: 9–48.

Singh V. 1964. Morphological and anatomical studies in Helobiae: I. Vegetative anatomy of some members of Potamogetonaceae. Proc. Indian Acad. Sci. 60B: 214–231.

Singh V. 1965. Morphological and anatomical studies in Helobiae. V. Vascular anatomy of the flower of *Lilaea scilloides* (Poir.) Hamm. Proc. Indian Acad Sci. B, 61: 316–535.

Soros-Pottruff C and U Posluszny. 1994. Developmental morphology of reproductive structures of *Phyllospadix* (Zosteraceae). Int. J. Plant Sci. 155: 405–420.

Soros-Pottruff C and U Posluszny. 1995. Developmental morphology of reproductive structures of *Zostera* and a reconsideration of *Heterozostera* (Zosteraceae). Int. J. Plant Sci. 156: 143–158.

Sorsa P. 1988. Pollen morphology of *Potamogeton* and *Groenlandia* (Potamogetonaceae) and its taxonomic significance. Ann. Bot. Fenn. 25: 179–199.

Soueges R. 1943. Embryogenie des Scheuchzeriacées: Developpement de l'embryon chez le *Triglochin maritimum* L. Compt. Rend. Hebd. Seances Acad. Sci. 216: 746–748.

Stenar H. 1935. Embryologische Beobachtungen fiber *Scheuchzeria palustris* L. Bot. Not. 1935: 78–86.
Stewart JG and L Ludenberg. 1980. Microsporocyte growth and meiosis in *Phyllospadix torreyi*, a marine monocotyledon. Am. J. Bot. 67: 949–954.
Sun K, J-K Chen, and Z-Y Zhang. 2001. Pollen morphology of Najadaceae and Zannichelliaceae. Acta Phytotax. Sinica 39: 31–37 (in Chinese with English summary).
Takaso T and F Bouman. 1984. Ovule ontogeny and seed development in *Potamogeton natans* L. (Potamogetonaceae), with a note on the campylotropous ovule. Acta Bot. Neerl. 33: 519–533.
Talavera S, P Garcia-Murillo, and J Herrera. 1993. Chromosome numbers and a new model for karyotype evolution in *Ruppia* L. (Ruppiaceae). Aquatic Bot. 45: 1–13.
Taylor ARA. 1957a. Studies of the development of *Zostera marina* L.: 1. The embryo and seed. Canad. J. Bot. 35: 477–499.
Taylor ARA. 1957b. Studies of the development of *Zostera marina* L.: 2. Germination and seedling development. Canad. J. Bot. 35: 681–695.
Terekhin ES. 1985. Potamogetonaceae, Ruppiaceae. In: A Takhtajan, ed. Comparative seed anatomy, vol 1, pp. 51–55. Nauka, Leningrad (in Russian).
Terechin ES and SI Chubarov. 1991. The embryological and carpological investigation of *Althenia jiliformis* (Zannichelliaceae). Bot. Zhurn. 76: 226–236 (in Russian with English summary).
Terechin ES and GV Shibakina. 1985. Zosteraceae. In: A. Takhtajan, ed. Comparative seed anatomy, vol 1, pp. 62–64. Nauka, Leningrad (in Russian).
Tomlinson PB and U Posluszny. 1976. Generic limits in the Zannichelliaceae (sensu Dumortier). Taxon 25: 273–279.
Tomlinson PB and U Posluszny. 1978. Aspects of floral morphology and development in the sea grass *Syringodium filiformis* (Cymodoceaceae). Bot. Gaz. 139: 333–345.
Tomlinson PB and U Posluszny. 2001. Generic limits in the seagrass family Zosteraceae. Taxon 50: 429–437.
Tutin TG. 1938. The autecology of *Zostera marina* in relation to its wasting disease. New Phytol. 37: 50–71.
Vijayraghavan MR and A Vidya Kumari. 1974, Embryology and systematic position of *Zannichellia palustris* L. J. Indian Bot. Soc. 53: 292–302.
Waycott M and DH Less. 1996. An integrated approach to the evolutionary study of seagrasses. In: J Kuo, RC Phillios, DI Walker, and H Kirkman, eds. Seagrass biology. Proceedings of an International Workshop, Rottnest Island, Western Australia. 25–29 January 1996, pp. 71–78. Perth.
Weigleb G. 1988. Notes on pondweeds, outline for a monographical treatment of the genus *Potamogeton* L. Feddes Repert. 99: 249–266.
Yamashita T. 1972. Eigenartige Wurzelanlage der Embryos bei *Ruppia maritima* L. Beitr. Biol. Pfl. 48: 157–170.
Yamashita T. 1973. Über die Embryo- und Wurzelentwicklung bei *Zostera japonicus* Aschers. et Graebn. J. Fac. Sci. Univ. Tokyo III, Bot. 11: 175–193.
Yamashita T. 1976. Über die Pollenbildung bei *Halodule pinifolia* und *H. uninervis*. Beitr. Biol. Pfl. 52: 217–226.
Zapata O and C McMillan. 1979. Phenolic acids in seagrasses. Aquatic Bot. 7: 307–317.

Superorder ARANAE

Order 5. ARALES

Very small to giant herbs with creeping or tuberous rhizomes or scrambling, slender shrubs, often climbers with aerial roots and true epiphytes, rarely free-floating aquatics. Roots mycorrhizal, without root hairs. Silica bodies lacking, but calcium oxalate crystals occur. Vessels in roots and seldom also in rhizome and in stem, always with scalariform perforations. Simple or uniseriate laticifers present in some genera of the Araceae. Resin ducts sometimes present as well as rows of mucilage cells. Sieve-element plastids P2c-type, or rarely (Pistiaceae) S-type. Leaves alternate, spirally arranged or rarely distichous, unifacial (Acora ceae) or bifacial, usually with distinct, basally sheathing petiole and expanded, entire to variously cleft or perforate or even compound lamina, variously parallel-veined or pinnately or palmately net-veined, usually developing acropetally (like typical dicotyledons). Often a row of intravaginal squamules occurs in the axils of the leaf sheaths or in the axils of the cataphylls. Stomata paracytic, less often tetracytic, or cyclocytic, sometimes anomocytic. Flowers small, usually numerous, in terminal, unbranched spike with fleshy axis, the spadix (extremely reduced in Lemnaceae) generally subtended by ovate and colored or rarely green, mostly rather large spathe (lacking in *Acorus*, *Gymnostachys*, *Orontium*, and in some Lemnaceae), which encloses the spadix in the juvenile stage, bisexual or unisexual (monoecious or rarely dioecious), actinomorphic, bractless. Perianth of 4 or 6(8) free or connate segments in two cycles, in unisexual flowers generally much reduced or wanting or always wanting. Stamens (1-)4 or 6, or 8–12, free or sometimes more or less connate and sometimes forming synandria; filaments mostly short and broad; anthers tetrasporangiate or disporangiate, basifixed, extrorse or very rarely (*Zamioculcas, Acorus*) introrse, opening by apical pores or slits or longitudinally. Tapetum plasmodial or (Acoraceae) secretory. Microsporogenesis successive. Pollen grains 2-celled or 3-celled, exceptionally variable, tectate-columellate to tectate-granular or atectate, 1-colpate and boat-shaped to globose and inaperturate, with various types of ornamentation. Gynoecium of (1-)3(-15) carpels; style terminal and generally short and thick to conical, or the stigmas sessile; ovary superior, plurilocular or unilocular, seldom pseudomonomerous, with one to many ovules per locule. Ovules pendulous, or horizontal, or ascending, anatropous, hemitropous, or orthotropous, sometimes campylotropous, bitegmic, very rarely unitegmic (*Symplocarpus* – Barabe, Froget, Chrétien 1987 and *Gymnostachys* – Buzgo 2001), crassinucellate or tenuinucellate, sometimes with endothelium, with or without parietal cell. Female gametophyte of *Polygonum*-type or rarely (as in most Lemnaceae) of *Allium*-type. Endosperm cellular, or helobial, or nuclear, with chalazal haustorial process. Fruits usually baccate, more rarely rather dry and firmly carnose or leathery, densely packed into a cylindric infructescence. Seeds one to many, sometimes fleshy, sometimes with operculum; seed coat formed by both integuments, predominantly by the inner one; embryo generally large, green, mostly linear, straight or sometimes curved, usually well differentiated, sometimes macropodous (having swollen hypocotyl or radicle); endosperm often present and then usually copious, with aleurone, simple starch grains, and oil or endosperm wanting; perisperm present only in *Pistia* (Mercado-Noriel and Mercado 1978; Vyshenskaya 1985). Plants contain flavone C-glycosides, flavonols and cyanogenic glucoide triglochinin.

The Arales are traditionally associated with Arecales and Cyclanthales mainly because of their small flowers arranged in dense, spadiciform inflorescence subtended by large principal bract or spathe. However, the accumulated data have shown that they are much closer to the alismatanae and share with them many taxonomically important features (Gow 1913; Jussen 1929; Huber 1977; Dahlgren and Clifford 1982; Dahlgren et al. 1985; Tillich 1985; Grayum 1991), including the presence of intravaginal squamules (but only in few taxon of the Arales, including some *Philodendron* species), the occurrence of both a spathe and spicate inflorescences, of laticifers, an amoeboid tapetum as well as of the cyanogenic substance triglochinin in Araceae and certain alismatanae (Nahrstedt 1975), and *rbc*L sequence (Duvall et al. 1993). However, the Arales differ markedly from the alismatanae in many respects.

Key to Families

1 Plants terrestrial or epiphytic or sometimes more or less aquatic, but not free-floating.

2 Leaves ensiform, anatomically unifacial. Perennial herbs with creeping, rather thick, aromatic, sympodial rhizomes. Have ethereal oil cells, but do not possess calcium oxalate crystals in any form, and no raphids. Vessels in roots and in rhizome, with scalariform perforations. Leaves tufted, long, linear, acuminate, equitant, parallel-veined, not narrowed or scarcely narrowed below, sheathing and surrounding the neck of rhizome at base. Stomata paracytic. Inflorescences scapiflorous (the scape keeled, with double vascularization, taken to represent adnation of the peduncles with the sheath of the subtending leaf; "solitary, lateral spadices, tepering acropetally to a blunt tipe, covered with tightly packed flowers, espatheate (in that the leafy point which terminates the scape above the manifestly lateral spadix seems to represent the blade of a terminal leaf, rather than a true spathe" – Watson and Dallwitz 2006). Flowers small to minute, bisexual, 3-merous. Perianth segments six, in two cycles, free, thick, incurved and truncate at apex. Stamens six; filaments linear; anthers tetrasporangiate, opening longitudinally. Tapetum secretory. Pollen grains 1-colpate, boat-shaped, without foot layer (Tarasevich, personal communication); the apertural exine is subpsilate (Grayum 1984). Gynoecium of (2-)3(-4) carpels; style very short, with small stigma; ovary superior, oblong, 2–3-locular. Ovules 2–4(-5), pendulous, orthotropous, bitegmic, tenuinucellate. Placentation axile to apical. Endosperm cellular, without haustorial processes. Fruits berries with thin, leathery pericarp. Seeds 1(-5)-9, oblong, with copious endosperm and single layer of radially elongated cells of perisperm; embryo straight, green, starchy, testa thick, without phytomelan. Contain proanthocyanidins. Basic chromosome number n = 12. . . . . 1. ACORACEAE.

2 Leaves with a more or less expanded, bifacial blade. Plants often cormous, or rhizomatous, or tuberous, sometimes climbing (stems twiners or scrambling, some very large). Laticifers present (sap clear, milky, or dark and taniniferous), rarely absent. Leaves sometimes distichous, simple, basal or cauline, rarely solitary (as in *Dracontium*), small to very large, sheathing (the sheath membranous, sometimes deciduous); blades often oblong, cordate, sagitate to hastate, sometimes perforated. Flowers small, very numerous, often fragrant, or malodorous, in a spadix that is sub-

tended by well developed and often colored spathe; plants usually monoecious, except *Arisaema,* which is dioecious). Stamens 1–4 or 6–12, free or united into synandria; anthers extrorse, tetrasporangiate or bisporangiate, opening by longitudinal or short slits, or by apical pores; staminodia sometimes present, free or united into synandria. Gynoecium of 1–8(-15) carpels, monomerous or syncarpous; ovary superior, 1–3-locular, rarely plurilocular; ovules one to many per locule, pendulous, or horizontal, or ascending, often with funicle, less often sessile. Fruits indehiscent (usually baccate, drupaceous, sometimes juice, often colorful, or nuts) or rarely capsular, capsules splitting irregularly (Watson and Dallwitz 2006). Seeds one to many; endosperm copious, the embryo straight, located in center of endosperm, or embryo curved, when endosperm lacking; embryo chlorophyllous or rarely (*Arisaema amurense*) achlorophyllous. Usually contain saponins, proanthocyanidins (commonly, sometimes very abundantly), sometimes flavonols (kaempferol and quercetin), cinnamic acids, cyanogenic glucosides, alkaloids and amines (Mayo et al. 1998); n = from 7 up to 84. . . . . 2. ARACEAE.

1 Free-floating or submerged aquatics. Lactifers wanting. Inflorescence very reduced, consists of one female flower and one or two male flowers.

3 Ovules tenuinucellate. Seeds with copious perisperm and scanty endosperm. Seed coat with very large hilum and an operculum. Fruits with many seeds. Small, stoloniferous, floating herbs, rarely anchored by their long, feathery, hanging roots; primary root absent. Branching sympodial, but internodes remain short. Leaves in rosettes, very shortly petiolated, entire, obovate to spatulate-oblong, truncate to emarginate at the apex, long-cuneate at base, with flabellate venation, covered with short depressed uniseriate hairs, moving together at night. Stomata anomocytic. Inflorescence small, inconspicuous, enclosed in small, short, greenish spathe that is tubular below, bisexual, with two perianthless flowers, male flower above the female one. Male flower with 4–6 stamens united into a synandrium. Pollen grains tectate-granular, 1(2)-colpate, boat-shaped, longitudinally strongly ridged; ridges (from 20 to 25) formed by the outer layer of the exine; the inner layer of exine forms beaklike thickenings. Female flowers naked, consisting of 1-locular carpel, partly attached to the spathe; style conic, with discoid stigma; ovary with numerous, basal-subparietal, orthotropous ovules. Micropyle formed by the outer integument. Female gametophyte *Polygonum*-type. Fruits with membranous, partly transparent pericarp, few-seeded. Seeds oblong or obovoid, testa ultimately rugose; embryo minute, apical, cuneiform; n = 7. . . . . 3. PISTIACEAE.

3 Ovules crassinucellate. Seeds without perisperm but with endosperm. Seed coat with operculum, but hilum is not so large. Small to minute, floating or submerged annual, much reduced aquatic herbs with thalloid plate-like bodies (fronds); thallus small to minute, globular, flat or linear, contain mucilage cells, with or without raphides. Roots simple, solitary or several per frond or lacking. Vascular system much reduced, the plants lacking both vessels and tracheids, or sometimes with tracheids in roots. Inflorescence very reduced, enclosed in membranous spathe or naked, consists of one female and one or two male flowers. Perianth lacking. Stamen solitary; anthers tetrasporangiate (*Lemna*) or disporangiate (*Wolffia*), opening by apical pores or short or longitudinal slits. Pollen grains 3-celled, tectate-columellate, 1-porate, echinate. Gynoecium of one carpel; style short, stigma funneliform; ovary 1-locular, with 1–7 basal, anatropous, hemitropous, or orthotropous ovules. Micropyle formed by the inner integument. Female gametophyte of *Allium*-type. Endosperm cellular. Fruits 1–6-seeded utricles. Seeds oblong-elliptical, smooth; embryo straight, rather large, well differentiated, but without primary root (Lodkina 1985), endosperm present or absent; n = 5, 8, 10, 11, 21. . . . . .4. LEMNACEAE.

1. ACORACEAE

Martynov 1820. 1/4. Holarctic and paleotropical: temperate Eurasia from Norway to Eastern Asia, Sulawesi, New Guinea; western and central North America.

Acorus.

Related to the Araceae, but they differ from them in the absence of raphides, presence of the ethereal cells, ensiform, unifacial leaves, presence of dermal perisperm, From the majority of Araceae (except *Gymnostachys*) Acoraceae differ also in the absence of the spathe. From the Araceae as well as from all other monocotyledons they markedly differ in the DNA sequences of the plastid *rbc*L, which according to Duvall et al. (1993) shows that *Acorus* "represents the most ancient surviving lineage of the ancestral monocotyledons." This hypothesis based mainly on the genetic constitution of the endosymbiotic organism, the chloroplast, is most improbable and strongly contradicts all other available data.

2. ARACEAE

A.L. de Jussieu 1763 (including Arisaraceae Rafinesque 1838; Caladiaceae Salisbury 1866; Callaceae Reichenbach ex Bartling 1830; Cryptocorynaceae J. Agardh 1858; Dracontiaceae Salisbury 1866; Lentiscaceae Horaninow 1843; Orontiaceae Bartling 1830; Pothoaceae Rafinesque 1838). 103/2500–3000. Predominantly tropical, only about 10 genera occur in temperate regions of the Northern Hemisphere.

2.1 GYMNOSTACHYDOIDEAE

Laticifers and spicular cells lacking. Leaves epetiolate, grasslike, bifacial. Prophylls foliar. Two to six more or less pendulous spadices are situated in a cluster on the 1–2 m high, unifacial peduncle. Spathe lacking. Flowers bisexual and 2-merous. Pollen grains 1-colpate, boat-shaped, stigma without papillae and possibly dry, very little secretion in the gynoecium, ovule unitegmic, integument is shorter than the nucellus (Buzgo 2001), n = 24. – *Gymnostachys* (1). Australia.

2.2 ORONTIOIDEAE

Laticifers present (*Orontium*) or absent. Swamp plants. Stems stout, erect. Leaves spirally arranged, oblong-elliptic or cordate, with acrodromous leaf venation. Flowers bisexual, 2–3-merous. Anthers terminal, opening by longitudinal slits. Ovary inferior, ovules anatropous, n = 13–15. – ORONTIEAE: *Orontium*; SYMPLOCARPEAE: *Lasichiton, Symplocarpus.*

2.3 POTHOIDEAE

Laticifers and spicular cells usually lacking. Leaves spirally arranged or distichous, often with winged petiole. Lateral veins of second and third orders connected by veinlets to form a reticulum. Flowers usually bisexual, with perianth. Ovules anatropous to hemitropous, n = (10, 12) 14 (15). – POTHEAE: *Pothos, Pedicellarum, Pothoidium, Anthurium*; MONSTEREAE: *Holochlamys, Spathiphyllum, Rhodospatha, Stenospermation, Scindapsus, Rhaphidophora, Anadendrum, Monstera, Alloschemone, Epipremnum, Amydrium, Heteropsis.*

2.4 LASIOIDEAE

Laticifers absent or the African genera often have laticifers or ducts, and biforines or biforine-like raphides. Leaves often sagittate, often deeply lobes and fenestrate; primary lateral veins of major lamina divisions pinnate to arcuate-parallel, order venation reticulate. Flowers bisexual, usually with perianth. Pollen grains colpate or inaperturate. Endosperm usually present, forming a thin but distinct layer, n = 13. – *Cyrtosperma, Lasiomorpha, Podolasia, Lasia, Anophyllum, Urospatha, Anaphyllopsis, Pycnospatha, Draconitum, Draconitoides.*

2.5 CALLOIDEAE

Laticifers present. Leaves distichous, cordate-pinnate, venation parallel-pinnate; petiole with type three aerenchyma. Flowers usually bisexual, without perigone. Pollen grains 2-colpate. Ovules anatropous. Endosperm copious, n = 18, 27. – *Calla.*

2.6 PHILODENDROIDEAE

Lactifers and ducts present in roots, stems, leaves and inflorescences, or absent. Flowers unisexual, without perigone. Ovules anatropous or orthotropous. Endosperm copious or rarely absent. – PHILODENDREAE: *Montrichardia, Anubias, Furtadoa, Philodendron, Homalomena*; ZANTEDESCHIEAE: *Zantedeschia, Callopsis;* STYLOCHAETONEAE: *Stylochaeton;* ZAMIOCULCADEAE: *Gonatopus, Zamioculcas*; AGLAONEMATEAE: *Nephthytis, Anchomanes, Pseudohydrosme, Aglaonema, Aglaodorum;* CULCASIEAE: *Culcasia, Cercestis;* SPATHICARPEAE: *Bognera, Dieffenbachia, Spathantheum, Gorgonidium, Synandrospadix, Gearum, Spathicarpa, Asterostigma, Mangonia, Taccarum.*

2.7 SCHISMATOGLOTTIDOIDEAE

Lactifers non-anastomosing, in all organs or absent from leaves. Inflorescence solitary, rarely two per sympodium spathe tube connate, persistent, spadix with unisexual flowers with perigone absent. Fruit a berry or syncarp. Seeds with endosperm, often copious. – CRYPTOCORYNEAE: *Cryptocoryne, Lagenandra*; SCHISMATOGLOTTIDEAE: *Phymatarum, Schisma-toglottis, Aridarum, Heteroaridarum, Hottarum, Piptospatha, Bucephalandra.*

2.8 AROIDEAE
Laticifers present, rarely absent, straight or anastomosing. Spathe usually differentiated into lower, convolute tube and upper, gaping blade. Flowers unisexual, without perianth. Pollen grains inaperturate. Ovules anatropous or orthotropous. Seeds usually with endosperm. – THOMSONIEAE *Amorphophallus, Pseudodracontium*; CALADIEAE: *Hapaline, Syngonium, Xanthosoma, Chlorospatha, Ulearum, Filarum, Zomicarpella, Caladium, Scaphispatha, Jasarum, Zomicarpa*; ARISAREAE: *Ambrosina, Arisarum*; PELTANDREAE: *Peltandra, Typhonodorum, Colletogyne, Carlephyton, Arophyton*; ARISAE-MATEAE: *Arisaema, Pinellia*; AREAE: *Typhonium, Sauromatum, Theriophonum, Biarum, Arum, Eminium, Dracunculus, Helicodiceros*; COLOCASIEAE: *Ariopsis, Alocasia, Remusatia, Colocasia, Steudnera, Protarum.*

3. PISTIACEAE

Richard ex C. Agardh 1822. 1/1. Tropical and subtropical regions.

Pistia.

Differ from the Araceae in habit, absence of primary root, leaf, and inflorescence morphology, absence of laticifers, anomocytic stomata, unique exine structure (Kuprianova and Tarasevich 1984; Tarasevich 1990), and especially in their perispermous seeds (see Mercado-Noriel and Mercado 1978; Vyshenskaya 1985) and in sieve-element plastids of S-type. According to Behnke (1995), the type of plastids clearly corroborate the isolated position of *Pistia.* Lindley (1847, 1853) included *Ambrosina* together with the genera comprising the modern family Lemnaceae in the Pistiaceae, whereas Buscalioni and Lanza (1935) included in the Pistiaceae only *Pistia* and *Ambrosina.* However, the genus *Pistia* differs markedly from *Ambrosina* and all other members of the Aroideae in its connate stamens, epetiolate leaves, the absence of tuberous stock, and secretion files, in addition to the characters mentioned above. The Pistiaceae originated probably from some ancient Aroideae-like ancestor.

4. LEMNACEAE

Martynov 1820 (including Wolffiaceae Bubani 1902). 4/c.35. Cosmopolitan.

4.1 LEMNOIDEAE
Raphides present. Roots present. Each frond has two flattened, budding pouches, one basal and one lateral on either side of the axis. Inflorescence is developed in one of the budding pouches and consists of one female and two male flowers enclosed by membranous spathe. Anther bilocular, opening transversely. Female flower is lateral to the male one. Seeds with longitudinal ribs. – *Spirodela* (including *Landoltia*), *Lemna.*

4.2 WOLFFIOIDEAE
Raphides and roots lacking. Each frond has only one, median budding pouch. Inflorescence is developed in a dorsal cavity, lacks a spathe and has one female and one male flower. Anthers unilocular, opening apically. Seeds nearly smooth. – *Wolffia, Wolffiella* (including *Pseudowolffia* and *Wolffiopsis*).

The Lemnaceae have some similarities with the Pistiaceac but differ markedly in pollen morphology (Kuprianova and Tarasevich 1984; Tarasevich 1990), crassinucellate ovules, the absence of perisperm, as well as in extremely reduced and modified vegetative body. Both the Pistiaceae and Lemnaceae evolved from Araceae – Aroideae. However, *Pistia* is not a link between the Araceae and Lemnaceae, but those two groups are not at all close in molecular data.

Bibliography

Amelunxen VF and G Gronau. 1969. Untersuchungen an den Gerbstoffzellen der Niederblatter von *Acorus calamus* L. Cytobiologie 1: 58–69.

Arber A. 1919. The vegetative morphology of *Pistia* and the Lemnaceae. Proc. Roy. Soc. Lond., Ser. B, Biol. Sci. 91: 96–103.

Barabe D and S Forget. 1988. Anatomie des fleurs fertiles et steriles de *Zamioculcas* (Araceae). Bull. Mus. Natl. Hist. Nat. B Adansonia 10: 411–419.

Barabé D, S Forget, and S Chrétien 1986. Sur les gynécées pseudo-monomeres: Cas de *Symplocarpus.* Compte Rendu Acad. Sci. Paris, sér.3, 302: 429–434.

Barabé D, S Forget, and S Chrétien. 1987. Organogénese de la fleur de *Symplocarpus foetidus* (Araceae). Canad. J. Bot. 65: 446–455.

Barabe D, A Bruneau, F Forest, and C Lacroix. 2002. The correlation between development of atypical bisexual flowers and phylogeny in the Aroideae (Araceae). Plant Syst. Evol. 232: 1–19.

Barabé D, C Lacroix, A Bruneau, A Archambault, and M Gibernau. 2004. Floral development and phylogenetic position of *Schismatoglottis* (Araceae). Int. J. Plant Sci. 165: 173–189.

Behnke H-D. 1995. P-type sieve-element plastids and the ssystematics of the Arales (*sensu* Cronquist 1988) – with S-type plastids in *Pistia*. Plant Syst. Evol. 195: 87–119.

Beppu T and A Takimoto. 1981. Geographical distribution and cytological variation of *Lemna paucicostata* Hegelm. Jpn. Bot. Mag. (Tokyo) 94: 11–20.

Blanc P. 1977. Contribution a l'etude des aracees. II. Remarques sur la croissance sympodiale chez l'*Anthurium scandens* Engl., le *Philodendron fenzlii* Engl. Et le *Philodendron speciosum* Schott. Rev. Gen. Bot. 84: 319–331.

Blanc P. 1980. Observations sur les flagelles des Araceae. Adansonia II 20: 325–338.

Blodgett F-H. 1923. The embryo of *Lemna*. Am. J. Bot. 10: 336–342.

Bogner J. 1979. A critical list of the aroid genera. Aroideana 1: 63–73.

Bogner J. 1987. Morphological variations in aroids. Aroideana 10(2): 4–16.

Bogner J and A Hay. 2000. Schismatoglottideae (Araceae) in Malesia. II – *Aridarum, Bucephalandra, Phymatarum* and *Piptospatha*. Telopea 9: 179–222.

Bogner J and M Hesse. 2005. Zamioculcadoideae, a new subfamily of Araceae. Aroideana 28: 3–20.

Bogner J and SJ Mayo. 1998. Acoraceae. In: K Kubitzki, ed. The families and genera of vascular plants, vol 4, pp. 7–11. Springer, Berlin/Heidelberg/New York.

Bogner J and DH Nicolson. 1991. A revised classification of the Araceae with dichotomous keys. Willdenowia 21: 35–50.

Bown D. 1988. Aroids – plants of the *Arum* family. Century, London.

Bown D. 2000. Aroids: Plants of the *Arum* family. Timber Press, Portland, OR.

Boyce PC. 1993. The genus *Arum*. Royal Botanic Gardens, Kew.

Boyce PC and A Hay. 2001. A taxonomic revision of Araceae tribe Potheae (*Pothos, Pothoidium* and *Pedicellarum*) for Malesia, Australia and the tropical western Pacific. Telopea 9: 449–571.

Buell MF. 1935. Seed and seeding of *Acorus calamus*. Bot. Gaz. 96: 758–795.

Buell MF. 1938. Embryology of *Acorus calamus*. Bot. Gaz. 99: 556–568.

Buscalioni L and D Lanza. 1935. Le basi morfologiche, anatomiche, teratologiche della nuova famiglia delle Pistiaceae (Buscalioni e Lanza) rappresentate dai duegeneri *Pistia* ed *Ambrosinia*. Malpighia 34: 103–180.

Buzgo M. 1994. Inflorescence development of *Pistia stratiotes* (Araceae). Bot. Jahrb. Syst. 115: 557–570.

Buzgo M. 2001. Flower structure and development of Araceae compared with alismatids and Acoraceae. Bot. J. Linn. Soc. 136: 393–425.

Buzgo M and PK Endress. 1998. Floral development of *Acorus calamus* (Acoraceae) in comparison with other basal monocots. In Monocots II, p. 13 (abstract). Sydney.

Buzgo M and PK Endress. 1999. The gynoecium of *Gymnostachys* (Araceae) and *Potamogeton* (Potomagetonaceae). In XVI International Botanical Congress: Abstracts, p. 240. St. Louis, MO.

Buzgo M and PK Endress. 2000. Floral structure and development of Acoraceae and its systematic relationships with basal angiosperms. Int. J. Plant Sci. 161: 23–41.

Campbell DH 1899. Notes on the structure of the embryo sac in *Sparganium* and *Lysichiton*. Bot. Gaz. 27: 153–166.

Campbell DH. 1900. Studies on the Araceae. 1. Ann. Bot. 14: 1–15.

Carlquist S and EL Schneider. 1997. Origins and nature of vessels in Monocotyledons. 1. *Acorus*. Int. J. Plant Sci. 158: 51–56.

Carlquist S and EL Schneider 1998. Origin and nature of vessels in monocotyledons. 5. Araceae subfamily Colocasioideae. Bot. J. Linn. Soc. 128: 71–86.

Chao Y and JD Palmer. 1999. Multiple acquisitions via horizontal transfer of a group I intron in the mitochondrial *cox*I gene during evolution of the Araceae family. Mol. Biol. Evol. 16: 1155–1165.

Chen YY, Li DZ, and H Wang. 2002. Infrageneric phylogeny and systematic position of the *Acoraceae* inferred from ITS, 18S and *rbc*L sequences. Acta Bot. Yunn. 24: 699–706.

Chouteau M, D Barabé, and M Gibernau. 2006. Pollen-ovule ratios in some Neotropical Araceae and their putative significance. Plant Syst. Evol. 257: 147–157.

Crawford DJ, E Landolt, DH Les, and RT Kimb. 2001. Allozyme studies in Lemnaceae: variation and relationships in *Lemna* sections *Alatae* and *Biformes*. Taxon 50: 987–999.

Crawford DJ, E Landolt, DH Les, and RT Kimball. 2006. Speciation in duckweeds (Lemnaceae): phylogenetic and ecological inferences. In: JT Columbus, EA Friar, JM Porter, LM Prince, and MG Simpson, eds. Monocots: comparative biology and evolution (excluding Poales), pp. 231–242, Rancho Santa Ana Botanical Garden, Claremont.

Croat TC. 1990. A comparison of aroid classification systems. Aroideana 13: 44–63.

Croat TC. 1998. History and current status of systematic research with Araceae. Aroideana. 21: 26–145.

Daubs EN. 1965. A monograph of Lemnaceae. Illinois Biological Monographs 34.

Den Hartog C and F van der Plas. 1970. A synopsis of the Lemnaceae. Blumea 18: 355–368.

Duvall MR. 2001. An anatomical study of anther development in *Acorus* L.: phylogenetic implications. Plant Syst. Evol. 228: 143–152.

Duvall MR, MT Clegg, MW Chase, WD Clark, WJ Kress, HG Hims, LE Eguiarte, JF Smith, BS Gaut, EA Zimmer, and GH Learn, Jr. 1993a. Phylogenetic hypotheses for the monocotyledons constructed from rbcL sequence data. Ann. Missouri Bot. Gard. 80: 607–619.

Duvall MR, GH Leaen, Jr., LE Eguiarte, and MT Clegg. 1993b. Phylogenetic analysis of *rbc*L sequences identifies *Acorus calamus* as the primal extant monocotyledon. Proc. Natl. Acad. Sci. USA 90: 4641–4644.

Engler A. 1876. Vergleichende Untersuchungen über die morphologischen Verhaltnisse der Araceae: I. Naturliches System der Araceae. Nova Acta Acad. Caes. Leop.-Carol. German. Nat. Cur. 39: 133–155.

Engler A. 1884. Beitrage zur Kenntnis der Araceae: V. 1, 2. Über den Entwicklungsgang in der Familie der Araceen und über die Blutenmorphologie derselben. Bot. Jahrb. Syst. 5: 141–188, 287–336.

Erti PO. 1932. Vergleichende Untersuchungen über die Entwicklung der Blattnervatur der Araceen. Flora 126: 115–248.

Evstatieva LN, MN Todorova, IV Ognyanov, and LV Kuleva. 1996. Chemical composition of the essential oil in *Acorus calamus* L. (Araceae). Fitologija (Bulgaria) 48: 19–23.

Eyde RH, DH Nicolson, and P Sherwin. 1967. A survey of floral anatomy in Araceae. Am. J. Bot. 54: 478–497.

Fox MG and JC French. 1988. Systematic occurrence of sterols in latex of Araceae: Colocasioideae. Am. J. Bot. 75: 132–137.

French JC. 1986a. Patterns of stamen vascularity in the Araceae. Am. J. Bot. 73: 434–449.

French JC. 1986b. Ovular vasculature in Araceae. Bot. Gaz. 147: 478–495.

French JC. 1986c. Patterns of endothecial wall thickenings in Araceae: subfamilies Colocasioideae, Aroideae, and Pistioideae. Bot. Gaz. 147: 166–179.

French JC. 1987a. Systematic occurrence of a sclerotic hypodermis in roots of Araceae. Am. J. Bot. 74: 891–903.

French JC. 1987b. Structure of ovular and placental trichomes of Araceae. Bot. Gaz. 148: 198–208.

French JC. 1987c. Systematic survey of resin canals in roots of Araceae. Bot. Gaz. 148: 360–371.

French JC. 1988. Systematic occurrence of anastomosing laticifers in Araceae. Bot. Gaz. 149: 71–81.

French JC. 1997. Vegetative anatomy. In: SJ Mayo, J Bogner, and PC Boyce. The genera of Araceae, pp. 9–24. Royal Botanic Gardens, Kew.

French JC and CT Kessler. 1989. Molecular systematics of the Araceae: are *Acorus* and *Gymnostachys* aroids? Am. J. Bot. 76(Suppl.): 242.

French JC and PB Tomlinson. 1981a. Vascular patterns in stems of Araceae: subfamily Pothoideae. Am. J. Bot. 68: 713–729.

French JC and PB Tomlinson. 1981b. Vascular patterns in stems of Araceae: subfamily Monsteroideae. Am. J. Bot. 68: 1115–1129.

French JC and PB Tomlinson. 1981c. Vascular patterns in stems of Araceae: subfamilies Calloideae and Lasioideae. Bot. Gaz. 142: 366–381.

French JC and PB Tomlinson. 1981d. Vascular patterns in stems of Araceae: subfamily Philodendroideae. Bot. Gaz. 142: 550–563.

French JC and PB Tomlinson. 1983. Vascular patterns in stems of Araceae: subfamilies Calocasioideae, Aroideae and Pistioideae. Am. J. Bot. 70: 756–771.

Gonçalves EG, Élder AS Paiva, and MA Nadruz Coelho. 2004. A preliminary survey of petiolar collenchyma in the Araceae. Ann. Missouri Bot. Gard. 91: 473–484.

Goremykin VM, B Holland, KI Hirsch-Ernst, and FH Hellwig. 2005. Analysis of *Acorus calamus* genome and its phylogenetic implications. Mol. Biol. Evol. 22: 1813–1222.

Gow JE. 1913. Phylogeny of the Araceae. Proc. Iowa Acad. Sci. 20: 161–168.

Govaerts R and DG Frodin. 2002. World checklist and bibliography of Araceae (and Acoraceae). Royal Botanic Gardens, Kew.

Grayum MH. 1984. Palynology and phylogeny of the Araceae. Ph.D. dissertation, University of Massachusetts, Amherst, MA.

Grayum MH. 1985. Evolutionary and ecological significance of starch storage in pollen of the Araceae. Am. J. Bot. 72: 1565–1577.

Grayum MH. 1986. Phylogenetic implications of pollen nuclear number in the Araceae. Plant Syst. Evol. 151: 145–161.

Grayum MH. 1987. A summary of evidence and arguments supporting the removal of *Acorus* from the Araceae. Taxon 36: 723–729.

Grayum MH. 1990. Evolution and phylogeny of the Araceae. Ann. Missouri Bot. Gard. 77: 628–697.

Grayum MH. 1991. Systematic embryology of the Araceae. Bot. Rev. 57: 167–203.

Grayum MN. 1992. Comparative external pollen ultrastructure of the Araceae and putatively related taxa. Monogr. Syst. Bot. Missouri Bot. Gard. 43: 1–167.

Grob GB, B Gravendeel, MCM Eurlings, and WLA Hetterscheld. 2002. Phylogeny of the tribe Thomsonieae (Araceae) based on chloroplast *mat*K and *trn*L intron sequences. Syst. Bot. 27: 453–467.

Gupta BL. 1935. Studies on the development of the pollen grain and embryo sac of *Wolffia arrhiza*. Curr. Sci. 4: 104–105.

Haccius B and KK Lakshmanan. 1966. Vergleichende Untersuchung der Entwicklung von Kotyledon und Sprofischeitel bei *Pistia stratiotes* und *Lemna gibba:* Ein Beitrag zum Problem der sogenannten terminalen Blattorgane. Beitr. Biol. Pfl. 42: 425–443.

Hartog C den and F von der Plas. 1970. A synopsis of the Lemnaceae. Blumea 18: 355–368.

Hay A. 1992. Tribal and subtribal delimitation and circumscription of the genera of Araceae tribe Lasieae. Ann. Missouri Bot. Gard. 79: 184–205.

Hesse M. 2002. The uniquely designed pollen aperture in *Lasioideae* (*Araceae*). Aroideana. 25: 51–59.

Hesse, M. 2006a. Pollen wall ultrastructure in Araceae and Lemnaceae in relation to molecular classifications. In: JT Columbus, EA Friar, JM Porter, LM Prince, and MG Simpson, eds. Monocots: comparative biology and evolution (excluding Poales), pp. 204–208. Rancho Santa Ana Botanical Garden, Claremont.

Hesse M. 2006b. Reasons and consequences of the lack of sporopollenin ektexine in Aroideae (Araceae). Flora 201: 421–428.

Hesse M, M Weber, and H Halbritter. 1998. Pollen wall stratification: its possible role in Araceae systematics. In Monocots II, p. 28. Sydney.

Hesse M, J Bogner, H Halbritter, and M Weber. 2001. Palynology of the perigoniate *Aroideae: Zamioculcas, Gonatopus* and *Stylochaeton* (*Araceae*). Grana 40: 26–34.

Hotta M. 1971. Study of the family Araceae – general remarks. Jpn. J. Bot. 20: 269–310.

Jussen FJ. 1929. Die Haploidgeneration der Araceen und ihre Verwertung fur die Systematik. Bot. Jahrb. Syst. 62: 155–283.

Il'ina GM. 1990. Lemnaceae. In: TB Batygina and MS Yakovlev, eds. Comparative embryology of flowering plants. Monocotyledons, pp. 279–286. Nauka, Leningrad (in Russian).

Ivanova IE. 1973. On the systematics of the family Lemnaceae. Bot. Zhurn. 58: 1413–1428 (in Russian).

Kaplan DR. 1970. Comparative foliar histogenesis of *Acorus calamus* and its bearing on the phyllode theory of monocotyledonous leaves. Am. J. Bot. 57: 331–361.

Kaplan DR. 1973. Comparative developmental analysis of heteroblastic leaf series of axillary shoots of *Acorus calamus* L. Cellule 69: 253–290.

Keating RC. 2000. Collenchyma in Araceae: trends and relation to classification. Bot. J. Linn. Soc. 134: 203–214.

Keating RC. 2002. Leaf anatomical characters and their value in understanding morphoclines in the Araceae. Bot. Rev. 68: 510–523.

Keating RC. 2003a. Acoraceae and Araceae. In M Gregory and D Cutler, eds. The anatomy of the Monocotyledons, vol. 9, pp. 1–327. Oxford University Press, Oxford.

Keating RC. 2003b. Leaf anatomical characters and their value in understanding morphoclines in the Araceae. Bot. Rev. 68: 510–523.

Keating RC. 2004a. Vegetative anatomical data and its relationship to a revised classification of the genera of Araceae. Ann. Missouri Bot. Gard. 91: 485–494.

Keating RC. 2004b. Systematic occurrence of raphide crystals in Araceae. Ann. Missouri Bot. Gard. 91: 495–504.

Korobova SN and NA Zhinkina. 1990. Araceae. In: TB Batygina and MS Yakovlev, eds. Comparative embryology of flowering plants. Monocotyledons, pp. 275–279. Nauka, Leningrad (in Russian).

Kozhevnikov DA. 1878. On the history of development of flower in the fam. Araceae. Nauka, Moscow (in Russian).

Kulkarni AR, D Dosi, and VM Manoj. 1990. Fruit and seed structure in Araceae. Proc. Indian Acad. Sci. 100B: 61–70.

Kuprianova LA and VF Tarasevich. 1984. The ultra-structure of the surface of pollen grain wall in some genera of the family Lemnaceae and the related genera of the family Araceae. Bot. Zhurn. 69: 1656–1661 (in Russian with English summary).

Landolt E. 1986. The family of Lemnaceae – a monographic study. Vol. 1: Morphology; karyology; ecology; geographic distribution; systematic position; nomenclature; descriptions. Veroff. Geobot. Inst. Eidg. Tech. Hochsch. Stift. Rubel Zuer. 71: 566.

Landolt E. 1998. Lemnaceae. In: K Kubitzki, ed. The families and genera of vascular plants, vol 4, pp. 264–270. Springer, Berlin/Heidelberg/New York.

Landolt E and R Kandeler. 1987. The family of Lemnaceae – a monographic study. Vol. 2: Phytochemistry; physiology; application; bibliography. Veroff. Geobot. Inst. Eidg. Tech. Hochsch. Stift Rubel Zuer. 95: 638.

Lawalree A. 1952. L'embryologie des Lemnaceae. Observations sur *Lemna minor*. Cellule 54: 305–326.

Lawalree A. 1961. La polinisation de *Lemna minor* L. Nat. Belg. 42: 164–165.

Lemon GD and U Posluszny. 2000a. Shoot development and evolution in *Pistia stratiotes* (Araceae). Int. J. Plant Sci. 161: 721–732.

Lemon GD and U Posluszny. 2000b. Comparative shoot development and evolution in the Lemnaceae. Int. J. Plant Sci. 161: 733–748.

Les DH and DJ Crawford. 1999. *Landoltia* (Lemnaceae) a new genus of duckweeds. Novon 9: 530–533.

Les DH, E Landolt, and DJ Crawford. 1994. Molecular systematics of the Lemnaceae. Am. J. Bot. 81: 168–169.

Les DH, E Landolt, and DJ Crawford. 1997. Systematics of Lemnaceae (duckweeds), inferences from micromolecular and morphological data. Plant Syst. Evol. 204: 161–177.

Les DH, DJ Crawford, E Landolt, JD Gabel, and RT Kimball. 2002. Phylogeny and systematics of Lemnaceae, the duckweed family. Syst. Bot. 27: 221–240.

Lodkina MM. 1985. Lemnaceae. In: A Takhtajan, ed. Comparative seed anatomy, vol 1, pp. 275–280. Nauka, Leningrad (in Russian).

Maheshwari SC. 1954. The embryology of *Wolffia*. Phytomorphology 4: 355–365.

Maheshwari SC. 1956. The endosperm and embryo of *Lemna* and systematic position of Lemnaceae. Phytomorphology 6: 51–55.

Maheshwari SC. 1958. *Spirodela polyrrhiza*: the link between the aroids and the duckweeds. Nature 181: 1745–1756.

Maheshwari SC and RN Kapil. 1963a. Morphological and embryological studies on the Lemnaceae. I. The floral structure and gametophytes of *Lemna paucicostata*. Am. J. Bot. 50: 677–686.

Maheshwari SC and RN Kapil. 1963b. Morphological and embryological studies on the Lemnaceae. II. The endosperm and embryo of *Lemna paucicostata*. Am. J. Bot. 50: 907–914.

Maheshwari SC and PP Khanna. 1956. The embryology of *Arisaema wallichianum* Hook. f. and the systematic position of the Araceae. Phytomorphology 6: 379–388.

Maheshwari SC and N Maheshwari. 1963. The female gametophyte, endosperm and embryo of *Spirodela polyrrhiza*. Beitr. Biol. Pflanz. 39: 179–188.

Mayo SJ, J Bogner, and PC Boyce. 1995. The Arales. In: PJ Rudall, PJ Cribb, DF Cutler, CJ Humphries, eds. Monocotyledons: systematics and evolution, pp. 277–286. Royal Botanic Gardens, Kew, London.

Mayo SJ, J Bogner, and PC Boyce. 1997. The genera of Araceae. Royal Botanic Gardens, Kew.

Mayo SJ, J Bogner, and PC Boyce. 1998. Araceae. In: K Kubitzki, ed. Families and genera of vascular plants, vol 4, pp. 26–74. Springer, Berlin/Heidelberg/New York.

Mayo SJ, L Cabrera, G Salazar, and MW Chase. 2003. Aroids and their watery beginnings. Ms.

McClure JW and RE Alston. 1966. A chemotaxonomic study of Lemnaceae. Am. J. Bot. 53: 849–860.

Mercado-Noriel LR and BT Mercado. 1978. Floral anatomy and seed morphology of water lettuce (*Pistia stratiotes*). Philipp. Agric. 61: 281–290.

Mucke M. 1908. Über den Bau und die Entwicklung der Fruchte und über die Herkunft von *Acorus calamus* L. Bot. Zeit. 66: 1–123.

Nahrstedt A. 1975. Triglochinin in Araceen. Phytochemistry 14: 2627–2628.

Nicolson DH. 1984. Suprageneric names attributable to Araceae. Taxon 33: 680–690.

Nicolson DH. 1988. History of Araceae systematics. Aroideana 10: 23–30.

Oganezova GG and NA Barsegyan. 1999. Some peculiarities of the generative organs of *Acorus calamus* L. from Armenian populations. Flora Rastitelnost Rast. Res. Armenia 12: 39–41 (in Russian).

Pan YH, KM Liu, and LG Lei. 2002. Advances in the systematics of *Acorus* L. and the re-establishment of Acoraceae. Bull. Bot. Res. (China). 22: 417–421.

Petersen G. 1989. Cytology and systematics of Araceae. Nord. J. Bot. 9: 116–166.

Ray TS. 1987. Leaf types in the Araceae. Am. J. Bot. 74: 1359–1372.

Ray TS. 1988. Survey of shoot organization in the Araceae. Am. J. Bot. 75: 56–84.

Remizova M and D Sokoloff. 2003. Inflorescence and floral morphology in *Tofieldia* (Tofieldiaceae) compared with *Araceae, Acoraceae* and *Alismatales* s.str. Bot. Jahrb. Syst. 124: 255–271.

Riaz M, S Qamar, and FM Chaudhary. 1995. Chemistry of the medicinal plants of the genus *Acorus* (family Araceae). Hamdard Med. 38(2): 50–62.

Rost LCM. 1979. Biosystematic inversigations with *Acorus*. 4. Communication: a synthetic approach to the classification of the genus. Planta Med. 37: 289–307.

Rostowzew S. 1905. Biology and morphology of duckweeds. Nauka, Moscow (in Russian).

Rothwell GW, MR Van Atta, HE Ballard Jr, and RA Stockey. 2004. Molecular phylogenetic relationships among Lemnaceae and Araceae using the chloroplast *trn*L-*trn*F intergenic spacer. Mol. Phylogenet. Evol. 30: 378–385.

Rudall PJ and CA Furness. 1997. Systematics of *Acorus*: ovule and anther. Int. J. Plant Sci. 158(5): 640–651.

Schneider EL and S Carlquist. 1998. Origin and nature of vessels in monocotyledons. 4. Araceae subfamily Philodendroideae. J. Torrey Bot. Soc. 125: 253–260.

Scribailo RW and PB Tomlinson. 1992. Shoot and floral development in *Calla palustris* (Araceae-Calloideae). Int. J. Plant Sci. 153: 1–13.

Seubert E. 1993. Die Samenmerkmale der Araceen und ihre Bedeutung für die Gliederum der Familie. Koeltz, Koenigstein.

Seubert E. 1997a. The sclereids of Araceae. Flora 192: 31–37.

Seubert E. 1997b. A comparative study of the seeds of Lasieae (Araceae). Bot. Jahrb. Syst. 119: 407–426.

Shadowsky AF. 1931. Einige Angaben fiber die Embryogenie von *Pistia stratiotes* L. Ber. Deutsch. Bot. Ges. 49: 350–356.

Silva CJ da. 1981. Observacoes sobre a biologia repro-dutiva *Pistia stratiotes* L. (Araceae). Acta Amazonica 11: 487–504.

Soukup A, JL Seago Jr, and O Votrubová. 2005. Developmental anatomy of the root cortex of the basal Monocotyledon, *Acorus calamus* (Acorales, Acoraceae). Ann. Bot. 96: 379–385.

Stevenson DW, JI Davis, JV Freudenstein, CR Hardy, MP Simmonds, and CD Specht. 2000. A phylogenetic analysis of the monocotyledons based on morphological and molecular character sets, with comments on the placement of *Acorus* and Hydatellaceae. In: KL Wilson and DA Morrison, eds. Monocots: systematics and evolution, pp. 17–24. CSIRO, Collingwood.

Su_XH, GM He, KL Sun, DL Bi, and M Wang. 2004. Study of comparative anatomy on structure of seedlings of *Acorus tatarinowii* and *Zantedeschia aethiopica*. Acta Bot. Bor. Occid. Sinica 24(3): 504–509.

Tam S-M, PC Boyce, TM Upson, D Barabé, A Bruneau, F Forest, and JS Parker. 2004. Intergeneric and infrafamilial phylogeny of subfamily Monsteroideae (Araceae) revealed by chloroplast *trnL*-F sequences. Am. J. Bot. 91: 490–498.

Tarasevich VF. 1989. Pollen grain ultrastructure in the genus *Anthurium* (Araceae) in connection with its systematics. Bot. Zhurn. 74: 314–324 (in Russian with English summary).

Tarasevich VF. 1990. Palynological evidence on the position of the Lemnaceae family in the system of flowering plants. Bot. Zhurn. 75: 959–965 (in Russian with English summary).

Tillich H-J. 1985. Keimlingsbau und verwandtschaftliche Bezeihungen der Araceae. Gleditschia 13: 63–73.

Tillich H-J. 2003. Seedling diversity in Araceae and its systematic implications. Feddes Repert. 114: 454–487.

Van der Ham RWJM, WLA Hetterscheid, and BJ Van Heuven. 1998. Notes on the genus *Amorphophallus* (Araceae) – 8 Pollen morphology of *Amorphophallus* and *Pseudodracontium*. Rev. Palaeobot. Palynol. 103: 95–142.

Vyshenskaya TD. 1985. Araceae. In: A Takhtajan, ed. Comparative seed anatomy, 1: 264–275. Nauka, Leningrad (in Russian).

Wang HZ, YG Chen, and CS Fan. 1998. Review of studies on chemical constituents and pharmacology in genus *Acorus* in China. Acta Bot. Yunn. Suppl. 10: 96–100.

Wang HZ, WL Li, ZJ Gu, and YY Chen. 2001. Cytological study on *Acorus* L. in southwestern China, with some cytogeographical notes on *A. calamus*. Acta Bot. Sinica 43: 354–358.

Wang PL and H Li. 1998. Report of pollen morphology of Araceae. Acta Bot. Yunn. Suppl. 10: 41–42.

Wang W and NX Zhao. 2002. Epidermal characters of leaves in Araceae. J. Wuhan Bot. Res. 20: 343–349.

Watling JR, SA Robinson, and RS Seymour. 2006. Contribution of the alternative pathway to respiration during thermogenesis in flowers of the Sacred Lotus. Plant. Physiol. 140: 1367–1373.

Weber M, H Halbritter, and M Hesse. 1999. The basic pollen wall types in Araceae. Int. J. Plant Sci. 160: 415–423.

Williams NH, JB Harborne, and SJ Mayo. 1981. Anthocyanin pigments and leaf flavonoids in the family Araceae. Phytochemistry 20: 217–234.

Wilson KA. 1960. The genera of the Arales in the southeastern United States. J. Arnold Arbor. 41: 47–72.

Zennie TM and JW McClure. 1977. The flavonoid chemistry of *Pistia stratiotes* L. and the origin of the Lemnaceae. Aquatic Bot. 3: 49–54.

Zhu ZY. 1985 Some new taxa of *Acorus* (Araceae) from Sichuan. Acta Bot. Bor.-Occid. Sinica 5: 118–121.

Subclass II. LILIIDAE

Perennial or annual herbs or more or less arborescent plants often with secondary growth of monocotyledonous type, often with well-developed subterranean storage organs, such as rhizomes, bulbs, corms, and tubers. Vessels present or absent, with scalariform or simple perforations. Leaves both basal and cauline or only basal, alternate, opposite, or verticillate, mostly with parallel venation. Stomata of various types. Flowers from very small to rather large in various kinds of inflorescences or solitary, bisexual or unisexual, actinomorphic or zygomorphic, with double or simple perianth, entomophilous or anemophilous. Tapetum usually secretory. Microsporogenesis mostly successive. Pollen grains 2-celled or 3-celled, 1-colpate or of various derived types. Gynoecium apocarpous or syncarpous. Ovules mostly anatropous, commonly bitegmic, usually crassinucellate. Female gametophyte mostly of *Polygonum*-type. Endosperm helobial or nuclear. Seeds mostly with more or less copious endosperm (lacking in Orchidales).

The most archaic liliids are characterized by some very primitive features, especially in their carpel and pollen morphology. However, some members of the Liliidae, including Orchidaceae, have reached a very high level of specialization.

Bibliography

Ambrose JD. 1980. A re-evaluation of the Melanthioideae (Liliaceae) using numerical analyses. In: CD Brickell et al., eds. Petaloid Monocotyledons, pp. 65–81. Academic, London.

Cheadle VI and H Kosakai. 1971 (1972). Vessels in Liliaceae. Phytomorphology 21: 320–333.

Chupov VS. 1994. Phylogeny and systematics of the Liliales and Asparagales. Bot. Zhurn. 79(3): 1–12 (in Russian with English summary).

Conover MV. 1983. The vegetative morphology of the reticulate-veined Liliiflorae. Telopea 2: 491–412.

Conover MV. 1991. Epidermal patterns of the reticulate-veined Liliiflorae and their parallel-veined allies. Bot. J. Linn. Soc. 107: 295–312.

Conran JG. 1989. Cladistic analysis of some net-veined Liliiflorae. Plant Syst. Evol 168: 123–141.

Fay MF, PJ Rudall, S Sullivan, KL Stobart, AY de Bruijn, G Reeves, F Qamaruz-Zaman, W-P Hong, J Joseph, WJ Hahn, JG Conran, and MW Chase. 2000. Phylogenetic studies of Asparagales based on four plastid DNA regions. In: KL Wilson and DA Morrison, eds. Monocots: systematics and evolution, pp. 360–371. CSIRO, Collingwood.

Goldblatt P. 1995. The status of R. Dahlgten's orders Liliales and Melanthiales. In: PJ Rudall, PJ Cribb, DF Cutler, and CJ Humphries, eds. Monocotyledons: systematics and evolution, vol. 1, pp. 181–200. Royal Botanic Gardens, Kew.

Huber H. 1969. Die Samenmerkmale und Verwandtschaftsverhaltnisse der Liliifloren. Mitt. Bot. Staatssamml. München 8: 219–538.

Kauff F, PJ Rudall, and JG Conran. 2000. Systematic root anatomy of Asparagales and other monocotyledons. Plant Syst. Evol. 223: 139–154.

Komar GA. 1978. Arils and aril-like formations in some Liliales. Bot. Zhurn. 63: 937–955 (in Russian).

Kosenko VN. 2001. Palynological data on the systematics of the superorder Lilianae. Bot. Zhurn. 86(8): 1–17 (in Russian with English summary).

Nair PKK and M Sharma. 1965. Pollen morphology of Liliaceae. J. Palyn. (Lucknow) 1: 39–61.

Oganezova GH. 2000. Systematic position of the Trilliaceae, Smilacaceae, Herreriaceae, Tecophilaeaceae, Dioscoreaceae families and the volume and phylogeny of the Asparagales (based on the seed structure). Bot. Zhurn. 85(9): 9–25 (in Russian with English summary).

Radulescu D. 1973. Liliiflorae: Discussions et consideradons phylogenetiques a l'aide de quelques recherches morphologiques. Acta Bot. Horti Bucurest. 1972–1973: 249–283.

Rudall PJ, KL Stobart, W-P Hong, JG Conran, CA Furness, G Kite, and MW Chase. 2000. Consider the Lilies: Systematics of Liliales. In: KL Wilson and D Morison, eds. Monocots: systematics and evolution, pp. 347–359. CSIRO, Collingwood.

Rudall PG and RM Bateman. 2002. Roles of synorganisation, zygomorphy and heterotopy in floral evolution: the

gynostemium and labellum of orchids and other lilioid monocots. Biol. Rev. 77: 403–441.
Satô D. 1942. Karyotype alteration and phylogeny in Liliaceae and allied families. Jpn. J. Bot. 12: 57–161.
Sen S. 1975. Cytotaxonomy of Liliales. Feddes Repert. 86: 255–305.
Shamrov II. 1999. The ovule and seed development in some representatives of the orders Liliales and Amaryllidales. Bot. Zhurn. 84(2): 13–33 (in Russian).
Slob A, B Jekel, and E Schlatmann. 1975. On the occurrence of tuliposides in the Liliiflorae. Phytochemistry 14: 1997–2005.
Vijayavalli B and PM Mathew. 1990. Cytotaxonomy of the Liliaceae and allied families. Continental Publishers, Kerala, India.
Vinersten A and K Bremer. 2001. Age and biogeography of major clades in Liliales. Amer. J. Bot. 88: 1695–1703.
Watson S. 1879. Contributions to American botany: I. Revision of the North American Liliaceae. Proc. Am. Acad. Arts Sci. 14: 213–288.
Williams CA, JB Harborne, and B Mathew. 1988. A chemical appraisal via leaf flavonoids of Dahlgren's Liliiflorae. Phytochemistry 27: 2609–2629.
Wunderlich R. 1936. Vergleichende Untersuchungen von Pollenkornern einiger Liliaceen und Amarylidaceen. Oesterr. Bot. Z. 85: 30–55.
Zomlefer WB. 1999. Advances in angiosperm systematics: examples from the Liliales and Asparagales. J. Torrey Bot. Soc. 126: 58–62.

Superorder LILIANAE

Order 5. MELANTHIALES

Perennial or rarely annual herbs, commonly with rhizomes, rarely cormlike, and sometimes with a bulblike base but without nutrient bulb scales. Raphides often present. Roots fibrous. Vessels only in roots, with scalariform perforations. Aerial stem erect, simple, stout to slender, with foliage or scaly leaves, sometimes basally thickened. Leaves deciduous or evergreen, alternate, all cauline (*Veratrum* spp.), or with large basal and small cauline leaves (*Xerophyllum*, *Zigadenus* spp.), or all leaves in a basal rosette (*Heloniopsis*), bifacial, distichous, sheathing at the base, linear, lanceolate, or rarely ovate, up to 50 cm long. Stomata anomocytic. Inflorescences terminal, often racemes, sometimes panicles, spikes or umbel-like, glabrous or pubescent. Flowers small, bracteate or ebracteate, without bracteoles, bisexual or very rarely unisexual, 3-merous, actinomorphic or rarely (*Chionographis*) zygomorphic. Perianth segments six in two cycles, generally similar, free or sometimes more or less basally connate, in some genera with basal nectaries. Stamens usually six in two cycles; filaments free or rarely adnate to the base of the perianth segments; filaments filiform to subulate; anthers tetrasporangiate, basifixed or dorsifixed-hypopeltate, mostly extrorse, opening longitudinally, Tapetum secretory. Microsporogenesis successive. Pollen grains 2-celled, tectate-columellate or less often (Heloniadeae and Chionographideae) intectate, mostly 1-colpate, or 4-porate (Chionographideae). Gynoecium of three carpels, free or more or less united and with free or connate stylodia; ovary superior or less often semi-inferior; each carpel, each locule and each parietal placenta with two to numerous ovules. Ovules anatropous or sometimes campylotropous and epitropic, bitegmic, crassinucellate to tenuicellate, with parietal cell. Female gametophyte of Polygonum-type or Allium-type. Endosperm helobial. Fruits septicidal Chionographideae), loculicidal (Xerophylleae and Heloniadeae), or ventricidal capsules, conical to globular, often 3-lobed. Seeds generally provided with terminal appendages or wings; seed coat essentially testal; the testal part has few cell layers, lack phytomelan but sometimes has phlobaphene; the tegminal part is thin and collapsed, red brown, or yellowish, though flattened; endosperm usually consists of thin-walled cells containing aleurone and fatty oils, sometimes also rounded starch grains; embryo small (1/5–1/9 of the length of endosperm), basal, globose or ovoid, or rather large and almost equal to the length of the seed (as in species of *Melanthium* and *Zigadenus*) and axial and linear. Steroidal saponins and chelidonic acid occur in many genera. Veratrum and some other related genera contain alkaloids derived from steroidal precursors; n = 6 (*Chionographis* – Zomlefer 1997) n = 8, 10–12, 15–17, 21–22.

Melanthiales, especially genera with more or less free carpels, are the most archaic members of the Liliidae and probably of the monocotyledons in general (see Lotsy 1911; Takhtajan 1959, 1966, 1980, 1987; Eames 1961; Radulescu 1973d).

1. MELANTHIACEAE

Batsch 1802 (including Chionographidaceae Takhtajan 1996; Heloniadaceae J. Agardh 1858; Veratraceae Salisbury 1807; Xerophyllaceae Takhtajan 1996). 13/100.

Temperate Eurasia, North America; *Schoenocaulon* (24) New Mexico, Texas and Florida through Mexico (c.20) south to Venezuela, southern Peru (1, *S. officinale*).

XEROPHYLLEAE: *Xerophyllum*; MELANTHIEAE: *Veratrum, Melanthium, Stenanthium, Amianthium, Schoenocaulon, Zigadenus, Anticlea, Toxicoscordion*; HELONIADEAE: *Helonias* (including *Heloniopsis* – Tanaka 1998), *Ypsilandra*; CHIONOGRAPHIDEAE: *Chamaelirium, Chionographis*.

Bibliography

Alison B, P Whiting, SD Sarker, L Dinan, E Underwood, V Sik, and HH Rees. 1997. 20-Hydroxyecdysone 2-B-D-glucopyranoside from the seeds of *Xerophyllum tenax*. Biochem. Syst. Ecol. 25: 255–261.

Ambrose JD. 1975. Comparative anatomy and morphology of the Melanthioideae (Liliaceae). Ph.D. dissertation, Cornell University. Ithaca, NY.

Ambrose JD. 1980. A re-evaluation of the Melanthioideae (Liliaceae) using numerical analyses. In: CD Brickell, DF Cutler and M Gregory, eds. Petaloid monocotyledons, pp. 65–81, pl. 1–2. Academic, London.

Badawi A. 1986. The main taxonomic view points on the intra- and the interrelationships of Melanthioideae (Liliaceae). Phytologia 61: 346–350.

Baillon H. 1893. L'organisation et les affinites des Campynemees. Bull. Mens. Soc. Linn. Paris 2: 1105–1109.

Behnke H-D. 2000. Forms and sizes of sieve-element plastids and evolution of the monocotyledons. In: Wilson KL and DA Morrison, eds. Monocots: systematics and evolution, pp. 163–188. CSIRO, Collingwood.

Behnke H-D. 2002 (2003). Sieve-element plastids and evolution of monocotyledons with emphasis on Melanthiaceae sensu lato and Aristolochiaceae-Asaroideae, a putative dicotyledon sister group. Bot. Rev. 68: 524–544.

Buxbaum F. 1925. Vergleichende Anatomic der Melan-thioideae. Repert. Spec. Nov. Reg. Veget. 29: 1–80.

Buxbaum F. 1927. Nachtrage zur vergleichenden Anatomic der Melanthioideae, part 1. Beih. Bot. Centralbl. 44: 255–263.

El-Hamidi A. 1952. Vergleichend-morphologische Untersuchungen am Gynoeceum der Unterfamilien Melanthioideae and Asphodelioideae der Liliaceae. Arbeit. Inst. allgem. Bot., Univ. Zürich, ser. A, 4: 1–50.

Frame DM. 2001. Chromosome studies in *Schoenocaulon* (Liliaceae: Melanthieae) a relict genus. An. Inst. Biol. Univ. Nac. Auton. Mex., Bot. 72: 123–129.

Frame DM, A Espejo, and AR López-Ferrari. 1999. A conspectus of Mexican Melanthiaceae including a description of new taxa of *Schoenocaulon* and *Zigadenus*. Acta Bot. Mexicana 48: 27–50.

Fuse S and MN Tamura 2000. A phylogenetic analysis of the plastid *mat*K gene with emphasis on Melanthiaceae sensu lato. Plant Biol. 2: 415–427.

Gates RR. 1918. A systematic study of the North American Melanthiaceae from a genetic standpoint. Bot. J. Linn. Soc. 44: 131–172.

Goldblatt P. 1995. The statis of R. Dahlgren's orders Liliales and Melanthiales. In: PJ Rudall, PJ Cribb, DF Cutler, and CJ Humphries, eds. Monocotyledons: systematics and evolution, pp. 181–200. Royal Botanic Gardens, Kew.

Gray A. 1837. Melanthacearum Americae Septentrionalis Revisio. Ann. Lyceum Nat. Hist. New York 4: 104–140.

Hara H. 1968. A revision of the genus *Chionographis* (Liliaceae). J. Jpn. Bot. 43: 257–267.

Kosenko VN. 1988. Pollen morphology in Chiono-graphideae, Uvularieae, Tricyrtideae, Scoliopeae, Anguillarieae, Iphigenieae, Glorioseae, Colchiceae (Melanthiaceae). Bot. Zhurn. 73: 172–185 (in Russian with English summary).

Kupchan SM, JH Zimmerman, and A Afonso. 1961. The alkaloids and taxonomy of *Veratrum* and related genera. Lloydia 24: 1–22.

Oganezova GG. 1984. Morphological and anatomical specific features of seed and fruit in some representatives of the subfamily Melanthioideae (Liliaceae) in relation with their systematics and phylogeny. Bot. Zhurn. 69: 772–781 (in Russian with English summary).

Oikawa K. 1961. The embryo sac of *Chionographis japonica* Maxim. Sci. Rep. Tohoku Imp. Univ., 4th ser. (Biol.) 2: 155–158.

Ono T. 1920. Embryologie der Liliaceae, mit besonderer Rücksicht auf die Endospermbildung. I. Melanthioideae und Aletroideae. Sci. Rep. Tohoku Univ. Biol. 4: 381–393.

Ono T. 1926. Embryologische Studien an *Heloniopsis breviscapa*. Sci. Rep. Tohoku Imp. Univ., 4th ser. (Biol) 2: 93.

Ono T. 1928. Endosperm development in Liliaceae. Bot. Mag. (Tokyo) 42: 335–440. (In Japanese).

Schulze W. 1978. Beiträge zur Taxonomie der Liliifloren. IV. Melanthiaceae. Wiss. Z. Friedrich-Schiller-Univ. Jena. Math.-Naturwiss. Reihe 27: 87–95.

Stenar H. 1928. Zur Embryologie der *Veratrum* und *Anthericum* Gruppe. Bot. Not. 1928: 357–378.

Sterling C. 1978. Comparative morphology of the carpel of the Liliaceae: Hewardieae, Petrosavieae, and Tricyrteae. Bot. J. Linn. Soc. 77: 95–106.

Sterling C. 1980. Comparative morphology of the carpel in the Liliaceae: Helonieae. Bot. J. Linn. Soc. 80: 341–356.

Sterling C. 1982. Comparative morphology of the carpel in the Liliaceae: Veratreae. Bot. J. Linn. Soc. 84: 57–77.

Takahashi M and S Kawano. 1989. Pollen morphology of the *Melanthiaceae* and its systematic implications. Ann. Missouri Bot. Gard. 76: 863–876.

Takhtajan AL. 1994. Six new families of flowering plants. Bot. Zhurn. 79(1): 96–97 (in Russian).

Takhtajan AL. 1994 (1995). New families of the monocotyledons. Bot. Zhurn. 79(12): 65–66 (in Russian).

Takhtajan AL. 1996. Validization of some formerly established families of flowering plants. Bot. Zhurn. 81(2): 85–86 (in Russian).

Tamura MN. 1998a. Melanthiaceae. In: K Kubitzki, ed. The families and genera of vascular plants, vol. 3, pp. 369–380. Springer, Berlin/Heidelberg/New York.

Tanaka NY. 1997, 1998. Phylogenetic and taxonomic studies on *Helonias, Ypsilandra* and *Heloniopsis*. I. Comparison of character states. II. Evolution and geographical distribution.

III. Taxonomic revision. J. Jpn. Bot. 72: 286–292, 329–336; 73: 102–115.
Tanaka NY. 1997. Taxonomic significance of some floral characters in *Helonias* and *Ypsilandra*. J. Jpn. Bot. 72: 110–116.
Tanaka NY and N Tanaka. 1977, 1979, 1980. Chromosome studies in *Chionographis* (Liliaceae). I. On the holokinetic nature of chromosomes in *Chionographis japonica* Maxim. II. Morphological characteristics of the somatic chromosomes of four Japanese members. III. The mode of meiosis. Cytologia 42: 753–763; 44: 935–949; 45: 809–817.
Utech RH. 1978a. Comparison of the vascular anatomy of *Xerophyllum asphodehides* (L.) Nutt. and *X. tenax* (Pursh) Nutt. (Liliaceae-Melanthioideae). Ann. Carnegie Mus. 47: 147–167.
Utech FH. 1978b. Vascular floral anatomy of *Helonias bullata* L. (Liliaceae-Helonieae) with a comparison to the Asian *Heloniopsis orientalis*. Ann. Carnegie Mus. 47: 169–191.
Utech FH and S Kawano. 1981. Vascular floral anatomy of the east Asian *Heloniopsis orientalis* (Thunb.) C. Tanaka (Liliaceae-Helonieae). Bot. Mag. Tokyo 94: 295–311.
Zomlefer WB. 1997. The genera of Melanthiaceae in the southeastern United States. Harvard Pap. Bot. 2: 133–177.
Zomlefer WB and WS Judd. 2002. Resurrection of segregates of the polyphyletic genus *Zigadenus* s.l. (Liliales: Melanthiaceae) and resulting new combinations. Novon 12: 299–308.
Zomlefer WB, NH Williams, WM Whitten, and WS Judd. 2001. Generic circumscription and relationships in the tribe Melanthieae (Liliales, Melanthiaceae), with emphasis on *Zigadenus*: evidence from ITS and *trn*L-F sequence data. Am. J. Bot. 88: 1657–1669.
Zomlefer WB, WM Whitten, NH Williams, and WS Judd. 2003. An overview of *Veratrum* s.l. (Liliales, Melanthiaceae) and an infrageneric phylogeny based on ITS sequence data. Syst. Bot. 28: 250–269.
Zomlefer WB, WM Whitten, NH Williams, and WS Judd. 2006. Infrageneric phylogeny of *Schoenocaulon* (Liliales: Melanthiaceae) with clarification of cryptic species based on ITS sequence data and geographical distribution. Am. J. Bot. 93: 1178–1192.

Order 7. TRILLIALES

Perennial herbs with long or short (sometimes tuberous), horizontal or vertical, monopodial rhizome and mostly one erect, unbranched stem. Oxalate raphides present in stem and leaves. Vessels only in roots, with scalariform perforations. Sieve-element plastids of P2c-type (Behnke 2002). Leaves in one verticil of three (*Trillium*) or 4–22 (other genera) at the top of stem, sessile or often petiolate, linear to ovate, elliptic, rounded or rhomboid, entire, sometimes more or less cordate at base, venation parallel to arcuate-pinnate with more or less reticulate secondary veins. Stomata tetracytic. Flowers always solitary and terminal, commonly rather large, mostly on long pedicel, mostly 3-merous or 4-merous, rarely (some species of *Daiswa*) 5-merous, bisexual, actinomorphic, with double perianth or sometimes apetalus. Perianth segments free. Sepals 3–9(10), mostly green, commonly lanceolate; petals 3–9(10), filiform to ovate, of various colours, rarely rudimentary. Stamens 3–24 (very rarely up to 34 in some species of *Daiswa*), mostly 6–12 (in *Trillium* usually 6), in 2–6 cycles, persistent, adnate to base of perianth segments; filaments usually short, slender; anthers basifixed, linear or oblong, often produced into well-developed apical appendage (up to 12–16 mm in *Daiswa thibetica*), dehiscent longitudinally, extrorse to introrse. Tapetum secretory. Microsporogenesis successive. Pollen grains 2-celled, 1-colpate and ellipsoidal or (most species of *Trillium*) inaperturate and spheroidal, tectate or intectate, reticulate, gemmate or spinulose. Septal nectaries present. Gynoecium of 3–12 carpels (in *Trillium* always three); stylodia free or more or less connate into deeply branched style that is usually separated from ovary by transverse rim; ovary superior, roundish or angular, sometimes depressed at the top, 3(-12) locular, 1-locular or 3(-12)-locular below and 1-locular above, with numerous anatropous ovules per locule or on intrusive parietal placentas. Ovules bitegmic, crassinucellate, with parietal cell. Crystal cells containing raphide bundles are common in the chalaza. Female gametophyte is bisporic of *Allium*-type. Endosperm helobial (*Trillium*) or nuclear (*Paris*). Fruits more or less fleshy, berry-like or more or less capsule-like (loculicidal in *Daiswa*). Seeds light to dark brown, ellipsoidal or almost globose, smooth or reticulate-foveolate, in *Daiswa* surrounded by juicy, orange or dark-red sarcotesta and in *Trillium* may bear an oil-rich elaiosome developed from the raphe and hilum; embryo minute, globose to ovoid, undifferentiated; seed coat has parenchymatous testa and phlobaphene containing tegmen; endosperm copious, contains aleurone, lipids, and starch grains. Contain steroidal saponins, flavonols (kaempferol, quercetins), and chelidonic acid. Alkaloids lacking, n = 5. Chromosomes usually very large (40 μm).

Trilliales have definite similarities with the Melanthiales, including free stylodia, helobial endosperm, the presence of parietal cell and crystal raphides, as well as the combined analysis of molecular and non-molecular data (Chase, Stevenson, Wikkin and Rudall 1995). According to Tamura (1998), possibly the Trilliaceae have recently been derived from the stock of the Melanthiaceae and have greatly changed their characteristics. However, from the Melanthiaceae the Trilliaceae differ in basic chromosome number, very large chromosomes, presence of septal nectaries, pollen morphology.

1. TRILLIACEAE

Chevallier 1827 (including Paridaceae Dumortier 1829). 5/70. Europe, Mediterranean, northeastern Anatolia, Caucasus, Siberia, Himalayas, northeastern India, northern Burma, eastern Asia, Indochina, North America.

Trillium (including *Pseudotrillium*), *Trillidium, Kinugasa, Daiswa, Paris.*

Bibliography

Berg RY. 1958. Seed dispersal, morphology, and phylogeny of *Trillium*. Skr. Nor. Videnstc-Akad. Oslo, n.s., 1: 1–36.

Berg RY. 1962. Contribution to the comparative embryology of the Liliaceae: *Scoliopus, Trillium, Paris,* and *Medeola*. Skr. Nor. Videnslc-Akad. Oslo, n.s., 4: 1–64.

Farmer SB and EE Schilling. 2002. Phylogenetic analyses of Trilliaceae based on morphological and molecular data. Syst. Bot. 27: 674–692.

Fukuda I. 2001a. The origin and evolution in *Trillium*. 1. The origin of the Himalayan *Trillium govanianum*. Cytologia 66: 106–111.

Fukuda I. 2001b. The origin and evolution in *Trillium*. 2. Chromosome variation of *Trillium undulatum* in North America. Cytologia 66: 319–327.

Gates RR. 1917. A systematic study of the North American genus *Trillium*, its variability, and its relation to *Paris* and *Medeola*. Annals Missouri Bot. Gard. 4: 43–93.

Geitler L. 1938. Weitere cytogenetische Untersuchungen an natürlichen Populationen von *Paris quadrifolia*. Zeitschr. Indukt. Abst. Verergsl. 75: 161–190.

Heatley M. 1916. A study of the life history of *Trillium cernuum* L. Bot. Gaz. 1: 425–429.

Howe TD. 1940. Development of the embryo sac in *Trillium grandiflorum*. Am. J. Bot. Suppl. 27: 11.

Jeffrey EC. 1939. The production of unfertilized seeds in *Trillium*. Science 90: 81–82.

Kato H, S Kawano, R Terauchi, M Ohara, and FH Utech. 1995. Evolutionary biology of *Trillium* and related genera (Trilliaceae). I. Restriction site mapping and variation of chloroplast DNA and its systematic implications. Plant Spec. Biol. 10: 17–29.

Kato H, R Terauchi, FH Utech, and S Kawano. 1995. Molecular systematics of the Trilliaceae sensu lato as inferred from *rbc*L sequence data. Mol. Phylogenet. Evol. 4: 184–193.

Kawano S and H Kato. 1995. Evolutionary biology of *Trillium* and related genera (Trilliaceae). II. Cladistic analyses on gross morphological characters, and phylogeny and evolution of the genus *Trillium*. Plant Spec. Biol. 10: 169–183.

Kazempour Osaloo S, FH Utech, M Ohara, and S Kawano. 1999. Molecular systematics of Trilliaceae I. Phylogenetic analysis of *Trillium* using *mat*K gene sequences. J. Plant Res. 112: 35–49.

Kazempour Osaloo S and S Kawano. 1999. Molecular systematics of Trilliaceae II. Phylogenetic analyses of *Trillium* and its allies using sequences of *rbc*L and *mat*K genes of cpDNA and internal transcribed spacers of 18s–26s nrDNA. Plant Spec. Biol. 14: 75–94.

Kotseruba VV. 2001. Intraspecific polymorphism of karyotype in *Daiswa hainanensis* subsp. *vietnamensis* Takht. Cytologia 43(11): 1075–1079.

Li H. 1984. The phylogeny of the genus *Paris*. Acta Bot. Yunn. 6: 351–362.

Li H. 1986. A study on taxonomy of the genus *Paris* L. Bull. Bot. Res. Kunming 6: 109–114 (in Chinese).

Li H. 1998. The genus *Paris* (Trilliaceae). Science Press, Beijing.

Li H, Z Gu and H Na. 1988. Cytogeographic study of the genus *Paris*. Acta Phytotax. Sinica 26: 10–21.

Naumova TN. 1990. Trilliaceae. In: TB Batygina and MS Yakovlev, eds. Comparative embryology of flowering plants: Monocotyledons, pp. 151–159. Nauka, Leningrad (in Russian).

Ohara M. 1989. Life history evolution in the genus *Trillium*. Plant Spec. Biol. 4: 1–28.

Punina EO, Yu A Myakoshina, AM Efimov, and AV Rodionov. 2000. Chromosome maps of Trilliaceae plants: heterochromatin nucleotide composition and mapping of 18S-26S rRNA genes in *Paris quadrifolia* L. Russ. J. Genet. 36: 546–549.

Smith MC and R Ingram. 1986. Heterochromatin banding in the genus *Paris*. Genetica 71: 141–145.

Spangler RC. 1925. Female gametophyte of *Trillium sessile*. Bot. Gaz. 79: 217–221.

Swamy BGL. 1948–49. On the post-fertilization development of *Trillium undulatum*. La Cellule 52: 5–14.

Takahashi M. 1982. Pollen morphology of North American species of *Trillium*. Am. J. Bot. 69: 1185–1195.

Takahashi M. 1984. Pollen morphology in *Paris* and its related genera. Bot. Mag. (Tokyo) 97: 233–245.

Takhtajan A. 1983. A revision of *Daiswa* (Trilliaceae). Brittonia 35: 255–270.

Tamura MN. 1998. Trilliaceae. In: K Kubitzki, ed. The families and genera of vascular plants, vol. 3, pp. 444–452. Springer, Berlin/Heidelberg/New York.

Trifonova VI. 1985. Trilliaceae. In: A Takhtajan, ed. Comparative seed anatomy, vol. 1, pp. 130–132. Nauka, Leningrad (in Russian).

Utech FH and S Kawamo. 1980. Vascular anatomy of the Japanese *Paris tetraphylla* A. Gray (Liliaceae-Parideae). J. Phytogeogr. Taxon. 28: 17–23.

Warmke HE. 1937. Cytology of the Pacific Coast *Trillium*. Am. J. Bot. 24: 376–383.

Wei ZX. 1995. Pollen morphology of *Trillium*. Acta Bot. Yunn. 17: 317–324.

Zomlefer WB. 1996. The Trilliaceae in the southeastern United States. Harv. Pap. Bot. 9: 91–120.

Order 8. LILIALES

Perennial or rarely annual herbs with rhizomes (tuberous in *Medeola*), tunicated corms or bulbs. Raphides lacking or rarely (Campynemataceae) present. Vessels only in roots, or both in roots and stems, with scalariform perforations. Leaves both cauline and basal, or only basal, alternate or rarely verticillate, sometimes distichous, sessile or sometimes petiolate, linear to ovate, usually parallel-veined, often sheathing at the base. Stomata anomocytic. Flowers small, medium-sized, or large, in various types of inflorescences or sometimes (*Tulipa* spp.) solitary, bisexual or very rarely unisexual, actinomorphic or rarely weakly zygomorphic (*Fritillaria* spp.). Perianth of six segments, free from each other, in two cycles, generally more or less similar or rarely (as in *Nomocharis*) different in the two whorls, with nectaries on the tepal bases (perigonal or less often androecial nectaries), or without nectaries (Mediolaceae). Stamens six, in two cycles, or rarely (Scoliopaceae) three, free from each other or epipetalous; anthers basifixed or more often dorsifixed, extrorse or latrorse, sometimes introrse, tetrasporangiate, opening longitudinally. Tapetum secretory. Microsporogenesis successive. Pollen grains 2- or 3-celled, 1-colpate or rarely with 2–3 distal colpi or inaperturate, often more or less reticulate. Gynoecium of three united carpels; stigma on short or long style, 3-lobed or with three stigmatic crests, papillate; ovary superior, 3-locular or sometimes 1–2-locular, with several to numerous ovules. Ovules anatropous or rarely campylotropous, usually bitegmic, crassinucellate or rarely weakly crassinucellate or (in some genera of Colchicaceae) tenuinucellate, without parietal tissue, often with funicular obturator. The archesporial cell functions directly as the megaspore mother cell without cutting off a parietal cell. Female gametophyte tetrasporic, mostly of *Fritillaria*-type, or *Polygonum*-type. Endosperm nuclear. Fruits septicidal or loculicidal capsules, or baccate (Streptopeae and *Medeola*). Seeds of various shape and size, seed coat is formed by both integuments, but in general it lacks mechanical strength; the testa is thin, sometimes only 2-cell-layered; without phytomelan, but usually with phlobaphenes; embryo mostly little differentiated; endosperm contains aleurone and fatty oils, sometimes with traces of starch. Flavonol glycosides based on kaempferol and quercetin present. Steroidal saponins and steroidal alkaloids occur in some genera.

Related to the Melanthiales.

Key to Families

1 Ovary superior.
  2 Plants rhizomatous.
    3 Stamens six in two cycles.
      4 Fruits capsular. Herbs with usually stoloniform rhizome, and mostly decaying every year, rarely thickened and surviving for a few years. Stem erect, mostly simple. Vessels both in roots and stem, with scalariform perforations which may have more than 40 bars. Leaves alternate, sessile, all cauline, oblong to ovate, sometimes clasping or perfoliate, reticulate-veined. Stomata surrounded by four stomatal contact cells, two cells lateral to the guard cells and one at each pole. Flowers rather large, terminal or axillary, solitary or in cymes or clusters, bisexual, actinomorphic. Perianth segments free to base, recurved above, caducous, the outer ones basally provided with globose, nectariferous pouches or less often with short spurs, more or less purple-spotted inside, the inner ones flat. Filaments slightly flattened, more or less connate below, spreading above; anthers oblong, dorsifixed, extrorse. Pollen grains 1-colpate, reticulate. Ovary superior, oblong, 3-angled, 3-locular, with numerous ovules. Stylodia connate into a columnar style with three spreading, bifurcate stigmas, tuberculate on inner side. Capsules narrowly oblong to linear, many-seeded. Seeds minute, reticulate, ovate to orbicular in outline, flat, piled upon each other (as in several genera of Liliaceae); testa contains lignin, tegmen totally collapsed. Have large, round crystals of calcium carbonate like those in *Erythronium* and *Lilium* (Goldblatt et al. 1984). Possesses quercetin and often kaempferol. n = 13, rarely 12. . . . . 3. TRICYRTIDACEAE.
      4 Fruits baccate. Plants with subterranean rhizome-like horizontal storage stem tuber decaying every year or short rhizome surviving for several years. Leaves cauline, in two pseudoverticillate groups (morphologically in dense spirals), one near the middle of stem and consists of (5)6–9(-11) oblanceolate to obovate, acuminate, and basally

attenuate leaves with three strong, convergent primary veins and a large number reticulate secondary veins, one at the apex consists of 3(-5) smaller leaves of different shape, short-petioled or sessile, mostly ovate, acuminate, basally obtuse to more or less truncate, with similar venation (*Medeola*) or leaves crowded at the base of stem. The basal parts of young leaves and stems are covered with flocculent and deciduous wool consisting of extremely long hairs that develop from epidermal cells by mere elongation. Inflorescence is a terminal and sessile, umbel-like cluster (synflorescence) of usually three sympodially arranged and contracted monochasia (bostryxes according to Berg 1962a) each of which is subtended by one of the leaves of the upper whorl (*Medeola*) or (*Clintonia*) inflorescence often a raceme, sometimes an umbel or 1-flowered. Flowers small and greenish; pedicels deflexing at anthesis and straightening after anthesis (*Medeola*) or erect to horizontal (*Clintonia*); perianth petaloid, with free, valvate segments. Nectaries lacking. Filaments filiform; anthers small, pseudobasifixed, extrorse. Pollen grains 1-colpate. Ovary 1-locular, with three deeply protruding parietal placentas that come quite close together (*Medeola*) or 2- or 3-locular (*Clintonia*); style very short, deciduous, but with three long, filiform, finely papillate, longitudinally grooved all along the inner side, stigmata branches are divaricate to finally deflexed-recurved over the anther (*Medeola*) or (*Clintonia*) trilobate, not deciduous. Fruits globose purplish-blue berry with 12–15 closely packed seeds. Seeds without appendages. Testa palisadlike. Embryo minute. n = 7 (usually), 8. . . . . 7. MEDEOLACEAE.

3 Stamens three, the inner stamen cycle lacking. Herbs with short, slender rhizome and short, subterranean stem, which at the soil surface bears two (rarely three or four) leaves. Leaves alternate, distichous, elliptic to oblong, sessile-clasping to subsessile, parallel veined with numerous transverse veinless, more or less mottled with purple spots. Inflorescences nearly sessile, 2–12-flowered, umbel-like, consists of axillary fascicles of elongated, 3-angled, twisting or recurving pedicels; each fascicle might be either a contracted bostryx or a contracted cincinnus (Berg 1959) or a contracted raceme (Utech 1992). Flowers rather large. Perianth segments free, dissimilar; outer segments petaloid, many-veined, striped, spreading, with basal nectary glands; inner segments linear, 3-veined, converging over the gynoecium. Stamens opposite and adnate basally to outer segments; filaments free, filiform-subulate; anthers oblong, dorsifixed, extrorse. Pollen grains 1-colpate, spheroidal or somewhat oblong, rather large, tectate-columellate, with rugulate arrangement of fine reticulum or foveolate. Ovary sessile, strongly 3-angled (the dorsal regions laterally flattened or compressed, and the ventral regions occupy the corners of ovary), 1-locular, with numerous ovules on parietal placentas; ovules with very long funicle; style short, with three linear, horizontally spreading to recurved stylar branches that are channeled on the upper stigmatic surface and bear small stigma at the apex. Fruits strongly 3-angled and thin-walled capsule terminated by persistent style with atypical irregular opening caused by parenchymatic cell degeneration of pericarp tissue between the six dorsal and ventral veins in the upper part of the fruit. Seeds oblong and slightly curved, longitudinally sulcate-striate, with an appendage (elaiosome) produced by the raphe that forms ventral excrescence along the seed, extending from near the micropyle to nearly across the chalazal end; embryo minute, approximately one tenth of the length of the endosperm, n = 7, 8. . . . . 4. SCOLIOPACEAE.

2 Plants with corm or bulb.

5 Plants with starchrich corms, in some genera stoloniferous, somewhat (*Gloriosa*) rhizomatous. Perennial herbs with a basal aggregation of leaves. Vessels only in roots. Stems erect or climbing (crimbers stem twiners and tendril climbers). Leaves alternate, spiral (usually) or distichous, simple, entire, linear or lanceolate, sheathing, parallel-veined. Flowers medium-sized to large, in bracteate raceme or spicate, bractless inflorescences, or solitary. Perianth segments more or less similar, free from each other or basally connate, in *Colchicum* and

related genera forming long tube. Nectaries perigonal (at the bases of perianth segments) or (Colchicoideae) androecial. Stamens inserted at the base of segments; filaments narrow or basally broad; anthers mostly short, ovate, basifixed or dorsifixed, extrorse or latrorse, very rarely almost introrse. Pollen grains tectate-columellate, 1-colpate or 2–4-(-many)-porate, reticulate, foveolate-reticulate or striate. Stylodia free or united into deeply 3-brachiate style, Ovary 3-locular, with several to many anatropous to weakly campylotropous, pseudocrassinucellate ovules per locule. Fruits septicidal or less often loculicidal capsules. Seeds globose or rarely (*Onixotis*) ovoid, sometimes with more or less well expressed raphal appendage (strophiole); testa mostly brown or red; embryo relatively short, elongate, or rarely subglobose; seedlings have conspicuously coleoptilelike cotyledon; endosperm oily. Lack true raphides but have crystal sand. Most genera contain colchicine-type (and related compounds) alkaloids with a tropolone ring, as well as many other alkaloids without a tropolone ring. n = 5, 7–12, 19. . . . . . 2. COLCHICACEAE.

5 Plants with tunicate bulbs.

6 Fruits septicidal capsules. Perennial, erect, often slender herbs with tunicated bulb having membranous or fibrous-reticulate coat. Vessels only in roots. Leaves alternate, flat, sessile, sheathing, simple, entire, linear to lanceolate, few or sometimes only one basal leaf. Flowers large, solitary or few together in a false umbel, with long pedicels, situated in the axils of green, lanceolate, or linear bracts. Perianth differentiated into two cycles; outer segments lanceolate to ovate, generally glabrous, and either green and sepaloid or similarly coloured as the inner segments; inner segments larger and broader, petaloid, cuneate to clawed, distinctly colored and usually marked with large or small spots, streaks or other patterns (nectar guides), more or less hairy inside and with an often densely bearded, depressed nectary gland or spot near the base. Stamens six, inserted at the base of the inner segments; filaments subulate and basally dilated; anthers pseudobasifixed (surrounding the tip of the filament with a tubular part of the connective), latrorse. Pollen grains 1-colpate, reticulate. Style very short or obsolete and three stigmatic branches are situated on the apex of the ovary; ovary 3-locular, with numerous anatropous ovules per locule. Fruits 3-angled or -winged, septicidal capsules. Seeds trigonous, tetragonous, or more or less irregularly shaped or flattened and lenticular; embryo slender, but well differentiated; endosperm containing fat and aleurone. Alkaloids sometimes present in the seeds; n = 7–10. . . . . 5. CALOCHORTACEAE.

6 Fruits loculicidal capsules. Plants with tunicated or nontunicated bulbs that have from one (*Gagea*) to numerous fleshy nutrient scales. Roots contractile. Leaves basal or cauline, alternate or rarely verticillate, linear to ovate-lanceolate, sessile or sometimes petiolate, parallel-veined, often sheathing at the base. Inflorescence cymose or rarely racemose and may form a raceme or (*Gagea*) be umbel-like, or sometimes reduced to single terminal flower. Flowers actinomorphic or rarely weakly zygomorphic; perianth segments similar or rarely (*Nomocharis*) dissimilar. Nectaries present at the tepal bases. Anthers epipeltate (as in *Lilium, Notholirion,* and *Nomocharis*) when part of the anther with its connective extends below its point of attachment, or pseudobasifixed when the filament tip is surrounded by the tubular connective. Pollen grains large, mostly monocolpate or less often with 2–3 morphologically distal colpi, rarely inaperturate or 1–3-porate, often coarsely reticulate. Ovary 3-locular; style short or long, apically 3-lobed or with three stigmatic crests; stigmatic surface papillate; placentas and margins of not completely fused carpels provided with stigmatoid tissue. Fruits loculicidal capsules. Seeds usually flat and disc-shaped, closely piled on top of each other, often provided with marginal rim, sometimes with raphal elaiosome; exotesta contain phlobaphenes, n usually = 12. . . . . 6. LILIACEAE.

1 Ovary inferior to semi-inferior. Slender herbs with very short rhizome and thick adventitious roots.

Vessels (very primitive) only in roots, but sometimes absent. Leaves and probably also other parts contain calcium oxalate raphide bundles in special idioblasts. Leaves alternate, sessile, few, one basal or several basally clustered and progressively smaller on flowering stem, linear to linear-lanceolate, sheathing at base, in *Campynemanthe* basal leaf apically 3-dentate. Stomata anomocytic. Flowers small, in terminal, bracteate few-flowered panicle, several-flowered pseudoumbel (*Campynemanthe*), or solitary, bisexual, actinomorphic. Perianth segments six, free, inserted near the top of the ovary, similar, ovate or broadly lanceolate, greenish, sometimes dotted-striated, without spots or spotted, persistent, enlarging after fertilization, provided with perigonal nectaries (*Campynemanthe*). Stamens six, inserted at base of perianth segments; filaments slender; anthers in *Campynema* sagittate and dorsifixed and extrorse, in *Campynemanthe* basifixed to subbasifixed and latrorse or weakly extrorse (*Campynema*). Tapetum secretory, Microsporogenesis successive. Pollen grains 1-colpate (*Campynema*) or inaperturate (*Campynemanthe*), reticulate. Gynoecium of three united carpels; stylodia free (*Campynema*) or connate into apically 3-lobed style (*Campynemanthe*); ovary inferior or semi-inferior (*Campynemanthe*), 3-locular, or 1-locular (*Campynema*), with few to many ovules; placentation when unilocular parietal, when trilocular –axile. Ovules anatropous, bitegmic, weakly crassinucellate. The parietal cell is formed from the archesporial cell and gives rise to parietal tissue. Fruits capsules dehiscing by decay of the lateral or dorsal walls or indehiscent. Seeds pale red-brown, in *Campynemanthe* nearly globose to angular and with long persistent funicles, in *Campynema* flattened and with spongy outer coat; seed coat formed (at least in *Campynemanthe*) by both integuments and both the testal and tegmic parts are 2-celled layer thick; the endotegmic layer contains phlobaphene; embryo minute; endosperm copious. Contain saponins; n = 11. . . . . .1. CAMPYNEMATACEAE.

1. CAMPYNEMATACEAE

Dumortier 1829. 2/4. New Caledonia (*Campynemanthe*) and Tasmania (*Campynema*).

Campynemanthe, Campynema.

The affinities of this rather isolated family is uncertain. Although it is often connected with or even included in the Melanthiaceae s. l., the nuclear development of endosperm (Lowry et al. 1987) makes it closer to the colchicalean families. However, Campynemataceae markedly differ from all other families of the Liliales by the presence of crystal raphides and almost inferior ovary.

2. COLCHICACEAE

A.P. de Candolle 1804 (including Burchardiaceae Takhtajan 1996; Merenderaceae Mirbel 1804; Uvulariaceae A. Gray ex Kunth 1843). 23/95. Africa (especially South Africa), Eurasia, Australia, New Zealand.

2.1 UVULARIOIDEAE

Plants with creeping or short rhizome. Stamens six, free; filaments expanded at base; anthers, extrorse or (*Kuntheria*) introrse. Pollen grains 1-colpate or (*Uvularia*) usually 2-colpate, tectate-columellate, foveolate-verrucate. Fruits somewhat fleshy loculicidal capsules or (*Uvularia*) septicidal capsules or (Streptopeae) baccate. Seeds without elaiosome. Contain calcium oxalate in the form of raphides, sometimes also crystal sand. Colchicin absent. n = 6, 7, 8, 9. – UVULAREAE: *Uvularia, Schelhammera, Tripladenia, Kuntheria*; STREPTOPEAE: *Prosartes, Disporum, Streptopus*; BURCHARDIEAE: *Burchardia.*

2.2 COLCHICOIDEAE

Plants with starch corms, in some genera stoloniferous. Stamens inserted at the base of perianth segments. Nectaries perigonal (at the bases of perianth segments) or androecial (Colchiceae) Pollen grains 1-colpate 2- or 4- to many-porate, reticulate, or striate. Most genera contain colchicines-type alkaloids. – WURMBEEAE (ANGUILLARIEAE): *Neodregea, Onixotis, Wurmbea, Baeometra;* IPHIGENIEAE: *Iphigenia, Camptorrhiza, Ornithoglossum, Hexacyrtis*; GLORIOSEAE: *Gloriosa, Littonia, Sandersonia*; COLCHICEAE: *Androcymbium, Merendera, Bulbocodium, Colchicum.*

3. TRICYRTIDACEAE

Takhtajan 1997 (including Compsoaceae Horaninow 1834). 1/18. Eastern Himalayas, northeastern India (Khasia Hills), China, Taiwan, Japan.

Tricyrtis.

Both chemical evidence (Williams et al. 1988) and morphological data support the separation of *Tricyrtis* from the Colchicaceae where it is usually placed.

4. SCOLIOPACEAE

Takhtajan 1996. 1/2. Western North America along the Pacific coast from Santa Cruz Mountains in California to the Cascade Mountains in Oregon.

Scoliopus.

The taxonomic position of the genus *Scoliopus* has been debated for a long time. At different times it has been included in Melanthiaceae s. l., Colchicaceae, Medeoleae of the Liliaceae s. l., and Trilliaceae. Dahlgren et al. (1985: 231) concludes that *Scoliopus* "is perhaps best placed in Uvulariaceae." Takhtajan (1987) placed the genus in a separate tribe Scoliopeae within Melanthiaceae-Melanthioideae, Tamura (1998) in Calochortaceae-Tricyrtidoideae, and Thorne (2006) includes it in Liliaceae. However, *Scoliopus* differs markedly from both the Melanthiaceae s. str., Uvulariaceae and Calochortaceae in many important features, including peculiar vegetative morphology, inflorescence, reduced androecium, 1-locular ovary with nonintruding parietal placenta, unique type of the fruit dehiscence, and seeds provided with strophiole.

5. CALOCHORTACEAE

Dumortier 1829. 1/65. Temperate western North America from British Columbia and Dakota south to Guatemala with the greatest concentration of species in California.

Calochortus.

Calochortaceae are rather isolated within the order, but embryologically they are nearest to the Scoliopaceae ("The greatest embryological similarity is found between *Scoliopus* and *Calochortus,*" Berg [1962: 51]).

6. LILIACEAE

A.L. de Jussieu 1789. (Erythroniaceae Martynov, 1820; Fritillariaceae Salisbury, 1866; Liriaceae Batsch ex Borkhausen 1797; Tulipaceae Batsch ex Borkhausen 1797). 13/550. Temperate and subtropical regions of the Northern Hemisphere, especially western, Central Asia, Himalayas, and eastern Asia.

LLOYDIEAE: *Gagea, Lloydia*; LILIEAE: *Cardiocrinum, Lilium, Notholirion, Nomocharis, Fritillaria Korolkowia, Rhinopetalum*; TULIPEAE: *Tulipa, Amana, Holongia, Erythronium.*

7. MEDEOLACEAE

Takhtajan 1987. 2/6. Eastern Asia and North America.

CLINTONIEAE: *Clintonia*; MEDEOLEAE: *Medeola.*

Related to the Liliaceae but differ from them markedly in basic chromosome number, type of the subterranean storage organ, the absence of nectaries, style morphology, crassinucellate ovules, and baccate fruits.

Bibliography

Baker JG. 1879. A synopsis of Colchicaceae and the aberrant tribes of Liliaceae. Bot. J. Linn. Soc. 17: 405–510.

Beal JM and M Ownbey. 1943. Cytological studies in relation to the classification of the genus *Calochortus*, part 3. Bot. Gaz. 104: 553–562.

Berg RY. 1959. Seed dispersal, morphology, and taxonomic position of *Scoliopus,* Liliaceae. Skr. Nor. Vidensk-Akad. Oslo 4: 1–56.

Berg RY. 1960. Ovary, ovule, and endosperm of *Calochortus amabilis*. Nytt. Mag. Bot. 8: 189–206.

Berg RY. 1962a. Morphology and taxonomic position of *Medeola,* Liliaceae. Skr. Nor. Vidensk.-Akad. Oslo, n.s., 3: 1–55.

Berg RY. 1962b. Contribution to the comparative embryology of the Liliaceae: *Scoliopus, Trillium, Paris,* and *Medeola,* Skr. Nor, Vidensk-Akad. Oslo 4: 1–64.

Björnstad IN. 1970. Comparative embryology of Asparagoideae-Polygonateae, Liliaceae. Nytt Mag. Bot. 17: 169–207.

Buxbaum F. 1937. Die Entwicklungslinien der Lilioideae. Bot. Arch. 38: 213–293, 305–398.

Buxbaum F. 1959. Beiträge zur Morphologic der Gattung *Tricyrtis*. Beitr. Biol. Pfl. 35: 55–75.

Cave MS. 1941. Megasporogenesis and embryo sac development in *Calochortus*. Am. J. Bot. 28: 390–394.

Cheadle VI and H Kosakai. 1971. Vessels in Liliaceae. Phytomorphology 21: 320–333.

Chupov VS. 1984a. The position of Liliaceae s. str. (subfamily Lilioideae of the family Liliaceae s. l.) in the system: Serological study. Bot. Zhurn. 69: 762–771 (in Russian with English summary).

Chupov VS. 1984b. The position of the family Liliaceae s. str. (subfamily Lilioideae of the family Liliaceae s. l.) in the system: An analysis of characters. Bot. Zhurn. 69: 1451–1461 (in Russian with English summary).

Dahlgren R and A-M Lu. 1985. *Campynemanthe* (Campynemataceae): morphology, microsporo-genesis, early ovule ontogeny, and relationships. Nord. J. Bot. 5: 321–330.

Eunus AM. 1951a. Contribution to the embryology of the Liliaceae: V. Life history of *Amianthium muscaetoxicum* Walt. Phytomorphology 1: 73–79.

Eunus AM. 1951b. Development of the embryo sac and fertilization in *Fritillaria pudica* Spring. Pakistan J. Sci. Res. 3: 106–113.

Fukuhara T and ZK Shinwari. 1994. Seed coat anatomy in Uvulariaceae (Liliales) of the Northern Hemisphere: Systematic implications. Acta Phytotax. Geobot. 45: 1–14.

Goldblatt P. 1986. Systematics and relationships of the bigeneric Pacific family Campynemataceae (Liliales). Bull. Mus. Natl. Hist. Nat. Paris, Sér. 4, 8: 117–132.

Goldblatt P, JE Henrich, and P Rudall. 1984. Occurrence of crystals in Iridaceae and allied families and their phylogenetic significance. Ann. Missouri Bot. Gard. 71: 1013–1020.

Haque A. 1951. The embryo sac of *Erythronium americanum*. Bot. Gaz. 112: 495–500.

He HP, FC Liu, L Hu, and HY Zhu. 1999. Alkaloids from the flowers of *Colchicum autumnale*. Acta Bot. Yunn. 21: 364–368.

Hong W-P, J Greenham, SL Jury, and GA Williams. 1999. Leaf flavonoid patterns in the genus *Tricyrtis* (Tricyrtidaceae *sensu stricto*, Liliaceae *sensu lato*). Bot. J. Linn. Soc. 130: 261–266.

Hruby Ch. 1938. Embryo sac development in *Erythronium dens canis*. Chronica Bot. 4: 20–21.

Joshi AC. 1940. Development of the embryo sac of *Gagea fascicularis* Salisb. Bull. Torrey Bot. Club 67: 155–158.

Kosenko VN. 1988. Pollen morphology in Chiono-graphideae, Uvularieae, Tricyrtideae, Scoliopeae, An-guillarieae, Iphigenieae, Glorioseae, Colchiceae (Melanthiaceae). Bot. Zhurn. 73: 172–185 (in Russian with English summary).

Kosenko VN. 1991. Palynomorphology of the family Liliaceae s. str. Bot. Zhurn. 76: 1696–1710 (in Russian with English summary).

Kosenko VN. 1992. Pollen morphology and systematic problems of the Liliaceae family. Bot. Zhurn. 77(3): 1–15 (in Russian with English summary).

Kubitzki K. 1998. Campynemataceae. In: K Kubitzki, ed. The families and genera of vascular plants, vol. 3, pp. 173–175. Springer, Berlin/Heidelberg/New York.

Lee NS and SH Yeau. 1990. A palynological study of *Streptopus ovalis* (Ohwi) Wang et Y.C. Tang and the relative species (tribe Poygonateae, Liliaceae). Korean J. Plant Taxon. 20: 81–94 (in Korean with English summary).

Lowry PP, P Goldblatt, and H Tobe. 1987. Notes on the floral biology, cytology, and embryology of *Campynemanthe* (Liliales: Campynemataceae). Ann. Missouri Bot. Gard. 74: 573–576.

Maheshwari P. 1946. The *Fritillaria* type of embryo sac: A critical review. J. Indian Bot. Soc. (M.D.P. Lyengar Comm. Vol.): 101–119.

Membrives N, J Martin, J Caujape Castells, and J Pedrola Monfort. 2002. Pollen morphology and biometry of the genus *Androcymbium* (Colchicaceae) in southern Africa: taxonomic and biogeographic considerations. Bothalia 32: 91–96.

Membrives N, J Pedrola Monfort, and J Caujape Castells. 2003a. Leaf morphology and anatomy of genus *Androcymbium* (Colchicaceae) in southwest Africa. Collect. Bot. 26: 83–99.

Membrives N, J Pedrola Monfort, and J Caujape Castells. 2003b. Morphological seed studies of southwest African *Androcymbium* (Colchicaceae). Bot. Macaronesica 24: 87–106.

Nawa N. 1928. Some cytological observations in *Tricyrtis*, *Sagittaria* and *Lilium*. Bot. Mag. (Tokyo) 42: 33–36.

Ness BD. 1989. Seed morphology and taxonomic relationships in *Calochortus* (Liliaceae). Syst. Bot. 14: 495–505.

Nordenstam B. 1998. Colchicaceae. In: K Kubitzki, ed. The Families and genera of vascular plants vol. 3, pp. 175–185. Springer, Berlin/Heidelberg/New York.

Oganezova GG. 1984. Morphologo-anatomical peculiarities of the fruit and seed in some representatives of the subfamily Wurmbaeoideae (Liliaceae) in connection with systematic and phylogeny. Bot. Zhurn. 69: 1317–1327 (in Russian with English summary).

Oganezova GG. 2000. Anatomy and systematics of some *Colchicum* species from Armenia. Bot. Chronika 13: 217–227 (in Russian).

Oganezova GG. 2002. Anatomy of Caucasian species of the genus *Merendera* (*Colchicaceae*) in context of their systematics. Bot. Zhurn. 87(2): 19–33 (in Russian with English summary).

Ogura H. 1964. On the embryo sac of two species of *Tricyrtis*. Sci. Rep. Tohoku Univ. Ser. IV (Biol.)30: 219–222.

Oikawa K. 1937. A note on the development of the embryo sac in *Cardiocrinum cordatum*. Sci. Rep. Tohoku Imp. Univ., 4th ser., 11: 303–306.

Oikawa K. 1940. The embryo sac of *Erythronium japonicum*. Bot. Mag. (Tokyo) 54: 366–369 (in Japanese).

Ownbey M. 1940. A monograph of the genus *Calochortus*. Ann. Missouri Bot. Gard. 27: 371–560.

Ownbey M. 1969. *Calochortus*. Univ. Wash. Publ. 17(1): 765–779.

Patterson TB and TJ Givnish. 2002. Phylogeny, concerted convergence, and phylogenetic niche conservatism in the core Liliales: Insights from *rbc*L and *ndh*F sequence data. Evolution 56: 233–252.

Petrova TF. 1977. Cytoembryology of the Liliaceae subfamily Lilioideae. Nauka, Moscow (in Russian).

Romanov ID. 1936. Die Embryosackentwicklung in der Gattung *Gagea* Salisb. Planta 25: 438–459.

Romanov ID. 1939. Two new forms of embryo sac in the genus *Tulipa*. Doklady Acad. Sci. URSS 22: 139–141.

Rudall PJ, KL Stobart, W-P Hong, JG Conran, CA Furness, GC Kite, and MW Chase. 2000. Consider the lilies: systematics of Liliales. In: KL Wilson and DA Morrison, eds. Monocots: systematic and evolution, pp. 347–359. CSIRO, Collingwood.

Sargant E. 1896. The formation of the sexual nuclei in *Lilium martagon:* I. Oogenesis. Ann. Bot. 10: 445–477.

Schnarf K. 1949. Der Umfang der Lilioideae im nature-lichen System. Oesterr. Bot. Z. 95: 257–269.

Schulze W. 1975. Beiträge zur Taxonomie der Liliifloren: II. Colchicaceae. Wiss. Z. Friedrich-Schiller-Univ. Jena, Math-Naturwiss. Reihe, 24: 417–428.

Shinwari ZK, R Terauchi, FH Utech, and S Kawano. 1994. Recognition of the New World *Disporum* section *Prosartes* as *Prosartes* (Liliaceae) based on the sequence data of the *rbc*L gene. Taxon 43: 353–366.

Shurukhina EA. 1994. Anatomical structure and ultra-structure of the seeds of Campynemataceae. Bot. Zhurn. 79(5): 58–62 (in Russian).

Sterling C. 1972. Comparative morphology of the carpel in the Liliaceae: Neodraegeae. Bot. J. Linn. Soc. 65: 163–171.

Sterling C. 1973a. Comparative morphology of the carpel in the Liliaceae: Wurmbaeae. Bot. J. Linn. Soc. 66: 75–82.

Sterling C. 1973b. Comparative morphology of the carpel in the Liliaceae: Colchiceae (*Colchicum*). Bot. J. Linn. Soc. 66: 213–221.

Sterling C. 1973c. Comparative morphology of the carpel in the Liliaceae: Colchiceae (*Androcymbium*). Bot. J. Linn. Soc. 67: 149–156.

Sterling C. 1974a. Comparative morphology of the carpel in the Liliaceae: *Baeometra, Burchardia,* and *Walleria*. Bot. J. Linn. Soc. 68: 283–290.

Sterling C. 1974b. Comparative morphology of the carpel in the Liliaceae: Iphigenieae. Bot. J. Linn. Soc. 68: 283–290.

Sterling C. 1975. Comparative morphology of the carpel Liliaceae: Glorioseae. Bot. J. Linn. Soc. 70: 341–349.

Sterling C. 1977. Comparative morphology of the carpel in the Liliaceae: Uvularieae. Bot. J. Linn. Soc. 74: 345–354.

Sterling C. 1978. Comparative morphology of the carpels in the Liliaceae: Hewardieae, Petrosavieae, and Tricyrteae. Bot. J. Linn. Soc. 77: 95–106.

Stewart RN and R Bamford. 1942. The chromosomes and nucleoli of *Medeola virginiana*. Am. J. Bot. 29: 301–303.

Takahashi H. 1980. A taxonomic study on the genus *Tricyrtis*. Sci. Rep. Fac. Educ., Gifu Univ. (Nat. Sci.) 6: 583–635.

Takahashi H. 1984. The floral biology of *Tricyrtis latifolia* Maxim. (Liliaceae). Bot. Mag. (Tokyo) 97: 207–217.

Takahashi H. 1987. A comparative floral and pollination biology of *Tricyrtis flava* Maxim., *T. nana* Yatabe and *T. ohsumiensis* Masamune (Liliaceae). Bot. Mag (Tokyo) 100: 185–293.

Takahashi Hr. 1994. Floral biology of *Tricyrtis macropoda* Miq. (Liliaceae). Acta Phytotaxon. Geobot. 45: 33–40.

Tamura MN. 1998a. Calochortaceae. In: K Kubitzki, ed. The families and genera of vascular plants, vol. 3, pp. 164–172. Springer, Berlin/Heidelberg/New York.

Tamura MN. 1998b. Liliaceae. In: K Kubitzki, ed. The families and genera of vascular plants, vol. 3, pp. 343–353. Springer, Berlin/Heidelberg/New York.

Tamura MN and EH Utech. 1992. Biosystematic studies in *Disporum* (Liliaceae-Polygonateae). IV. Karyotype analysis of some Asiatic and North American taxa with special reference to their systematic status. Plant Spec. Biol. 7: 103–120.

Utech FH. 1978a. Floral vascular anatomy of *Medeola virginiana* L. (Liliaceae-Parideae = Trilliaceae) and tribal note. Ann. Carnegie Mus. 47: 13–28.

Utech FH. 1978b. Somatic karyotype analysis of *Uvularia floridana* Chapman (Liliaceae). Cytologia 43: 671–678.

Utech FH. 1992. Biology of *Scoliopus* (Liliaceae): I. Phytogeography and systematics. Ann. Missouri Bot. Gard. 79: 126–142.

Vinnersten A and J Manning. 2006. A new classification of Colchicaceae. Taxon 56: 171–178.

Vinnersten A and G Reeves. 2003. Phylogenetic relationships within Colchicaceae. Amer. J. Bot. 90: 1455–1462.

Zou X, DW Fountain, and ER Morgan. 2001. Anatomical and morphological studies of seed development in *Sandersonia aurantiaca* (Hook.). South Afr. J. Bot. 67: 183–192.

Order 9. BURMANNIALES

Mostly small, perennial or annual mycorrhizal herbs, achlorophyllous and mycotrophic or less often autotrophic. The perennial autotrophic members with elongate or sometimes tuberous rhizome or tubers. Vessels with scalariform perforations present in roots, stems, and leaves of autotrophic species of *Burmannia*. Sieve-element plastids of P2c-type (Behnke 2002). Leaves alternate, basally sheathing, in mycotrophic species more or less reduced to scales, in autotrophic species linear to lanceolate and mostly concentrated at the base of the stem. Stomata anomocytic. Flowers terminal, in cymose inflorescences (usually a bifurcate cincinnus) or solitary, bisexual, actinomorphic to more or less zygomorphic, 3-merous. Perianth of 6(-8) segments in two cycles, more or less connate into a tube, sometimes the inner cycle absent. Stamens six in two cycles or (Burmanniaceae and rarely in Thismiaceae) only three and then opposite the inner segments (in contrast to the Iridales, where the stamens alternate with the inner segments); filaments usually short, attached to the perianth tube or (Corsiaceae) free from the perianth; anthers mostly with short filaments, tetrasporangiate, opening longitudinally or laterally, extrorse or introrse,. Tapetum secretory. Microsporogenesis successive. Pollen grains mostly 3-celled, tectate-columellate, 1-colpate or 1–2-porate, sometimes polyforaminate or inaperturate, psilate, undulate, folded or finely reticulate. Gynoecium of three united carpels, with filiform or short, simple or apically 3-brachiate style; ovary inferior, 3-locular to 1-locular, with numerous ovules per locule or per parietal placenta. Septal nectaries present or absent. Ovules minute, anatropous, bitegmic, tenuinucellate, without parietal cell. Female gametophyte of *Polygonum*-type or sometimes (in some species of *Burmannia*) of *Allium*-type. Endosperm helobial or rarely (some species of *Thismia*) cellular. Fruits usually capsular and dehisce longitudinally or transversely by slits or by valves, with terminal aperture or irregularly, rarely indehiscent. Seeds numerous and very small, with minute and undifferentiated, mostly globular embryo that often consists of 3–8 cells; endosperm usually consists of few cells; testa usually reticulate.

According to Dahlgren et all (1985: 216) "The closest relatives are perhaps the Melanthiales, in which Campynemataceae takes an intermediate position between the orders". Rudall and Eastman (2002) also came to the conclusion, that there are range of morphological similarities between *Corsia* and Campynemataceae, which in their opinion tends to support preliminary inferences from molecular data that they are closely related. Thus, Burmanniales are most probably closely related to Campynemataceae, which in my system is placed in Liliales.

Key to Families

1 Fertile stamens three. Autotrophic or saprophytic, annual or perennial herbs. Vessels, when present, in roots and stems, with scalariform perforations. Leaves well developed, or much reduced, or absent, basal or with neither basal nor terminal aggregations of leave, alternate, spiral or distichous, sessile, simple, entire, linear to ovate. Plants without silica bodies. Flowers solitary or in terminal cymes. Perianth generally actinomorphic, with straight or rarely curved tube and often very short lobes; often both perianth tube and ovary 3-angled or 3-winged. Stamens opposite the inner segments (when present); filaments short, inserted in the perianth tube; anthers introrse to latrorse, with thick and broad, sometimes transversely extended connective, opening transversely. Pollen monoporate or inaperturate, psilate. Nectaries septal at the top of ovary or dispersed on its side. Style apically divided into three branches, each of which bears stigmatic surface; ovary 3-locular and with protruded axile placentas or 1-locular with parietal placentas, which may emerge centrally at the ovary base. Ovary anatropous, tenuinucellate. Fruits capsules, loculicidal or rarely valvular. Seeds with testa extended into long wing at each end; embryo rudimentary at the time of seed release, testa without phytomelan; n = 6–8, much and high polyploidy. . . . 1. BURMANNIACEAE.

1 Fertile stamens usually six. Plants always mycotrophic. Nectaries absent.

2 Stamens inserted in perianth tube, usually six, but in *Oxygyne* three, recurved to pendent; thecae longitudinally dehiscent and united (separate when forming an anther tube), introrse. Small, often somewhat succulent, perennial, rhizomatous herbs; rhizome sometimes branched and coral-like. Leaves alternate or opposite, small, distichous, sessile, simple, entire. Mesophyll containing calcium oxalate crystals. Flowers in a few-flowered cincinnus or more often solitary, small, zygomorphic or actinomorphic. Perianth segments similar in the two whorls, or different (the lobes equal, or those of the inner whorl longer). Anthers basifixed, appendages (with apical enlargement of the connective. Tapetal cells unicellular, pollen porate. Ovary 1-locular, with three parietal placentas; styles one, apical, stigmas three, placentation parietal. Ovules numerous, anatropous. Fruits circumscissile capsules. Seeds numerous, minute, embryo rudimentary at the time of seed release, tecta without phytomelan; endosperm lacking; n = 6–8, 11–13. . . . . 2. THISMIACEAE.

2 Stamens free from the perianth. Mycotrophic herbs, erect or (*Corsia*) creeping with rhizomes or tubers. Plants without silica bodies. Leaves alternate, sessile, simple, entire, reduced to rather large scales, spiral to 2-ranked. Flowers solitary, terminal, bisexual or (*Corsiopsis*) dioecious or monoecious. Perianth strongly zygomorphic, one of the segments (the posterior of the outer cycle) large and brightly coloured (red-violet), adaxial; the other linear-filiform and more or less reflexed. Stamens six; filaments short; anthers dorsifixed, extrorse. Pollen grains sulcate or ulcerate. Ovary inferior, 1-locular, with three much intruded and bifurcate placentas. Stylodia free or connate into a style; stigmas thick. Ovules numerous, anatropous, tenuinucellate. Endosperm helobial. Fruits capsules dehiscing by three vertical valves down to the base or with terminal aperture. Seeds numerous, minute or small, with reticulate testa prolonged into a wing at each end and testa without phytomelan; embryo rudimentary at the time of seed release; endosperm multicellular (*Corsia*) or reduced to few cells (*Arachnitis*), oily or not oily; n = 9. . . . . 3. CORSIACEAE.

1. BURMANNIACEAE

Blume 1827. 10/95. Widely distributed in the tropics and subtropics.

Burmannia, Campylosiphon, Hexapterella, Dictyostega, Miersiella, Cymbocarpa, Gymnosiphon, Apteria, Marthella, Geomitra.

2. THISMIACEAE

J. Agardh 1858. 5/40. Southeastern Asia, Africa, America (mainly Brazil), Australia, and New Zealand.

Oxygyne, Afrothismia, Haplothismia, Thismia, Scaphiophora.

3. CORSIACEAE

Beccari 1878 (including Arachnitidaccae Munoz 1966). 3/30. China (*Corsiopsis*), New Guinea, Solomon Islands, and Australia (*Corsia,* 27) and Bolivia, Chile, Patagonia and Falkland Islands (*Arachnitis,* 2).

Corsia, Corsiopsis, Arachnitis.

Close related to the Burmanniaceae.

Bibliography

Chakrapani P and B Raj. 1971. Pollen morphology studies in the Burmanniaceae. Grana Palynol. 11: 161–179.

Colloza A. 1910. Contributo allo studio anatomico delle Burmanniaceae. Boll. Soc. Ital. 1910: 106–115.

Caddick LR, PJ Rudall, P Wilkin, TAJ Hedderson, and MW Chase. 2002. Phylogenetics of Dioscoreales based on combined analyses of morphological and molecular data. Bot. J. Linn. Soc. 138: 123–144.

Cribb PJ. 1985. The saprophytic genus *Corsia* in the Solomon Islands. Kew Mag. 2: 320–323.

Cribb PJ, P Wilkin, and M Clements. 1995. Corsiaceae: a new family for the Falkland Island. Kew Bull. 50: 171–172.

Ibisch P, C Neinhuis, and NP Rojas. 1996. On the biology, biogeography, and taxonomy of *Arachnitis* Phil. nom. cons. (Corsiaceae) in respect to a new record from Bolivia. Willdenowia 26: 321–332.

Jonker FP. 1938. A monograph of the Burmanniaceae. Meded. Bot. Mus. Herb. Rijks Univ. Utrecht 51: 1–279.

Kores P, DA White, and LB Thien. 1978. Chromosomes of *Corsia* (Corsiaceae). Am. J. Bot. 65: 584–585.

Larsen K. 1987. Thismiaceae. In: T Smitinand and K Larsen, eds. Flora of Thailand, vol. 5, part 1, pp. 124–126. Chutima Press, Bangkok.

Maas-van de Kamer H. 1998. Burmanniaceae. In: K Kubitzki, ed. The families and genera of vascular plants, vol. 3, pp. 154–164. Springer, Berlin/Heidelberg/New York.

Maas PJM, H Maas-van de Kamer, J van Bentham, HCM Snelders, and T Rübsamen 1986. Burmanniaceae. Flora Neotrop. Monogr. 42: 1–189.

Merckx V, P Schols, H Maas-van de Kamer, P Maas, S Huysmans, and E Smets. 2006. Phylogeny and evolution of Burmanniaceae (Dioscoreales) based on nuclear and mitochondrial data. Am. J. Bot. 93: 1684–1698.

Minoletti ML. 1986. *Arachnitis uniflora* Phil. una curiosa monocotiledonea de la flora Chilena. Bol. Soc. Biol., Concepcion, Chile 57: 7–20.

Neinhuis C and PL Ibisch. 1998. Corsiaceae. In: K Kubitzki, ed. The families and genera of vascular plants, vol. 3, pp. 198–201. Springer, Berlin/Heidelberg/New York.

Neyland R. 2002. A phylogeny inferred from large-subunit (26S) ribosomal DNA sequences suggests that Burmanniales are polyphyletic. Aust. Syst. Bot. 15: 19–28.

Neyland R and M Hennigan. 2003. A phylogeny inferred from large-subunit (26S) ribosome DNA sequences suggests that the Corsiaceae are polyphyletic. N. Z. J. Bot. 41: 1–11.

Pai RM. 1966. Studies in the floral morphology and anatomy of the Burmanniaceae: I. Vascular anatomy of the flower of *Burmannia pusilla* (Wall. ex Miers) Thw. Proc. Indian Acad. Sci. 63B: 301–308.

Rao VS. 1969. Certain salient features in the floral anatomy of *Burmannia, Gymnosiphon,* and *Thismia.* J. Indian Bot. Soc. 48(1–2): 22–29.

Rasmussen FN. 1995. Relationships of Burmanniales and Orchidales. In: P Rudall, PJ Cribb, DF Cutler, and CJ Humphries, eds. Monocotyledons: systematics and evolution, pp. 227–241. Royal Botanic Gardens, Kew.

Rübsamen T. 1983. Nectaries of the Burmanniaceae (Burmannieae). Acta Bot. Neerl. 32: 351.

Rübsamen T. 1986. Morphologische, embryologische, und systematische Untersuchungen an Burmanniaceae und Corsiaceae (Mit Ausblick auf die Orchidaceae-Apostasioideae). Diss. Bot. 92: 1–310.

Rudall P and S Morley. 1992. Embryo sac and early postfertilisation development in *Thismia* (Burmanniaceae). Kew Bull. 47: 625–632.

Rudall PJ and A Eastman. 2002. The questionable affinities of *Corsia* (Corsiaceae): evidence from floral anatomy and pollen morphology. Bot. J. Linn. Soc. 138: 315–324.

Terekhin ES. 1985. Burmanniales. In: A Takhtajan, ed. Comparative seed anatomy, vol. 1, pp. 138–141. Nauka, Leningrad (in Russian).

Wood CE, Jr. 1983. The genera of Burmanniaceae in the southeastern United States. J. Arnold Arbor. 64: 293–307.

Zhang D-X. 2000. Addition to the Flora Reipublicae Popularis Sinicae: the family Corsiaceae. Acta Phytotax. Sinica 38: 578–581.

Zhang DX. 2001. Phylogenetic reconstruction of *Burmannia* L. (Burmanniaceae): a preliminary study. Acta Phytotax. Sinica 39: 203–223.

Zhang D-X, Saunders RMK, and CM Hu. 1999. *Corsiopsis chinensis* gen. et sp. nov. (Corsiaceae): first record of the family in Asia. Syst. Bot. 24: 311–314.

Order 10. ALSTROEMERIALES

Erect or sometimes climbing rhizomatous perennial herbs or vines with fibrous and often also fusiform and fleshy tuberlike roots containing starch. Stems leafy, those of climbing species up to 4–6 m long. Calcium oxalate raphides present. Vessels both in roots (mostly with scalariform perforations) and stems (only with scalariform perforations). Axial parenchyma contains starch. Leaves alternate, linear to oblong, glabrous or pubescent on both leaf surfaces; epicuticular wax sometimes present. Stomata anomocytic. Inflorescences terminal or axillary, consisting of helicoid cymes, umbel-like or capitate, rarely flowers solitary. Flowers usually few to numerous (up to 100 in some *Bomarea*), bisexual, more or less actinomorphic or zygomorphic

(*Alstroemeria*), 3-merous. Perianth segments six, in two cycles, usually free. Stamens six in two cycles; filaments free, narrow; anthers basifixed, elongate, introrse or extrorse, tetrasporangiate, opening longitudinally, rarely by apical pores (Luzuriagaceae). Tapetum secretory. Microsporogenesis successive. Pollen grains 2-celled, large, 1-colpate, usually plano-convex. Gynoecium of three united carpels; style filiform, more orless3-lobed;ovaryinferiororsuperior(Lazuriagaceae), 3-locular or 1-locular. Ovules anatropous, bitegmic, tenuinucellate or (*Petermannia*) crassinucelate Female gametophyte of *Polygonum*-type. Endosperm nuclear. Fruits loculicidal capsules, or berries. Seeds globose to ellipsoid, embryo straight, cylindrical or linear; seeds lacking phytomelan; endosperm copious, contains hemicellulose, aleurone, and fatty oils; Steroidal saponins present or absent; contain chelidonic acid, flavonol glycosides; n = 5, 8, 9, 10.

Most probably related to the Liliaceae, which is also supported by the accumulation of appreciable amounts of tuliposide A (Slob et al. 1975). Hallier (1912) derived (with a question mark) Alstroemeriaceae from the Liliaceae.

Key to Families

1 Ovary superior.

2 Steroidal saponins present. Slender shrubby perennials from short rhizome. Calcium oxalate raphides present. Roots fibrous. Rhizome with monocotyledonous secondary thickening. Stems thin, branching, erect or twisting. Vessels with simple or scalariform perforations. Leaves alternate, distichous, shortly petiolate, ovate, parallel-veined, with many cross-veins. Stomata anomocytic. Flowers in axillary reduced cymes (bostryx), or solitary, functionally dioecious, small, hypogynous, actinomorphic, articulated on the pedicels with a short pericladium. Perianth segments six, not twisted after anthesis, united about two thirds of their length into a campanulate tube, persistent. Stamens six in two whorls; filaments adnate to the tepals; anthers 2-locular, tetrasporangiate, basal-dorsifixed, introrse. In female flowers stamens adnate to the perianth segments and reduced in size. Pollen grains obovate and sulcate, with reticulate exine. Gynoecium of three united carpels; style with 3-lobed, wet stigma; ovary 3-locular; ovules 2–3 per locule, anatropous, tenuinucellate; in male flowers ovary is shorter, lacks loculi, but has well-developed septal nectaries. Fruits berries; seeds angular, without phytomelan; testa and tegmen thin-walled; endosperm copious, storing aleurone and lipids; embryo clavate, about two thirds as long as the seed. . . . . 2. BEHNIACEAE.

2 Steroidal saponins absent. Terrestrial or epiphytic perennials with thin woody stems and thin rhizomes (*Drymophila*) or shrubs with short rhizomes (*Luzuriaga*). Stems thin, branching or simple, erect or scrambling. Vessels in the roots and stems, with simple or scalariform perforations. Calcium oxalate raphides and druses present in *Drymophila.* Leaves alternate, distichous, sessile or slightly petiolate, simple, linear to ovate, with numerous parallel veins and very slight reticulate or transverse venation, lamina inverted; mesophyll containing mucilage cells and calcium oxalate crystals. Flowers bisexual, in axillary, monobrachial cincinnus or solitary, actinomorphic, articulated on the pedicels (*Luzuriaga*) or at base of ovary (*Drymophila*). Perianth segments six or rarely eight in *Drymophila,* free almost to base, nectariferous at the base. Stamens six in two cycles; filaments free; anthers basifixed or dorsifixed, introrse or extrorse, opening longitudinally or sometimes by apical pores. Pollen grains 1-colpate, reticulate, or (*Geitonoplesium*) trichotomosulcate. Gynoecium of three united carpels; style with deeply 3-lobed (*Drymophila*) or capitate stigma; ovary superior 3-locular or rarely 1-locular, with several anatropous to hemitropous ovules. Fruits berries or berry-like, fleshy capsules. Seeds few to numerous, small, testal, in *Eustrephus* with fleshy strophiole. Endosperm copious, storing fat and aleurone and also hemicellulose (except *Geitonoplesium*); embryo straight, usually rather long, or curved (*Geitonoplesium, Eustrephus*). n = 10. . . . . 1. LUZURIAGACEAE.

1 Ovary inferior.

3 Fruits capsules or rarely (some sp. *Bomarea*) fruits berrylike and indehiscent. Erect or sometimes climbing rhizomatous perennial herbs or vines (most *Bomarea*), with fibrous and often also fusiform and fleshy tuberlike roots containing starch; very rarely (*Alstroemeria graminea*) annuals without storage roots. Stems unbranched, leafy, those of climbing species up to 4 m long or even longer. Calcium oxalate raphides present. Vessels both in

roots (mostly with scalariform perforations) and stems (only with scalariform perforations). Axial parenchyma contains starch. Leaves alternate, linear to oblong, narrowing at the base, generally twisted at the base by 180 degrees and reversing the surfaces, glabrous or pubescent on both leaf surfaces; epicuticular wax sometimes present. Stomata anomocytic. Inflorescences terminal, consisting of helicoid cymes, umbel-like or (*Leontochir*) capitate, rarely unifloral. Flowers usually few to numerous (up to 100 in some *Bomarea*), rather large, subtended by relatively large leaf-like bracts, bisexual, more or less actinomorphic or zygomorphic (*Alstroemeria*), 3-merous. Perianth segments six, in two cycles, usually free, more or less similar to conspicuously different in two cycles (the inner often longer, of different colour and more variegated than the outer one), in all genera the base of all or two of the inner tepals is canaliculate and in *Leontochir* even closed to form a pocket and contains nectar from succulent trichomes (Bayer 1998). Stamens six in two cycles; filaments free, narrow; anthers pseudobasifixed, elongate, sagittate, tetrasporangiate, opening longitudinally, latrorse. Tapetum secretory. Pollen grains 2-celled, large, 1-colpate, usually plano-convex, semitectate. Gynoecium of three united carpels; style filiform, more or less 3-lobed; ovary inferior, 3-locular or 1-locular (*Leontochir* and one species of *Alstroemeria* by abortation late in development), with numerous ovules per locule or per parietal placenta; distinct decurrent borders of the outer perianth segments are seen on the surface of the ovary as well as a circular scar around the top of the fruit left after the shedding of the perianth. Placentation axile (*Alstroemeria, Bomarea*) or parietal (*Leontochir, Alstroemeria pygmaea*). Ovules anatropous, tenuinucellate, with funicular obturator; no parietal cell cut off from the archesporial cell. Fruits more or less truncate capsules, opening loculicidally from the top to the base or only on the top, in *Alstroemeria* mostly explosively, rarely (some species of *Bomarea*) fruits berrylike and indehiscent. Seeds globose or globose to ellipsoid, mostly tuberculate, with long, narrow raphe and well-expressed hilum; testa of few to many cell layers, dry or (*Bomarea*) fleshy, with small amount of phlobaphenes in endotesta (at least in *Alstroemeria*), the inner integument reduced to thin membrane; endosperm copious, consists of cells with large pores, contains hemicellulose, aleurone, and fatty oils; embryo straight, cylindrical, about one-third to two-thirds the length of the endosperm. Produce steroidal saponins and chelidonic acid as well as flavonol glycosides and tuliposides (in which the order agrees with the Liliaceae), but alkaloids are lacking, n = 8 (*Alstroemeria*) or 9 (*Leontochir* and *Bomarea*). Chromosomes generally large. ...... 3. ALSTROEMERIACEAE.

3 Fruits berries. Perennial, rhizomatous, woody lianas with fibrous irregularly distributed adventitious roots and more or less prickly, climbing or scrambling stems to 6 m long. Calcium oxalate raphides present in all parts. Branching sympodial and some of morphologically terminal inflorescences transformed into branched tendrils. Vessels in roots and stem, with scalariform perforations. Leaves alternate, shortly petiolate, broadly lanceolate, acuminate, pinnate-striate veined with reticulate finer venation. Flowers in leaf-opposed or terminal branched cymes, bisexual, not articulated, ebracteate. Perianth segments six, subequal, spreading, or reflexed, basally nectariferous. Stamens six, free, inserted at the base of perianth; filaments filiform; anthers oblong, basifixed, tetrasporangiate, extrorse. Tapetum plasmodial. Pollen grains 1-colpate, with small spinules, contains starch grains at anthesis (Conran and Clifford 1986). Gynoecium of three carpels; style slender, with capitate, slightly 3-lobed stigma; ovary 1-locular, with numerous anatropous, crassinucellate ovules on parietal placentas. Berries bright red, with thick, fleshy pericarp. Seeds numerous more or less globose but somewhat angular, rugulose, dark brown, with thickened exotesta, several-layered mesotesta and thick-walled endotesta; tegmen consists of collapsed thin-walled cells; embryo linear, medium-sized; endosperm copious. Steroidal saponins and alkaloids lacking, n = 5. .... 4. PETERMANNIACEAE.

1. LUZURIAGACEAE

J.P. Lotsy 1911 (including Eustrephaceae Chupov 1994, including Geitonoplesiaceae R. Dahlgren ex Conran 1994). 4/7. South America from Peru to Tierra del Fuego, and Falkland Islands; Southern Africa, Malesia from Java to New Guinea, Australia (New South Wales to Tasmania), New Zealand, New Caledonia.

LUZURIAGEAE: *Drymophila, Luzuriaga*; EUSTREPHEAE: *Eustrephus*, *Geitonoplesium*

2. BEHNIACEAE

Conran, Rudall et Chase 1997. 1/1. Southern and South-Eastern Africa.

Behnia.

Taxonomic position of monotypic genus *Behnia* is debatable. Schlittler (1953) included this genus in Luzuriagoideae (Liliaceae s.l.). The molecular studies of Chase et al. (1995) found instead that *Behnia* occupied a basal position to clade containing the Agavaceae, Anthericaceae.

3. ALSTROEMERIACEAE

Dumortier 1829. 3/280. From central Mexico and West Indies to Chile and Argentina, mostly in the Andes.

Leontochir, Bomarea, Alstroemeria (including *Schickendantzia* and *Taltalia*).

Related to the Luzuriagaceae (Bayer 1998; Meerow 2004), which is also supported by the presence of colchicines alkaloids, flavonol glycosides (and absence of flavones) and the accumulation of appreciable amounts of tuliposide A (Hegnauer 1963; Slob et al. 1975), as well as by the molecular and morphological data (Chase et al. 1995; Rudall et al. 2000; Vinnersten and Bremer 2001; Soltis et al. 2006).

4. PETERMANNIACEAE

Hutchinson 1934. 1/1. Eastern Australia.

Petermannia.

Closely related to the Luzuriagaceae (Stevenson and Loconte (1995)

Bibliography

Aagesen L and AM Sanso. 1998. Phylogeny of the Alstroemeriaceae. In: Monocots II, p. 61 (abstract). Sydney.

Aagesen L and AM Sanso. 2003. The phylogeny of the Alstromeriaceae, based on morphology, *rps*16 Intron, and *rbc*L sequence data. Syst. Bot. 28: 47–69.

Aizen MA and A Basilio. 1995. Within and among flower sex-phase distribution in *Alstroemeria aurea* (Alstroemeriaceae). Canad. J. Bot. 73: 1984–1994.

Aker S and W Healy. 1990. The phytogeography of the genus *Alstroemeria*. Herbertia 45: 76–87.

Arroyo SC and BE Leuenberger. 1988. Leaf morphology and taxonomic history of *Luzuriaga* (Philesiaceae). Willdenowia 17: 159–172.

Baker JG. 1888. Handbook of the Amaryllidaceae including the Alstroemerieae and Agaveae. George Bell, London.

Bayer E. 1987. Die Gattung *Alstroemeria* in Chile. Mitt. Bot. Staatssamml. Münch. 24: 1–362.

Bayer E. 1988. Beitrag zur Cytologie der Alstroemeriaceae. Mitt. Bot. Staatssamml. Münch. 27: 1–6.

Bayer E. 1998. Alstroemeriaceae. In: K Kubitzki, ed. The families and genera of vascular plants, vol. 3, pp. 79–83. Springer, Berlin/Heidelberg/New York.

Buxbaum F. 1951. Die Grundachse von *Alstroemeria* und Einheit ihres morphologischen Typus mit den echten Liliaceae. Phytomorphology 1: 170–184.

Buxbaum F. 1954. Morphologic der Blüte und Frucht von *Alstroemeria* und der Anschluss der Alstroemerioideen bei den echten Liliaceen. Oesterr. Bot. Z. 101: 337–352.

Chupov VS. 1987. Taxonomic position of the genera *Geitonoplesium* and *Simethis*. Bot. Zhurn. 72: 904–908 (in Russian with English summary).

Clifford HT and JG Conran. 1987. *Drymophila*. Flora of Australia 45: 156–158. Canberra.

Conran JG. 1987. A phenetic study of the relationships of *Drymophila* R. Br. within the reticulate-veined Liliiflorae. Aust. J. Bot. 35: 283–300.

Conran JG. 1988. Embryology and possible relationships of *Petermannia cirrhosa* (Petermanniaceae). Nord. J. Bot. 8: 13–17.

Conran JG. 1989. Cladistic analyses of some net-veined Liliiflorae. Plant Syst. Evol. 168: 123–141.

Conran JG. 1998. Behniaceae. In: K Kubitzki, ed. The families and genera of vascular plants, vol. 3, pp. 146–138. Springer, Berlin/Heidelberg/New York.

Conran JG. 1999. Anatomy and morphology of *Behnia* (Behniaceae) and its relationships within Lilianae: Asparagales. Bot. J. Linn. Soc. 131: 115–129.

Conran JG and HT Clifford. 1998a. Luzuriagaceae. In: K Kubitzki, ed. The families and genera of vascular plants, vol. 3, pp. 365–368. Springer, Berlin/Heidelberg/ New York.

Conran JG and HT Clifford. 1998b. Petermanniaceae. In: K Kubitzki, ed. The families and genera of vascular plants, vol. 3, pp. 406–408. Springer, Berlin/Heidelberg/New York.

Conran JG, PJ Rudall, and MW Chase. 1997. Two new monocotyledon families: Anemarrhenaceae and Behniaceae (Lilianae, Asparagales). Kew Bull. 52: 995–999.

Goldblatt P. 1995. The status of R. Dahlgren's orders Liliales and Melanthiales. In: PJ Rudall, PJ Cribb, D. Cutler, and CJ Humphries, eds. Monocotyledons: systematics and evolution, pp. 181–200. Royal Botanic Gardens, Kew.

Hofreiter A and OB Lyshede. 2006. Fucntional leaf anatomy of *Bomarea* Mirb. (Alstroemeriaceae). Bot. J. Linn. Soc. 152: 73–90.

Hunziker AT. 1973. Notas sobre Alstroemeriaceae. Kurtziana 7: 133–135.
Hunziker JH. 1991. Protandry in *Alstroemeria psittacina* (Alstroemeriaceae). Polish Bot. Studies 2: 195–198.
Hunziker JH and CC Xifreda. 1990. Chromosome studies in *Bomarea* and *Alstroemeria* (Alstroemeriaceae). Darwiniana 30: 179–183.
Kosenko VN. 1994. Pollen morphology of the family Alstroemeriaceae. Bot Zhurn. 79(8): 1–8 (in Russian with English summary).
Lyshede OB. 2002. Comparative and functional leaf anatomy of selected Alstroemeriaceae of mainly Chilean origin. Bot. J. Linn. Soc. 140: 261–272.
Meerow AW. 2004. Alstroemeriaceae. In: N Smith, SA Mori, A Henderson, DW Stevenson, and SV Heald, eds. Flowering plants of the Neotropics, pp. 409–410. The New York Botanical Garden. Princeton University Press, Princeton.
Nemirovich-Danchenko EN. 1985. Alstroemeriaceae. In: A Takhtajan, ed. Comparative seed anatomy, vol. 1, pp. 77–78. Nauka, Leningrad (in Russian).
Oganezova GG. 1990. Seed and fruit anatomy of some Amaryllidaceae in connection with their systematics and phylogeny. Bot. Zhurn. 75: 615–630 (in Russian with English summary).
Reveal JL. 1997. Behniaceae (Magnoliophyta), a new family of Asteliales (Liliopsida). Phytologia 82: 273–274.
Rodriguez R and C Marticorena. 1987. Las especies del genera *Luzuriaga* R. et P. Gayana Bot. 44: 3–15.
Sanso AM. 1996. El género *Alstroemeria* (Alstroemeriaceae) en Argentina. Darwiniana 34: 349–382.
Sanso AM. 2002. Chromosome studies in Andean taxa of *Alstroemeria* (Alstroemeriaceae). Bot. J. Linn. Soc. 138: 451–459.
Sanso AM and JH Hanziker. 1998. Karyological studies in *Alstroemeria* and *Bomarea* (Alstroemeriaceae). Hereditas 129: 67–74.
Sanso AM and CC Xifreda. 1999. The synonymy of *Schickendantzia* with *Alstroemeria* (Alstroemeriaceae). Syst. Geogr. Plants 68: 315–323.
Sanso AM and CC Xifreda. 2001 Generic delimitation between *Alstroemeria* and *Bomarea* (Alstroemeriaceae). Ann. Bot. 38: 1057–1069.
Schlittler J. 1949. Die systematische Stellung der Gattung *Petermannia* E V. Muell. und ihre phylogenetische Beziehung zu den Dioscoreaceae Lindl. Vierteljahrsschr. Naturf. Ges. Zürich 1: 1–28.
Schlittler J. 1965 (1966). Sind die Luzuriagoideen wir-kliche Liliaceen oder haben die Ericales und Fern-stroemiales organo-phylogenetisch und stammesg-eschichtlich Beziehungen zur primitiven Liliefloren? Ber. Schweiz. Bot. Ges. 75: 96–109.
Slob A, B Jekel, and E Schlatmann. 1975. On the occurrence of tuliposides in the Liliiflorae. Phytochemistry 14: 1997–2005.
Stenar AHS. 1952. Notes on the embryology and anatomy of *Luzuriaga latifolia* Poir. Acta Horti Berg. 16: 219–232.
Tomlinson PB and ES Ayensu. 1969. Notes on the vegetative morphology and anatomy of the Petermanniaceae. Bot. J. Linn. Soc. 62: 17–26.
Vinnersten A and G Reeves. 2003. Phylogenetic relationships within Colchicaceae. Am. J. Bot. 90: 1455–1462.

Order 11. SMILACALES

Rhizomatous herbs or shrubs. Rhizomes with or without secondary thickening. Roots fibrous. Oxalate raphides present. Vessels in roots, in roots and stem, or sometimes also in leaves, with scalariform, simple, or scalariform and simple perforations. Leaves alternate, distichous, opposite or verticillate, sessile or petiolate, mostly without a sheathing base, mostly coriaceous, with several prominent longitudinal veins, often with usually prominent reticulate or transverse venation; midrib or main veins prominent. Stomata anomocytic. Flowers in terminal or axillary inflorescences or rarely solitary, mostly articulated on the pedicels, bisexual (Philesiaceae, and Ripogonaceae) or unisexual and dioecious, actinomorphic, 3-merous. Perianth segments six, usually more or less similar, free or connate into a tube. Stamens six, rarely (some Smilacaceae) three or nine (up to 18); filaments usually free or sometimes (*Heterosmilax*) more or less connate into a tube or column, sometimes adnate to the base of perianth segments or to the perianth tube; anthers basifixed or dorsifixed, tetrasporangiate or disporangiate, extrorse, latrorse or introrse, usually opening longitudinally. Tapetum secretory or (*Ripogonum*) plasmodial. Microsporogenesis successive. Pollen grains 2-celled, 1-colpate and with reticulate ornamentation or inaperturate and spinulose. Nectaries often occur at the bases of perianth segments or stamens. Gynoecium of three united carpels; stylodia free or only basally connate or connate into a style with 3-lobed or capitate stigma; ovary superior 3-locular or 1-locular, with one or two to numerous ovules per locule or per parietal placenta. Septal nectaries absent. Ovules anatropous, hemitropous, campylotropous, or orthotropous, bitegmic, crassinucellate, with parietal cell. Female gametophyte of *Polygonum*-type. Endosperm nuclear. Fruits berries. Seeds without phytomelan crust, with straight and sometimes very small embryo and copious endosperm containing aleurone, lipids, and sometimes hemicellulose, rarely (*Ripogonum*) starch. Lacking alkaloids.

Related to the Alstroemeriales and probably derived from the same stock.

Key to Families

1 Plants without tendrils.
  2 Leaves alternate, usually with inverted lamina. Erect (*Philesia*) or climbing (*Lapageria*) shrubs

from short woody rhizomes. Vessels in roots and stem, with scalariform perforations. Leaves sessile or petiolate, more or less sheathing at base, ovate to lanceolate or broadly linear and twisted at base, but lamina not inverted, with one (*Philesia*) or 3–5 (*Lapageria*) prominent parallel veins and distinctly reticulate or transverse venation. Stomata anomocytic (in *Lapageria* transversely oriented in relation to parallel veins). Flowers solitary in leaf axils or terminal on short branchlets; pedicels not articulated. Perianth segments free, in *Philesia* the outer cycle short and nearly sepaloid; inner segments always large, petaloid and overlapping to form a campanulate perigone. Filaments free (*Lapageria*) or connate half-way up (*Philesia*), anthers subbasifixed, extrorse or introrse, opening longitudinally. Pollen grains inaperturate and spinulose. Style linear, with 3-lobed or capitate stigma; ovary 1-locular, with numerous anatropous or amphitropous ovules on intrusive parietal placentas. Seeds globose; endosperm copious and consists of cells with rather thin, nonpitted walls; embryo 1/2 as long as the seed. Steroidal saponins present. n = 15, 19. . . . . 1. PHILESIACEAE.

2 Leaves opposite, decussate or verticillate. Lamina noninverted. Commonly climbing shrubs with glabrous or pubescent and often prickly stems, sometimes up to 20 m tall. Vessels only in roots and stem, with scalariform perforation. Leaves with twisted petiole, 3–5 parallel veins and reticulate transverse venation. Flowers in axillary racemes or spikes or in terminal racemose panicles, usually bisexual. Perianth segments free, similar, spreading. Filaments free, very short; anthers basifixed or subbasifixed, tetrasporangiate, opening longitudinally, latrorse. Pollen grains 1-colpate, reticulate and slightly foveolate. Style very short and thick or absent; stigma minutely 3-lobed; ovary 3-locular, with (1)2 anatropous ovules per locule. Seeds brown; seed coat derived mainly from tegmen and chalazal layers, whereas fleshy testa are represented by cuticular membrane that includes obliterated remnants of outer integument; endosperm copious, very hard, consists of radially elongated, thick-walled cells with scalariform-pitted walls, contains starchy grains and hemicellulose; embryo very small, linear. n = 15. . . . . .2. RIPOGONACEAE.

1 Plants mostly with paired tendrils on petiole. Woody or herbaceous creeping, climbing, or scrambling plants with rhizomatous or tuberous rootstock and often prickly stems and branches. Vessels in all parts, with scalariform perforations. Leaves alternate, lower ones reduced to scales, upper ones lanceolate to elliptic or cordate-orbiculate, petiolate, entire or slightly lobed, 3–9-veined, reticulate-veiny between parallel veins. Flowers rather small, in axillary, usually peduncled umbels or fascicles that are solitary or arranged on elongated axis, dioecious. Perianth segments free (*Smilax*) or united into dentate tube. Stamens six (rarely up to 15) or less often nine (*Pseudosmilax*) or only three (*Heterosmilax*) free or (*Heterosmilax*) united into a column; anthers basifixed, disporangiate, introrse. Pollen grains in monads, globose, inaperturate or indistinctly 1-colpate, spinulose. Male flowers without pistillodia. Stylodia short, thick, and spreading, free or rarely basally shortly connate, papillate; ovary 3-locular, with 1–2 pendulous, orthotropous ovules. Staminodia in female flowers present. Fruits berries. Seeds brown, shiny; when ripe, testa is obliterated and only tegmen remains, phlobaphenes; endosperm copious, consists of radially elongated, thick-walled cells and contains hemicellulose, aleurone, and fatty oils, but starch is lacking; embryo small, linear, straight. n = 13, 14, 15, 16. . . . . 3. SMILACACEAE.

1. PHILESIACEAE

Dumortier 1829. 2/2. Southern Chile.

Philesia, Lapageria.

Closely related to the Smilacaceae (see Conran 1989). Molecular studies also supported a close relationship between the Philesiaceae and the Smilacaceae (Chase et al. 1995; Soltis et al. 2006).

2. RIPOGONACEAE

Conran et H. Clifford 1985. 1/6. New Guinea, eastern Australia, New Zealand, and Stewart and Chatham Islands.

Ripogonum.

Related to the Philesiaceae and in some respects occupying an intermediate position between them and the Smilacaceae.

3. SMILACACEAE

Ventenat 1799. 2/320–370. Mainly tropical and subtropical regions, but also in warm-temperate regions.

Smilax, Heterosmilax (*Pseudosmilax*).

Close to the Ripogonaceae and the Philesiaceae. All this families have unique spiny pollen (Rudall et al. 2000).

Bibliography

Arber A. 1920. Tendrils of *Smilax*. Bot. Gaz. 60: 438–442.

Cameron KM and C-X Fu. 2006. A nuclear rDNA phylogeny of *Smilax* (Smilacaceae). In: JT Columbus, EA Friar, JM Porter, LM Prince, and MG Simpson, eds. Monocots: comparative biology and evolution (excluding Poales), pp. 598–605. Rancho Santa Ana Botanical Garden, Claremont.

Cave MS. 1966. The female gametophytes of *Lapageria rosea* and *Philesia magellanica*. Guyana Bot. 15: 25–31.

Chen S-C, Y-X Qiu, A-L Wang, and C-X Fu. 2006a. A phylogenetic analysis of the Smilacaceae based on morphological data. Acta phytotax. Sinica 14: 113–125.

Chen S-C, X-P Zhang, S-F Ni, C-X Fu, and KM Cameron. 2006b. The systematic value of pollen morphology in Smilacaceae. Plant Syst. Evol. 259: 19–37.

Conover M. 1983. The vegetative morphology of the reticulate veined Liliiflorae. Telopea 2: 401–412.

Conover M. 1991. Epidermal patterns of the reticulate-veined Liliiflorae and their parallel-veined allies. Bot. J. Linn. Soc. 107: 295–312.

Conran JG. 1989a. Cladistic analyses of some net-veined Liliiflorae. Plant Syst. Evol. 168: 123–141.

Conran JG. 1998b. Smilacaceae. In: K Kubitzki, ed. The families and genera of vascular plants, vol. 3, pp. 417–422. Springer, Berlin/Heidelberg/New York.

Conran JG and HT Clifford. 1985. The taxonomic affinities of the genus *Ripogonum*. Nord. J. Bot. 5: 215–219.

Conran JG and HT Clifford. 1998. Philesiaceae. In: K Kubitzki, ed. The families and genera of vascular plants, vol. 3, pp. 409–411. Springer, Berlin/Heidelberg/New York.

Conran JG, PJ Rudall, and MW Chase. 1997. Two new monocotyledon families: Anemarrhenaceae and Behniaceae (Lilianae, Asparagales). Kew Bull. 52: 995–999.

Fu C. 1998. Studies on systematics and evolution of *Smilax* and *Heterosmilax* (Smilacaceae). In Monocots II, p. 67 (abstract). Sydney.

Judd WS. 1998. The Smilacaceae in the southeastern United States. Harvard Pap. Bot. 3: 147–169.

Komar GA. 1985. Smilacaceae. In: A Takhtajan, ed. Comparative seed anatomy, vol. 1, pp. 132–133. Nauka, Leningrad (in Russian).

Koyama T. 1980. Materials toward a monograph of the genus *Smilax*. Quart. J. Taiwan Mus. 8: 1–62.

Koyama T. 1984. A taxonomic revision of the genus *Heterosmilax* (Smilacaceae). Brittonia 36: 184–205.

Patterson TB and TJ Givnish. 2002. Phylogeny, concerted convergence, and phylogenetic niche conservatism in the core Liliales: Insights from *rbc*L and *ndh*F sequence data. Evolution 56: 233–252.

Schlittler J. 1951. Die Gattung *Eusterphus* R. Br. ex Sims und *Geitonoplesium* (R. Br.) A. Cunn.: Morphologisch-anatomische Studie mit Berücksichtigung der systematischen, nomenklatorischen, und arealgeo-graphischen Verhältnisse. Ber. Schweiz. Bot. Ges. 151: 175–239.

Schulze W. 1931. Beiträge zur Taxonomie der Liliifloren. VIII. Wiss. Z. Friedrich-Schiller-Univ. Jena 31: 285–289.

Schulze W. 1982. Beiträge zur Taxonomie der Liliifloren. VII. Philesiaceae. Wiss. Z. Friedricg-Schiller-Univ. Jena 31: 277–283.

Simpson PG and WR Philipson. 1969. Vascular anatomy in vegetative shots of *Rhipogonum scandens* Forst. (Smilacaceae). N. Z. J. Bot. 7: 3–29.

Order 12. ORCHIDALES

Perennial herbs, terrestrial or epiphytic, usually mycotrophic or occasionally saprophytic, rhizomatous or tuberous. Silicia bodies enclosed in special cells, calcium oxalate raphides usually present. Vessels mostly only in roots, sometimes also in the stem and also in leaves, mostly with scalariform perforations. Leaves alternate or rarely opposite, parallel-veined. Stomata mostly paracytic or sometimes tetracytic. Flowers in various types of inflorescences, usually bisexual, mostly zygomorphic. Perianth segments six, free or more or less connate. Stamens six, three, two or usually one. Anthers dorsifixed or basifixed, introrse or seldom latrorse, tetrasporangiate, opening longitudinally. Tapetum secretory or rarely plasmodial. Microsporogenesis simultaneous or successive. Pollen grains mostly 1-colpate, tectate-columellate. Gynoecium of three united carpels; ovary superior or more often inferior, 3-locular or rarely unilocular, ovules of various number but mostly numerous, anatropous, very rarely hemitropous or campylotropous, crassinucellate or more often tenuinucellate, usually bitegmic. Female gametophyte of *Polygonum*-type or rarely of *Allium*-type, very rarely tetrasporic. Endosperm mostly nuclear. Fruits mostly capsular. Seeds usually numerous. Endosperm reduced or more often lacking, very rarely (*Lanaria*) copious. Embryo well developed or more often reduced and undifferentiated.

Very old group, which probably originated from the melanthianean ancestor.

Key to families

1 Ovary superior.

2 Seeds brown, clothed with colourless hairs often appearing white. Perennial herbs with more or less tuberous, short rhizome and fibrous roots. Leaves alternate, distichous, usually concentrated at the base of stem, linear, with prominent midrib and rough edges, sheathing slightly at base; cauline leaves small, bracteal. Stomata anomocytic. Flowers large, pendulous, in terminal racemes, bisexual, actinomorphic, campanulate, each subtended by a bract and two smaller bracteoles. Perianth segments six, petaloid, connate into a tube for about four-fifths of their length. Stamens six in two cycles, inserted in perianth tube; anthers dorsifixed, tetrasporangiate, dehiscing latrorsely by slits. Tapetum secretory. Microsporogenesis simultaneous. Pollen grains 1-colpate, with granular exine. Ovary superior, 3-locular, tapering below into a gynophore, stipitate, with 40–50 ovules per locule, inserted in two rows; style short, not distinct from ovary, with small 3-grooved stigma. Ovules anatropous, bitegmic, crassinucellate. Septal nectaries present as deep exterior grooves. Female gametophyte of *Polygonum*-type. Endosperm helobial. Fruits septicidal capsules. Seeds 5–6 mm long, oblong-falcate, angulate, with phytomelan in testa and phlobaphene in tegmen, with remnants of nucellar tissue; endosperm copious; embryo linear, slightly curved. n = 17, 27. . . . . . 1. BLANDFORDIACEAE.

2 Seeds black, glabrous. Perennial herbs (dwarf to large), shrubs or trees (*Cordyline*) to 10 m tall, often rhizomatous or tuberous, occasionally epiphytic (on trees). Raphides present. Secondary thickening anomalous (*Cordyline*) or absent. Vessels only in roots, with scalariform perforations. Leaves alternate, spiral or distichous, mesomorphic to leathery and stiff, linear to broadly lanceolate, mostly keeled, sheathing at base, usually densely pubescent, at least when young; the trichomes scalelike or stellate, multicellular and arising from multicellular base; mucilage canals present in leaves of some species of *Astelia* and *Collospermum*. Stomata paracytic. Flowers in one terminal and several lateral bracteate racemes or spikes, trimerous or (*Neoastelia*) penta- to heptamerous, bisexual, polygamo-dioecious or dioecious, actinomorphic. Perianth segments six in two cycles, free or basally connate, similar. Stamens six in two cycles, free or inserted on the perianth in *Milligania;* anthers basifixed to dorsifixed, opening longitudinally, introrse, unappendaged. Female flowers with staminodia. Floral nectaries present Microsporogenesis successive, or (*Milligania*) simultaneous. Pollen grains 1-colpate or (*Collospermum*) both 1-colpate and trichotomocolpate, more or less echinate. Gynoecium of 3(-7) united carpels; stylodia connate into short style (in *Neolitsea* stylar branches elongated with decurrent stigmas) or stigmas more or less sessile; in some species of *Astelia* there is a gradual transition between ovary and style. Ovary superior, 3-locular (placentation axile) or 1-locular (placentation parietal), with four to many anatropous ovules per locule; mucilage secreting intraovarian trichomes often present; septal nectaries present. Ovules, anatropous or campylotropous, bitegmic, crassinucellate. Fruits berries or (*Milligania*) loculicidal capsules splitting from the apex. Seeds ovate and often angular, black, more or less shiny, with mucilaginous hairs; exotesta encrusted with phytomelan; endosperm copious, containing aleurone and lipids; embryo straight or slightly bent, one third to half the length of endosperm. Saponins/sapogenins present; n = 8, 30, 35?. . . . . 2. ASTELIACEAE.

1 Ovary inferior

3 Stamens and style fused into gynostemium. Strongly mycotrophic or occasionally saprophytic perennial herbs, terrestrial or more often epiphytic, sometimes climbing or scandent, very rarely completely subterranean or rheophytic. Terrestrial species with fibrous or thickened tuberous or cordlike roots and usually rhizomatous, the epiphytic species often with one or several internodes modified into water storage organs (pseudobulbs) and often with aerial hanging roots covered by a thick layer of dead and usually spongy water-absorbing tissue derived from epidermis (velamen) that are also present in most terrestrials. Stems sympodial or (in certain Orchidoideae) monopodial. Silica bodies enclosed in special cells, calcium oxalate raphides, and mucilage cells generally present. Vessels mostly in roots, less often also in stem, very

rarely also in leaves, mostly with scalariform perforations. Leaves alternate, mostly distichous, rarely opposite or verticillate, simple and usually entire, sometimes plicate, linear to orbicular, sometimes pseudopetiolate, usually parallel-veined, often fleshy, sheathing basally with the sheaths generally closed and enveloping the stem, in saprophytic species reduced into small scales or rarely entirely absent. Stomata paracytic, less often anomocytic, sometimes tetracytic. Flowers in racemes, spikes, or racemose panicles, sometimes solitary, bracteate, bisexual or rarely unisexual (monoecious or dioecious), usually strongly zygomorphic, 3-merous, mostly resupinate at anthesis. Perianth segments six in two cycles; sepals green or more often coloured and petaloid, all similar or the median one (morphologically abaxial) differing in size and colour, mostly imbricate; the median petal (morphologically adaxial) commonly much larger, often highly modified and differing in colour from laterals forming the labellum, which often projects basally into a spur or sac with or without nectar, the laterals usually more or less like sepals. Stamens three (Apostasioideae), two (Cypripedioideae), or more often only one (Orchidoideae), all morphologically abaxial (but apparently adaxial), opposite labellum, when one united with the style forming gynostemium; anthers tetrasporangiate, opening longitudinally, introrse. Tapetum secretory. Microsporogenesis simultaneous. Pollen grains 2-celled, solitary (Apostasioideae and part of Cypripedioideae), in tetrads, or agglutinated into waxy or mealy masses (pollinia), tectate-columellate, or sometimes tectate-granular or even exineless, 1-colpate, 2-colpate, 1-porate or inaperturate, rarely (*Vanilla*) 3-porate, with different types of ornamentation. Gynoecium of three united carpels; style more or less apically inflexed and terminated by a 3-lobed stigma; all three stigma lobes may be fertile, but most commonly the two lateral ones are fertile and the median lobe is modified into a small sterile outgrowth, the rostellum; ovary inferior, 3-locular or generally 1-locular, with very numerous and minute ovules per locule or per parietal placenta. Ovules anatropous, usually bitegmic, tenuinucellate, without parietal cell. Female gametophyte of *Polygonum*-type or (mainly Cypripedioideae) of *Allium*-type, very rarely tetrasporic. Endosperm, when it develops, is nuclear, but usually endosperm formation is arrested very early or wholly omitted. Fruits usually capsules dehiscing by 3(6) hygroscopically sensitive valves, rarely fleshy and indehiscent. Seeds extremely numerous, commonly minute, mostly dustlike, often fusiform, usually germinating only in presence of appropriate fungus; seed coat membranous, formed mainly by the outer layer of the outer integument and generally consists of one layer of cells; the inner integument may persist as a thick layer; endosperm extremely reduced and mostly lacking; embryo usually not differentiated into distinct organs, and the cotyledon more or less develops only in a few species; on germination the embryo forms a tubercle (protocorm) with basal rhizoids but without radicle and usually without cotyledon. Produce flavone C-glycosides and flavonols as major leaf constituents, n = 6–29 +. . . . . 5. ORCHIDACEAE.

3 Style and stamens free.

4 Capsules many-seeded. Herbs with more or less tuberous rhizome or a corm often covered with membranous or fibrous sheaths of former leaves. Calcium oxalate raphides generally present, silica bodies always absent. Sieve-element plastids of form-P2c. Leaves basal, mostly tristichous, linear to lanceolate, prominently parallel-veined, with mucilage canals, often clothed with long, uniseriate hairs, but branched, multicellular hairs also present. Stomata paracytic or rarely tetracytic. Flowers in axillary racemes, spicate, corymbose or umbellate, sometimes solitary, bisexual (rarely unisexual in *Curculigo*), actinomorphic, 3-merous. Perianth segments six in two cycles (rarely four in two cycles), spreading, almost of equal size and similar in colour, free or less often connate into a sometimes long and narrow tube on top of the ovary, more or less persistent. Stamens six in two cycles or rarely (*Pauridia*) only three and then opposite the inner segments; filaments subulate to filiform, short to almost lacking; arising from the base of perianth segments or from the mouth of the perianth tube anthers elongate, tetrasporangiate, basifixed to dorsifixed, latrorse or introrse, opening

longitudinally. Tapetum plasmodial (amoeboid). Microsporogenesis successive. Pollen grains 2-celled, tectate-columellate, 1-colpate, finely reticulate. stylodia connate into short style that is apically divided into three branches; ovary inferior, 3-locular to 1-locular (*Empodium*). Ovules often numerous, anatropous, hemitropous, or campylotropous, bitegmic, crassinucellate or tenuinucellate, without parietal cell. Female gametophyte of *Polygonum*-type or rarely of *Allium*-type. Endosperm helobial or less often (*Pauridia* and some species of *Hypoxis* section *Ianthe*) nuclear; chalazal cell of helobial endosperm somewhat haustorial. Fruits either a capsule with loculicidal or circumscissile dehiscence, or sometimes more or less fleshy and indehiscent or irregularly dehiscent. Seeds small, brown or black, usually more or less globose, tuberculate to spinulose, with a prominent raphe, often with an elaiosome; seed coat is formed by both integuments, but the most prominent part is the outer epidermis of testa, the outer cells of which develop characteristic protuberances; exotesta is characterized by prominent, thick, black phytomelan crust; the pigments mainly being stored in the cell walls of the palisadlike exotesta, phlobaphenes present also in tegmen; endosperm composed of thin-walled, isodiametric cells storing aleurone and fatty oils; embryo cylindrical, straight, usually more or less underdeveloped, occupies two thirds or one half of the length of seed. Alkaloids and steroidal saponins absent but chelidonic acid present, n = 6–9, 11. . . . . .4. HYPOXIDACEAE.

4 Capsules 1-seeded. Tough herbs with relatively short vertical rhizome with upper part bearing several leaves and below them a coat of densely set, fibrous old leaf sheaths. Stem with white, plumose hairs above, increasingly dense on distal parts and on the inflorescences, this appearing white and woolly. Leaves linear, stiff, basal, sheathing, glabrous. Stomata paracytic. Flowers in much branched, dense, bracteate corymbose panicle with rhipidial lateral units. Perianth densely plumose, pubescent abaxially, glabrous adaxially, its segments connate into a cylindrical-infundibular tube. Stamens all fertile, inserted at the mouth of the tube; anthers dorsifixed, versatile, introrse, basally sagittate. Pollen grains 1-colpate, tectate-columellate. Style filiform, apically 3-lobed; ovary inferior or slightly semi-inferior, with two ovules per locule, with septal nectaries. The fertile locule dehiscing loculicidally. Seeds black, shiny; seed coat mainly of outer integument, with thick-walled, palisadlike epidermal cells encrusted with phytomelan, about six layers of testa below the epidermis and strongly compressed two cell layers of inner integument; endosperm copious, nonstarchy; embryo slightly curved, about as long as endosperm. n = 9. 3. LANARIACEAE.

1. BLANDFORDIACEAE

R. Dahlgren et H. Clifford 1985. 1/4. Eastern Australia.

Blandfordia.

Probably the most archaic member of the order.

2. ASTELIACEAE

Dumortier 1829. 5/55. Mascarenes, New Guinea, Australia, Tasmania, New Zealand, New Caledonia, Fiji, Samoa, and Polynesia to Hawaii and Chile.

Astelia, Collospermum, Cordyline (including *Cohnia*), *Neoastelia, Milligania.*

Shows some affinities with Blandfordiaceae and Hypoxidaceae (Bayer et al. 1998; Rudall et al. 1998).

3. LANARIACEAE

H. Huber ex R. Dahlgren 1988. 1/1. Cape Province of South Africa.

Lanaria.

Possible related to Asteliaceae and Hypoxidaceae (Chase et al. 1995).

4. HYPOXIDACEAE

R. Brown 1814. 9/100–120. Africa (mainly South Africa), Mascarenes, Seychelles, tropical and subtropical regions of Asia (in north to eastern Himalayas, southwestern China, and Japan). New Guinea, Australia, Tasmania, New Zealand, America from Atlantic part of the USA to West Indies and Uruguay.

Hypoxis, Rhodohypoxis, Spiloxene, Pauridia, Curculigo, Molineria, Hypoxidia, Empodium, Saniella.

Hypoxidaceae have probably some affinity with the Asteliaceae (Rudall et al. 1997; Nordal 1998). Although Hypoxidaceae and Asteliaceae differ from each other in several respects, including successive microsporogenesis, inferior ovary and pollen sculpturing, there are some unusual anatomical characters linking the two families, notably branched trichomes and mucilage canals in the leaves (Rudall et al. 1998).

5. ORCHIDACEAE

A.L. de Jussieu 1789 (including Apostasiaceae Lindley 1833; Cypripediaceae Lindley 1833; Limodoraceae Horaninow 1847; Neottiaceae Horaninow 1834; Neuwiediaceae R. Dahlgren ex Reveal et Hoogland 1991; Vanillaceae Lindley 1835). 750–800/20000–25000. Subcosmopolitan, chiefly in tropical and subtropical Asia, where representatives of all subfamilies and tribes occur, and tropical America.

5.1 APOSTASIOIDEAE

Terrestrial, sometimes (*Apostasia*) rhizomatous herbs. Vessels predominantly with simple perforations. Stems woody at the base. Leaves spirally arranged, entire, plicate. Stomata paracytic or tetracytic. Flowers slightly zygomorphic, resupinate in *Neuwiedia*. Labellum shallow. Stamens three (*Neuwiedia*), representing the median (adaxial) of the outer cycle and the lateral ones of the inner cycle, or two (*Apostasia*), representing lateral ones (the median stamen staminodial or missing); filaments only partly fused with the style in the gynostemium. Pollen grains free, 1-colpate with an operculum, reticulate. Style with 2- or 3-lobed stigma; stigma lobes equal and similar. Ovary 3-locular. Fruits thin-walled or fleshy capsules, disintegrating or (*Neuwiedia veratrifolia*) dehiscing loculicidally by three valves. Seeds ovoid or elliptic, dark brown, usually alveolate to reticulate, often sticky, sometimes with long funicular appendage or with a prominent appendage at both ends, n = 24. Southeastern Asia and from the Himalayas and Ceylon to New Guinea and northern Queensland. – *Neuwiedia, Apostasia.*

5.2 VANILLOIDEAE

Plant sympodial or monopodial. Vessels with scalariform perforation and the mostly opposite pitting of the lateral walls. Stomata tetracytic. Staminodes two. Anthers incumbent (by massive expansion of the apical column/connective). Pollen grains in tetrads. Seeds fusiform, crustose, n = 9, 10, 12, 14–16, 18. – *Cleistes, Isotria, Pogonia, Pogoniopsis, Duckeella, Erythrorchis, Cyrtosia, Galeola, Pseudovanilla, Clematepistephium, Dictyophyllaria, Epistephium, Eriaxis, Vanilla, Lecanorchis.*

5.3 CYPRIPEDIOIDEAE

Terrestrial or less often epiphytic herbs with fleshy or fibrous roots on short rhizomes. Vessels in roots or rarely in stems, with simple or scalariform perforations. Leaves spirally arranged or distichous, thin and plicate or fleshy and conduplicate. Stomata anomocytic or paracytic. Flowers strongly zygomorphic, resupinate. Two lateral sepals connate. Lateral petals often considerably longer and narrower than the others, median petal (labellum) deeply saccate, characteristically slipper-shaped. Two lateral stamens of the inner cycle always present, and the median stamen of the outer cycle is modified into a characteristic shield-like staminodium. Filaments largely united with style, forming a thick, inflexed gynostemium. Anthers subglobose and latrorse. Pollen grains in tetrads, usually free, 1-colpate, ulcerate or porate, with various ornamentation; pollen more or less viscid, but pollinia occur only in *Phragmipedium* longifolium and *Selenipedium* chica. Style thick, the stigma large and domelike. Ovary 3-locular or 1-locular. Fruits capsules. Seeds subspheric with hard testa (*Selenipedium*) or minute with membranous testa, n = 9 or more. Eurasia, America. – *Cypripedium, Selenipedium, Mexipedium, Paphiopedilum, Phragmipedium.*

5.4 NEOTTIOIDEAE

Mostly terrestrial herbs without pseudobulbs, often facultative geophytes. Fertile stamen one or rarely (*Diplandorchis*) two, the median one of both cycles. Anther movably attached to gynostemium through a connective, commonly persistent and soon withering, the thecae usually closely approximate, parallel. Pollinia granular, sectile or occasionally mealy, 2. – EPIPACTIDEAE: *Cephalanthera, Sinorchis, Tangtsima, Epipactis, Limodorum, Epipogium,* etc; NEOTTIEAE: *Listera, Neottia, Diplandrorchis, Goodyera,*

Ludisia, Anoectochilus, Zeuxine, Spiranthes, Cranichis, Ponthieva, etc.

5.5 EPIDENDROIDEAE. (including Vandoideae)
Mostly facultative epiphytes. Fertile stamen always 1. Anther movably attached to gynostemium through a connective, commonly deciduous. Pollinia waxy or cartilaginous, rarely mealy, 2, 4, 6, or 8. – ARETHUSEAE: *Arethusa, Sobralia, Elleanthus, Phaius, Calanthe, Bletia,* etc; GASTROIDEAE: *Nervilia, Gastrodia,* etc.; EPIPOGIEAE: *Epipogium, Stereosandra;* COELOGYNEAE: *Coelogyne, Pleione,* etc; MALAXIDEAE: *Liparis, Malaxis, Hammarbya, Oberoma,* etc.; CALYPSOEAE: *Calypso, Yoania;* EPIDENDREAE: *Epidendrum, Encyclia, Laelia, Cattleya, Eria, Pleurothallis, Stelis, Masdevallia, Restrepia, Dracula, Dendrobium, Flickingeria, Bulbophyllum,* etc.; POLYSTACHYEAE: *Polystachya,* etc.; CYMBIDIEAE: *Eulophia, Ansellia, Cymbidium, Grammatophyllum, Catasetum, Stanhopea, Gongora, Oncidium, Odontoglossum, Miltonia, Brassia,* etc.; MAXILLARIEAE: *Corallorrhiza, Zygopetalum, Lycaste, Maxillaria,* etc.; VANDEAE: *Phalaenopsis, Aerides, Vanda, Arachnis, Renanthera, Cleisostoma, Aerangis, Taeniophyllum, Microcoelia, Angraecum,* etc.

5.6 ORCHIDOIDEAE
Usually terrestrial herbs with spirally arranged leaves. Vessels in roots, with scalariform perforation. Commonly with storage organ, the tuberoid, which is a root structure around a core of stem tissue with an apical bud that will grow into a new shoot in the next season, with one of the axillary buds forming a new tuberoid. Fertile stamen one. Anther firmly adnate to gynostemium, not deciduous or withering. Thecae often widely spaced, parallel, or converging. Pollinia rather soft, with caudicles, the caudicles directed toward base of anther. – ORCHIDEAE: *Orchis, Dactylorhiza, Ophrys, Anacamptis, Platanthera, Gymnadenia, Habenaria, Cynorkis, Peristylus,* etc; CRANICHIDEAE: *Cyanicula, Aspidogyne, Moerenhoutia, Zeuxine, Beloglottis, Spiranthes, Cranichis, Manniella,* etc.; DIURIDEAE: *Acianthus, Corybas, Cyrtostylis, Bipinnula, Chloraea, Gavilea, Pterostylis, Stigmatodactylus, Adenochilus, Caladenia, Elythranthera, Eriochilus, Glossodia, Leptoceras, Ptaecoxanthus, Coilochilus, Diuris, Orthoceras, Caleana, Chiloglottis, Drakaea, Leporella, Lyperanthus, Megastylis, Paracaleana, Pyrorchis, Rumicola, Spiculaea, Genoplesium, Microtis, Thelymitra, Rhizanthella, Prasophyllum, Calochilus*; DISEAE: *Disa, Satyrium, Disperis*, etc.

A rather taxonomically isolated group, which is characterized by a number of unique characters, including miniaturization of the ovules and microspermy, silica bodies, and mesophyll idioblasts with helical thickening. They show connections with the Hypoxidaceae (Rolfe 1909, 1910; Hallier 1912; Gobi 1916; Hutchinson 1934, 1959, 1973; Garay 1960; Rao 1969; Takhtajan 1980, 1987, 1997; Goldblatt 1995), especially with the genera *Hypoxis* and *Curculigo,* there are indeed many similarities between Orchidaceae and Hypoxidaceae, and *Curculigo orchioides,* for example, strongly resemble the Orchidaceae. Also, chromosomes of the Hypoxidaceae resemble those of the Orchidaceae (Sharma 1969). Evidence from *rbc*L sequences also indicates that orchids are most closely related to Hypoxidaceae (Cameron et al. 1994). The connecting link between Hypoxidaceae and Orchidaceae is the most archaic subfamily the Apostasioideae. This intermediate position of apostasiads was already mentioned by Lindley (1833) in his *The Vegetable Kingdom* ("The order seems as if connecting Orchids with Hypoxids"). However, Orchidaceae could not be derived directly from the Hypoxidaceae.

Bibliography

Abe K. 1972. Contributions to the embryology of the family Orchidaceae: VII. A comparative study of the orchid embryo sac. Sci. Rep. Tohoku Imp. Univ., 4th ser. (Biol), 36: 179–201.

Ackermann JD and NH Williams. 1980. Pollen morphology of the tribe Neottieae and its impact on the classification of the Orchidaceae. Grana 19: 7–18.

Ames O. 1946. The evolution of the orchid flower. Am. Orchid. Soc. Bull. 14: 355–360.

Ames O. 1948. Orchids in retrospect: a collection of essays on the Orchidaceae. Botanical Museum **of** Harvard University, Cambridge, MA.

Arditti J. 1992. Fundamentals of orchid biology. Wiley, New York.

Atwood JT, Jr. 1984. The relationships of the slipper orchids (subfamily Cypripedioideae, Orchidaceae). Selbyana 7: 129–247.

Atwood JT, Jr. 1986. The size of the Orchidaceae and the systematic distribution of epiphytic orchids. Selbyana 9: 171–186.

Averyanov LV. 1990. The systems of Orchides (Orchidaceae) of the flora of Vietnam. The subfamilies Apostasioideae,

Cypripedioideae, Neottioideae, and Orchidoideae. Bot. Zhurn. 75(7): 1013–1028 (in Russian).

Averyanov LV. 1991. Main trends of the morphological evolution of the Orchidaceae family. Bot. Zhurn. 76: 921–935 (in Russian with English summary).

Averyanov LV, P Cribb, PK Loc, and NT Hiep. 2003. Slipper Orchids of Vietnam. Royal Botanic Gardens, Kew.

Baker JG. 1878. Synopsis of Hypoxidaceae. Bot. J. Linn. Soc. 17: 93–126.

Barthlott W. 1976. Morphologic der Samen von Orchideen im Hinblick auf taxonomische und funktionelle Aspekte. In: K Senghas, ed. Proc. 8th World Orchid Conf., pp. 438–443. Frankfurt.

Bayer C, O Appel, and PJ Rudall. 1998. Asteliaceae. In: K Kubitzki, ed. The families and genera of vascular plants, vol. 3, pp. 141–145. Springer, Berlin/Heidelberg/New York.

Benzing DH and JT Atwood, Jr. 1984. Orchidaceae: ancestral habitats and current status in forest canopies. Syst. Bot. 9: 155–165.

Burns-Balogh P and P Bernhardt. 1985. Evolutionary trends in the androecium of the Orchidaceae. Plant Syst. Evol. 149: 119–134.

Burns-Balogh P and P Bernhardt. 1988. Floral evolution and phylogeny in the tribe Thelymitreae (Orchidaceae: Neottioideae). Plant Syst. Evol. 159: 19–47.

Burns-Balogh P and V Funk. 1986. A phylogenetic analysis of the Orchidaceae. Smithsonian Contr. Bot. 61: 1–79.

Cameron KM. 2002. Intertribal relationships within Orchidaceae as inferred from analyses of five plastid genes. In Botany 2002: Botany in the Curriculum. Abstracts, p. 116. Madison, WI.

Cameron KM. 2004. Utility of plastid *psa*B gene sequences for investigating intrafmilial relationships within Orchidaceae. Mole. Phylogenet. Evol. 31: 1157–1180.

Cameron KM and MW Chase. 1996. Systematic investigations of the vanilloid orchids: evidence from DNA sequences, anatomy, and morphology. Am. J. Bot. 83: 143–144.

Cameron KM and MW Chase. 1998a. Systematic of Vanilloideae (Orchidaceae). In Monocots II, pp. 13–14 (abstract). Sydney.

Cameron KM and MW Chase. 1998b. Seed morphology of vanilloid orchids (Vanilloideae: Orchidaceae). Lindleyana 13: 148–169.

Cameron KM and MW Chase. 2000. Nuclear 18S rDNA sequences of Orchidaceae confirm the subfamilial status and circumscription of Vanilloideae. In: KL Wilson and DA Morrison, eds. Monocots: systematic and evolution, pp. 457–464. CSIRO, Collingwood.

Cameron KM, D Jarrell, and MW Chase. 1994. Evidence from rbcL sequences and phylogenetic relationships of major lineages within Orchidaceae. Am. J. Bot. 81(6): 145 (abstract).

Cameron KM, MW Chase, WM Whitten, PJ Kores, DC Jarrelll, VA Albert, T Yukawa, HG Hills, and DH Goldman. 1999. A phylogenetic analysis of the Orchidaceae: evidence from *rbcL* nucleotide sequences. Am. J. Bot. 86: 208–224.

Carlquist S and EL Schneider. 2006. Origins and nature of vessels in Monocotyledons: 8. Orchidaceae. Am. J. Bot. 93: 963–971.

Carlsward BS, WM Whitten, NH Williams, and B Bytebier. 2006. Molecular phylogenetics of Vandeae (Orchidaceae) and the evolution of leaflessness. Am. J. Bot. 93: 770–786.

Chase MW, JF Freudenstein, and KM Cameron. 2003. DNA data and Orchidaceae systematics: a new phylogenetic classification. In: KW Dixon, SP Pell, RL Barrett, and PJ Cribb, eds. Orchid Conservation, pp. 69–89. Kota Kinabalu, Sabah.

Clements MA, DL Jones, IK Scarma, ME Nightingale, MJ Garratt, KJ Fitzgerald, AM Mackenzie, and BPJ Molloy. 2002. Phylogenetics of Diuridae (Orchidaceae) based on the internal transcribed spacer (ITS) regions of nuclear ribosomal DNA. Lindleyana 17: 135–171.

Clifford HT and JG Conran 1998. Blandfordiaceae. In: K Kubitzki, ed. The families and genera of vascular plants, vol. 3, pp. 148–150. Springer, Berlin/Heidelberg/New York.

Clifford HT and PS Lavarack. 1974. The role of vegetative and reproductive attributes in the classification of the Orchidaceae. Bot. J. Linn. Soc. 6: 97–110.

Cribb P and T Cox 1998. Phylogeny of the Cypripedioideae. In Monocots II, p. 17 (abstract). Sydney.

Darwin C. 1862. On the various contrivances by which British and foreign orchids are fertilized. Murray, London.

De Vos MP. 1948. The development of the ovule and seed in the Hypoxideae: 1. *Ianthe* Salisb. South Afr. J. Bot. 14: 159–169.

De Vos MP. 1949. The development of the ovule and seed in the Hypoxideae: II. The genera *Pauridia* Harv. and *Forbesia* Ecldon. South Afr. J. Bot. 15: 13–22.

Di Fulvio TE and MS Cave. 1964. Embryology of *Blandfordia nobilis* Smith (Liliaceae) with special reference to its taxonomic position. Phytomorphology 14: 487–499.

Dodson CH. 1962. The importance of pollination in the evolution of the orchids of tropical America. Am. Orchid Soc. Bull. 31: 525–534, 641–649, 731–735.

Dora G and JM Edwards. 1991. Taxonomic status of *Lanaria lanata* and isolation of a novel biflavone. J. Nat. Prod. 54: 796–801.

Douzery JP, AM Pridgeon, PJ Kores, H Kurzweil, P Linder, and MW Chase. 1999. Molecular phylogenetics of Diseae (Orchidaceae): a contribution from nuclear ribosomal ITS sequences. Am. J. Bot. 86: 887–899.

Dressler RL. 1961. The structure of the orchid flower. Missouri Bot. Gard. Bull. 49: 60–69.

Dressler RL. 1981. The orchids: natural history and classification. Harvard University Press, Cambridge, MA/London.

Dressler RL. 1983. Classification of the Orchidaceae and their probable origin. Telopea 2: 413–424.

Dressler RL. 1986. Recent advances in orchid phylogeny. Lindleyana 1: 5–20.

Dressler RL. 1987. Cladistic analysis of the Orchidaceae: a commentary. Lindleyana 2: 66–71.

Dressler RL. 1990a. The Neottieae in orchid classification. Lindleyana 5: 101–109.

Dressler RL. 1990b. The Spiranthoideae: grade or subfamily? Lindleyana 5: 110–116.

Dressler RL. 1993. Phylogeny and classification of the orchid family. Dioscorides Press, Portland, OR.

Dressler RL and MW Chase. 1995. Whence the orchids? In: PJ Rudall, PJ Cribb, DF Cutler, and CJ Humphries, eds. Monocotyledons: systematics and evolution, pp. 217–226. Royal Botanic Gardens, Kew.

Dressler RL and CH Dodson. 1960. Classification and phylogeny in the Orchidaceae. Ann. Missouri Bot. Gard. 47: 25–68.

Fay MF, Rudall PJ, Sullivan S, Stobart KL, de Bruijn AY, Reeves G, Qamaruz-Zaman F, Hong WP, Joseph J, Hahn WJ, Conran JG, and Chase MW. 2000. Phylogenetic studies of Asparagales based on four plastid DNA regions. In: KL Wilson, and DA Morrison, eds. Monocots: systematics and evolution, pp. 360–371. CSIRO, Collingwood.

Freudenstein JV. 1991. A systematic study of endothecial thickenings in the Orchidaceae. Am. J. Bot. 78: 766–781.

Freudenstein JV and Chase MW. 2001. Analysis of mitochondrial nad1b-c intron sequences in Orchidaceae: Utility and coding of length-change characters. Syst. Bot. 26: 643–657.

Freudenstein JV and FN Rasmussen. 1997. Sectile pollinia and relationships in the Orchidaceae. Plant Syst. Evol. 205: 125–146.

Freudenstein JV and FN Rasmussen. 1999. What does morphology tell us about orchid relationships? – a cladistic analysis. Am. J. Bot. 86: 225–248.

Freudenstein JV, DM Senyo, and MW Chase. 2000. Mitochondrial DNA and relationships in the Orchidaceae. In: KL Wilson and DA Morrison, eds. Monocots: systematic and evolution, pp. 421–429. CSIRO, Collingwood.

Freudenstein JV, EM Harris, and FN Rasmussen. 2002. The evolution of anther morphology in orchids: incumbent anthers, superposed pollinia, and the vandoid complex. Am. J. Bot. 89: 1747–1755.

Freudenstein JV, C van den Berg, DH Goldman, PJ Kores, M Molvray, and MW Chase. 2004. An expanded plastid DNA phylogeny of Orchidaceae and analysis of jackknife branch support strategy. Am. J. Bot. 91: 149–157.

Garay LA. 1960. On the origin of the Orchidaceae, part 1. Bot. Mus. Lean. 19: 57–96.

Garay LA. 1972. On the origin of the Orchidaceae, part 2. J. Arnold Arbor. 53: 202–215.

Geerinck D. 1969. Genera des Haemodoraceae et des Hypoxidaceae. Bull. Jard. Bot. Nat. Belg. 39: 47–82.

Goldblatt P. 1995. The status of R. Dahlgren's orders Liliales and Melanthiales. In: PJ Rudall, PJ Cribb, DF Cutler, and CJ Humphries, eds. Monocotyledons: systematics and evolution, pp. 181–200. Royal Botanic Gardens, Kew.

Govindappa DA. 1967. Contribution to the embryology of *Hypoxis aurea* Lour. J. Indian Bot. Soc. 46: 193–198.

Govindappa DA and K Shamakumari. 1957. Development of embryo in *Hypoxis aurea* Lour. J. Indian Bot. Soc. 36(3): 324–327.

Hillard OM and BL Burtt. 1978. Notes on some plants of Southern Africa, chiefly from Natal. Part 7 (Hypoxidaceae). Notes Roy. Bot. Gard. Edinb. 36: 43–76.

Jain S, V Gupta, and MR Vijayaraghavan 1986. Structure and histochemistry of raphide idioblasts in *Apostasia wallichii* (R. Br.). Curr. Sci. 55: 932–934.

Johansen B and S Frederiksen. 2002. Orchid flowers: development and evolution. In: Cronk QCB, RM Bateman, and JA Hawkins, eds. Developmental genetics and plant evolution, pp. 206–219. Taylor & Francis, London.

Judd WS. 2000. The Hypoxidaceae in the Southeastern United States. Harvard Pap. Bot. 5: 79–98.

Judd WS, WL Stern, and VI Cheadle. 1993. Phylogenetic position of *Apostasia* and *Neuwiedia* (Orchidaceae). Bot. J. Linn. Soc. 113: 87–94.

Kocyan A and PK Endress. 2001a. Floral structure and development and systematic aspects of some "lower" Asparagales. Plant Syst. Evol. 229: 187–216.

Kocyan A and PK Endress. 2001b. Floral structure and development of *Apostasia* and *Neuwiedia* (Apostasioideae) and their relationships to other Orchidaceae. Int. J. Plant Sci. 162: 847–867.

Kocyan A, Y-L Qiu, PK Endress, and E Conti. 2004. A phylogenetic analysis of Apostasioideae (Orchidaceae) based on ITS, *trn*L-F and *mat*K sequences. Plant Syst. Evol. 247: 203–213.

Kores PJ, KM Cameron, M Molvray, and MW Chase. 1997. The phylogenetic relationships of Orchidoideae and Spiranthoideae (Orchidaceae) as inferred from *rbc*L plastid sequences. Lindleyana 12: 1–11.

Kores PJ, PH Weston, M Molvray, and MW Chase. 2000. Phylogenetic relationships within the Diurideae (Orchidaceae): inferences from plastid *mat*K DNA sequences. In: KL Wilson and DA Morrison, eds. Monocots: systematic and evolution, pp. 449–456. CSIRO, Collingwood.

Kores PJ, M Molvray, PH Weston, SD Hopper, AP Brown, KM Cameron, and MW Chase. 2001. A phylogenetic analysis of Diurideae (Orchidaceae) based on plastid DNA sequence data. Am. J. Bot. 88: 1903–1914.

Kosenko VN. 1994a. Morphology of pollen grains of the families Phormiaceae, Blandfordiaceae, and Doryanthaceae. Bot. Zhurn. 79: 1–12 (in Russian with English summary).

Kosenko VN. 1994b. Pollen morphology of the families Phormiaceae, Blandfordiaceae, and Doryanthaceae. Bot. Zhurn. 79(7): 1–12 (in Russian with English summary).

Kristiansen KA, FN Rasmussen, and HN Rasmussen. 2001. Seedlings of Neuwiedia (Orchidaceae subfamily Apostasioideae) have typical orchidaceous mycotrophic protocorms. Am. J. Bot. 88: 956–959.

Kumar M and KS Manilal. 1988. Floral anatomy of *Apostasia odorata* and the taxonomic status of apostasioids (Orchidaceae). Phytomorphology 38: 159–162.

Kurzweil H. 1987a. Developmental studies in orchid flowers. I. Epidendroid and vandoid species. Nord. J. Bot. 7: 427–442.

Kurzweil H. 1987b. Developmental studies in orchid flowers. II. Orchidoid species. Nord. J. Bot. 7: 443–451.

Kurzweil H. 1988. Developmental studies in orchid flowers. III. Neottioid species. Nord. J. Bot. 8: 271–282.

Kurzweil H. 1993. Developmental studies in orchid flowers. IV. Cypripedioid species. Nord. J. Bot. 13: 423–430.

Kurzweil H. 1998. Floral ontogeny of orchids: a review. Beitr. Biol. Pflanz 71: 45–100.

Kurzweil H. 2000. The value of early floral ontogeny in the systematics of Orchidaceae. In: KL Wilson and DA Morrison, eds. Monocots: systematic and evolution, pp. 436–440. CSIRO, Collingwood.

Kurzweil H and A Kocyan A. 2002. Ontogeny of orchid flowers. In: J Arditti and T Kull, eds. Orchid biology: reviews and perspectives. VIII, pp. 83–138. Kluwer, Dordrecht.

Lindley J. 1830–1840. The genera and species of Orchidaceous plants. Ridgways, London.

Lodkina MM. 1985. Orchidaceae. In: A Takhtajan, ed., Comparative seed anatomy, vol. 1, pp. 142–150. Nauka, Leningrad (in Russian).

Molvray M and P Kores. 1995. Character analysis of the seed coat in the Spiranthoideae and Orchidoideae, with special reference to the Diurideae (Orchidaceae). Am. J. Bot. 82: 1443–1454.

Molvray M, P Kores, and MW Chase. 2000. Polyphyly of mycoheterotrophic orchids and functional influences on floral and molecular characters. In: KL Wilson and DA Morrison, eds. Monocots: systematic and evolution, pp. 441–448. CSIRO, Collingwood.

Nelson E. 1965. Zur organophyletischen Natur des Orchideenlabellums. Bot. Jahrb. Syst. 84: 175–214.

Nelson E. 1967. Das Orchideenlabellum ein Homologon des einfachen medianen Petalums der Apostasiaceen oder ein zusammengesetztes Organ? Bot. Jahrb. Syst. 87: 22–35.

Nemirovich-Danchenko EN. 1985. Hypoxidaceae. In: A Takhtajan, ed. Comparative seed anatomy, vol. 1, 117–119. Nauka, Leningrad (in Russian).

Newton GD and NH Williams. 1978. Pollen morphology of the Cypripedioideae and the Apostasioideae (Orchidaceae). Selbyana 2: 169–182.

Neyland R and LE Urbatsch. 1996a. Evolution in the number and position of fertile anthers in Orchidaceae inferred from *ndh*F chloroplast gene sequences. Lindleyana 11: 47–53.

Neyland R and LE Urbatsch. 1996b. Phylogeny of subfamily Epidendroideae (Orchidaceae) inferred from *ndh*F chloroplast gene sequences. Am. J. Bot. 83: 1195–1206.

Nishimura G and M Tamura. 1993. Seed coat formation in *Apostasia nipponica*. J. Jpn. Bot. 68: 219–223.

Nordal I. 1998. Hypoxidaceae. In: K Kubitzki, ed. The families and genera of vascular plants, vol. 3, pp. 286–295. Springer, Berlin/Heidelberg/New York.

Oganezova GG. 1995. On the systematic position of the families Haemodoraceae, Hypoxidaceae and Taccaceae. Bot. Zhurn. 80: 12–25 (in Russian with English summary).

Okada H. 1988. Karyomorphological observations of *Apostasia nuda* and *Neuwiedia veratrifolia* (Apostasioideae, Orchidaceae). J. Jpn. Bot. 63: 344–350.

Oliviera VC and MG Sajo. 1999. Anatomia foliar de especies de Orchidaceae. Rev. Brasil. Bot. 22: 365–374.

Poddubnaya-Arnoldi VA. 1967. Comparative embryology of the Orchidaceae. Phytomorphology 17: 312–320.

Prakash N, and Ramsey M. 2000. Embryological development in Blandfordia and Neoastelia with comments on their systematic position. In: KL Wilson and DA Morrison, eds. Monocots: systematics and evolution, pp. 214–217. CSIRO, Collingwood.

Pridgeon AM and MW Chase. 1995. Subterranean axes in tribe Diurideae (Orchidaceae): morphology, anatomy, and systematic significance. Am. J. Bot. 82: 1473–1495.

Pridgeon AM, WS Stern, and DH Benzing. 1983. Tilosomes in roots of Orchidaceae: morphology and systematic occurrence. Am. J. Bot. 70: 1365–1377.

Pridgeon AM, PJ Cribb, and MW Chase. 1999. Genera orchidacearum, vol. 1. General Introduction, Apostasioideae, Cypripedioideae. Oxford University Press, Oxford.

Pridgeon AM, R Solano, and MW Chase. 2001a. Phylogenetic relationships in Pleurothallidinae (Orchidaceae): combined evidence from nuclear and plastid DNA sequences. Am. J. Bot. 88: 2286–2308.

Pridgeon AM PJ Cribb, and FN Rasmussen FN, eds. 2001b. Genera orchidacearum, vol. 2, Pt. 1. Orchidoideae. Oxford University Press, Oxford.

Pridgeon AM, Cribb PJ, and FN Rasmussen, eds. 2003. Genera orchidacearum, vol. 2, Pt. 2. Orchidoideae, Vanilloideae. Oxford University Press, Oxford.

Rao VS. 1969. The floral anatomy and relationships of the rare Apostasias. J. Indian Bot. Soc. 68: 374–385.

Rao VS. 1974. The relationships of the Apostasiaceae on the basis of floral anatomy. Bot. J. Linn. Soc. 68: 319–327.

Rasmussen F. 1982. The gynostemium of the neottioid orchids. Opera Bot. 65: 7–96.

Rasmussen F. 1986. Ontogeny and phylogeny in Orchidaceae. Lindleyana 1: 114–124.

Rasmussen FN. 1995. Relationships of Burmanniales and Orchidales. In: P Rudall, J Cribb, DF Cutler, and CJ Humphries, eds. Monocotyledons: systematic and evolution, pp. 227–241. Royal Botanic Gardens. Kew.

Rasmussen FN. 2000. Ins and outs of Orchid phylogeny. In: KL Wilson and DA Morrison, eds. Monocots: systematic and evolution, pp. 430–435. CSIRO, Collingwood.

Roife RA. 1909, 1910. The evolution of the Orchidaceae. Orchid Rev. 17: 129–132, 193–196, 289–292, 353–356, 1909; 18: 33–36, 87–99, 129–132, 162–166, 289–294, 321–325, 1910.

Rosso SW. 1966. The vegetative anatomy of the Cypripedioideae (Orchidaceae). Bot. J. Linn. Soc. 59: 309–341.

Rudall PJ. 1998. Lanariaceae. In: K Kubitzki, ed. The families and genera of vascular plants, vol. 3, pp. 340–342. Springer, Berlin/Heidelberg/New York.

Rudall PJ, MW Chase, DF Cutler, J Rusby, and AY De Bruijn. 1998. Anatomical and molecular systematics of Asteliaceae and Hypoxidaceae. Bot. J. Linn. Soc. 127: 1–12.

Scharf W. 1892. Beiträge zur Anatomic der Hypoxideen und einiger verwandter Pflanzen: *Lanaria plumosa* Ait. Beih. Bot. Centralbl. 52: 242–243.

Schill R. 1978. Palynologische Untersuchungen zur systematischen Stellung der Apostasiaceae. Bot. Jahrb. Syst. 99: 353–352.

Schlechter R. 1926. Die System der Orchidaceen. Notizbl. Bot. Gart. Berlin-Dahlem 9: 563–591.

Schlechter R. 1992. Die Orchideen, ed. 3, vol. 1A (FG Brieger, F Butzin, and K Senghas, eds). Paul Parey, Berlin.

Schlechter R. 1996. Die Orchideen, ed. 3, vol. 1B (FG Brieger, F Butzin, and K Senghas, eds). Paul Parey, Berlin.

Schlechter R. 2003. Die Orchideen, ed. 3, Literaturverzeichnis und Register zu Band I/A, B und C (K Senghas, ed.). Paul Parey, Berlin.

Sharma AK. 1969. Evolution and taxonomy of monocotyledons. In: CD Darlington, ed. Chromosomes Today, 2: 241–249.

Skottaberg C. 1960. *Astelia* on Mauritius. Sven. Bot. Tidskr. 54: 477–482.

Skottsberg C. 1934a. Studies in the genus *Astelia* Banks et Solander. K. Vetenscapakad. Handl. III. 14: 1–106.

Skottsberg C. 1934b. *Astelia* and *Pipturus* of Hawaii. Bernice P. Bishop. Mus. Bull. 117: 1–77.

Sood SK and PR Mohana Rao. 1988. Studies in the embryology of the diandrous orchid *Cypripedium cordigerum* (Cypripedieae, Orchidaceae). Plant Syst. Evol. 160: 159–168.

Stern WL 1993. Comparative vegetative anatomy and systematics of Spiranthoideae (Orchidaceae). Bot. J. Linn. Soc. 113: 161–197.

Stern WL and BS Carlsward. 2004. Vegetative constants in the anatomy of epiphytic orchids. Orchid Rev. 112: 119–122.

Stern WL and BS Carlsward. 2006. Comparative vegetative anatomy and systematics of the Oncidiinae (Maxillarieae, Orchidaceae). Bot. J. Linn. Soc. 152: 91–107.

Stern WL and WS Judd. 2000. Comparative anatomy and systematics of the orchid tribe Vanilleae excluding *Vanilla*. Bot. J. Linn. Soc. 134: 179–202.

Stern WL and WS Judd. 2001. Comparative anatomy and systematics of Catasetinae (Orchidaceae). Bot. J. Linn. Soc. 136: 153–178.

Stern WL and WM Whitten. 1999. Comparative vegetative anatomy of *Stanhopeinae* (Orchidaceae). Bot. J. Linn. Soc. 129: 87–103.

Stern WL, VA Cheadle, and J Thorsch. 1993. Apostasiads, systematic anatomy, and the origins of Orchidaceae. Bot. J. Linn. Soc. 111: 411–455.

Stern WL, MW Morris, WS Judd, AM Pridgeon, and RL Dressler. 1993. Comparative vegetative anatomy and systematics of *Spiranthoideae* (Orchidaceae). Bot. J. Linn. Soc. 113: 161–197.

Stern WL, WS Judd, and BS Karlsward. 2004. Systematic and comparative anatomy of *Maxillarieae* (Orchidaceae), sans Oncidiinae. Bot. J. Linn. Soc. 144: 251–274.

Swamy BGL. 1948. Vascular anatomy of orchid flowers. Bot. Mus. Leafl. 13: 61–95.

Szlachetko DL. 1995. Systema Orchidalium. Fragmenta Florist Geobot. 3 (Suppl.): 1–152.

Szlachetko DL and NB Margonska HB. 2002. Gynostemia orchidalium II. Orchidaceae (Epidendroideae). Acta Bot. Fenn. 173: 1–275.

Szlachetko DL and Rutkowski P. 2000. Gynostemia orchidalium I. Apostasiaceae, Cypripediaceae, Orchidaceae (Thelymitroideae, Orchidoideae, Tropidioideae, Spiranthoideae, Neottioideae, Vanilloideae). Acta Bot. Fenn. 169: 1–380.

Terekhin ES and OP Kamelina. 1969. Endosperm of the Orchidaceae. Bot. Zhurn. 54: 657–666 (in Russian).

Thompson MF. 1976, 1978, 1970. Studies in the Hypoxidaceae. I. Vegetative morphology and anatomy. II. Floral morphology and anatomy. III. The genus *Pauridia*. Bothalia 12: 111–117, 429–435, 621–625.

Tohda H. 1986. Seed morphology in Orchidaceae. III. Tribe Neottieae. Sci. Report Tohoku Univ. 4th ser, 39: 103–119.

Van den Berg C, DH Gioldman, JV Freudenstein, AM Pridgeon, KM Cameron, and MW Chase. 2005. An overview of the phylogenetic relationships within Epidendroideae inferred from multiple DNA regions and recircumscription of Epidendreae and Arethuseae (Orchidaceae). Am. J. Bot. 92: 613–624.

Vermuelen P. 1955. The rostellum of the Orchideae. Am. Orchid Soc. Bull. 24: 239–245.

Vermuelen P. 1959. The different structure of the rostellum in Ophrydeae and Neottieae. Acta Bot. Neerl. 8: 338–355.

Vermuelen P. 1966. The system of the Orchidales. Acta Bot. Neerl. 15: 224–253.

Vinogradova TN and EV Andronova. 2002. Development of orchid seeds and seedlings. In: J Arditti and T Kull, eds. Orchid biology: reviews and perspectives, vol. 8, pp. 167–234. Kluwer, Dordrecht.

Vogel EF De. 1969. Monograph of the tribe Apostasieae (Orchidaceae). Blumea 17: 313–350.

Wheeler JM. 1966. Cytotaxonomy of the large Asteliads (Liliaceae) of the North Island of New Zealand. N. Z. J. Bot. 4: 95–113.

Williams NH. 1979. Subsidiary cells in the Orchidaceae: Their general distribution with special reference to development in the Oncidieae. Bot. J. Linn. Soc. 78: 41–66.

Wilson KL and DA Morrison, eds. 2000. Monocots: systematics and evolution. CSIRO, Collingwood.

Yam TW, EC Yeung, XL Ye, SY Zee, and J Arditti. 2002. Orchid embryos. In: J Arditti and T Kull, eds. Orchid biology: reviews and perspectives, vol. 8, pp. 287–385. Kluwer, Dordrecht.

Order 13. IRIDALES

Mainly perennial herbs, generally with sympodial rhizomes or corms, less often with bulbs, occasionally shrubs with woody caudex and anomalous secondary growth, very rarely annuals (*Sisyrhinchium* spp.). Roots often mycorrhizal, without root hairs. Vessels usually only in roots (except *Sisyrhinchium*), with scalariform or simple perforations. Raphides lacking, but styloid calcium oxalate crystals present in vascular bundles and other tissues (absent only in *Sisyrhinchium* and closely related genera). Leaves basal or basal and cauline, generally distichous, often ensiform, linear or sometimes lanceolate or filiform, rarely (Doryanthaceae) very large and long (to 3 m long), rarely (*Geosiris*) scalelike, sheathing and equitant at the base, parallel-veined. Stomata anomocytic or paracytic. Flowers in terminal, cymose inflorescences or sometimes solitary, usually bracteate, mostly bisexual, actinomorphic or often more or less zygomorphic. Perianth of six petaloid segments in two cycles, all alike or the two cycles or the members of a single cycle more or less conspicuously different in shape and colour, often all connate below into a tube; occasionally the inner cycle reduced or lacking. Nectaries perigonal (Iridoideae), septal (Ixioideae and Nivenioideae) or sometimes lacking (e.g. *Isophysis*, most *Aristea*). Stamens six in two cycles or three (two in *Diplarrhena*); filaments free or sometimes connate basally or entirely; anthers basifixed to subbasifixed, or centrifixed, tetrasporangiate, extrorse to latrorse, usually opening longitudinally, rarely opening by apical pores. Tapetum secretory. Microsporogenesis mostly simultaneous. Pollen grains mostly 2-celled, tectate-columellate, usually

1-colpate, rarely 2- or 3-colpate, spiraperturate, or inaperturate, usually reticulate. Gynoecium of three united carpels; style terminal, subulate to filiform, usually 3-branched above, the branches often again divided, sometimes expanded and strongly petaloid with the stigma on the outer side of the branch rather than at the top; ovary inferior or very rarely (*Isophysis*) superior, 3-locular or very rarely (*Hermodactylis*) 1-locular with parietal placentation, usually with numerous ovules per locule, rarely with few or only one ovule. Ovules anatropous or campylotropous, bitegmic, crassinucellate, with parietal tissue and sometimes (*Iris pseudacorus*) with endothelium and in some genera with funicular obturator. Female gametophyte of *Polygonum*-type. Endosperm helobial (*Doryanthes, Ixiolirion, Lanaria, Odontostomum, Geosiris* and *Isophysis*) or nuclear. Fruits loculicidal capsules, rarely indehiscent. Seeds sometimes arillate; embryo rather small, cylindrical, mostly straight; both testa and tegmen are usually present; the outer epidermis is mostly pigmented with phlobaphene; endosperm hard, with thickened, pitted cell walls that contain reserves of hemicellulose; cells also contain aleurone and lipids but only rarely starch; embryo straight and fusiform and is usually 1/3–2/3 as long as the endosperm. The cotyledon of the seedling is coleoptilelike. Produces flavone C-glycosides and flavonols as major leaf constituents.

The distinctness of the Iridales is well supported by several characters, one of which are unique – calcium oxalate crystals in the form of styloids (Goldblatt et al. 1984). Probably related to the Melanthiaceae s. l. (Takhtajan 1959, 1966, 1969, 1980, 1997; Goldblatt 1990), and to the Campynemataceae. Embryologically the closest to the Melanthiales is *Geosiris* (Goldblatt 1990).

Key to Families

1 Microsporogenesis successive. Endosperm helobial.

2 Herbs with short vertical rootsock and large leaves. Leaves in dense basal rosette, from 1.3 to 3 m long and 20 cm wide, tough, swordlike, those on the stem much reduced, each with characteristic brown tubular tip. Mesophyll containing calcium oxalate crystals. Stomata paracytic. Inflorescences tall (up to 5 m long) terminal scapes, oblong thyrses or global compound racemes of numerous flowers, which may be substituted by bulbils. Flowers large (5–10 cm long), bisexual, actinomorphic or slightly zygomorphic. Perianth red, segments nearly free or basally connate into very short tube, long and falcate. Filaments adnate to perianth segments for about one-half of their length; anthers elongate, basifixed (the apex of filament inserted into a pit at base of connective), opening longitudinally. Microsporogenes successive. Pollen grains 1-colpate. Ovary inferior, 3-locular, with 40–50 anatropous ovules per locule; styles two long, tapering, with triangular stigma. Septal nectaries present. Endosperm helobial. Fruits large, turbinate, loculicidal capsules. Seeds light brown, more or less flattened, elliptic to rhomboid, with one-sided, broad, papery wing, nonlignified exotestal cell walls of *Doryanthes palmeri* impregnated with phlobaphene; embryo linear, from one-third to almost as long as the seed and with broad cotyledon. Testa without phytomelan. Proanthocyanidins and steroid saponines present. n = 17, 18, 22, 24, with clearly detectable differences in chromosome sizes. . . . . .5. DORYANTHACEAE.

2 Small, achlorophyllous, mycotrophic herbs with mycorrhizal roots. Vessels in roots, with scalariform perforations. Leaves small, scalelike, alternate, membranous, simple, entire. Stomata anomocytic. Flowers in terminal cymes, bracteate, small, sessile, actinomorphic. Perianth segments six in two whorled (the outer imbricate, the inner contorted), petaloid, blue; perigone tube short. Stamens three, opposite the outer perianth segments, very short; anthers basifixed, extrorse, opening longitudinally. Pollen grains 1-colpate. Septal nectaries absent. Ovary inferior, 3-locular; style one, trilobed, stigmas three; ovules anatropous, placentation axile. Fruits capsules or capsular-indehiscent, capsules triangular-obconical. Seeds minute, testa without phytomelan.. . . . . 6. GEOSIRIDACEAE.

1 Microsporogenesis simultaneous. Endosperm mostly nuclear.

3 Stamens mostly six, all fertile or 2–3 staminodial, attached to segments.

4 Ovules without chalazosperm.

5 Anthers opening longitudinally. Glabrous herbs with leafy stem and bulb-shaped tunicated corm that develops lateral corms at some distance from its base. Vessels in roots with scalariform perforations. Leaves linear, not stiff, cylindrical-subulate at the apex.

Flowers in thyrses with helicoid cymes as partial inflorescences, often reduced to few-flowered terminal false umbels. Perianth glabrous, its segments free, outer ones (rarely also inner ones) apiculate-mucronate. Stamens all fertile, in two unequal cycles; filaments adnate the lower part of perianth segments; anthers oblong-linear to ovoid, basifixed, introrse-latrorse. Pollen grains 1-colpate, reticulate. Style 3-brachiate; ovary inferior, with many superposed ovules per locule. Fruits dehiscing at the top, with numerous seeds. Seeds small, ovoid oblong, slightly curved, reticulate, black; testa consists of elongate, hexagonal epidermal cells with phytomelan and several cell layers with thin reddish brown walls; tegmen thin and membranous; endosperm cells thin-walled, walls pitted, containing aleurone, fats, hemicellulose, and also some starch in cells surrounding embryo; embryo straight, fusiform, nearly as long as endosperm. Alkaloids lacking, n = 12, chromosomes varying considerably in size. . . . . 1. IXIOLIRIACEAE.

5 Anthers opening apically by narrow pore or by an apical pore continuing in a short slit.

6 Ovary largely superior. Ovules generally campylotropous. Rootstock naked, fleshy, lobate, potatolike tuber with fibrous roots emerging from each lobe, rich in starch, with apical growing point. Basal underground part of stem with one or more scars after cataphylls. Aerial stem leafy, with cauline leaves successively larger toward the top, rigid, scabrid from inconspicuous hooks on longitudinal ridges. Leaves mesomorphic, sessile, rarely almost amplexicaul, linear to broadly lanceolate or ovate, the midrib on lower side generally scabrid, with recurved hooks, apically almost to distinctly cirrhous. Flowers on long, scabrid pedicels, solitary in leaf axils. Perianth segments basally connate into short tube. Stamens all fertile, inserted in perianth tube; anthers long, sometimes laterally coherent, basifixed. Pollen grains 1-colpate, elliptic, irregularly rugulose, with densely and minutely perforate tectum. Style with punctiform stigma; ovary with 4–12 ovules per locule; septal nectaries absent or (*Walleria nutans*) represented by small septal glands. Fruits globose, slightly 3-lobate capsules with 2–3 seeds. Seeds rounded to oblong, dark brown, warty to papillate, testa multicellular, without phytomelan, with small hair tuft crowning each wart or papilla; embryo relatively large, slightly curved, filling up two-thirds of the seed lumen, n = 12............. 2. WALLERIACEAE.

6 Ovary semi-inferior. Ovules anatropous. Perennial glabrous herbs. Rootstock tunicated corm with fibrous-reticulate covering. Stems partly leafy, leaves concentrated at the base. Vessels in roots, with scalariform perforations. Leaves alternate, spiral, or distichous, sessile, simple linear to lanceolate, sheathing at the base, glabrous, not cirrhous. Stomata amonocytic. Flowers in simple or compound racemes, bracteate. Perianth segments free or shortly connate at the base; perigone tube short or absent. Stamens all fertile or 2–3 staminodial; anthers mostly basifixed (peltate in *Tecophilaea*), sometimes connivent, with the connective sometimes prolonged at both ends. Pollen operculate. Style erect or slightly curved, with punctiform stigma; ovary 3-locular, with several to many ovules per locule. Fruits apically loculicidal capsules with 2–3 to many seeds. Seeds elongated to globose, smooth and black; testa encrusted with phytomelan; embryo straight or slightly curved; endosperm well developed, its cells with thick walls and numerous pores, n = 12. . . 3. TECOPHILAEACEAE.

4 Ovules with the tissue in chalazal part enlarged to form chalazosperm in which cells are swollen and filled with compound starch grains. Perennial herbs with laterally constricted nontunicated corm with roots emerging directly from the sides. Stems basally with scale leaves and basally concentrated 1–3 foliage leaves. Leaves alternate, distinctly petiolate with cordate-ovate lamina with arching main veins connected by transverse veinlets. Schizogenous secretory spaces with oil-like secretion present in leaves; mesophyll containing calcium oxalate crystals, or not.

Stomata paracytic. Flowers in racemes situated in the axils of bractlike or spathelike sheathing cataphylls. Perianth segments six in two cycles, blue, segments connate into short tube. Stamens inserted at the mouth of perianth tube, all fertile; filaments short, often S-shaped; anthers basifixed, elongate, opening by apical pore or short slit. Pollen grains 1-colpate or sometimes trichotomocolpate, nonoperculate, finely reticulate. Style more or less gynobasic, with punctiform stigma; ovary semi-inferior, with three centrally situated, septal nectary grooves, 3-lobed and 3-locular, with two basal ovules per locule; ovules anatropous, basally inserted, with a funicular obturator. Only one seed in one of three locules normally develops, and therefore the fruit may be classified as a mericarp of a schizocarp (Dahlgren et al. 1985). Seeds large, pearlike, deeply pitted, yellow to brown, completely lack phytomelan; embryo with large, terminal cotyledon containing starch grains (Nemirovich-Danchenko 1985) and sunken, lateral plumule; endosperm lacking (almost completely used during seed development), n = 11, 12. . . . . 4. CYANASTRACEAE.

3 Stamens three, rarely two, opposite the outer perianth segments. Perennial herbs or rarely shrubs, rhizomotous, or cormous, or rarely bulbaceous. Vessels in roots and stems, mostly with simple perforations, sometimes with scalariform and simple perforations. Leaves evergreen or deciduous, alternate, usually distichous, flat or terete, sessile or petiolate, simple, entire, sometimes foetid, lamina with secretory cavities, which containing mucilage; mesophyll with calcium oxalate crystals. Stomata anomocytic. Flowers in terminal panicles, cymes, spikes, umbels and corymbs, or solitary, small to large, actinomorphic or zygomorphic, 3-merous. Perianth segments six in two cycles, the inner sometimes much smaller, commonly spotted; perigone tube long or short. Stamens opposite the outer perianth lobes; filaments usually separate from one another; anthers basifixed, extrorse, opening longitudinally. Pollen grains usually reticulate, 1-colpate, rarely 2-, or 3-colpate. Ovary inferior and 3-locular, or rarely (*Isophysis*) superior and 1-locular. Styles 1, 3-lobed, stigmas papillate. Ovules (1-)2–50 per locule, anatropous. Fruits loculicidal capsules. Seeds testal and temic, phlobaphene present, endotesta with lipids, testa without phytomelan; embryo small to large, straight, chlorophyllous or achlorophyllous, cotyledon coleoptile-like; endosperm thick-walled, oily. Alkaloids present or absent, proanthocyanidins present or absent, flavonols present or absent; n = 3 –19, 22, 25, but mostly 7 and 10. . . . . 7. IRIDACEAE.

1. IXIOLIRIACEAE

Nakai 1943. 1/4–5. Egypt, Palestine and Irano-Turanian region to western China.

Ixiolirion.

Only a few botanists recognize a separate family Ixioliriaceae, and the genus is usually included in the Amaryllidaceae. However, it differs markedly from the Amaryllidaceae in its tunicated corms, leafy stem, leaf anatomy, inflorescence, simultaneous microsporogenesis, stamen and pollen morphology, and the absence of alkaloids. According to Dahlgren et al. (1985), Ixioliriaceae are closely related to some Alliaceae, to Tecophilaeaceae, or Phormiaceae. In my opinion the nearest relative is the family Tecophilaeaceae.

2. WALLERIACEAE

Takhtajan 1995. 1/3. Tropical and South Africa and Madagascar.

Walleria.

Walleria differs from the Tecophilaeaceae in so many features that it definitely deserves family rank (Huber 1969). Unlike the Tecophilaeaceae with prominent operculate pollen grains, *Walleria* has no typical operculum. According to Simpson (1985: 91), *Walleria* has an apertural membrane with an outer ridge of exinous deposits situated median and parallel to the aperture. In his opinion the apertural exine of *Walleria* is probably homologous with the operculum. In addition its flowers are solitary in cauline leaf axils, somewhat bean-like seeds are unique in being warty with a group of apical cells on each wart continuing each into a thin hair (Dahlgren et al. 1985).

3. TECOPHILAEACEAE

F. Leybold 1862. 6/20 (including Conantheraceae Pfeiffer 1873; Cyanellaceae Salibsbury 1866). Africa (*Cyanella, Kabuyea*) Chile (*Conanthera, Tecophilaea, Zephyra*), and California (*Odontostomum*).

Cyanella, Conanthera, Odontostomum, Tecophilaea, Zephyra.

Related to the Ixiolariaceae, these two families share corms, a leafy inflorescence, and, often, a nearly capitate inflorescence (Soltis et al. 2006). Endlicher (1841: 98) placed *Tecophilaea* after Iridaceae as "Genus Iridis affine," and Lindley (1853) included it in the Iridaceae between *Sisyrhinchium* and *Libertia.* Serological data supported a relationship between the Tecophilaeaceae and Iridaceae (Shneyer 1983). The Tecophilaeaceae are a sister group of the Iridaceae and share a common origin. Some members of Tecophilaeaceae, such as *Tecophilaea cyanocrocus,* resemble some Iridaceae even externally.

4. CYANASTRACEAE

Engler 1900. 1/3. Tropical Africa.

Cyanastrum (including *Kabuyea*).

Rather isolated within the order. The most automorphic characters are chalazospermous seeds, the presence of starch both in the storage tissue and embryo, and secretory spaces in the leaves. The habit is also quite different from that of Lanariaceae and Tecophilaeaceae and "indicates secondary adaptation to tropical forest habitats, e.g., in the amplified leaf lamina" (Dahlgren and van Wyk 1988: 80).

5. DORYANTHACEAE

R. Dahlgren et H. Clifford 1985. 1/2. Eastern Queensland and New South Wales.

Doryanthes.

P. Rudall (2003) suggested a close morphological relationship between Iridaceae and Doryanthaceae.

6. GEOSIRIDACEAE

Jonker 1939. 1/1. Eastern Madagascar.

Geosiris.

Closely related to Iridaceae, especially to Nivenioideae, but differ in achlorophyllous, reduced to small scales leaves, successive microsporogenesis, and helobial, starchy endosperm.

7. IRIDACEAE

A.L. de Jussieu 1789 (including Crocaceae Vest 1818; Galaxiaceae Rafinesque 1836; Gladiolaceae Rafinesque 1838; Isophysidaceae F.A. Barkley 1948; Ixiaceae Horaninow 1834). 78/1800. Widely distributed in tropical, subtropical, and temperate regions, but especially numerous in Africa south of the Sahara, in the eastern Mediterranean, western and eastern Asia, and Central and South America.

7.1 ISOPHYSIDOIDEAE

Ovary superior. Microsporogenesis simultaneous. Chlorophyllous plants with a thick rhizome and distichous, ensiform leaves. Flowers solitary, bisexual, actinomorphic. Perianth segments free, lanceolate, subequal. Nectary absent. Style short, with simple, recurved stigmatic branches. Fruits without an annulus at the summit. Biflavonoids present. – *Isophysis* (Tasmania).

7.2 NIVENIOIDEAE

Evergreen rosette plants with a rhizome or woody caudex, sometimes woody and with strongly developed secondary growth of the monocotyledonous type, often with persistent leaves. Vessels with scalariform perforation. Inflorescences bracteate panicles with units usually composed of paired and partly fused rhipidia, each enclosed in opposed leafy spathes. Flowers actinomorphic; perianth fugacious, usually blue, perianth segments more or less free or connate into a tube. Nectaries, when present, septal, rarely perigonal. Filaments free. Style slender, 3-lobed, the lobes often fringed. – *Aristea, Nivenia, Klattia, Witsenia, Patersonia.*

7.3 IRIDOIDEAE

Herbs with rhizomes or tunicated corms, rarely (in the tribe Irideae) with bulbs. Vessels usually only in roots, mostly with simple perforations. Inflorescence consists of one or more rhipidia enclosed in opposed leafy bracts. Flowers often large, generally actinomorphic. Perianth segments free or connate, similar or strongly different. Nectaries perigonal. Filaments free to basally connate. Style slender or more often short, branches undivided or more often apically bifurcate, and in a number of genera they are flat and petaloid. – SISYRINCHIEAE: *Libertia, Orthrosanthus Tapeinia, Solenomelus, Sisyrhinchium, Olsynium;* MARICEAE: *Trimezia, Pseudotrimezia, Neomarica;* TIGRIDEAE: *Cypella, Phalocallis, Larentia, Hesperoxiphion, Cipura, Calydorea, Eleutherine, Onira, Gelasine, Sphenostigma, Ennealophus* (including *Eurynotia*), *Herbertia, Kelissa, Mastigostyla, Cardenanthus, Nemastylis, Ainea, Alophia, Tigridia, Rigidella, Fosteria, Sessilanthera, Cobana;* IRIDEAE: *Dietes, Diplarrhena, Bobartia, Iris, Hermodactylis, Pardanthopsis, Belamcanda, Ferraria, Moraea* (including *Barnardiella, Galaxia, Gynandriris, Hexaglottis, Homeria, Rheome, Roggeveldia*).

7.4 IXIOIDEAE

Herbs with corms. Vessels only in roots, mostly with simple perforations. Inflorescences of sessile flowers arranged in a spike, which may become paniclelike, or flowers solitary. Flowers sessile, subtended by paired opposed bracts, actinomorphic or zygomorphic. Perianth segments connate into a tube. Perigonal nectaries lacking, but septal nectaries always present at the style base. Style slender, divided into three filiform branches. – PILLANSIEAE: *Pillansia;* WATSONIEAE: *Watsonia, Thereianthus, Micranthus, Lapeirousia, Anomatheca, Savannosiphon*; IXIEAE: *Tritoniopsis, Anapalina, Babiana, Antholyza, Zygotritonia, Gladiolus, Radinosiphon, Xenoscapa, Geissorhiza, Hesperantha* (including *Schizostylis*), *Melasphaerula, Romulea, Syringodea, Crocus, Freesia, Dierama, Ixia, Sparaxis, Tritonia, Duthiastrum, Crocosmia, Devia, Chasmanthe.*

Bibliography

Arroyo SC. 1982. Anatomia vegetativa de Ixiolirion Fisch. ex Herb. (Liliales) y su significado taxonomia. Parodiana 1: 271–286.

Arroyo S. 1986. Leaf anatomy in the Tecophilaeaceae. Bot. J. Linn. Soc. 93: 323–328.

Baker JG. 1878. Systema Iridearum. Bot. J. Linn. Soc. 16: 61–180.

Brummitt RK, H Banks, MAT Johnson, KA Docherty, K Jones, MW Chase, and PJ Rudall. 1998. Taxonomy of Cyanastroideae (Tecophilaeaceae): a multidisciplinary approach. Kew Bull. 53: 769–803.

Carter S. 1962. Revision of *Walleria* and *Cyanastrum* (Tecophilaeaceae). Kew Bull. 16: 190–200.

Cheadle V. 1969. Vessels in Amaryllidaceae and Tecophilaeaceae. Phytomorphology 19(1): 8–16.

Cheadle VI. 1963. Vessels in Iridaceae. Phytomorphology 13: 245–248.

Clausen RT. 1940. A review of Cyanastraceae. Gentes Herb. 4: 293–304.

Clifford HT. 1998. Doryanthaceae. In: K Kubitzki, ed. The families and genera of vascular plants, vol. 3, pp. 336–338. Springer, Berlin/Heidelberg/New York.

Dahlgren R and AE van Wyk. 1988. Structures and relationships of families endemic to or centered in southern Africa. Monograph Syst. Bot. Missouri Bot. Gard. 25: 1–94.

De Vos M. 1963. Studies on the embryology and relationships of South African genera of the Haemodoraceae: *Lanaria* Ait. South Afr. J. Bot. 29: 79–90.

Donato R, C Leach, and G Conran. 2000. Relationships of *Dietes* (Iridaceae) inferred from ITS2 sequences. In: KL Wilson and DA Morrison, eds. Monocots: systematics and evolution, pp. 407–413. CSIRO, Collingwood.

Goldblatt P. 1971. Cytological and morphological studies in the southern African Iridaceae. South Afr. J. Bot. 37: 317–460.

Goldblatt P. 1979. Preliminary cytology of Australasian Iridaceae. Ann. Missouri Bot. Gard. 66: 851–855.

Goldblatt P. 1981. Systematics, phylogeny, and evolution in *Dietes* (Iridaceae). Ann. Missouri Bot. Gard. 68: 131–152.

Goldblatt P. 1982. Chromosome cytology in relation to suprageneric systematics of Neotropical Iridaceae. Syst. Bot. 7: 186–198.

Goldblatt P. 1990. Phylogeny and classification of Iridaceae. Ann. Missouri Bot. Gard. 77: 607–627.

Goldblatt P. 1993. The woody Iridaceae: *Nivenia, Klattia,* and *Witsenia:* Systematic biology and evolution. Timber Press, Portland, OR.

Goldblatt P. 1998. Reduction of *Bamardiella, Galaxia, Gynandriris, Hexaglottis, Homeria,* and *Roggeveldia* in *Moraea* (Iridaceae: Irideae). Novon 8: 371–377.

Goldblatt P. 2002. Phylogeny and classification of the *Iridaceae* and the relationships of *Iris*. Ann. Bot. n.s. (Italy). 1(2): 13–28.

Goldblatt P and JC Manning. 1989. Chromosome number in *Walleria* (Tecophilaeaceae). Ann. Missouri Bot. Gard. 76: 925–926.

Goldblatt P and JC Manning. 2006. Radiation of pollination systems in the Iridaceae of sub-Saharan Africa. Ann. Bot. 97: 317–344.

Goldblatt P and P Rudall. 1993. Leaf anatomy and systematics of the Homeriinae (Iridaceae). Bot. J. Linn. Soc. 111: 379–397.

Goldblatt P and M Takei. 1997. Chromosome cytology of Iridaceae. Patterns of variation, determination of ancestral base numbers, and modes of karyotype change. Ann. Missouri Bot. Gard. 84: 285–304.

Goldblatt P, JE Henrich, and P Rudall. 1984. Occurrence of crystals in Iridaceae and allied families and their phylogenetic significance. Ann. Missouri Bot. Gard. 71: 1013–1020.

Goldblatt P, P Rudall, VL Cheadle, LJ Dorr, and CA Williams. 1987. Affinities of the Madagascan endemic *Geosiris*, Iridaceae or Geosiridaceae. Adansonia 9: 239–248.

Goldblatt P, P Rudall, and JE Henrich. 1990. The genera of the *Sisyrhinchium* alliance (Iridaceae-Iridoideae): Phylogeny and relationships. Syst. Bot. 15: 497–510.

Goldblatt P, JC Manning, and A Bari. 1991. Sulcus and operculum structure in the pollen grains of Iridaceae subfamily Iridoideae. Ann. Missouri Bot. Gard. 78: 950–961.

Goldblatt P, JC Manning, and P Rudall. 1998. Iridaceae. In: K Kubitzki, ed. The families and genera of vascular plants, vol. 3, pp. 295–333. Springer, Berlin/Heidelberg/New York.

Hartwell J, AV Cox, KM Cameron, ER Caddick, and MW Chase. 1994. Molecular systematics of the "lower" asparagoid lily families. Am. J. Bot. 81(6): 160 (abstract).

Jonker EP. 1939. Les Geosiridacees: Une nouvelle famille de Madagascar. Rec. Trav. Bot. Neerl. 36: 473–479.

Kenton A and CA Heywood. 1984. Cytological studies in South American Iridaceae. Plant Syst. Evol. 146: 87–104.

Kosenko VN. 1994. Pollen morphology of the families Phormiaceae, Blandfordiaceae, and Doryanthaceae. Bot. Zhurn. 79(7): 1–12 (in Russian with English summary).

Kubitzki K. 1998. Ixioliriaceae. In: K Kubitzki, ed. The families and genera of vascular plants, vol. 3, pp. 334–335. Springer, Berlin/Heidelberg/New York.

Lakshmanan KK and VJ Phillip. 1971. A contribution to the embryology of Iridaceae. Proc. Indian Acad. Sci. 73: 110–116.

Larsen PO, ET Sorensen, E Wieczorkowska, and P Goldblatt. 1981. Meta-carboxy-substituted aromatic amino acids and t-glutamyi peptides: chemical characters for classification in the Iridaceae. Biochem. Syst. Ecol. 18: 575–579.

Le Thomas A, M Suarez, and P Goldblatt. 2002. Pollen of Nivenioideae and its phylogenetic implications. Ann. Bot. n.s. (Italy). 1(2): 67–72.

Lewis GJ. 1954. Some aspects of the morphology, phylogeny, and taxonomy of the South African Iridaceae. Ann. S. Afr. Mus. 40: 15–113.

Manning JC and P Goldblatt. 1991. Systematic and phylogenetic significance of the seed coat in the shrubby African Iridaceae, *Nivenia, Klattia,* and *Witsenia*. Bot. J. Linn. Soc. 107: 387–404.

Nemirovich-Danchenko EN. 1985. Tecophilaeaceae, Cyanastraceae. In: A Takhtajan, ed. Comparative seed anatomy, vol. 1, pp. 106–108. Nauka, Leningrad (in Russian).

Newman FV. 1928, 1929. The life history of *Doryanthes excelsa*: I. Some ecological and vegetative features on spore production, Proc. Linn. Soc. N. S. W. 53: 499–538. II. The gametophytes, seed production, chromosome number, and general conclusions. Proc. Linn. Soc. N. S. W. 54: 411–435.

Nietsch H. 1940. Zur systematischen Stellung von *Cyanastrum*. Oesterr. Bot. Z. 90: 31–52.

Oganezova GG. 1981. Anatomical and morphological study in *Ixiolirion tataricum* ssp. *montanum*. Bot. Zhurn. 66: 702–713 (in Russian with English summary).

Oganezova GG. 1997a. The seed structure of some Iridaceae in connection with their systematics, geography and phylogeny. I. Subfamilies Nivenioideae and Irudoideae. Bot. Zhurn. 82(2): 14–29 (in Russian with English summary).

Oganezova GG. 1997b. The seed structure of some Iridaceae in connection with their systematics, geography and phylogeny. II. Subfamily Ixioideae. Bot. Zhurn. 82(3): 7–22 (in Russian with English summary).

Ornduff R. 1979. Chromosome numbers in *Cyanella* (Tecophilaeaceae). Ann. Missouri Bot. Gard. 66: 581–583.

Reeves G, MW Chase, P Goldblatt, P Rudall, MF Fay, AV Cox, B Lejeune, and T Souza-Chies. 2001. Molecular systematics of Iridaceae: evidence from four plastid DNA regions. Am. J. Bot. 88: 2074–2087.

Reeves G, P Goldblatt, PJ Rudall, and MW Chase. 2002. Molecular systematics of Iridaceae: a combined analysis of four plastid DNA sequence matrices. Ann. Bot. n.s. (Italy) 1(2): 29–42.

Rübsamen-Weustenfeld T, V Muldelka, and U Hamann. 1994. Zur Embryologie, Morphologic, und systematischen Stellung von *Geosiris aphylla* Baillon (monocotyledoneae-Geosiridaceae /Iridaceae) mit einigen embryologischen Daten zur Samenanlage von *Isophysis tasmanica* (Hook.) T. Moore (Iridaceae). Bot. Jahrb. Syst. 115: 475–545.

Rudall P. 1984. Taxonomic and evolutionary implications of rhizome structure and secondary thickening in Iridaceae. Bot. Gaz. 145: 524–534.

Rudall P. 1986. Taxonomic significance of leaf anatomy in Australasian Iridaceae. Nord. J. Bot. 6: 277–289.

Rudall P. 1991. Leaf anatomy of Tigridieae (Iridaceae). Plant Syst. Evol. 175: 1–10.

Rudall P. 1993. Leaf anatomy and systematics of Mariceae (Iridaceae). Kew Bull. 48: 151–160.

Rudall P. 1994. Anatomy and systematics of Iridaceae. Bot. J. Linn. Soc. 114: 1–21.

Rudall P. 1995. Iridaceae. In: DF Cutler and M Gregory, eds. Anatomy of the Monocotyledons, vol. 8. Clarendon, Oxford.

Rudall P. 2003. Unique floral structures and iterative evolutionary themes in Asparagales: Insights from a morphological cladistic analysis. Bot. Rev. 68: 488–509.

Rudall P and P Burns. 1989. Leaf anatomy of the woody South African Iridaceae. Kew Bull. 44: 525–532.

Rudall P and P Goldblatt. 1991. Leaf anatomy and phylogeny of Ixioideae (Iridaceae). Bot. J. Linn. Soc. 106: 329–345.

Rudall P and P Goldblatt. 1993. Leaf anatomy and systematics of Homeriinae (Iridaceae). Bot. J. Linn. Soc. 111: 379–397.

Rudall PJ and P Goldblatt. 2002. Floral anatomy and systematic position of *Diplarrhena* (Iridaceae): a new tribe Diplarrheneae. Ann. Bot. n.s. (Italy) 1(2): 59–66.

Rudall P and A Wheeler. 1988. Pollen morphology in Tigriidieae (Iridaceae). Kew Bull. 43: 693–701.

Schulze W. 1971. Beiträge zur Pollenmorphologie der Iridaceae und ihre Bedeutung für die Taxonomie. Feddes Repert. 82: 101–124.

Schulze W. 1984. Beitrage zur Taxonomie der Liliifloren: 11. Tecophilaeaceae und Cyanastraceae. (Contributions to the taxonomy of the Liliiflorae: 11. Tecophylaeaceae and Cyanastraceae.) Wiss. Zeitschr. Friedrich-Schiller Univ. Jena, Mat. Naturwiss. Beitr. Phytotax., 32(6): 957–964.

Shneyer VS. 1983. The relationship between Iridaceae s.l. as revealed by the serological analysis of seed proteins. Bot. Zhurn. 68: 49–54 (in Russian with English summary).

Simpson MG. 1985. Pollen ultrastructure of the Tecophilaeaceae. Grana 24: 77–92.

Simpson MG and P Rudall. 1998. Tecophilaeaceae. In: K Kubitzki, ed. The families and genera of vascular plants, vol. 3, pp. 429–436. Springer, Berlin/Heidelberg/New York.
Souza-Chies TT, G Bittar, S Nadot, L Carter, E Besin, and B Lejeune. 1997. Phylogenetic analysis of Iridaceae with parsimony and distance methods using the plastid gene *rps*4. Plant Syst. Evol. 204: 109–123.
Tillich H-J. 1995. Früchte, Samen und Keimpflanzen bei den Cyanastraceae Engler 1900 and einiger vermuteten Verwandten. Feddes Repert. 106: 483–493.
Tillich H-J. 2003. Seedling morphology in Iridaceae: indications for relationships within the family and to related families. Flora 198: 220–242.
Traub HP. 1943. The Ixiolirion tribe. Herbertia 9: 53–59.
Wilson CA. 2003. Phylogenetic relationships in *Iris* series *Californicae* based on ITS sequences of nuclear ribosomal DNA. Syst. Bot. 28: 39–46.
Williams CA, JB Harborne, and P Goldblatt. 1986. Correlations between phenolic patterns and tribal classification in the family Iridaceae. Phytochemistry 25: 2135–2154.
Zavada MS and G Scott. 1993. Pollen morphology of *Cyanella* species (Tecophilaeaceae). Grana 32: 189–192.

Order 14. AMARYLLIDALES

Perennial or sometimes biennial herbs or secondary arborescent plants with rhizomes, bulbs, or corms, or occasionally pachycaul trees with leaves in terminal rosettes. Roots often contractile, wiry or thick, tuberous or slightly succulent. Calcium oxalate in the form of cubes, raphides and sand is widespread. Vessels mostly only in roots, with scalariform or simple perforations. Leaves alternate, spirally arranged or distichous, usually keeled, mostly with sheathing base. Stomata usually anomocytic. Flowers small to large, in various kinds of inflorescences or solitary, bisexual or rarely unisexual, actinomorphic or more or less zygomorphic, 3-merous, articulated (Themidaceae, Anemarrhenaceae) or more often not articulated. Perianth segments six in two usually more or less similar cycles, free or more or less connate. Stamens six in two cycles, or rarely three or two, free or attached to the perianth; anthers basifixed to dorsifixed, usually introrse, tetrasporangiate, mostly opening longitudinally. Tapetum secretory. Microsporogenesis simultaneous or successive. Pollen grains 2-celled, mostly 1-colpate, reticulate or trichotomocolpate. Gynoecium of three united carpels; style with 3-lobed, capitate or punctiform stigma; ovary superior to inferior, 3-locular or less often 1-locular, with one or more often two to many ovules per locule, usually with septal nectaries. Ovules anatropous or less often campylotropous, usually bitegmic, crassinucellate or (some Alliaceae, Asphodelaceae) tenuinucellate, usually with parietal tissue. Female gametophyte mostly of *Polygonum*-type. Endosperm helobial or nuclear. Fruits mostly loculicidal capsules or sometimes nuts or berries. Seed coat formed mainly or only by the outer integument, mostly encrusted with phytomelan; embryos of various sizes, straight or less often more or less curved; endosperm copious or scarce, sometimes (*Yucca* in Agavaceae) substituted by perisperm. Often producing anthraquinones, contain steroidal saponins (except Amaryllidaceae) and chelidonic acid.

Probably derived from some Melanthiaceae-like ancestor.

Key to Families

1 Microsporogenesis simultaneous.
  2 Seeds without arils or ariloids.
    3 Pollen grains 1-colpate, reticulate.
      4 Flowers slightly to strongly zygomorphic, large, in one or two double helicoid cymes, on short pedicels. Perennial, glabrous herbs with short rhizome and often thickened and fusiform roots. Vessels only in roots, with scalariform perforations. Leaves all basal, alternate, distichous, flat or folded, linear, elongate, keeled. Stomata anomocytic. Flowers in terminal, scapiflorous cymes, actinomorphic to strongly zygomorphic, 3-merous. Perianth segments basally connate into a narrow tube, often recurved. Stamens slightly curved up, making flowers zygomorphic; filaments markedly unequal, free of one another, long, inserted in perianth tube; anthers dorsifixed, versatile, often twisted. Gynoecium of three carpels; style long, slender, slightly curved up, with punctiform, capitate stigma; ovary 3-locular, with many anatropous ovules per locule. Parietal tissue lacking. Endosperm nuclear. Fruits 3-angled, leathery, transversely rugulose, loculicidal capsules. Seeds black, shiny; endosperm contains aleurone and lipids; embryo about the same size as or smaller than endosperm; testa smooth, encrusted with phytomelan. Contain kaempferol and quercetis, steroidal saponins, and arthroquinones detected; n = 11. . . . . 1. HEMEROCALLIDACEAE.

4 Flowers actinomorphic. Robust, tufted, stemless perennials with short, erect rhizome hidden in massive fibrous remains of old leaves. Leaves basal, distichous, broadly ensiform, unifacial, equitant and isobilaterally flattened, rigid; sheathing base long, thick and fleshy. Inflorescence a terminal, brushlike spike, peduncle long, with several sheathing bracts. Perianth segments 20 mm long, free, more or less equal, at firs erect, becoming reflexed and twisted and so persisting with the entangled staminal filaments until the capsule opens. Anthers versatile. Style cylindric, with narrow stigma; ovary stipitate, 3-locular. Constricted above the middle, with several ovules per locule. Fruits capsules, more or less trigonous, broader below than above the equatorial constriction, the valves usually widely only in upper part. Seeds more or less spiny-papillose. n = 17, 18. . . . . 3. XERONEMATACEAE.

3 Pollen grains trichotomocolpate, rounded triangular in polar view. Tall, perennial, tufted herbs or subshrubs with short, stout rhizome and fascicled fibrous roots. Rhizome contains wax and leaves contain hemicellulose-like polysaccharide (Phormioideae) or wax and polysaccharide absent (Dianelloidea). Raphides present (Dianelloideae) or absent. Vessels only in roots, with simple and scalariform perforations. Leaves basal, alternate, long, distichous, sheathing, linear-ensiform, strongly keeled, very tough (with strong bast fibers). Stomata anomocytic. Flowers in tall, terminal panicle, bisexual, more or less zygomorphic, articulated just below flowers. Perianth segments connate at base, three inner ones longer. Stamens six in two cycles, inserted in the tube at base of perianth segments; filaments filiform, free of one another or (*Excremis*) basally connate into a ring, swollen apically and covered with nectariferous pappilae (*Dianella*) or with dense, woolly-tomentose beard often reduced to a tuft under the anther (*Stypandra*); anthers elongate, basifixed or dorsifixed, introrse, openings longitudinally or by apical pores. Ovary superior or (*Pasithea*) half-inferior, completely or incompletely 3-locular, with numerous anatropous ovules per locule; ovules crassinucellate or tenuinucellate (*Dianella*); style declinate, with capitate stigma. Endosperm helobial. Fruits long, loculicidal or septicidal capsules, rarely (*Dianella*) berries (sometimes blue). Seeds flattened, almost winged (*Phormium*), black or brown, shining, exotestal cells encrusted with phytomelan; embryo linear, long; endosperm oily; n = 8 (karyotype bimodal). . 2. PHORMIACEAE.

2 Seeds generally with arils or sometimes with ariloids. Mostly perennial, rhizomatous, succulent or non-succulent herbs, subshrubs, shrubs, or arborescent (some of woody forms with trunks up to several meters). Roots with velamen, vessels in roots with simple perforation. Secondary thickening absent, or anomalous; xylem without vessels or rarely with vessels (with scalariform perforations). Leaves small to very large, commonly basal, alternate, spiral (usually) or distichous, simple, entire, serrate, or dentate, and often with an apical spine. Stomata anomocytic, or tetracytic, or rarely paracytic. Flowers in terminal racemes or in spikes, actinomorphic or zygomorphic, small to large. Perianth segments six in two cycles, sepaloid and petaloid, white, to yellow or red, purple and brown (not blue or violet). Stamens six, in two cycles, free of one another, equal or markedly unequal; anthers dorsifixed. Styles apical; ovary superior, ovules 2–40 per locule, hemianatropous, anatropous, or nearly orthotropous (in *Aloe* and *Asphodelus*). Embryo helobial. Fruits loculicidal capsules. Seeds winged or wingless; embryo straight, endosperm oily; testa usually encrusted with phytomelan. Arthroquinones detected; n = 6–11, mostly n = 7, with four long and three short chromosomes. . . . . 4. ASPHODELACEAE.

1 Microsporogenesis successive.

5 One ovule per carpel. Pollen spiraperturate. Perennial tufted herbs with short sympodial rhizome emitting fibrous roots. Secondary growth occurs in rhizome. Vessels only in roots, with mostly simple perforations. Leaves reduced to long, reddish-brown, scarious radical sheaths and only rarely bear a short vestigial lamina. Stomata deeply sunken, anomocytic. Flowers in small, terminal, condensed, capitate inflorescences, consisting of 1–2(3) flowers borne slightly obliquely, sessile among imbricate hyaline bracts; each flower enclosed by one or two

free and five basally connate scales, indicating that the inflorescence is very reduced; flowers bisexual, actinomorphic. Perianth segments basally connate into a tube. Stamens six in two cycles, adnate to the base of perianth; filaments filiform; anthers dorsifixed. Pollen grains spiraperturate, echinate. Style filiform, shortly 3-brachiate, with papillate stigmatic surfaces; ovary 3-locular, with septal nectaries. Ovules anatropous to slightly campylotropous, bitegmic, crassinucelate, with parietal cell. Fruits membranous loculicidal capsules. Seeds ovoid, slightly flattened, black, rugulose, with crustaceous testa thinly encrusted with phytomelan; embryo linear and as long as endosperm; endosperm contains aleurone and lipids. Contain flavonoles, flavonoids and proanthocyanins and rhizome contains steroidal saponins, saponogen, wax alcohols. n = 8. . . . . . . 15. APHYLLANTHACEAE.

5 Two or more ovules per carpel. Pollen not spiraperturate.

6 Stamens usually six in two cycles (3 + 3). Xeromorphic arborescent plants with secondary growth derived from a secondary thickening meristem to acaulescent herbs with thick, woody, short rhizome. Roots contractile. Vessels in roots and leaves, with simple perforations in roots and scalariform ones in leaves. Leaves highly xeromorphic, spirally arranged, linear, firm, with thickened resiniferous bases and may be more than 1 m long, borne in dense tufts at the tips of branches. Stomata paracytic. Calcium oxalate raphides are widespread. Flowers usually small, in complex, dense, cylindrical, spikelike, and multiflorous inflorescences, sessile, bisexual, actinomorphic, supplied with relatively large bract and bracteole. Perianth segments six in two cycles, persistent, free, segments of the outer cycle chartaceous or scarious, those of the inner cycle membranous. Stamens six in two cycles, free and longer than the perianth; filaments flattened; anthers elongate, dorsifixed, opening longitudinally, introrse. Microsporogenesis successive. Pollen grains bean-shaped to ellipsoidal, 1-colpate, reticulate. Gynoecium of three united carpels; style long, subulate, tapering into 3-lobed or capitate to punctiform stigma; ovary superior, 3-locular, with several anatropous ovules per locule. Septal nectaries well developed. Ovules bitegmic, markedly crassinucellate, with parietal cell. Endosperm nuclear. Fruits woody or cartilaginous loculicidal capsules with one or two seeds per locule. Seeds flattened, black, testa with thick phytomelan crust; embryo fusiform, more or less curved; endosperm store aleurone and lipids. Stem and leaf bases produce copious acaroid resin which contains a major part of polymerized derivatives of cinnamic acids, n = 11. . . . . 5. XANTHORRHOEACEAE.

6 Stamens three in one cycle. Perennial, shortly rhizomatous herbs with thickened roots and short erect stems. Vessels with simple perforations occur in roots. Leaves alternate, basal, numerous, simple, entire, linear and grasslike, few-nerved and whit slightly sheathing bases. Stomata anomocytic. Inflorescences terminal, erect, pedunculate, leafy, bracteate, spike-like panicles. Flowers bisexual, small, greenish-white to brownish-purple, the pedicel is articulated. Perianth segments six, in two similar cycle, persistent; tepals free, connivent basally into a short tube with linear, spreading limbs. Stamens three, attached to the inner tepals; filaments free, short, flat; anthers linear, basifixed, introrse, dehiscent longitudinally. Pollen grains 2-celled, sulcate. Septal nectaries present. Gynoecium of three carpels; style short, filiform, stigma small, capitate; ovary superior, 3-locular, with two ovules per locule. Fruits loculicidal capsules. Seeds black, spindle-shaped, carinate; testa thick, phytomelan encrusted; embryo large, cylindric, strongly curved; endosperm fleshy. Contain saponins and the xanthone C-glycoside (Henauer 1863, Sato et al. 1994); n = 11. . . . . 7. ANEMARRHENACEAE.

7 Inflorescence simple or compound raceme, spike, panicle or thyrse.

8 Capsule septicidal, anthers basifixed. Perennials with usually tuberous rhizome and erect or climbing stems armed with prickles, sometimes stemless. Calcium oxalate raphides present in all parts of the plant. The stem of some species of *Herreria* have secondary thickening. Vessels only in roots, with scalariform perforations. Leaves in lateral or basal clusters, linear to lanceolate, sessile,

somewhat coriaceous, rather rigid, phylloclades, with closed parallel venation. Stomata anomocytic. Flowers small, in slender axillary racemes or panicles, bisexual, actinomorphic, with articulated pedicel. Perianth segments six in two cycled, free, all similar, spreading; in *Herreriopsis* saccate nectaries present at the base of perianth segments. Stamens six in two cycles, free, inserted at the base of perianth segments. Pollen grains 1-colpate. Gynoecium of three united carpels. Ovary superior, 3-locular, with intrusive placentas and septal nectaries (*Herreria*); style short to elongate, with 3-lobed stigma. Ovules many (*Herreriopsis*) or 3–6 (*Herreria*) per locule, anatropous. Seeds black, flattened, with thin, membranous wing all around; testa multilayered, exotestal cells encrusted with phytomelan; tegmen collapsed, represented by thin membrane; endosperm copious, containing aleurones and lipids, but lacking starch; embryo small, nearly straight. Saponines are present. n = 27, karyotype bimodal, with one large and 26 small chromosomes. . . . . . . 14. HERRERIACEAE.

8 Capsule loculicidal, anthers dorsifixed.

9 Sympodial perennials, often with thickened roots, flowers often articulated. Perennial herbs with short, often vertical rhizomes, and fibrous, fleshy or tuberous roots, rarely (*Chlorophytum suffruticosum*) with a woody stems. Stems erect or climbing. Calcium oxalate raphides present. Vessels in roots with scalariform or simple perforations, sometimes in the aerial stem and leaves (*Arthropodium*) with simple or scalariform perforations. Leaves alternate, spirally arranged, sessile or pseudopetiolate, sheathing at base, linear, linear-lanceolate to broadly ovate, sometimes (*Laxmannia* and *Sowerbaea*) with short ligul-like projection above the sheath, sometimes much reduced (in some Australian genera – Watson and Dallwitz, 2006), mucilaginous or not mucilaginous. Stomata anomocytic. Flowers in various kinds of inflorescences, usually pedicellate, subtended by single to numerous bracts, not articulated or articulated along the pedicels, bisexual, actinomorphic or zygomorphic. Perianth segment of six in two more or less similar cycles, petaloid, membranous, scarious or papery, often with fimbriate margins, persistent, free or connate. Stamens six or less often three, all fertile or three staminodial; filaments free or (*Echeandia*) basally connate; anthers basifixed or dorsifixed, dehiscing longitudinally or by apical pores, introrse or rarely latrorse or extrorse. Pollen grains 2-celled, 1-colpate or rarely trichotomocolpate or tetrachotomocolpate. Septal nectaries present. Gynoecium of three carpels, sessile or (*Chlorophytum*) stipitate. Style filiform, rarely (*Tricoryne*) gynobasic, with small, capitate or 2–3-lobed stigma; ovary superior, 3-locular; ovules single to numerous per locule. Fruits loculicidal or septicidal capsules, nuts or schizocarps (*Tricoryne*). Seeds testal, sometimes arillate, endosperm fleshy, embryo cylindric, straight to curved inwards; testa encrusted with phytomelan. Chelidonic acid, saponins/sapogenins present; n = 6–8, 10, 11, 13–15, 19, 33. . . . .6. ANTHERICACEAE.

9 Plants rosulate – herbs, shrubs or arborescent (often 'rosette trees'), flowers not articulated. Large, usually rhizomatous, often woody, or even arborescent plants that may have stout trunk; secondary growth absent, or anomalous (*Agave, Yucca, Furcraea*). Vessels only in roots, with scalariform or simple perforations. Leaves usually in rosettes, often basal, more or less thickened, lanceolate, linear, or subulate. Stomata of various types, generally deeply sunken. Flowers mostly in complex, much branched panicle, the lateral components of which consist of monochasia, or in racemes or spikes, mostly bisexual, actinomorphic or slightly zygomorphic, bracteate. Perianth segments free or more or less

connate into a tube. Stamens three or six, inserted at the base of perianth segments or in perianth tube; filaments filiform, inserted in perianth throat; anthers usually epipeltate, dorsifixed, unappendaged. Microsporogenesis successive. Pollen grains 1(2)-colpate, reticulate, in *Agave* occasionally in tetrads. Style short or rather long, with 3-lobate, capitate, or punctiform stigma; ovary superior to inferior, 3-locular, with several to many anatropous ovules per locule. Female gametophyte of *Polygonum*-type. Endosperm helobial or (as in *Furcraea* and some species of *Agave*) nuclear. Fruits loculicidal or sometimes septicidal (*Yucca* sect. *Chaenocarpa*) capsules, or indehiscent and dry, or baccate. Seeds often more or less compressed, sometimes with narrow wing, black. Epidermis of testa with thin crust of phytomelan; embryo small, cylindrical, surrounded by fleshy endosperm containing aleurone and fatty oils. Rich in steroidal saponins and also contains chelidonic acid; contain kaempferol. Crystal raphides lacking. n = 30, with five long chromosomes. . . . . 9. AGAVACEAE.

7 Inflorescence usually represents one or more contracted helicoid cymes, rarely (as in *Milula*) spikelike, subtended by involucre of (1) 2 or more membranous or scareous spathial bracts, free or basally united. Rootstock tuberous, erect rhizome.

10 Ovary inferior. Perennial herbs with tunicated bulbs with thick, fleshy scales, or rarely with a rhizome (as in *Scadoxus* and *Clivia*). Roots contractile. Vessels only in roots, with scalariform perforations. Leaves basally concentrated, linear to mostly orbicular, sheathing at the base, parallel-veined, with mucilage cells or elongate sacs with raphides. Stomata anomocytic. Inflorescence umbel-like, consists of one to several helicoid cymes with suppressed axes and subtended by 2–8 (rarely one) usually membranous bracts that are free or basally connate. Flowers usually large, bisexual, actinomorphic to slightly (very rarely strongly) zygomorphic, 3-merous. Perianth segments six, in two generally similar cycles, free or more or less connate, sometimes (as in *Narcissus*) with corona ("paracorolla"). Stamens six in two cycles (rarely three in *Zephira* and up to 18 in *Gethyllis*); filaments free or basally connate, inserted at the base of perianth segments or in the tube; anthers tetrasporangiate, generally elongate, epipeltate or sometimes basifixed, opening longitudinally or rarely (Galantheae) by apical pores. Tapetum secretory or (species of *Galanthus*) plasmodial. Pollen grains 2-celled, 1-colpate or less often (Amaryllideae) 2-colpate. Gynoecium of three united carpels; style slender, with punctiform, capitate, or 3-lobed stigma; ovary 3-locular, with several to many ovules per locule. Nectaries septal or (Galantheae) perigonal (secreted from the distal part of the inner perianth segments). Ovules anatropous or hemitropous, bitegmic or rarely (*Amaryllis* and *Nerine*) unitegmic or (*Crinum*) ategmic, crassinucellate or rarely (*Crinum, Narcissus, Zephyranthes*) tenuinucellate, mostly with parietal cell. Female gametophyte of *Polygonum*-type, rarely of *Adoxa*-type. Endosperm helobial or less often (Galantheae and Narcisseae) nuclear, in *Crinum* and related genera and in Eucharideae both types of endosperm development occur. Fruits loculicidal capsules or baccate. Seeds of various sizes, shapes, and colours, glabrous, usually shiny, sometimes with elaiosomes; testa consists of several cell layers, and usually its outer layers are incrusted with phytomelan, which is sometimes lacking (Dahlgren et al. 1985); tegmen is collapsed into a thin film; endosperm copious, fleshy, or rarely bony, contains aleurone and fatty oils, sometimes also hemicellulose stored in the cell walls, and often small

quantities of starch (more abundant at the early stages of development); embryo straight or slightly curved, generally more than half as long as the endosperm. Steroidal saponins lacking, but various particular alkaloids as well as chelidonic acid and norbelladine alkaloids are widely distributed. n = 5–15, 23, 27, 29, mostly 11.......13. AMARYLLIDACEAE.

10 Ovary superior. Plants rhizomatous. Roots fleshy, provided with a multiple velamen. Laticifers present. Vessels only in roots, mostly with scalariform perforations. Calcium oxalate raphides present. Leaves rosulate, distichous, linear, sheating at base. Flowers in a pseudo-umbel subtended by two involucral spathial bract, bisexual, zygomorphic, pedicels subtended by threadlike bracts. Perianth segments subequal, basally more or less connate; corona lacking. Stamens inserted on the perianth tube; filaments of unequal length, declinate; anthers dorsifixed. Pollen grains 1-colpate, reticulate. Style slender, hollow, declinate, with punctiform stigma; ovary 3-locular, with inner septal nectaries. Ovules numerous, campylotropous, crassinucellate. Parietal cell present. Fruits loculicidal capsules. Seeds black, angulate, elongated, winged; endosperm stores aleurone, lipids and reserve cellulose; embryo well developed, 4/5 of the length of endosperm Various steroidal saponins present. n = 15, rarely 16. ... 11. AGAPANTHACEAE.

11 Inflorescences monopodial racemes or spike. Rootstock mostly a bulb. Perennial herbs with bulbs or rarely (*Schoenolirion* and *Chlorogalum*) rhizomes; bulbs generally with membranous tunic and free or coalescent scales. Roots sometimes thick and commonly contractile. Vessels only in roots, with scalariform and/or simple perforations. Leaves concentrated at the base, linear to lanceolate, rarely elliptic to orbicular, sessile, sheathing at base. Inflorescences monopodial raceme or spike. Flowers not articulated, generally bisexual, actinomorphic or rarely zygomorphic. Perianth segments free or more often connate. Stamens six in two cycles, with sometimes one sterile cycle; filaments inserted at base of perianth segments or on the perianth tube; often broad and flat; sometimes connate at the base, rarely forming a paracorolla; anthers epipeltate. Pollen grains, 1-colpate. Style simple, with punctiform or sometimes distinctly 3-lobed stigma; ovary superior or rarely (*Bowiea*) semi-inferior, 3-locular, with septal nectaries (except for *Autonoe*), with two to numerous anatropous or campylotropous ovules per locule. Female gametophyte of various types, but mostly of *Polygonum*-type. Endosperm helobial or nuclear. Fruits loculicidal capsules. Seeds small to large, ovoid to obovate, with elaiosomes; outer epidermis of the testa consists of several layers and usually has thick to rather thin phytomelan crust (absent in *Chionodoxa, Pushkinia,* and a few species of *Scilla,* where outer epidermis collapses); inner integument forms thin membrane; endosperm usually copious, consists of pitted cells and contains aleurone and fatty oils and very rarely starch grains; embryo more or less cylindrical, usually straight. Crystal raphides contained in mucilage cells or canals are widely distributed. Producing steroidal saponins, chelidonic acid, polyhydroxyalkaloids, flavones, flavone C-glycosides. n = 2–15, 17, 19, 21, 27, 29......... 8. HYACINTHACEAE.

11 Inflorescences contracted cymes, rarely spikelike or umbellate.

12 Rootstock a bulb. Leaves spiral. Corona present. Alliaceous odour usually present. Rootstock bulbs or bulb-like corms with membranous or fibrous scales, more rarely rhizome. Articulated laticifers

present in leaves and bulb scales. Vessels only in roots, with scalariform or/and simple perforations. Leaves basally concentrated, filiform-linear, lanceolate or rarely ovate, sheathing at base, rarely constricted into a pseudopetiole. Flowers usually in one or more contracted cymes, rarely (as in *Milula*) spikelike, subtended by involucre of (1) 2 or more membranous spathial bracts, free or basally united, generally bisexual, actinomorphic or zygomorphic, not articulated. Perianth segments free or often connate. Functional stamens six or sometimes three or two with several staminodia; filaments inserted at the base of perianth segments or in perianth tube, more or less flat; anthers mostly elongate, epipeltate. Pollen grains, 1-colpate. Style simple, with 3-lobed or capitate stigma; ovary superior, 3-locular, with two to several anatropous, hemitropous, or campylotropous, crassinucellate (as in *Milula*) or tenuinucellate (as in *Allium*, *Nothoscordum*, *Tulbaghia*) ovules. Septal nectaries present. Female gametophyte of *Polygonum*- or *Allium*-type. Endosperm helobial. Fruits loculicidal capsules. Seeds of various sizes and shapes, with testal epidermis usually covered by rather thick crust of phytomelan and inner layers of testa compressed or collapsed, as is the tegmen; endosperm cells with rather thick and pitted walls and contain aleurone and fatty oils; embryo straight or more or less curved and usually more than half the length of endosperm. Contain flavonoids, steroidal saponins and strong-smelling allyl disulfides, propyl sulfides and vinyl sulfides that participate in essential oils causing the onion odour. n = 5–16. . . . . 12. ALLIACEAE.

12 Rootstock a starch-storing corm; a new corm is forming every year on top of the corm of the previous year. Lacticifers present. Calcium oxalate raphides present or absent. Vessels with scalariform and simple perforations. Leaves spirally arranged, linear, flat, forming closed sheath. Flowers in umbellate inflorescence, rarely reduced to single flower; pedicels often articulated, subtended by a small bract (two bracts when flower is solitary). Perianth segments usually more or less basally connate. A corona, scales or appendices often present between perianth segments and stamens. Fertile stamens three, the missing ones often transformed into staminodia; anthers more or less basifixed, introrse. Pollen grains 1-colpate, reticulate. Style with 3-lobed or capitate stigma; ovary sometimes on a gynophore, 3-locular, with two or several anatropous ovules per locule. Endosperm helobial (*Muilla* and *Triteleia*) or nuclear (*Brodiaea*). Fruits loculicidal capsules. Seeds angular; epidermis of testa with a crust of phytomelan; embryo short and thick; endosperm copious, containing aleurone and lipids. Alliaceous odour lacking. Steroidal saponins present or absent. n = 5–12, 14, 15, 21, 25, 27.. . . . . 10. THEMIDACEAE.

1. HEMEROCALLIDACEAE

R. Brown 1810. 1/ about 15. Central Europe (1) to Japan and China, mainly in China.

Hemerocallis.

The least advanced member of the order.

2. PHORMIACEAE

J. Agardh 1858. (including Dianellaceae R.A. Salisbury 1866). 10/30–35. Mediterranean (*Simethis*) Madagascar, Mascarenes, Seychelles, tropical East Africa, southern, eastern and southeastern Asia, Malesia, Hawaii, South Pacific Islands, Fiji, New Caledonia, New Zealand, Australia, Chile (*Pasithea*), and one monotypic genus *Excremis* in Andes of Ecuador, Colombia and Peru; *Phormium* – New Zealand (N and S Isl., Stewart Isl., Chatham Isl., Auckland Isl.) and Norfolk Island.

2.1 PHORMIOIDEAE
Rhizome contains wax and leaves contain hemicellulose-like polysaccharide. Raphides absent. Seeds winged. Proanthocyanidins absent. – *Phormium*.

2.2 DIANELLOIDEAE
Wax in rhizome and hemicellulose-like poly-saccharides in leaves absent. Raphides present Seeds wingless. Proanthocyanidins present (*Dianella*). – *Agrostocrinum, Pasithea, Herpolirion, Thelionema, Simethis, Stypandra, Rhuacophila, Dianella, Excremis*.

Related to the Hemerocallidaceae.

3. XERONEMATACEAE

M.W. Chase, Rudall and Fay, 2000. 1/2. New Caledonia, New Zealand.

Xeronema.

Possibly related to the Phormiaceae.

4. ASPHODELACEAE

A.L. de Jussieu 1789 (including Aloaceae Batsch 1802). 15/1000. Temperate, subtropical regions of the Old World, Alooideae – Africa (especially South Africa), Madagascar, Mascarenes, Arabia.

4.1 ASPHODELOIDEAE
Leaves generally not conspicuously succulent. Stomata generally anomocytic. Vessels with scalariform perforations, only occasionally present in the stem (species of *Asphodelus*). Typical caps of aloine cells are not formed at the phloem poles of the vascular bundles of leaves. Herbaceous plants. n = 6–11 – ASPHODELEAE: *Eremurus, Asphodelus, Asphodeline, Bulbine, Bulbinella, Jodrellia, Trachyandra, Paradisea*; KNIPHOFIEAE: *Kniphofia*.

4.2 ALOOIDEAE
Leaves more or less strongly succulent, often prickly along the margins. Stomata mostly sunken, more or less distinctly tetracytic. Large cap of aloine cells present at the phloem pole of most vascular bundles of leaves. Sparsely branched trees up to several meters tall to small rosette herbs, n mostly = 7, with four long and three short chromosomes. – *Aloe, Poellnitzia, Lomatophyllum, Gasteria, Haworthia, Chortolirion*.

5. XANTHORRHOEACEAE

Dumortier 1829. 1/30. Australia and Tasmania.

Xanthorroea.

6. ANTHERICACEAE

J. Agardh 1858 (including Boryaceae Rudall, Chase et Conran 1997; Johnsoniaceae Lotsy 1911; Laxmanniaceae Bubani 1902). 27/c.350. Africa, Madagascar, Mascarenes, Eurasia, Malesia, Australia, Pacific Islands, North, Central and South America. The largest genera is *Chlorophytum* (about 300).

ANTHERICIEAE: *Anthericum, Comospermum, Chlorophytum, Dichopogon, Hagenbachia, Diamena, Diora, Eremocrinum, Leucocrinum, Echeandia*; JOHNSONIEAE: *Tricoryne, Hodgsoniola, Caesia, Corynotheca, Arnocrinum, Hensmania, Johnsonia, Stawellia*; LAXMANNIEAE: *Arthropodium, Murchisonia, Thysanotus, Trichopetalum, Laxmannia, Sowerbaea, Chamaescilla*; BORYEAE: *Alania, Borya*.

7. ANEMARRHENACEAE

Conran, MW Chase and Rudall 1997. 1/65. Northern China and Korea.

Anemarrhena

8. HYACINTHACEAE

Batsch ex Borkhausen 1797 (Eucomidaceae Salisbury 1866; Lachenaliaceae Salisbury 1866; Ornithogalaceae Salisbury 1866; Scillaceae Vest 1818). 67/800–1000. Very widely distributed in both hemispheres, but especially diversified in South Africa, Mediterranean, and Irano-Turanian region.

8.1 OZIROËOIDEAE

Bulb leaves imbricate, mucilaginous. Foliage leaves few, thick and grooved. Scape with prominent bracts, 1–2(-3) flowered. Pedicels straight, pointing obliquely upwards. Stamens basally connate and adnate to the corolla. Seeds rounded; embryo as long as seed; n = 15, 17. – *Oziroë*.

8.2 ORNITHOGALOIDEAE

Bulb leaves usually lasting only 1–2 vegetation periods. Bracts large, prophylls lacking. Stamens three; filaments flat, often appendaged. Seeds often compressed and edged, ridged, or carinate. Cardenolides partly present; n = 2–10. – DIPCADIEAE: *Stellarioides, Coilonox, Pseudogaltonia, Dipcadi, Galtonia, Zahariadia, Melomphis, Cathissa, Eliokarmos, Loncomelos, Honorius*; ORNITHOGALEAE: *Ornithogalum, Albuca*.

8.3 URGINEOIDEAE

Bulb leaves usually imbricate. Scapes usually solitary, bracts spurred. Filaments not appendaged. Seeds flattened, sometimes winged; testa brittle, not tightly adherent to endosperm. Bufodienolids present; n = 6, 7, 10. – *Bowiea, Schizobasis, Igidia, Urgineopsis, Rhadamanthus, Litanthus, Rhadamanthopsis, Thuranthos, Tenicroa* (including *Sypharissa*), *Drimia Urginea*.

8.4 HYACINTHOIDEAE

Bulb leaves lasting 2–3 vegetation periods. Bracts usually small, rarely large or lacking. Prophylls present or absent. Seeds globose, drop-shaped or ellipsoid; elaiosomes formed from various parts of the testa; nucleus lacking protein crystals. Homoisoflavones present; n = 4–10. – PSEUDOPROSPEREAE: *Pseudoprospero*; MASSONIEAE: *Ledebouria, Resnova, Drimiopsis, Lachenalia, Veltheimia, Eucomis, Massonia, Namophila, Whiteheadia, Periboea, Polyxena, Androsiphon, Amphisiphon, Neobakeria, Daubenya*; HYACINTHEAE: *Hyacinthus, Merwilla, Schizocarpus, Fortunatia, Barnardia, Oncostema, Hyacynthoides, Chionodoxa, Puschkinia, Hyacinthella, Brimeura, Tractema, Alrawia, Prospero, Othocallis, Pfosseria, Nectaroscilla, Bellevalia, Muscari, Muscarimia, Leopoldia, Pseudomuscari, Scilla, Rhodocodon, Daubenya, Neopatersonia*.

Close related to the Anthericaceae and Agavaceae.

9. AGAVACEAE

Dumortier 1829 (including Chlorogalaceae Doweld et Reveal 2005; Funkiaceae Horaninow 1834; Hesperocallidaceae Traub 1972; Hostaceae Mathew 1988; Yuccaceae J. Agardh 1858). 15/300. Tropical and subtropical regions of America and West Indies, concentrated mostly in Mexico and southwestern and southern United States; primarily in dry areas.

9.1 CHLOROGALOIDEAE

Bulb leaves usually present, short-lived; scale and foliage leaves with little mucilage. Ovary superior, with septal nectaries; nectar ducts tubular or covered by hairs. – *Schoenolirion, Hastingsia, Camassia, Chlorogalum*.

9.2 YUCCOIDEAE

Caespitose or abrorescent plants. Ovary superior. – *Yucca, Hesperaloe*.

9.3 AGAVOIDEAE

Ovary inferior. – *Agave, Beschorneria, Furcraea, Littaea, Manfreda, Polianthes, Prochnyanthes*.

9.4 HESPEROCALLIDOIDEAE

Rootstock bulbous (*Hesperocallis*) or not bulbous (*Hosta*). Ovary superior. – *Hosta. Hesperocallis*.

10. THEMIDACEAE

R.A. Salisbury 1866. 12/60. Southwestern Canada, western United States and Mexico, with one species of *Milla* reaching Guatemala.

Muilla, Bloomeria, Dandya, Milla, Bessera, Petronymphe, Triteleia, Jaimehintona, Triteleiopsis, Brodiaea (*Themis*), *Dichelostemma, Androstephium*.

Related to the Hyacinthaceae and Alliaceae.

11. AGAPANTHACEAE

Voigt 1850. 1/9. South Africa from the Cape Peninsula to the south of Limpopo River.

Agapanthus.

12. ALLIACEAE

Borkhausen 1797 (including Cepaceae Salisbury 1866; Gilliesiaceae Lindley 1826; Milulaceae Traub 1972; Tulbaghiaceae Salisbury 1866). 18/650. Widely distributed in both hemispheres except for tropical regions, Australia, and New Zealand.

12.1 TULBAGHIOIDEAE
Plant often rhizomatous or bulb with thick roots and alliaceous smell; rhizomes or bulb containing starch. The closed sheaths at base of leaves very short. Flowers bracteate, in umbellate inflorescences. Perianth segments strongly connate, corona massive, lobes connate or not. Embryo straight or slightly curved, phytomelan crust rather thin; n = 6. – *Tulbaghia.*

12.2 ALLIOIDEAE
Mostly with tunicated bulbs and with alliaceous odour; bulbs lacking starch. Leaves more or less unifacial. Corolla basally connate, without corona. Stamens both basally connate and adnate to the corolla, often with winged filaments. Embryo curved; n = (7) 8 (9). *Allium, Caloscordum, Nectaroscordum, Milula, Nothoscordum, Ipheion, Tristagma, Garaventia.*

12.3 GILLIESIOIDEAE
Corona absent or of small scales. Stamens 2–3, variously connate and adnate; staminodes present. Embryo short; n = 6, 7, 9–10, 12. *Schickendantziella, Speea, Leucocoryne, Latace, Miersia, Gilliesia, Solaria* (including *Ancrumia, Gethyum*), *Erinna, Trichlora.*

Close to the Hyacinthaceae. According to Dahlgren et al. (1985), most of characters found in Alliaceae are found in the Hyacinthaceae although its inflorescence is racemose and a parietal cell is cut off in the nucellus, a condition rare in Alliaceae.

13. AMARYLLIDACEAE

J. Saint-Hilaire 1805 (incluing Brunsvigiaceae Horaninow 1834; Crinaceae Vest 1818; Cyrtanthaceae Salisbury 1866; Galanthaceae G. Meyer 1836; Gethyllidaceae Rafinesque 1838; Haemanthaceae Salisbury 1866; Leucojaceae Batsch ex Borkhausen 1797; Narcissaceae A.L. de Jussieu 1789; Oporanthaceae Salisbury 1866; Pancratiaceae Horaninow 1847; Strumariaceae Salisbury 1866; Zephyranthaceae Salisbury 1866). 61/850. Tropics and subtropics, especially America, Mediterranean, and Africa.

Classification after A.W. Meerow and D.A. Snijman (1998) and A.W. Meerow et al. (2000, 2002).

AMARYLLIDEAE: *Amaryllis, Boophone, Crinum, Ammocharis, Cybistetes, Crossyne, Strumaria* (including *Bokkeveldia, Gemmaria* and *Tedingea*), *Carpolyza, Nerine, Hessea, Namaquanula, Brunsvigia*; CYRTANTHEAE: *Cyrtanthus*; HAEMANTHEAE: *Clivia, Cryptostephanus, Scadoxus, Haemanthus*; CALOSTEMMATEAE: *Calostemma, Proiphys*; GETHYLLIDEAE: *Apodolirion, Gethyllis*; LYCORIDEAE: *Lycoris, Ungernia*; PANCRATIEAE: *Pancratium* (including *Chapmanolirion*); NARCISSEAE: *Narcissus, Braxireon, Sternbergia*; GALANTHEAE: *Galanthus, Leucojum, Vagaria, Lapiedra, Hannonia*; HIPPEASTREAE: *Hippeastrum, Worsleya, Griffinia, Rhodophiala, Zephyranthes* (including *Cooperia*), *Habranthus, Sprekelia, Pyrolirion, Placea, Traubia, Phycella*; EUCHARIDEAE: *Eucharis, Caliphruria, Plagiolirion, Urceolina*; HYMENOCALLIDEAE: *Hymenocallis, Leptochiton, Pamianthe, Ismene*; STENOMESSEAE: *Stenomesson, Phaedranassa, Rauhia, Eucrosia, Mathieua*; CLINANTHEAE: *Pseudostenomesson, Paramongaia, Pucara*; EUSTEPHIEAE: *Chlidanthus, Castellanoa, Eustephia, Hieronymiella.*

Related to the Alliaceae and differing from them mainly in their inferior ovary. Although they differ from the Alliaceae also in their unique alkaloids and in not containing steroidal saponins, Amaryllidaceae contain the same kind of mannose-specific lectins, which have similar molecular structures, agglutination properties, and amino acid composition and are serologically identical (Van Damme et al. 1991). The monophyly of Alliaceae and Amaryllidaceae is supported by *rbc*L sequences (Fay et al. 1994). It is interesting to note that in Amaryllidaceae plastids form are P2c, except *Rauhia* and *Worsleya*, which contain form-P2cf.

14. HERRERIACEAE

Endlicher 1841. 3/12. Temperate and subtropical South America (*Herreria, Clara*) and Madagascar (*Herreriopsis*).

Herreria, Clara, Herreriopsis.

Probably related to the Agavaceae.

15. APHYLLANTHACEAE

Burnett 1835. 1/1. Western Mediterranean.

Aphyllanthes

A wide variety of families have been proposed for this taxonomically rather isolated taxon, including Eriocaulaceae, Xanthorrhoeaceae, Liliaceae-Asphodeloideae, and Anthericaceae. *Aphyllanthes* shares a number of features with the Anthericaceae (including septal nectaries, helobial endosperm, seeds with phytomelan, and steroidal saponins).

Affinity is not clear, but probably belong to the Amaryllidales.

Bibliography

Althoff DM, KA Segraves, J Leebens-Mack, and O Pellmyr. 2006. Patterns of speciation in the yucca moths: parallel species radiations within the *Tegeticula yuccasella* species complex. Syst. Biol. 55: 398–410.

Alvarez A and E Kohler. 1987. Morfologia del polen de las Agavaceae y algunos generos afincs. Grana 26: 25–46.

Arroyo SC and DF Cutler. 1984. Evolutionary and taxonomic aspects of the internal morphology in Amaryllidaceae from South America and Southern Africa. Kew Bull. 39: 467–498.

Artyushenko ZT. 1989. Aspects of research on Amaryllidaceae Jaume. Herbertia 45: 131–137.

Bastide J and F Viladomat. 2002. Alkaloids of *Narcissus*. In: GR Hanks, ed. Narcissu and Daffodil, pp. 141–214. Taylor & Francis, London.

Beaumont J, DF Cutler, T Raynolds, and JG Vaughan. 1985. The secretory tissue of aloes and their allies. Israel J. Bot. 34: 265–282.

Berg RY. 1978. Development of ovule, embryo sac, and endosperm in *Brodiaea* (Liliales). Nord. J. Bot. 25: 1–7.

Berg RY. 1998. Development of ovule, embryo sac, and endosperm in *Dipterostemon* and *Dichelostemma* (Alliaceae) relative to taxonomy. Am. J. Bot. 83: 790–801.

Berg RY. 2003. Development of ovule, embryo sac, and endosperm in *Triteleia* (Themidaceae) relative to taxonomy. Am. J. Bot. 90: 937–948.

Berg RY and JR Maze. 1966. Contribution to the embryology of *Muilla*, with a remark on the taxonomic position of the genus. Madroño 18: 143–151.

Blunden G and K Jewers. 1973. The comparative leaf anatomy of *Agave, Beschorneria, Doryanthes,* and *Furcraea* species (Agavaceae: Agaveae). Bot. J. Linn. Soc. 66: 157–179.

Bogler DJ and BB Simpson. 1993. Molecular systematics of the Agavaceae: Evidence from sequencing the rDNA internal transcribed spacer region. Amer. J. Bot. 80(6): 133 (abstracts).

Bogler DJ and BB Simpson. 1995. A chloroplast DNA study of the Agavaceae. Syst. Bot. 20: 191–205.

Bogler DJ and BB Simpson. 1996. Phylogeny of Agavaceae based on its *r*DNA sequence variation. Am. J. Bot. 83: 1225–1235.

Bogler DJ, JC Pires and J Francisco-Ortega. 2006. Phylogeny of Agavaceae based on *ndhF*, *rbc*L, and ITS *r*DNA: Implications of molecular data for classification. In: JT Columbus, EA Friar, JM Porter, LM Prince, and MG Simpson, eds. Monocots: comparative biology and evolution (excluding Poales), pp. 313–328. Rancho Santa Ana Botanical Garden, Claremont.

Bouvier W. 1915. Beiträge zur vergleichenden Anatomie der Asphodeloideae (Tribus Asphodeleae und Hemerocallideae). Akad. Wiss. Wien Math.-Naturwiss. Kl. Denkschr. 91: 539–577.

Buchner L. 1948. Vergleichende embryologische Studien an Scilloideae. Oesterr. Bot. Z. 95: 428–451.

Cave MS. 1948. Sporogenesis and embryo sac development of *Hesperocallis* and *Leucocrinum* in relation to their systematic position. Am. J. Bot. 35: 343–349.

Cave MS. 1955. Sporogenesis and the female gametophyte of *Phormium tenax*. Phytomorphology 5: 247–253.

Cave MS. 1964. Cytological observations on some genera of the Agavaceae. Madroño 17: 163–170.

Cave MS. 1970. Chromosomes of California Liliaceae. Univ. Calif. Publ. Bot. 57: 1–48.

Cave MS. 1975. Embryological studies in *Stypandra* (Liliaceae). Phytomorphology 25: 95–99.

Chase MW, PJ Rudall, and JG Conran. 1996. New circumscriptions and a new family of asparagoid lilies: Genera formerly included in Anthericaceae. Kew Bull. 51: 667–680.

Chase MW, PJ Rudall, MF Fay, and KL Stobart. 2000. Xeronemataceae, a new family of asparagoid lilies from New Caledonia and New Zealand. Kew Bull. 55: 865–870.

Chase MW, A de Bruijn, G Reeves, AV Cox, PJ Rudall, MAT Johnson, and LE Equiarte. 2000. Phylogenetics of Asphodelaceae (Asparagales): an analysis of plastid *rbc*L and *trn*L-F DNA sequences. Ann. Bot. (London) 86: 935–956.

Chakroun S and Ch Hebant. 1983. Developmental anatomy of *Aphyllanthes monspeliensis,* a herbaceous monocotyledon with secondary growth. Plant Syst. Evol. 141: 231–241.

Cheadle VI. 1969 (1970). Vessels in Amaryllidaceae and Tecophilaeaceae. Phytomorphology 19: 8–16.

Chen ZK, FH Wang, and F Zhou. 1988a. On the origin, development and ultrastructure of the orbicules and pollenkit in the tapetum of *Anemarrhena asphodeloides* (Liliaceae). Grana 27: 273–282.

Chen ZK, FH Wang, and F Zhou. 1988b. The ultrastructural aspects of tapetum and Ubisch bodies in the *Anemarrhena asphodeloides*. Acta Bot. Sinica 30: 1–15 (in Chinese).

Chen ZK, F Zhou, FX Wang, and FH Wang. 1988c. Investigation on the development of male gametophyte in *Anemarrhena asphodeloides*. Acta Bot. Sinica 30: 569–573 (in Chinese).

Chen ZK, FH Wang, and ZH Li. 1990. Investigation on embryology of *Anemarrhena asphodeloides*. Acta Phytotaxon. Sinica 28: 223–227 (in Chinese).

Chung M-G and SB Jones, Jr. 1989. Pollen morphology of *Hosta* Tratt. (Funkiaceae) and related genera. Bull. Torrey Bot. Club 116: 31–44.

Chupov VS. 1987. Taxonomic position of the genera *Geitonoplesium* and *Simethis*. Bot. Zhurn. 72: 904–908. (in Russian with English summary).

Chupov VS and NG Kutiavina. 1978. The comparative immunoelectrophoretic investigations of seed proteins of Liliaceae.

Bot. Zhurn. 63: 473–493 (in Russian with English summary).

Chupov VS and NG Kutiavina. 1981. Serological studies in the order Liliales: II. Bot. Zhurn. 66: 408–418 (in Russian with English summary).

Clifford HT 1998. Xanthorroeaceae. In: K Kubitzki, ed. The families and genera of vascular plants, vol. 3, pp. 467–470. Springer, Berlin/Heidelberg/New York.

Clifford HT and JG Conran. 1998. Johnsoniaceae. In: K Kubitzki, ed. The families and genera of vascular plants, vol. 3, pp. 336–340. Springer, Berlin/Heidelberg/New York.

Clifford HT, RJF Henderson, and JG Conran 1998. Hemerocallidaceae. In: K Kubitzki, ed. The families and genera of vascular plants, vol. 3, pp. 245–252. Springer, Berlin/Heidelberg/New York.

Conran JG. 1998a. Anthericaceae. In: K Kubitzki, ed. The families and genera of vascular plants, vol. 3, pp. 114–121. Springer, Berlin/Heidelberg/New York.

Conran JG. 1998b. Aphyllanthaceae. In: K Kubitzki, ed. The families and genera of vascular plants, vol. 3, pp. 122–124. Springer, Berlin/Heidelberg/New York.

Conran JG. 1998c. Boryaceae. In: K Kubitzki, ed. The families and genera of vascular plants, vol. 3, pp. 151–154. Springer, Berlin/Heidelberg/New York.

Conran JG. 1998d. Herreriaceae. In: K Kubitzki, ed. The families and genera of vascular plants, vol. 3, pp. 253–255. Springer, Berlin/Heidelberg/New York.

Conran JG. 1998e. Lomandraceae. In: K Kubitzki, ed. The families and genera of vascular plants, vol. 3, 354–365. Springer, Berlin/Heidelberg/New York.

Conran JG and PJ Rudall. 1998. Anemarrhenaceae. In: K Kubitzki, ed. The families and genera of vascular plants, vol. 3, pp. 111–114. Springer, Berlin/Heidelberg/New York.

Conran JG and A Temby. 2000. Embryology and affinities of the Boryaceae (Asparagales). In: KL Wilson and DA Morrison, eds. Monocots: systematics and evolution, pp. 401–406. CSIRO, Collingwood.

Conran JG, MW Chase, and PJ Rudall. 1997. Two new monocotyledon families: Anemarrhenaceae and Behniaceae (Lilianae: Asparagales) Kew Bull. 52: 995–999.

Fahn A. 1954. The anatomical structure of the Xanthorroeaceae Dumort. Bot. J. Linn. Soc. 55: 158–184.

Fahn A. 1961. The anatomical structure of the Xanthor-roeaceae Dumort. and its taxonomic position. In Recent advances in botany, pp. 155–160. Taylor & Francis, Toronto.

Fay MF and MW Chase. 1996. Resurrection of Themidaceae for the *Brodiaea* alliance, and recircumscription of Alliaceae, Amaryllidaceae and Agapanthaceae. Taxon 45: 441–451.

Fay MF, J Hartwell, LR Caddick, A Cox, and MW Chase. 1994. A molecular evaluation of the monophyly of Alliaceae and Amaryllidaceae. Am. J. Bot. 81(6): 154 (abstracts).

Fay MF, PJ Rudall, S Sullivan, KL Stobart, AY de Bruijn, G Reeves, F Qamaruz-Zaman, W-P Hong, J Joseph, WJ Hahn, JG Conran, and MW Chase. 2000. Phylogenetic studies of Asparagales based on four plastid DNA regions. In: KL Wilson and DA Morrison, eds. Monocots: systematics and evolution, pp. 360–371. CSIRO, Collingwood.

Fisher JB and PB Tomlinson. 1971. Morphological studies in *Cordyline* (Agavaceae): I. Introduction and general morphology. J. Arnold Arbor. 52: 459–478.

Flory WS. 1977. Overview of chromosomal evolution in the Amaryllidaceae. Nucleus 20: 70–88.

Friesen N, RM Fritsch, and FR Blattner. 2006. Phylogeny and new infrageneric classification of *Allium* (Alliaceae) based on nuclear ribosomal DNA ITS sequences. Aliso 22: 372–395.

Fritsch RM and M Keusgen. 2006. Occurrence and taxonomic significance of cysteine sulphoxides in the genus *Allium* L. (Alliaceae). Phytochemistry 67: 1127–1135.

Good-Avila SV, V Souza, BS Gaut, and L Eguiarte. 2006. Timing and rate of speciation in *Agave* (Agavaceae). Proc. Natl. Acad. Sci. USA 103: 9124–9129.

Granick EB. 1944. A karyosystematic study of the genus *Agave*. Am. J. Bot. 31: 283–343.

Guaglianone ER and S Arroyo-Leuenberger. 2002. The South American genus *Oziroë* (Hyacinthaceae-Oziroëoideae). Darwiniana 40: 61–76.

Henderson RJF. 1991. Studies in *Dianella* Lam. ex Juss. (Phormiaceae): 2. Austrobaileya. 3(3): 473–480.

Henderson RFJ and HT Clifford. 1984. A recircumscription of the Phormiaceae Agardh. Taxon 33: 423–427.

Hoover RF. 1939. A definition of the genus *Brodiaea*. Bull. Torr. Bot. Club 66: 161–166.

Hoover RF. 1941. A systematic study of *Triteleia*. Am. Midland Naturalist 25: 73–100.

Huang S-M and C Sterling. 1970. Laticifers in bulb scales of *Allium*. Am. J. Bot. 57: 1000–1002.

Huynh KL. 1971. Etude de l'arrangement du pollen dans la tetrade chez les angiospermes sur la base de donnees cytologiques: 3. Le pollen trilete du genre *Dianella* Lam. (Liliaceae). Beitr. Biol. Pfl. 47(2): 277–286.

Ito M, A Kawamoto, Y Kita, T Yukawa, and S Kurita. 1999. Phylogeny of Amaryllidaceae based on *mat*K sequence data. Jpn. J. Plant Res. 112: 207–216.

Jin X-B. 1985. The chromosomes of *Hemerocallis* (Liliaceae). Kew Bull. 41: 379–391.

Keighery GJ. 1984. The Johnsonieae (Liliaceae): biology and classification. Flora 175: 103–108.

Klercker JEF. 1883. Recherches sur la structure anatomique de l'*Aphyllanthes monspeliensis* Lin. Bib. K. Svensk. Vetensk. Akad. Handl. 8(6): 1–23.

Kocyan A and PK Endress. 2001. Floral structure and development and systematic aspects of some 'lower' Asparagales. Plant Syst. Ecol. 229: 187–216.

Komar GA. 1976. The ultrastructure of seed appendages (elaiosomes) in *Scilla sibirica, Scilla mischtschenkoana,* and *Chionodoxa gigantea* (Liliaceae). Bot. Zhurn. 61: 332–341 (in Russian with English summary).

Komar GA. 1985. Alliaceae. In: A Takhtajan, ed. Comparative seed anatomy, vol. 1, pp. 78–82. Nauka, Leningrad (in Russian).

Kong H. 2001. Study on the seeds micro-morphological characteristics of *Hemerocallis* and its taxonomic significance. Acta Bot. Bor.-Occid. Sinica 21(2): 373–376.

Kosenko VN. 1994. Pollen morphology of the families Phormiaceae, Blandfordiaceae, and Doryanthaceae. Bot. Zhurn. 79(7): 1–12 (in Russian with English summary).

Kosenko VN and OY Sventorzhetskaya. 1999. Pollen morphology in the family Asphodelaceae (Asphodeleae, Kniphofieae). Grana. 38: 218–227.

Kubitzki K. 1998. Hostaceae. In: K Kubitzki, ed. The families and genera of vascular plants, vol. 3, pp. 256–260. Springer, Berlin/Heidelberg/New York.

Kubitzki K. 1998. Agapanthaceae. In: K Kubitzki, ed. The families and genera of vascular plants, vol. 3, pp. 58–60. Springer, Berlin/Heidelberg/New York.

Lebatha P, MH Buys, and B Stedje. 2006. *Ledebouria, Resnova* and *Drimiopsis*: a tale of three genera. Taxon 55: 643–652.

Lledó MD, AP Davis, MB Crespo, MW Chase, and MF Fay. 2004. Phylogenetic analysis of *Leucojum* and *Galanthus* (Amaryllidaceae) based on plastid *mat*K and nuclear ribosomal spacer (ITS) DNA sequences and morphology. Plant Syst. Evol. 246: 223–243.

McKelvey SD and K Sax. 1933. Taxonomic and cytological relationships of *Yucca* and *Agave*. J. Arnold Arbor. 14: 76–80.

McPherson MA, MF Fay, MW Chase, and SW Graham. 2004. Parallel loss of a slowly evolving intron from two closely related families in Asparagales. Syst. Bot. 29: 296–307.

Maekawa F and K Kaneko. 1968. Evolution of karyotype in *Hosta* (Liliaceae). J. Jpn. Bot. 43: 132–140 (in Japanese with English summary).

Manning JC, P Goldblatt, and MF Fay. 2004. A revised generic synopsis of Hyacintheaceae in sub-Saharan Africa, based on molecular evidence, including new combinations and the new tribe Pseudoprospereae. Edinb. J. Bot. 60: 533–568.

Marais W and J Reilly. 1978. *Chlorophytum* and its related genera (Liliaceae). Kew Bull. 32: 653–663.

Meerow AW. 1984. Karyotype evolution in the Amaryllidaceae. Herbertia 40: 139–154.

Meerow AW. 1985. The evolutionary significance of pancratoid floral morphology in the Amaryllidaceae. Am. J. Bot. 72(6): 962 (abstract).

Meerow AW. 1995. Towards a phylogeny of Amaryllidaceae. In: PJ Rudall, PJ Cribb, DF Cutler and CJ Humphries, eds. Monocotyledons: systematics and evolution, pp. 169–179. Royal Botanic Gardens, Kew.

Meerow AW and JR Clayton. 2004. Generic relationships among the baccate-fruited Amaryllidaceae (tribe Haemantheae) inferred from plastid and nuclear non-coding DNA sequences date. Plant Syst. Evol. 244: 141–155.

Meerow AW and B Dehgan. 1988. Pollen morphology of the Eucharideae (Amaryllidaceae). Am. J. Bot. 75: 1857–1870.

Meerow AW and DA Snijman. 1998. Amaryllidaceae. In: K. Kubitzki, ed. The families and genera of vascular plants, vol. 3, pp. 83–110. Springer, Berlin/Heidelberg/New York.

Meerow AW and DA Snijman. 2001. Phylogeny of Amaryllidaceae tribe Amaryllideae based on nrDNA sequences and morphology. Am. J. Bot. 88: 2321–2330.

Meerow AW, MF Fay, CL Guy, Q-B Li, FQ Zaman, and MW Chase. 1999. Systematics of Amarillidaceae based on cladistic analysis of plastid *rbc*L and *trn*L-F sequence data. Am. J. Bot. 86: 1325–1345.

Meerow AW, Fay MF, Chase MW, Guy CL, Li QB. 2000. The new phylogeny of the Amaryllidaceae. Herbertia. 54: 180–203.

Meerow AW, CL Guy, Q-B Li, and S-L Yang. 2000. Phylogeny of the American Amaryllidaceae based on *nr*DNA sequences. Syst. Bot. 25: 708–726.

Meerow AW, MF Fay, MW Chase, CL Guy, Q Li, D Snijman, and S-L Yang. 2000. Phylogeny of the Amarillidaceae: molecules and morphology. In: K Wilson and D Wallace, eds. Monocots: systematics and evolution, pp. 368–382. CSIRO, Collingwood.

Meerow AW, CL Guy, Q-B Li, and JR Clayton. 2002. Phylogeny of the tribe Hymenocallideae (Amaryllidaceae) based on morphology and molecular characters. Ann. Missouri Bot. Gard. 89: 400–413.

Meerow AW, J Francisco-Ortega, DN Kuhn, and RJ Schnell. 2006. Phylogenetic relationships and biogeography within the Eurasian clade of Amaryllidaceae based on plastid *ndh*F and nrDNA ITS sequences: lineage sorting in a reticulate area? Syst. Bot. 31: 42–60.

Müller-Doblies U and D Müller-Doblies. 1978. Zum Bauplan von *Ungernia*, der einzigen endemischen Amaryllidaceen-Gattung Zentralasiens. Bot. Jahrb. Syst. 99: 249–263.

Müller-Doblies D and U Müller-Doblies. 1985. De Liliifloris notulae: 2. De taxonomia subtribus Strumariinae (Amaryllidaceae). Bot. Jahrb. Syst. 107: 17–47.

Müller-Doblies D and U Müller-Doblies. 1996. Tribes and subtribes and some species combinations in Amaryllidaceae J. St-Hil. emend R. Dahlgren et al. 1985. Feddes Repert. 107: 1–9.

Müller-Doblies U and D Müller-Doblies. 1997. A partial revision of the tribe Massonieae (Hyacinthaceae). Feddes Repert. 108: 49–96.

Nandi S. 1974a. Chromosome characteristics and their correlation with the phenotypic and ecological variants in *Chlorophytum, Iphiopogon* and *Dianella*. Bull. Bot. Soc. Bengal. 28(1–2): 117–122.

Nandi S. 1974b. Chromosome characteristics and their correlations with the phenotypic and ecological variants in *Chlorophytum*. Bull. Bot. Soc. Bengal 28: 117–122.

Nemirovich-Danchenko EN. 1985. Agavaceae. In: A Takhtajan, ed. Comparative seed anatomy, vol. 1, pp. 88–92. Nauka, Leningrad (in Russian).

Nordal I, TE Eriksen, and M Fosing. 1990. Studies in the generic delimitation of Anthericaceae. Mitt. Staatsinst. Allg. Boy. Hamb. 23b: 535–559.

Oganezova GG. 1982. On the anatomical structure of fruit and seed of some Liliaceae in relation to systematics of the family: 2. Scilloideae. Bot. Zhurn. 67: 729–742 (in Russian with English summary).

Oganezova GG. 1986. Morphological and anatomical characters of seed and fruit in some members of the subfamily Allioideae (Liliaceae) in relation to their systematics and phylogeny. Bot. Zhurn. 71: 300–310 (in Russian with English summary).

Oganezova GG. 1987. Systematic position of some disputable genera for Asphodeloideae (Liliaceae) based on anatomical structure of their fruits and seeds. Bot. Zhurn. 72: 1009–1020 (in Russian with English summary).

Oganezova GG. 1990. Seed and fruit anatomy of some Amaryllidaceae in connection with their systematics and phylogeny. Bot. Zhurn. 75: 615–630 (in Russian with English summary).

Oganezova GG. 2000. Systematic position of the *Trilliaceae, Smilacaceae, Herreriaceae, Tecophilaeaceae, Dioscoreaceae* families and the volume and phylogeny of the *Asparagales* (based on the seed structure).) Bot. Zhurn. 85(9): 9–25 (in Russian with English summary).

Pellmyr O. 2003. Yuccas, yucca moths, and coevolution: a review. Ann. Missouri Bot. Gard. 90: 35–55.

Pellmyr O, JN Thompson, JM Brown, and RG Harrison. 1996. Evolution of pollination and mutualism in the yucca moth lineage. Am. Nat. 148: 827–847.

Pfosser M and F Speta. 1999. Phylogenetics of Hyacinthaceae based on plastid DNA sequences. Ann. Missouri Bot. Gard. 86: 625–875.

Pfosser M, W Wetschnig, S Ungar, and G Prenner. 2003. Phylogenetic relationships among genera of Massonieae (Hyacintheaceae) inferred from plastid DNA and seed morphology. J. Plant Res. 116: 115–132.

Pires JC and KJ Sytsma. 2002. A phylogenetic evaluation of a biosystematic framework: *Brodiaea* and related petaloid monocots (Themidaceae). Am. J. Bot. 89: 1342–1359.

Pires JC, MF Fay, WS Davis, L Hufford, J Rova, MW Chase, and KJ Sytsma. 2001. Molecular and phylogenetic analyses of Themidaceae (Asparagales). Kew Bull. 56: 601–626.

Pires JC, MF Fay, WS Davis, L Hufford, J Rova, MW Chase, and KJ Sytsma. 2001. Molecular and morphological phylogenetic analyses of Themidaceae (Asparagales). Kew Bull. 56: 601–626.

Pires JC, IJ Maureira, JP Rebman, GA Salazar, LI Cabrera, MF Fay, and MW Chase. 2004. Molecular data confirm the phylogenetic placement of the enigmatic *Hesperocallis* (Hesperocallidaceae) with *Agave*. Madroño 51: 307–311.

Rahn K. 1998a. Alliaceae. In: K Kubitzki, ed. The families and genera of vascular plants, vol. 3, pp. 70–78. Springer, Berlin/Heidelberg/New York.

Rahn K. 1998b. Themidaceae. In: K Kubitzki, ed. The families and genera of vascular plants, vol. 3, pp. 436–440. Springer, Berlin/Heidelberg/New York.

Raju MUS. 1957. Some aspects of the embryology of *Dianella nemorosa*. J. Indian Bot. Soc. 36: 223–226.

Reynolds T, ed. 2004. Aloes, the Genus *Aloe*. CRC Press, Boca Raton, FL.

Rudall P. 1994. The ovule and embryo sac in Xanthorraeaceae sensu lato. Flora 189: 335–351.

Rudall PJ. 1999. Flower anatomy and systematics of *Comospermum* (Asparagales). Syst. Geogr. Plants 68: 195–202.

Rudall PJ. 2003. Unique floral structures and iterative evolutionary themes in Asparagales: Insights from a morphological cladistic analysis. Bot. Rev. 68: 488–509.

Rudall PJ, MW Chase, and JG Conran. 1996. New circumscriptions and a new family of asparagoid lilies: genera formerly included in Anthericaceae. Kew Bull. 51: 667–680.

Rudall PJ, MW Chase, and JG Conran. 1997. Boryaceae. Kew Bull. 52: 416.

Rudall PJ, EM Engleman, L Hanson, and MW Chase. 1998. Embryology, cytology and systematics of *Hemiphylacus, Asparagus* and *Anemarrhena* (Asparagales). Plant Syst. Evol. 211: 181–199.

Satô D. 1935. Analysis of the karyotypes in *Yucca, Agave,* and related genera with special reference to the phylogenetic significance. Jpn. J. Genet. 11: 272–278.

Satô D. 1938. Karyotype alteration and phylogeny: IV. Karyotypes in Amarylidaceae with special reference to the SAT-chromosome. Cytologia 9: 203–242.

Schaffer WM and MV Schaffer. 1977. The reproductive biology of the Agavaceae: I. Pollen and nectar production in four Arizona agaves. Southwestern Naturalist 22: 157–167.

Schlimbach H. 1924. Beiträge zur Kenntnis der Samen-aniagen und Samen der Amaryllidaceen mit Berück-sichtigung des Wassergehaltes der Samen. Flora 117: 41–54.

Schlittler J. 1945. Untersuchungen über den Bau der Blütenstände in Bereich des *Anthericum*-typus (Asphodelinae-Anthericinae-Dianellinae). Mitt. Bot. Mus. Univ. Zürich. 174: 200–239.

Schnarf K und R Wunderlich. 1939. Zur vergleichender Embryologie der Liliaceae-Asphodeloideae. Flora 33: 297–327.

Schulze W. 1975. Beiträge zur Taxonomie der Liliifloren. I. Asphodelaceae. Wiss. Z. Friedrich-Schiller Univ. Jena 24, 4: 403–415.

Schulze W. 1982. Beiträge zur Taxonomie der Liliifloren: IX. Anthericaceae. Wiss. Z. Friedrich-Schiller Univ. Jena, Math.-Naturwiss. Reihe, 31: 291–307.

Schmid WG. 1991. The genus *Hosta*. Timber Press, Portland, OR.

Smith GF and EMA Steyn. 2004. Taxonomy of Aloaceae. In: T Reynolds, ed. Aloes, the Genus Aloe, pp. 15–36. CRC Press, Boca Raton, FL.

Smith GF and A-E Van Wyk. 1991. Generic relationships in the Alooideae (Asphodelaceae). Taxon 40: 557–581.

Smith GF and A-E van Wyk. 1998. Asphodelaceae. In: K Kubitzki, ed. The families and genera of vascular plants, vol. 3, pp. 130–140. Springer, Berlin/Heidelberg/New York.

Schnarf K und R Wunderlich. 1939. Zur vergleichenden Embryologie der Liliaceae-Asphodeloideae. Flora 133: 297–327.

Schuize W. 1983. Beiträge zur Taxonomie der Liliifloren: 14. Der Umfang der Amaryllidaceae. Wiss. Z. Friedrich-Schiller-Univ. Jena, Math.-Naturwiss. Reihe, 32: 985–1003.

Shah GL and BV Gopal. 1970. Structure and development of stomata on the vegetative and floral organs of some Amaryllidaceae. Ann. Bot. 34: 737–749.

Singh V. 1972. Floral morphology of the Amaryllidaceae: I. Subfamily Amaryllidoideae. Canad. J. Bot. 50: 1555–1565.

Snijman DA. 2000. Growth periodicity, flowering and phylogeny of the African *Amaryllideae* (Amaryllidaceae). In: JR Timberlake and S Kativu, eds. African plants: biodiversity, taxonomy and uses: proceedings of the 1997 AETFAT congress, Harare, Zimbabwe, pp. 389–404. Royal Botanic Gardens, Kew.

Snijman DA and HP Linder. 1996. Phylogenetic relationships, seed characters, and dispersal system evolution in Amaryllideae (Amaryllidaceae). Ann. Missouri Bot. Gard. 83: 362–386.

Sobotik M and F Speta. 1997. Beitrag zur Wurzelanatomie der Hyacinthaceae. Stapfia 50: 339–357.

Speta F. 1998. Hyacinthaceae. In: K Kubitzki, ed. The families and genera of vascular plants, vol. 3, pp. 261–285. Springer, Berlin/Heidelberg/New York.

Stedje B. 2000. The evolutionary relationships of the genera *Drimia, Thuranthos, Bowiea* and *Schizobasis* discussed in the light of morphology and chloroplast DNA sequence data. In: KL Wilson and DA Morrison, eds. Monocots: systematics and evolution, pp. 414–417. CSIRO, Collingwood.

Stedje B. 2001a. The generic delimitation within Hyacinthaceae, a comment on works by F. Speta. Bothalia 31: 192–195.

Steje B. 2001b. Generic delimitation of Hyacinthaceae, with special emphasis on sub-Saharan genera. Syst. Geogr. Plants 71: 449–454.

Stenar H. 1933. Zur Embryologie der *Agapanthus*-Gruppe. Bot. Not. 1933: 520–530.

Stenar H. 1951. Zur Embryologie von *Haemanthus Katherinae* Bak., Erorterungen über das helobiale Endosperm in den Amaryllidaceae und Liliaceae. Acta Horti Berg. 16: 57–72.

Sterling C and S-M Huang. 1972. Notes on the laticifers of *Allium, Caloscordum, Nothoscordum, Tristagma,* and *Tulbaghia*. Plant Life 28: 43–46.

Stevens PF. 1978. Generic limits in the Xeroteae (Liliaceae sensu lato). J. Arnold Arbor. 59: 129–155.
Stevenson DW and J Grimes. 1997. Herreriaceae do not have cladodes. Am. J. Bot. 84(6): 236 (abstract).
Sýkorová E, J Fajkus, M Mezníková, KY Lim, K Neplechová, FR Blattner, MW Chase, and AR Leitch. 2006. Minisatellite telomeres occur in the family Alliaceae but are lost in *Allium*. Am. J. Bot. 93: 814–823.
Tomita K. 1931. Über die Entwicklung des nackten Embryos von *Crinum latifolium* L. Sci. Rep. Tohoku Imp. Univ., 4th ser. (Biol.), 6: 163–169.
Tomlinson PB. 1965. Notes on the anatomy of *Aphyllanthes* (Liliaceae) and comparison with Eriocaulaceae. Bot. J. Linn. Soc. 59: 163–173.
Tomlinson RB and JB Fisher. 1971. Morphological studies in *Cordyline* (Agavaceae): I. Introduction and general morphology. J. Arnold Arbor. 52(3): 459–478.
Traub HP. 1957. Classification of the Amaryllidaceae: Subfamilies, tribes, and genera. Plant Life 13: 76–83.
Traub HP. 1963. The genera of Amaryllidaceae. La Jolla, Calif. The American Plant Life Soc, pp. 15–43.
Traub HP. 1972. The order Alliales. Plant Life 28: 129–138.
Treutlein J, GF Smith, B-E van Wyk, and M Wink. 2003. Phylogenetic relationships in Asphodelaceae (subfamily Alooideae) inferred from chloroplast DNA sequences (*rbc*L, *mat*K) and from genomic finger-printing (ISSR). Taxon 52: 193–207.
Vaikos NP and RM Pai. 1982. The floral anatomy of *Kniphofia uvaria* Hook. (Liliaceae: Kniphofieae). Proc. Indian Acad. Sci. 91: 351–356.
Vaikos NP and RM Pai. 1986. The floral anatomy of *Bowiea volubilis* Harv. J. Indian Bot. Soc. 65: 516–518.
Vaikos NP, SK Markandeya, and RM Pai. 1978. The floral anatomy of the Liliaceae: The tribe Aloineae. Indian J. Bot. 1: 61–63.
Vaikos NP, SK Markandeya, and RM Pai. 1981. The floral anatomy of the Liliaceae: The tribe Hemerocallideae. J. Indian Bot. Soc. 60: 222–231.
Van Damme EJM, IJ Goldstein, and WJ Peumans. 1991. A comparative study of mannose-binding lectins from the Amaryllidaceae and Alliaceae. Phytochemistry 30: 509–514.
Van Wyk B-E, GF Smith, AM Viljoen, J Treutlein, and M Wink. 2005. Secondary metabolites in relation to molecular phylogenies in *Aloe* and related genera (family Asphodelaceae). In Abstracts of XVII International Botanical Congress, p. 36. Vienna.
Verhoek S. 1998. Agavaceae. In: K Kubitzki, ed. The families and genera of vascular plants, vol. 3, pp. 60–70. Springer, Berlin/Heidelberg/New York.
Weichhardt-Kulessa K, T Börner, J. Schmitz, U Müller-Doblies, and D Müller-Doblies. 2000. Controversial taxonomy of Strumarinae (Amaryllidaceae) investigated by nuclear rDNA (ITS) sequences. Plant Syst. Evol. 223: 1–13.
Wetschnig W, M Pfosser, and G Prenner. 2002. Zur Samenmorphologie der Massonieae Baker 1871 (Hyacinthaceae) im Liche phylogenetisch interpretierter molekularer Befunde. Stapfia 80: 349–379.
Whitaker TW. 1934. Chromosome constitution in certain monocotyledons. J. Arnold Arbor. 15: 135–143.
Wunderlich R. 1950. Die Agavaceen Hutchinson's im Lichte ihrer Embryologie, ihres Gynözeum-, Staubblatt- und Blattbaues. Oesterr. Bot. Z. 97: 437–502.
Xiong Z-T and S-C Chen. 1998. Numerical cytotaxonomic studies of Hemerocallis (Liliaceae) from China. Acta Phytotax. Sinica 36: 206–215.
Xiong Z-T, S Chen, D Hong, and Y Luo. 1998. Pollen morphology and its evolutionary significance in *Hemerocallis* (Liliaceae). Nord. J. Bot. 18: 183–189.
Zomlefer WB. 1998. The genera of Hemerocallidaceae in the southeastern United States. Harvard Pap. Bot. 3: 113–145.

Order 15. ASPARAGALES

Perennial rhizomatous or tuberous (Eriospermaceae and some Asparagaceae) herbs or less often secondary arborescent plants often with extensive secondary thickening growth. Oxalate raphides generally present. Vessels only in roots or also in leaves and/or stem with scalariform and/or simple perforations. Leaves both basal and cauline or only basal, sessile or sometimes petiolate, alternate, occasionally distichous or tristichous, linear to ovate or cordate or reduced and scale-like, usually parallel-veined, often scleromorphic or succulent. Stomata anomocytic or less often paracytic or tetracytic. Flowers in various kinds of inflorescences, sometimes solitary, bisexual or less often unisexual, generally actinomorphic, usually 3-merous. Perianth usually of six segments in two cycles, similar or less often dissimilar, free or more or less connate. Stamens usually six, rarely four or eight, in two cycles; filaments linear, inserted at the base of perianth segments or the perianth tube, free from each other or connate into a column (Ruscaceae) or only basally (*Peliosanthes*); anthers tetrasporangiate, ovate to elongate, basifixed or dorsifixed, introrse or extrorse, opening longitudinally. Tapetum secretory. Microsporogenesis successive. Pollen grains 2-celled or rarely 3-celled, 1-colpate, rarely inaperturate. Gynoecium of three united carpels; style short or long, with 3-lobed, capitate or punctiform stigma; ovary superior or sometimes semi-inferior or inferior, 3-locular to less often 1-locular, rarely (some Convallariaceae) 4-or 2-locular, with 1–2 to many ovules per carpel. Septal nectaries mostly present. Ovules anatropous, hemitropous, campylotropous, or sometimes orthotropous, bitegmic, crassinucellate or tenuinucellate, mostly with a parietal cell. Female gametophyte of *Polygonum*-type or sometimes of *Allium*-type. Endosperm helobial or nuclear. Fruits capsular, baccate, or dry and indehiscent. Seeds of various sizes, shapes, and colours, with or without phytomelan; seed coat formed mainly by the outer integument; embryo small to large, straight to curved; endosperm storing aleurone, starch, and lipids. Usually produce chelidonic acid steroidal saponins. Alkaloids lacking.

Probably derived from some Colchicaceae – Uvularioideae-like ancestor.

Key to Families

1 Leaves well developed.

2 Ovules 2–4(-6) per carpel.

3 Plants not woody. Rhizomatous or tuberous herbs with fibrous or sometimes swollen and fleshy roots. Calcium oxalate raphides present. Vessels mainly in roots, with scalariform or reticulately perforated and scalariform perforations, and rarely in stems with scaraliform perforations. Leaves basal or cauline, alternate, opposite or verticillate, sessile or sometimes petiolate, simple, entire, linear, lanceolate, oblong, or ovate, sometimes cordate, sheathing or not sheathing. Stomata anomocytic. Flowers in terminal or axillary racemes, spikes, panicles, or flowers solitary, axillary, bisexual, actinomorphic, mostly 3-merous. Perianth segments six or four, rarely six or nine, free or more often connate into an urceolate or campanulate tube. Stamens six or rarely 4 or 8, 10, 12, in two cycles, inserted at the base of perianth segments or in the tube; filaments free or rarely (*Ophiopogon*) basally connate into a ring, sometimes nearly absent; anthers basifixed or dorsifixed, introrse. Pollen grains almost always 2-celled, 1-colpate or rarely (as in *Aspidistra, Tupistra*) inaperturate, tectate-columellate. Gynoecium of (2)3(4) carpels; style columnar or filiform, with lobed or capitate stigma, rarely stigma sessile; ovary superior or rarely (*Peliosanthes*) semi-inferior, 3-locular or rarely 4- or 2-locular, with two to many ovules per locule. Septal nectaries often present. Ovules 2–4 anatropous, campylotropous, or nearly orthotropous, generally crassinucellate, usually with parietal cell. Female gametophyte of *Allium*-type or less often of *Polygonum*-type, or *Drusa*-type, or *Scilla*-type. Endosperm nuclear or rarely helobial. Fruits berries, fleshy dehiscent capsule (*Gonioscypha*), drupe (*Tricalustra*) or dry, leathery, 1–3-seeded structures, rupturing in early development and exposing seeds. Seeds globose to ovoid, of various colour; the testal layer is almost completely obliterated during maturation, and a phytomelan crust is never formed; the tegmic layer collapses, but the cellular layer is retained. Endosperm consists of cells with thick pitted rugose walls and store aleurones and lipids; embryo usually more than half as long as the endosperm. Flavonols present (kaempferol and quercetin) or absent; steroidal sapinins in abundance); n = 9–20. . . . . . 1. CONVALLARIACEAE.

3 Woody, generally large, arborescent plants with a stout, simple or sparingly branched trunk up to a few meters tall, with terminal leaf rosettes. Secondary thickening anomalous. Vessels present in roots (with simple perforations) and in leaves (with scalariform perforations). Leaves alternate, linear, tough and hard or leathery, sessile, simple, entire, serrulate, or (*Dasylirion*) margins usually armed with curved spines. Stomata anomocytic, sunken, with oily contents. Inflorescences panicles, often large and many-flowered. Flowers polygamo-dioecious or dioecious, actinomorphic, articulated on their pedicels, bracteate. Perianth segments six, all similar, free, scarious. Stamens six in two whorls, free, inserted at base of perianth segments, vestigial in female flowers; anthers dorsifixed, introrse. Pollen grains 1-colpate, finely reticulate. Ovary 1–3-locular, with two anatropous ovules per carpel, or 3–6 in a single cavity in *Dasylirion*, reduced in male flowers; ovules tenuinucellate. Septal nectaries present. Style short, with 3-lobed stigma. Fruits dry, 3-winged samaras, or 3-lobed flatted capsules, or 3-ribbed nuts. Seeds without phytomelan, with cylindrical straight embryo. Flavonols (quercetin) present n = 18 and more often 19, chromosomes without marked difference in size. . . . . 3. NOLINACEAE.

2 Ovule one per carpel. Usually shrubby or treelike plants (*Dracaena draco* – the "Dragon Blood Tree" up to 15 m or more), sometimes scandent, or sometimes rhizomatous geophytes (*Sansevieria*); secondary thickening anomalous. Stems marked with transverse leaf scars or plants stemless or nearly stemless. Vessels in roots (mainly with simple perforations) and in leaves (with scalariform perforations). Leaves basal or

more often crowded on the tops of branches, sheathing at base, small to very large, linear-oblong to ovate or obovate, strap-shaped or ensiform, sometimes more or less distinctly petiolate, thin and flexible to thick and rigid, sometimes conspicuously succulent. Stomata anomocytic. Flowers numerous, in compound terminal or axillary, paniculate, spicate, or capitate racemes, bisexual, actinomorphic, bracteate and with one bracteole, articulated on the pedicel. Perianth segments six in two cycles, equal, connate into a short or long tube. Stamens six in two cycles, inserted on the tube; anthers dorsifixed, versatile, introrse. Pollen grains 1-colpate. Ovary superior, with one anatropous ovule per locule; septal nectaries present; style one, apical. stigma 3-lobed to capitate. Fruits 1–3-seeded, red or orange berries, but sometimes capsular-indehiscent, hard and woody. Seed coat has a three to four-layered outer integument, the outer epidermis of which is thick walled and lacks phytomelan, while the cells of the inner integument have collapsed and form a reddish brown membrane (Dahlgren et al. 1985); embryo small, endosperm oily, containing aleurone and lipids. Flavonols (kaempferol) present (*Dracaena*); n = 19–21. . . . . 2. DRACAENACEAE.

1 Leaves much reduced.

4 Seeds with phytomelan in testa and collapsed tegmen. Subshrubby or herbaceous perennials with erect or scandent stem growing from sympodial rhizome or rarely a tuber. Phylloclades perennial or deciduous, fasciculate or solitary, flat, and leaflike or more often slender and needlelike to filiform, apex mucronate, rarely (*Asparagus densiflorus*) bearing minute, scaly leaves, in some species (e.g., *Asparagus turkestanicus*) phylloclades lacking. Vessels in roots with scalariform or simple perforations, in stems only with scalariform perforations. Leaves more or less scalelike, scarious, with a spiny or soft basal spur. Stomata anomocytic. Flowers small and inconspicuous, solitary or in umbel-like or racemelike inflorescences, bisexual or unisexual; pedicels articulate, usually bracteolate at the base. Perianth segments six, free and spreading or basally connate and then forming a cup or tube. Stamens six, free from each other, inserted at the base of perianth segments; filaments filiform or flattened, rarely spurred at base; anthers more or less basifixed, sagittate, tetrasporangiate, dehiscing longitudinally, introrse. Microsporogenesis successive. Pollen grains 1-colpate. Gynoecium of three united carpels; ovary superior, sometimes slightly stipitate, 3-locular, style short, with 3-lobed or capitate stigma. Septal nectaries present. Ovules 2–12 per locule, hemitropous or almost orthotropous, crassinucellate. Female gametophyte of *Polygonum*-type. Fruits often red, smooth, or wrinkled berry, rarely nut, or (*Hemiphylacus*) loculicidal capsules. Seeds black; endosperm with distinctly pitted cell walls, contains aleurone, lipids and hemicellulose; embryo almost reach in the length of endosperm, slightly curved. n = 10, 56 (*Hemiphylacus*).. . . . . 5. ASPARAGACEAE.

4 Seeds without phytomelan.

5 Shrubs or subshrubs, often thicket forming, or lianas; secondary thickening absent. Phylloclades lanceolate to ovate, acute or pungent-pointed. In *Ruscus* vessels only in roots and with scalariform and simple perforations in other genera also in stems and with scalariform perforations. Leaves reduced to small, scarious scales; mesophyll sometimes containing mucilage cells and calcium oxalate crystals. Stomata anomocytic. Phylloclades bear on upper or lower surface racemelike inflorescence (*Ruscus*), umbel-like inflorescence on their margins (*Semele*), or short terminal racemes are free from the phylloclades (*Danae*). Flowers small, inconspicuous, bisexual or (*Ruscus*) unisexual (monoecious or dioecious), articulated with their pedicels. Perianth segments six, free or partly connate, in the latter case with a fleshy corona. Stamens six or (*Ruscus*) three; filaments connate into a column; anthers extrorse. Pollen grains 1-colpate or (*Semele*) inaperturate. Stigma sessile or subsessile, simple or more or less lobed. Ovary superior, 3-locular with two ovules per locule or (*Ruscus*) 1-locular with 1–4 ovules, without septal nectaries. Ovules hemitropous or orthotropous, tenuinucellate and have a large funicle. Female gametophyte of *Allium*-type. Fruits 1–4-seeded red or orange-red berries. Seeds relatively large, pale; testa lacks phytomelan and disintegrates during development, but tegmen is well developed and consists of thick-walled cells; endosperm store aleurones, lipids and hemicellulose; embryo is about ½ or 1/3 as

long as endosperm. Anthroquinones detected (*Ruscus*), present (*Danae*) kaempferol and quercetin, steroidal saponins and chelidonic acid;. n = 20; karyotypes of *Danae* and *Ruscus* are bimodal and consist of six large and 14 small chromosomes.......... 4. RUSCACEAE.

5 Herbs with generally globose hypocotyledonary tuber or a proliferation of tubers and stolons, usually with one to several slender sheathed necks. Leaves basally concentrated, solitary or several, basally sheathing, linear or lanceolate to ovate or broadly cordate and then prostrate, if broad basally constricted into pseudopetiole; leaves often with unique complex, pubescent, more or less dissected enations from the upper part of relatively small lamina. Stomata anomocytic. Inflorescence simple scapose raceme, the scape subtended by a basal cataphyll. Flowers ebracteolate. Perianth segments free, two cycles similar or dissimilar. Stamens six, all fertile; filaments basally attached to the perianth segments; anthers dorsifixed, epipeltate, elongate, opening longitudinally. Tapetum secretory. Microsporogenesis successive. Pollen grains 1-colpate. Style filiform, with punctiform stigma; ovary superior, 3-locular, with few to several ovules per locule. Fruits smooth loculicidal capsules. Seeds testal-tegmic, oval to comma-shaped, without phytomelan, covered with long, unicellular, air-filled, white, or later rust-colored hairs; embryo large, narrowly conical to cylindrical, and fills up most of the seed volume; mature seed endosperm is strongly reduced (consumed by developing embryo), but nucellar tissue divides to form a sheath of perispermatic tissue, which encloses the embryo either completely or only for the chalazal 2/3; the peripheral layer of the perisperm stores aleurone and fatty oils. n = 5–7, 10, 12 and 21 (the basic number appears to be x = 7, see Dahlgren and van Wyk 1988)................. 6. ERIOSPERMACEAE.

1. CONVALLARIACEAE

Horaninow 1834 (including Aspidistraceae Endlicher 1841; Ophiopogonaceae Endlicher 1841; Peliosanthaceae R.A. Salisbury 1866; Polygonataceae R.A. Salisbury 1866; Tupistraceae Schnitzlein 1846). 18/170. Eurasia (south to Bali and Lombok) and North and Central America.

1.1 CONVALLARIOIDEAE

Fruits berries. Seeds not berrylike. Ovary superior. n = 9–20. – POLYGONATEAE: *Polygonatum, Heteropolygonatum, Disporopsis, Maianthemum, Smilacina*; CONVALLARIEAE: *Theropogon, Speirantha, Convallaria, Reineckea, Rohdea*; ASPIDISTREAE: *Tricalistra, Aspidistra, Gonioscypha, Tupistra* (including *Campylandra*).

1.2 OPHIOPOGONOIDEAE

Fruits leathery or berrylike capsules soon dehiscent and deciduous after anthesis exposing seeds. Seeds berrylike, with fleshy seed coat. Ovary superior (*Liriope*) or semi-inferior. n = 18. – OPHIOPOGONEAE: *Ophiopogon, Liriope*; PELIOSANTHEAE: *Peliosanthes*.

Have many similarities with the Colchicaceae – Uvularioideae.

2. DRACAENACEAE

Salisbury 1866 (including Sansevieriaceae Nakai 1936). 2/100–160. Mainly tropical and subtropical regions of the Old World with only a few species in Hawaii, Central America and Cuba.

Dracaena, Sansevieria.

Close to the Convallariaceae, which is supported both by serotaxonomic (Chupov and Kutiavina 1978) and cytotaxonomic (Sharma and Chaudhuri 1964) data.

3. NOLINACEAE

Nakai 1943. 4/50. Southwestern United States, Mexico and northern Central America.

Nolina, Calibanus, Dasylirion, Beaucamea.

Closely related to the Convallariaceae and Dracaenaceae (Bogler and Simpson 1995, 1996; Bogler 1999). The rare flavonol 3-0-methyl-8-C-methyl-quercetin, has been found in all genera of the Nolinaceae and also occurs in Xanthorrhoeaceae (Williams et al. 1988).

4. RUSCACEAE

Sprengel 1826. 3/9–11. Macaronesia, western and central Europe, Mediterranean, western Asia.

Danae, Ruscus, Semele.

Closely related to the Convallariaceae.

5. ASPARAGACEAE

A.L. de Jussieu 1789 (including Hemiphylacaceae Doweld 2005). 2/210. Widely distributed in the Old World, mainly in arid and subarid regions.

Asparagus (? including *Myrsiphyllum* and *Protasparagus*), *Hemiphylacus* (5, Mexico).

According to Rudall et al. (1998) *Hemiphylacus* must be included in the Asparagaceae, but from this family it differs in fused perianth segments and chromosome number (n = 56).

6. ERIOSPERMACEAE

Endlicher 1841. 1/102. Africa south of the Sahara; with concentration of species in the southwestern Cape Province.

Eriospermum.

A rather isolated taxon, which affinity is still obscure. It shares some synapomorphies with the Cyanastraceae (Tecophileales), such as nuclear formation of endosperm, chalazosperm, successive microsporogenesis and large female gametophyte, but sharply differs in successive microsporogenesis, pollen ultrastructure, structure of the seed coat, embryo shape. According to Lu (1985), of the genera with which *Eriospermum* has been compared, *Bowiea* of the Hyacinthaceae may be the closest, but the differences are so obvious that a very close relationship is very unlikely. Recent molecular data support close relationships with the Asparagaceae (Chase et al. 1995, 2000; Fay et al. 2000).

Bibliography

Arber A. 1924a. *Danaë, Ruscus,* and *Semele:* a morphological study. Ann. Bot. 38: 229–260.

Arber A. 1924b. *Myrisphyllum* and *Asparagus*. Ann. Bot. 38: 635–659.

Bjornstad I. 1970. Comparative embryology Aspara-goideae-Polygonatae. Nytt. Mag. Bot. 17: 169–207.

Baijnath H and P Perry. 1980. Preliminary observations of leaf surface structures in *Eriospermum* Jacq. Proc. Electr. Microsc. Soc. S. Afr. 10: 39–40.

Baker JG. 1875. Revision of the genera and species of Asparagaceae. J. Linn. Soc. Bot. 14: 508–632.

Baker JG. 1877. Revision of the genera and species Anthericeae and Eriospermeae. Bot. J. Linn. Soc. 15: 253–363.

Björnstad I. 1970. Comparative embryology of Asparagoideae-Polygonateae, Liliaceae. Nytt. Mag. Bot. 17: 160–207.

Bogler D. 1998. Nolinaceae. In: K Kubitzki, ed. The families and genera of vascular plants, vol. 3, pp. 392–397. Springer, Berlin/Heidelberg/New York.

Bogler DJ and BB Simpson. 1995. A chloroplast DNA study of the Agavaceae. Syst. Bot. 20: 191–205.

Bogler DJ and BB Simpson. 1996. Phylogeny of Agavaceae based on ITS rDNA sequence variation. Am. J. Bot. 83: 1225–1235.

Bos JJ. 1998. Dracaenaceae. In: K Kubitzki, ed. The families and genera of vascular plants, vol. 3, pp. 238–240. Springer, Berlin/Heidelberg/New York.

Bugnon F. 1960. Contributions a l'etude de quelques problemes d'anatomie vegetale: V. Vascularisation de l'eperon foliar chez l'*Asparagus sprengeri* Regel. Bull. Soc. Bourg. 20: 65–71.

Conran JG. 1987. A phenetic study of the relationships of the genus *Drymophila* R. Br. within the reticulate-veined Liliiflorae. Aust. J. Bot. 35: 283–300.

Conran JG. 1989. Cladistic analysis of some net-veined Liliiflorae. Plant Syst. Evol. 168: 123–141.

Conran JG and MN Tamura. 1998. Convallariaceae. In: K Kubitzki, ed. The families and genera of vascular plants, vol. 3, pp. 186–198. Springer, Berlin/Heidelberg/New York.

Cooney-Sovetts C and R Sattler. 1986. Phylloclade development in the Asparagaceae: an example of homeosis. Bot. J. Linn. Soc. 94: 327–371.

Cutler DF. 1992. Vegetative anatomy of Ophiopogoneae (Convallariaceae). Bot. J. Linn. Soc. 110: 385–419.

Dai LK and SY Liang. 1991. Epidermal features of leaves and their taxonomic significance in subfamily Ophiopogonoideae (Liliaceae). Acta Phytotax. Sinica 29: 335–346.

Dickson A. 1886. On the occurrence of foliage leaves in *Ruscus* (*Semele*) *androgynus* with some structural and morphological observations. Trans. Proc. Bot. Soc. Edinb. 16: 130–149.

Duthie AV. 1940. Contribution to our knowledge of the genus *Eriospermum*. Ann. Univ. Stellenbosch, Reeks A, Wiss.-Natuurk. 18: 1–64.

Eguiarte LE et al. 1994. The systematic status of the Agavaceae and Nolinaceae and related Asparagales in the monocotyledons: an analysis based on the *rbcL* gene sequence. Bol. Soc. Bot. Mex. 54: 35–56.

Gu Z, Q Yang, and K Kondo. 1990. A karyomorphological study on *Disporopsis* Hance in China. La Kromosomo 57: 1916–1925.

Hernandez L. 1995. Taxonomic study of the Mexican genus *Hemiphylacus* (Hyacinthaceae). Syst. Bot. 20: 546–554.

Hirsch A. 1977. A developmental study of the phylloclades of *Ruscus aculeatus* L. Bot. J. Linn. Soc. 74: 355–365.

Hong D-Y and X-Y Zhu. 1990. Report on karyotypes of 6 species in 4 genera of Poligonateae from China. Acta Phytotax. Sinica 28: 185–198 (in Chinese with English summary).

Hong D-Y, LM Ma, and T Chen. 1987. A discussion on the karyotype and evolution of the tribe Convallarieae s. 1. (Liliaceae). In: DY Hong, ed. Plant chromosome research. Nishiki Print Co., Hiroshima.

Jang CG and M Pfosser. 2002. Phylogenetics of *Ruscaceae* sensu lato based on plastid *rbcL* and *trnL*-F DNA sequences. Stapfia. 80: 333–348.

Joyeux L. 1928. Valeur morphologique du cladode chez les Ruscees. Mem. Acad. Royale Belgique, Bruxelles 9: 1–94.

Judd WS. 2003. The genera of Ruscaceae in the Southeastern United States. Harvard Pap. Bot. 7: 93–149.

Ko SC, YO Kim, and YS Kim. 1985. A cytotaxonomical study of the tribe Ophiopogoneae in Korea. Korean J. Plant Taxon. 15: 111–125.

Kubitzki K and PJ Rudall. 1998. Asparagaceae. In: K Kubitzki, ed. The families and genera of vascular plants, vol. 3, pp. 125–129. Springer, Berlin/Heidelberg/New York.

Lazarte JE and BF Falser. 1979. Morphology, vascular anatomy, and embryology of pistillate and staminate flowers of *Asparagus officinalis*. Am. J. Bot. 66: 753–764.

Lu A-M. 1985. Embryology and probable relationships of *Eriospermum* Jacq. (Eriospermaceae). Nord. J. Bot. 5: 229–240.

Lü H-L, S-A Wu, J Yang, and G-Y Rao. 2000. Systematic study on the tribe Polygonateae (Liliaceae s.l.) with the evidence from leaf abaxial epidermis and seed coat. Acta Phytotax. Sinica 38: 30–42 (in Chinese with English summary).

Ma L-M and D-Y Hong. 1990. Pollen morphology and epidermal characters of leaves in Convallarieae (s.l.). Acta phytotax. Sinica 28: 228–236.

McPherson MA, MF Fay, MW Chase, and SW Graham. 2004. Parallel loss of a slowly evolving intron from two closely related families in Asparagales. Syst. Bot. 29: 296–307.

Nakai T. 1936. Subdivision of Convallariaceae Link. Jpn. J. Bot. 12: 145–150.

Oganezova GH. 2000. Fruit and seed structure of some Asparagaceae *s.l.* in connection with the volume and phylogeny of the family. Bot. Zhurn. 85(8): 14–31. (in Russian with English summary).

Perry PL. 1994. A revision of the genus *Eriospermum* (Eriospermaceae). Contrib. Bolus Herb. 17: 1–320.

Perry PL and PJ Rudall. 1998. Eriospermaceae. In: K Kubitzki, ed. The families and genera of vascular plants, vol. 3, pp. 241–244. Springer, Berlin/Heidelberg/New York.

Pires JC, IJ Maureira, TJ Givnish, KJ Sytsma, O Seberg, G Petersen, JI Davis, DW Stevenson, PJ Rudall, MF Fay, and MW Chase. 2004. Phylogeny, genome size, and chromosome evolution in Asparagaceae. In: JT Columbus, EA Friar, CW Hamilton, JM Porter, LM Prince, and MG Simpson, eds. Monocots: Comparative biology and evolution, vol. 1, pp. 287–304. Rancho Santa Ana Botanical Garden, Claremont.

Rao G-Y and K-Y Pan. 1994. Pollen morphology of the Polygonateae and its systematic significance. Cathaya 6: 75–91.

Rao RP and A Kaur. 1979. Embryology and systematic position of *Ophiopogon intermedius*. Proc. Indian Natl. Sci. Acad. 45: 175–187.

Rudall PJ and DF Cutler. 1995. Asparagales, a reappraisal. In: P Rudall, PJ Cribb, DF Cutler and CJ Humphries, eds. Monocotyledons: systematics and evolution, pp. 157–168. Royal Botanic Gardens, Kew.

Rudall PJ and G Campbell. 1998. Flower and pollen structure of Ruscaceae in relation to Aspidistreae and other Convallariaceae. Flora 194: 201–214.

Rudall PJ, CA Furness, MW Chase, and FF Michael. 1997. Microsporogenesis and pollen sulcus type in Asparagales (Lilianae). Canad. J. Bot. 75: 408–430.

Rudall PJ, EM Engleman, L Hanson, and MW Chase. 1998. Embryology, cytology and systematics of *Hemiphylacus, Asparagus* and *Anemarrhena* (Asparagales). Plant Syst. Evol. 211: 181–199.

Rudall PJ, JG Gonran, and MW Chase. 2000. Systematics of Ruscaceae/Convallariaceae: a combined morphological and molecular investigation. Bot. J. Linn. Soc. 134: 73–92.

Sarkar AK, N Datta, and U Chatterjee. 1981 (1982). Cytology of *Peliosanthes* Andr. (Liliaceae) as an aid to taxonomy. Caryologia 34: 467–472.

Schlitder J. 1953. Blütenartikulation und Phyllokladien der Liliaceae organophylogenetisch betrachtet, part 1 and 2. Feddes Repert. 55: 154–258.

Schlitter J. 1960. Die Asparageenphyllokladien erweisen sich auch ontogenetisch als Blätter. Bot. Jahrb. Syst. 79: 428–446.

Schulze W. 1982. Beiträge zur Taxonomie der Liliifloren X. Asparagaceae. Wiss. Z. Friedrich Schiller Univ. Jena 31: 309–330.

Sharma AK and M Chaudhuri. 1964. Cytological studies as an aid in assessing the status of *Sansevieria, Ophiopogon,* and *Curculigo*. Nucleus 7: 43–58.

Sharma AK and PC Datta. 1960. Chromosome studies in species of *Dracaena* with special reference to their means of speciation. J. Genet. 57: 43–76.

Sharma AK and I Ghosh. 1968. Cytotaxonomy of *Dracaena*. J. Biol. Sci. 11: 45–55.

Shinwari ZK, H Kato, R Terauchi, and S Kawano. 1994. Phylogenetic relationships among genera in the Liliaceae-Asparagoideae-Polygonatae s. 1. inferred from *rbc*L gene sequence data. Plant Syst. Evol. 192: 263–277.

Stützel T, U Reck, and D Müller-Doblies. 1991. Morphologische Studien zur Systematik der Convallariaceae. In: G Wagenitz, ed. 10th Symposium Morphologie, Anatomie und Systematik, p. 74. Göttingen.

Tamura MN. 1993. Biosystematic studies on the genus *Polygonatum* (Liliaceae) III. Morphology of staminal filaments and karyology of eleven Eurasian species. Bot. Jahrb. Syst. 115: 1–26.

Tamura MN. 1995. A karyological review of the orders Asparagales and Liliales (Monocotyledoneae). Feddes Repert. 106: 83–111.

Tamura MN, M Ogisu, and J-M Xu. 1997a. *Heteropolygonatum*, a new genus of the tribe Polygonatae (Convallariaceae). Kew Bull. 52: 949–956.

Tamura MN, AE Schwarzbach, S Kruse, et al. 1997b. Biosystematic studies on the genus *Polygonatum* (Convallariaceae) IV. Molecular phylogenetic analysis based on restriction site mapping of the chloroplast gene *trn*K. Feddes Repert. 108: 159–168.

Utech FH. 1979. Floral vascular anatomy of the Himalayan *Theropogon pallidus* Maxim. (Liliaceae-Convallarieae). Ann. Carnegie Mus. 48(3): 25–41.

Utech FH and S Kawano. 1976. Floral vascular anatomy of *Convallaria majalis* L. and *C.keiskei* Miq. (Liliaceae – Convallariinae). Bot. Mag. (Tokyo) 89: 173–182.

Vaikos NP, SK Markandeya, and RM Pai. 1989. Floral anatomy of the Liliaceae: tribe Convallarieae. Proc. Indian Acad. Sci. Plant Sci. 99: 91–95.

Velenovsky J. 1903. Zur Deutung der Phyllokladien der Asparageen. Beih. Bot. Centralbl. 15: 257–268.

Wu S-A, H-L LU, J Yang, G-Y Rao, R-L You, S Ge, and Y Zhong. 2000. Molecular systematic studies on the tribe Polygonateae (s.l.) in China based on RFLPs data of PCR-amplified chloroplast DNA fragments. Acta Phytotax. Sinica 38: 97–110 (in Chinese with English summary).

Yamashita J and MN Tamura. 2000. Molecular phylogeny of the Convallariaceae (Asparagales). In: KL Wilson and DA Morrison, eds. Monocots: systematics and evolution, pp. 387–400. CSIRO, Collingwood.

Yamashita J and MN Tamura. 2004. Phylogenetic analysis and chromosome evolution in Convallarieae (Ruscaceae sensu lato), with some taxonomic treatments. J. Plant Res. 117: 363–370.

Yang D-Q and K-F Zhu. 1990. Studies on karyotypes of 5 species of *Rohdea* and *Tupistra*. Acta Phytotax. Sinica 28: 199–206 (in Chinese with English summary).

Yang Y, H Li, X Liu, and K Katsuhiko. 1990. Karyotype study on the genus *Ophiopogon* in Yunnan. Acta Bot. Yunn. Suppl. 3: 94–102.

Yeo PF. 1968. A contribution to the taxonomy of the genus *Ruscus*. Notes Roy. Bot. Gard. Edinb. 28: 237–264.

Yeo PF. 1998. Ruscaceae. In: K Kubitzki, ed. The families and genera of vascular plants, vol. 3, pp. 412–416. Springer, Berlin/Heidelberg/New York.

Zimmermann MH and PB Tomlinson. 1969. The vascular system in the axis of *Dracaena fragrans* (Agavaceae): 1. Distribution and development of primary strands. J. Arnold Arbor. 50: 370–383.

Zimmermann MH and PB Tomlinson. 1970. The vascular system in the axis of *Dracaena fragrans* (Agavaceae): 2. Origin and distribution of secondary tissues. J. Arnold Arbor. 51: 478–491.

Superorder PANDANANAE

Order 16. PANDANALES

More or less arborescent plants, sometimes (*Freycinetia*) lianas with clasping, aerial roots, sometimes epiphytic; prop-roots usually at base of stem. Stems simple or sympodially branched, woody, but have no secondary growth or secondary vascular tissue and increase of trunk diameter is the result of primary thickening growth alone. Silica absent, but calcium oxalate occurs both as raphides and other kinds of crystal. Sieve-element plastids of form-P2cf. Vessels in all vegetative organs, in roots with very primitive scalariform perforations. Leaves spirally arranged, usually aggregated at the ends of the branches, tristichous or (*Sararanga*) tetrastichous, but the ranks run in well-marked spirals owing to spiral growth of the stem, usually linear-ensiform, long, usually rigid, coriaceous, sheathing at base, parallel-veined. Stomata more or less distinctly tetracytic. Flowers minute, numerous, much reduced, condensed in terminal or sometime lateral or axillary, usually unisexual inflorescences, subtended by bracts; lower bracts foliaceous, upper bracts variously coloured. Sometimes fragrant and slightly sweet; individual flowers strictly unisexual (most *Pandanus*), or with vestiges of the other sex, occasionally bisexual in some species of *Freycinetia,* without perianth or sometimes (*Sararanga*) with perianth-like cupule. Male flowers have many stamens (9–300); filaments fleshy, free or often connate; anthers basifixed, tetrasporangiate, sometimes apiculate. Tapetum plasmodial or secretory. Microsporogenesis successive. Pollen grains 2-celled, 1-porate, echinate, psilate, reticulate. Female flowers free or many connate into clusters. Gynoecium of one to several more or less united carpels; style short or wanting; stigmas reniform or hippocrepiform, sometimes elongated ovary superior, 1-locular, with one ovule in *Pandanus* and *Sararanga* or several or numerous in *Freycinetia*. Ovules anatropous, bitegmic, crassinucellate, with parietal cell. Female gametophyte of *Polygonum*-type or *Allium*-type. Endosperm nuclear. Fruits baccate or drupes, in heads. Seeds rather small, straight or variously curved; seed coat thin and membranous or (*Sararanga*) thickened; embryo small and straight; endosperm copious, oily or (*Freycinetia*) starchy. n = 25, 28, 30, 60.

Related to the Cyclanthales.

1. PANDANACEAE

R. Brown 1810 (including Freycinetiaceae Brongniart ex Le Maout et Decaisne 1868). 4/800–900. Tropical regions of the Old World, especially in Malesia, Melanesia, and Madagascar, with a few species in temperate regions (China, Japan, New Zealand).

1.1 PANDANOIDEAE
Trees generally supported basally by aerial roots. Flowers in repeatedly branched panicles (*Sararanga*) or in spikes or racemes of spikes (*Pandanus*). Ovary with one ovule. Fruits monodrupes or polydripes depending on the number of carpels (*Pandanus*) or polypyrenoid berries (*Sasraranga*). Endosperm oily. – *Sararanga, Pandanus*.

1.2 FREYCINETIOIDEAE
Woody lianas with clasping roots and linear to ovate or obovate leaves. Flowers in pseudoumbels (rarely racemes) of spikes. Ovary multiovulate. Fruits fleshy or ligneous berries. Endosperm starchy. – *Freycinetia, Martellidendron*.

Bibliography

Callmander MW, P Chassot, Küpfer, and PP Lowry. 2003. Recognition of *Martelliodendron*, a new genus of Pandanaceae, and its biogeographic implications. Taxon 52: 747–762.

Cox PA. 1981. Bisexuality in the Pandanaceae: new findings in the genus *Freycinetia*. Biotropica 13: 195–198.

Cox PA. 1990. Pollination and the evolution of breeding systems in Pandanaceae. Ann. Missouri Bot. Gard. 77: 816–840.

Cox PA, K-L Huynh, and BC Stone. 1995. Evolution and systematics of Pandanaceae. In: PJ Rudall, PJ Cribb, DF Cutler, and CJ Humphries, eds. Monocotyledons: systematics and evolution, vol. 2, pp. 663–684. Royal Botanic Gardens, Kew.

Fagerlind F. 1940. Stempelbau und Embryosackentwicklung bei einigen Pandanazeen. Ann. Jard. Bot. Buitenzorg 49: 55–78 + Taf. 7–13.

Furness CA and PJ Rudall. 2006. Comparative structure and development of pollen and tapetum in Pandanales. Int. J. Plant Sci. 167: 331–348.

Huynh K-L. 1974. La morphologic microscopique et la taxonomie du genre *Pandanus*. Bot. Jahrb. Syst. 94: 190–256.

Huynh K-L. 1991. The flower structure in the genus *Freycinetia*, Pandanaceae (part 1) – Potential bisexuality in the genus *Freycinetia*. Bot. Jahrb. Syst. 112: 295–328.

Huynh K-L. 1992. The flower structure in the genus *Freycinetia*, Pandanaceae (part 2) – Early differentiation of the sex organs, especially of the staminodes, and further notes on the anthers. Bot. Jahrb. Syst. 114: 417–441.

Huynh K-L. 2001. Contribution to the flower structure of *Sararanga* (Pandanaceae). Bot. J. Linn. Soc. 136: 239–245.

Huynh K-L and PA Cox. 1992. Flower structure and potential bisexuality in *Freycinetia reineckei* (Pandanaceae), a species of the Samoa Islands. Bot. J. Linn. Soc. 110: 235–265.

Jarzen DM. 1983. The fossil pollen record of the Pandanaceae. Gard. Bull. 36: 163–175.

Nambudiri EM and WD Tidwell. 1978. On probable affinities of *Viracarpon* Sahni from the Deccan Intertrappean flora of India. Paleontographica 166: 30–43.

North CA and AJ Willis. 1971. Contributions to the anatomy of *Sararanga* (Pandanaceae). Bot. J. Linn. Soc. 64: 411–421.

Pijl L van der. 1956. Remarks on pollination by bats in *Freycinetia, Duabanga,* and *Haplophragma,* and on chiropterophily in general. Acta Bot. Neerl. 5: 135–144.

Poppendieck H-H. 1987. Monoecy and sex changes in *Freycinetia* (Pandanaceae). Ann. Missouri Bot. Gard. 74: 314–320.

Stone BC. 1968a. Morphological studies in Pandanaceae: I. Staminodia and pistillodia of *Pandanus* and their hypothetical significance. Phytomorphology 18: 498–509.

Stone BC. 1968b. Materials for a monograph of *Freyci-netia* Gaud.: IV. Subdivision of the genus, with fifteen new sections. Blumea 16: 361–372.

Stone BC. 1970. Materials for a monograph of *Freycinetia* Gaud. (Pandanaceae): V. Singapore, Malaya, and Thailand. Gardens' Bulletin. Straits Settlements 25: 189–207.

Stone BC. 1972a. Materials for a monograph of *Freyci-netia* Gaud. (Pandanaceae): XV. The Sumatran species. Federat. Mus. J., n.s. 15: 203–207.

Stone BC. 1972b. A reconsideration of the evolutionary status of the family Pandanaceae and its significance in monocotyledon phylogeny. Quart. Rev. Biol. 47: 34–45.

Stone BC. 1990. New evidence for the reconciliation of floral organization in Pandanaceae with normal angiosperm patterns. In: P Baas, K Kalkman, and R Geesink, eds. The plant diversity of Malesia, pp. 33–35. Kluwer, Dordrecht.

Stone BC, K-L Huynh, and H-H Poppendieck. 1998. Pandanaceae. In: K Kubitzki, ed. The families and genera of vascular plants, vol. 3, pp. 397–404. Springer, Berlin/Heidelberg/New York.

Tomlinson PB. 1965. A study of the stomatal structure in Pandanaceae. Pac. Sci. 19: 38–54.

Vaughan RE and PO Wiehe. 1953. The genus *Pandanus* in the Mascarene Islands. Bot. J. Linn. Soc. 55: 1–32.

Zimmermann MH, PB Tomlinson, and J LeClaire. 1974. Vascular construction and development in the stems of certain Pandanaceae. Bot. J. Linn. Soc. 68: 21–41.

Order 17. CYCLANTHALES

Palm-like terrestrial or sometimes epiphytic perennial herbs or less often herbaceous vines or lianas. Stem very short to long and slender, rhizomatous to aerial, usually more or less lignified. Commonly with raphide sacs and sometimes also with other types of calcium oxalate crystals; silica bodies lacking. Watery or milky juice present in all organs. Mucilage canals occur in the vegetative parts of all genera except *Cyclanthus*; in this genus there are laticiferous vessels. Sieve-element plastids of form-P2c. Vessels in roots and leaves, with scalariform perforations. Leaves large, spiral, spirodistichous, or orthodistichous, sheathing at base, mostly petiolate; lamina very deeply bilobed to bifid or more rarely flabelliform-divided or entire, sometimes plicate, with parallel or parallel-pinnate venation and with cross-veins. Stomata tetracytic. Flowers small, in axillary or terminal spadices, sometimes screwlike, subtended by 2–4 (to 11) conspicuous, lanceolate-ovate, variously coloured bracts, the spathes, which enclose the spadix when young; individual flowers much reduced, unisexual, densely crowded, the female ones more or less embedded in axis, male and female flowers in same spadix. Male flowers without perianth or with symmetrical or asymmetrical perianth; perianth segments vary in shape and may have an adaxial secretory glandule. Stamens mostly numerous; filaments basally more or less connate and sometimes adnate to the perianth; in most genera swollen below, possessing a basal bulb, anthers basifixed, tetrasporangiate, usually opening longitudinally, sometimes with an apical secretory glandule. Tapetum secretory. Microsporogenesis successive. Pollen grains 2-celled, usually boat-shaped, tectate with a thick foot-layer and no endexine, 1-colpate or less often 1–2-porate or (*Dianthoveus* and *Evodianthus*) inaperturate, usually foveolate. Female flowers with simple perianth consisting of four free or connate segments, four opposite staminodia and gynoecium of four united carpels alternating with staminodia; stylodia free or basally connate into a short style; often stylodia are very short or wanting and then four stigmas nearly sessile; ovary superior, 1-locular, with numerous parietal, subapical or apical ovules. Ovules anatropous, bitegmic, weakly crassinucellate or almost tenuinucellate, with parietal cell. Female gametophyte of *Polygonum*-type. Endosperm helobial. Fruits many-seeded, baccate, often coalescent into multiple fruit. Seeds with generally straight, linear, small to medium-sized embryo and copious endosperm containing oil, aleurone, often hemicellulose, *Dicranopygium* rich in starch, which is also present in embryo, n = 9, 15, 16.

This order has features, especially in the plicate leaves, that are similar to those of the palms (Dahlgren et al. 1985), but it differs sufficiently in predominantly herbaceous habit, absence of silica, morphology of inflorescence, constant syncarpous gynoecium, absence of septal nectaries, constant successive microsporogenesis, helobial endosperm, starch-rich endosperm and embryo in *Dicranopygium,* as well as in the epicuticular wax without crystalloids. Probably related to the Pandanaceae. According to Harling (1958), *Freycinetia* is probably the nearest to the Cyclanthales. However, the order differs sufficiently from the Pandanales in secretory tapetum, helobial endosperm, more differentiated and plicate leaves, the non spiral growth of the stem.

1. CYCLANTHACEAE

Poiteau ex A. Richard 1824. 12/235. Tropical America, West Indies.

1.1 CARLUDOVICOIDEAE
Inflorescence with flowers in spirally arranged groups. Placentas four or one. Latex absent. Leaves apically bifid, fanlike, or entire, plicate. Fruits baccate, syncarpous or not. n = 9, 15, 16. – *Carludovica, Dianthoveus, Evodianthus, Aplundia, Thoracocarpus, Schultesiophytum, Dicranopygium, Ludovia, Sphaeradenia, Stelestylis, Chorigyne*.

1.2 CYCLANTHOIDEAE
Male and female flowers in separate, alternating cycles or flat spirals. Placentas numerous. Latex present. Leaves deeply cleft with a forked main rib, non plicate. Fruits dry, syncarpous. n = 9. – *Cyclanthus*.

Bibliography

Dahlgren RMT. 1982. Cyclanthaceae. Monocot Newslett. 2: 7–32.

Eriksson R. 1989. *Chorigyne*, a new genus of the Cyclanthaceae from Central America. Nord. J. Bot. 9: 31–45.

Eriksson R. 1993. Systematics of the Cyclanthaceae, especially *Sphaeradenia* and *Chorigyne*. Goteborg.

Eriksson R. 1994. Phylogeny of the Cyclanthaceae. Plant Syst. Evol. 190: 31–47.

Furness CA and PJ Rudall. 2006. Comparative structure and development of pollen and tapetum in Pandanales. Int. J. Plant Sci. 167: 331–348.

French CH, K Klancy, and PB Tomlinson. 1983. Vascular patterns in stems of the Cyclanthaceae. Am. J. Bot. 70: 1386–1400.

Gottsberger G. 1991. Pollination of some species if the Carludovicoideae, and remarks on the origin and evolution of the Cyclanthaceae. Bot. Jahrb. Syst. 113: 221–235.

Hammel BE and GJ Wilder. 1989. *Dianthoveus*: A new genus of Cyclanthaceae. Ann. Missouri Bot. Gard. 76: 112–123.

Harling G. 1946. Studien über den Blütenbau und die Embryologie der Familie Cyclanthaceae. Svensk Bot. Tidskr. 40: 257–272.

Harling G. 1958. Monograph of the Cyclanthaceae. Acta Horti Berg. 18: 1–428.

Harling G, GL Wilder, and R Eriksson. 1998. Cyclanthaceae. In: K Kubitzki, ed. The families and genera of vascular plants, vol. 3, pp. 202–215. Springer, Berlin/Heidelberg/ New York.

Surange KR. 1949. A contribution to the morphology and anatomy of the Cyclanthaceae. Trans. Natl. Inst. Sci. India, Calcutta 3(4): 159–209.

Tomlinson PB and GJ Wilder. 1984. Systematic anatomy of Cyclanthaceae (Monocotyledoneae) – an overview. Bot. Gaz. 145: 535–549.

Wilder GJ. 1976. Structure and development of leaves of *Carludovica palmata* (Cyclanthaceae) with reference to other Cyclanthaceae and Palmae. Am. J. Bot. 63: 1237–1256.

Wilder GJ. 1981a. Structure and development of *Cyclanthus bipartitus* Poit. (Cyclanthaceae) with reference to other Cyclanthaceae. II. Adult leaf. Bot. Gaz. 142: 96–114.

Wilder GJ. 1981b. Morphology of adult leaves in the Cyclanthaceae (Monocotyledoneae). Bot. Gaz. 142: 564–588.

Wilder GJ. 1984. Anatomy of noncostal portions of lamina in the Cyclanthaceae (Monocotyledoneae), part 5. Bot. Mus. Leafl. 30: 103–133.
Wilder GJ. 1985. Anatomy of noncostal portions of lamina in the Cyclanthaceae (Monocotyledoneae), parts 1–4. Bot. Gaz. 146: 82–105, 213–231, 375–394, 545–563.
Wilder GJ. 1986. Anatomy of first-order roots in the Cyclanthaceae (Monocotyledoneae), parts 1 and 2. Canad. J. Bot. 64: 2622–2644, 2848–2864.
Wilder GJ. 1987. Contributions to taxonomy and morphology of *Schultesiophytum chorianthum* Harl. and *Dicranopygium mirabile* Harl. (Cyclanthaceae). Opera Bot. 92: 277–291.
Wilder GJ. 1988. Inflorescence position as a taxonomic character in the Cyclanthaceae. Bot. Gaz. 149: 110–115.
Wilder GJ. 1989. Morphology of *Dianthoveus cremnophilus* (Cyclanthaceae). Canad. J. Bot. 67: 2450–2464.
Wilder GJ. 1992a. Comparative morphology and anatomy of absorbing roots and anchoring roots in three species of Cyclanthaceae (Monocotyledoneae). Canad. J. Bot. 70: 38–48.
Wilder GJ. 1992b. Orthodistichous and dorsiventral symmetry on adult shoots of *Cyclanthus bipartitus* (Cyclanthaceae, Monocotyledoneae). Canad. J. Bot. 70: 1388–1400.
Wilder GJ and DH Harris. 1981. Laticifers in *Cyclanthus bipartitus* Poit. (Cyclanthaceae). Bot. Gaz. 143: 84–93.

Order 18. TRIURIDALES

Usually perennial, commonly small, achlorophyllous, mycotrophic herbs with slender system of underground rhizomes often covered with scales. Raphides and silica bodies lacking. Roots filiform, with a cortex consisting of 1–3 layers of parenchyma cells containing mycorrhizal hyphae. Stems usually slender, erect. Vascular system weakly developed, with the vascular bundles in a single ring. Vessels lacking. Sieve-element plastids of P2c-type (Behnke 2002). Leaves reduced to alternate scales, have a single vascular strand. Stomata present or absent. Flowers very small, in terminal racemes (rarely reduced to a solitary flower), bracteate, bisexual (spp. of *Sciaphila*) or more often unisexual, monoecious or dioecious), actinomorphic. Perianth of 3–6 (-10) segments in a single or two cycles; segments mostly more or less basally connate, often reflexed and may form a starlike configuration, often conspicuously caudate apically, persistent. Stamens 2–6 (-10), often some staminodial, filamented, sessile or immersed in the conical androphore; anthers tetrasporangiate or less often 2–3-sporangiate, more or less extrorse, dehiscing longitudinally or more often transversely, typically extrorse, sometimes connective extends into a long, slender, terminal appendage. Tapetum plasmodial or secretory, with intermediate forms occurring within the Sciaphileae (Rubsamen-Weustenfeld 1991; Furness et Rudall 2006). Microsporogenesis successive. Pollen grains 3-celled, globose, inaperturate, with very thin exine which shows more or less spiny-gemmate, granulate or verrucate (Furness and Rudall 2006). Gynoecium of ten to many free or basally connate carpels, each with one basal ovule and nearly terminal or more often with more or less lateral or gynobasic stylodium. Ovules anatropous, bitegmic, tenuinucellate, without parietal cell, nucellar cap absent. Female gametophyte of *Polygonum*-type, but in *Triuris* very probably of *Fritillaria*-type (Rubsamen-Weustenfeld 1991). Endosperm nuclear. Fruits of small, thick-walled, achenelike follicles or rarely of achenes. Seeds small, straight or slightly curved, with small, undifferentiated embryo consisting of a globose, embryo proper and a more or less short suspensor; endosperm copious, containing hemicellulose, protein, oil, and exclusively in the Triurideae starch that disappears continuously during seed maturation; seed coat formed by the outer integument only, especially by its inner layer; middle cuticle very strongly thickened, directly adjacent to the endosperm tissue (Rubsamen-Weustenfeld 1991). n = 9, 11, 12 (mostly), 13, 14, 15, 16.

According to Chase et al. (2000), the Triuridaceae related to the Pandanaceae and Cyclanthaceae.

1. TRIURIDACEAE

Gardner 1843 (including Lacandoniaceae E. Marttnes et C.H. Ramos 1989). 10–11/45–50. Tropical regions of America, Africa, and Asia. The largest genus *Sciaphila* (30) is concentrated in the Old World.

SCIAPHILEAE: *Sciaphila, Seychellaria, Andruris, Hyalisma, Soridium;* TRIURIDEAE: *Triuris* (including *Lacandonia*), *Triuridopsis, Peltophyllum.* KUPEAEAE: *Kupea, Kihansia.*

In a variation of *Triuris*, described as a new genus *Lacandonia*, with central stamens and peripheral carpels, is probably no more than a very interesting case of homeotic mutation (Takhtajan 1997).

Bibliography

Ambrose BA, S Espinosa-Matís, S Vázquez-Santana, J Márquez-Guzmán, and ER Alvarez-Buylla. 2006. Comparative developmental series of the Mexican triurids support a euanthial interpretation for the unusual reproductive axes of *Lacandonia schismatica* (Triuridaceae). Am. J. Bot. 93: 15–35.

Cheek M. 2003. Kupeaeae, a new tribe of Triuridaceae from Africa. Kew Bull. 58: 939–949.

Cheek M, Williams SA, and Etuge M. 2003. *Kupea martinetugei*, a new genus and species of Triuridaceae from western Cameroon. Kew Bull. 58: 225–228.

Davidse G and SE Martinez. 1990. The chromosome number of *Lacandonia schismatica* (Lacandoniaceae). Syst. Bot. 15: 635–637.

Green PS and O Solbrig. 1966. *Sciaphila dolichostyla* (Triuridaceae). J. Arnold Arbor. 47: 266–269.

Furness CA and PJ Rudall. 2006. Comparative structure and development of pollen and tapetum in Pandanales. Int. J. Plant Sci. 167: 331–348.

Furness CA, PJ Rudall, and A Eastman. 2002. Contribution of pollen and tapetal characters to the systematics of Triuridaceae. Plant Syst. Evol. 235: 209–218.

Gandolfo MA, KC Nixon, and WL Crepet. 2002. Triuridaceae fossil flowers from the Upper Cretaceous of New Jersey. Am. J. Bot. 89: 1940–1957.

Igersheim A, M Buzgo, and PK Endress. 2001. Gynoecium diversity and systematics in basal monocots. Bot. J. Linn. Soc. 136: 1–65.

Imhof S. 1998. Subterranean structures and mycotrophy of the achlorophyllous *Triuris hyaline* Miers (Triuridaceae). Canad. J. Bot. 76: 2011–2019.

Imhof S. 2003. A dorsiventral mycorrhizal root in the achlorophyllous *Sciaphila polygyna* (Triuridaceae). Mycorrhiza 13: 327–332.

Imhof S. 2004. Morphology and development of the subterranean organs of the achlorophyllous *Sciaphila polygyna* (Triuridaceae). Bot. J. Linn. Soc. 146: 295–301.

Maas PJM and T Rubsamen. 1986. Triuridaceae. Flora Neotropica 40: 1–55. New York.

Maas van de Kramer H. 1995. Triudiflorae – Gardener's delight? In: PJ Rudall, PJ Cribb, DF Cutler, and CJ Humphries, eds. Monocotyledons: systematics and evolution, pp. 287–301. Royal Botanic Gardens, Kew.

Maas van de Kamer H and PJM Maas. 1994. *Triuridopsis*, a new monotypic genus in Triuridaceae. Plant Syst. Evol. 192: 257–262.

Maas van der Kamer H and T Rübsamen-Weustenfeld. 1998. Triuridaceae. In: K Kubitzki, ed. The families and genera of vascular plants, vol. 3, pp. 452–458. Springer, Berlin/Heidelberg/New York.

Marquez-Guzman J, S Vazquez-Santana, ME Engleman, A Martinez-Mena, and E Martinez. 1993. Pollen development and fertilization in *Lacandonia schismatica* (Lacandoniaceae) [Triuridaceae]. Ann. Missouri Bot. Gard. 80: 891–897.

Martinez E and CH Ramos. 1989. Lacandoniaceae (Triuridales): Una nueva familia de Mexico. Ann. Missouri Bot. Gard. 76: 128–135.

Meerendonk JPM van der. 1984. Triuridaceae. In: CGGJ van Steenis, ed. Flora Malesiana, ser. 1, 10: 109–121. Dodrecht.

Rübsamen-Weustenfeld T. 1991. Morphologische, embryologische, und systematische Untersuchungen an Triuridaceae. Bibl. Bot. 149: 1–113.

Rudall P. 2003. Monocot pseudanthis revisited: floral structure of the mycoheterotrophic family Triuridaceae. Int. J. Plant Sci. 164: S307–S320.

Sahashi N, T Ohmoto, K Uehara, M Ikuse, and C Chuma. 1991. Pollen morphology of *Andruris japonica* (Triuridaceae). Grana 30: 597–600.

Tomlinson PB. 1982. *Helobieae* (Alismatidae). In: CR Metcalfe, ed. Anatomy of the monocotyledons, 7: 1–522. Clarendon, Oxford.

Vergara-Silva F, S Espinosa-Matías, BA Ambrose, S Vázquez-Santana, A Martínez-Mena, J Márquez-Guzmán, E Martínez, EM Meyerowitz, and ER Alvarez-Buylla. 2003. Inside-out flowers characteristic of *Lacandonia schismatica* evolved at least before its divergence from a closely related taxon, *Triuris brevistyla*. Int. J. Plant Sci. 164: 345–357.

Order 19. VELLOZIALES

Usually arborescent or shrubby perennials of various sizes (sometimes over 6 m tall in South America) with simple or sparsely dichasially branched woody stems. Tanniniferous cells sometimes present. Commonly secreting resin or gum. Sieve-element plastids of Pf, -Pcf, -Pcfs, -Pfs, -P2c-types (Behnke et al. 2000). Vessels in roots with simple perforations and in leaves with scalariform perforations. Leaves xeromorphic, tristichous or spirotristichuosly arranged, crowded at the end of stem or its branches when new, linear, often pungent-pointed, parallel-veined, entire or often dentate-spinulose along the margins, sheathing at the base. Stomata paracytic or sometimes tetracytic. Adventitious roots alternating with the leaves. Flowers solitary, bisexual or (*Barbaceniopsis*) functionally unisexual, actinomorphic, 3-merous, in terminal, 1-many-flowered inflorescences. Perianth segments six, in two cycles, usually similar, petaloid, spreading, free or usually more or less connate into a short or long tube; perianth tube often with six appendages that are free or united into corona. Stamens six in two cycles or often, in most species of *Vellozia,* numerous (up to 76) and in six bundles; anthers tetrasporangiate or disporangiate, exceptionally 10-sporangiate (*Xerophyta schnizleiniana*), usually linear and long, basifixed to dorsifixed, opening longitudinally, latrorse to introrse, rarely extrorse. Tapetum mostly secretory. Microsporogenesis successive. Pollen grains 2- or 3-celled, solitary or (species of *Vellozia*) in tetrads, tectate-columellate, 1-colpate or rarely (*Vellozia*) inaperturate, finely verrucate to reticulate (Furness and Rudal 2006). Gynoecium of three united carpels; style slender, generally widening apically into a

3-lobed, capitate or clavate stigma; ovary variously pubescent on the outside, inferior or rarely semi-inferior (almost superior in *Vellozia burle-marxii*), 3-locular, with numerous ovules on stalked intrusive placentas. Ovules anatropous, bitegmic, weakly crassinucellate to tenuinucellate, with funicular obturator, without parietal cell. Female gametophyte of *Polygonum*-type. Endosperm helobial or (*Acanthochlamys*) nuclear (Li and Gao 1993). Fruits capsular, loculicidal or opening by pores, often flat or concave on the top, crowned with the scar of the perianth, or 6-toothed, sometimes spiny, with numerous seeds. Seeds rather small, black; seed coat testal-tegmic, formed by both integuments, without phytomelan but with phlobaphene in exotesta; endosperm copious, in its outer layers containing aleurone and lipids and in the inner layers plenty of globose starch grains; embryo small, cylindrical or narrowly ovoid, n = 7, 8, 17, 19, 24.

Possibly related to the Bromeliaceae (Huber 1969, 1977; Dahlgren et al. 1985; Takhtajan 1987), with which Velloziaceae share such common characters, as the para- and tetracytic stomata, helobial endosperm formation, testal-tegmic seed coat, and starchy endosperm. Affinity with the Bromeliaceae is also supported by the presence of the same kind of epicuticular waxes (Frölich and Barthlott 1988), but the lack of the UV fluorescence of the cell walls is at variance with this allocation (Kubitzki 1998). As regards the placement of the family together with the Cyclanthaceae, Pandanaceae, Stemonaceae and *Acanthochlamys* in the *rbc*L trees (Chase et al. 1995), I agree with Kubitzki (1998) that it "does not seem to be accompanied by morphological characters: at best it indicates the strong isolation of the family".

Key to Families

1 Anthers tetrasporangiate. Perennial herbs and sparsely branched shrubs. Stems are covered with persistent imbricate sheaths of fallen leaves on the upper part and on the persistent adventitious roots, piercing the withered leaf bases on the lower part. Raphides sometimes present. Leaves in a basal rosette or apical tufts, sometimes up to 1 m long, simple, linear-acuminate, with midrib, sometimes with a saw-toothed margin. Flowers solitary to 12–15 on each scape, pedicellate, terminal, bisexual, rarely functionally unisexual and the plants dioecious (*Barbaceniopsis*), actinomorphic, often large. Perianth segments six in two whorls, free or basally connate, sometimes (*Barbacenia*) with a small corona behind the anthers. Stamens 6–48, up to 76 (in *Vellozia*) in six bundles when more than six, lacerated staminal appendages often present in *Vellozia*; filaments epigynous (*Talbotia*) or adnate to the perianth, sometimes flattened; anthers long; pollen grains 1-sulcate or nonaperturate. Septal nectaries conspicuous. Ovary inferior or rarely semi-inferior, stigma large, erect or spreading. Fruits loculicidal capsules, or dehiscent by apical, basal, or longitudinal slits. Seed small, numerous; exotesta thickened or not (tegmen tanniniferous – *Pleurostima*); embryo small, endosperm copious, hard, with hemicellulose, protein, oil and starch. Biflavonoids present. n = 7, 8, 17, 24 (x = 12?). . . . . 1. VELLOZIACEAE.

1 Anthers disporangiate. Perennial herbs with short rhizome and thin, long fibriform roots. Stems without mantle of adventitious roots. Raphides and tannin cells absent. Sieve-element plastids form-P2c. Vessels with simple perforations and helical thickenings present in the root, scape and leaves. Leaves acerose, dorsiventral, ventrally subsemiorbicular and 2-canaliculate, dorsally flattened and 1-canaliculate, sheathed at the base. Midrib is composed of two "back-to-back" vascular bundles. Stomata paracytic. Inflorescence a compound capitulum on a scape arising from the rhizome, at the base usually surrounded by three leaflike aristate bracts, the peduncle bearing 5–8 few-flowered capityla, the flowers subtended by aristate bractlets. Flowers bisexual, actinomorphic, shortly pedicellate, perianth segments six in two cycles, connate into tube. Stamens six in two cycles, borne upon the corolla lobes; filaments short; anthers oblong, dorsifixed, dehiscent by longitudinal slits. Pollen grains 2- or 3-nucleate, spheroidal, 1-colpate, finely verrucate to reticulate. Septal nectaries absent. Fruits trigonous capsules. Seeds oblong, brown, tegmic, but the mechanical strength is provided mainly be the thickened periclinal walls of the cell layers immediately beneath the seed coat; this layer contains aleurone and free of starch; the rest of the endosperm contains starch in the form of compound starch grains; embryo large (Kao 1989; Kao and Kubitzki 1998). Steroid saponins may be present. n = 19. . . . . 2. ACANTHOCHLAMYDACEAE.

1. VELLOZIACEAE

Hooker 1827 (including Barbaceniaceae Arnott 1842). 8/250. America from Panama to Argentina (especially in southeastern Brazil), Madagascar, tropical and southern Africa, southwestern Arabia.

Talbotia, Xerophyta, Barbaceniopsis, Barbacenia (including *Aylthonia*), *Pleurostima, Burlemarxia, Nanuza, Vellozia.*

2. ACANTHOCHLAMYDACEAE

P.C. Kao 1989. 1/1. Subalpine xerophytic valley zone of southeastern Tibet, southwestern China and western Sichuan.

Acanthochlamys

According to Kao and Kubitzki (1998), a possible relationship of the Acanthochlamydaceae should be considered with all those families that share with them the possession of epigyny and starch accumulation in the endosperm, viz. Bromeliaceae, Haemodoraceae and Velloziaceae. They find it significant that these three families agree with *Acanthochlamys* in having starch grains composed of relatively few elements in the endosperm. Among these families, the Bromeliaceae differ too much from *Acanthochlamys* in their perianth differentiation into calyx and corolla and their epidermal cells containing silica bodies. The remaining two families agree with *Acanthochlamys* in the collapsed testa and 2-layered tegmen, the Velloziaceae also in the disporangiate anthers, and additionally the Haemodoraceae in the thickened outer periclinal wall of the aleurone layer. However, the absence of cell wall-bound ferulate is shared only with the Velloziaceae. Therefore, Kao and Kubitzki favour the affinity with the Velloziaceae, although in their opinion the relationship "may not be very close". According to Behnke et al., their "combined results, i.e. small form-P2c plastids without loosely packed crystals, clear distance in Neighbor Joining and strong bootstrap support, favour a monotypic family whose closest affinities are with Velloziaceae, and with no close affinities to Amaryllidaceae" (2000: 117).

Bibliography

Ayensu ES. 1968. The anatomy of *Barbaceniopsis:* a new genus recently described in the Velloziaceae. Am. J. Bot. 55: 399–405.

Ayensu ES. 1969. Leaf-anatomy and systematics of Old World Velloziaceae. Kew Bull. 23: 315–335.

Ayensu ED. 1972. Studies on pollen morphology in the Velloziaceae. Proc. Biol. Soc. Washington 85(40): 469–480.

Ayensu ES. 1973a. Biological and morphological aspects of Velloziaceae. Biotropica 5: 135–149.

Ayensu ES. 1973b. Phytogeography and evolution of the Velloziaceae. In: BJ Meggers, ES Ayensu, and WD Duckworth, eds. Tropical forest ecosystems in Africa and South America: a comparative review, pp. 105–119. Smithsonian Institution Press, Washington, DC.

Ayensu ES. 1974. Leaf anatomy and systematics of New World Velloziaceae. Smithsonian Contr. Bot. 15: 1–125.

Ayensu ES and JJ Skvarla. 1974. Fine structure of Velloziaceae pollen. Bull. Torrey Bot. Club 101: 250–266.

Baochun G. 1987. The sociological characteristics and pollen morphology of *Acanthochlamys*. Acta Bot. Yunn. 9: 401–405 (in Chinese with English summary).

Behnke H-D, J Treutlein, M Wink, K Kramer, C Schneider, and PC Kao. 2000. Systematics and evolution of Velloziaceae, with special reference to sieve-element plastids and *rbc*L sequence data. Bot. J. Linn. Soc. 134: 93–129.

Chen SC. 1981. Acanthochlamydoideae – a new subfamily of Amaryllidaceae. Acta Phytotaxon. Sinica 19: 323–329 (in Chinese with English summary).

Coetzee H. 1974. Anatomy of the leaves of the Velloziaceae in South Africa and South West Africa and a key based on leaf anatomy. Dinteria 10: 19–33.

Coetzee H, HP Schijff van der, and E Steyn. 1973. External morphology of the species of the South African Velloziaceae including a key based on external morphological characteristics. Dinteria 9: 3–21.

Dutt BSM. 1970. Velloziaceae. In Symposium on comparative embryology of angiosperms. Bull. Indian Nad. Sci. Acad. 41: 373–374.

Furness CA and PJ Rudall. 2006. Comparative structure and development of pollen and tapetum in Pandanales. Int. J. Plant Sci. 167: 331–348.

Gaff DE. 1971. Desiccation-tolerant flowering plants of southern Africa. Science 174: 1033–1034.

Gao B-C and P Li. 1993. Studies on the morphology and embryology of *Acanthochlamys bracteata* I. Morphological and anatomical studies on vegetative organs. J. Sichuan Ubiv. (Science ed.). 30: 534–537 (in Chinese with English summary).

Gao B and P Li. 1995. Studies on the morphology and embryology of *Acanthochlamys bracteata*: morphological and anatomic studies on vegetative organs. J. Sichuan Univ. Nat. Sci. Ed. 32 (special issue Feb.): 37–44 (in Chinese).

Greves S. 1921. A revision of the Old World species of *Vellozia*. J. Bot. 59: 273–284.

Goldblatt P and ME Poston. 1988. Observations on the chromosome cytology of Velloziaceae. Ann. Missouri Bot. Gard. 75: 192–195.

Ibisch PL, C Nowicki, R Vásquez, and K Koch. 2001. Taxonomy and biology of Andean Velloziaceae: *Vellozia andina* sp.nov. and notes on *Barbaceniopsis* (including *Barbaceniopsis castillonii* comb. nov.). Syst. Bot. 26: 5–16.

Kao PC. 1989. Acanthochlamydaceae – a new monocotyledon family. In: PC Kao and Z-M Tan, eds. Flora Sichuanica, 9: 483–507.

Kao PC and P Li. 1995. Studies on the morphology and embryology of *Acanthochlamys bracteata:* morphological and anatomic studies on vegetative organs. J. Sichuan Univ. 32: 37–44.

Kao PC and K Kubitzki 1998. Acanthochlamydaceae. In: K Kubitzki, ed. The families and genera of vascular plants, vol.3, pp. 55–58. Springer, Berlin/Heidelberg/New York.

Kubitzki K. 1998. Velloziaceae. In: K Kubitzki, ed. The families and genera of vascular plants, vol. 3, pp. 459–467. Springer, Berlin/Heidelberg/New York.

Li P and B-C Gao. 1993. Studies on morphology of *Acanthochlamys bracteata.* III. The investigation on double fertilization, embryogenesis and endosperm development of *Acanthochlamys bracteata.* J. Sichuan Univ. (Science ed.). 30: 260–263 (in Chinese with English summary).

Li P, P-C Gao, F Chen, and HX Luo. 1992. Studies on morphology and embryology of *Acanthochlamys bracteata.* II. The anther and ovule development. Bull. Bot. Res. 12: 389–395 (in Chinese with English summary).

Melo NF, M Guerra, AM Benko-Iseppon, and ML Menezes. 1997. Cytogenetics and cytotaxonomy of Velloziaceae. Plant Syst. Evol. 204: 257–273.

Mello-Silva R de. 1991. The infra-familial taxonomic circumscription of the Velloziaceae: a historical and critical analysis. Taxon 40: 45–51.

Mello-Silva R de. 2000. Partial cladistic analysis of *Vellozia* and characters for the phylogeny of Velloziaceae. In: KL Wilson and DA Morrison, eds. Monocots: systematic and evolution, pp. 505–522. CSIRO, Collingwood.

Mello-Silva R de. 2005. Morphological analysis, phylogenies and classification in Velloziaceae. Bot. J. Linn. Soc. 148: 157–173.

Menezes NL de. 1970. Aspectos anatomicos e a taxonomia da familia Velloziaceae. Ph.D. dissertation, University of Sao Paulo.

Menezes NL de. 1973. Natureza dos apendices petaloides em Barbacenioideae (Velloziaceae). Boletim de Zoologia e Biologia Marinha, n.s., 30: 713–755.

Menezes NL de. 1975. Presenca de traqueides de transfusao e bainha mestomatica em Barbacenioideae (Velloziaceae). Bol. Univ. São Paulo, Bot. 3: 29–60.

Menezes NL de. 1976. Megasporogenese, megagametogenese, e embriogenese em Velloziaceae. Bol. Univ. São Paulo, Bot. 4: 41–60.

Menezes NL de. 1980. Evolution in Velloziaceae, with special reference to androecial characters. In: CD Brickell, DE Cutler, and M Gregory, eds. Petaloid Monocotyledons: horticultural and botanical research, pp. 117–139. Academic, London.

Menezes NL de. 1988. Evolution of the anther in the family Velloziaceae. Bol. Univ. São Paulo, Bot. 10: 33–41.

Menezes NL de and J Semir 1990. New considerations regarding the corona in the Velloziaceae. Ann. Missouri Bot. Gard. 77: 539–544.

Menezes NL de, R de Mello-Silva, and SJ Mayo. 1993. A cladistic analysis of the Velloziaceae. Kew Bull. 49: 71–92.

Nemirovich-Danchenko EN. 1985. Velloziaceae. In: A Takhtajan, ed. Comparative seed anatomy, vol. 1, pp. 119–121. Nauka, Leningrad (in Russian).

Noher de Halac RI. 1969. Nuevos datos sobre la morfologia floral de *Barbaceniopsis bolivensis,* con especial referenda a la sexualidad. Kurtziana 5: 293–296.

Perrier H. 1946. Au sujet de la systematique des Vellosiacees et du genre *Xerophyta* Juss. Notul. Syst. (Paris) 12: 146–148.

Salatino A. 1999. Main results from *trn*L-F sequencing of Velloziaceae and allied taxa. Anais Acad. Brasil. Ciên. 71: 203–206.

Salatino A, MLF Salatino, R Mello-Silva, and I Duerholt-Oliveira. 1991. An appraisal of the plasticity of alkanes profiles of some species of Velloziaceae. Biochem. Syst. Ecol. 19: 241–248.

Salatino MLF, A Salatino, NL Menezes, and R de Mello-Silva 1989. Alkanes of foliar epicuticular waxes of Velloziaceae. Phytochemistry 28: 1105–1114.

Salatino MLF, A Salatino, R de Mello-Silva, M-A van Sluys, DE Giannasi, and RA Price. 2001. Phylogenetic inference in Velloziaceae using chloroplast *Trn*L-F Sequence. Syst. Bot. 26: 92–103.

Sazima M. 1979. Biologia floral de especies de Vellozia-ceaena Serra do Cipo, Minas Gerais. Ph.D. dissertation, University of São Paulo.

Smith LB. 1962. A synopsis of the American Velloziaceae. Contributions from the United States National Herbarium 35: 215–292.

Smith LB and ES Ayensu. 1974. Classification of Old World Velloziaceae. Kew Bull. 29: 181–205.

Smith LB and ES Ayensu. 1976. A revision of American Velloziaceae. Smithsonian Contr. Bot. 30: 1–172.

Warming E. 1893. Note sur la biologic et l'anatomie de la feuille des Vellosiacees. Oversight over del kongelige Danske Videnskabernes Selskabs. Forhandlingen: 57–100.

Williams CA, JB Harborne, J Greenham, and J Eagles. 1994. Differences in flavonoid patterns between genera within the Velloziaceae. Phytochemistry 36: 931–940.

Williams CA, JB Harborne, and NL Menezes. 1991. The utility of leaf flavonoids as taxomonic markers in the subfamily and generic classification of the Velloziaceae. Biochem. Syst. Ecol. 19: 483–495.

Order 20. STEMONALES

Perennial rhizomatous herbs with erect stems or trailing or twining herbs with fasciculate tuber. Raphides present or absent. Vessels only in roots or also in stem and even leaves, with scalariform perforations. Leaves alternate, opposite or verticillate, petiolate or sessile, lanceolate to ovate, with parallel, arcuate-striate or pinnate-striate venation. Stomata anomocytic or (Pentastemonaceae) tetracytic. Flowers solitary (axillary or terminal) or in few-flowered cymes or racemes, bisexual or rarely unisexual, actinomorphic, 2-merous, 3–4-merous, or rarely (*Pentastemona*) 5-merous. Perianth segments free or sometimes more or less connate, similar or more or less dissimilar. Stamens isomerous with the perianth segments; filaments free or basally connate, elongate or short; anthers tetrasporangiate, basifixed, opening longitudinally, introrse, often produced in sometimes rather long appendages. Tapetum secretory. Microsporogenesis successive. Pollen grains 1-colpate or (*Pentastemona* and *Stichoneuron*) inaperturate, reticulate, spiny-gammate or psilate (Furness and Rudal 2006). Gynoecium of two or 3–6(-8) or more united carpels; stylodia free or more or less connate; ovary superior to inferior, 3–6(-8)-locular or 1-locular, with various types of placentation. Ovules few or numerous, anatropous or hemitropous, bitegmic, crassinucellate, with parietal placentation. Female gametophyte or *Polygonum*-type. Endosperm nuclear. Fruits capsules (sometimes fleshy and berry-like) or less often berries. Seeds ellipsoidal to globose, usually with various kind of appendages; seed coat derived mainly from testa; endosperm copious, contains aleurone, fat, and starch grains; embryo usually small and nearly undifferentiated. Subterranean parts of *Stemona* and *Croomia* contain insecticidal alkaloids known only from these genera.

Key to Families

1 Ovary superior or semi-inferior. Flowers 2-merous. Twining or trailing, glabrous herbs with short rhizome and often tuberous roots or erect herbs with horizontally creeping rhizome and with scale leaves. Stems vascular bundles in one or two rings. Leaves alternate, opposite or verticillate, simple, linear, lanceolate, ovate, or triangular, petiolate or subsessile, cuneate to cordate or hastate at base, acuminate; venation parallel to arcuate-striate or arcuate-pinnate, with parallel transverse veinlets. Flowers large to rather small, in axillary few-flowered cymes or solitary, bisexual or rarely (*Stichoneuron*) unisexual, articulated from pedicel. Perianth segments four in two cycles, acuminate, free, sepaloid or petaloid. Stamens four in two cycles; filaments short, inserted at base of perianth; anthers in *Stemona* with appendaged connective. Pollen grains 1-colpate, tectate-columellate to granular, 1-colpate or (*Stichoneuron*) inaperturate, with microretivulate or reticulate (*Croomia*) ornamentation. Gynoecium of two carpels; stigma sessile; ovary semiinferior, 1-locular, with few to many, basal (*Stemona*) or apical ovules. Fruits 2-valved capsules. Seeds longitudinally ridged, with elaiosome of juicy, uniseriate or vesicular hairs from hilum raphe or micropyle; testa several-layered, ridges many cells high, tanniniferous; endosperm stores aleurone, lipids and some starch. n = 9, 12. . . . . 1. STEMONACEAE.

1 Ovary inferior, flowers 5-merous. Small, succulent monopodial herbs with scale leaves on short vertical rhizome; hairs uniseriate. Stem prostrate or slightly ascending, unbranched, fleshy, juicy, loosely attached to substratum with slender adventitious roots. Leaves alternate, dispersed, long-petiolate, shortly sheathing at the base, cordate or rounded at base, acute, subacuminate, with arcuate- pinnate venation and transverse parallel veinlets; petioles and sheathing leaf bases ciliate. Stomata paracytic, or tetracytic to encyclocytic. Flowers small, in axillary, simple or compound racemes, wholly or partly functionally unisexual, bracteate and with large dorsal bracteole; pedicels not articulated. Perianth segments five, in one cycle, almost free or partially connate into long to short tube, with five rounded imbricate lobes. Stamens five, in one cycle, attached to base of perianth tube; filaments very short, united into short, fleshy ring; anthers subsessile; connective broad, their extensions contacting stigmas. Pollen grains intectate, inaperturate, scabrate. Gynoecium of three carpels; style short and thick, with five apparently nectariferous pouches at its base inside staminal tube; ovary inferior, provided with ten longitudinal flanges or ribs, 1-locular, with numerous anatropous ovules on intrusive placentas. Fruits berries, distinctly ribbed, containing many seeds. Seeds oblong, endotestal. with 10–13 faint longitudinal ridges, provided with inflated funicular arillode covering about one third of the seed, with sarcotestalike hyaline exotesta; endotesta with massive U-shaped thickenings; endosperm copious, contains aleurone and lipids as well as starch grains mainly around embryo and near chalaza; embryo minute. n = 7. . . . . 2. PENTASTEMONACEAE.

1. STEMONACEAE

Caruel 1878 (including Roxburghiaceae Wallich 1832; Croomiaceae Nakai 1937). 3/30. From India, southern China, and Japan through Malesia to Australia and southeastern United States (one species of *Croomia*).

Stemona, Croomia, Stichoneuron.

2. PENTASTEMONACEAE

Duyfjes 1992. 1/2. Sumatra.

Pentastemona.

Related to the Stemonaceae, but markedly differ in 5-merous perianth and androecium, inferior ovary, prominent stigmatic lobes, parietal placentation, berrylike fruits, several-layered testa, intectate pollen grains, alternate leaves, leaf venation, compound midrib, tetracytic or encyclocytic stomata.

Bibliography

Ayensu ES. 1968. Comparative vegetative anatomy of the Stemonaceae (Roxburghiaceae). Bot. Gaz. 129: 160–165.

Bouman F and N Devente. 1992. A comparison of the structure of ovules and seeds in *Stemona* (Stemonaceae) and *Pentastemona* (Pentastemonaceae). Blumea 36: 501–514.

Conover M. 1991. Epidermal patterns in the reticulate-veined Liliiflorae and their parallel-veined allies. Bot. J. Linn. Soc. 107: 295–312.

Duyfjes BEE. 1991. Stemonaceae and Pentastemonaceae; with miscellaneous notes on members of both families. Blumea 36: 239–252.

Duyfjes BEE. 1992. Formal description of the family Pentastemonaceae with some additional notes on Pentastemonaceae and Stemonaceae. Blumea 36: 551–552.

Fedotova TA. 1985. Stemonaceae. In: A Takhtajan, ed. Comparative seed anatomy, vol. 1, pp. 128–130. Nauka, Leningrad (in Russian).

Furness CA and PJ Rudall. 2006. Comparative structure and development of pollen and tapetum in Pandanales. Int. J. Plant Sci. 167: 331–348.

Ham RWJM van der. 1991. Pollen morphology of the Stemonaceae. Blumea 36: 127–159.

Heel RWJM van der. 1992. Floral morphology of Stemonaceae and Pentastemonaceae. Blumea 36: 481–499.

Holm Th. 1905. *Croomia pauciflora*, an anatomical study. Am. J. Sci. 20: 50–54.

Kubitzki K. 1998a. Pentastemonaceae. In: K Kubitzki, ed. The families and genera of vascular plants, vol. 3, pp. 404–406. Springer, Berlin/Heidelberg/New York.

Kubitzki K. 1998b. Stemonaceae. In: K Kubitzki, ed. The families and genera of vascular plants, vol. 3, pp. 422–425. Springer, Berlin/Heidelberg/New York.

Lachner-Sandoval V. 1892. Beitrag zur Kentnis der Gat-tung *Roxburghia*. Bot. Centralbl. 50: 65–70, 97–104, 129–135.

Meijer W and J Bogner. 1983. *Pentastemona* (Stemonaceae): The elusive plant. Nature Malaysiana 8(1): 26–27.

Oginuma K, K Horiuchi, and T Fukuhara. 2001. Karyomorphology of two genera in Stemonaceae. Acta Phytotax. Geobot. 52: 57–63.

Rogers GK. 1982. The Stemonaceae in the southeastern United States. J. Arnold Arbor. 63: 327–336.

Rudall PJ, J Cunniff, P Wilkin, and LR Caddick. 2005. Evolution of dimery, pentamery and the monocarpellary condition in the monocot family Stemonaceae (Pandanales). Taxon 54: 701–711.

Steenis CGGJ van. 1982. *Pentastemona,* a new 5-merous genus of monocotyledons from North Sumatra (Stemonaceae). Blumea 28: 151–163.

Swamy BGL. 1964. Observation on the floral morphology and embryology of *Stemona tuberosa* Lour. Phytomorphology 14: 458–468.

Tomlinson PB and ES Ayensu. 1968. Morphology and anatomy of *Croomia pauciflora* (Stemonaceae). J. Arnold Arbor. 49: 260–277.

Wright CH. 1896. On the genus *Stemona* Lour. Bot. J. Linn. Soc. 32: 490–496.

Superorder DIOSCOREANAE

Order 21. DIOSCOREALES

Perennial herbs with twining or trailing or less often erect shoots arising from a fleshy, starchy rhizome or a tuber arising by hypertrophy of the internodes above the cotyledon (epicotyl) or/and by a lateral hypertrophy of the hypocotyl; the tuber with or without secondary growth. Roots commonly mycorrhizal, without root hairs. Tannin cells mostly present or (Trichopodaceae) lacking. Raphides in mucilaginous idioblasts generally present. Vessels in roots and sometimes also in stem, with scalariform perforations. Sieve-element plastids of P2c- or P1/2c-type (Behnke 2002). Leaves all basal, alternate, opposite or verticillate, sessile or more often petiolate, entire to compound, with various kinds of venation; petiole mostly with a pulvinus at each end. Ptyxis conduplicate or (Taccaceae) plicate. Stomata anomocytic. Flowers in various kind of inflorescences or solitary, bisexual or more often unisexual, actinomorphic, 2-merous, 3-merous. Perianth segments four, six in two cycles, more or less similar, usually basally connate into a short tube. Stamens isomerous with perianth segments; inner ones sometimes staminodial or obsolete; filaments free or adnate to the perianth segments; anthers basifixed, tetrasporangiate, introrse or sometimes extrorse, opening longitudinally, often with apically projected connective. Tapetum secretory. Microsporogenesis simultaneous or (*Trichopus*) successive. Pollen grains 2-celled, 1-, 2- or 4-colpate (sulcate) or (*Avetra*) 4(5)-porate and spinulose. Gynoecium of three united carpels; stylodia free or connate; ovary superior, semi-inferior or mostly inferior, 1- or 3-locular, with 1–2 or numerous ovules per locule. Ovules anatropous, bitegmic, crassinucellate or (Trichopodaceae) tenuinucellate. Female gametophyte of *Polygonum*-type or very rarely (in some species of *Dioscorea*) tetrasporic of *Drusa*-type. Endosperm nuclear. Fruits mostly capsules, rarely indehiscent and samaroid (*Trichopus* and *Rajania*) or berries. Seeds of various shapes, often (*Dioscorea* and *Stenomeris*) winged; seed coat formed by both the integuments (endotestal and exotegmic), lack phytomelan but often, especially those in capsular fruits, contain phlobaphene; endosperm copious, contains aleurone, lipids, some starch, and commonly also hemicellulose in form of cell-wall thickening (absent in *Stenomeris*); embryo small, but well differentiated, with subterminal plumule and broad, flat lateral cotyledon, or minute, undifferentiated. Often produce steroidal saponins, chelidonic acid, and sometimes (species of *Dioscorea*) alkaloid dioscorine. Flavones apparently lacking, although flavone C-glycosides may be present in the Trichopodaceae (Williams et al. 1988).

Closely related to the Stemonales and have a common origin with them.

Key to Families

1 Ptyxis plicate. Embryo minute, undifferentiated, basal. Acaulescent perennial herbs from starchy, more or less tuberous-thickened, globose or elongate rhizome containing alkaloids. Raphide cells present in stem and leaves. Vessels only in roots, with scalariform perforations. Leaves all basal, with basally more or less widened and often long petioles and entire to much dissected laminas; venation arcuate-striate, palmate-striate, or pinnate-striate with reticulate veinlets. Midrib compound. On both sides of leaves occur peculiar trichomes consisting of a short row of cells bearing a multicellular body on which is another cell row at the apex. Mesophyll with scattered mucilage cells. Stomata anomocytic or surrounded by one cell. Inflorescences umbellate, apparently cymose, on long peduncle, the bracts forming an involucre, the inner ones narrower and often filiform; flowers bisexual, actinomorphic, 3-merous. Perianth segments six in two cycles, imbricate, with a short, broad tube and six lobes, these biseriate, somewhat petaloid, the inner ones usually longer than the others. Stamens six in two cycles; filaments short, inserted on perianth tube, with inflexed margins, apically forming, together with the broad connectives, hoods over anthers; anthers tetrasporangiate, opening longitudinally, introrse. Pollen grains 2-celled, 1-colpate, tectate-columellate, verrucate to striate. Gynoecium of three united carpels; style short, with three broad stigmas often petaloid and reflexed over the style; ovary 6-ribbed, inferior, 1-locular, with numerous pendulous ovules on more or less intruded parietal placentas. Septal nectaries present in several species. Ovules anatropous, apotropous, with parietal cell. Fruits with a fleshy pericarp, 6-ribbed, berry-like, irregularly disintegrating, rarely loculicidal capsules (*Tacca* subgenus *Schizocapsa*). Seeds of various shapes, mostly provided with longitudinal ridges, sometimes with thin fleshy aril and fleshy raphe; seed coat formed by both the integuments (both 2-layered as in *Pentastemona*) which are well developed and take part in the mechanical function (restricted in the endotesta and exotegmen) and contains phlobaphene; endosperm copious, contains aleurone and lipids, but not starch; embryo minute, basal, undifferentiated, globose or ovoid, with laterally inserted cotyledon (as in *Dioscorea*). $n = 15$. . . . . 1. TACCACEAE.

1 Ptyxis conduplicate. Embryo small, but well differentiated, with subterminal plumule and broad flat cotyledons.

2 Flowers bisexual. The connective produced into a long appendage.

3 Fruits many-seeded loculicidal capsules. Tall climbers with short, subterranean horizontal rhizome. Stem tough, twining to the left. Leaves alternate, cordate, acuminate, arcuate-striate veined with many primary veins connected by more or less transverse commissural veinlets. Flowers in lax axillary panicles or solitary, pedicillate, pedicels bracteolate. Perianth segments connate into urn-shaped, 6-lobed tube. Stamens six, inserted in one series near the mouth of tube; filaments short, deflexed; anthers with broad connective that is apically prolonged into an elongated appendage with spathulate, horned apex. Style short, columnar, ending in three bifid stigmas; ovary with many ovules in each locule. Capsules linear and much elongated, 3-winged, with thin pericarp. Seeds flat, triangular, with large membranous wing on one side at the top; cells of mesotesta intensively coloured by phlobaphenes; cells of endosperm thin-walled. . . . . . 2. STENOMERIDACEAE.

3 Fruits indehiscent or opening by irregular ruptures, dry or slightly fleshy, 1- to 6-seeded. Erect herbs with short rhizome. Leaves entire, petiolate, ovate-elliptic to cordate-sagittate, 3–7-veined; primary veins palmate, secondary venation reticulate. Flowers solitary and axillary or in terminal or axillary short racemes or fascicles, with long pedicels. Perianth more or less campanulate. Stamens six, inserted at base of perianth segments; filaments very short; anthers with broad connective that is apically prolonged into long pointed appendage, introrse. Pollen grains 1-colpate, biconvex and not spinulose (*Trichopus*) or 4(5)-porate, spheroidal, more or less spinulate (*Avetra*). Style short, columnar, with 3-lobed stigma, the lobes bifid; ovary with two ovules per locule. Ovules tenuinucellate (at least in *Trichopus*). Fruits three wings or keels. Seeds more or less compressed, with deeply furrowed seed coat derived from both testa and tegmen; exotegmic cells elongated and differentiated as a mechanical tissue; endosperm copious, strongly ruminate; the outer wall layers are hard and thick owing to heavy deposition of hemicellulose; embryo minute, linear, straight. n = 14. . . . . 4. TRICHOPODACEAE.

2 Flowers unisexual, dioecious. Mostly lianous herbs or rarely shrubby plants with usually tuberous rhizome or woody rootstock that sometimes have secondary growth, rarely rhizome. Aerial stem usually climbing or trailing. Leaves alternate or rarely opposite, entire or lobed and often cordate or palmately compound; venation arcuate-striate or palmate-striate with 3–13 primary veins and reticulate veinlets; bulbils of cauline nature occur in leaf axils of many species of *Dioscorea*. Flowers small, in spikes, racemes, cymes, or panicles, pedicellate or sessile, often ebracteolate. Perianth segments usually basally connate into short tube. Nectaries commonly present. Fertile stamens six or three, very rarely one, erect or spreading; filaments free or sometimes shortly connate, attached to perianth tube; anthers not appendaged, rarely apiculate, dorsifixed, introrse or sometimes extrorse. Pollen grains 1-colpate, 2-colpate or occasionally 3-colpate. Pistillodia present or absent. Stylodia free or connate and apically 3-lobed or 3-brachiate; ovary with two to many ovules per locule. Staminodia often present in female flowers. Fruits usually loculicidal capsules, generally 3-angular or 3-winged, rarely samaras (*Rajania*) or berries (*Tamus*). Seeds winged or not, flat or scarcely compressed, rarely (*Tamus*) globose, not ribbed. Endotesta is mostly crystalliferous and in each cell a single crystal of calcium oxalate is embedded in a moiety of phlobaphene and sclerotic exotegmen consists of cells with pitted or sculptured walls (Huber 1998). n = 8, 9, 10, 12.. . . . . 3. DIOSCOREACEAE.

1. TACCACEAE

Dumortier 1829. 1/13 (or 35–50 ?). Pantropical, but best developed in Southeast Asia and Polynesia.

Tacca.

2. STENOMERIDACEAE

J. Agardh 1858. 1/2. Peninsular Malaya, northern Borneo and the Philippines.

Stenomeris.

The most archaic member of the order.

3. DIOSCOREACEAE

R. Brown 1810 (including Tamaceae Martynov 1820; Tamnaceae J. Kickx f. 1826). 6/650. Pantropical, a few species in subtropical and warm-temperate regions; *Tamus* (4–5), the only genus with baccate fruits, distributed in Macaronesia, Europe, Mediterranean region, and western Asia.

Dioscorea, Borderea, Testudinaria, Epipetrum, Rajania, Tamus.

Closely related to the Stenomeridaceae.

4. TRICHOPODACEAE

Hutchinson 1934 (including Avetraceae Takhtajan 1997). 2/2. India (southwestern Deccan Peninsula), Sri Lanka, Malay Peninsula (*Trichopus*) and eastern Madagascar (*Avetra*).

Trichopus, Avetra.

Bibliography

Al-Shehbaz IA and BG Schubert. 1989. The Dioscoreaceae in the southeastern United States. J. Arnold Arbor. 70: 57–95.

Ayensu ES. 1966. Taxonomic status of *Trichopus:* anatomical evidence. Bot. J. Linn. Soc. 59: 425–430.

Ayensu ES. 1969. Aspects of the complex nodal anatomy of the Discoreaceae. J. Arnold Arbor. 50: 124–137.

Ayensu ES. 1970. Analysis of the complex vascularity in stems of *Dioscorea composita*. J. Arnold Arbor. 51: 228–240.

Ayensu ES. 1972. Dioscoreales. In: CR Metcalfe, ed. Anatomy of the monocotyledons, vol. 6. Clarendon, Oxford.

Behnke H-D. 1965. Über das Phloem der Dioscoreaceen unter besonderer Berucksichtigung ihrer Phloembecken, parts 1 and 2. Z. Pflanzenphysiol. 53: 149–155, 214–244.

Behnke H-D. 1984. Plant trichomes – structure und ultrastructures: general terminology, taxonomic applications, and aspect of trichome-bacteria interaction in leaf tips of *Dioscorea*. In: E Rodrigues, PL Healy and I Mehta, eds. Biology and chemistry of plant trichomes, pp. 1–21. Plenum, New York.

Bharathan G, L Raz, and P Wilkin. 2001. The true yams, Dioscorea (Dioscoreaceae): Phylogenetic analysis of chloroplast nucleotide sequences. In Botany 2001: Plants and People. Abstracts, p. 154. Albuquerque.

Bouman F. 1995. Seed structure and systematics in Dioscoreales. In: PJ Rudall, PJ Cribb, DF Cutler, and DF Humphries, eds. Monocotyledons: systematics and evolution, vol. 1, pp. 139–156. Royal Botanic Gardens, Kew.

Burkill I. H. 1960. The organography and the evolution of Dioscoreaceae. Bot. J. Linn. Soc. 56: 319–412.

Caddick LR and P Wilkin. 1998. A revision of the genus *Stenomeris* (Dioscoreaceae). Kew Bull. 53: 703–712.

Caddick LR, CA Furness, KL Stobart, and PA Rudall. 1998. Microsporogenesis and pollen morphology in Dioscoreales and allied taxa. Grana 37: 321.

Caddick LR, PJ Rudall, and P Wilkin. 2000a. Floral morphology and development in Dioscoreales. Feddes Repert. 111: 189–230.

Caddick LR, PJ Rudall, P Wilkin, and MW Chase. 2000b. Yams and their allies: systematics of Dioscoreales. In: KL Wilson and DA Morrison, eds. Monocots: systematic and evolution, pp. 475–487. CSIRO, Collingwood.

Caddick LR, PJ Rudall, P Wilkin, TA Hedderson, and MW Chase. 2002a. Phylogenetics of dioscoreales based on combined analyses of morphological and molecular data. Bot. J. Linn. Soc. 138: 123–144.

Caddick LR, P Wilkin, PJ Rudall, TAJ Hedderson, and MW Chase. 2002b. Yams reclassified: a recircumscription of Dioscoreaceae and Dioscoreales. Taxon 51: 103–114.

Caddick LR, P Wilkin, PJ Rudall, TAJ Hedderson, and MW Chase. 2002c. Yams reclassified: a recircumscription of Dioscoreaceae and Dioscoreales. Taxon 51: 102–114.

Cheadle VI and H Kosakai. 1976. Vessels in Dioscoreales. Phyta 1: 41–53.

Drenth E. 1972. A revision of the family Taccaceae. Blumea 20: 367–406.

Drenth E. 1976. Taccaceae. In: CGGJ van Steenis, ed. Flora Malesiana, ser. 1, 7(4): 806–819. Leyden.

Huber H. 1998a. Dioscoreaceae. In: K Kubitzki, ed. The families and genera of vascular plants, vol. 3, pp. 216–235. Springer, Berlin/Heidelberg/New York.

Huber H. 1998b. Trichopodaceae. In: K Kubitzki, ed. The families and genera of vascular plants, vol. 3, pp. 441–444. Springer, Berlin/Heidelberg/New York.

Kale NN and RM Pai. 1979. The floral anatomy of *Trichopus zeylanicus* Gaertn. Proc. Indian Acad. Sci. B Plant Sci. 88: 63–67.

Kubitzki K. 1998. Taccaceae. In: K Kubitzki, ed. The families and genera of vascular plants, vol. 3, pp. 425–428. Springer, Berlin/Heidelberg/New York.

Ling P-P. 1981. Stomatal studies in Chinese Taccaceae with a discussion of its taxonomical significance. Bull. Nanjing Bot. Gard., Mem. Sun. Yat. Sen. 1981: 20–24.

Nagaraja Rao C. 1955. Embryology of *Trichopus zeylanicus* Gaertn. J. Indian Bot. Soc. 34: 213–221.

Oganezova GG. 1995. On the systematical position of the families Haemodoraceae, Hypoxidaceae and Taccaceae (data on the seed structure). Bot. Zhurn. 80(7): 12–25 (in Russian with English summary).

Perrier de la Bathie H. 1924. Un nouveau genre de Dioscoreacées. 71: 25–28.

Petrova LP and LK Safina. 1985. Dioscoreaceae. In: A Takhtajan, ed. Comparative seed anatomy, vol. 1, pp. 133–135. Nauka, Leningrad (in Russian).

Prieto CA, JL Fernandez Alonso, and RL Liesner. 2000, Notas sobre la familia Taccaceae (Liliidae) y su presencia en Colombia. Caldasia 22(2): 265–270.

Rao NA. 1953. Embryology of *Dioscorea oppositifolia* L. Phytomorphology 3: 121–126.

Rao NA. 1955. Embryology of *Trichopus zeylanicus* Gaertn. J. Indian Bot. Soc. 34: 213–221.

Rao VS. 1969. The vascular anatomy of *Tacca pinnatifida*. J. Univ. Bombay 38: 18–24.

Schols P, CA Furness, P Wilkin, S Huysmans, and E Smets. 2001. Morphology of pollen and orbicules in some *Dioscorea* species and its systematic implications. Bot. J. Linn. Soc. 136: 295–311.

Schols P, CA Furness, P Wilkin, E Smets, V Cielen, and S Huysmans. 2003. Pollen morphology of *Dioscorea* (*Dioscoreaceae*) and its relation to systematics. Bot. J. Linn. Soc. 143: 375–390.

Schols P, CA Furness, V Merckx, P Wilkin, and E Smets. 2005a. Comparative pollen development in Dioscoreales. Int. J. Plant Sci. 166: 909–924.

Schols P, P Wilkin, CA Furness, S Huysmans, and E Smets. 2005b. Pollen evolution in yams (*Dioscorea*: Dioscoreaceae). Syst. Bot. 30: 750–758.

Sivarajan VV, P Pushpangadan, and PK Ratheesh Kumar. 1990. A revision of *Trichopus* (Trichopodiaceae). Kew Bull. 45: 353–360.

Watson EV. 1936. A study of the anatomy of *Trichopus zeylanicus* Gaertn. Notes Roy. Bot. Gard. Edinb. 19: 135–156.

Wilkin P, P Schols, MW Chase, K Chayamarit, CA Furness, S Huysmans, F Rakotonasolo, E Smets, and C Thapyai. 2005. A plastid gene phylogeny of the yam genus, *Dioscorea*: roots, fruits and Madagascar. Syst. Bot. 30: 736–749.

Xifreda CC. 2000. Evaluation of pollen and vegetative characters in the systematics of South American species of *Dioscorea* (Dioscoreaceae). In: KL Wilson and DA Morrison, eds. Monocots: systematic and evolution, pp. 488–496. CSIRO, Collingwood.

Subclass III. ARECIDAE

The subclass Arecidae is one of the most ancient phyla of the liliopsids and probably had a common origin with the Liliidae. Both subclasses derived from a hypothetical extinct common ancestor, which most likely was a terrestrial rhizomatous perennial herb of more or less wet habitats without vessels and with primitive 1-colpate, boat-shaped, pollen grains (2-celled when shed) and primitive free, conduplicate carpels with decurrent stigmas. In both subclasses there are still members with a number of primitive characters, including apocarpous gynoecium and primitive carpels. During the evolution of the Arecidae a progressive reduction and specialization of flowers has taken place, which has been compensated by the origin of very specialized compound inflorescences subtended by a usually well-developed bract, the spathe.

Superorder ARECANAE

Order 22. ARECALES

Usually large, sometimes very large, rarely rather small (only 2 cm in diameter and less than 25 cm tall in some species of *Chamaedorea*), more or less treelike or shrubby plants, branched or more often unbranched, erect or sometimes climbing, rarely stemless, sometimes with creeping rhizomes, but never producing bulbs or corms; lacking cambium but sometimes increasing in diameter by primary growth. Silica bodies common. Calcium oxalate usually occurs as raphides, rarely as isolated crystals or as crystal sand. Vessels usually in all organs, in roots usually with simple perforations, in stem and leaves mostly with scalariform perforations (always scalariform in *Nypa*). Leaves spirally arranged, usually in terminal rosettes, scattered along the stem in climbing species, rarely basal, often large or very large, usually clearly differentiated into petiole and lamina, with broad, clasping sheath at base; lamina usually palmately or pinnately divided, less often entire or bipinnate (*Caryota*), undivided and densely plicate in bud, with a stout central axis (costa or rachis), which is short or absent in truly palmate leaves, better developed in costapalmate leaves, and prominent in the pinnate leaves; some laminas are divided so that the segments (of palmate leaves) and pinnae (of pinnate leaves) appear to be either V-shaped (termed induplicate) while the others are so divided that they appear to be like an inverted V (termed reduplicate). Stomata tetracytic. Epicuticular wax of *Strelitzia* type. Flowers usually small, numerous, in usually large, compound or rarely simple spikes or panicles, sometimes in dichasia, rarely in heads, subtended by one or more bracts, which often enclose the young inflorescence; individual flowers sessile and often even embedded in the axis or rarely shortly pedicellate, bisexual or much more often unisexual (monoecious or less often dioecious) or sometimes polygamous, actinomorphic or less often weakly zygomorphic, commonly 3-merous, usually bracteate. Perianth segments six or sometimes four, in two more or less dissimilar cycles or less often up to ten and spirally arranged, free or more or less connate, often scale-like, rarely reduced or lacking; outer segments (sepals) usually smaller, imbricate or open in bud; inner segments (petals) valvate in male flowers and imbricate in female flowers. Stamens mostly six in two cycles, but seldom, as in *Nypa*, one of two cycles lacking, or sometimes numbers range from seven to eight to mostly nine to many (120 to over 250 in *Phytelephas,* up to 521 in *Ammandra* and up to 954 in *Palandra*), filaments free

A. Takhtajan, *Flowering Plants*,

or more or less connate or/and adnate to the inner perianth segments; anthers tetrasporangiate, basifixed or dorsifixed, latrorse, opening usually longitudinally or rarely (some species of *Areca*) by apical pores. Tapetum secretory. Microsporogenesis simultaneous or sometimes successive (reported for *Nypa fruticans*, *Phoenix sylvestris*, and *Pinanga disticha*, but needs confirmation). Pollen grains tectate-columellate, 1(2)-colpate, trichotomocolpate, zonocolpate or 1(2)-porate, predominantly reticulate. Staminodia often present. Gynoecium of (1)3(4–10) free or more often united carpels, sometimes pseudomonomerous; stylodia free or connate; ovary superior, in syncarpous forms usually 3-locular, with one erect or pendulous ovule per locule or per carpel. Septal nectaries often present. Ovules anatropous or less often hemitropous, campylotropous, or orthotropous, bitegmic, crassinucellate, with parietal cell. Female gametophyte usually of *Polygonum*-type, rarely of *Allium*-type. Endosperm nuclear. Fruits dry or fleshy drupe with the endocarp usually attached to the seed or less often baccate, rarely more or less dehiscent. Seeds with very small, cylindric, or conical embryo embedded apically, laterally or basally in copious endosperm; endosperm usually bony, sometimes ruminate, contains fats and oils, aleurone and hemicellulose deposited in cell walls, but without starch. n = 8–10, 12–19, especially 14, 16, 18. In *Voanioala* 2n = 606 ±3 (Johnson 1989; Roser 1994).

Arecales are one of the most ancient branches of the monocotyledons and are among the first families of magnoliophytes that are definitely recognizable in the fossil record. They are also very heterobathmic. There are 16 arecaceous genera with the apocarpous gynoecium, and, what is even more important, in some of them the carpels are conduplicate and stipitate, with open sutures and laminar or sublaminar placentation (see Uhl and Moore 1971; Moore 1973). In *Trachycarpus fortunei* trichomes are present along and in some extent within the unsealed suture, as in some taxa of archaic dicotyledons (Uhl and Moore 1971; Moore and Uhl 1982). Stigmas are sessile or subsessile. In Arecales septal nectaries developed independently, and as Eames (1953) says "the story here parallels that in the Liliaceae; septal glands have developed independently, perhaps several times."

The Arecales resemble the commelinalean and poalean families in possessing epicuticular wax of the *Strelitzia*-type and cell walls containing UV-fluorescent compounds. Besides, the Arecales resemble the Joinvilleaceae, Flagellariaceae, and Poaceae in plicate ptyxis of leaves (found in *Joinvillea* and some grasses), silica bodies, the occurrence of vessels in the stems, the pronounced development of ligules and ligulelike structures, the conspicuously similar flavonoid compounds, such as tricin and sulfonated flavonoids, and abundant cyanogenesis (Dahlgren and Rasmussen 1983; Dahlgren et al. 1985). But in contrast to the Poales, septal nectaries are common in palms, and, what is phylogenetically more important, some of them still have such plesiomorphic characters as apocarpous gynoecium of very primitive carpels and simultaneous microsporogenesis. In general, palms are much nearer to the archaic Liliidae and Alismatidae than to the Poanae and related groups. However, they occupy a very isolated taxonomic position, and evidently originated independently and directly from the earliest monocotyledons.

1. ARECACEAE

Schultz-Schultzenstein 1832 or Palmae A. L. de Jussieu 1789 (nom. altern.) (including Acristaceae O. F. Cook 1913; Borassaceae Schultz-Schultzenstein 1832; Calamaceae Kunth ex Perleb 1838; Caryotaceae O. F. Cook 1913; Ceroxylaceae O. F. Cook 1913; Chamaedoraceae O. F. Cook 1913; Cocosaceae Schultz-Schultzenstein 1832; Coryphaceae Schultz-Schultzenstein 1832; Geonomataceae O. F. Cook 1913; Iriarteaceae O. F. Cook et Doyle 1913; Lepidocaryaceae C. Martius 1838; Malortieaceae O. F. Cook 1913; Manicariaceae O. F. Cook 1910; Nypaceae Brongniart ex Le Maout et Decaisne 1868; Phoenicaceae Burnett 1835; Phytelephantaceae Martius ex Perleb 1838; Pseudophoenicaceae O. F. Cook 1913; Sabalaceae Schultz-Schultzenstein 1832; Sagaceae Schultz-Schultzenstein 1832; Saginaceae Berchtold et J. Presl 1820; Synechanthaceae O. F. Cook 1913). 188/2000–2700. Pantropical, with some subtropical and a few temperate species.

Classification after J. Dransfield et al. 2005 and M.V. Norup et al. 2006.

1.1 CALAMOIDEAE

Erect or climbing, often spiny. Leaves palmate, costapalmate, or, more often, pinnate, reduplicate. Flowers almost always in dyads or dyad derivatives, bisexual, polygamous, monoecious or dioecious, with syncarpous gynoecium. Fruits 1–3-seeded, with thin or thick, fleshy or spongy mesocarp, covered with reflexed imbricate scales. – EUGEISSONEAE: *Eugeissona*; LEPIDOCARYEAE: *Oncocalamus, Eremospatha, Laccosperma, Raphia, Lepidocaryum, Mauritia, Mauritiella*; CALAMEAE: *Korthalsia, Eleiodoxa, Salacca, Metroxylon, Pigafetta, Plectocomia, Myrialepis, Plectocomiopsis, Calamus* (including *Calospatha*), *Retispatha, Daemonorops, Ceratolobus, Pogonotium*.

1.2 NYPOIDEAE

Mangrove palm with dichotomously branched, creeping stem. Leaves paripinnate, reduplicate. Flowers unisexual, monoecious, with androecium of only three stamens and apocarpous gynoecium; carpels large, asymmetrical, have a vascular system different from that of all other palms. Fruits with fibrous mesocarp. – *Nypa*.

1.3 CORYPHOIDEAE

Dwarf or creeping to large fan palms with solitary to multiple trunks. Leaves palmate or costapalmate, rarely entire or pinnate, almost always induplicate. Flowers solitary or in cincinni, bisexual or unisexual, with apocarpous or syncarpous gynoecium. Septal glands occur in syncarpous genera. Fruits usually with fleshy mesocarp. – SABALEAE: *Sabal*; CRYSOPHILEAE: *Schippia, Trithrinax, Zombia, Coccothrinax, Hemithrinax, Thrinax, Chelyocarpus, Cryosophila, Itaya*; PHOENICEAE: *Phoenix*; LIVISTONEAE: *Chamaerops, Guihaia, Trachycarpus, Rhapidophyllum, Maxburretia, Rhapis, Livistona, Licuala, Johannesteijsmannia, Pholidocarpus, Pritchardiopsis, Acoelorraphe, Brahea, Colpothrinax, Copernicia, Pritchardia, Serenoa, Washingtonia*; CHUNIOPHOENICEAE: *Chuniophoenix, Nannorrhops, Kerriodoxa*; CARYOTEAE: *Caryota, Arenga, Wallichia*; CORYPHEAE: *Corypha*; BORRASEAE: *Bismarckia, Satranala, Hyphaene, Medemia, Latania, Lodoicea, Borassodendron, Borassus*.

1.4 CEROXYLOIDEAE

Small to tall. Leaves pinnate or entire and pinnately ribbed, induplicate. Flowers unisexual (monoecious or dioecious) or rarely bisexual, with syncarpous gynoecium. – CYCLOSPATHEAE: *Pseudophoenix*; CEROXYLEAE: *Ceroxylon, Oraniopsis, Juania, Ravenea*; PHYTELEPHEAE: *Phytelephas, Aphandra, Ammandra*.

1.5 ARECOIDEAE

Very small to large, from erect to repent. Leaves pinnate, pinnately ribbed or bipinnate, induplicate or reduplicate. Flowers always unisexual, monoecious or dioecious, sessile and embedded in axis, with syncarpous, often pseudomonomerous gynoecium. – IRIARTEEAE: *Dictyocaryum, Iriartella, Iriartea, Socratea Wettinia*; CHAMAEDOREEAE: *Hyophorbe, Wendandiella, Synechanthus, Chamaedorea, Gaussia*; PODOCOCCEAE: *Podococcus*; ORANIEAE: *Orania*; SCLEROSPERMEAE: *Sclerosperma*; ROYSTONEAE: *Roystonea*: REINHARDTIEAE: *Reinhardtia*; COCOSEAE: *Beccariophoenix, Jubaeopsis, Voanioala, Allagoptera* (including *Polyandrococos*), *Attalea, Butia, Cocos, Jubaea, Lytocaryum, Syagrus, Parajubaea, Acrocomia, Barcella, Elaeis, Gastrococos, Aiphanes, Bactris, Desmoncus, Astrocaryum*; MANICARIEAE: *Manicaria*, EUTERPEAE: *Hyospathe, Euterpe, Prestoea, Neoniholsonia, Oenocarpus*: GEONOMEAE: *Pholidostachys, Welfia, Calyptonoma, Calyptrogyne, Asterogyne, Geonoma*; LEOPOLDINIEAE: *Leopoldinia*; PELAGODOXEAE: *Sommieria, Pelagodoxa*; ARECEAE: *Archontophoenix, Chamebeyronia, Kentiopsis, Actinokentia, Actinorhytis, Nenga, Pinanga, Areca, Alloschmidia, Cyphophoenix, Campecarpus, Basselinia, Cyphosperma, Veillonia, Burretiokentia, Physokentia, Neoveitchia, Carpoxylon, Satakentia, Brongniartikentia, Clinosperma, Cyphokentia, Moratia, Lavoixia, Dypsis, Lemurophoenix, Masoala, Marojejya, Calyptrocalyx, Linospadix, Laccospadix, Howea, Deckenia, Acanthophoenix, Oncosperma, Tectiphiala, Drymophloeus, Carpentaria, Veitchia, Balaka, Normanbya, Wodyetia, Ptychosperma, Ponapea, Adonidia, Solfia, Ptychococcus, Brassiophoenix, Hedyscepe, Rhopalostylis, Verschaffeltia, Roscheria, Phoenicophorium, Nephrosperma*.

Unplaced members of Arecaceae: *Alsmithia, Bentinckia, Clinostigma, Cyrtostachys, Dictyosperma, Heterospathe, Hydriastele* (including *Siphokentia, Gronophyllum, Gulubia*), *Iguanura, Lepidorrhachis, Loxococcus, Rhopaloblaste*.

Genus dubious: *Dransfieldia*.

Bibliography

Arber A. 1922. On the development and morphology of the leaves of palms. Proc. Roy. Soc. London, Ser. B., Biol. Sci. 93: 249–261.

Asmussen CB. 1999. Towards a chloroplast DNA phylogeny of the Geonomeae (Palmae). In: A Henderson and F Borchsenius, eds. Evolution and classification of palms. Mem. New York Bot. Gard. 83: 121–129.

Asmussen CB and MW Chase. 2001. Coding and non-coding plastid DNA in palm systematics. Am. J. Bot. 88: 1103–1117.

Asmussen CB, WJ Baker, and J Dransfield. 2000. Phylogeny of the palm family (Arecaceae) based on *rps*16 intron and *trn*L-*trn*F plastid DNA sequences. In: KL Wilson and DA Morrison, eds. Monocots: systematic and evolution, pp. 525–535. CSIRO, Collingwood, VI.

Asmussen CB, J Dransfield, V Deickmann, AS Barfod, J-C Pintaud, and WJ Baker. 2006. A new subfamily classification of the palm family (Arecaceae): evidence from plastid DNA phylogeny. Bot. J. Linn. Soc. 151: 15–38.

Barfod A. 1991. A monographic study of the subfamily Phytelephantoideae (Arecaceae). Opera Bot. 105: 5–73.

Baker WJ and J Dransfield. 2000. Towards a biogeographic explanation of the Calamoid palms. In: KL Wilson and DA Morrison, eds. Monocots: systematic and evolution, pp. 545–553. CSIRO, Collingwood, VI.

Baker WJ and AHB Loo. 2004. A synopsis of the genus *Hydriastele* (Arecaceae). Kew Bull. 59: 61–68.

Baker WJ, J Dransfield, MM Harley, and A Bruneau. 1999a. Morphology and cladistic analysis of subfamily Calamoideae (Palmae). In: A Henderson and F Borchsenius, eds. Evolution and classification of palms. Mem. New York Bot. Gard. 83: 307–324.

Baker WJ, CB Asmussen, SC Barrow, J Dransfield, and TA Hedderson. 1999b. A phylogenetic study of the palm family (Palmae) based on chloroplast DNA sequences from the *trn*L —*trn*F region. Plant Syst Evol. 219: 111–126.

Baker WJ, J Dransfield, and TA Henderson. 2000a. Phylogeny, character evolution, and a new classification of the Calamoid Palms. Syst. Bot. 25: 297–322.

Baker WJ, TA Hedderson, and J Dransfield. 2000b. Molecular phylogenetics of subfamily Calamoideae (Palmae) based on mrDNA ITS and cpDNA *rps*16 intron sequence data. Molec. Phylogen. Evol. 14: 195–217.

Baker WJ, TA Hedderson, and J Dransfield. 2000c. Molecular phylogenetics of *Calamus* (Palmae) and related rattan genera based on 5S nrDNA spacer sequence data. Molec. Phylogen. Evol. 14: 218–231.

Barfod AS. 1991. A monographic study of the subfamily Phytelephantoideae (Arecaceae). Opera Bot. 105: 1–73.

Bayton RP. 2005. *Borassus* L. and the Borassoid palms: systematics and evolution. Ph.D. thesis, University of Reading.

Belin-Depoux M and M Hering de Queiroz. 1972. Remarques sur le developpement des feuilles des palmiers: Rapprochement avec d'autres Monocotyledons. Phytomor-phology 21: 337–353.

Blombery A and T Rodd. 1982. Palms. Angus & Robertson, Sydney.

Bosch E. 1947. Blutenmorphologische und zytologische Untersuchungen an Palmen. Ber. Schweiz. Bot. Ges. 57: 37–100.

Corner EJH. 1966. The natural history of plants. Weidenfeld & Nicolson, Berkeley.

Dransfield J and NW Uhl. 1998. Palmae. In: K Kubitzki, ed. The families and genera of vascular plants, vol. 4, pp. 306–389. Springer, Berlin/Heidelberg/New York.

Dransfield J, IK Ferguson, and NW Uhl. 1990.The Coryphoid palms: Patterns of variation and evolution. Ann. Missouri Bot. Gard. 77: 802–815.

Dransfield J, NW Uhl, CB Asmussen, WJ Baker, MM Harley, and CE Lewis. 2005. A new phylogenetic classification of the palm family, Arecaceae. Kew Bull. 60: 559–569.

Eames AJ. 1953. Neglected morphology of the palm leaf. Phytomorphology 3: 172–189.

Ferguson IK. 1986. Observations on the variation in pollen morphology of Palmae and its significance. Canad. J. Bot. 64: 3079–3090.

Ferguson IK and MM Harley. 1993. The significance of new and recent work on pollen morphology in the Palmae. Kew Bull. 48: 205–243.

Fisher JB and J Dransfield. 1977. Comparative morphology and development of inflorescence adnation in rattan palms. Bot. J. Linn. Soc. 75: 119–140.

Fisher JB and HE Moore, Jr. 1977. Multiple inflorescences in palms (Arecaceae): their development and significance. Bot. Jahrb. Syst. 98: 573–611.

Gunn B. 2004. The phylogeny of the *Cocoeae* (Arecaceae) with emphasis on *Cocos nucifera*. Ann. Missouri Bot. Gard. 91: 505–522.

Haccius B and VJ Philip. 1979. Embryo development in *Cocos nucifera* L.: a critical contribution to a general understanding of palm embryogenesis. Plant Syst. Evol. 132: 91–106.

Hahn WJ. 2002a. A molecular phylogenetic study of the Palmae *(Arecaceae)* based on *atpB*, *rbcL*, and 18S nrDNA sequences. Syst. Biol. 51: 92–112.

Hahn WJ. 2002b. A phylogenetic analysis of the Arecoid Line of palms based on plastid DNA sequence data. Molec. Phytogen. Evol. 23: 189–204.

Harborne JB and CA Williams. 1991. Distribution and evolution of flavonoids in the Palmae and related monocotyledonous families. Bot. Jahrb. Syst. 113: 237–254.

Harley MM. 1990. Occurrence of simple, tectate, monosulcate, or trichotomosulcate pollen grains within the Palmae. Rev. Palaeobot. Palynol. 64: 137–147.

Harley MM. 1999a. The fossil record and palm pollen apertures. In XVI International Botanical Congress: Abstracts, p. 282. St. Louis.

Harley MM. 1999b. Palm pollen: overview and examples of taxonomic value at species level. In: A Henderson and F Borschenius, eds. Evolution, variation, and classification of Palms, Mem. New York Bot. Gard. 83: 95–120.

Harley MM, MH Kurmann, and IK Ferguson. 1991. Systematic implications of comparative morphology in selected Tertiary and extant pollen from the Palmae and the Sapotaceae. In: S Blackmore and SH Barnes, eds. Pollen and spores: patterns of diversification, vol. 44, pp. 335–238. Clarendon Press, Oxford.

Harley MM and WJ Baker. 2001. Pollen aperture morphology in Arecaceae: application within phylogenetic analysis, and a summary of the fossil record of palm-like pollen. Grana. 40: 45–77.

Harley MM and J Dransfield. 2003. Triporate pollen in the Arecaceae. Grana 42: 3–19.

Henderson A. 1986. A review of pollination studies in the palms. Bot. Rev. 52: 221–259.

Henderson A. 2006. Traditional morphometrics in plant sytematics and its role in palm systematics. Bot. J. Linn. Soc. 151: 103–111.

Henderson A and F Borschenius, eds. 1999. Evolution, variation, and classification of Palms. Mem. New York Bot. Gard. 83: 1–324.

Henderson FM. 2006. Morphology and anatomy of palm seedlings. Bot. Rev. 72: 273–329.

Imchanitzkaja NN. 1985. Palms. Nauka, Leningrad (in Russian).

Johnson MAT. 1989. An unusually high chromosome number in *Voaniola gerardii* (Palmae: Arecoideae: Cocoeae: Butiinae). Kew Bull. 44: 207–210.

Klotz LH. 1978a. Form of the perforation plates in the wide vessels of metaxylem in palms. J. Arnold Arbor. 59: 105–128.

Klotz LH. 1978b. The number of wide vessels in petiolar vascular bundles of palms: an anatomical feature of systematic significance. Principes 22: 64–69.

Klotz LH. 1978c. Observations on diameters of vessels in palms. Principes 22: 99–106.

Langlois AC. 1976. Supplement to palms of the world. Gainesville.

Lewis CE and JJ Doyle. 2001. Phylogenetic utility of the nuclear gene malate synthase in the palm family (Arecaceae). Molec. Phylogen. Evol. 19: 409–420.

Lewis CE and JJ Doyle. 2002. A phylogenetic analysis of tribe Areceae (Arecaceae) using two low-copy nuclear genes. Plant Syst. Evol. 236: 1–17

Mason CT. 1999. Arecaceae palm family. J. Ariz. Nev. Acad. Sci. 32: 22–23.

Moore HE, Jr. 1973. The major groups of palms and their distribution. Gentes Herb. 11: 27–141.

Moore HE, Jr and NW Uhl. 1973. Palms and the origin and evolution of monocotyledons. Quart. Rev. Biol. 48: 414–436.

Moore HE, Jr and NW Uhl. 1982. Major trends of evolution in palms. Bot. Rev. 48: 1–69.

Norup MV, J Dransfield, MW Chase, AS Barfod, ES Fernando, and WJ Baker. 2006. Homoplasious character combinations and generic delimitation: a case study from the Indo-Pacific arecoid palms (Arecaceae: Areceae). Am. J. Bot. 93: 1065–1080.

Periasamy K. 1962. Morphological and ontogenetic studies in palms: I. Development of the plicate condition in the palm leaf. Phytomorphology 12: 54–64.

Periasamy K. 1977. Morphological and ontogenetic studies in palms: VI. On the ontogeny of plication in the palm leaf. Proc. Indian Acad. Sci. 85B: 269–273.

Read RW and LJ Hickey 1972. A revised classification of fossil palm and palmlike leaves. Taxon 21: 129–137.

Roncal JJ, J Francisco-Ortega, CB Asmussen, and CE Lewis. 2005. Molecular phylogenetic of tribe Geonomeae (Arecaceae) using nuclear DNA sequence of phosphoribulokinase and RNA polymerase II. Syst. Bot. 30: 275–283.

Röser M. 1994. Pathways of karyological differentiation in palms (Arecaceae). Plant Syst. Evol. 189: 83–122.

Röser M. 2000. DNA amounts and qualitative properties of nuclear genomes in palms (Arecaceae). In: KL Wilson and DA Morrison, eds. Monocots: systematic and evolution, pp. 538–544. CSIRO Collingwood.

Sannier J, S Nadot, A Forchioni, M Harley, and B Albert. 2006. Variations in the microsporogenesis of monoculcate palm pollen. Bot. J. Linn. Soc. 151: 93–102.

Sato D. 1949. Karyotype alterations and phylogeny: VI. Karyotype analysis in Palmae. Cytologia 14 (3–4): 174–186.

Sowunmi MA. 1972. Pollen morphology of the Palmae and its bearing on taxonomy. Rev. Palaeobot. Palynol. 13: 1–80.

Stauffer FW and PK Endress. 2003. Comparative morphology of female flowers and systematics in Geonomeae (Arecaceae). Plant Syst. Evol. 242: 171–203.

Thanikaimoni G. 1970a. Les palmiers: palynologie et systematique. Trav. Sect. Sci. Tech. Inst. Fr. Pondichery, 2: 1–286.

Thanikaimoni G. 1970b. Pollen morphology, classification, and phylogeny of Palmae. Adansonia, ser. 2, 10: 347–365.

Tomlinson PB. 1960–1962. Essays on the morphology of palms. Principes 4: 55–61, 140–143, 1960; 5: 8–12, 46–53, 83–89, 117–124, 1961; 6: 44–52, 122–124, 1962.

Tomlinson PB. 1961. Palmae. In: CR Metcalfe, ed. Anatomy of the monocotyledons, vol. 2. Clarendon Press, Oxford.

Tomlinson PB. 1962. The leaf base in palms, its morphology and mechanical biology. J. Arnold Arbor. 43: 23–50.

Tomlinson PB. 1979. Systematics and ecology of the Palmae. Ann. Rev. Ecol. Syst. 10: 85–107.

Tomlinson PB. 2006a. The uniqueness of palms. Bot. J. Linn. Soc. 151: 5–14.

Tomlinson PB. 2006b. Stem anatomy or climbing palms in relation to long-distance water transport. In: JT Columbus, EA Friar, JM Porter, LM Prince, Mc Simpson, eds. Monocot: comparative biology and evolution. Excluding Poales, pp. 265–277 Rancho Santa Ana Bot. Garden. Rancho Santa Ana Botanic Garden, Claremont, CA.

Tomlinson PB and JR Vincent. 1984. Anatomy of the palm *Rhapis excelsa*. X. Differentiation of stem conducting tissues. J. Arnold Arbor. 65: 191–214.

Uhl NW. 1972. Inflorescence and flower structure in *Nypa fruticans* (Palmae). Am. J. Bot. 59: 729–743.

Uhl NW and J Dransfield. 1987. Genera palmarum: a classification of palms based on the work of Moore, H.E. Jr and Bailey, L.H. Hortorium and the International Palm Society, Allen Press, Kansas.

Uhl NW and HE Moore, Jr. 1971. The palm gynoecium. Am. J. Bot. 58: 945–992.

Uhl NW and HE Moore, Jr. 1977. Centrifugalstamen initiation in phytelephantoid palms. Am. J. Bot. 64: 1152–1161.

Uhl NW, J Dransfield, JI Davis, MA Luckow, KS Hansen, and JJ Doyle. 1995. Phylogenetic relationships among palms: cladistic analyses of morphological and chloroplast DNA restriction site variation. In: PJ Rudall, PJ Cribb, DF Cutler, and CJ Humphries, eds. Monocotyledons: systematics and evolution 2: 623–661. Kew.

Williams CA and JB Harborne. 1973. Negatively charged flavones and tricin as chemosystematic markers in the Palmae. Phytochemistry 12: 2417–2430.

Zimmermann MH and JS Sperry. 1983. Anatomy of the palm *Rhapis exelsa*. IX. Xylem structure of the leaf insertion. J. Arnold Arbor. 64: 599–609.

Zimmermann MH and PB Tomlinson. 1974. Vascular patterns in palm stems: variations of the *Rhapis principle*. J. Arnold Arbor. 55: 402–424.

Zona S. 1997. The genera of Palmae (Arecaceae) in the southeastern United States. Harvard Pap. Bot. 11: 71–107.

Zona S. 2004. Raphides in palm embryos and their systematic significance. Ann. Bot. N.S. 93: 415–421.

Subclass IV. COMMELINIDAE

A large group of orders derived from the Liliidae-Lilianae, most likely from the Amaryllidales-like ancestor. The Commelinidae are characterized by such synapomorphies as endosperm with copious starch, cell walls containing UV-fluorescent compounds (ferulic and coumaric acids), silicon dioxide bodies in leaves, stomatal complexes with well differentiated subsidiary cells, epicuticular waxes of the *Strelitzia*-type. The commelinids are rather advanced, mostly with very specialized inflorescences and often reduced and anemophilous flowers. They constitute a monophyletic group, which is supported both by cladistic analysis (Dahlgren and Rasmussen 1983) and *rbc*L sequence data (Duvall et al. 1993).

Bibliography

Clark WD, BS Gaut, MR Duvall, and MT Clegg. 1993. Phylogenetic relationships of the Bromeliflorae-Commeliniflorae-Zingiberiflorae complex of monocots based on *rbc*L sequence comparisons. Ann. Missouri Bot. Gard. 80: 987–998.

Givnish T J, TM Evans, JC Pires, and KJ Sytsma. 1999. Polyphyly and convergent morphological evolution in Commelinales and Commelinidae: Evidence from *rbc*L sequence data. Molec. Phylogen. Evol. 12: 360–385.

Graham SW, RG Olmstead, and SCH Barett. 2002. Rooting phylogenetic trees with distant outgroups: a case study from the commelinoid monocots. Molec. Biol. Evol. 19: 1769–1781.

Superorder BROMELIANAE

Order 23. BROMELIALES

Mostly short-stemmed epiphytic perennial herbs or sometimes terrestrial xerophytes. Many species of epiphytic genus *Tillandsia* have elongate stem, and species of *Puya* have more or less woody stem, which in *P. raimondii* reaches a height of 10 m; in some species of *Pitcairnia* stem is climbing. In epiphytic species roots often serve as climbing roots. Sacs with bundles of calcium oxalate raphides occur in all parts as well as small, spherical silica bodies in the inner periclinal walls of the epidermal cells of stem and leaves. Sieve-element plastids of form-P2c. Vessels only in roots, in roots and stem, or also in leaves, with scalariform or less often simple perforations. Leaves spirally arranged or rarely (some species of *Tillandsia*) distichous, mostly in basal rosettes, less often cauline, often rigid and xeromorphous, with broad, sheathing base, simple, entire or spinose-derrate, linear, broadly subulate, lanceolate, or rarely ovate, parallel-veined, generally with internal water-storing mesenchymatous tissue and with characteristic water-absorbing, peltate scales with uniseriate stalk, and multicellular, stellate to platelike shield which is often pressed against the epidermis, often forms a vase-shaped rosette in which water accumulates and from which it is absorbed. Stomata tetracytic (with two narrow lateral and two short, terminal subsidiary cells) or sometimes with six subsidiary cells. Flowers in simple or compound racemes, spikes, heads, or panicles, rarely solitary, commonly with well-developed, brightly colored, spathaceous bracts, bisexual or sometimes functionally unisexual, actinomorphic or rarely somewhat zygomorphic, almost always 3-merous. Perianth segments six in two more or less unequal cycles; outer segments generally much smaller, often hyaline or greenish, free or more or less connate; inner segments petaloid and usually brightly colored, free or basally connate into short tube, often basally with ligular appendages. Stamens six in two cycles; filaments narrow, inserted at the base of perianth segments when they are free, and adnate to them when they are connate;

anthers often long and linear, tetrasporangiate, basifixed or dorsifixed, opening longitudinally, introrse. Tapetum plasmodial. Microsporogenesis successive. Pollen grains generally 2-celled, 1-colpate, inaperturate or 2-polyporate (restricted to Bromelioideae), subtectate, mostly reticulate. Gynoecium of three united carpels; style usually long, slender, more or less 3-lobed, with different kinds of stigma (conduplicate-spiral or less often convolute-bladed, simple-erect, cupulate and coralliform, see Brown and Gilmartin 1984, 1988); ovary superior to inferior, 3-locular, usually with more or less numerous ovules. Septal nectaries present. Ovules anatropous or rarely campylotropous, very rarely (*Hohenbergia*) orthotropous, bitegmic, crassinucellate, with parietal cell. Female gametophyte of *Polygonum*-type. Endosperm development of a special subtype of helobial type, in which cell walls are formed sooner in the chalazal chamber than in the micropylar, and often the chalazal chamber lacks starch grains which are copious in the micropylar endosperm (Dahlgren et al. 1985). Fruits usually septicidal capsules or more often berries. Seeds usually small and appendaged, rarely naked; seed coat formed by the inner integument with the various degree of participation of the outer integument; embryo small or sometimes rather large, cylindrical, mostly lateral, with the radicle next to the hilum; endosperm copious, mealy, consisting of large cells containing simple or compound starch grains; peripheral layer of endosperm represented by smaller cells containing aleurone. Flavonoles are present. n = 25, rarely 16, 17, 21, 24, 17, 27, 36, chromosomes very small.

Bromeliaceae are set apart from other monocotyledons in many respects. Their synapomorphies include testal-tegmic seeds (Danilova et al. 1995), unique stellate or scalelike multicellular trichomes, solitary silica bodies, a conspicuously sepaloid and petaloid perianth, and the unusual conduplicate, spiral stigmas, common in all three subfamilies (Gilmartin and Brown 1987; Smith and Till 1998). They are relatively closest to the Velloziaceae (see Huber 1977; Gilmartin and Brown 1987), and both of them are most probably related to the Lilianae one hand and to the Commelinanae (especially Rapateaceae, see Smith 1934) on the other. Similarities between the Bromeliales and Lilianae include the presence of raphides, steroidal saponins and flavonoids, mostly scalariform perforations of the root vessels, the presence of septal nectaries, and helobial endosperm. Also, there are definite similarities between the pollen grains of the Bromeliaceae and those of certain families of the Lilianae, including Hypoxidaceae (Erdtman and Praglowski 1874) and Agavaceae (particularly *Agave*) (Takhtajan 1980). However, from most of the Lilianae the Bromeliaceae differ in their stomatal complex (see Tomlinson 1969) and the presence of silica bodies. In addition, it is important to mention, that members of the Bromeliales show UV-fluorescence in the cell walls, indicating the presence of the same compounds as are found in the Arecanae, Commelinanae and Zingiberanae (Harris and Hartley 1980). Dahlgren et al. (1985) presume that the ancestors of the Bromelianae belonged to the same branch as did the ancestors of both Commelinanae and Zingiberanae (possibly also those of the Arecanae). Unlike most of the Commelinanae they had retained the entomophilous syndrome and in this they are still "liliifloran".

1. BROMELIACEAE

A. L. de Jussieu 1789 (including Tillandsiaceae Wilbread 1834). 57/2700 or more. Tropical and partly subtropical regions of America from Virginia to Patagonia, 1 species of *Pitcairnia* in tropical West Africa.

1.1 PITCAIRNIOIDEAE

Nearly always terrestrial. Roots present and largely functional. Leaves nearly always spinose-serrate. Ovary superior (or largely so in most genera) to rarely inferior or semi-inferior (*Pitcairnia*). Ovules with a chalazal appendage. Fruits generally septicidal capsules with an additional loculicidal apical dehiscence in some genera or rarely indehiscent. Germination epigeal. – BROCCHINIEAE: *Brocchinia;* PITCAIRNIEAE: *Navia, Ayensua, Steyerbromelia, Lindmania, Cottendorfia, Connellia, Fosterella, Pepinia, Pitcairnia;* PUYEAE: *Brewcaria, Deuterocohnia, Puya, Encholirium, Dyckia, Hechtia.*

1.2 TILLANDSIOIDEAE

Mainly epiphytic. Roots often functioning as holdfasts, sometimes lacking. Leaves generally with spineless margin. Petals appendages present or absent. Ovary superior or nearly so, rarely (*Glomeropitcairnia*) semi-inferior. Ovules with or without chalazal appendages. Fruits septicidal capsules. – CATOPSIDEAE: *Gatopsis*; GLOMEROPITCAIRNIEAE: *Glomeropitcairnia*; VRUESEEAE:

Alcanterea, Vriesea, Werauhia; TILLANDSIEAE: *Mezobromelia, Racinaea, Guzmania, Tillandsia, Viridantha.*

1.3 BROMELIOIDEAE

Mainly epiphytic, usually acaulescent. Roots often functioning as holdfasts. Leaves usually spinose-serrate or serrulate. Ovary inferior or nearly so. Ovules usually lack chalazal appendages. Fruits baccate, but often dry. – *Orthophyllum, Cryptanthus, Araeococcus, Fascicularia, Greigia, Fernseea, Ochagavia, Ursulaea, Bilbergia, Neoglaziovia, Bromelia, Deinacanthon, Chevalliera, Pseudananas, Ananas, Acanthostachys, Hohenbergia, Aechmea, Quesnelia, Canistrum, Wittrockia, Portea, Pseudaechmea, Hohenbergiopsis, Androlepis, Lymania, Nidularium, Neoregelia, Streptocalyx, Ronnbergia, Disteganthus.*

Bibliography

Barfuss M, MR Samuel, and W Till. 2004. Molecular phylogeny in subfamily Tillandsioideae (Bromeliaceae) based on six cpDNA markers: An update. J. Brom. Soc. 54: 9–17.

Barfuss MHJ, MR Samuel, W Till, and TF Stuessy. 2005. Phylogenetic relationships in subfamily Tillandioideae (Bromeliaceae) based on DNA sequence data from seven plastid regions. Am. J. Bot. 92: 337–351.

Benzing DH. 1976. Bromeliad trichomes: Structure, function, and ecological significance. Selbyana 1: 330–348.

Benzing DH. 1980. The biology of the Bromeliads. Mad-River Press, Eureka, CA.

Benzing DH, ed. 2000. Bromeliaceae: Profile of an adaptive radiation. Cambridge University Press, Cambridge.

Benzing DH, K Henderson, B Kessel, and J Sulak. 1976. The absorptive capacities of bromeliad trichomes. Am. J. Bot. 63: 1009–1014.

Benzing DH, J Seemann, and A Renfrew. 1978. The foliar epidermis in Tillandsioideae (Bromeliaceae) and its role in habitat selection. Am. J. Bot. 65: 359–365.

Benzing DH, T Givnish, and D Bermudes. 1985. Absorptive trichomes in *Brochhinia reducta* (Bromeliaceae) and their evolutionary and systematic significance. Syst. Bot. 10: 81–91.

Bohme S. 1988. Bromelienstudien: III. Vergleichende Untersuchungen zu Bau, Lage, und systematischer Verwertbarkeit der Septalnektarien von Bromeliaceen. Trop. Subtrop. Pflanzenwelt 62: 1–154.

Brown GK and AJ Gilmartin. 1984. Stigma structure and variation in Bromeliaceae: Neglected taxonomic characters. Brittonia 36: 364–374.

Brown GK and AJ Gilmartin. 1988. Comparative ontogeny of bromeliaceous stigma. In: P Leins, SC Tucker, and PK Endress, eds. Aspects of floral development, pp. 191–204.

Brown GK and AJ Gilmartin. 1989a. Stigma types in Bromeliaceae: A systematic survey. Syst. Bot. 14: 110–132.

Brown GK and AJ Gilmartin. 1989b. Chromosome numbers in Bromeliaceae. Am. J. Bot. 76: 657–665.

Brown GK and RG Terry. 1992. Petal appendages in Bromelliaceae. Am. J. Bot. 79: 1051–1071.

Budnowski A. 1922. The septal glands of the Bromeliaceae. Bot. Arch. 1: 47–80.

Cheadle VI. 1955. Conducting elements in the xylem of the Bromeliaceae. Bull. Bromeliad Soc. 5: 3–7.

Crayn DM, RG Terry, JAC Smith and K Winter. 2000. Molecular systematic investigations in Pitcairnioideae (Bromeliaceae) as a basis for understanding the evolution of crassulacean acid metabolism (CAM). In: KL Wilson and DA Morrison, eds. Monocots: systematic and evolution, pp. 569–579. CSIRO, Collingwood.

Crayn DM, K Winter, and JAC Smith. 2004. Multiple origins of crassulacean acid metabolism and the epiphytic habit in the neotropical family Bromeliaceae. Proc. National Acad. Sci. USA 101: 3703–3708.

Ehler N and R Schill. 1973. Die Pollenmorphologie der Bromeliaceae. Pollen Spores 15: 13–49.

Erdtman G and J Praglowski. 1974. A note on pollen morphology. In: LB Smith and RJ Downs. Bromeliaceae. Flora Neotropica 14: 28–33.

Gilmartin AJ and GK Brown. 1987. Bromeliales, related monocots, and resolution of relationships among Bromeliaceae subfamilies. Syst. Bot. 12: 493–500.

Gilmartin AJ, GK Brown, GS Varadarajan, and M Neighbours. 1989. Status of *Glomeropitcairnia* within evolutionary history of Bromeliaceae. Syst. Bot. 14: 339–348.

Gitai J, R Horres, and AM Benko-Iseppon. 2005. Chromosomal features and evolution of Bromeliaceae. Plant Syst. Evol. 253: 65–80.

Givnish TJ, KC Millan, and KJ Systma. 2003. Origin adaptative radiation, and biogeographic diversification of Bromeliaceae inferred from *dnh*F sequences. In: Monocots III Abstracts, p. 32. Claremont.

Givnish TJ, KJ Systsma, and JF Smith. 1990. A reexamination of phylogenetic relationships among bromeliad subfamilies using cpDNA restriction site variation. Suppl. Am. J. Bot. 77(6): 133 (Abstract).

Givnish TJ, JC Pires, SW Graham, MA McPherson, LM Prince, TB Patterson. 2006. Phylogeny, biogeography, and ecological evolution in Bromeliaceae: insights from *ndh*F sequences. In: JT Columbus, EA Friar, CW Hamilton, JM Porter, LM Prince, and MG Simpson, eds. Monocots: comparative biology and evolution, vol. 2. Rancho Santa Ana Botanical Garden, Claremont.

Gross E. 1988. Bromelienstudien IV. Zur Morphologie der Bromeliaceen-Samen unter Berücksichtung systematisch-taxonomischer Aspekte. Trop. Subtrop. Pflanzenwelt 64: 1–215.

Halbritter H. 1992. Morphologic und systematische Bedeutung des Pollens der Bromeliaceae. Grana 31: 197–212.

Horres R, G Zizka, G Kahl and K Weising. 2000. Molecular systematics of Bromeliaceae: evidence from *trn*L (UAA) intron sequences of the chloroplast genome. Plant Biol. 2: 306–315.

Marchant CJ. 1967. Chromosome evolution in the Bromeliaceae. Kew Bull. 21: 161–168.

Nemirovich-Danchenko EN. 1983. Structure of the seed coat in the representatives of the Bromeliaceae. Bot. Zhurn. 68(8): 1094–1101 (in Russian with English summary).
Ortlieb U and S Winkler. 1977. Ökologische Differenzierungsmuster in der Evolution der Bromeliaceen. Bot. Jahrb. Syst. 97: 586–602.
Owen TP and WW Thomson. 1991. Structure and function of a specialized cell wall in the trichomes of the carnivorous bromeliad *Brocchinia reducta*. Canad. J. Bot. 69: 1700– 1706.
Owen TP, DH Benzing, and WW Thomson. 1988. Apoplastic and ultrastructural characterizations of the trichomes from the carnivorous bromeliad *Brocchinia reducta*. Canad. J. Bot. 66: 941–948.
Ranker TA, DE Soltis, PS Soltis, and AJ Gilmartin. 1990. Subfamilial phylogenetic relationships of the Bromeliaceae, evidence from chloroplast DNA restriction site variation. Syst. Bot. 15: 425–434.
Rauh W. 1981. Bromelien. 2nd ed. Verlag Eugen Ulmer, Stuttgart.
Reinert F, CAM Russo and LO Salles. 2003. The evolution of CAM in the subfamily Pitcairnioideae (Bromeliaceae). Bot. J. Linn. Soc. 80: 261–268.
Sajo MG, CJ Prychid, and PJ Rudall. 2004a. Structure and development of the ovule in Bromeliaceae. Kew Bull. 59: 261–267.
Sajo MG, PJ Rudall, and CJ Prychid. 2004b. Floral anatomy of Bromeliaceae, with particular reference to the evolution of epigyny and septal nectaries in commelinid monocots. Plant Syst. Evol. 247: 215–231.
Sajo MG, CA Furness, CJ Prychid, and PJ Rudall. 2005. Microsporogenesis and anther development in Bromeliaceae. Grana 44: 65–74.
Schulte K, R Horres, and J Zizka. 2005. Molecular phylogeny of Bromelioideae and its implications on biogeography and the evolution of CAM in the family. Sencken. Biol. 85: 113–125.
Sharma AK and I Ghosh. 1971. Cytotaxonomy of the family Bromeliaceae. Cytologia 36: 237–247.
Simpson MG. 1988. A critique of "Bromeliales, related monocots, and resolution of relationships among Bromeliaceae subfamilies". Syst. Bot. 13: 610–614.
Smith JAC. 1989. Epiphytic bromeliads. In: U Lüttge, ed. Vascular plants as epiphytes, evolution and ecophysiology, pp. 108–138. Springer, Berlin.
Smith LB. 1934. Geographical evidence on the lines of evolution in the Bromeliaceae. Bot. Jahrb. Syst. 66: 446–468.
Smith LB. 1955. The Bromeliaceae of Brazil. Smithson. Misce. Collect. 126: 1–290.
Smith LB. 1988. New key to the genera of Bromeliaceae. Beitr. Biol. Pflanzen 63: 403–411.
Smith LB and RJ Downs. 1974, 1977, 1979. Bromeliaceae: 1. Pitcairnioideae. 2. Tillandsioideae. 3. Bromelioideae. Flora Neotropica 14, 1974; Monogr. 14, part 2, 1977; Monogr. 14, part 3, 1979.
Smith LB and W Till. 1998. Bromeliaceae. In: K Kubitzki, ed. The families and genera of vascular plants, vol. 4, pp. 74–99. Springer, Berlin/Heidelberg/New York.
Smith LB and CE Wood. 1975. The genera of Bromeliaceae in the southeastern United States. J. Arnold Arbor. 56: 375–397.
Terry RG, GK Brown, and RG Olmstead. 1997a. Examination of subfamilial phylogeny in Bromeliaceae using comparative sequencing of the plastid locus *ndh*F. Am. J. Bot. 84: 664–670.
Terry RG, GK Brown, and RG Olmstead. 1997b. Phylogenetic relationships in subfamily Tillandsioideae (Bromeliaceae) using *ndh*F sequences. Syst. Bot. 22: 333–345.
Tomlinson G. 1969. Bromeliaceae. In: CR Metcalfe, ed. Anatomy of the monocotyledons, vol. 3, pp. 193–294. Clarendon Press, Oxford.
Varadarajan GS and AJ Gilmartin. 1987. Foliar scales in the subfamily Pitcairnioideae (Bromeliaceae). Syst. Bot. 12: 452–571.
Varadarajan GS and AJ Gilmartin. 1988a. Phylogenetic relationships of groups of genera within the subfamily Pitcairnioideae (Bromeliaceae). Syst. Bot. 13: 283–293.
Varadarajan GS and AJ Gilmartin. 1988b. Taxonomic realignments within the subfamily Pitcairnioideae (Bromeliaceae). Syst. Bot. 13: 294–299.
Varadarajan GS and AJ Gilmartin. 1988c. Seed morphology of the subfamily Pitcairnioideae and its systematic interpretation. Am. J. Bot. 75: 808–818.

Superorder ZINGIBERANAE

Order 24. ZINGIBERALES (CANNALES)

Perennial herbs with sympodial, very short to elongate, often tuberous-thickened and starchy rhizomes or arboreal with a simple, unbranched trunk lacking secondary growth (some Strelitziaceae). Silica bodies in bundle sheath. Vessels only in roots, sometimes present also in the shoot, with simple or scalariform perforations. Leaves distichous or spirally arranged, small to very large, with an opened or closed sheath, usually a distinct petiole, and an entire, simple blade, that is rolled up one side to the other in bud. Stomata paracytic or more seldom tetracytic or hexacytic. Flowers usually in terminal, racemose inflorescences, zygomorphic or asymmetrical, bisexual or unisexual. Sepals three, free or connate to a tube, unlike the petals. Petals three, sometimes dissimilar, free or connate, sometimes joined to the sepals. Functional stamens one or five, rarely six (*Ravenala* and some *Ensete*), usually one to several staminodes are present. Anthers basifixed or rarely dorsifixed (some Zingiberaceae), opening longitudinally. Tapetum secretory or rarely (some Zingiberaceae and Cannaceae) amoeboid. Microsporogenesis successive. Pollen grains usually 2-celled, inaperturate or more rarely 1-colpate (*Zingiber* and *Dimerocostus*), spiraperturate (*Tapeinochilos*) and

5–6-porate (*Costus, Monocostus*). Gynoecium of three united carpels, or one or two of them are more or less reduced or even obsolete. Ovary superior or semi-inferior to inferior, with numerous ovules on the intrusive placentas. Septal nectaries commonly present. Ovules anatropous or rarely (some Marantaceae) campylotropous or (some Zingiberaceae) orthotropous, bitegmic, crassinucellate. Female gametophyte *Polygonum*-type or rarely (Costaceae) *Adoxa*-type. Endosperm nuclear or rarely (Zingiberaceae and Costaceae) helobial. Fruits loculicidal capsules, sometimes schizocarpous, nuts or berries. Seeds often operculate next to the radicle, often arillate, seed coat derived from the outer integument, embryo straight or curved, endosperm copious with compound starch grains, or rarely (Maranthaceae) absent, perisperm usually present, rarely (Strelitziaceae) absent. Contain raphides and chelidonic acid.

Related to Lilianae and Bromelianae and probably have a common origin with Bromeliales from ancient lilealean stock.

Key to Families

1 Functional stamens six or five.

2 The median perianth segment of the inner cycle not larger than the others. Giant or medium-sized more or less arborescent plants with erect, woody in texture trunks or (most species of *Strelitzia* except *S. nicolai*) herbs with short and cormlike stem, generally with a sympodial starchy rhizome with branches terminating in aerial shoots. Roots with anomalous stele filled with wide medullary vessels and scattered phloem islands with a single sieve tube in each strand. Starch grains ellipsoidal to spherical or (*Ravenala*) markedly flattened. Silica cells associated with vascular bundles in all parts of the plant except the roots, silica bodies druselike or spherical. Raphide sacks present in the ground tissue of all organs. Vessels in root and woody stem or (*Ravenala*) only in roots, with scalariform and sometimes also simple perforations. Leaves commonly very large, more or less petiolate (petiole long to often indistinct or even lacking), with short basal sheath; lamina with coarse midrib and closely set parallel, slightly sigmoid lateral veins interconnected by tertiary cross-veins and fusing near margin. Stomata generally tetracytic to hexacytic, often with sunken guard cells. Flowers in axillary or (*Phenacospermum*) terminal thyrse with 1 to several lateral, distichously arranged, few-flowered, compact, monochasial cymes, each one in the axil of large, stiff, brightly colored, persistent, spathaceous bract; individual flowers bi-sexual, more or less zygomorphic, subtended by carinate bracteoles, sweet-smelling. Perianth segments all more or less petaloid, outer segments free from each other and more or less equal, the inner ones slightly or more often strongly dissimilar, the lateral being more or less fused with each other and in *Strelitzia* form a stiff, arrow-shaped keel enclosing functional stamens and style, the median free or only basally connate with the others. Stamens six (*Ravenala*) or five functional and one (belonged to the inner cycle) lacking; filaments elongate, rigid; anthers in *Strelitzia* with slightly protruded connective. Style filiform, often dilated distally; stigmas 3-lobed, long turbinate. Ovary 3-locular, semi-inferior, with conspicuous solid extension above the locules. Ovules numerous in each carpel. Fruits loculicidal capsules. Seeds with brightly colored, hairy arils, with mostly rudimentary operculum (well developed only in *Ravenala*) but without micropylar collar, copious starch- and aleurone-storing endosperm and 1-layered, starchless perisperm; embryo straight or curved. Phenylphenalenones (cyanidin) and flavonols present (*Strelitzia*). n = 7, 8, 11. . . . . 1. STRELITZIACEAE.

2 The median (adaxial) perianth segment of the inner cycle developed as a large labellum. Medium-sized or small perennial herbs with creeping sympodial rhizome rich in compound starch grains and bearing scalelike leaves and very short, sometimes branched stem and rather thick roots. Hat-shaped silica bodies occur in connection with the vascular bundles, and raphide sacs occur in all parts. Vessels only in roots, with scalariform perforations with numerous bars. Leaves of moderate size, distinctly petioled, with broadly lanceolate or elliptic lamina with arching and apically convergent longitudinal lateral veins emerging in regular intervals from the prominent midvein and regularly connected by fine transversal veinlets; palisade tissue lacking, mesophyll of irregularly arranged large and small cells. Stomata paracytic, with deeply sunken asymmetrical guard cells. Inflorescences are

terminal on short leafy branches of the stem; the partial inflorescences consist of sympodially connected 1-flowered monochasial cymes subtended by scalelike bracts; individual flowers bisexual, strongly zygomorphic and superficially orchid-like, bracteate, malodorous. Perianth segments six in two cycles, outer segments subequal, narrow, basally connate into tube with linear-lanceolate and pointed lobes, inner segments strongly dissimilar, the median (adaxial) one large and transformed into variously shaped labellum, the two lateral ones smaller and overlap the stamens. Stamens five, the median (adaxial) one of the inner cycle (opposite the labellum) missing or sometimes staminodial; filaments short; anthers with shortly subulate connective appendage. Style erect, slender, apically widens into 3-lobed, asymmetric, flattened, marginally laciniate-fimbriate stigmatic lobes. Stigma monosymmetric, dorsiventrally flattened, the base part of adaxial side contain secretory tissue (viscidium). Ovary has a solid apical extension, and with numerous ovules per locule. Fruits many-seeded elongate capsules with papery pericarp terminated by a beak formed by the proximal part of the ovary extension. Seeds globose to pyriform, smooth or hairy, with a large lacerate aril and an operculum in front of the radicle. Testa vascularized, exotesta and next two layers lignified, endotesta of radially elongated sclereids; testa encrusted with phytomelan; perisperm slight (Wen et al. 1997). Embryo cylindrical, endosperm copious, with isodiametric starch grains, mostly only one cell layer thick and at the chalazal part several cell layers thick. n = 9. . . . 2. LOWIACEAE.

3 Flowers bisexual. Commonly very tall herbs with creeping sympodial rhizomes, nearly acaulescent or with slender, unbranched aerial stems. Roots with normal polyarch stele or with occasional internal phloem strands. Starch grains cylindrical or ellipsoidal but never flattened. Laticifers lacking. Silica cells (stigmata) internal, adjacent to vascular bundles, with unequally thickened walls, each including a small rectangular silica body, each with a deep central depression. Vessels only in roots, with scalariform perforations. Leaves distichous, with long, basal sheath and petiole (sometimes indistinct or lacking). Stomata tetracytic or less often paracytic, with narrow deep subsidiary cells adjacent to guard cells. Flowers in terminal flattened thyrse consisting of distichously arranged, dense, monochasial cymes (cincinni), each one in the axil of large or rather small, usually brightly colored, keeled or more often boat-shaped bract; individual flowers subtended by carinate bracteoles, strongly zygomorphic. Perianth segments all petaloid, dissimilar, all fused mutually and with the filament bases to form a basal tube; above this the median posterior (adaxial) one of the outer cycle and the filaments are distinct, whereas the other five segments are further connate into a 5-dented or 5-lobed, boat-shaped upper lip. Functional stamens five, with the filaments adnate to the tubular part of the perianth, the outer median stamen staminodial, inserted on the top of the perianth tube opposite the median sepal. Ovules solitary in each locule. Fruits schizocarps with a rather thin pericarp splitting into (2)3 usually blue, fleshy mericarps each of which contains a single seed. Seeds exarillate, with an operculum formed by the tissue of pericarp, whereas only the rudimentary operculum if any are formed by the outer integument (Plisko 1985); perisperm represented by thin aleurone-storing layer, but endosperm copious, starch- and oil-storing; embryo straight or curved. n = 11, 12. .4. HELICONIACEAE.

3 Flowers unisexual. Large to gigantic herbs with short, massive underground rhizome. Root stele anomalous in possessing numerous scattered vessels and phloem islands, each phloem strand with several wide sieve tubes. Starch grains flattened and eccentric. Articulated laticifers occur in association with vascular bundles in all parts except roots. Silica present in all parts except roots as small, more or less rectangular, troughlike bodies in cells (stigmata) with unevenly thickened walls. Vessels in roots, with scalariform and simple perforations. Leaves spirally arranged, with well-developed, coarse, tubular sheath and easily frayed lamina with coarse, compound midvein and closely set, parallel, slightly sigmoid lateral veins interconnects by tertiary cross-veins and fusing near margin. Stomata tetracytic. Flowers in terminal

thyrse which consists of spirally arranged lateral dense and compact monochasial cymes (cincinni) each in the axil of large, firm, keeled or boat-shaped bracts; individual flowers are subtended by hyaline, recurved bracts and unisexual, either female (basally in thyrse), with nonfunctional stamens, or male (apically in thyrse), with nonfunctional gynoecium. Perianth segments petaloid, dissimilar, the three segments of outer cycle and two segments of the inner cycle united into usually 5-dented or 5-lobed, initially tubular structure that soon splits on one side, the third (adaxial) segment of the inner cycle is distinct from the others and directed downward. Functional stamens 6 (most species of *Ensete*) or the median (adaxial) of the inner cycle missing. Ovary 3-locular, inferior; ovules 10–50 in each locule. Fruits berries. Seeds exarillate, with well-developed operculum and micropylar collar; peri-sperm represented by thin layer, but endosperm copious, starchy and mealy; embryo straight or slightly curved (*Ensete*). Alkaloids (indole), and flavonols (kaempferol and quercetin) present or absent, phenylphenalenones (cyanidin, or cyanidin and delphinidin) present; n = 7, 9–14, 16, 17. .... 3. MUSACEAE.

1 Functional stamen one.

4 Leaves distichous. Aromatic herbs. Cells containing ethereal oils present in the parenchyma of all parts of the plant and very common in the abaxial leaf epidermis. Tannin cells in ground parenchyma are common in all parts. Silica commonly restricted to epidermis of lamina, either as silica sand or as solitary, irregularly spherical bodies. Starch grains often flattened. Vessels chiefly in roots and stems, with scalariform, or occasionally with simple perforations. Leaves have long open or rarely closed sheath, with or without ligule. Two sterile lateral stamens of the inner cycle united to form the petaloid labellum: two lateral staminodia of the outer cycle petaloid or inconspicuous, flanking the stamen or adnate to the labellum: the median anterior member of the outer cycle always reduced; anthers usually opening longitudinally, occasionally dehiscing by pores. Pollen grains inaperturate or rarely (*Zingiber*) 1-colpate; exine very thin, hardly resistant to acetolysis, but the intine well developed and lamellated. Septal nectaries (2, sometimes 3 or 1) of various shape present at the style base. Ovary at first always 3-locular but when fully developed either incompletely 3-locular or 1-locular and with parietal or basal-parietal placentation; ovules numerous to many, anatropous or orthotropous. The chalazal chamber of the endosperm degenerates quite early, often as early as the two-celled stage. Fruits mostly loculicidally dehiscing capsules, sometimes indehiscent or fleshy, berrylike. Seeds with thin, generally lobed or laciniate aril (except *Caulokaempferia coenobiales*) and are filled up by starchy perisperm (with mostly compound starch grains) surrounding the thinner endosperm that may contain aleurone grains and sometimes also starch grains; chalazosperm poorly developed or absent; embryo straight (except in *Alpinia*), usually reaching the extension of the perisperm, rarely shorter; testa encrusted with phytomelan. Phenylpropanoids and related curcumins, ethereal oils present. n = 8–21 and 25, mostly 12. . . . 5. ZINGIBERACEAE.

4 Leaves spirally or four-ranked arranged. Nonaromatic herbs with multicellular hairs. Oil and tannin cells lacking. Silica bodies stellate, druse-shaped, restricted to internal cells next to vascular bundles. Starch grains not flattened, often large and irregular. Vessels mainly in roots and stems, usually only with scalariform perforations. Leaves with short, closed tubular sheath; ligule present. Stomata paracytic or tetracytic. Flowers solitary or in terminal or axillary inflorescences, bracteate and sometimes bracteolate. Perianth segments with distinct calyx and corolla; calyx imbricate Fertile stamen often petaloid. All staminodia (2 of the inner cycle and 3 of the outer one) fused into a conspicuous labellum which is united with the basal part of the stamen into a papillate tube. Pollen grains very large, distinctly aperturate, 1–2-colpate, spiraperturate or 5–16-porate, resistant to acetolysis. Ovary 3-locular (*Costus*) or 1–2-locular; ovules numerous per locule, anatropous The chalazal chamber of the endosperm is coenocytic and remains at the time when the embryo is differentiated. Septal nectaries two, immersed in the apical part of the ovary. Fruits capsules dehiscing loculicidally or rarely explosively or

irregularly, occasionally indehiscent (nuts or achene-like). Seeds arillate, filled with strongly developed perisperm containing copious starch composed of simple concentric starch grains and well-developed chalazal tissue; endosperm is little developed and contains starch; embryo cylindrical, rather short, has a cotyledonary sheath and contains fat; testa encrusted with phytomelan. Benzoquinones and steroidal saponins present. n = 8, 9, 14. . . .6. COSTACEAE.

5 Pollen grains spinulose. Perennial rhizomatous herbs, sometimes up to 5 m tall. Stem and rhizome with conspicuous mucilage canals or cavities. Vessels only in root, with simple perforations, or roots without vessels. Starch grains flattened. Leaves distichous to spirally arranged, with arching-convergent lateral veins and without pulvinus; the mesophyll containing calcium oxalate crystals. Stomata paracytic, tetracytic or hexacytic. Flower bisexual, in terminal spikes or in panicles (thyrses), usually composed of 2-flowered cincinni, pairs not enantiomorphic. Inner staminodia petaloid. The single fertile stamen bearing only one half-anther; anther introrse, disporangiate, unilocular. Style flat and fleshy-petaloid, with terminal stigma. Ovary 3-locular; ovules 20–50 per locule, anatropous. Fruits warty capsules, opening by collapse of the pericarp. Seeds exotestal, exarillate, with starch, without operculum but with distinct germination slit and micropylar collar; perisperm without canals; embryo straight; endosperm consisting of a thin, starchy layer. Present cyanidin, flavonols (kaempferol and quercetin), chelidonic acid and aromatic resin; n = 9. .7. CANNACEAE.

5 Pollen grains psilate. Few-flowered herbs, rosulate or with aerial stem of variable length, sometimes lianescent and highly-climbing, rarely epiphytic. Mucilage canals usually lacking (but present in *Thalia*). Vessels in stems and roots with simple or scalariform perforations. Starch grains ellipsoidal and slightly eccentric, rarely flattened or angular. Leaves generally distichous, with arching lateral veins that fuse marginally, some of them at intervals being thicker than the others; petiole with swollen pulvinus at or near the summit that controls the orientation of lamina. Flower pairs are in mirror image (enantiomorphic). The single fertile stamen with a pollen sac on one edge). Two inner staminodia (cucullate and callous-thickened) modified into structures for the explosive release of pollen. Septal nectaries present. Style not petaloid, often expanded and lobed at the summit, with lateral stigma in depression between lobes. Ovary 3-locular, but two often empty or absolete; ovules solitary in fertile locule, anatropous or anacampylotropous Fruits commonly 1- to 3-seeded capsules with dry to fleshy pericarp, sometimes baccate or (*Thalia*) caryopsis-like. Seeds rather large, endotestal, usually with a basal aril (lacking or transformed into a pulp in indehiscent fruits) and with an operculum; testa encrusted with phytomelan; embryo horseshoe-shaped; perisperm with simple, laterally dilated or variously branched canals that originates during seed maturation by an intrusion of the chalazal tissue into the nucellus (Grootjen 1983). Proanthocyanidins (cyanidin) and flavonols (kaempferol and quercetin) present or absent; n = 4–14, 21, 23. . . . . 8. MARANTACEAE.

1. STRELITZIACEAE

Hutchinson 1934. 3/7. Tropical South America (*Phenakospermum*), East coast of South Africa from Cape Province to Transvaal (*Strelitzia*), and Madagascar (*Ravenala*).

RAVENALEAE: *Ravenala;* STRELITZIEAE: *Phenakospermum, Strelitzia.*

According to Tomlinson (1962, 1969), the Strelitziaceae may be considered to represent the nearest approach to the ancestral stock from which the whole order Zingiberales s.l. has evolved. However, Strelitziaceae are a very heterobathmic group that combines some primitive characters (such as complete or nearly complete androecium) with some very advanced characters (such as very specialized perianth and inferior ovary). I agree with Kress (1990: 716) that "the arborescent nature of the Strelitziaceae is a derived character and not a primitive feature as (Tomlinson 1962) suggested."

2. LOWIACEAE

Ridley 1924. 1/16. Southern China (Kwangtung province and Hainan), Indochina, Malay Peninsula, Sumatra, Java, Borneo, and Brunei.

Orchidantha (Lowia).

A very distinct taxon characterized by many unique characters that isolate it from other zingiberids (see Kress 1990). However, it definitely belong to the "banana group" of the Zingiberanae (Kress 1995) and has many common features with the Musaceae (including operculate seeds) and especially with the Strelitziaceae (including capsular fruits, arillate seeds and pollen grains resembling much of *Ravenala*) (Larsen 1998).

3. MUSACEAE

A. L. de Jussieu 1789. 3/60. Tropical Africa (from the Cameroon to Ethiopia and Transvaal), India, southeastern Asia, Malesia, New Guinea, northeastern Australia, and the Pacific Islands.

Musa, Ensete, Musella.

4. HELICONIACEAE

Nakai 1941. 1/200 +. Southern Central America and South America (especially northern parts of the Andes), with a few species in, Melanesia.

Heliconia.

Markedly differ from the related Musaceae in the absence or articulated laticifers, root anatomy, distichous leaves, bisexual flowers, adaxially placed median member of the outer cycle of the perianth, the presence of a single staminodium opposite the unpaired adaxial member of the outer cycle of perianth, plasmodial tapetum, and especially the unique type of the schizocarpous fruits.

5. ZINGIBERACEAE

Martynov 1820 (including Alpiniaceae Link 1821). 53/1350. Madagascar, Mascarenes, Seychelles, tropical and South Africa, southern, eastern, and Southeast Asia, Malesia, Pacific Islands, New Guinea, and northern and eastern Australia (mainly Queensland), with some representatives in Neotropics. The center of diversity is in tropical Asia. *Tamijia* – Kalimantan, *Siphonochilus* – Africa and Madagascar.

Classification after Kress, Prince and Williams (2002).

5.1 TAMIJIOIDEAE

1/1. Plants evergreen with fibrous rhizomes; ovary unilocular with parietal placentation. Plane of distichy of leaves perpendicular to rhizome. Lateral staminodes well developed and fused to labellum. – *Tamijia.*

5.2 SIPHONOCHILOIDEAE

1/15. Plants with seasonal dormancy period and fleshy rhizomes; ovary trilocular with parietal placentation. Plane of distichy of leaves perpendicular to rhizome. Lateral staminodes well developed and fused to labellum. – *Siphonochilus.*

5.3 ALPINIOIDEAE

Lateral staminodes reduced or absent. Plane of distichy of leaves perpendicular to rhizome. – ALPINIEAE: *Aframomum, Alpinia, Amomum, Autolandra, Cyphos-tigma, Elettaria, Elettariopsis, Etlingera, Geocharis, Geostachys, Hornstedtia, Leptosolena, Paramomum, Plagiostachys, Renealmia, Vanoverberghia*; RIEDELIEAE: *Burbidgea, Pleuranthodium, Riedelia, Siamanthus.*

Insertae sedis: *Siliquamomum* (1, southern China, Indochina).

5.4 ZINGIBEROIDEAE

Plane of distichy of leaves parallel to rhizome. – ZINGIBEREAE: *Boesenbergia, Camptandra, Cautleya, Cornukaempferia, Curcuma, Curcumorpha, Distichochlamys, Haniffia, Haplochorema, Hedychium, Hitchenia, Kaempferia, Laosanthus, Nanochilus, Paracautleya, Parakaempferia, Pommereschea, Pyrgophyllum, Rhynchanthus, Roscoea, Scaphochlamys, Smithatris, Stadiochilus, Stahlianthus, Zingiber*; GLOBBEAE: *Gagnepainia, Globba, Hemiorchis, Mantisia.*

Insertae sedis: *Caulokaempferia* (10, Himalayas, Southeast Asia).

6. COSTACEAE

Nakai 1941. 4/120. The largest genus *Costus* (95) pantropical, *Monocostus* (1, Peru) and *Dimerocostus* (3–5) neotropical, *Tapeinochilus* (18) in East Malesia to northeastern Australia.

Costus, Dimerocostus, Monocostus, Tapeinochilos. Close related to the Zingiberaceae.

7. CANNACEAE

A. L. de Jussieu 1789. 1/10-25. Tropical and subtropical regions of America, especially of western South America.

Canna.

Related to the Marantaceae. "Differences between the two families including the positioning of pollen on the ventral face in Cannaceae, the dorsal face in Marantaceae, indicate that both are different lineages" (Kubitzki 1998).

8. MARANTACEAE

R. Brown 1814. 31/550. Nearly pantropical, but especially numerous in tropical America and absent in Australia.

Phrynium, Ataenidia, Monophrynium, Phacelophrynium, Stachyphrynium, Galathea, Sanblasia, Pleiostachya, Ischnosiphon, Monotagma, Donax, Schumannianthus, Hypselodelphys, Trachyphrynium, Megaphrynium, Sarcophrynium, Maranta, Marantochloa, Afrocalathea, Monophyllanthe, Koernickanthe, Myrosma, Saranthe, Hylaeanthe, Ctenanthe, Stromanthe, Halopegia, Thalia, Thaumatococcus, Cominsia, Haumania.

Bibliography

Andersson L. 1981. The neotropical genera of Marantaceae: circumscription and relationships. Nord. J. Bot. 1: 218–245.

Andersson L. 1998. Heliconiaceae, Marantaceae, Musaceae, Strelitziaceae. In: K Kubitzki, ed. The families and genera of vascular plants, vol. 4, pp. 226–230, 278–293, 296–300, 451–455. Springer, Berlin/Heidelberg/New York.

Andersson L and MW Chase. 2001. Phylogeny and classification of Marantaceae. Bot. J. Linn. Soc. 135: 275–287.

Beltran IC and K Kiew. 1984. Cytotaxonomic studies in the Zingiberaceae. Notes Roy. Bot. Gard. Edinb. 41: 541–559.

Box MS and PJ Rudall. 2006. Floral structure and ontogeny in *Globba* (Zingiberaceae). Plant Syst. Evol. 258: 107–122.

Burn BL and RM Smith. 1972. Tentative keys to the subfamilies, tribes, and genera of Zingiberaceae. Notes Roy Bot. Gard. Edinb. 31: 171–176.

Burtt BL. 1972. General introduction to papers on Zingiberaceae. Notes Roy. Bot. Gard. Edinb. 31: 155–165.

Burtt BL and RM Smith. 1972. Key species in the taxonomic history of Zingiberaceae. Notes Roy. Bot. Gard. Edinb. 31: 177–227.

Carlquist S and EL Schneider. 1998. Origins and nature of vessels in monocotyledons. 3. Lowiaceae, with comments on rhizome anatomy. Blumea 43: 219–224.

Cheesman EE. 1947. Classification of the bananas. Kew Bull. 2: 97–117.

Chen Z-Y. 1989. Evolutionary patterns in cytology and pollen structure of Asian Zingiberaceae. In: LB Holm-Nielsen, IC Nielsen, and H Balslev, eds. Tropical forests, pp. 185–191. Academic Press, London.

Fahn A. 1953. The origin of the banana inflorescence. Kew Bull. 1953: 299–306.

Fahn A and P Benouaiche. 1979. Ultrastructure, development, and secretion in the nectary of banana flowers. Ann. Bot. 44: 85–93.

Fisher JB. 1978. Leaf-opposed buds in *Musa*: their development and a comparison with allied Monocotyledons. Am. J. Bot. 65: 784–791.

Friedrich WL and F Strauch. 1975. Der Arillus der Gattung *Musa*. Bot. Not. 128: 339–349.

Graven P, CGT Koster, JJ Boon, and F Bouman. 1996. Structure and macromolecular composition of the seed coat of the Musaceae. Ann. Bot. (UK). 7(2): 105–122.

Graven P, CG De Koster, JJ Boon, and F Bouman. 1997. Functional aspects of mature seed coat of the *Cannaceae*. Plant Syst. Evol. 205: 223–240.

Gris A. 1860. Note sur l'origine et le mode de formation des canaux perispermiques dans la graine des Marantees. Ann. Sci. Nat. Bot, ser. 4, 13: 97–102.

Grootjen CJ. 1983. Development of ovule and seed in Marantaceae. Acta Bot. Neerl. 32: 69–86.

Grootjen CJ and E Bouman. 1981. Development of ovule and seed in *Costus cuspidatus* (N. E. Br.) Maas (Zingiberaceae), with special reference to the formation of the operculum. Bot. J. Linn. Soc. 83: 27–39.

Grootjen CJ and F Bouman. 1988. Seed structure in Cannaceae: taxonomic and ecological implications. Ann. Bot. 61: 363–371.

Hesse M and M Waha. 1983. The fine structure of the pollen wall in *Strelitzia reginae* (Musaceae). Plant Syst. Evol. 141: 285–298.

Holttum RE. 1950. The Zingiberaceae of the Malay Peninsula. Gard. Bull. Straits Settlem. 13: 1–249.

Holttum RE. 1951. The Marantaceae of Malaya. Gard Bull. Singapore 13: 254–296.

Holttum RE. 1970. The genus *Orchidantha* (Lowiaceae). Gard. Bull. Singapore 25: 239–246.

Holttum RE. 1974. A commentary on comparative morphology in Zingiberaceae. Gard. Bull. Singapore 27: 155–165.

Humphrey JE. 1896. The development of the seed in the Scitamineae. Ann. Bot. 10: 1–40.

Jain RK. 1963. Studies of Musaceae: I. *Musa cardiosperma* sp. nov., a fossil banana fruit from the Deccan Intertrappean series, India. Paleobotanist 12: 45–58.

Jaramillo MA and WJ Kress. 1997. Phytlogenetic relationships of the genera of the family Costaceae. Bull. *Heliconia* Soc. Int. 9: 5–8.

Johansen LB. 2005. Phylogeny of *Orchidantha* (Lowiaceae) and the Zingiberales based on six DNA regions. Syst. Bot. 30: 106–117.

Kamelina OP. 1990. The development of male and female embryonic structures in the Marantaceae family. Bot. Zhurn. 75: 480–483 (in Russian with English summary).

Kamelina OP and PH Raven 1997. Embryological features of the Lowiaceae family. Bot. Zhurn. 82: 1–6 (in Russian with English summary).

Kennedy H. 1978. Systematics and pollination of the "closed flowered" species of *Calathea* (Marantaceae). Univ. Calif. Publ. Bot. 71: 1–90.

Kirchoff BK. 1983. Floral organogenesis in five genera of the Marantaceae and in *Canna* (Cannaceae). Am. J. Bot. 70: 508–523.

Kirchoff BK. 1988a. Floral ontogeny and evolution in the ginger group of the Zingiberales. In: P Leins, SC Tucker, and PK Endress, eds. Aspects of floral development, pp. 45–56. J. Cramer, Berlin.

Kirchoff BK. 1988b. Inflorescence and flower development in *Costus scaber* (Costaceae). Canad. J. Bot. 66: 339–345.

Kirchoff BK. 1991. Homeosis in the flowers of the Zingiberales. Am. J. Bot. 78(6): 833–837.

Kirchoff BK. 1992. Ovary structure and anatomy in the Heliconiaceae and Musaceae (Zingiberales). Canad. J. Bot. 70: 2490–2508.

Kirchoff BK. 1997. Inflorescence and flower development in the Hedychieae (Zingiberaceae): *Hedychium*. Canad. J. Bot. 75: 581–594.

Kirchoff BK and H Kennedy. 1985 Foliar, nonstructural nectaries in the Marantaceae. Canad. J. Bot. 63: 1785–1788.

Kirchoff BK and H Kunze. 1995. Inflorescence and floral development in *Orchidantha maxillarioides* (Lowiaceae). Int. J. Plant Sci. 156: 159–171.

Kress WJ. 1984. Systematics of Central American *Heliconia* (Heliconiaceae) with pendant inflorescences. J. Arnold Arbor. 65: 429–532.

Kress WJ. 1986. Exineless pollen structure and pollination system of tropical *Heliconia* (Heliconiaceae). In: S Blackmore and IK Ferguson, eds. Pollen and spores: form and function, pp. 329–345. Linnean Soc., London.

Kress WJ. 1990a. The phylogeny and classification of the Zingiberales. Ann. Missouri Bot. Gard. 77: 698–721.

Kress WJ. 1990b. The taxonomy of Old World *Heliconia* (Heliconiaceae). Allertonia 6: 1–58.

Kress WJ. 1993. Morphology and floral biology of *Phenakospermum* (Strelitziaceae), an arborescent herb of the Neotropics. Biotropica 25: 290–300.

Kress WJ. 1995. Phylogeny of the Zingiberanae: morphology and molecules. In: PJ Rudall, PJ Cribb, DF Cutler, and CJ Humphries, eds. Monocotyledons: systematics and evolution, vol. 2, pp. 443–460. Royal Botanic Gardens, Kew.

Kress WJ. 1998. The Zingiberales: phylogenetic relationships among the families – one last time. In: Monocots II, p.34 (Abstract). Sydney.

Kress J and K Larsen. 2001. *Smithatris*, a new genus of Zingiberaceae from Southeast Asia. Syst. Bot. 26: 226–230.

Kress WJ and DE Stone. 1982. Nature of the sporoderm in monocotyledons, with special reference to the pollen grains of *Canna* and *Heliconia*. Grana 21: 129–148.

Kress WJ and DE Stone. 1983. Morphology and phylogenetic significance of exineless pollen of *Heliconia* (Heliconiaceae). Syst. Bot. 8: 149–167.

Kress WJ and CD Specht. 2005. Between Cancer and Capricorn: phylogeny, evolution and ecology of the primarily tropical Zingiberales. Biol. Skr. 55: 459–478.

Kress WJ and CD Specht. 2006. The evolutionary and biogeographic origin and diversification of the tropical monocot order Zingiberales. In: JT Columbus, EA Friar, CW Hamilton, JM Porter, LM Prince, and MG Simpson, eds. Moconots: comparative biology and evolution, pp.619–630. Clermont. [Aliso 22: 619–630].

Kress WJ, DE Stone and SC Sellers. 1978. Ultrastructure of exineless pollen: *Heliconia* (Heliconiaceae). Am. J. Bot. 65: 1064–1076.

Kress WJ, LM Prince, WJ Hahn and EA Zimmer. 2001. Unraveling the evolutionary radiation of the families of the Zingiberales using morphological and molecular evidence. Syst. Biol. 50: 926–944.

Kress WJ, LM Prince and KJ Williams. 2002. The phylogeny and new classification of the gingers (Zingiberales); evidence from molecular data. Am. J. Bot. 89: 1682–1696.

Kronestedt E and B Walles. 1986. Anatomy of the *Strelitzia reginae* flower (Strelitziaceae). Nord. J. Bot. 6: 307–320.

Kubitzki K. 1998. Cannaceae. In: K Kubitzki, ed. The families and genera of vascular plants, vol. 4, pp. 103–106. Springer, Berlin/Heidelberg/New York.

Kunze H. 1984. Vergleichende Studien an Cannaceen- und Marantaceenblüten. Flora 175: 301–318.

Kunze H. 1985. Die Infloreszenzen der Marantaceen und ihr Zusammenhang mit dem Typus der Zingiberales-Synfloreszenz. Beitr. Biol. Pfl. 60: 93–140.

Kunze H. 1986. Infloreszenz- und Blütenmorphologie von *Orchidantha maxillarioides* (Ridl.) K. Schum. (Lowiaceae). Beitr. Biol. Pfl. 61: 221–234.

Lane IE. 1955. Genera and generic relationships in Musaceae. Mitt. Bot. Staatssamml. München 2(13): 114–131.

Larsen K. 1966. Chromosome cytology and relationship of the Lowiaceae. Nat. Hist. Bull. Siam. Soc. 21: 21–24.

Larsen K. 1998a. Costaceae. In: K Kubitzki, ed. The families and genera of vascular plants, vol. 4, pp. 128–132. Springer, Berlin/Heidelberg/New York.

Larsen K. 1998b. Lowiaceae. In: K Kubitzki, ed. The families and genera of vascular plants, vol. 4, pp. 275–277. Springer, Berlin/Heidelberg/New York.

Larsen K and J Mood. 1998. *Siamanthus*, a new genus of Zingiberaceae from Thailand. Nat. Hist. Bull. Siam Soc. 45: 217–221.

Larsen K and T Jenjittikul. 2001. *Laosanthus*, a new genus of Zingiberaceae from Laos. Nord. J. Bot. 21: 135–138.

Larsen K, JM Lock, H Maas, and PJM Maas. 1998. Zingiberaceae. In: K Kubitzki, ed. The families and genera of vascular plants, vol. 4, pp. 474–495. Springer, Berlin/Heidelberg/New York.

Liao J, Y Wen, and Q Wu. 1998 Studies on vascular system anatomy of the flower of *Orchidantha chinensis* T. L. Wu. J. Trop. Subtrop. Bot. 6: 275–282.

Liao J-P and Q-G Wu. 2000. A preliminary study of the seed anatomy of Zingiberaceae. Bot. J. Linn. Soc. 134: 287–300.

Liao J-P, Y-J Tang, X-L Ye, and Q-G Wu. 2004. Seed anatomy of species in banana families of Zingiberales and its systematic significance. J. Trop. Subtrop. Bot. 12: 291–297 (in Chinese.).

Long H and Y Wen. 1997. Pollen morphology of Lowiaceae from China. J. Trop. Subtrop. Bot. 5(3): 6–9.

Maas PJM. 1972. Costoideae (Zingiberaceae). Flora Neotropica, Monograph 8. Hafner, New York.

Maas PJM. 1977. *Renealmia* (Zingiberaceae-Zingiber-oideae) Costoideae (Additions) (Zingiberaceae). Flora Neotropica, Monograph 18. New York.

Mangaly JK and J Nayar. 1990. Palynology of South Indian Zingiberaceae. Bot. J. Linn. Soc. 103: 351–366.

Manning JC and P Goldblatt. 1989. Chromosome number in *Phenacospermum* and *Strelitzia* and the basic chromosome number in Strelitziaceae (Zingiberales). Ann. Missouri Bot. Gard. 76: 932–933.

Mauritzon J. 1936. Samenbau und Embryologie einiger Scitamineen. Acta Univ. Lund. 31(9): 1–31.

Mood J and K Larsen. 1997. *Cornukaemperia*, a new genus of Zingiberaceae from Thailand. Natural History Bulletin of the Siam Society 45: 217–221.

Nakai T. 1948. A new attempt in the classification of the Strelitziaceae. Bull. Tokyo Sci. Mus. 22: 19–24.

Nam TV. 1975. Costaceae et Zingiberaceae, leurs appareils ligulaires. Adansonia, ser. 2, 14: 561–570.

Newman SWH and BK Kirchoff. 1992. Ovary structure in the Costaceae (Zingiberales). Int. J. Plant Sci. 153: 471–487.

Ngamriabsakul C, MF Newman, and QC Cronk. 2004. The phylogeny of the tribe Zingibereae (Zigiberaceae) based on ITS (nrDNA) and *trn*L-F (cpDNA) sequences. Edinb. J. Bot. 60: 483–507.

Nur N. 1976. Studies on pollination in Musaceae. Ann. Bot. 40: 167–177.

Olatunji OA. 1980. The structure and development of stomata in some Zingiberales. Notes Roy. Bot. Gard. Edinb. 38: 499–516.

Pai RM. 1965. The floral anatomy of *Elettaria cardamomum* Maton. A reinvestigation. New Phytol. 64: 187–204.

Pedersen LB. 2004. Phylogenetic analysis of the subfamily Alpinioideae (Zingiberaceae), particularly *Etlingera* Giseke, based on nuclear and plastid DNA. Plant Syst. Evol. 245: 239–258.

Pedersen LB and B Johansen. 2004. Anatomy of the unusual stigma in *Orchidantha* (Lowiaceae). Am. J. Bot. 91: 299–305.

Plisko MA. 1985. Zingiberales. In: A Takhtajan, ed. Comparative seed anatomy, vol. 1, pp. 206–235. Nauka, Leningrad (in Russian).

Prince LM and WJ Kress. 2006. Phylogenetic relationships and classification in Marantaceae: insights from plastid DNA sequence data. Taxon 55: 281–296.

Pugialli HRL, MAC Kaplan and OR Gottlieb. 1994. Evolucao flavonoidica em *Zingiberales*. An. Acad. Bras. Ci. 66: 491–494.

Punt W. 1968. Pollen morphology of the American species of the subfamily *Costoideae* (Zingiberaceae). Rev. Palaeobot. Palynol. 7: 31–43.

Rao VS. 1963. The epigynous glands of *Zingiberaceae*. New. Phytol. 62: 342–349.

Raynolds PK. 1927. The banana. Houghton Mifflin Company, Cambridge, MA.

Rao VS, H Karnik, and K Gupte. 1954. The floral anatomy of some *Scitamineae*. I. J. Indian Bot. Soc. 33: 118–147.

Rogers GK. 1984. The Zingiberales (Cannaceae, Maranthaceae and Zingiberaceae) in the southeastern United States. J. Arnold Arbor. 65: 5–55.

Rowley JR and JJ Skvarla. 1986. Development of the pollen grain wall in Canna. Nord. J. Bot. 6: 39–65.

Rudall P and RM Bateman. 2003. Evolution of zygomorphy in monocot flowers: iterative patterns and developmental constraints. New Phytol. 162: 25–44.

Sakai S and H Nagamasu. 2000. Systematic studies of Bornean Zingiberaceae: III. *Tamijia*: a new genus. Edinb. J. Bot. 57: 245–255.

Schachner J. 1924. Beiträge zur Kenntnis der Blüten- und Samenentwicklung der Scitamineen. Flora 117: 16–40.

Scutch AF. Anatomy of the axis of the banana. Bot. Gaz. 93: 233–258.

Searle RJ and TAJ Hedderson. 2000. A preliminary phylogeny of the Hedychieae tribe (Zingiberaceae) based on ITS sequences of the nuclear rRNA cistron. In: KL Wilson and DA Morrison, eds. Monocots: systematics and evolution, pp. 710–718. CSIRO Collinwood.

Simao DG and VL Scatena. 2001. Morphology and anatomy in *Heliconia angusta* Vell. and *H. velloziana* L. Emygd. (*Zingiberales: Heliconiaceae*) from the Atlantic forest of southeastern Brazil. Rev. Brasil. Bot. 24(4): 415–424.

Simão DG, VL Scatena, F Bouman. 2006. Developmental anatomy and morphology of the ovule and seed of *Heliconia* (Heliconiaceae, Zingiberales). Plant Biol. 8: 143–154.

Simmonds NW. 1962. The evolution of the bananas. Longman, London.

Skvarla J and JR Rowley. 1970. The pollen wall of *Canna* and its similarity to the germinal apertures of other pollen. Am. J. Bot. 57: 519–529.

Smith JE, WJ Kress, and EA Zimmer. 1993. Phylogenetic analysis of the Zingiberales based on *rbc*L sequences. Ann. Missouri Bot. Gard. 80: 620–630.

Song J-J, J-P Liao, Y-J Tang, and Z-Y Chen. 2004. Chromosome numbers in *Orchidantha* (Lowiaceae) and their biogeographic and systematic implications. Ann. Bot. Fenn. 41: 429–433.

Song J-J, YJ Tang, J-P Liao, X-X Huang and Z-Y Chen. 2003. Chromosome numbers of *Orchidantha (Lowiaceae)*. Acta Bot. Yunn. 25: 609–612 (in Chinese).

Specht CD. 2005. Phylogenetics, floral evolution and rapid radiation in the tropical monocot family Costaceae (Zingiberales). In: AK Sharma and A Sharma, eds. Plant genome: biodiversity and evolution, pp. 29–60. Science Publishers, Enfield, NH.

Specht CD. 2006. Systematics and evolution of the tropical monocot family Costaceae (Zingiberales): a multiple dataset approach. Syst. Bot. 31: 89–106.

Specht CD and DW Stevenson. 2006. A new phylogeny-based generic taxonomy for the monocot family Costaceae (Zingiberales). Taxon 55: 153–163.

Specht CD, WJ Kress, DW Stevenson and R DeSalle. 2001. A molecular phylogeny of Costaceae (Zingiberales). Molec. Phylogen. Evol. 21: 333–345.

Stone DE, SC Sellers, and WJ Kress. 1979. Ontogeny of exineless pollen in *Heliconia,* a banana relative. Ann. Missouri Bot. Gard. 66: 701–730.

Stone DE, SC Sellers, and WJ Kress. 1981. Ontogenetic and evolutionary implications of a neotenous exine in *Tapeinochilos* (Zingiberales: Costaceae) pollen. Am. J. Bot. 68: 49–63.

Tanaka N. 2001. Taxonomic revision of the family Cannaceae in the New World and Asia. Makinoa, n.s. 1: 1–74.
Tang YJ and JP Liao. 2001. Anatomy and histochemistry of *Paramomus* (sic) *petaloideum* seeds. Acta. Bot. Yunn. 23: 321–327.
Tang YJ, JP Liao, YQ Wen, and QG Wu. 2000 Studies on vascular system anatomy of *Strelitzia nicolai* flower. Acta Bot. Yunn. 22: 291–297.
Tang Y-J, Z-Y Xie, J-P Liao, and Q-C Wu. 2005. Anatomy and histochemistry of the seeds in ginger families of Zingiberales and their systematic significance. Acta Bot. Bor.-Occid. Sinica 25: 343–354. (In Chinese).
Thellade I and J Thellade. 1996. Ontogeny of pollen grains of *Zingiber spectabile* (Zingiberaceae). Grana 35: 162–170.
Tiwari SC and BES Gunning. 1986. Development of tapetum and microspores in *Canna* L.: an example of an invasive but non-syncytial tapetum. Ann. Bot. 57: 557–463.
Tomlinson P. B. 1956. Studies in the systematic anatomy of the Zingiberaceae. Bot. J. Linn. Soc. 55:547–592.
Tomlinson PB. 1959. An anatomical approach to the classification of the Musaceae. Bot. J. Linn. Soc. 55: 779–809.
Tomlinson PB. 1960. The anatomy of *Phenakospermum* (Musaceae). J. Arnold Arbor. 41: 287–297.
Tomlinson PB. 1961. Morphological and anatomical characteristics of the Marantaceae. Bot. J. Linn. Soc. 58: 55–78.
Tomlinson PB. 1961. The anatomy of *Canna*. Bot. J. Linn. Soc. 56: 467–473.
Tomlinson PB. 1962. Phylogeny of the Scitamineae Morphological and anatomical considerations. Evolution 16: 192–213.
Tomlinson PB. 1969. Classification of the Zingiberales (Scitamineae) with special reference to anatomical evidence. In: CR Metcalfe, ed. Anatomy of the monocotyledons, vol. 3, pp. 295–302. Clarendon Press, Oxford.
Van de Venter HA. 1976. Notes on the morphology of the embryo and seedling of *Strelitzia reginae* Ait. South Afr. J. Bot. 42: 63–69.
Wen Y and J Liao. 1999. Floral morphology and anatomy of *Orchidantha chinensis* (Lowiaceae). J. Trop. Subtrop. Bot. 7(4): 329–336.
Wen YQ and JP Liao. 2001. Embryological study on *Orchidantha chinensis* (Lowiaceae). J. Trop. Subtrop. Bot. 9(4): 301–305.
Wen Y, J Liao, and Q Wu. 1997. Anatomy and histochemistry of the seed of *Orchidantha sinensis* T.L. Wu. Guihaia 17: 235–241.
Williams KJ, WJ Kress and PS Manos. 2004. The phylogeny, evolution, and classification of the genus *Globba* and tribe Globbeae (Zingiberaceae): appendages do matter. Am. J. Bot. 91: 100–114.
Winnell S, Newman H and Kirchoff BK. 1992. Ovary structure in the Costaceae (Zingiberales). Int. J. Plant Sci. 153: 471–487.
Wood TH, WM Whitten, and NH Williams. 2000. Phylogeny of *Hedichium* and related genera (Zingiberaceae) based on ITS sequence data. Edinb. J. Bot. 57: 261–270.
Wu Felin. 1964. Lowiaceae, a family new to the flora of China. Acta Phytotax. Sinica. 9: 335–343.
Wu T-L, Q-G Wu, and Z-Y. Chen. 1996. Proceedings of the second symposium on the family Zingiberaceae. Zhongshan University, China.

Superorder COMMELIANAE

Order 25. COMMELINALES

Perennial or rarely annual, mostly terrestrial herbs with various types of underground storage organs, including rhizomes, root tubers or, rarely, bulbs. Raphides (calcium oxalate crystals) are present in raphide sacs or raphide canals; silica bodies present in some genera of Commelinaceae and *Hanguana*. Vessels in roots, sometimes also in stems and leaves, with scalariform of simple perforations. Leaves alternate, spirally arranged or distichous, petiolate or sessile, basal or cauline, simple, entire, usually with parallel venation. Stomata usually tetracytic or paracytic. Flowers in various kinds or terminal or axillary inflorescences, sometimes solitary, bisexual or rarely unisexual, actinomorphic to zygomophic, usually 3-merous, bracteate or not. Perianth segments six in two cycles, rarely (Philydraceae) of four perianth segments. Stamens basically six in two cycles, rarely stamen solitary (Philydraceae), or three stamens, with or without staminodia; filaments free or basally connate, glabrous or sometimes pubescent; anthers basifixed or dorsifixed, introrse to extrorse, sometimes versatile, tetrasporangiate, usually opening longitudinally. Tapetum secretory, amoeboid. Microsporogenesis successive. Pollen grains generally 2-celled and 1-colpate, sometimes inaperturate. Gynoecium of three united carpels; style with capitate, 3-lobed stigma; ovary mostly 3-locular, superior or rarely inferior, with solitary to numerous ovules per locule. Ovules anatropous, hemitropous, orthotropous, or rarely campylotropous, bitegmic, crassinucellate to tenuinucellate. Female gametophyte of *Polygonum*-type or rarely *Allium*-type. Fruits mostly loculicidal capsules, rarely indehiscent and than sometimes baccate. Seeds usually small, of various shapes and sculpturing, rarely arillate; embryo small, usually straight, well-differentiated to undifferentiated; endosperm copious, with starch grains and aleuron bodies. Containing flavonoids, ferulic and diferulic acids.

Have many similarities with the Bromelianae as well as with the Zingiberanae. They are linked with the Bromelianae and Zingiberanae by such conspicuous synapomorphies as epiticular wax of the *Strelitzia*-type, cell walls containing UV-fluorescent compounds, and endosperm with copious starch (Dahlgren and Rasmussen 1983).

Key to Families

1 Vessels in all organs. Perennial or sometimes annual, terrestrial or rarely epiphytic herbs with rhizomes, root tubers or, rarely, bulbs. Roots fibrous or tuberous. Calcium oxalate crystals are present in raphide sacs forming articulated raphide-canals (lacking only in *Cartonema*). Contain silica bodies. Starch, when present in vegetative parts, consists of simple, excentric grains. Vessels usually in all organs, with simple perforations. Three-celled glandular microhairs present in most genera. Leaves usually alternate, spirally arranged or distichous, simple, entire, commonly somewhat succulent, often narrowed into a false petiole, parallel-veined; leaf-sheath closed and well-defined. Stomata tetracytic, hexacytic, or sometimes paracytic. Flowers in various kinds of terminal or axillary, basically cymose inflorescences, sometimes solitary, bisexual or sometimes polygamous, actinomorphic or more or less zygomorphic, 3-merous, without nectaries. Perianth segments six in two distinct cycles. Sepals green or sometimes petaloid and colored, usually free, imbricate, generally similar. Petals ephemeral, free and sometimes clawed or occasionally basally connate, imbricate. Stamens basically six in two cycles, but often there are only three to one fertile stamens and two to four staminodia, rarely three or fewer fertile stamens without staminodia; filaments slender, free or rarely basally connate, glabrous or some or all bearded with brightly colored long hairs; anthers basifixed or dorsifixed, sometimes versatile, introrse to extrorse, opening longitudinally or rarely by apical pores or porelike slits. Tapetum amoeboid. Pollen grains generally 2-celled, mostly 1-colpate, rarely with two additional apertures. Gynoecium of three united carpels; style with capitate, 3-lobed or penicillate stigma; ovary sessile or stipitate, 3-locular or rarely 2-locular (1 locule being underdeveloped or even suppressed), with 1 to ca. 20 ovules per locule. Ovules anatropous, hemitropous, orthotropous, or rarely campylotropous, crassinucellate to tenuinucellate, with or without parietal cell. Female gametophyte of *Polygonum*-type or rarely *Allium*-type. Endosperm nuclear. Fruits mostly loculicidal capsules, rarely indehiscent and then sometimes baccate. Seeds generally small, hard, of various shapes and sculpturing, rarely arillate; seed coat testal-tegmic, formed by endotesta and exotegmen and in some degree also endotegmen (Vyshenskaya 1985), with disclike or conic operculum (embryotega) covering the embryo and surrounded by a micropylar collar except in *Cartonema*, where it still can be identified as a bump; endotesta silicified; endosperm copious, starchy, contains compound starch grains and a small amount of aleurone; embryo small, straight, lateral, well differentiated except in *Cartonema*, situated at the opposite end to the hilum beneath the operculum. 6-hydroxyflavonoids are present. n = 4 to 29. . . . . .2. COMMELINACEAE.

1 Vessels only in roots and sometimes in stems.

2 Vessels with scalariform perforation

3 Ovules crassinucellate

4 Tapetum secretory, its cells 2–4-nucleate and always contain raphids. Perennial pubescent herbs with rhizomes or corms and fibrous roots. Small glandular hairs and longer, uniseriate hairs often present. Tanniniferous cells and idioblasts with styloids of calcium oxalate widespread in all organs, but silica bodies lacking in the vegetative parts. Vessels in roots and in stems (Schnaider and Carlquist 2005), with scalariform perforations. Leaves sessile, unifacial, ensiform to linear or terete, mostly basal, basal and lower cauline leaves distichous, with well-developed sheath, the rest smaller and spirally arranged; venation parallel. Stomata paracytic, but in *Helmholtzia* and *Orthothylax* have two to four lateral and two less distinct polar subsidiary cells. Flowers in terminal simple or paniculately branched spikes, sessile in the axils of rather large spathaceous bracts, bisexual, strongly zygomorphic, 3-merous. Perianth petaloid, of four free or basally connate members; the large (adaxial) member, which is frequently 3- or 2-cuspidate, consists of 3 fused perianth segments – 2 segments of the outer cycle and the reduced median (adaxial) segment of the inner cycle; two other lateral segments of the inner cycle are very small, resemble staminodia, and are basally more or less adnate to the stamen; the large abaxial member of the perianth represents the only free segment of the

outer cycle (Hamann 1966, 1998, 1998). Stamen solitary, opposite the abaxial perianth segment, corresponds to the median stamen of the outer cycle of the ancestral androecium which is otherwise suppressed; the inner cycle of stamens is always lacking; filament strongly flattened, inserted at the base of the perianth segment; anthers basifixed to dorsifixed, opening longitudinally, straight and introrse, or curved (*Phillydrella*), or helically twisted (*Philydrum*). Tapetum secretory, its cells 2–4-nucleate and always contain raphids. Pollen grains 2-celled, in monads or (*Philydrum*) tetragonal tetrads, 1-colpate, tectate-columellate, with distinctive lamellae inside the homogeneous foot layer, reticulate; the aperture wall consists of a thick, 2-layered intine with exine absent or composed of scattered deposits (Simpson 1985). Gynoecium of 3 united carpels; style persistent, with a capitate, slightly 3-lobed stigma; ovary superior, 3-locular or for the greater part 1-locular with deeply intruded and bibrachiate parietal placentas, with numerous ovules on each placenta. Septal nectaries or other nectariferous glands absent. Ovules anatropous, weakly crassinucellate, with parietal cell. Female gametophyte of *Polygonum*-type. Endosperm formation of a peculiar subtype of the helobial type, in which the small chalazal endosperm chamber becomes cellular earlier than the large micropylar chamber; only the micropylar chamber serves as storage tissue (Hamann 1966, 1998, 1998). Fruits loculicidal capsules or rarely berrylike. Seeds small to minute, generally very numerous, reddish-brown, ellipsoidal to ovoid or linear-cylindrical, spirally striate-tuberculate (*Philydrum*), reticulate-rugose (*Philydrella*) or silky-striate (*Helmholtzia*), with generally conspicuous red-colored chalazal cap and an operculum (later pushed off by the germinating embryo) formed by the micropylar region of both layers of the inner integument; seed coat testal-tegmic; endosperm copious, rich in starch (consisting of large single grains), crystalline aleuron bodies and also some oil; embryo small, linear, straight, weakly differentiated. Ferulic and diferulic acids present in the cell wall, also contain proanthocyanins. n = 8, 16, 17. . . . . 3. PHILYDRACEAE

4 Tapetum amoeboid; cells 1–2-nucleare without raphids. Perennial or rarely (*Hydrothrix*) annual, rhizomatous or stoloniferous, swamp or aquatic herbs ranging from amphibious through floating-leaved and free-floating to almost completely submerged. Branching of the main axis usually sympodial. Trichomes are absent on vegetative organs. Tanniniferous idioblasts have been reported from the petioles. Oxalate raphides are widely present and cells with one or two prismatic crystals have been found in some genera. Vessels in roots and often also in stem, with scalariform perforations. Leaves both basal and cauline, distichous or spiro-distichuous along stem, usually with considerable sheath enveloping stem, with distinct and sometimes inflated petiole and lanceolate to ovate or broadly cordate, rarely linear or even; filiform undifferentiated lamina, sometimes reduced. Stomata paracytic. Flowers in terminal racemes, spikes, or panicles or solitary, often subtended by two spathelike bracts, bisexual, actinomorphic to zygomorphic, 3-merous, bracteate or not, sometimes triheterostylous and with trimorphic pollen grains. Perianth of 6 (in *Scholleropsis* mostly 4) usually petaloid segments in two cycles, nearly free or more often basally connate into a tube. Stamens mostly six in two cycles or only three stamens (sometimes unequal), rarely (*Scholleropsis*) three stamens and one staminodium, or (*Hydrothrix*) only one stamen and two staminodia; filaments slender, attached to the perianth tube; anthers basifixed or peltate, opening longitudinally and introrse or opening by apical pores. Tapetum amoeboid, cells 1–2-nucleate. Pollen grains 2-celled or rarely 3-celled, with 1–3 distal or subequatorial

colpi, exine verrucate, scabrate to finely reticulate. Gynoecium usually of three united carpels (in *Pontederia* pseudomonomerous); style slender, with capitate or 3-lobed stigma; ovary superior, mostly 3-locular, with several or numerous ovules or less often (*Pontederia*) with one apical pendulous ovule. Septal nectaries mostly present. Ovules anatropous, crassinucellate, with parietal cell. Female gametophyte of *Polygonum*-type. Endosperm helobial, with a small chalazal chamber where cell wall formation begins much earlier than in the large micropylar chamber. Fruits many-seeded loculicidal capsules or (*Pontederia* and *Reussia*) 1-seeded nutlets surrounded by postfloral-modified, hardened, rigid perianth bases. Seeds small, usually longitudinally furrowed, with relatively large, cylindrical embryo surrounded by copious endosperm consisting of elongate cells containing starch and aleurone layer; cotyledon linear, bifacial; seed coat exotestal; exotestal cells box-like, elongated, endotestal cells thin, elongated transversely; the inner one is represented by strongly compressed layer containing a reddish-brown pigment. n = (7)8(-13). . . . . 4. PONTEDERIACEAE

3 Ovules tenuinucellate. Robust perennial herbs growing in humid forests, lowland swamps or along slowly moving rivers and lakes (sometimes floating), with creeping rhizomes and long stolons arising from the base of the erect stems and forming dense colonies. Short multicellular branched hairs common on stem and leaf of young plants, each with a deeply sunken, uniseriate base. Conspicuous tannin cells present. Vessels only in roots, with scalariform perforations. Leaves rolled unilaterally in bud (ptyxis supervolute), mostly basal, spirally arranged, linear to lanceolate, acute, pseudopetiolate, with strong midrib, parallel longitudinal veins are linked with numerous close-set thin cross veinlets, sheathing at base, usually with silica bodies; the sheat and petiolar region contain large mucilage channels. Stomata tetracytic. Flowers small, in large terminal panicles composed of spike-like partial inflorescences, with broad, large bracts, dioecious, actinomorphic, 3-merous. Perianth segments six in two cycles, shortly connate at base, persistent, the inner cycle red-dotted, larger and vaulted. Male flowers with rudimentary gynoecium which is surrounded by a lobed, disc-like structure Stamens six, inserted on the base of perianth segments; filaments subulate with broader base; anthers small, basifixed, tetrasporangiate, opening longitudinally, introrse; female flowers with six staminodia. Tapetum amoeboid, with a periplasmodium. Pollen grains more or less spheroidal, inaperturate, with thin, hyaline exine provided with small spinules. Gynoecium of three united carpels; ovary superior, 3-locular, with one ovule per locule; stigma subsessile or sessile, broadly 3-lobed. Characteristic mucilage-secreting hairs present within the ovary locule (Rudallet al. 1999). Ovules almost orthotropous, probably tenuinucellate, with funicular obturator. Fruits fleshy thick-walled 1–3-seeded drupes. Seeds bowl-shaped, meso-endotestal, without phytomelan or phlobaphene; endosperm copious, clearly divided into several (up to 7) separate segments which possibly are the remnants of nucellus (Nemirovich-Danchenko 1985) and enclosed by a single peripheral later of aleurone cells, with simple and compound starch grains of various size and form (Nemirovich-Danchenko); embryo small, globose. n = 12, 18, and more than 45; chromosomes very small.. . . . . 1. HANGUANACEAE

2 Vessels with simple or rarely scalariform perforation. Perennial rhizomatous and stoloniferous or less often cormose to bulbous herbs mostly with various kinds of trichomes on both the vegetative and reproductive parts. Roots fasciculate and fibrous, sometimes tuberous. Oxalate raphides present in all parts, but silica bodies lacking. Sieve-element plastids of form-P2c. Vessels usually only in roots but sometimes (*Lachnanthes*) also in stem, with simple or (*Dilatris* and *Tribonanthes, Anigozanthos flavidus* – Conostylidaceae) scalariform perforations. Leaves mostly basal or subbasal, distichous, lanceolate, narrowly linear or acicular, unifacial, sheathing at base, parallel-veined, sometimes (*Barberetta* and *Wachendorfia*) plicate. Stomata paracytic. Flowers in terminal, generally cymose

inflorescences or solitary, pedicellate or subsessile, bisexual, actinomorphic to strongly zygomorphic, 3-merous, conspicuously hairy or glabrous. Perianth segments six, persistent, membranous to coriaceous, either composed of three outer and three inner imbricate, free or basally connate similar segments or (in *Anigozanthos, Blancoa,* and *Conostylis*) perianth connate into a short or long, straight or curved tube with six monocyclic and valvate lobes. Stamens six in two cycles or three opposite the inner perianth segments or (*Pyrrorhiza*) one opposite the inner anterior perianth segment and two staminodia opposite the inner latero-posterior segments; filaments free or adnate to perianth tube, equal or unequal in length and/or position; staminodia sometimes present; anthers generally oblong, tetrasporangiate, basifixed to dorsifixed, opening longitudinally, introrse, straight or arcuate, sometimes with a connective appendage. Tapetum amoeboid; radially oriented, lignified cell wall thickenings occur in the endothecial and middle layer cells (Simpson 1998b). Pollen grains 2-celled, with 1–3-layered atectate exine, 1-colpate or 2–8-porate. Gynoecium of three united carpels, all fertile or (*Barberetta*) only the posterior carpel fertile at maturity; style slender, straight to strongly curved to one side of the flower, with capitate or 3-lobed stigma; ovary superior to inferior, 3-locular or (*Phlebocarya*) basally 3- and apically 1-locular, with one to numerous ovules per carpel. Septal nectaries present or rarely (as in *Xiphidium*) absent. Ovules anatropous or orthotropous, crassinucellate, with parietal cell. Female gametophyte of *Polygonum*-type. Endosperm helobial, with early cell formation in the chalazal chamber. Fruits loculicidal to apically poricidal capsule, sometimes indehiscent (in *Macropidia* dehiscing into three single-seeded mericarps with a hard endocarp). Seeds of various shapes, glabrous or hairy; seed coat formed by both integuments, either predominantly by the outer integument (in Haemodoroideae) or mainly by the inner integument (Conostylidoideae); embryo globose to ovoid or kidney-shaped, lateral, in Conostylidoideae minute and undifferentiated, in Haemodoroideae larger and more differentiated; endosperm copious, differentiated into peripheral aleurone and lipid layer, in which cell walls store cellulose, and an inner layer containing simple or compound starch grains. Flavones, fructans and arylphenalenones are present. The order is chemically unique in possessing perinaphthenone pigments which are responsible for the red coloration prominent in the roots and rootstocks of the Haemodoraceae s. str. Chelidonic acid widely distributed in the order. Species of *Wachendorfia* and *Anigozanthos* have cell wall-bound p-coumaric and diferulic acids.. . . 5. HAEMODORACEAE

1. HANGUANACEAE

Airy Shaw 1965. 1/5 or more. Sri Lanka, Indochina, Malesia, Palau Islands, Australia.

Hanguana.

Bentham and Hooker (1883), Baillon (1892), Lotsy (1911), Hutchinson (1934, 1973) and many others included *Hanguana* in the Flagellariaceae. However, it differs markedly from the Flagellariaceae in many respects (Erdtman 1952; Smithson 1956; Tomlinson 1969). Endlicher (1841) included *Hanguana* in Astelieae, which he considered "Genera Juncaceis affinia," and the affinity to the Asteliaceae has been accepted by Dahlgren et al. (1985) and Takhtajan (1987). However, there are many important differences between Asteliaceae and Hanguanaceae, including tetracytic stomata, inaperturate pollen grains, presence of funicular obturator, bowl-shaped seeds, and peculiar segmented and starchy endosperm. According to Stevenson and Loconte (1995), and Rudall, Stevenson and Linder (1999) the Hanguanaceae are related to the Zingiberales. According to molecular data (Chase et al. 2005) Commelinaceae and Hanguanaceae are sister taxa. Both families contain silica bodies in leaves and ferulic acid in cell walls (Harris and Hartley 1980; Rudall and Caddick 1994).

2. COMMELINACEAE

Mirbel 1804 (including Cartonemataceae Pichon 1946, Ephemeraceae Batsch 1802, Tradescantiaceae Salisbury 1807). 40/650. Tropical and subtropical regions, especially Africa, southern Asia, Mexico, and northern Central America with relatively few representatives in temperate parts of eastern Asia, southern North America, and Australia. Subfamily Cartonematoideae is confined to the Southern Hemisphere.

Classification after T.M. Evans, R. B. Faden, M. G. Simpson, and K. J. Sytsma (2000).

2.1 CARTONEMATOIDEAE
Stem collenchyma absent. Glandular microhairs lacking. Raphide canals lacking or only next to the veins. Flowers actinomorphic – CARTONEMATEAE: *Cartonema;* TRICERATELLEAE: *Triceratella.*

2.2 COMMELINOIDEAE
Stem collenchyma present. Glandular microhairs nearly always present. Raphide canals present and mostly between the veins of the lamina. Flowers actinomorphic to zygomorphic — TRADESCANTIEAE: *Palisota, Streptolirion, Spatholirion, Aetheolirion, Cyanotis* (including *Amischophacelus*), *Belosynapsis, Coleotrype, Porandra, Amischotolype, Dichorisandra, Siderasis, Geogenanthus, Cochliostema, Plowmanianthus, Thyrsanthemum, Gibasoides, Tinantia, Elasis, Matudanthus, Weldenia, Gibasis, Tradescantia, Callisia, Tripogandra, Sauvallea*; COMMELINEAE: *Stanfieldiella, Floscopa, Buforrestia, Murdannia, Anthericopsis, Tricarpelema, Pseudoparis, Polyspatha, Dictyospermum, Pollia, Aneilema, Rhopalephora, Commelina.*

3. PHILYDRACEAE

Link 1821. 4/5–6. Eastern and Southeast Asia, Guam, Andaman Islands, New Guinea, Australia (all 4 genera and 5 species occur in Australia).

Helmholtzia, Orthothylax, Philydrum, Philydrella.

Philydraceae are related to the Pontederiaceae. The common features of the Philydraceae and Pontederiaceae include the ovary position, embryology (embryo and endosperm in the ripe seed), vegetative anatomy (paracytic stomata, styloid crystal idioblasts, etc.) and similar proanthocyanins (see Hamann 1966, 1998).

4. PONTEDERIACEAE

C. Kunth 1816 (including Heterantheraceae J. Agardh 1858, Unisemataceae Rafinesque 1837). 9/33. Widespread in tropics, with a few species extending into warm-temperate areas; the centre of diversity is tropical and subtropical America.

EICHHORNIEAE: *Eichhornia, Monochoria*; PONTEDERIEAE: *Pontederia, Reussia;* HETERANTHEREEAE: *Heteranthera, Zosterella, Hydrothrix, Scholleropsis, Eurystemon.*

Related to the Philydraceae, but even closer to the Haemodoraceae (see Dahlgren and Rasmussen 1983 and Simpson 1987). According to Simpson (1987: 113), the exine architecture of the non-tectate-columellate genera of the Pontederiaceae is strikingly similar (and presumably homologous) to that of members of the family Haemodoraceae. The UV-fluorescent ferulate in the cell walls, oxalate raphides and styloids, helobial endosperm with a diminutive chalazal chamber and plasmodial tapetum suggest relationships with both the Philydraceae and Haemodoraceae (Cook 1998). In addition the seedlings of the deriaceae is very similar to those of Philydraceae (Tillich 1995).

5. HAEMODORACEAE

R. Brown 1810 (including Conostylidaceae Takhtajan 1987, Dilatridaceae M. Roemer 1840, Wachendorfiaceae W. Herbert 1837, Xiphidiaceae Dumortier 1829). 14/10. Southern Africa, New Guinea, Australia, Tasmania, eastern and southeastern North America and Nova Scotia, and tropical America.

5.1 HAEMODOROIDEAE
Mostly with red coloration in roots and rootstocks. Distinctive tannin cells absent. Starch grains in vegetative parts simple. Outer cycle of androecium lacking or (*Schiekia*) represented by two vestiges of outer cycle, inner cycle represented by three fertile stamens or (*Pyrrorhiza*) by one fertile stamen and two staminodia. Pollen grains boat-shaped, colpate, heteropolar, with verrucate or (*Schiekia*) foveolate exine sculpturing. Ovary superior or inferior. n = 7, 8, 12, 15, 19–21. – *Haemodorum, Lachnanthes, Dilatris, Schiekia, Pyrrorhiza, Xiphidium, Barberetta, Wachendorfia.*

5.2 CONOSTYLIDOIDEAE
Roots and rootstocks without red coloration. Distinctive tanniniferous idioblasts present in various organs, especially in placental tissues. Starch grains in vegetative parts compound, but those in endosperm simple. In

Phlebocarya silica bodies are present. Stamens six; staminodia absent Pollen grains, with distinctive rugulate (brainlike) exine sculpturing and 2–8 circular apertures, either fusiform with two apertures (*Anigozanthos, Blancoa, Conostylis* spp., *Macropidia*), cylindric with two apertures (*Phlebocarya*), triangular with three apertures (*Conostylis* spp.), or globose with 7–8 apertures. Ovary mostly inferior (semi-inferior in *Blancoa*). n = 4–8, 11, 14, 16, 21, 28 (mostly 6 or 8). – *Phlebocarya, Tribonanthes, Conostylis, Blancoa, Anigozanthos, Macropidia.*

Related to the Pontederiaceae (Dahlgren and Rasmussen 1983), which is supported by palynological data (Simpson 1987, 1990, 1998), and to the Philydraceae, supported by such common features as ensiform leaves, paracytic stomata, the pubescent flowering stems, features of helobial endosperm formation and its starchy contents and plasmodial tapetum (Hamann 1998), and to both the Pontederiaceae and Philydraceae in possessing p-coumaric and diferulic acids (Simpson 1998). In addition, members of Philydraceae and Conostylidoideae and most members of Pontederiaceae have tannin idioblasts scattered in various plant organs (Simpson 1998).

Bibliography

Adams LG. 1987. Philydraceae. Flora Australia 45: 40–46.

Anderberg AA and P Eldenas. 1991. A cladistic analysis of *Anigozanthos* and *Macropidia* (Haemodoraceae). Austral. J. Bot. 4: 655–664.

Banerji I and S Haldar. 1942. A contribution to the morphology and cytology of *Monochoria hastaefolia* Presl. Proc. Indian Acad. Sci. 16B: 91–106.

Barett SCH. 1988. Evolution of breeding systems in *Eichhornia* (Pontederiaceae): a review. Ann. Missouri Bot. Gard. 75: 741–760.

Barrett SCH and SW Graham. 1997. Adaptive radiation in the aquatic plant family Pontederiaceae: insight from phylogenetic analysis. In: YJ Givnish and KJ Sytsma, eds. Molecular evolution and adaptive radiation, pp. 225–228. Cmabridge University Press, Cambridge.

Bayer C, O Appel, and PJ Rudall. 1998. Hanguanaceae. In: K Kubitzki, ed. The families and genera of vascular plants, vol. 4, pp. 223–225. Springer, Berlin/Heidelberg/New York.

Brenan JPM. 1966. The classification of Commelinaceae. Bot. J. Linn. Soc. 59: 340–370.

Brückner G. 1926. Beiträge zur Anatomie, Morphologie und Systematik der Commelinaceae. Bot. Jahrb. Syst. 61, Beiblatt 137: 1–70.

Burton E and MG Simpson. 1996. Floral anatomy and phylogeny of the Pontederiaceae using new and revised morphological characters and outgroup information. Am. J. Bot. 83: 143 (Abstract).

Chikkannaiah PS. 1963. Embryology of some members of the family Commelinaceae: *Commelina subulata* Roth. Phytomorphology 13: 174–184.

Chub VV and EV Mavrodiev. 2001. Morphological characters of the leaf series in the Commelinaceae family with special emphasis on the number of prophylls and their homology in monocots. Bot. Zhurn. 86(4): 1–11 (in Russian with English summary).

Coker WC. 1907. The development of the seed in the Pontederiaceae. Bot. Gaz. 44: 293–301.

Cook CDK. 1998. Pontederiaceae. In: K Kubitzki, ed. The families and genera of vascular plants, vol. 4, pp. 395–403. Springer, Berlin/Heidelberg/New York.

Cooke RG and JM Edwards. 1981. Naturally occurring phenalenones and related compounds. Fortschr. Chem. Org. Naturst. 40: 158–190.

Daumann E. 1965. Das Blütennektarium bei den Pontederiaceen und die systematische Stellung dieser Familie. Preslia 37: 407–412.

Dellert R. 1933. Zur systematischen Stellung von *Wachendorfia*. Oesterr. Bot. Z. 82: 335–345.

De Vos MP. 1956. Studies on the embryology and relationships of South African genera of the Haemodoraceae: *Dilatris* Berg. and *Wachendorfia* Burm. South Afr. J. Bot. 22: 41–63.

De Vos MP. 1961. On the embryology and relationships of the South African genera of the Haemodoraceae. In: Recent Adv. Bot. 1: 194–698.

Eckenwalder JE and SCH Barnett. 1986. Phylogenetic systematics of Pontederiaceae. Syst. Bot. 11: 373–391.

Evans TM, RB Faden, and KJ Sytsma. 2000a. Homoplasy in the Commelinaceae: a comparison of different classes of morphological characters. In: KL Wilson and DA Morrison, eds. Monocots: systematic and evolution, pp. 557–565. CSIRO, Collingwood.

Evans TM, RB Faden, MG Simpson, and KJ Sytsma. 2000b. Phylogenetic relationships in the Commelinaceae: I. A cladistic analysis of morphological data. Syst. Bot. 25: 668–691.

Evans TM, KJ Sytsma, RB Faden, and TJ Givnish. 2003. Phylogenetic relationships in the Commelinaceae: II. A cladistic analysis of *rbc*L sequences and morphology. Syst. Bot. 28: 270–292.

Faden RB. 1991. The morphology and taxonomy of *Aneilema* R. Brown (Commelinaceae). Smithsonian Contr. Bot. 76: 1–166.

Faden RB. 1998. Commelinaceae. In: K Kubitzki, ed. The families and genera of vascular plants, vol. 4, pp. 109–128. Springer, Berlin/Heidelberg/New York.

Faden RB and DR Hunt. 1991. The classification of the Commelinaceae. Taxon 40: 19–31.

Faden RB and Y Suda. 1980. Cytotaxonomy of Commelinaceae: chromosome numbers of some African and Asiatic species. Bot. J. Linn. Soc. 81: 301–325.

Geerinck D. 1968. Considerations taxonomiques au sujet des Haemodoraceae et des Hypoxidaceae (Monocotyledones). Bull. Soc. Bot. Belg. 101: 265–278.

Geerinck D. 1969a. Genera des Haemodoraceae et des Hypoxidaceae. Bull. Jard. Bot. Nat. Belg. 39: 47–82.

Geerinck D. 1969b. Le genre *Conostylis* R. Br. (Haemodoraceae d'Australie). Bull. Jard. Bot. Nat. Belg. 39: 167–177.

Geerinck D. 1970. Revision du genre *Anigozanthos* Labill. (Haemodoraceae d'Australie). Bull. Jard. Bot. Nat. Belg. 40: 261–276.

Givnish TJ, TM Evans, and KJ Sytsma. 1995. The Commelinales: an extreme example of convergent evolution in monocots. Am. J. Bot. 82(6): 132 (Abstract).

Givnish TJ, TM Evans, JC Pires and KJ Sytsma. 1999. Polyphyly and convergent morphological evolution in Commelinales and Commelinidae: evidence from *rbc*L sequence data. Molec. Phylogen. Evol. 12: 360–385.

Graham SW and SCH Barrett. 1995. Phylogenetic systematics of Pontederiales: implications for breeding -system evolution. In: PJ Rudall, PJ Cribb, DF Cutler and CJ Humphries, eds. Monocotyledons: systematics and evolution, vol. 2, pp. 415–441. Royal Botanic Gardens, Kew.

Graham SW, JR Kohn, BR Morton, JE Eckenwalder, and SCH Barrett. 1998. Phylogenetic congruence and discordance among one morphological and three molecular data sets from Pontederiaceae. Syst. Biol. 47(4): 545–567.

Green JW. 1959. The genus *Conostylis* R. Br.: I. Leaf anatomy Proc. Linn. Soc. N. S. W. 84: 194–206.

Green JW. 1961. The genus *Conostylis* R. Br.: II. Taxonomy. Proc. Linn. Soc. N. S. W. 85: 334–373.

Hamann U. 1962. Über Bau und Entwicklung des Endosperms der Philydraceae und über die Begriffe "mehliges Nährgewebe" und "Farinosae." Bot. Jahrb. Syst. 81: 397–407.

Hamann U. 1963. Die Embryologie von *Philydrum lanuginosum* (Monocotyledoneae-Philydraceae). Ber. Deutsch. Bot. Ges. 76: 203–208.

Hamann U. 1966. Embryologische, morphologisch-anatomische, und systematische Untersuchungen an Philydraceen. Willdenowia 4: 1–178.

Hamann U. 1998. Philydraceae. In: K Kubitzki, ed. The families and genera of vascular plants, vol. 4, pp. 389–394. Springer, Berlin/Heidelberg/New York.

Hardy CR and DW Stevenson. 2000. Development of gametophytes, flower and floral vasculature of *Cochliostema odoratissimum* (Commelinaceae). Bot. J. Linn. Soc. 134: 131–157.

Heime NA and HP Linder. 1992. Morphology, evolution, and taxonomy of *Wachendorfia* (Haemodoraceae). Bothalia 22: 59–75.

Hofreiter A and HJ Tillich. 2002. Root anatomy of the Commelinaceae (Monocotyledoneae). Feddes Repert. 113(3–4): 231–255.

Hopper SD and NA Campbell. 1977. A multivariate morphometric study of species relationships in kangaroo paws (*Anigozanthos* Labill. and *Macropidia* Drumm. ex Harv.: Haemodoraceae). Austral. J. Bot. 25: 523–544.

Hopper SD, MF Fay, M Rossetto, and MW Chase. 1999. A molecular phylogenetic analysis of the bloodroot and kangaroo paw family, Haemodoraceae: taxonomic, biogeographic and conservation implications. Bot. J. Linn. Soc. 131: 285–299.

Horn CN. 1998. Pontederiaceae pickerelweed family. J. Arizona-Nevada Acad. Sci. 30(2): 133–136.

Jones K and C Jopling. 1972. Chromosomes and the classification of the Commelinaceae. Bot. J. Linn. Soc. 65: 129–162.

Kapil RN and K Walia. 1965. The embryology of *Philydrum lanuginosum* Banks ex Gaertn, and the systematic position of the Philydraceae. Beitr. Biol. Pfl. 41: 381–404.

Keighery GJ. 1981. Pollination and the generic status of *Blancoa canescens* Lindl. (Haemodoraceae). Flora 171: 521–524.

Larsen K. 1972. Flagellariaceae; Hanguanaceae. Flora of Thailand. 2(2): 162–166. Bankok, Applied Scientific Research Corporation of Thailand, [for flora of Thailand]

Larsen K. 1983. Hanguanaceae. In: JF Leroy, ed. Flore du Cambodge, du Laos et du Vietnam. Part 20: 129–132. Paris.

Lee DW, Yap Kim Pin, and Liew Foo Yew. 1975. Serological evidence on the distinctness of the monocotyledonous families Flagellariaceae, Hanguanaceae and Joinvilleaceae. Bot. J. Linn. Soc. 70: 77–81.

Maas PJM and H Maasvan de Kamer. 1993. Neotropical Haemodoraceae. Flora Neotropica 61: 1–44. New York.

MacFarlane TD, SD Hopper, RW Purdie, AS George, and SJ Patrick. 1987. Haemodoraceae. Flora Australia 45: 55–148. Canberra.

Maheshwari SC and B Baldev. 1958. A contribution to the morphology of *Commelina forskalaei* Vahl. Phy-tomorphology 8: 277–298.

Maury MP. 1888. Sur les affinites du genre *Susum*. Bull. Soc. Bot. France 35: 410–417.

Murty YSN, NP Saxena, and V Singh. 1974. Floral morphology of the Commelinaceae. J. Indian Bot. Soc. 53: 127–136.

Nemirovich-Danchenko EN. 1985a. Hanguanaceae. In: A Takhtajan, ed. Comparative seed anatomy, vol. 1, pp. 98–99. Nauka, Leningrad (in Russian).

Nemirovich-Danchenko EN. 1985b. Haemodoraceae. In: A Takhtajan, ed. Comparative seed anatomy, vol. 1, pp. 113–117. Nauka, Leningrad (in Russian).

Nemirovich-Danchenko EN. 1985c. Pontederiaceae. In: A Takhtajan, ed. Comparative seed anatomy, vol. 1, pp. 121–122. Nauka, Leningrad (in Russian).

Oganezova GG. 1995. On the systematical position of the families Haemodoraceae, Hypoxidaceae and Taccaceae (data on the seed structure). Bot. Zhurn. 80(7): 12–25 (in Russian with English summary).

Ono T. 1928. Embryologische Studien an einigen Pontederiaceen. Sci. Rep. Tohoku Imp, Univ., 4th ser. (Biol.) 3: 405–415.

Ornduff R. 1979. Chromosome numbers and relationships of certain African and American genera of Haemodoraceae. Ann. Missouri Bot. Gard. 66: 577–580.

Pichon M. 1946. Sur les Commélinacées. Notul. Syst. (Paris) 12: 217–242.

Poole MM and DR Hunt. 1980. Pollen morphology and the taxonomy of the Commelinaceae: an exploratory survey. Vol. 8 of American Commelinaceae. Kew Bull. 34: 639–660.

Prychid CJ, CA Furness and PJ Rudall. 2003. Systematic significance of cell inclusions in Haemodoraceae and allied families: silica bodies and tapetal raphides. Ann. Bot. 92: 571–580.

Qiong Y and Q-E Yang. 2006. Chromosomes of four species in three genera of Commelinaceae from China and their systematic implications. Bot. J. Linn. Soc. 152: 399–403.

Radulescu D. 1973. La morphologie du pollen chez quelques *Haemodoraceae*. Lucr. Gard. Bot. Bucuresti 1972–1973: 123–132.

Robertson KR. 1976. The genera of Haemodoraceae in the southeastern United States. J. Arnold Arbor. 57: 205–216.

Rohweder O. 1963. Anatomische und histogenetische Untersuchungen an Laubsprossen und Blüten der Commelinaceen. Bot. Jahrb. Syst. 82: 1–99.

Rohweder O. 1969. Beiträge zur Blütenmorphologie und anatomic der Commelinaceen mit Anmerkungen zur Begrenzung und Gliederung der Familie. Ber. Schweiz. Bot. Ges. 79: 199–220.

Rosatti TR. 1987. The genera of Pontederiaceae in southeastern United States. J. Arnold Arbor. 68: 35–71.

Rudall PJ, DW Stevenson, and HP Linder. 1999. Structure and systematics of *Hanguana*, a monocotyledon of uncertain affinity. Austral. Syst. Bot. 12: 311–330.

Schneider EL and S Carlquist. 2005a. Orogin and nature of vessels in Monocotyledons. 7. Philydraceae and Haemodoraceae. J. Torrey Bot. Soc. 132: 377–383.

Schneider EL and S Carlquist. 2005b. Origins and nature of vessels in monocotyledons. 6. *Hanguana* (Hanguanaceae). Pacific Sci. 59: 393–398.

Schwartz O. 1927. Zur Systematik und Geographie der Pontederiaceen. Bot Jahrb. Syst. 61, Beibl. 139: 28–50.

Simpson MG. 1983. Pollen ultrastructure of the Haemodoraceae and its taxonomic significance. Grana 22: 79–103.

Simpson MG. 1985. Pollen ultrastructure of the Philydraceae. Grana 24: 23–31.

Simpson MG. 1987. Pollen ultrastructure of the Pontederiaceae: evidence for exine homology with the Haemodoraceae. Grana 26: 113–126.

Simpson MG. 1988. Embryological development of *Lachnanthes caroliniana* (Haemodoraceae). Am. J. Bot. 75: 1394–1408.

Simpson MG. 1989. Pollen wall development of *Xiphidium coeruleum* (Haemodoraceae) and its systematic implications. Ann. Bot. 64: 257–269.

Simpson MG. 1990. Phylogeny and classification of the Haemodoraceae. Ann Missouri Bot. Gard. 77: 722–784.

Simpson MG. 1993. Septal nectary anatomy and phylogeny in the Haemodoraceae. Syst. Bot. 18: 593–613.

Simpson MG. 1998a. Reversal in ovary position from inferior to superior in the Haemodoraceae, evidence from floral ontogeny. Int. J. Plant Sci. 159: 466–479.

Simpson MG. 1998b. Haemodoraceae. In: K Kubitzki, ed. The families and genera of vascular plants, vol. 4, pp. 212–232. Springer, Berlin/Heidelberg/New York.

Simpson MG and WC Dickison. 1981. Comparative anatomy of *Lachnanthes* and *Lophiola* (Haemodoraceae). Flora 171: 95–113.

Singh V. 1962. Vascular anatomy of the flower of some species of the Pontederiaceae. Proc. Indian Acad. Sci. B 56: 339–353.

Skottsberg C. 1932. Bemerkungen über die Philydraceen. Bot. Jahrb. Syst. 65: 253–274.

Skottsberg C. 1948. Philydraceae. In: CGGJ van Steenis, ed. Flora Malesiana, ser. 1, 4: 5–7. Leyden.

Smith RW. 1898. A contribution to the life history of the Pontederiaceae. Bot. Gaz. 25: 324–337.

Smith RW. 1908. Endosperm of *Pontederia*. Bot. Gaz. 45: 338–339.

Smithson E. 1956. The comparative anatomy of the Flagellariaceae. Kew Bull. 9: 491–501.

Steinecke H und U Hamann. 1989. Embryologisch-Systematische Untersuchungen an Haemodoraceen. Bot. Jahrb. Syst. 111: 247–262.

Stenar H. 1927. Zur Entwicklungsgeschichte der Gattung *Anigozanthos* Labill. Bot. Not. 1927: 104–114.

Stenar H. 1938. Die systematische Stellung der Gattung *Xiphidium*. Svensk Bot. Tidskr. 32: 274.

Strange A, PJ Rudall and CJ Prychid. 2004. Comparative floral anatomy of Pontederiaceae. Bot. J. Linn. Soc. 144: 395–408.

Tillich H-J. 1994. Untersuchungen zum Bau der keimpflanzen der Philydraceae und Pontederiaceae (Monocotyledonae). Sendtnera 2: 171–186.

Tillich H-J. 1996. Seeds and seedling in Hanguanaceae and Flagellariaceae (Monocotyledons). Sendtnera 3: 187–197.

Tillich H-J und E Sill. 1999. Systematische Studien zur Morphologie und Anatomie von *Hanguana* Blume (Hanguanaceae) und *Flagellaria* L. (Flagellariaceae), mit der Beschreibung einer neuen Art *Hanguana bogneri* spec. nov. Sendtnera 6: 215–238.

Tomlinson PB. 1966. Anatomical data in the classification of Commelinaceae. Bot. J. Linn. Soc. 59: 371–395.

Tomlinson PB. 1969. Commelinales-Zingiberales. In: CR Metcalfe, ed. Anatomy of the monocotyledons, vol. 3, pp. 78–81. Clarendon Press, Oxford.

Vyshenskaya TD. 1985. Commelinaceae. In: A Takhtajan, ed. Comparative seed anatomy, vol. 1, pp. 170–177. Nauka, Leningard (in Russian).

Order 26. XYRIDALES

Perennial or annual herbs, terrestrial or rarely aquatic or epiphytic, stemless or caulescent, commonly rhizomatous, rarely bulbous. Vessels in root and stem, or in all parts (Xyridaceae), with scalariform or more often with simple perforations. Leaves mostly all at the base, distichous or spirally arranged, usually with a sheath, sometimes ligulate, parallel-veined. Stomata paracytic, anomocytic, seldom tetracytic, hexacytic or with several subsidiary cells. Flowers solitary or in different type of inflorescences, sometimes much reduced, bisexual or sometimes (Eriocaulaceae, Hydatellaceae) unisexual, actinomorphic or zygomorphic, 3-merous, with or without nectaries. Perianth of six segments (rarely 4 – Eriopcaulaceae) in two cycles, differentiated into calyx and corolla, rarely (Hydatellaceae) perianth lacking. Sepals usually free or basally connate, imbricate or rarely valvate. Petals free or connate, imbricate. Stamens 6 or reduced to 1–3 fertile, sometimes staminodia present; filaments short (Xyridaceae) or slender (Mayacaceae, Hydatellaceae), adnate to the petals or hypogynous. Anthers mostly tetrasporangiate and dithecal, or sometimes tetrathecal, or seldom disporangiate and dithecal, basifixed or (Eriocaulaceae) dorsifixed, opening longitudinally or by lateral slits (Hydatellaceae) or apical pores. Tapetum secretory. Microsporogenesis successive or (Rapateaceae) simultaneous. Pollen grains 2-celled or seldom 3-celled, 1-colpate or seldom inaperturate, 3-aperturate, spiraperturate, 2-colpate, zonocolpate, trichotomocolpate, with a pair of short colpae in addition to the principal one or irregularly colpate, Gynoecium of three united carpels, seldom one or two carpels more or less reduced or even suppressed, with simple or lobed stigma. Placentation central or parietal. Ovary superior, 1–3-locular, sometimes stipitate (Eriocaulaceae, Hydatellaceae). Ovules solitary to numerous in each locule, anatropous to hemitropous, orthotropous or campylotropous, bitegmic, crassinucellate or tenuicellate. Female gametophyte of *Polygonum*- or *Allium*-type. Endosperm nuclear or helobial, rarely (Hydatellaceae) cellular. Fruits dehiscent (loculicidal or rarely septicidal capsules), sometimes indehiscent. Seeds small, sometimes with embryotega (Mayacaceae); endosperm usually copious, containing starch and often also protein, and seldom – oil; embryo small, mostly indifferentiated. Flavonoids and anthraquinones, ferulic and diferulic acids present.

Related to the Commelinales and share a common origin from the commelinalean ancestor.

Key to Families

1 Stamens 6 or 2–3.
  2 Flowers solitary or in different types of simple inflorescences, but not in globose or cylindrical heads.
    3 Leaves mostly all at the base.
      4 Leaf base symmetric, silica bodies lacking. Perennial or less often annual herbs, mostly of wet habitats (usually marshes, rarely aquatic), stemless or caulescent. Rootstocks commonly rhizomes, rarely bulbous. Roots fibrous, diffuse, mostly filiform, less often fleshy. Raphides are lacking, but small crystals of calcium oxalate present. Vessels in all parts, with simple perforations often in oblique or very oblique end walls. Leaves mostly basal, distichous or less often spirally arranged, ligulate or eligulate, with short, open sheath, narrow, flat to terete, parallel-veined. Stomata paracytic or anomocytic. Inflorescence usually an elongate or head-like spike, borne at the summit of a usually nude scape or rarely leafy stem. Individual flowers small, bisexual, actinomorphic to zygomorphic, 3-merous, without nectaries, in the axils of stiff or coriaceous, spirally arranged, closely imbricate bracts. Perianth of six segments in distinct two cycles, differentiated into calyx and corolla; sepals usually free, imbricate, hyaline or chaffy-scarious, subequal or more often the outer (anterior) large, more or less enclosing the rest of flower, but pushed away by the development of corolla, sometimes more or less reduced or even obsolete, the lateral ones small, boat-shaped, keeled; corolla colored, usually ephemeral, petals free or connate, sometimes to form a rather long, slender tube, imbricate. Fertile stamens only three and opposite petals in all genera (with the exception of one *Xyris*, which has 6), three stamens of the outer cycle staminodial or very rarely absent; filaments short, adnate to the petals or hypogynous; anthers tetrasporangiate, basifixed, often elongate, latrorse or introrse, opening longitudinally. Microsporogenesis successive. Pollen grains 2-celled or 3-celled, 1-colpate, 2-colpate, irregularly colpate or (*Achlyphila, Abolboda, Aratitiyopea,* and *Orectanthe*) inaperturate and intectate. Gynoecium of 3 united carpels; style slender, with simple or 3-lobed stigma, sometimes with 2–3 basal glandular appendages; ovary 3-locular to 1-locular, with several or numerous or rarely only one ovule per locule or per parietal placenta. Ovules anatropous, orthotropous or weakly campylotropous, weakly crassinucellate, without parietal cell. Female gametophyte of *Allium*- or *Polygonum*-type. Endosperm nuclear or (*Abolboda*) helobial. Fruits small loculicidal capsules enclosed by the persistent perianth. Seeds small, numerous, with strong longitudinal ridges and inner cross-lines; seed coat formed by both integuments, but only the tegmic layers, which are filled with a resinlike tanninlike substance, provide mechanical strength; endosperm copious, rich in aleurone, compound starch

grains, and sometimes also fatty oils; embryo small, more or less lens-shaped, little differentiated, lying in the micropylar part. Anthraquinoses are present. n = 8?, 9, 13, 14, 16, 17 (*Xyris*)..... 2. XYRIDACEAE.

4 Leaf base asymmetric, epidermal cells containing solitary or aggregated silica bodies. Perennial herbs often of wet habitats, rarely epiphytic, with short, thick, vertical or prostrate rhizome and short stem. Trichomes nonglandular uniseriate slime-producing hairs, mucilage cells present. Raphides and calcium oxalate crystals lacking. Vessels usually in root and stem, with scalariform (Saxofriedericioideae) or scalariform and simple perforations (Rapateoideae). Leaves crowded in basal rosette, unifacial or bifacial, distichous or less often spirodistichuosly arranged, rarely petiolate, with folded open sheath at base, linear to lanceolate, often rather firm, parallel-veined; ligules rarely present. Stomata paracytic or less often tetracytic. Inflorescence axillary or terminal, on leafless scapes, mostly with an involucre of one, two, or several large, coriaceous, basally broad bracts subtending the head or one-sided raceme of spikelets; each spikelet with a single terminal flower and several imbricate, spirally arranged sterile bracts beneath it; individual flowers 3-merous, bisexual, actinomorphic or nearly so, with or without nectaries. Perianth segments six in two cycles strongly differentiated into calyx and corolla; sepals firm, hyaline at base, free or basally connate, imbricate; petals ephemeral, usually ba-sally connate into a short or long tube with imbricate lobes. Stamens six in two cycles; filaments commonly short, usually attached to the corolla tube and often basally connate; anthers tetrasporangiate, basifixed, linear, usually very long, sometimes with short, apical connective appendage, dehiscing by four, two, or one apical or subapically lateral pores, extrorse or latrorse. Microsporogenesis simultaneous. Pollen grains 2-celled, oblong to elliptic, tectate-columellate, with a thick foot-layer and granular endexine, 1-colpate or rarely 2–3-colpate or zonocolpate. Gynoecium of three united carpels; style with punctiform or small, capitate and more or less papillate stigma; ovary 3-locular or rarely incompletely 3-locular, with one to several ovules per locule. Ovules anatropous, crassinucellate, sometimes with conspicuous funicular obturator, with parietal cell. Female gametophyte of *Polygonum*-type. Endosperm nuclear. Fruits septicidal capsules often with only one or two fertile locules. Seeds of different shape and sculpturing, with large hilar scar and a chalazal appendage (caruncle); seed coat endotestal-exotegmic, formed by the outer layer of the inner integument and the innermost of the outer one; endosperm copious, mealy, rich in simple starch grains and with a proteinaceous outer layer; embryo small, lenticular, undifferentiated, situated at the micropylar end, with flat side appressed to endosperm. Accumulate a considerable amount of aluminium. n = 11 (*Maschalocephalus*), 26...3. RAPATEACEAE.

3 Leaves well distributed along the stem. Small perennial herbs of shallow fresh water or very wet habitats with adventitious roots. Raphides, calcium oxalate crystals, and silica lacking. Longitudinal septate air canals present in all vegetative organs. Vessels in root, stem and leaves, with long scalariform, indistinctly perforate end walls. Leaves spirally arranged, multiranked, not sheathing, commonly with bidentate apex, 1-veined. Stomata paracytic. Flowers solitary in the axils of sheathing bracts, bisexual, actinomorphic, 3-merous, without nectaries. Perianth segments six, free, in two distinct cycles; sepals green, valvate to subvalvate; petals white, imbricate, shortly clawed. Stamens three, represent the outer staminal cycle; filaments slender, glabrous; anthers tetrasporangiate, basi-fixed, opening by apical pores or porelike slits, sometimes by pores at the end of tubular apical appendage. Microsporogenesis successive. Pollen grains 2-celled, 1-colpate, columellate, finely reticulate. Gynoecium of three united carpels; style filiform, with capitate or shortly trifid stigma; ovary 1-locular, with several ovules on each parietal placenta. Ovules orthotropous, tenuinucellate; micropyle formed by both integuments,

the inner one closing over the nucellus to form a narrow endostome. Female gametophyte of *Polygonum*-type. Endosperm nuclear. Fruits loculicidal capsules. Seeds small, ovoid to globose, minutely beaked and scrobiculate; seed coat exotestal, mainly formed by the outer layer of the outer integument, the inner layer remaining undifferentiated; the inner layer of the inner integument is tanniniferous, the outer layer is squeezed flat (Venturelli and Bouman 1986); the exotestal cells greatly increase in size, especially in radial direction and accumulate starch grains, the lateral and tangential walls become strongly thickened, are pitted and contain lignin; these exotestal cells can hold air and thus instrumental in keeping the mature seed afloat for some time (Venturelli and Bouman 1986); the integumentary layers in the micropylar region are transformed into a characteristic micropylar beak, which by many authors (including myself) are erroneously interpreted as an operculum; it is formed by the inner layer of the outer integument, the outer layer of the inner integument and to a lesser extent also by the inner layer of the inner integument, all of which undergo an appreciable radial elongation to close the exo- and endostome; only the cells of the inner layer of the outer integument have slightly thickened walls (Venturelli and Bouman 1986); the micropylar tissue probably disintegrates before germination and provides a channel for the emergence of the seedling (Hamann 1961; Venturelli and Bouman 1986); endosperm copious, the cells of central endosperm are mainly filled with closely-packed simple starch grains, the outermost layer of the endosperm is present as an aleurone layer which stores spherical protein bodies and lipids; embryo small, depressed-ovoid, undifferentiated, apical, lying opposite micropyle, with flat side facing the endosperm, n = 8. . . . . 1. MAYACACEAE.

2 Inflorescence globose to cylindrical, dense, centripetally flowering small heads (headlike spikes) borne on thin peduncles generally supplied with closed sheath at the base formed by a single leaf, or composed of many (up to more than 1,000) such heads: heads subtended by an involucre consisting of more or less numerous, imbricate, veinless bracts different in size and shape from bracts subtending flowers, although they connected by intermediates. Small (sometimes very small) to medium-sized perennial or rarely annual herbs of various types of habitats – poor sandy soils, marshy or boggy habitats, rivers, ponds, or temporarily flooded places, sometime even on rocks, adapted both to short dry periods and to submerged growth. Perennials with erect or horizontal, often thick rhizomes. Roots fibrous. Trichomes multicellular, uniseriate, with smooth surface or glandular, terminal cell simple or bibranched (malpighiaceous). Raphides lacking, but calcium oxalate crystals and rarely also silica bodies present. Vessels in all vegetative organs, with simple perforations in roots and simple and scalariform perforations in stems and leaves. Leaves usually basal, spirally arranged or rarely distichous, commonly linear, parallel-veined, more or less grasslike, distinctly or indistinctly sheathed at base. Stomata paracytic. Individual flowers small or very small, numerous, sessile or shortly pedicillate, unisexual (monoecious or rarely dioecious) or very rarely bisexual, actinomorphic or slightly zygomorphic, 3-merous or less often 2-merous, without nectaries (except for the nectariferous glands within the tip of the petal in *Eriocaulon* and *Mesanthemum*). Perianth of six or four segments in two cycles, scarious, often white, differing in shape and texture but not clearly differentiated into calyx and corolla; outer segments membranous-hyaline, free or less often more or less connate, sometimes with basal tube or forming spathe; in male flowers inner segments usually form 3- or 2-lobed tube, in female flowers they are free, often very small or reduced to hairs. Stamens usually three or two, opposite the petals, or (*Eriocaulon* and *Mesanthemum*) six or four in two cycles, rarely (*Syngonanthus amazonicus*) solitary; filaments free or rarely connate; anthers tetrasporangiate or rarely disporangiate, basifixed or dorsifixed, introrse, opening longitudinally. Microsporogenesis successive. Pollen grains usually 3-celled, spheroidal to ellipsoid, spiraperturate, echinate to echinulate and may have spinulae of different size. Male flowers often with pistillodium. Gynoecium of three or two united carpels; style often appendaged, 2–3-lobed, the lobes often bifid at the apex; ovary often stipitate, 3- or 2-loc-

ular, with one ventral-apical, pendulous ovule per locule. Ovules orthotropous, tenuinucellate, without parietal cell. Female flowers with conspicuous staminodia (*Rondo-nanthus*) or with inconspicuous staminodia or without staminodia. Female gametophyte of *Polygonum*-type. Endosperm nuclear. Fruits small, thin-walled loculicidal capsules. Seeds small, ellipsoidal to subglobose; seed coat formed by both integuments, but the protecting layer formed primarily by the inner epidermis of the inner integument; endosperm copious, mealy, rich in compound starch grains; embryo very small, lens-shaped or bell-shaped, undifferentiated, situated at the micropylar end. Cell wall-bound ferulic, p-coumaric, and/or diferulic acids present. n = 8, 9, 15, 20, 25.. .4. ERIOCAULACEAE.

1 Stamen solitary. Very small aquatic or subaquatic, annual herbs with numerous unbranched roots and reduced vascular system. Stem very short, branching to form a leafy tuft, often with minute multicellular hairs in leaf axils. Oxalate raphides and silica bodies lacking. Vessels in roots and stem, with scalariform perforations. Leaves all basal, submerged and emergent, alternate, glabrous, linear to filiform, lack a distinct sheath or ligules, 1-veined. Stomata anomocytic (*Trithuria*), in submerged species of *Hydadella* apparently lack stomata (Hamann 1998). Inflorescence terminal, sessile or scapose head with an involucrum of 1–4 pairs of opposite, membranous, 1-veined bracts often forming an apparent whorl; flowers numerous, minute, unisexual, lacking perianth and bracteoles, hydrophilous or autogamous. Filaments filiform; anthers tetrasporangiate, basifixed, linear to elliptic, opening longitudinally by lateral slits. Pollen grains 2-celled, tectate-columellate, 1-colpate and spinulose (*Trithuria*) or with indistinct porelike aperture. Gynoecium monomerous or more probably pseudo-monomerous (according to Hamann, 1998, the three prominent ribs on the ovary wall in *Trithuria* and the 3-valved fruit in *T. submersa* indicate a basically 3-merous gynoecium); stigma terminal, sessile, with 2–10 hairs, each a single row of distended cells; ovary superior, shortly stipitate, 1-locular, with one pendulous ovule. Ovules anatropous, crassinucellate or pseudocrassinucellate. Endosperm very probably cellular (Hamann 1998). Fruits small, non-fleshy, dehiscent or indehiscent, 3-valvular capsules (*Trithuria*), or achenelike (*Hydatlla*). Seeds minute, ovoid, with an operculum formed by the apical cells of the inner layer of the inner integument; seed coat includes both testa and tegmen, but the protective layer represented by derivatives of the outer epidermis of the outer integument (seeds exotestal); most part of endosperm replaced by starchy perisperm, the rest of endosperm occupies small zone close to embryo and consists of a few cells devoid of starch; embryo minute, peripheral, lens-shaped, very weakly differentiated. . . . . .5. HYDATELLACEAE

1. MAYACACEAE

Kunth 1842. 1/4 or up to 10. Southeastern United States, West Indies, Central and South America, 1 species in Africa (Angola, Zaire, Zambia).

Mayaca.

They share more characters with the Xyridaceae. According to Stevenson (1983, 1998), Venturelli and Bouman (1986) and Johri et al. (1992), embryologically Mayacaceae resemble the Xyridaceae more than the Commelinaceae and can be placed near Xyridaceae. However they markedly differ from the Xyridaceae in aquatic habit, anatomy of vegetative organs, dispersed, 1-veined leaves without distinct sheath, solitary flowers, apically dehiscing anthers, pollen morphology.

2. XYRIDACEAE

C. Agardh 1823 (including Abolbodaceae Nakai 1943). 5/325. Widespread in tropical and subtropical regions, especially of America, with relatively few species extending into temperate regions. All five genera have representatives on the Guiana Highlands, which might be the primary centre (Kral 1998).

Very close to the Mayacaceae and share a common origin from the commelinalean ancestor.

2.1 XYRIDOIDEAE

Leaves more or less clearly distichous, unifacial, often with an abrupt transition fro sheath to lamina. Petals commonly free, yellow. Pollen grains smooth and 1-colpate (occasionally 2-colpate). Staminodia typically present and bifid (lacking in *Achlyphila)*. Styles symmetric and unappendaged. Ovary 1-locular or incompletely 3-locular, with parietal or free basal placenta. – *Achlyphila, Xyris.*

2.2 ABOLBODOIDEAE
Leaves polystichous, bifacial, without abrupt transition from sheath to lamina. Petals united, blue or white. Pollen grains spinulose and inaperturate. Staminodia reduced to filaments or lacking. Styles mostly asymmetric and appendaged. Ovary 3-locular, with axile placentas. – *Abolboda, Orectanthe, Aratitiyopea.*

3. RAPATEACEAE

Dumortier 1829. 16/80. Mainly tropical South America, especially Guyana Highlands, with one genus (*Epidryos*) extending into Panama and one monotypic genus (*Maschalocephalus*) in tropical West Africa (Sierra Leone and Liberia).

3.1 SAXOFRIDERICIOIDEAE
Carpels with several ovules, on axil or on septa. Seeds prismatic or pyramidal. – SAXOFRIDERICEAE: *Saxofridericia, Phelpsiella, Stegolepis, Amphiphyllum, Marahuacaea, Epidryos;* SPHOENOCEPHALEAE: *Kunhardtia, Guacamaya, Schoenocephalium.*

3.2 RAPATEOIDEAE
Carpels with one ovule inserted at or near the base. Supernumerary antipodals are common. Seeds oval or oblong, with flattened and papillate apical appendages. – RAPATEAE: *Rapatea, Cephalostemon* (including *Duckea*), *Spathanthus*; MONOTREMEAE: *Monotrema, Potarophytum, Windsorina, Maschalocephalus.*

A very heterobathmic family with relatively primitive vessels in many genera (Carlquist 1966, 1969) and a rather specialized inflorescence and flowers. The Rapateaceae are usually placed near the Xyridaceae (see Hallier 1912; Hutchinson 1934, 1959, 1973; Engler's Syllabus 1964; Takhtajan 1966, 1980, 1987; Cronquist 1980, 1988; Dahlgren et al. 1985; Thorne 1992a, b). However, there are many important differences, including asymmetric leaf bases, the presence of silica bodies in the epidermal cells and tannin-bearing cells, the peculiar chlorenchyma cells, and the nature of the vascular core of the roots (Tomlinson 1969), poricidal dehiscence of anthers, pollen morphology, simultaneous microsporogenesis, as well as morphology of inflorescences, fruits, and seed ana-tomy, and accumulation of aluminum. Tiemann (1985) is of the opinion that the Rapateaceae should be placed closer to the Cyperaceae and Juncaceae to which they are related because of numerous common embryological features. Also in Tiemann's opinion flower morphology and anatomy support a relationship with the Thurniaceae, which she considers a connecting link. However, the Rapateaceae differ markedly from the Juncales in their pollen grains and seed coat structure as well as in their perianth morphology and some anatomical features. Many years ago Pilger (1930) and Smith (1934) suggested the relationship of the Rapateaceae with the Bromeliaceae. However, morphologically the Rapateaceae are nearer to the families of the Commelinanae, especially to the Xyridaceae.

4. ERIOCAULACEAE

Martynov 1820. 10/1000-1500. Mainly tropical regions, especially of America; a few species of *Eriocaulon* extend into temperate regions of Europe, eastern Asia and North America and most species *Lachnocaulon* occur in United States.

4.1 ERIOCAULOIDEAE
Submerged and floating, rarely terrestrial plants. Roots and leaves with aerenchyma. Petals free, with black tips; stamens adnate to the corolla. – *Eriocaulon.*

4.2 PAEPALANTHOIDEAE
Usually terrestrial plants. Petals often connate. Stylar appendages, stigmas commissural. – *Rondonanthus, Mesanthemum, Lachnocaulon, Syngonanthus, Philodice, Leiothrix, Tonina, Paepalanthus, Blastocaulon.*

A distinct and rather advanced family that has some similarities with the Xyridaceae, including many vegetative characters, compact inflorescences, echinate pollen, and stylar appendages like those in some Xyridaceae. However, the Eriocaulaceae differ from the Xyridaceae in unique type of inflorescences, spiraperturate pollen grains, solitary orthotropous pendulous ovule per locule, endotegmic seeds.

5. HYDATELLACEAE

Hamann 1976. 2/10. Temperate Australia and Tasmania, 1 species of *Hydatella* in New Zealand and 1 species of *Trithuria* in India.

Hydatella, Trithuria.

A very specialized and isolated family which related to Xyridaceae and Mayacaceae (Bremer 2002; Soltis 2006).

Bibliography

Arekal GD and SN Ramaswamy. 1980. Embryology of *Eriocaulon hookerianum* Stapf and the systematic position of Eriocaulaceae. Bot. Not. 133: 295–309.

Benko-Iseppson AM and MGL Wanderley. 2002. Cytogenetic studies on Brazilian *Xyris* species (Xyridaceae). Bot. J. Linn. Soc. 138: 245–252.

Boubier A-M. 1895. Remarques surranatomie systématique des Rapateacées et des families voisines. Bull. Herb. Boiss. 3(2): 115–120.

Carlquist S. 1960. Anatomy of Guyana Xyridaceae: *Abolboda, Orectanthe,* and *Achlyphila.* Mem. New York Bot. Gard. 10(2): 65–117.

Carlquist S. 1961. Pollen morphology of Rapateaceae. Aliso 5: 39–66.

Carlquist S. 1966. Anatomy of Rapateaceae: roots and stems. Phytomorphology 16 (I): 17–38.

Carlquist S. 1969. Rapateaceae. In: CE Metcalfe, ed. Anatomy of the monocotyledons, vol. 3, pp. 130–145. Clarendon Press, Oxford.

Cheadle V and H Kosakai. 1982. Occurrence and specialization of vessels in Xyridales. Nord J. Bot. 2: 97–109.

Cooke DA. 1987. Hydatellaceae. In: AS George, ed. Flora of Australia, vol. 45, pp. 1–5. Australian Government Publishing Service, Canberra.

Giulietti AM, WR Monteiro, SJ Mayo, and J Stephens. 1988. A preliminary survey of testa sculpture in Eriocaulaceae. Beiträge Biol. Pflanzen 62: 189–209.

Givnish TJ, TM Evans, and KJ Sytsma. 1994. Molecular evolution and adaptive radiation in South American elements of the plant family Rapateaceae. Am. J. Bot. 81(6): 445 (Abstract).

Givnish TJ, TM Evans, KJ Sytsma, TB Patterson, and MI Zjhra. 1998. Molecular evolution, adaptive radiation, and origin of the amphiatlantic distribution of the family Rapateaceae. In: Monocots II, p. 22 (Abstract). Sydney.

Govil GM and S Lavania. 1982. Floral vasculature of *Eriocaulon* L. J. Indian Bot. Soc. 61: 371–376.

Govindappa DA (= Arekal GD). 1956. Embryological studies in *Xyris pauciflora* Willd. Proc. Indian Acad. Sci. 42B: 47–57.

Grootjen CJ. 1983. Development of ovule and seed in *Cartonema spicatum* R. Br. (Cartonemataceae). Austral. J. Bot. 31: 297–305.

Hamann U. 1961. Merkmalsbestand und Verwandschaftsbeziehungen der Farinosae: Ein Beitrag zum System der Monocotyledonen. Willdenowia, 2, 5: 639–763.

Hamann U. 1962. Weiteres über Merkmalsbestand und Verwandtschaftsbeziehungen der "Farinosae". Willdenowia 3: 169–207.

Hamann U. 1975. Neue Untersuchungen zur Embryologie und Systematik den Centrolepidaceae. Bot. Jahrb. Syst. 96: 154–191.

Hamann U. 1976. Hydatellaceae: a new family of Monocotyledoneae. New Zealand J. Bot. 14: 193–196.

Hamann U. 1998. Hydatellaceae. In: K Kubitzki, ed. The families and genera of vascular plants vol. 4, pp. 231–234. Springer, Berlin/Heidelberg/New York.

Hamann U, K Kaplan, and T Rübsamen. 1979. Über die Samenschalenstruktur der Hydatellaceae (Monocotyledoneae) und die systematische Stellung von *Hydatella filamentosa.* Bot. Jahrb. Syst. 100: 555–563.

Hohendorff U. 1981. Embryologische Untersuchungen an Eriocaulaceen. Thesis, University of Bochum.

Horn af Rantzien H. 1946. Notes on the Mayacaceae of the Regnelian Herbarium in the Riksmuseum Stockholm. Svensk Bot. Tidskr. 40: 405–424.

Kircher E. 1977. Embryologische Untersuchungen an *Xyris capensis* Thunb. Thesis, University of Bochum.

Körnicke F. 1873. Monographic der Rapateaceen. Linnaea 37: 417–494.

Kral R. 1966a. *Xyris* (Xyridaceae) of the continental United States and Canada. SIDA 2: 177–260.

Kral R. 1966b. Eriocaulaceae of continental North America north of Mexico. SIDA 2: 285–332.

Kral R. 1972. A treatment of American Xyridaceae exclusive of *Xyris.* Ann. Missouri Bot. Gard. 79: 819–895.

Kral R. 1983. The Xyridaceae in the southeastern United States. J. Arnold Arbor. 64: 421–429.

Kral R. 1989. The genera of Eriocaulaceae in the southeastern United States. J. Arnold Arbor. 70: 131–142.

Kral R. 1992. A treatment of American Xyridaceae exclusive of *Xyris.* Ann. Missouri Bot. Gard. 79: 819–885.

Kral R. 1998. Xyridaceae. In: K Kubitzki, ed. The families and genera of vascular plants, vol. 4, pp. 461–469. Springer, Berlin/Heidelberg/New York.

Krauss JE, MG Sajo, CL Dias Leme, MGL Wanderley. 1994. Aspectos morfologicos do desenvolvimento pos-seminal em especies de *Xyris* L. (Xyridaceae). Hoehnea 21: 29–38.

Lourteg A. 1952. Mayacaceae. Notul. Syst. (Paris) 14: 234–248.

Maguire B and JJ Wurdack. 1958. The botany of the Guyana Highland, part 3. Mem. New York Bot. Gard. 10: 19–49.

Maguire B and JJ Wurdack. 1960. Xyridaceae. In: B Maguire et al. The botany of the Guyana Highland, part 4. Mem. New York Bot. Gard. 10(2): 11–15.

Maguire B and JJ Wurdack. 1965. The botany of the Guyana Highland, part 6. Mem. New York Bot. Gard. 12: 69–102.

Maguire B and LB Smith. 1963. Xyridales. In: B Maguire et al. The botany of the Guyana Highland, part 5. Mem. New York Bot. Gard. 10(5): 7–37.

Maguire B. 1979. Additions to the Rapateaceae. Acta Amazonia 9: 267–269.

Malrne G. 1925. Xyridologische Beiträge. Arkiv Bot. 19(13): 1–8.

Malrne G. 1933. Beiträge zur Kenntnis der südamerikanischen Xyridaceen. Arkiv Bot. 25(12): 1–18.

Ramaswamy SN and GD Arekal. 1982. Embryology of *Eriocaulon xeranthemum* Mart. (Eriocaulaceae). Acta Bot. Neerl. 31: 41–54.

Ramaswamy SN and MVS Raju. 1982. The embryo sac of *Xyris schoenoides* Mart. (Xyridaceae). Bull. Torrey Bot. Club 109(3): 325–329.

Roisa MM and VL Scatena. 2003. Floral anatomy of *Eriocaulon elichrysoides* and *Syngonanthus caulescens* (Eriocaulaceae). Flora 198: 188–199.

Rosa MM and VL Scatena. 2007. Floral anatomy of Paepalanthoideae (Eriocaulaceae, Poales) and their nectariferous structures. Ann. Bot. 99: 131–139.

Saarela JM, HS Rai, JA Doyle, PK Endress, S Mathews, AD Marchant, B Briggs, and SW Graham. 2007. Hydatellaceae identified as a new branch near the base of the angiosperm phylogenetic tree. Nature 446: 312–315.

Sajo MG. 1992. Estudos morfoanatomicos em orgais foliares de *Xyris* L. (Xyridaceae). Bol. Bot. Univ. São Paulo 13: 67–86.

Sajo MG and PJ Rudall. 1999. Systematic vegetative anatomy and ensiform leaf development in *Xyris* (Xyridaceae). Bot. J. Linn. Soc. 130: 171–182.

Scatena VL and F Bauman. 2001. Embryology and seed development of *Paepalanthus* sect. *Actinocephalus* (Koern.) Ruhland (Eriocaulaceae). Plant Biol. 3: 341–350.

Scatena VL, AM Giulietti, EL Borba, and C van den Berg. 2005. Anatomy of Brazilian Eriocaulaceae: correlation with taxonomy and habitat using multivatiate analyses. Plant Syst. Evol. 253: 1–22.

Splett S, W Barthlott, T Stützel, and MAG Barros. 1993. Leaf anatomy of Brazilian Eriocaulaceae and its diagnostic significance. Flora 188: 399–411.

Stevenson DW. 1983. Systematic implications of the floral morphology of the Mayacaceae. Am. J. Bot. 70(5, pt.2): 32.

Stevenson DW. 1998. Mayacaceae. In: K Kubitzki, ed. The families and genera of vascular plants, vol. 4, pp. 294–296. Springer, Berlin/Heidelberg/New York.

Stevenson DW, M Colella and B Boom. 1998. Rapateaceae. In: K Kubitzki, ed. The families and genera of vascular plants, vol. 4, pp. 415–424. Springer, Berlin/Heidelberg/New York.

Stevenson DW, JI Davis, JV Freudenstein, CR Hardy, MP Simmonds, and CD Specht. 2000. A phylogenetic analysis of the monocotyledons based on morphological and molecular character sets, with comments on the placement of *Acorus* and *Hydatellaceae*. In: KL Wilson and DA Morrison, eds. Monocots: systematics and evolution, pp.17–24. CSIRO, Collingwood.

Steyermark JA. 1984. Flora of the Venezuelan Guyana, part 1. Ann. Missouri Bot. Gard. 71: 297–340.

Stützel T. 1984. Blüten- und infloreszenzmorphologische Untersuchungen zur Systematik der Eriocaulaceae. Diss. Bot. 71: 1–108.

Stützel T. 1985. Die epipetalen Drüsen der Gattung *Eriocaulon* (Eriocaulaceae). Beitr. Biol. Pflanz. 60: 271–276.

Stützel T. 1988. Untersuchungen zur Wurzelanatomie der Eriocaulaceen. Flora 180: 223–239.

Stützel T. 1990. "Appendices" am Gynoeceum der Xyridaceen. Beitr. Biol. Pflanzen 65: 275–299.

Stützel T. 1998. Eriocaulaceae. In: K Kubitzki, ed. The families and genera of vascular plants, vol. 4, pp. 197–207. Springer, Berlin/Heidelberg/New York.

Stützel T and N Gansser. 1995. Floral morphology of North American Eriocaulaceae and its taxonomic implications. Feddes Repert 106: 495–502.

Stützel T and F Weberling. 1982. Untersuchungen über Verzweigung und Infloreszenzaufbau von Eriocaulaceen. Flora 172: 105–112.

Suessenguth K and R Beyerle. 1936. Über die Xyridaceengattung *Abolboda* Humb. et Bonpl. Bot. Jahrb. Syst. 67(2): 132–142.

Thanikaimoni G. 1965. Contribution to the pollen morphology of Eriocaulaceae. Pollen Spores 7: 181–191.

Thieret JW. 1975. The Mayacaceae in the southeastern United States. J. Arnold Arbor. 56: 248–255.

Thorsch JA and VI Cheadle. 1996. Vessels in Eriocaulaceae. IAWA J. 17: 183–204.

Tiemann A. 1985. Untersuchungen zur Embryologie, Blütenmorphologie, und Systematik der Rapateaceen und der Xyridaceen-Gattung *Abolboda* (Monocotyledoneae). Diss. Bot. 82: 1–202.

Tomlinson PB. 1964. Notes on the anatomy of *Triceratella* (Commelinaceae). Kirkia 4: 207–212.

Uphof JCT. 1924. The physiological anatomy of *Mayaca fluviatilis*. Ann. Bot. 38: 389–393.

Venturelli M and F Bouman. 1988. Development of ovule and seed in Rapateaceae. Bot. J. Linn. Soc. 97: 267–294.

Venturelli M and F Bouman. 1986. Embryology and seed development in *Mayaca fluviatilis* (Mayacaceae). Acta Bot. Neerl. 35: 497–516.

Vyshenskaya TD. 1985a. Rapateaceae, Xyridaceae. In: A Takhtajan, ed. Comparative seed anatomy, vol. 1, pp. 167–169. Nauka, Leningrad (in Russian).

Vyshenskaya TD. 1985b. Mayacaceae, Eriocaulaceae. In: A Takhtajan, ed. Comparative seed anatomy, vol. 1, pp. 177–183. Nauka, Leningrad (in Russian).

Vyshenskaya TD. 1985c. Hydatellaceae. In: A Takhtajan, ed. Comparative seed anatomy, vol. 1, pp. 190–191. Nauka, Leningrad (in Russian).

Weinzieher S. 1914. Beiträge zur Entwicklungsgeschichte von *Xyris indica* L. Flora 106: 393–432.

Order 27. DASYPOGONALES

Mainly xeromorphic, rhizomatous perennial herbs or shrubby or arborescent plants sometimes with secondary growth. Vessels in root, with simple or scalariform (*Baxteria, Kingia*) perforations. Raphides present or (Calectasiaceae) absent. Sieve-element plastids of P2c-type (Lomandraceae). Leaves small to large, alternate, leathery or modified into spines, linear or strap-shaped, more or less V- or U-shaped in transection, often distichous or crowded in a rosette, rarely scattered along the stem, usually with broad, often sheathing base that persists after lamina falls or is burnt off. The mesophyll containing calcium oxalate crystals or without calcium oxalate crystals. Stomata anomocytic or (*Baxteria*) paracytic/tetracytic. Flowers in various kinds of inflorescences or solitary, bisexual or dioecious, actinomorphic, 3-merous, bracteate. Perianth segments six in two cycles, mostly dry and glumaceous, similar, more or less free or united basally or

(*Calectasia*) into a tube, persistent. Stamens six in two cycles, usually attached to base of perianth segments; anthers tetrasporangiate, basifixed or (*Dasypogon*) dorsifixed, opening longitudinally or by apical pores, introrse or rarely extrorse; the endothecial thickenings spiral, or girdling (*Lomandra*). Pollen grains mostly 1-colpate and microreticulate or psilate or (*Baxteria*) the exine is divided into geometrically defined areas, a nearly unique type among flowering plants (Chanda and Ghosh 1976; Clifford et al. 1998). Gynoecium of three united carpels; stylodia more or less completely connate; ovary superior or inferior, 3-locular or 1-locular (*Dasypogon*), with 1 ovule per carpel, with or without septal nectaries. Ovules anatropous or rarely campylotropous, bitegmic, crassinucellate. Female gametophyte of *Polygonum*-type. Endosperm nuclear. Fruits dehiscent (capsules or capsular-indehiscent) or more often indehiscent, enclosed in persistent perianth. Seeds more or less globose or oblong, with pale yellow testa; seed coat formed predominantly by the outer integument; embryo lens-shaped and broad; endosperm copious, without starch, with a mealy, very thin-walled parenchyma (at least in *Dasypogon*). Testa without phytomelan. Cell wall-bound ferulic acid is present in all genera (Rudall and Caddick 1994); rarely (only in one species of *Dasypogon*) is present chelidonic acid; in *Lomandra* detected arthroquinones; n = 7, 8, 9.

A rather isolated group, which, according to Clifford et al. (1998), is possibly near the Rapateaceae; probably belong to the Commelinanae.

Key to Families

1 Anthers opening longitudinally.

2 Fruits capsules.

3 Fruits explosively dehiscent capsules; dehiscence septifragal, the valves also splitting into two valves, each half with the inner wall forming a plate held under tension and springing upward to eject the seed. Perennial stemless herbs with short, thick rhizomes and thick roots. Vessels in roots, with scalariform perforations. Leaves crowded, linear, pungent, with sheathing bases. Flowers large, 7–8 cm long, solitary, terminal, sessile, surrounded by several bracts, scented like rotten meat. Perianth segments free, almost similar, indurated, persistent. Stamens six, attached to the bases of the perianth; anthers apiculate, introrse. Pollen grains of unique pantocolpate type: the surface divided by the apertures into geometrically defined areas; the aperture membrane formed by coarse pieces, the surface pattern of which is essentially the same as that of the interaperturate areas; it seems as if the ends of the colpi of pantocolpate grains are united to make a continuous aperture that encloses several zones of geometrical pattern; ornamentation microreticulate (Chanda and Ghosh 1976: 538). Ovary 3-locular, with one erect ovule per locule; style subulate, more or less as long as perianth, with small 3-lobed stigma. Seeds globose. n = 7.. . . . . 1. BAXTERIACEAE.

3 Fruits loculicidal capsules, ovoid or trigonous. Perennial tufted herbs with generally short rhizomes and wiry roots or (*Lomandra*) herbs or shrubs with thick, woody rhizomes with secondary thickening and fibrous roots. Stems elongated (erect or procumbent) or very short. Vessels only in roots with simple and scalariform perforations or (*Acanthocarpus*) also in leaves, with scalariform perforations. Leaves alternate, scleromorphic, perennial or (*Xerolirion*) deciduous, generally distichous or spirodistichuous, sheathing at base, linear, sometimes prickly or scarious. Stomata generally anomocytic, although paracytic and tetracytic in some *Lomandra* species. Flowers in usually contracted cymose or racemose inflorescences or solitary, subtended by single to several bracts, bisexual or (*Lomandra* and *Xerolirion*) unisexual and dioecious, usually pedicillate, pedicels articulated or (*Xerolirion*) not articulated. Perianth segments free or shortly united basally, fleshy, petaloid or sometimes chartaceous, persistent. Stamens six in two cycles, more or less inserted on the perianth segments; anthers dorsifixed. Seeds testal; present napthoquinones (*Lomandra*); n = 7, 8. . . . . .2. LOMANDRACEAE.

2 Fruits indehiscent, within persistent perianth. Perennial herbs or secondary arborescent plants with short rhizome or stout caudex of fibrous pith surrounded by persistent packed leaf bases. Roots thick or fibrous. Stems sometimes (*Kingia*) with a cover of aerial roots penetrating the persistent leaf bases. Vessels in roots, with scalariform (*Kingia*) or simple (*Dasypogon*) perforations. Leaves linear, imbricate at base,

sheathing. Inflorescence a globose head on terminal peduncle bearing scattered bracts (*Dasypogon*) or axillary headlike, their peduncles covered by imbricate sheathing bracts. Flowers small, sessile; perianth segments narrow and dry; outer segments free or (*Dasypogon*) united above the middle and thickened; inner segments free or almost free. Stamens inserted on perianth; anthers dorsifixed, versatile (*Dasypogon*) or basifixed, introrse (*Kingia*). Pollen grains 1-colpate, finely reticulate or punctate. Gynoecium of three carpels; style subulate, stigma small; ovary sessile, 3-locular, with one ovule per locule; style long, with punctiform stigma. Septal nectaries present as shallow basal grooves. n = 7.. . . . . .3. DASYPOGONACEAE.

1 Anthers opening by two apical pores. Xeromorphic, rhizomatous perennial herbs, shrublets or shrubs, with wiry roots. Branches stiff, covered by old leaf sheaths or leaves. Vessels in roots, with simple perforations. Leaves alternate, small, linear or lanceolate, simple, entire, pungent, the base indurated and closely sheathing stem. Flowers bisexual, solitary, terminal on the ends of short branchlets, surrounded by many scarious, small imbricate leaves. Perianth segments connate into a short tube, the lobes spreading, stiffly scarious, lilac-blue or purple, variously pubescent outside. Stamens six inserted at the throat of perianth lobes, all equal to markedly unequal, free of one another. Anthers basifixed, terminally dehiscing via pores to dehiscing via short slits. Pollen grains 1-colpate, reticulate. Gynoecium of three carpels; style long, slender, attenuate from the ovary, much longer than the ovary. Ovary 1-locular, with three basal, anatropous ovules; style filiform, with punctiform stigma. Septal nectaries absent. Fruits nuts, enclosed in persistent faded perianth. Seeds elongate, with thin membranous testa, n = 9.. . . . 4. CALECTASIACEAE.

1. BAXTERIACEAE

Takhtajan 1996. 1/1. Southwestern Australia.
Baxteria.
A rather isolated family characterized by very large solitary flowers, unique type of multiaperturate pollen grains, and septifragal capsules.

2. LOMANDRACEAE

Lotsy 1911. 5/80 Australia, New Guinea, New Caledonia.
Lomandra, Romnalda, Chamaexeros, Xerolirion, Acanthocarpus.
Closely related to the Dasypogonaceae.

3. DASYPOGONACEAE

Dumortier 1829 (including Kingiaceae Endlicher 1838). 2/4. Southwestern Australia.
Dasypogon, Kingia.

4. CALECTASIACEAE

Endlicher 1836. 1/3. Southwestern and southeastern Australia.
Calectasia.
Probably closely related to the Dasypogonaceae (both are characterized by large chalazosperm, see Rudall [1994]), but differ from them markedly in anthers opening by apical pores, and solitary flowers on the ends of short branchlets.

Bibliography

Anway JC. 1969. The evolution and taxonomy of *Calectasia cyanea* R. Br. (Xanthorrhoeaceae) in terms of its present day variation and cytogenetics. Austral. J. Bot. 17: 147–159.

Barrett RL and Dixon KW. 2001. A revision of the genus *Calectasia* (Calectasiaceae) with eight new species described from south-west Western Australia. Nuytsia 13: 411–448.

Bedford DJ, AT Lee, TD MacFarlane, RJF Henderson, and AS George. 1986. Xanthorrhoeaceae. Fl. Austral. 46: 88–171.

Briggs BG. 1986. Chromosome numbers in *Lomandra* (Dasypogonaceae). Telopea 2: 741–744.

Chanda S and K Ghosh. 1976. Pollen morphology and its evolutionary significance in Xanthorrhoeaceae. In: IK Ferguson and J Muller, eds. The evolutionary significance of the exine, pp. 527–559. Linn. Soc. Symposium, No. 1. London/New York.

Chanda S, B Lugardon, and G Thanikaimoni. 1978. On the ultrastructure of pollen apertures in *Calectasia* R.Br. (Xanthorrhoeaceae). Pollen Spores 20: 351–356.

Clifford HT, GJ Keighery, and JG Conran. 1998. Dasypogonaceae. In: K Kubitzki, ed. The families and genera of vascular plants, vol. 4, pp. 190–194. Springer, Berlin/Heidelberg/New York.

Conran JG. 1998e. Lomandraceae. In: K Kubitzki, ed. The families and genera of vascular plants, vol. 3, pp. 354–365. Springer, Berlin/Heidelberg/New York.
Fahn A. 1954. The anatomical structure of the Xanthorrhoeaceae Dumort. Bot. J. Linn. Soc. 55: 158–184.
Keighery GJ. 1983. Balistochory (explosive seed dispersal) in *Baxteria* R. Br. (Xanthorrhoeaceae). W. Austral. Nat. 15: 163–166.
Neyland R. 2002. A phylogeny inferred from large-subunit (26S) ribosomal DNA sequences suggests that the family Dasypogonaceae is closely aligned with the Restionaceae allies. Austral. Syst. Bot. 15: 749–754.
Rudall P. 1994. The ovule and embryo sac in Xanthorrhoeaceae *sensu lato*. Flora 189: 335–351.
Rudall P and MW Chase. 1996. Systematics of Xanthorrhoeaceae *sensu lato*, evidence for polyphyly. Thelopea 6, 629–647.

Superorder JUNCANAE

Order 28. JUNCALES

Perennial, often rhizomatous herbs or less often annuals, rarely shrubs or lianas, mostly of wet habitats, rarely aquatic, very rarely (*Prionium*) rather large shrubby plants with hard persistent trunk (homologous to the rhizome of other genera) without secondary growth. Rhizomes underground, sympodial, sometimes tuberous, starchy. Stems commonly trigonous, rarely compressed, very rarely articulated, mostly solid, but sometimes hollow or transversely septate. Raphides absent, but silica bodies commonly present. Vessels usually in all vegetative organs, with scalariform or/and simple perforations. Leaves alternate, tristichous or more rarely distichous or polystichous, basal or cauline, usually sheathing at base, sheats usually closed, sessile or rarely pseudopetiolate, mostly linear, parallel-veined, rarely reduced or absent; ligule present or absent. Stomata paracytic or sometimes tetracytic. Flowers small and inconspicuous, sessile in axils of spirally arranged or distichous bracts, forming spikes or spikelets that are sometimes solitary and terminal, but usually arranged in secondary inflorescences, bisexual or often unisexual (monoecious or very rarely dioecious). Perianth segments six (rarely 4) in two cycles, free, glumaceous or herbaceous, equal or subequal, sometimes replaced by bristles or numerous hairs or lacking; perianth in some genera (such as *Oreobolus*) of six scales in two cycles but usually strongly reduced into bristles or hairs, or wanting. Stamens mostly three (1 medio-anterior, 2 lateral), less often reduced to two or one, rarely six or more, in two cycles, the inner cycle sometimes lacking; filaments generally free from each other; anthers tetrasporangiate, basifixed, introrse, opening longitudinally, often apiculate. Tapetum secretory. Microsporogenesis simultaneous. Pollen grains 3-celled, tectate-columellate, 1-porate or less often up to 7-porate, in monads (Mapanioideae) (Tarasevich 1986) or more often borne in cryptotetrads (the 4 microspores being retained within a common wall, but three of them soon degenerate so that at maturity pollen grain appears to be single), or (in Juncaceae and Thurneraceae) usually in tetrahedral tetrads that are surrounded by the common tetrad wall consisting of sporopollenin, atectate, with one reduced, poroid, distal aperture. Gynoecium of three or less often two, rarely four, united carpels; stylodia free or more often connate into short, linear style, often swollen at base, with three, two, or very rarely eight filiform, linear, or lanceolate stigmatic branches. Ovary superior, 3-locular or less often 1-locular, with one to many basal ovules per carpels. Ovules anatropous, bitegmic (outer integument very thin), crassinucellate, with parietal cell. Female gametophyte of *Polygonum*-type. Endosperm helobial or (Cyperaceae) nuclear. Fruits commonly trigonous, lenticular or globose nuts, rarely drupelike, or three to many-seeded loculicidal capsules or sometimes circumscissile. Seeds small, free from the pericarp; seed coat testal-tegmic, thin and delicate, consists of two layers formed by outer epidermis of the outer integument and inner integument of the inner integument; embryo small, basal, usually more or less differentiated, surrounded by abundant endosperm containing lipids, aleurone, and simple starch grains. Aleurones, flavonoid sulphates, flavone C-glycosides, kestose and isokestose storage oligosaccharides present. Chromosomes often with diffuse centromere. n = 3-13 +.

Related to the Commelinanae, and possibly to the Xyridales.

Key to Families

1 Plants without silica bodies. Annual or perennial rhizomatous or tuberous herbs or rarely shrubs (a few with woody trunks, esp. *Prionium*). Plants lacking crystal oxalate raphides, but accumulated starch exclusively "pteridophyte type". Vessels in roots, stems and leaves, with simple, or scalariform and simple perforations. Leaves evergreen, alternate,

nearly always tristichous, or rarely (*Distichia*) distichous, folded, or rolled, terete or flat, leathery or membranous and sometimes reduced to membranous sheaths; sessile, simple; *Prionium* has leaves in terminal rosette, linear, tapering, flat or canaliculate, serrate. Stomata paracytic. Inflorescences terminal or pseudolateral, compound, of open cymose panicles, or often congested and head- or spike-like, or flowers sometimes solitary and terminal (*Rostkovia*, *Marsippospermum*) or lateral (*Distichia*, *Oxychloe* and *Patosia*). Flowers with one or more spathal bracts, bracteolate or ebracteolate, small, actinomorphic, bisexual or unisexual. Perianth segments six in two cycles, or rarely three, similar or sometimes very different in size, free, glumaceous. Stamens 6 or rarely 2–3; filaments filiform of flat and widened at the base; anthers introrse or latrorse, appendages or unappendaged. Pollen ulcerate with granular exine, in tetrahedral and cross tetrads. Ovary 1-locular or 3-locular; styles one or three, free of partially joined; stigmas two, brushlike. Ovules 3 (*Luzula*), or 7–100 per locule, ascending. Fruits mostly loculicidal capsules, but in *Distichia*, *Oxychloe* and *Patosia* less well defined and sometimes circumscissile (Baslev 1996). Seeds numerous or only three (*Luzula*), ovoid or rarely fusiform, globose or angular, not hispid, usually not pointed at ends, with (mucilaginous) exotesta and endotegmen; testa without phytomelan; embryo straight, small; endosperm starchy. Free flavones present (quercetin); n = 3–42, mostly 6, 12, 18, 24. 2. JUNCACEAE.

1 Plants with silica bodies.

2 Ovary 3-locular. Large, tough herbs with large, subterranean rhizomes. Silica bodies spheroidal, present in some of the cells of parenchyma and epidermis. Vessels in roots, stems and leaves, mostly with scalariform perforations. Leaves all basal, alternate, fourranked, or tristichous, flat or canaliculate, leathery, sessile, sheathing, simple, entire; leaf bundles in pairs above each other in a leaf transaction, the lower (and smaller) bundle of a pair with the phloem on top (adaxial), facing the phloem of the upper bundle. Stomata paracytic, but sometimes some of them tetracytic. Inflorescences of one or more dense, racemose, globose to ellipsoid heads subtended by spreading, leafy bracts on an otherwise leafless, bluntly triangular or quadrangular peduncle. Flowers bisexual, actinomorphic, small, with short, swollen, puberulous pedicles. Perianth segments six, free, similar, brown, spotted, persistent. Stamens six, in two cycles, adnate to the base of the perianth segments and free of one another; anthers introrse. Styles united and very short, attenuate from the ovary; stigmas three, elongate. Ovules 1–7 per locule, ascending, with a zig-zag micropyle, placentation basal. Fruits loculicidal capsules. Seeds spindle-shaped, hispid, subulate-pointed at both ends; testa without phytomelan; embryo small, straight, cylindrical; endosperm copious, mealy, starchy. . . . . . . 1. THURNIACEAE.

2 Ovary 1-locular. Small to tall perennial or annual herbs, rarely dwarf shrubs, or lianas, terrestrial or helophytic, rarely aquatic; the perennials tufted, rhizomatous, stoloniferous, tuberous. Stems with solid internodes to with spongy internodes, or with hollow internodes. Vessels with scalariform, or simple, or scalariform and simple perforations. Sieve-element plastids of P-type. Leaves all basal or one to few cauline, rarely all cauline, alternate, distichous or tristichous, rarely spiral; flat or folded, sheathing, simple, entire, but commonly with prickle hairs. Inflorescences terminal, more rarely pseudolateral, very rarely reduced to a single spikelet. Flowers bracteate or ebracteate, bracteolate, usually small to minute, bisexual or (functionally) unisexual, and then spikelets bisexual or unisexual, rarely dioecious. Perianth segments vestigial (represented by bristles or hairs) and than 6, free, sometimes more or less sepaloid, or absent. Stamens usually 3 (1 medio-anterior, 2 lateral), but sometimes reduced to 1–2 or 4–6, and even more numerous (to 22); filaments sometimes strongly elongating after anthesis, rarely connate; anthers introrse or latrorse, appendaged or unappendaged. Gynoecium of 2, or 3(-4) carpels; styles 2–3, free to partially joined; stigmas (1-)2, or 3(-15). Ovule solitary, basal, bitegmic, outer integument not contributing to the micropyle. Fruits achene-like or rarely drupaceous. Seeds with thin testa (phytomelan lacking), raphe and chalaza usually conspicuous; embryo small, basal, surrounded by abundant mealy or oily endosperm. Alkaloids and proanthocyanidins present or absent; when present – cyanidin and delphinidin; flavonols present (quercetin) or absent; n = 5 or more. . . . . 3. CYPERACEAE

1. THURNIACEAE

Engler 1907. 1/3. Tropical South America (Guyana Highlands and Brazil).

Thurnia.

Close related to the Juncaceae, but vessels mostly with scalariform perforations and silica bodies present.

2. JUNCACEAE

A.L de Jussieu 1789 (including Prioniaceae Munro et Linder 1998). 8/400-440. The genera *Juncus* and *Luzula* distributed mostly in the cold and temperate regions of the Northern Hemisphere, the rest in the Southern Hemisphere; *Prionium* – Cape Province of South Africa.

Marsippospermum, Rostkovia, Distichia, Patosia, Oxychloe, Luzula, Juncus, Prionium.

The taxonomic position of *Oxychloe* (7, Andes) is uncertain. Some authors prefer to place it in Cyperaceae (Plunkett et al. 1995; Thorne 2000), but morphological traits of *Oxychloe* are typically juncaceous (Drabkova et al. 2003). The genus *Prionium* starts apart in many characters, especially in its woody trunk and leaf anatomy (Cutler 1969; Munro and Linder 1997, 1998), but in floral morphology and pollen tetrads is juncaceous.

3. CYPERACEAE

A. L. de Jussieu 1789 (including Caricaceae Burnett 1835, Elynaceae Reichenbach 1828, Kobresiaceae Gilly 1952, Mapaniaceae Shipunov 2003, Papyraceae Burnett 1835, Scirpaceae Batsch ex Borkhausen 1797). 108/5300. Cosmopolitan, but mainly in cold and temperate regions.

Classification after P. Goetghebeur (1998).

3.1 MAPANIOIDEAE

Pollen grains in monads. Spike-lets cymose with one terminal female flower and several axillary male flowers. Intraspicular prophylls two, always opposite, scalelike or rarely saclike. – HYPOLYTREAE: *Scirpodendron, Hypolytrum, Principina, Mapania, Mapaniopsis, Paramapania, Diplasia*; CHRYSITRICHEAE: *Capitularina, Exocarya, Chorizandra, Lepironia, Hellmuthia, Chrysitrix.*

3.2 CYPEROIDEAE

Pollen grains in cryptomonads. Spikelets without terminal flower, never prophyllate, many-flowered. Flowers subtended by floral scale, generally bisexual. – SCIRPEAE: *Scirpus, Eriophorum, Phylloscirpus, Oreobolopsis, Amphiscirpus, Trichophorum* (including *Baeothryon*); FUIRENEAE: *Fuirena, Pseudoschoenus, Bolboschoenus, Actinoscirpus, Schoenoplectus*; ELEOCHARIDEAE: *Eleocharis, Egleria, Websteria*; ABILDGAARDIEAE: *Fimbristylis, Dichostylis, Crosslandia, Bulbostylis, Abildgaardia, Nemum, Nelmesia*; CYPEREAE: *Ficinia, Desmoschoenus, Isolepis, Scirpoides, Oxycaryum, Kyllingiella, Androtrichum, Cyperus, Juncellus, Mariscus, Torulinum, Courtoisina, Remirea, Sphaerocyperus, Ascopholis, Alinula, Ascolepis, Pycreus, Queenslandiella, Kyllinga, Volkiella, Lipocarpha*; DULICHIEAE: *Dulichium, Sumatroscirpus, Blysmus*; SCHOENEAE: *Arthrostylis, Actinoschoenus, Trichoschoenus, Trachystylis, Rhynchospora, Pleurostachys, Schoenus, Gymnoschoenus, Mesomelaena, Ptilothrix, Cyathochaeta, Oreobolus, Carpha, Trianoptiles, Tetraria, Cyathocoma, Neesenbeckia, Epischoenus, Costularia, Gahnia, Morelotia, Reedia, Evandra, Caustis, Cladium, Rhynchocladium, Machaerina, Lepidosperma, Tricostularia.*

3.3 SCLERIOIDEAE

Spikelets with few glumes, one to few of which each subtend a strictly unisexual flower. A perianth often present. – CRYPTANGIEAE: *Lagenocarpus, Didymiandrum, Everardia, Cephalocarpus*; TRILEPIDEAE: *Trilepis, Afrotrilepis, Coleochloa, Microdracoides*; SCLERIEAE: *Scleria*; BISBOECKELEREAE: *Becquerelia, Diplacrum, Bisboeckelera, Calyptrocarya.*

Insertae sedis: *Exochogyne, Koyamaea.*

3.4 CARICOIDEAE

Pollen grains in cryptomonads. Spikelets without terminal flower, female flowers subtended by strongly metamorphosed prophyll, which is bottle-shaped or spathaceous; flowers always unisexual, without perianth. – *Kobresia, Schoenoxiphium, Uncinia, Vesicarex, Carex, Cymophyllus.*

Have many similarities with the Juncaceae (see especially Shah 1967; Simpson 1995). According to Simpson characters linking Juncaceae and Cyperaceae are the following: leaves usually tristichous, stomata usually paracytic, calcium oxalate raphides absent, chromosomes holocentric, occurrence of postreductional meiosis, perianth segments bracteal, microsporogenesis simultaneous, developing pollen grains in tetrads, pollen grains ulcerate, embryogeny onagrad, presence of luteolin 5-methyl ether, and parasitism by similar rust and smut species. However, the Cyperaceae differ from the Juncaceae in the presence of silica bodies, Cyperaceae-type simultaneous microsporogenesis, nuclear endosperm development, pollen in pseudomonads, solitary ovule per locule, indehiscent fruits, presence of yellow aurone pigments and flavone tricin (Simpson 1995).

The Cyperaceae could either derived from some ancient juncalean ancestor or more probably had a common origin. It is important to note that according to Cheadle (1955) no species in the Juncaceae has vessels quite primitive enough throughout the plant to permit use of the Juncaceae as the source of the most primitive members of the Cyperaceae. Besides, the pollen grains of the Cyperaceae, especially those of Mapanioideae, are less specialized.

Bibliography

Balslev H. 1996. Juncaceae. Flora Neotropica Monograph. New York Botanical Garden, New York.

Balslev H. 1998. Juncaceae. In: K Kubitzki, ed. The families and genera of vascular plants, vol. 4, pp. 252–269. Springer, Berlin/Heidelberg/New York.

Barnard C. 1958. Floral histogenesis in monocotyledons: III. The Juncaceae. Austral. J. Bot. 6: 285–298.

Blaser NW. 1941, 1944. Studies in the morphology of the Cyperaceae: I. Morphology of flowers: A. Scirpoid genera. B. Rhynchosporoid genera. II. The prophyll. Am. J. Bot. 28: 542–551, 832–838, 1941; 31: 53–54, 1944.

Bruhl J. 1995. Sedge genera of the world: relationships and a new classification of the Cyperaceae. Austral. Syst. Bot. 8: 125–305.

Bruhl J, L Watson, and MJ Dallwitz. 1992. Genera of Cyperaceae: Interactive identification and information retrieval. Taxon 41: 225–234.

Cheadle VI. 1955. The taxonomic use of specialization of vessels in the metaxylem of Gramineae, Cyperaceae, Juncaceae, and Restionaceae. J. Arnold Arbor. 36: 141–157.

Cheadle VI and H Kosakai. 1972. Vessels in the Cyperaceae. Bot. Gaz. 133: 214–223.

Cheadle VI and H Kosakai. 1973. Vessels in Juncales: I. Juncaceae and Thurniaceae. Phytomorphology 23: 80–87.

Cutler DF. 1965. Vegetative anatomy of Thurniaceae. Kew Bull. 19: 431–441.

Cutler DF. 1969. Juncales. In: CR Metcalfe, ed. Anatomy of the monocotyledons, vol. 4. Clarendon Press, Oxford.

Drábková K, J Kirschner, O Seberg, G Petersen, and Č Vlček. 2003. Phylogeny of the Juncaceae based on *rbc*L sequences, with special emphasis on *Luzula* DC and *Juncus* L. Plant Syst. Evol. 240: 133–147.

Dunbar A. 1973. Pollen development in the *Eleocharis palustris* group (Cyperaceae): I. Ultrastructure and ontogeny. Bot. Not. 126: 197–254.

Eiten LT. 1976. Inflorescence units in the Cyperaceae. Ann. Missouri Bot. Gard. 63: 81–112.

Goetghebeur P. 1986. Genera Cyperacearum. Masters thesis, University of Ghent.

Goetghebeur P. 1998. Cyperaceae. In: K Kubitzki, ed. The families and genera of vascular plants, vol. 4, pp. 141–198. Springer, Berlin/Heidelberg/New York.

Holttum RE. 1948. The Spikelets in Cyperaceae. Bot. Rev. 14: 525–541.

Kirpes CC, LG Clark, and NR Lersten. 1996. Systematic significance of pollen arrangement in microsporangia of Poaceae and Cyperaceae, review and observations on representative taxa. Am. J. Bot. 83: 1609–1622.

Kirschner J. 2002a. Species plantarum: Flora of the World, Part 6. Juncaceae 1: *Rostkovia* to *Luzula*. Australian Biological Resources Study, Canberra.

Kirschner J. 2002b. Species plantarum: Flora of the World, Part 7. Juncaceae 2: *Juncus* subgenus *Juncus*. Australian Biological Resources Study, Canberra.

Kirschner J. 2002c. Species plantarum: Flora of the World, Part 8. Juncaceae 3: *Juncus* subgenus *Agathryon*. Australian Biological Resources Study, Canberra.

Kovtonjuk N. 1999. Systematic significance of seed surfaces in some *Juncaceae* and *Caryophyllaceae*. In: MH Kurmann, AR Hemsley, eds. The evolution of plant architecture, pp. 367–374. Royal Botanic Gardens, Kew.

Koyama T. 1961. Classification of the family Cyperaceae, part 1. J. Fac. Sci. Univ. Tokyo, Sect. 3, Bot. 8: 37–148.

Kristiansen KA, M Cilieborg, L Drábková, T Jørgensenm, G Petersen, O Seberg. 2005. DNA taxonomy – the riddle of *Oxychloë* (Juncaceae). Syst. Bot. 30: 284–289.

Kubitzki K. 1966. Untersuchungen über den Blütenbau von *Oreobolus* R. Br. Bot. Jahrb. Syst. 85: 80–87.

Kubitzki K. 1998. Thurniaceae. In: K Kubitzki, ed. The families and genera of vascular plants, vol. 4, pp. 455–457. Springer, Berlin/Heidelberg/New York.

Kukkonen I. 1984. On the inflorescence in the family Cyperaceae. Ann. Bot. Fenn. 21: 257–264.

Laurent M. 1904. Recherches sur le développement des Joncées. Ann. Sci. Nat. Bot. 19: 97–194.

Lye KA. 2000. Achene structure and function of structure in Cyperaceae. In: KL Wilson and DA Morrison, eds. Monocots: systematic and evolution, pp. 615–628. CSIRO, Collingwood.

Makde KH. 1982. Pollen development in the Cyperaceae. J. Indian Bot. Soc. 61: 242–249.

Makde KH and SM Bhuskute. 1984. Trends of specialization in endosperm of Cyperaceae. Current Science 53: 504–506.

Metcalfe CR. 1971. Cyperaceae. In: CR Metcalfe, ed. Anatomy of the monocotyledons, vol. 5. Clarendon Press, Oxford.

Michalska A. 1953. Cytological investigations on *Luzula*. Acta Soc. Bot. Polon. 22: 169–186.

Muasya AM, JJ Bruhl, DA Simpson, A Culham, and MW Chase. 2000. Suprageneric phylogeny of Cyperaceae: a combined analysis. In: KL Wilson and DA Morrison, eds. Monocots: systematic and evolution, pp. 593–601. CSIRO, Collingwood.

Muasya AM, DA Simpson, MW Chase, and A Culham. 1998. An assessment of the suprageneric phylogeny in Cyperaceae using *rbc*L DNA sequences. Plant Syst. Evol. 211: 257–271.

Munro SL and HP Linder. 1997. The embryology and systematic relationships of *Prionium serratum* (Juncaceae: Juncales). Am. J. Bot. 84: 850–860.

Munro SL and HP Linder. 1998. The phylogenetic position of *Prionium* (Juncaceae) within the order Juncales based on morphological and *rbc*L sequence data. Syst. Bot. 23: 43–55.

Nemirovich-Danchenko EN. 1985. Thurniaceae. In: A Takhtajan, ed. Comparative seed anatomy, vol. 1, pp. 161–162. Nauka, Leningrad (in Russian).

Plunkett GM, DE Soltis, PS Soltis and RE Brooks. 1995. Phylogenetic relationships between Juncaceae and Cyperaceae, insights from *rbc*L sequence data. Am. J. Bot. 82: 520–525.

Reihmann D. 1977. Vergleichende Embryologie und systematische Stellung der Juncaceae mit neuen Beiträgen zur Samenentwicklung von *Juncus* und *Luzula*. Thesis, University of Bochum.

Richards J, JJ Bruhl, and KL Wilson. 2005. Understanding floral morphology of mapaniid Cyperaceae: development of reproductive structures in *Exocarya scleroides*. In: Botany 2005. Learning from plants (Abstracts), pp. 37–38.

Roalson EH. 2005. Phylogenetic relationships in the Juncaceae inferred from nuclear ribosomal DNA internal transcribed spacer sequence data. Int. J. Plant Sci. 166: 397–413.

Schneider M. 1932. Untersuchungen über die Embryobildung und Entwicklung der Cyperaceen. Beih. Bot. Centralbl. 49: 649–674.

Schultze-Motel W. 1959. Entwicklungsgeschichtliche und vergleichend-morphologische Untersuchungen im Blütenbereich der Cyperaceae. Bot. Jahrb. Syst. 78: 129–170.

Shah CK. 1967. A taxonomic evaluation of the families Cyperaceae and Juncaceae. Bull. Nation. Inst. Sci. India 34: 248–256.

Shah CK. 1968. Development of pericarp and seed coat in the Cyperaceae. Naturaliste Canad. 95: 39–48.

Shah CK. 1974. Morphology and embryology of the family Cyperaceae. Adv. Plant Morph. 1972: 102–112.

Simpson D. 1995. Relationships within Cyperales. In: Rudall PJ, PJ Cribb, DF Cutler, and CJ Humphries, eds. Monocotyledons: systematics and evolution, vol. 2, pp. 497–509. Royal Botanic Gardens, Kew.

Simpson D. 1998. Mapaniaceae – fact or fiction? In Monocots II, p. 50 (Abstract). Sydney.

Simpson DA, CA Furness, TR Hodkinson, AM Muasya, and MW Chase. 2003. Phylogenetic relationships in Cyperaceae subfamily Mapanioideae inferred from pollen and plastid DNA sequence data. Am. J. Bot. 90: 1971–1086.

Smith DL and JS Faulkner. 1976. The inflorescence of *Carex* and related genera. Bot. Rev. 42: 53–81.

Snell RS. 1936. Anatomy of the spikelets and flowers of *Carex, Kobresia,* and *Uncinia*. Bull. Torrey Bot. Club. 63: 277–295.

Starr JR, RJ Bayer, and BA Ford. 1999. The phylogenetic position of *Carex* section *Phyllostachys* and its implications for phylogeny and subgeneric circumscription in *Carex* (Cyperaceae). Am. J. Bot. 86: 563–577.

Starr JR, SA Harris, and DA Simpson. 2004. Phylogeny of the unispicate taxa in Cyperaceae tribe Cariceae I: generic relationships and evolutionary scenarios. Syst. Bot. 29: 528–544.

Starr JR, Teoh V, E Roalson, AM Muasya, and DA Simpson. 2006. Towards a phylogenetic classification of sedges (Cyperaceae): chloroplast (*rbc*L, *mat*K, *Ndh*F) and nuclear (ADC) data. In: Botany 2006 – Looking to the future, conserving the past (Abstracts), pp. 258–259.

Tarasevich VF. 1986. On the position of the family Cyperaceae in the system of flowering plants in connection with the palynological data. In: VN Tikhomirov, ed. Sources of information in the phylogenetic systematics of plants, pp. 75–76. Nauka, Moscow (in Russian).

Tejavathi DH. 1987. Seed development in some members of Cyperaceae. Beitr. Biol. Pflanz. 62: 43–55.

Thimm U. 1985. Zur Embryologie, Blüten- und Fruchtanatomie der isolierten Juncales-Gattungen *Prionia* und *Thurnia*. Diploma Thesis Bochum. University of Bochum, Division of Biology.

Tucker GC. 1987. The genera of Cyperaceae in the SE United States. J. Arnold Arbor. 68: 361–445.

Van der Veken P. 1965. Contribution a embryographie systematique des Cyperaceae-Cyperoideae. Bull. Jard. Bot. Brux. 35: 285–354.

Van Wichelen J, K Camelbeke, P Chaerle, P Goetghebeur, and S Huysmans. 1999. Comparison of different treatments for LM and SEM studies and systematic value of pollen grains in Cyperaceae. Grana 38: 50–58.

Vovk AG. 1985a. Juncaceae. In: A Takhtajan, ed. Comparative seed anatomy, vol. 1, pp. 158–159. Nauka, Leningrad (in Russian).

Vovk AG. 1985b. Cyperaceae. In: A Takhtajan, ed. Comparative seed anatomy, vol. 1, pp. 163–166. Nauka, Leningrad (in Russian).

Vrijdaghs A, P Caris, P Goetghebeur, and E Smets. 2004. The bristles of *Dulichium* (Cyperaceae), a floral ontogenetic study. In: Botany 2004. Alpine diversity: adapted to the peaks (Abstracts), p. 164.

Vrijdaghs A, P Caris, P Goetghebeur, and E Smets. 2005a. Floral ontogeny in *Scirpus*, *Eriophorum* and *Dulichium* (Cyperaceae), with special reference to the perianth. Ann. Bot. 95: 1199–1209.

Vrijdaghs A, P Goetghebeur, AM Muasya, P Caris, and E Smets. 2005b. Floral ontogeny in *Ficinia* and *Isolepis* (Cyperaceae), with focus on the nature and origin of the gynophore. Ann. Bot. 96: 1247–1264.

Wilczek E. 1892. Beiträge zur Kenntnis des Baues der Frucht und des Samens der Cyperaceen. Bot. Centralbl. 51: 129–138, 193–201, 225–233, 257–265.

Williams CA and JB Harborne. 1975. Luteolin and daphnetin derivatives in the Juncaceae and their systematic significance. Biochem. Syst. Ecol. 3: 181–190.

Wulff HD. 1939. Die Pollenentwicklung der Juncaceen nebst einer Auswertung der embryologischen Befunde hinsichtlich einer Verwandtschaft zwischen den Juncaceen und Cyperaceen. Jahrb. Wiss. Bot. 87: 533–556.

Yen AC and RG Olmstead. 2000. Molecular Systematics of Cyperaceae Tribe Cariceae based on two chloroplast DNA

regions: *ndh*F and *trn*L intron-intergenic spacer. Syst. Bot. 25: 479–494.
Zhang X, KL Wilson, and JJ Bruhl. 2004. Sympodial structure of spikelets in the tribe Schoeneae (Cyperaceae). Am. J. Bot. 91: 24–36.

Superorder POANAE

Order 29. TYPHALES

Perennial herbs of shallow-water and wet habitats, sometimes wholly submerse and floating, with creeping, sympodial rhizomes rich in starch (starch grains pteridophyte-type). Raphide bundles are common, calcium oxalate crystals present. Silica bodies sometimes present. Cells with mucilage common. Vessels in all vegetative organs, with scalariform perforations. Leaves alternate, distichous, with strongly sheathing base, long, linear, flat or keeled, with parallel venation, possess a prominent system of aerenchymatous channels. Stomata paracytic. Flowers small or minute, numerous, basically 3-merius or 2-merous, unisexual, in compound, basally female and apically male racemose inflorescences consisting of complex globular heads or very dense cylindric spikes. Perianth of one to several inconspicuous, membranous scales or more or less numerous thin bristles. Stamens 1–8, mostly 3; filaments free or often basally connate; anthers tetrasporangiate, basifixed, extrorse. Tapetum plasmodial. Microsporogenesis successive. Pollen grains 2-celled, free or in tetrads, tectate-columellate, 1-porate (ulcerate), finely reticulate. Gynoecium mostly pseudomonomerous, in Typhaceae seemingly monomerous while in Sparganiaceae with one more or less adaxial fertile locule and one abaxial empty locule, but some have gynoecia with two or three fertile locules and correspondingly with two or three stigmatic branches; style short or long with shortly decurrent stigma; ovary superior, with one pendulous ovule in the fertile locule (or fertile locules). Ovules anatropous, apotropous, bitegmic, crassinucellate, with parietal cell. Female gametophyte of *Polygonum*-type. Endosperm helobial. Fruits small, dry, indehiscent or dorsally dehiscent. Seeds fill up the fruit locule, with thin, membranous seed coat, straight, cuneate or fusiform, well-differentiated embryo, copious, mealy endosperm containing starch, aleurone, and small quantity of lipids, and thin perisperm containing starch, aleurone, and lipids (*Sparganium*) or only starch (*Typha*). The unlignified cell walls contain bound ferulate. n = 15.

Taxonomic position is disputable. The order Typhales is enigmatic, having many automorphic features, but few characters which links it clearly to other orders (Linder and Kellogg 1995). They are associated either with the commelinids or more traditionally with the Arales and Pandanales. They resemble the Arales in starchy rhizome, floral morphology of *Sparganium*, the abundance of crystal raphides, plasmodial tapetum, successive microsporogenesis, starchy endosperm (only in certain Araceae). Seeds of the Typhales have perisperm, like those of Pistiaceae and Acoraceae. Besides, *Sparganium* and *Acorus* share the same rust parasite, *Uromices sparganii* (Parmelee and Savile 1954; Savile 1979). There are also many similarities to the Pandanales, including vessel characteristics (Wagner 1977), close resemblances between the branching in *Sparganium* and that in *Pandanus*, and striking resemblances in the arrangement of flowers (see Rendle 1953). As Rendle points out, the inflorescence of *Sparganium* recalls that of *Pandanus* on a smaller scale. According to Müller-Doblies (1969), there is "die frappierende Ahnlichkeit der kopfehenformigen Partialinfloreszenzen von *Sparganium* mit jenen mancher Pandani (z. B. *Pandanus sparganioides* Bak.)." The affinity between these two orders is supported also by cytological data (Sharma 1964; Malick and Sharma 1966). In both of them n = 15. "Moreover, in the morphology of the chromosomes too homogeneity is evident, all being characterized by very small chromosomes, with nearly identical types of restrictions. It is very difficult to distinguish the three genera on the basis of their karyotypes as they look very similar to each other" (Sharma 1964). Therefore Sharma even suggests the inclusion of all of them under one order Pandanales. Chemical data also support the affinity between Typhales and Pandanales. Both of them have caffeic and p-coumaric acids (Gibbs 1974), and in both of them similar flavonoids, simple phenols, raphides, steroids, tannins, and triterpenoids (Radhakrishnaiah et al. 1984) are present. Finally, a close phylogenetic relationship between Typhales and Pandanales is supported by serological study (Bergner and Jensen 1989). However, Typhales are herbaceous, silica bodies sometimes present, cells with mucilage absent, leaves not aggregated at the end of branches, distichous, with prominent system of aerenchymatous channels compartimentized by diaphragms, stomata paracytic, endosperm formation helobial (probably of particular type with small chalazal chamber), fruits dry, seed coat very reduced, seeds with perisperm and always rich in starch, and the unlignified cell walls are fluorescent under UV and contain bound ferulic acid. In addition, the epicuticular wax is consistently of *Strelitzia* type. Dahlgren et al. (1985) place the Typhales in the vicinity of Pontederiaceae, Haemodoraceae and Philydraceae, but don't exclude the possibility that they also come near the ancestors of Cyperales or Pandanales. According to Linder and Kellogg (1995) Typhales are the sister-group to Poales s. l. I agree with Kubitzki (1998: 459), that a close relationship to Poales s. l. (Poanae sensu mihi) at present is the best supported hypothesis.

Key to families

1 Inflorescence axillary, of dense globose units, each comprising twice-thrice-compound heads; the inflorescence system basally female, apically male. March or aquatic perennial herbs. Leaves emergent and floating, alternate, distichous, sessile, sheathing simple, entire. Mesophyll sometimes containing mucilage cells (with raphides) and calcium oxalate crystals. Flowers bracteate (female flowers) or ebrancheate (male flowers), small. Floral receptacle developing an androphore. Perianth of the female flowers of (1-)3–4(-6) small segments. Stamens (1-)3(-6). Gynoecium of 1–2 carpels, monomerous or syncarpous (and then pseudomonomerous, with one locule empty). Ovary 2-locular. Fruits 1-seeded, drupaceous or nucular, hydrochorous or endozoochorous, sessile or nearly so; seeds starchy, epicarp consists of three layers: firm exocarp, spongy mesocarp and thick and stony endocarp; embryo small, endosperm oily. . . . . 1. SPARGANIACEAE.

1 Inflorescence a terminal, dense, elongate, cylindrical spike. Aquatic herbs; leaves emergent, alternate, leathery, sessile, sheathing, simple. Silica bodies present or absent. The mesophyll crystals raphides, or druses, or solitary-prismatic, or styloids. Perianth of the female flowers mostly of numerous capillary bristles. Stamens (1-)3(-5); anthers appendaged (via apical projection of the connective). Gynoecium of one carpel. Fruits anemochorous, a follicle (but tiny and achene-like before dehiscence), long-stipitate, epicarp consists of two layers: thin exocarp of thin-walled cells and thin endocarp of thick-walled cells. . . . . .2. TYPHACEAE.

1. SPARGANIACEAE

Hanin 1811. 1/14. Extratropical regions of the Northern Hemisphere, mountains of Sumatra and New Guinea, and southeastern Australia and New Zealand.

Sparganium.

2. TYPHACEAE

A. L. de Jussieu 1789. 1/10-15. Nearly cosmopolitan, but mainly in Eurasia and North America.

Typha (including *Rohrbachia*).

Very close to the Sparganiaceae. The differences between them are clearly related to the switch from hydrochory to anemochory (Kubitzki 1998).

Bibliography

Asplund I. 1972. Embryological studies in the genus *Typha*. Svensk Bot. Tidskr. 66: 1–17.

Bergner I and U Jensen. 1989. Phytoserological contribution to the systematic placement of the Typhales. Nord. J. Bot. 8: 447–456.

Cook CDK and MS Nicholls. 1986, 1987. A monographic study of the genus *Sparganium* (Sparganiaceae). Part I. Subgenus *Xanthosparganium* Holmberg. Bot. Helv. 96: 213–267, 1987; Part II. Subgenus *Sparganium*. Bot. Helv. 97: 1–44, 1987.

Gibbs RD. 1974. Chemotaxonomy of flowering plants, pp. 1–4. McGill-Queen's University Press, Montreal.

Graef PE. 1955. Ovule and embryo sac development in *Typha latifolia*. Am. J. Bot. 42: 806–809.

Haines RW and KA Lye. 1972. Studies in African Cyperaceae: VII. Panicle morphology and possible relationships in Sclerieae and Cariceae. Bot. Not. 125: 331–343.

Lee DW and DE Fairbrothers. 1972. Taxonomic placement of the Typhales within the monocotyledons: preliminary serological investigation. Taxon 21: 39–44.

Malik R and AK Sharma. 1966. Chromosome studies in Indian Pandanales. Cytologia 31: 402–410.

Mavrodoev EV. 2001. *Rohrbachia*, a new genus of the Typhaceae. Bot. Zhurn. 86, 9: 120–124 (in Russian with English summary).

Meyer FJ. 1933. Beiträge zur vergleichenden Anatomie der Typhaceae (Gattung *Typha*). Beitr. Bot. Centralbl. 51: 335–376.

Müller-Doblies U. 1969. Über die Blütenstände und Blüten sowie zur Embryologie von *Sparganium*. Bot. Jahrb. Syst. 89: 459–450.

Müller-Doblies D. 1970. Über die Verwandtschaft von *Typha* und *Sparganium* im Infloreszenz- und Blütenbau. Bot. Jahrb. Syst. 89: 451–562.

Parmelee J and BDO Savile. 1954. Life history and relationship of rusts of *Sparganium* and *Acorus*. Mycologia 46: 823–836.

Radhakrishnaiah M, G Nageshwar, and LL Narayana. 1984. Chemotaxonomy of *Pandanus* and *Typha*. Curr. Sci. 53: 759–760.
Rendle A. 1953. The classification of flowering plants. Vol. 1. Gymnosperms and monocotyledons, 2nd ed. Cambridge University Press, Cambridge
Rowlatt U. 1992. Architecture of the leaf of the greater reed mace, *Typha latifolia* L. Bot. J. Linn. Soc. 110: 161–170.
Savile DB0. 1979. Fungi as aids in higher plant classification. Bot. Rev. 45: 377–503.
St. John H. 1941. Teratological *Typha*. Rhodora 43: 85–91.
Thieret JW. 1982. The Sparganiaceae in the southeastern United States. J. Arnold Arbor. 63: 341–355.
Wagner P. 1977. Vessel types of monocotyledons: a survey. Bot Not. 130: 383–402.
Zubkova IG and LK Shabes. 1985. Typhaceae. In: A Takhtajan, ed. Comparative seed anatomy, vol. 1, pp. 262–263. Nauka, Leningrad (in Russian).

Order 30. RESTIONALES

Perennial or less often annual herbs, sometimes lianous, with solid or fistulose stems. Silica bodies often present. Oxalate raphides mostly present (absent in most Restionaceae and Centrolepidaceae). Vessels in roots and stems, with simple or scalariform perforations. Leaves alternate, commonly distichous, entire, with open or closed sheath and narrow, parallel-veined lamina that is sometimes reduced or suppressed. Stomata paracytic, of graminoid type. Flowers small and inconspicuous, in various kinds of inflorescences, often in spikelets, bisexual or more often unisexual, mostly 3-merous. Perianth small, membranous, in two cycles or in one cycle, sometimes lacking. Stamens 6 in 2 cycles or 3(4) in 1 cycle, sometimes only 2 or even 1; anthers basifixed to dorsifixed, tetrasporangiate or less often disporangiate, latrorse or introrse, opening longitudinally. Tapetum secretory. Microsporogenesis simultaneous (Flagellariaceae) or more often successive. Pollen grains 3-celled or less often 2-celled, tectate-columellate, 1-porate, often with annulate graminoid pore. Gynoecium commonly of three united carpels, less often two carpels; in Centrolepidaceae of one carpel; stylodia more or less connate into style; ovary superior, 3-locular or by reduction of one or two carpels 2-locular or even 1-locular, with one pendulous ovule per locule. Ovules orthotropous, bitegmic, crassinucellate (Flagellariaceae) or more often tenuinucellate, without parietal cell. Female gametophyte of *Polygonum*- or *Allium*-type. Endosperm nuclear. Fruits loculicidal capsules or nuts, in Flagellariaceae drupaceous. Seeds small, globose or ovoid, with or without operculum; seed coat formed by both integuments; endosperm copious, starchy, with aleurone layer; embryo small, lens-shaped or obovate, weakly differentiated. Bound p-coumaric, ferulic and diferulic acids present. n = 6–13, 18, 19.

Related to the Typhales and Juncales and have a common origin from a commelinalean ancestors (Linder and Kellogg 1995).

Key to Families

1 Anthers disporangiate or rarely (Restionaceae p.p.) tetrasporangiate.

2 Stamens three; gynoecium generally of 3 carpels. Usually xeromorphic perennial herbs, tufted or with an elongate rhizome with overlapping scales, glabrous or densely woolly with multicellular hairs. Stem with a continuous cylinder of sclerenchyma ring inside the chlorenchyma. Silica bodies present or absent; plants accumulated starch exclusively 'pteridophyte type'. Vessels in stem with scalariform, or with simple and/or scalariform, or with simple perforations; in the rhizomes with exclusively simple perforations and in roots with scalariform perforations. Stems with solid or spongy, or hollow internodes. Leaves alternate, mostly distichous, usually reduced to an open sheath with scarcely any lamina, usually not ligulate, stipulate (in the form of membranous lobes at the tops of the sheath margins), or estipulate. Stomata, when present, paracytic. Flowers in spicate or paniculate inflorescences, aggregated into spikelets subtended by a single bract or less often solitary and mostly are also subtended by bracteoles, unisexual, dioecious or rarely monoecious; in some species of *Lepyrodia* and in one species of *Restio* bisexual flowers may occur. Perianth segments six in two series, rarely four in two cycles or only three or even absent. Stamens opposite the inner perianth segments; filaments sometimes connate into a column; anthers dorsifixed, unilocular, introrse or latrorse, disponagiate or tetrasporangiate, appendaged or unappendages. Pollen grains 2-celled or 3-celled, scrobiculate, pores with annulus or without annulus. Staminodia present or absent. Pistillodia may be present in the male flowers. Stylodia free or connate below, with elongate, papillate, or shortly plumose stigma on their inner sides; ovary 3- to 1-locular, sessile to

stipitate. Hypostase present. Ovules 1–3, tenuinucellate, rarely (*Alexgeorgia*) crassinucellate. Large starch bodies are surrounding polar nuclei. Female gametophyte of *Polygonum*-type. Fruits 1–3-locular loculicidal capsules or small nuts. Seeds endotegmic, with copious, mealy, starchy endosperm capped by minute, lenticular or obovate, weakly differentiated embryo. Contain flavonols (kaempferol, quercetin, and myricetin), glycoflavones, and proanthcyanidins. n = 7, 8, 11, 12, 13, 16, 20. . . . . 3. RESTIONACEAE.

2 Stamen one; gynoecium of one carpel. Small, tufted, annual or perennial and rhizomatous herbs of wet habitats. Raphides and silica bodies lacking. Multicellular, branched or unbranched filamentous hairs sometimes present. Roots usually without differentiated pericycle. Stems solid. Vessels with scalariform perforations in root, stems and leaves. Leaves alternate, basal or less often imbricately inserted along the stem, linear, lanceolate or setaceous, basally with distinct, membranous, open sheath. Flowers minute, in highly reduced, terminal, compound, spikelike or headlike inflorescences subtended by one to several leafy bracts and bearing two to several distichous, glumelike bracts; each bract encloses a group of male flowers or one to several female flowers, or, most frequently, one or more bisexual synanthia; each synanthium is a spikeletlike cyme of one or two male flowers and one to several superposed or collateral female flowers and often has one to three small, hyaline bracts or bracteoles. Filaments filiform, glabrous; anthers 1-locular, dorsifixed, versatile. Microsporogenesis successive. Pollen grains 3-celled, tectate-granular, 1-porate (ulcerate), with poorly defined, aperturate margins, verrucate, tectum perforated by small punctae (scrobiculi). Stylodium filiform, with elongate, ventrally decurrent, papillate stigma; ovary superior, with one apical, pendulous ovule. Female gametophyte of *Polygonum*-type. Fruits small, membranous, 1-seeded, indehiscent (*Aphelia*) or more often dehiscing longitudinally on a dorsal side, which gives an appearance of paired valves in the fruiting pseudanthia of *Gaimardia* and *Centrolepis* (Cooke 1998); collateral dehiscing fruits are generally united into a capsulelike collective fruit. Seeds small; seed coat rather thin, membranous, formed mainly by the inner epidermis of the inner integument, the outer integument being very thin; endosperm copious, mealy, rich in compound starch grains, without aleurone layer; embryo minute, weakly differentiated, globose, obconical, or bell-like, peripheral, lying on the micropylar side. n = 10–13. . . . . 6. CENTROLEPIDACEAE.

1 Anthers tetrasporangiate.

3 Perennial, high climbing lianas with solid stem arising from sympodial rhizomes, supported by means of long, coiled tendrils terminating the leaves and representing the extension of the thickened midrib. Leaf blades larger than sheaths and ending in a tendril. Silica bodies present. Scattered secretory cells present in leaf and stem. Vessels in leaves and stems, with scalariform and simple perforations. Leaves alternate, circinately enrolled in bud; lamina lanceolate, constricted at base; leaf sheaths closed. Flowers in more or less branched terminal panicles, bisexual or very rarely unisexual, bracteate. Perianth segments five, in two cycles, free. Stamens six in two cycles, occasionally reduced to staminodia; filaments very short; anthers sagittate basifixed, latrorse. Microsporogenesis simultaneous. Pollen grains 2-celled. Cynoecium of three carpels; styles three, sometimes connate basally, free or partially joined. Ovary 2-locular, ovules one per locule, pendulous. Female gametophyte of *Allium*-type. Fruits fleshy, red or black drupes. Seeds endotestal; embryo minute, lenticular; endosperm copious, starchy. Flavonoles (kaempferol) present; alkaloids present or absent; n = 19. . . . . 1. FLAGELLARIACEAE.

3 Reed-like, coarse, unbranched, erect herbs to 5 m tall. Stems terete, hollow, arising from a short sympodial rhizomes, without scabrid margins, unbranched below the inflorescence. Silica bodies and calcium oxalate crystals present. Secretory cells absent. Vessels in all vegetative organs, with scalariform and simple perforations. Leaves large (the lamina up to 1 m long), alternate, distichous, flat, sessile, with tubular, open, sheathing bases, the mouths of sheaths with diminutive ligules, the lamina abruptly narrowed at base and articulated to the sheath, long, conspicuously plicate, grasslike, bearing short prickly hairs and branched filamentous hairs. Flowers in long,

much branched terminal panicles, bisexual, bracteate and have caducous bracteoles. Perianth segments six, in two cycles, free to joined, sepaloid (bractlike), sometimes the outer members much larger, persistent. Stamens six; anthers basifixed (safittate) latrorse. Pollen grains 3-celled. Ovary 3-locular; styler branches or sessile stigmatic areas three; stigmas covered with short papillae. Ovules one per locule, pendulous. Fruits drupaceous, red, yellow or black. Seeds with copious, mealy, starchy endosperm capped by minute, lenticular, undifferentiated, discoid embryo. n = 18. . . . . 2. JOINVILLEACEAE.

4 Monoecious plants with male and female flowers together in the same spikelet. Stamens six or four. Perennial evergreen herbs with slender cylindrical solid photosynthetic stems arising from rhizomes. Silica bodies (as silica sand) present in the culm mesophyll. Vessels in stems and in roots, with scalariform and simple perforations. Leaves reduced to 1–4 tightly convoluted sheaths up to 10 cm long; sheath split to the base, usually with a mucro which may be developed into a small awn. Flowers in single- or several-flowered spikelets aggregated into terminal conical or cylindrical dense spikelike heads, subtended by a glossy, dark brown or black, rigid, ovate bract longer than the flower, dorsiventrally flattened. Perianth of six, unequal, glumaceous segments. Female flowers with three staminodes. Stamens six or four; filaments free, slender; anthers basifixed, tetrasporangiate, latrorse. Pollen grains annulate and with a plug; interapertural wall smooth, and lacks scrobiculi. Pistillodia present. Gynoecium with 2-locular ovary; styles two, free, attenuate from the ovary, apical; stigmas two; ovules one per locule, pendulous Female gametophyte tetrasporic, most similar to the *Drusa*-type. Fruits a 3-locular capsule or a nut. Seeds with minute embryo and copious endosperm. n = 32, 33 and ca. 24. . . . . 5. ECDEIOCOLEACEAE.

4 Dioecious plants; culms not striate or with striations formed by narrow ridges of sclerenchyma not be deep crypts; stamens three. Perennial herbs with solid or fistulose stems arising from short, rhizomes, which covered with overlapping, glabous scales; stems simple or branched, sometimes very compressed. Vessels with scalariform perforations. Silica bodies lacking, but oxalate crystals occur in some species. Leaves linear to filiform, laterally flattened, equitant, with a small ligule at the transition to the open sheath. Flowers in few to many flowers paniculate inflorescence with each branch subtended by elongated leaflike bract, pedicellate or sessile, each with one or two short bracts. Perianth of six more or less similar and glumaceous segments in two cycles. Stamens three; filaments free connate; anthers dorsifixed, tetrasporangiate, latrorse. Pollen grains 3-celled, scrobiculate, pores annulate. Gynoecium of three carpels; style free, stigmas decurrent ovary 3-locular with solitary ovule per locule. Female gametophyte of *Polygonum*-type. Fruits 1-seeded angular capsules dehiscing at the angles or nutlets. Seeds with minute, lenticular embryo and copious mealy, starchy endosperm. Flavonol glycosides present. n = 11.. . . . . 4. ANARTHRIACEAE

1. FLAGELLARIACEAE

Dumortier 1829. 1/4. Tropical Africa, Southeast Asia, Malesia, Micronesia, New Caledonia, northern Australia, and eastward to Samoa and Niue.

Flagellaria.

2. JOINVILLEACEAE

Tomlinson et A. C. Smith. 1970. 1/2. From western Malesia eastward to Samoa and Hawaii, but absent from southern and eastern Indonesia, New Guinea, and Australia.

Joinvillea.

3. RESTIONACEAE

R. Brown 1810 (including Elegiaceae Rafinesque 1838, Hopkinsiaceae B. G. Briggs et L. A. S. Johnson 2000; Lyginiaceae B. G. Briggs et L. A. S. Johnson 2000). 57/520. Tropical and South Africa, Madagascar, Southeast Asia, Malesia, New Guinea, Australia, Tasmania, New Zealand, and southern South America;

concentrated mostly in southwestern Africa and extratropical regions of Australia and Tasmania.

Lyginia, Hopkinsia, Staberoha, Ischyrolepis, Elegia, Chondropetalum, Dovea, Askidiosperma, Platycaulos, Restio, Calopsis, Thamnochortus, Rhodocoma, Ceratocaryum, Cannomois, Nevillea, Hydrophilus, Anthochortus (including *Anthochortus*), *Mastersiella, Hypodiscus, Willdenowia, Lepyrodia, Sporadanthus, Calorophus, Winifredia, Empodisma, Coleocarya, Desmocladus, Harperia, Onychosepalum, Catacolea, Kulinia, Lepidobolus, Baloskion, Curingalia, Acion, Saropsis, Alexgeorgia, Chordifex, Dielsia, Eurychorda, Platychorda, Tremulina, Melanostachya, Cytogonidium, Loxocarya* (including *Megalotheca*), *Taraxis, Tyrbastes, Pseudoloxocarya, Leptocarpus, Hypolaena, Meeboldina, Stenotalis, Chaetanthus, Dapsilanthus, Apodasmia, Pseudoloxocarya.*

4. ANARTHRIACEAE

D. F. Cutler et Airy Shaw 1965. 1/7. Southwestern Australia.

Anarthria.

Closely related to the Restionaceae, differing mainly in the culm anatomy.

5. ECDEIOCOLEACEAE

D. F. Cutler et Airy Shaw 1965. 2/2. Southwestern Australia.

Ecdeiocolea, Georgeantha.

6. CENTROLEPIDACEAE

Endlicher 1836. 4/35. Southern China (Hainan), Southeast Asia, Malesia, New Guinea, Australia, Tasmania, New Zealand, and subantarctic South America (Tierra del Fuego and Falkland Islands).

Centrolepis (including *Pseudalepyrum*), *Aphelia, Brizula, Gaimardia.*

Closely related to the Restionaceae, but differ from them markedly in very reduced and specialized inflorescence with unisexual flowers clustered into synanthia, minute and perianthless flowers, male flowers usually with one stamen, disporangiate anthers, "centrolepidoid" pollen grains, and monocarpellate gynoecium. Also in most species of all genera except *Gaimardia,* they have an unusual type of epidermal cell arrangement whereby one cell overlaps the end of the next in a file (a unique feature in monocotyledons), they have no pericycle in the root, and their root hairs arise in an abnormal fashion from one side of root epidermal cells (Cutler 1969: 96). Already Dahlgren (1975) proposed a separate order Centrolepidales although in his later publications he included the Centrolepidaceae in the Restionales. Linder and Ferguson (1985: 75) suggested that the Centrolepidaceae be excluded from the orders Restionales and Poales. According to Hamann (1975), the Centrolepidaceae could have a neotenous origin from the Restionaceae.

Bibliography

Appel O and C Bayer. 1998. Flagellariaceae. In: K Kubitzki, ed. The families and genera of vascular plants, vol. 4, pp. 208–211. Springer, Berlin/Heidelberg/New York.

Baker CA. 1951. Flagellariaceae. In: CGGJ van Steenis, ed. Flora Malesiana, ser. 1, 4: 245–250. Leyden.

Baranova MA. 1975. Stomatographical studies of the family Flagellariaceae. Bot. Zhurn. 60: 1690–1697 (in Russian).

Bayer C and O Appel. 1998. Joinvilleaceae. In: K Kubitzki, ed. The families and genera of vascular plants, vol. 4, pp. 249–251. Springer, Berlin/Heidelberg/New York.

Borwein B, ML Goetsee, and S Krupko. 1949. Development of the embryo sac of *Restio dodii* and *Elegia racemosa*. South Afr. J. Bot. 15: 1–11.

Briggs BG and LAS Johnson. 1998. New combinations arising from a new classification of non-African Restionaceae. Telopea 8: 21–33.

Briggs BG and LAS Johnson. 1999. A guide to a new classification of Restionaceae and allied families. In: KA Meney and JS Pate, eds. Australian rushes, biology, identification and conservation of Restionaceae and allied families, pp. 25–56. University of Western Australia Press, Perth.

Briggs BG and LAS Johnson. 2000. Hopkinsiaceae and Lyginiaceae, two new families of Poales in Western Australia, with revisions of *Hopkinsia* and *Lyginia*. Telopea 8: 477–502.

Briggs BG, AD Marchant, S Gilmore, and CL Porter. 2000. A molecular phylogeny of Restionaceae and allies. In: KL Wilson and DA Morrison, eds. Monocots: systematics and evolution, pp. 661–671. CSIRO, Collinwood.

Chanda S. 1966. On the pollen morphology of the Centrolepidaceae, Restionaceae, and Flagellariaceae with special reference to taxonomy. Grana Palynol. 6: 355–415.

Chanda S and S Rowley. 1967. Apertural types in pollen of Restionaceae and Flagellariaceae. Grana Palynol. 7: 16–36.

Cheadle VI. 1955. The taxonomic use of specialization of vessels in the metaxylem of Gramineae, Cyperaceae, Juncaceae, and Restionaceae. J. Arnold Arbor. 36: 141–157.

Cheadle VI and H Kosakai. 1975. Vessels in Juncales, II. Centrolepidaceae and Restionaceae. Am. J. Bot. 62: 1017–1026.

Cooke DA. 1998. Centrolepidaceae. In: K Kubitzki, ed. The families and genera of vascular plants, vol. 4, pp. 106–109. Springer, Berlin/Heidelberg/New York.

Cutler DF. 1969. Juncales. In: CR Metcalfe, ed. Anatomy of the monocotyledons, vol. 4. Clarendon Press, Oxford.

Cutler DF and HK Airy Shaw. 1965. Anarthriaceae and Ecdeiocoleaceae: Two new monocotyledonous families, separated from the Restionaceae. Kew Bull. 19: 489–499.

Eldenäs PK and HP Linder. 2000. Congruence and complementarity of morphological and *trn*L-*trn*F sequence data and the phylogeny of the African Restionaceae. Syst. Bot. 25: 692–701.

Hamann U. 1962. Beitrag zur Embryologie der Centrolepidaceae mit Bemerkungen über den Bau der Blüten und Blütenstande und die systematische Stellung der Familie. Ber. Deutsch. Bot. Ges. 75: 153–171.

Hamann U. 1963. Über die Entwicklung und den Bau des Spaltöffnungsapparates der Centrolepidaceae. Bot. Jahrb. Syst. 82: 316–320.

Hamann U. 1975. Neue Untersuchungen zur Embryologie und Systematik der Centrolepidaceae. Bot. Jahrb. Syst. 96: 154–191.

Harborne JB, CA Williams, BG Briggs, and LAS Johnson. 2000. Flavonoid patterns and the phylogeny of the Restionaceae. In: KL Wilson and DA Morrison, eds. Monocots: systematics and evolution, pp. 661–671. CSIRO, Collinwood.

Hochuli PA. 1979. Ursprung und Verbreitung der Re-stionaceen. Vierteljahrsschr. Naturf. Ges. Zürich 124: 109–131.

Johnson LAS and BG Briggs. 1981. Three old southern families Myrtaceae, Proteaceae, and Restionaceae. In: A Keast, ed. Ecological biogeography of Australia, pp. 427–469. W. Junk, Utrecht.

Keighery GJ. 1998. Systematics and biology of the Centrolepidaceae. In: Monocots II, p. 30 (Abstract). Sydney.

Kircher P. 1986. Untersuchungen zur Blüten- und Infloreszenzmorphologie, Embryologie, und Systematik der Restionaceen im Vergleich mit Gramineen und verwandten Familien. Diss. Bot. 94: 1–219.

Krupko S. 1962. Embryological and cytological investigations in *Hypodiscus aristatus* Nees (Restionaceae). South Afr. J. Bot. 28: 21–44.

Krupko S. 1963. Macrosporogenesis and embryo sac development in *Chondropetalum hookerianum* (Mast.) Pillans (Restionaceae). Acta Soc. Bot. Polon. 32: 17–190.

Krupko S. 1966. Some loose embryological and cytological observations on members of Restionaceae family. Bull. Soc. Amis Sci. Poznan, ser. D, 7: 59–67.

Ladd PG. 1977. Pollen morphology of some members of the Restionaceae and related families, with notes on the fossil record. Grana Palynol. 16: 1–14.

Lee DW, YK Pin, and LF Yew. 1975. Serological evidence on the distinctness of the monocotyledonous families Flagellariaceae, Hanguanaceae, and Joinvilleaceae. Bot. J. Linn. Soc. 70: 77–81.

Linder HP. 1984. A phylogenetic classification of the genera of the African Restionaceae. Bothalia 15: 11–76.

Linder HP. 1985. Conspectus of the African species of Restionaceae. Bothalia 15: 387–503.

Linder HP. 1987. A hypothesis on the evolutionary history of the Poales/Restionales. Kew Bull. 42: 297–318.

Linder HP. 1991. A review of the African Restionaceae. Contrib. Bolus Herb. 13: 209–264.

Linder HP. 1992a. The structure and evolution of the female flower of the African Restionaceae. Bot. J. Linn. Soc. 109: 401–425.

Linder HP. 1992b. The gynoecia of Australian Restionaceae: morphology, anatomy and systematic implications. Austral. Syst. Bot. 109: 227–245.

Linder HP. 2000. Vicariance, climate change, anatomy and phylogeny of Restionaceae. Bot. J. Linn. Soc. 134: 159–177.

Linder HP and LR Caddick. 2001. Restionaceae seedlings: morphology, anatomy and systematic implications. Feddes Repert. 112(1–2): 59–80

Linder HP and IK Ferguson. 1985. On the pollen morphology and phylogeny of the Restionales and Poales. Grana 24: 65–76.

Linder HP and PJ Rudall. 1993. The megagametophyte in *Anarthria* (Anarthriaceae, Poales) and its implications for the phylogeny of the Poales. Am. J. Bot. 80: 1455–1464.

Linder HP, BG Briggs, and LAS. Johnson. 1998a. Anarthriaceae. In: K Kubitzki, ed. The families and genera of vascular plants, vol. 4, pp. 19–21. Springer, Berlin/Heidelberg/New York.

Linder HP, BG Briggs, and LAS Johnson. 1998b. Ecdeiocoleaceae. In: K Kubitzki, ed. The families and genera of vascular plants, vol. 4, pp. 195–197. Springer, Berlin/Heidelberg/New York.

Linder HP, BG Briggs, and LAS Johnson. 1998c. Restionaceae. In: K Kubitzki, ed. The families and genera of vascular plants, vol. 4, pp. 425–445. Springer, Berlin/Heidelberg/New York.

Linder HP, BG Briggs, and LAS Johnson. 2000. Restionaceae – a morphological phylogeny. In: KL Wilson and DA Morrison, eds. Monocots: systematics and evolution, pp. 653–660. CSIRO, Collinwood.

Manning JC and HP Linder. 1990. Cladistic analysis of patterns of endothecial thickenings in the Poales/Restionales. Am. J. Bot. 77: 196–210.

Meney KA, KW Dixon, and JS Pate. 1999. Seed reproduction and germination ecology in Restionaceae. In: KA Meney and JS Pate, eds. Australian rushes, biology, identification and conservation of Restionaceae and allied families. pp. 97–108. University of Western Australia Press, Perth.

Michelangeli FA, JI Davis, and DW Stevenson. 2002. Phylogenetic relationships among Poaceae and related families as inferred from morphology, chloroplast structure, and sequence data from the mitochondrial and plastid genomes. In: Botany 2002: Botany in the curriculum (Abstracts), p. 139. Madison.

Michelangeli FA, JI Davis and DW Stevenson. 2003. Phylogenetic relationships among Poaceae and related families as inferred from morphology, inversions in the plastid genome, and sequence data from the mitochondrial and plastid genomes. Am. J. Bot. 90: 93–106.

Newell TK. 1969. A study of the genus *Joinvillea* (Fla-gellariaceae). J. Arnold Arbor. 50: 527–555.

Neyland R. 2002. A phylogeny inferred from large-subunit (26S) ribosomal DNA sequences suggests that the family Dasypogonaceae is closely aligned with the Restionaceae allies. Austral. Syst. Bot. 15: 749–754

Pin YK and LF Yew. 1975. Serological evidence on the distinctness of the monocotyledonous families Flagellariaceae, Huanguanaceae and Joinvilleaceae. J. Linn. Soc. Bot. 70: 77–81.

Prakash N. 1970. The floral development and embryology of *Centrolepis fascicularis*. Phytomorphology 19: 285–291.

Ronse Decraene LP, PH Linder and EF Smets. 2001. Floral ontogenetic evidence in support of the *Willdenowia* clade of South African Restionaceae. J. Plant Res. 114: 329–342.

Ronse Decraene LP, HP Linder, and EF Smets. 2002. Ontogeny and evolution of the flowers of South African Restionaceae with special emphasis on the gynoecium. Plant Syst. Evol. 231: 225–258.

Rudall P. 1990. Development of the ovule and megagametophyte in *Ecdeiocolea monostachya*. Austral. Syst. Bot. 3: 265–274.

Rudall P and HP Linder. 1988. Megagametophyte and nucellus in Restionaceae and Flagellariaceae. Am. J. Bot. 75: 1777–1786.

Rudall P, W Stuppy, J Cunniff, EA Kellogg, BC Briggs. 2005. Evolution of reproductive structures in grasses (Poaceae) inferred by sister-group comparison with their putative closest living relatives, Ecdeiocoleaceae. Am. J. Bot. 92: 1432–1443.

Sack FD. 1994. Structure of the stomatal complex of the monocot *Flagellaria indica*. Am. J. Bot. 81: 339–344.

Smith AC. 1978. *Flagellaria*. Allertonia 1: 341–344.

Smithson E. 1956. Comparative anatomy of the Flagellariaceae. Kew Bull. 11: 491–501.

Subramanyam K and HS Narayana. 1972. Some aspects of the floral morphology and embryology of *Flagellaria indica* L. In: YS Murty, BM Johri, HY Mohan Ram, and TM Varghese, eds. Advances in plant morphology, pp. 211–217. Sarita Prakashan, Meerut, India.

Tillich HJ. 1996. Seeds and seedlings in Hanguanaceae and Flagellariaceae (Monocotyledons). Sendtnera 3: 187–197

Tillich H-J und E Sill. 1999. Systematische Studien zur Morphologie und Anatomie von *Hanguana* Blume (Hanguanaceae) und *Flagellaria* L. (Flagellariaceae), mit der Beschreibung einer neuen Art *Hanguana bogneri* spec. nov. Sendtnera 6: 215–238.

Tomlinson PB and AC Smith. 1970. Joinvilleaceae, a new family of monocotyledons. Taxon 19: 887–889.

Tomlinson PB and U Posluszny. 1977. Features of dichotomizing apices in *Flagellaria indica* (Monocotyledones). Am. J. Bot. 64: 1057–1065.

Vyshenskaya TD. 1985a. Flagellariaceae. In: A Takhtajan, ed. Comparative seed anatomy, vol. 1, pp. 184–185. Nauka, Leningrad (in Russian).

Vyshenskaya TD. 1985b. Centrolepidaceae. In: A Takhtajan, ed. Comparative seed anatomy, vol. 1, pp. 187–189. Nauka, Leningrad (in Russian).

Whipple CJ, and RJ Schmidt. 2006. Genetics of grass flower development. Adv. Bot. Res. 44: 385–424.

Williams CA, JB Harborne, J Greenham, BG Briggs, and LAS Johnson. 1997. Flavonoid evidence and the classification of the Anarthriaceae within the Poales. Phytochemistry 45: 1189–1196.

Williams CA, JB Harborne, J Greenham, BG Briggs, and LAS Johnson. 1998. Flavonoid patterns and the revised classification of Australian Restionaceae. Phytochemistry 49: 529–552.

Order 31. POALES

Perennial or less often annual or biennial herbs or with more or less woody stems. Raphides and crystals of calcium oxalate lacking, but silica bodies present. Internodes solid or more often hollow. Vessels commonly in all vegetative organs, with simple or sometimes scalariform perforations. Sieve-element plastids of P2c-type (Behnke 2002). Leaves alternate, distichous, differentiated into open or rarely closed sheath and usually narrow and elongate, but in some tropical members broad and in *Phyllorachis* even sagittate, in many Bambusoideae with more or less distinct petiole (up to 25 cm in *Anomochloa*); venation parallel or rarely (*Leptaspis* and *Pharus*) with pinnately disposed lateral veins; a membranous ligule generally present at the juncture of the sheath and lamina on the adaxial side, sometimes reduced or represented by a row of hairs or even wholly suppressed. Stomata paracytic, with dumbbell-shaped guard cells. Flowers small, very reduced, bisexual or sometimes unisexual (monoecious or dioecious), arranged in panicles or sometimes in spikelike or racemelike inflorescences, which consist of one to to many-flowered spikelets; spikelets usually with a pair of subopposite bracts lacking axillary flowers (subtending glumes) and one to several florets often on more or less zigzag axis (rachilla); each floral axis arises in the axil of a bract (lemma) and usually bears a two-keeled prophyll (palea), (1)2 or (*Stipa* and Bambusoideae) 3 much smaller and inconspicuous scales (lodicules) above these, and the androecium and gynoecium; upper (adaxial) scale (palea) typically has 2 main veins and interpreted as 2 connate segments of the outer cycle of the ancestral 3-merous perianth (Čelakovsky 1889), the third segment nearly always suppressed; lodicules, which sometimes, as in *Arundinaria*, are relatively large and well developed, apparently represent the inner cycle of the ancestral perianth (Čelakovsky 1889; Butzin 1965). Stamens in many Bambusoideae six in two cycles, but mostly three, less often one, two, four, or (in some Bambusoideae) more than six; filaments elongate, filiform; anthers tetrasporangiate, elongate, basifixed but deeply sagittate so as to appear versatile, opening longitudinally. Tapetum secretory. Microsporogenesis successive. Pollen grains 3-celled, tectate-columellate with a thick foot-layer, 1-porate, operculate, annulate, psilate, spinulose or minutely scabrate. Gynoecium pseudomonomerous, consists of two or (Bambusoideae) three carpels; stylodia free or more or less connate into a short style; ovary superior, 1-locular, with one ovule attached to the wall. Ovules anatropous, hemitropous, campylotropous, or orthotropous, bitegmic or rarely (e.g., in *Melocanna)* unitegmic, crassinucellate, without parietal cell. Female gametophyte of *Polygonum*-type. Endosperm nuclear. Fruits mostly a caryopsis with a thin pericarp usually adhering to the seed or sometimes (*Crypsis, Sporobolus, Eleusine*) utricle-like with a free, membranous or gelatinous pericarp, rarely nutlike (some members of the Bambusoideae) or baccate (*Melocanna baccifera*). Seeds with copious, starchy endosperm usually with proteinaceous tissue and sometimes also oily, rarely (*Melocanna* and related genera) lacking; embryo straight, varying in size from very small (Festuceae and most of temperate tribes) to as long as the fruit (*Spartina,* etc.), placed on the abaxial side of the fruit, strongly differentiated and more or less completely enveloped by the edges of cylindrical coleoptile (the first leaf modified as a soil-penetrating sheath for the plumule), radicle enclosed in coleorhiza (interpreted by Brown [1965] as the remainder of the base of the proembryo) and well-developed and strongly modified cotyledon (scutellum) appressed to the endosperm, n = 2–23 +.

A rather distinct order characterized by unique spikelets, fine channels in tectum, and especially by its highly differentiated and specialized embryo unique among monocotyledons. The Poales are closely related to and derived from the Restionales, most probably from some extinct ancestor of the type of the living genus *Joinvillea*. Hallier (1902, 1912) included grasses in Enantioblastae and placed them between Flagellariaceae and Restionaceae, and Wettstein (1908, 1935) derived them from Enantioblastae. Various morphological, anatomical, palynological, and embryological studies confirm the close affinity between Poales and Restionales. According to Campbell and Kellogg (1987) and Rudall et al. (2005), Joinvilleaceae and Poaceae are sister groups based on such a strong and in angiosperms unique synapomorphy as large, solitary silica bodies in epidermal short cells, and in reproductive structures. Besides, there are such common features as the ligules, distichy, open leaf sheath, hollow internodes, vessel perforations, 3-celled pollen grains, exine structure, morphology of integuments, binucleate antipodal cells, seed anatomy, Asterad-type of embryogenesis. Relationships are especially close between *Joinvillea* and archaic grasses, such as *Streptochaeta.* I therefore agree with Campbell and Kellogg (1987: 218) that the synapomorphy uniting the Poaceae and Joinvilleaceae is more convincing than that linking the grasses to the Restionaceae.

1. POACEAE

Barnhart 1895 or Gramineae A.L. de Jussieu 1789 (nom. altern.). (including Aegilopaceae Martynov 1820, Agrostidaceae Burnett 1835, Alopecuraceae Martynov 1820, Andropogonaceae Martynov 1820, Anomochloaceae Nakai 1943, Arundinaceae Hochstetter 1850, Arundinellaceae Herter 1940, Avenaceae Martynov 1820, Bambusaceae Burnett 1835, Chloridaceae Herter 1940, Eragrostidaceae Herter 1940, Festucaceae Sprengel 1825, Hordeaceae Burnett 1835, Lepturaceae Herter 1940, Melicaceae Martynov 1820, Miliaceae Burnett 1835, Nardaceae Martynov 1820, Oryzaceae Burnett 1835, Panicaceae Voight 1845, Pappophoraceae Herter 1940, Parianaceae Nakai 1943, Phalaridaceae Burnett 1835, Pharaceae Herter 1940, Saccharaceae Martynov 1820, Spartinaceae Burnett 1835, Stipaceae Burnett 1835, Streptochaetaceae Nakai 1943, Triticaceae Hochstetter 1848, Zeaceae Kerner 1891). 670-850/10500-11000. Cosmopolitan.

1.1 ANOMOCHLOOIDEAE

Perennial, rhizomatous, herbaceous grasses. Leaves with either a distichous or spiral phyllotaxis; outer ligule absent; inner ligule a short fringe of cilia or absent, not membranous; pseudopetioles short to very long, with dark, turgid swellings at both ends or only at the summit; blades with parallel venation. Inflorescences terminal, stalked, spike-like, erect when young, but arching and drooping at maturity, with complicated branching patterns; the bracts within the spikelet equivalents with distichous or spiral phyllotaxis; lodicules absent or replaced by a ring of short brownish cilia borne on a low membranous ring; stamens 4–6; ovary glabrous, stigmas 1–3, hispid. Fruit a caryopsis, large, fusoid to oblong-rectangular, embryo large, the hilum linear, shallow and inconspicuous; endosperm mealy. n = 12 or 18 (Clark and Judziewicz 1996). – STREPTOCHAETEAE: *Streptochaeta*; ANOMOCHLOEAE: *Anomochloa*.

1.2 PHAROIDEAE
Perennial, rhizomatous, monoecious, herbaceous grasses. Rhizomes sympodial. Culms hollow or solid. Leaves distichous, the pseudopetiole prominent and twisted, the blade resupinate; outer ligule absent, the inner ligule present, membranous; blades with lateral nerves diverging obliquely from the midnerve and running straight to the margins. Inflorescence terminal, open, paniculate, the rachis and branches disarticulating or not, covered with uncinate macrohairs, bracts outside of the spikelets absent. Spikelets unisexual, one-flowered, mostly in male-female pairs on short branchlets, or some female spikelets solitary. Female spikelets large, short-pedicelled; glumes two, shorter than the floret; lemma indurate, tubular or inflated, covered wholly or in part by macrohairs; palea narrow, bicarinate; lodicules absent; ovary glabrous, style one, stigmas three. Fruit a caryopsis, the hilum extending its full length, embryo small, endosperm mealy. Male spikelets small, short to long-pedicelled, membranous; glumes two, shorter that the floret; lodicules three or none, if present then minute, elliptic, glabrous, and nerveless; stamens 6. n = 12. – *Pharus, Leptaspis, Scrotochloa.*

1.3 PUELIOIDEAE
Perennial. Calms herbaceous, usually unbranched above; internodes hollow. Leaves distichous, not basally aggregated; outer ligule absent (*Guaduella*) or present (*Puelia*); blades pseudopetiolate, large and broad; sheaths non-auriculate. Synflorescences racemose or paniculate, sometimes with small bracts subtending the branches. Spikelets with two glumes, several florets and elongated rachilla internodes, the 1–3 basal florets male, the next several florets female-fertile and bisexual, with apical incomplete florets (*Guaduella*), or the basal 3–6 florets male or neuter with the single apical floret female (*Puelia*), disarticulating above the glumes, and between the florets (*Guaduella*) or not (*Puelia*); stamens six, filaments free (*Guaduella*) or monadelphous (*Puelia*); ovary glabrous or hairy, an apical appendage present or not; stigmas 2–3. Caryopsis with a long-linear hilum; embryo small. n = 12. – PUELIEAE: *Puelia*; GUADUELLEAE: *Guaduella*.

1.4 BAMBUSOIDEAE
Perennials (rarely annual) rhizomatous herbaceous or woody. Culms hollow or solid. Leaves almost always petiolate, articulate-deciduous or less often persistent; laminas usually lanceolate to lanceovate, very rarely linear. Spikelets (or spikelets proper of the pseudospikelets) bisexual (Bambuseae) or unisexual (Olyreae), often with an indefinite number of flowers, not infrequently branched, often with more than two glumes at the base; lemmas and paleas usually with a large and indefinite number (up to 25) of veins, lodicules usually three, less often two, more than three, or absent. Stamens usually six, less often three, sometimes two, four, or more than six (up to 120); filaments not or rarely connate into a tube. Stigmas 2 or 3, less often 1 or 4–6, not rarely on a long style. Caryopsts often with the pericarp thickened in the upper part, often nutlike or drupaceous. Meristem under the base of the plumule absent. Chromosomes small; n = 9–12, seldom 7. – BAMBUSEAE: *Arthrostylidium, Bambusa, Chusquea, Dinochloa, Merostachys, Neurolepis,* etc.; OLYREAE: (including BUERGERSIOCHLEAE and PARIANEAE): *Buergersiochloa, Lithachne, Olyra, Raddia, Eremitis, Pariana,* etc.

1.5 EHRHARTOIDEAE
Plants annual or perennial (rhizomatous or stoloniferous), herbaceous to suffrutescent. Culms hollow or solid. Leaves distichous; abaxial ligule absent; adaxial ligule a fringed or unfringed membrane, or a fringe of hairs; blades rarely basally cordate or sagittate (Phyllorachideae), somewhat broad to usually narrow, sometimes pseudopetiolate; sheaths sometimes bearing auricles. Inflorescences paniculate or racemose, bracts outside of the spikelets rarely present. Spikelets bisexual or unisexual, with glumes 2 (absent in some Oryzeae), sterile florets 0 to 2, and female-fertile floret 1; palea well developed; lodicules 2, membranous or rarely fleshy; stamens usually 3 or 6 (sometimes 1, 2, or 4); ovary glabrous, apical appendage absent; styles 2, free, fused basally of for their full length (*Zizaniopsis*), close; stigmas 2. Caryopsis with the hilum long-linear; endosperm hard, without lipid, containing compound starch grains (rarely simple); embryo small, embryonic leaf usually with overlapping margins (meeting in *Potamophila*). n = 12 (10 in *Microlaena*, 15 in *Zizania*). – ORYZEAE: *Leersia, Luziola, Oryza, Zizania,* etc.; PHYLLORACHIDEAE: *Humbertochloa, Phyllorachis*; EHRHARTEAE: *Ehrharta, Microlaena,* etc.

1.6 POOIDEAE
Perennial or annual herbs, very seldom weakly ligneous. Culms hollow (rarely solid). Leaves without a

petiole, not deciduous, very seldom petiolate; laminas usually linear, seldom lanceolate or lanceolate. Spikelets with a definite or indefinite number of flowers, not branched, provided with two glumes at the base, seldom without glumes or with only one; lemmas usually with (1-) 3–9(-11) veins; paleas with 2 veins forming keels; lodicules two, seldom three, of various types, or sometimes wanting. Stamens usually three, very seldom one, two, four, or six; filaments free. Stylodia two, very seldom three or one, very rarely basally connate into a short style. Caryopsis always with a thin pericarp, this generally adnate to the seed, more rarely free from it, sometimes saclike. Meristematic layer present under the coleoptile, as a result of which the base of the plumule can elongate during germination and produce adventitious roots. Chromosomes small or large; n = 2, 4–12 and more. – BRACHYPODIEAE: *Brachypodium;* TRITICEAE: *Aegilops, Agropyron, Elymus, Elytrigia, Eremopyrum, Henrardia, Hordelymus, Hordeum, Leymus, Psathyrostachys, Secale, Triticum,* etc.; BROMEAE: *Bromus, Boissiera,* etc.; POEAE: *Agrostis, Aira, Anthoxanthum, Arrhenatherum, Avena, Briza, Calamagrostis, Catabrosa, Cinna, Colpodium, Cutandia, Cynosurus, Dactylis, Deschampsia, Eremopoa, Festuca, Helictotrichon, Hierochloe, Holcus, Koeleria, Lolium, Millum, Polypogon, Puccinellia, Trisetum, Vulpia,* etc.; MELICEAE: *Glyceria, Melica, Lophochlaena,* etc; BRYLKINIEAE: *Brylkinia*; DIARRHENEAE: *Diarrhena*; BRACHYELYTREAE: *Brachyelytrum*; AMPELODESMEAE: *Ampelodesmos*; STIPEAE: *Oryzopsis, Piptatherum, Stipa,* etc.; LYGEEAE: *Lygeum*; NARDEAE: *Nardus*; PHAENOSPERMATEAE: *Phaenosperma.*

1.7 ARISTIDOIDEAE

Plants annual or perennial, caespitose, herbaceous, xerophytic or less commonly mesophytic. Culms solid or hollow. Leaves distichous. Inflorescences paniculate, bracts outside of the spikelets absent. Spikelets with bisexual florets, glumes two, female-fertile floret one, and no rachilla extension, cylindrical or laterally compressed, disarticulating above the glumes; lemma with three awns, the awns separate from each other, or fused below into a twisted column; palea short, less than half the lemma length; lodicules present or rarely absent, when present two, free, membranous, glabrous, heavily vascularized; stamens one to three; ovary glabrous, apical appendage absent, haustorial synergids absent, styles two, free, close; stigmas two. Caryopsis with the hilum short or long-linear; endosperm hard, without lipid, containing compound starch grains; embryo small (*Sartidia*) or large (*Aristida, Stipagrostis*), epiblast absent, scutellar cleft present or absent (*Sartidia*). n = 11, 12. – ARISTIDEAE: *Aristida, Sartidia, Stipagrostis.*

1.8 DANTHONIOIDEAE

Perennial (caespitose, rhizomatous or stoloniferous) or less commonly annual, herbaceous or rarely suffrutescent plants. Culms solid or very rarely hollow. Leaves distichous; abaxial ligule usually absent (sometimes present in *Cortaderia, Karroochloa,* and *Pentaschistis*); adaxial ligule a fringe of hairs or a fringed membrane; blades relatively narrow, without a speudopetiole; sheats not auriculate except in *Pentameris thuarii.* Inflorescences paniculate or less commonly racemose or spicate, bracts outside of the spikelets absent. Spikelets bisexual (but sometimes without bisexual florets in *Cortaderia*) or unisexual (*Cortaderia, Lamprothyrsus*); glumes 2 and usually equal, female-fertile florets 1–6 (to 20), with apical reduction and a rachilla extension usually present, laterally compressed; lemma lacking uncinate macrohairs, awn single and from a sinus; palea well developed, sometimes relatively short; lodicules 2, free (rarely joined), fleshy or rarely with an apical membranous flap, glabrous or ciliate, often with microhairs, sometimes heavily vascularized; stamens 3; ovary glabrous or rarely with apical hairs (*Pentameris*), apical appendage absent, haustorial synergids present, only weakly developed in a few taxa; styles 2, the bases usually widely separated, stigmas 2. Caryopsis with the hilum short or long-linear; endosperm hard, containing compound starch grains (simple in *Prionanthium*); embryo large or small, epiblast absent, scutellar cleft present. n = 6, 7, 9. – DANTHONIEAE: *Pentaschistis, Pentameris, Prionanthium, Chionochloa, Cortaderia, Lamprothyrsus, Pseudopentameris, Chaetobromus, Rytidosperma, Notodanthonia, Austrodanthonia, Joycea, Karroochloa, Schismus, Tribolium, Merxmuellera, Danthonia, Notochloe, Plinthanthesis.*

1.9 ARUNDINOIDEAE

Annuals or perennials, tufted, rhizomatous or stoloniferous, herbaceous to somewhat woody. Culms hollow or less commonly solid. Leaves narrow and the ligule is membranous or cartilaginous or may form a rim of cilia; microhairs usually present, arm cells sometimes occur, fusoid cells always absent. Inflorescences usually paniculate, rarely spicate or racemose, bracts outside of the spikelets absent. Spikelets with bisexual florets, glumes two, female-fertile florets one to several, apical reduction usually present; pales usually well developed; lodicules two, free (rarely joined at the base), fleshy, glabrous or infrequently ciliate; stamens 1–3; ovary glabrous, apical appendage absent, haustorial synergids absent; styles two, usually free, close; stigmas two. Caryopsis with the hilum short or long-linear; endosperm hard, without lipid, containing compound starch grains; embryo large or small. n = 6, 9, 12. – ARUNDINEAE: *Arundo, Amphipogon, Dregeochloa, Hakonechloa, Molinia, Phragmites, Triodia,* etc.

1.10 MICRAIROIDEAE

Annual or perennial, terrestrial and rarely aquatic, rhizomatous, matforming, sometimes decumbent herbs. Culms solid or hollow. Leaves distichous or sometimes spirally arranged; abaxial ligule occasionally present as a line of hairs, adaxial ligule a fringe of hairs. Inflorescences panicles, racemes or spikes, terminal or axillary. Spikelets bisexual, with two glumes (one in *Sphaerocaryum*); glumes tardily or rarely early deciduous. Pales well developed, lodicules two or sometimes absent, cuneate, free, fleshy. Stamens 2–3; ovary usually glabrous, apical appendage absent; styles 2, free, stigmas 2. Fruits caryopsis with the hilum short linear to long linear; endosperm hard, containin simple starch grains; embryo small; n = 10. – ERIACHNEAE: *Eriachne, Pheidochloa*; ISACHNEAE: *Coelachne, Heteranthoecia, Isachne, Limnopoa, Sphaerocaryum*; MICRAIREAE: *Micraira.*

1.11 CHLORIDOIDEAE

Annual or perennial herbs, tufted or rhizomatous (stoloniferous) and sometimes robust (rarely woody). Culms solid or hollow. Leaves mostly narrow and the ligule is hair-fringed or reduced to a rim of hairs. Silica bodies are mostly saddle-shaped and microhairs are always present. Distal cell usually inflated and sometimes glandular. Mesophyll lacks arm cells and fusoid cells and has a Kranz type PS anatomy. Inflorescences paniculate, paniculate with spicate branches, racemose, or spicate, bracts outside of the spikelets absent. Spikelets bisexual or sometimes unisexual (if so the plants dioecious or monoecious), with one to several florets and the rachilla usually prolonged; pales well developed, lodicules two or absent, fleshy, glabrous. Stamens 1–3, stigmas 2; ovary glabrous, apical appendage absent, haustorial synergids absent. Fruits caryopsis or utricle, hilum short; endosperm hard, without lipid, containing simple or compound starch grains; embryo large or rarely small. n = 7, 8, 9 or 10. – CYNODONTEAE (CHLORIDEAE): *Bouteloua, Soderstroimia, Griffithsochloa, Pringleochloa, Buchloe, Cathestecum, Cyclostachya, Opizia, Pentarrhaphis, Melanocenchris, Schaffnerella, Lintonia, Pommereulla, Pleuraphis, Chloris, Arachne, Afrotrichloris, Apochiton, Crypsis, Cynodon, Diplachne, Enteropogon, Eustachys, Trichloris, Astrebla, Brachyachne, Cypholepis, Eleusine, Tetrapogon, Ochthochloa, Saugetia, Tragus, Aegopogon, Decaryella, Dignathia, Farrago, Hilaria, Lepturus, Zoysia, Perotis, Lophiolepis, Lycurus, Tetrachaete,* etc.; ERAGROSTIDEAE: *Coelochyrum, Eleusine, Eragrostis, Acamptoclados, Ectrosia, Harpachne, Heterachne, Panichloa, Steirachne, Viguierella, Entoplocamia, Desmostachya, Myriostachya, Tetrachne, Distichlis, Jouvea, Psilolemma, Crypsis, Urochondra, Allolepis, Dasyochloa, Erioneuron, Munroa, Dinebra, Trichoneura, Tridens, Aeluropus, Muhlenbergia, Tripogon, Oropetium, Craspedorhachis, Kengia, Bewsia, Ctenium, Sporobolus, Monodia, Plectrachne, Fingerhuthia, Uniola, Neostapfia, Orcuttia, Pappophorum, Spartina,* etc.;

1.12 CENTOTHECOIDEAE

Annuals or rhizomatous or /and stoloniferous perennials, herbaceous or reedlike. Calms solid or hollow, Leaves typically petiolate with relatively broad tessellate blades; ligules membranous or ciliate, or membranous with ciliate margins. Inflorescences racemiform or paniculiform; bracts outside of the spikelets absent. Spikelets bisexual or unisexual, often compressed laterally, (1-)2 to many-flowered with reduction either below or above the fertile florets; lemma lacking uncinate macrohairs, if awned, the awn single; palea usually well developed, sometimes relatively short; lodicules 2 or 0, more or less cuneate; stamens (1-)2–3; stigmas 2; ovary glabrous, apical appendage absent. Caryopsis with the hilum basal, usually flattened laterally; endosperm hard, without lipid, containing simple or compound starch grains; embryo small or large. n = 12 (Tenório 1976; Clayton 1978, Soderstrom 1981). – CENTOTHECEAE: *Centotheca, Zeugites, Bromuniola, Calderonella, Chasmanthium, Chevalierella, Lophatherum, Megastachya, Orthoclada*; THYSANOLAENEAE: *Thysanolaena.*

1.13 PANICOIDEAE

Annual or perennial herbs, either tufted or rhizomatous and sometimes with robust woody culms. Leaves broad or narrow and the ligule is a ciliate membrane or a rim of hairs, sometimes it is absent. The silica bodies cross- or dumb-bell-shaped and microhairs are always present. Neither arm cells nor fusoid cells occur in the mesophyll and the leaf anatomy is usually of Kranz Type, both the PS and MS subtypes occurring. The flowers usually with 2 fleshy cuneate lodicules; stamens 1–3(-6); stigmas 2. – ARUNDINELLEAE: *Arundinella, Loudetia*, etc.; HUBBARDIEAE: *Hubbardia*; PANICEAE: *Cenchrus, Pennisetum, Digitaria, Leptocoryphium, Melinis, Acroceras, Axonopus, Brachiaria, Cenchrus, Echinochloa, Eriochha, Ichnanthus, Mesosetum, Oplismenus, Panicum, Paspalum, Sacciolepis, Setaria, Spinifex*, etc.; ANDROPOGONEAE: *Andropogon, Chrysopogon, Arthraxon, Coix, Cymbopogon, Hyparrhenia, Dimeria, Erianthus, Heteropogon, Rottboellia, Imperata, Miscanthus, Saccharum, Leptatherum, Germainia, Eiclasta, Spathia, Trachypogon, Agenium, Homozeugos, Sorghum, Sorghastrum, Dichanthium, Themeda, Tripsacum, Zea, Chionachne, Polytoca, Sclerachne, Trilobachne*, etc.; STEYERMARKOCHLOEAE: *Steyermarkochloa*.

Bibliography

Andre JP. 1998. A study of the vascular organization of bamboos (Poaceae-Bambuseae) using a microcasting method. IAWA J. 19: 265–301.

Anton AM and AE Cocucci. 1984. The grass mega-gametophyte and its possible phylogenetic implications. Plant Syst. Evol. 146: 117–121.

Arber A. 1934. The Gramineae: a study of cereal, bamboo, and grass. Cambridge University Press, Cambridge.

Avdulov NP. 1931. Karyosystematische Untersuchungen der Familie Gramineen. Bull. Appl. Bot. (Suppl.) 44: 1–428 (in Russian with German summary).

Barker NP and HP Linder. 1999. Sequences of the grass-specific insert in the chloroplast *rpo*C2 gene elucidate generic relationships of the Arundinoideae (Poaceae). Syst. Bot. 23: 327–350.

Barker NP, HP Linder, and EH Harley. 1995. Polyphyly of the Arundinoideae (Poaceae): evidence from *rbc*L sequence data. Syst. Bot. 20: 423–435.

Barker NP, HP Linder, and EH Harley. 1999. Sequences of the grass-specific insert in the chloroplast *rpo*C2 gene elucidate generic relationships of the Arundinoideae (Poaceae). Syst. Bot. 23: 327–350.

Barker NP, CM Morton and HP Linder. 2000. The Danthonieae: generic composition and relationships. In: SWL Jacobs and J Everett, eds. Grasses: systematics and evolution, pp. 221–229. CSIRO, Collingwood.

Barker NP, LG Clark, JI Davis, MR Duvall, GF Guala, C Hsiao, EA Kellogg, HP Linder, RJ Mason-Gamer, SY Mathews, MP Simmons, RJ Soreng, RE Spangler (GPWG, Grass Phylogeny Working Group). 2001. Phylogeny and subfamilial classification of the grasses (Poaceae). Ann. Missouri Bot. Gard. 88: 373–457.

Barkworth ME. 2000. Changing perceptions of the Triticeae. In: SWL Jacobs and J Everett, eds. Grasses: systematics and evolution, pp. 110–120. CSIRO, Collingwood.

Batygina TB and MS Yakovlev. 1990. Poaceae. In: TB Batygina and MS Yakovlev, eds. Comparative embryology of flowering plants: Monocotyledons, pp. 217–234. Nauka, Leningrad (in Russian).

Baum BR and BO Savile. 1985. Rusts (Redinales) of Triticeae: evolution and extent of coevolution, a cladistic analysis. Bot. J. Linn. Soc. 91: 367–394.

Baum BR, JR Estes and PK Gupta. 1987. Assesment of the genomic system of classification in the Triticeae. Am. J. Bot. 74: 1338–1395.

and ecology. Longman, London.

Bhanwra RK. 1988. Embryology in relation to systematics of Gramineae. Ann. Bot. 62: 215–233.

Bhanwra RK, N Kaur, and A Garg. 1991. Embryological studies in some grasses and their taxonomic significance. Bot. J. Linn. Soc. 107: 405–417.

Birch WR. 1963. Epiblast in Gramineae. Nature 198 (4877): 304.

Brown WV. 1958. Leaf anatomy in grass systematics. Bot. Gaz. (Crawfordsville) 119: 170–178.

Brown WV. 1959. The epiblast and coleoptile of the grass embryo. Bull. Torrey Bot. Club 86: 13–16.

Brown WV. 1960. The morphology of the grass embryo. Phytomorphology 10: 215–323.

Brown WV. 1965. The grass embryo: a rebuttal. Phytomorphology 15: 274–284.

Brown WV and SC Johnson. 1962. The fine structure of the grass guard cell. Am. J. Bot. 49: 110–115.

Butzin F. 1965. Neue Untersuchungen über die Blüte der Gramineae. Ph.D. dissetration., University of Berlin.

Cai LB. 2002. A preliminary discussion on the taxonomic values of the main characters of reproductive organ of Poaceae and their ranks suitable for differentiating taxa. Bull. Bot. Res. 22: 278–284.

Calderon CE and TR Soderstrom. 1980. The genera of Bambusoideae (Poaceae) of the American continent: keys and comments. Smithsonian Contr. Bot. 44: 1–27.

Campbell CS. 1985. The subfamilies and tribes of Gramineae (Poaceae) in the southeastern United States. J. Arnold Arbor. 66: 123–299.

Campbell CS and EA Kellogg. 1987. Sister group relationships in the Poaceae. In: TR Soderstrom et al., eds. Grass systematics and evolution, pp. 217–224. Smithsonian Institution Press, Washington.

Caro JA. 1982. Sinopsis taxonomica de las gramineas Argentinas. Dominguezia 4: 1–51.

Carolin RC, SWL Jacobs and M Vesk. 1973. The structure of the cells of the mesophyll and parenchymatous bundle sheath of the Gramineae. Bot. J. Linn. Soc. 66: 269–273.

Čelakovsky L. 1889. Über den Archenbau der brasilianischen Grasgattung *Streptochaeta* Schrad. Sitzungsber. Königl. Böhm. Ges. Wiss. Prag. Math.-Naturwiss. Kl. 1: 14–42.

Čelakovsky L. 1897. Über die Homologien des Grasembryos. Bot. Z. 55: 141–174.

Chandra N. 1963. Morphological studies in the Gramineae: III. On the nature of the gynoecium in the Gramineae. J. Indian Bot. Soc. 42: 252–259.

Cheadle VI. 1955. The taxonomic use of specialization of vessels in the metaxylem of Gramineae, Cyperaceae, Juncaceae, and Restionaceae. J. Arnold Arbor. 36: 141–157.

Clark LG and EJ Judziewicz. 1996. The grass subfamilies Anomochlooideae and Pharoideae (Poaceae). Taxon 45: 641–645.

Clark LG, W Zhang and JF Wendel. 1995. A phylogeny of the grass family (Poaceae) based on *ndh*F sequence data. Syst. Bot. 20: 463–460.

Clark LG, M Kobayashi, S Mathews, RE Spangler, and EA Kellogg. 2000. The Puelioideae, a new subfamily of Poaceae. Syst. Bot. 25: 181–187.

Clayton WD. 1978. Gramineae. In: VW Heywood et al., eds. Flowering plants of the world, pp. 285–290. Mayflower Books, New York.

Clayton WD. 1981. Evolution and distribution of grasses. Ann. Missouri Bot. Gard. 68: 5–14.

Clayton WD and SA Renvoize. 1986. Genera Grammum. Kew Bull., additional ser., 13: 1–389.

Clifford HT. 1961. Floral evolution in the family Gramineae. Evolution 15: 455–460.

Conert HJ. 1987. Current concepts in the systematics of the Arundinoideae. In: TR Soderstrom, KW Hilu, CS Campbell and ME Barkworth, eds. Grass systematics and evolution, pp. 239–250. Smithsonian Institution Press, Washington.

Coro JA. 1982. Sinopsis taxonómica de las gramneas Argentina. Dominguezia 4: 1–4.

Davis JI and RJ Soreng. 1993. Phylogenetic structure in the grass family (Poaceae) as inferred from chloroplast DNA restriction site variation. Am. J. Bot. 80: 1444–1454.

Doebley J, M Durbin, E Golenberg, M Clegg, and D Ma. 1990. Evolutionary analysis of the large subunit of carboxylase (*rbcL*) nucleotide sequence among the grasses (Gramineae). Evolution 44: 1097–1108.

Duvall MR, JD Noll, and AH Minn. 2001. Phylogenetics of Paniceae (Poaceae). Am. J. Bot. 88: 1988–1992.

Ellis RP. 1987. A review of comparative leaf blade anatomy in the systematics of the Poaceae: the past twenty-five years. In: TR Soderstrom et al., eds. Grass systematics and evolution, pp. 3–10. Smithsonian Institution Press, Washington.

Esen A and KW Hilu. 1989. Immunological affinities among subfamilies of the Poaceae. Am. J. Bot. 76: 196–203.

Esen A and KW Hilu. 1991. Electrophoretic and immunological studies of prolamins in the Poaceae: II. Phylogenetic affinities of the Aristideae. Taxon 40: 5–17.

Francis A. 1990. The Tripsacinae: an interdisciplinary review of maize *(Zea mays)* and its relatives. Acta Bot. Fenn. 140: 1–51.

Gaut BS, LP Tredway, C Kubik, RL Gaut, and W Meyer. 2000. Phylogenetic relationships and genetic diversity among members of the *Festuca-Lolium* complex (Poaceae) based on ITS sequence data. Plant Syst. Evol. 224: 33–53.

Ge S, A Li, BR Lu, SZ Zhang, and DY Hong. 2002. A phylogeny of the rice tribe Oryzeae (Poaceae) based on *mat*K sequence data. Am. J. Bot. 89: 1967–1972.

Gillespie LJ and RJ Soreng. 2005. A phylogenetic analysis of the bluegrass genus *Poa* based on *cp*DNA restriction site data. Syst. Bot. 30: 84–105.

Giussani LM, JH Cota Sanchez, FO Zuloaga, and EA Kellogg. 2001. A molecular phylogeny of the grass subfamily Panicoideae (Poaceae) shows multiple origins of C4 photosynthesis. Am. J. Bot. 88: 1993–2012.

Gomez-Martinez R and A Culham. 2000. Phylogeny of the subfamily Panicoideae with emphasis on the tribe Paniceae: evicence from the chloroplast *trn*L-F cpDNA region. In: SWL Jacobs and J Everett, eds. Grasses: systematics and evolution, pp. 136–140. CSIRO, Collingwood.

Gould FW and RB Shaw. 1983. Grass systematics, 2nd ed. Texas A&M University Press, College Station.

Grass Phylogeny Working Group (GPWG). 2001. Phylogeny and subfamilial classification of the grasses (Poaceae). Ann. Missouri Bot. Gard. 88: 373–457.

Grebenstein B, M Roser, W Sauer, and V Hemleben. 1998. Molecular phylogenetic relationships in Aveneae (Poaceae) species and other grasses as inferred from ITS1 and ITS2 rDNA sequences. Plant Syst. Evol. 213: 233–250.

Guala GF. 2000. The relation of space and geography to cladogenic events in *Agenium* and *Homozeugos* (Poaceae: Andropogoneae) in South America and Africa. In: SWL Jacobs and J Everett, eds. Grasses: systematics and evolution, pp. 159–166. CSIRO, Collingwood.

Guignard JL. 1961. Recherches sur l'embryogénie des Graminées: Rapports des Graminées avec les autres Monocotylédones. Ann. Sci. Nat. Bot., 12th ser., 2: 491–610.

Gundel PE, PH Maseda, MM Vila-Aiub, GM Chersa, and R Benech-Arnold. 2006. Effects of *Neotyphodium* fungi on *Lolium multiflorum* seed germination in relation to water availablity. Ann. Bot. 97: 571–577.

Guo Y-L, and S Ge. 2005. Molecular phylogeny of Oryzeae (Poaceae) based on DNA sequences from chloroplast, mitochondrial, and nuclear genomes. Am. J. Bot. 92: 1548–1558.

Hartley W. 1958, 1973. Studies on the origin, evolution, and distribution of the Gramineae: II. The tribe Paniceae. V. The subfamily Festucoideae. Austral. J. Bot. 6: 343–357, 1958; 21: 201–234, 1973.

Hartley W. 1964. The distribution of the grasses. In: C Bernard, ed. Grasses and grasslands, pp. 29–46. Macmillan, London.

Hilu KW. 1985. Trends of variation and systematics of Poaceae. Taxon 34: 102–114.

Hilu KW. 2000. Contribution of prolamin size diversity and structure to the systematics of the Poaceae. – In: SWL Jacobs and J Everett, eds. Grasses: systematics and evolution, pp. 241–247. CSIRO, Collingwood.

Hilu KW and LA Alice. 1999. Evolutionary implications of *matK* indels in Poaceae. Am. J. Bot. 86: 1735–1741.

Hilu KW and LA Alice. 2000. Phylogenetic relationships in subfamily Chloridoideae (Poaceae) based on *matK* sequences: a preliminary assessment. In: SWL Jacobs and J Everett, eds. Grasses: systematics and evolution, pp. 173–179. CSIRO, Collingwood.

Hilu KW and LA Alice. 2001. A Phylogeny of Chloridoideae (Poaceae) based on *matK* sequences. Syst. Bot. 26: 386–405.

Hilu KW and A Esen. 1988. Prolamin size diversity in the Poaceae. Biochem. Syst. Ecol. 16: 457–465.

Hilu KW and A Esen. 1990. Prolamins in systematics of Poaceae subfam. Arundinoideae. Plant Syst. Evol. 173: 47–70.

Hilu KW and A Esen. 1993. Prolamin and immunological studies in the Poaceae: III. Subfamily Chloridoideae. Am. J. Bot. 80: 104–113.

Hilu KW and K Wright. 1982. Systematics of Gramineae: cluster analysis study. Taxon 31: 9–36.

Hilu KW, LA Alice, and H Liang. 1999. Phylogeny of Poaceae inferred from *mat*K sequences. Ann. Missouri Bot. Gard. 86: 835–851.

Hodkinson TR, MW Chase, MD Lledo, N Salamin, and SA Renvoize. 2002. Phylogenetics of *Miscanthus*, *Saccharum* and related genera (Saccharinae, Andropogoneae, Poaceae) based on DNA sequences from ITS nuclear ribosomal DNA and plastid *trn*L intron and *trn*L-F intergenic spacers. J. Plant Res. 115: 381–392.

Holttum RE. 1956. The classification of the bamboos. Phytomorphology 6: 73–90.

Hsiao C, SWL Jacobs, Barker NP, and NJ Chatterton. 1998. A molecular phylogeny of the subfamily Arundinoideae (Poaceae) based on sequences of rDNA. Austral. Syst. Bot. 11: 41–52.

Hsiao C, SWL Jacobs, NJ Chatterton, and KH Asay. 1999. A molecular phylogeny of the grass family (Poaceae) based on the sequences of nuclear ribosomal DNA (ITS). Austral. Syst. Bot. 11: 667–688.

Hubbard C. E. 1954. Grasses. Penguin Books, London.

Irish EE. 1998. Grass spikelets: a thorny problem. BioEssays 20: 789–793.

Jacobs SWL, J Everett, M Barkworth, and C Hsiao. 1998. Relationships within the stipoid grasses (Gramineae). Monocots II, pp. 29–30 (Abstract). Sydney.

Judziewicz EJ. 1987. Taxonomy and morphology of the tribe Phareae (Poaceae, Bambusoideae). Ph.D. Thesis. Madison.

Judziewicz EJ and TH Soderstrom. 1989. Morphological, anatomical, and taxonomic studies in *Anomochloa* and *Streptochaeta* (Poaceae: Bambusoideae). Smithsonian Contr. Bot. 68: 1–51.

Judziewicz EJ, RJ Soreng, G Davidse, PM Peterson, TS Filgueiras, and FO Zuloaga. 2000. Catalogue of New World grasses (Poaceae): 1. Subfamilies Anomochlooideae, Bambusoideae, Ehrhartoideae and Pharoideae. Smithsonian Institution Press, Washington.

Kellogg EA. 2000. Molecular and morphological evolution in the Andropogoneae. In: SWL Jacobs and J Everett, eds. Grasses: systematics and evolution, pp. 149–158. CSIRO, Collingwood.

Kellogg EA and CS Campbell. 1987. Phylogenetic analyses of the Gramineae. In: TR Soderstrom et al., eds. Grass systematics and evolution, pp. 310–322. Smithsonian Institution Press, Washington.

Kellogg EA and HP Linder. 1995. Phylogeny of Poales. In: PJ Rudall, PJ Cribb, DF Cutler, and CJ Humphries, eds. Monocotyledons: systematics and evolution, vol. 2, pp. 511–542. Royal Botanic Gardens, Kew.

Kellogg EA and L Watson. 1993. Phylogenetic studies of a large data set: I. Bambusoideae, Andropogonoideae, and Pooideae (Gramineae). Bot. Rev. 59: 273–343.

Kircher P. 1986. Untersuchungen zur Blüten- und Infloreszenzmorphologie, Embryologie, und Systematik der Restionaceen im Vergleich mit Gramineen und verwandten Familien. Diss. Bot. 94: 1–219.

Kirpes CC, LG Klark, and NR Lersten. 1996. Systematic significance of pollen arrangement in microsporangia of Poaceae and Cyperaceae, review and observations on representative taxa. Am. J. Bot. 83: 1609–1622.

Le Roux LG and EA Kellogg. 1999. Floral development and the formation of unisexual spikelets in the Andropogoneae (Poaceae). Am. J. Bot. 86: 354–366.

Liang H and KW Hilu. 1996. Application of the *matK* gene sequences to grass systematics. Canad. J. Bot. 74: 125–134.

Linder HP and IK Ferguson. 1985. On the pollen morphology and phylogeny of the Restionales and Poales. Grana 24: 65–76.

Linder HP and PJ Rudall. 2005. Evolutionary history of Poales. Ann. Rev. Ecol. Syst. 36: 107–124.

Liu Q, N-X Zhao, and G Hao. 2005. The phylogeny of the Chloridoideae (Gramineae): a cladistic analysis. J. Trop. Subtrop. Bot. 13: 432–442.

Macfarlane TD. 1987. Poaceae subfamily Pooideae. In: TR Soderstrom et al., eds. Grass systematics and evolution, pp. 265–276. Smithsonian Institution Press, Washington.

Macfarlane TD and L Watson. 1982. The classification of Poaceae subfamily Pooideae. Taxon 31: 178–203.

Malcomber ST, JC Preston, R Reinheimer, J Kossuth, and EA Kellogg. 2006. Developmental gene evolution and the origin of grass inflorescence diversity. Adv. Bot. Res. 44: 425–481.

Mant JG, RJ Bayer, MD Crisp, and JWH Treuman. 2000. A phylogeny of Triodieae (Poaceae: Chloridoideae) based on the ITS region of *nr*DNA: testing conflict between anatomical and inflorescence characters. In: SWL Jacobs and J Everett, eds. Grasses: systematics and evolution, pp. 213–217. CSIRO, Collingwood.

Mathews S, RC Tsal, and EA Kellog. 2000. Phylogenetic structure in the grass family (Poaceae): evidence from the nuclear gene Phytochrome B. Am. J. Bot. 87: 96–107.

Mathews S, RE Spangler, RJ Mason-Gamer, and EA Kellog. 2002. Phylogeny of Andropogoneae inferred from phytochrome B, GBSSI, and NDHF. Int. J. Plant Sci. 163: 441–450.

Mathews S, RC Tsal, and EA Kellog. 2001. Phylogenetic structure in the grass family (Poaceae): the wild type and the gynomonoecious fsf1 mutant. Am. J. Bot. 88: 363–381.

Mejia-Saulés T and FA Bisby. 2000. Preliminary views on the tribe Meliceae (Gramineae: Pooideae). In: SWL Jacobs and J Everett, eds. Grasses: systematics and evolution, pp. 83–88. CSIRO, Collingwood.

Mensah JK, and LS Gill. 1997. Cuticular and leaf blade anatomical studies of the tribe Sporoboleae (Poaceae) from West Africa. J. Plant Anat. Morphol. 7: 72–81.

Metcalfe CR. 1960. Gramineae. In: CR Metcalfe, ed., Anatomy of the monocotyledons, vol. 1. Clarendon Press, Oxford.

Michelangeli FA, JI Davis, and DW Stevenson. 2003. Phylogenetic relationships among Poaceae and related families as inferred from morphology, inversions in the plastid genome, and sequence data from the mitochondrial and plastid genomes. Am. J. Bot. 90: 93–106.

Negby M and D Koller. 1962. Homologies in the grass embryo: a reevaluation. Phytomorphology 12: 289–296.

Ortiz-Diaz J-J and A Culham. 2000. Phylogenetic relationships of the genus *Sporobolus* (Poaceae: Eragrostideae) based on nuclear ribosomal DNA ITS sequences. In: SWL Jacobs and J Everett, eds. Grasses: systematics and evolution, pp. 184–188. CSIRO, Collingwood.

Page VM. 1947. Leaf anatomy of *Streptochaeta* and the relation of this genus to the Bamboos. Bull. Torrey Bot. Club 74: 232–239.

Page VM. 1951. Morphology of the spikelet of *Streptochaeta*. Bull. Torrey Bot. Club 78: 22–37.

Peterson PM, RD Webster, and J Valdés-Reyna. 1995. Subtribal classification of the New World Eragrostideae (Poaceae, Chloridoideae). SIDA 16: 529–544.

Peterson PM, RD Webster, and J Valdés-Reyna. 1997. Genera of New World Eragrostideae (Poaceae: Chloridioideae). Smithsonian Contr. Bot. 87: 1–50.

Peterson PM, RJ Soreng, G Davidse, TS Filgueiras, FO Zuloaga, and EJ Judziewicz. 2001. Catalogue of New World grasses (Poaceae): II. Subfamily Chloridoideae. Smithsonian Institution Press, Washington.

Petersen G, O Seberg, M Yde, and K Berthelsen. 2006. Phylogenetic relationships of *Triticum* and *Aegilops* and evidence for the origin of the **A**, **B**, and **D** genomes of common wheat (*Triticum aestivum*). Molec. Phylogen. Evol. 39: 70–82.

Petrova LR and NN Tsvelev. 1974. On the evolution of inflorescence in Poaceae: On the nature and functions of lodicules. Bot. Zhurn. 59: 1713–1720 (in Russian with English summary).

Petrova LR, TA Fedotova, and TC Nikolaevskaya. 1985. Poaceae. In: A Takhtajan, ed. Comparative seed anatomy, vol. 1, pp. 192–205. Nauka, Leningrad (in Russian).

Philipson WR. 1985. Is the grass gynoecium monocarpellary? Am. J. Bot. 72: 1954–1961.

Pilger R. 1954. Das System der Gramineae. Bot. Jahrb. Syst. 76: 281–384.

Piperno DR and H-D Sues. 2005. Dinosaurs dined on grass. Science 310: 1126–1128.

Pizzolato TD. 2000. A systematic view of the development of vascular systems in culms and inflorescences of grasses. In: SWL Jacobs and J Everett, eds. Grasses: systematics and evolution, pp. 8–28. CSIRO, Collingwood.

Poddubnaya-Arnoldi VA. 1978. Cytoembryological characteristics of the Poaceae. Bull. Main Bot. Gard. 109: 57–60 (in Russian).

Prat H. 1936. La systématique des Graminées. Ann. Sci. Nat. Bot., 10th ser., 18: 165–258.

Prat H. 1960. Vers une classification naturelle des Graminées. Bull. Soc. Bot. France 107: 32–79.

Reeder JR. 1953. The embryo of *Streptochaeta* and its bearing on the homology of the coleoptile. Am. J. Bot. 40: 77–80.

Reeder JR. 1957. The embryo in grass systematics. Am. J. Bot. 44: 756–768.

Reeder JR. 1962. The bambusoid embryo: A reappraisal. Am. J. Bot. 49: 639–641.

Reimer E and JH Cota-Sanches. 2007. An SEM survey of the leaf epidermis in Danthonoioid grasses (Poaceae: Danthonioideae). Syst. Bot. 32: 60–70.

Renvoize SA. 1981. The subfamily Arundinoideae and its position in relation to a general classification of the Gramineae. Kew Bull. 36: 85–102.

Renvoize SA. 1985. A survey of leaf-blade anatomy in grasses. VII. Pommereulleae, Orcuttieae and Pappophoreae. Kew Bull. 40: 737–744.

Rondeau R, C Rouch, and G Besnard. 2005. NADP-malate dehydrogenase gene evolution in Andropogoneae (Poaceae): Gene duplication folllowed by sub-functionalization. Ann. Bot., n.s. 96: 1307–1314.

Roth I. 1955. Zur morphologischen Deutung des Gra-sembryos und verwandter Embryotypen. Flora 142: 564–600.

Roshevits R Yu. 1937. Grasses. Nauka, Moscow/Leningrad (in Russian).

Rudall PJ. 2005. Evolution of reproductive structures in grasses (Poaceae) inferred by sister-group comparison with their putative closest living relatives, Ecdeiocoleaceae. Am. J. Bot. 92: 1432–1443.

Sanchez-Ken JG and LG Clark. 2001. Gynerieae, a new neotropical tribe of grasses (Poaceae). Novon 11: 350–352.

Sanchez-Ken JG, LG Clark, EA Kellogg, and EE Kay. 2007. Reinstatement and emendation of subfamily Micrairoideae (Poaceae). Syst. Bot. 32: 71–80.

Savile DBO. 1990. Relationships of Poaceae, Cyperaceae, and Juncaceae reflected by their fungal parasites. Canad. J. Bot. 68: 731–734.

Schuster J. 1910. Über die Morphologic der Grasblute. Flora (Jena) 100: 213–266.

Seberg O and S Frederiksen. 2001. A phylogenetic analysis of the monogenomic Triticeae (Poaceae) based on morphology. Bot. J. Linn. Soc. 136: 75–97.

Skvortsov AK. 1977. Once more on the morphological nature of the parts of embryo and seedling. Bull. Moscow Soc. Naturalists 82(5): 96–111 (in Russian with English summary).

Soderstrom TR. 1981. The crass subfamily Centostecoideae. Taxon 30: 614–616.

Soderstrom TR. 1987. Some evolutionary trends in the Bambusoideae (Poaceae). Ann. Missouri Bot. Gard. 68: 15–47.

Soderstrom TR and RP Ellis. 1987. The position of bamboo genera and allies in a system of grass classification. In: TR Soderstrom, KW Hilu, CS Campbell, and ME Barkworth, eds. Grass systematics and evolution, pp. 225–238. Washington.

Soreng RJ and JI Davis. 1998a. Phylogenetics and character evolution in the grass family (Poaceae): simultaneous analysis of morphological and chloroplast DNA restriction site character sets. Bot. Rev. 64: 1–85.

Soreng RJ and JI Davis. 1998b. A cladistic analysis of Poaceae subfamily Pooideae. In: Monocots II, p. 52 (Abstract). Sydney.

Soreng RJ and JI Davis. 2000. Phylogenetic structure in Poaceae subfamily Pooideae as inferred from molecular and morphological characters: misclassification versus reticulation. In: SWL Jacobs and J Everett, eds. Grasses: systematics and evolution, pp. 61–74. CSIRO, Collingwood.

Soreng RJ, JI Davis, and JJ Doyle. 1990. A phylogenetic analysis of chloroplast DNA restriction site variation in Poaceae subfam. Pooideae. Plant Syst. Evol. 172: 83–97.

Stanley KE. 1999. Evolutionary trends in the grasses (Poaceae): a review. Mich. Bot. 38: 3–12.

Stebbins GL. 1956. Cytogenetics and evolution of the grass family. Am. J. Bot. 43: 890–905.

Stebbins GL. 1981. Coevolution of grasses and herbivores. Ann. Missouri Bot. Gard. 68: 75–86.

Tateoka T. 1962. Starch grains of endosperm in grass systematics. Bot. Mag. (Tokyo) 75(892): 377–383.

Tateoka T, S Inoue, and S Kawano. 1959. Notes on some grasses. IX. Systematic significance of bicellular microhairs of leaf epidermis. Bot. Gaz. 21: 80–91.

Tenório E. 1976. The subfamily Centostecoideae (Gramineae). Ph.D. thesis. University of Maryland.

Terrell EE and PM Peterson. 1993. Caryopsis morphology and classification in the Triticeae (Pooideae, Poaceae). Smithsonian Contr. Bot. 83: 1–24.

Tulloch AP. 1981. Composition of epicuticular waxes from 28 genera of Gramineae: Differences between subfamilies. Canad. J. Bot. 59: 1213–1221.

Tsvelev NN. 1989. The system of grasses (Poaceae) and their evolution. Bot. Rev. 55: 142–204.

Van Den Borre A and L Watson. 1997. On the classification of the Chloridoideae. Austral. Syst. Bot. 10: 491–531.

Van Den Borre A and L Watson. 2000. On the classification of the Chloridoideae: results from morphological and leaf anatomical data analyses. In: SWL Jacobs and J Everett, eds. Grasses: systematics and evolution, pp. 180–183. CSIRO, Collingwood.

Van Tieghem P. 1897. Morphologie de l'embryon et de la plantule chez les Graminees et les Cyperacees. Ann. Sci. Natur. Bot. Biol. Veget. 8: 259–309.

Vegetti AC. 1998a. The structure of the paracladial zone in Luziolinae (Oryzeae-Poaceae). Beitr. Biol. Pfl. 70: 101–106.

Vegetti AC. 1998b. Estudio fenetico de la inflorescencia en Andropogoneae (Poaceae). Kurtziana 26: 145–163.

Vegetti AC. 2000. Typology of synflorescences in Oryzeae (Poaceae). Phyton (Austria) 40: 71–88.

Vegetti AC. 2002. Caracterizacion de los sistemas de ramificacion en especies de Oryzeae (Poaceae). Candollea 57: 251–260.

Vegetti C and AM Anton. 1995. Some evolution trends in the inflorescence of Poaceae. Flora 190: 225–228.

Vegetti C and AM Anton. 2000. The grass inflorescence. In: SWL Jacobs and J Everett, eds. Grasses: systematics and evolution, pp. 29–31. CSIRO, Collingwood.

Watson L, HT Clifford, and MJ Dallwitz. 1985. The classification of the Poaceae: Subfamilies supertribes. Austral. J. Bot. 33: 433–484.

Watson L and MJ Dallwitz. 1992. The grass genera of the world. CAB International. Wallingford, UK.

Webster RD. 1988. Genera of the North American Paniceae (Poaceae: Panicoideae). Syst. Bot. 13: 576–609.

Whipple CJ and RJ Schmidt. 2006. Genetics of grass flower development. Adv. Bot. Res. 44: 385–424.

Yakovlev MS. 1948. Morphological types of embryo and grass embryo. Doklady of Armenian Akad. Sci. 8: 127–134 (in Russian).

Yakovlev MS. 1950. Endosperm and embryo structure of grasses as a taxonomic feature. Trudy Komarov Bot. Inst. Acad. Sci. USSR, 7th sen, 1: 121–218 (in Russian).

Zhang W-P and LG Clark. 2000. Phylogeny and classification of the Bambusoideae (Poaceae). In: SWL Jacobs and J Everett, eds. Grasses: systematics and evolution, pp. 35–42. CSIRO, Collingwood.

Zuloaga FO, O Morrone, and LM Giussani. 2000. A cladistic analysis of the Paniceae: a preliminary approach. In: SWL Jacobs and J Everett, eds. Grasses: systematics and evolution, pp. 123–135. CSIRO, Collingwood.

Index

D

J

N

O

MIX
Papier aus verantwortungsvollen Quellen
Paper from responsible sources
FSC® C105338

If you have any concerns about our products,
you can contact us on
ProductSafety@springernature.com

In case Publisher is established outside the EU,
the EU authorized representative is:
Springer Nature Customer Service Center GmbH
Europaplatz 3, 69115 Heidelberg, Germany

Printed by Libri Plureos GmbH
in Hamburg, Germany